Management of Animal Care and Use Programs in Research, Education, and Testing

SECOND EDITION

Management of Animal Care and Use Programs in Research, Education, and Testing

SECOND EDITION

Edited by

Robert H. Weichbrod

Gail A. (Heidbrink)

Thompson John N. Norton

CRC Press
Taylor & Francis Group
Boca Raton London New York

CRC Press is an imprint of the
Taylor & Francis Group, an **informa** business

An Association of American Publishers (AAP) 2019 PROSE Award Finalist in Biological Science honors scholarly works published in 2018.

Note: Since 1976, the Association of American Publishers Awards for Professional and Scholarly Excellence (PROSE Awards) have recognized publishers who produce books, journals, and digital products of extraordinary merit that make a significant contribution to a field of study each year.

American College of Laboratory Animal Medicine

CALAS | ACSAL

eclam European College of Laboratory Animal Medicine

iaclam

iat Institute of Animal Technology

JCLAM

KCLAM KOREAN COLLEGE OF LABORATORY ANIMAL MEDICINE

LAMA

Cover design by Bonnie Hamalainen, MFA, Baltimore, Maryland.

CRC Press
Taylor & Francis Group
6000 Broken Sound Parkway NW, Suite 300
Boca Raton, FL 33487-2742

Printed on acid-free paper

International Standard Book Number-13: 978-1-4987-4844-5 (Hardback)

Library of Congress Cataloging-in-Publication Data

Names: Weichbrod, Robert H., editor. | Thompson, Gail A., editor. | Norton, John N., editor.
Title: Management of animal care and use programs in research, education, and testing / editors, Robert H. Weichbrod, Gail A. Thompson, John N. Norton.
Other titles: Management of laboratory animal care and use programs
Description: Second edition. | Boca Raton : Taylor & Francis, 2018. |
Preceded by Management of laboratory animal care and use programs / edited by Mark A. Suckow, Fred A. Douglas, Robert H. Weichbrod. c2002. | Includes bibliographical references and index.
Identifiers: LCCN 2016059771 | ISBN 9781498748445 (hardback : alk. paper)
Subjects: | MESH: Animal Welfare--organization & administration | Animals, Laboratory | Laboratories--organization & administration
Classification: LCC SF406 | NLM QY 50 | DDC 636.088/5--dc23
LC record available at https://lccn.loc.gov/2016059771

Visit the Taylor & Francis Web site at
http://www.taylorandfrancis.com

and the CRC Press Web site at
http://www.crcpress.com

For Product Safety Concerns and Information please contact our EU representative: GPSR@taylorandfrancis.com
Taylor & Francis Verlag GmbH, Kaufingerstraße 24, 80331 München, Germany

Contents

Section VII Husbandry

Section VIII Animal Health and Care

Section IX Conclusion

Foreword

I am pleased and honored to have been asked to write the Foreword to the second edition of *Management of Animal Care and Use Programs in Research, Education, and Testing.* I am especially grateful to Dr. Weichbrod, Dr. Norton, and Ms. Thompson for allowing me to convey my personal views on the importance and value of providing exceptional laboratory animal care and management and its vital link to performing outstanding biomedical science.

Today, in 2017, science in general, and biomedical science in particular, is under attack from those who believe that science and the funding of science should not be a federal, state, or local priority. There are those who believe that breakthroughs in health care come merely from studies that solely use computer models, cell-based studies, bioinformatics, and device engineering. The scientific community and the lay public have discussed these issues for many years. The reality is that no advances in health care can be achieved with just the use of these adjuncts; the use of animals and animal models of disease remains imperative, lest we do great harm to humans and animals. The competition for research funding is at its highest level in over 20 years. Everyone is aware, and most would agree, that science is synonymous with a search for truth. Acquiring believable facts is the only way biomedical science will provide therapies that work and cures to both human and animal diseases. Importantly, federal agencies, universities, industry, and journal publishers have increased their emphasis on issues related to rigor and reproducibility in science so that more science can be trusted and will contribute to understanding basic biological processes that lead to new therapies and cures. As part of the animal kingdom, the quality of our lives, however, matters. The quality of laboratory animals' lives and human lives is indelibly intertwined. In many ways, it is within this context that *Management of Animal Care and Use Programs in Research, Education, and Testing* is such an important body of work.

When I started my career as a graduate student in 1966, the guidelines for carrying out experiments on large and small animals were generally left to the discretion of the investigator. Even laboratory animal facilities were rather lax in both oversight of investigators and their own husbandry practices. Certainly, this was an era well before the advent of Institutional Animal Care and Use Committees (IACUCs) and/or oversight bodies. It is now clear that the evolution of governmental, local, and institutional guidelines has changed the landscape and certainly the quality of the science produced. In 1976, I became the chair of the University of Nebraska Medical Center's (UNMC) advisory committee on animal resources, a precursor to the UNMC's IACUC formed in 1985, a position I held until 1988 when I became chair of the Department of Physiology. I became a student member of the American Physiological Society (APS) in 1968, and in 1985, I served and was the chair of the APS's Animal Care and Experimentation Committee until 1988. I became chair of the APS Public Affairs Committee in 2001. I then went on to be elected councilor and the president of the APS in 2008. I also served on the council of the Society for Experimental Biology and Medicine from 1998 to 2002. I bring these service appointments up because in every discussion as well as on every scientific review panel, the issue of animal care and associated regulations was discussed and continues to this day. It's appropriate to discuss these issues because without continual refinement, the science may suffer.

Given that the readers of this book agree that animal models of human disease are necessary and given that there are no perfect animal models of human disease, it behooves investigators to maximize the science that is acquired from these models. Unless a study is proposed to evaluate the mechanisms of stress per se, it is not beneficial to perform studies on stressed animals, either under anesthesia or in the conscious state. I'm confident that potential life-saving information has been lost in experiments carried out on animals where optimal husbandry, maintenance of consistent and appropriate environment conditions, and the successful alleviation of stress were not maximized effectively.

The community of biomedical science is vast and includes principal investigators, administrators, laboratory facility managers, animal caregivers, veterinarians, physicians, and allied trade representatives.

It cannot be emphasized enough that individual science is a thing of the past. Today, team science is the norm. We need to all recognize that working and using animals in biomedical research or testing is a privilege, not a right. This concept comes out loud and clear in this book. The editors have put together a text that is not only for those directly involved in the care and use of experimental animals but also for all of us that need the support of the animal care team to do good and believable science. The book focuses on every issue and concern from the culture of caring to technical issues related to animal husbandry. Importantly, it also reviews issues related to compliance and regulation. How much regulation is too much? What is reasonable and how to comply with federal guidelines without impairing the ability of investigators to carry out good science are just some of the important topics covered in this text. The book makes it clear that as new techniques are developed, issues relating to animal welfare change accordingly. Finally, the future of animal oversight and the impact of pending legislation are discussed.

The editors and contributors have taken the second edition to a new level. Compiling the second edition was a real *tour de force* and I congratulate the editors for this outstanding text. This book should be used by anyone who cares about biomedical science: principal investigators, members of IACUCs or other oversight bodies, institutional administrators, policy makers (including legislators), as well as animal facility personnel. I would also recommend that trainees make this book part of their library, as important as a text on physiology, biochemistry, genetics, or any other biomedical discipline. It is important that only well-trained individuals carry out procedures on animals for the benefit of science. Those of us who have worked with animals for many years have an obligation to pass on high-quality practices to those who will do science in years to come. The culture of science will be changed by this edition, which should be a required reading for all those contemplating a career in biomedical science.

Irving H. Zucker, PhD, FAHA, FAPS
Omaha, Nebraska

Preface to the Second Edition

Since the publication of the first edition, there have been many notable advances in biomedical research that have resulted in a better understanding of complex biological systems. This has led to the development of new therapeutic modalities that have improved the quality of life for people and animals across the world.

The importance of continuing biomedical research with animals remains critical to these developments. Therefore, it is important for administrators, managers, and scientists working with and responsible for laboratory animal resources to have the practical experience, knowledge, and understanding of the topics concerning humane animal care and use.

The *Management of Animal Care and Use Programs in Research, Education, and Testing,* Second Edition, is an expanded and revised edition to the original text published in 2001. Like the first edition, this book is meant to serve as a first-line management resource, provide a strong advocacy reference for advancing quality animal welfare and science worldwide, and continue as a valuable seminal reference for those engaged in all types of programs involving animal care and use. The book has been greatly expanded to provide a more thorough overview of the current breadth and depth of the field, with applicability to an international audience.

Over the 16 years since the publication of the first edition of the book, *Management of Laboratory Animal Care and Use Programs*, a multitude of critical regulatory guidance documents have been updated and expanded. Among these are the

- *Guide for the Care and Use of Laboratory Animals*, Eighth Edition (National Research Council 2011)
- *Guide for the Care and Use of Agricultural Animals in Research and Teaching* (Federation of Animal Science Societies 2010)
- European Convention for the Protection of Vertebrate Animals Used for Experimental and Other Scientific Purposes (European Treaty Series 123)
- *Biosafety in Microbiological and Biomedical Laboratories*, Fifth Edition (U.S. Department of Health and Human Services 2009)
- AVMA Guidelines on Euthanasia of Animals (American Veterinary Medical Association 2013)

Additionally, there have been updated regulatory and policy guidelines and direction from the

- Office of Laboratory Animal Welfare, the NIH office overseeing compliance with
  - U.S. Public Health Service Policy on Humane Care and Use of Laboratory Animals
  - Animal welfare assurances
  - Statements of compliance
- U.S. Department of Agriculture (USDA), with policy directives on
  - Animal Welfare Act of 1966 (Public Law 89-544) and subsequent amendments, as promulgated in USDA regulations 9 CFR Chapter 1, Subchapter A, Animal Welfare, Parts 1–3
- AAALAC International, with
  - More than 980 accredited units in 44 countries
  - Applicable position statements and reference resources
  - Relevant governmental legislation and regulatory updates

This second edition has been broadly expanded to address changes and provide a more thorough overview of the current complexity and extensiveness of the field. It is timely that the material in the second edition be expanded and updated to reflect the advancements and new information both within and outside the United States.

The new features and expanded scope of the book will be highly beneficial to readers and practitioners as well as provide strong rationale for advancing the quality of animal welfare and science worldwide. The editors intend this book to be of the highest quality and surpass the much-deserved reputation of the first edition as a first-line management resource, and to continue to be a valuable seminal work for those engaged in all types of programs involving the management and administration of animal care and use programs.

The book consists of 36 chapters written by individuals with notable experience and expertise in the pertinent topical areas. They bring insights into emerging technologies and an appreciation for the needs of an international audience. Each chapter has been peer-reviewed and contains the latest information, resources, and reference materials from the scientific literature. Key revisions to the book include a new section on environment and housing, which contains chapters focused on the management considerations of housing and enrichment delineated by species; an enhanced narrative throughout the book of critical topics in program management, physical plant, animal health, and husbandry; and expanded coverage of regulatory compliance, assessment, and assurance processes. These topics are addressed and information is provided about the impact of globalization and efforts to harmonize and build on cultural awareness initiatives around the world. Emphasis is given in new chapters on the development of a collaborative "culture of care" within an animal care and use program, and how behavioral management through animal training can play an integral role in a veterinary health program.

The overall goal of the book is to improve the reader's ability to recognize, interpret, and adapt in a more complex and dynamically changing environment, as well as to provide a better understanding of engineering principles, the application of performance-based standards, and the valuable use of professional judgment. Ultimately, the book represents foundational information for managers and administrators to bridge the connection between quality science and quality animal care to the international audience. While the application of the information and principles contained within this book may vary among different managers, organizations, or specific countries, the editors hope the book provides strong advocacy for advancing and connecting animal welfare and science of the highest quality worldwide.

The editors acknowledge and thank all the contributing authors and reviewers of the chapters for sharing their time and expertise. We offer our sincerest thanks to Bonnie Hamalainen, MFA, for contributing her superb art design talents in creating the second edition text cover.

And a special tribute is given to Tim Allen who worked with the USDA's Animal Welfare Information Center in Beltsville, MD for many years. Tim, who passed away suddenly at the end of 2016, was a remarkable human being and he will be truly missed. He thoroughly embraced and gave valuable sage advice in the development of the second edition text. Tim provided the tools and expertise in providing high-quality comprehensive literature searches for many of our second edition authors. Over the course of his career, he provided valuable information on improved animal welfare practices to numerous individuals on an international basis. Tim left an indelible mark on many people in the way they approached their research in the field of animal care and use.

Royalties due the editors of *Management of Animal Care and Use Programs in Research, Education, and Testing*, Second Edition, will be donated to the following professional societies who have agreed to copublish the book:

American College of Laboratory Animal Medicine (ACLAM)
Canadian Association for Laboratory Animal Science (CALAS)
European College of Laboratory Animal Medicine (ECLAM)
International Association of Colleges of Laboratory Animal Medicine (IACLAM)
Institute for Animal Technology (IAT)
Japanese College of Laboratory Animal Medicine (JCLAM)
Korean College of Laboratory Animal Medicine (KCLAM)
Laboratory Animal Management Association (LAMA)

Editors

Robert H. Weichbrod is the chief animal program administrator with the National Eye Institute of the National Institutes of Health in Bethesda, Maryland. Before this position, Dr. Weichbrod was responsible for managing laboratory animal resources for the U.S. Department of Defense's Uniformed Services University of the Health Sciences. Dr. Weichbrod received his bachelor of science in zoology from the University of Maryland, master of business administration from Marymount University, and doctor of philosophy in public administration and policy from Walden University. His dissertation, providing an analysis of the care and use of laboratory animals in Department of Defense activities, is a seminal work that has been frequently cited during congressional hearings. Dr. Weichbrod earned his laboratory animal technologist certification from the American Association for Laboratory Animal Science (AALAS) and was a charter class graduate of AALAS's Institute for Laboratory Animal Management.

Dr. Weichbrod's distinguished career in animal care and use spans over 35 years. He has worked in positions ranging from an entry-level animal care technician (age 24) to animal program administrator. Dr. Weichbrod has served in a wide variety of leadership roles during his career, including AALAS president in 2000, member of the American Association for Accreditation of Laboratory Animal Care (AAALAC) International's Council on Accreditation from 1997 to 2009, on its board of trustees from 2010 to 2016, and currently as a member organization delegate for AAALAC International. Dr. Weichbrod has served as a vice president for the Institute of Animal Technology in England since 2002.

Dr. Weichbrod has authored more than 70 scientific, managerial, and technical articles pertaining to programs involving the effective care and use of animals in biomedical research, training, and education. He was the founding editor in chief and publisher of the Laboratory Animal Management Association's (LAMA) journal the *LAMA Review* and coeditor of the book *Management of Laboratory Animal Care and Use Programs* (CRC Press 2001). He has served on numerous scientific advisory review committees and journal editorial boards. Dr. Weichbrod is an experienced and dedicated advocate for the welfare and responsible use of animals in research. He is committed to maximizing the outcomes of the biomedical research team, enabling increased knowledge of complex biological systems, resulting in health benefits for animals and people.

Among Dr. Weichbrod's awards are AALAS's Joseph J. Garvey Award and George R. Collins Award; LAMA's William O. Umiker Memorial Award, U. Kristina Stephens Award, and Charles River Medallion; the Purina LabDiet Animal Technician of the Year Award; and the Award of Excellence from the U.S. Secretary of Defense.

Gail A. (Heidbrink) Thompson has been active in the laboratory animal science community since beginning her career at the University of Minnesota in 1966. She has held research positions at the University of Minnesota, at Emory University/Yerkes Regional Research Primate Center, and as director of animal resources at National Jewish Health in Denver, Colorado. She was a principal and cofounder of Britz-Heidbrink, Inc. (a manufacturing company for animal research facilities and zoological parks). She was the founder, owner, and president of Peak Animal Resources+, Inc. until her retirement in December 2014.

Along with her long career in laboratory animal science and zoological park facilities, Gail has been a dedicated volunteer in several associations that strive for the humane care of laboratory and captive animals through facility and housing improvement, education and training, and oversight and management. The associations include but are not limited to AAALAC International, trustee (2000–2016), delegate and board of directors (current); the American Association for Laboratory Animal Science (AALAS) since 1973; the Laboratory Animal Management Association (LAMA), formed in 1983, founding member; the Institute of Animal Technology United Kingdom since 1989; and the Mile High Branch of AALAS since 1977, founding member. Additional organizations in which she has participated include the American

Society of Primatologists, Canadian Association for Laboratory Animal Science, Scientists Center for Animal Welfare, Scandinavian Society for Laboratory Animal Science, International Conference on Environmental Enrichment, Association of Zoos and Aquariums, American Association of Zoo Keepers, and several branches of AALAS.

Significant activities or positions held include the honor of serving as president of AALAS in 2004 and holding committee appointments and leadership roles on most AALAS committees over the years. In 1991, Gail initiated and developed the Institute for Laboratory Animal Management as a joint program for AALAS and LAMA. She served as a director and chair of the board of regents of the Institute for Laboratory Animal Management for 6 years and also as the editor of *Tech Talk*. Gail has been involved in the formation and continuation of the AAALAC International Fellowship Award since its conception in 2005. She has served as chair of the award selection committee since 2005. Additionally, she serves as a vice president of the Institute of Animal Technology and member of the President's Advisory Group. She has served as president, treasurer, and program chair for LAMA.

Gail has authored, coauthored, or presented in more than 100 publications, platform presentations, and workshops. Several have been international, with presentations in the United Kingdom, Denmark, Brazil, Canada, and Australia, and for the Federation of European Laboratory Animal Science Associations.

She is a recipient of several awards, including Ralston Purina Animal Technician of the Year, George Collins Award (AALAS), Joseph Garvey Award (AALAS), Ulla Kristina Stephens Award (LAMA), Charles River Medallion, Wm. O. Umiker Award (LAMA), Roland Tibbetts Award (U.S. Small Business Administration), Manufacturing Legacy of the Year (University of Wyoming), and life memberships for AALAS, LAMA, and the Mile High Branch of AALAS.

Beyond her career activities, she has been active in community events, including serving as a mentor for science fair students; a board member on Wyoming's Science, Technology and Energy Authority; and a distinguished lieutenant governor and club president for the Rocky Mountain District of Kiwanis International.

John N. Norton is the director of the Division of Laboratory Animal Resources, is a professor of pathology, and serves as the attending veterinarian for all animal care and use activities at Duke University in Durham, North Carolina. He also serves as an adjunct professor of clinical sciences at North Carolina State University College of Veterinary Medicine. Prior to returning to academia, Dr. Norton served as a toxicologist and directed a laboratory animal resources organization in the private sector.

Dr. Norton received his bachelor of science degrees and doctor of veterinary medicine degree from North Carolina State University and his doctor of philosophy in pharmacology from Vanderbilt University. He is board certified through the American College of Laboratory Animal Medicine and the American Board of Toxicology.

Dr. Norton's distinguished career in both toxicology and laboratory animal medicine spans over 25 years. He has managed preclinical discovery and development projects, prepared regulatory dossiers for both domestic and international submissions, designed and managed complex animal facilities, and strived to optimize research outcomes while ensuring regulatory compliance of animal programs. In his roles in drug and biomedical device development, he has served as study director and/or manager in more than 150 preclinical pharmacology and safety studies in both the academic and private sectors. In his current position, he has developed a preclinical core capable of performing a wide variety of discovery and developmental studies, including those requiring performance under good laboratory practice regulations.

Dr. Norton has served on numerous committees of professional organizations, such as the American College of Laboratory Animal Medicine, and he currently serves as a board member for the National Association for Biomedical Research. In addition, he serves on the board of directors for the North Carolina Association for Biomedical Research, including current service as the chair, and formerly served on the board of directors for the Texas Society for Biomedical Research. In addition, Dr. Norton served as a member of AAALAC International's Council on Accreditation from 2004 to 2016, and as its president during 2013 to 2015. He currently serves as a council member emeritus for AAALAC International.

Dr. Norton's collaborative research focuses on extrinsic factors that may influence the animal research model, specifically in the area of noise and vibration. He has published via scientific, managerial, and technical articles and book chapters on a variety of topics, and he is a frequent reviewer of scientific articles and grant proposals. Dr. Norton is an advocate of ensuring animal welfare while focusing on quality research outcomes and translational research involving novel therapeutics and biomedical devices.

Contributors

E. Douglas Allen, BS, LATG
University of Maryland School of Medicine
Baltimore, Maryland

Lynn C. Anderson, DVM, DACLAM
Covance Laboratories
Madison, Wisconsin

Diana P. Baumann, BSc (Hons), RLATG, CMAR
Stowers Institute for Medical Research
Kansas City, Missouri

Joseph N. Benoit, PhD
Burrell College of Osteopathic Medicine
Las Cruces, New Mexico

Valerie K. Bergdall, DVM, DACLAM
The Ohio State University
Columbus, Ohio

Jan Bernal, DVM
Pfizer
Groton, Connecticut

Amanda Bettis, BS, LVT
Texas A&M University
College Station, Texas

M. Elizabeth Blackburn, CPIA
University of Nebraska Medical Center
Omaha, Nebraska

Mollie A. Bloomsmith, PhD
Yerkes National Primate Research Center
Atlanta, Georgia

Kenneth R. Boschert, DVM, DACLAM
Washington University Medical School
St. Louis, Missouri

Ronald L. Bowman, BS, RLATG, CMAR
National Biodefense Analysis and Countermeasures Center
Frederick, Maryland

John F. Bradfield, DVM, PhD, DACLAM
University of Nebraska Medical Center
Omaha, Nebraska

Natalie A. Bratcher, BS, MS
AbbVie
North Chicago, Illinois

Marilyn J. Brown, DVM, MS, DACLAM, DECLAM
Charles River Laboratories
Wilmington, Massachusetts

Cindy A. Buckmaster, PhD, CMAR, RLATG
Baylor College of Medicine
Houston, Texas

Ronald L. Carter, RLATG, CMAR
Memorial Sloan Kettering Cancer Center
Weill Cornell Medicine and the Hospital for Special Surgery
New York, New York

Larry T. Chapin, MS
Michigan State University
East Lansing, Michigan

Randall R. Clevenger, BS, LATG
National Institutes of Health
Bethesda, Maryland

Sylvie Cloutier, BSc, MSc, PhD
Canadian Council on Animal Care
Ottawa, Canada

Laura A. Conour, DVM, DACLAM
Princeton University
Princeton, New Jersey

James D. Cox, RLATG, MLAS
Howard Hughes Medical Institute
Ashburn, Virginia

Edgar F. Czarra, BS, MBA
National Institutes of Health
Bethesda, Maryland

William E. Dale, PhD
Oregon Health & Science University
Portland, Oregon

Louis DeTolla, VMD, MS, PhD, DACLAM
University of Maryland School of Medicine
Baltimore, Maryland

Robert S. Dixon, DVM, MS, DACLAM
University of Nebraska Medical Center
Omaha, Nebraska

John C. Donaho, BS, CMAR
University of Texas Health Science Center
Houston, Texas

Lynell M. Dupepe, BS, RLATG, CMAR
Tulane University
New Orleans, Louisiana

James J. Elliott, DVM, DACLAM, DACVPM
Texas A&M University
College Station, Texas

Kim Froeschl, BS, RLATG, CMAR, CPLP
Pfizer Inc.
Pearl River, New York

Janet C. Garber, DVM, PhD
Garber Consulting
Pinehurst, North Carolina

Teresa R. Gleason, BS, LVT, SRS, LATG
Charles River Laboratories
Ashland, Ohio

Javier Guillén, DVM
AAALAC International
Pamplona, Spain

Sander O. Hacker, VMD, DACLAM
University of Texas Health Science Center
at San Antonio
San Antonio, Texas

James A. Hagarman, PhD
The Jackson Laboratory
Bar Harbor, Maine

George J. Haluska, PhD
Oregon Health & Science University (Retired)
Portland, Oregon

F. Claire Hankenson, DVM, MS, DACLAM
Michigan State University
East Lansing, Michigan

Debra L. Hickman, DVM, MS, DACLAM, DACAW
Indiana University
Indianapolis, Indiana

Judy M. Hickman-Davis, DVM, PhD, DACLAM
The Ohio State University
Columbus, Ohio

Margaret C. Hogan, RLATG, CMAR
Duke University
Durham, North Carolina

Rebecca K. Holt, DVM, MPH
National Biodefense Analysis and
Countermeasures Center
Frederick, Maryland

Diane (Dee) Horne, CVT, CMAR
Oregon Health & Science University (Retired)
Portland, Oregon

Robert F. Hoyt, Jr., DVM, DACLAM
Leidos Biomedical Research
Bethesda, Maryland

Michael J. Huerkamp, DVM, DACLAM
Emory University
Atlanta, Georgia

Eric Hutchinson, DVM, DACLAM
Johns Hopkins University
Baltimore, Maryland

Stephen T. Kelley, DVM, MS, DACLAM
University of Washington
Seattle, Washington

Bruce W. Kennedy, MS, RLATG, CMAR, CPIA, EdD
California State Polytechnic University
Pomona, California

William W. King, DVM, PhD, DACLAM
University of Louisville
Louisville, Kentucky

Hilton Klein, VMD, MS, DACLAM, DECLAM
Fox Chase Cancer Center
Philadelphia, Pennsylvania

Susan P. Lambeth, BS, LATG
The University of Texas MD Anderson Cancer Center
Bastrop, Texas

George Langan, DVM, DACLAM
The University of Chicago
Chicago, Illinois

Christian Lawrence, MS
Boston Children's Hospital
Boston, Massachusetts

Jori K. Leszczynski, DVM, DACLAM
University of Colorado, Denver
Denver, Colorado

Jan Linkenhoker, DVM, MSPH, DACLAM
National Institutes of Health
Bethesda, Maryland

Neil S. Lipman, VMD, DACLAM
Memorial Sloan Kettering Cancer Center
Weill Cornell Medicine and the Hospital for Special Surgery
New York, New York

Kerith Luchins, DVM, DACLAM
The University of Chicago
Chicago, Illinois

Elizabeth R. Magden, DVM, MS, DACLAM
The University of Texas MD Anderson Cancer Center
Bastrop, Texas

Carolyn M. Malinowski, DVM, MS, RLATG, CMAR, CPIA
University of Florida
Gainesville, Florida

David Mallon, RLATG, CMAR, BS
National Institutes of Health
Bethesda, Maryland

Mary Catherine Mareno, BA
The University of Texas MD Anderson Cancer Center
Bastrop, Texas

Leticia V. Medina, DVM, DACLAM
AbbVie
North Chicago, Illinois

Thomas R. Meier, DVM, MBA, DACLAM
Mayo Clinic
Rochester, Minnesota

Charles T. Miller, PhD
The Jackson Laboratory
Bar Harbor, Maine

Norman Mortell, BA (Hons), MIAT, RAnTech
Agenda Life Sciences
Hull, East Riding of Yorkshire, United Kingdom

Christian E. Newcomer, VMD, MS, DACLAM
AAALAC International
Brookville, Maryland

Marek Niekrasz, DVM, DACLAM
The University of Chicago
Chicago, Illinois

John N. Norton, DVM, PhD, DABT, DACLAM
Duke University
Durham, North Carolina

Dorcas P. O'Rourke, DVM, MS, DACLAM
East Carolina University
Greenville, North Carolina

Yasmina A. Paramastri, Drh, DACLAM
National University of Singapore
Singapore, Singapore

Gerald Percifield, NCARB
HDR
Atlanta, Georgia

Kathy A. Perdue, DVM, DACLAM
National Institutes of Health
Baltimore, Maryland

Jaine E. Perlman, RLATG, BA
Yerkes National Primate Research Center
Atlanta, Georgia

Jessica Peveler, BA, RVT, RLATG
Indiana University School of Medicine
Indianapolis, Indiana

Theodore Plemons, MBA, CMAR, RLATG
National Institutes of Health
Bethesda, Maryland

Stephen J. Pomeroy, MHA, RLATG
Duke University (Retired)
Durham, North Carolina

Ernest D. Prentice, PhD
University of Nebraska Medical Center
Omaha, Nebraska

James M. Raber, DVM, MS, PhD
National Institutes of Health
Bethesda, Maryland

Lisa A. Reamer, MRes
The University of Texas MD Anderson Cancer Center
Bastrop, Texas

Randall P. Reynolds, DVM, MS, DACLAM
Duke University
Durham, North Carolina

Tracie Rindfield, LAT, SRS
Rindfield Consulting
Lawrence, Michigan

Gordon Roble, DVM, DACLAM
New York University
New York, New York

John J. Sancenito
Information Network Associates
Harrisburg, Pennsylvania

George E. Sanders, DVM, MS, AFS/FHS
University of Washington
Seattle, Washington

Steven J. Schapiro, PhD
The University of Texas MD Anderson Cancer Center
Bastrop, Texas

James M. Schmitt, MD, MS
National Institutes of Health
Bethesda, Maryland

Mark Sharpless, CMAR, RLATG, MBA, JM
Yerkes National Primate Research Center
Atlanta, Georgia

Richard C. Simmonds, DVM, MS, DACLAM
Nevada System of Higher Education (Retired)
Reno, Nevada

Christine P. Sivula, DVM, MS, DACLAM
University of Minnesota
Minneapolis, Minnesota

Michael Stephens, BA, MSc
Medical Research Council
London, United Kingdom

Daniel T. Stimson, PhD, JD
National Institutes of Health
Bethesda, Maryland

Pamela A. Straeter, MS, RLATG, CMAR, CPIA
Princeton University
Princeton, New Jersey

Mark A. Suckow, DVM, DACLAM
University of Minnesota
Minneapolis, Minnesota

Melissa Swan, BA, RVT, RLATG
Purdue University
West Lafayette, Indiana

Janice C. Swanson, MS, PhD, AABS
Michigan State University
East Lansing, Michigan

James R. Swearengen, DVM, DACLAM
AAALAC International
Frederick, Maryland

Camellia Symonowicz, BS, MS, RLATG, CMAR
Middletown, Connecticut

Jamie Tackett, CVT, RLATG, BS
Abbvie
North Chicago, Illinois

Michael Talcott, DVM, DACLAM
Washington University
St. Louis, Missouri

Suzette D. Tardif, PhD
Southwest National Primate Research Center
San Antonio, Texas

James F. Taylor, DVM, MS, DACLAM
National Institutes of Health (Retired)
Derwood, Maryland

Joseph D. Thulin, DVM, MS, DACLAM
Medical College of Wisconsin
Milwaukee, Wisconsin

Michelle Wallace-Fields, BS, RLATG, CMAR
University of Colorado, Denver
Denver, Colorado

Robert H. Weichbrod, RLATG, MBA, PhD
National Institutes of Health
Bethesda, Maryland

Carole Wilson, BSc (Hons), MIAT, RAnTech
University College London
London, United Kingdom

Deborah E. Wilson, DrPH, CBSP
National Institutes of Health
Bethesda, Maryland

Jeffrey D. Wyatt, DVM, DACLAM
University of Rochester
Rochester, New York

Irving H. Zucker, PhD, FAHA, FAPS
University of Nebraska Medical Center
Omaha, Nebraska

Reviewers

Christian R. Abee, DVM, MS, DACLAM
The University of Texas MD Anderson Cancer Center
Bastrop, Texas

Tim Allen, BS
U.S. Department of Agriculture
Beltsville, Maryland

Lynn C. Anderson, DVM, DACLAM
Covance Laboratories
Madison, Wisconsin

Matthew R. Bailey
National Association for Biomedical Research
Washington, DC

Philippe Baneux, DVM, DECLAM
Cornell University
Ithaca, New York

Kathryn Bayne, MS, PhD, DVM, DACLAM, DACAW, CAAB
AAALAC International
Frederick, Maryland

B. Taylor Bennett, DVM, PhD, DACLAM, DACAW
Management Consultant
Hinsdale, Illinois

Joseph N. Benoit, PhD
Burrell College of Osteopathic Medicine
Las Cruces, New Mexico

Rudolf P. Bohm Jr., DVM, DACLAM
Tulane National Primate Research Center
Covington, Louisiana

Gary L. Borkowski, DVM, MS, DACLAM
Eli Lilly & Company
Indianapolis, Indiana

Thomas M. Butler, DVM, MS, DACLAM
Consultant
Fair Oaks Ranch, Texas

Michelle D. Calkins, LVT, BS, RLATG
Duke University
Durham, North Carolina

Arthur L. Caplan, PhD
NYU Langone Medical Center
New York, New York

Calvin Carpenter, DVM, DACLAM
Alcon Research
Fort Worth, Texas

Bart Carter, DVM, MS, DACLAM
University of Texas Southwestern Medical Center
Dallas, Texas

Daniel Castranova, RLATG, MS
National Institutes of Health (Charles River)
Bethesda, Maryland

Randall R. Clevenger, BS, LATG
National Institutes of Health
Bethesda, Maryland

J. G. Collins, PhD
Yale University School of Medicine
New Haven, Connecticut

William Cullen
National Institutes of Health
Bethesda, Maryland

Edgar F. Czarra, BS, MBA
National Institutes of Health
Bethesda, Maryland

William E. Dale, PhD
Oregon Health & Science University
Portland, Oregon

David DeGrazia, BA, MStud, PhD
National Institutes of Health
Bethesda, Maryland

Stephen L. Denny, DVM, MS, DACVPM, DACLAM
National Institutes of Health
Bethesda, Maryland

Helen E. Diggs, MEd, DVM, DACLAM
Oregon State University
Corvallis, Oregon

Deb Donohue, BA, RLATG
Medical College of Wisconsin
Milwaukee, Wisconsin

Paige A. Ebert, RLATG, CMAR, MBA
AbbVie
North Chicago, Illinois

Gaylen L. Edwards, DVM, PhD
University of Georgia
Athens, Georgia

Craig A. Fletcher, DVM, PhD, DACLAM
University of North Carolina–Chapel Hill
Chapel Hill, North Carolina

Patricia L. Foley, DVM, CPIA, DACLAM
Georgetown University
Washington, DC

Pamela Fontenot
Priority One Services
Alexandria, Virginia

Elizabeth Ford, DVM, MPVM, DACLAM
The Scripps Research Institute
La Jolla, California

James G. Fox, DVM, DACLAM
Massachusetts Institute of Technology
Cambridge, Massachusetts

Lisa J. Garrett, MS
National Institutes of Health
Bethesda, Maryland

Bradford S. Goodwin Jr., DVM, DACLAM
University of Texas Health Science Center–Houston (Retired)
Houston, Texas

Andrew W. Grady, DVM, MS, DACLAM
University of Mississippi Medical Center
Jackson, Mississippi

Molly Greene BA, CPIA
Michigan State University
East Lansing, Michigan

Javier Guillén, DVM
AAALAC International
Frederick, Maryland

Anna L. Hampton, DVM, DACLAM
Duke University
Durham, North Carolina

Ruth Hanessian, BS
Rockville Science Center
Rockville, Maryland

F. Claire Hankenson, DVM, MS, DACLAM
Michigan State University
East Lansing, Michigan

John J. Hasenau, DVM, DACLAM
Lab Animal Consultants
Sparks, Nevada

J. R. Haywood, PhD
Michigan State University
East Lansing, Michigan

Tanya L. Herzog, DVM, DACLAM
Houston Methodist Research Institute
Houston, Texas

Melanie Hoar, PhD
PMI Nutrition International
St. Louis, Missouri

Robert Franklin Hoyt Jr., DVM, MS, DACLAM
Leidos Biomedical Research
Bethesda, Maryland

Richard B. Huneke, DVM, MPH, DACLAM
Drexel University
Philadelphia, Pennsylvania

Elisa A. Hurley, PhD
Public Responsibility in Medicine and Research (PRIM&R)
Boston, Massachusetts

Todd A. Jackson, DVM, DACLAM
Oklahoma State University
Stillwater, Oklahoma

Michael D. Kastello, DVM, PhD, DACLAM, DECLAM
Hampton, Virginia

Christopher King, DVM, DACLAM
University of Georgia
Athens, Georgia

William W. King, DVM, PhD, DACLAM
University of Louisville
Louisville, Kentucky

Hilton Klein, VMD, MS, DACLAM, DECLAM
Fox Chase Cancer Center
Philadelphia, Pennsylvania

Gail E. Laule, MA
Active Environments
Lompoc, California

Kirk Leech, BA, MA
European Animal Research Association
London, United Kingdom

Morag Mackay, VN, CertEd, RLATG, SRS
La Jolla Institute for Allergy and Immunology
San Diego, California

David Mallon, RLATG, CMAR, BS
National Institutes of Health
Bethesda, Maryland

Ronald M. McLaughlin, DVM, MS, DACLAM
University of Missouri–Columbia
Fulton, Missouri

John G. Miller, DVM, DACLAM
Frederick, Maryland

Scott A. Mischler, DVM, PhD, DACLAM
Pfizer
Pearl River, New York

Guy B. Mulder, DVM, MS, MBA, DACLAM
Charles River Laboratories
Wilmington, Massachusetts

Christian E. Newcomer, VMD, MS, DACLAM
AAALAC International
Brookville, Maryland

Steven M. Niemi, DVM, DACLAM
Harvard University
Cambridge, Massachusetts

John N. Norton, DVM, PhD, DABT, DACLAM
Duke University
Durham, North Carolina

Peter Nowlan, MVB, MSc, MRCVS, CertLas
Trinity College Dublin
Dublin, Ireland

Bryan E. Ogden, DVM, DACLAM
SingHealth Experimental Medicine Centre
Singapore, Singapore

James O'Malley, DVM, MPH, DACLAM
National Institutes of Health
Bethesda, Maryland

Lauren Pandolfo, MS, CMAR
National Institutes of Health
Bethesda, Maryland

Christine M. Parks, DVM, PhD, DACLAM
Parks Veterinary Services
Waterloo, Wisconsin

Cynthia Pekow, DVM, DACLAM
Veterans Affairs Puget Sound Health Care System
Seattle, Washington

Marky Pitts, CPIA
IACUC Consultant
Joseph, Oregon

Ernest D. Prentice, PhD
University of Nebraska Medical Center
Omaha, Nebraska

Robert H. Quinn, DVM, DACLAM
SUNY Upstate Medical University
Syracuse, New York

Randall P. Reynolds, DVM, MS, DACLAM
Duke University
Durham, North Carolina

Mary A. Robinson, DVM, DACLAM
University of Texas Health Science Center at Houston
Houston, Texas

Gordon Roble, DVM, DACLAM
New York University
New York, New York

Alan C. Rosenquist, PhD
University of Pennsylvania
Philadelphia, Pennsylvania

Clay D. Rouse, DVM
Duke University
Durham, North Carolina

John J. Sancenito
Information Network Associates
Harrisburg, Pennsylvania

Kim E. Saunders, DVM, DACLAM
Oregon Health & Science University
Portland, Oregon

Diana G. Scorpio, DVM, MPH, DACLAM
National Institutes of Health
Bethesda, Maryland

James T. Sheets, DVM, MPH, DACLAM, DACVPM
University of Iowa
Iowa City, Iowa

Susan Brust Silk, MS
National Institutes of Health
Bethesda, Maryland

Richard C. Simmonds, DVM, MS, DACLAM
Nevada System of Higher Education (Retired)
Reno, Nevada

Margaret Snyder
National Institutes of Health
Bethesda, Maryland

D. M. Stark, DVM, PhD, DACLAM
Laboratory Animal Science Professional (Retired)
New York, New York

Mark D. Stevens, DPA
National Institutes of Health
Bethesda, Maryland

Daniel T. Stimson, PhD, JD
National Institutes of Health
Bethesda, Maryland

James G. Strake, DVM, DACLAM
Zoetis
Kalamazoo, Michigan

John Stubblefield, MSc
Institute of Marine and Environmental Technology
Baltimore, Maryland

Mark A. Suckow, DVM, DACLAM
University of Minnesota
Minneapolis, Minnesota

James R. Swearengen, DVM, DACLAM
AAALAC International
Frederick, Maryland

Douglas K. Taylor, DVM, DACLAM
Emory University
Atlanta, Georgia

James F. Taylor, DVM, MS, DACLAM
National Institutes of Health (Retired)
Derwood, Maryland

Wayne R. Thomann, DrPH
Duke University
Durham, North Carolina

Gail A. (Heidbrink) Thompson, RLATG
Peak Animal Resources+ (Retired)
Wheatland, Wyoming

Peggy T. Tinkey, DVM, DACLAM
The University of Texas MD Anderson Cancer Center
Houston, Texas

Patricia V. Turner, MS, DVM, DVSc, DACLAM, DABT, DECAWBM (WSEL)
University of Guelph
Guelph, Canada

Daniel Uhlrich, PhD
University of Wisconsin School of Medicine and Public Health
Madison, Wisconsin

Wendy Underwood, DVM, MS
Eli Lilly & Company
Indianapolis, Indiana

Fausto Vela
National Institutes of Health
Bethesda, Maryland

Ted Wasky
Envigo
Indianapolis, Indiana

James L. Weed, PhD
Centers for Disease Control and Prevention
Atlanta, Georgia

Robert H. Weichbrod, RLATG, MBA, PhD
National Institutes of Health
Bethesda, Maryland

Mark T. Whary, DVM, PhD, DACLAM
Massachusetts Institute of Technology
Cambridge, Massachusetts

Kyha D. Williams, DVM, DACLAM
Duke University
Durham, North Carolina

Marc Willis, LATG, BS, MBA
Lab Products
Decatur, Illinois

Margaret A. Wittaker
Active Environments
Lompoc, California

Irving H. Zucker, PhD, FAHA, FAPS
University of Nebraska Medical Center
Omaha, Nebraska

Section I

Introduction/Historical Overview

1

Evolution of Laboratory Animal Program Management

James F. Taylor

The use of animals for research and teaching began many hundreds of years ago, wherein animal dissection provided education and training for scientists, medical students, and physicians. Such animal use coupled with human's ownership and subsequent treatment of domesticated species eventually led to the creation of societies for the prevention of cruelty to animals; the Royal Society for the Prevention of Cruelty to Animals was the first one, created in the United Kingdom in 1824. The first national law addressing animal experimentation, the Cruelty to Animals Act, was passed in Britain in 1876.

Since that time, the presence of those societies and antivivisection and animal welfare organizations, and the passage of associated animal anticruelty laws throughout the hemispheres eventually formed the societal posture upon which laws, regulations, and standards evolved that now form the regulatory and oversight environment we now work under in our pursuit of knowledge through the humane and responsible care and use of animals in biomedical research, education, and testing.

In 1947, the Laboratory Animal Bureau was formed in the United Kingdom. The initial directors were R. E. Glover and W. Lane-Petter. They recognized that there was no standardized education system for laboratory animal care providers, and without a standard education and training program, the quality of animal care and research studies would be inconsistent and variable. The bureau organized the first of several conferences for animal care personnel on April 20, 1948, at the Royal Veterinary College of London. Subsequent conferences were held all around the United Kingdom. The organizational meeting for the new Animal Technician Association (ATA) was held on August 27, 1949 (renamed the Institute of Animal Technology [IAT] in 1965). Under the chairmanship of Dr. W. Lane-Petter, the association was ratified on March 30, 1950. The actions to establish a certification program and branches, appoint journal editors, and elect officers were also ratified.

In the meantime, in the United States, the Animal Care Panel (ACP) was formed in the late 1950s by a group of veterinarians to facilitate the sharing of information regarding the management of research animal colonies. The ACP published several species standards, and in 1963 published the *Guide for Laboratory Animal Facilities and Care* (the first *Guide*). In conjunction with that publication, revised in 1965, the ACP's Animal Facilities Certification Committee, later renamed the Animal Facilities Accreditation Board, became the American Association for Accreditation of Laboratory Animal Care (AAALAC) International. The ACP later became the American Association for Laboratory Animal Science (AALAS) in 1966. Thus, by 1965 we had in place the seminal nonregulatory entities (the *Guide*, AAALAC International, and AALAS) that continue to set the majority of the standards that are followed in the United States in pursuit of using animals in biomedical research, teaching, and testing.

Internationally, the United Kingdom passed the Medicines Act in 1968, largely in response to the thalidomide tragedy. Subsequently, the Animals (Scientific Procedures) Act (ASPA) was passed in 1986, and revised to comply with EU Directive 2010/63/E2 in 2012. The ASPA is enforced by the Home Office, which must issue specific licenses to individuals who design and perform animal procedures. Enforcement is based on establishment, project, and individual licenses; training requirements; ethical review of projects; and annual or more frequent inspections. Each establishment must also have at least one Named Animal Care and Welfare Officer (NACWO). The NACWO acts as an advocate for and provides advice on the welfare of animals being used under the project licenses.

The theft of pet dogs from backyards by animal dealers for sale to research laboratories led to startling articles in *Sports Illustrated* and *Life* magazines in 1965 and 1966, respectively, that served as the final stimulus for the U.S. Congress to vote on the Animal Welfare Act (AWA). Congressional hearings on the humane treatment of animals in experiments were first held in 1962, subsequent hearings were held in 1965 and 1966, and the legislation passed in August 1966. The AWA was the first U.S. federal law that specifically regulates the care and use of animals in research; enforcement of the AWA was assigned to the U.S. Department of Agriculture (USDA).

The original law covered dogs and cats, nonhuman primates, rabbits, guinea pigs, and hamsters held by dealers or by research facilities prior to the conduct of studies. Animal dealers delivering dogs or cats across state lines to research facilities had to be registered. Research facilities were required to register only if they received federal funding.

The 1970 amendment to the AWA expanded species coverage to all warm-blooded animals, removed the requirement for interstate transport of the animals, added the need for the appropriate use of anesthetics and tranquilizers in experiments, and added minimal standards of care for research animals. The 1976 amendment revised the standards for the transport of animals and expanded the definition of a carrier. Also, the U.S. Freedom of Information Act (FOIA), passed in 1966, received major amendments in 1974.

The next major amendment occurred in 1985. The USDA's jurisdiction was enlarged to cover animal testing, as well as animals used in teaching. Regulations were established for exercising dogs and providing for the psychological well-being of nonhuman primates, designing studies to minimize pain and distress, and requiring scientists to consider alternatives to procedures causing pain or distress. Practices that were considered to be painful were defined, and animals could not be used in more than one recoverable operative procedure; however, exceptions were presented. The amendment mandated the creation of Institutional Animal Care and Use Committees (IACUCs) and listed their roles, composition, and responsibilities. The amendment also created the Animal Welfare Information Center (AWIC).

The 1990s' amendments created a minimum holding period for dogs and cats in animal shelters destined for sale to dealers and further required dealers to provide written background information about those animals to the buyers. Space and environmental condition requirements were also revised for guinea pigs, hamsters, rabbits, horses and other farm animals, and elephants, and dealers selling pocket pets were brought under the Animal and Plant Health Inspection Service's (APHIS) jurisdiction.

Amendments since the 1990s have added housing and care requirements for marine mammals held in captivity, and have specified that only birds bred for research were to be covered under the AWA. The National Institutes of Health (NIH) adopted standards for housing chimpanzees held in federally funded sanctuaries and placed restrictions on the importation of dogs destined for research to only dogs with import permits.

USDA Animal Care (AC), a component of APHIS of the USDA, administers the AWA, and its inspectors are required to inspect registered research facilities on an annual basis. They base their findings on the Animal Welfare Regulations (AWR). Per the AWA, the secretary (USDA) shall promulgate standards (AWR) to govern the humane handling, care, treatment, and transportation of animals by research facilities (as well as exhibitors and dealers). AC also publishes the *Animal Care Policy Manual* and the *Animal Welfare Inspection Guide*, which augment and specify how the AWR shall be enforced. Inspectors make unannounced visits to research facilities and inspect and review facilities, practices, and documentation regarding all regulated aspects of animal care and use at those facilities. The results of those visits, as well as facility annual reports, are now posted on the USDA website. AC also operates the Center for Animal Welfare and the Animal Care Emergency Programs, as well as enforces the Horse Protection Act.

The U.S. Public Health Service (PHS), a component of the Department of Health and Human Services (DHHS), through the NIH, the Food and Drug Administration (FDA), and the Centers for Disease Control and Prevention (CDC), provides the largest U.S. investment in biomedical research through the award of grants and contracts. In 1971, the NIH issued a policy on the humane care of laboratory animals. That policy required institutions using warm-blooded animals in NIH-funded projects to evaluate their facilities regarding the acceptable standards for the care, use, and treatment of those animals. In 1985, the Health Research Extension Act (HREA) provided the statutory mandate for the publication of the PHS Policy on Humane Care and Use of Laboratory Animals (PHS Policy). The secretary of the Department of Health, Education, and Welfare (now the DHHS), through the director of the NIH, via the NIH Office for Protection from Research Risks (OPRR) (now the NIH Office of Laboratory Animal Welfare [OLAW]), implemented the HREA. The act requires the proper care and treatment of the research animals, directing the appointment of an IACUC to semiannually review and document the institution's program of animal care and treatment, identify any violations, and report those findings annually to the OPRR.

During that same year, in parallel with the release of the International Guiding Principles for Biomedical Research Involving Animals, the U.S. government adopted the Interagency Research Animal Committee publication "U.S. Government Principles for the Utilization and Care of Vertebrate Animals Used in Testing, Research, and Training" (U.S. Government Principles). The PHS Policy, published in 1986, endorses the U.S. Government Principles and requires institutions receiving PHS funding for animal activities to use the *Guide* (last revised by the National Research Council [NRC] of the U.S. National Academy of Sciences in 2011; *Guide for the Care and Use of Laboratory Animals* [NRC Guide]) in developing and implementing their animal care and use program. The PHS Policy describes the assurance process; defines the roles and responsibilities of the IACUC, to include animal program and scientific protocol reviews; and requires proper training and access to a health program for personnel having animal contact. It also requires compliance with applicable USDA regulations, and describes the annual reporting requirements. Compliance with PHS Policy is administered by OLAW through a self-reporting, self-correcting program, with a minimum of for-cause inspections by OLAW staff. Institutions are required to inform the OLAW of noncompliant issues, with subsequent written reports when the institution has implemented corrective actions that are designed to preclude such incidents from occurring in the future.

The *Guide for Laboratory Animal Facilities and Care*, published by the ACP in 1963, was first revised in 1965 and renamed the *Guide for the Care and Use of Laboratory Animals* (NRC Guide). The 1965 revision, and subsequent editions, was published by the Institute for Laboratory Animal Resources (ILAR) (now known as the Institute for Laboratory Animal Research), a component of the NRC of the National Academy of Sciences, in conjunction with the NIH. Pursuant to National Academy practices, panels of laboratory animal science experts were appointed to committees to create revisions to the NRC Guide. In 1985, the ILAR published the sixth edition of the *Guide*. That particular edition reflected the significant regulatory changes that were brought about by the 1985 AWR amendments and the newly published PHS Policy. The *Guide* was last revised in 2011, resulting in the eighth edition of that publication. The last two editions reflected the growing practice of the application of performance standards (in comparison with the previous use of engineering standards), as well as recognition of professional judgment by providers and users to achieve compliance with the recommendations enunciated in the *Guide*. The *Guide* has become an internationally recognized reference on the care and use of animals in biomedical research.

AAALAC International is a private, nonprofit accrediting organization that provides peer-reviewed assessments of institutional animal care and use programs. AAALAC International was founded in 1965. Accreditation by AAALAC International is voluntary and thus is not a regulated endeavor. AAALAC International began as the American Association for Accreditation of Laboratory Animal Care and became international in the mid-1990s. There are now more than 980 institutions accredited in greater than 44 countries. AAALAC International uses the NRC Guide, the Federation of Animal Science Societies' *Guide for the Care and Use of Agricultural Animals in Research and Teaching* (Ag Guide), and the European Convention for the Protection of Vertebrate Animals Used for Experimental and Other Scientific Purposes, Council of Europe (European Treaty Series 123) as its three primary standards to evaluate animal care and use programs. The PHS Policy recognizes AAALAC International accreditation as evidence that an assured institution is in compliance with the recommendations enunciated in the NRC Guide. AAALAC International performs site visits to accredited units every 3 years to reevaluate their programs and reassure their compliance with the standards, as applicable, mentioned above.

Training of research, veterinary, and animal care staff members is a requirement of the regulations and standards. The USDA AWR require training programs that, at a minimum, include instruction on humane animal care and experimentation; research and testing methods that minimize or eliminate the use of animals or limit animal pain or distress; use of the information service at the National Agricultural Library (AWIC), established under a subsection of the regulations; and methods wherein deficiencies in animal care and treatment should be reported.

The *Guide* calls for personnel to receive training or have the experience to complete the tasks for which they are responsible, as well as to regularly participate in continuing education activities. AALAS provides the AALAS Learning Library (ALL), an online educational resource with curriculum topics for technicians, veterinarians, IACUC members, and research personnel. AALAS, in conjunction with the Laboratory Animal Management Association (LAMA), has offered laboratory animal management education through the Institute for Laboratory Animal Management (ILAM) since 1992. The LAMA is an international membership organization "... dedicated to advancing the quality of management and care of laboratory animals throughout the world." LAMA provides a quarterly management publication and training sessions, and holds an annual meeting. Laboratory Animal Welfare Training Exchange (LAWTE) is a membership organization that generates training materials for use by trainers and sponsors seminars and webinars for animal care and use staff members. In addition, there are a number of college-level programs that offer 2-year associate of science degrees or 4-year bachelor of science degrees in veterinary technology. The Scientists Center for Animal Welfare (SCAW) is a membership organization that provides educational resources in support of humane animal care and use in the research environment. Those resources include conferences, workshops, publications, and training materials.

AALAS has developed certification programs for laboratory animal technicians that confer recognition of an individual's experience and knowledge: the three levels of certification include the assistant laboratory animal technician, laboratory animal technician, and laboratory animal technologist. AALAS sponsors a voluntary registry continuing education program wherein certified technicians or technologists are recognized as having stayed current in the field of animal science. AALAS also sponsors (in conjunction with the Institute for Certified Professional Managers) a certification program for managers in laboratory animal science positions, wherein they can become a Certified Manager of Animal Resources. In addition, AALAS and LAMA sponsor an on-site educational program for technicians, technologists, and managers entitled the Institute for Laboratory Animal Management. That program offers instruction in management concepts and provides communications, team building, and networking opportunities for attendees.

Internationally, the oldest (founded in 1950) education and training program offered for animal care and/or technical personnel is provided by the IAT. The IAT offers programs, certifications, and registry for several levels of technicians, technologists, and NACWO licensees. Additionally, they offer a government-recognized 4-year college education program. The IAT education and certification programs are used and recognized by several European and Scandinavian countries and India.

The American Veterinary Medical Association (AVMA) is a professional membership organization for veterinarians engaged in private, corporate, government, academic, and uniformed services practices. The AVMA is governed by a house of delegates, which includes 70 members from state, territorial, and allied veterinary organizations. Among those allied organizations is the American Society of Laboratory Animal Practitioners (ASLAP). ASLAP is the only organization that represents laboratory animal veterinarians within the AVMA. ASLAP provides support for educational activities and advocacy for veterinary involvement in humane animal care and use in research endeavors. The AVMA's American Board of Veterinary Specialties (ABVS) recognizes specialty organizations that meet ABVS standards for board certification of veterinarians (diplomates) in a specialized field of veterinary medicine. Among the 22 AVMA-recognized specialties is the American College of Laboratory Animal Medicine (ACLAM). ACLAM currently has greater than 975 active diplomates engaged in support of the responsible and appropriate care and use of animals in research. Those diplomates provide clinical and management expertise to the myriad species used in research, participate in IACUC deliberations of research protocols, and serve as subject matter experts to both institutional officials and research investigators.

In the United Kingdom, the Laboratory Animals Veterinary Association (LAVA), founded in 1963, formerly the British Laboratory Animal Veterinary Association, is made up of veterinary surgeons and students from the United Kingdom and elsewhere who are interested in laboratory animal medicine and science. Many members act as named veterinary surgeons under ASPA in a full- or part-time capacity. Vets working in academic, drug discovery, commercial, and contract research environments are represented as clinicians or in other aspects of the care and use of animals in biomedical research. LAVA promotes best practice and the dissemination of new technologies through its biannual meetings and other avenues. It represents laboratory animal veterinarians at the national and international level when needed for discussions and decision making on laboratory animal law, welfare, ethics, transport, techniques, and disease.

Internationally, the Japanese College of Laboratory Animal Medicine was formed in 1999; the European Society of Laboratory Animal Veterinarians and the European College of Laboratory Animal Medicine were formed in 1996 and 2000, respectively; and in 2006, the Korean College of Laboratory Animal Medicine was formed. Their missions and functions are similar to those described above for ASLAP and ACLAM. The International Association of Colleges of Laboratory Animal Medicine provides a communication platform for those colleges and promotes the responsible use of laboratory animals through the certification of veterinary specialists.

In 1959 in the United Kingdom, W. M. S. Russell and R. L. Burch published *The Principles of Humane Experimental Technique*. That publication arose from an initiative generated in the mid-1950s by the Universities Federation for Animal Welfare (UFAW). That book espoused the principles of replacement, reduction, and refinement (the 3Rs) as steps scientists could apply to improve the welfare of research animals.

The creation of IACUCs brought a regulatory and oversight presence down to the local (institutional) level. The IACUCs are empowered by both regulatory (AWR and PHS Policy) and standards-driven (*Guide*) requirements to proactively oversee the care and use of animals in research, testing, and education on a continuing basis. The members, to include (at a minimum) an individual not affiliated with the institution, a scientist involved in animal studies, and a veterinarian with delegated program authority, have defined responsibilities: review and approval of research protocols (prior to the conduct of studies); assessment of the training and occupational safety and health of personnel involved in animal activities; at least semiannual review of the institutions' animal care and use program components, as well as physical inspection of the animal facilities and animal activity areas where animal procedures are performed; investigation of alleged noncompliant actions by staff involved in animal activities; and communication of their activities to institutional officials. The presence of the IACUC brought oversight and implementation of compliance monitoring and enforcement to the very local level. Such IACUC oversight and review activities permitted assessment and judgment to be applied by individuals with direct familiarity and understanding of the animal activities performed by the research staff, enabling the application of professional judgment and performance standards enunciated in the *Guide*.

There are a few international organizations that have maintained prominence in global laboratory animal science.

The Council for International Organizations of Medical Sciences (CIOMS) is an international, nongovernmental, nonprofit organization established jointly by the World Health Organization (WHO) and the United Nations Educational, Scientific and Cultural Organization in 1949. Starting in the early 1980s, the CIOMS developed the International Guiding Principles for Biomedical Research Involving Animals. In 1985, the executive secretary of CIOMS stated, "In elaborating and publishing the International Guiding Principles the objective of CIOMS is not to duplicate such national regulations or voluntary codes as already exist but to provide a conceptual and ethical framework, acceptable both to the international biomedical community and to moderate animal welfare groups, for whatever regulatory measure each country or scientific body chooses to adopt in respect of the animals used for scientific purposes." At the time of their release in 1985, those principles were fully endorsed by the European Medical Research Councils and the WHO Advisory Committee on Medical Research.

The International Council for Laboratory Animal Science (ICLAS) is a scientific membership organization promoting the ethical care and use of laboratory animals. Membership categories include national (countries), scientific and union, institutional, associate, and affiliate members. ICLAS was formed in 1956 (initially known as the International Committee on Laboratory Animals; renamed ICLAS in 1979). ICLAS collects and disseminates information on laboratory animal science, fosters harmonization of policies and practices, and promotes international collaboration of animal care and use.

In 2012, an ad hoc committee cochaired by CIOMS and ICLAS revised the International Guiding Principles. In 2013, the OLAW ruled that institutions outside the United States that receive PHS funding are required to have a foreign assurance that commits the institutions to following the revised guiding principles.

The Federation of European Laboratory Animal Science Associations (FELASA) is a membership organization open to laboratory animal science organizations of Europe and beyond. FELASA has published numerous internationally recognized guidelines, recommendations, and position statements and continues to host triennial international congresses. FELASA also sponsors an accreditation program for education and training programs for individuals who perform any of the four functions (animal care, carrying out animal procedures, designing animal procedures and projects, and killing animals) outlined in EU Directive 2010/63.

While the practices and equipment related to animal care and use have become extremely sophisticated and significantly refined, there remain detractors to the animal research enterprise. Those organizations continue to voice their opposition to the use of animals in research through public demonstrations, lobbying efforts to support legislation opposing or limiting the scope of animal research, antivivisection advertisements, and attempting to use internal institutional document material (obtained through FOIA requests for federally funded organizations or state sunshine laws) to embarrass or demand regulatory investigations at institutions where noncompliant activities were alleged to have occurred. Members of a small number of these organizations conduct illegal acts of violence or terrorism against institutional facilities or individuals to demonstrate their objections to the use of animals in research. Several arrests of individuals committing acts objecting to the use of animals in research have been made, in accordance with the 2006 amendment to the 1992 U.S. Animal Enterprise Terrorism Act. Among the more active detractors are the Animal Legal Defense Fund (ALDF), the Animal Liberation Front (ALF), Cruelty Free International (previously called the British Union for the Abolition of Vivisection [BUAV]), the Humane Society of the United States (HSUS), In Defense of Animals (IDA), the National Anti-Vivisection Society (NAVS), the New England Anti-Vivisection Society (NEAVS), People for the Ethical Treatment of Animals (PETA), the Physicians Committee for Responsible Medicine (PCRM), and Stop Animal Exploitation Now (SAEN).

The relative explosion of electronic technology over the past couple decades has amplified and made possible almost instantaneous communications and data sharing among individuals and organizations. The Internet, and the associated hardware and software supporting it, has literally erased international borders and brought knowledge and social interchange to practically all the world's population. While the overwhelming majority of that interchange is positive and productive, persons or organizations that either disagree with or oppose the use of animals in research can easily reach a practically unlimited number of recipients. Posting information and opinions and/or photo and video on the Internet (or unsolicited e-mails) is a practically unhindered activity, as there are essentially no "filters" or *a priori* reviews of that information, other than establishing a log-in name and password. The social media websites enjoying the greatest popularity at present include Facebook, Google, Instagram, Twitter, and YouTube.

Conclusion

As noted above, the care and use of animals in biomedical research has a long and somewhat complex and involved history. The training and experience of animal care staff and the veterinary specialists engaged in supporting these research endeavors has grown in leaps and bounds to keep up with the explosive growth of sophistication in the technology of facilities and equipment paralleling the advancement of scientific techniques and procedures. The remaining chapters of this book will greatly expand on the management and implementation of the myriad activities carried out on behalf of the biomedical research enterprise.

Section II

Developing a Collaborative Culture of Caring

2

Culture of Care: Organizational Responsibilities

Marilyn J. Brown, Camellia Symonowicz, Leticia V. Medina, Natalie A. Bratcher, Cindy A. Buckmaster, Hilton Klein, and Lynn C. Anderson

CONTENTS

Introduction

Animal use in research has contributed significantly to advances in science and medicine, and the role of laboratory animal professionals in this process is pivotal (AALAS 2001; Medina 2008). While it is desirable to use alternatives to live animals for this process, the use of animals continues to be necessary to protect human and animal health and the environment (EU 2010). To preserve the privilege to use animals in research, a strong program of animal care and use becomes important for several reasons: regulatory compliance, quality of scientific results, addressing public sensitivities, managing staff sensitivities, and moral obligations to the animals themselves. Regulations impacting on the care and use of research animals are covered in greater detail elsewhere in this text. The most commonly referenced regulatory standards include the *Guide for the Care and Use of Laboratory Animals* (*Guide*) (recognized internationally as setting standards for animal care and use), European Union (EU) Directive 2010/63/EU, and the World Organization for Animal Health (OIE). The *Guide* states that "all who care for, use, or produce animals for research, testing or teaching must assume responsibility for their well-being," and that "both researchers and institutions have affirmative duties of humane care and use" of research animals, which is later defined as "those actions taken to ensure that laboratory animals are treated according to high ethical and scientific standards" (NRC 2011). The *Guide* further states that "it is the institution's responsibility to put into place policies, procedures, standards, organizational structure, staffing, facilities, and practices to ensure the humane care and use of laboratory animals throughout the institution" (NRC 2011). The EU Directive states that animals have intrinsic value that must be respected and that "animal welfare considerations should be given the highest priority … that each use is carefully evaluated," and that principles of replacement, reduction, and refinement (the 3Rs) should be considered systematically when using animals in research (EU 2010). The OIE, comprised of more than 170 member countries, has eight guiding principles on animal welfare outlined in its Animal Health Code. These principles also support incorporation of the 3Rs and state "that the use of animals carries with it an ethical responsibility to ensure their welfare to the greatest extent practicable" (OIE 2008).

Biomedical progress depends, fundamentally, on scientific excellence, which is dependent on quality animal care (Ad Hoc Committee to Revise the International Guiding Principles 2012; Friese 2013). The provision of excellent care also addresses some of the ethical and moral concerns of the general public regarding the use of animals in research. Animal care and use carries with it the responsibility to ensure that high ethical and scientific standards (NRC 2011) are met, and the public is reassured by knowing how much effort is expended by animal caregiving staff on behalf of the animals, to adhere to the intent as well as the scope of the laws that protect research animals (Medina 2008; EU 2010; Coleman 2011). Strong animal care and use programs also address the sensitivities of staff, who often choose careers in animal research because of their love and compassion for animals (Coleman 2011; Davies and Horst 2015).

Institutional culture influences the productivity and performance of many enterprises (Simone 2009; Ng'ang'a and Nyongesa 2012; Uddin et al. 2013), and cultures that promote caring for the animals and people supporting animal care and use programs can provide a basis for an exceptional animal care and use program. This culture, often referred to as the "culture of caring" or "culture of care," promotes compassion and respect for laboratory animals and the people who work with them. In discussing a "strong culture" at a successful large technology company, Kunda focuses on the "self-conscious and tireless celebration of the company's strong culture—one in which employees are creative, committed, entrepreneurial, independent, and moral." This involves not only employees' intellectual skills and physical presence, but also their emotions, moral sense, and personal loyalties (Kunda 2006). Care is less something to be rigidly defined than a style of thinking. It "directs attention to what was once rendered invisible within scientific research … as opposed to the calculable and controllable" (Mol et al. 2010). Davies and Horst (2015) write about the relationship between "craft" (or skills) and "care" and reflect on the potential implications of the promotion of a culture of care in a research setting. They propose a model of craft as a caring practice "which brings together skill, a focus on utility or purpose and a particular emotional orientation (care, passion and commitment)." In their analysis of numerous research labs globally, they found that "a happy group was understood as a productive one," and the strongest leaders accommodated different individuals and viewed treating people well as vital, "both because it is the right thing to do and because it is, ultimately, good for science."

A culture of care goes beyond being compliant with applicable rules and regulations and strives to meet the full intent of established rules and regulations—excellent animal welfare and reproducible scientific results. Many of the laws and guidelines surrounding animal care and use allow for the use of professional judgment (Klein and Bayne 2007). This should not be interpreted to support a minimalistic approach that just meets the letter of the law, but instead should be applied to working with animals in a manner that strives to provide the best possible care for the animals, thus producing the highest-quality scientific results (Medina 2008). A culture of care often starts with an institutional mission and value statement that clearly states the institution's commitment to the humane care and use of animals (Phanuel Kofi Darbi 2012). This mission statement frequently refers to the advancement of knowledge, the development of life-saving procedures and drugs, improving the quality of life for humans and animals, or some similar goal. The corresponding value statement, often referred to as "core values," articulates the institution's commitment to animal welfare, the humane care and use of laboratory animals, and/or the implementation of the 3Rs. Examples include

- "[Our Institution] is committed to the humane care of the research animals we produce and work with in all of our activities" (http://www.criver.com/about-us/humane-care/best-practices).
- "We are committed to reducing our reliance on animal testing methods, and promoting the development, validation and use of non-animal testing models. [The Institution] requires

that where animals have been or may be used for research or testing, that we abide by the principles of the 3Rs of animal research" (http://www.bms.com/sustainability/environmental_performance/Pages/product_stewardship.aspx).

- "[Our Institution] is committed to ensuring the humane care and use of laboratory animals in the company's research and development programs. We recognize that high quality science and humane animal care are inseparable. In addition to complying with applicable legislation and regulations, [Our Institution's] laboratory animal research programs and facilities aim to exceed regulatory agency standards" (http://www.abbvie.com/responsibility/transparency-policies/home.html#).

A culture of care usually includes

- Strong institutional commitment to provide the resources and leadership necessary, such as ongoing communication from management that reinforces the commitment to animal welfare for all institutional stakeholders (scientists, technicians, shareholders, and the public)
- Creation of an environment where staff feel empowered to come forward with any concerns or suggestions they have to improve the animal care and use program and that respects and nurtures staff compassion
- Mechanisms to support open communications on all aspects of the program
- A well-defined program of training on aspects of animal care and use, including ethics for all employees (from animal care technicians to top research scientists) and mechanisms to ensure competency
- Programs that recognize excellence in animal care and use
- Empowerment of animal welfare oversight committees, such as the Institutional Animal Care and Use Committee (IACUC), Ethics Committees (ECs), and Animal Welfare Bodies (AWBs)
- Commitment to, and proactive implementation of, the 3Rs

The productivity of any enterprise is, ultimately, dependent on the culture established to drive its success (Kunda 2006). Biomedical advances in a research culture are a significant aspect of their measure of productivity, and as a result, these advances continue to improve and save human and animal lives. There are still many unmet medical needs for both people and animals. Therapeutic discoveries are necessary to address these various diseases and disorders. The research community cannot provide the cures and treatments needed without collecting scientific data in both preclinical animal studies and human clinical trials, which both must adhere to high scientific and ethical standards as described in good laboratory practice (GLP) and good clinical practice (GCP) regulations, respectively (FDA 2001). The miracles of tomorrow depend on ongoing innovations in biomedical research progress today (Brouwers et al. 2011). Caring for research animals can present a variety of emotional challenges for research and laboratory animal professionals (AALAS 2013); however, a strong culture of care that supports the overall well-being of all the animals and people involved in biomedical discovery may drive productivity in unprecedented ways.

This chapter elaborates on the important components described above, providing examples and suggestions for how a culture of care can be incorporated into any animal care and use program, regardless of size or scientific mission.

Organizational Commitment

A commitment to animal welfare starts with the institutional leadership. Institutional leaders include officers of a company, academic administrators, managers, faculty, and any others having organizational leadership responsibilities. Leaders of programs in which animals are used must be able to articulate the commitment to animal welfare and be able to discuss it with both internal and external stakeholders. Institutional leaders should also seek opportunities to reinforce the messages supportive of a culture of care, for example, through staff meetings, "all hands" meetings, and video messaging. If institutional leaders stress the importance of animal welfare as part of the overall institutional culture they support, other employees, be they management or technicians, will understand the importance of demonstrating this commitment in their work at the institution. Institutional leaders should also convey that activities that are not in keeping with the institution's values and culture of care will not be tolerated (Albanese et al. 2015). Based on relevant animal care and use job responsibilities, it is good management practice for an employee's commitment to animal welfare to be assessed as part of his or her performance review, so the importance of this commitment is emphasized and formally recognized as part of the institutional culture (Meehan et al. 2008; Phanuel Kofi Darbi 2012).

A commitment to animal welfare should be applied to all species of animals, regardless of their intended use for research, teaching, testing, or production or inclusion in regulations and standards. This commitment is often articulated in a short value statement, mentioned earlier, which should be something that resonates with all employees (including management) involved in animal work, as well as those employees who may be involved in other aspects of the institution's work (ancillary support staff such as receptionists, administrative assistants, janitorial and grounds staff, and facilities engineers). This value statement should be something that all employees can refer to when discussing their employment with others. The message of the value statement can be reinforced by displaying it prominently in areas, such as the institutional website, institutional handbooks, and facilities where animals are housed. This display makes the institution's commitment clear not only to employees but also to visitors and regulators.

An institution can demonstrate its commitment to animal welfare and a culture of care in a variety of ways, depending on the size and scope of the institution. Some large institutions employ staff specifically dedicated to the ongoing support and enhancement of the culture of care (Albanese et al. 2015; Bratcher and Reinhard 2015). Support for committees that work to enhance environmental enrichment and normal animal behavior is another way institutions can demonstrate their commitment to animal welfare. Committees, dedicated personnel, or other initiatives that foster implementation of alternatives, also known as the 3Rs (covered later in this chapter), are also important aspects of a culture of care (James et al. 1995; Medina 2008). Strong support for training initiatives, including the recruitment and development of highly competent trainers and providing adequate time for initial training and reinforcement training, is another key aspect of institutional support. Institutional support for training should include not only internal training but also support for continuing education at professional meetings for various stakeholders, such as technicians, veterinarians, IACUC members, and managers.

The IACUC, EC, AWB, or comparable internal animal welfare oversight body can also influence the culture of care and help ensure animal welfare, sound science, implementation of the 3Rs, and regulatory compliance (Coleman 2011). As such, this committee should be strongly supported by the institution. Other ways an institution can show its support for a quality animal care program and a culture of care include involvement in professional organizations such as the American Association for Laboratory Animal Science (AALAS) or similar organizations in the institution's country, involvement in AAALAC International, and supporting internal research efforts and publications in the areas of animal welfare and the 3Rs. Support for recognition programs can also be powerful and should start at the top and include all levels of management.

Having key institutional leaders involved in recognition events is just one way of demonstrating an institutional commitment to a culture of care (Saunderson 2004). In addition, institutions should provide support for staff experiencing difficulty, which can be inherent when compassionate individuals are involved in tasks such as euthanasia (Herzog 2002). Management should ensure that staff who work with animals are given high regard in the institutional framework, to acknowledge the importance of the challenging work they do in support of ethical biomedical research. Last, since senior leaders (institutional officials, IACUC chairs, and others) play critical roles in establishing organizational culture, consideration should be given to succession planning that will help ensure future leaders who will maintain a culture of care (Valentine 2012).

There are several internationally accepted guidelines and principles that can serve as a foundation when developing or assessing an organization's culture of care. One example is the CIOMS-ICLAS International Guiding Principles (Table 2.1) (Ad Hoc Committee to Revise the International Guiding Principles 2012). Another reference is the Five Freedoms (Table 2.2) created by the Bramble Commission in the United Kingdom in 1955 (Mellor 2016).

TABLE 2.1

CIOMS-ICLAS Principles: International Guiding Principles for Biomedical Research Involving Animals, December 2012

The following principles should be used by the international scientific community to guide the responsible use of vertebrate animals in scientific and/or educational activities.

I. The advancement of scientific knowledge is important for improvement of human and animal health and welfare, conservation of the environment, and the good of society. Animals play a vital role in these scientific activities and good animal welfare is integral to achieving scientific and educational goals. Decisions regarding the welfare, care, and use of animals should be guided by scientific knowledge and professional judgment, reflect ethical and societal values, and consider the potential benefits and the impact on the well-being of the animals involved.

II. The use of animals for scientific and/or educational purposes is a privilege that carries with it moral obligations and responsibilities for institutions and individuals to ensure the welfare of these animals to the greatest extent possible. This is best achieved in an institution with a culture of care and conscience in which individuals working with animals willingly, deliberately, and consistently act in an ethical, humane and compliant way. Institutions and individuals using animals have an obligation to demonstrate respect for animals, to be responsible and accountable for their decisions and actions pertaining to animal welfare, care and use, and to ensure that the highest standards of scientific integrity prevail.

III. Animals should be used only when necessary and only when their use is scientifically and ethically justified. The principles of the Three Rs—Replacement, Reduction and Refinement—should be incorporated into the design and conduct of scientific and/or educational activities that involve animals. Scientifically sound results and avoidance of unnecessary duplication of animal-based activities are achieved through study and understanding of the scientific literature and proper experimental design. When no alternative methods, such as mathematical models, computer simulation, *in vitro* biological systems, or other nonanimal (adjunct) approaches, are available to replace the use of live animals, the minimum number of animals should be used to achieve the scientific or educational goals. Cost and convenience must not take precedence over these principles.

IV. Animals selected for the activity should be suitable for the purpose and of an appropriate species and genetic background to ensure scientific validity and reproducibility. The nutritional, microbiological, and general health status as well as the physiological and behavioral characteristics of the animals should be appropriate to the planned use as determined by scientific and veterinary medical experts and/or the scientific literature.

V. The health and welfare of animals should be primary considerations in decisions regarding the program of veterinary medical care to include animal acquisition and/or production, transportation, husbandry and management, housing, restraint, and final disposition of animals, whether euthanasia, rehoming, or release. Measures should be taken to ensure that the animals' environment and management are appropriate for the species and contribute to the animals' well-being.

VI. The welfare, care, and use of animals should be under the supervision of a veterinarian or scientist trained and experienced in the health, welfare, proper handling, and use of the species being maintained or studied. The individual or team responsible for animal welfare, care and use should be involved in the development and maintenance of all aspects of the program. Animal health and welfare should be continuously monitored and assessed with measures to ensure that indicators of potential suffering are promptly detected and managed. Appropriate veterinary care should always be available and provided as necessary by a veterinarian.

VII. Investigators should assume that procedures that would cause pain or distress in human beings cause pain or distress in animals, unless there is evidence to the contrary. Thus, there is a moral imperative to prevent or minimize stress, distress, discomfort, and pain in animals, consistent with sound scientific or veterinary medical practice. Taking into account the research and educational goals, more than momentary or minimal pain and/or distress in animals should be managed and mitigated by refinement of experimental techniques and/or appropriate sedation, analgesia, anesthesia, nonpharmacological interventions, and/or other palliative measures developed in consultation with a qualified veterinarian or scientist. Surgical or other painful procedures should not be performed on unanesthetized animals.

VIII. Endpoints and timely interventions should be established for both humane and experimental reasons. Humane endpoints and/or interventions should be established before animal use begins, should be assessed throughout the course of the study, and should be applied as early as possible to prevent, ameliorate, or minimize unnecessary and/or unintended pain and/or distress. Animals that would otherwise suffer severe or chronic pain, distress, or discomfort that cannot be relieved and is not part of the experimental design, should be removed from the study and/or euthanized using a procedure appropriate for the species and condition of the animal.

(*Continued*)

TABLE 2.1 (CONTINUED)

CIOMS-ICLAS Principles: International Guiding Principles for Biomedical Research Involving Animals, December 2012

IX. It is the responsibility of the institution to ensure that personnel responsible for the welfare, care, and use of animals are appropriately qualified and competent through training and experience for the procedures they perform. Adequate opportunities should be provided for ongoing training and education in the humane and responsible treatment of animals. Institutions also are responsible for supervision of personnel to ensure proficiency and the use of appropriate procedures.

X. While implementation of these Principles may vary from country to country according to cultural, economic, religious, and social factors, a system of animal use oversight that verifies commitment to the Principles should be implemented in each country. This system should include a mechanism for authorization (such as licensing or registering of institutions, scientist, and/or projects) and oversight which may be assessed at the institutional, regional, and/or national level. The oversight framework should encompass both ethical review of animal use as well as considerations related to animal welfare and care. It should promote a harm–benefit analysis for animal use, balancing the benefits derived from the research or educational activity with the potential for pain and/or distress experienced by the animal. Accurate records should be maintained to document a system of sound program management, research oversight, and adequate veterinary medical care.

Source: Ad Hoc Committee to Revise the International Guiding Principles, CIOMS International Guiding Principles for Biomedical Research Involving Animals, 2012, retrieved April 4, 2016, from https://grants.nih.gov/grants/olaw/Guiding_Principles_2012.pdf.

TABLE 2.2

Five Freedoms, December 1979

1. Freedom from thirst, hunger and malnutrition—by providing ready access to fresh water and a diet to maintain full health and vigour.
2. Freedom from discomfort and exposure—by providing an appropriate environment including shelter and a comfortable resting area.
3. Freedom from pain, injury or disease—by prevention or rapid diagnosis and treatment.
4. Freedom from fear and distress—by ensuring conditions and treatment which avoid mental suffering.
5. Freedom to express normal behaviour—by providing sufficient space, proper facilities and company of the animal's own kind.

Source: Mellor, D.J., *Animals (Basel)*, 6(10): (Article 59) 1–7, 2016.

Designing a Culture of Care

In addition to a succinct institutional value statement, how an institution embraces the importance of animal-based research, embodies a commitment to compassion and respect, and shares this both within the organization and outside the organization also defines its culture of care. The structure of a program helps define responsibility and expectations and can create an environment where teamwork toward a common goal—that of animal welfare and quality science—can thrive. This environment creates a supportive workplace where compassion, recognition of the importance of the human–animal bond, and respect are a focus. Caring environments support open communication about all aspects of the culture of care, and ensure that all staff are appropriately trained and competent, and that excellence is recognized and rewarded. In this way, the core values of the culture are reinforced continually, ensuring ongoing assessment of activities and supporting improvements, including opportunities for more fully implementing the principles of the 3Rs.

Key Characteristics of a Culture of Care

Structure

Many individuals contribute to the creation of a culture of care. As previously stated, it is very useful to have informed, engaged senior leadership set the tone for the commitment to animal welfare (Albanese et al. 2015). Additional key staff include managers, veterinary staff, husbandry caregivers, research technicians, scientists, and the local animal welfare oversight body (IACUC, ECs, AWBs, etc.). These positions have unique and shared responsibilities for ensuring animal welfare. Clearly understood expectations, open communication, and teamwork are key to functionality. For example, the establishment and implementation of setting humane endpoints illustrates how this can work. During review of a protocol, the animal welfare oversight body discusses the impacts of the proposed study on animal welfare, including strategies that will be used to minimize pain and/or distress. Through interaction with site veterinary staff and the scientist, plans of action are developed. Including research and husbandry technicians as part of the team in this discussion, since they are often at the forefront of identifying adverse effects on animals and also often the ones who will carry out many of the treatment and supportive care strategies, helps promote a robust plan. Team meetings, prior to the start of a study, help ensure that the plan is well understood by all who will be involved in the study. Intervention points, be they treatment, supportive care, or euthanasia, can be discussed and understood so as to avoid unnecessary delays in the implementation of the intervention plan. Committees may also be used to establish policies and procedures relating to animal welfare since different perspectives can lead to a more informed process and encourage buy-in by the various people involved in the success of the project.

Developing, growing, and fostering a culture of care program can be enhanced through committees and working groups, dedicated positions, and volunteer programs—all of which can be adapted to any research setting. Internal committees and working groups, such as alternatives committees, enrichment committees, and other welfare or project-focused working groups, bring together expertise, potentially globally, with varied influence and experience (Bratcher et al. 2012; Albanese et al. 2015). Working together as a team, across areas of primary responsibility, fosters a sense of the common goal of animal welfare. In addition to internal committees and working groups, staff can be encouraged to be involved in external consortia aimed at welfare advances (https://iqconsortium.org/initiatives/leadership-groups/3rs; https://www.nc3rs.org.uk/). Such involvement can provide dividends to an institution as it speaks well of the institution's commitment to animal welfare and may also result in staff bringing learned ideas for improvement back to the home institution.

Human–Animal Bond and Staff Empowerment

The key attributes of a culture of care are compassion and respect for both the animals and the people who work with them (Buckmaster 2012; Brown 2014; Albanese et al. 2015). The majority of caregivers in research laboratories have chosen the field because of a strong interest in animals (Chang and Hart 2002; Coleman 2011). Caregivers work closely with animals and often develop strong relationships with them. These relationships can bring joy to caregivers, referred to as compassion satisfaction (Figley and Roop 2006; Mehelich 2011; Neff 2012). These positive relationships can minimize the impacts of stressors on animals, such as alterations of the hypothalamic–pituitary–adrenal axis, cardiovascular function (Von Holst 1998; Gerber et al. 2002), and reproductive and immunological functions (Rogers et al. 1999; Bethea et al. 2008). Stress also increases experimental variability (Schapiro et al. 2000; Weed and Raber 2005). Positive interactions with caregivers can reduce abnormal behavior, increase species-appropriate behavior, and promote coping skills that help mitigate stress reactivity toward novel objects or situations (Rennie and Buchanan-Smith 2006). Engaging in positive interactions with the animals leads to increased morale and job satisfaction in caregivers, which leads to better care and improved animal well-being (Waitt et al. 2002). Positive human–animal relationships facilitate research by allowing human observers to approach the animals easily and safely, thereby facilitating daily observations and health checks (Lehman 1992). Training animals using positive reinforcement training (PRT) can create situations where the animals cooperate for common husbandry, health care, and research procedures. Although this training requires an investment in time and a level of skill on the part of the caregiver, PRT can reduce fear and stress in the animal and reduce the need for pharmacological restraint, thus reducing stress- and drug-associated variability in the data (Crockett et al. 2000; Bassett et al. 2003; Laule et al. 2003; Lambeth et al. 2006). PRT can also decrease potential animal or human injury that can result from struggling and can also decrease the technician time needed to perform husbandry and research procedures. One potential complication of the positive human–animal bond can be the development of favoritism—when one or more select, favorite animals receive extra attention and perhaps treats. This can have unintended consequences, such as weight gain or disruption of normal conspecific social relationships (Coleman 2011). Researchers and caregivers should be mindful of the impact of their interactions and how their relationships may be perceived by the animals in their care (Shyan-Norwalt 2009).

In a strong culture of care, an institution develops mechanisms to celebrate the compassion, caring, and respect their staff exhibit toward animals, and will be prepared with resources to address issues if such care creates personal conflict. This conflict can result in a condition known as compassion fatigue. This condition has been recognized in various health care fields (Carmack and Becker 1988; Baran et al. 2009). Because of the emotional investment on the part of research staff, illness and death of research animals may lead to compassion fatigue in caregivers (Coleman 2011). The "cost of caring" in the animal research setting has been recognized since the 1980s and has resulted in numerous publications and guides (Herzog 2002; Overhulse 2002; Russow 2002; Kelly 2015). The AALAS publication "The Cost of Caring" provides an introduction to this topic (AALAS 2013). It stresses that "close contact with animals affords personnel intense feelings of satisfaction in knowing they are not only providing essential needs such as food, water, and clean bedding but also affection" (AALAS 2013). It also recognizes that personnel may experience grief and mourning with the death of an animal and describes the steps that may be experienced in the grieving process (AALAS 2013).

Moral and emotional consequences of caring for animals should not be ignored or minimized (Coleman 2011). "Institutions and caregivers benefit when grief associated with the loss of animals is acknowledged and supported" (Coleman 2011). An institution can develop strategies to reduce the moral conflict that may be present in research staff. This can start by acknowledging that these moral conflicts exist. Scientists can explain the importance of their studies and the reasons for the procedures used and provide caregivers with a voice in the ethical decisions (e.g., "brown bag" lunches, staff training, and service on the IACUC) so they can feel that their opinions and moral concerns are being considered (Herzog 2002). Engaging individuals at all levels of animal use in the review process for animal activities (IACUC, ECs, and AWBs) so different perspectives are heard, and staff feel an ownership of decisions made about animal care and use, can enhance staff empowerment. Helping staff understand the nature of the research, expected clinical signs, and humane endpoints (and the rationale for those endpoints) can help them to cope with animal illness and death. Consideration should be given to allow staff to opt out of situations that cause angst, such as euthanasia of an animal to which they have become particularly attached. Having programs for rehoming animals, when euthanasia is not a necessary part of the study, is also a proactive way to help staff cope. Such programs can also utilize employees who are not caregivers to assist in socialization and other steps that must be taken before adoption, thus making such a program a high-profile aspect of an institution's animal care and use program (Bratcher 2014; Albanese et al. 2015). Individuals who cannot cope with the loss of animal life that is inherent in most animal research programs may need professional support. Access to professional grief counselors is a benefit offered by many institutions. Their guidance and support can be invaluable in individual and group settings. However, informal gatherings that allow caregivers to share their emotional experiences with each other are often just as powerful for healing. Discussing feelings about the loss of animals and celebrating their lives and contributions can also be deeply comforting and strengthen the caregivers' connections with one another (Coleman 2011; AALAS 2013). An increasing number of institutions have developed programs of remembrance or paying tribute to laboratory animals to help staff handle the emotional consequences of their compassion (James et al. 1995; Herzog 2002; Iliff 2002). Such activities may include a plaque, organized "ceremonies," or creation of a dedicated quiet place, such as a garden, where staff can go to reflect. Such activities allow staff a way to recognize the loss, as well as the important contributions animals have made to research. This may also give people an opportunity to discuss and share their feelings in a supportive environment (Lynch and Slaughter 2001; Iliff 2002; Coleman 2011; Nishikawa and Morishita 2012; Dickens 2013; Wenting 2016).

Every animal research institution should create an environment where staff feel empowered to raise concerns about animal welfare. In fact, reporting animal welfare issues should be an expectation of all employees. The *Guide* clearly states that "the institution must develop methods for reporting and investigating animal welfare concerns, and employees should be aware of the importance of and the mechanisms for reporting animal welfare concerns.... Mechanisms for reporting concerns should be posted in prominent locations.... Multiple points of contact, including senior leadership are recommended.... The process should include a mechanism to report anonymously and nondiscrimination against the concerned/reporting party" (NRC 2011).

Individuals involved in the mechanisms for reporting, such as veterinary staff, supervisors, or IACUC, EC, and AWB members, should know how to receive such concerns. They should also know how to initiate the review process, which should highlight root causes and the development of sustainable corrective and preventive actions. The last step of the review process should include a method that provides feedback to staff members sharing a concern. Without this last step, staff may become bitter and disengaged because they feel their concerns and suggestions are not being heard (Albanese et al. 2015). When concerns are raised, an opportunity is created for employees to collaborate on solutions to a situation that results in better animal welfare, better understanding among stakeholders of the issues and goals, and improved staff morale. This proactive approach can minimize the potential for reactive responses to whistle-blower concerns.

Communication

Communication begins with ensuring the visibility of the culture of care program throughout the institution. As previously noted, the value statement should be prominently displayed. Another option is to post signs in or near the animal facility that recognize the purpose of animal research. For example, if cancer research is being conducted, a sign could note, "We Are Curing Cancer Here." The concept and goals of a culture of care can be shared throughout the organization. Nonanimal users may be asked about their place of work, and having an understanding of the value placed on animal welfare may help this employee speak proudly about the organization's commitment. The institution should develop multiple mechanisms to communicate details of the culture of care both internally and externally. This is most often done via meetings, newsletters, websites, bulletin boards, and e-mail announcements. Involvement by human resources and communications professionals in preparing these communications tools is an excellent way to engage management and to increase management awareness of these issues. Posters, particularly those that have a visual feature, such as a logo, to clearly link them to the culture of care, are another way of getting the word out. Of course, the mechanisms to report concerns and suggestions for improvement and the program's commitment to recognize excellence are also key opportunities for open sharing of information.

Communication that promotes a culture of care at every level of an organization is necessary for ensuring a consistent approach to facilitating the sharing of best practices in large, multisite, and global organizations. Larger organizations may benefit from having dedicated personnel, such as a central coordinator or overarching senior management champion, for the culture of care. However, activities are often overseen locally by the IACUC, EC, and AWB (Brown 2014). It is desirable to have some type of mechanism to ensure communication between these bodies throughout the organization (often organized by the group of dedicated personnel or champion mentioned earlier). Such communication can occur through telecommunications and face-to-face meetings. These should occur on a regular basis and often enough that ideas and momentum are not lost.

An additional opportunity for enhancing communication about the culture of care is to hold a "Culture of Care Day" where researchers present the work that they are doing and how the care for the animals is integral to the success of their work. Participation by all employees in the Culture of Care Day celebration will help them gain insight for the work that is accomplished and meet the people who take care of the animals. A Culture of Care Day is also an opportunity to recognize the animals who have enabled discoveries (the idea of remembrance tributes is covered in more detail elsewhere in this chapter). An invitation to attend this celebration can be sent to all employees involved in the animal care and use program and also be extended to the facilities staff who are responsible for the animal environment and repairs. It is important for them to understand their role in ensuring optimal animal care and use. Engaging others in the culture of care program, and having consistent reminders of the program until it matures and is embedded in the work culture, will help to ensure the success of the program.

Public outreach and educational opportunities help to gain public trust in the care and use of research animals. The research and laboratory animal care staff will benefit from instruction that helps them to talk about the work that they do and the impact that it has on both human and animal life. Various organizations provide resource materials to help with public outreach efforts. Although it is not possible to provide an exhaustive list, some examples include the Foundation for Biomedical Research (http://fbresearch.org/), Americans for Medical Progress (www.amprogress.org), Understanding Animal Research (www.understandinganimalresearch.org.uk), Speaking of Research (speakingofresearch.com/), and States United for Biomedical Research (www.statesforbiomed.org/).

Training

Training, and the topics that must be covered, is mandatory in most countries (EU 2010; NRC 2011; USDA 2013) and is covered in greater detail elsewhere in this text. Therefore, this section highlights the aspects of training that directly support a culture of care. There are two major categories of training for individuals involved in an animal care and use program. The first is technical training. Ensuring that staff have adequate technical training to perform necessary tasks with animals prior to actually working with animals is paramount and mandatory for a successful program (EU 2010; NRC 2011; USDA 2013). Such training should include an assessment of each person's technical competency (NRC 1991, 2011; EU 2010; Brown et al. 2013). Refresher training and periodic confirmation of competency is also part of a strong training program. During technical training, gentle handling, careful observation, and recognition of abnormalities (both physical and behavioral) are topics that will support a culture of care. The second part of a training program should include topics central to a culture of care that impact attitudes, such as understanding why animals are used in research, the ethics of such use, and animal welfare (Brown 2013). Training must take into consideration different local cultural characteristics. This training must also include the mechanism to report concerns or suggestions for improvement in animal welfare and inform employees of their protection from retribution for coming forward with a concern. This training should be required for all employees who have animal handling responsibilities (EU 2010; NRC 2011; USDA 2013) and can be either mandatory or optional for ancillary staff who provide support for the animal care and use staff. Human resources staff should have a good understanding of the purpose of animal research and the need for the animal care and use staff to perform their jobs with technical proficiency, as well as compassion and sensitivity. Every employee is a representative of the institution in the public sphere and is thus in a position to convey the institution's strong commitment to animal welfare and a culture of care. Since the use of animals in research can be a sensitive topic, training can be provided to help them talk with their friends, family, and neighbors about the job that they do. Such training helps them better understand the issues, encouraging pride in their contributions to biomedical research.

Training on why animals are necessary in research and teaching, how we work with them, and the importance of animal welfare can be delivered easily in lectures or self-learning materials. Training that instills compassion, respect, and sensitivity may be done most effectively face-to-face, allowing staff to gain a comfort level for these discussions. Training on the availability of support programs for employees who are having problems related to the nature of their work should also be considered.

Recognition

Programs of recognition help demonstrate that an organization or group values individuals and teams who have exhibited excellence or made exceptional contributions in areas of importance to that organization or group. Recognition can be an extremely powerful employee motivational tool (McConnell 1997). Awards have been shown to encourage higher standards of performance (Shuaib et al. 2015). Recognition can be both internal and external to an organization. A program of recognition within an organization that has a strong culture of care may recognize staff who demonstrate caring by providing ideas for improvements to animal care and use, creating solutions to problems, going "above and beyond" their job description to enhance animal welfare, advancing the implementation of the 3Rs, or serving as role models in exemplifying the behaviors and characteristics consistent with a culture of care. While some employees may be motivated by tangible rewards, others may find the act of being recognized and appreciated, particularly when done in a group setting by someone high up in the organization, to be even more rewarding (Saunderson 2004). Examples of tangible awards that have been used by some research organizations include special pins or mugs, gift cards, certificates for time off, a special parking spot, certificates of achievement, and monetary awards. In addition to rewarding employees who have achieved something particularly special, such a program can also serve to motivate others who would like to be recognized "next time" (Shuaib et al. 2015).

Internal recognition programs may have an "open nominating process" wherein any employee may nominate a deserving individual. Other programs may limit who can submit a nomination to individuals in more senior positions. Some programs may have both types of nomination processes. To help foster a team approach and the idea that caring values are shared by all employees, an organization may choose to have the winners selected by an awards committee, consisting of individuals who represent various types of positions—caregivers, research technicians, scientists, veterinary staff, and others. Generally, a recognition event is held at least annually, but some organizations have found that smaller, quarterly events help keep the momentum going and keep the values of their culture front and center in everyone's thinking.

As mentioned previously, there are also external awards that recognize excellence in animal welfare, laboratory animal science and medicine, and the 3Rs. Creating a strong nomination packet for an employee or colleague, even if they are not selected for the actual award, is a form of recognition in itself. Examples of just some of the organizations having such awards include AALAS, AAALAC International, the American Veterinary Medical Association (AVMA), the National Centre for the Replacement, Refinement & Reduction of Animals in Research (NC3Rs), Charles River Laboratories, F. Hoffmann La Roche, and the World Congress on Alternatives and Animals in the Life Sciences. Information on these awards can be found on the organizations' websites.

Auditing and Continuous Improvement

An effective culture of care program requires the commitment of employees at all levels of the organization. Broadly distributed institutional policies or statements that codify the importance and priority of animal care and welfare in research are a powerful way to promote this commitment. Compliance with the relevant federal, state, and local regulatory requirements is just the starting point for a culture of care. Compliance with established industry standards, such as those described in the eighth edition of the *Guide* (NRC 2011), the *Guide for the Care and Use of Agricultural Animals in Research and Teaching* (*Ag Guide*) (FASS 2010), the European Convention for the Protection of Vertebrate Animals Used for Experimental and Other Scientific Purposes (ETS 123) (Council of Europe 1986), and the AVMA Guidelines for the Euthanasia of Animals (AVMA Panel on Euthanasia 2013), as well as clearly defined internal performance standards and operating procedures, is also central to a culture of care. In addition, there are several journals and online resources available that feature successful strategies for improving animal welfare in research settings. Some of the most popular include AALAS's *Laboratory Animal Science Professional*, the *AALAS Journal* and *Comparative Medicine*, Nature's *Lab Animal* magazine, the *ILAR Journal*, the Institute of Animal Technology's *Animal Technology and Welfare*, and NC3Rs' website. Making time to review these and related resources regularly is an excellent way to stay current on the latest developments for improving animal welfare, an important component for maintaining a strong and progressive culture of care.

Internal Audits and AAALAC Accreditation

Regular assessments or audits of the animal care and use program are a critical feature of a culture of care. Required IACUC audits can be combined with additional internal audits conducted by researchers, study directors, members of the husbandry and veterinary teams, and others that have a professional investment in the program, to measure its success. Soliciting feedback in this manner also encourages involvement and suggestions for improvement that foster alignment with the culture of care that supports established animal care and use policies. A variety of performance measures and trends can be tracked and evaluated to inform and guide programmatic adjustments. Some examples include the number and types of substantiated compliance concerns reported, rodent breeding performance, success rates for socially housed animals, number and types of abnormal animal behaviors documented, technical proficiency measures, and employee retention rates.

Programmatic reviews by external consultants or inspectors are also helpful, and some are required by federal law, for example, the Animal Welfare Act and Animal Welfare Regulations (USDA 2013). In addition, many organizations voluntarily request assessment every 3 years by AAALAC International (www.aaalac.org). AAALAC's mission is to enhance the "quality of research, teaching, and testing by promoting humane, responsible animal care and use." AAALAC site visits involve an intense and detailed programmatic review by a variety of industry experts from all over the world. Animal care and use programs are evaluated using the three standards mentioned earlier (*Guide*, *Ag Guide*, and ETS 123), and formal accreditation is only granted to programs that meet or exceed them. AAALAC has been assessing and accrediting animal care and use programs for more than 50 years and has accredited more than 950 research institutions in 41 countries. Suggestions offered by the collective wisdom and global expertise of AAALAC site visitors are unique and unparalleled. As such, triennial AAALAC reviews can be especially powerful for improving a culture of care, providing continual guidance and opportunities to raise the bar for animal care and well-being in research settings.

Reinforcement

Having a variety of mechanisms in place to reinforce the culture of care across the various stakeholder groups is helpful to continuously highlight the institutional commitment to provide excellent care for research animals. One element is to create a written "commitment to humane care and use of animals" that all employees who work with animals read and sign upon initial employment and on an ongoing basis thereafter, for example, after regularly scheduled animal welfare training (Albanese et al. 2015). In addition to reminding employees of their personal commitment to animal welfare, the use of a written pledge to the culture of care enables the IACUC, EC, AWB, and managers to reinforce the institutional commitment to uphold high standards for research animal welfare. Another way to reinforce this culture of care is for the IACUC, EC, and AWB members, including the committee chair, to attend research department and animal support staff meetings. They can then provide updates regarding the culture of care, including the need to maintain sensitivity when working with living animals and the need to respond quickly when animal welfare issues are identified. These visits from the animal welfare oversight body leadership help to reinforce one of the reasons the IACUC, EC, or AWB exists: to provide oversight to ensure that all individuals within a program do their part to maintain a high-quality animal care and use program (EU 2010; NRC 2011; USDA 2013; Brown 2014). Visual features, such as a logo that represents the culture of care, that convey the commitment to animal welfare can be placed on training materials, name badges, lanyards, awards certificates, and posters and be posted throughout the animal facilities and beyond (Brown 2003; Albanese et al. 2015). Such posters can be created internally or can be obtained from a variety of sources, such as http://www.criver.com/aboutus/humane-care/humane-care-posters, http://www.aalasfoundation.org/outreach/resources/posters, https://www.aaalac.org/commerce/bookstore.cfm, and http://oacu.od.nih.gov/posters/index.htm. Finally, creating animal welfare–focused programs that reach out beyond the research community to employees who are not involved in the animal program can greatly enhance a culture of care. An example of this is establishing a program where employees from around the institution can be trained to interact with research dogs through play, exercise, and caring attention (Bratcher et al. 2012).

The 3Rs

The 3Rs, as stated by Russell and Burch in *The Principles of Humane Experimental Technique*, are supported by laboratory animal research institutions in various ways (Russell and Burch 1959). Concepts of the 3Rs are stated in the Animal Welfare Act, the Public Health Service Policy on Humane Care and Use of Laboratory Animals, the *Guide*, the EU Directive, and various other national and regional guidelines. Institutions can further enhance their commitment by defining the expectations of the 3Rs in institutional policies and guidance documents (Bratcher and Reinhard 2015). By incorporating the 3Rs into the company culture, many other aspects of an optimal animal welfare program will fall into place. The regulations and guidelines help programs to build a foundation of animal care and use that is based on the 3Rs, but each institution can have a variety of different approaches for furthering its culture of the 3Rs based on specific program needs and resource availability. There are many approaches to promoting a culture of care. One example is the establishment of an alternatives committee that draws support from a broad base of scientists, veterinarians, animal care technicians, and the IACUC, ECs, and AWBs.

This broad-based approach to the 3Rs, with ideally one or more key leaders to coordinate efforts, ensures that champions of alternatives are empowered to advance the most ethical science and animal welfare through the development of annual goals and coordinated cross-institutional initiatives—which are a mutually beneficial and sustainable way to promote a culture of care (James et al. 1995; Bratcher and Reinhard 2015). The IACUC, ECs, and AWBs or alternatives committee can also send out a regular newsletter, for example, quarterly or semiannually, that highlights advances in the 3Rs to help reinforce the message that we are continuously seeking ways to refine, reduce, and replace animals as part of our culture of care. Hosting a journal club for the 3Rs or "lunch-n-learns" on hot topics about animal welfare can also strengthen a culture of care.

Conclusion

A strong culture of care supports the humane care of animals, which in turn supports quality research, compliance with regulatory requirements, improved public trust in the process of biomedical discovery, and our moral obligations to the staff and animals. A culture of care requires strong and visible commitment from organizational leadership, clearly defined responsibilities, and open communication fostering a team approach to animal welfare. The program recognizes and, indeed, celebrates compassion and respect for animals and the people who work with them. The organization ensures that all staff working with animals are appropriately trained and competent and has programs for recognition of excellence to encourage continued employee commitment to a culture of care. Implementation of the 3Rs of animal use is found throughout the program.

Achieving a culture of care is not the end—this program will require ongoing support and enhancement as conditions change, new scientific information is available, and new technologies arise. A culture of care is served by embracing innovation and evolution. Advances in alternatives will require the reassessment of current practices and an openness to incorporate new technologies, be they *in vitro* opportunities; new animal models, such as fish or invertebrates; or new genetically engineered models that may present their own challenges with creation and maintenance. As new information becomes available in the area of animal welfare science, an institution may need to reassess its practices of animal care and use to ensure the highest standard of animal welfare.

ACKNOWLEDGMENTS

The authors would like to acknowledge the efforts of Judy Murray (Charles River Laboratories), Tim Allen (National Agriculture Library), and Toni Albanese (Bristol-Myers Squibb) for their valuable assistance with references for this chapter.

REFERENCES

AALAS (American Association for Laboratory Animal Science). (2001). Cost of caring: Recognizing human emotions in the care of laboratory animals. Memphis, TN: AALAS.

AALAS (American Association for Laboratory Animal Science). (2013). Cost of caring: Recognizing human emotions in the care of laboratory animals. Memphis, TN: AALAS.

Ad Hoc Committee to Revise the International Guiding Principles. (2012). CIOMS International Guiding Principles for Biomedical Research Involving Animals. Retrieved April 4, 2016, from https://grants.nih.gov/grants/olaw/Guiding_Principles_2012.pdf.

Albanese, T. L., M. J. Brown, J. L. S. Lofgren, L. V. Medina and C. M. Symonowicz. (2015). *Culture of Care: Elements for Success*. Phoenix, AZ: National AALAS.

AVMA Panel on Euthanasia. (2013). *AVMA Guidelines for the Euthanasia of Animals*. 2013 ed. Schaumburg, IL: American Veterinary Medical Association.

Baran, B. E., J. A. Allen, S. G. Rogelberg, C. Spitzmuller, N. A. Digiacomo, J. B. Webb, N. T. Carter, O. L. Clark, L. A. Teeter and A. G. Walker. (2009). Euthanasia-related strain and coping strategies in animal shelter employees. *J Am Vet Med Assoc* 235(1): 83–88.

Bassett, L., H. M. Buchanan-Smith, J. McKinley and T. E. Smith. (2003). Effects of training on stress-related behavior of the common marmoset (*Callithrix jacchus*) in relation to coping with routine husbandry procedures. *J Appl Anim Welf Sci* 6(3): 221–233.

Bethea, C. L., M. L. Centeno and J. L. Cameron. (2008). Neurobiology of stress-induced reproductive dysfunction in female macaques. *Mol Neurobiol* 38(3): 199–230.

Bratcher, N., J. Cohan, T. Von Sande, S. Morgan and S. Medina. (2012). P109 enhanced dog socialization and adoption program: Improved animal welfare and improved employee morale. *JAALAS* 51(5): 680.

Bratcher, N. A. (2014). Adoption of research beagles—Considerations for creating a sussessful adoption program. Presented at MSMR Enrichment Symposium, Newton, MA.

Bratcher, N. A. and G. R. Reinhard. (2015). Creative implementation of 3Rs principles within industry programs: Beyond regulations and guidelines. *J Am Assoc Lab Anim Sci* 54(2): 133–138.

Brouwers, C.-A., A. Garrison and P. Ponce de León Baridó. (2011). Emerging biopharmaceutical companies: Ensuring a favorable environment for continued innovation. Retrieved March 31, 2016, from http://phrma.org/sites/default/files/pdf/bcg_emerging_biopharmaceutical_companies_white_paper.pdf.

Brown, M. (2013). Ethics and animal welfare. In *Laboratory Animal Welfare*, ed. K. Bayne and P. Turner. New York: Elsevier, 7–14.

Brown, M. (2014). Creating a culture of care. NC3Rs News Blog. London: National Centre for the Replacement, Refinement & Reduction of Animals in Research. Here Dr. Brown describes the culture of care at Charles River and gives practical tips for other institutions wishing to follow their lead.

Brown, M. J. (2003). Strategies for enhancing animal welfare in the commercial production of rodents. Australian & New Zealand Council for the Care of Animals in Research and Teaching. August. Retrieved March 4, 2016, from http://anzccart.org.nz/wp-content/uploads/2014/08/2003-conf-session6.pdf.

Brown, S., P. Flecknell, I. Jackson, M. Leggett, V. Robinson and M. Yates. (2013). Summary of the Brown Report: Independent Investigation into Animal Research at Imperial College London. http://brownreport.info/lay-summary/.

Buckmaster, C. (2012). Shifting the culture of lab animal care. *Lab Anim (NY)* 41(7): 205.

Carmack, B. J. and J. Becker. (1988). Staff stress. *Lab Anim* 17: 21–27.

Chang, F. T. and L. H. Hart. (2002). Human-animal bonds in the laboratory: How animal behavior affects the perspectives of caregivers. *ILAR J* 43(1): 10–18.

Coleman, K. (2011). Caring for nonhuman primates in biomedical research facilities: Scientific, moral and emotional considerations. *Am J Primatol* 73(3): 220–225.

Council of Europe. (1986). ETS 123: European Convention for the Protection of Vertebrate Animals Used for Experimental and Other Scientific Purposes. Retrieved April 4, 2016, from http://www.coe.int/en/web/conventions/full-list/-/conventions/treaty/123.

Crockett, C. M., M. Shimoji and D. M. Bowden. (2000). Behavior, appetite, and urinary cortisol responses by adult female pigtailed macaques to cage size, cage level, room change, and ketamine sedation. *Am J Primatol* 52(2): 63–80.

Davies, S. R. and M. Horst. (2015). Crafting the group: Care in research management. *Soc Stud Sci* 45(3): 371–393.

Dickens, D. (2013). Russian scientists build monument to honor lab rats. BuzzFeed. Retrieved March 4, 2016, from http://www.buzzfeed.com/donnad/russian-scientists-build-monument-to-honor-lab-rats#.jhOvwr60GL.

EU (European Union). (2010). Directive 2010/63/EU. The European Parliament and the Council of the European Union. *Off J Eur Union* L 276/233.

FASS (Federation of Animal Science Societies). (2010). *Guide for the Care and Use of Agricultural Animals in Research and Teaching.* Champaign, IL: FASS.

FDA (Food and Drug Administration). (2001). Guidance for FDA staff. In Bioresearch Monitoring Good Laboratory Practice Compliance Program 7348.808 (Nonclinical Laboratories). Silver Spring, MD: FDA. Retrieved March 31, 2016, from http://www.fda.gov/iceci/enforcementactions/bioresearchmonitoring/ucm133789.htm.

Figley, C. R. and R. G. Roop. (2006). The definition of terms and the nature of compassion. In *Compassion Fatigue in the Animal-Care Community.* Washington, DC: Humane Society Press, 7–28.

Friese, C. (2013). Realizing potential in translational medicine: The uncanny emergence of care as science. *Curr Anthropol* 54(S7): S129–S138.

Gerber, P., C. R. Schnell and G. Anzenberger. (2002). Behavioral and cardiophysiological responses of common marmosets (*Callithrix jacchus*) to social and environmental changes. *Primates* 43(3): 201–216.

Herzog, H. (2002). Ethical aspects of relationships between humans and research animals. *ILAR J* 43(1): 27–32.

Iliff, S. A. (2002). An additional "R": Remembering the animals. *ILAR J* 43(1): 38–47.

James, M. L., L. A. Mininni and L. C. Anderson. (1995). Establishment of an animal alternatives committee. *Contemp Top Lab Anim Sci* 34(3): 61–64.

Kelly, H. (2015). Overcoming compassion fatigue in the biomedical lab. *ALN Magazine*, July/August.

Klein, H. J. and K. A. Bayne. (2007). Establishing a culture of care, conscience, and responsibility: Addressing the improvement of scientific discovery and animal welfare through science-based performance standards. *ILAR J* 48(1): 3–11.

Kunda, G. (2006). *Engineering Culture: Control and Commitment in a High-Tech Corporation.* Philadelphia: Temple University Press.

Lambeth, S. P., J. Hau, J. E. Perlman, M. Martino and S. J. Schapiro. (2006). Positive reinforcement training affects hematologic and serum chemistry values in captive chimpanzees (*Pan troglodytes*). *Am J Primatol* 68(3): 245–256.

Laule, G. E., M. A. Bloomsmith and S. J. Schapiro. (2003). The use of positive reinforcement training techniques to enhance the care, management, and welfare of primates in the laboratory. *J Appl Anim Welf Sci* 6(3): 163–173.

Lehman, H. (1992). Scientist-animal bonding: Some philosophical reflections. In *The Inevitable Bond: Examining Scientist-Animal Interactions*, ed. H. Davis and D. Balfour. Cambridge: Cambridge University Press, 383–396.

Lynch, J. and B. Slaughter. (2001). Recognizing animal suffering and death in medicine. *West J Med* 175(2): 131–132.

McConnell, C. R. (1997). Employee recognition: A little oil on the troubled waters of change. *Health Care Superv* 15(4): 83–90.

Medina, L. V. (2008). Building a culture of animal welfare: Past, present and future. *ARBS Annu Rev Biomed Sci.* São Paulo: São Paulo State University, Abbott Laboratories.

Meehan, P., D. Rigby and P. Rogers. (2008). Creating and sustaining a winning culture. *Harvard Management Update.* Retrieved March 30, 2016, from https://hbr.org/2008/02/creating-and-sustaining-a-winn-1.

Mehelich, C. (2011). Compassion fatigue: Emotional burnout in the animal care field. *Bella Dog*, September.

Mellor, D. J. (2016). Moving beyond the "Five Freedoms" by Updating the "Five Provisions" and Introducing aligned "Animal Welfare Aims". *Animals (Basel)* 6(10): (Article 59) 1–7.

Mol, A. M., I. Moser and J. Pols, eds. (2010). *Care in Practice: On Tinkering in Clinics, Homes and Farms.* Bielefled, Germany: Transcript Verlag.

Neff, K. D. (2012). The science of self-compassion. In *Compassion and Wisdom in Psychotherapy*, ed. C. Germer and R. Siegel. New York: Guilford Press, 79–92.

Ng'ang'a, M. J. and W. J. Nyongesa. (2012). The impact of organisational culture on performance of educational institutions. *Int J Bus Soc Sci* 3(8): 211–217.

Nishikawa, T. and N. Morishita. (2012). Current status of memorial services for laboratory animals in Japan: A questionnaire survey. *Exp Anim* 61(2): 177–181.

NRC (National Research Council). (1991). *Education and Training in the Care and Use of Laboratory Animals: A Guide for Developing Institutional Programs.* Washington, DC: National Academy Press.

NRC (National Research Council). (2011). *Guide for the Care and Use of Laboratory Animals.* Washington, DC: National Academies Press.

OIE (Organization for Animal Health). (2008). Introduction to the recommendation for animal welfare. Paris: World Organization for Animal Health. Retrieved March 22, 2017, from http://www.oie.int/index.php?id=169&L=0&htmfile=chapitre_aw_introduction.htm.

Overhulse, K. (2002). Coping with lab animal morbidity and mortality: A trainer's role. *Lab Anim* 31(6): 39–42.

Phanuel Kofi Darbi, W. (2012). Of mission and vision statements and their potential impact on employee behavior and attitudes: The case of a public but profit-oriented tertiary institution. *Int J Bus Soc Sci* 3(14): 95–109.

Rennie, A. and H. Buchanan-Smith. (2006). Refinement of the use of non-human primates in scientific research. *Anim Welf* 15: 203–213.

Rogers, C. J., C. S. Brissette-Storkus, W. H. Chambers and J. L. Cameron. (1999). Acute stress impairs NK cell adhesion and cytotoxicity through CD2, but not LFA-1. *J Neuroimmunol* 99(2): 230–241.

Russell, W. M. S. and R. L. Burch. (1959). *The Principles of Humane Experimental Technique.* London: W. M. S. Russell.

Russow, L. M. (2002). Ethical implications of the human-animal bond in the laboratory. *ILAR J* 43(1): 33–37.

Saunderson, R. (2004). Survey findings of the effectiveness of employee recognition in the public sector. *Public Pers Manage* 3(3): 255–275.

Schapiro, S. J., P. N. Nehete, J. E. Perlman and K. J. Sastry. (2000). A comparison of cell-mediated immune responses in rhesus macaques housed singly, in pairs, or in groups. *Appl Anim Behav Sci* 68(1): 67–84.

Shuaib, W., A. M. Saeed, H. Shahid, N. Hashmi, R. Alweis, M. Ahmad and L. R. Sanchez. (2015). Award incentives to improve quality care in internal medicine. *Ir J Med Sci* 184(2): 483–486.

Shyan-Norwalt, M. R. (2009). The human-animal bond with laboratory animals. *Lab Anim* 38: 132–136.

Simone, J. V. (2009). Institutional culture. *Oncol Times* 31(5): 5–6.

Uddin, M. J., R. H. Luva and S. M. M. Hossian. (2013). Impact of organizational culture on employee performance and productivity: A case study of telecommunication sector in Bangladesh. *Int J Bus Manag* 8(2): 63–77.

USDA (U.S. Department of Agriculture). (2013). Animal Welfare Act and Animal Welfare Regulations. Title 7—Agriculture, Chapter 54—Transportation, sale, and handling of certain animals. Washington, DC: USDA.

Valentine, D. (2012). Maintaining organization culture through leadership succession planning. *Franklin Bus Law J* (3): 130.

Von Holst, D. (1998). The concept of stress and its relevance for animal behavior. In *Stress and Behavior*, ed. A. Moller, M. Milinski and P. Slater. San Diego: Academic Press, 1–109.

Waitt, C., H. M. Buchanan-Smith and K. Morris. (2002). The effects of caretaker-primate relationships on primates in the laboratory. *J Appl Anim Welf Sci* 5(4): 309–319.

Weed, J. L. and J. M. Raber. (2005). Balancing animal research with animal well-being: Establishment of goals and harmonization of approaches. *ILAR J* 46(2): 118–128.

Wenting, Z. (2016). 'We have a responsibility to be kind' to lab animals: Expert. *China Daily.* Retrieved March 4, 2016, from http://usa.chinadaily.com.cn/china/2016-01/17/content_23122942.htm.

3

Fostering Collaborative Roles and Responsibilities for Members of an IACUC or Oversight Body

Ernest D. Prentice, M. Elizabeth Blackburn, and Robert S. Dixon

CONTENTS

Introduction

One of the key components of an animal care and use program (hereafter referred to as the "Program") is the Institutional Animal Care and Use Committee (IACUC). In the United States, the establishment of an IACUC is mandated by the Public Health Service (PHS) Policy on Humane Care and Use of Laboratory Animals (Office of Laboratory Animal Welfare 2015) and the Animal Welfare Regulations, which apply to U.S. Department of Agriculture (USDA)–regulated species (APHIS 2013). The overall purpose of an IACUC is to provide assessment and oversight of the Program. The eighth edition of the *Guide for the Care and Use of Laboratory Animals* (hereafter referred to as the *Guide*) also uses the term *IACUC*. The *Guide*, which is an internationally recognized reference, describes the IACUC as being "responsible for assessment and oversight of the institution's Program, components and facilities" (NRC 2011). In the international setting, there are other equivalent oversight bodies (OBs) that serve the same function as an IACUC but have a different label. However, in order to ensure both consistency and clarity in this chapter, the term *IACUC* is used to designate any committee or OB responsible for the Program as described above. The term *institution* refers to a university, medical center, corporation, or other organization that has a Program, and *researcher* means a person who uses animals for research, testing, or teaching.

The primary purpose of this chapter is to address how best to foster collaborative roles and responsibilities for members of the IACUC, with particular emphasis on the IACUC chair, IACUC administrator, and attending veterinarian (AV). In addition, the position of the institutional official (IO) will be addressed, although the IO is usually not a member of the committee. There is, however, no intent to address in any detail the roles and responsibilities of any of the above. Indeed, much has been written about IACUC functions, such as review of protocols that propose to use animals for research, testing, and teaching (hereafter referred to as "activities" when textually appropriate); semiannual program evaluations; and postapproval monitoring. Much has also been written about the AV's responsibility for providing adequate veterinary care for animals. Therefore, the focus will instead be on a "fostering process" and desirable management skills related to enhancing the functionality of the committee within the Program.

We believe our ideas about fostering collaborative roles and responsibilities for members of IACUCs that are presented in this chapter will help an institution develop and sustain a successful Program that necessarily requires an effective committee. This is a committee that is structured and resourced and operates in a way that best serves the institution, its researchers, and the animals. Furthermore, it is our hope that both institutions and their IACUCs will recognize that while animal well-being and compliance go hand in hand, undue regulatory burden is widely viewed as a major problem (this topic is covered in Chapter 10 of this text). Clearly, a bureaucratically driven overemphasis on compliance does not help the animals, the institution, or its researchers, and is, in effect, tantamount to compliance micromanagement. The IACUC should not lose sight of its charge and obligation to ensure humane animal care and use while facilitating valuable research. This charge is reinforced by the congressional findings for the 1985 amendment of the Animal Welfare Act, where Congress stated "the use of animals is instrumental in certain research and education for advancing knowledge of cures and treatment for diseases and injuries which afflict both humans and animals" (Agriculture 2011). Indeed, the core of the Program is the committee's review and approval of worthwhile research and other activities involving animals for societal benefit.

The Program

The *Guide* provides the following insight related to the Program: "The primary oversight responsibilities in the Program rest with the IO, the AV, and the IACUC. Their roles fit in a defined organizational structure where the reporting relationships, authorities, and responsibilities of each are clearly defined and transparent. Together they establish policies and procedures, ensure regulatory compliance, monitor Program performance, and support high-quality science and humane animal use. A Program that includes these elements and establishes a balance among them has the best chance of efficiently using resources while attaining the highest standards of animal wellbeing and scientific quality" (NRC 2011). A well-defined and balanced organizational structure with detailed policies and procedures, quality assessments, and transparency is, however, insufficient without the right culture: a culture of compliance, conscience, and respect. This tripartite culture forms the ethical foundation of the Program and facilitates the creation and operation of an effective IACUC.

Establishing a Culture of Compliance, Conscience, and Respect

Every member of the IACUC, the researchers, and the personnel in the facility who provide daily care for the animals should recognize they are providing stewardship for the animals used in activities approved by the committee. Indeed, all personnel involved in the Program in any capacity should strive to work collaboratively and responsibly in the interest of the animals that are primarily used for human health advances, advances in animal health, and other important societal benefits. This is, in effect, "self-regulation," which is far more beneficial to the research community and society at large than increased federal involvement in oversight and compliance enforcement.

The *Guide* addresses the key concept of self-regulation by stating, "Both researchers and institutions have affirmative duties of humane care and use [of animals] that are supported by practical, ethical and scientific principles" (NRC 2011). Successful self-regulation, however, requires establishment of an institutional culture of compliance, conscience, and respect. This kind of culture means that everyone from the chief executive officer (CEO) on down recognizes the importance of humane animal research, testing, and teaching carried out in accordance with the highest ethical standards. Since ethics goes hand in hand with compliance, activities involving animals must also comply with all applicable requirements of the country, region, and local government. To put it simply, "everyone does the right thing because it is the right thing to do," and relationships are built on trust.

A culture based on the concept of "doing what is right" allows self-regulation to flourish, contributes to animal well-being, minimizes noncompliance, and ultimately reduces costly regulatory burden. Indeed, effective self-regulation helps minimize the ever-present possibility that more stringent animal welfare laws and regulations will be promulgated. The question, of course, that must be addressed is this: How does an institution establish a culture of compliance, conscience, and respect? The answer can be found both at the top and at the bottom of the personnel pyramid. The CEO, with proactive buy-in from the senior administration and other key employees, needs to formulate an unequivocal mission statement that incorporates the concept that the use of animals is a privilege granted by society to the institution, and the institution, in turn, grants this privilege to its researchers after review and approval by the IACUC.

The mission statement should clearly articulate and reinforce the institution's commitment to ensure that the humane care and use of animals meets all regulatory requirements, as well as the universal standards reflected in the *Guide* (NRC 2011). In other words, the ethical use of animals should be "job 1," reflecting the values of the institution. The mission statement should be directly disseminated to all employees, and also be posted in laboratories and animal care facilities where personnel can receive daily reminders of their obligations. Indeed, a compelling case can be made for requiring all researchers and animal care personnel to sign their understanding and intended compliance with the mission statement, at least initially upon employment and perhaps even yearly. This can be easily accomplished using an online educational tracking system (Blackboard 2016) or other similar systems.

The institutional culture described above must emphasize a value approach where all individuals involved in the Program are valued no matter what their assigned role. All personnel should have an open and collaborative relationship with full and proactive participation with each other in pursuit of the institution's mission. Critical elements include the clarification of expectations, information gathering, continuing education, the free flow of ideas, and respect for each team member's input and their role in ensuring the ethical use of the animals and provision of humane care. There must also be trust, as mentioned previously. Trust in the integrity of the researchers and their staff is essential.

The bottom line that should be conveyed to all personnel, and strongly reinforced by supervisors, is a message that reflects both commitment and accountability, for example, "We want you, we need you, we value you. We expect you to live up to the institution's commitment to conduct activities involving animals that are both scientifically and ethically justified. Humane care of all animals will be provided in accordance with the highest possible standards. We hold you accountable for living in a culture of compliance, conscience, and respect where actions are guided by what is right."

The IACUC

It is common knowledge that the time has long passed when researchers could freely use animals without any prior review and approval of a protocol by the IACUC or being subject to any ongoing oversight of their animal use. Indeed, one can reasonably argue that an effective Program is highly dependent on a knowledgeable, dedicated, proactive, and adequately resourced committee that acts as a prominent and empowered agent of the institution, working on behalf of both the animals and the researchers. The members of this important agent of the institution should understand their responsibilities, as well as appreciate the ethical foundation upon which the IACUC operates. Indeed, from an ethical standpoint it is not just a matter of regulatory oversight of the Program by the committee. It is also a shared responsibility and partnership with researchers and other involved personnel for the humane care and use of animals during the pursuit of scientific advancement and other societal benefits.

The responsibilities of the IACUC extend far beyond the review and approval of protocols in a partnership with researchers. For example, the IACUC should regularly evaluate the entire Program. This includes a review of the training provided to researchers and animal care personnel, which should fit the needs of the institution. The IACUC should assess the institution's disaster plan, which should be tailored to the geographical location of the institution and the related risk from events such as flooding or tornadoes. Evaluation of the effectiveness of the occupational health and safety program is a responsibility of the IACUC, as well as inspection of the institution's animal facilities. The IACUC should respond to concerns involving the care and use of the animals at the institution and ensure that changes in the Program are implemented promptly when they are needed. Finally, the IACUC should evaluate the functionality of the committee itself. Indeed, this should be an ongoing process. An IACUC that is not knowledgeable, engaged, committed, and proactive may not be operating at an optimal level.

The IACUC, which is usually appointed by the CEO or the IO under the delegated authority of the CEO, typically includes a chair; one veterinarian trained or experienced in laboratory animal science, commonly referred to as the AV; one or more scientists; one or more nonscientists; and most importantly, one or more community members who are unaffiliated with the institution. It is not unusual for medium to large institutions to have an IACUC with well over 12 members, which may include one or more clinical veterinarians, in addition to the AV. Considering the broad range of research conducted at many institutions, it is obviously advantageous to have a larger committee that collectively has a greater range of scientific expertise and experience. It may also be useful to have alternate members who can serve when the primary member is unavailable. In addition, expert consultants should be utilized when necessary to help the IACUC complete protocol reviews, provide advice regarding the resolution of problems impacting the program, and assist in identifying ways to enhance the program.

The IACUC should operate in accordance with quorum rules that require the presence of a simple majority of the membership. The utilization of alternate members, when necessary, can help alleviate quorum problems that may arise and require cancellation of a scheduled meeting or termination of a meeting in progress. The IACUC should always be cognizant of the fact that researchers cannot initiate or continue their research without approval by the committee. Therefore, the IACUC has an obligation to remember that the researchers are its customers and take steps to avoid unnecessary delays. Committee activities should be completed in a timely manner and not include bureaucratic impediments. For example, protocol review letters should not suffer from a lack of clarity, contain irrelevant questions, and require modifications that are unjustifiably inconsistent with previous reviews. In addition, researchers should be encouraged to question any requirement they do not understand or think is unreasonable.

Given the threat of various animal rights groups combined with a largely unfounded fear of regulatory inspections, such as those conducted in the United States by the USDA, it is not surprising that one of the greatest contributors to regulatory burden is the IACUC itself. For example, at some institutions, protocol submission forms have unnecessarily increased in both length and complexity, institutional policies and procedures related to animal care and use have proliferated, and inflexible engineering standards have too often become the norm. Too many IACUCs have become overly risk aversive and overzealous in their pursuit of absolute compliance. There has also been a concomitant erosion of trust in the integrity of the researcher, which is very alarming. It would seem the traditional "trust but verify" has, unfortunately, become "verify then trust." Self-regulation simply does not work without reliance on the integrity of the researchers and their staff.

Service on the IACUC should be viewed as a moral responsibility for stewardship of the animals that is shared with the institution's researchers. Since many protocols involve some degree of animal pain and discomfort, and almost all protocols result in euthanasia of the animal subjects, committee members and the researchers should balance the science with the ethics. Scientific need should not prevail if animal welfare will be compromised beyond an acceptable limit. Therefore, the goal of the IACUC in full partnership with the researcher is to ensure there is an acceptable harm–benefit relationship where the potential scientific or societal benefit of the activity outweighs, or at least balances, the harm to which animals will be subjected. A valid assessment of the harm–benefit relationship depends on the researcher submitting a carefully written protocol and the members of the IACUC engaging in a thoughtful review and follow-up discussion with the researcher as necessary. Finally, it is important for each committee member to be cognizant of their role as society's gatekeeper in order to ensure the use of animals is fully justified. Considering the fact that service on the IACUC is a labor-intensive endeavor that has both societal and institutional ramifications, each committee member should be acknowledged and valued by the institution.

The Institutional Official

The IO, as defined by the *Guide*, is "the individual who, as a representative of senior administration, bears ultimate responsibility for the Program and is responsible for resource planning and ensuring alignment of Program goals with the institution's mission" (NRC 2011). In the United States, the PHS Policy similarly defines the IO as "an individual who signs, and has the authority to sign the institution's assurance, making a commitment on behalf of the institution that requirements of this Policy will be met" (Office of Laboratory Animal Welfare 2015), while the Animal Welfare Regulations describe the IO as "the individual at a research facility who is authorized to legally commit on behalf of the research facility that the requirements of 9 CFR parts 1, 2, and 3 will be met" (APHIS 2013). At most institutions, the IO is the person to whom the IACUC reports and is also responsible for appointing its members.

Regardless of which definition of an IO is applicable for an institution, it is clear that this individual should be a senior administrator or manager, such as a vice president for research, a vice chancellor, a vice provost, or even a CEO. The IO does not have to be an experienced researcher or know all the requirements that govern the care and use of animals. Indeed, most IOs logically rely on the IACUC chair, the IACUC administrator, the AV, and of course, the committee itself. However, the IO must understand and fully support the Program, including the autonomy of the IACUC. This requires an involved IO who meets regularly with the chair, the administrator, and the AV but does not engage in micromanagement. The IO should also periodically meet with the IACUC but refrain from regularly attending meetings in order to ensure committee discussions are not inhibited.

When situations arise where the IACUC and a researcher are in an unresolved disagreement, it is incumbent upon the IO to act as an objective arbitrator in order to help resolve the problem. This can be challenging since the committee must function as an independent agent of the institution and not be subjected to inappropriate pressure from the IO, other senior administrators, or researchers. It is certainly reasonable for the IO to meet with upset or dissatisfied researchers and ask the IACUC for further justification of any given decision when there are legitimate questions or concerns. However, it is not acceptable for the IO to pressure the committee in any way to change a decision. On the other hand, it is possible that an IACUC could become unfettered to Program goals or regulatory requirements. This, in turn, could compromise achievement of the institution's mission. If this situation were to occur, it is incumbent upon the IO to take all necessary steps to resolve the problem up to and including replacement of committee members.

The IACUC Chair

The IACUC chair is responsible for ensuring that the committee, which operates under the chair's direction, fulfills all its responsibilities mandated by federal and other requirements. A successful committee is one that operates under a chair who is a strong leader of process, as well as a respected and skilled leader of people. This can be particularly difficult in an institutional environment where the researchers and other personnel are bright, creative, and independent. In general, the chair should be a person who is well respected within the organization and occupies an upper-level position within the institution, for example, a tenured full professor or senior manager. However, there are community members, such as veterinarians, retired scientists, and individuals from a variety of other professions, who successfully serve as IACUC chairs. The commonly expressed rationale for appointing a chair who has no affiliation with the institution is the absence of any real or perceived conflict of interest.

The IACUC chair should, obviously, be knowledgeable about all applicable regulations and other requirements that govern the care and use of animals. It is also advantageous for the chair to have a successful animal research or testing background. Activities involving animals often require complex protocol design and highly technical procedures. In addition, there is species variability, occupational health and safety hazards, and myriad other factors important to the committee's review. However, a background in animal research, husbandry, or other animal-related activities is certainly not an absolute prerequisite for achieving success as a chair. Some IACUCs have attorneys or other compliance professionals serving as the chair who may themselves lack scientific experience but know how to effectively use the expertise of the committee members.

Regardless of whether the IACUC chair has a scientific background or is affiliated with the institution, the chair must be able to run a thorough and efficient convened meeting of the committee. This is, undoubtedly, the most difficult task that a chair faces, and therefore it deserves particular attention. The chair must ensure that all relevant scientific and animal welfare issues are identified and discussed, and a satisfactory resolution is achieved. During committee deliberations, it is not uncommon for very contentious issues to arise that impact the science, animal welfare, or both. Indeed, it is not uncommon for heated debate to take place where committee members express strong polarized opinions. In these situations, it is imperative that the chair be able to bring the committee to a resolution that considers the opinions of all members, addresses the scientific needs of the researchers, and ensures the humane use of animals.

It is important for the chair to avoid dominating the committee or allowing other members to dominate the discussion during the meeting or at other committee functions. When an IACUC is controlled by a vocal minority, this will likely stifle participation by other committee members. This, in turn, can create a sense of disenfranchisement, which is obviously counterproductive. The chair should therefore encourage all committee members to fully engage in the process. When a viewpoint expressed by a member is irrelevant or even wrong, a skilled chair can guide the discussion so that no one feels intimidated or insulted. Diplomacy is the key to appropriately handling personality differences or lack of knowledge factors that can impact the effectiveness of the committee.

Whether an IACUC is large or small, not all committee members may feel equal. For example, the scientific members may include both well-known and highly funded researchers and junior researchers who are still striving to achieve a successful career in science. This can create a situation of scientific inequality where junior members of the IACUC may be reluctant to express views that oppose those of senior members. This perception of inequality in status may be particularly prevalent in the minds of the nonscientist member and the nonaffiliated (community) member. Members who do not have a scientific background may struggle with the complexity of protocols and other issues that are integral parts of a Program. Some of their questions could be inappropriately characterized as naïve by other committee members. However, from a voting standpoint all IACUC members are equal. Each member has the power of their vote. It is incumbent upon the chair to ensure that the aforementioned perception of inequality does not compromise anyone's participation in the voting process. No member of the IACUC should feel devalued or experience a sense of not belonging. Every member should feel free to independently cast their vote in favor of or against a motion on the table and abstain as necessary without pressure or undue influence. It should also be remembered that the IACUC is acting as society's gatekeeper. Therefore, one can argue that the nonscientist members and the nonaffiliated members are, in many ways, more important gatekeepers than the scientific members.

Finally, the IACUC chair should develop an amicable relationship with the researchers and avoid being characterized as a bureaucrat whose only role is to enforce compliance. The chair is not a policeman, and certainly the IACUC is not a police force or a punitive body. The goal of the chair and the committee should be to help the researcher both understand and apply applicable requirements in order to maintain compliance and ensure that activities involving animals are conducted in accordance with the highest ethical standards. This, in turn, will help the animals, facilitate worthwhile animal usage, and reduce the risks to the institution for failing to maintain compliance. This is not an easy goal to achieve where researchers are under intense pressure to obtain funding and publish their results in peer-reviewed journals. Failure to maintain sufficient scientific productivity may result in a researcher being unable to achieve that next promotion or even facing the possible loss of his or her position. Thus, the IACUC should work with the scientists to facilitate and improve the quality of science at the institution.

The IACUC Administrator

Medium and large-sized institutions normally provide the IACUC chair with the assistance of an administrator and other support staff. The IACUC administrator should possess an up-to-date working knowledge of all applicable regulations and institutional requirements. He or she should have the ability to maintain a clear and consistent framework while working hand in hand with the chair, the committee, and the researchers. The most important characteristics for an IACUC administrator to possess are analytical thinking skills, organizational ability, written and verbal communication proficiency, and personality traits that allow him or her to work diplomatically with strong-willed, and sometimes difficult, researchers and committee members.

The administrator should be viewed as a professional. Ideally, he or she should obtain certification such as that offered by the Public Responsibility in Medicine and Research (PRIM&R 2016). Certification validates the individual's mastery of the knowledge necessary to serve as an effective administrator, which may be valuable to the Program. It is important to stress the fact that a certified administrator is a professional who may know the regulations in greater detail than the chair or some other members of the committee. Therefore, the chair, the IACUC, and the researchers should recognize the administrator as a valued committee resource.

To best achieve a balance between the regulatory requirements and the concerns and needs of the IACUC and the researchers, we believe the administrator should remain neutral and not serve as a voting member of the committee. To this end, he or she must possess the ability to remove himself or herself from expressing strong opinions, listen to the concerns and needs of all parties (including the researcher who may not be present), help the chair apply the regulations, and suggest solutions that best meet the concerns of the committee and the needs of the researcher while ensuring regulatory compliance. Achieving this balance may well be the most important aspect of the administrator's role.

The Attending Veterinarian

The PHS Policy, USDA regulations, and the *Guide* clearly indicate that the AV is responsible for the health and well-being of all laboratory animals used in research, testing, and teaching at the institution (Agriculture 2011; APHIS 2013; Office of Laboratory Animal Welfare 2015). Fulfillment of this responsibility requires the institution to have a program of veterinary care that meets the American College of Laboratory Animal Medicine (ACLAM) Guidelines for Adequate Veterinary Care or equivalent standards (American College of Laboratory Animal Medicine 2016). The AV is, accordingly, an integral member of the IACUC who provides the necessary expertise in laboratory animal medicine and husbandry. Indeed, depending on the size of the institution, the AV may be the only member who has such expertise. In addition to having the necessary training and experience, the AV should also have a mind-set that focuses on the needs of the science combined with animal welfare, and certainly should keep abreast of rapid advances in the development and use of animal models.

It is critical for the institution to ensure that the AV has the authority and the resources to carry out the above-mentioned responsibility. The AV's judgment regarding adequate veterinary care should never be influenced by institutional interests that are contrary to the provision of such care. It is also important for researchers to recognize that the standards of veterinary care for laboratory animals continue to evolve, and it is therefore essential for the AV to work closely with them in the design and conduct of their research, testing, and teaching activities. It is equally important for members of the IACUC to rely on the expertise of the AV during the review and approval of protocols and in other committee functions, such as program and animal facility review.

The AV, as a laboratory animal veterinarian, cannot always function within the Program in the way a typical clinical veterinarian in private practice would. This, in turn, requires the AV to assume a very challenging role. On the one hand, animals enrolled as subjects in activities approved by the IACUC may experience unanticipated pain and discomfort associated with injuries or diseases that are unrelated to the activities. In this situation, the AV holds the sole authority to treat the animals, and therefore the animal must be either cared for in accordance with current veterinary standards or removed from the study if no longer eligible to participate in the protocol. Neither action requires prior approval by the committee. This authority needs to be communicated to all researchers as part of their orientation to using animals at the institution.

On the other hand, some protocols are designed to test the toxicity of a new agent; produce a disorder, such as heart failure; induce an injury, such as a brain concussion; or cause another health anomaly. In these circumstances, the protocol for scientific reasons may prohibit treatment of pain and discomfort. In such cases, the AV is no longer the sole authority or decision maker regarding pain-related issues. It is, nevertheless, the responsibility of the AV to carefully consider the scientific necessity to withhold pain-relieving agents, evaluate the potential effect on the animals, and advise the IACUC accordingly. Since the AV is only one member of the committee with one vote, ultimately the committee decides on the approvability of the protocol, regardless of whether the AV disagrees with the committee or casts a negative vote. It is therefore important for the AV to recognize that his or her role on the IACUC is different than his or her role within the Program in general. At times, these roles may create an ethical dichotomy that the AV must understand and handle appropriately.

Regardless of the nature of the protocol in question, it is imperative that the AV and all other members of the IACUC ensure that any pain and discomfort the animals may experience is minimized to the greatest extent possible, consistent with legitimate, literature-supported scientific constraints. To that end, it is particularly important for the AV and the committee to be satisfied that the protocol specifies valid experimental and humane endpoints. The experimental endpoint is reached when the study objectives have been achieved. The humane endpoint is the point at which pain will be relieved, usually by euthanasia. In consideration of the principle of refinement, in many protocols the experimental endpoint should approximate the humane endpoint as much as possible (Wallace 2000). The IACUC should refrain from approving any protocol that does not specify an early humane endpoint compatible with the scientific objectives. It is important for the researcher to consult with the AV during the planning stages to design a study that best addresses any animal welfare concerns prior to review by the committee.

Finally, it is important that the IACUC not be run by the AV, despite the fact that the AV is undoubtedly one of the more important members of the committee and, at times, the most critical member. As mentioned previously, the goal of the IACUC is to ensure that there is an acceptable harm–benefit relationship that justifies the use of animals. A valid assessment of this relationship clearly requires the expertise of the AV, who can help the committee perform the harm–benefit assessment, which can be quite challenging.

Conclusion

Institutions where animal activities involving research, testing, and teaching are conducted have an obligation to develop and maintain a Program where self-regulation without undue regulatory burden works for the optimal benefit of the institution, its researchers, and the animals. This requires a culture where trusted researchers and their staff operate on the basis of doing what is right and there is not a bureaucratically imposed and punitive emphasis on absolute compliance. Such a culture allows self-regulation to flourish, minimizes noncompliance, and ultimately, decreases the cost of the use of animals for societal purposes. The IACUC, which has primary oversight responsibilities of the Program, together with the IO and the AV, should work collaboratively within this culture to ensure regulatory compliance and the humane care and use of animals that meets the highest possible scientific and ethical standards. That is the essence of serving as society's gatekeeper. We can and should do no less.

REFERENCES

Agriculture: Transportation, sale, and handling of certain animals—Congressional findings for 1985 amendment. 2011. 7 USC §2131. https://www.gpo.gov/fdsys/pkg/USCODE-2011-title7/html/USCODE-2011-title7-chap54-sec2131.htm.

American College of Laboratory Animal Medicine. 2016. http://www.aclam.org.

APHIS (Animal and Plant Health Inspection Service), Department of Agriculture. 2013. Animal welfare: Parts 1–3. Code of Federal Regulations, Title 9: 7-123. https://www.gpo.gov/fdsys/pkg/CFR-2012-title9-vol1/pdf/CFR-2012-title9-vol1-chapI-subchapA.pdf.

Blackboard. 2016. Blackboard.com. http://www.blackboard.com/.

NRC (National Research Council). 2011. *Guide for the Care and Use of Laboratory Animals.* Washington, DC: National Academies Press.

Office of Laboratory Animal Welfare. 2015. Public Health Service policy on humane care and use of laboratory animals. Bethesda, MD: National Institutes of Health.

PRIM&R (Public Responsibility in Medicine and Research). 2016. Certified Professional IACUC Administrator. http://www.primr.org/Subpage.aspx?id=1587.

Wallace, J. Humane endpoints and cancer research. *ILAR Journal* 41, no. 2 (2000): 87–93.

4

Bioethics and Animal Use in Programs of Research, Teaching, and Testing

Richard C. Simmonds

CONTENTS

Introduction

The use of nonhuman animals (hereinafter "animals") for the benefit of humans and other animals is a contentious subject throughout many of the world's countries. Cultural morals and traditions, the status of animals in various religions, individual and cultural ethical values, and the diversity of concern for specific species (including the perceptions about the "warm and fuzzies" [e.g., dogs and cats] versus the "creepy-crawlies" [e.g., mice, rats, and reptiles]) all contribute significantly to the complexity of addressing moral (rightness or wrongness) and ethical (what ought to be) issues involved when considering the use of animals in biomedical research, teaching, and testing activities (hereinafter "biomedical activities"). Further, it is most certainly difficult, if not impossible, to define what the moral status of sentient animals (species that can feel pain and suffer), creatures that are not persons yet not mere things, should be that would be universally accepted. These factors also contribute to the complexity of the regulatory climate in the numerous countries in which there is some sort of oversight of animal use in biomedical activities (see Attachment I for examples of U.S. laws, regulations, and guidelines and the list of worldwide regulations [excluding the United States] compiled by AAALAC International) (AAALAC International 2015).

Readers of this chapter will not find a definitive answer to the question "Is it ethical or moral to use animals in biomedical activities?" There are no black or white answers, only many shades of gray. The intent of this chapter is for readers to gain a basic understanding of the philosophical, ethical, and moral basis for the various possible responses to this question. Those wishing to examine the issues involved in the use of animals in biomedical activities will find a plethora of germane literature available. Furthermore, in the interest of full disclosure, I am not a philosopher or ethicist by training (nor will you be after reading this chapter); rather, I am a veterinarian who has had an incredible variety of management positions in the specialty of laboratory animal medicine for more than 50 years and has had a personal interest in fostering the moral, ethical, and legal use of animals in biomedical activities. My personal position regarding the use of animals in biomedical activities may best be described as welfarist or utilitarian.

Implementation of ethical programs for the care and use of animals in biomedical activities requires buy-in from all members of an institution, from the highest-ranking administrators to the animal care and research staffs. The achievement of an ethical, legally compliant, and high-quality animal care and use program is in the self-interest of the institution, as well as those individuals who care for and use the animals. High-quality animal care and use will maximize the quality of the research results, which, in turn, will maximize the prestige of the institution, reputation of the animal users, and likely job satisfaction for the animal care and use staffs.

A tangential usefulness of the material presented herein is that institutional officials and managers of animal care and use programs who understand the origins and foundations for public perceptions regarding the use of animals in biomedical activities will likely be better able to organize their programs to minimize the possibilities for criticism and maximize their ability to respond in a positive manner to any such criticisms.

Finally, all individuals involved in biomedical activities should, at least, believe in their own minds that what they are doing meets their own sense of what is right, moral, and ethical, and it is hoped that the material presented in this chapter may be of help in their ability to make this determination.

Relevant Historical Philosophical Approaches Regarding Animal Use

Philosophical Concepts

The foundation for the philosophy of animal rights may be said to have been laid down by the Greek philosopher and mathematician Pythagoras of Samos circa mid-fifth century BC. Quotes attributed to him include

- "Animals share with us the privilege of having a soul. Alas, what wickedness to swallow flesh into our own flesh, to fatten our greedy bodies by cramming in other bodies, to have one living creature fed by the death of another?"

- "As long as man continues to be the ruthless destroyer of lower living beings he will never know health or peace."
- "For as long as men massacre animals, they will kill each other" (Pythagoras of Samos 2015).

A somewhat contrasting view regarding animals having souls was expressed by the thirteenth-century Saint Thomas Aquinas, as he was reported to believe that animals did not have souls and thus were not worthy of moral concern, but that cruelty to animals was unacceptable because cruel behavior toward animals would progress to cruelty to other humans (Aquinas 1956).

From Pythagoras's time until about the time of the Industrial Revolution (ca. late eighteenth century through the mid-nineteenth century), public interest in animal welfare, let alone animal rights, was largely nonexistent, as the average person throughout the world was more interested in personal survival. Two examples of the emerging interest in animal welfare were

1. The establishment of the first regulations pertaining to the welfare of farm animals in America found in the Massachusetts Body of Liberties issued in 1641, which contained the following regarding animal care:
   a. "Article 92. No man shall exercise any Tirranny or Crueltie towards any bruite Creature which are usuallie kept for man's use."
   b. "Article 93. If any man shall have occasion to leade or drive Cattel from place to place that is far of, so that they be weary, or hungry, or fall sick, or lambe, It shall be lawful to rest or refresh them, for competant time, in any open place that is not Corne, meadow, or inclosed for some peculiar use" (Massachusetts Body of Liberties 1641).
2. During the mid-nineteenth century, the emerging animal rights and welfare movement in England caught the attention of Queen Victoria and led to her recognizing the efforts of the Society for the Prevention of Cruelty to Animals (SPCA) (organized in 1824) to foster animal welfare and bestowing upon them the authority to add "Royal" to their name, indicating her support for their efforts (RSPCA) (Royal Society for the Prevention of Cruelty to Animals 2015).

Returning to the subject of whether animals have souls, the seventeenth-century French philosopher and mathematician Rene Descartes postulated that animals did not have souls; therefore, animals were not subjects of moral concern. In simplistic terms, Cartesian philosophy would allow humans to do whatever they wish with and to animals, without concern for any possibility of the animals suffering therefrom. Without souls, Descartes believed that animals could be viewed as being nothing more than equivalent to machines, and any vocalizations that an observer could possibly interpret as the animals being in pain should be of no more moral concern than a squeaky wheel on a wagon (Rene Descartes 2015). While the Cartesian philosophy might have been the philosophy of choice by biomedical scientists conducting surgical research with animals before 1846, when the first successful use of ether as an anesthetic was demonstrated (Robinson and Toledo 2012), it is unlikely that contemporary biomedical scientists would consider themselves as accepting of this philosophy.

Another ethical position, utilitarianism, was emphasized by eighteenth- and nineteenth-century English philosopher Jeremy Bentham, who published the following:

> The day may come when the rest of the animal creation may acquire those rights which never could have been withholden from them by the hand of tyranny. The French have already discovered that the blackness of the skin is no reason why a human being should be abandoned without redress to the caprice of a tormentor. It may one day come to be recognized that the number of legs, the villosity of the skin, or the termination of the os sacrum are reasons equally insufficient for abandoning a sensitive being to the same fate. What else is it that should trace the insuperable line? Is it the faculty of reason, or perhaps the faculty of discourse? But a full-grown horse or dog is beyond comparison a more rational, as well as a more conversable animal, than an infant of a day or week or even a month, old. But suppose they were otherwise, what would it avail? *The question is not, Can they reason? nor Can they talk? but, Can they suffer?* (emphasis added)

This quote is given in its entirety to provide context. The last three phrases in the paragraph are frequently quoted by animal rights advocates and are generally considered to be the modern foundation for the philosophy of animal rights, as it effectively introduces suffering or, more accurately, sentience, that is, having the capacity to feel pain and suffer, into the moral and ethical equation. Simply, sentient animals should have the right not to suffer (Bentham 2007).

Even though Bentham's quote would seem to imply that he anticipated the contemporary animal rights movement, and certainly opposed gratuitous mistreatment of animals, he did not oppose the use of animals by humans, including the use of animals in biomedical activities. In a letter to the editor of the *Morning Chronicle* (March 4, 1825), he stated, "I never have seen, nor ever can see, any objection to the putting of dogs and other inferior animals to pain, in the way of medical experiment, when that experiment has a determinate object, beneficial to mankind, accompanied with a fair prospect of the accomplishment of it. But I have a decided and insuperable objection to the putting of them to pain without any such view." This stresses the foundation of the utilitarian view that weighs benefits and costs and focuses on the overall good an activity provides. Also, even though Bentham's quote in the preceding paragraph is often quoted by vegetarianism advocates, there does not seem to be any reference to him being a vegetarian himself.

Philosopher Peter Singer, another utilitarian often quoted by animal rights activists, defines *speciesism* as "a prejudice or attitude of bias in favor of the interests of members of one's own species against those of members of other species" (Singer 1975). In "A Philosophical Self-Portrait," he elaborates that speciesism is an "essential philosophical view ... [that] is simple but revolutionary. Species is, in itself, as irrelevant to moral status as race or sex. Hence all beings with interests are entitled to equal consideration: that is, we should not give their interests any less consideration than we give to the similar interests of members of our own species. Taken seriously, this conclusion requires radical changes in almost every interaction we have with animals, including our diet, our economy, and our relations with the natural environment." He goes on to indicate that his "ethical position is a form of preference-utilitarianism" (Singer 1997).

Since the philosophy of *utilitarianism* and the "ethical position of preference-utilitarianism" have been introduced at this point, it is worthwhile to provide an abbreviated explanation of these terms here. Utilitarianism is a form of philosophical thought that the "consideration of right conduct should be the usefulness of its consequences; specifically: a theory that the aim of action should be the largest possible balance of pleasure over pain or the greatest happiness of the greatest number" or "the belief that a morally good action is one that helps the greatest number of people" (Merriam-Webster Online Dictionary 2015). *Preference utilitarianism* may be defined as the equal consideration of the preferences of all individuals (both humans and animals) involved (my condensation of various definitions found). A more detailed discussion of the philosophy of utilitarianism is included in a separate section below.

Representing the animal rights view, philosopher Bernard Rollin makes the case that there are no "morally relevant" reasons for excluding animals as "objects of moral concern"; thus, by extension, they are deserving of "rights." In his book titled *Animal Rights and Human Morality*, he states, "From a strictly philosophical point of view, I think that we must draw a startling conclusion: If a certain sort of research on human beings is considered to be immoral, a *prima facie* case exists for saying that such research is immoral when conducted on animals." It seems to me that if one were to accept this point of view, all research on animals would be immoral, and thus unethical, since the research subjects could not give informed consent (although, possibly, a court appointed "guardian" for the animal subjects could give consent and owners of animals could give consent for their animals to participate in clinical therapeutic trials). Further in the same paragraph, he states that this "criterion would effectively curtail the vast majority of research" in basic biological and applied basic biomedical research, the development of drugs and therapeutic chemicals and biologics, the testing of consumer goods, the use of animals in education (demonstration, dissection, surgery practice, high school science projects, etc.), and the extraction of products from animals (e.g., serum, musk, and blood). He does appear to accept the premise that animals can be used in biomedical activities under the same conditions that permit the use of human subjects; for example, the specific animals used will potentially benefit from the specific study (Rollin 1981). In a more recent report, Professor Rollin seems to make the case that there are no moral grounds to support the use of sentient animals in most, if not all, biomedical activities (Rollin 2012).

Philosopher David DeGrazia argues in his book titled *Taking Animals Seriously: Mental Life and Moral Status* that "equal consideration for animals is more reasonable than its denial, given the failure of opponents of equal consideration to meet their burden of proof ... those who doubt that animals should be extended equal consideration should note that a commitment to giving animals serious consideration would be enough to support most of the foregoing conclusions" (referring to the cases made in the previous content of the book). Dr. DeGrazia sums up his position by providing the reader with 15 ethical conclusions regarding our treatment of animals, which include (the verbiage in parentheses is my suggested examples) "Don't cause unnecessary harm" (e.g., meat eating is unnecessary for individuals who have viable alternative food options), "Make every reasonable effort not to provide financial support for institutions that cause or support unnecessary harm" (e.g., don't support circuses or movies that include animal acts), "Don't kill sentient animals unnecessarily" (e.g., for food or for most research), and "There is a presumption against disabling sentient animals (that is, damaging their ability to function in ways that significantly interfere with their ability to live a good life) and, if they are non-dangerous, the presumption is virtually absolute" (e.g., research involving any animal impairment would be unacceptable) (DeGrazia 1996).

From a more moral–legal perspective, attorney and professor of law Gary L. Francione makes the case that only by incrementally destroying the legal concept of animals as property can true animal rights ever be achieved. Late in his book titled *Rain without Thunder: The Ideology of the Animal Rights Movement*, he states, "In a sense, we are really only talking about one right—the right not to be treated as property.... As long as animals are regarded as property, we cannot really talk about their rights. That property cannot have rights follows from what it is to be property.... We can be responsible for property, but not to property" (Francione 1996). The content of this 1996 book explains in detail what was meant by the statements in the "Point/Counterpoint" article cowritten by Professor Francione and Dr. Regan, discussed below.

As this chapter is being written, there are at least two current efforts attempting to reclassify animals from merely "property" to "sentient beings," one in New Zealand (Library of Congress 2015) and another in the National Assembly in Quebec City, Canada (Anonymous 2015c), as well as continuing efforts in the United States to obtain "personhood" status for chimpanzees (Slate Website 2015). Interestingly, in a blog post copyrighted in 2006, Professor Francione made the case that the campaign to gain personhood for chimpanzees is "problematic because [it] suggest[s] that a certain species of nonhuman is 'special' based on similarity to humans. That does not challenge the speciesist hierarchy—it reinforces it—in at least two ways" (Francione 2006). This position is completely consistent with his contention in his book.

Melding together the ethical and legal points, Dr. Regan and Professor Francione published an article in which they state, "Many animal advocates hold that there really is no difference between animal welfare and animal rights. Others claim that while there is a difference, advancing animal welfare is a necessary prerequisite to advancing animal rights. Given either assumption, many conscientious activists conclude that we must support welfarist means in our march toward animal rights ends. ... We believe these views are mistaken. *Not only are the philosophies of animal rights and animal welfare separated by irreconcilable differences*, and not only are the practical reforms grounded in animal welfare morally at odds with those sanctioned by the philosophy of animal rights, but *also the enactment of animal welfare measures actually impedes the achievement of animal rights.* ... The goal of the animal rights movement is nothing less than the total liberation of nonhuman animals from human tyranny" (Regan and Francione 1992, emphasis added). By the use of the phrase "irreconcilable differences," it would seem to me these authors are clearly saying that there is no philosophical way to justify the use of animals in any way that benefits humans and, presumably, other animals. It also seems that they are saying that the enactment of animal welfare laws will impede the ultimate goal of no animal use, presumably because if the general public is convinced that animal "welfare" is achieved, they will not support the more radical implementation of animal rights as envisioned by strident animal rights advocates.

Philosopher Tom Regan also makes the case that animals are "moral patients" deserving of rights. It appears that he bases his argument on an expansion of Jeremy Bentham's "Can they suffer?" test. Moral "agents" have the capacity to speak for themselves, for example, a fully mentally competent adult human.

Moral patients are members of a moral community that cannot speak for themselves and must have someone speak for them, for example, an adult human in a coma or a newborn infant. Regan advances the proposition that sentient animals are deserving of moral "concern" and, by virtue of this fact alone, are deserving of being considered moral patients, leading to the necessity for someone to speak for them and, presumably, protect them from use in biomedical activities (Regan 1983).

Providing a slight contrast to the modern animal rights viewpoint, philosopher Michael Allen Fox states, "The conclusion [that] ... so far as it pertains to animals ... [their lack of the] various degrees the possession of capacities on which moral autonomy or agency depends, animals fail to meet the conditions specified for full membership in the moral community and likewise fail to qualify for having rights" (Fox 1986). Interestingly, approximately a year after his book was published, Fox published a retraction and indicated that he had changed his mind and no longer supported the case that he made in the book (Fox 1987). Regardless of Fox's recanting his arguments for supporting the use of animals in biomedical activities, what he presented in the book in support of the quote cited above would seem to be valid and would support the necessary and ethical use of animals in biomedical activities, at least in activities with a high benefit-to-risk ranking.

In making the case for animals having moral rights, Professor Jerry Tannenbaum has postulated a number of "myths of the animal rights movement" that are germane to our discussion (Tannenbaum 1995):

- Myth 1: "One must choose between animal rights and animal welfare." He discusses the reasons that aspects of concern for animal welfare might support the conclusion that animals have moral rights to certain levels of care and to limitations on how they may be used or treated; thus, there is nothing to support the position that animals do not have rights. He goes on to make the case that animals have *moral* rights that devolve from the fact that they are sentient and have interests. Since the intended audience for his book, titled *Veterinary Ethics: Animal Welfare, Client Relations, Competition and Collegiality*, is veterinarians, Professor Tannenbaum goes on to make the case that veterinarians (and by extension, those of us involved in biomedical studies with animals?) should accept his premise that animals have moral rights so that they are not marginalized in the public arena of debate over animal rights. That said, one could make the case that these terms are not synonymous; that is, animal welfare is ensuring responsible care and use of animals however they are being used (in biomedical activities, as pets or service animals, in agriculture, in zoos and aquatic parks, etc.), while animal rights, as defined by the strident animal rights movement, would require "nothing less than the total liberation of nonhuman animals from human tyranny" (Regan and Francione 1992).
- Myth 2: "Animals do not yet have legal rights." He makes the case that the various statutes that exist to protect animals from cruel treatment can be considered sufficient to conclude that animals should be considered as having legal rights. In this regard, he also states, "If one believes, as many people surely do, that animals should have some legal rights, one is not thereby committed to the demand of the animal rights movement that animals should be given legal standing to sue their owners, their veterinarians, or other people for money or other kinds of relief. The concept of legal rights for animals, like the concept of moral rights for animals, does not entail the platform of the animal rights movement." While Professor Tannenbaum makes a compelling case here, another possible case could be made that animal welfare statutes do not convey to animals any rights; rather, they simply codify human obligations to animals similarly to many of the statutes that govern the protection of endangered species of flora (certainly such protected plants would not be considered as having rights).

Presenting yet another viewpoint on the use of animals in biomedical activities, Sir William Paton, an eminent British pharmacologist, makes the case that the use of animals in biomedical activities to advance knowledge in medicine [both human and veterinary], biology, and the basic sciences is the *most* ethical use of animals of all the ways humans use animals (emphasis added). His reasoning is that any knowledge gained from such studies would live on in perpetuity, while, the benefits, whatever there are of using animals for other purposes, are time limited (Paton 1984).

Utilitarianism

As noted previously, utilitarianism is a form of philosophical thought that "the determining consideration of right conduct should be the usefulness of its consequences; *specifically*: a theory that the aim of any action should be the largest possible balance of pleasure over pain or the greatest happiness of the greatest number" or "the belief that a morally good action is one that helps the greatest number of people [or animals?]" (Merriam-Webster Online Dictionary 2015).

While it appears that animal rights has emerged as a recognized system of philosophical thought regarding the status of animals, utilitarianism is most likely a philosophy that would be a better fit for most people's views. In the "Summary and Conclusion" of the 1986 Office of Technology assessment report about the use of animals in biomedical activities, it is stated, "Because it extends the scope of moral concern to animals without committing itself to a vulnerable theory of animal rights, Utilitarianism has become the theory of choice among those who would press for more constraints on humans' treatment of animals" (U.S. Congress 1986).

This report from a government committee composed of representatives from many stakeholder perspectives is one of the few attempts to present an unbiased assessment of the pros and cons, or need and value, of the use of animals in biomedical activities—thus the conclusion that the most appropriate philosophical approach to such use is utilitarianism, simplistically, the greatest good for the most individuals. That said, even this approach is fraught with problems since it is unclear who or what should be considered within the term *individuals*, and animals cannot speak for themselves and cannot articulate an "informed decision" to participate as a subject in biomedical activities. Besides, when considering the "utility" of any specific proposed biomedical activity, further questions are evoked, including, what criteria will be used to balance the "good" versus "bad" of the activity? Unfortunately, there are no universally accepted guidelines for how to calculate the utility of any particular use of animals (see also the "Risk (Harm) versus Benefit Assessment" section that follows later).

Another source, philosopher Carl Cohen, says that "utilitarianism appeals to many people—it is practical and concrete and seems to make sense in daily life. Utilitarianism does not say using animals for research is wrong; what it does say is that to decide on the moral rightness of an action you need to look at whether that research might promote an aggregate good for a greater number of people than not doing the research. Some would include animals in this equation since animals do benefit from research. For the Abolitionist, animal research would be wrong since it is morally wrong to use an animal merely as a means, even … to a good end" (Cohen 1986).

Morally Relevant Difference between Humans and Animals

One can make the claim that in order for a living entity to claim rights, it must be a member of a moral community (North Carolina State University 2015), that is, a community in which individual members can make moral decisions and recognize right from wrong. Also, with rights come obligations that members of the community must be willing to accept, that is, obligations to recognize the rights of others. It is unlikely that animals can either recognize obligations to other animals or knowingly accept them. If animals had rights, they would have them even if the human species did not exist, and any prey animal would be able to exercise its right not to be eaten by a predator (an obviously absurd situation). However, to the best of our knowledge, only humans possess the intellectual capacity to make moral decisions and accept reciprocal obligations; therefore, the logical conclusion is that only humans have rights. An objection to this point of view is that there are individual humans who are not able or competent to make moral decisions (e.g., infants, persons in a coma, and persons with dementia); however, an argument can be made that such persons either have the potential to make claims for their rights (e.g., the infant who grows up to be a mentally competent adult) or previously had such capabilities but no longer do (e.g., the person in an irreversible coma). Thus, all persons should be considered members of the moral community.

That said, and even if this conclusion is correct, the fact that only humans have rights does not give us the unhindered right to do whatever we want to, or with, animals. Since all humans are members of a moral community, mostly capable of recognizing right from wrong and accepting the obligations that come with rights, and it is generally recognized that it is wrong to mistreat animals, we are thus obligated to treat animals humanely. However, this obligation is to us, to our own humanity, and not to the animals per se, although one could say that this obligation also imposes on humans a moral duty to not mistreat animals.

While animals may not have rights, nations and states can and do confer on them legal protections and impose legal restrictions on how they may be used and treated, for example, animal abuse and control laws and international wildlife protection treaties. These legal protections may severely restrict how we may use animals and what we can do to animals. Furthermore, enforcement of these protections may even result in severe penalties for violating the laws. The laws do not convey rights to animals; rather, they simply codify our moral obligations to animals as defined by our various societies and cultures. While there are a number of highly respected philosophers who have made the case that animals have moral rights based on their sentience, one can still contend that in order to have rights, individuals need to be members of a moral community, and no animals currently share this status with humans.

Our body of laws pertaining to animal protection form the foundation for animal welfare within our society. Most importantly in this regard is the generally high degree of public concern for animal welfare that is found throughout most contemporary societies and cultures, including the vast majority of individuals working in various biomedical activities. We should keep in mind that animal welfare as a concept is not synonymous with the philosophy of animal rights, especially as this latter term is defined by the most radical animal rightists. An animal welfarist believes in responsible pet ownership, responsible animal agriculture, humane animal care, and so forth. A true animal rightist demands no human use of animals, including but not limited to pet ownership and agriculture, and no animal care (since animals should not be property requiring care). In the utopian world of the true animal rightist, no animal will be subjugated to human welfare, need, or desire.

One argument made by proponents of rights for animals is that there are no morally relevant differences between a mature animal (see discussion on Regan above), with its full mental faculties, and a severely mentally deficient human, for example, a brain-dead person. However, a case can be made that there is, in fact, at least one significant, morally relevant difference between humans and animals, and that is morality itself. So far as we know and can determine at present, the human species is the only species that attempts to interact with its environment (e.g., protect the rain forest) and other species in some moral way (e.g., save endangered species). Obviously, there may be some variability from culture to culture with regard to the definition of what actions are and are not moral. That said, we are the only species that has ever worried about the fate of other species. It would seem to me that this characteristic of humans in general should certainly be considered a morally relevant difference, thus supporting the conclusion that only humans have rights.

Guidelines and Principles

The 3Rs

Replacement, reduction, and refinement (the 3Rs), proposed by Russell and Burch in their 1959 book titled *The Principles of Humane Experimental Technique*, have become foundational planks for many laws, regulations, guidelines, and local policies governing oversight of biomedical activities involving the use of animals throughout the world, and should continue to be such into the future. While the widespread implementation of these principles most certainly has resulted in more humane animal use, and in some reduction in the numbers of animals used (especially "higher" vertebrate mammals), it is unlikely that the total numbers of individual animals have decreased significantly when all species are counted. The growing importance of zebrafish and genetically modified rodents in biomedical activities may actually lead to an increase in the absolute number of animals being used in biomedical activities in the future.

- *Replacement*: In the original 1959 book, this term was meant to mean actual replacement of animals with "insentient material" (i.e., absolute replacement). Over time, the meaning has evolved to include the substitution of an animal with another animal species "lower" on the phylogenic scale (i.e., relative replacement).

- *Reduction*: Originally defined to mean "reduction in the numbers of animals used to obtain information of a given amount and precision." Some of the ways to implement the reduction of animals are to ensure that proposed activities are well designed and statistically valid, that one is using the best animal model available, that the animals are in the best health possible (including free of confounding microorganisms), and that the animals receive the best care possible to minimize induced variables that can negatively affect the animals, resulting in the need for more animals to obtain statistically valid study results. Interestingly, this may actually require the use of more animals in a given activity if it were needed to ensure statistical validity of the activity, thus negating the need for repetitive studies, which would ultimately reduce the numbers needed to obtain valid results.
- *Refinement*: Originally defined to mean "any decrease in the incidence or severity of inhumane procedures applied to those animals which still need to be used." Today, this R includes using procedures that will minimize pain, discomfort, and distress. One concept for implementing refinement is especially promising: whenever possible, make the animals willing members of the research team. For example, nonhuman primates can be taught to present an arm for a blood draw in return for receiving a desired food treat.

  Another example of refinement is to select or train dogs through positive reinforcement to voluntarily run on a treadmill if required in a project studying cardiovascular function. Unfortunately, this approach is generally reserved for use in long-term studies, as it requires time and enhanced human interaction. Positive human interaction can also improve animal welfare but will likely increase human attachment to the animals. This should be encouraged, as it will benefit the animal; however, for many studies the animals must be euthanatized at the end. In respect for human caretakers, procedures should be in place to minimize the emotional impact on the people who have become attached to the animals, perhaps by having other individuals not so attached conduct the euthanasia and collection procedures. Other, short-term examples of refinement include the use of multimodal analgesia for procedures that historically use no or only one analgesic, or use of nonpharmaceutical therapies (e.g., environmental enrichment, heating pad, and soft bedding) if painful procedures are being performed and analgesics must be withheld for scientific purposes.

  Tannenbaum and Bennett (2015a) have recently published in the *Journal of the American Association for Laboratory Animal Science* an excellent and exhaustive discussion about the 3Rs, including their original meaning and how the meanings and intent have evolved since first proposed. In addition, a letter to the editor, authored by Carbone (2015), along with a rebuttal by Tannenbaum and Bennett (2015b), further illuminates some of the contemporary aspects of the 3Rs. These three publications are well worth reading for a more thorough understanding of the importance of the 3Rs with regard to contemporary efforts to establish ethical programs of animal care and use in biomedical activities.

In Europe, "Directive 2010/63/EU revising Directive 86/609/EEC on the protection of animals used for scientific purposes ... is firmly based on the principle of the Three Rs, to replace, reduce and refine the use of animals used for scientific purposes" (European Directive 2010/63/EU).

U.S. Government Principles for the Utilization and Care of Vertebrate Animals Used in Testing, Research, and Training

Although intended to be applicable to biomedical activities accomplished with funds provided by the U.S. government, these principles would likely be just as applicable to all biomedical activities involving animals conducted anywhere in the world.

> The development of knowledge necessary for the improvement of the health and well-being of humans as well as other animals requires *in vivo* experimentation with a wide variety of animal species. Whenever U.S. Government agencies develop requirements for testing, research, or training procedures involving the use of vertebrate animals, the following principles shall

be considered; and whenever these agencies actually perform or sponsor such procedures, the responsible Institutional Official shall ensure that these principles are adhered to:

I. The transportation, care, and use of animals should be in accordance with the Animal Welfare Act (7 U.S.C. 2131 et. seq.) and other applicable Federal laws, guidelines, and policies.

II. Procedures involving animals should be designed and performed with due consideration of their relevance to human or animal health, the advancement of knowledge, or the good of society.

III. The animals selected for a procedure should be of an appropriate species and quality and the minimum number required to obtain valid results. Methods such as mathematical models, computer simulation, and *in vitro* biological systems should be considered.

Proper use of animals, including the avoidance or minimization of discomfort, distress, and pain when consistent with sound scientific practices, is imperative. Unless the contrary is established, investigators should consider that procedures that cause pain or distress in human beings may cause pain or distress in other animals.

V. Procedures with animals that may cause more than momentary or slight pain or distress should be performed with appropriate sedation, analgesia, or anesthesia. Surgical or other painful procedures should not be performed on unanesthetized animals paralyzed by chemical agents.

VI. Animals that would otherwise suffer severe or chronic pain or distress that cannot be relieved should be painlessly killed at the end of the procedure or, if appropriate, during the procedure.

VII. The living conditions of animals should be appropriate for their species and contribute to their health and comfort. Normally, the housing, feeding, and care of all animals used for biomedical purposes must be directed by a veterinarian or other scientist trained and experienced in the proper care, handling, and use of the species being maintained or studied. In any case, veterinary care shall be provided as indicated.

VIII. Investigators and other personnel shall be appropriately qualified and experienced for conducting procedures on living animals. Adequate arrangements shall be made for their in-service training, including the proper and humane care and use of laboratory animals.

IX. Where exceptions are required in relation to the provisions of these Principles, the decisions should not rest with the investigators directly concerned but should be made, with due regard to Principle II, by an appropriate review group such as an institutional animal care and use committee. Such exceptions should not be made solely for the purposes of teaching or demonstration.

NASA Sundowner Principles

In October 1996, the National Aeronautics and Space Administration (NASA) convened a meeting of a group of diverse individuals to "implement bioethics policies for animal experimentation" sponsored by or involving NASA. The group consisted of a public policy expert, three bioethicists, and representatives from the animal welfare community, AAALAC International, NASA, and other federal agencies involved in animal research. This diverse group concluded that (Rowan 2000)

> Among the basic moral principles generally accepted in our culture, three are particularly relevant to the ethics of research using animals: respect for life, societal benefit, and non-maleficence.
>
> - Respect for Life—Killing entails moral costs:
>
> Living creatures deserve respect. This principle requires that animals used in research should be of an appropriate species and health status and should involve the minimum number required to obtain valid scientific results [reduction]. It also recognizes that the use of different species may raise different ethical concerns. Selection of appropriate species should consider cognitive capacity and other morally relevant factors. Additionally, methods such as mathematical models,

computer simulation, and *in vitro* systems should be considered and used whenever possible [replacement].

- Societal Benefit—Advancing knowledge and health is a strong justification for research: The advancement of biological knowledge and improvements in the protection of the health and well-being of both humans and other animals provide strong justification for biomedical and behavioral research. This principle entails that where animals are used, the assessment of the overall ethical value of such use should include consideration of the full range of potential societal goods, the population affected, and the burdens that are expected to be borne by the subjects of the research.
- Non-Maleficence—Minimization of distress, pain, and suffering is a moral imperative: Vertebrate animals are sentient. This principle entails that the minimization of distress, pain, and suffering is a moral imperative [refinement]. Unless the contrary is established, investigators should consider that procedures that cause pain or distress in humans may cause pain and distress in other sentient animals.

Because of the diversity of this group, representing a broad range of philosophical thought and beliefs, these three principles resulting from the group's deliberations should be considered foundational components of any institution's program for the ethical use of animals in biomedical activities.

Religion

Most of the religions of the world include tenants that encourage concern for animals and their welfare, and some even revere certain species of animals and give them special status within their culture. A few religions so revere all animal life that they will go to great extremes to live lives without harming even the most humble of creatures, such as insects (e.g., Jainism [Anonymous 2015a]). The Reverend Professor Andrew Linzey has also made a case that animals have rights from a theistic perspective, that is, animal "rights [can be] based on the inherent worth of creatures possessing the property of being elected by God in love" (Cahill 2016). Therefore, on a global basis there may be situations where a proposed biomedical activity using animals may be morally acceptable within some segment of a local culture, but the range of animals acceptable for use might be limited by consideration for regional or local religious concerns; for example, dogs are generally considered to be "dirty" in Islam but can be kept as pets with some reservations (Banderker 2015). Persons involved in biomedical activities should be aware of any such concerns and design proposed activities and staffing plans accordingly.

Contemporary Issues

Biomedical Activities Are Not Monolithic

As illustrated in Figure 4.1, the term *biomedical activities* (or more generally referred to as "biomedical research") encompasses a very broad range of activities involving the use of animal subjects, and this diversity presents another very large layer of complexity to ethical issues involved therein.

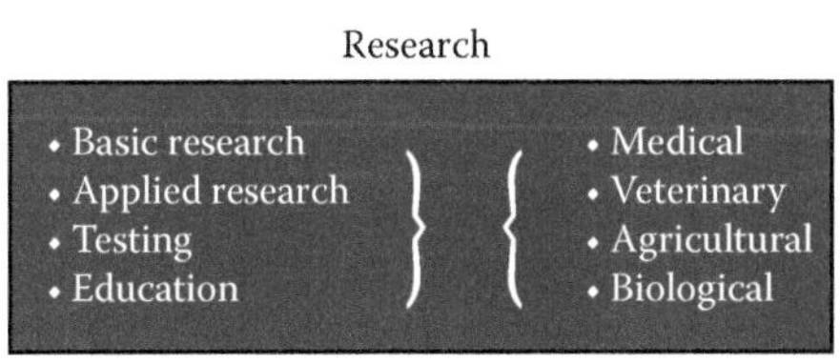

FIGURE 4.1 Illustration of the interrelationships between the types of biomedical activities on the left and the different biomedical disciplines involving the use of animal subjects on the right; that is, all four types of biomedical activities are found in each of the four disciplines.

Often, the public discourse regarding the use of animals in biomedical activities focuses on the benefits that "might" be derived only for humans. However, a significant proportion of such activities are undertaken with the direct or tangential intent of benefitting animals (e.g., development of a vaccine for an emerging disease, such as parvovirus in dogs). Also, in many cases the goal of an activity involving animal subjects is simply the advancement of knowledge without definitive knowledge of its benefit for either humans or animals, that is, basic research activities.

If it is decided that it is "unethical" to use animals in biomedical activities intended to benefit humans, the ethical principle of *justice*, which would include equal consideration of the interests of all (sentient?) animal subjects, would dictate that it is also unethical to use animals in biomedical activities intended to benefit other animals (e.g., endangered species captive propagation research or research to develop a vaccine for a newly discovered disease of dogs or horses), that is, "what's good for the goose is good for the gander." In contrast, one might even be able to make the case that the principle of *beneficence*, that is, the obligation to do good and avoid harm, would dictate that efforts to save an endangered species should be performed, even if it meant that some animals of the same species or another similar species might die during this effort. Another example of this conundrum is the use of some members of an animal species, deliberately infected with a newly identified pathogen resulting in overt disease in the recipients, to develop a vaccine for the newly discovered fatal disease in order prevent the disease in future populations of the species (e.g., Potomac horse fever in horses).

As an example of how this conundrum might apply to, and impact on, public opinion on a global scale, consider the widely accepted concern for endangered species. There are 181 parties currently signed on to the provisions of the Convention on International Trade in Endangered Species of Wild Fauna and Flora (CITES 2015), and in the United States, there is strong public and political support for the Endangered Species Act (1973). If there were laws prohibiting the use of sentient animals in biomedical activities, then the critical research needed to protect endangered species, including field studies and captive breeding for reintroduction efforts, would have to cease since such activities would likely involve using animal subjects.

A recent report noted the possibility that research for a vaccine against the Ebola virus using chimpanzees will result in a vaccine that likely can be used to protect the remaining and endangered wild chimpanzees from this fatal disease (Choi 2014). If this were to be the result of current vaccine development efforts, it would be another example of a discovery involving the use of animals that turns out to be critically important to animals. As an aside, Dr. Francis Collins, director of the National Institutes of Health (NIH), announced on November 19, 2015, that the NIH will permanently stop using chimpanzees for biomedical testing and is retiring its remaining chimpanzee population, thus potentially decreasing the probability of development of a successful vaccine (St. Fleur 2015).

Basic research often can seem like we are looking for keys to locks unknown. Because of this level of general ambiguity, basic research is often poorly understood by the general public; thus, what may be perceived as "unnecessary" or "trivial" may lead to public outcry about the cruelty of any use of animals in such pursuits. Senator William Proxmire (Democrat, Wisconsin, 1957–1981) would periodically issue "golden fleece awards." These awards were targeted at federally funded scientific research that he considered wasteful. However, the creation of this award reflected the fundamental misunderstanding of how science works, and how such research can turn out to be extremely important regardless of whether it makes sense to nonscientists. Indeed, such research can have a major impact on society. The nature of scientific research is that its impact is hard to predict (Anonymous 2015b).

However, without the foundation of basic research, followed by applied research, biological and medical advances would be impossible. When speaking to public or lay audiences and describing this point, I have frequently been asked, "How much basic research ultimately turns out not to have been of any particular significance?" My response, "Probably a significant percentage!" often surprises the questioner. I then go on to say, "However, the problem is that no one can, a priori, tell which basic research endeavors will fall into this status." Two examples that illustrate this point involve discoveries in the then emerging discipline of genetics in the 1930s and 1940s:

- During the 1930s and 1940s, there were two investigators conducting basic genetic research. One was Dr. George Snell, who was studying the genetics of mice at the Jackson Laboratory in Bar Harbor, Maine. The other was Dr. Barbara McClintock, who was studying the genes in corn at Cornell University's College of Agriculture and the Department of Botany at the University of Missouri–Columbia. It is highly unlikely that either Dr. Snell or Dr. McClintock, in the 1930s and 1940s, could have predicted that their basic research would lead to genetic discoveries that proved to be critically important in what is now known as "medical genetics" and would lead to their receipt of Nobel Prizes in Physiology or Medicine in 1980 (shared with two others) and 1983, respectively. Yet, their discoveries proved to be significant foundational blocks in the development of today's medical genetics.

- Another example unrelated to direct human interests that seems to be enlightening for laypersons is the captive breeding programs that have successfully allowed for the reintroduction of the endangered whooping cranes, California condors, and black-footed ferrets into their historic habitat ranges. All these species were on the brink of extinction when the captive programs began. Without captive breeding research efforts involving similar species (sand hill cranes, Andean condors, and nonthreatened ferret species, respectively), these three recovery efforts likely would not have succeeded.

Finally, it is critical to be realistic regarding the speed with which research goes from the benchtop to the bedside. This fact is poorly understood by the general public, which, in turn, can lead to opposition to some basic research studies involving animals when the ultimate value of the activities may not be known for decades. Finding disease cures can take up to a century—from some initial basic science discovery, a single brick in an allegorical huge pyramid, to the final "useful" product at the pyramid's apex is very hard for lay individuals to comprehend. The authors of one analysis, titled "From Scientific Discovery to Cures: Bright Stars within a Galaxy," state, "new treatments depend upon a broad base of scientific knowledge plus special contributions from a few exceptional scientists." Working backwards through the published medical literature, the analysis team looked at the step-by-step advances that were necessary to lead to the development of two drugs. According to this analysis, more than 7000 researchers from 5700 different institutions, working in succession over 100 years, were needed to develop a cancer drug. In support of this concept, the development of a cystic fibrosis drug was also daunting: 2900 scientists with ties to 2500 different institutions, laboring for 60 years. The bottom line, according to study coauthor Alexander Pico, is that "it takes contributions from a surprisingly large and complex network of individual scientists working in many locales to reach a cure" (Williams et al. 2015). Misunderstanding of this aspect of discovery can lead laypersons to oppose, what seems to them, to be excessive or frivolous use of animals in biomedical activities.

Changing the Legal Status of Animals

Recently, some animal rights groups have mounted campaigns to change references to animal "owners" in laws, statutes, and regulations to read animal "guardians." At least three U.S. cities have adopted amendments to their animal welfare statutes to replace the term *animal owner* with the term *animal guardian*: Boulder, Colorado; West Hollywood, California; and Berkeley, California. While the sponsors of this movement allege that their motive is simply to raise the public's awareness of their responsibilities to their animals, it is likely that their real agenda is to give legal "standing" to animals as things other than property (since the word *guardian* has very specific legal meaning and is defined as "a person lawfully invested with the power, and charged with the duty, of taking care of the person" (Garner 2009). It would not be much of a stretch to have this definition revised to substitute the word *animal* for the word *person*.

Such a change could theoretically lead to someone seeking legal guardianship of an animal that is not theirs based on a perception that the existing guardian (e.g., an animal in a research institution) is unfit. Should such a court case be filed, and even if the original guardian prevails in court, the potential legal fees could be significant. Further, if an animal needed expensive veterinary care that the guardian (in this example a private individual) could not afford, he or she might be liable for charges of animal endangerment for not providing the care and could be brought into court by someone suing on the animal's behalf.

While it is highly unlikely that these efforts to change the legal status of animals will lead to such draconian results, it is highly likely that there will be a good deal of turmoil created by this seemingly insignificant change in verbiage among the general public, animal owners, legislators, and most significantly, lawyers and veterinarians.

One development over the last several years that might indicate that these concerns about the possibility of legal turmoil over this issue, as well as other animal-related issues, such as personhood for chimpanzees, are valid is the burgeoning number of law schools that offer or have offered credit for animal law courses or seminars (at least 119 schools) (NABR 2015a) and that have received millions of dollars to support such courses (NABR 2015b).

Perhaps if the efforts to use a term other than *owner* for persons who "own" animals were to advocate for use of the term *caretaker*, or some similar term not fraught with potential legal implications, the goal of recognizing that animals are not just things may advance more rapidly. In regard to concerns that "unfit" owners, that is, those that abuse or treat animals cruelly, ought to be sanctioned, it is worthwhile to remember the many laws, regulations, and international treaties that exist to protect animals include serious sanctions for both individuals and institutions for violations thereof.

Other Contemporary Issues

When attempting to justify the value of the use of animals in biomedical activities, it may be necessary to include consideration of a subset of contemporary cultural and ethical issues:

- Is it morally acceptable to use animals for *any* human purpose? Could the use of any captive or domestic animal for any purpose whatsoever be equivalent to forced human slavery? It is unlikely that this represents the majority ethical viewpoint for most cultures. However, it is possible to say that some individuals might take this ethical position. If one should conclude that the answer to this question is no, further discussion about the value or necessity of using animal subjects in biomedical activities would be purely academic.
- There are different cultural perceptions regarding the moral acceptance of animal use, which can encompass the way the animal is used, the purpose of the use, or even the species being used. It is likely that most cultures and societies accept the use of some species of animals as pets. There are species of animals that are revered in one culture that may be eaten in another (e.g., cows and dogs). There are species that have a significant place in the cultural heritage of various native peoples, and may still be hunted even though they are nationally recognized as endangered (e.g., whales and eagles). These varied perceptions further complicate efforts to resolve the moral and ethical aspects of animal use in biomedical activities.
- What is the "necessity" of any particular use of animals? As will be discussed in more detail in the "Risk (Harm) versus Benefit Assessment" section that follows, what is necessary with regard to biomedical activities is a prominent consideration in numerous laws, regulations, policies, and guidelines that pertain to animal use in biomedical activities. Is it necessary to use an animal for a specific research proposal, and is it necessary to use a specific species? Is it necessary to genetically modify an animal model to do the research, even if it means that the animal may develop a disease that might be painful? Unfortunately, there is no universally accepted definition of what *necessary* means when used in this context. For example, in a study designed to develop a new treatment for some major emerging human disease, would it be necessary that the potential treatment be such as to save a minimum of 1000 human lives, or would the saving of 10 lives be an acceptable "cost" in terms of animals used? Even the eminent medical missionary who fostered a "reverence for [all] life" philosophy, Albert Schweitzer, has been quoted as saying that "it is *necessary* for the advancement of medical understanding" (emphasis added) when asked about his views on the use of laboratory animals for biomedical research (Pittman 1990).
- Problematic words. Many of the terms that we use in discussing the ethical use of animals lack precise or universally accepted definitions. With regard to the use of the term *obscenity*, there

is an old adage; it can be hard to define, but "we know it when we see it." Well, this colloquial expression can apply equally well to the terms *well-being, distress*, and *suffering*; we cannot exactly define them, but we know them when we see them.

General well-being: To assess an animal's general well-being, one may use physiological criteria, behavioral criteria, or functional criteria, or a combination of all three, in an effort to arrive at an approximation of well-being.

- Physiological criteria: Animals must be able to maintain internal homeostasis in order to survive. Most physiological parameters can be measured, many by non- or minimally invasive procedures (e.g., body temperature or blood cortisol levels). "Normal" ranges for the parameters can be established, and animals with values within those parameters may be said to have an appropriate level of physiological well-being.
- Functional criteria: *Functional* means can the animal function normally within the circumstances of its life; for example, can it obtain and utilize adequate nutrients to maintain itself, can it engage in adequate reproductive behavior to propagate its species at a relatively normal rate, and/or can it engage in sufficient grooming behavior to prevent injury to itself?
- Behavioral criteria: Many persons involved in trying to assess animal well-being advocate that behavioral criteria are more reliable than are either physiological or functional criteria. They base their position on the proposition that psychological and subclinical physiological stress or distress will manifest in abnormal behavior. The problem with this approach is that there is no uniform agreement on what is the normal behavioral baseline against which to judge whether a particular behavior is abnormal. For example, should we use the normal behavior of tigers living in the wild as the baseline for assessing the behavior of a captive-bred tiger living in a zoo when the cat has never known what life in the wild is like and appears to be physiologically and functionally doing well even though it repeatedly paces in its enclosure (mimicking a tiger's travels over its range?) without injury?

Psychological well-being: The U.S. Federal Animal Welfare Act implementation regulations (Part 3, "Standards," Subpart D, §3.81) require that institutions provide "environment enhancement to promote [the] *psychological* well-being" of nonhuman primates (Federal Animal Welfare Act and Animal Welfare Regulations 2013, emphasis added). While one may be able to assess the general well-being of an animal by using the three criteria above, assessing its psychological well-being may not be so simple. "To assess an animal's 'psychological' state ('well-being' or otherwise), requires [lingual] communication of feelings"; thus, "when evaluating an animal's well-being, the most parsimonious [simplistic] approach is to use behavioral indices. To say that 'abnormal' behaviors indicate abnormal 'psychology' is to make a great leap and say that you are able to get inside the mind of animals" (Rasmussen 2000). Therefore, we can truly assess psychological well-being only in conscious, lingual humans. Even with a nonverbal human patient, psychological well-being cannot be assessed. For example, persons in comas do not receive antidepressant drugs since it cannot be determined that they are depressed. There can be no psychological intervention with a person in a coma since their psychological state cannot be determined.

In truth, there is probably no universally accepted way to assess the psychological well-being of any nonhuman animal, even though it is required in the U.S. regulations for nonhuman primates. To meet this requirement, behavioral criteria appear to be acceptable to ensure compliance with the law.

Distress: Distress may be defined as physical or mental anguish or suffering; however, as has been stated many times, much of the difficulty in achieving a broadly accepted approach to categorizing, and then addressing, pain and distress is due to the absence of a concise definition. From a behavioral perspective, this inability to arrive at a *Webster's Dictionary* type of definition is due in part to the fact that (1) pain and distress are not discrete states, but are a continuum of experience; (2) signs differ between species, and most animals hide signs of pain because such a sign of weakness may provoke an attack from predators or subordinate members of the group; (3) there is a lack of specific behavioral indicators of pain and distress; (4) in the course of identifying distress, interobserver variability can be large; and (5) there is a tendency to anthropomorphize, which is encouraged by U.S. government principle IV. That principle states that "unless the contrary is established, investigators should consider that procedures that cause pain or distress in human beings may cause pain or distress in other animals" (Institute for Laboratory Animal Research 2011). Note that this principle seems to associate distress with pain, which is commonly, but not universally, the case. Distress may be caused by factors other than pain, for example, noise, crowded enclosures, pheromones from other species, and environmental temperatures. As with psychological well-being discussed above, distress is often assessed by the animal's behavior. Physiological and functional criteria can also be used, with behavior, to identify distress.

Suffering: One could define suffering as an unpleasant emotional state associated with pain or distress. However, this definition would not seem to provide practical guidance when attempting to determine whether an animal is suffering when obvious gross indications are absent. Trying to define suffering in a manner that would be acceptable to all, which covers all the possible situations that may be associated with animal use in biomedical activities, is very difficult and, possibly, impossible. The previous discussion regarding well-being would seem to be applicable here, as if well-being is adequately addressed in an institution's animal care and use program, animal suffering would be minimized or eliminated to the maximum extent possible.

Alternatives: The term *alternatives* is most frequently interpreted as meaning the replacement of animals in biomedical activities with nonanimal technology. Tissue cultures (both human and animal), computer modeling and simulation, *in vitro* techniques, and retrospective statistical analyses are only a few examples of techniques that can replace the use of animals in some cases. Several organizations have been established to promote and develop nonanimal alternatives (e.g., the U.S. Department of Agriculture Animal Welfare Information Center [2015], the Interagency Coordinating Committee on the Validation of Alternative Methods [2015], the Center for Alternatives to Animal Testing [2015], FRAME [2015], and the National Centre for the Replacement, Refinement & Reduction of Animals in Research [2015]).

Is there an acceptable alternative methodology that does not require the use of animals? There are a few circumstances where a nonanimal alternative can replace the use of an animal. For example, for some compounds the Corrositex test method for dermal corrosivity can be used as a stand-alone assay for evaluating corrosivity and replace the use of animals (Corrositex 2015). It is likely that as more is learned about cellular biology and function, as well as molecular biology, the development of more nonanimal methodologies will be validated and replace animals used in some research activities. However, the "horns of our dilemma" are that the primary reason animals are used in biomedical research activities is the exquisite complexity of intact living organisms. It is highly unlikely that alternative methodologies will replace most animal use any time soon. However, what is certainly safe to say is that validated, nonanimal methodologies will be welcomed by most individuals responsible for biomedical activities currently involving the use of live animal subjects, if for no other reasons than a simple aversion to the use of sentient animals for biomedical activities and, from a purely pragmatic calculation, the alternative may be less expensive and faster than using animals. Few researchers would prefer the complexity and burden of animal experimentation if there were accurate alternatives.

As is true of general science advancements today, potential alternative technologies are emerging rapidly. For example, recent reports have been published describing the development of a "brain" in a dish (cell culture) that may replace animals in studies of brain cell function in some cases (Erickson 2015). At the 2015 annual meeting of the American Association for Laboratory Animal Science, the keynote speaker, Dr. Uwe Marx, presented an exciting summary of current efforts to develop human "multiorgan chips" or micro test cells. Other presentations elaborated on some of the organ test systems in existence that can simulate one or two organs that can be used in toxicological testing in lieu of live animals. While these systems are largely in the validation stage, Dr. Marx believes that within 20 years, there will be validated systems that will consist of up to 10 different "organoids" on a single chip that will actually be capable of interacting with each other, thus providing a cell culture–based alternative to a living animal for screening tests of candidate pharmaceutical agents or toxicological studies.

Another relatively new nonanimal alternative to live animals has been developed for teaching veterinary students rectal palpation of cattle and horses using haptic (sense of touch) virtual reality simulators. Using these simulators, veterinary students gain the ability to conduct such exams before they conduct them on live animals, significantly reducing the need for live animals in the students' early efforts to learn how to accomplish these diagnostic examinations (Farminguk Web Page 2015). Along with high-fidelity mechanical manikins, these simulators are reducing the use of live animals in many medical and veterinary school training programs.

As an aside, most individuals proposing to use animal subjects in biomedical activities are usually well informed about the latest *validated* alternative (to animals) technology that would be appropriate for use in the proposed activities and invariably will have already incorporated them into their activities. After all, as already noted, it is in their own self-interest to use the best, and least expensive, technology to obtain the highest-quality data. Besides, most scientists and research technicians will likely be emotionally delighted to use a nonanimal alternative, rather than live animals, so long as the alternative methodology provides equivalent results.

- Where to draw the line?

  If society were to decide that at least sentient animals have moral or legal rights, where along the taxonomic scale should the line be drawn? Today, most laws and regulations governing how we treat animals apply only to vertebrate animals and, in some cases, only certain vertebrate species. The theory for selecting these taxonomic criteria is that only vertebrates have the anatomical and physiological capacity to be considered sentient and thus experience pain. However, it has been well demonstrated that at least some invertebrate species, for example, octopuses, have a level of intelligence that would infer sentience (Borell 2015). Even earthworms produce endorphins, which are hormones that are known to function to reduce pain in humans (DeGrazia 1996), perhaps indicating that earthworms can feel pain and thus might be sentient. It might be concluded that trying to define a firm line dividing animals into those that are deserving of rights and those that are not, based on sentience, would be extremely difficult, if not impossible. Thus, the best ethical approach would be to provide the best, least aversive care for all animals used in biomedical activities.

Public Perceptions about the Use of Animals in Biomedical Activities

As documented in the two references cited, a majority of the public support the use of animals in biomedical activities provided the animals are "humanely" cared for and used and the activity is "necessary" (Cox and Montrose 2016; NABR 2016).

- Many in the general public seem to believe that many animals used in biomedical activities "suffer," at least to some extent. If one were to ask a friend or acquaintance who has nothing to do with biomedical activities if it is okay to use animals in biomedical activities, it is likely that the response would be something like, "Well, it's a shame that they have to suffer, but it's okay if [followed with some qualifying statement]." Control, or preferably elimination of pain and/or distress, seems to be a major concern of many individuals. Thus, control or elimination of pain and/or distress should be a major commitment of all animal care and use programs, and provisions for minimizing pain and/or distress should be well documented in proposals for animal use and in actual application during the accomplishment of animal use activities. When talking to lay audiences about the pain issue, one may use the following example regarding toothache pain to illustrate how pain can be addressed in biomedical activities:

> Ask those present to recall the last time they had a toothache and follow with the question, "When you first became aware of the pain, did you immediately make an appointment with a dentist or, more likely, did you think to yourself, 'Oh, it will go away'?" Then note that most likely the answer for most in the audience would be the latter; that is, they waited until the pain became intolerable to consult a dentist. Then point out that the first inclination of pain would represent the "perception threshold," and the intolerable point would be the "tolerance threshold," and studies involving some level of minimum pain likely could be considered acceptably humane if the level was only at or near the perceptual threshold and well under the tolerance threshold. This approach seems to be received well by most members of audiences. In addition, a discussion of the availability and use of modern analgesics to control pain in animals used in biomedical activities also seems to be well received by lay audiences.

- Also, it seems that many in the public believe that the use of animals in biomedical activities is, at best, poorly regulated—in spite of the fact that in the United States alone, there are approximately 42 laws and regulations that can govern how animals are cared for and used in biomedical activities, depending on the nature of the studies and the source of funding (see Attachment I for some selected examples). The number and variety of regulations that govern animal use in biomedical activities worldwide is truly impressive (AAALAC International 2015). Unregulated, hardly!

Media Portrayal of Animal Use in Biomedical Activities

The cover art of the December 26, 1988 *Newsweek* magazine issue depicted an illustration of a black and white striped cat sitting in a completely bare, wire cage just slightly larger than the animal. The expression on the cat's face staring at the observer could be anthropomorphically interpreted as "apprehensive." The cover presents multiple subliminal messages that could sway a layperson's perception about biomedical activities involving the use of animals. To begin with, the animal shown is a "warm and fuzzy" cat, not a "creepy crawly" rat or mouse. In addition, the photo implies that the cat lives in a sterile cage that is obviously way too small for it and lacks food, water, a litter box, or a resting shelf. Then, finally, what might be the most powerful subliminal message is in the title and, most importantly, the subtitle, i.e., "The Battle over Animal Rights: A Question of **Suffering** and Science" [emphasis added]—that is, animal suffering is the norm in research using animals (with the "truth" of this statement being "proven" by the obvious "cruelty" of how the cat is being housed)!

Much of the media material about the use of animals in biomedical activities over the last five or so decades appears to be biased toward the animal rights point of view, frequently equating *animal rights* and *animal welfare* as synonyms. These apparent biases are frequently a result of the reporters or commentators not being well versed about the material they are presenting. Another significant reason for this misreporting is the research community's apparent reluctance to present the facts about our biomedical activities. For example, an institution gets a multi-million-dollar, multiyear grant for a project involving the use of animal subjects, but the announcement of the grant emphasizes the dollar amount and alludes to the importance of the project in terms of human health or scientific discovery, but rarely mentions the animals to be used. Announcements of some new, allegedly significant discovery usually laud the institutional support and the scientist making the discovery but almost never includes any acknowledgment of the critical role the use of animals played in making the discovery. One can make the case that emphasizing the use of animals in our biomedical activities carries the risk that those opposed to such animal use may target institutions that are forthcoming. However, by having highly ethical caring and legally compliant animal care and use programs that can withstand public scrutiny, along with being proactive and forthcoming with our public stakeholders, including the media, we should be able to ultimately persuade the general public of the validity of our activities.

Value of Biomedical Activities Involving Animals

If you Google "advances in medicine and science attributed to animal research," you will find more than 7.5 million results. These include numerous lists of advances in biomedical disciplines, as well as numerous sites refuting the "facts" of such claims, using emotional arguments and examples of some misdeeds by some scientists, to denigrate those using animals in biomedical activities.

As is the case with most human activities, there have been examples of inappropriate animal care or use in some biomedical activities at some institutions. However, if there are a few examples of children being mistreated in childcare centers, all childcare centers are not closed; rather, the guilty parties are punished and the management of their specific centers corrected or the centers closed. Thus, because there have been some (likely a very small percentage of individuals or institutions involved in conducting biomedical activities) instances of inappropriate or illegal violations of contemporary standards for the care, use, or welfare of animals in biomedical activities, caution should be exercised not to paint the entire biomedical community with the same paintbrush. In addition, those institutions and individuals that are found to be violating standards, policies, and laws are frequently subject to severe penalties if discovered.

Former U.S. Surgeon General C. Everett Koop, MD, has been quoted as stating, "Virtually every major medical advance for both humans and animals has been achieved through biomedical research using animal models to study and find a cure for a disease and through animal testing to prove the safety and efficacy of a new treatment" (Anonymous 2001). Not only is Dr. Koop's statement true, but also it is likely that the majority of medical advancements and achievements that we take for granted today have been the result of activities using animal subjects.

Some Specific Examples of the Value of Animal-Based Medical Research

In the 1960s, thalidomide (generic name rofecoxib, trade name Vioxx), a drug used in sleeping pills and sedatives, was discovered to help alleviate morning sickness in pregnant women and was widely used in Europe and other countries. However, due to one Food and Drug Administration (FDA) inspector who believed that there were insufficient animal data to support its use in pregnant women, and in spite of much pressure to approve the drug for use in the United States, thalidomide was not approved for sale in the United States for use by pregnant women. Within several years of thalidomide being first prescribed for pregnant women, it became obvious that the drug was causing limb malformations in newborn infants and its sale was halted worldwide. The FDA inspector who held up approval of the drug in the United States was Dr. Frances Kelsey, a Canadian working at the FDA. In 1962, Dr. Kelsey was awarded the President's Award for Distinguished Federal Civilian Service by President John F. Kennedy, largely in recognition of her efforts to keep thalidomide out of the U.S. market (Wikipedia Web Page 2015). In June 2015, she was named to the Order of Canada, again largely in recognition of her efforts to keep thalidomide off the market, but also for being "an instrumental figure in shaping and enforcing drug licensing protocols."

Dr. Albert Sabin, recognized as the developer of the oral polio vaccine, in a personal letter to this author, dated September 28, 1991, noted thusly, "My own experience of over 60 years in biomedical research amply demonstrated that without the use of animals and of human beings, it would have been impossible to acquire the important knowledge needed to prevent much suffering and premature death not only among humans but also among animals." Also, in a paper published in the *Journal of the American Medical Association*, Dr. Sabin noted that "during the preceding four years approximately 9,000 monkeys, 150 chimpanzees, and 133 human volunteers have been used thus far in the quantitative studies of the various characteristics of different strains of polioviruses" (Sabin 1956). Even though use of the oral polio vaccine has been discontinued in many of the world's developed countries (due to the reported incidence of 1 in 2.4 million cases of induced polio in recipients of the vaccine and the increased safety of the injectable vaccine), it is still used in many developing countries due to the ease of administration and some cultural resistance to the injectable vaccine.

Dr. Helen Taussig was the codeveloper (with Drs. Alfred Blalock and Vivien Thomas) of the Blalock–Taussig–Thomas shunt for correcting the malposition of cardiac vessels with or without ventricle perforations in newborn infants (commonly called the blue baby syndrome, which, uncorrected, invariably led to the death of the affected infants). In a personal conversation with this author (ca. mid-1960s), she indicated that the use of dogs was absolutely critical in the development of the shunt, and because of the results of the dog studies, early surgeries on human infants were largely successful. On September 28, 1962, in testimony before the congressional house hearing of the Subcommittee of the Committee on Interstate and Foreign Commerce, Dr. Taussig affirmed that the use of dogs had been critical for the development of the Blalock–Taussig–Thomas shunt and noted that the procedure had "saved thousands of lives throughout the world. It opened up the field of pediatric cardiac surgery" (Taussig 1962).

Dr. Emanuel Grunberg, while not as well known by the public as Drs. Sabin and Taussig, was a member of the team that discovered the antituberculosis drug isoniazid. In the early 1950s, Dr. Grunberg was working at Roche Pharmaceuticals, where there was a major program to find an antibiotic that would be effective against *Mycobacterium tuberculosis*, the causative agent of human tuberculosis. Prior to the discovery of isoniazid, the only antibiotic that was available to treat tuberculous was streptomycin, but it was more bacteriostatic than bactericidal and treatment could take years and long-term use of the drug could lead to some serious potential side effects, such as vertigo and allergic reactions. Over dinner at his son's house one evening (ca. mid-1980s), Dr. Grunberg told this author that testing of isoniazid in mice was a critical factor in discovery of the drug, as *in vitro* testing with it indicated only marginal effectiveness against *M. tuberculosis*. However, when tested in mice it proved to be highly effective against the bacterium, an effect ultimately determined to be a result of a metabolic product produced by the animals that was the active agent. Isoniazid is still considered to be a first-line medication in the prevention and treatment of tuberculosis.

Is History Prologue?

In a 2011 Hastings Center Report, Dr. Larry Carbone wrote, "History is informative, but not conclusive. To say that dogs were vital to the discovery of the role of the pancreas in diabetes in the 1920s is not to conclude that other approaches could not have worked then, or that the dog studies would be necessary in the twenty-first century" (Carbone 2011). Dr. Carbone's implied caution about using the criticality of some past use of animals as justification for continued use of animals can be problematic when viewed through the lens of contemporary technological knowledge.

However, conscientious application of the 3Rs should maximize the probability that future use of animals in biomedical activities will prove to have been valid in the context of our knowledge today. Contemporary knowledge may still require an animal model for studies of diabetes, but it may be a genetically altered mouse hosting human pancreas cells rather than dogs. More importantly, critical well-designed research, based on well-documented rationale and that can best be completed using animals today, should not be delayed on the hope that a better alternative might be discovered sometime in the future.

Impact of Emerging Technologies

All too often scientific advances occur at a rate greatly outpacing the ability of philosophers, ethicists, politicians, and the general public to address issues that arise therefrom. The cloning of animals, the creation of chimeric animals that host human cells, the genetic manipulation of germ cell DNA to permanently eliminate a genetic disease in future offspring, and so forth, have raised serious ethical issues in the past, are doing so today, and most certainly will continue to do so in the future as yet undiscovered biomedical technologies advance. As a single example, recall the ethical uproar that arose about the use of human embryonic stem cells when early studies required obtaining cells from human embryos or umbilical cord blood. Now, at least in some cases, technology allows us to regress one's own fat stem cells to their pluripotential state and use them to treat some medical conditions, largely negating most of the previous concerns about human stem cell research.

It is impossible to predict here what kinds of new biological or biomedical technologies will emerge in the future. With regard to the use of animals in biomedical activities, review committees need to be informed about any potential adverse effects that a new technology may have on the welfare of any animals involved in a proposed activity. For example, if the proposed activity involves creating a novel genetically modified strain of mouse, the responsible applicant should be required to provide (1) the best possible prediction regarding adverse impacts on the animals' welfare, (2) the criteria for determining whether any adverse impacts on the animals' welfare occur, (3) a description of how any adverse impacts on the animals' welfare will be addressed when or if they occur, and (4) the criteria that will be used to determine when to terminate the use of an affected animal due to welfare concerns (humane endpoint). The same four aspects of any new technology involving animal subjects should also be addressed in active proposals.

Can Animal Use Be Prohibited?

Could we do away with the use of animals in biomedical activities? Of course we *can*, but at what cost? Abolishment of the use of animals (just vertebrates or all sentient animals) in biomedical activities would most certainly impede advances in human and veterinary medicine, impede foundational basic research, and probably significantly slow advances in knowledge that would alleviate future human and animal suffering. What it would not do is significantly minimize human exploitation or use of animals since the single thing that could be done as a society to mitigate animal use would be to mandate by law vegetarianism or vegan lifestyles, which would immediately "save" billions of future animals that would be produced for consumption.

Application of Ethical Concerns of Oversight Bodies Such as Institutional Animal Care and Use Committees and Ethics Review Committees

In the various countries with regulations regarding the use of animals in biomedical activities, there are instances where committees charged with reviewing the use of animals are specifically expected to consider the ethical aspects and implications of the proposed activities during their deliberations. That said, and considering that references to "ethics" or "biomedical ethics" are frequently absent from regulatory or policy documents, it is proposed here that all such committees, as well as institutional managers and officials, have at least an implied responsibility to consider the ethical aspects of any proposed use of animals, whether or not required by law, regulations, or guidelines. In cases where there may be serious ethical concerns about a particular activity, strong justifications for approving the animal use should be well documented by the individual or unit requesting such use.

For instance, members of an Institutional Animal Care and Use Committee (IACUC), as constituted by the U.S Federal Animal Welfare Act (Federal Animal Welfare Act and Animal Welfare Regulations 2013) and/or the U.S. Public Health Service Policy on Humane Care and Use of Laboratory Animals (U.S. Department of Health and Human Services 2015) (and their respective regulations and policies), should consider the "scientific merit" of a proposed activity. The argument against consideration of merit by these review committees is usually presented something like this: "Since funding agencies base their awards largely on the merit of proposals, and have their own review procedures to specifically evaluate the merit of applications, local review committees should not review proposals for merit." However, if an IACUC member believes that there truly is *no* merit to a proposed animal use activity, only a disapproval vote would make ethical sense. Thus, each IACUC member must be convinced that there is at least some minimal level of merit to vote for approval. Obviously, the acceptable level would be an individual decision. While "merit" has not been discussed heretofore as a component of an "ethical" review, it would seem that approval of an activity that is believed to be truly without merit would be unethical.

In addition, if the members of a review committee determine that there is sufficient merit to consider a request for animal use, they should then determine whether the proposal for the activity has, at a minimum and to the greatest extent possible, (1) addressed the 3Rs, (2) documented that the proposed animal model is the best possible one for use, (3) shown that all care and use procedures have been well addressed so that the welfare of the animals is ensured to the greatest extent possible, and (4) demonstrated that there are adequate resources available to care for the animals. Only then should a vote for approval be received.

While not always available, effort should be made to identify a person trained in bioethics to be a member of all review committees.

Risk (Harm) versus Benefit Assessment

Generally, cost–benefit assessments are made in regard to the financial costs of a project versus the potential income to be expected or, in the case of some projects, such as a new interstate highway, the potential for fewer vehicular accidents and/or the more efficient movement of traffic, all with reasonably well-established criteria and methodologies for accomplishing the necessary data analyses.

Even risk–benefit assessments with regard to new or potential medical treatments in human medicine can be accomplished using reasonably well-defined criteria; however, integral to these assessments is input and informed consent from the mentally competent patient, human research subject, or competent human guardian for the patient or research subject (Pinkertom et al. 2002). In all these cases, the potential for benefit (to the patient or subject) must outweigh the potential risks or harm.

In regard to the use of sentient animals in biomedical activities, such assessments suffer from four very significant problems: (1) the animals cannot participate in the discussions, nor can they give informed consent; (2) in at least some cases, the animals will be subjected to situations that, even with conscientious application of the 3Rs, will result in at least some discomfort, distress, or [hopefully] minimal pain; (3) in many if not most cases, the animals will be euthanatized at the end of the activity for purposes of collection of organs, tissues, or other biological samples; and (4) consideration must be given to who will approve the animal use. The decision to use animals in biomedical activities may depend on the principal investigator, animal review committee members (including the public members), institutional administrators (such as in-house grants supervisors), representatives of regulatory or funding agencies, or ethicists. Thus, these problems make it imperative that review procedures are structured to be as objective and transparent as possible.

In a recent issue of *Lab Animal*, Drs. Kinter and Johnson (2015) provided an excellent discussion on the use of the terms *risk–benefit* or *harm–benefit* with regard to assessing proposals for the use of animals in biomedical activities. They make an excellent case for maintaining the term *risk–benefit* as found in the American *Guide for the Care and Use of Laboratory Animals* (Institute for Laboratory Animal Research 2011) versus the term *harm–benefit* as found in Article 38 of European Directive 2010/63/EU. The authors conclude by recommending "that scientists, IACUCs, and groups that accredit animal research programs maintain the language and concept of 'risk-benefit analyses' for assessing the risk that research animals might experience pain, distress, or injury. 'Risk-benefit analysis' transparently and unambiguously describes and addresses the necessary considerations of animal research, and its terminology agrees with the specific language in the *Guide for the Care and Use of Laboratory Animals*" (Institute for Laboratory Animal Research 2011). That is, risk (of pain, physical harm, etc.) can be more reasonably predicted and quantified than can the harm, although, some harm may occur even if the risks are low (e.g., euthanasia for collection of tissue samples).

In a 2010 article published in the *Archives of Biological Science*, a case is made for the use of the term *cost–benefit* rather than *risk–benefit*, in that "the term 'cost' defines the expected harm, pain and distress that is likely to be experienced by the animals." However, as noted above, the authors failed to provide objective criteria for quantifying the relative costs (to the animals) versus the benefits to be derived from the studies (Kostomitsopoulos and Durasevic 2010).

Role of Animal Care and Use Administrators, Directors, and Managers in Fostering Ethical and Compliant Programs

As noted in the introduction to this chapter, institutional officials and managers of animal care and use programs who understand the origins and foundations for the variety of public perceptions regarding the use of animals in biomedical activities will likely be better able to organize their programs to minimize the possibilities for criticism and maximize their ability to respond in a positive manner to any such criticisms. Even though this knowledge of history and the various ethical concepts and philosophies about animal use in biomedical activities will enable managers to establish operational policies and procedures of a truly ethical program, there will never be a program totally acceptable to all the public stakeholders.

Animal care and use program managers should insist on the necessity of quality animal care and use programs and be vocal advocates for the animals *and* the users. These are not contradictory roles. Strongly advocating for quality animal care and at the same time assisting the animal users in designing legally compliant and ethically appropriate proposed activities are complementary and will maximize the probability of quality science.

A major role for managers is to ensure that all members of the animal care and use staff are fully aware of the reasons animals are used in biomedical activities and of the efforts in place to ensure that they are used in the most ethical and humane manner possible. Hopefully, successful managers will help all staff personnel understand that their institution's programs for animal care and use are ethical, necessary, and of high moral quality and will ensure that those beliefs are based on factual knowledge about the institution's programs.

In some institutions, directors and managers of animal care and use programs may be asked to participate with public affairs officers in responding to inquires about animal use since they may be the staff members who are most familiar with the institution's animal use program. In such cases, the institution would be remiss if it did not provide professional media training to the director or manager.

Application of Ethical Concerns by Other Institutional Individuals

While it should be obvious, it is in the best interest of both all institutions and all individuals within an institution to conduct all animal care and use activities in an ethical manner, and it is intrinsically important to do so. Because of the controversies involved in the use of animals in biomedical activities, it is especially important that all personnel, at all levels, be supportive of efforts to foster implementation of ethical and legally compliant animal care and use programs.

The emphasis on an institution's commitment to having an ethical and legally compliant animal care and use program should start with the chief executive officer issuing an unambiguous statement indicating that the institution believes that the appropriate use of animal subjects in biomedical activities is necessary, and that there is a firm commitment to conduct such activities in the most ethical and compliant manner possible. This statement and the commitment it documents should be prominently displayed throughout the institution and included in all in-house training materials. As part of this commitment, there should be well-documented support for quality programs of animal care and use, and this should be prominently advertised through the use of posted policies encouraging the reporting of any concerns about animal care and use, with assurances that persons reporting concerns will not be penalized for coming forward, and listing the institutional officials to whom concerns should be reported.

To maximize the quality of an institution's animal care and use program, and minimize to the extent possible any unethical or noncompliant situation, all ancillary support personnel must be included in the institution's training program so that they are knowledgeable about the requirements for the use of animals in biomedical activities that apply to their respective areas of responsibilities.

Other individuals who may be overlooked as needing information regarding an institution's commitment to, and implementation of, a quality animal care and use program are the members of other institutional review committees that have oversight responsibilities for certain biomedical activities, such as biosafety, occupational health, radiological safety, and laboratory safety, all activities that may involve the use of animal subjects.

Value of External Review of Animal Care and Use Programs

As good as an animal care and use program may be, it never hurts to have a review by a disinterested individual or organization that is not regulatory in nature. Today, the "gold standard" for such external review is accreditation by AAALAC International. While AAALAC International accreditation is not inexpensive or easy to obtain and keep, it is well worth it. The post–site visit report that the institution receives is of tremendous value in determining the quality of the institution's animal care and use program and will highlight any deficiencies discovered. One of the best aspects of this external review is that it is conducted by highly qualified individuals, who are thoroughly knowledgeable about animal use in biomedical activities, and the results of the review are completely confidential. These external review panels can sometimes identify "problems" that could lead to, or be interpreted as, impacting the ethical status of an institution's animal care and use program, which can then be addressed and corrected if necessary.

Conclusion

As noted in the introduction, there are no black and white answers regarding the bioethics of animal use in programs of research, teaching, and testing, and it depends on a person's experience, knowledge, understanding of the issues, and personal philosophy. Philosophically, none of the various philosophies presented earlier in the chapter, including animal rights, are necessarily wrong, nor are any of them universally accepted. Thus, we may perceive that we are wallowing in a philosophical quagmire. However, a conscientious effort and commitment on our part to establish and maintain animal care and use programs of the highest quality, including the fullest implementation possible of the 3Rs and responsible use of animals, will enable us to be comfortable that we are providing the most ethical and morally acceptable program possible.

Hopefully, the material presented herein will be helpful to new and experienced administrators and managers of animal care and use programs involved in biomedical activities in developing their personal understanding of the complexity of the ethical and moral issues involved in our profession, and in instilling that understanding in those they work with and for at their institutions.

ATTACHMENT I

Selected Examples of the Laws and Regulations That May Apply to Biomedical Activities Conducted in the United States and Canada

- *Federal Animal Welfare Act and Regulations* (http://www.aphis.usda.gov/).
- *Public Health Service Policy* (http://grants.nih.gov/grants/olaw/olaw.htm) and the associated *Guide for the Care and Use of Laboratory Animals* (the *Guide*), published by the National Research Council, and most recently revised in 2011 (https://grants.nih.gov/grants/olaw/Guide-for-the-Care-and-use-of-laboratory-animals.pdf).
- *Good Laboratory Practice Act.* This act and its implementing regulations are enforced by the federal FDA, and the provisions of the act may impact how the care and use of animals is documented when animals are used in biomedical activities resulting in data submitted in support of applications for approval of a new pharmaceutical agent or medical device (http://www.21cfrpart11.com/files/library/pred_rules/mcdowall_glp_annotate.pdf).
- *Good Manufacturing Act.* As with the Good Laboratory Practice Act, this act and its implementing regulations are enforced by the federal FDA and the provisions of the act may impact how the care and use of animals is documented when animals are used in activities involving the manufacturing of pharmaceutical products (e.g., to test efficacy) (http://www.fda.gov/Drugs/DevelopmentApprovalProcess/Manufacturing/ucm169105.htm).
- *Lacey Act.* Under the Lacey Act, it is unlawful to import, export, sell, acquire, or purchase fish, wildlife, or plants that are taken, possessed, transported, or sold (1) in violation of U.S. or Indian law or (2) in interstate or foreign commerce involving any fish, wildlife, or plants taken possessed, or sold in violation of state or foreign law. The law covers all fish and wildlife and their parts or products, plants protected by the CITES, and those protected by

state law. Commercial guiding and outfitting are considered to be a sale under the provisions of the act (http://www.fws.gov/international/laws-treaties-agreements/us-conservation-laws/lacey-act.html).

- *Convention on International Trade in Endangered Species of Wild Fauna and Flora* (https://www.cites.org/). Depending on the species of animals being sought for biomedical activities, CITES provisions may preclude using them and/or may require special permits and waivers to obtain them.

Other Import and Export of Animals

- *Animal and Plant Health Inspection Service (APHIS)* (https://www.aphis.usda.gov/wps/portal/aphis/ourfocus/importexport)
- *Centers for Disease Control and Prevention (CDC)* (http://www.cdc.gov/importation/bringing-an-animal-into-the-united-states/)
- *U.S. Fish and Wildlife Service* (http://www.ncbi.nlm.nih.gov/books/NBK19632/).
- Interstate transport—APHIS

State and Local Regulations

Institutions using animals in research, teaching, or testing may be subject to additional state and local laws. State and local legislatures should be consulted for more details. For example, in the state of Nevada, the Nevada Department of Wildlife requires permits for trapping wild native animals for research and for having and maintaining certain prohibited species, such as African clawed frogs.

Guidelines

Many professional societies publish "best-practice" guidelines for use of animals in biomedical activities. While such guidelines may not be legally binding, they may affect public perceptions about an institution's animal care and use program, as well as a program's ability to obtain AAALAC International accreditation. Some examples of such guidelines are

- *Guide for the Care and Use of Agricultural Animals in Research and Teaching* (*Ag Guide*). The *Ag Guide* is published by the Federation of Animal Science Societies (FASS) and was most recently updated in 2010 (http://aaalac.org/about/Ag_Guide_3rd_ed.pdf).
- *Guidelines of the American Society of Mammalogists for the Use of Wild Mammals in Research* (http://www.mammalsociety.org/uploads/Sikes%20et%20al%202011.pdf).
- The Ornithological Council's *Guidelines to the Use of Wild Birds in Research* (http://www.nmnh.si.edu/BIRDNET/guide/).
- The American Society of Ichthyologists and Herpetologists' *Guidelines for the Use of Fishes in Research (2013)* (http://www.asih.org/sites/default/files/documents/publications/asf-guidelines-use-of-fishes-in-research-2013.pdf) and *Guidelines for Use of Live Amphibians and Reptiles in Field and Laboratory Research* (http://www.asih.org/sites/default/files/documents/resources/guidelinesherpsresearch2004.pdf).

REFERENCES

AAALAC International. 2015. International regulations and resources. http://www.aaalac.org/resources/internationalregs.cfm (accessed September 19, 2015).

Anonymous. 2001. The importance of animal research to medical discovery. *UCSD Perspectives*. Summer 2001. https://michaeldabney.files.wordpress.com/2010/03/animal.pdf (accessed March 20, 2017).

Anonymous. 2015a. Why animals matter: A religious and philosophical perspective: Janism. http://www.think-differently-about-sheep.com/Why_Animals-Rights_Matter_Jainism_Ahimsa_and_Animal_Rights.htm (accessed September 19, 2015).

Anonymous. 2015b. Senator William Proxmire's golden fleece awards. http://www.goldengooseaward.org/history/ (accessed September 19, 2015).

Anonymous. 2015c. Quebec Animal Welfare Bill hearings begin at National Assembly. *CBC News*, September 14. http://www.cbc.ca/news/canada/montreal/bill-54-quebec-animal-welfare-national-assembly-1.3226924 (accessed March 20, 2017).

Aquinas, T. 1956. *On the Truth of the Catholic Faith, Summa Contra Gentiles*. Doubleday, Garden City, NY.

Banderker, A.M. 2015. Status of dogs in Islam. http://www.islamicconcern.com/dogs.asp (accessed September 19, 2015).

Bentham, J. 2007. *Introduction to the Principles of Morals and Legislation (1823)*. Dover, Mineola, NY, 2007.

Borrell, B. 2009. Are octopuses smart? *Scientific American*, February 27. https://www.scientificamerican.com/article/are-octopuses-smart/ (accessed March 20, 2017).

CAAT (Center for Alternatives to Animal Testing). 2015. http://caat.jhsph.edu/ (accessed September 19, 2015).

Cahill, J.M. 2016. Grounded in love: A theistic account of animal rights. *J. Anim. Ethics* 6(1), 67–80.

Carbone, L. 2011. The utility of basic animal research. *Hastings Cent. Rep.* 42(Suppl.) S12–S15.

Carbone, L. 2015. Letter to the editor; Tannenbaum, J. and Bennett, B.T. 2015b. Response letter to the editor. *J. Am. Assoc. Lab. Anim. Sci.* 54(4), 351–352.

Choi, C.Q. 2014. Ebola vaccine shows promise in chimp study apes. *Live Science*. http://www.livescience.com/45890-ebola-vaccine-protects-chimps.html (accessed September 19, 2015).

CITES (Convention on International Trade in Endangered Species of Wild Fauna and Flora). 2015. https://www.cites.org/ (accessed September 19, 2015).

Cohen, C. 1986. The case for the use of animals in biomedical research. *N. Engl. J. Med.* 315(14), 865–870.

Corrositex: An in vitro test method for assessing dermal corrosivity potential of chemicals. 2015. https://ntp.niehs.nih.gov/iccvam/docs/dermal_docs/corprrep.pdf (accessed October 25, 2015).

Cox, L. and Montrose, T. 2016. How do human-animal emotional relationships influence public perceptions of animal use? *J. Anim. Ethics* 6(1), 44–53.

DeGrazia, D. 1996. *Taking Animals Seriously: Mental Life and Moral Status*. Cambridge University Press, Port Chester, NY.

Endangered Species Act. 1973. http://www.fws.gov/endangered/laws-policies/ (accessed September 19, 2015).

Erickson, B.E. 2015. "Brain in a dish" could replace toxic animal tests. *Scientific American*. http://www.scientificamerican.com/article/brain-in-a-dish-could-replace-toxic-animal-tests/ (accessed October 25, 2015).

European Directive 2010/63/EU. 2010. The protection of animals used for scientific purposes. http://eur-lex.europa.eu/legal-content/EN/ALL/?uri=CELEX:32010L0063 (accessed September 25, 2015).

Farminguk Web Page. 2015. Arizona University installs state of the art "haptic cow and horse." http://www.farminguk.com/News/Arizona-university-installs-state-of-the-art-Haptic-Cow-and-Horse-_37175.html (accessed October 25, 2015).

Federal Animal Welfare Act and Animal Welfare Regulations. 2013. https://www.aphis.usda.gov/animal_welfare/downloads/Animal%20Care%20Blue%20Book%20-%202013%20-%20FINAL.pdf (accessed September 19, 2015).

Fox, M.A. 1986. *The Case for Animal Experimentation*. University of California Press, Berkeley.

Fox, M.A. 1987. Animal experimentation: A philosopher's changing views. *Between Species* 3(2), 55.

FRAME. 2015. http://www.frame.org.uk/ (accessed September 19, 2015).

Francione, G.L. 1996. *Rain without Thunder: The Ideology of the Animal Rights Movement*. Temple University Press, Philadelphia.

Francione, G.L. 2006. The great ape project: Not so great. Blog post, December 20. http://www.abolitionistapproach.com/the-great-ape-project-not-so-great/#.WJ5HlqOZN7M (accessed September 19, 2015).

Garner, B.A., ed. 2009. *Black's Law Dictionary*. 9th ed. West Group, Eagan, MN.

Institute for Laboratory Animal Research, National Academy of Sciences. 2011. *Guide for the Care and Use of Laboratory Animals*. 8th ed. Washington, DC: National Academy of Sciences, Appendix B, 199.

Interagency Coordinating Committee on the Validation of Alternative Methods. 2015. http://iccvam.niehs.nih.gov/home.htm (accessed September 19, 2015).

Kinter, L.B. and Johnson, D.K. 2015. A defense of "risk-benefit" terminology. *Lab. Anim.* 46(10), 403–407.

Kostomitsopoulos, N.G. and Durasevic, S.F. 2010. The ethical justification for the use of animals in biomedical research. *Arch. Biol. Sci. (Belgrade)* 62(3), 783–789.

Library of Congress. 2015. Global Legal Monitor: New Zealand: Animal welfare legislation recognizes animals as sentient, bans cosmetic testing. May 19. http://www.loc.gov/law/foreign-news/article/new-zealand-animal-welfare-legislation-recognizes-animals-as-sentient-bans-cosmetic-testing/ (accessed September 19, 2015).

The Massachusetts Body of Liberties. 1641. https://history.hanover.edu/texts/masslib.html (accessed September 14, 2015).

Merriam-Webster Online Dictionary. 2015. Utilitarianism. http://www.merriam-Webster.com/dictionary/utilitarianism (accessed September 19, 2015).

NABR (National Association for Biomedical Research). 2015a. Animal law courses. http://www.nabranimallaw.org/law-schools/animal-law-courses/ (accessed November 21, 2015).

NABR (National Association for Biomedical Research). 2015b. Bob Barker's donations to law schools. http://www.nabranimallaw.org/law-schools/bob-barkers-donations-to-law-schools/ (accessed November 21, 2015).

NABR (National Association for Biomedical Research). 2016. Poll: American public support for animal research on the rise. Washington, DC: NABR. http://www.nabr.org/poll-american-public-support-for-animal-research-on-the-rise/ (accessed April 18, 2016).

NC3Rs (National Centre for the Replacement, Refinement & Reduction of Animals in Research). 2015. www.nc3rs.org.uk (accessed September 19, 2015).

North Carolina State University. 2015. Animal subjects in research. http://www.ncsu.edu/grad/preparing-future-leaders/rcr/modules/module_4.doc (accessed October 27, 2015).

Paton, W.D.M. 1984. *Man and Mouse: Animals in Medical Research.* Oxford University Press, Oxford.

Pinkertom, S.D., Johnson-Masotti, A.P., Derse, A. and Layde, P.M. 2002. Ethical issues in cost-benefit analysis. *Eval Program Plan* 25, 71–83.

Pittman, J.A. 1990. Letter to the editor. *New York Times.* http://www.nytimes.com/1990/05/26/opinion/l-dr-schweitzer-on-the-use-of-lab-animals-734490.html (accessed November 21, 2015).

Pythagoras of Samos. Pythagoras quotes about animals. http://www.azquotes.com/author/11949-Pythagoras/tag/animal (accessed November 22, 2015).

Rasmussen, C.T. 2000. Personal communication. Professor and chair emeritus, University of Nevada, Las Vegas, Department of Psychology.

Regan, T. 1983. *The Case for Animal Rights.* University of California Press, Berkeley.

Regan, T. and Francione, G.L. 1992. A movement's means create its ends. *The Animals' Agenda*, January/February, 40–45.

Rene Descartes. 2015. http://www.animalethics.org.uk/descartes.html (accessed November 22, 2015).

Robinson, D.H.I. and Toledo, A.H. 2012. Historical development of modern anesthesia. *J. Invest. Surg.* 25(3), 141–149.

Rollin, B.E. 1981. *Animal Rights and Human Morality.* Prometheus Books, Buffalo, NY.

Rollin, B.E. 2012. The moral status of invasive animal research. *Hastings Cent. Rep.* 42(6), S4–S6.

Rowan, A.N. 2000. Ethical principles for animal research and the Sundowner principles. In A.L. Kraus and D. Renquist, eds., *Bioethics and the Use of Laboratory Animals: Ethics in Theory and Practice.* American College of Laboratory Animal Medicine, Gregory C. Benoit Publishing, Dubuque, IA, 15–28.

Royal Society for the Prevention of Cruelty to Animals, United Kingdom. 2015. http://www.rspca.org.uk/utilities/aboutus/history (accessed September 19, 2015).

Russell, W.M.S. and Burch, R.L. 1959. *The Principles of Humane Experimental Technique.* Methuen & CO LTD, London.

Sabin, A.B. 1956. Present status of attenuated live-virus poliomyelitis vaccine. *JAMA* 162(18), 1589–1596.

Singer, P. 1975. *Animal Liberation.* New rev. ed. Avon Books, New York.

Singer, P. 1997. A philosophical self-portrait. In T. Mautner, *The Penguin Dictionary of Philosophy.* Penguin Books, London, 521–522.

Slate Website. 2015. Judge denies legal personhood to chimps—For now. July 31. http://www.slate.com/articles/news_and_politics/jurisprudence/2015/07/habeas_corpus_and_personhood_for_chimps_judge_barbara_jaffe_gives_liberty.html (accessed September 19, 2015).

St. Fleur, N. 2015. N.I.H. to end backing for invasive research on chimps. http://www.nytimes.com/2015/11/20/science/nih-to-end-backing-for-invasive-research-on-chimps.html?_r=0 (accessed November 20, 2015).

Tannenbaum, J. 1995. *Veterinary Ethics: Animal Welfare, Client Relations, Competition and Collegiality*. Mosby, St. Louis, MO.

Tannenbaum, J. and Bennett, B.T. 2015a. Russell and Burch's 3Rs then and now: The need for clarity in definition and purpose. *J. Am. Assoc. Lab. Anim. Sci.* 54(2), 120–132.

Taussig, H.B. 1962. Humane treatment of animals used in research. Congressional House Hearing of the Subcommittee of the Committee on Interstate and Foreign Commerce, September 28–29.

U.S. Congress, Office of Technology. 1986. *Assessment, Alternatives to Animal Use in Research, Testing and Education*. OTA-BA-273. Washington, DC: U.S. Government Printing Office, 83.

USDA (U.S. Department of Agriculture), National Agricultural Library, Animal Welfare Information Center. 2015. https://awic.nal.usda.gov/ (accessed September 19, 2015).

U.S. Department of Health and Human Services, National Institutes of Health. 2015. Public Health Service policy on humane care and use of laboratory animals. Bethesda, MD: National Institutes of Health. https://grants.nih.gov/grants/olaw/references/phspol.htm (accessed September 25, 2015).

Wikipedia Web Page. 2015. Frances Oldham Kelsey. https://en.wikipedia.org/wiki/Frances_Oldham_Kelsey (accessed September 25, 2015).

Williams, R.S., Lotia, S., Holloway, A.K. and Pico, A.R. 2015. From scientific discovery to cures: Bright stars within a galaxy. *Cell* 163(1), 21–23.

5

Behavioral Management Programs to Promote Laboratory Animal Welfare

Mollie A. Bloomsmith, Jaine E. Perlman, Eric Hutchinson, and Mark Sharpless

CONTENTS

Introduction to Behavioral Management

Animal care in research, teaching, and testing facilities has undergone a remarkable transformation in the past two decades, and there is a heightened emphasis on the psychological status of animals living in these institutions. Today, providing for the psychological well-being of research animals is an integral part of animal care programs and is key to promoting a "culture of caring" within our facilities. This change occurred as the scientific community realized that animals have many behavioral needs, which, if not met, can adversely affect their behavior, physical health, and research utility. Programs to address the well-being of laboratory animals by providing them with more complex and interesting environments are now common across all taxa, including primates, dogs, cats, pigs, sheep, rodents, rabbits, birds, and fish.

The term *behavioral management* refers to a comprehensive approach to improving the welfare of captive animals by employing social housing, environmental enrichment, animal training, facility design, and the assessment of behavior and behavioral problems. Behavioral management builds from a foundation of understanding the behavior of the target species (or of closely related species) in natural conditions, with the aim of improving animal care and enhancing animal welfare. An important premise of behavioral management is that the individual components of enrichment, animal training, social housing, and behavioral assessment can be integrated to achieve behavioral goals for captive animals better and more completely than any one technique applied in isolation. Behavioral management practices that improve psychological well-being and reduce stress for research subjects are important refinements in animal care.

Social housing of laboratory animals is fundamental to effective behavioral management programs since the most commonly used laboratory animals are social species. The *Guide for the Care and Use of Laboratory Animals* (hereafter referred to as the *Guide*) states, "Appropriate social interactions among members of the same species (conspecifics) are essential to normal development and well-being" (National Research Council 2011, p. 64). Social housing in compatible groupings or pairs offers the animals opportunities for engaging in a variety of species-typical behaviors, such as grooming, affiliation, agonism, play, and rearing of offspring. Social housing also reduces the occurrence of abnormal behaviors in many species, and it generally is especially important to socially house young animals.

Environmental enrichment enhances the well-being of animals by providing them with species-specific opportunities for exercise and manipulating objects, and cognitive challenges (*Guide*, p. 53). Enrichment is classified into broad and overlapping categories, including four types of nonsocial enrichment: feeding, physical, sensory, and cognitive and occupational enrichment (Keeling et al. 1991). When used together, these different categories of enrichment can improve many different facets of the animals' behavioral repertoire.

Animal training, and especially positive reinforcement training (PRT), is another component of behavioral management programs. PRT is a form of conditioning in which the subjects are given positive reinforcement (e.g., food treat or social interaction) following their performance of a desired behavior. PRT relies on the voluntary cooperation of animals, rather than on compelling them to comply through manual restraint or chemical restraint. PRT is generally seen as preferable to negative reinforcement training, in which the subject performs a desired behavior to avoid or escape from a negative stimulus (e.g., loud sound). While PRT has been used most often with nonhuman primates, it is now being applied to a variety of other laboratory animal species.

Behavioral assessment and monitoring are important aspects of behavioral management programs, as they are needed to identify behavioral problems, determine needed treatments for those problems, and assess whether treatments are effective. Behavioral assessment can be based on quantitative behavioral data collection and formal studies of behavioral change, or on a systematic application of less rigorous methods.

Specialized positions in behavioral management are becoming more common as research facilities are employing more individuals with expertise in behavior and animal welfare. When this type of focused position is not in place, others in the organization should be designated to take on the behavioral responsibilities, and this should be included as part of their job descriptions; performance of these responsibilities should be measured in their performance reviews; and they should be given the resources to learn about behavioral management.

Behavioral management programs (sometimes referred to as environmental enrichment programs or psychological well-being programs) are required by some regulatory agencies and are emphasized in professional standards of research animal care across the world. Multinational organizations such as the Council for International Organizations of Medical Sciences and the International Council for Laboratory Animal Science have set out principles of responsibility and oversight for countries with animal research and teaching programs, and these include the principle that the animals' environment and management are species appropriate and contribute to their well-being (iclas.org/wp-content/uploads/2013/03/CIOMS-ICLAS-Principles-Final1.pdf). Some countries have established federal legislation that addresses these programs. For example, in the United States, the Animal Welfare Act (AWA) regulations direct that research institutions must "develop, document, and follow an appropriate plan for environment enhancement adequate to promote psychological well-being of nonhuman primates" (9 CFR 3.81; U.S. Department of Agriculture [USDA] 1991, www.nal.usda.gov/awic/final-rules-animal-welfare-9-cfr-part-3-0). This environmental enhancement plan should be an Institutional Animal Care and Use Committee (IACUC)–approved policy or standard operating procedure (SOP), and it must be made available to the Animal and Plant Health Inspection Service (APHIS) and any federal funding agency upon request. The AWA also requires exercise for dogs (Section 13(a)(2)(B) of the AWA [7 USC 2143]). Live vertebrate animal research being conducted with federal funding from the U.S. Public Health Service (notably including funding from the National Institutes of Health) must comply with the *Guide*.

The *Guide* stipulates that enrichment should be provided for all laboratory animals. In addition, those research facilities that choose to become accredited by AAALAC International must also comply with the *Guide* or with the *Guide for the Care and Use of Agricultural Animals in Research and Teaching* (Federation of Animal Science Societies 2010), which also describes the provision of enrichment, and says most forms of enrichment should be provided. All relevant legal and professional standards should be followed when behavioral management programs are developed.

Constructs to Achieve with Behavioral Management

Behavioral management programs have general goals of increasing species-appropriate behavior, decreasing abnormal behavior, increasing resiliency, and reducing distress within the laboratory environment. In the *Guide*, there are also several other general tenets that can be met through behavioral management: offering novelty, providing opportunities for choice and control, habituating and preparing animals for procedures, and socially housing social species.

Offering animals novelty can be done through exposing them to novel objects, spaces, or activities that are stimulating to captive animals and give them practice in dealing with new challenges. This is helpful because if animals have been maintained in a completely consistent environment (i.e., lacking novelty), they may respond with fear or avoidance when novel stimuli are encountered. Novelty can be offered by rotating enrichment items, by offering new types of resources, by moving animals into larger activity caging, and by using PRT to teach animals new behaviors (i.e., the learning of new tasks is a stimulating experience). Social housing provides tremendous novelty with unpredictable and complex interactions with other animals. There is an important caveat to note with regard to novelty. Making changes in animal environments too often or too quickly may be stressful for some individuals. For example, individual animals who are rated as having highly inhibited temperaments are less likely to approach new objects; novel activities or objects may actually increase stress for such individuals without proper preparation (Roma et al. 2006; Coleman 2012).

Another principle is to give animals control and choice in their environments: "Well-conceived enrichment provides animals with choices and a degree of control over their environments, which allows them to better cope with environmental stressors" (*Guide*, p. 53). For example, different types of physical enrichment (e.g., visual barriers, elevated shelving, and protected contact panels) may allow animals to control their proximity to others, and this may reduce conflict and stress. Giving mice adequate nesting material also allows them to control their temperature and avoid stress associated with cooler temperatures. PRT relies on the animals' voluntary cooperation, and allows them to exercise choice in whether to participate in training sessions (Laule et al. 2003).

Requiring animals to be habituated to and prepared for research procedures is another broad principle in modern standards for animal care. The *Guide* states that "dogs, nonhuman primates, and many other animals can be trained, through use of positive reinforcement techniques, to cooperate with research procedures.… Animals to be placed in restraint devices should be given training (with positive reinforcement) to adapt to the equipment and personnel … [and] … animals that do not adapt to necessary restraint systems should be removed from the study" (*Guide*, p. 29). Animal training programs based in positive methods are needed to meet these standards.

A final general tenet that behavioral management programs most directly address is that social species should be socially housed. "Single housing of social species should be the exception and justified based on experimental requirements or veterinary-related concerns about animal well-being. In these cases, it should be limited to the minimum period necessary" (*Guide*, p. 64), and social housing is the "default" method for social species. For some species there are few difficulties or dangers in social housing, but others require skilled staff members to form social groups and monitor animals in social settings.

Personnel Involved in Behavioral Management

Successful behavioral management programs vary widely in design and implementation, but all require the involvement of people with different backgrounds and roles in a research facility working together as a team. The complexity of a program and the part each person will play depend on many facility-specific factors, such as vivarium size and staffing, species being used, and the presence of dedicated behavior staff. In a small colony of mice, for instance, a successful behavioral management program may consist of a relatively static enrichment plan overseen and periodically updated by a single attending veterinarian in concert with the rest of the IACUC, which would review the program regularly, and may be implemented by animal care or research staff who would provide feedback to the veterinarian on predefined performance measures. In a larger and more complex facility, however, one can imagine that an ideal behavioral management program would include expanded roles for many layers of personnel.

Animal care staff members see the animals multiple times daily and are often the most familiar with individual animals or strain temperaments and traits. Thus, a member of the care staff trained to distinguish normal from abnormal behaviors may identify animals experiencing early or mild psychological distress or, in some cases, physiological disease (Gaskill et al. 2013). The recognition of abnormal behaviors (both the presence and absence of) is crucial because this is one of the primary performance measures available to gauge whether a program is successful, as outlined in the *Guide*. Having the care staff provide enrichment, such as novel foods, also offers an opportunity for daily positive human contact and the development of a positive relationship between animals and the humans they interact with most. In facilities where there is a dedicated behavior staff, having care staff be the primary providers of enrichment resources may preserve the ability of behavioral specialists to remain neutral observers and view animals in their most typical state. Care staff may also train animals for routine husbandry tasks, allowing for another positive human contact opportunity and also allowing the training of many animals (Adams et al. 2004).

In many ways, the role of research staff parallels that of care staff. Researchers are often in frequent contact with their animals and may notice subtle signs of (di)stress due to their familiarity with the individuals or changes in their experimental measures. Collaboration between research staff and the rest of the behavioral management team is critical because many behavioral management techniques have the potential to impact research outcomes, and many research procedures have the potential to impact animal behavior. Open communication regarding experimental procedures, socialization attempts, abnormal behaviors, and so forth, is key to striking a balance between animal well-being and successful research outcomes.

The broad training and expertise of laboratory animal veterinarians in matters of animal health and research model design gives them a valuable perspective in ensuring balance between behavioral management, animal safety, and research integrity. Especially in facilities with dedicated behavior staff, there are several areas of frequent interface between veterinarians and behavioral management. In facilities with a robust socialization program, veterinarians will necessarily be involved in treating wounds caused by social aggression. Veterinarians have an increasing role in treating abnormal behavior via psychoactive medications, sometimes in combination with behavioral techniques. The use of positive reinforcement to train animals for cooperation with common veterinary techniques, such as physical exams and blood collection, also requires the interaction of veterinarians and behavioral management staff.

As behavioral management programs have become more complex, the need for staff dedicated to these programs has become more apparent, and indeed, such an arrangement has become the norm in many large primate programs. While American veterinarians are given legal responsibility for environmental enhancement plans for nonhuman primates by the AWA (USDA 1991), 48% of primate facilities reported having a behavioral scientist (with a PhD or master's degree) rather than a veterinarian providing primary oversight for behavioral management (Baker 2016). As discussed below, there are many necessary facets of successful behavior programs, and to expect care staff or veterinarians to manage these while keeping abreast of new developments in the large and evolving field of animal behavior, in addition to their normal duties, can be overwhelming to them.

Many other entities and individuals within a research organization may also contribute to behavioral management. The IACUC, notably, should include review of the behavioral management program during program reviews. Individual IACUC members should be trained to recognize abnormal behaviors or other signs of distress and note these when seen in animals during facility inspections, and the IACUC as a whole should work with operations staff and veterinarians to development optimal enrichment strategies for the species within a program. Facility managers and purchasing officers can provide valuable input into the cost–benefit analysis of new enrichment devices or procedures, while safety officials may have ideas for how to ensure personnel safety during implementation. Facility maintenance or a mechanical shop may be involved in the design and construction of enrichment devices or enhanced enclosures. Information technology staff may design ways to monitor the performance of a device. Regardless of which individuals in a facility have formal responsibility within the behavioral management program, all entities must work together toward the common goal of increased animal welfare and enrichment to achieve success.

Social Housing Program

Social housing serves a basic biological need for many laboratory animals. As indicated in the *Guide*, and by AAALAC International (http://www.aaalac.org/accreditation/positionstatements.cfm#social), social species should be housed in stable pairs or groups of compatible individuals unless precluded by experimental reasons, clinical issues, or behavioral incompatibility. Individually housed animals that are members of social species require scientific justification for single housing in the IACUC protocol to which they are assigned. Intermittent social or pair housing has been applied in some situations, typically to facilitate research needs. For most social species, group living is the most species-appropriate and desirable form of housing, although monogamous species (e.g., prairie vole and marmosets) are an obvious exception. If living in groups is not feasible, living with one other partner may be. For some species, particularly some domestic animals, interactions with humans is also an important means of supplying social stimulation.

Social housing programs should include an outline of the social introduction and monitoring processes to be used, the criteria used to select animals for introduction to one another (e.g., sex, age, and temperament), and the methods by which they are introduced. Some species, such as primates, may require a stepwise introduction process, which includes observations of interactions when animals are provided with first visual access, then protected contact via a divider of mesh or small holes, and finally full contact, while for other large animal species, such as livestock and dogs, introductions typically proceed from visual and olfactory contact directly to supervised full access in spaces large enough to allow escape in the case of an inappropriately aggressive interaction. For most rodents, typical management practices lead to the pair or group housing of littermates or young, sexually immature adults, which may reduce the necessity for closely observed introductions, or breeding pairs introduced via immediate full access contact.

Regardless of the species or method of introduction, socially housed animals should be regularly monitored for compatibility. Observations may be more intense in the days or weeks following introductions, but should continue at some level for all socially housed animals. Observations for compatibility should be conducted during routine cleaning, feeding, and research activities. Monitoring can be conducted by a trained observer in the room, via video recorder, or by remote viewing systems, such as Internet-secure Internet protocol (IP) cameras. Communication about the social introduction should be directed to all affected staff, if that information might impact their daily operations.

Environmental Enrichment Program

In an effort to address application of the term *environmental enrichment* to any change in an animal's environment, regardless of measurable benefit for the animal, Newberry (1995) defined it as "an improvement in the biological functioning of captive animals resulting from modifications to their environment." The scientific literature and regulatory documents have expanded on this definition, referring to environmental enrichment as features of a captive animal's environment or husbandry that increase that animal's number and range of normal behaviors, decrease abnormal behaviors, increase interaction with the environment, and enhance its ability to cope with behavioral and physiological challenges (Federation of Animal Science Societies 2010, p. 30). An environmental enrichment program consists of using a variety of techniques and providing a variety of resources to animals with the goal of modifying their behavior in these ways. Most of the nonsocial approaches used to improve the environments of research animals can be categorized into four general types: feeding, physical, sensory, and cognitive and occupational (Keeling et al. 1991). Food enrichment is consumable items that are either different in type from the animals' standard diets or delivered in such a way as to encourage natural foraging behaviors. Physical enrichments are elements of an enclosure that allow animals to experience different microenvironmental conditions, such as shelters, perches, or bedding. Sensory resources are those that engage an animal's senses without necessarily requiring interaction, such as television, music, or smells. Cognitive and occupational resources are those engaging animals in interaction with their environment in such a way as to be mentally stimulating, including the provision of puzzles, activities, or PRT. Of course, many techniques for improving animal environments could fall into more than one of these categories, such as puzzle feeders qualifying as both cognitive and food enrichment, and swimming pools qualifying as physical and sensory resources, but these general categories are useful for considering which aspects of an animal's experience could be enhanced.

In general, the design and selection of resources for the enrichment program is made to encourage appropriate behaviors that mimic those seen for a species in their natural environment. In this way, enrichment resource design and management should rely heavily on the natural history of a species and not on what might look interesting or colorful to a human observer. Natural histories describe the environment and context in which species live naturally. These remain relevant to the welfare of laboratory animals despite sometimes thousands of generations of removal from the "wild" state, as laboratory strains released into the wild consistently revert to naturalistic patterns of behavior (Hutchinson et al. 2005). Some important elements of natural histories with particular relevance to enrichment resource design include diet and food gathering behaviors, including what animals eat, and whether they live near a readily available food source or spend many hours a day finding food; sleep habits, including when in the day an animal sleeps and whether it seeks or builds shelter; and social structure, such as whether animals form long-lasting dominance relationships or are territorial. For example, wild mice instinctively build elaborate nests in which they sleep during the day; this could be used to infer that mice might benefit from the provision of nesting material and minimizing disturbances during the day (Olsson and Dahlborn 2002; Hutchinson et al. 2005). Likewise, macaques spend a great deal of time covering wide ranges in the search of food, and techniques to increase foraging time have been shown to have some beneficial psychological effects in captive primates (Bennett et al. 2010; Gottlieb et al. 2011).

Implementing any new enrichment technique or resource requires a team approach with input from a variety of personnel. Any new enrichment should be approved prior to distribution to any research animals. This includes cases of "one-time" or limited enrichment resources that are often chosen as much for their effect on human morale as for animal benefit (i.e., seasonally themed novel foods) and those with the intent to be implemented broadly or on a repeat basis. A behavioral specialist should evaluate potential for efficacy or harmful effects with the help of peer-reviewed scientific literature. A veterinarian should assess the enrichment for any health concerns, safety risks, or potential confounding effects on the research. The animal care and facility management staffs should determine a method of sanitation and logistical aspects related to the provisioning of the enrichment. The goal should be to provide maximum benefit and minimum harm to animals, at a minimum cost in labor and money (Bennett et al. 2010). In facilities where a behavioral scientist heads the behavioral management program, she will likely conduct the initial review and design or selection. In facilities without such personnel, resource selection may fall to a designated individual or committee. It should be noted that information may be found outside the literature search databases typically used for biomedical research and may be found in information provided by a manufacturer or published in non-peer-reviewed trade journals. An example of the importance of enrichment resource preapproval is the known interaction between grapefruit and many pharmaceutical compounds commonly used in research facilities (Bailey et al. 2013). This interaction is likely to be known to veterinarians, but perhaps not to a well-intentioned enrichment committee seeking to implement new food resources.

The construction and purchase of enrichment resources will depend on the funds, supplies, and means available. Some facilities have dedicated mechanical shops that construct enrichment devices, while others rely on dedicated behavioral specialists or enlist volunteer care staff for their design, construction, and repair. Regardless, any new enrichment device should undergo pilot testing prior to widespread adoption. Considerations prior to and during this testing phase include the potential impact on the research project utilizing the "test subjects," which will necessitate involvement of at least one willing primary investigator. During testing, devices should be monitored routinely to ensure they do not pose any safety risk to personnel or animals (e.g., certain plastics may be chewed and quickly develop sharp points). An assessment should be made of the efficacy of the enrichments, although the extent of this may depend on the resources and personnel available at the institution. Such a review would, at a minimum, determine whether the animals actually utilize the new device or resource, but optimally would include measuring effects on normal and abnormal behaviors or physiology in an objective manner. The ability to sanitize the enrichment device should also be reviewed as part of the testing phase. Whether this consists of visual inspection or more objective measures by such tools as ATPase testing or cultures will depend on the institution, existing SOPs, the vulnerability of the animals or personnel to potential fomite-transmitted pathogens, and the opinion of the design team and facility management.

Once an enrichment device has been tested, a team approach to program implementation should continue and determine whether the item is worth implementing more broadly. Again, this team should include the behavioral specialist (as available), veterinarian, and facility management and/or animal care staff. It may also include a purchasing or financial officer, to discuss the cost–benefit of implementation. All investigators whose animals will be impacted by the new enrichment device should be consulted, or at a minimum informed of the change, since enrichment devices and resource may impact important research variables. For enrichment resources that require manual distribution, such as novel foods and manipulanda, a system of documenting this provision should be developed to ensure complete implementation. After an enrichment device or resource is in use, there should also be a system by which concerns are reported by animal care staff or other personnel and where the device can be rereviewed for these reported concerns. Any injuries or negative health outcomes involving a device should be reported first to a veterinarian and, based on the severity and context of the incident, to relevant regulatory bodies as required by law. Such an incident should prompt a review of that device. If such a review is completed, it is important to consider the injury severity and rate against the total use and benefit when determining whether to remove or alter a device; a rarely used device with a high rate of injury would be a more obvious candidate for discontinuation than one used broadly with a low rate of minor injuries or mishaps. A periodic review of enrichment device efficacy should be conducted. As animals are exposed repeatedly to devices, some lose their novelty. This is especially true of puzzle-type devices provided to nonhuman primates (Meehan and Mench 2007). Literature regarding a device should also be monitored. An example of how consensus regarding enrichment programs evolves over time can be found in the consideration of mouse enrichment resources over the past few decades. By the late 1990s, a consensus had emerged that nesting material was a superior enrichment resource for rodents (Olsson and Dahlborn 2002). Since then, however, this consensus has developed significant nuance, with current literature suggesting the type of nesting material provided and subsequent form of the nest is a significant variable in determining how much benefit accrues to the animals (Hess et al. 2008; Gaskill et al. 2013).

Animal Training Program

Animal training is an integral component of behavioral management, and PRT is a powerful tool to gain cooperation and compliance by laboratory animals. Animal training occurs each time there is an interaction with an animal. When looking at the daily operations of a facility through the lens of animal training and operant conditioning, one can see animals responding to environmental cues all the time. For example, an animal caregiver approaches a group of animals with food, and the animals move toward the caregiver and are given the food. The animals are positively reinforced for moving toward the animal caregiver, and they are likely to do it again to gain food. An animal has learned and is now more likely to approach a person, as it previously received something it wanted (food) following that behavior. An animal caregiver spraying water with a hose approaches a group of animals who then move into an adjacent cage. Once they move, the spray from the hose ceases. These animals have been negatively reinforced for moving away from the animal caregiver and the next time will likely move quickly away from the caregiver spraying water. The behavior to move from one location to another is motived by avoiding something they find aversive (spraying water). These are examples of two different experiences for the animal, and some literature shows the use of negative reinforcement is more stressful for the animal than positive reinforcement (as measured by physiological and behavioral means, and response to humans). Personnel should be trained to recognize what types of negative reinforcement are used, to use it in the smallest amount possible to get the needed behavior, and to offer positive reinforcement following the animal's cooperation.

Initiating a PRT program (sometimes called "clicker training") is valuable and meets recommendations of the *Guide* (Perlman et al. 2012). Animal training techniques and some average required training times for a variety of behaviors are well documented in the literature (Schapiro et al. 2003; Coleman et al. 2008; Veeder et al. 2009; McMillan et al. 2014; Bloomsmith et al. 2015; Bliss-Moreau and Moadab 2016). These include conscious blood collection, urine collection, semen collection, restraint behaviors, presenting for injection, movement between enclosures, and managing aggressive behavior (see Schapiro et al. in Chapter 32 of this volume for a thorough review). When trainers talk to one another, they often describe individual differences in animals' willingness to train. Coleman (2012) has described how temperament can impact training success in primates; in one study, animals with exploratory temperaments were more successful training candidates than those with inhibited temperaments.

As a manager of an animal program, it is critical to understand that animal training should be applied by people who have been educated in animal training techniques. It is also important to note that progress is measured in terms of the animal advancing through the training steps, not just in the time the trainer has devoted to training sessions. To that point, McMillan et al. (2014) described a refinement in training primates for chair restraint in which PRT techniques were applied for several months. For those animals not progressing through the training steps by a set time point prior to initiation of the study requiring the chair restraint, the lowest magnitude of negative reinforcement required was applied and then was followed immediately by a food reinforcer. Minimal stress was caused, and subjects were trained for the behavior in the needed time frame. Once an animal is fully trained, the behavior needs to be maintained if it is to remain useful. Managers should schedule maintenance training for their staff to employ as needed; the frequency will vary with behaviors and individual animals.

Behavioral Assessment Program

Behavioral assessment is a critical component of an effective behavioral management program. A solid understanding of the natural behavior of a species is necessary to identify species-appropriate behaviors, provide an effective enrichment program to promote a broad repertoire of species-appropriate behaviors, and identify abnormal behavior so that appropriate treatments can be applied. For example, ferrets and squirrel monkeys exhibit scent-marking behavior, so providing enrichment to promote this behavior is important. Most primates spend much of their day searching for, processing, and consuming food, so developing ways for them to express these behaviors in captivity is an essential part of the enrichment program for primates. Understanding behaviors that are fundamental to a species' "lifestyle" and supplying the means for them to express them is an important consideration for their captive welfare. Per the *Guide*, "personnel responsible for animal care and husbandry should receive training in the behavioral biology of the species they work with to appropriately monitor the effects of enrichment as well as identify the development of adverse or abnormal behaviors" (p. 53). Education about the behavior of the species with which an individual works should begin at hire and be supplemented, reinforced, and updated throughout the career of a laboratory animal professional. This training should include the behavioral biology of the species, understanding behaviors that could be of concern, the mechanism by which to report behaviors of concern, and appropriate staff–animal interactions.

Since there are many different personnel (veterinary, animal care, and research staff) who work with any one animal, behavioral assessment programs should include a system of reporting and responding to behaviors of concern, such as stereotypic behaviors, fearful or aggressive behavior, hair plucking, barbering, hyperactivity, anxiety, and self-injurious behaviors. For example, animal caregivers may observe social aggression during routine feeding, veterinarians might observe an animal behaving fearfully when collecting biological samples, and research team members may observe an animal exhibiting self-directed abnormal behavior during testing. When such behaviors are observed, they should be reported to personnel responsible for behavioral care. This may be done verbally, especially if the animal is injured or in danger and immediate action is required, or in written form, such as on log sheets posted outside of each animal housing room. These log sheets should be reviewed regularly by personnel responsible for behavioral care who should assess the problem and provide viable treatments if appropriate. If treatments are applied, they should be documented and evaluated for efficacy.

A behavioral assessment program can be used to evaluate the presence, absence, rates, and durations of selected behaviors. These data can be recorded by using different observation and recording methodologies (Tables 5.1 and 5.2). As one example, the one-zero method will indicate whether individuals are expressing or not expressing particular behaviors during a time interval. Abnormal behaviors, fear-related behaviors, and other behaviors of concern are noted during brief observation periods using a check sheet–type format. Then if an individual meets predetermined thresholds for problematic behaviors, a more in-depth focal behavioral assessment can be completed. Based on this assessment, treatments are then provided and evaluated. Possible treatments may include social changes (such as a new social partner), animal training applications (such as desensitization to a stimulus causing fearful behavior), additional enrichment applications (such as additional foraging devices, destructibles, and visual barriers), and operational changes (such as cleaning the animal's stall first, or reducing the amount of time an animal is separated from its social group for a procedure). If the behavioral problem is reduced or ameliorated, that treatment is typically maintained. Hair coverage scores are also sometimes completed as part of a behavioral assessment program for some species prone to self-induced alopecia. These scores are useful to identify individuals with pronounced or increasing hair loss, and objective evaluations of hair regrowth following treatment (see Figure 5.1 as an example of a hair scoring diagram and scoring system based on Luchins et al. 2011).

TABLE 5.1

Methods of Observation of Animal Behavior

Method	Definition	Advantages	Disadvantages	Common Uses
Ad libitum	Observer writes down anything that seems relevant	Easy Allows for description of rare behaviors	Biased toward most conspicuous behaviors or individuals	Preliminary observations Recording rare events
Focal sampling	Observe one individual, recording all instances of behavior for specified amount of time	Detailed observation/ data collection possible	Difficult in field because animals go out of sight	Can be used with continuous, instantaneous, or one-zero sampling (see below)
Scan sampling	Whole group observation—behaviors of all individuals recorded at set intervals	Allows study of more subjects	Can record only small number of general behaviors/categories Biased toward more conspicuous behaviors and individuals	Determining species activity budgets Must use instantaneous sampling Sometimes combined with focal sampling
Behavior sampling	Observe whole group, record each occurrence of a particular behavior (focus is on behavior rather than individual)	Catches some rare behaviors that focal or scan sampling can miss	Biased toward conspicuous occurrences	Studying rare but important behavior, such as fighting/mating Can be combined with other techniques

Source: Based on Martin, P., and Bateson, P., *Measuring Behavior: An Introductory Guide*, Cambridge University Press, Cambridge, MA, 1993.

TABLE 5.2

Methods of Recording Animal Behavior

Method	Definition	Advantages	Disadvantages	Common Uses
Continuous recording	"All-occurrences" recording—observer writes down frequencies and durations for behaviors throughout observation interval	Produces a more exact record of behaviors Allows behavioral sequences to be recorded Less biased	Demanding data collection Limit to how many categories of behavior can be recorded at once Interobserver reliability can be difficult	Used in focal animal sampling
One-zero recording	Observation period divided into short intervals; on beep, observer notes whether a behavior has occurred at any time during the previous interval	Easier data collection, high interobserver reliability	Does not provide information such as how often behavior occurred during interval or for how long Biased data—usually overestimates durations and underestimates frequencies	Used in focal animal or scan sampling
Instantaneous sampling	Observation period divided into short intervals; on beep, observer records whether behavior is currently occurring	Good when intervals are short and behavior durations are relatively long Easier data collection—can take data on many categories of behavior Usually good interobserver reliability	If behaviors are short in duration, they may be missed Subject to observer bias if an interesting behavior occurs close to, but not on, the beep	Used in focal animal and scan sample techniques Good choice when behavior occurs rapidly, making continuous sampling difficult

Source: Based on Martin, P., and Bateson, P., *Measuring Behavior: An Introductory Guide*, Cambridge University Press, Cambridge, MA, 1993.

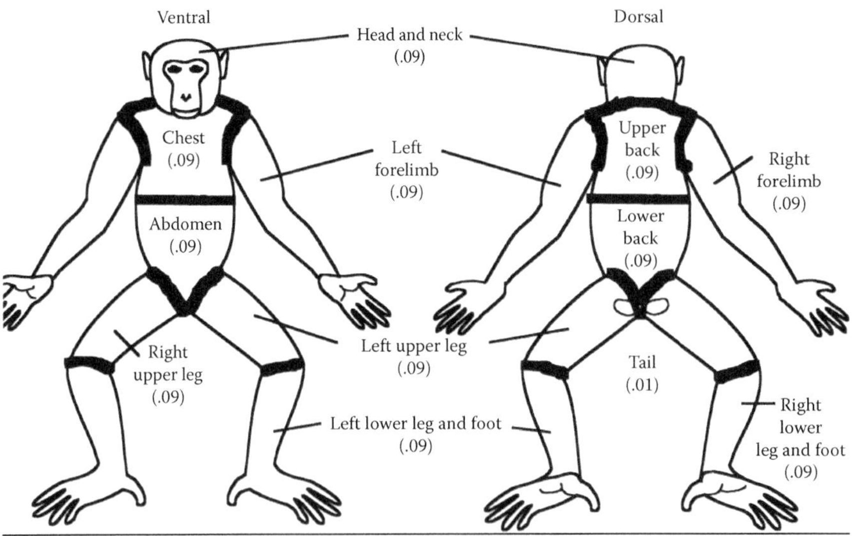

Alopecia score	% Body surface affected
0	0
1	>0 to 10
2	>10 to 25
3	>25 to 50
4	>50 to 75
5	>75 to 100

FIGURE 5.1 Example of hair score template for nonhuman primates. (From Luchins, K.R. et al., *J. Am. Assoc. Lab. Anim. Sci.*, 50(6), 926–938, 2011.)

Facility and Housing Design

The design of laboratory animal housing is a vital aspect of promoting animal well-being, as the physical environment can have a major influence on animal behavior. Animal housing should be designed to facilitate the expression of species-typical locomotor behavior, resting patterns, and activities. For example, arboreal species should have an adequate enclosure height and structures for climbing and resting off of the ground, and more terrestrial species should have more floor space available. Providing animals with areas for escape from others may reduce behavioral problems such as aggression. It is also important to keep in mind husbandry and care practices, policies, and requirements when designing animal rooms. Construction materials must be durable, sanitizable, and relatively easy to maintain. In addition, there are often regulatory requirements regarding room environmental parameters (e.g., airflow, temperature, and humidity) that can make providing the appropriate environment for a particular species challenging (e.g., nude, severe combined immunodeficiency [SCID], aged, or immunocompromised rodents many need to be housed in rooms with higher temperatures). Animal area and room doors should also be wide and tall enough to accommodate cage racks and equipment. With newer types of caging intended to accommodate social pairs or groups of animals, and with the growing use of larger "activity cages" to provide more space (sometimes on a rotating basis), rooms and doorways should be of sufficient size to allow their use. Room doors with viewing windows that have shutter-type options are preferred. Animal rooms with windows to the outside may be beneficial for some species. Another option is incorporating specialized lighting features and control mechanisms into the facility design that accommodates the natural behaviors of the animals, as well as allowing for husbandry and care procedures. For example, some nocturnal animals may be housed under reverse light cycle conditions such that husbandry and monitoring are done using red lights. Some other creative facility design ideas or concepts, such as a trickling water feature, may provide some background "white noise" (DeGenova and Abee 2006).

Sound abatement and use of materials that decrease sound transmission help reduce disruption and facilitate better behavioral management. Animal housing rooms and areas should be separated from high-traffic areas, as well as from cage wash or cage processing areas and areas where there is loud activity. Sound abatement measures should be considered, especially when housing sensitive species (e.g., rodent breeding colonies, gerbils, and rabbits) or particularly noisy species (e.g., dogs, swine, and nonhuman primates). It may also be important for animal housing areas and rooms to be in close proximity to testing and clinical procedure areas.

Constructing in-room pens or other enclosures instead of having caging in rooms has become more popular to maximize the space available to the animals, and to allow group housing of a variety of species (e.g., rabbits, cats, and farm animals). Such housing also allows the use of substrates (e.g., wood shavings, hay, and excelsior) on the floor of enclosures, which provides softness and a good opportunity to scatter food to promote foraging or rooting in the substrate. When caging is used, it should be flexible enough to allow introductions of unfamiliar animals (e.g., panels between cages that allow varying degrees of tactile contact between animals, and tunnels between cages). Features that add complexity and variety to the caging, such as "balconies," "porches," and "viewing ports," have been reported to be well used by some species.

It is possible to appropriately house some laboratory animals outdoors (e.g., nonhuman primates and dogs); this should be done whenever possible, as the complexity of the outdoor environment is believed to be beneficial. Outdoor enclosures must be appropriately designed and constructed for the specific species to be housed so that the environment is safe and secure. Outdoor areas must include provisions for shade and protection from wind and other weather elements, as well as provide visual barriers from other animals in the group, and opportunities for species-typical behaviors and activities. Visual barriers and shade structures can also increase the complexity of the environment, and should be constructed to allow the research team and other personnel to clearly view the animals. Ground cover or other substrate may be needed to provide the animals with a comfortable living surface. Selection of materials will depend on the species as well as the population density. A suitable ground cover can assist with erosion control, but the ground cover needs to be a material that can be regularly maintained. Grass, sod, and small gravel, as well as wood chips, hay, and excelsior, are all possible choices that should be evaluated and considered by the veterinary, animal care, facility maintenance, and behavioral management staffs prior to use. The ability to incorporate enrichment items (e.g., small foods) in the substrates should be evaluated. Rotating these substrates or combining them may also be an option that would add variety to the environment. Outdoor areas may also include climbing and play structures. These outdoor areas and equipment should be regularly cleaned and sanitized. They should also be inspected for wear and tear or other potential hazards, and in such case, the damage should be immediately repaired or the equipment replaced. In some cases, indoor living animals can spend brief periods of time outside, and this can be beneficial.

When animals are housed in large indoor enclosures, in large groups, or in outdoor enclosures, animal training programs become more important to achieve reliable access to the animals. It is desirable to design enclosures in such a way that the animals move toward people to shift for cleaning or to enter a transport box, for example, rather than moving away from people, so that PRT can be used for the training. The design of the area into which the animal is moved should be considered carefully. For example, tall animals are unlikely to readily move into a transport box that requires an unnatural or uncomfortable body position. Thus, poor equipment design can thwart the benefits of PRT. A transport box or enclosure design that provides adequate space can make a tremendous difference on the training outcome (e.g., moving the animal) and the methods necessary to complete the task (positive or negative reinforcement).

To determine features of caging systems, considerations must be made for personnel and animals, including species-specific characteristics, socialization potential, research needs and applications, provisions for enrichment devices and distribution methods, husbandry routines, and animal accessing and training. These factors also play a sizable role in the preparation of facility budgets and allocating funds and resources, including personnel and materials costs. There may be financial implications to staffing and supply budgets as well, as new demands on infrastructure to accommodate additional space for preparation and storage of enrichment program materials arise. However, there are also some creative ways to minimize and reduce costs. For example, some facilities incorporate the use of recycled cardboard or paper materials for bedding. In some cases, modifications to husbandry or care practices are implemented (e.g., the interval between husbandry practices may be extended, thereby decreasing personnel time devoted to cleaning and disturbing animals less frequently), with appropriate approvals and while remaining within the relevant recommendations and requirements (e.g., *Guide*, Office of Laboratory Animal Welfare [OLAW], and AAALAC). In such situations, managers and supervisors will need to evaluate caregiver time for activities, such as preparing, distributing, collecting, and cleaning enrichment materials and devices, and SOPs will need to be reviewed, modified, and approved by the IACUC to expand the approved cleaning methods.

Coordination of Behavioral Management with Research

Behavioral management programs should be carefully coordinated with biomedical and behavioral research in which the animals are involved. Finding the best way to balance the needs of laboratory animals with the needs of the research project in which they are subjects is a complex task. With increasing attention being paid to the psychological well-being of animals, the process must also include this consideration. Behavioral management plans must be harmonized with the goals of the research so that the animals benefit from social housing, enrichment, and animal training, all applied in such a way that the research is not unnecessarily compromised.

The IACUC review process is one forum for this negotiation between investigators and those responsible for behavioral management. The review of research proposals by the IACUC should involve careful consideration of how the research might involve deviation from the standard animal care program, including any requested changes in enrichment or social housing. Preferably, there should be at least one individual serving on the IACUC who has expertise in behavioral management. IACUC protocol forms should include sections for describing plans for social housing, environmental enrichment, and animal training. If an IACUC proposal indicates that social species are to be housed alone, scientific justification must be provided to explain why the research cannot be accomplished with socially housed subjects. Any time that must be spent alone or in "protected contact" (i.e., housing with limited tactile interaction possible between two animals) must be minimized, and additional enrichment should be considered. Each protocol should include information on how animal training will be used within the study so that it is clear what procedures the research staff will train or habituate the animals to, and how that training will be accomplished. There should be a plan for what will be done if subjects do not habituate to research procedures during the expected time frames.

As IACUC protocols are reviewed, investigative staff members may be asked to make changes or accommodations to their studies that they have not made in the past. With new techniques available to improve well-being and with new expectations for standards of care, investigators are essentially being asked to recalibrate some of their studies, and this may take time. Clear communication is important, and instruction, including working examples from other facilities and reviewing published literature with investigators as they attempt to make accommodations, will be helpful. Investigators can be challenged if their funding is not adequate to support these changes, which may bring about increased time that animals are assigned to a study and time that staff is needed to work on the study. In the future, the researcher will be able to incorporate new techniques (e.g., time to train animals to cooperate with research procedures) into funding proposals.

Some routine behavioral management activities may need to be coordinated with ongoing research. For some studies, it is best to have the social environment consistent throughout the study duration, whereas other studies may accommodate social housing only during some phases, so social introductions and separations need to be coordinated with the research staff as the study progresses. Approaches such as intermittent social housing (e.g., animals are housed together during portions of each day or each week) may be useful. Providing enrichment resources, especially foods, may need to be coordinated with research staff, particularly if the research involves any restriction of access to foods or fluids. In some cases, food enrichment resources may need to be provided only after certain research procedures are completed, and scheduling may need to be determined. Low or no-calorie foods may be given to some animals that have restricted access to foods for research purposes. Animal training procedures to help animals habituate to research procedures are typically performed by the research staff members, but behaviors that are trained by others (e.g., animal care, behavioral management, or veterinary personnel) should be transferred to all who might be able to apply the same training methods. For example, if a monkey is trained to cooperate with an injection of a substance for research purposes, the veterinary staff should be taught the process so that they can use the same training methods when accessing the monkey for a physical examination.

Behavioral management can augment research goals by improving the quality of the research. Animal training can make research procedures less stressful for the animals and more efficient for the research team. Handling that reduces animal stress also reduces the potential confounding influence of distress on the study's dependent measures. Human interactions, such as tickling rats, may help to reduce fear and avoidance of humans, and assist the rats in recovering more quickly from research procedures (Cloutier et al. 2015). Social housing can reduce the stressful influence of some research procedures, as the presence of a familiar companion can "buffer" the animal from these experiences (Gilbert and Baker 2011). Good science depends on animal models that are healthy both physically and psychologically, and good behavioral management programs can be an important means of maintaining healthy research animals.

Coordination of Behavioral Management with Veterinary and Animal Care Personnel

Teamwork, communication, and coordination of behavioral management activities with animal care and veterinary medicine are paramount to optimize the impact on animal welfare and to provide a holistic system of care. Setting up a system describing each group's responsibility for communication and action is important for reducing confusion and ensuring that behavioral problems are addressed properly. As an example, when animals show excessive fearful behavior, the people who observe the behavior should communicate the incident to staff providing the behavioral care. Animal care staff may report an animal was expressing fearful behavior when they were passing out food, veterinary staff might see fearful behavior when they are restraining an animal, and research staff may report fearful behavior when they are exposing an animal to new research equipment. The staff responsible for the behavioral care should assess the behavior and its environmental triggers, implement treatments such as positive human interaction and animal training, and assess these treatments for effectiveness. If appropriate, treatments should be communicated and coordinated for implementation with other staff as needed. A report of excessive fear may be discussed with veterinary staff and result in evaluation as to whether psychoactive medications might assist in reducing the behavior. Any such treatment should be communicated with staff responsible for behavioral care and researchers, as they will be evaluating any increase or decrease in the behavior or there may be an impact on the study.

Record Keeping and Documentation

In general, the behavioral management program should be reviewed on a regular basis by behavioral management staff, the attending veterinarian, and the IACUC to ensure it is beneficial to animal welfare and that it reflects current scientific knowledge (*Guide*, p. 53). Record keeping is important for developing behavioral management programs and for monitoring and improving established programs. Records are useful to determine whether the program is running as designed, whether assigned responsibilities are completed, which interventions are effective, and to report on program progress. Development of meaningful documentation systems leads to objective evaluation systems that can be used for identifying programmatic strengths and weaknesses, internal reporting, regulatory reporting, and in some cases, scientific reporting in journals and professional conferences. All aspects of behavioral management (behavior, enrichment, socialization, and animal training) should be documented. This is consistent with recommendations in the *Guide*, which states, "Records of rearing and housing histories, mating histories, and behavioral profiles are useful for the management of many species, especially nonhuman primates" (p. 75).

Behavioral assessment records allow examining the frequency, duration, or severity of behaviors of concern, the impact of treatments on the problem, and determining whether a different course of action is needed. This information can also be used to evaluate the population's behavioral health, such as the percentage of a colony expressing particular behaviors, and can depict behavioral responses to programmatic changes over time.

Record keeping should reflect that staff is following institutional SOPs for feeding, supplying enrichment devices, and treatments. Documentation should inform staff about what animals receive which enrichment resource, and should reflect enrichment device cleaning and sanitation outcomes. Records for social housing might include outcomes of each social introduction, reasons for separations, and the total number of socially and singly housed animals in the colony. If nonhuman primates are singly housed, the attending veterinarian is required to review the reasons for individual housing every 30 days. Maintaining such information allows programmatic evaluation and generation of progress reports. These are useful for assessments by outside groups (AAALAC International, U.S. Department of Agriculture [USDA], and Canadian Council on Animal Care [CCAC]) and IACUCs, as well as for advancing the program and comparing it with other programs and publications.

Animal training programs should have established communication and documentation systems, including clear communication of programmatic training goals, applications of training at the facility, and expectations for personnel implementation. These programs provide opportunities for consistency in the way in which animals are worked (i.e., to move the animals to where they are needed, all animals are called to the staff person, not sent away from a staff person), and communication among trainers and staff implementing or maintaining animal training is essential to its success. Records should be kept showing which animals have been trained and for what behaviors, the methods by which animals were trained (via positive or negative or a blend of the reinforcements), the cues the animals know, the expected animal response for each cue, a measurement of consistency in response, and the maintenance schedule (Perlman et al. 2012).

Selecting New Employees and Employee Training on Behavioral Management

In many programs, the animal caregivers are responsible for feeding and cleaning and also for preparing and distributing the environmental enrichment, moving animals from one location to another to facilitate social housing, and managing the feeding of animal groups. They should be trained on the importance of these daily responsibilities and that their interactions with animals can impact the animal's welfare. With these daily human–animal interactions in mind, the institution should provide training on essential elements of behavioral management: behavior, enrichment, animal training, social group, and pair housing.

First and foremost, staff education is needed to identify appropriate social and nonsocial behaviors and behaviors of concern (e.g., self-injurious behaviors, hair plucking or barbering, and fearful behaviors) for each species, as well as how and to whom concerning behaviors should be reported. Behavioral education provides a caregiver with information on how the animal is responding to its environment and to the events taking place. As an example, being able to identify when animals are showing signs of social incompatibility during feeding times can alert the caregiver that tension is mounting and a different feeding technique may help and/or to contact a behavioral specialist so the problem can be evaluated.

Socialization programs can be complex and require significant time investment, hands-on training, and detailed documentation (Truelove et al. 2015). To complete their role, animal caregivers must be able to identify social behaviors such as aggression or play, apply feeding techniques involving PRT, and use training and reinforcement for social separation to facilitate research.

Caregivers should be trained on requirements for environmental enrichment, such as their role in distribution, documentation, sanitization, and monitoring for safety. This includes enrichment basics such as methods to load and clean enrichment devices; schedules for using different types of foraging devices and other forms of enrichment; how to identify when enrichment devices, structures, or items are damaged and in need of repair; and understanding that damaged devices can cause harm to animals and need to be removed, replaced, and repaired. Other instruction could include how to rotate devices or toys, learning where clean enrichment is stored, and knowing with whom to communicate if enrichment supplies are needed. In addition to these responsibilities, instruction should be given on the outcomes of providing enrichment, such as increasing species-typical behavior, such as nesting, enhanced social interaction, manipulation, and play, as well as the potential for the expression of fear and anxiety to novel objects, and what to do when this is observed.

Caregivers should be trained on how to positively reinforce animal behavior and how to recognize the beneficial impact that it has to their daily work, such as giving food reinforcement or praise to a pig that calmly walks down the hallway to a scale for routine weighing. It is important to discuss how human actions can impact animal responses, and how to appropriately and safely behave around a particular species. As an example, loud noises can startle and increase stress for the animals. Quietly placing items on metal tables or quietly closing a door can greatly reduce noise and disturbance for many research animals.

Upon hire, there should be some instruction on basic behavioral management, and then in the months and years following, continuing educational opportunities. Continuing education in aspects of behavioral management is essential for all staff within an organization as the program changes and is enhanced. Supervisors and managers need to provide leadership and support, as well as work together to create a culture that reflects the importance and significance of the behavioral management program. These opportunities may include further instruction in formalized staff meetings, specialized hands-on training sessions from behavioral management personnel or consultants, webinars, or attending workshops or conferences. In facilities with behavioral management staff on site, animal care and behavioral management cross-training is highly encouraged. These interactions promote the exchange of ideas, understanding of job responsibilities and challenges, and opportunities to teach behavioral management techniques relevant to assigned responsibilities, such as providing food reinforcement immediately following handling or movement from one location to another.

As with all jobs, there are particular personality traits that may lead to more successful personnel and program outcomes. For those working closely with animals in a research environment, managers should be looking for individuals with compassion for animals, willingness to cooperate and collaborate with others, thoroughness with attention to detail, and evidence of being accountable and responsible. Young and Cipreste (2004) describe characteristics of good animal trainers and say these factors are important in the success of the animal training program. These characteristics include patience, a calm demeanor, and the ability to be both consistent and analytical of their own behavior.

For those working in behavioral management programs, learning how to correctly and thoroughly implement behavioral management techniques requires training, time, and practice. Many of the skills needed for behavioral management staff can be acquired through hands-on experience, classroom instruction, workshops, reading published literature, or visiting behavioral management consultants. Especially when training opportunities are limited within a facility, there are many professional organizations, workshops, conferences, web-based resources, and other reading material that will be useful for staff training and continuing education in behavioral management (Tables 5.3 and 5.4).

TABLE 5.3

Examples of Professional Organizations Whose Scope Relates to the Behavioral Management of Captive Animals

Professional Organizations	Website
American Association for Laboratory Animal Science	www.aalas.org
International Council for Laboratory Animal Science	www.iclas.org
Laboratory Animal Welfare and Teaching Exchange	www.lawte.org
Animal Welfare Institute	www.awionline.org
Laboratory Animal Management Association	www.lama-online.org
Veterinary Support Personnel Network	www.vspn.org
The Animal Behavior Management Alliance	www.theabma.org
American Society of Primatologists	www.asp.org
Academy of Veterinary Behavior Technicians	www.avbt.net

TABLE 5.4

Examples of Workshops, Conferences, and Other Learning Resources Related to Behavioral Management

Workshops, Conferences, and Other Learning Resources	Website
Symposium on Social Housing of Laboratory Animals	caat.jhsph.edu/programs/workshops/social_housing.html
International Conference on Environmental Enrichment	www.enrichment.org
Enrichment Extravaganza Conferences	
Primadaption Workshops	www.primateproducts.com/blog/2012/04/19/fall-2012-primadaption-workshops
Animal Training Applications in Zoo and Aquarium Settings	www.aza.org/animal-training-applications-zoo-aquarium-settings
Primate Behavioral Management Conference	www.mdanderson.org/education-and-research
AALAS Learning Library	www.aalaslearninglibrary.org
AAALAC Resources on Regulations and Standards Worldwide	www.aaalac.org/resources/internationalregs.cfm
Enrichment Record	www.enrichmentrecord.com
Mouse Ethogram	http://web.stanford.edu/group/compmed/cgi-bin/index.php
Lab Animal Science Professional	www.aalas.org/publications/las-pro#.Vk42v3arRmM
Shape of Enrichment	www.enrichment.org
Primate Enrichment Forum	www.primate.wisc.edu (to apply for listserv)
Workshop on Macaque Pair Housing	www.yerkes.emory.edu/education/2016%20Behavioral%20Management%20Workshop.html
Karen Pryor Academy for Animal Training and Behavior	https://www.karenpryoracademy.com/

Structure of Behavioral Management Programs

Behavioral management programs typically fall into three categories: project based, section-wide, and facility-wide (Perlman et al. 2012). One institution may implement multiple approaches simultaneously, and programs may evolve from one structure to another. A project-based approach includes a few animals and a limited number of staff members. Often, this approach is used within a research laboratory or with a small group of animals, and relies on a motivated individual and may not persist when that individual is not present. The benefits include increased welfare of animals and a positive impact on job satisfaction of those implementing the program. Drawbacks may include limited resources and support, the efforts can be viewed as expendable when time is restricted, often there may be less emphasis on safety for animals and staff, and limited training of staff and supervision of activities. Lastly, inconsistent implementation of behavioral management programs across an institution may be problematic for oversight organizations such as USDA, IACUC, and AAALAC.

A section-wide approach includes a greater number of animals and sometimes more staff members than the project-based approach. This approach often has the support of a manager and is implemented within a unit or department. The benefits include improved safety and staff training, and improved animal management. With more animals and staff involved, there may be a greater chance that additional changes will follow. Drawbacks include limited support from other departments that may interface with the same animals, limited opportunities for the continuing education of staff (such as attending conferences or workshops), and the behavioral management activities may not persist beyond the manager's tenure.

A facility-wide approach is implemented and supported by all departments throughout the entire facility and impacts the greatest number of animals. The benefits include support for continuing education and training of staff members, and allocation of resources toward the effort, such as dedicated staff and equipment. The program and goals are clearly defined, which includes both what is provided for the animals (e.g., all animals receive foraging opportunities daily) and who is responsible and accountable for providing those opportunities. Drawbacks include the cost of increased personnel and educational opportunities, but these are minimized through proper communication and education. Defining returns on investments should be part of the overall plan.

Characteristics of Successful Behavioral Management Programs

There are a number of hallmarks of research, teaching, and testing facilities that have superior behavioral management programs. When the psychological aspects of animal care programs are viewed as essential to daily work rather than as luxuries that will only be carried out when resources allow, then these programs flourish. Although upper-level management and administrative staff members may not engage directly in behavioral management programs, their support and interest are vital. Upper-level managers may determine the resources that will be devoted to behavioral management. Financial support is needed for personnel, supplies, equipment, storage space, and office space. Funds to send staff members to attend conferences and workshops on behavioral management are important for ensuring that they have the knowledge and skills to carry out the work. This is especially important in facilities that do not have individuals on staff that are trained in species-specific animal behavior. The most effective behavioral management programs also have forward-thinking individuals with a passion for improving animal welfare, and an organizational culture that embraces change and seeks improvement.

Behavioral management programs require constant evaluation and revision, especially as the animal population changes (e.g., new species are acquired), as needs of the animals change (e.g., new research procedures begin), and as new behavioral management techniques are employed (e.g., an animal training program is started). If behavioral problems are becoming more evident or more widespread (e.g., a certain abnormal behavior is observed more frequently or among a larger portion of the animal colony), then new plans should be devised to address and possibly prevent this problem behavior.

As with other elements of animal care programs, effective and standardized methods of communication with intentional redundancy are needed. Communication between those directing the behavioral management activities and animal care staff, veterinary staff, research staff, and shop or maintenance staff will be needed to carry out the program. For example, typically animal care staff members deliver enrichment items to the animals, but representatives of all these groups should be involved in determining the appropriateness of various types of enrichment for research animals.

Like other areas of laboratory animal care, behavioral management programs should be supported by SOPs and other written documents to describe central practices. Examples include SOPs for the enrichment program, for training animals for restraint procedures, and for social introduction procedures. Other written guidelines may include a schedule or calendar for delivering enrichment resources, detailed animal training plans for common procedures, descriptions of behaviors to look for among socially housed animals, and a written process for reporting behavioral problems. Written policies to promote the psychological well-being of nonhuman primates and to exercise dogs are required under the AWA.

Conclusions and Summary

Behavioral management is a critical part of a well-functioning animal research program, and is receiving increasing support and emphasis from the public, peer review and accreditation groups, and regulatory agencies. Behavior management programs are based on the goal of increasing animal well-being and species-appropriate behavior while decreasing abnormal behavior and distress. These goals are achieved by offering animals novelty, choice, and control within their environments (via environmental enrichment programs, good facility design, and animal training), habituating them to the procedures they will undergo (via collaboration with researchers, management, and veterinary staff), and allowing animals maximum social contact with conspecifics (via a species-appropriate socialization program). Each of these methods requires a team approach, with the investment of time, resources, and commitment from a variety of individuals within a facility, including behavior specialists, veterinarians, facility management, animal care staff, researchers, and the IACUC.

Successful behavioral management programs will necessarily evolve as new literature and technologies become available. With the increasing ease of use and decreasing cost of touchscreen devices, it is easy to anticipate these devices may be more commonly used to allow animals to interact with their environment in increasingly sophisticated ways, such as having switches to turn on or off sensory enrichment devices, such as television or music, or by completing simple mazes or puzzles to obtain a treat or other reward. With the advent of electronic record-keeping and analysis software built to handle "big data," the review of a behavior management strategy's efficacy may become much more routine, but may also introduce new context to conventional wisdom, such as with the increasing nuance regarding mouse nesting materials. Video and image analysis software are already used for automated behavior coding for mouse phenotyping purposes and to analyze nonhuman primate alopecia, and as these technologies become more reliable and cheaper, they could be used more routinely for behavioral assessment. Regardless of what changes in specific techniques may come, the characteristics of a well-functioning behavioral management team will remain constant: open communication with all relevant staff, adequate training in animal behavior according to each person's role, the establishment and analysis of valid performance measures, and regular self-evaluation.

ACKNOWLEDGMENTS

We greatly appreciate the collaboration of the entire Animal Resources Division at the Yerkes National Primate Research Center. Funding support was from the National Institutes of Health, Office of Research Infrastructure Programs (OD P51) OD011132. All aspects of management and research use at the Yerkes Center conformed to the Emory University IACUC and U.S. federal regulations and guidelines. We also appreciate the support of the National Institutes of Health Division of Veterinary Resources and its behavior unit. We are grateful for the assistance of Kristina Adams from the Animal Welfare Information Center for collecting literature resources. The Yerkes Center and the National Institutes of Health are fully accredited by the Association for Assessment and Accreditation of Laboratory Animal Care International.

FURTHER READINGS

Bayne, K., and P.V. Turner. 2014. *Laboratory Animal Welfare*. New York: Elsevier.
Young, R.J. 2003. *Environmental Enrichment for Captive Animals*. Oxford: Wiley-Blackwell.

REFERENCES

Adams, K.M., A.M. Navarro, E.K. Hutchinson, and J.L. Weed. 2004. A canine socialization and training program at the National Institutes of Health. *Lab Animal* 33: 32–36.

Bailey, D.G., G. Dresser, and J.M. Arnold 2013. Grapefruit-medication interactions: Forbidden fruit or avoidable consequences? *CMAJ: Canadian Medical Association Journal* 185(4): 309–316.

Baker, K.C. 2016. Survey of 2014 behavioral management programs for laboratory primates in the United States. *American Journal of Primatology* 78(7): 780–796.

Bennett, A.J., C.A. Corcoran, V.A. Hardy, L.R. Miller, and P.J. Pierre. 2010. Multidimensional cost–benefit analysis to guide evidence-based environmental enrichment: Providing bedding and foraging substrate to pen-housed monkeys. *Journal of the American Association for Laboratory Animal Science* 49(5): 571–577.

Bliss-Moreau, E., and G. Moadab. 2016. Variation in behavioral reactivity is associated with cooperative restraint training efficiency. *Journal of the American Association for Laboratory Animal Science* 55(1): 41–49.

Bloomsmith, M.A., K. Neu, A. Franklin, C. Griffis, and J. McMillan. 2015. Positive reinforcement methods to train chimpanzees to cooperate with urine collection. *Journal of the American Association for Laboratory Animal Science* 54(1): 66–69.

Cloutier, S., K.L. Wahl, J. Panksepp, and R.C. Newberry. 2015. Playful handling of laboratory rats is more beneficial when applied before than after routine injections. *Applied Animal Behaviour Science* 164: 81–90.

Coleman, K. 2012. Individual differences in temperament and behavioral management practices for nonhuman primates. *Applied Animal Behaviour Science* 137: 106–113.

Coleman, K., L. Pranger, A. Maier, S.P. Lambeth, J.E. Perlman, E. Theile, and S.J. Schapiro. 2008. Training rhesus macaques for venipuncture using positive reinforcement techniques: A comparison with chimpanzees *Journal of the American Association for Laboratory Animal Science* 47(1): 37–41.

DeGenova, R.L., and C.R. Abee. 2006. Creating a naturalistic research environment. *ALN Magazine*. www.alnmag.com/article/2006/04/creating-naturalistic-research-environment.

Federation of Animal Science Societies. 2010. *Guide for the Care and Use of Agricultural Animals in Research and Teaching*. Champaign, IL: Federation of Animal Science Societies.

Gaskill, B.N., A.Z. Karas, J.P. Garnerand, and K.R. Pritchett-Corning. 2013. Nest building as an indicator of health and welfare in laboratory mice. *Journal of Visualized Experiments: JoVE* (82): 51012.

Gilbert, M.H., and K.C. Baker. 2011. Social buffering in adult male rhesus macaques (*Macaca mulatta*): Effects of stressful events in single vs. pair housing. *Journal of Medical Primatology* 40(2): 71–78.

Gottlieb, D.H., S. Ghirardo, D.E. Minier, N. Sharpe, L. Tatum, and B. McCowan. 2011. Efficacy of 3 types of foraging enrichment for rhesus macques (Macaca malatta). *Journal of the American Association for Laboratory Animal Science* 50(6): 888–894.

Hess, S.E., S. Rohr, B.D. Dufour, B.N. Gaskill, E.A. Pajorand, and J.P. Garner. 2008. Home improvement: C57BL/6J mice given more naturalistic nesting materials build better nests. *Journal of the American Association for Laboratory Animal Science* 47(6): 25–31.

Hutchinson, E., A. Avery, and S. Vandewoude. 2005. Environmental enrichment for laboratory rodents. *ILAR Journal* 46: 148–161.

Keeling, M.E., P.L. Alford, and M.A. Bloomsmith. 1991. Decision analysis for developing programs of psychological well-being: A bias-for-action approach. In M.A. Novak and A.J. Petto (eds.), *Through the Looking Glass* (pp. 57–65). Washington, DC: American Psychological Association.

Laule, G.E., M.A. Bloomsmith, and S.J. Schapiro. 2003. The use of positive reinforcement training techniques to enhance the care, management, and welfare of primates in the laboratory. *Journal of Applied Animal Welfare Science* 6: 163–174.

Luchins, K.R., K.C. Baker, M.H. Gilbert, J.L. Blanchard, D.X. Lui, L. Myers, and R.P. Bohm. 2011. Application of the diagnostic evaluation for alopecia in traditional veterinary species to laboratory rhesus macaques (*Macaca mulatta*). *Journal of the American Association for Laboratory Animal Science* 50(6): 926–938.

Martin, P., and P. Bateson. 1993. *Measuring Behavior: An Introductory Guide*. Cambridge, MA: Cambridge University Press.

McMillan, J.L., J.E. Perlman, A. Galvan, M.A. Bloomsmith, and T. Wichmann. 2014. Refining the pole and collar method of restraint: Emphasizing the use of positive training techniques with rhesus macaques. *Journal of the American Association for Laboratory Animal Science* 53: 61–68.

Meehan, C.L., and J.A. Mench. 2007. The challenge of challenge: Can problem solving opportunities enhance animal welfare? *Applied Animal Behaviour Science* 102(3–4): 246–261.

National Research Council. 2011. *Guide for the Care and Use of Laboratory Animals*. 8th ed. Washington, DC: National Academies Press.

Newberry, R.C. 1995. Environmental enrichment: Increasing the biological relevance of captive environments. *Applied Animal Behaviour Science* 44: 229–243.

Olsson, A.I., and K. Dahlborn. 2002. Improving housing conditions for laboratory mice: A review of environmental enrichment. *Laboratory Animals* 36(3): 243–270.

Perlman, J.E., M.A. Bloomsmith, M.A. Whittaker, J.L. McMillan, D.E. Minier, and B. McCowan. 2012. Implementing positive reinforcement animal training programs at primate laboratories. *Applied Animal Behaviour Science* 137: 114–126.

Roma, P.G., M. Champoux, and S.J. Suomi. 2006. Environmental control, social context, and individual differences in behavioral and cortisol responses to novelty in infant rhesus monkeys. *Child Development* 77: 118–131.

Schapiro, S.J., M.A. Bloomsmith, and G.E. Laule. 2003. Positive reinforcement training as a technique to alter nonhuman primate behavior: Quantitative assessments of effectiveness. *Journal of Applied Animal Welfare Science* 6(3): 175–187.

Truelove, M.A., A.L. Martin, J.E. Perlman, J.S. Wood, and M.A. Bloomsmith. 2015. Pair housing of macaques: A review of partner selection, introduction, techniques, monitoring for compatibility, and methods for long-term maintenance of pairs. *American Journal of Primatology* 79(1): 1–15.

USDA (U.S. Department of Agriculture). 1991. www.nal.usda.gov/awic/final-rules-animal-welfare-9-cfr-part-3-0.

Veeder, C.L., M.A. Bloomsmith, J.L. McMillan, J.E. Perlman, and A.L. Martin. 2009. Positive reinforcement training to enhance the voluntary movement of group-housed sooty mangabeys (*Cercocebus atys atys*). *Journal of the American Association for Laboratory Animal Science* 48(2): 192–195.

Young, R.J., and C.F. Cipreste. 2004. Applying animal learning theory: Training captive animals to comply with veterinary and husbandry procedures. *Animal Welfare* 13: 225–232.

6

Education and Outreach Programs

Daniel T. Stimson

CONTENTS

Animal Research: Why Is It Necessary?

Animal research has played a vital role in virtually every major medical advance of the last century—for both human and animal health. Antibiotics, blood transfusions, chemotherapy, dialysis, insulin, organ transplantation, heart bypass surgery, joint replacement, vaccinations, and a host of other treatments to limit disease, pain, and suffering are based on knowledge attained through research with lab animals.

The use of animals in research began in antiquity, and gathered force in the eighteenth century, when philosophers and scientists shifted from studying an ideal, often spiritual world to studying the natural world. The nineteenth century saw the birth of physiology—the study of living organisms and their parts—with deepening exploration of human and animal bodies through the study of human cadavers and living animals. Two other developments during that time would have a lasting impact on biomedical research and the place of animals within it. In 1847, William Morton discovered the anesthetic properties of ether, allowing researchers to study animal physiology while reducing animal suffering. And in 1859, Charles Darwin published *The Origin of Species*, in which he proposed that all humans and nonhuman animals evolved from a common source, "having been originally breathed into a few forms or into one" (Darwin 1859). This shared origin of humans and animals—and the shared physiologies we have inherited—provided a strong scientific basis for studying animals as a proxy for humans.

In the modern era, genetic research has further strengthened this rationale. Despite obvious differences between human and mouse, the mouse genome is about 92% identical to ours. Even the fruit fly genome shares about 44% identity with the human genome. Together, our similarities to animals—in physiology and genetics, as well as behavior, development, and molecular biology—provide a rationale not only for studying animals to gain insights into human health, but also for treating them humanely.

Insights drawn from animal research have been critically important in efforts to improve human and animal health. Since 1900, the average human life expectancy in the United States has increased by about 30 years. In 2015, infant mortality in the United States—a key indicator of the nation's health—was measured at 5.8 deaths per 1000 live births, compared with 55.7 deaths per 1000 live births in 1935 (National Center for Health Statistics).

In the early twentieth century, infectious diseases were the leading cause of death in the United States. Today, we have antibiotics to treat potentially fatal bacterial infections, largely thanks to research on rodent models. Research conducted on animals has also led to vaccines against deadly viruses such as polio, rabies, mumps, rubella, and hepatitis.

Animal research has also played an important role in developing drugs for common disorders, such as heart failure, stroke, asthma, diabetes, and arthritis. The lipid-lowering effects of statins, which can reduce the risk of heart attack and stroke, were first discovered in animal models (Tobert 2003). Insulin, the pancreatic cells that produce it, and its ability to lower the spikes in blood sugar associated with diabetes were all discovered in animals (White 2014). (For more examples, see "Animals behind the Top 25 Prescribed Drugs in the United States (2014)" in Appendix 6.1.)

Life-saving medical procedures and surgeries have also been developed through animal research. Blood transfusions provide a powerful example. Each year in the United States, more than 14 million blood transfusions are given to support blood volume during surgery, to compensate for acute blood loss, or to treat severe burns or infections (CDC 2016b). Knowledge of blood flow and its role in transporting oxygen and nutrients throughout the body is derived from animal studies that began in the seventeenth century. Animal studies also enabled the earliest transfusions, and contributed to the understanding of different blood types and their compatibility (Giangrande 2000).

Many have questioned whether animal studies are necessary to advance medical research. Here, it is important to trace the development of new therapies from discovery to first-in-human testing, and to consider the alternatives to using animals during this process.

Developing a new method to prevent or treat disease begins with understanding disease mechanisms; human studies might provide clues, but animal studies are typically used to show cause and effect. Again, research on blood helps illustrate the point. Until the 1960s, as many as 10,000 infants died annually in the United States from a condition called hemolytic disease of the newborn. Human case studies pointed to blood-typing proteins as the likely culprit. But it was a series of blood transfusion studies on rhesus monkeys and other animals that pinpointed the cause—a protein called Rh factor that, when present in fetal blood, could cause a pregnant mother's immune system to attack her child. Risk of this reaction can now be determined by a simple blood test and prevented with a drug called RhoGAM (Stockman and de Alarcon 2001). Discovery of the Rh factor is just one example of a vast body of Nobel Prize–winning work made possible through animal research (see Appendix 6.1).

A mass poisoning in American history illustrates how tragedy can unfold when a new treatment is introduced for human use without the benefit of animal studies. In 1937, in response to high demand for strep throat medications, the S. E. Massengill Company began making a new liquid antibiotic. Before marketing, the drug was tested for taste, appearance, and odor—but not for toxicity—which was not legally required at the time. Unknown to the company's chief chemist, the liquid solvent was poisonous, and ultimately killed hundreds of people. Public outrage led to the passage of the Food, Drug, and Cosmetic Act, which expanded the powers of the U.S. Food and Drug Administration (FDA) to ensure drug safety (Ballentine 1981). Today, FDA rules generally require toxicity studies in animals before new drugs can move to human testing (Food and Drug Administration 2016).

There are useful methods to complement studies of humans and animals. These include computer modeling, noninvasive imaging, and laboratory-grown cells. Such cellular models are becoming increasingly sophisticated; there are now many tissue and organ "chips" that combine cells with other organic and artificial components to model parts of the human body. Researchers have developed these systems in recognition that animal models cannot answer every question in biomedical research, and in deference to the 3Rs. These guiding principles call for researchers to work toward *reduction* in the number of animals used in research, *refinement* of methods to reduce animals' pain and suffering, and *replacement* of animals with alternative models (Russell and Burch 1959). But despite advances in alternative models, it remains the case that for many biomedical research questions, animals provide the closest approximation to how our bodies function in health and disease.

Recent Animal Research Achievements

Alzheimer's Disease

Alzheimer's is the sixth leading cause of death in the United States and currently affects more than 5 million Americans. In 2015, the cost of treating Alzheimer's and associated dementias in the United States was $226 billion. As Alzheimer's progresses, a protein fragment called beta amyloid accumulates in the brain. Studies of rodent models suggest that beta amyloid plays a causal role in the disease. Researchers are thus pursuing beta amyloid and functionally related proteins as targets for therapy. In several ongoing clinical trials, scientists are testing the effects of an approach called immunotherapy, which involves triggering an immune reaction to destroy the target proteins (Alzforum 2016).

Cancer

Thanks in part to diagnostic tests and therapies based on animal research, cancer deaths have been on a continuous decline in the United States since the 1990s. Advances in treating childhood leukemia provide one of the most dramatic success stories in cancer history. In the 1940s, acute lymphocytic leukemia (ALL) was uniformly fatal. But thanks to the development of effective chemotherapy—tested in part with mice—the cure rate for children with ALL is now about 80% (Simone 2008). Animal research has also been critical for developing a new class of treatment called targeted therapy. While chemotherapeutic drugs work by poisoning cancer cells—and can poison healthy cells too—targeted therapies attack unique molecules found in cancer cells. The FDA has approved many targeted therapies for diverse types of cancer (National Cancer Institute 2016) (Figure 6.1).

Cystic Fibrosis

Cystic fibrosis (CF) is a genetic disease in which mucus builds up in the lungs, blocking the airways and creating a welcome environment for bacteria. Decades of research with mouse models helped establish that the disease is caused by defects in a protein needed to clear fluid from the lungs (Keiser and Engelhardt 2011). Advances in supporting ventilation and preventing infection are helping people with CF live longer and healthier lives, but the average life span for adults with CF is still less than 40 years (National Library of Medicine 2016). Since 2012, the FDA has approved two drugs that target the protein defects at the root of the disease (Pollack 2015). Gene therapy for CF has also shown promise in a clinical trial (Alton et al. 2015).

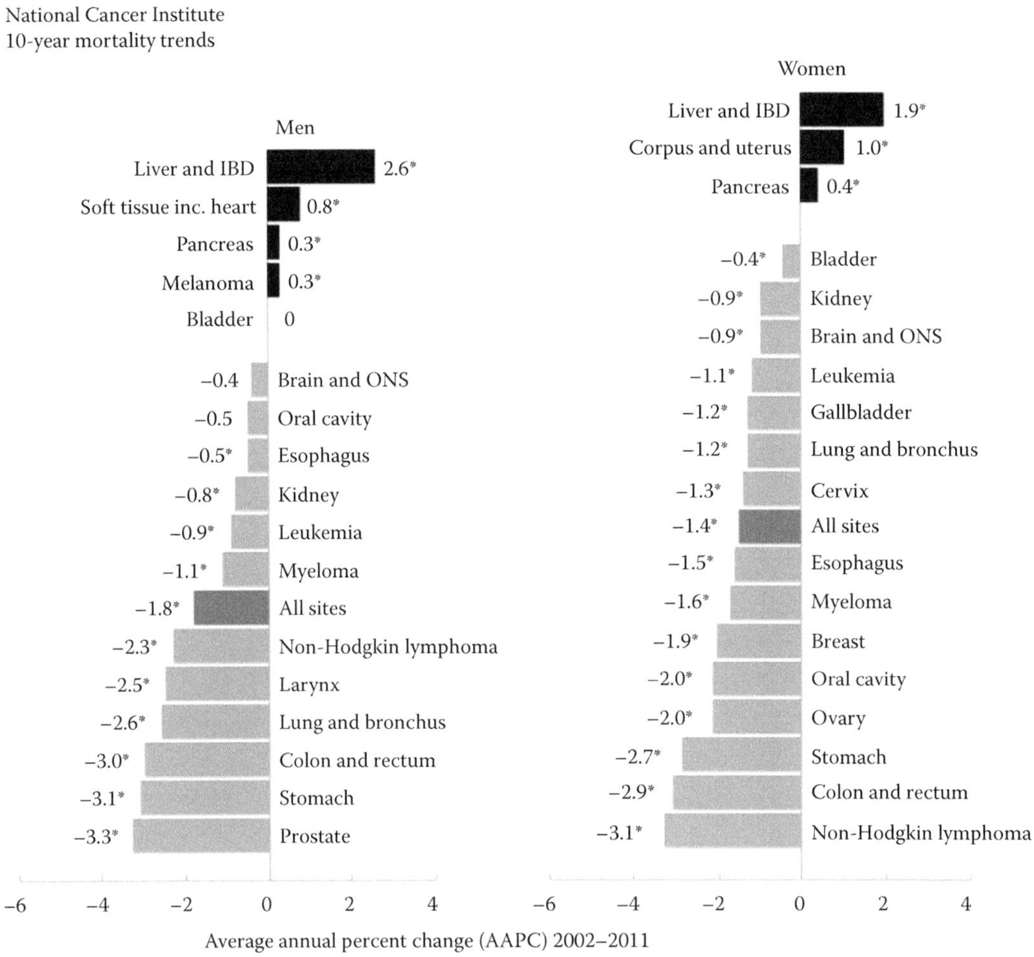

FIGURE 6.1 Ten-year mortality trends for the most common cancers, from 2002 to 2011. During this time, death rates increased for some types of cancers (black bars), but decreased for most (light gray bars). Overall cancer death rates declined by about 1.8% per year among men and by 1.4% per year among women (gray bars). *AAPC is significantly different from zero ($p < .05$). (Annual Report to the Nation on the Status of Cancer 1975–2011, National Cancer Institute, Bethesda, MD.)

Ebola

The 2014 Ebola outbreak claimed more than 11,000 lives in West Africa. Ebola causes severe hemorrhagic fever in humans and other mammals (CDC 2016a). Survivors may continue to suffer long-term effects even after the infection has cleared. Because Ebola evolved from primate viruses, animal research has been essential in prevention and treatment efforts. More than a dozen candidate vaccines have shown promise in animal studies, and four have moved into clinical trials, all at record-breaking speed (World Health Organization 2016a).

Epilepsy

There are now more than 20 antiseizure medications available to the 2.3 million adults and 450,000 children living with epilepsy in the United States (National Institute of Neurological Disorders and Stroke 2016a). None of these drugs would have been approved for human use unless their safety had been tested first in animals. Since 1975, the Epilepsy Therapy Screening Program at the National Institutes of Health (NIH) has used rodent models to rapidly test new candidate epilepsy drugs. The program has helped develop some of the most commonly prescribed epilepsy drugs (National Institute of Neurological Disorders and Stroke 2016b).

Heart Attack and Stroke

Animal research has contributed to dozens of drugs that help prevent heart attack and stroke through various means, such as controlling blood clots, blood pressure, and diabetes. Among the more recent additions to this arsenal are cholesterol-lowering drugs called statins. High cholesterol was identified as a potential risk factor for heart attacks and strokes in the 1960s. By the late 1970s, researchers began to develop and test statins in animals (including rabbits, dogs, and monkeys). Today, nearly 30% of adults over age 40 have taken a cholesterol-lowering medication within the past 30 days (Tobert 2003).

Hepatitis

Viruses are the most common cause of hepatitis, an inflammation of the liver than can lead to cirrhosis, end-stage liver disease, and liver cancer. More than half a billion people worldwide are infected with hepatitis B, C, and D each year (National Institute of Allergy and Infectious Diseases 2016a). Thanks to animal studies, a vaccine for hepatitis B became available in the 1980s, and since then, the number of people infected with the virus each year in the United States has dropped from about 250,000 to about 70,000 (Centers for Disease Control and Prevention 2002). Although more elusive, two vaccines for hepatitis C have moved into clinical trials thanks to encouraging tests in nonhuman primates (White et al. 2014). A variety of antiviral drugs can help reduce viral load and manage symptoms in people with chronic hepatitis infections. Animal research has helped lead to a new generation of antivirals for hepatitis C that are more effective and easier to tolerate than older drugs (Feeney and Chung 2014).

HIV

More than 1.2 million Americans are currently living with HIV (CDC 2016c). Over the course of the global HIV/AIDS epidemic, more than 70 million people have been infected and about 35 million people have died (World Health Organization 2016b). When the epidemic began in the 1980s, HIV/AIDS was universally fatal. Today, there are 31 FDA-approved antiretroviral drugs available to treat HIV infection, and with proper treatment, many people with HIV are able to live a normal life, to have children, and to watch them grow up (National Institute of Allergy and Infectious Diseases 2016b). An HIV vaccine, once thought impossible due to the virus's capacity for rapid mutation, is now in clinical trials thanks in part to animal research (National Institute of Allergy and Infectious Diseases 2016c).

Influenza

Certain strains of influenza can be severely debilitating and even deadly, especially for children, pregnant women, and the elderly. Each year, influenza kills as many as 500,000 people worldwide. As many as 160 million doses of influenza vaccine are distributed annually in the United States to prevent flu and reduce its impact (CDC 2016d). But infectious disease experts must still take their best guess about the most dangerous and prevalent strains of influenza from year to year. To eliminate this guesswork, researchers are working on a universal flu vaccine with the help of animal models. Historically, ferrets, mice, and guinea pigs have been important animal models in influenza research (Bouvier 2015).

Joint Disorders

The majority of patients who undergo successful hip and knee replacements each year no longer require the use of wheelchairs and experience less pain when walking. Companion pets, most commonly dogs, are also the recipients of successful joint replacements, which can greatly improve their quality of life (Ohio State University 2016).

Kidney Disease

Prior to the 1960s, kidney failure was a leading cause of death among young adults (Couser 2016). Today, people with kidney disease are living longer, healthier lives thanks to improvements in dialysis (in which the blood is filtered of waste) and kidney transplantation. Research on sheep was essential for developing the shunts that are used in dialysis, and for the drugs used to prevent immune rejection of kidney transplants.

Organ Failure

The lives of hundreds of thousands of kidney, liver, and heart transplant recipients have been prolonged and enhanced thanks to surgical advances and the development of effective immunosuppressive drugs that prevent organ rejection. In 2014 alone, more than 29,000 people received organ transplants in the United States (Health Resources and Services Administration). Moreover, life-saving procedures, such as open-heart surgery, coronary artery bypass, valve replacement, and the repair of congenital heart defects—surgeries developed with the help of lab animals—are now common practice. For a decade, two types of artificial heart—both tested in calves—have been providing heart failure patients the critical time they need to find suitable heart donors (O'Brien 2015).

Malaria

Malaria is a chronic, sometimes fatal disease caused by a parasite transmitted by mosquitoes. In 2015, an estimated 214 million people worldwide were diagnosed with malaria, and 438,000 died. A new generation of drugs has been developed to fight the most severe forms of this disease, and researchers are working to develop a malaria vaccine with the help of rodents and other animal models.

Paralysis

About 6 million people in the United States are living with paralysis (Reeve Foundation 2016). Scientists are studying rodents and other mammals to develop potential new therapies to spur neurons in the spinal cord to grow and create new connections, enabling recovery of sensations and motor functions. An approach that combines spinal stimulation with supported locomotion on a treadmill has even helped people with paralyzing spinal injuries recover some ability to move their legs (Angeli et al. 2014). Other researchers have used cats to understand and treat facial paralysis (Iritani et al. 1991).

Parkinson's Disease

Each year, about 50,000 Americans are diagnosed with Parkinson's disease, which causes tremors, slower movements, poor balance, and muscle rigidity (National Institute of Neurological Disorders and Stroke 2016c). Nobel Prize–winning work with rabbits taught us much about the drug L-DOPA, which helps control these symptoms and is a cornerstone of treatment for Parkinson's disease (see Appendix 6.1). People who become resistant to L-DOPA can benefit from deep brain stimulation (DBS), which was developed in part through research with monkeys and approved for treating Parkinson's in the 1990s. In DBS, a very fine wire is implanted into the brain and used to stimulate motor control areas. The stimulation is powered by a battery pack implanted near the collarbone. The technique can bring about dramatic and long-lasting improvement (Noonan 2014).

Vision Loss

More than 10 million Americans have age-related macular degeneration (AMD), the leading cause of vision loss in the United States. AMD destroys central vision, which is needed for everyday tasks like reading, driving, or simply recognizing the faces of loved ones (American Macular Degeneration 2016).

A class of drugs called vascular endothelial growth factor (VEGF) inhibitors was developed through animal research, and can slow vision loss from the most common form of advanced AMD (called wet or neovascular AMD). A recent study found that after 5 years of treatment with VEGF inhibitors, half of patients retained 20/40 or better vision, good enough to drive or read (Maguire et al. 2016). State-of-the-art therapies are under development for other vision disorders. A gene therapy approach—first studied in mice—has been shown to restore at least some sight to young people blinded by the genetic disease Leber congenital amaurosis (National Eye Institute 2016).

Animal Research for Animals

Animal research has improved health care not only for humans, but also for animals, including companion animals, farm animals, wildlife, and endangered species. Our genetic and physiological similarities to animals means that we are susceptible to many of the same health problems, and that we can benefit from similar interventions (Figure 6.2). We share vulnerability to some infectious diseases, including rabies and Lyme disease. Studies of dogs, in particular, were helpful in developing antibiotic treatment regimens and diagnostic blood tests that are used for both canine and human patients with Lyme disease (Krupka and Straubinger 2010).

In addition to infectious diseases, animals and humans are susceptible to common diseases and conditions that depend on factors such as aging, diet, physical activity, and genetics. Examples of shared treatments for diseases that are common to humans and animals include

- Anti-inflammatory drugs for arthritis
- Chemotherapy for cancer
- Eye drops for glaucoma
- Insulin for diabetes
- Pacemakers for an abnormal heartbeat
- Surgeries for traumatic injury
- Transplants to repair damaged organs and tissues (bone, cornea, and kidney transplants are common among companion animals)

According to a 2005 report from the National Academies exploring challenges in veterinary medicine and research, the level of public funding devoted to a particular animal health issue tends to correlate with its relevance to human health. The report notes that animal diseases that also threaten human health have been funded at "munificent levels" by the NIH, and that increases in the NIH budget from 1998 to 2003 have led to significant benefits for animal health (Committee on Assessing the Nation's Framework for Addressing Animal Diseases 2005).

Animal Models

About 95% of all lab animals are mice, rats, fish, and birds (Hastings Center 2012). Mice, in particular, have become the preeminent animal model for a host of diseases. Its small size and ability to reproduce rapidly have made the mouse an attractive choice to biologists for nearly 100 years (Jackson Laboratory). We now know that the mouse genome contains essentially the same complement of genes found in the human genome. So studying how a particular gene works in mice can shed light on that gene's role in human health and disease (National Human Genome Research Institute 2002). Mouse genetics has opened many doors to new therapies not only for rare genetic diseases, but also for common diseases, including many forms of cancer. Scientists also are able to breed mice with genetic mutations that cause human diseases. In the past 25 years, it has also become possible to turn mouse genes on and off at specific times during the life span and in specific tissues and organs. These conditional mutations allow

FIGURE 6.2 Because of our shared physiology, animals and humans also have a shared vulnerability to diseases such as diabetes and cancer. Studying animals to develop and test new interventions for these diseases has benefited humans and animals alike. (Courtesy of the Foundation for Biomedical Research.)

researchers to probe exactly when and where in the body a gene is needed, thus supporting progress toward therapies that act with similar specificity, without unwanted side effects.

Mice are not always the best model for answering questions about human health and disease. To seek an understanding of fundamental, shared biology or search for new genetic pathways, researchers often turn to cell models, or fruit flies or other invertebrates (National Human Genome Research Institute 2002). On the other hand, some research questions are more effectively addressed in studies of large mammals. For example, nonhuman primates have been essential for studying basic brain function, certain brain disorders, and infectious diseases that preferentially affect primates, such as Ebola and HIV (Lankau et al. 2014). Because their organs are similar in size to ours, pigs have been important for the development of life-saving surgeries, including organ transplants (Dehoux and Gianello 2007). Research on the visual systems of dogs and cats—which are similar to ours—has led to treatments for visual disorders and blindness (Wiesel 1982).

While nonhuman primates, pigs, dogs, and cats have made clear contributions to medicine, they make up less than one-half of 1% of animals used in research. As described below, federal laws and regulations impose important limits on research involving animals, especially large mammals.

Oversight

In the United States, the use of animals in research is regulated by the Animal Welfare Act (AWA) and Public Health Service (PHS) policy. The AWA has requirements for housing, feeding, ventilation, and other care that is appropriate for the animal species. The law also requires researchers to minimize animals' exposure to pain, distress, and discomfort. Researchers must provide exercise for dogs, and ensure psychological well-being for primates (U.S. Department of Agriculture 2016a). Although the AWA does not apply to mice, rats, or birds that are bred for research, PHS policy applies AWA requirements to the use of these animals in federally funded research (Office of Laboratory Animal Welfare 2015; National Science Foundation 2016).

Under the AWA and PHS policy, research institutions are required to establish an Institutional Animal Care and Use Committee (IACUC) to oversee all work with animals. IACUCs review animal research projects from the planning stage through completion. They require researchers to justify their need for animals, select the most appropriate species, explain why the research cannot be accomplished without an alternative model, and use the fewest number of animals possible to answer a specific question. They enforce the AWA's requirement that researchers take steps to minimize pain, and otherwise ensure animal welfare to the extent that they can without affecting the science. All IACUCs include at least one veterinarian and one community representative, unaffiliated with the institution. These committees oversee and monitor every potential experiment to help ensure optimal animal care. They have the authority to reject any research proposal and stop any project they believe has failed to meet proper standards.

IACUCs are required to conduct regular inspections of animal research facilities, and to investigate any complaints regarding animal care (U.S. Department of Agriculture 2016b). If an IACUC finds that a facility is not complying with its preapproved research plan, it must document the noncompliance and recommend corrective action. Failure to correct a problem that affects animal health, safety, or well-being must be reported to any federal agency that supports the research, and can lead to a revocation of funds (Office of Laboratory Animal Welfare 2016).

Violations of the AWA are the rare exception, not the rule. The research community advocates the highest quality of animal care and treatment for two key reasons. First, the use of animals in research is a privilege, and those animals that are helping us unlock the mysteries of disease deserve our respect and the best possible care. Second, a well-treated animal will provide more reliable scientific results, which is the goal of all researchers.

Laws and policies regulating animal research have evolved with changes in science and public opinion. The AWA itself was passed in 1966 in response to public concerns about the use of dogs and cats in research, bolstered by a high-profile case in which a companion dog was abducted and sold to a lab (Adams and Larson 2016). (Federal rules now require strict documentation regarding the source of laboratory animals, and the NIH will not fund research on dogs or cats obtained from small dealers or pounds [National Institutes of Health 2013].) In 2010, the NIH commissioned the Institute of Medicine (IOM) (now the National Academy of Medicine) to examine the need for chimpanzees in NIH-funded research. The IOM concluded that "while the chimpanzee has been a valuable animal model in the past, most current biomedical research use of chimpanzees is not necessary" (Institute of Medicine and National Research Council of the National Academies 2011). Following the IOM's report, the NIH announced in 2013 that it would significantly reduce the use of chimpanzees in research. In 2015, the agency announced that it would end support for chimpanzee research (Collins 2015) and retire all NIH-owned and supported chimpanzees to a sanctuary called Chimp Haven (2016), which occupies 200 acres of forest in southern Louisiana.

Public Support

Until recently, U.S. public support for the use of animals in biomedical research remained steady at around 70%, but according to the Foundation for Biomedical Research, it now ranges from 45% to 57% in various surveys. If public support for animal research continues to decline, this could have serious implications for investments by the NIH, Centers for Disease Control and Prevention, National Science Foundation, and other publicly funded agencies that support biomedical research. The scientific community must be prepared to convey the value of animals in research and our commitment to their humane treatment—to students, health professionals, policy makers, government leaders, philanthropists, media, and the general public.

Animal rights extremists often make the false claim that animal research has contributed little to nothing in advancing clinical or veterinary care. For the last 30 years, as animal rights extremism has grown exponentially in both scope and size, the biomedical research community has mostly remained silent, assuming that logical minds would prevail, and that people would see the connection between animal research and improvements in health. This has not been the case, and animal research critics have controlled the dialogue (Figure 6.3).

Communication Tips

Communications to the public should focus on the results of animal research—disease mysteries solved, cures discovered, lives saved, and stories of people who might not be here today if it were not for animal

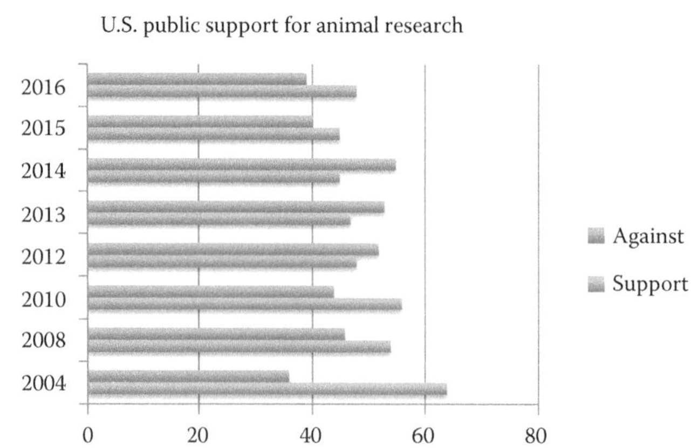

FIGURE 6.3 (See color insert.) Surveys by the Foundation for Biomedical Research show that U.S. public support for the use of animals in medical research has declined from about 70% in 2004 to about 45% in recent years. Survey results for 2015 and 2016 included respondents who answered "unsure," 5% and 13%, respectively. Those responses are not included in the chart. (Courtesy of the Foundation for Biomedical Research.)

research. For people who doubt the value of animal research, it is worth asking them to consider the alternative. Would they want a loved one to receive a new, potentially hazardous drug or procedure that had never been tested in another live being? The fact that animal research helps animals, and not just humans, can also be a powerful message; it is new information for many people. Here are a few important tips to consider:

- Contact your institution's communications or press office for advice on how to talk about animal research—before a crisis or potential crisis occurs. They can help you develop a plan to anticipate and counter uninformed challenges from animal rights extremists.
  - They may recommend designating one or more spokespersons to talk about animal research. Most communications offices offer media training.
  - They can prepare materials that describe research at your organization, the role of animals, and your institution's animal welfare policies.
  - The advocacy groups in the next section can offer additional support and resources.
- Familiarize yourself with biomedical breakthroughs made possible because of animals, like the ones described in this chapter. Be ready to share them.
- Learn more about the leading causes of death and disability, and how animal research is supporting the development of new tests and treatments—for humans and animals. How many lives could be saved? How many disabilities could be prevented or overcome? Provide examples from your lab or organization, where applicable.
- Find people and families affected by disease in your community who are willing to speak out on behalf of the need for animal research. Most individuals and families affected by diseases like cancer and Alzheimer's understand the need for animal research.
- Stress that the majority of lab animals are purpose-bred mice and rats. Nonhuman primates, dogs, cats, and pigs make up less than one-half of 1% of all animals used in research.
- Emphasize that humane treatment of lab animals is the law and the common practice at almost all institutions. Allegations of lab animal neglect and abuse represent the exception. These allegations are investigated by the sponsoring institutions and by government agencies, and can lead to revocation of research funding.

Domestic and International Advocacy Groups

- Alliance for Biomedical Research in Europe (BioMed Alliance): This is a unique initiative representing 21 leading research-oriented medical societies that include more than 400,000 researchers across Europe. The BioMed Alliance is committed through its actions to promote excellence in European biomedical research and innovation. Visit biomedeurope.org for more information.
- American Association for Laboratory Animal Science (AALAS): AALAS is a membership association of professionals employed around the world in academia, government, and private industry who are dedicated to the humane care and treatment of laboratory animals, as well as the quality research that leads to scientific gains that benefit people and animals. Visit aalas.org for more information.
- Americans for Medical Progress (AMP): AMP is a U.S.-based nonprofit that nurtures public understanding of and support for the humane, necessary, and valuable use of animals in medicine. Through various specialty publications, outreach initiatives, and the media, AMP informs the public about the facts of animal-based research. Visit amprogress.org for more information.
- AnimalResearch.info (2016): This website aims to provide reliable, detailed information about when and why it is appropriate to use animals, and the history of this area of research. An international collaboration of scientists and expert contributors provide source information about

the contribution of animals to scientific advances, and the rationale for why they are used. Visit animalresearch.info for more information.

- European Animal Research Association (EARA): This a communications and advocacy organization supporting biomedical research and health care development across Europe. EARA facilitates open discussion about animal research, and provides accurate, evidence-based information about its benefits. Visit eara.eu/en for more information.
- European Federation of Pharmaceutical Industries and Associations (EFPIA): EFPIA brings together more than 70 pharmaceutical industry associations and leading companies in Europe undertaking research, development, and the manufacture of medicinal products for human use. EFPIA supports transparent dialogue about the use of animals in research. Visit efpia.eu/topics /innovation/animal-welfare for more information.
- Foundation for Biomedical Research (FBR): FBR is the oldest and largest nonprofit organization in the United States dedicated to improving human and animal health by promoting public understanding and support for biomedical research. FBR's mission is to educate people about the essential role of humane and responsible animal research in the quest for medical advancements, treatments, and cures for both people and animals. Using video and educational and media campaigns to raise public awareness, FBR illustrates how animal research improves health for both people and animals. Visit fbresearch.org for more information.
- National Association for Biomedical Research (NABR): NABR provides a unified voice for the scientific community on legislative and regulatory matters affecting laboratory animal research. NABR works to safeguard the future of biomedical research on behalf of public and private universities, medical and veterinary schools, teaching hospitals, voluntary health agencies, professional societies, pharmaceutical and biotech industries, and other animal research–related firms. Visit nabr.org for more information.
- Speaking of Research (2016): This is an international advocacy group that provides accurate information about the importance of animal research in medical and veterinary science. Speaking of Research aims to challenge animal rights extremists by participating in talks and debates on campuses across the country and by utilizing online tools to provide encouragement, information, and support to all who care about medical progress. Visit speakingofresearch.com for more information.
- States United for Biomedical Research (SUBR): SUBR is a network of nonprofit associations that have joined forces to promote health through science and education. SUBR is committed to building collaborative networks of organizations, institutions, businesses, and individuals to promote public understanding of and support for biomedical research, including the humane care and use of research animals. Visit statesforbiomed.org for more information.
- Understanding Animal Research: This UK-based organization aims to achieve broad understanding and acceptance of the humane use of animals in biomedical research in the United Kingdom in order to advance science and medicine. Visit understandinganimalresearch.org.uk for more information.

DISCLAIMER

This chapter should not be interpreted as representing the viewpoint of the U.S. Department of Health and Human Services, the National Institutes of Health, or the National Heart, Lung, and Blood Institute. The author and editors extend their appreciation to the Foundation for Biomedical Research (FBR), Washington, DC, for providing expertise in the development of this chapter.

APPENDIX 6.1

Nobel Prizes Based on Animal Research (Foundation for Biomedical Research 2016b)

Nearly every Nobel Prize in Medicine awarded in the last three decades was dependent on data from animal models. Overall, 83% of Nobel Prizes awarded for outstanding contributions to medicine have involved animal research since the awards program was founded in 1901.

Year	Nobel Laureate	Animal Model	Contribution to Modern Medicine
2015	William C. Campbell, Satoshi Ōmura, and Youyou Tu	Mouse, nematode (roundworm)	Novel therapies against roundworm infections and malaria
2014	John O'Keefe, May-Britt Moser, and Edvard I. Moser	Rat	Discoveries of cells that constitute a positioning system in the brain (an inner GPS)
2013	James E. Rothman and Thomas C. Südhof	Hamster, mouse	Discoveries of machinery regulating vesicle traffic, a transport system in our cells
2012	Sir John B. Gurdon and Shinya Yamanaka	Frog, mouse	Discovery that mature cells can be reprogrammed to behave like embryonic stem cells
2011	Bruce A. Beutler, Jules A. Hoffmann, and Ralph M. Steinman	Fruit fly, mouse	Discoveries concerning the activation of innate immunity
2010	Robert G. Edwards	Rabbit	Development of in vitro fertilization
2009	Carol W. Greider, Elizabeth H. Blackburn, and Jack W. Szostak	Frog, mouse	Discovery of how chromosomes are protected by telomeres and the enzyme telomerase
2008	Harald zur Hausen, Françoise Barré-Sinoussi, and Luc Montagnier	Cow, chimpanzee, hamster, monkey, mouse	Discovery of human papilloma viruses causing cervical cancer
2007	Mario R. Capecchi, Sir Martin J. Evans, and Oliver Smithies	Chick, mouse	Principles of gene modification in mice through the use of embryonic stem cells
2006	Andrew Z. Fire and Craig C. Mello	Nematode	Discovery of RNA interference—gene silencing by double-stranded RNA
2005	Barry J. Marshall	Piglet	Discovery of the bacterium *Helicobacter pylori* and its role in gastritis and peptic ulcers
2004	Richard Axel and Linda B. Buck	Fruit fly, mouse	Discovery of odorant receptors and the organization of the olfactory system
2003	Paul C. Lauterbur and Sir Peter Mansfield	Chimpanzee, clam, dog, frog, mouse, pig, rabbit, rat	Discoveries concerning magnetic resonance imaging (MRI)
2002	H. Robert Horvitz, Sydney Brenner, and John E. Sulston	Nematode	Genetic regulation of organ development and programmed cell death
2001	Leland H. Hartwell, Tim Hunt, and Sir Paul M. Nurse	Clam, frog, rabbit, sea urchin	Discoveries of key regulators of the cell cycle
2000	Arvid Carlsson, Paul Greengard, and Eric R. Kandel	Mouse, sea slug	Discoveries in signal transduction in the nervous system
1999	Günter Blobel	Dog, mouse, rat	Discovery that proteins have intrinsic signals that govern their transport and localization
1998	Robert F. Furchgott, Louis J. Ignarro, and Ferid Murad	Rabbit	Regulation of blood pressure by nitric oxide (NO)
1997	Stanley B. Prusiner	Hamster, mouse	Discovery of prions, a new biological principle of infection

(*Continued*)

Year	Nobel Laureate	Animal Model	Contribution to Modern Medicine
1996	Peter C. Doherty and Rolf M. Zinkernagel	Mouse	Recognition of virus-infected cells by the immune system
1995	Edward B. Lewis, Christiane Nüsslein-Volhard, and Eric F. Wieschaus	Fruit fly	Genetic control of early embryonic development
1994	Alfred G. Gilman and Martin Rodbell	Cow, guinea pig, rabbit, rat, turkey	G-proteins and their role in signal transduction in cells
1993	Richard J. Roberts and Phillip A. Sharp	Mouse, rat	Discoveries of split genes
1992	Edmond H. Fischer and Edwin G. Krebs	Rabbit, rat	Discoveries concerning reversible phosphorylation (to regulate proteins)
1991	Erwin Neher and Bert Sakmann	Frog	Chemical communication between cells
1990	Joseph E. Murray and E. Donnall Thomas	Dog	Organ transplantation techniques
1989	Harold E. Varmus and J. Michael Bishop	Chicken	Discovery of the cellular origin of retroviral oncogenes
1988	Sir James W. Black, Gertrude B. Elion, and George H. Hitchings	Cat, dog, guinea pig, monkey, mouse, rabbit, rat	Discoveries of important principles for drug treatment
1987	Susumu Tonegawa	Mouse	Discovery of genetic principle behind antibody diversity
1986	Stanley Cohen and Rita Levi-Montalcini	Chick, mouse, snake	Nerve growth factor and epidermal growth factor
1985	Michael S. Brown and Joseph L. Goldstein	Rat	Regulation of cholesterol metabolism
1984	Niels K. Jerne, Georges J. F. Köhler, and César Milstein	Mouse	Techniques of monoclonal antibody formation
1982	Sune K. Bergström, Bengt I. Samuelsson, and John R. Vane	Guinea pig, rabbit, rat	Discovery of prostaglandins
1981	Roger W. Sperry	Cat, monkey	Functional specialization of the cerebral hemispheres
1981	David H. Hubel and Torsten N. Wiesel	Cat, monkey	Information processing in the visual system
1980	Baruj Benacerraf, Jean Dausset, and George D. Snell	Guinea pig, mouse	Histocompatibility (self) antigens and their mechanism of action
1979	Allan M. Cormack and Godfrey N. Hounsfield	Pig	Development of computer-assisted tomography (CAT) scan
1977	Roger Guillemin and Andrew V. Schally	Pig, sheep	Peptide hormones produced in the brain
1977	Rosalyn Yalow	Pig, sheep	Development of radioimmunoassays of peptide hormones
1976	Baruch S. Blumberg and D. Carleton Gajdusek	Chimpanzee	New mechanisms for the origin and dissemination of infectious diseases
1975	David Baltimore, Renato Dulbecco, and Howard Martin Temin	Chicken, horse, monkey, mouse	Interaction between tumor viruses and genetic material
1974	Albert Claude, Christian de Duve, and George E. Palade	Chicken, guinea pig, rat	Structural and functional organization of cells
1973	Karl von Frisch, Konrad Lorenz, and Nikolaas Tinbergen	Bee, bird, fish	Organization of social and behavior patterns in animals
1972	Gerald M. Edelman and Rodney R. Porter	Guinea pig, rabbit	Chemical structure of antibodies
1971	Earl W. Sutherland Jr.	Dog, rat	Mechanism of the actions of hormones
1970	Sir Bernard Katz, Ulf von Euler, and Julius Axelrod	Cat, rat	Mechanism of storage and release of nerve transmitters
1968	Robert W. Holley, Har Gobind Khorana, and Marshall W. Nirenberg	Cat, frog, guinea pig, rat, snake, tortoise	Interpretation of the genetic code and its function in protein synthesis

(Continued)

Year	Nobel Laureate	Animal Model	Contribution to Modern Medicine
1967	Ragnar Granit, Haldan Keffer Hartline, and George Wald	Chicken, crab, fish, rabbit	Primary physiological and chemical processes of vision
1966	Peyton Rous	Chicken, rabbit, rat	Discovery of tumor-inducing viruses
1966	Charles Brenton Huggins	Chicken, rabbit, rat	Hormonal treatment of prostate cancer
1964	Konrad Bloch and Feodor Lynen	Rat	Regulation of cholesterol and fatty acid metabolism
1963	Sir John Carew Eccles, Alan Lloyd Hodgkin, and Andrew Fielding Huxley	Cat, crab, frog, squid	Ionic mechanisms involved in nerve cell activity
1961	Georg von Békésy	Guinea pig	Mechanism of stimulation within the cochlea (inner ear)
1960	Sir Frank Macfarlane Burnet and Peter Brian Medawar	Rabbit	Discovery of acquired immune tolerance
1957	Daniel Bovet	Dog, rabbit	Production of compounds that act on blood vessels and muscle (including antihistamines)
1955	Axel Hugo Theodor Theorell	Horse	Nature and mode of action of oxidation enzymes
1954	John Franklin Enders, Thomas Huckle Weller, and Frederick Chapman Robbins	Monkey, mouse	Culture of poliovirus that led to development of vaccine
1953	Hans Adolf Krebs	Pigeon	Discovery of the citric acid cycle
1953	Fritz Albert Lipmann	Pigeon	Discovery of coenzyme A and its importance in metabolism
1952	Selman Abraham Waksman	Guinea pig	Discovery of streptomycin, the first effective antibiotic against tuberculosis
1951	Max Theiler	Monkey, mouse	Development of yellow fever vaccine
1950	Edward Calvin Kendall, Tadeus Reichstein, and Philip Showalter Hench	Cow	Antiarthritic role of adrenal hormones
1949	Walter Rudolf Hess	Cat	Functional organization of the brain as a coordinator of internal organs
1949	Antonio Caetano de Abreu Freire Egas Moniz	Cat	Therapeutic value of leucotomy (frontal lobotomy) in certain psychoses
1947	Carl Ferdinand Cori and Gerty Theresa Cori (née Radnitz)	Dog, frog, toad	Catalytic conversion of glycogen to glucose
1947	Bernardo Alberto Houssay	Dog, frog, toad	Role of pituitary gland in sugar metabolism
1945	Sir Alexander Fleming, Ernst Boris Chain, and Sir Howard Walter Florey	Mouse	Discovery of penicillin and its curative effect in various infectious diseases
1944	Joseph Erlanger and Herbert Spencer Gasser	Cat	Specific functions of nerve cells
1943	Henrik Carl Peter Dam and Edward Adelbert Doisy	Chick, dog, mouse, rat	Discovery of vitamin K function
1939	Gerhard Domagk	Mouse, rabbit	Antibacterial effects of prontosil
1938	Corneille Jean François Heymans	Dog	Role of the sinus and aortic mechanisms in regulation of respiration
1936	Sir Henry Hallett Dale and Otto Loewi	Bird, cat, frog, reptile	Chemical transmission of nerve impulses
1935	Hans Spemann	Frog, newt	Organizer effect in embryonic development
1934	George Hoyt Whipple, George Richards Minot, and William Parry Murphy	Dog	Liver therapy for anemia
1932	Sir Charles Scott Sherrington and Edgar Douglas Adrian	Cat, dog	Function of neurons

(*Continued*)

Year	Nobel Laureate	Animal Model	Contribution to Modern Medicine
1930	Karl Landsteiner	Monkey	Discovery of human blood types
1929	Christiaan Eijkman and Sir Frederick Gowland Hopkins	Chicken	Discovery of antineuritic and growth-stimulating vitamins
1928	Charles Jules Henri Nicolle	Guinea pig, monkey, mouse, rat	Pathogenesis of typhus
1924	Willem Einthoven	Dog	Mechanism of the electrocardiogram
1923	Frederick Grant Banting and John James Richard Macleod	Dog, fish, rabbit	Discovery of insulin and mechanism of diabetes
1922	Archibald Vivian Hill	Frog	Discovery relating to the production of heat in muscle
1920	Schack August Steenberg Krogh	Frog	Discovery of capillary motor-regulating mechanism
1919	Jules Bordet	Guinea pig, horse, rabbit	Mechanisms of immunity
1913	Charles Robert Richet	Dog, rabbit	Mechanisms of anaphylaxis
1912	Alexis Carrel	Dog	Surgical advances in the suture and grafting of blood vessels
1910	Albrecht Kossel	Bird	Knowledge of cell chemistry through work on proteins
1908	Ilya Ilyich Mechnikov and Paul Ehrlich	Bird, fish, guinea pig	Immune reactions and functions of phagocytes
1907	Charles Louis Alphonse Laveran	Bird	Role of protozoa as cause of disease
1906	Camillo Golgi and Santiago Ramón y Cajal	Dog, horse	Characterization of the central nervous system
1905	Robert Koch	Cow, sheep	Pathogenesis of tuberculosis
1904	Ivan Petrovich Pavlov	Dog	Physiology of digestion
1902	Ronald Ross	Pigeon	Understanding of malaria life cycle
1901	Emil Adolf von Behring	Guinea pig	Development of diphtheria antiserum

Animals behind the Top 25 Prescribed Drugs in the United States (2014)

Rank	Drug	Drug Usage	Species
1	Synthroid	Hypothyroidism	Dog, mouse, rat
2	Crestor	High cholesterol	Cat, dog, monkey, mouse, rabbit, rat
3	Nexium	Heartburn, acid reflux	Dog, mouse, rabbit, rat
4	Ventolin HFA	Bronchospasm	Dog, rabbit, rat
5	Advair Diskus	Asthma, chronic obstructive pulmonary disease (COPD)	Dog, guinea pig, hamster, monkey, mouse, rabbit, rat
6	Lantus Solostar	Diabetes	Dog, guinea pig, rabbit, rat
7	Cymbalta	Fibromyalgia, major depression, general anxiety disorder	Dog, mouse, rabbit, rat
8	Vyvanse	Attention deficit hyperactivity disorder (ADHD), severe binge eating	Dog, guinea pig, rat
9	Lyrica	Fibromyalgia, epilepsy	Monkey, mouse, rabbit, rat
10	Spiriva Handihaler	COPD	Dog, mouse, rabbit, rat
11	Lantus	Diabetes	Dog, guinea pig, monkey, rat
12	Celebrex	Arthritis, ankylosing spondylitis	Dog, monkey, mouse, rat
13	Abilify	Schizophrenia, bipolar disorder	Dog, monkey, rabbit, rat
14	Januvia	Diabetes	Dog, monkey, rabbit, rat
15	Namenda	Dementia	Dog, mouse, rat
16	Viagra	Erectile dysfunction	Dog, mouse, rabbit, rat

(Continued)

Rank	Drug	Drug Usage	Species
17	Cialis	Erectile dysfunction	Dog, mouse, rat
18	Zetia	High cholesterol	Dog, mouse, rat
19	Nasonex	Nasal allergies	Dog, rabbit, rat
20	Suboxone	Opioid addiction	Baboon, dog, mouse, rabbit, rat
21	Symbicort	Asthma	Dog, mouse, rat
22	Bystolic	Hypertension	Dog, mouse, rat
23	Namenda	Dementia	Dog, mouse, rat
24	Flovent HFA	Asthma	Dog, rabbit, rat
25	Oxycontin	Severe pain	Rabbit, rat

Source: Foundation for Biomedical Research. 2016. Animals behind top drugs. https://fbresearch.org/medical-advances/animal-testing-and-research-achievements/animal-research-behind-top-drugs/ (accessed December 6, 2016).

REFERENCES

Adams, B., and J. Larson. 2016. Legislative history of the Animal Welfare Act: Introduction. https://www.nal.usda.gov/awic/legislative-history-animal-welfare-act-introduction (accessed December 5, 2016).

Alton, E. W. F. W., D. K. Armstrong, D. Ashby, K. J. Bayfield, D. Bilton, E. V. Bloomfield, A. C. Boyd et al. 2015. Repeated nebulisation of non-viral CFTR gene therapy in patients with cystic fibrosis: A randomised, double-blind, placebo-controlled, phase 2b trial. *Lancet Respir Med* 3 (9):684–691.

Alzforum. 2016. Alzheimer's disease therapeutics. http://www.alzforum.org/therapeutics (accessed December 4, 2016).

American Macular Degeneration Foundation. 2016. What is macular degeneration? https://www.macular.org/what-macular-degeneration (accessed December 4, 2016).

Angeli, C. A., V. R. Edgerton, Y. P. Gerasimenko, and S. J. Harkema. 2014. Altering spinal cord excitability enables voluntary movements after chronic complete paralysis in humans. *Brain* 137 (Pt 5):1394–1409.

AnimalResearch.Info. 2016. Malaria. http://www.animalresearch.info/en/medical-advances/diseases-research/malaria/ (accessed December 4, 2016).

Ballentine, C. 1981. Sulfanilamide disaster: Taste of raspberries, taste of death. *FDA Consumer*. http://www.fda.gov/AboutFDA/WhatWeDo/History/ProductRegulation/SulfanilamideDisaster/default.htm.

Bouvier, N. M. 2015. Animal models for influenza virus transmission studies: A historical perspective. *Curr Opin Virol* 13:101–108.

Centers for Disease Control and Prevention. 2002. Achievements in public health: Hepatitis B vaccination—United States, 1982–2002. https://www.cdc.gov/mmwr/preview/mmwrhtml/mm5125a3.htm?mobile=nocontent.

Centers for Disease Control and Prevention. 2016a. 2014 Ebola outbreak in West Africa—Case counts. http://www.cdc.gov/vhf/ebola/outbreaks/2014-west-africa/case-counts.html (accessed December 3, 2016).

Centers for Disease Control and Prevention. 2016b. Blood safety basics. http://www.cdc.gov/bloodsafety/basics.html (accessed December 4, 2016).

Centers for Disease Control and Prevention. 2016c. HIV in the United States: At a glance. http://www.cdc.gov/hiv/statistics/overview/ataglance.html (accessed December 4, 2016).

Centers for Disease Control and Prevention. 2016d. Seasonal influenza vaccine total doses distributed. http://www.cdc.gov/flu/professionals/vaccination/vaccinesupply.htm (accessed December 4, 2016).

Central Intelligence Agency. 2016. *The World Factbook*. https://www.cia.gov/library/publications/the-world-factbook/geos/us.html (accessed December 3, 2016).

Chimp Haven. 2016. http://www.chimphaven.org/ (accessed December 5, 2016).

Christopher and Dana Reeve Foundation. 2016. Prevalence of paralysis in the United States. https://www.christopherreeve.org/living-with-paralysis/stats-about-paralysis (accessed December 4, 2016).

Collins, F. 2015. NIH will no longer support biomedical research on chimpanzees. https://www.nih.gov/about-nih/who-we-are/nih-director/statements/nih-will-no-longer-support-biomedical-research-chimpanzees (accessed December 5, 2016).

Committee on Assessing the Nation's Framework for Addressing Animal Diseases. 2005. *Animal Health at the Crossroads: Preventing, Detecting, and Diagnosing Animal Diseases*. Washington, DC: National Research Council of the National Academies.

Couser, W. 2016. Animal research essential in fight against kidney disease. https://www.kidney.org/news/newsroom/positionpaper01.

Darwin, C. 1859. *On the Origin of Species by Means of Natural Selection.* 1st ed. London: John Murray.

Dehoux, J. P., and P. Gianello. 2007. The importance of large animal models in transplantation. *Front Biosci* 12:4864–4880.

Feeney, E. R., and R. T. Chung. 2014. Antiviral treatment of hepatitis C. *BMJ* 349:g3308.

Food and Drug Administration. 2016. Code of Federal Regulations Title 21: Food and drugs. Part 312: Investigational new drug application. 21 CFR 312. Silver Spring: Food and Drug Administration.

Foundation for Biomedical Research. 2016a. Animals behind top drugs. https://fbresearch.org/medical-advances/animal-testing-and-research-achievements/animal-research-behind-top-drugs/ (accessed December 6, 2016).

Foundation for Biomedical Research. 2016b. Nobel Prizes in Medicine. http://fbresearch.org/education/nobel-prize/ (accessed December 4, 2016).

Giangrande, P. L. F. 2000. The history of blood transfusion. *Br J Haematol* 110 (4):758–767.

Hastings Center. 2012. Fact sheet: Animals used in research in the U.S. http://animalresearch.thehastingscenter.org/facts-sheets/animals-used-in-research-in-the-united-states/ (accessed December 5, 2016).

Health Resources and Services Administration. 2016. Organ Procurement and Transplantation Network: Need continues to grow. https://optn.transplant.hrsa.gov/need-continues-to-grow/ (accessed December 4, 2016).

Institute of Medicine and National Research Council of the National Academies. 2011. *Chimpanzees in Biomedical and Behavioral Research: Assessing the Necessity.* Washington, DC: National Academies Press.

Jackson Laboratory. 2016. Why mouse genetics? https://www.jax.org/genetics-and-healthcare/genetics-and-genomics/why-mouse-genetics# (accessed December 5, 2016).

Iritani, H., Y. Nishimura, and T. Minatogawa. 1991. Neurophysiology of ischemic facial nerve paralysis in an animal model. *Acta Otolaryngol* 111 (5):934–942.

Keiser, N. W., and J. F. Engelhardt. 2011. New animal models of cystic fibrosis: What are they teaching us? *Curr Opin Pulmon Med* 17 (6):478–483.

Krupka, I., and R. K. Straubinger. 2010. Lyme borreliosis in dogs and cats: Background, diagnosis, treatment and prevention of infections with *Borrelia burgdorferi sensu stricto. Vet Clin North Am Small Anim Pract* 40 (6):1103–1119.

Lankau, E. W., P. V. Turner, R. J. Mullan, and G. G. Galland. 2014. Use of nonhuman primates in research in North America. *J Am Assoc Lab Anim Sci* 53 (3):278–282.

Maguire, M. G., D. F. Martin, G.-S. Ying, G. J. Jaffe, E. Daniel, J. E. Grunwald, C. A. Toth, F. L. Ferris III, and S. L. Fine. 2016. Five-year outcomes with anti-vascular endothelial growth factor treatment of neovascular age-related macular degeneration. *Ophthalmology* 123 (8):1751–1761.

National Cancer Institute. 2016. Cancer treatment: Targeted therapy. https://www.cancer.gov/about-cancer/treatment/types/targeted-therapies (accessed December 4, 2016).

National Center for Health Statistics. Age-adjusted death rates and life-expectancy at birth. Centers for Disease Control and Prevention. https://data.cdc.gov/NCHS/NCHS-Age-adjusted-death-rates-and-life-expectancy-/w9j2-ggv5#column-menu (accessed March 21, 2017).

National Eye Institute. 2016. The research road: Gene therapy for Leber congenital amaurosis. https://nei.nih.gov/lca/timeline (accessed December 4, 2016).

National Human Genome Research Institute. 2002. Background on mouse as a model organism. https://www.genome.gov/10005834/background-on-mouse-as-a-model-organism/.

National Institute of Allergy and Infectious Diseases. 2016a. Hepatitis. https://www.niaid.nih.gov/diseases-conditions/hepatitis (accessed December 4, 2016).

National Institute of Allergy and Infectious Diseases. 2016b. HIV/AIDS: Treatment. https://www.niaid.nih.gov/diseases-conditions/hiv-treatment (accessed December 4, 2016).

National Institute of Allergy and Infectious Diseases. 2016c. HIV/AIDS: Vaccine development. https://www.niaid.nih.gov/diseases-conditions/hiv-vaccine-development (accessed December 4, 2016).

National Institute of Neurological Disorders and Stroke. 2016a. The epilepsies and seizures: Hope through research. http://www.ninds.nih.gov/disorders/epilepsy/detail_epilepsy.htm#3109_29 (accessed December 3, 2016).

National Institute of Neurological Disorders and Stroke. 2016b. Epilepsy Therapy Screening Program. http://www.ninds.nih.gov/research/asp/index.htm (accessed December 4, 2016).
National Institute of Neurological Disorders and Stroke. 2016c. Parkinson's disease: Hope through research. http://www.ninds.nih.gov/disorders/parkinsons_disease/detail_parkinsons_disease.htm (accessed December 4, 2016).
National Institutes of Health. 2013. Notice regarding NIH plan to transition from use of USDA Class B dogs to other legal sources. https://grants.nih.gov/grants/guide/notice-files/NOT-OD-14-034.html.
National Library of Medicine. 2016. Medline Plus: Cystic fibrosis. https://medlineplus.gov/ency/article/000107.htm (accessed December 4, 2016).
National Science Foundation. 2016. Proposal preparation instructions: Proposals involving vertebrate animals. https://www.nsf.gov/pubs/policydocs/pappguide/nsf16001/gpg_2.jsp#IID7 (accessed December 5, 2016).
Noonan, D. 2014. Inside the science of an amazing new surgery called deep brain stimulation. http://www.smithsonianmag.com/innovation/inside-science-amazing-new-surgery-called-deep-brain-stimulation-180951170/.
O'Brien, A. 2015. How to mend a broken heart. *Huffington Post.* http://www.huffingtonpost.com/2015/06/08/artificial-heart_n_7515526.html (accessed December 4, 2016).
Office of Laboratory Animal Welfare, National Institutes of Health. 2015. Public Health Service policy on humane care and use of laboratory animals. https://grants.nih.gov/grants/olaw/references/phspol.htm (accessed December 5, 2016).
Office of Laboratory Animal Welfare, National Institutes of Health. 2016. Health Research Extension Act of 1985. http://grants.nih.gov/grants/olaw/references/hrea1985.htm (accessed December 5, 2016).
Ohio State University Veterinary Medical Center. 2016. FAQ: Total hip replacement. http://vet.osu.edu/vmc/companion/our-services/orthopedic-surgery/faq-total-hip-replacement (accessed December 4, 2016).
Pollack, A. 2015. Orkambi, a new cystic fibrosis drug, wins F.D.A. approval. *New York Times.* http://www.nytimes.com/2015/07/03/business/orkambi-a-new-cystic-fibrosis-drug-wins-fda-approval.html.
Russell, W. M. S., and R. L. Burch. 1959. *The Principles of Humane Experimental Technique.* London: Methuen.
Simone, J. V. 2008. Curing pediatric lymphoblastic leukemia. In *50 Years in Hematology.* Washington, DC: American Society of Hematology.
Speaking of Research. 2016. U.S. statistics. https://speakingofresearch.com/facts/statistics/ (accessed December 5, 2016).
Stockman, J. A., and P. A. de Alarcon. 2001. Overview of the state of the art of Rh disease: History, current clinical management, and recent progress. *J Pediatr Hematol Oncol* 23 (6):385–393.
Tobert, J. A. 2003. Lovastatin and beyond: The history of the HMG-CoA reductase inhibitors. *Nat Rev Drug Discov* 2 (7):517–526.
U.S. Department of Agriculture. 2016a. Animal Welfare Act. https://www.nal.usda.gov/awic/animal-welfare-act (accessed December 5, 2016).
U.S. Department of Agriculture. 2016b. A quick reference of the responsibilities & functions of the Institutional Animal Care & Use Committee (IACUC) under the Animal Welfare Act. https://www.nal.usda.gov/awic/animal-welfare-act-quick-reference-guides#Q15 (accessed December 5, 2016).
White, B., A. Madden, M. Prins, M. Hellard, H. Wand, G. J. Dore, K. Page, and L. Maher. 2014. Assessing the feasibility of hepatitis C virus vaccine trials: Results from the Hepatitis C Incidence and Transmission Study-community (HITS-c) vaccine preparedness study. *Vaccine* 32 (42):5460–5467.
White, J. R. 2014. A brief history of the development of diabetes medications. *Diabetes Spectr* 27 (2):82–86.
Wiesel, T. N. 1982. The postnatal development of the visual cortex and the influence of environment. *Biosci Rep* 2 (6):351–377.
World Health Organization. 2016a. Ebola vaccine trials. http://www.who.int/medicines/ebola-treatment/emp_ebola_vaccines/en/ (accessed December 4, 2016).
World Health Organization. 2016b. Global health observatory data: HIV/AIDS. http://www.who.int/gho/hiv/en/ (accessed December 4, 2016).

Section III

Compliance, Assessment, and Assurance

7

Compliance and Regulatory Programs

William W. King, Yasmina A. Paramastri, and Javier Guillén

CONTENTS

Disclaimer: The authors have made every attempt to anglicize the titles and references of works originating from non-English-speaking countries and regret taking any liberties not intended by the original authors. The authors also apologize for any inadvertent omission of specific regulations impacting any country, region, commonwealth, state, territory, or other municipality not mentioned in this work.

Introduction

The past several decades have witnessed dramatic changes in the regulatory environment involving the use of animals in research and education. A cursory review of citations using the keywords "animal model or animal models" in PubMed (http://www.ncbi.nlm.nih.gov/pubmed) indicates a steady increase in manuscripts involving the use of animals (Figure 7.1). As research findings enhanced collective knowledge and experimental techniques and goals became more sophisticated, scientists demanded greater control of experimental variables. Concerns about variability inherent to animal models, such as housing environment, disease status, procedure-related pain and distress, nutritional stabilization, and sanitation practices, dictated adaptation of recognized *standards* (Box 7.1). As science continues to globalize, reproducibility and comparability provided by the application of accepted standards become increasingly important.

The use of animals for research and teaching has consistently polarized public opinion, and governments have recognized a need for *regulations* to address the ethical responsibility for acknowledging the needs of the animals used for societal purposes (Schwindaman 1999). Regulations therefore provide a means of reassuring the public that animal use is limited to situations of strict necessity and conforms to their demands for providing species-appropriate daily care, reducing avoidable pain and distress, eliminating unnecessary duplication, and preventing illegal activities, such as animal theft.

Therefore, an animal compliance program is simply an oversight system designed to address two fundamental objectives: conformity to established standards as dictated by scientific rigor and assurance of humane animal treatment as demanded by the public (Box 7.2).

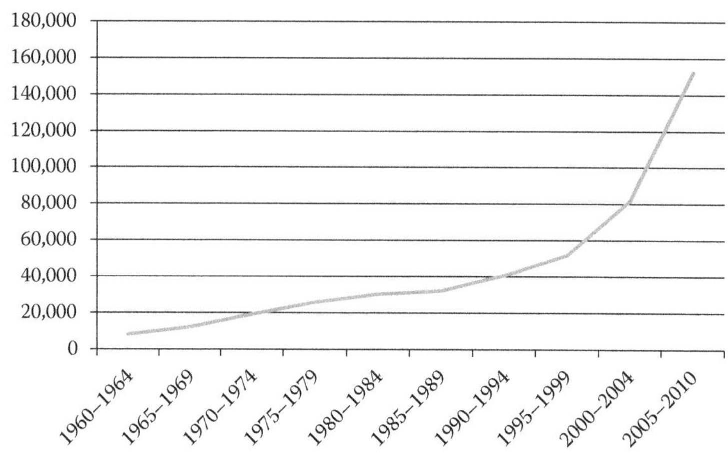

FIGURE 7.1 Citations involving the keywords "animal model or animal models" in PubMed.

BOX 7.1

Standard: A level of quality, achievement, etc., that is considered acceptable or desirable.
Standardization: To change (things) so that they are similar and consistent and agree with rules about what is proper and acceptable.

Source: Merriam-Webster, 2016, https://www.merriam-webster.com.

BOX 7.2

Because much of the funding for biomedical research originates from federal sources, scientists are accountable to the public and the public's wishes. (Steneck 2007)

Oversight is part of the social contract between scientists and funding agencies that is implicit with the privilege of performing research, working with research subjects, and using public facilities and funds. (Haywood and Greene 2008)

Those with management responsibilities for animal care and use programs play an important role in ensuring regulatory compliance and in serving as a primary conduit of communication between the animal users and regulatory authorities. Thus, it behooves the manager to develop a keen awareness and understanding of regulatory oversight and the responsibilities of various institutional and external entities involved, and a basic understanding of pertinent laws, regulations, principles, guidelines, and policies at local, state, national, and international levels. This is no small feat; regulatory complexities have been compared with Russian matryoshka dolls with seemingly unending layers of intricacy (Bayne and Garnett 2008). Knowing where to seek additional information is often the most crucial step. The goal of this chapter is to introduce concepts concerning animal program compliance, describe the entities that are commonly involved, and provide a brief summary of pertinent sources of guidance.

Typical Compliance Programs

Salient Features

Hierarchy of Guidance

To establish a framework for understanding the components of a compliance program, one must be a student of the published work that serves as guidance, including regulations (i.e., codes, rules, decrees, laws, orders, and statutes) and guidelines (e.g., guides, principles, manuals, handbooks, or other sets of instructions). Figure 7.2 depicts a hierarchy of such information in order of obligation.

- Arguably, an individual's or society's *ethical principles* provide a foundation upon which all other sources of guidance are based. Because simply abiding by regulations does not equate to being ethical (Pimple 2002), these underlying precepts contribute to what is considered acceptable behavior and conduct.
- *Regulations* represent a government's attempt to express the will of the public they serve. Regulatory application may be limited, not only geographically, but also in scope, species, intended purpose, and source of financial support; however, adherence to regulatory requirements is usually not optional.

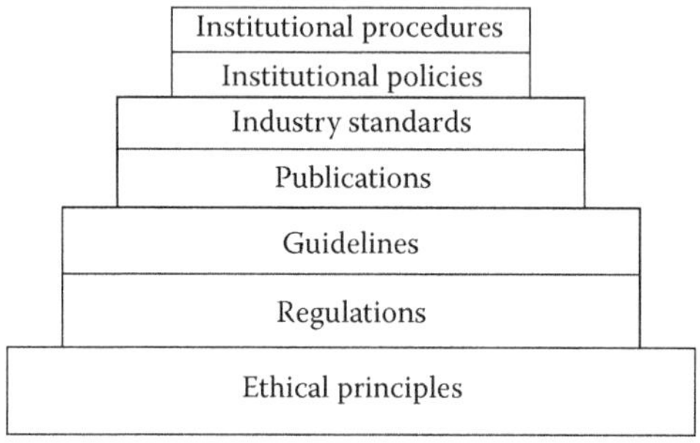

FIGURE 7.2 Hierarchy of guidance used in compliance programs.

- *Guidelines* typically provide additional breadth and detail to the concepts introduced in regulations. Frequently, guidelines are considered recommendations; however, when adopted by regulatory bodies, they may become the equivalent of regulations.
- Peer-reviewed *publications* present a nimble source of guidance based on new knowledge or novel interpretation, and frequently yield additional information for specific circumstances. Their application is generally not compulsory, although program managers have an obligation to remain familiar with current information (ILAR 2011a).
- *Industry standards* are relatively subjective and not easily defined, but add useful direct experience and anecdotal evidence, especially in the absence of other guidance.
- *Institutional policies* afford a valuable means of interpreting and applying guidance to specific circumstances. Once established, institutional policies are often supplemented with *practices* outlined in step-by-step *procedures* (e.g., standard operating procedures [SOPs]).

The relevance of the various forms of guidance can be protean. In fact, determining which *must* be used and which *can* be used presents a formidable challenge to the program manager. Applicability is frequently dictated by location and mission of the institution, species utilized, and funding source for the specific project. For example, a U.S.-based pharmaceutical company in Boston using only mice with private funds may be required to follow municipal, state, and some but not all federal regulations. Because of their wider applicability, this chapter focuses primarily on national regulations and guidelines. This should not be interpreted to be a dismissal of other types of guidance when managing a compliance program.

Regulations cannot address all experimental circumstances and may be silent on some issues impacting animal welfare in a complex research environment. This reality requires that program managers maintain at least a cursory knowledge of the guidance typically applied to other institutions or situations. For example, many institutions choose to voluntarily apply a given federal agency's regulations program-wide to enhance consistency, for example, the Public Health Service Policy on Humane Care and Use of Laboratory Animals (PHS Policy), regardless of the funding source (see description below). Similarly, U.S. regulations lack specific recommendations on primary enclosure space for ferrets and numerous other species; the descriptions presented in the European standards (CoE 1986) may provide a useful baseline. While presenting some obstacles, benefits to global harmonization of standards include the use of data to support federal applications to multiple countries, collaborative scientific efforts, and an increase in international transfer of animal models (e.g., genetically engineered mice), which has both transportation and health status implications (ILAR 2004, 2011b, 2012; Guillén 2014; see also Chapter 8).

Engineering, Performance, Practice, and Professional Standards

When reviewing regulations and guidelines, one should acknowledge the difference between *engineering*, *performance*, *practice*, and *professional standards* (ILAR 2011a), which reflect some of the themes presented in the hierarchy of guidance. Engineering standards are rigid, firmly established, prescribed criteria that represent minimally expected values or ranges of values for a given circumstance. Applying and assessing compliance with such standards is relatively straightforward because deviation beyond the defined units is generally considered unacceptable. Performance standards are goal oriented and defined by a specified outcome, regardless of the means of achieving that outcome. A benefit of a performance standard is its flexibility; that is, as long as the desired result is clearly defined, the user may determine the path to achieve the goal. Note that an important and often overlooked precondition for employing performance standards is objectively measuring the outcome to ensure intended impact (Klein and Bayne 2007).

Historically, regulations have tended to rely heavily on engineering rather than performance standards, although that trend has shifted in recent years. Regardless, contemporary animal compliance programs should incorporate both. Engineering standards are used when required or to provide a baseline; performance standards may be adopted when engineering standards are unavailable, or when acceptable substitutes are identified and, through careful evaluation, prove to be preferable.

Underlying all sources of guidance is the *professional judgment* of the program manager, veterinarian, and other specially trained personnel used to identify and interpret guidance and its application. When evaluating the suitability of standards, the utilization of generally accepted "practice" or "industry standards" and incorporating the experience of competent and qualified scientists and veterinarians becomes indispensable, especially in the absence of established specifications.

Self-Regulation and Trust

Society expects its members to acknowledge their responsibility to obey established rules of order. Similarly, regulatory agencies presuppose a willingness to comply with established standards. Evidence suggests that promoting a culture of compliance by an organization is more effective in determining personal behavior than the existence of rules and regulations (Brønstad and Berg 2011). Therefore, regulatory authorities expect self-regulation by both the institution and individuals within the program (Box 7.3).

In most countries, regulatory authority to use animals in research and education is granted to the institution, not the individual scientist. Similarly, penalties for noncompliance are addressed to the institution, not the individual responsible for the infraction. Thus, to protect the interests of all animal-using scientists, institutions are compelled to develop a comprehensive system of institutional oversight based on a spirit of self-regulation. In some ways, this is purely practical—no regulatory body could sustain the resources required to police all activities. On the other hand, the fact that self-regulatory systems function effectively illustrates the research community's acknowledgment of its responsibilities to the public.

Self-regulation can only succeed if all parties deliberately and faithfully engage and respectively acknowledge each other's roles. Therefore, the first obligation to the compliance program is to ensure cooperation among those entities involved. For the principal investigator (PI), the nature of modern science dictates a comfort with self-regulation. The processes of grant application, presentation, and publication based on peer review are inherently self-monitoring (Steneck 2007). However, scientists also face a difficult balance pursuing original discovery within regulatory and financial constraints

BOX 7.3

The scientific research enterprise, like other human activities, is built on a foundation of trust. Scientists trust that the results reported by others are valid. Society trusts that the results of research reflect an honest attempt by scientists to describe the world accurately and without bias. The level of trust that has characterized science and its relationship with society has contributed to a period of unparalleled scientific productivity. But this trust will endure only if the scientific community devotes itself to exemplifying and transmitting the values associated with ethical scientific conduct. (NAS 1995)

(Bayne and Garnett 2008). Compliance with animal welfare regulations is particularly challenging if those who are regulated simply view compliance programs as another level of bureaucracy, especially if the rationale behind the regulations appears capricious or counterproductive (Dress 2007).

This underscores the need to promote not only a culture of compliance, but also adoption of scientifically based standards. Science-based regulations provide the best means of counteracting concerns over regulatory burden, a topic that has absorbed increasing bandwidth in recent years (Decker et al. 2007; see also Chapter 10). There is no doubt that regulations have benefitted both animal welfare and research data. However, there remains growing concern over the negative impact of excessive regulation. As stated by Thulin and colleagues (2014), regulations "will not be rescinded, and, if history tells us anything, these requirements are likely to become even more stringent."

Although momentum from some initiatives to rectify conflicting regulations and duplicative reporting requirements has abated (Klein and Bayne 2007), U.S. federal agencies acknowledge the benefits of reducing regulatory burden. For example, the National Institutes of Health's (NIH) Office of Laboratory Animal Welfare (OLAW) maintains memoranda of understanding outlining cooperation between the U.S. Department of Agriculture's (USDA) Animal and Plant Health Inspection Service (APHIS) and Food and Drug Administration (FDA) (OLAW 2012), the National Science Foundation (NSF) (OLAW 2015b), and the Department of Veterans Affairs (VA) (OLAW 2013).

Fortunately, the concepts of performance- and science-based standards are correlative. Performance-based standards incorporate developing knowledge without depending on preestablished engineering standards that may not apply to an institution, species, or circumstance (Klein and Bayne 2007). Unfortunately, an unintended side effect may be the creation of overly intricate mechanisms of oversight extending beyond the intent of regulations (DeHaven 2002). If motivated by absolute risk aversion, institutions may inadvertently or willingly exaggerate the burden of compliance without fully examining or realizing that there may be a lack of benefit to animal welfare, thereby introducing unfortunate opportunities for alienating those who are regulated (Bayne and Garnett 2008; Haywood and Greene 2008; Thulin et al. 2014).

The escalating financial requirements for research should also be considered (CLS 1988). The use of resources for unnecessary compliance programs may hinder the ability to identify improvements in animal care. Reduction in animal use should be directed by increased availability of better alternatives, not simply because of fiscal constraints. Science-based performance standards, which focus on outcomes, can be an important means of improving research cost-effectiveness (Klein and Bayne 2007).

Institutions must bear these concepts in mind. As long as standards can be seen to directly improve animal welfare and research data, regulatory compliance may be viewed as a professional virtue comparable with scientific integrity (DuBois 2004).

Constituents and Their Interaction

Ensuring appropriate oversight of animal care and use involves numerous institutional entities, including not only the PI, but also the institutional official (IO), Institutional Animal Care and Use Committee (IACUC) or equivalent ethical oversight body (EOB), attending veterinarian (AV), and their respective staffs (Box 7.4). Compliance programs also routinely involve analogous oversight of hazardous agent use, such as biological and radiation safety officers and committees, chemical hygienists, and occupational health professionals and emergency preparedness personnel. The organizational structure of

BOX 7.4

There is growing recognition that the care of laboratory animals is an institutional responsibility as well as the responsibility of individual investigators. The animal care programs of most large institutions are based increasingly on this partnership of responsibility; the recommendations in the *Guide* assume it. (ILAR 1965) animal compliance programs implies a system of checks and balances, drawing benefits from the unique perspectives and expertise of each component. A crucial characteristic of such arrangements is the effective interaction among the above-mentioned parties. An awareness of and commitment to synergy and a sense of common purpose is paramount (see also Chapter 3).

Role of the Principal Investigator

As previously stated, the primary role of the PI in the regulatory program is to recognize the benefits of viewing the compliance program as a partnership and to remain committed to that relationship. The responsibility for actual study conduct obligates the PI to both the funding agency or sponsor and the public through regulatory agencies via the IACUC or EOB. PIs must also acknowledge animal welfare as an ethical and scientific obligation. Given the complexities involved in devising and executing scientific experimentation and managing a laboratory, a team approach utilizing the abilities and skills of professionals with specific expertise is crucial.

Role of the Institutional Official

Accountability for the animal care and use program rests with the IO (ILAR 2011a). The IO is the primary agent committing the institution to abide by the regulations and enforce institutional policies across departmental and divisional boundaries (Potkay and DeHaven 2000; Brown and Shepherd 2014). To meet this responsibility, the IO must maintain adequate authority to not only ensure adequate financial resources for program needs, but also impart a spirit of cooperation and commitment to humane care and use "from the top down." Given his or her status in the organization, the IO is rarely directly involved in day-to-day management decisions, but appropriately depends on the self-monitoring of the PI, the professional judgment of the AV, and the oversight of the IACUC or EOB. However, the IO cannot completely relinquish operational duties. Lowman (2008) describes the critical elements of the IO as (1) an ability to understand and predict future programmatic research needs and ensure that the animal care and use program is suited to meet those needs, (2) a commitment to and means of assessing both quality and integrity, (3) dynamic engagement in planning and resource development, and (4) an emphasis on accountability.

Role of the Institutional Animal Care and Use Committee or Ethical Oversight Body

To the IACUC or EOB belongs the primary task of overseeing animal welfare compliance. Pertinent guidance agrees that a "jury of peers" is the best entity to (1) compare the likely beneficial outcomes of a study or exercise with the possible induction of unrelieved pain or distress to animal subjects, that is, potential "harm versus benefit analysis"; (2) assess the oversight, management, and ultimate use of animals in the institution; and (3) investigate and resolve alleged instances of noncompliance. The diverse roles and responsibilities of this important committee have been the subject of numerous texts, to which the reader is referred (Podolsky and Lucas 1999; ARENA and OLAW 2002; Silverman et al. 2014).

A mainstay of IACUC or EOB responsibility is the review of proposed animal use, often summarized in documents referred to as animal use protocols. Written protocols provide the committee membership, which usually consists of other scientists, veterinarians, administrators, and community representatives, with a detailed description of anticipated animal use. This includes not only animal-related procedures, but also the rationale behind the need for the procedures specified, the species and animal models selected, the number of animals requested, and an assurance of addressing and incorporating the "three Rs" (see description below). Committee endorsement provides the PI with the knowledge that, if conducted in accordance with the IACUC- or EOB-approved protocol, all relevant regulatory requirements are met or exceptions are adequately justified.

The IACUC or EOB generally assists the institution in ensuring that personnel have or obtain appropriate training in the tenets of animal care and use and are proficient in the specific procedures they are expected to perform. There are numerous references available to assist the IACUC or EOB in this task (ILAR 1991; Kennedy 2002; Pritt et al. 2004; Dobrovolny et al. 2007; Greene et al. 2007; Medina et al. 2007; Van Zutphen 2007; Rush and Dyson 2014; see also Chapters 5 and 12).

Inadequate composition and function of the IACUC have been a primary source of citations from regulatory authorities in the United States (Potkay and DeHaven 2000). The IACUC or EOB must remain informed and engaged with the animal care and use program; otherwise, the system of institutional self-regulation is threatened. Comprehensive programmatic oversight demands the IACUC's or EOB's intimate familiarity with the daily operation of the animal care and use facilities. Protocol review must be thorough, requiring adequate detail to fully grasp the potential experiences of animals assigned to each experimental group. That said, while expected project outcomes should be readily apparent, the IACUC or EOB should avoid limiting animal-based projects or experiments to those that appear to directly lead to immediate therapeutic findings. The utility of study results may only be apparent in a distant future when viewed as one component of an overwhelmingly complex process (CLS 1988).

Role of the Attending Veterinarian

While international regulations vary in the emphasis assigned to the institutional entities involved in overseeing veterinary care, that is, the veterinary staff, research personnel, or ethics committees, provision of adequate veterinary care is universally recognized as a fundamental element of an appropriately operating animal care and use program (Zurlo et al. 2009). Veterinary professional organizations worldwide acknowledge the importance of the attending or designated veterinarian (Poirier et al. 2015).

The American College of Laboratory Animal Medicine (ACLAM) maintains a position statement on adequate veterinary care (ACLAM 1996). This document outlines key components of a veterinary care program, including a reliance on the professional judgment of specially trained and qualified veterinarians directing programs of disease detection, surveillance, prevention, diagnosis, treatment, and resolution; handling, restraint, the use of anesthetics and analgesics, and euthanasia; surgical and perioperative care; and overall animal well-being. In its position statement, AAALAC International summarizes the *Guide for the Care and Use of Laboratory Animals* (the *Guide*) (ILAR 2011a; see description below) and its interpretation of the AV's role, which also includes assurance of emergency medical care, integration within the IACUC and protocol review, and involvement in ensuring personnel training and proficiency (AAALAC 2016a). These and many other references confirm the need for the AV to retain adequate authority to oversee or assist in the oversight of the overall animal care and use program and to intervene on the behalf of animals in pain or distress beyond that explicitly approved by the IACUC or EOB.

Role of the Program Manager

In many institutions, the role of the manager closely approximates that of the AV, and in fact, these functions may be provided by the same person. Depending on the size and complexity of the program, however, the attention of the AV may be directed primarily to veterinary medical issues and IACUC or EOB responsibilities. This may require the manager to direct centrally provided animal facility and support services, and thereby implement much of the program of daily animal care approved by the IACUC or EOB. As content experts, program managers may also be asked to serve as voting or *ex officio* members of the IACUC or EOB, or at least provide input on committee decisions. Regardless, because all activities that potentially impact animal health and well-being are the purview of the IACUC or EOB, a healthy working relationship is advisable.

Team Approach to Compliance

Once again, the importance of collegial cooperation among these constituents cannot be overstated. This is best accomplished by generating a "culture of care," which requires an IO with appropriate authority and involvement, an IACUC or EOB willing and able to share its expectations with the research staff, veterinary staff that are seen as colleagues in the research team, and training programs to ensure uniform distribution of these tenets (Klein and Bayne 2007).

Inherent to any compliance system is a potential for conflicts of interest (COIs). COIs, both real and perceived, must be avoided, as their prevalence will erode public confidence in the oversight structure.

That said, some COIs are inevitable. In fact, the composition of the IACUC or EOB presents several opportunities for COIs, for example, instances when the IO, AV, or IACUC chair also serves as a PI. Of course, even the AV's or scientists' service on the IACUC or EOB presents a potential for COIs. Notably in academia, PIs are expected to publish their own results while training and collaborating with other scientists, and even reviewing potential competitors' manuscripts and grant applications (Steneck 2007). Although complete elimination is not feasible (Thulin et al. 2014), continued efforts to address avoidable COIs will bolster public perception of a compliance program. Examples may include preventing IO service on the IACUC or EOB, or the AV serving as IACUC or EOB chair, and IACUC or EOB member recusal from voting on protocols for which he or she has a vested interest.

Van Sluyters (2008) illustrates productive cooperation between the IO, AV, and IACUC or EOB via a "three-legged stool"; effective programs depend on the appropriate level of pressure and control by each. Figure 7.3 offers an alternate representation that includes the program manager and PI, depicting the realms of concern and influence as spheres. Each entity involved in the compliance program has its own area of expertise and bias; successful programs encourage these constituents toward central, common objectives.

Program Assessment

The increasing quantity and complexity of regulations present opportunities for noncompliance, especially if the rationale behind certain standards has less obvious direct benefits to science and/or animal well-being. Performing studies or procedures not explicitly described and approved by the IACUC or EOB is considered among the most serious of noncompliance. Mechanisms to detect and prevent noncompliance are paramount, but institutions should also regularly assess the interaction between the PI, IACUC or EOB, and other constituents to minimize weaknesses that may lead to noncompliance. Conditions of grant awards include regulatory compliance, which provides another motivation for ongoing evaluation. Considerable financial ramifications are associated with an animal study out of compliance with IACUC or EOB approval, institutional policies, or regulations (Bayne and Garnett 2008). Advisory committees and confidential surveys can gauge the success of IACUC or EOB oversight (Ingham et al. 2000; Van Sluyters 2008). Using methods frequently adopted by continuous quality improvement programs may also engender PI responsiveness by improving both animal health and data integrity, as well as protocol approval time (Nolte et al. 2008). There are several other means of assessing these working relationships, which include regulatory oversight, voluntary accreditation, and periodic evaluation by the IACUC or EOB, which are often mandated by regulations.

Postapproval monitoring (PAM) is a term gaining in popularity for addressing the institutional responsibility in ongoing oversight of study conduct. The duties of the IACUC or EOB include such monitoring, accomplished via periodic rereview of protocols, routine inspections of facilities and laboratories, and regular evaluation of the overall animal care and use program. Special attention to other IACUC or EOB practices, such as *de novo* protocol and amendment review, investigating reports of animal welfare concerns, and assessing unexpected outcomes, augments the IACUC's or EOB's responsibility for continuing review (Plante and James 2008). Results of evaluations by regulatory authorities, and especially by accrediting agencies (e.g., AAALAC International), provide exceptional assessment of program effectiveness. Given their expertise and intimate familiarity with ongoing studies, the program manager and other support personnel provide important sources of "real-time" information.

To enhance or at least verify self-regulation, many programs have found it necessary or advisable to institute additional study conduct monitoring via specially trained personnel (Banks and Norton 2008; Collins 2008; Dale 2008; Lowman 2008). As discussed in Chapters 10 and 15, PAM can take myriad shapes and complexities and should be tailored to the needs of the program. Regardless of mechanisms used, ensuring open lines of communication and conducting PAM in a collegial manner is crucial to its success. Any PAM must avoid the appearance as a regulatory mandate without value to each individual researcher and demonstrate to all parties involved, including researchers, the IACUC or EOB, veterinarians, program managers, and institutional safety officers, that the PAM is a resource profiting both animal welfare and scientific results (Dale 2008).

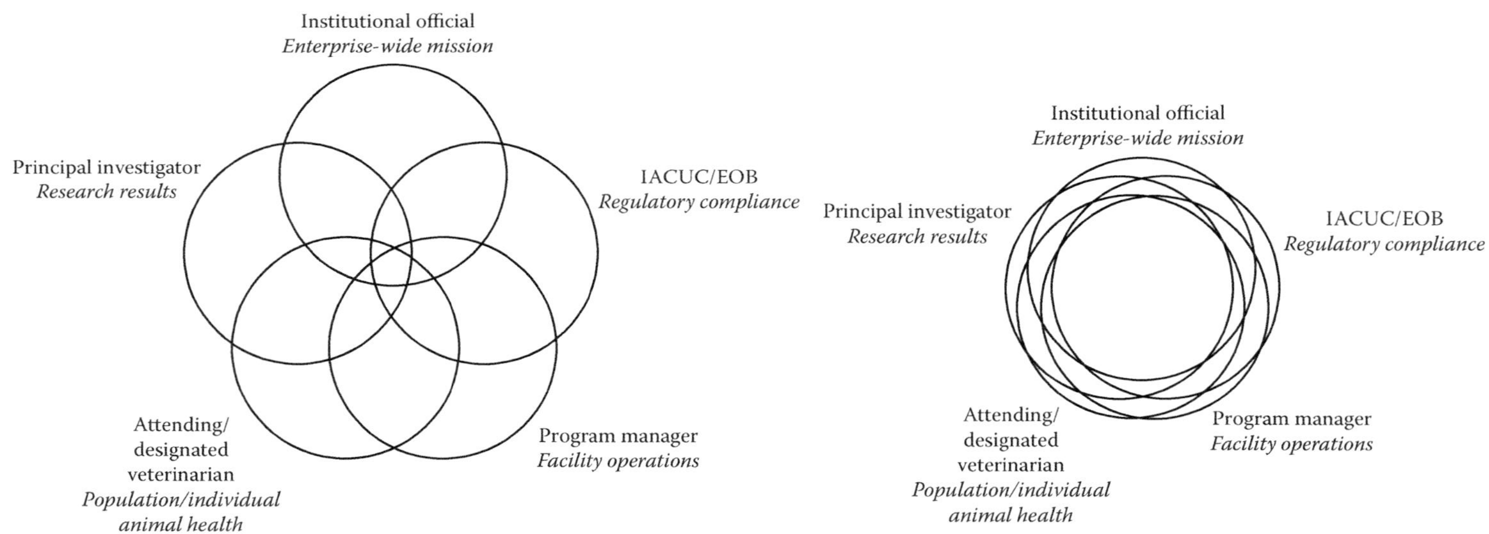

FIGURE 7.3 Realms of concern for the constituents of an animal research compliance program. The left figure illustrates a small amount of success in achieving a central goal when each constituent places excessive attention on its own bias. The right figure demonstrates greater cooperation and thereby a greater chance for success.

Special Considerations for the Program Manager

As previously stated, the program manager must balance several obligations. As depicted in Figure 7.3, the primary duty is daily operational management, often with a focus on cost-effectiveness. Program managers are also frequently recruited for routine animal health and research-related technical procedures. Furthermore, the program manager is charged with implementing many guidance-driven issues, such as environmental enrichment, social housing, monitoring for pain and distress, and reducing husbandry-related research variables, all of which require careful and continuous attention. This frequently involves compiling uniform practices outlined in SOPs, which encourage consistency. Review of SOPs by the IACUC or EOB also provides an efficient and effective means of augmenting committee oversight (DeHaven 2002).

By definition, research involves change. PIs request new species, models, and methods. Novel situations present opportunities for achieving outcomes. Similarly, innovation in addressing pain and distress evolves. Advances in techniques and methods emerge regularly; programs and program managers must remain current and apply enhancements whenever feasible. The judicious manager benefits from interacting with colleagues and sharing "what works and why." Fortunately, there are a number of comparative medicine and laboratory animal science societies, associations, and other means of communicating with colleagues with which the manager should be familiar. These associations frequently host continuing education events and often play integral roles in developing guidelines, especially in regions without national regulations. AAALAC International maintains a list of advocacy and educational groups that provides a useful resource (AAALAC 2016b).

A hallmark of adopting performance and professional standards is the ability to apply contemporary and creative processes. Utilization of "best practices" is sensible. However, it is important to recognize that simply because something *can* be done does not mean that it *should* be done. Scientists are appropriately wary of change; change must be driven by superior results and approached cautiously so as to not introduce unexpected variables to ongoing studies. In compliance programs, imprudent fixation on best practices can also lead to regulatory "creep," or imposing a practice that was deemed necessary for one circumstance or facility onto another. Alternative practices should not simply do no harm or have negligible impact, they should demonstrably improve animal welfare.

The spirit of self-regulation also requires a willingness to self-report. Regulatory authorities expect some level of reporting noncompliance and adverse events to the IACUC or EOB and the governmental agency. In the United States, failure to self-report is among the more common sources of noncompliance cited by regulatory agencies (Potkay and DeHaven 2000). A reluctance to report deviations or departures may result in a loss of credibility or worse—allowing conditions to negatively influence animal well-being or research results. Conversely, a reputation for excessive reporting beyond regulatory requirements risks loss of the collegial relationship required of compliance programs and the flexibility intrinsic to application of performance standards. When serving as an extension of the IACUC or EOB, this requires careful attention to diplomacy for the program manager. Ensuring that all parties maintain attention to their ultimate shared goals, research standardization and regulatory compliance, will assist in determining the most effective response plan to reported concerns.

Introduction to Regulations Impacting Animal-Based Research

The reader is strongly encouraged to review the source documents for regulations and guidelines that impact their program directly. No summary can provide adequate detail for the issues raised and discussed in these documents. The intent of the remainder of the chapter is to provide an introduction to the extant regulations and guidance documents around the globe and to identify common themes. Note that AAALAC International maintains an impressive collection of links to U.S. and international regulations (AAALAC 2016c).

An introduction to regulations and guiding documents would be remiss without a brief discussion of Russell and Burch's *The Principles of Humane Experimental Technique* (1959). This historic text is divided into two parts. Part One, "The Scope of Humane Technique," describes concepts of inhumanity and underlying themes of removing inhumanity. Part Two, "The Progress of Humane Technique," introduced what has since been referred to as the three Rs: replacement, reduction, and refinement. Numerous works describe and expand the basic maxims of the three Rs (e.g., Goldberg et al. 1996; Guhad 2005; Brønstad and Berg 2011; Curzer et al. 2013) and encourage their implementation in theory (e.g., Van Zutphen and Van Der Valk 2001; Lloyd et al. 2008; Jennings et al. 2010; Slob 2014; Bratcher and Reinhard 2015; Kramer and Font 2015) and in practice (e.g., Agelan et al. 2008; Kim et al. 2012; Michelini et al. 2014; Lilley et al. 2015), or reflect their influence (e.g., Carlsson et al. 2004; Fenwick et al. 2009; Bayne 2011; Törnqvist et al. 2014).

Briefly, *replacement* strategies include the use of tissue culture, microorganisms, and other models in lieu of live animals. Appropriate study design and analysis of sources of variation address the *reduction* in numbers of animals required. *Refinement* may take many paths, but involves consideration of alternatives to potentially painful procedures and other means of minimizing animal distress. The vast majority of, if not all, regulations, guidelines, and other references to animal use in research, teaching, and testing refer to the fundamental concepts of the three Rs, which have been cited as the primary benchmarks for ensuring public confidence and avoiding unnecessary pain and distress (ILAR 2008).

The Americas

There are several excellent summaries of U.S. regulations, including within the previous edition of this text (Gonder 2002; see also Anderson 2002; Cardon et al. 2012; Silk et al. 2013; Bradfield et al. 2014).

United States

Animal Welfare Act and Regulations

Stimulated at least in part by animal welfare organizations and the general public, as exemplified by a cover article in *Life* magazine investigating pet theft (Wayman 1966), the U.S. Congress passed the Laboratory Animal Welfare Act in 1966 (Public Law [P.L.] 89-544), which primarily addressed the acquisition, transportation, and basic veterinary care of dogs, cats, nonhuman primates, rabbits, guinea pigs, and hamsters (LAWA 1966; Loew and Cohen 2002).

The original 1966 act focused on the acquisition of dogs and cats, and required licensing of dealers and registration by research facilities using these two species. The use of these companion species in research remains a polarizing topic (ILAR 2009). The title of the act was pared to the Animal Welfare Act in 1970 (P.L. 91-579), reflecting an expansion of the species covered; this amendment also broadened the entities requiring licensure and introduced a requirement for annual reports summarizing activities and providing assurance of appropriate animal care (AWA 1970). Amendments to the act in 1976 (P.L. 94-279) broadened the oversight of animal transportation (AWA 1976). In 1985, the Food Security Act (P.L. 99-198) included the Improved Standards for Laboratory Animals Act, which compelled oversight of animal research by an IACUC (AWA 1985). The Pet Theft Act of 1990 (AWA 1990) reemphasized the public concerns over the acquisition of companion animals in research (Anderson 2002). Amendments in 2002 and 2008 clarified the species included in the definition of *animal* and fines levied for violations (AWA 2002, 2008; Cardon et al. 2012).

The text of the Animal Welfare Act is primarily a collection of principles that empower the USDA to develop and implement specific standards within regulations. The regulations are established and enforced by the Animal Care section of APHIS; public input is sought by circulating proposed revisions to standards, followed by accepted revisions, in the *Federal Register*. The regulations are subsequently published in the Code of Federal Regulations (CFR), Title 9, "Animals and Animal Product," Subchapter A, "Animal Welfare," Parts 1–4. For use in biomedical research and education, the species covered by the USDA regulations are homeothermic vertebrates excepting traditional farm animals used for production research and rats of the genus *Rattus*, mice of the genus *Mus*, and birds bred specifically for research. Other bird species are covered; however, standards for their use have not been established (USDA 2015a).

The Animal Welfare Regulations are published along with the act in a document internally referred to as the "Animal Care Blue Book" (USDA 2015b). The regulations are divided into four parts: (1) definition of terms, (2) regulations, (3) standards, and (4) rules of practice governing proceedings under the Animal Welfare Act. Part 2 contains 9 subparts. Subpart C pertains specifically to research facilities; other subparts that impact or mention research facilities include (E) identification of animals, (F) stolen animals, (G) records, (H) compliance with standards and holding period, and (I) miscellaneous. Part 3 outlines "specifications for the humane handling, care, treatment, and transportation" in six subparts for (A) dogs and cats, (B) guinea pigs and hamsters, (C) rabbits, (D) nonhuman primates, (E) marine mammals, and (F) other warm-blooded species.

The regulations contain specific requirements for facility registration; IACUC membership and operation, including review of proposed animal use and semiannual program assessment and facility inspections; investigating animal-related concerns; personnel training and qualifications; the role of the AV and adequate veterinary care; recordkeeping; and annual reports. Emphasis includes avoiding unnecessary pain, distress, and duplicative experiments; considering alternatives to procedures likely to cause pain or distress; and special consideration for multiple major survival surgical procedures. The USDA regulations also oblige institutions to develop means by which animal welfare concerns may be reported and investigated, with stipulations for anonymity and freedom from repercussion, also known as "whistleblower" protection.

Inspectors for the USDA, termed either veterinary medical officers (VMOs) or animal care inspectors (ACIs), are charged with conducting annual inspections of licensees and registrants (USDA 2015a). In 2013, the USDA released the *Animal Welfare Inspection Guide* (AWIG) (USDA 2013), which replaced the earlier *Research Facility Inspection Guide* and the *Consolidated Inspection Guide* (Bennett et al. 2014). The purpose of the 424-page AWIG "is to provide an aid for APHIS Animal Care personnel when inspecting USDA licensed and registered facilities." Information specifically related to the inspection of research facilities is included in the first seven of the eight chapters of the AWIG. Chapter 9 contains 20 Animal Care policies, also published separately as the *Animal Care Policy Manual* (USDA 2015c), which incorporated the former 2011 *Animal Care Resource Guide*. These documents, along with "fact sheets" covering topics such as a summary of the act, methods of conducting compliance inspections, and appealing inspection findings, may be found at the USDA APHIS Animal Welfare website (USDA 2015d). Note that none of these texts are considered regulations, but are intended to "improve the quality and uniformity of inspections, documentation, and enforcement of the Animal Care Program."

Violations identified during USDA inspections are outlined as noncompliant items in publicly available inspection reports; severe or ongoing violations may result in fines to the institution. Several authors outline effective means of managing the USDA inspection process (e.g., Ingham et al. 2004; Cardon et al. 2012). All emphasize the importance of ensuring that institutional representatives are intimately familiar with not only the regulations, but also the additional guidance available and used by VMOs and ACIs when conducting inspections.

Public Health Service Policy

For decades, participation in the NIH intra- or extramural research program required recipients to follow the recommendations of the *Guide* (ILAR 2011a; see description below) and maintain either accreditation by AAALAC International or a system of internal oversight by an institutional committee (ILAR 1978). In 1985, the Health Research Extension Act (P.L. 99-158) authorized the secretary of Health and Human Services to work with the director of the NIH and develop guidelines for the appropriate use of animals in biomedical research (HREA 1985). Incorporating and supplementing the U.S. Government Principles for the Utilization and Care of Vertebrate Animals Used in Testing, Research, and Training (IRAC Principles) (IRAC 1985), the subsequent PHS Policy (OLAW 2002) was published in 1986 and revised in 2002 and 2015 (NIH 2015) (Box 7.5).

The PHS Policy applies to any use of live vertebrate animals conducted or supported by PHS agencies, including the NIH, FDA, Centers for Disease Control and Prevention (CDC), and Health Resources and Services Administration (HRSA). Through a memorandum of understanding, institutions supported by the NSF must also employ PHS Policy (OLAW 2015b). Implementation of the policy requires negotiation of an "animal welfare assurance" between OLAW and an institution wishing to conduct PHS-supported activities. The assurance details the institutional animal care and use program and methods used to comply with the PHS Policy. Features such as organizational structure, personnel training and qualifications,

BOX 7.5 U.S. GOVERNMENT PRINCIPLES FOR THE UTILIZATION AND CARE OF VERTEBRATE ANIMALS USED IN TESTING, RESEARCH, AND TRAINING

The development of knowledge necessary for the improvement of the health and well-being of humans as well as other animals requires *in vivo* experimentation with a wide variety of animal species. Whenever U.S. Government agencies develop requirements for testing, research, or training procedures involving the use of vertebrate animals, the following principles shall be considered; and whenever these agencies actually perform or sponsor such procedures, the responsible Institutional Official shall ensure that these principles are adhered to:

I. The transportation, care, and use of animals should be in accordance with the Animal Welfare Act (7 U.S.C. 2131 et seq.) and other applicable Federal laws, guidelines, and policies.*

II. Procedures involving animals should be designed and performed with due consideration of their relevance to human or animal health, the advancement of knowledge, or the good of society.

III. The animals selected for a procedure should be of an appropriate species and quality and the minimum number required to obtain valid results. Methods such as mathematical models, computer simulation, and *in vitro* biological systems should be considered.

IV. Proper use of animals, including the avoidance or minimization of discomfort, distress, and pain when consistent with sound scientific practices, is imperative. Unless the contrary is established, investigators should consider that procedures that cause pain or distress in human beings may cause pain or distress in other animals.

V. Procedures with animals that may cause more than momentary or slight pain or distress should be performed with appropriate sedation, analgesia, or anesthesia. Surgical or other painful procedures should not be performed on unanesthetized animals paralyzed by chemical agents.

VI. Animals that would otherwise suffer severe or chronic pain or distress that cannot be relieved should be painlessly killed at the end of the procedure or, if appropriate, during the procedure.

VII. The living conditions of animals should be appropriate for their species and contribute to their health and comfort. Normally, the housing, feeding, and care of all animals used for biomedical purposes must be directed by a veterinarian or other scientist trained and experienced in the proper care, handling, and use of the species being maintained or studied. In any case, veterinary care shall be provided as indicated.

VIII. Investigators and other personnel shall be appropriately qualified and experienced for conducting procedures on living animals. Adequate arrangements shall be made for their in-service training, including the proper and humane care and use of laboratory animals.

IX. Where exceptions are required in relation to the provisions of these Principles, the decisions should not rest with the investigators directly concerned but should be made, with due regard to Principle II, by an appropriate review group such as an institutional animal care and use committee. Such exceptions should not be made solely for the purposes of teaching or demonstration.†

* For guidance throughout these principles, the reader is referred to the *Guide for the Care and Use of Laboratory Animals*, prepared by the Institute of Laboratory Animal Resources, National Academy of Sciences.

† Published in the *Federal Register*, May 20, 1985, vol. 50, no. 97, by the Office of Science and Technology Policy (FR Doc. 85-12059).

Source: Interagency Research Animal Committee, *Federal Register*, 50, 97, 1985. https://grants.nih.gov/grants/olaw/references/phspol.htm.

and occupational health and safety for personnel are included in the description. While not requiring accreditation by AAALAC International, OLAW categorizes institutions based on accreditation status. Special mechanisms are also in place to ensure compliance with PHS Policy at institutions with no animal care and use program and foreign institutions receiving PHS funding.

In addition to the IRAC Principles, the PHS Policy requires compliance with the Animal Welfare Act and other federal regulations, as applicable; the *Guide* (ILAR 2011a); and the *AVMA Guidelines for the Euthanasia of Animals* (AVMA 2013). OLAW requires that non-U.S. institutions receiving PHS support pledge compliance with the Council for International Organizations of Medical Sciences (CIOMS) and International Council for Laboratory Animal Science (ICLAS) International Guiding Principles for Biomedical Research Involving Animals (CIOMS and ICLAS 2012; see description below).

The PHS Policy outlines the membership requirements and major functions of the IACUC, including semiannual facility inspections, programmatic reviews, and reports to the IO; approval of PHS-conducted or -supported research projects; and review of animal-related concerns; and otherwise advises the IO. Assured institutions are required to submit an annual report, notify OLAW of protocol suspensions and instances of noncompliance with the policy or *Guide*, and periodically renegotiate the assurance. Through OLAW, PHS performs site visits to assured institutions both "for cause" and on a random basis (Potkay and DeHaven 2000). Federal funding may not be used to support noncompliant activities; failure to abide by PHS Policy is considered a breach of contract. Thus, the impact of noncompliance may range from a single project to the entire institution's ability to use or receive PHS awards.

The OLAW website contains links to these references, as well as a considerable amount of guidance, including, at the time of writing, 36 NIH Guide Notices and 13 "Dear Colleague" letters, 84 frequently asked questions (FAQs), 20 articles penned by OLAW staff, 65 published commentaries on protocol review, and 3 letters to editors. Educational resources include archives of 34 webinars and podcasts, a tutorial on the PHS Policy, training videos, and other resources (OLAW 2015a).

Other Regulations

There are numerous other regulations impacting the use of animals in research and/or employees working in vivaria in the United States. For example, the Food, Drug, and Cosmetic Act, originally legislated in 1938 with significant amendments in 1962 and 1976, regulates the development of drugs, biological products, and medical devices for use in humans and animals (FDA 2015a). This act requires that the FDA ensures that research supporting, or intending to support, applications for such products will be conducted in accordance with the good laboratory practice (GLP) standards (21 CFR, Part 58; Anderson 2002; Gonder 2002; FDA 2015c). The FDA also includes provisions termed "the Animal Rule," for limited market approval of drugs completing animal trials but for which human clinical trials are not ethical or feasible (FDA 2015b). Similarly, research supporting, or intending to support, applications to the Environmental Protection Agency (EPA) for new pesticides must also abide by GLP standards (Anderson 2002).

Other regulations affecting the use of animals in research include the Occupational Safety and Health Act, designed to protect the health of personnel; the Controlled Substances Act, regulating the use of opioids and other substances overseen by the Drug Enforcement Administration; the Atomic Energy Act, controlling the use of radioactive materials; the Radiation Control for Health and Safety Act, regulating the use of medical devices emitting radiation; and the Endangered Species Act, monitoring the use of threatened species (Anderson 2002; Gonder 2002).

As previously mentioned, there are agreements between OLAW and the USDA, FDA, NSF, and VA (OLAW 2012, 2013, 2015b) for sharing compliance reports and other information. Note that in addition to distribution within federal agencies, most compliance reports become available to the general public through the Freedom of Information Act (FOIA) (Potkay and DeHaven 2000).

Numerous states and other municipalities also have laws impacting the use of animals in research, including the use of animals in education, acquisition from humane shelters, and direct oversight, including inspection by local authorities. Many laws also have been promulgated to protect animal-using enterprises. The National Association for Biomedical Research website maintains a list of such statutes (NABR 2016).

The Guide

First published in 1963 as the *Guide for Laboratory Animal Facilities and Care* by the U.S. Public Health Service (Animal Care Panel 1963) and now in its eighth edition, the *Guide* (ILAR 2011a) has become an internationally recognized resource for assisting "institutions in caring for and using animals in ways judged to be scientifically, technically, and humanely appropriate," with recommendations "based on scientific principles and on expert opinion and experience with methods and practices that have proved to be consistent with high quality care." From its inception, the authors of the *Guide* acknowledged the "scientific community's ethical commitment to provide the best possible care for animals used in the service of [hu]man[s] and animals" and highlighted the need for professionally qualified leadership, trained and experienced animal care personnel, and efforts to enhance animal health and well-being through appropriate facility construction and operation (Animal Care Panel 1963). Subsequent prefaces confirmed the widespread acceptance of the *Guide* by the scientific community and an ongoing commitment to the adoption of high standards. The 1968 *Guide* stated that it intended to extend recently adopted federal legislation by "defining humane care in professional terms and describing the facilities that provide humane care" (ILAR 1968). All editions reconfirm the notion of the shared responsibility between individual scientists and the institution, as well as the concept that nothing should interfere with the conduct of appropriately designed experiments. The latest edition states, "The *Guide* is created by scientists and veterinarians for scientists and veterinarians to uphold the scientific rigor and integrity of biomedical research with laboratory animals as expected by their colleagues and society at large" (ILAR 2011a).

As described previously, PHS Policy requires compliance not only with the IRAC Principles, Animal Welfare Act and Regulations, and other federal regulations, but also with the *Guide* (NIH 2015). The *Guide* is also considered a reference by USDA inspectors (USDA 2013), and is one of the three primary standards used by AAALAC International in evaluating animal care and use programs (AAALAC 2016d). Furthermore, numerous other countries, research sponsors and funding sources, scientific societies, and journal editorial boards require adherence to the recommendations of this text.

The eighth edition of the *Guide* contains five chapters and four appendices. Chapter 1, "Key Concepts," defines *laboratory animal* as "any vertebrate animal (i.e., traditional laboratory animals, agricultural animals, wildlife, and aquatic species) produced for or used in research, testing, or teaching." While the authors state that the *Guide* does not specifically address the use of invertebrates, wildlife used in field studies, or agricultural research or teaching, it "establishes general principles and ethical considerations that are also applicable to these species and situations." This section reconfirms the three Rs and incorporates not only personnel but "policies, procedures, standards, organizational structure, staffing, facilities, and practices put into place" into the institutional animal care and use program.

Chapter 2, "Animal Care and Use Program," outlines the roles and responsibilities of the IO, AV, and IACUC or EOB. It emphasizes the importance of personnel training and education, occupational health and safety, means of investigating and reporting concerns regarding animal welfare, and institutional preparedness for natural disasters and other emergencies.

Recommendations regarding the maintenance of animal housing environments and husbandry practices, such as environmental enrichment, illumination, noise and vibration, and population management, are separated into terrestrial and aquatic animal sections in Chapter 3, "Environment, Housing, and Management." While this section provides some recommended engineering standards that institutions may find useful, the *Guide* repeatedly endorses performance standards, especially for such considerations as available space within the primary enclosure.

Chapter 4, "Veterinary Care," expands on the bulk of the 1963 and subsequent editions of the *Guide* and provides a detailed review of concepts inherent to both individual animal and colony health. The subject matter includes animal procurement and transportation; preventive medicine practices, such as quarantine, acclimation, and health surveillance; and clinical care, including routine and emergency care. The largest section of the chapter is devoted to surgery, outlining recommendations from surgeon training and presurgical planning to postoperative care. This chapter also describes approaches to managing pain and distress via anesthesia, analgesia, and euthanasia.

Recommendations for the construction and maintenance of animal facilities provided the greater part of earlier *Guide* editions and are now included as the final chapter, "Physical Plant." The eighth edition expands discussions on special housing facilities, such as barriers, aquatics, and hazard containment, and special procedural areas for surgery, behavioral observation, imaging, and irradiation.

The Ag Guide

The *Guide for the Care and Use of Agricultural Animals in Agricultural Research and Teaching* (the *Ag Guide*) was first published by the Federation of Animal Science Societies (FASS) in 1988 and revised in 1999. In the latest edition (FASS 2010), the title was changed to *Guide for the Care and Use of Agricultural Animals in Research and Teaching* to reflect a broader application of its recommendations, stating, "Farm animals have certain needs and requirements and these needs and requirements do not necessarily change because of the objectives of the research or teaching activity." The *Ag Guide* endorses the IRAC Principles (1985). The *Ag Guide* is used as a reference for USDA inspectors (USDA 2013) and is one of the three primary standards used by AAALAC International in its accreditation program (AAALAC 2016d).

The first five chapters of the *Ag Guide* outline key features of institutional animal care and use programs. Chapter 1, "Institutional Policies," describes the roles and responsibilities of the IACUC, and highlights the importance of developing written operating procedures, ensuring biosecurity, and providing attention to personnel qualifications and occupational health. Chapter 2, "Agricultural Animal Health Care," considers veterinary health issues, such as medical care, surgery, restraint, and euthanasia; monitoring for zoonotic diseases and drug residues; and special considerations for genetically modified and cloned animals. Chapters 3 and 4, "Husbandry, Housing, and Biosecurity" and "Environmental Enrichment," respectively, discuss standard practices and environmental management, emphasize biological security and containment, and delineate species-specific recommendations. Chapter 5, "Animal Handling and Transport," provides background information and practical guidance for transporting, moving, and restraining animals to minimize the potential for animal and personnel distress and injury. The remaining six chapters expand the concepts outlined in Chapters 3 through 5 for beef cattle, dairy cattle, horses, poultry, sheep and goats, and swine.

Canada

The *Guiding Principles on the Care of Experimental Animals* was first published in 1961 by the Canadian Federation of Biological Societies (Griffin and Baar 2014). In 1968, the Canadian Council on Animal Care (CCAC) was established with the primary responsibility to create, maintain, and oversee the implementation of high ethical and quality assurance standards for the care and use of experimental animals throughout Canada. The CCAC is a nonprofit, national, peer review organization funded primarily by the Canadian Institutes of Health Research (CIHR) and the Natural Sciences and Engineering Research Council (NSERC), and through participating institutions (CCAC 2016a).

Species under the scope of CCAC oversight include all nonhuman vertebrates and cephalopods. The CCAC establishes recommended practices by its standards program. Policy statements provide basic requirements for animal care and use (CCAC 2016b). The *Guide to the Care and Use of Experimental Animals* (*Guide*) outlines more specific standards and species-specific recommendations (CCAC 1993). Guidelines address specific aspects of program oversight, such as humane endpoints, antibody production, euthanasia, animal procurement, protocol review, personnel training, and use of wildlife, fish, and transgenic, farm, and marine mammals (CCAC 2016c).

A CCAC Certificate of Good Animal Practice (GAP) is earned by institutions that participate in the Assessment and Certification Program (CCAC 2016b,c,d). Using the policy statements, guidelines, and the *Guide*, the Assessment and Certification Committee awards certificates based on reviews conducted by an assessment panel every 3 years.

All Canadian provinces have adopted some legislation regarding the use of animals in research and teaching. For example, in Ontario, animal use is governed by the Animals for Research Act (Griffin and Baar 2014). Although participation in the CCAC certification program is intended to be voluntary, most provinces reference CCAC standards, and several require it. Furthermore, certain sponsors require an institutional certificate as a condition of funding.

Establishing an institutional Animal Care Committee (ACC) is a prerequisite for CCAC eligibility. The ACC is expected to oversee all aspects of animal care and use, including protocol review and approval. ACCs are expected to meet at least twice annually, conduct facility inspections at least annually, and otherwise ensure institutional compliance with veterinary, personnel training, occupational health and safety, and facility maintenance standards.

Latin America

Latin America is an enormous and heterogeneous area with the Spanish language as the main common denominator, except for Brazil, where Portuguese is spoken. Only three countries in Latin America have developed specific laws protecting animals used for research: Brazil, Mexico, and Uruguay. In others, for example, Peru, Venezuela, and Costa Rica, general animal welfare laws contain brief references to research animals. Other countries have no such regulations; the level of care and use in research is left exclusively to institutional and personal commitment.

The Brazilian regulation is known commonly as the "Arouca law" and was issued in 2008 (Brazil 2008). This law mandates the creation of the National Council to Control Animal Experimentation (CONCEA), composed of representatives from several prestigious public institutions and charged with establishing a framework for overseeing animal care and use. In Mexico, the specific regulation is the Mexican Official Norm for the Production, Care, and Use of Laboratory Animals (NOM-062-ZOO-1999) (Mexico 1999). The law in Uruguay is entitled Use of Animals in Experiments, Teaching and Research Activities Law (No. 18.611) (Uruguay 2009). Similar to Brazil, a National Commission for Animal Experimentation (CNEA) is established to control the use of research animals. These three pieces of legislation mandate a prospective evaluation of all research proposals to be performed by an institutional ethical oversight committee.

In these and other countries lacking specific research animal welfare regulations, professional national and supranational associations and societies play a vital role in developing standards, training personnel, and promoting new legislation. Examples include the Federation of South American Societies for Laboratory Animal Science (FESSACAL) (http://www.fessacal.com/) and the Federation of Hispanic Societies and Associations for Laboratory Animal Science of North America, Central America and Caribbean (FESAHANCCCAL) (http://www.fesahancccal.com/).

Europe

European Union

To understand the regulatory framework in Europe, it is essential to understand the organization of the European countries. The Council of Europe (CoE) comprises most of the European countries (47 of currently 51 independent states), including the 28 European Union (EU) member states. The CoE is intergovernmental, has no legislative powers, seeks voluntary cooperation, and may issue recommendations, agreements, and conventions. One of these conventions is the European Treaty Series 123 (ETS 123) (CoE 1986), which established the foundations for the current research animal protection legal framework across Europe. However, conventions must be voluntarily ratified and signed by CoE members, by which they commit to incorporating them into their national legislation. But as a voluntary process, signature and ratification does not always occur. The current chart of signatures (26) and ratifications (21) is available at http://www.coe.int/en/web/conventions/full-list/-/conventions/treaty/123/signatures.

On the other hand, the supranational EU retains policy-making and legislative powers. The EU issues regulations, requiring direct compliance by member states, and *directives*, which must be transposed into national legislation within a certain time period. The current piece of legislation concerning the protection of animals used for scientific purposes is Directive 2010/63/EU (EP and CEU 2010). This directive has already been adopted by all EU member states, and establishes common requirements for breeders, suppliers, and users in the EU, which must be authorized by, and registered with, the "competent authority" in each member state. However, during the transposition process, member states can implement in a different manner some of the requirements (see example described below). Importantly, although requirements established prior to implementation of the directive may be maintained, member states cannot extend new stipulations beyond those established by the original charge.

Directive 2010/63/EU is based explicitly on the principles of the three Rs (Article 4), and applies to all vertebrates, including independently feeding larval forms and mammalian fetuses from the last third of normal development, as well as live cephalopods. Annex I lists the species that must be purpose bred, which includes traditional laboratory species and others, such as zebrafish. Special attention is paid to the use of nonhuman primates, which is limited to specific purposes (Article 8).

A feasibility study by the European Commission to be finalized before November 10, 2017, will be used to investigate an intent to limit the use of nonhuman primates to the second generation only, that is, the offspring of animals bred in captivity or from self-sustaining colonies. The use of great apes is banned, although a safeguard clause could permit their use in cases of emergency (Article 55).

All procedures must be classified as "nonrecovery," "mild," "moderate," or "severe" on a case-by-case basis using the assignment criteria set out in Annex VIII. The classification of "severe" has implications concerning the reuse of animals (Article 16) and the need to perform a retrospective assessment (Article 39).

Directive requirements for animal care and use (Annex III) are based on ETS 123 Appendix A, guidelines for accommodation and care of animals (CoE 2006). Of particular impact, adoption in the directive transforms recommendations within Appendix A for cage sizes into obligations for EU member states, significantly increasing required minimum cage sizes for many species by comparison with previous standards. Nonetheless, institutions had to comply with the directive by January 1, 2017.

Other important concepts in the directive include the competence of personnel, who must be adequately educated and trained prior to (1) designing procedures and projects, or performing (2) animal-related procedures, (3) animal care, or (4) euthanasia (Article 22). For the latter three functions, a period of direct supervision to demonstrate competency is required; that is, training itself is not sufficient. In fact, institutions must designate one or several persons responsible for ensuring that staff are trained and supervised until demonstrating competency, and receive continuing education. Note that authority for identifying specific training requirements remains with the member states, and Annex IV lists acceptable methods of performing species-specific euthanasia.

All institutions must also have a designated veterinarian with expertise in laboratory animal medicine charged with advisory duties related to animal treatment and well-being (Article 25). The article allows substitution for the designated veterinarian by a "suitably-qualified expert." This controversial clause was intended to provide flexibility for institutions using only rare species; however, the loophole may be used to justify a lack of an institutional veterinarian. That said, the responsibilities and authority assigned to the designated veterinarian beyond animal medical care are mainly advisory, which includes providing input to the Animal Welfare Body (AWB) (see below) and advice concerning reuse or other nonterminal disposition.

Oversight and ethical review is distributed among several parties, including one or more designated persons bearing the responsibility for overseeing animal care and welfare for the institution. Additionally, institutions must establish an AWB composed of a scientist and the designated responsible individuals for animal care and welfare. The designated veterinarian, or other expert, also provides input to the AWB. Minimum AWB functions include advising on the care and use of animals and implementing the three Rs, assisting in internal operational processes, following the development and outcome of projects, and recommending rehoming schemes (Article 27).

A critical difference compared with oversight systems in other areas of the world is that the project evaluation required for authorization is not assigned to the institutional AWB, but to the "competent authority." This is generally a public entity; however, member states retain some flexibility in identifying others to perform the mandatory project evaluation. This may include designation of the AWB or other institutional ethics committees, or an external body (Guillén et al. 2015). In any case, project evaluation must be transparent and performed in an impartial manner, and must include, in addition to the customary items (listed in Annex VI), a "harm–benefit analysis," and assessment of the severity classification. Designation of the body responsible for project review exemplifies the flexible implementation and transposition into national legislation from the common framework provided by the directive.

Following evaluation, projects may be authorized for up to 5 years. Final authorization, which generally must not take longer than 40 working days, or 55 in case of great complexity, is granted only at the competent authority level, except in Belgium, where authorization may be granted by the institutional body performing the evaluation. To enhance transparency, nontechnical project summaries produced by the applicant are published anonymously by the competent authorities for general public access.

Competent authorities are required to inspect establishments. Inspections are to be carried out on at least one-third of the user institutions each year in accordance with a risk analysis. However, breeders, suppliers, and users of nonhuman primates must be inspected at least annually.

To facilitate harmonization in the implementation of the directive, the European Commission developed a number of Expert Working Groups (EWGs) that have produced important consensus documents (European Council 2016). These texts, addressing such areas as AWB function, genetically altered animals, inspections and enforcement, nontechnical summaries, project evaluation, and severity assessment, may be adopted by member states and others affected by the directive. Consensus documents involving education and training continue to significantly foster harmonization of practices across member states. Note that ancillary topics such as occupational health and safety and animal transportation are not addressed in Directive 2010/63/EU, but may be presented in other independent pieces of legislation at the EU or national level (Guillén et al. 2014).

Other European Countries

Many European countries not subjected to the EU Directive have specific legislation. For example, Switzerland has a well-established regulation, including an Animal Welfare Act (FASF 2005) with amendments and ordinances addressing the care and use of research animals. The act includes both cephalopods and decapods, and abides by the principles adopted by the EU. Project evaluation is performed at the competent authority level, with committees established at the regional (Kanton) level.

Norway also has specific regulations (NMAF 2015) that closely adhere to the EU Directive. Given a relatively high use of fish in research, Norwegian legislation typically expects involvement of a veterinarian, or "fish health specialist" to serve the role of a veterinarian, concepts similar to those in Article 25 of the EU Directive.

Another country that belongs to the CoE but not to the EU, and has neither signed nor ratified ETS 123, is the Russian Federation. Lacking specific legislation, the Russian Laboratory Animal Science Association (Rus-LASA) continues to promote modern animal welfare regulations.

Laboratory animal professionals in Israel have close relationships with European colleagues and participate in European professional organizations, such as the Federation of European Laboratory Animal Science Associations (FELASA) and European Society for Laboratory Animal Veterinarians (ESLAV). Research animal care and use has been legislated in this country since 1994 with the prevention of cruelty to animals law and prevention of cruelty to animals rules (Assembly of Israel 2005; Kalman et al. 2014). The law in Israel is based on a national council appointed by the Minister of Health and the establishment of institutional animal care and use programs overseen by an IACUC appointed by the council. The council can issue rules and guidance; the latest edition of the *Guide* (ILAR 2011a) is used to define the legal framework. As such, animal research oversight closely resembles that in the United States, with the IACUC retaining primary authority and supported by an AV.

Africa

Little can be said on specific legislation on the protection, care, and use of research animals in Africa (Hau 2014). Only general laws concerning animal welfare can be found in some countries, such as Kenya (1983), South Africa (1962), and Tanzania (2008). The most recent, Tanzanian Animal Welfare Act, contains a general reference to the "Five Freedoms" (FAWC 1979) and, more importantly, explicit recognition of the three Rs when addressing animal experimentation. In South Africa, where more laboratory animal research is conducted, there are other less official codes of conduct concerning the care and use of laboratory animals. For example, the South African Medical Research Council published ethical guidelines on the use of animals for research and training (MRC 2004). The South African national standard for the care and use of animals for scientific purposes (SABS 2008) is a voluntary national standard. Minimum standards for research animal facilities are also described within the rules related to the practice of veterinary medicine in this country (DAFF 2015). Personnel who complete training by the British Institute of Animal Technology (IAT), with additional practical training at institutions recognized by the South African Veterinary Council (SAVC), can be registered as laboratory animal technologists.

Given humanitarian needs on the African continent, it is not surprising that laboratory animal legislation has not been prioritized; however, initiatives are underway. Continuing education courses are routinely organized in the Magreb (Tunisia, Algeria). The association ACURET.ORG (2016) is "a multinational, interdisciplinary non-governmental not-for-profit organization, incorporated in Nigeria for the purpose of promoting humane animal care and use for scientific purposes in developing countries, in particular, in Africa" that "values and implements the Principles of the 3Rs as the basis for its promotion."

Asia

People's Republic of China

Regulation of the People's Republic of China for Administration of Laboratory Animals

The Regulation of the People's Republic of China for Administration of Laboratory Animals was approved in 1988 by the state council under Decree No. 2 Order of the State Science and Technology Commission of the People's Republic of China (SSTC 1988). The regulation consists of 8 chapters and 35 articles. It addresses general requirements for the quarantine and control of infectious disease, personnel training and certification, and the import, export, and use of laboratory animals. The term *laboratory animals* under Chapter I, Article 2 of the regulation is defined as "artificially raised and bred animals with controlled microbes and parasites and definite genetic background and clear sources that are used in scientific research, teaching, production, examinations, and other scientific experiments." This definition does not further limit the definition of species covered, although sections discussing quarantine specifically address handling procedures for wild-caught species before use as laboratory animals. The regulation is applicable to any institution and individual maintaining, raising, breeding, supplying, or using animals for experimental purposes.

Other Standards, Guidelines, and Local Regulations

The Ministry of Science and Technology (MOST) is the government agency that oversees laboratory animal science development in China, and establishes general requirements and guidelines. Detailed requirements and specific topics on animal care and use are described in National Standards of the People's Republic of China, including Laboratory Animals—Requirements of Environment and Housing Facilities (GB 14925–2010) (NSPRC 2010a) and Laboratory Animals—Microbiological Standards and Monitoring (GB 14922.2–2010) (NSPRC 2010b). Specific standards for animal facilities are described in Architectural and Technical Codes for Laboratory Animal Facilities (GB 50447–2008) (NSORC 2008) and, more recently, Guidelines of Humane Treatment of Laboratory Animals (MOST 2006).

National regulations require institutions using animals to also comply with local regulations. The Municipal/Provincial Science and Technology Commission oversees the care and use of laboratory animals at the local level (e.g., state or province), which includes conducting an annual inspection. MOST's Management of Laboratory Animal License Systems authorizes the local administration office to issue individual and institutional registrations and licenses to conduct animal-based scientific activities. For example, the Beijing Administrative Office of Laboratory Animals (BOALA) administers the Legal Administration Documents for Laboratory Animals in People's Republic of China and Beijing Municipality, which issues licenses in Beijing (BAOLA 2000). Prerequisites to apply for licensure include records of personnel training, certification, and medical evaluation, and quality assurance monitoring. The license is valid for five calendar years.

MOST requires self-monitoring to be performed by an institutional ethical committee, for example, an IACUC (Bayne and Wang 2014). Detailed requirements for ethical committee function and protocol review are regulated by the local authority. MOST guidelines also address animal transportation, with additional guidelines provided in the local regulation. Permits from the Ministry of Forestry are also required for breeding, transporting, and export of nonhuman primates or nonhuman primate tissues (Bayne and Wang 2014).

Taiwan, Republic of China

Animal Protection Act and Enforcement Rules of Animal Protection

The Animal Protection Act was promulgated in 1998 by the Taiwan's Legislative Yuan and last amended in 2011 (COA 2011). The act, which also covers animal transport in general, consists of 6 chapters and 40 articles; implementation of the act is the authority of the Council of Agriculture (COA), to which institutions must submit annual reports. According to the act, *animal* is defined as "a dog, a cat, [or other] vertebrate that is raised and kept by people, including economic animals, experiment[al] animals, pets, and other animals." The scope of scientific application for experimental animals includes "for the purpose of teaching, experiments, manufacturing biological preparations, experimental products, drug or toxic substance, and organ transplantation." Chapter 2 captures humane aspects of animal use. Chapter 3, Articles 15–18, details the scientific application of animals, including the establishment of a care and use committee or panel to oversee the scientific aspects of animal use. The local competent authority's inspector may inspect facilities to review the implementation of the act.

The Enforcement Rules of Animal Protection were promulgated in 2000 and amended in 2013, and include specifications for the types of institutions conducting animal-based scientific activities (COA 2013).

Guidebook and Other Resources

A Guidebook for the Care and Use of Laboratory Animal (*Guidebook*) published by COA in 2010 addresses husbandry and enrichment, veterinary care, physical plant, breeding, occupational health and safety programs, disaster and emergency planning, and species-specific information (Kurosawa et al. 2014). The standards outlined in the *Guidebook* are used by inspection teams for annual evaluation of facilities selected by the COA.

Training courses for IACUC and program management are offered regularly, sponsored by the Chinese Association for Laboratory Animal Science (CALAS), Chinese Taipei Society for Laboratory Animal Science (CSLAS), and National Laboratory Animal Center (NLAC) (Chen 2008).

Japan

Law of Humane Treatment and Management of Animals

The Law of Humane Treatment and Management of Animals No. 105 was promulgated in 1973; a 2005 amendment endorses the principles of the three Rs (Kurosawa et al. 2014). The law addresses the proper treatment of animals, such as the responsibilities of the owner and registration and inspection requirements for animal-handling businesses. Although articles under the law pertain to all persons engaged in a business involving animals, the term *animal* is limited to mammals, birds, and reptiles.

Standards Relating to the Care and Management of Experimental Animals

The Standards Relating to the Care and Management of Experimental Animals: Experimental Animal Regulation was published in 1980 (Prime Minister's Office 1980). In 2006, the Ministry of Environment released a revision that included specific guidance for the relief of pain (MEJ 2006). The standards define *experimental animal* as "a mammal or bird reared or kept at a facility (including animal being transported to a facility) for use in experiments." The term *experiment* is defined as an "education purpose, experimental research, or manufacture of biotics or other scientific application." The standards address general considerations for transport, basic care, and the health and safety of animals (Kurosawa et al. 2014).

Other Laws and Resources

Additional guidelines are dependent on the jurisdiction and funding agencies applicable to the institution. For example, the basic policy for the conduct of animal experimentation is pertinent for research under the Ministry of Health, Labor and Welfare (MHLW) and the Fundamental Guidelines for Proper Conduct of Animal Experiments (Fundamental Guidelines) are applicable to organizations conducting animal experiments for the Ministry of Education, Culture, Sports, Science and Technology (MEXT 2006).

According to the Fundamental Guidelines, the term *animal* is defined as a mammalian, avian, or reptilian species, for experimentally related activities conducted in academic research institutions. The Fundamental Guidelines address the responsibility of the research institution and the establishment of institutional regulations and an animal experimental committee. The role and responsibility of the animal experimental committee includes protocol review and approval to ensure concurrence with the law and standards, education, training, and self-inspection.

The Guidelines for Proper Conduct of Animal Experiments were developed by the Science Council of Japan in 2006, and serve as an additional support document for activities under both MEXT and MHLW (SCJ 2006). These guidelines define *animal experiment* as "utilization of animals for education, testing, research, manufacture of biological products or other scientific purposes." The guidelines provide additional details regarding IACUC responsibilities and function; animal health, care, and management; and personnel training and safety. Laboratory animals covered under this guideline includes mammalian, avian, or reptilian species.

Use of genetically modified animals is regulated by the Law Concerning the Conservation and Sustainable Use of Biological Diversity through Regulation on the Use of Living Modified Organism (LCCSU 2007).

South Korea

Animal Protection Law

The Korean Animal Protection Act (APA) was first issued by the Ministry of Agriculture and Forestry (MAF) in 1991, and recently amended under the Ministry of Agriculture, Food, and Rural Affairs (Republic of Korea 2014). A broader definition of *animal* has been reduced in the current law to mean "mammals, birds, reptiles, amphibia, and fish, animals that are prescribed by Presidential Decree after discussion between the head of a relevant central administrative agency and the Minister of Agriculture, Food and Rural Affairs."

The revised law contains 47 articles, including establishment of a central Animal Welfare Committee within the ministry (Article 5), animal transportation (Article 9), and requirements for registration (Article 12). Chapter III (Articles 23–28) discusses animal experimentation, focusing primarily on minimizing animal pain and suffering; considerations for replacing animal models; qualifications of personnel conducting experiments; and reducing the number of animals used. Article 26 mandates the establishment of an ethics committee, termed the Animal Experimentation Ethics Committee, to ensure ethical treatment, and describes the committee's appointment, composition, and other functions. At least one member of the committee is selected by the ministry.

Laboratory Animal Act

The Laboratory Animal Act (LAA), which is overseen by the Ministry of Food and Drug Safety (MFDS), first issued in 2008 and amended in 2010 (LAA 2010), enhances ethical consideration of laboratory animal use and animal testing. Consisting of seven chapters, the LAA mandates that activities comply with the APA, and specifically applies to testing activities involved in "development, safety control and quality control of foods, functional health foods, medical and pharmaceutical products, non-medical and pharmaceutical products, biomedicines, medical appliances, and cosmetics" and "safety control and quality control of narcotics."

The LAA defines *laboratory animal* as a "vertebrate used or raised for the purpose of animal testing." The term *animal testing* means "testing conducted on laboratory animals, or scientific procedures for scientific purposes, such as education, testing, research and production of biological medicines or such."

Chapter II of the LAA describes education of personnel, consideration of alternatives to animal testing, personnel safety, and the establishment of a Laboratory Animal Management Committee to ensure the ethics, safety, and reliability of animal testing. Other chapters cover animal testing facilities, supply of laboratory animals, safety control, and disclosure of records to the public.

India

Prevention of Cruelty to Animals Act

The Prevention of Cruelty to Animals Act of 1960 (Act 59 of 1960) was the first law to legislate animal welfare in India, and was amended by the Central Act 26 of 1982 (PCAA 1982). The term *animal* is defined in the act as "any living creature other than a human being." Legal provision of experimentation on animals is stipulated in Chapter IV, applying to an "animal for the purpose of advancing new discovery of physiological knowledge for saving or prolong[ing] life, alleviating suffering or combating disease of human beings, animals, or plants." With the aim of preventing unnecessary animal pain or suffering, the act (Chapter IV, Section 15) dictates central oversight of animal experimentation by the Committee for the Purpose of Control and Supervision of Experiments on Animals (CPCSEA). CPCSEA is authorized to supervise and control experimental activity, including conducting inspections (Chapter IV, Section 18).

Breeding of and Experiments on Animal (Control and Supervision) Rules

In 1998, the Indian Parliament, through the Ministry of Social Justice and Empowerment, and later the Ministry of Environment and Forestry, reinforced CPCSEA by enacting the Breeding of and Experiments on Animal (Control and Supervision) Rules, which were later amended in 2001 and 2006 (MSJE 1998, 2001; MEF 2006). The definition of *experimental activity* under the 2001 amendments was expanded to include "use of animals for the acquisition of knowledge of a biological, psychological, ethological, physical or chemical nature; and includes the use of animals in the production of reagents and products such as antigens and antibodies, routine diagnostics, testing activities, and establishment of transgenic stock for the purpose of saving or prolonging life, or alleviating suffering, or for combating any disease whether on human being or animals."

The rules (Section 4) mandate registration of animal facilities conducting breeding and/or experimentation. The registrant is subject to inspection by individuals nominated and authorized by CPCSEA. The requirements include establishment of an ethics committee, known as the Institutional Animal Ethics Committee (IAEC), to review and approve the experimental proposals. The composition of IAEC membership, requirements for records, and qualifications for personnel assuming supervisory responsibilities are described in the rules.

The 2001 revision also required compliance with the Indian National Science Academy (INSA) Guidelines for Care and Use of Animals in Scientific Research (Guidelines). First published in 1992, the INSA Guidelines describe ethical responsibilities and committee review; housing and environment; breeding, genetics, and transgenic animals; animal husbandry and disease control; personnel training; animal transportation; and veterinary care, including anesthesia and euthanasia (INSA 2000).

Recently, CPCSEA released additional guidelines related to the use of animals for experiments, including the CPCSEA Guidelines for Laboratory Animal Facility (CPCSEA Guidelines) (CPCSEA 2003) and CPCSEA Standard Operating Procedures for Institutional Animal Ethics Committees (CPCSEA SOPs for IAEC) (CPCSEA 2010). The CPCSEA Guidelines contain eight annexures addressing requirements for veterinary care programs, responsibilities of the veterinarian, and anesthesia and euthanasia; personnel training and occupational health and safety; animal restraint, husbandry, and social environment; animal environment control and physical plant; transportation; transgenic animal use; and special procedures for using biological hazards. The CPCSEA SOPs for IAEC were issued in 2010 to ensure quality, consistency, and compliance with the rules and act in the function of the IAEC and its review mechanisms. This comprehensive document stipulates the IAEC objectives, role and function, membership composition, appointment, authority, and requirements for quorum, record keeping, and reporting. Functions of the IAEC include the review and approval of proposed animal use prior to implementation; conducting facility inspections, and submission of reports to CPCSEA. The IAEC is responsible for monitoring research activities and ensuring their compliance with the rules and act.

CPCSEA establishes an official animal experiment proposal form in the "Application for Permission for Animal Experiments (Form B)" and "Checklist for Protocol." While the IAEC is authorized to approve experimentation involving rodents and rabbits, use of other species requires additional approval by the CPCSEA National Committee, based on a positive recommendation from the IAEC.

IAEC membership includes at least one nonaffiliated member appointed by CPCSEA, also known as the nominee of CPCSEA. All IAEC decisions occur after discussion in a meeting that meets quorum; the presence of the nominee of CPCSEA member is mandatory.

The CPCSEA also provides a "Checklist for Inspection" to assist in conducting facility reviews, and an experimental animal health record describing maintenance of animal receipt, medical and experimental procedures, and rehabilitation efforts. Annual reports to CPCSEA include copies of IAEC minutes and facility inspection reports. Animal procurement, including importation of rodents for research purposes, requires certificates issued by CPCSEA after approval by the IAEC.

The 2006 revision of the rules requires investigators to consider rehabilitation rather than euthanasia for animals used in nonterminal studies, excluding those resulting in an inability to resume natural function or persistent pain, or those involving exposure to hazardous agents posing a risk to humans and other animals. Aftercare and/or rehabilitation is specifically described in the Recommendations of the Sub-Committee on Rehabilitation of Animals after Experimentation Set Up by CPCSEA (CPCSEA 2006).

Thailand

The National Research Council of Thailand (NRCT) has issued regulations applicable for animal care and use for scientific purposes in the *Ethical Principles and Guidelines for the Use of Animals for Scientific Purposes* (*Ethical Guidelines*), the *Fundamental Principles of Designing Animal Care and Use Facilities for Scientific Work* (*Fundamental Principles*), and the *Standards for Institutional Animal Care and Use Committees* (*IACUC Standards*) (NRCT 2006).

The *Ethical Guidelines* reflect revision of the *Ethics for Animal Experimentation* of 1999, which were developed primarily based on the CIOMS International Guiding Principles for Biomedical Research Involving Animals of 1985 (Gettayacamin et al. 2014). The *Ethical Guidelines* define the term *animal* as wildlife or vertebrate laboratory animals. *Laboratory animal* is defined as "animals that are procreated and nurtured in confinement, and are used by human[s] for the benefit of any branch of science and technology," and is applicable to both animal users and breeders. The document expands the concepts of the three Rs, in five ethical principles: addressing alternatives to animal models, minimizing animal numbers, conserving wildlife, avoiding unnecessary pain and distress, and compliance with animal experimental protocols. General precepts in animal transportation, facilities, husbandry, and health; personnel qualifications; and reliability of equipment are described.

Monitoring implementation of the ethical principles is to be conducted at both the institutional and national level. Institutions using animals in research, testing, the production of biological materials, and teaching must establish an ethics committee to oversee and ensure adherence to the guidelines. Committee responsibilities include establishing SOPs, review and approval of projects involving animal use, monitoring animal use activities, and providing training and continuing education of personnel.

The NRCT appoints the National Committee for Research Animal Development (NCRAD) to monitor and promote the ethical use of animals in scientific activities. In addition, editorial boards of academic journals are advised to request detailed information for research paper submission, including genetic background, justification of animal numbers, animal care, and protocol approval by the institutional ethics committee. In 2009, the NRCT also issued *Fundamental Principles*, which provides recommended standards for facility design, physical plant, animal environment, and personnel safety. *IACUC Standards* addresses IACUC membership and function, and emphasizes IACUC responsibility in accordance with the *Ethical Guidelines* (Gettayacamin et al. 2014).

Singapore

Animals and Birds Act and Animals and Birds (Care and Use of Animals in Scientific Purposes) Rules

The Animals and Birds Act originally enacted under Ordinance 3 of 1965 covered the use of any mammal (other than humans), bird, fish, other living creature, or those categorized under the class of animal, including genetically modified animals (ABA 2002). The Animals and Birds (Care and Use of Animals in Scientific Purposes) Rules, passed by the Ministry for National Development of Singapore, limited the definition of *animal* to "any live vertebrate, including any fish, amphibian, reptile, bird, and mammal [other than] human beings" (AVA 2004).

Prerequisites for transporting, holding, or using animals for scientific purposes include an AV, an appropriately constituted IACUC, a process for approving animal-related projects, and licenses issued by the Agri-Food and Veterinary Authority (AVA). The appointment, membership, and function of the IACUC are stipulated in Rules 7 and 8, and include semiannual review of the animal care and use program, annual inspection of the research facility, investigating reports of noncompliance, and PAM. Rule 11 specifies requirements for the AV and adequate veterinary care, including specifications for training and experience. Licensees submit an annual report to the AVA and are required to immediately report suspected disease outbreaks or unusual mortality levels.

National Advisory Committee for Laboratory Animal Research Guidelines

The AVA also requires compliance with the National Advisory Committee for Laboratory Animal Research (NACLAR) Guidelines on the Care and Use of Animals for Scientific Purposes (NACLAR Guidelines) (NACLAR 2004) for facilities housing or using vertebrates for teaching, field trials, environmental studies, research, diagnostics, product testing, and production of biological products. The NACLAR Guidelines are comprised of (1) the Guiding Principles for the Care and Use of Animals in Scientific Purposes (Guiding Principles), which address humane care and use, including appendices for fish and nonhuman primates, (2) Guidelines for the Institutional Animal Care and Use Committee (IACUC Guidelines); and (3) training guidelines.

The Guiding Principles detail animal housing and management, animal procurement and transportation, veterinary care, and IACUC responsibilities (Chapter 8). An AV licensed by the AVA must be part of the program and has authority to oversee animal care and use.

Additional functions of the IACUC are described in the IACUC Guidelines. Oversight function at the institutional level is maintained by the IACUC. Their responsibilities include protocol review and approval of proposed animal activities, semiannual program review, annual inspection of animal housing and procedure areas, and annual reporting to AVA ensuring compliance with and describing IACUC-approved exemptions to the Guiding Principles.

The training guidelines outline recommendations for all personnel involved in the animal care and use program, including researchers, animal facility personnel, and IACUC members. IACUC members are required to attend the Responsible Care and Use of Laboratory Animals Course within 12 months of appointment. Occupational health and safety is also addressed, and must comply with biosafety standards issued by the Ministry of Health and the Ministry of Manpower.

Other Guidelines

Supported by the Genetic Modification Advisory Committee (GMAC) guidelines, the GMAC is responsible for overseeing the transportation and use of genetically modified organisms (GMOs) (GMAC 2013).

Malaysia

As of this writing, the proposed Malaysian Code of Practice for the Care and Use of Animals for Scientific Purposes has been developed by the Laboratory Animal Science Association of Malaysia (LASAM) and adopted by many research institutions and the National Bioethics Council (Gettayacamin et al. 2014; LASAM 2016).

The term *animal* under the proposed code is applied to any member of the animal kingdom and includes "any mammal (other than [hu]man), bird, reptile, amphibian, fish, mollusk, arthropod, and other vertebrate or invertebrate, whether alive or dead, and the egg, young or immature from thereof." The scope envelops all aspects of animal care and use in scientific activities, "in medicine, biology, agriculture, veterinary or other animal science, industry and teaching," and in "research, teaching, field trials, product testing, diagnosis, the production of biological products and environmental studies."

The proposed code endorses the three Rs and stipulates an institutional ethics committee. It describes the investigator responsibilities; animal housing, transportation, and veterinary care; and personnel qualification and training; and outlines principles and recommended engineering standards for animal space and environmental control. Section 2 provides a comprehensive description of IACUC appointment, responsibility, membership, proposal review processes, monitoring, and reporting functions.

Indonesia

Animal Welfare Law

The Law of Republic Indonesia No. 6, issued in 1967, specifies animal welfare in Article 22 (LRI 1967). The law was replaced in 2009 by the Law of the Republic Indonesia No. 18, which includes sections devoted to veterinary public health and animal welfare (Chapter VI), and veterinary authority in laboratory animal and comparative medicine (Chapter VII, Article 74). The term *animal* under this law (Chapter VI, Part 2, Article 66) includes both vertebrates and invertebrates. It governs the housing, care, handling and restraint, transportation, use, euthanasia, and humane treatment of animals (LRI 2009).

Regulation and Guidelines for Care and Use of Animals for Scientific Purposes

Principles of the ethical treatment of animals in comparative medicine are described in the government of the Republic of Indonesia Regulation No. 95 (GRIR 2012). Detailed guidelines for laboratory animal care and use in biomedical research supported by the Ministry of Health were developed by the Health Research Ethics Committee, and consist of (1) National Guidelines on Health Research Ethics (HREC 2011a) and (2) *Teaching Guide Book for Ethics on Health Research* (*Teaching Guide Book*) (HREC 2011b). The guidelines and *Teaching Guide Book* address the three Rs and the "Five Freedom Principles" (hunger and thirst; discomfort; pain, injury, and disease; fear and distress; and expression of natural behaviors) (FAWC 1979), and other aspects of an animal care and use program.

The guidelines describe appointment and oversight responsibilities of a veterinarian, competency and training for personnel involved in animal care and use, behavior and environmental management, and occupational health and safety. Mandated oversight by the Health Research Ethics Committee is limited to governmental institutions under the Ministry of Health; however, most other academic and private research institutions have voluntarily established ethics committees in accordance with the guidelines.

Philippines

Animal Welfare Act

The Animal Welfare Act (Republic Act No. 8485) was enacted by Congress in 1998 (AWA 1998) and amended in 2013 (AWA 2013). Animal experimentation is briefly addressed under Section 6.

Rules and Regulation on the Conduct of Scientific Procedures Using Animals

Following the act, the Department of Agriculture Administrative Order No. 40 (AO No. 40) Series of 1999 requires the registration of institutions using animals for research and scientific procedures, excluding animal clinical tests for evaluating veterinary products. A formal animal care and use program overseen by an IACUC is required to obtain authorization. Additionally, the Rules and Regulations on the Conduct of Scientific Procedures Using Animals govern the use of animals for scientific purposes (RPDA 1999).

The rules define *animal* as "any live vertebrate (domestic or wild) that is used or intended for use in scientific purpose[s]." Pertinent activities include biomedical research, teaching and instruction, product testing, and the production of antisera or other biologicals. The rules describe requirements for authorization, and detail appropriate euthanasia procedures for commonly used species and agents. Annexes include specific guidance for the IACUC, including composition, obligations, protocol review, and inspection processes (Annex A); the application for authorization (Annex B); and a description of the animal care and use program (Annex C), which includes veterinary medical care, husbandry, physical environment, personnel qualification, and physical plant; and a standardized animal care and use statement/protocol review form.

Other Regulations and Guidelines

The Code of Practice for the Care and Use of Laboratory Animals (COP) developed by the Philippine Association of Laboratory Animal Science (PALAS), is applicable to animals used in biomedical research, teaching, testing, and the production of biological products. It references and affirms both the CIOMS International Guiding Principles for Biomedical Research Involving Animals (CIOMS and ICLAS 2012) and the *Guide* (ILAR 2011a). The COP is endorsed by local regulatory agencies and includes standards for husbandry, occupational health and safety, training, and the ethics committee; transportation, housing, and environmental enrichment; and veterinarian responsibilities, veterinary care, and euthanasia methods (Gettayacamin et al. 2014). Authorization for use of animals is granted by the director of the Bureau of Animal Industry (BAI) under the Department of Agriculture, and presumes adherence to PALAS COP and ethics committee approval (Gettayacamin et al. 2014).

Cambodia

Chapter 7.1 of the OIE *Terrestrial Animal Health Code* (OIE 2015) is generally adopted by government and academic research institutions in Cambodia. Permits to export nonhuman primates are authorized by the Ministry of Agriculture, Forestry and Fisheries (MAFF), and enforced by the Forestry Administration (FA) and the Department of Animal Health and Production (DAHP). Regular inspections of nonhuman primate breeding and holding facilities are conducted by the veterinary authority of FA and DAHP (Gettayacamin et al. 2014).

Australasia

Australia

Australian Code for the Care and Use of Animals for Scientific Purposes

The *Australian Code for the Care and Use of Animals for Scientific Purposes* (the *Code*) was first developed in 1969 to promote the ethical, humane, and responsible care and use of research animals. The most recent revision of the *Code*, the eighth edition, was released in 2013 (ARC 2013). Although the *Code* serves as guidance, adherence to the *Code* is mandated for institutions receiving funding from the National Health and Medical Research Council (NHMRC) (Schofield et al. 2014). The *Code* applies to all scientific activities involving animals, defined as "all activit[ies] conducted with the aim [of] acquiring, developing, or demonstrating knowledge or techniques in all area[s] of science." This includes the use of animals for teaching, field studies, the generation of animal lines with undefined genetic impact on animal well-being, disease diagnosis, biological agent production, and product testing, noting that the latter is precluded from NHMRC funding (NHMRC 2016). The definition of *animal* according to the *Code* includes "any live nonhuman vertebrate (that is fish, amphibians, reptiles, birds, mammals, encompassing domestic animals, purpose-bred animals, livestock, wildlife), and cephalopods." Some territories and states may include other species under local legislation.

Awareness of and compliance with the *Code* and acceptance of specified responsibilities is required of institutions, including the Animal Ethics Committee (AEC) and all personnel involved in any aspect of the animal care and use program. The *Code* applies throughout the involvement of the animal in the scientific activities, from acquisition, transport, and breeding until the ultimate disposition, and requires that all involved in the care and use of animals maintain compliance with other relevant and applicable commonwealth, state, and territory legislations. The *Code* is comprised of six sections: (1) governing principles, (2) responsibilities, (3) animal well-being, (4) care and use of animals for the achievement of educational outcome in science, (5) complaints and noncompliance, and (6) independent external review of the operation of institutions.

Section 2 of the *Code* stipulates the responsibilities of the institution, investigators, animal care personnel, and AEC, which is given primary responsibility for ensuring that all animal care and use activities are conducted in accordance with the *Code*. The AEC must review applications for proposed animal care and management, monitor animal care and use activities, take appropriate action regarding unexpected adverse events and noncompliance, and establish institutional guidelines for animal care and use. The *Code* specifies activity monitoring methods, such as inspection of animals, animal housing areas, and procedures, and review of records and reports. The AEC is expected to determine the frequency and timing of inspections in which the nonaffiliated member should participate. This section also outlines institutional responsibilities regarding occupational health and safety related to the animal care and use program.

Animal importation, discussed in Section 3, requires authorization by the Australian Department of Agriculture, Fisheries and Forestry, and Quarantine Inspection Service. Air transportation of genetically modified and exotic animals is subject to commonwealth, state, and territory regulation as well. Although the *Code* does not enumerate training, qualifications, or experience requirements for veterinary expertise, the availability of veterinarians to advise and oversee the veterinary care program is specifically outlined.

Section 4 of the *Code* describes standards for primary enclosures, husbandry, environmental control and maintenance, and emergency action plans. Additional standards for housing, including environmental enrichment, are generally specified by state and territory animal welfare legislation.

The *Code* mandates an independent external review to be conducted at least every 4 years. Details on the coordination, membership requirements, and duties of the review panel are specified in Section 6 of the *Code* (ARC 2013).

NHMRC Guidelines

Other guidelines have been developed by the NHMRC to supplement the *Code*. Like the *Code*, these documents are mandatory for institutions receiving funding from the NHMRC.

The Guidelines to Promote the Well-Being of Animals Used for Scientific Purposes: The Assessment and Alleviation of Pain and Stress in Research Animals (NHMRC 2008a) provides background material, basic strategies, and fact sheets to advance the well-being of animals use for scientific purposes. Parts of this document also discuss appropriate means of euthanasia, anesthesia, and analgesia, and use of anxiolytics. An additional reference frequently cited for euthanasia is *Euthanasia of Animals Used for Scientific Purposes* by the Australian and New Zealand Council for the Care of Animals in Research and Teaching (ANZCCART 2001).

Other relevant references developed by NHMRC include *A Guide to the Care and Use of Australian Native Mammals in Research and Teaching* (NHMRC 2014), Policy on the Use of Non-Human Primates for Scientific Purposes (NHMRC 2003), Guidelines for the Generation, Breeding, Care and Use of Genetically Modified and Cloned Animals for Scientific Purposes (NHMRC 2007), Guidelines on the Use of Animals for Training Interventional Medical Practitioners and Demonstrating New Medical Equipment and Techniques (NHMRC 2009a), Guidelines for Monoclonal Antibody Production (NHMRC 2008a), and Guidelines on the Care of Dogs Used for Scientific Purposes (NHMRC 2009b).

Local Legislation

While compliance with the *Code* is only required for institutions receiving funding from NHMRC, animal welfare regulations impacting the use of animals in research exist for states and territories, which adhere to its principles and guidelines. For example, the New South Wales Animal Research Review Panel Guidelines encompass several aspects of animal care and use, such as species-specific care and housing standards, acquisition of dogs and cats, use of feral animals in research, and production of monoclonal antibodies. Institutions in Victoria are referred to the Code of Practice for Housing and Care of Laboratory Mice, Rats, Guinea Pigs, and Rabbits and the Code of Practice for the Use of Animals from Municipal Pounds in Scientific Procedures (Schofield et al. 2014).

Ethical review and oversight functions by an AEC, as outlined in the *Code*, are required by all states and territories. Facility licensing and registration is also required in all states and territories for institutions that use, breed, supply, or hold certain species for scientific purposes. An annual report of compliance with the *Code* is required.

Training and education for animal users, care staff, and AEC members are described in both the *Code* and regional legislation. Resources for AEC members are specifically available from the ANZCCART and animal welfare units in certain states, such as Queensland, New South Wales, and Victoria. Training for new or inexperienced personnel working with nonhuman primates must be arranged in consultation with NHMRC National Nonhuman Primate Breeding Colonies.

Other Guidelines

GMOs, including laboratory animals, are regulated under the Commonwealth of Australia Gene Technology Act of 2000, which addresses licensing, certification, accreditation, advisory and ethics committees, and inspections (GTA 2014); Gene Technology Regulations 2001, which include standards for containment and monitoring activities (GTR 2001); and the Guidelines for the Transport, Storage and Disposal of GMOs (DHA 2011).

New Zealand

Animal Welfare Act

New Zealand's Animal Welfare Act was promulgated in 1999 (Public Act No. 142) and has been amended several times, most recently in 2015 (NZAWA 1999, 2015). The term *animal* under the act is defined as "any live member of the animal kingdom," that is, a mammal, bird, reptile, amphibian, or fish; octopus, squid, crab, lobster, or crayfish; any other member as specified by the governor general, by order in council; and any mammalian fetus, or avian or reptilian prehatched young that is in the last half of its period of gestation or development. Use in research includes any work involving the manipulation of animals in investigative, experimental, diagnostic, toxicity, or potency testing; in producing antisera or other biological products; and in teaching. The act specifically prohibits the use of animals in any research, testing, or teaching for the purpose of developing, making, or testing cosmetics, or an ingredient that is intended exclusively for use in cosmetics.

Part 6 of the act, entitled "Use of Animals in Research, Testing, and Teaching," dictates that any person wishing to use animals in research is required to both have a project approval from the IAEC and maintain an approved code of ethical conduct (CEC) document. The CEC is authorized by the director general of the Ministry of Agriculture and Forestry (MAF), based on the recommendation from the National Animal Ethics Advisory Committee (NAEAC), and is issued for up to 5 years (NZAWA 1999). The CEC holder is generally the chief executive officer of an institution. To assist applicants, the NAEAC released the *Guide to the Preparation of Codes of Ethical Conduct* (NAEAC 2012). The AEC functions on behalf of the CEC holder, reviewing applications for project approval and monitoring compliance with the conditions of the project and with CEC. Part 6 also details the AEC membership composition and reporting of noncompliance. The act specifies restrictions to and classification of surgical procedures, including conditions applied when personnel other than veterinarians perform surgical procedures.

Good Practice Guide for the Use of Animals in Research, Testing, and Teaching

The most recent edition of the *Good Practice Guide for the Use of Animals in Research, Testing, and Teaching* (*Practice Guide*) was published by the NAEAC in 2010 (NAEAC 2010). The *Practice Guide*, which is intended to expand and supplement the act, applies to all aspects of the care and use of animals in scientific activities in medicine, biology, agriculture, veterinary and other animal science, industry, and education. The *Practice Guide* describes standards for animal acquisition and transportation; environmental enrichment and facility management, including special considerations for farm animals; recommendations for personnel qualifications, training, and staffing levels; hazard communication; and the responsibilities of investigators and instructors using animals.

Other Guidelines and Standards

The *Practice Guide* provides general guidance and principles related to animal euthanasia; detailed guidelines are published in *Euthanasia of Animals Used for Scientific Purposes* (ANZCCART 2001). Similarly, expanded information regarding animal transport is described in International Air Transport

Association (IATA) regulations and "Transport of Animals within New Zealand" by the National Animal Welfare Advisory Committee (2011).

Genetically modified animal use is overseen by the Hazardous Substances and New Organism Act of 1996 and the Biosecurity Act of 1993. Requirements for facility construction, operation, and semiannual internal review, and periodic MAF audits are specified in MAF Biosecurity Authority Standard 154.03.03: Containment Facilities for Vertebrate Laboratory Animals (MAF 2002). Authorization from the Ministry of Primary Industries is required to transfer genetically modified animals (Schofield et al. 2014).

Multinational Guidance

AAALAC International

The American Association for the Accreditation of Laboratory Animal Care (AAALAC) was conceived by the Animal Care Panel, which became the American Association for Laboratory Animal Science (AALAS) as a means of supporting high-quality science by ensuring humane animal care and use, and thereby assuring the general public that research was conducted appropriately (AAALAC 2016e). The organization was incorporated in 1965 as a not-for-profit association conducting voluntary reviews of animal care and use programs. In 1996, the name was changed to the Association for the Assessment and Accreditation of Laboratory Animal Care International (AAALAC International), to reflect a rapidly growing interest in the benefits of accreditation worldwide. Most recently, the association discontinued defining the acronym to avoid overinterpretation of the term *laboratory* and uses "AAALAC International" as its official title (AAALAC 2016e). The AAALAC International mission statement reads, "AAALAC International is a voluntary accrediting organization that enhances the quality of research, teaching, and testing by promoting humane, responsible animal care and use. It provides advice and independent assessments to participating institutions and accredits those that meet or exceed applicable standards" (AAALAC 2016f).

Governing the organization is a Board of Directors appointed by delegates from more than 60 member organizations that share AAALAC's core values. The board appoints internationally recognized experts in laboratory animal medicine, science, and use as members of the Council on Accreditation, who conduct triennial reviews leading to a determination of accreditation status. These reviews include on-site evaluation of institutions for initial or continuing accreditation, using three primary standards: the *Guide*, the *Ag Guide*, and ETS 123. Augmenting these standards are numerous reference resources acknowledged by the council (AAALAC 2016g), eight position statements summarizing the council's interpretation of high-impact issues (AAALAC 2016h), and at the time of this writing, 58 frequently asked questions (AAALAC 2016i).

Accreditation is discussed in more detail in Chapter 9. While not a regulatory organization, accreditation provides an extraordinarily valuable means of program assessment.

Terrestrial Animal Health Code

Another organization seeking to advance common animal welfare principles is the World Organization for Animal Health (OIE). Headquartered in Paris, the OIE is recognized by the World Trade Organization, represents 180 member countries, and is "the intergovernmental organization responsible for improving animal health worldwide" (OIE 2016). Among its contributions are research animal welfare standards described in the *Terrestrial Animal Health Code* (OIE 2015), which specifically endorses the three Rs and a system of oversight by independent institutional, regional, or national ethical review entities (see also Chapter 8).

International Guiding Principles for Biomedical Research Involving Animals

Two additional associations with similar interests include the CIOMS (2016) and the ICLAS (2016). In 2012, these organizations published the International Guiding Principles for Biomedical Research Involving Animals (CIOMS and ICLAS 2012). This document shares many of the same principles as the

OIE *Terrestrial Animal Health Code*, such as an affirmation and application of the three Rs in ethical reviews of proposed animal use, including minimizing opportunities for pain and distress and incorporation of humane endpoints, and the importance of personnel competency and provision of veterinary care (see also Chapter 8).

Summary

Although it is tempting to justify practices "because the regulations require it," managers are urged to view compliance as a by-product of the partnership between and pursuit of common goals by the PI, IO, AV, and IACUC or EOB, and thereby serve as a *tool for validating* their efforts (see also Chapter 10). Applying accepted standards to animal care and use encourages consistency and minimizes the potential for animal pain and distress, thereby meeting both scientific and ethical obligations. Oversight, including that provided by external reviewers, such as regulators and accrediting bodies, and the outcome of their scrutiny should be used to foster and confirm the existence of a well-managed program with those two symbiotic objectives.

REFERENCES

AALAC International. 2016a. The attending veterinarian and veterinary care. http://www.aaalac.org/accreditation/positionstatements.cfm#vetcare (accessed February 11, 2016).

AAALAC International. 2016b. Advocacy and education. http://www.aaalac.org/resources/links.cfm#advocate (accessed February 11, 2016).

AAALAC International. 2016c. International regulations and resources. http://www.aaalac.org/resources/internationalregs.cfm (accessed February 12, 2016).

AAALAC International. 2016d. About AAALAC. http://www.aaalac.org/about/guidelines.cfm (accessed February 12, 2016).

AAALAC International. 2016e. About AAALAC, history. http://www.aaalac.org/about/history.cfm (accessed February 11, 2016).

AAALAC International. 2016f. AAALAC international mission statement. http://www.aaalac.org/about/mission.cfm (accessed February 11, 2016).

AAALAC International. 2016g. AAALAC's reference resources. http://www.aaalac.org/accreditation/resources.cfm (accessed February 11, 2016).

AAALAC International. 2016h. Position statements. http://www.aaalac.org/accreditation/positionstatements.cfm (accessed February 11, 2016).

AAALAC International. 2016i. Frequently asked questions. http://www.aaalac.org/accreditation/faq_landing.cfm (accessed February 11, 2016).

ABA (Animals and Birds Act). 2002. The statutes of the Republic of Singapore. http://statutes.agc.gov.sg/aol/download/0/0/pdf/binaryFile/pdfFile.pdf?CompId:b39eb414-ccf3-415e-8b4d-541340b82590 (accessed February 13, 2016).

ACLAM (American College of Laboratory Animal Medicine). 1996. Adequate veterinary care. https://www.aclam.org/education-and-training/position-statements-and-reports (accessed January 20, 2016).

ACURET.ORG. 2016. https://www.linkedin.com/in/acuret-org-5a4215b2 (accessed February 15, 2016).

Agelan, A., A.S. Braverman, G.E. Dean, and M.R. Ruggieri. 2008. Refinement in the management of the denervated canine urinary bladder using an abdominal vesicostomy. *ILAR J* 49(4): E8–E14.

Anderson, L. 2002. Laws, regulations, and policies affecting the use of laboratory animals. In *Laboratory Animal Medicine*, ed. J.G. Fox, L.C. Anderson, F.M. Loew, and F.W. Quimby, 19–33. 2nd ed., American College of Laboratory Animal Medicine Series. San Diego: Academic Press/Elsevier Science.

Animal Care Panel. 1963. *Guide for Laboratory Animal Facilities and Care*. Washington, DC: U.S. Government Printing Office.

ANZCCART (Australian and New Zealand Council for the Care of Animals in Research and Teaching). 2001. *Euthanasia of Animals Used for Scientific Purposes*. 2nd ed. https://www.adelaide.edu.au/ANZCCART/publications/Euthanasia.pdf (accessed February 13, 2016).

ARC (Australian Research Council), National Health and Medical Research Council, Australian Government. 2013. *Australian Code for the Care and Use of Animals for Scientific Purposes*. 8th ed. https://www.nhmrc.gov.au/_files_nhmrc/publications/attachments/ea28_code_care_use_animals_131209.pdf (accessed February 13, 2016).

ARENA and OLAW (Applied Research Ethics National Association and the Office of Laboratory Animal Welfare). 2002. *Institutional Care and Use Guidebook*. 2nd ed. Bethesda, MD: Office of Laboratory Animal Welfare.

Assembly of Israel. 2005. Prevention of cruelty to animals laws (experiments on animals; 5754–1994) and the prevention of cruelty to animals rules (experiments on animals; 5761–2001). www.weizmann.ac.il/vet/Law.htm (accessed February 13, 2016).

AVA (Agri-Food and Veterinary Authority of Singapore). 2004. Animals and birds (care and use of animals for scientific purposes) rules. http://www.ava.gov.sg/docs/default-source/legislation/animals-and-birds-act/19web_ab_careanduseofanimalsforscientificpurposesr.pdf (accessed February 13, 2016).

AVMA (American Veterinary Medical Association), Panel on Euthanasia. 2013. *AVMA Guidelines for the Euthanasia of Animals*. 2013 ed. https://www.avma.org/KB/Policies/Documents/euthanasia.pdf (accessed February 13, 2016).

AWA (Animal Welfare Act). 1970. P.L. 91-579, United States. Approved December 24, 1970.

AWA (Animal Welfare Act). 1976. Amendments. P.L. 94-279, United States. Approved April 22, 1976.

AWA (Animal Welfare Act). 1985. Food Security Act. P.L. 99-198, United States. Approved December 23, 1985.

AWA (Animal Welfare Act). 1990. Food, Agriculture, Conservation, and Trade Act. P.L. 101-624, United States. Approved November 28, 1990.

AWA (Animal Welfare Act). 1998. Republic Act No. 8485, Republic of Philippines. http://www.paws.org.ph/uploads/2/1/2/7/21275318/__ra8485_animal_welfare_act.pdf (accessed February 13, 2016).

AWA (Animal Welfare Act). 2002. Farm Security and Rural Investment Act. P.L. 107-171, United States. Approved May 13, 2002.

AWA (Animal Welfare Act). 2008. Food, Conservation, and Energy Act. P.L. 110-246, United States. Approved June 18, 2008.

AWA (Animal Welfare Act). 2013. Amendment. Republic Act No. 10631, Republic of Philippines. http://www.gov.ph/2013/10/03/republic-act-no-10631/ (accessed February 13, 2016).

Banks, R.E., and J.N. Norton. 2008. A sample postapproval monitoring program in academia. *ILAR J* 19(4): 402–418.

BAOLA (Beijing Administrative Office of Laboratory Animals). 2000. Legal administration documents for laboratory animalS in People's Republic of China and Beijing municipality. English translation. http://www.aaalac.org/intlRefs/IntRegs/China/P.R.%20China/Legal%20Administration%20Documents%20for%20Lab%20Animals.PDF (accessed February 8, 2016).

Bayne, K. 2011. The IACUC's impact on advancing best practices and the three R's. *ILAR J* 52(Suppl): 436–439.

Bayne, K., and N. Garnett. 2008. Mitigating risk, facilitating research. *ILAR J* 49(4): 369–371.

Bayne, K., and J. Wang. 2014. Oversight of animal research in China. In *Laboratory Animals: Regulations and Recommendations for Global Collaborative Research*, ed. J. Guillén, 244–266. San Diego: Academic Press/Elsevier.

Bennett, B.T., A.D. Cardon, and M.R. Bailey. 2014. USDA *Animal Welfare Inspection Guide*. *Lab Anim (NY)* 43(6): 191.

Bradfield, J.F., B.T. Bennett, and C.S. Gillett. 2014. Oversight of research animal welfare in the United States. In *Laboratory Animals: Regulations and Recommendations for Global Collaborative Research*, ed. J. Guillén, 5–59. San Diego: Academic Press/Elsevier.

Bratcher, N.A., and G.R. Reinhard. 2015. Creative implementation of the 3Rs principles within industry programs: Beyond regulations and guidelines. *J Am Assoc Lab Anim Sci* 54(2): 133–138.

Brazil (Presidência da República, Casa Civil, Subchefia para Assuntos Jurídicos Brazil). 2008. Lei No. 11.794 de 8 de Outubro, 2008 (Arouca law). http://www.planalto.gov.br/ccivil_03/_ato2007-2010/2008/lei/l11794.htm (accessed February 11, 2016).

Brønstad, A., and A.G. Berg. 2011. The role of organizational culture in compliance with the principles of the 3Rs. *Lab Anim* 40(1): 22–26.

Brown, M., and M.J. Shepherd. 2014. Reporting lines of the IACUC. In *The IACUC Handbook*, ed. J. Silverman, M.A. Suckow, and S. Murthy, 39–43. 3rd ed. Boca Raton, FL: CRC Press.

Cardon, A.D., M.R. Bailey, and B.T. Bennett. 2012. The Animal Welfare Act: From enactment to enforcement. *J Am Assoc Lab Anim Sci* 51(3): 301–305.

Carlsson, H.E., J. Hageline, and J. Hau. 2004. Implementation of the 'three Rs' in biomedical research. *Vet Rec* 154(15): 467–470.

CCAC (Canadian Council on Animal Care). 1993. *Guide to the Care and Use of Experimental Animals*. Vol. 1, 2nd ed. Ottawa. http://www.ccac.ca/Documents/Standards/Guidelines/Experimental_Animals_Vol1.pdf (accessed February 8, 2016).

CCAC (Canadian Council on Animal Care). 2016a. About the CCAC. http://www.ccac.ca/en_/about (accessed February 8, 2016).

CCAC (Canadian Council on Animal Care). 2016b. Policy statements. http://www.ccac.ca/en_/standards/policies (accessed February 8, 2016).

CCAC (Canadian Council on Animal Care). 2016c. Guidelines. http://www.ccac.ca/en_/standards/guidelines (accessed February 8, 2016).

CCAC (Canadian Council on Animal Care). 2016d. CCAC certification. http://www.ccac.ca/en_/assessment/certification (accessed February 8, 2016).

Chen, H.H. 2008. The past to present animal use and current animal protection law in Taiwan. Japanese Society for Alternative in Animal Experiments. *AATEX J* 14: 743–748.

CIOMS (Council for International Organizations of Medical Sciences). 2016. About us. http://www.cioms.ch/index.php/2012-06-07-19-16-08/about-us (accessed February 2016).

CIOMS and ICLAS (Council for International Organizations for Medical Sciences and International Council for Laboratory Animal Science). 2012. International guiding principles for biomedical research involving animals. http://iclas.org/wp-content/uploads/2012/10/CIOMS-ICLAS-Principles-Final.pdf (accessed February 15, 2016).

CLS (Commission on Life Sciences), Committee on the Use of Laboratory Animals in Biomedical and Behavioral Research, National Research Council. 1988. *Use of Laboratory Animals in Biomedical and Behavioral Research*. Washington, DC: National Academies Press.

COA (Council of Agriculture), Executive Yuan Republic of China (R.O.C. Taiwan). 2011. Taiwan Animal Protection Act. http://law.coa.gov.tw/GLRSnewsout/EngLawContent.aspx?Type=E&id=120 (accessed March 28, 2017).

COA (Council of Agriculture), Executive Yuan Republic of China (R.O.C. Taiwan). 2013. Enforcement rules of animal protection. http://law.coa.gov.tw/GLRSnewsout/EngLawContent.aspx?Type=E&id=151 (accessed March 28, 2017).

CoE (Council of Europe). 1986. European Convention for the Protection of Vertebrate Animals Used for Experimental and Other Scientific Purposes. European Treaty Series 123. Strasbourg, France: CoE.

CoE (Council of Europe). 2006. Appendix A of the European Convention for the Protection of Vertebrate Animals Used for Experimental and Other Scientific Purposes (ETS No. 123). Guidelines for accommodation and care of animals (Article 5 of the convention). Approved by the multilateral consultation. Strasbourg, France: CoE.

Collins, J.G. 2008. Post approval monitoring and the Institutional Animal Care and Use Committee (IACUC). *ILAR J* 49(4): 388–392.

CPCSEA (Committee for the Purpose of Control and Supervision of Experiments on Animals). 2006. Recommendations of the Sub-Committee on Rehabilitation of Animals after Experimentation Set Up by CPCSEA. http://hhcro.org/down/Rehabilitation%20of%20animals%20Guidelines.pdf (accessed February 13, 2016).

CPCSEA (Committee for the Purpose of Control and Supervision of Experiments on Animals). 2003. CPCSEA guidelines for laboratory animal facility. *Indian J Pharmacol* 35: 257–274.

CPCSEA (Committee for the Purpose of Control and Supervision of Experiments on Animals). 2010. Standard operating procedures for Institutional Animal Ethics Committees (IAEC). http://cpcsea.nic.in/WriteReadData/userfiles/file/SOP_CPCSEA_inner_page.pdf (accessed February 13, 2016).

Curzer, H.J., M.C. Wallace, G. Perry, P.J. Muhlberger, and D. Perry. 2013. The ethics of wildlife research: A nine R theory. *ILAR J* 54(1): 52–57.

Dale, W.E. 2008. Postapproval monitoring and the role of the compliance office. *ILAR J* 49(4): 393–401.

Decker, R.S., L. Wimsatt, A.G. Trice, and J.A. Konstan. 2007. A profile of federal-grant administrative burden among federal demonstration partnership faculty. A report of the Faculty Standing Committee of the Federal Demonstration Partnership. http://sites.nationalacademies.org/cs/groups/pgasite/documents/webpage/pga_054586.pdf (accessed January 22, 2016).

DeHaven, W.R. 2002. Best practices for Animal Care Committees and animal use oversight. *ILAR J* 43(Suppl): S59–S62.

DAFF (Department of Agriculture, Forestry, and Fisheries). 2015. Commencement of the Veterinary and Para-Veterinary Professions Amendment Act (Act Number 15 of 2012). *Government Gazette* 29380, November 9, 2015.

DHA (Department of Health and Ageing), Office of the Gene and Technology Regulator. 2011. Guidelines for the transport, storage and disposal of GMOs. http://www.ogtr.gov.au/internet/ogtr/publishing.nsf/Content/tsd-guidelines-toc/$FILE/tsd-guidelines.pdf (accessed February 13, 2016).

Dobrovolny, J., J. Stevens, and L.V. Medina. 2007. Training in the laboratory animal science community: Strategies to support adult learning. *ILAR J* 48(2): 75–89.

Dress, N.K. 2007. The realities of conducting laboratory animal research. Psychological Science Agenda. http://www.apa.org/science/about/psa/2007/01/dess.aspx (accessed January 8, 2016).

DuBois, J.M. 2004. Is compliance a professional virtue of researchers? Reflections on promoting the responsible conduct of research. *Ethics Behav* 14(4): 383–395.

EP and CEU (European Parliament and the Council of the European Union). 2010. Directive 2010/63/EU of the European Parliament and of the council of 22 September 2010 on the protection of animals used for scientific purposes. *Off J Eur Union* 2010: L276/33–79.

European Council. 2016. Consensus documents. http://ec.europa.eu/environment/chemicals/lab_animals/interpretation_en.htm (accessed June 23, 2016).

FASF (Federal Assembly of the Swiss Federation). 2005. Animal Welfare Act.

FASS (Federation of Animal Science Societies). 2010. *Guide for the Care and Use of Agricultural Animals in Research and Teaching*. 3rd ed. Champaign, IL: FASS.

FAWC (Farm Animal Welfare Council). 1979. Five freedoms. http://webarchive.nationalarchives.gov.uk/20121007104210/http:/www.fawc.org.uk/freedoms.htm (accessed February 15, 2016).

FDA (Food and Drug Administration). 2015a. Federal Food, Drug, and Cosmetic Act. Title 21, Chapter 9. http://www.fda.gov/RegulatoryInformation/Legislation/default.htm (accessed December 16, 2015).

FDA (Food and Drug Administration). 2015b. Product development under the animal rule: Guidance for industry. http://www.fda.gov/downloads/drugs/guidancecomplianceregulatoryinformation/guidances/ucm399217.pdf (accessed December 16, 2015).

FDA (Food and Drug Administration). 2015c. Good laboratory practice for nonclinical laboratory studies. http://www.ecfr.gov/cgi-bin/text-idx?SID=ea9665b648fb85cd46b99bbd0e5337ff&mc=true&node=pt21.1.58&rgn=div5 (accessed December 16, 2015).

Fenwick, N., G. Griffin, and C. Gauthier. 2009. The welfare of animals used in science: How the "three Rs" ethic guides improvements. *Can Vet J* 50(5): 523–530.

Gettayacamin, M., R. Grant, I. Winoto, D. Sajuthi, Y. Paramastri, J. Khoo, P. Chatikavanij, J. Villano, and A.R. Mutalib. 2014. Laws, regulations, guidelines, and principles pertaining to laboratory animals in South East Asia. In *Laboratory Animals: Regulations and Recommendations for Global Collaborative Research*, ed. J. Guillén, 296–332. San Diego: Academic Press/Elsevier.

GMAC (Genetic Monitoring Advisory Committee). 2013. Singapore biosafety guidelines for research on genetically modified organisms (GMOs) http://www.gmac.gov.sg/pdf/Research/Singapore%20Biosafety%20Guidelines%20for%20GMO%20Research_Jan%202013.pdf (accessed February 13, 2016).

Goldberg, A.M., J. Zurlo, and D. Rudacille. 1996. The three Rs and biomedical research. *Science* 272: 1403.

Gonder, J.C. 2002. Regulatory compliance. In *Management of Laboratory Animal Care and Use Programs*, ed. M.A. Suckow, F.A. Douglas, and R.H. Weichbrod, 163–185. Boca Raton, FL: CRC Press.

Greene, M.E., M.E. Pitts, and M.L. James 2007. Training strategies for Institutional Animal Care and Use Committee (IACUC) members and the institutional official (IO). *ILAR J* 48(2): 131–142.

Griffin, G., and M. Baar. 2014. Canada's oversight of animal use in science. In *Laboratory Animals: Regulations and Recommendations for Global Collaborative Research*, ed. J. Guillén, 61–93. San Diego: Academic Press/Elsevier.

GRIR (Government of the Republic of Indonesia). 2012. Regulation No. 95. Veterinary Public Health and Animal Welfare.

GTA (Gene Technology Act). 2014. Australian government. https://www.comlaw.gov.au/Details/C2014C00587 (accessed February 13, 2016).

GTR (Gene Technology Regulation). 2001. Australian government. https://www.comlaw.gov.au/Details/F2011C00732 (accessed February 13, 2016).

Guhad, F. 2005. Introduction to the 3Rs (refinement, reduction, and replacement). *Contemp Top Lab Anim Sci* 44(2): 58–59.

Guillén, J., ed. 2014. *Laboratory Animals, Regulations and Recommendations for Global Collaborative Research*. San Diego: Academic Press/Elsevier.

Guillén, J., J.B. Prins, D. Smith, and A.-D. Degryse. 2014. The European framework on research animal welfare regulations and guidelines. In *Laboratory Animals, Regulations and Recommendations for Global Collaborative Research*, ed. J. Guillén, 117–188. San Diego: Academic Press/Elsevier.

Guillén, J., S. Robinson, T. Decelle, C. Exner, and M. Fentener van Vlissingen. 2015. Approaches to animal research project evaluation in Europe after implementation of Directive 2010/63/EU. *Lab Anim* 44(1): 23–31.

Hau, A.R., F.A. Guhad, M.E. Cooper, I.O. Farah, O. Souilem, and J. Hau. 2014. Animal experimentation in Africa: Legislation and guidelines. In *Laboratory Animals, Regulations and Recommendations for Global Collaborative Research*, ed. Javier Guillén, 117–188. San Diego: Academic Press/Elsevier.

Haywood, J.R., and M. Greene. 2008. Avoiding an overzealous approach: A perspective on regulatory burden. *ILAR J* 49(4): 426–434.

HREA (Health Research Extension Act). 1985. P.L. 99-158. Approved November 20, 1985.

HREC (Health Research Ethics Committee). 2011a. National guidelines on health research ethics. Ministry of Health, Republic of Indonesia.

HREC (Health Research Ethics Committee). 2011b. *Teaching Guide Book for Ethics on Health Research*. Ministry of Health, Republic of Indonesia.

ICLAS (International Council for Laboratory Animal Science). 2016. Mission. http://iclas.org/about-iclas (accessed February 15, 2016).

ILAR (Institute for Laboratory Animal Research), National Research Council. 1965. *Guide for Laboratory Animal Facilities and Care*. Washington, DC: U.S. Government Printing Office.

ILAR (Institute for Laboratory Animal Research), National Research Council. 1968. *Guide for Laboratory Animal Facilities and Care*. Washington, DC: U.S. Government Printing Office.

ILAR (Institute for Laboratory Animal Research), National Research Council. 1978. *Guide for the Care and Use of Laboratory Animals*. Washington, DC: U.S. Government Printing Office.

ILAR (Institute for Laboratory Animal Research), National Research Council. 1991. *Education and Training in the Care and Use of Laboratory Animals: A Guide for Developing Institutional Programs*. Washington, DC: National Academy Press.

ILAR (Institute for Laboratory Animal Research), National Research Council. 2004. *The Development of Science-Based Guidelines for Laboratory Animal Care: Proceedings of the November 2003 International Workshop*. Washington, DC: National Academies Press.

ILAR (Institute for Laboratory Animal Research), National Research Council. 2008. *Recognition and Alleviation of Distress in Laboratory Animals*. Washington, DC: National Academies Press.

ILAR (Institute for Laboratory Animal Research), National Research Council. 2009. *Scientific and Humane Issues in the Use of Random Source Dogs and Cats in Research*. Washington, DC: National Academies Press.

ILAR (Institute for Laboratory Animal Research), National Research Council. 2011a. *Guide for the Care and Use of Laboratory Animals*. 8th ed. Washington, DC: National Academies Press.

ILAR (Institute for Laboratory Animal Research), National Research Council. 2011b. *Animal Research in a Global Environment: Meeting the Challenges: Proceedings of the November 2008 International Workshop*. Washington, DC: National Academies Press.

ILAR (Institute for Laboratory Animal Research), National Research Council. 2012. *International Animal Research Regulations: Impact on Neuroscience Research: Workshop Summary*. Washington, DC: National Academies Press.

Ingham, K.M., K. Field, and J. Mackta. 2004. Managing inspections: Guidance to biomedical research facilities. *Contemp Top Lab Anim Sci* 43: 60.

Ingham, K.M., J.A. Goldberg, H.J. Klein, R.G. Johnson, and M.D. Kastello. 2000. A novel approach for assessing the quality and effectiveness of IACUC oversight in investigator compliance. *Contemp Top Lab Anim Sci* 39(1): 28–31.

INSA (Indian National Science Academy). 2000. Guidelines for care and use of animals in scientific research. https://www.scribd.com/document/216973822/INSA-Guidelines-for-Care-and-Use-of-Animals-for-Research (accessed March 28, 2017).

IRAC (Interagency Research Animal Committee). 1985. U.S. government principles for the utilization and care of vertebrate animals used in testing, research, and training. *Federal Register* 50: 97. https://grants.nih.gov/grants/olaw/references/phspol.htm (accessed February 12, 2016).

Jennings, M., D.B. Morton, E. Charton, J. Cooper, C. Hendriksen, S. Martin, M.C. Pearce et al. 2010. Application of the three Rs to challenge assays used in vaccine testing: Tenth report of the BVAAWF/FRAME/RSPCA/UFAW Joint Working Group on Refinement. *Biologicals* 38(6): 684–695.

Kalman, R., A. Harmelin, E. Ziv, and Y. Fischer. 2014. Israeli legislation and regulation on the use of animals in biological and medical research. *In Laboratory Animals, Regulations and Recommendations for Global Collaborative Research,* ed. J. Guillén, 189–204. San Diego: Academic Press/Elsevier.

Kennedy, B.W. 2002. Management of education and training programs. In *Management of Laboratory Animal Care and Use Programs*, ed. M.A. Suckow, F.A. Douglas, and R.H. Weichbrod, 15–77. Boca Raton, FL: CRC Press.

Kenya. 1983. Prevention of Cruelty to Animals Act.

Kim, S.H., S.B. Kim, G.H. Yang, and C.H. Rhee. 2012. Mouse compound muscle action potential assay: An alternative method to conduct the LD50 botulinum toxin type A potency test. *Toxicon* 60(3): 341–347.

Klein, H.J., and K.A. Bayne. 2007. Establishing a culture of care, conscience, and responsibility: Addressing the improvement of scientific discovery and animal welfare through science-based performance standard. *ILAR J* 48(1): 3–11.

Kramer, M., and E. Font. 2015. Reducing sample size in experiments with animals: Historical controls and related strategies. *Biol Rev Camb Philos Soc* 92(1): 431–445.

Kurosawa, T.M., J.H. Park, and C.C. Hong. 2014. Laws, regulations, guidelines and principles pertaining to laboratory animals in Far East Asia. In *Laboratory Animals: Regulations and Recommendations for Global Collaborative Research*, ed. J. Guillén, 268–294. San Diego: Academic Press/Elsevier.

LAA (Laboratory Animal Act), Republic of South Korea. 2010. http://www.mfds.go.kr/eng/eng/index.do?nMenuCode=16&searchKeyCode=131&page=1&mode=view&boardSeq=67031 (accessed February 10, 2016).

LASAM (Laboratory Animal Science Association of Malaysia). 2016. Proposed Malaysian code of practice for the care and use of animals for scientific purposes. http://www.lasam.org.my/sites/default/files/resources/document/Code%20of%20Practice.%20-%20Proposed%20for%20Malaysia.pdf (accessed February 10, 2016).

LAWA (Laboratory Animal Welfare Act). 1966. P.L. 89-544, United States. Approved August 24, 1966.

LCCSU (Law Concerning the Conservation and Sustainable Use of Biological Diversity through Regulation on the Use of Living Modified Organism). 2007. English translation. https://www.env.go.jp/en/laws/nature/law_ccsubdrlmo.pdf (accessed February 10, 2016).

Lilley, E., R. Armstrong, N. Clark, P. Gray, P. Hawkins, K. Mason, N. López-Salesansky, A.K. Stark, S.K. Jackson, C. Thiemermann, and M. Nandi. 2015. Refinement of animal models of sepsis and septic shock. *Shock* 43(4): 304–316.

Lloyd, M.H., B.W. Foden, and S.E. Wolfensohn. 2008. Refinement: Promoting the three Rs in practice. *Lab Anim* 42(3): 284–293.

Loew, F.M., and B.J. Cohen. 2002. Laboratory animal medicine: Historical perspectives. In *Laboratory Animal Medicine*, ed. J.G. Fox, L.C. Anderson, F.M. Loew, and F.W. Quimby, 1–17. 2nd ed., American College of Laboratory Animal Medicine Series. San Diego: Academic Press/Elsevier Science.

Lowman, R.P. 2008. The institutional official and postapproval monitoring: The view from 10,000 feet. *ILAR J* 49(4): 379–387.

LRI (Law of Republic Indonesia) No. 6. 1967. Article 22.

LRI (Law of Republic Indonesia) No. 18. 2009. Chapters VI–VII.

MAF (Ministry of Agriculture and Forestry), New Zealand. 2002. Containment facilities for vertebrate laboratory animals. Biosecurity Authority Standard 154.03.03. https://www.mpi.govt.nz/ (accessed February 13, 2016).

Medina, L.V., K. Hrapkiewicz, M. Tear, and L.C. Anderson. 2007. Fundamental training for individuals involved in the care and use of laboratory animals: A review and update of the 1991 NRC core training module. *ILAR J* 48(2): 96–108.

MEF (Ministry of Environment and Forests). 2006. Breeding of and experiments on animals (control and supervision) rules. http://envfor.nic.in/legis/awbi/so1818(e).pdf (accessed February 10, 2016).

MEJ (Ministry of Environment of Japan). 2006. Standards relating to the care and management of laboratory animals and relief of pain. Notice No. 88.

Merriam-Webster. 2016. https://www.merriam-webster.com/ (accessed February 13, 2016).

Mexico (Secretaría de Agricultura, Ganadería, Desarrollo Rural, Pesca y Alimentación). 1999. Especificaciones Técnicas para la Producción, Cuidado y Uso de Animales de Laboratorio. Diario Oficial de la Federación 22 de agosto de 2001 Mexico. NOM-062-ZOO-1999. http://www.fmvz.unam.mx/fmvz/principal/archivos/062ZOO.PDF (accessed February 8, 2016).

MEXT (Ministry of Education, Culture, Sport, Science and Technology of Japan). 2006. Fundamental guidelines for proper conduct of animal experiment. English translation. http://www.clar.med.tohoku.ac.jp/data/kitei/10th/hoi1-10th-E-monkashou.pdf (accessed February 10, 2016).

Michelini, E., L. Cevenini, M.M. Calabretta, D. Calabria, and A. Roda. 2014. Exploiting in vitro and in vivo bioluminescence for the implementation of the three Rs principle (replacement, reduction, and refinement) in drug discovery. *Anal Bioanal Chem* 406(23): 5531–5539.

MOST (Ministry of Science and Technology), People's Republic of China. 2006. Guidelines of humane treatment of laboratory animals.

MRC (Medical Research Council), South Africa. 2004. Guidelines on ethics for medical research: Use of animals in research and training. http://www.kznhealth.gov.za/research/ethics3.pdf (accessed February 15, 2016).

MSJE (Ministry of Social Justice and Empowerment), Indian Parliament. 1998. Breeding of and experiments on animals (control and supervision) rules. http://envfor.nic.in/legis/awbi/awbi10.html (accessed February 10, 2016).

MSJE (Ministry of Social Justice and Empowerment), Indian Parliament. 2001. Breeding of and experiments on animal (control and supervision) rules. http://envfor.nic.in/legis/awbi/awbi11.html (accessed February 10, 2015).

NABR (National Association for Biomedical Research). 2016. State government. http://www.nabranimallaw.org/research-animal-protection/state-government/ (accessed February 12, 2016).

NACLAR (National Advisory Committee for Laboratory Animal Research), Singapore. 2004. The guidelines on the care and use of animals for scientific purposes. http://www.ava.gov.sg/docs/default-source/tools-and-resources/resources-for-businesses/Attach3_AnimalsforScientificPurposes.PDF (accessed February 10, 2016).

NAEAC (National Animal Ethics Advisory Committee), Ministry of Agriculture and Forestry, New Zealand. 2010. *Good Practice Guide for the Use of Animals in Research, Testing, and Teaching*. http://www.mpi.govt.nz/protection-and-response/animal-welfare/animals-in-research-testing-teaching/resources/ (accessed February 13, 2016).

NAEAC (National Animal Ethics Advisory Committee), Ministry of Agriculture and Forestry, New Zealand. 2012. *Guide to the Preparation of Codes of Ethical Conduct*. http://www.mpi.govt.nz/protection-and-response/animal-welfare/animals-in-research-testing-teaching/resources/ (accessed February 13, 2016).

NAS (National Academy of Sciences). 1995. *On Being a Scientist*. http://www.nap.edu/readingroom/books/obas/preface.html (accessed February 13, 2016).

NAWAC (National Animal Welfare Advisory Committee). 2011. Transport of animals within New Zealand by transportation. https://www.mpi.govt.nz/protection-and-response/animal-welfare/codes-of-welfare/ (accessed February 13, 2016).

NHMRC (National Health and Medical Research Council). 2003. Policy on the use of non-human primates for scientific purposes. https://www.nhmrc.gov.au/_files_nhmrc/publications/attachments/ea14.pdf (accessed February 13, 2016).

NHMRC (National Health and Medical Research Council). 2007. Guidelines for the generation, breeding, care and use of genetically modified and cloned animals for scientific purposes. http://www.nhmrc.gov.au/guidelines-publications/ea17 (accessed February 13, 2016).

NHMRC (National Health and Medical Research Council). 2008a. Guidelines to promote the well-being of animals used for scientific purposes: The assessment and alleviation of pain and stress in research animals. https://www.nhmrc.gov.au/_files_nhmrc/publications/attachments/ea18.pdf (accessed February 13, 2016).

NHMRC (National Health and Medical Research Council). 2008b. Guidelines for monoclonal antibody production. http://www.nhmrc.gov.au/guidelines-publications/ea19 (accessed February 13, 2016).

NHMRC (National Health and Medical Research Council). 2009a. Guidelines on the use of animals for training interventional medical practitioners and demonstrating new medical equipment and techniques. http://www.nhmrc.gov.au/guidelines-publications/ea26 (accessed February 13, 2016).

NHMRC (National Health and Medical Research Council). 2009b. NHMRC guidelines on the care of dogs used for scientific purposes. http://www.nhmrc.gov.au/guidelines-publications/ea25 (accessed February 13, 2016).

NHMRC (National Health and Medical Research Council). 2014. *A Guide to the Care and Use of Australian Native Mammals in Research and Teaching*. https://www.nhmrc.gov.au/_files_nhmrc/publications/attachments/ea29_guide_aus_native_mammals_141003.pdf (accessed June 23, 2016).

NHMRC (National Health and Medical Research Council). 2016. Use of animals in the testing of cosmetics. http://www.nhmrc.gov.au/health-ethics/animal-research-ethics/use-animals-testing-cosmetics (accessed February 10, 2016).

NIH (National Institutes of Health). 2015. Public Health Service policy on humane care and use of laboratory animals. http://grants.nih.gov/grants/OLAW/references/phspol.htm (accessed December 15, 2015).

NMAF (Norwegian Ministry of Agriculture and Food). 2015. Regulation on the use of animals in research.

Nolte, K.B., D.M. Stewart, K.C. O'Hair, W.L. Gannon, M.S. Briggs, M. Barron, J. Pointer, and R.S. Larson. 2008. Speaking the right language: The scientific method as a framework for a continuous quality improvement program within academic medical research compliance units. *Acad Med* 83(10): 941–948.

NRCT (National Research Council of Thailand). 2006. *Ethical Principles and Guidelines for the Use of Animals for Scientific Purposes*. https://www.aaalac.org/resources/Ethical_Principles_and_Guidelines_for_the_Use_of_Animals_for_Scientific_purposes.pdf (accessed February 13, 2016).

NSPRC (National Standards of the People's Republic of China). 2008. Architectural and technical codes for laboratory animal facilities (GB 50447–2008).

NSPRC (National Standards of the People's Republic of China). 2010a. Laboratory animals—Requirements of environment and housing facilities (GB 14925–2010).

NSPRC (National Standards of the People's Republic of China). 2010b. Laboratory animals—Microbiological standards and monitoring (GB 14922.2–2010).

NZAWA (New Zealand Animal Welfare Act). 1999. http://www.legislation.govt.nz/act/public/1999/0142/latest/whole.html#DLM49664 (accessed February 13, 2016).

NZAWA (New Zealand Animal Welfare Act). 2015. Amendment (No. 2). http://www.legislation.govt.nz/act/public/2015/0049/latest/DLM5174807.html (accessed February 13, 2016).

OIE (World Organization for Animal Health). 2015. *Terrestrial Animal Health Code*. http://www.oie.int/international-standard-setting/terrestrial-code/ (accessed February 15, 2016).

OIE (World Organization for Animal Health). 2016. About us. http://www.oie.int/en/about-us/ (accessed February 15, 2016).

OLAW (Office of Laboratory Animal Welfare). 2012. Memorandum of understanding among the Animal and Plant Health Inspection Service, U.S. Department of Agriculture and the Food and Drug Administration, U.S. Department of Health and Human Services, and the National Institutes of Health, U.S. Department of Health and Human Services, concerning laboratory animal welfare. APHIS Agreement No. 11-6100-0027-MU. MOU No. 225-06-4000. http://grants.nih.gov/grants/olaw/references/finalmou.htm (accessed February 12, 2016).

OLAW (Office of Laboratory Animal Welfare). 2013. Memorandum of understanding between the Office of Laboratory Animal Welfare, National Institutes of Health, U.S. Department of Health and Human Services, and the Office of Research Oversight and the Office of Research and Development, Veterans Health Administration, U.S. Department of Veterans Affairs, concerning laboratory animal welfare. http://grants.nih.gov/grants/olaw/references/mou_olaw.htm (accessed February 12, 2016).

OLAW (Office of Laboratory Animal Welfare). 2015a. Guidance. http://grants.nih.gov/grants/olaw/olaw.htm (accessed 15 December 2015).

OLAW (Office of Laboratory Animal Welfare). 2015b. Memorandum of understanding between NIH and NSF. http://grants.nih.gov/grants/olaw/references/mou_nsf.htm (accessed June 22, 2016).

OLAW (Office of Laboratory Animal Welfare), National Institutes of Health. 2002. Public Health Service policy on humane care and use of laboratory animals. Bethesda, MD: OLAW.

PCAA (Prevention of Cruelty to Animals Act). 1982. Ministry of Environment and Forest. Government of India. http://envfor.nic.in/legis/awbi/awbi01.html (accessed February 10, 2016).

Pimple, K.D. 2002. Six domains of research ethics: A heuristic framework for the responsible conduct of research. *Sci Eng Ethics* 8: 191–205.

Plante, A., and M.L. James. 2008. Program oversight enhancements (POE): The big PAM. *ILAR J* 49(4): 419–425.

Prime Minister's Office. 1980. Standards relating to the care and management of experimental animals: Experimental Animal Regulation. Notification No. 6. English translation. http://www.alive-net.net/english/en-law/viv-kijun.html (accessed February 10, 2016).

Podolsky, M.L., and V.S. Lucas, eds. 1999. *The Care and Feeding of an IACUC: The Organization and Management of an Institutional Animal Care and Use Committee.* Boca Raton, FL: CRC Press.

Poirier, G.M., C. Bergmann, D.G. Denais-Lalieve, I.A. Dontas, N. Dudoignon, H. Ehall, J.M. Fentener van Vlissingen et al. 2015. ESLAV/ECLAM/LAVA/EVERI recommendations for the roles, responsibilities and training of the laboratory animal veterinarian and the designated veterinarian under Directive 2010/63/EU. *Lab Anim* 49(2): 89–99.

Potkay, S., and W.R. DeHaven. 2000. OLAW and APHIS: Common areas of noncompliance. *Lab Anim (NY)* 29(5): 32–37.

Pritt, S., P. Samalonis, L. Bindley, and A. Schade. 2004. Creating a comprehensive training documentation program. *Lab Anim (NY)* 33(4): 38–41.

Republic of Korea, Ministry of Agriculture, Food, and Rural Affairs. 2014. Animal Protection Act. http://elaw.klri.re.kr/eng_service/lawView.do?hseq=32052&lang=ENG (assessed February 14, 2016).

RPDA (Republic of Philippines Department of Agriculture). 1999. Rules and regulations on the conduct of scientific procedures using animals. http://www.paws.org.ph/uploads/2/1/2/7/21275318/awa_ao40-rules_and_regulations_on_the_conduct_of_scientific_procedures_using_animals.pdf (accessed February 13, 2016).

Rush, H.G., and M.C. Dyson. 2014. Personnel training. In *The IACUC Handbook*, ed. J. Silverman, M.A. Suckow, and S. Murthy, 483–502. Boca Raton, FL: CRC Press.

Russell, W.M.S., and R.L. Burch. 1959. *The Principles of Humane Experimental Technique.* London: Methuen and Co.

SABS Standards Division. 2008. South African national standard. The care and use of animals for scientific purposes.

Schofield, J., D. Noonan, Y. Chen, and P. Penson. 2014. Laboratory animal regulations and recommendations for global collaborative research: Australia and New Zealand. In *Laboratory Animals: Regulations and Recommendations for Global Collaborative Research*, ed. J. Guillén, 333–376. San Diego: Academic Press/Elsevier.

Schwindaman, D.F. 1999. The history of the Animal Welfare Act. In *50 Years of Laboratory Animal Science*, ed. C.W. McPherson and S.F. Mattingly, 147–151. Memphis, TN: American Association for Laboratory Animal Science.

SCJ (Science Council of Japan). 2006. Guidelines for proper conduct of animal experiments. English translation. http://www.scj.go.jp/ja/info/kohyo/pdf/kohyo-20-k16-2e.pdf (accessed February 10, 2016).

Silk, S.B., L.L. Hampton, and P.A. Brown. 2013. What investigators need to know about the use of animals. *ILAR J* 54(3): 324–328.

Silverman, J., M.A. Suckow, and S. Murthy, eds. 2014. *The IACUC Handbook.* 3rd ed. Boca Raton, FL: CRC Press.

Slob, W. 2014. Benchmark dose and the three Rs: Getting more information from the same number of animals. *Crit Rev Toxicol* 44(7): 557–567.

South Africa. 1962. Animal Protection Act.

SSTC (State Science and Technology Commission of the People's Republic of China). 1988. Regulation of the People's Republic of China for Administration of Laboratory Animals. English translation. http://www.aaalac.org/intlRefs/IntRegs/China/P.R.%20China/Legal%20Administration%20Documents%20for%20Lab%20Animals.PDF (accessed February 8, 2016).

Steneck, N.H. 2007. *Introduction to the Responsible Conduct of Research*. Washington, DC: Office of Research Integrity, Department of Health and Human Services.

Tanzania. 2008. Animal Welfare Act, Part V.

Thulin, J.D., J.F. Bradfield, V.K. Bergdall, L.A. Conour, A.W. Grady, D.L. Hickman, J.N. Norton, and J.M. Wallace. 2014. The cost of self-imposed regulatory burden in animal research. *FASEB J* 28: 3297–3300.

Törnqvist, E., A. Annas, B. Granath, E. Jalkesten, I. Cotgreave, and M. Öberg. 2014. Strategic focus on 3R principles reveals major reductions in the use of animals in pharmaceutical toxicity testing. *PLoS One* 9(7): e101638.

Uruguay (Senado y Cámara de Representantes de la República Oriental del Uruguay). 2009. Utilización de Animales en Actividades de Experimentación, Docencia e Investigación Científica. Ley No. 18.611. http://www.cic.hc.edu.uy/images/stories/ley_18.611-_investigacin_con_animales.pdf (accessed March 28, 2017).

USDA (U.S. Department of Agriculture). 2013. *Animal Welfare Inspection Guide*. https://www.aphis.usda.gov/animal_welfare/downloads/Animal-Care-Inspection-Guide.pdf (assessed March 24, 2017).

USDA (U.S. Department of Agriculture). 2015a. Animal Welfare Act. https://www.aphis.usda.gov/aphis/ourfocus/animalwelfare/sa_awa/ct_awa_program_information (accessed December 13, 2015).

USDA (U.S. Department of Agriculture). 2015b. Animal Welfare Act and Animal Welfare Regulations. https://www.aphis.usda.gov/animal_welfare/downloads/Animal%20Care%20Blue%20Book%20-%202013%20-%20FINAL.pdf (accessed December 13, 2015).

USDA (U.S. Department of Agriculture). 2015c. *Animal Care Policy Manual*. https://www.aphis.usda.gov/animal_welfare/downloads/Animal%20Care%20Policy%20Manual.pdf (assessed December 13, 2015).

USDA (U.S. Department of Agriculture). 2015d. Publications, forms and guidance documents. https://www.aphis.usda.gov/aphis/ourfocus/animalwelfare/sa_publications/ct_publications_and_guidance_documents (accessed December 13, 2015).

Van Sluyters, R.C. 2008. A guide to risk assessment in animal care and use program: The metaphor of the 3-legged stool. *ILAR J* 49(4): 372–378.

Van Zutphen, B. 2007. Education and training for the care and use of laboratory animals: On overview of current practices. *ILAR J* 48(2): 72–74.

Van Zutphen, L.F.M., and J.B.F. Van der Valk. 2001. Development on the implementation of the three Rs in research and education. *Toxicol In Vitro* 15: 591–595.

Wayman, S. 1966. Concentration camps for dogs. *Life* 60(5): 22–29.

Zurlo, J., K. Bayne, and J.M. Clark. 2009. Adequate veterinary care for animals in research: A comparison of guidelines from around the world. *ILAR J* 50(1): 85–88.

8

Harmonizing International Animal Care and Use Programs

John F. Bradfield, Javier Guillén, and Lynn C. Anderson

CONTENTS

Global Principles of Research Animal Welfare

Introduction

The 3R principles of replacement (to avoid or replace the use of animals), reduction (to minimize the number of animals used), and refinement (to minimize animal suffering and increase animal well-being) were first proposed in 1959 (Russell and Burch 1959), but their adoption by the scientific community has been gradual. Now, they are a key component of regulatory and guidance documents across the globe, creating a basic international framework for animal care and use. There are differences in the way these principles are implemented, as the care and use of research animals is not and cannot be standardized completely. However, the international acceptance of these basic principles has made them global, allowing a level of harmonization of animal care and use practices worldwide. Harmonization is achieved by following the same principles regardless of the way they are implemented to obtain the same desired outcome. As an example described later in the chapter, there may be different legal systems for protocol or project evaluation, carried out by government authorities, institutional bodies, or a combination of both, and all are valid methods if the evaluation is thoroughly performed in an impartial manner from the scientific point of view.

Principles in the Legislation

The same set of basic welfare principles can be found in most of the regulations on the protection and care and use of research animals worldwide (Guillén 2014). The 3Rs are part of most regulations, although in different levels of explicitness. For example, in the United States principles such as the consideration of alternatives, minimization of pain and distress, qualification of personnel, and use of the minimum number of animals are embedded in the Animal Welfare Act (AWA 1966); the Public Health Service (PHS) Policy (OLAW 2002); the U.S. Government Principles for the Utilization and Care of Vertebrate Animals Used in Testing, Research, and Training (OLAW 2015); the *Guide for the Care and Use of Laboratory Animals* (*Guide*) (NAS 2011); and the *Guide for the Care and Use of Agricultural Animals in Research and Teaching* (*Ag Guide*) (FASS 2010). The legal framework in the European Union is more explicit in its Directive 2010/63/EU (European Parliament and Council of the European Union), where in addition to several articles developing some of these concepts, there is a full article focusing on the requirement to apply the 3R principles. The European Convention for the Protection of Vertebrate Animals Used for Experimental and Other Scientific Purposes, Council of Europe, European Treaty Series 123 (ETS 123) served as the first harmonization tool in Europe (Council of Europe 1986, 2006) and was adopted by the European Union and many countries of the Council of Europe. The convention includes all basic principles, and most of the standards for care and accommodation contained in its Appendix A (revised in 2007) were later partially adopted by Directive 2010/63/EU. The principles are not exclusive of the U.S. and European frameworks, but other regulations around the world apply them as well. For example, countries in Latin America (Brazil, Mexico, and Uruguay), or Asia (India, Singapore, and South Korea) have enacted legislation based on the same principles (see Chapter 7). Another concept used, especially in some of the Asian countries, is the Five Freedoms (from hunger and thirst; from discomfort; from pain, injury, or disease; to express normal behavior; and from fear and distress), which was defined for the farm animal environment as early as in 1965 for the Farm Animal Welfare Council (Brambell 1965) and was later adapted for research animals (Mellor and Reid 1994). Unfortunately, there are a number of countries and geographical areas lacking specific legislation on research animals. It is in these areas where guidance provided by global and other professional organizations can play a more fundamental role serving as the main tool for harmonization.

Principles of Global Organizations

Several organizations that have global representation are contributing to the promotion of global principles and harmonization of research animal welfare. The World Organization for Animal Health (OIE), representing 178 countries and territories, developed welfare standards for animals used in research and education that were included in the OIE *Terrestrial Animal Health Code* in 2012 (WHO 2012). This code refers explicitly to the 3Rs and describes standards according to these principles for the main areas of animal care and use. As an example of how the document follows a harmonization approach while respecting different options for methods of implementation, it states that "ethical review may be undertaken by regional, national or local ethical review bodies or committees" considering impartiality and independence of evaluators, and indicates the minimum expertise needed for such body or committee.

The Council for International Organizations of Medical Sciences (CIOMS) and the International Council for Laboratory Animal Science (ICLAS) published the International Guiding Principles for Biomedical Research Involving Animals (CIOMS and ICLAS 2012). The principles also refer to the 3Rs and address many of the same concepts as the OIE document. Principles shared between CIOMS/ICLAS and the OIE include proposal for oversight of animal use (including ethical evaluation), assurance of training and competency of personnel (all categories of personnel working with animals), provision of veterinary care (highlighting the important role of veterinarians), implementation of humane endpoints, and alleviation of pain and distress. The OIE document offers additional recommendations on the source of animals, physical facility and environmental conditions appropriate to the species, and good husbandry, with emphasis on animal behavior. Both documents share the harmonization concepts as described above for the OIE. CIOMS/ICLAS Principle X also acknowledges that "while implementation of these principles may vary from country to country according to cultural, economic, religious and social factors, a system of animal use oversight that verifies commitment to the principles should be implemented in each country."

Also, at a global level in the private sector, a number of pharmaceutical and biotechnology companies have established a not-for-profit organization named the International Consortium for Innovation and Quality in Pharmaceutical Development (IQ Consortium) (https://iqconsortium.org/), with the mission of advancing science and technology to augment the capability of member companies to develop transformational solutions that benefit patients, regulators, and the broader research and development community. One of the actions undertaken by the IQ Consortium has been the promotion of the 3Rs. The IQ Consortium has launched, with the cooperation of AAALAC International, an annual global 3Rs award that recognizes significant innovative contributions toward the 3Rs of animal research to advance ethical science.

Role of Professional Associations

Under the umbrella provided by the legal frameworks and the global organizations, a number of professional associations help to advance the practical implementation of research animal welfare standards. These associations, whose members may include all types of personnel involved in research animal care and use, are typically organized nationally and/or regionally, and focus on the development and implementation of standards coming from international principles. Professional associations promote the education and training of personnel through scientific meetings, courses, online resources, or scientific publications, and collaboration with government competent authorities for the establishment of regulations. They also educate the public on the benefits of animal research, how animals are provided with humane care, and the regulations protecting animals used in research. The American Association for Laboratory Animal Science (AALAS) (www.aalas.org), with more than 60 years of history, has provided one of the most powerful tools for the education and training of personnel within the United States, and its educational materials are available and used by the international research animal community. AALAS also publishes two of the most recognized scientific journals in the field: the *Journal of the American Association for Laboratory Animal Science* (JAALAS) and *Comparative Medicine.* The Canadian Council on Animal Care (CCAC) has also published guidelines on the training of personnel working with animals in science (CCAC 2015). In Europe, the Institute of Animal Technology (IAT) (http://www.iat.org.uk/), founded in 1950 in the United Kingdom, has played a leading role in the training of animal technologists, especially in the United Kingdom, although its training is also recognized in distant areas, such as South Africa. Also registered in the United Kingdom, Laboratory Animals Limited (LAL) (http://www.lal.org.uk/) is a charity whose aims are to promote education and training in laboratory animal science. LAL funds a variety of initiatives (such as bursaries for training) and publishes the journal *Laboratory Animals*, which is the official journal of several laboratory animal associations in Europe. Most of the national associations in Europe belong to the Federation of European Laboratory Animal Science Associations (FELASA) (www.felasa.eu). FELASA has published numerous recommendations on a variety of areas (Guillén 2012), which are widely accepted, especially those concerning education and training, and animal health monitoring. FELASA plays a significant role in the establishment of the training schemes across Europe, as evidenced by the FELASA leadership of the recently created Education and Training Platform for Laboratory Animal Science (http://www.etplas.eu). One important step toward global harmonization was taken with the establishment of the AALAS-FELASA liaison body that works in areas of mutual interest; several joint AALAS-FELASA working groups are producing recommendations that may be accepted on both sides of the Atlantic, including those for health monitoring of rodents for animal transfer (Pritchett-Corning et al. 2014), harm–benefit analysis (Brønstad et al. 2016; Laber et al. 2016), and animal transportation (in press). More recently, the Asian Federation of Laboratory Animal Science Associations (AFLAS) (http://www.aflas-office.org/), the Federation of South American Societies for Laboratory Animal Science (FESSACAL) (http://www.fessacal.com/), and the Federation of Hispanic Societies and Associations for Laboratory Animal Science of North America, Central America and Caribbean (FESAHANCCCAL) (http://www.fesahancccal.com) have started to improve communication, collaboration, and harmonization in their respective geographic areas.

ICLAS is playing an important role in the harmonization of genetic and microbiological control thanks to the Genetic Monitoring Reference Program (http://iclas.org/animal-quality-network/iclas-genetic-monitoring-reference-program) and the Performance Evaluation Program for Diagnostic Laboratories (http://iclas.org/animal-quality-network/performance-evaluation-program-for-diagnostic-laboratories-pep), programs designed for self-assessment of the genetic background of strains and health diagnostic tools, respectively. Before the establishment of this Genetic Monitoring Reference Program, the ICLAS Working Group on Harmonization had already published guidelines on the production, care, and use of genetically altered animals (Turner et al. 2015).

At the laboratory animal medicine level, professionals have gathered in laboratory animal medicine colleges in several geographical areas, such as America (American College of Laboratory Animal Medicine [ACLAM]; https://www.aclam.org/), Europe (European College of Laboratory Animal Medicine [ECLAM]; http://eslav-eclam.org/); Japan (Japanese College of Laboratory Animal Medicine [JCLAM]; http://plaza.umin.ac.jp/~jclam/_AAAAAA_/index.html), and South Korea (Korean College of Laboratory Animal Medicine [KCLAM]; http://www.kclam.org/), which together form the International Association of Colleges of Laboratory Animal Medicine (IACLAM) (http://www.iaclam.org/). IACLAM assists in achieving the shared objectives of all these laboratory animal medicine specialty organizations to promote the welfare and responsible use of laboratory animals through the certification of veterinary specialists, education, the dissemination of information relevant to the field, and serving as research partners. IACLAM and ICLAS have also collaborated in the promotion of the 3Rs globally (Rose et al. 2013).

A unique tool for global harmonization is the accreditation program of AAALAC International (www.aaalac.org). This organization, with the contribution of independent professionals from North America, Europe, and the Pacific Rim (the Council on Accreditation), has accredited more than 950 animal care and use programs in 41 countries across the world, applying the standards of the *Guide*, the *Ag Guide*, and ETS 123, which are considered the primary standards (http://www.aaalac.org/about/guidelines.cfm), and other guidance documents, considered reference resources (http://www.aaalac.org/accreditation/resources.cfm). The impact of AAALAC International on global harmonization is more significant since its spreading across the globe, and more especially in the Pacific Rim area, where in some countries regulations on the use of animals in research are not very strict. The accreditation establishes a common framework with the *Guide* as the baseline.

Why Harmonization Matters

Globalization of Science and the Core Ethical Principles That Result from Harmonization

Science and the scientific method are universal in the pursuit to understand our natural world. Biomedical science is the universal quest to understand and improve health and well-being. When science involves animals as human surrogates, there is an ethical imperative to minimize harm to animals and maximize the benefits of their use. This ethical imperative is as universal as the science itself. Scientists have generally agreed on the principles that comprise appropriate scientific methods, so it follows that the ethics of animal use in science also be comprised of universally held tenets to which all agree.

The conduct of science is done on an ever-increasing global scale, and it is logical to assume that the breadth and scope of the scientific enterprise will only increase. Scientific collaboration occurs unencumbered by geography, culture, or politics, and this collaboration creates the need to consider a harmonized approach to animal care and use to ensure a fundamental universal regard for research animals and their care. The need to harmonize standards for animal care and use in the overall scientific enterprise is for the benefit of animals and their welfare, but is also important to enhance the quality of scientific outcomes. In fact, recent concerns regarding the reproducibility of animal-based data have highlighted the need for high-quality, universal standards in study methods (Kilkenny et al. 2010). The animal component of research should not be a limiting factor in the universal application of sound scientific principles, so the evolving harmonization of ethical animal use has become an integral aspect of global science. The veracity of animal-based scientific data is essential, and harmonized standards of animal care and use are an integral component of sound study design. Further, although the 3Rs do not specifically address harmonization per se, their consistent application in study design can result in a more harmonized approach, which results in benefits such as reduction in animal use. Törnqvist et al. (2014) report that coordinating projects within a company, coupled with increased collaboration among test sites, resulted in a significant reduction in the number of animals used. Although harmonized global principles of animal use ethics and the organizations that promote them are developing, there are inevitably differences in how various countries and cultures implement these ethical principles. These differences in implementation are likely a necessary aspect to ensure their relevance and practicality. Yet, the fundamental tenets of ethical animal-based research translate across many boundaries:

- In the context of the scientific aims of the research, animal care and use programs must ensure animal well-being. An efficient, well-run program of animal care that is managed by well-trained, competent personnel is of paramount importance. Harmonized standards augment quality performance by ensuring consistent, effective outcomes.
- Application of the 3Rs is essential, not only in study design but also throughout an effective and efficient animal care and use program (e.g., properly sized and managed breeding colonies that ensure that animals are not overproduced).
- In 1965, the concept of the Five Freedoms emerged from a report in the United Kingdom that considered five basic tenets of livestock welfare (Brambell 1965). Application of the Five Freedoms in research animal welfare remains important today; these are freedom from hunger or thirst; from discomfort; from pain, injury, or disease; to express normal behavior; and from fear and distress. These are essential aspects of sound animal care and use practices. Within the needs of the specific research, study design should uphold these principles and husbandry programs must ensure the Five Freedoms are met.
- Ensure animal use is justified based on sound scientific principles, using appropriate methodology, and endorsed by a favorable harm–benefit analysis.
- When one of the basic tenets of ethical animal use cannot be met because of the scientific aims of the study (e.g., the study of pain), it is particularly important to conduct a rigorous harm–benefit analysis. In such circumstances, a harmonized, team approach to animal care translates to better overall care, and minimization of animal pain and distress. A standardized approach to implementing humane endpoints is paramount in such studies.

Consideration and implementation of these basic tenets impacts all aspects of the animal care and use program and will form the framework for the husbandry and management program, the veterinary care program, the responsibilities and function of the animal care and use committee (or ethics committee or oversight body), and the design and function of animal facilities. Therefore, harmonizing animal care and use to the extent practical will lead to a more uniform and efficient standard of animal care that promotes animal welfare and high-quality science.

Operational Efficiency, Teamwork and High Standards, and Scientific Quality

Given that a more harmonized set of global standards is developing, institutions are increasingly considering the practical benefits of harmonizing animal care within the institution itself. In particular, institutions with large, diverse programs that may have many facilities in separate locations, and perhaps locations in several countries, have harmonized practices throughout the program. Even smaller animal care and use programs can benefit from institution-wide harmonized practices as they tend to provide a uniform high standard of care that offers several advantages.

Operational Efficiency

One of the main reasons institutions strive for harmonized animal care practices is to achieve operational efficiency. Several of the benefits of implementing harmonized methods of operation in the animal care program include a streamlined personnel training and performance program, greater latitude in the utilization of the workforce, streamlined use of equipment and supplies, standardized performance expectations, help in creating a unified management team, and more accurate budgeting and cost control.

Teamwork and High Standards

In order for a diverse animal care and use program to become a harmonized program, a uniform consensus about the appropriate standard of care must be established. Developing consensus brings together many stakeholders throughout the institution, which might include upper-level administration, facility operations management, animal caregivers, the Institutional Animal Care and Use Committee (IACUC) veterinary care personnel, and investigators. These stakeholders establish the institutional expectations of animal care that meet the appropriate regulatory guidelines, set any additional institution-specific guidelines, and agree on practices that meet requirements, and which are efficient and practical. This consensus-building process produces high standards of animal care that are clearly delineated, understood, and consistently implemented.

Scientific Quality

At all times, but especially when scientific inquiry involves animals, it is imperative to ensure that studies are well designed, controlled, and well executed in order to minimize confounding variables that could jeopardize scientific outcomes. It is this fundamental aspect of animal-based research that provides the impetus for harmonization of the animal care and use program. Practices that are harmonized across the animal care program ensure a uniform standard of care that promotes quality science. All aspects of the husbandry program, animal environment and management, veterinary care, and physical plant should ensure the same level of care, irrespective of facility, staff, location, or specific program, within the institution. The implementation of a performance-based approach to animal care ensures a uniform standard of care while still allowing for a variety of methods to achieve the harmonized standard. Such a harmonized program allows greater collaboration among scientists and promotes accurate and reliable data that are more effective and impactful. When harmonized standards are employed, differences in scientific outcomes are more likely due to study variables, rather than confounding variables, thereby improving the accuracy of the science.

Strategies for Effective Harmonization

Engineering and Performance Standards

Engineering standards are usually found in the regulatory requirements, which specify both the characteristics and technical details required in order to meet the standard. An engineering standard not only specifies what the standard or outcome must be, but also how it is achieved. Engineering standards dictate the methods that must be used in order to achieve the standard or outcome. The benefit of these standards is that they ensure a uniformity of methods across the animal program, or among different programs. It is usually straightforward to ensure that engineering standards are followed, as they tend to be very specific. However, engineering standards allow little or no flexibility in methods, which is sometimes helpful when implementing practices for unique programs.

Performance standards are designed to allow flexibility in methods while still ensuring that specific outcomes are achieved. Performance standards identify the specific goal, standard, or outcome that is expected, but do not specify *how* the goal is achieved, thereby allowing latitude about the best, most effective manner to meet the standard given the unique aspects of each program. A variety of methods might all result in an outcome that meets the specified goal or standard. The performance standard approach is a key element of the eighth edition of the *Guide*, which describes many specific outcomes that are necessary for appropriate animal care and use. Yet the *Guide* provides no information as to how exactly those outcomes are attained. Three basic steps are required to ensure a performance standard is clearly identified and achievable:

- Identify a specific and precise definition of the standard.
- Establish the assessment criteria to determine that the standard is achieved.
- Develop and implement methods for ongoing evaluation.

An inherent part of establishing appropriate performance-based standards is the concept of professional judgment, or the collective judgment of a profession. Professional judgment must be an essential component of implementing performance-based methods to ensure that they satisfy relevant regulations or guidelines, are appropriate for the species and scientific aims, promote animal welfare or well-being, are tailored to the institution's unique circumstances or needs, and are achievable and not overly burdensome. Once a performance standard is identified, implemented, and validated, it should be uniformly followed throughout the animal care program to ensure an unvarying, harmonized approach to animal care and use.

Management Structure and Commitment

Harmonization within research institutions that have animal facilities in multiple, global locations can be especially challenging, due to differences in applicable animal welfare regulations, societal expectations, and cultures. However, with visible commitment from the institution's most senior leadership, institutions are able to set expectations and emphasize the importance of high animal welfare standards and a culture of caring across all sites, regardless of location. This is often accomplished in the form of an institutional policy that prioritizes and articulates the institution's commitment to the humane treatment of animals, consideration and implementation of the 3Rs, personnel training to ensure high-quality science and animal well-being, and compliance with applicable laws and regulations.

Global institutional oversight of animal welfare can be achieved through a management structure that includes the various stakeholders in the animal care and use programs at all locations. For example, some global institutions have formed animal welfare councils whose members have global responsibility for the scientific enterprise, animal care and facility operations, veterinary services, environmental and occupational health and safety, public relations, security, and legal concerns. These councils are often charged with identifying the overarching strategies and projects that support the institution's policy to ensure and advance animal welfare. They are not intended, nor should they be expected to, replace the IACUCs or other equivalent oversight bodies, such as animal welfare bodies or ethics committees at each site.

Irrespective of the structure of the ethical review and oversight committee or the processes they use, the following activities must be effectively performed by the welfare councils to help ensure animal welfare:

- Providing oversight of animal care and use at their site
- Maintaining animal welfare compliance
- Helping to define global animal welfare standards
- Promoting the sharing of best practices among the sites
- Engaging and inspiring employee commitment to the 3Rs
- Ensuring that all animal use is fully justified
- Offering practical advice and support for refinements
- Providing clear communication channels among the stakeholders

Challenges of Harmonization

Although the existence of global principals mentioned above and the implementation of performance standards help harmonize many aspects of animal care and use, there are also a number of factors that present challenges to harmonization. Perception of animals may vary in countries with different religious beliefs and traditions. When the most widespread religions were born, there was no animal research, so religions are adapting their tenets to the global principles of animal care and use (Pontifical Academy for Life 2004; Mehdi Naderi et al. 2012).

The level of societal concern regarding the use of animals for research also varies across geographical areas, and appears related to the history of animal research and the knowledge of it by society. More commonly, in areas with a longer tradition of animal research, societal concerns and influence are higher, resulting in stricter regulatory framework and institutional practices and control, thus challenging international harmonization. This challenge is intensified by the increase in research being conducted in areas with less strict or underdeveloped animal protection regulations, probably caused by a general growth of the economy in these areas and also, in some cases, by pragmatic reasons due to the financial and legal facilities offered.

There are also a number of socioeconomic factors that impact animal welfare. Factors such as limited resources to acquire quality, purpose-bred animals; employ or train qualified professional staff; or provide adequate animal housing or caging systems, husbandry, and facilities create variation in the ability to harmonize standards across the animal care programs in different regions or countries. The challenge is posed by this variable context where institutions have to implement harmonized animal care and use programs, a context that makes careful planning and application of performance standards paramount to achieve harmonization.

Implementation of an Institutionally Harmonized Program

Effective global harmonization depends, in part, on achieving institutional agreement on the interpretation of global standards, such as the *Guide*. Institutional policies and guidelines can then be developed to document the institution's expectations for various aspects of its animal care and use programs, regardless of location. The policies and guidelines must also allow for the differences in country-specific regulations and flexibility in the application of site-specific procedures.

Harmonization of various aspects of the animal care and use program will help to provide consistency across sites and minimize variables during studies that may be conducted at more than one location. Activities that lend themselves to global harmonization can be broadly categorized into four areas: personnel management, animal care and veterinary medical management, animal welfare bodies (such as IACUCs), and facility operations. Some specific targets for harmonization include animal welfare training, use of personal protective equipment, biosecurity, rodent health monitoring, social housing, environmental enrichment, acclimation to procedures and restraint, blood and dosing volumes and techniques, water quality testing, and heating, ventilation, and air-conditioning performance.

The harmonization process begins with identification of a standard of care for a particular activity or outcome for which there can be agreement across multiple sites, in the form of policies or guidelines. While the specific processes or method to achieve the standard may vary slightly from one site to the next, a guidance document can be developed to establish performance expectations and provide general direction for the activity.

As an illustration of this process, consider a large, global research institution that needs to harmonize expectations (or standards) for animal drinking water quality across all sites. The first step would be to form a working group with representatives knowledgeable about their site-specific water quality standards and testing practices, and with one or more subject matter experts. Members of the team might include veterinarians, facility managers, or environmental health and safety personnel. This project team's first task would be to clarify and align their goals. Using this example, the team's charter might look like this:

1. Objective: Harmonize animal drinking water quality requirements across all sites
2. Project scope
    a. In scope
        i. Harmonization of parameters tested

        ii. Harmonization of criteria used to qualify acceptable tests
        iii. Harmonization of testing frequency
    b. Out of scope
        i. Water source (municipal or wells)
        ii. On-site water treatment (filters, ultraviolet light, chlorine)
3. Deliverable: Publication of a global guideline for water quality standards
4. Activities
    a. Collect information on current site practices for animal drinking water analysis
    b. Review relevant references
    c. Determine what parameters must be tested
    d. Determine the minimum standards for each parameter
    e. Determine minimum testing frequency
5. Benefits by functional area
    a. Consistent water quality standards
    b. Meet animal welfare standards
    c. Minimize variables to support quality science
6. Critical success factors: Engagement of the project team members and active participation in team meetings
7. Stakeholders: Research operations, facilities management, veterinary services, and investigators at each site

The first project team activity is to collect baseline data from each site. In this example, each site would provide information on the water quality tests it conducts and the frequency of testing. A full test profile would likely include an assessment of heavy metals, organic and inorganic solvents, pesticides, and bacteria. Water analysis using the full test profile may be conducted once or twice a year, while microbial analysis may be done monthly, quarterly, or semiannually.

Next, the regulatory basis and standards for each site's practices should be identified. The *Guide* states that animals should have access to potable, uncontaminated drinking water, but acknowledges that the definition of potable water can vary with locality. The *Guide* also recommends periodic testing for pH, hardness, and microbial or chemical contamination. In the United States, the Environmental Protection Agency establishes drinking water standards for human consumption, which can be used as the basis for testing animal drinking water. Many other countries and the World Health Organization have also established water quality standards.

Having collected the baseline data and reviewed the applicable references, the project team should then conduct a gap analysis. In this case study, factors they should examine include (1) the tests that are common to all sites, (2) the tests that are only done at certain sites and the rationale for those extra tests, (3) the frequency of testing, and (4) the various references cited. This gap analysis should drive discussion leading to best practices and establishment of the parameters that must be tested to ensure water quality and the minimum standards for each parameter based on published references, and the minimum testing frequency. The exact methods or processes used to collect and test water quality are likely to differ somewhat from site to site, depending on the water source, how it is delivered to the site, the animal facility water distribution system, and the individual cage design.

Having made those decisions, the project team can draft a document that provides the guidance on testing water quality. This draft should be circulated to stakeholders for comment, and revised as needed. The final guidance document should be dated and a proposed review date provided, as standards and best practices may vary over time.

The final step is to implement training for those individuals who are responsible for the water analysis process at each site. Failure to provide adequate training will most likely result in failure to meet the internal guidelines established by the project team. Lack of training will also lead to inconsistent testing approaches across multiple sites. The goal of harmonization should be to minimize variables to support quality science, reliable data, and animal health and welfare.

Summary

Globalization of science and animal use entails a universal approach to animal care to ensure quality, ethically based research. Efforts to globally harmonize standards for the humane care and use of research animals have been underway for some time, and there are a number of nations and professional organizations that strive to implement and promote these standards. In spite of significant cultural diversity and many other challenges across the globe, there are emerging regulations and guidelines that are recognized as important benchmarks in harmonization that employ core ethical principles that are universal. At the same time that global standards of animal care and use are developing, institutions are increasingly realizing the benefits of harmonizing standards across the entire animal care and use program itself. Implementing a harmonized approach to institutional animal care and use requires a performance-based system that allows flexibility in methods while still achieving appropriate, high-level standards. Strategies for institutional harmonization include identifying those practices that are appropriate to harmonize, defining their scope, determining outcomes, outlining specific activities needed to develop a harmonized goal, engaging the right project team members, and ensuring that all stakeholders are considered and valued. The benefits of such a process are clearly identified goals, streamlined efficiency, cost-effectiveness, improved animal welfare, and quality science.

REFERENCES

AWA (Animal Welfare Act). 1966. Public Law 89-544 and subsequent amendments. U.S. Code, vol. 7, secs. 2131–2157 et seq.

Brambell, R. 1965. *Report of the Technical Committee to Enquire into the Welfare of Animals Kept under Intensive Livestock Husbandry Systems*. December. London: HMSO.

Brønstad, A., Newcomer, C.E., Decelle, T., Everitt, J.I., Guillen, J., and Laber, K. 2016. Current concepts of harm–benefit analysis of animal experiments—Report from the AALAS–FELASA Working Group on Harm–Benefit Analysis. Part 1. *Lab Anim* 50(1S): 1–20.

CCAC (Canadian Council on Animal Care). 2015. CCAC guidelines on training of personnel working with animals in science. www.ccac.ca.

CIOMS and ICLAS (Council for International Organizations of Medical Sciences and International Council for Laboratory Animal Science). 2012. International guiding principles for biomedical research involving animals. www.iclas.org.

Council of Europe. 1986. European Convention for the Protection of Vertebrate Animals Used for Experimental and Other Scientific Purposes. European Treaty Series 123. Strasbourg, France: Council of Europe.

Council of Europe. 2006. Guidelines for accommodation and care of animals (Article 5 of the Convention). Appendix A of the European Convention for the Protection of Vertebrate Animals Used for Experimental and Other Scientific Purposes. European Treaty Series 123. Strasbourg, France: Council of Europe.

European Parliament and Council of the European Union. 2010. Directive 2010/63/EU of the European Parliament and of the council of 22 September 2010 on the protection of animals used for scientific purposes. *Off J Eur Union* L 276: 33–79.

FASS (Federation of Animal Science Societies). 2010. *Guide for the Care and Use of Agricultural Animals in Research and Teaching*. 3rd ed. Champaign, IL: FASS.

Guillén, J. 2012. FELASA guidelines and recommendations. *J Am Assoc Lab Anim Sci* 51(3): 311–21.

Guillén, J. 2014. *Laboratory Animals: Regulations and Recommendations for Global Collaborative Research*. London: Academic Press.

Kilkenny, C., Browne, W.J., Cuthill, I.C., Emerson, M., and Altman, D.G. 2010. Improving bioscience research reporting: The ARRIVE guidelines for reporting animal research. *PLoS Biol* 8(6): e1000412.

Laber, K., Newcomer, C.E., Decelle, Y., Everitt, J. I., Guillen, J., and Brønstad, A. 2016. Recommendations for addressing harm–benefit analysis and implementation in ethical evaluation—Report from the AALAS–FELASA Working Group on Harm–Benefit Analysis. Part 2. *Lab Anim* 50(1S): 21–42.

Mehdi Naderi, M.M., Sarvari, A., Milanifar, A., Borjian Boroujeni, S.A., and Mehdi Akhondi, M.M. 2012. Regulations and ethical considerations in animal experiments: International laws and Islamic perspectives. *Avicenna J Med Biotechnol* 4(3): 114–120.

Mellor, D., and Reid, C. 1994. Concepts of animal well-being and predicting the impact of procedures on experimental animals. In Baker R.M., Jenkin G., and Mellor D.J. (eds.), *Improving the Well-Being of Animals in the Research Environment.* Adelaide: Australian and New Zealand Council for the Care of Animals in Research and Teaching, pp. 3–18.

National Academy of Sciences (NAS). 2011. *Guide for the Care and Use of Laboratory Animals.* Washington, DC: National Academy Press.

OLAW (Office of Laboratory Animal Welfare), National Institutes of Health, Public Health Service. 2002. Public Health Service policy on humane care and use of laboratory animals. Rockville, MD: OLAW.

OLAW (Office of Laboratory Animal Welfare). 2015. U.S. Government Principles for the Utilization and Care of Vertebrate Animals Used in Testing, Research, and Training. Rockville, MD: OLAW. http://grants.nih.gov/grants/olaw/references/phspol.htm#USGovPrinciples.

Pontifical Academy for Life. 2004. Concluding communiqué on the "Ethics of Biomedical Research for a Christian Vision." Presented at Proceedings of the Ninth Assembly of the Pontifical Academy for Life, Vatican City, February 24–26, 2003.

Pritchett-Corning, K.R., Prins, J.B., Feinstein, R., Goodwin, J., Nicklas, W., and Riley, L. 2014. AALAS/FELASA Working Group on Health Monitoring of Rodents for Animal Transfer. *J Am Assoc Lab Anim Sci* 53(6): 633–640.

Rose, M., Everitt, J., Hedrich, H., Schofield, J., Dennis, M., Scott, E., and Griffin, G. 2013. ICLAS Working Group on Harmonization: International guidance concerning the production care and use of genetically-altered animals. *Lab Anim* 47(3): 142–152.

Russell, W.M.S., and Burch, R.L. 1959. *The Principles of Humane Experimental Technique.* Special edition. Potters Bar, England: Universities Federation for Animal Welfare.

Törnqvist, E., Annas, A., Granath, B., Jalkesten, E., Cotgreave, I., and Öberg, M. 2014. Strategic focus on 3R principles reveals major reductions in the use of animals in pharmaceutical toxicity testing. *PLoS One* 9(7): e101638.

Turner, P.V., Pekow, C., Clark, J.M., Vergara, P., Bayne, K., White, W.J., Kurosawa, T.M., Seok, S.H., and Baneux, P. 2015. Roles of the International Council for Laboratory Animal Science (ICLAS) and International Association of Colleges of Laboratory Animal Medicine (IACLAM) in the global organization and support of 3Rs advances in laboratory animal science. *J Am Assoc Lab Anim Sci* 54(2): 174–180.

WHO (World Health Organization). 2012. *Terrestrial Animal Health Code.* http://www.oie.int/international-standard-setting/terrestrial-code/access-online/.

9

Assessment and Accreditation Programs for Research Animal Care and Use

Christian E. Newcomer and Sylvie Cloutier

CONTENTS

Introduction

External peer review as a means for ensuring and advancing the welfare of animals used in scientific activities in research, teaching, testing, and production, and concomitantly the quality of science conducted with animals, has been in existence for more than 50 years and is now widely accepted globally. Its proponents now encompass diverse audiences, including animal care and use professionals that provide the support systems for research animal activities; scientists using animals in exploratory research, testing, and teaching; organizations representing the interests of scientists, animal welfare, and the beneficiaries of science; research administrators; and public officials charged with research animal oversight. In some instances, sponsoring organizations and entities have adopted the position that research programs must participate in the assessment and accreditation or certification process as a condition of receiving funding. However, in the vast majority of instances, organizations participating in assessment and accreditation programs are internally motivated to participate in a rigorous and objective external peer review process as a test of their program's quality and to affirm their stewardship and ethical consideration of research animal subjects.

As of 2016, there are two organizations recognized for their roles in the independent, expert, peer review assessment of research animal programs: AAALAC International, an assessment and accreditation program with global participation, and the Canadian Council on Animal Care (CCAC), an independent, nonprofit organization that assesses and certifies animal ethics and care in science programs in Canada, whose reference documents on topics in research animal care and use are also widely utilized in other countries.

History and Governance Structures for the AAALAC International Accreditation and CCAC Certification Programs

AAALAC International

During the expansion in biomedical research in the years following World War II, scientific research involving animals evolved quickly. Care of research animals was typically the exclusive domain and responsibility of individual researchers. However, in the 1950s and 1960s, the responsibility began to shift to the institution, as the size and complexity of research activities became more challenging. During this period, a small group of forward-thinking veterinarians from the Chicago area convened to share information on the care of laboratory animals so that their institutions could improve the conditions of the animals under their care to promote the validity of experimental results. These pioneers eventually founded a national organization known as the Animal Care Panel (ACP).

In 1950, the ACP began to meet annually, and by the end of the decade had taken the lead in developing standards for animal care and use and encouraging the education and certification of animal care and use professionals. In 1960, the ACP created the Animal Facilities Certification Board (AFCB) to develop a framework of professional standards in anticipation of eventually offering an accreditation program. Their work was greatly influenced by the publication of the *Guide for Laboratory Animal Facilities and Care* in 1963 (U.S. Public Health Service 1963), the forerunner of the *Guide for the Care and Use of Laboratory Animals* (hereinafter referred to as the *Guide*), which, in its eighth edition (ILAR 2011), remains a centerpiece of AAALAC International's three primary standards (TPSs). The AFCB became the Animal Facilities Accreditation Board (AFAB), which formally proposed a voluntary accreditation process to help institutions achieve high standards for animal care and use. In 1964, a National Advisory Committee was convened with representatives from several influential scientific and medical associations. This was a key factor in the future success of the accreditation program—bringing scientists and researchers to the table to ensure that all interests were well represented in an endorsement of the AFAB's findings. AAALAC International's predecessor, the American Association for Assessment and Accreditation of Laboratory Animal Care, was finally established by the AFAB in 1965 with 14 charter member organizations to guide its efforts. Remaining anchored in science today, AAALAC International continues to operate as an independent, nonprofit organization providing a voluntary, confidential, peer review accreditation program to promote animal welfare and improve the quality of research conducted with animals.

The governance structure of AAALAC International today reflects the diverse and robust involvement of animal welfare organizations and scientific organizations globally. Each member organization appoints a representative to serve as a delegate to AAALAC International. The approximately 70 current organizations cluster into three general categories according to their interests: scientific associations whose members are actively engaged in animal research, veterinary specialty and interest groups and/or other professionals focused primarily on the care and welfare of research animals, and patient or industry and issue advocacy groups working to advance animal welfare in the conduct of science. The AAALAC Board of Directors (BOD) is vested with the responsibility for establishing the general rules and standards of accreditation and ensuring the effective and efficient operation of AAALAC International's administration. The BOD also provides some oversight of the appointment of members of the AAALAC International Council on Accreditation (COA) and COA emeriti, who are charged with leading program evaluations, and of the COA's decisions pertaining to accreditation outcomes. In the case of adverse accreditation outcomes (i.e., revoke or withhold accreditation), the AAALAC International BOD serves as the second-level appellate body. Other responsibilities of the AAALAC International BOD can be found in AAALAC International's bylaws at AAALAC International (2016a).

CCAC

The phenomenal research growth period of the 1950s and 1960s also witnessed a growing public concern over the use of animals in Canadian science. The scientific community became increasingly aware that this was a sensitive issue that raised serious ethical questions, not the least of which was responsibility for animal ethics and care in science. In 1963, the Medical Research Council (MRC), now known as the Canadian Institutes of Health Research (CIHR), decided the matter warranted further study; the following year, it requested that the National Research Council (NRC) of Canada establish a committee to investigate the ethical oversight and humane care of experimental animals in Canada. The report of the Special Committee on the Care of Experimental Animals, published in 1966, recommended the creation of a voluntary control program exercised by scientists at each institution, subject to peer judgment and committed to implementing the guiding principles developed by an independent advisory organization (CCAC 2016a).

A feasibility study was undertaken in 1967, and as a result, universities and government departments performing animal-based scientific activities agreed to support the formation of the CCAC. Consequently, the CCAC was established in 1968 as a standing committee of the Association of Universities and Colleges of Canada (AUCC), now known as Universities Canada, until its incorporation as an independent, nonprofit organization in 1982 (CCAC 2016a).

The CCAC is a registered not-for-profit organization responsible for establishing, maintaining, and overseeing the implementation of high standards for animal ethics and care in science throughout Canada (CCAC 2015). The CCAC's mission is to ensure that animal-based science in Canada occurs only when necessary and that the animals in the studies receive optimal care according to high-quality, research-informed standards (CCAC 2015).

The CCAC's mandate is to act in the interests of the Canadian people in advancing animal ethics and care in science by (CCAC 2015)

- Developing research-informed standards that incorporate expert opinion, the values of Canadians, and strategies to reduce the need for, and potential harm to, animals in science, while promoting their well-being

- Encouraging the implementation of the highest standards of ethics and care for animals in science in collaboration with the animal care community and scientists across Canada
- Administering assessment and certification programs that empower scientific institutions to achieve high standards of animal ethics and care
- Providing education, training, and networking opportunities to support individuals, animal care committees, and institutions

In Canada, animals studied in the wild, on farms, and in research facilities for scientific purposes are included in the CCAC's mandate. The CCAC programs also cover animal-based scientific activities in research, teaching, testing, and production (of animals or biologics). All proposed animal-based activities must be clearly described in a protocol to be approved by an institutional animal care committee before the activity commences. Only if animals are an essential component will their involvement be permitted (CCAC 2016b).

In Canada, there is no federal legislation on the use of animals in science. Animal welfare is a provincial responsibility, as defined under the Canadian Constitution, and eight provinces refer to CCAC standards in their regulations (CCAC 2006, 2016c). The guidelines and policies established by the CCAC therefore serve as national reference standards.

The CCAC currently comprises 22 member organizations drawn from both the Canadian scientific and animal welfare communities. The diverse composition of these organizations reflects a wide range of interests, concerns, and objectives regarding animal-based science in Canada. Member organizations elect the CCAC board of directors, which is charged with providing direction for the organization in accordance with CCAC's mandate (CCAC 2016d). In the pursuit of this objective, the board of directors works in collaboration with four advisory standing committees (Standards, Assessment and Certification, Public Affairs and Communications, and Governance and Nominations) and is supported by the CCAC secretariat staff (CCAC 2016d). Committee members are selected from the representative of member organizations, as well as from community representatives with complementary expertise. The purpose of the Standards Committee is to provide for the development and renewal of guidelines, policies, and evidence-based practices regarding animal ethics and care in research, teaching, and testing (CCAC 2016e). The Assessment and Certification Committee oversees and supports the Assessment and Certification Program for animal-based research, teaching, and testing programs in academic, government, and private institutions. The committee provides for the development or renewal of policies related to the assessment and certification process of animal ethics and care programs and has oversight responsibilities on the appropriate implementation of these policies and guidelines by animal care committees across Canada. It also reviews reports and recommendations outcomes for assessment and certification of participating organizations (CCAC 2016e). The Public Affairs and Communications Committee promotes the CCAC's role in advancing animal ethics and care in science. To achieve this purpose, it advises the board of directors on initiatives to increase public awareness of the CCAC, its mission, and its ongoing impact, and on the effectiveness of its communications with all its stakeholders (CCAC 2016e). The Governance and Nominations Committee advises the board of directors regarding policies; terms of references describing the purpose, structure, and functioning of committees and projects; and procedures that will guide the functioning of the board and its standing committees. It is also responsible for recruiting future members of the board, as well as committees, working groups, and task forces (CCAC 2016e).

The CCAC delivers its mandate through three complementary and mutually supportive programs: Standards Setting and Maintenance, Assessment and Certification, and Training and Networking (CCAC 2016d). Thus, the system used by the CCAC integrates the delivery of these three types of programs under one organization. The Standards Setting and Maintenance Program is responsible for developing reference standards, promoting the principles of Russell and Burch's three Rs (replacement, reduction, and refinement), achieving international harmonization, and establishing national and international credibility (CCAC 2016f). The Assessment and Certification Program serves as a quality assurance system through the assessment and certification of animal-based research, teaching, and testing programs in individual institutions. It is peer based, and involves scientists, veterinarians, community representatives, administrators, and animal health technicians working together to ensure that animal-based science activities are appropriate (CCAC 2016f). The Training and Networking Program provides opportunities and resources for members of CCAC-certified institutions to learn more about CCAC guidelines, policies, and evidence-based practices that underpin the ethics and care of animal-based science (CCAC 2016g). Last, two service units (Public Affairs and Communications, and Operations) provide additional administrative support to these programs (CCAC 2016g).

The CCAC is financed by the CIHR and Natural Sciences and Engineering Research Council (NSERC), with additional contributions from federal science-based departments and agencies, as well as annual program participation fees paid by CCAC-certified institutions (CCAC 2016h).

Demographics of Programs Participating in AAALAC International Accreditation and CCAC Certification

As of 2016, AAALAC International's accreditation program includes more than 950 accredited units in 40 countries, and has sustained an average annual growth in participating units of greater than 2% since 1990. The number of organizations accredited internationally began to climb in the late 1990s in response to AAALAC International's active engagement of international participants as a strategic goal. International expansion has been especially marked in the past decade. AAALAC International had 45 accredited units outside of North America in 2005, surging to 255 by 2015. Participation among European countries has almost tripled during this interval (from 28 to 75), and in the Pacific Rim, it has increased more than 10-fold (from 14 to 149 accredited programs). During this same interval, the net gain in number of programs participating in accreditation from North America increased by only 50. In addition, since 2005 programs from 16 new countries have successfully entered the accreditation program. Accreditation encompasses all types of organizations conducting animal-based scientific activities in research, teaching, testing, and production, which AAALAC International tracks by category. The most recent category profile for participating programs is given in Table 9.1.

In AAALAC International's data from 2000, the percentage of commercial programs and the number of university programs (campus-wide and limited combined) were approximately equal, at 31% each. However, with the collaborative global commercialization of science and more frequent and rapid exchange of research animal data internationally in recent years, the impetus for the commercial sector to participate in AAALAC International accreditation has soared, as reflected in these data. Although participation in every sector of the industry has expanded since 2000, it is evident that growth potential for AAALAC International remains fertile across the globe.

The CCAC certifies Canadian-based institutions within three broad sectors: academia (i.e., research centers, colleges, and universities), governmental research and testing centers, and private organizations conducting research, testing, and production of animals or biologics for scientific purposes (CCAC 2008). The size and complexity of institutional animal ethics and care programs vary widely, from small colleges limited to teaching programs to large universities with several affiliated research centers. The total number of CCAC-certified institutions has been increasing over the years. As an example, between 2012 and 2015 there were between 167 and 198 CCAC-certified institutions (CCAC 2013, 2014, 2016i).

TABLE 9.1

Total Number of Accredited Units by Category (2015 Data)

Category	Accredited Units	% Accredited Units
Commercial	415	44
Government	128	14
Nonprofit	99	10
Hospital	44	5
University/campus-wide	145	15
University/limited	119	12
	950	100

Other Quality Assurance Programs

Some programs involved in research animal care and use, particularly those involving animal production and/or commercial product development using animals, have pursued certification by the International Organization for Standardization (ISO) 9000 family of quality management systems standards to provide assurance to customers and other stakeholders that they are consistently meeting quality, statutory, and regulatory requirements related to their product (US TAG to ISO/TC 176 2015; Wikipedia 2016). Experts certified in the ISO 9000 process lead assessments focused on seven core management principles. When ISO 9000 assessments are applied to research animal care and use activities, technical experts in research animal care and use assist assessors in assuring that *in vivo* outcomes and quality parameters provided by the organization meet regulatory and industry standards. However, the individuals conducting the actual site assessment submit their findings to a separate body that makes the determination if the site warrants certification. ISO certification has not attracted wide participation from the research animal industry, likely because most organizations have concluded that the excessive effort required to obtain and sustain certification does not substantially contribute to the delivery of high-quality animal care and use or meaningful animal welfare outcomes. For an anecdotal comparison, AAALAC International has published the reflections of individuals who have experience with both ISO 9000 certification and AAALAC International accreditation (AAALAC International 2016b).

Standards Used in the Assessment and Certification Process: Selection and Application

AAALAC International's Standards and Guidance

Programs participating in AAALAC International accreditation are expected to conform to all local, regional, national, and supranational laws and guidelines that are applicable to their operations, and to comply with recommendations of AAALAC International's TPSs, as applicable to circumstances. These standards are

- The *Guide for the Care and Use of Laboratory Animals*, published by the NRC (ILAR 2011)
- The *Guide for the Care and Use of Agricultural Animals in Research and Teaching* (*Ag Guide*), published by the Federation of Animal Science Societies (FASS 2010)
- The European Convention for the Protection of Vertebrate Animals Used for Experimental and Other Scientific Purposes, by the Council of Europe (ETS 123) (Council of Europe 2016)

The adoption of these three documents by AAALAC International's BOD as primary standards in 2011 signifies the importance of these performance-based guidelines in the accreditation process. The *Guide* has been used by AAALAC International since the inception of the accreditation program as the overarching framework for the comprehensive assessment of participating research animal care and use programs. The scope of applicability of ETS 123 is limited to the member countries of the Council of Europe that have voluntarily ratified the convention, and thus is not a prevailing standard in the United States or other countries. The *Ag Guide* is a multiauthored document with more than 60 expert contributors and substantial international participation offering extensive and detailed guidelines on the care and use of animal subjects typically employed in agricultural or production research.

AAALAC International encourages institutions to make animal care and use determinations based on regulatory and funding requirements, overlaid with a performance approach that enhances animal welfare and quality science. Programs in countries with no underlying regulatory requirement may elect to adopt one or more of the TPSs. For example, ETS 123 and local guidelines may be used in programs located outside of Europe, and similarly, the *Guide* in programs located outside of the United States. In some situations, it may be useful to consider AAALAC International's collation of performance standards from the *Guide*, *Ag Guide*, and Appendix A of ETS 123 to address particular programmatic needs. Each guidance document provides sound recommendations regarding housing environments and cage or pen space availability for a variety of species. As an example, since the *Guide* does not provide specific recommendations on appropriate housing for ferrets, the excellent guidance provided in ETS 123 for this species could be useful for institutions in the United States and globally.

Since 1975, AAALAC International has also referred to other specialty publications to supplement the information within the TPSs regarding practices, procedures, or techniques that may be useful to address particular programmatic needs. These specialty publications are designated as reference resources and cover the following topics: general considerations, biosafety, education, euthanasia, animal health monitoring, occupational health and safety, research considerations, and species-specific information.

References included in this reference resources list have been formally reviewed and adopted by AAALAC International's COA as guidance for accredited units. These texts may also be utilized by the AAALAC International site visitors and during COA deliberations when discussing issues identified during site visits. When applicable, reference-specific notes provide additional information on any exclusions or clarification in the reference. Prospective new reference resources are generally identified through review of the literature and/or policy documents performed regularly by members of the AAALAC International COA or BOD. However, AAALAC International invites the nomination of documents from all sources. AAALAC International's reference resources can be reviewed in detail at AAALAC International (2016c).

AAALAC International's Rules of Accreditation

AAALAC International operates the accreditation program according to the rules of accreditation that are established by its BOD. Organizations contemplating application for accreditation should thoroughly familiarize themselves with the rules (AAALAC International 2016d). The highlights of the nine sections of rules of accreditation are offered. Section 1 defines the types of animals included in the evaluation, the activity criteria that must be met by the unit seeking accreditation, and the administrative and organizational arrangements and alignment of authority that are compatible with eligibility for application. All animals used or to be used in research, teaching, or testing at accreditable units are to be included and evaluated in accordance with AAALAC International's standards. This may include invertebrate species where they are relevant to the unit's mission. Organizations seeking accreditation must have fully operational animal care and use programs functioning at a reasonable activity level in relation to the unit's capacity for conducting research animal activities. Programs that do not have sufficient activity are ineligible for assessment or accreditation, and accredited programs that cease animal activities are required to notify AAALAC International and reinstate activities within 12 months to remain eligible for accreditation. Readers should consult Section 1 of the rules to understand how AAALAC International evaluates whether an institution is proposing an acceptably structured and demarcated accredited unit. Importantly, once an appropriate accreditable unit or program is defined, all the animals *owned* by that unit will be included as part of the program of research animal care and use in AAALAC International's evaluation, regardless of the site of animal housing and use.

Section 2 of the rules describes the standards used in AAALAC International's assessment and accreditation activities. A key provision is the requirement that accredited units submit an annual report updating the continuity and any changes within the program. In addition, accredited units are expected to promptly alert AAALAC International of any adverse events related to the animal care and use program. These should include investigations performed by regulatory entities, as well as reports of serious incidents or concerns that negatively impacted animal well-being.

Section 3 of the rules establishes the fee requirements for participation in the accreditation program and provides for their routine periodic adjustment according to economic conditions. It also allows for the levy of supplemental charges to units involved in protracted accreditation evaluations resulting from programmatic deficiencies and correction efforts.

Section 4 of the rules concerns general information about site visits and site visitors. Site visits are typically conducted at 3-year intervals, although AAALAC International reserves the right to alter the interval to ensure animal welfare and the integrity of the accreditation program. Site visit teams comprise at least two site visitors, except in very special circumstances (e.g., a small or remote facility or activity); the lead visitor is a member of the COA or a COA emeritus (see Section 3 below), and assistants are selected from the pool of ad hoc consultants or specialists appointed by the COA. AAALAC International strives to avoid conflicts of interest in the assignment of site visitors; correspondence from the AAALAC

International administrative office to the accredited unit, as well as the unit's site visit planning with the COA member, also engages the unit to ensure a fair and objective review. Similar provisions for avoiding conflicts of interest permeate all of AAALAC International's practices through the rendering of a final accreditation decision. Site visitors' COA may only conduct AAALAC International business following formal assignment by AAALAC International, and all materials utilized and findings identified during the site visit may not be disclosed to any person or agency accept AAALAC International. Strict confidentiality is a cornerstone of AAALAC International's contract with programs participating in accreditation.

Sections 5–7 of the rules encompass provisions for granting or denying accreditation, accreditation status and considerations for programs that are in less than full accreditation status, and the process for hearings and appeals if COA's decision is to withhold or revoke accreditation. The topic of accreditation status is covered later in this chapter in some detail. However, it is important to note here that the rules provide for two levels of appeal when the COA declares a withhold or revoke action. Accredited units or new applicants for accreditation are given the opportunity to review the reasons for the COA's denial of accreditation and have 30 days to file for an appeal hearing and/or submit material evidence to the COA on why the adverse decision is unjustified. Should the appeal to the COA be unsuccessful, the unit may then exercise a similar process and appeal the COA's decision before the AAALAC International BOD.

AAALAC International's Position Statements

In addition to the TPSs, AAALAC International has also promulgated several position statements to clarify key issues and highlight its interpretation and expectations for the treatment of certain topics within the TPSs. The topics addressed are deemed highly significant and impactful; nonconformance with these position statements generally precludes an institution from achieving full accreditation. Like any new accreditation standard, the position statements are developed collaboratively by the COA and BOD and are adopted by the BOD after a period of public circulation, comment, and revision if necessary.

Currently, AAALAC International has eight position statements, six of which have a bearing on accreditation decisions. The topics covered include defining the expectations of the attending veterinarian and the program of veterinary care, standards for cage and pen space, attention to social housing, safety considerations for walk-in cage or rack washers and bulk sterilizers, and occupational health and safety and management considerations for *Macacine herpesvirus* 1. Two other position statements clarify AAALAC International's use of the term *laboratory animals* and provide guidance to organizations involved with agricultural animals used in research and the selection of standards of care and use. The full position statements can be found at AAALAC International (2016d).

AAALAC International's FAQs

AAALAC International offers responses to more than 50 frequently asked questions (FAQs) to provide organizations with an understanding of the applicability, nature, and administrative requirements of the assessment and accreditation process, and guidance on both general and specific topics pertaining to accreditation. New FAQs are added (or deleted) in response to AAALAC International's adoption of new standards. Others may result from information gleaned during site visits or other communication from the accredited community suggesting that an interpretive summary of AAALAC International's view might help organizations sharpen and expedite their approach to important areas of interest. For example, 15 new or revised FAQs were issued following the adoption of TPSs by AAALAC International in 2011. The summary of topics covered in the FAQs by broad category include AAALAC International's assessment process; animals included in the AAALAC International accredited "unit"; institutional responsibilities; animal environment, housing, and management; veterinary medical care; physical plant; administration; the accreditation process; and maintaining accreditation. The FAQs can be found at AAALAC International (2016e), and some of these topics are discussed in other sections of this chapter.

As stated in AAALAC International's rules of accreditation, "the accreditable unit shall observe any and all statutes and governmental regulations which bear upon animal care and use including, but not limited to, the prevailing standards of sanitation, health, labor and safety of the jurisdiction(s) in which it is located." AAALAC International has no direct connection or affiliation with regulatory entities, nor does it have any interest in functioning as an ersatz regulator. However, research animal welfare laws and regulations express the public and governmental expectations specific to the locality, and adherence to these requirements is important for ethical, legal, and public relations reasons. For this reason, the AAALAC International site visit process includes review of the institution's documents concerning regulatory interactions, and institutions are asked to confirm that corrective actions have been taken in instances of regulatory noncompliance.

AAALAC International's accreditation findings are conveyed to organizations in the following consistent format:

1. *What* did the site visitors observe that indicated program quality was potentially compromised?
2. *Why* is the finding regarded as a potentially negative and/or serious impact on the quality of the research animal care and use program, that is, which AAALAC International reference or standard is relevant?
3. *What* is the organization expected to do to address the finding successfully?

The importance of the finding and the urgency and criticality of corrective actions required are communicated in the *why* and second *what* statements above through the use of the verbs *must* or *should*. This long-standing AAALAC International precedence has been modeled after the use of these terms in the *Guide* and is explained in the FAQ on this subject (AAALAC International 2016f). Given the fundamental importance of the recommendations in the *Guide* that are prefaced with a *must*, AAALAC International's COA typically categorizes site visit findings that do not conform with a *must* statement in the *Guide* as mandatory items for correction. In AAALAC International's parlance, a mandatory item is a serious deviation from the recommendations of the *Guide* and/or other AAALAC International standards, which has to be corrected to achieve or continue full accreditation. This judgment is based on the COA's assessment of the potential for the program deficiency to adversely affect the health, well-being, or safety of animals or humans. Other terms connoting the imperative to act are discussed in the FAQ referenced above.

Other findings identified by AAALAC International during the on-site assessments of animal care and use programs, and deliberated by COA in convened sessions, are considered "suggestions for improvement" (SFIs). These are recommendations that the COA feels are desirable to enhance an already acceptable or even commendable program. SFIs are used to draw attention to recommendations that are typically denoted as *should* statements in the *Guide*. AAALAC International considers the offering of SFIs to be an element of the peer review process that is designed to assist accredited programs and new applicants by sharing the cumulative knowledge and experience of the COA in light of contemporary literature. It should be noted that there is no obligation for institutions to make program changes based on SFIs; implementation of suggestions is, however, one means of promoting a high-quality animal care and use program. Also, an SFI does not automatically become a mandatory item for correction during the next site visit cycle if the same situation (procedure, practice, etc.) is observed. However, if an issue is identified that is a *should* statement in the *Guide*, but is one of numerous issues noted within the same program area that collectively signal a broader problem, then it may be included within a mandatory item for correction.

CCAC's Standards and Guidance

In Canada, the guidelines and policies developed and established by the CCAC serve as national reference standards. Thus, the delivery of the CCAC's mandate is based on guidelines and policies that provide clear guidance to institutional animal ethics and care programs. Assessments of institutional programs are also based on the CCAC guidelines, policies, and other relevant documents (CCAC 2016j).

CCAC Guidelines

The CCAC guidelines are intended to provide assistance implementing evidence-based practices to achieve Russell and Burch's "three Rs" tenets for animal-based science (replacement, reduction, and refinement). Throughout CCAC guidelines, the terms *should* and *must* are also used. The term *should* is used to indicate an obligation for which exceptions must be justified and approved by a properly constituted animal care committee. The term *must* is used for mandatory requirements, for which no lower standard of practice could be accepted or where regulatory requirements are involved (CCAC 2016k).

Guidelines are developed and revised in response to the current and emerging needs of the research community, advances in animal care, and the needs of the Assessment and Certification Program. This process, overseen by the Standards Setting and Maintenance Program, is led by subcommittees of experts working to produce texts that are based on sound scientific evidence and expert opinion, and receive extensive peer and public review. Strong peer involvement ensures that guidelines are aligned with advances in science and animal care, nationally and internationally, and are suited to the needs of Canadian institutions and investigators (CCAC 2016l). The CCAC currently has guidelines pertaining to specific types of animals (farm animals, fish, laboratory animals, marine mammals, and wildlife), and others covering procedures or management (transgenic animals, procurement of animals, antibody production, endpoints, euthanasia, laboratory animal facilities, protocol review, and training of personnel). Typically, the animal category guidelines cover topics such as acquisition, transportation, handling and restraint, housing and husbandry, health care, procedures, euthanasia, facility design and management, and occupational health and safety (CCAC 2016k).

CCAC Policies

The CCAC publishes policies that establish the ground rules and basic requirements for the oversight of the institutional animal ethics and care programs within the Canadian system (CCAC 2016m). Policies can be grouped in three categories: (1) general policies for animal-based science (e.g., categories of invasiveness, and pedagogical merit of live animal–based teaching and training); (2) policies for institutional animal ethics and care programs (provide guidance for institutions' administrators and animal care committees); and (3) policies governing the CCAC Assessment and Certification Program (provide guidance on the assessment and certification process).

Other CCAC-Recognized Standards

In addition to CCAC guidelines and policies, the following four distinct sets of documents are also used in CCAC assessments (CCAC 2016n):

1. The Canadian Association for Laboratory Animal Medicine (CALAM) *Standards of Veterinary Care*
2. The *Codes of Practice for the Care and Handling of Farm Animals*, published by the National Farm Animal Care COA
3. The *Canadian Biosafety Standards and Guidelines*, Second Edition, published by the Public Health Agency of Canada and the Canadian Food Inspection Agency
4. The *Containment Standards for Facilities Handling Aquatic Animal Pathogens*, First Edition, published by the Canadian Food Inspection Agency

Three Rs

The tenets of the three Rs are embedded in the conduct of animal-based science in Canada and are at the heart of all CCAC policies and guidelines. Resources for researchers, veterinarians, animal care staff, and animal care committees, such as the CCAC three Rs microsite (http://3rs.ccac.ca/en/), provide complementary information needed to fully implement the guidelines and to achieve high standards of animal welfare (CCAC 2016o).

Use of Local or National Laws in the Certification Process

As stated previously, there is no federal legislation in Canada regulating the use of animals in science, as this is of provincial jurisdiction, as defined in the Canadian Constitution. Animal welfare and animal-based science are provincial responsibilities, with the guidelines and policies set by the CCAC serving as national reference standards (CCAC 2006). The provincial legislations in eight Canadian provinces (Alberta, Manitoba, New Brunswick, Newfoundland and Labrador, Nova Scotia, Prince Edward Island, Quebec, and Saskatchewan) have been amended to directly reference CCAC standards. Ontario has its own Animals for Research Act and regulations (CCAC 2006, 2016c).

Institutions must follow provincial, federal, and municipal requirements (where applicable), and in particular those regarding occupational health and safety programs and the use of dangerous compounds when working with animals, including (CCAC 2008)

- The federal Health of Animals Act (http://laws-lois.justice.gc.ca/eng/acts/H-3.3/FullText.html), which governs the control of animal diseases and toxic substances
- The Workplace Hazardous Materials Information System (WHMIS) (http://hc-sc.gc.ca/ewh-semt/occup-travail/whmis-simdut/index-eng.php), which requires that each employer provide safe working conditions and that employees be informed about all hazards that they will face
- The federal Controlled Drugs and Substances Act and its related regulations (http://laws-lois.justice.gc.ca/eng/acts/C-38.8/FullText.html), which governs the use of narcotics and other controlled substances
- Federal regulations for the use of radioactive material
- Provincial occupational health and safety acts for the use of x-rays

Professionals and Professionalism Involved in Conducting Assessment and Certification Visits

AAALAC International's Professionals on the Site Visit Team

AAALAC International has more than 450 volunteer professionals who are eligible to participate on the teams conducting the assessment and accreditation site visits. These individuals are not employees of AAALAC International, but have been selected through a competitive process and represent a broad spectrum of backgrounds and expertise. The general criteria for selection include educational, professional, and scholarly accomplishments; a strong understanding of pertinent aspects of animal care and use programs; and a demonstrated ability to interact professionally and independently with integrity, objectivity, neutrality, and collegiality. The professionals involved in AAALAC International site visits fall into two general categories: COA members or COA emeritus members and ad hoc consultants and specialists.

COA members or COA emeriti are the individuals that lead AAALAC International site visit teams, and site visits are not conducted without their participation. The COA members are individuals that have distinguished themselves during their volunteer service to AAALAC International as ad hoc consultants and specialists based on standard criteria used by the COA site visit team leaders to assess the performance of the team. As the size of the COA expands or COA members complete their term limits or resign, a COA nominating committee selects a pool of top candidates among the ad hoc consultants and specialists for a competitive election of new COA members. Newly elected COA members must then be ratified by the AAALAC International BOD. As of 2016, the COA consisted of 63 members who can serve a maximum of four 3-year terms (12 years total). Of the current 63 COA members, 9 are nonveterinarians with a doctoral degree in another specialty area of science or medicine. The remaining individuals predominantly have a doctoral degree in veterinary medicine, and many also have other clinical certifications or advanced scientific degrees. There is usually some turnover on the COA membership annually, allowing the selection of new members and enabling the COA to alter its expertise and composition according to needs within the accredited community.

Due to the high workload of the COA, AAALAC International developed a program to retain COA emeritus members as site visit leaders. The program has evolved over time, and the COA emeriti must meet defined criteria in contemporary site visit experience and continuing education, are subject to competitive term reappointment, and have limited terms. COA emeriti serve as the team leaders of the site visits they are assigned. They also work in collaboration with a partner on the current COA and participate in the COA deliberations on the accreditation status of the units they have site visited. There are currently 28 COA emeriti supporting the AAALAC International accreditation program.

AAALAC International has a pool of ad hoc consultants and specialists from which the COA and COA emeritus team leaders can choose to assemble a professional team with an array of talent complementing the needs of the organization being visited. There are currently approximately 363 ad hoc consultants and specialists participating in AAALAC International's peer review activities, representing a majority of the 43 countries with AAALAC International accredited programs. This pool is composed of approximately 250 veterinarians, 133 other doctorates, 18 individuals who are certified professional Institutional Animal Care and Use Committee (IACUC) administrators, and 10 who are certified managers of animal resources qualified. The consultant designation connotes an individual with extensive general qualifications, whereas a specialist designates an individual with demonstrated expertise in a specific area that may be especially helpful for the review of programs encompassing that interest. Ad hoc consultants and specialists also have term appointments and limits. Due to the size of this pool, many positions typically turn over or are reappointed annually in a competitive process. Those interested in learning more about applying for consideration as an ad hoc consultant or specialist should consult the AAALAC International website at AAALAC International (2016g).

CCAC's Professionals on the Assessment Site Visit Team

Visits to institutions participating in CCAC programs are conducted by an assessment panel. The panels for regular assessment visits are composed of a (CCAC 1999, 2006, 2016p)

- Community representative
- Scientist
- Veterinarian
- CCAC associate director of assessment (ADA), as an *ex officio* member

Assessment panels are an integral part of the CCAC Assessment and Certification Program. The CCAC ADA, present at every assessment visit as an *ex officio* member, serves as a steward of the assessment process to ensure consistency. The ADA organizes the visit, selects panel members, works to ensure consistency between panels, and assists panel members in undertaking their duties, in clarifying issues, and in formulating recommendations.

Other panel members are volunteers, selected for their experience and ability to be actively involved in the process, and to communicate clearly and constructively with their peers on the panel and with representatives of the institution to be assessed. One of the members is identified as the panel chair, generally a scientist or a veterinarian; the chair leads discussions and works toward achieving consensus among the panel members.

The community representative is generally selected from a humane society in the local area of the institution or from the pool of community representatives on local animal care committees (other than the committees of the institution to be visited), in order to provide a local, public perspective. It allows members of the public to be involved in the ethical decision-making process regarding animal-based research, provide scientific institutions and the CCAC with an external perspective, and furnish a means of identifying and addressing public concerns.

The scientific representative of the panel is selected according to his or her expertise with reference to the nature of the research, teaching, or testing at the institution to be assessed. The veterinarian, who has expertise in experimental animal medicine, is also selected according to his or her specific expertise and experience.

The role of an assessment panel is to review all aspects of the animal-based research, teaching, and testing programs within an institution; to assess the procedures and facilities used in the implementation of animal-based programs; and to note and comment on any matters that may not be in accord with the

CCAC's guidelines and policies. They also promote implementation of high standards of animal ethics and care, which have been defined by peers through CCAC guidelines and policies. Panel members act as unbiased, informed advisors to the institution, not as inspectors.

Components of the Assessment, Accreditation, and Certification Processes

Preparing for an AAALAC International Site Visit

Organizations can determine if their programs are eligible to participate in accreditation by reviewing the "rules of accreditation" available online at AAALAC International (2016d). Eligible institutions may request an application package from the AAALAC International office, or download one from AAALAC International's website. The application begins with a two-page "application for accreditation" form that collects information on an institution's key contacts and other administrative information (AAALAC International 2016h). The application also contains instructions aiding preparation of the program description (PD), the key document used by AAALAC International's COA to evaluate an animal care and use program. Other information in the package guides the prospective applicant to the TPSs used by the COA to evaluate programs, a number of international guidelines, and a list of additional resources that may be used during the assessment process as appropriate to the location of the program (AAALAC International 2016c).

Organizations seeking and maintaining accreditation must pay a one-time application fee and subsequent annual fees. Both of these fees are based on a sliding scale that generally correlates with the size of an institution's animal facility or facilities. AAALAC International determines the group classification during the review of the application. The fee schedule, established and approved by the AAALAC International BOD, may be amended as necessary to reflect the costs of operating the accreditation program. Revocation of accreditation is automatic if a unit is 12 months in arrears for payment of fees. Refer to the fee schedule at AAALAC International (2016i).

The PD is a critical document for AAALAC International's COA to review an institution's animal care and use program. In addition, the effort, engagement, integration, and teamwork required by the institution to create this comprehensive and detailed document may help the institution identify and address weaknesses, resulting in enhanced animal well-being and improved conditions for sound science from research animal studies. The narrative portion of the PD addresses the following broad areas:

- Animal care and use program
- Animal environment, housing, and management
- Veterinary care
- Physical plant

AAALAC International provides a template that guides the institution through the creation of the PD. In addition to the narrative portion of the PD, appendices are used for the collection of other information vital to AAALAC International's evaluation. The PD must be submitted in English, with the exception of some appendices. Full information on the preparation of a PD is available at AAALAC International (2016j).

Completed application packages undergo administrative review to ensure that applicant programs meet the criteria listed in the rules of accreditation. Based on the information in the PD and correspondence from the applicant program, a site visit team, made up of two or more AAALAC International COA members and ad hoc consultants or specialists, is assigned to conduct the review. The AAALAC International office provides the applicant with the names and affiliations of the site visit team members. This permits identification of potential conflicts of interest not previously detected during the site visitor assignment process. The COA member contacts the organization to schedule the visit at their discretion at a mutually agreeable time.

AAALAC International's accreditation program is strictly confidential. Site visit teams and the AAALAC International office manage all institutional information accordingly, which obviates a need for any additional confidentiality agreements with participating programs. However, if an organization believes an additional confidentiality agreement is required, this must be submitted to the AAALAC International corporate office for review.

AAALAC International Site Visit Process

Typically, the site visit team meets the evening before the site visit to discuss the institution's PD, results of the previous site visit, and any other issues related to the visit. The site visit leader will answer any questions that other site visitors may have about conduct of the site visit, and describe how the site visit will proceed. Site visits begin with an "in-briefing," during which institutional representatives meet with the site visit team to discuss the accreditation process. After the in-briefing, the site visit team interacts with a smaller group from the institution composed of key personnel (e.g., attending veterinarian, IACUC chair and/or administrator, facility managers, physical plant personnel, and health and safety personnel), and others deemed helpful by the institution to review and clarify any aspects of the PD, as required. Although the site visit team thoroughly reviews the PD before the visit, this gives the team members an opportunity to ask specific questions or request additional documents to ensure their understanding of the program.

Following completion of the PD review, the site team conducts a tour of the animal facilities inclusive of animal support areas, outdoor acreage if applicable, and at least a representative sample of laboratories used in animal care and use activities. This review may take as little as a few hours in a very small program to several days in larger programs with more complex research missions or disseminated facilities. Generally, site visitors work in two-person teams throughout this process; however, visitors occasionally split to review multiple rooms containing the same species housed under standardized conditions to increase site visit efficiency. Site visitors may also occasionally employ virtual technology to accomplish the review, for example, mobile video applications, of remote animal facilities.

Site visitors also typically reserve a time dedicated to a discussion with the members of the IACUC or comparable oversight body to learn more about pertinent IACUC issues and exchange information about the AAALAC International accreditation process. Traditionally, this meeting is held over a lunch break to promote conversation and enhance the efficiency of the visit. Based on the PD review, discussions with the IACUC, and observations of the site visit team during the facilities and laboratory tour, the team may request additional information on specific protocols or procedures. This enables the site visitors to evaluate the soundness and consistency of protocol review practices and ensure that studies were performed according to the conditions of protocol approval with satisfactory animal care and use outcomes.

At the conclusion of the facility and laboratory review, the site visit team meets alone in an "executive session." During this session, they typically review previously requested documents and may wish to have additional information at their disposal to address aspects of the program about which they still have questions. During this session, they discuss their observations and decide what information should be communicated at the conclusion of the visit.

Finally, an "exit briefing" is held for the institution's representatives and any others, as deemed appropriate by the institution's leadership. During the briefing, the team leader shares the key findings, including both commendations and issues of concern, and conveys what the team intends to recommend to the COA regarding the institution's accreditation status. This phase of the visit continues to afford the institution opportunity to correct any misperceptions or factual errors that the site visit team may have. Following the visit, the site visitors document their findings and recommendations in an official site visit report, which is used in the COA deliberations on accreditation status. It is important to note that regardless of the experience of the site visitors and their efforts to construct a collegial, lucid, and insightful exit briefing, the recommendations offered during the exit briefing are only preliminary. Site visitors should affirm that issues discussed during the exit briefing, including their expected categorization as "mandatory findings" or SFIs, represent their preliminary judgment. The COA reserves the authority and responsibility for the site visit outcome, including any specific findings. Therefore, in response to these tentative findings highlighted by an AAALAC International site visit team, organizations are encouraged to only make program or facility changes based on their own evaluation of and concurrence with the basis of the team's findings. AAALAC International discourages organizations from making large financial expenditures to address any findings until receiving the final letter from AAALAC International containing the COA's decision following deliberation.

Institutions are invited to formally respond to the site visit team's initial findings through what is called "post–site visit communication" (PSVC). The PSVC is a letter and accompanying documentation that are submitted to address any misperceptions, or explain how the institution has already responded to concerns raised during the exit briefing. PSVC should provide a direct, cogent, definitive, and well-documented response to a finding; correspondence explaining partial and preliminary actions or proffering nonspecific approaches to later resolution of problems should be avoided. Additional details on the exit briefing and the PSVC process are available at AAALAC International (2016k).

COA Process for Evaluation and Conferral of Accreditation Status

The COA makes all final decisions regarding accreditation. The COA meets three times a year (January, May, and September) to consider site visits from the preceding trimester. Prior to the meeting, at least four additional COA members not present at the site visit review and comment on the site visit report. During the COA deliberations, the COA member who led the site visit describes the institution's program, shares the site visit team's observations, and presents the institution's perspective on findings under discussion or dispute. In the ensuing discussions, AAALAC International uses the performance standards approach of the *Guide* in its assessments. The COA members evaluate whether institutions are achieving specified program outcomes (e.g., adequate sanitation), without being prescriptive regarding the exact manner in which to achieve the outcome or goal (e.g., frequency or methods used for sanitation). While engineering standards specify in detail a method or technique for achieving a desired outcome, they do not provide for interpretation or modification of the technique should an alternative be deemed equally acceptable. Alternatively, performance standards define the outcome in detail and provide measurable criteria for assessing whether the outcome is achieved. For more details on AAALAC International's use and endorsement of performance standards, see AAALAC International (2016l). Based on the COA's discussions, a letter is drafted that explains the accreditation status. After the COA meeting, the letter is further reviewed, edited, and approved by members of the COA leadership and AAALAC International staff. In most instances, this allows official notification of accreditation status within 4–8 weeks of the COA meeting.

There are a number of different categories of accreditation status that are described in more detail at AAALAC International (2016m). Briefly, new applicants for accreditation are originally accredited only when designated award full accreditation or award full accreditation with conditions. A condition is a finding that requires corrective action but for which implementation is underway on a defined, easily achievable timetable. New applications that are awarded provisional status must correct a mandatory finding within a 24-month time limit to be considered accredited. Withhold accreditation is awarded to programs with more mandatory corrections required than can be reasonably achieved. Therefore, these programs must reapply for consideration in the accreditation program.

Following subsequent triennial reviews of accredited programs, the optimal outcome is an award of continued full accreditation. Programs that have mandatory findings, depending on the severity and complexity of the findings, are awarded either deferred accreditation or probation; deferred accreditation entails a simpler, less resource-demanding corrective action, and probation involves a more complicated and potentially costlier organizational response. In both cases, the organization is still accredited, but the mandated corrections must be completed within a specified time frame to attain continued full accreditation. If they are not, the programs may proceed to an action of revoke accreditation. In the revoke accreditation status, organizations remain accredited through two levels of appeal, first to the COA and then to the BOD, after which they are removed from accreditation or restored to more favorable status if substantial, definitive progress on mandatory items has been achieved.

AAALAC International's Program Status Evaluation Program

AAALAC International offers a confidential Program Status Evaluation (PSE) review that helps assess the quality of all aspects of an animal research program, including animal husbandry, veterinary care, institutional policies, and the facilities where animals are housed and used. The decision to offer assessment services (in addition to the accreditation program) was prompted by a number of requests from nonaccredited institutions for a "pre–AAALAC International site visit." These institutions, particularly those outside of the United States, are typically less familiar with the accreditation process and want to find out how their programs compare with AAALAC International standards—before participating in the formal accreditation program.

The objective of the PSE service is twofold. First, it is meant to assist institutions by identifying weaknesses and suggesting ways to improve or correct them. Second, it is meant to familiarize institutions with the AAALAC International accreditation process and encourage them to participate. Therefore, the PSE is conducted in a style mimicking a regular AAALAC International site visit, although led by a senior member of the AAALAC International staff, accompanied by covisitors who are former members of the COA. The other salient hallmark of a PSE is that the recommendations for correcting programmatic deficiencies are consultative in nature, providing the organization with directions on the path forward. The occurrence and outcome of the PSE visit are unknown to the COA in the event that the institution subsequently applies for accreditation. More information about AAALAC International's PSE program may be found at AAALAC International (2016n).

CCAC Assessment Process

The various types of assessment visits use the CCAC guidelines, policies, and associated documents when conducting an assessment of the structure and resources of the animal ethics and care program; the composition, functioning, and effectiveness of the animal care committee; and the appropriateness of animal care, animal-based practices, procedures, and facilities (CCAC 2016q).

- *Orientation visits* are scheduled for institutions wishing to join the CCAC program. They are conducted by an ADA (CCAC 2016r).
- *Regular assessment visits* are conducted by assessment panels of peers for institutions that have already joined the CCAC program. These visits are conducted every 6 years (CCAC 2016q).
- *Interim assessment visits* are scheduled between full visits, 3 years after the regular visit. Interim visits are conducted by an ADA (either alone or with a peer reviewer) (CCAC 2016q).
- *Special visits* to an institution may be undertaken between assessments if conditions at an institution warrant it, or following a request by the institution. Special visits are generally conducted by an ADA without a panel (CCAC 2016q). For example, the Assessment and Certification Committee could request an ADA to visit an institution in probation to assess the progress made in addressing issues. An institution could request a special visit prior to an assessment to verify that a new or renovated animal facility meets the CCAC guidelines.

Institutions that are assessed and certified by the CCAC must pay an annual program participation fee. The CCAC secretariat ensures that the institution to be visited remains in good standing with regard to its annual fees.

Preparing for the CCAC Visit

The CCAC contacts an institution approximately 3–6 months prior to a visit to confirm the date of the visit. Once an assessment visit by the CCAC is confirmed with the institution, the ADA selects panel members, and the CCAC secretariat forwards all relevant forms and documents (also available on the CCAC website) to be completed and submitted to the CCAC prior to the visit. Those documents are (CCAC 2016s)

- *Agenda of visit*: Two months before the visit, an agenda drafted by the institution is sent to the ADA. Assessment visits may be scheduled for one or more days, depending on the size of the institution's program. The agenda may be modified to accommodate institutional or CCAC needs (CCAC 2016t).

- *Animal ethics and care program review form (PRF)*: Three to four weeks before the assessment visit, the PRF is sent to the CCAC secretariat and the panel members by the institution to be assessed. The PRF allows the institution to review its own program thoroughly and to provide details on aspects of its animal ethics and care program to the CCAC assessment panel members selected for the visit. The PRF includes seven sections (and appendices) covering the different aspects of the animal ethics and care program: (1) general information about the program, activities involving animals, and species of animals; (2) animal care committee functioning; (3) animal use protocol forms; (4) veterinary or animal care program; (5) continuing education and training for veterinary and care personnel and users, occupational health and safety, and crisis management; (6) animal facilities; and (7) summary, which includes a description of the strengths and weaknesses of the program based on the institution self-review. The PRF is used as the basis for all assessments by CCAC panels (CCAC 2016u).

Panel members prepare for the assessment visit by reviewing all relevant preassessment documentation, which includes the agenda, the PRF, and oftentimes the assessment and implementation reports from the previous visit, and familiarizing themselves with pertinent CCAC policies and guidelines (CCAC 2006, 2016v).

A preassessment meeting of panel members is held prior to the visit (often the evening before the visit), to review the goals of the assessment, the institution's preassessment documentation, the results of the previous assessment visit, and any other issues related to the visit. The panel chair and ADA use this time to answer any questions that the panel members may have, and to build consensus on how the assessment panel will function and how the assessment will take place. If needed, a list of specific animal-based protocols or standard operating procedures (SOPs) that panel members want to review, or specific scientists or teachers that they would like to meet during the assessment, are prepared and identified during the preassessment meeting (CCAC 1999, 2006, 2016v).

CCAC Site Visit Process

The following is an overview of the elements of a typical, 1-day assessment visit (CCAC 1999, 2016w):

- Initial meeting with CCAC panel and institutional representatives, including the animal care committee
- Review of protocols and SOPs
- Tour of any animal facilities
- Panel meeting in camera
- Summary meeting with institutional representatives

A visit typically starts with an initial meeting to briefly describe the program and to complete a thorough review of the institution's animal ethics and care program, focusing on the structure and administration of the program, animal care committee function, veterinary and animal care services, continuing education and training, and occupational health and safety. The agenda of the initial meeting is generally as follows (CCAC 1999, 2006):

- Welcome from the senior administration of the institution and presentation of the institutional animal ethics and care program, including any significant changes or plans since the previous assessment and for the foreseeable future
- Update on the CCAC program from the ADA, and introduction of the panel members
- Questions by the panel on the structure and administration of the animal ethics and care program, and then on the PRF, as completed by the institution
- Concerns of the institution (e.g., regarding animal care or CCAC programs)

Upon their arrival at the institution, CCAC assessment panel members request a few selected animal research protocols, with associated correspondence (complete protocol files), SOPs, and other relevant documents, to review following the initial meeting.

During the site visit, the panel members tour the animal facilities, including animal holding rooms, procedure rooms (which may include investigators' laboratories), and support areas (washing areas, storage areas, etc.). The panel may be accompanied by the chair of the animal care committee, senior veterinary and/or animal care representatives, and/or other animal care committee or institutional representatives. It is expected that as many investigators, study directors, course instructors, and caretakers as possible are available to meet with the assessment panel during the tour of the facilities (CCAC 1999, 2006).

Following the site visit, and prior to the final meeting with institutional representatives, the assessment panel meets (in camera) to discuss any concerns resulting from the visit, prepare a summary of the institution's strengths and weaknesses, and finalize the recommendations and commendations that will be presented verbally at the final meeting (CCAC 1999, 2006).

During the summary meeting with the institutional animal care committee and administrative representatives, the assessment panel verbally presents a summary of its observations, commendations, and preliminary recommendations by priority. All recommendations and commendations are contained in the final written report sent to the institution after the visit. The final meeting also provides an opportunity for institutional representatives to discuss with the panel members the most appropriate means to address any concerns raised (CCAC 2006).

CCAC Process for Evaluation and Conferral of Certification Status

Following the assessment visit to an institution, the CCAC ADA, in collaboration with the assessment panel members, prepares a detailed report of the visit based on the panel's analysis of information from the PRF and other relevant documents provided by the institution, and the panel's findings during the visit (CCAC 2016x).

This draft report is then circulated electronically to the panel members and to the members of the CCAC Assessment and Certification Committee for comments and approval. Once the assessment report is finalized, it is forwarded to the senior administrator responsible for the institution's animal ethics and care program, usually within 10 weeks of the assessment visit. Assessment reports support institutions in achieving good practices in animal ethics and care, highlight strengths and weaknesses of institutional programs, identify deficiencies, and where necessary, include recommendations to help institutions improve their animal ethics and care program and meet CCAC guidelines and policies (CCAC 2006, 2016x). Excellent conditions, practices, or personnel are recognized in formal commendations (CCAC 2016x,y).

Recommendations are categorized as follows (CCAC 2016y):

- *Major*: An immediate and significant threat to animal health or welfare found during a CCAC visit for which the institution must take immediate appropriate action. Failure to take such action will result in the removal of the institution's CCAC Certificate of Good Animal Practice® (GAP).
- *Serious*: Significant or long-standing weaknesses in the animal ethics and care program. The measures taken and planned in response to these recommendations must be provided to the CCAC, typically within 3 months of the institution receiving the written recommendations.
- *Regular*: Other weaknesses in the animal ethics and care program. The measures taken and planned in response to these recommendations must be provided to the CCAC within 6 months of the institution receiving the written recommendations.
- *Commendatory*: Excellent conditions, practices, or personnel in an animal ethics and care program.

Answers to the recommendations are reviewed by the ADA, panel members, and the Assessment and Certification Committee. Answers must be satisfactory in order for the institution to be certified (CCAC 2016x).

Two types of certificates can be awarded by the CCAC following an assessment: CCAC Certificate of GAP and CCAC Probationary Certificate of GAP. A CCAC Certificate of GAP is issued to institutions that have been found by the panel and by the CCAC Assessment and Certification Committee to have an animal-based research, teaching, or testing program in compliance with CCAC guidelines and policies. A Probationary Certificate of GAP is issued to (1) institutions having significant, unresolved deficiencies in their animal ethics and care program, or (2) new institutions joining the program (see below). The CCAC Certificate of GAP is valid for a period of 3 years, whereas the probationary certificate is valid for a period of 1 year (CCAC 2016z).

An institution issued a probationary certificate for unresolved deficiencies is given specific deadlines for responding to the CCAC's recommendations within the 1-year probationary period. The CCAC Assessment and Certification Committee evaluates the progress made by the institution at each deadline. The Assessment and Certification Committee may remove the Probationary Certificate of GAP if serious problems are not resolved within the given time frames. The committee may grant a Certificate of GAP to the institution before or at the end of the probationary period if it considers that the responses are satisfactory and the issues have been resolved. A thorough review of the institution's progress is undertaken at the end of the probationary period (CCAC 2016z).

Canadian federal granting agencies (CIHR and NSERC) require all institutions using animals for research to hold valid CCAC certification (either the Certificate of GAP or the Probationary Certificate of GAP) in order to be eligible to receive agency research funds (CCAC 2015), as described in the 2013 federal granting agency Agreement on the Administration of Agency Grants and Awards by Research Institutions, and in particular Section 3.5, "Research Involving Animals." Loss of CCAC certification constitutes a breach of the agreement, which would be addressed as specified in Section 4.4, "Default and Remedies of the Agency Agreement." The CCAC notifies the federal granting agencies of any changes in the certification status of an institution, if applicable, and removes the name of the institution from the published CCAC list of certificate holders (CCAC 2016z).

CCAC's Requirements for Initial Certification

All institutions wishing to obtain an initial Certificate of GAP are required to pay an annual program participation fee to the CCAC and must have the following elements in place before their first assessment visit (CCAC 2016z):

- Compliance with provincial legislation and regulations.
- An animal care committee whose composition, authority, responsibilities, and functions are defined in written terms of reference based on the most recent version of the CCAC policy statement on terms of reference for animal care committees.
- An animal use protocol form or forms, according to the guidance contained in the CCAC policy statement on terms of reference for animal care committees.
- Complete protocols submitted by animal users for all planned animal-based work, with the animal care committee having reviewed and approved any protocols planned for the near future. The protocol reviews should be based on relevant guidance (as found in Section 3e of the CCAC policy statement on terms of reference for animal care committees) and documented in the minutes of one or more animal care committee meetings.
- A formal agreement or agreements for veterinary services, based on the main elements of the CALAM *Standards of Veterinary Care.*
- Trained, qualified personnel in sufficient numbers to provide daily care for all animals.
- Other programs based on CCAC guidance for
    - The training of animal users
    - Postapproval monitoring
    - Occupational health and safety, to cover all animal project-related risks
    - Crisis management

- Where animal facilities are needed, either they must meet relevant CCAC guidance or a detailed plan with timelines must be in place describing how they will be improved to meet the guidance.
- The animal care committee must have visited all facilities designated for animal care or use, and must have approved of their use in one or more written site visit reports.

Institutional Steps to Maintain or Restore Accreditation or Certification

Developing and Communicating a Compelling Corrective Action Plan in Response to AAALAC International's Findings

The value and integrity of the AAALAC International accreditation program require that participating institutions integrate the principles outlined in AAALAC International's TPSs and other pertinent guidance into their daily operations to develop and sustain a durable and resilient animal care and use program. Organizations that are unable to incorporate these goals ultimately encounter problems that result in COA's issuance of a status of less than continued full accreditation. For programs failing to meet standards in multiple areas or with a chronic history of not meeting their obligations and the corrective actions they have previously proffered, the COA may elect to revoke accreditation. This action requires the program to critically reevaluate its interest in accreditation and to adopt a new approach, if not a new organizational culture, team, and commitment of resources, if it intends to reenter the accreditation program. Organizations that attempt to use triennial accreditation visits to intensify their programs for evaluation during peak performance, and then allow their programs to languish between visits, quickly realize that such an approach is unsustainable.

Institutions are required to respond when the AAALAC International COA identifies a mandatory finding. The institutional responses (IRs) should reply directly to the COA's findings with a detailed, definitive, comprehensive, and cogent account of the actions taken. The IRs should not only communicate the specific corrective actions the institution has taken (or is preparing to take) for the specific examples cited in the letter from COA, but also encompass, by extension, any other potential manifestations of the circumstances identified. As a general guideline, the IR should cover the key parameters important in journalism: who, what, why, when, where, and how. Specifically, who is the individual or group responsible and authorized to take corrective action, and who is the target audience of the action if it involves people? What are the specific corrective actions and the relevant outcome measure for success? Why, in the context of the program or institution, should COA believe that the action will be effective? When will the corrective action be implemented or completed? Where in the program does the corrective action apply? And, how do these steps in aggregate create an enduring solution? It is very common that additional documents are requested, along with the IR, to demonstrate facets of the corrective action and validate the extent of progress achieved.

In many cases, a single well-written IR resolves all the COA's concerns allowing restoration of full accreditation. However, the AAALAC International COA realizes that some mandatory findings are very difficult to address due to programmatic complexity and extensive resources that may be required; thus, the resolution of a single issue may require multiple IRs. In some cases involving extensive or complicated corrective actions, the COA may elect to revisit the program to ensure that quality has been restored.

Drop-In Visits from AAALAC International

Normally, AAALAC International site visits are conducted on 3-year intervals. While quite rare, drop-in visits "for cause" or "not for cause" may occur between the routinely scheduled triennial visits. In both cases, relatively short notice (one to a few days) of the pending drop-in visit is provided. The scope of the drop-in visit is typically very focused; more rarely, the entire animal care and use program may be reviewed during a drop-in visit. Costs for the drop-in visit are absorbed by AAALAC International.

Occasionally, a site visit team will conduct a drop-in visit to an institution in a geographic area where a routinely scheduled visit is occurring at another organization. These visits may be prompted by significant programmatic changes that have occurred at the institution subsequent to the routine site visit, such as critical organizational changes or a reversal of a commitment made to COA (e.g., reopening an animal housing area that had been closed in response to a site visit observation). Such not-for-cause visits provide the COA an update as to the status or functioning of the institution. Drop-in visits for cause may be conducted following a meeting of the COA, during which the COA has determined that an issue identified during the regularly scheduled site visit was of sufficient seriousness that immediate follow-up by the COA, in the form of an additional on-site assessment, was necessary. Occasionally, a drop-in visit for cause may be initiated by the executive office following receipt of a verifiable allegation related to the animal care and use program at an accredited institution.

Reports of observations made during either type of drop-in visit are discussed by the COA. The observations made during a not-for-cause drop-in generally result in no action being taken by the COA. However, the report may prompt the COA to require additional information regarding the matter from the institution, or the COA may elect to schedule a revisit prior to the typical 3-year interval. Responses to observations made during a drop-in for-cause visit range from no change to the accreditation status of the institution to less than full accreditation pending correction of the issue.

Other Actions Required for AAALAC International Accreditation

A requirement for continued accreditation is the submission of an annual report that provides updates on basic information about an accredited program. A form is made available online in mid-December; however, there are no specific due dates for submitting an annual report. An institution may choose from a variety of reporting periods (e.g., university fiscal year, calendar year, federal government fiscal year, or government oversight body reporting period) one that is best suited to their needs, as long as the period covered is continuous with previous reports to eliminate any gaps.

The information provided in the annual reports also includes notification of significant compliance issues, such as protocol violations, animal use not approved by IACUC, and protocol suspensions; revisions to facility size, location, name, or other administrative data; and changes in key personnel or other aspects of the animal care and use program. AAALAC International also requests the number of animals used by species during the year to gauge the level of research animal activity and assist in the selection of the next site visit team.

AAALAC International also expects accredited units to promptly report adverse events, such as unexpected animal deaths; natural disasters; significant animal rights activities; inappropriate euthanasia techniques and/or failure to confirm euthanasia; allegations, complaints, or reports regarding animal welfare concerns; inadequate veterinary care; and outcomes of regulatory investigations (e.g., Office of Laboratory Animal Welfare [OLAW], CCAC, UK Home Office). In some cases, the correspondence addressed to the regulatory entities may be sufficient for AAALAC International's purposes, and AAALAC International can simply be copied to ensure a prompt update [AAALAC International 2016o]). AAALAC International requests information on any changes in unit contact person (name, including degree, title, address, phone and fax numbers, and e-mail) or in facility size, location, or name for a site visit scheduled prior to the next annual report.

Measures to Ensure Success in Continuing CCAC Certification

Institutions wishing to maintain CCAC certification must ensure that (CCAC 2016z)

- They are compliant with provincial legislation and regulations
- Their animal care committee remains active and functional, meeting at least twice every year, visiting all animal facilities at least once every year, and fulfilling all the responsibilities described in the CCAC policy statement on terms of reference for animal care committees, including postapproval monitoring of animal ethics and care in science

- Their veterinary and animal care services continue to meet institutional needs and CCAC standards
- Their training, occupational health and safety, and crisis management programs are relevant, complete, up to date, and in line with CCAC guidance
- Their facilities (if required) meet institutional needs and CCAC standards
- They are in good standing with program participation fees

Implementation Report: Response of the Institution to the CCAC Recommendations

The senior administrator in charge of the animal ethics and care program at the institution must submit one or more implementation reports responding to recommendations in the CCAC assessment report. The implementation reports must comprehensively address the recommendations to ensure certification. As described above, the timeline for response reflects the type of recommendation (major, serious, and regular), with major recommendations requiring an immediate response (CCAC 2006, 2016x). Institutions are typically given 3 months to respond to a serious recommendation and 6 months to respond to a regular recommendation.

The implementation reports are reviewed by the ADA, the panel members who participated in the assessment, and the Assessment and Certification Committee. Institutions may be asked to submit additional information to clarify or complete their implementation report, or to provide an update on work in progress at the time of the original response. Special visits may be conducted if necessary. Unsatisfactory IRs or the absence of a response can lead to a probationary certification, and eventually to certificate removal (CCAC 2016x).

Institutions are expected to consider the following points when preparing an implementation report (CCAC 2016x):

1. Implementation reports should be reviewed and sent by the senior administrator responsible for the entire animal care and use program, following consultation with the animal care committee, the animal care services, and any other interested parties. This is to ensure that the report reflects the position of the institution as a whole, and to demonstrate the institution's commitment to its animal care and use program.
2. The CCAC cannot accept an implementation report without evidence that either all recommendations have been fully implemented or a sound process with timelines has been established to appropriately address them.
3. If a recommendation is to be implemented gradually (e.g., facility renovation or construction, or writing or reviewing SOPs), institutions are expected to provide a timeline with dates for the various stages of the process and include supporting documentation.
4. To assist the assessment panel and Assessment and Certification Committee members in reviewing the implementation report, the recommendation should be quoted preceding the response. Answers should be specific to each part of all recommendations.

The CCAC works closely with institutions to resolve difficulties in implementing CCAC recommendations. The CCAC Assessment and Certification Committee may recommend at any time that a special visit of an institution be undertaken to evaluate the implementation of recommendations. Special visits, conducted by an ADA without a panel (CCAC 2016r), may also follow a request of the institution. The findings of a special visit may be used by the Assessment and Certification Committee to modify the status of an institution (CCAC 2010). An institution's CCAC Certificate of GAP is removed if (CCAC 2016z)

1. The institution does not take immediate, appropriate action after being issued a major recommendation during a CCAC visit
2. After being assigned a probationary certificate, the institution fails to respond to the Assessment and Certification Committee's satisfaction to serious recommendations contained in CCAC reports, despite being given a specified period in which to do so
3. The institution is no longer "in good standing" with the CCAC, that is, fails to remit its annual program participation fee

Benefits of Accreditation and Certification

Both AAALAC International accreditation and CCAC certification claim comparable and significant benefits. The processes used in accreditation and certification are well described, readily accessible, efficient, and well coordinated and integrated. The CCAC develops standards (standards component), provides support to learn about them and how to apply them (education and three Rs components), and supports their implementation (assessment and three Rs components). AAALAC International primarily has adopted standards developed by other experts and authorities and issues position statements, clarifications, and FAQs to inform the community of how to interpret and use the standards successfully in accreditation. The CCAC ethical review system is designed to integrate the needs of scientists, animals, and the community at the local level, and to set standards for animal ethics and care at the national level. The AAALAC International assessment and accreditation process also incorporates the needs of the scientific community and animals, and the interests of diverse public communities globally by ensuring that all national and local laws governing animal research are met and by ensuring the appropriate composition of the site visit teams.

Both AAALAC International accreditation and CCAC certification require institutions to first perform a comprehensive self-evaluation that helps catalyze institutional introspection, critical analysis, teamwork, and consensus building. Subsequently, a team of peers provides an in-depth, confidential, on-site evaluation of the institution's animal care and use program. This independent, rigorous peer review provides a collaborative and potentially formative approach that supports the institution's program in meeting all applicable standards.

Accreditation or certification also engages scientists, veterinarians, managers, and administrators in an independent, rigorous assessment of their institution's animal care and use program—an assessment that ultimately results in improved animal welfare, cooperation in programmatic planning to meet the needs of an evolving research mission, and better research practices and outcomes.

Accreditation or certification fosters appropriate care and husbandry programs, which encourages species-specific expression of physiological, behavioral, and psychological needs, and improves the reliability of scientific results.

Accreditation or certification confirms that an organization is compliant with relevant legislation and regulations through the review of regulatory documents and through the review of IACUC or animal care committee records and discussions with committee members. In the case of the CCAC, the certification also allows institutions to receive research funding from the CIHR, NSERC, National Institutes of Health (NIH), and other funders, and access government contracts in Canada. AAALAC International accreditation confers Category I status on an institution under the Public Health Service (PHS) Policy on Humane Care and Use of Laboratory Animals (revised in 2015), which reduces the reporting requirements for organizations that receive PHS funding (Public Health Service 2016). Because it promotes animal well-being and ensures high-quality animal care, accreditation or certification complements other quality assurance programs, such as the ISO or the good laboratory practice standards.

AAALAC International accreditation and CCAC certification are recognized internationally for acknowledging high-quality programs. In the scientific community, accreditation or certification demonstrates that an institution is committed to establishing, achieving, and maintaining high standards for animal welfare in science. Moreover, both the conditions of animal care and use and an assurance of sound ethical review of the research plan and conduct are important factors considered during the review of manuscripts for publication in internationally recognized scientific journals; accreditation or certification broadly responds to these concerns.

In today's world, research institutions increasingly partner or contract with other research entities at the national and international levels to pursue multidisciplinary collaborative research and enhance the efficiency of resource utilization. Accreditation or certification promotes harmonization of care within and between institutions and can be used as a way to gauge the quality of a particular program, and provide assurance to diverse stakeholders. Institutions, companies, and organizations are also held to very high levels of accountability by their own constituents and the general public. Thus, a program achieving accreditation or certification serves as a beacon of competence that may aid other institutions seeking to improve their animal care and use program. In addition, voluntary participation in an accreditation or certification program demonstrates to the public that the institution is dedicated to achieving the highest standards of animal care and use and the ethical oversight of animal-based research.

Both AAALAC International accreditation and the CCAC Assessment and Certification Program are collaborative, peer processes that stimulate the pursuit and exchange of information and knowledge and encourage the growth of professionalism, expertise, cooperation, and teamwork within the institution's animal care and use team. This invariably leads to continuous improvement in the animal care and use program. Earning and maintaining certification or accreditation requires institutions to remain current and to incorporate contemporary practices that are proven scientifically sound and are pertinent and appropriate to the organization's research mission. Accreditation or certification provides an optimal means of illustrating a true commitment to humane animal care and use benefitting both animal well-being and quality science.

ACKNOWLEDGMENTS

Sylvie Cloutier is thankful to Dr. G. Griffin, Dr. M. Baar, P. Belleau, and E. Verlinden for their input and assistance with the preparation of the chapter section on the CCAC.

REFERENCES

AAALAC International. 2016a. Bylaws. http://www.aaalac.org/accreditation/AAALAC_Bylaws_Approved_06-04-15.pdf (cited August 2, 2016).

AAALAC International. 2016b. *AAALAC International Connection*. http://www.aaalac.org/publications/connection/winter_spring_2001.pdf (cited August 3, 2016).

AAALAC International. 2016c. AAALAC International's reference resources. http://www.aaalac.org/accreditation/resources.cfm (cited August 3, 2016).

AAALAC International. 2016d. Rules of accreditation. http://www.aaalac.org/accreditation/rules.cfm (cited August 3, 2016).

AAALAC International. 2016e. Frequently asked questions. https://www.aaalac.org/accreditation/faq_landing.cfm (cited August 3, 2016).

AAALAC International. 2016f. "Should" vs. "must." https://aaalac.org/accreditation/faq_landing.cfm#G5 (cited August 3, 2016).

AAALAC International. 2016g. Apply to become an AAALAC ad hoc consultant. http://www.aaalac.org/about/adhoc_app.cfm (cited August 3, 2016).

AAALAC International. 2016h. Application for accreditation. http://www.aaalac.org/accreditation/Application_for_Accreditation_t_08_11.pdf (cited August 3, 2016).

AAALAC International. 2016i. Fees and deadlines. http://www.aaalac.org/accreditation/fees.cfm (cited August 3, 2016).

AAALAC International. 2016j. Download a copy of the AAALAC program description. http://www.aaalac.org/programdesc/index.cfm (cited August 3, 2016).

AAALAC International. 2016k. AAALAC i-brief. www.aaalac.org/publications/ibrief.cfm (cited August 3, 2016).

AAALAC International. 2016l. AAALAC International's application of performance standards. http://www.aaalac.org/accreditation/faq_landing.cfm#performance (cited August 3, 2016).

AAALAC International. 2016m. Categories of accreditation. http://www.aaalac.org/accreditation/categories.cfm (cited February 12, 2017).

AAALAC International. 2016n. Assessement: The Program Status Evaluation (PSE) program. http://www.aaalac.org/programstatus/index.cfm (cited August 3, 2016).

AAALAC International. 2016o. Annual reports. http://aaalac.org/accreditation/faq_landing.cfm#H1 (cited August 3, 2016).

CCAC (Canadian Council on Animal Care). 1999. CCAC assessment panel policy. Ottawa: CCAC.

CCAC (Canadian Council on Animal Care). 2006. *CFHS-FSCAA Manual for Community Representatives*. Ottawa: CCAC.

CCAC (Canadian Council on Animal Care). 2008. CCAC policy statement for senior administrators responsible for animal care and use programs. Ottawa: CCAC.

CCAC (Canadian Council on Animal Care). 2010. CCAC policy on certification of animal care and use program. Ottawa: CCAC.

CCAC (Canadian Council on Animal Care). 2013. Advancing the ethical use and welfare of animals in Canadian science: Annual report 2012–2013. Ottawa: CCAC.

CCAC (Canadian Council on Animal Care). 2014. Hearing what Canadians have to say: Annual report 2013–2014. Ottawa: CCAC.

CCAC (Canadian Council on Animal Care). 2015. Working together to enhance animal ethics and care in science. Strategic plan 2015–2020. Ottawa: CCAC.

CCAC (Canadian Council on Animal Care). 2016a. History. http://www.ccac.ca/en_/about/history (cited February 23, 2016).

CCAC (Canadian Council on Animal Care). 2016b. Scope of CCAC programs. http://www.ccac.ca/Documents/Standards/Policies/The_Scope_of_CCAC_Programs.pdf (cited February 23, 2016).

CCAC (Canadian Council on Animal Care). 2016c. National recognition. http://www.ccac.ca/en_/standards/recognition/in-canada (cited February 23, 2016).

CCAC (Canadian Council on Animal Care). 2016d. Governance. http://www.ccac.ca/en_/about/governance (cited February 23, 2016).

CCAC (Canadian Council on Animal Care). 2016e. Standing committees. http://www.ccac.ca/en_/about/governance/standing-committees (cited February 23, 2016).

CCAC (Canadian Council on Animal Care). 2016f. Programs and operations. http://www.ccac.ca/en_/about/programs-and-operations (cited February 23, 2016).

CCAC (Canadian Council on Animal Care). 2016g. Secretariat structure. http://www.ccac.ca/Documents/About/CCAC_Secretariat-structure.pdf (cited February 23, 2016).

CCAC (Canadian Council on Animal Care). 2016h. About the CCAC. http://www.ccac.ca/en_/about (cited February 23, 2016).

CCAC (Canadian Council on Animal Care). 2016i. Where are these animals studies? http://www.ccac.ca/en_/facts-and-figures/animals-in-canadian-science/where (cited August 15, 2016).

CCAC (Canadian Council on Animal Care). 2016j. Standards and policies. http://www.ccac.ca/en_/standards (cited February 23, 2016).

CCAC (Canadian Council on Animal Care). 2016k. Guidelines. http://www.ccac.ca/en_/standards/guidelines (cited February 23, 2016).

CCAC (Canadian Council on Animal Care). 2016l. Development of peer-based guidelines. http://www.ccac.ca/en_/standards/guidelines/development (cited February 23, 2016).

CCAC (Canadian Council on Animal Care). 2016m. Policy statements. http://www.ccac.ca/en_/standards/policies (cited February 23, 2016).

CCAC (Canadian Council on Animal Care). 2016n. Other CCAC-recognized standards. http://www.ccac.ca/en_/standards/other_standards (cited February 23, 2016).

CCAC (Canadian Council on Animal Care). 2016o. Three Rs. 4. http://www.ccac.ca/en_/assessment/a_panels (cited February 23, 2016).

CCAC (Canadian Council on Animal Care). 2016p. Assessment panels. http://www.ccac.ca/en_/assessment/a_panels (cited February 23, 2016).

CCAC (Canadian Council on Animal Care). 2016q. Asssessment visits. http://www.ccac.ca/en_/assessment/visits (cited February 23, 2016).

CCAC (Canadian Council on Animal Care). 2016r. *Advancing the Ethical Use and Welfare of Animals*. Annual Report 2012–2013. Ottawa: CCAC.

CCAC (Canadian Council on Animal Care). 2016s. Before the visit. http://www.ccac.ca/en_/assessment/visits/preparing (cited February 23, 2016).

CCAC (Canadian Council on Animal Care). 2016t. Agenda of visits. http://www.ccac.ca/en_/assessment/visits/preparing/agenda (cited February 23, 2016).

CCAC (Canadian Council on Animal Care). 2016u. Animal care and use program review form. http://www.ccac.ca/en_/assessment/visits/preparing/prf (cited February 23, 2016).

CCAC (Canadian Council on Animal Care). 2016v. CCAC assessment panel member—A crucial role. http://www.ccac.ca/en_/assessment/a_panels/crucial-role (cited February 23, 2016).

CCAC (Canadian Council on Animal Care). 2016w. During the visit. http://www.ccac.ca/en_/assessment/visits/during (cited February 23, 2016).
CCAC (Canadian Council on Animal Care). 2016x. After the visit. http://www.ccac.ca/en_/assessment/visits/after (cited February 23, 2016).
CCAC (Canadian Council on Animal Care). 2016y. Recommendations made in CCAC assessment reports. http://www.ccac.ca/Documents/Standards/Policies/Recommendations_made_in_CCAC_assessment_reports.pdf (cited February 23, 2016).
CCAC (Canadian Council on Animal Care). 2016z. CCAC policy statement on the certification of animal ethics and care in science programs. Ottawa: CCAC.
Council of Europe. 2016. Appendix A of the European Convention for the Protection of Vertebrate Animals Used for Experimental and Other Scientific Purposes. http://www.aaalac.org/about/AppA-ETS123.pdf (cited August 3, 2016).
FASS (Federation of Animal Science Societies). 2010. *Guide for the Care and Use of Agricultural Animals in Research and Teaching*. 3rd ed. Champaign, IL: FASS.
ILAR (Institute for Laboratory Animal Research), Division of Earth and Life Sciences, National Research Council. 2011. *Guide for the Care and Use of Laboratory Animals*. Washington, DC: National Academies Press.
Public Health Service, National Institutes of Health Office of Extramural Research. 2016. Office of Laboratory Animal Welfare—Public Health Service policy on humane care and use of laboratory animals. http://grants.nih.gov/grants/olaw/references/phspol.htm (cited August 3, 2016).
U.S. Public Health Service. 1963. *Guide for Laboratory Animal Facilities and Care*. Washington, DC: U.S. Public Health Service.
US TAG to ISO/TC 176. 2015. 5. https://www.iso.org/files/live/sites/isoorg/files/archive/pdf/en/qmp_2012.pdf (cited August 3, 2016).
Wikipedia. 2016. ISO 9000. https://en.wikipedia.org/wiki/ISO_9000 (cited August 3, 2016).

10

Facilitating the Research Process: Limiting Regulatory Burden and Leveraging Performance Standards

Joseph D. Thulin, Valerie K. Bergdall, and John F. Bradfield

CONTENTS

Introduction

Animal care and use program administrators and managers are responsible for both promoting animal well-being and facilitating the research. However, since animal welfare laws and regulations primarily articulate edicts intended to mitigate harm to animals, it can be difficult to see how an institution's regulatory compliance activities can both protect animal subjects and serve the research enterprise. While it can be a challenge to balance these seemingly competing goals, one should not conclude that ensuring high program achievement in terms of compliance precludes concomitant highly productive research.

In this chapter, we explore how the expanding environment of compliance—driven by both legal requirements and the associated institutional regulatory processes—drives the development of programs for oversight of animal care and use and impacts the work of scientists using animals. We describe some opportunities and strategies that program administrators and managers can use to facilitate the research enterprise without compromising compliance demands. Two general approaches are presented: (1) alleviating unnecessary regulatory burden, which conserves institutional and researcher resources, and (2) selecting appropriate standards and methods for assessing outcomes, which promotes program flexibility. It is worth noting at the outset that whatever the approach, it is crucial for the program manager or administrator to engage with the institutional animal care and use committee (IACUC) and other institutional oversight entities, such as the responsible institutional official (IO) and attending veterinarian (AV), as well as other program stakeholders. (Note: The term *institutional animal care and use committee*, or IACUC, is used in a generic sense to denote the institutional body, regardless of name, charged with oversight and evaluation of the animal care and use program.)

Impact of Regulatory Compliance Burden

Regulatory Burden

National and regional laws and regulations pertaining to the use and care of animals in research vary considerably with respect to the compliance responsibilities required of the regulated institutions. For example, in the United States, the federal regulatory system relies heavily on self-enforcement at the institutional level, while in countries of the European Union, there is greater reliance on direct enforcement of compliance by the state. Regardless of the explicit local regulatory environment, however, each well-managed institutional program will not only design animal care and use processes that conform to regulatory standards, but also exercise some degree of internal oversight to ensure those standards are being met. Thus, regulatory burden can be considered the total effort expended and concomitant cost incurred to achieve and maintain compliance. With regard to animal research, regulatory burden encompasses, but is not limited to, the activities and expenses of the IACUC, committee administrative support, government interface (hosting inspections, preparing and filing required applications and reports, etc.), and providing animal care procedures and facilities that meet standards. Likewise, the researchers themselves devote considerable effort to fulfilling compliance requirements. These include securing and maintaining licensures, preparing animal use protocols for IACUC review, maintaining IACUC approvals, hosting laboratory audits, fulfilling program orientation and training requirements, and similar activities. Quite obviously, satisfying compliance requirements comes at considerable cost to the institution.

The scientific community clearly accepts, as one of many social obligations, the regulatory burden necessary for safeguarding proper treatment and care of research animals. Without question, institutions recognize that maintaining their privilege to conduct research with animals depends on their ability to ensure compliance with regulatory standards. But as necessary as compliance may be, when regulatory burden becomes excessive, it can impinge upon research productivity while failing to yield any further enhancements to animal welfare. The impact of expanding regulatory requirements on research in the United States has been described (Federal Demonstration Partnership 2014; National Science Board 2014; Federation of American Societies for Experimental Biology 2015; Committee on Federal Research Regulations and Reporting Requirements et al. 2016). For example, in a recent survey of principal investigators of federally sponsored research projects in the United States, it was reported that they spend more than 40% of their time on administrative tasks associated with compliance (Federal Demonstration Partnership 2014). Global data on regulatory burden are not available, but it is reasonable to assume it has significant impact on research output in many countries. So the total cost of regulatory burden is limited not only to resource outlay, but also to lost productivity. As such, program managers should implement processes that effectively ensure regulatory compliance while minimizing regulatory burden so that research output is protected and unnecessary costs are contained.

Self-Imposed Compliance Burden

Haywood and Greene (2008) considered three sources of regulatory burden: explicit legal and regulatory requirements, additional interpretive guidance given by regulatory agencies, and internal, self-imposed burden. For the purposes of this discussion, however, we consider just two: external requirements imposed by law, regulation, or regulatory agency, and internal, self-imposed requirements not explicitly required by law, regulation, or regulatory guidance. In other words, we differentiate between compulsory and elective standards and processes. This distinction is important because program administrators and managers have relatively little opportunity to alleviate the burden stemming from compulsory standards, while they can have a significant influence on diminishing burden originating from elective institutional requirements.

Examples of self-imposed burdens adopted by institutions and opinions regarding how these burdens arise have been described (Haywood and Greene 2008; Plante and James 2008; Thulin et al. 2014;

Pritt et al. 2016). Those facets of the program commonly augmented by elective practices include, but are not limited to, animal use protocol review, institutional policies, investigator training, documentation, and monitoring of laboratories. Maintaining accreditation by AAALAC International is another common example of voluntary program enhancement. All these result in burden beyond that stemming from the minimum external requirements, and most probably originate from both a desire for excellence and aversion to the risk of noncompliance.

Internal, elective standards are not always without merit—many may be worth the additional effort they create. This is especially true when they address gaps in the regulatory standards or promote program quality in ways not specifically addressed by the regulations. For example, while participation in the AAALAC International accreditation program is voluntary and requires considerable effort, many institutions find that the value of the periodic peer reviews outweighs the costs. Nonetheless, it is likely the case that some institutions have implemented various elective standards and processes without ever thoroughly vetting them for benefit and cost. Consequently, the compliance programs at many of these institutions may unnecessarily stymie research activities without enhancing animal well-being or even reducing institutional risk.

Minimizing Regulatory Compliance Burden

Regulatory agencies view compliance oversight as a means to ensure rules are being followed, and institutions often view this as a process to help ensure they remain compliant and therefore avoid "trouble" with regulatory agencies or the public. Typically, the research community views compliance as a hurdle rather than a facilitative process. *The challenge for leadership in the animal care and use program at an institution, therefore, is to promote a culture in which compliance truly does facilitate the research process while at the same time protecting the institution from noncompliance events that may threaten its privilege to conduct animal research and secure associated grant funding* (Bayne and Garnett 2008). To promote such a culture, it is imperative that stakeholders understand the goals of compliance processes, and that the processes are clear and without unnecessary and burdensome steps. In addition, flexibility in achieving compliance goals should be maintained when feasible, and compliance outcomes assessed routinely to improve the process as needed over time. Imposing new processes without input from stakeholders, that result in additional work, are inflexible, and do not have obvious benefits will not be embraced. This can lead to erosion of the relationship between the research community and institutional compliance efforts.

Perhaps the most important way compliance programs can become more facilitative of the research is to minimize the regulatory burden placed on the researchers. A basic approach to reducing regulatory burden involves determining first whether a compliance requirement or process is mandated or elective, then eliminating elective practices based on benefit–cost analysis, and finally, determining if remaining mandated or elective practices can be done more efficiently (Haywood and Greene 2008). With thorough knowledge of the regulatory standards, establishing whether a process is mandatory should be relatively straightforward. But weighing the benefits and costs of elective procedures will be more complex. A rigorous assessment will require questions such as

- What are the intended benefits of the policy, procedure, or requirement?
- What entities, for example, animals, researchers, IACUC members, or institution, benefit from the policy, procedure, or requirement?
- Are the intended benefits being realized, and can the achievement be measured?
- What are the costs, for example, staffing, researcher time (correlates to time that cannot be devoted to research), space, or equipment, of the policy, procedure, or requirement?
- Can the costs be measured?

The answers to these questions will provide the data or measures from which objective decisions can be made about the value of policies and procedures and whether they should be retained or retired.

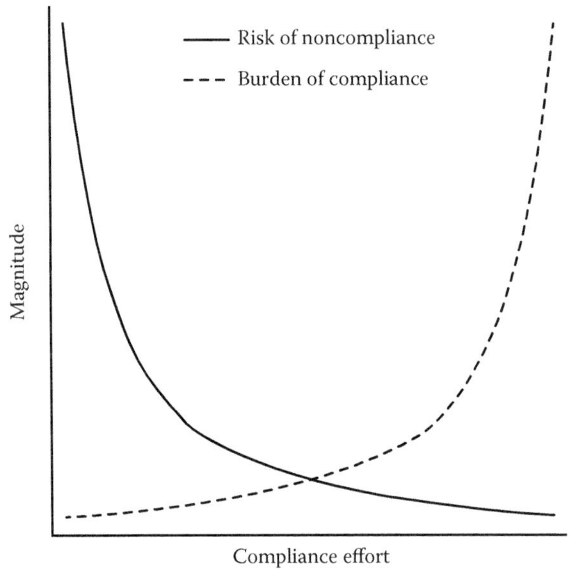

FIGURE 10.1 Relationship of overall compliance effort to risk of noncompliance and compliance burden. With little or no compliance effort, the risk of noncompliance is high, while the compliance burden is low. As the effort increases, risk decreases while burden increases. However, there likely is a point, perhaps at the theoretical intersection of the two curves, when the amount of risk reduction to be gained, combined with the amount of added burden, does not support increasing the effort. (Reprinted from Thulin, J.D. et al., *FASEB J.*, 28, 3297–3300, 2014. With permission.)

Finding more efficient ways to perform the remaining compliance procedures can be a significant challenge. Adopting a standardized approach to process improvement can be helpful in this regard. One such approach is "Lean thinking," a method by which the value of processes (or steps within processes) is assessed from the perspective of defined customers, for example, the researchers or the animals, and waste within processes identified, thus facilitating process improvements that achieve desired outcomes (Kim et al. 2009). *With respect to compliance, it is important to understand that more compliance effort does not necessarily lead to "better" compliance outcomes; rather, there is a point at which added effort likely results in diminishing returns and wasted resources.* (Figure 10.1 illustrates the effect of compliance effort on risk noncompliance and regulatory burden.) As we will point out later, inspections are not the best way to achieve desired outcomes. With this in mind, program managers and oversight bodies should strive for that balance between too little and too much compliance effort.

Assessing Outcomes

It is imperative to "measure" the effectiveness of the activities associated with animal care and research procedures to ensure animal well-being and promote scientific integrity. It is not sufficient to simply design a program that ought to work well. An institution should also have some means to verify that animal activities have appropriate outcomes. Yet, institutions should not fall prey to the notion that measuring outcomes alone achieves quality. Program quality is achieved by implementing well-designed and effective procedures, while outcome measures simply provide the information needed to assess effectiveness and drive improvement where needed. Developing appropriate outcome metrics requires a thorough understanding of the different types of standards that apply to the animal care and use program.

Engineering Standards

Engineering standards are often described in the regulations and specify both the characteristics and technical details required in order to meet the standard. An engineering standard not only specifies what the standard or outcome must be, but also how it is achieved. They dictate the methods to be used in order to achieve the standard or outcome. The benefit of such an approach is that there is uniformity of methods across the animal program and among animal programs. It is easier to assess compliance with engineering standards because they require fixed methods for achieving outcomes, but they do not allow flexibility to adopt methods that may be more effective for a particular program. An engineering standard leaves little or no room for interpretation about what must be done and, as a result, avoids the onus of careful and accurate interpretation that is required when implementing performance standards.

An example of a commonly used engineering standard is the space recommendations in the eighth edition of the *Guide for the Care and Use of Laboratory Animals* (*Guide*) specified in the tables 3.2 through 3.6 (NRC 2011). These tables provide specific floor space metrics required to meet *Guide* standards, and it is a straightforward process to determine whether these standards are achieved. A process to establish outcome assessment is unnecessary.

Performance Standards

In many ways, performance standards are considered the opposite of engineering standards in that they often have qualitative outcomes, whereas engineering standards usually have quantitative outcomes. "Performance standards" or the "performance-based approach" is a concept that has been part of animal research management for many years. The term *performance approach* became a prominent feature of the seventh edition of the *Guide* (NRC 1996). However, in previous editions of the *Guide*, as early as the fourth edition (NRC 1972), the term *professional judgment* was used in the context of how the recommendations are to be used and interpreted. Professional judgment is a key element of performance standards, so the fundamental concepts of the performance-based approach are not new. It has been said that professional judgment is the collective judgment of a profession, and not simply the judgment of one professional. This is an important distinction as one develops performance standards.

The key advantage of the performance-based approach is the programmatic latitude it affords. A performance standard identifies a specific goal or outcome that is to be achieved, but does not specify *how* the goal is achieved. The intent of such a standard is to allow latitude about the best, most effective manner to meet the standard given the unique aspects of each program. A variety of methods might all result in an outcome that meets the specified goal or standard. The performance standard approach is a key element of the eighth edition of the *Guide*. The *Guide* provides much detail about what constitutes appropriate standards (goals) of animal care and use. It is very clear on what *must*, or *should*, be done, but it is virtually silent about how the standards are to be achieved. This was done intentionally to allow flexibility in methods, which is considered to be a positive, less prescriptive approach. However, with this flexibility comes greater institutional responsibility to ensure that the standards, or goals, are correctly and accurately interpreted. When using performance standards in the animal program, it is critical to neither under- nor overinterpret what a particular standard or goal is.

When considering performance-based standards, three basic steps are required to ensure a standard is clearly identified and achievable:

1. Identify a specific and precise definition of the standard.
2. Establish the assessment criteria to determine that the standard is achieved.
3. Develop and implement methods for ongoing evaluation.

An example of commonly used performance standards is the criteria described in the *Guide* regarding adequate cage space. In contrast to the *Guide's* tables 3.2 through 3.6 that specify engineering metrics for cage space, there is much language in the *Guide* that describes the qualitative, performance-based aspects of the provision of adequate cage space (e.g., the need to accommodate postural adjustments, the importance of cage volume, the provision of environmental enrichment, and space considerations for social grouping and breeding). The *Guide* highlights the need for professional judgment when assessing the performance-based criteria for adequate cage space. Because there are a number of ways to achieve these performance standards, it is important that institutions develop specific criteria to assess whether the performance standards are met: (1) precisely defining adequate cage space characteristics (relying on

Guide parameters), (2) developing assessment criteria to benchmark adequacy of space, and (3) implementing methods for ongoing assessments of cage space provided.

This kind of approach embodies the performance-based concept of operational latitude while still satisfying the requirements specified in the *Guide*. Of course, the definition of the standard and criteria for monitoring and methods of assessment for an entire animal program will be much more complex, but the above example illustrates the utility of the performance-based approach. When considering this approach, however, one should be mindful of the fact that procedures deemed acceptable for achieving a performance standard in one program may not be the most effective in another program. *In addition, it is important to note that implementation of performance standards may require more effort than a corresponding engineering standard.*

Practice Standards

In addition to describing performance and engineering standards, the *Guide* also refers to "practice standards," a term that has recently been used to denote another aspect of benchmarking practices. *Practice standard* is a term that designates a time-proven, effective practice that is the result of collective professional judgment that usually evolves from practical experience and information in peer-reviewed scientific literature and textbooks. An example of a practice standard is the common practice of a cage-change interval once every 2 weeks for ventilated mouse cages. This is a widely accepted practice that has arisen from a combination of scientific literature, cage manufacturer information, and years of practical experience across a variety of institutions and research programs.

Because the field of laboratory animal science is continuously evolving, widespread practices emerge and become commonly accepted by the profession as sound, effective, and beneficial. Practice standards are usually improved as a natural result of scientific progress, improvements in technology, and advances in management rather than by regulatory requirement.

Institutions may implement practice standards that have evolved over time because the practices have proven to be effective in the context of their animal care and use program, and are based on the specific needs of the animals and experimental requirements. It is important to validate practice standards to ensure they are effective and promote animal well-being and good science. Furthermore, this validation will ensure that these practice standards are not simply based on tradition and convenience. Validation of institution-specific practice standards can be done in much the same way that performance standards are assessed, although practice standards, having been time proven, typically require less rigorous analysis of the outcomes.

Selection of Appropriate Standards

Most animal care and use programs employ a combination of engineering, performance, and practice standards. Each type of standard has its benefit, and the choice to use one versus another is best made after careful consideration of the regulations and guidelines required by external agencies, the scientific needs of the program, and the overall management and organization of the animal care and use program. Engineering standards typically result from regulatory requirements. Performance and practice standards are often employed to allow appropriate flexibility, often required in diverse, dynamic, and complex research programs.

For engineering standards, the institution must simply ensure that the methods and standards imposed by the regulatory statutes are fulfilled as directed, a relatively straightforward process. Other guidelines, such as the *Guide*, expect programs to implement performance and practice standards that meet the specified outcomes. Finally, there may be institutional guidelines and standards added to those required by external agencies, and these can be engineering, performance, or practice standards. The latitude afforded by implementing performance-based standards comes with choosing the best means to achieve the goal. However, once the most appropriate methods are established, adherence to them is important to ensure that standards are achieved.

The implementation of performance-based standards requires a well-reasoned, thoughtful approach to achieve high-quality animal care and use, while at the same time ensuring that performance-based methods are not overly cumbersome and onerous. It is imperative that the methods and practices employed promote animal welfare and high-quality science, and are straightforward so that meeting the standards is an inherent part of daily animal care and use. Conversely, overreliance on a compliance monitoring program to "enforce" the standards is usually less effective, inefficient, and tedious.

Evaluation of Outcomes

Once appropriate outcome metrics have been established (i.e., what to measure), the next step is to determine how the outcomes will be measured. The process for measuring how well an institution meets animal program standards requires routine, ongoing commitment and attention to detail. It can also require much time and effort. However, the benefit to a well-implemented and efficient program of monitoring provides valuable information on animal well-being, animal use activities, and institutional commitment. Effective monitoring will provide information that can be used to validate strengths and address weaknesses while avoiding unnecessary, burdensome processes that have little or no impact on animal care and use. Although monitoring is a necessary part of attaining quality, emphasis should be on implementing procedures that effectively achieve the outcome.

Personnel in each area of the animal program will likely have a role in the evaluation of outcome metrics. However, the IACUC plays a central role and has the responsibility to evaluate all aspects of the program. The IACUC may rely primarily on its periodic reviews of the program and facilities for evaluation or, in larger, more diverse and complex programs, delegate monitoring to others. In recent years in the United States, employing dedicated personnel to monitor the program on behalf of the IACUC has become common (see below), although there is no requirement that such personnel or formal monitoring program be implemented.

Alternatively, the IACUC can rely on other program personnel to help with monitoring activities. Investigators play a key role in monitoring animal use activities. Ongoing monitoring of animal activities throughout the experiment is part of the scientific process, and this level of monitoring by investigators also aids the IACUC in ensuring compliance with regulations and guidelines. Many institutions have advisory committees comprised of scientists to help provide information and feedback to the IACUC and animal program personnel. The IACUC should also rely on veterinarians to monitor aspects of animal health and well-being, as well as some aspects of animal use, such as surgery or emerging and unexpected health issues. Facility managers and operational staff can supply objective information on metrics involving the animal environment and facilities, husbandry, sanitation, and day-to-day animal care. Maintenance staff are also important and can provide assessments of critical mechanical systems, such as heating, ventilation, and air-conditioning (HVAC), electrical power, and lighting. Institutional security personnel are instrumental in providing information on physical and personal security and regulating appropriate access to animal facilities. Occupational health and safety professionals are integral for providing information on the occupational health and safety program (OHSP) for personnel with animal contact, and they should provide ongoing feedback to the IACUC regarding the overall functionality of the OHSP. Finally, careful tracking of issues and error rate analysis can highlight common areas of concern that may emerge and help to focus institutional resources in the appropriate areas.

Application of Performance Standards to Oversight of the Animal Care and Use Program

For most animal care and use programs, the application of performance standards across all components of the program is a mainstay for day-to-day function. The IO, IACUC, and AV are responsible for ensuring that performance standards satisfy regulatory requirements, promote animal welfare, are tailored to the institution's needs, and are not overly burdensome or costly. This responsibility creates a significant workload, but when done effectively, everyone benefits: the animals, animal caregivers, investigators, management, and IACUC itself.

Animal Care Program

The main components of the animal care and use program include institutional administration and management, the IACUC (protocol review and postapproval monitoring [PAM]), the program of veterinary care, animal environment and management, facilities and physical plant, and the OHSP. Although the IACUC does not perform or implement all the program components, the IACUC must ensure that performance standards in all these component areas are in place and functional. Regardless of the specific component of the program being evaluated, the collection of objective, relevant information, and error rate analysis are critical. Remember that part 3 of developing a performance standard involved developing methods for ongoing assessment of performance criteria. It is that ongoing evaluation of criteria that provides objective data as to whether a performance standard has been met, and provides the IACUC with key information needed for overall evaluation of the program.

It is beyond the scope of this chapter to detail how the IACUC provides oversight of all program components; however, some examples are provided that illustrate practical methods for the IACUC to ensure that performance-based standards are appropriate.

1. *Institutional administration and management*: The institution must provide sufficient resources to the animal care and use program to ensure that the research enterprise is viable and high standards of animal welfare are in place. A straightforward method for the IACUC to evaluate the overall institutional support of the program involves a careful review of several semiannual reports. It is the responsibility of the IACUC to identify problematic trends and work with management to ensure corrections are made. When several annual or semiannual reports are reviewed over time, concerning trends may become evident that indicate that the institution's administration may need to augment the resources allocated to the program. If problematic items are repeatedly identified by the IACUC without resolution, this may be a good indication that administrative support may be insufficient. On the other hand, if issues identified by the IACUC are consistently resolved, administrative support of the program is likely sound.
2. *Program of veterinary care*: The program of veterinary care must ensure that healthy animals are acquired and maintained. Programs that meet current veterinary practice standards must be in place for animal clinical care and preventive medicine, emergency care, health monitoring and biosecurity, anesthesia and analgesia, surgery, pathology, and euthanasia. Metrics for ensuring an adequate veterinary care program often include evaluating the timeliness of veterinary clinical care for sick or injured animals; the monitoring and control of endemic animal pathogens; the procurement of high-quality, healthy animals; adequate veterinary support for services such as surgery and pathology; and the programs and strategies for control of animal pain or distress. By simply tracking and collecting data regarding these, and other veterinary activities, the effectiveness of the program of veterinary care can be assessed.
3. *Animal environment and management*: Standards for animal environment and management include species-appropriate housing that promotes psychological and physical well-being and minimizes experimental variables that confound research data. The needs for socialization, environmental enrichment, quality feed and water, husbandry, and sanitation are important to ensure animal welfare. Metrics to evaluate these aspects of the program are often part of the semiannual review process. Considerations when evaluating the program of day-to-day care of animals often include the assessment of the overall adequacy of animal facility operations management: careful review of husbandry records to ensure key elements of daily care are regularly performed; review of sanitation intervals for cages and pens and the facility, as well as records and data relative to verification of sanitation effectiveness; review of the implementation and effectiveness of programs for social housing and environmental enrichment; review of feed storage and feeding practices; review of methods to provide quality drinking water and reports of water quality assessments; evaluation of standard operating procedures (SOPs) that impact animal well-being and assess work practices to ensure that SOPs are followed; assessment of staffing needs and the levels of expertise required; and evaluating events impacting animal well-being or research integrity.

4. *Animal facilities*: The animal facilities must support the research activities involving animals, be well designed to promote a healthy environment for animals, have properly designed and functioning mechanical systems, maintain biosecurity, be clean and well maintained, and provide a safe working environment for personnel. They should contain adequate space for animal housing, procedural areas, specialized areas (e.g., surgery, necropsy, and quarantine), bio- and hazard containment as necessary, and adequate support areas (e.g., storage and cage-washing facilities). Evaluation of the animal facilities often is done concurrently with assessment of the animal environment and management, as these two components of the program are linked and interdependent. Evaluation of the facilities is relatively straightforward and involves ongoing assessment for needs relative to overall space, design, and maintenance in support of the institution's research enterprise; regular evaluation of critical mechanical systems, such as HVAC, lighting, electrical power, and disaster planning safeguards; careful monitoring of the animal environment to ensure stable environments for research integrity, as well as animal safety; review of environmental monitoring data to ensure stable temperature and humidity according to the needs of the animals and scientific goals; review of adequacy of caging and pen systems to support animal well-being and the scientific needs of the program; and operation of animal facilities to foster animal and human safety.
5. *Protocol review and approval processes*: Discussions of performance standards typically focus on how the IACUC or other oversight body can leverage them in the evaluation of the animal care and use program and facilities. However, the performance-based approach can also be applied to the processes used to support what is arguably the most important responsibility of the IACUC—protocol review. Researchers have indicated that animal care and use protocol approval and revision are among the most significant regulatory burdens associated with animal-based research (Federal Demonstration Partnership 2014). As such, careful design and ongoing evaluation of these processes are imperative not only for ensuring regulatory compliance, but also for facilitating the conduct of research.

   There are myriad regulatory standards governing animal use protocol review and approval. In some cases, the regulations are noncommittal or even silent on the processes that must be used. Under these circumstances, institutions have tremendous opportunity for the application of performance-based standards in planning, evaluating, and improving protocol review processes. Unfortunately, many institutions fail to take full advantage of the opportunity afforded here because protocol review processes tend to be developed with a singular focus on regulatory standards and relatively little consideration for efficiency for either the IACUC or investigator. As a result, the institution may achieve its regulatory compliance objectives, but with greater burden than is necessary. To address this, institutions should adopt outcome-based standards for protocol review that not only give credence to the regulatory requirements and ethical responsibilities to the animal subjects, but also do so as efficiently as possible. This means careful examination of processes and eliminating waste by modifying or purging activities or requirements that do not add value.

   Haywood and Greene (2008) provide a lengthy list of self-imposed burdens associated with protocol review. Considering these suggested items, most programs will readily identify areas where processes can be improved. For purposes of illustration here, we explore two common problem areas in the process: overly complex protocol forms and excessive need for revisions to protocol submissions. These two interrelated facets are both sources of undue waste in the process.

   Completing the protocol form, which is the primary tool used by most IACUCs for review of animal activities, can be a daunting task for researchers. While the regulatory standards usually specify the content or specific considerations (e.g., identification of species, description of animal use, alternatives to painful procedures, and number of animals used) that must be addressed in animal use protocols, the forms at many institutions require information far in excess of the regulatory standards, or that the information be provided in a more complex format than the regulatory standards require. For instance, a standard required by regulation might be for the protocol to include a complete description of animal use; however, protocol forms rarely ask simply for a complete description of animal use. Instead, forms often attempt to illicit a complete description of animal use through the use of many questions and form fields, such as overview of the animal procedures, description of nonsurgical procedures, description of survival surgical procedures, description of postsurgical care, description of nonsurvival surgical procedures, table for frequency of volumes of blood collected, table of drugs administered, and listing of facility locations where animal procedures will be performed. If done well, some parsing of the base content requirements may actually make it easier for the investigator to provide the necessary information. But if done poorly, the parsing, along with the accompanying form field instructions, can result in needlessly lengthy and complex forms that investigators find difficult to complete. This in turn directly contributes to the need for revisions of submissions, which creates extra work for submitters and reviewers alike. For these reasons, it is advisable for the IACUC to evaluate the benefit and cost of the protocol form fields in a manner similar to the method described earlier.

In the authors' collective experience, it is relatively rare for protocols to wend their way through the review process without revisions and reworks being required. When reworks are commonly required in any process, the process itself becomes suspect. For example, if in the manufacturing of widgets 90% of them had to go back for rework before they were acceptable for market, one would consider the manufacturing process to be seriously flawed. Of course, preparation of an animal use protocol is not the same as the fabrication of a product in a factory, but when a high percentage of protocol submissions require revision, it still is indicative of underlying flaws in the process.

For years, quality management specialists have been exhorting the recognition that inspection does not improve the quality of a product or service because it is retrospective in nature—the product or service, whether of the desired quality or not, has already been produced or performed (Deming and Hollnagel 2000). Not to diminish the importance of the protocol review process, but much of it is indeed mere inspection. Is the form complete? Is the correct information provided? Are the descriptions sufficiently detailed? Are the proposed activities consistent with applicable regulations and policies? Is the literature search current? Imagine the time and effort that could be saved for both protocol authors and reviewers if protocols were completed to the IACUC's satisfaction at the time of their original submission. More or better protocol reviews are not likely to lead to a significant reduction in the number of protocol reworks required; rather, only changes to the way protocols are produced or authored will accomplish this.

Investigation into the root causes of inadequate protocol submissions may help uncover elements of the protocol authoring and submission process that can be revised in order to reduce the likelihood of revision requests. The causes may include unclear or overly complicated form instructions, language barriers, or poorly trained authors. In any case, identification of root causes will unveil those process elements that can be targeted for improvements and, subsequently, whether the improvements effect a decreased rate of protocol revisions.

Ensuring the protocol form is easy to complete and does not ask for superfluous information, and changing the protocol authoring and submission process to reduce the need for revisions are just two ways whereby improved overall efficiency can be garnered. Of course, there are many other aspects of the protocol review process that can and should be evaluated using a performance-based standard of optimized efficiency. One payoff for this approach is reduced compliance burden for the researchers, which frees more time that can be devoted to the conduct of research. In addition, these types of process improvements can free the IACUC to concentrate more on the matters, such as harm–benefit analysis, that are of greatest import to animal welfare and research success.

6. *PAM*: The investigator has responsibility to ensure activities are being conducted in congruence with the IACUC-approved protocol. In the United States, the IACUC is further tasked with "continued oversight" of approved activities. Historically, this monitoring has been achieved

primarily via the semiannual program review and facilities inspections, communications with the animal care staff, and external agency inspections. More recently, many U.S. institutions have implemented a PAM program in which a dedicated staff is tasked with assessing congruency between the IACUC-approved activities and actual activities in the animal care and use program. In some cases, this dedicated monitoring program includes oversight of the animal care operations, in addition to research activities. Regardless of how the PAM program is conducted, it is critical that the program identifies problem areas and addresses them in a timely fashion to ensure a quality program. Instituting a PAM system without using the information gained to refine the program, and thereby prevent noncompliance from happening, is wasteful and directs resources away from more beneficial components of the animal care program (Plante and James 2008).

When performance standards (outcomes focus) are applied to PAM, the leadership of the animal care and use program can more effectively determine how noncompliance items are typically identified at their institution, the types of noncompliance identified by the various monitoring programs, and the approximate associated costs. This approach allows the program to minimize those non-value-added components and minimize the regulatory burden when possible.

Semiannual facility evaluations, which are a regulatory requirement in the United States, typically identify physical facility issues associated with the animal care and use program. The quality of this inspection process can vary greatly, and represents an opportunity for the IACUC to self-assess its effectiveness. One possible outcome measure or performance standard for semiannual inspections would be to identify with a high level of accuracy direct animal care deficiencies present in the animal housing areas. For example, at Institution A, the AAALAC site visitors identified facility problems that were not noted in the IACUC semiannual inspections. Upon introspection, Institution A noted that the semiannual inspections appeared to be cursory in nature, as the time required to complete the inspection was minimal. To address this concern, Institution A refined its inspection process to ensure that the inspection team included one of the veterinarians, and provided additional education to the IACUC members regarding their role in the semiannual inspection and items to look for as they conduct inspections. Over the course of the next year, the IACUC then analyzed the effectiveness of this change by scrutinizing the number of findings on the semiannual inspection, as well as the type of finding. Overall, findings increased by 200%, and furthermore, the findings themselves more frequently involved a direct animal care concern or animal welfare issue. The cost of this effort was strictly personnel in nature, and included additional veterinary time, preparation time, and IACUC member time. Institution A's IACUC determined that this additional time cost was worth the effort since it identified findings that directly impacted the animals at a higher rate than before the change was implemented. Most importantly, however, Institution A was able to use the information gained on the inspections to leverage facility repairs and enhance preventive maintenance procedures in the vivarium to minimize problems from occurring in the first place.

Another example is Institution B, whose IACUC recognized a concern with their oversight of investigator lab areas as identified by an external accreditation agency. In response to the agency findings, Institution B implemented a change to include all animal use investigator areas that utilize non–U.S. Department of Agriculture (USDA)–covered species in the 6-month facility inspections. Note that regulations require those areas to be inspected "regularly" rather than "every 6 months" to ensure adequate IACUC oversight. Since many of the lab findings related to rodent surgery procedures, additional training was implemented to target aseptic technique and postoperative care. During the first several years of this change, the IACUC noted that findings in these investigator areas were common; however, after about the third year, it was uncommon to identify a problem in these locations. Based on this outcome, the IACUC at Institution B redesigned the inspection process to lower the frequency of inspections for areas that had no findings on multiple inspections, and maintained the 6-month frequency for laboratory areas that had findings. The use of this continuous process improvement approach as applied to the semiannual inspection process has allowed both institutions to achieve an effective and efficient program of oversight for animal care and use.

The animal care and veterinary staff are an integral component of the PAM program, and frequently are the individuals who identify serious noncompliance issues that impact animal welfare. The IACUC should ensure that animal care and veterinary staffs are familiar with the research protocols and activities and know how to report problems to the IACUC as applicable. On an international level, it is the animal care and veterinary staff to whom monitoring and compliance activities often are delegated (Whittaker 2014).

Institutions that are considering use of a dedicated PAM specialist should determine what areas of the animal care and use program may require this additional oversight, and develop a feedback mechanism to facilitate correction of the root causes of the noncompliance. The qualifications and reporting structure of the specialist may vary depending on the institution's oversight needs. Investing in dedicated PAM personnel can direct resources away from direct animal care and research support. Therefore, institutions should make an informed decision on the cost–benefit of this nonrequired activity for their institution (Thulin et al. 2014).

As Haywood and Green (2008) advocated, the IACUC should consider the following three questions: Why do we do this? Does it help the animals? Can the end be achieved in a more efficient, cost-effective manner? As the PAM program is structured, it is useful to consider these three items to ensure efforts are directed toward the most important areas for their institution.

Organizational Considerations

Institutions provide support of the animal care and use program utilizing a management structure in which all support activities are centralized or a structure in which support is distributed among various units at the institution. The approach to management varies between academics and private industry based on a survey done in 2013, with private industry having a higher predominance of centralized support, with all functional areas of the animal care program being provided from the same administrative unit. In academia, the predominant administrative structure in support of the animal care and use program is a distributed model with functional areas (animal care administration and IACUC support) utilizing separate but parallel reporting structures (Bradfield and Thulin 2013). Pinson (2012) suggested that this separation may weaken the programmatic authority and fiscal stability in the animal care program. When studying the impact of organizational culture on the animal care and use program, it becomes easy to understand how communication in a distributed administrative structure can be especially challenging. Specifically, each administrative group is responsible for separate yet complementary functional components of the animal care program (Ellenberger and Corning 1999). On one hand, the animal care program provides administrative support for direct animal care, while on the other, the IACUC office provides administrative support for regulatory documentation associated with animal use. The latter function is typically removed from direct animal care, and the focus becomes more regulatory, or bureaucratic in nature. The consequence of this organizational structure can be a disconnect of the IACUC support staff from both the direct animal care program and the researcher, and lead to disharmony within the functional components of the animal care program (Thulin et al. 2014). Regardless of which organizational model is utilized, the institutional culture and personnel involved (Bronstad and Tronsdal Berg 2011) will ultimately determine how well the organizational structure functions at an individual institution.

Summary

Ensuring the animal care and use program conforms to the regulatory standards does indeed come at considerable cost and effort to the institution and the investigators conducting the science. The challenge for program managers and oversight bodies is to conduct the compliance activities in a way that not only ensures proper animal welfare, but also is no more restrictive to research than is necessary. Facilitation then is about minimizing compliance burden and better accommodating science through flexibility—not just about making sure the institution retains its research privileges. Reduction in compliance burden can be achieved through careful examination of the attendant processes to eliminate those that are unnecessary or do not add value, and to find more efficient ways to perform all others. Embracing performance standards paired with careful evaluation of outcomes is important for achieving program flexibility. Programs that successfully realize these approaches will certainly be viewed as facilitating both animal welfare and the research enterprise.

REFERENCES

Bayne, K.A., and Garnett, N.L. 2008. Mitigating risk, facilitating research. *ILAR J* 49:369–371.

Bradfield, J.F., and Thulin, J.D. 2013. Institutional oversight: The role of the attending veterinarian. Presented at the American College of Laboratory Animal Medicine Forum, Williamsburg, VA, April 14–17.

Bronstad, A., and Tronsdal Berg, A. 2011. The role of organizational culture in compliance with the principles of the 3Rs. *Lab Anim* 40:22–26.

Committee on Federal Research Regulations and Reporting Requirements: A New Framework for Research Universities in the 21st Century; Committee on Science, Technology, and Law; Board on Higher Education and Workforce; Policy and Global Affairs; and National Academies of Sciences, Engineering, and Medicine. 2016. *Optimizing the Nation's Investment in Academic Research: A New Regulatory Framework for the 21st Century*. Washington, DC: National Academies Press. https://www.nap.edu/download/21824 (accessed September 14, 2016).

Deming, W.D., and E. Hollnagel. 2000. *Out of the Crisis*. Cambridge, MA: MIT Press.

Ellenberger, M.A., and Corning, B.R. 1999. The Animal Care and IACUC offices: United or divided? *Lab Anim* 28:44–47.

Federation of American Societies for Experimental Biology. 2015. Sustaining discovery in biological and medical sciences: A framework for discussion. http://www.faseb.org/Portals/2/PDFs/opa/2015/10.23.15%20Sustaining%20Discovery%20for%20print%2031Aug15.pdf (accessed September 14, 2016).

Federal Demonstration Partnership. 2014. 2012 faculty workload survey: Research report. http://sites.nationalacademies.org/cs/groups/pgasite/documents/webpage/pga_087667.pdf (accessed September 14, 2016).

Haywood, J.R., and Greene, M. 2008. Avoiding an overzealous approach: A perspective on regulatory burden. *ILAR J* 49:426–434.

Kim, C.S., Spahlinger, D.A., and Billi, J.E. 2009. Creating value in health care: The case for Lean thinking. *J Clin Outcomes Manag* 16:557–562.

National Research Council. 1972. *Guide for the Care and Use of Laboratory Animals*. 4th edition. Washington: US Government Printing Office.

National Research Council. 1996. *Guide for the Care and Use of Laboratory Animals*. 7th edition. Washington: National Academy Press.

National Science Board. 2014. Reducing investigators' administrative workload for federally funded research. March 10. http://www.nsf.gov/pubs/2014/nsb1418/nsb1418.pdf (accessed September 18, 2015).

NRC (National Research Council). 2011. *Guide for the Care and Use of Laboratory Animals*. 8th ed. Washington, DC: National Academies Press.

Pinson, D. 2012. A training primer for institutional officials. *Lab Animal* 41:198–203.

Plante, A., and James, M.L. 2008. Program oversight enhancements (POE): The big PAM. *ILAR J* 49:419–425.

Pritt, S., McNulty, J.A. Greene, M., Light, S., and Brown, M. 2016. Decreasing institutionally imposed regulatory burden for animal research. *Lab Anim* 45(8):297–300.

Thulin, J.D., Bradfield, J.F., Bergdall, V.K. et al. 2014. The cost of self-imposed regulatory burden in animal research. *FASEB J* 28:3297–3300.

Whittaker, A. 2014. Animal research regulation in Australia—Does it pass the test of robustness? *Glob J Anim Law* 1:1–14.

Section IV

Program Management and Stewardship of Resources

11

Human Care

Pamela A. Straeter, Carolyn M. Malinowski, and Laura A. Conour

CONTENTS

Introduction

The daily roles a manager fulfills within an organization are numerous and diverse. Although certain skills, such as emotional intelligence and communication, are innate for some managers, many, including staff management and oversight of personnel performance issues, must be learned. In this chapter, one of the most important roles of a manager is explored, the supervision and care of humans. As managers of human resources, we need to recruit, develop, and retain people with the knowledge and skills necessary to function at a level that will aid in achieving organizational goals (Jones et al. 2010a).

What Is a Manager?

The *Merriam-Webster Dictionary* defines *management* as "the act or skill of controlling and making decisions about a business, department, etc." (Merriam-Webster 2015b), whereas the *Oxford English Dictionary* defines *management* as "the process of dealing with or controlling things or people" (Oxford Dictionaries 2015). Yet another definition was coined by Mary Parker Follett, a pioneer in the field of organizational behavior in the early 1900s. She defined *management* as "the art of getting things done through people" (Reddy and Tripathi 2008). Although there are slight differences in the definitions above, the central theme of accomplishing tasks is pervasive. Based on these definitions, one is able to intuitively assume that a manager is someone with responsibilities in both personnel and task management. In this chapter, the focus will remain on the personnel aspect of a manager's job.

Central to managing people is the theory of emotional intelligence. This concept is "the ability to perceive emotions, to access and generate emotions so as to assist thought, to understand emotions and emotional knowledge, and to reflectively regulate emotions so as to promote emotional and intellectual growth" (Swijtink 2016). In layperson's terms, managers should be able to perceive, understand, use, and manage emotions through self-awareness, self-regulation, social skills, empathy, and motivation (Swijtink 2016). While emotional intelligence in inherent for some managers, it may be a learned trait for others.

In addition to emotional intelligence, there are many other traits that are desirable in a manager, including diplomacy, tact, and fairness. These traits, along with personal attitudes, behaviors, and actions, determine a manager's individual style. No single recipe for the perfect manager exists. Management style is often influenced by existing infrastructure, such as the location, size, and type of institution (e.g., academic, government, military, pharmaceutical, or contract research); the type of labor force (e.g., contract or union); and the goals of the organization. Regardless of infrastructure type, a manager who is able to establish a positive working environment, empower and encourage employees, provide for the development of staff, and acknowledge and appreciate staff contributions will realize a sizable return on investment in terms of efficiency, effectiveness, and loyalty. However, several caveats exist: What is effective at a large pharmaceutical firm may not be the case at a large university. Loyalty may be evident with a union labor force, but not with contracted, temporary staff. What is efficient at a large institution may not be practical in a smaller one. Each manager must determine what works for him or her and which style abides by the culture of his or her workplace environment.

An important delineation must be made between the management of groups and individuals, as the two are often handled very differently. To ensure uniform treatment of staff, most institutions have developed policies, guidelines, or procedures for handling tasks, such as recruitment, discipline, and performance reviews. Often, these procedures are in place to ensure consistent treatment of all staff. However, the application of established protocols should be adapted to both the individual and the situation. For example, one employee may find directness intimidating, while another finds it refreshing. Thus, a manager must learn to effectively interact with individuals while treating all employees equally and fairly.

Management Activities

The activities in which managers engage are too numerous to individually identify in a single book chapter. This section breaks these activities down into core concepts and how the resulting activities may be measured.

According to the Institute for Certified Professional Managers (ICPM), there are four main activities in which managers engage: planning, organizing, leading, and controlling. Although these activities may occur individually under certain circumstances, they often occur in the order listed above. When planning, managers identify strategies, goals, and courses of action. Additionally, they allocate resources to aid in achieving set goals and objectives. When organizing, managers work to establish good working relationships between coworkers, enabling them to cooperate and collaborate. Leading occurs when managers motivate employees and teams. Finally, when controlling, managers evaluate, measure, monitor, maintain, and improve the performance of outlined objectives (Jones et al. 2010b).

Two key, recurring management concepts are efficiency and effectiveness, both of which may be used to measure performance in the attainment of goals. Efficiency focuses on how well resources are used, while effectiveness focuses on both the appropriateness and degree of achievement of organizational goals (Jones et al. 2010c). These are common concepts in popular management strategies, including Lean Management™, Six Sigma™, and ISO 9000™, all of which emphasize continuous improvement in value, efficiency, and quality.

Training for Managers

In the field of laboratory animal care, it is not unusual for employees to advance through the ranks without formal training in the skills necessary to be successful managers. However, it is essential that managers receive training to support success in fulfilling their new roles and responsibilities. One of the most common sources of training is through the American Association for Laboratory Animal Science (AALAS), where various certifications may be obtained. These levels ascend in hierarchy based on length of time in the field, education, and experience. They include the assistant laboratory animal technician (ALAT), laboratory animal technician (LAT), laboratory animal technologist (LATG), and Certified Manager of Animal Resources (CMAR). Another source of management training offered by

AALAS is the Institute for Laboratory Animal Management (ILAM), a 2-year program designed with a focus on management in the laboratory animal field. Additionally, the Laboratory Animal Management Association (LAMA) offers management training through workshops and continuing education opportunities at its annual conference.

In 2008, AALAS conducted a survey of LATGs and CMARs to identify key knowledge and tasks essential for competent job performance. Several of the survey sections addressed the topic of management. The majority of CMAR respondents and 38% of LATGs classified themselves as either managers or supervisors. However, 86% of CMARs and 55% of LATGs reported that supervising or managing employees was part of their job, despite not officially classifying themselves as managers or supervisors. More than 80% of both LATGs and CMARs reported that their job responsibilities include the following tasks: employee hiring, performance appraisal, corrective actions, and employee termination—again, with a notable percentage of both groups lacking a formal title of "manager." Both LATGs and CMARs ranked conflict management as one of the top five competencies to possess in human resource management, while CMARs also listed management principles and techniques, including interview skills, performance appraisals, employee supervision, and career development, as the remaining highly ranked competencies. The results of this survey show that management is an important aspect of the job for many LATGs and CMARs, regardless of whether their position title references "manager." Additionally, this survey emphasizes the importance of management skills in everyday work duties and the need to ensure staff are provided with the necessary training and education to fulfill their roles and responsibilities (AALAS 2008).

An alternative informal way in which mangers are able to learn necessary skills is through the development of a local or institutional teaching group, in which managers gather together to share experiences. This allows managers to learn from both the successes and failures of others, as well as how to facilitate the development of compassion, understanding, and objectivity that allows them to function within the diversity of culture and inherently human situations. Learning by understanding the consequences of actions taken is one of the most powerful tools available.

Employee Composition

A significant amount of information has been shared about attributes of an effective manager, yet it is equally important to understand the composition of employees. An employee, as defined by the *Merriam-Webster Dictionary*, is "one employed by another usually for wages or salary and in a position below the executive level" (Merriam-Webster 2015a). Aside from this definition, employees can also be defined based on the method by which they are paid, such as exempt (not eligible for overtime) or nonexempt (eligible for overtime), or by the manner in which they are employed (e.g., unionized or contract staff). While these labels describe the type of employee, every attempt should be made to ensure that employees are treated equally and fairly. Understanding appropriate ways to handle different types of employees can be the key to success for an effective manager.

Unionized or contract employees may require a different management approach than nonunionized, permanent staff. Union leaders, through collective bargaining, negotiate wages, working conditions, personnel safety, hours, and other benefits on behalf of their union members. Unless a manager is at the collective bargaining table during contract negotiations, he or she will have little influence on the wages, job descriptions, and promotional criteria for his or her staff. Similarly, contract staff are paid at a negotiated rate between the company and staffing service based on organizational need and the desired skill set of the employees, limiting a manager's ability to financially motivate and developmentally enrich his or her employees. These challenges are not, however, impossible to surmount. A talented manager can develop unique opportunities for all types of staff without violating rules or contracts.

Contract employees are not permanent employees of a company, but rather temporary personnel hired for a specific prearranged period. Contract employees may not be required to participate in office meetings and organizational events. The use of contractors can permit an organization to quickly and nimbly adjust the size of the workforce to match the labor needs of the organization. However, contract employees do not receive employer-paid benefits, such as group health insurance, paid vacation and sick leave, or

401(k) programs, which may lead to a lack of commitment to the organization and potential for divisive interactions with noncontract employees (Dale 2015).

In contrast, some contract staffing companies have full-time permanent employees that are assigned to specific government, academia, and pharmaceutical companies for long- and short-term periods. These employees participate in the staffing company's group medical plans, 401(k) plans, and all company-wide benefits and events. Accrued paid vacation, sick leave hours, and paid holidays are a standard as well. These employees maintain seniority within the company regardless of where they are assigned or relocated.

In addition to the various types of employment, it is equally vital to understand the importance of diversity within staff groups. Staffing diversity can create challenges due to differences in views and cultural beliefs. For example, there are cultures that consider and value the role of animals in society and use in research differently, resulting in dissimilar interpretations of animal welfare relative to pain and distress, as well as the use of positive versus negative reinforcement when training animals. It is extremely important to understand a potential employee's ethical views in these areas while also conveying the organization's philosophy and values for the use of animals in research.

An organization must maintain a culture that is welcoming and hospitable to all employees regardless of background or differences; otherwise, costly turnovers will continue as talent leaves for more favorable work environments. It is for this reason that attracting a workforce with various backgrounds is the first step in creating a strong and heterogeneous workplace. Leading and maintaining that diversity in the workforce is much more than managing staff; it is enriching the organization's culture with diversity programs and embracing differences. Diversity training is an essential part of building a cohesive work environment steeped in awareness (Holt 2015). Often, there are many diversity training workshops available both within and external to an organization. Some courses may include promoting age, racial, cultural, sexual orientation, gender diversity, and discrimination awareness. The combination of a welcoming and diverse workplace culture, maintained through frequently offered programs and courses, ensures employees at all levels feel valued and are motivated to productively work for any organization. With these differences in mind, it is important for managers to recognize and appreciate the web of relations encompassing a diverse workforce and to appreciate the value it brings.

Recruitment

It is often said that a good manager is only as good as the team he or she builds. However, it is rare that an individual walks into a management role with a highly effective and functional team already in place, making it necessary to bring in new team members to complete or enhance an existing team. This makes the ability to recruit highly qualified applicants a valuable tool for any manager.

Recruiting staff can be a long and exhaustive process, and if a team is already understaffed and overwhelmed, the sense of urgency to accomplish this may be strong. According to a 2013 CareerBuilder survey, 66% of U.S. employers reported that bad hires lowered their company's productivity, affected worker morale, and even resulted in legal issues. Although tempting, a manager should not rush into a hasty hiring decision in order to fill an empty staffing position. The more thoroughly candidates are vetted, the less likely they will be a poor match for the team. The right balance of job fit, personality, and experience is desired in staffing hires. Job fit is an approach that determines if an applicant's strengths, needs, and experience match the requirements of a particular job and work environment (Heathfield 2015). An individual with an excellent skill set is not always the right fit for the position.

When considering candidates, the work site or facility "culture" fit is almost as important as the required experience level. This concept is often used to define the harmony of a work group, where finding the right fit and/or balance is essential. The resulting harmonious work group will have strong lines of communication, culminating in an efficient group of employees that meets the program's goals. The right fit may be as simple as like-minded skill sets, work ethic, or interest in the same hobbies or beliefs. Some qualities, such as a soft-spoken demeanor or an outgoing personality, are simple to observe, while other, equally important qualities require a little more effort to discover. The balance between the required skill set, level of education, and type of personality are all important components in finding the right fit for vacant positions. The candidate with higher learning credentials and key words on their curriculum vitae (CV) or résumé is not necessarily the best cultural fit for the group. If the candidate possesses the essential job requirements, then a manager who is open-minded and a good listener can learn more about his or her disposition through effective dialogue.

Whether a position is open due to a vacancy or the addition of new head count, managers should invest some time in planning their recruitment strategy. Some of the more widely used recruitment methods include placing advertisements on industry-related national and local websites, in newsletters, on list-servs, and on company Internet sites; publishing internal announcements; using personal referrals; and enlisting employment or executive recruitment agencies. Progressive managers may also use social and new recruitment media strategies to target the best candidates for their staffing needs. Managers may also need to adapt their recruitment strategies according to the level of the position. To attract more senior candidates, managers should tap into their own networks, as well as those of their employees, and might also consider using an executive search firm.

With the number of talented individuals in the field of laboratory animal care, a hiring manager must consider several factors when selecting a team member to join the organization. The importance of recruiting for and promoting workplace diversity is essential. According to many organizations, diversity is typically defined as valuing and accepting the differences among people with respect to age, class, gender, ethnicity, sexual orientation, and mental and physical capabilities (Murphy 2015). To aid in the recruitment of individuals from diverse backgrounds, the following suggestions should be considered:

- Develop specific strategies to increase the flow of applicants from a range of backgrounds. For example, advertise in a variety of publications, on Internet sites, at international job fairs, and at colleges and universities with a diverse student population.
- Establish internships to bring in individuals from a variety of backgrounds to enable interested individuals to gain on-the-job experience and skills.
- Promote the hiring of people from nontraditional backgrounds, and implement support systems for their transition to the team and workplace.
- Seek referrals from existing employees to find promising candidates within the organization or laboratory animal community (Wood 2015).

The larger the pool of applicants a manager can select from, the greater the likelihood that the best and right fit for the organization and team will be identified. It is additionally critical that a hiring manager utilizes the appropriate tools needed to sift through many applications to ensure applicants not only have the right experience and skill set, but also possess employment aspirations that are genuine. The recruiting process can include several phases, depending on the size and type of organization and the hiring manager's preference. These may include an initial review of applications, easily eliminating those that do not meet the requirements outlined in the job description; telephone interviews to screen candidates' genuine interest and salary expectations; and in-person interviews, potentially involving a first round with a larger pool of interviewers and then a final round with a small number of interviewers from within and outside of the department. Frequently, an institution's human resources department (HRD) can assist a manager in initiating the recruiting process based on organizational procedures.

Once the position has been advertised and the applicant pool begins to grow, managers will sift through the cover letters, résumés, and applications to identify those candidates that meet the qualifications of the position. Regardless of profession or job title, managers seek a résumé highlighting not only the candidate's skills required to identify, anticipate, prevent, and solve the typical problems that arise daily in that position, but also the accomplishments related to previous efforts (Wood 2015). Therefore, managers should focus on applicants with experience, an understanding of the job's deliverables, the tools to perform the job duties, and results-based achievements evident within their résumé or application. Candidates who do not qualify can be effectively eliminated from the applicant pool when these defined criteria are used as part of the selection process.

Following the initial review of applicants, most hiring managers will set up telephone interviews or Skype™ calls in order to conduct initial screens prior to arranging in-person interviews. Some typical screening questions are provided in Table 11.1 (Doyle 2015).

TABLE 11.1

Typical Screening Questions for Telephone Interviews

What is the previous employment history, including the names of employers, job titles and descriptions, and dates of employment?
What are/were your roles and responsibilities? Describe your experiences working directly with animals?
What major challenges and problems did you face? How did you handle them?
How do you handle stress and pressure?
What motivates you?
What is the difference between animal rights and animal welfare?
What type of work environment do you prefer?
How would this position fulfill your short-term and long-term goals?
What are your compensation expectations?

These preliminary screening questions are just a small sampling of inquiries that can be made, particularly when individuals are new to the industry or have limited experience. It is equally important to take the time during the screening process to convey the nature of the work the position entails, particularly if the applicant has no prior experience in the field of laboratory animal science. Transparency about the less appealing responsibilities of the job, such as standing to perform cleaning and cage-changing tasks, wearing personal protective equipment, and handling soiled caging, is essential.

Once a manager has selected the top candidates (typically three to five at the most), the next step in the recruitment process is to initiate in-person interviews. Interviews should include key stakeholders who will interact regularly with the candidate (Gebelin et al. 2004). For example, when hiring a veterinary technician, an interview panel may include a veterinarian, facility supervisor, principal investigator, and peer veterinary technician. Depending on the level and complexity of the position, a manager can choose to either combine the stakeholders in teams for the interview or arrange one-on-one interactions tailoring the type of questions that may be asked. When coordinating the time allotted for the interview, the manager may schedule 1–2 hours depending on the number of stakeholders involved and the configuration of the interview sessions. Higher-level positions will require proportionally increased interview time and stakeholder numbers. For example, an interview for a veterinary technician may only last 2–4 hours, whereas the time needed to interview for a supervisor may extend to a full day. Another critical step is to coordinate interview questions to ensure the same questions are not repeated throughout the interview process. Questions should revolve around the job candidate's personality, strengths, weaknesses, knowledge, skills, abilities, and talents that account for on-the-job performance. This type of questioning is called competency-based interviewing, and it is the most effective method used (Types of interviews 2015). This method is also referred to as targeted selection interviewing, used to evaluate candidates' past performance and behaviors as predictors of future performance. Furthermore, the use of open-ended questions with this approach is most effective, such as

- Describe the best boss you have reported to.
- Describe motivations and frustrations on the job.
- Tell me about the toughest job-related decision you have ever made.

When using this approach, the candidate may talk about past lessons learned and reveal details about his or her personality, diplomacy skills, and ability to work as part of a team.

Providing an opportunity to tour the applicants through the work environment to give them a firsthand account of the type of work they will be expected to do is an effective approach to include in the interview process. Additionally, a manager may pair them with a seasoned staff member performing that job so they can speak informally about the advantages and challenges of working in that role. It is prudent to ensure that the staff member conducting the tour understands the ultimate goal of having a dependable coworker, and that he or she is representing the department and organization to a prospective employee.

This exercise also helps to convey job expectations to a potentially overqualified applicant. Any method the hiring manager employs to ensure that candidates fully understand the job responsibilities will minimize the potential for employee turnover and failed recruitment efforts.

Once the candidate pool has been narrowed, the selection process becomes more difficult. A manager may be left with two to three ideal candidates warranting a second round of interviews with a different set of interviewers using a battery of more detailed inquiries about the applicant, his or her qualifications, and his or her ability to perform for the organization. During the second interview, discussion may ensue around such topics as terms of employment, working conditions, benefits, and pay. Managers should be prepared to field questions but avoid committing to specific terms relating to pay wages and benefits unless they have the approval to negotiate specific terms.

Once the interviews have been completed, the interview panel should provide their feedback directly to the hiring manager. Managers may find it beneficial to set up a meeting with the interview panel to assess the extent to which each candidate met their selection criteria; however, it is important to ensure interviewers are not swayed by others into changing their rankings and opinions. Ideally, the candidates will be ranked based on qualifications, ability to perform the job, and the responses from the targeted selection method of gathering past performance experiences. At this point, the manager typically transitions the hiring process to his or her HRD representative, who will make the actual job offer, to include the terms of employment, such as wages, benefits, background checks, and a mutually agreed upon start date. A best practice following receipt of the feedback on candidates should involve collecting and retaining notes from the interviews conducted. Most organizations require the retention of interview documents, as they are important in protecting the institution against an accusation of unfair hiring practices.

A critical part of the hiring process, and one that smaller institutions in particular may overlook, is a background check on the chosen candidate. Employers within the biomedical research community are vulnerable to individuals from animal extremist groups looking for opportunities to infiltrate organizations, potentially stealing, vandalizing, and misrepresenting research activities at the institution. It is highly recommended that hiring managers conduct searches using publicly available information available through Internet searches, social media, and professional association organizations. For example, does a candidate maintain local or national AALAS membership, which might serve as an indication of his or her commitment to the profession of laboratory animal science? Any background information gleaned from these various sources could be useful in determining if the applicant is ideal for a position involving highly sensitive responsibilities, such as caring for and working with research animals. It is worth the investment to use organizations that provide background and security checks for a fee.

Selection and Staff Onboarding

Once the preferred candidate has been identified, terms of employment agreed upon, and a successful background check completed, the hiring process is complete. The next step is to initiate the onboarding of the new staff member—a process equally as important as recruitment. There are many functions (e.g., animal care, veterinary services, and business office administration) within laboratory animal care programs that require customized orientation and training. Managers are responsible for orientating new employees to the institution and the expectations of their new role. Even if an organization has a formal institution-wide onboarding program, the manager remains the most important person in the onboarding process (Vernon 2013). Table 11.2 contains typical activities that a manager may incorporate into the process of orientation and onboarding of the new employee (Purdue University 2015).

When considering how to develop an onboarding strategy, there are three broad categories of information a new employee needs: general information for all employees about the organization, department-specific information, and job-specific information. While each manager may have a unique approach to new employee onboarding, it must be ensured that new employees have all the tools necessary to be successful productive members of the team.

TABLE 11.2

Typical Onboarding Activities to Be Conducted by the Manager/Management Team

Timing	Objective	Sample Activities to Be Done for the New Hire
Prior to the first day	Create the employee's first impression.	• Send out introduction and welcome letter to colleagues. • Create an onboarding schedule. • Gather requisite forms to be completed. • Work with the HRD to schedule organizational orientation. • Identify workspace. • Order supplies/equipment needed to work.
First day	Welcome the employee! This is a day of discovery, administration, and introductions.	• Schedule meetings with the business office to complete new hire paperwork. • Schedule tours of the department and workspace. • Arrange a meeting with the manager to discuss job expectations and management style.
First week	Help the employee get acquainted with the organization and new role.	• Have employee work with a colleague for on-the-job training. • Set up initial manager check-in meeting.
First month	The employee becomes a productive member of the organization.	• Schedule meetings with other departments for the employee to learn their functions. • Facilitate opportunities to learn more about the organization's services and benefits. • Ensure that the employee is immersing in organizational culture. • Schedule a meeting to create a development plan.
Future actions	The employee continues orientation. Onboarding is an ongoing learning process.	• Identify any additional formal training needed. • Hold ongoing check-in meetings. • Perform 30-, 60-, and 90-day performance reviews and annual evaluation.

Following the initial orientation and onboarding activities, managers should continue to focus their attention on

- Providing the appropriate balance of structure and independence to promote the success of new hires in their roles and responsibilities
- The use of coaching conversations to diffuse anxiety, establish responsibilities and expectations, and provide actionable feedback to guarantee skill development and achievement of performance goals
- Adapting best practices for managing a diverse workforce and personal work styles
- Providing appropriate opportunities for leadership whenever possible

Managers who appropriately structure new employees' onboarding activities will convey to them that they are a priority to the organization. It is critical to welcome new employees, thus affirming their choice to join the organization, motivating them to continuously contribute to the success of the business, and exciting them with the prospects of a new job (Gebelin et al. 2004).

Retention

The recruitment of individuals with the desired skills and knowledge to be successful in their job is but the first step in staff management. An equal challenge is how to retain these individuals within the organization.

Regardless of the employee type or background, most individuals will look to their managers for opportunities to grow within the organization and further develop their knowledge, skills, and abilities in the field of animal care and use. Managers can assist staff by sharing a potential positions diagram (an example is shown in Figure 11.1), discussing how the person can prepare for different types of jobs

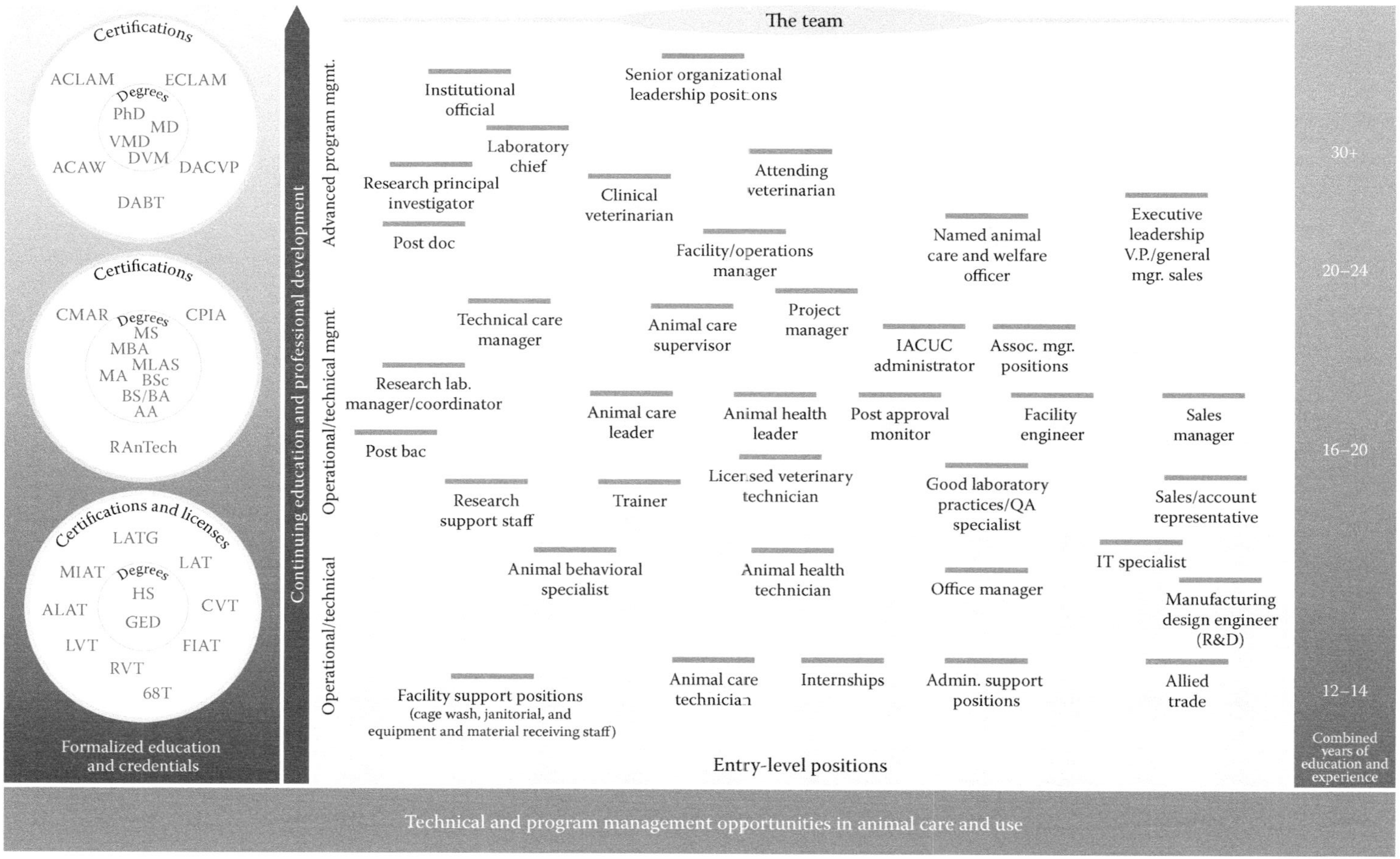

FIGURE 11.1 **(See color insert.)** Example of a team that shares the responsibility to work closely together in programs of animal care and use around the world to ensure quality animal well-being and the responsible conduct of research, education, and testing with animals. This is not intended as representative of any particular organization or institution. (See abbreviation key on the next page.) (Developed by R. H. Weichbrod and K. McKibbin with Graphics Prints, Inc., Herndon, VA.)

KEY TO FIGURE 11.1

CERTIFICATIONS

American Association for Laboratory Animal Science (AALAS) (https://www.aalas.org/)

ALAT	Assistant Laboratory Animal Technician
LAT	Laboratory Animal Technician
LATG	Laboratory Animal Technologist

AALAS in partnership with the Laboratory Animal Management Association (LAMA) (http://www.lama-online.org/) **and Institute for Certified Professional Managers (ICPM)** (https://www.icpm.biz/)

CMAR	Certified Manager of Animal Resources

Public Responsibility in Medicine and Research (PRIM&R) (http://www.primr.org/)

CPIA	Certified Professional Institutional Animal Care and Use Committee (IACUC) Administrator

Institute of Animal Technology (IAT) (http://www.iat.org.uk/)

MIAT	Member of the IAT
RAnTech	Registered Animal Technologist
NACWO	Named Animal Care and Welfare Officer
FIAT	Fellow of the IAT

LICENSES

RVT	Registered Veterinary Technician
CVT	Certified Veterinary Technician
LVT	Licensed Veterinary Technician

MILITARY OCCUPATIONAL SPECIALITY

68T	Animal Care Specialist (United States Army)

BOARD CERTIFICATIONS

ACLAM	American College of Laboratory Animal Medicine
ECLAM	European College of Laboratory Animal Medicine
ACAW	American College of Animal Welfare
DABT	Diplomate, American Board of Toxicology
DACVP	Diplomate, American College of Veterinary Pathology

FORMALIZED EDUCATION

HS	High school diploma
GED	Graduate equivalency degree
AA	Associate of arts
BA	Bachelor of arts

BS	Bachelor of science
BSc	Bachelor of science (university degree)
MA	Master of arts
MBA	Master of business
MS	Master of science
MLAS	Master of laboratory animal science
DVM	Doctor of veterinary medicine
VMD	Veterinary medical doctor
PhD	Doctor of philosophy
MD	Medical doctor

REFERENCES

American Association for Laboratory Animal Science. 2016. Careers in biomedical research: Accept the challenge to care. https://www.aalas.org/public-outreach/resources#.WKP1KqOZN7M. Fields of biomedical research and related careers.

Eisenstein, M. 2005. Profiles in laboratory animal science: 11 paths to the top of the field. *Lab Animal* 34(6): 100–105.

Institute of Animal Technology. 2016. The career pathway for animal technologists. http://www.iat.org.uk/#!pathway/clq0. Interactive career pathway, clicking on buttons and text for further details, such as position title, brief description of duties, likely qualifications, and salary.

in the field, and what qualifications may be required to qualify for the various opportunities. It is important to stress to employees that there are occasions where an employee has fulfilled all the qualifications for advancement, but a particular position within an organization is not available or an opportunity for promotion does not exist at that time.

Managers use coaching, motivation, performance reviews, and discipline to tailor employee efforts and contributions progressing toward achievement of individual performance and organizational goals. Before a manager can engage in any of the activities listed above, he or she must first have an understanding of the employee as an individual. This includes understanding the employee's professional and personal goals, desires, and aspirations or lack thereof.

Managers serve as coaches by engaging with employees on a daily basis to enhance and strengthen their presence within an organization. Coaching is an informal process by which the manager provides insight, guidance, and encouragement to aid employees in performance improvement and goal attainment. In some circumstances, managers may also act as mentors for employees, although more commonly, mentors are not the employee's direct supervisor. A mentor–protégé relationship is one that typically develops between a high-ranking person with a great deal of experience and a junior person. The mentor may fulfill many roles for the protégé, including that of role model, teacher, confidant, counselor, coach, and advisor. These functions fall into two main categories, career and psychosocial. In the career category, a mentor engages in activities such as coaching, sponsorship, providing exposure and visibility, challenging perceived limitations, and providing career and organizational advancement advice that help to enhance protégé career progression. In the psychosocial category, mentors participate in interpersonal aspects that give the protégé a sense of identity, competence, confidence, and effectiveness in personal as well as professional development. Whether coaching or mentoring, it is the function of the manager to provide support to help guide the employee to meet development goals and performance standards and, if desired, attain new heights in his or her career (Ragins and Cotton 1999).

Keeping employees motivated to perform at high levels can prove challenging, as motivation differs for each employee, frequently changes with personal circumstances, and is often dependent on the stage of the employee's career. To help understand the source of motivation, the manager should first determine if the behaviors exhibited by the employee are either intrinsically or extrinsically motivated. Those that are intrinsic are evident when the employee derives a sense of accomplishment or satisfaction from completing assigned tasks or projects and helping the organization attain its goals. Comparatively, extrinsically motivated behavior is performed either to acquire something (e.g., recognition, bonus, or award) or avoid punishment (Jones et al. 2010c). There are many resources available to help guide managers in understanding what motivates their employees. A few of the major theories proposed that drive motivation are described below (Jones et al. 2010c):

- Expectancy theories, such as the one developed by Victor Vroom in the 1960s, state that if workers believe that high effort levels lead to high performance, which in turn leads to a desired outcome (e.g., bonus or promotion), then motivation will be high.
- Needs theories, such as Maslow's hierarchy of needs, focus on five basic needs that people must meet. In this theory, the lowest level of unmet need drives motivation. Needs starts with physiological demands (e.g., food, water, and shelter) and safety (e.g., security and stability), progresses to the need for belonging (e.g., friendship and love) and esteem (e.g., feeling good about oneself), and ends with self-actualization and the realization of one's full potential.
- Equity theories are based on the employee's perception that there is equity within an organization and the effort, skills, and knowledge that they contribute, as well as the outcomes they receive, are equal to those of their peers.

When offering feedback, managers assess and evaluate employee behaviors, contributions, and performance in comparison with set expectations and standards, in either a formal or informal manner (Jones et al. 2010a; Taylor 2015). Feedback is usually given at a set interval in the form of a performance review, but it may also occur at variable frequencies, including immediately at the completion of a task or project, at set intervals throughout the year, or annually. Feedback may be presented either formally or informally, depending on the situation. Depending on the institution, formal performance reviews may occur either semiannually or annually, while informal reviews occur as the manager deems necessary. The goal of feedback is to reshape or reinforce behaviors and strive for continuous improvement and development (Taylor 2015). Feedback is deemed to be effective if the following elements are met: specific, work related, nonpersonal, descriptive, timely, frequent, documentable, purposeful, nonprescriptive, constructive, balanced, occurring in an appropriate setting, and interactive (Jones et al. 2010a). As a manager, it is desirable not only to be able to give feedback, but also to seek and receive it from others for both employees' and the manager's own reviews.

In addition to providing feedback during performance reviews, managers also assess future programmatic needs and set goals for the employee to accomplish. A common acronym used when setting goals is SMART, where goals that are specific, measurable, attainable, realistic, and time-bound are made (Table 11.3). Goals may focus on attaining a certain level of performance, mastering a task, or demonstrating skills or knowledge (Doran 1981).

Assessment of goals may be based on how the employee performs his or her job, performance results, and accomplishments. In addition, the skills, abilities, and personality characteristics that directly pertain to the job may be included in the review. Individual goals should be aligned with those of the institution and should give employees opportunities for growth and development. By setting SMART goals, the manager establishes concrete criteria against which to evaluate an employee's performance. These criteria may then be used to provide objective information for decisions regarding raises, bonuses, and promotions, as well as to identify any gaps that may exist in knowledge or skills (Jones et al. 2010c). It is important to consider including feedback from others at the institution, such as peers, customers, subordinates, or often the employees themselves (e.g., self-assessment), as this may contribute to a more well-rounded and complete performance assessment.

Recognition, incentives, rewards, bonuses, and promotions are key aspects of motivation and performance-based reviews. While bonuses and promotions are self-explanatory, it is necessary to distinguish between other key motivational aspects based on subtle differences. Recognition can be an award, verbal praise, or public announcement to recognize an accomplishment made by an employee. Rewards are benefits given to an employee for an achievement, an effort made, or a service performed, the desirability of which is contingent on what the reward is, the value the employee places on the reward, how it is given, and how receiving an award is perceived socially. Incentives are something desired by the employee that he or she works toward attaining. This could be in the form of a bonus or reward. These are all approaches that may be employed by managers to help boost performance.

TABLE 11.3

SMART Criteria for Goal Setting

Acronym	Criteria	Description
S	Specific	*What*: What does one want to accomplish? *Why*: Specific reasons, purpose, or benefits of accomplishing the goal. *Who*: Who is involved? *Where*: Identify a location. *Which*: Identify requirements and constraints.
M	Measurable	How much? How many? When will it be accomplished? Indicators should be quantifiable.
A	Achievable	How can the goal be accomplished? How realistic is the goal based on other constraints?
R	Realistic	Does this seem worthwhile? Is this the right time? Does this match other efforts/needs? Is this the right person for the job?
T	Time-bound	When? What can be done 6 months from now? What can be done 6 weeks from now? What can be done today?

Issues with performance may arise for a plethora of reasons. Conflict is one type of performance problem that occurs often in the workplace and can arise due to a wide variety of reasons, including communication or interpersonal misunderstandings; discrepancy of opinions; clashes of opposing wishes, goals, wants, or needs; scarcity of resources; overlapping authority; feelings of unfairness or underappreciation; differing rewards or evaluation systems; perceived inconsistencies or inequalities; and job dissatisfaction (Jones et al. 2010c). Conflicts may occur at any level within an organization, between individuals, or within or between groups or teams. When approaching conflict, a manager should consider the following (Jones et al. 2010c):

- Identify the problem and focus only on the problem.
- Use peer mediation to let the involved parties solve the conflict themselves.
- Negotiation may be necessary with the use of mediators as facilitators or arbitrators to settle a dispute.
- Compromise is essential, and any sense of unfairness will perpetuate the situation.
- Encourage collaboration and cooperation.

Unresolved conflict is one of the many issues that frequently leads to disciplinary problems. When an employee requires discipline, first identify the problem and determine if the action or behavior was intentional or unintentional. Some behaviors or actions may be grounds for immediate dismissal from the institution. Behaviors that fall into this category are often outlined by the HRD during the employee orientation process. Dealing with disciplinary issues is often very uncomfortable for many managers. Table 11.4 describes considerations for managers when dealing with disciplinary issues.

As a manager, your approach to discipline will greatly influence how the employee perceives the event. For employees, a sense of fairness ranks high on the scale of importance with regards to the outcome of a disciplinary event. Research shows that once the discipline has occurred, many employees only remember the negative aspect of the event, even if behaviors or actions change, while managers tend to focus on the positive outcomes. These perceptions often stem from the emotions of feeling reprimanded and the social stigma of peer awareness of the event. Discipline may also involve financial or legal ramifications, such as ineligibility for bonuses or performance-based increases, and may also include a final written

TABLE 11.4
Approaches to Disciplinary Actions

Be sensitive to personal and cultural differences. The same approach may not work for everyone.
Make sure discipline is done in private.
Try to keep emotion out of the discipline event.
Use a counseling approach, not a legal one.
Adopt positive nonverbal clues, such as making eye contact, nodding in understanding, keeping the body relaxed, and using a soft tone of voice.
Make sure employees understand the policies and discipline procedure.
Provide a clear explanation of the behavior or problem by being specific and identifying the details of the incident.
Make clear that the purpose of the discipline is to help the employee improve or change.

warning leading up to dismissal (Atwater et al. 2007). It is imperative to ensure that there is written documentation of all disciplinary events; this documentation is essential for tracking recurring undesirable events that may progress to employee termination. It is important for a manager to keep records of all disciplinary action performed, whether informal or formal. A convenient method of documentation to employ for informal actions is for managers to send e-mails to themselves with event details, including conversations and actions taken. Informal records help the manager remember incidences that, while not egregious on their own, may be part of a larger problem. Should termination become necessary, these e-mail records will prove invaluable, avoiding vague recollections at a later date when attempting to recall specific events. In the event of a termination, it is advisable to seek guidance from the HRD to ensure that the dismissal is conducted in a legal manner.

Disciplinary action is a very sensitive area and often presents an emotionally charged situation. It is essential that managers not only are familiar with the institutional forms and processes, but also work in close consultation with the HRD. Confidentiality and consistency are of utmost importance both in the handling of the situation and during the documentation process.

Development

Employee development reduces costs due to decreased personnel turnover, thus yielding a larger return on employee investment. Organizational investment in an employee's career improves morale, boosts productivity, and increases organizational efficiency (Jehanzebi and Bashiri 2013). Employee development comes in many forms and targets two overall areas. While these two areas sound similar and are often mistakenly used interchangeably, employee development (also referred to as staff development) and career development are distinctly different. Staff development typically refers to a company's efforts to train and develop an employee for internal benefits. Career development is generally used to describe the efforts that an employee takes to learn new job skills to enhance current career opportunities or to enter a different field altogether. In order to understand the manager's role in employee development and where the greatest impact lies for the organization, as well as the employee, it is important to understand the contrasts between staff and career development, as well as where these areas may overlap.

Staff development tends to focus on job-related skills that contribute directly to the function of the organization (Johnston 2015). To offer staff development, a manager must first identify operational areas where productivity or efficiency needs to be improved. A manager can then find or develop training programs to address specific skills that will immediately improve job performance. For example, if a manager identifies an employee with supervisory potential, he or she may recommend a management course offered either internally or externally. Alternatively, an employee may struggle when dealing with fellow staff that have varying opinions or with different personality types, making a workshop on conflict resolution an appropriate developmental training session to be recommended by the manager. These are just a couple of examples of staff development that can help an employee's overall work performance.

The identification of learning and development needs, in addition to the implementation of staff development, is a shared responsibility between management and employees. The organization provides the policies and organizational directions that often set the stage for staff development. The manager aligns the development of the individual with the directions of the organization while encouraging individual staff members to be responsible for monitoring their own skill levels with regard to job requirements and development. An example of this can be seen when an organization institutes new core values and implements training initiatives in support of this effort. The manager works with staff to ensure that training is completed, staff understand how the change impacts the manner in which their daily work is carried out, and they are responsible for training compliance.

Career or professional development tends to focus on career-related skills, some of which may benefit the organization immediately and directly, while others contribute to the person's professionalism and competency in more general terms. Professional development should be considered a benefit offered, much like health insurance. This benefit may not address training directly specific to job-related tasks, but it can help model personality and professionalism in an employee. For example, AALAS certification may not be required for a position, but once obtained, the employee has demonstrated not only competency for the job but also a professional commitment to the field. Industry certifications will also provide opportunities for the professional to relocate or gain employment within global organizations.

Professional development that seeks to provide broad instruction in relation to careers must focus on specific competencies that can actually improve performance within the organization. Using the example above, an individual's status in the profession is enhanced once he or she is certified as an ALAT; however, by virtue of the knowledge required to attain the certification, the certification may actually improve the day-to-day work performance within the organization.

The responsibility for career development primarily resides with the employee, thus placing the manager in the role of facilitator and counselor. Employees should be actively involved in their own career planning by formulating career goals and developing a plan to reach those goals. Career planning requires a conscious effort on the part of the employee; specifically, it requires time, research, and effort. An employee should devote time to reflect on likes, dislikes, and how they pertain to the career of interest. This may even involve looking beyond an employee's current job for transferable skills. Career planning is difficult work; it does not happen automatically. A major component of career planning involves setting short- and long-term career goals, which will assist in making the process more manageable as career plans progress or change. Furthermore, the manager can assist by providing encouragement and guidance to the employee, but the drive must come from within the individual (Byars and Rue 2006). Moreover, managers are also responsible for assisting staff with setting goals and identifying tools to help with career development. A manager can facilitate employee understanding of available career opportunities within the organization as they relate to employees' abilities and interests. For example, a person might strongly desire to advance to the position of department manager until discovering that there is only one manager role in the department, and that this position requires advanced education and various certifications, such as CMAR. Most importantly, a manager must recognize that career planning or development is not something one person can do for another; it has to come from the individual. Only the individual knows what she or he really wants out of a career, and certainly these desires vary appreciably from person to person.

One of the most effective approaches for achieving career aspirations is for staff to share their career plans and goals with their manager to ensure that they are not only realistic, but also attainable. The creation of a timeline for this plan is equally important using incremental milestones and scheduled checkpoints to discuss progress. Plans and timelines often require adjustments, as unexpected work or life events impede progress.

In the laboratory animal care field, there are many opportunities for individuals to develop and advance their careers. Many institutions offer tuition reimbursement packages for those who wish to further their educational endeavors. Most opportunities for advancement to a manager or supervisor level require a degree in higher education, such as a bachelor's or master's degree in the sciences. In the past, it was difficult to attend college-level courses while working full-time. Today, alternative options exist for those interested in obtaining a master in laboratory animal science. One option includes "distance learning," a term used to describe an education that is received at an off-site location, primarily with the use of computer-based programs developed for virtual teacher and student classroom interactions. Previously, students who participated in a distance learning program received their education through correspondence courses, but today's technology makes many other options possible. Whether it is Drexel University, Eastern Virginia Medical School, Newcastle University, or the Institute of Animal Technology (IAT), accredited distance learning programs offer graduate-level courses, which can be taken through online courses anytime and anywhere in the world.

Institutions often require one of several industry-related certifications as a means to not only demonstrate a commitment to continued education in the field of laboratory animal science, but also serve as an authoritative endorsement of an individual's level of knowledge and competency in laboratory animal technology. A global professional body in the field of animal technology founded in 1950, IAT in the United Kingdom offers continued education opportunities that encourage animal technologists to develop their knowledge, skills, and attitudes. Additionally, the AALAS-sponsored technician certification designations of ALAT, LAT, and LATG are well known and widely used throughout the varied fields of laboratory animal science. It is worth noting that these certifications begin at the entry level for a technician and provide for a natural progression as experiential skills are attained. Many institutions offer incentives for taking the certification exams, which include reimbursement of exam fees, provision of study materials or courses, employer-paid time to study, promotions, bonuses, or salary increases with successful completion of a certification exam. As a manager, it is important to emphasize the importance of these certifications, which are examples of where staff and career development overlap: employers may require AALAS certification for a particular job title, but the employee also benefits professionally from an industry perspective. Employees may also demonstrate their dedication to continuing education by maintaining their certification in the AALAS Registry, which requires submission of continued education units (CEUs) every 2 years to keep certification current and active.

Managers and supervisors also have the opportunity to further their own developmental endeavors through the AALAS-sponsored CMAR program, which is designed to increase competency and professionalism in the field of laboratory animal resources management. In addition to the CMAR program, AALAS offers the ILAM educational program. This program was originally established by LAMA and was adopted by AALAS as a cosponsor. ILAM was developed to provide instruction in management concepts applicable to the laboratory animal science field, as well as to enhance communication, team building, and networking among colleagues with mutual interests. Additionally, many organizations have internal courses, potentially as part of a management certificate program, which offer the basic components of people management, such as conflict resolution or diversity training, often including company-specific core values and behaviors. Managers are encouraged to take advantage of all the resources available to them within their organization, as well as those offered externally.

Workforce Diversity and Career Development

Career development is not uniform across a diverse workforce. Depending on an individual's background, developmental needs can differ greatly, and an effective manager understands that one size does not fit all. Members of a diverse workforce identify by attributes such as culture, ethnicity, gender, race, religion, and multigenerational identification. Understanding differences between the generations is fundamental to building a successful multigenerational workplace. Each generation has particular needs that include specific preferences, expectations, beliefs, and work style. A brief description of each generation and their values and views toward development and career planning is included in Table 11.5 (Symonowicz and Straeter 2008).

A key to effective management lies in understanding the types and attributes of these individuals. Much of this knowledge is founded in the understanding of the generation to which these individuals belong.

Another subset of a diverse workforce includes staff with special needs or disabilities. The Americans with Disabilities Act of 1990 (ADA 1990) was established to mandate ending discrimination in employment, as well as opening society to the disabled for most situations in everyday life (ADA 1990). *Disability* is defined as a physical or mental impairment, which substantially limits one or more major life activities (Suckow et al. 2001). These could include such activities as hearing, seeing, speaking, walking, performing manual tasks, caring for oneself, learning, or working. According to the ADA, people with disabilities must be qualified to perform the essential functions or duties of a job, with

TABLE 11.5

Generational Categories

Generation	Values	Development Approach
Traditionalists (Born 1925–1945)	• Change averse; not very risk tolerant; respect authority and hard work • Build a legacy, a lifetime career with one company	• Most retired; if not, provide low-stress and stable environment • Not looking for advancement; will work hard for wages
Baby boomers (Born 1946–1964)	• Lifestyle revolves around work; balance is a quaint idea but not really a possibility • Build a perfect career and excel • Team players and loyal, but do not adapt well to change	• Interested in exciting, challenging projects • Strive to make a difference
Generation X (Born 1965–1980)	• Develop behaviors of independence, resilience, and adaptability; work to live and view the world with a little cynicism and distrust • Strong team players and collaborators • Build a transferable career; variety of skills and experiences • Entrepreneurial thinking, but rank low on executive presence	• Opportunities to learn and grow • Desire manager's acknowledgment of job well done
Generation Y—Millennials (Born 1981 and 2000)	• Constant experience in the networked world (grew up on computers and the Internet); computer access has had a profound impact on their style in approaching problem-solving situations • Build several parallel careers; have several jobs simultaneously • Generally, not strong team players	• Desire flexible work schedules and a more rounded work–life balance
Generation Z—Nexters (Born after 2000)	• Just beginning to enter the workforce as part-time high school students	• Behaviors are expected to follow those of generation Y

or without reasonable accommodation, in order to be protected from job discrimination. The individual must satisfy the employer's requirements for the job, such as education, employment experience, skills, or licenses. The individual must also be able to perform the essential functions of the job with or without reasonable accommodation. "Reasonable accommodation" is any change or adjustment to a job or work environment that permits a qualified applicant or employee with a disability to participate in the job application process, to perform the essential functions of a job, or to enjoy benefits and privileges of employment equal to those enjoyed by employees without disabilities (U.S. Equal Employment Opportunity Commission 2015). For example, reasonable accommodation may include provision or modification of equipment or devices; restructuring of job tasks or responsibilities; modification of work schedules; reassignment to a vacant position; adjustment or modification of examinations, training materials, or policies; or provision of readers and interpreters to make the workplace readily accessible and usable by people with disabilities. Some of the activities mentioned above are critical to the essential job requirements, making reasonable accommodations unrealistic and difficult. In addition to the ADA, the U.S. Department of State addresses the global concerns around disabilities through the Special Advisor for International Disability Rights (SADR), which leads the U.S. comprehensive strategy to promote and protect the rights of persons with disabilities internationally. The United States, as part of the foreign policy, works to remove barriers and create a world in which disabled people enjoy dignity and full inclusion. Whether within country or outside, managers should make every effort to familiarize themselves with not only U.S. federal and state regulations, but also those rules and policies that may impact managing individuals with disabilities in a global environment.

An employer is required to provide a reasonable accommodation to a qualified applicant or employee with a disability unless the employer can show that the accommodation would create an undue hardship, requiring significant difficulty or expense. Managers and organizations can benefit greatly by the inclusion of these individuals in their workplace, potentially enriching the lives of those with special needs. Individualized training is emphasized and is a key factor in ensuring a successful integration of the individual into an organization (Weichbrod 2015).

A unionized or bargaining unit staff is another type of work group diversity that may require a different managerial approach, which should be in alignment with the collective bargaining agreements (CBAs) in place. Managing unionized staff can sometimes have additional challenges, particularly when it comes to hiring and recognition practices. However, regardless of the type of staff, managing with consistency, fairness, and respect will lead to a more successful work environment. Additionally, understanding and respecting the terms of the CBA or union contract is an important aspect of successful workforce management. Most agreements contain job descriptions that may or may not be detailed and specific, creating challenges to provide individualized developmental and career progression plans. For example, if a highly qualified and skilled union employee were to be given a leadership or growth opportunity that was not offered to other union employees, there is a possibility that the manager will be grieved by the union as a result of a violation to the CBA. However, there are many ways to create unique opportunities for the development of unionized staff. One example is through the use of self-directed work teams (SDWTs), which allows for staff autotomy and the ability to have control and input on work processes without compromising quality and workplace policies. SDWTs are one aspect of the self-managed approaches developed for unionized staff that has proved effective for Taco Bell®, Walmart®, and Costco®, as well as a 3M™ manufacturing facility in the early 1990s. More specifically, an SDWT is a group of people, usually employees in a company, who combine different skills and talents to work toward a common purpose or goal without the usual managerial supervision (Williams 1995). In order for an SDWT to succeed, the company or organization must provide a meaningful mission statement to the team, empower the team to make important decisions, establish boundaries through the provision of rules and company policies, and train members in the skills and knowledge needed to accomplish their purpose. Similarly, a high-involvement workplace is another scenario where employees exercise control over daily tasks, cited as one of the most important work characteristics to produce happiness on the job (Reinhard 2015). High-involvement work teams improve productivity and job satisfaction by giving workers more autonomy and encouraging cooperative work efforts. With these tools, the team is held accountable for the success or failure of a project (Williams 1995). In both examples, work teams are focused on learning and continuous improvement. The team's mandate is not just to get a job done, but to use competitive intelligence, market information, and internal statistics to find better ways to perform its job responsibilities (Reinhard 2015). To optimize the effectiveness of high-performance work teams, managers must focus on employee continuing education and career development.

Conclusion

In today's competitive workplace, companies and institutions must offer more than a competitive salary to attract and retain excellent employees. Employee loyalty is at a reportedly all-time low, and individuals focus more on maintaining a portable skill set in order to move from opportunity to opportunity, rather than forging a long-term commitment to a company (K@W 2012). According to a 2011 Careerbuilder.com report (http://business.time.com/2012/05/11/declining-employee-loyalty-a-casualty-of-the-new-workplace/), 76% of full-time workers, while not actively looking for a new job, would leave their current workplace if the right opportunity came along. Other studies show that each year, the average company loses from 20% to 50% of its employee base. Following a MetLife study in 2011, the vice president of MetLife's U.S. business made a profound statement: "Businesses are understandably focused on expenses, but they're taking their eye off the ball with human capital issues, notably what drives employee satisfaction and loyalty" (Petrecca 2011). Given the current uncertain climate, managers are forced to learn motivational strategies to attract and retain the best workers at the risk of losing them to a competitor who has strong motivational retention strategies. Unlike previous generations, today's generation of employees work for different reasons and look for different job-related experiences. Historically, a weekly paycheck served as a sufficient motivator, but now is only one in a myriad of factors that motivate employees. Some people desire a social connection with their colleagues, some work to fulfill their self-worth, and yet others work to move up the corporate ladder. Managers today need to do more than give the occasional pat on the back; they need to develop a dynamic and diverse workforce spanning several generations that requires a number of management tools to address individual staff characteristics, needs, and issues. It starts from the beginning, with recruiting the right team members, implementing the appropriate developmental plan for each individual, and demonstrating the desire to retain the talent. Although not all-inclusive, this chapter is designed to provide some of the important tools that will equip the manager for success. The field of human resources is not always born out of instinct, and managers are warned not to become overly confident in their skills, but instead to continually invest in their self-growth.

REFERENCES

AALAS (American Association for Laboratory Animal Science). (2008). Job analysis survey—LATG/CMAR. Collierville, TN: AALAS.

ADA (Americans with Disabilities Act). (1990). P.L. 101-336. Amended September 28, 2008.

Atwater, L., J. Brett, and A. Charles. (2007). Workplace discipline: Lessons learned. *Organizational Dynamics* 36(4): 392–403.

Byars, L., and L. Rue. (2006). *Human Resources Management*. New York: McGraw-Hill Irwin.

Dale, R. (2015). Advantages and disadvantages of being a contract employee. http:http://www.ehow.com/info_10074329_advantages-disadvantages-being-contract-employee.html (retrieved October 13, 2015).

Doran, G. (1981). There's a S.M.A.R.T. way to write management's goals and objectives. *Management Review (AMA FORUM)* 70(11): 35–36.

Doyle, A. (2015). Phone interview questions and answers. http://jobsearch.about.com/od/interviewquestionsanswers/a/phoneintervquest.htm (retrieved September 13, 2015).

Gebelin, S., B. Davis, and K. Nelson-Neuhaus. (2004). *Successful Manager's Handbook*. Minneapolis: Personnel Decisions International.

Heathfield, S. (2015). Assess job fit when you select employees. http://humanresources.about.com/od/glossaryj/g/job-fit.htm (retrieved September 13, 2015).

Holt, M. (2015). Training in workplace diversity. *Houston Chronicle*. http://smallbusiness.chron.com/training-workplace-diversity-3026.html (retrieved September 13, 2015).

Jehanzebi, K., and N. Bashiri. (2013). Training and development program and its benefits to employee and organization. *European Journal of Business and Management* 5(2): 243–244.

Johnston, K. (2015). The differences between staff development & professional development. http://smallbusiness.chron.com/differences-between-staff-development-professional-development-37039.html (retrieved September 13, 2015).

Jones, G. R., J. M. George, S. C. Dejanasz, K. O. Dowd, and B. Z. Schneider. (2010a). *Management Skills II: Planning and Organizing*. New York: McGraw-Hill.

Jones, G. R., J. M. George, J. M. Ivancevich, R. Konopaske, M. T. Matteson, W. G. Nickels, J. M. McHugh, and S. M. McHugh. (2010b). *Management Skills 1: Foundations of Management*. New York: McGraw-Hill.

Jones, G. R., J. M. George, J. M. Ivancevich, R. Konopaske, M. T. Matteson, W. G. Nickels, J. M. McHugh et al. (2010c). *Management Skills 3: Leading and Controlling*. New York: McGraw-Hill.

K@W. (2012). Declining employee loyalty: A casualty of the new workplace. http://knowledge.wharton.upenn.edu/article/declining-employee-loyalty-a-casualty-of-the-new-workplace/ (retrieved September 13, 2015).

Merriam-Webster. (2015a). Employee. http://merriam-webster.com/dictionary/employee (retrieved August 22, 2015).

Merriam-Webster. (2015b). Management. http://merriam-webster.com/dictionary/management (retrieved August 22, 2015).

Murphy, B. (2015). 7 things great leaders always do (but mere managers always fear). http://www.inc.com/bill murphy-jr/7-things-great-leaders-always-do-but-mere-managers-always-fear.html (retrieved September 13, 2015).

Oxford Dictionaries. (2015). Management. http://www.oxforddictionaries.com/definition/english/management (retrieved August 22, 2015).

Petrecca, L. (2011). Employee loyalty is at a three-year low. http://usatoday30.usatoday.com/money/workplace/2011-03-26-employees-less-loyal.htm (retrieved September 13, 2015).

Purdue University. (2015). Supervisor's guide to effectively onboarding a new employee. https://www.purdue.edu/hr/pdf/supervisorGuideOnboarding.pdf.

Ragins, B., and J. Cotton. (1999). Mentor functions and outcomes: A comparison of men and women in formal and informal mentoring relationships. *Journal of Applied Psychology* 84(4): 529–550.

Reddy, P., and P. Tripathi. (2008). *Principles of Management*. New Delhi: Tata McGraw-Hill.

Reinhard, C. (2015). Characteristics of a high involvement work team. http://smallbusiness.chron.com/characteristics-high-involvement-work-team-15916.html (retrieved September 13, 2015).

Suckow, M. A., F. A. Douglas, and R. H. Weichbrod, eds. (2001). *Management of Laboratory Animal Care and Use Programs*. New York: CRC Press.

Swijtink, Z. (2016). Daniel Goleman's five components of emotional intelligence. https://sonoma.edu/users/s/swijtink/teaching/philosophy_101/paper1/goleman.htm (retrieved August 12, 2016).

Symonowicz, C., and P. Straeter. (2008). Managing the generational divide. *LAMA Review* 20(1–4): (1)26; (22)10; (23)10–11; (24)10–11.

Taylor, B. (2015). The integrated dynamics of motivation and performance in the workplace. *Performance Improvement* 54(5): 28–37.

Types of interviews. (2015). http://hr.berkeley.edu/hr-network/central-guide-managing-hr/managing-hr/recruiting-staff/employment/interviewing/types (retrieved September 13, 2015).

U.S. Equal Employment Opportunity Commission. (2015). The ADA: Your employment rights as an individual with a disability. https://www.eeoc.gov/facts/ada18.html (retrieved September 13, 2015).

Vernon, A. (2013). Managers role in onboarding webinar. https://www.td.org/Digital-Resources/Webcasts/TD/2013/03/Managers-Role-in-Onboarding (retrieved September 13, 2015).

Weichbrod, R. H., S. Blanks, M. D. Plemons, J. Chinchilla, N. S. Linde, J. B. Drake, and J. M. Raber. (2015). Untapped potential: Animal care and use programs adopt alternative training and recruitment strategies for individuals with intellectual and developmental disabilities. *Lab Anim Sci Prof* 3(4): 12–20.

Williams, R. (1995). Self-directed work teams: A competitive advantage. http://qualitydigest.com/magazine/1995/nov/article/self-directed-work-team-15916.html (retrieved September 13, 2015).

Wood, H. (2015). Needs & benefits of a diverse workforce. http://smallbusiness.chron.com/needs-benefits-diverse-workforce-21703.html (retrieved September 13, 2015).

12

Education and Training

Bruce W. Kennedy and Kim Froeschl

CONTENTS

Each research facility shall provide for the training of scientists, animal technicians, and other personnel involved with animal care and treatment in such facility.

AWAR 2013

The U.S. Government Principles, Health Research Extension Act of 1985, and the PHS Policy repeatedly refer to appropriately trained, qualified, and experienced personnel, and availability of instruction and training. The institution is responsible for the training of its staff.

Maloney 2013

Introduction

The two quotes opening this chapter attest to the continued importance of and purpose and requirement for training in laboratory animal science. Moreover, evidence of the support for training in the conduct of research, teaching, and testing at institutions where animals are used has escalated. Since the publication of the first version of this chapter (Kennedy 2002a), the position of laboratory animal trainer has been defined (Kennedy 2002b) and is included on the organizational charts of many facilities. The current edition of the *Guide for the Care and Use of Laboratory Animals* (*Guide*) (ILAR 2011) mentions more than 150 different instances of the word *training*. Many lab animal societies have incorporated more training offerings into their membership benefits and conferences, most notably the American Association for Laboratory Animal Science (AALAS) and the Institute of Animal Technology (IAT). The European lab animal community, represented by the Federation of European Laboratory Animal Science Associations (FELASA 2015), have proposed revisions and recommendations to enhance their accredited training programs. Several other international laboratory animal organizations—including universities, national societies, and research entities—have modeled their own training programs on these prominent groups.

Increasingly, animal research oversight bodies around the world mandate that individuals who work with animals in research and training settings provide evidence of their knowledge and skill at whatever task they do. This is aligned with the principles of refinement from the 3Rs (replacement, reduction, and refinement), where knowledge obtained through training addresses animal welfare, recognition of pain and distress, and appropriate housing (NC3Rs 2016a). And, whereas the institutional animal care and use committees (IACUCs*) today have the ultimate responsibility, it is typically the management of the program that is charged with and implements the training. To be clear, as stated in the *Guide* (ILAR 2011), "the IACUC (or institutional equivalent) is responsible for assessment and oversight of the institution's program components and facilities." Further, "the IACUC is responsible for providing oversight and for evaluating the effectiveness of the training program" (Foshay and Tinkey 2007).

This chapter is written to be both a practical and a conceptual guide for the laboratory animal manager engaged directly or indirectly in training. It is based on and extends the principles of managing training presented in the original chapter (Kennedy 2002a). The practical portion comes first and builds on the train-the-trainer (TtT) structures used to establish a training program, adding passages on the manager's role in developing both cultures and programs of training. In support of practice, educational theories and concepts related to training and applied to training management in laboratory situations are also presented. Using a training metaphor, like learning to drive a car, the immediate want is to get in and go. That may seem to be practical, but it helps to know a little about the car (concepts), even if the driver (manager) is experienced. Thus, with both driving and training, an awareness of several factors can be beneficial: what the drivers need to operate, how the "ride" can be improved, how to avoid accidents, what compliance with the rules of the road means, and so forth.

In 2002, the notion of a facility trainer was novel (Kennedy 2002b). While training itself had been occurring for many decades, the more formalized practice of training in lab animal science was young and developing, having previously been largely a part-time responsibility of someone in the leadership of the lab animal facility (Pritt and Clifford 2014). As these authors noted, lab animal training has evolved in a relatively short time from basic concepts of content to considerations of learning. Previously, the focus was on topics pertaining to "what" should be taught and reiterations of the regulatory requirements to the "hows," meaning educational techniques and ideas for presenting and then evaluating the effectiveness of training. Training now proceeds with the inclusion of all facility staff, from repair personnel to institutional officials, rather than only the select members of the husbandry team. At today's laboratory animal conferences, there are more topics related to educating laboratory animal staff, and increasingly, they are focused on the details of competency and assessment tools, electronic learning management systems (LMSs), English language learners (ELLs), and unique organizational (institutional) models for training.

Alongside this evolution has been the growth of professional training organizations like the Laboratory Animal Welfare Training Exchange (LAWTE), which started in 1994 (Pritt and Clifford 2014; Kennedy 2015). Laboratory animal managers are working side by side with designated trainers, fulfilling training objectives as expressed in the LAWTE mission statement of "expanding animal welfare and

* In this chapter, the term *IACUC* will be used to designate the committee or oversight body responsible for oversight of an institution's animal care and use program.

enhancing public understanding through effective training and education of animal research professionals" (LAWTE 2015). Doubtless, the many lab animal organizations and programs around the world have contributed to and fostered the concepts of training in vivaria. The reader is referred to the abundance of listings in the appendices.

One perspective on what it means today to be involved with training in a lab animal setting comes from the website of the American College of Laboratory Animal Medicine (ACLAM 2015). While written expressly for veterinarians, the following passage conveniently summarizes the responsibilities for all who are engaged in training, especially facility managers.

> As laboratory animal experts, ACLAM Diplomates often teach and train others. This teaching may range from formal classroom instruction to informal training in a laboratory or animal facility setting. Just as the type of teaching can embrace a wide spectrum, so too can the categories of people instructed. Individuals may be fellow scientists, research technicians, or students who are learning techniques of animal experimentation or about the regulations and requirements of an animal care and use program; they may be animal technicians who are learning about the care and use of laboratory animals; or they may be veterinarians, veterinary students, or undergraduate students who are studying laboratory animal medicine, science or management.
>
> Another aspect of teaching involves informing the public about the use of animals in research. Elected representatives, the media, and the general public need to be educated concerning the laws and regulations which protect research animals, the excellent care that is given to laboratory animals, and the great contribution animal research can make to the life, health, and well-being of humans and animals. Because of their expertise, Diplomates have credibility with the public and are, therefore, important members of the institution's public information team.

Having a Foundation in Lab Animal Training

As mentioned, this chapter builds on the preceding version (Chapter 2) from the book *Management of Laboratory Animal Care and Use Programs* (Suckow et al. 2002). That chapter is in the process of being archived with LAWTE for access and reference in the future. It was written with the theme of TtT underlying the sections because the typical lab animal manager had limited background in training, yet most were obliged to train their own staff and frequently others in the research environment. Managerial skill sets a few decades ago were usually based primarily on animal husbandry, facility management, and human resources (HR)—not training. Over time, new directions and roles involving training have emerged for managers, paralleling additional animal-focused responsibilities. Today, many lab animal facility managers are directly responsible for implementing the training program as part of their duties; they supervise trainers or work closely with a training entity, such as a consultant or institutional training department.

Consequently, this chapter presents training topics more from the perspective of managing a training program and less so from being the actual trainer. Nevertheless, familiarity with TtT topics as a foundation—related to laboratory animal training—will be helpful and potentially abbreviate the manager's learning curve. The following topics are the highlights from the original chapter on training for managers, many of which are updated in this chapter:

- What constitutes training sessions, recognizing that they can be constructed around the 6Ws tool (who, what, when, where, how, and why)
- What the objectives of training are, including regulatory requirements and fulfillment of the principles of the 3Rs (Russell and Burch 1959)
- The ABCs of training, meaning how people act (B for behavior) when they learn something (C for cognition) and feel afterward about what they learned (A for attitude)
- The concept of adult learning, that is, andragogy, as studied and articulated by Professor Malcolm Knowles

- The psychology of learning, in terms of WIIFM (what's in it for me) and Professor Abraham Maslow's hierarchy of needs, which addressed the intrinsic and extrinsic motivating factors applied to learning
- The numerous sources for developing training curriculum, adapted from several related disciplines within education
- The resources available from laboratory animal science organizations to aid trainers, the types and amounts of which have since increased greatly
- An introduction to computer-based training (CBT), much of which has been eclipsed by today's electronic technologies, known more commonly as e-learning
- The relationships, separations, crossovers, and distinctions between training practices and HR management
- Case studies that explore and introduce training principles as applied to managing the laboratory animal facility

These practices and concepts are still quite valid and functional for use with training in today's lab animal facility. In this chapter, they are made relevant to the current training situations occurring in facilities.

Roles and Expectations of the Facility Manager in the Training Environment

The lab animal facility manager is sometimes held directly responsible for the outcomes of training within the facility. For example, the manager could be the supervisor of one or more trainers, could manage the budget of the training department, or could be a part- or full-time trainer. In these capacities, here is a list of items, any of which the manager could anticipate doing. The reader is encouraged to add others and make a personal list.

- Being engaged with the day-to-day operations regarding training
- Reviewing training records and compiling reports to submit to regulatory agencies or upper management
- Maintaining an LMS, including collection and entry of learner data and information
- Engaging personally in training activities
- Establishing and contributing to the culture of training within smaller departments or for the whole institution
- "Walking the talk" about training, sharing its importance and significance
- Conducting training sessions
- Designing curriculum
- Anticipating and identifying training needs
- Recognizing that not all persons learn the same way, nor do all trainers teach the same way
- Recognizing that training subjects are wide and varied (e.g., consider the different needs for a highly secured lab, a traditional barrier, and wildlife habitats)
- Directly managing the training, which is an active process, or supporting it more passively, as in training management
- Supporting trainers, which includes providing resources, committing time, and helping with coordination
- Facilitating learning, including arranging work time for study, mentoring, and planning
- Enabling learning in both formal (e.g., classrooms and professional conferences) and informal (e.g., spontaneous hallway conversations and e-mailing materials) settings

A facility manager shows leadership in many aspects of laboratory animal operations. Leadership (qualities, characteristics, and actions) may vary depending on specific circumstances and objectives.

Sergiovanni (2007), a scholar of educational leadership, has formulated a list of roles, what he calls the *five forces of leadership*, and how they are applied in environments of educating and training. Here, they are further interpreted for situations in managing lab animal training.

1. The *technical force* leader in training has the role of management engineer or coordinator. This aspect involves planning, organizing, and scheduling training, which are analogous to the 6Ws tool (Kennedy 2002a, 2016b) of answering the *where* and *when* questions. It may be the manager who needs to arrange a place and time for training or offers to host an investigator presentation for husbandry staff or an AALAS activity like a webinar or conference.
2. The *humanistic force* leader needs to address the personal and human relations issues associated with training. In this role, the manager provides support to trainers and trainees* to meet the training objectives of the institution. The manager may be motivating a group of technicians studying for a certification examination or may be building morale and clarifying for investigators who question the need for required training. The manager who ignores the training needs of colleagues is likely to have a harder time gaining respect.
3. The *educational force* leader appears in the manager when asked to be a trainer (or a teacher; see the "Is It Education, Teaching, or Training?" section of Appendix 12.1) to share professional and technical experience. Doing so demonstrates being part of the team and an aspect that is not managing all the time (staff may say, "I didn't know she knew how to do that"). Such a manager demonstrates an intimate knowledge of the institution, its equipment, its people, and its purpose. An educational force manager has a different, yet complementary, function to that of a pedagogical instructor in conveying knowledge.
4. The *symbolic force* leader exemplifies the importance of training. The manager speaks to the impact (e.g., welfare and reproducible data) for both animals and personnel and, when joined with the humanistic force role, the rewards that come from training. Attending celebrations for persons who achieve a training award, participating in training sessions, suggesting and promoting the purchase of training aids, and calling out the *teachable moment* are all examples of symbolic leadership that is visible to others.
5. The *cultural force* leader often sets the tone for training, ensuring that learning is part of the culture of the institution. These are examples that show support of a learning culture: (a) emphasizing the need to attend training, despite the day being busier than expected; (b) writing into the budget a request to purchase new training materials; and (c) advocating for the goals to have more staff become credentialed, whether Registered Veterinary Technician (RVT), laboratory animal technologist (LATG), diplomate of the American College of Laboratory Animal Medicine (DACLAM), Certified Professional IACUC Administrator (CPIA), and so forth.

Culture of Training Set by the Manager

Imagine a day before you knew much about working with animals, whether it was your pet to whom you taught a trick or a class you took about laboratory animal science or your first job in the research field. Unknowing, unsure, and unskilled are probably appropriate descriptors. Those situations—and more, as will be discussed—are exactly why training in laboratory animal science is necessary, critical, and warranted.

Despite being well meaning in what we do in animal research, as humans we tend to forget what we have learned, we may be unaware of new regulations or techniques, and we can lose our skills without retraining or practice. Training, offered and supported by institutional management, is often the

* The terms *learners*, *students*, and *trainees* will be used synonymously in this chapter. In practice, *students* may refer to those in a typical classroom setting, and *trainees* may refer to those completing residencies and internships. All are learners being trained.

mechanism to refresh and enhance our capabilities. An improved and strengthened animal care program results, something that all stakeholders—researcher, trainer, administrator, public, and most of all, the animals—can appreciate.

The following is a list of several ways by which the manager, applying the five educational forces, can contribute to and create a culture of training:

- Suggest topics for training sessions.
- Set competency criteria to support husbandry and technical skills.
- Become a member of an organization that provides training (see Appendix 12.5). Offer memberships to staff as well.
- Follow up on training sessions. Ask what happened, what was learned, and how it will be used in the facility. Seek input from those who received the training.
- Send people to off-site training with the expectation of a "report" at the next group meeting. The report can take the form of a written page for distribution, a 5-minute talk, a prepared presentation with slides, a discussion of an item or topic, and so forth.
- Review training records for being current, complete, and relevant to what is needed for the institutional animal program.
- Facilitate the various learning styles of those receiving training; likewise, facilitate evaluation styles.
- Engage in postapproval monitoring (PAM) activities that involve training components.
- Smile at the opportunities, rather than frown, when training obligations are to be met and costs are to be expended.
- Train while wandering around (TWWA), which is a cousin to manage by walking around (MBWA).
- Include training—either attending or presenting sessions—as part of the annual performance reviews for members of staff.
- Consider facility inspections not as a burden, but rather as learning and training opportunities to make improvements.
- Negotiate with staff to get work done so that training can occur (e.g., come in earlier than usual).
- Talk about the training done as a manager, acknowledging that learning is a lifelong process.

The manager often sets the tone, consciously or not, for much in the laboratory animal facility, including the culture of training. By scheduling time into the day for training and making it clear that training is a priority, the manager shows commitment to the program. But, it is not solely in the hands of management on behalf of the institution, IACUC, and others. It is incumbent upon employees to be engaged to identify the opportunities (available time, types of programs, gaps in knowledge, etc.) when they can bring learning into their own work for the benefit of animal research and their organization or company. Further, staff create value when they train themselves in ways that are mutually beneficial with the institution. By engaging with the training culture and processes, staff contribute to the overall mission and provide opportunities for others to gain proficiency and confidence, and contribute to animal welfare. They do not just do their jobs; they participate in teaching, too. Overall, training is part of the quality of a team, its reputation, its capability, and its accomplishments. When management responds to staff training requests and feedback, the culture of training can be enhanced.

Outreach Training

A particular means to develop a culture of learning is to create opportunities and encourage others to be involved in outreach training, which is providing information to those who might not otherwise have access through a training experience. Examples include teaching or talking about laboratory animal

FIGURE 12.1 Mirror images of the terms teach and learn.

science with other departments in the institution, in local K–12 school systems, in veterinary and vet tech college programs, and with civic organizations and similar entities (Benjamin 2016; Kennedy 2016a).

While it is instructing others, outreach training provides a manager or trainer with the learning experience of conveying information to a different audience, as well as instructional practice. As Figure 12.1 shows, teaching and learning are reflections of each other. In fact, it is a Latin proverb, *docendo discimus*, which literally translates to "We learn by teaching" (https://www.merriam-webster.com/dictionary/docendo%20discimus), but commonly it is understood as "By learning you will teach; by teaching you will learn." Training knowledge and skills can be practiced with outreach training.

An additional benefit of outreach training is to provide school systems support in the STEM disciplines (science, technology, engineering, and mathematics). By conducting outreach in schools, a manager is also reaching out to the potential future workforce (Benjamin 2016).

Training Yourself, the Manager

The manager who attends continuing education (CE) sessions and brings the information back to the institution is sending a message about training—that it is an important way to spend one's work time. Participation in groups geared toward training and managerial development (such as the Laboratory Animal Management Association [LAMA]; see Appendix 12.5), as well as the presentation of posters, case studies, and presentations at regional and national meetings, is another way in which managers can lead by example. Laboratory animal science is a fast-moving field, and the facility manager must prioritize or commit to training, staying abreast of developments in research of all kinds: HR in terms of law and cultures, the technologies of both learning and working, and organizational systems in order to be efficient and capable.

A new manager should plan a training ladder for himself or herself. It is like planning a career ladder that includes self-evaluations to help determine what is needed to progress in the field. It should be recognized as a different paradigm to do training management, compared with the more typical management of people, outcomes, animals, projects, and so forth. Training oneself through resources like a 360° evaluation and books such as *The First 90 Days: Proven Strategies for Getting Up to Speed Faster and Smarter* by Michael D. Watkins will likely prove fruitful. The 360° feedback evaluation, or multisource feedback, is an exercise taken from industrial psychology and organizational development (Armstrong et al. 2000), where evaluations are received from an individual's work circle. Used in business, it has been applied to the training environment (Armstrong et al. 2000), where the educational leader or trainer is evaluated by the circle of stakeholders that includes students or trainees, peer trainers, and superiors concerned with managing the learning environment. The direct stakeholders involved should be comprised of animal care providers, veterinary staff, technicians, cage wash staff, investigators, administrators, and oversight committee members. Others perhaps interested in training outcomes resulting from 360° evaluations could include regulatory oversight and accrediting agencies, commercial clients, the public, and vendors or suppliers.

Being Engaged with Training

A buzzword associated with establishing a culture, whether for training or another purpose, is the term *engaged*. Engaged means involvement in something to a high level of commitment. Part of knowing that training has worked is seeing the trainee engaged with what was learned, which can be observed as a demonstration of knowledge or skill, behaving in a way that reflects the training, and expressing internalization of the training. After being engaged with training, a manager should observe that the trainee knows what to do, and does it as expected and with the proper attitude. See also the "Assessment: Measuring Competency and Evaluating Training" section in this chapter.

One of the challenges about managing staff is that not everyone "buys into the corporate culture" or wants to be deeply engaged (Dinkel 2013). This struggle has been especially described for millennials, those born in the 1980s and 1990s (Caprino 2014). In the highly regulated, technologically quick moving, and often emotionally charged discipline of animal research, stakeholders who do not adopt training and accept the importance of training as a continuous process could become problematic for facility managers. Using an approach that can apply to animal care personnel of all ages, Caprino recognized the challenge, saying, "Quit trying to engage millennials." Among her pieces of advice to enhance adoption of training are avoid using the direct term of *being engaged*, share the obligation of training with the trainees, incorporate computing devices as training tools, and encourage individual thinking on the team, which includes how each person wants to learn. Her words resonate with the learner-focused principles of adult learning too (Knowles et al. 2012).

Training Collaboration between Facility Management and the Oversight Board

The oversight board, herein referred to as the IACUC, of a research animal facility has the responsibility to evaluate the effectiveness of the institutional training program. The management in collaboration with the IACUC ensures that appropriate training (and, as found to be necessary, any remedial or retraining) occurs within the institution as prescribed in the regulations. Communication on such matters is critical for compliance and animal welfare. Members of the IACUC might not themselves conduct the actual training, ensuring instead it is done by other entities, such as the IACUC staff, veterinary staff, managers and supervisors, or a group of trainers. Facility management often does and should assist the IACUC with regard to training in several ways:

- Maintaining records of training that the staff receive and entering relevant data into an LMS
- Keeping training up to date through CE opportunities
- Participating in and initiating training needed by the institution
- Matching training to the activities proposed in the animal study protocol and identifying needs for additional or remedial training
- Informing the IACUC of opportunities made available for training, for example, through announcements received from the organizations, societies, and training partners, as listed in the appendices of this chapter
- Hosting opportunities for training of IACUC members, such as webinars and industry speakers
- Appreciating that training is not only a regulatory obligation that the institution and IACUC are fulfilling, but also one that addresses the welfare of animals, the 3Rs, and the quality of scientific results
- Evaluating the actual performance of various tasks, including surgery and animal care, in the facility and asking one-on-one questions when observing on-the-job training opportunities

- Supporting and demonstrating the IACUC's institutional approach to the competent performance and refinement of animal-related tasks
- Facilitating the learning process by engaging with users of the animal facility

Education Related to Lab Animal Training

Traditional education in the workplace came in the form of the relationship between journeyman and apprentice. This approach was also typical in lab animal training where the experienced manager or senior technician taught the junior technicians. Typical examples could be a session in the break room of the facility for a biology lecture about one of the animals used in the facility or instruction alongside a large piece of equipment in the cage wash area. Such one-on-one educational formats were prevalent for decades, until the advent of CBT, which has become widespread in education—for example, the AALAS Learning Library (ALL 2016) and the Collaborative Institutional Training Initiative program (CITI 2016), and others as listed in Appendices 4 and 5.

The beginnings of lab animal training are traced to the early twentieth century and attributed to veterinarian Simon Brimhall, recognized as the father of laboratory animal medicine. In 1915, Dr. Brimhall joined the Mayo Clinic (Rochester, Minnesota), filling the first position devoted to laboratory animal medicine in an American medical research institution (AALAS 1999; Moreland 1999; Fox and Bennett 2015). Being a new discipline, there was little existing training for anyone. As Moreland wrote (1999), "During the early years these veterinarians relied on their basic education in veterinary medicine, on what they could learn from other biomedical scientists, from one another, and by 'trial-and-error.'" Stephens (1999), writing about animal caretakers, said that they too developed their knowledge while on the job—in other words, practical experience. Eventually, for many in the field, "the advantage of training soon became evident." And not long afterwards (in 1950), the forerunner of AALAS, that is, the Animal Care Panel, was founded in part to inform and train those engaged in laboratory animal research.

Today, there are many organizations throughout the world devoted to training (see Appendices 2 through 5). The first recognized lab animal training society is the IAT in the United Kingdom, established in in 1949, which preceded the founding of AALAS by a year (Kennedy 2015). Today, there are laboratory animal science programs at many colleges and universities that confer degrees, certification programs offered by laboratory animal societies, and professional educational programs to train trainers, whether a designated trainer, manager, or training manager.

Since the early part of the twentieth century, various educational movements have been proposed. Among them, an educational theorist named John Dewey emerged as a proponent of the concepts of progressive education (Dewey 1938), which has been widely adopted. Dewey held to the ideas of students realizing their potential, fulfilling their own capacity to learn, and developing abilities from within themselves. This notion was different than the traditional top-down training, that is, from journeyman to apprentice. And it is evident in laboratory animal science, where several programs embody these concepts to promote career advancement closely linked to accumulated, demonstrated, and recognized knowledge and skills.

- Certification for technicians as offered, for example, by AALAS (assistant laboratory animal technician [ALAT], laboratory animal technician [LAT], and laboratory animal technologist [LATG]) (AALAS 2016b)
- Training for managers with the Institute for Laboratory Animal Management [ILAM] (AALAS 2016c) and certification as a Certified Manager of Animal Resources (CMAR) (AALAS 2016b)
- Credential for laboratory animal veterinarians from the American College of Laboratory Animal Medicine as a diplomate (DACLAM) (ACLAM 2015)
- Certifications for animal surgeons, including surgical research anesthetist (SRA), surgical research technician (SRT), and surgical research specialist (SRS), from the Academy of Surgical Research (ASR 2016)
- Credential for IACUC staff known as Certified Professional IACUC Administrator (CPIA), from Public Responsibility in Medicine and Research (PRIM&R 2016)

As a result of the pioneers in laboratory animal training, the varieties and choices for educating and training processes should be greatly appreciated by today's manager of a laboratory animal facility. The people who responsibly work, research, and teach with animals now are products of those processes. The future of training in laboratory animal science was recognized by Stephens (1999), saying that animal care providers "must be on a constant learning curve. We need to make sure that our knowledge and skills develop as our field becomes more technical and specialized, our equipment becomes more sophisticated, and become knowledgeable about additional animal models." Also needing to be addressed in managing training programs are divergent learning styles (e.g., kinesthetic, auditory, and visual), learning challenges (e.g., English learners and hazardous environments), competency evaluations, generational experiences, and international agreements about credentials. National and international credentials and certifications are especially useful in assessing competencies of employees and potential employees. Hiring managers can get a general idea of a candidate's knowledge and skills as related to laboratory science by the credentials that the candidate holds. In addition, in our global industry, knowing what certifications equate from one organization to another is critical. For example, if a technician gained a certification in Europe but is now working in the United States, does his training match the training that would be provided in America? Knowing this can help a manager in assessing what additional training will be needed when moving staff from one location to another. Overall, the entire discipline of—and the various roles of persons engaged in—laboratory animal science must be part of training management in this century.

Andragogy and Learning Styles Revisited

The concepts of adult learning or andragogy were introduced in the first edition of this chapter. It is the recognition that adults learn differently than children. Whereas teachers in the typical K–12 school system are "pushing" content on their students—known as pedagogy—a trainer in an adult environment is facilitating the learning of the content based on the motivation of the students (Kennedy 2014b). In pedagogical models, the learner is dependent on the teacher, who takes most of the responsibility for what is learned and when it is learned. Typically, the teacher has more experience or knowledge of the topic. With andragogy, the learner—generally an adult, but not always—is at the center of the educational processes (see point 4 about orientation in Table 12.1 and Figure 12.2). Both approaches are important in lab animal training, to the extent that managers should recognize when it is probably better to use one before the other.

Dr. Malcolm Knowles (the son of a veterinarian and a lifelong engaged learner) is regarded as the father of andragogy (Knowles et al. 2012). Knowles's ideas of adult learning, first proposed in the 1970s, included just four assumptions or principles contributing to the theory. But as it developed, Knowles later decided to include the aspects of motivation and needing to know (assumptions 5 and 6 in Table 12.1). Ironically, many managers might think these two should have been his first thoughts about adult learning. Figure 12.1 shows the theory graphically. The table summarizes Knowles's key assumptions about andragogy.

TABLE 12.1
Assumptions about Adult Learning (Andragogy) Based upon Knowles

1. **Role of the learner's experiences or foundation:** Experience, including making mistakes, provides a strong basis for learning activities with adults.
2. **Learner's self-concept:** Adults need to be responsible for their decisions when being educated. They want a part in planning and evaluating their instruction.
3. **Readiness to learn:** Adults are most interested in learning that which has immediate relevance to their work and/or personal lives.
4. **Orientation to learning:** Adult learning is problem-centered rather than content-oriented (but not exclusively).
5. **Motivation to learn:** Adults often respond better to internal versus external motivators.
6. **Need to know:** Adults want to understand the reasons for learning something.

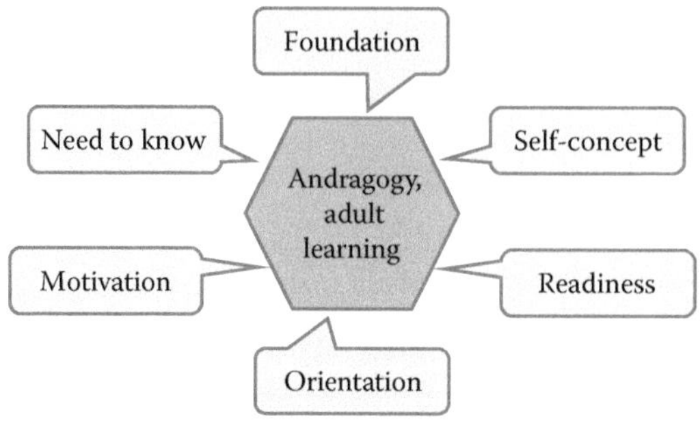

FIGURE 12.2 Six assumptions or principles of Knowles's andragogy.

Andragogy is different from pedagogy because the assumptions focus more on the learner than on the content. For a trainer, that does not mean they are mutually exclusive. Rather, they should be thought of as an instructional continuum, and both may be utilized in the same training session. For example, when teaching about safety, personal protection, and animal biosafety levels (ABSLs), it is entirely appropriate for the trainer to be content based, emphasizing facts and evaluating the rote memorization of such principles. When the training addresses gowning procedures in ABSL 2–4 environments, the trainer probably should allow the student to experience the processes independently, before being corrected to use safer practices (reference Knowles's assumption 1 related to the foundation in Table 12.1).

Another example of the instructional continuum is the use of case studies of animal diseases. In this training, the instructor passes out a report from a relevant journal for the class participants to read and contemplate. These directions are pedagogical in nature, but discussion of the health concerns, diagnostic tests, differentials, and so forth, is based on problem-solving models, and thereby demonstrates andragogy and higher-order cognitive application.

Knowles's assumption 5 about motivation in learning reflects a specific meaning and purpose in lab animal training. Recall that regulations (AWAR 2013) and guidelines (ILAR 2011) associated with conducting animal research require training. A question that both trainers and managers of training need to ask is, what kind of motivation is influencing the adult learner? Is it the mandate of the administration demanding training, the fear of a regulatory citation, the wish to be promoted, a new technology with an animal, an interest in the research, achieving certification or licensure, or something else? The answers should be considered by management when developing and evaluating training sessions in order to make the best use of resources dedicated to training. And the answers should be placed in the context of two other concepts of what motivates adults to learn, which are Maslow's hierarchy of needs and WIIFM. These are respectively models of *where* in terms of life is the learner at the time of learning and *what* will the learner obtain from the learning. The manager who ignores these concepts will probably find it more difficult to instill a culture of training and to achieve acceptance of training from individuals. Instead, and more effectively, the manager should be using these management concepts as contributions to the structuring of training around the needs and abilities of the trainees, as well as those of the institution.

There are actually many learning styles and strategies that trainers and managers can use to engage learners (Hawk and Shah 2007). It is critical when managing training to recognize individual learning preferences. A common one is the kinesthetic, visual, and aural style, probably best known as feeling or touching, seeing, and listening. Another is the notion of active and passive learning (Tweed 2014), which recognizes the variable style of both the learners and the trainers (Figure 12.3). On some occasions, an active learner may become passive with an active trainer because of subtle personality or cultural conflicts. In another learning environment, the passive learner may be drawn into greater activity because of the types of active learning tools used by the trainer (Figure 12.3). Dobrovolny et al. (2007), Kennedy (2002a), and Knowles et al. (2012) describe other styles and provide additional resources to create learning environments and promote learning, which the manager can share with the individual primarily responsible for training. Utilizing most or all these tools in a training program will increase learning and learner interest. It is well recognized from educational research (Handelsman et al. 2007) that active

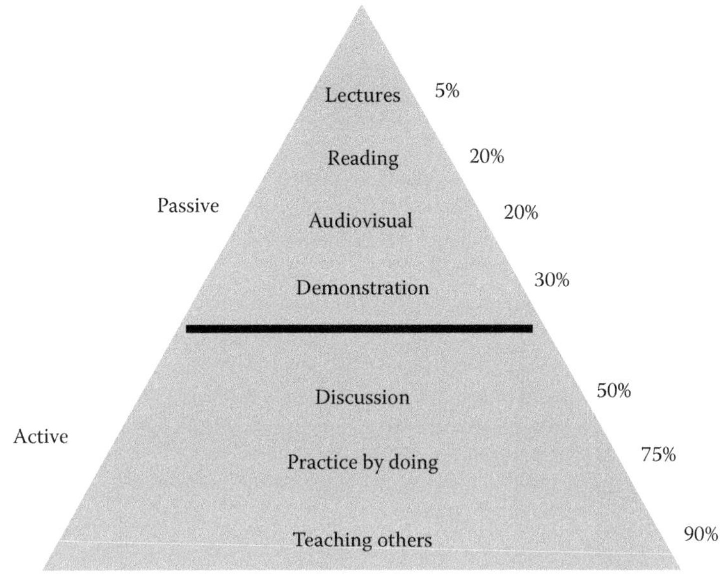

FIGURE 12.3 Tasks associated with active and passive learning, which build upon themselves.

learners take more responsibility for their learning when engaged in problem solving, group work, mapping (hand drawing), and other activities, like how we participated and learned as children (pedagogy).

Appreciating Cultural Differences and How We Learn

Many different species are represented in the vivarium, being used as research models for multitudes of purposes. Yet only one species, *Homo sapiens*, has the responsibility and task to care for them, and we are just as diverse. As much as the mouse and monkey have different husbandry requirements, there are differences among the cultures of vivarium personnel that also need to be recognized and appreciated by those responsible for training them.

Cultural competence: Pritt et al. (2014) described the concept of cultural competence—or cultural intelligence—as applied to laboratory animal science. The terms mean the "set of behaviors, knowledge, and attitudes among professionals that foster effective work in cross-cultural situations." To embody cultural competence "requires a willingness to learn about others' traditions, characteristics, language, behavior, and other attributes in order to interact effectively with others at work" (Pritt et al. 2014). With awareness of such differences in culture, the manager and the trainer may need to modify how the evaluations of performance are conducted to assess outcomes in training situations. For example, the trainer should recognize that some trainees demonstrate their knowledge better verbally than through written summaries.

Diversity: Alworth and colleagues (2010) addressed the challenges presented by the variety of persons working in laboratory animal science, which heretofore has received only minimal attention. Specific examples were cited: Cultural and religious attitudes on multiple topics may vary among personnel and affect interactions. Attitudes regarding animals can differ, such as what level of respect they deserve and whether certain species cannot be handled. Different attitudes about the appropriate way to interact with coworkers based on their sex, age, race, sexual orientation, and other attributes may cause friction among personnel. Their article included several suggestions and lab animal industry programs to increase awareness and provide solutions. Many of the concepts are crossovers between and combinations of traditional management and training, about which the facility manager should be cognizant.

TABLE 12.2
Tips When Training ESL Students

1. Don't address the pause in conversation; allow for silence and mental processing.
2. Limit use of idioms.
3. Check frequently for understanding.
4. Appreciate the differences in English and where the staff member learned it.
5. Teach how to use automated grammar and style checkers for e-mails, health reports, and the like.
6. Review training materials to provide another perspective on understanding.
7. Learn something yourself in another language.

Source: Adapted from Hughes, L., *Lab. Anim.*, 43, 101, 2014.

Generational: Those individuals born between 1977 and 1994, also known as millennials, gen Y or generation Y, are described (Dinkel 2013; Caprino 2014) as having always learned in the digital age, being savvy with computers, smartphones, and other technological devices. For them, information is obtained and delivered easily and quickly from electronic sources, such as the Internet or web. Their expectations for training differ from those of their predecessors—the baby boomers and the Xers. Managing their training includes utilizing educational games, long-distance and connected learning, tailored instruction, and training on the fly.

Scientific or research language: English has emerged as the primary language for science and business. Yet while walking through the corridors of a vivarium or research lab, scores of other languages often will be heard. Journal articles are still published in many other languages, even though there are electronic means to instantly translate them. At conferences, many languages will be heard and papers are often presented in native languages. Moreover, the vernacular of science is almost another language, where words and phrases can be confusing if not defined and explained. Trainers must pay attention to this possibility and evaluate the three components of a language—what is read, what is said, and what is heard—in their training materials. There is so much brain activity while translating, it could be a distraction to comprehension for the learner.

English as a second language (ESL): Nonnative English speakers (NNESs) work in many roles at laboratory animal facilities, where they need to learn both the job and the language. The manager will need to address training topics with them. Only recently have various authors begun to analyze this challenge in laboratory animal settings, despite the efforts with ESL students in schools over several decades. Hughes (2014) provides tips to consider, as shown in Table 12.2, in the context of managing NNES staff, which led her to analyze the situation and develop successful approaches (Hughes 2013; Sherly 2014). Pritt et al. (2014) made many of the same suggestions and, in echoing Hughes's point 1, say that adjusting the pace of training may be effective for some NNES.

Developing a Training Program

Perhaps you are a manager tasked with developing a training program for your animal care and use program. Maybe you find yourself inaugurating or creating one from scratch. It is conceivable that a decision was made to enhance or expand the existing program because of a recommendation by an outside entity (e.g., AAALAC International or a regulatory agency). Either way, you are faced with designing the learning specifications unique to your situation and institution; fortunately, there are resources to help.

Modifying or starting a new program can be challenging. The first suggestion is to obtain the "buy-in" of other managers and administrators as appropriate in the groups to be supported. The IACUC should also be on board with the proposed training program, making sure there is an understanding of the regulations requiring training and the need to ensure that anyone who works with animals is properly trained and competent in the skills they are performing. Related to this approach is to ensure that all recognize how the training program will help them to be in compliance. Keep in mind that the manager is often there to manage the process.

The Committee on Educational Programs in Laboratory Animal Science published a guide in 1991 with a section (V) on how to develop, deliver, and evaluate an educational program. While dated, it will be useful to initiate the process and compel the answering of training-related questions pertinent to your institution. The publication noted that "the IACUC and the course coordinator are responsible for developing clear objectives for each phase of the training program." So, the manager is not alone in these efforts. Furthermore, it is emphasized that "these objectives must incorporate both federally mandated and institutional requirements" (Committee on Educational Programs in Laboratory Animal Science 1991). Accordingly, the combined effort depends on others to help define what is needed in the training. Finally, "the methods for presenting the material will depend on the audience, the objectives that have been set, the nature of the content and the resources available" (Committee on Educational Programs in Laboratory Animal Science 1991). Thus, there are some boundaries to be considered and set for a training program.

The literature on laboratory animal technology has started to include accounts and case studies of training program designs. Heenan (2010) described a program developed with the objective to standardize training given to rodent handlers. The concept included identifying a person from investigator laboratories to be responsible for coordinating animal activities within that research group. Lockworth et al. (2011) found a benefit and improvement in animal care after providing specialty training to veterinary care technicians and husbandry staff. Dyson and Rush (2012) reported on their success involving lab animal veterinarians, who designed, implemented, and provided training sessions for research personnel.

As mentioned throughout this chapter, several lab animal organizations, like LAWTE, LAMA, IAT, and AALAS, provide information about developing training programs particular to the animal perspective. For general training ideas, a Google search with the terms "Developing a training program" will produce thousands of hits that can be scanned and searched for ideas and details. Well known is the organization Association for Talent Development (ATD), which is the former American Society for Training and Development (ASTD), which has assisted trainers in many disciplines, some of which align to laboratory animal science. Among these resources, the manager will be able to find the information needed in order to feel capable to take on training tasks. Nevertheless, additional help in the form of a checklist, based on a user's actual experience, and considerations for hiring a training company on a consulting basis are provided in this chapter.

Checklist for the Training Manager

While conceptualizing a training program for an institution, one facility manager devised the scheme shown in Box 12.1, categorizing the various concerns of the stakeholders with questions that needed to be answered in order to fulfill the specific objectives of the animal care and use program. Is it complete? No. An initial plan seldom is, because it will be modified with new information that reflects design changes and implementations as they get done. Nevertheless, it is a list of questions that can be utilized by any training manager, whether beginning or modifying an established training program.

Summary of Who Needs Training and What They Need

Lest the lab animal manager feel overwhelmed by the responsibilities in training, recall that institutions are responsible for providing appropriate resources to support personnel training (Anderson 2007), and the IACUC is responsible for providing oversight and for evaluating the effectiveness of the training program (Foshay and Tinkey 2007). All program personnel training should be documented (ILAR 2011). There are many aspects to training that fall within the realm of the IACUC; thus, a full appreciation of all the regulatory needs must be gained by consulting other references.

Maloney (2013) described during an Office of Laboratory Animal Welfare (OLAW) webinar who needs training in laboratory animal research, training, and testing.

> Minimum requirements under the PHS Policy indicate those who are responsible for housing, feeding and care of the animals used, as well as those who will be conducting procedures on live animals, all require training. The Policy also discusses proper constitution of the [Institutional]

BOX 12.1 TRAINING MANAGER'S CHECKLIST

INITIAL ASSESSMENT: WHAT DO WE NEED?

- What regulations, guidelines, policies (state, federal, and granting agencies) is the institution bound by?
- What is the scope of the program (e.g., species, research procedures, models, and objectives)?
- What is the purpose of the program (e.g., basic research, safety testing, or multifaceted)?
- Who are the stakeholders (e.g., research personnel, institutional representatives, regulatory committees, safety, and operational staff)?
- What are our goals and priorities (e.g., building a program, grant funding, AAALAC International accreditation, U.S. Department of Agriculture [USDA] registration, or culture of improvement)?
- What is the level of institutional support for this initiative?

PLANNING: WHAT IS OUR APPROACH, OUR TIME FRAME, AND OUR STRATEGY?

- What are our specific goals (e.g., training types, formats, and subject matter)?
- How will we meet them (e.g., instructed training, online programs, hands-on training, and qualification processes)?
- What processes or resources do we have in place and how can we leverage them (e.g., trainers, advocates, infrastructure, and resources)?
- What preparations are needed (e.g., budgeting, data compilation, program outline, stakeholder input, or IACUC approval)?
- What is our rollout plan (e.g., beta phase, pilot phase, targeted rollout, or complete rollout)?
- What is our communication plan (e.g., staged communications, controlled feedback, improvements, training, implementation dates, and preparations)?

IMPLEMENTATION: HOW WILL OUR PLAN BE PUT INTO PLACE?

- What is our timeline?
- What steps are needed?
- Who will be responsible for implementation?
- How will we roll out and test our plan for implementation?
- Do we have all necessary institutional support and approval?

REASSESSMENT: HOW IS OUR PLAN WORKING?

- What is working well?
- What can be improved?
- Are we meeting all requirements?
- What are our opportunities for improvement (stakeholder feedback)?

Animal Care and Use Committee and the general training and experience required for each type of member. This includes: the veterinarians and other professional staff such as pathology, imaging, and behavioral professionals; animal care personnel such as the clinicians and husbandry staff; the research team including principal investigators, study directors, technicians, post-doctoral fellows, students, and visiting scientists; and last but not least, the Institutional Animal Care and Use Committee itself.

Other key personnel, such as the institutional official and occupational health and safety professionals, may require training in order to succeed in their roles. Numerous other types of personnel may be involved as well—either directly or indirectly—in the animal program and it would be best practice to ensure these people receive appropriate training. Other personnel may include: administrative staff such as the IACUC support team; those involved in animal transportation both at the shipping and receiving ends; facilities personnel including those responsible for design and renovation of animal facilities; engineering staff; maintenance and custodial staff; and security personnel. And when considering the requirement for a contingency or disaster plan, institutions may find it beneficial to include emergency management personnel in the training program—and this may involve both institutional and local fire and law enforcement officials. And finally, program administrators might consider including their institutional human resources management and legal professionals in the training roster.

What Topics Should Managers Include in Training?

Of course it is not practical, but the answer should be everything. Another possible way of phrasing the question is, what should a trainer leave out? That is equally difficult and challenging to answer, but the response could be "it depends on the needs of the institution." And from a practical standpoint, there are general requirements as described in the Animal Welfare Act and Regulations (AWAR) and *Guide* that must always be included. The *Guide* has it right where it says (ILAR 2011) that "the number and qualifications of personnel required to conduct and support a Program depend on several factors, including the type and size of the institution, the administrative structure for providing adequate animal care, the characteristics of the physical plant, the number and species of animals maintained, and the nature of the research, testing, teaching, and production activities." Furthermore, the *Guide* advises, "Educators and trainers can use the *Guide* as a document to assess both the scope and adequacy of training programs supported by the institution."

The traditional curriculum for training in laboratory animal sciences would include, at a minimum, the following: occupational health, surgery, anatomy, physiology, regulations, species-specific characteristics, health and disease, sanitation, genetics, and more. Every lab animal program will add to and subtract from this (and the managers reading this chapter should do the mental exercise), as they address what is relevant and appropriate for the conduct of research, testing, and teaching with animals. And that is the point: animal facility managers are among those who have the responsibility to identify the needs specific to their programs—as such, it is part of the culture of training.

What is stated in the *Guide* yields another definition of lab animal curriculum:

> The institution should provide appropriate education and training to members of research teams—including principal investigators, study directors, research technicians, postdoctoral fellows, students, and visiting scientists—to ensure that they have the necessary knowledge and expertise for the specific animal procedures proposed and the species animal care and use program used (Conarello and Shepard 2007). Training should be tailored to the particular needs of research groups; however, all research groups should receive training in animal care and use legislation, IACUC function, ethics of animal use and the concepts of the Three Rs, methods for reporting concerns about animal use, occupational health and safety issues pertaining to animal use, animal handling, aseptic surgical technique, anesthesia and analgesia, euthanasia, and other subjects, as required by statute.

While the previous passage would seem all-encompassing, there are several topics that are not typically included in a lab animal science curriculum. Some might increase in importance during the next decade and deserve more attention. Today's lab animal manager should be aware of the following:

- *Animal behavior*: Operant conditioning with animals to gain their participation in procedures has been demonstrated in a number of species. Samples and data can be obtained with much stress for both the animals and investigators. Training must be obtained in order to perform "positive reinforcement training (PRT) methods which reward desired behavior," which has "become a valuable tool for the humane care and use of laboratory animals" (NC3RS 2016b).
- *Communication with the public*: The ability to present to the public what is discovered in research is critical. The public often supports the work financially, yet is challenged to understand it. Science moves quickly and transparency has become a buzzword. Dissemination of research findings comes in the form of traditional publications and presentations, as well as webinars, websites, videos, and town hall meetings. Depending on institutional policies, it may be incumbent for managers to communicate what we do in schools, governmental hearings, and other venues for public outreach. Furthermore, whereas English is the primary language of science, cultural shifts and employment practices indicate the need for both digital communication and non-English languages.
- *Animal welfare*: In the last decade, the American Veterinary Medical Association (AVMA), the Royal College of Veterinary Surgeons, and the Australian and New Zealand College of Veterinary Scientists have established distinct credentials to recognize those persons specializing in animal welfare. According to the AVMA website (www.avma.org), the American College of Animal Welfare (www.acaw.org) offers certification to veterinarians successfully completing advanced education and training to become animal welfare specialists.
- *Ethics*: Indeed, ethics have taken an even greater foothold in the animal facility (Dinkel 2012). Training embodies the ethical responsibility we have when conducting animal research, testing, and teaching (King 2004). Animal ethics, as broad a topic as that is, is explored in depth elsewhere in this book. The future animal facility managers will need to be trained and conversant, from Aristotle to the 3Rs to the emerging legislation resulting from law school students studying in the newly emerging discipline of animal ethics.
- *Robotics*: More and more equipment in the facility is being operated or monitored by robots and computerized devices, supported by computer-controlled interfaces and, to some degree, artificial intelligence. In a way similar to the integration of trainers in vivaria, personnel with an understanding of computers and mechanics will be (are) needed in the cage wash area, at the research bench, and with data management systems.
- *The 3Rs*: Education and training in the laboratory animal facility has increasingly embraced the principles of refinement, reduction, and replacement as outlined by Russell and Burch (1959) in *The Principles of Humane Experimental Technique* (Nevalainen 2004; Guhad 2005; Franco and Olsson 2014). The Johns Hopkins University Center for Alternatives to Animal Testing advocates that "the concept of recognizing, minimizing, and eliminating pain and distress in laboratory animals should be included in training programs for all persons involved in the care and use of laboratory animals" (Zurlo et al. 2016). The application of the 3Rs, especially as applied to refinement, is demonstrated by more frequent training agendas at conferences that share improvements on techniques involving animals and, by a wider choice, of inanimate models to replace live animals in training sessions. Training and the Rs have become part of the institutional culture of using laboratory animals. However, King (2004) noted the knowledge gap of "the lack of existing research data comparing the educational value of alternative, with traditional animal-based, instruction methods." This presents a challenge to trainers and managers in laboratory animal science.
- *Veterinary care technicians and husbandry staff development*: Pritt and Duffee (2007) advocated that animal care providers receive "a strong training program on diverse topics according to staff duties." Their recommendations for topics included ethics, regulatory compliance,

and "species-specific biology and behavior, animal facility equipment and operations, animal health procedures, animal research policies, occupational health and safety equipment and practices, computer usage, training, and management." Lockworth et al. (2011) presented a case report where the roles and responsibilities of veterinarians, veterinary care technicians, and husbandry staff were modified to accommodate the expansion of a rodent program. With the backing of management, a training program was designed for technicians to become more engaged, from simply identifying health issues to actually making decisions for treating and euthanizing rodents. This training program empowered (following the principles of andragogy) all team members and resulted in a staff that could provide consistent, high-quality veterinary care more efficiently.

- *Those not trained*: It has been mentioned that the facility manager will often be called on to assist the IACUC in the institutional obligation to train. Benoit and Bayne (2005) addressed those persons who, for various reasons, may not be "captured" (called the *forgotten* by them) through traditional channels as needing training. Specifically, it may be the visiting scientist or summer student who has not been identified but suddenly appears one day working in the facility. The manager will need to decide how to handle the situation—for example, by engaging another approved user from the same laboratory group, curtailing the animal activity, contacting the IACUC, or documenting and reporting the incident—all the while managing the lack of training incident from a HR perspective.

What Does the Training Cost?

High among the concerns of a manager is how to pay for products and services, which includes training. Conversely, the question needs to be asked by managers, what would, or could, it cost if there is no training? Phrasing that differently, the cost of doing it wrong is just too much—and reflects poorly on facility management.

Managers acknowledge that money is needed for the purchase of training materials (e.g., AALAS 2016a), attending conferences, subscribing to journals and webinars, holding memberships with training organizations, and so forth. The cost of these items can add up quickly. Other costs may be less apparent, but they are still quite visible under examination. For example, they include the salary for a trainer and trainee, the items used in developing curriculum, and the rooms where training occurs.

Some hidden costs may not even be apparent, but nevertheless have an impact on the overall animal use and care program when training is inadequate. Consider these scenarios:

- The extra time taken by a PI to write a protocol or by the IACUC to review one that is not well written and is not complete
- The losses from procrastination or personal grief (recall the ABCs of training) associated with a task that is not well learned and consequently is not conducted properly or perhaps not at all
- The health effects resulting to either an animal or staff member that manifest long after an exposure because the user had not learned correctly how to use personal protective equipment initially
- The issues, expenses, and cost of public relations following the revelation that an animal has been mistreated or an untreated hazard has left the facility
- The cost to institutional prestige and reputation following preventable losses, and consequently grant proposals being awarded elsewhere

The facility manager should consider that training is protective and should be bought as a precaution, much as we do with insurance. DeOrnellis (2013) wrote, "As such, trained individuals may be likened to an insurance policy, 'purchased' by the institution, to mitigate some of the risk associated with *in vivo* research. But an important distinction between insurance and training needs to be made. An insurance policy never improves an individual's position; it can only protect the individual from disaster. Insurance equals risk mitigation. A strong training program also mitigates risk, but unlike an insurance policy, training can improve an institution's animal care and use program."

The cost of doing it wrong is just too much! This expression needs to be repeated in lab animal training sessions. Trainers, those responsible for PAM, and facility administration should look for ways whereby training can help to improve animal well-being and increase efficiencies in facility activities. Good managers have a direct line of communication with most aspects and stakeholders of the animal care and use program, and therefore they are a unique resource for issues where training could be helpful to correct a problem.

At times, it is appropriate for managers to calculate the return on investment (ROI) for training. Training is an expense item that can be measured. How much benefit came back when a training session was held, whether it was at a conference to which staff traveled or it was a trainer hired to do on-site training? How much contact time was there? Did staff learn "after hours" during the conference? Were staff talking after the contracted trainer left about what was learned? Such a calculation involves factors other than just dollars or euros paid.

If the institution is fortunate to have a dedicated training center, is it also considered a cost center in terms of administration and budgeting? Costs associated with training are in most cases better tracked when there is a cost center, so that training activity is accurately measured, and it has associated line items in the facility or institutional budget. Besides accounting for the money spent, the training coordinator can track staff members trained, store training materials, survey users, and provide on-demand training.

The concept of cost accounting in relationship to training should not be regarded only from a negative perspective. Consider these concepts, each of which contributes to the overall culture of training evolving at the research institution:

- The ROI means staff perform work more efficiently.
- Facts are retained, or it is known where they can be looked up.
- Through discussions in and after class and training sessions, better animal-related practices are suggested.
- Full training records, meaning large files with frequent training encounters, elicit kudos from visitors (including regulatory inspectors and AAALAC International) to the animal facility.
- Trainers become acquainted with members of the research team, and their relationship is positive, not confrontational.
- The newly recognized skills and knowledge of lab animal staff can be utilized by the training staff.

Methods That Managers Use to Accomplish Training In-House

Educational methods include storytelling, discussion, teaching, training, and directed research. In school, we expect to learn from teachers, who often practice *pedagogy*, which is the giving of information based on established traditions and norms of a broader scope of education. Traditionally, the process of instruction or teaching has been conducted through rote learning. Managers and trainers in laboratory animal science should be asking themselves when and where this model is appropriate. Probably, it can work well in preparation for a certification examination, but will it suffice for tasks like surgery and cage washing?

Training is said to be the education one receives when specializing in a field. Moreover, it is the practical application of knowledge acquired (Paschal 2013), based on the needs of the student or learner, which distinguishes training from teaching. Because training is often associated with what adults want or need to learn, it has frequently been referred to as adult-learning education, or *andragogy* (Knowles et al. 2012).

Knowledge is the factual information or the skill set obtained through experiential learning (e.g., on the job) or a formal educational process (e.g., in a classroom setting) (Oxford 2015b). Appropriate use of training or educational methods is often the goal of the specific audience and in-house training content.

E-Learning in Laboratory Animal Science

Of the many advances in managing animal facilities, those in information technology are among the most impressive. Great amounts of data can be collected and stored without paper. Information can be sent and received within a single centralized facility or exchanged with colleagues elsewhere in the world within seconds. This has benefitted training as well through the discipline of e-learning.

The social media of the twenty-first century started in lab animal science as electronic blackboards. Credit is given to Dr. Ken Boschert at Washington University (St. Louis, Missouri) for his vision of working with AALAS to establish a means by which lab animal professionals could exchange information—thereby training one another. This became CompMed (AALAS 2016d)—a unique community of users providing electronic sharing of knowledge, coverage of comparative medicine and laboratory animals, resources, opinions, and so forth.

Webinars are available at all times, both at the first showing and in the archives of the provider. They are a means to obtain focused information on many topics pertinent to operating a laboratory animal facility. Often complimentary, vendors share their latest products and federal agencies inform about the latest regulatory changes via webinars. Members of lab animal societies can watch webinars for free as a member benefit or at a discounted cost. Refer to the appendices at the end of this chapter to find organizations providing these e-training opportunities.

The libraries and curricula of the electronic "classrooms" devoted to laboratory animal science, such as the ALL (www.aalaslearninglibrary.org) and CITI (www.citiprogram.org), have increased significantly in the last decade. Both represent worthy subscription resources for learning about and obtaining training in laboratory animal science disciplines. Lab animal managers must be cautioned not to simply direct staff to take the classes, however. This information is primarily knowledge-based content with minimal demonstrations of skills. Competency and capability with the application of facts and utilization of skills should be assessed by the institution. Recalling the driving metaphor, because an individual has watched a video about driving a car does not mean he or she will be able to proceed down the road without incident!

The advantage to training material archived on the web is that it can be accessed any hour of the day, it can be watched any number of times, and it can give time to the trainer for personalized training instead of conducting rote training sessions. Unlike the libraries of videotapes that facilities used to buy and keep in a static manner, media on the web can be immediately updated to current knowledge. The astute facility manager involved with training will be aware of and evaluate such benefits.

Another feature of e-technology is the development of LMSs. This is software that accumulates into a database the training records and experiences of staff. Commercial examples include Moodle, Canvas, Blackboard, Google Classroom, and Compliance Wire. Selecting, installing, and maintaining an LMS may require the assistance of the institution's computer and information technology group.

An LMS is informative for reminders about training that needs to be periodically retaken and to quickly identify who among staff are experienced with a technique or an animal model. Managers will find an LMS useful in several ways. Imagine one that extracts from the résumé of a new hire all of her capabilities and puts them into a personnel record, onto which all additional training is recorded, so that all of her training or expertise is available to the manager. Imagine the LMS that ties into an electronic protocol system, such that entering the name of Technician Tom automatically populates a document (e.g., a protocol or grant proposal) and informs the IACUC that he is qualified to work with the specific animals on that study. Imagine the institutional library that can be built within an LMS for PowerPoint presentations, conferences attended, guest speakers heard, webinars attended, certifications achieved, and so forth. With an LMS, the *Guide* (ILAR 2011) requirement that "all program personnel training should be documented" may be more easily satisfied.

Choosing an LMS takes its own set of skills and project planning. Knowing exactly what the organization needs or wants the LMS to do is critical from the outset, to which the lab animal manager should contribute. A simple program that lists training and competencies is easy to manage and easy to install. A more complex system that documents standard operating procedures (SOPs) review, hands-on training, classroom training, CBT, and so forth, takes much more time both to set up and to manage. A dedicated trainer or training staff will need to have time in their schedule to manage the LMS.

The benefits of a more complex system, though, can be immeasurable. It can improve both compliance and standardization of training across large organizations. A well-organized and thought-out LMS can be set up with curricula for many different job roles. These jobs can send reminders to learners when training is due, store CBT modules, alert managers when training is delinquent, generate transcripts for animal program reviewers (e.g., inspectors), and more. Be aware that one can only get out of the LMS what one is willing to put into it. Organizations that cannot dedicate personnel to manage the training program and LMS would probably be better served with a simpler LMS.

E-learning is one of the buzzwords of training now. Bound paper books to read, plastic animal models to manipulate, and hands-on physical labs to conduct still have a function. But, the online Portable Document Format (PDF), computerized body systems, and virtual labs of almost all procedures embody the 3Rs for training purposes. The workforce (the millennials or generation Y) entering biomedical research in 2017 have had these learning experiences since kindergarten. Trainers and managers who do not integrate e-technologies will be behind the e-learning curve.

There are many pieces of educational software that can aid trainers in conveying information about laboratory animal science. The fast developments in the field preclude mentioning extensive software systems. Nevertheless, laboratory animal facility managers and trainers might use the two below as examples for fostering learning:

- Poll Everywhere (www.polleverywhere.com) is a tool to engage audiences from large to small in a training session (Kennedy 2014c). It can be used as an icebreaker to open a training session or as an assessment tool for longer training sessions. Using either a mobile phone or a computer, the student participants respond to questions either posed in the application or embedded in a PowerPoint. Results are displayed instantaneously, bringing "quiet" or passive learners into the presentation.
- Padlet (www.padlet.com) is similar to Poll Everywhere. Its capability is that it can collect postings made to a preset web page established by the trainer. For example, an assignment, consisting of five standard questions, can be given during a training session to characterize a particular lab animal research model. The learners—singularly or in groups—can complete the tasks in real time and post their work for all to see and compare with oversight by the trainer.

Managers and trainers need to be cognizant of the Americans with Disabilities Act (ADA) requirements with e-learning materials to provide access to information resources. Compliance with this law is important for training a diverse staff. An example standard is Voluntary Product Accessibility Template (VPAT), used in the California State University system (CSU 2015).

Position of Trainer

In the last century, the position of a trainer in a laboratory animal facility was a somewhat rare entity. Certainly, there were persons who trained, like managers, but they tended to have other primary responsibilities. Their jobs were not devoted entirely to training, and so they had to carve out the time each day and week to educate both users of the facility and themselves. Many benefitted from the educational opportunities provided by national and regional AALAS meetings. LAMA, established in 1984, addressed many of the training responsibilities of managers, while LAWTE, established in 1994, was created to help empower staff with their training responsibilities.

Momentum has built for "named" trainers in facilities. In 2002, Kennedy published a model position description for a lab animal trainer. More trainers and more facilities with trainers are visible in the laboratory animal field. Although there is no database of trainers, it is reasonable to think that there is at least one trainer, part- or full-time, in every laboratory animal facility. Calculating an estimate for the United States, it is possibly 1500–2000 trainers, based on almost 1000 USDA-registered facilities, 1300 Public Health Service (PHS)–assured institutions, and more than 900 AAALAC-accredited institutions (like a Venn diagram, the numbers will overlap). This estimate is supported by the number of persons participating with LAWTE, which in 2015 had a global membership exceeding 300 persons and more than 800 subscribers to its complimentary listserv, which shares training topics (Kennedy 2015). The Laboratory

Animal Facility Compensation Survey conducted by AALAS in 2014 reported that 50 individuals held a position where training was their primary task (AALAS 2014).

Today's laboratory animal facility trainer could be a member of the vivarium staff, the training department, the veterinary medicine department, or the IACUC or equivalent oversight ethics committee. Directing the individuals and their appropriate efforts with the institutional program in mind is often the role of the laboratory animal science manager.

The Manager Who Trains

Sometimes the lab animal facility manager is also the trainer, particularly in smaller facilities. In that circumstance, there are advantages, such as knowing the capabilities and capacities of the staff, knowing what specific experience (knowledge) or skill is needed for a project, and knowing what resources are available for training. A special benefit is finding *teachable moments*, which is when a manager spontaneously recognizes a situation as an opportunity to train on something being discussed for other institutional purposes.

While the dual role of manager–trainer is efficient in terms of institutional resources, the circumstance has the potential to decrease the resource availability of that person. This dual role may undermine individual capabilities as a trainer because of more pressing responsibilities as a manager. For example, consider the scenario of assigning, as the manager, the task of processing a delivery of animals instead of conducting a scheduled training session. Another is the awkwardness that arises when a manager addresses a difficult HR issue with an employee, which upsets both, and later the disciplined staff member attends a training session led by the manager.

Contracting Training

In recent years, it has become possible to hire trainers and training on a contractual basis. Several companies have made it part of their business to offer training on a fee-for-service basis. In other situations, training is a line item among many services provided in a contract by a full-service animal care company, such as veterinary care, cage changing, and cage washing. As with other contracts in the animal facility, the manager often negotiates the specifics and should sort the details of training services as well. Recalling the 6Ws process, they can be defined. Who will receive the training—staff only or all employees of the institution? Where will training occur—at the site of the contracted company or the institution? What will the training consist of—basic husbandry, surgical skills, regulations, and so forth? Why do this training—both contracted company and institution need to know? How much training is to be provided—for only basic internal needs or CE? When will training happen—for each new employee, on an as-needed basis, once per year?

As discussed elsewhere in this chapter and with regard to cost accounting in this book, the matter of charges, payments, per diems, and similar cost items will need to be determined. Costs can be absorbed as an institutional support service item under a general operating budget. They may fall into the laboratory animal resources component or possibly the IACUC, or even an established training cost center.

Assessment: Measuring Competency and Evaluating Training

Determining whether someone "knows his stuff" has taken prominence in lab animal training, as in other fields—and rightly so. Much is at stake when an animal is involved. Because investigators and the IACUC, through the protocol review process, have determined that it is ethically appropriate to use animals in a procedure, those persons actually conducting the procedure must be deemed competent at what they do.

Four terms are commonly associated with evaluating training. *Assessment* is the evaluation or estimation of the nature, ability, or quality of something. *Qualifying* is the demonstration of capability at something in order to be qualified or accomplished, enabling someone suitable for a particular job or activity—one is said to have qualifications. *Competence* is the ability to do something successfully or efficiently. And, *proficiency* is a high degree of competence or skill—one is said to have expertise.

The purposes of assessment in lab animals are many (Popham 2010; Kennedy et al. 2016). They include

- Comparing actual learner performance to the goals of the instruction
- Helping learners make decisions about their next action since staff often work alone
- Monitoring ongoing progress with, for example, complicated surgical procedures
- Assessing the teaching methods as conducted by the institutional trainer
- Revising and improving the institutional training program
- Providing information about capabilities to other stakeholders (e.g., clients of contract research organizations [CROs], AAALAC international site visitors, and USDA veterinary medical officers)
- Qualifying an individual's background knowledge and experience (e.g., during the hiring process or to determine whether a person can opt out of additional training)
- Determining learner satisfaction with training sessions (i.e., fulfilling the principles of WIIFM or motivation according to andragogy)
- Developing self-assessment procedures

The lab animal manager will recognize that being qualified or competent is a demonstration of training and a learner's ability that exceeds basic or elementary copying and regurgitation of educational material. Many organizations throughout the world recognize achievements, accomplishments, and expertise in various disciplines within laboratory animal science; see Appendices 2 through 5. A simple explanation of a new concept (fact or technique) does not equal knowledge or an ability to apply it.

The styles used to assess or evaluate are different from learning styles. In the same way that language, science vernacular, learning disabilities, anxiety, and so forth, affect learning, an assessment of an individual may be challenged by these factors. The facility manager may need to assist in accommodating both, applying one or more of the educational leadership forces presented previously in this chapter.

The facility manager needs to know basic characteristics of assessment processes. They include being simple and fair, which can be said to be transparent; being purposeful and related to job skills and tasks; matching the priorities of the institution; utilizing appropriate testing methods; allowing for recording mistakes in order to inform others (e.g., the IACUC and institutional official) and to identify retraining needs; and offering continuous improvement. The manager should critically assess how evaluations are done. Evaluations should not be assessing what is easiest and what is not relevant. The manager should spot when there is a potential conflict of interest as well.

Examples of what teachers and trainers use for assessment are the following: check-off sheets, adherence to SOPs, concept understanding forms, assessment of previous experience, tests with and without psychometric analysis applied, competency standards coming from the industry (AALAS and PRIM&R apply them for their certification tests), observations while performing tasks, analysis of outcomes and performance (e.g., numbers of cages changed, record-keeping errors, and morbidity and mortality statistics), and rubrics (see further in this chapter). The tools can utilize technology or homegrown products. What is critical is that assessment tools are reliable and validated. In order to accomplish this, the competency criteria should be established ahead of time and communicated to the learner before or during training. Additional information and descriptions can be found in educational references such as Popham (2010) and from the ATD (see Appendix 12.3).

Many members of a research institution can be responsible for and engaged in the various processes of assessing competency. At minimum, they, who are called assessors by Singleton (Kennedy et al. 2016), would include trainers, auditors, veterinarians, vet techs, researchers, managers, and animal care staff. The facility manager is arguably toward the top of the list and should work with the IACUC and trainers to ensure that individuals performing animal activities are capable to the greatest extent possible. The following items are a summary of several management and evaluation approaches the manager can consider:

- Certification and certificates. Managers should appreciate the variety of ways that individuals earn recognition for their training. Generally, a certification indicates a study of a body of knowledge with an evaluative test at the end. A certificate may indicate attendance at a single training session or completion of several criteria in order to show participation. Several institutions and organizations provide training in laboratory animal science, which have been accumulated in the appendices at the end of this chapter.

- Attempts have been made to harmonize training among countries and grant reciprocity for completion of similar study and accumulated experience. Managers should evaluate and appreciate for institutional needs the accomplishments of individuals at other institutions and animal programs. Some of them are described in the appendices.
- Evaluative educational research indicates that persons demonstrate their knowledge and skills in different ways. One person may test better, whereas another may show a skill better, and both would be assets to the animal research team. Managers should recognize that human resource requirements in job descriptions that are too stringent may restrict the pool of good candidates. The terms *book learned* and *street smarts* imply different ways to gain knowledge and apply it.
- Postapproval monitoring. The concepts of PAM and the institutional means by which it is conducted are intended to ensure compliance with animal welfare and regulations. As well, PAM activities can potentially be used to evaluate animal users in real circumstances on what they have learned in training sessions and subsequently apply when working in the facility. Additional training may be a suggested outcome of a PAM session.
- Capability in one area may prepare a learner for something else. Educational assessment incorporates formative (as something develops) with summative (at the point of completion) determinations. In another way of managers being motivators of learning, they can utilize the concept of training mixed with "have to know, good to know, fun to know" topics. Training managers can explain, for example, the animal model in an interesting way (fun) and then train on the requirements (have to know). Both of these processes relate to the aforementioned principles of andragogy.
- Indicate that "going to training" is still "going to work." The learner should return from a session and share what was learned, which serves as a measure of knowledge gained. The manager is then assessing the value of a training opportunity, the fact that the learner attended it, and what was presented. This process relates to the concept of ROI in training.
- Challenging trainees. Veteran trainers know students informally categorized as the newbie, the prima donna, and the experienced new hire. Each requires a different training management approach. The newbie is generally the neediest in terms of training because the entire curriculum is new and the starting point may be challenging to find. The prima donna comes across as knowing it all and objects to being trained again; an option is to test or evaluate the individual for proficiency and use this information to determine training requirements. The experienced new hire challenges the trainer by saying, "At my old job, we did it this way," which can be used to consider as a point of evaluation by the institution or be turned back to say, "Thanks, good to know; here, we want to be consistent in our practices."
- The "oops!" factor of making new mistakes (MnMs). Mistakes will happen in research. Einstein said, "If we knew what it was we were doing, it would not be called research, would it?" These sentiments are not granting permission to make mistakes, but are supportive of the R for refinement to practice before doing. MnMs in a controlled learning environment with *in vitro* methods can be a satisfactory approach. Use mouse models, then mouse carcasses, and then the real mouse for the actual procedure in order to manage a training process. Increasingly, there are publications in the lab animal science literature describing methods to train using alternatives to animals.
- Protocol documentation of training. The possibility of having abbreviated résumés or training transcripts for each staff member in a database (i.e., LMS), kept up to date and accessible by the institution, could assist in completing the training and experience section of animal use protocols.
- Continuous professional development (CPD). The manager offers training opportunities on a continual basis as part of the institutional and leadership culture about training to be as proficient, competent, and updated as possible. Many job categories in laboratory animal science require professional credentials, and persons must maintain their knowledge through CE. For some, credentials are maintained in a registry (e.g., the AALAS RALAT, RLAT, and RLATG technician levels, where the R prefix indicates being registered). More and more, training

programs are receiving Registry of Approved Continuing Education (RACE) approval by the American Association of Veterinary State Boards (AAVSB) (www.aavsb.org/race) for CE to help maintain state licensure for veterinarians and veterinary technicians. RACE is seen as a measure of quality control in training.

- Rubrics. A rubric is a tool to conduct an evaluation, by scoring ability, whether it is stating a fact or performing a task. There are three elements in a rubric (Popham 2010): (1) evaluative criteria to judge, (2) quality distinctions on how to judge, and (3) an application strategy for the score. For example, a score of zero means unable, 1 for capable, and 2 for well capable. The scoring scale is usually expanded to allow for fewer absolute distinctions, but there is still an acceptable cutoff score based on objective criteria. "Technical skills seem best suited for this numeric metric evaluation," opined Waldis (2013), who, along with Clifford et al. (2013), has published examples of rubrics for laboratory animal science. The facility manager can work with the facility trainer to help establish the validity and reliability of institutionally based rubrics.
- In pedagogical schooling, where the passing of information is focused on the student, the teacher often conducts assessments (commonly quizzes, tests, and practicums) that are formative and summative. Formative assessments occur during training, as knowledge develops or forms. Summative assessments, on the other hand, tend to occur at the end of a unit of training or perhaps as an overall cumulative examination or demonstration of knowledge. Both have their value for the learners to be informed about their progress and accomplishment of course objectives. Additionally, the manager can use these assessments to evaluate trainers, to inform them about their impact on trainees. Assessments can be powerful for all the stakeholders to gauge learning. Keeping records of assessments is a management tool to measure learning and progress over time.
- Self-assessment is another kind of evaluation, but it is more personal. The individual learner is asked to compare knowledge and skills gained over a period of time. It is a reflection of what has been achieved. A self-assessment can be structured with specific questions, tied into learning objectives, or left to be free form. It can take the structure known as pre- and posttesting, where existing knowledge is measured at the beginning and end of a training session using the same evaluative tool, like a 10-question quiz.
- Training metrics. While not a measurement of learning per se, metrics used to evaluate a training program may be part of the manager's job. The data might include the numbers of persons trained and retrained, numbers certified, types of training, costs associated with training both internally and externally, and numbers of hours spent training. One perspective on what can be measured is available as a webinar entitled "Metrics Applied to Training in Lab Animal Sciences" (Kennedy 2014d).

Summary

As presented throughout this book, the duties, expectations, and responsibilities of the manager of a laboratory animal facility have generally expanded and become more complex since the publication of the first edition. Part of that increase has been the manager's tasks associated with training. Concurrently, for many reasons, the training environment has changed. Whereas before the manager's position description probably included training as one more task to address, today the breadth of training responsibilities is far greater. Regulations, animal welfare, procedural refinements, societal ethics, and more are all demanding increased effort at training. Thus, any of the following are possibilities related to training and what the manager might be doing:

- To be the manager of the training department in a large animal care and use program
- To be a manager alongside one or more trainers
- To be the manager supervising one or more trainers
- To be responsible for implementing or revising a training program
- To maintain the database of training records (and a composite of knowledge) for presentation to regulatory agencies or clients using a CRO

- To work with the public relations department of an institution to inform the public, providing outreach programs to community organizations and schools
- To assist with the education of students, from elementary to graduate, on the STEM disciplines, because predictions are that a science and technology educated workforce will be greatly needed in the next decade and beyond
- To work with the IACUC in managing teaching and training protocols, as well as research and testing protocols

In other words, as would be expected, there is no "one-size-fits-all" approach that describes what it means to manage training. Each institution will have its own needs and outcomes. Accordingly, this chapter has first familiarized—as a foundation—the lab animal manager with concepts of developing training programs, what many would label as TtT. There exist now many more resources for that information. Assuming that base level of knowledge, the topics in this revised chapter have moved to focusing more on the management concepts of running a training program. With that have been bits of educational theory for the manager to appreciate.

The manager of a laboratory animal facility plays a very important role in facilitating training within an institution. One should appreciate the interesting play on words here—the verb *facilitate* and the noun *facility*—which originate from the word *facilis* in Latin-based languages, meaning "to help something run more smoothly and effectively." It is the responsibility of the institution's ethical oversight committee—the IACUC—to ensure that appropriate training is accomplished. The manager can make that more *facile*, meaning "with ease," for all the stakeholders.

The facility manager is aided by various organizations and modalities to ensure training in the care and use of laboratory animals. In today's learning environments, distance and time do not limit the availability of training. And there are numerous teaching methods to assist learners with the advent of technology and studies of cognition. Regardless, the traditional humanistic approaches to instill the culture of training—the "people" part—are where the facility manager can and should contribute.

APPENDIX 12.1: EDUCATIONAL THEORY AND CONCEPTS APPLIED TO LABORATORY ANIMAL SCIENCE

Introduction

Typically, the manager of an organization is charged with "getting the job done." In response, the commitment to and actions taken may be directed toward those goals. While this approach is practical and generally efficient, it also prudent to be familiar with theories and concepts behind the practices of managing a particular process. This chapter has focused on the practical and applied steps related to managing training. It turns now to some of the reasoning and history that support them.

In the spirit of learning—being a lifelong learner—the topics in this section are meant to enrich the background of those who manage training in some respect. Broader training is seldom a bad idea, but a topic may not be relevant and efficient to learn "at the moment."

Is It Education, Teaching, or Training?

Usage of the terms *education* and *training* is more often a philosophical conversation by those (epistemologists) who study knowledge and acquiring it. Likely, it has minimal impact on the daily operations in the animal facility. However, a manager will be better informed when discussing the distinctions and nuances of curriculum and competency with trainers and educators, who of course use the specific vocabulary of their profession, as do scientists and architects, with whom the facility manager may be more familiar.

One definition of *education* is "the process of receiving or giving systematic instruction, especially at a school or university" (Oxford 2015a), which is synonymous with the concept of formal education as reflected in academic degrees (BS, MS, PhD, etc.). But, an education comes from many sources, as managers know from reviewing résumés and hiring staff.

Irrespective of the terms, the learning strategies—as used by those training in lab animal facilities—are dependent on the purpose and type of instruction. The word *instruction* comes from the Latin *instruere*, for "building in" (Shakeshaft et al. 2013). So, whether educating, teaching, or training, the objective is to build in information, by instructing the student, first imparting the knowledge and later assessing its application. When the student is, for example, husbandry staff, veterinary staff, scientist, IACUC member, or facility repair person, that is the point where terms like *education* and *training* become relevant for the lab animal manager.

Relationship between Training and Human Resources

The facility manager should recognize what are the responsibilities of HR and what is correctable through training. HR does assist in the traditional duties of hiring new staff and disciplining existing members. From perhaps a more positive aspect, the terms *human capacity* and *human capital* have emerged in HR vocabulary to recognize the development of knowledge, skills, and attitudes (KSA) by employees. Training can and should be used by managers to build capacity and capital in their employees. Such actions can be applied to specific individuals for their career development, as well as to various sized groups to improve the overall production of the institution. An astute manager will identify where training—as a tool—can be used for this purpose and take advantage of the potential (Dinkel 2015). Further, it is why managers need to engage with their institutional training programs.

The two leadership forces of technical and humanistic are generic and applicable to all forms of management. The other three forces—educational, symbolic, and cultural—can be seen as prevalent characteristics of exemplary training programs. For example, a proactive manager demonstrates these educational leadership traits when considering the wasted resources represented by a training room sitting empty or staff not attending a local training course because they do not value learning or feel there is no personal benefit. On the other hand, mixed leadership messages are sent when the manager turns down a staff member willing to pay the transportation costs herself to attend a local training session because she asked for approval of a day—a "training day"—to attend.

APPENDIX 12.2: SCHOOLS OFFERING TRAINING IN LAB ANIMAL SCIENCE

Since the early days of Dr. Brimhall at the Mayo Clinic, when learning about lab animal science was an on-the-job training effort, various schools have developed curriculum to teach about animal research.

From an informal poll of trainers on the LAWTE listserv conducted in 2015, it was learned that there are scores of educational institutions offering extensive coursework in laboratory animal science topics throughout the world. By extensive, it is meant that there is a strong and specific focus to teach about laboratory animal science, but a degree in the subject is not necessarily conferred by the institution. What this body of training means to laboratory animal facility managers is that informed groups of students will enter the animal research field already possessing skill sets and an animal welfare mind-set for research, testing, and teaching, and opportunities for collaboration between research and educational institutions.

Veterinary Technician Programs

Possessing coursework in general animal health and procedures is valuable in staff who work in laboratory animal facilities. Veterinary technician programs are available in most states in America, where there are both 2- and 4-year programs, and in many provinces in Canada, where the programs are 2 and 3 years in length. Many programs have been accredited by committees of the individual countries. In America and Canada, accreditation is conducted by the Committee on Veterinary Technician Education and Activities (CVTEA) of the AVMA. There are several distance education programs as well. In an accredited veterinary technician program, part of the curriculum is the requirement for content in "laboratory animal procedures"; see Section 8 of the task list at https://www.avma.org/ProfessionalDevelopment/Education/Accreditation/Programs/Pages/cvtea-pp-appendix-i.aspx.

Bachelor's Programs

A few schools offer a degree in animal health science to train veterinary technicians for a variety of animal fields, including animal research. The curriculum at California State Polytechnic University, Pomona has been published (Kennedy 2014a), which indicates one approach to satisfy the breadth of material needed to prepare for a career in lab animal care. Several other schools, as shown in the table, offer an emphasis in laboratory animal science in their programs.

Master's Programs

Drexel University, when it was known as Hahnemann, was the first school to develop the master of laboratory animal science (MLAS) degree. Starting as a face-to-face program in 1987, since 2009 it has conducted an online program as well. See http://catalog.drexel.edu/graduate/schoolofbiomedicalsciences/laboratoryanimalsciences and Kennedy (2014e) for more information. Another program was launched at Eastern Virginia Medical School in 2014. Similar to Drexel, it "allows maximum flexibility for the students … who may be currently working within the field of laboratory animal science [or] looking for opportunities in research laboratories but do not want to relocate to attend traditional master's programs." See www.evms.edu/education/masters_programs/las for more information.

Doctorate-Level Programs

Well known are the residency programs to train veterinarians in laboratory animal science in order to become diplomates of the American College of Laboratory Animal Medicine (DACLAM). One combines the veterinary medicine degree with a lab animal master's degree (Tufts). More information can be obtained from the website of the American Society of Laboratory Animal Practitioners (ASLAP) (www.aslap.org), which maintains a voluntary listing of residency programs, and ACLAM, which lists all ACLAM-recognized training programs (www.aclam.org/education-and-training/training-programs). Additionally, the AVMA maintains lists of the veterinary colleges (www.avma.org/ProfessionalDevelopment/Education/Foreign/Pages/ECFVG-world-colleges-degrees.aspx) and the types of veterinary medical degrees (www.avma.org/ProfessionalDevelopment/Education/Foreign/Documents/ecfvg12.pdf) granted throughout the world.

Representative Schools in the World with an Emphasized Curriculum in Lab Animal Science

School	Location	Program	General Comment
Aachen University	Germany	Master in laboratory animal science	http://msc-lab-animal.com/program
Alamance Community College	North Carolina	Associate's degree, 2 year	
Becker College	Massachusetts	Animal science/veterinary technology degree with a laboratory animal emphasis, 4 year	
Bel-Rae Institute of Animal Technology	Colorado	2-year associate's degree in applied science in veterinary technology	
California State Polytechnic University, Pomona	California	Bachelor of science, animal health science	One of several veterinary technology programs accredited by the CVTEA of the AVMA
Delaware Valley University	Pennsylvania	Bachelor of science, 4 year	www.delval.edu/academics/undergraduate/agriculture-environmental-sciences/animal-science/animal-science

(*Continued*)

School	Location	Program	General Comment
Drexel University	Pennsylvania	Master of laboratory animal science	http://catalog.drexel.edu/graduate/schoolofbiomedicalsciences/laboratoryanimalsciences
Eastern Virginia Medical School	Virginia	Master of laboratory animal science	www.evms.edu/education/masters_programs/las
FRAME (Fund for the Replacement of Animals in Medical Experiments)	United Kingdom, Central Europe, and Scandinavia	Training schools in experimental design and statistics done in association with universities	www.frame.org.uk/training-schools
Harcum College	Pennsylvania	Veterinary technology, 2 year	1972, AVMA accredited
Individual institutions	Mexico	Permanent training program for staff	Mexican legislation; NOM-062—technical specifications for the production, care, and use of laboratory animals
Individual institutions	Argentina, Brazil, Chile, Colombia, Cuba, Uruguay, and Venezuela	Coursework for local staff	
Johns Hopkins School of Medicine	Maryland	Research fellowship in molecular and comparative pathobiology	Residency, including ACLAM
Kerala Veterinary and Animal Sciences University	India	Certificate in laboratory animal medicine	Doctor of veterinary medicine (DVM) oriented, in affiliation with the University of Guelph, Canada; online course with a practical component in India, www.kvasu.ac.in and www.open.uoguelph.ca
Massachusetts Institute of Technology (MIT) Division of Comparative Medicine	Massachusetts		Residency, including ACLAM
Mesa College	California	Associate's degree program in veterinary technology	
Metropolitan Community College—Maple Woods	Missouri	Associate's degree in applied animal science; veterinary technology, 2 year	
Michigan State University	Michigan	Bachelor's for technologist, 4 year	
Norfolk County Agricultural High School	Massachusetts	High school	Not accredited; by arrangement with local biotechnology companies
North Carolina A&T University	North Carolina	Bachelor's and master's in lab animal science	
Norwegian Veterinary Institute	Norway	Courses for researchers and technicians	http://oslovet.norecopa.no/dokument.aspx?dokument=74
Pennsylvania State University	Pennsylvania	Bachelor's, 4 year; graduate program	http://vbs.psu.edu/ and http://animalscience.psu.edu/
Pierce College	Washington	Veterinary technology, 2 year	
Purdue University	Indiana	Associate's and bachelor's, veterinary technology	Bachelor of science includes laboratory animal management (regulatory issues and facilities design/management)

(*Continued*)

School	Location	Program	General Comment
State University of New York, Delhi	New York	2-year associate's degree in applied science in veterinary technology	www.delhi.edu/academics/majors/vet-sci/index.php
Tufts University	Massachusetts	Combined DVM and lab animal master's	
Universitat Autonoma de Barcelona	Spain	Master of laboratory animal science and welfare	http://pagines.uab.cat
University of California, Merced	California	Academic credit	In development
University of Cincinnati	Ohio	Master's degree program in biomedical research	In development
University of Copenhagen	Denmark	Postgraduate education for designated veterinarians according to EU criteria	http://labveteurope.ku.dk/
University of Guelph	Canada	Certificate in laboratory animal medicine	DVM oriented, online course with a practical component in Canada, www.lam-cdn.opened.uoguelph.ca
University of Guelph, Canada	United States	Certificate in laboratory animal medicine	DVM oriented, in affiliation with the University of Guelph, Canada; online course with a practical component in the United States, www.lam.OpenEd.uoguelph.ca
University of Guelph, Canada	Mexico	Certificate in laboratory animal medicine	DVM oriented, in affiliation with the University of Guelph, Canada; online course with a practical component in Mexico, www.open.uoguelph.ca
University of Maryland, College Park	Maryland	Bachelor of science, laboratory animal management, 4 year	
Wayne County Community College/ Wayne State University	Michigan	Veterinary technician	Joint program to offer the student an associate's degree or a bachelor's degree
Weill Cornell Medical College participates with Memorial Sloan Kettering Cancer Center and Rockefeller University	New York	Residency program for veterinarians	Tri-institutional training program in laboratory animal medicine and science; NIH supported

Note: This compilation of schools is subject to change and does not include all with a curriculum in lab animal science.

APPENDIX 12.3: PROGRAMS FOR TRAINERS

Learning is an everyday, all-the-time practice. Sometimes the trainer has to go back to school, both to learn to train and to learn the subject material. For formalized learning, the lab animal trainer has numerous options, including the following.

Association for Talent Development

What has been known as the American Society for Training and Development (ASTD) rebranded itself in 2014 to become the Association for Talent Development (ATD), "a professional membership organization supporting those who develop the knowledge and skills of employees in organizations around the world" (www.td.org). ATD intends to "support the talent development profession by providing trusted content in the form of research, books, webcasts, events, and education programs."

The association hosts several conferences each year. ATD also offers an internationally recognized credential program (Certified Professional in Learning and Performance [CPLP]) that covers the entire talent development profession. Eligible candidates must complete both a knowledge and a skills application exam (SAE).

Association of Veterinary Technician Educators

The Association of Veterinary Technician Educators (AVTE) was established in 1973 to promote quality education in veterinary technology. Focused on motivating and preparing teachers in the classroom for "the next generation of professional veterinary technicians" (www.avte.net), there is a mutual pedagogy and interest in animal welfare with laboratory animal science. The AVTE maintains a listserv and offers a biennial symposium. Related information about veterinary technology can also be found at https://www.aavsb.org/VTNE.

Colleges and Universities

Many institutions of higher learning offer classes in association with their education or teacher training departments for members of the community. Topics like curriculum development, educational technology, and educational leadership are among the offerings. It is possible to obtain the bachelor of arts (BA), master of education (MEd), doctorate in education (EdD), and philosophy doctorate (PhD) degrees. The website www.educationdegree.com has a compilation of more than 20,000 teacher education degree and certificate programs. Some programs are offered online.

Laboratory Animal Welfare Training Exchange

Established in 1994, LAWTE has become known internationally for its purpose of training trainers in the specific domain of laboratory animal science. The mission statement is clear: "expanding animal welfare and enhancing public understanding through effective training and education of animal research professionals" (www.lawte.org). LAWTE has had conferences every two years since its inception, presents on training subjects at other animal meetings, and presents webinars on training topics.

APPENDIX 12.4: RECOGNIZED LABORATORY ANIMAL SCIENCE CREDENTIALS

Education and training programs vary around the world. How individuals are recognized for completing their training to establish their capabilities and credentials to use laboratory animals in research, testing, and teaching also shows great variation. This table is a compilation of the designations, licenses, registrations, registries, certifications, and other institutional or organizational terms used for that purpose. It should help the lab animal manager to recognize and confirm the credentials from other regions and to find more information. Because programs evolve, omissions are likely.

Organization	Credentials	Website for More Information
AALAS American Association for Laboratory Animal Science	Assistant laboratory animal technician (ALAT) Laboratory animal technician (LAT) Laboratory animal technologist (LATG) Certified Manager of Animal Resources (CMAR)	www.aalas.org/certification
ACAW American College of Animal Welfare	Diplomate, DACAW	www.acaw.org

(Continued)

Organization	Credentials	Website for More Information
ACLAM American College of Laboratory Animal Medicine	Diplomate, DACLAM	www.aclam.org/certification
ANZLAA Australian and New Zealand Laboratory Animal Association	There is not a formal credential, but employers recognize the diploma of animal technology offered by providers of technical and further education (TAFE) training	www.anzlaa.org
Argentina	Curso sobre diseño experimental y tamaño de la muestra en trabajos con animales de laboratorio XI edición (Course on experimental design and sample size in work with laboratory animals, XI edition)	
ASR Academy of Surgical Research	Surgical research anesthetist (SRA) Surgical research technician (SRT) Surgical research specialist (SRS)	http://surgicalresearch.org
AVMA American Veterinary Medical Association	22 AVMA-recognized veterinary specialty organizations comprising 40 distinct specialties	www.avma.org/ProfessionalDevelopment/Education/Specialties/Pages/default.aspx
CALAS/ACSAL Canadian Association for Laboratory Animal Science	Associate Registered Laboratory Animal Technician (ARLAT) Associate Registered Laboratory Animal Technician Administrative (ARLATA) Registered Laboratory Animal Technician (RLAT) Registered Master Laboratory Animal Technician (RMLAT)	http://calas-acsal.org/registry-exams www.ccac.ca/en_/assessment
Chile	Recognizes the category C program from FELASA	
CVTEA AVMA Committee on Veterinary Technician Education and Activities	Accredited veterinary technology programs, more than 200 in America: Animal health technician, AHT Registered Veterinary Technician, RVT Licensed Veterinary Technician, LVT Certified Veterinary Technician, CVT *Note*: These programs are not specific to laboratory animal science, but do have a required component about the discipline	www.avma.org/professionaldevelopment/education/accreditation/programs/pages/cvtea-about.aspx
FELASA Federation of European Laboratory Animal Science Associations	Functions (not the same as categories): A: Persons caring for animals B: Persons performing animal experiments C: Persons responsible for directing animal experiments D: Laboratory animal science specialist Functions A through D are as defined by EU Directive, Article 23, as well as for specialists in laboratory animal science in Articles 24–26. These are recommendations—at the time of this writing—and are subject to change. The research institutes of the member countries give courses that are aligned in part with the FELASA recommendations, as well as country law	www.researchgate.net/publication/285582574_FELESA_2015_Recommendations_for_the_Accreditation_of_Education_and_Training_Courses_in_Laboratory_Animal_Science
NAVTA National Association of Veterinary Technicians in America	See CVTEA	www.navta.net/?page=specialties

(*Continued*)

Organization	Credentials	Website for More Information
PRIM&R Public Responsibility in Medicine and Research	Certified Professional IACUC Administrator (CPIA)	www.primr.org/Subpage.aspx?id =1587
United Kingdom	Registered Veterinary Nursing Program and advanced diploma in veterinary nursing	
United Kingdom Home Office	Examples of named positions: Named animal care and welfare officer (NACWO) Named training and competency officer (NTCO) Named information officer (NIO)	www.gov.uk/government/uploads /system/uploads/attachment_data /file/291350/Guidance_on_the _Operation_of_ASPA.pdf
United Kingdom IAT Institute of Animal Technology	Level 2: Diploma in laboratory animal husbandry Levels 2–3: Further education diplomas in laboratory animal science and technology Levels 4–6: Higher education (degree-level) diplomas in laboratory animal science and technology Registered Animal Technologist (RAnTech) Member of the Institute of Animal Technology (MIAT) Fellow of the Institute of Animal Technology (FIAT)	www.iat.org.uk/education www.iat.org.uk/rantech www.iat.org.uk/#!qualifications /c1zu6 www.iat.org.uk/#!grades/cld3 www.iat.org.uk/#!rantech/c5ni
United Kingdom Royal Society of Biology	Society of Biology (RSciTech)	www.rsb.org.uk/careers-and-cpd /registers/rscitech

APPENDIX 12.5: ORGANIZATIONS WITHIN LABORATORY ANIMAL SCIENCE OFFERING TRAINING, COURSES, AND MORE

Training is a lifelong journey, and in research we constantly are learning something new. Two terms describe this concept for laboratory animal managers: *continuous professional development* (CPD), often used in Europe, and *continuing education* (CE) or *continuing education units* (CEUs), accumulated from participation in a training event. Various groups in laboratory animal science, including professional societies, government agencies, and vendors, provide CPD and CE training. The LAWTE organization has designated vendors as CTPers, which stands for commercial training partners (Kennedy 2013), who, besides product literature, provide online resources, have trainers on staff, and host webinars to educate both generally and specifically on their services, products, and technologies.

By no means could a listing like this be complete, and fortunately for the field, it is ever expanding. This list is provided as a start for managers who find themselves needing training resources; from there, it is a journey through links. Many of these are based in America, but are connected internationally. It is an indication of the network of trainers, teachers, and educators in all disciplines throughout the world. It also demonstrates the dedication of both people and organizations to the notion of training in laboratory animal science.

The selection includes organizations that offer training in laboratory animal science that the facility manager can consider for staff and personal CPD or CE. Much of the information is quoted directly from the organization's websites, links to which are provided. Besides the websites, many organizations have social media accounts associated with Facebook, Twitter, and LinkedIn, which facilitate the exchange of information and provide resources for managing all aspects of the laboratory animal facility.

AAALAC International, www.aaalac.org

AAALAC International, formerly known as the Association for Assessment and Accreditation of Laboratory Animal Care, is a "private, nonprofit organization that promotes the humane treatment of animals in science through voluntary accreditation and assessment programs." It does so by offering a variety of training opportunities throughout the world. These include webinars, presentations at conferences, and the AAALAC International Education and Outreach Program. For its accreditation mandate, it works with managers of animal facilities. Through individualized programs, it offers "education and outreach modules designed to provide the information you need to proactively manage animal care and use issues in ways that meet AAALAC International standards."

AALAS, www.aalas.org

Besides supporting the largest conference for laboratory animal personnel in the world, the American Association for Laboratory Animal Science (AALAS) provides many options for training, including the AALAS Learning Library (ALL 2016); the Institute for Laboratory Animal Management (AALAS 2016c); several listservs (AALAS 2016d), including TechLink, CompMed, and the IACUC Forum; manuals for the certification examinations for ALAT, LAT, and LATG; webinars; and educational resources, including audiovisual aids. Products are available at https://www.aalas.org/education/educational-resources.

Additionally, local branches (chapters) of AALAS produce training products for use by their members. Modeled for the AALAS technician certification exams, they are varied in their approach to training, being online, scheduled group sessions, developed with the aid of curriculum designers; self-paced, with and without assignments; mentored by experienced technicians and managers; and more. Participation can be used for the AALAS Registry Program.

ACLAM, www.aclam.org

The goals of the American College of Laboratory Animal Medicine (ACLAM) are today as when defined in 1958: "to encourage education, training and research in laboratory animal medicine; to establish standards of training and experience for qualification of specialists in laboratory animal medicine; and to further the recognition of such qualified specialists by suitable certification and other means." The members of ACLAM produce textbooks and educational materials. The college promotes the training of veterinarians to become board certified in laboratory animal medicine, known by the designation of diplomate.

APV, www.primatevets.org

The Association of Primate Veterinarians (APV) focuses on education and training for those working with nonhuman primates (NHPs). Each year, the APV meets prior to the AALAS conference to discuss matters concerned with the management of NHPs. It is not restricted to veterinarians. The APV website provides training materials from its past meetings and many links for other sources of NHP information.

CALAS/ACSAL, www.calas-acsal.org

The Canadian Association for Laboratory Animal Science (CALAS)/Association Canadienne pour la Science des Animaux de Laboratoire (ACSAL) is "a national association dedicated to providing high quality training and educational resources to animal care professionals across Canada. We believe animal research, when necessary, must be conducted professionally, ethically and compassionately." CALAS/ACSAL has training and certification programs that are voluntary and support the veterinary group.

CCAC, www.ccac.ca

The Canadian Council on Animal Care (CCAC) (Conseil Canadien de Protection des Animaux), often working with CALAS, offers web-based seminars, training modules, workshops, and other educational materials in both English and French. The CCAC develops different educational and reference materials to support the specialized training needs of all animal users (including investigators, graduate students, postdoctoral fellows, and research staff), animal health professionals, animal care committee members, and institutional officials (senior administrators) directly responsible for animal care and use programs. CCAC's guidelines can be found at www.ccac.ca/Documents/Standards/Guidelines/CCAC_Guidelines_on_Training_of_Personnel_Working_With_Animals_in_Science.pdf.

CITI, www.citiprogram.org

The Collaborative Institutional Training Initiative (CITI) provides a subscription service of online training modules, available 24/7, in many languages, maintaining a database of learners and e-mail transcripts. The CITI mission statement is "to promote the public's trust in the research enterprise by providing high quality, peer reviewed, web based, research education materials to enhance the integrity and professionalism of investigators and staff conducting research." The curriculum includes many disciplines pertaining to training persons for work in an animal facility: Animal Care and Use (ACU), Biosafety and Biosecurity (BSS), Export Control (EC), Good Clinical Practice (GCP), Human Subjects Research (HSR), Information Privacy and Security (IPS), and Responsible Conduct of Research (RCR), including the Conflicts of Interest (COI) course. As taken from the CITI website, "the ACU materials cover general principles of ethical care and use of animals in research, training, and testing, as well as focusing on the care and use of particular animals. Content is designed to meet U.S. Department of Agriculture (USDA) and Office of Laboratory Animal Welfare (OLAW) requirements for basic training in the humane care and use of animals."

CLATR, www.clatr.org

The Committee for Laboratory Animal Training and Research (CLATR) (formerly known as ACLAD) is an organization that was created to aide laboratory animal and comparative medicine trainee career development by providing research mentorship and networking opportunities, facilitating sharing of training resources among training programs, and promoting interactions of trainee-focused committees of national organizations.

CRL, www.criver.com/customer-service/education-training/educations

Charles River Laboratories (CRL) "hosts frequent courses, seminars, and symposiums dedicated to advancing attendees' knowledge of their respective fields" around the world. CRL is well known for its annual short course held in the Boston area, which covers both contemporary and traditional animal issues in its curriculum, designed for all position levels in the facility, including the facility manager. It also issues booklets, posters, and other training, much of it free of charge.

ETPLAS, http://etplas.eu

The Education and Training Platform for Laboratory Animal Science (ETPLAS), or simply The Platform, provides "a forum for exchanging information on laboratory animal science education and training therefore helping to establish mutually recognised training courses. Successful completion of such a harmonised course enables working with laboratory animals and promotes free movement of competent personnel in the EU." It started in 2015 with its information exchange.

FBR, http://fbresearch.org

"Established in 1981, the Foundation for Biomedical Research (FBR) is the nation's oldest and largest organization dedicated to improving human and animal health by promoting public understanding and support for humane and responsible animal research." FBR does not directly host conferences or issue credentials, but its frequent posts and e-mail are a source of information useful for training on a broad number of subjects, from animal models to regulations. Managers and trainers can subscribe and share awareness about laboratory animal research, testing, and teaching beyond what otherwise might be provided.

FGB, www.fondazioneguidobernardini.org/en/foundation/aboutus.aspx

The Fondazione Guido Bernardini (FGB) was established in 2009 in Milano, Italy, as an independent, non-profit organization in memory of Guido Bernardini. "FGB is devoted to continuing education and training of professionals involved in the care, welfare and use of laboratory animals; the overall mission is to support the humane and responsible use of animals in science and to promote the quality of biomedical research by encouraging high standards of knowledge and competence in scientific and technical staff."

FLAIRE Learning, https://flairelearning.com

This website was developed by Professor Paul Flecknell and colleagues at Newcastle University. It focuses on training in anesthesia, pain, and distress. Much of the curriculum is aligned with the learning objectives of the European Union and is offered gratis.

Humane Endpoints, www.humane-endpoints.info/en

The Humane Endpoints website helps scientists identify humane endpoints in rodents. The website offers pictures, videos, and training modules, which can be used in training programs for CPD.

IAT, www.iat.org.uk

Started in 1949, but officially recognized a year later, the Institute of Animal Technology (IAT) in the United Kingdom is the oldest laboratory animal science professional society (Kennedy 2015). As stated at its website, "the IAT's purpose is to advance knowledge and promote excellence in the care and welfare of animals in science and to enhance the standards and status of those professionally engaged in the care, welfare and use of animals in science." The IAT is internationally regarded for its training programs for animal care personnel, which offers many levels of credentials (see Appendix 12.4), and its annual congress. Its website is also rich with educational and welfare materials.

iCARE Project, https://grants.nih.gov/grants/olaw/interagency_icare.htm, www.primr.org/icare

In 2017, the NIH, in partnership with PRIM&R and others, developed the Interagency Collaborative Animal Research Education (iCARE) Project and began offering Train the Trainers Institute (TTI) programs. The objective of iCARE is to use active learning pedagogy to train IACUC members and institutional and animal program personnel to meet their responsibilities for animal welfare oversight.

ICLAS, http://iclas.org

The mission of the International Council for Laboratory Animal Science (ICLAS) education committee is to promote and harmonize education and training in laboratory animal science, particularly in regions of the world where such opportunities are lacking or few. Among the council's goals relating to laboratory animal management and training are to assess needs of ICLAS members with regard to education and training; provide assistance to regions for setting up training and certification programs for laboratory animal care staff, research staff, veterinary staff, and laboratory animal specialists; and decide what role the education committee should play in promoting public education and understanding of laboratory animal science.

ILAM, www.aalas.org/education/ilam

Founded in 1991 by LAMA in cooperation with AALAS, the Institute for Laboratory Animal Management (ILAM) is now supported by AALAS. The focus and emphasis is on both historical and emerging issues in laboratory animal management. ILAM is an "educational program developed to provide instruction in management concepts that is applicable to the laboratory animal science industry and to enhance communication, team building, and networking among colleagues with mutual interests." ILAM is a 2-year program conducted in Memphis, Tennessee, with classes of about 50 students networking and learning from knowledgeable persons in their disciplines.

JAX, www.jax.org/education-and-learning

The Jackson (JAX) Laboratory not only is a provider of thousands of different kinds of genetically engineered mice, but also has a mission to teach. The opening page for training on its website is very telling: "Education and Learning. From high school summer programs to conferences that further the education of practicing scientists and professionals, we advance science and improve health through our commitment to education." The JAX Lab offers webinars, training workshops, conferences at its laboratories and research institutions, on-demand teaching materials, and more.

LAMA, www.lama-online.org

The Laboratory Animal Management Association (LAMA) is recognized as the premier organization to provide training on managing laboratory animal facilities. "LAMA is an association dedicated to advancing the quality of management and care of laboratory animals throughout the world. Since its establishment in 1984, LAMA has grown to over 700 members residing in geographical locations as widespread as Asia, Australia, Europe and Canada. The membership continues an active role in AALAS and the career field by providing leadership to numerous committees and organizations on local, state and national levels."

LAWTE, www.lawte.org

The Laboratory Animal Welfare Training Exchange is the consummate organization for materials related to training in laboratory animal science in general. The association offers a listserv, which is maintained by AALAS. The website has an archive of training materials for members. Its members are international (in 2015, a dozen countries were represented), and they host a biennial conference.

NAS, www.nasonline.org

The National Academy of Sciences (NAS) was founded in 1863 to support science. Today, the three academies for science, engineering, and medicine address pressing issues of society and "solve complex problems and inform public policy decisions," many of which concern the work of animal research facilities, for example, pain and distress, occupational health, and the compilation of the *Guide*. Through workshops, publications, and conferences, the teaching mission of the NAS is reflected in this sentence quoted from the mission statement: "[the] Academies also encourage education and research, recognize outstanding contributions to knowledge, and increase public understanding in matters of science, engineering, and medicine."

NAVTA, www.navta.net

The National Association of Veterinary Technicians in America (NAVTA) represents veterinary technicians and assistants in all disciplines of veterinary health care, many of whom work in laboratory animal science as well. Specifically, NAVTA has partnered with the Society of Laboratory Animal Veterinary Technicians (www.slavt.net) "to develop a network of professional veterinary technicians dedicated to the advancement of responsible and humane laboratory animal care and use to benefit humans and animals, to exchange information and expertise in the care and use of laboratory animals and to advance by actively seeking continuing education opportunities for the members." Several CE resources are maintained in a tab on the NAVTA website.

NC3RS, www.nc3rs.org.uk

The NC3Rs is a UK-based scientific organization dedicated to replacing, refining, and reducing the use of animals in research and testing (the 3Rs). Various "projects aim to provide training through web-based tutorials and other resources on best practice in the refinement of animal experiments."

NIH/OLAW, http://grants.nih.gov/grants/olaw/educational_resources.htm

While the U.S. federal Office of Laboratory Animal Welfare is primarily regarded for its role in monitoring compliance with PHS policy in "assured" institutions, it does so because of the educational programs that it supports and provides to the global laboratory animal science community to ensure the humane care and use of animals. The website is an information-rich resource for both managers and trainers to facilitate training, which includes regional workshops, societal conferences, online seminars, archives of presentations, and more.

NORINA, www.oslovet.norecopa.no/NORINA

The Norwegian Inventory of Alternatives (NORINA) is an English language database of more than 3500 audiovisual aids useful as alternatives or supplements in education and training. It is affiliated with Norecopa, which is Norway's National Consensus Platform for the advancement of the 3Rs. Norecopa was founded in 2007.

PRIM&R, www.primr.org

Public Responsibility in Medicine and Research (PRIM&R) offers conferences, webinars, regional meetings, and archives of training materials. Since 2007, PRIM&R has maintained the CPIA program. The organization's mission statement is to "advance the highest ethical standards in the conduct of biomedical, behavioral, and social science research. We accomplish this mission through education, membership services, professional certification, public policy initiatives, and community building."

Purina LabDiet, www.labdiet.com

Purina offers the PMI LabDiet Laboratory Animal Care Course, which was inaugurated circa 1960 to provide an introduction to the laboratory animal field. It is a correspondence course designed for entry-level technicians and valuable for anyone new to laboratory animal science and wishing to progress to AALAS certification. The course is offered in the classic print form or in a new interactive online form. A certificate of completion is provided.

SCAW, www.scaw.com

The Scientist Center for Animal Welfare (SCAW) supports training objectives through conferences and educational materials at its website and regional conferences around the United States. "By addressing animal research challenges directly through education and training, SCAW serves to facilitate open discussion and helps craft solutions as well as promote best practices. Through its outreach to the broader research community, SCAW demonstrates its commitment to excellence in animal care and science."

SLAVT, http://slavt.net

Training managers can support the education of their facility veterinary technicians through the Society of Laboratory Animal Veterinary Technicians (SLAVT). "The Society of Laboratory Animal Veterinary Technicians is an emerging group focused on national recognition through training and certification for veterinary technicians in the field of laboratory animal medicine." It holds meetings in conjunction with AALAS and its branches.

USDA/APHIS, www.aphis.usda.gov/wps/portal/aphis/ourfocus/animalwelfare

The Animal Care division of the U.S. Department of Agriculture/Animal and Plant Health Inspection Service (USDA/APHIS) is responsible for upholding and enforcing the Animal Welfare Act. Representatives frequently participate in conferences with AALAS, National Institutes of Health (NIH)/OLAW, AAALAC International, PRIM&R, and other organizations concerned with the training of animal care providers. The experience and training of individuals engaged with animals must be documented in applications proposed for their use in research, testing, and training.

The Animal Welfare Information Center (AWIC) within USDA (https://awic.nal.usda.gov/about-awic) was established in 1985 to provide "information products, services, and activities to help the regulated community with employee training" related to animal use. It participates in several outreach services available to facility managers, including "conducting workshops both at [the National Agricultural Library] and outside facilities for researchers concerned with meeting the requirements of the AWA, exhibiting at animal-related scientific meetings, giving presentations on AWA-related topics, and publishing the *Bulletin*."

Various Biomedical Advocacy Groups

States United for Biomedical Research (SUBR) (www.statesforbiomed.org) is "the network of nonprofit associations who have joined forces to promote health through science and education." Across the United States (see their website for the coverage), they promote science education and provide regional conferences about laboratory animal science, often associated with AALAS branch meetings. The topics vary; for example, the Massachusetts Society for Medical Research (MSMR) (www.msmr.org) offers an online course for the unaffiliated member of the animal care and use committee. The California Society for Biomedical Research (CSBR) (www.ca-biomed.org/csbr) works with the state legislature and writes science curriculum. Other states and associations are represented in the graphic available at http://statesforbiomed.org/about-us/members-funders.

Journals

Several journals publish articles on topics related to training and the management of training. The following are representative:

- *Animal Lab News* (www.alnmag.com), including the "Tools for Training" column
- *Animal Technology and Welfare* (www.iat.org.uk), the official journal of the IAT and European Federation of Animal Technologists
- *ILAR Journal* (http://ilarjournal.oxfordjournals.org/)
- *Journal of the American Association for Laboratory Animal Science* (JAALAS) and *Laboratory Animal Science Professional* (LAS Pro) (www.aalas.org)
- *Journal of Veterinary Medical Education* (http://jvme.utpjournals.press)
- *Lab Animal* (American and European editions) (www.labanimal.com), including the "Fruits of Education" column
- *Laboratory Animals* (Sage Publications) (http://lan.sagepub.com), the official journal for several laboratory animal societies, including the following:
    - AFSTAL (Association Française des Sciences et Techniques de l'Animal de Laboratoire)
    - ESLAV (European Society of Laboratory Animal Veterinarians)
    - FELASA (Federation of European Laboratory Animal Science Associations)
    - GV-SOLAS (Gesellschaft für Versuchstierkunde)
    - ILAF (Israeli Laboratory Animal Forum)
    - LASA (Laboratory Animal Science Association)
    - NVP (Nederlandse Vereniging voor Proefdierkunde)
    - SECAL (Sociedad Española para las Ciencias del Animal de Laboratorio)
    - SGV (Schweizerische Gesellschaft für Versuchstierkunde)
    - SPCAL (Sociedade Portuguese de Cièncias em Animas de Laboratório)
- *Scandinavian Journal of Laboratory Animal Science* (http://sjlas.org/index.php/SJLAS)

Additional Training Groups

There are many organizations and societies throughout the world with an interest in training personnel who work with animals in some capacity, whether research, teaching, zoo, wildlife, welfare, alternatives, managing, or another environment. Regardless of the language, the words *education* and *training* stand out. A few are listed here to be representative.

- Academy of Surgical Research (ASR), http://surgicalresearch.org
- Alternatives to Laboratory Animals (ATLA), www.atla.org.uk
- Asian Federation of Laboratory Animal Science Associations (AFLAS), www.aflas-office.org
- Association of Zoos and Aquariums (AZA), www.aza.org
- Australian and New Zealand Laboratory Animal Association (ANZLAA), www.anzlaa.org
- Belgian Council for Laboratory Animal Science (BCLAS), www.bclas.org
- Center for Alternatives to Animal Testing (CAAT), http://caat.jhsph.edu/about/index.html
- French Association of Animal Science and Technology Laboratory (AFSTAL), L'Association Française des Sciences et Techniques de l'Animal de Laboratoire, www.afstal.com
- [German] Society of Laboratory Animals (GV-SOLAS), Die Gesellschaft für Versuchstierkunde, www.vtk-online.de
- Italian Association for Laboratory Animal Science (AISAL), Associazione Italiana per le Scienze degli Animali da Laboratorio, www.aisal.org
- Japanese Association for Laboratory Animal Science (JALAS), http://www.jalas.jp/english/en_about_jalas.html
- SCANBUR Academy, www.scanbur.com/academy; courses are available for animal technicians, researchers, veterinarians, facility managers, and others employed in laboratory animal science
- Singapore Association for Laboratory Animal Science (SALAS), www.salas.sg/v1
- Sociedad Española para las Ciencias del Animal de Laboratorio (SECAL), http://secal.es
- Swiss Laboratory Animal Science Association (SGV), Schweizerische Gesellschaft für Versuchstierkunde, www.naturalsciences.ch/organisations/sgv/education_commitee
- Zebrafish Husbandry Association, www.zhaonline.org

CONTRIBUTORS AND COLLABORATORS

The contributions of the following groups and individuals are gratefully acknowledged by the authors. Data on the various lab animal programs around the world were shared. Others read early drafts of the chapter and made valuable improvements.

LAWTE members and survey responders
Shannon Reynolds
Laura Pearson
Jussi Helppi
Alistair Reid
Rubina Calderia
Octavio Villanueva
Cindy Pekow
Catalina Guerra
Janet Trombley
Roy Hoglund

Leah Schmidt
Lee Shuman
Amanda Dillard
Stacy Pritt
Regina Correa-Murphy
Nirah Shomer
Julie R. Kramer
Tim Allen
Mark Sharpless
Pam Grabeel
Richard Cluck
Steven Cubitt
Morag Mackay
Michelle Calkins
Dennis Stark
Anonymous reviewers
All those who have trained and managed us over the years

REFERENCES

AALAS (American Association for Laboratory Animal Science). (1999). *50 Years of Laboratory Animal Science*. Hanover, PA: Sheridan Books.

AALAS (American Association for Laboratory Animal Science). (2014). 2014 laboratory animal facility compensation survey. https://www.aalas.org/store/detail?productId=5814027.

ALL (American Association for Laboratory Animal Science Learning Library). (2016). Animal care and use in research and education. https://www.aalaslearninglibrary.org/ (accessed February 20, 2016).

AALAS. (2016a). Bookstore. https://www.aalas.org/store (accessed February 20, 2016).

AALAS. (2016b). Certification. https://www.aalas.org/certification (accessed February 20, 2016).

AALAS. (2016c). ILAM. https://www.aalas.org/education/ilam (accessed February 20, 2016).

AALAS. (2016d). Listservs. https://www.aalas.org/get-involved/listservs (accessed February 20, 2016).

ACLAM (American College of Laboratory Animal Medicine). (2015). Teaching and training. http://www.aclam.org/certification/aclam-diplomates.

Alworth, L., Ardayfio, K. L., Blickman, A., Greenhill, L., Hill, W., Sharp, P., Talmage, R., Plaut, V. C., and Goren, M. J. (2010). Diversity in laboratory animal science: Issues and initiatives. *J Am Assoc Lab Anim Sci* 49(2): 138–146.

Anderson, L. C. (2007). Institutional and IACUC responsibilities for animal care and use education and training programs. *ILAR J* 48(2): 90–95.

Armstrong, T., Blake, S., and Piotrowski, C. (2000). The application of a 360-degree feedback managerial development program in higher education: The Florida model. *Education* 120(4): 691.

ASR (Academy of Surgical Research). (2016). Academy of Surgical Research: Certifications http://surgicalresearch.org/requestPage.asp?ID=15 (accessed February 21, 2016).

AWAR (Animal Welfare Act and Regulations). (2013). Training of scientists, animal technicians, and other personnel involved with animal care and treatment at research facilities. Section 2143(d). https://www.aphis.usda.gov/animal_welfare/downloads/Animal%20Care%20Blue%20Book%20-%202013%20-%20FINAL.pdf (accessed February 20, 2016).

Benjamin, K. (2016). Sharing our work with pride. Outreach column. *Lab Anim* 45(5): 151–152.

Benoit, J. N., and Bayne, K. (2005). Training the trainee: The institution's responsibility to the often forgotten. *Lab Anim* (34): 46–50.

Caprino, K. (2014). Quit trying to 'engage' millennials. May 1. http://www.forbes.com/sites/kathycaprino/2014/05/01/quit-trying-to-engage-millennials/.

CITI (Collaborative Institutional Training Initiative). (2016). The Collaborative Institutional Training Initiative (CITI Program) at the University of Miami. https://www.citiprogram.org (accessed February 21, 2016).

Clifford, P., Melfi, N., Bogdanske, J., Johnson, E., Kehler, J., and Baran, S. (2013). Assessment of proficiency and competency in laboratory animal biomethodologies. *J Am Assoc Lab Anim Sci* 52(6): 711–716.

Committee on Educational Programs in Laboratory Animal Science. (1991). *Education and Training in the Care and Use of Laboratory Animals: A Guide for Developing Institutional Programs*. Washington, DC: Commission on Life Sciences, Institute for Laboratory Animal Research, Division on Earth and Life Studies, National Research Council.

Conarello, S. L., and Shepherd, M. J. (2007). Training strategies for research investigators and technicians. *ILAR Journal* 48(2): 120–130.

CSU (California State University). (2015). CSU guide to completing the Voluntary Product Accessibility Template (VPAT). www.calstate.edu/Accessibility/EIT_Procurement/APPENDIX.C.VPAT_GUIDE.doc.

DeOrnellis, D. (2013). An insurance policy (and more) for the animal care and use program. Fruits of Education column. *Lab Anim* 42(11).

Dewey, J. (1938). *Experience and Education*. New York: Touchstone (Simon & Schuster).

Dinkel, A. M. (2012). Ethics training, Part 2: Strategies for exploring common moral dilemmas. *ALN Magazine*. http://www.alnmag.com/articles/2012/02/ethics-training-part-2-strategies-exploring-common-moral-dilemmas (accessed September 11, 2015).

Dinkel, A. M. (2013). Leadership training for millennials. *ALN Magazine*. http://www.alnmag.com/articles/2013/09/leadership-training-millennials (accessed September 11, 2015).

Dinkel, A. M. (2015). Career focused learning. *ALN Magazine*. http://www.alnmag.com/articles/2015/04/career-focused-learning (accessed September 11, 2015).

Dobrovolny, J., Stevens, J., and Medina, L. V. (2007). Training in the laboratory animal science community: Strategies to support adult learning. *ILAR J* 48(2): 75–89.

Dyson, M. C., and Rush, H. G. (2012). Institutional training programs for research personnel conducted by laboratory-animal veterinarians. *J Vet Med Educ* 39(2): 160–168.

FELASA (Federation of European Laboratory Animal Science Associations). (2015). Recommendations for the accreditation of education and training courses in laboratory animal science. https:/www.researchgate.net/publication/285582574_FELASA_2015_Recommendations_for_the_Accreditation_of_Education_and_Training_Courses_in_Laboratory_Animal_Science. doi: 10.13140/RG.2.1.3799.1765.

Foshay, W. R., and Tinkey, P. T. (2007). Evaluating the effectiveness of training strategies: Performance goals and testing. *ILAR J* 48(2): 156–162.

Fox, J. G., and Bennett, B. T. (2015). Laboratory animal medicine. In *Historical Perspectives in Laboratory Animal Medicine*, ed. L. C. Anderson, G. Otto, K. R. Pritchett-Corning, and M. T. Whary. 3rd ed. London: Academic Press.

Franco, N. H., and Olsson, I. A. S. (2014). Scientists and the 3Rs: Attitudes to animal use in biomedical research and the effect of mandatory training in laboratory animal science. *Lab Anim* 48(1): 50–60.

Guhad, F. (2005). Introduction to the 3Rs (refinement, reduction and replacement). *Contemp Topics Lab Anim Sci* 44(2): 58–59.

Handelsman, J., Miller, S., and Pfund, C. (2007). *Scientific Teaching*. London: WH Freeman and Co.

Hawk, T. F., and Shah, A. J. (2007). Using learning style instruments to enhance student learning. *Decision Sci J Innov Educ* 5: 1–19.

Heenan, T. (2010). A program for standardized training in rodent handling at a large academic institution. *Lab Anim* 39(4): 113–117.

Hughes, L. (2013). Training initiatives for non-native English speakers in laboratory animal care. *Lab Anim* 42: 101.

Hughes, L. (2014). Tips for training non-native English speakers. *Lab Anim* 43: 101.

ILAR (Institute for Laboratory Animal Research). (2011). *Guide for the Care and Use of Laboratory Animals*. 8th ed. Washington, DC: National Academies Press.

Kennedy, B. W. (2002a). Management of education and training programs. In *Management of Laboratory Animal Care and Use Programs*, M. A. Suckow, F. A. Douglas, and R. H. Weichbrod. Boca Raton, FL: CRC Press.

Kennedy, B. W. (2002b). Creating a training coordinator position. *Lab Anim.* 31(6): 34–38.

Kennedy, B. W. (2013). Vendors assisting in training programs, parts 1 and 2. Fruits of Education column. *Lab Anim* 42(2, 4): 101, 177.

Kennedy, B. W. (2014a). Lab animal management syllabus, parts 1 and 2. Fruits of Education column. *Lab Anim* 43(4–5).

Kennedy, B. W. (2014b). Facilitate, don't dictate. Fruits of Education column. *Lab Anim* 43(6): 251.

Kennedy, B. W. (2014c). Engaging students to participate. Fruits of Education column. *Lab Anim* 43(7): 291.

Kennedy, B. W. (2014d). Metrics applied to training in lab animal sciences. https://www.youtube.com/watch?v=I4hoqP2vYjw. LabRoots, https://www.labroots.com/webinar/measuring-teaching-metrics-applied-to-training-in-lab-animal-sciences.

Kennedy, B. W. (2014e). The virtual fruits of education. Fruits of Education column. *Lab Anim* 43(9): 373.

Kennedy, B. W. (2015). LAWTE turns 21. Fruits of Education column. *Lab Anim* 44(11): 449.

Kennedy, B. W. (2016a). Outreach: Managers need to do it. Fruits of Education column. *Lab Anim* 45(5): 149.

Kennedy, B. W. (2016b). Tools for organizing training, part 1. Fruits of Education column. *Lab Anim* 45(7): 269.

Kennedy, B. W., Kilpatrick, A., and Singleton, W. (2016). Assessing and ensuring qualifications, competence, and proficiency. Presented at the 2016 PRIM&R IACUC Conference, Bellevue, WA. http://www.primr.org/iacuc16/.

King, L. (2004). Ethics and welfare of animals used in education: An overview. *Anim Welf* 13: S221–S227.

Knowles, M. S., Holton, E. F., and Swanson, R. A. (2012). *The Adult Learner: The Definitive Classic in Adult Education and Human Resource Development*. 7th ed. New York: Routledge.

Latin proverb quotes. (n.d.). http://www.quotes.net/quote/16678 (accessed February 22, 2016).

LAWTE (Laboratory Animal Welfare Training Exchange). (2015). Mission statement. www.lawte.org.

Lockworth, C. R., Craig, S. L., Liu, J., and Tinkey, P. T. (2011). Training veterinary care technicians and husbandry staff improves animal care. *J Am Assoc Lab Anim Sci* 50(1): 84–93.

Maloney, S. (2013). Training issues. OLAW Online Seminar, June 20. https://www.youtube.com/watch?v=31p2aVy4yRE&feature=youtu.be. Transcript and slides, June 20, 2013, http://grants.nih.gov/grants/olaw/130620_Maloney_slides.pdf.

Moreland, A. F. (1999). Development of training programs in laboratory animal medicine. In *50 Years of Laboratory Animal Science*, chap. 14. Hanover, PA: Sheridan Books.

Nevalainen, T. (2004). Training for reduction in laboratory animal use. *Altern Lab Anim* 32(Suppl. 2): 65–67.

NC3Rs (National Centre for the Replacement, Refinement & Reduction of Animals in Research). (2016a). The 3Rs. https://www.nc3rs.org.uk/the-3rs (accessed May 7, 2016).

NC3Rs (National Centre for the Replacement, Refinement & Reduction of Animals in Research). (2016b). Training animals. https://www.nc3rs.org.uk/training-animals (accessed May 7, 2016).

Oxford. (2015a). Education. http://www.oxforddictionaries.com/us/definition/american_english/education.

Oxford. (2015b). Knowledge. http://www.oxforddictionaries.com/us/definition/american_english/knowledge.

Paschal, H. (2013). Teaching vs training. http://www.correlationsllc.com/teaching-vs-training/ (accessed January 14, 2014).

Popham, W. J. (2010). *Everything School Leaders Need to Know about Assessment*. Thousand Oaks, CA: Corwin/Sage.

PRIM&R (Public Responsibility in Medicine and Research). (2016). Certified Professional IACUC Administrator (CPIA). http://www.primr.org/subpage.aspx?id=1587 (accessed February 21, 2016).

Pritt, S., and Clifford, P. (2014). Happy 20th birthday to the Laboratory Animal Welfare Training Exchange (LAWTE)! *ALN Magazine*, December.

Pritt, S., and Duffee, N. (2007). Training strategies for animal care technicians and veterinary technical staff. *ILAR J* 48(2): 109–119.

Pritt, S., Hill, W., Whitcomb, T., Stegmeyer, L., Blesh, R., and Sharp, P. (2014). Cultural competence in laboratory animal science. *Lab Anim Sci Prof* 2(1): 31–34.

Russell, W. M. S., and Burch, R. L. (1959). *The Principles of Humane Experimental Technique*. London: Methuen.

Sergiovanni, T. J. (2007). *Rethinking Leadership: A Collection of Articles*. 2nd ed. Thousand Oaks, CA: Sage.

Shakeshaft, N. G., Trzaskowski, M., Mcmillan, A., Rimfeld, K., Krapohl, E., Haworth, C. M. A., Dale, P. S., and Plomin, R. (2013). Strong genetic influence on a UK nationwide test of educational achievement at the end of compulsory education at age 16. *PLoS One* 8(12): e80341.

Sherly, M. (2014). Letter to the editor. *Lab Anim Sci Prof* 2(1): 8.

Stephens, U. K. (1999). The development of the technician's role, technician training, and certification. In *50 Years of Laboratory Animal Science*, chap. 12. Hanover, PA: Sheridan Books.

Suckow, M. A., Douglas, F. A., and Weichbrod, R. H. (2002). *Management of Laboratory Animal Care and Use Programs*. Boca Raton, FL: CRC Press.

Tweed, J. (2014). The differences between active and passive learning. http://www.coracleonline.com/article/differences-between-active-and-passive-learning (accessed September 8, 2015).

Waldis, H. (2013). Competency matters: Establishing standardization in your training program. *Lab Anim Sci Prof* 1(4): 24–25.

Zurlo, J., Rudacille, D., and Goldberg, A. M. (2016). The three R's: The way forward. http://caat.jhsph.edu/publications/Articles/3r.html (accessed May 9, 2016).

13

Fiscal Management

Stephen J. Pomeroy and Theodore Plemons

CONTENTS

Introduction

A sound financial footing is essential for a well-run animal care program, and it is important that program directors and managers have a working knowledge of program finances. Animal care programs are being forced to do more with less while operational expenses continue to climb or remain flat at best. Given this new reality, informed and relentless financial management is more important than ever to ensure that programs meet their various stakeholders' needs. Innovative programs are adapting a proactive budgeting and tracking process in which every cost is challenged with the intent to reduce or eliminate it eventually, and embedding this process in a general management culture that is committed to unyielding continuous improvement through eliminating unnecessary work while providing researchers and animals with faster, better, cheaper, safer services.

Effective and inventive fiscal management is of interest to many other laboratory animal constituencies, including

- The *parent organization*, be it a university, governmental agency, or company looking for a quality program whose cost is reasonable and competitive. In addition, internal institutional groups (internal audit, budget offices, etc.) evaluate whether the program is fiscally sound.
- *Researchers* want a well-run program that is affordable. Most universities and some private companies charge daily housing (also known as per diems) and other fees—researchers pay attention to these rates and "comparison shop"—often without appreciating that variances between institutions' per diem rates are due to differential institutional subsidies rather than operating costs. Daily housing fees may not be as prevalent at governmental agencies or companies, but researchers still want to ensure that the animal care program provides necessary resources.

  Attractive wages and employee benefits are critical to maintaining animal care program staff. It is also important that fiscal managers embrace change and prepare for constant improvement and updating of resources needed by the animal program staff, such as supplies, equipment, and repairs and renovations.

- *Funding agencies*, such as the National Institutes of Health (NIH) and foundations, require the animal care program to properly charge expenses according to funding agency guidelines. Billing rates should be based on actual costs.
- *Inspection and accrediting agencies* (e.g., the U.S. Department of Agriculture [USDA] and Association for Assessment and Accreditation of Laboratory Animal Care [AAALAC]), although not interested in program finances, focus on whether necessary staffing, equipment, repairs and renovations, and so forth, have been funded and provided. These agencies help assure private and funding sources that animal use is justified and humane, and can have a favorable impact on an investigator's proposal review. It is critical that animal program management ensure adequate financial resources to support these needs.

Sound financial management should tell the financial story in clear terms. It should meet the needs and format of the institution being served, and include acceptable accounting practices, sound budget planning, and frequent monitoring and adjustments. It should constantly examine practices and expenses, striving to reduce or eliminate wasteful spending and practices. It is very important to remember that unlike most supplies and services used by the investigators, animal care and maintenance is usually a monopoly and therefore will be viewed critically. The more transparent and fiscally sound the accounting practices, the easier it will be for the manager to tell the program's story.

Financial management comprises four processes: *planning*, *budgeting*, *monitoring*, and *cost accounting* (Alford 2001). Each of the four steps should be performed on a regular (monthly is recommended) basis for both short-term and long-term management. To ignore any steps truly jeopardizes fiscal management in any organization.

Program Planning

The planning process is the prequel to populating budgets with figures and is the opportunity for animal facility management to express their vision and plan for the future. At this juncture, program managers should strategize where the program should head, receiving input from investigators, program staff, and administration. Decisions should be made as to predicted animal populations, services to be added or eliminated, staffing needs, technical service levels to be delivered, and so forth. It is particularly important to have historical data available to make sure any projections and forecasts have a solid foundation. Animal population trends, revenue by various sources, technical services provided, expense details, and so forth, should be accessible for multiple years. If past spending involved wrong or wasteful expenses, the related historical data and methodology must be scrutinized heavily and modified. Questions to be answered about the upcoming year include

- What are the organization's purposes, goals, and objectives?
- What should the organization be doing in the next and future years?
- What core activities are essential?
- Were there instances of previous spending that proved wrong or wasteful, and if so, was the historical data it was based on wrong?
- What additional programs might be taken on if cost were not an object?
- What one-time expenditures might be considered, such as upgrading equipment, improving physical plant, or reducing operational expenses?
- What peer programs are attractive for comparison, and how does your program compare with these organizations? Is there benchmarking information to evaluate, and is that data applicable to my program?

Input and involvement from as many stakeholders as possible should be solicited to understand true needs and problems, as a wide variety of information and input help planners make better decisions (National Research Council 2000).

Budgeting

The budgeting process should always be related to the planning process noted above. Planning for the future requires a clear sense of available resources. Conversely, sensible budgets and financial plans cannot be put together without knowing the direction the organization will take in the future. Simply allocating money is not enough. Instead, proper and strategic use of resources is key to budgeting.

Simply put, budgeting is predicting how much it will cost to carry out the activities identified in the planning process after the requisite details are compiled. The budget is completed after negotiations with the parent organization and approval. Most animal programs will utilize the budgeting mechanisms of the parent organization's financial systems to develop and track budgeting information. In some cases, these systems may not be detailed enough, especially for monitoring multiple categories of expenses, and the program may have to augment the financial system output to further categorize and monitor expenses. This level of detail should help program financial managers. On the other hand, financial managers should strive to simplify financial output for easier sharing and understanding within the program.

The animal program should be tailored to a structure that best tells its financial story in clear terms and logically tracks expenses and revenues. Examples of possible structures and budget formatting decisions to be considered are

- Individual financial statements or budgets arranged by program or function (i.e., a large animal program could create a separate budget covering all barrier rodents, another for all large animals, still another for veterinary technical services, etc.).
- Financial statements or budgets established for each animal facility on campus.
- Level of expense and revenue detail for each budget. For instance, is a budget line for supplies sufficient, or should supplies be tracked further to include cleaning supplies, personal protective equipment (PPE), feed, bedding, and so forth? More detail is always better if one wants to understand where the money is going and is motivated to reduce or eliminate costs.
- Budgeting capital equipment purchases (e.g., equipment whose acquisition cost is $5000 or greater). Capital acquisitions will typically be addressed in a capital budget, separate from the operating budget, that projects and programs for large expenditures over a multiyear period. Capital purchases should always be accompanied by a justification as, well as a return on investment (ROI) whenever possible. In addition, it is a display of good management when big-ticket items are prioritized strategically.
- An amortization process should be considered for capital purchases. Amortization in the simple form basically refers to (1) depreciating the purchase cost of capital items over time and (2) charging the program's operating budget the annual depreciation cost for the capital items and transferring those funds to an amortization fund. For example, 20 years ago ABC Animal Facility purchased a cage wash machine for $100,000. ABC estimates that the machine had a useful life of 20 years, so it transferred $5000 annually from operating expenses to its amortization account for the past 20 years. ABC had an epic cage wash failure and the cage wash needs to be replaced. Through the use of the amortization process, ABC has the funding available to purchase its next cage wash. Depreciation periods are often shorter than the expected lifetime of items, often in the 5- to 10-year time frame.

The organization's financial system must have sufficient flexibility to track according to desired categories. In many cases, the budgeting process will be *incremental*, that is, relies heavily on information contained in the previous year's actual expenses and income, with some adjustments. Some organizations utilize a *zero-based budgeting* approach, which requires each program and line item of a budget to be calculated anew. Staff members are told that any item in the budget will be zero unless they can provide a full justification for a new budget figure. Zero-based budgeting represents a proactive budgeting and tracking process in which every cost is challenged and works well with a general management culture that is committed to unyielding continuous improvement through eliminating unnecessary work and relentless financial management. Zero-based budgeting requires a different mind-set than traditional budgeting methodology, but it can be a more useful tool during times of doing more with less, providing a more innovative and exacting analysis.

Once the budget framework has been finalized, it is time to populate the budget based on projections made in the planning process. Salary and nonpersonnel expenses will be entered into the budgeting tool (usually provided by your organization) according to the various budget tracking categories the animal program management has agreed upon. Estimates should be made of predicted revenues based on activity levels (animal census, technical services provided, etc.) produced in the planning process. In many instances, the parent organization may provide budget subsidies and the subsidies may be applied equally across the entire program or may be allocated appropriately with guidance from the institution.

Pharmaceutical, other for-profit companies, and government facilities will typically utilize an annual budget that is developed and approved based on projected costs without recapturing any charges to investigators. Examples of annual costs that should be considered are

- Animal facility staffing, to include full-time employees (FTEs) and/or contractors (note that when dealing with staffing costs, do not forget to include fringe benefits)
- Training costs associated with new staff, as well as continual learning activities for existing employees
- Space, utilities, and other infrastructure charges
- Food, bedding, and supplies
- Sentinel programs
- Veterinary care
- Dedicated facility space and administrative costs assessed by the parent institution
- Animal facility-related equipment and preventive maintenance agreements, as well as contingencies for repairs

New programs will be especially challenging to estimate, as there is little historical data and many unknowns. The tendency is to underestimate costs and overstate revenues, but an accurate approximation should be attempted with a caveat regarding the circumstances of the estimate. Budget writers should also be sure to anticipate the cost that new programs will have on core staff, supervision, space, and equipment.

Once all expenses and revenues are entered, it is time to evaluate the bottom line of the various budgets. Are they acceptable, or do they need adjustment? Do projected expenses and revenues echo the organization's purposes, goals, and objectives? Are the planned activities central to the organization? Do they help the organization 1–5 years down the road? Comparing the estimated expenses with the estimated income is usually very revealing. It may be clear in the process that some activities will have to be given up if the projected budget does not balance. In fact, if management discovers that income is adequate to cover all expenditures (less institutional subsidies) at this point in the budgeting process, perhaps the organization is not reaching far enough or being ambitious enough in its planning. Alternatively, any excess income could be applied to lower per diem rates.

Adjustments may be needed to bring budgets into expectations, particularly the expectancies of the parent organization. For instance, the parent organization may have provided a subsidy amount for the new fiscal year, but the animal program's preliminary budget shows a need for a greater subsidy. The animal program then has the options of making the case for a greater subsidy, reexamining revenues and expenses to reduce the animal program budget, or increasing rates to reduce the deficit. Rate increases can certainly help close the deficit, but the animal program management team must be sure that users can absorb such an increase. Rate increases that are too precipitous can reduce utilization and actually increase deficits. Negotiations among the animal program management team could help clarify and lead to adjustments that serve all. It is essential that all parties agree to the final budget, including the parent organization, researchers, and animal program management staff.

Monitoring

The budget monitoring process is just as critical as the budget development phase. As the fiscal year progresses, timely budget reports should be shared and reviewed by all stakeholders. How do actual expenses and revenues compare to what was predicted? Are there problem areas that stand out and need to be modified? The animal care program management team should feel ownership of the budget and be willing to monitor and react to issues. Potential overages should be evaluated as to root causes, and affected parties should work on resolutions. Above all, upper administration should be aware of any issues and planned responses as soon as possible.

Special Topic: Budget Cuts and Shortfalls

As financial times have gotten harder, institutions have taken steps that impact the animal care program. Most animal care programs have been asked to do more with less, particularly in response to the Great Recession of 2008. Many animal care programs have seen subsidies provided by the parent organization decreased, and have been asked to increase their animal housing charges, as well as other fees, to help offset costs. However, animal programs have to be very careful in increasing rates, as researchers also have been hit by static grant budgets, and cannot absorb increases easily. Per diem increases that are too drastic may lead to decreased animal populations and may start a downward program slide. Any reaction to budget cuts or shortfalls should be carefully thought through. A recent study by Baker and Kearney analyzed whether increasing per diem charges had a negative effect on animal census. Utilizing techniques from econometrics, a field of economics that applies mathematical and statistical methods to economic data, they found a correlation between higher per diem prices and decreased census, particularly for large animals. The authors caution that their analysis is preliminary, but their use of econometrics is a first for the laboratory animal field and hopefully will encourage programs to carry out more robust studies of other economic questions facing laboratory animal research (Baker and Kearney 2015).

In reaction to budget cuts of shortfalls, animal program management should go back to the planning process and review key elements: What are the organization's purposes, goals, and objectives? What should the organization be doing in the next and future years? What core activities are essential?

Answering these questions may help identify nonessential activities that could be curtailed or eliminated. Creative approaches can help reduce expenses. For instance, sharing program responsibilities between functional areas may reduce salary and other expenses. Recruitment for vacant positions can be suspended and functions absorbed by existing employees. Nonessential purchases can be delayed, and the animal program can continually evaluate approaches to improve operational efficiencies within the facility that will not compromise research programs. These may include streamlining animal husbandry procedures or changing types or sources of animal feed, bedding, and supplies, or reevaluating the animal program's PPE policies. Marketing of services to researchers may increase revenues. Involving a representative animal program core in decision making during difficult fiscal times may help mitigate the pain and create buy-in to the steps taken.

Cost Accounting

Budget planning, development, and monitoring are the financial processes that guide the operation of the animal care program for a defined period of time. Another financial process, cost accounting or cost analysis, analyzes the animal program's budget details to determine accurate and defensible charges for services provided by the program. A cost accounting accumulates and evaluates actual expenses for a completed fiscal year, categorizing and matching expenses to the type of service or activity provided. A cost accounting represents an effort to set reasonable rates for such activities as animal housing, technical services, and other charges based on those actual expenses. In addition, this exercise is required for institutions that receive federal funds, with a prescribed methodology for assigning and calculating those costs. The organization's accounting system logs, classifies, and summarizes the financial activity of the animal care program. The cost analysis manipulates this historical information from the accounting system, allocating expenses to cost centers and arriving at a per item cost for providing services to researchers (Silverman 2009).

Cost accounting activities for an animal program are guided by the following principles:

- Billing rates must be based on actual animal program costs.
- Billing rates for identifiable services that involve significant activities of the animal care program must be established.
- Billing units should logically represent the service provided.
- All rates and the overall animal program should be operated as closely as possible to a "break-even" basis. Adjustments should be made to compensate for surpluses or deficits (variances).
- All costs associated with providing an animal service should be included in the total cost of each service.
- The costs must be treated consistently as either direct or support costs.
- The assignment of costs to cost centers and the allocation of support costs to direct cost centers should be based on beneficial relationships.
- Animal programs cannot charge a "profit" to federal programs; that is, they cannot continually charge more for services than the cost to provide.
- All users should be charged consistently at full rates (or the revenue should be imputed).
- Revenue and costs should be compared at least annually to identify surpluses and deficits for each service.

Federal guidelines do not require the animal program to break even each year as long as overall adjustments are made to maintain rates as close to costs as possible, and an institution can subsidize rates; that is, it can use institutional funds to lower rates below actual costs.

Cost accounting is especially relevant to those institutions receiving federal funds for research involving animals, as these institutions are required to perform a periodic cost analysis of their animal program. A document published in 2000 by the NIH, the "Cost Analysis and Rate Setting Manual for Animal Research Facilities," provides guidance to animal programs for carrying out such a cost accounting and determining reasonable rates to charge researchers for services. Note that such rates are subject to federal audit if federal grants or contracts are involved. Therefore, close collaboration between an institution's animal program and it grants and contracts office is encouraged to make sure all intramural parties are involved in setting and assessing those rates.

A key component of the cost analysis process is the cost center, which is defined as identified activity areas to which expenses are allocated. Cost centers can be direct activities that are chargeable to researchers (barrier mouse cage per diem, veterinary technical services, etc.) or indirect support activities, such as cage wash, animal health care, and feed and bedding. Both personnel and nonpersonnel expenses should be carefully evaluated for proper allocation to the appropriate cost center.

The effort survey asks staff to track their activities for a period of time during the fiscal year to identify which cost centers their efforts are devoted to, that is, which species they perform animal husbandry activities for, when they provide animal health care, when they devote time to administrative tasks, and so forth. This provides a snapshot of personnel activity in the cost accounting process and allows personnel expenses to be allocated to the various cost centers. Ideally, the animal care program should sample employee efforts multiple times during the fiscal year so that a true picture of staff effort is recorded. Effort surveys can be administered in many ways; an employee can keep track of effort by a simple paper worksheet, or the animal program may develop an online application easily accessible to all employees. Whatever the method, the intent is to understand where employee effort is devoted and use the results to rationally allocate personnel expenses to the correct cost center.

Nonpersonnel expenses for the fiscal year in question must also be allocated to the identified cost centers. Ideally, the animal program's financial system permits the categorization of expenses when the cost is incurred. Otherwise, someone will have to go through all expenses retrospectively and categorize them under the appropriate cost center, which can be a time-consuming effort.

An alternative cost analysis approach may be more appropriate if the program is engaged in continuous improvement of its operations in order to identify and eliminate unnecessary activities and expenses while improving quality. In such programs, staff activities may change, perhaps even several times in the same year, as efficiencies are realized in some tasks while additional time and emphasis are assigned to others, like more staff training or the introduction of a new animal species. Consequently, time-and-motion or other activity studies, such as how many cages each technician changes or washes on average every day, are less reliable for consistent costing because they may not be performed in a consistent manner. In this case, programs divide annual cost category totals by the expected census for each animal unit (e.g., mouse cage) to arrive at a per diem rate for each species and husbandry scenario.

One challenging area of expenses, particularly for university-based and government animal programs, is indirect costs. Indirect costs are identified as those costs incurred for common or joint purposes that cannot be specifically associated with an activity. Indirect costs usually refer to central overhead expenses incurred by the parent organization, and include utilities, central administrative offices (human resources, procurement, etc.), and building maintenance. Organizations receiving federal funds are reimbursed for these indirect, or facilities and administrative (F&A) costs, in addition to direct reimbursement for awarded grants. It is important that the animal program follow its organization's guidelines and not include any costs in the cost analysis that may be already included in the organization's overall F&A costs. Lang, in a 2009 article, notes the challenge of categorizing indirect and direct costs for the animal program and makes a case for including a number of animal program activities that are difficult to associate with a specific activity (e.g., regulatory issues and overall veterinary care) in the organization's F&A calculations (Lang 2009). Activities related to regulatory compliance, as well as animal procurement, are specifically prohibited from cost-setting exercises establishing per diem rates under federal policy. Conversely, the NIH Rate Setting Manual does allow for internal support costs within an animal program. Due to budget constraints, upper management may be looking to spread more allowable indirect space costs across functional groups, such as the animal program.

Once all personnel and nonpersonnel expenses are allocated to the cost centers, the internal support cost centers (cage wash, animal health care, feed and bedding, etc.) must be further allocated to the chargeable cost centers via a reasonable method.

For instance, total costs of cage wash may be allocated to the various species based on a study of cage wash activities. If 20% of all cage wash effort is devoted to rat cage per diem, then 20% of cage wash expenses will be allocated to the rat cage per diem.

An accurate census-taking system is key to any charge-back program. This may be manual or automated (bar coding or radio frequency identification [RFID] tags) and must involve a frequent animal or cage count by species, type of housing, and investigator.

Total costs for the chargeable cost centers are then divided by the number of service units or other activities provided during the fiscal year. This will produce the actual per unit cost for providing daily animal care to the mix of species in the program, or the hourly cost for items such as veterinary technical services. An example of the final step of the cost analysis process is as follows:

Total expenses allocated to the mouse barrier cost center	$2,300,000
Total average daily census of mouse barrier cages	4,500
Total annual care days for mouse barrier cages	1,642,500
(4,500 average daily census × 365 days annually)	
Daily housing cost for mouse barrier cage	$1.40
(Total expenses $2,300,000 divided by 1,642,500 annual care days)	

These calculations show that it costs this animal program $1.40 per cage per day to house mice in a barrier cage. The animal care program should compare this developed cost to its current rates and make adjustments accordingly. Perhaps central administration feels this rate is too high for researchers and may provide a subsidy to reduce the per diem. Whatever decisions are made, both the animal program and parent organization should remember that billing rates must be based on actual costs, all rates and the overall animal program should be operated as closely as possible to a break-even basis, and a profit cannot be charged to federal programs; that is, they cannot continually charge more for services than the cost to provide.

Special Topic: Benchmarking and Quality Improvement

Many animal programs receiving federal funds for research involving animals are keenly aware of how their per diem and other rates compare with those of their peer institutions. Researchers will often discuss rates with colleagues throughout the country, and lower animal housing rates can be an excellent recruiting tool to attract new faculty. Animal program management should be aware of how their service rates stand up to those of other organizations, both regionally and nationally. One source of information is the Animal Resources Cost Survey periodically sponsored by Yale University's Animal Resources Center. Participation is open to academic and nonprofit academically oriented institutions. The survey aggregates responses from institutions throughout the country and provides a great deal of data, including information on average and median housing rates. The survey also asks questions regarding institutional subsidies, staffing, and many other key animal program topics.

The American Association for Laboratory Animal Science (AALAS), American College of Laboratory Animal Medicine (ACLAM), and American Society of Laboratory Animal Practitioners (ASLAP) periodically perform laboratory animal compensation surveys that are helpful for benchmarking animal program salaries to regional and national averages. Many institutions will also perform their own ad hoc surveys comparing their animal program with those of their peers for such topics as animal program salaries, institutional subsidies, and animal program responsibilities. Such benchmarking and surveys are very helpful for providing perspective for the animal program's operation and performance. Comparisons should be considered carefully, however, as program structure, institutional funding and subsidies, program resources provided, and so forth, can contribute to differences between animal programs.

Regarding quality improvement, the Vivarium Operational Excellence Network (VOE-N) is a consortium of animal care facility experts who have agreed to share best practices for continuous improvement, with the ultimate goals of improving quality and efficiency and reducing operating costs. Members participate in a free-flowing exchange of ideas, share the results of individual improvement efforts, and provide feedback and support. The VOE-N also provides educational opportunities designed to train the next generation of animal care facility leaders.

Grants, Other Sources of Revenue, and New Services

Some animal care programs are reacting to challenging fiscal times by seeking nontraditional sources of funding. The federal government, as well as nongovernmental agencies, offers grants and other awards that can help upgrade facilities or support research into laboratory animal issues. For instance, the NIH, through its Office of Research Infrastructure Programs, provides a limited number of animal facility improvement grants to upgrade animal facilities and equipment.

The funding agency usually requires a detailed application requiring significant details of proposal, as well as information about the institution's research environment. Animal program management should be willing to devote significant time to an application, as grant writing is an intensive effort. Once the grant is submitted, the funding agency will review proposals and choose those that best respond to the agency's grant criteria. Once awarded, the animal care program must follow the funding agency's requirements and timelines closely, making sure funds are expended as outlined in the grant proposal and within the time period of the grant. Many institutions have significant grant administration programs, and animal care management should utilize these resources to prepare the grant application and properly administer the grant if the animal program is lucky enough to receive an award.

Animal care programs may also provide services to outside organizations for supplemental sources of income. The animal care program should ensure there is capacity within the program and that all institutional users' needs are met first, before considering outside collaboration. The animal care program should also review institutional guidelines regarding collaborations with external organizations before providing services to outside groups. Many research institutions are considered not-for-profit, and unauthorized outside agreements could jeopardize that status, particularly if the arrangements do not meet the institution's research or education missions. Nevertheless, outside agreements that meet guidelines can be fruitful for both the animal program and the outside group. The animal program's parent institution may have contract and grants specialists that can guide management through the correct steps for establishing such a relationship.

The animal care program should be cognizant of the needs of the researchers it serves and be willing to provide new services. This can both help the research community and provide a source of increased support for the program. More researchers are utilizing genetically altered mice models, and some newer animal users do not have the time or experience to carry out all the needed research steps. Many programs have realized that there are unmet needs and have added a number of new areas of expertise, including breeding services, technical and surgical assistance, transgenic cores, and gnotobiotic facilities.

The animal program should develop a business plan before embarking on any new initiatives. This business plan should include an evaluation of the interest within the research community and possible usage; a survey of other organizations already providing such services, getting an idea of pricing and organization; a proposed structure of the new program, and how it will be initiated and developed; and projected expenses and revenues over a 3- to 5-year period. New programs will be especially challenging to estimate, as there may be little historical data and many unknowns. Costs should be estimated on the high side, and revenues should not be overly optimistic. Any new initiatives should be reviewed and approved by the parent organization before proceeding.

Technology

Program managers are tasked with ensuring that their animal program has a solid financial footing. To accomplish this, they must understand and address business and research issues that impact their fiscal needs. Increasingly, technology has empowered problem solving and enables new opportunities and approaches to modernize and streamline all key management functions, allowing the highest level of productivity and effectiveness in managing and tracking fiscal resources.

The unique capability of emerging technologies is already in use in some aspects of animal care and use. There are many of these tools that are changing the way we manage animal facilities. Two of the technologies are RFID and global positioning system (GPS). Both are currently available within the health care field and appear capable of decreasing operating costs but may require significant initial capital investment. The RFID application in the animal research facility is readily being used for animal census, whereas GPS technology might be more beneficial for tracking equipment throughout the facility, such as racks and other accountable equipment (McGrady et al. 2010).

Technology is constantly evolving and can be applied within vivaria. Technology is available to remotely connect to an animal room and use video cameras to check the health status of animals. Sensors to notify robots that a cage is wet or has an elevated ammonia level may become the norm and could dispatch an automated robot to change the cage. Globalization, software, and the Internet are having a synergistic effect on how we all live and work. Software is providing more services via the Internet every day, and the actual physical location of those servers and software matter less and less. It is estimated that by 2026, nearly all data needed by the end user will be accessible at all times. Smartphones and other highly portable devices with a web browser can access this data in the cloud. The ability for the animal facility to store its data in the cloud and have it accessible via a web browser can help improve facility overall operations. This type of technology is enabling animal facilities and program management to work seamlessly together in tracking expenses (Doughman 2016).

Program managers will need to keep their fingers on the pulse of this technology and determine its cost-effectiveness. Managers should develop a systematic approach in investigating the full range of emerging technologies, see their potential possibilities, understand their uncertainties, compare the technologies, and select the ones to focus on and continually review new information and choices. Thoroughly evaluating those possibilities is key to making good choices about where and when to invest (Evans 2009; see also https://grants.nih.gov/grants/policy/air/rate_setting_manual_2000.pdf).

GLOSSARY OF FINANCIAL TERMS

allocation: assignment or distribution of costs or expenses to cost centers as part of the cost accounting process

benchmarking: comparing an organization's programs, strategies, and so forth, to peer organizations

budget: an estimate of costs, revenues, and resources over a specified period, reflecting a reading of future financial conditions and goals

capital equipment: items of considerable value and durability that are used in the animal care program. Accounting rules usually consider objects costing more than $5000 and having an extended lifetime as capital equipment

cost accounting or cost analysis: the accumulation, examination, and manipulation of cost data for the purpose of setting animal program per diem and other rates

cost center: a defined area to which costs are allocated in the cost analysis process. Includes direct activities that are chargeable to researchers (mouse per diem) or indirect support activities, such as cage wash

effort survey: asks staff to track their activities for a period of time during the fiscal year to identify which cost centers their efforts are devoted to, that is, which species they perform animal husbandry activities for, when they provide animal health care, when they devote time to administrative tasks, and so forth

financial or fiscal management: the planning, directing, monitoring, organizing, and controlling of an organization's monetary resources

fiscal year: the 12-month period an organization defines as its budget year. The U.S. federal government's fiscal year is October 1 to September 30, while many organizations use a July 1 to June 30 year

goals: an observable and measurable desired end result to be achieved during a defined time frame

planning: the identification of goals and objectives to be achieved, along with a strategy to achieve them

return on investment: the time required to recoup through reduced operating costs the investment in a major capital item

revenues or support: payments or fees that help support the expenses of an animal care program. Sources can include per diem charges, fees for technical services, grant funds, and institutional subvention

zero-based budgeting: a method for preparing budgets that starts from scratch each year with no pre-authorized funds. Each activity must be justified on a basis of cost–benefit, no present commitments existing, and there being no balances to carry forward

REFERENCES

Alford, J. 2001. Fiscal management. In *Management of Laboratory Animal Care and Use Programs*, ed. M. Suckow, 243–255. Boca Raton, FL: CRC Press.

Baker, D., Kearney, M. 2015. The need for econometric research in laboratory animal operations. *Lab Animal* 44 (6): 217–220.

Doughman, E. 2016. What will vivarium technology look like in 2026? *Animal Laboratory News*. http://www.alnmag.com/articles/2016/03/what-will-vivarium-technology-look-2026.

Evans, N., Ralston, B., Broderick, A. 2009. Strategic thinking about disruptive technologies. *Strategy & Leadership* 37 (1): 23–30.

Lang, C. 2009. The cost of animal research. *Lab Animal* 38 (10): 335–338.

McGrady, E., Conger, S., Blanke, S., Landry, B., Zalucki, P. 2010. Emerging technologies in healthcare: Navigating risks, evaluating rewards. *Journal of Healthcare Management* 55 (5): 353–364; discussion 364–365.

National Institutes of Health, National Center for Research Resources. 2000. Cost analysis and rate setting manual for animal research facilities. https://grants.nih.gov/grants/policy/air/rate_setting_manual_2000.pdf.

National Research Council. 2000. *Strategies That Influence Cost Containment in Animal Research Facilities.* Washington, DC: National Academies Press. http://www.nap.edu/catalog/10006/strategies-that-influence-cost-containment-in-animal-research-facilities.

Silverman, J. 2009. *Managing the Laboratory Animal Facility.* Boca Raton, FL: CRC Press.

14

Occupational Safety and Health

James M. Schmitt, Deborah E. Wilson, and James M. Raber

CONTENTS

Background

Occupational safety and health considerations are primary concerns of animal care and use program management regardless of the size of the program or facility. In order to appropriately address the associated safety and health concerns, a concerted effort of hazard identification must be undertaken. Each activity or job should be analyzed to identify the potential hazards associated with the performance of each activity. Once the hazards are identified and the associated risks assessed, they must of course be mitigated in some fashion to prevent accidents and injuries. The job hazard analysis (JHA) or job safety analysis (JSA) and subsequent risk assessment approach provide a proactive path toward injury and illness prevention associated with the animal care and use program. The JHA or JSA is a standard tool used in the practice of occupational safety and health, and information regarding its use is broadly available elsewhere (OSHA 2002; Blair 2014). The JHA or JSA focuses on the relationships among the worker, the task, the tools, and the work environment. It is also used as a tool to identify and mitigate hazards that have been associated with an accident, injury, or work-related illness. The mitigation of identified hazards should follow the broadly accepted approach using the "hierarchy of controls" strategy (OSHA 2005) (Figure 14.1). The hazard controls in the hierarchy are, in order of decreasing effectiveness,

- Elimination or substitution of the hazard
- Use of effective engineering controls
- Appropriate and effective administrative controls
- Use of personal protective equipment (PPE) appropriate for the job or task at hand

While this approach sounds simple, a team is often required and recommended. A team approach can help personnel to fully understand the hazards that may be present, regulations and policies that may apply, support and training required, and institutional policies and liabilities that may impact the interventions or mitigation strategies selected and implemented. Although all jobs and tasks supporting an animal care and use program should eventually undergo a JHA, jobs that may entail the highest potential risks should be prioritized and assessed first in order to reduce the likelihood of injury or illness. A historical review of Occupational Safety and Health Administration (OSHA) accident logs,

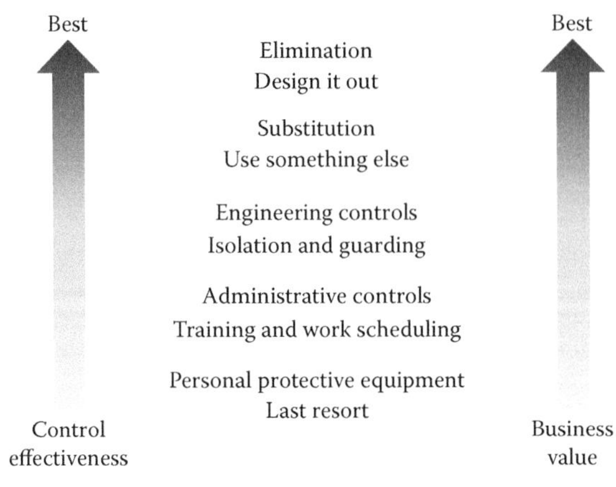

FIGURE 14.1 Hierarchy of controls.

near-miss reports, and facility injury reports can all be helpful in prioritizing your approach. It is recommended that the previous 3 years of reports and data be reviewed in order to set priorities (Blair 2014). Areas or tasks where injuries have previously occurred, or have been associated with a high frequency of injury, illness, environmental harm, or equipment damage, should be given the highest priority. In order to fully understand and address the safety issues associated with the conduct of animal research, a safety team should be constituted with representation from a variety of disciplines. In a larger, more complex program, the team may be composed of a wide variety of individuals with a range of specialized expertise. For example, safety and health professionals, veterinarians, animal facility managers or supervisors, environmental compliance personnel, occupational medicine advisors, research scientists, facility or maintenance engineering personnel, and financial and human resources personnel, as deemed necessary, are required to fully understand the hazards and risks present, the potential mitigations, and the resources required to bring the issues to successful closure. In smaller programs, a safety professional working with a facility manager, a veterinarian, and an engineering representative may suffice. Regardless of the program size, safety is a team sport.

There are a number of excellent references that can help guide institutions in establishing occupational safety and health requirements that meet the needs of the animal care and use program (NRC 1997, 2003, 2011a). Additionally, there are national, regional, or provincial and European Congress regulations and directives establishing occupational safety and health requirements. Based on the animal facility location, these requirements must also be met when establishing the occupational safety and health program.

Roles and Responsibilities

Successful animal care and use programs share certain characteristics. Most importantly, there is a strong, institutional commitment to the safe and compassionate conduct of research. Typically, the institutional official exemplifies this commitment. Model programs typically take a team approach to assessing needs and developing consensus approaches to meet those needs. At a minimum, the occupational safety and health management team is composed of a veterinarian with experience and training in the species used within the program, a facility manager, and a safety and health specialist with knowledge relevant to biomedical research and animal care and use. Successful collaboration as team members requires a cooperative approach and effective communication skills. The animal care and use committee or oversight body is responsible for ensuring that an effective occupational safety and health program is in place. A detailed description of the roles and responsibilities of these various parties is addressed in Chapter 3.

Occupational Safety

As discussed in the previous section, the responsibility for the overall design and execution of a program that protects the safety, health, and welfare of all personnel rests in the hands of a team of key individuals. It is the responsibility of the institutional official to ensure that the accountability for the day-to-day operation of the program is clearly delineated and the responsible individuals are knowledgeable of both the safety issues and daily program requirements. It is the responsibility of the occupational safety and health management team to identify potential safety hazards, conduct risk assessments, and ensure that appropriate safety measures are in place to protect all personnel. Hazard identification and risk assessment is an ongoing process within an animal care and use program. The *Guide for the Care and Use of Laboratory Animals* (*Guide*) (NRC 2011a) states that the risks associated with the experimental use of animals and research materials must be identified and reduced to minimal and acceptable levels.

As outlined in the following sections, hazards within the animal program fall in several categories, ranging from environmental and animal hazards to research hazards, such as biohazards and radiation. Hazard identification and risk assessment are the first steps in developing a comprehensive program that is both safety conscious and efficient. Both the severity and seriousness of the potential consequences of the hazard must be weighed, in combination with the prevalence or probability of the hazard occurring.

These factors can help the team rank or scale potential hazards in the order of their importance for prevention measures and resource allocation. The selection of safeguards to minimize or remove potential risk factors and the assessment of the effectiveness of the chosen safeguards are primary functions of the management team. Through this risk assessment procedure, the team should be able to describe the decision-making process used and the rationale for the selection or development of specific safeguards, policies, procedures, and practices.

Clear, concise, and timely communication at all levels is critical for the development, implementation, and assessment of a dynamic occupational safety and health program. Personnel at all levels should be encouraged to share their ideas, problems, and experiences related to workplace safety and hazards without punitive consequences. The development and implementation of a training program that is both proactive and responsive to changing program requirements is critical in order to equip individuals with the knowledge and skills needed to work safely in a research environment with animals. Training should be reoccurring and documented. Training should be directed at identified problem areas and those areas assessed for effectiveness on a regular basis.

Engineering versus Procedural Controls

The creation of a safe working environment starts with the physical features of the facility and equipment used in the work environment. Engineering or physical controls are preferred over procedural controls because they are not dependent on human actions to minimize hazardous exposures.

Barriers can be created between an individual and a potential hazard, which can prevent physical injury or contact with the hazardous agent. Barriers can be defined in many ways and take many forms. Solid barriers can protect an equipment operator from pinch or crush hazards, and directional airflow can protect adjacent rooms or hallways from contamination.

Good facility design is often the first step in creating an effective engineered barrier control program (Lipman and Leary 2015). Understanding the functional use of a room, including the potential for using hazardous agents in the area, can help in deciding if the room air pressure should be positive or negative to adjacent areas. For example, animal rooms or areas in which infectious material will be handled are commonly maintained with a negative air pressure to the adjacent areas, whereas "clean" rooms (surgery suites, clean cage wash, etc.) are routinely maintained with a positive room air pressure to adjacent areas. Architectural barriers that control personnel access and flow patterns can further isolate an area and provide protection for surrounding locations. Air locks and/or anterooms have been used in some facilities to provide additional protection for animal biosafety level 3 and 4 (ABSLs 3 and 4) areas.

Local exhaust fume hoods, biological safety cabinets, and downdraft tables are devices used to prevent personnel exposure to hazardous or flammable gases, aerosols, vapors, or dusts. These devices should ideally be located in low-traffic areas, away from room entrance doors and air supply and exhaust ducts, which can create air turbulence and disrupt the normal functioning of the device. These devices should be evaluated on a regular basis according to manufacture recommendations or regulatory guidance and, if possible, certified to ensure adequate air velocities and filter performances, as appropriate. These devices are typically maintained with a face velocity airflow of 80–100 ft/min (NRC 1995). Ideally, the device should include a continuous airflow monitor and alarm to aid in ensuring the functionality of the unit when working with hazardous materials. Care should be taken to understand the safe operating distance and requirements (such as front window sash height) of the unit. This is especially critical when working with downdraft tables.

Biological safety cabinets are one of the most effective primary containment devices for working with infectious agents and other hazards with potential for aerosol production. Proper air balance and inward airflow are critical to the function of these units. High-efficiency particulate air (HEPA) is brought into the cabinet, and the air is again HEPA filtered when leaving the cabinet. For infectious work, biological safety cabinets should be certified in accordance with the National Sanitation Foundation Standard 49 (NSF 2014). Ideally, the unit should be recertified annually and whenever the cabinet is moved. Individuals using a biological safety cabinet must be trained on its use, functionality, and limitations. Like fume hoods, inward airflow is critical to the functionality of the device; containment can be compromised by interruption of the inward airflow pattern by rapid movements of the operator's arms, changes in room ventilation patterns, creation of air turbulence outside of the cabinet, or disruption of the inward airflow by an abundance of items located inside the cabinet. Certain biological safety cabinets can also be used to protect the clean status of an animal or product within the cabinet. When used in this capacity, the biological safety cabinet not only protects the animal or product from contamination, but also protects personnel and prevents environmental contamination from aerosols and animal allergens.

Laminar flow change hoods are not biological safety cabinets and not designed for work with infectious agents. It is critical that personnel understand the difference between a laminar flow change hood and a biological safety cabinet. Laminar flow hoods are designed primarily for the protection of animals or products used within the hood and not for the protection of personnel or the environment. These mass air displacement (MAD) Class 100–type devices flush the inside of the cabinet with HEPA-filtered air but are not guaranteed to recapture and refilter all the air prior to releasing it back into the room. Although some laminar flow hoods may act like a biosafety cabinet by redirecting some of the air leaving the cabinet through a HEPA filter prior to releasing it back into the room, the efficiency of these units is questionable and they should not be confused with a biosafety cabinet. Therefore, positive-pressure, laminar flow change hoods or cleaning stations are not recommended because of their tendency to circulate animal allergens out of the hood onto the user and into the room (Lipman and Leary 2015).

Administrative Controls

In addition to engineering controls for controlling hazards in the animal facility, administrative controls are another effective technique for hazard control. Administrative controls are often used when engineering controls are impractical or prohibitively expensive. They may also be used in conjunction with engineering controls. Administrative controls are used when available and provide a plethora of tools for use in hazard control and mitigation, through establishment of policies (e.g., rest breaks), personnel practices (e.g., required training), management (e.g., rotating work assignments), monitoring, limiting worker exposure, measuring performance, training and education, housekeeping, maintenance, and purchasing. Use of and rigorous adherence to standard operating procedures is also a highly effective administrative control. For instance, the "two-man" rule or "buddy system" in nonhuman primate facilities is an example of an administrative control, as is having a policy of substituting "green products" for use where possible, eliminating more hazardous and environmentally damaging cleaning products.

Hazard Communication and Signage

Warning or hazard signage is an important method of hazard communication and control that is administrative in nature. Recent efforts to globally harmonize chemical hazard warning symbols (Global Harmonization System [GHS]) emphasize the importance of signage in communicating hazards and its appropriate use in the animal facility. See examples of GHS warning pictograms in Figure 14.2. In the United States, OSHA requires safety data sheets (SDSs) under its hazard communication standard (29 CFR 1910.1200). The OSHA published a rule (77 *Federal Register* 17574–17896, March 26, 2012) that aligns its hazard communication standard and SDS requirements with the GHS.

In addition to potentially hazardous chemicals, radioactive and/or biohazardous materials may be used or present in the animal facility. Universal warning symbols for these hazards have also been established and should be used on warning signage when these potentially hazardous materials are present. Examples of the international symbols for biohazards and radioactive materials are provided in Figure 14.3. These symbols should be incorporated into appropriate warning signage.

Personnel Protective Equipment

PPE is often worn in an animal facility to minimize exposure to serious workplace injuries and illnesses. Injuries and/or illnesses may result from exposure to chemical, biological (blood body fluids, infectious agents, and allergens), radiological, physical, electrical, mechanical, or a host of other animal facility workplace hazards. These hazards should be identified for mitigation. Mitigation procedures include (1) elimination or substitution of a less hazardous substance or method, (2) the use of

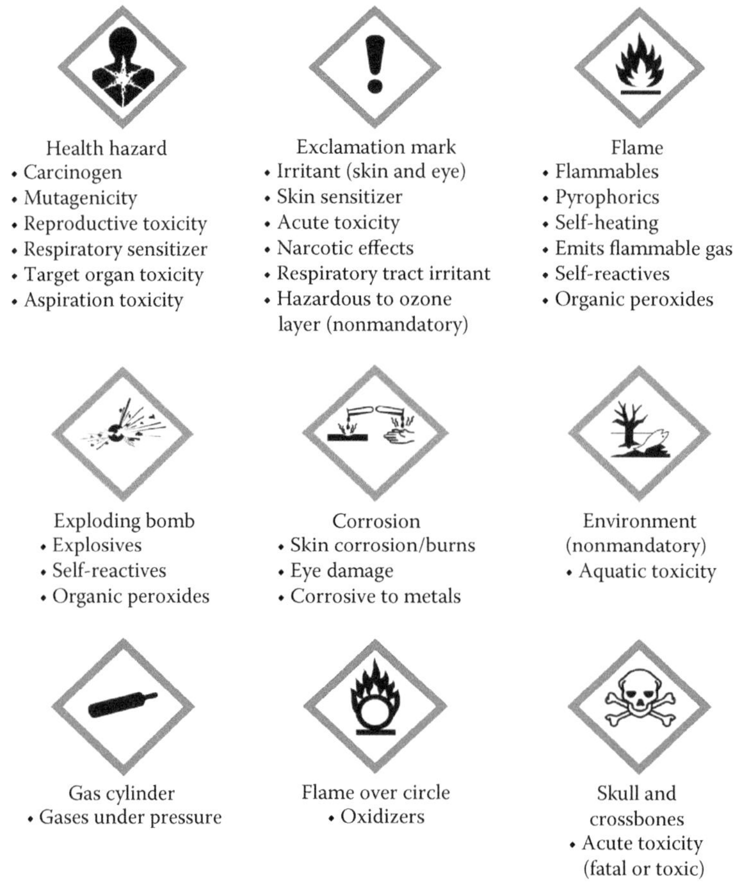

FIGURE 14.2 Examples of GHS warning pictograms.

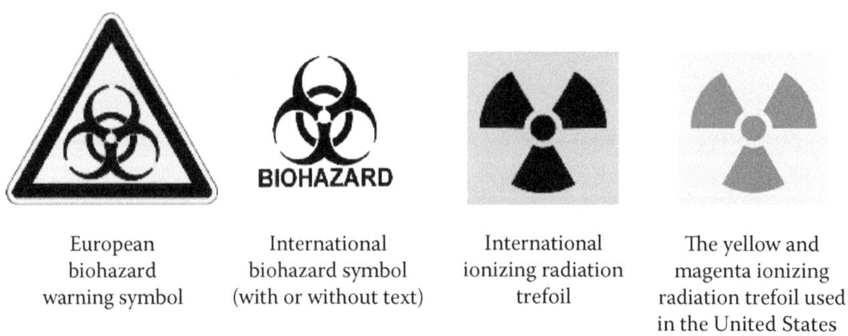

FIGURE 14.3 **(See color insert.)** International warning symbols for biohazardous and radioactive materials.

engineering controls to reduce potential exposures, (3) administrative or procedural controls, and (4) the use of PPE—in that order. This is often referred to as the "hierarchy of controls." PPE should be utilized when the hazards associated with the work cannot be eliminated or reasonably reduced using the other controls established in the hierarchy to provide a safe work environment for employees. PPE may include items such as gloves, safety glasses and shoes, earplugs or muffs, respirators, eye and mucous membrane protective devices, and coveralls and full-body suits. PPE is used also in animal facility management strategies to protect the health of the animals. This discussion, however, is limited to PPE used for the protection of employees only.

PPE programs are often mandated by law. In the United States, the use of PPE is covered by the U.S. Department of Labor OSHA standards. In general, 29 CFR 1910 Subpart I—PPE governs the use of these devices. Additionally, there are a number of other OSHA standards relating to specific hazards that require the use of PPE, and many require that PPE be provided free of charge to employees by the employer. In Canada, similar mandates for federal operations are found under the Canada Labor Code and in the corresponding Canada Occupational Health and Safety Regulations, specifically Part XII—Safety Materials, Equipment, Devices and Clothing. Additional provincial regulations may also apply. In the European Union (EU), Directive 89/656/EEC—Use of PPE provides guidance and protections. If PPE is to be used, these standards have similar requirements—programs should address the hazards present; the selection, maintenance, and use of PPE; the training of employees; medical support; and monitoring of the program to ensure its ongoing effectiveness.

PPE should be selected that will protect against the hazard identified, the activities being performed, and the animal species involved in the work. Table 14.5 to Table 14.15 provide examples of hazards associated with activities necessary for providing adequate husbandry of selected animal species and recommendations for PPE to be used during the conduct of these activities.

PPE selection should be done through a risk assessment process with full consideration of the human factor. Wearing PPE can impact human performance and may increase some risks to the wearer by reducing dexterity, impeding vision, reducing ability to communicate, and increasing heat stress possibility. A balanced approach must be used when selecting the appropriate PPE for a given tack and environment. When selecting PPE, it is important to remember that more is not always better. Specific descriptions of the types of PPE recommended for various species and the activities being performed in the animal facility are provided in Table 14.5 to Table 14.15.

Clothing Protection and Contamination Control

Contaminants of all types may be spread on clothing, compromising the facility, other animals, and even coworkers and family members should the contaminant be carried back to an administrative area or home on clothing soiled at work. After a careful risk assessment, contamination may be addressed through the use of uniforms, laboratory coats, jumpsuits, and hair and shoe coverings. Appropriate donning, doffing, and disposal of work clothing or clothing coverings and shoe coverings is important in contamination control and personal protection. Incorrect removal of the protective garments can lead to exposures to the agents that the PPE was intended to protect the personnel from. Training on the appropriate and effective removal and disposal process is critical to ensure personnel safety.

Gloves

Selection of glove type and material is based on the type of exposure and nature of the hazard. Some chemicals can easily penetrate gloves that work very well for other chemicals. The following should be considered when choosing gloves necessary for work in the animal facility:

- Chemical type
- Temperature extremes and cryogenic properties
- Physical hazards (sharps and piercing objects)
- pH

- Toxicity
- Infectious potential
- The procedure to be conducted

Additionally, one should consider whether the gloves are for incidental or extended contact with a hazardous agent. Incidental contact may involve little or no direct exposure to the hazardous material, and uses may include activities such as accidental spills or splashes, overspray from a dispensing device, handling of infectious agents that require barrier protection, and preventing contamination of materials (product protection), or during sterile or aseptic procedures. Disposable, surgical-type gloves are appropriate for incidental contact, as are nitrile gloves. Often, nitrile gloves are preferable to latex gloves because of better chemical resistance and the tendency to visibly tear when punctured, and in situations where the worker has a latex allergy or to protect against development of a latex allergy. When the choice is made to use disposable gloves, check for tears or punctures before use, remove and replace gloves immediately if overtly contaminated, and avoid touching common objects (door knobs, keyboards, etc.) while wearing gloves. Never wash or reuse disposable gloves.

Extended glove contact to hazardous materials may include situations such as handling highly contaminated materials, submerging hands in or prolonged contact with a chemical or other hazardous substance, and the need for physical protection from temperature extremes or sharp objects. More substantial gloves are generally required for extended use, and many of these gloves are reusable. Reusable gloves must be inspected for tears or punctures before and after each use and for evidence of prior contamination and/or degradation. Extended-use, reusable gloves should be washed and air-dried after removal. Table 14.1 will aid the reader in proper glove selection.

TABLE 14.1

Proper Glove Selection in the Animal Facility

Glove Type	Use	Advantages and Disadvantages
Latex (natural rubber)	Incidental	• Good for biological and water-based materials • Poor for organic solvents • Little chemical protection • Hard to detect punctures/holes • May cause or trigger latex allergies
Nitrile	Incidental contact (disposable exam glove) Extended contact (thicker reusable glove)	• Excellent general use glove • Good for solvents, oils, greases, and some acids and bases • Clear indication of tears and breaks • Good alternative for those with latex allergies
Butyl rubber	Extended contact	• Good for ketones and esters • Poor for gasoline and aliphatic, aromatic, and halogenated hydrocarbons
Neoprene	Extended contact	• Good for acids, bases, alcohols, fuels, peroxides, hydrocarbons, and phenols • Poor for halogenated and aromatic hydrocarbons • Good for most hazardous chemicals
Polyvinyl chloride (PVC)	Specific use	• Good for acids, bases, oils, fats, peroxides, and amines • Good resistance to abrasions • Poor for most organic solvents
Polyvinyl alcohol (PVA)	Specific use	• Good for aromatic and chlorinated solvents • Poor for water-based solutions
Cut-resistant gloves (stainless steel, Kevlar, leather)	Specific use	• Sleeves are available to provide protection to wrists and forearms • If there is potential for biological or chemical contamination, wear appropriate disposable gloves underneath cut-resistant gloves and discard disposables after use

Mucous Membrane Protection

Mucous membrane protection refers to protection of the eyes, nose, and mouth from splash and splatter of potentially contaminated materials. This nomenclature originally referred to protective devices used to prevent occupational transmission of human immunodeficiency virus (HIV) in the health care workplace via mucous membrane exposure of the workers. The use of the term has evolved and is now more broadly applicable. This type of personal protection includes goggles, face masks, and face shields, and specific devices should be chosen after careful risk assessment of the task at hand. Ideally, appropriate mucus membrane protection should be selected with the guidance of a trained occupation safety specialist. Figure 14.4 provides additional guidance on the selection of face masks.

Respiratory Protection

Respirators are devices fitted to the individual that can protect workers against insufficient oxygen environments, harmful dusts and other particulates, fogs, smokes, mists, gases, vapors, and sprays (http://www.cdc.gov/niosh/docs/2005-100/pdfs/2005-100.pdf). Respirators protect the user in two basic ways. The first is by the removal of contaminants from the air. Respirators of this type include particulate respirators, which filter out airborne particles, and powered air-purifying respirators (PAPRs) with cartridges or canisters that filter out chemicals and gases. Other respirators protect by supplying clean breathable air from another source. Respirators that fall into this category use compressed breathing air from a remote source (such as those used in maximum containment facilities—ABSL-4) and self-contained breathing apparatus (SCBA), which is a dedicated air supply (tank air). Workers should use respirators for protection from contaminants in the air only if other hazard control methods are not practical or possible under the circumstances. Respirators should not be the first choice for respiratory protection in workplaces. They should only be used

- When following the "hierarchy of controls" is not possible (elimination, substitution, engineering, or administrative controls)
- While engineering controls are being installed or repaired
- When emergencies or other temporary situations arise (e.g., maintenance operations)

Workers wearing respirators to accomplish work tasks must be enrolled in an appropriately constituted respiratory protection program as defined by national or regional authorities, consult an industrial hygienist or other safety and health professional in selection and fit testing of the appropriate respiratory protective device, and be medically cleared to use a respirator.

Hearing Protection

Animal facilities have high-noise areas, such as mechanical spaces, cage wash areas, and housing areas where animals may loudly vocalize. Exposure to high levels of noise can cause permanent hearing loss (https://www.osha.gov/SLTC/noisehearingconservation/). Neither a hearing aid nor surgery can correct this type of hearing loss. Short-term exposure to loud noise can also cause a temporary change in hearing (ears may feel "stuffed up" or clogged) or a ringing may be present (tinnitus). These short-term problems may go away within a few minutes or hours after leaving the noisy area. However, repeated exposures to loud noise can lead to permanent tinnitus and/or hearing loss.

Loud noise can also create physical and psychological stress, reduce productivity, interfere with communication and concentration, and contribute to workplace accidents and injuries by making it difficult to hear warning signals or instructions. Animal facilities with noisy areas should establish a hearing conservation program (HCP) to identify and monitor areas where high noise levels exist and to subsequently identify employees whose job duties may expose them to noise levels exceeding the 8-hour time-weighted average. The U.S. OSHA (29 CFR 1910.95) sets the 8-hour time-weighted average action limit (or threshold limit value [TLV]) at or above 90 dBA (decibels measured on the A scale, slow response). Other authorities, such as the American Conference of Governmental Industrial Hygienists (ACGIH)

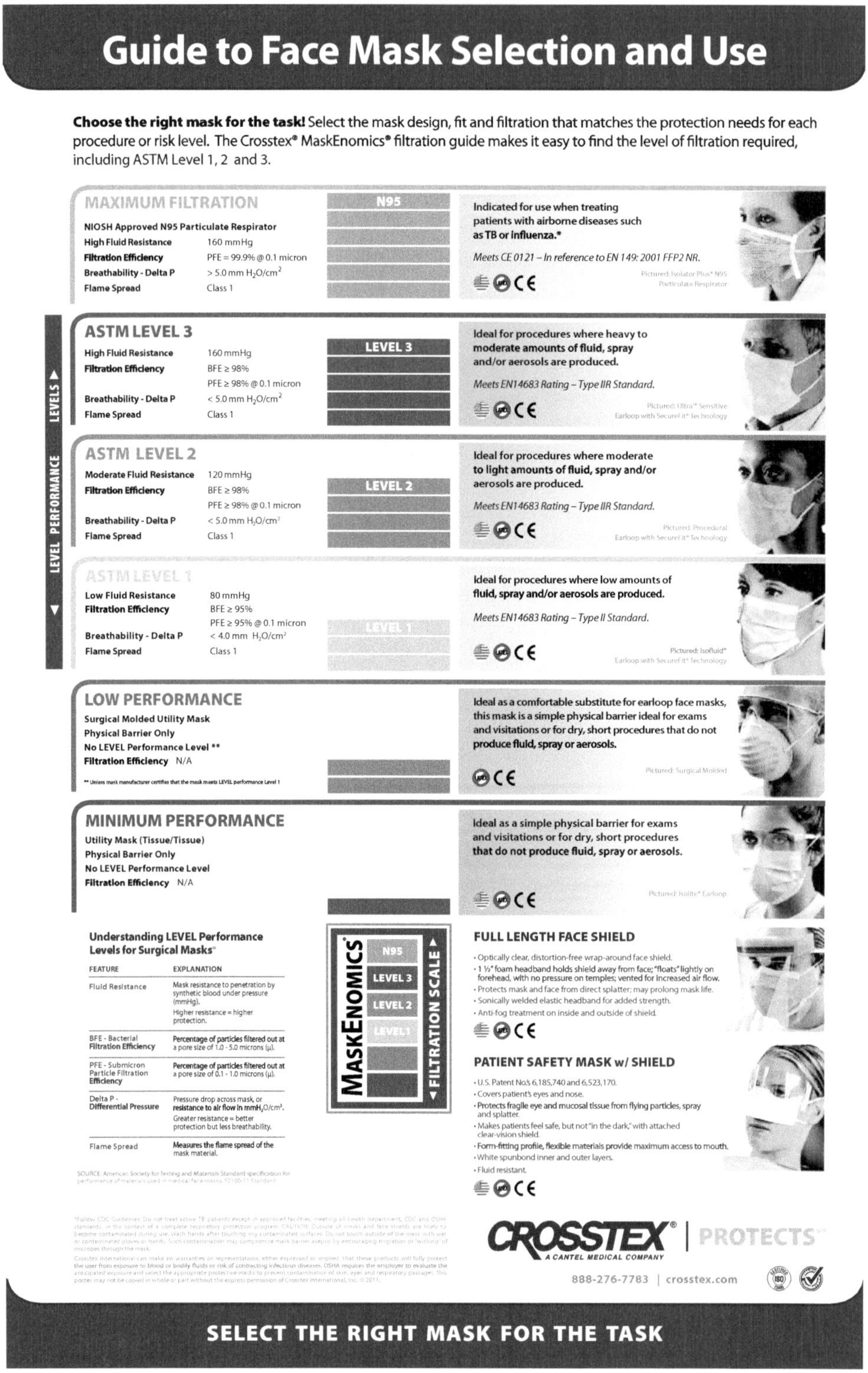

FIGURE 14.4 (See color insert.) Guidelines for the selection and use of face masks. (From CROSSTEX, Hauppauge, NY, http://www.crosstex.com/home.asp.)

and the EU Directive 2003/10/EC on workplace noise, set more conservative 8-hour exposure limits at 85 and 80 dB(A), respectively.

An HCP should address (1) high-noise-area identification and monitoring, (2) employee monitoring (personal noise dosimetry), (3) audiometric monitoring, (4) noise control (engineering and administrative controls), (5) hearing protection devices (HPDs), (6) training, (7) medical monitoring, and (8) record-keeping requirements. Personnel working in high-noise areas of the animal facility should have appropriate hearing protective devices available to them. An HPD is a personal safety product that is worn to reduce the harmful auditory and/or annoying effects of noise. HPDs should be viewed as a last resort, when other means, such as engineering and administrative controls, are not practical or economical. All HPDs have an associated noise reduction rating (NRR). Aural inserts (earplugs) fit directly into the ear canal. Earplugs in both formable and premolded versions are available in various sizes. Circumaural protectors (ear muffs), which are plastic domes that cover the ears and are connected with a spring band that fits on top of the head or is attached to a hard hat, are also an option. The HPD chosen must have an NRR that reduces the worker's noise exposure to the established level.

Occupational Medicine

Designing Occupational Medical Support Services

Because the medical services provided for the workforce should match the potential health hazards in the workplace, the design of the offering logically begins with a risk assessment. This approach enhances the medical provider's understanding of the work environment and should result in more effective services. Failure to tailor the medical services to meet the workers' likely needs risks compromising the quality of the medical care and the workers' respect for it. The risk assessment requires a team approach and is an ongoing process. Initially, safety specialists, veterinarians, managers, and principal investigators (PIs) identify potential health risks in the workplace. The medical provider occasionally must modify the offering to address new hazards and discontinue services for hazards that are eliminated from the workplace. Developing and maintaining constructive relationships with the partners that helped with the initial design of the program is critically important for subsequently identifying changes in workplace hazards in a timely fashion. The medical provider ideally shares draft versions of the proposed medical support services and subsequent modifications with the safety officials, veterinarians, managers, and the investigators involved in the risk assessment. Encouraging their meaningful involvement in the design process enhances the quality of the offering and likely will encourage their support of it.

Medical support for animal-based research should not be referred to as "medical surveillance" because it does not meet the definition of the term. By definition, medical surveillance involves ongoing evaluation or testing of workers at risk for an adverse health consequence from a workplace hazard. The monitoring tool must be used appropriately: sensitive enough to identify an injury before the worker would seek medical attention, specific enough to identify the consequence of a particular workplace injury, and acceptably affordable and palatable to the worker. Moreover, data generated from the testing must be analyzed in a timely fashion and shared with those tested, and authorities capable of making adjustments in the workplace. Unfortunately, other than testing the hearing of workers who are exposed to noise above the OSHA-permissible exposure level, there are no tests that meet this definition for workers who support animal-based research. It is more accurate to describe the offering as medical support services tailored to address workplace health hazards. The services often include an occupational history, a targeted review of relevant personal medical details, work-related health counseling (verbal and in writing), and immunizations and laboratory testing, if indicated by both occupational tasks and clinically warranted.

It is convenient to divide the occupational medical support services as occurring in three separate time frames. The initial medical evaluation should occur prior to the individual's starting work with animals and is referred to as a preplacement medical evaluation. The second possible contact would be routine periodic medical evaluations. Routine periodic medical evaluations in most situations are not clinically warranted for most individuals working with animals. The third contact option is medical evaluation following occupational injuries, potential work-related illnesses, or other general job-related health concerns.

Preplacement Medical Evaluation

This visit provides an opportunity for the health care provider to assess the individual and determine the physical demands and environmental hazards of the proposed position. Ideally, human resources or the direct supervisor provide the work site information as part of the paperwork for the evaluation. The health care provider gauges the worker's understanding of the type of contact the individual will have with animals, other potential workplace health hazards, and the nature of the research. With those details in mind, the health care provider obtains a history of current medical conditions and their treatment, a personal history for allergies, and a history for immunizations that are relevant for the proposed work. Not infrequently, during this initial evaluation newly hired workers acknowledge symptoms suggestive of allergies to animals related to prior employment. If the worker will need to utilize a respirator, the health care provider should administer the OSHA Initial Medical Questionnaire for Respirator Use (https://www.osha.gov/pls/oshaweb/owadisp.show_document?p_table=STANDARDS&p_id=9783). A physical exam generally is not warranted and should not be included in the preplacement medical evaluation. Rarely, the health care provider may be concerned that a worker's preexisting medical condition may interfere with his or her ability to perform the duties of the proposed position effectively and safely (e.g., musculoskeletal issues). When that occurs, the most effective approach to resolve the matter is for the health care provider to review copies of the worker's relevant personal medical records.

Potentially, the most valuable portion of the preplacement medical evaluation is the opportunity to provide individualized counseling. The health care provider discusses potential health hazards, such as animal proteins, musculoskeletal injuries, animal bites and scratches, and exposure to chemical and biological agents. The health care provider emphasizes that all work-related injuries and illnesses must be reported at the workplace and to the medical provider for prompt attention. Finally, the health care provider describes what constitutes proper first aid and the steps the worker should take to access emergency medical support. Additionally, detailed medical information should be provided for high-consequence human pathogens involved in the research (ABSL-3, BSL-3AG, and ABSL-4 agents) or potentially harbored by the animal used in the research (e.g., herpes B-virus (*Macacine herpesvirus* 1, formerly *Cercopithecine herpesvirus* 1 [CHV-1]). The counseling helps focus the worker's attention on potential health hazards at the workplace, how best to avoid them, and the steps to take to reduce the risk of significant injury following an accident. Informational handouts help keep the message consistent and may provide a useful reference for the worker later.

The health care provider updates the worker's immunizations as occupationally indicated and clinically warranted, and performs relevant testing (e.g., testing for rubeola and tuberculosis (TB) for individuals handling nonhuman primates, and rabies titers if there is an elevated occupational risk for exposure to rabies) (www.cdc.gov/rabies/specific_groups/doctors/serology.html). Pulmonary function testing typically is not needed to medically clear a worker to use a respirator. There is very little, if any, value associated with the practice of storing a worker's serum for possible future reference (Lehner et al. 1994). The practice is not warranted.

Routine Periodic Medical Evaluations

Individuals working with laboratory animals should periodically be reminded of the health hazards in their work environment and the actions they should take in the event of a possible occupational injury or illness. These reminders may be provided as part of annual training or in a targeted e-mail message. Some programs opt to meet this need by requiring periodic medical evaluations. However, as stated above, routine periodic medical evaluations are in most situations not clinically warranted and may be cost-prohibitive. Those advocating for periodic medical evaluations may contend that the visits permit the health care provider an opportunity to test the worker for evidence of an allergy to animal proteins. However, available testing strategies for allergies are of dubious clinical value. Elevated immunoglobulin E levels do not reliably help predict who will become sensitized to laboratory animal proteins. Similarly, skin testing for allergies to animal proteins is not helpful. Positive skin tests do not correlate well with clinical findings. In addition, 13% of workers with positive skin tests are asymptomatic (possible false positives) and another 6% will have negative results in the presence of symptoms suggestive of allergic reactions (false negatives) (Bush and Stave 2003). Use of enzyme-linked immunosorbent assays and radioallergosorbent testing (RAST) has been suggested; however, results from these tests correlate even less well than skin test results. Allergic reactions to animal proteins usually begin within 6 months of exposure and rarely occur after 2–3 years of work with laboratory animals (Gross 1980). A more practical approach to determining who is having a problem is to provide detailed counseling regarding allergic reactions to animals before the individual begins work with laboratory animals, and then periodically remind those workers of the risk and their need to report symptoms suggestive of an allergic reaction promptly. Once an allergic reaction is detected, the best approach to controlling the condition is reducing exposure to the proteins and symptomatic treatment. Others advocate periodic medical evaluations to determine whether workers are immunocompromised or pregnant. In neither case is a positive response likely to alter the individual's medical clearance to perform the duties of his or her assigned position. As is the case with workers who develop allergic reactions to animal proteins, the additional medical history does not significantly alter the advice provided during the preplacement medical evaluation. Individuals working with nonhuman primates are routinely recalled for periodic testing for subclinical infections with *Mycobacteria tuberculosis* to minimize the risk for inadvertent infection of the research animals. The interval questionnaire for respirator fit testing and training may be administered by the person performing the testing or training. Workers with positive responses or an interest in speaking with a health care provider should be referred back to the medical support staff.

Medical Evaluation for Work-Related Injuries and Illnesses

Medical support for work-related injuries and illnesses is critically important. Injury reports permit insight into the circumstances that lead to occupational injuries and opportunities for corrective actions to prevent further injury. The organization must take all reasonable steps to facilitate these reports, including ensuring that the process is not perceived as punitive. If contract staff is involved in the work, the language for their contract should include injury reports as a deliverable item in the contract. Physical and financial barriers to accessing health care services in a timely fashion should be minimized.

The majority of injuries reported from animal care facilities can be classified as musculoskeletal trauma, although cuts, bites, and scratches are also frequently reported. Table 14.2 provides a summary of 194 occupational injuries and illnesses reported to the National Institutes of Health (NIH) Occupational Medical Service in a 12-month period ending in 2014 from animal care areas. Not surprisingly, the majority of the injuries reported occurred while the worker was handling an animal. Interestingly, only 1% of the work-related injuries and illnesses reported that year involved an allergic reaction to animal proteins. A few factors may explain this finding. A significant majority of workers with allergic reactions to animal proteins are discovered during the preplacement medical evaluations. Second, safety measures currently employed to reduce occupational exposures to animal proteins are effective.

The health care provider, in collaboration with biosafety professionals and researchers, should develop a detailed response plan for potential exposures to all high-consequence biologic and chemical hazards. The plan should describe the emergency medical response, including who needs to be notified of the

TABLE 14.2

Summary of Animal-Related Occupational Injuries and Illnesses Reported to the NIH Occupational Medical Service in a 12-Month Period Ending in 2014

- 56 (34%) occurred during the feeding, handling, or restraining of an animal
- 46 (24%) occurred during the performance of a medical intervention
- 50 (26%) involved work with, or in proximity to, a cage
- 30 (15%) were sprains, strains, repetitive motion injuries, punctures, or lacerations not involving a research animal or a cage
- 2 (1%) were new reports of work-related allergies

event. In rare instances, the results of an exposure can be severe and occur so quickly that the medical response should be available almost immediately. Examples of such incidents include exposures to the neurotoxin 1-methyl-4-phenyl-1,2,3,6-tetrahydropyridine (MPTP) and concentrated hydrofluoric acid. In these cases, the antidote (sublingual selegiline and intradermal injection of calcium gluconate, respectively) should be maintained in the workplace. Although the likelihood of actual occupational illness will likely be much lower, a similar argument may be used for storing a starter dose of valacyclovir in work areas with Old World macaques. If such an approach is warranted, responsibility for properly storing and maintaining the medications must be established and treatment should only be initiated at the direction of the responsible health care provider. Regardless of whether medical attention for a potential exposure will be needed immediately or urgently, the medical support team should be intimately familiar with the response plan and available to provide assistance, regardless of the day or week or time of day. Drills testing the plan can help familiarize the medical support staff with the details of the plan and identify potential weaknesses in the strategy. Common findings in these drills include the importance of having someone record events as they occur and the need for clear guidance on communication requirements.

In assessing the significance of a potential exposure to a biological hazard, medical providers should consider separately the risk of exposure (RoE) occurring and the risk of disease (RoD) following an exposure. This approach should improve communications and reduce misunderstandings. For a worker to be at risk for an exposure to a biological agent, two conditions must be met. First, a viable biological agent must be present. Second, the worker's PPE or innate protection (e.g., intact skin) must be breached, resulting in a potential exposure (e.g., by percutaneous injury, cutaneous and mucous membrane contamination, inhalation, or ingestion) to the biological hazard. The worker, PI, biosafety specialist, and the medical provider cooperatively determine the RoE. The risk is stratified into one of four categories: no risk; low but not zero (LBNZ) risk; minimal risk; and elevated risk (Table 14.3).

Occasionally, the circumstances of an accident warrant an estimate of the RoE for more than one biological agent. For example, a percutaneous injury that occurs while dissecting neurologic tissue from a macaque known to be infected with B-virus and experimentally infected with a biological hazard would warrant a RoE for B-virus and a second RoE for the biological hazard. Similarly, a bite injury involving an animal infected experimentally with a biological agent would warrant an assessment for the RoE to the animal's oral flora and a second estimate for the RoE to the biological hazard.

The RoD is an estimate of the probability that an exposure will result in an infection or an estimate of the probability that an illness has been caused by a biological agent under investigation. The medical

TABLE 14.3

Risk-Level Descriptors

Risk of Exposure Level	Description
No risk RoE = 0	1. Either the biological hazard was not present *or* 2. The worker's PPE was not breached or it is not possible that the worker could have sustained an exposure to the biological hazard. 3. There is no conceivable RoE.
Low but not zero risk RoE = 1	1. The biological hazard may have been present *and* 2. The worker's PPE was breached and, although the RoE is too small to quantify, it cannot be excluded that the worker sustained an exposure to the biological hazard. 3. Although theoretically possible, the risk is extremely unlikely.
Minimal risk RoE = 2	1. The biological hazard may have been present *and* 2. The worker's PPE was breached and the worker may have sustained an exposure to the biological hazard. 3. The risk is unlikely.
Elevated risk RoE = 3	1. The biological hazard likely was present *and* 2. The worker's PPE was breached and the worker sustained an exposure to the biological hazard. 3. The risk is plausible.

TABLE 14.4

Factors Modifying the Risk of Disease

1. Virulence: Is the biological hazard a known human pathogen (e.g., SIV, a close relative of HIV-2, or known to infect humans without causing illness)? How likely is it to cause disease in humans (e.g., only 10% of people with latent TB infection develop TB)? How likely is it to cause severe health consequences (e.g., case fatality rates of influenza strains differ significantly)? Has the organism been modified to enhance or diminish the parent strain's pathogenic potential (e.g., nonvirulent Sterne strain of *Bacillus anthracis*)?
2. Volume and concentration: What was the estimated dose of exposure (e.g., decreased size of inoculum if the needle goes through gloves first or syringe plunger was not depressed during injury)? How much viable pathogen may have been introduced (e.g., was source material chemically or physically inactivated prior to the incident)? What is the agent's minimum infectious dose in a healthy host? What is the toxin's lethal dose?
3. Route of exposure: The clinical significance of the event may be impacted by the route of exposure (e.g., the risk of a percutaneous exposure to HIV is greater than a mucosal exposure, and the risk of a respiratory exposure to influenza is likely higher than a percutaneous exposure).
4. First aid: How much time passed from exposure to starting to cleanse the affected body part (immediate cleansing of agent may reduce disease risk)? Was first aid delivered appropriately (e.g., adequate duration and volume, and use of appropriate disinfectant)?
5. Preexposure protection: Was the individual vaccinated against the biological hazard? Does the individual have protective antibody titers? How effective is the vaccine? (Prior vaccination may lower the disease risk.)
6. Worker's medical conditions and treatment: Does the individual have an illness or take medications that predispose for higher risk for disease (e.g., TB) or complications (e.g., influenza)?
7. Postexposure medical countermeasures: Are pharmacologic or immunologic agents available that are known to be effective against the biological hazard (e.g., antibiotics, antivirals, or specific immunoglobulin G [IgG] therapy)? Are there any investigational medical treatments available (e.g., serum from survivors of a disease in cases of high-risk exposure to the same disease agent)?

provider and infectious disease specialists cooperatively determine the RoD. Several factors influence estimates of the likelihood an exposed worker will be at risk for developing the disease associated with the biological hazard. These factors include, but are not restricted to, the virulence of the biological hazard, the volume (size of inoculum) and concentration of the biological hazard in fluid, the route of exposure, the adequacy of the first aid (e.g., timeliness, technique, duration, and agent used), the worker's protection to the biological hazard from prior immunization or infection, the worker's medical conditions and treatments, and the adequacy of postexposure chemoprophylaxis (e.g., timeliness and effectiveness). The RoD is usually the same as the RoE. However, the RoD may be higher or lower based on any of the factors described in Table 14.4.

Serum stored at the time of an exposure to a biological agent is occasionally clinically useful in assessing whether the injury resulted in a subsequent laboratory acquired infection (LAI). That specimen is referred to as an "acute sample." If that specimen is tested, a "convalescent sample" acquired 10 or more days following the injury should be tested simultaneously. Medical providers should exercise caution when attributing clinical significance to test results from a diagnostic study that is not commercially available.

Reporting, Record Keeping, and Quality Assurance

Given that workers who have access to animals used in research are required to receive medical support services, the medical provider must have a mechanism for communicating who has received the preplacement medical evaluation and any required periodic medical evaluations (e.g., testing workers with access to nonhuman primates for recent infection with *M. tuberculosis*). As mentioned above, injury reports provide valuable information. The health care provider promptly should share the circumstances of work-related injuries with their counterparts in safety and management. This approach permits a timely review of incidents and ideally the identification of potential corrective actions that may prevent recurrences. The data is also needed for generating an OSHA 300 log. The health care provider should be familiar with the workers' compensation application process and facilitate workers' completion of the reporting forms. Reviewing injury data for a specified period may help identify critical patterns in the injury data (e.g., a disproportionate share of the injuries involve a particular process or piece of equipment).

The medical provider must comply with its country's regulations governing the generation and maintenance of all clinical records. For example, in the United States a medical provider must comply with all state and federal regulations.

Finally, the health care provider should have an active quality assurance program that utilizes specific, meaningful indicators that test the extent to which the medical services conform to established expectations. For instance, the provider may elect to track whether routine and emergency medical care is provided in a timely fashion or whether the care provided matches the description in standard operating procedures.

Unique Hazards Encountered in the Care and Use of Laboratory Animals

An animal care and use program presents many challenges in the identification, assessment, and control of occupational hazards. Program hazards can be divided into three broad categories: (1) environmental, (2) animal, and (3) research related. Although some hazards are common for all programs, many of the potential hazards are unique to a given program or facility, and are dependent on the experimental research being conducted and animal models being used. Therefore, the unique and dynamic nature of each laboratory animal care and use program requires that hazard identification and risk assessment be a customized and ongoing process. The hazard identification and risk assessment process must include individuals with the training and experience needed to fully assess the possible range of potential program hazards and the risks. This process commonly includes managers, veterinarians, and safety specialists. Although it is impossible to capture all potential hazards and risks that can be encountered in a laboratory animal program in one short chapter, the next section attempts to provide a general outline of common hazards that may be encountered.

Environmental Hazards

Surface Hazards

The need for moisture-impermeable, highly sanitizable surfaces in the animal facility, especially for flooring, often creates slick and slippery working conditions. These situations can at times be reduced or eliminated by selecting textured materials, for example, the addition of sand or other textured additives to epoxy flooring. In addition, careful attention to moisture on the floor, the selection of facility shoes with slip-resistant soles, and/or the use of slip-resistant shoe covers helps reduce the problem.

Ergonomic Hazards

Attention to good ergonomic design can reduce or eliminate the occurrence of musculoskeletal disorders and repetitive motion injuries while improving health, safety, and productivity. It takes forethought to design and arrange a work environment that is user-friendly, efficient, and safe. Consideration must be given to the mental, physical, and organizational aspects of the work, as well as the setting in which the work is conducted. The goal is to reduce, minimize, or eliminate factors that can result in injury, pain, or discomfort.

Although potential ergonomic hazards can be identified in many of the jobs related to animal care and use, these injuries are most common in production areas where the tasks require repetitive movements or the movement of heavy objects. The purposeful placement of handholds in natural positions, coupled with attention to body position and back support, can help to eliminate many hand, wrist, shoulder, and back injuries. One example is the use of back supports coupled with the use of heavy-duty, removable quick-release suction cup handles (e.g., Harbor Freight Tools). Their use has significantly reduced ergonomic injuries related to the movement of large, heavy primate cage racks (NIH Program Data). Care must be used when recommending the use of back supports, as they may provide a false sense of security to the user. It is for this reason that some programs have stopped recommending their use. When used with appropriate training, back supports serve as a reminder to use good body posture, mechanics, and lifting procedures.

When space permits, powered equipment, such as "tugs" and forklifts, has been used to move heavy and bulky caging. The selection of smaller, lighter animal racks can also reduce ergonomic issues, but this is not always possible when trying to maximize housing densities and reduce housing costs. Another area where we have been able to improve ergonomics is the selection of uniquely designed biosafety cabinets and change hoods. Many vendors now manufacture adjustable-height cabinets and hoods that allow the employee to adjust the height of the unit to meet his or her specific requirements. Other examples of ergonomic improvements that have benefitted our employees are (1) the use of antifatigue mats and runners in areas where personnel are required to stand for prolonged periods of time; (2) the procurement of smaller, lighter bags of food and bedding; (3) the use of self-opening doors; (4) the use of lighter fiberglass cage pans versus heavier metal units; and (5) the use of ramps or other floor-leveling strategies to smooth the transition from one work area to another.

Machinery and Equipment Hazards

Any piece of machinery or equipment that has moving parts can pose a significant hazard to the operator. It is critical that the equipment is well designed and in good working order, and that precautions are taken to protect the operator from potential pinch, crush, and entrapment areas (e.g., rack washers and bulk autoclaves). Engineering controls such as strategic placement of finger guards or other barriers to minimize accidental exposure of fingers and hands are often beneficial in preventing these types of injuries. For example, retrofitting a metal guard plate at the interface of the tunnel washer conveyor belt and the cage loading area has provided additional protection for the equipment operators.

When purchasing any piece of machinery or equipment, the product should be evaluated not only for functionality, but also for safety and ergonomic design. Care should be taken to evaluate the safety of the equipment not only during routine operation, but also during equipment cleaning and maintenance. Consideration should be given to the presence and location of emergency shutoffs, signage related to potential dangers, and safety features.

It is important to evaluate not only the machinery or equipment for safety, but also the required standard operating procedures and training practices for the safe operation of the equipment. Engineering safety controls or procedural safety measures are preferred; however, they are not always possible. Therefore, well-thought-out standard operating procedures, coupled with a well-designed and -documented training program, are often required.

Heat, Steam, Cryogens, and Pressure Hazards

Heat, steam, and pressure hazards are of primary concern in the facility cage wash area, as well as other areas using heat or steam to sanitize and sterilize equipment and instruments. The use of cryogens (e.g., liquid nitrogen and dry ice) are also common in the modern laboratory animal facility. Cryogen boil-off can lead to asphyxiation of both personnel and animals. Rooms housing magnetic resonance (MR) scanners in which cryogen gases are stored must be equipped with oxygen sensors and a method for increasing room ventilation to exhaust inert gases during cryogen filling (Klaunberg and Davis 2008).

The appropriate selection and use of PPE (e.g., aprons, gloves, safety glasses, and goggles) is normally sufficient to protect personnel when they are coupled with appropriate standard operating procedures and training. It is important to ensure that gloves are both moisture and heat resistant when working in wet cage wash areas or with surfaces that may be hot, as well as moist. In addition, appropriate ventilation must be present to prevent steam and fog buildup from creating fogged glasses, goggles, and an unsafe work environment. Steam release safety interlocks must be present and in good working order in equipment where increased internal pressures are present and steam is allowed to build up (e.g., steam autoclaves). Larger pieces of walk-in equipment, for example, rack washers and autoclaves, should ideally be equipped with internal emergency shutdown or de-energizing and egress mechanisms to prevent personnel from becoming trapped inside. When working with cryogens, care must be taken to ensure that the work area is well ventilated and that the gloves selected are appropriate for the temperature ranges of the agents. When working with liquid nitrogen, eye protection should also be worn.

Lighting, Electrical, and Electromagnetic Hazards

Lighting can be a source of hazard secondary to the type of lighting fixture and amount of light provided. Fixtures that are not properly installed, in good repair, grounded, and water resistant can pose a shock hazard, especially in a wet environment. Ceiling fixtures in the animal facility are often covered and sealed with a gasket to prevent moisture penetration and bulbs from being accidently broken during room sanitation. Ideally, lighting fixtures should be easily accessible for bulb changing without having to clear the room.

Some animals are purposefully reared in dark, low-light-intensity conditions, or under red lighting. These lighting conditions can present potential safety hazards for individuals working and navigating in the area. Standard operating procedures for individuals working in low-light areas often require an acclimation period to prevent personnel injury or task performance errors. Rearing animals in the dark also requires the use of night vision goggles, which must be of sufficient quality and fit to allow routine tasks to be conducted adequately, including animal health checks. It is also important to remember that more than one pair of goggles will be required to ensure continuity of care in case one pair becomes damaged.

Electrical outlets and switches throughout high-water-use areas, such as cage washing areas and high moisture holding rooms (e.g., aquatics, flushed runs, and rooms that are hosed down), should be ground fault interrupted (GFI) and have moisture-resistant covers. In addition, grounded (three-prong) plugs should be standard for all equipment throughout the program.

The use of lasers and ultraviolet (UV) light is becoming more common in the laboratory and animal facility. Both areas require controlled access to the area when the laser or UV light is in use, and appropriate signage to warn personnel entering of the potential danger. Depending on the power and nature of the laser or intensity and wavelength of the UV light, appropriate eye and skin protection is required.

With the development of magnetic resonance imaging (MRI), electromagnetic hazards have become prevalent in animal care and use programs. It must be understood that the MRI magnet is *always* on. Because of the strong magnetic field created by the equipment, any ferromagnetic object brought into the scanner room or critical zone of the equipment will become magnetized and can become a dangerous projectile that could injure personnel and damage equipment. These high magnetic fields can also cause a life-threatening situation in individuals with implanted pacemakers, stents, and aneurysm clip implants. Body piercing jewelry can move when brought within the magnetic field and, in some cases, generate heat and cause burns. This is also true for some transdermal patches, tattoos, and foreign bodies, which may contain metallic components. Individuals who may be pregnant should consult with their health physicist or safety officer before working in close proximity to potential magnetic field hazards. Here again, appropriate signage and training is critical to the management of these electromagnetic hazards.

Chemical Hazards

Potential chemical hazards are numerous, but manageable, in the modern research laboratory and animal facility. It is important to always check with your safety specialist on the appropriate use and storage of potentially hazardous chemicals. Depending on the chemical hazard, chemical showers and/or eye wash stations may be required in the area. In addition to the product labeling, the material safety data sheets (MSDSs), or what are now commonly called safety data sheets (SDSs), are a good source of information and should be readily accessible to staff within the area using the material.

Acids, bases, and caustic and corrosive chemicals should be stored on shelving or in cabinets that are low to the floor. Ideally, these items should also be placed in a secondary container, such as a spill tray. The secondary containment should be big enough to contain the spill of the largest container being stored. Ideally, acids and bases should be stored away from each other, in separate secondary containment areas. Since vapors may escape a container, the cabinet or area where chemicals are stored should be ventilated or located near a ventilation system, such as a fume hood. Never store corrosive chemicals under a sink where the vapors may cause corrosion of plumbing fixtures. Ideally, shelves used to store any chemical should be designed with "endcaps" to prevent containers from accidently being pushed off the end.

In all cases, appropriate signage should be present where chemical hazards are present. Care should be taken to ensure that standard operating procedures outline the appropriate use and storage of the chemicals, required PPE, and disposal requirements.

Fire Hazards

Many types of fire hazards and flammable materials are present in the animal laboratory and facility. The National Fire Protection Association (NFPA) categorizes potential fire hazards by class, where each class indicates the nature of the flammable or combustible material. Class A hazards are ordinary combustibles, Class B flammable liquids, and so forth. Most combustibles within an animal facility fall into Classes A and B. The NFPA further classifies the area in which fire hazards are stored into three classifications: light (low) hazard, ordinary (moderate) hazard, or extra (high) hazard. Light (low) hazard areas include offices, classrooms, and meeting areas. Ordinary (moderate) hazard areas are areas where the quantity and combustibility of material is moderate, such as light manufacturing areas, research operations, parking garages, workshops, or maintenance and service areas. Extra (high) hazard areas are areas that store or use high volumes of flammable material.

The proper storage of Class B liquids is determined by the material's flash point, the lowest temperature required for the liquid to produce a vapor that can propagate a flame. Programs must evaluate the MSDSs or SDSs of each substance to ensure that their handling and storage are in compliance with all OSHA laws and regulations. Flammable liquids are common in necropsy, surgery, and diagnostic areas of an animal facility. Smaller quantities of Class B liquids can be stored openly in the laboratory or facility; however, larger quantities will require storage in explosion-proof cabinets or refrigerators.

Noise Hazards

As stated above under the "Hearing Protection" section, animals and the equipment used for their care may produce sufficient noise to be considered a hearing hazard. Small rodent species rarely pose a noise hazard, larger species (e.g., dogs, pigs, and nonhuman primates) can create enough noise during routine husbandry to pose a serious threat. This is also true for many areas of an animal facility housing noise-generating equipment (e.g., cage washers, pressure washers, and vacuums). The magnitude of the problem is equated with the number of animals present, the acoustics of the room they are housed in, and the required personnel exposure time.

Depending on the acoustics of the room and the nature of the sound generated by the animals or equipment, engineering controls should be considered prior to requiring hearing protections in the area. The magnitude of sound generated can often be dampened by modifying the room acoustics by varying the room surfaces or other structural relationships. Unfortunately, many common sound-dampening surfaces are unable to be sanitized and are permeable to moisture and dirt. The net result is that they are unusable in the animal facility. That being said, removable and sanitizable acoustic panels (e.g., Soundbreak™) have been used successfully in animal facilities to decrease noise levels by up to 10 dB.

Sharps Hazards

Sharps hazards can be found throughout the animal laboratory and facility environment. Inappropriately handled needles, syringes, pipettes, scalpels, broken glass, and other sharp equipment (e.g., damaged caging) can result in traumatic tissue injuries. In addition to the tissue trauma, these injuries can also carry a risk of contamination from pathogenic, zoonotic, and chemical agents. Although minimizing the use of sharps is the idea method for reducing this hazard, it is critical that personnel be trained on how to safely handle and dispose of sharps. The disposal of sharps must be in compliance with all local regulations and policies. The use of appropriately labeled puncture-resistant, leak-proof containers for the disposal of sharps is imperative. It is recommended that needles not be recapped, bent, cut, or removed from syringes, and one should never attempt to compact the contents of the container. Many products are marketed today to aid in the prevention of needle sticks and sharp hazards. For example, needle and scalpel shields, which are integral to the needle or scalpel itself, can shield the point or edge to protect personnel once the procedure has been completed. If recapping a needle or removal of the scalpel blade from the scalpel handle is necessary, personnel should be trained on the appropriate method to be used, for example, using a one-handed technique to hold the syringe with attached needle while picking up the needle cap from a flat counter surface. Scalpel blades should never be removed from the handle by hand. The use of a long needle holder or hemostat to grasp and remove the blade is a safer alternative.

Animal-Related Hazards

Bites, Scratches, and Other Animal-Related Physical Injuries

Although mostly preventable, potential hazards related to animal bites, scratches, kicks, and crushing injuries are ever present in the laboratory animal environment. Animal-related hazards fall into two broad categories, physical and biological. Physical injuries are the result of the mechanical damage to tissue caused by an animal's bite, scratch, or direct trauma. Biological injuries are secondary to the inoculation of the wound site with microbial pathogens or zoonotic agents present in or on the animal's mouth, skin, or nails (e.g., B-virus, rabies virus, and methicillin-resistant *Staphylococcus aureus*). Most animal-related injuries are the result of direct human animal contact during periods of animal transfer, restraint, or other procedure.

The proper handling and restraint of an animal is the most effective way to protect both the personnel and animals. Chemical restraint is often used to provide a safe working environment for both personnel and animals. When chemical restraint is not possible or practical, consideration should be given to acclimatizing the animal prior to the procedure using positive reinforcement training and desensitization techniques. Personnel working with awake animals must be trained in the normal behavior of the species, as well as their behavior under distressful situations. Training on the appropriate restraint techniques for the species, coupled with appropriate PPE, will help prevent physical injury. When working with animals, it is important to control as many of the environmental variables as possible, while insuring the humane handling of the animal. Ideally, the environment should be quiet and free from distractions. All required restraint equipment should be in good working order and readily accessible prior to working with the animal. Animals not previously acclimated to being restrained often become defensive and aggressive. Personnel should remain calm and move smoothly with confidence and purpose, while trying to anticipate the animal's reaction and movements.

Various bite-resistant gloves manufactured out of Kevlar® and/or stainless steel mesh are on the market today. Many of the gloves can be worn over moisture-impermeable gloves or under other protective gloves to reduce punctures secondary to animal bites. When working with some species (e.g., nonhuman primates), arm-length, bite-resistant gloves, usually manufactured of leather or a similar material, should be used to protect both the hand and forearms. It should be noted that protective gloves may not prevent the animal from biting or causing injury, but they are effective in preventing the bite from breaking the skin. The main disadvantage of most gloves is that they greatly reduce the tactile ability and mobility of the user.

All animal bites, scratches, and other related injuries should be taken seriously, and a determination should be made by trained occupational medical personnel whether subsequent treatment of the injury is required. Animal-related injuries should be logged, tracked, and periodically reviewed by program management and safety personnel. This review should identify predisposing factors and possible training gaps and then stimulate program modification to improve injury awareness and prevention efforts. Retraining of current personnel may also be required to prevent reoccurrences.

Allergens

Workers may develop allergic reactions to an animal protein, also referred to as an allergen. These allergens are found in dander, hair, urine, and saliva. Some personnel working with animals become sensitized over time, whereas many never become sensitized. Those who develop allergic reactions often do so within the first 12 months of contact with an animal at home or in the workplace. In rare situations, individuals may not display signs of developing an allergy for several years. The earliest symptoms typically consist of upper airway and skin complaints, such as nasal congestion or stuffiness, a "runny" nose with nasal drainage, sneezing, red and irritated eyes, skin itching, and hives. More advanced cases display lower airway findings, such as coughing, wheezing, shortness of breath, asthma, and in rare circumstances, life-threatening anaphylaxis. The proposed mechanism for these reactions is an immunological and physiological sensitization to specific animal proteins (allergens). Animal allergens can be carried through the air and potentially contaminate all surfaces within an animal facility or animal laboratory. Allergens tend to stick to fur, dander, bedding, and dust. Individuals working directly with animals or in an area contaminated by animal allergens can carry the allergens out of the area on their hair and clothing. Exposure to antigens can be through inhalation of airborne particles, ingestion, or direct contact with the skin or eyes.

Although all animal proteins can be allergenic, some species may pose more of a problem than others. Rabbit and cat allergies are common, whereas allergic reactions to nonhuman primates and aquatic species are less common (NRC 1997). Several factors are associated with an individual's risk of developing an allergic reaction: the individual's immune system, and the intensity, frequency, and route of exposure to the animal allergens. Some activities have a higher RoE than others, for example, cage changing. The worker at greatest risk for developing an allergic reaction to laboratory animal proteins is the worker who has a history of allergic reactions to household pets. Individuals with a personal history for asthma, seasonal sinus problems, eczema, and other allergies are also at increased risk for developing an allergic reaction to laboratory animal proteins.

Individuals entering a laboratory animal program should be evaluated for existing animal-related allergies and educated on protective measures. The goal in the workplace is to minimize the chances that workers will inhale or have skin contact with animal-related allergens. Allergen risks can be minimized or eliminated by the use of appropriate engineering controls, such as air containment devices (e.g., biological safety cabinets and chemical fume hoods) and waste handling equipment (e.g., HEPA-filtered bedding dump stations). Respiratory exposures can also be further controlled by housing animals in filter-topped cages and working in a well-ventilated area, such as a room with nonrecirculating room air or near a fume hood. The *Guide* states that PPE should be used to supplement but not to replace engineering and process controls.

In animal facilities and laboratories using animals, signage that identifies hazards should be clearly posted (Figure 14.5). Policies and procedures should be developed to minimize the environmental burden of the allergens through the decontamination of work areas and, where possible, working with animals in a biological safety cabinet or fume hood. The use of containment caging systems such as microisolator caging has also helped to reduce the environmental burden of allergens. To prevent allergens from

FIGURE 14.5 Example of animal hazard signage.

leaving the facility or animal laboratory, consideration should be given to working in a negative-air-pressure room and ensuring that PPE (e.g., lab coats) are not worn outside of the area. Ideally, lab coats should be disposable. If nondisposable coats are worn, they should be commercially laundered on a frequent basis.

Standard operating procedures should promote the frequent washing of hands and forearms with soap and water while working and prior to leaving the area. Although alcohol-based hand rubs can be used to reduce skin pathogens, they are not effective for allergens and are not a substitution for hand washing with soap and copious rinsing with free-flowing water.

Most individuals do not require respiratory protection when working with or around animals, although a dust or mist face mask may be helpful to reduce exposure to antigens and larger particles, such as hair and dust. In situations where respiratory protection from particulates is required, the use of a disposable filtering face respirator, such as an N-95, or other negative-pressure respirator, such as a PAPR, can provide the needed protection. Individuals using an N-95 respirator or other respirator should be enrolled in the institutional respiratory protection program. Contact allergies can be prevented by protecting eyes and exposed skin with appropriate PPE (e.g., goggles, long sleeves, and gloves).

Zoonotic Diseases

Many of the animals used in the laboratory can harbor potentially zoonotic infections. The likelihood of an animal possessing a zoonotic disease is greatest in wild-caught or random-source animals, where historical information and health data are minimal or lacking. Today, many animals can be procured from captive breeding colonies with a long history of effective programs of disease detection, diagnosis, treatment, and prevention. Using animals from colonies with a known history and disease status, coupled with the development of appropriate standard operating procedures for PPE and animal handling, has helped to make the transmission of zoonotic diseases in the modern laboratory animal facility a rare occurrence.

Although the transmission of zoonotic diseases is rare, some animals must be obtained from the wild or from random sources, and some zoonotic diseases are hard to detect or eliminate from captive-bred animals. A few notable examples are B-virus in Old World macaques, Q fever and contagious ecthyma in small ruminants, cat scratch fever in felines, and mycobacteriosis in fish and frogs. In addition, when working with animals believed to be free of zoonotic diseases, care must be taken not to introduce potential zoonotic agents through the use of contaminated tumors, cell lines, biological agents, or other products. Disease-free colonies have also been contaminated by the inadvertent introduction of feral animals or food or bedding that has been contaminated prior to use.

To prevent the transfer of zoonotic agents, wound care of animal injuries is of critical importance. Depending on the agent in question, programs should develop standard operating procedures for the first aid of wounds and in some cases mucous membrane contaminations. These procedures should be developed in consultation with both medical and safety specialists who are knowledgeable of the potential zoonotic risks associated with each species used within the program. The procedures should address both the immediate care to be provided at the time of the injury and any required follow-up care.

Detailed information on potential zoonotic agents associated with laboratory animals can be found in many publications and reports (NRC 1997; Fox et al. 2015). The following provides a brief overview of the most important zoonotic pathogens associated with common laboratory animals. In addition, Table 14.5 to Table 14.15 provide suggested guidelines for procedures and PPE used within the program at the NIH in Bethesda, Maryland. It must be noted that the suggested PPE outlined in the next sections represents only the guidance provided as a framework for the establishment of best practices. The guidelines were compiled by a team consisting of laboratory animal veterinarians, facility managers, and safety and occupational medical specialists. The guidelines were subsequently adopted by the NIH Animal Research Advisory Committee.

Nonhuman Primates

Although humans and nonhuman primates share many of the same diseases (Table 14.5), the pathogenesis of an organism can vary widely between humans and other primates. An organism can be relatively

TABLE 14.5

Relevant Zoonotic Diseases in Laboratory Nonhuman Primates

Zoonosis	Agent	Route of Transmission
Diarrhea; gram-negative sepsis	Enterobacteriaceae: *Salmonella* spp., *Shigella* spp., *Campbylobacter* spp., *Yersenia* spp.	Fecal-oral
Protozoal diarrhea	*Entamoeba histolytica*, *Giardia* spp., *Balantidium coli*, *Cryptosporidium* spp.	Fecal-oral
Tuberculosis	*Mycobacteria tuberculosis*, *Mycobacteria bovis*	Splash, spray, inhalation of aerosols
B-virus meningoencephalitis	*Macacine herpesvirus* 1	Bite, scratch, splash exposure of mucous membranes, or blood/tissue exposure to contaminated instruments, needles, and equipment (e.g., cages)
Hepatitis	Hepatitis A virus, hepatitis E virus	Fecal-oral
Rabies	Rabies virus (*Lyssavirus*)	Wound or bite, contact with saliva or brain
Measles	Rubeola virus	Splash/spray, aerosols, contact with contaminated fomites
Foamy virus	Spumavirus	Direct blood/tissue contact with infected tissue or contaminated materials
Herpes simplex	Herpes simplex	Direct contact
Helminths	*Oesophagostonum* spp., *Strongyloides* spp.	Fecal-oral
Dermatomycosis (ringworm)	*Trichophyton* spp.	Direct contact

Source: National Institutes of Health—Animal Research Advisory Committee, Guidelines for personnel protection in animal facilities, 2016, https://oacu.oir.nih.gov/sites/default/files/uploads/arac-guidelines/ppe.pdf.

asymptomatic in one species, while having devastating consequences in another. For example, B-virus can produce a life-threatening disease in humans, while in Old World macaques it produces a mild clinical disease similar to human cold sores secondary to human simplex virus. While most nonhuman primates are extensively tested in quarantine or originate from captive-bred colonies, these animals can still harbor zoonotic diseases. In addition, the nonfixed tissues from nonhuman primates can carry potential pathogens and require special handling. Therefore, it is critical for a program to develop a dynamic training program to educate personnel on the potential hazards and steps required to protect their safety and health.

TB can be a devastating disease in both humans and nonhuman primates. Ideally, a program should be implemented for the routine testing of both the individuals working with nonhuman primates and the nonhuman primates used within the program. The prevalence of TB in both humans and animals is high in many parts of the world. Recently, the United States has experienced the smallest decline in more than a decade in the number of TB cases, as well as an increase in the number of multiple-drug-resistant TB strains (http://www.cdc.gov/tb/publications/factsheets/statistics/TBTrends.htm). Although vaccination is used in some parts of the world in both animals and humans, it does not prevent infection, but only the growth of the organism and development of clinical disease. Therefore, vaccination is not recommended or used as a method of control in the United States. The foundation of most TB control programs is the testing of both humans and animals using a tuberculin sensitivity skin test. Programs should develop standard operating procedures for the testing of staff and animals, to include the methodology, test frequency, test interpretation, and handling of positive animals. Because false-positive skin tests are possible, consideration should be given to retesting using the same or a different site and/or methodology.

Measles (rubeola) is a highly communicable disease in humans, New World, and Old World nonhuman primates (Fox et al. 2015). Outbreaks arise in nonhuman primates from contact with humans infected with the virus or humans who have carried the virus from home on their body or clothing. It has been reported that the measles virus can survive in the external environment for approximately 4 days (Walther and Ewald 2004). Therefore, prevention programs should stress the importance of vaccinating

TABLE 14.6

Personnel Protective Equipment Guidelines for Personnel Working in Nonhuman Primate Facilities: Risk Assessment and Suggested PPE

Requirements: The appropriate protection for specific work depends on the degree of risk involved. In general, work activities can be characterized as low, moderate, or high risk, as detailed in the table. If a particular activity is not listed, use the example that provides the nearest match. *Donning a street clothes covering (e.g., disposable lab coat/jumpsuit or dedicated lab coat/uniform) is required to enter a nonhuman primate facility.*[a] Thorough washing of hands is recommended when exiting any animal facility or laboratory. These areas are not designated and posted as a hazardous work area.[b] The following provides a framework for the establishment of best practices for PPE.

Activity Risk Level	Description
Low	• **Working environment:** *Does not support the generation of splashes or droplets.* • **Animal activities:** Slight risk or no risk of direct contact with restrained, sedated, or anesthetized Old World nonhuman primates (NHPs) or their unfixed tissues. • **Equipment contact:** Slight risk or no risk of contact with equipment and surfaces that have been in contact with Old World NHPs, or their unfixed tissues. • **Proximity risk:** Very low (personnel can stay out of the proximity [>3 ft] of the face of restrained, sedated, or anesthetized Old World NHPs, as well as unfixed Old World NHP tissues or body fluids where very low or no splash risk exists).
Moderate	• **Working environment:** *Supports the generation of splashes, but not aerosols.* Work in proximity (<3 ft) of an awake macaque, the face of a sedated or anesthetized macaque, an unfixed Old World NHP brain and spinal tissue, or cerebrospinal fluid (CSF) where a splash potential is present. • **Animal activities:** Direct contact, proximity (<3 ft), or the possibility of direct contact with an awake Old World NHP or work within the proximity of the face of a sedated or anesthetized macaque. • **Equipment contact:** Activities may generate splashes from equipment and surfaces that have been in contact with a macaque, the unfixed brain, spinal tissue, or CSF from an Old World NHP. • **Proximity risk:** Personnel cannot always stay out of the proximity (≥3 ft) of the face, restraining device, or cage of an Old World NHP, as well as the unfixed tissues or body fluids where a splash risk exists.
High	• **Working environment:** *Supports the generation of both splashes and aerosols.* • **Animal activities:** Direct contact with live (i.e., awake, sedated, or anesthetized) or dead Old World NHPs or their unfixed tissues and body fluids. • **Equipment contact:** Activities that have the potential to generate aerosols from equipment and surfaces that have been in contact with live or dead Old World NHPs or their unfixed tissues and body fluids.

[a] Dedicated or disposable long-sleeved uniform/scrubs or lab coat; jumpsuit or coveralls required to cover exposed skin on the arms or legs.

[b] ABSL-3, ABSL-4, and radioactive work environments may require additional PPE; NHP work areas are considered ABSL-2.

Code	Suggested PPE	Protection of	From
B	Specialized arm-length, bite-resistant gloves	Personnel	Bite/scratch
G	Moisture-impermeable gloves	Personnel	Scratch
M	Mucous membrane protection (goggles + surgical face mask, face shield)	Personnel	Splash and droplets
R	Respiratory protection (respirator, N-95 respirator, PAPR) (Note: Most PAPRs provide concurrent eye protection)	Personnel	Aerosols
H	Hair bonnet or lab coat protection of shoulder length or longer hair (required if in an animal holding room or in proximity of an awake NHP)	Personnel	Mechanical injury
S	± Shoe covers or facility-dedicated shoes as defined by the facility/program standard operating procedures	Personnel	Soiling and contamination
E	Eye protection (goggles, approved safety glasses)	Personnel	Splash and droplets

(*Continued*)

TABLE 14.6 (CONTINUED)

Personnel Protective Equipment Guidelines for Personnel Working in Nonhuman Primate Facilities: Risk Assessment and Suggested PPE

Example Activities	Street Clothes Covering plus PPE Codes
Facility corridor activities	S
Entering an NHP animal room containing *New World NHPs* with no direct contact with the animals/caging (Note: Requires face mask for animal protection)	G, S
Entering an NHP animal holding room containing *Old World NHPs* with no direct contact with the animals/caging	G, H, M, S
Entering a room containing a restrained, sedated, or anesthetized Old World NHP with no proximity contact (≥3 ft) of an animal or equipment	Nothing additional
Contact with an *enclosed* restraint chair holding an awake, "non-head-fixed" Old World NHP	G, M
Proximity (≤3 ft) to the *front* of an enclosed restraint chair holding an awake, "head-fixed" Old World NHP	G, M
Proximity (≤3 ft) to the *rear* of an enclosed restraint chair holding an awake, head-fixed Old World NHP	G
Proximity (≤3 ft) to a *nonenclosed* restraint, transfer device, or cage holding an awake macaque	G, H, M, S
Transfer of an alert macaque using a stand-off method, such as pole/collar technique or transfer cage	G, H, M, S
Transferring alert macaques into clean caging or transport cages	G, H, M, S
Hand transfer of an awake New or Old World primate	G, H, M, S, B
Hand transfer of an anesthetized Old World primate	G, H, M, S
Hand transfer of an Old World primate that is awake or lightly sedated (e.g., recovering from anesthesia)	G, H, M, S, B
Enclosed cart transport of an anesthetized Old World primate (Note: Mucous membrane protection must be available)	G
Open cart transport of an anesthetized Old World primate within an animal facility, staying out of the proximity of the animal's face (Note: Mucous membrane protection should be available)	G, S
Minor procedures on a restrained or sedated animal (suture removal, venipuncture, physical exam, anesthesia monitoring, etc.) where the individual *cannot* always stay out of the proximity of the face of the Old World NHP	G, M
Procedures on an anesthetized Old World NHP staying away from the face (e.g., suture removal, venipuncture, or physical exam)	G
Intubation of an anesthetized Old World NHP	G, M
Evaluating or placing electrodes in an implant cylinder from *behind* a chaired awake macaque with its head fixed	G
Cleaning the cranial cylinders of an awake macaque with its head fixed with flush solutions, etc.	G, M
Physiology or behavior lab activities conducted remotely while the animal is fully enclosed in a chair, separate room, or test box	No additional PPE required
Operating room procedures not in proximity (≤3 ft) of the anesthetized macaque and not touching equipment or surfaces previously in contact with the animals, a non-aerosol-forming environment	S
Operating room or dissection procedures that that *do not produce splashes or droplets* in a macaque	G, S
Operating room or necropsy procedures with the *potential to produce splashes, but not aerosols* in a macaque (Note: When using an operating microscope, the scope can replace ocular splash protection at the time of use)	G, M, S

(Continued)

TABLE 14.6 (CONTINUED)

Personnel Protective Equipment Guidelines for Personnel Working in Nonhuman Primate Facilities: Risk Assessment and Suggested PPE

Operating room or necropsy procedures with the *potential to produce aerosols* in a macaque (striker saw, etc.)	G, E, R, S
Dental or oral surgery procedures that *do not produce aerosols*	G, M
Dental or oral surgery procedures that *do produce aerosols* (e.g., Cavitron® or high-speed rotary drill)	G, E, R
Proximity contact with an infant macaque	G, M, S
Mop sanitizing of an Old World NHP room/wipe-down of contaminated cages or other equipment	G, M, S
Sanitizing an NHP room with a pressure washer or hose/hose-down of soiled caging or other equipment	G, E, R, S
Experiments in laboratories that involve *fixed* Old World NHP tissue or body fluids	G
Experiments in laboratories that involve *nonfixed Old World NHP tissue or body fluids other than brain, spinal cord, or CSF*	G
Experiments that involve nonfixed nervous tissue or CSF from an Old World NHP with the *potential to produce splashes, but not aerosols*	G, M
Experiments that involve nonfixed nervous tissue or CSF from an Old World NHP that have a *potential to produce aerosols*	G, E, R

Source: National Institutes of Health—Animal Research Advisory Committee, Guidelines for personnel protection in animal facilities, 2016, https://oacu.oir.nih.gov/sites/default/files/uploads/arac-guidelines/ppe.pdf.

TABLE 14.7

Relevant Zoonotic Diseases of Rodents and Rabbits

Zoonosis	Agent	Species	Route of Transmission
Rat bite fever	*Streptobacillus moniliformis*, *Spirillum minus*	Rodents	Bites, fecal-oral (*S. moniliformis*) Bites (*S. minus*)
Lymphocytic choriomeningitis	LCM virus	Rodents	Aerosol, bites, direct contact, fecal-oral
Hantavirus pulmonary syndrome	Hantavirus	Rodents	Aerosol
Cheyletiellosis	*Cheyletiella parasitivorax*	Rabbit	Direct contact
Dermatophytosis (ringworm)	*Trichophyton* sp., *Microsporum* sp.	Rodent, rabbit	Direct contact
Tapeworm	*Hymenolepis nana*	Rodents	Fecal-oral
Tularemia	*Francisella tularensis*	Rabbit	Ingestion/aerosol

Source: National Institutes of Health—Animal Research Advisory Committee, Guidelines for personnel protection in animal facilities, 2016, https://oacu.oir.nih.gov/sites/default/files/uploads/arac-guidelines/ppe.pdf.

the workforce and educating individuals of the risk they pose to nonhuman primates if they have come in contact with an individual who has measles. Depending on the availability of a vaccine, programs should also consider the vaccination of nonhuman primates. Whether vaccinating humans or animals, it is important to determine if a protective titer had been reached through subsequent serology testing.

B-virus is enzootic among monkeys of the genus *Macaca* and causes minimal morbidity in its natural host. In contrast, infections in humans, New World, and other Old World monkeys are often fatal and present with rapidly ascending encephalomyelitis. Although this infection remains an uncommon result of macaque-related injuries, the increase in the use of macaques for research on simian immunodeficiency virus (SIV) and other potentially stressful conditions has expanded the number of opportunities for human exposure. Like the human herpes simplex virus, B-virus can reside in the nervous system, where it becomes latent and is only shed when the animal becomes stressed or the immune system

TABLE 14.8

Protective Clothing Requirements for Personnel in Rodent and Rabbit Facilities: Risk Assessment and Suggested PPE

Considerations: In rodent and rabbit facilities, PPE functions to reduce staff exposure to allergens and protect animals from infectious agents. The type of PPE needed depends on multiple factors, including the use of allergen-reducing equipment, such as ventilated racks and biosafety cabinets, the susceptibility of the animal colony being housed, and the activity being performed. *Donning a street clothes covering (e.g., disposable lab coat/jumpsuit or dedicated lab coat/uniform) is required to enter a rodent and rabbit facility or room at the NIH.* Thorough washing of hands is recommended when exiting any animal facility. The following provides a framework for the establishment of best practices for PPE.

Activity Risk Level	Description
Low risk	Entering area with no anticipation of physical exposure to animals or soiled caging
Moderate risk	Exposure to animals, animal allergens, or nonbiohazardous soiled caging
High risk	Potential exposure to biohazardous material or zoonotic agents

Code	Suggested PPE	Protection of	From
G	Moisture-impermeable gloves	Personnel	Animal allergens; zoonotic, biologic, and chemical agents
H	± Hair bonnet/covering as defined by facility standard operating procedures	Animals, personnel	Dust and hair, carried animal pathogens, animal allergens
R	Surgical face mask, dust/mist face mask, N-95 respirator	Animals, personnel	Animal pathogens, animal allergens, airborne particulates
C	Respirator with appropriate cartridge	Personnel	Infectious aerosols, chemical vapors

Example Activity	Street Clothes Covering plus PPE Code (s)[a]
Corridor activities	H
Enter animal holding room for brief visual inspection without opening a cage	H
Contact with primary and rodent enclosures	H
Opening an animal cage	H, G, R (recommended)
Direct contact with animals	H, G, R (recommended)
Cage change in biosafety cabinet	H, G, R (recommended)
Cage change on cart within a holding room	H, G, R (dust/mist face mask, recommended)
Cage change using sterilant-level disinfection (200 ppm)	H, G, C
Biohazardous and radioactive studies	As required by study

Source: National Institutes of Health—Animal Research Advisory Committee, Guidelines for personnel protection in animal facilities, 2016, https://oacu.oir.nih.gov/sites/default/files/uploads/arac-guidelines/ppe.pdf.

[a] Additional mucous membrane protection may be required by individuals with known sensitivities to various species. Individuals are encouraged to consult with their safety representatives for further information.

suppressed. It is also important to remember that B-virus infection can also occur subsequent to exposure to nonfixed macaque tissues and products (e.g., brain, spinal cord, nerve, and cell cultures).

B-virus can produce oral or urogenital lesions in macaques, and the virus has been found in ocular secretions and saliva. Transmission is commonly through direct bites and scratches, but airborne droplets can pose a splash hazard through the exposure of mucous membranes. It is also important to remember that injuries with inanimate objects contaminated with the virus (e.g., needles, scalpels, and cages) can also pose a hazard risk. Although serologically B-virus-negative macaques can be procured today, as many are bred in virus-negative captive breeding colonies, it is critical to remember that *all* macaques must be handled as though they are potentially infected because viral shedding is intermittent and viral serology may not accurately reflect the animal's viral status, even with repeated testing.

TABLE 14.9

Relevant Zoonotic Diseases of Laboratory Carnivores

Zoonosis	Agent	Species	Route of Transmission
Bite and scratch bacterial agent	*Capnocytophaga canimorsus*	Dog, cat	Direct contact
Cat scratch disease	*Bartonella henselae* (Cat scratch fever)	Cat	Bite, scratch
Pasturellosis	*Pasturella multocida*	Dog, cat	Bite, scratch
Rabies	Rabies virus (*Lyssavirus*)	All	Wound or bite, contact with saliva or brain
Dermatomycoses (ringworm)	*Microsporum* sp., *Trichophyton* sp.	Dog, cat	Direct contact
Acariasis	*Sarcoptes scabei*	Dog, cat	Direct contact

Source: National Institutes of Health—Animal Research Advisory Committee, Guidelines for personnel protection in animal facilities, 2016, https://oacu.oir.nih.gov/sites/default/files/uploads/arac-guidelines/ppe.pdf.

TABLE 14.10

Protective Clothing Requirements for Personnel in Carnivore Facilities: Risk Assessments and Suggested PPE

Considerations: In carnivore facilities, PPE functions to reduce staff exposure to allergens and noise and to protect staff from infectious agents. *Shoe covers or facility-dedicated footwear and hearing protection are required to enter a carnivore facility at the NIH.* Thorough washing of hands is recommended when exiting any animal facility. The following provides a framework for the establishment of best practices for PPE.

Activity Risk Level	Description
Low risk	Entering area with no anticipation of physical exposure to animals or soiled caging
Moderate risk	Exposure to animals, animal allergens, or nonbiohazardous soiled caging
High risk	Potential exposure to biohazardous material or zoonotic agents

Risk Code	Suggested PPE	Protection of	From
G	Moisture-impermeable gloves	Personnel	Scratch
E	Eye protection (goggles, approved safety glasses)	Personnel	Splash and droplets
M	Mucous membrane protection (goggles + surgical face mask, face shield)	Personnel	Splash and droplets
R	Respiratory protection (respirator, N-95 respirator, PAPR) (Note: Most PAPRs provide concurrent eye protection)	Personnel	Aerosols

Example Activity	Shoe Cover and Hearing Protection plus Codes
Entry into animal holding areas	None additional
Direct contact with animals	G
Cleaning animal holding areas (indoor or outdoor)	G, M
Contact with animals with biohazardous agent	G, M, R (when indicated)

Source: National Institutes of Health—Animal Research Advisory Committee, Guidelines for personnel protection in animal facilities, 2016, https://oacu.oir.nih.gov/sites/default/files/uploads/arac-guidelines/ppe.pdf.

The development of appropriate bite, scratch, and splash exposure care policies, procedures, and training on the handling of bites, scratches, and splashes of body fluids from macaque monkeys or other injuries from equipment contaminated with body fluids is critical to ensuring the health and safety of individuals working with these animals. The adequacy and timeliness of wound decontamination procedures are the most important factors determining the risk of infection after exposure to B-virus (Cohen et al. 2002). Thorough cleaning within 5 minutes of injury or exposure is the best means of preventing B-virus contamination from progressing to actual infection. Studies have demonstrated that B-virus may

TABLE 14.11

Relevant Zoonotic Diseases of Ungulates

Zoonosis	Agent	Species	Route of Transmission
Q fever	*Coxiella burnetti*	Sheep, cattle, goats	Aerosol, or direct contact, especially when dealing with the products of conception
Contagious ecthyma (Orf)	Pox virus	Sheep, goats	Direct contact
Tuberculosis	*Mycobacterium bovis*, *Mycobacterium avium*, or *Mycobacterium tuberculosis*	Swine, sheep, goats	Aerosol, or direct contact
Campylobacteriosis	*Campylobacter jejuni*	Swine, sheep, cattle	Fecal-oral
Dermatomycoses (ringworm)	*Trichophyton* or *Microsporum* spp.	Cattle, sheep, goats, swine	Direct contact
Bovine spongiform encephalopathy	Prion	Cattle	Direct blood/tissue contact with infected tissue or contaminated material (e.g., brain, spinal cord, or ocular/retina)

Source: National Institutes of Health—Animal Research Advisory Committee, Guidelines for personnel protection in animal facilities, 2016, https://oacu.oir.nih.gov/sites/default/files/uploads/arac-guidelines/ppe.pdf.

TABLE 14.12

Protective Clothing Requirements for Personnel in Ungulate Facilities: Risk Assessments and Suggested PPE

Considerations: In ungulate facilities, PPE functions to reduce staff exposure to allergens and protect staff from infectious agents. *Shoe covers or facility-dedicated footwear is required to enter an ungulate facility at the NIH. In addition, hearing protection is required for some swine holding areas.* Thorough washing of hands is recommended when exiting any animal facility. Steel-toed safety shoes should be worn when working with larger ungulates. The following provides a framework for the establishment of best practices for PPE.

Activity Risk Level	Description
Low risk	Entering area with no anticipation of physical exposure to animals or soiled caging
Moderate risk	Exposure to animals, animal allergens, or nonbiohazardous soiled caging
High risk	Potential exposure to biohazardous material or zoonotic agents

Risk Code	Suggested PPE	Protection of	From
G	Moisture-impermeable gloves	Personnel	Scratch
E	Eye protection (goggles, approved safety glasses)	Personnel	Splash and droplets
M	Mucous membrane protection (goggles + surgical face mask, face shield)	Personnel	Splash and droplets
R	Respiratory protection (respirator, N-95 respirator, PAPR) (Note: Most PAPRs provide concurrent eye protection)	Personnel	Aerosols

Example Activity	Shoe Cover plus Codes
Entry into indoor animal holding areas	None additional
Entry into outdoor pens, runs, stables, etc.	None additional
Direct contact with animals	G
Cleaning animal holding areas (indoor or outdoor)	G, M
Contact with pregnant sheep, goats, or cattle during parturition, their birth products, or bedding and other wastes	G, M, R (when indicated)
Contact with animals with biohazardous agent	G, M, R (when indicated)

Source: National Institutes of Health—Animal Research Advisory Committee, Guidelines for personnel protection in animal facilities, 2016, https://oacu.oir.nih.gov/sites/default/files/uploads/arac-guidelines/ppe.pdf.

TABLE 14.13

Relevant Zoonotic Diseases of Laboratory Fish and Frogs

Zoonosis	Agent	Species	Route of Transmission
Mycobacteriosis	*Mycobacterium xenopi*, *Mycobacterium fortuitum*, *Mycobacterium marinum*, *Mycobacterium chelonei*, *Mycobacterium ulcerans*	Fish, frogs	Breaks in skin surface
Salmonellosis	*Salmonella* spp.	Frogs	Breaks in skin surface
Vibriosis	*Vibrio vulnificus*	Fish	Breaks in skin surface

Source: National Institutes of Health—Animal Research Advisory Committee, Guidelines for personnel protection in animal facilities, 2016, https://oacu.oir.nih.gov/sites/default/files/uploads/arac-guidelines/ppe.pdf.

TABLE 14.14

Protective Clothing Requirements for Personnel in Fish and Frog Facilities

Considerations: In aquatic facilities, exposure to potential zoonotic agents or allergens occurs through direct contact with the fish or frogs or indirect contact through exposure to system water. Personnel safety includes the wearing of close-toed shoes with nonskid soles as the work environment includes wet floors. Waterproof gloves should be worn when holding or manipulating animals. Many husbandry tasks involve exposure to water without direct contact with animals. Use of gloves can limit exposure in some of these tasks; however, many routine tasks may involve immersing the hands into system water. In these situations, the gloves can trap system water against the skin, causing prolonged exposure and no opportunity for the skin to dry. Thus, some husbandry tasks involving contact with system water may be more appropriately performed without gloves. As with other animal facilities, thorough washing of hands when exiting is essential. Hand wash or hand sanitation stations should be available after exiting the aquatic facility. The following provides a framework for the establishment of best practices for PPE.

Activity Risk Level	Description
Low risk	Entering area with no anticipation of physical exposure to animals or system water
Moderate risk	Exposure to animals, system water, or dirty tanks
High risk	Potential exposure to biohazardous or chemically hazardous material

Risk/Code	Suggested PPE	From
G	Moisture-impermeable gloves	Zoonosis/allergen
M	Face shield (mucous membrane protection)	Zoonosis/allergen

Example Activity	Suggested PPE Codes
Corridor activities	None additional
Enter animal holding room for brief visual inspection without entering exposure to system water	None additional
Contact with system water (reaching into tank, reservoir)	G/none additional
Netting fish/frogs	G
Direct contact with fish/frog (manual collection of sperm/ova)	G
Scrubbing tanks	G
Biohazardous and chemically hazardous studies	Based on risk assessment
Splash hazards (spraying with system water)	M

Source: National Institutes of Health—Animal Research Advisory Committee, Guidelines for personnel protection in animal facilities, 2016, https://oacu.oir.nih.gov/sites/default/files/uploads/arac-guidelines/ppe.pdf.

enter host cells within 5 minutes of exposure. Therefore, sponge scrubbing the wound with povidone-iodine or chlorhexidine surgical scrub under running water for 15 minutes is recommended. After the 15-minute scrub, the wound should be rinsed thoroughly with water. For splash exposures to the mucous membranes of the eyes, mouth, or nose, the exposure site should be flushed with water or saline for 15 minutes. Following the cleaning or flushing of the exposed area, personnel must report to a health care professional who is knowledgeable on B-virus and the associated hazards.

TABLE 14.15

Protective Clothing Requirements for Personnel in Cage Wash Areas: Risk Assessments and Suggested PPE

Requirements: The appropriate protection for cage wash–related activities depends on the degree of risk involved. If a particular activity is not listed, use the example that provides the nearest match. *Unless otherwise noted, a disposable street clothes covering, facility-dedicated uniform, or lab coat, and shoe covers or facility-specific steel-toed safety shoes must be worn.* Thorough washing of hands is recommended when exiting any animal facility. The following provides a framework for the establishment of best practices for PPE.

Activity Risk Level	Description
Low risk	Entering area with no anticipation of physical exposure to caging or equipment
Moderate risk	Procedures performed in wet areas or may involve exposure to high volumes of water; physical exposure to caging, bedding, water, and equipment
High risk	Procedures that may actively aerosolize waste/fluids or generate potentially contaminated fluids at either a high velocity or high volume; exposure to large volumes of steam; exposure to chemicals; risk to mucous membranes

Codes	Hazards	Suggested Additional PPE
B	Mechanical injury	Safety belt and safety training
C	Chemical	Appropriate gloves, respirator with appropriate cartridge
D	Dust	Dust/mist face mask
E	Eye protection	Goggles/face shield
G	Contamination	Gloves
G1	Heat/burn	Heat-resistant gloves
M	Mucous membrane	Goggles and face mask, face shield, PAPR
N	Noise	Hearing protection (earplugs, ear muffs, etc.)
R	Respiratory	Respirator, N-95, PAPR
S	Splash	Rubber boots, waterproof apron

Example Activity	Street Clothes Covering and Shoe Cover plus PPE Codes
Transporting cages/equipment to and from cage wash area	G, B
Dirty Side	
When machinery is running, all activities	N
Dumping cages	M, S, G
Dumping water bottles	S, G
Dumping acidified/chemically treated water bottles	S, G
Loading equipment on tunnel washer and/or rack washer	M, S, G
Dumping chemicals	C, S, G
Handling nonhuman primate caging/equipment	M, G
Handling dumping nonhuman primate pans	M, S
Loading autoclave before operating	M, G
Loading ABSL-3 autoclave	E, R, S, G
Emptying autoclave after operating	G1, S
Emptying autoclave of large amounts of hot liquids	E, M, S
Cleaning/hosing down area	E, M, S
Routine maintenance on cage wash machines	G1, S
Visitors to dirty side ACTIVE	See functions above
Visitors to dirty side PASSIVE	–
Maintenance personnel to dirty side ACTIVE	See functions above
Maintenance personnel to dirty side PASSIVE	–
Clean Side	
When machinery is running, all activities	N
Offloading equipment from tunnel and/or rack washer while operational	G1, S
Removing equipment from autoclave	G1

(*Continued*)

TABLE 14.15 (CONTINUED)

Protective Clothing Requirements for Personnel in Cage Wash Areas: Risk Assessments and Suggested PPE

Removing large amounts of hot liquids from autoclave	E, M, S
Automatically bedding caging	E, D
Manually bedding caging	E
Preparing acidified water	C, E, S
Preparing chemicals	C, E, S
Cleaning/hosing down area	E, M, S
Routine maintenance on cage wash machines	G1, S
Visitors to clean side ACTIVE (within 6 ft of above functions)	See functions above
Visitors to clean side PASSIVE	–
Maintenance personnel to clean side ACTIVE (within 6 ft of above functions)	See functions above
Maintenance personnel to clean side PASSIVE	–

Source: National Institutes of Health—Animal Research Advisory Committee, Guidelines for personnel protection in animal facilities, 2016, https://oacu.oir.nih.gov/sites/default/files/uploads/arac-guidelines/ppe.pdf.

It is recommended that a program using Old World nonhuman primates develop a "macaque bite, scratch, and splash kit." These kits should be located in convenient areas throughout Old World nonhuman primate work areas. The kits should be portable, self-contained units that include a list of instructions on how to use the kit, as well as the following items: (1) iodine and/or chlorhexidine surgical scrub brushes, (2) emergency room directions and instructions for emergency physicians, (3) a small quantity of oral antiviral medication to be taken as directed by a physician, and (4) viral culture swabs and media to be used by emergency personnel to test the wound. Following any human exposure, it is recommended that a complete physical be conducted on the involved animal by trained veterinary personnel, including serological testing.

Measures to prevent B-virus exposures center on the development of standard operating procedures to prevent macaque bites, scratches, and splashes through appropriate animal handling and PPE. A program should develop guidelines for ranking the potential hazard risk associated with various activities related to the husbandry and use of macaques in research (Table 14.6). Each risk level should be associated with the use of appropriate PPE for each activity. It should be noted that mucous membrane protection is defined as a device or combination of devices, which protect the mouth, nose, and eyes from splash or droplets, such as an approved full-face shield, approved safety glasses or protective glasses plus surgical face mask, approved surgical face mask and eye shield combination, form-fitting goggles plus a surgical face mask, or a PAPR. Approved full-face shields provide sufficient protection above the shield to prevent droplets from running down into the eyes and adequate side protection to prevent material from splashing into the eyes or mouth (Figure 14.6). Approved safety glasses or protective glasses must also meet the same requirement to protect the eyes from running droplets and side splashes. Because of individual variations in the shape of each person's head and face, face shields and safety glasses may need to be evaluated and approved on an individual basis. In activities where aerosols may be formed and respiratory protection required, protection is defined as a device or combination of devices that protects the mouth, nose, eyes, upper airways, bronchi, and lungs from splashes, droplets, and aerosols. Respiratory protection includes the use of an approved full-face shield plus a surgical face mask or other approved respirator, form-fitting goggles plus a surgical face mask or other approved respirator, or a PAPR. The use of an N-95 respirator or form-fitting respiratory other than a PAPR should be approved through the program occupational medical service due to the potential risk when used by individuals with certain respiratory or cardiac conditions.

Rodents and Rabbits

Zoonotic diseases associated with rodent and rabbit use in the modern animal research laboratory and animal holding facility (Table 14.7) are uncommon due the availability and use of commercially available specific pathogen-free animals. In addition, the development of facility programs and policies for colony health monitoring and the importation and testing of noncommercially available animals and

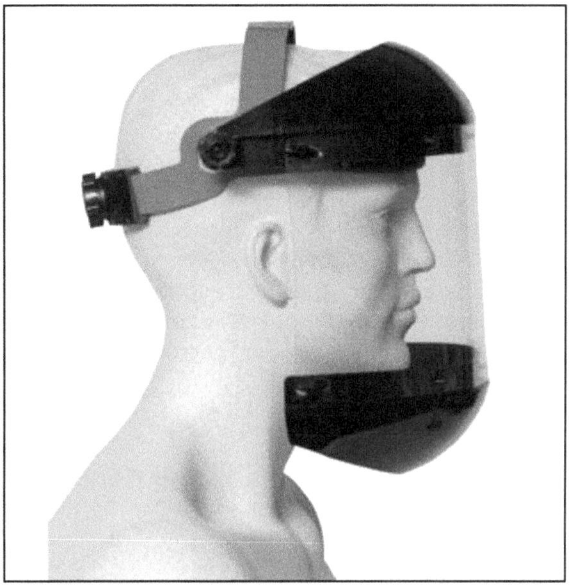

FIGURE 14.6 Example of a ratcheted full-face shield with chin guard that provides mucous membrane splash protection from all angles. (From Life Science Products, Chestertown, MD, http://lspinc.com/.)

animal products has also helped to prevent the introduction of zoonotic agents into research colonies. In many situations, the inadvertent introduction of feral animals or material contaminated by feral animals (e.g., feed and bedding) represents one of the more common routes of introduction of pathogens.

It should be noted that the PPE recommended for rodents and rabbits (Table 14.8) provides personnel protection not only from potential zoonotic agents, but also from potential allergens, as outlined in the "Allergens" section above. The use of hair bonnets, lab coats, and gloves may also potentially protect specific pathogen-free animals from diseases that personnel may carry on their cloths, hair, or hands from outside of the facility. That being said, it has been recently identified that the donning of shoe covers was a potential source of contamination and that wearing shoe covers did not significantly impact the health of the colony (Hickman-Davis et al. 2012). In addition, not donning hair protection was found to result in the transport of rodent allergens on the hair of personnel working with rodents (Krop et al. 2007).

Carnivores

The most common carnivores used in research are dogs, cats, and ferrets. The most common zoonotic agents seen in laboratory carnivores in the United States are rabies, enteric pathogens, and parasites (Table 14.9). Many of the zoonotic diseases of carnivores can be prevented by vaccination (e.g., rabies and leptospirosis) and appropriate health management (e.g., parasite checks). Here again, the PPE (Table 14.10) serves to prevent the transmission of disease in two directions, from animals to humans and from humans to animals. It is common for personnel to keep dogs, cats, and ferrets as pets at home, and their health status may vary from that of the animals maintained at work. Therefore, program management should be aware of the pets kept at home by their staff and develop policies and procedures that protect the health and welfare of the research colony.

Ungulates

Zoonotic disease associated with ungulates (Table 14.11) used in research is often dependent on the source of the animals and policies and procedures used to manage the research herd. Closed herds with appropriate vaccination and herd health programs may present a minimal risk, whereas herds maintained outside in unprotected open areas may present a higher risk. Q fever is prevalent in sheep throughout the United States, and sheep are the species most associated with outbreaks of the disease in laboratory animal populations. The PPE used with ungulates (Table 14.12) will vary with the risk level, husbandry style, and nature of contact.

Fish and Frogs

Zoonotic diseases associated with fish and frogs used in research are infrequent (Table 14.13), but can occur. The availability of commercially available animals of known health status has helped to decrease the zoonotic risk. Mycobacteria infections of the skin are one of the more common zoonotic diseases seen in the laboratory environment. Care should be taken to use gloves (Table 14.14), especially if there are any breaks in the skin. The thorough washing of hands and arms when leaving the aquatic holding area is recommended. Although splash hazards to the eyes and mucous membranes are not normally problematic, when occurring, the area should be thoroughly flushed with running water for 5 minutes.

Cage Wash Areas

The facility cage wash area can be particularly problematic due to the potential for exposure to allergens, zoonotic enteric agents, and physical hazards. Table 14.15 summarizes the protective clothing requirements for the area based on the hazard risk and activity.

Additional Hazards

Hazards Associated with Animal Study Protocols

All animal study protocols (ASPs) should be closely examined to identify and address potential hazards associated with the protocol. There may be multiple classes of hazards (infectious, radiologic, chemical, and physical) associated with each ASP, and they may not be fully recognized by the investigator. Each ASP should be closely reviewed by a health and safety professional prior to final Institutional Animal Care and Use Committee (IACUC) approval to ensure that appropriate interventions, mitigations, and disposal procedures are fully addressed and understood by all parties.

Biological Agents with Human Pathogen Potential

There are special considerations for working with experimentally infected animals housed in indoor research environments. Institutional management must provide facilities, equipment, staff, and established practices that reasonably ensure appropriate levels of environmental quality, safety, security, and care for the laboratory animal while protecting workers, the environment, and the public from the infectious agents themselves. The vivarium is considered a type of specialized research laboratory, and biosafety levels have been described for working in these areas with animals that have been experimentally infected. Biosafety levels are appropriate combinations of facility features, safety equipment, and operations and procedures that are used to ensure necessary containment of the infectious agent during the conduct of the research. In the animal facility, biosafety levels are referred to as ABSLs 1–4 because the specialized nature and complexities of animal research are addressed in conjunction with the need for biological containment. Biosafety levels increase based on the infectious agents in use and the risk assessment of the procedures to be performed. An overview of these requirements is presented in Table 14.16. For a full description of ABSLs, the reader should consult the current edition of Centers for Disease Control and Prevention (CDC)/NIH publication entitled *Biosafety in Microbiological and Biomedical Laboratories*.

These recommendations assume that laboratory animal facilities, operational practices, and the quality of animal care meet applicable standards and regulations (e.g., the *Guide* [NRC 2011a] and laboratory Animal Welfare Regulations [AWAR 2013]) and that appropriate species and numbers of animals have been selected for animal experiments.

In Canada, similar containment levels (CLs) are employed but are more heavily regulated by the government. The reader is referred to *Canadian Biosafety Standard* (CBS), Second Edition, 2015

TABLE 14.16

Summary of Recommended Animal Biosafety Levels for Activities in Which Experimentally or Naturally Infected Vertebrate Animals Are Used

BSL	Agents	Practices	Primary Barriers and Safety Equipment	Facilities (Secondary Barriers)
1	Not known to consistently cause diseases in healthy adults	Standard animal care and management practices, including appropriate medical support services	As required for normal care of each species • PPE: Laboratory coats and gloves; eye and face protection, as needed	Standard animal facility • No recirculation of exhaust air • Directional airflow recommended • Hand washing sink is available
2	• Agents associated with human disease • Hazard: Percutaneous injury, ingestion, mucous membrane exposure	ABSL-1 practice plus • Limited access • Biohazard warning signs • Sharps precautions • Biosafety manual • Decontamination of all infectious wastes and animal cages prior to washing	ABSL-1 equipment plus primary barriers • Containment equipment appropriate for animal special procedures • PPE: Laboratory coats; gloves; face, eye, and respiratory protection, as needed	ABSL-1 plus • Autoclave available • Hand washing sink available • Mechanical cage washer recommended • Negative airflow into animal and procedure rooms recommended
3	Indigenous or exotic agents that may cause serious or potentially lethal disease through the inhalation route of exposure	ABSL-2 practice plus • Controlled access • Decontamination of clothing before laundering • Cages decontaminated before bedding is removed • Disinfectant foot bath as needed • Baseline serum sample obtained	ABSL-2 equipment plus • Containment equipment for housing animals and cage dumping activities • Class I, II, or III Biological Safety Cabinets (BSCs) available for manipulative procedures (e.g., inoculation and necropsy) that may create infectious aerosols • PPE: Appropriate respiratory protection	ABSL-2 facility plus • Physical separation from access corridors • Self-closing, double-door access • Sealed penetrations • Sealed windows • Autoclave available in facility • Entry through anteroom or air lock • Negative airflow into animal and procedure rooms • Hand washing sink near exit of animal or procedure room
4	• Dangerous/exotic agents that post high risk of aerosol transmitted laboratory infections that are frequently fatal, for which there are no vaccines or treatments • Agents with a close or identical antigenic relationship to an agent requiring BSL-4 until data is available to redesignate the level • Related agents with unknown risk of transmission	ABSL-3 practices plus • Entrance through change room where personal clothing is removed and laboratory clothing is put on; shower on exiting • All wastes are decontaminated before removal from the facility	ABSL-3 equipment plus • Maximum containment equipment (i.e., Class III BSC or partial containment equipment in combination with full-body, air-supplied, positive-pressure suit) used for all procedures and activities	ABSL-3 facility plus • Separate building or isolated zone • Dedicated supply and exhaust, vacuum, and decontamination systems • Other requirements outlined in the text

Source: Adapted from Centers for Disease Control and Prevention and National Institutes of Health, *Biosafety in Microbiological and Biomedical Laboratories*, 5th ed., U.S. Government Printing Office, Washington, DC, 2009.

(http://canadianbiosafetystandards.collaboration.gc.ca/cbs-ncb/index-eng.php), which is a harmonized national standard for the handling and/or storing of human and terrestrial animal pathogens and toxins based on pathogen risk groups 1–4 (RGs 1–4) established by the World Health Organization (WHO) (Table 14.17).

Based on risk assessment, CLs 1–4 are assigned. Small and large animal zones may be included within laboratories meeting CLs 1–4. The reader should consult the *Canadian Biosafety Handbook* (CBH), Second Edition, 2015. The CBH is a companion document to the CBS and provides core information and guidance to aid in achieving the biosafety and biosecurity requirements outlined in the CBS.

Working with Recombinant or Synthetic Nucleic Acid Molecules

Guidance for working with recombinant or synthetic nucleic acids in animals can be found in the *NIH Guidelines for Research Involving Recombinant or Synthetic Nucleic Acid Molecules* (NIH Guidelines 2016) (http://osp.od.nih.gov/sites/default/files/resources/NIH_Guidelines.pdf). These guidelines must be consulted when planning work involving recombinant or synthetic acid molecules. The use of the NIH Guidelines is also risk assessment based, relying on the WHO RGs as a starting point. Using the *Biosafety in Microbiological and Biomedical Laboratories* and the NIH Guidelines as companion documents can be problematic because nomenclature for CLs has not been harmonized. For example, when experiments involve whole animals in which the animal's genome has been altered or which involve viable microbes containing recombinant DNA (other than viruses that are only vertically transmitted) that are tested on whole animals, a minimum containment of BL2 or BL2-N is required. This nomenclature, in practice, equates to BSL-2 and ABSL-2. However, in cases where the introduction of recombinant DNA into animals might lead to the creation of novel mechanisms or increased transmission of a recombinant pathogen or production of undesirable traits in a host animal, higher-level containment conditions should be implemented. When experimenting with animals that contain sequences from viral vectors that do not lead to transmissible infection as a result of complementation or recombination in the host animal, BL1 or BL1-N containment may be used (BSL-1 and ABSL-1, respectively). The Institutional Biosafety Committee (IBC), in each institution, serves as the governance and review body for the oversight of recombinant work and should be consulted prior to beginning this type of work involving animals.

A number of agencies within the U.S. government oversee different aspects of work with recombinant nucleic acids and human pathogens. These include

- NIH Guidelines
- EPA Regulations
- U.S. Department of Agriculture (USDA) Animal Plant Health Inspection Service (APHIS) Regulations
- FDA Regulations

TABLE 14.17

World Health Organization Pathogen Risk Groups

WHO Risk Group	Definition
Risk group 1 (no or low individual and community risk)	A microorganism that is unlikely to cause human disease or animal disease.
Risk group 2 (moderate individual risk, low community risk)	A pathogen that can cause human or animal disease but is unlikely to be a serious hazard to laboratory workers, the community, livestock, or the environment. Laboratory exposures may cause serious infection, but effective treatment and preventative measures are available and the risk of spread of infection is limited.
Risk group 3 (high individual risk, low community risk)	A pathogen that usually causes serious human or animal disease but does not ordinarily spread from one infected individual to another. Effective treatment and preventive measures are available.
Risk group 4 (high individual and community risk)	A pathogen that usually causes serious human or animal disease and that can be readily transmitted from one individual to another, directly or indirectly. Effective treatment and preventive measures are not usually available.

- Commerce Department Regulations
- Select Agent Rules
- HHS Synthesis Screening Guidance for Providers of Synthetic Double-Stranded DNA

The CBS refer to the NIH Guidelines for review and risk assessment assistance regarding recombinant nucleic acid work. In the EU, the following directives are salient regarding genetically modified microorganisms (GMMs) and should be consulted:

- Directive 90/219/EEC on contained use of GMMs
- Directive 2001/18/EC on deliberate release into the environment of GMMs

Use of Radionuclides in Animal Research

Typically, very small quantities of radioactive materials are used in animal research, and the risk to an employee's health is low. This is not to say that radiation protection in these situations is inconsequential; however, it is typically a lesser concern than the other risks associated with animal research. In all cases, the animal study proposal must be reviewed by the institution's radiation protection office or health physicist and should specifically address the radioactive conditions permitted, personal hygiene precautions to be taken by animal care personnel, and instructions for cage cleaning and the collection, labeling, and disposal of radioactive wastes.

Personal protection used with radioactive materials in animal research is usually similar to that used routinely in animal care. The use of gloves, laboratory coats, and other protective clothing minimizes the chances for the ingestion or absorption of radioactive materials when working with animals. The protocols should specify the training that must be provided to all personnel regarding the specific radiation safety requirements for the tasks to be performed. This training must be directly related to the duties of the individual, and commensurate with the risks.

Use of Hazardous Chemicals and Drugs in Animal Research

Potentially hazardous chemicals and drugs are commonly found in animal research and should be specifically addressed in animal study proposals. The ASP must identify potential hazards of the chemicals and drugs to be used, communicate information concerning the hazards to those involved in the conduct of the research, and describe the appropriate protective measures to employees. Alternatively, standard operating procedures can also be used to communicate the required information. Chemical safety in research involving animals is governed by numerous local, state, and federal regulations, as well as numerous "standards of practice." Standards of practice are guidelines used to determine what should or should not be done in a given situation. In the case of safety and health practice, this guidance may be issued by other than regulatory agencies, such as the National Institute for Safety and Health (NIOSH), CDC, and the National Research Council. These guidelines may also be published and widely distributed in authoritative texts such as *Prudent Practices in the Laboratory: Handling and Management of Chemical Hazards* (NRC 2011b). These accepted standards of practice may be considered regulatory under OSHA's "General Duty" clause that requires a workplace free of recognized hazards (Thomann 2003). Some chemicals may be explicitly regulated by OSHA in a specific health standard, such as formaldehyde or ethylene oxide. Most chemicals in laboratories and animal facilities are regulated more generally through 29 CFR 1910.1450—Occupational Exposure to Hazardous Chemicals in Laboratories. This regulation specifically addresses the unique and variable aspects of the laboratory environment, including the animal facility. In any case, it is the employer's responsibility to minimize exposures to hazardous chemicals in the workplace.

In preparing an ASP, pertinent chemical safety information may be found on MSDSs or SDSs, pharmaceutical inserts or information, the Registry of Toxic Effects of Chemical Substances (RTECS), or similar services.

Certain common and ubiquitous chemical agents may be overlooked when considering usage in an animal facility and may not be addressed directly in an ASP. Consider the use of disinfectants and sterilants. By definition, these compounds are "poisons" and must be handled responsibly. They are used to destroy or irreversibly inactivate pathogenic microorganisms on inanimate surfaces (Van Swearingen and Shoaf 2001). Many different classes of disinfectants and sterilants are found and used in animal care facilities, and each has associated potential health effects, standards and recommendations for safe use, and methods for exposure control. Care should be taken during use of these compounds.

Anesthetic gases are another potential chemical hazard often present in animal facilities and during veterinary procedures. The anesthetic gases and vapors that leak into the surrounding room during medical procedures are considered waste anesthetic gases. Some potential effects of exposure to waste anesthetic gases are nausea, dizziness, headaches, fatigue, and irritability, as well as sterility, miscarriages, birth defects, cancer, and liver and kidney disease, among operating room staff or their spouses (in the case of miscarriages and birth defects). Employers and employees should be aware of the potential effects and be advised to take appropriate precautions. Scavenging devices should be used to capture waste gases, limiting occupational exposure. Routine surveys or leak tests of each anesthetic breathing circuit (machine, scavenging device, tubing, etc.) or location should be performed and documentation retained for presentation to AAALAC International site visitors, safety specialists, and so forth, as necessary. Records of certification for other safety equipment, including fume hoods, biosafety cabinets, downdraft tables, and necropsy tables, should also be kept available for inspection.

Disposal of Hazardous Materials

Disposal of hazardous materials or waste can be a complicated matter. Most animal programs or facilities will generate hazardous wastes related specifically to the work being performed. Each animal program should have a hazardous materials management plan (HMMP) and a designated hazardous materials manager. The HMMP addresses proper labeling, packaging, and manifesting of biological, chemical, and radioactive waste in compliance with the Nuclear Regulatory Commission, Department of Transportation, Federal Resource Conservation and Recovery Act, and specific state or local requirements or those of a particular jurisdiction. Strict compliance with these regulations ensures the waste is managed, transported, and disposed of properly while reducing potential liability to the animal program or parent organization. In most cases, the institution with which the animal program or facility is associated will have established an HMMP that, if well done, will specifically address the needs of the animal program and will arrange to provide hazardous waste services to the animal facility. Each animal study proposal should address potential hazardous waste accumulation and the proper disposal of the waste.

In general, wastes produced from hazardous materials used in animal research or in animal facilities can be classified as chemical, radioactive, medical, or pathological, or multihazardous waste. They are classified as such based on the characteristics of the waste. Characteristics defining hazardous materials will differ by country, or by consensus of member nations in the case of the EU (Commission Regulation [EU] No. 1357/2014). The program or facility management must be cognizant of local, state, and federal requirements for the disposal and transport of hazardous wastes, and that regulatory authorities (environmental, public health, transportation, etc.) may each have regulations to which the animal program must adhere. "Cradle-to-grave" management of hazardous waste (from generation to final disposal) is an expectation when hazardous materials are used in research.

When planning research projects, it is important to consider the type and amount of waste that may be produced. Planning should incorporate methods or techniques that will limit or reduce the hazardous materials to be used, as well as limiting waste accumulation. The following principles will assist in reaching this goal:

- Consider use of alternative materials.
- Order only what you need.

- Use only what you need.
- Reuse what you can.

Each waste stream is handled in a specific way and therefore requires proper segregation. Segregation of wastes makes waste disposal safer, more efficient, and more cost-effective. The animal program must make every effort to understand and comply with the institution's HMMP.

Removal of waste from the site depends on what arrangements have been made for transport and final disposal. It is generally safer and more cost-effective to retain a professional contract service to provide removal and disposal services. Disposal of waste usually involves filling out a hazardous waste manifest that indicates what kind of waste is in the container, who produced the waste, the name of the transporter, and the manner of disposal. By signing these forms, the hazardous waste generator, transporter, and disposal site all share some responsibility for the waste; however, governmental authorities tend to view the waste as belonging to the generator (cradle) until it has reached final disposal (grave). This manifest tracks the chain of custody for hazardous waste shipments.

Summary

A dynamic occupational safety and health program is a critical aspect of any animal care and use program, regardless of its size or complexity. A quality occupational safety and health program is a team event that requires the active, ongoing participation of all personnel. An effective program includes the assessment of facility design and equipment, as well as the development of work-specific training plans. The establishment of appropriate standard operating procedures and PPE, coupled with ongoing monitoring and compliance with established procedures, further strengthens a program. From the development of a JHA to the assessment of work-related injuries, the program must adapt to the changing demands of the work environment while ensuring the safety and health of all employees.

REFERENCES

AWAR (Animal Welfare Act and Regulations). 2013. Code of Federal Regulations, Vol. 1, Subchapter A, Title 9. http://awic.nal.usda.gov/government-and-professional-resources/federal-laws/animal-welfare-act.

Blair, E. 2014. Improving safety performance with behavior-based safety. In *Fundamentals of Occupational Safety and Health*, ed. M. A. Friend and J. P. Kohn. New York: Bernan Press.

Bush, R. K., and G. M. Stave. 2003. Laboratory animal allergy: An update. *ILAR J* 44(1):28–51.

Cohen, J. I., Davenport, D. S., Stewart, J. A., Deitchman, S., Hilliard, J. K., Chapman, L. E., and the B Virus Working Group. 2002. Recommendations for prevention of and therapy for exposure to B virus (Cercopithecine herpesvirus 1). *Clin Infect Dis* 35(10):1191–1203.

Fox, J. G., Otto, G., and L. A. Colby. 2015. Selected zoonoses. In *Laboratory Animal Medicine*, ed. J. G. Fox, L. C Anderson, G. Otto, K. R. Pritchett-Corning, and M. T. Whary, 1543–1597. 3rd ed. Amsterdam: Elsevier/Academic Press.

Gross, N. J. 1980. Allergy to laboratory animals: Epidemiologic, chronological and physiologic aspects, and a trial of cromolyn in its management. *J Allergy Clin Immunol* 66:158–165.

Hickman-Davis, J. M., Nicolaus, M. L., Petty, J. M., and V. K. Bergdall. 2012. Effectiveness of shoe covers for bioexclusion within an animal facility. *J Am Assoc Lab Anim Sci* 51(2):181–188.

Klaunberg, B. A., and J. A. Davis. 2008. Consideration for laboratory animal imaging center design and setup. *ILAR J* 49(1):4–16.

Krop, E. J., Doekes, G., Stone, M. J., Aalverse, R. C., and J. S. van der Zee. 2007. Spreading of occupational allergens: Laboratory animal allergens on hair-covering caps and in mattress dust of laboratory animal workers. *Occup Environ Med* 64(4):267–272.

Lehner, N. L., Huerkamp, M. J., and D. L. Dillehay. 1994. Reference serum revisited. *Contemp Topics Lab Anim Sci* 33(6):61–63.

Lipman, N., and S. Leary. 2015. Design and management of research facilities. In *Laboratory Animal Medicine*, ed. J. G. Fox, L. C Anderson, G. Otto, K. R. Pritchett-Corning, and M. T. Whary, 1543–1597. 3rd ed. Amsterdam: Elsevier/Academic Press.

National Institutes of Health—Animal Research Advisory Committee. 2016. Guidelines for personnel protection in animal facilities. https://oacu.oir.nih.gov/sites/default/files/uploads/arac-guidelines/ppe.pdf.

NIH Guidelines (Department of Health and Human Services, National Institutes of Health). 2016. NIH guidelines for research involving recombinant or synthetic nucleic acid molecules. Available at https://auth.osp.od.nih.gov/sites/default/files/resources/NIH_Guidelines.pdf.

NRC (National Research Council). 1997. *Occupational Health and Safety in the Care and Use of Research Animals*. Washington, DC: National Academies Press.

NRC (National Research Council). 2003. *Occupational Health and Safety in the Care and Use of Nonhuman Primates*. Washington, DC: National Academies Press.

NRC (National Research Council). 2011a. *Guide for the Care and Use of Laboratory Animals*. Washington, DC: National Academies Press.

NRC (National Research Council). 2011b. Prudent Practices in the Laboratory: Handling and Management of Chemical Hazards. Washington, DC: National Academies Press.

NSF (National Sanitation Foundation). 2014. Biosafety cabinetry: Design, construction, performance, and field certification. NSF/ANSI 49-2014. Ann Arbor, MI: NSF International.

OSHA (Occupational Safety and Health Administration), U.S. Department of Labor. 2002. Job hazard analysis. OSHA 3071. https://www.osha.gov/Publications/osha3071.pdf.

OSHA (Occupational Safety and Health Administration), U.S. Department of Labor. 2005. Small business handbook. OSHA 2209-02R. https://www.osha.gov/Publications/smallbusiness/small-business.pdf.

Thomann, W. R. 2003. Chemical safety in animal care, use, and research. *ILAR J* 44(1):13–19.

Van Swearingen, F. L., and A. R. Shoaf. 2001. Biological safety: Emergency response and decontamination procedures. In *Handbook of Chemical Health and Safety*, ed. A. J. Alaimo, 343–344. Washington, DC: American Chemical Society.

Walther, B. A., and P. W. Ewald. 2004. Pathogen survival in the external environment and the evolution of virulence. *Biol Rev* 79:849–869.

15

Program Documentation and Monitoring

William E. Dale, George J. Haluska, and Diane (Dee) Horne

CONTENTS

Introduction

Animal care and use programs are complex, and not surprisingly, extensive documentation is required by regulation and policy to ensure that appropriate standards are achieved. This chapter presents four areas of an animal use program for which specific documentation is necessary: the Institutional Animal Care and Use Committee/Oversite Body (IACUC/OB), the animal resources and facility operations, good laboratory practice (GLP), and select agents.

Institutional Animal Care and Use Committee/Oversite Body

In the United States, the Animal Welfare Act (AWA) and Animal Welfare Act and Regulations (AWAR) (USDA 2015) provide standards for ensuring the humane care and use of certain animal species in research facilities. Similarly, the U.S. Health Research Extension Act of 1985 and the Public Health Service Policy on Humane Care and Use of Laboratory Animals (PHS Policy) (NIH 2015) establish guidelines for the proper care and use of all vertebrate animal species in PHS-supported research activities. In Europe, the European Treaty Series 123 (ETS 123) (Council of Europe 1986) establishes the legal foundation for the protection of research animals in 47 different countries. The Directive 2010/63/EU (European Parliament and the Council of the European Union 2010), adopted by all European Union member countries, is the current legislation regarding protection of animals used for research purposes. Regulatory requirements for the use of animals in research vary between countries in Asia, and this topic is discussed in greater detail in Chapter 8. However, as an example, the Law of Humane Treatment and Management of Animals No. 105 was adopted in Japan to address the proper treatment of animals, including mammals, birds, and reptiles. Nevertheless, for all regulations, whether country specific or international, there is a requirement for some type of ethical review of the proposed use of animals. The IACUC/OB is usually the primary entity overseeing animal use activities in research or teaching. Not uncommonly, the IACUC/OB will conduct a review of any proposed research project based on specific information that is presented in an animal use protocol.

Animal Use Protocols

Animal use protocol review is an important aspect of every animal care and use program in research and represents the initial step toward obtaining approval for conducting animal-based research. The

U.S. requirements for review of animal activities are implemented by federal agencies, including the U.S. Department of Agriculture/Animal and Plant Health Inspection Service (USDA/APHIS), which is responsible for ensuring conformance with the AWA and AWAR, and the National Institutes of Health (NIH) Office of Laboratory Animal Welfare (OLAW), which is responsible for ensuring implementation of PHS Policy. The standards document used by OLAW is the *Guide for the Care and Use of Laboratory Animals* (i.e., *Guide*), published by the Institute for Laboratory Animal Resources, National Research Council (NRC 2011), and it presents expectations for the approval of animal use by the IACUC/OB in terms of protocol review. The *Guide* is an internationally recognized document, and numerous countries require review of animal use activities in accordance with the standards presented therein. As such, AAALAC International utilizes the *Guide* as a primary standards documents, along with the *Guide for the Care and Use of Agricultural Animals in Research and Teaching* (*Ag Guide*) (FASS 2010) and ETS 123 (Council of Europe 1986).

The IACUC/OB should consider information provided in the animal use protocol and evaluate legitimacy of the proposed study based on regulatory requirements. Animal use information presented to the IACUC/OB is typically on forms developed by the institution, and in the past, paper documents were used for this purpose. However, more recently, electronic systems are commonly used for protocol review, as these can provide a more manageable alternative by allowing for easy storage and retrieval of these important documents. The animal use protocol should include the rationale for performing the research project, as well as a description of the research in lay language that an average newspaper reader would understand. For species covered by the U.S. AWAR and in many other countries, investigators must also document that they have considered alternatives to pain and distress by performing a search of relevant databases, documenting the keywords used in the search and the databases searched, and the dates covered by the search. The results of the search are typically incorporated into consideration of the replacement, reduction, and refinement (three Rs) of animal use. A description of how the search results may impact development of the research plan may also be required, and the researcher should account for the number of animals to be used for the research. This should be justified statistically whenever possible. Other items that should be documented in the protocol include:

- Verification that the research proposed does not unnecessarily duplicate research already performed. The necessity to provide additional proof of the validity of previous results should be scientifically justified (USDA 2015).
- Appropriate use of anesthetics and analgesics. Use of these agents must be addressed, and withholding them must be justified for scientific reasons (USDA 2015). Proper use of anesthetics and analgesics is an ethical and scientific imperative (NRC 2011).
- Housing standards that will be used, according to species, as defined in the AWAR (USDA 2015) and in the *Guide* (NRC 2011). Deviations from these standards should be justified in the protocol.
- Surgical procedures that will be used. An animal should not undergo more than one major survival surgical procedure unless there is scientific justification for doing so.
- Euthanasia methods that will be used. U.S. institutions and many others worldwide refer to the American Veterinary Medical Association (AVMA) *Guidelines for the Euthanasia of Animals* (AVMA 2013) for guidance on methods of euthanasia. Deviations from the recommendations in the AVMA *Guidelines* should be scientifically justified in the animal use protocol.
- Personnel qualifications. All individuals involved in the use of animals must be qualified to perform the animal-related tasks to which they are assigned, and staff should be provided instruction and training on procedures to be performed. Training and/or experience with the animal species and techniques should be documented, and this information may be retained in the protocol. Additionally, many countries, including the United States (NRC 2011; USDA 2015), require that research personnel receive training in regard to relevant legislation, IACUC/OB function, the ethics of animal use, the three Rs principle, reporting of animal welfare concerns, and occupational health and safety, in addition to animal handling, surgery techniques, and anesthesia and analgesia. Accessibility of training records may be important for review by the IACUC/OB or government regulators, and electronic systems may be useful in this regard.

Special Considerations for IACUC/OB Review

Not uncommonly, special considerations will arise during protocol review, and these should be considered carefully by the IACUC/OB because of their potential for causing unrelieved pain or distress, or other animal welfare concerns (NRC 2011). Refinements should be considered in order to minimize pain and distress while accomplishing the required scientific objectives. The IACUC/OB should also weigh the objectives of the study against the potential harm to the research animal and animal well-being. Special considerations listed in the *Guide* (NRC 2011) include:

- *Experimental and humane endpoints.* The experimental endpoint of a study occurs when the scientific aims and objectives have been reached. The humane endpoint is the point at which pain or distress in an experimental animal is prevented, terminated, or relieved. The use of humane endpoints is particularly important when the animal may experience pain or distress, or may become physically impaired, such as during tumor growth or toxicology studies. Information in the protocol should include a clear description of the humane endpoints, along with the required frequency of animal observation, details of the training provided to those responsible for recognizing humane endpoints, and the actions to be taken when these are observed. The veterinarian should be involved in defining humane endpoints, and these should be clearly stated in the protocol. Not uncommonly, a table or chart that helps ascertain humane endpoints using a quantifiable approach can be generated and included in the protocol for ease of use by the researcher. Further discussion regarding humane endpoints may be found in a previous issue of the *ILAR Journal* addressing this topic (ILAR 2000).
- *Unexpected outcomes.* It is difficult to anticipate all unexpected outcomes that may occur during the course of a scientific investigation. Nevertheless, it is recognized that breeding experiments that result in new genotypes may produce offspring with abnormal phenotypic characteristics. Depending on the phenotypic alteration, there can be impacts on the animals' well-being, ability to achieve normal physiological and behavioral function, or survivability. It is important that the offspring of genetically modified animals be monitored closely for abnormal characteristics, and if these are observed, the animal use protocol should be modified to address the number of animals that may be required or any special care that might be needed. Humane endpoints should be considered for genetically modified animals that are severely affected.
- *Prolonged physical restraint.* Limiting an animal's normal movement by manual or mechanical means may be necessary during scientific studies. Restraint can be relatively brief and last only a matter of minutes, or it may be prolonged and last for hours. Regardless of the time length, a description of the restraint in the animal use protocol is needed. Scientific justification for the use of restraint should be included, along with a description of the training that will be provided to the animals so they can adapt to the restraint device. Along with a description of the training should be an assurance that animals that fail to adapt to the restraint will be removed from the study. Other aspects of restraint that should be included in the protocol include the time period of restraint, the observation frequency during the period of restraint, and a description of untoward effects (e.g., lesions, behavioral changes, or illness) that will necessitate removal of an animal from the restraint device and possible elimination from the study.
- *Multiple survival surgeries.* Survival surgeries are commonly classified as major or minor, and many institutions rely on the definition for these presented in the *Guide* (NRC 2011). That is, major survival surgery penetrates and exposes a body cavity, produces substantial impairment of physical or physiologic functions, or involves extensive tissue dissection or transection. Minor survival surgery does not expose a body cavity and causes little or no physical impairment. Whether major or minor, the need to perform multiple surgical procedures on a single animal should be identified in the protocol for evaluation by the IACUC/OB. Justification for performing multiple survival surgeries should be provided, including an explanation of the necessity to perform these procedures in order to achieve the scientific aims. Multiple survival

surgeries may also be necessary for unforeseen experimental or clinical reasons, for example, replacement of a cranial head post that was dislodged or emergency surgery to address an unanticipated health concern. If these are anticipated, they should be included in the animal use protocol as multiple survival surgical procedures that may occur. Importantly, using a single animal for multiple survival surgeries on different protocols is discouraged and, when proposed, should be reviewed carefully. For animals covered under the AWA in the United States, a written request and supporting data must be sent to USDA/APHIS for review and approval if any additional major survival surgical procedure is required on a different protocol.

- *Food and fluid restriction*. Animal studies across several disciplines (e.g., behavior, neuroscience, and physiology) may require restriction of food or fluid to achieve the desired endpoint. Such studies require careful consideration, and the animal use protocol should contain enough information to ensure that only the minimal amount of restriction will be incurred, that methods are in place to ensure animal health and well-being, and that potential negative impacts on the animal will be addressed (NRC 2003). There is no standard "one-size-fits-all" approach to food or fluid studies, as the species, strain, gender, age, and level of restriction, among a host of other factors, may influence the impact on the animal (Heiderstadt et al. 2000; Rowland 2007). Consequently, food or fluid restriction studies are best considered on a case-by-case basis, and the animal use protocol should include a thorough description of this aspect of the study. Monitoring parameters, frequency of body weight assessment, and written record requirements during the period of food or fluid restriction are important considerations. Also, it should be indicated in the protocol that written daily records will be maintained and will include documentation of consumption of food and/or water, physical and clinical observations, and behavioral changes that are observed. A description of endpoints used to remove an animal from the food- or fluid-restricted aspect of the study should also be included.
- *Non-pharmaceutical-grade drugs or chemicals*. Pharmaceutical-grade drugs and chemicals help provide assurance that toxic or undesired side effects will not be introduced into the animal. Both the *Guide* (NRC 2011) and USDA/APHIS (USDA 2016) indicate that pharmaceutical-grade drugs or chemicals should be used whenever available. Should non-pharmaceutical-grade drugs or chemicals be needed, their use should be described and justified in the animal use protocol. A higher cost for pharmaceutical-grade drugs is not considered to be adequate justification for use of nonpharmaceuticals, although OLAW has indicated that recent exorbitant cost increases of pentobarbital have placed it logistically into the unavailable category (OLAW 2012). However, pentobarbital from a reagent or analytical-grade powder, properly prepared by a pharmacist or other knowledgeable individual (e.g., chemist, veterinarian, or researcher), with assurance of appropriate storage and handling, and approval by the IACUC/OB, is acceptable. New experimental drugs or compounds are rarely, if ever, available in pharmaceutical grade, and consequently, their use as a non-pharmaceutical-grade substance can be justified, assuming there is relevance to the study objectives. Nevertheless, the protocol should contain a description of how the purity, sterility, pH, pyrogenicity, osmolality, and stability of such compounds will be assured (NIH 2008).
- *Field investigations*. Studies using wild animals usually require approval from relevant agencies at the local, state, or country level, and the animal use protocol should contain verification that the relevant permits or licenses have been obtained. As with any study, veterinary input will be necessary to ensure appropriate handling or treatment for wild animals that will be captured, held, anesthetized, or used in any way that may compromise their well-being. Nevertheless, other resources, such as a qualified wildlife biologist or relevant professional wildlife biology society publications (e.g., Braun 2005), may be referred to in the protocol as a valid source of information for these studies.
- *Agricultural animals*. The use of farm animals (bovine, equine, ungulate, etc.) may occur for either biomedical- or agricultural-based research, or for studies that overlap both areas. Regardless of the research objectives, the IACUC/OB is obligated to provide oversight of these studies to ensure appropriate animal care and use. The *Guide* (NRC 2011) does not focus on

the use of agricultural animals for studies of food and fiber production; however, the *Ag Guide* (FASS 2010) is an excellent resource in this regard. For agricultural studies, the protocol should address all aspects of animal oversight, including minimization of pain and distress. For agricultural animals used in biomedical research, the standards of the *Guide* should apply and be reflected in the protocol even in a farm setting (NRC 2011).

- *Recombinant DNA and hazardous agent use with animals.* Recombinant DNA and hazardous agents may pose a threat to animal and plant health or public health and safety, and applicable regulations (local, state, and country) should be considered in the animal use protocol. The protocol should list all such agents that are administrated to research animals so that appropriate review regarding use, containment, and disposal may occur. The Institutional Biosafety Committee (IBC) or similar oversight body, along with research biosafety personnel, typically review the use of such agents, and the IACUC/OB should avoid approval of an animal use protocol prior to its review. In the United States, certain records are required to be maintained related to recombinant DNA research depending on the biosafety level (NIH 2016). Such records may include documentation of all animals moved in and out of the facility, a list of experiments that are currently in progress, documentation of the time and date of personnel entering and leaving the facility, temperature records for decontamination and incineration, a permanent record of all animals used in an experiment, and occupational health records related to personnel exposure to spills or accidents.
- *IACUC/OB standard operating procedures (SOPs).* At many institutions and among various scientific disciplines, certain procedures carried out in animal research are routinely identical (e.g., standard blood draws, vascular cannulation, and intraperitoneal injections). As such, SOPs detailing these methods may be preapproved by the IACUC/OB and used, as appropriate, in any protocol. The use of IACUC/OB SOPs can eliminate the need for the scientist to "reinvent the wheel" with each new protocol submission and for redundant review of commonly used procedures by veterinary staff and IACUC/OB members. Thus, IACUC/OB SOPs can save considerable time and effort for both the scientist and those performing protocol review. IACUC/OB SOPs should be reviewed, updated, and reapproved by the committee on a regular basis to ensure that the most up-to-date methods are being utilized.

Electronic Systems for Managing Protocol Review

The availability of electronic protocol review systems has greatly reduced the need to use paper protocols at many institutions. Electronic systems provide a number of inherent advantages, including ease of records accessibility and retention, and almost instantaneous retrieval capabilities for specific parameters (e.g., which protocols are approved for multiple surgeries, and how many protocols utilize non-pharmaceutical-grade drugs?). These advantages may be especially important at big institutions that have a large number of approved animal protocols. Electronic protocol review systems are not perfect, however, and disadvantages may include cost and the need for periodic software updates. There is no shortage of electronic systems available on the market and a web search for "IACUC protocol software" (or similar terms) will generate numerous commercial options. Some institutions have chosen to develop an electronic system in-house to manage protocol review, and this approach may be convenient in terms of customizability and control. Nevertheless, in-house systems mandate considerable investment in developer time, with the realization that local software development expertise will always be needed to fix software "bugs" or make system changes to reflect regulatory requirements. Given the many options for electronic software systems, consultation with other institutions may be helpful in determining which would be best for your organization.

IACUC/OB Meeting Minutes

Documentation of IACUC/OB meeting activities should be maintained in the form of minutes. The AWAR (USDA 2015) and PHS Policy (NIH 2015) contain identical requirements for meeting minutes at U.S. institutions, and these are typical for other countries also. Namely, meeting minutes should include records of attendance, IACUC/OB activities, and descriptions of IACUC/OB deliberations such that major issues and outcomes are provided in enough detail for an outsider to ascertain the nature of the discussions and the conclusions reached. Importantly, meeting minutes should not contain details that might place the institution or any individual at risk, or which could compromise privacy or proprietary information (Public Health Service 2013). These expectations are applicable to discussions regarding policy or SOP development, inspection reports, incidents of noncompliance, and reports of animal welfare concern, in addition to discussions regarding animal use activities (i.e., protocol review and approval).

Retention Requirements

Regulatory requirements for retaining documents relevant to animal use will vary internationally. For example, the Directive 2010/63/EU (European Parliament and the Council of the European Union 2010) indicates that inspection records shall be kept for 5 years, but in the United States, it is mandated that protocols, meeting minutes, and records of deliberations be retained for 3 years after completion of the animal activity (NIH 2015; USDA 2015). For U.S. institutions, this includes retention of a copy of the approved Animal Welfare Assurance, the protocol application and associated significant changes, semiannual reports to the institutional official (IO), and records of the outcome of site visits performed by AAALAC International or another accrediting agency. Regardless of the national requirements, all such records should be retained as a hard copy or electronically in a central location for ease of access when needed.

IACUC/OB Policies

Development and adoption of formal written policies by the IACUC/OB can serve to set program standards for investigators, administrators, and the institution. By creating policies before difficult issues actually arise, the committee gives itself time for careful thought and consideration of complex questions without the pressure associated with an immediate crisis. Policies can also help the IACUC/OB achieve consistency while attending to the myriad responsibilities related to protocol review, inspections, and addressing animal-related concerns. Although formal IACUC/OB policies are potentially of benefit, international requirements for their development can vary, and there is no regulatory requirement in the United States that animal program policies be developed. Nevertheless, many institutions, both globally and in the United States, have adopted formal IACUC/OB policies on topics that typically impact animal use programs, such as euthanasia standards, environmental enrichment requirements, and cage density constraints, to name just a few. The timing of periodic review of policies by the IACUC/OB varies, but many institutions revise these documents at least every 3 years or whenever regulatory requirements change. Policies can be maintained on paper or electronically, but regardless of the medium, it is critically important that investigators have access to the most recent version. At larger institutions, it is recommended that electronic dissemination via e-mail or web page be used to ensure that the latest version of any policy is available to all laboratories.

Facility Inspection and Program Review

Regular facility inspections and review of the animal care program by the IACUC/OB can help ensure compliance with recommended standards. In performing these tasks, the IACUC/OB may identify deficiencies or program weaknesses that should be addressed to ensure that a high level of quality animal care is provided. Requirements for performing facility inspections and reviewing the animal care program can vary from one country to the next. For example, the Directive 2010/63/EU (European Parliament and the Council of the European Union 2010) mandates that member states shall ensure that the competent authorities carry out regular inspections of all animal users, with the frequency of inspections based on a risk analysis for each institution. In comparison, the *Guide* (NRC 2011) states that facility inspections and program review should occur at least annually, or more often if necessary. Given the international diversity in conducting this type of review, the reader should refer to specific regulatory requirements for the country of interest. The information presented here focuses on U.S. regulatory requirements.

The AWAR (USDA 2015) and PHS Policy (NIH 2015) both require that the IACUC/OB inspect all animal facilities every 6 months, document its findings by indicating if an item is significant or minor, and develop a timeline for correction. Information gathered during inspections should be recorded, and some inspectors have started entering the information onto tablets or laptop computers that are carried with them. In the United States, if there is a significant deficiency identified during program review or inspections involving AWAR-covered species, a date for correction of the deficiency must be identified and completed within 15 days of the correction date. A written notice to the USDA is required if correction of a significant deficiency does not occur within the 15-day window. The results of both the facility inspection and the program review should be documented, signed by a majority of the committee members, and sent to the IO. Minority views of any IACUC/OB member related to the inspection and program review should be included, along with any changes to the animal use program and any departures from the *Guide* or PHS Policy, or the AWAR.

Animal Welfare Concerns and Noncompliance

Many countries have regulatory requirements for responding to animal welfare concerns and noncompliance, and the reader is referred to Chapter 8 for more detailed discussion of global regulations. In the United States, the AWAR (USDA 2015) indicate that the IACUC/OB must review and as necessary investigate any public complaints or noncompliance reports from personnel related to the use of animals at research facilities. Adequate methods for receiving complaints confidentially are expected to be in place. The *Guide* (NRC 2011) indicates that mechanisms for reporting concerns should be documented and posted in the facility. The posting should have multiple points of contact and should include a mechanism for anonymous reporting, and indication of compliance with whistle-blower policies and protection from reprisals. The response to animal-related concerns or investigations of noncompliant activities should be well documented, and software systems designed to track such activities are available (Capterra 2016). Retention of associated documents should be defined by institutional policy or regulatory requirements.

Postapproval Monitoring

Postapproval monitoring (PAM) programs vary from institution to institution, and the type and intensity of a PAM program often are dependent on institutional size and complexity (Haywood and Greene 2008). Certain responsibilities of the IACUC/OB, including periodic facility inspections, program review, and continuing protocol review, should be considered as part of a PAM program, and the documentation required for these activities has been discussed. More formal PAM programs will require additional documentation to ensure that all relevant parties (i.e., PI, laboratory staff, and IACUC) are made aware of observations and outcomes (Dale 2008). Regardless of the intensity of the PAM program, documentation of observations is evidence of a well-established program of care that is focused on animal well-being.

Animal Resources and Facility Operations

Documentation is an integral part of a laboratory animal care and use program, but the records that are required will vary in scope as the program dictates. We will not discuss the how-to and why of each topic, as that information is covered in other chapters and in referenced publications. The following sections describe a general overview of each area of a laboratory animal care and use program that should be documented.

General

Laboratory animal programs should have a written program of veterinary care outlining the provision of adequate medical care for all animals, including delivery of both routine and emergency services.

Institutions that are applying for accreditation by AAALAC must complete a program description that describes all aspects of the veterinary care program. Other documents that may be required include a PHS-approved Animal Welfare Assurance statement when funding is received from the PHS or other government agencies.

Animal Resource and Facility Standard Operating Procedures

Animal resource and facility SOPs are documents that present step-by-step instructions and responsibilities for tasks to be performed by individuals working in laboratory animal care and use programs. Specific SOPs should be identified by each program and may include categories such as personnel, husbandry, animal health and well-being, canine exercise programs, primate psychological well-being programs, facility maintenance, equipment operation, and personnel safety. Animal resource and facility SOPs should be written in a thorough and standardized way so that all personnel can follow and execute the activities consistently. Written SOPs are essential in many circumstances, but even when they are not required, they are a useful tool to ensure consistency within the program.

Animal Resource and Facility SOP Authorship and General Organization

One or more individuals, such as an animal resource manager or a committee of knowledgeable staff members, are typically charged with authorship of animal resource and facility SOPs pertinent to their area of expertise. Animal facility SOPs should be logically organized with separations between major sections, such as husbandry, sanitation, animal health, and security. The number of SOPs and the level of detail will vary depending on program organization and complexity.

Authorization of Animal Resource and Facility SOPs

Animal resource and facility management should approve written SOPs, and the IACUC/OB has a responsibility to review these (NRC 2011). Review and approval of SOPs help confirm that an appropriate method is being used for a given procedure and that there is standardization of procedural performance. Also, management and the IACUC/OB can help convey the SOP to all those who may be impacted.

Animal Resource and Facility SOP Format

The format of SOPs varies by institution. Typically, animal resource and facility SOPs include an edition number or code, serial number, title, author signature, and date, along with an approval signature effective date. The format chosen is typically followed without variation between each SOP.

Animal Resource and Facility SOP Content

Animal resource and facility SOPs should include the purpose, scope, references, definitions, procedure, and any other related documentation. The purpose section outlines the objective or intent of the document. The scope outlines the department, group, or personnel to which the procedure applies. The references section details other documents that have a bearing on the activities within the procedure. The definitions section explains any word not easily understood. The procedure section details the actions involved in the activity by stating who does what and how, where, when, and if needed, the reasons why the activity is carried out. The documentation section includes any other documents referred to within the procedure or that may be generated as a result of procedural implementation. Documents may be attached as appendices and published literature may serve to supplement an SOP document.

Use and Management of Animal Resource and Facility SOPs

Each laboratory or animal use area should have immediately available animal resource and facility SOPs that relate to the procedures being performed. The location and content of relevant SOPs should be common knowledge for laboratory or animal use area personnel. A robust SOP training program can help achieve this objective. An historical file of SOPs that includes revisions and the dates the document was revised should be maintained. Distribution and receipt of SOPs should be documented. Electronic copies should be controlled and managed by institutional policy to ensure that the most recent version is available.

Animal Resource and Facility SOP Summary

SOPs can be a useful tool to help ensure consistency within an animal care and use program. Not only are animal resource and facility SOPs a valuable asset to the institution, but also they bring the opportunity to clarify working routines and areas of responsibility. A collection of good SOPs that are reviewed and updated on a regular basis also shows management's commitment to the production of quality documentation and work practices. Importantly, SOPs are only successful if they are read, understood, and followed by those who are carrying out the work.

Staff Education and Training

Staff education and training are critical to ensure quality animal care in any laboratory animal use program. Basic methods of documenting education and training will be presented here. However, Chapter 12 of this volume presents the necessary elements of a thorough and inclusive education and training program that meets regulatory requirements.

Training Records

A central location should be identified for maintaining records of education and training. Documents should always be retrievable and available for review by oversite bodies during inspections, site visits, or investigations. Hard copies and a secure file system may be adequate for storing these documents, but most programs rely on electronic storage of all educational and training information.

Staff Qualifications

Applications and curriculum vitae of animal care personnel should be accessible so they can be referenced when needed in order to verify the qualifications of staff providing for the care and use of animals. Curriculum vitae should be updated as necessary to reflect newly acquired certifications and job skills.

Continuing Education

Continuing education and training should be documented and maintained. All programs are required to provide ongoing training, although the frequency, topics, and requirements will vary depending on the scope of the program. Documentation of relevant classroom and hands-on training sessions should be maintained, and this may be done easily by retaining a signed attendance log for those attending the training session. Participation in off-site learning opportunities, such as veterinary technology programs, local branch symposia, and national and international professional meetings, should be documented and retained. Certificates of continuing education are typically presented to attendees at such meetings, and these should be retained centrally so they can be retrieved when needed.

On-the-job training is an important aspect of continuing education, and regular review of procedures, changes in rules and regulations, new animal resource and facility SOPs, and other pertinent information should be documented. Also, documentation should be retained for completion of refresher courses that are required when working with chemical, biological, or radiological hazards.

Research Staff Training

Documentation of research staff training is typically included in the institutional animal use protocol and should be shared with the animal care program veterinarians and managers. Written documentation may be suitable for smaller programs, but many large programs utilize electronic systems specifically designed for animal protocol review. These electronic systems commonly incorporate a searchable database that simplifies document retrieval. Documentation of research staff training should be kept current and updated commensurate with instruction as it is provided.

Husbandry Assignments and Completion of Activities

Completion of assigned animal husbandry duties should be documented. These activities may be conducted by one animal care person or by many individuals, depending on the size of the program. A simple written daily log of activities may be appropriate in small programs, but large programs with multiple animal care personnel often develop more robust documentation systems.

Documentation of daily assignment schedules and completion of activities is a great resource for both animal care personnel and those at the managerial level. Many organizations utilize electronically managed programs, such as a spreadsheet to distribute workload. A well-maintained electronic spreadsheet or management program can be a valuable tool to oversee husbandry assignments when dealing with fluctuating animal populations and husbandry staff. The information that is maintained usually includes assignable tasks with time allocations for each chore. Programs can be set up to calculate times for all assigned work, including, but not limited to, changing animal cages, checking animal health, feeding, cleaning, stocking supplies, and housing newly received animals.

Completion of husbandry tasks, as well as daily observations of animals for health and well-being, assurance of access to feed and water, and verification of a safe and appropriate environment, should be documented. Completion of cleaning tasks outlined in animal resource and facility SOPs should also be documented. A simple calendar sign-off sheet maintained in the animal housing area may suffice for documentation of completed basic husbandry tasks, room environmental parameters, and verification of animal well-being. Portable or wall-mounted electronic documentation devices are now more commonly used at many institutions.

Acquisition and Disposition of Animals

Records regarding the acquisition and disposition of all animal species should be maintained. The USDA also requires specific documentation regarding the acquisition, transportation. and disposition of cats and dogs (USDA 2015). This information must be maintained for a reasonable period of time, and most institutions retain these records for a minimum of 3 years following the final disposition of the animal.

Identification, Cage Cards, and Census and Billing

All animals held for research must have appropriate identification, which starts with their acquisition and receipt. Animal records can vary from limited information on a cage card for rodents, birds, and fish, to detailed computerized records for individual animals. Means of identification include room, rack, pen, stall, and cage cards with written, bar-coded, or radio frequency identification (RFID) information. Identification cards should include the source of the animal, strain or stock, names and contact information for the responsible investigators, pertinent dates (e.g., arrival date and birth date), and protocol number when applicable. Genotype information, when applicable, should also be included, and should be consistent.

Large animals such as dogs, cats, nonhuman primates, and farm animals typically have an identification tag attached to their collar or a legible tattoo. Some large animals within a primary enclosure (e.g., penned cattle) may be identified by coat color and pattern differences. Records containing basic descriptive information are essential for the management of colonies of large long-lived species and should be maintained for each animal (NRC 1979; Dyke 1993; Suckow and Doerning 2007). These records often include species, animal identifier, sire and/or dam identifier, sex, birth or acquisition date, source, exit date, and final disposition. Such animal records are essential for genetic management and historical assessments of colonies. Records of rearing and housing histories, mating records, and behavioral profiles are useful for the management of many species, especially nonhuman primates

(NRC 1979). Relevant recorded information should be provided when animals are transferred between institutions.

Documenting and tracking animal census is important for historical assessment of populations and is used for appropriate billing. This can be conducted using a traditional hand count of animals. More sophisticated electronic programs are available that use bar codes or chips for scanning into a tracking program. RFID systems are now being used for fast input of information for all animals housed within an entire room.

Environmental Enrichment

A formal, written plan for environmental enrichment is standard practice at most facilities for all species, with the primary aim of enhancing animal well-being. The enrichment plan should describe the structural elements of primary enclosures that may enhance the well-being of animals, along with the nonstructural provisions that promote species-typical activity patterns. Records should be maintained for tracking the actual enrichment provided for each species and for individual animals, including any rotation of specific devices. The environmental enrichment plan should also describe the strategy for social housing of all social species and should describe the steps that will be taken with isolated or individually housed animals to compensate for the absence of conspecifics.

For dogs, facilities should have an appropriate plan to provide them with the opportunity for exercise. The plan should include written standard procedures to be followed in providing the opportunity for exercise and must be approved by the attending veterinarian. Exercise provided or justification for not providing the exercise must be documented to meet requirements of the AWA (USDA 2015).

Emergency (Disaster) Plan

A comprehensive emergency plan is an expectation of the AWA and the *Guide* and is intended to protect the health and safety of humans and animals. Facilities should develop an appropriate plan to provide for the humane handling, treatment, transportation, housing, and care of animals in the event of an emergency or disaster. The facility should identify emergency situations that would trigger the need for implementation of a contingency plan. Examples of emergencies include electrical outages, faulty heating, ventilation, and air-conditioning (HVAC) systems, fires, animal escapes, and natural disasters that the facility is most likely to experience. The plan should outline specific tasks and procedures for emergency animal evacuation, shelter in place, provision of food and water, sanitation, ventilation, bedding, and veterinary care. The plan should identify a chain of command and who (by name or by position title) will be responsible for fulfilling responsibilities of emergency tasks. The plan also should address how response and recovery will be handled in terms of materials, resources, and training.

Emergency training of all personnel should be conducted and documented on a regular basis. It is advisable to review emergency plans at least annually, and any changes should be communicated to employees through documented training within 30 days.

Animal Health Surveillance

Maintaining an established health status for animals and colonies, as well as keeping unwanted pathogens out of facilities, is a primary concern for any animal care program. It is important to document known pathogens and verify new pathogens as they emerge. An important first step in accomplishing this is to document which pathogens are to be excluded for each species and every housing area. All incoming animals should have health data reviewed before admission into the facility or transfer to other locations, both internal and external. Animals routed through quarantine or isolation typically have further evaluation and health screening that is documented. Health surveillance documentation may be kept as a part of the individual, herd, or specific group of animals within a central filing system or electronic program. Electronic health surveillance and biosecurity programs are very useful and may be used to generate reports for review and action. Ongoing health data information should be collected, reviewed, and maintained centrally, and this may be done using electronic health surveillance programs or other custom computer software. The frequency and type of pathogen testing must be determined and included in the program. Ongoing test results should be archived in the program and reviewed regularly.

Animal Health Records

Animal health records are considered critical for documenting animal well-being and tracking animal care and use at a facility (NRC 2011). The American College of Laboratory Animal Medicine (ACLAM) has issued a public statement indicating that medical records are considered to be a key element of a program of adequate veterinary care for animals used in research, teaching, and testing (Field et al. 2007). Individual health records should be maintained for animals that receive regular, individual health evaluations as deemed appropriate by the institution. Documentation may range from detailed records for each individual animal to limited information recorded in group or herd records. Many facilities use electronic record keeping, while others continue to use handwritten paper records. Typically, dogs, cats, nonhuman primates, marine mammals, and other larger species have their own individual health record, while other species, such as rodents, birds, zebrafish, and other small animals, may have records that document health issues within a selected group, room, or herd of animals.

Health record information should include information associated with the management of clinical disease, diagnostic or therapeutic procedures performed, and preventative medical procedures. Health records may also be useful for recording husbandry information, dietary issues, and necropsy findings where applicable.

Notations in the health record should be made by individuals who have administered treatment, or made direct observations or evaluations of the animals or their diagnostic results. Individual health records should follow the animal throughout its life. A copy of the health record should be transferred with the animal if it is relocated. Individual or group health records can be maintained by the program indefinitely for historical purposes. Otherwise, these documents can be maintained for a reasonable period of time, but typically for a minimum of at least 3 years beyond the final disposition of each animal.

Environmental Health and Safety

It is essential that institutions have in place effective programs for controlling hazards and minimizing occupational risks of injury and illness in the workplace. Programs may have a functional environmental health and radiation safety (EHRS) office available to oversee the institution, while others may have other technical resources available with expertise to oversee specific activities within the program. All critical areas that are overseen by EHRS must function so as to minimize risks, and oversight of these activities should be documented. Documentation should include administrative procedures, facility design and operation, exposure control methods, engineering controls, work practices, personal protective equipment, education and training, equipment performance, information management, emergency procedures, and program evaluation.

Every person who works with animals should be provided an opportunity to enroll and participate in an occupational health and safety program. This should include students, visiting scientists, volunteers, maintenance and custodial personnel, vendors, and other nonemployees. A tracking mechanism should be in place to document enrollment. Documentation of occupational exposures, safety training, medical surveillance, and work-related injury and illness is important and is useful for evaluating the effectiveness of the occupational health and safety program for the institution. This documentation is confidential and should only be accessible to oversight personnel.

Written work practices should be in place in the form of SOPs and policies to help prevent accidents, exposures, and injuries. Special practices should be in writing and followed for specific hazards, such as chemicals, radiation, or infectious agents. These specific hazards should have written procedures (typically posted on the entrance where used) outlining the unique hazards and steps that should be taken to prevent exposure and ensure containment. All staff working with these types of hazardous agents should receive specific training, and this should be documented.

Specialized equipment and work areas should be clearly posted with safety precautions, and staff should be continuously trained concerning safe work practices. Some areas requiring special attention and documentation for use of potentially dangerous equipment include cage washers, sterilization equipment, heat sources, pallet jacks, forklifts, and transport vehicles.

Environmental Monitoring

The animal resource facility environment is an essential component of animal well-being, quality research, and production of laboratory animals, as well as providing conditions that promote the health and safety of personnel who work in these facilities. Active monitoring of the environmental parameters of an animal facility should be in place and the information must be retrievable. Whenever animals are exposed to environmental changes or any unplanned events, the events should be documented and shared with the affected research staff, veterinarians, and the IACUC/OB.

The parameters of the microenvironment immediately surrounding the animals should be monitored and documented. This typically includes temperature and humidity, although concentrations of gases (e.g., ammonia) and particulate matter within the primary enclosure may also be measured using specialized equipment. A baseline of the parameters within the microenvironment is essential to know and document before husbandry practices such as cage type, bedding and enrichment choice, number of animals allowed per enclosure, and cage change frequency are selected. Once the baseline has been established, ongoing checking and documentation of the microenvironment is essential to ensure environmental consistency.

The macroenvironment surrounding the primary enclosure should be documented continuously whenever possible, but at a minimum, daily observations should be recorded. Manual systems and devices such as high and low thermometers, hygrometers, light timers, water supply systems, data loggers, and light meters are useful in this regard. Many institutions have electronic programs that continuously monitor these parameters with built-in alarming and remote notification to key personnel who respond 24/7 when preset parameters are breached. Most electronic programs store collected information that can be reviewed by oversight bodies and research staff for verification of environmental parameters. Nevertheless, even the most sophisticated monitoring program should periodically be checked to validate accuracy.

Verification of air quality, pressure differentials, and ventilation should be documented periodically to ensure that filters and systems are performing as intended, valves are functioning, air quality is as designed, and air exchange rates are within the desired parameters. Periodic air balancing and pressure differential reports should be documented. Any problems discovered should be addressed and the resolutions recorded.

Noise and vibration produced by animals and husbandry activities are inherent in the operation of an animal facility. Routine practices should be in place to reduce noise exposure to animals and humans alike. Unfortunately, unplanned events related to construction, accidents, and natural events will happen. Any event, planned or unplanned, that has the potential to affect animal well-being should be documented and reviewed. Actions put into place to reduce the potential for future negative impacts should also be documented.

Equipment Function

All equipment within the animal facility should be kept in good operating condition. Most large pieces of equipment should have well-documented preventative maintenance programs, including dates of routine maintenance or repair. Documentation should include any testing results verifying proper equipment function. The frequency of functional testing of equipment should be decided by each facility. Equipment in the preventative maintenance and functional performance testing program typically include cage and bottle washers, autoclaves, anesthetic vaporizers, room sterilization equipment, individually ventilated cage systems, biological safety cabinets, fume hoods, laminar flow animal transfer stations, automated water systems, humidifiers, and environmental monitoring systems.

Pest Control

Pest control programs should be well documented. This includes documentation of how structural penetrations are sealed, how vermin entry is prevented, and methods of monitoring and protecting outside areas from vermin nesting or congregation. Additional documentation should include monitoring that is in place to trap unwanted vermin and what is done if insects or evidence of wild animal entry is found. If any traps or chemicals are employed, these should be described in the pest control program with information related to the protection of research animals when using physical or chemical deterrents. The pest control program should indicate that researchers will be notified when vermin control measures are to be implemented.

Security and Access Control

Security of the laboratory animal facility is essential to the welfare of the animals and to research studies that are being conducted. A central paper or electronic file should be kept identifying each individual who receives access to animal facilities. The rooms, areas, and other information concerning access should be kept in these files. Ideally, a method for reviewing access records will be in place with the ability to terminate access when appropriate. Such control may not be practical, however, for decentralized or agricultural facilities.

Water Quality Monitoring

Water is an essential component of the health and well-being for both terrestrial and aquatic animals. Documentation must be in place describing the components of any water system, and this should include a description of the water source, treatment or purification process, and how water is provided to the animals. Quality control should include monitoring for contaminants and other parameters, including pH, alkalinity, hardness, and other parameters listed in the *Guide* (NRC 2011). Routine test results should be documented and reviewed periodically. If automatic water delivery systems are used, a description or SOP for maintenance, testing, and sanitization is advisable.

Sanitation Monitoring

Sanitation monitoring should be documented on a regularly scheduled basis as determined by each facility. Items that may be included in the sanitation monitoring program are cages, housing racks, room surfaces, transport equipment, enrichment devices, and research or behavioral testing equipment or other hard-surface devices used within the animal facility. Monitoring information should be collected routinely and documented, and reviewed by knowledgeable staff. Documentation of contamination correction plans should be retained for future reference.

Controlled Drugs

Drug security and control are the responsibility of each facility that receives, maintains, or distributes any controlled drugs. Most countries have established their own enforcement agencies similar to the U.S. Department of Justice Drug Enforcement Administration. Drugs, substances, and certain chemicals used to make drugs are classified into five distinct categories or schedules depending on the drug's acceptable medical use, and abuse or dependency potential. Schedule I drugs are considered the most dangerous, with a high potential for abuse. As the drug schedule changes—Schedule II, Schedule III, and so forth—so does the abuse potential; Schedule V drugs have the least potential for abuse.

Detailed records are required, and documentation of receipt and dispensing of controlled drugs must be maintained. A controlled substance logbook is typically kept with the name of the drug, identification number, and date the drug was received. When a controlled substance is dispensed, the date, amount used, patient or project, and identification of the person who dispensed the drug should be recorded in the controlled drug log. Records retention may vary by country, but in the United States, all records pertaining to controlled substances must be kept for 2 years and be made available for inspection by duly authorized officials of the Drug Enforcement Administration (CFR 2016a). In the case of loss of controlled substances, the regional Drug Enforcement Administration and local police department must be notified. A theft report form (DEA-106) must be completed and sent to the nearest office of the Drug Enforcement Administration.

Inspections and Site Visits

In the United States, whenever animals are held that are covered under the AWA, institutions are inspected without notification on a regular basis by a USDA/APHIS inspector. Inspectors will review all documentation associated with these animals to ensure that the facility, husbandry, veterinary care, and IACUC/OB practices are in compliance with the AWAR (USDA 2015). It is essential that all documentation related to the animal care and use program be current, accurate, and available. In the United Kingdom and most of Europe, a Named Animal Care and Welfare Officer (NACWO) is appointed and charged with oversite and reporting back to the Home Office (Council of Europe 1986; European Parliament and the Council of the European Union 2010).

During any USDA/APHIS inspection, it is important to provide relevant and required records. Active records kept in animal rooms, surgery, and support spaces must be made available, and historic files concerning the care and use of all covered animals must be accessible. Typically, animal care and use files are maintained for at least 3 years after the completion of the activity and inspection reports from USDA/APHIS, and all related correspondence should be maintained for at least 3 years. Whenever the USDA/APHIS administrator notifies a research facility in writing that specified records shall be retained pending completion of an investigation or proceeding under the AWA, the research facility shall hold those records until their disposition is authorized in writing by the administrator.

AAALAC International is a private, nonprofit organization that promotes the humane treatment of animals in science through a voluntary assessment and accreditation program. Research programs that participate in the AAALAC accreditation process demonstrate that they meet the minimum standards required by law, and are also going the extra step to achieve excellence in animal care and use. An important document related to the AAALAC accreditation process is the program description, which provides a description of the animal care and use program in specific detail. AAALAC site visitors will review all aspects of an animal care and use program (typically every 3 years), and they will verify that all operations are as described in the program description. Key representatives of the animal program will be interviewed, and facilities will be toured while practices and procedures are observed. Also, documentation related to animal care and use will be reviewed to ensure compliance with regulatory requirements and high standards of animal care. Facilities should document the highlights of the site visit and record any comments presented at the exit briefing. Follow-up activities and correspondence will be documented according to the AAALAC accreditation process (AAALAC International 2016).

Good Laboratory Practice

Preclinical studies conducted in the United States that demonstrate the safety of a drug, biologic, or medical device are regulated by the U.S. Food and Drug Administration (FDA) and are typically referred to as GLP for nonclinical studies (CFR 2016b). FDA GLP regulations were implemented in 1979 with the intent of ensuring the reliability, validity, and reproducibility of test results in order to establish a dependable foundation for assessing the risks and benefits of the product being tested. GLP studies typically utilize animals, and FDA regulations mandate specific documentation responsibilities. Reference is made to Subpart J in CFR (2016b) for specific information regarding requirements for storage, retrieval, and retention of GLP records. Although the focus of this chapter is GLP, mention is made here also of good manufacturing practices (GMPs), which is another set of regulations found in CFR 21, Part 211 that are governed by the FDA for ensuring the safety and integrity of drugs (CFR 2016c). The GMPs are applied for products that are developed for use by human beings, while GLPs are applied for nonclinical laboratory studies. The areas that come under GMP are facilities and buildings, equipment, production, process control, packaging and labeling, laboratory controls, and returned or salvaged drug products. Additional information regarding GMP requirements may be found in Oechslein (2015).

GLP documentation requirements may be met using electronic systems, and commercial software packages are available for this purpose. The regulations contain specific requirements for electronic records and signatures, and the reader is referred to "Bioresearch Monitoring Good Laboratory Practice" (FDA 2016) for additional information.

Organization and Personnel

Employee Files

Personnel who supervise or conduct a GLP study must have the education, training, and experience to perform their job duties. This information must be retained for each employee.

Management Responsibilities

The testing facility management is responsible for several assurances, and in regard to documentation, it must ensure that any deviations from the regulations are recorded. Other documentation responsibilities at the management level include

- The study director has overall responsibility for study documentation and must ensure that all experimental data is accurately recorded and verified. Additionally, the study director is responsible for documenting unforeseen circumstances and corrective actions that are taken, along with ensuring that all study documentation (including raw data, protocols, and final reports) is archived at the close of the study.
- The quality assurance unit must maintain a copy of a master schedule sheet containing general information regarding all GLP studies conducted at the facility. Also, the responsibilities and procedures of the quality assurance unit, along with inspection records and the method of indexing inspection documents, must be written down. Copies of protocols, study status reports, and the final study report must be documented.

Standard Operating Procedures

A GLP testing facility must have written SOPs. Deviations from SOPs must be documented and significant changes must be authorized in writing by management. SOPs must be established for at least the following animal study activities:

- Animal room preparation
- Animal care
- Receipt, identification, storage, handling, mixing, and method of sampling test and control articles
- Test system (i.e., animal) observations
- Laboratory tests
- Handling of animals found moribund or dead during a study
- Necropsy of animals or postmortem examination of animals
- Collection and identification of specimens
- Histopathology
- Data handling, storage, and retrieval
- Maintenance and calibration of equipment
- Transfer, proper placement, and identification of animals

SOPs must be immediately available in the laboratory area and an historical file of each, including revisions and corresponding dates, must be maintained.

Equipment

Written SOPs that define the methods, materials, and schedules used in inspection, cleaning, maintenance, testing, calibration, and/or standardization of equipment are required. Specific personnel (generally identified by job title) who are responsible for performing each operation must be indicated. SOPs should also specify actions to be taken in the event of failure or malfunction of the equipment.

Written records of equipment operations and maintenance must be maintained. These records must contain the date of the operation and whether the maintenance operations were routine and followed the written SOPs. Nonroutine repairs performed as a result of equipment failure and malfunction require written records that document the nature of the defect, how and when the defect was discovered, and any remedial action taken.

Animal Care

SOPs are required for the housing, feeding, handling, and care of animals. Any clinical treatment must be documented and these records retained. Feed and water must be analyzed periodically and the raw data must be maintained. Finally, the use of pest control measures must be documented. FDA inspectors may ask to see a written SOP for pest control, along with documentation of implementation of these procedures.

Select Agents

Biological agents and toxins that have the potential to pose a severe threat to public health and safety, animal or plant health, or animal or plant products are referred to as select agents and toxins (or simply select agents). Oversight of select agents occurs through the Federal Select Agent Program (FSAP), which is implemented jointly by the Centers for Disease Control and Prevention's (CDC) Division of Select Agents and Toxins (DSAT), and the USDA/APHIS, Agricultural Select Agent Services (AgSAS). Registration with the FSAP and adherence to the select agent regulations (CDC and USDA/APHIS 2014a) is required for any person or entity that possesses, uses, and/or transfers a select agent or toxin. Part of this responsibility includes appointing a responsible official (RO) who is accountable for compliance with the regulations and who has authority to act on behalf of the institution. There are a total of 65 select agents (CDC and USDA/APHIS 2014b), and extensive documentation is required to meet the regulatory requirements and ensure that the institution is prepared to manage their use appropriately. Records must be accurate, comprehensive, and secure. All records must be retained for at least 3 years. Records and documentation may be maintained on paper and/or electronically, and it is advisable that backup copies be kept.

Initial Registration and Other Forms

The Application for Registration for Possession, Use and Transfer of Select Agents and Toxins (APHIS/CDC Form 1) must be completed and sent to the DSAT (human select agents), to AgSAS (animal or plant select agents), or to either agency if overlap agents (i.e., those that are hazardous to both humans and animals) will be used. The appropriate sections of Form 1 are to be submitted anytime there is a change at the institution. Any correspondence associated with an application or an amendment should be retained. Other APHIS/CDC forms and the reasons for their use are:

- Form 2: To request approval to transfer select agents from one registered entity to another registered entity. Shipping invoices and related materials should be included with Form 2. Note that approval is not required to transfer toxins in quantities below a specified amount listed in the regulations, although these may only be transferred after documenting that the recipient has a legitimate need to use and possess the toxin.
- Form 3: To be submitted within 7 calendar days upon discovery of a theft, loss, or release of a select agent and toxin.
- Form 4: To report identification of a select agent or toxin isolated from a clinical specimen or environmental sample (Form 4A), or from a proficiency (comparative) sample test (Form 4B).

- Form 5: To request an exemption from the select agent regulations for an investigational product.
- FD 961: To obtain a security risk assessment (SRA) for every individual listed on the registration form who requires access to select agents; resubmitted for each person every 3 years.

Biosafety Plan

A written biosafety plan commensurate with the risk of the select agents that are possessed must be developed and implemented (CDC and USDA/APHIS 2014a). The plan must include an occupational health program for individuals with access to Tier 1 agents (i.e., a subset of select agents that pose a severe threat to public health and safety) and to severe acute respiratory syndrome (SARS). Additional information regarding the occupational health program requirements may be found at "Occupational Health Program Guidance Document for Working with Tier 1 Select Agents and Toxins" (CDC and USDA/APHIS 2013a). The biosafety plan must be specific to each laboratory where the select agents are used. Alternatively, an institutional plan may be developed with an attached plan for each laboratory. The plan must provide instructions on the risks and ways to mitigate them.

The elements of the biosafety plan can only be determined by performing an institution-specific risk assessment, although guidance for what may be included has been provided in the *Responsible Official Resource Manual* (CDC and USDA/APHIS 2013b). The presence of animals at a facility necessitates a biosafety plan for their care and handling, and the components of the plan will depend on the animal species, agents used, procedures conducted, and facilities. SOPs may be used, and if so, these should be referenced and where they can be accessed in the biosafety plan noted.

Electronic accessibility of the biosafety plan, via web page or otherwise, is becoming more common, as this format enhances accessibility and editorial convenience. Staff should be kept apprised of changes when they are made to the electronic document. Written biosafety plans should be subject to document tracking to ensure that when changes are made, all plans are updated accordingly.

Security Plan

A security plan, based on a site-specific security risk assessment, is required for every registered entity. Each security plan is unique; however, the following components must be included in all plans:

- A description of procedures for controlling the long-term select agent inventory
- Procedures for ensuring information security (e.g., inventory access logs, passwords, and entry logs)
- A description of how access to animals exposed (or possibly exposed) to select agents and their safeguarding is controlled, along with escaped animal procedures
- Security procedures when the laboratory must be cleaned or repaired, including information regarding granting and recording access and escort provisions for maintenance personnel, and securing select agents during cleaning or repair
- Procedures for identifying and removing unauthorized individuals
- A description of access procedures (e.g., identify who authorizes and provides keys or access cards or codes, and who removes access and collects keys or cards), along with the response procedures when keys, cards, or codes are lost
- Response procedures following theft or loss of select agents, or when inventory records are altered
- A description of requirements (e.g., security drills and exercises) that will ensure that all personnel with select agent access are aware of security policies and procedures

The security plan should undergo annual review or whenever changes, such as facility renovation or addition of new projects, occur.

Incident Response Plan

A written incident response plan is required with the intent of defining measures that would be taken to mitigate the impact of natural and man-made threats. The plan must be based on a site-specific risk assessment, and it should be developed in harmony with institution- or community-wide plans. Notably, for institutions using animals, the plan must include how animals infected with select agents will be handled during emergency situations, and this must include a description of the response to the release of infected animals into the laboratory, building, or community. Also, if the institution works with animals and it possesses select agent plant pathogens, the impact of the release of the plant pathogens on animals and animal products must be considered. Other required parts of the incident response plan include a full description of the response to:

- Theft, loss, release, or inventory discrepancy for a select agent
- Security breaches
- Severe weather and natural disasters given the geographic location
- Workplace violence, bomb threats, or suspicious packages
- Fire, gas leak, explosion, and power outage

Contact information (home or cell phone, work phone, etc.) for key individuals should be included in the plan, including those for the RO and alternate RO, biosafety officer, animal care and veterinary staff, building manager and facilities management, and physical security official for the building. Contact information for local public safety entities (e.g., police, fire department, and hazardous materials management), along with that for the regional Federal Bureau of Investigation Weapons of Mass Destruction office, should also be listed. Other information that should be in the plan in case of an incident includes a description of each responder's role in the response, information on procedures for rescue and provision of medical care, a list of personal protective equipment and its location, security and evacuation procedures, a description of emergency medical treatment and first aid, and a description of decontamination procedures for impacted areas.

For institutions with Tier 1 agents (i.e., a subset of select agents that pose a severe threat to public health and safety), the incident response plan must also describe the response to a failure of intrusion detection systems such as locks, cameras, or alarms. Additionally, the plan must contain a description of how federal, state, or local law enforcement will be notified of possible criminal activity related to the institution, its personnel, or its select agents. Examples include unusual interest by outsiders in an institution's select agent work, and unauthorized removal of select agents from designated areas.

Training

The RO has the responsibility to ensure that training is conducted and documented (CDC and USDA/APHIS 2014a). Training should be appropriate for the work that an individual is performing; for example, a laboratory technician who conducts animal research with plague bacteria will require extensive biosafety training to ensure that he or she knows how to work in a safe manner with this agent. Security training regarding how to protect select agents from theft or loss, and incident response training on how to react to emergencies also should be provided. Note there is a special training requirement for institutions with Tier 1 select agents in that insider threat awareness briefings, which relate to identification and reporting of suspicious behaviors, must be provided annually to individuals with an SRA (CDC and USDA/APHIS 2014a).

Initial and annual refresher training must be provided and documented for each individual who has access to select agents, that is, for those individuals with a security risk assessment. The RO must ensure that each training record includes:

- Name of the individual
- Date of training

- Description of the training
- Means used to verify that the employee understood the training (e.g., by testing to ascertain the employee's comprehension, or by asking the employee to sign a verification that he or she understood the training)

Individuals who do not have an SRA (e.g., IACUC/OB members during semiannual inspections) may go into select agent areas with an escort. However, there must be a training record that these individuals have received instructions and signed that they received and understood the instructions prior to escorted entry. Recurring visitors (i.e., individuals who enter a select agent area on a regular basis) must receive training once per year, and there must be documentation that they received and understood the training.

It is the RO's responsibility to ensure that all training records are maintained for at least 3 years, even if the individual leaves the institution before the end of that time. There is no required way to document training, although an individual file for each trainee may help ensure accurate oversight. Paper records are acceptable, as are computer-based systems, which are available commercially.

Other Documentation

Further details on the documentation listed below may be found in the *Responsible Official Resource Manual* (CDC and USDA/APHIS 2013b).

- Access records: The RO must ensure that there is a record of who has access to rooms where select agents are used and stored. This should include a record of who has been assigned codes, keys, combinations, or other security mechanisms that are needed for access.
- Audits
  - Internal: Annual internal audits are to be conducted, and a copy of these, including documentation of the deficiencies identified and their resolution, are to be maintained.
  - External: This includes inspections, reviews, certifications, or audits by other agencies, accreditation organizations, or groups. There is no regulatory requirement to maintain these documents; however, they could be useful in responding to FSAP inquiries.
- IBC minutes: FSAP inspectors may ask to review IBC minutes (although there is no requirement under the select agent regulations for this committee).
- List of select agents and toxins: The RO should maintain a record of the select agents for which the institution is registered (these will be identified in the initial registration and amendments), where the select agents are located, and a current inventory for each. Electronic inventory systems are available that can facilitate this requirement.
- List of personnel with an SRA: For each individual, the list should include:
  - Name
  - Date of submission request for SRA
  - Date of SRA approval
  - Date when access to select agents was allowed
  - SRA expiration date (3 years after approval date)
- Health-related entry requirements: These include personal records of immunizations, respiratory evaluations, baseline sera, or health evaluations required for individuals working with select agents. This information is often confidential and is kept by the occupational health service.
- FSAP inspection reports: These reports, including the response to deficiencies and all follow-up documentation, should be maintained by the RO.
- Drills and exercises: These interchangeable terms refer to activities conducted by an institution that ensure that the biosafety, incident response, and security plans are adequate. Documentation of such activities, although not required, provides evidence that the institution has engaged in these drills or exercises.

- Commissioning and verification: Commissioning is a process that ensures that all systems in a facility perform as designed. The original commissioning document for animal biosafety levels (ABSLs) 3 and 4 facilities and subsequent annual verification documentation should be maintained.
- Equipment records: Maintenance records for biological safety cabinets, autoclaves, digesters, liquid effluent decontamination systems, high-efficiency particulate air (HEPA) filters, exhaust fans, and other lab equipment should be maintained. Also, sterilization verification should be kept for autoclaves, and records of decontamination verification should be maintained for digesters and liquid effluent decontamination systems.
- Internal transfer records: Transfer of select agents within an institution (e.g., from one laboratory to another) should be maintained. An internal transfer form may be developed for this purpose (note that APHIS/CDC Form 2 is not required for internal transfers).
- Restricted experiments: These are experiments that require review and approval by FSAP before they may be conducted. Specific information regarding these types of experiments may be found in Section 13 of the select agent regulations (CDC and USDA/APHIS 2014a) and in the *Responsible Official Resource Manual* (CDC and USDA/APHIS 2013b). All documentation, including the letter requesting the FSAP review, supportive documents, and related correspondence, should be maintained.

REFERENCES

AAALAC International. 2016. Available at http://www.aaalac.org/ (accessed July 29, 2016).

AVMA (American Veterinary Medical Association). 2013. *AVMA Guidelines for the Euthanasia of Animals: 2013 Edition*. Available at https://www.avma.org/KB/Policies/Documents/euthanasia.pdf (accessed July 22, 2016).

Braun CE, ed. 2005. *Techniques for Wildlife Investigations and Management*. 6th ed. Bethesda MD: The Wildlife Society.

Capterra. 2016. Top compliance software products. Available at http://www.capterra.com/compliance-software/ (accessed July 29, 2016).

CDC (Centers for Disease Control and Prevention) and USDA/APHIS (U.S. Department of Agriculture/Animal and Plant Health Inspection Service). 2013a. Occupational Health Program guidance document for working with Tier 1 select agents and toxins. Available at http://www.selectagents.gov/resources/Occupational_Health_Program_Guidance_v3-English.pdf (accessed July 30, 2016).

CDC (Centers for Disease Control and Prevention) and USDA/APHIS (U.S. Department of Agriculture/Animal and Plant Health Inspection Service). 2013b. *Responsible Official Resource Manual*. Available at http://www.selectagents.gov/resources/RO_Manual_2014.pdf (accessed July 30, 2016).

CDC (Centers for Disease Control and Prevention) and USDA/APHIS (U.S. Department of Agriculture/Animal and Plant Health Inspection Service). 2014a. Federal Select Agent Program. Title 42, Chapter 1, Subchapter F, Part 73. Available at https://www.ecfr.gov/cgi-bin/retrieveECFR?gp=&SID=8a4be60456973b5ec6bef5dfeaffd49a&r=PART&n=42y1.0.1.6.61 (accessed April 7, 2017).

CDC (Centers for Disease Control and Prevention) and USDA/APHIS (U.S. Department of Agriculture/Animal and Plant Health Inspection Service). 2014b. Federal Select Agent Program, select agents and toxins list. Available at http://www.selectagents.gov/SelectAgentsandToxinsList.html (accessed July 30, 2016).

CFR (Code of Federal Regulations). 2016a. Maintenance of records and inventories. Title 21, Part 1304, 1304.04. Available at http://www.deadiversion.usdoj.gov/21cfr/cfr/1304/1304_04.htm (accessed July 29, 2016).

CFR (Code of Federal Regulations). 2016b. Food and drugs: Good laboratory practice for nonclinical laboratory studies. Title 21, Part 58. Available at https://www.accessdata.fda.gov/scripts/cdrh/cfdocs/cfcfr/CFRSearch.cfm?CFRPart=58 (accessed July 29, 2016).

CFR (Code of Federal Regulations). 2016c. Food and drugs: Current good manufacturing practice for finished pharmaceuticals. Title 21, Part 211. Available at https://www.accessdata.fda.gov/scripts/cdrh/cfdocs/cfcfr/CFRSearch.cfm?CFRPart=211 (accessed July 29, 2016).

Council of Europe. 1986. European Convention for the Protection of Vertebrate Animals Used for Experimental and Other Scientific Purposes. European Treaty Series 123. Strasbourg, France: Council of Europe.

Dale WE. 2008. Postapproval monitoring and the role of the compliance office. *ILAR J* 49:393–401.

Dyke B. 1993. Basic data standards for primate colonies. *Am J Primatol* 29:125–143.

European Parliament and the Council of the European Union. 2010. Directive 2010/63/EU of the European Parliament and of the Council of 22 September 2010 on the protection of animals used for scientific purposes. *Off J Eur Union* L276/33–79.

FASS (Federation of Animal Science Societies). 2010. *Guide for the Care and Use of Agricultural Animals in Research and Teaching*. 3rd ed. Champaign, IL: FASS.

FDA (Food and Drug Administration). 2016. Bioresearch monitoring good laboratory practice. Compliance Program 7348.808 (Nonclinical Laboratories), Attachment A, Computerized Systems. Available at http://www.fda.gov/downloads/ICECI/EnforcementActions/BioresearchMonitoring/UCM133765.pdf (accessed July 29, 2016).

Field K, Bailey M, Foresman LL, Harris RL, Motzel SL, Rockar RA, Ruble G, Suckow MA. 2007. Medical records for animals used in research, teaching, and testing: Public statement from the American College of Laboratory Animal Medicine. *ILAR* J 48:37–41.

Haywood JR, Greene M. 2008. Animal use and oversight: Balancing risk management with burden. *ILAR J* 49 (4):426–434.

Health Research Extension Act. 1985. P.L. 99-158. Approved November 20, 1985.

Heiderstadt KM, McLaughlin RM, Wright DC, Walker SE, Gomez-Sanchez CE. 2000. The effect of chronic food and water restriction on open-field behaviour and serum corticosterone levels in rats. *Lab Anim* 34:20–28.

ILAR (Institute for Laboratory Animal Research), National Research Council. 2000. Humane endpoints for animals used in biomedical research and testing. *ILAR J* 41:59–123.

NIH (National Institutes of Health). 2008. Guidelines for the use of non-pharmaceutical grade compounds in laboratory animals. Bethesda, MD: Animal Research Advisory Committee, Office of Animal Care and Use, NIH. Available at http://oacu.od.nih.gov/ARAC/documents/Pharmaceutical_Compounds.pdf (accessed July 24, 2016).

NIH (National Institutes of Health). 2015. Public Health Service policy on humane care and use of laboratory animals. Available at http://grants.nih.gov/grants/OLAW/references/phspol.htm (accessed July 24, 2016).

NIH (National Institutes of Health). 2016. NIH guidelines for research involving recombinant or synthetic nucleic acid molecules (NIH guidelines). Department of Health and Human Services. Available at http://osp.od.nih.gov/sites/default/files/NIH_Guidelines.html (accessed July 24, 2016).

NRC (National Research Council). 1979. *Laboratory Animal Records*. Washington, DC: National Academy Press.

NRC (National Research Council). 2003. *Guidelines for the Care and Use of Mammals in Neuroscience and Behavioral Research*. Washington, DC: National Academies Press.

NRC (National Research Council). 2011. *Guide for the Care and Use of Laboratory Animals*. 8th ed. Washington, DC: National Academy Press.

Oechslein C. 2015. *GMP Fundamentals—A Step-by-Step Guide for Good Manufacturing*. Schopfheim: Maas & Peither AG GMP Publishing.

OLAW (Office of Laboratory Animal Welfare). 2012. Use of non-pharmaceutical-grade chemicals and other substances in research with animals. Available at http://grants.nih.gov/grants/olaw/120301_seminar_transcript.pdf (accessed July 24, 2016).

Public Health Service. 2013. Policy on humane care and use of laboratory animals—Frequently asked questions. IACUC Composition, Functions and Authority, Question No. B.7. Washington, DC: U.S. Department of Health and Human Services.

Rowland NE. 2007. Food or fluid restriction in common laboratory animals: Balancing welfare considerations with scientific inquiry. *Comp Med* 57:149–160.

Suckow MA, Doerning BJ. 2007. Assessment of veterinary care. In Silverman J, Suckow MA, Murthy S, eds., *The IACUC Handbook*. 2nd ed. Boca Raton, FL: CRC Press.

USDA (U.S. Department of Agriculture). 2015. Animal Welfare Act and Animal Welfare Regulations. Available at https://www.aphis.usda.gov/animal_welfare/downloads/Animal%20Care%20Blue%20Book%20-%202013%20-%20FINAL.pdf (accessed July 22, 2016).

USDA (U.S. Department of Agriculture). 2016. Veterinary care (March 14, 2014). APHIS Policy #3. Available at https://www.aphis.usda.gov/animal_welfare/downloads/Animal%20Care%20Policy%20Manual.pdf (accessed July 24, 2016).

16

Existing and Emerging Information Technology

Thomas R. Meier and Kenneth R. Boschert

CONTENTS

Introduction

Information technology (IT) is the technology involving the development, maintenance, and use of computer systems, software, and networks for the processing and distribution of data. IT management is the discipline in various forms of how IT is managed in accordance with priorities and needs of the business or organization. IT can be used to support many aspects of animal care and use programs, including the Institutional Animal Care and Use Committee (IACUC) and animal care departments responsible for husbandry and veterinary care.

IT in the IACUC Office

As animal care programs, regulatory expectations, and IACUC processes have become increasingly complex, the utilization of IT to track, manage, and report on programmatic elements has become commonplace. Institutions typically purchase a commercial software package or utilize their own IT department to develop a custom application to support the animal care program. Commercially available IT solutions may require customization to meet the individual program needs. Institutions are encouraged to evaluate both in-house and external options when selecting an animal program management software suite.

Several IACUC office functions benefit from an electronic interface. Protocol authorship, modification, and submission through electronic systems have become quite common. Depending on the size of the program and volume of protocols, the utilization of IT in the protocol review and submission process may range from very simple to extremely complex. For example, a small program with relatively few protocols may be able to leverage the typical suite of office tools (Word®, Excel®, and Adobe Acrobat®), with e-mail communication to effectively manage the program. Advanced editing features, such as tracking changes and adding comments to documents, can become a component of the protocol review process. The disadvantage of this simple approach is that the manual efforts required to move protocols through the different phases of review are not automated. In addition, the content and classification of particular studies required for reporting to external agencies become burdensome due to the manual efforts of tracking information.

In larger, more complex programs, the utilization of a system to manage this information is critical to help reduce administrative burden. Commercially available and in-house developed programs can provide a secure, web-based method of submitting, modifying, reviewing, and approving IACUC submissions. Furthermore, data within the protocols can be summarized in reports (e.g., studies involving hazards, locations in which animals are utilized, and U.S. Department of Agriculture [USDA] pain categories). A well-constructed protocol form and reporting system can assist the animal care program in tracking and generating metrics pertaining to the program. It can also expedite routine tasks for the IACUC office, such as generating meeting minutes, excerpts, approval letters, and inspection reports.

IACUC protocols may include the use of biohazardous, environmental, or radiation hazards, and these often require external review by a related safety committee or officer. It may be advantageous to expand the IACUC protocol review process system to allow this outside review to occur in the same system. For example, linking the IACUC protocol to the Institutional Biosafety Committee (IBC) review process can reduce delays in review (e.g., even allow for simultaneous review by the IACUC and the particular external review group). It also helps to ensure congruency between submissions.

The IACUC is responsible for ensuring that personnel are adequately trained or have experience in the work being proposed. Regardless of whether the training is in a hands-on, web-based, or lecture format, tracking the completion of training or documenting competencies electronically facilitates the review of training and competency by the IACUC. The system used to track training does not necessarily need to be inside the IACUC protocol, if the two can be tied together through IT. The Collaborative Institutional Training Initiative (CITI) and American Association for Laboratory Animal Science (AALAS) Learning Library are both commercial electronic training solutions that support data migration and linkage to IACUC systems. The data feed requires some additional programming and internal support by the institution's IT department. This provides the IACUC the ability to review personnel qualifications during protocol review, as well as assign and record training by staff.

Additional activities that can benefit from linking electronic systems include protocol–grant congruency and participation in the occupational health program for personnel. Institutions conducting federally funded projects are expected to review the grant to ensure that the associated IACUC protocol is congruent. This is facilitated if the grant can be associated with the protocol electronically. Last, while personnel medical records are confidential, the IACUC protocol can help identify personnel that may require additional evaluation by the occupational health and safety department.

The ability to run reports and summarize the animal care program is essential. Organizations are required to report information about the animal care program to outside regulatory and accrediting agencies on a regular basis. A well-constructed protocol form with logically developed reports can generate an organization's annual USDA report, appendices, and tables for AAALAC program descriptions, and facilitate the inspection process by identifying areas in the program that require semiannual inspections. Specific reporting requirements for animal care programs are covered in Chapter 7.

Emerging IT in support of the IACUC function includes the use of tablets and mobile database software to facilitate semiannual inspections.

IT in the Animal Care Department

As technology improves, the utilization of IT-based solutions for tracking information is growing in animal care departments. Animal program management software used by the IACUC often includes options for animal ordering, census, billing, and even electronic medical records.

The use of radio frequency identification (RFID) microchips or bar codes as a method of "counting" cages or animals for census or billing purposes has become commonplace. These methods reduce human error and can reduce the manual effort in collecting billing information. In today's environment, animal care programs may have tens of thousands of rodent cages assigned to various investigators, under different IACUC protocols that are funded by different sources. Electronic identification of each cage helps ensure that the correct principal investigator (PI) and funding source is charged for services. Electronic census, as this process is commonly referred to, is integrated into the animal facility management software for billing purposes. In addition, the animal care department can use this information to evaluate volumes of cages and project growth in various housing types and across areas of the vivarium.

Electronic census still requires staff to perform census counts and does not replace the need for periodic manual reconciliation of cage counts. Electronic census has the advantages of improving the accuracy of billing, expediting the counting process, and reducing human error versus hand-counting cages. However, it does not necessarily reduce the number of staff required to support census-related activities.

Another common component of animal facility management software facilitates the animal ordering process. Web-based or software applications allow research staff to order animals for a particular protocol. The application also gives investigators control over the quantity and time that specific animals are delivered to a chosen animal holding room in a particular facility. Depending on the complexity of the ordering system, investigative staff can choose the particular strain, vendor, age, and/or weight of animals for their study. This information is typically relayed to the animal care department, which supports the procurement process. Current electronic ordering systems are capable of facilitating communication by sending automated messages to study staff and facility managers when animals are ordered and received. This provides facility managers the opportunity to verify that adequate housing exists prior to approval or shipment of the order.

Electronic medical record systems provide a systematic method of documenting clinical observations, treatments, surgical procedures, and preventive medicine activities. Electronic medical records can hold diagnostic information, such as radiograph or clinical images, as well as clinical notes by the veterinary and research staff. The systems also permit multiple users to view a record from different locations simultaneously. The emergence of tablets and mobile devices further expands the range of the utility of electronic medical records, as a desktop computer is no longer a prerequisite for access.

Electronic records also bring the advantages of not occupying physical space for storage and archiving of records. Institutions using paper-based medical records may consider electronic archiving of old records through document scanning. Desktop scanners are currently capable of scanning more than 40 double-sided pages per minute (in color), making this a reasonable consideration for archiving paper records of any type. Scanning software is also capable of optical character recognition (OCR) to recognize text in documents, thus making the contents of these documents searchable for easier retrieval. OCR is available through scanning software and Portable Document Format (PDF) applications, such as Adobe Acrobat Professional®. It should be noted that OCR is designed to recognize type-written text versus an individual's handwriting; therefore, its application is best aimed toward printed fonts versus handwritten material.

While not technically a component of the animal's medical record, documentation of animal enrichment and social housing records represents an emerging need in animal care programs. The authors expect that the ability to document social housing groupings and pairings and enrichment provisions has already been, or will soon be, incorporated into electronic medical record systems developed to support the current animal research facility. Paper-based records for fish and rodent colonies have always posed a challenge for animal care programs to manage effectively. The use of wireless Internet-enabled mobile devices to access and update medical records eliminates many of the challenges with large volumes of records.

Complex breeding records for transgenic and knockout mice may be tracked and maintained via several animal colony management databases (Allwood et al. 2015). These systems may help researchers improve efficiency and reduce animal numbers. Various software packages, either free or commercially available, may provide real-time information and track animal events, as well as provide searching and other tools for the maintenance of cage cards, pedigree, mating, and litter information.

In today's environment, standard operating procedures (SOPs) are typically in an electronic format. These documents are often based on a Microsoft Word format and can be converted to PDF to prevent unauthorized editing.

Content management systems are applications (often web based) used to "manage the content" of a variety of different online resources. Content management systems are utilized in support of websites to keep information organized and current. They can allow authorized users access to publish information to the website for instantaneous updates to the site's content. In addition, content management systems can be used to create fully searchable libraries of SOPs. Such systems can trigger electronic notifications when SOPs are due for regular review, and they often include a review process to ensure proper document control.

Essentially every event in the animal facility is subject to being tracked or measured by some means. Simple examples utilized by the authors include monitoring the incidences of automated watering valve failure and managing a schedule for health monitoring testing. By putting information into spreadsheets, databases, and other systems, the organization of data becomes easier to manage. In turn, the summation of large amounts of data becomes easier to accomplish.

Communication and Collaboration in the Current Environment

A server is essentially one or more computers linked together on a network that are capable of "serving up" client requests (i.e., requests for information) or storage information on a remote and secure drive. Web servers, for example, allow for the hosting of web pages on an internal or external site (i.e., hosting a website). Data servers often consist of entire rooms of linked computers that are used to store data. One primary advantage of saving data on a server is that the institution's IT department may perform regular backup of this information. Access to servers or network storage is managed by setting individual user permissions to determine what data is accessible. Servers are also considered to be fairly secure since the server is usually only available through the institution or a virtual private network (VPN) connection to the institution. The centralization of data creates the necessity for safeguards to be in place to protect this information. This can be achieved by managing which user accounts have access to what data, whether their access is read only versus read/write, as well as ensuring that there is an adequate backup plan in case of system failure.

Hard drives and USB drives are considered "local" storage. The data accessed on the hard drive or USB drive is only available to the individual using the computer. The use of local storage modalities for secure data is not recommended due to the fact that the risk of theft of the drive and/or failure of the drive can result in total data loss. USB drives with encryption and password protection are available to address data security; however, this does not help the user in the event of loss or theft.

Cloud storage (e.g., Google Drive®, Dropbox®, and Microsoft OneDrive®) represents existing and emerging methods of storing and sharing data "in the cloud." In all cases, the data is stored on remote servers and accessible through integrated software or web-based applications (Aharony 2014; Goldsborough 2015). Google's approach to cloud-based computing is entirely web based. Google Drive, Docs, and Sheets represent web-based applications for online document creation, editing, sharing, and collaboration. Users can create a document and share with others for collaboration through granting-sharing permissions. Institutions should look to their IT department for guidance on whether cloud storage solutions are recommended or supported. Any product that allows users to share information with others is at risk of inadvertent dissemination of private data.

Dropbox differs from Google Drive in the respect that a local version of the files in the cloud drive is maintained on the user's computer in a folder called "Dropbox." The Dropbox application runs in the background and monitors the Dropbox folder and subfolders for changes. If a file is added, removed, or changed, the respective changes are uploaded to the Dropbox server and subsequently synchronized to any other computer the account owner has authorized to use the application. Dropbox users can share documents and folders within their Dropbox with other users to facilitate everything from document collaboration to sharing photos with family.

OneDrive is Microsoft's solution for online cloud storage. It functions very similar to Dropbox in that the user can maintain their OneDrive folder on their local computer, which synchronizes with the cloud and other devices with OneDrive installed. The competitive advantage of OneDrive is simple integration into Microsoft Word, Excel, PowerPoint®, and other Microsoft applications that can save directly to the cloud server without the use of a local folder.

These are but a few of the more common and popular cloud-based solutions for file storage and collaboration. All offer free accounts with a limited amount of storage, with options to increase storage for a fee. In addition, all these cloud-based methods of storage are commonly integrated into mobile device applications, giving the user access to their Dropbox, Google Drive, or OneDrive from their smartphone or tablet. There are several factors that impact end users' adoption to cloud computing, many of which are end user characteristics. These include perceived ease of use, personal innovativeness, threat or challenge, self-efficacy, and openness to experience.

The acceleration of smartphone and tablet technology has revolutionized the way information is gathered, shared, and communicated among peers. These devices facilitate calendaring, web access, task management, electronic books, and instant messaging in a device once limited to voice calls. Current smartphones can provide location information using global positioning system (GPS), capture quality images, record high-definition video and audio, and transmit this data through Bluetooth, Wi-Fi, or cellular data connections. Mobile devices also represent a viable videoconferencing option. The possibilities for utilization of mobile devices are still growing, as the catalog of available apps is continually expanding.

The use of videoconferences and teleconferences has existed for some time. Videoconferences can include live video or audio of participants and the ability to share the view of one person's computer screen with others over a secure connection. The combination of broadcasting the audio and screen of a presenter to one or many viewing participants is the definition of a webinar, where a presentation is hosted and transmitted to many viewers. Webinars give the user an opportunity to participate in a lecture or presentation from the comfort of their office. Several oversight and educational bodies, such as the Office of Laboratory Animal Welfare (OLAW), National Association for Biomedical Research (NABR), AAALAC, AALAS, and USDA, host webinars targeting animal care and use programs for the dissemination of information on timely topics, regulatory updates, and other helpful information for animal care program managers.

Information Technology Security

Keeping data secure and protected is a field that is evolving as fast as IT itself. We refer the reader to Chapter 19 for further discussion of IT security.

Social Media

Social media, by a variety of electronic tools and platforms, serves as a means of communication and collaboration. Social media forums provide space for users to have conversations, share information, and create web content. There are many platforms for social media, including blogs, microblogs, wikis, social networking sites, instant messaging, photo-sharing sites, video-sharing sites, podcasts, and virtual worlds. Earlier Internet tools, such as e-mail and LISTSERV® mailing lists, continue to provide solid communication venues for laboratory animal professionals. We live in a time of great innovation, and undoubtedly, different methods will arrive on the scene to connect us in new and exciting ways.

Many people are interested in developing and maintaining a social media presence in personal and professional capacities. When used appropriately, social media can be a useful adjunct to a department's or individual's communications approach. On an individual level, social media can contribute to productive task-oriented and relationship-building behavior. Using social media makes it possible to share information and make connections around the campus or around the world. On a personal level, social media allows you to communicate with family and friends, learn new concepts, develop your interests, or be entertained. On a professional level, social media can expand your knowledge in a particular field and allow you to connect with other professionals that have common interests. At a company level, social media allows you to have a conversation with your audience, gain customer feedback, recruit new employees, and elevate your brand. You can also use social media to control your message or provide correct information quickly.

Unfortunately, social media can also support deviant behavior (Carlson et al. 2016). Social media allows for users to share photos, videos, thoughts, and feelings with a large networked audience in ways that have not existed before. The repercussions for some actions, whether malicious or accidental, can be significant and nearly impossible to control. Responsibility for what is posted by a user is a personal responsibility. Institutions are encouraged to create social media policies or guidelines for staff.

In academic settings, some IT departments make internal tools or common platforms available for their users (Sharepoint®, Wordpress®, and Wordsphere®). If you wish to establish a social media presence in the public domain, the choices available are numerous, although not all social media sites will be worthwhile or appropriate for your department's brand or goals.

Before launching an official account on a new social media site for your department, you may want to consider trying it on a personal level. Create an account for yourself and use it. Observe how others use the site and the frequency, type, and popularity of content being generated. Look at how your department would fit in. An effective strategy would be to master one or two sites and avoid diluting your social media efforts. Choose and use a platform that allows you to share your content with the appropriate audience.

Presently, some of the more common web-based and mobile app social media platforms include Facebook®, Tumblr®, Twitter®, LinkedIn®, and YouTube®.

Facebook (http://www.facebook.com) is a social networking site that allows people from around the world to network with friends, companies, and organizations. In the authors' opinion, departments looking to build overall brand awareness should consider using Facebook. You can post new content, including videos, photos, events, and links to news stories.

Tumblr (http://www.tumblr.com), a blogging site, allows users to post text, images, videos, links, quotes, and audio. It is especially useful for pictures, videos, and longer written content.

Twitter (http://www.twitter.com), a microblogging site, allows people to post messages ("tweets") limited to 140 characters or less. Those looking to engage an audience at a high frequency and have the resources to respond promptly should consider using Twitter. It is useful for posting information about events, research, news stories, and more.

LinkedIn (http://www.linkedin.com), a business-related social networking site, is used mainly for professional networking. It is very useful to connect with current employees or alumni via creation of a LinkedIn group. Most often, it is used for job postings, describing academic programs and research projects, and announcing meetings and other events.

Several lab animal–related organizations maintain an online presence on LinkedIn, including AALAS, the American College of Laboratory Animal Medicine (ACLAM), the American Society of Laboratory Animal Practitioners (ASLAP), and the Academy of Surgical Research. Informal discussion groups are also present, including several relevant to lab animal professionals and veterinarians. These are easily discoverable via a group search on the LinkedIn main site (Baran 2012).

YouTube (http://www.youtube.com) is a massive video-sharing site run by Google. Those producing a large volume of video content should consider creating a YouTube channel to publish their availability.

With all social media efforts, keep in mind the sensitive nature of animal research. While you can maintain a presence on any social media platform, it is important to maintain discretion when deciding what might be posted for public consumption.

Other social media services that may appeal to specialty audiences include

- ResearchGate (http://www.researchgate.net/), a service that connects researchers and allows them to share and access scientific output, knowledge, and expertise
- Academia (https://www.academia.edu/), which provides a method of sharing scientific papers and improving citation relevance

E-Mail

Before the presence of web and mobile applications, e-mail was the predominant method of communication. E-mail arguably remains the most common method to communicate among lab animal professionals and those in science. E-mail distribution servers (i.e., LISTSERV) have long been available whereby groups with a common interest can send and receive e-mail from subscribers to that group. Some remain very popular and have thousands of subscribers.

AALAS (http://www.aalas.org) sponsors several mailing lists of interest for lab animal professionals, including Compmed, Techlink, IACUC-Forum, and Laboratory Animal Welfare Training Exchange (LAWTE). Other popular listserv mailing lists include ProMED (http://www.promedmail.org), Gnotobiotics (https://listserv.uab.edu/scgi-bin/wa?A0=GNOTOBIOTICS), MGI-List (http://www.informatics.jax.org/mgihome/lists/lists.shtml), Rat Community Forum (http://mailman.mcw.edu/mailman/listinfo/rat-forum), Embryo Mail (http://embryomail.net), and VetProf (http://www.vetprof.com).

A number of groups provide informative e-mail newsletters covering a wide variety of medical, scientific, and management topics, including Americans for Medical Progress (AMP) (http://www.amprogress.org), NABR (http://www.nabr.org), Foundation for Biomedical Research (FBR) (http://www.fbresearch.org), American Veterinary Medical Association (AVMA) (http://www.avma.org), AAALAC (http://www.aaalac.org), Davis Thompson Foundation (http://www.cldavis.org), Jackson Laboratory (http://www.jax.org), Charles River (http://www.criver.com), Taconic (http://www.taconic.com), Envigo (http://www.envigo.com), Covance (http://www.covance.com/), and ALN (http://www.alnmag.com) and *Lab Animal* (http://www.labanimal.com) magazines. Several government agencies (https://www.govdelivery.com) also provide e-mail notifications of their activities, including OLAW (http://grants.nih.gov/grants/olaw/references/list.htm), the Food and Drug Administration (FDA) (http://www.fda.gov/), the National Institutes of Health (NIH) (https://www.nih.gov/), the Centers for Disease Control and Prevention (CDC) (http://www.cdc.gov/), and the USDA (http://www.usda.gov/).

Useful Websites for Laboratory Animal Professionals

There are seemingly an infinite number of websites, which come and go each day, on the World Wide Web. While browser software applications such as Google® or Bing® provide great starting points for web searches, a number of specialty websites and databases may prove useful for laboratory animal professionals.

AALAS Learning Library (https://www.aalaslearninglibrary.org): Provides training for technicians, veterinarians, managers, IACUC members, and investigators

ACVP—American College of Veterinary Pathologists (http://www.acvp.org): Promotes the advancement and sharing of knowledge, lifelong learning, and professional competency through its certifying examinations, educational programs, and journal

AFLAS—Asian Federation of Laboratory Animal Science Associations (http://www.aflas-office.org): Promotes through international cooperation Asian Congresses on Laboratory Animal Science for the purpose of reviewing scientific, technical, and educational problems in laboratory animal science

Agricola (http://agricola.nal.usda.gov): USDA database that provides citations to agricultural literature

Altweb (http://altweb.jhsph.edu): Clearinghouse for information on alternatives to animal testing

AnimalResearch.Info (http://www.animalresearch.info): An international collaboration of scientists and researchers to provide reliable, detailed information about when and why it is appropriate to use animals, and the history of this area of research

ANZLAA—Australian and New Zealand Laboratory Animal Association (http://www.anzlaa.org): Association of professional experts involved with the care and welfare of animals used in research in Australia and New Zealand

APV—Association of Primate Veterinarians (https://www.primatevets.org): International organization concerned with the health, care, and welfare of nonhuman primates

ASR—Academy of Surgical Research (http://surgicalresearch.org): Promotes the advancement of professional and academic standards, education, and research in the arts and sciences of experimental surgery

BioGRID—Biological General Repository for Interaction Datasets (http://thebiogrid.org): Public database that archives and disseminates genetic and protein interaction data from model organisms and humans

CALAS—Canadian Association for Laboratory Animal Science (http://calas-acsal.org): Association dedicated to providing high-quality training and educational resources to animal care professionals across Canada

CCAC—Canadian Council on Animal Care (http://www.ccac.ca): National peer review organization responsible for setting, maintaining, and overseeing the implementation of high standards for animal ethics and care in science throughout Canada

CDC—Centers for Disease Control and Prevention (http://www.cdc.gov): Conducts critical science and provides health information to protect against health threats

Digires—Digital Resources for Veterinary Trainers (http://www.digires.co.uk): Produces digital material for those involved in the training of research workers, veterinarians, and others who work with animals

EARA—European Animal Research Association (http://eara.eu): Communications and advocacy organization whose mission is to uphold the interests of biomedical research and health care development across Europe

eMICE (http://emice.nci.nih.gov): Provides information, communication, and education about the wide variety of animal cancer models

Enrichment Record (http://enrichmentrecord.com): Engages the lab animal community in an informed discussion about all aspects of enrichment for laboratory animals

EurekAlert! (http://www.eurekalert.org): Online global news service operated by the American Association for the Advancement of Science (AAAS)

FELASA—Federation of European Laboratory Animal Science Associations (http://www.felasa.eu): Represents common interests in the furtherance of all aspects of laboratory animal science in Europe

FLAIRE Learning—Flecknell Laboratory Animal Interactive Resources (https://flairelearning.com): Interactive resources to support the training of laboratory animal research workers and to continue the professional development of others who work with laboratory animals

FOAVet—Free open access veterinary knowledge sharing and collaboration (http://www.dvm.com): Promotes free knowledge sharing and collaboration for veterinary professionals

IAA—IACUC Administrators Association (http://iacucaa.org): Organization that facilitates opportunities for IACUC administrative professionals to share successful processes and ideas, as well as providing a venue to discuss emerging problems

IACLAM—International Association of Colleges of Laboratory Animal Medicine (http://www.iaclam.org): Provides a common platform at the global level for communication and representation of these colleges and their diplomates

IACUC Central (http://iacuc.org): Sponsored by AALAS, a comprehensive repository for all things IACUC

ICLAS—International Council for Laboratory Animal Science (http://iclas.org): International scientific organization dedicated to advancing human and animal health by promoting the ethical care and use of laboratory animals in research worldwide

ILAR—Institute for Laboratory Animal Research (http://dels.nas.edu/ilar): A unit in the Division on Earth and Life Studies of the National Research Council, National Academy of Sciences; evaluates and reports on scientific, technological, and ethical use of animals and related biological resources, and of nonanimal alternatives in nonfood settings, such as research, testing, education, and production of pharmaceuticals

ILAR Journal (http://ilarjournal.oxfordjournals.org): Publication of the ILAR, which provides timely information for all who use, care for, and oversee the use of animals in research

International Mouse Strain Resource (http://www.findmice.org): Searchable online database of mouse strains, stocks, and mutant embryonic stem (ES) cell lines available worldwide, including inbred, mutant, and genetically engineered strains

Interspecies Differences Database (https://www.interspeciesinfo.com): Helps researchers make an optimal choice of species and strain of animal model

IVIS—International Veterinary Information Service (http://www.ivis.org): Provides information to veterinarians, veterinary students, technicians, and animal health professionals worldwide with free access to electronic books, proceedings of veterinary meetings, journals, short courses, and an international calendar of veterinary events

JoVE—*Journal of Visualized Experiments* (http://www.jove.com): A PubMed-indexed video journal with the goal of increasing the productivity of scientific research by publishing scientific research in a visual format

KOMP Repository—Knockout Mouse Project (https://www.komp.org): Trans-NIH initiative that aims to generate a comprehensive and public resource comprised of mouse ES cells containing a null mutation in every gene in the mouse genome

Laboratory Animals Limited (http://www.lal.org.uk): Registered charity that promotes education and training in laboratory animal science; publishes *Laboratory Animals* journal

LAMA—Laboratory Animal Management Association (http://www.lama-online.org): Association dedicated to advancing the quality of management and care of laboratory animals throughout the world

LAWTE—Laboratory Animal Welfare Training Exchange (http://www.lawte.org): Global organization that is expanding animal welfare and enhancing public understanding through effective training and education of animal research professionals

Macaque website (http://www.nc3rs.org.uk/macaques): Sponsored by the National Centre for the Replacement, Refinement & Reduction of Animals in Research (NC3Rs) and covers the life history of rhesus and cynomolgus macaques, their behavior, captive management, and tools for welfare assessment

Merck Veterinary Manual (http://www.merckvetmanual.com): Contains guidelines for the diagnosis, treatment, and prevention of animal disorders and diseases

MMRRC—Mutant Mouse Resource & Research Centers (https://www.mmrrc.org): Distributes and cryopreserves scientifically valuable, genetically engineered mouse strains and mouse ES cell lines with potential value for the genetics and biomedical research community

Monarch Initiative (https://monarchinitiative.org): Integrates, aligns, and redistributes cross-species gene, genotype, variant, disease, and phenotype data

MRB (Mouse Resource Browser) (http://bioit.fleming.gr/mrb): Resource management project that provides a dynamic and interactive view of worldwide available mouse resources

National Academies Press (http://www.nap.edu): Publishes more than 200 books a year on a wide range of topics in science, engineering, and medicine, providing authoritative information on important matters in science and health policy

Netvet Veterinary Resources (http://netvet.wustl.edu): Large collection of links to veterinary medical and animal-related resources

NIH ORIP—Office of Research Infrastructure Programs (http://dpcpsi.nih.gov/orip): Provides the research infrastructure and related research programs, and coordinates NIH's science education efforts

Norina (http://oslovet.norecopa.no): Provides information on laboratory animal science and alternatives to the use of animals in research, teaching, and school dissection classes

One Health Initiative (http://www.onehealthinitiative.com): Worldwide strategy for expanding interdisciplinary collaborations and communications in all aspects of health care for humans, animals, and the environment

PRIM&R—Public Responsibility in Medicine and Research (http://www.primr.org): Community of ethics-minded research administration and oversight personnel that also provides educational and professional development opportunities

Primate Info Net (http://pin.primate.wisc.edu): Covers the field of primatology, providing original content and links to resources about nonhuman primates in research, education, and conservation

PRRS—Primate Resource Referral Service (http://prrs.wanprc.org): Provides a communications and database network for efficient acquisition and sharing of existing captive primates and primate-related resources

PubMed (http://www.ncbi.nlm.nih.gov/pubmed): Database containing more than 25 million citations for biomedical literature from MEDLINE®, life science journals, and online books

RGD—Rat Genome Database (http://rgd.mcw.edu): Site for genetic, genomic, phenotype, and disease data generated from rat research

SCAW—Scientists Center for Animal Welfare (http://www.scaw.com): Research professionals dedicated to balancing animal welfare and excellence in basic and applied scientific inquiry

ScienceDaily (http://www.sciencedaily.com): Popular science news website, updated frequently and freely accessible

SLAVT—Society of Laboratory Animal Veterinary Technicians (http://www.slavt.com): Network of professional veterinary technicians dedicated to the advancement of responsible and humane laboratory animal care and use to benefit humans and animals

Speaking of Research (http://speakingofresearch.com): Provides information about the importance of animal research and animal testing in medical and veterinary science

SUBR—States United for Biomedical Research (http://www.statesforbiomed.org): Network of nonprofit associations who promote health through science and education

Translational Microbiome Research Forum (http://www.translationalmicrobiome.org): Online resource for scientists engaged in translational microbiome research to access current, topical information and to provide a platform to exchange knowledge and ideas

UFAW—Universities Federation for Animal Welfare (http://www.ufaw.org.uk): Works with the animal welfare science community worldwide to develop and promote improvements in the welfare of farm, companion, laboratory, and captive wild animals

USDA Animal and Plant Health Inspection Service (https://www.aphis.usda.gov): Protects and promotes U.S. agricultural health, regulating genetically engineered organisms, administering the Animal Welfare Act, and carrying out wildlife damage management activities

USDA Animal Welfare Information Center (https://awic.nal.usda.gov): Provides information for improved animal care and use in research, testing, and teaching

VBI—Veterinary Bioscience Institute (http://vetbiotech.com): Provides experimental and veterinary surgical and biomethodology training

VetScite (http://vetscite.org): International current awareness journal for veterinary and related research scientists

VIN—Veterinary Information Network (http://www.vin.com): Large communications and information resource for the veterinary profession

ZFIN (Zebrafish Model Organism Database) (http://zfin.org): Online database of information for zebrafish researchers

REFERENCES

Aharony, N. 2014. Cloud computing: Information professionals' and educational technology experts' perspectives. *Library Hi Tech* 32 (4):645–666.

Allwood, M. A., D. Griffith, C. Allen, J. Reed, Q. H. Mahmoud, K. R. Brunt, and J. A. Simpson. 2015. Lennie: A smartphone application with novel implications for the management of animal colonies. *Laboratory Animals* 49 (3):258–262.

Baran, S. W. 2012. Fostering a virtual community for lab animal scientists. *Lab Animal* 41 (2):49.

Carlson, J. R., S. Zivnuska, R. B. Harris, K. J. Harris, and D. S. Carlson. 2016. Social media use in the workplace: A study of dual effects. *Journal of Organizational and End User Computing* 28 (1):15–31.

Goldsborough, R. 2015. The changing world of cloud storage. *Tech Directions* 75 (2):12.

17

Emergency Response and Management

Lynell M. Dupepe, John C. Donaho, and Gordon Roble

CONTENTS

Introduction

Emergency response and business continuity (BC) planning is an essential component of the research organization's program. A well-developed plan ensures the humane care and treatment of laboratory animals in a crisis situation. Animal facility management must contemplate the following questions: If our facility is faced with a crisis—natural or man-made—are we prepared? Are the local authorities aware of specialized facility needs? Does our response plan integrate with that of city, county, and state emergency plans? Are our essential personnel equipped to respond? Can we recover after the disaster? Taking responsibility for preparedness and response is vital to reduce potential damage and hasten recovery from any adverse event, large or small. Following a disaster, business resumption is the priority of the institution. The goal of comprehensive crisis planning is to create resiliency within an organization and increase the chance of a successful outcome when a disaster or emergency occurs.

Although there are distinctions between crisis, emergency, and disaster, for the purposes of this chapter, planning for any of the above comes under the heading *crisis*. This chapter uses the definition of *crisis* as provided by the Canadian Council on Animal Care (CCAC) (2016). Out of many, varied definitions, it most succinctly encompasses what the authors felt a crisis management plan should consider. The CCAC defines a crisis as "any unplanned event which triggers a real, perceived, or possible threat to the health and safety of animals or personnel or to the Institution's credibility" (Canadian Council on Animal Care 2016).

To delineate differences between an emergency and a disaster, it is helpful to consider the scope of response required for the event. A disaster is considered a calamitous event that requires a response beyond that available locally, such as state or federal assistance (International Federation of Red Cross and Red Crescent Societies). An emergency can be resolved with local resources. For example, an emergency is when a water supply line bursts, flooding one or more animal rooms, but cleanup and repair take place in a reasonably short period of time. This situation requires an immediate response, but it usually does not result in business disruption or relocation of staff. While a broken water line might be an emergent situation handled at the room or building level, a broken dam upstream from a town would be a disaster, as flood waters take out utilities, businesses, and homes. Because the scope of a disaster can be sizable, planning for crisis management and BC becomes more important.

Many administrators lump crisis planning and BC planning into one document. Although this is a practical solution, a distinction should be made between these two entities. Mortell and Nicholls (2013) describe crisis preparedness as the actions needed "to mitigate the impact of threats that can reasonably be predicted." Predicting the impact of these threats comes from appropriate crisis preparedness, as described later in this chapter. They further define BC planning as "the actions taken by an organization to ensure that critical business functions will be sustained in the event of a disaster." It may be easiest to consider the crisis response and recovery plan as actions taken at the time of the adverse event, as compared with the longer-term goals of a BC plan. BC planning expands on the crisis management recovery phase goals to ensure that business continues through the crisis and onward with minimal interruption. BC planning is discussed in further detail at the end of this chapter. Regardless of whether an institute specifically delineates these terms, the planning for both has significant overlap, and appropriately created plans will account for all details discussed in Chapter 18.

As described in the *New England Journal of Medicine*, the global biomedical research and development (R&D) expenditures in 2012 alone were $268.4 billion (Chakma et al. 2014). Emergency and BC planning will mitigate the escalating costs of recovery. Laboratory animal professionals have a moral and fiscal responsibility to protect the staff, animals, and research efforts at their respective organizations. Directors, veterinarians, facility managers, supervisors, and researchers are vital in creating a culture of preparedness within the animal facility. The goal of this chapter is to assist managers, veterinarians, supervisors, or other personnel in mitigating these crises with a well-thought-out plan and a well-prepared team. This chapter examines the concepts of risk assessment, mitigation, preparedness, response and recovery, and BC. A crisis is not the time to develop a plan; it is the time to execute a carefully constructed and *rehearsed* plan. It is not a question of if; it is a matter of when! When disaster comes knocking, will the institution be ready?

Crisis Management

Crisis management is a community affair, and for it to work well, private, local, state, and federal entities must execute their respective emergency management responsibilities. It is important to understand the organization of crisis response on a local, state, and federal level to integrate institutional plans with them effectively. Globally, emergency response and management systems are not standardized and differ widely from country to country. In the United States, for instance, emergency response is a state and local government responsibility and the federal government acts in a supportive manner, while the contrary is true in the United Kingdom. The European Union provides preplanning and disaster coordination among the member nations. While some Latin American and Caribbean countries rely almost exclusively on civil defense, the vast majority have adopted a decentralized approach, increasing the role of local governments. Although organizationally dissimilar, a common thread in all global systems is the need for comprehensive strategies that include prevention, mitigation, preparedness, and provisions for reconstruction and rehabilitation activities (Heath 1999).

Crisis Response Coordination

Customarily, entities such as institutional emergency response groups act as the liaison between the local agencies and the organization. The emergency response department of an institution may utilize an incident command model, as discussed later in this chapter, allowing communication between institutional departments and leadership, and local command centers. Depending on the level of the crisis, local and state plans may be enacted in tandem with the organizational or facility plan. The institutional plans should be shared with local police, fire, and other relevant municipal departments so that if the need arises, they may assist with recovery efforts (National Research Council 2011). For example, in the event of a mandatory evacuation of a city, consider the possibility of a municipal evacuation protocol where traffic flow of all interstate highways reverts to outbound traffic only. How will contraflow traffic patterns impact essential personnel trying to report to work or evacuation of an animal facility?

State governments are responsible for protecting communities and citizens within each state and carrying out statewide crisis management activities. They are also the liaison between resources available at the federal level and local level. Both state and local governments develop emergency plans for hazards that threaten their communities. These plans generally follow the format recommended by the Federal Emergency Management Agency (FEMA) and include mitigation activities, preparedness plans, response to emergencies, and recovery operations (FEMA 1996).

The Department of Homeland Security (DHS) and FEMA are federal agencies and play significant roles in disaster response. The DHS was created in response to the September 11, 2001, terrorist attacks in New York and Washington, DC, to protect U.S. territories from terrorist attacks and man-made and natural disasters. FEMA was created in 1979 and tasked with the mission to "lead America to prepare for, prevent, respond to and recover from disasters with a vision of 'A Nation Prepared'" (FEMA 2015), and is the principal federal agency responsible for responding to disasters. FEMA was placed under the umbrella of the DHS in 2003 and given the added responsibility to ensure that first responders are trained and equipped to deal with weapons of mass destruction. In 2006, the Post-Katrina Emergency Reform Act was signed into law, significantly reorganizing FEMA. Under this act, FEMA was given substantial new authority to remedy gaps that became apparent in the response to Hurricane Katrina in August 2005, thus giving the agency a more robust preparedness mission. In addition to emergency response and recovery activities, FEMA provides training programs and information on up-to-date mitigation measures, reviews and coordinates state emergency plans, and provides subsidies to state and local offices of emergency management for emergency management programs.

Animal Facility Crisis Response Planning

Regulatory Compliance

Common sense suggests that all animal facilities need an emergency plan. However, all institutions that maintain a written Animal Welfare Assurance describing compliance with the Public Health Service (PHS) policy (National Institutes of Health 2002) or AAALAC International accreditation, both in the United States and abroad, are required to develop and maintain current, updated emergency response plans (National Research Council 2011). The *Guide for the Care and Use of Laboratory Animals* (*Guide*) first introduced the importance of emergency planning in the seventh edition (1996) and expanded this requirement in the eighth edition (2011) as a mandatory recommendation. Office of Laboratory Animal Welfare (OLAW) guidance for developing an emergency plan can be found on the OLAW website (http://grants.nih.gov/grants/olaw/olaw.htm). These new standards are in direct response to the increase in worldwide catastrophes.

Developing the Plan

There are four phases to consider while constructing a crisis plan:

1. Risk assessment and mitigation
2. Preparedness
3. Response
4. Recovery

The first and most important is risk assessment and mitigation. Risk assessment is the foundation for everything else that follows because it identifies areas of weakness. Risk mitigation eliminates or reduces those areas of weakness or risks. The preparedness phase encompasses the planning and action to ensure human and animal safety and the capability of facilities. This is accomplished by putting people and materials in place to provide essential services and protections required during times of crisis. The

FIGURE 17.1 Emergency management circle of life.

response portion of the plan determines how quickly damage can be assessed, repaired, or replaced to allow recovery. Response is the phase during and immediately after a crisis. The response portion of the plan is critical but short-lived compared with the recovery phase. Planning for recovery and subsequent BC, the fourth phase, is often an afterthought. Both should have been considered and developed in the risk assessment and mitigation stage. Having a BC plan in place can ensure a much faster return to normal business operations. Any delay in recovery can spell failure in the long term (Figure 17.1).

Risk Assessment

Risk assessment considers the spectrum of potential internal and external human and naturally occurring impacts to a facility. It is a process where hazards and hazard effects are identified, the risks and vulnerabilities associated with those hazards are evaluated and scored, and mitigations to reduce the risks are sought. Hazards are anything that can cause harm to animals, people, or business operations. Hazards include obvious things like wind, fire, and flood. Other hazards might include disruption of normal animal care activities as a result of a pandemic or security breaches, such as break-ins or data loss. The list can be exhaustive, but identifying hazards is of utmost importance for the mitigation and preparedness phases (Figure 17.2).

Planning considers the results of a hazard, not necessarily the specific hazard. A fire or storm in a nearby neighborhood resulting in widespread power outage or evacuations due to hazardous fumes could disrupt operations in the animal facility. The real hazard in this case is loss of power or forced evacuation of the area, not the fire or storm. By considering the result and not the cause, the number of scenarios can be simplified because the response is the same whether the loss of power was caused by fire or flood. The goal is to manage the impact from the loss of power.

Understanding the differences between hazard, risk, and vulnerability is critical when building a plan. In simple terms, hazards are something out of your control. Hazards, such as an earthquake or tsunami, cannot be stopped, but we can plan for the results of those events. Facility management needs to know and understand what threats exist for their location in order to assign risk. Risk, as opposed to hazards, can be managed. Risk is assigning a value or probability of a hazard happening at a specific location, along with the calculated damage that the hazard may cause (Lundberg and Willis 2015). Another way of looking at this is how vulnerable or how much exposure a facility has to a specific hazard. Vulnerability is a measure of physical, social, economic, and environmental factors that increase susceptibility to the impact of a hazard (Palliyaguru et al. 2014). Vulnerability is often directly tied to geography. Facilities located along rivers or coastlines might have a higher vulnerability to flooding and hurricanes. Exposure

	Natural	Blizzard, hurricane, tornado, lightning, wildfire, earthquake, windstorm, drought
	Technology	Data loss or theft, hardware failure, hacking, viruses, telecommunications, cyber-attack, unauthorized access
	Facilities	Loss of utilities, fire, explosion, HVAC failure, power failure, security failure, loss of access, hazardous spill
	Personnel	Loss of key personnel, strikes, pandemics, human error, single point of failure
	Operational	Loss of funding, regulatory issues, financial issues, negative publicity
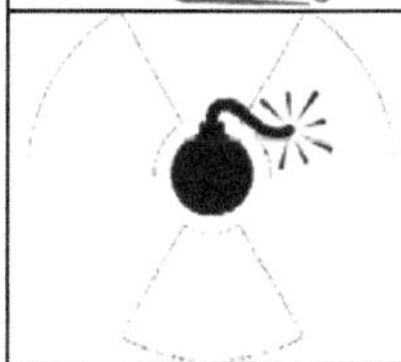	Social	Infiltration, break-in, civil unrest, riots, bomb threats, terrorism

FIGURE 17.2 Threat types.

is another component of risk, and refers to that which is affected by natural disasters, such as people and property. Risk is also a factor of the resilience of the facility in a time of crisis. Resilience is the ability of people to plan for, respond to, and recover from a crisis. As discussed later in the chapter, developing a crisis plan, drilling the plan, and having resources available for recovery enhances organizational resilience. The publication by Arms and Van Zante (2010) speaks to the importance of crisis planning and organizational resilience. These factors were fundamental in their successful and timely evacuation of an animal facility during the threat of California wildfires.

The Venn diagram in Figure 17.3 shows how hazards, vulnerability and exposure, and resilience interact. The intersection where all three factors overlap is the area of highest risk. For each hazard, the level of vulnerability and resilience will vary, increasing or decreasing the relative amount of risk for that hazard. Wildfire hazard provides an example for facilities located within certain arid regions. In this case, well-maintained perimeter fire breaks would minimize the hazard by decreasing the amount of dry fuel easily ignited. Installing smoke filters in air intakes would provide another level of safety, reducing risk. That is what mitigation is all about. Mitigation addresses risks and lowers vulnerability to reduce impacts on business operations.

There are several good risk assessment tools available to determine hazards and vulnerability, such as that from Kaiser Permanente (Kaiser Foundation Hospitals and Health Plan 2001). It can be as simple as a high or low designation, but assigning risk for hazards helps establish priorities (Figure 17.4). In most, the hazards are categorized by type, including naturally occurring, human-related, mechanical-technological, and hazardous materials events. Tornadoes, floods, and earthquakes are considered naturally occurring events. Terrorism, bomb threats, break-ins, and civil disturbances are considered human-caused hazards. Mechanical and technological events include equipment failure, loss of

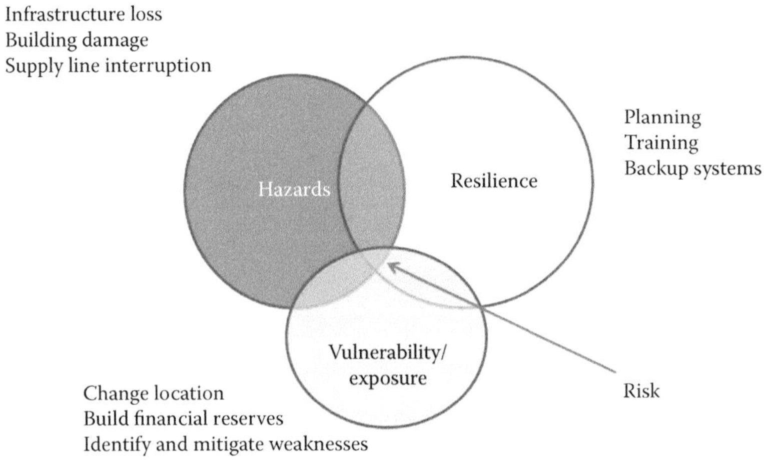

FIGURE 17.3 This diagram represents the relationship between risk vulnerability and resilience.

utilities, information systems outages, and data breaches or supply shortages. Hazardous material events encompass chemical spills and radiologic and biohazard exposures. Creating an exhaustive list of all possible hazards is not realistic.

For risk assessment, facility management can and should anticipate those hazards most likely to impact a facility based on location and past history. Consider possible hazards that could damage facility operations either directly or indirectly. An electrical fire within an animal room is an example of direct damage to operations. There could be loss of animals and research, and perhaps injury to staff members. A truck carrying "just-in-time" animal diets involved in a serious crash hundreds of miles away is an example of indirect damage. Both impact how business will be conducted, and both could result in loss of research. Determining assessment priorities becomes a factor when allocating funding for mitigation and response. Facilities located along coastlines or in low areas near rivers might be assigned a high risk for flooding. Flooding could be the result of a hurricane, tsunami, or a large water main break in the street adjacent to a facility. Flooding is the hazard that requires a risk assignment.

Ultimately, vulnerability is what is left in the risk box when all the mitigation, planning, and preparation are done. A facility located inland away from a major waterway might have a higher vulnerability to an earthquake or tornado. Vulnerability is also tied to what mitigation has occurred at a particular location. A facility in a known floodplain with levees, flood walls, or submarine door installations is less vulnerable to flooding than a facility with no mitigation. Assessing vulnerability requires a thorough knowledge of location and operations to provide a complete picture of risk. Planning and preparation enhance mitigation techniques to reduce risk and the overall vulnerability of a facility.

Mitigation

Mitigation involves addressing vulnerabilities identified during the risk assessment to reduce or eliminate their impact. In the long run, mitigation is often far less expensive than rebuilding after a crisis. However, funding is often an issue delaying or preventing mitigation. A commonly cited example is cryopreservation. This is a proven strategy to protect valuable strains of animals, yet this important expenditure is often not included as a line item in research or departmental operating budgets. Even though this is an up-front expenditure, the costs associated with not having cryopreserved strains in terms of redevelopment and lost research time can be significantly more (Thorat et al. 2013). Further considerations for research continuity include repositioning emergency power or backing up vital records at accessible, off-site locations. Organizational leadership must resolve critical issues when forming mitigation plans, such as how much damage is an organization willing to take in the event of a crisis and how much damage can be prevented by mitigation? How much loss of animal lives and research is acceptable? Mitigation costs compared with the costs of catastrophic loss are calculable. Hurricanes Katrina and

Hazard and vulnerability assessment tool naturally occurring events

KAISER PERMANENTE.

Event	Probability	Severity = (magnitude – mitigation)						Risk
		Human impact	Property impact	Business impact	Preparedness	Internal response	External response	
	Likelihood this will occur	*Possibility of death or injury*	*Physical losses and damages*	*Interruption of services*	*Preplanning*	*Time, effectiveness, resources*	*Community/ mutual aid staff and supplies*	*Relative threat**
Score	*0 = N/A* *1 = Low* *2 = Moderate* *3 = High*	*0 = N/A* *1 = Low* *2 = Moderate* *3 = High*	*0 = N/A* *1 = Low* *2 = Moderate* *3 = High*	*0 = N/A* *1 = Low* *2 = Moderate* *3 = High*	*0 = N/A* *1 = High* *2 = Moderate* *3 = Low or none*	*0 = N/A* *1 = High* *2 = Moderate* *3 = Low or none*	*0 = N/A* *1 = High* *2 = Moderate* *3 = Low or none*	*0–100%*
Hurricane								0%
Tornado								0%
Severe thunderstorm								0%
Snowfall								0%
Blizzard								0%
Ice storm								0%
Earthquake								0%
Tidal wave								0%
Temperature extremes								0%
Drought								0%
Flood, external								0%
Wildfire								0%
Landslide								0%
Dam inundation								
Volcano								0%

FIGURE 17.4 (See color insert.) Example of hazard and vulnerability assessment tool. (From Kaiser Foundation Hospitals and Health Plan, Kaiser hazard vulnerability assessment (HVA), Kaiser Foundation Hospitals and Health Plan, Oakland, CA, 2001.)

Ike impeded or ended long, successful research careers due to the loss of data and animals, extension of project timelines, or institutional budget cuts (Willyard 2010).

Very few institutions have the resources to respond to and recover from crises without outside help. Facility management should develop a network with surrounding institutions, vendors, and emergency response agencies to aid in emergency response and recovery. For example, development of an evacuation plan to preserve irreplaceable animals is a good insurance policy (Willyard 2010). Initiating a memorandum of understanding (MOU) with other institutions to accept evacuated laboratory animals can save many irreplaceable colonies. Consider working with facilities in different geographic areas, as they are less likely to be impacted by the same event. Communicate often with the receiving institutions so that all parties are apprised of real-time inventories, housing availability, and staffing needs. Space utilization fluctuates from month to month, and space available today for emergency housing at another institution might not be available at the time required. The receiving institution should ensure that it can be prepared within a 96-hour notice to receive animals should conditions warrant that type of action. This constant stream of information will assist in preparedness operations for both institutions (Figure 17.5).

Who makes the decision on evacuation or when to evacuate? What guidelines are used to make the decision? Will the plan include a priority list for evacuation? These questions must be answered as part of the planning process, especially as evacuation of some facilities, those with very large animal populations, could take upwards of 2–3 days. Facility management should evaluate both the time and coordination metrics for evacuation and transport within a variety of time constraints—immediate, short-term (within 12–24 hours), and long-term (within a week). The biosecurity of research animals and humans is another consideration in evacuation. What techniques will be used to ensure biosecurity for rodents? One method used successfully is to wrap complete racks with installed cages (water bottles turned upside down) with plastic pallet wrap. The racks can be quickly moved from building to building or loaded on trucks for transport to a safe location. Certain large animals, such as nonhuman primates, have human health risks, making coordination of their evacuation and transport in a short period of time difficult. Ultimately, there may also have to be a decision to euthanize some or all animals.

Both the *Guide* and Directive 2010/63/EU of the European Parliament and of the council of September 22, 2010 (European Commission 2010), on the protection of animals used for scientific purposes contain text regarding euthanasia of laboratory animals during emergency situations. Euthanasia guidelines under emergency conditions must be evaluated during the planning stages, as a great deal of thought is required to develop appropriate euthanasia methods for instances when animals cannot be relocated. Specific personnel need to be trained and psychologically prepared to undertake this difficult task, as disaster response may require euthanasia of large populations of animals. Accommodations for mass euthanasia of rodents must also be considered. For example, multiple carbon dioxide euthanasia stations for rodents may be necessary to euthanize large colonies of rodents in a short time frame. The standard single euthanasia station would require extended periods of time to complete this task, rendering it ineffective in rapid response (Roble et al. 2010). Sufficient quantities of carbon dioxide may also be unavailable for larger rodent facilities. Facilities with larger species of animals or aquatics will face similar challenges in euthanasia.

Some facilities may be able to provide safe housing for responders on campus. How will food, water, and bath facilities be provided? Your plan should also include alternate housing arrangements for responders. When assessing risk, safe areas outside the expected impact area can be identified that are less likely to sustain damage from a natural event. Contracts can be negotiated with hotels both inside and outside of predicted impact zones months in advance to provide housing on short notice of 48 hours or less. The contract should guarantee a certain number of rooms to be available. Other items to consider in negotiating alternate housing are geographic location, backup power for lights and heating, ventilation, and air-conditioning (HVAC), Internet connectivity, and the ability to provide meals to responders. Are vendors available to provide catering or should facilities maintain stocks of minimally perishable foods, such as meals ready to eat (MREs) (Roble et al. 2010)? Other considerations are security and the ability to travel back into the disaster zone. Special arrangements developed before catastrophes will identify and provide clearance for responders to reenter disaster areas. Examples include creation of essential personnel lists, special identification badges, and vehicle placards. These prearrangements do

FIGURE 17.5 Evacuation of irreplaceable animals.

not guarantee entry, as lapses in communication between essential facility personnel and local or non-local law enforcement or other officials may occur and impede entry of essential personnel. Discussing and providing a solution for this contingency will hopefully eliminate its occurrence. Additional options include using a tiered system to categorize responders. Animal health responders are often in a high tier, just behind fire, police, and medical personnel. Other institutions find that establishing an escort plan may be effective in some situations. Essential personnel rendezvous at a predetermined location and are escorted to the facility by emergency personnel postevent.

Preparedness

Crisis Preparedness and Response Training Just as BC planning increases a successful return to normal operations in the aftermath of an emergency or disaster, training enhances the successful execution of a crisis plan. Ejeta et al. (2015) recognized that although awareness and incidence of disasters are increasing, effective preparedness is not increasing concurrently. Gowan et al. (2015) further categorized training needs to the level of behavioral preparedness—ensuring that individual personnel have the resilience to handle crisis.

The aim of training is to familiarize all staff members with crisis response procedures so that they become an automatic reaction in an emergent situation. This requires frequent personnel training sessions. Training sessions are enhanced by the input from all levels of laboratory animal personnel to most faithfully reproduce emergent situations. Interactive web-based training modules offered by FEMA's Emergency Management Institute or local emergency management courses are helpful resources. Ideally, these training exercises will validate the emergency plan, develop staff competency, and test established procedures.

Emergency response plans are fluid, living plans and should be reviewed and revised yearly and after each incident. When plans become stagnant, they are no longer effective. Facility management needs to nurture plans through drills. With repeated simulations, areas of weakness can be identified and addressed. When weaknesses are identified, the plan should be updated to document changes. Emergency preparedness is a multifaceted objective, and there is not a set way to achieve this goal appropriately. There are many types of exercises for use with personnel in crisis preparedness training. These all extend the utility of the written plan. Each component carries equal weight; one type of exercise does not substitute for another. The remainder of this section focuses on several of these exercises, such as tabletop exercises, facility drills, and specific personnel development.

Drilling the Plan Far too often, the existence of a written plan promotes a false sense of security. Plans will fail due to a lack of staff preparedness. Facilities need to develop instruction methods to introduce the plan to personnel. This will provide the foundation for emergency response. Personnel who do not have at least a familiarity with plans will not respond appropriately or timely when required. Training exercises can be divided into two categories: discussion-based exercises, including seminars, workshops, and tabletops, or more complex operations-based exercises, including drills and full-scale exercises. An excellent outline and training program is available through FEMA's Emergency Management Institute (2015). Several training types will be briefly outlined here. Ultimately, training exercises will test procedures and increase staff confidence to respond appropriately in the stress of a calamitous event.

Types of Exercises

1. *Seminars and workshops.* Seminars and workshops serve to orient staff to response plans. These are opportunities to teach toward the plan or allow for discussion or development of plans or portions of plans. These planning sessions often result in the creation of emergency operating procedures (Swearengen et al. 2010). Lecture and discussion instruction are the recommended first steps in developing a training exercise program (Cabinet Office and National Security and Intelligence 2014). These can be tailored for distribution as lectures, small groups, or one-on-one training. Plan dissemination or discussion may differ depending on the job category for involved personnel. Focus and emphasis should be given to plan sections or topics that will be relevant to a given audience. For example, computer security may not be relevant

to certain personnel, but a critical item for others. These orientation techniques give essential personnel the opportunity to discuss the plans and learn their proposed role and the roles of others.

2. *Tabletop exercises.* Tabletop or scenario exercises are based on hypothetical, yet realistic events. The success of the exercise can be enhanced if all participants are familiar with the crisis plans prior to the exercise. The exercise usually takes place around a table with participants under the guidance of a facilitator. The group is given a scenario to discuss and formulate a course of action in response to the crisis. These events allow for deliberate, well-considered problem solving, thus strengthening their use in the identification of weaknesses in emergency plans. It provides an opportunity to validate and update the plan as necessary. Tabletop exercises may be conducted at both the department and interdepartmental levels and are usually a planned event. Careful attention to detail in preparing the scenario is crucial to a successful tabletop exercise. Mortell and Nicholls (2013) suggest that tabletop exercises may provide a false sense of security regarding preparation since these exercises are done with advanced preparation and notice, without the element of stress.
3. *Live practice drills.* Live practice drills are a rehearsal for an event. These can focus on training to a specific procedure or can be expanded to a full-scale exercise that involves the entire facility staff and may extend to outside response agencies, like the local police. These test the logistics of execution, lines of communication, and physical capabilities of the facilities (FEMA Emergency Management Institute 2015). This type of exercise helps participants build confidence by simulating the crisis. The exercise also serves to identify potential weaknesses in the response, such as unanticipated supply needs or lack of communication, teamwork, or leadership capabilities.

Training raises awareness of the types of crises the staff may encounter and, consequently, how to effectively prepare and respond. Training also builds confidence in the staff's ability to respond quickly and without hesitation or confusion. Some emergencies and disasters can be very overwhelming; however, training mitigates the stress of dealing with each situation. All staff considered essential personnel, or who are responsible for planning, training, or crisis preparedness and response, should participate. Inclusion of personnel from other key departments (principal investigators, facility services, environmental health and safety, biosafety, etc.) or outside agencies (i.e., fire and police departments) is essential and should be considered. The FEMA Emergency Management Institute (https://training.fema.gov/is/) provides excellent training courses on all the above topics.

Nontechnical Skill Training

Personnel are the most critical aspect of crisis response. Without personnel, no plan can be implemented. Furthermore, no two crises will be the same, so no plan can account for all contingencies, often necessitating creativity from responders. Most facilities assume that personnel will be available to implement a crisis plan, but facilities must recognize the reasons that people will not respond to a crisis (Davidson et al. 2009; Adams and Berry 2012). Major reasons include fear, other real or perceived responsibilities, uncertainty, inability to reach the work site, or not understanding the importance of their role. Human behavior will impact all responses. Many authors recognize that preparedness training must account for and focus on human behavior to create an empowered employee during the response (Donaho 2014; Ejeta et al. 2015; Gowan et al. 2015). This employee then has practiced skills, bestowed through training. Ultimately, a responder must be able to remain calm and assess the situation rapidly and make a decision. If personnel are not prepared for the psychological aspects inherent in crisis response, they may not respond appropriately, no matter how well they know the plan. As suggested by Crichton and Flin (2001), laboratory animal program leaders should invest appropriate time increasing the nontechnical skills of their employees to enhance their response capabilities.

These skills, such as decision making, leadership, teamwork, and stress management, are as important as the knowledge of the crisis response plan or on-the-job skills (Crichton and Flin 2001). Increasing this skill set has a twofold purpose: first, it may assist in identifying personnel with strong nontechnical skills who could be assigned leadership roles during an emergency response, and second, it will increase the confidence of responding personnel. A majority of polled responders following a disaster stated they looked for guidance from upper management before acting (Pullium et al. 2014). This could lead to serious delays if upper management is not readily available. Thus, the need for leadership and decision making in response is imperative. There may not be time or a way to communicate effectively in a crisis, and personnel should have the training and confidence to make a decision or lead a response. Personnel may also have to work and communicate effectively with unfamiliar persons during crisis. These skills need to be learned and strengthened prior to the crisis.

Personnel do not necessarily fear for their own safety during a response; however, it is likely that they will fear for the safety of another. Davidson et al. (2009) found that families and pets are the oft-cited reasons that people do not come to work during or following a disaster. This holds true for persons considered essential personnel. Facility management must account for this attrition and determine the minimum number of effective responders required to maintain an operation. Facility management can mitigate attrition by determining what factors reduce willingness to work, and creating a contingency plans for these. Considerations must include

- *Care for dependents*: Can children or pets be provided safe location on site with parents, or can other assured care be provided while parents are at the disaster site?
- *Transportation*: Can people get to work, and what methods can be used to facilitate this?
- *Personnel feeling that they will be cared for on site*: Will there be food, shelter, and bedding provided, or perhaps a supervisor or other personnel who can provide psychological support?

Facility management must verify that personnel understand the importance of their jobs and what the facility will do to protect and support them and their families. Staff will likely have significant internal struggles regarding what to do during most catastrophic events. Preparation to assuage these concerns should begin early in crisis response planning to increase likelihood success.

A regular training program will aid all staff, whether serving as an introduction for new staff or as a refresher course for long-term staff. At the end of the training exercises, all staff should be familiar with their potential roles in crisis response, either individually or as a team; have a good working knowledge of the response plan; know who the key decision makers are; and be able to identify methods of communication before, during, and after an emergency.

Response

Response takes place during or immediately following a crisis. This phase is focused primarily on the safety of animals and humans and attempting to minimize damage to infrastructure or equipment. In principle, the initial response to a crisis will be coordinated by local government's emergency services, that is, first responders such as police, fire departments, and emergency medical services. These responders will likely exist within the hierarchy of an incident command system (ICS). Nearby municipalities and state and volunteer agencies may assist as available. The scope of danger in a facility may not become clear until later in response, and often external authorities may control access to the facility until the danger has been assessed and resolved.

The first responder's mission is to take action to save lives and prevent further damage. Wingfield et al. (2010) recognized that it cannot be expected that these responders will have the capability to handle animals in a laboratory animal setting. For a lab animal facility, the appropriate response will depend on the essential personnel. As described previously, the safety, well-being, and effectiveness of the response team depends on their previous preparation. The safety of personnel is the first priority in a crisis; the animals may become secondary. Often, the initial response is a damage assessment to determine if areas or buildings are safe to enter. Hazards have to be identified and corrected before animal care can commence. For a lab animal facility, the appropriate response will depend on the essential personnel available.

Related to external control of disaster sites, a frequent problem during the response is the inability of personnel to travel to or enter the animal holding facilities. As mentioned earlier, essential personnel must carry identification providing them access to the site. In addition, first responders must recognize these forms of identification or must be informed of the entry requirements for facility essential personnel. Unsafe circumstances may prevent access; however, first responders and essential personnel must be able to communicate and recognize one another to engage appropriately.

Once on site, the exact actions needed during a response are unpredictable. Essential personnel may find themselves in unfamiliar locations, stymied by not having appropriate equipment due to restricted access points and stuck in the dark. Responders must rely on their training and adhering to facility standard operating procedures (SOPs) as closely as possible, but anticipate the need for improvisation and creativity. Three of the most recent large-scale animal facility disasters involved flooding—all responded in a different manner, emphasizing the individual nature of disaster (Schub 2002; Goodwin and Donaho 2010; Pullium et al. 2014).

Incident Command Bigley and Roberts (2001) cite an ICS as a best-practice system for emergency management. The concept of ICS grew out of management studies of the disastrous 1970s' California wildfires, which found numerous holes in emergency response and sought to create a standardized response system to emergency and disaster (Jensen and Thompson 2016). The ICS established a common, flexible framework allowing multiple crisis response agencies to interact effectively and efficiently (Jensen and Thompson 2016). FEMA provides a description of the use of an ICS as an integrative management tool, combining multiple agencies or departments into an effective response structure (FEMA Emergency Management Institute 2015). Briefly, at a local, state, or national level, the ICS structure consists of an incident commander, command staff, and general staff (Smith and Kuldau 2014). The incident commander has overall response leadership, and the command staff is composed of safety, public information, and liaison officers (Smith and Kuldau 2014). The liaison officer coordinates activities among the various agencies involved. At the next level of ICS is the general staff. Within general staff are section chiefs representing operations, planning, logistics, and finance and administration (Smith and Kuldau 2014). This hierarchy enhances information exchange and rapid decision making. Using this modular system, ICS can be tailored and scaled to adapt to each situation. Laboratory animal facilities are recommended to develop their own ICS structures, which generally exist as an extension of the general staff. Descriptions of ICS structures for laboratory animal facilities can be found in Vogelweid (1998) and Roble et al. (2010). These descriptions provide outlines for the integration of the laboratory animal facility responders within a local, state, or national ICS and provide guidance on the personnel who should be included within facility ICS. These personnel include, but are not limited to, directors, veterinarians and technicians, husbandry staff, and administrative staff. Successful use of an ICS requires the appropriate training and understanding of the system for all individual participants (Jensen and Waugh 2014). The concept of the ICS must be included within training sessions—both discussion based and operations based. The same skills required to manage the response to crisis are required to work within an ICS: leadership, communication ability, situational awareness, capable decision making, and creativity (Crichton et al. 2005). These skills enhance an ICS and aid in creating calm and order from chaos. Having a dedicated animal facility representative within the incident command structure improves communication and enhances the likelihood of decisions that benefit the animal facility.

Communication Methods Coordination of the crisis response cannot happen without appropriate communication. This includes the ability to communicate the risk of and response to the emergency among those at the site, those in the command center, and the general public (McCormick 2015). Satellite phones are most frequently used when landline infrastructure is damaged (McCormick 2015). Experience shows that these and other forms of communication, including cell phones, may fail in disaster areas as infrastructure is damaged or service is overwhelmed. Satellite phones prove unreliable due to weather or limited link availability, as first responders use the same satellite systems. The traditional public safety communications systems using two-way radio (often called walkie-talkies if handheld) can also become limited during disasters. Portable radio repeaters to extend handheld walkie-talkie ranges can be rented to establish a radio communications network in an emergency. This can be preplanned with contracts ready when needed. Similarly, ham radio (also two-way radio) provides an ad hoc communications system when all else fails (Reid 2016). Licensed ham radio groups commonly set up emergency networks during disasters. A new option is the portable broadband antenna system capable of moving large amounts of data to support video, voice, and text communications. These systems deploy quickly and provide private networking virtually anywhere. Institutions must not forget the value of pen and paper as well. Establishing known meeting points and regular meeting times as part of crisis planning allows for safety checks and information dissemination. Simple systems, such as writing instructions on black- or whiteboards or paper and easels, give responders directions and updates. Regular town hall meetings during the recovery phase are also highly recommended, as these can be conducted in the absence of power. However, as McCormick (2015) states, more and more reliance is falling on social media and mobile-based technology.

Superstorm Sandy signified a shift in the use of social media for disaster response and information gathering (Cohen 2013). National organizations such as FEMA and private citizens and institutions alike put forth information on the superstorm via social media platforms such as Instagram, Twitter, Facebook, and LinkedIn. Other technologies, such as Short Message Service (SMS) (text messaging) and websites, also provided a ready spread of information. FEMA reached 6 million Twitter users with one tweet and increased its Facebook users to 300,000 (from 12,000) on the day Hurricane Sandy made landfall (Cohen 2013). Bonnan-White et al. (2014) exemplified the use of crisis informatics during a snow disaster and defined this as a new method of harnessing communication for crisis management. This real-time access to information is being channeled by government and private agencies to improve situational awareness. Social media represents enormous potential in crisis response when information is conveyed accurately. Personnel empowered to present and receive real-time data will make significantly improved decisions and likely improve response. The real-time power of social media may be exemplified as well in horrific crisis situations, such as active shooting or other domestic terror threats where communications are not blocked, but access to devices other than handheld ones may be limited by the danger. The earliest or only information obtained in these instances may come from social media sites (Gillard 2011).

The DHS Science and Technology Directorate's Virtual Social Media Working Group has published five documents providing guidance on the use of social media in emergency response (U.S. Department of Homeland Security Science and Technology 2015). These documents provide an excellent starting point for developing a strategy for use of social media. This includes determining the right technology; determining security measures regarding the intended audience for the information, either institutional personnel or the general public; and understanding legal ramifications of information spread. An active social media policy also allows institutions to provide information ahead of the inevitable rumors that swirl around disasters. Misinformation can spread very quickly. Accurate information disseminated quickly establishes credibility and trust, but it can be difficult to determine the veracity of information rapidly. For example, during Hurricane Ike in 2008, news reports of escaped tigers on Galveston Island circulated widely when, in fact, it was rumor (KVUE Staff 2008). Institutions should be prepared to have personnel actively monitoring social media sites and correcting misinformation as soon as possible. There are a number of ways to facilitate information verification by including embedded hashtags (#) to identify topic threads, keywords, URLs, photos, or videos to provide greater information (U.S. Department of Homeland Security Science and Technology 2015). Keywords and hashtags allow rapid scanning of messages for relevant content. Determining appropriate hashtags and keywords in advance of crises is a useful exercise and should be part of the planning process. This can also be used to crowdsource information and aggregate greater information or understanding about the situation from the general public posting information on what it witnesses. This information gathering and analysis will likely be done by outside agencies such as FEMA or other responders, as this requires a complex coordination and intelligence gathering framework (U.S. Department of Homeland Security Science and Technology 2015). Establishing a social media response team within the command center is vital in today's interconnected world. This is especially true to ensure that the proper information is given to the general public regardless of the size or scope of the crisis.

Recovery

In the recovery phase, actions are taken to return to a normal or safer situation following an emergency or disaster. This will be different for all facilities, and this chapter aims to provide considerations for this phase. The process of recovery can be the most difficult part. Recovery begins after the initial response is completed or winding down. While working to assess the damage or impact on facilities, equipment, and animals, recovery teams may be strained by less than ideal conditions. Work may be physically taxing and emotionally stressful, especially if there was catastrophic injury to animal populations. Crisis recovery should be considered in dual parts: institutional repairs, including facility restoration and animal replacement required to return to normal work, and emotional repairs, required to return employees to normal work.

Many excellent references exist on mechanisms of recovery (Schub 2002; Goodwin and Donaho 2010; Swearengen et al. 2010; Ikeda 2012). These provide direction on steps that all institutions should include as part of their planning to facilitate their recovery. An accurate description of facility inventory (animal and equipment) should be prepared and maintained for insurance purposes. This should include all specialty equipment associated with an animal facility, including caging, sanitation, and sterilization devices, and specialized research and surgical instrumentation (Goodwin and Donaho 2010).

Personnel's mental and physical well-being is of paramount importance as well. Facilities should provide trauma and grief counseling for all impacted employees, especially if animals were lost or damage necessitates euthanasia during the recovery phase. Posttraumatic stress disorder is well recognized in survivors of disasters (Galea et al. 2005). Mental health professionals must be available to provide this therapeutic service. Early intervention and debriefing include the use of psychological techniques such as critical incident stress debriefing (CISD), followed by further therapy as required (McNally et al. 2003). CISD should be done within the first 24 hours of a disaster and may significantly improve therapeutic recovery from anxiety or depression (Donaho 2014). Costs for these services should be included by institutions in planning stages. Many facilities have also held memorial services for those animals lost, as part of the healing process (Goodwin and Donaho 2010; Ikeda 2012). Employee care may also require provisions for furlough or relocation as personnel requirements are reduced because of limited day-to-day operations.

Recovery also requires reporting details of the crisis to the appropriate government funding and regulatory agencies and accrediting agencies such as OLAW, the U.S. Department of Agriculture (USDA), the Centers for Disease Control and Prevention (CDC), and AAALAC International (Schub 2002; Swearengen et al. 2010; Ejeta et al. 2015). These will be dependent on whether institutions hold assurances, registrations, or accreditation with these organizations. The crisis response plans should contain contact information for accrediting agencies. Each of these organizations, as well as the American Veterinary Medical Association, provides excellent guidance for both reporting structure and crisis response, as outlined below. Notification is a task that is best completed as soon as possible following the crisis for two reasons. First, this fulfills reporting requirements, and second, it protects the facility against negative or inappropriate press. Activists against biomedical research involving animals will often twist these events to meet their agenda. Appropriate documentation and evaluation by regulatory agencies can provide protection against exaggerated claims. As the Institutional Animal Care and Use Committee (IACUC) and veterinarians will play an instrumental role in assisting management with reporting, developing mechanisms for IACUC continuance is also essential (Roble et al. 2010).

Concepts of Business Continuity Management

BC planning is important, yet frequently overlooked, in the overall scheme of crisis planning. BC planning provides the framework to return to business after a disrupting event, large or small. In the laboratory animal setting, BC means reestablishing the ability to conduct research. As discussed previously, a BC plan is a natural continuation to a crisis management plan. Generally, these plans should be created independently, but a properly constituted crisis response plan may eliminate the need for a BC plan if all measures effectively mitigate damage. However, a BC plan may be necessary in noncrisis situations. Snyder (2013) provides insight into this concept through the example of a leaky roof. A "leaky roof is not a crisis situation, but the leak may impact equipment storage or an employee location. In either case, alternate work location plans should be implemented so that critical business processes may continue" (Snyder 2013).

Similarly to a crisis response plan, a BC plan accounts for both internal and external factors that disrupt business. Internal factors include items like computer failure or loss of a piece of equipment due to mechanical failure, which may not be in the scope of a crisis management plan. External factors are items such as a strike at a manufacturing plant that disrupts the supply chain (Snyder 2013). Each of these problems may not be considered a crisis, but they assuredly have the potential to impact research. The BC plan should cover the most to least likely events an animal facility may face. As with crisis planning, BC planning is program specific and should be developed by the individuals involved in the emergency response. The animal facility plan should be coordinated with the overall organizational BC plan and all pertinent internal departments and external entities. Effective animal facility administration requires expertise in personnel management, resource management, education, regulatory compliance, and fiscal management. To effectively integrate these aspects into a BC plan, the planning team should include personnel from departments like veterinary resources, human resources, payroll, benefits, security, information technology, and environmental health and safety. Continuity planning should detail how the organization will handle failures of varying significance. For example, initial stage planning must provide methods to ensure supply chain continuity in the event that electronic databases or ordering systems are unavailable. Maintaining a secured supply of currency on site is recommended for emergent use. Having a basic response strategy for the facility both during and after a crisis will help employees minimize disruption. For example, does the BC plan include a mechanism to pay and support employees through each phase of the crisis? Security in the knowledge that one can take care of personal issues helps employees focus on the recovery and continued success of the business.

Think of BC planning as proactive insurance, a method to return to function with the least amount of disruption. Facilities with an appropriate, rehearsed BC plan more effectively and promptly respond to and recover from emergencies.

Benefits of Crisis and Business Continuity Planning: Resilience

A resilient organization is able to rapidly respond and move forward with operations despite conditions that disrupt normal routines. An excellent review described both crisis management and BC as the pillars supporting a resilient organization, further emphasizing the importance of these two concepts (Snyder 2013). A resilient organization must be able to withstand all hazards and sustain its mission. The DHS Risk Lexicon (September 2008) defines resilience as the "ability to resist, absorb, recover from or successfully adapt to adversity or a change in conditions." As described throughout this chapter, resilience is built through planning for and mitigation against known risks, through leadership and training, and by incorporating trained personnel throughout the organization into the design and execution of the plan. The ability to adapt quickly to disruptions, while maintaining continuous business operations and safeguarding people, assets, and reputation, promotes flexibility. Planning for the future, whatever it brings, instills resilience. Five key considerations for developing a crisis and BC plan for the laboratory animal research facility include

1. In the wake of several natural disasters, research organizations are now required to develop emergency plans as per the OLAW. The eighth edition of the *Guide* requires animal facilities to develop and maintain a current crisis plan. OLAW and Directive 2012/63/EU of the European Parliament and of the council of September 22, 2010, on the protection of animals used for scientific purposes and AAALAC accreditation require a crisis plan document.
2. These plans aid in mitigating loss of important research assets and help generate documentation needed for insurance claims and assistance through government agencies like FEMA or funding organizations.
3. Crisis and BC plans aid in acquiring the supplies necessary to care for animals and personnel during a disaster.
4. The process of crisis and BC planning will aid in identifying weaknesses in the organization and improves understanding of facility function.
5. Crisis and BC planning protects the organization's image and reputation. A good reputation is important for the prospects of securing grants and personnel and gaining public support.

Conclusion

Those who have experienced a disaster understand the turmoil of emotion that accompanies it. There is confusion and urgency in needing to ensure the humane treatment of animals, the safety of personnel, and the return to normalcy as soon as possible. Crisis and BC planning not only helps to safeguard the organization from catastrophe; it also prepares the animal care staff and other key departments for any disaster. Several sources rich in advice on developing continuity plans and emergency management are included in the references for further review, study, and application. Recovery does not end when the lights come back on and the cleanup is done. A major disaster can take years of work to return to the same status as prior to the disaster. This chapter is dedicated to the brave men and women who have given, and will give of themselves, unselfishly to protect the laboratory animals in their charge.

APPENDIX 17.1: RESOURCES FOR CRISIS PLAN DEVELOPMENT

AAALAC: Disaster preparedness and response for veterinarians. http://www.aaalac.org/accreditation/faq_landing.cfm#H.

Alliance for Biomedical Research in Europe (EBRA): www.biomedeurope.org.

American Association for Laboratory Animal Science: Checklist of disaster planning expectations in the *Guides* and Animal Welfare Regulations. https://www.aalas.org/iacuc/iacuc_resources/disaster-preparedness#.Vq18ML8oqvU.

Americans for Medical Progress: www.amprogress.org.

American Veterinary Medical Association: A collection of topical papers regarding disaster preparedness for animals. https://www.avma.org/news/journals/collections/pages/avma-collections-disaster-preparedness-and-response.aspx.

Animal Welfare Information Center: U.S. Department of Agriculture Agricultural Research Service National Agricultural Library. https://awic.nal.usda.gov/.

Canadians for Health Research: www.chrcrm.org/.

CDC: http://emergency.cdc.gov/.

DHS: BC plan. http://www.ready.gov/sites/default/files/documents/files/BusinessContinuityPlan.pdf.

FEMA: Online training module to increase awareness and preparedness among animal owners about typical hazards, how animals are affected, and what can be done to reduce the impact of disasters. https://training.fema.gov/emi.aspx.

FEMA Emergency Management Institute: Online training module to guide emergency management officials, animal owners, and industries in preparing community disaster plans.

Filling the Ark: Animal Welfare in Disasters (2009), by Leslie Irvine: http://www.temple.edu/tempress/titles/1977_reg.html.

Foundation for Biomedical Research (FBR): fbresearch.org.

Institute for Laboratory Animal Research, National Academy of Sciences: dels.nas.edu/ilar/.

National Alliance of State Animal and Agricultural Emergency Programs Best Practices Library: Partners in disaster response. Includes lists of animal emergency response courses and reference materials. http://nasaaep.org/.

Office for Human Research Protections, U.S. Department of Health and Human Services: www.hhs.gov/ohrp/.

OLAW: Provides links to resources about planning for and responding to natural and other disasters. https://grants.nih.gov/grants/olaw/disaster_planning.htm.

Resources for Crisis Management in Zoos and Other Animal Care Facilities (1999), edited by S. D. Chan, W. K. Baker, and D. L. Guerrero: http://www.amazon.com/Resources-Crisis-Management-Animal-Facilities/dp/1929672020.

"Sandy Destroys Years of Medical Research": http://www.cnn.com/2012/11/08/health/sandy-research-time/index.html?eref=mrss_igoogle_cnn.

Scientists Center for Animal Welfare (SCAW): www.scaw.com.

Society for Neuroscience: Guidelines for crisis management. Includes supporting institutions for assistance when research is questioned by activists. www.sfn.org.

REFERENCES

Adams, L. M., and D. Berry. 2012. Who will show up? Estimating ability and willingness of essential hospital personnel to report to work in response to a disaster. *Online J Issues Nurs* 17 (2):8.

Arms, M. M., and J. D. Van Zante. 2010. Wildfire evacuation: Outrunning the witch's curse—One animal center's experience. *ILAR J* 51 (2):158–163.

Bigley, G. A., and K. H. Roberts. 2001. The incident command system: High-reliability organizing for complex and volatile task environments. *Acad Manage J* 44 (6):1281–1299.

Bonnan-White, J., J. Shulman, and A. Bielecke. 2014. Snow tweets: Emergency information dissemination in a US county during 2014 winter storms. *PLoS Curr* 6.

Cabinet Office and National Security and Intelligence. 2014. Emergency planning and preparedness: Exercises and training. London: Cabinet Office and National Security and Intelligence. https://www.gov.uk/guidance/emergency-planning-and-preparedness-exercises-and-training (cited January 20, 2016).

Canadian Council on Animal Care. 2016. Crisis management program 2016. Ottawa: Canadian Council on Animal Care. http://www.ccac.ca/en_/assessment/acc/resources_acc/crisis (cited January 28 2016).

Chakma, J., G. H. Sun, J. D. Steinberg, S. M. Sammut, and R. Jagsi. 2014. Asia's ascent—Global trends in biomedical R&D expenditures. *N Engl J Med* 370 (1):3–6.

Cohen, S. E. 2013. Sandy marked a shift for social media use in disasters. *Emergency Management*. http://www.emergencymgmt.com/disaster/sandy-social-media-use-in-disasters.html.

Crichton, M., and R. Flin. 2001. Training for emergency management: Tactical decision games. *J Hazard Mater* 88 (2–3):255–266.

Crichton, M. T., K. Lauche, and R. Flin. 2005. Incident command skills in the management of an oil industry drilling incident: A case study. *J Contingencies Crisis Manage* 13 (3):116–128.

Davidson, J. E., A. Sekayan, D. Agan, L. Good, D. Shaw, and R. Smilde. 2009. Disaster dilemma: Factors affecting decision to come to work during a natural disaster. *Adv Emerg Nurs J* 31 (3):248–257.

Donaho, J. 2014. Building the human component into contingency plans. *Lab Anim (NY)* 43 (1):27–32.

Ejeta, L. T., A. Ardalan, and D. Paton. 2015. Application of behavioral theories to disaster and emergency health preparedness: A systematic review. *PLoS Curr* 7.

European Commission. 2010. Directive 2010/63/EU of the European Parliament and the council of 22 September on the protection of animals used for scientific purposes. Brussels: European Commission.

FEMA (Federal Emergency Management Agency). 1996. *Guide for All-Hazard Emergency Operations Planning*. Washington, DC: FEMA.

FEMA (Federal Emergency Management Agency). 2015. About the agency. Washington, DC: FEMA. http://www.fema.gov/about-agency.

FEMA Emergency Management Institute. 2015. ICS resource center. Washington, DC: FEMA. http://training.fema.gov/EMIWeb/IS/ICSResource/index.htm (cited December 24, 2015).

Galea, S., A. Nandi, and D. Vlahov. 2005. The epidemiology of post-traumatic stress disorder after disasters. *Epidemiol Rev* 27:78–91.

Gillard, E. 2011. Virginia Tech shooting reinforces the power of social media. *Daily Press*. http://articles.dailypress.com/2011-12-08/sports/dp-nws-tech-social-media-1209–20111208_1_twitter-and-facebook-twitter-website-power-of-social-media.

Goodwin, B. S., Jr., and J. C. Donaho. 2010. Tropical storm and hurricane recovery and preparedness strategies. *ILAR J* 51 (2):104–119.

Gowan, M. E., J. A. Sloan, and R. C. Kirk. 2015. Prepared for what? Addressing the disaster readiness gap beyond preparedness for survival. *BMC Public Health* 15 (1):1139.

Heath, S. E. 1999. *Animal Management in Disasters*. St. Louis, MO: Mosby.

International Federation of Red Cross and Red Crescent Societies. What is a disaster? Geneva: International Federation of Red Cross and Red Crescent Societies. http://www.ifrc.org/en/what-we-do/disaster-management/about-disasters/what-is-a-disaster/.

Ikeda, T. 2012. Crisis management and recovery from the damage to the laboratory animal production facility due to the Great East Japan Earthquake. *Exp Anim* 61 (1):1–11.

Jensen, J., and S. Thompson. 2016. The incident command system: A literature review. *Disasters* 40 (1):158–182.

Jensen, J., and W. L. Waugh. 2014. The United States' experience with the incident command system: What we think we know and what we need to know more about. *J Contingencies Crisis Manage* 22 (1):5–17.

Kaiser Foundation Hospitals and Health Plan. 2001. Kaiser hazard vulnerability assessment (HVA). Oakland, CA: Kaiser Foundation Hospitals and Health Plan.

KVUE Staff. 2008. Busting the myths and rumors around Ike. http://www.kvue.com/story/news/local/2014/05/16/2176120/.

Lundberg, R., and H. Willis. 2015. Assessing homeland security risks: A comparative risk assessment of 10 hazards. *Homeland Security Affairs* 11:1–23.

McCormick, S. 2015. New tools for emergency managers: An assessment of obstacles to use and implementation. *Disasters* 40 (2).

McNally, R. J., R. A. Bryant, and A. Ehlers. 2003. Does early psychological intervention promote recovery from posttraumatic stress? *Psychol Sci Public Interest* 4 (2):45–79.

Mortell, N., and S. Nicholls. 2013. Practical considerations for disaster preparedness and continuity management in research facilities. *Lab Anim (NY)* 42 (10):F18–F24.

National Institutes of Health, Office of Laboratory Animal Welfare. 2002. Public Health Service policy on humane care and use of laboratory animals. Bethesda, MD: National Institutes of Health, Office of Laboratory Animal Welfare.

National Research Council. 2011. *Guide for the Care and Use of Laboratory Animals*. 8th ed. Washington, DC: National Academies Press.

Palliyaguru, R., D. Amaratunga, and D. Baldry. 2014. Constructing a holistic approach to disaster risk reduction: The significance of focusing on vulnerability reduction. *Disasters* 38 (1):45–61.

Pullium, J. K., G. S. Roble, and M. A. Raymond. 2014. Emergency planning: Be prepared. *Nature* 514 (7523):430.

Reid, K. 2016. Where to go during a communications shutdown. *Emergency Management*, February 3. http://www.emergencymgmt.com/disaster/Where-to-go-During-a-Total-Communications-Shutdown.html?utm_medium=email&utm_source=Act-On+Software&utm_content=email&utm_campaign=How%20Worried%20Should%20we%20be%20About%20the%20Zika%20Virus&utm_term=Where%20to%20go%20During%20a%20Total%20Communications%20Shutdown.

Roble, G. S., N. M. Lingenhol, B. Baker, A. Wilkerson, and R. J. Tolwani. 2010. A comprehensive laboratory animal facility pandemic response plan. *J Am Assoc Lab Anim Sci* 49 (5):623–632.

Schub, T. 2002. The year of the flood: Tropical Storm Allison's impact on Texas Medical Center. *Lab Anim (NY)* 31 (8):34–39.

Smith, J. S., and G. A. Kuldau. 2014. Methods of instruction of the incident command system and related topics at US veterinary schools. *Disaster Med Public Health Prep* 8 (6):505–510.

Snyder, H. 2013. The Wikipedia of business resilience. *Disaster Recovery Journal*, April 22. http://www.drj.com/articles/online-exclusive/thewikipedia-of-business-resilience.html.

Swearengen, J. R., K. J. Vargas, M. K. Tate, and N. S. Linde. 2010. Disaster preparedness in biocontainment animal research facilities: Developing and implementing an incident response plan (IRP). *ILAR J* 51 (2):120–126.

Thorat, R., S. Ahire, and A. Ingle. 2013. Re-establishment of a breeding colony of immunocompromised mice through revival of cryopreserved embryos. *Lab Anim (NY)* 42 (4):131–134.

U.S. Department of Homeland Security Science and Technology. Virtual social media working group 2015. Washington, DC: U.S. Department of Homeland Security. http://www.firstresponder.gov/SitePages/Technology/Profile.aspx?s=Technology&itemID=16.

Vogelweid, C. M. 1998. Developing emergency management plans for university laboratory animal programs and facilities. *Contemp Top Lab Anim Sci* 37 (5):52–56.

Willyard, C. 2010. When natural disasters strike, tragedy can unfold in the lab. *Nat Med* 16 (6):616–617.

Wingfield, W. E., B. E. Rollin, and R. A. Bowen. 2010. You have a disaster plan but are you really prepared? *ILAR J* 51 (2):164–170.

Section V

Physical Plant

18

Facility Design, Planning, and Renovation

Michael J. Huerkamp, David Mallon, and Gerald Percifield

CONTENTS

Introduction

Building or renovating an animal research facility (ARF) is a daunting task. Aside from meeting all the regulations and guidelines for quality animal care and use and employee safety, a facility project requires accommodating the variety of research protocols ongoing at the institution and projecting the possible protocols and species requirements in the future. Due to physical, financial, and philosophical concerns, restrictions and compromises inevitably affect the design of every new ARF. For example, the available footprint and/or laboratory configuration may dictate the location of load-bearing columns and possible ARF room sizes. The potential layout and organization may be impacted by factors dictating column placement. The objectives in designing an ARF are to achieve efficiency and flexibility, both in operations and in animal housing capacity (species and volume), and control of the environment. The facility must enable the maintenance of research animals free of unwanted and unacceptable diseases, and chemical and biological contaminants, while fostering a steady-state existence in which they are subjected to minimal stress and environmental fluctuations. The degree to which these objectives are met is largely dependent on the design, as well as the management, of the ARF. These responsibilities fall on the animal resources program, a team of veterinarians, and management and technical experts, in providing the facilities, equipment, personnel, and programs to promote high-quality research with animals.

Constituents: Animals, Investigators, Staff, and Stakeholders

As a program manager or director of a major operation, it is essential to have a deep understanding of the constituents. In animal research programs, this includes a very diverse group of people, needs, and requirements. Constituents include researchers, care and operational staff, and the animals themselves. All have needs that must be accommodated. This chapter identifies those needs and how a successful leader can accommodate those needs and requirements.

The first priority of any animal facility is to maintain the health and safety of the animals placed in the care of the animal research program. Most developed countries have legislated laws and regulations regarding the use of some or all animal species in research. Such legislation places significant responsibility on institutions and their people, and provides for significant legal penalties for violations of the laws. Additionally, nonlegislated standards have been established to facilitate "best practices." The *Guide for the Care and Use of Laboratory Animals* (*Guide*) (ILAR 2011) provides guidance on many aspects of vivarium operations in support of the legislative requirement and by providing guides to best practices. An understanding of the applicable legal requirements and best practices standards is crucial to providing the best and appropriate care for all constituents to meet the desired timeless design and construction standards.

The second priority of the ARF is the health and safety of the occupants. Most standard facility health and safety requirements are prescribed by a wide variety of building, life safety, and fire codes, and institutional requirements. Specific types of hazardous protocols may require animal biosafety containment, which may require additional safety features, such as those outlined in the *Biosafety in Microbiological and Biomedical Laboratories* (BMBL) (CDC 2009).

The ARF is a vital extension of the research laboratory and should be situated as close to the research laboratories as possible. It should be designed to meet and adapt to the needs of research conducted at the institution. Access to the ARF should be enabled and unencumbered for authorized personnel, yet hardened to prevent interference by unauthorized individuals. The facility must promote conformance and compliance with standard operating procedures (SOPs) and desired behaviors. It should also enable nonconformance to be identified in order to allow correction through training, retraining, or other measures. The facility should allow for the logical control, allocation, and distribution of space and be built and operated to function at minimal cost, without sacrificing quality.

Another key user group served by the facility is the animal resources team. The facility should be centralized to the extent reasonable to enable efficient staff operations and provide for shared support elements (NRC 2000). It should be easy to maintain and manage where operations and procedures can be standardized, be ergonomically sound, and provide for a working environment and amenities that promote staff retention. The facility design should account for physical plant durability, reliability, and maintainability, while retaining the flexibility to adapt to unpredictable needs. Generic, general, and modular design concepts best meet these needs in the majority of these cases. Secure, restricted access promotes a safe environment, free of undesirable disruption, culminating in controlled environmental conditions. Other critical constituents are the physical plant maintenance personnel. They have a stake in the systems operating correctly, efficiently, and safely. Other stakeholders that must be considered are legal, regulatory, accreditation, and funding agencies and the public. This group provides the funding and oversight of the conditions for the research environment. The institutional administrators and leadership are ultimately responsible for the operations and the animals. The facility should promote safe, comfortable, consistent, wholesome, and appropriately enriched housing for the animals.

Institutional leadership relies on a design that is durable, easy to maintain, and economical to operate. The facility must meet all legal and regulatory requirements, protect the public trust and reputation of the institution, meet the research needs of investigators, provide for the needs of the animals, ensure their good care, and enable the program to meet all moral and ethical obligations of all parties.

Significance of Animal Housing

The validity and reliability of research data is no greater than the least reliable element in the process used to derive the data. Research involving experimental animals focuses on the measurement of their biological processes as the primary determinant of data. These biological processes are expressions of their genetic makeup and environmental influences, at every point from conception to death. It is in the control of environmental variables where ARFs play a vital role in the validity of nonconfounded research. The salient environmental variables to consider are

- Physical
  - Temperature
  - Relative humidity
  - Draftiness
  - Lighting
  - Sound
- Psychosocial
  - Population density
  - Transportation, movement, and handling
  - Group composition
  - Opportunities for stimulation and enrichment
  - Isolation
  - Pheromones
- Chemical
  - Contaminants of air, water, food, and bedding
  - Diet stability
  - Unsanitary conditions
  - Caging materials
  - Chemicals used in the facility
- Microbial
  - Disease-producing microbes
  - Viruses
  - Bacteria
  - Related organisms, fungi, and parasites
  - Nonpathogenic organisms that adventitiously affect biological systems

All these variables are addressed in two manners; one is in the operational protocols and SOPs, and the other is in the design of the facility. Both components are critical in creating safe and effective research environments.

Regulatory Requirements

The design, construction, and operation should meet all applicable codes, regulations, and standards (e.g., AAALAC International, the National Research Council *Guide*, the CCAC [2003] "Guide," Directive 2010/63/EU, the Animal Welfare Act and Regulations [AWAR], National Institutes of Health [NIH] construction standards, the BMBL, the American National Standards Institute [ANSI], the Occupational Safety and Health Administration [OSHA], the National Institute for Safety and Health [NIOSH], and the Americans with Disabilities Act [ADA]). These specialized vivarium-focused regulations are in addition to the building codes, zoning regulations, life safety codes, and individual institutional requirements that greatly influence the design of all animal facilities. Many of the vivarium-specific documents listed above are guidelines for the design and operation of animal facilities and are not necessarily regulatory requirements. It is sometimes required that international research work meet the requirements of the international community. Exploration of sources of research funding or exporting of research data may drive additional design criteria. It is the responsibility of vivarium management staff and institutional management to determine where these guidelines should be applied and where certain operational procedures override specific recommendations. An example is the general recommendation for locating a hand sink in all animal holding rooms. While in most facilities this is a desired location, if SOPs require the wearing of more than one layer of gloves and the doffing of the outer glove is to occur outside of the holding room, then a hand sink should not be necessary.

Interface between Physical Plant and Vivarium Operations

The physical plant must be designed to support the vivarium operations. The architect selected to design the animal facility must have experience and a deep understanding of vivarium operations in order to successfully integrate operational requirements into the physical plant. It is critical to the success of the ARF that the entire design team understand all the procedures and protocols that are to be implemented so that the facilities will be able to support those activities. Procedures for entering and exiting the facility, posted signs, cage changing processes, and other protocols must be well understood in order for the design to support these operations. The development of "flow diagrams" that outline a variety of vivarium operations are very useful in understanding the design requirements. (See the sections entitled "Modular Scheme Concepts" and "Functional and Zonal Relationships and Allocations" in this chapter that discuss in more detail the use of flow diagrams to drive the facility design.)

The Building

Mission, Goals, Operational Requirements, and "Basis of Design"

The facility should be flexible and adaptable with sealed, unadorned, durable surfaces. Beyond that, there are a number of considerations in outfitting and operating the facility and accommodating research including, but not limited to

- *Casework*: Where required, it should be easy to maintain and clean, and otherwise minimally built out.
- *Drains*: Although optional, functionality depends on the species. Incorporating drains, capped when not in use, enhances adaptability. Trough drains should be considered if large animals might be housed. Additionally, drains should be included if there is the prospect for housing aquatic species and also in the cage wash areas (when allowed by local ordinances). The capacity of the drain system must be adequate for all the facility's liquid discharge.
- *Procedural areas*: The provision of sufficient procedural space within the ARF will help to minimize or eliminate animal work outside of the facility.
- *HVAC*: Heating, ventilation, and air-conditioning (HVAC) and electrical systems should be designed with sufficient redundancy to maintain environmental parameters consistently within desired ranges within both the animal housing and study rooms and in consideration of potential system components failing. Ideally, the design should allow for the maintainability of mechanical, electrical, and building automation systems located outside animal housing rooms.
- *Animal entry*: A quality facility allows for sensible options for animals to enter the facility from approved sources, destined for quarantine and/or to be rederived. For rodents, a combination of active and passive design concepts should be employed to provide or maintain barrier conditions. Entry designs should be appropriate to encourage reasonable compliance to SOPs by personnel.
- *Life cycle costs*: Life cycle costs should be evaluated relative to initial purchasing costs to provide for the most cost-effective construction and operations evaluations and to provide true value engineering. It is at the time of facility design that sustainable economic operations and low per diem rates (NRC 2000) have the highest probable result.
- *Working environment*: Other important attributes for the ARF are to provide a safe and healthy working environment for all personnel, and provide amenities that assist in recruiting and retaining staff, and result in maintenance, appearance, and operations that instill confidence in the research community that their animals are receiving quality care.

Phases of Construction and Renovation

The four phases of renovation or new construction are planning, programming, design, and construction. These phases represent progressive stages of ever-greater detail in defining and delivering the final product. The planning phase begins with a vision that includes the identification of goals and assignment of subsequent work to task groups to achieve those goals. Programming is a process bringing ever greater granularity to the vision. It is a step where architecture/engineering (A/E) are intensively engaged, and it is highly, although not solely, reliant on the expert participation of members of the animal resources team. The high-paced, intense design stage culminates with the issuance of detailed construction drawings. Finally, construction and renovation represents the physical act of transforming drawings into a three-dimensional structure, and subsequently continues with "punch lists" and validation procedures.

Planning

Planning begins with the identification of a need, with the genesis usually at the level of top management. It may be driven by growth of programs, increased research staff, changes in research programs, evolving science or the needs of science, demands for greater efficiency, and/or a need for more sophisticated capability. The impetus for top management to instigate this process may ascend from a number of sources, such as strategic planning, goals for recruiting or acquiring new programs, desires of a substantial donor, or program growth driving the need to expand.

The next step in planning is to appoint the appropriate planning groups and then to develop a statement of work (SOW).

Statement of Work

A SOW is the product of a visionary task group of key leaders. It results in a preliminary set of general drawings and should be facilitated by the participation of animal resources personnel, but not always. A well-crafted SOW will

1. Identify the intent and objectives of the project
2. Define the requirements for accomplishing it, including the performance schedule
3. Set the budget amount or limitations
4. Project for the number of investigators, type of research to be served, and species to be housed
5. Establish the size of the project and whether it will be a stand-alone structure or relate to other buildings and adjacencies
6. Determine the basic and special operational requirements for the facility
7. Ascertain the sustainability objectives and potential environmental impacts
8. Stipulate the deliverables expected of the A/E specialists, including construction drawings
9. Prepare the more specific program requirements

Task Groups

The next step in the process is to form the various task groups that will drive a task or have a liaison relationship with the project. One such critical group is the *end users*. This group consists of those who will use, operate, and maintain the facility. It includes researchers, animal resources specialists, top management, physical plant management, engineering personnel, and occupational health and safety specialists. Their task is to review and evaluate proposed designs from an operational, performance, and scientific needs perspective. They may gather considerable amounts of information by questionnaires or surveys.

The *user group* is the most important group during the planning phase. The *design task group* is second in importance during planning, but assumes a role of primacy thereafter. This group serves as the hub of the design effort and provides central management throughout the process. It is authorized to make final decisions about the design and minimally consists of animal resources personnel, technical experts, safety representatives (including security), a budget manager, and a project manager. This group is also often responsible for resolving conflicts or referring them to the oversight group. Animal resources personnel include more than the director and may include husbandry management, cage wash personnel, the veterinary staff, and other appropriate representatives of the end user stakeholders.

The *oversight group*, composed of senior leadership, is the final arbitrator for unresolved issues. It has the authority to change the budget, commit funds, and provide organizational direction, and is responsible for ensuring that the final design meets the original goals. A fourth group is that of *legal counsel*, which is advisory to the oversight group on legalities, contracts, and the like.

Programming after the SOW Is Defined

More detailed planning occurs in the programming period after the SOW is defined. In this process, the operational and user needs are further refined and brought into focus. This step is augmented by the A/E group working with other applicable groups, which most commonly involve the project manager, representative scientists, animal resources personnel, and top management. Needs and requirements are identified, such as

- Surgery, imaging, behavioral, or other specialized procedural space
- Specialized housing requisites, such as for aquatics, large animals, or potential hazards vital to research (e.g., radiation, dangerous chemicals, and infectious agents)
- How the physical plant staff will relate to and maintain the facility

The type and size of the animal or animals to be studied, the degree of pathogen exclusion desired, and the level of biocontainment form the fulcrum on which the entire facility programming and then design must balance. Mice are the species typically requiring the greatest allocation of space in modern ARFs (Figure 18.1). At this time, additional information is collected either by getting all the critical people together and asking them for projections of future needs or by polling the key stakeholder groups. Animal resources will be called on to provide the historical context of census growth and ratios of animals to investigators. They will also orient the A/E firm to the housing, equipment, and procedures preferred in animal care processes.

Animal resources is expected to refine goals and objectives and then clearly communicate them to the A/E firm and provide ever greater detail in the qualitative and quantitative descriptions of programs and

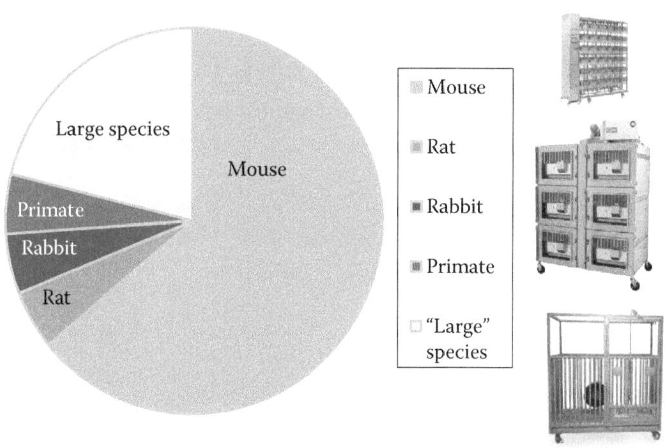

FIGURE 18.1 **(See color insert.)** Relative proportion of space required by species used in research in 2006. (Figure and data compliments of HDR, Inc., Atlanta, GA.)

activities to be served by the end product. The means used to house animals and types of them, determination of the various types of space and square footage accorded, and how traffic and work will flow in the facility are elucidated. It is at this stage that conflicts first arise between competing needs and priorities of the various end users in light of the limitations of the budget and unforeseen constraints emerge bearing on any of the aforementioned areas.

Program of Requirements

Programming culminates in the development of a program of requirements (POR) and the design intent document. In the design community, this is commonly referred to as the basis of design (BOD). The POR or BOD is a comprehensive document consisting of a detailed narrative and diagrams, and it also serves as the starting point for commissioning and validation efforts. Here, the goals and objectives have been refined. The POR will provide the spatial allocations between nonprogram and program components, including administrative, mechanical, animal housing, and animal care support. Space functions and quantities, the functional adjacencies, and staffing projections will be defined by this time. Following this process, the comfort and ease of research for the human occupants should be a priority. Conditions allowing flexibility will have been considered. Cost will come into focus. Code constraints and existing conditions at the site and other impediments will have been identified and addressed. Critical infrastructure requirements will be defined, including the quality of construction; mechanical, electrical, and plumbing (MEP); quality requirements for materials; redundancy (e.g., power, fans, and elevators); security and traffic control; and types and location of fixed equipment. At this time, the schedule also takes shape and, in the case of renovation or disruption posed to nearby activities, any phasing and relocation plans if the facility is to remain occupied. Programming concludes with a detailed budget, precise drawings, and a tentative construction schedule. The design intent document describes the minimum functional requirements of the facility, defines the commissioning process, and states the methods of performance testing and assessment of functionality.

Errors in Planning and Programming

The errors that commonly plague planning and programming are

- Questionnaires and interviewer biases (such as collecting opinions or unsubstantiated and wildly enthusiastic projections instead of facts)
- Collecting too much or too little information
- Catering to a dominant or domineering user or need in lieu of basic requirements
- Failing to establish true priorities and attending to wishes rather than needs
- Perpetuating design flaws
- Ignoring occupancy needs for fixed equipment and for postwarranty preventive maintenance of expensive equipment

Design flaws arise from inexperienced design teams and conclude in the failure to meet the needs of both the research and animal care end users. Design flaws occur when the team overlooks essential activities, such as

- Where feed or bedding will be stored and how these materials will be received, processed, and moved to animal rooms or staging areas
- How people will enter rooms and where any personal protective equipment (PPE) might be kept
- How animals will enter or exit and the level of biosecurity
- The methods and paths of waste removal, dock capacity, and traffic flow, which are particular risks

In medical center facilities, careful consideration should be given when entering into projects that are enveloped within, or adjacent to, patient care areas, as there will almost invariably be complications regarding dock access, traffic flow, maintenance, and renovation that must be addressed.

The needs for nonfixed equipment are considerable and expensive, and should be considered for inclusion within the scope of work. This may include the caging (or a portion thereof) and support equipment to enable "turnkey" use of the facility upon acceptance of ownership. Also to be considered are biosafety cabinets, lab chairs made of sanitizable material for use in procedure rooms, and office furniture and hardware. Ignoring the needs for nonfixed equipment is often the consequence of establishing the budget prematurely, excluding these important line items, or the surreptitious pirating of the appropriation toward other needs. Oversight in ensuring sustained preventive maintenance of vital, expensive equipment is a simple error that must be prevented in order to ensure continuity of operations and prevent animal resources from taking on unforeseen financial burdens. It is important to remember that whatever equipment is installed must eventually be removed, and the design must take this into consideration.

Design

Step three, the design phase, starts when the POR is approved, and it ends with the award of the construction contract. Exertions devoted to design by all parties may account for about one-third of the total project cost (Wilson and Palmer 2005), so it is nothing to be trivialized. This phase requires close and regular work between the A/E and design task groups. It is a period of iteration and reiteration, drawing and redrawing, and back-and-forth exchange of concepts that requires prioritization, extra devotion (often beyond normal working hours), and thick skin. Design transforms the POR data into biddable and constructible documents. It follows a sequential process of conceptual design to design development to review of construction documents to review of "prefinal" construction documents to review of the final bid documents.

Concept Development

Conceptual (schematic) design is the preliminary phase and includes conversion of the spatial relationships and block diagrams of the POR into line drawings, floor plans, and elevations. Several concepts are usually developed. New constraints may begin to exert an effect, such as in an asymmetrical footprint allocation or unusual dimensions of the structure or certain areas or components. This phase is influenced by many factors, and there is no "one way" to do it. Conceptual development ends with the selection of a single concept and proceeds to design development.

Design Development

The product of design development is a written report on major design objectives, how they will be achieved, and the rationale for the design. From that report, drawings, schematics, and floor plans are developed that show details of corridors, rooms, doors, elevators, shafts, and the overall exterior. As progress is made, the construction documents are spawned, giving rise to drawings showing room titles; numbers and dimensions; door sizes, types, swings, and dimensions; fixed and portable equipment; and plumbing fixtures. Special facilities are drawn in detail. The cost estimates and projections become clearer and more meticulous. The prefinal review set of documents is, in essence, a complete set of drawings, product specifications, and budget estimates to be submitted for approval. These drawings are complete, including colors, finishes, furnishing, and signage. It is at this point that there may be finalization of commissioning design intention and specifications. Once this step has been completed, significant changes cannot be made easily. After approval, the final bid documents are created. These specify the quality of items and systems, insurance and bonds, locations of the bid opening, bidding dates, and conditions of the bid.

Bid Conditions

There are generally two conditions of the bid. The general contractor approach is where the construction company provides a total cost and budget for the entire project and accrues any savings in time, materials, and the like that occur along the way. The benefit to the project owner is knowing the total project costs, but the risk is for inferior substitution or work. It is possible under these conditions for savings to be split in a negotiated fashion between the owner and contractor. An experienced project team, with adequate administrative support, serves as a deterrent to inferior substitutions. The construction management approach allows room for negotiation during the course of the project, but under the constraints of the budget. For example, if the cost of flooring materials is less than budgeted, the savings could be diverted by the owner to other upgrades (as in better flooring or new doors). This approach is more time-consuming to manage, but allows for balance of cost and quality and sometimes for upgrades of quality.

Errors in Design

The common errors inherent to design arise from many factors, such as

- Budget imbalances between mechanical systems, equipment, and finishes
- Problems related to the selection of inferior interior finishes
- Selecting the wrong or inadequate equipment
- Not considering how large equipment will enter the facility for installation (or removal and replacement) or if surfaces have adequate weight-bearing properties
- Ignoring redundancy and fail-safe systems that protect animals and experiments
- Inadequate security
- Failing to plan for coordination among the building trades as they are brought into the process
- Suboptimal storage capacity
- Vermin control during and after construction
- Disregarding the possibilities of flexibility and less intrusive future expansion
- Not accounting for noise transmission and abatement
- Overlooking codes, safety factors, and emergency procedures

Construction

Construction begins with the notice to proceed and ends with the occupancy of the facility. It accounts for about 67% of the total cost of a project. It is important to appreciate that a building does not arise from the design documents, but rather from shop drawings developed from the design documents. Shop drawings include schematics, drawings, parts lists, catalog cuts, mechanical and electrical diagrams, and product samples. They are developed for almost every component of the facility: electrical, plumbing, ventilation, and the like. As product or other substitutions are often necessary, proposed, or undertaken during this phase, it is critical that the design group, including the animal resources staff, define which specific submittals they must review as part of the approval process. Economic realities often dictate that the end users do not get everything they want or that was originally intended, but the facility user should have some voice in "value engineering" decisions and an ability to make cost–benefit decisions. The A/E and construction groups drive the construction phase.

It is very important that the animal resources staff stay engaged in the process, because there is arguably no group that is more invested and interested in the outcome of the facility. Animal resources should have representation at all regular meetings and be authorized to take ad hoc walk-throughs and evaluations of the site at will in order to ensure that everything is being done according to the original plan, and that all equipment, installation, and construction meet specifications. Being regularly on the site allows for mistakes to be addressed immediately and prevents surprises at the end. While this requires additional effort and documentation, it prevents costly errors. Another important element is to have zero tolerance for eating and drinking within the ARF confines during construction in order to prevent attraction of vermin and possible contamination. It is at this stage where vermin prevention interventions, such as boric acid installation in cavities, should be implemented.

Project Management

Project management is integral to this phase. This is usually provided by an institutional employee skilled and knowledgeable in the areas of design, construction, specialized ARF requirements, interacting with contractors and the trades, managing the budget, scheduling, and representing the ultimate owner and not the contractor. Project managers may arise from different backgrounds, but they must be skilled in sensibly balancing end user needs with top management desires for on-time schedule completion and budget control. Project managers typically are operating at the point of tension between controlling up-front costs for top management and delivering a product that allows for low operating costs, maximum efficiency, and more rapid payback to the institution in savings, investigator retention, and scientist recruiting. If project managers are overcommitted to the end user at the expense of top management, they risk getting fired. If they are overly loyal to the interests of top management at the expense of the end user, the end user may be victimized with uninformed and unsanctioned "value engineering" decisions or left with an unfinished facility that will remain incomplete or require finishing with funds from other sources. These realities make it obvious that the sympathies of the project manager may ultimately lie elsewhere than with the needs of the end users. At an operative level, the project manager should take the lead with regard to notifying other institutional shareholders, such as hospitals or imaging sites, of construction-related impacts, such as road closures, utility shutdowns, and any other noise- or vibration-generating activities. During a renovation project, the ARF should take the lead to ensure that construction-related impacts are minimized and communicated to the users.

Construction Monitoring

It is critical for animal resources personnel to understand that their regular inspections of the facility under construction best ensure that the final product conforms to the design documents. While the primary responsibility for this rests with the A/E group, periodic multidiscipline reviews must be made of the construction. As part of this monitoring, the A/E group and representatives of the owners are responsible for documenting accepted work so that payments can be made. During the construction phase, meetings are regularly held by the contractor and progress reports are made by the A/E to the owner to keep the project on schedule and verify the construction in place.

Change Orders

"Change orders" and their management are another vital area requiring understanding. Change orders result in alterations to the original project and derive either from unforeseen developments or as a consequence of poor planning, ill-defined programming, design errors, or impulse. Change orders should be avoided because of the associated attendant high cost and scheduling delays (Figure 18.2). This reinforces the importance of thorough programming, planning, and design. For example, it is significantly

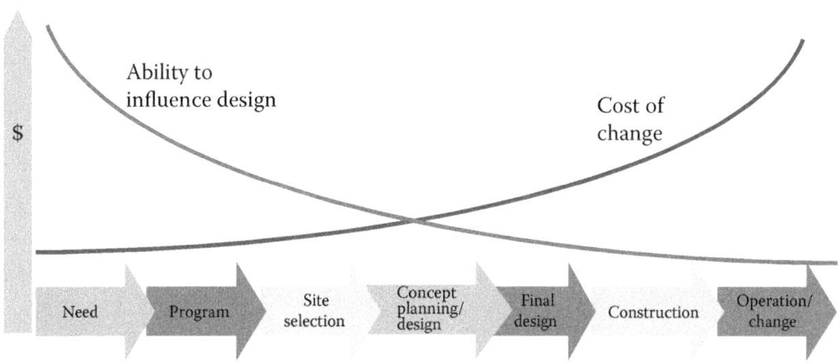

FIGURE 18.2 General representation of the cost of changes relative to the progress of a construction project. The cost of changes and of change orders escalates and the ability to influence design declines as a project advances.

less expensive to install extra conduit (to possibly electronically secure some or all animal rooms in the future) during the scheduled phase of construction than to retrofit conduit and surface penetrations after the structure has been mostly completed. It is prohibitively expensive to add plumbing and drains to a facility constructed without them.

Project Closeout

Project closeout is a process to ensure that all aspects of the contract have been met and installation is acceptable. At this time, all systems and equipment should be tested. A separate "commissioning agent" is often employed by the owner to verify that the equipment and systems operate as specified. A step beyond commissioning is to engage in validation, whereby the work, materials, and design are constantly evaluated every step of the way and where all are logged, tagged, and approved. Optimally, this involves additional engineers or "validation consultants" and more up-front expense, but it may pay off in an immediately functional and highly efficient facility. In reality, this service often is restricted to the project manager and animal resources team. A punch list is developed by the A/E group and representatives of the owner during final inspections to document required corrections or omissions that must be addressed as a condition of acceptance. The animal resources staff must take an active part in developing this list, since it is their workplace that is at stake. All changes during construction should be documented in "as-built" drawings for future reference. Accurate as-built documents are essential, as modifications and renovations years in the future are inevitable and having the as-built conditions is a critical legacy.

Final payment is made to the contractor after a "release of claims" is obtained, which ensures that all subcontractors and suppliers have been paid. With respect to occupancy, it is best not to take possession until all punch list items have been corrected. Otherwise, the owner may be responsible for any damage or unattended items not on the punch list once occupancy occurs. Most items and construction are warrantied for 1 year. Often overlooked is the "11-month walk-through," where deficiencies are identified and corrected before exhaustion of the warranty. Warranties should be in place and negotiated to become effective and start at the time of occupancy or first use, not when first delivered to the site. Equipment requiring preventive maintenance contracts beyond the first year should be identified and contracts arranged. All complaints should be collected by the users and passed on to the A/E group during the first year. This is important, as the A/E group and contractor are no longer on site and will be unaware of conditions if not notified.

Errors in Construction

The most common management and leadership mistakes that occur in association with construction are not having a plan for conflict resolution and expectations for the schedule. Another error is ambiguity in assignments and charges to the task groups, including responsibilities and lines of authority. Other confounders at this stage include ill-defined program requirements, which predispose the process to frequent change orders or inferior construction and a loss of focus, resulting in failures to work around problems or address them immediately to maintain momentum. Imbalances of scheduling and budget with respect to the functionality of the facility risk overweighting decisions in favor of conserving "up-front" (or "first") costs and not balancing these with operating ("life cycle") costs. For example, a finish that is cheap to apply may require expensive maintenance or premature refinishing down the road compared with a more expensive, lasting, and durable product. Other risks include letting aesthetics take priority over needs and ignoring users simply to remain on budget or schedule. Insufficient user input and fidelity to user needs can result in a new facility that is underutilized or requires immediate renovation. Enslavement to the budget may also predispose to poor quality control (QC), whereby budget demands converging with a hurried pace and compressed schedule leave insufficient time to monitor and act to correct deficiencies.

The Design Team with the A/E Group

The A/E group formally joins the project during programming or at the beginning of the design phase and consists of experts from a qualified firm. It is critical that the firm have experience in ARF design, particularly with HVAC engineering, as good HVAC engineers are a rare commodity. They should also be knowledgeable of hardware and finishes appropriate for any ARF and able to project costs, savings, and payback relative to certain designs, space utilization, animal housing systems, staffing levels, census figures, and automation options. The selection of the A/E group may be made by top management or in a more egalitarian fashion, with the institution inviting firms to bid on the project with subsequent steps of developing a short list, considering references, and scheduling presentations and interviews. It is important to look past the firm itself and to the credentials of the specific A/E group proposed for the project.

Design/Renovation

Due to changing requirements or inherent imperfection, renovation to some degree is intrinsic in the life cycle of some or all of a facility. The usual drivers leading to renovation are

- To correct deficiencies and defects, save money or energy, or enhance sustainability
- Provide redundancy
- Enhance security
- Improve operations logistics and efficiency
- Address safety
- Allow for greater flexibility
- Permanently repurpose space in response to a change in activities

The scale of renovation can range from major, grand, all-encompassing projects where walls are ripped down and infrastructure is gutted to more focused applications leaving room dimensions essentially intact, such as to add automated drinking water delivery, switching from static to ventilated cages, refurbishing marred surfaces, or replacing failing equipment. There is also the simple fact that structures, systems, and equipment eventually begin to show their age and fail to hold up. The ideal is to avoid or minimize the need for renovation, as it is costly, disruptive of adjacent activities, and often dusty and noisy, thus placing the onus on proper construction of adaptable resources the first time. Considering the original cost of the space, adding on the expense of demolition and waste disposal plus rebuilding in a confined area, the effects on the availability of utilities, and the value of the lost use of the space and forsaken productivity, the outlays for renovation can be staggering. If only a few existing conditions substantially penalize a renovation project, such as the expensive, time-consuming, and hazardous removal of asbestos-containing material, new construction can quickly become more economically attractive than renovation. On the positive side, the principles of renovation design are essentially those of construction, except with the element of demolition often added.

The overall complexity of renovating in the midst of preexisting activity places an even greater premium on planning than new construction and on managing and sustaining research through the alteration. The impact of noise and vibration on the care and maintenance of animals undergoing experiments during facility renovations can be substantial and extends well beyond the traditional zone of demolition and reconstruction. There is some evidence referencing the need to mitigate noise and vibration during construction projects (Rasmussen et al. 2009). Phasing the renovation is a way of potentially limiting the impact of noise on the animals. Constructing the facility properly and with principles of adaptability the first time is also a way to mitigate (i.e., prevent) renovation.

The cost of renovation may exceed construction on a per gross square foot basis (Wilson and Palmer 2005). Just as with construction, the MEP components are generally substantial in renovation, and with resource conservation and environmental effects having considerable weight. Renovations of substance almost invariably require either new air handling systems or significant upgrades and modifications. Beyond the sophistication and capacity of equipment exerting cost-effectiveness, complex building systems are usually concealed behind the sometimes cramped, constrained, and congested walls and ceilings of the ARF. Adequate floor-to-floor height is an important factor in this regard, making top floors and sometimes basements often best suited as the sites for renovation. The infrastructure, for example, includes the weight-bearing capability of the floors that must be significantly robust, as animal research sites and their occupants are particularly hard on their environment and also susceptible to vibrations from power sources, machinery, and traffic. Another cost factor occurs in low-rise and midrise animal facilities where verticality adds costs and complexity. This increase is due to the challenge and higher cost of installing and maintaining more complex HVAC systems in multistory buildings and managing the needs of occupants on adjacent floors. Payback periods may vary by institution and circumstance, but projects in most cases usually are seen favorably and deemed worthwhile if the payback is 5 or fewer years.

Establishing the Budget

The approximate parameters of the budget are usually established at or before the start of planning, sometimes resulting from feasibility studies. There is a danger in setting the budget prematurely before completion of the planning and programming stages. Premature estimates may result in having to adjust or eliminate needed facets within a low, premature cap. Budgetary constraints and space limitations frequently cause tension between the end users focused on safety and low operating costs, scientists envisioning alternative uses for space, and the administration and members of the project management team whose objectives are low up-front costs, finishing under budget, and moving on to tackle the next project. These sorts of differences are not unnatural, however. Cost justification, the weighing of the expenses and benefits of an initiative or perceived improvement, is a normal and customary business practice. It allows for optimal monetary investment and uses dollars as the unit and basis of comparison. Consequently, when competing for the same allocation of money, the relative merit of needs of the animal resources program must be subjected to the same intense scrutiny as all other potential applications. Animal facilities, in general, are some of the most expensive facilities to construct and operate. These high costs are associated with the continuous, 24/7/365 operational requirements, the robustness of the engineered systems, the level of finishes that are required, and the equipment required to operate a vivarium facility. It is very important that the vivarium management participate in the initial stages of the development of a project to ensure that there is adequate funding to construct the specific facility required. The cost can range from \$400 per square foot to more than \$700 per square foot, depending on the complexity, location, and size of a given facility.

Construction Delivery Process

The selection of the delivery process, or the method by which the contractual approach to the construction of the facility is accomplished, can be a large cost driver. The most traditional approach is termed "design–bid–build," which describes the process where the design of the facility is 100% completed by the selected A/E firm and then the construction documents are issued to construction companies, who then bid on the project based on their estimated costs. This is typically a very competitive process, where the lowest price is typically awarded the project. While cost is always a factor when considering a contract, a "value-added" approach may be specified. This enables the contract evaluation team to consider intangible factors (i.e., experience of the A/E firm with regard to vivarium design) versus having to accept the lowest bid. The contractors who bid on the project can be selected by the institution or invited bidders, or the project can be made open to any qualified contractor. At most public institutions, open, competitive bidding is often required, whereas private institutions are able to select the contractor or contractors that they want to bid on the project. The approach taken by the public institutions is seen as delivering the most competitive pricing; however, this process requires that the construction documents be very thorough, as any missing elements or mistakes in these documents will lead to additional costs required to cover the deficiencies in the documents.

A second approach is called "design–build"; this approach teams an A/E firm with a construction firm to offer turnkey services to the customer. This approach has not been widely accepted by institutional clients, as they and many others see benefits in controlling both the A/E firm and construction firm and their contracts to ensure that the goals are met for each party. In this approach, the A/E firm is typically selected by the contractor, and it is this entity that holds the contract for the A/E services. The drawback of this arrangement is that the A/E firm may be prevented from acting as design and engineering advocates for the client because the client has no control over the A/E, as it works for the contractor. This approach, however, is widely used in the private sector, particularly on industrial-type projects. The process is seen to provide the fastest schedule from design to occupancy. The design–build team works together to prepare the design and construction documents, often looking for the lowest-cost approach, and then presents both the design and the costs in one complete package for the owner to approve. If the design–build approach is used, the animal resources group should carefully review all drawings and submittals to avoid value engineering approaches that may require costly renovations or lead to an inefficient use of part or all of the facility.

In the last 10 years or so, a third approach has become very popular, particularly with institutional clients, called the construction manager (CM) at risk approach. This process begins much like the traditional design–bid–build approach, with the client engaging the A/E firm directly to begin the design process. At some point, typically early in the design process, a contractor is selected to provide the actual construction, without having priced the entire project. The CM then works with both the A/E firm and the client to provide interim cost estimates as the design progresses, guiding the overall costs of the project. This is often seen as a benefit to both the client and the A/E firm in that the project costs are estimated earlier in the process and changes required to manage an established budget can be implemented in early design phases rather than later, causing major changes to the construction documents. This approach seems to provide owners with a more secure sense that their project will be completed with the available funding and that cost overruns can be avoided. However, due to the ongoing cost estimates and the contractor's need to eliminate risks of cost overruns, this approach is typically the most expensive, possibly adding 10%–15% to the costs of a given project. This premium is often seen by the client as the cost associated with minimizing the risk of overbudget projects.

There are several iterations to each of these approaches, but the above describes the basics of the three most widely used delivery methods. Each client must evaluate what is most important to it in selecting the appropriate delivery process—overall costs, schedule, and/or quality. Each approach has its pros and cons, but the selection of a delivery method should be made with a thorough understanding of the risks and benefits of each approach.

Cost Considerations

A vivarium is one of the most expensive types of facilities to construct and operate. The actual products of construction visible to the end user (surfaces, equipment, and interior accoutrements) constitute less than 50% of the investment in the project; approximately 50% of the cost is that of the MEP components (Figure 18.3). Cost factors, when meeting the needs of the end users (typically research staff), are a primary criterion. Therefore, the size and types of spaces provided must be considered with the ultimate costs of the space and equipment. There is a minimum quality standard that must be provided to meet the requirements

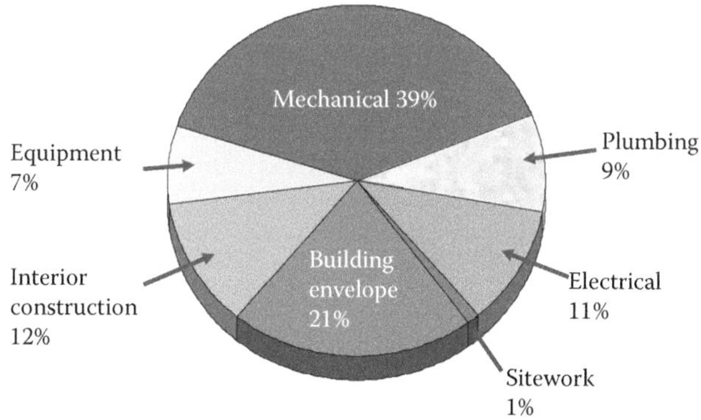

FIGURE 18.3 General proportions of construction costs by system and/or type of work illustrating the high combined costs of MEP components.

of the regulatory bodies, which will typically require a minimum cost per square foot to deliver. This minimal quality level must be achieved to successfully operate a facility that is safe and protects the health of all the occupants. The only other factor that can be changed to meet an established budget is the reduction in the overall size and scope of the facility. Constructing a below-standard facility should never be considered, as human and animal health and safety could be at risk. It is easier to downscale size and develop additional space in the future, whereas upgrading a poor-quality facility is much more difficult and costly. The vivarium management must be diligent during the design processes to ensure that the minimum quality standards are built into the design, which will then be implemented during construction. The ARF management team should be involved in the establishment of the initial scope and budget for any proposed new or renovation project, to ensure that there is adequate funding to meet the scope and quality required.

Scheduling matters can also affect cost. Procurement times, construction phasing, and local factors affecting the availability of trade specialists require planning that takes these into consideration. In particular, local factors, such as several large construction projects in a given market, may drive the availability of trade professionals and price of construction materials much higher than historical averages, both of which lead to increased costs. Similarly, after large-scale devastating events, such as hurricanes or earthquakes, construction costs and contractors will be at a premium and may be diverted away from ongoing local and regional projects.

Site

Site selection for a vivarium facility often does not engage the vivarium management staff with the responsibility left with the design team. However, there are a few critical issues that should be considered when reviewing a potential site for an animal facility that require input from the vivarium staff. These include the accessibility to the loading dock area, the delivery of supplies and materials, and the removal of waste products, which will be a daily occurrence. The "grade" of the land should also be considered. For example, a loading dock area that is not even can provide significant access challenges in locales that receive snow or ice. The availability of highly reliable utilities should also be an important consideration. In some locations, this may require utilizing well water, and percolation test results that will support the intended facility must be obtained. Obvious issues of floodplain locations and geotechnical challenges should not be overlooked. Other important criteria may include the ability to expand the facility and if the site is sufficient for expansion.

Permanent and Temporary Structures

A simple, but fundamental early decision in the design and construction process is to determine whether to create a new building, renovate an existing one, or buy a prefabricated structure. Specific to our industry, there are fast-track, fully operational, highly integrated, turnkey-ready, modular ARFs that can be delivered rapidly, at competitive prices and situated in a number of sites, including on top of existing buildings. If desired, the modular building can be preequipped with caging, cage washers, and mechanical equipment included and installed. If one elects to go prefab, then the designer will require certain information, such as the species; equipment needs; requirements for the flow of personnel, materials, and animals; preferences on room sizes and layouts; space allocation for program needs; and overall square footage required.

If one elects to build or renovate, questions to be addressed are

- Will the facility allow for centralization of activities or must there be some degree of decentralization?
- Will there be single or multiple stories?
- How will maintenance personnel access MEP?
- What will be the number and arrangement of corridors?

Stratification of the ARF onto multiple floors has its pros and cons (Table 18.1). The best advantage is convenient access to ARFs by researchers. A significant drawback is that these facilities are highly elevator

TABLE 18.1

Comparative Assessment of Single versus Multilevel Facilities

Criteria	Single Level	Multilevel
Security	●	◒
Management/supervision	◒	○
Construction costs	●	◒
Operating costs	●	◒
User access	◒	●
Service access	●	◒
Disease control	◒	●
Noise control	◒	●
Environmental control	●	◒
Flexibility	◒	◒
Vertical transportation	●	○
Travel distance	◒	●
Renovation	◒	●
Expansion/contraction	◒	●
Net to gross	◒	◒

Note: ● = optimal; ◒ = acceptable; ○ = suboptimal.

dependent, thus adding considerable up-front construction costs, making access and servicing by the animal care team subject to bottlenecks, and creating the risk or reality of use by nonanimal users. Another downside is that the piston-like movement of the elevator cars in the shaft may result in the injection of air from the animal facility onto other floors and vice versa. Multiple floors may also spawn investigator and departmental territorialism and a reluctance to share space beyond a related cohort. Typically, multilevels are most viable for the largest of facilities (e.g., those encompassing greater than 20,000 net square feet [NSF]).

Corridors are important considerations. Will the rooms be accessed for all purposes from a single corridor? Will there be multiple corridors for a clean-to-dirty traffic flow pattern? Or, will the corridors open into suites containing one or more of a variety of rooms?

Types of Space

The programmed types of space critical to an ARF segregate along the lines of what is mandatory and what components are optional, depending on how the facility will be used. Animal holding rooms and certain support spaces are required elements for even the most meager satellite-type facilities. Additional spaces may be needed if the ARF is to be sufficiently large to operate with cost-effective efficiency, requiring continuous staffing during business hours, and usually with an associated cage wash resource regularly in operation. In a facility optimally staffed with a supervisor and six to eight animal care and cage wash technicians, beyond animal holding rooms, the following types of space are usually included:

- Procedure rooms (including euthanasia)
- Cage wash with related support spaces
    - Soiled items staging
    - Gross wash-down
    - Mechanical equipment space
    - Work areas for the processing of soiled and clean materials
    - Clean holding (10%–20% of caging and accessories in use at 100% capacity)
    - Sterilization and sterile holding
    - Utilities service and access

- Spaces and amenities to facilitate related administrative functions
- Spaces and amenities to support veterinary clinical activities
  - Pharmacy
  - General procedure room for emergency minor surgery and other clinical interventions
  - Record-keeping and communications area in lieu of using mobile technology
  - Isolation, hospitalization, major surgery, and substantive diagnostic procedures as separate areas or incorporated into a clinical hub
  - Necropsy and veterinary diagnostic lab
  - Space and storage suitable for the processing of clinical and research samples
- Support spaces
  - Dock access for receiving and shipping
    - Determine if both clean and soiled docks are needed
  - Clean storage, such as feed and bedding
  - Idle equipment
  - Gas cylinders
  - Expendables, such as PPE and including cleaning and other chemicals
  - Holding and staging areas for trash, soiled bedding, other waste, and carcasses
- Other spaces to consider depending on size of facility and need
  - Feed preparation
  - Custodial closets
  - On-site waste processing and disposition (e.g., chemical digester and incinerator)
- Vivarium staff areas (and research staff while in the ARF)
  - Offices to enable confidential interactions
  - Technical workstations
  - Diverse communications resources
  - Break rooms
  - Locker rooms
- Administration areas
  - Senior staff offices
  - Reception
  - Conference rooms
  - Training room
  - Library
  - File room or archives
  - Control room for servers, environmental monitoring, and security hardware

If the program has an active research component, wet laboratories for faculty purposes may also be necessary. Large, complex facilities may be outfitted with select animal research labs, large animal surgery resources, and other modalities. Specialized laboratory facilities that may be needed include those for the generation and cryopreservation of genetically engineered animals; quarantine and health rederivation; behavioral and physiologic phenotyping; acoustical, photoperiod, and other environmental manipulations; and imaging and irradiation. A large animal surgical resource will require one or more operating rooms, a site for preoperative animal preparation, an operating room prep, a surgeon scrub, a recovery or intensive care unit (possibly shared with the animal prep area), a gas cylinder closet, a janitor closet, equipment storage (C-arms, cautery, tables, monitor stations, and the like when not in use), controlled substance storage, a pharmacy, an office, dressing rooms with lockers, and laundry storage and processing.

It is important to understand the concepts of net square footage and gross square footage, as they contribute to many facilities that start out tracking the needed space against the budget. The budget and associated costs are based on the total facility, or the gross square feet (GSF) of the building. This includes every component of the building and what is needed to support the operations and includes usable spaces, connecting corridors, all MEP areas, stairwells and elevators, duct and utility shafts, information technology (IT) and service closets and rooms, structural columns, and even the space occupied by the thickness of the walls and partitions. The net square footage typically refers to the spaces listed above, which represent the usable areas of the building. The building efficiency is the ratio between the net or usable square footage and the gross or total square footage, and in a typical ARF, the efficiency will be between 50% and 60%. The cost of the facility is always based on the gross or total building square footage, so in the early planning stages, before floor plans are even prepared, it is essential to apply the efficiency factor to the required space requirements, net square footage, when preparing early cost estimates.

Modular Scheme Concepts

The types of spaces desired for the ARF are then organized into programmed zones, although these may not be strictly segregated and certain components of zones may be comingled (Figure 18.4). The usual zones that are considered are standard animal housing and procedures, specialized animal housing and

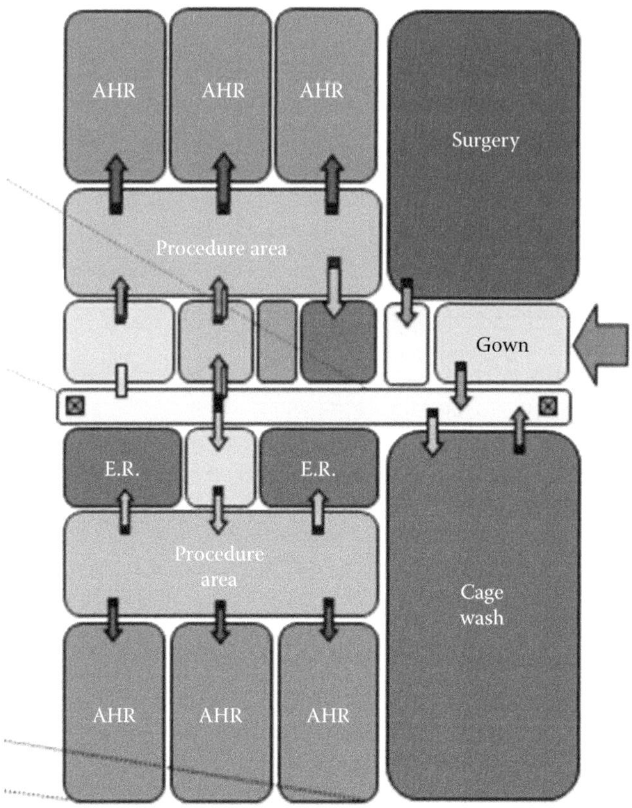

FIGURE 18.4 **(See color insert.)** Modular scheme relationship and flow concept at the design stage of a construction or renovation project. AHR, animal holding room; E.R., equipment room.

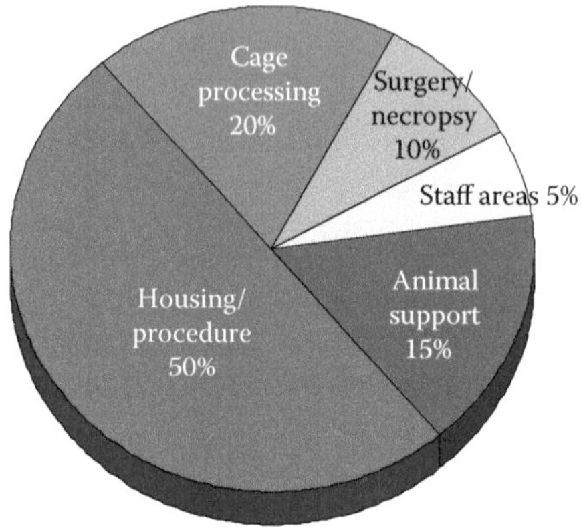

FIGURE 18.5 Demonstration of the typical programmable space allocations by proportion in a typical animal research facility.

procedures (e.g., barriers and hazard containment), cage processing, logistics, personnel, administrative, and traffic (corridors, elevators, and stairwells) (Figure 18.5). The standard animal housing procedure zone is usually organized with three primary considerations:

1. Will the zone be organized around corridors or use a suite system?
2. A suite system typically constitutes 50% or more of the program space.
3. The standard zone must be organized and built to isolate and insulate species and research procedures sensitive to noise from noise-generating species and activities.

Special animal housing and procedure zones are usually built to contain hazards, particularly infectious agents, but may also have applicability in certain cases for dangerous chemicals and radiation. In some cases, barrier exclusion facilities may be constructed, but it is important to consider that these are costly to build and maintain. They are also highly reliant on user compliance, which can overcome and render ineffectual the most superb design. The cage processing zone is fairly explanatory in that it consists of all washing staging and preparation of caging and sanitizable equipment, along with related storage and aspects of distribution. The logistics zone addresses receiving, waste holding, general supply storage, bedding storage, and food holding with components sited near the dock and cage processing zone, and with smaller satellite components distributed throughout the facility. The personnel zones are usually a mix of a proportion centralized in a contiguous arrangement, but with some dispersal. This zone constitutes locker and dressing rooms, toilets, showers and other amenities, air showers, supervisor offices and clerical space, break rooms, and anterooms. The administrative zone may consist of offices for senior leadership and clerical and secretarial staff, office equipment, a conference room or library, a kitchenette, a mail room, a systems room, office supplies, and one or more waiting rooms. Corridors are critical for the flow of material and personnel and segregation of "clean" from "soiled" or contaminated areas. Corridors also may serve to differentiate public spaces (e.g., the administrative zone) from animal activity or other restricted areas, and identify transitions to different types of space (e.g., locker rooms or surgery), and may be used to separate pathogen-free areas (animal housing or clean cage wash) from contaminated or "dirty" zones (soiled cage wash).

Functional versus Zonal Relationships, Including Space Allocations

The ARF consists of program space and nonprogram space and is typically quantified in terms of both GSF and NSF. A term closely related to the latter is net assignable square feet (NASF). The GSF of the facility comprises all the program and nonprogram space and is the sum of space of all floors out to the exterior wall surfaces—it essentially amounts to all the spaces and surfaces that must be constructed

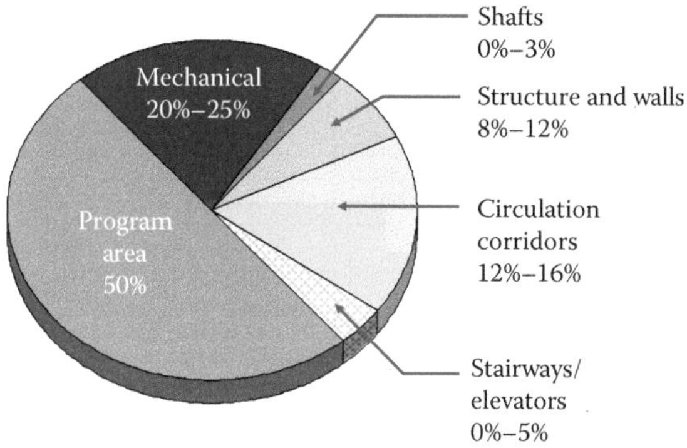

FIGURE 18.6 Illustration of the typical distribution of space between that which is programmable and the other types of structural and support features.

and maintained and is the figure used in calculating construction costs for a facility (Figure 18.6). A/E groups use NSF to address all the internal components that must be designed. NASF is the area that is suitable for occupancy for specific activities. It excludes shared and nonprogrammed space and is often used to distribute costs to activities. The nonprogram spaces are the general areas that cannot be assigned to specific activities and generally amount to those required for circulation, such as corridors, stairwells, lobbies, and elevator shafts; those needed for support responsibilities, such as utility closets and mechanical rooms; and those needed for architectural features, such as columns. As an estimate and by rough order of magnitude (ROM), a facility is about 50:50 program to nonprogram. With respect to the NSF that is programmable, animal housing usually occupies 40%–50%, with support space constituting the remainder (Figure 18.5). Through repetitive, reiterative analysis and reconfigurations, it is possible to squeeze additional program space from the footprint, possibly exceeding 60%.

A number of factors may converge to dictate the size of the ARF. For large research-specific structures, a good starting point is to plan for an ARF that is 10%–20% GSF of the total research space. Relative to this allocation, the larger the research facility is and the higher the animal density housing (i.e., the greater the ratio of rodents and fish to other species), the smaller the overall percent of the ARF can be. Where animal work may be prohibited or minimized in laboratories, the overall program space may need to be 25% larger (i.e., 12.5%–25% of the GSF) to account for additional procedural space. Space requirements also may be projected based on factors such as the number of investigators multiplied by the average GSF or NSF used per principal investigator (PI) at the institution or, for mice, using the average number of mouse cages per investigator and projecting for about one mouse cage per NASF. Overall space and its allocation may be mitigated by special programs or requirements, such as a transgenic mouse facility, a behavioral core, a large animal surgery, an infectious disease biocontainment suite, a fully contained shower-in barrier, isolation suite labs, imaging facilities, a gnotobiology facility with isolators, or teaching and classroom space. For example, an MRI unit may have an allocation of 1500 NASF per device. When using the ROM approach, it is best to use multiple methods to project the facility size. If the results are concordant, this should be reassuring that forecasting is on the right track. Where there is divergence, the situation should be analyzed further.

Room Detail Sheets

The room detail sheet is an import planning and design tool used by many architects and planners to document the requirements of each individual space in the facility. It can also be used as a facility management tool, informing the team of the parameters around which each space is designed (Figure 18.7). As a facility manager, the input to and the review of these data sheets is important to ensure that the individual spaces will meet their intended function. The data should include everything that is required of the space, including the size, the environmental requirements, the required utilities, the finishes, and the equipment that is to be included. Equipment should include not only the built-in features, but also the removable

Detailed room requirements

Facility name:
Room name:
Contact:

Date:
Produced by:

Utilization
- Hours of use
- Hours of operation
- Days per week

Architectural
- Function
- Min. floor area
- Fire rating
- Sound transmission coefficient
- Adjacencies
  - Primary
  - Secondary
- Floor finish
- Partitions
- Crash rails
- Base
- Ceiling
  - Height
- Varmin control
- Interior borrowed light
- Door 1
  - Primary leaf size
  - Secondary leaf size
  - Door material
  - Frame material
  - Rating
  - Vision panel
  - Access control
  - Lock type
  - Door jamb protection
- Door 2
  - Primary leaf size
  - Secondary leaf size
  - Door material
  - Frame material
  - Rating
  - Vision panel
  - Access control
  - Lock type
  - Interlocked
  - Door jamb protection

Special construction
- Floor loading
- Load shielding
- Acoustical
- Other (note)

Audio/visual
- Overhead projector
- Projection screen
- Video monitors
- Smart walls/devices

Mechanical
- Temperature
  - Other:
- Humidity
  - Winter 30%±10%
  - Summer 50%±10%
  - Range
  - Uncontrolled
- Quality
  - Min. air charges/hr.
    - Occupied
    - Unoccupied
  - 100% cubicle air
  - Recirculated air
  - Directional airflow
  - Air supply
  - Air exhaust
  - HEPA filtration/supply
  - HEPA filtration/exhaust
  - Clean room class
  - Rough-in filter
  - Mech noise criteria (NC)
  - Low slotted exhaust
  - Heat generating equip.
- Clean steam
- Steam/condensate return
- Chilled water (CHW S/R)
- Other:

Casework (linear mm)
- Base cabinets
- Wall cabinets
- Standing countertop
- Seabed countertop
- Open shelving
- Tech station
- Cart/storage
- Tall glassware case
- Reagent rack
- Special needs (identify)

Lab equipment characteristics
- Floor vibration criteria
- Light sensitive
- Concentrated loading
- Vibration producing
- Heat producing
- Noise producing
- Animal species
- Number of cages

Electrical
- Lighting type:
- Lighting level at 750 mm AFF
- General lighting level
- Dimmer
- Timer
- Safe light
- Nightlight
- Zoned lighting
- Special lighting
- Standby lighting
- Standby power
- UPS
- Receptacles
  - Special equip. outlet
  - (See notes below if checked)
- Communications:
  - Telephone
  - Data
  - Intercom
  - TV outlet
  - Speakers
- Special equipment

Security/monitor
- Information database system
- Badge reader
- Special monitoring
- Alarm/strobe
- Panic button
- Standby power
- Camera
- Other:

Chemicals
- Bases
- Acids
- Solvents
- Carcinogens/regulated

Storage misc. (linear mm)
- Bookcases
- Tack board
- Chalk/marker board
- Lockers/cubicles
- Lab cost hacks

Plumbing
- Laboratory natural gas (G)
- Laboratory vacuum (V)
- Laboratory air (A)
- Portable hot water (HW)
- Portable cold water (CW)
- High purity water
- Lab animal watering
- Nitrogen gas (N_2)
- Carbon dioxide (CO_2)
- Special gas
- Cylinder gases
  - Inert
  - Flammable
  - Tools
- Sinks
  - Cap (100×180×155)
  - Hand (230×300×150)
  - Medium (350×400×150)
  - Large (400×550×250)
  - Service (400×700×250)
  - Wall sink
- Floor drain (FD)
- Floor sink (FS)
- Safety shower/eyewash
- Drench hose (DH)
- Hose bibb
- Hot/cold wash down
- Other:

Fire protection
- Sprinkle system
- Extinguisher
- Fire alarm:
  - Horn
  - Strobe
- Other:

Hazardous storage
- Flammable
- Chemical
- Radioisotope storage
- Biological storage
- Controlled substances
- Safety closet
- Other:

Hoods	Size	Services (see plumbing)	Count
Chemical fume hood			
Radioisotope hood			
Perchloric hood			
Distillation hood			
Walk-in hood			
Laminar flow hood			
Canopy hood			
Snorkel			
Low slotted exhaust			
Biological safety cabinet			
Type:			
Type:			
Special enclosures			
Type:			

Notes and remarks

FIGURE 18.7 Room detail sheet example.

equipment, such as caging systems, cage change equipment, food and trash receptacles, and any other components that could impact the proper circulation, access, and operational protocols. These data sheets should be reviewed and approved by all stakeholders prior to the start of the actual facility design. A helpful asset in addressing room detail sheets from the institutional perspective is for the institution to have predefined all applicable design, construction, materials, and equipment performance standards required. It is ideal for the institution to have established standards for such items as architectural features, MEP, environmental control, monitoring and alarms, communication, security, animal drinking water (ADW), acoustics, life safety, and emergency power. Specifications should be clear down to the level of items installed or mounted in various spaces, including case work, drains, implement hangers, and even hand soap and paper towel dispensers. These standards can then be compared with the content of the room detail sheet. This is often presented as a checklist of choices for each specific type of space (e.g., cage wash, four-rack room, or pen room), and the appropriate selections and entries can then be accomplished.

Architecture

The architecture of the vivarium includes many components, from the exterior design of the building to the selection of appropriate finishes for all the interiors. Generally, these are comprised of floors, walls, ceilings, doors, penetrations, and appurtenances. It is important to select an architect who has significant experience in the design and planning of vivarium facilities, as they are very complex, and only through experiences and lessons learned can the architect successfully complete your facility. Animal facilities are typically not built as stand-alone facilities, but are usually part of a larger, research-focused facility. The intermixing of multiple functions with differing criteria poses many challenges to the successful design. Everything from the column bay spacing and corridor widths to the MEP systems needs to be carefully integrated to ensure that they support all the functions required in a given facility. The actual building design should be reviewed for compliance with the room data sheets, accessibility to loading functions, accessibility to the research laboratories, and other collaborative functions. In a multistory facility, the vertical circulation can have a major impact on the facilities operation, and the location of stairwells versus service and passenger elevators is critical for both operational requirements and security control (see the "Security Systems" section for more specific information). In multiuse facilities, the vivarium is often designed as a stand-alone component within the facility due to the specialized nature of this type of space. Dedicated HVAC, plumbing, and electrical systems are often included to isolate the vivarium and allow it to operate on different schedules from the rest of the facility. The shape of the building must also be considered in that many contemporary designs involve curves, complex exterior designs, and unusual angles. The efficiency of the animal facility can be greatly impacted, as the typical rooms and individual spaces and their associated equipment require simpler, rectilinear spaces to function properly.

Mechanical Systems

The mechanical systems are typically one of the most important components of any animal facility and generally the single largest component of the construction costs. These systems must be sufficiently robust to support continuous operations over many years without shutdown or failure. Only high-performance HVAC should be considered. The two main components of the HVAC system are its supply air side and exhaust air side. Both supply and exhaust sides should have 100% redundancy (backup units of sufficient size for both supply and exhaust) built in. The supply air side must provide adequate conditioned air to meet the loads of every space, within the tolerances established by the design requirements. The loads that are included in determining the volume and temperature of the supply air include internal heat loads generated by the occupants of the space, both human and animal loads, the heat generated by the lighting system, and the equipment within the room.

Two delivery methods are available to furnish the supply and exhaust air systems. One is a constant volume (CV) system that, as its name implies, provides a constant, fixed rate of both supply and exhaust air based on the loads described above. The room pressure or directional airflow is established by a fixed differential between the supply and exhaust air systems. The other approach is a variable air volume (VAV) system that uses a series of automatic dampers, variable-speed fans, and digital controls to modulate both the supply and exhaust air systems to respond to changes in the space. Examples would be fume hoods or other exhaust devices cycling on and off. The CV system is less costly and easier to maintain, while the variable volume system provides more accurate control of the space and saves energy, but requires higher levels of sophisticated controls. In addition, spaces located along an exterior wall will need to include thermal loads created by the external environment. The total heating and cooling loads are a major component in sizing the HVAC systems; however, other requirements may generate loads that exceed all thermal loads. These could include institutional standards for minimum ventilation rates established for specific operations. These ventilation loads are typically measured in air changes per hour, and are typically established by the institution and the regulatory guidelines outlined elsewhere in this chapter. The established or required ventilation rates can, and typically do, exceed the internal heat loads, and then become the driver for sizing the HVAC systems.

In some specialized facilities, a third component can add additional requirements to the sizing of the HVAC system, and that is the exhaust requirements of operational and/or research equipment. Fume hoods, ducted biological safety cabinets, and specialized research tools can require ventilation rates that exceed the other two components of the loads on any given space. The HVAC engineer must calculate each of these components, and it is the largest of the three that will finally size the HVAC system.

It is considered a best practice to require these systems to use 100% of outside air, prohibiting the recirculation of air from one space to another. This practice greatly increases the size and loads on the mechanical equipment, as outside air temperatures can range from well below 0° to well over 100°. It is important for the facility management to understand the loads and which component is the driver for the system sizing because over time, changes within the individual spaces will occur and reanalysis of the system's capacity must be completed. It is a common approach in a multiuse facility to have dedicated HVAC equipment for the vivarium component due to the continuous operational requirements, and the more stringent environmental conditions that are required to be maintained. The location and accessibility of the HVAC systems is critical to allow for proper maintenance, repair, and eventual replacement of the equipment.

The exhaust air side of the HVAC systems is an integral part of the overall systems and is used to remove thermal loads, odors, and fumes from the facility, and to provide the required ventilation rates. Every room or space should be served with an exhaust component, which, in addition to removing odors, heat load, and airborne contaminants, works in concert with the supply side to provide differential air pressure between any adjacent spaces and from the building to the exterior environment. Downdraft tables, ducted biosafety cabinets, and fume hoods should also be considered part of the exhaust system. This differential pressure is created by the difference between the supply air volume and the exhaust air volume; the greater the difference between these two, the greater the pressure differential will be. This differential is critical because it dictates the "directional airflow" between two adjacent spaces, creating either positive or negative pressurized spaces. This directional airflow is used to move air from the cleanest spaces to the dirtier spaces in the building and between the building and the exterior environment. In general, the building itself should be positive to the exterior environment to prevent the intrusion of outside air, moisture, and airborne contaminants. Individual rooms can be either positively or negatively pressurized based on their function and the requirements for room air cleanliness. For example, an operating room would be positively pressurized to reduce incoming contaminants, while typical quarantine holding rooms would be negatively pressurized to prevent pathogens from permeating the rest of the facility. It is a critical component of the room detail sheets to document the requirements of each space, including a requirement for being either positively or negatively pressurized. It is also important to provide rooms that are adaptable, that is, moving from being positively pressurized to negatively pressurized. This is achieved by the use of room air valve systems that provide a range of air volumes that can be adjusted up or down to change the pressure differential. Obviously, this adaptability comes at a cost, as the adjustable valves are more expensive. Most systems use sensors and a feedback mechanism to maintain the temperatures and pressures of the air supplied and exhausted. The location of the sensors should be such that they are accessible for repair and replacement. For the ARF, it may be worth considering the installation of differential airflow sensors with visible indicators in areas where the correct differential is critical, such as in operating rooms, quarantine or isolation rooms, and cage wash.

The location of the HVAC and other mechanical equipment must also be considered. This equipment, as with all mechanical systems, requires maintenance and access, so clearances around the equipment are important for regular maintenance, repair, and over time, eventual replacement. The location and ease of access to the ductwork and room air valve systems should also be considered. Most rooms will require regular "air balance" testing to affirm the differential pressures. Accessing the air valve systems from a hallway or service corridor is preferable to having service personnel having to access areas above the ceiling in an animal room, and with careful planning to prevent overcrowding, this can be accomplished. This means considering how much interstitial space between floors is necessary so that all supply and exhaust ductwork is accessible for repair and replacement. For some facilities, it may be advantageous to consider a large enough interstitial space to allow for full standing height, enabling maintenance personnel to access HVAC, IT, and even lighting equipment foregoing the need to enter the ARF. The more difficult this equipment is to service, the more likely that regular maintenance will be reduced. The large equipment also requires large floor space, and planning of a vivarium must take into consideration that more than 40% of the facility will be devoted to support space required to operate the facility. In most instances, vivaria are designed to be supplied with 100% outside air systems—*no recirculation*—which requires large volumes of outside air to be brought into the mechanical air handling units (AHUs), so their location in relation to the exterior is important. The intake air grills must be located away from external contaminants, such as loading docks, trash areas, other building exhaust systems, and other areas where contaminants will be drawn into the AHUs. Conversely, the location of the exhaust fan system must also be carefully selected to avoid any reentrainment of the exhaust air back into the supply air system of the facility or surrounding facilities. In dense, urban areas, it is sometimes required to have a wind–wake analysis performed to understand how the exhaust and supply air components are impacted by the local environment, and to verify that the proposed locations of the systems do not create reentrainment of exhausted air. Location of the supply and exhaust air within each room should also be carefully considered. Refer to the room diagrams and the planned caging to avoid having an air supply (draft) directly in line with cages. A position in the middle of the room is generally preferred. The location of the exhaust ducts is also critical so that their placement in the room is not later blocked by caging or equipment, impeding the airflow. The locations for both the supply air diffusers and exhaust grills are critical to maximize the performance of the HVAC systems and ensure that no "dead zones" in the rooms are created and that odors and contaminants are removed from all parts of the space.

The preferred arrangement is to supply animal facilities with 100% fresh (i.e., nonrecirculated) air, despite the fact that recycled air provides substantial energy savings. Recycled air brings with it the risk of accompanying odors, allergens, and airborne pathogen cross-contamination. If air is recirculated between spaces, it should not exceed 50% of the supply. Unless a specific risk assessment dictates otherwise, the air must be recycled strictly to the room of origin after conditioning with activated charcoal and subjected to high-efficiency particulate air (HEPA) filtration to remove volatile pollutants and airborne particles. Considerable design input is merited in these cases, as the extent and efficiency of filtration should be proportional to the estimated risk. Assessment must account for the amount of fresh air necessary to ensure air quality and thermal and humidity requirements of the species, animal population, and/or activities in the space. Although the temptation to reduce energy costs is imposing, the reality is that the risks associated with recycling air and the costs of maintaining the necessary attendant filtration systems may be too great.

Humidification of the outside air may also be considered part of the HVAC equipment and mechanical systems. Achieving the *Guide*'s recommended range of 30%–70% relative humidification for the most commonly used laboratory animals can be a serious challenge for some facilities, particularly in some locales or during some seasons. Neotropical species of nonhuman primates housed indoors may benefit from humidification of air. Humidification systems should be sized so that they can maintain the facility at 50% relative humidity during the least humid times of the year based on the locale, taking into account the relatively high amounts of air changes per hour that occur in a vivarium. This should allow the ARF to maintain relative humidity within the recommended range throughout the year.

Plumbing Systems

The plumbing systems include not only the potable water and sanitary drainage systems required for an ARF, but also all the specialty piping systems, including ADW systems and the specialty gases used to support the research. These gases can include natural gas, compressed air, vacuum, and other specialty gases. (Refer to the "Special Utilities" section for more detail.)

Incoming water can be treated by a variety of different means. Sand bed filters can be used for removing particulates. Charcoal filters can be used to remove chemicals, although they do require regular maintenance. Reverse osmosis (RO) filters remove many types of particulates and ions from the water, but again require regular maintenance. Deionized (DI) water has almost all the minerals removed from the water. Ultraviolet (UV) light may also be used in a supplement to filtration to disinfect water by killing bacteria. UV bulbs must be periodically changed, as they lose their "kill" effectiveness over time. Water softeners may be useful in reducing scale buildup if the source water is hard (i.e., high in calcium or magnesium cations). Consideration should be given to the order in which the various components are employed; for example, water softeners that add salts will not be beneficial if installed upstream of an RO filter. Source water should be checked throughout the year, as seasonal shifts may occur (e.g., higher salt content in winter months and higher cold water temperatures in summer months).

Domestic hot water should be available for hand sinks, slop sinks, hose connections, scrub sinks, kitchenettes, and showers, generally supplied at ~110°F. Domestic cold water should be available for all the same functions, to also include eyewash stations and safety showers, at whatever temperature the municipality, or well, supplies. Chilled water lines (40°F–50°F) may be used in the HVAC system or required for some steam sterilizers to operate properly if the domestic cold water is not cold enough for vacuum pumps to achieve enough vacuum. This would be a consideration in areas with generally high year-round temperatures or if the piping insulation is insufficient. "Laboratory" hot water (>140°F) should be available to operate cage washers to avoid causing undo strain on the equipment.

The design of the sanitary drainage system in animal housing rooms is generally based on the animal species to be housed in the room. However, if adaptability is a critical requirement of the facility, the drainage system should be designed to support a variety of species. Rooms supporting larger animals are designed for "wash-down" procedures, and typically require continuous trench drains running the length of the room. The floors are sloped, to quickly remove the water used in hosing down the room. In rooms that are dedicated to rodents or if flexibility is not required, a small floor drain, located under a hand sink, should be considered, but even here, it may not be required. In considering drains, keep in mind that it is overall vastly less expensive to install them as a component of the original construction and cap them when in disuse than to undergo an expensive and disruptive renovation or retrofit later. Consider a trap or garbage disposal unit for each drain line where large animals may be housed. Drain lines in an ARF will get clogged, so clean-outs should be accessible.

The cage processing areas, often the spaces with the largest plumbing requirements, must be carefully designed to support all aspects of the cage processing (washing) functions, from initial wash-down of heavily soiled cages and equipment to the final sanitization cycles. Large cage washing equipment often requires pits in order for caging systems to be rolled directly into the washers without requiring ramps. Once located, this equipment is difficult and expensive to relocate, so careful planning, ideally based on process flow diagrams, must be implemented. The sizing of water lines to meet all the dynamic pressure and flow requirements for cage washing equipment is also important, as is the need for adequate hot water supplies. Booster water heaters may also be required to provide 180° water to the equipment. These booster heaters can be integrated with the equipment or can be furnished as part of the building infrastructure. Sanitary waste piping must also be sized to adequately remove large quantities of water released in short periods of time without backing up, and the building plumbing engineer must select the appropriate drain system. Drainpipes should be at least 4 inches (10.2 cm) in diameter proceeding from the animal facility sources to the municipal main sewer line. Larger diameters (>6 inches) of pipe, however, should be considered, as they promote greater flexibility in the use and adaptation to a wide range of activities (ILAR 2011). Consider the municipality's requirements with regard to discharge water (temperature, antibiotics, pH, and contaminants) and whether a system may be needed to meet the discharge requirements.

The ADW system is one of the most critical systems in the ARF. The first decision that must be made in developing this system is determining the type and quality of the drinking water to be delivered to the animals. The *Guide* and other ARF operational recommendations typically require that clean, potable water be provided. However, each veterinary team can and must determine the type and quality of water to be used from the many options available. RO water is one common type of water used in the ARF.

Other options and upgrades to the basic water are available and can be applied to potable or RO water. Dechlorinated, hyperchlorinated, acidified, and other treatments can be selected, some based on specialized animal needs, but more often the selection is based on the experience of the veterinarian team. The researchers can also participate in the selection of the type and quality of the drinking water system.

Once the type and quality of the drinking water to be used are selected, the method of distributing that water to the animals must be determined. The two approaches are an automated distribution system that delivers the drinking water directly to the animal cages and one that employs the use of individual water bottles that are installed in the cages. Within the last 10–15 years, the automated delivery of drinking water, often called autowatering, has become more widely used for many reasons. These include

- Reduced labor costs associated with the use of water bottles
- Belief that this system delivers a more consistent quality of water
- Autowatering provides a continuous fresh supply
- Water bottles provide a limited supply

Most autowatering manufacturers "design and install" their system. These systems require space for the pumps, and possibly chlorination or acidification systems. They also require access to a drain (i.e., function specific, sink, or floor) for periodic system flushes. There are several reasons why water bottles can be seen as the best option, and these include

- Elimination of the possibility that the automated water distribution systems could run continuously and flood the cage
- Elimination of the possibility of system error causing a failure to provide water
- Medications or other materials cannot be delivered to the animals with an automated system

Delivering drinking water through water bottles requires equipment to process and clean or sanitize the bottles and additional equipment to fill the individual bottles. New systems that upgrade the traditional water bottle include a disposable or one-use pouch or sack system that eliminates the need to wash and sanitize reusable bottles. However, these one-use products still require labor to fill and deliver to the cages. The autowatering system is not maintenance or labor-free, as the sipper tube inside the cage must be cleaned and serviced on a regular basis. The building distribution system must be maintained to prevent the buildup of biological contaminants inside the piping. The ADW should be monitored for quality on a daily basis, and the watering manifolds on each rack should be sanitized and flushed each time they are rotated, or more frequently for rats and mice, as they do not drink that much water. Further details on watering systems may be found in Chapter 28.

Special Utilities

The special utilities for an ARF can vary, depending on the types of animals to be housed and the equipment to be used, but more importantly, on the types of research and procedures that must be accommodated. Steam is often provided by boilers serving the entire building or may come from a central plant facility serving many buildings. The quantity and quality of the steam required to meet all the requirements within the ARF must be reviewed to determine the best approach in supplying steam to the ARF. The quality of the steam supplied from central plants or even building-specific boiler systems must be reviewed, as the chemicals used in many systems to prevent corrosion of the piping can impact ARF processes, including potential damage to certain thermoplastics and the buildup of scale on cages and equipment. In addition, many systems require a period of shutdown for maintenance, which is typically not acceptable in the ARF. It is not uncommon for the ARF to have a stand-alone boiler system to control the quality and availability of this critical system. Steam traps should be properly installed to remove condensate from the steam lines. Most cage washers and autoclaves need to be plumbed to a condensate line or condensate return system.

The majority of special utilities can operate from portable gas cylinders or compressors, or they can be plumbed into the facility. If the facility will be plumbed for any of the gases, consider the location of the bulk tanks and how easy it will be to have them refilled. Portable gas cylinders may require special storage and handling (CFR, Title 29, Section 1910.101) depending on the locality. If any gases are plumbed, the type of wall fitting used should match the existing or planned equipment. Several different types of connections and fittings (e.g., Chemetron and Ohmeda) are available for the different gases; which type should be specified during the design phase.

- *Carbon dioxide*: Carbon dioxide (CO_2) is used in most facilities, whether it is for euthanasia or the operation of incubators. Having centrally plumbed CO_2 in specific areas can minimize the need for portable cylinders, although a portable units should still be available. A dynamic pressure of approximately 50 psi is suitable for most facilities. Consideration should be given to automated monitoring of tank farms to prevent complete exhaustion of the inventory, particularly in the middle of a procedure.
- *Compressed air*: Compressed air is used to operate pneumatic devices, commonly found on washers and sterilizers. Compressed air should be supplied so that it is clean and dry. Portable air compressors are available if the facility is not plumbed to supply compressed air, although there is an additional noise factor to consider, along with ensuring that the air being supplied is of high enough quality (i.e., oil-free and dry). Dynamic supply is typically 80–120 psi so that it can be downregulated as needed for specific applications.
- *Natural gas*: In a vivarium, natural gas is often used to operate Bunsen burners and may be plumbed to laboratory or procedure areas or directly to biological safety cabinets, although this latter arrangement is often discouraged by safety experts. The dynamic pressure may range from 0.25 to 2.00 psi.
- *Nitrogen*: Nitrogen is typically used to operate surgical equipment. A dynamic pressure of ~175 psi can be downregulated at the point of use as required.
- *Medical air*: Medical air may be needed for certain respiratory applications, either for patients or for medical devices. Typically maintained at ~55 psi, medical air is subject to specific regulations (NFPA 2015).
- *Nitrous oxide*: Nitrous oxide is typically supplied to surgical suites for use as an anesthetic. It will typically be in a portable tank with dynamic pressures at ~50 psi.
- *Oxygen*: Oxygen is used primarily in surgical suites or locations where patients may require supplemental oxygen. A dynamic pressure of ~55 psi can be downregulated at the point of use as required.
- *Vacuum*: Medical vacuum can be used to eliminate waste anesthetic gas (WAG) from the immediate environment. Medical vacuum can also be used to provide suction through suction canisters that contain any aspirate and may filter the vacuumed air. Portable suction units with compressors are available if the facility is not plumbed for suction, although there is an additional noise factor to consider with the compressor-driven units. A continuous vacuum of ~75 kPa should be expected from either plumbed or compressor-driven vacuum lines.

Electrical Systems

All areas should be wired with an anticipation of the maximum amperage draw that will be needed in the space. A backup generator, powered by an alternative source (e.g., diesel), should be in place to allow at least some functionality within the vivarium during electricity loss. Backup generators should be maintained on a regular basis. Animal holding rooms should be wired with "emergency" electrical connections that draw power from a backup generator, with the possible addition of "regular" receptacles. It would be ideal to wire animal holding rooms so that the entire room is not on the same circuit, so that when a breaker is blown there are some functional outlets and lights in the room. Some individually ventilated cage (IVC) systems have supply and exhaust units that sit on top of the rack, so receptacles in the ceiling or high up on the walls may be advantageous for cord management. During the design phase, consider what connections may be needed in certain areas in the future; for example, a dryer may need a

40 A circuit (as opposed to a 20 A circuit for general receptacles) and could require additional electrical work after the facility is up and running. Any electrical work requires completed "schedules"; a map of what breaker serves each receptacle, light, and device in each electrical box; and a master book for the facility, which will be invaluable in the future. Waterproof covers should be used wherever there is a chance of a splash hazard. Covered and gasketed electrical receptacles should be used in animal holding rooms. Areas where a lot of water is generally present (e.g., cage wash areas and aquatics areas) should use weatherproof receptacle covers.

Lighting

Lighting requirements within an ARF should be carefully considered to provide a species-specific environment that also meets the research and husbandry needs of the animals and the facility. There are multiple considerations with regard to lighting, most notably the intensity of the light, spectrum of the light, and cycle of the light. Additionally, ultrasound emitted from fluorescent lights may be a confounding factor for certain types of acoustical, behavioral, breeding, and other types of experiments. Commercially available octave band analyzers or bat detectors (Jennings et al. 1998) have proven useful in identifying ultrasonic noises and their sources in areas where animals are housed or studied.

The *Guide* advocates automated control of the lighting cycle and, as a general rule, illumination of 325 lux measured 1 m from the floor. Most rodent species can thrive at a lower intensity level, in the 40-lumen vicinity (NASA 1988). Albino strains of rodents may suffer retinal damage when exposed to light above 130 lumens (ILAR 2011). When deciding on an operating intensity, the needs of the people working with the animals should be balanced against the ideal setting for the animals. It may be useful to maintain the light intensity in the room at a level comfortable for the species, but with some lights that can be turned on via a timer to accommodate the needs of the people working in the room. Red lights, in a range not visible to rodents, can provide enough light for people to work in a room while maintaining a dark cycle for the rodents. Night vision goggles may also prove useful when working with any species during a "dark" cycle.

Spectrum of illumination may be important for certain species under unique circumstances. With sunlight exposure, cholesterol is converted to vitamin D3 in the ultraviolet B (UVB)–exposed skin. However, this is addressed for laboratory animals in that diets formulated for them are routinely fortified with vitamin D3. Providing this is the case, special illumination is not ordinarily required. Animals housed indoors and without exposure to natural sunlight, particularly marmosets, tamarins, nursing nonhuman primate infants, and poultry, may benefit from a supplemental source of UVB radiation (NRC 2003). In light of the continued imprecise understanding of the minimal requirements and safe upper limits of vitamin D3 in the diet and the effects of certain experimental manipulations on food consumption and nutritive state (NRC 2003), some designers and facility managers find it prudent to provide supplemental natural or artificial UVB radiation in animal housing areas. This necessitates a design that allows either unimpeded exposure to solar radiation, the installation of windows or skylights fabricated with UVB-transparent panes, or artificial light sources for room illumination that emit substantial UVB energy at appropriate wavelengths (i.e., 290–315 nm) (Ullrey and Bernard 1999). Natural lighting via window or skylight may present a security concern and add unneeded research variables with regard to light cycle.

Zebrafish and some other aquatic species seem to thrive more when provided with a dusk–dawn lighting cycle (Helfman 1981). Other species may require a light cycle that is constantly adjusted to account for their natural rhythm. Ideally, the lighting in an ARF will be easily controlled via electronic means. Light cycles should be programmable for each individual room, and the programs easy to adjust as needed. Owing to the importance of photoperiod consistency and regularity, and its impact on virtually every biological response, light cycle monitoring via a system with a sensor that provides on–off time reports, and possibly intensity levels, will be labor saving and help ensure animal well-being and protect experiments. Undetected lighting malfunctions arise most commonly from technological failures and human activity (i.e., activating lights during the dark cycle). Consequently, the cycling of the photoperiod should be nominally monitored using electrical current sensors for each lighting level in each room and/or with photocell sensors reporting via the building automation software. If manually operated lights are to be used in any animal room, the switches should be of a timer type to avoid unnecessary light exposure. Each room should be on an individual circuit (breaker) whenever possible to avoid lighting issues with multiple rooms when a problem occurs in one.

Additional considerations are

- Ceiling-mounted surgical lighting may require a mounting plate in the ceiling.
- Light ballasts should have gaskets and be sealed on the room side of the ballast to facilitate pest management.
- In high-humidity environments, such as cage wash or aquatics areas, consideration should be given to light ballasts that are completely sealed, not just on the room side.
- The type of light ballasts chosen should be weighed with regard to desired functionality and how easy it will be to change out the bulbs.
- Emergency lighting powered by a backup system should be of a level to allow people to make their way safely through the facility.

IT, Data, and Communication Systems

IT, data, and communication systems are becoming increasingly complex and important to the operation and management of modern ARFs. Traditional communication systems include facility-wide intercoms and hardwired telephone systems, which are rapidly being replaced with wireless systems, which allow for communications within the facility and also connect the ARF to outside communication links. It should be noted that wireless systems in the ARF must be more robust than in other types of facilities due to the separation of spaces, often with heavy masonry constructions, which reduces the effectiveness of the system. Routers and/or additional repeaters are required to reach and service all areas of the ARF and must be carefully located. Data systems are also more widely used in the day-to-day management of the ARF, documenting daily census counts and health status of the animals and documenting and tracking the wide variety of research protocols that are in effect throughout the facility. Researchers, veterinary staff, and animal care technicians must all have access to this type of information to ensure that all procedures, health issues, and other concerns are quickly and efficiently distributed to all those that need it.

A fairly recent development in ARF management is the advent of radio frequency identification (RFI) tagging, or the use of electronic markings and telemetry tools and infrastructure to monitor animal census counts in "real time." This allows for immediate monitoring of the animals, and serves to support the collection of per diems as required by most funding agencies. These systems are replacing the requirements for paper "cage cards" that have been used to document specific protocols in place for the animals in each cage, the frequency and cycles for cage cleaning schedules, and the health status of these animals. This is a time-consuming process with many opportunities for errors and is rapidly being replaced by the new RFI systems. An expert in this area should be consulted in selecting the best system for each specific facility for both the type of system to be employed and the unique installation requirements for the RFI infrastructure.

Access to all the critical data associated with the information required to successfully maintain and operate an ARF is important, and the location of computer workstations and access ports is an important component in the design process. Typically, workstations are provided for animal care staff, veterinary staff, and researchers, outside of the animal areas, where gowning and other personnel protocols are reduced. A segregated area for access and input into ARF data systems is often a desired feature in the ARF, but data input and access areas immediately adjacent to animal holding rooms and procedure areas should be provided to allow for immediate access of daily reports and condition assessments. In addition, certain data systems storing critical and/or long-term data should have a backup system located away from the ARF in case of lost data due to computer malfunctions or, in the worst case, data systems that are destroyed or damaged in a fire or other disaster.

Life Safety Systems

There are numerous regulatory requirements that document the "life safety" requirements for all facilities, both for new construction and for renovated facilities. For the most part, these deal with the safety of the human occupants of the facility, and include requirements for exiting a facility in an emergency, fire protection, and lighting and emergency communication systems. In the early stages of the design process, the type and fire rating of the building structure, separating floors and walls, will be determined based on the occupancy of the entire building. Once those have been determined, the specific requirements are then set by the regulatory building codes and must be adhered to in the design of the facility. Fire separation of exit ways, stairwells, and exit corridors will be determined and then must be maintained throughout the life of the facility. The fire rating of these exit ways is an important consideration, as any penetrations, including windows and doorways, are limited and require fire ratings equal to that required for the exit pathway. These are particularly important to understand during any renovations to the facility, as the required exit ways must always be maintained.

Current building regulations typically require sprinkler systems to be installed in all new construction, and their operation should be understood by the users of the facility, as accidental actuation of a wet pipe sprinkler system can cause significant damage to equipment and impact the health of any exposed research animals. Sprinkler heads require a certain amount of "clearance" between the ceiling and the highest objects in the room, and this should be considered when deciding on ceiling height or the installation of vertically integrated structures (e.g., shelves). The fire sprinkler system is a component of the overall fire protection system, which includes the fire alarms. The fire alarm system includes both audio and visual signals that must be located throughout the ARF so that all areas are covered. The ARF poses unique challenges to these systems because the sound level of the audio alarms and the flashing lights of strobes can have a negative impact on the research animals. There are audio systems available that limit the sound levels to those less disturbing to the animals, but these must be reviewed and approved by the authorities having jurisdiction (AHJs). The impact of the flashing strobe lights can be mitigated by the careful location of the strobes themselves, isolating them away from direct lines of sight into animal holding rooms. Again, the design and installation of these critical systems must be approved by the AHJ.

In addition to the building fire rating requirements and the fire protection systems, certain components of the facility are required to operate in the case of an emergency. These include lighting systems that provide adequate illumination to support safe exiting of the facility and certain communication systems that are used by first responders as they enter and evaluate the condition of the facility and search for any left-behind occupants. The emergency lighting systems can be battery operated or, more commonly in an ARF, powered by the emergency generator that is typically available in many ARFs. It should be mentioned here that there is different terminology regarding this topic, which may be important to understand. The building codes and fire protection requirements refer to *emergency power* as a system that supports the mandatory life safety systems discussed above. In the ARF, critical systems are required to operate due to the nature of the facility and special occupants housed in the facility. Those systems are supported by *stand-by* or *backup* power, differentiated from emergency power required by the building codes. While this can be seen as semantics, it is important to understand the difference. Life safety is a mandatory requirement and must be provided for, while critical ARF functions are required by best practices and animal use guidelines. Many ARFs combine these requirements and support them on the same emergency generator. However, building codes require that these two services be distinct and separated by transfer switches and subpanels. In rare instances, it is possible that generator power could be limited to those needs for life safety, and other functions may be bypassed. In these and all cases of emergency, an emergency or disaster plan should document these differences, locate these critical switches, and be prepared to implement other strategies, should generator power not be available to support critical ARF functions.

Emergency eyewash stations and showers are an OSHA requirement where the eyes or body of any person may be exposed to "injurious corrosive materials" (CFR, Title 29, Section 1910.151(c)). The ANSI standard uses the definition of *hazardous material*, which broadens the spectrum to include "compounds that have the capability of producing adverse effects on the health and safety of humans" (ANSI 2014). In addition, the eyewash stations must be readily available when conducting any work greater than biosafety level 1. The placement of safety systems should be such that they are easily accessed in the case of an exposure. Both portable and plumbed systems exist. If plumbed systems are used, they should use cold (or room temperature) water. Eyewash stations should have covers for the spray jets to prevent dust or dirt accumulation, and these covers should be self-removing when the unit is activated.

Security Systems

Security systems in the ARF are a critical component in the overall operational management of the facility, requiring a balance between effectiveness and encumbrance and encompassing a number of possibilities (Figure 18.8). Institutions typically have specific security requirements for specific functional areas, such as an ARF. The overall security of the ARF should protect the animals and research from the well-meaning but unauthorized, as well as those that may be ill intentioned. The system capabilities should take into consideration any number of noncriminal factors that may result in a breach as related to human behavior (e.g., propping doors), HVAC imbalances (e.g., causing doors to fail to close completely), power outages, and the consequences of technology (e.g., insufficient IT capacity). As such, expert consultants are often best at assisting with the responsibility of security.

Physical security may include a fence around the perimeter of the facility or institution, keyed doors, and security guards. Fences are generally a one-time cost that require minimal upkeep; however, they may not be practical for a facility or fit within an institution's overall image. All access doors into the ARF should be covered by some secured device, even if it is a simple keyed lock. Keys are generally used as a backup to other access methods, although they may be the primary means for certain restricted or less accessed areas. Software programs exist for tracking key numbers, assignments, and access. Advanced security devices include punch-code hardware, card reader devices, and even biometric readers that use fingerprints, palm prints, or retinal scans as identifiers for approved entry. Security guards are a continuous cost to an ARF, in both space and salaries, so their inclusion in a security plan should be carefully considered. Although security guards can be a deterrent, most security guards do not have weapons and often act only as a first alert via a phone call to the police or responding authorities.

Electronic security may consist of alarms when an inappropriate entry is attempted, "card" readers, or keypads that require a personnel identification number (PIN). These methods do not always allow for individual identification of the user, as keys, cards, and keypad codes can be inappropriately shared with an untrained or unauthorized individual. Card keys may also be poorly secured when not in use, which creates additional risk for misappropriation. Closed-circuit television (CCTV) or video surveillance systems, which transmit to a location with a recorder and monitors, may prove valuable for monitoring key areas of an ARF. The use of CCTVs or combined (card and code) systems may reduce the amount of "borrowing" that may occur. The deployment of cards that have multipurposes or enable access to multiple areas (e.g., facility gymnasium or library) also reduces the risk of lending or inadequate security. As the technology improves, biometric (i.e., fingerprint or retinal) scanners are slowly becoming more popular because they are specific to an individual. Fingerprint scanners can

Access control: Card readers
- Biometrics or a guard are options

Magnetic locks/electric strikes
High-security locks and key controls
Closed-circuit TV (CCTV) with recording capability
Alarm systems
Photoidentification system
Intercoms
Instructional signage
Emergency power

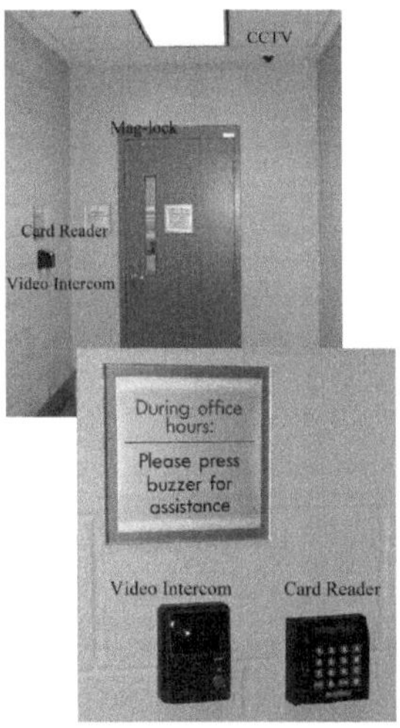

FIGURE 18.8 Pictorial depiction and list of the attributes of a security system.

prove troublesome in some areas of an ARF, where it may be common to wear gloves when attempting to access a particular room.

The design of the security systems should be based on a risk assessment undertaken by the ARF leadership, along with the institutional groups that have security responsibilities. Risks associated with certain locations or regions, types of research, species on census, past security issues, and any special components of the ARF may dictate a heightened security concern. The inclusion of high levels of biosecured research and procedures is one example of a programmatic requirement that may drive the risk assessment to require higher levels of external and internal security. Locations that have a history of protests from the community may also drive the need for more advanced security systems.

It is likely that a combination of physical security and one or more types of secure access systems will be required. At the building level, any system put in place should allow for 24/7/365 individual identification at all ARF entrances and exits. It is preferable that each room, at least each animal holding room, have its own secure entrance requirement. If an electrical system (card reader, biometric scanner, or keypad) will be used, it is important to understand how the system works during an electrical failure. During an electrical failure, do the doors remain locked or do they fail open, and does this apply to both sides of the door? Ideally, the ARF would have the option to allow some doors to fail open, and some doors to fail closed. Any door that fails closed should have the means to be opened from the inside. At a minimum, the ARF staff should have control of, or be able to monitor, the systems allowing access to the individual animal rooms. It is helpful for the ARF staff to have access to the entry logs from electronic systems and videos from CCTV when resolving procedural, not security, concerns.

Commissioning and Validation

Commissioning and validation are practices, procedures, and protocols that are implemented to support an overall QC and quality assurance (QA) program. The commission process starts with a review of the BOD report prepared early in the planning and design phases that documents all the critical and required operational and programmatic requirements of the facility. These include all environmental parameters that must be maintained, the levels of flexibility that are expected, the capacities and species of animals that are projected to be housed, the hours of operation, and all other critical elements of the facility that are required to be met for the ARF to be successful. The commissioning process then includes visual observations, physical testing, and measuring to verify that all the required parameters are met. The commissioning process is typically performed by independent firms that specialize in this process to provide an independent review and testing of the required systems. These firms should also exhibit significant experience in the commissioning of ARF, to ensure a deep understanding of the specialized needs of these facilities. Many of these tests typically occur toward the end of the construction phase; however, there are many benefits of bringing the commissioning team on board early to provide "constructability" reviews during the various design phases of the project. These reviews can recognize design and engineering issues prior to the start of construction, when changes to the design documents are more difficult, and raise the overall quality of the ARF. The final commissioning report can also serve as a valuable, long-term management tool, in that it documents both the design requirements and the capabilities of the constructed systems, which can be invaluable during future renovation projects. Design reviews will also verify that equipment and/or systems are specified to meet the requirements of recognized industry standards, such as Underwriters Laboratories (UL), ANSI, American Society of Heating, Refrigerating, and Air-Conditioning Engineers (ASHRAE), and Factory Mutual (FM).

Validation and *commissioning* are often terms used interchangeably; however, they can have significantly different requirements and procedures. The term *validation* is used extensively in the pharmaceutical and other highly process-driven industries, and is heavily used in testing the quality of process equipment. A typical validation process includes three components: the installation qualifications (IQs), the operational qualifications (OQs), and the performance qualifications (PQs). IQ begins at the time of selection of equipment and systems and concludes at installation, where evaluation demonstrates that assembly and installation have been done to specification and where all supporting services, such as utilities, are provided and properly connected. The OQ component reviews and documents that the system and/or equipment operates correctly and delivers the technical requirements and capacities that it was designed and engineered for. Finally, the PQ component reviews and documents how the system and/or equipment works and delivers the capacities and technical requirements, when combined in an overall system. An example of this process follows: An AHU is selected to provide heat and cooling to a variety of spaces. The IQ component would include the review of the manufacturer's technical data and other documents to ensure that the selected unit is appropriate for its intended purpose. The OQ component would include the testing of the AHU, once it is installed. It would be tested to confirm that the volume and temperature of the supplied air are aligned with the design requirements. The PQ component, the final step in the process, will measure and test that the rooms that the AHU serves, as part of the overall HVAC system, actually maintain the environmental parameters that the system was designed to achieve.

The validation process can be seen as a detailed component of an overall commissioning and QC and QA program. These processes are typically applied to all critical ARF systems and can include HVAC systems; exhaust systems, including fume hoods and biological safety cabinets; lighting systems; emergency power systems; plumbing systems; and equipment, including cage processing devices and autoclaves. The fume hoods and biological safety cabinets are required to have additional testing, as they are required by regulatory agencies to be certified that they perform as designed and installed. These devices require annual certification to ensure that they continue to perform as required.

Some aspects of validation are best delegated to the animal resources program, or at least involve it. A well-established example is the completion of the punch list, the list of owner-identified tasks, repairs, and small details that must be attended to as the facility nears completion to satisfy the terms of the construction project. Beyond a mere checklist of the acceptability of architectural features and finishes and major components of MEP, validation should also include demonstration of the proper function of electrical outlets, drains, lighting intensity and control, and hot and cold water at faucets; intactness of the drinking water distribution system; environmental monitoring, including simulating alarm conditions; and emergency power.

Sustainable Design

Sustainable design typically refers to the utilization of nonrenewable resources and design, and engineering and operational solutions that minimize the use of those nonrenewable resources. To these categories need to be added design, engineering, and operational solutions that can be maintained with the available expertise when the facility is occupied and operating. Designing with the use of "appropriate technologies" is crucial to the long-term operation and maintenance of these demanding facilities. The focal points of sustainability when it comes to the ARF are energy and water conservation and efficiency, waste and disposal cost reduction, efficient use of materials and resources in construction, recycling and increased use of products with recycled content, and interior improvements leading to increased productivity. Energy usage in these facilities is high, as they are required to operate 24/7/365 and to maintain fairly stringent and reliably consistent environmental conditions.

Generally, when it comes to being "green," the focus is on equipment, particularly with respect to energy savings and water use minimization, and how design can facilitate processes that preserve energy and resources and reduce waste. One way to best ensure sustainability is to aspire to the Leadership in Energy and Environmental Design (LEED) program standards established by the U.S. Green Building Council (USGBC), which also includes a commissioning obligation and ensures that buildings incorporate sustainability and operate as efficiently as possible. In design, whole-building software enables the simulation of conditions and analysis of assumptions that maximize attributes of sustainability before the permanency of bricks and mortar. At the level of the ARF, the specific technologies that promote sustainability are dry heat sterilizers (instead of autoclaves), washers and autoclaves that conserve water, recirculating drinking water systems instead of single-pass flushing, means to capture and reuse gray water, processing biowaste via rotoclaves or alkaline hydrolysis digesters instead of by incineration or to landfill, and the use of IVC systems for rodents. The latter allow for room air quality to be maintained with reduced air changes per hour through rooms not only by reducing energy usage, and the resultant reduced utility costs, but also by lowering the amount and capacities of air handling equipment and its related capital equipment costs (Geertsema and Lindsell 2015). A design that allows air to pass through the IVC and then into the building exhaust system saves the cost of reconditioning that room air by removing motor and animal BTUs from the environment, and with the additional benefit of ducting odors and allergens out and promoting workplace comfort. The use of IVCs oftentimes allows for extending the cage change interval from 7 to 14 days (or more, depending up circumstances), which affords a substantial reduction in the energy required for washers, autoclaves, and steam production; in the water used for washing and steam; and in cage wash detergents and chemicals, and results in a longer useful cage life.

While ARFs operate with requirements for precise and unwavering control of the environment, there are gains that can be had in realms that ordinarily cannot be varied, such as lighting. For example, when illumination in intermittent use, nonanimal areas, such as storage and corridors, is controlled by occupancy sensors, energy use can be reduced without affecting experiments. Likewise, provisions can be designed and implemented for controls to enable the ventilation rates and precision of temperature control of unoccupied sites to be reduced either manually or via occupancy (CO_2) sensors. Another means to reduce the carbon footprint is through the composting of soiled animal bedding. This practice is wholly reliant on a commitment during programming and planning to material handling, traffic, and waste management features, such as dedicated dumpsters, their loading and access, the equipment, and/or the procedures used to convey waste. It is also dependent on local regulations with respect to the pretreatment of waste generated from animal care activities. In some cases, much of what can be accomplished will be outside the animal facility or behind the scenes in the form of key infrastructure, building-wide MEP, insulation, condenser operations, boiler economizers, energy-efficient HVAC equipment, and heat recovery systems. Some operational parameters can contribute to reducing energy consumption, but can also have adverse consequences. This speaks to the value of ARF representation and participation in the planning of the whole building, not just the ARF. For example, a zoning approach to facility layout may allow access to daylight via windows or other means at office and in nonsensitive support spaces associated with the ARF. This sort of implementation, however, may create security vulnerabilities that must be considered. Heat can also be conserved and used in heat recovery systems, but if these leak, there may be disastrous consequences for odor and allergen dissemination beyond the ARF. The obvious dependence of sustainable operations on design reinforces that sustainability is maximized at and around project conception, rather than via changes or retrofits made at its conclusion. It is important to note that even after implementing all cost-effective energy reduction strategies, the ARF will remain a substantial energy consumer. Be that as it may, only the limits of imagination, time, and money and the important attribute of timing, in concert with new construction, dictate possible sustainability impacts. Rather than wasting water suitable for consumption, might reclaimed water that is mostly clean, but not potable and not required to be suitable for drinking, be useful for preliminary rinse cycles of soiled cages? At this time, regulations continue to plague in many cases what might be possible.

Adherence to Local Restrictions and Possible Future Restrictions

Responsible research institutions have been properly processing waste generated from research for years. This includes such interventions as the inactivation of infectious agents, the sequestering and safe post-use processing of chemicals and radioactivity, the dispersal by dilution of contaminated air, and the neutralization of wastewater from washing equipment. As communities and local governments become more sustainability conscious and aware of the public service costs and environmental impacts associated with research, undoubtedly scientific institutions will come under increasing scrutiny. Several practices are subject to scrutiny and possible future restriction or abolishment. They include

- The consumption of water and the temperature, impurities, pH, and biochemical oxygen demand of wastewater from washing equipment
- The disposal of waste such as soiled bedding via landfill or the sanitary sewer
- The use of incineration as a means of pathological waste destruction
- The disposition of hazards

From the inside, high utility bills paid by the institution might also guide senior leadership in the establishment of restrictions or programs that change behaviors and practices.

Ergonomics

Ergonomics is an applied science focused on the study of the relationship of the work environment to workers and their use of tools with the goal of maximizing efficiency and safety in the workplace. There are many sources of hazard that can be managed with ergonomic sensibility in the ARF (Figure 18.9). Integrating ergonomic principles into design, construction, and the selection of equipment pays off in the enduring safe and economical operation and use of animal facilities. The critical advocacy for this rests with the ergonomics-attuned animal care specialists involved with the project. Failing this responsibility risks a final product that does not optimize animal care and research efficiency, may lead to increased workers' compensation costs, and may require premature, expensive corrections to the facility.

Optimal ergonomic features of a facility enabled through proficient design are

- Short, flat, and straight distances to move heavy equipment and navigate between key sites, including at docks.
- Seamless flooring that combines slip and skid resistance with relative smoothness.
- Doors, openings, and passages of adequate dimension, particularly in challenging areas, such as docks and bulk food and bedding storage areas, and with respect to enabling future replacement of large, fixed pieces of equipment.
- Corridors of at least 7-foot width after the application of wall guards and in light of other projections.
- Ceilings of 9-foot and sometimes 10-foot height.
- Hoist, overhead, pneumatic, and drag conveyor systems to move bulks materials, such as clean and soiled bedding.
- Washing and sterilizing equipment of sufficient capacity to manage the full census and daily throughput during normal operating hours.
- Automation ranging from the simple and standard, such as automated ADW systems, cage bedding stations integrated into wash lines, hydraulic lifts at docks, and sliding doors, to

Hazards related to animals
Trauma
Allergens
Zoonoses

Work environment hazards
Ergonomic (e.g., lifting and slips/falls)
Machinery related
Heat, steam, and pressure
Ionizing radiation
Flammable materials
Sharp objects
Noise
Light and electromagnetic

Research-related risks
Diverse and protocol related
Microbial agents of all types
Recombinant DNA
Radioactive agents
Chemical substances
Nanoparticles

FIGURE 18.9 (See color insert.) Presentation of various types of safety signage and a list of hazards common to animal research facilities.

the complex and special, such as fully or partly roboticized material handling in cage wash. Automation is particularly valuable where a task is accomplished at a high rate or with force.

- Easy access to service utilities and fixtures accomplished outside of animal holding rooms. Choosing light fixtures, vented duct covers, and paper towel dispensers that are accessed using simple latches, rather than keys, screws, or nuts and bolts, can offer efficiency and safety.
- Ample space for support activities and storage, including PPE and stowing and charging powered devices, and the collection of contaminated outerwear.

Work areas and stations must be designed, installed, and used in ways that promote linear movements, minimize twisting, facilitate short reaches, and are adjustable to enable processes to be completed at a height between the shoulders and the waist. Dumpsters are not always appreciated as a workstation, but if soiled bedding is hauled manually to a dumpster, the design and transportation of the waste bin and the relationship to the bulk receptacle should be ergonomically appropriate.

The common normal practices of using single-pass air handling systems, providing negative differential airflow for animal housing rooms, ventilating the animal housing and procedure rooms separately from laboratory and office spaces, using inward or downward HEPA-filtered airflow workstations, and providing surfaces that can be readily washed down all enable allergen and hazard containment and management.

Adequate management of temperature, relative humidity, light, noise, and vibration in the work environment, alone and in combination, is important to optimize efficiency and prevent worker discomfort, fatigue, distraction, and/or injury (Huerkamp et al. 2008). However, this can usually be realized by meeting established guidelines for animal facility design and function (ILAR 2011).

A sound approach as construction begins, particularly for substantial projects, is to require a mockup room to be built out to allow for the evaluation of key dimensions and architectural features, such as floors and walls, before construction proceeds unencumbered.

Given, in theory, that any technology in the market is available to us all, the differentiating factor in productivity and originality between programs comes down to the quality and continuous engagement of its people, with the latter highly dependent on workplace safety. Applying the principles of ergonomics at the time of facility construction and renovation offers the best opportunity to ensure worker welfare and enable premium animal care.

Disaster Planning

Disaster planning is most often thought of as preparing for natural disasters, floods, earthquakes, and other natural occurrences. However, two of the most common causes of significant disruption to animal care and research are failure of infrastructure (primary and backup utilities, and access and egress routes) and shortage of personnel. The former prominently includes major equipment failures, HVAC system failures, and power outages. The *Guide* requires that all vivarium facilities have a disaster plan in place covering any system failures, regardless of the cause, natural or man-made. These plans must address the procedures that would be implemented should a disaster occur (ILAR 2011). The most critical component of these plans must be for the health and safety of all staff associated with the facility, closely followed by caring for the health and safety of the research animals. With this in mind, the disaster resistance of the facility and its ease of maintainability form the foundation of preparedness for emergencies and disasters and in the prevention or minimization of any effects. After considering the adverse events that arise from natural causes and are indigenous to a location, in design it is then critical to assess the vulnerability of each type of animal husbandry system and research dependency, the building's automated infrastructure, and the suitability of construction materials. Design and construction based on a vulnerability assessment where the adverse consequences of an event are recognized, the means to minimize impact are identified, and potential losses are quantified enables disaster resistance based on sensible prioritization. Consider the types of animals and their husbandry needs. Rank animal care and research units according to their dependence on backup utilities and intensity of care. Species that live in controlled environments depend on a reliable source of heating and cooling, and the lack of backup utilities to sustain their environment makes these animals and related research vulnerable to disasters. Typical examples of facilities that are susceptible to catastrophic loss include buildings that do not meet current standards of construction to withstand likely regional geophysical hazards, such as earthquakes, floods, and hurricanes; sites without fire suppression systems; and locations that can only be serviced via elevators. Construction standards and materials chosen in light of vulnerabilities enable resistance to earthquakes, strong winds, floods, hurricanes, tornados, and fires. It is also important to appreciate that not all disaster events begin as a regional cataclysm, but some may start on a small scale and/or with a limited scope. For example, the loss of an elevator serving multiple facility floors will have increasing impact as time progresses that would be mitigated by budgeting and designing for multiple elevators. Where only single fans are in place for air supply and exhaust, there are heightened risks of the loss of ventilation or differential airflow, with the latter being especially an issue in association with high-containment or exclusion areas. Beyond elevators and ventilation, redundancy must be considered for drinking water supply, washing and sterilizing, data storage, storage of critical spare parts on site, and environmental monitoring. With respect to drinking water, think through how systems will be replenished, if necessary, in the event of loss of the municipal water supply. Are there filling stations for emergency water, and has the storage of emergency water been considered? Another critical element of redundancy through design is in backup generator power and fuel reserves, and the most important related decisions are in what to power and what this will cost relative to risk. Minimally, backup power should maintain fresh air supply, life safety lighting, and power to ventilated racks and drinking water distribution systems. Beyond that, expensive decisions will lie ahead in whether to enable undiminished photoperiod maintenance, sustained electrical supply to chillers, certain degrees of heating, and operation of wash centers. Ideally, chillers should have design features to easily accept water from a tanker. Also important in the continuity of operations are water reserves in reservoirs, bottles, or bags and stored feed. In some situations in design, there are trade-offs. Ground-floor and low-level locations of ARFs may create risks for flooding in the event of a deluge, but also offer fortune in the face of a water loss event, as water will settle downward in a building, with the animals among the last to lose supply. Multiple entry and egress points increase security challenges, but also afford more opportunities to enter or evacuate a damaged structure. In light of the reality that responders will not be permitted to enter a damaged structure until it is deemed safe by a qualified engineer, systems and setups, such as ventilation and automated watering, must be able to function automatically and independently of human involvement for a time. Beyond these attributes, ease of maintenance is another feature of design and construction that hardens a facility to the effects of an untoward event. A useful early step in mitigating disasters through effective design is to review and identify historical causes of disruption and to consider the vulnerabilities of each component of the infrastructure. Identify risks for burst pipes, clogged drains, worn wiring, and vulnerable plumbing. Ensure that fire exits can be kept clear, fire hazards are eliminated, and storage capacity is sufficient to prevent hallways from being used. When disaster strikes, what were plainly trivial or minor annoyances in function escalate to impede response and recovery efforts. If facility conditions are regularly monitored, repaired, and improved, many emergency situations will be eliminated and others will have reduced effect. Take, for example, where there are redundant fans by design that are each to manage 50% of the ventilation load, but one is broken and the other running at 100%. When that second fan malfunctions in the middle of a summer heat wave, an emergency will likely ensue. The same can be true for compressors, steam, and other mechanical components of the facility. It is in the elimination of existing and common causes of everyday disruptions, in combination with a proactive preventive maintenance program, that the threshold is increased whereby events may lead to major disruptions. This makes the physical plant maintenance personnel and their ability to efficiently do their jobs key contributors to any program of disaster reduction. In the event of an emergency, it is critical that they have ease of access so that their efforts may quickly return services back to normal. Design that takes into account the use of temporary equipment and a ready inventory of critical spare parts is also key in effective intervention by physical plant maintenance experts. Already having a well-considered and continuously maintained disaster and emergency preparedness plan is an imperative to facilitate design and construction through the benefit of the a priori contemplation of cause, effect, and response in association with various scenarios. Refer to Chapter 17, where this topic is covered in depth.

Equipment

Housing Systems

The caging systems that will be used should be identified during the planning phase of the process. If at least some existing caging is to be utilized, from a management perspective, it is preferable to use the same type throughout the facility. The "caging" systems may include individual cages, tanks, racks, pens and runs, enclosures, or even the room. The specific type utilized will be driven primarily by the animals being studied. Cages need to be of sufficient size to meet the applicable legal and regulatory requirements and should meet those of the *Guide*. Any caging system used needs to be safe, sanitizable, and durable, and allow for the observation of the animals, while promoting species-specific behaviors. Some caging systems may be built into the animal holding room, generally when the research protocol requirement for a long timeline is well known. An ARF requiring more flexibility may select mobile racks or modular pens and runs and enclosures, to allow for their positioning in the most efficient manner possible within the space available. Mobile racks and the largest piece of a modular unit should fit through all the doorways (elevators included) leading to and from the cage wash area, including the rack washer doors and inside the unit itself. Mobile racks should have wheels and casters that will allow for easy maneuvering of a "full" rack even after being washed and autoclaved multiple times, but should not be injurious to the flooring system used in the ARF.

There are many manufacturers and styles of both static cages and IVCs for rodents. The initial cost for static caging is cheaper, because they are generally placed on a shelf rack that costs far less than an IVC rack. Because air is supplied to and exhausted from each cage, the bedding in an IVC unit stays drier and ammonia takes longer to accumulate at the cage level, allowing for a longer cage change duration, which in turn decreases the labor per cage. Most IVCs, when used with the proper technique, will minimize disease transmission within a given rodent colony. Some rodent cage manufacturers now market disposable static cages and IVCs, eliminating the need for a cage processing area and the costly cage washing equipment. Disposable caging may require some additional storage area and careful consideration with regard to the delivery frequency, how it will be moved within the facility, and the disposal method. The utility (e.g., electrical, exhaust, LAN, or Wi-Fi) requirements and thermal load for each rodent rack need to be accounted for in the planning stage to ensure that the proper types and number of connections are present in each room. IVC racks differ in how they respond during an electrical failure. They may become negative to the room, positive to the room, or static. Hermetically sealed IVCs and disposable cages have only a limited number of hours during an electrical failure before air quality at the cage level becomes a concern, which means a short-term power loss may need to be included in the disaster plan. The type of caging that best fits the program needs should be determined and planned for in the design process.

Caging systems for larger animals are usually premised on either a "dry" or "wet" system. The type of caging system, dry versus wet or a combination of the two, should be determined in the design phase. A dry system usually involves regular changing and washing of excreta pans with or without bedding or liners. In addition to the labor required for regular changing and cleaning of dry pans, this system requires transport to and from the animal holding rooms and storage space for the pans that will be used as a change-out. A wet system usually involves cleaning or flushing the excreta pan *in situ*. Wet systems involve more plumbing to flush and drain the system within the room, and could require additional PPE if aerosols will be created when cleaning a macaque room. In choosing a caging system, one should also consider the means of enrichment for the species; this may include shelves, ledges, or perches built into the cage or as part of the run. The ability to introduce a new animal to an existing group in stages should also be a part of housing systems for larger species. A means to house more than one animal, while allowing for the isolation of a specific individual, will allow for greater flexibility in a group housing system.

Special types of caging systems may include those used for enrichment and exercise, continuous infusion, intensive care or recovery, telemetry, and inhalation and metabolic studies. Because of their specialized functionality, most of these caging systems have a larger footprint than the type of systems regularly used for the species. Enrichment cages that are built into the facility may offer a larger and more dynamic space for the animals, possibly doing double duty as a group housing area, but at the cost of reducing the housing capacity within the facility. Modular or mobile enrichment cages are also available, although generally smaller than a built-in cage. They do afford a space allowing for more movement and exercise. Intensive care and recovery cages generally require utilities—electricity at a minimum, but these could include water and oxygen—to operate as designed. If these cages will be regularly used, a specific location should be designed for them, allowing for the extra utility requirements and the increased thermal load the cage systems will add to the room. Most telemetry caging and some metabolic caging systems will require additional electrical receptacles, possible Wi-Fi or LAN connections, and some metabolic cages may require oxygen. Certain research procedures using animals with minimal or no immune systems (i.e., germ-free and gnotobiotic animals) require unique caging systems, which often require extensive sterilization to maintain animal health status. This can complicate cage processing procedures and add significantly to the process loads for the autoclaving equipment. Careful consideration and understanding of these requirements should be undertaken in the initial design and planning of a facility to prevent a major bottleneck at the autoclaves.

For most facilities, an ADW system for larger species will be an advantage. Most racking or modular systems will accommodate connections to an ADW system, and the ADW equipment that can be used in pens or runs is easy to purchase and install as needed. Rooms should be designed to accommodate the racks or modular systems that are chosen so that there is adequate space to allow for ancillary equipment (i.e., feed bins, medical records, and enrichment equipment). Rooms should also be of sufficient size to allow for safe husbandry activities (e.g., cleaning or changing of excreta pans and racks and cages), while allowing for researchers to enter the room and work with their animals.

Aquatic Systems

Aquatic systems operate based on two general principles, flow-through or recirculating systems. Within a range, most aquatic species can adapt to slow changes in water quality. Aquatic systems will require an input of water, which is then purified (e.g., RO) and pushed to a reservoir. The water from the reservoir must be salinized and of the appropriate pH before it is pumped to the systems. Water added to any aquatics system should be free of chlorine and other compounds and chemicals deleterious to the species housed. Water hardness should be considered and addressed through means of conditioning equipment. Maintaining the temperature of an aquatics system is critical to the well-being of the animals. Appropriate heaters and/or chillers should be in place to maintain the water temperature. The pumps that move the water will add noise and heat to the room and may be better placed in a separate room. Aquatics units should be powered from the backup power source.

Flow-through systems push conditioned water through the tanks and then to the drain. These systems tend to be more resource-intensive because fresh "conditioned" water is needed on a reoccurring basis and this water is not reused. A flow-through system may be less costly with regard to materials and provide better biosecurity, as the water is not shared among tanks, so the health status in each tank may be preserved.

In contrast, recirculating systems use conditioned water at a predetermined rate; only a portion (e.g., 10%) of the entire system is turned over at a time. Because only a small percentage of the water is removed from a recirculating system, this greatly reduces the need for conditioned water. Conversely, recirculating systems require more maintenance with regard to the filters (biological, physical, chemical, and UV) that remove deleterious items from the recirculating water, and because the water is shared among all tanks, these systems encourage a "herd" health mentality for the animals housed in them.

Monitoring of appropriate water quality parameters (e.g., temperature, pH, conductivity, total gas pressure (TGP), and flow) should be considered prior to installation. Phone or LAN lines may be needed to ensure that remote monitoring or notification capabilities are in place. Depending on the location of the aquatics systems and the sensitivity of nearby areas on the same level or the floor below, floor sensors for water detection and/or catch basins may help prevent, contain, and minimize water damage to surrounding areas in the event of a leak. Automatic sump pumps that periodically run and pump to a drain can also help mitigate the risk of potential water damage if any part of an aquatics system ends up leaking.

Cage Processing

The processing of cages and racks is a costly (e.g., labor, space, equipment, chemicals, and utilities) function in an ARF. All facilities typically require some modicum of cage and accessory processing support. Depending on the needs of the facility, this could be a relatively small space with a cabinet washer or could include multiple tunnel and rack washers for larger facilities. If racks, from larger than rodent species, will be employed in the facility, a spray-down or pretreatment area should be part of the cage washing area. Adequately sized floor drains with traps and accessible clean-outs should be employed in any cage processing area, including the pretreatment area. Depending on the municipality's effluent water requirements, it may be beneficial to have chemically treated water from a cage processor pumped to a holding tank where the water can be neutralized and tempered (cooled) prior to discharge to the sewer. The flooring in the cage processing area should be extremely durable and able to withstand a lot of heavy traffic, standing water, and overflowing chemicals, both acidic and alkaline. Walls and ceilings should be waterproof and have adequate protection from all the heavy equipment. Cage processing areas are generally noisy, so an engineering standard noise abatement should be included in the design plans. Supply and exhaust airflow requirements in a cage processing area are generally much higher than those in any other location within an ARF because of the heat and moisture associated with the equipment and the need to prevent the contamination of clean areas as a consequence of directional airflow. The majority of the overall footprint of the cage processing equipment will be in the dirty cage processing area, which will drive the total size of the space required. The clean cage processing areas should have differential air pressure positive to contiguous areas to minimize the potential contamination of clean supplies. The dirty cage processing areas should have negative air pressure relative to their surroundings to contain allergens and odors. Given that even clean areas can be a source of odors, careful mapping of the entire facility should be undertaken to understand how the airflows impact all areas. Most cage washers employ a combination of chemicals and heated water to thoroughly sanitize the caging and ancillary equipment. An adequate, safe (spill protection, safety showers, etc.) space for the bulk storage of the chemicals to be used should be included in the design plans. Bulk chemical storage space will need to be accessible to the chemical supplier and within a reasonable range of the cage processing equipment. Measuring bulk chemical use may require electrical receptacles, and Ethernet or Wi-Fi availability in the chemical storage area.

Prior to purchasing any cage processing equipment, a realistic idea of the maximum number of racks and cages to be processed per day should be calculated, so that the appropriately sized cage processing equipment can be procured. The maximum number should be calculated based on projected needs when the facility is at full operational capacity, and depending on the projected labor, it would be wise to calculate the maximum during a holiday week, or 4-day week. Once the throughput needs are decided, the additional requirements of the equipment can be determined. It cannot be overemphasized that the calculations need to take into consideration cage changes to be done on a regular schedule, but also the spot changing that regularly occurs and that can amount to up to 30% more of the census over the interval between scheduled cage changes.

Tunnel washers can continuously process rodent cage supplies or other items that fit the width of the belt and height of the opening. Tunnel washers are composed of prewash, wash, rinse, and final rinse sections; on average, each section is 5–8 feet in length. Dryer sections are optional and will generally add another 5–8 feet to the length of the machine. This overall length needs to be considered in the room layout of the design phase. Additional space will be required for operators and possibly ancillary equipment at the load and unload ends. Newer model tunnel washers have controls for belt speed that can be used to increase throughput, providing that adequate decontamination occurs, which will be utility dependent to some extent. Some units are designed to deliver "sanitized" cages and require a final 180° final rinse. This must be considered when designing the hot water delivery system to ensure that adequate capacity is available. With regard to throughput calculations, it may take 10 minutes for an object to travel the length of a belt (with dryer) on a low-speed setting, but on a higher setting it may be done in 4 minutes. The width of the belt can then be determined based on the throughput needs. Utility requirements will vary based on the unit, but steam, condensate return, exhaust connections, hot water, cold water, and drains will all need to be available. Some units may require compressed air, from either a house line or a compressor. An Internet connection may allow for remote connection to the equipment, which can be useful for off-site troubleshooting and software updates and upgrades.

Rack washers are cycle-based cabinet washers, similar to a dishwasher without a dry cycle. Specific racks can be purchased so that rack washers can be used to process rodent caging and pans, in addition to washing racks. Rack washer dimensions, length, width, and height, should be considered with regard to the largest items that will be washed. A longer rack washer may accommodate multiple racks, which can increase throughput. Rack washer cycles are programmed by the end user to accommodate different cleaning needs. Cycle times will depend on utilities and dynamic supplies at adequate pressures and temperatures, and will generally be slightly slower the larger the unit because of the increased volumes of water to be heated and moved. Cycle times may range from 10 minutes for a fast 180° rinse cycle to more than 60 minutes for heavily soiled large animal equipment and under conditions when demand from competing consumers of hot water (e.g., rack and tunnel washers running at the same time) slows cycles if there is not sufficient hot water. Safety features that are critical to walk-in rack washers to prevent operator entrapment are explosion-relief door latches allowing ease of opening from inside the chamber, an interior emergency shutoff mechanism accessible from anywhere in the machine, and a lockout capability where a restart requires deliberate disengagement of the emergency stop. Some units are provided with see-through doors that enable visual monitoring of the interior and the cycle. Some manufacturers have the capability for the rack washer to be combined with different sterilization methods to turn the rack washer cabinet into a liquid sterilization chamber. Rack washers may require a "pit" to sit in so that the floor of the rack washer is even with the rooms. The pit will need to be sloped to either a drain or a low point where the water can be pumped out. Raised areas, pads, will be required in machine-specific locations within a pit. Utility requirements will vary based on the unit, but steam, condensate return, an exhaust connection, hot water, cold water, and a drain will all need to be available. Some units may require compressed air, from either a house line or a compressor. An Internet connection may allow for remote connection to the equipment, which can be useful for off-site troubleshooting and software updates and upgrades.

Many cage processing areas include a steam sterilizer, the specifics of which are discussed in the "Sanitization, Sterilization, and Decontamination" section. For planning purposes, the footprint of the sterilizer should be considered with regard to overall space available and proximity to the cage washing equipment and required utilities.

Bedding removal systems and HEPA-filtered cage dumping stations should also be considered when designing a cage wash area. HEPA-filtered cage dumping stations help contain allergens and dust. They are available from multiple manufacturers and differ little in their functionality but more in their footprint and the ergonomics of their use. Most bedding removal systems operate as a vacuum, to remove the soiled bedding from the cage or pan. The soiled bedding is then transported directly to a dumpster, which is either inside or outside of the facility and regularly emptied. The way the soiled bedding is moved to the dumpster can vary by manufacturer and may not be compatible with some bedding types. Water-based systems are also available for bedding removal. These units are common in the restaurant and cafeteria industry. In these systems, the soiled bedding is run through an industrial-sized garbage disposal to grind the pieces to a very fine pulp, which is then pumped in the water over to an auger that extracts the bedding. The extracted bedding is then moved to the dumpster, while the water is reused. Garbel-type systems, where the bedding is ground, mixed with water, agitated in a vortex, and dumped directly into the sanitary sewer system, have been employed in many facilities, but they are losing favor due to the high water consumption and impacts on the local treatment plants. Careful review of any bedding removal and waste management system is required to ensure compliance with local regulations.

Bedding dispensers, positioned in the clean cage processing area, can reduce the need to manually add bedding to cages or pans, minimizing ergonomic concerns and increasing reproducibility with regard to the amount of bedding added. Some units can be connected to the unload end of the tunnel washer so that rodent cages remain on the belt, are flipped, and have bedding added, all automatically. Other bedding dispensers can be positioned in the clean cage processing area, as needed, to provide an efficient workflow. Bedding dispensers can be adjusted so that different amounts of bedding can be dispensed, allowing for cage size variation. There are types of bedding that cannot be effectively used in bedding dispensers, and this should be considered when evaluating the need for a unit and the specific products available. Some dispensers can be connected to an automated filling system, which can be located in a different area of the facility, eliminating the need for the operators to stop and refill the bedding.

Facilities that do not utilize automated drinking water may want to consider water filling stations. These stations are generally located in the clean cage processing area so that water bottles can be refilled after they are washed, but there are point-of-use models available. Water bottle filling stations generally fill all the bottles in a bottle basket during one cycle and can be adjusted for various sizes and configurations. Other water filling units fill plastic bags or sacks, which can be placed into rodent cages with a sipper tube, either already in place or added. The used plastic bags can then be recycled. Water filling stations generally require an operator to load and unload. Each unit will also require source water of some type, possibly domestic cold or treated water, and 120 VAC; for these reasons, the units are generally not mobile.

Facilities that do use automated watering should have a means to sanitize the watering manifolds attached to the racks and the hoses used for connection to the system. These units may be placed in the dirty or clean cage processing area. Rack and hose flush stations allow for the connection of racks or hoses for a sanitization cycle. The cycle generally consists of exposure to a chlorine solution, followed by a freshwater rinse and then blowing the water out of the hose or manifold using compressed air. The need for source water, compressed air, and power will require these stations to be fixed in place, so their location should be considered during the design phase.

The ARF will have to decide how clean caging will be prepared and moved to the animal holding rooms during the design phase of the process. If rodent caging is to be assembled in the clean cage processing area, then sufficient space to allow for the storage of all the clean caging and appropriate assembly locations should be designed. If rodent cages will be assembled at the room level, then a smaller staging area where the cages are placed onto flatbeds, racks, or carts for transport should be part of the clean cage processing area or located in close proximity. Pans for larger species will also require a staging space, as they are placed onto carts or racks for transport to the animal holding rooms.

Sanitization, Sterilization, and Decontamination

This section could cover an entire chapter if it addressed every aspect of sanitization, sterilization, and decontamination, but instead, the focus is on areas of cage processing and animal holding room sanitation. The definitions of *sanitization*, *sterilization*, and *decontamination* are first required, as they can be interpreted very differently and a common understanding is critical. Sanitization is to make sanitary (as by cleaning) or to make more acceptable by removing unpleasant or undesired features. Washing with soap and water is an example of sanitization. Sterilization is the destruction of all organisms on an object by treating it with chemicals or subjecting it to high heat or radiation. It may also refer to procedures that result in infertility. Decontamination (or disinfection) is the process of reducing the number of pathogenic organisms on an object to a harmless level. Harmless types or amounts of organisms may persist after decontamination procedures. These definitions are similar, but in the ARF, the processes that deliver each of these conditions are operationally different, depending on the requirements and the items or spaces that are being treated.

In the ARF, the cages and caging systems are the most frequently cleaned components, and even here, different applications and measures may be applied. Rodent cages are changed and cleaned on a regular basis, typically established by the veterinary and animal care staff, and dependent on the health status required, the cleaning process will vary. In some facilities, cleaning and sanitization is achieved through normal cage washing equipment that delivers a 180° pasteurizing final rinse cycle. The exposure to this temperature water for a certain duration, in many instances, is considered to deliver a sanitized cage that is acceptable for reuse. In barrier facilities, populated with immune-suppressed animals requiring a higher health status, cages are often sterilized by the use of autoclaves or through the use of dry heat. In the ARF, this is typically the highest level of cleanliness required. The choice of sanitization through hot water exposure or sterilization through a second step is elected by the veterinary and animal care staff, with input from the researchers whose animals may be required to have a higher level of protection to maintain a certain health status.

In the ARF, the term *decontamination*, as it applies to animal cages, usually implies the need to remove harmful pathogens, and the autoclaving process is typically the method of choice. For larger cages or caging systems that are not portable or do not easily lend themselves to be put through a cage processing system and autoclaving, decontamination is achieved through the use of chemical or gaseous materials that are known to be effective, while leaving no residue that could be harmful to the animals after cleaning.

Steam sterilizers use a combination of steam, pressure, and time to sterilize. They often have reinforced external "jackets" around the chamber to withstand the pressure differentials. This extra reinforcement requires more space and, in larger units, adds considerable weight, such that their location should be planned in advance to account for the overall weight. Benchtop steam sterilizers are available for the sterilization of surgical instruments or other small items. The placement in the facility of larger, bulk-sized steam sterilizers should also be planned based on access to the area. These units are commonly used to bulk sterilize loads of bedding, food, caging, water, and contaminated cages. Most manufacturers assemble and weld the chamber into one piece at the factory before it is sent to a facility. This may require that the unit is put in place before walls or ceilings are finished to accommodate the size of the chamber. Some manufacturers have "sectional" chambers where multiple pieces are assembled at the factory and shipped to the facility. This may allow for easier final placement at the facility but also requires on-site welding of the sections and overall testing and certification of the chamber.

Whether the unit will have one door or two and be used as a "pass-through" sterilizer is another decision to be made during the design phase of the project. Pass-through units with a bioseal are a must for removing items from a high-containment room. Each cycle of a newly installed unit will need to be validated, as outlined in the "Commissioning and Validation" section of this chapter. Manufacturers may include validation as part of the purchase package or offer the service if desired. Steam sterilizers will require steam (clean) and condensate return lines, along with a drain and cold water supply to run the vacuum pump. Hot water supply and compressed air may be required by particular units. An Internet connection may allow for remote connection to the equipment, which can be useful for off-site troubleshooting and software updates and upgrades.

Dry heat sterilizers operate similarly to an oven, using much higher temperatures than a steam sterilizer to sterilize items. Although benchtop models are available for instruments, because these units do not require the same pressure reinforcements that a steam sterilizer does, a larger sterilization chamber can be accommodated in the same space footprint. Exposure to these higher temperatures is not suitable for all items, but may work well for sterilizing rodent caging or certain medical instruments. As dry heat sterilizers can be assembled in sections without a need for chamber certification, they may fit into areas that a steam sterilizer would not. In addition, dry heat sterilizers only require electricity to operate, which gives greater flexibility in the placement within a facility.

Ethylene oxide (EtO) sterilization can be used for items that are heat, moisture, or pressure sensitive. EtO sterilizers operate in the 50°C–60°C range. The sterilizers are explosion-proof because EtO is highly flammable in air. The cycle times are generally 16–18 hours, and items must be well aerated prior to use. OSHA has special handling guidelines for EtO. The exhaust requirements, storage areas for EtO, and footprint of the unit should all be considered during the design phase.

Hydrogen peroxide vapor (HPV) sterilization is also available for items that are heat or moisture sensitive. Some systems use HPV and plasma. Some HPV systems use vapor injected into the cabinet. Paper and some wrapping materials (e.g., woven cloth and cotton) are not compatible with HPV sterilization. HPV sterilization cycle times are also much faster than EtO. Plasma unit cycle times are typically 45–70 minutes when operating in the 45°C–50°C range. Vapor units typically have a cycle time of 2 hours but operate in the 30°C–40°C range.

Peracetic acid liquid sterilizers are available for items that are submersible. Peracetic acid sterilizers operate at the 50°C–55°C range and cycle times are fast, less than 30 minutes. Any device with a lumen must be appropriately connected for sterilization to occur. The final part of the cycle involves a sterile water rinse, leaving items ready for use immediately after the cycle concludes.

Alternative methods of liquid sterilization may include using chlorine dioxide or ozone within a chamber. Both can be corrosive, so their use should be carefully considered. Sterilization via x-ray or gamma irradiation is generally not practical for an ARF. High-level chemical disinfection can be achieved using ortho-phthaladehyde (OPA), a safer replacement for the commonly used glutaraldehyde solutions. Label claims in the United States require 12 minutes of exposure time at 20°C. Any item disinfected with OPA should be copiously rinsed. Direct skin contact with OPA should be avoided.

Robotics and Automation

A key but expensive aspect of ergonomics is in the sensible application of automation to animal care. This technology, well-established in medical technology and extensively used in manufacturing, is only emerging in animal resources. Machines are most suitable for preprogrammed activities, high-throughput repetition, applying ballistic forces, working in hazardous areas, and process consistency; hence, the opportunities for automation in the ARF are numerous. Current applications include soiled contact bedding disposal, clean contact bedding dispensation, cage and water bottle washing (including capping and uncapping), ADW supply, and feed delivery. Automation in these cases may reduce stress; eliminate many potential injuries and disorders associated with the overuse of muscles, bad posture, and repeated tasks (by reducing the risk of injury from inattentiveness brought on by the boredom of repetitive tasks); and reduce exposure to allergens. It allows for human resources to be allocated to animal care and science support activities rather than the monotonous repetition of inanimate material handling processes.

From an ergonomic perspective, handling large volumes of rodent cages, water bottles, and accessories can be a highly repetitive activity. When engaged in such work, there is risk for a combination of bad postures occurring at an excessive frequency and sometimes involving considerable force. This may cause regular microtrauma and subsequent musculoskeletal injury. For example, in the dumping of soiled bedding and inverting solid-bottom rodent cages onto the conveyor of a tunnel washer, finger pinch grips and presses, radial (lateral) and ulnar (medial) deviations at the wrist, inward rotation, full extension of the forearm, and twisting at the waist—all inappropriate postures or movements—may occur at excessive frequency and predispose employees to a number of injuries, including carpal tunnel syndrome, ganglion cysts, wrist or elbow tendonitis, and back strain. The handling of water bottles has been ranked as one of the highest-risk procedures for work-related musculoskeletal disorders (WMSDs), followed by various scenarios of handling rodent cages.

The most dramatic and innovative application of ergonomics to ARF operations has been in cage washing. The range of options for cage washing begins with washing by hand and extends all the way to all-in and all-out robotic handling of rodent cages and water bottles. Even the hazard for WMSDs presented by the water bottle uncapping and capping process can be automated. Despite such impressive applications, however, one must remain mindful that automation itself may introduce the novel risk of injury where none existed before. For example, persons working with robots are at risk of being struck by arms programmed to move regularly and not capable of sensing a human obstruction. As such, any staff member involved in the automation area must be educated on equipment performance and safety procedures to avoid such injuries. The automation of any component of the ARF processes must include a review of the financial benefits of employing these systems, as they remain expensive and are typically not seen as providing a reliable return on investment (ROI) at smaller facilities. Even though automation has been shown to significantly reduce operating costs and repetitive stress injuries, program administrators still demand careful cost control on every project.

Operational Support

Space is usually at a premium within an ARF, particularly the non-revenue-generating space that is needed to support ongoing operations. Office or desk space will be required for veterinary, management, training, and administrative staff. This should include workstations for frontline supervisors and technicians, as they often require computer access. Conference space that can be used for meetings or training should also be available. A cafeteria or breakroom will be needed by the staff at mealtimes but could also double as a meeting or training space. Locker rooms where the staff can change into scrubs or other work uniforms, and securely store personal items, may be appropriate for some facilities. Showering and toilet facilities are usually included with the locker rooms. In barrier facilities, it may be appropriate that the locker rooms are the primary means of entrance to the ARF. Storage space for PPE, cleaning supplies, replacement filters and other parts, caging, disaster supplies, and carts, flatbeds, and racks should be designed into the facility. A temperature-controlled feed and bedding storage area should also be identified. Even if rooms are designed with drains, centralized areas with slop sinks should be considered in the design, if only to help with the mitigation of potential leaks. Exercise or play areas may be appropriate for some species; although the space may not have to be designed into the plans, it should be considered.

Sufficient procedural space should be available to perform veterinary care, treatment, or euthanasia of the species housed within the facility. A space for the medications that will be used should also be set aside. This could range from a cabinet for the storage of commonly used rodent treatments to an entire pharmacy for larger facilities with a wide range of species. Facilities housing larger animals should have some kind of aseptic operating space available, whether dedicated operating rooms or a designated room that fulfills all appropriate regulatory requirements. Facilities that anticipate supporting research that involves surgery on larger species should consider additional support space, such as preop preparation, instrument cleaning and preparation, diagnosis, recovery, and necropsy.

Research Support

As more scientists appreciate the value of collecting data and using equipment where or near where their animals are housed, greater pressure is exerted on the facility to support and enable the use of various research devices and forms of equipment. As science continues to progress, the frontier is almost limitless and somewhat unpredictable as to what may be needed to accommodate the types, functions, purposes, numbers, and sizes of these constantly evolving and improving items. The enduring doctrines of creating general spaces that are designed and outfitted for flexibility (e.g., drains and convertible sources of power) and for ease of renovation allow for research needs to be met. Within this context, storage space devoted to research equipment that is used with regular periodicity should be allocated by design or enabled through flexibility from design. A caution is that ARF storage can become exhausted, particularly with the abandonment by users of obsolete, neglected, or broken equipment. The postoccupancy principles of encouraging the development of cores, thus reducing the unnecessary and wasteful redundancy of research equipment, accompanied by a clearly articulated policy on the responsibilities for storage (i.e., lab vs. ARF), will promote operational efficiency.

ROI Calculations and Modeling

While ROI has a number of applications worldwide in finance, marketing, business management, environmentalism, home improvement, and other venues, from the perspective of the design and construction of the ARF, it is a decision tool used as a means to determine if an investment in technology, materials, or other features intended to improve safety or economy provides rapid payback. Will it subsequently contribute the greatest ongoing impact to efficient operations? The obvious application for this is in the financial justification for labor-saving, "high-ticket" up-front items, such as automated watering systems (instead of cheaper bottles), high-density rodent caging systems, robotics, mechanized waste removal systems, and high-volume, validated, programmed euthanasia systems. For example, ROI can be used to provide the analytical basis for cage and rack system selection, with their impact on space needs and construction costs balanced with system purchase cost and an ultimate comparison with operating costs to determine payback time in years. All information to do this sort of modeling should be extractable from the A/E and ARP groups. This approach logically aids in demonstrating comparisons to promote sensible decision making. Table 18.2 illustrates a pro forma ROI matrix that can be used to compare the costs of two alternatives. For example, system 1 might be a ventilated caging system with automated watering to be compared with system 2, nonventilated (static) filter-top cages with water bottles. Static cages require more frequent cage changes, increasing the labor cost per cage. IVCs have a higher initial cost than static cages and require periodic maintenance for the supply and exhaust air components and the racks themselves. Operationally, the ROI approach is also useful for evaluating the monetary effect of comparative work practices, such as cage wash throughput and waste processing.

TABLE 18.2

Construction and Equipment Cost Pro Forma ROI Matrix

Cost Center	System 1	System 2
Equipment	a	e
Space	b	f
Personnel (cost per year)	c	g
Maintenance (cost per year)	d	h
First cost total, equipment + space	a + b = i	e + f = j
First cost difference	i – j	
Annual cost total, personnel + maintenance	c + d = k	g + h = l
Annual cost total difference		k – l
Simple payback (years)		(i – j)/(k – l) = m

Operations

The ARF is a very complex building type that requires continuous operational and maintenance support. This section reviews three components of the running of an ARF: operations, maintenance, and housekeeping. These three elements are completely interconnected, work to support each other, and are critical to a successful program. For the purposes of this section, we define each of these three elements as follows:

- *Operations*: This component refers to the use and care of the animals and associated research activities that are housed in the ARF.
- *Maintenance*: This component refers to the daily, regularly scheduled, and emergency repairs of the building, its systems, and their associated components.
- *Housekeeping*: This component refers to the daily cleaning and trash removal required for this type of facility.

The responsibility for the operation of the ARF normally falls to the organization established by the institution to oversee the care and use of animal models required by their research enterprise. These organizations have different names at different institutions, but their functions and responsibilities are the same and the types of staff required are all very similar. This includes veterinarians, animal care technicians, veterinarian technicians, and a variety of support and administrative staff. A complete chapter or even a separate book could be written on the processes and requirements of operating an ARF, but from the perspective of the impacts on the facility and its design, there are several elements that must be considered.

First and foremost, several key members of the operations team *must* be engaged in the initial programming and planning of the ARF, as each organization will have specific procedures and protocols that they have implemented, that will impact the design of the facility.

SOPs

SOPs may not be required for all ARF activities; however, they do facilitate the transfer of knowledge and standardization of procedures. Each facility will have different needs with regard to the numbers and types of SOPs generated. Facilities that will conduct studies according to good laboratory practices (GLPs) at a minimum must have SOPs to cover the specifics outlined (CFR, Title 21, Part 58). SOPs may address multiple daily routines, such as standard administrative and record-keeping tasks, husbandry for each species, veterinary care, regularly performed technical procedures, controlled drugs and pharmaceuticals, receipt and storage of feed and bedding, sanitization, disinfection, sterilization, waste disposal, and hazard containment. As opposed to operating from instruction manuals, maintaining SOPs for the various pieces of equipment in an ARF can allow for a user to have a quick reference for how to perform common procedures, as well as providing for the care and maintenance of the equipment.

Program management should agree on a standard template, generation, and revision process if SOPs are to be used. A table of contents that groups SOPs into logically related tasks, with some type of numbering system, will assist in the quick referencing of SOPs. Multiple companies offer document control software and shareware systems that can assist with the organization, standardization, and review processes. SOPs should be readily available for all staff members to reference; however, master versions should have restricted access. SOPs should be "living" documents, such that they require constant refinement and updating. As such, a method of version control should be implemented, in addition to a means of archiving previous versions for later reference. Generally, someone familiar with the procedure should generate the draft SOP for review by others. A well-written SOP should allow someone new to the facility to perform the procedures as outlined.

Process Flows

Thoughtfully considered process flows will maximize the efficiency within the ARF and preserve and protect biosecurity for the species housed within the facility. Process flows should be considered during the design phase of a new project or during the planning phase of a renovation project. Some key process flows include movement of staff (research and support) between different functional and holding areas, movement of animals into and within the facility, cage and rack movement between housing areas and cleaning areas, movement of supplies into the facility, and removal of waste.

Some ARFs may be more vertical in design, requiring elevators (at least two for redundancy) for movement of supplies and people, versus horizontal designs, which may require much greater travel times. Process flows should be considered based on these designs. Appropriate wheels and casters for the flooring type, expected weights, and type of transport will greatly aid in maintaining the facility while safely and easily allowing for the transport of caging and supplies.

Movement of Personnel

Personnel entry into the ARF should occur in prescribed locations. These locations might not correspond with emergency exit doors, which are required by building codes. Each location should be secured 24/7/365. Close to the entrance, appropriate PPE should be available to protect the people entering the facility and the animals within. Many animal facilities require the support staff to wear scrubs, or facility-dedicated clothing, which is donned in a locker room adjoining the facility. Research staff may be required to don additional PPE if wearing "street" clothes. The barrier status of the facility, or area, may require all personnel to enter through a locker room, shower, and don facility-specific garb prior to entering the facility.

All personnel need to realize that once inside the facility, there will probably be a specific order of entry for different rooms. This order is maintained to protect the health of the animals and may be dictated by the health status of the animals in the rooms (i.e., general holding prior to quarantine) or by the species in the rooms (i.e., New World nonhuman primates prior to Old World nonhuman primates). Facilities with a cage wash area should have engineering (i.e., separate locker rooms) and procedural standards (i.e., shower before entry) to avoid having personnel travel from dirty to clean areas.

Movement of Animals

Animals arriving at the facility should not enter through a personnel entrance but rather a designated area that is temperature controlled. Accompanying paperwork should be verified prior to moving the animals to a holding room. There should be systems in place so that the support staff has foreknowledge of animal deliveries, and there should be systems in place for how to handle the unannounced deliveries. Some modified or true barrier facilities may only allow rederived animals into the facility or certain housing areas. Facilities that require rederivation "in" should have separate areas for dirty animals and the rederived animals while awaiting a clean health status. Clinical or research needs may require the temporary removal and return of an animal from the facility. Appropriate procedures should be put in place within the facility and at the other location to maintain the health of the animal, the safety of personnel, and the biosecurity of the other animals within the facility.

It is likely that animals will be moved within the facility, whether they are transported to procedure areas, clinical support locations, or research laboratories; the means of transport for intrafacility movement should be specified and appropriate for the species. Procedures should be in place to determine if the movement is short-term (i.e., an animal is having an x-ray taken and will be returning) or long-term (i.e., an animal is being moved to another holding room), which should require documentation. Facilities that house animals that are not easy to handle should have equipment (i.e., nets and catch poles) and procedures in place so that if an animal does get "loose," it can be returned to its home cage.

Movement of Supplies

To aid in the receipt of supplies, most facilities will need a loading dock with a standard height platform and the means to accommodate smaller box trucks, whether with a lift or a ramp. Once the supplies are offloaded from the transport vehicle, but prior to entry into the facility, it may be necessary to decontaminate them. Spraying or misting can be accomplished with spray bottles, pump sprayers, misting tunnels, and so forth. The exterior of delicate items that are not suitable for spraying or misting may need to be carefully wiped off with a suitable agent. Some facilities may require access to a pass-through sterilizer to sterilize the supplies "into" the facility. Once within the facility, all supplies will require suitable storage space; for animal feed, this may require a temperature- and humidity-controlled space. The amount of supplies that can be maintained in the storage space available should be considered with regard to contingency planning. The location of specialized storage containers (i.e., flammable or combustible) should be planned to allow for safe containment and accessibility. Every facility should have a secure means to safely store and, using purpose-designed carts, transport gas cylinders. Most supplies will generally be moved within the facility using carts, flatbeds, transport trucks, or pallets. When designing the facility, ensure that whichever methods to be used will have the proper clearance through door frames and around corners.

Movement of Caging

The movement of caging within an ARF is something that will occur on a regular basis and, as such, should be planned in advance. Most facilities will need to move caging from a cleaning area, or receiving area if disposable cages are used, to the animal holding room or cage changing area. Facilities with a cage wash will need to transfer cages to an area where they can be prepared (i.e., bedding, feeders, and enrichment added) for use and/or transport. Most small animal caging will fit on carts or transport trucks, which can be used to transfer cages from the cleaning area to either an assembly or staging area. Cage assembly areas should have space for staging unassembled, work-in-process, and completed cages. If small animal caging is to be assembled in the room, the transfer of clean caging and the necessary supplies (e.g., feed, bedding, and feeders) needs to be coordinated and staged as well. Staging areas where cages are kept prior to transport to the animal rooms or cage changing stations should be large enough to accommodate the number of carts and/or transport trucks used; this may include complete cage setups; feed, bedding, and enrichment; cage components (feeders and lids); and water bottles (if used). Animal holding racks may not require an assembly area, but will require some type of staging area that is at least large enough to accommodate the greatest number of racks expected to be staged at any given time. Animal holding racks in facilities using a "dry" system will probably need staging space for the excreta pans and their means of transport (i.e., cart, flatbed, or baker's rack).

Soiled caging will need to be transported to a processing area and then either a disposal or cleaning area. To lengthen the life of the rack washer, soiled animal holding racks should be manually cleaned prior to washing; this will require a dirty staging area, preferably with substantial drainage and supply water. Excreta pans may also be dumped and manually cleaned in the same or similar type of area if a bedding disposal system is not used. Some large animal racks and pans may also be best maintained by the application of prewashing agents (i.e., foam). Once the prewash is applied, the equipment will have to stand for a time period before it should be washed, which may require a prewash staging area.

Small animal cages may have the soiled bedding disposed of at the room level, area level, or facility level. Consideration for allergen exposure and waste removal should be factored into the disposal locations. Processing the soiled bedding at the facility level may warrant investment in a bedding disposal system, most of which operate via vacuum, sucking the soiled waste from the cages and transporting it to a dumpster. There are limits to the vertical and horizontal distances these vacuum systems can transport soiled bedding, in which case a water-based system may make more sense. Any type of automated system will greatly reduce the need for waste transport from the cleaning area to a dumpster. Disposable small animal caging may also need to have the waste removed prior to transport from the facility and will require a staging area for pickup.

The type of research being conducted may require the sterilization of cages and/or racks prior to cleaning. This may need to occur at the room level using a pass-through sterilizer that allows for the safe retrieval of the racks from the uncontaminated side. It may be more practical for a small number of cages to seal them in bags or place them in an enclosed cart suitable for steam sterilization, prior to transporting them to a centralized sterilizer.

Removal of Waste

Waste in an animal facility should be segregated with holding locations for each type (i.e., recyclable, trash, biological, chemical, and radioactive). Some institutions have programs in effect that specify the types and methods of waste disposal. Harmful waste should not enter the refuse stream unless precautions are taken with regard to packaging and the disposal mechanism. The type of waste may require a staging area, prior to disposal. This area needs to securely store the waste while being accessible to those who will be dropping off and picking up the waste.

Recycling stations should be conveniently placed to encourage use, but be accessible for those who will be emptying them. Animal facilities generate a lot of trash, and most of it typically comes from the soiled bedding, so the dirty side of the cage wash should be in close proximity to at least one dumpster. Because of the volume of trash generated, it may be worth considering the use of dumping carts and whether they would work with a platform or lift placed in conjunction with the dumpster. Biological waste may include carcasses, requiring a separate area for storage prior to removal from the facility, whether it is a freezer or a dedicated cold room. Chemical waste staging areas should be well ventilated and allow for the appropriate storage, based on the type of chemicals. Chemical disposal services can package and remove both liquid and solid chemical waste.

To maximize containment and minimize contamination, the transport of radioactive waste should be limited in scope to what is absolutely necessary. Ideally, radioactive waste will stay in the area of use until it is no longer radioactive, which is only an option for short-lived isotopes, or until it is ready for transport and removal from the facility. Containment of the waste during transport through the facility must be considered; the methods used will be isotope dependent. No regular traffic should be permitted through the areas the radioactive waste is being transported, until the waste is removed and the areas have been verified to be free of contamination. After the radioactive waste has left the facility, the transport carts and/or containers should be decontaminated, along with the temporary radioactive waste staging areas.

Staffing

The staffing needs for each facility will differ based on the size of the facility and complexity of the research supported. One staff requirement is an attending veterinarian, and although he or she does not have to be part of the facility staff, most programs will require at least one attending veterinarian whom is responsible for program oversight and veterinary care. One of the challenges every facility faces is finding staff members dedicated to the support of the animals and research every day of the year, including weekends, holidays, and even in disaster situations. Full- or part-time staff should be the standard because of the training involved to perform the various tasks within an ARF. Temporary employees are rarely cost-effective or able to provide the level of support necessary. The geographic location of the ARF, and its proximity to similar facilities, will also have an effect on the pool of available and qualified potential employees. Although the

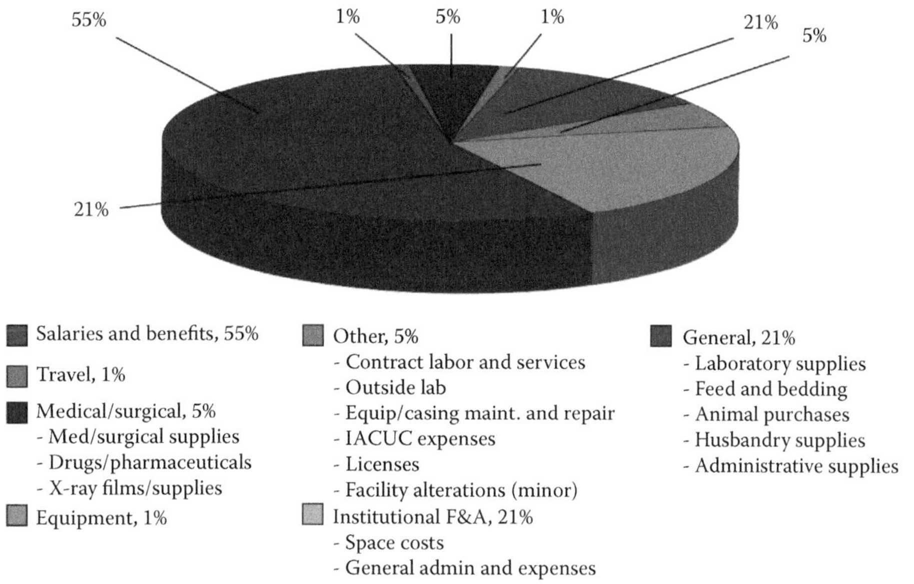

FIGURE 18.10 **(See color insert.)** Representative operating costs of a functioning animal research facility. IACUC, Institutional Animal Care and Use Committee; F&A, facilities and administrative.

challenges are the same, each institution will have its own requirements with regard to the means of staffing the facility, whether they are institutional employees, contract employees, or union employees.

Depending on the size of the facility and scope of the research being conducted, multiple types of positions may be necessary. These positions may vary and overlap by program, but generally consist of loading dock support, cage wash, animal care, veterinary care, technical support, specialized research support, surgical support, administrative support, and management.

Financials

Once constructed and occupied, the greatest expense in the operation of the ARF is salaries, which can be 55% or more of the overall operating costs (Figure 18.10). Consequently, it is incumbent on design and construction to facilitate cost-effective operations. Cost-effective operations translate into greater proportions of funding applied to actual research rather than the maintenance of animals or the physical plant.

Maintenance

Continuous

Building maintenance, after the owner takes control of the site, is often thought of in terms of the large-scale, engineered, and centralized components, often in penthouses or appended to buildings and distributed through chases and the overhead interstitium, and maintained by a separate professional staff with expertise in the maintenance of complex systems. These specialists provide preventive maintenance, where the building and equipment are regularly scheduled for upkeep to prevent failures that may range from the annoying to the disastrous; operational maintenance, where form and function are honed and fine-tuned during use to ensure optimized performance; and corrective maintenance, where repair is required in response to failure. This serves as reinforcement that those who maintain the physical plant and MEP should have someone at the table during the design process and on through construction. The ideal is for the project manager to meet this obligation and to clearly maintain productive relations with these teammates afterward.

If one subscribes to the belief that those of us in the animal care business participate in every experiment by providing a wholesome and consistent environment where animals reside, experiments are conducted, and data is collected, then it is easy to appreciate the hand-in-glove relationship and invaluable partnership between those who manage and operate the ARF and those who maintain it and provide steam, normal and emergency power, conditioned air, water, and the plumbing. A productive relationship with the facilities maintenance team is critical after construction is completed and the facility is occupied.

The ARF typically has an array of highly specialized equipment that usually is beyond the capability of the institutional mechanics, electricians, plumbers, and carpenters to maintain. Autoclaves, dry heat sterilizers, mechanical washers, tunnel washers, bottle fillers, automated waste disposal systems, robots, biosafety cabinets, ventilated caging systems, water treatment and distribution systems, euthanasia chambers, imaging devices, and laboratory diagnostic apparatuses constitute a fairly representative, albeit not exhaustive, list of these modalities requiring specialized maintenance service. Often, this need is accomplished via renewable contract arrangements with contractors, but in some cases, internal expertise may be developed. An integral mechanic function is financially feasible and desirable for large programs in order to allow for expedient and regular maintenance, modification, and repair of pens, cages, caging systems, powered machines, motorized devices, and nonfixed material handling equipment; salvage of parts; completion of small improvements and installations; conduct of simple and routine fleet maintenance procedures; and attention to complex decentralized and dispersed MEP analogs, such as blowers on ventilated caging systems and the pumps, filters, and distribution system for automated watering. In this case, design and construction must account for space and outfitting needs for both desk and mechanical work. Postconstruction, as the warranties on new equipment expire, there will need to be a renewable budget allocation for preventive maintenance contracts and for covering the expenses related to normal wear and tear, as well as those for damage from accidents and abuse.

Conclusions

The changes required of animal facilities will be driven by the nature of research into the human condition, the safeguarding of our ecosystem, and the sensible use of natural and/or finite resources, particularly people. These obligations of stewardship will be enabled by new technologies that improve the conduct of experiments and the efficiency and safety of operations, with ramifications for existing facilities, driving renovation, and new construction projects. A premium will continue to be placed on the machinery used in laboratory animal care to conserve water and energy. Other improvements undoubtedly will materialize in affordable automation, including dimensions that allow for its reach to extend far into the facility, wireless and other communications, research data acquisition and streaming, lighting, and HVAC. Great gains remain possible and will be attainable in terms of energy-efficient ventilation. Needs to conserve will drive environmental control to the room level, making zonal control concepts obsolete. At the room level, relative occupancy sensors will allow vacant rooms to automatically revert to diminished standby air turnover rates, and for variable ventilation in adapting to the existing biomass present while maintaining the stability of temperature and relative humidity. That does not mean, however, that energy savings will reign absolute. Significant, energy-consuming supply and/or discharge air treatment and filtration requirements may be necessary to protect the endogenous microbial profile of certain animals, to enable studies of the microbiome not dependent on isolators, possibly for enhanced GLP studies, and/or where hazards are involved. Regulatory action is likely to exert effects on animal care waste management practices, possibly imperiling disposal by landfill, incineration, or sewer in some locales, and spelling the doom of the massive water-consuming chillers integral to so many institutional air-conditioning systems today. Where 9 feet 0 inches typically has been the peak height of facility rooms, minimum clearances of 10 feet 0 inches may be necessary to accommodate or convert to alkaline hydrolysis digestion units for rendering of certain organic wastes. Design and construction best serve this possibility by providing the capacity for pits at some sites or expandability of some ceilings to 10 feet 0 inches or more. Higher ceilings also allow for the deployment of overhead gantry systems in cage wash and other areas.

On the science side, the human conditions of aging, the neurosciences, and infectious diseases will continue to influence the facility in terms of the capability to enable behavioral research, imaging, and the containment of hazards. For behavior, the focus will be on minimizing interruptions to animal behavior imposed by noise, vibration, other environmental disruptions, and human intrusions of any sort. Improvements of wireless remote technologies and in the field of biometrics will allow for remote sensing and recording, hands-off animal experiments, and a broadened range of investigator-independent animal experiments, and at the same time, it will challenge traditional animal care practices. Automated physiologic phenotyping, recording, and analysis of behavior, and even health monitoring down to the level of the cage, may progress toward the norm.

Perhaps the greatest need and effect that could be realized is in automated drinking water supply systems. While these have been a boon for worker safety and efficiency, and even lauded as a tremendous animal care development as early as 1959 (Russell and Burch 1959), the experience of many facilities is that the consequences of cage floods and cases of dehydration in association with these systems remain the greatest source of preventable losses of rodents. Might breakthroughs at the cage sensing and reporting level of high moisture and/or abnormal water flow provide the tipping point to achieve elusive rodent safety via prompt, real-time flood and dehydration detection, thus allowing for rapid emergency intervention? The bottom line is that no one can fully predict the future, making our responsibility to enable it rather than to predict or dictate it. To accomplish that means that we must design and build with flexibility and with any number of possibilities, including the incredible, in mind.

For the present, it is critical that the animal resources personnel who occupy and use a facility know and understand it, and therefore the design principles and construction standards used to deliver the product. When faced with building or renovating, it is vital to have and enable an absolutely clear vision, and agreement among all end users, on what the new facility needs to accomplish. Insufficient user input risks a new facility that is underutilized or requires immediate renovation. Never assume someone else will anticipate or provide for your needs. The stakeholders working in the facility will only be served by representation at the table from the time of programming on, but also when facility management maintains a continuous hands-on approach through the completion of construction. As it has been said, if you are not at the table, then you will be on the menu. It is critical to stay involved from the start of planning to the completion of construction and to visit the construction site regularly. As such, you should own a hard hat, a safety vest and glasses, and an architect's scale ruler. Understand that this is a team game and the object is not to have winners and losers, but the best outcome. To that end, not everyone will get everything they want. Be prepared to establish priorities, trade or design for future adaptability, and eye accommodations that will allow for lesser needs that cannot be immediately funded to be met in time. Additionally, do not think small, but program with future growth in mind. Encourage flexibility in layout and design whenever possible to consider future research needs. Design is a detail-oriented responsibility, and one can never drill down to minute detail enough or pore over a plan sufficiently. Pay close attention to trouble areas that are sometimes overlooked, such as floor and ceiling finishes, cage wash areas, and the flow and throughput of work. Serve as an advocate to ensure that needs take priority over design and first costs, and that there is investment in oversight systems and management practices to ensure that the project is done correctly the first time. If you are approaching this responsibility from the perspective of a novice, seek out and work with people who really know ARFs and understand the importance of meeting the needs of the end user and learn from them.

REFERENCES

ANSI (American National Standards Institute). Eyewash standard. Z358.1-2014. Washington, DC: ANSI.

CCAC (Canadian Council on Animal Care). Guidelines on: Laboratory animal facilities—Characteristics, design, and development. Ottawa: CCAC, 2003.

CDC (Centers for Disease Control and Prevention). *Biosafety in Microbiological and Biomedical Laboratories (BMBL)*. 5th ed. Atlanta, GA: CDC, 2009. http://www.cdc.gov/biosafety/publications/bmbl5/index.htm.

CFR (Code of Federal Regulations). Good laboratory practice for nonclinical laboratory studies. Title 21, Part 58. Washington, DC: U.S. Food and Drug Administration. http://www.accessdata.fda.gov/scripts/cdrh/cfdocs/cfCFR/CFRSearch.cfm?CFRPart=58&showFR=1.

CFR (Code of Federal Regulations). Compressed gases (general requirements). Title 29, Section 1910.101. Washington, DC: U.S. Government Publishing Office. http://www.ecfr.gov/cgi-bin/text-idx?SID=2a82e93e8d104a96f4b04b54816791fd&mc=true&node=se29.5.1910_1101&rgn=div8.

CFR (Code of Federal Regulations). Occupational health and safety standards: Medical devices and first aid. Title 29, Section 1910.151(c). Washington, DC: U.S. Government Publishing Office. http://www.ecfr.gov/cgi-bin/text-idx?SID=2a82e93e8d104a96f4b04b54816791fd&mc=true&node=se29.5.1910_1151&rgn=div8.

Directive 2010/63/EU. On the protection of animals used for scientific purposes. Brussels: European Parliament and the Council of the European Union, 2010.

Geertsema RS, Lindsell CE. Effect of room ventilation rates in rodent rooms with direct-exhaust IVC systems. *J. Am. Assoc. Lab. Anim. Sci.* 54(5) (2015): 521–6.

Helfman GS. Twilight activities and temporal structure in a freshwater fish community. *J. Fisheries Aquatic Sci.* 38 (1981): 1405–20.

Huerkamp MJ, Gladle M, Mottet M, Forde K. Ergonomic considerations and allergen management. In *Planning and Designing Research Animal Facilities*, ed. JR Hessler, NDM Lehner. Kidlington, UK: Elsevier, 2008, 115–28.

ILAR (Institute for Laboratory Animal Research), Committee for the Update of the Guide for the Care and Use of Laboratory Animals. *Guide for the Care and Use of Laboratory Animals.* 8th ed. Washington, DC: National Research Council, National Academies Press, 2011.

Jennings M, Batchelor GR, Brain PF, Dick A, Elliott H, Francis RJ, Hubrecht RC et al. Refining rodent husbandry: The mouse. *Lab Anim.* 32 (1998): 233–59.

NASA (National Aeronautics and Space Administration). Summary of conclusions reached in workshop and recommendations for lighting animal housing modules used in microgravity related projects. In *Lighting Requirements in Microgravity: Rodents and Nonhuman Primates*, ed. DC Holley, DM Winget, HA Leon. NASA Technical Memorandum 101077. Moffett Field, CA: Ames Research Center, 1988, 5–8.

NFPA (National Fire Prevention Association). Health care facilities code. NFPA-99. Quincy, MA: NFPA, 2015, paragraph 5.1.3.5.2.

NRC (National Research Council). *Strategies That Influence Cost Containment in Animal Research Facilities.* Washington, DC: National Academies Press, Committee on Cost of and Payment for Animal Research, 2000.

NRC (National Research Council). *Nutrient Requirements of Nonhuman Primates.* 2nd rev. ed. Washington, DC: National Academies Press, Committee on Animal Nutrition, Ad Hoc Committee on Nonhuman Primate Nutrition, 2003, 116–22.

Rasmussen S, Glickman G, Norinsky R, Quimby FW, Tolwani RJ. Construction noise decreases reproductive efficiency in mice. *J. Am. Assoc. Lab. Anim. Sci.* 48(4) (2009): 363–70.

Russell WMS, Burch RL. *The Principles of Humane Experimental Technique.* Special ed. South Mimms, Potters Bar, Herts, England: Universities Federation for Animal Welfare, 1959, 165.

Ullrey DE, Bernard JB. Vitamin D: Metabolism, sources, unique problems in zoo animals, meeting needs. In *Zoo and Wild Animal Medicine: Current Therapy*, ed. ME Fowler, RE Miller. Vol. 4. Philadelphia: Saunders, 1999, 63–78.

Wilson M, Palmer D. Facility construction. *ALN*, August 31, 2005.

19

Special Security Considerations for Protecting Programs That Use Animals

John J. Sancenito, Norman Mortell, Michael Stephens, and Robert H. Weichbrod

CONTENTS

Introduction

Security breaches have serious consequences and, in some cases, may be catastrophic. In addition to jeopardizing personal safety, some additional potential outcomes might involve research program disruption, loss of critical scientific data, biohazard release, and/or the theft of research animals. The protection of assets within any program of animal care and use requires considerable planning, coordination, education, and resources. Security is everyone's responsibility, regardless of whether there is a dedicated department or if the duties are less centralized. A safe and secure environment creates the opportunity for a more productive workforce, reduces liability, and allows employees to focus on the organization's mission. All organizations have a responsibility to protect people, data, and property; however, animal care and use programs have unique security challenges that must be addressed. In many ways, the best security for an institution is good animal welfare.

Security concerns can generally be divided into two categories, external threats and insider threats. External threats originate outside the company, government agency, or institution. Examples include campaigns from animal rights extremists, cyberattacks, and general crime. Internal threats can involve trusted individuals, including existing or former employees, vendors, and subcontractors. Some insider threats are malicious, while others put the organization in jeopardy through negligence. Successful organizations are able to implement effective security programs that do a thorough job of assessing the risk to the organization based on both types of threats and then put measures into place to mitigate their impact.

Security Awareness and Preparedness

Security and convenience are inversely proportionate to one another. The stricter the security measures are, the more problems they create. Most people will accept the fact that we need locked doors to maintain privacy and keep unauthorized people out. The number of locks the door has, however, will dictate the threshold that is considered to be inconvenient, which then could lead to the door being propped open. If security measures are perceived to be too difficult, employees will find ways to bypass them (Sancenito 2014). The relationship between convenience and security is illustrated in Figure 19.1. A delicate balance must be achieved to help ensure security protocols are accepted and followed.

There are four steps that should be taken to develop a comprehensive security plan. These steps are to conduct a risk analysis, conduct a physical security review, conduct emergency planning, and develop a system for educating all employees on security issues. At each step, the organization should evaluate if it has qualified personnel internally to perform the tasks necessary or if an outside consultant will be necessary. There are pros and cons to both sides. Internal personnel are more familiar with the organization's function and operations, while an independent contractor may be more objective and not as easily influenced by internal politics.

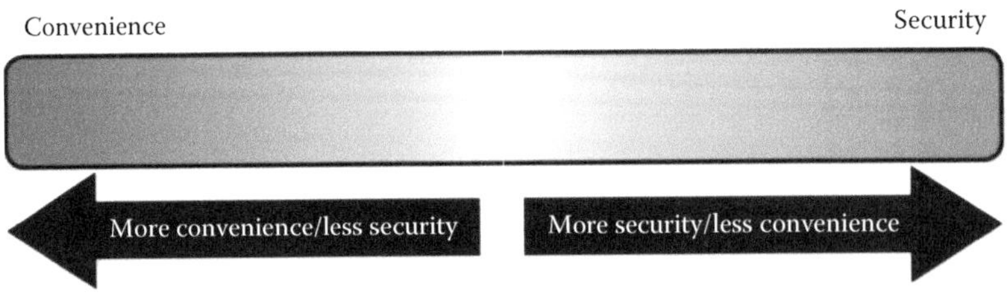

FIGURE 19.1 **(See color insert.)** Relationship between convenience and security.

Step 1: Conduct a Risk Analysis

The first step is to conduct a risk analysis. This process involves making a list of possible events that could affect operations at the facility. Possible risks should be measured using a risk evaluation matrix chart, as depicted in Figure 19.2, in which the probability of an event occurring is evaluated, along with the impact the event would have on business operations.

By measuring risk in this manner, the organization can more effectively determine the amount of resources that should be applied to the risk. In addition to assessing the potential for major downgrading events, other things should be considered when conducting a risk analysis. They include what needs to be protected and how critical it is to operations, what is around the facility that may bring increased risks, and what has been the history of impactful events at the facility.

- *Critical*: Priority and resources should be given to these risks and countermeasures put in place. In addition to actions included in the high category, risks in this category may also require additional insurance to protect the organization.
- *High*: In addition to detailed policies and procedures, detailed contingency plans should exist for these risks. Drills should be conducted on a regular basis. A considerable amount of time should be spent planning for these events.
- *Medium*: Policies and procedures should be in place. A moderate amount of attention should be provided to this category.
- *Low*: Events in this category are not worthy of putting time and attention into them, as they may be easily managed. They are covered by policies and procedures put in place for higher-risk categories.

Each phase is cumulative, to include the actions in the previous one.

For instance, if the geographic area of a laboratory rarely has tornadoes, the likelihood may be considered low, while the impact would be high. Tornadoes would receive a "medium" risk designation. The result would be to have a written plan to deal with the possibility. Tornado drills are probably unnecessary. If a laboratory is located in an area that has frequent and severe tornadoes, the risk may be categorized as "critical." The proper response for this scenario would be to have written policies and procedures, conduct regular emergency drills, and ensure that this type of disaster has proper insurance coverage.

When evaluating possible risks, it is important to prepare a plan for any possible incident that may disrupt business operations. This "all-hazards" approach allows for a series of guidelines to be put in place

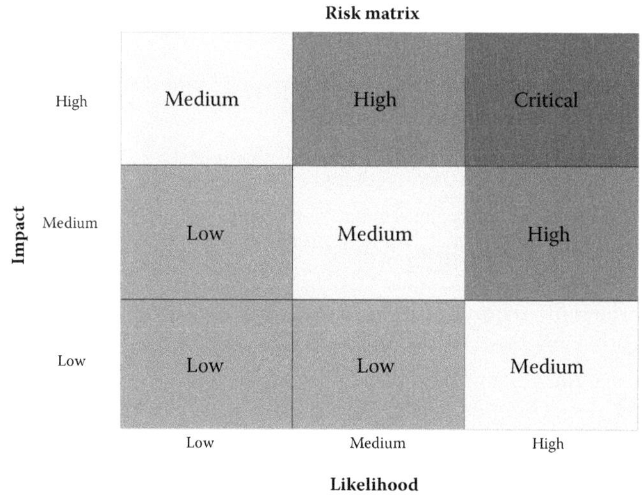

FIGURE 19.2 (See color insert.) Risk evaluation matrix.

for dealing with multiple types of crisis situations, instead of just focusing on a single type of threat. An effective plan is a guide for managing situations that include

- *Natural disasters*: Flooding, earthquakes, wildland fires, extreme wind damage, and severe winter weather
- *Accidents*: Hazardous material spill on a nearby roadway, chemical leak from a neighboring business, explosions (combustible material, high-pressure steam, etc.), fire, or an extended power interruption
- *Man-made disasters*: Terrorism, sabotage, theft of animals, destruction of research, arson, or a malicious computer virus

Special risk factors accompany animal facilities and can directly influence the risk profile of an organization. The species of animals used is a major consideration. Regulated species require registration, animal use reporting, and routine government inspections. These activities create documentation that can be obtained through use of the Freedom of Information Act. Reports filed with the U.S. Department of Agriculture's (USDA) Animal and Plant Health Inspection Service (APHIS) are often used by activist groups. Nonhuman primates, dogs, and cats tend to draw much more attention from animal rights activists than does the use of rodents and fish. The amount of Animal Welfare Act violations (negative regulatory actions) a facility has received may also affect risk. Other factors can include the nature of the research (invasive or noninvasive) and the quantity of animals used in research programs.

Step 2: Conduct a Physical Security Review

Physical security measures deter, detect, and delay an individual from gaining access to a building or restricted area. It is important to conduct a thorough review of existing security measures and evaluate their effectiveness. Some physical security measures can be costly to install and maintain. It is therefore important to evaluate which measures will provide the greatest return on investment.

A physical security review is most effective when performed by an independent person removed from the organization. The review should be conducted by a person or team who is knowledgeable of recognized standards, best practices, and industry benchmarks. There are many aspects to physical security, including crime prevention by environmental design, perimeter security, building architecture, critical infrastructure, disaster planning, electronic security systems, access control, security patrols, and information technology (IT) security. Some measures may involve significant capital expense and may require budgeting for them over a period of time, and others may be implemented with little or no expenditure. Measures to consider include the following

- Have a clearly defined perimeter. Post "No Trespassing" signs around the property that clearly define boundaries.
- Have adequate lighting on the outside of the buildings, particularly near doors and windows. Lighting creates a psychological deterrent and enables detection of intruders and hazards.
- Maintain all vegetation and shrubbery near buildings to less than 4 feet tall to prevent the possibility of someone hiding in them. Nothing higher than 2 feet tall should be planted within 20 feet of building entrances.
- Have exterior doors that are of solid construction and rust-free (including hinges) and shut properly. Exterior doors should have automatic returns on nonremovable hinges.
- Ensure that all exterior windows and glass doors are break resistant. Ideally, no windows on the first floor should open. Animal vivarium and procedure rooms should be located along interior walls, if possible, and should not have exterior windows.
- Animal receiving areas should not be visible from outside the property.
- Combustible material, refuse containers, and dumpsters should not be located within 50 feet of the building or near air intake vents.

- Video surveillance systems should have both monitoring and recording capability. Video surveillance systems (closed-circuit television [CCTV]) should be on closed networks and air gapped on the organization's network to ensure that only authorized personnel are monitoring. It is recommended that two factor authentication procedures are used and function in low-light situations.
- Regularly test access control, video, fire detection and suppression systems, and exterior lighting. If timers are used, ensure that they are adjusted for daylight savings time and/or seasonal changes to daylight hours.
- Conduct annual audits of physical keys and access control badges. Limit access whenever possible and evaluate if employees' access rights are appropriate for their job function and work hours. Access control audits should be conducted at least quarterly to determine if there are unusual patterns, inconsistent with an individual's or group's work routine.
- Secure all roof access points and skylights.
- Have barriers around buildings and utilities that boarder vehicle travel or parking areas.

Physical security measures are only one part of a security program and are useless if they are bypassed or ignored. Ensuring security measures are utilized requires that policies and procedures are closely followed. The goal is to have security measures utilized in daily routines.

Policies and Procedures

Policies and procedures provide guidelines to employees on what is expected of them. They are the cornerstone of any effective security program, but are worthless if they are not followed and enforced. Policies and procedures should exist for everything—visitor control, opening and closing procedures, information security, and preemployment screening. Key security-related policies and procedures should include

- Disaster and crisis management plans that include checklists and up-to-date points of contact.
- Protest and demonstration action plans for managing protests outside the building, as well as disruptive persons in the building (see Appendix 19.1).
- Visitor and vendor handling and escort requirements that define the visitor procedures and the management of vendors working in sensitive areas.
- Bomb threat checklist that includes questions to ask of the caller (see Appendix 19.2).
- Suspicious phone call handling procedures (see Appendix 19.3).
- Suspicious mail and package handling procedures.
- Workplace violence procedures, including lockdown and active shooter protocols.
- Photography policies to prohibit unauthorized photographs in animal vivarium areas. The policy should also address where digital photographs may be stored and managed on IT networks.
- Operational and information security policies. A policy should be in place to address the handling, storing, destruction, and management of data and documents.
- Good access control policies prohibiting the loaning of badges, allowing others to enter without credentials, and propping open unattended doors.

In order to create a culture of security consciousness, it is important to have top-down support for security practices from the highest level of the organization. Security policies and procedures must apply to everyone, regardless of their position or status. If upper management does not follow or show respect for security measures, employees will feel the measures are not important. A duplicitous system undermines the credibility of a security program.

Step 3: Emergency Planning

Managing and quickly recovering from a crisis or disaster does not just happen; it is the result of meticulous advanced planning. Business continuity is defined by ASIS International as "a comprehensive managed effort to prioritize key business processes, identify significant threats to normal operation, and plan mitigation strategies to ensure effective and efficient organizational response to the challenges that surface during and after a crisis" (ASIS International 2005).

Most organizations at some point will experience an unforeseen incident that causes business interruption. The difference between an organization bouncing back or shutting down is dependent on how well it is prepared. It is therefore important not only to have a business continuity plan, but also to repeatedly test it against evolving threats and real-world situations. A business continuity plan helps an organization plan for situations and assists in identifying weaknesses in its operational resiliency. Facilities that may not be directly affected by severe weather or flooding situations may be indirectly affected by long periods of utility shutdowns or road closures. Emergency plans need to be thought out in advance and include plans for relocating animals and critical materials (e.g., preserved specimens in ultra-low-temperature equipment) to alternate facilities.

Security Committees

Security responsibilities may vary greatly depending on the size and type of the organization and the nature of biohazards studied there. Some organizations have dedicated security departments, while in other organizations security duties fall to the facilities department. Regardless of where security responsibilities are placed in the organizational chart, it is important to have a security committee to provide ongoing input to safety and security matters. A well-rounded committee with representatives from all critical departments can offer a perspective of operational and logistical considerations of proposed security measures and their impact. Involving departments in the planning phase can make implantation and compliance much easier.

An effective team will have representatives from each of the following areas:

- Management
- Facilities
- Security or safety
- Human resources
- Public relations
- Finance
- Animal program and scientific or research staff

This group should meet on a regular basis (quarterly at a minimum) to discuss security issues and to evaluate institutional risk and concerns raised in the physical security review. Specific roles of individuals during an emergency should be identified. Succession planning should also be addressed in the event that key decision makers are unavailable or incapacitated during a crisis. Priorities should be decided and major improvements budgeted for over a period of time, if necessary. Some of the committee responsibilities include

- Development of standard operating procedures and internal policies
- Evaluation of security preparedness
- Holding regular meetings and conducting tabletop exercises or drills
- Educating others within the organization on risks, procedures, and policies
- Managing crisis situations
- Conducting postincident reviews and identifying what worked well and what did not

Tabletop Exercises

A tabletop exercise is a meeting to discuss a simulated emergency situation. During an exercise, a facilitator presents a situation or series of situations and asks the group to explain how they would respond. The exercise allows institutions to test their emergency plans and disaster preparedness in a low-stress environment. Tabletop exercises can be an effective tool in evaluating security preparedness and identifying weaknesses in incident management plans. A well-organized tabletop exercise can prepare organizations for real-world threats and help promote improved communications and a better working relationship between departments.

Crisis management plans are not effective if they are just put on the shelf. Unfortunately, during a crisis situation, institutions often find that their plan is not effective and was not properly updated and communicated to key personnel. A tabletop exercise provides an excellent opportunity for an organization to apply, test, and refine its preparedness. Vulnerabilities or flaws may be identified and addressed long before a situation occurs.

A tabletop exercise can only be effective if it involves critical decision makers, involves all divisions of an institution, and has the unwavering support of executive management. Organizations must determine if they have the resources in-house or if they should hire an outside consultant to facilitate the exercise. There are pros and cons to both scenarios. An outside consultant may not be familiar with the corporate culture, policies, and procedures, but will bring fresh ideas and a true understanding of how other companies have managed similar incidents. An outsider may also be more effective as his or her opinions may be perceived as being more objective. In some areas, local emergency responder organizations (e.g., law enforcement, fire departments, agricultural authorities, and the military) may be willing to participate in or help organize tabletop exercises for institutions. Establishing cohesive relationships with emergency response agencies can be extremely beneficial. Having a memorandum of understanding or service level agreement is a vehicle that can be used to institutionalize expected response protocols.

Step 4: Employee Awareness Campaigns

Security awareness is one of the most important but commonly overlooked parts of a successful security and risk management program. It involves training employees on security policies and procedures, educating them on suspicious activity to be aware of, and empowering them to report violations and suspicious behavior. Employees should be trained to be aware of their surroundings and know what to do if they see someone taking pictures from the perimeter, suspicious vehicles in the parking lot, strangers asking for information about animals or what the building is used for, or fellow employees who are exhibiting strange behaviors. Physical security measures, policies and procedures, and employee awareness campaigns are dependent on each other to be effective (Figure 19.3).

Employees should be provided security awareness training at the time of hire and then periodically throughout their employment. Topics to be emphasized include

- The importance of operational, information, and IT security
- Recognizing, challenging, and reporting suspicious people, behavior, and persons
- Social engineering tactics (both existing and emerging) and policies regarding the release of information

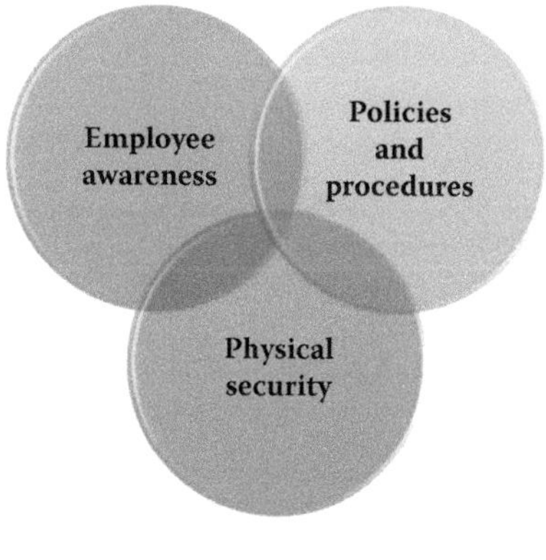

FIGURE 19.3 Connections between physical security, policies and procedures, and employee awareness.

- The process for reporting company policy and security violations
- Workplace violence awareness
- IT and cybersecurity awareness and restrictions
- Evacuation and shelter-in-place procedures
- Employees' individual roles during emergencies

Finding a creative, humorous, or informative way to present security issues, such as posters around the facility, increases the possibility that employees will retain the key concepts. Decide which key security concepts you would like employees to remember and focus on those areas.

Risk from Animal Rights Extremism

The use of animals in research, teaching, and testing is a highly controversial issue that transcends both the scientific and public arenas. According to a 2014 public opinion poll conducted by the Pew Research Center, 50% of U.S. adults say they oppose the use of animals in research, while 46% say they support it (Funk and Rainie 2015). It should be noted that there is a difference between those who support animal welfare and those who support animal rights. Animal welfare supporters believe in the humane care and treatment of animals, and the elimination of unnecessary pain and suffering. Animal rights supporters believe animals are sentient beings and should not be used for food, clothing, research, or entertainment.

Some animal rights supporters feel morally compelled to take action to stop what they perceive to be the suffering of animals. Animal rights activists have a deep philosophical opposition to the use of animals in research, teaching, and testing. Most animal rights activists express their views in lawful ways, such as adopting a vegan lifestyle or engaging in peaceful activities to promote their views. A small number of these individuals, however, believe change will only occur through pressure campaigns and intimidation. An even smaller subset has adopted militant tactics to further their cause.

The term *direct action* is widely utilized by both the animal rights community and law enforcement to describe the variety of actions that social justice activists utilize to impact their intended targets. Examples of nonviolent direct action (often called nonviolent resistance or civil resistance) include strikes, workplace occupations, sit-ins, sabotage, graffiti, and cyber-crime (Boundless 2015).

Examples of violent direct action include arson, bombings, and assault. The use of violent direct action by extremist groups has had a significant impact on animal-related research, not because of a high number of occurrences, but because of the fear it invokes. Activists often refer to past actions in an attempt to intimidate researchers. While there have been some heinous actions committed in the name of animal rights, the reality is that these incidents occur infrequently.

Animal Rights Movement's Use of Direct Action

Data concerning direct actions was collected by Information Network Associates, Inc., an international security consulting firm, over the 4-year time period from 2011 to 2014. This information was categorized by factors including attack method and target industry in order to provide a historical statistical analysis of animal rights criminality as it relates to the animal research industry. This data provides the basis for studying trends of animal rights criminality that can impact the research industry and provides an opportunity to forecast future risks.

The majority of the data collected for the study came from activist self-reporting. Although in some cases this self-reporting is dubious, much of the criminal actions reported by activists include evidentiary documentation. This documentation included photographs and video of activists conducting the criminal actions. In many cases, the actions claimed by activists are reported in the mainstream media, which provides additional validity that the action reported by activists occurred. In some cases, the mainstream media accounts are the primary or only source that can confirm an attack by animal rights activists happened.

However, there remain obvious limitations when studying criminal activity and violence by animal rights activists, such as individuals not sharing information about their actions in order to avoid prosecution. Animal rights criminality remains unique from common law breaking in that the purpose for activists engaging in criminal actions is to ensure they receive attention. Activists will sometimes send anonymous e-mails claiming responsibility for criminal actions to mainstream media outlets, but more often use alternative or independent news sources, websites, and social media. This reporting allows activists to publicize their cause and instill fear and intimidation against the target. It is particularly troublesome for those organizations that possess similar profiles. While it is possible that some of the activist self-reporting is simply fabricated, it is just as likely that more serious criminal actions or failed attempts are never publicized or claimed.

A total of 1750 animal rights–related direct action attacks were identified internationally between 2011 and 2015. These incidents represent all animal rights–related attacks, regardless of targeted industry. The annual number of attacks ranged from 203 to 429 over this 5-year period; however, animal research was targeted in only 11% of these incidents. A breakdown of these attacks by industry is identified in Table 19.1.

An analysis of the 195 attacks against the research industry reveals that the most actively targeted sector was laboratory animal transportation. This sector is part of a relatively new campaign focusing on animal transportation (airlines, shipping, ferries, and trucking companies). Contract research organizations (CROs) are the next heavily targeted, followed closely by academic research institutions. Pharmaceutical companies and biotech or nonprofit agencies trail behind, as depicted in Figure 19.4.

TABLE 19.1

International Direct Action Attacks by Targeted Industry, 2011–2015

Industry	Direct Action Attacks	Percentage of Attacks
Entertainment (bullfighting, circus, zoo, horse/dog racing, etc.)	130	7
Research (government, academic, CRO, pharma, biotech/nonprofit organizations, lab animal breeders, transportation, etc.)	195	11
Clothing (fur industry, leather products, etc.)	245	14
Food (restaurants, farms, slaughterhouses, etc.)	507	29
Other (hunting, pest control, non-research-related animal breeders/suppliers, etc.)	673	38
Total	1750	100

Source: INA (Information Network Associates), Data collected by INA, Inc. Harrisburg, PA, 2016, www.ina-inc.com.

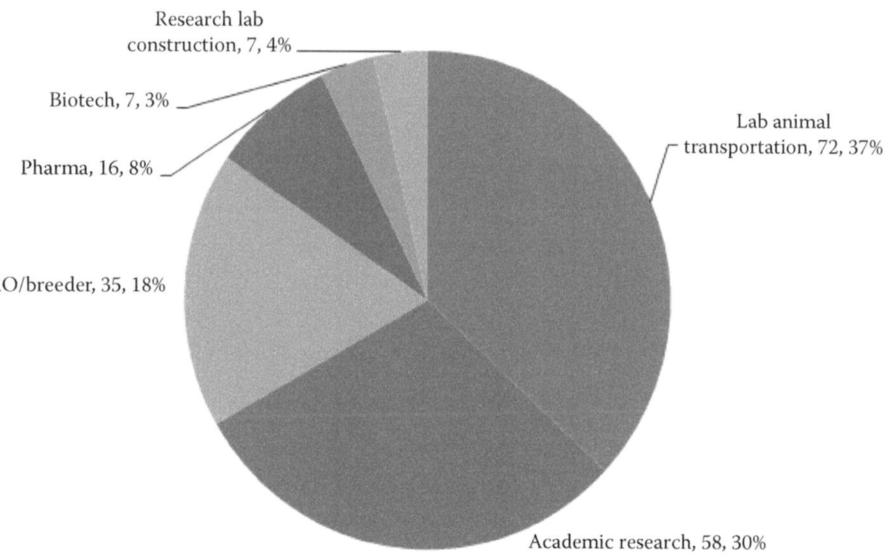

FIGURE 19.4 (See color insert.) Breakdown of 195 direct action attacks against research-related industries from 2011 to 2015. (From INA [Information Network Associates], Data collected by INA, Inc. Harrisburg, PA, 2016, www.ina-inc.com.)

An international look at direct actions against the research industry shows that the animal rights movement is predominantly a western European and U.S. phenomenon. Spikes in other countries, such as Turkey, occur mainly because of an active cell of activists committing a crime spree in a particular area. The relatively low number of incidents can make the movement seem more influential in certain areas than it really is. Figure 19.5 shows the number of direct action attacks against animal research institutions by country.

Activists often target organizations they perceive to be softer targets that may be easily influenced by pressure campaigns. Some extremists are willing to use extreme methods to inflict economic damage or violence if it is necessary to achieve their goals. While many actions are considered acts of terrorism because they are willing to commit criminal acts to further a social ideology, there are major differences between animal rights–related incidents and international terrorism. International terrorists target the general population, wish to inflict mass casualties, are willing to physically confront opposition, and are usually prepared to sacrifice their own lives in the process of achieving their objective. Those committing terrorist acts in the name of animal rights target organizations they feel benefit from what they perceive to be animal suffering, want to avoid detection and their capture if at all possible, and typically try to avoid hurting humans and animals in the process. This is not to say that people have not been seriously injured in animal rights direct actions, as they most certainly have. In 2001, the managing director of Huntington Life Sciences, a British CRO, sustained head injuries that required stitches after he was assaulted outside his home by three masked men carrying baseball bats (Kelso 2001). A series of letter bombs sent to various animal-related industries in Great Britain resulted in a woman being blinded in one eye and a 6-year-old girl being badly burned on the lower parts of her body (Stokes 2001). On August 28, 2003, two bombs exploded at the Chiron Corporation in Emeryville, California. Less than a month later, another bomb exploded outside the Shaklee Corporation in Pleasanton, California (Rodriquez et al. 2003). While no one was injured in the two California bombings, the incident was claimed in support of animal rights. Thankfully, however, these attacks are committed infrequently and have been declining in recent years.

The vast majority of criminal activity against research-related industries from 2011 to 2015 has been minor vandalism, as shown in Figure 19.6.

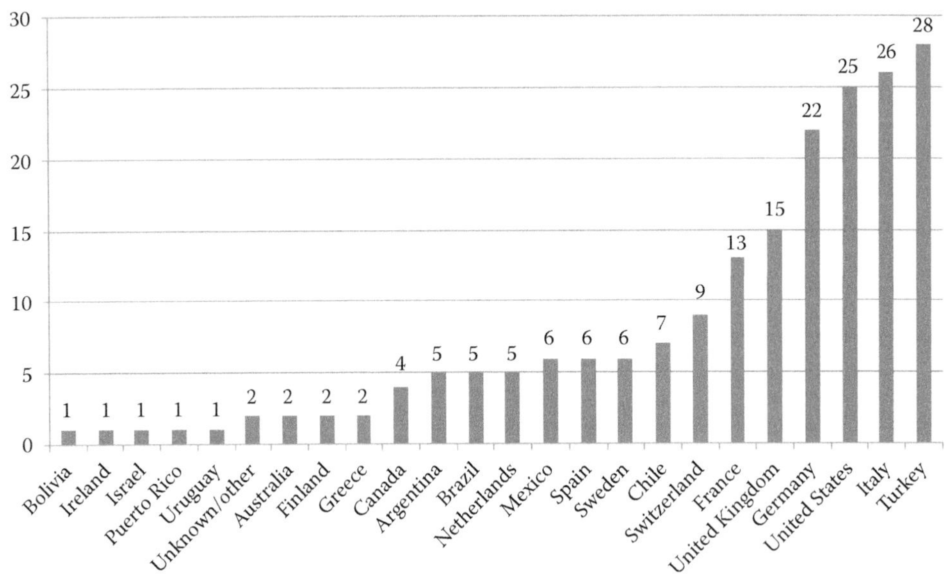

FIGURE 19.5 Breakdown of 195 direct action attacks against research-related industries by country from 2011 to 2015. (From INA [Information Network Associates], Data collected by INA, Inc., Harrisburg, PA, 2016, www.ina-inc.com.)

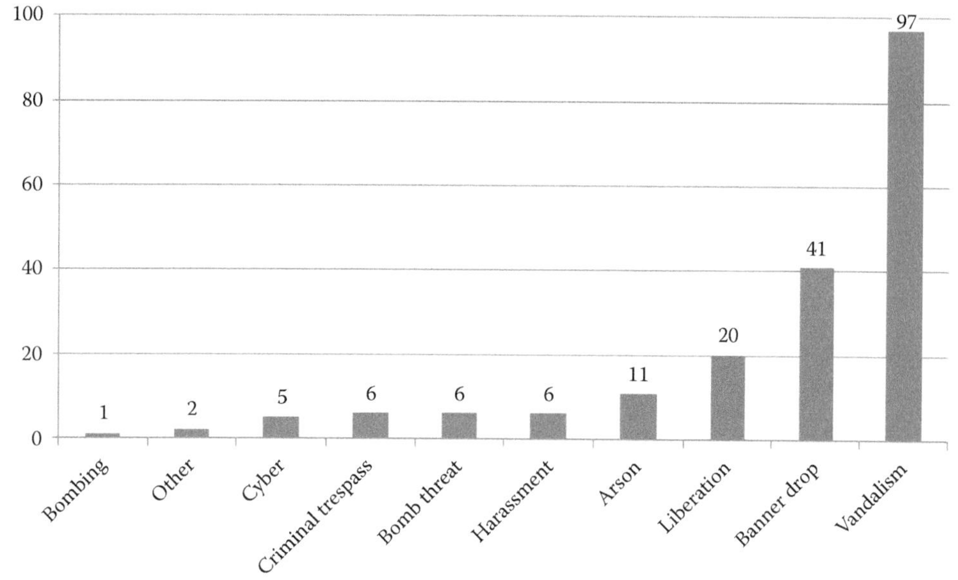

FIGURE 19.6 Type of direct action attacks against research-related targets from 2011 to 2015.

More serious direct actions during this time period have included 1 bombing, 6 bomb threats, 11 arsons, 5 cyberattacks, and 1 claim of product contamination against consumer products.

Direct action peaked in 2008 when a total of 700 international direct action incidents were reported, 119 of which were research related. The direct action then dropped suddenly before leveling out. The reason for the decline in direct action is related to several factors, which include

- Increased law enforcement attention
- Improvements to laws and increased penalties for violations thereof, such as the Animal Enterprise Terrorism Act in the United States and the Anti-Social Behaviour Act in the United Kingdom
- Arrests of key radical leaders
- Increased influence by more moderate mainstream activist groups

The animal rights movement has adapted and evolved over time. While current trends are focused on legal tactics, there are those who reject the focus on animal welfare and support an abolitionist approach. Radical groups operate in small anonymous cells, which can make their actions difficult to predict. The threat posed by animal rights activists will continue to be a risk as long as the use of animals is a necessary and essential part of scientific, medical, and environmental research. It is imperative, therefore, that institutions with animal use and care programs take special precautions and security measures to manage this increased risk.

Insider Threats: Risk of Infiltration

Animal rights extremists use a wide variety of tactics to target organizations with animal use and care programs. One of these tactics is to gain employment with the targeted facility in order to clandestinely gather information to be used in an attempt to further their cause. More than 65 known infiltrations have occurred at research laboratories since 1981. In each case, covert video or photographs were taken from inside the facility, and in some cases, confidential documents and data were stolen. The purpose of these infiltrations was to further an animal rights agenda and obtain images they used to influence public opinion, bring regulatory scrutiny, and motivate activists. Images they obtained during the infiltration were also used in fund-raising campaigns. Each infiltration was followed by complaints made to the USDA demanding action be taken for perceived Animal Welfare Act violations.

Video produced during undercover investigations is captured by hidden cameras worn by operatives over many months. The accuracy and credibility of some pictures and videos can be called into question as having been edited, taken out of context, staged, or due to the negligent actions of the undercover operative. In some cases, however, videos obtained by undercover activists have depicted employees mistreating animals and acting in a manner inconsistent with accepted and very prevalent animal care and use standards. These unfortunate incidents are exploited by activists on social media sites and are used to recruit new activists to their cause.

The vast majority of undercover videos depict legitimate and industry-accepted scientific methods. The dilemma, however, is that the majority of the public does not understand the necessity for animal-based research or the scientific or medical advancements produced. The videos are intended to elicit the maximum emotional response from the viewer. The videos are sometimes narrated by celebrities to give them credibility and appeal to a wider audience.

The impact of infiltrations can be devastating to an organization and should not be underestimated. The potential severity of animal rights clandestine actions must be made clear to the research staff, from the most "prominent" scientist to the "lowliest" technician, so they realize that buy-in is most definitely in their own interest.

In 1981, the Institute for Behavior Research in Silver Spring, Maryland, was infiltrated by someone working with People for the Ethical Treatment of Animals (PETA). Local law enforcement arrested Dr. Edward Taub and seized the research animals after the undercover employee made accusations of animal cruelty. Dr. Taub was convicted in state court, but his conviction was overturned on an appeal based on a ruling that state laws did not apply to federal research programs (*Taub v. State*, 296 Md. 439, 463 A.2d 819 [1983]).

Professional Laboratory and Research Services, Inc., a North Carolina research laboratory, was forced to shut down and surrender hundreds of animals as a result of an undercover PETA investigation in 2010 (Clay 2010). Four employees were also arrested and charged with animal cruelty related offenses. Many others have had to downsize their workforce due to organizations cancelling or withdrawing research studies because of negative publicity. A laboratory's reputation can be forever tarnished and the recruitment of personnel or ability to obtain grants and funding could be affected.

Even organizations with model high-quality animal care and use programs are not immune from employees who do not follow established procedures and highly scrutinized and approved animal study protocols. The risk of infiltration cannot be eliminated, but it can managed by having effective preemployment screening programs, proper supervision of employees, and physical security measures in place to limit access to data and sensitive areas.

Recruiting and screening qualified employees to work within an animal care and use program can be challenging. Nothing can be more important than hiring employees with the proper qualifications, experience, and temperament to care for the animals entrusted to their care. Recruiting and selecting good candidates is difficult, time-consuming, and expensive. Despite this fact, due diligence should never be compromised for expediency. Effective hiring practices ultimately save time and money by avoiding costly hiring mistakes.

An effective hiring process includes a process to validate genuine information at each stage and address concerns or red flags that may arise. Each step provides you with a little more confidence that the applicant is genuine, has the right qualifications for the job, and fits in with your culture and values. There are several stages in an effective recruitment process, which may include those shown in Figure 19.7.

Following the steps in the hiring confidence pyramid moves the employer from a position where they know nothing about a person to a point where the job applicant is seen as a viable candidate for the position and potentially a future trusted employee. Each organization will have a different process, but the key elements are the same. Filtering out unsuitable candidates at each phase provides greater confidence in those selected to go to the next level.

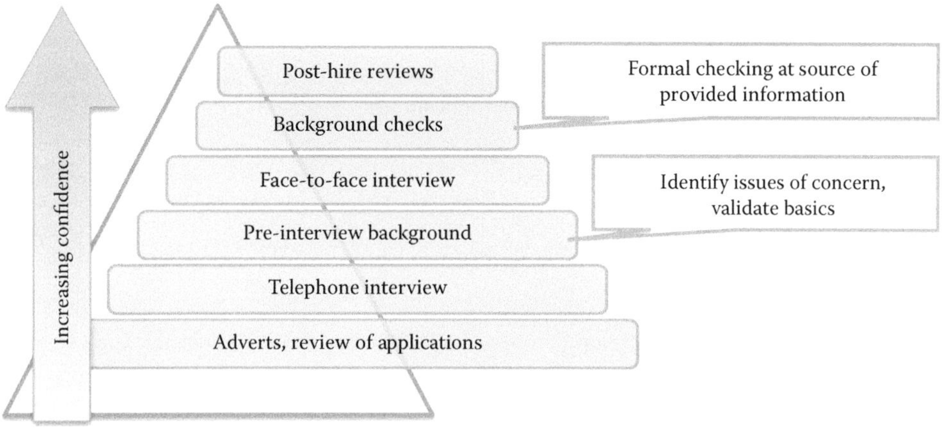

FIGURE 19.7 Hiring confidence pyramid diagram.

Preemployment Screening

Traditional background searches, such as criminal history checks, are not completely effective in identifying bad hires or potential infiltrators. Organizations need to use specialized background checks that include social media searches in order to identify individuals with disingenuous motives. Activist groups meticulously lay out backstories for undercover operatives and train them to blend in. They will, however, almost always use their own names and personal identifying information. Social media should be considered, as information uncovered in this media has successfully identified true motivations and potential infiltrators.

The aim in all cases is to protect the organization's interests while maintaining compliance with all applicable employment laws. Legal counsel should review and approve each step of the hiring process, as laws vary greatly from one area to another. A release from the applicant must be signed authorizing the employer (or a third party, if applicable) to conduct a preemployment background check. The form should include an honesty statement where the applicant verifies that all information is accurate and any false or misleading information will be grounds for disqualification or dismissal, regardless of when it is discovered.

Employers should verify that the information provided to them by an applicant is accurate. This involves verification *at a minimum* of the following basic information:

- The person's identity
- His or her right to work in the country or location
- His or her current address and all others within the past 5 years
- His or her employment history (typically most recent employment, plus the previous 3–5 years)
- His or her education or qualifications and memberships appropriate to the job

Employment that involves working with animals deserves additional scrutiny and due diligence during each phase of the process. Additional checks, such as open-source or Internet mining, antiterrorist and/or money laundering databases (these are often free to access), criminal records, driver's license, and credit history, may be required depending on the position being recruited for and job duties. Contacting references directly through their place of work also helps to validate each reference's identity.

Interviewers should have sufficient training in how to conduct interviews in a fair, consistent, and legal manner. The ability to ask probing questions to get to the core values of the person is essential, particularly about the applicant's view on the use of animals in research. Circumstances that warrant additional scrutiny include

- Relocation from long distances for entry-level positions
- Gaps in employment history
- Prior addresses that are post office boxes
- Nonverifiable previous employers
- A desire to only work with larger species of animals or in certain laboratories or functional areas

Questions should be asked concerning an applicant's personal beliefs about the use of animals in research. It is important not to just accept a statement in his or her responses, such as "society accepts that it is necessary." Interviewers should probe further to ascertain the individual's core beliefs on the subject. If a telephone interview was conducted earlier, the interviewer could refer to the applicant's initial response or ask the question in a different way. It is important to have a template of questions for consistency, but red flags and follow-ups from the telephone interview should also be addressed. As an added security measure, consideration should be given to avoiding telephone interviews, but conducting Skype or FaceTime interviews and saving a screenshot of the interviewee to ensure that follow-up interviews are given by the same individual.

Interviewers should always use open-ended questions that require a narrative answer and not simply a yes or no answer. Questions should also be asked that may cause someone who provided inaccurate information to react, such as "Our firm has a standard policy of conducting background checks on all hires before an offer is made or finalized. You have already signed a release form. Do you have any concerns about that?" and "Tell me about any unexplained gaps in your employment history" (Rosen 2004). Both of these questions offer the applicant the opportunity to provide additional information and the interviewer an opportunity to spot any negative body language or changes in responses that could cause some concern.

Interviewers should carefully note verbal and nonverbal behaviors during the interview process. Signs of nervousness are typical during the interview process; however, sudden changes in body language in response to a particular topic or line of questioning may be a sign of stress or deception.

Things to be aware of should include changes in speech (stuttering, pauses, pitch, or volume), posture (crossing or uncrossing arms or legs, holding hand over the mouth, or tapping fingers), and grooming gestures (twirling hair, rubbing their arms or legs, wringing their hands, and touching the nose). These behaviors are typical signs of nervousness and increased stress. The behaviors typically occur in clusters. They should be considered only if they occur repeatedly when a topic is discussed.

Consider the following scenario:

> A candidate is being asked routine questions during a preemployment interview. He seems relaxed and calm and answers the questions in a sincere manner. The interviewer then asks if the person has any objections to the use of animals in research. The candidate suddenly breaks from his normally relaxed posture to one that is more rigid. He sits up straighter, crosses his arms, and rubs his hands together, and his foot begins to tap on the floor. The person stutters for the first time and responds in a higher than normal pitch that he has no objection and recognizes that animals are a necessary part of new drug discovery.

The interviewer may consider changing the subject to see if the person's body language returns to the relaxed state. The animal use topic could then be reasked to determine if the interviewee's behavior changes again in response to the question.

While body language alone does not necessarily indicate deception, it can be a useful tool in determining which areas deserve further inquiry. The interviewee in the above scenario may be deceptive about his true feelings concerning the use of animals in research, or there may be another reason for the increased anxiety, such as a fear of handling certain types of animals.

Background screening typically raises internal objections based on the fact that recruiting can be a long, costly, and difficult process. An organizational process should be in place to prevent internal pressure to quickly fill positions and supersede a proper and thorough background check. A bad hire is far more expensive than an effective and proven background screening program.

Postemployment Supervision

Animal care and use program staff at all levels have a fundamentally important role in identifying suspicious behavior among existing employees. Following every infiltration, management or coworkers almost always say they thought there was something odd about the individual and his or her behavior. To the extent permitted by law and local policies and contracts, new hires should receive close scrutiny, both openly and "quietly," for at least the first year after employment.

Training managers, supervisors, and other members of the animal care and use program to identify and report suspicious behavior is an essential part of any security program. Some suspicious behavior to be aware of can include

- Employees who repeatedly disregard and violate established security procedures and protocols
- Employees found in areas not associated with their job
- Employees wearing bulky clothing or odd accessories
- Employees excessively questioning and having an exaggerated interest in other departments
- Employees changing their clothes alone and guarding their locker contents (being secretive)
- Employees consistently offering to work late or work alone
- Employees arriving to work at unusual times
- Employees taking part in unusual copying or computer searches
- Employees perceived to be "loners"

Employees with less than 1 year of employment require additional scrutiny and monitoring. Evaluation of employees should continue after they are hired, and input from multiple sources (supervisors, fellow employees, and other staff) should be considered. Communication and documentation is an important part of this process. Some procedures to consider include

- Have periodic meetings with staff to introduce new employees and discuss any concerns. Lab managers, supervisors, and veterinarians should be present during these meetings.
- Monitor the activities of new employees and, when practical, make sure they are paired with an experienced employee.
- Conduct periodic unannounced inspections.
- Have different badges for employees, contractors, and visitors.
- Restrict unnecessary access to specific locations and information systems.
- Monitor IT network activity and badge swipes.

Contract staff should be held to the same standard of background checks that is applicable to permanent staff. Vendors or staffing agencies should be contractually obligated to perform the background checks prior to access being granted. Obtain assurance from the agency that the appropriate screening has been undertaken and conduct audits to confirm compliance.

Policies and Procedures

Written security policies and procedures should exist for access control, handling intellectual property, and reporting suspicious behavior. Each employee's access should be restricted to locations and times required.

A policy restricting cellular phones and cameras in animal areas is advisable. It should be understood that covert cameras can be disguised as nearly anything. While this policy cannot be effective in preventing someone from wearing a covert camera, it does force the individual to intentionally violate a documented security policy. Photography policies also help control legitimate employees from taking and distributing unauthorized images.

Visitors, contractors, and housekeeping personnel should always be escorted in animal areas. A copy of key company policies, including safety precautions and photography restrictions, should be acknowledged by them prior to accessing the facility.

Written policies should also exist that represent the organization's culture of respect for animals. Employees should be directed not to yell or curse at the animals, or to speak or act in any degrading way toward them. In many undercover videos, employees are recorded making derogatory comments about the animals they work with. Even if their actions do not rise to the level of their comments, such behavior is unprofessional and inconsistent with the reverent attitude that should be fostered for those working with research animals.

Not only should animal care and use programs have a zero animal mistreatment policy, but also they should have a policy that mandates the reporting of such behavior by coworkers. A clear policy should exist that obligates an employee witnessing animal mistreatment, neglect, or unprofessional conduct to immediately report it through internal channels. All complaints regarding animal mistreatment should be thoroughly investigated. All organizations using animals should make their best efforts to ensure that if an individual did wear a covert video and audio device into the facility, the only thing he or she would document is high-quality humane animal treatment and the best scientific practices.

Workplace Violence

Workplace violence is a major concern for all employers, and history has shown us that these incidents can happen anywhere. The Occupational Safety and Health Act defines workplace violence as any act or threat of physical violence, harassment, intimidation, or other threatening disruptive behavior that occurs at the work site. It ranges from threats and verbal abuse to physical assaults and even homicide. It can affect and involve employees, clients, customers, and/or visitors.

The General Duty Clause, Section 5(a)(1) of the Occupational Safety and Health Act of 1970, requires employers to provide their employees with a place of employment that is free from recognizable hazards that are causing or likely to cause death and serious harm to employees.

Media reports of workplace violence incidents make it increasingly difficult for organizations to claim that these incidents were unforeseeable.

There are currently no specific standards for employers to follow other than a general directive to have a workplace violence prevention program. A successful program involves physical security controls, procedural controls, and training.

- *Physical security controls and environmental design*: Physical security controls to consider in the plan include windows and doors, access control measures, exit accessibility, lighting, communication systems, and security or police response.
- *Administrative controls*: Procedural controls include internal policies, procedures and standards such as preemployment background checks, workplace violence zero tolerance, disciplinary actions, hotlines, employee assistance programs, and procedures for evaluating threats or potentially violent individuals.
- *Training*: Training includes the training of human resources and supervisory personnel on the warning signs of potentially violent individuals and de-escalation techniques, and general workplace violence training for staff.

There are many types of workplace violence incidents and motivations for them. Violence may be committed by complete strangers, customers or clients, and employees on one another, or in personal relationships. An effective workplace violence plan addresses each of these possibilities and the steps necessary to prevent them. Some potential situations that pose a threat include

- Former employees who have been laid off or terminated
- Current employees who are being disciplined
- Disgruntled customers or visitors
- Rejected suiters, separation, or divorce situations (particularly those with a history of violence or legal protection order)
- Individuals who have exhibited stalking or obsessive behavior
- Employees with alcohol or drug problems, financial difficulties, or emotional problems
- Persons with an intent to commit a crime, such as robbery or sexual assault
- Threats of violence to the facility or individuals working there

Early intervention is an essential part of workplace violence prevention. It is important to train supervisors on the signs of a potentially violent person. Some of the individual warning signs include

- Intimidating or threatening words or behavior.
- Holding a grudge and blaming others for their mistakes or problems. Feeling victimized and talking of getting even with others for a perceived wrong.
- Excessive talk about workplace violence incidents.
- Preoccupation with violence or an extreme interest or obsession with weapons.
- Marked changes in mood or behavior. May include a sudden decline in personal hygiene or work performance.
- Depressed or expressed hopelessness.
- Demonstrated violence toward animals or inanimate objects.

Identifying at-risk employees and taking proactive measures can substantially reduce the likeliness of a violent incident. All employers should have a plan and train supervisors to recognize warning signals and make all employees aware of how to report potentially violent behavior.

Managing and Protecting Data

Research organizations generate not just research data but also sensitive information, including intellectual property, financial, and personnel records. The failure to adequately secure organizational networks, IT systems, and/or other electronic assets will be detrimental to the organization (Mortell and Nicholls 2013). All information has value to an organization and should be managed effectively to safeguard it:

- *Confidentiality (disclosure)*: Data and information can only be disclosed to those that have the appropriate permissions and reasons.
- *Integrity (authenticity)*: Data is safe and cannot be manipulated.
- *Availability (access)*: Only staff with the appropriate permissions can access the data quickly and easily to conduct their legitimate work.

Most data security measures attempt to stop someone from hacking into the network to steal electronically stored information. The reality, however, is that insider threats account for the vast majority of data and intellectual property theft. A global survey conducted by Symantec indicated that half of employees who left or lost their jobs in the last 12 months kept confidential corporate data, and 40% of them plan to use it in their new jobs (Symantec 2013). Despite this risk, however, most organizations do not have effective methods to control access to data or audit which files were accessed or downloaded.

In order to evaluate the appropriate level of data security, it must first be evaluated and classified based on its value, risk, and obligation to safeguard it. There is no one system appropriate for all organizations.

Each organization must define and implement a system appropriate for its individual programs and requirements. The sample data classification system included in Table 19.2 is an example of such a system created from combining multiple open-source systems.

Compliance with privacy laws can be difficult, as laws governing data collection, storage retention, and use differ greatly from one jurisdiction to another. National, regional, and local laws may have special (sometimes more restrictive) requirements.

Data should be classified and protected based on its importance and sensitivity, regardless of its storage medium. Large collections of data in spreadsheets and databases may require additional control measures, even if they do not clearly fit into the classification table.

Classification of information is not enough on its own. The journey of information through an organization should be a documented process with an auditable trail. For example, data for a study might start in the animal's holding room; pass through a level of technicians, supervisors, and managers before going to the study director; and ultimately end up in a records archive area. Printed and electronic storage of information should be considered. Each aspect of data storage and transfer has vulnerabilities that should require appropriate security measures. Each station along the route may have separate security controls, access control systems, and storage mediums. Mapping the process data flows through the organization helps identify potential weaknesses and provides opportunities to designate the appropriate and necessary safeguards that should be applied. A sample data flow process is represented in Figure 19.8.

In addition to data collected in the research setting, there is also confidential information, including occupational health and safety, personnel, and security data. The measures in place to protect data must be measured against the importance of the data to the organization, as well as regulatory requirements to safeguard it.

Human error plays a role in almost all data breaches. Most employees do not intentionally put information at risk, but do so because of ignorance, not being vigilant, or because they do not follow established protocols. All employees should understand their role in managing information security. Some preventive measures individuals can take at their desks or workstations include

- Getting into the habit of locking computer screens before leaving them unattended
- Having a clear-desk policy, using data classification to ensure confidential and/or sensitive information is properly stored
- Properly destroying confidential waste paperwork

Ongoing training is necessary to reinforce good information security practices. The human element cannot be eliminated, so it is necessary to supplement it with procedural and technical solutions.

TABLE 19.2

Sample Data Classification System

Classification	Description	Examples
Public	Nonconfidential or publicly available information	Press releases, job postings, information on publicly accessible websites, and marketing material
Internal	Benign information that an organization chooses to keep confidential	Employee lists, business partnerships, client lists, capacity, employee salaries, internal memos, or operating expenses
Confidential	Sensitive data that could cause harm if disclosed; access should be restricted only to those with a need to know	Individual employment files, employee salaries, legally privileged information, information covered under Family Educational Rights and Privacy Act (FERPA), and financial records covered by the Gramm–Leach–Bliley Act or similar regulations or policies
Restricted use	Extremely sensitive information that is obligated to be safeguarded in the most stringent manner; includes information that would trigger notification to individuals, clients, or regulatory agencies if compromised	Data controlled by U.S. government, classified data, financial account numbers covered by the Payment Card Industry Data Security Act (PCI-DSS), medical records covered under the Health Insurance Portability and Accountability Act (HIPAA), contractual obligations, and IT system passwords

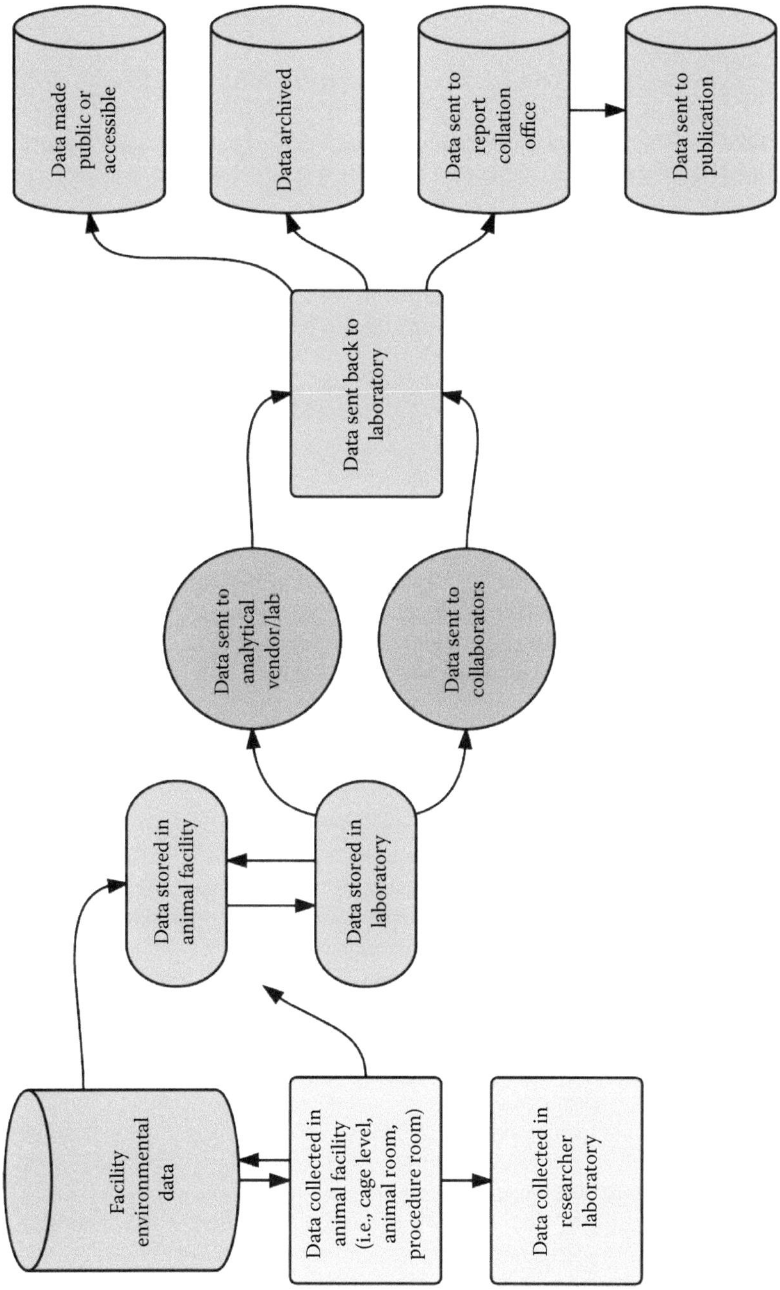

FIGURE 19.8 Example research laboratory data flowchart.

Social Engineering

The human element of security is by far the weakest link in the security chain. Physical security systems and advanced technology have helped strengthen security operations, but these controls can be easily defeated in the face of human manipulation and persuasion. Social engineering is a process by which a person attempts to obtain privileged or confidential information that he or she is not privileged to have by the use of deception or trickery.

Social engineering principles rely on the notion that human trust can be manipulated, or engineered, to retrieve desired information. People generally have a natural inclination toward helping others and usually assume trust to avoid conflict. Furthermore, people will reveal information they believe to be innocuous. It is therefore important to grade the information on its value and properly train employees on how to safeguard it.

Social engineers target the research industry for a variety of reasons, including recruiting, unsolicited sales, competitive intelligence, insider trading, fraud, and investigative journalism. Animal rights and other social justice activists also engage in social engineering to gather information to further their cause. The social engineering process is usually engaged in over the phone by way of a pretext phone call, by e-mail, or by social media. These methods afford anonymity and little chance of detection for the information seeker, aiding in the success of the social engineering process.

Pretext Interactions

Establishment of a pretext, or seemingly legitimate use, is usually part of the social engineering process. Activists may make pretext phone calls to the organization in an attempt to gain information valuable to their campaign. Information of value to activists may relate to employee personal and contact information, types of research programs and projects, animal study models, company business associations, company meeting information, third-party contracts, facility addresses, facility layouts, and security standards. Activists may make pretext phone calls to receive tacit confirmation of an idea they believe to be true. For example, an individual may assume a certain facility houses nonhuman primates. This individual may call the research organization and, by asking a benign question about primates, may be able to persuade the person answering to make some sort of admission or confirmation.

The first step in the pretext phone call is the establishment of the pretext, or the invented scenario, that affords the caller the legitimacy needed to persuade the recipient of the call to release information. Pretext callers typically research their target entity, so they may act knowledgeably about the company or related subject matter. The social engineer can engage in a number of simple tactics to collect what seems to be "inside" information, including names of company employees and key executives, company events, operations, protocol, and other pieces of information that can help establish legitimacy. Internet research, information from former employees, dumpster diving for documents and evidence, and surveillance are all means an individual may use toward gaining information that may be used to establish credibility and gain the confidence of his or her target. Pretext callers may hold themselves out to be a legitimate employee; an employee from a different department, facility, or office within the same organization; a representative of a known company affiliate or contractor; a legitimate third-party inquirer; or a government regulator.

Risk is spread company-wide and must be controlled at all levels of the organization. Employees should have a clear understanding of what type of information needs to remain confidential and secure, and how to maintain it. Security orientation and training should be held upon hire and periodically throughout employment to reiterate security standards. When it comes to an animal use and care program, information that may seem benign and harmless to an employee may provide a gold mine of information for activists. A policy clearly defining information that should only be given out in person to trusted, designated individuals would reduce the chance that sensitive information is provided to an unauthorized party. As a general rule, employees should provide publicly relatable information or refer the caller to a management official responsible for public affairs.

Addressing Suspicious Phone Calls

Organizations should establish a strict policy regarding the disclosure of sensitive information over the phone. For example, it is a good practice to have a policy that prohibits the disclosure of sensitive information over the phone to individuals that one has even the slightest reservations regarding their identity. If a caller requests sensitive information, establish a policy whereby these inquiries are handled via a return phone call. In this way, a caller's contact information can be taken and the request forwarded to an appropriate party. If the caller declines to provide contact information or hangs up, the call may have been a social engineering attempt. In the event of a suspicious phone call, the following are useful tips in dealing with the caller:

- Remain calm and do not become intimidated by the caller. The caller should not be permitted to sway the recipient of the call into providing information by way of appeals to authority or threats of supervisory action.
- Listen carefully and take notes during the call. Take care to include information regarding the likely age, gender, or any identifiable accent of the caller. This will provide an opportunity for a staff member to identify the voice later, if the same individual attempts additional phone calls. One should also listen for background noises that may help to identify the point of origin of the call.
- Be polite and show interest in the caller's stated concerns. Doing so will likely keep the caller calm and talking.
- It is preferred that any requests for information about the organization be deferred to allow time for the legitimacy of the call to be verified. Advise the caller that permission must be given from a senior staff member to provide the information the caller is seeking. Ask for the caller's name, address, and phone number, and carefully document this and other information the caller may supply during the conversation.
- Assess the caller's response to answers given. If they are unreasonably agitated by requests for standard information, such as their name and contact information, it could be an indication that the suspicions are well founded.
- At the conclusion of the conversation, carefully record the date and time of the call and initial any notes for future reference.

Suspected social engineering attempts should be followed up on in accordance with established company guidelines. Callers may attempt to place multiple calls to different employees in search of the information they desire. Quick action could thwart this type of subversion. A sample suspicious caller form is included in Appendix 19.3.

Phishing

Phishing is a form of social engineering that is done via e-mail. The phisher may request information or may ask the user to take an action, such as follow a link or open an attachment. The link may take the user to a website that resembles a legitimate website in an attempt to get the user to divulge personal information, passwords, or other confidential information. E-mail attachments can direct the user to a bogus website or download a malicious code, such as a virus, virus detection hoaxes, ransomware, or keystroke loggers.

Employee awareness campaigns should be conducted to reinforce prevention. It is not just sufficient to train new employees at the time of hire. Threats change and security topics should be constantly reinforced to be effective. E-mail reminders and posters around the facility are one way of continually reminding employees of potential risks. Topics that should be conveyed to employees include

- Be suspicious of all e-mail attachments, including those sent from friends or coworkers. Scan all attachments for malware prior to access, including opening or executing the attachment.
- Never open attachments or click on hyperlinks in e-mails sent from unknown or untrusted senders.
- Do not plug unknown media, such as jump drives and CDs, into computers.
- Report suspicious e-mails or activity to appropriate personnel.

Malware

Malware is often present in e-mail attachments, or can be automatically downloaded and installed on computers when hyperlinks are clicked on. The e-mail may appear to come from a legitimate source, including a colleague or supervisor. The consequences are not always immediately apparent, as the malware could be set to activate at a certain date and time, or may work in the background.

This is particularly true when it comes to animal rights–related websites. Company or personal computers should not be used unless special precautions are undertaken. Many of these websites have a malicious code on them and track the IP addresses of anyone visiting their site. They often post the names of companies who did so. Online monitoring of activist activity should not be done without careful consideration. It is suggested that filters be put in place to block activist websites from organizational computers, as these sites may contain malware. Employees should be trained to be suspicious of requests to download applications or to install "add-ons," such as ActiveX components or strange multimedia players. Never download or install software from unknown or untrusted websites.

Cyber-Squatting

Cyber-squatting is the registering, selling, or using of a domain name with the intent of profiting from the goodwill of someone else's trademark. The cyber-squatter may register domain names containing variations of popular trademarks. This method relies on the high probability that individuals will make typographical errors when entering domain names into their web browsers.

The motivation for cyber-squatting may be financial gain, identity theft, or gathering confidential information. An individual may register a bogus site that is similar to a trusted website domain of a company or trade association. The individual may even make the fake website look exactly like the one in question. He or she will then attempt to capture confidential information from visitors who attempt to log in to a private part of the website or who attempt to make online purchases. Unsuspecting individuals may enter their usernames and passwords, believing they are on the website they intended to visit (INA 2012). E-mail domains are also registered so that anyone mistyping an e-mail address could accidently e-mail important documents to the scammers. Animal rights activists have used this tactic to gain information about animal studies and research programs.

- Be cautious when following links within e-mails to make sure you are not redirected. Type it manually if you are suspicious.
- Be cautious when typing website URLs or e-mail addresses.
- Check variations of your organization's web address to ensure other variations of the website have not been registered for nefarious purposes.
- Purchase similar domains to your organization's domain to ensure no one else does.
- Take legal action, if necessary, to protect your organization's name or brand identity.

Social Media

Social media is a wonderful tool for people to remain connected. It is also a wonderful tool for those with malicious intent to gather information. Activists monitor social media and use it as a tool to develop information. Not using privacy restrictions or accepting a friend request from an unknown or unverified individual could have serious consequences.

Organizations should have a social media policy that limits what information employees may post on social media. Employers cannot control everything employees post on personal social media pages, but they can dictate that proprietary and confidential information not be disclosed. The policy should prohibit employees from divulging confidential information, which includes the types of animals used in research studies, clients' names, and security protocols. The posting of photographs taken within the facility should be strictly prohibited. It is important not to make this policy overreaching, as legal decisions allow employees to discuss the terms and conditions of their employment. Legal council should review social media policies before implementation.

Malware is increasingly spread through social networking sites by installing nefarious third-party add-on applications or by clicking on hyperlinks in messages. Additionally, there seems to be a false sense of security when using social media. Social media awareness programs should remind employees to

- Utilize security settings to privatize their social media accounts
- Have a strong password and change it periodically
- Never accept a friend request from an unknown and unverified profile
- Only install third-party social networking applications that are approved by IT
- Never click on hyperlinks in messages from unknown or untrusted contacts, and avoid clicking on message hyperlinks sent from trusted contacts unless they are certain where doing so leads

The possibility of a data breach is substantially reduced in companies who embrace a strong information security culture. Such a program can only be successful if it is adopted by staff at all levels and becomes a routine part of operations. Management must embrace information sharing rules and not make requests that would induce an employee to violate them. Once information is revealed, it can never be retracted and can be detrimental to an individual or organization. Efforts must be made to mitigate the security risk that humans pose to even the most advanced security programs.

Conclusion

Security is an ongoing process, and security adaptation to emerging risk is not something that is permanently achieved. Risk factors will change over time, and an effective program must evolve and adapt. A risk-based approach should be taken to ensure the maximum return on investment. Risk mitigation programs should include physical security measures, effective policies and procedures, and ongoing employee awareness.

The unpredictable human element of an organization will never be fully eliminated, but efforts can be made to strengthen it. A program that includes continuous employee education is an organization's best defense.

Security is often looked at as a burden or cost center within organizations. It is impossible to measure or quantify the deterrence value or return on investment for security measures. Increasing security awareness is essential to gain support for security initiatives. It is also important to gain buy-in at all levels of the organization, especially from scientists and research staff, who frequently balk at the perceived "excessive" emphasis on security. All organizations have a duty to protect people, property, and data. Animal care and use programs have an increased responsibility—that of safeguarding the animals entrusted to them.

APPENDIX 19.1: GUIDELINES FOR MANAGING PROTESTS

- Lock or monitor lobby and exterior doors.
- Notify local police and security immediately.
- Do not attempt to engage or communicate with activists for any reason.
- Conceal your company ID, if visible.
- Avoid leaving the building during the demonstration, if possible.
- Postpone any deliveries, if possible.
- Instruct employees not to watch or videotape the protest.
- Expect that you are being videotaped.

APPENDIX 19.2: SAMPLE BOMB THREAT FORM

Be calm. Be courteous. Listen. Do not interrupt the caller. Notify your supervisor/security department by a prearranged signal while the caller is on the line, if possible.

Date: ____________ **Time:** ____________

Exact words of person placing call:

Ask the following questions:

1. When is the bomb going to explode?
2. Where is the bomb right now?
3. What kind of a bomb is it?
4. What does it look like?
5. Why did you place the bomb?

Try to determine the following: (Circle as appropriate)

Caller's identity:	Male	Female	Adult	Juvenile	Age ____ years		
Voice:	Loud	Soft	High pitch	Deep raspy	Pleasant	Intoxicated	
	Other__________						
Accent:	Local	Not local	Foreign	Region ____________________			
Speech:	Fast	Slow	Distant	Distorted	Stutter	Nasal	
	Slurred	Lisp					
Language:	Excellent	Good	Fair	Poor	Foul	Other ______________	
Manner:	Calm	Angry	Rational	Irrational	Coherent	Incoherent	Deliberate
	Emotional	Righteous	Laughing	Intoxicated			
Background noises:	Bedlam	Trains	Animals	Music	Quiet	Voices	
	Airplanes	Traffic	Party	Atmosphere	Office machines	Factory machines	

Caller ID information: ________________________________

Additional information:

Notify your supervisor and talk to no one about the incident until they instruct you to do so.

________________________ ________________________

Receiving telephone number Person receiving call

APPENDIX 19.3: SAMPLE SUSPICIOUS PHONE CALL FORM

Be calm. Be courteous. Listen. Do not interrupt the caller. Notify your supervisor/security department by a prearranged signal while the caller is on the line, if possible.

Date: ____________ **Time:** ____________

The following response guidelines are suggested in the event of a suspicious phone call:

- Remain calm and do not become intimidated by the caller. Do not let the caller sway you into providing information by way of appeals to authority or threats of supervisory action.
- Listen carefully and take notes during the call. Take care to include information regarding the likely age, gender, and any identifiable accent, if possible.
- Advise the caller that you must seek permission from a senior staff member to provide the information that he or she is seeking. Ask for the caller's name, address, and telephone number, and carefully document this and other information that the caller provides.
- Immediately notify a supervisor. It is not unheard of for callers to place multiple calls to different employees in search of the information they desire.

What did the caller ask for or say?

Try to determine the following: (Circle as appropriate)

Caller's identity:	Male	Female	Adult	Juvenile	Age ____ years		
Voice:	Loud	Soft	High pitch	Deep raspy	Pleasant	Intoxicated	
	Other__________						
Accent:	Local	Not local	Foreign	Region ____________________			
Speech:	Fast	Slow	Distant	Distorted	Stutter	Nasal	
	Slurred	Lisp					
Language:	Excellent	Good	Fair	Poor	Foul	Other ______________	
Manner:	Calm	Angry	Rational	Irrational	Coherent	Incoherent	Deliberate
	Emotional	Righteous	Laughing	Intoxicated			
Background noises:	Bedlam	Trains	Animals	Music	Quiet	Voices	
	Airplanes	Traffic	Party	Atmosphere	Office machines	Factory machines	

Caller ID information: ________________________________

Additional information:

Notify your supervisor and talk to no one about the incident until they instruct you to do so.

_________________________ _______________________________

Receiving telephone number Person receiving call

USEFUL RESOURCES

Industry Resources

Americans for Medical Progress: www.amprogress.org
National Association for Biomedical Research: www.nabr.org
States United for Biomedical Research: www.statesforbiomed.org

Disaster Preparedness

www.ready.gov
www.dhs.gov
www.fema.gov

Risk Assessment Help

www.cabinetoffice.gov.uk/content/risk-assessment

Insider Threat Best Practice

Federal Bureau of Investigation: https://www.fbi.gov/investigate/counterintelligence; www.cert.org/insider-threat/best-practices
The Common Sense Guide to Mitigating Insider Threats, Fourth Edition: www.sei.cmu.edu/reports/12tr012.pdf
Centre for the Protection of the National Infrastructure—Insider Data Collection Study: https://www.cpni.gov.uk

Lighting

Illuminating Engineering Society: www.ies.org

Removing Personal Details Online and Data Protection

https://ico.gov.uk

Check Your Online Impression

www.pipl.com
www.zabasearch.com

Check Your Password Strength

www.passwordmeter.com

Workplace Violence

https://leb.fbi.gov/2011/january/workplace-violence-prevention-readiness-and-response
http://www.bjs.gov/index.cfm?ty=pbse&sid=56
https://www.osha.gov/SLTC/workplaceviolence

REFERENCES

ASIS International. *Business Continuity Guideline: A Practical Approach for Emergency Preparedness, Crisis Management, and Disaster Recovery.* Alexandria, VA: ASIS International, 2005.

Boundless. Forms of disagreement. Boundless, July 21, 2015. https://www.boundless.com/political-science/textbooks/boundless-political-science-textbook/public-opinion-6/public-opinion-44/forms-of-disagreement-255-8025 (accessed January 15, 2016).

Clay, D. North Carolina animal research facility closes due to PETA investigation. *The Examiner*, September 16, 2010. http://www.examiner.com/article/north-carolina-animal-research-facility-closes-due-to-peta-investigation (accessed February 23, 2016).

Funk, C., and L. Rainie. Americans, politics and science issues. Washington, DC: Pew Research Center, July 1, 2015. http://www.pewinternet.org/2015/07/01/americans-politics-and-science-issues/.

INA (Information Network Associates). Domain cybersquatting. White paper. Harrisburg, PA: INA, October 15, 2012. www.ina-inc.com.

INA (Information Network Associates). Data collected by INA. Harrisburg, PA: INA, 2016. www.ina-inc.com.

Kelso, P. Animal lab boss attacked. *The Guardian*, February 2, 2001. http://www.theguardian.com/uk/2001/feb/24/paulkelso1 (accessed February 23, 2016).

Mortell, N., and S. Nicholls. Practical considerations for disaster preparedness and continuity management in research facilities. *Lab Animal*, 42 (10), F18–F24, 2013.

Rodriquez, M., J.-R. Chong, and G. Krikorian. Suspect is sought in bombings. *Los Angeles Times*, October 10, 2003. http://articles.latimes.com/2003/oct/10/local/me-warrant10 (accessed February 23, 2016).

Rosen, L. S. *The Safe Hiring Manual: The Complete Guide to Keeping Criminals, Imposters and Terrorists Out of the Workplace.* 1st ed. Tempe, AZ: Facts on Demand Press, 2004.

Sancenito, J. Social engineering: Are you the weakest link? *Laboratory Animal Science Professional*, 3 (4), 2014.

Stokes, P. Animal rights bomber admits terror campaign. *The Telegraph*, July 28, 2001. http://www.telegraph.co.uk/news/uknews/1335581/Animal-rights-bomber-admits-terror-campaign.html (accessed February 23, 2016).

Symantec. What's yours is mine: How employees are putting your intellectual property at risk. White paper. Mountain View, CA: Symantec, 2013.

Section VI

Environment and Housing

20

Environmental Factors: Macroenvironment versus Microenvironment

Margaret C. Hogan, John N. Norton, and Randall P. Reynolds

CONTENTS

Introduction

Animals used in research, teaching, and testing are complex living creatures that respond to environmental parameters. Whether living in a tightly controlled internal environment or an outdoor environment subject to weather changes, animals, through their physiological and behavioral processes, will respond to various stressors in their environment (Hale 1969; Sossinka 1982; Price 1984). Even subtle changes in the living and experimental environment can lead to confounding and variable research outcomes. Although the effects of the environment may not be readily noticeable to the investigator, variations may occur, resulting in research variability and potentially erroneous conclusions (Roe 1965; Magee 1970; van der Touw et al. 1978; Chvedoff et al. 1980; Kokolus et al. 2013). Arguably, the degree to which environmental variables influence research outcomes may not be appreciated by many researchers (Vesell and Lang 1976; Clough 1982; Siegel 2011; Bustin 2014), when their focus is on controlling direct experimental variables inherent in their research. Recently, there has been renewed emphasis regarding reproducibility in animal research outcomes (Siegel 2011; Bustin 2014), whether research is conducted between different research facilities or within the same facility, and an increased emphasis on the reporting of conditions that can cause variability when animals are used in research (Kilkenny et al. 2009, 2010).

The use of animals, whether in research, teaching, or testing, is governed by regulations, policies, and guidelines (Bayne and Anderson 2015). These documents commonly describe the care and use expectations for various species and delineate the common environmental variables, such as housing requirements, sanitation, feed, water, lighting, and temperature. Animal well-being directly correlates to the appropriateness of the many environmental variables, whether physical, nutritional, or social enrichment (Yousef 1985; Curtis 1986; Baker and Lipman 2015). Environmental conditions should minimize stress, illness, mortality, injury, and behavioral problems. Controlling environmental variables (Figure 20.1), along with the provision of appropriate husbandry, is crucial for the appropriate use and well-being of animals.

When attempting to control and minimize the effect of environmental variables on laboratory animals, the design and management of the research facility is paramount (Hessler 1999). The modern research facility, with specialized housing rooms and caging systems, provides an environment where surrounding variables are minimized regardless of species, thus creating an optimal setting and preventing perturbations of the data. In contrast, natural environments commonly used for agricultural, aquatic, and traditional nonrodent species and field studies are inherently prone to variance in environmental conditions, thus producing perturbations of generated data. Additionally, natural environments are well suited for colonies of breeding animals or holding animals until needed for use. Natural environments can subject animals to weather events, such as summer heat, winter storms, and noise. The animals need to acclimate, and appropriate shelter and wind breaks may be needed to ensure animal well-being. Regardless of the type of environment, proper housing and management are essential to ensure animal well-being and the quality of research data obtained from the animals. Qualified and dedicated personnel who adhere to well-conceived operating procedures for the facility create an environment that results in high-quality animal care.

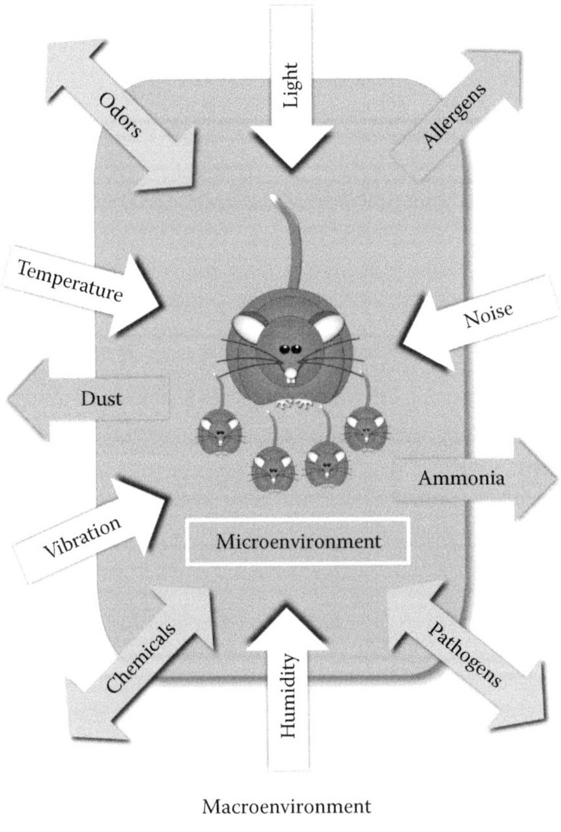

FIGURE 20.1 Environmental factors that can influence the health, well-being, and research outcomes in animals.

The animal's environment consists of both the micro- and the macroenvironment. The microenvironment refers to an animal's primary enclosure, which includes the immediate area surrounding the animal, either in an aquarium, cage, pen, or stall, and is the environment in direct contact with the animal (FASS 2010; NRC 2011). Animals are exposed to the conditions within the microenvironment, such as temperature, humidity, noise, vibration, and air composition. The physical conditions surrounding the microenvironment are defined as the macroenvironment, which is composed of the room, barn, or pasture (NRC 2011). In many situations, the microenvironment is identical to the macroenvironment due to open caging systems or natural housing environments. In contrast, the microenvironment can be substantially dissimilar to the macroenvironment due to the design of the primary enclosure. Ventilated caging systems permit different microenvironments to exist within the same macroenvironment. For example, by providing increased high-efficiency particulate air (HEPA)–filtered airflow to and from each cage, the cages on one rack could be positive pressure, relative to the macroenvironment, to prevent entry of infectious agents, and the cages on another rack could be negative pressure to prevent the escape of infectious agents. The microenvironment and macroenvironment should be appropriate for the genetic background and age of the animals and the purpose for which they are being used.

The following sections discuss some considerations of the macro- and microenvironment related to research animals.

The Macroenvironment

The macroenvironment must be maintained in a manner that will provide stable, comfortable, and clean living conditions for the animals. Some factors that must be addressed are lighting, air quality (e.g., airflow, temperature, and humidity), minimal noise and vibration, and proper sanitization practices.

Lighting

Appropriate lighting is crucial to an animal's physiology and behavior (Dauchy et al. 2013a, 2013b), and alterations in light cycle, intensity, or spectrum may lead to stress (Stoskopf 1983). In addition to duration, intensity, and wavelength, the species and strain of the animals are also factors to consider (NRC 2011) for the macroenvironment.

Lighting should be diffused evenly throughout the macroenvironment to provide for the animal's specific lighting requirements and provide sufficient light to perform husbandry, sanitation, and animal health evaluation procedures. Lighting should be at appropriate levels regardless of the animal's location in the room (NRC 2011). In the United States, most animal research facilities follow the recommendation of 325 lux (30 foot-candles) approximately 1 m (3.3 ft) above the floor for routine animal housing (NRC 2011). This lux level has been deemed appropriate for accommodating husbandry, veterinary, and research activities without causing retinal damage to albino rodents (Schofield and Brown 1996; Faith and Huerkamp 2009). European guidelines further stipulate that darker areas for withdrawal should be available within the animal enclosures (Council of Europe 2006).

To control light intensity and provide appropriate lighting for a variety of species, most modern animal research facilities are designed to include two-stage lighting systems. With two-stage lighting, lux can be set at a low level to accommodate albino rodents or at higher levels for animals with normally pigmented eyes. The lighting can also be adjusted for specific research needs or to accommodate husbandry activities. Some of the more sophisticated systems that control two-stage lighting are computer based and can be administered from the desktop or by remote access by way of a variety of mobile devices. Preprogrammed illumination override switches are often available at the room level for short-term use. These systems help to eliminate the possibility of unauthorized room-level alteration of light intensity and may even generate alarms if lighting parameters are out of prescribed ranges.

Research facilities are designed to provide a controlled light environment that prevents outside environmental influences, such as natural lighting, from creating variances from programmed photoperiods. Light cycles should ensure that animal health is not compromised and should provide for normal diurnal and circadian rhythms. Light cycles that provide 12–14 hours of light daily are appropriate for most laboratory animals (Lipman 2007; Pritchett-Corning et al. 2011).

Since most human activity in a research facility normally occurs during the day, the most widely used photoperiod in animal research facilities is illumination during the day and dark at night. During this cycle, animal holding room lights are on during the vivarium's normal business hours and at an acceptable illumination level to permit routine husbandry, veterinary care, and research activities. However, in some situations researchers choose reverse light cycles, where lights are off during the day and on during the night. Since rodents are nocturnal animals, their highest level of activity is during the day when housed under a reverse light cycle. This type of photoperiod is convenient for researchers to observe nocturnal activities and behavior during daytime hours, but presents challenges for the daily operation of the animal facility. Red spectrum lighting and night vision goggles can alleviate some of the human vision challenges and allow staff to safely carry on with daily activities. Red spectrum lighting works because of a difference in the cells that make up the retina of the rodent and human eye (Faith and Huerkamp 2009; Pritchett-Corning et al. 2011; NIGMS 2012); rodents cannot perceive red light, while humans can see this wavelength of light. Wavelength filtering materials, such as commercially available red film over light sources, will provide a perceived dark environment for rodents, but humans can continue daily operations in the red light illuminated environment. Red wavelength illumination can be generated by door window tinting, fluorescent bulb sleeves, or tube guards. Although red spectrum lighting is widely recognized for allowing a safe working environment for humans, emerging literature points to evidence that continuous exposure to red spectrum illumination is not without negative implications for laboratory rodents. A recent study showed evidence of marked changes in circadian hormone rhythms and plasma corticosterone in nude rats that were housed in red-tinted cages (Dauchy et al. 2013b). As an alternative to using red spectrum lighting, Faith and Huerkamp described the benefits of sodium (vapor) lamps, which have wavelengths at the margins of what rodents see, but in the visual field of humans. They provide sufficient light levels for humans but dull the light for rodents to a level that allows nocturnal behavior (Faith and Huerkamp 2009).

Advanced electronic lighting systems, such as the Lutron System™ (Lutron Lighting Systems, Inc., Cooperstown, Pennsylvania), are capable of phototransition. This is the gradual change in a photoperiod by slowly increasing or decreasing light levels, creating a dusk-to-dawn effect instead of a sudden change in illumination. Sudden changes in illumination intensity or light cycle can elicit a startle response in animals. For example, birds not accustomed to programmed photoperiods may take flight and collide with walls or other obstacles when unexpectedly subjected to darkness (Faith and Huerkamp 2009). Some aquatic species require transitional lighting to stimulate feeding, cleaning, and comfort behaviors (Stoskopf 1983).

It is widely recognized that appropriate light cycles are extremely important for maintaining natural circadian rhythm, which determines when the body should be awake or asleep. Deviations from natural photoperiods cause disruptions in circadian rhythm and can have detrimental effects on the brain, body function, and behavior (Lipman 2007). When mice were subjected to a 20-hour photoperiod (10 hours light and 10 hours dark) for 6–8 weeks, they exhibited profound changes in cognition and physiology, increased body temperature, disruption of normal hormone levels, and increased weight gain (Karatsoreos 2009). Other studies have shown that photoperiod changes resulted in behavioral alterations (e.g., increased aggression), as well as changes in reproduction and increased susceptibility to cancer and infectious disease (Kennaway 2005; Dauchy et al. 2011). Since unscheduled disruptions in photoperiods can be devastating to research outcomes, lighting should be electronically controlled and regularly monitored to ensure that cycles remain consistent. If cycle timers fail or override switches are not functioning properly, lights can remain on or off for long periods of time outside the scheduled photoperiod. The lighting parameters should be controlled and documented by the building environmental monitoring system and alarm if room parameters deviate outside of a specified range. During the light "on" period, task lighting in rodent rooms should be provided by override for a time-limited period before returning to the programmed lighting levels and schedule. During the lighting "off" period, the room lighting should be provided via override in the red spectrum for a time-limited period. Care should also be taken to eliminate disruption caused by light contamination during the dark period, caused by light leaking around and under door frames or unprotected windows. Digital displays on ventilated rack blowers or other equipment maintained in the animal room may also emit enough light to disrupt a dark cycle (Dauchy et al. 2011).

Historically, animal research facilities have used overhead fluorescent lighting fixtures to provide illumination in corridors, as well as in animal holding rooms. Generally, these are ballast–lamp systems holding cool white fluorescent (CWF) tubes that emit sufficient light levels in an appropriate wavelength to enable the human eye to adequately evaluate and work with the animals. In addition, most CWF fixtures produce the wavelength and spectral pattern of natural light (Hessler 2009). Fluorescent fixtures are also available in a wide variety of colors, and there is growing research interest in evaluating the effects of different light spectrums on animal physiology. In addition to color variety, the environmental impact of lighting fixtures is now a consideration in vivarium lighting. One of the newest environmentally friendly concepts is the introduction of light-emitting diodes (LEDs), which, at appropriate levels, have been shown not to interfere with circadian rhythms or cause phototoxic effects (Hessler 2009). Although LED lighting is usually more expensive initially, it consumes less power and operates longer than traditional light fixtures. Another option for room lighting is augmentation with natural lighting through windows or sky lights. This is infrequently recommended in animal research facilities because of the inability to strictly control photoperiods, but can be part of an environmental enrichment program for higher-level species, such as nonhuman primates or farm animals housed in research facilities.

Heating, Ventilation, and Air-Conditioning

Ensuring consistent, species-specific temperature and humidity in an animal room is critical for animal thermoregulation, to maintain normal physiology, and to promote natural behavioral tendencies. Species-specific room-level, dry-bulb temperature tolerance ranges for common laboratory species are listed in the *Guide for the Care and Use of Laboratory Animals* (*Guide*) (NRC 2011). (The dry-bulb temperature is the temperature of the air measured by an ordinary thermometer.) Temperatures in the *Guide* for rodents are typically set lower than their thermoneutral zone (the temperature range where the animals do not need to gain or lose heat) to prevent heat stress during periods of activity. The addition of nesting material is important to keep mice warm when at rest. Since wide fluctuations in temperatures can be detrimental to animal health and well-being, temperatures should be maintained at a set point toward the middle of the range and within 2° of the set point. For example, if the desired temperature for a rodent room is 70°F, the room temperature should remain between 68°F and 72°F. Wide fluctuations in temperature can produce metabolic and behavioral responses that can confound research results or alter an animal's response to research procedures (Hessler 2009). Temperature extremes can cause a reduction in breeding, a decrease in food intake, increased disease susceptibility, and disturbances in sleep–wake patterns (Laber and Gonder 2007; Lipman 2007; Faith and Huerkamp 2009). Macroenvironmental fluctuations and high rates of intracage ventilation can be mitigated at the cage level by providing appropriate nesting materials or nesting boxes that can aid in thermoregulation. These materials provide a choice for animals to build nests, take shelter in a box, or burrow into bedding in response to environmental conditions inside the cage (Baumans et al. 2002; Gaskill et al. 2009, 2012).

Relative humidity (RH) is another component of the macroenvironment that could have detrimental effects on animal health if not maintained within tolerable ranges. There is a wider range of control for humidity (30%–70%) than temperature, but extreme variations from recommended RH should be avoided (NRC 2011). Although these values are based on tolerable ranges for most mammalian species, maintaining consistency in RH is important for all animals used in biomedical research. Extreme fluctuations have been shown to impede an animal's rate of heat loss, impact normal activity, and cause changes in normal amounts of food intake (Faith and Huerkamp 2009; Rosenbaum et al. 2010; NRC 2011). Certain species can develop dermatitis or flaky skin, and rats can develop ringtail from prolonged periods of exposure to humidity below 30% (Schofield and Brown 1996; NRC 2011). RH that is extremely high can also influence conditions in the microenvironment, such as increased moisture levels in the bedding, cage wall condensation, higher cage temperatures, food spoilage, and bacterial generation of ammonia (Burn and Mason 2005; NRC 2011).

Considering the ramifications of uncontrolled environmental conditions, temperature and humidity should be monitored and recorded on a regular basis. Equipment as simple as a room-level hygrothermometer or a sophisticated, facility-wide electronic environmental monitoring system can ensure that animals are not subjected to extended periods of extreme temperature and humidity fluctuations.

Air Quality

The quality of the air at the room level not only impacts animal health but also is an important component of creating a safe environment for personnel working in animal facilities. One of the most common occupational hazards for animal care staff or others working in animal facilities is the development of animal-related allergies or increased sensitivity to allergens (NRC 1997; Harrison 2001). The quality of the animal room air is affected by the particulate load of a variety of particles, such as animal dander and dust generated from animal bedding, by gases such as ammonia vapors and by infectious agents from both animals and humans. Most research institutions control the quality of room air by exchanging the entire volume of the air with 100% outside air at a set rate per hour, based on the type of caging or species housed within the room. The number of times the room air is replaced per hour is referred to as air changes per hour (ACH). Exchanging the volume of the room air can also aid in reducing heat load, decreasing RH, and replacing carbon dioxide generated by animals and humans with oxygen. Ten to fifteen constant volume ACH are recommended in most animal housing rooms, but these recommendations do not take into account the variations in heat load; the size, species, and numbers of animals; the room design; and the microenvironment (NRC 2011). Setting of ACH at the higher end of the suggested range may be unnecessary and a waste of energy. Conversely, at the low end, a room that is heavily populated or housing a significant number of large animals could be underventilated at 10–15 ACH, leading to increases in temperature, RH, particulate counts, and noxious odors. In order to ensure a higher level of air quality, modern facilities frequently consult with engineering specialists to include computational fluid dynamics (CFD) analysis when designing heating, ventilation, and air-conditioning (HVAC) systems for new construction. Computerized CFD software effectively evaluates how air will enter, circulate, and exit the room based on volume of air and dynamics created by the air entering the room, in conjunction with the placement of supply and exhaust registers (Norton and Brouwer 2009). This analysis utilizes various types and amounts of caging planned for each room, as well as variations in species. Obvious flaws will be identified, and adjustments can be made to provide the best possible macroenvironment for animals and humans.

Noise and Vibration

Noise and vibration have been shown to affect many behavioral and physiological parameters in animals and can be a confounding variable in research studies (Anthony and Harclerode 1959; Buckley and Smooker 1970; Turner et al. 2005, 2007; Small and Dietrich 2007). As a result, the current edition of the *Guide* has placed emphasis on taking steps to control the effects of these factors (NCR 2011). Although both sound and vibration can adversely affect animals, it is not a simple matter to determine what magnitudes and frequencies of sound or vibration are problematic. The information below summarizes the relevant scientific properties of these factors and what is currently known about the effects of sound or vibration on laboratory animals.

Sound and vibration are forms of energy that travel in waves, with sound being perceived by what we hear and vibration by what we feel. These waves have both amplitude and frequency (Figure 20.2).

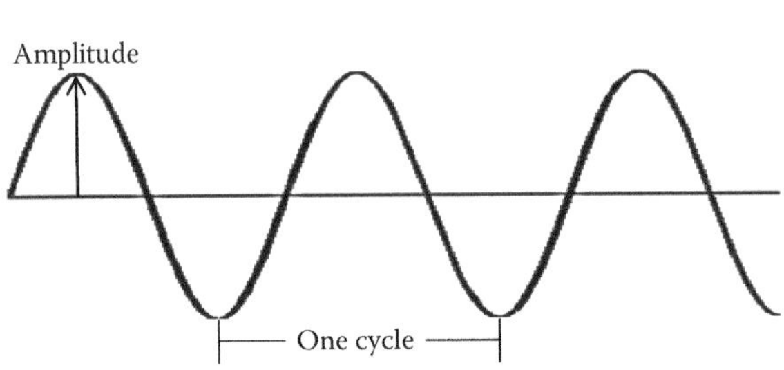

FIGURE 20.2 Depiction of sound or vibration wave demonstrating the amplitude (intensity) and wave cycle of sound or vibration.

The amplitude indicates the intensity of the sound or vibration and is represented by how far the peak of the wave moves past the neutral position. The frequency is the amount of time that it takes to complete one cycle from a point on one wave to the same point on the next wave. The term *hertz* is used as a unit of measure for frequency and is the number of cycles per second. One hertz is one cycle per second (Crocker 2007). As discussed in the text below, both the amplitude and frequency have to be considered when determining the effects of sound or vibration on laboratory animals.

Noise

The amplitude of sound can be measured in decibels. As a sound wave moves through the air, it creates pressure. The measurement of this pressure relative to the sound pressure at the hearing threshold can be expressed as "decibels (sound pressure level)" or "dB (SPL)" (Crocker 2007). In Figure 20.3, the hearing threshold curves in dB (SPL) at various frequencies are approximated for the human, dog, and mouse. The rabbit and rhesus monkey have similar hearing profiles as the dog (Turner et al. 2005). Each curve, or audiogram, has a similar shape in that it takes a higher sound pressure ("louder" sound) for hearing to occur at lower frequencies than in the middle of the frequency range, and then higher-level sound is required for hearing to occur as the frequencies continue to increase.

As noted in Figure 20.3, both dogs and mice hear higher-frequency sound than humans. Studies have shown that husbandry practices and common equipment found in animal facilities produce sound at frequencies above what humans can hear but within the hearing range of animals (Milligan et al. 1993; Small and Dietrich 2007). Conversely, sound at frequencies 1000 Hz would be audible to humans, but animals, such as the mouse, may not be able to hear sound at these frequencies. Although it has been shown that a ventilated rack and animal transfer station both produced sound pressure levels above the ambient level within the human hearing range, the sound pressure levels within the mouse hearing range did not increase above ambient noise from either noise source. In the same study, when various types of construction equipment were used adjacent to the ventilated rack, the sound pressure level within the mouse hearing range was increased, but to a lesser degree for each implement than were the sound pressure levels within the human hearing range. At more distant locations within the animal facility, sound pressure levels from a large jackhammer within the mouse hearing range decreased much more rapidly than did those in the human hearing range, indicating that less of the sound is perceived by mice than by humans (Reynolds et al. 2010). Therefore, it cannot be assumed that animals hear sounds that humans hear or that humans hear sounds that are audible to animals.

In controlled studies where mice are exposed to prolonged high-decibel sound at frequencies well within their hearing range, noise has caused teratogenic and reproductive effects. Noise may also cause changes in stress hormones and sleep, and behavioral disturbances in laboratory animals, as well as

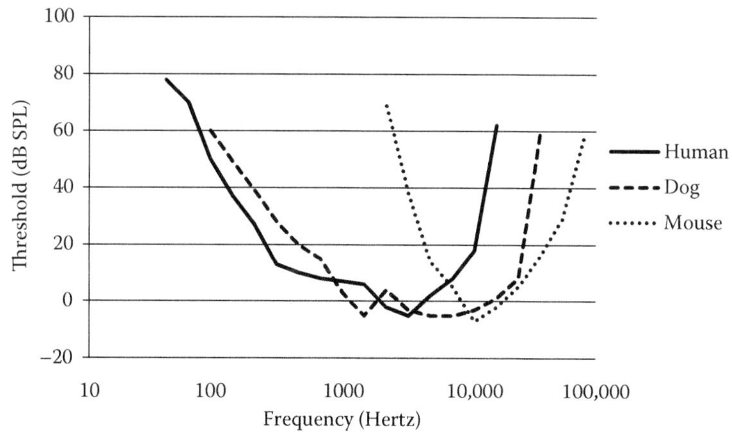

FIGURE 20.3 Approximation of audiograms for the human, dog, and mouse.

changes to the gastrointestinal, cardiovascular, immune, reproductive, and nervous systems (Pfaff 1974; Zakem and Alliston 1974; Fletcher 1976; Peterson 1980; Nawrot et al. 1981; Murata and Takigawa 1989; Turner et al. 2005; Rabat 2007).

Vibration

Similar to sound, the effects of vibration depend not only on the amplitude but also the frequency at which it occurs. The resonance frequency is the frequency of applied external vibration that causes an object to more readily vibrate, and even amplify the vibration, in comparison with other frequencies. Every object or part of the body has a resonance frequency (F_n), which is calculated by the formula $F_n = 1/(2\pi)*\sqrt{(k/m)}$, where k is the stiffness constant and m is the mass. Knowledge of resonance frequency is important because vibration at this frequency will be perceived more strongly and ultimately will induce more adverse effects (Griffin 1996). Because the mass is in the denominator of the equation, the heavier the object, the lower the resonance frequency, assuming that the stiffness constant is the same. There is evidence that the rat and mouse have a resonance frequency range of 31–50 and 41–60 Hz, respectively, as a whole-body composite (Rabey et al. 2014). Vibration at frequencies outside of these ranges may also be perceived by animals (Norton et al. 2011) and could contribute to distress and alter research. Because of the varying sensitivities to vibration, different frequencies, as well as amplitudes, need to be considered when evaluating the effects on laboratory animals.

Anecdotal reports of the effects of vibration include a reduction in mouse breeding efficiency in rodent breeding colonies, reductions in food intake and weight gain, and behavioral modifications (Faith and Miller 2007). Low-level whole-body vibration in mice caused a decrease in fat production, lowered liver triglyceride levels (Rubin et al. 2007), and caused an increase in bone volume and/or bone formation (Xie et al. 2006, 2008). Whole-body vibration in mice has also been shown to cause an increase in heart rate and blood pressure (Li et al. 2015). Whole-body vibration in rats caused an increase in plasma stress hormone and brain serotonin levels, as well as a decrease in gastric emptying time, decreased organ weight, and increased adrenal weight (Sackler and Weltman 1966; Toraason et al. 1980; Ariizumi and Okada 1983). In other animals, low-level, whole-body vibration has been shown to cause cardiovascular effects in dogs and swine (Edwards et al. 1972), avoidance behavior in poultry (Abeyesinghe et al. 2001), and behavioral changes and an increase in stress-related hormones in swine (Perremans et al. 2001).

Controlling Noise and Vibration in the Laboratory Animal Facility

Some sources of noise and vibration are inherent in daily operations (e.g., movement of equipment, cleaning procedures, and animal vocalization) (Rozema 2009). To reduce operational noise, all casters on equipment should be in good working condition so that noise and vibration are not generated. Animal cage change procedures should be performed as quietly as possible, including adding feed to hoppers and the watering of animals. Animal transport should be performed in a manner where the animals will not experience noise or vibration, such as from a cart on an uneven floor. The researchers should be made aware of the effects of noise and vibration on study animals and trained in how to minimize them. Dogs, swine, and primates should be housed in separate wings of a building from other species or in quarters designed to provide auditory separation from other species sensitive to noise and vibration. Loud or sudden noises may negatively impact rodent breeding, behavioral studies, and sensitive species, such as rabbits and guinea pigs. Therefore, they should be housed in areas where minimal noise is present (e.g., away from noisy species like dogs or nonhuman primates and away from cage wash operations).

Sources of noise or vibration that are generated from the physical plant include the ventilation system, light fixtures, ventilated racks, cage change stations, cage washers, autoclaves, and computers. Supply and return fans that reduce noise and the incorporation of silencers for these fans are important measures for noise reduction of the ventilation system (Rozema 2011). Light fixtures should not have a buzz or hum sound, and computers should have limited use in the animal room unless they are enclosed in a soundproof enclosure. As described above, light fixtures and computers can generate noise that animals can hear, but is inaudible to humans. Low-noise racks and transfer stations are now available and should be incorporated into the animal facility whenever possible. Animal breeding rooms should not be placed next to the cage wash area or autoclaves, or near other noise- or vibration-generating equipment, such as pumps and compressors. Fire alarms can be another disturbance in the laboratory animal environment and should emit sound less than 400 Hz, or a visual alert system may serve as an alternative to sound-generating alarms (Zoontjens 2012).

Sanitation

Sanitation is the maintenance of environmental conditions in a manner that will limit microbial levels. Most animal facilities include both cleaning and disinfection as components of a sanitation program. Cleaning removes gross debris and organic materials that shield microorganisms to allow for effective disinfectant penetration (NRC 2011). Disinfection eliminates or significantly reduces microbial concentrations. Good sanitation programs start with conscientious planning to ensure that appropriate surface materials are used to allow for ease of sanitation of both the macro- and microenvironment. Surfaces should be smooth and impervious and composed of materials capable of withstanding frequent, vigorous applications of degreasers, detergents, and corrosive disinfectants. Animal rooms should not have unsealed penetrations, cracks, or seams that could house vermin or microorganisms (Rollin and Kesel 1990). To prevent cross-contamination between rooms, each room should have its own dedicated cleaning implements. Mops, buckets, brooms, and so forth, should not be shared between rooms and corridors, as they may be fomites for the spread of pathogens.

Frequency of sanitation is usually determined by the species housed, the type of animal enclosures, and the amount of debris generated by the animals (NRC 2011). Rooms housing large species such as dogs, pigs, and nonhuman primates in open enclosures or kennels require more frequent room sanitation intervals than rooms with rodents housed in microisolator caging (NRC 2011). In general, floors are usually mopped or hosed on a daily basis in the large animal rooms and complete room sanitation is performed every 1–2 weeks, depending on the institutional sanitation program. For small animal rooms, surfaces are cleaned daily and mopped at intervals ranging from daily to once a week (Casebolt 2009). Based on institutional performance standards, rodent rooms are usually sanitized at least monthly and when an empty room is reopened for occupation. In rooms where detergents and disinfectants are applied by a hose or a foamer, proper protective gear, such as goggles or a face shield and a mask, should be worn to protect against aerosolized chemicals, fecal material, or infectious agents. Eye protection should also be worn when applying detergents or disinfectants on walls and ceilings. Care should be taken to ensure that all surfaces are thoroughly rinsed of detergents or disinfectants and that the animals are not sprayed with water or cleaning chemicals during sanitization activities.

Sanitization Chemicals

Careful consideration should be given when choosing cleaning and disinfectant chemicals for use in animal facilities. Obvious concerns are the effectiveness of detergents to remove gross organic soil and the ability of disinfectants to aid in the control of the spread of pathogens or experimental infectious agents. Combination detergent–disinfectants are commonly used for sanitation of the macroenvironment and are effective against a wide variety of microorganisms (Ingraham et al. 2013). Examples of disinfectants commonly found in combination formulas are detergents plus a quaternary ammonium and/or phenolics or iodophors (Small and Dietrich 2007). These are considered low-level disinfectants and are not effective against spores and some fungi or viruses. Intermediate-level disinfectant–detergents contain chemicals that can kill *Mycobacterium tuberculosis* var. *bovis*, but cannot kill spores. Spore-killing combinations may contain chlorine dioxide, hydrogen peroxide, or peracetic acid. It is extremely important to follow the manufacturer's directions regarding the dilution of the agent and the contact time required to achieve appropriate surface disinfection (Small and Dietrich 2007). Chemicals must be thoroughly rinsed from all surfaces to prevent chemical burns or dermatitis as a result of residue exposure to footpads or skin. Residual odors should also be avoided. Chemicals that are designed to cover odors or cleaning solutions that are scented should not be used in animal facilities (NRC 2011). Perfumes or volatile chemicals may cause changes in normal physiology, behavior, or metabolism (Castelhano-Carlos and Baumans 2009).

After completion of an infectious disease study or in response to a pathogen outbreak, it may be necessary to decontaminate an entire room or enclosure. Agents such as vaporized hydrogen peroxide (VHP) or chlorine dioxide gas are extremely effective when complete sterilization is needed (Small and Dietrich 2007). For this application, chemicals are vaporized and distributed by an external generator attached to ports or custom penetrations outside a sealed animal room or enclosure. Both VHP and chlorine dioxide have antimicrobial properties and are effective against spores. Neither is corrosive in the vapor or gaseous form. Chlorine dioxide can also be used in a liquid form for surface decontamination when sprayed on equipment such as animal change stations, biological safety cabinets, or countertops.

Confirming Sanitation Effectiveness

To ensure an environment that contributes to the health and well-being of research animals, it is important to monitor the effectiveness of the cleaning and disinfection program. Regular evaluation of the effectiveness of sanitation processes with methods such as visual inspection, monitoring water temperature, and microbiologic testing is important (NRC 2011). For microbiologic testing, RODAC™ (Replicate Organism Detection and Counting) agar plates, culture swabs, and adenosine triphosphate (ATP) bioluminescence meters are commonly used tools in laboratory animal facilities to determine the presence or absence of microorganisms or organic materials (Schondelmeyer et al. 2006; Turner et al. 2010). In each instance, it is critical that the surface to be tested is free of residual disinfectants, which could alter colony counts or confound ATP readings.

The swabbing method involves swiping a sterile swab that has been immersed in lecithin broth over a surface that has been disinfected. The swab is then inserted into a broth tube and incubated for 24 hours prior to streaking on appropriate agar media. The plate is incubated and microscopically evaluated for microbial growth at defined periods of time over the course of up to 3 days (Small and Dietrich 2007). RODAC plating is a similar process, except that the agar plate is pressed directly on a disinfected surface (Ingraham et al. 2013). The plate is immediately covered, incubated, and observed for the number of bacterial colonies. Both methods take a few days for visible growth. The third and much faster method, ATP bioluminescence testing, also utilizes a swab technique but delivers results in a matter of seconds using a portable luminometer (Patel 1994; Turner et al. 2010). A prepackaged swab is swiped across a disinfected surface and then immersed in a tube containing an ATP amplifying solution. The tube is inserted into the luminometer, which analyzes the luminescence of the sample and detects the presence of parasites, such as pinworm eggs, organic materials, or microorganisms commonly found in a laboratory animal facility. Since microbiologic monitoring with swabs or plates will not detect all types of organisms and ATP testing is limited in its detection of gram-negative bacteria, using a combination of the two in areas where a high level of cleanliness is required would be beneficial (Turner et al. 2010).

Pest Control

A pest control program should prevent, control, or eliminate pest infestations in the animal environment (NRC 2011) due to their potential for disease transmission and contamination of feed and bedding. Pest control plans should include prevention of attraction and access to the facility by insects and vermin from the outside, as well as a vigorous sanitation program on the inside. The implementation, control, and monitoring of the plan should be documented.

How the exterior of a facility is designed can play a major role in preventing vermin excursions into the interior of the facility. For example, the type of landscaping and type of light fixtures, doors, and finishes, if chosen properly, can enhance an external pest control program. The National Institutes of Health (NIH) *Design Requirements Manual* recommends using landscaping plans that do not include ground cover–type plantings in close proximity to the building that could provide nesting and harborage for pests (NIHOM 2013). Conduit and lighting fixtures should be sealed and flush with exterior surfaces to prevent a nesting area for insects or wild birds. All exterior wall penetrations should be sealed, and doors should have door sweeps. The use of chemical pesticides around the facility exterior should be carefully considered for safety and effectiveness and be in compliance with state and federal regulations. If exterior trapping devices are used, they should be regularly checked and documented.

Like the exterior of the building, pest control inside the facility begins with good design features and a robust sanitation program. Interior surfaces should be smooth and of a material that can withstand repeated exposure to cleaners and disinfectants. All junctions and cracks should be sealed and drains fitted with screens or covered in order to prevent infestation by insects or vermin. Light fixtures should be sealed and water resistant to enable adequate sanitation. If "drop-down" ceilings are used in animal areas, the grid system should be tightly gasketed and the tiles made of a washable material. Routine sanitization of animal areas at a frequency appropriate to the species housed is key to the deterrent of pests. The use of chemical pesticides should be avoided whenever possible in animal housing areas due to potential adverse effects on animals (NRC 1997, 2011). Chemical pesticides should not be used prior to review by the veterinarian, operations manager, and investigators. Only humane rodent traps should be used, and they should be checked daily. "Sticky boards" that could entrap rodents should not be used for humane reasons.

Environmental Monitoring

In addition to monitoring the disease prevention and sanitation programs in the animal facility, other environmental influences, such as physical plant functionality, must be monitored and documented as well. Consistency in temperature, ACH, and humidity and light cycles are critical to maintaining normal metabolism and normal behavioral activities in research animals. Automated monitoring systems are currently available that are capable of sampling, recording, and with enhanced options, controlling room conditions. Preferably, an automated system should trigger alarms and send alerts via e-mail, pager, or phone when excursions outside of set parameters occur or when malfunctions are detected in automatic watering systems. In many cases, automated systems designed for laboratory animal facilities serve as redundant systems for building automation systems.

The Microenvironment

Terrestrial Animals

For terrestrial animals, the immediate surrounding area serves as the primary enclosure or microenvironment and may be a cage, pen, stall, or kennel. Enclosures should be provided that are designed to fit the needs of the species occupying the space. Modern technology offers the ability to provide environments that significantly stabilize the conditions within the microenvironment. Until the advent of individually ventilated caging (IVC) systems, the static microisolator cage was the predominant enclosure for microisolation rodent housing. When compared with static caging, mechanically ventilated mouse isolator cages at approximately 60 ACH have been shown to have dramatically lower levels of ammonia, carbon dioxide, and RH (Memarzadeh et al. 2004). Although IVC improves air quality within the cage, there are aspects of IVC systems that may have an unintended negative impact. Baumans et al. (2002) illustrated that high air velocities (60 ACH and higher) may cause stress to the animals, which was manifested in the study through avoidance behavior. In their study, when given the freedom to move between adjoining mechanically ventilated cages and static caging, mice showed a preference for the static cages and avoided air inlets that were placed low in the IVC systems. The study reinforced the importance of providing appropriate amounts of nesting material in IVC since when provided with huts or nesting material, avoidance behavior decreased and nests were built by mice in IVC. Mice in this study also demonstrated a preference for the type of IVC systems where forced air entered from the cage lid instead of the cage wall (Baumans et al. 2002). Along with controlling the flow dynamics inside the cage, most modern IVC systems allow for changing the directional flow of air through the cages from positive to negative and vice versa relative to the room. Caging in a barrier setting would normally have a positive directional flow in relation to the macroenvironment, and in most situations, caging being used for control of pathogens or experimental infectious agents would incorporate a negative airflow setting (Lipman 2009). Advancing technology also provides methods for monitoring the microenvironment in IVC. Many vendors can provide optional specialized caging features that will sample and report on environmental conditions, such as temperature, humidity, and ammonia concentration inside of the cage. Newer models of exhaust blowers on IVC systems will display cage exhaust temperature digitally. Caging systems can also be equipped with wireless technology capable of monitoring, reporting, and storing data on environmental conditions. This technology can also generate alarms for excursions outside of desired parameters or power loss. An emergency power supply, to ensure that the cages are properly ventilated in the event of a power failure, is common when using IVC systems. IVC racks are portable equipment, so it is important that consideration be given to using twist-lock plugs or some other method of preventing inadvertent disconnection from the electrical source in the event that the IVC rack is moved (NRC 2011).

In direct contrast to a controlled, predictable micro- and macroenvironment, research animals such as dogs, nonhuman primates, and agricultural animals have often been housed in outdoor or indoor–outdoor enclosures. Frequently, animals with access to indoor–outdoor housing are group housed in an effort to provide added enrichment. Although animals can benefit from this type of setting, group housing could present species-specific challenges in regard to enclosure density and temperament or social rank of the animals. This type of environment can be distressful to some animals if not carefully controlled (Novak and Suomi 1989; Overhall and Dyer 2005). Indoor–outdoor and outdoor environments can also confound research results due to natural influences, such as seasonal changes in temperature, humidity and light cycles, an inability to control nutrient intake in grazing animals, and the possible presence of naturally occurring pathogens (NRC 2011).

Along with controlling and monitoring the microenvironment, we must also consider sanitation practices as an important component of providing for the health and well-being of the animals. Cage changing frequency and effective sanitation of the primary enclosure both contribute to the quality of the microenvironment. Enclosures and accessories should generally be sanitized every 2 weeks. Some types of caging systems or enclosures may require more or less frequent sanitization based on the cage's size, animal holding density, or type of air ventilation system (NRC 2011; Horn et al. 2012). The process of sanitation or disinfection may necessitate the use of chemical cleaners or disinfectants. Products used for sanitizing the microenvironment should be free of perfumes or odors and should be thoroughly rinsed from all surfaces. Studies evaluating routine distressful environmental conditions in the animal facility report that the presence of perfumes or odors from volatile chemicals can elicit physiological reactions (Castelhano-Carlos and Baumans 2009). A variety of cleaning methods may be effective, such as mechanical cage washers or manual sanitization; however, the method should be appropriate for the materials being used (NRC 2011). Recommended methods for ensuring and validating the effectiveness of the sanitization process can include visual inspection, monitoring water temperature, and microbial and ATP testing. These methods are appropriate for both the macro- and the microenvironment.

Lighting is another component of the microenvironment that usually differs significantly from the macroenvironment. Type of caging and location of the cage in the room or on the rack can influence the amount of light reaching the animals within the enclosure. Studies have shown that light intensity can vary as much as 80-fold from the top to the bottom of the rack in transparent cages and up to 20-fold inside the cage (NRC 2011). Cage material can affect the amount of light that penetrates the enclosure, with clear translucent caging allowing the highest light intensity. There is evidence that light levels that are too low can negatively impact normal physiological responses in some strains of rodents (Faith and Huerkamp 2009). Room lighting should be such that consideration is given to the well-being of the species to be housed at the cage level, in addition to the ability of staff to safely visualize the animals (NRC 2011).

Aquatics and Semiaquatics

The use of aquatics, especially zebrafish, has increased in biomedical research for a variety of reasons. It is relatively easy to manipulate their genome, they produce many offspring, and they are more easily housed in a natural environment than rodents. Because of the nature of their environment, aquatic species require somewhat different considerations regarding their husbandry. Water quality is of the upmost importance, as well as what types of materials are used for their housing. Lighting, temperature, and cage sanitization are also critical components of aquatic species husbandry.

Water quality is obviously an essential concern when housing aquatic species. The pH of the water is one important consideration, with a pH between 6.8 and 8.0 commonly used (Lawrence 2007; Green 2010). The temperature of the water is species dependent, with temperatures generally maintained within the range of their natural habitat. If changes in temperature or pH are needed, these parameters should be changed very slowly at a rate that will be tolerated by the species. Other water quality parameters to be considered are total dissolved ions (ionized minerals), total water hardness (amount of calcium and magnesium), ammonia, nitrite, nitrate, and chlorine (Astrofsky et al. 2002). The species-specific water parameters should be established through information acquired from the vendor or the literature prior to housing aquatics.

The water used for aquatics should be treated to remove chlorine and chloramines, reduce suspended sediments and gases, and eliminate pathogens. Chlorine and chloramines can be removed from a municipal water supply by activated charcoal or sodium thiosulfate. The type of pipes used in the water system can introduce toxicants into the water. Black iron or plastic pipes are generally preferable for aquatic species (Hodson and Spry 1985).

The ventilation system should also be tailored to the species' humidity needs. The air exchange rate in an aquatic room may be required to provide sufficient ventilation to reduce humidity levels, in the case of fish, for example, or may need to be reduced to allow sufficient humidity for amphibians (O'Rourke and Schultz 2002). Some amphibians may benefit from a water-soaked sponge or foam placed in the primary enclosure to provide additional humidity (Astrofsky et al. 2002).

A defined and stable photoperiod is important to the health and well-being of aquatic species. A typical light cycle will consist of 12–16 hours of light per day (Astrofsky et al. 2002). As with rodents housed in cages, attention must be given to ensure that light is not too intense or insufficient depending on the position of the aquaria on a rack system. A range of 54–324 lux at the surface of the water has been suggested to be appropriate for zebrafish (Matthews et al. 2002), with the lower intensities limiting the growth of algae in tanks. The placement of lights directly over tanks should be avoided to prevent algae growth.

Primary enclosures often consist of a transparent material, such as glass, polycarbonate, acrylic, or Plexiglas, as well as stainless steel (Matthews et al. 2002; O'Rourke and Schultz 2002). The material should be impermeable to moisture and easily sanitized. Lids should be nonabrasive due to the propensity for aquatic species to jump. All materials that will contact the water or the animals (e.g., pipes, tubing, and connectors) should be made out of a material that will not leach toxic compounds into the water (Brand et al. 2002). Hiding places are beneficial to many aquatic species. For example, polyvinyl chloride pipe provides a good area of seclusion for frogs. Semiaquatic species should be provided with a floor that slopes so that they can rest when emerged from the water.

Housing density of aquatic species is an important factor in their health and well-being. The number of fish that can be housed per liter of water varies with the species, as well as the water quality. Depending on the species, both too few and too many animals in an enclosure can have detrimental effects. The vendor for the species or the literature should be consulted on proper housing densities for each species. Adult *Xenopus* densities range from one frog per 3 L to four frogs per 5–10 L (O'Rourke and Schultz 2002), and five adult zebrafish per liter is common (Reed and Jennings 2011).

All electrical fixtures should be protected from moisture, and any outlets should be ground fault interrupted. Due to the electrical needs for aeration, filtration, and lighting, emergency power is critical to prevent disastrous results. A few hours without these life-support systems could result in high mortality in some species (Astrofsky et al. 2002).

Cleaning and disinfection of cages requires precautions to prevent chemical residue that could harm aquatic species. Disinfectant or detergent residues can cause illness or death. Tanks can be washed in a dish or cage washer, but chemicals or detergents should not be used or should be designed for aquatic washing systems (e.g., IWT Tecniplast). The use of alcohol or 5% acetic acid, followed by 3% hydrogen peroxide in 0.1% sodium hydroxide, to wipe the tanks has been successful. The tanks are then rinsed several times with clean, dechlorinated water (Brand et al. 2002).

Summary

Animals are complex living creatures that will respond to environmental conditions within their micro- and macroenvironments. Understanding and controlling these variables will likely result in better-controlled experiments involving animals and in promoting animal welfare. Additional information concerning environmental and housing considerations for species is provided in Chapters 21 through 25.

REFERENCES

Abeyesinghe, S.M., C.M. Wathes, C.J. Nicol et al. 2001. The aversion of broiler chickens to concurrent vibrational and thermal stressors. *Appl Anim Behav Sci* 73:199–215.

Anthony, A., and J.E. Harclerode. 1959. Noise stress in laboratory rodents. II. Effects of chronic noise exposures on sexual performance and reproductive function of guinea pigs. *J Acoust Soc Am* 31:1437–40.

Ariizumi, M., and A. Okada. 1983. Effect of whole-body vibration on the rat brain content of serotonin and plasma corticosterone. *Eur J Appl Physiol Occup Physiol* 52:15–9.

Astrofsky K.M., R.A. Bullis, and C.G. Sagerström. 2002. Biology and management of zebrafish. In *Laboratory Animal Medicine*, ed. J.G. Fox, L.C. Anderson, F.M. Loew, and F.W. Quimby, 861–83. San Diego: Academic Press.

Baker, D.G., and N.S. Lipman. 2015. Factors that can influence animal research. In *Laboratory Animal Medicine*, ed. J.G. Fox, L.C. Anderson, G.M. Otto, K.R. Pritchett-Corning, and M.T. Whary, 1441–95. San Diego: Elsevier.

Baumans, V., F. Schlingmann, M. Vonck et al. 2002. Individually ventilated cages: Beneficial for mice and men? *Contemp Top Lab Anim Sci* 41:13–9.

Bayne, K., and L.C. Anderson. 2015. Laws, regulations, and policies affecting the use of laboratory animals. In *Laboratory Animal Medicine*, ed. J.G. Fox, L.C. Anderson, G.M. Otto, K.R. Pritchett-Corning, and M.T. Whary, 23–42. San Diego: Elsevier.

Brand, M., M. Granato, and C. Nüsslein-Volhard. 2002. Keeping and raising zebrafish. In *Zebrafish—A Practical Approach*, ed. C. Nüsslein-Volhard and R. Dahm, 12–15. Oxford: Oxford University Press.

Buckley, J.P., and H.H. Smooker. 1970. Cardiovascular and biochemical effects of chronic intermittent neurogenic stimulation. In *Physiological Effects of Noise*, ed. B.L. Welch and A.S. Welch, 75–84. New York: Plenum Press.

Burn, C.C., and G.J. Mason. 2005. Absorbencies of 6 different rodent beddings: Commercially advertised absorbencies are potentially misleading. *Lab Anim* 39:68–74.

Bustin, S.A. 2014. The reproducibility of biomedical research: Sleepers awake! *Biomol Detect Quantif* 2:35–42.

Casebolt, D.B. 2009. Facilities for dogs, swine, sheep, goats and miscellaneous species. In *Planning and Designing Research Animal Facilities*, ed. J.R. Hessler and N.D.M. Lehner, 313–21. San Diego: Elsevier.

Castelhano-Carlos, M.J., and V. Baumans. 2009. The impact of light, noise, cage cleaning and in-house transport on welfare and stress of laboratory rats. *Lab Anim* 43:311–27.

Chvedoff, M., M.R. Clarke, E. Irisarri et al. 1980. Effects of housing conditions on food intake, body weight and spontaneous lesions in mice: A review of the literature and results of an 18-month study. *Food Cosmet Toxicol* 18:517–22.

Clough, G. 1982. Environmental effects on animals used in biomedical research. *Biol Rev* 57:487–523.

Council of Europe. 2006. European Convention for the Protection of Vertebrate Animals Used for Experimental and Other Scientific Purposes. European Treaty Series 123. Guidelines for the Accommodation and Care of Animals. Article 5, Appendix A. Strasbourg: Council of Europe.

Crocker, M.J. 2007. Fundamental of acoustics, noise and vibration. In *Handbook of Noise and Vibration Control*, ed. M.J. Crocker. Hoboken, NJ: John Wiley & Sons.

Curtis, S.E. 1986. The case for intensive farming of food animals. In *Advances in Animal Welfare Science*, ed. M.W. Fox and L.D. Mickley, 245–55. Washington, DC: Humane Society of the United States.

Dauchy, R.T., E.M. Dauchy, J.P. Hanifin et al. 2013a. Effects of spectral lighting transmittance through standard laboratory cages on circadian metabolism and physiology in nude rats. *J Am Assoc Lab Anim Sci* 52:146–56.

Dauchy, R.T., L.M. Dupepe, T.G. Ooms et al. 2011. Eliminating animal facility light-at-night contamination and its effects on circadian regulation of rodent physiology, tumor growth, and metabolism: A challenge in the relocation of a cancer research laboratory. *J Am Assoc Lab Anim Sci* 50:326–36.

Dauchy, R.T., M.A. Wren, E.M. Dauchy et al. 2013b. Effect of spectral transmittance through red-tinted rodent cages on circadian metabolism and physiology in nude rats. *J Am Assoc Lab Anim Sci* 52:745–55.

Edwards, R.G., E.P. McCutcheon, and C.F. Knapp. 1972. Cardiovascular changes produced by brief whole-body vibration of animals. *J Appl Physiol* 32:386–90.

Faith, R.F., and M.J. Huerkamp. 2009. Environmental considerations for research animals. In *Planning and Designing Research Animal Facilities*, ed. J.R. Hessler and N.D.M. Lehner, 59–83. San Diego: Elsevier.

Faith, R.F., and S.J. Miller. 2007. The need for sound and vibration standards in US research animal rooms. *ALN Magazine*, July/August, 31–8.

FASS (Federation of Animal Science Societies). 2010. *Guide for the Care and Use of Agricultural Animals in Research and Teaching*. 3rd ed. Champaign, IL: FASS.

Fletcher, J.L. 1976. Influence of noise on animals. In *Control of the Animal House Environment: Laboratory Animal Handbook*, ed. T. McSheehy, 51–62. Huntingdon, UK: Laboratory Animals Ltd.

Gaskill, B.N., C.J. Gordon, E.A. Pajor et al. 2012. Heat or insulation: Behavioral titration of mouse preference for warmth or access to a nest. *PLoS One* 7:e32799ne.

Gaskill, B.N., S.A. Rohr, E.A. Pajor et al. 2009. Some like it hot: Mouse temperature preferences in laboratory housing. *Appl Anim Behav Sci* 116:279–85.

Green, S.L. 2010. *The Laboratory* Xenopus *sp.* Boca Raton, FL: CRC Press/Taylor & Francis.

Griffin, M.J. 1996. Whole-body vibration and health. In *Handbook of Human Vibration*, ed. M.J. Griffin, 27–42. San Diego: Elsevier Academic Press.

Hale, F.B. 1969. Domestication and the evolution of behaviour. In *The Behaviour of Domestic Animals*, ed. E.S.E. Hafez, 22–42. Baltimore: Williams & Wilkins.

Harrison, D.J. 2001. Controlling exposure to laboratory animal allergens. *ILAR J* 42:17–36.

Hessler, J.R. 1999. The history of environmental improvements in laboratory animal science: Caging systems, equipment, and facility design. In *Fifty Years of Laboratory Animal Science*, ed. C.W. McPherson and S.F. Mattingly, 92–120. Memphis: American Association for Laboratory Animal Science.

Hessler, J.R. 2009. Electrical: Special considerations. In *Planning and Designing Research Animal Facilities*, ed. J.R. Hessler and N.D.M. Lehner, 455–60. San Diego: Elsevier.

Hodson, R.G., and D.J. Spry. 1985. The effect of sulfite dechlorination on the accumulation of lead by fish in aqueous bioassays. *Can J Fish Aquat Sci* 42:841–44.

Horn, M.J., S.V. Hudson, L.A. Bostrom et al. 2012. Effects of cage density, sanitation frequency, and bedding type on animal wellbeing and health and cage environment in mice and rats. *J Am Assoc Lab Anim Sci* 6:781–8.

Ingraham, A., F.E. Lynch, and K.B. Shapiro. 2013. Sanitation chemicals for laboratory animal science. *Animal Lab News*, January, vol. 6(1), 10–5. http://www.alnmag.com/articles/2013/01/sanitation-chemicals-laboratory-animal-science.

Karatsoreos, I. 2009. Disruption of circadian rhythms affects both brain and body, mouse study finds. *Science Daily*, October 28. http://www.sciencedaily.com/releases/2009/10/091026225744.htm.

Kennaway, D.J. 2005. The role of circadian rhythmicity in reproduction. *Hum Reprod Update* 11:91–101.

Kilkenny, C., W.J. Browne, I.C. Cuthill et al. 2010. Improving bioscience research reporting: The ARRIVE guidelines for reporting animal research. *PLoS Biol* 8(6):e1000412.

Kilkenny, C., N. Parsons, E. Kadyszewski et al. 2009. Survey of the quality of experimental design, statistical analysis and reporting of research using animals. *PLoS One* 4(11):e7824.

Kokolus, K.M., M.L. Capitano, C.T. Lee et al. 2013. Baseline tumor growth and immune control in laboratory mice are significantly influenced by subthermoneutral housing temperature. *Proc Natl Acad Sci USA* 110:20176–81.

Laber, K., and J.C. Gonder. 2007. A renewed look at laboratory rodent housing and management. *ILAR J* 48:29–36.

Lawrence, C. 2007. The husbandry of zebrafish (*Danio rerio*): A review. *Aquaculture* 269:1–20.

Li, Y., K.N. Rabey, D. Schmitt et al. 2015. Characteristics of vibration that alter cardiovascular parameters in mice. *J Am Assoc Lab Anim Sci* 54:372–7.

Lipman, N.S. 2007. Design and management of research facilities for mice. In *The Mouse in Biomedical Research: Normative Biology, Husbandry and Models*, ed. J.G. Fox, S.W. Barthhold, M.T. Davisson, C.E. Newcomer, F.W. Quimby, and A.L. Smith, 271–319. Burlington, MA: Academic Press.

Lipman, N.S. 2009. Rodent facilities and caging systems. In *Planning and Designing Research Animal Facilities*, ed. J.R. Hessler and N.D.M. Lehner, 265–88. San Diego: Elsevier.

Magee, P.N. 1970. Tests for carcinogenic potential. In *Methods in Toxicology*, ed. G.E. Paget, 158–96. Oxford: Blackwell.

Matthews, M., B. Trevarrow, and J. Matthews. 2002. A virtual tour of the *Guide* for zebrafish users. *Lab Anim* 31:34–40.

Memarzadeh, F., P.C. Harrison, G.L Riskowski et al. 2004. Comparison of environment and mice in static and mechanically ventilated isolator cages with different air velocities and ventilation designs. *Contemp Top Lab Anim Sci* 43:14–20.

Milligan, S.R., G.D. Sales, and K. Khirynkh. 1993. Sound levels in room housing laboratory animals: An uncontrolled daily variable. *Physiol Behav* 53:1067–76.

Murata, M., and H. Takigawa. 1989. Teratogenic effects of noise in mice. *J Sound Vib* 132:11–18.

Nawrot, P.S., R.O. Cook, and C.W. Hamm. 1981. Embryotoxicity of broadband high-frequency noise in the CD1 mouse. *J Toxicol Environ Health* 8:151–7.

NIGMS (National Institute of General Medical Sciences). 2012. Circadian rhythms fact sheet. http://www.nigms.nih.gov/Education/Pages/Factsheet_CircadianRhythms.aspx.

NIHOM (National Institutes of Health Office of Management). 2013. Integrated pest management. In *Design Requirements Manual*, chap. 1, sect. 1-11. http://orf.od.nih.gov/PoliciesAndGuidelines/BiomedicalandAnimalResearchFacilitiesDesignPoliciesandGuidelines/DRMHTMLver/Pages/DesignRequirementsManualTableofContents.aspx.

Norton, J.N., and A.B. Brouwer. 2009. The planning, design and construction process. In *Planning and Designing Research Animal Facilities*, ed. J.R. Hessler and N.D.M. Lehner, 17–44. San Diego: Elsevier.

Norton, J.N., W.L. Kinard, and R.P. Reynolds. 2011. Comparative vibration levels perceived among species in a laboratory animal facility. *J Am Assoc Lab Anim Sci* 50:653–9.

Novak, M.A., and S.J. Suomi. 1989. Psychological well-being of primates in captivity. *ILAR J* 31:5–15.

NRC (National Research Council). 1997. *Occupational Health and Safety in the Care and Use of Research Animals*. Washington, DC: National Academies Press.

NRC (National Research Council). 2011. *Guide for the Care and Use of Laboratory Animals*. Washington, DC: National Academies Press.

O'Rourke, D.P., and T.W. Schultz. 2002. Biology and diseases of amphibians. In *Laboratory Animal Medicine*, ed. J.G. Fox, L.C. Anderson, F.M. Loew, and F.W. Quimby, 793–826. San Diego: Academic Press.

Overhall, K.L., and D. Dyer. 2005. Enrichment strategies for laboratory animals from the viewpoint of clinical veterinary behavioral medicine: Emphasis on cats and dogs. *ILAR J* 46:202–16.

Patel, P. 1994. *Rapid Analysis Techniques in Food Microbiology*. Exeter, Devon, UK: Chapman & Hall.

Perremans, S., J.M. Randall, G. Rombouts et al. 2001. Effect of whole-body vibration in the vertical axis on cortisol and adrenocorticotropic hormone levels in piglets. *J Anim Sci* 79:975–81.

Peterson, E.A. 1980. Noise and laboratory animals. Find all citations in this journal (default). Or filter your current search. *Lab Anim Sci* 30:422–39.

Pfaff, J. 1974. Noise as an environmental problem in the animal house. *Lab Anim* 8:347–54.

Price, E.O. 1984. The behavioral aspects of animal domestication. *Q Rev Biol* 59:1–32.

Pritchett-Corning, K.R, S.T. Chou, L.A. Conour, and B.J. Elder. 2011. Biology and reproductive biology. In *Guidebook on Mouse and Rat Colony Management*, 14–44. Wilmington, MA: Charles River Laboratories.

Rabat, A. 2007. Extra-auditory effects of noise in laboratory animals: The relationship between noise and sleep. *J Am Assoc Lab Anim Sci* 46:35–41.

Rabey, K.N., Y. Li, J.N. Norton et al. 2014. Vibrating frequency thresholds in mice and rats: Implications for the effects of vibrations on animal health. *Ann Biomed Eng* 43:1957–64.

Reed, B., and M. Jennings. 2011. Guidance on the housing and care of zebrafish, *Danio rerio*. Southwater, West Sussex, UK: Royal Society for the Prevention of Cruelty to Animals, Research Animal Department. https://www.aaalac.org/accreditation/refresources/zebrafishreference.pdf (accessed June 17, 2016).

Reynolds, R.P., W.L. Kinard, J.J. Degraff et al. 2010. Noise in a laboratory animal facility from the human and mouse perspectives. *J Am Assoc Lab Anim Sci* 49:592–7.

Roe, F.J.C. 1965. Spontaneous tumors in rats and mice. *Food Cosmet Toxicol* 3:707–20.

Rollin, B.E., and M.J. Kesel. 1990. *The Experimental Animal in Biomedical Research*. Vol. 1. Boca Raton, FL: CRC Press.

Rosenbaum, M.D., S. VandeWoude, J. Volckens et al. 2010. Disparities in ammonia, temperature, humidity and airborne particulate matter between the micro and macro-environments of mice in individually ventilated caging. *J Am Assoc Lab Anim Sci* 49:177–83.

Rozema, R. 2009. Noise and vibration considerations for the animal lab environment. *ALN Mag*, April, vol. 8, 23–7.

Rozema, R. 2011. Acoustic and vibration levels in research animal facilities—What is required? *ALN Mag*, April, vol. 10(3), 25–9.

Rubin, C.T., E. Capilla, Y.K. Luu et al. 2007. Adipogenesis is inhibited by brief, daily exposure to high-frequency, extremely low-magnitude mechanical signals. *Proc Natl Acad Sci USA* 104:17879–84.

Sackler, A.M., and A.S. Weltman. 1966. Effects of vibration on the endocrine system of male and female rats. *Aerosp Med* 37:158–66.

Schofield, J.C., and M.J. Brown. 1996. Animal care and use: A nonexperimental variable. In *Essential for Animal Research: A Primer for Research Personnel*, ed. B.T. Bennet, M.J. Brown and J.C. Schofield, 18–26. Darby, PA: Diane Publishing Company.

Schondelmeyer, C.W., D.L. Dillehay, S.K. Webb et al. 2006. Investigation of appropriate sanitization frequency for rodent caging accessories: Evidence supporting less-frequent cleaning. *J Am Assoc Lab Anim Sci* 45:40–3.

Siegel, V. 2011. Reproducibility in research. *Dis Model Mech* 4:279–80.

Small, J.D., and T. Dietrich. 2007. Environmental and equipment monitoring. In *The Mouse in Biomedical Research*, ed. J.G. Fox, S.W. Barthold, M.T. Davisson, C.E. Newcomer, F.W. Quimby, and A.L. Smith, 409–36. San Diego: Academic Press.

Sossinka, R. 1982. Domestication in birds. In *Avian Biology*, ed. D.S. Famer, J.R. King, and K.C. Parkes, 373–403. Vol. VI. New York: Academic Press.

Stoskopf, M.K. 1983. The physiological effects of psychological stress. *Zoo Biol* 2:179–90.

Toraason, M.A., D.W. Badger, and G.L. Wright. 1980. Gastrointestinal response in rats to vibration and restraint. *Environ Res* 23:341–7.

Turner, D.E., E.K. Daugherty, C. Altier et al. 2010. Efficacy and limitations of an ATP-based monitoring system. *J Am Assoc Lab Anim Sci* 49:190–5.

Turner, J.G., C.A. Bauer, and L.P. Rybak. 2007. Noise in animal facilities and why it matters. *J Am Assoc Lab Anim Sci* 46:10–3.

Turner, J.G., J.L. Parrish, L.F. Hughes et al. 2005. Hearing in laboratory animals: Strain differences and non-auditory effects of noise. *Comp Med* 55:12–23.

Van der Touw, J., S.J. Thrower, and J. Olley. 1978. Non-specific neural stimuli and metabolic rhythms in rats. *Physiol Bohemos* 27:501–4.

Vesell, E.S., and C.M. Lang. 1976. Symposium on environmental and genetic factors affecting laboratory animals: Impact on biomedical research. *Fed Proc* 35:1123–65.

Xie, L., J.M. Jacobson, E.S. Choi et al. 2006. Low-level mechanical vibrations can influence bone resorption in the growing skeleton. *Bone* 39:1059–66.

Xie, L., C. Rubin, and S. Judex. 2008. Enhancement of the adolescent murine musculoskeletal system using low-level mechanical vibrations. *J Appl Physiol* 104:1056–62.

Yousef, M.K. 1985. *Stress Physiology in Livestock*, vol. I, *Basic Principles*. Boca Raton, FL: CRC Press.

Zakem, H.B., and C.W. Alliston. 1974. The effects of noise level and elevated ambient temperatures upon selected reproductive traits in female Swiss–Webster mice. *Lab Anim Sci* 24:469–75.

Zoontjens, L. 2012. Notes on the acoustical design of animal holding rooms within medical research facilities. *Proc Acoustics (Fremantle)* 21–23:1–6.

21

Small Animal Enclosures and Housing

Debra L. Hickman, Judy M. Hickman-Davis, Jessica Peveler, and Melissa Swan

CONTENTS

Introduction

The selection of small animal housing enclosures for laboratory facilities is dictated by a multitude of factors. Design of the actual facility and the configuration of the caging system can significantly affect the health and well-being of the rodents used in research, as well as the research results. In general, the preferred housing environment is comfortable for the animals and allows them to engage in species-specific behaviors, while allowing scientific study. This chapter reviews the primary considerations in the selection of an appropriate housing environment for small animals, predominantly rodents. Additionally, it reviews the importance of selection of appropriate bedding, feed, and environmental enrichments, with considerations for operational efficiencies and animal well-being.

Regulatory Requirements and Considerations

The regulatory requirements for small animal enclosures and housing in the United States focus primarily on animal health and well-being. Regulatory issues for animal housing differ according to species and are intended to account for physical, physiologic, and behavioral needs. Legal requirements for research animal housing are captured in the Animal Welfare Act (AWA) (Public Law 89-544) (APHIS 2002) and the Health Research Extension Act of 1985 (Public Law 99-158) (United States 1985). Additional guidance can be found in the *Guide for the Care and Use of Laboratory Animals* (*Guide*) (NRC 2011).

The AWA was established in 1966 and has been amended many times. The AWA confers authority for enforcement of the law to the U.S. Department of Agriculture (USDA) Animal Plant Health Inspection Service (APHIS). These regulations cover care, handling, housing, and transportation for rabbits, guinea pigs, and hamsters, but specifically exclude the commonly used laboratory rats of the genus *Rattus* and mice of the genus *Mus*. In general, primary enclosures for covered species are required to be structurally sound, in good repair, and constructed of materials such that animals are safe and dry. Animals housed in solid-bottom cages should be provided with bedding, and for mesh-bottomed cages, construction should protect feet and legs from injury while minimizing animal contact with waste. Space requirements are provided for each species, but all mandate that primary enclosures must allow "each animal to make normal postural adjustments with adequate freedom of movement" with "convenient access to clean food and water" (APHIS 2002).

The Health Research Extension Act requires compliance with the Public Health Service (PHS) Policy on Humane Care and Use of Laboratory Animals in order for institutions to compete for or receive federal funds. The Office of Laboratory Animal Welfare (OLAW) is responsible for ensuring compliance with the PHS Policy, utilizing the *Guide* as the basis for evaluating animal care programs. The *Guide* represents the current state of knowledge surrounding best practices for the care and maintenance of laboratory animals. The *Guide* also acts as the primary reference document for AAALAC International, which provides a voluntary assessment for accreditation of institutional animal care programs. These documents define laboratory animal as *any* vertebrate animal used in research, teaching, or testing, and therefore including rats and mice. The eighth edition of the *Guide* reaffirms a dedication to the promotion of animal well-being; however, in relation to the primary enclosure, increased emphasis has been placed on provision of enrichment and increased space to account for an animal's species-specific needs (NRC 2011).

Internationally, there is a growing focus on the need to enhance laboratory animal well-being. Guidelines and regulations regarding small animal enclosures have been adopted by several countries, including Canada (Canadian Council on Animal Care 2003), Europe (Council of Europe 1986), Australia (National Health and Medical Research Council 2013), and Singapore (Singapore Government National Advisory Committee for Laboratory Animal Research 2004). Cultural and religious differences between countries may impact animal treatment and the definition of animal welfare. A discussion of the international issues and a comparison of trends in findings by AAALAC International provide an interesting perspective on the attitude toward animal environment and housing worldwide (Bayne et al. 2014).

Caging Systems

Rodent caging systems can be quite variable, but most fit a standard design of a simple box with modifications to provide environmental enhancements for the animals, to meet experimental objectives, and to ensure bioprotection or containment of pathogens or allergens. Although the actual space needs for rodents are not well defined (e.g., see Patterson-Kane et al. 2004; Nicholson et al. 2009; Whittaker et al. 2012) and can vary greatly by species and strain and gender, the majority of regulatory agencies provide minimum floor space and height requirements in their guidance documents (Table 21.1).

TABLE 21.1

Space Requirements for Rodents According to International Sources

Cage Space Requirements for Mice in Groups According to the *Guide for the Care and Use of Laboratory Animals*[a]

Body Weight, g	Floor Area, in.^{2} (cm^2)	Height, in. (cm)
<10	6 (38.7)	5 (12.7)
Up to 15	8 (51.6)	5 (12.7)
Up to 25	12 (77.4)	5 (12.7)
>25	>15 (>96.7)	5 (12.7)

Cage Space Requirements for Rats in Groups According to the *Guide for the Care and Use of Laboratory Animals*[a]

Body Weight, g	Floor Area, in.^{2} (cm^2)	Height, in. (cm)
<100	17 (109.6)	7 (17.8)
Up to 200	23 (148.35)	7 (17.8)
Up to 300	29 (187.05)	7 (17.8)
Up to 400	40 (258.0)	7 (17.8)
Up to 500	60 (387.0)	7 (17.8)
>500	≥70 (≥451.5)	7 (17.8)

Cage Space Requirements for Mice in Groups According to the FELASA Guidelines and Recommendations[b]

Body Weight, g	Floor Area, in.^{2} (cm^2)	Height, in. (cm)
Up to 20	9.3 (60)	4.7 (12)
>20 to 25	10.85 (70)	4.7 (12)
>25 to 30	12.4 (80)	4.7 (12)
>30	15.5 (100)	4.7 (12)

Cage Space Requirements for Rats in Groups According to the FELASA Guidelines and Recommendations[b]

Body Weight, g	Floor Area, in.^{2} (cm^2)	Height, in. (cm)
Up to 200	31 (200)	7.1 (18)
>200 to 300	38.75 (250)	7.1 (18)
>300 to 400	54.25 (350)	7.1 (18)
>400 to 600	69.75 (450)	7.1 (18)
>600	93 (600)	7.1 (18)

Note: FELASA, Federation of European Laboratory Animal Science Associations.

[a] NRC (National Research Council), *Guide for the Care and Use of Laboratory Animals*. 8th ed. Washington, DC: National Academies Press, 2011.

[b] European Union, Directive 2010/63/EU of the European Parliament and of the council of 22 September 2010 on the protection of animals used for scientific purposes. *Off J Eur Union* 276:33–79, 2010.

Materials

As the body of knowledge in animal care increases, the selection of materials for the construction of rodent caging has adapted to provide increased bioprotection, balanced with improvements to animal well-being.

Stainless steel is a common cage construction component. Historically, suspended and shoe box–style caging have been manufactured from stainless steel due to their longevity and sturdiness (Figure 21.1). However, these cages were heavy to manipulate and created a cool, dark housing environment for the rodents, potentially interfering with normal physiology. In the modern animal facility, stainless steel is a

FIGURE 21.1 (See color insert.) Example of a stainless steel, mesh-bottomed caging for rats. The food bin is located on the inside of the cage. Water bottles would be placed on the outside (where the springs are located). Note the white plastic resting boards on the floors of the cage to provide the rats with a solid-bottom surface for comfort. (Photo by Deb Hickman.)

common component of the feed and water holders for shoe box caging. The vast majority of racks, including those for larger rodents, such as guinea pigs and chinchillas, are also constructed from stainless steel, although generally with plastic molded floors (Figure 21.2). Aluminum is utilized in the construction of some individually ventilated caging (IVC) units to minimize the weight associated with them, but care must be taken with these units because of their increased susceptibility to physical damage.

Modern caging for small rodents is generally constructed of plastics, with the type of plastic utilized having changed over time. With the addition of ventilation of rodent enclosures, caging has moved from opaque plastics (generally constructed from polypropylene) (Figure 21.3) to translucent plastics (such as polycarbonate or polysulfone) that afford better visibility of the cage inhabitants (Figure 21.4). Cages that allow for easier observation of the animals without removal of the cage from the rack are especially important in IVC systems where the interval between cage changes may be 14 days or more. Additionally, the need for biosecurity and sterility has increased the demand for plastics that can withstand repeated cleaning with laboratory-grade detergents and steam sterilization without deterioration. In general, modern caging plastics can be divided into two broad categories: thermoplastics intended for repeated use and disposable plastics intended for single-use applications. Current thermoplastics include polycarbonate, polysulfonate, polythermide, and polypropylene. Caging manufacturers use injection molding techniques with these thermoplastics to create a caging unit with minimal variability and even strength.

One of the first thermoplastics used in the construction of rodent caging was polypropylene (Lipman 1999). This thermoplastic made an opaque cage that hid the animals from view, and as mice and rats are

(a)

FIGURE 21.2 Examples of housing for (a) guinea pigs. Note the molded plastic bottom with bedding. *(Continued)*

burrowing animals in the wild, it is possible that rodents felt more secure in the opaque caging. However, these cages interfered with the ability of technical staff to perform the mandated daily health checks of animals without disturbing the cage. They also deteriorated when subjected to the high temperatures and pressures associated with autoclaving, so the advent of static microisolation and IVC resulted in decreased use of caging manufactured from this thermoplastic, although these cages can still be purchased commercially.

(b)

(c)

(d)

FIGURE 21.2 (CONTINUED) Examples of housing for (b) chinchillas or rabbits. Note the molded plastic bottom with bedding. Additional enrichment items can be placed, such as the house (c) or the shelter and ball (d). The perforated plastic panel behind the chinchilla in (d) is a barrier that can be removed to provide additional floor space for the animals, facilitating group housing of stable social groups. (Photos by Judy Hickman-Davis.)

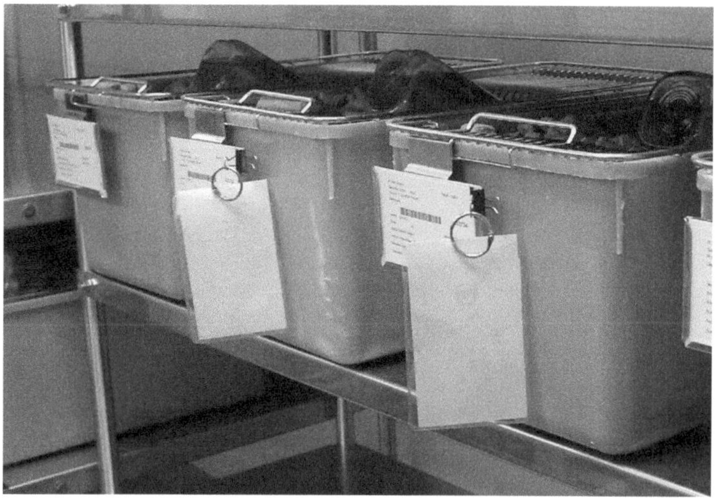

FIGURE 21.3 Example of rat caging made from polypropylene. Note the opaque nature of the cage and the presence of a wire lid to hold food and water. This caging is open to the room environment. (Photo by Deb Hickman.)

FIGURE 21.4 Example of mouse caging made from polysulfonate. Note that it is possible to visualize the mouse from outside of the cage. This configuration has added a lid with a filter top for enhanced bioprotection of the mouse within the cage. (Photo by Deb Hickman.)

Polycarbonate was the next commonly used thermoplastic. This plastic is clear, can withstand temperatures of 121°C (250°F), and generally has a life span of approximately 1 year under conditions of normal use. It can be tinted, which could provide additional visual protections for the animals (Dauchy et al. 2013). Rodents do not have good visual acuity associated with wavelengths of light in the infrared end of the spectrum. Therefore, there is speculation that use of caging that has been tinted red or yellow may decrease the intensity of the light perceived by rodents, perhaps even to the extent of appearing opaque. Even if this is not the case, protecting rodents from exposure to bright light decreases general stressors (Dauchy et al. 2013) and physical changes, such as retinal degeneration (LaVail et al. 1987).

Polysulfones have also been used in caging construction. These thermoplastics have the advantage of being able to withstand temperatures of 134°C (273°F), with a prolonged life span of approximately 3 years under conditions of normal use. Naturally colored amber or yellow, these plastics could also provide some visual protection from light intensities for the rodents. Most recently, polyethermide (PEI) has been utilized for the construction of rodent caging. This thermoplastic is able to withstand temperatures of 134°C (274°F) and has a life span of approximately 5 years under conditions of normal use, making it a sound financial investment for the animal facility. These cages are also deep amber to yellow, possibly providing some visual protection to the animals.

Disposable caging, manufactured from polyethylene terephthalate (PET), can also be used for the housing of small rodents. These cages provide superior bioprotection or containment opportunities, as they can be used once and disposed of through either the normal waste stream or recycling, if available. In addition to their value for high-level barrier or biocontainment situations, their use can also increase operational efficiency by circumventing the need for a cage wash space. If this space is not needed, the facility can use the footprint for additional housing, while resulting in a cost savings because there is no purchase of costly equipment with significant ongoing utility costs.

Types

The different types of caging that are currently available provide an interesting look at the developmental progress of the laboratory animal science field and how the priorities of bioprotection and animal well-being have shaped modifications to existing designs.

Conventional housing is defined as housing where the housing environment is not isolated from the animal housing room. Although this style of housing is seen much less commonly than containment caging, such as static microisolation and IVC, in the modern animal facility, it can still be an appropriate housing method for some commercial and experimental needs. There are generally two styles of conventional caging: suspended wire bottom caging (Figure 21.1) and shoe box caging with a wire top feeder (Figure 21.3). The major concern with both of these methods is that there is no bioprotection of the animals or the personnel working with the animals.

Suspended wire bottom caging was the historically preferred method of housing rats and mice used in research, as these caging systems required significantly less maintenance than those used today. However, there are health and well-being concerns associated with the use of this caging for those species. For example, older, heavier rats are predisposed to pododermatitis when housed for prolonged periods of time in these cages (Blair 2013), and neonates quickly become chilled if a secure nesting location is not provided. The use of this type of caging now requires scientific justification and approval of the animal ethics body. When used, modern caging includes modifications such as resting boards and provision of other enrichment devices, such as houses or manipulanda on which to gnaw. Generally, the racks have pans with paper or bedding beneath them ("indirect bedding"), which are changed at least once a week, while the actual animal housing cages are washed less frequently.

The conventional shoe box cage with a wire top feeder is the more common configuration of conventional rodent housing seen in the modern vivarium. The cage is generally made of plastic with a solid bottom, into which contact bedding is placed and directly bedded. Feed (and sometimes water) is provided using a stainless steel wire bar lid that holds the food above the animals (Figure 21.3).

A modification of the conventional shoe box caging system was the addition of a cage lid with filtration capabilities (Figure 21.4). These caging systems allowed the housing environment of the cage to be isolated from the animal housing room—especially when used in conjunction with laminar flow workstations of biosafety cabinets. These lids are sufficiently porous enough to allow the exchange of gases and moderation of temperature, but prevent the escape of potentially infectious or allergenic compounds. However, because the exchange of gases and heat is not as efficient as a system without the filtration top, these cages often require changes that are much more frequent than the conventional shoe box caging system.

IVC systems were created in response to the accumulation of gases that were noted in the static filter-top systems. In an IVC system, the lidded cages are placed on a rack that forces air through the caging, creating air changes within the housing environment to remove the potentially noxious gases (Figure 21.5). There are many different designs of IVC systems (Figure 21.6). In the original designs, air

(a)

(b)

FIGURE 21.5 (See color insert.) (a) Example of an IVC rack. There are no cages, but the rack can hold up to 160 mouse cages. The metal wall in the center of the rack provides forced air to the cages on the rack. (b) Close-up of IVC with empty bays next to the cages. On the metal wall, there are two round blue and white circles. These are the air vents that deliver and exhaust air from the cage. The silver nozzle below the air vents is an automatic watering system that provides water on demand to the animals in the cage. The yellow levers hold the cage in place, to ensure a tight seal for efficient air exchanges within the cage. (Photos by Deb Hickman.)

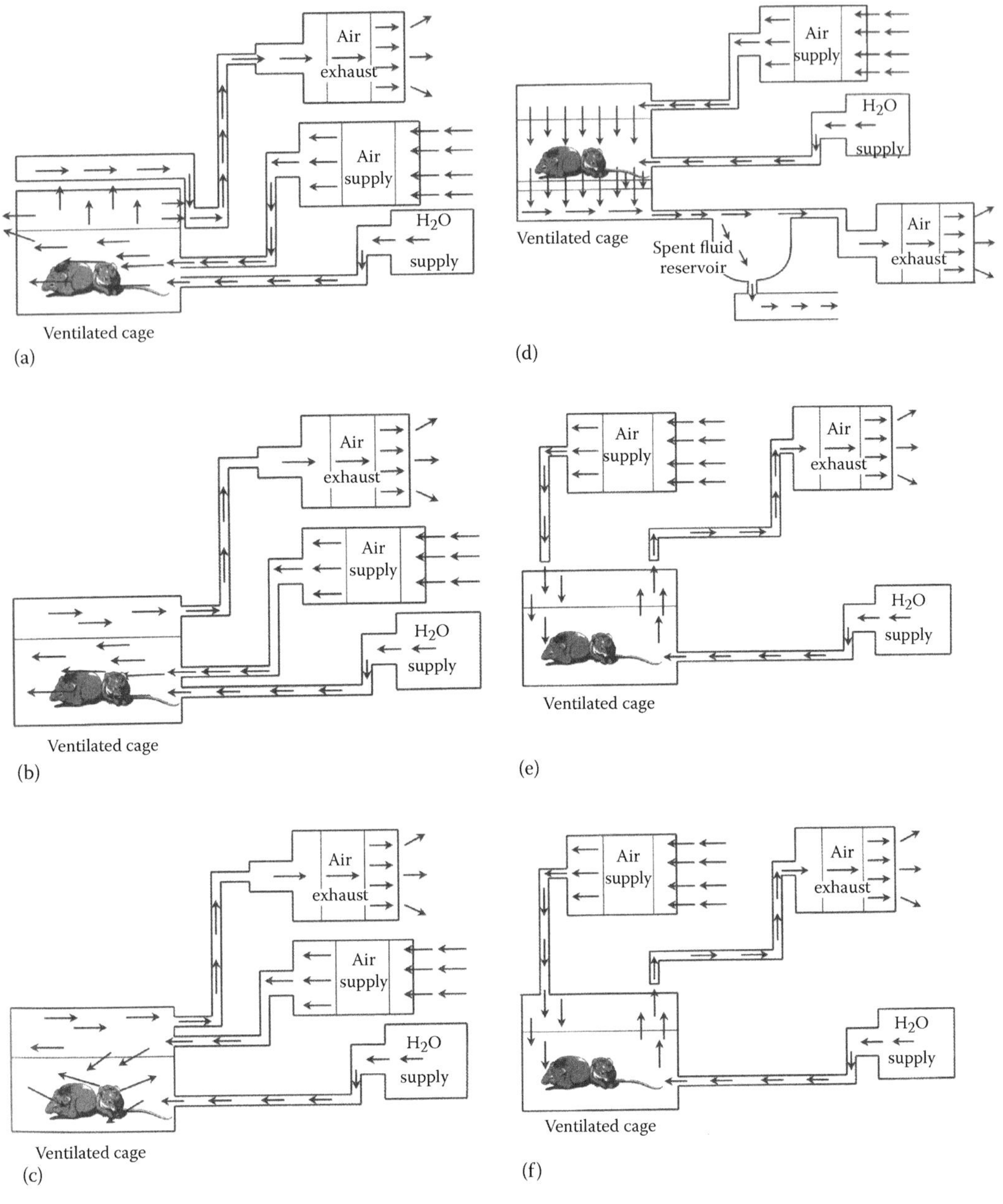

FIGURE 21.6 Schematic representations of types of commercially available IVC. All systems are shown with automatic watering. Numbers within brackets reflect the three commercially available versions of the intracage supply–intracage exhaust ventilated caging systems. (a) Intracage supply–perimeter capture. (b) Intracage supply–intracage exhaust (direct) [1]. (c) Intracage supply–intracage exhaust (direct) [2]. (d) Intracage supply–intracage exhaust (direct) [3]. (e) Intracage supply–intracage exhaust (indirect). (f) Intracage supply–intracage exhaust (combinaton). (From Lipman, N. S., *Contemp. Top. Lab. Anim. Sci.*, 38(5), 9–17, 1999.)

was brought into the cage from the room, without filtration aside from that provided by the filter-top lid, and was discharged from the cage by exhaust blowers. Now, these units are generally designed so that air is brought in from the animal room, passed through a high-efficiency particulate air (HEPA) filter, delivered to the cage, and then HEPA filtered out, although there are many configurations available from the various vendors (Figure 21.6). High-level biocontainment studies utilize specialized IVC systems where a gasket is present between the lid and the cage to ensure a tight seal with no leakage of any pathogens of concern from the cage. Filtration of both the supply and exhaust air produces superior bioprotection for the animals and decreases particulate contamination of the animal housing room. In addition to providing a controlled housing environment, another significant advantage to the use of IVC systems is that it is possible to house more cages within an existing footprint.

Isolators are specialized caging systems that are used in cases where bioprotection of the animals needs to be very high. Examples include very valuable immunocompromised animals, germ-free animals, and gnotobiotics. In an isolation system, sterile caging with wire top lids are introduced into a sterile environment (although the equipment can be placed prior to sterilization). Rodents are generally introduced after caesarean section to allow for control of the biome. The entire environment is maintained aseptically. Items such as food and bedding are sterilized prior to introduction into the isolator. Caging equipment can be washed within the isolator or removed for cleaning, but then is sterilized prior to reintroduction into the isolator. In the strictest protocols, nothing (aside from waste) is removed from the isolator until the colony is removed, and the unit is broken down and cleaned and resterilized prior to its next use.

Metabolic cages allow the researcher to collect information about the animal by collecting samples from the cages. The most commonly used caging designs house the rat or mouse in a Plexiglas or glass cylinder with a wire mesh floor. All feces and urine are collected over a period of time, generally 24 hours. A funnel system that is incorporated into the floor of the cage allows the collection of urine and feces separately from each other, facilitating *in vitro* analysis. More recently, metabolic cages have been designed that can measure more of the metabolic output of the rodent than just the urine and feces. In this style of metabolic chamber, the rodent is placed in a sealed cage that is provided with fixed amounts of air. The computerized system is able to measure multiple variables, such as activity, body weight, feeding, drinking, urine collection, sleep detection, body temperature, heart rate, and carbon dioxide production.

Integration

The racks and caging systems that are used in an animal facility drive some of the specialized infrastructure requirements in the facility. Structurally, doors need to be sufficiently large to allow personnel to move racks, with or without their exhaust and supply blowers, in and out of rooms. This is needed as the functions of rooms may change over time, in addition to the regular sanitation requirements.

The facility heating, ventilation, and air-conditioning (HVAC) system plays a significant role in the success of the housing system selected. When using conventional caging systems, high levels of air changes, such as the 10–15 air changes per hour recommended by the *Guide*, reduce the odors and allergens in the animal rooms (Korpi et al. 2007). The use of IVC systems and isolators may remove the requirement for the high level of air changes for the control of odors and dander, especially if the racks exhaust directly into the building exhaust. However, the motors associated with these units also add a heat load to the room, which often requires that the air changes per hour remain high to prevent the animal housing room from warming.

Control of temperature at the animal housing room level is targeted to be within 1°–2° of the established set point. Recently, more consideration has been given to the effect of the IVC on the housing environment of the cage. As air changes per hour at the cage level can be rather high (upward of 60 air changes per hour in some models), there is a possibility of drafts and chilling of the cage inhabitants. For this reason, it has been recommended that the lowest air changes per hour required to remove gases be utilized and that nesting material be provided to rodents, to allow them to thermoregulate. All this is very dependent on the system that is in use, which is why it behooves the laboratory animal facility personnel to be familiar with how the air flows in and out of the cages and to be observant of how their animals are using the space and enrichments provided.

When working with IVC, isolators, and the modern computerized metabolism systems, emergency power is absolutely critical. Because these systems require forced air, if the power to the fans is disrupted, the levels of ammonia, carbon dioxide, and relative humidity, and the temperature in the housing environment of the cages can rise very quickly, resulting in the death of animals (Huerkamp et al. 2003). Conventional housing does not have this risk, although emergency power is still recommended to ensure that the HVAC and light cycle for the animal housing room are consistent during extended outages.

Sanitation

The *Guide* is a primary resource for animal facility managers and Institutional Animal Care and Use Committees (IACUCs) when selecting and approving standard operating procedures for maintenance and sanitation of primary enclosures. Housing density, cage change frequency, bedding material, and the use of ventilated caging affect the housing environmental conditions (Reeb-Whitaker et al. 2001; Rosenbaum et al. 2009). The frequency of bedding changes and cleaning and disinfection of the cage is a primary consideration for the maintenance of environmental conditions appropriate for animal health and well-being (NRC 2011). Fluctuations in these parameters can cause alterations in metabolic or physiologic processes, as well as disease susceptibility. The *Guide* utilizes performance standards such as temperature, humidity, and concentration of gases and particulate matter to support the selected frequency and intensity of cleaning and disinfection. The frequency and intensity of sanitation also depend on specific physiologic and behavioral characteristics of the species being housed, the use and type of bedding material, and the rate and degree of soiling of the enclosure surface.

The accepted practice for sanitation of solid-bottom cages is once per week, with sanitation of equipment and accessories such as cage lids once every 2 weeks (NRC 2011). The use of IVC has increased the sanitation interval for solid-bottom cages to 2 weeks, as these cages receive frequent HEPA-filtered air changes. Increased intracage ventilation removes excess humidity and waste gases from cages, reduces disease transmission, and maximizes the efficient use of space and personnel resources (Reeb et al. 1998; Carty 2008; Rosenbaum et al. 2009). Recommended cage sizes for rodents are primarily based on the weight of the animal and the numbers of animals per cage; however, as cage density increases, cage change frequency must also increase to prevent accumulation of waste gases (Gonder and Laber 2007; Rosenbaum et al. 2009, 2010). Standard practices in many rodent breeding facilities permit less frequent cage changes to prevent disruption of newborn pups; however, studies with C57BL/6NTac mice and NTac:NIH-Whn rats indicate that breeding performance may not be impacted by disruption of neonates during the cage change-out (Sanderson et al. 2010).

In terms of sanitation, bedding provided in rodent cages can act to absorb moisture, minimize bacterial growth, limit animal contact with excreta, and reduce the accumulation of intracage ammonia (NRC 2011). Several bedding materials are available for rodents, including corncob, wood chips, paper products, and grass fiber pellets. Bedding should be evaluated for absorbency, biodegradability, toxicity, dust, palatability, comfort, cost, availability, damage to cage washers, animal preference, and effect on research (Wirth 1983; Ras et al. 2002; Smith et al. 2004; Domer et al. 2012). Soiled bedding should be changed "as often as necessary" to keep ammonia levels low and animals clean and dry. Intracage ammonia concentration is influenced by temperature, humidity, ventilation, urine, and bacteria. Humidity may be influenced by ventilation or bedding material, with highly absorbent materials and increased airflow decreasing intracage humidity levels (Lipman et al. 1992). Intracage ammonia is produced by the combination of urine and urease-producing bacteria found in feces and processed bedding material. The initial bacterial content in bedding can be controlled by autoclaving prior to use (Reeb et al. 1998). Disadvantages of autoclaving bedding include personnel time, autoclave energy consumption and maintenance costs, and loss of absorptive capacity of the bedding material. Although a correlation between cage appearance and animal health has not been documented, human perception of cage cleanliness remains a significant factor in the determination of cage change frequency (Rosenbaum et al. 2009; Domer et al. 2012).

Selection of the appropriate plastic materials is an important consideration for the maintenance and sanitation of cages (Demorotski 2008). Frequency of cage change-out, type of detergents, autoclaving, and water quality and temperature will impact cage longevity. Deterioration of plastic and residue deposition from the sanitation process may cause cages to become opaque. Opaque cages make visualization of the animal difficult and increase the time required for animal health observations. Hard water and calcium buildup on cages may be removed by manual scraping, which may damage the plastic, or by use of chemical rinses, which may damage expensive cage wash equipment (Chipps et al. 2012). Routine examination and treatment or removal of damaged or opaque plastic should be performed to ensure continued effective sanitation and clear observation of the animal. Likewise ventilated caging systems that utilize air nozzles that penetrate the cage should routinely be examined for occlusion by hair, bedding, food particulates, or debris to ensure that appropriate airflow is maintained to the cage. Occluded nozzles may result in increased humidity and accumulation of ammonia within the cage (Creamer et al. 2014), necessitating more frequent cage change-outs.

Water

Animals must have access to potable, uncontaminated drinking water (NRC 2011). Water entering an animal research facility is typically supplied by a local source and meets the standard necessary for human consumption (Hessler 2011). Most groundwater in the United States contains significant levels of calcium, magnesium, or iron, which may cause mineral scale to develop within pipes or sipper tubes. Water treatment with ultrafiltration or reverse osmosis (RO) provides high-quality drinking water and reduces the potential for scale to form on equipment or within pipes. Further treatment of animal drinking water by hyperchlorination (12–15 ppm) or acidification (pH 2.5–3) has been used to limit the growth of opportunistic microorganisms that may have a negative impact on transgenic or immunosuppressed rodents (Hessler and Lehner 2009). The choice of water delivery system is determined by species, facility, and type of water available. Water may be supplied to cages by water bottles, pouches (i.e., Hydropac®), or automatic water systems (Figure 21.5b). A detailed discussion of water treatment and the pros and cons of the different cage delivery systems is provided by Allen et al. in Chapter 28 of this book.

The use of closed-top or specialized caging systems to maintain barrier classifications for rodent housing has implications for water delivery and consumption. Water intake by rodents has been shown to be affected by intracage humidity (Hoyt et al. 2007) and type of caging (Memarzadeh et al. 2004; Rosenbaum et al. 2010). The increased use of ventilated racks for housing rodents has raised concerns about the potential for higher airflow to cause dehydration (Bekkevold et al. 2013). Conversely, higher intracage humidity has been reported to be associated with decreased water consumption by mice housed in closed-top ventilated and static microisolator cages (Memarzadeh et al. 2004; York et al. 2012). Wire bar or smooth plastic cage lids are molded to accommodate water bottles or bags as a separate component of the rodent cage. Cage lids are designed to allow the water bottle or bag to be added either externally, that is, outside the closed cage lid, or internally, that is, completely captured within the bonnet of the cage lid. The definition of barrier status at the level of the cage may dictate the need for a particular style of cage lid.

Automatic water systems are a common mechanism for the delivery of potable water to a variety of small laboratory animal species. Ventilated rodent racks may support automatic water systems with water valves permanently attached to the rack, removable valves attached to the rack, or valves attached permanently to the inside of the cage. Water valves permanently attached to the inside of the cage provide challenges for sanitation, requiring more attention for removal of bedding that may become caught up around the inverted stem. Water valves that remain affixed to the rack provide the opportunity for cross-contamination of cages if cage locations are inadvertently switched during cage change-out or during experimental procedures. Recommendations for sanitation and maintenance of automatic watering systems have been proposed for certain predefined conditions (Gonzalez et al. 2011; Molk et al. 2013); however, unique standard operating procedures should be developed to meet the individual facility and equipment conditions.

Changing Stations and Biological Safety Cabinets

When working with static microisolators or IVC, a changing station is a critical tool to ensure that the cage-level containment is not compromised. Most changing stations are built with a laminar flow design, where HEPA-filtered air is forced across the work surface, providing protection to the animals in the cage, and then captured and exhausted out through an additional HEPA filter to provide allergen protection to the animal housing room and the personnel working with the caging. However, the placement of HEPA filters is not consistent, and the design of the unit should be evaluated to ensure that it is performing in a way that meets the intended need. The design can have the air flowing from the back of the unit, across the work surface, to the front of the unit, or it can have a top-down configuration, where the air is forced from the ceiling of the cabinet down onto the work surface, and collected to be exhausted at the work surface (Figure 21.7). Both of these configurations provide excellent bioprotection for the majority of rodent pathogens and allergens, although they do not provide appropriate guarantees of protection of the worker from possibly infectious zoonotic pathogens or from exposure to chemical contaminants, such as isoflurane.

When working with agents that are listed at biosafety level 2 or higher, a changing station will not provide sufficient personnel protection and a biological safety cabinet (BSC) should be used. Class I biosafety cabinets are generally not advised for use in an animal facility. These units provide personnel and environmental protection, but no product (animal) protection, thus compromising the biocontainment established with the use of static filtration or IVC. Class II cabinets protect the animals, the animal housing room, and the personnel. There are four subtypes of Class II cabinets. Types A1 and A2 blow the air over the work surface from the top of the cabinet. The air is exhausted at the bottom of the cabinet and returned to the top, where 70% of the air recirculates through the system HEPA and 30% is exhausted through the exhaust HEPA. These are normally "thimble" ducted, but they do not have to be ducted to operate properly. The difference between Type A1 and A2 cabinets is in the inflow velocity of the supply air. Type B1 and B2 cabinets must be hard-ducted to an exhaust system (Figure 21.8). In these cabinets, 60% of the air is exhausted and only 40% is recirculated. These cabinets can be used with chemical hazards, in addition to biological hazards, due to their configuration. Class II cabinets are the style that is commonly used in most laboratories. Class III cabinets are generally reserved for maximum containment laboratories.

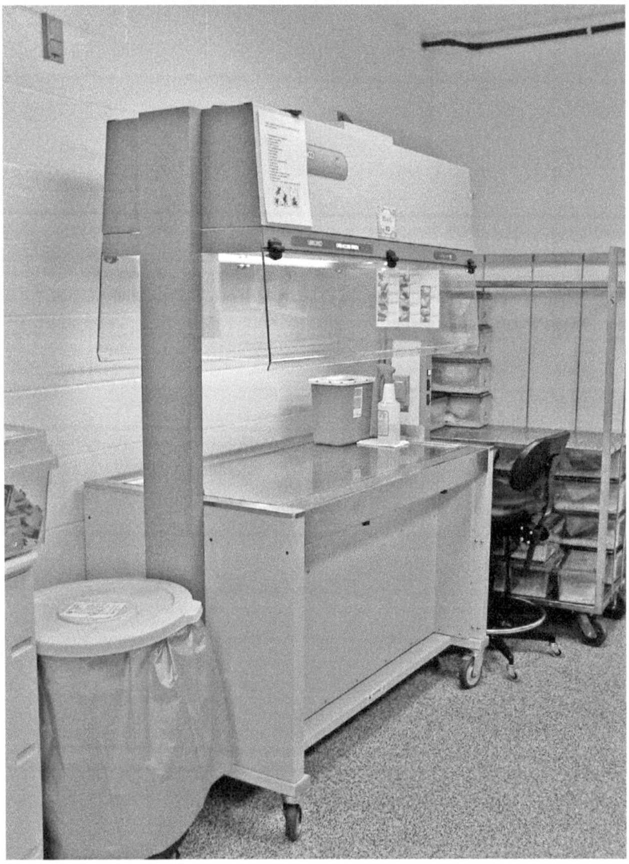

FIGURE 21.7 Example of a laminar flow workstation. In this configuration, air is HEPA filtered in the top of the unit and forced down over the work surface. Exhaust ports on the edges of the work surface collect the air for additional filtration before the air is exhausted into the room. (Photo by Deb Hickman.)

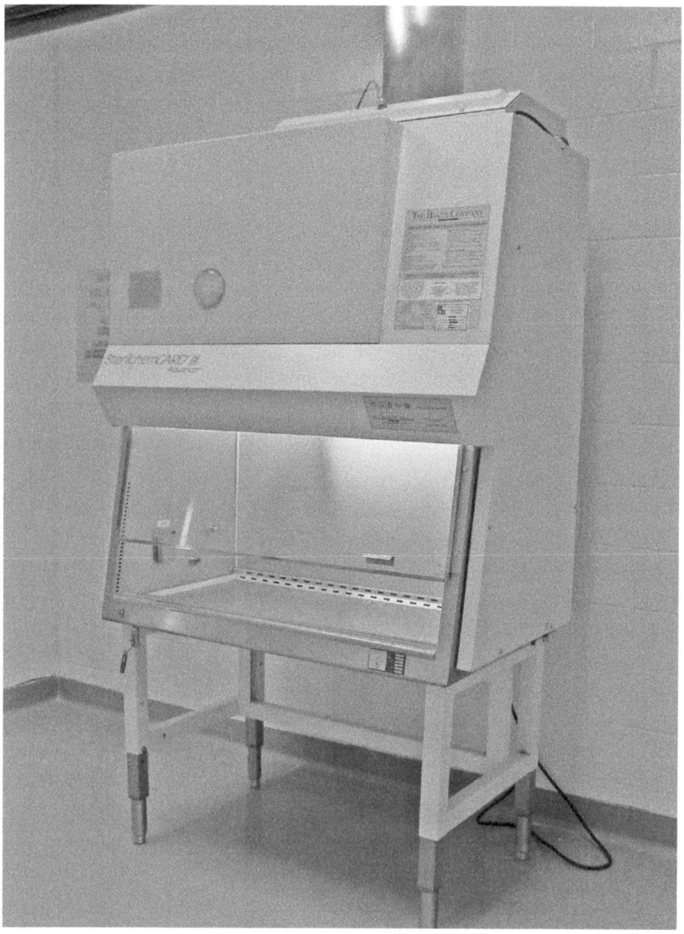

FIGURE 21.8 Example of a BSC suitable for use with biological hazards. (Photo by Deb Hickman.)

Barrier Housing

The use of barrier housing for rodents has become the standard for housing research and breeding animals (Hessler 2011). Barrier facilities are classified on the basis of operational criteria that are used to protect animals from unwanted or excluded organisms (Committee on Infectious Diseases of Mice and Rats 1991). The need for barrier housing has expanded with the use of immunocompromised, fragile, or extremely valuable transgenic animals. The barrier may exist at the cage, room, or facility level. In addition, specialized caging and equipment, such as ventilated racks, flexible film, or hard-sided isolators, may be utilized to create a barrier within a specific area of an animal facility (Hessler 2011). The criteria for classification of a barrier system consist of a number of factors that have not changed over the years (Committee on Infectious Diseases of Mice and Rats 1991): (1) quality, number, and source of animals; (2) frequency and method of animal introduction; (3) processing of materials into the barrier; (4) entry of personnel into the barrier; (5) environmental systems (HVAC, temperature, and lighting); and (6) monitoring practices.

The effective maintenance of a barrier requires the development of standard operating procedures that are understood and supported by all personnel. Designation of entry order based on the criteria listed above may be applied to the closed microisolator cage, room, or facility access. Establishment of the barrier at the cage level requires specialized handling procedures, sanitized microisolator caging, and sterile water, food, and bedding (Baker et al. 2014). Barrier facilities may be managed at different levels of microbiological control, which in turn dictates the way in which supplies and personnel enter the facility. Movement of materials and personnel should always travel from the "cleanest" to the "dirtiest" areas, with barrier-specific requirements for personnel to shower and/or change clothes prior to reentry from a dirtier to a cleaner barrier area. Materials or supplies that are cleaned or set up outside of the barrier may be wrapped or autoclaved for transport into the barrier. The use of pass-through autoclaves may eliminate the need for wrapping and unwrapping of supplies transported into the barrier (Hessler 2011). Conversely, the utilization of irradiated food and bedding may decrease employee time spent autoclaving supplies. Laminar flow rooms, biosafety cabinets, and change stations allow for animals to be handled outside of the barrier of the cage and still maintain strict barrier status.

The use of personal protective equipment (PPE) is an integral component of the barrier management practices designed to maintain animals free of excluded organisms. Animal care personnel are often required to enter through a vestibule with interlocking doors, where sterile outer garments or uniforms are donned prior to entry into the barrier facility (Hessler 2011). In addition to a sterile lab coat or uniform, typical PPE for entry into a barrier includes hair bonnet, face mask, shoe covers, and gloves (ACLAM 2010). The general consensus within the laboratory animal community is that PPE is important; however, recent discussion regarding PPE has focused on determining what amount of PPE is essential for disease control versus what is worn as part of historical practices (ACLAM 2010). The increased use of microisolator and individually ventilated rodent caging systems has been shown to be very effective at controlling pathogen transmission between cages within the same room, and even on the same rack, for a variety of disease outbreaks (Clough et al. 1995; Hasegawa et al. 2003; Bohr et al. 2006). A shift in the definition of the barrier unit from the facility or room to the cage level may allow for decreased use of some PPE and still maintain barrier status (Hickman-Davis et al. 2012; Baker et al. 2014).

Enrichment

All species evolved in a rich world of experiences, but in order to reduce variability in studies, small rodent housing environments can be quite barren, resulting in aberrant behavior and physiology in some animals (Olsson et al. 2003; NRC 2011; Weary 2013). Although it has been assumed that a standardized environment results in standardized animals, the concerns with this assumption are demonstrated repeatedly in colonies of genetically identical rodents where there are behavioral differences in incidence of maternal cannibalism, fighting, and self-destructive compulsive behaviors. Therefore, a more appropriate animal model would be one raised and living in an enriched environment (Smith and Corrow 2005; Arranz et al. 2010; Konkle et al. 2010). Continual improvements in understanding and technology should be used to reduce and refine animal use (Russell and Burch 1959). Enrichment needs to be a committed part of every institution's budget and management plan (Patterson-Kane 2011). This is necessary to nurture not only the animals under our care, but also the staff that provide the care. When unsupported by the institution, caring personnel may enrich animals on their own, resulting in a loss of personal resources for staff and the very real possibility of unaccounted for variables for animal studies (Patterson-Kane 2011).

There are many ways to enrich animals because animals have a variety of behaviors to fulfill. Enrichment strategies can be broken down into several general categories: social, bedding, nesting, shelters, oral, exercise, scent, and sound. Unfortunately, there is no one perfect "best enrichment" practice or formula. Allowable and ideal enrichment strategies change depending on facility operation, study design and intent, animal strain, and individual animal preferences (Baumans 2005). When strategizing a complete enrichment plan, many considerations must be taken under advisement. For some institutions, it is preferable to invest in durable devices that can be sanitized and reused; others may find it better to use disposable devices. When reusable items are employed, they should be considered a potential vector, with impacts on cleaning and quality control monitoring (Smith and Hargaden 2001). Understanding of the species and strain needs is also critical. For example, an enrichment device can be a resource to be guarded and result in escalated aggression, especially in some group-housed male mice (Howerton et al. 2008).

Using a positive psychology approach to environmental enrichment, the first step is defining the ideal quality of life for each species (Olsson et al. 2003; Timmins et al. 2007). An international panel of 12 experts determined that *welfare* usually refers to observable and measurable experiences of an animal, whereas *quality of life* describes animals' mental state (Timmins et al. 2007). Often, the picture of quality of life is established using the concept of these animals in the wild or as pets. The goal then should be to continually move toward that standard.

Social

The *Guide* recommends housing social animals with conspecifics as a default (NRC 2011). It is through play with other young and interactions with adults that animals learn important social skills, such as fighting, appeasement, mating calls, and parenting (Panksepp and Beatty 1980; Varlinskaya et al. 1999; Burgdorf et al. 2008; Viegas 2010; Kulesskaya et al. 2011). A social partner can be a soothing presence during times of stress, a playmate for exercise, and with the rise of IVC, a partner to share body heat. Mice in isolation show higher heart rates and disruptions in sleep patterns than pair-housed mice (Spani et al. 2003). The importance of social housing is illustrated in a study that demonstrated that a mouse will choose to be with a historically aggressive mouse over solitary housing (Van Loo et al. 2003).

Social enrichment may also be in the form of positive interactions between animals and their handlers. Although it is impractical in many facilities due to the large number of animals, gently handling animals does have its rewards. This can be done with young animals, for a few minutes a day, or even with adult animals that will be handled for procedures. Habituation is the simplest form of learning, and it has the benefit of reducing the novelty, and accompanying fear, of human handling (Varty et al. 2000). Handlers should keep in mind that every interaction with an animal is a training session; therefore, calm, gentle, respectful handling should be practiced during every encounter.

Bedding

In general, bedding is a standard addition to cage bottoms for the purpose of absorbing waste. There are many bedding options, including various wood, corn, and paper products. As burrowing creatures in the wild, some rodents prefer deep bedding (Freymann et al. 2014). Unfortunately, this practice is not practical for widespread institutional use, because it would add significant cost and ergonomic risk due to increased weight of the bedded cages.

Nesting

A common mistake is to assume that bedding and nesting in a cage are the same. Bedding can be considered the litter pan of the cage, and serve a similar purpose (unless provided in such volume as to also be able to be used for burrowing). Animals have been shown to give off separate pheromones in relation to these two different purposes (Van Loo et al. 2000). Nesting materials are used by rodents to build areas for warmth, sleep, and relaxation (Van Loo et al. 2004; Winnicker 2012; David et al. 2013; Gaskill et al. 2013b). Nesting material is likely the best single enrichment device for mice, coming also in a variety of options, most commonly facial tissue, paper strips, and compressed cotton (Hutchinson et al. 2005, 2012; Smith and Corrow 2005; Kulesskaya et al. 2011; Narver 2013). Recently, there has been a rise in the popularity of paper strips, as they allow mice to weave complex nests, as seen in the wild (Hess et al. 2008; Gaskill et al. 2012, 2013a). As always, selection of the best nesting material will depend on strain, study, and institution. Nude or hairless mice can get conjunctivitis from fibers of nesting material. Weak mice may lack the constitution required to make use of compressed cotton or even paper strips. Some nesting materials may strangulate or stick to neonates, causing injury or death (Smith and Corrow 2005). Rats tend to use nesting material to form bedded resting areas rather than enclosed nests, but it has been shown to be a valid enrichment option that rats will work to acquire (Manser et al. 1998; Van Loo and Baumans 2004) (Figure 21.9).

FIGURE 21.9 Example of nesting material use in rats. In this case, the rat was provided with two styles of nesting material—cotton batting and shredded paper. Note how the rat has created a shelter, which it is climbing out of, and how the cotton is used to line the bottom of the cage. (Photo by Deb Hickman.)

Shelters

In-cage shelters provide both mice and rats respite from the drafts of ventilated caging, the ability to withdraw from bright lights, a place of security or escape from cage mates, and a place to hide from perceived predators. Rats will stay in a nest box for the duration of a room's light cycle, and when a shelter is not available, will retreat to a darkened cage corner (Manser et al. 1998). Although it conceals them from husbandry staff, rats prefer opaque shelters with enclosed corners (Manser et al. 1998) (Figure 21.10).

Diets, Treats, and Gnawing

Foraging is a positive time expense for many species; in fact, mice prefer to search for food, even when it is readily available (Baumans 2005; Pritchett-Corning et al. 2013). In the absence of adequate nesting material, mice can compensate for the thermal stress of IVC by food grinding for higher-energy parts of food (Pritchett-Corning et al. 2013).

Rats demonstrate a clear preference for gnawing enrichment (Smith and Corrow 2005). Gnawing is an activity that should be understood to be different from eating. It is reasonable to expect some small pieces of gnawing enrichment to make their way through the gastrointestinal tract; however, these small pieces generally tend to be innocuous (Smith and Hargaden 2001), assuming that the devices have been assessed to ensure they are nontoxic (Figure 21.11).

Exercise

An exercise wheel is the most common tool utilized to stimulate exercise within a cage. The challenge in providing this for lab animals is finding a design that will fit and remain stable within a typical shoe box–style cage. When available, rodents do make use of them with resulting beneficial effects (Leggio et al. 2005; Neuman et al. 2013). Other options for in-cage exercise opportunities can be provided via a trapeze or hook hanging from the hopper. Additionally, expanding the vertical space of cages (Figure 21.12) (Wheeler et al. 2014) or including shelves to increase the environmental complexity has been shown to increase the quality of life of laboratory rodents.

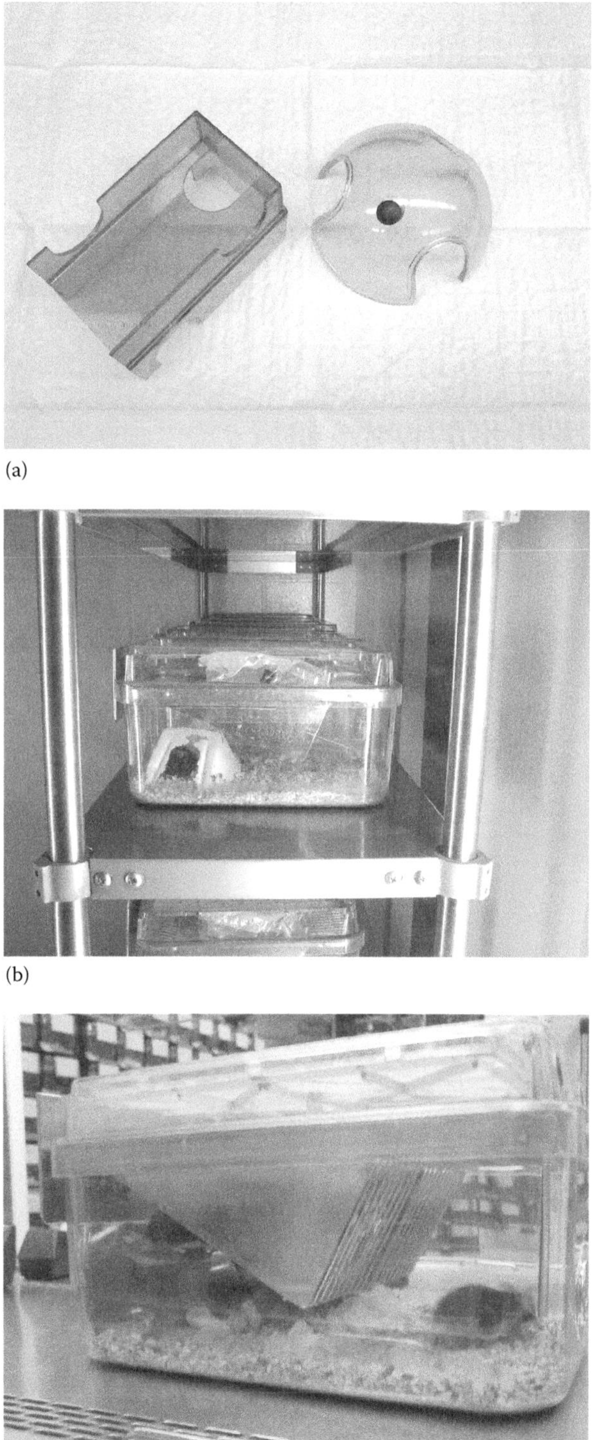

(a)

(b)

(c)

FIGURE 21.10 Examples of intracage shelters. These come in a variety of styles and are fabricated from a variety of materials, including plastic (a, c) and paper (b). (Photo by Deb Hickman.)

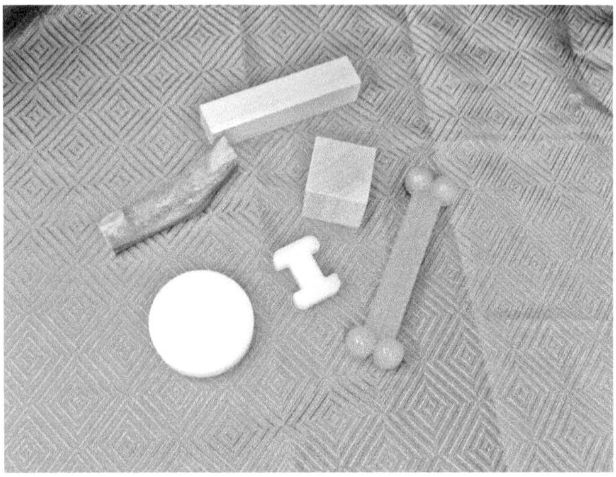

FIGURE 21.11 Examples of gnawing devices. The top devices are fabricated from wood. The bottom devices are fabricated from a variety of plastics. The variety in firmness can create novel gnawing experiences for the rodents. (Photo by Deb Hickman.)

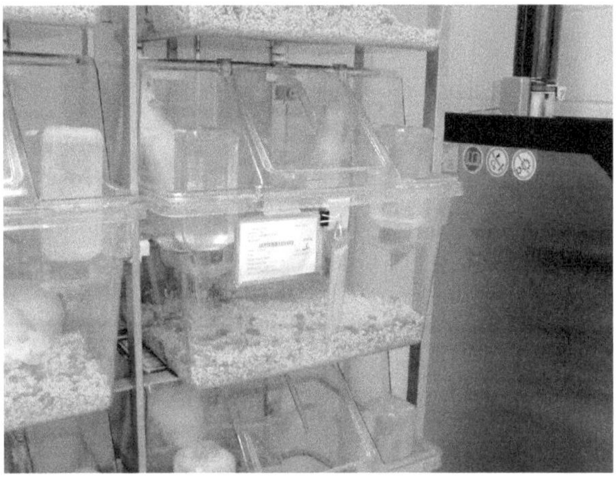

(a)

(b)

FIGURE 21.12 Example of a multilevel caging system for rodents. (a) Note the rats utilizing the upper shelf, while (b) provides a detailed view of the shelf that bisects the cage. The shelf and space to stand provide a complex environment for the rodents.

Sound and Scent

Noise stress (such as a fire alarm, motor vibration, and loud in-room human activity) has been demonstrated to have an adverse effect on breeding and stress levels (Castelhano-Carlos and Baumans 2009; Alworth and Buerkle 2013). Although music has been shown to have a beneficial effect on captive animals (Whiteman 2001; Wells et al. 2002; Boone and Quelch 2003; Hutchinson et al. 2005; Menon and Levitin 2005; Ying et al. 2007; Leeds and Wagner 2008; Davila et al. 2011; Kogan et al. 2012; Alworth and Buerkle 2013), it is unclear if this is a direct effect of the music itself. It is possible that it is an indirect effect resulting from the possibilities that the music enrichment serves to mask other unpleasant vivarium noises, has a calming effect on caretakers that extends to the animals, or serves as mental stimulation in the otherwise monotonous cage (Burgdorf et al. 2008; Leeds and Wagner 2008; Alworth and Buerkle 2013). If employed, calming music should be used, at a conversational (or lower) level, and it should be turned off at the end of each workday (NRC 2011). Music after a noise stress or restraint has been shown to reduce stress levels in mice, and rats exposed to music composed by Mozart had a significant reduction in blood pressure as long as 2 hours postexposure (Alworth and Buerkle 2013).

Scent is a critical component of rodent communication. The role of pheromones in the synchronization of the reproductive cycle (Brennan 2010), social facilitation (Stockley et al. 2013; Semple et al. 2014; MacRae et al. 2015), and signaling of distress (Inagaki et al. 2010; Voznessenskaya 2014) has been described, but is still not well understood. As nesting material can contain "appeasement" pheromones, movement of the nest to a new cage at the time of cage change may help to decrease aggression between mice (Whary et al. 2015). Additional studies are ongoing to determine how scents and pheromones can be used to improve the quality of life for laboratory rodents.

Emerging Technologies

Improvements to the operational efficiencies of animal facilities represent a rapidly progressing arena. By the time this book is published, it is a certainty that there will be new technologies available to improve the quality of life of animals and the efficiencies of the animal facilities. This section addresses some currently emerging technologies as examples of the opportunities that are coming.

Automated Census

The accurate collection of census data is critical for the vast majority of animal facilities that rely on per diems to provide the fiscal support for their operation. Bar code scanning has been a significant improvement over manual counting of cages, but it still requires that each cage be scanned individually to be counted. The use of wireless technology and radio frequency identification (RFID) has provided an opportunity for significant improvements in operational efficiencies. These systems allow the cages to essentially self-report their presence, fully automating the census-taking operation.

Alarms

Technology is also providing opportunities to intervene in cases of equipment malfunction at the level of the cage. Automated systems that alert a user when a cage is not engaged properly to IVC can help detect in real time that the cage is receiving appropriate air changes or access to an automatic watering system. Some manufacturers have developed IVC racks that integrate with the facility animal housing room monitoring systems to alert facility staff when there is a concern with a particular rack. Ultimately, technology will allow alerts at the housing environment level as well.

Thermoregulation at the Cage Level

Recognition that the animal housing room is generally kept at temperatures that are comfortable for humans, but could be cool for rodents, has become another area for potential refinement. The potential cooling of rodents is further complicated by the relatively high air changes per hour that are present in an IVC housing environment. Additionally, there are some studies that show that cold stress of the rodent model can negatively affect research results, such as tumor growth (Kokolus et al. 2010, 2014). To ameliorate this issue, provision of nesting material has been recommended, as this allows the mice to behaviorally thermoregulate (Gordon et al. 2014; Maher et al. 2015). To support some tumor growth studies, recommendations have been made to increase the temperature of the animal housing room. Additionally, some caging manufacturers have developed caging systems that will allow the housing environmental temperature of the cage to be elevated without affecting the room temperature.

Research Impact and Considerations

The effect of the housing environment on experimental variables is an area of ongoing active evaluation between scientists and the animal facility staff. The use of filtration tops for rodent caging was driven by the understanding that exposure to pathogens resulted in confounding variables in a wide variety of studies, including studies that are affected by challenges to the immune system, such as tumor growth (Baker 1998). Additionally, scientists who are concerned about the development and functioning of the brain often have concerns about changes in the housing environment that may affect the development or behavior of the animals that they are studying (Toth et al. 2011; Toth 2015). Inappropriate social groups and provision of environmental enrichment that might create distress for the animals are also confounding variables that must be evaluated in consideration of selection of the appropriate housing environment to be provided (Toth et al. 2011). Light contamination can result in alterations of tumor growth (Dauchy et al. 2013), as can changes in the temperature of the macro- and housing environment (Kokolus et al. 2014). Some strains breed well in IVC, while others do better in static filtration or conventional situations. Because there is so much variation in the range of scientific evaluations that are performed on a daily basis, laboratory animal professionals are highly encouraged to work closely with their scientific colleagues when evaluating the options that are available for housing environments.

REFERENCES

ACLAM (American College of Laboratory Animal Medicine). ACLAM 2010 Forum—PPE questionaire. https://www.aclam.org/Content/files/files/Public/Active/forum2010_ppe_surveyresults.pdf.

Alworth, L. C., and S. C. Buerkle. 2013. The effects of music on animal physiology, behavior and welfare. *Lab Anim (NY)* 42 (2):54–61.

APHIS (Animal and Plant Health Inspection Service). 2002. Animal Welfare Act and animal welfare regulations, 2131–5. Washington, DC: U.S. Department of Agriculture, APHIS.

Arranz, L., N. M. De Castro, I. Baeza, I. Mate, M. P. Viveros, and M. De la Fuente. 2010. Environmental enrichment improves age-related immune system impairment: Long-term exposure since adulthood increases life span in mice. *Rejuvenation Res* 13 (4):415–28.

Baker, D. G. 1998. Natural pathogens of laboratory mice, rats, and rabbits and their effects on research. *Clin Microbiol Rev* 11 (2):231–66.

Baker, S. W., K. A. Prestia, and B. Karolewski. 2014. Using reduced personal protective equipment in an endemically infected mouse colony. *J Am Assoc Lab Anim Sci* 53 (3):273–7.

Baumans, V. 2005. Environmental enrichment for laboratory rodents and rabbits: Requirements of rodents, rabbits, and research. *ILAR J* 46 (2):162–70.

Bayne, K., A. C. D. Bayvel, and V. Williams. 2014. Laboratory animal welfare: International issues. In *Laboratory Animal Welfare*, ed. K. Bayne and P. Turner, pp. 77–93. San Diego: Elsevier.

Bekkevold, C. M., K. L. Robertson, M. K. Reinhard, A. H. Battles, and N. E. Rowland. 2013. Dehydration parameters and standards for laboratory mice. *J Am Assoc Lab Anim Sci* 52 (3):233–9.

Blair, J. 2013. Bumblefoot: A comparison of clinical presentation and treatment of pododermatitis in rabbits, rodents, and birds. *Vet Clin North Am Exot Anim Pract* 16 (3):715–35.

Bohr, U. R., M. Selgrad, C. Ochmann, S. Backert, W. Konig, A. Fenske, T. Wex, and P. Malfertheiner. 2006. Prevalence and spread of enterohepatic *Helicobacter* species in mice reared in a specific-pathogen-free animal facility. *J Clin Microbiol* 44 (3):738–42.

Boone, A., and V. Quelch. 2003. Effects of harp music therapy on canine patients in the veterinary hospital setting. *Harp Ther J* 8 (2):4–5, 15.

Brennan, P. A. 2010. Pheromones and mammalian behavior. In *The Neurobiology of Olfaction*, ed. A. Menini, pp. 167–9. Boca Raton, FL: CRC Press.

Burgdorf, J., R. A. Kroes, J. R. Moskal, J. G. Pfaus, S. M. Brudzynski, and J. Panksepp. 2008. Ultrasonic vocalizations of rats (*Rattus norvegicus*) during mating, play, and aggression: Behavioral concomitants, relationship to reward, and self-administration of playback. *J Comp Psychol* 122 (4):357–67.

Canadian Council on Animal Care. 2003. *Laboratory Animal Facilities—Characteristics, Design and Development.* Ottawa, ON: Canadian Council on Animal Care.

Carty, A. J. 2008. Opportunistic infections of mice and rats: Jacoby and Lindsey revisited. *ILAR J* 49 (3):272–6.

Castelhano-Carlos, M. J., and V. Baumans. 2009. The impact of light, noise, cage cleaning and in-house transport on welfare and stress of laboratory rats. *Lab Anim* 43:311–27.

Chipps, J., V. K. Bergdall, N. Hanson, and D. Hadziselimovic. 2012. Increasing the usable lifespan of clouded rodent caging via implementation of a novel cage wash sanitation process. *J Am Assoc Lab Anim Sci* 51:637.

Clough, G., J. Wallace, M. R. Gamble, E. R. Merryweather, and E. Bailey. 1995. A positive, individually ventilated caging system: A local barrier system to protect both animals and personnel. *Lab Anim* 29 (2):139–51.

Committee on Infectious Diseases of Mice and Rats, Institute of Laboratory Animal Resources, Commission on Life Sciences, National Research Council. 1991. *Infectious Diseases of Mice and Rats.* Washington, DC: National Academy Press.

Council of Europe. 1986. European Convention for the Protection of Vertebrate Aimals Used for Experimental and Other Scientific Purposes. Strasbourg, France: Council of Europe. http://conventions.coe.int/Treaty/en/Treaties/Html/123.htm.

Creamer, M. A., J. Petty, T. Martin, V. K. Bergdall, and J. M. Hickman-Davis. 2014. Implications of natural occlusion of ventilated racks on ammonia and sanitation practices. *J Am Assoc Lab Anim Sci* 53 (2):174–9.

Dauchy, R. T., E. M. Dauchy, J. P. Hanifin, S. L. Gauthreaux, L. Mao, V. P. Belancio, T. G. Ooms et al. 2013. Effects of spectral transmittance through standard laboratory cages on circadian metabolism and physiology in nude rats. *J Am Assoc Lab Anim Sci* 52 (2):146–56.

David, J. M., S. Knowles, D. M. Lamkin, and D. B. Stout. 2013. Individually ventilated cages impose cold stress on laboratory mice: A source of systemic experimental variability. *J Am Assoc Lab Anim Sci* 52 (6):738–44.

Davila, S. G., J. L. Campo, M. G. Gil, M. T. Prieto, and O. Torres. 2011. Effects of auditory and physical enrichment on 3 measurements of fear and stress (tonic immobility duration, heterophil to lymphocyte ratio, and fluctuating asymmetry) in several breeds of layer chicks. *Poult Sci* 90 (11):2459–66.

Demorotski, D. 2008. Selecting plastic animal cages. *ALN.* http://www.alnmag.com/articles/2008/02/selecting-plastic-animal-cages.

Domer, D. A., R. L. Erickson, J. M. Petty, V. K. Bergdall, and J. M. Hickman-Davis. 2012. Processing and treatment of corncob bedding affects cage-change frequency for C57BL/6 mice. *J Am Assoc Lab Anim Sci* 51 (2):162–9.

European Union. 2010. Directive 2010/63/EU of the European Parliament and of the council of 22 September 2010 on the protection of animals used for scientific purposes. *Off J Eur Union* 276:33–79.

Freymann, J., P. P. Tsai, H. Stelzer, and H. Hackbarth. 2014. The amount of cage bedding preferred by female BALB/c and C57BL/6 mice. *Lab Anim (NY)* 44 (1):17–22.

Gaskill, B. N., C. J. Gordon, E. A. Pajor, J. R. Lucas, J. K. Davis, and J. P. Garner. 2012. Heat or insulation: Behavioral titration of mouse preference for warmth or access to a nest. *PLoS One* 7 (3):e32799.

Gaskill, B. N., A. Z. Karas, J. P. Garner, and K. R. Pritchett-Corning. 2013a. Nest building as an indicator of health and welfare in laboratory mice. *J Vis Exp* (82):51012.

Gaskill, B. N., K. R. Pritchett-Corning, C. J. Gordon, E. A. Pajor, J. R. Lucas, J. K. Davis, and J. P. Garner. 2013b. Energy reallocation to breeding performance through improved nest building in laboratory mice. *PLoS One* 8 (9):e74153.

Gonder, J. C., and K. Laber. 2007. A renewed look at laboratory rodent housing and management. *ILAR J* 48 (1):29–36.

Gonzalez, D. M., S. J. Graciano, J. Karlstad, M. Leblanc, T. Clark, S. Holmes, and J. D. Reuter. 2011. Failure and life cycle evaluation of watering valves. *J Am Assoc Lab Anim Sci* 50 (5):713–8.

Gordon, C. J., C. Aydin, E. A. Repasky, K. M. Kokolus, G. Dheyongera, and A. F. Johnstone. 2014. Behaviorally mediated, warm adaptation: A physiological strategy when mice behaviorally thermoregulate. *J Therm Biol* 44:41–6.

Hasegawa, M., S. Kagiyama, M. Tajima, K. Yoshida, Y. Minami, and T. Kurosawa. 2003. Evaluation of a forced-air-ventilated micro-isolation system for protection of mice against *Pasteurella pneumotropica*. *Exp Anim* 52 (2):145–51.

Hess, S. E., S. Rohr, B. D. Dufour, B. N. Gaskill, E. A. Pajor, and J. P. Garner. 2008. Home improvement: C57BL/6J mice given more naturalistic nesting materials build better nests. *J Am Assoc Lab Anim Sci* 47 (6):25–31.

Hessler, J., and N. Lehner. 2009. *Planning and Designing Research Animal Facililties*. San Diego: Academic Press.

Hessler, J. R. 2011. Laboratory animal facilities and equipment for conventional, barrier, and containment housing systems. In *Handbook of Laboratory Animal Science*, ed. S. J. Schapiro and J. Hau, pp. 74–5. Boca Raton, FL: CRC Press.

Hickman-Davis, J. M., M. L. Nicolaus, J. M. Petty, D. M. Harrison, and V. K. Bergdall. 2012. Effectiveness of shoe covers for bioexclusion within an animal facility. *J Am Assoc Lab Anim Sci* 51 (2):181–8.

Howerton, C. L., J. P. Garner, and J. A. Mench. 2008. Effects of a running wheel-igloo enrichment on aggression, hierarchy linearity, and stereotypy in group-housed male CD-1 (ICR) mice. *Appl Anim Behav Sci* 115 (1–2):90–103.

Hoyt, R., J. Hawkins, M. St. Claire, and M. Kennett. 2007. Mouse physiology. In *The Mouse in Biomedical Research*, ed. W. Barthold, S. Fox, M. T. Davisson, F. Newcomer, W. Quimby, and A. L. Smith, 23–90. San Diego: Elsevier.

Huerkamp, M. J., W. D. Thompson, and N. D. Lehner. 2003. Failed air supply to individually ventilated caging system causes acute hypoxia and mortality of rats. *Contemp Top Lab Anim Sci* 42 (3):44–5.

Hutchinson, E., A. Avery, and S. Vandewoude. 2005. Environmental enrichment for laboratory rodents. *ILAR J* 46 (2):148–61.

Hutchinson, E. K., A. C. Avery, and S. Vandewoude. 2012. Environmental enrichment during rearing alters corticosterone levels, thymocyte numbers, and aggression in female BALB/c mice. *J Am Assoc Lab Anim Sci* 51 (1):18–24.

Inagaki, H., Y. Kiyokawa, Y. Takeuchi, and Y. Mori. 2010. The alarm pheromone in male rats as a unique anxiety model: Psychopharmacological evidence using anxiolytics. *Pharmacol Biochem Behav* 94 (4):575–9.

Kogan, L. R., R. Schoenfeld-Tacher, and A. A. Simon. 2012. Behavioral effects of auditory stimulation on kenneled dogs. *J Vet Behav* 7 (5):268–75.

Kokolus, K. M., C. C. Hong, and E. A. Repasky. 2010. Feeling too hot or cold after breast cancer: Is it just a nuisance or a potentially important prognostic factor? *Int J Hyperthermia* 26 (7):662–80.

Kokolus, K. M., H. M. Spangler, B. J. Povinelli, M. R. Farren, K. P. Lee, and E. A. Repasky. 2014. Stressful presentations: Mild cold stress in laboratory mice influences phenotype of dendritic cells in naive and tumor-bearing mice. *Front Immunol* 5:23.

Konkle, A. T., A. C. Kentner, S. L. Baker, A. Stewart, and C. Bielajew. 2010. Environmental-enrichment-related variations in behavioral, biochemical, and physiologic responses of Sprague-Dawley and Long Evans rats. *J Am Assoc Lab Anim Sci* 49 (4):427–36.

Korpi, A., S. Lappalainen, E. Kaliste, P. Kalliokoski, K. Reijula, and A. L. Pasanen. 2007. A multi-faceted approach to risk assessment of laboratory animal allergens at two facilities. *Am J Ind Med* 50 (2):127–35.

Kulesskaya, N., H. Rauvala, and V. Voikar. 2011. Evaluation of social and physical enrichment in modulation of behavioural phenotype in C57BL/6J female mice. *PLoS One* 6 (9):e24755.

LaVail, M. M., G. M. Gorrin, and M. A. Repaci. 1987. Strain differences in sensitivity to light-induced photoreceptor degeneration in albino mice. *Curr Eye Res* 6 (6):825–34.

Leeds, J., and S. Wagner. 2008. *Through a Dog's Ear*. Boulder, CO: Sounds True, Inc.

Leggio, M. G., L. Mandolesi, F. Federico, F. Spirito, B. Ricci, F. Gelfo, and L. Petrosini. 2005. Environmental enrichment promotes improved spatial abilities and enhanced dendritic growth in the rat. *Behav Brain Res* 163 (1):78–90.

Lipman, N. S. 1999. Isolator rodent caging systems (state of the art): A critical view. *Contemp Top Lab Anim Sci* 38 (5):9–17.

Lipman, N. S., B. F. Corning, and M. A. Coiro Sr. 1992. The effects of intracage ventilation on microenvironmental conditions in filter-top cages. *Lab Anim* 26 (3):206–10.

MacRae, M., W. M. Kenkel, and A. C. Kentner. 2015. Social rejection following neonatal inflammation is mediated by olfactory scent cues. *Brain Behav Immun* 49:43–8.

Maher, R. L., S. M. Barbash, D. V. Lynch, and S. J. Swoap. 2015. Group housing and nest building only slightly ameliorate the cold stress of typical housing in female C57BL/6J mice. *Am J Physiol Regul Integr Comp Physiol* 308 (12):R1070–9.

Manser, C. E., D. M. Broom, P. Overend, and T. H. Morris. 1998. Investigations into the preferences of laboratory rats for nest-boxes and nesting materials. *Lab Anim* 32 (1):23–35.

Memarzadeh, F., P. C. Harrison, G. L. Riskowski, and T. Henze. 2004. Comparison of environment and mice in static and mechanically ventilated isolator cages with different air velocities and ventilation designs. *Contemp Top Lab Anim Sci* 43 (1):14–20.

Menon, V., and D. J. Levitin. 2005. The rewards of music listening: Response and physiological connectivity of the mesolimbic system. *Neuroimage* 28 (1):175–84.

Molk, D. M., C. L. Karr-May, E. D. Trang, and G. E. Sanders. 2013. Sanitization of an automatic reverse-osmosis watering system: Removal of a clinically significant biofilm. *J Am Assoc Lab Anim Sci* 52 (2):197–205.

Narver, H. L. 2013. Care and monitoring of a mouse model of melanoma. *Lab Anim (NY)* 42 (3):92–8.

National Health and Medical Research Council. 2013. Australian code of practice for the care and use of animals for scientific purposes. Canberra: National Health and Medical Research Council. https://www.nhmrc.gov.au/guidelines-publications/ea16.

Neuman, J. C., K. A. Albright, and K. L. Schalinske. 2013. Exercise prevents hyperhomocysteinemia in a dietary folate-restricted mouse model. *Nutr Res* 33 (6):487–93.

Nicholson, A., R. D. Malcolm, P. L. Russ, K. Cough, C. Touma, R. Palme, and M. V. Wiles. 2009. The response of C57BL/6J and BALB/cJ mice to increased housing density. *J Am Assoc Lab Anim Sci* 48 (6):740–53.

NRC (National Research Council). 2011. *Guide for the Care and Use of Laboratory Animals*. 8th ed. Washington, DC: National Academies Press.

Olsson, I. A. S., C. M. Nevison, E. G. Patterson-Kane, C. M. Sherwin, H. A. Van de Weerd, and H. Würbel. 2003. Understanding behaviour: The relevance of ethological approaches in laboratory animal science. *Appl Anim Behav Sci* 81 (3):245–64.

Panksepp, J., and W. W. Beatty. 1980. Social deprivation and play in rats. *Behav Neural Biol* 30 (2):197–206.

Patterson-Kane, E. P. 2011. Horizons of enrichment: The history, accomplishments, obstacles and aspirations of environmental enrichment. *Enrichment Rec* 7 (April):12–14.

Patterson-Kane, E. P., M. Hunt, and D. Harper. 2004. Short communication: Rat's demand for group size. *J Appl Anim Welf Sci* 7 (4):267–72.

Pritchett-Corning, K. R., R. Keefe, J. P. Garner, and B. N. Gaskill. 2013. Can seeds help mice with the daily grind? *Lab Anim* 47 (4):312–5.

Ras, T., M. van de Ven, E. G. Patterson-Kane, and K. Nelson. 2002. Rats' preferences for corn versus wood-based bedding and nesting materials. *Lab Anim* 36 (4):420–5.

Reeb, C., R. Jones, D. Bearg, H. Bedigan, D. Myers, and B. Paigen. 1998. Microenvironment in ventilated animal cages with differing ventilation rates, mice populations, and frequency of bedding changes. *Contemp Top Lab Anim Sci* 37 (2):43–49.

Reeb-Whitaker, C. K., B. Paigen, W. G. Beamer, R. T. Bronson, G. A. Churchill, I. B. Schweitzer, and D. D. Myers. 2001. The impact of reduced frequency of cage changes on the health of mice housed in ventilated cages. *Lab Anim* 35 (1):58–73.

Rosenbaum, M. D., S. VandeWoude, and T. E. Johnson. 2009. Effects of cage-change frequency and bedding volume on mice and their microenvironment. *J Am Assoc Lab Anim Sci* 48 (6):763–73.

Rosenbaum, M. D., S. VandeWoude, J. Volckens, and T. Johnson. 2010. Disparities in ammonia, temperature, humidity, and airborne particulate matter between the micro- and macroenvironments of mice in individually ventilated caging. *J Am Assoc Lab Anim Sci* 49 (2):177–83.

Russell, W. M. S., and R. L. Burch. 1959. *The Principles of Humane Experimental Technique*. London: Methuen.

Sanderson, A. E., H. M. Multari, J. J. Lohmiller, and S. R. Boutin. 2010. Effect of cage-change frequency on rodent breeding performance. *Lab Anim (NY)* 39 (6):177–82.

Semple, B. D., L. J. Noble-Haeusslein, Y. Jun Kwon, P. N. Sam, A. M. Gibson, S. Grissom, S. Brown et al. 2014. Sociosexual and communication deficits after traumatic injury to the developing murine brain. *PLoS One* 9 (8):e103386.

Singapore Government National Advisory Committee for Laboratory Animal Research. 2004. Animals and birds act guidelines on the care and use of animals for scientific purposes. Singapore: Singapore Government National Advisory Committee for Laboratory Animal Research.

Smith, A. L., and D. J. Corrow. 2005. Modifications to husbandry and housing conditions of laboratory rodents for improved well-being. *ILAR J* 46 (2):140–7.

Smith, E., J. D. Stockwell, I. Schweitzer, S. H. Langley, and A. L. Smith. 2004. Evaluation of cage microenvironment of mice housed on various types of bedding materials. *Contemp Top Lab Anim Sci* 43 (4):12–17.

Smith, M. M., and M. Hargaden. 2001. Developing a rodent enrichment program. *Lab Anim (NY)* 30 (8):36–41.

Spani, D., M. Arras, B. Konig, and T. Rulicke. 2003. Higher heart rate of laboratory mice housed individually vs in pairs. *Lab Anim* 37 (1):54–62.

Stockley, P., L. Bottell, and J. L. Hurst. 2013. Wake up and smell the conflict: Odour signals in female competition. *Philos Trans R Soc Lond B Biol Sci* 368 (1631):20130082.

Timmins, R. P., K. D. Cliff, C. T. Day, B. L. Hart, L. A. Hart, R. C. Hubrecht, K. F. Hurley et al. 2007. Enhancing quality of life for dogs and cats in confined situations. *Anim Welf* 16 (2):83–7.

Toth, L. A. 2015. The influence of the cage environment on rodent physiology and behavior: Implications for reproducibility of pre-clinical rodent research. *Exp Neurol* 270:72–7.

Toth, L. A., K. Kregel, L. Leon, and T. I. Musch. 2011. Environmental enrichment of laboratory rodents: The answer depends on the question. *Comp Med* 61 (4):314–21.

United States. 1985. Health Research Extension Act of 1985. Public Law 99-158. *U.S. Statutes at Large* 99 (Title IV, Sections 1–12), 99–158.

Van Loo, P. L., and V. Baumans. 2004. The importance of learning young: The use of nesting material in laboratory rats. *Lab Anim* 38 (1):17–24.

Van Loo, P. L. P., C. L. J. J. Kruitwagen, L. F. M. Van Zutphen, J. M. Koolhaas, and V. Baumans. 2000. Modulation of aggression in male mice: Influence of cage cleaning regime and scent marks. *Anim Welf* 9 (3):281–95.

Van Loo, P. L. P., E. Van der Meer, C. L. J. J. Kruitwagen, J. M. Koolhaas, L. F. M. Van Zutphen, and V. Baumans. 2004. Long-term effects of husbandry procedures on stress-related parameters in male mice of two strains. *Lab Anim* 38 (2):169–77.

Van Loo, P. L. P., L. F. M. Van Zutphen, and V. Baumans. 2003. Male management: Coping with aggression problems in male laboratory mice. *Lab Anim* 37:300–13.

Varlinskaya, E. I., L. P. Spear, and N. E. Spear. 1999. Social behavior and social motivation in adolescent rats. *Physiol Behav* 67 (4):475–82.

Varty, G. B., M. P. Paulus, D. L. Braff, and M. A. Geyer. 2000. Environmental enrichment and isolation rearing in the rat: Effects on locomotor behavior and startle response plasticity. *Biol Psychiatry* 47 (10):864–73.

Viegas, J. 2010. Male mice sing ultrasonic love songs. March 5. http://news.discovery.com/animals/zoo-animals/mouse-courtship-songs-ultrasonic.htm.

Voznessenskaya, V. V. 2014. Influence of cat odor on reproductive behavior and physiology in the house mouse: (*Mus musculus*). In *Neurobiology of Chemical Communication*, ed. C. Mucignat-Caretta, pp. 403–5. Boca Raton, FL: CRC Press.

Weary, D. M. 2013. What is suffering in animals? 23. http://awp.landfood.ubc.ca/files/2013/06/Weary_suffering.pdf.

Wells, D. L., L. Graham, and P. G. Hepper. 2002. The influence of auditory stimulation on the behaviour of dogs housed in a rescue shelter. *Anim Welf* 11 (4):385–93.

Whary, M. T., N. Baumgarth, J. G. Fox, and S. W. Barthold. 2015. Biology and diseases of mice. In *Laboratory Animal Medicine*, ed. L. C. Anderson, G. Otto, K. R. Pritchett-Corning, and M. T. Whary. San Diego: Elsevier.

Wheeler, R. R., M. P. Swan, and D. L. Hickman. 2014. Effect of multilevel laboratory rat caging system on the well-being of the singly-housed Sprague Dawley rat. *Lab Anim* 49 (1).

Whiteman, B. 2001. 'Moosic study' reveals way of increasing milk yields. University of Leicester. http://www.le.ac.uk/press/press/moosicstudy.html (accessed October 16, 2014).

Whittaker, A. L., G. S. Howarth, and D. L. Hickman. 2012. Effects of space allocation and housing density on measures of wellbeing in laboratory mice: A review. *Lab Anim* 46 (1):3–13.

Winnicker, C. 2012. Mice no longer out in the cold. January 10. http://www.criver.com/about-us/eureka/blog/january-2012/mice-no-longer-out-in-the-cold.

Wirth, H. 1983. Criteria for the evaluation of laboratory animal bedding. *Lab Anim* 17 (2):81–4.

Ying, H., X. Lan-Wen, Y. Fei and Y. Ping. 2007. The effects of enrichment with music or colorful light on the welfare of restrained mice. *Lab Anim Comp Med* 27 (2):71–6.

York, J. M., A. W. McDaniel, N. A. Blevins, R. R. Guillet, S. O. Allison, K. A. Cengel, and G. G. Freund. 2012. Individually ventilated cages cause chronic low-grade hypoxia impacting mice hematologically and behaviorally. *Brain Behav Immun* 26 (6):951–8.

22

Large (Nonagricultural) Animal Enclosures and Housing

Jeffrey D. Wyatt

CONTENTS

Introduction

The trend to provide larger, social, more complex and enriching, naturalistic housing for research animals is driven by growing expectations or requirements from the public, changes in the scientific knowledge regarding animal welfare needs, animal care and research staff, funding and accreditation agencies, and national or regional animal welfare regulatory entities. All species of laboratory animals have ancestral or wild counterparts, which may provide relevant information specific to housing and management practices that promote well-being and natural behaviors. The concept that the research environment should allow the animal to live in a manner for which it was designed or has evolved in nature is beginning to bridge a gap between research animal ethics and animal welfare science. A focus on the species-specific "telos," how the animal is hardwired in nature, be it gregarious, nocturnal, arboreal, a flyer, or a hopper, assists our approach to fostering the best welfare experience possible within the practical constraints of a laboratory facility (Fraser 1999). A similar need exists to balance the species-specific telos and the needs of research and available resources to support those needs.

The objective of this chapter is to provide scientific information for the facility manager, researcher, or veterinarian specific to opportunities for housing and managing four large laboratory animal species: rabbits, nonhuman primates, cats, and dogs. The intended goal is to meet oversight agency requirements while considering their natural history (telos) and the practical husbandry and operational needs to promote good science.

Priorities

Oversight Organizations

International research animal welfare regulatory agencies considered in this chapter are United States and European centric. They are the U.S. Department of Agriculture (USDA) Animal Welfare Act Regulations (AWA Regulations) and Appendix A of the European Convention for the Protection of Vertebrate Animals Used for Experimental and Other Scientific Purposes, including Article 5 of the convention, Guidelines for Accommodation and Care of Animals (ETS 123) (Council of Europe 2006). These are considered minimum standards, and some institutions may want to meet or even exceed the minimum regulatory standard. These choices are not intended to minimize utility of other countries' or regions' animal welfare regulations. Canadian animal welfare legislation for research animals falls under provincial jurisdiction, where most provinces have adopted the Canadian Council on Animal Care (CCAC) guidelines, which are currently under revision (CCAC 1993). There is impressive growth in Asian research animal care programs pursuing AAALAC International accreditation. A draft proposal of an animal protection law, released in 2009 in China, remains yet to be adopted by the legislature. A Prevention of Cruelty to Animals (PCA) Act was established in India in 1960, with subsequent legislation requiring the Committee for the Purpose of Control and Supervision of Experiments on Animals (CPCSEA) to regulate experimentation on animals following "Breeding of and Experiments on Animals (Control and Supervision) Rules, 1998," as amended in 2001 and 2006 (Bayne et al. 2015). Federal laws designed to promote the humane care and use of research animals exist in many other countries, including Brazil, Australia, and New Zealand (Mellor and Bayvel 2008; Trez 2010). Since all the regulations governing research animal welfare listed above include housing space metrics and management practices within the ranges promulgated by the USDA AWA Regulations and ETS 123, the information in this chapter remains globally relevant.

The *Guide for the Care and Use of Laboratory Animals*, Eighth Edition (*Guide*) (NRC 2011) describes guidelines for vertebrate laboratory animal housing and care performance standards to be met and followed as a condition for institutions awarded research funding by the National Institutes of Health (NIH). The Association for Assessment and Accreditation of Laboratory Animal Care International (AAALAC International 2015), the only global organization performing peer review, assessment, and accreditation of institutions using animals in research, testing, and teaching, uses the *Guide* as a primary reference, in addition to relevant federal or territorial regulations. The *Guide* will also be used in this chapter for referencing housing and management expectations for nonhuman primates, rabbits, dogs, and cats.

Welfare

Animal welfare, the state of an animal's quality of life, is measured on a continuum of poor to good. The concept of "Five Freedoms" has been considered and widely discussed and applied to the care of farmed livestock to assess animal welfare (Brambell 1965). The Five Freedoms are freedom (1) from hunger and thirst; (2) from discomfort; (3) from pain, injury, or disease; (4) to express normal behaviors; and (5) from fear and distress. A more positive interpretation of the Five Freedoms has been proposed for zoo animals as Five Opportunities to thrive (1) receiving nutritionally complete diets resulting in natural feeding response and behaviors, (2) affording living experiences with choice and control, (3) experiencing good physical health, (4) providing spaces to live in appropriate social groups promoting natural behavior, and (5) experiencing healthy stress responses, challenges, and coping skills, avoiding distress and chronic stress (Vicino 2015). These same principles may be applied to laboratory animals across all species. The space (floor, vertical, and cubic area) and features designed for environmental enrichment (rearing, hopping, climbing, swinging, swimming, etc.) with respect to the natural history of the species are discussed. In addition to the physical aspects of housing, such as substrates, perches, shelving, and privacy nest boxes, behavioral enhancement and enrichment strategies, including choice between items, foraging opportunities, social experiences, reproduction, and positive human interaction, are considered for each of the four species.

Husbandry

The practical husbandry aspects of laboratory animal housing must also be considered. The increased complexity and size of animal housing choices may pose challenges to effective sanitization and have a major impact on the capacity of the facility. While the *Guide* offers flexibility in frequency of bedding changes and waste removal from daily to weekly and primary enclosure sanitation done at least at 2-week intervals, the required staffing numbers will vary widely depending on many variables, including housing density, types of surfaces, style of movable or fixed caging or pens, and the availability of swing space. Animal and personnel safety, including ergonomics and zoonotic hazard exposure, must be considered with the selection of housing options (NRC 2003). Provision of food or water on different schedules (restricted amounts, intermittent schedules, or *ad lib*) in a variety of housing types, from caging to pens, with animals housed singly or in groups, may create difficulties for animal care and research staff. Appropriate management of environmental factors, such as heating, ventilation, and air-conditioning (HVAC); light intensity, spectrum, and cycle; noise and vibration; and usage of visual stimulation, provides opportunities to enhance the animals' experiences (NRC 2011). For all these housing and management considerations, the cost or risk, for example, labor, materials, inconvenience, or animal incompatibility, must be weighed against the benefits, such as improvement in animal health and welfare, staff morale, and scientific outcomes.

Supporting Research and Emerging Technologies

The animal housing and management styles should not interfere with or compromise the scientific goals of the research. Ideally, they should enhance or provide a refined or more efficient approach to the research. Accessing animals for research activities may be facilitated with switch gates, doors, and vestibules. Operant conditioning and training promoting target training, stationing, switching, and crating provides a win–win, nondistressful, and safe experience for both the animal and handler. Giving the animals, especially nonhuman primates, a choice to participate in neuroscience research with cage- or pen-mounted touchscreens or joysticks provides more data over time. Use of the home cage negates the need to manage the animal with a pole and collar and/or chair restraint with subsequent transport to the lab, resulting in less stress to the animal and greater safety for personnel. Telemetry devices, internally implanted or externally protected in a jacket, provide 24/7 monitoring of unrestrained animals for physiological parameters, including blood pressure, temperature, heart and respiratory rate, and activity using waterproof receivers mounted inside home cages and pens.

Cost–Benefit

With limited research funding, regulatory burden, and ever-increasing animal and animal care costs, the reality of cost-effective operations significantly, and understandably, influences choices for animal housing and management (Haywood and Greene 2008). While the up-front capital cost of movable or compartmentalized caging or pens, which may be mechanically sanitized consistently at a high water temperature in a rack washer, may be staggering, the return on that investment may be offset in the long run by savings in labor associated with meticulous hand washing and sanitizing fixed pens or cages. Moving animals safely away from the spray, detergents, and disinfectants as required by the *Guide* and AWA Regulations while hand sanitizing fixed pens or caging requires swing space or large enclosures. Occupational health issues in all aspects of cage washing, including transporting soiled caging to cage wash, the potential for exposure to zoonoses and chemicals, and other occupational health and safety issues, like exposure to allergens and repetitive motion injury during hand washing, and concern regarding entrapment in a rack washer remain occupational health risk assessment priorities. Regular evaluation of sanitation effectiveness may yield more consistent results with mechanized cage washing systems than hand sanitization (Bayne and Davis 1983; Lloyd-Evans et al. 1986; Wardrip et al. 2000; Trachoo and Frank 2002; Luchins et al. 2011).

The rationales for selecting particular animal housing and management systems, including improved animal welfare and health, the promotion of natural behaviors, making enrichment biologically relevant, the management of stereotypies, enhancing staff morale, and increasing research efficiencies, should be considered when performing a cost– or risk–benefit analysis (Newberry 1995; Baumans 2005; Tarou and Bashaw 2007).

Nonhuman Primates

Welfare

Space

Primatologists and behaviorists for all species of nonhuman primates argue that using one factor, body weight, to determine minimum cage size is insufficient and provides no assurance that the animals' physical, behavioral, and social needs are being met because of the complexities of their behavioral repertoires. Nonhuman primates with overlapping body weights, such as crab-eating or long-tailed macaques (*Macaca fascicularis*) and rhesus macaques (*Macaca mulatta*), have adapted to very different niches in the wild. This evolutionary divergence, regardless of similar body masses, requires more attention to providing an arboreal cage or pen experience with three-dimensional climbing opportunities and perches for crab-eating macaques than for the more terrestrial rhesus macaque. Attention to the animal's natural history during cage design promotes improved psychological well-being, as evidenced by crab-eating macaques demonstrating less self-injurious behavior in vertically designed caging (Buchanan-Smith et al. 2004). In another study, individually housed rhesus and crab-eating macaques preferred to spend the majority (68%) of their time in the upper tier of a one-over-one cage, regardless of illumination levels (MacLean et al. 2009). While body weight provides a simplistic and easily legislated starting point for determining minimum cage space, the species-typical behaviors and adaptations for survival in nature must be considered.

Larger housing spaces for nonhuman primates often, but not always, promote good welfare. The required minimum floor and cubic space available to all nonhuman primates is significantly greater in the ETS 123 European standards (Table 22.1) than in the USDA AWA Regulations and the *Guide* (Table 22.2). Common marmosets (*Callithrix jacchus*) displayed significantly higher levels of locomotion after only increasing the vertical dimension of caging (Kitchen and Martin 1996). The behavioral benefits

TABLE 22.1

Minimum Cage Space for Nonhuman Primates per ETS 123

	Floor Space, m^2	Volume per Animal, m^3	Height, m	Enclosure Volume, m^3
Marmosets	0.5 for 1–2 adults plus offspring up to 5 months	0.2 per additional animal >5 months	1.5 if top of enclosure is at least 1.8 m from floor	
Tamarins	1.5 for 1–2 adults plus offspring up to 5 months	0.2 per additional animal >5 months	1.5 if top of enclosure is at least 1.8 m from floor	
Squirrel monkeys	2.0 for 1–2 adults	0.5 per additional animal >6 months	1.8	
Macaques and vervets <3 years	2.0 for up to 3 animals	1.0	1.8	3.6
Macaques and vervets ≥3 years	2.0 for up to 2 animals	1.8	1.8	3.6
Macaques and vervets, breeding	2.0 includes offspring up to 2 years of age	3.5 for up to one mother with offspring up to 2 years of age	2.0	3.6
Baboons <4 years	4.0 for up to 2 animals	3.0	1.8	7.2
Baboons ≥4 years	7.0 for up to 2 animals	6.0	1.8	12.6
Baboons, breeding	7.0, includes offspring up to 2 years of age	12.0, includes offspring up to 2 years of age	2.0	12.6

TABLE 22.2

Comparison of Minimum Cage Space Requirements for Nonhuman Primates per AWA Regulations and the *Guide*

AWA Regulations Groups	Body Weight, kg	AWA Regulations Minimum Space: Floor Height, m^2 (cm)	*Guide* Groups across All Nonhuman Primate Species	*Guide* Minimum Space: Floor Height, m^2 (cm)
Marmosets, tamarins, and infants across species <6 months	<1	0.15 (50.8)	Up to 1.5 kg	0.2 (76.2)
Capuchins, squirrel monkeys, and juveniles across species (6 months to 3 years)	1–3	0.28 (76.2)	Up to 3 kg	0.28 (76.2)
Macaques and African species	3–10	0.40 (76.2)	Up to 10 kg	0.40 (76.2)
Male macaques and large African species	10–15	0.56 (81.28)	Up to 15 kg	0.56 (81.3)
Baboons and nonbrachiating species >15 kg	15–25	0.74 (91.44)	Up to 20 kg	0.74 (91.4)
Great apes >25 kg and brachiating species	>25	2.33 (213.36)	Up to 25 kg	0.93 (116.8)
			Up to 30 kg	1.40 (116.8)
			>30 kg	≥2.32 (152.4)

displayed by nonhuman primates housed in naturalistic exhibits in zoos using soft architecture, such as pools, tree branches, leafy browse, ropes, and hammocks, instead of metallic or stainless steel surfaces and climbing structures were predicted to in the future afford the same benefits in nonhuman primate research institutions (Maple and Finlay 1989). Such foreshadowing has been realized in research facilities providing pools of water for macaques (Robins and Waitt 2011), rope swings and hammocks for many species of nonhuman primates (Fillman-Holliday and Landi 2002; Maier et al. 2004), branched leafy tree browse for macaques (Strazzeri et al. 2014), and deep litter (straw and wood wool) floor substrate for Wolf's guenons (*Cercopithecus wolfi*) (Fuller et al. 2010). Crab-eating macaques, singly housed in cages meeting AWA Regulations minimum floor and height dimensions, demonstrated increased exploratory behavior and elimination of stereotypies when placed individually for even short periods of time in a highly enriched playpen measuring approximately 4 × 5 × 3 ft (1.2 × 1.5 × 1.05 m). When returned to the smaller home cage, stereotypies recurred (Bryant et al. 1988). Considering floor space alone compromises space utilization, as shown when pen-housed rhesus macaques chose to spend 90% of their time on elevated structures approximately 1–4 ft (40–130 cm) above the floor. The more dominant animals chose the higher structures, whereas juveniles were more likely to use the movable swings and Ferris wheel, emphasizing the benefit for animals to make choices and demonstrate species-typical behavior in captive housing (Reinhardt 1992a).

Another cage feature, recognized to be desirable for pair-housed female rhesus monkeys, was a privacy panel inserted inside the cage where one animal could choose time alone behind a visual barrier while still sharing the cage. The paired macaques spent 75% of the time next to each other with a privacy panel available, compared with 60.8% of the time when a privacy panel was not available. Even though macaques are highly social in nature, it appears that intermittent control over being alone promotes more time in close proximity and a better cage-mate relationship (Reinhardt and Reinhardt 1991). The double-wide QUAD-style macaque caging used commonly for pairing macaques, with a sliding door connecting the passageway between cage compartments, accomplishes the same outcome with more vertical and floor space available. Placing a large viewing window in the animal room door promotes an enriching experience as staff walk by, allowing animals to make choices to ignore or interact (author's personal experience).

Enrichment

The behavior analysis tool involving preference assessment has been used in food animal management programs, as well as with humans with impaired communication, and has been suggested for nonhuman primates housed in research settings to determine which environmental enrichment is preferred (Sumpter et al. 2002; Staal et al. 2003; Schapiro and Lambeth 2007). Preference assessment testing identifies choice for otherwise nonverbal experimental subjects. When making a choice by touching a cage-mounted screen, Japanese macaques (*Macaca fuscata*) consistently selected movies with animation or humans, and preferred to watch movies not previously viewed instead of viewing the same movie over and over again (Ogura and Matsuzawa 2012). Common marmosets (*Callithrix jacchus*), when given control over a supplemental light and heat source, made choices to adjust the light. More importantly, the study demonstrated that even though the marmosets preferred supplemental light and heat, the opportunity for control more importantly improved animal welfare, as measured by calmer activity patterns and lower scent marking (Buchanan-Smith and Badihi 2012). When given a choice, common marmosets preferred to be fed in a bowl located high in the cage, most likely because of their arboreal existence in the wild (Buchanan-Smith et al. 2002). Providing choice and measuring welfare impact (e.g., behavior and activity score, social interaction, and mentation) of environmental enrichment and housing designs preferred by the nonhuman primate will advance how effectively we promote a multisensory environment.

Social Housing

The natural history, including dominance style, of the species of nonhuman primate should be considered when implementing a social housing program. Even though rhesus and crab-eating macaques are highly social in nature, the rhesus, being the more aggressive or "despotic" species, may need more attempts and space for pairing than the considerably more passive "egalitarian" crab-eating macaque (Buchanan-Smith et al. 2004). The standard of practice to acclimate macaques for pairing is a double-cage design in accordance with AWA Regulations minimum space requirements that allows two isosexual macaques to first see and possibly touch each other through a protected contact panel with a mesh or larger spaces, such as twelve 2 × 3 in. (5.1 × 8 cm) openings. This approach results in impressive animal welfare benefits in spite of a minority of animal incompatibilities (Baker et al. 2012). While AAALAC International considers protected contact to be a social "experience" not fully meeting the definition of social housing (AAALAC International 2015), the welfare benefits of providing two macaques "grooming or protected contact" through a panel with vertical bars as far apart as 2 in. (5.16 cm) has been well demonstrated (Crockett et al. 1997; Lee et al. 2012). Pregnancy prevention panels have been created by using two panels with bars providing 2 in. spacing but with an additional 2.5 in. (6 cm) gap between cages, allowing grooming contact for paired male and female crab-eating macaques. This strategy resulted in more successful pairing outcomes than with isosexual pairs (Crockett et al. 1997). Experiment-driven, scientifically justified, and Institutional Animal Care and Use Committee (IACUC)–approved (e.g., neurotoxin-dosed, impaired primates; good laboratory practice [GLP] studies; and some infectious disease studies), or veterinary-required or incompatibility-based exemptions from social housing are permitted for NIH-funded research (NRC 2011). Enhanced enrichment should be offered in these cases.

Husbandry

Husbandry-related variables highlighted by Reinhardt (2004), such as understimulation due to lack of suitable enrichment or social deprivation, adversely affected animal welfare and research results, including an increase in stress sensitivity and disease progression, immunosuppression, stress-induced anxiety as evidenced by increased plasma prolactin concentrations, abnormal brain function, and physiological evidence of depression. Manual restraint of nonhuman primates is distressful since they experience this as a predator–prey relationship. Stress-sensitive physiological parameters show that nonhuman primates do not acclimate to repeated, brief, or prolonged manual restraint (Reinhardt 2004). Positive reinforcement training in contrast to forcing mechanical or manual restraint for husbandry, medical, or research purposes alleviates animal anxiety and distress. Otherwise, the simple entry of research, animal care, or veterinary staff into the animal room becomes a daily intrinsic stressor due to previous negative conditioning experiences (Reinhardt 2004). Lower-tier caging provides reduced quality and quantity of light and a more restricted view of incoming staff by the nonhuman primate. It also impairs observation of those animals by animal care staff compared with primates housed in upper-tier caging. Since most primates escape terrestrial threats in nature by moving vertically, the animals housed in lower-tier caging may feel insecure and distressed. These extraneous husbandry-related variables should be addressed to minimize confounders and variability in research results and to promote good animal welfare (Reinhardt 2004).

Providing animals with choices associated with routine husbandry practices promotes welfare. Primate biscuits provided simultaneously in an open-top food hopper adjacent to an externally mounted food puzzle feeder presented a simple choice for rhesus macaques in a study by Reinhardt (1994). After first consuming some of the biscuits from the open hopper, the animals spent a 55-fold increase in time (11 minutes total at 1 minute per biscuit) retrieving biscuits from the puzzle feeder compared with the open-top feeder (32 seconds total at 1.1 second per biscuit while discarding most of the biscuits), indicating that the macaques preferred to forage or work at retrieving their food, as they do in nature. Positive reinforcement training allowing nonhuman primates control and choice to cooperatively enter a switch or transport crate through two guillotine doors without squeezing from the home cage, avoiding distress associated with mechanical squeeze or manual restraint (Reinhardt 1992b). Other management techniques, from cage changing to weighing, demonstrate the shift to using positive reinforcement training to benefit research results and nonhuman primate welfare across species (Bassett et al. 2003; Laule et al. 2003; McKinley et al. 2003; Prescott et al. 2003; Schapiro et al. 2003; Prescott and Buchanan-Smith 2005).

Cost–Benefit

The array of primate caging available presents the facility manager and veterinarian with an overwhelming number of suitable choices. The decision to purchase new caging (movable cages or compartmentalized pens), to repurpose existing caging for pair housing, or to build wall-mounted pens with relatively inexpensive commercial stainless steel or chain-link fencing is not simple. The answer is situational, depending on what type of resources are primarily available, such as sufficient capital to purchase relatively expensive, portable, social housing pens or cages possible to clean in an automated cage washer or an adequate staffing level for labor-intense sanitization of less expensive wall-mounted pens. The advantages of choosing movable caging are their durability and ease of cleaning. The consistently validated cleaning and sanitizing methods in an automated, mechanized cage rack washer far outweigh the disadvantages of labor-intense, potentially hazardous, inconsistently effective hand scrubbing, power washing, sanitizing, and rinsing of nonmovable pens. Per the *Guide*, the effectiveness of sanitation of equipment must be assessed with special attention to hand-washed surfaces that directly contact animals. The main disadvantage of purchasing movable caging and compartmentalized pens that may go through cage wash is the up-front capital cost, ranging from $18,000 for a QUAD cage that may house two pairs of macaques to a higher expense for modular pens housing four or more macaques (author's personal experience). The advantage of hiring a fencing company to custom-build two to three floor-to-ceiling, 8 ft wide by 5 ft deep, pens is the lower cost of $10,000 total for all three larger pens in one room (author's personal experience). If there is no swing space to move the primates from the pens, the cleaning process of occupied pens poses significant health and welfare risks to both primates and personnel. While wall-mounted pens provide larger space and lower up-front cost, this author prefers to make the capital investment in purchasing movable caging to house all nonhuman primates in pairs and recover the costs over the 20-year life of the caging in renewal and replacement charge-backs to the per diem. The less expensive, yet highly enriched, floor-to-ceiling mounted pens have their place in a program by serving as exercise or playpens through which primates may rotate, depending on their research use. When facing the reality that purchase of new exercise pens or caging is not feasible, a resourceful institutional team of animal care, veterinary, and facilities staff should include a science-based approach and emphasize that as well. Staff may modify outdated stainless steel dog cages or one-over-one macaque cages. The end result will be cost-effective and high-functioning macaque exercise or pair housing cages, which improve animal welfare, recycle outdated caging, and cost less than 10% of the purchase price of new caging (Storey et al. 2000; Martin et al. 2002).

Enrichment strategies impacting husbandry that are developed and implemented by all stakeholders, including the finance administrator, the facility manager, and animal care, research, and veterinary staff, become achievable and sustainable. The best example of such teamwork occurred as an institution determined the feasibility and cost versus benefit for providing wood shavings on floors of primate pens. The welfare benefits of softwood shavings as natural substrates over bare floors, including reduction in aggression, stereotypies, and overgrooming, and promotion of foraging and exploratory behavior, were well documented in peer-reviewed publications. Questions regarding labor costs for implementing the strategy and for physical plant modifications remained unanswered. Animal care staff first recognized that the floor drain covers needed redesigning to prevent clogging. Next, staff designed and implemented a daily spot cleaning strategy of the shavings with removal of all shavings at 2-week intervals, followed by chemical disinfection of the floor and replacement of the shavings. Costs, including animal care staff labor, disinfectants, bedding, drain modifications, and other materials, were calculated and compared with the previous housing strategy, which required daily hosing and disinfection of bare floors. In addition to the anticipated animal welfare benefits, an overall 9% decrease in husbandry-associated cost was achieved, as well as a reduction in water and disinfectant. This type of collaborative and evidence-based evaluation of an enrichment strategy impacting husbandry could be replicated as a "best practice" when considering the cost and benefit of novel environmental enhancement or housing practices across all species (Dean 1999; Bennett et al. 2010).

Supporting Research and Emerging Technologies

Accessing nonhuman primates from large pens or even smaller cages may be difficult, distressful, and dangerous for both the animal and the handler. Reinhardt (1992b) demonstrated that without the use of a squeeze apparatus or chemical immobilization, 298 single-housed, adult rhesus macaques were trained by animal care staff to voluntarily jump into a hand-carried transport box (36 cm long, 28 cm wide, and 33 cm high) and rewarded with a food treat after being weighed and returned to their home cage. A positive relationship with animal care staff, who interact with the primates daily, is the key to success with other operant conditioning techniques, such as target training or stationing. Giving nonhuman primates a choice to cooperate with venipuncture for routine blood collection avoids the stress-related effect from manual restraint or drug effect of anesthetics on physiologic and serum chemistry and hematology profiles (Reinhardt 2003; Lambeth et al. 2006; Coleman et al. 2008).

Positive reinforcement training techniques for pole-and-collar restraint first habituating animals to the equipment and shaping behavior over 17 weeks with 85 training sessions (mean 1085 total training minutes per macaque) using fruit rewards enhanced welfare and research (McMillan et al. 2014). Additional examples of using positive reinforcement to train nonhuman primates for pole-and-collar restraint, removal from home cage without collars, and home cage training without food or water restriction underscore the trend away from more negative techniques forcing the desired behavior (Scott et al. 2003; Franklin 2016). A "smart chair" designed to automatically train macaques to enter a nonhuman primate chair for water reward without a human in the animal room decreased training person-hours per day while avoiding resistance by animals traditionally trained using a more negative and relatively unsafe technique of leash pulling (Ponce et al. 2016). These and other trained behaviors also assist research staff to overcome the challenge of retrieving primates from social housing in cages or pens.

Research on human and nonhuman primate gaze tracking traditionally involves a coil incorporated in a contact lens for humans or surgically implanted subconjunctivally in the sclera of the nonhuman primate. The tracking of eye movement and visual orienting is then typically measured by taking the chair-restrained nonhuman primate to the laboratory and rewarding successful trials with fluid or food. Shepherd and Platt (2006) designed and implemented a helmet-mounted eye camera apparatus allowing telemetric infrared video gaze tracking in two ring-tailed lemurs (*Lemur catta*) while pair housed and navigating their naturalistic home pen. Several months of animal training and habituation resulted in the ability to provide a unique opportunity to track eye movement while the animals were socially interacting and foraging, without requiring chair restraint or food or fluid restriction, as is more traditionally used in the macaque model. Such technology is potentially scalable across other primate species as equipment is refined and collaborative relationships between research, animal care, and facility management staff provide more opportunities to consider emerging research technologies. Performing more research activities within the animal housing facility benefits the animal and research staff by eliminating the need to remove and transport the animal from the familiar home environment to the research lab. However, the facility manager and space planners must recognize the reduction in animal housing capacity as more specialized research space is expanded within the vivarium.

A more efficient use of animal facility space provides the opportunity to perform all research activities within the home cage or pen. A cage-mounted joystick and game board apparatus attached to a Lexan-enclosed computer mounted to the animal room wall allowed macaques to have 24-hour access to perform a battery of 18 computerized tasks. The animals received fruit-flavored chow pellets, automatically delivered with each successful trial. The 18 tasks began first with relatively simple shaping tasks, while subsequently advancing to more complex learning, problem-solving, and memory-testing tasks. This home cage paradigm, which engaged macaques with choices for participation, proved to be enriching and resulted in the animals learning the tasks more quickly, with progression through all tasks in as few as 6 months, compared with several years with previously studied animals, which performed the same tasks after being removed from the home cage only during weekday work hours (Washburn and Rumbaugh 1992). Rhesus macaques also participated in home cage cognitive testing using a cage-mounted, computer-automated training and testing system with touchscreen technology for delayed matching-to-sample (DMTS) studies, which are useful in Alzheimer's disease research (Buccafusco et al. 2002). Similar research in cognitive science using cage-mounted joysticks or touchscreens is possible in socially housed nonhuman primates, including in capuchins (*Cebus apella*), which are notoriously difficult to remove from the home cage in a social housing setting. The capuchin pen caging was modified with a bank of test enclosures high in the cage, catering to their arboreal lifestyle. Individual capuchin monkeys were trained over weeks to enter the testing vestibule and eventually accept the door being closed to separate them from their conspecifics while being rewarded with highly preferred food treats. Eventually, computerized test sessions with detachable apparatuses mounted to the cage front were performed and lasted 90–120 minutes. Even though the capuchins only had up to 2 hours of daily access to the cage-mounted computer, the time to complete experiments (2 days to 4 weeks) with up to 1500 trials in 120 minutes compared well with that of the macaques with 24/7 access to the same computerized testing apparatus (Evans et al. 2008). Group-housed macaques also participated in very similar joystick and touchscreen computerized experiments, accomplished simply by separating the test animal inside their home cage only during the trials (author's personal experience). This type of cage-side research requires significant collaboration between facility managers, animal care staff, and researchers. One disadvantage with this technology is the complication posed for adequately sanitizing caging without damaging research equipment (author's personal experience). The advantages include savings in research staff labor and training; elimination of animal stress from being removed from the home cage; less handling of a potentially dangerous, fully awake nonhuman primate; increased research data in relatively less time; and enhanced animal welfare due to animal choice and control while participating in the research.

Telemetry technology provides opportunities to continuously monitor a wide range of physiologic parameters in social or single-housed animals with surgically implanted devices and multifrequency data loggers in the animal room (Hawkins et al. 2004). Examples of physiologic parameters monitored with telemetry equipment include brain function (electroencephalograms), eye movement (electro-oculograms), body temperature, blood pressure, intraocular pressure and heart rate (Lane et al. 1999; Reinhart et al. 2002; Downs et al. 2011; Zhdanova et al. 2011). Modifications to animal housing or the room to accommodate radio frequency identification (RFID) and data loggers are necessary to use this technology.

Rabbits

Welfare

Space

The minimum floor space recommended per rabbit and internal cage height required by ETS 123 are much greater than the dimensions indicated in the *Guide* or AWA Regulations (Table 22.3).

Preference assessments, as described for people and nonhuman primates and considered for rabbits, assist the caregiver with providing an environment most enriching to the animal or human subject. When given a choice of cage heights, 112 socially housed 5-week-old weanling rabbits in cage blocks of 12 or 16 rabbits per square meter preferred 40 cm when active and 20 cm when resting, while completely avoiding open-top cages. Rabbit ear lesions in this same study, which are a reflection of interrabbit aggression and poor welfare, were minimal in cages with a 30 cm height compared with a 20 cm height (Princz et al. 2008a). Sixty-three percent of group-housed rabbits from 5 to 13 weeks old preferred housing with walls half-covered with mirrors, irrespective of stocking densities of either 12 or 16 rabbits per square meter. Seventy-two percent of single-housed rabbits in the same age range also preferred caging with mirrors on walls, presumably promoting comfort and good welfare by giving the rabbits a choice (Dalle Zotte et al. 2009). Platforms inside the cage enhanced natural exploratory behaviors, with rabbits spending half of their time on the raised platform compared with the floor (Dalle Zotte et al. 2009). Benefits to animal welfare were demonstrated in a study in which single-housed, female New Zealand White rabbits in cages

TABLE 22.3

Minimum Cage Space for Rabbits per the *Guide*, AWA Regulations, and ETS 123

	Guide Floor Height, m² (cm)	AWA Regulations Floor Height, m² (cm)		ETS 123 Floor Height, m² (cm)
<2 kg	0.14 (40.5)	0.14 (35.56)	7–10 weeks	0.12 (40)
2–4 kg	0.28 (40.5)	0.28 (35.56)	<3 kg	0.35 (45)
>4–5.4 kg	0.37 (40.5)	0.37 (35.56)	3–5 kg	0.42 (45)
>5.4 kg	≥0.46 (40.5)	0.46 (35.56)	>5 kg	0.54 (60)

enriched with raised platforms showed much fewer stress behaviors, such as bar gnawing, restlessness, and excessive grooming, than rabbits in conventional cages (Hansen and Berthelsen 2000). Formally evaluating preferences and publishing outcomes will assist the industry by providing scientific evidence to promote best practices in the design of rabbit housing. Floor space provided to socially housed rabbits in low stocking densities (6.2 rabbits per square meter) in groups of 10, 10–15, or 40 from 60 to 80 days of age resulted in medium to severe wounding in 7%, 6.5%, and 21%, respectively, regardless of gender composition (Trocino 2006). However, no pathologies or lesions were observed in another study where rabbits were housed at a stocking density of 15 rabbits per square meter in groups of 6 per cage, 10 per pen, or 60 per pen if euthanized at 72 days of age (Postollec et al. 2003). Rabbit age advanced beyond 72 days (10.3 weeks), not floor space, was the primary risk factor for aggression-related injuries for socially housed rabbits. Floor space alone is, however, positively correlated with rabbit welfare indices. Rabbits individually housed in caging with floor spaces of 0.88, 1.68, and 3.35 m^2 were more active and interacted with environmental enrichment in cages providing greater floor space (Dixon et al. 2010).

Enrichment

The New Zealand White rabbit (*Oryctolagus cuniculus*) is the only domestic animal whose behavior may be assessed by the behavior of its wild counterpart, specifically the European wild rabbit (*O. cuniculus*) (Lehmann 1991; Trocino and Xiccato 2006). While extrapolation of European wild rabbit behavioral traits directly to the domesticated New Zealand White rabbit may not be accurate for every situation, especially in laboratory caging (Princz et al. 2008b), the social behavioral similarities specific to aggression are striking (Myers and Poole 1959, 1961; Mykytowycz 1959, 1960). Extrapolation from a different genus or species of wild rabbits (e.g., the Eastern cotton tail [*Sylvalagus floridanus*] and the pygmy rabbit [*Brachylagus idahoensis*]) would be less appropriate. Considering the wild rabbit's natural space preferences, social choices, foraging, and courtship behaviors when designing enrichment strategies in the research laboratory is reasonable (Trocino 2006). An awareness of the species' natural history, or telos, as described in the introduction of the chapter, well positions the facility manager for success when designing housing and management practices. Foraging, a natural behavior, in hay or hay cubes not surprisingly decreases stereotypies in single or socially housed rabbits (Trocino 2006). A similar successful pattern was seen where single-housed male rabbits most preferred hay for enrichment, followed by hay cubes as a second choice, promoting more activity and fewer stereotypies, such as bar biting and excessive fur licking, than in rabbits with chew sticks and controls with no enrichment (Lidfors 1997). In contrast, rabbits housed alone in barren cages all demonstrated boredom behaviors, represented by being inactive 56% of the time and otherwise showing stereotypies, such as hair and bar chewing, head swaying, and pawing (Gunn and Morton 1995). Ninety-six single-housed rabbits equally distributed between two cage systems with identical floor space (3.5 m^2), but with one cage enhanced with 80 cm vertical height and a nest box at one end, showed a preference and improved welfare in the enhanced cage (Hansen and Berthelsen 2000). Adding chewable toys such as cardboard rolls, cardboard rings, and rubber balls with a bell inside all had equally positive behavioral effects in single-housed male New Zealand White rabbits, with time spent chewing on toys instead of displaying stereotypies, as seen with the control rabbits (Poggiagliolmi et al. 2011). Frequent handling (three times a day) of individually caged or pen-housed rabbits by a familiar person over only 5 days demonstrated a clear benefit, with a reduction of fear in the rabbit, while strengthening the human–animal bond (Podberscek et al. 1991a). Offering enrichment is never completely risk-free, as evidenced by a report of a plastic whiffle ball becoming lodged between the lower and upper incisors, causing gingival damage and requiring removal (Shomer et al. 2001). In choosing enrichment and management strategies, one must assess the risk for deleterious events to occur and consider the potential for avoiding psychological harm and negative effects on research by trying novel management and enrichment practices.

Social Housing

The ETS 123 provisions offer a global exemption from social housing for intact, adult male rabbits with each other. The guidelines emphasize that young littermates of either gender and juvenile to adult females may be group housed in pens, with a caveat that adult females need to be observed carefully and managed for aggression (ETS 123). Aggression is reduced, but not eliminated, when rabbits have opportunities to isolate or hide from each other (Reinhardt 2004). The *Guide* and AAALAC International indicate that social housing for socially compatible animals should be the default if they are communal in nature (NRC 2011; AAALAC International 2015). Immature rabbits less than 12 weeks old, especially littermates, are almost always compatible in social groups. For this reason, meat production rabbitries socially house New Zealand White rabbits in mixed gender groups only up to 80 days (11.5 weeks) to make the most of the fattening market weight while minimizing premature culling due to aggression (Postollec 2003; Princz et al. 2008b; Trocino 2006). Adult wild European rabbits (*O. cuniculus*) of either gender are not often socially compatible and demonstrate aggression and severe fighting even if housed in multiacre outdoor enclosures (Mykytowycz 1959, 1960), possibly explaining the unpublished successes reported of pairing antibody-producing female rabbits over years in cages or pens. Cases of severe aggression causing abandonment of further attempts to socially house same-sex, adult rabbits of either gender, even in large, complex pens, underscore the importance of respecting the species' behavior in the wild (Rommers et al. 2014; author's personal experience). Neutering male rabbits that are socially housed has reduced but not eliminated aggression (Raje and Stewart 1997). Once hierarchies are established in wild populations in large areas, aggression is infrequent (Trocino 2010). In conclusion, domestic rabbits behave similarly to their ancestral wild counterparts. If given the amount of space typically available in nature, they situationally fight and/or retreat, respecting the dominance social hierarchy. This survival adaptation of retreat from aggression is not possible in confined laboratory pens or cages (Albonetti et al. 1990; Princz et al. 2008a), justifying single housing of rabbits of either gender greater than 80 days old. The welfare benefit of social or pair housing juvenile rabbits less than 12 weeks old is clear. After 12 weeks, all male rabbits should be exempt from social housing due to incompatibility, as indicated by the ETS 123 provisions. Rabbits of either gender greater than 12 weeks of age should be exempt from social housing as practiced by the meat industry to prevent trauma from aggression (Postollec et al. 2003; Trocino 2003). At best, adult female rabbits in the wild may mutually tolerate each other (Myers and Poole 1959, 1961) possibly explaining the unpublished successes reported of pairing antibody producing, female rabbits over years in cages or pens. For this reason, select, mature female rabbits may appear mutually tolerant of each other over years (author's personal experience), but identifying compatible pairs comes with a welfare cost of failed attempts; thus, a harm–benefit assessment should be made by the stakeholders, including the research staff, manager, attending veterinarian, animal care staff, and IACUC. Single-housed rabbits benefit from enriched cages (Lidfors 1997) and, if feasible, time alone locomoting in a floor pen, promoting activity and guarding against osteoporosis and bone abnormalities seen in inactive rabbits (Chu et al. 2004).

Husbandry and Cost–Benefit

Social housing of six rabbits of either gender less than 12 weeks old in a pen that provides approximately 4–6 ft^2 per rabbit (concurrent with 0.35 m^2 per Appendix A, ETS 123) promotes desirable, affiliative, species-typical behaviors, such as rearing, hopping, and nuzzling (author's personal experience). As beneficial as pen housing immature rabbits has proven to be, the labor associated with spot cleaning bedding and complete replacement of bedding and sanitation of the room at 2-week intervals is 10-fold greater per rabbit compared with cage housing, even in pairs, due to inefficient hand cleaning of pen surfaces and the opportunity cost associated with low-density housing in a pen room compared with three-tier rabbit caging (author's personal experience). A cost-effective alternative to housing immature rabbits in pens is pair housing in commercial cages, which may be sanitized in a mechanical cage washer. Rabbits thumping their hind feet in metal and plastic cages generate surprisingly loud, high-intensity sounds up to 78 dB (10–40 kHz) and 60 dB, respectively (Milligan et al. 1993). Attempts to lessen these sound levels may provide the benefit of decreased stress, especially in kit rearing settings, and should be considered a potential experimental variable (Suckow et al. 2010).

Supporting Research and Emerging Technologies

Radiotelemetry technology provides continuous monitoring of physiological data in many laboratory animal species, including rabbits. Conventional tonometry (intraocular pressure measurement) in rabbits traditionally requires removing the animal from its home cage and placing a tonometer on the surface of the topically anesthetized cornea. A more refined technology allows continuous monitoring of intraocular pressure for up to a year using a surgically implanted pressure transducer. The intraocular pressures are broadcast by amplitude modulation to one or more receivers inside the animal cage (McLaren et al. 1996). Similar monitoring of blood pressure, heart rate, and oxygen levels using surgically implanted transmitter sensors is possible (Sato et al. 1995; Ward et al. 2002). The advantage of this emerging telemetry technology is that it allows continuous monitoring without removing the animal from its home cage, but it requires specialized broadcast technologies or RFID systems in the animal room.

Canines

Welfare

Space

Animal welfare is promoted when choices and control by the animal are possible (Sumpter et al. 2002). Concurrent preference assessments or choice studies, as done with primates and rabbits specific to housing preferences, need to be further explored for dogs (Hetts 1991). Floor space restriction to 1.7 m^2 for 6 weeks with no visual contact with other dogs produced progressively increased salivary and urinary cortisol levels throughout the 6 weeks after being housed outdoors for 7 weeks in groups in large (36 m^2) grassy pens (Beerda et al. 1999a). Behavioral indices of stress, including lower postures, autogrooming, paw lifting, vocalizing, coprophagy, and repetitive behavior, were increased for the spatially restricted and socially isolated dogs compared with when these same dogs were socially housed outdoors in spacious pens (Beerda et al. 1999b). These findings suggest that both space restriction and social isolation create a state of chronic stress and poor welfare in dogs. The trend of the European standards to require more floor and vertical space holds true for dogs, as with the other species covered in this chapter. Providing more housing space is intuitively sensible as a tool to enhance animal welfare, but scientific studies evaluating dog preferences specific to housing design and novel housing strategies are needed (Tables 22.4 through 22.6).

Enrichment

Humans have thousands of years of experience interacting with dogs, often in mutually enriching ways. Social experiences with canine conspecifics and humans are widely accepted as being enriching for both humans and dogs. Socialization and obedience training programs of dogs performed by animal care staff enrich the dogs and staff and result in better research models (Adams et al. 2004). Inanimate options,

TABLE 22.4

Minimum Cage Size for Dogs per the *Guide*

Weight, kg	Floor Area, m^2	Height
<15	0.74	Stand erect with feet on floor
Up to 30	1.2	Stand erect with feet on floor
>30	≥2.4	Stand erect with feet on floor

TABLE 22.5

Minimum Cage Size for Dogs per AWA Regulations

Floor Space, ft^2	Height
Mathematical square of sum of length of dog in inches (tip of nose to base of tail) plus 6 in., with product divided by 144	6 in. higher than head when standing normally

TABLE 22.6

Minimum Cage Size for Dogs per ETS 123

Weight, kg	Minimum Enclosure Size, m^2	Minimum Floor Area per Dog, m^2	Minimum Height, m
Up to 5	4	0.5	2
>5–10	4	1.0	2
>10–15	4	1.5	2
>15–20	4	2	2
>20	8	4	2

including olfactory and auditory stimuli, chew toys, platforms, and time outdoors, are also considered to be possible ways to reduce boredom and encourage play. Toys that are chewable and make noise are preferred by laboratory dogs. Suspending these toys not only reduces labor needed for cleaning but also reduces possessive aggression between dogs (Wells 2004). A more complex, three-dimensional design of the cage or pen with shelving, elevated platforms, and ramps provides choice and an expanded view of their surroundings, reducing patterned jumping to presumably see over solid cage walls (Hubrecht 1993, 2002). Pens or caging should allow dogs in socially compatible groups to sleep away from urine and feces, preferably on a platform or bed, and see activities in the room (Hubrecht 1995; Dean 1999). Simple recycled, fleece-lined, polyethylene barrels have been preferred by older beagles over traditional benches (Eisele 2001). Particular types of auditory stimulation can be beneficial to laboratory-housed canines. For example, playing 4 hours of classical music was associated with a calming effect, in contrast to heavy metal rock music, which encouraged barking (Wells et al. 2002). Similarly, olfactory enrichment with lavender and chamomile resulted in a calming effect compared with rosemary and peppermint (Graham et al. 2005). In addition to the use of animate enrichment with people or other dogs, there will always be a need for novel and effective inanimate enriching alternatives in the kennel (Taylor and Mills 2007). Dogs have been enrolled in interesting concurrent preferences assessment studies demonstrating, for example, that they far prefer being petted to being vocally praised (Feuerbacher and Wynne 2015), but show a greater preference for being offered food over being petted (Feuerbacher and Wynne 2014). These two specific preference assessment studies provide insight for targeting impactful management to promote laboratory dog welfare.

The AWA Regulations require that the research institutions develop an exercise program approved by the attending veterinarian (Kulpa-Eddy et al. 2005). The opportunity for exercise may be accomplished by providing groups of compatible dogs access to pens or runs with 100% minimum floor space per dog or always providing individually housed dogs with twice the minimum space required by USDA AWA Regulations or providing access to an exercise area prescribed by the attending veterinarian. ETS 123 also recognizes the value of offering dogs the opportunity for exercise by encouraging staff to take dogs to an area separate from their housing for exercise in compatible groups with human interaction. When given a choice, research beagles spent 32% of their day in an outdoor yard voluntarily moving between the inside and outside 103 times daily, resulting in increased activity and presumably improved welfare (Spangenberg et al. 2006). The welfare impact of exercise (20 minutes, 3 days per week for 12 weeks) alone in caged (1.12 m^2) beagles was evaluated by measuring behavioral and physiological parameters. Four groups of dogs were exercised individually, with another dog taken to the exercise room and immediately returned to the cage or nonexercised as a cage control. All groups showed similar trends in parameters measured over time, supporting a conclusion of no effect of exercise on well-being with one exception. Both groups of exercised dogs barked more often on nonexercise days than nonexercised dogs (Clark et al. 1997). This pattern of barking may suggest that the exercised dogs were soliciting the opportunity with a preference to leave the cage to the exercise pen alone or with another dog.

Social Housing

As with the other species in this chapter, the consideration of the ancestral counterpart's natural history, in the dog's case the southern wolf (*Canis lupus pallipes*), provides insight into beneficial housing or management strategies. With 14,000 years of domestication, the social nature of the dog with other dogs and humans is intrinsic (Hubrecht 1995). ETS 123 encourages social housing of compatible dogs. Efforts to evaluate social housing and enrichment opportunities for laboratory animals, including dogs, consistent with the goals of toxicological studies, which historically prohibited such practices, are underway (Dean 1999; Bayne 2003; Turner et al. 2003). Overnight social housing from 3:00 p.m. to 8:00 a.m. promotes desirable social behaviors in canine subjects separated only for dosing and collecting clinical data for approximately 4 hours during the workday (Mack et al. 2003). Socially housing research dogs in pairs and larger groups either continuously or intermittently is the accepted best welfare and research practice (Hetts et al. 1992; Hubrecht et al. 1992; Beerda et al. 1999a, 1999b; Bayne 2003). Dogs group housed in a shelter led to a significant reduction in barking, displayed a friendlier interaction with people, and had fewer behavioral problems and stereopathies than individually housed dogs (Mertens and Unshelm 1996). Social isolation and space restriction produced chronic stress and poor animal welfare, as described above in the "Space" section (Beerda et al. 1999a, 1999b).

Husbandry

Incorporating operant conditioning and training of dogs benefits the research, promotes good animal welfare, and strengthens the bond of cooperation between personnel and dogs. Training dogs for experimental manipulations, including oral or intravenous dosing, blood collections, and stationing for imaging or body weights, results in a much more positive, easily repeatable experience (Hubrecht 2002). Separation of socially housed dogs for feeding or movement to another area of the pen or run may be accomplished by creating flexibility in the pen or cage design with interconnecting passages, which may be opened or closed depending on the circumstances. All such pens or caging should be able to accommodate at least two dogs (Hubrecht 2002).

Barking may cause prolonged sound levels in excess of 100 dB, requiring animal care personnel to wear hearing protection, as well as possibly distressing or causing hearing damage in some dogs (Hubrecht 1995). As described under the social housing section, a benefit realized with group housing dogs in a shelter was significantly less barking than single-housed dogs (Mertens and Unshelm 1996), an assessment that should be made in the research setting.

Cost–Benefit

When comparing environmental enrichment techniques, including social housing, a rotating selection of chew toys, and 30 seconds of daily human grooming and handling, Hubrecht (1993) determined that the dogs preferred the alternating selection of chew toys. Data from preference studies may provide direction for cost-effective enrichment strategies, especially for large facilities. As with nonhuman primates, wall- or floor-mounted dog pens with connecting swing gates are relatively inexpensive to build when contracted to local fencing companies. Such pens, however, pose sanitation challenges, as they require much more labor to manually clean, as well as more scrutiny on microbiological monitoring for assessing the effectiveness of sanitation compared with movable, compartmentalized pens, which may be disassembled and passed through a mechanical cage washer (author's personal experience). The ATP bioluminescent tests serve as reliable alternatives to bacterial culturing (e.g., on RODAC plates) of sanitized surfaces, providing immediate results for interpretation of bacterial contamination after sanitation (Ednie et al. 1998; Turner et al. 2010). Institutions should project ongoing labor costs associated with sanitizing fixed or wall- or floor-mounted pens before deciding against purchasing the more costly, but more easily sanitizable and durable movable pen units. Outdoor pens or walking dogs outdoors provides novel exercise opportunities for dogs, as well as a morale booster for staff (Hubrecht 2002).

Supporting Research and Emerging Technologies

Continuous telemetric monitoring of physiological parameters in dogs has emerged as a refinement to avoid the confounders, labor, and potential stress associated with otherwise removing the animals from their housing environment to gather recordings. The challenge to modify pens or caging to accommodate water-resistant, dog-proof yet removable receivers has been easily overcome with advanced technology.

Collecting data simultaneously from pair-housed dogs has been challenging, with concerns of data being confounded by pair housing, in addition to technical complications related to recording from the same frequency from two dogs with existing radiotelemetry equipment. In these situations, paired dogs would have to be separated during telemetry recordings. However, modification of commercially available digital radio transmitters (PhysioTel™) allowed cardiovascular data to be collected from paired dogs simultaneously using different frequencies, producing results consistent with measurements taken in single-housed dogs (Prior et al. 2015). Research priorities, including sleep apnea, hypertension, long-term blood pressure measurements, and recording of cardiovascular parameters during pharmacologic studies, have all been met by optimizing animal housing, producing high-quality data collection, and upholding animal welfare in both single- and socially housed dogs (Brooks et al. 1996, 1997a, 1997b; Klumpp et al. 2006).

Felines

Welfare

Space

The space afforded to cats, especially when socially housed, greatly assists with conflict avoidance. Provision of 10 m^2 per cat successfully promoted compatibility in a group of 14 neutered individuals (Bernstein and Strack 1996). While this floor space is 27- and 13-fold greater than the minimum required by the AWA Regulations and ETS 123 Appendix A, as shown in Table 22.7, much smaller spaces may promote compatibility when choices to avoid each other are available (Rochlitz 2000). Abundant horizontal resting platforms and surfaces at different levels promoted the natural behavior of cats preferring to rest alone (Podberscek et al. 1991b; Rochlitz 2000). Providing socially housed cats with a minimum distance of 1–3 m from each other in all directions horizontally and vertically is ideal (Ellis 2009; Herron and Buffington 2010). Sharing the same spots in an enclosure or pen at different times, sometimes only within specific groups, has been observed (Bernstein and Strack 1996). Exceeding a group density of 0.6 animals per square meter (1 cat per 1.6 m^2) produced high stress levels in cats housed in boarding catteries or shelters, emphasizing the importance of acclimating the cats to each other before allowing free contact (Kessler and Turner 1999).

Enrichment

Successful enrichment provides cats with opportunities to demonstrate natural behaviors (Herron and Buffington 2010). Vertical substrates, scratching posts, and items to pick up, toss, smack, chase, and pounce on are good examples of enrichment. Offering food in puzzle feeders or balls that release dry food pellets when manipulated promotes an enriching experience simulating hunting (Ellis 2009; Herron and Buffington 2010). Group-housed cats do not demonstrate aggression or competition when presented with enrichment toys but instead perform group exploratory behaviors (Dantas-Divers et al. 2011). Visual enrichment, such as watching prey species (e.g., birds) from a window in situations where photoperiod and security are not critical, has been replicated with television screens playing videos of birds, mice prey species, or cats playing together (Ellis 1990). Olfactory enrichment, including catnip, which 50%–70% of cats are genetically sensitive to, has been used in research animal care and use

TABLE 22.7

Minimum Cage Size for Cats per the *Guide*, AWA Regulations, and ETS 123

Guide Weight, kg	*Guide* Floor, m^2	*Guide* Height, cm	AWA Floor, m^2	AWA Height, cm	ETS 123	ETS 123 Floor, m^2	ETS 123 Height, cm	ETS 123 Shelves, m^2
≤4	0.28	60.8	0.28	60.96	1 adult	1.5	200	0.5
>4	≥0.37	60.8	0.37	60.96	Each additional	0.75	n/a	0.25

programs (Ellis 1990). Commercially available pheromones such as Feliway, a synthetic analog of the F3 feline facial pheromone (FFP) secreted by chin, lip, and vibrissae glands, reduced stress, fear, anxiety, and aggression in cats when sprayed on examination tables and kennels in unfamiliar venues, such as a veterinary practice (Rodan et al. 2011; Hewson 2014; Pereira et al. 2015). Another synthetic feline facial pheromone (Felifriend F4) produced the same calming state in cats in a rescue shelter (Patel et al. 2010). These pheromones have been used to enrich cats in research housing (Ellis 1990). Placing a large viewing window in the cat room door promotes tremendously enriching interaction between staff and cats (Ellis 2009; author's personal experience). However, the corridor light cycle must be in synchrony with the animal room's cycle.

Social Housing

With an understanding of the natural history of the cat's evolution from the African wildcat (*Felis sylvestris*), informed decisions may be made how to best meet housing and management conditions in the laboratory environment (Rochlitz 2000; Jongman 2007). Even though the African wildcat is solitary, the domestic cat may be maintained in a social environment if enough space is provided to promote a wide range of natural behaviors. These behaviors include playing, chasing, resting, sleeping, exploring, and the choice to hide. Neutered male and female cats that were pair housed in households demonstrated a wide range of individual relationships with no gender-specific differences in aggressive or affiliative behaviors. However, male pairs spent more time in close proximity, while female pairs were never observed to allorub (i.e., chin or head bump) each other (Barry and Crowell-Davis 1999). Lessons learned from cat behavior in household settings may be useful when establishing social housing of laboratory animals. Cats evade conflict by avoiding each other, which requires additional space and visual barriers (Herron and Buffington 2010). Stress-related behavior, specifically prolonged autogrooming, persists in social settings where a cat experiences long-lasting uncontrollable distress (van den Bos 1998). Stress in socially housed cats, which may be objectively scored, is driven more by familiarity and unrelatedness, especially in rescue shelters, which identify higher levels of stress in communal rather than discrete-unit housing (Ottway and Hawkins 2003).

Husbandry

Enough space (both horizontal and vertical) must be afforded, especially in a social environment for distribution of food, water, and litter pans; climbing structures; and platforms, so cats may choose to be far from each other to avoid conspecific conflict and aggression (Rochlitz 2000). Socially housed cats should be weighed often to confirm equal opportunity for food consumption. Five basic systems of the cat's housing environment influence welfare: space, nutrition, defecation and urination, social, and behavioral (Herron and Buffington 2010). A key feature of successful space planning for cats involves choice and control over perching or hiding options. Feeding socially housed cats in separate bowls hidden from view of each other promotes a more natural predatory behavior. Separate litterboxes should be provided for each socially housed cat plus one additional box, all preferably out of sight of each other, scooped out daily, and completely cleaned weekly. Interactions with animal care staff significantly influence urinary cortisol levels, a physiological parameter for stress. When animal care staff stopped talking to and petting single-housed laboratory cats, as well as cleaned and fed them on a very irregular schedule over 3 weeks, elevated urinary cortisol levels confirmed chronic stress compared with a control colony of cats. Stressed cats demonstrated reduced play and exploratory behavior and increased withdrawal and hiding behavior (Carlstead et al. 1993). The daily schedule and interactions between laboratory cats and care staff may significantly confound research outcomes and affect animal welfare. However, socially housed cats in a laboratory setting demonstrated more exploratory behavior with an unfamiliar caretaker on the first day, and then on subsequent days displayed diminished interest unless the person interacted with the cats in a playful way (Podberscek et al. 1991a). Early socialization of cats with humans influences successful interactions with humans later in life (Ellis 2009). The daily impact that animal care staff has on all species of laboratory animals is validated scientifically in such studies.

Cost–Benefit

The husbandry cost efficiencies or disadvantages associated with cleaning a room housing free-ranging cats compared with movable caged (individually or socially housed) cats will vary depending on swing space to move all animals out and types of movable caging transported to a mechanical cage washer. The most cost-effective process that promotes the best animal welfare involves free-ranging cats in a room with sanitizable or disposable surfaces or materials (e.g., plastic furniture and stretched canvas platforms) combined with standard stainless steel cages with open doors and litter pans, which may be taken to a cage wash machine at 2-week intervals. This strategy provides cats with choices and control while maximizing efficient and effective sanitation of movable enclosures. With this housing method, cats may be separated temporarily, if needed, singly, or in pairs inside a traditional cage for experimental reasons and otherwise freely exercise in a highly enriched room. The cost of operation of this housing option is no different than continually housing the cats in the cages since the cages are taken to cage wash at 2-week intervals and the floor is mopped as needed with cats placed inside the cages (author's personal experience). The ideal room design for socially housed cats includes a safety gate or vestibule at the entrance to prevent animals from escaping when the door is opened.

REFERENCES

AAALAC International. AAALAC position statement and frequently asked questions (FAQ) on social housing. https://www.aaalac.org/accreditation/faq_landing.cfm#C6 and https://www.aaalac.org/accreditation/positionstatements.cfm#social (accessed December 1, 2015).

Adams, K., A. M. Navarro, and E. K. Hutchinson. A canine socialization and training program at the National Institutes of Health. *Lab Animal* 33, no. 1 (2004): 32.

Albonetti, M. E., F. Dessi-Fulgheri, and F. Farabollini. Intrafemale agonistic interactions in the domestic rabbit (*Oryctolagus cuniculus L.*). *Aggressive Behavior* 16 (1990): 77–86.

Baker, K. C., M. A. Bloomsmith, B. Oettinger, K. Neu, C. Griffis, V. Schoof, and M. Maloney. Benefits of pair housing are consistent across a diverse population of rhesus macaques. *Applied Animal Behaviour Science* 137, no. 3 (2012): 148–156.

Barry, K. J., and S. L. Crowell-Davis. Gender differences in the social behavior of the neutered indoor-only domestic cat. *Applied Animal Behaviour Science* 64, no. 3 (1999): 193–211.

Bassett, L., H. M. Buchanan-Smith, J. McKinley, and T. E. Smith. Effects of training on stress-related behavior of the common marmoset (*Callithrix jacchus*) in relation to coping with routine husbandry procedures. *Journal of Applied Animal Welfare Science* 6, no. 3 (2003): 221–233.

Baumans, V. Environmental enrichment for laboratory rodents and rabbits: Requirements of rodents, rabbits, and research. *ILAR Journal* 46, no. 2 (2005): 162–170.

Bayne, K., G. S. Ramachandra, E. A. Rivera, and J. Wang. The evolution of animal welfare and the 3Rs in Brazil, China, and India. *Journal of the American Association for Laboratory Animal Science* 54, no. 2 (2015): 181–191.

Bayne, K. A. Environmental enrichment of nonhuman primates, dogs and rabbits used in toxicology studies. *Toxicologic Pathology* 31, no. 1 (Suppl.) (2003): 132–137.

Bayne, K. A. L., and R. T. Davis. Susceptibility of rhesus monkeys (*Macaca mulatta*) to the Ponzo illusion. *Bulletin of the Psychonomic Society* 21, no. 6 (1983): 476–478.

Beerda, B., M. B. H. Schilder, J. A. Van Hooff, H. W. De Vries, and J. A. Mol. Chronic stress in dogs subjected to social and spatial restriction. I. Behavioral responses. *Physiology & Behavior* 66, no. 2 (1999a): 233–242.

Beerda, B., M. B. H. Schilder, W. Bernadina, J. A. Van Hooff, H. W. De Vries, and J. A. Mol. Chronic stress in dogs subjected to social and spatial restriction. II. Hormonal and immunological responses. *Physiology & Behavior* 66, no. 2 (1999b): 243–254.

Bennett, A. J., C. A. Corcoran, V. A. Hardy, L. R. Miller, and P. J. Pierre. Multidimensional cost–benefit analysis to guide evidence-based environmental enrichment: Providing bedding and foraging substrate to pen-housed monkeys. *Journal of the American Association for Laboratory Animal Science* 49, no. 5 (2010): 571–575.

Bernstein, P. L., and M. Strack. A game of cat and house: Spatial patterns and behavior of 14 domestic cats (*Felis catus*) in the home. *Anthrozoös* 9, no. 1 (1996): 25–39.

Brambell, R. *Report of the Technical Committee to Enquire into the Welfare of Animals Kept under Intensive Livestock Husbandry Systems*. London: Her Majesty's Stationery Office, 1965, 1–84.

Brooks, D., R. L. Horner, L. F. Kozar, C. L. Render-Teixeira, and E. A. Phillipson. Obstructive sleep apnea as a cause of systemic hypertension. Evidence from a canine model. *Journal of Clinical Investigation* 99, no. 1 (1997a): 106–109.

Brooks, D., R. L. Horner, L. F. Kozar, C. L. Render-Teixeira, and E. A. Phillipson. Obstructive sleep apnea as a cause of systemic hypertension. Evidence from a canine model. *Journal of Clinical Investigation* 99, no. 1 (1997b): 106.

Brooks, D., R. L. Horner, L. F. Kozar, T. K. Waddell, C. L. Render, and E. A. Phillipson. Validation of a telemetry system for long-term measurement of blood pressure. *Journal of Applied Physiology* 81, no. 2 (1996): 1012–1018.

Bryant, C. E., N. M. J. Rupniak, and S. D. Iversen. Effects of different environmental enrichment devices on cage stereotypies and autoaggression in captive cynomolgus monkeys. *Journal of Medical Primatology* 17, no. 5 (1988): 257–269.

Buccafusco, J. J., A. V. Terry, and P. B. Murdoch. A computer-assisted cognitive test battery for aged monkeys. *Journal of Molecular Neuroscience* 19, no. 1–2 (2002): 179–185.

Buchanan-Smith, H. M., and I. Badihi. The psychology of control: Effects of control over supplementary light on welfare of marmosets. *Applied Animal Behaviour Science* 137, no. 3 (2012): 166–174.

Buchanan-Smith, H. M., M. J. Prescott, and N. J. Cross. What factors should determine cage sizes for primates in the laboratory? *Animal Welfare* 13 (2004): S197–S201.

Buchanan-Smith, H. M., C. Shand, and K. Morris. Cage use and feeding height preferences of captive common marmosets (*Callithrix j. jacchus*) in two-tier cages. *Journal of Applied Animal Welfare Science* 5, no. 2 (2002): 139–149.

Carlstead, K., J. L. Brown, and W. Strawn. Behavioral and physiological correlates of stress in laboratory cats. *Applied Animal Behaviour Science* 38, no. 2 (1993): 143–158.

CCAC (Canadian Council on Animal Care). *Guide to the Care and Use of Experimental Animals*. Ottawa: Canadian Council on Animal Care, 1993.

Chu, L.-R., J. P. Garner, and J. A. Mench. A behavioral comparison of New Zealand White rabbits (*Oryctolagus cuniculus*) housed individually or in pairs in conventional laboratory cages. *Applied Animal Behaviour Science* 85, no. 1 (2004): 121–139.

Clark, J. Derrell, Dawn R. Rager, Sharon Crowell-Davis, and Donald L. Evans. Housing and exercise of dogs: Effects on behavior, immune function, and cortisol concentration. *Comparative Medicine* 47, no. 5 (1997): 500–510.

Coleman, K., L. Pranger, A. Maier, S. P. Lambeth, J. E. Perlman, E. Thiele, and S. J. Schapiro. Training rhesus macaques for venipuncture using positive reinforcement techniques: A comparison with chimpanzees. *Journal of the American Association for Laboratory Animal Science* 47, no. 1 (2008): 37–41.

Crockett, C. M., R. U. Bellanca, C. L. Bowers, and D. M. Bowden. Grooming-contact bars provide social contact for individually caged laboratory macaques. *Journal of the American Association for Laboratory Animal Science* 36, no. 6 (1997): 53–60.

Council of Europe 2006. Appendix A of the European Convention for the Protection of Vertebrate Animals Used for Experimental and Other Scientific Purposes (ETS No. 123), including Article 5 of the convention, Guidelines for Accommodation and Care of Animals. https://www.aaalac.org/about/AppA-ETS123.pdf (accessed December 1, 2015).

Dalle Zotte, A., Z. Princz, Z. Matics, Z. Gerencsér, S. Metzger, and Z. Szendrő. Rabbit preference for cages and pens with or without mirrors. *Applied Animal Behaviour Science* 116, no. 2 (2009): 273–278.

Dantas-Divers, L. M. S., S. L. Crowell-Davis, K. Alford, G. Genaro, J. M. D'Almeida, and R. L. Paixao. Agonistic behavior and environmental enrichment of cats communally housed in a shelter. *Journal of the American Veterinary Medical Association* 239, no. 6 (2011): 796–802.

Dean, S. W. Environmental enrichment of laboratory animals used in regulatory toxicology studies. *Laboratory Animals* 33, no. 4 (1999): 309–327.

Dixon, L. M., J. R. Hardiman, and J. J. Cooper. The effects of spatial restriction on the behavior of rabbits (*Oryctolagus cuniculus*). *Journal of Veterinary Behavior: Clinical Applications and Research* 5, no. 6 (2010): 302–308.

Downs, J. C., C. F. Burgoyne, W. P. Seigfreid, J. F. Reynaud, N. G. Strouthidis, and V. Sallee. 24-Hour IOP telemetry in the nonhuman primate: Implant system performance and initial characterization of IOP at multiple timescales. *Investigative Ophthalmology and Visual Science* 52, no. 10 (2011): 7365–7375.

Ednie, D. L., R. P. Wilson, and C. M. Lang. Comparison of two sanitation monitoring methods in an animal research facility. *Journal of the American Association for Laboratory Animal Science* 37, no. 6 (1998): 71–74.

Eisele, P. H. A practical dog bed for environmental enrichment for geriatric beagles, with applications for puppies and other small dogs. *Journal of the American Association for Laboratory Animal Science* 40, no. 3 (2001): 36–38.

Ellis, S. L. H. Environmental enrichment: Practical strategies for improving feline welfare. *Journal of Feline Medicine and Surgery* 11, no. 11 (2009): 901–912.

Evans, T. A., M. J. Beran, B. Chan, E. D. Klein, and C. R. Menzel. An efficient computerized testing method for the capuchin monkey (*Cebus apella*): Adaptation of the LRC-CTS to a socially housed nonhuman primate species. *Behavior Research Methods* 40, no. 2 (2008): 590–596.

Feuerbacher, E. N., and C. D. L. Wynne. Most domestic dogs (*Canis lupus familiaris*) prefer food to petting: Population, context, and schedule effects in concurrent choice. *Journal of the Experimental Analysis of Behavior* 101, no. 3 (2014): 385–405.

Feuerbacher, E. N., and C. D. L. Wynne. Shut up and pet me! Domestic dogs (*Canis lupus familiaris*) prefer petting to vocal praise in concurrent and single-alternative choice procedures. *Behavioural Processes* 110 (2015): 47–59.

Fillman-Holliday, D., and M. S. Landi. Animal care best practices for regulatory testing. *ILAR Journal* 43, no. (Suppl. 1) (2002): S49–S58.

Franklin, E. G. The training of nonhuman primates with pole and collar for chair work using positive reinforcement. May 2016. http://arizona.openrepository.com/arizona/bitstream/10150/612942/1/azu_etd_mr_2016_0078_sip1_m.pdf (accessed July 31, 2015).

Fraser, D. Animal ethics and animal welfare science: Bridging the two cultures. *Applied Animal Behaviour Science* 65, no. 3 (1999): 171–189.

Fuller, G., L. Sadowski, C. Cassella, and K. E. Lukas. Examining deep litter as environmental enrichment for a family group of wolf's guenons, *Cercopithecus wolfi*. *Zoo Biology* 29, no. 5 (2010): 626–632.

Graham, L., D. L. Wells, and P. G. Hepper. The influence of olfactory stimulation on the behaviour of dogs housed in a rescue shelter. *Applied Animal Behaviour Science* 91, no. 1 (2005): 143–153.

Gunn, D., and D. B. Morton. Inventory of the behaviour of New Zealand White rabbits in laboratory cages. *Applied Animal Behaviour Science* 45, no. 3 (1995): 277–292.

Haywood, J. R., and M. Greene. Avoiding an overzealous approach: A perspective on regulatory burden. *ILAR Journal* 49, no. 4 (2008): 426–434.

Hansen, L. T., and H. Berthelsen. The effect of environmental enrichment on the behaviour of caged rabbits (*Oryctolagus cuniculus*). *Applied Animal Behaviour Science* 68, no. 2 (2000): 163–178.

Hawkins, P., D. B. Morton, R. Bevan, K. Heath, J. Kirkwood, P. Pearce, L. Scott, G. Whelan, and A. Webb. Husbandry refinements for rats, mice, dogs and non-human primates used in telemetry procedures. *Laboratory Animals* 38, no. 1 (2004): 1–10.

Herron, M. E., and C. A. T. Buffington. Environmental enrichment for indoor cats. *Compendium: Continuing Education for Veterinarians* 32, no. 12 (2010): E4.

Hetts, S. Psychologic well-being: Conceptual issues, behavioral measures, and implications for dogs. *Veterinary Clinics of North America: Small Animal Practice* 21, no. 2 (1991): 369–387.

Hetts, S., J. D. Clark, J. P. Calpin, C. E. Arnold, and J. M. Mateo. Influence of housing conditions on beagle behaviour. *Applied Animal Behaviour Science* 34, no. 1 (1992): 137–155.

Hewson, C. Evidence-based approaches to reducing in-patient stress. Part 2. Synthetic pheromone preparations. *Veterinary Nursing Journal* 29, no. 6 (2014): 204–206.

Hubrecht, R. Dogs and dog housing. In *Environmental Enrichment Information Resources for Laboratory Animals: 1965–1995: Birds, Cats, Dogs, Farm Animals, Ferrets, Rabbits and Rodents*. Washington, DC: U.S. Department of Agriculture, 1995, 43–62.

Hubrecht, R. Comfortable quarters for dogs in research institutions. *Comfortable Quarters for Laboratory Animals* (2002): 56–64.

Hubrecht, R. C. A comparison of social and environmental enrichment methods for laboratory housed dogs. *Applied Animal Behaviour Science* 37, no. 4 (1993): 345–361.

Hubrecht, R. C., J. A. Serpell, and T. B. Poole. Correlates of pen size and housing conditions on the behaviour of kennelled dogs. *Applied Animal Behaviour Science* 34, no. 4 (1992): 365–383.

Jongman, E. C. Adaptation of domestic cats to confinement. *Journal of Veterinary Behavior: Clinical Applications and Research* 2, no. 6 (2007): 193–196.

Kessler, M. R., and D. C. Turner. Effects of density and cage size on stress in domestic cats (*Felis silvestris catus*) housed in animal shelters and boarding catteries. *Animal Welfare* 8, no. 3 (1999): 259–267.

Kitchen, A. M., and A. A. Martin. The effects of cage size and complexity on the behaviour of captive common marmosets, *Callithrix jacchus jacchus*. *Laboratory Animals* 30, no. 4 (1996): 317–326.

Klumpp, A., T. Trautmann, M. Markert, and B. Guth. Optimizing the experimental environment for dog telemetry studies. *Journal of Pharmacological and Toxicological Methods* 54, no. 2 (2006): 141–149.

Kulpa-Eddy, J. A., S. Taylor, and K. M. Adams. USDA perspective on environmental enrichment for animals. *ILAR Journal* 46, no. 2 (2005): 83–94.

Lambeth, S. P., J. Hau, J. E. Perlman, M. Martino, and S. J. Schapiro. Positive reinforcement training affects hematologic and serum chemistry values in captive chimpanzees (*Pan troglodytes*). *American Journal of Primatology* 68, no. 3 (2006): 245–256.

Lane, M. A., D. K. Ingram, and G. S. Roth. Calorie restriction in nonhuman primates: Effects on diabetes and cardiovascular disease risk. *Toxicological Sciences* 52, no. suppl 1 (1999): 41–48.

Laule, G. E., M. A. Bloomsmith, and S. J. Schapiro. The use of positive reinforcement training techniques to enhance the care, management, and welfare of primates in the laboratory. *Journal of Applied Animal Welfare Science* 6, no. 3 (2003): 163–173.

Lee, G. H., J. P. Thom, K. L. Chu, and C. M. Crockett. Comparing the relative benefits of grooming-contact and full-contact pairing for laboratory-housed adult female *Macaca fascicularis*. *Applied Animal Behaviour Science* 137, no. 3 (2012): 157–165.

Lehmann, M. Social behaviour in young domestic rabbits under semi-natural conditions. *Applied Animal Behaviour Science* 32, no. 2 (1991): 269–292.

Lidfors, L. Behavioural effects of environmental enrichment for individually caged rabbits. *Applied Animal Behaviour Science* 52, no. 1 (1997): 157–169.

Lloyd-Evans, N., V. S. Springthorpe, and S. A. Sattar. Chemical disinfection of human rotavirus-contaminated inanimate surfaces. *Journal of Hygiene* 97, no. 1 (1986): 163–173.

Luchins, K. R., K. C. Baker, M. H. Gilbert, J. L. Blanchard, and R. P. Bohm. Manzanita wood: A sanitizable enrichment option for nonhuman primates. *Journal of the American Association for Laboratory Animal Science* 50, no. 6 (2011): 884–887.

Mack, P. A., R. M. Bell, B. L. Tubo, J. A. Ashline, and K. L. Smiler. Validation study of social housing of canines in toxicology studies. *Journal of the American Association for Laboratory Animal Science* 42, no. 2 (2003): 29–30.

MacLean, E. L., S. R. Prior, M. L. Platt, and E. M. Brannon. Primate location preference in a double-tier cage: The effects of illumination and cage height. *Journal of Applied Animal Welfare Science* 12, no. 1 (2009): 73–81.

Maier, J. X., J. G. Neuhoff, N. K. Logothetis, and A. A. Ghazanfar. Multisensory integration of looming signals by rhesus monkeys. *Neuron* 43, no. 2 (2004): 177–181.

Maple, T. L., and T. W. Finlay. Applied primatology in the modern zoo. *Zoo Biology* 8, no. S1 (1989): 101–116.

Martin, D. P., T. Gilberto, C. Burns, and H. C. Pautler. Nonhuman primate cage modifications for environmental enrichment. *Journal of the American Association for Laboratory Animal Science* 41, no. 5 (2002): 47–49.

McKinley, J., H. M. Buchanan-Smith, L. Bassett, and K. Morris. Training common marmosets (*Callithrix jacchus*) to cooperate during routine laboratory procedures: Ease of training and time investment. *Journal of Applied Animal Welfare Science* 6, no. 3 (2003): 209–220.

McLaren, J. W., R. F. Brubaker, and J. S. Fitzsimon. Continuous measurement of intraocular pressure in rabbits by telemetry. *Investigative Ophthalmology and Visual Science* 37, no. 6 (1996): 966–975.

McMillan, J. L., J. E. Perlman, A. Galvan, T. Wichmann, and M. A. Bloomsmith. Refining the pole-and-collar method of restraint: Emphasizing the use of positive training techniques with rhesus macaques (*Macaca mulatta*). *Journal of the American Association for Laboratory Animal Science* 53, no. 1 (2014): 61–68.

Mellor, D. J., and A. C. D. Bayvel. New Zealand's inclusive science-based system for setting animal welfare standards. *Applied Animal Behaviour Science* 113, no. 4 (2008): 313–329.

Mertens, P. A., and J. Unshelm. Effects of group and individual housing on the behavior of kenneled dogs in animal shelters. *Anthrozoös* 9, no. 1 (1996): 40–51.

Milligan, S. R., G. D. Sales, and K. Khirnykh. Sound levels in rooms housing laboratory animals: An uncontrolled daily variable. *Physiology & Behavior* 53, no. 6 (1993): 1067–1076.

Myers, K., and W. E. Poole. A study of the biology of the wild rabbit, *Oryctolagus cuniculus* (L.), in confined populations. I. The effects of density on home range and the formation of breeding grounds. *Wildlife Research* 4, no. 1 (1959): 14–26.

Myers, K., and W. E. Poole. A study of the biology of the wild rabbit, *Oryctolagus cuniculus* (L.), in confined populations. II. The effects of season and population increase on behaviour. *Wildlife Research* 6, no. 1 (1961): 1–41.

Mykytowycz, R. Social behavior of an experimental colony of wild rabbits. II. First breeding season., *Wildlife Research* 4 (1959): 1–13.

Mykytowycz, R. Social behaviour of an experimental colony of wild rabbits, *Oryctolagus cuniculus* (L.). III. Second breeding season. *Wildlife Research* 5, no. 1 (1960): 1–20.

Newberry, R. C. Environmental enrichment: Increasing the biological relevance of captive environments. *Applied Animal Behaviour Science* 44, no. 2 (1995): 229–243.

NRC (National Research Council). Risk management: The principles underlying the design and implementation of an occupational health and safety plan. In *Occupational Health and Safety in the Care and Use of Nonhuman Primates*. Washington, DC: National Academies Press, 2003, chap. 7.

NRC (National Research Council). *Guide for the Care and Use of Laboratory Animals*. 8th ed. Washington, DC: National Academies Press, 2011.

Ogura, T., and T. Matsuzawa. Video preference assessment and behavioral management of single-caged Japanese macaques (*Macaca fuscata*) by movie presentation. *Journal of Applied Animal Welfare Science* 15, no. 2 (2012): 101–112.

Ottway, D. S., and D. M. Hawkins. Cat housing in rescue shelters: A welfare comparison between communal and discrete-unit housing. *Animal Welfare* 12, no. 2 (2003): 173–189.

Patel, G., S. Heath, K. Coyne, and A. C. German. Pilot study to investigate whether a feline pheromone analogue reduces anxiety-related behavior during clinical examination of cats in a rescue shelter. *Journal of Veterinary Behavior: Clinical Applications and Research* 5, no. 1 (2010): 33.

Pereira, J. S., S. Fragoso, A. Beck, S. Lavigne, A. S. Varejão, and G. da Graça Pereira. Improving the feline veterinary consultation: The usefulness of Feliway spray in reducing cats' stress. *Journal of Feline Medicine and Surgery* (2015): 1–6.

Podberscek, A. L., J. K. Blackshaw, and A. W. Beattie. The effects of repeated handling by familiar and unfamiliar people on rabbits in individual cages and group pens. *Applied Animal Behaviour Science* 28, no. 4 (1991a): 365–373.

Podberscek, A. L., J. K. Blackshaw, and A. W. Beattie. The behaviour of laboratory colony cats and their reactions to a familiar and unfamiliar person. *Applied Animal Behaviour Science* 31, no. 1 (1991b): 119–130.

Poggiagliolmi, S., S. L. Crowell-Davis, L. C. Alworth, and S. B. Harvey. Environmental enrichment of New Zealand White rabbits living in laboratory cages. *Journal of Veterinary Behavior: Clinical Applications and Research* 6, no. 6 (2011): 343–350.

Ponce, C. R., M. P. Genecin, G. Perez-Melara, and M. S. Livingstone. Automated chair-training of rhesus macaques. *Journal of Neuroscience Methods* 263 (2016): 75–80.

Postollec, G., E. Boilletot, R. Maurice, and V. Michel. Influence de l'apport d'une structure d'enrichissement (plate-forme) sur les performances zootechniques et le comportement des lapins élevés en parcs. In *10èmes Journées de la Recherche Cunicole*, Paris, November 2003, 173–176.

Prescott, M. J., V. A. Bowell, and H. M. Buchanan-Smith. Training laboratory-housed non-human primates, part 2: Resources for developing and implementing training programmes. *Animal Technology and Welfare* 4, no. 3 (2005): 133.

Prescott, M. J., and H. M. Buchanan-Smith. Training nonhuman primates using positive reinforcement techniques. *Journal of Applied Animal Welfare Science* 6, no. 3 (2003): 157–161.

Princz, Z., I. Radnai, E. Biró-Németh, Z. Matics, Z. Gerencsér, I. Nagy, and Z. Szendrő. Effect of cage height on the welfare of growing rabbits. *Applied Animal Behaviour Science* 114, no. 1 (2008a): 284–295.

Princz, Z., A. D. Zotte, I. Radnai, E. Bíró-Németh, Z. Matics, Z. Gerencsér, I. Nagy, and Z. Szendrő. Behaviour of growing rabbits under various housing conditions. *Applied Animal Behaviour Science* 111, no. 3 (2008b): 342–356.

Prior, H., R. Billing, I. Wallace, M. South, K. Oldman, J. Moors, M. Skinner, N. Edmunds, and A. Milne. Pair-housed dog telemetry: Animal welfare refinement with early indications of similar study sensitivity. *Journal of Pharmacological and Toxicological Methods* 75 (2015): 182.

Raje, S., and K. L. Stewart. Group housing for male New Zealand White rabbits. *Lab Animal* 26, no. 4 (1997): 36–38.

Reinhardt, V. Space utilization by captive rhesus macaques. *Animal Technology* 43, no. 1 (1992a): 11–17.

Reinhardt, V. Transport-cage training of caged rhesus macaques. *Animal Technology* 43 (1992b): 57–61.

Reinhardt, V. Caged rhesus macaques voluntarily work for ordinary food. *Primates* 35, no. 1 (1994): 95–98.

Reinhardt, V. Working with rather than against macaques during blood collection. *Journal of Applied Animal Welfare Science* 6, no. 3 (2003): 189–197.

Reinhardt, V. Common husbandry-related variables in biomedical research with animals. *Laboratory Animals* 38, no. 3 (2004): 213–235.

Reinhardt, V., and A. Reinhardt. Impact of a privacy panel on the behavior of caged female rhesus monkeys living in pairs. *Journal of Experimental Animal Science* 34 (1991): 55–58.

Reinhart, G. A., L. C. Preusser, S. E. Burke, J. L. Wessale, C. D. Wegner, T. J. Opgenorth, and B. F. Cox. Hypertension induced by blockade of ETB receptors in conscious nonhuman primates: Role of ETAreceptors. *American Journal of Physiology—Heart and Circulatory Physiology* 283, no. 4 (2002): H1555–H1561.

Robins, J. G., and C. D. Waitt. Improving the welfare of captive macaques (*Macaca* sp.) through the use of water as enrichment. *Journal of Applied Animal Welfare Science* 14, no. 1 (2011): 75–84.

Rochlitz, I. Recommendations for the housing and care of domestic cats in laboratories. *Laboratory Animals* 34, no. 1 (2000): 1–9.

Rodan, I., E. Sundahl, H. Carney, A.-C. Gagnon, S. Heath, G. Landsberg, K. Seksel, and S. Yin. AAFP and ISFM feline-friendly handling guidelines. *Journal of Feline Medicine and Surgery* 13, no. 5 (2011): 364–375.

Rommers, J. M., B. J. F. Reuvekamp, H. Gunnink, and I. C. de Jong. Effect of hiding places, straw and territory on aggression in group-housed rabbit does. *Applied Animal Behaviour Science* 157 (2014): 117–126.

Sato, K., F. Chatani, and S. Sato. Circadian and short-term variabilities in blood pressure and heart rate measured by telemetry in rabbits and rats. *Journal of the Autonomic Nervous System* 54, no. 3 (1995): 235–246.

Schapiro, S. J., M. A. Bloomsmith, and G. E. Laule. Positive reinforcement training as a technique to alter nonhuman primate behavior: Quantitative assessments of effectiveness. *Journal of Applied Animal Welfare Science* 6, no. 3 (2003): 175–187.

Schapiro, S. J., and S. P. Lambeth. Control, choice, and assessments of the value of behavioral management to nonhuman primates in captivity. *Journal of Applied Animal Welfare Science* 10, no. 1 (2007): 39–47.

Scott, L., P. Pearce, S. Fairhall, N. Muggleton, and J. Smith. Training nonhuman primates to cooperate with scientific procedures in applied biomedical research. *Journal of Applied Animal Welfare Science* 6, no. 3 (2003): 199–207.

Shepherd, S. V., and M. L. Platt. Noninvasive telemetric gaze tracking in freely moving socially housed prosimian primates. *Methods* 38, no. 3 (2006): 185–194.

Shomer, N. H., S. Peikert, and G. Terwilliger. Enrichment-toy trauma in a New Zealand White rabbit. *Journal of the American Association for Laboratory Animal Science* 40, no. 1 (2001): 31–32.

Spangenberg, E. M. F., L. Björklund, and K. Dahlborn. Outdoor housing of laboratory dogs: Effects on activity, behaviour and physiology. *Applied Animal Behaviour Science* 98, no. 3 (2006): 260–276.

Staal, J. A., L. Pinkney, and D. M. Roane. Assessment of stimulus preferences in multisensory environment therapy for older people with dementia. *British Journal of Occupational Therapy* 66, no. 12 (2003): 542–550.

Storey, P. L., P. V. Turner, and J. L. Tremblay. Environmental enrichment for rhesus macaques: A cost-effective exercise cage. *Journal of the American Association for Laboratory Animal Science* 39, no. 1 (2000): 14–16.

Strazzeri, J. M., J. J. Hunter, B. D. Masella, L. Yin, W. S. Fischer, D. A. DiLoreto, R. T. Libby, D. R. Williams, and W. H. Merigan. Focal damage to macaque photoreceptors produces persistent visual loss. *Experimental Eye Research* 119 (2014): 88–96.

Suckow, M. A., V. Schroeder, and F. A. Douglas. *The Laboratory Rabbit.* Boca Raton, FL: CRC Press, 2010.

Sumpter, C. E., T. M. Foster, and W. Temple. Assessing animals' preferences: Concurrent schedules of reinforcement. *International Journal of Comparative Psychology* 15, no. 2 (2002): 107–126.

Tarou, L. R., and M. J. Bashaw. Maximizing the effectiveness of environmental enrichment: Suggestions from the experimental analysis of behavior. *Applied Animal Behaviour Science* 102, no. 3 (2007): 189–204.

Taylor, K. D., and D. S. Mills. The effect of the kennel environment on canine welfare: A critical review of experimental studies. *Animal Welfare* 16, no. 4 (2007): 435–447.

Trachoo, N., and J. F. Frank. Effectiveness of chemical sanitizers against *Campylobacter jejuni*-containing biofilms. *Journal of Food Protection* 65, no. 7 (2002): 1117–1121.

Trez, T. A. Refining animal experiments: The first Brazilian regulation on animal experimentation. *Alternatives to Laboratory Animals* 38, no. 3 (2010): 239–244.

Trocino, A., and G. Xiccato. Animal welfare in reared rabbits: A review with emphasis on housing systems. *World Rabbit Science* 14, no. 2 (2006): 77–93.

Turner, D. E., E. K. Daugherity, C. Altier, and K. J. Maurer. Efficacy and limitations of an ATP-based monitoring system. *Journal of the American Association for Laboratory Animal Science* 49, no. 2 (2010): 190–195.

Turner, P. V., K. L. Smiler, M. Hargaden, and M. A. Koch. Refinements in the care and use of animals in toxicology studies—Regulation, validation, and progress. *Journal of the American Association for Laboratory Animal Science* 42, no. 6 (2003): 8–15.

USDA (U.S. Department of Agriculture). Animal Welfare Act and Animal Welfare Regulations. https://www.aphis.usda.gov/animal_welfare/downloads/Animal%20Care%20Blue%20Book%20-%202013%20-%20FINAL.pdf (accessed December 1, 2015).

van den Bos, R. Post-conflict stress-response in confined group-living cats (*Felis silvestris catus*). *Applied Animal Behaviour Science* 59, no. 4 (1998): 323–330.

Vicino, G. From prevention of cruelty to optimizing welfare: Opportunities to thrive. In *Proceedings of World Association of Zoos and Aquariums 70th Annual Conference*, Al Ain, UAE, October 11–15, 2015, 23. http://www.waza.org/files/webcontent/1.public_site/7.get_involved/waza_annual_conferences/Al%20Ain%20Conference%20Proc.pdf (accessed July 30, 2016).

Ward, W. K., M. D. Wood, and E. P. Slobodzian. Continuous amperometric monitoring of subcutaneous oxygen in rabbit by telemetry. *Journal of Medical Engineering & Technology* 26, no. 4 (2002): 158–167.

Wardrip, C. L., J. E. Artwohl, J. Oswald, and B. T. Bennett. Verification of bacterial killing effects of cage wash time and temperature combinations using standard penicylinder methods. *Journal of the American Association for Laboratory Animal Science* 39, no. 4 (2000): 9–12.

Washburn, D. A., and D. M. Rumbaugh. Testing primates with joystick-based automated apparatus: Lessons from the Language Research Center's computerized test system. *Behavior Research Methods, Instruments, & Computers* 24, no. 2 (1992): 157–164.

Wells, D. L. A review of environmental enrichment for kenneled dogs, *Canis familiaris*. *Applied Animal Behaviour Science* 85, no. 3 (2004): 307–317.

Wells, D. L., L. Graham, and P. G. Hepper. The influence of auditory stimulation on the behaviour of dogs housed in a rescue shelter. *Animal Welfare* 11, no. 4 (2002): 385–393.

Zhdanova, I. V., K. Masuda, C. Quasarano-Kourkoulis, D. L. Rosene, R. J. Killiany, and S. Wang. Aging of intrinsic circadian rhythms and sleep in a diurnal nonhuman primate, *Macaca mulatta*. *Journal of Biological Rhythms* 26, no. 2 (2011): 149–159.

23

Agricultural Animals

Janice C. Swanson, Larry T. Chapin, and F. Claire Hankenson

CONTENTS

Introduction

Distinguishing the Guidance

The purpose of this chapter is to provide an overview of the important considerations associated with the care and use of agricultural animals regarding their environment and housing and to highlight the available resources to assist program managers, veterinarians, and research staff. In the United States, two guidelines may be utilized when managing programs engaged in research, testing, and teaching with agricultural animals: the Federation of Animal Science Societies (FASS) *Guide for the Care and Use of Agricultural Animals in Research and Teaching* (Ag Guide) (FASS 2010) and the National Research Council (NRC) *Guide for the Care and Use of Laboratory Animals* (NRC Guide) (NRC 2011). In most developed countries, there are laws and/or standards in place for the use of animals in research. For example, in Europe the EU 2010/63/EU Directive is used (European Commission 2016), and in Canada the Canadian Council on Animal Care (CCAC) guidelines and the Canadian Codes of Practice are referenced. Care should be taken to check for laws and guidance documents operating within a country if your animal research program is international in scope. For the purpose of providing a range of resources to draw guidance, we include references to guidelines and training materials from other countries that may be useful auxiliary resources.

The NRC Guide has evolved through numerous editions and provides instruction on requirements for all animals used in biomedical endeavors that are funded by federal programs administered through the Public Health Service (PHS) and National Science Foundation (NSF). The Ag Guide was first developed in 1988 for institutional programs using agricultural animals to conduct agricultural research related to food and fiber production. However, it too has had to evolve over time to include more guidance with respect to the use of agricultural animals in biomedical or agricultural research, largely because farm animals and agricultural species' fundamental needs do not change based on research assignment. Therefore, depending on the research objectives and housing situation, both the NRC Guide and Ag Guide may be employed simultaneously. There is remarkable harmonization between the NRC Guide and Ag Guide with respect to program recommendations and animal care; however, there are also notable differences, such as requirements for the composition of the Institutional Animal Care and Use Committee (IACUC), some disagreement on cage and pen space requirements for certain species, and environmental temperature recommendations.

In the United States, the Animal Welfare Act (AWA) (7 USC 2131 et seq.) and its regulations (Title 9 CFR Subchapter A, Parts 1–3) also come into play when it has been determined by the IACUC that approved protocols represent biomedical research, testing, or training, or include human medical and veterinary student education. Any animals used for these purposes are "regulated" or "covered," as are the spaces, facilities (i.e., surgery suite), and housing areas in which they are maintained. Further clarification regarding aspects of agricultural animal use that are regulated versus exempt from U.S. Department of Agriculture (USDA) coverage can be found in the Animal Care Resource Guide, specifically Policy 17, "Regulation of Agricultural Animals." Although species-specific standards for agricultural animals have not yet been developed under the AWA, the NRC Guide and Ag Guide provide strong supplemental support to meet the regulatory mandate for animal care. It will be important to remain current on any changes that occur between guidelines and state or federal regulations that may affect your program compliance.

Management Approaches Based on Animal Use Goals

Determining the Goals of Animal Use Based on Production, Research, and Teaching

Both the proper management of agricultural animals and related compliance with animal care obligations will necessitate special attention to the purpose and requirements of research using agricultural animals. The first step is to determine the purpose of the proposed agricultural animal use (Swanson 1998). The IACUC is the body that must make the determination of the type of use. Typically, specific attention is paid to the funding source and the justification of the experimental or educational goal. If the goal is to conduct science to improve the health of humans or to endeavor human teaching and training (whether veterinary or medical) for a similar health or medical purpose, then the use is likely to be categorized as biomedical (Stricklin and Mench 1994). The funding source often provides a strong clue. For example, if the research activity is funded by the National Institutes of Health (NIH), then the research would most likely be classified as biomedical; as such, the PHS Policy on Humane Care and Use of Laboratory Animals, the NRC Guide, and the AWA regulations are principle references. The Ag Guide will likely serve as an important source of species-specific guidance for areas not covered in depth by the other two documents.

Regardless of the type of research, all animals used in research, teaching, or testing, including breeding stock, should be covered by IACUC oversight. All personnel responsible for agricultural animal care and use must achieve an understanding of the applicable regulations and guidelines required to meet program compliance. It is highly recommended that agricultural animal facility and research managers carefully review and discuss approved IACUC protocols with the principal investigator (PI) and the attending veterinarian (AV) to ensure compliance is met.

Sustaining Multiple Missions within an Agricultural Program

With the current emphasis on transdisciplinary scientific endeavors, it is not unlikely that an institution may be simultaneously conducting biomedical and agricultural research and teaching courses using farm animals housed within a shared facility; as an example, horses may be used for investigating aspects of species reproduction, for biomedical procedures like investigation of hormones and antivenoms, and as models for student education and therapeutic wellness programs. The diversity of potential animal use requires that staff managing livestock, horses, and poultry at such facilities be conversant in, and knowledgeable of, the requirements for the care and housing of animals for the different categories of work. Planning for biomedical research to be conducted at an agricultural-type facility requires advance input from the PI, the veterinary staff charged with the provision of training and preventative and therapeutic care, and the veterinarian in charge of the institutional animal care program (typically the AV) to determine if the site is suitable to conduct the proposed activity. For example, if survival surgery is to be conducted on site, then presurgical, surgical, and postsurgical requirements must be met. Further clarification regarding veterinary care aspects of surgery, drugs, and materials for agricultural animal use can be found in the Animal Care Resource Guide, specifically Policy 3, Veterinary Care (USDA 2016c). Equipping facilities to meet the demands of dual-purpose activities should be undertaken as projects are approved and prior to the start of work.

Impact on Budgets, Personnel, and Facilities

As with all research endeavors using animals, there are essential investments in constructing and maintaining facilities and equipment, providing for qualified staff and their training, veterinary care for animals, and the day-to-day requirements for overseeing colonies, herds, and flocks of animals. In general, the infrastructure to maintain, handle, and conduct on-site research using agricultural animals is expensive. When dual-purpose (both food and fiber and biomedical) research is being conducted, there can be additional challenges in obtaining the considerable financial resources necessary to achieve regulatory compliance. For example, the environment and conditions under which to conduct survival surgery may carry different expectations for a biomedical protocol versus an agricultural one. Financially, this might require either an addition to a facility in order to meet biomedical standards or renovating an existing biomedical surgical room at another location. Although every attempt has been made to harmonize the guidelines and regulations covering both types of categorical research, it is important for managers of agricultural facilities to be acutely aware of the differences. An institution that is conducting biomedical and agricultural research using the same facilities should be prepared to ensure, or build into the facilities, the rooms, pens, and other spaces necessary to meet compliance for both types of research. Alternatively, animals can be transported from housing areas to remote facilities that meet the appropriate regulatory expectations for animal needs. Preplanning for meeting research objectives and ensuring the ability to secure resources when changes are required are important management strategies for averting excessive pressure on budgets, personnel, and facilities. This also requires clear communication of needs and issues within the institutional structure that oversees and budgets for animal research and programmatic growth.

Housing

Management of Biomedical Research Animals versus Agricultural Animals

Type of Facility

Agricultural animals can be housed in a variety of management systems, from extensive (e.g., pastures or range) to intensive (e.g., houses, pens, or cages). The physical accommodations are dependent on the husbandry needs of each species; the scope and type of research, teaching, or testing being conducted; and whether the work is biomedical or agricultural in nature. As explained in the various resource guidelines, a critical component is to have appropriately maintained facilities for proper management, secure housing, and husbandry and health support of animals (CCAC 2009; FASS 2010).

Housing Arrangements for Biomedical Research Animals (USDA-Regulated Species)

During a USDA inspection, the Animal and Plant Health Inspection Service (APHIS) designee (typically a veterinary medical officer [VMO]) should be given full access to all areas (pens, barns, lots, pastures, or facilities) where regulated animals are kept. During the inspection, all areas of care and treatment that are covered under the AWA should be reviewed. This includes observing regulated species on site or owned by the institution, as well as inspecting the quality of facilities, including enclosures, the integrity of the building and construction materials and space, and food storage areas (USDA 2012).

From a USDA perspective, nonregulated animals (typically laboratory mice, rats, and birds, along with those agricultural animals used for food and fiber research) should *not* be inspected or included on the inspection report, unless they exhibit some clear potential to impact the health or well-being of the regulated animals (USDA 2016b). This is an important point to keep in mind when both regulated and nonregulated animals are housed in the same facility. However, in other countries, where other laws and directives apply, this may not be the case.

To provide a rapid visual distinction between animals housed in a proximity that may be used for varying purposes, it may be useful to mark animals with a coded system. There are a variety of easily identifiable and distinguishable options to choose from, such as housing groups in distinct pens or cages, or using colored tags, colored bands, cards, electronic or physical identifiers, and/or markings that allow for identification and sorting of both individuals and groups of regulated and nonregulated animals. In addition, managers are required to maintain accurate and complete records of the sources, dates of acquisition and disposition, and proper identification of regulated animals on the premises (USDA 2012).

When Animal Research Involves Surgical Procedures

Surgical procedures in agricultural animals may be those that are required for experimental purposes or those that arise spontaneously due to husbandry, breeding, or natural events (e.g., surgical correction of colic in a horse or caesarean section for a sheep in dystocia). Animals that undergo invasive surgical procedures require appropriate perioperative support, regardless of the rationale for the procedure. Further, for those proposals that may preemptively suggest the potential for emergency surgery, the associated IACUC protocols should have written provisions for veterinary care and handling of these types of surgical cases, or make reference to a site-specific standard operating procedure (SOP) regarding emergency veterinary care measures.

No matter what the rationale for conducting surgery, standards of appropriate veterinary care should be applied. This standard should include sterile instruments and appropriate protective equipment (e.g., disposable gowns and gloves). Skin incision or biopsy sites should be disinfected appropriately prior to surgical operations. For major survival surgeries, it is expected that facilities are available to accommodate the species undergoing procedures and that every effort will be made to perform survival surgery under stringent conditions that maximize sanitation and minimize breaches to aseptic technique, according to the AWA regulations (USDA 2016c). The Canadian guidance on surgical expectations is explored in depth in Section 9, "Specialized Procedures Used in Research and Testing" (CCAC 2009), and other countries may have relevant guidance that should be followed.

Animal Procurement

Agricultural housing areas are often established to produce all or most of the animals required on site, with breeding and production as the goal. This closed-herd method can preserve biosecurity for improved herd health, assist with monitoring genetics, and serve as a key training area for students interested in agricultural practices. However, this approach will necessitate that the facility be designed to accommodate and manage a range of animal ages (e.g., birth to adulthood), sexes, and physiological states (e.g., neonate, growing, mature, cycling, breeding, pregnant, and lactating) for a particular species on a continuous basis.

Bringing animals into the existing group from outside sources allows for the acquisition of only the type of animals required for use at a particular time. This is a common practice for "all in–all out" facilities, where all animals are delivered or arrive in a group at one time and are removed at one time (e.g., poultry or swine facilities). Management approaches will be influenced by the cost and availability of the type of animals needed; each new animal that enters the herd, flock, or unit is a potential new point of contact for disease transmission, thus increasing the need for secure biocontainment across the facility. For newly introduced animals, domestic isolation and quarantine procedures are vital and should be designed to isolate animals from the main herd or flock in order to monitor their health status and ensure their compatibility with the resident population.

Background Characteristics of Animals Required for the Category of Work

A wide range of characteristics and requirements will need to be accommodated for biomedical or agricultural needs, including considerations of breed, line, strain, phenotype, age, sex, size, physiological state, offspring, genetic modifications, diseases, and/or parasitic infestations of animals to be used for research, teaching, or testing (Helke et al. 2015; Underwood et al. 2015).

In general, the standards of agricultural animal care and welfare will be the same for all research animals, regardless of research type. However, alterations in management and housing are often required for special strains; for specific genetic modifications or the exposure or introduction of hazardous materials to animals in order to accommodate animal and personnel welfare; for biological, chemical, and/or physical safety; and to abide by the related laws, regulations, and guidelines that pertain to these specific situations (CCAC 2009; FASS 2010; NRC 2011).

Research with Privately Owned Animals on Private Farms

Work Conducted on Private Farms

When research, teaching, or testing is to be conducted by institutional employees on private property (e.g., farms), conducted with privately owned animals on institutional property, or conducted with institutionally owned or leased animals on private property, it is essential that the IACUC reviews and approves this type of work. Institutions must develop policies for off-site activities in order to ensure appropriate and humane animal care and use (FASS 2010; NRC 2011). Owners of the animals and/or property must be informed of the expectations and approved procedures and must have agreed to the procedures through formalized consent.

The agreement of understanding may be documented as a contract, memorandum of understanding (MOU), consent form, or other arrangement. The document should be comprehensible to a layperson and realistic about potential risks and outcomes. The agreement should contain the following items: project title, description and purpose of the study, duration, animal species and number, procedures, responsibility for non-research-related health and medical issues, disposition of animals at project conclusion, voluntary participation and right to withdraw from the project, access to institutionally owned animals on private property, name of the principal investigator and primary contact persons, expectations of the owner, consequences and procedures if an owner decides to withdraw from the research, client consent statement to participate (acknowledgments and waivers), and stakeholders' signatures, printed names, titles, and agreement date. Consent agreements should be reviewed by the IACUC and/or the institution's legal counsel (Russow and Theran 2003; University of Illinois 2008; University of Kentucky 2013).

Provision of Off-Site Veterinary Care

The oversight of agricultural animal health for research, teaching, and testing is the responsibility of the institutional veterinarian (e.g., AV). This individual has direct or delegated authority and ultimately is responsible for animal care and use, husbandry and nutrition, disease control, sanitation and biosecurity, and preservation of animal well-being. The AV, or a qualified designee that is also a veterinarian, should be in routine communication about biomedical and agricultural animal health expectations.

For those agricultural settings that may be remote from a campus or from the centralized administrative unit, it is not uncommon to retain clinical veterinary practitioners that conduct their business or practice in closer proximity to farms. Consultant large animal veterinarians can be hired for routine and emergent health concerns and may more readily access animals for assessments and treatments than the AV located at a distant central campus site. These noninstitutional or "outside" veterinarians should receive an orientation on the required guidelines and reporting lines for the institution. If nonveterinary staff are appropriately trained, certain treatments may be administered according to dedicated SOPs (e.g., veterinary medical care guidelines) that have been reviewed and approved by the AV (or their designee) or reviewed and approved by the IACUC (CCAC 2009; FASS 2010; NRC 2011).

Planning for Adequate Space for Animals on Research Trials

Because of the larger genetic variation in farm species relative to traditional rodent models, thorough consideration should be given to experimental design, in order to address this issue. Increased variability may require an increased number of animals to detect treatment differences. To obtain meaningful data from research trials, it is imperative that appropriate experimental designs are developed to answer the specific study objectives while controlling for variation (CCAC 2016a,b). In order to accomplish this, it is essential to know what the experimental unit will be and how units will be grouped for analyses. This has a direct effect on the type of housing and number of animals utilized for research. PIs should consult with statistical experts during the planning stages of animal experimentation to determine correct study design and the required number of experimental units. Managers should gain an understanding of the experimental design and the required number of animals in order to plan for the provision of adequate housing and procedural space.

The facility itself should be designed for maximum flexibility based on the types of research trials to be run. One should consider how the treatments will be applied to the animals and if treatments will be applied to individual animals or groups of animals. For example, when the same treatment is applied to a group of animals in one enclosure (e.g., pen, cage, or pasture), the enclosure serves as the experimental unit, not the individual animal. In this example, one must plan for an appropriate number of enclosures to accommodate the statistical power required in the experimental design. If a treatment can be randomly applied to an individual animal in such a way to ensure that the treatment response in any one animal is not influenced by the response in the others, then the animal is the experimental unit (Johnstone 2013). And lastly, depending on the experimental design, the experimental unit could potentially be both group and individual (Festing and Altman 2000).

Building and Facility Redundancies for Continued Functionality

Agricultural animals living in enclosed housing facilities are dependent on the continuous operation of the heating and ventilation system for temperature and humidity control and air quality. In addition, electrical power is required for lighting and essential facility operations, such as fans, milking equipment, water wells, surgical and procedure room equipment, cold rooms and carcass freezers, and computers and security systems. As animal facilities become more technologically dependent, they become more vulnerable to equipment faults and power outages (European Union 2016). In closed housing systems with high animal densities (e.g., poultry or swine buildings), the animal environment quality can quickly deteriorate with a prolonged loss of ventilation or power.

Alarm systems or warning devices are essential to monitor environmental conditions (e.g., temperature and humidity), equipment faults, and power failures within animal facilities. The alarm system should notify a responsible person of the emergency condition so appropriate remedial action can be taken as quickly as possible (CCAC 2009; European Union 2016). For monitoring a gas such as carbon monoxide (CO), a CO meter or data logger allows the user to determine if there is a problem and locate the source of the emission (University of Georgia Cooperative Extension Service 2007). Emergency management SOPs must be developed that describe procedures to follow (e.g., alerting personnel to out-of-acceptable range) in the event of these alarm conditions (CCAC 2009). Existing facility alarm and security systems are available, either hardwired or with wireless capabilities, and include an array of sensors (such a CO monitors), cameras, recorders, and notification options. Due to the size and design of indoor agricultural animal housing systems and the nature of their environments (e.g., dusty, corrosive, humid, and harsh cleanings), a specialist experienced in agricultural facilities should be consulted in the selection, installation, and maintenance of commercial systems designed for these conditions.

Backup or emergency power in the form of generators or alternative power supplies should be able to maintain ongoing support functions in animal rooms, operating suites, and other essential areas until critical procedures are completed (e.g., surgery), or until alternative actions can be taken (e.g., repair or relocation) or the power is restored. Emergency SOPs should be developed that describe procedures for staff to follow in the event of a power outage.

If generators are to be delivered to the site as the backup power supply, consider having electrical switchgears preinstalled on critical power circuits to allow for more rapid generator placement and initiation. It is critical to consider the placement of emergency generators, electrical switchgears, and sufficient generator fuel supplies to avoid placement in areas subject to flooding (e.g., basements, floodplains, and storm surges). Work with institutional physical plant experts to post breaker box locations on equipment and to assess critical infrastructure needs prior to an emergency (Goodwin and Donaho 2010; Swearengen et al. 2010).

Alternate sources of backup power include battery-stored power, wind power (e.g., turbines), solar power (e.g., panels on rooftops), hydropower, and fuel cells to supply electricity; all these options should be evaluated for applicability, safety compliance, and accessibility (U.S. Department of Energy 2016). Other alternatives to consider are opening security doors, windows, and removable (breakout) panels to facilitate air exchange; the use of portable gasoline engines as temporary replacements for electric motors (e.g., to run feed augers); and the use of uninterruptible power supplies (UPSs) to run critical equipment (e.g., anesthesia ventilators, cautery units, and surgical lamps) until such procedures have been completed.

Environment

Enrichment of the Environment

The use of environmental enrichments can mitigate or prevent problems by providing for the expression of natural social and behavioral needs. Ultimately, the goal of any enrichment program is to prevent stereotypies and detrimental behavior from developing by improving the complexity of the animal's environment to encourage the expression of a normal range of behavior, encourage the animal to utilize the environment to its advantage, and to promote adaptation to the environment allowing the animal to cope with research activity (see Chapter 5 for more information on behavioral management strategies).

Agricultural versus Biomedical Guidance on Enrichment

There are important concepts and considerations for developing a successful enrichment program. First, knowledge of the animal's behavioral, physiological, and anatomical attributes, and the natural history of the species provides critical grounding for identifying and implementing suitable enrichment strategies. Second, an enrichment program requires trained personnel to deliver the enrichment, monitor the impact of the enrichment on the animal's behavior to determine its effectiveness, and make adjustments and changes when required.

The major guidance documents introduced in the beginning of the chapter address aspects of environmental enrichment. The Ag Guide provides detailed information with species-specific recommendations and guidance on the types and use of environmental enrichment. General guidance is provided by the CCAC Guide and the NRC Guide, and numerous web resources and texts (Young 2003) are also available on this subject.

Natural enrichment approaches typically favor housing animals in surroundings consistent with their evolved behaviors. For example, the daily provision of a grazing pasture with patches of forest and a flowing stream for a small herd of cattle would be an exceptionally enriched natural enrichment. For swine, provision of an outdoor lot with wallows and rooting areas would promote species-specific behaviors. However, the provision of habitats such as those described here must also meet applicable environmental regulations.

For those agricultural species that are housed in agricultural production facilities, integration of enrichment may be engineered into the environment in creative ways (brushes hung at head height for cattle, reflective mirrors positioned in sheep pens, and straw bedding and substrates for rooting in swine areas). Types of enrichments are mentioned in this section, and further descriptions of these enrichments can be found in the Ag Guide.

Conversely, for agricultural animals housed in biomedical settings (e.g., off pasture) and within fully enclosed buildings, enrichment devices should be fashioned to sustain routine washing and sanitization, and then rotated with other novel devices. Depending on the species and the research aim, enrichment in biomedical settings for agricultural animals should promote species-specific behaviors and may include structural additions, provision of bedding substrates (straw and shavings), toys that can be manipulated, hanging objects on short tethers that prevent entanglement, and nutritional supplements.

Under research conditions, there are challenges to meeting the behavioral and social needs of animals. All agricultural species can develop abnormal behaviors and stereotypies if housed in barren environments, intensive groups, or under conditions of social restriction, such a single housing or isolation. Stall weaving by horses, wool eating by sheep, tongue lolling by cattle, tail biting in pigs, and feather pecking and cannibalism in chickens are examples of stereotypies and detrimental behaviors that can cause injuries to the animal or others. In order to address the welfare needs of animals directly, the unique natures and needs of differing agricultural species must be considered in the conduct of housing and husbandry (Russow and Theran 2003). Welfare considerations for agricultural species are a critical aspect of research considerations and have been reviewed by experts in related publications (Underwood et al. 2014; Grandin 2015b).

Types and Sources of Enrichment

In brief, there are five types of enrichment used to satisfy animal needs: social, occupational, physical, sensory, and nutritional (Bloomsmith et al. 1991). Each type of enrichment requires extensive knowledge of animal behavior, as well as the species' natural history and biology, to ensure that sources of enrichment match the need that is to be addressed and will illicit the desired response and outcome.

1. Social enrichment is used under a variety of conditions, where animals may be grouped, individually housed, or isolated from their conspecifics. Social enrichment can be contact or noncontact in nature. Social contact enrichment typically employs various grouping strategies with other animals. Noncontact social enrichment engages the auditory, olfactory, or visual senses to prevent isolation or separation distress.
2. Occupational enrichment is used to meet a psychological and/or physical need of the animal. The manipulation of a device controlling the dispensation of feed or access to a paddock (for horses in particular) or a device such as a hot walker (e.g., treadmill) for horses to encourage exercise are examples of occupational enrichment (Jockheer-Sheehy and Houpt 2015). The Ag Guide provides species-specific examples, such as the use of rubbing and scratching devices for self-maintenance in cattle, toys such as pull chains or balls for exercise in pigs, and perches for roosting in the vertical space.
3. Physical enrichments are changes made to the physical environment in which the animal lives. Strategies can range from designing floor space to accommodate specific types of behavior, such as lying down, or dunging areas to prevent soiling in areas meant for resting.
4. Sensory enrichments engage the animal's senses, such as audio playback of sounds, tactile stimulation, and placement of windows or pleasant or calming odors.
5. Nutritional enrichments typically involve the use of food delivery mechanisms or toys with species-specific foods.

Future developments in enrichment will come from creative application of technology, efficient data collection, improved understanding of behavioral genetics and evolutionary relationships, sophisticated application of new findings in learning theory, and the incorporation of new methods for quantifying and understanding differing animal temperaments (Ha and Andrews 2016).

Alterations to the Physical Environment Require Acclimation

Species-specific behaviors need to be considered when applying enriching treatments that change or alter the physical or social housing arrangement. For example, acclimation of animals to relocation, new groups, temperature changes, or isolation for procedures must all be well planned. The fairly innocuous decision to change stalls for an animal, or to introduce new devices or toys, can trigger stress or fear instead of curiosity upon first exposure. In some cases, animals must learn how to use the enrichment. Careful planning and animal safety must be thoroughly considered when implementing and managing an enrichment program and species-specific behaviors; that is, dairy cattle tend to learn their new stall locations easier when the turns are the same and only the relative distance changes.

Boredom with any particular enrichment strategy requires a program to be dynamic and may require a periodic and routine cycling of the enrichment options. Whether it is a change in the housing environment or an interactive element, when the novelty of the enrichment is reduced through prolonged exposure, consider changing the strategy. During daily observations of animals, it is important to note animal interactions with, or the early signs of failure of, provided enrichment tools and devices.

Projects Requiring Confinement

Agricultural animals in close confinement projects or other intensive procedures requiring prolonged restraint (e.g., metabolism stall and stanchions) will experience less stress if they have been adapted to contact with people and are trained to cooperate voluntarily with procedures. Because of the degree and type of restraint in metabolism stalls, this type of stall should only be used in approved studies and not for routine or permanent housing. Animals should have the ability to maintain a comfortable posture and to eat, rise, and rest normally. Comfort can be enhanced in metabolism stalls with coated flooring and by making the front portion of the flooring solid. When possible, allow for visual, auditory, and olfactory contact with conspecifics (CCAC 2009). If necessary, animals not included in the study could be housed in the same room to ease the distress associated with isolation.

Because animals cannot typically make their own thermal adjustments in this type of housing system, there needs to be appropriate environmental control of temperature and ventilation for animal comfort. Personnel should observe animals frequently and monitor for signs of thermal stress, altered behaviors, and changes in appetite, urinary and fecal outputs, and the condition of skin, feet, and legs to determine if adjustments are required or if humane interventions are needed. The length of time that an animal remains in a metabolism stall or stanchion before removal for exercise is a welfare decision that should be based on professional judgment and experience and approved by the IACUC. When animals are turned out for exercise, the exercise area (e.g., pen or outside lot) should be safe for animals to move about with good footing and with no exposure to sharp edges or protrusions (CCAC 2009; FASS 2010; New South Wales Department of Primary Industries 2010; NRC 2011).

Monitoring Air Quality within Indoor Facilities

The environmental control system for animals housed indoors should provide an acceptable housing environment and appropriate air quality to support good animal health and well-being. Acceptable air quality can usually be achieved with proper ventilation, waste management, and routine husbandry. Performance specifications and ranges for ventilation and air quality are provided in the major guidance documents.

The primary air pollutants from indoor housing of agricultural species are ammonia, hydrogen sulfide (H_2S), CO, methane, and dust, both respirable (particles 5 μm or less) and total (FASS 2010). Gas detectors (e.g., bellows or pump type) can be used to spot-check levels of these gases, and dust particle size counters can be used when it is suspected that respirable dust levels (normally not visible) are high (CCAC 2009; Wheeler 2009).

Ammonia concentration is a common concern for agricultural animals. For example, the United Egg Producers welfare guidelines (2016) state that ammonia exposure for birds should ideally be less than 10 ppm and should rarely exceed 25 ppm, which is in agreement with the Ag Guide recommendations.

Agricultural housing systems with manure storage pits are a source of H_2S and methane gas. H_2S gas in high concentrations can seriously injure or even kill animals and people. Methane gas in high concentrations can explode if exposed to an electrical source or spark. This is of major concern when manure storage pits are located under floors of animal housing units. High concentrations of H_2S and methane are given off during manure agitation or removal. It is recommended that when manure pits are pumped out, there are no animals or humans in the barn, and there is appropriate ventilation to eradicate the gases, which then rise up through slotted floors. This is a danger not only in enclosed barns, but also in open-sided barns with short side walls or curtains on calm still days when there is little to no air movement. Based on the guidance documents, persons working in barns should be aware of these health concerns and may need to wear breathing apparati in areas where H_2S gas or ammonia concentration becomes excessively high due to failure of facility systems (Choinière 1993; CCAC 2009; FASS 2010; NRC 2011; Rozeboom and May 2012).

CO is an odorless, colorless, and toxic gas produced from incomplete combustion of carbon-based fuels. Sources of CO in confinement facilities include unvented or malfunctioning fuel-fired heaters and engine exhaust from equipment such as power washers, portable generators, or vehicles and tractors. To minimize CO exposure, heating equipment must be properly vented and maintained, and whenever possible, portable equipment with engines should be kept outside. If engines must be operated inside an enclosed facility, they must be vented to the outside or the facility must be adequately ventilated to remove exhaust fumes during use. Attention should also be given to enclosed trailers and trucks when transporting animals, especially during cold weather, to prevent vehicle engine exhaust from entering animal enclosures (National Farm Animal Care Council 2001; NRC 2006; University of Georgia Cooperative Extension Service 2007; FASS 2010).

Monitoring the Environment for Thermal Stress

Most agricultural animals are quite adaptable to the wide range of thermal environments that are typically found in the natural outdoor surroundings of various climatic regions of the continental United States. The thermal environment, or effective environmental temperature, entails the combined effects of both the physical environment and the physiological state of the animal. It is important to monitor both the animal and its thermal environment in order to assess the potential for thermal stress. Equipment can be used to monitor temperature, humidity, air speed, and air movement or drafts; examples include the electronic fast-response thermometer, infrared thermometer to measure surface temperature, minimum–maximum temperature thermometer, relative humidity hygrometer, air velocity meter, and smoke pencils (CCAC 2009; FASS 2010).

Animals at high risk of thermal stress include young animals, darkly pigmented animals, and animals that have been sick or have a previous history of respiratory disease. Heat stress tolerances also vary between and within species depending on hair coat type, skin pigmentation, breed origin, production level, coat color, body weight, and length of acclimation following an environmental or pen change (Agriculture Victoria 2016).

The temperature–humidity index (THI) is used to better characterize the influence of ambient temperature and humidity on the animal in moderate to hot conditions. The Livestock Weather Safety Index or Livestock Weather Hazard Guide (LWHG) is based on the THI and is used to describe categories or levels of heat stress associated with hot weather conditions. The ambient temperature and percent relative humidity are obtained from weather reports to determine the THI and the LWHG heat stress level. Ranges of THI values have been defined to indicate that the potential for heat stress in livestock exists all the way up to the emergency level, indicating when plans for handling livestock should be postponed, if at all possible. Differing references on calculating THI for livestock versus lactating dairy cows are available (Alkire 2009; FASS 2010; Collier et al. 2012; USDA 2016a). Heat stress in animals is cumulative, and managers should be aware of the clinical and behavioral signs associated with heat stress (Table 23.1). If night temperatures do not get low enough to cool animals, then animals cannot make a full physiological recovery before exposure to the next day's heat event (Dahlen and Stoltenow 2012).

TABLE 23.1

Clinical and Behavioral Signs Indicative of Thermal (Heat) Stress

Clinical and Behavioral Signs	Livestock	Poultry/Caged Layers	Horses
Group behaviors	Crowded around water tanks or shade	Standing erect with wings held away from body	Profuse sweating; dehydration with a prolonged skin tent of several seconds at the neck or shoulders
Activity	Lethargy, immobility or aimless wandering, staggering		Fatigue, droopy ears
Appetite	Poor	Poor	Poor
Respiratory effort	Increased respiratory rate and heart rate	Panting	Increased respiratory rate and heart rate
Temperature	Elevated rectal temperature		Rectal temperature greater than 103°F
Individual behaviors	Drooling, slobbering, open-mouthed breathing	Gular (throat) flutter	Lethargy
Emergency conditions	Collapse, nonresponsiveness, seizures, death	Collapse and death	Collapse and death

Source: Kerr, S., Livestock heat stress: Recognition, response, and prevention, Washington State University Extension Fact Sheet FS157E, Pullman: Washington State University, 2015, http://cru.cahe.wsu.edu/CEPublications/FS157E/FS157E.pdf; Hunton, P., and L.J. Weber, Poultry: Heat stress in caged layers, Fact sheet, Guelph: Ontario Ministry of Agriculture, Food and Rural Affairs, 1988, http://www.omafra.gov.on.ca/english/livestock/poultry/facts/88-111.htm; Martinson, K. et al., Managing horses during hot weather, Minneapolis: University of Minnesota Extension, 2006, http://www.extension.umn.edu/agriculture/horse/care/managing-horses-during-hot-weather/.

The wind chill effect takes into account the ambient air temperature and the wind speed. This was developed in order to arrive at an effective ("feels like" or "still air") temperature that accounts for the cooling effect of the air as it draws heat away from the body in cold temperatures. Cold stress occurs as the animal loses its ability to conserve and generate body heat to stay within its thermoneutral zone. Personnel and managers must be aware of normal behaviors of the animals under their care in order to identify abnormal behaviors related to thermal stress. Fact sheets are available from various extension sources that describe signs and mitigation strategies for thermal stress in agricultural livestock species, which also include cellphone applications. Refer to the following resources: Tarr (2007), Kerr (2015), Hunton and Weber (1988), Martinson et al. (2016). Heat stress applications (apps) for smart phones calculate heat stress conditions and provide mitigation suggestions. Apps are available from Google Play, BlackBerry World and the App Store (e.g., USDA Agricultural Research Service (HEAT STRESS), The University of Missouri (ThermalAid), The University of Guelph (Canada) and the Ontario Ministry of Agriculture and Food and the Ministry of Rural Affairs (HEAT STRESS IN LIVESTOCK AND POULTRY) as well as livestock feed companies).

Corrective Actions to Offset Temperature Effects

Before heat or cold or chill stress occurs, the management team should be aware of which animals, flocks, or herds are most susceptible and have a plan to alleviate the stressors.

To reduce potential for heat stress, managers should consider enhanced access to water, alternating feeding times to the coolest periods of the day, promoting air movement, cooling with sprinklers or fans, enhanced shade areas, and potentially providing bedding substrates that do not retain heat. Animals may also be provided with electrolyte supplementation if prone to fluid loss or dehydration, and handling activity should be minimized to conserve energy (Dahlen and Stoltenow 2012).

To reduce cold stress in animals, monitor the weather and adjust feed and energy for low critical temperatures, provide wind breaks and shelters, design structures to reduce stress from excessively cold temperatures and wind chills, keep animals clean and dry in well-bedded areas, and be sure to provide ready access to additional feed and ample unobstructed water (Hamilton 2006; Beef Quality Assurance 2016).

Oversight of Veterinary Care and Animal Welfare

Qualified Veterinary Care

All guidance and regulatory expectations state that a veterinarian with experience in the care of agricultural animals must be accessible to research investigators and instructors. Veterinarians in their role should be familiar with the national or federal and local regulatory guidance and any institutional affiliated animal care policies (see the introduction for reference to guidance documents) as part of their job requirements. The amount of time that the veterinarian contributes to the program is not mandated; however, the veterinarian must have the authority to ensure that aspects of the animal care program are met. Familiarity with agricultural practices, herd health, and husbandry nuances will foster the important relationship between veterinary and nonveterinary staff members; these relationships are critical, since direct, frequent, and routine communication must occur between this animal oversight group and the AV.

Veterinary Care by Farm Staff

No program has to rely on a sole veterinarian for clinical assistance and regulatory advisement; instead, the AV can delegate authority to other qualified veterinarians and further enlist subject matter and species' experts when making decisions about agricultural animal care. Ideally, the veterinarian that works with the nonveterinary team (e.g., barn or farm managers, dairy personnel, and husbandry teams) will have experience with the species of animals being assessed. Contracting with regional veterinary college faculty and staff and/or private practitioners for ruminants, equine, poultry, swine, and other specialties can facilitate provision of timely and species-appropriate veterinary care.

Written programs of veterinary care can involve veterinary medical instructions for trained nonveterinary staff members that are employed in farm, dairy, and production environments. These instructions or veterinary medical care guidelines (i.e., vet care SOPs) can include guidance on recognition of sick animals, assessment tools, surveillance techniques, treatment options, and administration of therapeutics. It is advisable to provide these instructions to farm managers and staff in order to explicitly outline what medical care can be offered promptly by the staff and when to contact the veterinary team on medical cases. These documents should be reviewed and updated on a routine basis by veterinary staff and discussed with the IACUC.

These veterinary medical care guidelines should be available at all farm sites for agricultural species used for agricultural research, teaching, and testing and ideally should contain information to include

- Emergency contact names and numbers (e.g., farm manager, veterinarian, and investigator staff)
- Expectations on animal identification, surveillance, and observations, and a checklist of abnormal conditions to document (attitude, appetite, locomotion, any bodily fluids or discharges noted, respiratory rate and effort, and fecal and urine output)
- Checklist of emergencies that require veterinary attention (e.g., inability to ambulate normally or stand, signs of pain or distress, respiratory distress, hemorrhage, or neurologic abnormalities)
- Instructions for how to initiate treatment with drugs, doses, route, dosing schedule, duration of use, and potential withdrawal times
- Reminders throughout the treatment protocols that if the condition of the animal worsens or the animal is nonresponsive to treatment, a clinical veterinarian should be contacted for consultation and further assessments

Related to the veterinary medical care guidelines, access to antibiotics and routine agricultural practice drugs (e.g., vaccines, antiparasitics, vitamins, and minerals) should be outlined so that no animal goes without treatment for a prolonged period of time for any reason, even for routine issues in farm settings. Decisions about essential veterinary procedures (which diagnostic tests may be needed, the decision to euthanize unhealthy animals, and approval for necropsies for individual assessments or herd health surveillance) should be done only under the direction of the AV or designee. Funding for these types of services, which can be costly for larger animals and/or multiple animals from a particular herd or flock, needs to be provided as part of the program of veterinary care and should be supported by the administrators of the animal care program.

If an establishment does not have a full-time AV, then a written program of care for USDA-covered studies must be documented on USDA APHIS Form 7002, "Program of Veterinary Care (PVC) for Research Facilities or Exhibitors/Dealers" or an equivalent format. This written program is to be reviewed on a routine basis and updated whenever necessary. By law, such programs must include regularly scheduled visits to the animal housing sites by the veterinarian to best monitor animal health and husbandry needs.

Use of Scoring Systems to Evaluate Animal Well-Being

Body condition scoring (BCS) is the term used to describe the relative fatness or thinness and overall nutritional status and health of animal species. This concept has provided a significant improvement to the provision of adequate veterinary care and husbandry. Scoring is often graded on a 5- or 9-point scale, where the central number on the scale (e.g., 3 on a BCS 5-point scale and 5 on a BCS 9-point scale) represents an animal that is in average flesh and appropriate appearance. Scoring is based on palpation of an animal and determination of visual assessments of bony protuberances, spinous processes, and conformation. Representative pictures of animals at the designated scores (from thin to obese) ideally will accompany scoring descriptions to ensure an objective and consistent assessment by any observer of the same animal. Individual animals can also be scored relative to the overall flock or herd score as a useful management tool, in particular to judge the overall fullness of muscling and amount of fat cover in animals, as well as indicators of potential reproductive success. Visual scoring can be augmented by the use of tape measurements and ultrasound images of back fat estimates in certain species, like pigs, sheep, horses and beef cattle (Mississippi State University Extension Service 2016; National Hog Farmer 2016; University of Arkansas at Pine Bluff 2016; University of Nebraska-Lincoln 2016). Layer hens and poultry can be assessed by palpating keel and breast muscles; a more well-fleshed bird will have a less prominent keel and plump, rounded musculature surrounding the bone.

Locomotion Assessments

Lameness in agricultural species is one of the more common veterinary care issues for those animals housed on pasture, as well as those housed on indoor surfaces and grating. Hooved animals are prone to foot rot, which can result in locomotion issues, chronic pain, and reduced productivity. Lameness in cattle placed in indoor confinement areas has been attributed to bacterial infections of the hoof, as well as injury to the horn of the claws of the foot.

Hoof trimmings should be part of scheduled veterinary care for both livestock and horses. Hoof trimming is also a particular concern for sheep and swine housed indoors in biomedical settings where the floor surfaces are not conducive to promoting hoof wear and tear. In addition, individual animals, as well as flocks, may be treated with antimicrobials and routine foot baths to combat the persistence of hoof infections.

For cattle that are housed indoors in confinement spaces for any period of time, veterinary care may need to be provided to alleviate hock lesions and joint swelling related to injury. Environmental contributors that can exacerbate limb lesions include the type of flooring, how wet or dry the flooring is at the time animals are ambulating, the design of stalls, the presence of manure, and stocking density that might promote animals stepping on herd mates. Anecdotally, provision of a softer substrate (i.e., sand), floor padding, or supplemental bedding materials may greatly reduce hock lesions, particularly in dairy cattle.

Expectations and Management of Sanitation

Cleanliness of agricultural animal facilities is encouraged as a particular element of any veterinary care program; importantly, this does not mean that the housing and flooring environment has to be sterile. Clean areas promote animal health and well-being, particularly in indoor environments with limited housing space for animals enrolled in biomedical studies. Keeping animal waste removed from the biomedical housing areas is a requirement of the NRC Guide; however, this daily removal of waste is challenging to achieve in a pasture or barn setting when multiple animals are cohoused in an agricultural field or similar outdoor setting. Many agricultural facilities may only be sanitized between successive groups of animals, and this may only be feasible semiannually. For ruminants, swine, and poultry, disinfection of a primary housing stall or barn enclosures will be the exception, not the rule, in agricultural research (Tillman 1994). In equine research environments, proper drainage and removal of excrement is essential in stalls, paddocks, and walkways for preservation of foot health; in addition, high amounts of bedding substrates will help to absorb urine and prevent backsplash and scalding (Jonckheer-Sheehy and Houpt 2015). Overall, a plan should be enacted to ensure that animals are maintained in reasonably clean and dry surroundings that promote the health of the animal species and also the health of the human workers that interact with the animals in the housing environment.

Animal Handling and Restraints

Safe handling and restraint of agricultural animals is essential for a successful research, teaching, or testing program or production facility. Animal welfare is strongly influenced by the relationship between animals and humans. A positive respectful attitude toward animals and an understanding of animal behavior and human interactions with animals will help to reduce stress, improve outcomes, and reduce potential research bias.

Agricultural animals present substantial challenges to animal caretakers with respect handling and restraint. They are significantly larger and are herd or flock oriented, which can lead to safety problems if personnel are not familiar with that particular species' norms of behavior. Excellent resources are available on handling and restraint in the Ag Guide (FASS 2010). Independently offered workshops are another resource for training personnel (see Charles River Laboratories, FASS online modules, CCAC farm animals stream training, etc.), and the majority of research institutions using animals typically offer training to personnel handling animals.

Agricultural animals should be given appropriate opportunities to acclimate to humans that will work with them. In order to accomplish this, personnel need to be trained and experienced in animal behavior, in how to approach animals in a calm slow manner, and how to capture, lift, move, or herd animals by understanding flight zones and how to use them. This can be done by staff consciously and slowly walking in a nonaggressive manner through pen and stall areas or down aisles where animals are housed. Avoidance of quick movements (an action signatory of predators), limiting direct eye contact, and assuming a calm demeanor helps to dampen fear (CCAC 2009).

In a production agriculture setting with large numbers of animals, the animals may have only limited contact with humans and may require commercial handling equipment. In a biomedical setting where animals are housed indoors or in small groups, there is the opportunity for frequent human contact and the need to habituate animals to the restraints and procedures required in order to reduce stress in both humans and animals. In this case, it may also be possible to use reward systems (positive reinforcement) to elicit the animal's cooperation with the procedures.

Handling and human–animal interactions are improved with properly designed facilities. Factors to consider include facility flooring, equipment design and maintenance, sanitation, animal conditioning, minimal restraint if possible, and limiting restraint time. Examples of acceptable restraint devices include equipment such as hobbles, squeeze chutes, stanchions, or Panepinto slings (Grandin 2016b; Grandin 2014; FASS 2010). With the use of appropriate moving devices, such as pig boards, flags, slappers, rattle paddles, streamers, or halters with lead ropes, animals can be guided with minimal manual manipulation through facility areas. Electric prods should be limited to only specific situations with appropriate training on their use (Grandin 2016a; National Farm Animal Care Council 2016).

Animal Transportation and Equipment

Livestock transportation is a process that involves handling, loading, moving, unloading, and changes in environment for the animal. This process can be further impacted by factors such as isolation, new social groups, unstable footing, changes in ambient temperature and humidity, exhaust fumes, withholding of feed and water, and potential exposure to pathogenic organisms. Whether transporting animals within the facility or across jurisdictions, the primary concern should be the safety and comfort of the animal. In addition to the Ag Guide (FASS 2010), the NRC Guidelines for the Humane Transportation of Research Animals (NRC 2006) also contains an overview and recommendations for livestock, with special tables addressing considerations important to safely transporting livestock, including critical temperature zones and an index for safe transport based on weather conditions.

Prior to transportation, the health and physiological status of the animal and the duration of the move should be assessed to determine if the animal should be transported. Transport vehicles (e.g., transport carts, containers, trailers, and trucks with loading ramps) should be sized to accommodate the species and should be clean and dry, have a nonslip floor, be well bedded when needed, and have proper ventilation. The transport vehicle or trailer should be inspected to ensure that doors, latches, welds, flooring, hitches, tires, lights, and brakes are in working condition. Consideration should be given to loading animals at stocking densities appropriate for their weight, sex, species, and trip duration. While most agricultural animals have a large range for their thermoneutral zone, they need to be protected from temperature extremes during loading and unloading and transport. For the health and welfare of the animal, minimize the time that livestock and poultry are on vehicles and minimize stationary time when ventilation depends on vehicle movement.

Transporting livestock or poultry requires a knowledge of, and compliance with, current local, state, and federal regulations and identification requirements. There are additional regulations, inspections, and health certificates when animals cross state lines or international borders, depending on species. For long-duration transport, there are regulations and considerations for feed, water, exercise, and lairage (resting areas). A transportation emergency plan should be developed; this plan should include responses for emergencies such as delays, vehicle damage, accidents, escapes, illness, segregation, and euthanasia, and contact information for trained and qualified persons to assist during an emergency (Canadian Agri-Food Research Council 2001; NRC 2006; NRC 2011; FASS 2010; North American Meat Institute 2013).

Medical Records

Medical records, as a means to document animal health observations, are critical for monitoring animal health issues that may arise as part of the experimental outcomes or through iatrogenic routes. The presence of medical records is widely recognized as a hallmark of a program of adequate veterinary care (Field et al. 2007), and the AV must have the authority to oversee the medical records program.

These medical records should contain, at minimum, the identification of the animal, any relevant medical history of the animal, vaccination records, research purposes and any related research use or interventions, surgical reports, treatments, withdrawal dates from medication, and documentation of resolution of concerns. Personnel that typically may be responsible for making medical record notations include veterinary and technical staff, animal care staff, and research groups.

Information on active medical cases should be reviewed daily, and subsequent entries on animal health observations should be chronologically provided, dated, and initialed by the observer. Records may be maintained in electronic database format, or may include legible paper files maintained on site within barns or housing areas. In summary, every facility should have a system of medical records sufficiently maintained to demonstrate adequate veterinary involvement.

Medical Record Maintenance

It may be prudent to forego individual animal records in lieu of group health records for animals kept in cohorts. Methods of how to design medical records should be made in consultation with the AV and/or veterinary designees and should reflect an appropriate level of communication about existing, ongoing, and resolved health concerns in the agricultural populations.

Individual health records may be preferable for those specific animals that receive routine independent health evaluations or that have ongoing health needs that require frequent rechecks, treatments, or observations for progression of clinical signs or improvements.

Medical Record Accessibility

Per the AWA and Animal Welfare Regulations expectations on veterinary care, "animal medical records must be kept and made available for APHIS inspection." These records should be maintained for the duration of the activity and kept for an additional 3 years after completion of the projects. All records shall be available for inspection and copying by authorized APHIS representatives within a reasonable time frame; original materials are not to be removed by the inspector from the premises and confidentiality of the information will be maintained by USDA inspectors. Should a particular animal medical record be requested by the public from the USDA (e.g., from a Freedom of Information Act [FOIA] request) or from the institution itself, legal cooperation may be necessary to address the planned response.

Records Reflecting Animals Undergoing Surgical and/or Invasive Procedures

Aspects of surgical or invasive procedures on individual or cohorts of agricultural animals should be recorded and essentially present an accurate depiction of the process and outcome of the surgery. The use of expired medical materials (fluids, sutures, anesthetics, etc.) during any survival surgery procedure on a regulated species is not acceptable and inconsistent with adequate veterinary care (USDA 2016c). Similarly, pharmaceutical-grade substances are expected to be used whenever they are available, including for acute terminal procedures; cost savings alone is not a sufficient justification for the use of non-pharmaceutical-grade items.

Procedures should be thoroughly described in detail for surgical events. If animals are to be reused for multiple procedures, particularly palpation laboratories for teaching of students, the relative number of palpations over a designated time frame should be recorded and reviewed by veterinary personnel. In particular, postoperative analgesia to relieve pain and/or distress is expected to be delivered, barring a worthy scientific justification to withhold such treatment. The AV has the authority to alter postoperative care measures if unexpected painful outcomes occur in animals. If an agricultural animal is taken off site (to a farm or clinic) for further postoperative assistance, then the site may also need to be inspected by federal, regional, or institutional staff to ensure compliance with programmatic expectations of veterinary care.

Biosecurity and Animal Disposition

Biosecurity for agricultural animals involves the security measures taken to prevent unintentional animal infections and infestations. Biosecurity is achieved through bioexclusion, which can be accomplished through appropriate facility design, training, and operational precautions. The degree of biosecurity depends on the animals' susceptibility to disease and the intended use of the animals. A mechanism to review, restrict, and/or approve animal acquisitions should be in place to ensure that adequate housing is available and appropriate veterinary quarantine procedures are established before shipping or arrival of animals.

Animals should be obtained from reputable sources with good health management, and who can provide a known health status. Some studies may require animals with specific criteria, such as specific pathogen-free (SPF), nutritional status, genotype, disease status, or production traits. Health requirements may vary for animals required for acute versus short-term studies versus long-term studies. Animals of unknown origin (e.g., livestock auctions) may pose a significant health risk and should be handled accordingly. When facilities bring in new animals, they must be acquired and transported legally in compliance with applicable international, federal, and state regulations and institutional procedures. As mentioned previously, upon arrival animals should undergo a domestic quarantine and acclimation period, including veterinary assessment (physical examination) and administration of any appropriate preventative or clinical treatments.

Biosecurity for personnel should include a training program about animal hazards, immunizations against zoonotic agents (e.g., influenza) that can infect both humans and animals, personal protective equipment (PPE), avoidance of hand-to-mouth or -face activities in animal facilities, and policies for working within livestock areas. Visitors should be informed of the risks of entering an animal facility, and a biosecurity risk assessment should be performed to determine the risks of allowing visitors into an animal facility. Based on the outcomes of the biosecurity risk assessment, visitation may be limited to days or weeks following any dedicated animal contact, with a requirement for PPE and shower-in for entry, or may be forbidden based on disease concerns in general agricultural populations (e.g., to avoid potential transmission of avian influenza into a closed flock.)

Vermin, predators, and other animals pose additional biosecurity risks to be evaluated, excluded, or controlled. Examples include other domesticated agricultural species, dogs, cats, wildlife, wild rodents, birds, insects, and feral livestock. These animals can bring in or carry out pathogenic and zoonotic diseases, adversely affect resident populations, and contaminate feed and water supplies.

From both a biosecurity and a facility security perspective, a perimeter fence around agricultural animal areas should be strongly considered. A perimeter fence can prevent or discourage animals from leaving facility grounds and reduce potential liability due to property damage or personal injury when animals enter roads or other unsafe areas. A fence can deter other animals or people from entering the animal facility grounds, while providing dedicated access points that can be controlled, monitored, or locked when required for biosecurity purposes (e.g., highly pathogenic outbreaks) or for facility security. Considerations for a perimeter fence include function; design; material; fence posts; gates, crossings, and guards; access controls; signage; alarms; monitoring devices (e.g., cameras); power requirements; proximity to buildings or storage areas; deliveries and load-outs; buffer zone with double fencing; life span; maintenance; boundary lines; and applicable laws, regulations, ordinances, or rules (Levis and Baker 2016; Worley 2016).

Additional points to consider for biosecurity risk assessment are facility access points, facility layout, fencing, alarm systems, aerosol transmission (wind and airstreams), signage, personnel traffic patterns within and between buildings, boot cleaning and disinfection, contamination in feed, impure water, vehicles (transport, delivery, and service), equipment, manure handling, and carcass disposal. Hazard analysis algorithms are available for use within the agricultural setting to assess overall biosecurity (e.g., Hazard Analysis Critical Control Point [HACCP] system) (Beef Quality Assurance 2016).

Disposition of animals that are no longer required for research, teaching, or testing includes abattoir, auction market, return to home farm, sale, gifting, loaning, transfer of ownership, adoption, or euthanasia and appropriate disposal. The type of disposition will depend on a health and welfare assessment of the animal for suitability to the method of disposition and should be in compliance with veterinary recommendations, such as those from the American Veterinary Medical Association (AVMA). Final decisions on disposition should be made by the veterinarian in conjunction with recommendations from the research team, and must follow applicable international, federal, and state regulations, as well as any institutional procedures (CCAC 2009; FASS 2010).

Animals and Personnel Safety

People need to have a realistic understanding of agricultural animals and their behavior to avoid putting themselves or others in situations that could cause stress, harm, or damage to people, animals, or facilities. People who are not properly trained need to be closely supervised until competence is demonstrated in their required areas of work (Ontario Ministry of Labour 2006; CCAC 2009; FASS 2010).

Facilities should be designed with appropriate flooring, adequate sanitation, secure enclosures, and chutes, alleys, or raceways that prevent turning around or wedging and with sides high enough to prevent jumping out. Ramps and platforms should facilitate walking and prevent slippage, while step heights and slopes should be no greater than recommended maximums. Livestock tend to be curious about structures and items in and around their environment. It is best to examine an animal's structural environment to correct any elements that would allow an animal to become stuck, caught, or hung up on design elements. It is also important to maintain cleaning practices and remove foreign objects from the animal's environment before livestock lick, chew, or swallow objects they should not ingest.

When it comes to working with livestock, size matters. Many species of livestock are larger, stronger, and quicker than humans. As a result, there is the potential for serious injury or death to humans or other animals. Social structures and instincts, such as herding, hierarchy, dominance, and mothering, can result in radical changes in behavior when individual animals are separated from the herd or flock, new animals are introduced to the herd or flock, or during breeding, birthing, or mothering. In addition, some individuals, breeds, strains, and sexes have unpredictable temperaments, making them more difficult or even dangerous to handle. Safe handling of animals requires emphasis on education and training, properly designed facilities, appropriate handling equipment, and proper PPE (Dogan and Demirci 2012; Grandin 2015a).

An occupational health and safety program is a requirement for individuals working with animals (CCAC 2009; FASS 2010; NRC 2011). It should be noted that when animals are reared on farms in agricultural settings, the National Safety Council ranks agriculture as one of the most hazardous industries in the nation (OSHA 2009; Wilkinson and Tilma 2016). Thus, the risks must be appropriately assessed by occupational health and safety specialists and be consistent with federal, state, and local regulations.

Common hazards for people in an agricultural animal setting include (OSHA 2016; Victoria State Government, Australia 2016)

- Animals: Bites, kicks, scratches; knocked, tossed, gored, crushed, rammed, trampled; zoonotic infection
- Chemicals: Burns, respiratory illness, poisoning
- Confined spaces: Poisoning, engulfment, suffocation, drowning
- Drugs and drug exposures

    - Veterinary: Human reproductive issues, bronchoconstriction in asthmatic individuals, animal or human death, residue in animals or animal products
    - Research: Nonpharmaceutical grade, unknown side effects, adverse reactions, animal death, residue in animals or animal products
- Electricity: Shocks or electrocution
- Ergonomic risks: Musculoskeletal injury
- Heights: Falls resulting in injury, death
- Machinery: Crashes, rollovers, falls, amputations, crushing, ejection, entanglement
- Noise pollution: Hearing loss
- Vehicles (motorized or animal): Crashes, falls
- Water: Slip and fall, drowning
- Weather: Sunburn, heat stroke, dehydration, frostbite, hypothermia, lightning strike

Special Research Considerations

Researchers and farm personnel should meet to discuss needs, expectations, and priorities prior to beginning a research project to achieve a mutual understanding on the following topics:

- Project and the expected timelines and outcomes
- Labor requirements
- Equipment and facility needs
- Scheduling of activities (husbandry efforts and experimental treatments)
- Special aspects or requirements of the project

Researchers will aim for as much consistency of activities as possible for animals on research projects (e.g., husbandry efforts to feed and clean at consistent times daily); this same effort for consistency should be a priority of the animal management group. Realistically, it is impossible for every project to be a first or top priority. Understanding and achieving consensus between the research teams and husbandry management on these details will help reduce potential conflicts.

Tracking Animals on Experiments and across Teaching Courses

Once research and teaching projects have been approved by the IACUC, it is imperative to maintain records of animals on these projects. These records facilitate tracking animal usage and decisions on the potential for planned reuse of animals. A variety of information can be maintained for each animal, including origin, medical record (health history), experimental use (e.g., number of palpations performed on the animal by trainees, blood draw descriptions, and physical examination practice), and ability to be utilized on differing projects to meet institutional objectives. The following list offers suggestions of data important to track:

- Animal medical history (e.g., health, lameness and hoof lesions, treatment, preventative care, anesthetic, surgical care, mortality, and necropsy)
- Potentially harmful products to which the animal has been exposed (e.g., drugs, chemicals, and biologicals)
- Individual animal and/or group of animals (including acquisition and disposition)
- Nutrition information
- Production records
- Breed or pedigree

- Experimental animal use on approved IACUC protocols
- Genetic manipulations
- Pesticide use
- Animals on hand (inventory)
- Attending or clinical veterinary visits
- Room or pen maintenance logs

Because large agricultural species are typically long-lived, within food and fiber production settings it is not uncommon for animals to be used in multiple projects (e.g., nutrition, production, and breeding). Tracking usage (and reuse) becomes critical in documenting an animal's complete medical history and making decisions about its suitability for a new project.

Record keeping and maintenance of the associated paperwork takes time and effort. The time requirement should be acknowledged as an expected part of the manager's job, and appropriate time must be allotted to accomplish these critical tasks (CCAC 2009; FASS 2010; NRC 2011; ; European Union 2016; USDA 2016b).

Effects of Research on Animals and Animal Products

Working with agricultural animals allows for the potential sale or harvest of the animals and their products in the form of meat, milk, eggs, fiber, and waste. To determine the actual cost of an activity and the appropriate disposition of both animals and animal products, it must first be determined whether the animals and/or their products are edible, adulterated, or contaminated. This information dictates whether the animals and/or their food products may be processed through normal food production channels or if they must not enter the human food chain. Both the Food and Drug Administration and the USDA are primary regulatory agencies that oversee food law with respect to animal products in the United States.

Sale or disposal will be determined by the applicable laws and regulations of the country, state, or locale. While income can be gained through the potential sale of healthy animals and/or their products, there can also be income losses and costs associated with the disposal of these animals and their products. Consultation with a veterinarian, qualified animal scientist, or appropriate agency official familiar with the species and products of interest can be useful in making these decisions (CCAC 2009; FASS 2010; NRC 2011).

Review of Projects and Communication about Project Requirements

Animal use projects, SOPs for routine animal care, and veterinary guidelines are reviewed as part of the IACUC oversight. However, institutional approval may not be the same thing as farm or facility approval. PIs or teaching instructors planning to use animals should contact facility or husbandry managers during the project planning phase and before protocols are submitted for IACUC approval so that animals, facilities, project needs, and scheduling are assured. Once an animal use protocol has been approved, written notification from the IACUC is sent to the PI or class instructor confirming that work may commence. It is the responsibility of the PI or instructor to inform the farm or facility personnel that a particular research or teaching protocol has been approved, to make arrangements for commencement of the project, and to provide the essential details that farm or facility requires to meet the request. There are any number of ways this notification of relevant stakeholders can take place; the next paragraphs describe forms and PI notifications as options.

Department or Unit Animal Use Form

After projects are approved by the IACUC, a form can be completed by the PI that provides a project overview and the specific farm or facility details required to conduct the project (e.g., special feed requirements, animal availability, appropriate housing space, and equipment and labor needs). This completed form is then sent to each responsible party (e.g., farm or facility manager, research coordinator, chairperson, and unit director) to review and sign off on the project, including agreement on other important issues, such scheduling and per diem charges. The document can then be kept for reference throughout the lifetime of the project.

PI Notification to Farm or Facility Manager

Alternatively, PI notification may only need to go to the farm or facility manager. It is imperative that farm or facility managers are fully informed about the expectations of the project, have enough space to house the species and animal numbers requested, and have indicated when the project can begin. Notifications may also be conducted through designated user groups (e.g., a farm or facility research committee) that include key stakeholders involved in animal projects.

Scheduling Projects to Run in Parallel within the Same Facility or Farm

In active animal facilities, the number of projects and the subsequent demand for animals can outpace the capacity of the facility. Prioritization is required to maximize use of both animals and facilities.

There are several approaches to prioritize these decisions depending on established relationships and arrangements within the institution or facility. The decision could be made by an individual (e.g., facility manager), small group (e.g., facility manager and research coordinator or veterinarian), or a committee (e.g., facility manager, research coordinator, scientists, and veterinarian) familiar with the projects or activities to be performed; these individuals should also have an understanding of the facilities, animal species, and farm area management.

Adverse Event Reporting

Procedures must be established to report problems such as animal injury, illness, or death, and mechanical breakdowns or malfunctions within the animal facility. A reporting structure must first be established to identify those people responsible for the health, safety, and welfare of the animals, facilities, and activities, for example, the facility manager, the AV or institutional veterinarian, and the investigator(s), if part of a research trial. Animal care management and research personnel must be aware of these procedures and the obligation to report such problems in a timely manner to the responsible individuals. A beneficial approach to reporting is "if you see something, say something." It is also important for all personnel to understand what to do when an unexpected adverse event occurs. An adverse event is an outcome that has a negative impact on an animal's welfare beyond what was described in an approved protocol or SOP. Such an event would not only be a detriment to animal health and welfare but also be a disruption of intended data collection or research goals. This can be a point of confusion in agricultural production systems when an adverse event occurs in the course of routine agricultural animal management and not on a research protocol. Whether the animals are on protocol or not, any unexpected adverse event should be reported with expediency (NRC 2011; Queensland Government, Department of Agriculture and Fisheries 2016).

Training

Training is a fundamental requirement that is mandated to ensure animal care workers are qualified for the tasks they perform. This is emphasized in all the guidance documents for agricultural and biomedical research, teaching, and testing. Training for agricultural species and work in agricultural settings requires nuanced topics due to the complexity and scope of typical farm operations.

The categories of training are similar to those for other laboratory animal care workers, which include training for species-specific clinically healthy appearance and behaviors, SOPs (e.g., for all aspects of farm operations, equipment maintenance, and husbandry), IACUC and regulatory expectations, environmental health and safety, and occupational health and safety.

Agricultural facilities are unique in their type, size, amount, and variety of animal housing; feeds and storage; manure handling and storage; equipment; noise; and dust. Within each broad training category listed above, there may be more specific training requirements based on the variety of activities in agricultural operations. For example, occupational health and safety training could include back safety training for lifting, confined space training for manure pits and grain bins, fall protection training for silos, hand and portable power tool training for workshops, heat stress training, ladder safety training, lockout and tag-out training for the servicing and maintenance of machines and equipment, nonregulated vehicle (utility vehicle and skid steer) training, hearing conservation programs and testing, respirator use and fit test training, right-to-know training for hazardous materials, and tractor safety training and performance evaluation. A thorough professional risk assessment will help establish what training is needed for working with agricultural species in specific facility settings.

IACUC Membership Should Include Individuals with Agricultural Background and Training

In the United States, blending of projects that use agricultural species for biomedical use necessitates that both the Ag and NRC Guides (FASS 2010; NRC 2011) are incorporated and that the IACUC ultimately concurs with the expectations for the care, housing, husbandry, and facilities for the animals. Enrolling agricultural species specialists, animal science experts, and large animal and food animal veterinarians into the membership of the IACUC is the best practice for enhancing the discussion and critical conversations necessary to provide training and education on species-specific needs and nonresearch areas, like behavior, husbandry, and standards of large animal veterinary practice. The advantage to a single institutional animal oversight committee is that the established guidelines and policies will apply across academic lines in an "unbiased" manner, compared with another scenario where two distinct committees review agricultural and biomedical work separately (Curtis 1994; Tillman 1994).

An institution that uses agricultural animals in research, teaching, and testing should have a constituted IACUC that includes members with agricultural expertise. These members will be able to assess and monitor agricultural animal care and use compared with more traditional laboratory animal (e.g., mouse and rat) care and use. Within the IACUC, one person may fulfill more than a single dedicated role. For agricultural animal research, teaching, and testing, the IACUC should include a scientist who has experience in agricultural research or teaching involving agricultural animals; an animal, dairy, or poultry scientist who has training and experience in the management of agricultural animals; and a veterinarian who has training and experience in agricultural animal medicine and who is licensed or eligible to be licensed to practice veterinary medicine (FASS 2010).

Public Health Service Assurance Statement for the Institution

In order to receive PHS funding, the federal Office of Laboratory Animal Welfare (OLAW) must first review and approve an institution's Animal Welfare Assurance document. This document describes how the institution will comply with the PHS Policy on Humane Care and Use of Laboratory Animals (OLAW 2015). How the institution's Animal Welfare Assurance statement is written and approved will determine the level of regulatory oversight and reporting, as well as which guidelines (NRC Guide or Ag Guide) must be applied for agricultural animal research and teaching versus biomedical animal research and teaching. As mentioned previously, under the AWA, the law specifically excludes the following agricultural animals: horses not used for research purposes, and other farm animals, such as, but not limited to, livestock or poultry used or intended for use as food or fiber, or livestock or poultry used or intended for use for improving animal nutrition, breeding, management, or production efficiency, or for improving the quality of food or fiber.

If an institution writes its Animal Welfare Assurance document for all live vertebrate animals, then all animals come under the NRC Guide and its attending oversight and reporting requirements. If the Animal Welfare Assurance document is written only for PHS or NSF-funded vertebrates, then only animals used in biomedical research, teaching, or testing will fall under the NRC Guide and its oversight and reporting requirements (Smith 2011).

Administrative Commitments to Agricultural Animal Research

Administrative commitments to academic agricultural animal research programs may be confounded by individual programs or departments in an academic setting, or even by virtue of which external sources of funding (if any) are available to support facilities. As an example, institutions may have differing expectations for care of animals in agricultural farm settings when compared with how the same species would be housed indoors for biomedical or surgical models of human disease; these differences will arise almost entirely due to the planned research objectives. The distinction is not insignificant, nor does it indicate that there is a failure of one level of oversight compared with another. In fact, the Ag Guide specifically addresses that activities for agricultural research should be appreciated for what they contribute to agricultural research and teaching, without the level of prescription that is entailed in the NRC Guide. The fact that agricultural animals used for agricultural research are not covered by the AWA can inherently set up a schism between the oversight expectations at a particular institution. Ultimately, there are certain practices (i.e., confinement housing) that, while common in animal agriculture, would not be permitted under the NRC Guide. Despite differences in practices, support for the ultimate research goals is essential and will functionally elevate animal welfare and data outcomes.

Recruiting New Biomedical Animal Users Who Use Agricultural Species

It is advised that the AV and nonveterinary facility and support staff be kept informed of planned initiatives for biomedical/agricultural research at the institution. At approximately the second return visit for a prospective research hire, determining the hire's animal needs, numbers, and research goals will be very helpful for planning their space and regulatory needs. Start-up packages with research funding, equipment needs, and other affiliated bonuses should attempt to accurately represent the location for animal housing and available institutional resources.

If agricultural animals are to be housed in novel settings (e.g., sheep for surgical biomedical research at an institution that has previously not housed agricultural animals indoors), critical discussions about surgical suite expectations, facility requirements, and applicable biosecurity procedures should happen early in the recruitment phase. Ideally, these discussions should involve the veterinary staff, architects, contractors, and facility and planning personnel from central administration, along with the research group that is hoping to establish their new program. There can be a considerable cost to renovations or new construction for enabling agricultural facilities to meet biomedical standards; therefore, identification of funding for such renovations and equipment needs (hydraulic tables, stainless steel caging, social housing space, etc.) should occur prior to hiring of new researchers.

Summary

This chapter has undertaken the topic of managing agricultural species (dairy, swine, sheep, horse, and poultry) for a multitude of research, teaching, or production purposes. While the background for the provided guidance is based on U.S. perspectives, topics of animal health, welfare, housing, enrichment, and care are applicable globally. A distinction between whether work is agricultural or biomedical in scope is critical for the application of appropriate policies and expectations for the animal environment. Frequent communication among stakeholders in the animal work (investigators, IACUC, and veterinary professionals) will assist with planning and preparedness for optimal facility, housing, and agricultural animal care needs.

REFERENCES

Agriculture Victoria. 2016. Caring for animals during extreme heat. Melbourne: Agriculture Victoria. http://agriculture.vic.gov.au/agriculture/animal-health-and-welfare/animal-health/animals-in-hot-conditions/caring-for-animals-during-extreme-heat (accessed April 20, 2016).

Alkire, D. 2009. Monitor and manage heat stress. In *Livestock Weather Hazard Guide*. Ardmore, OK: Samuel Roberts Noble Foundation. http://www.noble.org/ag/livestock/monitorheatstress/ (accessed April 20, 2016).

Beef Quality Assurance. 2016. *Beef Quality Assurance Manual*. Centennial, CO: Beef Quality Assurance. www.BQA.org or http://www.bqa.org/resources/manuals (accessed April 20, 2016).

Bloomsmith, M.A., L.Y. Brent, and S.J. Shapiro. 1991. Guidelines for developing and managing an environmental enrichment program for nonhuman primates. *Laboratory Animal Science* 41: 372–377.

Canadian Agri-Food Research Council. 2001. *Recommended code of practice for the care and handling of farm animals – Transportation*. Nepean, Ontario: Canadian Food Inspection Agency Animal Health & Production Division. https://www.nfacc.ca/pdfs/codes/transport_code_of_practice.pdf (accessed March 28, 2017).

CCAC (Canadian Council on Animal Care). 2009. *Guidelines On: The Care and Use of Farm Animals in Research, Teaching and Testing*. Ottawa: Canadian Council on Animal Care. http://www.ccac.ca (accessed March 28, 2017).

CCAC (Canadian Council on Animal Care). 2016a. Training module: Farm animals used in biomedical research. Companion notes. Ottawa: Canadian Council on Animal Care. http://www.ccac.ca/Documents/Education /Modules/Farm-animals/Biomedical_research/Companion-notes.pdf (accessed April 20, 2016).

CCAC (Canadian Council on Animal Care). 2016b. Training module on: Farm animals used in biomedical research. PowerPoint presentation. Ottawa: Canadian Council on Animal Care. http://www.ccac.ca/en _/training/niaut/farm/ethical_use_and_care_biomedical (accessed April 20, 2016).

Choinière, Y. 1993. Air quality inside livestock barns. Fact sheet. Guelph: Centre for Food and Animal Research, Agriculture and Agri-Food Canada, Ontario Ministry of Agriculture, Food and Rural Affairs. http://www.omafra.gov.on.ca/english/livestock/swine/facts/93-001.htm (accessed April 20, 2016).

Collier, R., L.W. Hall, S. Rungruan, and R.B. Zimbleman. 2012. Revised temperature humidity index for lactating dairy cows. http://dairy.ifas.ufl.edu/rns/2012/6CollierRNS2012a.pdf (accessed April 20, 2016).

Curtis, S.E. 1994. Commentary: Farm animal use in biomedical science—Melding the guidelines. *ILAR News* 36: 35–39.

Dahlen, C.R., and C.L. Stoltenow. 2012. Dealing with heat stress in beef cattle operations. Fargo: North Dakota State University Extension Service. http://www.ag..edu/pubs/ansci/beef/as1615.pdf (accessed April 20, 2016).

Dogan, K.H., and S. Demirci. 2012. Livestock-handling related injuries and deaths. In *Livestock Production*, ed. K. Javed. Rijeka, Croatia: InTech. http://www.intechopen.com/books/livestock-production/livestock -handling-related-injuries-and-deaths (accessed April 20, 2016).

European Commission. 2016. Directive 2010/63/EU on protection of animals used for scientific purposes: Education and training framework. Caring for animals aiming for better science. Brussels: European Commission. http://ec.europa.eu/environment/chemicals/lab_animals/pdf/guidance/education_training/en .pdf (accessed April 20, 2016).

European Union. 2016. Appendix A of the European Convention for the Protection of Vertebrate Animals Used for Experimental and Other Scientific Purposes (ETS 123). Guidelines for Accommodation and Care of Animals (Article 5 of the Convention). Strasbourg: Council of Europe. https://rm.coe .int/CoERMPublicCommonSearchServices/DisplayDCTMContent?documentId=090000168007a445 (accessed April 20, 2016).

FASS (Federation of Animal Science Societies). 2010. *Guide for the Care and Use of Agricultural Animals in Research and Teaching*. 3rd ed. Champagne, IL: FASS.

Festing, M.F.W., and D.G. Altman. 2002. Guidelines for the design and statistical analysis of experiments using laboratory animals. *ILAR Journal* 43: 244–258.

Field, K., M. Bailey, L.L Foreseman, R.L. Harris, S.L. Motzel, R.A. Rockar, G. Ruble, and M.A. Suckow. 2007. Medical records for animals used in research, teaching, and testing: Public statement from the American College of Laboratory Animal Medicine. *ILAR Journal* 48: 37–41.

Goodwin, B.S., Jr., and J.C. Donaho. 2010. Tropical storm and hurricane recovery and preparedness strategies. *ILAR Journal* 51(2): 104–119.

Grandin, T. 2014. *Livestock Handling and Transport*. 4th ed. http://www.grandin.com/inc/livestock.handling .transport.4th.html (accessed April 20, 2016). Wallingford, UK: CABI International.

Grandin, T. 2015a. Assessment of temperament in cattle and its effect on weight gain and meat quality and other recent research on hairwhorls, coat color, bone thickness, and fertility. http://www.grandin.com /behaviour/principles/assessment.temperament.html (accessed April 20, 2016).

Grandin, T. 2015b. *Improving Animal Welfare: A Practical Approach*. 2nd ed. Wallingford, UK: CABI International.

Grandin, T. 2016a. Recommended basic livestock handling safety tips for workers. http://www.grandin.com/behaviour/principles/principles.html (accessed April 20, 2016).

Grandin, T. 2016b. Website. http://www.grandin.com/ (accessed April 20, 2016).

Ha, J.C., and N. Andrews. 2014. The future of enrichment for lab animals. *The Enrichment Record*, p.6-7. Website. http://enrichmentrecord.com/wp-content/uploads/2014/07/FUTURE-OF-ENRICHMENT.pdf (accessed March 27, 2017).

Hamilton, T. 2006. Beef cattle's adjustment to cold weather. Fact sheet. Guelph: Production Systems, Ontario Ministry of Agriculture, Food and Rural Affairs. http://www.omafra.gov.on.ca/english/livestock/beef/news/vbn1006a4.htm (accessed April 20, 2016).

Helke, K.L., P.C. Ezell, Duran-Struuck, and M.M. Swindle. 2015. Biology and diseases of swine. In *Laboratory Animal Medicine*, ed. L.C. Anderson, O.G. Pritchett, M.T. Whary, and J.G. Fox, 695–769. 3rd ed. Cambridge, MA: Academic Press.

Hunton, P., and L.J. Weber. 1988. Poultry: Heat stress in caged layers. Fact sheet. Guelph: Ontario Ministry of Agriculture, Food and Rural Affairs. http://www.omafra.gov.on.ca/english/livestock/poultry/facts/88-111.htm (accessed April 20, 2016).

Johnstone, P. 2013. *Planning and Managing Agricultural and Ecological Experiments*. New York: Routledge.

Jonckheer-Sheehy, V.S.M., and K.A. Houpt. 2015. Management methods to improve the welfare of horses used in research. *Lab Animal* 44: 350–358.

Kerr, S. 2015. Livestock heat stress: Recognition, response, and prevention. Washington State University Extension Fact Sheet FS157E. Pullman: Washington State University. http://cru.cahe.wsu.edu/CEPublications/FS157E/FS157E.pdf (accessed April 20, 2016).

Levis, D., and R. Baker. 2016. Biosecurity of pigs and farm security. NebGuide EC289. Lincoln: University of Nebraska–Lincoln Extension. http://extensionpublications.unl.edu/assets/pdf/ec289.pdf (accessed May 2, 2016)

Martinson, K., M. Hathaway, C. Ward, and R. Johnson. 2016. Managing horses during hot weather. Minneapolis: University of Minnesota Extension. http://www.extension.umn.edu/agriculture/horse/care/managing-horses-during-hot-weather/ (accessed April 20, 2016).

Mississippi State University Extension Service. 2016. Body condition scoring in horses. Starkville: Mississippi State University Extension Service. http://msucares.com/livestock/equine/pdfs/p2465-horses.pdf (accessed April 20, 2016).

National Farm Animal Care Council. 2001. The code of practice for the care and handling of farm animals. Transportation. Lacombe, Alberta: National Farm Animal Care Council. http://www.nfacc.ca/codes-of-practice/transport (accessed April 20, 2016).

National Farm Animal Care Council. 2016. Codes of practice for the care and handling of farm animals. Lacombe, Alberta: National Farm Animal Care Council. http://www.nfacc.ca/codes-of-practice (accessed April 20, 2016).

National Hog Farmer. 2016. Sow body condition scoring guidelines. http://nationalhogfarmer.com/site-files/nationalhogfarmer.com/files/archive/nationalhogfarmer.com/posters/BodyConditionScorePoster.pdf (accessed April 20, 2016).

New South Wales Department of Primary Industries, Animal Welfare Branch. 2010. Animal Research Review Panel. Guideline 23: Guidelines for the housing of sheep in scientific institutions. Orange: New South Wales Department of Primary Industries. Emergencies and animal welfare, industry and investment animal ethics infolink: http://www.animalethics.org.au (accessed April 20, 2016); http://www.animalethics.org.au/policies-and-guidelines/animal-care (accessed March 28, 2017).

North American Meat Institute. 2013. *Recommended Animal Handling Guidelines & Audit Guide*. 2013 ed. Washington, DC: North American Meat Institute.

NRC (National Research Council). 2006. *Guidelines for the Humane Transportation of Research Animals*. Washington, DC: National Academies Press.

NRC (National Research Council). 2011. *Guide for the Care and Use of Laboratory Animals*. 8th ed. Washington, DC: National Research Council, National Academies Press.

OLAW (Office of Laboratory Animal Welfare). 2015. Obtaining an Assurance. Bethesda, MD: OLAW. https://grants.nih.gov/grants/olaw/obtain_assurance.htm (accessed March 30, 2017)

Ontario Ministry of Labour. 2006. Large animal handling: Occupational health and safety guidelines for farming operations in Ontario. Toronto: Ontario Ministry of Labour. http://www.labour.gov.on.ca/english/hs/pubs/farming/gl_animal.php (accessed April 20, 2016).

OSHA (Occupational Safety and Health Administration). 2009. Fact sheet. Farm safety. Washington, DC: OSHA. www.OSHA.gov (accessed April 20, 2016); https://www.osha.gov/OshDoc/data_General_Facts/FarmFactS2.pdf (accessed April 20, 2016).

OSHA (Occupational Safety and Health Administration). 2016. Agricultural operations hazards and controls Washington, DC: OSHA. https://www.osha.gov/dsg/topics/agriculturaloperations/hazards_controls.html (accessed April 20, 2016).

Queensland Government, Department of Agriculture and Fisheries. 2016. Reporting unexpected adverse events by investigators, teachers and animal facilities where animals are used for scientific purposes. https://www.daf.qld.gov.au/animal-industries/welfare-and-ethics/using-animals-for-scientificpurposes/animal-ethics-committees/reporting-unexpected-adverse-events. (accessed April 20, 2016).

Rozeboom, D., and J. May. 2012. Safety precautions for agitating and pumping manure. East Lansing: Michigan State University Extension. http://www.msue.msu.edu (accessed April 20, 2016); http://msue.anr.msu.edu/news/safety_precautions_for_agitating_and_pumping_manure (accessed April 20, 2016).

Russow, L.M., and P. Theran. 2003. Ethical issues concerning animal research outside the laboratory. *ILAR Journal* 44 (3): 87–190.

Smith, D. 2011. Public Health Service (PHS) assurance statement for the institution. In *Managing the Research University*, 228. Oxford: Oxford University Press.

Stricklin, W.R., and J.A. Mench. 1994. Oversight of the use of agricultural animals in university teaching and research. *ILAR News* 36: 9–14.

Swanson, J.C. 1998. Oversight of farm animals in research. *Lab Animal* 27: 2–5.

Swearengen, J.R., K.J. Vargas, M.K. Tate, and N.S. Linde. 2010. Disaster preparedness in biocontainment animal research facilities: Developing and implementing an incident response plan (IRP). *ILAR Journal* 51: 120–126.

Tarr, B. 2007. Cold stress in cows. Fact sheet. Guelph: Ontario Ministry of Agriculture, Food and Rural Affairs. http://www.omafra.gov.on.ca/english/livestock/beef/facts/07-001.htm (accessed April 20, 2016).

Tillman, P. 1994. Integrating agricultural and biomedical research policies: Conflicts and opportunities. *ILAR News* 36: 29–35.

Underwood, W.J., R. Blauwiekel, M.L. Delano, R. Gillesby, S. Mischler, and A. Schoell. 2015. Biology and diseases of ruminants (sheep, goats, and cattle). In *Laboratory Animal Medicine*, ed. L.C. Anderson, O.G. Pritchett, M.T. Whary, and J.G. Fox, 623–694. 3rd ed. Cambridge, MA: Academic Press.

Underwood W.J., J.J. McGlone, J. Swanson, K.A. Anderson, and R. Anthony. 2014. Agricultural animal welfare. In *Laboratory Animal Welfare*, ed. K. Bayne, 233–278. Amsterdam: Elsevier.

United Egg Producers. 2016. Animal husbandry guidelines for U.S. egg laying flocks: 2016 edition. Johns Creek, GA: United Egg Producers. http://www.unitedegg.org/information/pdf/UEP-Animal-Welfare-Guidelines2016.pdf or http://www.unitedegg.org/AnimalWelfare/ (accessed May 2, 2016).

USDA (U.S. Department of Agriculture), Animal and Plant Health Inspection Service. 2012. Animal care fact sheet. APHIS 41-05-012. Washington, DC: USDA. https://www.aphis.usda.gov/publications/animal_welfare/content/printable_version/fs_compliance_inspection.pdf (accessed March 22, 2017).

USDA (U.S. Department of Agriculture), Agricultural Research Service. 2016a. Cattle heat stress forecast. Washington, DC: USDA. http://www.ars.usda.gov/Main/docs.htm?docid=21306 (accessed April 20, 2016).

USDA (U.S. Department of Agriculture), Animal and Plant Health Inspection Service. 2016b. *USDA Animal Welfare Inspection Guide*. Washington, DC: USDA. https://www.aphis.usda.gov/animal_welfare/downloads/Animal-Care-Inspection-Guide.pdf (accessed March 22, 2017).

USDA (U.S. Department of Agriculture), Animal and Plant Health Inspection Service. 2016c. Animal Care Policy #3: Veterinary care. Washington, DC: USDA. https://www.nal.usda.gov/sites/default/files/Policy3.pdf (accessed March 27, 2017).

U.S. Department of Energy. 2016. Using backup generators: Alternative backup power options. Washington, DC: Office of Electricity Delivery & Energy Reliability. http://energy.gov/oe/community-guidelines-energy-emergencies/using-backup-generators-alternative-backup-power-options (accessed April 20, 2016).

University of Arkansas at Pine Bluff. 2016. Body condition scoring of sheep. Pine Bluff: University of Arkansas. http://www.uaex.edu/publications/pdf/FSA-9610.pdf (accessed April 20, 2016).

University of Georgia Cooperative Extension Service. 2007. Carbon monoxide measuring and monitoring. *Poultry Housing Tips* 19(3). https://www.poultryventilation.com/tips/vol19/n3 (accessed April 20, 2016).

University of Illinois. 2008. Guidelines for determining when IACUC oversight, veterinary authority, and facility inspections are required for research, teaching, demonstration and extension activities using animals. Approved by the Institutional Animal Care and Use Committee, July. Urbana-Champaign: University of Illinois. https://research.illinois.edu/files/upload/off-campus_animal_use.pdf (accessed April 20, 2016).

University of Kentucky. 2013. IACUC policies, procedures, and guidelines: Conducting research, teaching, or testing using privately owned animals on university property or on privately owned land. Amended and approved by the Institutional Animal Care and Use Committee, November 13. Lexington: University of Kentucky. http://www.research.uky.edu/ori/univet/resources/sop/IACUC_123_Off-Site_Animal_Use.pdf (accessed April 20, 2016).

University of Nebraska–Lincoln. 2016. Body condition scoring your beef cow herd. Lincoln: University of Nebraska. http://beef.unl.edu/learning/condition1a.shtml (accessed April 20, 2016).

Victoria State Government, Australia. 2016. Better health channel. Farm safety—Risks and hazards. https://www.betterhealth.vic.gov.au/health/healthyliving/farm-safety-risks-and-hazards (accessed April 20, 2016).

Wheeler, E.F. 2009. Ammonia monitoring in animal environments using simple instruments. Fact sheet G-110. University Park: Pennsylvania State University Cooperative Extension Service.

Wilkinson, R., and A. Tilma. 2016. Livestock handling and confinement safety. National Ag Safety Database. East Lansing: Michigan State University Extension. http://nasdonline.org/document/913/d000754/livestock-handling-and-confinement-safety.html (accessed April 20, 2016).

Worley, J. 2016. Fences for the farm (C 774). Athens: University of Georgia Cooperative Extension. http://extension.uga.edu/publications/detail.cfm?number=C774 (accessed May 2, 2016).

Young, R.J. 2003. *Environmental Enrichment for Captive Animals.* Wiley Blackwell Animal Welfare Series. Chichester, UK: Wiley Blackwell, Universities Federation for Animal Welfare.

24

Aquatics

Christian Lawrence, George E. Sanders, and Carole Wilson

CONTENTS

Introduction

This chapter provides the reader with a detailed overview of the basic requirements for housing aquatic animals used in research. The treatment of this subject in the chapter spans from the microenvironment (critical parameters of water quality and how various aquaculture systems and aquatic enclosures are designed and maintain this) to the macroenvironment (facility design and system functionality).

As is the case for any species of laboratory animal, whether aquatic or terrestrial, the overarching goal of housing is to provide appropriate living conditions that maintain environmental parameters within a range that promotes maximum welfare. It is also critical to consider that the conditions experienced by research animals are, in essence, the most basic component of the experiments in which they are utilized. Therefore, housing should promote stability and definition of the environment such that the myriad physical variables that may impact the animal are controlled and known. Not only is this approach the most ethical course of action, but it also promotes science of the highest quality and reproducibility.

At the same time, regulatory oversight governing the use of aquatic animals is becoming more stringent. Both in Europe and the United States, the most recent versions of the primary regulatory documents followed by laboratory animal programs, the European Convention for the Protection of Vertebrate Animals Used for Experimental and Other Scientific Purposes (ETS 123), European Directive 2010/63/EU, and the *Guide for the Care and Use of Laboratory Animals*, now contain specific information related to the care and use of aquatic species (COE 1991; European Union 2010; Institute of Laboratory Animal Research 2011; Short Guide 2014). This development, which has been mirrored elsewhere in the world (Perry 2013), compels veterinarians, facility managers, and researchers to increase their understanding of the multitude of variables that impact the care and use of aquatic species, none of which are more elemental than the environment in which they are maintained.

The Microenvironment

The central component of the microenvironment of an aquatic animal is water. The aqueous environment has numerous features, both physical and chemical in nature. These parameters interact with each other and the organisms that reside within it (both target and incidental) to form a dynamic setting that requires frequent control and monitoring to ensure that the conditions are maintained within a favorable range to support life. This landscape may be collectively referred to as water quality, and is central to the entire concept of housing aquatic animals. The following sections describe the physical and chemical components of water quality most critical in aquatic animal housing environments.

Physical Parameters

Temperature

Temperature exerts profound effects on aquatic animals, as well as on the biological and chemical processes that define their environments. Species that cannot maintain thermal homeostasis, poikilotherms (i.e., those that have a body temperature that varies with their surroundings) in particular, display varying degrees of tolerance to changes in temperature, as well as a more narrow optimum range in which they perform best (Kelsch and Neill 1990).

Temperature influences virtually every facet of the biology of a given aquatic species: activity, feeding, behavior, growth, and reproduction, to name a few. In addition, it is critical to consider that the nontarget organisms residing within aquatic systems, such as microbes, will also have varying temperature tolerances. This may impact the virulence of certain pathogens (Frans et al. 2011), as well as the efficiency of nitrifying bacteria in biological filters (Masser et al. 1999).

Temperature also impacts water chemistry in various ways. For example, temperature influences the level of dissolved gases in solution. In general, the solubility of gases increases as temperature decreases. It also plays a role in determining the relative toxicity of total ammonia nitrogen (TAN) in water; as temperature increases, so too does the fraction of ammonia that is toxic to aquatic animals

(Timmons et al. 2002). This means there is generally a greater risk of ammonia poisoning in warm water systems than in cold water systems.

Chemical Parameters

pH

The pH of an aqueous solution is determined by the relative concentration of acids and bases within it. One of the major goals of water quality management in aquatic animal housing is to maintain the water within a pH range that supports both the health of the animals and the functionality of the system itself. This is no simple proposition, since a number of processes, including animal metabolism and the breakdown of organic wastes in the water, all tend to drive the pH of the water downward. If the water is not buffered, the pH will typically first drop below the threshold required for biological filtration (see the "Chemical Parameters" section), resulting in an ammonia spike that will then pose a direct threat to the lives of the animals housed within. Accordingly, bases must be added to the water to offset the production of acids. This can be achieved via manual addition of buffers, usually sodium bicarbonate, to the system. Systems with automated pH monitoring and control options measure the pH continuously via in-line probes, and can be programmed to automatically dose the system with sodium bicarbonate solution when the pH level drops below a predetermined set point. When this happens, the system is dosed until the pH moves back above this target. At that point, the dosing stops and the system resumes normal operation until values drop below set points again.

Alkalinity

Alkalinity is the measurement of all bases present in an aqueous solution. In practice, alkalinity can be thought of as the buffering capacity of the solution; the higher the alkalinity, the more resistant it is to changes in pH. Conversely, solutions with low alkalinity values are highly susceptible to swings in pH. The most important components of alkalinity, which is expressed in milligrams per liter of $CaCO_3$, are carbonate $\left(CO_3^{2-}\right)$ and bicarbonate $\left(HCO_3^-\right)$ ions. A number of biological processes, including animal respiration and decomposition, tend to produce acids and drive pH values down in aquatic housing systems. The alkalinity of the water determines the extent to which the pH will actually move as a result of these factors. In general, it is advisable to maintain alkalinity values within the range of 50–150 mg/L $CaCO_3$ for fish (Wurts 2002) and frogs (Delpire et al. 2011). Requirements are likely to be similar for any aquatic animal. Waters with alkalinity values of less than 20 mg/L are considered unacceptably low and are inherently unstable (Wurts 2002). Alkalinity is also crucially important for efficient biological filtration (see below), as the microbial communities that metabolize (and therefore detoxify) nitrogenous wastes within aquatic housing systems require the bicarbonate portion of alkalinity for survival and growth (Yanong 2003b). When alkalinity is low, these bacteria may become stressed to the extent that they do not perform their function adequately, which may result in a rise in ammonia and/or nitrite that would compromise the health of the animals residing within the system.

Alkalinity is very much a dynamic value in that it will be impacted by the biomass of a given aquatic system. The more intensive the system is (higher stocking densities, elevated feed inputs, etc.), the greater the demand will be for alkalinity, and the faster it will become "used up." To combat this, a buffering agent must be regularly added to the water to maintain alkalinity. In systems with higher loads, this will have to happen more frequently. The most commonly used and readily available buffers are sodium bicarbonate (baking soda) and, to a lesser extent, crystalized forms of calcium carbonate, such as limestone, dolomite, or crushed coral.

Hardness

Water hardness measures the quantity of divalent ions, primarily calcium and magnesium, and, to lesser extent, iron and selenium, in water (Wurts 2002). Aquatic animals require these ions to support various processes, including ossification, muscle contraction, blood clotting, and osmoregulation, and for the activation of several key enzymes (Hossain and Yoshimatsu 2014). Hardness also may influence the pathology of certain diseases (CCAC 2005). Captive aquatic animals must be provided these minerals in the water and/or the diet.

Hardness is measured in units of $CaCO_3$, and can be measured by simple, readily available reagent test kits or, more accurately, by mass spectrometry. Water sources will vary in their hardness levels. When purified water sources, deionized or reverse osmosis (RO), are utilized as a source for aquatic animals (especially common in the laboratory environment), the water will contain minimal hardness, even after synthetic salts have been added back to them. It is possible to raise the hardness of the water by directly adding calcium and magnesium salts to the water, or by passing the water through beds or reactors of crystalized forms of $CaCO_3$ noted previously (Harper and Lawrence 2010).

Salinity

Salinity reflects the total concentration of all dissolved ions in water, and exerts a profound influence on animals residing within it. Aquatic animals living in freshwater are hyperosmotic to their environment, and thus tend to gain water and lose salts by diffusion across the skin and the gills in larval forms of amphibians and fish, and across the skin in adult amphibians. Consequently, they must maintain their internal water and salt balance by voiding copious amounts of dilute urine while actively transporting ions back into the blood via chloride cells on the gill epithelium or skin. Freshwater fish rarely, if ever, drink. Freshwater amphibians do not drink, and instead are able to obtain water through a specialized region in their skin called the ventral pelvic area. Aquatic animals living in salt water have the reverse problem; they are hypoosmotic to their environment, meaning that they tend to lose water and gain salts across permeable membranes. To maintain osmotic balance, marine fish drink copious amounts of water and actively excrete excess salts via chloride cells in the gills. Saltwater reptiles (turtles, snakes, crocodiles, etc.) have salt glands that perform the same function. Saltwater animals, especially fish, produce very small amounts of urine to help conserve water.

There is an energetic cost to osmoregulation that varies with the salinity of the external medium, and all aquatic species exhibit a preferred level of salinity where this cost is lowest. Maintaining animals above or below this optimum is possible (the degree to which depends on the particular species), but because the organism must expend more energy in doing so, it can impact growth, survival, and reproduction in a negative fashion.

Nitrogenous Wastes

The primary waste product of aquatic animal (fish, aquatic turtles, aquatic reptiles, and aquatic amphibians) metabolism is ammonia, which is also produced during the decomposition of decaying organic matter. Ammonia is toxic to aquatic animals, but these animals are able to excrete nitrogen in the form of ammonia into the external aqueous environment because doing so dilutes it to nontoxic levels within the body.

This mechanism presents a challenge as it relates to managing aquatic animals in laboratory environments because the dilution effect that renders ammonia levels harmless in the wild is less effective in the smaller water volumes typical of artificial or captive settings. In these scenarios, ammonia levels become an item of critical concern due to the greater potential of accumulation within these systems. Two forms of ammonia exist in equilibrium aquatic systems, ammonia (NH_3) and ammonium $\left(NH_4^+\right)$, the sum of which is referred to as TAN. The ratio of the highly toxic NH_3 ("un-ionized ammonia") to the nontoxic ammonium NH_4^+ ("ionized ammonia") increases with pH and temperature and, to a lesser extent, decreases as salinities are increased. Levels of NH_3 in excess of 0.02 mg/L are generally toxic to aquatic animals, and therefore must be eliminated from housing situations, as the levels of NH_3 will rapidly accumulate in closed systems. In recirculating aquaculture systems (RASs), ammonia is removed by nitrifying bacterial species that oxidize NH_3/NH_4 into nitrite (NO_2) and then nitrate (NO_3). The intermediate product of this conversion, nitrite, is also toxic, and can be problematic in freshwater systems at concentrations in excess of 1 ppm (Wheaton 2002). The final product of the oxidization process, nitrate, is relatively nontoxic to aquatic organisms. Most species of fish will tolerate up to 1000 mg/L

(Wheaton 2002), although perhaps a more conservative target would be 200 mg/L, a level shown to be toxic in zebrafish (Learmonth and Carvalho 2015). In RASs, nitrate levels can be controlled by regular exchange of a percentage of water that results in the maintenance of nitrate levels at or below the desired target value, or by absorption via the use of nitrogen-fixing plants. Ammonia, nitrite, and nitrate can all be removed by simply flushing the system, which is a hallmark of flow-through aquaculture configurations. These approaches for nitrogenous waste management will be expanded on in a subsequent section of the chapter.

Amphibians, aquatic reptiles, and some species of fish (sharks, rays, skates, and lungfish) also have the ability to excrete nitrogen as urea, which is much less toxic than ammonia. In amphibians and reptiles (and in lungfish), this occurs when the animals are out of water (Chew et al. 2003; Walls 2008; Rasmussen et al. 2011).

Dissolved O_2

Dissolved oxygen is a critical environmental parameter in aquaculture. Aquatic animals require oxygen for respiration, and the demand for this depends on a number of factors, most notably body size, feeding rate, activity levels, and temperature. The availability of dissolved oxygen in the water is determined by water temperature, salinity, and biological demand.

While the culture requirements for dissolved oxygen will vary greatly by species in accordance with their natural history, the safest management strategy is that levels should be maintained at or just under saturation to ensure continued health of the target animal. For informational purposes, the recommended minimum dissolved oxygen level for warm water fish is 4 mg/L (Wedemeyer 1996), while the minimum required for biological filtration is 2 mg/L (Wheaton 2002). Gas supersaturation occurs when the total pressure of gases dissolved in water is higher than the atmospheric pressure. This can occur when water temperature is rapidly increased, when water is pumped to the surface from deep wells, when leaky or cracked pipes suck air under pressure, when pumps cavitate, or when water enters a plunge pool and air is forced into solution via pressure (Weitkamp and Katz 2012). This situation can cause acute, subacute, or chronic mortality in aquatic animals exposed to this water, depending on the level of gas supersaturation.

CO_2

Carbon dioxide (CO_2) is produced in aquatic systems by the respiration of animals, by the photosynthesis of plants and phytoplankton, and during the breakdown of organic matter (Harper and Lawrence 2010). As with ammonia, challenges arise in closed aquatic systems as a result of accumulation of CO_2 due to high organic loads in reduced volumes. Levels below 20 mg/L free CO_2 are generally recommended for cultures of fish in RASs (Swann 1997). In these environments, excess CO_2 is removed by aeration or "off-gassing" (any process that increases air–water contact).

Water Treatment

Pretreatment

The source of the water used to house aquatic animals is of paramount importance. To put it simply, it must be safe and clean. The safety of the water source is most directly centered around contaminants, both chemical and biological. Chemical contaminants are any compounds that have the potential to be toxic to the target animal. Heavy metals, such as copper, lead, and zinc, are toxic to aquatic animals. So too are pesticides and herbicides that may at times be present in source water. Both types must be removed from the source water prior to its entry into housing systems (Yanong 2003a; Kent et al. 2012). Many municipalities will also treat their water supplies with chemical disinfecting agents, such as chlorines and/or chloramines, both of which can be highly toxic to aquatic species (Yanong 2003a; Kent et al. 2012). For these reasons, water sources should be tested prior to use in aquatic housing systems by a qualified water-testing laboratory and periodically thereafter.

Depending on the results of these analyses, different measures may be taken to remove problem compounds. In many cases, the source water can be treated by passing it through a bed of activated carbon, which will remove organic contaminants, particularly chlorines and to a lesser extent chloramines (Harper and Lawrence 2010). Water purification processes, like RO filtration or distillation (DI) will remove heavy metals, pesticides, and other dissolved solids from the water. When RO or DI is used to treat source water, it may be necessary to add specific minerals back to the water to "condition" it prior to its use in housing systems. This is typically done using commercial aquarium salt mixtures. When using the latter, it may be important to determine the profile of cations and anions in artificial seawater prior to use in marine facilities. Seawater composition can be an important determinant of fish health (Grguric et al. 1999). Some of the ions that are commonly quantified (in more comprehensive tests, usually by ion chromatography) include the cations sodium, magnesium, calcium, potassium, strontium, and lithium, and the anions chloride, sulfate, bromide, fluoride, and phosphate, as well as nitrate or nitrite. Typically, testing is done prior to first use of a new seawater formulation or when problems arise (e.g., fish health issues). Some facilities go so far as to assay every new batch of artificial salt water, but this level of testing is more common with aquariums and public exhibits than in research facilities.

Water sources may also contain biological or microbial contaminants. Bacteria, fungi, algae, protozoa, and viruses may all be found in some water supplies, and could pose a threat to the health of the target animals within the housing systems. Ultraviolet (UV) disinfection is an effective means to eliminate many microbial contaminants from the water (Summerfelt 2003). Ozone may also be used, either independently or in concert with UV, to reduce microbial burdens in water (Summerfelt 2003) When necessary, more advanced pretreatment systems will include measures for removing both chemical (activated carbon and RO) and microbial (UV and ozone) contaminants prior to use in housing systems. In all cases, source water should be regularly tested to ensure that the right strategies are in place to make it safe to use for aquatic animals.

Treatment within Aquatic Housing Systems

Once water is delivered to the housing system enclosures (e.g., tanks), there is an entirely new level of treatment that must take place. The primary concern is with solid (feces and uneaten feed) and chemical (ammonia) wastes produced by the animals residing within the system. Fish and aquatic life stages and species of amphibians and reptiles excrete these products directly into their external environment. If these toxic products are not removed from the water, the ability of the housing system to support life will effectively be nil. Therefore, their success will be defined by its ability to keep the levels of these waste products at a minimum.

Solids Removal

Organic solids are generated by fish and amphibian excreta and feed residue, and are the primary pollutants that need to be removed from the water prior to discharge into the environment or biological filtration in RASs (Cripps and Bergheim 2000). These substances will increase oxygen demand and ammonia production, and can harbor potentially pathogenic microorganisms. The three types of waste solids found in aquatic housing systems are settleable, suspended, and fine or dissolved solids (Masser et al. 1999). Settleable solids are those that will drop out of the water column. Suspended solids are those that are carried by the water column and do not settle. Fine or dissolved solids become incorporated or bound to water molecules and are not removed by filtration alone.

There are various mechanisms utilized to remove solids from aquatic housing systems. The simplest element of this may be incorporated into tank design and operation itself; the shape of the tank and flow of water into it can facilitate "automatic" removal of solids. For example, housing systems typically used for zebrafish and other small-bodied fish will incorporate sloped tank bottoms and a tank baffle and/or suction mechanism driven by flow into the tank from above to move solids along the bottom of the tank up behind the baffle and out into effluent gutters (Lawrence and Mason 2012). Tanks for larger species, such as trout, tilapia, bass, and minnows, may also be designed to function in a similar manner (Wheaton 2002). Once solids are removed from housing, they will subsequently be removed from the water by

(a) (b)

FIGURE 24.1 Solids filters for aquaculture. (a) Backwashing sand filter. (b) Rotating drum filter.

one of several means. The simplest is via gravity, where solids are allowed to fall out of suspension in a "settling basin," where they can be collected and subsequently removed. Solids may also be captured by filtration; wastewater is pumped through a media that traps the particles suspended within it and allows water to flow through. Examples of this would be sand filters, rotating drums, or bead filter units (Figure 24.1). Trapped particles are then periodically backwashed and sent to a drain (Lawrence and Mason 2012). Solids may also be fractionated out; vigorous aeration will be used to produce air bubbles that will trap solids that can be later removed.

Chemical Removal

There are three different types of aquaculture systems, all of which are defined by the mechanism they employ to remove chemical waste products (primarily ammonia) from water: static, flow-through, and recirculating systems (Figure 24.2).

Static Systems

Static aquaculture systems rely on the same principles as flow-through aquaculture; the primary difference is the duration between or frequency of flushing events. Depending on the species and application,

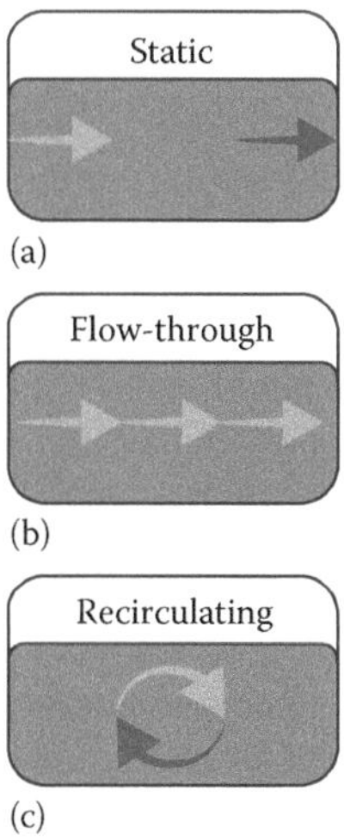

FIGURE 24.2 (See color insert.) Three primary modes of aquaculture. (a) Static. Water is exchanged periodically. (b) Flow-through. Water flows in a unidirectional flow, removing wastes. (c) Recirculating. Clean water flows into tanks, and effluent flows out and is treated to remove wastes before flowing back into tanks.

it may be days or even weeks between water changes. This is often necessary for species adapted to living in stagnant environments (e.g., *Kryptolebias marmoratus* [Mourabit et al. 2011]). These systems may or may not have plumbing and pumps to automatically deliver and drain water; some will require that water changes be done manually.

Flow-Through Systems

In flow-through aquaculture systems, ammonia is removed by flushing; essentially, the flow of water into and out of housing tanks is unidirectional. Clean water flows (is pumped) in, and effluent water (made "dirty" by the animals living within it) flows out. This exchange, which can be continuous, periodic (pulsed), or timed, serves to keep the levels of TAN in the system at or near zero.

Recirculating Systems

In RASs, which are commonly employed for small research fish like zebrafish and medaka, as well as for larger-bodied aquaculture species where discharge of wastewater to the environment needs to be limited, ammonia is removed by nitrifying bacterial species that oxidize ammonia to nitrite and then nitrate in a process referred to as biological filtration or biofiltration. These microbes, which are ubiquitous in air, soil, and water, naturally colonize and grow on the surface of all substrates within aquatic housing systems.

During normal operation of the RAS, effluent water is removed from housing tanks and is sent to a treatment zone within the system that contains a specially designed substrate with a high surface area, known as the biological filter, upon which very high concentrations of these nitrifying bacteria attach and grow. As the effluent water (containing ammonia) passes through this biological filter, certain species of bacteria (e.g., *Nitrosomonas* spp., *Nitrobacter* spp., and *Nitrospira* spp.) living on the filter oxidize the ammonia in the water to nitrite and then finally nitrate (Wheaton 2002). Nitrates, which are toxic to fish in only very high concentrations, can be controlled through removal by regular water changes or by the use of nitrogen-fixing plants.

The efficiency with which the bacteria operate is highly dependent on a number of chemical and physical parameters, especially alkalinity and dissolved oxygen, and to a lesser extent on salinity and temperature. Thus, biofiltration components of water treatment zones are designed to maximize gas exchange (high rates of aeration) in order to maximally support the health of the bacteria. Alkalinity is maintained in the system by addition of sodium bicarbonate (see earlier in chapter). Reductions in either of these parameters will negatively impact the biofilter and compromise the system's ability to support life (Lawrence and Mason 2012).

Disinfection

Any aquatic housing system will necessarily also support a diverse fauna of microorganisms, including bacteria, viruses, protozoa, algae, and fungi. For the most part, this is unavoidable, except in situations where great care is taken to ensure gnotobiotic conditions (Rawls et al. 2004). The problem presented by this natural phenomenon is that in a closed or small aquatic housing system, some of these organisms (e.g., pathogens) might accumulate to levels that could be harmful to the animals in the system. The situation is significantly exacerbated in recirculating systems. To keep the populations of these organisms in check, many aquaculture systems will incorporate a means of disinfecting water, either before it gets to animals in flow-through or static situations, or within the water treatment loop of RASs (Lawrence and Mason 2012). In RASs, the effluent water would be passed through a disinfection unit of some configuration, subsequent to solids removal and biological filtration. It is important to consider that whatever the method used, true sterilization is not achieved; rather, it is more accurate to define the goal of this level of treatment as a reduction in the microbial load of a given housing system.

UV disinfection is the most frequently applied method of disinfection in aquatic housing systems. In these units, water is passed through a tube (usually nested in stainless steel or polyvinyl chloride [PVC]) that contains a special light bulb, housed within a quartz sleeve to maximize transmittance, that emits light in the ultraviolet C (UVC) range to damage the DNA of organisms suspended within the

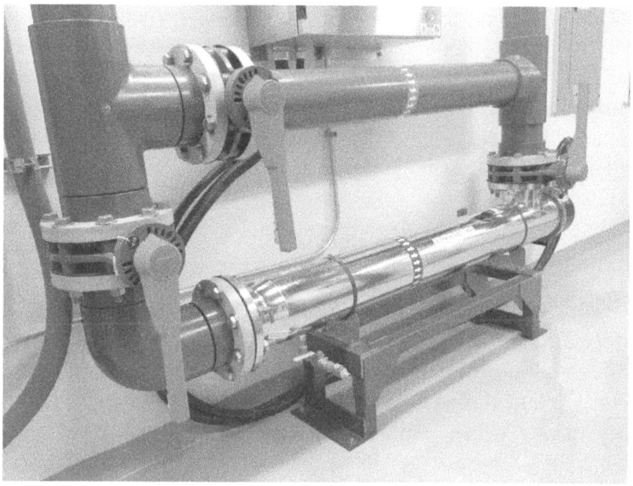

FIGURE 24.3 UV sanitization. An example of a UV sanitizing unit for aquaculture.

water column, rendering them noninfectious or inactive (Harper and Lawrence 2010) (Figure 24.3). The effectiveness of UV disinfection units is determined by a number of factors, including the wattage of the lamp (intensity), the ability of the UV light to penetrate the water sample (transmittance), the length of time the water sample is exposed to the light (contact time), and the size and biological complexity of the organism targeted (Yanong 2003b). These units must be serviced regularly to ensure optimal operation; of particular importance is the cleanliness and clarity of the quartz sleeve and life of the bulb.

Another method of water disinfection in aquaculture is ozone, although this is less commonly employed in research facilities, mostly because of human and animal safety concerns. Ozone (O_3) is an extremely reactive oxidant and, when used correctly, is very effective at eliminating bacteria and viruses (Summerfelt 2003). The most commonly used ozone system in aquatic animal housing is called a corona discharge generator, which produces ozone by generating a high-voltage spark that splits divalent oxygen into valent oxygen atoms that quickly bond to available oxygen molecules to form ozone (Summerfelt 2003). Like UV, its efficacy is impacted by contact time with the water, as well as the maintenance of adequate concentrations of the molecules in the water.

Water Quality Testing, Monitoring, and Control

Water quality in aquatic housing systems must be monitored and managed to ensure that it remains within target ranges (Lawrence and Mason 2012). The most elemental part of water quality management is regular testing of the parameters most critical for the species being maintained. Standard testing approaches should be developed for each species, based on their tolerance ranges.

There are a variety of testing methods that are readily available for common water quality parameters, discussed earlier in this chapter. Options range from the most simple, inexpensive, or rapid (test strips), to the more time-consuming and more accurate (colorimetric reagent kits), to the most accurate, expensive, and comprehensive equipment, such a spectrophotometer. For some parameters, handheld electronic meters are also an option.

In all cases, it is important to regularly cross-reference results with more than one testing method to ensure the highest accuracy of your results. It is also critical to make sure the chosen test kits are used before expiration dates, and are stored and handled properly. Meters and probes should also be serviced, calibrated, maintained, and replaced in accordance with manufacturer instructions and recommendations. Reliance on faulty testing methodology results is a recipe for disaster in any aquatic facility. In the case where probes are also automatically controlling a parameter (e.g., pH, salinity, and temperature or conductivity), it is especially critical to ensure that the equipment is working properly (Lawrence and Mason 2012).

Testing and monitoring is only part of the water quality management equation. For each species and housing system, facility managers should develop an overall strategy that includes not only the regular testing and recording of key parameters, but also standard protocols that are to be followed in the event that values move out of the target range. These approaches should be developed for each species being maintained. In some cases, there may be different testing regimens required for different life stages, management applications, and/or experimental procedures.

Housing

Tanks and Enclosures

There are a wide variety of tank and enclosure types used to house aquatic animals. The physical features of these enclosures that are most important for consideration include volume, shape, depth, and color (CCAC 2005), and each of these characteristics should be matched to the greatest possible extent with the natural history and behavior of the target species. It is also critical to consider that the selection of a tank or enclosure has the potential to influence the behavior of the animals, and that this may impact research endeavors. For example, it has been shown that the opacity of the housing tank in which zebrafish are maintained will influence their performance in standard behavioral tests, with animals being held in transparent tanks exhibiting a stronger depth preference than fish housed in opaque tanks (Blaser and Rosemberg 2012).

It is also important to consider the nature of support or bracing used to stabilize the tanks and enclosures. Water is extremely heavy (~28 kg/ft^3), as are some commonly used tank or enclosure materials (especially glass, fiberglass, and concrete), and so enclosures must be contained in a manner that will ensure that they do not fail or overwhelm the structural support of the room or space in which they are located. The seismic activity of the given geographic region should also be taken into account; in some regions with high seismic activity, specialized bracing is used to mitigate against movement, tipping, falling, or structural failure during earthquakes (Lawrence and Mason 2012).

Construction/Materials

Tanks and enclosures should be made of impervious, inert materials that are injection molded or sealed or joined in such a way that they contain no jagged or sharp edges that could injure the animals residing within. Suitable materials for tank or enclosure construction include glass, fiberglass, acrylic, concrete, or various plastics (polycarbonate, polysulfone, etc.). However, given the possibility that some plastics, such as polycarbonate, have the potential to leach small amounts of bisphenol A (BPA) over extended periods of use, it may be advisable to avoid these materials, especially if the animals are to be used in experiments where these compounds could unduly influence results (Lawrence and Mason 2012). BPA has been shown to be toxic to a wide variety of species at even relatively low concentrations (Lindholst et al. 2003; Duan et al. 2008; Chung et al. 2011; Saili et al. 2012; Tse et al. 2013). Special attention should also be paid to the nature of materials in plumbing, fixtures, and any sealants used in the construction of a housing system to ensure that they do not contain chemicals that pose a threat to the health of the animals. Corrosion of metallic components can become a significant problem in aquatic animal housing facilities. As such, care should be taken to ensure the proper selection of appropriate grades of stainless steel (e.g., 304, 316, or higher), the use of protective coatings (e.g., powder coat or epoxy paint), and regular cleaning to minimize the development of corrosion over time.

Maintenance and Sanitization

Tanks and enclosures should be cleaned on a regular basis to ensure the health of the animals living within them. The production of solid wastes, both from excreta and from uneaten feed and other biological materials, must be dealt with at the level of the enclosure to ensure maintenance of good water quality. Tanks and enclosures should be constructed in such a way to facilitate the removal of these materials. For example, commercially produced tanks used to house zebrafish and other small-bodied fish and/or amphibian species are often designed to be self-cleaning in this regard, by virtue of a baffle or siphoning apparatus that relies on an inflow of water into the tank to move solids along the bottom and up and out of the drain in the back of the tank, where they can be subsequently filtered from the water (Lawrence and Mason 2012). Many other tank designs incorporate circular flow patterns and conducive shapes (round, octagonal, hexagonal, and square with rounded corners) to promote the accumulation or settling of solids in the center bottom of the tank that can be removed manually by siphoning and/or placement of a perforated standpipe above an outflow drain (Masser et al. 1999). Any enclosure that does not possess continuous removal of solids will require regular siphoning of solids during water changes.

Tanks and enclosures housing aquatic animals will naturally accumulate biofilms, cyanobacteria, and algae over time. The rate at which these and other microorganisms colonize tank or enclosure surfaces will be accelerated by nutrient loading (more animals, feed, phosphates, etc.), flow rates of water in the tank, and lighting. There are two basic problems associated with this phenomenon. First, these microorganisms act as a "biological scaffold," providing a substrate upon which numerous other microorganisms will grow, some of which have the potential to be pathogenic to the resident animals being housed (Yanong et al. 2010). Second, if the growth of biofilms and algae or cyanobacteria becomes too excessive, it will limit the visibility of tanks, thereby making it more difficult for individuals to assess the health and well-being of target animals on a regular basis. Therefore, tanks and enclosures need to be regularly cleaned on an appropriate cycle to prevent this from happening. The frequency of cleaning will vary depending on several factors, including but not limited to tank or enclosure configuration and material type, stocking density, flow rate, effectiveness of solid or particulate matter removal, and feeding rate.

There are several different ways in which enclosures and tanks may be cleaned and sanitized. Cleaning tanks while animals are in them should generally be avoided, if at all possible, as materials liberated from the surface during scrubbing may pose a threat to the animals themselves. For example, it has been shown that the infection rates of *Mycobacterium chelonae* in zebrafish populations were higher when animals were not removed during tank cleaning (Murray et al. 2011). So when animals are held in smaller, more interchangeable housing units, the best approach may be a tank change (fish are moved to a new tank, and the old tank is cleaned and sanitized before reuse). In instances where enclosures are larger or more permanent in nature, the best option for cleaning may be to first remove the animals, followed by scrubbing, sanitization, and flushing of the tanks and enclosures before animals are reintroduced.

Sanitization of the tanks and enclosures can be done chemically, but this should be performed with caution. If detergents are used, it is advisable to use products that are free of surfactants to maximize safety and risibility to facilitate removal after product use. Bleach, hydrogen peroxide, and iodide-based products may also be used, but in all cases, it is critical to ensure that there are no residues of any of these agents remaining on the surfaces of the enclosures before animals are reintroduced. Heat (180°C) can also be used, either alone or in combination with chemical means. There are new commercial options available for cage washing small aquatic tanks that combine surfactant-free detergents with heat and high-pressure spraying with deionized water to sanitize tanks (Figure 24.4). Regardless of the cleaning and sanitization methods used, the methods should be verified on a regular basis and goals should be set for the degree of sanitization achieved via each cleaning and disinfection method (Garcia and Sanders 2011).

Environmental Enrichment

Although environmental enrichment techniques have been utilized extensively in the housing of mammalian laboratory animal species (Baumans 2005; Hutchinson et al. 2005), there is a general paucity of data concerning their application in the housing of fish and other aquatic animals (Williams et al. 2009). The reason for this has much to do with a limited knowledge of the natural history of many of these species. Furthermore, the presence of structural complexity in housing exerts profound effects on the physiology and behavior of animals (Brydges and Braithwaite 2009; Volpato 2009; Kistler et al. 2011; Lawrence 2011), and a limited understanding of these impacts has the potential to influence some studies in unexpected ways (Bayne 2005). Together, these two facts have limited the widespread adoption of these techniques for many laboratory aquatic animals.

Research on the most suitable applications is being actively conducted to address the information gap (Kistler et al. 2011; Wafer et al. 2015). However, environmental enrichment is employed on a limited

FIGURE 24.4 Tank sanitization. New tank washers are being developed specifically to safely sanitize aquatic housing tanks.

basis for some species (Williams et al. 2009), most commonly in the form of increasing structural complexity (e.g., plastic plants and flower pots) within enclosures, the selection and configuration of which are highly dependent on the species in question, as well as the type of research in which they are being utilized (Williams et al. 2009). It should be noted that any enrichment product should be evaluated for safety (does not break apart or degrade, is smooth without sharp edges, does not leach toxic chemicals or agents into the water, cannot be consumed by the animal during exposure, etc.) prior to use, and it should be able to be sanitized by prescrubbing in combination with immersion in a disinfecting solution and/or cage washer, as with any other component of housing. Indeed, the fact that added structural complexity within holding tanks makes sanitization cumbersome is another factor that has further limited adoption of environmental enrichment, at least within the laboratory research community.

Macroenvironment

The *Guide for the Care and Use of Laboratory Animals* defines the macroenvironment as being "the physical environment of the secondary enclosure (e.g., a room, a barn, or an outdoor habitat)"

(Institute of Laboratory Animal Research 2011). The following sections detail components of the macroenvironment for aquatic animals.

Room Design

General Characteristics

Rooms or space used to house aquatic animals must adhere to a general standard of design that centers on the concept of water handling and containment. In short, aquatic facilities are inherently "wet" spaces, and the room or space should be designed to account for that reality.

Structural Requirements

As mentioned previously, the weight of water (~28 kg/ft^3) should be accounted for in room design. The structural integrity of the underlying support of a space should be assessed prior to introduction of aquatic housing systems. In some cases, specialized bracing or support systems may be required to ensure that the load of a given aquatic operation can be properly supported.

Containment of water that is liberated from housing systems in the course of normal operation and maintenance or in the event of an accidental release or flood should also be accounted for in design. To this end, floor or trench drains are a critical component of any aquatic facility. These should be sited throughout a facility and be appropriately sized for the amount or volume of water to be discharged, and if at all possible, the floor should be sloped toward them to help contain water flowing on the floor within the room and to facilitate cleanup. If for some reason floor drains are not a possibility, adequate means for flood detection, for example, water detectors or physical barriers, should be included. For the reasons of structural integrity and water containment, in many instances, aquatic facilities are often situated in the basement or ground floors of buildings.

Electrical components and all outlets should be water sealed and ground fault interrupted (GFI). These should also be located outside the normal "splash zone" of any equipment or housing systems (CCAC 2005). Pumps and motors may need to be raised above floor level to prevent damage from overflows and standing water on floors.

Maintenance and Sanitization

Aquatic facility rooms should be designed to facilitate both management and sanitization. Aquatic housing systems will necessarily contain various mechanical components (filters, pumps, and monitoring equipment) that will require routine servicing and occasional replacement and repair. When possible, these components should be sited within discrete, compartmentalized spaces within a larger aquatic space (i.e., a physical plant) that provides physical separation between them and the animals. This practice serves to reduce disturbance (noise, vibration, and heat produced during normal operation) and to ensure that maintenance and repairs can be made quickly and with minimal disturbance to the animals and ongoing work. If such compartmentalization is not achievable, the goal of design should still be to separate these processes from housing areas to the greatest extent possible.

Sanitization of the space itself is also critical. All flooring should be sealed and slip resistant. Countertops and fixtures should be made from materials that are resistant to corrosion. Walls and ceilings should be made from or coated with materials that provide a measure of water resistance and resist damage from normal husbandry activities. Penetrations into the floor, walls, or ceiling should be sealed. All surfaces should be made to allow for thorough cleaning and disinfection.

Temperature

Temperature in aquatic rooms will exert profound effects on the microenvironment (discussed above), as well as on the well-being of the people working in the space. Although accepted room temperatures are generally around 73°F, with a relative humidity of 45%–55%, the relatively high volume of water present in aquatic facilities will also necessarily increase humidity in the rooms, and this will be exacerbated when warm water or tropical species are being housed. Excess humidity will promote the growth of unwanted microorganisms (bacteria, mold, or fungi) on surfaces in a facility. To combat this, aquatic facilities should be well ventilated so that humidity is controlled. This will also serve to minimize aerosol transfer between tanks and throughout the facility, a phenomenon that has been shown to transmit some certain pathogens between tanks in aquaculture facilities (Roberts-Thomson et al. 2006). As with all physical parameters, the most appropriate humidity levels will vary with species; for example, dendrobatid frogs and other amphibians may require elevated relative humidity in enclosures (de Vosjoli 1999; St. Claire et al. 2005).

Lighting and Illumination

Light is another physical parameter that, like temperature, impacts numerous facets of the biology of a given species. There are three primary facets of illumination that are critical to consider relative to the housing of captive animals: photoperiod, intensity, and spectrum. In all cases, preferences will very much be species specific and reflective of natural history. In terms of photoperiod, regular day–light cycles are an important determinant of circadian rhythms, the abrupt alteration or absence of which will negatively affect growth, survival, immune function, and reproduction, across species (Sadoglu 1979; Delgado et al. 1987; Bayarri et al. 2003; Leonardi and Klempau 2003; El-Sayed and Kawanna 2004; Almazán-Rueda et al. 2005; Ángeles Esteban et al. 2006; Valenzuela et al. 2012; Khattak et al. 2014). In some instances, particularly if the species being housed is crepuscular, it may be desirable to phase lighting on and off more gradually with programmable dimming switches (Lawrence 2011). Both the intensity and spectrum of lighting will exert major effects on the growth and development of a wide range of aquatic species (Boeuf and Le Bail 1999; Godinho et al. 2005; Villamizar et al. 2013). It should also be noted that illumination will influence nontarget organisms. For example, higher-intensity lighting will promote the growth of algae and cyanobacteria in aquatic housing systems (Koerber and Kalishman 2009).

Vibration

Many aquatic species, especially fish, are sensitive to noise and vibration, and in some cases, it may induce developmental problems (Vandenberg et al. 2012) and both acute and chronic stress (Smith 2004; Neo et al. 2015). Therefore, care should be taken in aquatic facilities to insulate the space from any potential sources of outside noise and vibration and to minimize these within the facility proper. More proximately, life-support-associated equipment—pumps, filters, and so forth—that produce noise and vibration should be sited to minimize their impact on housed animals. The best way to achieve this is to compartmentalize mechanical and housing space, but in instances where this is not possible, mechanical equipment should be located as far as possible from housing space. In particular, pumps and other vibrating components (such as air compressors) should be supported with sound- and vibration-dampening braces to minimize their impact. Noise arising from construction within or near aquatic facilities should be mitigated as well.

Additional Concerns

Isolation and Quarantine

As with any research animal, the management of aquatic animals in a facility requires strict pathogen control. A hallmark of health management concerns the oversight and control of incoming animals originating from outside sources. For most aquatic species, commercially available sources of certified or specific pathogen-free animals do not exist. Therefore, aquatic facilities should possess dedicated quarantine or isolation space and housing that allows animals to be brought in from the outside and isolated for a defined period of time before safely being introduced into the main housing area.

The key to quarantine in aquatics facilities is separation. If at all possible, quarantine or isolation housing should be completely separate from the main population. Ideally, this separation should be complete; quarantine should be in its own room, complete with its own housing, equipment, sanitization, and water treatment. Nothing should be shared with the main facility. This way, incoming animals may be received into this space, and housed there either until they pass a defined period of quarantine or until new animals can be safely derived from them as fertilized eggs or embryos that can be treated (i.e., surface disinfection [Sanders et al. 2012]) and/or screened for selected pathogens before introduction into the main population (Kent et al. 2011).

Sentinels and Health Monitoring

Another hallmark of health management in laboratory animal populations is the regular monitoring of existing populations and housing systems for the presence of pathogens (Lipman and Homberger 2003; Varga 2011; Willeberg 2012). Health monitoring programs for aquatic animals follow similar principles as other laboratory animals. In particular, the use of sentinel animals that are exposed to effluent water from an entire housing system or population (the equivalent of "dirty bedding" sentinels in rodent colonies) can be an effective way to monitor the health status of a given room or population (Varga 2011). A growing number of commercially available fish housing systems now offer specifically designed sentinel systems to meet this growing demand (Lawrence and Mason 2012) (Figure 24.5).

FIGURE 24.5 Sentinel tanks. Specialized housing tanks called sentinel tanks are designed to expose animals to effluent water to detect pathogens in the system. In this example (white arrow), the tanks are mounted on a special, designated rack on a RAS.

FIGURE 24.6 Breeding systems. Specialized group breeding tanks (for zebrafish) are used to generate large quantities of embryos for experiments. Inset: Small breeding tank/insert used for individual pairs or small groups (<10) of fish.

Feeding Systems

One of the primary advantages of some aquatic species, especially small-bodied species of fish like zebrafish or medaka, is their space efficiency, that is, the ability to house many thousands of animals in relatively small amounts of space. This space efficiency comes at a cost, though, as the feeding of hundreds of small tanks requires a significant amount of labor and supplies (Lawrence et al. 2012). This is also a concern in large-scale aquaculture operations with large populations of fish or other aquatic species.

One solution to this problem is automated feeding systems. The most basic automatic feeding systems can be mounted on individual tanks or enclosures. There are also more sophisticated options, such as robot feeding systems that are mounted to aquarium racks and are capable of feeding hundreds or thousands of tanks automatically (Lawrence and Mason 2012).

Breeding Systems

Many experimental applications in which aquatic animals are utilized are predicated on the production of embryonic or larval forms for use in various experiments or assays. This is particularly the case for fish species like zebrafish. While it may be possible to spawn aquatic animals directly within their housing systems, it may be necessary in some instances to remove them for spawning. In these cases, specialized spawning chambers are utilized (Figure 24.6). Commercial vendors have met these needs by designing integrated breeding units within housing systems that allow for manipulations or controls over the reproduction of target species. The same sanitation, durability, and material toxicity concerns that apply to racks and tanks apply to breeding equipment as well.

Summary

Aquatic animals are playing an increasingly prominent role in basic and biomedical research, and this trend is expected to continue well into the next decade and beyond. The growth and utility of these model organisms is highly dependent on the maintenance of favorable, stable, and defined environmental conditions in housing systems. While this is no different than it is for any animal model, terrestrial or otherwise, it is perhaps more challenging with fish, amphibians, and aquatic reptiles because of the nature of water and its properties as a universal solvent.

It is critical for managers, veterinarians, scientists, and technicians to possess a detailed understanding of the physical, chemical, and mechanical complexities of aquatic housing systems. This knowledge must be coupled with data on the target species in order to develop physical infrastructure and management strategies that promote the best possible welfare for these animals in laboratories. The information presented in this chapter provides readers with a comprehensive base from which to start and continue this process. Ultimately, these efforts will serve to allow aquatic models to fulfill their great promise and help to accelerate the pace of scientific discovery.

REFERENCES

Almazán-Rueda, P., A. T. M. Van Helmond, J. A. J. Verreth, and J. W. Schrama. 2005. Photoperiod affects growth, behaviour and stress variables in Clarias gariepinus. *Journal of Fish Biology* 67: 1029–39.

Ángeles Esteban, M., A. Cuesta, A. Rodríguez, and J. Meseguer. 2006. Effect of photoperiod on the fish innate immune system: A link between fish pineal gland and the immune system. *Journal of Pineal Research* 41: 261–66.

Baumans, V. 2005. Environmental enrichment for laboratory rodents and rabbits: Requirements of rodents, rabbits, and research. *ILAR Journal* 46 (2): 162–70.

Bayarri, M. J., L. Rodríguez, S. Zanuy, J. A. Madrid, F. J. Sánchez-Vázquez, and M. Carrillo. 2003. Effect of photoperiod manipulation on daily rhythms of melatonin and reproductive hormones in caged European sea bass (*Dicentrarchus labrax*). *Fish Physiology and Biochemistry* 28: 37–38.

Bayne, K. 2005. Potential for unintended consequences of environmental enrichment for laboratory animals and research results. *ILAR Journal* 46 (2): 129–39.

Blaser, R. E., and D. B. Rosemberg. 2012. Measures of anxiety in zebrafish (*Danio rerio*): Dissociation of black/white preference and novel tank test. *PLoS One* 7: e36931.

Boeuf, G., and P. Y. Le Bail. 1999. Does light have an influence on fish growth? *Aquaculture* 177: 129–52.

Brydges, N. M., and V. A. Braithwaite. 2009. Does environmental enrichment affect the behaviour of fish commonly used in laboratory work? *Applied Animal Behaviour Science* 118 (3–4): 137–43.

CCAC (Canadian Council on Animal Care). 2005. Guidelines on the care and use of fish in research, teaching, and testing. Ottawa: Canadian Council on Animal Care. http://www.ccac.ca.

Chew, S. F., T. F. Ong, L. Ho, W. L. Tam, A. M. Loong, K. C. Hiong, W. P. Wong, and Y. K. Ip. 2003. Urea synthesis in the African lungfish *Protopterus dolloi*—Hepatic carbamoyl phosphate synthetase III and glutamine synthetase are upregulated by 6 days of aerial exposure. *Journal of Experimental Biology* 206: 3615–24.

Chung, E., M. C. Genco, L. Megrelis, and J. V. Ruderman. 2011. Effects of bisphenol A and triclocarban on brain-specific expression of aromatase in early zebrafish embryos. *Proceedings of the National Academy of Sciences of the United States of America* 108 (43): 17732–7.

COE (Council of Europe). 1991. European Convention for the Protection of Vertebrate Animals Used for Experimental and Other Scientific Purposes. ETS 123. http://www.coe.int/en/web/conventions/full-list/-/conventions/treaty/123.

Cripps, S. J., and A. Bergheim. 2000. Solids management and removal for intensive land-based aquaculture production systems. *Aquacultural Engineering* 22: 33–56.

Delgado, M. J., P. Gutiérrez, and M. Alonso-Bedate. 1987. Melatonin and photoperiod alter growth and larval development in *Xenopus laevis* tadpoles. *Comparative Biochemistry and Physiology, Part A: Physiology* 86 (3): 417–21.

Delpire, E., K. B. Gagnon, J. J. Ledford, and J. M. Wallace. 2011. Housing and husbandry of *Xenopus laevis* affect the quality of oocytes for heterologous expression studies. *Journal of the American Association for Laboratory Animal Science* 50 (1): 46–53.

de Vosjoli, P. 1999. Designing environments for captive amphibians and reptiles. *Veterinary Clinics of North America: Exotic Animal Practice* 2 (1): 43–68, v–vi.

Duan, Z. H., L. Zhu, L. Y. Zhu, K. Yao, and X. S. Zhu. 2008. Individual and joint toxic effects of pentachlorophenol and bisphenol A on the development of zebraftsh (*Danio rerio*) embryo. *Ecotoxicology and Environmental Safety* 71 (3): 774–80.

El-Sayed, A. F. M., and M. Kawanna. 2004. Effects of photoperiod on the performance of farmed nile tilapia *Oreochromis niloticus*. I. Growth, feed utilization efficiency and survival of fry and fingerlings. *Aquaculture* 231: 393–402.

European Union. 2010. Directive 2010/63/EU of the European Parliament and of the council of 22 September 2010 on the protection of animals used for scientific purposes. *Official Journal of the European Union* 33–79.

Frans, I., C. W. Michiels, P. Bossier, K. A. Willems, B. Lievens, and H. Rediers. 2011. *Vibrio anguillarum* as a fish pathogen: Virulence factors, diagnosis and prevention. *Journal of Fish Diseases* 34 (9): 643–61.

Garcia, R. L., and G. E. Sanders. 2011. Efficacy of cleaning and disinfection procedures in a zebrafish (*Danio rerio*) facility. *Journal of the American Association for Laboratory Animal Science* 50: 895–900.

Godinho, L., J. S. Mumm, P. R. Williams, E. H. Schroeter, A. Koerber, S. W. Park, S. D. Leach, and R. O. Wong. 2005. Targeting of amacrine cell neurites to appropriate synaptic laminae in the developing zebrafish retina. *Development* 132 (22): 5069–79.

Grguric, G., J. A. Komas, and L. A. Gainor. 1999. Differences in major ions composition of artificial seawater from aquarium tanks at the New Jersey State Aquarium. *Aquarium Sciences and Conservation* 2: 145–59.

Harper, C., and C. Lawrence. 2010. *The Laboratory Zebrafish (Laboratory Animal Pocket Reference)*. Boca Raton, FL: CRC Press.

Hossain, M. A., and T. Yoshimatsu. 2014. Dietary calcium requirement in fishes. *Aquaculture Nutrition* 20: 1–11.

Hutchinson, E., A. Avery, and S. VandeWoude. 2005. Environmental enrichment for laboratory rodents. *ILAR Journal* 46: 148–61.

Institute of Laboratory Animal Research. 2011. *Guide for the Care and Use of Laboratory Animals*. 8th ed. Washington, DC: National Research Council, National Academies Press.

Kelsch, S. W., and W. H. Neill. 1990. Temperature preference versus acclimation in fishes—Selection for changing metabolic optima. *Transactions of the American Fisheries Society* 119 (4): 601–10.

Kent, M. L., C. Buchner, V. G. Watral, J. L. Sanders, J. Ladu, T. S. Peterson, and R. L. Tanguay. 2011. Development and maintenance of a specific pathogen-free (SPF) zebrafish research facility for *Pseudoloma neurophilia*. *Diseases of Aquatic Organisms* 95 (1): 73–79.

Kent, M. L., J. M. Spitsbergen, J. M. Matthews, J. W. Fournie, K. N. Murray, and M. Westerfield. 2012. Diseases of zebrafish in research facilities. In *ZIRC Health Services Zebrafish Disease Manual*. Eugene, OR: Zebrafish International Resource Center.

Khattak, S., P. Murawala, H. Andreas, V. Kappert, M. Schuez, T. Sandoval-Guzmán, K. Crawford, and E. M. Tanaka. 2014. Optimized axolotl (*Ambystoma mexicanum*) husbandry, breeding, metamorphosis, transgenesis and tamoxifen-mediated recombination. *Nature Protocols* 9 (3): 529–40.

Kistler, C., D. Hegglin, H. Würbel, and B. König. 2011. Preference for structured environment in zebrafish (*Danio rerio*) and checker barbs (*Puntius oligolepis*). 135: 318–27.

Koerber, A. S., and J. Kalishman. 2009. Preparing for a semiannual IACUC inspection of a satellite zebrafish (*Danio rerio*) facility. *Journal of the American Association for Laboratory Animal Science* 48 (1): 65–75.

Lawrence, C. 2011. Environmental enrichment and the zebrafish. *Enrichment Record*, April 2011, Volume 11, pp. 11–15.

Lawrence, C., J. Best, A. James, and K. Maloney. 2012. The effects of feeding frequency on growth and reproduction in zebrafish (*Danio rerio*). *Aquaculture* 368–369: 103–8.

Lawrence, C., and T. Mason. 2012. Zebrafish housing systems: A review of basic operating principles and considerations for design and functionality. *ILAR Journal* 53 (2): 179–91.

Learmonth, C., and A. P. Carvalho. 2015. Acute and chronic toxicity of nitrate to early life stages of zebrafish—Setting nitrate safety levels for zebrafish rearing. *Zebrafish* 12 (4): 305–11.

Leonardi, M. O., and A. E. Klempau. 2003. Artificial photoperiod influence on the immune system of juvenile rainbow trout (*Oncorhynchus mykiss*) in the southern hemisphere. *Aquaculture* 221: 581–91.

Lindholst, C., P. M. Wynne, P. Marriott, S. N. Pedersen, and P. Bjerregaard. 2003. Metabolism of bisphenol A in zebrafish (*Danio rerio*) and rainbow trout (*Oncorhynchus mykiss*) in relation to estrogenic response. *Comparative Biochemistry and Physiology, Part C: Toxicology and Pharmacology* 135: 169–77.

Lipman, N. S., and F. R. Homberger. 2003. Rodent quality assurance testing: Use of sentinel animal systems. *Lab Animal* 32 (5): 36–43.

Masser, M. P., J. Rakocy, and T. Losordo. 1999. *Recirculating aquaculture tank production systems: Management of recirculating systems. Stoneville*, MS: Southern Regional Aquaculture Center.

Mourabit, S., M. Edenbrow, D. P. Croft, and T. Kudoh. 2011. Embryonic development of the self-fertilizing mangrove killifish *Kryptolebias marmoratus. Developmental Dynamics* 240 (7): 1694–704.

Murray, K. N., J. Bauer, A. Tallen, J. L. Matthews, M. Westerfield, and Z. M. Varga. 2011. Characterization and management of asymptomatic mycobacterium infections at the Zebrafish International Resource Center. *Journal of the American Association for Laboratory Animal Science* 50 (5): 675–79.

Neo, Y. Y., L. Parie, F. Bakker, P. Snelderwaard, C. Tudorache, M. Schaaf, and H. Slabbekoorn. 2015. Behavioral changes in response to sound exposure and no spatial avoidance of noisy conditions in captive zebrafish. *Frontiers in Behavioral Neuroscience* 9: 28.

Perry, M. 2013. Revised Australian Code of Practice for the care and use of animals for scientific purposes. *Australian Veterinary Journal* 76 (4): 286.

Rasmussen, A. R., J. C. Murphy, M. Ompi, J. W. Gibbons, and P. Uetz. 2011. Marine reptiles. *PLoS One* 6 (11): e27373.

Rawls, J. F., J. I. Gordon, and B. S. Samuel. 2004. Gnotobiotic zebrafish reveal evolutionarily conserved responses to the gut microbiota. *Proceedings of the National Academy of Sciences of the United States of America* 101 (13): 4596–601.

Roberts-Thomson, A., A. Barnes, D. S. Fielder, R. J. G. Lester, and R. D. Adlard. 2006. Aerosol dispersal of the fish pathogen, *Amyloodinium ocellatum. Aquaculture* 257 (1–4): 118–23.

Sadoglu, P. 1979. A breeding method for blind *Astyanax mexicanus* based on annual spawning patterns. *Copeia* 1979: 369–71.

Saili, K. S., M. M. Corvi, D. N. Weber, A. U. Patel, S. R. Das, J. Przybyla, K. A. Anderson, and R. L. Tanguay. 2012. Neurodevelopmental low-dose bisphenol A exposure leads to early life-stage hyperactivity and learning deficits in adult zebrafish. *Toxicology* 291: 83–92.

Sanders, J. L., V. Watral, and M. L. Kent. 2012. Microsporidiosis in zebrafish research facilities. *ILAR Journal* 53 (2): 106–13.

Short Guide. 2014. *Code of Practice for the Housing and Care of Animals Bred, Supplied or Used for Scientific Purposes: Short Guide to Using the Code of Practice.* London: Home Office.

Smith, M. E. 2004. Noise-induced stress response and hearing loss in goldfish (*Carassius auratus*). *Journal of Experimental Biology* 207 (Pt. 3): 427–35.

St. Claire, M. B., M. J. Kennett, M. L. Thomas, and J. W. Daly. 2005. The husbandry and care of dendrobatid frogs. *Contemporary Topics in Laboratory Animal Science* 44 (6): 8–14.

Summerfelt, S. T. 2003. Ozonation and UV irradiation—An introduction and examples of current applications. *Aquacultural Engineering* 28: 21–36.

Swann, L. 1997. *A fish farmer's guide to understanding water quality in aquaculture.* Urbana, IL: Illinois-Indiana Sea Grant Program, Aquaculture Extension.

Timmons, M. B., J. M. Ebeling, J. M. Wheaton, S. T. Summerelt, and B. J. Vinci. 2002. *Recirculating Aquaculture Systems.* Ithaca, NY: Cayuga Aqua Ventures.

Tse, W. K. F., B. H. Y. Yeung, H. T. Wan, and C. K. C. Wong. 2013. Early embryogenesis in zebrafish is affected by bisphenol A exposure. *Biology Open* 2: 466–71.

Valenzuela, A., V. Campos, F. Yañez, K. Alveal, P. Gutiérrez, M. Rivas, N. Contreras, A. Klempau, I. Fernandez, and C. Oyarzun. 2012. Application of artificial photoperiod in fish: A factor that increases susceptibility to infectious diseases? *Fish Physiology and Biochemistry* 38: 943–50.

Vandenberg, L. N., C. Stevenson, and M. Levin. 2012. Low frequency vibrations induce malformations in two aquatic species in a frequency-, waveform-, and direction-specific manner. *PLoS One* 7 (12): e51473.

Varga, Z. M. 2011. Aquaculture and husbandry at the Zebrafish International Resource Center. *Methods in Cell Biology* 104: 453–78.

Villamizar, N., L. M. Vera, N. S. Foulkes, and F. J. Sánchez-Vázquez. 2013. Effect of lighting conditions on zebrafish growth and development. *Zebrafish* 11 (2): 173–81.

Volpato, G. L. 2009. Challenges in assessing fish welfare. *ILAR Journal* 50 (4): 329–37.

Wafer, L., J. Whitney, V. Behrana Jensen, R. Flores, T. Gomez, and B. S. Goodwin. 2015. Environmental enrichment influences fertility and fecundity of zebrafish (*Danio rerio*). Presented at *World Aquaculture Society, Aquaculture America Meeting*, New Orleans, LA, February 19–22.

Walls, S. C. 2008. The ecology and behavior of amphibians. *Quarterly Review of Biology.* 83 (4).

Wedemeyer, G. A. 1996. *Physiology of Fish in Intensive Culture Systems.* New York: Chapman & Hall.

Weitkamp, D. E., and M. Katz. 2012. A review of dissolved gas supersaturation literature. *Transactions of the American Fisheries Society* 109 (6): 659–702.

Wheaton, F. 2002. Recirculating aquaculture systems: An overview of waste management. In *Proceedings of the Fourth International Conference on Recirculating Aquaculture*, Blacksburg, VA, 57–68.

Willeberg, P. 2012. Animal health surveillance applications: The interaction of science and management. *Preventive Veterinary Medicine* 105 (4): 287–96.

Williams, T. D., G. D. Readman, and S. F. Owen. 2009. Key issues concerning environmental enrichment for laboratory-held fish species. *Laboratory Animals* 43 (2): 107–20.

Wurts, W. A. 2002. Alkalinity and hardness in production ponds. *World Aquaculture* 33: 16–7.

Yanong, R. 2003a. Fish health management considerations in recirculating aquaculture systems. Part 1: Introduction and general principles. Circular FA-120. Gainesville: University of Florida Institute of Food and Agricultural Services Cooperative Extension Service.

Yanong, R. 2003b. Fish health management considerations in recirculating aquaculture systems. Part 2: Pathogens. Circular FA-121. Gainesville: University of Florida Institute of Food and Agricultural Services Cooperative Extension Service.

Yanong, R. P. E., D. B. Pouder, and J. O. Falkinham. 2010. Association of mycobacteria in recirculating aquaculture systems and mycobacterial disease in fish. *Journal of Aquatic Animal Health* 22 (4): 219–23.

25

Nontraditional Species

Dorcas P. O'Rourke, James D. Cox, and Diana P. Baumann

CONTENTS

Introduction

Birds, reptiles, terrestrial amphibians, and invertebrates comprise a minority of species used in research, yet they appear with some regularity in animal facilities. The more commonly seen bird species include pigeons, chickens, ducks, finches, and parrots. Pigeons are typically used in psychology research and behavioral experiments involving cognition and learning. Chickens, quail, and other poultry are studied extensively in agricultural production-based research, vaccine development, and developmental toxicology studies. Chickens and ducks are also important species for infectious disease research. Passerines (songbirds) such as zebra finches are used as models for neurobiology of vocal learning, effects of substance abuse on learning and memory, ecotoxicology, cognition, and behavioral research (Bateson and Feenders 2010). Parrots, budgerigars, cockatiels, and related psittacine species are used primarily in cognition and behavior studies, as well as pharmacology and avian disease research. Caring for these highly intelligent and behaviorally complex animals can be quite interesting and challenging (Kalmar et al. 2010).

Among reptile species, turtles and lizards are most frequently kept in research facilities. Turtles have been extensively used in physiology research and teaching, including investigations of hypoxia and freeze tolerance (Packard and Packard 2004; Overgaard et al. 2007). Lizards are subjects of behavioral, endocrine, physiology, reproductive, toxicology, and tail regeneration studies (Hutchins et al. 2014). The effects of toxicants and environmental pollutants have been studied in alligators, as well as the antimicrobial activity of alligator blood extracts (Merchant et al. 2006). Snakes are commonly kept in herpetological teaching collections, used for venom research, and are studied in behavioral and physiological research (O'Rourke 2014). Snake venom has also been used to develop drugs for the treatment of hypertension, thrombosis, and cancer (Koh and Kini 2012).

Leopard frogs are the most common semiaquatic amphibian species used in physiology research and teaching. Along with toads and salamanders, they are an animal model for environmental toxicology and endocrine disruptor studies. Fruit flies (*Drosophila melanogaster*) and the roundworm (*Caenorhabditis elegans*) have traditionally dominated the invertebrate research field, contributing to knowledge gained in many areas of genetic and developmental research.

In addition to the aforementioned use as animal models, birds, reptiles, amphibians, and invertebrates are most commonly studied by biologists to better understand the natural history, ecology, behavior, and conservation of the animals themselves. While this work is frequently conducted in the field, it sometimes necessitates bringing wild-caught animals into the laboratory for holding and further research.

Regulatory Considerations

Regulatory considerations for nontraditional species differ somewhat from those for most laboratory animals. The Convention on International Trade in Endangered Species of Wild Fauna and Flora (CITES) (https://www.cites.org/) is applicable to any of the approximately 5600 species of animals (or parts of animals, such as skin, feathers, tissues, etc.) that are identified in one of the CITES appendices as species whose survival is threatened by international trade. Importation or exportation of these animals requires special approvals and permits. The U.S. Fish and Wildlife Service (USFWS) (http://www.fws.gov/endangered/) is the U.S. agency responsible for issuing CITES permits. USFWS also issues scientific permits for work with exotic birds through the Wild Bird Conservation Act (http://www.fws.gov/international/laws-treaties-agreements/us-conservation-laws/wild-bird-conservation-act.html).

Native amphibians, reptiles, and birds that are classified as threatened or endangered in the United States are protected under the Endangered Species Act. Permits to work with these animals are issued by USFWS (land-based sea turtle activities are issued by the National Marine Fisheries Service). The USFWS also issues permits through the Bald and Golden Eagle Protection Act and Migratory Bird Treaty Act for research with applicable species.

Both the Public Health Service (PHS) and the National Science Foundation (NSF) require that institutions have a PHS assurance to receive funding (http://grants.nih.gov/grants/olaw/references/mou_nsf.htm). The *Guide for the Care and Use of Laboratory Animals* (*Guide*) (NRC 2011) provides specific recommendations on cage size and temperature requirements for chickens, quail, and pigeons. More detailed information related to the care and use of poultry in agricultural research settings is contained in the *Guide for the Care and Use of Agricultural Animals in Research and Teaching* (*Ag Guide*) (FASS 2010). Guidelines for research involving wild birds are published by the Ornithological Council (2010). Similarly, the American Society of Ichthyologists and Herpetologists (ASIH) has published the *Guidelines for Use of Live Amphibians and Reptiles in Field and Laboratory Research* (ASIH 2004). Appendix A of the European Convention for the Protection of Vertebrate Animals Used for Experimental and Other Scientific Purposes (ETS 123) (Council of Europe 2006) provides specific information for chickens, ducks, quail, pigeons, zebra finches, amphibians, and reptiles, as does the *UFAW Handbook on the Care and Management of Laboratory and Other Research Animals* (Hubrecht and Kirkwood 2010).

Invertebrates

More than 1.3 million species of invertebrates have been identified (Brusca and Brusca 2003), comprising more than 98% of all known animal species (Ruppert et al. 2004). Occupying over 30 phyla (Cooper 2011; Lewbart 2011), they offer a tremendous diversity, and their presence in the laboratory animal arena is on the rise (Figure 25.1). Studies on the fruit fly (*D. melanogaster*) and nematode (*C. elegans*) have contributed significantly to research efforts, and these classical model species were chosen due to their amenability to laboratory research and their suitability for studying a wide range of biological problems (Wilson-Sanders 2011). Indeed, much of our knowledge of genetics, development, physiology, ecology, and evolution comes from the study of invertebrates (Pechenik 2015). However, technologies continue to advance, providing exciting new approaches and allowing previously unanswerable questions to be explored. The time and cost to sequence genomes has dropped considerably, and current techniques allow selective altering of gene expression patterns across a much wider variety of species. In particular, the emergence of next-generation sequencing (NGS) has opened a door of opportunities in contributing to unravelling the mysteries of the diversity of life. NGS is also known as high-throughput sequencing and describes a number of modern technologies that allow for faster, cheaper sequencing of DNA and RNA. The huge drop in cost has made it feasible to generate

FIGURE 25.1 Housing system for *Aplysia*, a species commonly used in neuroscience research. (Photo by D.P. Baumann.)

reference-quality genome sequences for almost any species, providing new material for comparative analysis studies. Additionally, sequencing allows examination of gene expression at an unprecedented depth, as well as mapping of transcription factor binding sites. Evolutionary developmental biology (evo-devo) is one example of a relatively new field of study that has taken off in a major way as a consequence of NGS.

Invertebrates recognized as emerging model organisms include animals as varied as snails, leeches, squid, butterflies, and crickets. Planarians, free-living flatworms, provide a powerful model for regeneration; ants provide a model for social organization and chemical communication; honeybees provide a model for learning memory and behavior; and comb jellies are furthering our understanding of evolution (*Emerging Model Organisms*, Volumes 1 and 2). Multiple species, including sea anemones and sea squirts, are used in comparative genomics to study the similarities and differences of evolutionary lineages, allowing origin, change, and loss of genomic features to be mapped. Additionally, many diseases of humans and animals are caused by certain invertebrates, leading to another area of interest in the biomedical research field (Pechenik 2015).

Laws, regulations, and guidelines involved with invertebrate animal studies vary by country. In the United States, there are currently no regulations regarding work on invertebrates. Various countries, including Canada, Australia, and New Zealand, have included cephalopods (cuttlefish, squid, octopuses, and nautiluses) in their national codes of practice and legislation covering research (Smith et al. 2013; Harvey-Clark 2011). In the European Union, Directive 2010/63/EU, on the protection of animals used for scientific purposes, gives cephalopods the same European Union legal protection as vertebrates (Council of Europe 2010). Other regulations may also apply, particularly for at-risk, threatened, or endangered species. These include whether the animal is listed under the CITES agreement. If the animal is determined to be an invasive species, this can require significant housing containment preparations, including on-site inspections. State and local laws, along with permits required to obtain, transport, and house the animals, must all be investigated. Even for nonregulated species, a best practice of having the veterinarian and the animal care and use committee oversee the work can provide significant contributions toward ensuring optimum animal welfare and results.

Sourcing invertebrates may take many avenues. The simplest cases involve animals specifically bred for research and/or commercial availability, such as various *Drosophila* stock centers or the University of Miami's National Resource for *Aplysia* facility (Harvey-Clark 2011). It is also worth investigating whether other labs working on that particular species can spare some from their colony. Other options include obtaining animals through the pet trade, paying a local organization to collect them, or undertaking wild capture. It is worthwhile spending time on species familiarization, and essential that proper and thorough training be provided to technicians who will actually be caring for the animals. Presentations on anatomy and natural history can assist with knowing what to expect in terms of husbandry and animal behavior. New species research may include consulting with zoos, other laboratories, natural history museums, aquariums, private breeders, and online pet forums, in addition to researching and reviewing the published literature.

Management of invertebrates in a laboratory setting will be totally dependent on the species. Being able to house a large colony in a relatively small space is advantageous in terms of costs, but the usual problems of monoculture exist on an elevated level. Close attention needs to be paid to housing and husbandry standards, methods of capture and restraint, quarantine, biosecurity, and food quality and variety. As with all species, good hygiene practices and careful documentation can prevent a multitude of problems (Cooper 2004). In situations where little is known about the natural history of that animal, trial and error may be necessary, and so sufficient animals should be procured to allow housing them in multiple set-ups. Guidelines will need to be developed, covering capture, transport, handling, housing, care, maintenance, health monitoring, humane anesthesia, analgesia, and euthanasia. Given the vast diversity of invertebrates, specific details of housing, husbandry, and species survival are impractical for this chapter; however, the value of appropriate choices must be recognized (Frye 1992). The most current information should be used to assess environmental conditions, and monitoring should be in place to assess and maintain required parameters. Location should be selected to avoid noise and vibration, while providing a safe, secure facility. Additional measures may need to be put in place to provide containment appropriate for the species.

Knowledge of the animal's life cycle and reproductive strategy is essential to colony maintenance.

Invertebrates exhibit a varied selection of patterns of reproduction (Brusca and Brusca 2003). Asexual mechanisms include fission and parthenogenesis, while sexual mechanisms include gonochoristic and hermaphroditic patterns, in addition to the ability to switch sex. Internal fertilization occurs with very few exceptions on land and in freshwater invertebrates, whereas external fertilization is much more common in the marine environment. Life spans are similarly varied, with some invertebrates living for many years, while others may live for less than a year. Within this existence, some invertebrates breed repeatedly and some only once in their lifetime (Olive 2002; Pechenik 2015).

Veterinary staff and caregivers will need to gain an understanding of normal behavior in the species to assess animal health. Both alterations in behavior and physical changes, such as lethargy, anorexia, lesions, and color changes, can indicate illness in invertebrates. A herd health approach is usually taken, and although some treatments exist, environmental review and even depopulation are often the necessary courses of action in disease outbreak situations. In terms of health and safety of staff, zoonoses, although uncommon, do exist, and allergic reactions can also be issues. Invertebrate defense mechanisms can provide challenges to animal handling, including urticating hairs, biting with mandibles, pinching, stinging, sharp spines, and emission of noxious substances. A variety of restraint devices and techniques assist with safe handling in these circumstances, particularly with venomous species. Other challenges of invertebrate colony management include animal identification (Figure 25.2), census tracking, and enrichment, which can take the form of social groupings, environmental complexity, and activities that allow species-typical behavior, such as catching live food.

Anesthesia is used to both immobilize animals for examination and perform procedures that have the potential to cause pain (Cooper 2001, 2011). Although the ability to feel pain remains a much debated topic (Elwood 2011), it has been shown that most invertebrates will respond to noxious stimuli (Lewbart 2011; Horvath et al. 2013). The *AVMA Guidelines for the Euthanasia of Animals* (AVMA 2013) states, "A conservative and humane approach to the care of any creature is warranted, justifiable, and expected by society." Methods of anesthesia and euthanasia chosen need to minimize or eliminate pain and distress, and will undoubtedly change over time as further research is carried out. Generalization of methods is not possible given the vast diversity of taxa; however, a two-step approach to euthanasia is generally

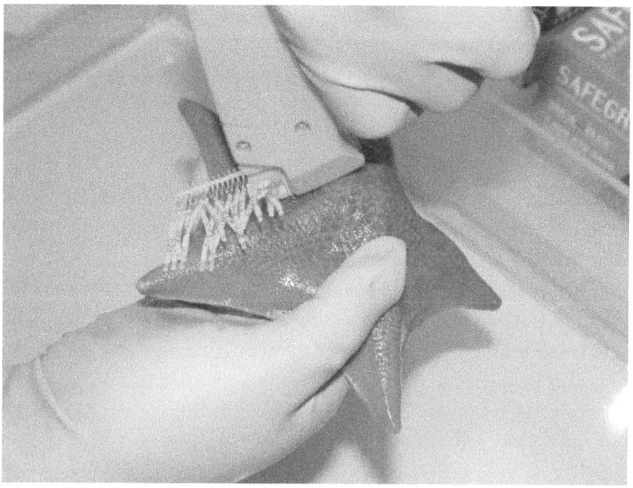

FIGURE 25.2 Tagging system used for starfish identification. (Photo by D.P. Baumann.)

preferred, with a chemical method to induce terminal anesthesia, followed by a physical or chemical second step to ensure death. For terrestrial species, inhalation anesthesia is commonplace, while immersion anesthesia is generally exercised for aquatic invertebrates (Lewbart 2011; AVMA 2013).

Birds (Songbirds, Pigeons, Chickens and Ducks, and Psittacines)

Class Aves is represented by more than 200 families of birds in approximately 40 orders. Within the families, species demonstrate remarkable physiological and behavioral adaptations to the wide variety of niches that they occupy. Commonly represented species in research environments include members of the orders Columbiformes (pigeons and doves), Galliformes (chickens and quail), Anseriformes (ducks and geese), Passeriformes (songbirds), and Psittaciformes (parrots and budgerigars). Biological and behavioral needs can vary significantly; therefore, it is imperative to consult relevant species-specific literature when designing housing and husbandry systems.

Primary enclosures should be large enough to permit species-typical behaviors. In most cases, this means cage sizes that permit birds to stand or perch and spread their wings without contacting walls or the cage top (Kalmar et al. 2010; NRC 2011). Cage dimensions should be adequate to permit flight if needed; therefore, horizontal space is critical. Many birds are highly social and do well when housed together, provided that adequate space and cover are available to escape aggressors. Most primary enclosures are made of wire bars or grid. Spacing between bars and grids must be sufficiently small to prevent escape. Corners and junctions where panels meet may also offer potential escape routes and must be carefully checked to ensure that there are no gaps. Parrots are highly intelligent and easily learn to open cage doors. Psittacine species will climb on and chew cage bars, and may develop zinc toxicity if they ingest oxidized zinc that has not been removed by prior thorough brushing and dilute acetic acid washing (Howard 1992; Lightfoot and Yeager 2008; Kalmar et al. 2010).

Chickens, quail, pigeons, and other species that spend a great deal of time on the ground should be provided with a substrate that encourages foraging behavior. Coarse hardwood shavings work well, are absorbent, and provide birds the ability to scratch and peck. Hardwood shavings can also be used in large aviaries for birds such as finches. Alternatively, paper should be used to line cage bottoms and has the advantage of permitting daily observations of droppings and ease of change for health assessment purposes. Abrasive surfaces, such as concrete, when used exclusively for waterfowl flooring, can be a contributing factor to the development of bumblefoot.

Nest boxes are critical components of hen enrichment (Mench et al. 2010), and hens should be provided with enclosed nest boxes containing appropriate nesting material. Nests provide refuges for other bird species as well, and are essential components of breeding aviaries. There should always be enough nests to accommodate all occupants of the cage or aviary, as some species can be quite territorial. Likewise, perches should be plentiful and accessible to the animals. In large cages and aviaries, perches should be placed in a fashion that encourages flight. The perches themselves should be species appropriate. Diameter should be sufficient to permit comfortable perching, with toes extending two-thirds around and perches placed in locations at heights preferred by the species being housed. Natural branches of varying diameters are frequently used for passerines and psittacines. Manzanita and other woods afford contoured nonabrasive surfaces that encourage foot exercise (Figure 25.3). Care should be taken to avoid abrasive materials such as sandpaper-covered perches, as these can predispose birds to developing foot lesions, and they should never be the highest perch in the cage. Dust baths should be made available to chickens (Duncan 2010) and other species that engage in this behavior (Matheson et al. 2015). Most chickens do not bathe in water and use the dust baths to maintain optimal feather condition (Duncan 2010). Most birds commonly kept in research facilities are prey species. Wild-caught animals, in particular, can become easily stressed by smaller enclosures and unfamiliar activities. It is therefore essential to provide appropriate cover to allow animals a safe retreat. Cover can be simple, such as vertical panels, or more naturalistic items, like artificial and nontoxic natural plants. Large psittacines such as parrots are highly intelligent and require complex enrichment strategies, such as provision of manipulanda, foraging opportunities, and objects that encourage chewing and shredding, in addition to human interaction.

Clean water must be easily accessible to all birds in the enclosure. Ducks and other waterfowl benefit from water, typically provided in pools, for keeping feathers healthy and for engaging in species-typical behaviors (Mench et al. 2010). Ponds or pools give these species opportunity to swim and engage in normal behaviors. Most waterfowl can be accommodated with a water depth of 2 feet (Flinchum 2006).

Other bird species will use shallow pans of water to bathe. Misting from above, especially for rain forest species, simulates rainfall and also encourages grooming behaviors (Figure 25.4).

Avian nutritional requirements (increasingly understood through research) and feeding habits vary widely among species. Many songbirds have omnivorous tendencies. Even species primarily thought of as seed eaters, such as finches and sparrows, ingest insects and other protein sources. Seeds, in fact, are recognized to be deficient in protein and high in fats, and feeding exclusive seed diets to many species is strongly discouraged. Commercially balanced, nutritionally complete granular or pelleted diets are available for chickens and quail, ducks, finches, budgerigars, and parrots. High-quality, clean, fresh seed mixes and millet sprays can be used as supplements for seedeaters, and mealworms, waxworms,

FIGURE 25.3 Finch aviary demonstrating multiple, spaced manzanita perches of varying diameters, hanging artificial vegetation for cover, and abundant nest boxes. (Photo by D.P. O'Rourke.)

FIGURE 25.4 **(See color insert.)** Gentle misting of parrots and other species provides "showers," which promote natural grooming behaviors and feather health.

and other invertebrates can be provided to species that also eat insects. Waterfowl will forage on grass or plants. Fresh vegetables and fruits should be made available to psittacines and other species, and hard-boiled eggs also make excellent supplements. Birds typically forage throughout the day, so offering a variety of foods allows them to engage in normal feeding behaviors. In all cases, clear knowledge of the dietary needs of each species housed is essential when determining proper diets for captive birds.

Birds should be housed in rooms or enclosures that provide adequate ventilation but are not drafty. Air exchange rates will vary depending on the species and number of animals housed. Standard animal room temperatures are adequate for many species, although animals from tropical environments may require warmer ambient temperatures. Care should be taken to avoid humidity extremes. Many bird species housed in outdoor aviaries can be gradually acclimated to tolerate seasonal climate changes, but animals should always be provided with shelter and a means to avoid temperature extremes.

Full-spectrum light is generally recommended for birds for behavioral, reproductive, and health reasons. The light source should be chosen carefully based on the species' physiology and behavior. There is evidence that chickens and pigeons can perceive flickering in certain frequencies of fluorescent lighting. This flickering can potentially impact the animal's health and behavior (Lisney et al. 2012). These factors should be taken into account when housing any bird species in conventional animal facilities, and should also be considered as fluorescent lighting is phased out and replaced with light-emitting diode (LED) lighting. Light cycles will vary, depending on species requirements and the study needs. Provision of dawn and dusk periods in the daily cycle permits birds to access roosts without experiencing sudden darkness. Alternatively, low-level lighting kept on throughout the dark cycle permits birds to move about more safely.

Sudden dramatic temperature changes can have negative impacts on a bird's health. Even species that have been acclimated to wider temperature ranges in outside aviaries require appropriate shelter from extremes of heat and cold. Many such aviaries have indoor access for animals to seek shelter in environmentally controlled areas. As for all animal facilities, bird housing areas should be equipped with emergency power.

The sanitation frequency of avian primary enclosures can be variable, depending on species, substrate, enclosure size, and number of inhabitants. In smaller cages, paper liners are typically changed daily. Larger aviaries with shavings are spot cleaned daily and changed weekly to biweekly. Nonporous surfaces for waterfowl and similar species are usually hosed once or twice daily. Sanitation should be weekly for small cages, and biweekly for floor pens and aviaries. It is critical to avoid exposure of birds to chemicals during *in situ* sanitation processes. Quaternary ammonium compounds and other standard chemicals, such as bleach, can be used. Phenolics should be avoided. All surfaces must be thoroughly rinsed before reintroducing birds.

Two important considerations when handling birds are minimizing stress and avoiding feather damage. Assembling all necessary items for transfer, examination, or procedures prior to capture is critical to avoid unnecessarily long restraint times. This is especially true for small passerines and psittacines that can die acutely when stressed. Darkening the room will calm the bird and make capture less stressful since most species do not see well in the dark. Small passerines can be gently picked up by hand. Pigeons, chickens, and ducks can be restrained by holding the wings against the body. Psittacines are often best restrained by wrapping in an appropriately sized towel, which will protect the feathers and cover the sharp toenails. All birds will defend themselves if threatened, and large psittacines, in particular, can deliver very painful bites. Therefore, in addition to wrapping, psittacines should always have their head restrained.

Birds have high metabolic rates, and lose body mass rapidly when not eating. They also mask signs of illness, so a bird that is not perching, sitting huddled with ruffled feathers, and not responding to stress of a visitor is likely quite ill. Supportive care includes providing a warm, quiet environment and ensuring adequate nutrition and water intake. Commercially available emergency diets are available for gavaging critically ill animals. Birds that are less ill can sometimes be provided warm palatable diets to encourage eating. In addition to providing a warm, quiet environment, perches should be placed on or near the cage bottom in close proximity to food and water bowls. Antibiotics, antifungals, and other therapeutic agents, if indicated, should be dosed according to species (Marx 2006).

Newly arrived birds should be quarantined away from the main colony to avoid introduction of viral, bacterial, parasitic, and fungal agents into the established population. Source (wild caught vs. colony reared) will help determine the duration of quarantine.

Birds can harbor zoonotic agents. Pigeons and psittacines, in particular, can be infected with *Chlamydiophila psittaci*, the causative agent of psittacosis or "parrot fever." Avian mycobacteriosis is another potential zoonotic disease. Both agents can be detected through fecal polymerase chain reaction (PCR).

Inhalant anesthetics such as isoflurane and sevoflurane work well for most avian anesthesia procedures. Birds can be induced by face mask or chamber and then intubated, if necessary, for maintenance. Injectable anesthetics, in general, are not as controllable and reliable across species as are inhalants. Analgesics, such as nonsteroidal anti-inflammatory agents, should always be provided when appropriate.

Euthanasia of avian species should be in accordance with the *AVMA Guidelines for the Euthanasia of Animals*. Intravenous barbiturate overdose is an acceptable method. Information on methods considered acceptable with conditions (inhalant anesthetic overdose, cervical dislocation, decapitation, etc.) and methods considered unacceptable can also be found in these guidelines (AVMA 2013).

Reptiles (Snakes, Lizards, Turtles, and Crocodilians)

The class Reptilia is represented by more than 9500 species of living reptiles. Lizards and snakes (squamates) comprise the majority of these animals (Pincheira-Donoso et al. 2013), with turtles and crocodilians making up the remainder of reptiles commonly seen in laboratory animal facilities. In addition to many species being studied by herpetologists and evolutionary biologists as research subjects in their own right, reptiles are also used as animal models. Among the more commonly encountered reptile species in animal facilities are red-eared sliders (*Trachemys scripta elegans*), box turtles (*Terrapene* spp.), green anoles (*Anolis carolinensis*), green iguanas (*Iguana iguana*), American alligators (*Alligator mississippiensis*), garter snakes (*Thamnophis sirtalis*), and rat snakes (*Pantherophis* spp.).

Selection of an appropriate cage or enclosure depends on the species needs and behavior. The most important aspect of preparing to house a new species of reptile is conducting a thorough literature review to determine habitat, food requirements, and normal behavior of that species. Many snakes are "sit and wait" ambush predators and will remain coiled up under a shelter regardless of the amount of space provided. In contrast, other snake species and many lizard species are quite active and require larger enclosures to engage in species-specific behaviors. Reptile housing options include glass or acrylic aquaria, stackable caging systems, fiberglass tanks, and other types of impervious primary enclosures. Sufficient height should be provided for species that climb or perch, such as iguanas, anoles, rat snakes, and corn snakes. Containers for aquatic and semiaquatic turtles and crocodilians must include a pond area for swimming. The water must be deep enough for the animal to right itself if it becomes turned over, and provision of a haul-out area for basking is very important. Lids do not need to be provided for tanks and cages housing turtles, crocodilians, and certain lizard species, provided that the tank sides are tall enough to prevent escape. In contrast, lids for snake and many lizard cages must be not only provided, but also tightly fitted and secured to prevent escape. Venomous species of reptiles should be kept in non-breakable cages that are lockable and completely secure (Figure 25.5). Newer cage designs for venomous species often include shift panels. These are particularly useful, as they allow the animal to be segregated away from the area being serviced, and decrease personnel risk by minimizing animal handling (O'Rourke and Lertpiriyapong 2015).

Substrates for enclosures include paper, hardwood shavings, indoor–outdoor carpet, or more natural substances. Selection of the most appropriate material depends on the needs of a given species. For example, shavings permit burrowing behaviors in many species. For water snakes and related species that spend much time in the water bowl, precut indoor–outdoor carpet keeps animals dry, allows traction for movement, and is easily sanitized and replaced. Natural substrates, while aesthetically pleasing, can harbor pathogens and must be disinfected prior to use; they are also more difficult and labor-intensive to maintain.

Clean, fresh water must be available at all times. Water bowl size is dictated by species behavior and number of occupants in the enclosure. All animals within the cage must be able to easily access the water, and the container must be large enough for animals to soak, since many snakes soak to soften and loosen skin prior to shedding. Box turtles and tortoises have difficulty climbing over the sides of tall water bowls, and should be provided water in large, shallow dishes that are easy to access and cannot be tipped over (Figure 25.6).

Some reptiles tend to be secretive; therefore, cages must be provided with refuges or visual barriers for animals to hide from view. Species that climb should also have branches or ledges on which to perch.

Some species of reptiles can be housed in pairs or groups. Examples include many turtles and tortoises, young crocodilians, and certain species of snakes and lizards. Exceptions include territorial male lizards, such as anoles, and cannibalistic snakes, such as king snakes.

The dietary needs of reptiles vary significantly, depending on species. Snakes feed on whole prey. Most species can be maintained on euthanized rodents. Unfortunately, individuals may not adapt and will have to be fed live prey. Other species feed on insects, fish, or frogs, and some snake species have evolved specialized, fastidious dietary requirements. If snakes are housed together, they should be separated for feeding. Crocodilians are strict carnivores; however, they can be transitioned from whole prey to a balanced, high-protein pelleted diet. Depending on species, lizards may be insectivores, carnivores, omnivores, or herbivores. Anoles can be fed crickets and other insects. Prey insects must be maintained on nutritious diets, to ensure that they are not nutritionally deficient when fed to the reptiles. Iguanas are

FIGURE 25.5 (See color insert.) Venomous snakes can be housed in locking shift cages to allow husbandry with minimal risk. The red X marks the acrylic panel, which is easily removed. (Photo by D.P. O'Rourke.)

FIGURE 25.6 Water bowls for reptiles should be large, sturdy, and easily accessible to the species being housed. (Photo by D.P. O'Rourke.)

herbivores, and feed on leafy greens, legumes, and other protein- and calcium-rich vegetables, and occasional fruit. Sliders and other aquatic turtles eat fish and aquatic vegetation. Box turtles eat animal-based protein, as well as mushrooms, berries, and fruit, whereas tortoises are more herbivorous. Nutritionally balanced, commercial diets are available for many reptile species.

Reptiles are ectothermic; that is, they cannot internally regulate their body temperature. Consequently, they must be provided with an environment that is warmed or allows them to behaviorally thermoregulate. Strategies include placing a focused heat source on part of the cage (which also accommodates basking behavior), use of heat strips below part of the cage, or simply warming the entire room. In all cases, care must be exercised to monitor temperatures so that animals will not be exposed to sudden excursions or temperature extremes. With a few exceptions, room humidity levels of 30%–70% are appropriate for most reptile species. In general, reduced airflow rates are acceptable for many reptiles.

Many lizards and turtles, particularly young growing animals, require exposure to ultraviolet (UV) light in the range of 290–320 nm as a necessary component of proper vitamin D and calcium metabolism (Gehrmann 1996). UV light sources should be placed within 24 inches of animals and changed at 6-month intervals to ensure that animals receive the proper wavelengths (Figure 25.7). The light should not pass through any materials that absorb critical wavelengths (Burger et al. 2007). It is advisable to test all light sources advertising specific wavelengths, to verify the accuracy of the wavelength claims.

Reptiles are very sensitive to sudden temperature excursions. Failure of electrical and heating, ventilation, and air-conditioning (HVAC) systems resulting in cold room temperatures can result in immune suppression, illness, and death. Similarly, sudden dramatic increases in room temperatures can be lethal. Therefore, as for other species, animal facilities housing reptiles should have backup emergency power.

Agents commonly used for sanitation of reptile cages include quaternary ammonium compounds and bleach solutions. Phenolics are toxic to reptiles and should not be used to sanitize cages or rooms. Sanitation frequency will vary depending on species. Many reptiles eat and defecate less frequently than mammals; therefore, cage cleaning intervals, especially if spot cleaning is implemented, can be less frequent. However, prolonged intervals between sanitation will lead to buildup of excrement and subsequent pathogen overgrowth. In general, snake and lizard cages can be changed at 1- to 2-week intervals, depending on dietary habits and level of soiling. Aquatic and semiaquatic species of turtles and crocodilians will need more frequent sanitation if pool areas are static and unfiltered.

When handling reptiles, care must be taken to support the animal's body. Snakes will bite if they perceive a threat, so they should be approached carefully. Many snakes can be moved by using a snake

FIGURE 25.7 **(See color insert.)** Many lizard species must have a basking site and access to UV light of the appropriate wavelength. (Photo by D. P. Baumann.)

hook to lift and transfer the animal. Turtles lack teeth, but many species will inflict a painful bite; they can be held by the sides of the shell. Crocodilians must have their head securely restrained. The jaws can be held shut and a tape muzzle applied to prevent the animal from opening its mouth. Crocodilians also have a powerful muscular tail that should be restrained. Lizards will bite, and many can slap with their tails. Additionally, many species will break off their tails as a predator avoidance mechanism, so tails should never be grabbed to capture or restrain a lizard.

Clear acrylic or plastic snake tubes can facilitate examination and treatment of snakes. The snake is encouraged to crawl into the tube, and then is prevented from backing out by holding the animal and the tube securely with the same hand (Figure 25.8). This method is particularly useful for manipulating venomous snakes. Any reptile that is anorectic, has a poor body condition, is having difficulty shedding, or is not behaving normally should be evaluated for potential illness. Prior to shedding, the scales covering the eyes of a snake (spectacles) will turn a milky bluish color. During this time, the animal will be more irritable and prone to defensive actions. The opacity clears within several days, when the skin loosens and is typically shed in one complete piece. Patchy shedding or retained spectacles can be indicative of poor environmental quality or a disease process. In contrast, lizard skin and turtle scales (scutes) are normally shed in pieces. Any reptile that is suspected of being ill should be observed in its enclosure for evidence of abnormal breathing, vomiting, or diarrhea. Reptiles are susceptible to the same types

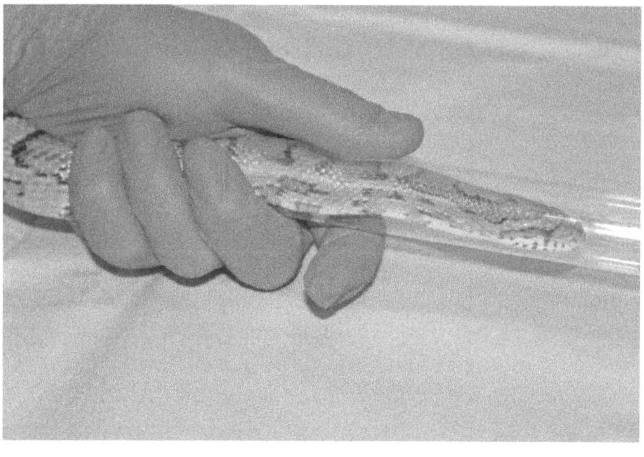

FIGURE 25.8 Snakes can be restrained in clear plastic tubes for ease of handling and safety. (Photo by D.P. O'Rourke.)

of diseases as mammals and birds, and can be treated accordingly. Therapeutic agents, dosages, and frequency of administration should be species specific whenever possible, due to the slower metabolic rates of reptiles. Formularies for reptiles are available (Carpenter et al. 2014). Supportive care includes provision of a warm environment and ensuring hydration. Tube feeding may also be indicated (O'Rourke and Lertpiriyapong 2015).

Newly acquired reptiles should be quarantined for several weeks, especially if they are wild caught. During this time, they should be evaluated for parasites and potential pathogens. The most common zoonotic disease associated with reptiles is *Salmonella*. Many species carry and shed the bacterium but exhibit no clinical signs. Proper hand washing and wearing gloves as indicated will help prevent transmission.

Inhalants such as isoflurane are the most reliable and controllable forms of anesthesia for reptiles. Animals can be induced in a chamber and then intubated if indicated. Some species, such as turtles and crocodilians, will breath-hold for very prolonged periods; these animals can be induced with an injectable agent and then intubated and maintained on inhalants. Analgesics used in reptiles include opioids and nonsteroidal anti-inflammatory agents.

Euthanasia methods for reptiles must be species appropriate. Injectable barbiturate and anesthetic overdose are acceptable. Inhalant anesthetic overdose is considered acceptable with conditions. Hypothermia is unacceptable for euthanasia, except for liquid nitrogen immersion of animals weighing less than 4 g (AVMA 2013).

Amphibians (Terrestrial and Semiaquatic Frogs, Toads, and Salamanders)

Amphibians comprise more than 6000 different species across three distinct order classifications: *Anura*, *Caudata*, and *Gymnophiona* (Pough 2007). The focus of this section is on *Caudata*, better known as salamanders, and *Anura*, which includes frogs and toads, as *Gynophiona* species, also known as caecilians, are rarely used in research. Together, these comprise the majority of all amphibian species in the world, with relatively few species used in research and teaching (Pough 2007). With so few amphibians being used in research, it is often challenging to obtain the animals necessary for the studies, as it requires either live capture or commercial acquisition through vendors in the pet trade. Live capture typically requires permits from the state in which collections will take place and may also require a federal permit. Some states may also require a permit to purchase the animals from vendors. Although captive-bred animals can be obtained for *Xenopus* and some *Ambystoma* species, the majority of amphibians obtained from commercial vendors are live captured for the pet trade or for use as a reptile food source. It is therefore important to assess the quality of the vendors by understanding how they capture, house, and ship animals.

Due to the diversity of species, habitats, social behaviors, and environmental cues, it is difficult to prescribe specific husbandry practices that can be standardized across all species and facilities, as is the case with many of the mammalian species maintained in laboratories. It is therefore essential to have an understanding of and appreciation for the natural environment in which the species resides and attempt to replicate that environment in the laboratory setting. The following descriptions of caging, husbandry and care, and macro- and microenvironments provide examples of what has worked for some of the more common amphibian species used in research, including *Xenopus* sp., *Rana* (*Lithobates*) sp., *Bufo* (*Anaxyrus*) sp., *Hymenochirus* sp., *Bombina* sp., *Ambystoma* sp., *Necturus* sp., and *Notophthalmus* sp. (DeNardo 1995; O'Rourke 2007).

As with all laboratory animals, the primary enclosure should be sized and contain the necessary substrates to provide comfort and safety, while allowing for species-typical behaviors. Many amphibian species are capable of jumping high distances or climbing smooth vertical surfaces, so it is important that the primary enclosure be escape-proof. This typically includes the use of a lid that also prevents escape of live foods, allows airflow, and permits the passage of UV light (Browne and Zippel 2007). The enclosure material should allow for easy sanitation, and both plastic and fiberglass (DeNardo 1995) have proven to be successful materials. Standard glass aquarium tanks can also be used. It is important that whatever material is chosen does not contain rough or sharp edges or surfaces, as these can damage the amphibian's skin, which is an important barrier for disease prevention. Clear walls on all sides may cause a problem, as it is sometimes difficult for amphibians to recognize that there is a physical barrier. Then during an attempt to escape, they could run into the tank wall, damaging their rostral area.

Inside the primary enclosure, it is important to provide a substrate and other objects that support the natural environment of the species. This could include coconut husks or even compressed paper bedding like ALPHA-dri® or Pure-oCel®, which promote burrowing behaviors by *Bufo* (*Anaxyrus*) and terrestrial *Ambystoma* species (Figure 25.9). For amphibian species that do not burrow, the use of butcher paper, newsprint, and artificial turf has proved successful (Beaupre et al. 2004). The use of cage and pan liners like those used with other lab animal species may be preferred, as these are typically free of any contaminants. It is likely that amphibians will ingest loose substrate in the enclosure when feeding, such that corncob, kitty litter, and pine shavings should be avoided due to their potential to swell, as well as cedar shavings due to their potential toxicity (Beaupre et al. 2004). In addition to the substrate, the use of hiding places is necessary, as many amphibian species spend a significant portion of the day in these areas (Pough 2007). These can include polyvinyl chloride (PVC) tubes, aquarium rock structures, and artificial logs.

Water is an essential and critical part of the environment for aquatic, semiaquatic, and terrestrial amphibians. Many amphibian species do not consume water, but rather absorb water through the skin. Because of this, the water source and quality are important. The principles and requirements for water that are used for zebrafish can be applied to amphibians.

Tap water can be used, but it is important to know the water treatment practices of the local water authority. Residual chlorine or other chlorinated compounds used in water treatment plants that are left untreated would be detrimental to amphibian populations. In addition, heavy metals and hardness can be toxic, and periodic changes in the treatment by municipalities can be disastrous to the amphibians (DeNardo 1995). The use of an activated charcoal filter is necessary if tap water is used. It is critical to ensure the effectiveness of the charcoal in removing contaminants, such as chlorine, and to replace the filter at regular intervals. Reverse osmosis (RO), distilled, and deionized water can all be used, but the pH and osmotic balance are critical. There are a variety of recipes that can be used to provide this balance, depending on the species and life stage of the amphibians being housed (DeNardo 1995).

It is necessary to conduct periodic monitoring of not only the source water and the treated water provided to the colonies, but also the water in the enclosures. Excess food and animal waste, along with insufficient cleaning practices, can all lead to poor water quality and potential disease (DeNardo 1995). The tank or cage water should be tested for the by-products of the nitrogenous waste cycle. The excess

FIGURE 25.9 (See color insert.) Proper housing for toads includes a substrate such as coconut husks, a water source, and a PVC shelter for hiding. Note the toad at the base of the shelter buried in the substrate. (Photo by J.D. Cox.)

food and animal waste generate ammonia, which is toxic to amphibians. Systems that recirculate the water through filtration mechanisms and back into the tank typically contain a biological filter with nitrifying bacteria to break down the ammonia buildup. One species of bacteria converts ammonia to nitrite, which is toxic, but less so than ammonia, and another species converts nitrite down to nitrate. Nitrates are even less toxic, and daily water changes will prevent the buildup of this compound. Nitric acid is another by-product of this process, which is why recirculating systems require the addition of a buffer like sodium bicarbonate. If the tanks do not have a recirculating system or the amphibians only require a shallow bowl of water, it may be necessary to perform a complete water change after feeding so as to avoid the buildup of toxic substances.

Due to the variety of amphibian species and the limited information available on captive husbandry, it is necessary to consider the natural environment when considering enrichment and what are the species-typical behaviors in those environments. Some species (*Ambystoma* sp.) like to burrow, while others (*Eurycea* sp.) like to hide under rocks and in crevices. Aquatic species may also desire a shelter, either at the water surface, like a lily pad, or a tunnel-like device, such as PVC tubes. It is important to ensure that any enrichment device provided is free of sharp edges due to the sensitive nature of amphibians' skin and the potential for disease when the skin is compromised (Figure 25.10).

Food can be another source of enrichment, especially for terrestrial species. The use of live food is necessary for most amphibians, as they require prey movement for feeding (DeNardo 1995). Crickets and mealworms are the most common prey species provided, but fruit flies, tubifex worms, brine shrimp, earthworms, cockroaches, waxworms, and even small rodents have been used (DeNardo 1995; Pough 2007). It is important that a balanced diet be provided to the prey species to ensure that the amphibian species is obtaining sufficient nutrition. In addition to providing a variety of prey species, you may also need to gut load or dust the prey with a vitamin and/or mineral supplement to ensure a balanced diet. Some aquatic amphibian species will consume pelleted diets. *Xenopus* species are the most common research model that receives a pelleted diet, which is commercially available. It is also important to note that different life stages may require different feeds. For example, tadpoles will eat standard fish flake food, spinach, and alfalfa pellets (DeNardo 1995).

Amphibian species are ectothermic, so it is paramount to ensure that the temperature in the housing room meets the requirements of the species housed. Understanding the temperature variation in the species' natural environment is necessary to determine the acceptable temperature range for laboratory-housed species. Because amphibians are more sensitive to stress associated with higher temperatures (DeNardo 1995), it is often best to set the acceptable temperature at the lower range of the natural environment. This is true for both aquatic and terrestrial species. The ability to provide temperature gradients through the use of heating devices under the cage or with heat lamps aids in digestion, development, and the ability to fight off diseases (Browne and Zippel 2007).

FIGURE 25.10 **(See color insert.)** Provision of both an aquatic environment and a terrestrial environment is essential to species such as the tiger salamander (*Ambystoma tigrinum*) as it undergoes metamorphosis. (Photo by J.D. Cox.)

It is also important to match the humidity in the cage to the natural environment of the species, and most require higher levels (>70%) than are typically found or can be achieved in a typical research animal facility. This requires providing supplemental humidity in the microenvironment, which can be achieved by misting or through the addition of a water feature, such as a waterfall (Wright and Whitaker 2001). Plants can also be added to the tank to increase humidity. As with temperatures, it is best to provide humidity gradients in the tank to allow the animals to find their desired microenvironment, depending on their physiological needs.

It is not necessary to maintain amphibian rooms at the same high level of air changes (10–15 air changes per hour [ACH]) as are typical in animal research facilities. High levels can be detrimental to amphibian species due to the potential for desiccation (DeNardo 1995). Room air changes of 1–2 per hour will provide sufficient ventilation and aid in maintaining the required humidity levels for the species housed (Wright and Whitaker 2001).

Although amphibians can be housed using standard lighting conditions used in a typical research animal facility (DeNardo 1995), depending on the species and the type of research, it may be necessary to make modifications to accommodate these needs. Just as mammals require ultraviolet B (UVB) radiation for the production of vitamin D_3, so do some species of amphibians (Wright and Whitaker 2001; Beaupre et al. 2004; Pough 2007; Browne and Zippel 2007). Because of this requirement, facilities should ensure that there is a lighting source that provides adequate UVB at the animal's level. Most plastics and glass block UVB penetration, so even though fluorescent lights at the room level produce UVB, the distance from the animals and the type of caging used may prevent the animals from obtaining any UVB exposure. Fluorescent, mercury vapor, and halogen lamps can all be used to provide the necessary UVB exposure (Browne and Zippel 2007), but should be placed no higher than 18 inches from the animal. It is also possible to use LED lights, but this will require one set of lights for the light–dark cycle and another set to provide the necessary UVB exposure. Regardless of how the UVB light is provided, be aware that too much UVB can be detrimental to all stages of the amphibian life cycle (Antwis and Browne 2009). It may also be necessary to provide a dawn-to-dusk light cycle, especially if there is a need to breed the species (Pough 2007). It is understood that amphibians predominantly rely on a visual stimulus to detect and capture prey (Ruhl and Dicke 2012; Borghuis and Leonardo 2015), such that amphibian species that are predominantly nocturnal may even require a low level of light, similar to moonlight, during the dark phase to ensure that they are able to detect prey (Pough 2007).

Amphibian housing facilities that experience power or HVAC failures should take into consideration principles similar to those for a typical research animal facility: temperature, airflow, and lighting levels and cycles. Ideally, the controls for these utilities will be maintained on generator power, such that if there is a power failure, this will only lead to a loss of function for no more than a few minutes. If the species is provided water that is on a recirculating filtration system, this becomes a major concern, as without flow, the water quality will quickly deteriorate. Within 24–48 hours, there will be an increase in toxic nitrogenous compounds, such as ammonia and nitrites. The biological filter that maintains these compounds in check will also begin to die off.

In addition to the husbandry steps already mentioned with regards to feeding, water quality, and environmental conditions, it is important to consider the necessary steps for maintaining a clean facility. This includes cleaning tanks and enrichment devices, which is typically a manual process, as well as floors and other horizontal surfaces. Due to the permeability of amphibian skin, soaps and detergents are typically not used. Historically, low concentrations of chlorine bleach (0.06% to 0.6% sodium hypochlorite) have been used. By using chlorine bleach, even if the rinse was not effective, any residual chlorine will dissipate by allowing the equipment to air out for 24 hours. With recent developments of tank cleaning chemicals for zebrafish, there may be other options in the future that can reduce the amount of manual labor necessary to clean tanks and enrichment devices.

When handling amphibians, it is important to do so in a way that causes minimal disruption to the skin. It is also important to protect the handler, as many amphibians have defensive secretions that are toxic. The use of powder-free gloves that have been rinsed with distilled water is a safe way to pick up and restrain amphibian species. Transferring the animal to a clear container to allow for a complete visual examination or for transport may be needed. While the animal is being evaluated, it is critical to provide moisture to minimize drying or disruption of the amphibian's skin.

Since there are few amphibian species bred in captivity, there is a high probability that animals will be wild caught, which presents challenges as this relates to animal health. The disease state of these wild-caught animals is unknown; thus, a strict quarantine program is critical. It is ideal to house newly arrived animals in a separate room, but at a minimum, there should be a separation of tanks and the use of new gloves, instruments, and cleaning equipment between tanks, with the quarantine tanks worked with last. The basic principles for a good animal quarantine program can be applied to amphibians as well, but more specifics on the program have been previously described (Browne and Zippel 2007).

Many of the animal health concerns associated with amphibian populations are due to stress from shipping or poor environmental conditions. As with most animal species, this stress leads to a suppressed immune response, allowing for the development of a disease state due to opportunistic pathogens (Wright and Whitaker 2001). There is also a risk of disease with wild-caught animals, depending on the time of year in which they were collected (O'Rourke 2007). Intestinal and respiratory parasites, as well as chytridiomycosis, are some of the more common concerns with wild-caught specimens (O'Rourke and Rosenbaum 2015).

Surgical procedures should be performed following standard practices for most other species. Although the skin may contain some antimicrobial compounds, it is important to start with sterile instruments and use aseptic techniques throughout the procedure (DeNardo 1995). Tricaine methanesulfonate (MS-222), isoflurane baths, and isoflurane gel are commonly used anesthetics for surgical procedures (Wright and Whitaker 2001). The use of a surgical scrub on the skin should be limited to 0.75% chlorhexidine- or benzalkonium chloride (2 mg/L)–soaked gauze, left on the surgical area for a minimum of 10 minutes (Wright and Whitaker 2001). It is important to maintain a moist skin surface due to the requirement for oxygen exchange through the skin. This can be achieved through the use of moist gauze, the circulation of freshwater over the skin, or the use of artificial slime (Wright and Whitaker 2001). The surgical site can be closed with tissue glue or suture (DeNardo 1995). The suture should be a monofilament nonabsorbable material that is placed in a simple interrupted or horizontal mattress pattern to prevent dehiscence (DeNardo 1995; Wright and Whitaker 2001). It is important to minimize the tightness of the suture and remove all ligatures after 4 weeks if they have not come off naturally (DeNardo 1995). Silk and gut suture should not be used, as silk causes tissue reaction and gut dehisces (Tuttle et al. 2006). As with any surgical procedure, there is an assumption that the animal will experience some pain during the postoperative recovery period and should be provided analgesia. Several analgesics have been used in amphibians (Koeller 2009; Coble et al. 2011; Wright et al. 2014). The method of euthanasia used is dependent on the needs of the research. The most common method is an overdose of the anesthetic MS-222 by intracoelemic injection or immersion. The use of ethanol, barbiturates, and benzocaine as an injectable euthanasia has also been described (DeNardo 1995; Wright and Whitaker 2001). With any of the injectable methods, it is critical to ensure death, which can be done by removal of the heart as a secondary method. Physical methods include decapitation, followed by pithing and blunt cranial trauma, but should only be performed by trained and experienced individuals. Freezing of small specimens that are not freeze tolerant is also acceptable (Wright and Whitaker 2001; AVMA 2013).

Summary

Care of nontraditional species in a laboratory animal facility carries a responsibility of investing extra time and effort to learn as much as possible about the biology, natural history, and behavior of these widely variable and unique creatures. Only when we understand their fundamental needs can we begin to provide appropriate environments for these remarkable groups of animals.

REFERENCES

Antwis, R.E., and R.K. Browne. 2009. Ultraviolet radiation and vitamin D3 in amphibian health, behavior, diet and conservation. *Comp Biochem Physiol A* 154:184–190.

ASIH (American Society of Ichthyologists and Herpetologists). 2004. *Guidelines for Use of Live Amphibians and Reptiles in Field and Laboratory Research.* 2nd ed. Lawrence, KS: ASIH.

AVMA (American Veterinary Medical Association). 2013. *AVMA Guidelines for the Euthanasia of Animals.* 2013 ed. Schaumburg, IL: American Veterinary Medical Association.

Bateson, M., and G. Feenders. 2010. The use of passerine bird species in laboratory research: Implications of basic biology for husbandry and welfare. *ILAR J* 51:394–408.

Beaupre, S.J., Jacobson, E.R., Lillywhite, H.B., and Zamudio, K. Herpetological Animal Care and Use Committee. American Society of Ichthyologists and Herpatologists. 2004. Guidelines for use of live amphibians and reptiles in field and laboratory research, 2nd ed. http://www.asih.org/sites/default/files/documents/resources/guidelinesherpsresearch2004.pdf.

Borghuis, B.G., and A. Leonardo. 2015. The role of motion extrapolation in amphibian prey capture. *J Neurosci* 35(46):15430–15441.

Browne, R.K., and K. Zippel. 2007. Reproduction and larval rearing of amphibians. *ILAR J* 48:214–234.

Browne, R.K., R.A. Odum, T. Herman, and K. Zippel. 2007. Facility design and associated services for the study of amphibians. *ILAR J* 48:188–202.

Brusca, R., and G. Brusca. 2003. *Invertebrates*. 2nd ed. Sunderland, MA: Sinauer Associates.

Burger, R.M., W.H. Gehrmann, and G.W. Ferguson. 2007. Evaluation of UVB reduction by materials commonly used in reptile husbandry. *Zoo Biol* 26:417–423.

Carpenter, J.W., E. Klaphake, and P.M. Gibbons. 2014. Reptile formulary and laboratory normals. In *Current Therapy in Reptile Medicine and Surgery*, ed. D.R. Mader and S.J. Divers, 382–410. St. Louis, MO: Elsevier.

Coble, D.J., D.K. Taylor, and D.M. Mook. 2011. Analgesic effects of meloxicam, morphine sulfate, flunixin meglumine, and xylazine hydrochloride in African-clawed frogs (*Xenopus laevis*). *J Am Assoc Lab Anim Sci* 50:355–360.

Cooper, J.E. 2001. Invertebrate anesthesia. *Vet Clin N Am Exot Anim Prac* 4:57–67.

Cooper, J.E. 2004. Invertebrate care. *Vet Clin N Am Exot Anim Prac* 7:473–486.

Cooper, J.E. 2011. Anesthesia, analgesia, and euthanasia of invertebrates. *ILAR J* 52:196–204.

Council of Europe. 2006. European Convention for the Protection of Vertebrate Animals Used for Experimental and Other Scientific Purposes. ETS 123. Strasbourg, France: Council of Europe.

Council of Europe. 2010. Directive 2010/63/EU of the European Parliament and of the council of 22 September 2010 on the protection of animals used for scientific purposes. Strasbourg, France: Council of Europe.

DeNardo, D. 1995. Amphibians as laboratory animals. *ILAR J* 37:173–181.

Duncan, I.J.H. 2010. The domestic fowl. In *UFAW Handbook on the Care and Management of Laboratory and Other Research Animals*, ed. R. Hubrecht and J. Kirkwood, 637–654. 8th ed. West Sussex, UK: Wiley-Blackwell.

Elwood, R.W. 2011. Pain and suffering in invertebrates? *ILAR J* 52:175–184.

Emerging Model Organisms: A Laboratory Manual. Vol. 1. 2009a. Cold Spring Harbor, NY: Cold Spring Harbor Laboratory Press.

Emerging Model Organisms: A Laboratory Manual. Vol. 2. 2009b. Cold Spring Harbor, NY: Cold Spring Harbor Laboratory Press.

FASS (Federation of Animal Science Societies). 2010. *Guide for the Care and Use of Agricultural Animals in Research and Teaching*. 3rd ed. Champaign, IL: FASS.

Flinchum, G.B. 2006. Management of waterfowl. In *Clinical Avian Medicine*, ed. G.J. Harrison and T.L. Lightfoot, 831–847. Vol. II. Palm Beach, FL: Spix Publishing.

Frye, F. 1992. *Captive Invertebrates: A Guide to Their Biology and Husbandry*. Malabar, FL: Krieger Publishing.

Gehrmann, E.H. 1996. Evaluation of artificial lighting. In *Reptile Medicine and Surgery*, ed. D.R. Mader, 463–465. Philadelphia: Saunders.

Harvey-Clark, C. 2011. IACUC challenges in invertebrate research. *ILAR J* 52:213–220.

Horvath, K., D. Angeletti, G. Nascetti, and C. Carere. 2013. Invertebrate welfare: An overlooked issue. *Ann Ist Sup Sanit* 49:9–17.

Howard, B.R. 1992. Health risks of housing small psittacines in galvanized wire mesh cages. *JAVMA* 200:1667–1674.

Hubrecht, R., and J. Kirkwood, eds. 2010. *UFAW Handbook on the Care and Management of Laboratory and Other Research Animals*. 8th ed. West Sussex, UK: Wiley-Blackwell.

Hutchins, E.D., G.J. Markov, W.L. Eckalbar et al. 2014. Transcriptomic analysis of tail regeneration in the lizard *Anolis carolinensis* reveals activation of conserved vertebrate developmental and repair mechanisms. *PLoS One* 9:e105004.

Kalmar, I.D., G.P.J. Janssens, and C.P.H. Moons. 2010. Guidelines and ethical considerations for housing and management of psittacine birds used in research. *ILAR J* 51:409–423.

Koeller, C.A. 2009. Comparison of buprenorphine and butorphanol analgesia in the eastern red-spotted newt (*Notophthalmus viridescens*). *J Am Assoc Lab Anim Sci* 48:171–175.

Koh, C.Y., and R.M. Kini. 2012. From snake venom toxins to therapeutics—Cardiovascular examples. *Toxicon* 59:497–506.

Lewbart, G.A. 2011. *Invertebrate Medicine*. 2nd ed. West Sussex, UK: Wiley-Blackwell.

Lightfoot, T.L., and J.M. Yeager. 2008. Pet bird toxicity and related environmental concerns. *Vet Clin Exot Anim* 11:229–259.

Lisney, T.J., B. Ekesten, R. Tauson, O. Hastad, and A. Odeen. 2012. Using electroretinograms to assess flicker fusion frequency in domestic hens *Gallus gallus domesticus*. *Vis Res* 62:125–133.

Marx, K.L. 2006. Therapeutic agents. In *Clinical Avian Medicine*, ed. G.J. Harrison and T.L. Lightfoot, 241–342. Vol. I. Palm Beach, FL: Spix Publishing.

Matheson, S.M., J. Donbavand, V. Sandilands, T. Pennycott, and S.P. Turner. 2015. An ethological approach to determining housing requirements of gamebirds in raised laying units. *Appl Anim Behav Sci* 165:17–24.

Mench, J., R. Newberry, S. Millman, C. Tucker, and L. Katz. 2010. Environmental enrichment. In *Guide for the Care and Use of Agricultural Animals in Research and Teaching*, 30–44. 3rd ed. Champaign, IL: Federation of Animal Science Societies.

Merchant, M., N. Leger, E. Jerkins, M.B. Pallansch, R.L. Paulman, and R.G. Ptak. 2006. Broad spectrum antimicrobial activity of leukocyte extracts from the American alligator (*Alligator mississippiensis*). *Vet Immunol Immunopathol* 110:221–228.

NRC (National Research Council). 2011. *Guide for the Care and Use of Laboratory Animals*. 8th ed. Washington, DC: National Research Council.

Olive, P.J.W. 2002. Reproduction and life cycles in invertebrates. In *Encyclopedia of Life Sciences* p. 6. Hoboken, NJ: John Wiley & Sons.

Ornithological Council. 2010. *Guidelines to the Use of Wild Birds in Research*. 3rd ed. Washington, DC: Ornithological Council.

O'Rourke, D.P. 2007. Amphibians used in research and teaching. *ILAR J* 48:183–187.

O'Rourke, D.P. 2014. Reptiles and amphibians in laboratory animal medicine. In *Current Therapy in Reptile Medicine and Surgery*, ed. D.R. Mader and S.J. Divers, 290–295. St. Louis, MO: Elsevier.

O'Rourke, D.P., and K. Lertpiriyapong. 2015. Biology and diseases of reptiles. In *Laboratory Animal Medicine*, ed. J. Fox, L. Anderson, G. Otto, K. Pritchett-Corning, and M. Whary, 967–1013. 3rd ed. London: Elsevier.

O'Rourke, D.P., and M.D. Rosenbaum. 2015. Biology and diseases of amphibians. In *Laboratory Animal Medicine*, ed. J. Fox, L. Anderson, G. Otto, K. Pritchett-Corning, and M. Whary, 931–965. 3rd ed. London: Elsevier.

Overgaard, J., H. Gesser, and T. Wang. 2007. Tribute to P.L. Lutz: Cardiac performance and cardiovascular regulation during anoxia/hypoxia in freshwater turtles. *J Exp Biol* 210:1687–1699.

Packard, G.C., and M.J. Packard. 2004. To freeze or not to freeze: Adaptions for overwintering by hatchlings of the North America painted turtle. *J Exp Biol* 207:2897–2906.

Pechenik, J. 2015. *Biology of the Invertebrates*. New York: McGraw-Hill Education.

Pincheira-Donoso, D., A.M. Bauer, S. Meiri, and P. Uetz. 2013. Global taxonomic diversity of living reptiles. *PLoS One* 8:e59741.

Pough, F.H. 2007. Amphibian biology and husbandry. *ILAR J* 48:203–213.

Ruhl, T., and U. Dicke. 2012. The role of the dorsal thalamus in visual processing and object selection: A case of an attentional system in amphibians. *Eur J Neurosci* 36:3459–3470.

Ruppert, E.E., R.S. Fox, and R.D. Barnes. 2004. *Invertebrate Zoology: A Functional Evolutionary Approach*. 7th ed. Boston: Brooks Cole.

Smith, J.A., P.R.L. Andrews, P. Hawkins, S. Louhimies, G. Ponte, and L. Dickel. 2013. Cephalopod research and EU Directive 2010/63/EU: Requirements, impacts and ethical review. *J Exp Mar Biol Ecol* 447:31–45.

Tuttle, A.D., J.M. Law, C.A. Harms, G.A. Lewbart, and S.B. Harvey. 2006. Evaluation of the gross and histological reactions to five commonly used suture materials in the skin of the African clawed frog (*Xenopus laevis*). *J Am Assoc Lab Anim Sci* 45:22–26.

Wilson-Sanders, S.E. 2011. Invertebrate models for biomedical research, testing, and education. *ILAR J* 52:126–152.

Wright, K., J.W. Carpenter, and R.S. DeVoe. 2014. Abridged formulary for amphibians. In *Current Therapy in Reptile Medicine and Surgery*, ed. D.R. Mader and S.J. Divers, 411–416. St. Louis, MO: Elsevier.

Wright, K.M., and B.R. Whitaker. 2001. *Amphibian Medicine and Captive Husbandry*. Malabar, FL: Krieger Publishing.

Section VII

Husbandry

26

Basic Animal Facility Management

Jori K. Leszczynski, Jamie Tackett, and Michelle Wallace-Fields

CONTENTS

Introduction

Managers of facilities are faced with many different responsibilities when overseeing the management of the animal care programs at their institutions. These will vary based on the type of facility that they are overseeing, as well as the species and type of research that is being conducted. In this chapter, managers will gain knowledge of basic animal facility management in areas such as staffing, husbandry, animal identification, working with hazards, equipment maintenance, maintaining safety equipment, animal procurement, and receiving and exporting animals into the facility.

Types of Facilities

There are many different types and sizes of facilities that house animals used for research. Facility design, size, and resources available depend on many factors, including the species being housed, the goals and types of research being performed, and the size of the larger research enterprise that the facility serves. Program types can range from colleges and universities, private nonprofit research institutes, and contract research organizations, to those serving industries such as pharmaceuticals. In general, a centrally managed animal research program that has as few independent facilities as possible is the most efficient and cost-effective method to run a program. However, due to research demands and available resources, many programs have multiple sites housing animals, and thus varying management structures. It is important that the management team of the institution take into consideration the unique needs of each facility, in conjunction with the applicable regulations and policies of the institution, as well as the local area that the program is located in, when determining management practices (Stark et al. 2010).

Satellite Facilities

An area that has come under increasing scrutiny by AAALAC International and, in the United States, the U.S. Department of Agriculture (USDA) and the Office of Laboratory Animal Welfare (OLAW) is how satellite facilities are managed and monitored to ensure compliance with the *Guide for the Care and Use of Laboratory Animals* (*Guide*) (NRC 2011) and Animal Welfare Regulations (AWR). Many satellite facilities are run by research investigator groups that will need to be trained appropriately regarding applicable rules, laws, and/or policies; animal care; facility maintenance; and documentation standards. Regardless of whether the satellite facility is under the management of a centralized animal program or run by research staff, it is strongly recommended that satellites strive to establish the same operating procedures that are in use in the main facilities, including, but not limited to, husbandry practices, veterinary care, documentation, training, and sanitation practices, for overall institutional program consistency.

Compliance and Monitoring of Satellite Facilities

Monitoring of a satellite facility for compliance is the role of the animal oversight body, usually called the Institutional Animal Care and Use Committee (IACUC); however, the veterinary staff should play a role in ensuring that animal care standards are appropriate for the species being housed. Oversight of the satellite can be performed as an extension of the IACUC or of the centralized management of the animal program. The individual monitoring the facility could be a veterinarian, a veterinary technician, a member of the animal husbandry team, or a member of the IACUC staff, depending on the complexity of the work within the satellite and how the animal program is established at the institution. IACUCs may have someone that oversees a formal postapproval monitoring program as well, which can add to the oversight of the satellite.

Equipment Sanitation in Satellite Facilities

Cage sanitation practices in a satellite facility must meet the requirements of the *Guide* and, if applicable, the AWR. However, many satellites are small and lack mechanical cage washing equipment. Therefore, those involved in the care of the facility should make a decision as to how these requirements can be met. Options include transporting dirty supplies to a centralized cage wash room in one of the centrally managed animal facilities; using a washer that is located near the satellite, such as a glassware washer; purchasing disposable caging; and hand sanitizing. The cost of using disposable cages should be weighed against the labor, supply, transport, and utility costs of hand washing or transporting equipment to a centralized animal washing facility. Large equipment, such as shelving racks and tables, is generally not able to be removed from the facility, and so a method and frequency of sanitizing these components in place should be developed. With any method that is used, a system to validate the method of cleaning to show that sanitation is appropriate must be set up according to the *Guide*.

Training of Research Personnel to Care for Animal Colonies

With any satellite that is run by research staff, training must be conducted to ensure that they are following all local, state, and federal regulations and policies with regards to care and use of animals. It is recommended that the training be conducted by members of both the IACUC and animal facility staff, as there must be a thorough understanding of all aspects of animal care prior to beginning work in the facility. Research staff must understand that animal care is a 365-day process, and proper documentation is essential for all aspects of the care of animals. Animal facility staff are also in the best position to share standard operating procedures (SOPs) and documentation forms to assist the research staff with setting up the facility.

Core Resources

As more expensive equipment is used to conduct research, institutions are increasingly moving toward shared equipment programs. It is important that shared equipment policies are established that include a fee structure to cover the costs of preventive maintenance, repairs, disposable supplies, and ultimately replacement costs. Skilled operator costs may also be factored into the equation. It is important to establish an individual or group who will be responsible for maintaining the equipment should there be problems. These responsibilities could fall to the animal program or to the head of a core service (Berns et al. 1996), depending on institutional management structures.

Staffing Models

As animal husbandry and management has become more complex, staffing for animal facilities has become more specialized, with many institutions having multiple categories of employees that work within them. The American Association for Laboratory Animal Science (AALAS) has recognized this and has increased the number of job titles that it collects compensation data on. For example, in 2010 there were 12 job categories, which grew to 18 categories in 2014 (AALAS 2010, 2014). The types of positions at a facility will depend on the types of species housed, the size of the institution, and the complexity of the organizational structure. As facilities grow, job specialization, where employees spend their time on a narrower group of tasks, tends to be more common in order to gain efficiencies in the operation. While repetitive motion injuries and employee burnout should be considered with all employees, those that are specializing in specific tasks should be monitored more closely and appropriate opportunities for cross-training should be developed. Managers should be encouraged to look at their program and determine the best staffing model for their specific situation (Boisvert and Morgan 2006; Schray et al. 2011).

Common job descriptions within an animal facility are listed in Table 26.1 (AALAS 2014).

An additional job description that is starting to become more prevalent is the enrichment coordinator or animal behaviorist. This position could be filled by a veterinarian, but many times it is filled by someone with an educational background in animal behavior. These positions are growing in number as the welfare of animals in laboratory animal facilities continues to improve and, along with it, the interest in ensuring appropriate species behavior. The AWR sections related to primates and dogs (CFR 2013) and the *Guide* both have language on providing species-specific environmental enhancement. Behavior specialists and enrichment coordinators, once only employed in large primate programs, are now employed by many facilities to meet these requirements for all species. These individuals may work independently to find and treat behavior disorders in individual animals or provide broader programmatic direction with regard to enrichment and species-specific behavior.

Training

Regardless of job description, a comprehensive training program should be employed for all staff members, including initial training and verification of the ability to perform duties, as well as regular opportunities to refresh training on the key components of the program. Time should also be allocated in employees' schedules to ensure that training required by other groups at the institution, such as environmental health and safety (EHS), who may oversee blood borne pathogen and chemical hygiene training, for example, is completed. Regardless of the training being performed, a method for documenting when training occurs, as well as retaining records of training, should be developed. These methods could be simple paper files, a spreadsheet, or a more complex training database system.

Personnel Considerations

Consideration should be given to personnel needs when working in an animal facility, such as dedicated clothing, locker rooms, hand washing sinks, laundry, offices, and break rooms. According to the *Guide* and AWR, personal protective equipment (PPE) provided should be based on a risk-based analysis of the work being conducted, to include hazards that are present in the work environment. Occupational health and EHS professionals should be consulted when developing the risk assessment. Common or basic PPE used in biomedical research facilities may include dedicated clothing, gowns, hair bonnets, surgical masks, shoe covers, dedicated shoes, eye protection, and gloves. Additional PPE can be added to the basic PPE to protect users against other hazards in the work environment, such as coveralls to provide full-body coverage, or N95 masks when working with respiratory hazards. PPE should be removed at a defined boundary within the facility, and showers may be required to control exposure of hazards to personnel or animals. Hearing protection should be provided when the maximum safe decibel level for continuous noise exposure, that is, currently 85 dB for 8 hours (NIOSH 2007), is exceeded, and should

TABLE 26.1
Common Job Descriptions

Job Title	Job Description
Director	Usually requires an advanced college degree in a related field, e.g., DVM, PhD, or specialty board certification. Ensures overall facility compliance with federal, state, and local regulations. Has responsibility for animal care program, policy, planning, and administration.
Assistant/associate director	Usually requires a college degree; may require a DVM and specialty board certification. Assists director in oversight of the animal care program, policy, planning, and administration. May have direct supervisory responsibilities for one or more components of facility operations.
Manager	May or may not require LATG or CMAR certification; usually requires a college degree in a related field. Provides direct supervision of supervisory personnel. Ensures standard operating procedures are current and accurate, and that occupational safety procedures for animal care are observed. Recommends personnel actions to include hiring, training, promotion, and discipline. May have responsibility for facilities management and planning.
Administrative/ business manager	May require a business degree and 3 or more years of experience in an office management position. Works with the director to prepare budgets and provide monthly reports. Oversees purchasing of laboratory animal facility supplies, develops bids and oversees contracts for large purchases, prepares and assists with the preparation of annual per diem cost analyses, develops and oversees facility business processes, and fulfills other administrative duties.
Supervisor	Usually requires LATG certification and/or college degree in a related field. Provides first-line direct supervision of laboratory animal care personnel. Prepares schedules for work assignments, and animal husbandry. Trains and instructs personnel in appropriate animal care and facility procedures.
Clinical veterinarian	Requires a DVM. Provides veterinary medical care for research animals. May plan, execute, and follow up on technical research projects. Develops, implements, and controls procedures related to the acquisition, maintenance, quarantine, and disposition of all research animals. Coordinates with, provides technical support for, and acts as a liaison to management and other divisions regarding animal health issues.
Facility compliance manager	Usually requires a minimum of a bachelor's degree or its equivalent. Manages regulatory compliance processes, including handling corrective and preventive actions; establishes and administers quality assurance programs to measure and track compliance with regulations; and assists with semiannual inspections.
IACUC coordinator	Oversees compliance with regulatory standards and accrediting agencies. Facilitates activities of the IACUC, including administration of the protocol review process, semiannual facility inspection and program review, and record keeping.
Training coordinator	May require LAT, LATG, or CMAR certification. May need a 2-year college degree or higher in a related field. Develops, coordinates, and provides training for animal care staff, scientists, and others who may be exposed to laboratory animals. Training provided may include topics such as biology, care, biomethodologies, and special techniques for a variety of laboratory animals, and may be provided via web, didactic, and hands-on methods. Also responsible for managing training documentation and supporting compliance with training mandates.
Research technician	May require college degree in a related field. Provides assistance to researchers by performing specific technical procedures of the research project. Collects research data and maintains accurate research records.
Animal health care technician	May require a certification, such as LATG, RVT, or CVT, or a licensure, such as LVT, and/or previous veterinary technology-related experience. Provides support to the veterinary staff in collecting biological samples, performs diagnostic and imaging laboratory procedures, provides pre- and postoperative care to animals, and assists in surgery. Operates and maintains equipment such as anesthesia machines, diagnostic laboratory equipment, and other specialized instruments. Administers medications and medical treatments. Maintains and stores inventories of medical and diagnostic supplies.
Senior-level laboratory animal technician	May require at least LAT certification. Provides husbandry of animals in more specialized areas of the animal facility, to include barrier and biocontainment areas. Maintains supplies, feed, and bedding inventories and storage. Assists researchers and veterinary staff when needed. Trains entry- and midlevel technicians.

(Continued)

TABLE 26.1 (CONTINUED)

Common Job Descriptions

Job Title	Job Description
Midlevel laboratory animal technician	May require at least LAT certification. Provides basic daily husbandry of laboratory animals, to include more technical procedures, such as animal identification, collecting biological samples, and administering medications. Maintains accurate animal records. In addition to basic sanitizing supplies and equipment, may use autoclaves, necropsy equipment, and laminar flow hoods. Receives and stores materials and supplies.
Entry-level laboratory animal technician	Usually no AALAS certification required upon entry, but may be required within a certain length of time on the job. Provides daily husbandry of laboratory animals, to include observation of animals, handling and restraint, feeding and watering, changing cages/racks, and sanitizing animal quarters and equipment. Operates and/or uses supplies and equipment necessary to perform the husbandry duties. Provides daily record keeping for animals and room environments.
Shipping coordinator	May require AALAS certification. Coordinates receipt of animals from and/or shipping of animals to other institutions/companies. Supervises the packing of animals for shipment and follows all regulations regarding transport of animals, including use of appropriate containers, feed, bedding, and water as needed. Coordinates shipping arrangements with other institutions and/or shipping companies. Maintains records of all deliveries and shipments.
Purchasing coordinator	Usually requires a minimum of a business-related associate's degree. Works with vendors to obtain competitive pricing bids, tracks deliveries of purchases, prepares purchase orders, and provides customer support to research staff or department on requested purchases.
Administrative support assistant	Assists manager with coordinating administrative support tasks, such as payroll, purchasing, and customer service.
Cage washer	Usually no AALAS certification required. Sanitizes and/or sterilizes animal cages, racks, and other related equipment; operates and/or uses rack washers and other sanitization and sterilization equipment; prepares and uses disinfecting and sanitizing solutions. Performs preventive maintenance and maintains service records on racks and caging equipment.

Source: Adapted from American Association for Laboratory Animal Science, 2014 laboratory animal facility compensation survey, Industry Insights, Inc., Dublin, OH, 2014.

Note: DVM, doctor of veterinary medicine; LATG, laboratory animal technologist; CMAR, Certified Manager of Animal Resources; LAT, laboratory animal technician; RVT, Registered Veterinary Technician; CVT, Certified Veterinary Technician; LVT, Licensed Veterinary Technician.

be part of an overall hearing conservation plan (OSHA 2002). An effort should be made in the engineering of spaces, as well as the use of appropriate equipment and appropriate PPE, in order to reduce personnel exposure to animal allergens (Gordon et al. 1997).

If additional respiratory protection is required, there are many different types of respirators that can be used to provide protection against the inhalation of harmful airborne contaminants, including N95 masks, half- or full-face chemical respirators, self-contained breathing apparatus (SCUBA), and personal purified air respirators (PAPRs) (Figure 26.1). It is recommended that occupational health and EHS professionals be involved with the decision on which of these methods of protection is appropriate for the hazards being faced, as well as to ensure that individuals are cleared medically and appropriately fit tested before their use and on an annual basis (NRC 1997). Eye protection, such as goggles or full-face shields, are important where splashes may occur (NRC 1997). In most cases, utilizing nitrile gloves instead of latex gloves provides more protection from chemicals, as well as reduces the possibility of inadvertently exposing an individual with a latex allergy to latex powders. Additional protective equipment can include arm protectors for chemical and direct contact allergen control and bite gloves when handling animals that may bite. The types of PPE that are used in the facility should be evaluated on a regular basis to ensure that all components are safe, as well as achieve the goals of protecting the humans, animals, or both (NRC 1997). Programs have reported large cost savings through reduction or elimination of components of the traditional PPE model that do not provide additional support of the programmatic goals of limiting disease spread potential and employee safety (Hickman-Davis et al. 2012; Baker et al. 2014).

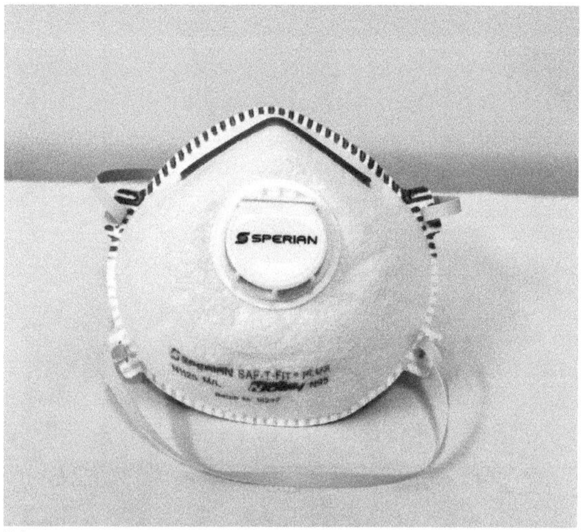

(a)

(b)

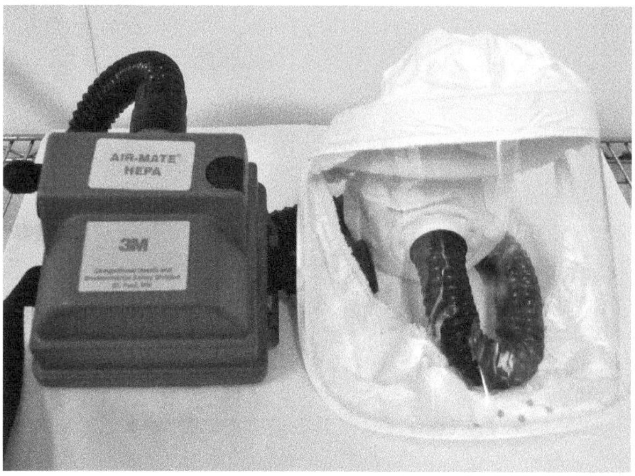

(c)

FIGURE 26.1 Types of respirators: (a) N95, (b) full-face, and (c) PAPR. (Photo by Michelle Wallace-Fields.)

Traffic Patterns

Traffic patterns and the movement of supplies are dependent on the facility design and the health status of the facilities within the animal program. Generally, facility design is centered on the cage washing facility, with equipment and personnel moving out from the clean side of the cage wash. From there, it is important to work with the veterinary staff to determine the best traffic pattern that will minimize cross-contamination between areas of the facility. Generally, this requires entry into the most disease-free rooms first (i.e., germ-free, strict isolation barrier, or other clean room), followed progressively to rooms that have a higher risk of disease, and ending with the rooms with the highest risk of disease (i.e., quarantine or biohazard rooms). Ultimately, caging and staff would end up in the dirty side of the cage wash. Proper facility design will increase efficiency in ensuring that these traffic patterns are followed, as design can be used to move people in the correct direction. It can also help to reduce the possibility of contaminants and prevent unnecessary exposure of personnel to animals and animal wastes. However, as research requirements change and as animal programs grow, facility design is generally only one component of this process. Therefore, it is important to establish workflows, SOPs, and facility policies that support these goals, to reduce the potential for disease outbreak and transmission of pathogens (Trott et al. 2004), as well as promote efficiency of staff and operations.

Movement of Specialized Staff between Facilities

Programs that have specialized staff that need to work across multiple facilities or within facilities with varied health status, such as behaviorists or veterinary staff, controlling for cross contamination between facilities or within a specific facility is more of a challenge. This is especially true when there may be only one individual performing a specific function that may require them to travel in between these areas within the same day. It is important that the facility management staff and veterinary staff discuss appropriate traffic patterns for the program for these specialized staff on a routine basis. Any individual that has access to the facility should be regularly trained to ensure that he or she knows the current order of entry into different areas. To assist personnel, signage could be employed that specifies the room entry order, or rooms could be grouped based on a similar health status, by facility, by color coding, or by numerical codes. Instructions and training should be provided to staff that need to work in areas of varying health statuses regarding room entry order, changes in PPE, shower requirements, or other procedures that should be used to prevent cross-contamination. To promote efficient use of staff time, it is best if individuals working in the facility schedule their workdays or workweeks in order to minimize the potential for cross-contamination. For example, individuals should move through their day or week starting at the most disease-free areas, and then move to the areas with a higher risk for disease. If an individual is required to enter a disease-free area after entering a lower health status room, specific procedures, such as showering and changing clothes or mandated time out of the facility, should be developed to prevent cross-contamination.

Basic Husbandry Methods

One of the most important steps for any animal facility is determining the type of caging, bedding, food, and delivery methods for food and water, for each species of animal housed. When determining what type of caging to utilize, the institution should take into account the species to be housed, the social housing needs of those species, the enrichment to be used, the maximum number of animals that will be allowed per cage or pen, whether the animals will be breeding, whether a ventilated or a static housed cage is most appropriate for the animals, the studies being conducted, whether cage floors should be solid or slotted metal, and cage washing equipment and capabilities. Cage and pen cleaning and changing frequencies are outlined in the *Guide* and the AWR. If the institution will be using a cage or pen cleaning and change frequency that is outside of the recommendations of the *Guide* or AWR, then the institution should develop criteria to determine if performance standards are being met. These "exceptions" to the *Guide* or AWR should be approved by the IACUC and appropriately noted.

Rodent Cage Cleaning Frequency

The frequency of cage unit changes and soiled bedding change is a matter of professional judgment of the animal care personnel based on consultation with the investigator, taking into consideration the standards of the *Guide*. Soiled bedding, cages, enrichment, and other cage structures (i.e., wire bar lids, water bottles, and microisolator lids) should be removed and replaced with fresh materials as often as necessary to keep the animals clean and dry, as well as to ensure appropriate conditions for the animals. The frequency of changing out each component will depend on the performance standards you document at your institution (Schondelmeyer et al. 2006).

The frequency of bedding and cage changing depends on such factors as the number and size of the animals in the primary enclosure, the size of the enclosure, the urinary and fecal output, the appearance and wetness of the bedding, and the type of bedding used. The *Guide* recommends that solid-bottom caging be sanitized at least once a week, but some types of cages might require less frequent sanitation, which includes individually ventilated cage (IVC) units. AAALAC International refers its accredited units to the *Guide* when establishing their interpretation of cage changing frequency and states that "for IVC cage change intervals of longer than 2 weeks, verification of microenvironmental conditions may include measurement of pollutants such as ammonia and CO_2, microbiologic load, observation of the animals' behavior and appearance, and the condition of bedding and cage surfaces. Also, as with any performance standard, there should be a system in place to monitor the outcome and report back to the IACUC." For example, Rosenbaum et al. (2009) found that as you increase bedding volume, there is a statistically significant decrease in the rate at which ammonia levels build in the microenvironment. In addition, cages with low bedding volumes have increased humidity, which is also linked to increased ammonia production. In order to stay within appropriate ammonia levels, they recommend a 14-day change cycle for a mouse cage containing five female adult mice when using 400 mL of aspen chip bedding. It is also important to consider species' preferences for structuring cage component changes. For example, rats prefer scent-marked material, such as enrichment, nesting material, or even a small amount of soiled bedding, to be moved to the new cage when cage changing (Meller et al. 2011), whereas there is evidence that mice do not (Rasmussen et al. 2010).

Lean management principles have also been adopted in the laboratory animal field. An example of Lean principles involves moving away from "scheduled" cage changes. These principles include changing cages on an "as needed" basis or identifying areas that can extend beyond the typical industry standard. For example, a single-housed CD-1 mouse in disposable, ventilated caging was evaluated for extending the needed cage change date beyond 14 days (Vogelweid et al. 2011). The study found that these single-housed mice could be extended to 28-day cage changes, and that the in-cage ammonia levels were found to still be below the industry standard of 25 ppm. The Center for Comparative Medicine at Massachusetts General Hospital and the Vivarium at Seattle Children's Research Institute (SCRI) have developed Lean management practices for the frequency of cage changes within their facilities. Both programs report large savings of technician time and bedding use. With these savings comes decreased interruption to the animals, and decreased use of cage wash equipment. At Massachusetts General, the average population is 27,000 cages. They found that with the Lean practices of cage changing, they will save 8000 hours a year in labor costs (Stahl 2012). SCRI found that Lean cage changing practices save 1.5 hours per day of cage washing and result in a 50% reduction in the cost of bedding (Carbasho 2013); both groups report a 30%–35% reduction in cage changes.

Nonrodent Mammals

Additional consideration should be given for nonrodent laboratory animal species. There are many types of caging options available for nonrodent mammals. Some caging options are transient, such as rabbit or primate caging that are on wheels (Figure 26.2), or portable runs for larger species, such as pigs or dogs, that are able to be broken down and sent to cage wash for processing. Other enclosures are permanently installed, such as permanently fixed dog or pig runs (Figure 26.3), and must be cleaned in place. All enclosures must be cleaned at an appropriate interval to meet standards of the *Guide* or other local, state, or national standards, such as the AWR. In general, primary enclosures must be cleaned of excreta and food waste daily, and under the enclosures is cleaned as necessary to prevent the soiling of the animals.

FIGURE 26.2 Rabbit caging on wheels to allow for easier movement to cage wash facilities for sanitization. (Photo by Michelle Wallace-Fields.)

In most cases, enclosures must be cleaned with a sanitizing agent or via a chemical cage washer at least every 2 weeks. If the animal is in a fixed enclosure, it is important that the animal is either removed or able to escape the area so that it does not get wet or harmed by the sanitizing agent.

Fish, Reptiles, and Amphibian Housing

Housing and sanitation of enclosures for nonmammalian species vary greatly by the species being housed (Figure 26.4). It is recommended that facilities consult with experts in the field when setting up enclosures and determining methods for cleaning. For example, when working with aquatic species, water quality is generally the most important aspect of the environment and should be monitored closely (the *Guide*; Major and Wassersug 1998). In addition, the use of chemicals is strongly discouraged, and hot water should be used to sanitize equipment. If chemicals are used, then extensive rinsing should be employed. Enclosure sanitation frequency will vary by species and performance standards. Taxon-specific guidelines, wildlife and zoo publications, and internal guidelines are all resources that can be used when determining the ideal housing and care of these species.

Food

Once cage type and bedding change frequency have been determined, each institution should also determine what feed vendor to utilize and how staff are going to handle, store, and transport food for those cages. The *Guide* states, "Animals should be fed palatable, non-contaminated, and nutritionally adequate

FIGURE 26.3 Permanently fixed large animal run for housing dogs, pigs, sheep, goats, or other larger laboratory animal species. Fixed runs require methods for sanitization in place. (Photo by Michelle Wallace-Fields.)

food daily or according to their particular requirements unless the protocol in which they are being used requires otherwise." Facilities are encouraged to consider manufacturers' and suppliers' procedures and practices for protecting and ensuring diet quality (e.g., storage, vermin control, and handling procedures). Some feed vendors provide feed in different volumes and bag sizes. Typically, the smaller the bag, the higher the cost per unit of weight; however, using lighter bags could reduce staff injuries. Diet typically comes in nonsterilized and sterilized form. The sterilized diet can be either irradiated or autoclaved. When considering the use of sterilized feed versus nonsterilized feed, the pathogen status of the animals in the facility should be considered, as well as the overall biosecurity. For example, there has been at least one report of unsterilized food as the apparent cause of a mouse parvovirus outbreak (Watson 2013). In all cases, the expiration date for the diet should be noted and, unless stated otherwise, the assumption should be that diets can be used until 6 months after the date of manufacture. If feed is transferred to a secondary container, the container should be sanitized and labeled with the type of diet and mill or expiration date.

Water

There are several options for administering drinking water to animals, including water bottles, automated waterers, bowls, and even pouches that can provide potable drinking water. Generally, the source

FIGURE 26.4 *Xenopus* sp. frog tanks housed on a rack with a flow-through water supply system. This type of system allows water to be conditioned prior to being introduced to the animals, but it is not recirculated. (Photo by Michelle Wallace-Fields.)

of water comes from the local municipal water supply, and so reviewing local water reports on a set interval is recommended in order to ensure that the water meets appropriate standards for research animals. In many facilities, water is further treated to ensure that chemical and microbial contaminants are minimized. The most common methods for reducing contaminants are to use progressive filters or reverse osmosis. For reduction in microbial load, which is important when working with immune-compromised animals, water can be treated with ultraviolet (UV) light, autoclaved, acidified, and/or hyperchlorinated (Hessler and Leary 2002). When deciding whether to sterilize the water via chemical means, take into consideration the possible effects on the research that is being conducted (Lipman and Perkins 2002). Major considerations for deciding on the type of delivery device that works best for your institution will be based on the species housed, the staging and storing capabilities of your facility, the water restrictions in the area the facility is located, and the ability to wash and/or sanitize these devices. For automatic watering systems, generally ensuring the water remains sanitary is accomplished via regular flush of the lines or through filtration and UV sterilization for recirculating systems (Schultz 2006). All methods of water delivery have advantages and disadvantages, so it is important to evaluate your processes on a routine basis to ensure that the systems in place perform optimally (Gonzalez et al. 2011; Gordon and Wyatt 2011). The water quality and source should also be evaluated on a routine basis to ensure that it is potable. In general, water that has been treated will remain potable and the risk of contamination is minimal, whether using bottles or automatic watering systems (Haist et al. 2004; Meier et al. 2008).

Environmental Enrichment

Environmental enrichment is important to allow animals to display species-specific behaviors. All laboratory animal facilities should have a program to provide environmental enrichment for all species according to the *Guide*. The AWR requires a program that provides for the psychological well-being of nonhuman primates and an exercise program for dogs. The *Guide* recommends that an environmental enrichment program be provided that promotes the well-being of all species through species-appropriate methods and that is routinely evaluated to ensure that it is meeting the needs of the animals in the program. When designing the program, the type of species housed and the type of research being performed should be evaluated. For example, social housing is one of the most common methods for achieving species-specific behaviors for social species, but it may cause stress when incompatible animals are placed together (Whary et al. 1993). Some of these stressors can be reduced by including visual barriers and houses. Manipulanda, such as toys, chew sticks, and nesting material (Gaskill et al. 2009), may also be used as part of an enrichment program. It is important for animal facility managers to work with the IACUC, as well as the research scientists, to ensure compliance with the regulations, as well as to ensure the integrity of the science being conducted. By working with the research scientists to determine the best way to enrich animals on study, facilities can help to control the possible variability that enrichment can bring to research results. This can include changes in physiologic parameters, such as hematology results, cardiovascular telemetry data, or food consumption (Andrews-Kelly 2014).

Health Checking

Performing a daily check of animal health and environment, or health checking, is another important aspect of every laboratory animal facility. According to the *Guide*, all animals should be observed for signs of illness, injury, or abnormal behavior by a person trained to recognize such signs. A proper health check should consist of evaluating food and water availability, the health status of the animal, and the condition of the microenvironment. In addition, animal facility personnel should also examine the macroenvironment, to include temperature, humidity, noise, light intensity, light cycle, functionality of equipment, cleanliness, and organization. The *Guide* recommends that daily observations be performed, but recognizes that some applications (i.e., large outdoor settings) may make daily observation difficult and allows for professional judgment to minimize risks to personnel and animals. Each institution should establish policies and procedures for the reporting of adverse conditions that could affect the animal's health and welfare, as well as illness and unexpected death.

Animal Identification

Importance of Genus and Species Recognition

As a manager, it is important to know what species will be housed in order to plan appropriate micro- and macroenvironments for the animals. Having the appropriate environment for the animals also will ensure scientific integrity and maximize staff and researcher safety. For example, when working with *Xenopus* frogs, it would be important to know if they are *Xenopus laevis* or *Xenopus tropicalis*, as the two different species require different environments, even though they are of the same genus (Major et al. 1998). The *Guide* provides information on appropriate temperature, humidity, and housing density for a variety of commonly used laboratory animals. It is also important to make sure that all regulations pertaining to the locality that the facility is located in are followed, such as the AWR in the United States.

Genetic Identification

Genetic identification of animals is very important, especially in the age of the genetically modified animal. Many labs work with multiple different lines of animals, mainly rodents, that range from commercially available strains and stocks to complex genetic lines with multiple knocked-out genes. While genetically modified animals of all species are being developed, mice and rats are by far the most common species involved. It is important for managers and their staff to have a basic understanding of the rules for nomenclature for these animals. The best resource is published by the International Committee on Standardized Genetic Nomenclature for Mice (2015) and can be found on the Internet at the Mouse Genome Informatics site (http://www.informatics.jax.org/mgihome/nomen/index.shtml).

As the use of genetically modified animals has increased, so has the amount of breeding that occurs at institutions. As a result, determining the genotype of an animal has become an important step in ensuring scientific integrity by not only ensuring that the genetic changes that are desired are being passed on to the next generation, but also ensuring that inadvertent breeding or genetic drift has not occurred. There are several methods to establish the genotype of an animal, all having to do with collection of a DNA sample from the animal. Methods to gain DNA depend on the species and developmental status of the animal. For rodents, these methods include taking a small tip of the tail in rodents (Hankenson et al. 2008), using the skin produced from an ear notch or snip, or using a toe clip (Paluch et al. 2014), saliva, skin biopsy, blood, or feces (Symonds 2013). For most other species, a small skin biopsy, blood, saliva, fin clip, or feces can be used. For any species, it should be noted that the taking of any tissue samples should be considered a potentially painful or distressful procedure, so noninvasive methods should be encouraged if possible. All sample collections should receive the appropriate IACUC approval and be appropriate for the species and age of the animal.

When managing breeding colonies, detailed records should be kept to ensure that appropriate genetics are maintained and animals are used judiciously (Ayadi et al. 2011). Systems can range from very simple, such as using hard copies and cage cards (Jude et al. 2012), to complex database systems. There are commercially available systems that can integrate into the institution's animal program management software, as well as several free or fee-for-service programs that are available to download from the Internet (Donnelly et al. 2010). Many research labs use homegrown systems that employ simple database software programs. Using these electronic programs, colony managers can consider the use of technology further by implementing systems that can be used at the cage side with smartphones and tablets. Other species, such as primates, cats, guinea pigs, and rabbits, can draw on similar colony management systems in order to efficiently track animals and ensure proper genetic makeup of the colony (Chikazawa et al. 2012).

Sex Differentiation

Employees should be trained on how to differentiate the sex of an animal as part of their basic duties. The most common method for most rodents is to look for a difference in anogenital distance, where the anogenital distance is greater in males than in females (Figure 26.5). For most other mammals, it is

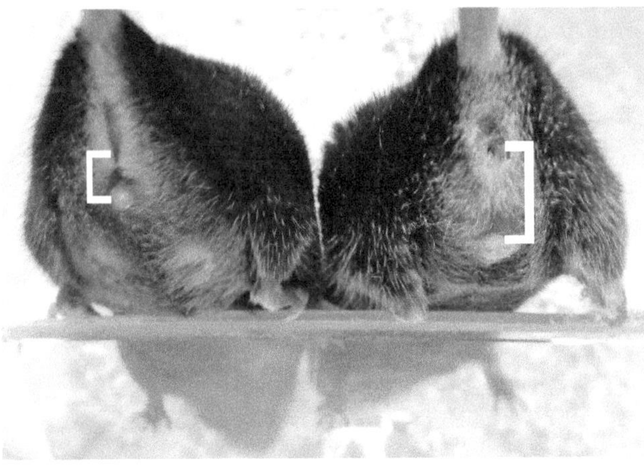

FIGURE 26.5 **(See color insert.)** Sex differentiation in rodents is generally achieved by looking at anogenital distance. Anogenital distance is greater in males (right) than females (left). (Photo by Michelle Wallace-Fields.)

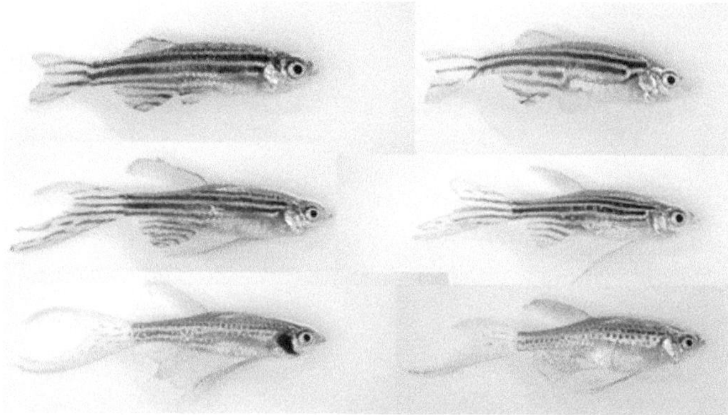

FIGURE 26.6 (See color insert.) Sex differentiation in zebrafish. Males have a more streamlined torpedo shape; females have a rounder abdomen. On wild-type fish, females have a dorsal fin that is more yellow. Females also have a cloaca (not pictured). (Photo by Morgan Singleton.)

typical to look for a penis and testicles in males or vulva in females. For nonmammalian species, generally there are other sexual dimorphic traits to look for, such as size, anatomic features, or color patterns (Figure 26.6).

Identification Techniques

While the *Guide* and the AWR both have the requirement for animals being identified, for most species the standards are not proscriptive and allow for flexibility to apply a method that works within an individual facility. Most institutions employ an identification card system that is either hand generated or generated via a computerized census system using a bar code or radio frequency identification device (RFID) scanner (Scher 2008; Gentile 2011). According to the *Guide*, these cards "should include the source of the animal, the strain or stock, names and contact information for the responsible investigator(s), pertinent dates, and protocol number when applicable." If the animal is genetically modified, appropriate nomenclature for the genotype should also be included.

As science becomes more precise, the need to individually identify animals becomes greater. There are many acceptable methods for identification of individual animals. For mammals, these include photos to identify unique color patterns, collars, ear tags (Randle et al. 2011), ear notching and punching, tattoos, RFID microchips implanted under the skin (Troyk 1999; Catarinucci et al. 2014), light-activated microtransponders (Gruda et al. 2010), and temporary or permanent hair dye (Figure 26.7). Neonatal animals, especially rodents, can be difficult to identify. Footpad tattoos can be a noninvasive method of identification for rodents at a young age, where 5–10 μL of tattoo ink is injected into the footpads of neonates less than 5 days old (Figure 26.8). Toe clipping for rodents (Paluch et al. 2014) is another option, but it is discouraged at most institutions due to the invasive nature and resultant potential for pain and distress to the neonate. However, it is sometimes scientifically necessary, especially if it is combined with genotyping. Some studies suggest that the adverse effects are minimized if it is performed at less than 7 days of age (Castelhano-Carlos et al. 2010) and appropriate anesthesia or analgesia is used (Hankenson et al. 2008). Regardless, it must be approved by the IACUC.

Reptiles can be identified using shell notching, tail notching, scale clipping, toe clipping, photos of skin color or patterns, RFID microchip transponders, and/or nontoxic dyes (O'Rourke and Schumacher 2002). Amphibians can be identified by their color pattern, toe clipping, tattooing, freeze branding, glass or plastic beads sewn into the muscle, and RFID microchip transponders. Because amphibians shed their skin rapidly or regenerate limbs, it is important to consider which method will be used, as many of these methods may be temporary (O'Rourke and Schultz 2002).

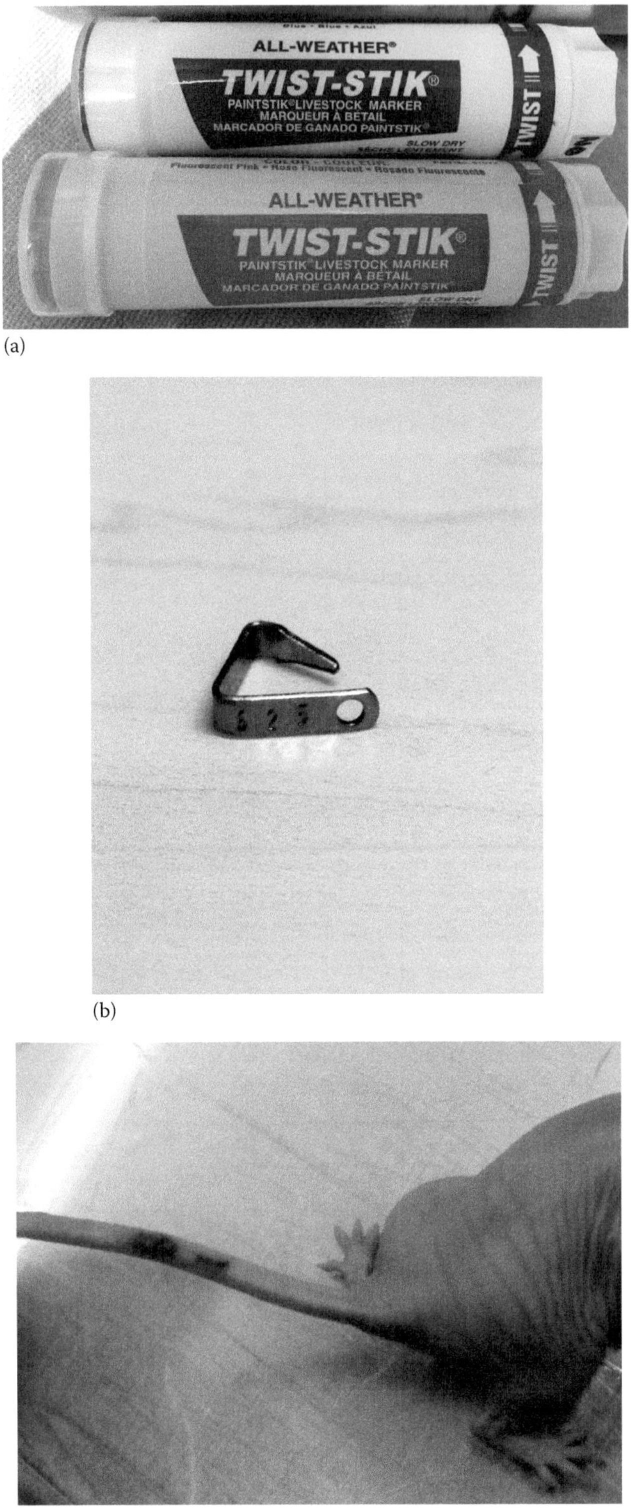

(c)

FIGURE 26.7 There are many identification techniques in animals, including the use of (a) paint sticks, (b) ear tags, and (c) tattoos. (Photo by Michelle Wallace-Fields.)

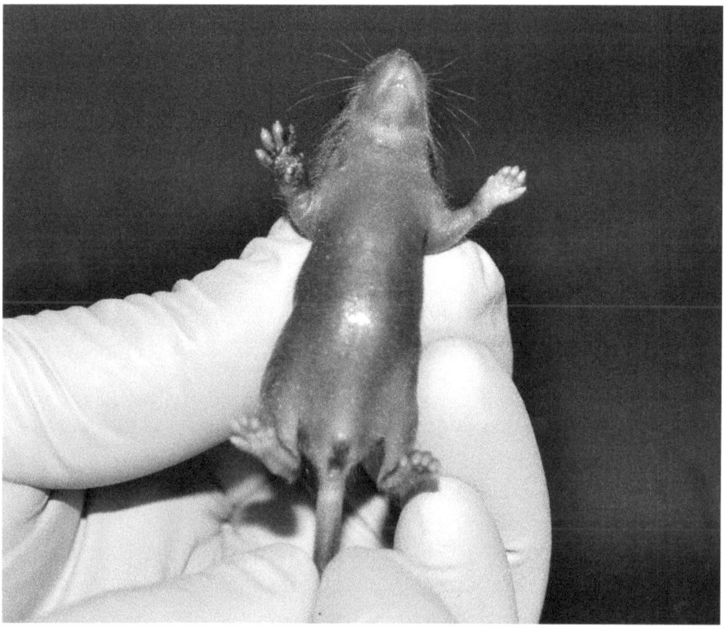

(a)

ID #	Footpad tattoo
0	○○ ○○
1	●○ ○○
2	○● ○○
3	●● ○○
4	○○ ●○
5	●○ ●○
6	○● ●○
7	●● ●○
8	○○ ○●
9	●○ ○●
10	○● ○●
11	●● ○●
12	○○ ●●
13	●○ ●●
14	○● ●●
15	●● ●●

(b)

(c)

FIGURE 26.8 (See color insert.) (a) Neonatal animals can be identified at less than 5 days old by using footpad tattoos. (b) Up to 16 animals in a litter can be identified using this method, where (c) the right front pad is identified as 1, etc. (Photo by Michelle Wallace-Fields.)

Supply Management and Rotation Policies

A challenge in all laboratory animal facilities is inventory management and establishing rotation policies and audit practices for all supplies located in the facility, not just those that are managed by the centralized animal program. For example, if research personnel are allowed to store items and/or equipment in the animal facility, methods to audit those items should also be devised to ensure that they remain within the established expiration date or are cleaned, sanitized, and/or calibrated on an appropriate schedule. The semiannual inspections of the animal facilities by the IACUC can also be used to help ensure that items are appropriately stored.

When establishing the audit system to ensure that appropriate rotation and supply management is being conducted, facilities should utilize local, state, and federal guidelines, as well as institutional standards, to ensure consistency. Audit forms can be helpful in standardizing the process, as well as ensuring the frequency, quality, and thoroughness of tasks when they are completed. Items that can be included on audit forms are checking for expiration dates on feed and medications, and ensuring that waste containers are replaced or sanitized on a standard time frame, unsecured or uncapped sharps are managed, proper labeling is used, hazard storage containment is appropriate, proper signage on caging and in rooms is used, research-specific items are sanitized and within date, and sanitation schedules within the facility are being complied with.

It is important for your institution to develop a process to verify the sanitation effectiveness and frequency of stationary equipment or equipment that is sensitive to wet sanitation applications (i.e., electronics and hand-washed equipment). This can be performed by using Replicate Organism Detection and Counting (RODAC) plates or bioluminescence technology; see the section on room maintenance for additional information on these technologies.

Rotation of supplies with a finite shelf life should ensure that the oldest products are used first to prevent waste. It is important to set up institutional SOPs or policies to ensure that expiration dates for items provide sufficient time to use the supplies before they expire, such as checking mill dates on feed bags on arrival. These same policies can be used to establish criteria to determine the shelf life of disposable goods that do not have mill or expiration dates. Some facilities incorporate Lean practices for the purchasing of supplies on an as needed basis rather than storing supplies for long periods of time prior to use. This method of supply use should be balanced with the need to ensure that sufficient supplies are available according to your institution's disaster plan.

Pest Control

An adequate pest control program is essential for managing all laboratory animal facilities. Pests can harbor disease, cause physical damage to facilities, or present as fomites that carry contamination from outside of the facility in or from one area within the facility to the another. Therefore, an integrated pest management (IPM) program should be developed (CDC-NIH 2009). The purpose of an IPM program is to develop a comprehensive program that integrates multiple factors, including facility design, monitoring, sanitation, facility maintenance, communication, and record keeping, along with pest control, in order to prevent, control, or eliminate pests from entering or infesting a facility. The goal of an IPM program is to minimize the use of pesticides that have the potential to impact research animals and therefore research results (Gunasekara et al. 2008). If pesticides are used, the decision should include the pest control specialist, facility management, veterinary staff, and research staff, as well as be in compliance with local, state, and federal regulations. Whenever possible, nontoxic means of pest control should be used, such as live traps for rodents, insect growth regulators (Donahue et al. 1989), amorphous silica gel, or sticky traps specifically designed for insects. It is important when using sticky traps for insects that only those designed to not capture rodents are used, as improper sticky traps that can capture rodents are considered inhumane. The use of nonchemical means is especially important when managing aquatic species due to the potential for water contamination (Rodney et al. 2013). If live traps are used (Figure 26.9), a system should be developed for monitoring them on a frequent basis, including a plan for humane euthanasia that is also safe for personnel, should an animal be found (Mason et al. 2003). If there are animals housed in outdoor facilities, the same principles should apply, and should focus on eliminating

FIGURE 26.9 Live rodent traps can be used in a facility to humanely catch wild and escaped research rodents. It is important to set up a system to monitor traps frequently and have a plan for disposition of the animals if one is caught by this method. (Photo by Michelle Wallace-Fields.)

or minimizing the potential risk associated with pests and predators, such as ensuring the proper storage of food (NRC 2011).

Room Maintenance

Room maintenance and sanitation practices are used to ensure that the environment is clear of debris and clutter, odors are managed, and potential contaminants that could spread disease, affect research, or harm research personnel, such as allergens, are minimized. Ensuring that equipment is in working order and maintained as part of these practices is important. For example, with proper airflow within the facility and the advent of filter paper tops, ventilated cages, and filtered workstations in rodent rooms, allergen concentrations in rooms can be decreased by more than 75% (Reeb-Whitaker et al. 1999; Gordon and Preece 2003).

Institutions should evaluate sanitation schedules that best fit their programmatic needs. Sanitation schedules can consist of daily, weekly, monthly, quarterly, semiannual, and annual tasks. Items that need to be addressed include walls, floors, drains, ceilings, doors, light fixtures, and so forth. Record keeping of the tasks performed is needed to document that these items are being completed. In order to set the frequency of tasks, it is important to evaluate performance standards. Methods that can verify whether items have been sanitized include the use of bioluminescence meters (Figure 26.10) and RODAC plates. RODAC plates are used to detect and quantify the microorganisms that are present on surfaces that you have sanitized to determine the effectiveness of your cleaning technique and schedule. RODAC plates require incubation of 3–5 days to achieve results. Bioluminescence works by detecting the level of organic matter that is present on a surface and can be used both before and after cleaning. It uses a natural enzyme that is found in fireflies that reacts with adenosine triphosphate (ATP), which is a component of organic matter, to produce a light reaction that can be measured. With bioluminescence, the more organic matter that is present, the higher the ATP reading, and therefore the higher the level of light detected. Because the reaction can be measured in a few minutes, the entire process is generally more efficient, faster, and more cost-effective than other methods, such as RODAC plates, which has contributed to its increase in popularity for determining sanitation effectiveness in animal facilities (Ednie et al. 1998). One thing to be aware of is that the technology can be limited, as it is not able to detect the difference between live and dead bacteria, as in both cases the cell wall that creates the ATP reaction is still present (Turner et al. 2010). Finally, due to the fact that every facility operates uniquely and each manufacturer's equipment is slightly different, it is important to validate the use of bioluminescence in your facility using other methods, such as RODAC plates, to ensure that proper parameters are established.

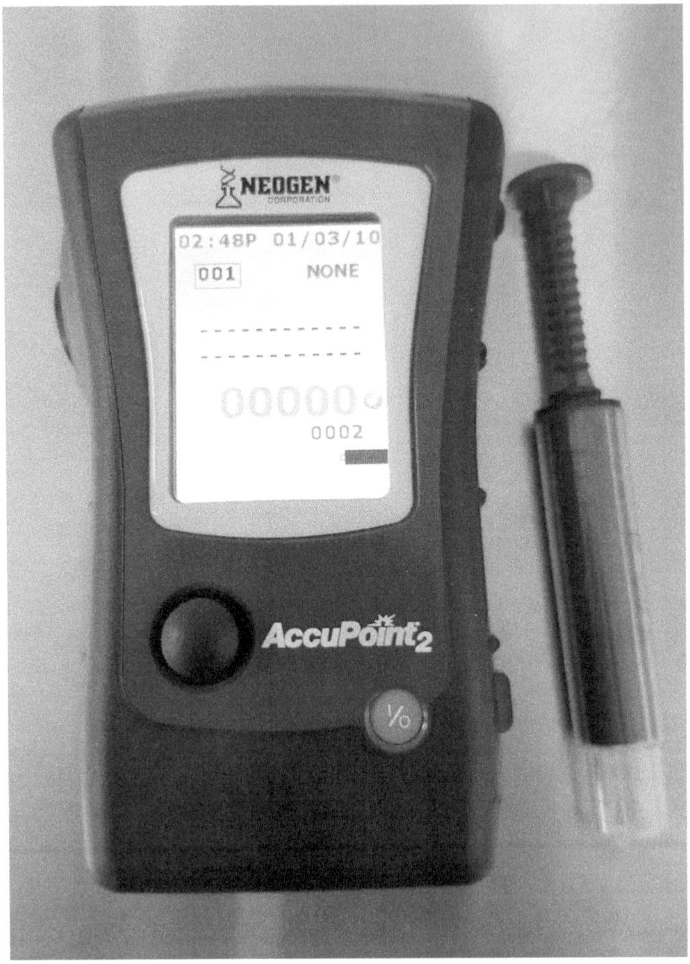

(a)

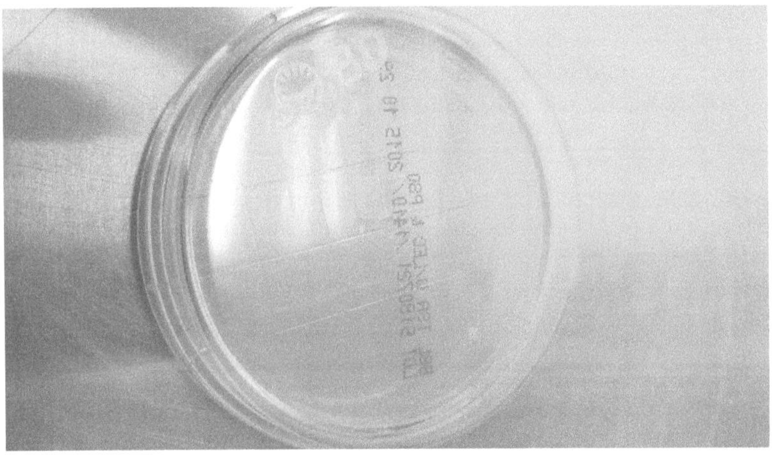

(b)

FIGURE 26.10 (a) Bioluminescence meters can be used to determine sanitation effectiveness in your program by reading the amount of ATP involved in a reaction, which indicates a level of organic matter on a surface. (b) RODAC plates can be used to test for microbial contamination and hygenic status of surfaces. (Photo by Michelle Wallace-Fields.)

Sanitation Principles

Principles

Following appropriate principles of sanitation for an animal facility is not only good management, but also required by the *Guide* and AWR. It is important to understand the differences between cleaning, disinfection, and sterilization (Table 26.2), as well as the type of facility and species that will be managed before determining which method of sanitation will be required. For example, a conventional facility housing rodents in static cages will likely follow different sanitation practices than a barrier facility housing animals in germ-free isolators, and a facility housing large animals in fixed cages will require different practices than an aquatic facility.

Facilities should work to ensure that appropriate sanitation is occurring on a regular basis by ensuring appropriate cleanliness of the facility via routine cleaning and removal of waste. Most surfaces, such as countertops, walls, floors, door handles, caging, and utensils, should undergo routine disinfection. The frequency of disinfection will depend on regulatory guidance (i.e., caging and cage furniture) or performance standards set by the facility. Disinfection eliminates nearly all recognized pathogenic microorganisms on inanimate objects, and the level of disinfection will depend on the established standard for the item and the facility. It is important that organic material is removed in order to ensure that appropriate disinfection can occur, as many disinfectants are neutralized when this type of material is present, as well as proper contact times and temperatures are used to ensure a high success rate. Large-scale decontamination of rooms may be conducted with formaldehyde vapor, hydrogen peroxide vapor (Krause et al. 2001), or chlorine dioxide gas (Lowe et al. 2013).

Depending on the requirement for the facility, certain equipment may be required to undergo a higher level of sanitation and be sterilized, in order to definitively destroy all living microorganisms and viruses, including spores. Sterilization methods include heat, ethylene oxide gas, hydrogen peroxide, plasma, ozone, and radiation. Beyond sterile surgical equipment, animal programs may choose to sterilize caging, cage furniture, food, bedding, and water, in order to enhance biosecurity for their animals. This is most commonly performed with barrier rodent colonies (CDC-NIH 2009).

There are several classes of sterilants and disinfectants (Table 26.3). The chosen method will depend on the item and desired level of sanitation that is required. Consideration of the effect on the animals that are housed should also be considered. For example, there have been reports that quaternary ammonium compounds can impair reproduction in mice when used to sanitize rodent cages (Harrington 2014).

Validation and Quality Assurance Programs

Regardless of the methods used for sanitation, an appropriate quality assurance program that has been validated should be established for an animal care program. Mechanical washers should be monitored daily to ensure that appropriate temperatures are being reached to achieve required sanitation standards. This can be performed via mechanical means or through the use of temperature indicators (Figure 26.11). Autoclaves should also be monitored daily to ensure that they are properly functioning. This can be achieved through steam indicator strips, autoclave tape, and/or biological indicators (Figure 26.11). Beyond the use of these items, all surfaces that have been sanitized, whether by mechanical means or

TABLE 26.2

Principles of Sanitation

Term	Definition
Cleaning	Mechanical removal of visible surface contaminants, such as dirt, soil, and marks, usually by washing, wiping, or brushing with soap and water.
Disinfection	Involves using chemicals to kill most microorganisms, except spores. Cleaning is generally required prior to disinfectant use, as dirt tends to deactivate the disinfectants.
Sterilization	Involves killing all forms of microbial life, including fungi, spores, bacteria, and viruses. Methods include autoclaving, hydrogen peroxide, and some chemical sterilants.

TABLE 26.3

Cleaning and Disinfection Compounds

Classes of Disinfectants	Application and Safety Profile	Examples
Alcohols	• Active against vegetative bacteria, fungi, and lipid-containing viruses	• 70% ethyl alcohol
	• Does not kill spores; variable against non-lipid-containing viruses	• Isopropyl
	• Best if used at 70% concentration in water; does not leave residue	
	• Volatile and flammable	
	• Use on skin, work surfaces, and benchtops	
Aldehydes	• When used as a gas, can kill all microorganisms and spores, but not prions	• Formaldehyde
	• Slow acting and requires high humidity	• Araformaldehyde
	• Use to decontaminate enclosed spaces (i.e., biosafety cabinets and rooms)	• Glutaraldehyde
	• Suspected carcinogen; mucous membrane irritant; use proper venting	• Cidex
Chlorine compounds	• Fast-acting oxidant; broad-spectrum germicide for general disinfection	• Bleach
	• Highly alkaline and corrosive to metal; use as gas or liquid	• Chlorine dioxide
	• Activity reduced by organic matter; activity decreases once mixed	• Clidox-S®
	• Concentration for use depends on level of disinfection required	• Vimoba®
	• Gas can be used for disinfection of enclosed spaces	
	• Chlorine gas is highly toxic, so caution should be used	
Iodophor	• Can stain surfaces and fabrics	• Betadyne®
	• Good antiseptics, especially for skin; use as a presurgical scrub	• Povidone®
	• Should not be used on aluminum or copper	
Oxidizing agents	• Broad-spectrum germicide; use at 3%–6% concentration for decontamination of work surfaces	• Hydrogen peroxide
	• At higher concentrations or vaporized hydrogen peroxide or peracetic acid can sterilize heat-sensitive medical equipment	• Peroxyacetic acid®
	• Used alone, it is slow acting, but combine with stabilizers to make less corrosive and work faster	• Virkon S®
	• Can be corrosive to metals; store away from heat and light	• Acidulate®
Phenolic compounds	• Active against vegetative bacteria and lipid-containing viruses	• Pine-Sol®
	• Does not kill spores; variable against nonlipid viruses/mycobacterium	• Lysol®
	• Decontamination of environmental surfaces	
	• Some formulations used as skin and mucous membrane sanitizers	
	• Usually must be diluted with deionized water	
Quaternary ammonium compounds (Quats)	• Active against some vegetative bacteria and lipid-containing viruses	• Roccal® • Quatricide®
	• Activity reduced by organic matter, water hardness, and anionic detergents	
	• Some are used as antiseptics	• Sani-Plex®
	• Good surface decontaminants, such as walls, floors, and counters	

Source: WHO (World Health Organization), *Laboratory Biosafety Manual*, 3rd ed., WHO, Geneva, 2004, http://www.who.int/csr/resources/publications/biosafety/WHO_CDS_CSR_LYO_2004_11/en/.

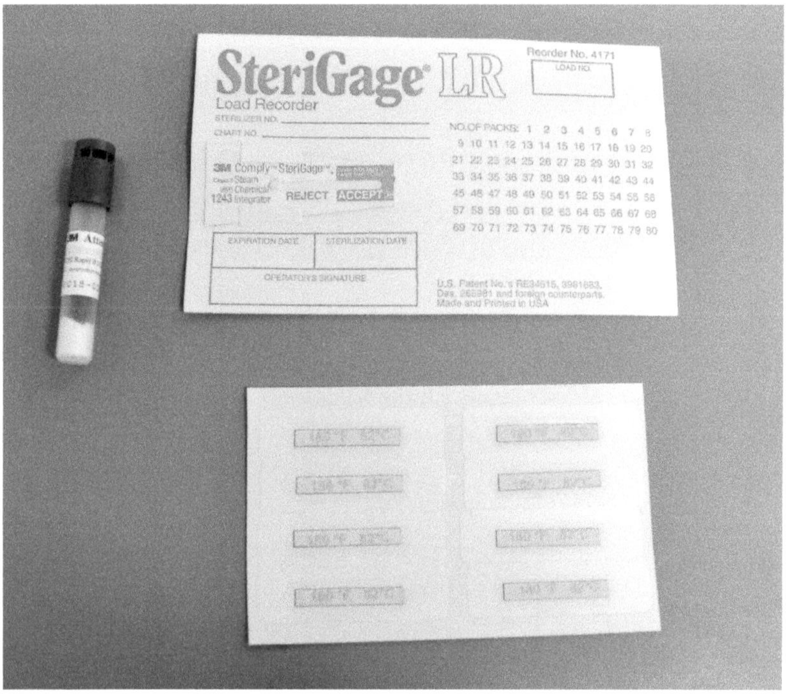

FIGURE 26.11 Quality assurance methods for the cage wash include the use of biological indicators (top left) and steam indicator strips (top right) for autoclaves and temperature indicator tapes for washers (bottom). (Photo by Michelle Wallace-Fields.)

by hand, should have the method validated for effectiveness, including investigator equipment, such as behavioral devices (Guide 2011). Methods that can be used are RODAC plates or ATP monitoring through bioluminescence.

Hazard Identification

Communication between the animal husbandry staff and the research staff is key when working with hazards in the animal facility. Institutions have a legal requirement to inform anyone that comes into contact with a container with a hazard or an animal that has been exposed to a hazard, about that hazard. Systems should be developed to ensure the appropriate communication of the hazard name, dose, date of exposure, method of disposal, personal precautions, and spill procedures and who to contact if a spill or an exposure occurs. At a minimum, basic information should be provided at the container or cage level (Figure 26.12) to direct husbandry and research staff to basic procedures and where to find additional information for more in-depth questions and safety procedures. The goal of good signage is to decrease exposure of personnel to hazardous agents, and to communicate what to do in an emergency. This communication should contain universal symbols for biohazardous, chemical, and radiologic waste, as well as appropriately direct the proper PPE and disposal procedures for waste.

Waste Management

Laboratory animal research programs generate a large amount of waste, and the regulations that govern waste disposal can sometimes be difficult to navigate. A good relationship and open lines of communication with your institution's EHS team will help you to determine appropriate waste streams for all waste generated and ensure the compliance of your facility and institution, as well as the safety of personnel (Hill 1999).

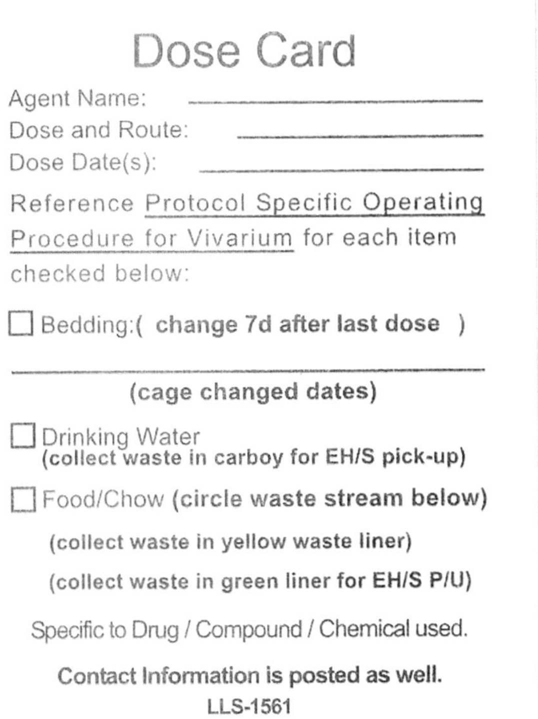
Dose Card

Agent Name: ____________

Dose and Route: ____________

Dose Date(s): ____________

Reference Protocol Specific Operating Procedure for Vivarium for each item checked below:

☐ Bedding:(change 7d after last dose)

____

(cage changed dates)

☐ Drinking Water
(collect waste in carboy for EH/S pick-up)

☐ Food/Chow (circle waste stream below)

(collect waste in yellow waste liner)

(collect waste in green liner for EH/S P/U)

Specific to Drug / Compound / Chemical used.

Contact Information is posted as well.

LLS-1561

FIGURE 26.12 Example of hazard signage that can be placed at the cage level to provide information as to the agent, dose, route, and date of administration. Additional information, such as how to handle bedding, food, and water, can be included, or staff can be directed to see additional information. (Photo by Michelle Wallace-Fields.)

The EHS personnel at your institution will have firsthand knowledge of local, state, and federal regulations that must be followed for your specific location. It is "best practice" to always follow the most stringent of the applicable regulations and policies that the facility falls under to ensure compliance. This starts with developing a team approach with the EHS personnel to ensure that any changes to the regulations are followed and that they are aware of any new or changing research projects that could impact how waste is handled at the institution. Strategies such as one-on-one meetings with research groups known to use high-risk reagents, signage related to proper disposal at key points in the facility, color-coding disposal containers (trash cans, sharps containers, etc.) for personnel to easily identify correct waste streams, and regular updates through electronic mechanisms and town hall meetings are some of the methods that can be considered to ensure compliance. The following categories of waste are addressed in this section: nonhazardous solid waste, carcasses, pharmaceutical waste, and infectious and biological wastes.

Nonhazardous Solid Waste

Nonhazardous solid waste can be defined as anything that can be placed into a sanitary landfill. In most cases, soiled bedding collected from healthy animals and most of the general trash that is generated in the facility can be classified under this category. Due to the volume of soiled bedding generated, some institutions have recently begun working with local composting companies to dispose of their waste. When considering this option, it is important to ensure practices within the facility that guarantee that only bedding from healthy animals is diverted to this disposal method. In addition, facilities that use disposable cages have successfully incorporated recycling into their programs. When handling soiled materials, containment precautions should be used to control allergenic exposures, such as dump stations or appropriate respirators. Users must be properly trained in the use of safety equipment and in the importance of wearing proper PPE. A waste pickup schedule should be created that can handle the volume of the buildup of materials being stored and keeps it to a minimum. Storage areas should be located outside of animal holding and clean equipment areas to decrease contamination, odor, vermin, and disease transmission risks (CFR 2013).

Waste containers must be available and located in the proximity of the work being performed. Waste containers must conform to the needs of the work based on size and ease of use. Containers should be leak-proof, have a lid (unless they are emptied daily), and be sanitized or replaced on a set schedule.

Carcasses

A method for carcass disposal should be developed that is appropriate for the facility and meets local, state, and federal requirements (Cornell Waste Management Institution 2015). Methods of carcass disposal include incineration, contracted carcass disposal companies, and chemical digestion through alkaline hydrolysis (Murphy et al. 2009). Institutions may also want to consider donating carcasses of healthy, nongenetically modified animals that have been euthanized via nonchemical means to zoos, raptor societies, and so forth. While there are some in existence, most facilities do not have the ability to incinerate carcasses directly on site and must contract with a licensed crematorium or biohazardous waste service. Some institutions have moved to chemical digestion of carcasses on site because of this. When considering this option, check with local wastewater officials to determine how the effluent from the digester will be handled. Some local wastewater regulations will not allow the effluent to be discharged into the local waste treatment system, and therefore require a separate system to capture and dispose of the effluent through the solid waste stream. In all cases, sufficient cold storage and containment procedures should be available to handle the volume of carcasses generated by the institution prior to scheduled disposal.

Pharmaceutical and Chemical Waste

Pharmaceutical and chemical waste, if improperly handled, can end up in the water supply and harm animals and plants that live in the water, as well as humans. As a result, it is important to develop procedures that ensure that these agents are appropriately collected and disposed of. Educating the research and animal facility personnel regarding the different classifications of waste (i.e., pharmaceutical, chemical, and cytotoxic) and collection methods for each type is essential to ensure that these agents are not disposed of through the sanitary sewer system or in the nonhazardous solid waste stream. This can include ensuring that chemicals used in the cage wash area are appropriately collected and/or neutralized prior to discharge into the sanitary sewer system. Methods to accomplish this include setting up collection points near where hazards are used, signage, and training courses developed in conjunction with your institution's EHS personnel, as some hazards require more stringent collection, disposal, and documentation than others. For staff safety, all chemical hazards used in the facility should be identified in the location of use during the time period that the hazard has the potential to be present. This includes notations at the cage level for hazards given to animals. Appropriate signage and procedures to follow in the event of an exposure should be easily located in the area that the hazard is being used.

Biohazardous Waste

When working with biohazardous waste, it is important to work with your institution's biosafety officer and EHS personnel to ensure that a risk analysis is conducted for each type of agent that will be used. The risk analysis should develop methods to ensure that the appropriate containment and waste handling methods for each type of hazard are present in the facility.

There are varying methods by which biohazardous waste may be handled, and they will depend on the type of biohazards, as well as the available resources at the institution. The best method of destroying waste is through incineration, as much of the bedding and supplies are combustible high-density waste. By incinerating the waste, it prevents the need for reuse of potentially contaminated supplies, as well as avoids the potential for the formation of infectious aerosols during handling and disposal of bedding and other wastes. In addition, large volumes of infectious waste, including animal carcasses, are able to be processed in a relatively short period of time. However, incinerators on site at an institution, especially those that are certified to handle hazardous waste, are becoming less common. Institutions that do not have this resource available must work with waste removal companies to move hazards off site, which will involve local, state, and federal approvals and permits, as well as special packaging, labeling, and handling before physically removing biohazardous materials from the institution. It is important when using this method that prior to moving the materials, all infectious materials have been decontaminated or are packaged in an approved manner for transport. It is important that all documentation is generated and maintained related to shipping of infectious or biological waste to meet all applicable regulations for your area. Consult with EHS personnel and/or the biosafety officer to ensure that correct methods are used.

Depending on local resources and the types of agents that are being worked with, incineration may not be practical or necessary, so other methods of decontamination may be considered, including chemical methods and sterilization through autoclaving. Many institutions use autoclaving to decontaminate bedding materials, excreta within cage units, and other forms of solid waste. Decontaminated solid waste can then, in most cases, be managed as general waste. It is important to work with the institution's biosafety officer and EHS personnel to ensure that waste handling is appropriate before ultimate disposal. In addition, appropriate signage and procedures to follow in the event of an exposure while handling biohazardous waste should be easily located in the areas where hazard work is occurring.

Preventive Maintenance and Equipment Repair Programs

Animal facilities use many pieces of equipment that require routine preventive maintenance. Animal programs should evaluate the available resources within their institution to determine what aspects of the program can be handled internally, such as routine cleaning and basic maintenance. For many programs, outside service groups are required to maintain equipment. Service contracts should be considered to ensure that required maintenance is performed on a predetermined schedule, especially on large capital pieces of equipment, such as biological safety cabinets (BSCs), cage and rack washers, and autoclaves, as well as other minor pieces of equipment, such as surgical monitoring equipment, imaging equipment, and watering valves. Service contracts also can help with budgeting, in that general maintenance costs can be generally known in advance, and can help to decrease the frequency of unanticipated downtime for equipment. The frequency of maintenance will depend on the service being conducted, the age of the equipment, the frequency of use of the equipment, and the manufacturer's guidance. The terms of the service contract will also depend on the location of the service provider in relation to the facility and how many pieces of equipment need to be maintained. It is important to include emergency service clauses in the contracts for heavily used equipment to ensure that essential equipment will be serviced in a timely manner.

Areas of special consideration for appropriate preventive maintenance are BSCs, animal change stations, and blowers for ventilated cage systems. All these types of units contain high-efficiency particulate air (HEPA) filters and are designed for varied protection against biological hazards, depending on the type of device that is being used. The HEPA filters can become loaded so that airflow cannot be maintained and thus need to be replaced. The function of these systems should be verified by an accredited technologist at installation and then recertified at least annually (NSF and ANSI 2004). Fume hoods provide needed protection from chemical hazards and should also be recertified annually to ensure appropriate negative airflow.

Maintaining Safety Equipment

The *Guide* requires that all safety equipment should be properly maintained and periodically validated to ensure that it is properly installed and operates correctly. Facilities should develop a defined schedule of maintenance for all safety equipment that is in compliance with specific guidelines and regulations. For all safety equipment, pathways to these emergency services should be checked on a weekly basis for obstructions and cleared if they are found present. All persons that may need to use a safety device should be trained in the proper operation of the said device. All devices should be checked at least annually to ensure that they meet the applicable standards. In the next sections are common types of safety equipment that may be found within an animal facility.

Portable Fire Extinguishers

Portable fire extinguishers are governed by the National Fire Protection Association (NFPA 2013) and OSHA standards (OSHA 29 CFR 1910.157). Extinguishers should be maintained and charged by certified personnel and be labeled indicating when they were certified. All extinguishers should be inspected when placed into service and then inspected at least monthly via visual or electronic means to ensure that the extinguisher is located in the correct location, is not obstructed, is full, and is in good condition. Annually, all extinguishers should be manually checked to ensure that the units do not have obvious physical damage, do have all seals, and the attachments are appropriate. Depending on the type of extinguishers in your facility, additional testing may need to be performed in 3- to 6-year intervals. It is best to check with a certified individual to ensure that fire extinguishers are appropriately maintained.

Eyewash Stations and Safety Showers

Eyewash stations (Figure 26.13) and safety showers should be available in all areas of the facility, especially areas where chemicals are used, including cage wash. Eyewash stations can range from as simple as commercially available sterile eyewash bottles, to those that are gravity fed, and ultimately to those that are plumbed with tempered water. Please note that sterile eyewash bottles generally do not provide enough water to meet the required flushing time of approximately 15 minutes, and so alternative methods, such as a plumbed station, should be available for continued flushing after using this method for the initial flush procedure. It is also important to ensure that bottles are kept within their expiration date at all times. Regardless of the methods used in the facility, a weekly inspection of devices and applicable expiration dates is recommended by the American National Standards Institute (ANSI). For piped-in systems, such as eyewash stations, drench hoses, and safety showers, ANSI/ISEA Z358.1–2014 includes specifics related to flow of water, temperature of water, location of units, and

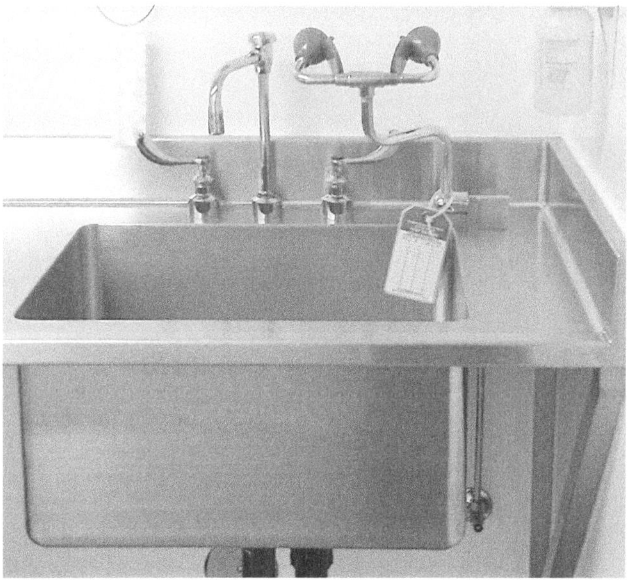

FIGURE 26.13 Eyewash station with card for documenting when station was tested. (Photo by Michelle Wallace-Fields.)

other important design considerations to ensure proper function. In addition, it is recommended that all plumbed systems are activated weekly to ensure that they are not obstructed, operate properly (ANSI and ISEA 2014), and have proper water temperature, as well as to prevent stagnation of water and an accumulation of particulates.

First Aid Kits

First aid kits should be available for minor injuries within the facility and should be checked on a routine schedule, generally monthly, for expired items and to ensure appropriate levels of supplies are present. At a minimum, basic bandage materials to handle cuts and abrasions should be available. More extensive first aid kits will depend on the hazards present in the facility and should involve the institution's occupational health provider's input. Staff should be appropriately trained to report all injuries and exposures that occur within the facility, as well as to complete appropriate documentation in accordance with their institution's procedures for reporting work-related injuries.

Bite and Scratch Kits

All animal facilities should have a designated process for dealing with bites and scratches. The species housed in the facility and agents being used in research will impact the materials that are gathered and maintained for a bite and scratch kit, as well as the reporting requirements and actions to be taken when an exposure occurs. When working with species with zoonotic potential, such as macaques that may be positive for herpes B, it is essential that the protocol to follow when a bite or scratch occurs is developed in collaboration with the occupational health provider for the program. This will involve ensuring that employees are well trained to report the type of exposure so that appropriate medical attention can be provided as soon as possible.

Sharps Containers

Sharps containers should be placed near the vicinity of the work being performed. Sharps produce a physical hazard to personnel, and so containers should be prominently labeled, be leak and puncture resistant, and be disposed of in a manner consistent with your institution's waste management plan. Appropriate training and signage should be employed instructing personnel to either use safety sharps or to not recap needles. For noncontaminated glass, sturdy corrugated cardboard boxes may be used.

Compressed Gas Alarms

Compressed gas alarm regulations require that access to panels must remain clear and void of obstruction (NFPA 99–1999 edition, Section 4-3.1.2.3(d)) and the placement of panels is to be in the corridor or on the floor it services. This allows for the gases to be turned off in the event of an emergency from a local source. Chapter 9 of the NFPA 99–2005 edition states that "periodic testing procedure for nonflammable medical gas and vacuum and related alarm systems shall be implemented and records of the tests for maintenance are kept until the next test is performed."

Compressed Gas Cylinders

Gas cylinders should be secured at all times to prevent tipping and transported on carts with the caps on (Figure 26.14). If bulk storage of tanks is required, an appropriate location that meets the applicable local code should be designed. This could be inside or outside of the facility, but should allow for the cylinders to be secured away from sources of ignition or heat.

FIGURE 26.14 Example gas tank–securing device. This type of cart allows the tank to be easily moved as well. (Photo by Michelle Wallace-Fields.)

Facility Administration

Animal Census

Determining how many animals a facility is taking care of, or the animal census of the facility, is an important aspect of the administration of an animal program. Determining animal census numbers can aid in calculating per diem charges, assigning work tasks, projecting the budget, allocating resources, allocating space for projects, and calculating the amount of assigned and unassigned space. It is also an important tool to monitor numbers of animals approved on protocols. In addition, the *Guide* and AWR dictate that IACUCs review and approve the number of animals required and that institutions track the use of animals used in research.

There are three primary methods to collect census information in a laboratory animal facility: bar coding, RFID technology (Figure 26.15), and hand counting of individual animals or cages. Each institution should determine which method works best for its programmatic needs and financial capabilities and at what frequency census should be performed. Of the three methods discussed, RFID is the least labor-intensive but most costly due to proprietary equipment and software. The RFID chip is placed either in the cage card holder or on an individual cage card and broadcasts signals that can be captured by static readers in the ceiling, a mobile reader on a cart, or a handheld reader. Bar coding is midlevel

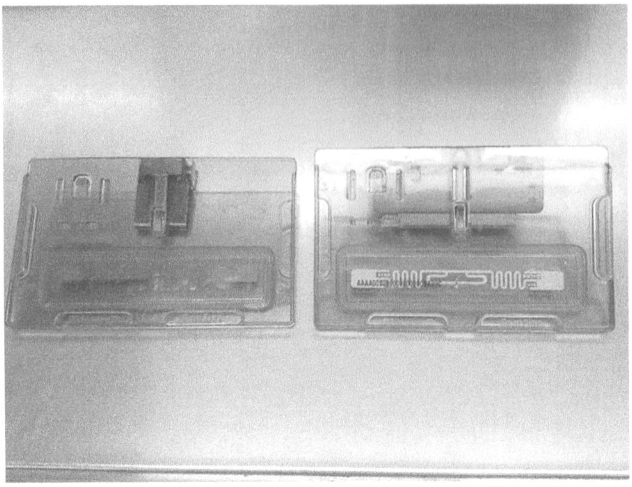

(a)

Cage Card #: 0000110963

PI: Doe, James	Requisition #: 16HSD278
Protocol Number: 123456(10)1D	Protocol Expiration Date: 10/16/2016
Strain: Hsd:ICR(CD-1) (Harlan)	Vendor: Envigo (Formerly Harlan)
Requestor: Scott, Mary	Location: RC2, 0334-RC2
Cage Type: *Mouse Ventilated*	DOB: 12/25/2015
Account: Doe, James - 62027731	
Special Caging:	
Comments:	

Cage Card #: 0000110963

Notes:

(b)

FIGURE 26.15 **(See color insert.)** Example methods to collect census and protocol information include (a) RFID technology, which can be placed into a cage card, and (b) bar-coded cage cards. Both require special scanning devices that can be uploaded into software programs. (Photo by Michelle Wallace-Fields.)

in cost, as well as the amount of labor required, as each bar code must be individually scanned using a handheld reader. Hand counting is the most labor- and time-intensive, but least expensive, due to needing to ensure that each cage is checked for the proper research investigator, protocol number, and in some facilities, billing account number. Both RFID and bar coding are generally a component of a larger electronic inventory management system. Uses of both of these technologies are intended to streamline processes by virtually eliminating data entry errors, collect large amounts of data with little time and effort, and improve the way information is tracked (Lucid and Lepidi 2010).

Rate Setting

Many animal programs are required by their institutions to derive at least a portion of their operating budget by collecting funds for taking care of animals and performing services for research groups. The daily rate that is charged on a per cage or per animal basis is called a per diem rate. To calculate per diem rates, all costs related to the care of an animal in a defined type of situation should be added together, such as all mice housed in static cages and rats housed in ventilated cages. Costs that should be considered for inclusion in the per diem rate include space charges, personnel, utilities, food, bedding, equipment purchase, equipment maintenance, and medical care. Once a total cost is determined, it should be divided by the total number of animals that you expect to care for, and again by the number of days of care anticipated over a period of time, such as a year. There are many ways to perform a cost analysis and rate setting for an animal program. For those institutions that are receiving funds from the National Institutes of Health (NIH), the Office of Management and Budget (OMB) within the NIH issued Circular A-21. The circular is directed at educational institutions that are receiving NIH funding in the form of grants, contracts, and/or other agreements. It refers to the *Cost Analysis and Rate Setting Manual for Animal Research Facilities* (NCRR 2000), which focuses specifically on laboratory animal research facilities and has several examples on how to perform a cost analysis.

Animal Procurement

Obtaining animals for research requires verification that the species, strain, and number of animals to be obtained are approved on an active animal care and use protocol and funds are in place to support their care. Most institutions have a defined animal procurement program and strict animal ordering requirements. It is the responsibility of the institution to ensure that animals are procured lawfully following local, state, national, and international regulations and standards. Vendors should be evaluated by veterinary staff on a routine basis to ensure the appropriate health status of their animals to enter the facility. The facility should also be prepared and capable of housing and managing the care for all species that will be procured in advance of the animals' arrival.

Vendor Evaluation

There are typically three methods for procuring animals for research: (1) in-house breeding, (2) purchasing from vendors or dealers that you have previously approved (routine sources), and (3) importing from other research institutions or vendors you have not previously approved (nonroutine sources). Animals should never be purchased from pet stores or pet distributors due to the unknown genetic makeup, origin of the animal, or health status. Obtaining animals from these sources increases the potential for introducing pathogens to the colony at large, or to personnel that care for them (Roble et al. 2012). Because the greatest risk of pathogen and parasite transmission is from the introduction of new animals into an established colony (Reuter and Dysko 2003), the facility veterinarian should be involved with the development of guidelines and recommendations on the overall biosecurity program for the facility in order to minimize the risk associated with the introduction of new animals into a colony or facility.

When selecting vendors for all species, it is best to determine if the vendors are AAALAC International accredited, as this will indicate that a specific level of animal care has been established. Depending on the country, there may be additional regulatory or accrediting agencies and rules that a vendor would be required to follow. For example, in the United States, vendors that provide animals other than rats, mice, birds, and agricultural animals used for agricultural research are required to be licensed by the USDA as either a Class A or Class B vendor. Most USDA-covered species come from Class A vendors, which indicates that they are licensed breeders that produce "animals that are bred and raised on the premises in a closed or stable colony and those animals acquired for the sole purpose of maintaining or enhancing the breeding colony" and are commonly called purpose bred (CFR 2013). Because these are purpose-bred animals, vendors should be able to provide routine health assessment and documentation regarding important information about the colony. Additional information that may be available and important for assessment of the animal is genetic screening and pathogen status. Some animals are procured via Class B vendors. Class B vendors are licensed by the USDA and indicate that the animals are not raised by the vendor, but instead are purchased and resold (AWR 2013). If a Class B vendor holds the animal for a period of time to ensure health status or treats the animal to ensure a specific health status, those animals are considered "conditioned." Information regarding USDA license and registration holders is available through the USDA. You can verify USDA reports by visiting the USDA web page (https://www.aphis.usda.gov/aphis/home/).

Beginning in September 2015, newly funded NIH grants must procure cats and dogs from Class A dealers or other legal sources, such as privately owned colonies. Investigators with existing grants or funding from nonfederal sources are permitted to use Class B dealers (NIH 2013). Regardless of the source, the institution should check for unique animal IDs upon arrival. For Class B animals, animals should be checked for other indicators that the animal was previously a pet, such as by using a microchip scanner. If there is evidence that the animal may have been a pet, measures to confirm ownership should be taken (NRC 2011).

If utilizing fish, amphibians, or reptiles in research, consult the appropriate government offices where the facility is located. For example, in the United States, the Fish and Wildlife Service should be contacted to help identify the status of threatened or endangered species animal populations that may be required for research. In Europe, the European Union habitats (European Union Habitat Directive 1992) and birds (European Union Bird Directive 2009) directives should be consulted. The Zebrafish International Resource Center (ZIRC) maintains a listing of stockfish that it carries that are bred specifically for research.

Additional verification of a vendor's status should include ensuring that the vendor's program for animal health and biosecurity conforms to the program's needs. Vendors should be able to provide the facility with a list of pathogens that they test for on a routine basis, the method they use (serology, polymerase chain reaction [PCR], or direct examination), and the frequency with which they are tested. It is important to compare this information to the list of pathogens that you exclude from your facility and make a determination of whether animals from this source will be accepted directly into the facilities animal colonies or will require quarantine or rederivation. This decision will be dependent on the species of animal that is being worked with and the resources available at the institution.

Procuring Animals from Nonroutine Sources

Due to the increased popularity of genetically modified animals and increasing cross-institutional collaborations, facilities are routinely asked to procure animals from nonroutine sources, such as universities, private research foundations, biotech companies, or vendors that have not been previously approved by the veterinary staff. Because the pathogen status of nonroutine sources is often unknown, does not have a sufficient history, or is unreliable, appropriate quarantine and biosecurity protocols should be employed when receiving animals from these sources. A veterinarian should oversee the program and perform a risk analysis of each shipment from nonroutine sources. Recent health status reports, generally from the previous year, should be reviewed. In addition, standardized questionnaires are helpful in ensuring that consistent information is captured for each shipment. Animals that do not meet the institution's health requirements should be refused, be rederived, or undergo an enhanced quarantine procedure, such as targeted testing of the animals prior to shipment and/or additional testing and monitoring once received at the institution.

Management of Quarantine

Because most institutions are dealing with limited space, it is not unusual for multiple groups of animals to undergo quarantine at the same time. Careful consideration of housing and handling of animals and equipment should be employed to prevent cross-contamination. If at all possible, animals from different shipments should be isolated from one another either at the room level, in isolation cubicles, or by employing specific traffic patterns in the room and aseptic technique. It is important to move through husbandry procedures by working with the veterinary staff to work from the lowest-risk animals to the highest-risk animals. Strict biosecurity protocols of changing gloves and cleaning work space in between quarantine groups should also be employed. Other considerations should include whether research staff will be allowed in quarantine areas and whether breeding and other research-related procedures should be limited due to space or biosecurity concerns.

It is vital that some level of documentation regarding which cohort the animals arrived with is maintained with the animals at the cage level in order to avoid mistakes. This could be accomplished by identifying individual quarantine groups by utilizing visual cues such as colored cage cards, special stickers, unique images, or numbers.

Management of Aquatic Animal Quarantine

Importing and/or quarantining of aquatic animals should employ precautions similar to those of terrestrial species. Animals should be placed in conditioned tanks that are isolated from the general population, ideally in a separate room in either isolation tanks or on a quarantine rack that receives water from the centralized system but does not send the water to be recirculated. This is especially important if the facility employs the use of a recirculating system to house colony animals, as that could result in contaminating the entire colony through the water system. The veterinarian should determine the length of time that animals should be placed on the rack and the conditions under which they can be placed with the regular population based on the pathogens that are of concern. For example, with zebrafish, it is ideal for adults to remain on the quarantine rack and only fry produced from washed embryos to be introduced onto a centralized life-support system. As with all quarantine systems, since the health status of quarantined animals is unknown, staff should be directed to work with existing colonies prior to working with colonies of unknown health status.

Receiving Animals

When receiving animals, it is important that the facility has procedures in place to maintain the health and pathogen status of incoming animals. This includes using an area that allows for the safe receipt of animals that is environmentally controlled, away from the public, and easily decontaminated. Facilities should ensure that routine source vendors are using vehicles that are decontaminated after each use, and that they are following protocols for the packaging, handling, and transport of the animals that prevent contamination during shipment. Nonroutine source shipments may arrive via transport that is not dedicated to animal shipments, but the same level of packaging, handling, and transport of animals to prevent contamination should be employed. Shipping containers for all species should be built to withstand the rigors of transport (Orcutt et al. 2001), have appropriate openings that support ventilation, and ensure that animals have access to both feed and water sources during transport. For animals that require a restricted pathogen status, the use of filtered coverings on the openings should be used to prevent exposure to pathogens during transport.

Upon receipt, the exterior of shipping containers should be disinfected prior to opening to prevent the spread of pathogens that may have been picked up on the outside of the container during transport (NRC 2006). Disinfectants should be chosen in consultation with the veterinary staff to ensure that they are safe to use around the species being received and effective against the pathogens of concern. Containers should be examined for damage upon arrival, and any breaks in integrity should be documented (Figure 26.16). Animals with specific pathogen-free status or that are immunocompromised may be particularly impacted, and the veterinary staff should be notified to assess whether animals will still be useful to research studies. Once the exterior of the shipping container has been determined to be intact and has been decontaminated, gloves should be changed after opening, but before handling the animals. When working with small animals, a dual system may be employed to decrease the chance of contaminating the animals with pathogens on the outside of the shipping container. In this situation, one person opens the shipping container, and a second, with clean gloves, handles and houses the animals, being careful not to touch the outside of the container (Figure 26.17).

FIGURE 26.16 When receiving animals, containers should be examined for damage. Any breaks should be documented and reported to the veterinary staff to determine the possible impact on the biosecurity of the facility. (Photo by Michelle Wallace-Fields.)

FIGURE 26.17 Dual system for housing rodents. One person handles the outside of the container only and opens the lid, while the second person only touches the inside contents to house the animals. (Photo by Michelle Wallace-Fields.)

As part of the housing procedure, it is important to document the number, sex, and other characteristics of the animals received. In addition, it is important to note if any animals appear sick or are found dead and immediately contact the veterinary staff to provide care or determine the cause of death. Oftentimes, vendors will replace sick or dead animals but require notification within a specific number of days after receiving the delivery in order to process the request for replacement animals. In addition, depending on the species involved, the circumstances, where the animals came from, and where the facility is located, death of animals during transport may require a report to a regulatory oversight body by either the facility, the vendor, or the party that transported the animal (NRC 2006). Once animals are received, they should, according to institutional policy, be quarantined or released into the general housing and allowed an appropriate acclimation period to allow homeostasis to develop (Obernier and Baldwin 2006; Fernstrom et al. 2008).

Transport

Interinstitutional Transport

At times, facilities need to send animals to other research institutions. The transportation of research animals may be covered by several different regulations, policies, or international treaties, depending on what species is being transported and to where (NRC 2006; IATA Live Animal Regulations). As regulations vary from area to area and can change without notice, it is best to check the regulations that pertain to the sending and receiving facilities to ensure compliance. Commercial shipping companies that specialize in the transport of research animals may be able to assist by providing information on what approvals and paperwork will be required in advance of shipping.

The *Guide for the Care and Use of Laboratory Animals*, Eighth Edition recommends that institutions consider the following prior to the shipment of research animals:

- Animal safety and well-being
- Appropriate level of biosecurity
- Minimization of zoonosis
- Microenvironment needs are met (feed, water, and temperature)

- Outside factors like noise
- Fluctuations in temperature
- Direct exposure to sunlight is minimized or managed
- Animals are housed for transport in the proper size carrier or container
- Animals are not overcrowded and have proper space for species-specific behaviors
- No physical, mental, or psychological trauma ensues

When shipping, transporting, and/or receiving animals, ensure that all applicable regulations and guidelines are followed and that all regulations pertaining to record retention are followed. In most cases, documentation should be retained for at least 3 years, and multiple copies may be helpful in the event that one agency removes critical paperwork.

Intrainstitutional Transport

If the animal program allows the movement of animals from one physical location to another within the same institution, such as between facilities or to laboratories, SOPs and institutional policies should be developed in conjunction with the facility management, veterinary staff, and IACUC. The reasons for ensuring that the movement of animals occurs appropriately include that animals of different health statuses are not intermingled, the sentinel health history of a room accurately reflects the animals housed, and the movement of potentially infected cages can be tracked in the event of an outbreak, as well as ensuring animal well-being and personnel and general public safety. Consideration should be given to the appropriate length of time the animals can be in transit; what container is best suited for transit to protect both the animal and the public, from a biosecurity and a visual standpoint; and what routes should be used. For small animals, because caging can be a significant investment, as well as a biosecurity hazard when returned to the main facility, many institutions provide disposable containers for transport out of the facility to laboratory spaces (Figure 26.18). When deciding on alternative containers for small animals, it is important to consider the following: the species being transported, the number of animals, the distance or length of time of the transport, the housing in the laboratory posttransport, the ability of the material to absorb urine, access to food and water, and the ability to shield the animals from public view. Some institutions use a combination of disposable transport containers for short-term experiments and disposable cages (Figure 26.19) for longer housing in the laboratory if the transport containers are not able to provide sufficient space, keep the animals dry, or allow for access to food and/or water for the period of time that the animal will be temporarily housed in the laboratory. These decisions should be made in conjunction with the IACUC. It is best if the institution develops a transportation

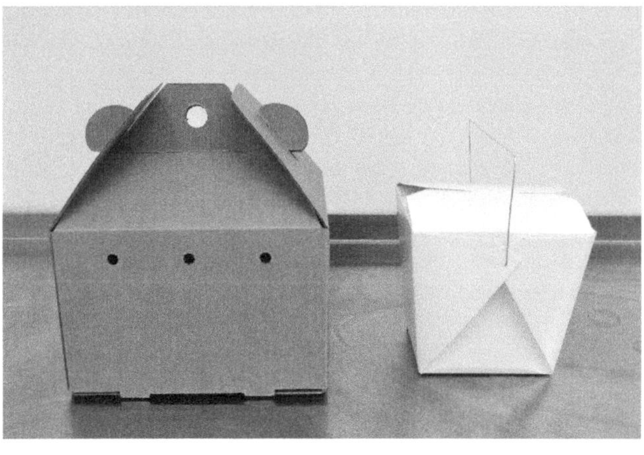

FIGURE 26.18 Examples of disposable transport containers used for intrainstitutional transport. (Photo by Michelle Wallace-Fields.)

FIGURE 26.19 Disposable cages can be used to hold animals in the laboratory. This method is best used when animals will need to be in the laboratory for longer than a few minutes, so that dry bedding, food, and water can more easily be provided than with disposable transport containers. (Photo by Michelle Wallace-Fields.)

policy for the animal program that addresses all the guidelines and applicable regulatory requirements. In addition, it is important to consider how animals will be shielded from view of the public and how occupational health and safety considerations will be addressed for situations such as rodent allergens and potential zoonotic diseases (e.g., Q fever) when animals are moved.

When moving animals between institutions, facilities, or even between rooms within the same facility, it is important that the animal be given time to acclimate to its new housing environment. Studies suggest anywhere from 2 to 5 days for animals to acclimate after being transported, as well as to allow for water and food consumption to normalize (Tuli et al. 1995; Conour et al. 2006). Therefore, the research staff, veterinarians, and IACUC should carefully consider the length of acclimation periods to be included in policies on a per species and research-type basis to ensure that animals are given adequate time to normalize prior to use.

Emerging Technologies

Laboratory animal management staff should always be on the alert for technologies that can help to enhance animal welfare, enhance ergonomics and employee health, increase customer service, and make operations more efficient and cost-effective. RFID chips are becoming more common in animal facilities for not only census and billing, but also for tracking of caging and supplies so that real-time inventories can be more easily performed. The use of the Internet and web-based applications that can be accessed through a smartphone or tablet computer will allow for greater communication between facility personnel and research staff, as well as reduce paperwork by allowing direct input of information into systems. Apps and social media may also be leveraged for performing functions like training and communication. In all cases, when using electronic systems it is important to ensure that the institution's information technology office is involved to ensure that appropriate digital safety and security methods are employed. Automation will become more prevalent in areas such as cage changing and cage wash processing to help decrease ergonomic injuries (Glass et al. 2004; Gobbi 2009; University of California Ergonomics Team 2014). Ultimately, those involved with the management of animal research programs should look not only within the laboratory animal science community, but across all industries, to learn how to best employ emerging technologies to enhance the functionality of their facilities.

REFERENCES

AALAS (American Association for Laboratory Animal Science). 2010. 2010 Laboratory Animal Facility Compensation Survey. Dublin, OH: Industry Insights, Inc.

AALAS (American Association for Laboratory Animal Science). 2014. 2014 Laboratory Animal Facility Compensation Survey. Dublin, OH: Industry Insights, Inc.

Andrews-Kelly, G. 2014. Laboratory animal enrichment: Bridging compliance with science. Presented at the *Public Responsibility in Medicine and Research Conference, Denver, Colorado.*

ANSI (American National Standards Institute) and ISEA (International Safety Equipment Association). 2014. Emergency eyewash and shower equipment: Selection, installation, and use guide. ANSI/ISEA Z358.1–2014. https://safetyequipment.org/wp-content/uploads/2015/05/EWS-Guide-Feb_2015.pdf (accessed August 12, 2015).

Ayadi, A., G. Ferrand, I.G. Cruz, and X. Warot. 2011. Mouse breeding and colony management. *Current Protocols in Mouse Biology* 1(1):239–264.

Baker, S.W., K.A. Prestia, and B. Karolewski. 2014. Using reduced personal protective equipment in an endemically infected mouse colony. *Journal of the American Association for Laboratory Animal Science* 53(3):273–277.

Berns, K., E.C. Bond, and F.J. Manning. 1996. Resource sharing in biomedical research. Washington, DC: Committee on Resource Sharing in Biomedical Research, Division of Health Sciences Policy, Institute of Medicine, National Academy Press.

Boisvert, C.M., and P.B. Morgan. 2006. The study support associate: An alternative to the traditional animal care position. *Lab Animal* 35(1):34–38.

Carbasho, T. 2013. Seattle Children's Research Institute adopts lean strategies for vivarium. https://www.tradelineinc.com/reports/2013-7/seattle-children%E2%80%99s-research-institute-adopts-lean-strategies-vivarium (accessed August 17, 2015).

Castelhano-Carlos, M.J., N. Sousa, F. Ohl, and V. Baumans. 2010. Identification methods in newborn C57Bl/6 mice: A developmental and behavioral evaluation. *Lab Animal* 4:88–103.

Catarinucci, L., R. Colella, L. Mainetti, L. Patrono, S. Pieretti, I. Sergi, and L. Tarricone. 2014. Smart RFID antenna system for indoor tracking and behavior analysis of small animals in colony cages. *IEEE Sensors Journal* 14(4):1198–1206.

CDC-NIH (Centers for Disease Control and Prevention—National Institutes of Health). 2009. *Biosafety in Microbiological and Biomedical Laboratories*. 5th ed. Washington, DC: U.S. Department of Health and Human Services.

CFR (Code of Federal Regulations). 2013. Title 9, Animals and animal products; Chapter 1, Animal and Plant Health Inspection Service, Department of Agriculture; Subchapter A, Animal welfare; Part 2, Regulations and Part 3, Standards. Washington, DC: Office of the Federal Register.

Chikazawa, S.A., A.D. Cerqueda, K.L. Cummings, J.A. Scarbrough, S.M. McKinney, C.N. Docampo, and M.M. Crane. 2012. Animal identification procedures aid in meeting colony management and research goals in rhesus macaque (*Macaca mulatta*) and sooty mangabey (*Cercocebus atys*) colonies. *American Journal of Primatology* 74(S1):42.

Conour, L.A., K.A. Murray, and M.J. Brown. 2006. Preparation of animal for research—Issues to consider for rodents and rabbits. *ILAR Journal* 47(4):283–293.

Cornell Waste Management Institution. 2015. Disposal of animal carcasses state laws. http://compost.css.cornell.edu/mapsdisposal.html# (accessed September 10, 2015).

Directive 2009/147/ECFernstrom, A.L., W. Sutian, F. Royo, K. Westlund, T. Nilsson, H.E. Carlsson, T. Paramastri, J. Pamungkas, D. Sajuthi, S.J. Schapiro, and J. Hau. 2008. Stress in cynomologus monkeys (*Macaca fascicularis*) subjected to long-distance transport and simulated transport housing conditions. *Stress—The International Journal on the Biology of Stress* 11(6):467–476.

Donahue, W.A., D.N. VanGundy, W.C. Satterfield, and L.J. Coghlan. 1989. Solving a tough problem. *Pest Control* 57:46–50.

Donnelly, C.J., M. McFarland, A. Ames, B. Sundberg, D. Springer, P. Blauth, and C.J. Bult. 2010. JAX Colony Management System (JCMS): An extensible colony and phenotype data management system. *Mammalian Genome* 21(3–4):205–215.

Ednie, D.L., R.P. Wilson, and C.M. Lang. 1998. Comparison of two sanitation monitoring methods in an animal research facility. *Contemporary Topics in Laboratory Animal Science* 37:71–74.

European Union Bird Directive. 2009. Directive 2009/147/EC. http://ec.europa.eu/environment/nature/legislation/birdsdirective/index_en.htm.

European Union Habitat Directive. 1992. Council Directive 92/43/EEC. http://ec.europa.eu/environment/nature/legislation/habitatsdirective/index_en.htm.

Gaskill, B.N., S.A. Rohr, E.A. Pajor, J.R. Lucas, and J.P. Garner. 2009. Some like it hot: Mouse temperature preferences in laboratory housing. *Applied Animal Behavior Science* 116:279–285.

Gentile, F. 2011. Improving efficiency and productivity by implementing a web-based Laboratory Animal Management Systems combined with RFID technology. *Journal of the American Association for Laboratory Animal Science* 50(5):766–767.

Glass, J.D., H.J. Klein, W. Terpeluk, and W.J. Pouch. 2004. The implementation of robotic technology in an animal facility: A case study. *Contemporary Topics in Laboratory Animal Science* 43(3):14–17.

Gobbi, A. 2009. New directions in cage processing automation: Expanding beyond the dump station. *Journal of the American Association for Laboratory Animal Science* 48(5):537. Presented at the *Abstracts of Scientific Presentations, 2009 AALAS National Meeting, Denver, Colorado.*

Gonzalez, D.M., S.J. Graciano, L. Karlstad, M. LeBlanc, T. Clark, S. Holmes, and J.D. Reuter. 2011. Failure and life cycle evaluation of watering valves. *Journal of the American Association for Laboratory Animal Science* 50(5):713–718.

Gordon, A., and J. Wyatt. 2011. The water delivery system affects the rate of weight gain in C57Bl/6J mice during the first week after weaning. *Journal of the American Association for Laboratory Animal Science* 50(1):37–40.

Gordon, S., and R. Preece. 2003. Prevention of laboratory animal allergy. *Occupational Medicine* 53:371–377.

Gordon, S., J. Wallace, A. Cook, R.D. Tee, and A.J. Newman Taylor. 1997. Reduction of exposure to laboratory animal allergens in the workplace. *Clinical and Experimental Allergy* 27(7):744–751.

Gruda, M.C., A. Pinto, A. Craelius, H. Davidowitz, W.M. Kopacka, J. Li, L. Qian, E. Rodriguez, E. Kuspiel, and W. Mandecki. 2010. A system for implanting laboratory mice with light-activated microtransponders. *Journal of the American Association for Laboratory Animal Science* 49(6):826–831.

Gunasekara, A.S., A.L. Rubin, K.S. Goh, F.C. Spurlock, and R.S. Tjeerdema. 2008. Environmental fate and toxicology of carbaryl. In *Reviews of Environmental Contamination and Toxicology*, ed. D.A. Whitacre, 95–121. Vol. 196. New York: Springer.

Haist, C.J., J.M. Cadillac, and R.C. Dysko. 2004. Assessment of bacterial contamination of drinking water provided to mice. *Journal of the American Association for Laboratory Animal Science* 43(6):8–13.

Hankenson, F.C., L.M. Garzel, D.D. Fischer, B. Nolan, and K.D. Hankenson. 2008. Evaluation of tail biopsy collection in laboratory mice (*Mus musculus*): Vertebral ossification, DNA quantity, and acute behavioral responses. *Journal of the American Association for Laboratory Animal Science* 47(6):10–18.

Harrington, M. 2014. Common disinfectants impair reproduction in mice. *Lab Animal* 43(10):335.

Hessler, J.R., and S.L. Leary. 2002. Design and management of animal facilities. In *Laboratory Animal Medicine*, ed. J.G. Fox, L.C. Anderson, F.M. Loew, and F.W. Quimby, 909–953. San Diego: Academic Press.

Hickman-Davis, J.M., M.L. Nicolaus, J.M. Petty, D.M. Harrison, and V.K. Bergdall. 2012. Effectiveness of shoe covers for bioexclusion within an animal facility. *Journal of the American Association for Laboratory Animal Science* 51(2):181–188.

Hill, D. 1999. Safe handling and disposal of laboratory animal waste. *Occupational Medicine—State of the Art Reviews* 14(2):449–468.

IATA (International Air Transport Association). Live animal regulations. Montreal: IATA. www.iata.org.

International Committee on Standardized Genetic Nomenclature for Mice. 2015. Rules for nomenclature of mouse and rat strains. http://www.informatics.jax.org/mgihome/nomen/index.shtml (accessed September 12, 2015).

Jude, L.L., N. Zielinski-Mozny, and S.I. Levin. 2012. Abstract: Rodent cage card notification system for colony management. *Journal of the American Association for Laboratory Animal Science* 51(5):677.

Krause, J., G. McDonnell, and H. Riedesel. 2001. Biodecontamination of animal rooms and heat-sensitive equipment with vaporized hydrogen peroxide. *Journal of the American Association for Laboratory Animal Science* 40(6):18–21.

Lipman, N.S., and S. Perkins. 2002. Factors that may influence animal research. In *Laboratory Animal Medicine*, ed. J.G. Fox, L.C. Anderson, F.M. Loew, and F.W. Quimby, 1143–1184. San Diego: Academic Press.

Lowe, J.J., S.G. Gibbs, P.C. Iwen, P.W. Smith, and A.L. Hewlett. 2013. Decontamination of a hospital room using gaseous chlorine dioxide: *Bacillus anthracis*, *Francisella tularensis*, and *Yersinia pestis*. *Journal of Occupational and Environmental Hygiene* 10(10):533–539.

Lucid, B.D., and K. Lepidi. 2010. The use of barcode and RFID technology in lab animal facilities. *ALN Magazine*, September 12. http://www.alnmag.com/articles/2010/09/use-barcode-and-rfid-technology-lab-animal-facilities (accessed September 10, 2015).

Major, N., and R.J. Wassersug. 1998. Survey of current techniques in the care and maintenance of the African clawed frog (*Xenopus laevis*). *Contemporary Topics* 37(5):57–60.

Mason, G., and K.E. Littin. 2003. The humaneness of rodent pest control. *Animal Welfare* 12:1–37.

Meier, T.R., C.J. Maute, J.M. Cadillac, J.Y. Lee, D. Righter, K.M.S. Hugunin, R.A. Deininger, and R.C. Dysko. 2008. Quantification, distribution, and possible source of bacterial biofilm in mouse automated watering systems. *Journal of the American Association for Laboratory Animal Science* 47(2):63–70.

Meller, A., I. Kasanen, O. Rukšėnas, N. Apanavičiene, Ž. Baturaitė, H.-M. Voipio, and T. Nevalainen. 2011. Refining cage change routines: Comparison of cardiovascular responses to three different ways of cage change in rats. *Laboratory Animals* 45:167–173.

Murphy, R.G.L., J.A. Scanga, B.E. Powers, J.L. Pilon, K.C. VerCauteren, P.B. Nash, G.C. Smith, and K.E. Belk. 2009. Alkaline hydrolysis of mouse-adapted scrapie for inactivation and disposal of prion-positive material. *Journal of Animal Science* 87(5):1787–1793.

NCRR (National Center for Research Resources). 2000. Cost analysis and rate setting manual for animal research facilities. https://grants.nih.gov/grants/policy/air/rate_setting_manual_2000.pdf.

NFPA (National Fire Protection Association) 10. 2013 edition. Standard for portable fire extinguishers. http://www.nfpa.org/codes-and-standards/all-codes-and-standards/list-of-codes-and-standards?mode=code&code=10.

NIH (National Institutes of Health). 2013. Notice regarding NIH plan to transition from use of USDA Class B dogs to other legal sources. NOT-OD-14-034. December 17. Bethesda, MD: NIH.

NIOSH (National Institute for Safety and Health). 2007. Code of Federal Regulations; Title 49, Transportation; Vol. 4, 2011-10-01; Part 227, Occupational noise exposure. Atlanta, GA: NIOSH.

NRC (National Research Council). 1997. *Occupational Health and Safety in the Care and Use of Research Animals*. Washington, DC: National Academies Press.

NRC (National Research Council). 2006. *Guidelines for the Human Transportation of Research Animals*. Washington, DC: National Academies Press.

NRC (National Research Council). 2011. *Guide for the Care and Use of Laboratory Animals*, ed. E.D. Olfert, B.M. Cross, and A.A. McWilliam. Washington, DC: National Academies Press.

NSF (National Science Foundation) and American National Standards Institute (ANSI). 2004. Class II (laminar flow) biosafety cabinetry. NSF/ANSI Standard 49-2007. Ann Arbor, MI: NSF/ANSI.

Obernier, J.A., and R.L. Baldwin. 2006. Establishing an appropriate period of acclimatization following transportation of laboratory animals. *ILAR Journal* 47(4):364–369.

OMB (Office of Management and Budget). OMB Circular A-21. Part 220 – Cost Principles for Educational Institutions. https://www.gpo.gov/fdsys/pkg/CFR-2012-title2-vol1/pdf/CFR-2012-title2-vol1-part220.pdf.

Orcutt, R.P., R.S. Phelan, and J.G. Geistfeld. 2001. Exclusion of mouse hepatitis virus from a filtered, plastic rodent shipping container during an in transit field challenge. *Contemporary Topics in Laboratory Animal Science* 40(4):32–35.

O'Rourke, D.P., and T.W. Schultz. 2002. Biology and diseases of amphibians. In *Laboratory Animal Medicine*, ed. J.G Fox, L.C. Anderson, F.M. Loew, and F.W. Quimby, 793–826. San Diego: Academic Press.

O'Rourke, D.P., and J. Schumacher. 2002. Biology and diseases of reptiles. In *Laboratory Animal Medicine*, ed. J.G Fox, L.C. Anderson, F.M. Loew, and F.W. Quimby, 827–860. San Diego: Academic Press.

OSHA (Occupational Safety and Health Administration), United States Department of Labor. Portable fire extinguishers. OSHA 29 CFR 1910.157.

OSHA (Occupational Safety and Health Administration), United States Department of Labor. Hearing conservation. OSHA 3074 2002 (Revised).

Paluch, L.R., C.C. Lieggi, M. Dumont, S. Monette, E.R. Riedel, and N.S. Lipman. 2014. Developmental and behavioral effects of toe clipping on neonatal and preweanling mice with and without vapocoolant anesthesia. *Journal of the American Association for Laboratory Animal Science* 53(2):132–140.

Randle, S.J., R. Vij, T. Bowman, G. Morrow, J. Greve, A. Ogasawara, V. Ramaswamy, E. Biewagen, P. Carter, and J. Hongo. 2011. High throughput identification of animals using miniature ID digital ear tags. *Journal of the American Association for Laboratory Animal Science* 50(5):784. Poster presented at *2011 National Meeting of the American Association for Laboratory Animal Science, San Diego, California*.

Rasmussen, S., M.M. Miller, S.B. Filipski, and R.J. Tolwani. 2010. Cage change influences serum corticosterone and anxiety-like behaviors in the mouse. *Journal of the American Association for Laboratory Animal Science* 50(4):479–483.

Reeb-Whitaker, C.K., D.J. Harrison, R.B. Jones, J.B. Kacergis, D.D. Myers, and B. Paigen. 1999. Environmental and occupational disorders: Control strategies for aeroallergens in an animal facility. *Journal of Allergy and Clinical Immunology* 103(1):139–146.

Reuter, J.D., and R.C. Dysko. 2003. Quality assurance/surveillance monitoring programs for rodent colonies. In *Laboratory Animal Medicine and Management*, ed. J.D. Reuter and M.A. Suckow. Ithaca, NY: International Veterinary Information Service. http://www.uff.br/animaislab/a12.

Roble, G.S., V. Gillespie, and N.S. Lipman. 2012. Infectious disease survey of *Mus musculus* from pet stores in New York City. *Journal of the American Association for Laboratory Animal Science* 51(1):37–41.

Rodney, S., T.R. Scott, and D.R.J. Moore. 2013. Estimating the toxicity of pesticide mixtures to aquatic organisms: A review. *Human and Ecological Risk Assessment* 19(6):1556–1575.

Rosenbaum, M.D., S. VandeWoude, and T.E. Johnson. 2009. Effects of cage-change frequency and bedding volume on mice and their microenvironment. *Journal of the American Association for Laboratory Animal Science* 48(6):763–773.

Scher, B. 2008. New solutions for management and husbandry: RFID census and wash cycle tracking. *Journal of the American Association for Laboratory Animal Science* 47(5):76.

Schondelmeyer, C.W., D.L. Dillehay, S.K. Webb, M.J. Huerkamp, D.M. Mook, and J.K. Pullium. 2006. Investigation of appropriate sanitization frequency for rodent caging accessories: Evidence supporting less-frequent cleaning. *Journal of the American Association for Laboratory Animal Science* 45(6):40–43.

Schray, C.L., V.A. Hill, M.H. Nowland, and H.G. Rush. 2011. Development of a technical services team for husbandry staff in a large university setting. *Journal of the American Association for Laboratory Animal Science* 50(5):761.

Schultz, C.C. 2006. Recirculating vs. flushing: Animal watering system alternatives. *ALN Magazine*, February 28. http://www.alnmag.com/articles/2006/02/re-circulating-vs-flushing-animal-watering-system-alternatives.

Stahl, N.Z. 2012. Mass General cuts costs in the vivarium with Lean Process Implementation. Pursuing the relentless elimination of everything that does not create value. *Tradeline Reports*. https://www.tradelineinc.com/reports/2012-2/mass-general-cuts-costs-vivarium-lean-process-implementation.

Stark, S., J. Petitto, and S. Darr. 2010. Animal research facility. Whole Building Design Guide, a program of the National Institute of Building Sciences. https://www.wbdg.org/design/animal_research.php (accessed September 10, 2015).

Symonds, E.L. 2013. Fecal DNA genotyping: A non-invasive approach to characterize mouse models for nutrigenomics cancer chemoprevention studies. *Current Pharmacogenomics and Personalized Medicine* 11(1):12–21.

Trott, K., B.A. Stacy, B.D. Lifland, H.E. Diggs, R.M. Harland, M.K. Khokha, T.C. Grammer, and J.M. Parker. 2004. Characterization of a *Mycobacterium ulcerans*-like infection in a colony of African tropical clawed frogs (*Xenopus tropicalis*). *Comparative Medicine* 54(3):309–317.

Troyk, P.R. 1999. Injectable electronic identification, monitoring, and stimulation systems. *Annual Review of Biomedical Engineering* 1:177–209.

Tuli, J.S., J.A. Smith, and D.B. Morton. 1995. Stress measurements in mice after transportation. *Lab Animal* 29(2):132–138.

Turner, D.E., E.K. Daugherty, C. Altier, and K.J. Maurer. 2010. Efficacy and limitations of an ATP-based monitoring system. *Journal of the American Association for Laboratory Animal Science* 49(2):190–195.

University of California Ergonomics Project Team. 2014. Ergonomics study of animal care positions at the University of California. https://ehs.ucsf.edu/sites/ehs.ucsf.edu/files/Ergonomics%20Study%20of%20Animal%20Care%20Positions%20at%20The%20University%20of%20California.pdf.

Vogelweid, C.M., K.A. Zapien, M.J. Honigford, L. Li, H. Li, and H. Marshall. 2011. Effects of a 28-day cage-change interval on intracage ammonia levels, nasal histology, and perceived welfare of CD1 mice. *Journal of the American Association for Laboratory Animal Science* 50(6):868–878.

Watson, J. 2013. Unsterilized feed as the apparent cause of a mouse parvovirus outbreak. *Journal of the American Association for Laboratory Animal Science* 52(1):83–88.

Whary, M., R. Peper, G. Borkowski, W. Lawrence, and F. Ferguson. 1993. The effects of group housing on the research use of the laboratory rabbit. *Lab Animal* 27(4):330–341.

WHO (World Health Organization). 2004. *Laboratory Biosafety Manual*. 3rd ed. Geneva: WHO. http://www.who.int/csr/resources/publications/biosafety/WHO_CDS_CSR_LYO_2004_11/en/ (accessed September 5, 2015).

27

Feed and Bedding

Ronald L. Carter and Neil S. Lipman

CONTENTS

Introduction

Providing for an animal's nutritional needs, as well as an appropriate environment to reside, is the cornerstone of animal husbandry. Feed and bedding can directly influence experimental outcomes, as their chemical and physical characteristics can affect the animal's physiology (Baker and Lipman 2015). The production and subsequent handling of feed and bedding, whether by the manufacturer or distributor, and continuing within the animal facility, can also affect the quality of the product in use. The selection, management, and provision of high-quality feed and bedding is essential to ensure the health and welfare of the animals and the integrity of the results obtained from them.

In this chapter, feed and bedding are reviewed separately. The importance of, and the methods by which, feed and bedding are selected and their impact on research are discussed. Processing, handling, and storage of feed and bedding is also reviewed. The reader is provided an overview of the concepts relevant to selecting and managing feed and bedding in the laboratory animal setting. As the scope of this topic is broad, the reader should consult the references for additional information.

Feed

The provision of high-quality feed to animals is essential to meet their physiologic needs, specifically growth, maintenance, and reproduction. There are numerous products, which differ in nutrient content, available in various formulations from a variety of feed manufacturers (Fox and Newberne 1980). A primary directive when formulating feed is to ensure sufficient content of the six classes of *nutrients*: water, carbohydrates, fats (lipids), proteins, minerals, and vitamins. While all animals require each of the six classes of nutrients, some species may require higher levels of specific nutrients than others. Additionally, nutrients are considered either essential or nonessential. Essential nutrients are those that an animal cannot synthesize or cannot synthesize in sufficient quantities to maintain health and must be obtained from an external source, that is, in the diet. It is important to recognize, when selecting a diet, that a particular nutrient may be essential (i.e., necessary in the diet or otherwise provided) for some species but not for others. Nonessential nutrients are those nutrients that can be produced by the animal or its microbial flora and are not a dietary requirement. Examples of nutrient categories are provided in the following paragraphs, with select examples of species that have species-specific nutrient requirements.

Fiber, a complex *carbohydrate*, provides bulk for the contents of the digestive tract and is important for efficient and effective digestion in most laboratory animal species, particularly rabbits, guinea pigs, some species of nonhuman primates, and ruminants. These species should be fed a high-fiber diet to ensure proper gastrointestinal (GI) physiology and to avoid GI obstruction. *Vitamins* are needed in small amounts to maintain health. Some vitamins may be produced in the body, but not in sufficient quantities. These vitamins must be supplied in the diet. For example, vitamin C cannot be synthesized by guinea pigs and some nonhuman primate species. These species are dependent on receiving their vitamin C requirement in their feed or, less commonly, in their water, or through the provision of dietary supplements, for example, fruit; therefore, their diets are specially formulated to include vitamin C, without which they may develop scurvy. Vitamin D is supplied to laboratory animals, in the form of cholecalciferol (D3), in their feed, although the absolute requirement for vitamin D is unknown for many species. Some nonhuman primate species, if not exposed to ultraviolet (sun) light, require vitamin D3 in their diet to help prevent rickets, as they cannot utilize vitamin D2 to meet their needs. Amino acids, the building blocks of *proteins*, are needed to carry out many important bodily functions. The amino acid taurine is essential for cats but not for other species. For this reason, cat foods are supplemented with taurine to avoid a deficiency that may lead to blindness and tooth decay. The primary functions of *fats* are to supply and store energy. High-fat diets (up to 11%) are at times used as a response to increased neonatal morbidity and mortality, or to support breeding in rodents, because their higher energy content can help the animal meet the energy demands of gestation and lactation. *Minerals*, in the appropriate ratios, are necessary to maintain normal physiology. A mineral deficiency or the provision of minerals in an inappropriate ratio can lead to disease. For example, a diet deficient in calcium can lead to rickets in young animals of many species, or in mature animals to osteoporosis, a disease in which the bones are fragile, making them susceptible to fracture.

Subcommittees of the National Research Council Committee on Animal Nutrition have prepared comprehensive reports for the nutrient requirements of commonly used laboratory animal species (NRC 1977, 1982, 1993, 1994, 1995, 1998, 2000, 2001, 2003a, 2003b, 2006a, 2006b, 2007); these publications also include information on quality assurance, contaminants and toxicants, bioavailability, and palatability.

Formulas and Types

Most laboratory diets are *natural ingredients* and nutritionally complete. Natural-ingredient diets are formulated from agricultural- and/or animal-based ingredients, such as processed whole grains and fish meal, and commodities subjected to limited refinement. Formulations of natural-ingredient diets differ depending on the species to which they are fed. Even within a species, there may be different formulations depending on the animals' use. For example, there are many formulations of rodent diets, differing principally with respect to protein and fat concentration for breeding versus nonbreeding animals and for growth and maintenance.

Most natural-ingredient diets are *"closed" formula*, in which the individual components of the diet are proprietary and are not specified by the manufacturer, although a guaranteed analysis is provided, including a list of ingredients and nutrients with their calculated values. The diet's ingredients may change in association with commodity prices or availability. Closed-formula diets typically contain ingredients such as ground corn, ground oats, alfalfa meal, soybean meal, and ground wheat. Vitamins, minerals, and fat are added to ensure nutritional adequacy. Categories of closed-formula diets include least-cost, fixed, and variable formula diets.

Least-cost formula is the practice of substituting one ingredient in a diet with a lower-cost ingredient at any given time, to be cost-effective. Of the various types of closed-formula diets, least cost introduces the most variables due to the possibility of constant and broad change in the diet's ingredients.

Fixed formula ensures the ingredient formulation does not change, unless the formula itself is updated by the manufacturer to change nutritional content. It is important to note that since the nutritional content of the dietary constituent can naturally vary with harvest location and across growing seasons, there will be some variability from batch to batch in a fixed formula diet.

The goal of a *variable formula* is to maintain the concentrations of known nutrients in the diet in response to the naturally changing nutritional content of the ingredients. The formula is altered as ingredients are assayed to ensure the diet remains nutritionally stable across batches (Knapka 1997). Variable formula diets, although likely to be more stable with regard to nutritional content than least-cost and fixed formula diets, may still vary over time, as the ingredients used do differ from place to place and season to season.

Although less commonly utilized, *"open"-formula* diets are manufactured in accordance with an established known ingredient formulation that is known to the purchaser. Open-formula diets were generated to reduce diet as a variable and may be natural ingredient, but most commonly are purified diets. For example, in the 1970s, the American Institute of Nutrition (AIN) initiated a program to standardize laboratory animal diets, formed a committee, and designed the open-formula AIN-76 rodent diet to be used as a standard reference diet in an effort to reduce some variability. AIN-76A was revised and improved in 1993, and two new formulations were derived: AIN-93G for growth, pregnancy, and lactation, and AIN-93M for adult maintenance. Open-formula diets offer certain advantages, including, but not limited to, the following: their quantitative ingredient formulations are available to the user, they facilitate control of potential research variables, they allow for repeatability in research, and the diets can be purchased from multiple vendors, encouraging competitive and quality incentives (Barnard et al. 2009).

Purified diets, also referred to as semisynthetic diets, are open-formula diets produced from purified components, such as carbohydrates (e.g., starch or sucrose), purified sources of protein (e.g., lactalbumin and casein), refined oils, and synthetic vitamins and minerals. Purified diets are typically used when altering the nutritional content of the diet or when compounding with additives. Purified diets allow for complete ingredient control. Specific ingredients can be added or excluded. The nutritional content of a purified diet is highly stable from batch to batch, as the specific ingredients and amounts of each are known. Purified diets are considerably more expensive than natural-ingredient diets.

Chemically defined diets are formulated from chemically pure nutrients, for example, specific carbohydrates, triglycerides, individual amino acids, essential fatty acids, vitamins, and minerals. Chemically defined diets are generally utilized when altering a specific nutritional dietary component. They are extremely expensive and highly labile.

It is important to note that natural-ingredient diets vary greatly from purified and chemically defined diets, and while each has its own advantages; it is not feasible to compare results when a natural-ingredient diet is used for a control study for cost savings, and a purified or chemically defined diet is used for the test diet.

Certified diets are natural-ingredient diets produced to meet the requirements of the governmental good laboratory practice (GLP) standards, requiring periodic feed analysis for environmental contaminants that may interfere with research studies (CFR 2004; EU 2004; OECD 2011). Feed samples are analyzed and certified to contain no more than the established maximum level of environmental contaminants, including heavy metals, chlorinated hydrocarbons, organophosphates, and aflatoxins. Many contaminants are found naturally in plant materials or are agricultural residues. Diets may also be contaminated during storage or formulation. A batch-specific certificate of analysis with a guarantee that the contaminants do not exceed the acceptable maximum limits is provided with the diet. For preclinical toxicology studies required to meet GLP standards, sufficient diet is commonly procured to ensure that the animals are fed diet from a single batch for the duration of the study.

Medicated and *compounded* diets have drugs or other chemicals added. They are routinely used in research facilities housing rodents. Many compounded diets are available "off the shelf" from commercial suppliers. Examples of medicated diets include feed containing the anthelmintic fenbendazole, as well as several antibiotics, including trimethoprim/sulfamethoxazole and doxycycline. Fenbendazole feed is commonly employed to treat mice for nematodes (e.g., *Syphacia* and *Aspiculuris* spp.), trimethoprim/sulfamethoxazole-compounded feed is used to control pneumocystosis in immunodeficient mouse strains, and doxycycline-containing feed is used to induce or inhibit transcription in conditional transgenic mice that contain components of the tetracycline transactivator system (Coghlan et al. 1993; Lewandoski

2001; Ryding et al. 2001). Compounded feed is manufactured in special manufacturing plants, distinct from those used to produce other diet types, to ensure there is no carryover of the added compound to noncompounded feed. Compounded feed can be manufactured using distinct dyes, permitting the diet to be differentiated from standard diets. Feed can also be pigmented with a small amount of a Food, Drug, and Cosmetic Act (FD&C) food color to distinguish different diets. Diets can also be compounded with various additives for specific needs. For example, at the authors' institution, feed compounded with vitamin E is used to treat ulcerative dermatitis in C57BL/6 (B6) mice and genetically engineered mice on a B6 background (Lawson et al. 2003). Medicated and compounded diets may have a shorter shelf life, as indicated by the manufacturer.

The majority of diets provided to rodents, and therefore used in animal research, are *pelleted* (Figure 27.1a and b), which provides the densest and highest energy content per unit weight, allowing efficient delivery. The effect of the pelleting process prevents ingredient separation during handling and feeding, minimizes waste, and decreases the need for storage space. Pelleting involves mixing, grinding, and exposing various dietary constituents, such as proteins, fibers, and minerals, to steam (typically 65°C–80°C); compressing the conditioned ingredients into a die through which the meal is forced, expelled, and cut to the desired pellet length; conveying the meal to a dryer, which reduces moisture content to levels that impede microbial growth, permitting a relatively long shelf life; and cooling and sieving before packaging (Tobin et al. 2007). Since pelleting subjects feed constituents to heat and pressure and then rapid cooling, the bacterial and fungal loads that may be found in unprocessed ingredients are reduced (Halls and Tallentire 1978). Pelleting ensures formulation stability, allows easy verification of animal access to feed, is dustless and prevents inhalation, and in rodents, provides a hard substrate to "wear" down incisors.

Extruded diet (Figure 27.1c and d) is produced by extrusion, a process in which ingredients are exposed to high heat and pressure. Initially, the feed is prepared as it is for pelleting; however, the ingredients are more finely ground, increasing gelatinization of starch. Dietary constituents are conditioned at ~80°C–95°C before being exposed to a temperature of ~150°C under high pressure when forced through the die. During the process, the material moves from an area of high pressure to one of lower pressure, in which superheated water vapor, trapped in the feed, cools and begins to expand and pop, introducing and trapping air into the feed, producing a product less dense than a typical pellet. The resulting

FIGURE 27.1 Various feed forms: (a) pellet (rodent), (b) pellet (rabbit), (c) extruded collet (nonhuman primate), (d) extruded collet (dog), (e) meal, and (f) powdered. (Courtesy of Envigo [a–d] and PMI LabDiet [f].)

collet (extruded pellet) is cut, dried, cooled, and sieved to remove fines (Tobin et al. 2007). Because extruded diets are produced at higher temperatures than pelleted diets, the microbiologic burden is less. Extruded diet is typically used for larger species, such as nonhuman primates, canine, and swine. While extruded diets are commonly fed to rodents in Europe, extruded rodent diets are infrequently fed to rodents in the United States, principally because of their low density and cost. Many of the advantages of pelleting also apply to extruded feeds, such as formulation stability, easy verification of animal access, and being dustless. Additional advantages of extruded diets are the ability to add high levels of fat to the diet, the diet is more digestible, and it performs better following autoclaving, with little to no clumping or excessive hardening, which commonly occurs when a pelleted diet is subjected to steam sterilization. However, extruded feeds are more expensive to manufacture and, being less dense, need to be fed more often and require more space for storage.

Meal diets (Figure 27.1e) consist of ground ingredients that are mixed together with no further processing, that is, pelleting or extrusion. Microbiologic levels are much higher than those found in pelleted or extruded feed due to the lack of heat used during processing, making meal more susceptible to rancidity and insect infestation (Eva and Rickett 1983). If the ingredients in the meal are different-sized particles, segregation of ingredients can occur. Also, some species may be able to isolate the particles they prefer, leaving the others behind, thus negating a "balanced" diet. Feeding meal mixes generally require the use of a container, such as a jar, trough, or tub.

Powdered or ground diet (Figure 27.1f) is a pelleted or extruded diet that is ground to a powder after production. Powdered diets are commonly used when needing to provide additives after formulation. However, waste is high with powdered diets, and caking commonly occurs when exposed to the microenvironment. Powdered diet requires special feeders, which can easily tip, spilling the feed, which is subsequently soiled with urine, feces, or saliva. Due to its considerably greater surface area, powdered diets can spoil much easier than pellets and can separate. Some powdered diets are formulated to be suspended in water. These liquid diets are uncommon and are most often used for alcohol studies or postsurgical recovery. The diet has to be prepared more often and needs to be monitored. Once in solution, the stability of nutrients is significantly shorted.

Autoclavable diets, which are formulated to be sterilized prior to provision, are enriched with heat-labile nutrients, including thiamin, vitamins A, B12 and E, pantothenic acid, and pyridoxine, whose concentrations are reduced during the autoclaving process. Steam autoclaving at 121°C for 15–20 minutes is frequently recommended for diet sterilization. Some diet formulations may be adversely affected at this temperature, and pasteurization is used in lieu of sterilization. Pasteurization is typically achieved by exposure to a temperature of 107°C for 15–20 minutes in an autoclave (Caulfield et al. 2008). Slightly lower temperatures and durations can be used to pasteurize with decreasing effects on the nutritional and microbial content of the diet (Faith and Hessler 2006). Unlike sterilization, pasteurization does not kill all microorganisms in the diet, but instead achieves a "log reduction" in the number of viable organisms. Autoclavable pelleted diets may be coated with silicon dioxide or calcium bentonite to reduce the likelihood of clumping and adherence, which occurs as a result of pellet swelling during steam sterilization. To avoid clumping postautoclaving, feed may be decanted into bags with additional space to accommodate swelling, or it may be sterilized on trays at a depth of ~3 inches. Sterilization and pasteurization cycles must be developed and verified to ensure the sterility or microbial load reduction of the feed. Pulsed vacuum sterilization, which removes air from the autoclave chamber, is preferred to ensure adequate steam penetration when autoclaving feed in the manufacturer's original packing materials (bag). Feed subjected to excessive sterilization may be depleted of essential nutrients and the protein quality may be reduced, but more likely it may become too hard—a result of the polymerization of select feed constituents—for some rodent strains to eat (Ford 1987). Validation of autoclave cycles poststerilization for sterility or postpasteurization for desired microbial reduction is recommended on a regular basis. Additional postautoclaving monitoring should include assurance that appropriate levels of heat-labile constituents are provided at suitable levels (Lipman 2007). Access to mass spectroscopy and other technologies is available to evaluate feed for contaminants if contamination is expected.

Gamma-irradiated diets have become commonplace and have replaced the use of autoclaved diets in many settings, as they require less processing after receipt and are not subject to the effects of heat and temperature that result from autoclaving. Most feed producers subject irradiated diet to between 10 and

40 kGy (1 and 4 Mrad) by exposing the bags to a cobalt source or electron beam. As feed bags are palletized and then irradiated, irradiation exposure differs, depending on a specific bag's location within the load or even among pellets or collets in a single bag. The irradiation dose is stated as the minimal dose exposure. Some diets will be exposed to greater amounts of irradiation. Although irradiated feed is not purported to be sterile, bacterial (cells and spores) and fungal loads are markedly reduced, to less than 100 bacteria or fungi per gram of feed (Cover and Belcher 1992). This is in contrast to standard diets whose bacterial loads fluctuate seasonally and can reach levels as high as 500,000 total bacteria per gram of feed. Irradiation is purported to be ineffective against some viruses, for example, mouse parvovirus (MPV), as the exposure dose is insufficient (W. Shek, personal communication, 2004; Lipman 2007); however, some laboratory animal veterinarians affirm that sporadic parvovirus outbreaks can be eliminated by using irradiated diet. In contrast to steam sterilization, irradiation has much less of an effect on nutritional quality. Irradiated and fortified autoclavable diets are commonly used for axenic and microbiologically defined rodents, and immunodeficient animals (NRC 2011).

Various *soft* and/or *moist* diets are commercially available. These are commonly available for dogs and cats. In rodents, they are used at weaning to ease the transition from lactation to a solid diet; at postsurgery to hasten recovery; to improve the nutrition of animals subject to the effects of experimental manipulation; for select mutants, such as those with dental deformities, that have difficulty ingesting hard feed; for animals having difficulty ambulating; and during shipping.

Supplemental fiber is frequently provided to ruminants, as well as rabbits and guinea pigs, in addition to a balanced commercial diet. Common plants used include timothy and alfalfa hay. Hay should be of good quality, contain few thick tough stems and weeds, and have no mold or dust. Irradiated, cubed hay is commercially available to provide a product free of parasites and microorganisms.

Live feed may be used for certain species, such as fish, amphibians, and reptiles. Examples of live feed include brine shrimp larvae, rotifers, paramecia, algae, protozoa, drosophila, crickets, waxworms, and mealworms. Some reptiles need live feed to be supplemented with a balanced commercial diet or calcium powder, augmented with vitamin D3 to assist in absorbing the calcium, sprinkled onto live prey at feeding time. Live feed is highly perishable and requires on-site equipment and processes to maintain daily production of rations. Standardized timing for collection of live feed ensures that the highest nutritional value is provided. If sourcing a live product, it is imperative to have a reliable vendor for procurement to avoid inconsistencies (e.g., hatching rates). Additionally, the nutritional values of various types of live feed differ and must be taken into account when selecting the type to be used for a particular species.

Flaked or *dry* diets are available and used for certain aquatics species. Generally, flaked diets are used for convenience and efficiency in large colonies, for example, zebrafish. Nutritional superiority of flaked or dry over live feed for aquatics species is under debate, although many facilities use a mixture of both live and flaked diets to ensure a nutritionally complete diet for the various stages of development (Lawrence et al. 2012). One advantage of flaked or dry feed is the ability to prevent the introduction of unwanted pathogens by treating the product with gamma-irradiation.

Influence of Feed on Experimental Results

A fundamental goal in scientific research is to eliminate variables. There are important considerations in choosing the right diet for an experiment. When deciding on what type of diet to use, it is best to look in the literature to see what diet has been used in similar studies. Additionally, food provided as enrichment should be factored into the overall caloric intake to ensure adequate nutrition and prevent undesired health or experimental effects. Feed constituents and formulations serve as research variables and, in specific research areas, require the use of purified and chemically refined diets. Some contaminants, such as heavy metals, including arsenic, cadmium, lead, and mercury, and pesticides, may be introduced from the environment. Phytoestrogens, although not necessarily negative in all circumstances, are a well-known example of a naturally occurring compound in plants that have estrogenic activity and which can interfere with behavior, reproduction, bone development, and metabolic activity. Therefore, feed manufacturers are producing diets that avoid protein sources known to contain isoflavones (Allred et al. 2001; Ju et al. 2001, 2002; Thigpen et al. 2001, 2002, 2003). The main subclasses of phytoestrogens found in ingredients used in research diets are isoflavones (found in soybean meal), coumestans (found in alfalfa), and lignans (mainly associated with plant fibers) (Tobin et al. 2007). To date, numerous studies in animals leave little doubt that isoflavones affect a variety of experimental endpoints (Baker and Lipman 2015). For example, rodents consuming isoflavones have been noted to have fewer tumors and/or a delay in tumor development in mammary, liver, colon, and prostate cancer (Leiter 2009). Although experimental results may vary depending on many factors, the evidence for isoflavones' role in cancer is mounting and must not be dismissed as insignificant in cancer research (Leiter 2009). Phytoestrogens have been reported to cause reproductive problems, including impaired ovarian function and reduced fecundity, in sheep and cattle (Adams 1995). Diets containing alfalfa, which fluoresce naturally, can affect image quality when performing fluorescence optical imaging (Inoue et al. 2008). Therefore, alfalfa-free imaging diets are available for rodents. Many other studies describe the experimental impact of various dietary constituents. It is impossible to know which yet-to-be-discovered compounds may have such an effect. Additionally, as noted earlier, even closed-formula diets cannot be exactly replicated batch after batch to ensure complete nutritional consistency. At the very least, the concept of feed as a variable in most research should be taken into account when planning studies.

Bedding

Bedding is an integral component of the husbandry provided to most terrestrial species. It is used to absorb, dilute, and/or limit the animal's contact with its excreta, and is used for nest building; provides insulation and therefore allows the animal to thermoregulate; can serve to provide environmental enrichment; minimizes the growth of microorganisms; and in some cases, reduces the accumulation of intracage ammonia (Perkins and Lipman 1995; Smith et al. 2004). Importantly, bedding can also influence experimental data. A variety of materials are utilized as both contact and noncontact bedding. By definition, contact bedding that which the animals have direct contact. Noncontact bedding is typically provided as a sheet or on a roll, lining a pan or cage, and does not typically come into physical contact with the animal; rather, it sits below a rack or cage to collect and absorb urine and feces. Bedding selection should be based on a variety of factors, the most important of which are animal preference and materials that minimally interfere with the investigations for which the animals are used. For example, mice exhibit a preference for large fibrous materials that they can manipulate and use to build a nest (Blom et al. 1996; Van de Weerd et al. 1997). Bedding that enables burrowing is encouraged for some species, such as mice and hamsters. Pigs naturally forage and explore, even if there are no obvious stimuli (Wood-Gush et al. 1993), so if bedding is provided, it should be of a nature that encourages and satisfies that behavior (Bollen and Ritskes-Hoiinga 2007). No type of bedding is ideal for all species under all management and experimental conditions. For example, in nude or hairless mice that lack eyelashes, fibers from some forms of cellulose bedding can result in periorbital abscesses (White et al. 2008). Vendors' manufacturing, monitoring, and storage methods are important in bedding selection, as bedding may be contaminated with toxins and environmental pollutants, as well as bacteria, fungi, and vermin (NRC 2011). Other considerations for bedding selection are product cost; availability; absorbency; palatability, or lack thereof; ease of handling, transportation, and storage, including packaging and product weight (dry and wet); the ability to sterilize and/or obtain gamma-irradiated product; disposability; the ability to control ammonia accumulation; and the amount of associated dust.

Types of Bedding and the Potential to Affect Research

Bedding is generally manufactured from plant materials such as wood, cotton, and corncob, which are subject to varying degrees of processing. Minimally processed wood is the most commonly used contact bedding. Soft- or hardwoods, devoid of bark, are chopped, shredded, or shaved, and then heated at temperatures up to 1200°F to reduce the bacterial and moisture content before packaging. Hardwood bedding (Figure 27.2a) is manufactured from aspen, beech, maple, and/or birch. Softwoods, such as pine or cedar, are generally avoided, as the volatile aromatic amines that give these materials their pleasant aroma alter hepatic microsomal enzyme concentrations and therefore xenobiotic processing (Ferguson 1966; Vessel 1966). Most wood bedding, especially shredded or shaved products, has excellent nest-building characteristics (Blom et al. 1996). Larger species, such as pigs, may be provided with shaved wood products, which encourage natural foraging, assist in body heat retention, and absorb urine (Figure 27.2).

Corncob (Figure 27.2b), produced from the woody-ring portion of the cob by processing with a hammer mill and roller mill and subsequently dried, is available in several pellet sizes. One-eighth inch, 1/4 inch, or a mixture of both are commonly used for rodents. Corncob has excellent characteristics with respect to inhibiting the accumulation of ammonia, and therefore is preferred when using static isolator caging (Perkins and Lipman 1995; Smith et al. 2004). The specific characteristics of corncob inhibiting ammonia accumulation are unknown, but they are unrelated to its absorbency. Corncob can be abrasive and has been associated with foot lesions in highly immunocompromised mouse strains (author's [N.S.L.] personal communication). Off-gassing of acetic acid has also been observed, presumably from the decay of residual organic matter (Perkins and Lipman 1995). The density of corncob limits nest building, and therefore it is frequently supplemented with nesting material or mixed with other bedding types. The authors recommend autoclaving or purchasing irradiated product, as the porosity of corncob leads to mold growth in unsterilized or nonirradiated product underneath the sipper tube, where spillover occurs. Corncob expands and adheres during autoclaving, requiring the need to dissociate the pellets after steam sterilization.

A variety of processed wood products, for example, cellulose (Figure 27.2c), both virgin and recycled, are available for use as both contact and noncontact bedding. Products differ in absorbency, color, shape, and size. They may also be blended with other products, such as corncob. Cellulose products are typically more expensive than wood and generally have good nest-building characteristics. In some countries, nontraditional bedding is used (e.g., in India, bedding composed of rice husks is commonly used); however, appropriate quality assurance practices should be in place.

Certified bedding, in which levels of specific toxic environmental contaminants are measured and determined to not exceed maximum concentrations, is available for use in studies that must meet GLP standards.

A variety of products, generally manufactured from cellulose, are available for use as noncontact bedding (Figure 27.2d). Noncontact bedding includes plastic-backed absorbent paper and cage board. Products are available in precut sheets, formed trays, and roll stock. Material used for noncontact bedding can be impregnated with antibiotics, for example, neomycin, to inhibit bacterial growth and the subsequent ammonia production.

(a) (b) (c) (d)

FIGURE 27.2 Various bedding types: (a) wood, (b) corncob, (c) cellulose, and (d) noncontact cage board. (Courtesy of P.J. Murphy [a], the Andersons, [b], and Shepherd Specialty Papers [c and d].)

Feed and Bedding Handling and Processing

The receipt, handling, storage, and distribution of feed and bedding are integral components of a biosecurity program, especially for barrier-maintained rodent species. These processes are heavily influenced by floor plan, the operational concepts employed in the facility, and the level of adventitious agent exclusion. Quality assurance programs for feed and bedding processes should include monitoring and validating sterilization or pasteurization and/or decontamination procedures.

Some facility designs provide distinct corridors to separate the distribution of feed and bedding (with or without associated caging) between the clean cage wash (CCW), the animal holding rooms (AHRs) and the soiled cage wash (SCW), in order to control cross-contamination between clean and soiled supplies. However, the use of dual-corridor facilities has become less common, as the space utilization is less efficient than single-corridor facilities, animal facility construction and operational costs have escalated, and the use of specific pathogen-free animals has become commonplace. Regardless, facility management must take into account the specific design of the facility to make important decisions in the interest of minimizing or eliminating contamination during the processing, storage, and handling of these materials.

The method of processing feed and/or bedding into CCW is of critical importance, as the bulk of the materials used in a vivarium originate from this area and are distributed throughout the facility. An operational failure in CCW, resulting in contamination, is likely to be widely spread. Feed may be autoclaved into the facility, but use of gamma-irradiated diet has become commonplace and has replaced the use of autoclaved diets in a variety of settings, as they require less processing after receipt and are not subject to the effects of heat and temperature that result from autoclaving. Validation of autoclave cycles poststerilization for sterility or postpasteurization for desired microbial reduction is recommended on a regular basis. Biological indicators are used to verify the sterility of bedding. Similar to bedding, additional postautoclaving monitoring should include assurance that appropriate levels of heat-labile constituents within feed are at suitable levels. Access to mass spectroscopy and other technologies is available to evaluate bedding for contaminants if contamination is expected.

Irradiated feed is available in a variety of packaging. To provide assurance that a particular bag or package of diet has been irradiated, the bags or packages are individually marked with indicator labels using polyvinyl chloride (PVC) impregnated with an acid-sensitive dye. Irradiation results in a color change caused by the release of hydrochloric acid within the PVC label (Tobin et al. 2007). The outer paper packaging of irradiated diets can be exposed to contaminants during shipping and storage and therefore should be sprayed with a disinfectant solution and/or carefully removed to reveal an inner plastic bag. The sealed inner plastic bag also permits decontamination with a liquid or spray disinfectant. This is generally done using a spray bottle, or alternatively, there are automated misting chambers available. Irradiated diet can also be obtained in small, watertight, vacuum-sealed plastic bags for use in isolators, change stations, or BSCs. Compromised packages are easily identified, as the vacuum seal is lost and the bag inflates. The use of vaporized hydrogen peroxide, as a method to decontaminate the external surface of feed bags, even those with an inner plastic bag, must be carefully considered, as the hydrogen peroxide gas may penetrate the feed (author's [N.S.L.] personal communication). Subjecting irradiated feed bags to an extremely short "flash" sterilization cycle has been employed as a method for surface decontamination prior to relocation of feed into a rodent barrier (Thurlow et al. 2007). Feed for nonrodent species is not typically irradiated or available in a formulation that permits autoclaving. These bags are generally surface decontaminated with a liquid disinfectant or sterilant prior to, or upon, entry into the facility. It may be necessary to ensure biosecurity that feed used for nonrodent species be stored and handled separately from feed used for barrier-maintained rodents.

Bedding comes in a variety of sizes and packaging and is generally packaged by cubic feet, as opposed to weight, and in autoclavable paper bags. Bulk totes of bedding can be procured for larger facilities to reduce handling by staff when used with a pneumatic dispensing system. Irradiated bedding products generally receive a minimum dose of 15 kGy, and each bag is marked with an indicator label confirming exposure. Smaller bags are vacuumed packaged and shipped in protective cartons. Certified bedding is packaged in lot-coded, certified stamped bags and arrives with a corresponding laboratory analysis report. Bedding is typically autoclaved into CCW or gamma-irradiated and surface decontaminated using a method similar to that used to process feed. During autoclaving, bedding can absorb moisture and, as a result, loose absorbency and support the growth of microorganisms. Therefore, if autoclaved, sufficient time should be provided for drying before stacking, or alternatively, gamma-irradiated bedding may be used.

Once treated, caging may be filled with bedding manually, for example, with a scoop, or using an automatic dispenser. Several types of automated dispensers are available. The most accurate is a dispenser into which cages are inserted and a metered amount of bedding is released, using a foot-activated or automated switch, into the cages (Figure 27.3). This dispenser type is frequently employed with articulated-arm or foundry robotic cage washing systems or in facilities processing limited numbers of rodent cages. In-line bedding dispensers are also available for tunnel washers (Figure 27.4). These dispensers continuously "rain" bedding from above. Cages are flipped as they exit the unload end of the washer and are subsequently filled as they pass under the dispenser. The amount of bedding dispensed is dependent on both the speed of the belt on which the cage moves and the volume of bedding continuously released from the dispenser. In-line dispensers are useful when processing large numbers of rodent cages. Dispensers can be filled manually, or automated delivery systems are available to fill the unit to reduce the labor-intensive task of handling the bedding. The latter is particularly valuable if the point of receipt, that is, loading dock, is a considerable distance away from the cage wash. Regardless of the method of distribution, given the significance of exposure to particulate, consideration must be given, in consultation with an occupational health expert, to the use of respiratory protection when handling loose bedding.

The flow of feed and bedding from the CCW or another postprocessing distribution point facility to the AHR, whether in cages or other containers, is also of critical importance. Some facilities steam sterilize all materials that come in contact with laboratory rodents, including all cage components, before leaving the CCW for distribution in order to ensure unwanted infectious agents are excluded. Other facilities limit treatment to feed and bedding (such as sterilization), as they pose the more likely risk of introducing excluded agents. In single-corridor facilities, the use of reusable or disposable equipment covers during transport is highly recommended to reduce the likelihood of cross-contamination during the transport of both clean and soiled equipment containing feed and bedding.

The SCW is the repository of all returning and potentially contaminated material. Therefore, procedures that ensure adequate containment are necessary. The potential for contamination resulting from an infectious disease outbreak is omnipresent, and caging containing soiled bedding from contaminated animals is commonly processed along with those from "clean" colonies. Soiled material sterilization is generally limited to situations when a known contaminant is present. Given the significance of animal allergens, prevention or minimization of exposure should be of high priority during the handling and disposal of bedding (Harrison 2001). A variety of solutions are available for this purpose, including, but not limited to, dust masks, N95 particulate respirators, Class I biological safety cabinets, and pneumatic waste disposal systems with integrated downdraft stations.

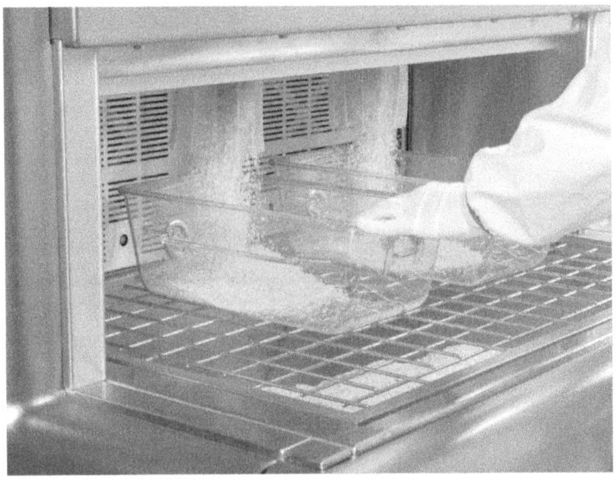

FIGURE 27.3 Automatic bedding dispenser. (Courtesy of Tecniplast.)

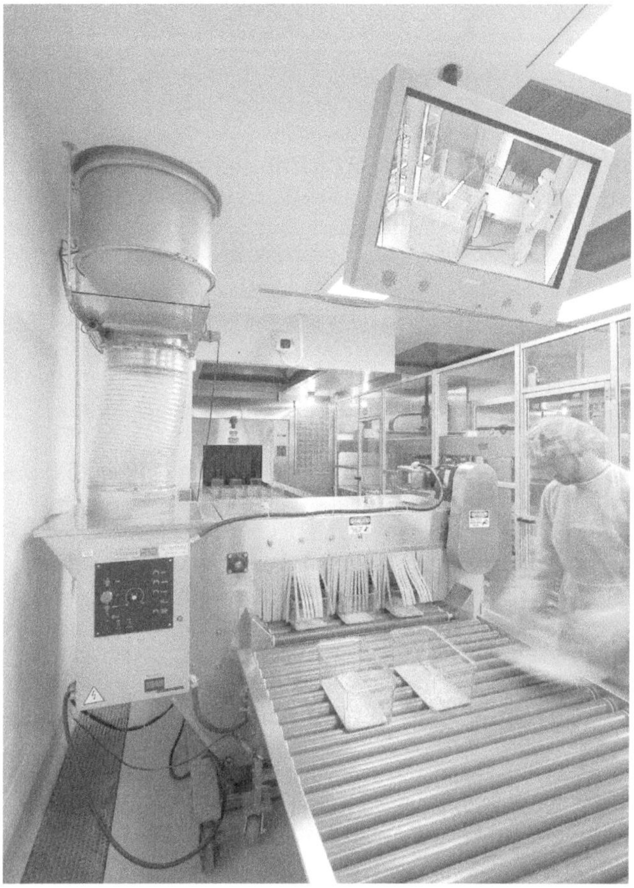

FIGURE 27.4 In-line bedding dispenser with integrated pneumatic delivery system. (Courtesy of MSKCC.)

There are several systems marketed to the laboratory animal industry for soiled bedding disposal. Systems differ based on whether the bedding is altered by adding water before transport and disposal. In dry systems, bedding is dumped into a collecting receptacle and transported by vacuum or auger to a waste receptacle, generally at a loading dock, for subsequent removal. The receptacles are commonly outfitted with a shredder or grinder to macerate larger materials that may be inadvertently introduced into the system. There are also several systems that convert bedding and cellulose materials, such as cage board, into a slurry by pulverizing them and adding water. The material can then be flushed into the sanitary waste system, if code permits, or the water can be extracted and the waste, which is reduced in volume, is containerized for disposal. The collecting receptacle used in all these systems is generally provided with downdraft, for personnel protection against aerosols. Foundry, or articulated-arm, robots can be used with a tunnel washer for waste disposal. On the soiled side of the cage wash, the robot's gripper picks up the uppermost row of soiled cages, which have been nested on specially designed pallets. The robot then dumps the cages and places them on the tunnel washer belt. After washing, the robot on the clean side collects the cages from the washer belt, inserts them into an automated bedding dispenser, and subsequently stacks the bedded cages onto pallets.

Feed and Bedding Receipt and Storage

Upon receipt, bags of feed and bedding should be checked for milling and expiration dates and examined to ensure that they are intact and unstained, demonstrating that their contents have not been damaged, potentially exposed to vermin, penetrated by liquids, or otherwise contaminated. Bags with defects should be rejected because of the possibility of contamination.

Inherent risks for feed and/or bedding storage are nutrient degradation, contamination, spoilage, and vermin infestation. It is important to know the manufacture date and other factors that affect the shelf life of a diet. Stale food or food transported or stored inappropriately can become deficient in nutrients. Stock should be rotated regularly upon receipt of each shipment to ensure that the oldest material is used first and to ensure that the freshest is always available. Feed and bedding should be stored off the floor on pallets, racks, or carts in a manner that facilitates sanitation and visual confirmation of the manufacture date. Placing materials a minimum of 6 inches away from the wall will help protect them against pests, permit air circulation, and allow for cleaning behind and under the stored material. Opened feed and bedding bags should be stored in vermin-proof containers, such as plastic lined cans with tight-fitting lids, to minimize contamination and avoid the potential spread of pathogens. Containers should be cleaned regularly and new liners placed before adding new material, with the expiration date of any materials marked on the container. Chemicals should not be stored in the same room as feed and bedding because of the possibility that bags could be contaminated by the chemicals.

Feed should be maintained under cool, dry conditions and in a well-ventilated area not exposed to direct sunlight. A constant cool temperature and low humidity in storage areas help to avoid spoilage. Daily minimum and maximum temperatures and humidity values should be closely monitored to ensure appropriate environmental conditions where feed and bedding are stored. Natural-ingredient diets should be stored at temperatures less than 70°F, and relative humidities of ~50% are ideal. Exposure of feed to elevated temperatures induces rancidity, in which unsaturated fats and lipids are oxidized and converted into hydroperoxides, which subsequently break down into volatile aldehydes, esters, alcohols, ketones, and hydrocarbons, giving the feed a disagreeable odor and taste. Most natural-ingredient, dry laboratory animal diets stored properly can be used up to 6 months, or longer in some cases, after manufacture. Specialty diets may have a shorter shelf life. Specialty diets, such as purified diets, containing high fat levels are subject to spoilage and must be kept in a 4°C cooler to extend the shelf life. Nonstabilized vitamin C in manufactured feeds generally has a shelf life of only 3 months, but commonly used stabilized forms can extend the shelf life of the diet to 6 months. Other perishable items, such as fruit and vegetables, used for larger species, should be stored in appropriate containers and kept at a lower temperature to avoid spoilage. Live food sources need to be maintained and managed to ensure a steady supply and the health and suitability of the organism as a food and should be stored in a type-appropriate manner to preserve nutritional content, minimize contamination, and prevent entry of pests. The shelf life for live food sources, such as brine shrimp eggs, or commercial (pellet or flake) diets for aquatics species should be determined based on manufacture recommendations or follow commonly accepted practices (NRC 2011). Unopened canned feed can typically be stored for up to 2 years or longer. General feed and bedding inventory quantities should be maintained such that they may be rotated into circulation and consumed before the set expiration date (for feed) but have sufficient on-hand stock in case of a delivery delay. It is important to recognize that some manufacturers ship directly to the end user, while others ship to distributors, who then send feed and bedding to the end user.

Pests such as cockroaches and wild rodents are potential carriers of disease-causing agents and should be prevented from any contact with materials used in the laboratory setting. Whether in the plant (vendor), distributor's warehouse, or the animal facility, a stringent pest control program should be in place with regular inspections forwarded for review and archive by the end user, for example, the facility manager or attending veterinarian. All storage locations should be vermin-proof. Walls and floors must be free of cracks and crevices. Pipelines, drains, and air filters should be well sealed and inspected frequently. Newly received supplies should be thoroughly inspected for evidence of pests. Inside a facility, pests seek food, water, and the protection of dark areas for shelter and breeding. In addition to using metal or plastic dunnage racks, carts or shelves should be used to keep materials off the ground and away from the walls, doors and entrances should have rodent guards or door sweeps to prevent entry of vermin, and insect screens may be used in specific locations. Movable equipment, cage racks, shelves, and drawers must be routinely moved and cleaned. Humane mechanical traps should be placed on the floor against walls, which is where rodents commonly travel, and insect pheromone traps at key locations throughout the area. Traps that catch pests live require frequent observation. Electrical insect trap lamps may be used to monitor insect activity, but should not be used as a primary method of control. Cleanliness and good housekeeping are the primary methods of vermin control for feed and bedding storage locations. If a facility is infested with insects, pesticides are undesirable, but may be necessary, in which case feed and bedding would need to be protected. Inappropriate use of pesticides in an animal facility can be dangerous to personnel and experimental animals. Exposure of research animals to pesticides can compromise research data. The introduction of pesticides makes controlling experimental variables more difficult—hence the importance of vermin-proofing storage areas and cleanliness prior to relying on traps or other means of ridding an area of pests.

Regular monitoring of the pest control program with inspections is a must for plant (vendor), distributor, and animal facility locations. Inspections should give assurance that the pest control program is being executed as designed and to monitor vermin activity levels, and action should be taken if deficiencies are found. Reports from the plant and distributor locations should be shared regularly with animal facility management.

Conclusion

Feed and bedding are the single most important commodities used in the maintenance of laboratory animals. They are also among the primary factors contributing to variables in the animals' environment and have the potential to impact the research studies in which the animals are utilized. Understanding the types of feed and bedding available and selecting those appropriate for the research to be conducted are critical decisions made by facility management and veterinarians, in concert with research staff. Ensuring that the procedures for receipt, storage, processing, and distribution of feed and bedding are well defined is an essential management responsibility to ensure animal health and welfare, and ultimately contribute to attaining high-quality research data.

REFERENCES

Adams, N.R. 1995. Detection of the effects of phytoestrogens on sheep and cattle. *J Anim Sci* 73:1509–1515.

Allred, C., Allred, K., Young, H. et al. 2001. Soy diets containing varying amounts of genistein stimulate growth of estrogen dependent (MCF-7) tumors in a dose-dependent manner. *Cancer Res* 61:5045–5050.

Baker, D.A., and Lipman, N.S. 2015. Factors that can influence animal research. In *Laboratory Animal Medicine*, ed. J.G. Fox, L.C. Anderson, G. Otto, K.R. Pritchett-Corning, and M.T. Whary, 1441–1496. 3rd ed. Oxford: Academic Press.

Barnard, D.E., Lewis, S.M., Teter, B.B., and Thigpen, J.E. 2009. Open- and closed-formula laboratory animal diets and their importance to research. *J Am Assoc Lab Anim Sci* 48:709–713.

Blom, H.J.M., Van Tintelen, G., and Van, V. 1996. Preferences of mice and rats for types of bedding material. *Lab Anim* 30:234–244.

Bollen, P., and Ritskes-Hoiinga, M. 2007. The welfare of pigs and minipigs. In *The Welfare of Laboratory Animals*, ed. E. Kaliste, 275–289. Vol. 2. New York: Springer Publishing.

Caulfield, C.D., Cassidy, J.P., and Kelly, J.P. 2008. Effects of gamma irradiation and pasteurization on the nutritive composition of commercially available animal diets. *J Am Assoc Lab Anim Sci* 47:61–66.

CFR (Code of Federal Regulations). 2004. Good laboratory practice regulations. 21 CFR §58.105-58.113.

Coghlan, L.G., Lee, D.R., Psencik, B., and Weiss, D. 1993. Practical and effective eradication of pinworms (*Syphacia muris*) in rats by use of fenbendazole. *Lab Anim Sci* 43:481–487.

Cover, C., and Belcher, L. 1992. Effect of an irradiated rodent diet on growth and food consumption: A comparative study. *Contemp Top Lab Anim Sci* 31:13–17.

EU (European Union). 2004. Directive 2004/10/EC of the European Parliament and of the council of 11 February 2004 on the harmonisation of laws, regulations and administrative provisions relating to the application of the principles of good laboratory practice and the verification of their applications for tests on chemical substances. *Off J Eur Union* L50:44–59.

Eva, J.K., and Rickett, M.J. 1983. Improvements in diet stability through processing. *Lab Anim* 17:1–6.

Faith, R.E., and Hessler, J.R. 2006. Housing and environment. In *The Laboratory Rat*, ed. M.A. Suckow, S.H. Weisbroth, and C.L. Franklin, 303–337. 2nd ed. Oxford: Academic Press.

Ferguson, H. 1966. Effect of red cedar chip bedding on hexobarbital and pentobarbital sleep time. *J Pharm Sci* 55:1142–1148.

Ford, D. 1987. Nutrition and feeding. In *The UFAW Handbook on the Care and Management of Laboratory Animals*, ed. T.B. Poole, 50–53. 6th ed. Essex, UK: Longman Scientific and Technical.

Fox, J.G., and Newberne, P.M. 1980. Nutritional adequacy and quality control of rodent diets. *Lab Anim Sci* 30:352–363.

Halls, N.A., and Tallentire, A. 1978. Effects of processing and gamma irradiation on the microbiological contaminants of a laboratory animal diet. *Lab Anim* 12:5–10.

Harrison, D.J. 2001. Controlling exposure to laboratory animal allergens. *ILAR J* 42:17–36.

Inoue, Y., Izawa, K., Kiriyu, S., Tojo, A., and Ohtomo, K. 2008. Diet and abdominal autofluorescence detected by in vivo fluorescence imaging of living mice. *Mol Imaging* 7:21–27.

Ju, Y., Allred, C., Allred, K. et al. 2001. Physiological concentrations of dietary genistein dose-dependently stimulate growth of estrogen-dependent human breast cancer (MCF-7) tumors implanted in athymic nude mice. *J Nutr* 131:2957–2962.

Ju, Y., Doerge, D., Allred, K. et al. 2002. Dietary genistein negates the inhibitory effect of tamoxifen on growth of estrogen-dependent human breast cancer (MCF-7) cells implanted in athymic mice. *Cancer Res* 62:2474–2477.

Knapka, J.J. 1997. Natural-ingredient diets: Managing the variation in dietary nutrient concentrations. *Lab Anim* 26:40–42.

Lawrence, C., Best, J., James, A., and Maloney, K. 2012. The effects of feeding frequency on growth and reproduction in zebrafish (*Danio rerio*). *Aquaculture* 368–369:103–108.

Lawson, K.A., Anderson, K., Menchaca, M. et al. 2003. Novel vitamin E analogue decreases syngeneic mouse mammary tumor burden and reduces lung metastasis. *Mol Cancer Ther* 2:437–444.

Leiter, L.M. 2009. Phytoestrogens—Uninvited, troublesome guests in scientific research. New Brunswick, NJ: Research Diets, Inc. http://www.researchdiets.com/system/resources/BAhbBlsHOgZmIi8yMDEyLzA0LzIwLzEzXzU2XzU0XzU5M19QaHl0b2VzdHJvZ2Vucy5wZGY/Phytoestrogens.pdf.

Lewandoski, M. 2001. Conditional control of gene expression in the mouse. *Nat Rev Genet* 10:743–755.

Lipman, N.S. 2007. Design and management of research facilities for mice. In *The Mouse in Biomedical Research*, Vol. III, *Normative Biology, Husbandry, and Model*, ed. J.G. Fox, M.T. Davisson, F.W. Quimby, S.W. Barthold, C.E. Newcomer, and A.L. Smith, 271–319. 2nd ed. Oxford: Academic Press.

NRC (National Research Council). 1977. *Nutrient Requirements of Rabbits*. 2nd rev. ed. Washington, DC: National Academy Press.

NRC (National Research Council). 1982. *Nutrient Requirements of Mink and Foxes*. 2nd rev. ed. Washington, DC: National Academy Press.

NRC (National Research Council). 1993. *Nutrient Requirements of Fish*. Washington, DC: National Academy Press.

NRC (National Research Council). 1994. *Nutrient Requirements of Poultry*. 9th rev. ed. Washington, DC: National Academy Press.

NRC (National Research Council). 1995. *Nutrient Requirements of Laboratory Animals*. 4th rev. ed. Washington, DC: National Academy Press.

NRC (National Research Council). 1998. *Nutrient Requirements of Swine*. 10th rev. ed. Washington, DC: National Academy Press.

NRC (National Research Council). 2000. *Nutrient Requirements of Beef Cattle*. 7th rev. ed. Washington, DC: National Academy Press.

NRC (National Research Council). 2001. *Nutrient Requirements of Dairy Cattle*. 7th rev. ed. Washington, DC: National Academy Press.

NRC (National Research Council). 2003a. *Nutrient Requirements of Nonhuman Primates*. 2nd rev. ed. Washington, DC: National Academies Press.

NRC (National Research Council). 2003b. *Guidelines for the Care and Use of Mammals in Neuroscience and Behavioral Research*. Washington, DC: National Academies Press.

NRC (National Research Council). 2006a. *Nutrient Requirements of Dogs and Cats*. Washington, DC: National Academies Press.

NRC (National Research Council). 2006b. *Nutrient Requirements of Horses*. 6th rev. ed. Washington, DC: National Academies Press.

NRC (National Research Council). 2007. *Nutrient Requirements of Small Ruminants: Sheep, Goats, Cervids, and New World Camelids*. Washington, DC: National Academies Press.

NRC (National Research Council). 2011. *Guide for the Care and Use of Laboratory Animals*. Washington, DC: National Academy of Sciences.

OECD (Organisation for Economic Co-operation and Development). 2011. OECD series on principles of good laboratory practice (GLP) and compliance monitoring. Paris: OECD. http://www.oecd.org/document/63/0,3746,en_2649_34381_2346175_1_1_1_1,00.html (accessed September 22, 2016).

Perkins, S.E., and Lipman, N.S. 1995. Characterization and qualification of microenvironmental contaminants in isolator cages with a variety of contact beddings. *Contemp Top Lab Anim Sci* 34:93–98.

Ryding, A.D., Sharp, M.G., and Mullins, J.J. 2001. Conditional transgenic technologies. *J Endocrinol* 171:1–14.

Smith, E., Stockwell, J.D., Schweitzer, I., Langley, S.H., and Smith, A.L. 2004. Evaluation of cage microenvironment of mice housed on various types of bedding materials. *Contemp Top Lab Anim Sci* 43:12–17.

Thigpen, J., Haseman, J., Saunders, H. et al. 2002. Dietary factors affecting uterine weights of immature CD-1 mice used in uterotrophic bioassays. *Cancer Detect Prev* 26:381–393.

Thigpen, J., Haseman, J., Saunders, H. et al. 2003. Dietary phytoestrogens accelerate the time of vaginal opening in immature CD-1 mice. *Comp Med* 53:477–485.

Thigpen, J., Locklear, J., Haseman, J. et al. 2001. Effects of the dietary phytoestrogens daidzein and genistein on the incidence of vulvar carcinomas in 129/J mice. *Cancer Detect Prev* 25:527–532.

Thurlow, R., Arriola, R., Soll, C., and Lipman, N.S. 2007. Evaluation of a flash disinfection process for surface decontamination of gamma-irradiated feed packaging. *J Am Assoc Lab Anim Sci* 46:46–49.

Tobin, G., Stevens, K.A., Russell, R.J. 2007. Nutrition. In *The Mouse in Biomedical Research*, Vol. III, *Normative Biology, Husbandry, and Model*, ed. J.G. Fox, M.T. Davisson, F.W. Quimby, S.W. Barthold, C.E. Newcomer, and A.L. Smith, 321–383. 2nd ed. Oxford: Academic Press.

Van de Weerd, H., Van Loo, P., Van Zutphen, L. et al. 1997. Preferences for nesting material as environmental enrichment for laboratory mice. *Lab Anim* 31:133–143.

Vessel, E. 1966. Induction of drug-metabolizing enzymes in liver microsomes of rats and mice by softwood bedding. *Science* 157:1057–1058.

White, W.J., Hawk, C.T., and Vasbinder, M.A. 2008. The use of laboratory animals in toxicology research. In *Principles and Methods in Toxicology*, ed. A.W. Hays, 1055–1101. 5th ed. Boca Raton, FL: CRC Press.

Wood-Gush, D.G.M., Vestergaard, K., and Petersen, H.V. 1991. The seeking of novelty and its relation to play. *Anim Behav* 42:599–606.

28

Water Quality and Water Delivery Systems

E. Douglas Allen, Edgar F. Czarra, and Louis DeTolla

CONTENTS

Introduction

Water is essential to sustain an animal's physiological and biochemical processes, and therefore necessary to sustain life. Water can be found within all tissues and cells and is the intracellular and extracellular medium where physiologic processes are carried out.

It is important to recognize the significance of the relationship of water to the health and well-being of laboratory animals. It is equally important to recognize the relevance of water quality and the reliability of the systems used to deliver water to laboratory animals used in biomedical research. Providing high-quality water to research animals will help to minimize experimental variables, and can have a positive effect on overall animal health. This chapter is intended to provide some relevant background information on water sources, the importance of water quality, and some guidance to animal resource management and animal care personnel on drinking water treatment and delivery systems. As water quality serves as the microenvironment for aquatic species, refer to Chapter 24 for additional information.

Sources of Water

Domestic water supplies originate from either surface water or groundwater. Surface water originates from streams, rivers, lakes, and reservoirs. Groundwater originates from surface water, which penetrates the earth's crust, is collected in aquifers, and is accessed through community water system wells. Approximately 70% of community water system users derive their drinking water from surface water sources, leaving the balance derived from ground sources (USEPA 2012).

Source water contaminants vary in concentration and generally fall under the following classifications: heavy metals, organic and inorganic chemicals, pesticides and herbicides, pathogens, and radionuclides. To address this issue, the U.S. Congress passed the Safe Drinking Water Act (SDWA) in 1974 (amended in 1986 and 1996). Per the Act, the SDWA is administered by the U.S. Environmental Protection Agency (EPA), and it was enacted to protect drinking water and its sources. The SDWA established the minimum quality standards for domestic water supplies. The SDWA authorizes the EPA to establish national health-based standards for drinking water and is applicable to all public water supplies.

The SDWA regulations call for meeting mandatory maximum contaminant levels (MCLs) and nonenforceable maximum contamination-level goals (MCGLs) (USEPA 2014, 2016b). There are six groups of standards for drinking water: microorganisms, disinfectants, disinfection by-products, inorganic chemicals, organic chemicals, and radionuclides. Municipalities are required annually to produce and provide to consumers the Consumer Confidence Report (CCR) under the EPA Consumer Confidence Rule, which may be found through the EPA's website (USEPA 2015) or your local water supplier.

The composition of source water, whether it is from surface water or groundwater sources, can vary considerably. The factors attributed to surface water and groundwater quality changes are complex in nature and vary from region to region and depending on the source of the drinking water. Anthropogenic, atmospheric deposition; seasonal changes; municipal and industrial discharge; brownfield sites (USEPA 2016a); urban-related runoff; and other major pollution sources have all been identified as contributors to water quality degradation. In 1991, the U.S. Geological Survey implemented the National Water-Quality Assessment Program (USGS 2014) to evaluate the condition of U.S. streams, rivers, groundwater, and aquatic systems to provide scientific information to national, regional, state, and local resource managers and policy managers for making sound decisions about the management of their water supply resources.

Municipal Water Treatment and Distribution

Drinking water providers treat source water in a variety of ways to produce a potable product before distribution through the municipal supply. Water treatment plants remove suspended particles and unwanted contaminants, and then filter and disinfect the source water to produce an acceptable product. Activated carbon filtration may be added to the filtration spectrum in order to remove odors, improve palatability, or remove chemical contaminants. The pH may be adjusted to reduce water distribution pipe corrosion and leaching of heavy metals (e.g., lead and copper) from the supply distribution system and to reduce the formation of alkaline metals (e.g., Ca^{2+} and Mg^{2+}), carbonate deposits that may lead to premature distribution system failure and/or decline in water quality. An additional final filtration step may be used to remove suspended particles and organic matter created during the treatment process. The final step in water treatment is disinfection of the water supply. Introduction of ozone, chlorine compounds, or potassium permanganate are among various means to initiate disinfection before distribution to water storage facilities and the municipal distribution system (MDS).

The MDS is composed of a network of pipes fabricated from a wide variety of materials. Cast iron, ductile and coated ductile iron, concrete and prestressed concrete, galvanized steel, polyvinyl chloride (PVC) and chlorinated polyvinyl chloride (CPVC), copper, and other materials are typically used. PVC compounds may contain unbound phthalates, a chemical used to make PVC and CPVC softer and more flexible. However, phthalates are known to be endocrine disruptors, mimicking naturally occurring hormones that can interfere with the endocrine system and produce adverse reproductive effects and developmental abnormalities (Colborn 2004; Mathieu-Denoncourt et al. 2015). The MDS transports water to the point of consumption by the end user. During transport, the treated municipal water supply is susceptible to uncontrolled chemical and biological reactors, which can contribute to water quality variability and an undesirable product. Although the conduit materials used in the MDS are durable and long lasting, they can also contribute to a decline in water quality. Iron corrosion and leaching of lead and copper from pipe walls and joints (Tchounwou et al. 2012) is a common water quality threat from older municipal systems. Other aspects of the MDS that can compromise water quality include the gradual decrease in residual disinfectants, bacterial regrowth and colonization, disruption of the water supply from water main failures, and formation of toxic disinfectant by-products (USEPA 2013).

Water Quality

Although the potable product delivered to research facilities may be of sufficient quality for human consumption, it may not be of adequate quality for laboratory animal consumption. The presence of disinfection by-products in sufficient concentrations can adversely affect animal health, by introducing undesirable and unintended variables that can confound animal research outcomes. Furthermore, compromised distribution systems can introduce other contaminants that can have the same deleterious effects. Microbial contamination (e.g., *Legionella* sp., *Giardia* sp., and *Cryptosporidium* sp.) of municipal supplies (Beer et al. 2015) are well known and may lead to drinking water–associated disease in humans. Microbial contamination of municipal water supplies also represents a significant health concern for laboratory animals.

From a regulatory perspective, water quality guidance and standards are defined and outlined in the animal welfare regulations (9 CFR §3) and the *Guide for the Care and Use of Laboratory Animals* (National Research Council 2011). For institutions conducting nonclinical studies under the Food and Drug Administration's good laboratory practices (21 CFR§58.90(g)), routine and periodic water quality assessment is required. The purpose of this regulatory requirement is to ensure that contaminants known to be capable of interfering with a study are not at or above levels that can interfere with and adversely affect the outcome of the study.

Considering the complex and variable nature of our water supply, evaluation of source water quality is a critical step in determining whether the water delivered to laboratory animals is of adequate quality and whether additional treatments are needed. Evaluation of CCRs from the water-supplying municipality may be considered a baseline assessment of water quality. However, additional facility-by-facility evaluation at the distribution point may also provide additional information on water quality and should be used for consideration of site-specific treatment options. The combined water quality information obtained through assessment should be jointly reviewed by animal resource administrators, laboratory animal veterinarians, and key representatives of the research community to formulate what treatment options are required to meet the institution's research objectives in order to mitigate the inherent risks associated with source water quality variability. There are a variety of water treatment options and water delivery systems available to laboratory animal research facilities (LARFs) housing research animals. In this chapter, the advantages and disadvantages of each of these systems are described in greater detail in order to provide the reader with the information necessary to select an animal drinking water system that is most appropriate based on the research needs of the facility, species and strain sensitivity, facility infrastructure, and facility cost constraints.

Water Treatment

There are various methodologies available to purify and treat water, with each solution having possible pros and cons dependent on a LARF's needs. These methodologies can be acquired and used as stand-alone treatment systems or can be combined for point of use, floor by floor, or building by building. Water purification systems are designed to remove particulates, as well as chemical and biological contaminants from water. Mechanical purification methods most commonly used to treat animal drinking water include filtration, irradiation, and steam sterilization. Filtration systems vary considerably in their design and function, with reverse osmosis (RO) systems producing a product that is far more pure—devoid of particulate and chemical and biological contaminants—than other filtration systems. Irradiation and steam sterilization are means of addressing biological contaminants but have no effect on the removal of chemical contaminants and are therefore likely to be used as an adjunct water treatment option. The same can be said about acidification and chlorination as a chemical approach to addressing biological contamination. Although chlorination and acidification are effective water disinfection options widely used in research facilities, the addition of chemical additives may be contraindicated for certain types of research. In addition, reactivity with certain water contaminants may introduce toxic disinfectant by-products that can be detrimental to animal health and introduce undesirable research variables (Komulainen 2004; Richardson et al. 2007). A "one-size-fits-all" approach may be applicable to certain specific types of research activities but not applicable to others. It is important to gain a full understanding of water treatment options and water delivery systems, in order to engage principal investigators on the type of research their laboratories are carrying out and discuss what water treatment options are available and what solution is most appropriate to meet their specific needs. Consultation with the animal breeder or source may yield valuable information with regard to species or strain sensitivity to water treatment options. Water treatment systems used in research animal facilities are discussed next.

Media Filtration

Media-type filters are commonly used in LARFs as either stand-alone systems or pretreatment modules in more complex water treatment systems. There are many types of media filters used. The number and type of filters to use is dependent on the volume of water to be treated, whether water must be filtered continuously or in batches, the amount and sizes of suspended particles in the source water, the concentration of organic compounds, and the acceptable amount of contaminants in the treated water. A comprehensive water analysis (Curran and Smart 2007) should be conducted on the facility's domestic water supply in order to determine the need for media filtration. Once the need has been determined, a water quality consultant or engineer can help determine the type of media filtration that would be best suited for removing particular water contaminants of concern. A water quality consultant or engineer will also be able to assist with the sizing of the media filtration and the correct order that the domestic water supply should pass through the media filtration units in order to achieve the facility's water quality goal.

Media filters are made from a wide variety of materials and can filter gross particulates from the domestic water supply down to particles as small as 0.1 μ. Media filters are designed to entrap suspended particles as the water passes through the media. Much like high-efficiency particulate air (HEPA) filters, as the media loads with particles, the flow of water through the media surface becomes restricted, resulting in a reduced flow rate and a reduced volume of treated water produced over time. With cartridge media filters, the filter membrane can be easily removed from the filter housing and replaced with a new cartridge. Sand filters, which are designed for the removal of gross particulates from the domestic water supply, require periodic back-flushing in order to maintain filtration efficacy. The periodic back-flushing, which is normally accomplished automatically with a timer–valve assembly, flushes the particulates from the sand media to a drain.

Activated carbon filtration is another type of media filtration that is commonly used to treat domestic water supplies. Activated carbon filtration is specifically designed to remove organic compounds. Activated carbon filtration differs from sand or cartridge-type filtration by utilizing the adsorptive properties of carbon rather than the particle entrapment process utilized with sand or cartridge-type filtration processes. Through the adsorptive process, organic compounds that are dissolved in the domestic water supply are attracted to the surface of the activated carbon particles and thereby removed from solution. There are many factors that influence the efficiency of this process, including the carbon particle size and pore structure, the amount of actual time that the incoming water is exposed to the carbon media, the incoming water flow rate, and the amount of activated carbon remaining in the filter media housing. Once carbon adsorption sites are occupied by contaminants removed from the incoming water supply, the ability of activated carbon to adsorb additional organic compounds is lost, resulting in "breakthrough." Breakthrough is the process by which previously bound organic contaminants are released back into the water by the activated carbon media. Once the breakthrough point has been reached, the filtering capacity of the activated carbon filtration media will be less than optimum (Dvorak and Skipton 2013). Similar to sand media filtration, periodic back-flushing to remove captured contaminants from the activated carbon media or replacement of the carbon media is required in order to maintain adequate filtration of dissolved organic compounds from the domestic water supply.

Reverse Osmosis and Ultrafiltration

In order to reduce the potential for experimental variables in research protocols, many LARFs have invested in water purification systems that reduce or eliminate the risk of using domestic water supplies that may have contaminants remaining in the drinking water after having been treated with simple media filtration. RO and ultrafiltration (UF) are two water purification methods that are commonly used to purify animal drinking water destined for laboratory animal consumption.

RO is a process that utilizes high-feed water pressure to overcome osmotic pressure and produce a product that is predominantly free of low-molecular-weight contaminants. RO membranes are capable of removing contaminants larger than 0.001 μ, including monovalent and multivalent ions, pyrogens, proteins, colloidal substances, and microorganisms. In order to understand how RO systems work, it is worthwhile understanding the theory of RO, the design and function of the RO membrane, and the related components that produce a purified water product that is suitable for consumption by laboratory animals.

The RO membrane is the primary element of an RO system. The RO membrane is semipermeable, allowing purified water to pass through the membrane when high pressure is applied to the incoming water. The purified water that passes through the semipermeable membrane and is suitable for laboratory animal consumption is aptly named the "permeate," whereas the water that remains on the inlet side of the membrane and is highly concentrated with dissolved ions and precipitated heavy metals is aptly called the "concentrate." There are several different RO membranes from which to choose. The choice should be based on the water quality goals of the LARF, the water quality test of the domestic water supply, and the operating environment of the RO system. The materials commonly used to make RO membranes include cellulose acetate, polyamide, polyether sulfone, polyacrylonitrile, and polyvinylidiene fluoride. Polyamide and cellulose acetate are the most common membrane materials used in laboratory animal drinking water applications. Polyamide membranes have a slightly higher salt rejection rate than cellulose acetate membranes. However, polyamide membranes will degrade and become ineffective at removing water contaminants if exposed to chlorine. If polyamide membranes are selected, chlorine must be removed from the domestic water supply prior to the RO process, and should be injected into the RO permeate water after the membrane of the LARF is interested in affording protection against possible biofilm formation.

RO membranes are very thin (≈0.2 μ), with a pore size as small as 0.0001 μ. These membranes are supported by a polymeric microporous thermoplastic layer that adds strength and protects the integrity of the membrane. The RO membrane is composed of several of these composite layers, which increases the filtration surface area available for the feed water.

Over the years, there have been many advances in the design of RO membrane modules. Today, RO membranes are available as hollow fiber modules, spiral wound modules, and plate or frame modules. Each module design has application-specific characteristics. Modules are assembled in special pressure vessels of various dimensions, diameters, and lengths and fitted with feed water, permeate, and rejection water fittings. Earlier RO membrane designs recovered less than 10% of the water used. With continued advances in RO membrane technology and the ability to interconnect multiple RO membrane modules in series, significantly higher recovery rates of greater than 90% (Stoughton et al. 2013) can be achieved while providing higher flow rates and flux to meet any demand.

Although a high capital investment, RO systems offer animal resources the distinct advantage of providing a consistent and pure source of drinking water regardless of changes in supply water quality. Commercial RO systems are available in a variety of different configurations and options that can provide variable permeate flow rates, percent recovery, and percent rejection. Depending on the specific needs of the institution, RO systems can be sized as a single centralized site plant with distribution to all campus animal resource centers or designed to handle the specific needs of a dedicated point of use.

UF can provide a cost-effective alternative to RO water purification. Similar to RO purification, UF has the capability to remove microorganisms, particulates, and macromolecules that are greater than 0.01 μ. Unlike RO membranes, UF membranes do not have the ability to remove low-molecular-weight organics and elemental ions such as calcium and sodium. UF membranes have greater pore sizes than their RO counterparts, and therefore require less applied pressure to overcome osmotic pressure. The lower applied feed water pressure, in combination with the greater pore size of the UF membrane, translates into a higher flux rate (>90%) of purified water (Pilutti and Nemeth 2003).

The domestic water supply that feeds an RO or UF system will likely contain concentrations of suspended solids, dissolved ions, and organic compounds that have the potential to prematurely foul RO or UF membranes. It is important to have a water quality test performed to determine what pretreatment methods should be used to reduce these water contaminants so that maximum membrane life is achieved, and to also ensure that the pretreated supply water is compatible with the intended membrane type. As the supply feed water interfaces with the semipermeable membrane barrier, contaminants from the supply feed water settle on the membrane surface and load the various microporous membrane layers and feed water channels. This process is referred to as membrane fouling, which can negatively impact the performance of an RO or UF system (Amy 2008). Under the right conditions, dissolved salts may also precipitate from the feed water stream and accumulate on the membrane surface in the form of scale.

To circumvent these conditions and prevent premature fouling of RO and UF membranes, pretreatment of the feed water is commonly used to improve the quality of feed water. As described in the previous section, media filtration may be used to remove suspended particles, colloidal materials, and microorganisms. Activated carbon media filters will remove organic compounds and chlorine. Residual chlorine used as a disinfectant in domestic water supplies may damage certain RO or UF membrane materials (polyamide). Newer membrane technology is far more tolerant of residual chlorine (Lee et al. 2011). To protect the RO or UF membranes from scale formation, water softeners may be used to remove Ca^{2+} and Mg^{2+} ions.

Although pretreatment provides some membrane fouling protection for RO and UF systems, over time the total volume of water passing through the membrane vessel will ultimately lead to membrane fouling. Periodic back-flushing of RO or UF membranes with detergents is an additional procedure that can be utilized to extend membrane life.

Most RO and UF systems cannot produce purified water at a rate that will keep up with the peak demand experienced in most LARFs. This is particularly true in facilities that utilize water bottles or disposable water pouches. If the LARF is utilizing an automated animal drinking water system, a smaller-volume tank may be sufficient, but the storage tank will need to be adequately sized to accommodate the daily flushing of the room distribution piping and the manifold piping on the individual animal holding racks. Large purified water storage tanks allow the RO and UF systems to spread production over time, yet still meet the immediate demand for animal drinking water during peak hours of facility operation. Large purified water storage tanks also provide the LARF with a purified water reserve, which can prove essential during disaster scenarios, during periods when the RO or UF system is down for maintenance, and for other contingency planning.

Posttreatment of RO or UF purified water is essential to maintaining animal drinking water that is free from microorganisms. If uncontrolled, bacteria will begin to proliferate in the purified water medium and will attach to the surfaces of the tank and to the water distribution piping and manifolds, forming biofilms (Molk et al. 2013). There are a number of ways to proactively approach protection, including the use of low-level chlorine or acid injection after RO or UF purification. When using purified water recirculating loops as a method of distributing the purified water to bottle filling systems or automated animal drinking water systems, ultraviolet C (UV-C) radiation, in conjunction with additional microfiltration, is often used to keep bacteria from proliferating and forming biofilms within the purified water holding tanks and distribution piping. However, in this application UV-C has limitations (described later) that should be given due consideration.

Monitoring RO or UF system performance is a critical function to ensure that water is always available and consistent in quality and to ensure that system performance is within specifications. Most modern RO and UF purification systems offer continuous monitoring and notification features that will monitor key parameters of the system and proactively notify designated institutional staff when parameters are out of bounds. It also serves as a means of timely remediation of potentially significant issues without compromise to continuation of service or system integrity.

Ultraviolet Irradiation

Used in water treatment and wastewater treatment facilities to disinfect and enhance water quality, ultraviolet (UV) irradiation is also a viable solution available for disinfecting laboratory animal drinking water in LARFs. As mentioned previously, UV irradiation is typically used as a treatment method for reducing the proliferation of bacteria and biofilms in recirculating purified water loops. UV radiation is electromagnetic radiation that lies between the visible light spectrum and x-rays and typically falls between 100 and 400 nm. UV-C radiation, with a wavelength falling between 245 and 285 nm on the UV spectrum, is the most effective radiation range for inactivating microorganisms in water systems. This wavelength is produced by special low-pressure mercury vapor lamps housed within a UV-C reactor that emit electromagnetic radiation within this wavelength. The UV-generating lamps do not come in contact with the water. The lamps are either surrounded by a quartz sleeve within the lumen of the UV-C reactor or arranged around UV-C-penetrable tubes. The purified water source must constantly pass through the reactor in order for the water to receive the UV-C treatment, thereby destroying microorganisms as they come into contact with the UV-C radiation. UV-C energy does not add anything to the water, nor does it alter water chemically or remove other water contaminants (Benjamin and Lawler 2013). UV systems used in conjunction with recirculated or single-pass systems do offer the benefit of reducing microbial loads; however, due to design limitations, these systems may not totally eliminate the threat of microbial growth. Biofilm-producing colonies at any point in the recirculating loop or in the water distribution system will be unaffected by UV unless these colonies break free and pass through the UV reactor, where they will be exposed to the UV-C radiation. Therefore, additional water treatment options should be considered depending on the water quality goals of the laboratory animal research program.

UV-C energy penetrates and inactivates microorganisms by photochemically transforming UV-sensitive nucleotides, particularly pyrimidines. When exposed to UV-C radiation, the RNA or DNA transcription and replication process within the microorganism is impaired, compromising the essential cellular processes that eventually lead to the death of the microorganism. Certain microorganisms have the ability to repair the photochemical damage caused by the UV-C radiation through a process known as photoreactivation or dark repair. The ability of certain microorganisms to survive UV-C exposure is usually due to the effect of radiation doses that are too low. It has been shown that photoreactivated microorganisms show a greater resistance to UV radiation than microorganisms that have not been exposed and survived (Hijnen 2010).

There are several factors that will determine the efficacy of UV-C irradiation as a reliable method for disinfection of laboratory animal drinking water systems. These factors include the UV sensitivity of microbial species, the UV-C radiation wavelength or dose, the presence of water contaminants that absorb UV-C radiation, the presence of suspended particles that can block UV-C penetration, the age of the UV-C lamps, the physical distance between the UV-C radiation source and the microorganisms, and the presence of radiation-blocking film on the interior surfaces of the quartz sleeves or tubes. These factors must be monitored for effect, and periodic maintenance must be performed on the system to maintain optimal performance.

In general, most viruses, yeasts, and bacteria are sensitive to UV-C irradiation and require much lower UV radiation doses than molds and protozoa. *Giardia* and *Cryptosporidium*, which are typical protozoa found in water, require much higher doses of radiation for effective treatment (Chen et al. 2006).

Dosage, intensity, and exposure time are interrelated and important factors of UV system efficacy. Dosage is the product of intensity and exposure time expressed in seconds. Intensity is the output energy from the UV-C lamp and expressed as microwatts per square centimeter ($\mu W/cm^2$). Organisms passing through a UV reactor at a faster and more direct rate will receive a much lower dose than those that take a more circuitous path and slower rate.

UV-C lamp intensity does not remain constant over the life of the lamp. New lamp intensity declines rapidly at first, but the rate of decline lessens over time. As intensity declines, a point is reached where lamps should be replaced. Replacement frequency will largely be dictated by the type of lamp used and the minimum intensity determined empirically. UV-C intensity meters are available to assist in this regard. Generally speaking, low-pressure mercury-type lamps have a life of roughly 1 year.

The presence of UV-absorbing contaminants like sulfites, nitrites, phenols, humic and fulvic acid, iron, and turbidity reduces the UV radiation intensity that is available for microbiological control. Suspended matter harboring microbes may block UV energy and also reduce UV system efficacy. Contaminant deposits on the internal surfaces of the reactor or lamp sleeves also affect system efficacy. To mitigate the affect water quality has on UV systems, establishing a program of assessing intensity, cleaning UV reactors, and replacing lamps is warranted to ensure optimal performance.

UV systems used in conjunction with recirculated or single-pass animal drinking water systems do offer the benefit of reducing microbial loads; however, they may not totally eliminate them. Biofilm that may be growing downstream of the UV reactor or at any point in the recirculating loop or water distribution system will be unaffected by UV-C radiation. Bacterial colonies that break away from biofilms can be neutralized as they pass through the UV reactor; however, the inactivated bacteria will provide a rich nutrient source for biofilms growing on surfaces downstream. Additional water treatment options should be considered depending on the water quality goals of the LARF.

Acidified Water

Inorganic acid has long been used in laboratory animal drinking water systems as a means of controlling bacterial contamination. Acidified water has been particularly effective at eliminating gram-negative opportunistic pathogens like *Pseudomonas aeruginosa*, a common organism found in domestic water supplies. Although effective against *P. aeruginosa* and other gram-negative microbes, some microorganisms (e.g., acid-resistant fungi) (Edstrom Industries 2003) can be unaffected and survive (Meltzer 1993).

Hydrochloric acid is the predominant type of acid used to treat water supplied to laboratory animals, although sulfuric acid provides an effective option. A pH of between 2.5 and 3.0 is the recommended concentration of acid in animal drinking water systems. A pH below 2.5 has been shown to affect weight gains and water consumption in male mice (Hall et al. 1980). The use of acidified water may have some undesirable side effects on laboratory animal research, which should be considered prior to use. In some instances, acidified water may react with water bottle stoppers and release undesirable substances (Kennedy and Beal 1991). Acidified water may also alter the excretion of phenol red, and result in lowered proteinuria and decreased urine volume in rats (Clausing and Gottschalk 1989). Acidified water has also been shown to alter the gut microbiome and the incidence and onset rate of diabetes (Sofi et al. 2014; Wolf et al. 2014).

Acids used to treat water are corrosive by nature, and can cause damage to animal drinking water treatment and delivery systems, and even injury to facility personnel. Care should be exercised in the handling and storing of acid reagents following facility standard operating procedures (SOPs), using appropriate personal protective equipment (PPE), and referencing the appropriate safety data sheet (SDS). Materials used in the design and manufacture of animal drinking water treatment and distribution systems should be resistant to the corrosive properties of these acids. Stainless steel and certain plastic piping materials will resist the corrosive nature of inorganic acids. Silicone elastomers and Teflon used to join and seal piping materials are also resistant to the corrosive effects of acid. Brass and other metallic materials should be avoided. For automated animal drinking water systems, type 316L stainless steel is the material of choice. Type 316L stainless steel is resistant to prolonged exposure to low pH drinking water; however, it is incumbent upon the LARF staff to monitor the pH for maintaining the acceptable range of 2.5–3.0. Severe and irreversible damage has occurred to systems using 316L stainless steel when acidified animal drinking water has been maintained at levels below 2.5 pH. The anticorrosive properties of 316L stainless steel are due primarily to its unique concentrations of nickel and chromium and the addition of up to 3% molybdenum. Type 316L stainless steel utilized in animal drinking water systems is commonly exposed to a "passivation" process during its manufacture. Passivation of stainless steel further enhances the corrosion-resistant properties of the material by removing free iron from the surface and leaving a protective chromium oxide layer. Damage to this chromium oxide layer can occur when acidified water is maintained below a pH of 2.5, allowing the acidified water exposure to the ferrous material underneath the protective layer. When this protective layer becomes breached, pitting of the stainless steel will occur, allowing for pinhole leaks and eventual failure of the material.

Disposal of acidified water into the municipal waste stream may require special handling (e.g., neutralization) prior to its release. Therefore, it is suggested that facility administrators give this point due consideration should acidification of the animal drinking water be implemented.

Chlorination

The use of chlorine compounds to treat animal drinking water is an effective and acceptable means of disinfection. Chlorine compounds added to the animal drinking water supply preserve the quality of the water by preventing microorganisms sensitive to chlorine from colonizing the water distribution system. Even in low concentrations, chlorine compounds in an aqueous solution have a profound effect on eliminating episodic microbial contaminations. Understanding that animal drinking water can become contaminated with microbes at any given time, consideration should be given to the use of chlorine compounds as an adjunct treatment of the water provided to laboratory animals.

There are many chlorine compounds available to treat potable water supplies. Chloramines, sodium hypochlorite (NaOCl), and chlorine dioxide (ClO_2) are among the most predominant compounds used today for water disinfection. At the correct concentrations, chlorine compounds are safe for human and animal consumption and may be a means for eradicating endemic chlorine-sensitive microorganisms from rodent colonies (Takimoto et al. 2013). The chlorine compound and final concentration used for biocidal activity, whether it is for preventative measures or treating a water supply, should be determined based on the biocidal results at a given concentration and understanding that certain contaminants in the source water can react with chlorine, making its biocidal characteristics less effective.

The chlorine compound used for treatment and the final concentration of chlorine used in animal drinking water should be given due consideration. Although somewhat variable depending on the contamination level of the water supply, concentrations of free chlorine at <10 ppm for a sufficient period of contact time are generally adequate to treat for most microbial contaminants. However, microbiological assessment of the water supply and determination of the optimal chlorine concentration are warranted. Although laboratory rodents can tolerate high concentrations in drinking water, previous studies carried out by the National Toxicology Program (NTP 1992) in F344/N male and female rats and B6C3F1 male and female mice indicated there was an inverse relationship between increasing concentrations of two chlorine compounds (up to 275 ppm available atomic chlorine and up to 200 ppm chloramine administered to both species) and decreased water consumption and body weights over a 2-year period.

Chlorine compounds used as water disinfectants are most effective in solution when water chemistry conditions are optimal. The effectiveness of chlorine compounds in solution is influenced by pH, the presence of organic matter, water hardness, temperature, and chlorine concentration. The pH of the water has the greatest influence on the antimicrobial properties of chlorine (Dychdala 2001). In the case of sodium hypochlorite, an increase in pH significantly decreases the disinfecting ability of chlorine, whereas a decrease in pH has the opposite effect. Sodium hypochlorite is the active ingredient in household bleach and is perhaps the least expensive source of sodium hypochlorite available for use as a water disinfectant. When added to animal drinking water supplies, sodium hypochlorite molecules react with the water molecules to form hypochlorous acid (HOCl) and sodium hydroxide (NaOH). HOCl dissociates in water to form hydrogen (H^+) and hypochlorite (OCl^-) ions. Of the two compounds, HOCl exhibits far more germicidal activity than OCl^-. The relationship of the HOCl and OCl^- ions in solution is pH dependent. As pH increases, the concentration of OCl^- increases. Conversely, as the pH decreases, the concentration of HOCl increases (Dychdala 2001). When used as a disinfectant in water supplies, sodium hypochlorite can react with organic compounds dissolved in the water, forming trihalomethanes. Trihalomethanes are suspected to be carcinogenic and may add an additional variable when consumed by animal research subjects under research protocols.

Chlorine dioxide is more frequently being used for drinking water disinfection in municipal water supplies. Unlike other chlorine compounds, chlorine dioxide is unaffected by high pH and does not form trihalomethanes in the presence of organic material. Chlorine dioxide has a higher oxidative capacity than sodium hypochlorite and is therefore much more effective as a biocide at lower concentrations.

Chlorine is a powerful oxidizing agent and corrosive by nature. Stainless steel is susceptible to the corrosive effects of chlorine; however, various grades of stainless steel are more resistant to the effects of chlorine corrosion than others. As is the case with acids, type 316L is more resistant to chlorine corrosion damage than other stainless steel grades commonly used in the industry.

Steam Sterilization

Steam sterilization is one of the oldest and most effective methods of treating laboratory animal drinking water. Under the right conditions, steam sterilization destroys all microorganisms present in the water medium, including viruses, bacteria, spores, fungi, molds, and protozoa. Steam sterilization is effective at deactivating these microorganisms both in the water and when they are attached to water storage and containment surfaces. Steam is a penetrating sterilization medium, able to overcome biofilms and the cellular defense mechanisms of the individual microorganisms.

Achieving sterility is a function of adequate exposure time to both temperature and pressure. The quality of the steam produced by the sterilizer is also very important. Dry saturated steam or superheated steam is preferred over wet steam. Superheated or "dry" steam has less entrained water droplets than wet steam and, as a result, can transfer more heat energy during the sterilization process. Transferrable heat energy is important, as it promotes sterilization efficiency and efficacy. Steam sterilization is a nontoxic approach to treating animal drinking water, is relatively inexpensive, and kills microorganisms rapidly.

Steam sterilization is typically used to sterilize animal drinking water bottles that are either filled or empty. The sterilizer chambers that are common on most steam sterilizers used in LARFs are significant in size, allowing the facility to sterilize hundreds of bottles per sterilization cycle. Steam sterilization of filled and empty bottles is handled in very different ways. Empty bottles can be sterilized very quickly using a high-vacuum sterilization cycle. In contrast, bottles filled with animal drinking water must utilize a slower processing cycle in order to prevent the drinking water from boiling out of the bottles. This process exchanges vacuum pulses for steam pulses to evacuate air from the sterilization chamber. A slow exhaust cycle is used to cool the chamber, rather than the dry cycle that is used for nonliquid materials. Purging air from the chamber with steam is a function of the density of the steam versus the air. Since air is denser than steam and therefore heavier, it settles to the bottom of the chamber and is forced to the sterilizer drain as steam is added to the chamber. The exhaust process involves slowly lowering the chamber temperature and water temperature in the bottles to a point where the water temperature falls below the boiling point so that the water remains in a liquid state and does not boil over. Many sterilizer manufacturers offer load probes as a means of confirming product temperature throughout the sterilization process. However, a load probe provides only one reference point within the chamber, so other means of verification may be required.

It is important to establish a process of evaluating sterilization efficacy and validation of sterilization cycle parameters after the sterilization cycle is complete. Validation can be accomplished through the use of biological indicators for liquid cycles. It is important to place biological indicators at multiple sites throughout the entire load to provide greater assurance that the load has been sterilized.

Steam sterilization is an effective means of sterilizing animal drinking valves that are an integral component of automated animal drinking water systems and some disposable water pouch systems (e.g., stainless steel animal drinking valves). Since animal drinking valves are the farthest point away in terms of distance from where the water was initially treated, and since these valves come in direct contact with the research animals, animal drinking valves can harbor microbial contaminants that can cross-contaminate through the cage replacement process. The stainless steel manifold pipes that are integral to automated animal drinking water systems used to supply water to individually ventilated caging (IVC) racks and other laboratory animal housing systems are designed to be sanitized either chemically or through the use of steam sterilization. Attention to the manufacturer's recommendations of cycle parameters, including temperature, vacuum depth, and dry cycle, is important in order to prevent damage to this equipment.

Water Delivery Systems

Water Bottles

Arguably one of the most versatile water delivery systems, water bottles have withstood the test of time and remain a viable option for water delivery to laboratory research animals. Water bottles also provide a means of customizing water treatments for subsets of laboratory animals for which it is much more difficult to provide using other water delivery systems. Water bottles come in a variety of shapes and volumes to meet the needs of the animals, husbandry schedules, and research protocols. Volume is a particularly important aspect, as it relates to its service life and recycle frequency. The bottle volume choice should be compatible with the animal housing type, with ample consideration given to the recycle (changing) frequency of the cage. Operational costs should also be considered in the selection of the bottle volume and materials. The change frequency of the cage and the bottle will affect the labor associated with the bottle handling process, so larger bottle volumes that will extend bottle change-out will reduce operational costs.

The materials used in the manufacture of animal drinking water bottles and sipper caps should also be considered. Water bottles are commercially fabricated from a variety of materials, including glass and various plastic polymers. Exposure of bottles and sipper caps to sanitizing chemicals and the steam sterilization process will degrade the material composition over time.

Glass is a fragile material long used in the industry, and although still used today, glass represents handling issues for husbandry and scientific staff due to its relative weight compared with plastic polymers, and due to operational and personnel injury costs due to frequent breakage. Research has shown that under certain circumstances, glass bottles may contribute silicon to the drinking water, which can introduce a research variable (Lohmiller and Lipman 1998).

Polymer-type bottles offer the advantage of being lightweight, durable, transparent, and resistant to heat and chemical degradation. They are available in a variety of volumes and shapes to suit the considerable variety of laboratory animal research species and the corresponding variety of animal housing systems. Although polymer-type bottles offer many positive features for use as a water containment vessel in LARFs, one type of plastic polymer, polycarbonate, has been associated with the leaching of bisphenol A (Howdeshell et al. 2003; Hunt et al. 2003), a synthetic organic compound used in the production of polycarbonates and other thermoplastics. Bisphenol A and analogs bisphenol S and bisphenol F (Rochester and Bolden 2015) are characterized as endocrine disruptors, exhibiting estrogen-like properties. In contrast, polypropylene, polyethylene, and polyethylene terephthalate (PET) are not made with bisphenols or other known endocrine disruptors, suggesting that these materials represent a viable alternative to the use of affected thermoplastics. The introduction of endocrine disruptors through the supplied drinking water can add a variable to research, and perhaps disturb normal breeding and reproductive cycles in laboratory animal species.

Stoppers or sipper caps that are affixed to the open end of water bottles are also available in a wide range of materials. Black rubber and neoprene stoppers are the predominant types. Of these two types, neoprene is more resistant to degradation from repetitive steam sterilization cycles; however, hardening of the stoppers, as a consequence of steam sterilization, occurs with both types. Hardening of the stoppers can lead to the breakdown and release of particles from the stopper into the drinking water. These particles will ultimately be consumed by research animals with potentially negative consequences. Rubber and neoprene stoppers are also known to release minerals and heavy metals (Kennedy and Beal 1991; Nunamaker et al. 2013) into the drinking water.

An alternative to conventional stopper and sipper tubes is capped bottles with a "pinhole" orifice. The design characteristic of this approach relies on the inherent surface tension of water to prevent it from leaking out of bottles while providing free access to water. However, accumulation of debris in the orifice can lead to water delivery failure. Therefore, careful attention to appropriate cleaning and inspection of the bottle and orifice is warranted.

Automated Watering Systems

Automated watering systems in LARFs have been in existence for more than 50 years. These systems provide a viable and reliable alternative to using water bottles. Automated watering is a centralized approach to delivering water to laboratory animals. These systems begin with centralized water purification and treatment, and then use a network of pressure reduction stations and distribution piping to deliver the purified and treated animal drinking water to the animal holding rooms within the LARF. Manifolds on large animal enclosures, or racks for smaller species can connect to the animal room drinking water distribution system to provide a quality drinking water product at the cage or enclosure level. Drinking valves have been designed to accommodate the different water volume requirements and natural water access instincts of most laboratory animal species. Modern automated watering systems are designed with control systems that allow for automated daily flushing of the room, rack, or enclosure distribution system. These systems may also be equipped with advanced monitoring capabilities to enable notification of facility staff in the event of a system failure or leak. Automated watering systems can be designed to address specific facility layouts and budgets, and to meet institutional water quality standards.

Animal drinking valves are adequately designed to reliably and repeatedly deliver high-quality water to laboratory animals with minimal effort and maintenance. With the wide adaptation of IVC systems for rodents, drinking valves were required to be located on the inside of the individual cages. The design of these new valves had to effectively prevent water leakage into the cage caused by particulates and other contaminants in the drinking water, or by the natural behavior of rodents pushing bedding into the frontal opening of the valve. Although the initial designs of these new rodent drinking valves did not perform well, the various manufacturers have refined and improved valve designs to the point where rodent automated drinking water systems are reliable, acceptable, and safe for these valuable research subjects.

Automated watering systems utilize two typical design approaches to distribute treated animal drinking water to the racks and kennels located in animal holding rooms. Room or rack flushing systems rely on a principle that once the supply water is treated and distributed to the room or rack level, the water should be either utilized for animal drinking or flushed to drain. This system design eliminates the potential of the water dissolving contaminants downstream at the room or cage level and bringing those contaminants back to the treated water storage tank. Since animal drinking volumes rarely provide enough water exchange to keep the water fresh and free from bacterial growth, room or rack flushing systems rely on facility staff to either manually flush the distribution lines and manifolds to drain or utilize a sequencing controller to automatically flush solenoids at the end of the distribution lines. To reduce the risk of microbial growth on the internal surfaces of the distribution system (biofilm), modern room and rack flushing systems utilize a proportioning system to inject low levels of chlorine or inorganic acid into the drinking water. Automated flushing provides a mechanism to flush at prescribed intervals, for a given duration and at prescribed times. Modern flush sequencing and monitoring control panels have the added capability of providing documentation that the flushing actually occurred. They can also monitor key system parameters and provide alarm notification to facility personnel in the event of a system failure.

In contrast, recirculating room and rack distribution systems rely on a continuous flow of water from the animal drinking water holding tank to the animal holding rooms and back to the holding tank as a means of providing clean, fresh water to the research animals. These systems rely on additional microfiltration and the use of a UV-C reactor to reduce the potential for microbial contamination. Recirculating systems offer the advantage of conserving water; however, they employ the use of one or more continuously running pumps to move the water. For the reasons noted above, recirculating systems are more easily contaminated with microorganisms when relying solely on UV-C radiation and filtration. For this reason, recirculating systems often have a proportioning system that will allow for the injection of inorganic acid into the drinking water as a means of controlling microbial growth. Chlorine is not normally considered a suitable residual biocide with recirculating systems since it is difficult to control the concentration of free chlorine when the water is recirculating. However, routine testing for the presence of free chlorine or the use of chlorine monitoring and reporting equipment can adequately address the use of chlorine in the water delivery scenario. UV radiation also has a tendency to dissipate free chlorine concentrations.

Animal drinking water manifolds that are integral to animal holding racks or kennels are connected to the room distribution system by way of one or more detachable recoil hoses. These hoses allow for the holding racks or kennels to be detached from the system to facilitate room distribution system sanitizing without exposing animals to sanitizing agents, allowing for rack manifold and recoil hose sanitizing and for storing housing systems when no longer needed. If rack and kennel manifolds are designed to be flushed manually by facility staff, these manifolds are attached to the room distribution pipe using a single flexible recoil hose. Flushing of the rack and kennel manifolds is accomplished by manually opening a drain valve at the end of the manifold, allowing the water contained in the manifold to flow either to a room drain or to a bucket or catch pan that can be emptied into a local drain. Although manual flushing is effective if performed properly and on a daily basis, it relies on facility staff to accomplish this task. If facility personnel overlook this important maintenance step, there is increased risk of microbial contamination.

If rack or kennel manifolds are designed for automated rack flushing, or are part of a recirculating system that allows the water to flow through the rack manifolds during the water recirculating process, manifolds utilizing this design are connected to the room distribution system using two flexible recoil hoses. In rack flushing systems, one of the hoses will be attached to the water supply distribution line and one of the hoses will be attached to the water distribution drain line. There is an electrically controlled solenoid on the drain line associated with each rack or kennel manifold that can be sequenced to open during a flush cycle. The sequencing panel will also validate that the flushing cycle was completed. In recirculating flow-through-the-rack systems, one recoil hose is attached to the water supply line, and one recoil hose is attached to the water return line. Water will continuously flow through the room distribution piping, and each of the manifold pipes as the water makes its journey to and from the supply tank and through the UV-C reactor and microfiltration systems.

Manifold design is an important consideration when automated watering systems are purchased and utilized within a LARF. The design of the manifold has a direct bearing on animal drinking water quality and on the ability of the research animals to access drinking water *ad libitum*. A properly designed rack manifold will not only ensure that each cage on the rack has continuous access to the treated drinking water, but also ensure that air bubbles that become entrapped in the system are reduced and flushed to drain so that the research animals receive water from the drinking valves rather than air. A properly designed rack manifold will also ensure that all of the water in the manifold can be easily flushed and exchanged with clean treated water during the daily manual or automated flush sequences.

Daily flushing of rack and kennel manifolds is an important process for maintaining quality animal drinking water. At the manifold level, the animal drinking water is under low pressure, and the only exchange of the water is derived from animal drinking. The low and intermittent flow of water from animal drinking activity is not enough to thoroughly exchange the water in the manifold. If the water is not thoroughly exchanged on a daily basis, microorganisms will begin to proliferate and bacterial biofilms will begin to develop on the surfaces of the manifold and drinking valve components. To thoroughly and efficiently exchange the water in a manifold, the manifold must be designed as one continuous pipe from cage to cage and rack shelf to rack shelf (S configuration). To ensure that entrapped air is adequately removed during the flushing and filling of the manifold, a reverse S manifold design is preferred over an S configuration. With a reverse S–designed manifold, the water inlet flow begins with the horizontal pipe on the bottom of the rack and flows through the pipe on each row of cages as the water courses through to the last cage position on the top row of the rack. For automated systems that flush the manifold piping on each individual rack, reverse S manifolds are typically used.

There are other manifold designs that are used on rodent housing racks and on large animal enclosures. These other manifold designs may require manual flushing by the animal husbandry staff. Other common manifold designs are named after the letter shapes that they resemble. H manifolds, I manifolds, U manifolds, and upside-down U manifolds are just some of the more common designs that are utilized for different species. Animal husbandry staff should be aware that these manifolds may work equally well in distributing animal drinking water to the individual cages as the S and reverse S manifold designs; however, greater care and attention must be devoted to these manifold types to ensure that the animal drinking water is thoroughly exchanged on a daily basis and that entrapped air has been removed during the manifold filling and flushing process.

Most LARFs that utilize automated animal drinking water systems have dedicated rack manifold and recoil hose flushing stations. These stations are typically located within the clean side of the cage wash facility and also serve as a solution for flushing rack manifolds after sanitizing. Although these flush stations can be purchased as either manual or with microprocessor controls for an automated flush, they have the ability to flush rack manifolds and recoil hoses with a hyperchlorinated solution for greater sanitizing potential. The automated flush stations have flush and soak cycles to provide adequate contact time with the chlorine solution, and a final rinse cycle to rid the manifold of any residual chlorine solution. In addition to the flush, soak, and rinse cycles, the automated recoil hose flush stations have a compressed air cycle to dry the internal surfaces of the recoil hoses so that they can be either put back into immediate use or placed in storage with the quick-disconnect ends connected.

Daily flushing, or the continuous flow of animal drinking water through an automated watering system, is an integral component of maintaining an institution's water quality standards. Periodic testing of the animal drinking water at the cage or enclosure level is considered a good practice to mitigate the risk of microbial contamination, and to ensure that the LARF operation is adhering to the institution's water quality standards.

If test results indicate that bacteria are present in the room distribution system, sanitizing the system with a hyperchlorinated water solution may be indicated. A hyperchlorinated water solution of 20 ppm of free chlorine and 30 minutes of contact time is recommended by the manufacturers of these systems. Concentrations higher than 20 ppm and in combination with longer contact times may actually damage the system materials and should be avoided. Portable sanitizing units for mixing and pumping hyperchlorinated water solutions through the room distribution system are available from the manufacturers of automated watering systems. It is advisable to disconnect rack manifolds from the automated watering system while performing a hyperchlorinated sanitation process to avoid introducing a research variable and, moreover, to avoid potential animal health-related consequences as a result of an atypical exposure to a higher level of chlorine. Animals must be moved to other housing areas or provided with another source of treated animal drinking water during a hyperchlorinated sanitation process. After exposing the distribution system to the hyperchlorinated water solution for 30 minutes of contact time, the distribution system should be completely flushed with the treated drinking water supply to completely rid the system of the hyperchlorinated water solution.

One of the more valuable features of automated watering systems is the integration of monitoring capabilities into the system. The current technologies not only provide real-time information on system performance and status, but also, more importantly, provide a feedback system when the system integrity is compromised. Remote notification keeps key personnel informed of system breaches at all hours of the day, facilitating immediate response and remediation by responsible facility staff.

Automated watering systems typically require a higher initial capital investment over water bottle systems; however, automated systems will usually recover the initial capital within 3–5 years due to low operating costs and labor savings.

Disposable and Semidisposable Systems

Perhaps the most recent developments in water delivery solutions are disposable and semidisposable systems. They complement and provide a viable alternative to existing solutions for water delivery. The fundamental properties of these systems are that they provide a defined volume, provide for a variety of water treatment options, and can be used with most animal housing systems with the use of specialized adapters or for housing systems designed to use them. The principal theory of their use is substituting the capital and maintenance costs of the various processing equipment, labor costs, and operational costs for refined maintenance, operational, and labor costs.

Disposable Solutions: Pouches and Water Bottles

Disposable animal drinking water solutions are single-use products that provide a fixed quantity of water to laboratory research animals. These systems are primarily used for rodent species, but they have been used to a limited extent with larger species utilizing specialized caging adapters. Water is gravity fed from the water container or "pouch" through a specially designed valve delivering water at a species-specific flow rate. The pouch collapses as the water is consumed. The pouch is completely disposable, and some types may be recycled. The drinking valve, depending on the design and material, can either be discarded after one or more changing cycles or be reused after sanitation or sterilization (Gianni and Willis 2005).

Modern disposable pouch systems are designed to offer a variety of water treatment options to meet institutional needs and requirements. These pouch filling systems take up minimal space within the facility and can be conveniently connected to purified water and other treated water sources, including media filtration, UV-C reactor, and chlorine–acid proportioner. Utility connections are reasonable, requiring common electrical voltages and common water pressure and flow rates. An added benefit of these disposable pouch systems is the ability to customize water treatment from production run to production run to meet the needs of discrete research projects. If medications are required to be used in the animal drinking water, a silicone patch can be applied to the surface of the pouch, providing a leak-resistant means of introducing medications into the drinking water via small-gauge needle and syringe injection. The low volume of the pouch allows for conserving medication costs and for those instances where treatments are of a relatively short duration.

Pouches can be either filled and sealed on site with purified or treated water, or purchased prefilled with purified or treated water. The pouch filling equipment has a relatively high capital cost, paralleling that of other cage washing, bottle washing, and automated watering system equipment. Pouch volumes are typically 8 or 13 oz. by default; however, volumes less than 8 oz. can be produced to meet unique species, research, or operational requirements. The pouch filling equipment that is currently available on the market can produce up to 700 or 1800 pouches per hour, depending on the equipment model. Adjusting for setup and shutdown time of the pouch filling machine, in a typical 8-hour workday one person can produce up to 11,700 or 4,550 pouches, respectively, and approximately 1,200 gallons or 462 gallons of reserve water, respectively.

The disposable pouch filling system is capable of producing a water product free from microbial contaminants. However, the exterior pouch surfaces should be tested periodically to reduce the risk of contamination. If testing reveals significant microbial contamination, facility personnel should use appropriate levels of PPE and disinfect the interior and exterior of the pouch containers with an appropriate chemical disinfectant, such as chlorine dioxide.

Disposable water bottles and sipper caps have become more popular with the advent of disposable caging. Made of recyclable PET, these disposable bottle systems are available prefilled with acidified water or empty so that the bottles may be custom-filled on site at the LARF. Currently, the sole manufacturer of these disposable bottle systems also offers irradiation as an option for both prefilled and empty bottles. Designed to fit this specific manufacturer's disposable caging system, the disposable nature of the product and exclusive use of this technology eliminate the need for the facility to invest in most of the bottle filling, washing, and sterilization equipment that is associated with a conventional LARF. In exchange for the initial capital investment required for most conventional cage wash facilities using conventional caging and bottles, facilities opting to use disposable caging and bottle solutions will have increased material, transportation, storage, and disposal costs and increased costs associated with inventory management.

Disposable water pouch systems can be purchased with disposable drinking valves made of plastic polymers, or with drinking valves made with stainless steel that can be reused after sanitizing or sterilizing. Filling systems are available that allow for the filling of disposable water pouches at the point of use in the animal holding and/or procedure rooms. These portable filling stations are typically located at an animal cage changing station and connected to the facility's automated watering distribution systems. Treated water pouches are filled on demand as individual cages are changed or manipulated by husbandry or research staff.

The disposable water delivery systems offer additional benefits. The footprint required to store water is relatively small, particularly if vertical space is used. The use of these systems provides for the opportunity to develop a strategy to store water to ensure a continuous supply during drinking water supply interruptions and during those times when contingency plans are evoked. The production and collection of pouches in stackable containers stored on transport carts provides for easy transfer and efficient storage to the animal holding and storage areas. From a contingency planning standpoint and during periods of unscheduled water outages, it is feasible to deliver animal drinking water stored in pouches to most laboratory animals maintained on other water delivery systems (Allen 2013; Joseph et al. 2016), including automated watering systems and water bottles. Water can easily be transferred from pouches to mobile bins, containers, or carboys and dispensed by way of gravity or with electrically driven submersible pumps. Evaluating the building or floor water distribution system layout and its connections to the water delivery system provides insight into how water can be recovered from disposable systems and delivered to other species by way of their typical water delivery system.

Gels

Used by commercial breeders and animal exporting institutions for hydration during animal transportation, gels have the unique characteristic of being largely composed of water available in a semisolid form. When provided in sufficient quantity during transportation, studies suggest that animals remain well hydrated and experience weight gains when provided with both hydration gel and diet (Fredenburg et al. 2009; Pruet et al. 2010).

Although water is the major component, incorporation of nutrients in a balanced manner allows for meeting both the water and nutritional needs of the animal. This approach is particularly useful in circumstances where conventional means of delivering water is not effective or where animals cannot physically access the water source. In a balanced combination with other feed nutrients, nutritive gels are useful for those circumstances where both water and conventional diets do not provide for the special needs of the animal. Nutritive gels are highly palatable and offer a solution for transitioning weanling animals to standard diets and, more importantly, where traditional methods of transitioning offspring to a water source may not be effective.

Gels may also be considered for use in those circumstances where supportive care is indicated in a research protocol (Khaing et al. 2012). Gels may also provide a convenient medium for postoperative analgesic delivery (Christy et al. 2014) or for continuous delivery of medications and compounds (Overk et al. 2011).

Consideration of the use of gels as an adjunct hydration method or nutritive supplement should take into account not only the benefits derived from their use, but also whether they are compatible with the objectives and predicted outcomes of existing and proposed research protocols. Discussion and coordination with prospective researchers is therefore highly recommended.

Conclusion

Water is an essential nutrient required to sustain life. However, water can introduce experimental variables to research studies and compromise animal health if the quality of the animal drinking water is not adequately assessed. Whether carrying out studies under GLP conditions or supporting a multidisciplinary research enterprise, the process for determining what type of water will be acceptable starts with assessing water quality.

There are a variety of methods presented here to address the treatment options for the water supply, the details of each water treatment option, and other hydration options to consider. The decisions to be made regarding water treatment methods are complicated. There is not necessarily only one option that will satisfy an institution's needs, and within that institution, there may be multiple approaches needed to meet the research and animal health needs. The methods used to treat the water supply, and the solutions available for delivering water to research animals, are all interrelated components of managing laboratory animal resources. The availability and diversity of the continually evolving water purification solutions, water treatment options, and delivery systems provide animal resource administrators numerous options for meeting the unique needs of the research community and should provide for the general and unique needs of laboratory animals in your charge. Careful consideration of the content of this chapter and familiarity with past and present trends in delivering water are paramount.

REFERENCES

Allen ED. 2013. Hydropac®'s emerging role in contingency planning. Presented at the 2013 Hydropac® Exchange, Baltimore, MD, October 27.

Amy G. 2008. Fundamental understanding of organic matter fouling of membranes. *Desalination* 231 (1–3): 44–51.

Beer KD, Gargano JW, Roberts VA, Hill VR, Garrison LE, Kutty PK, Hilborn ED, Wade TJ, Fullerton KE, and Yoder JS. 2015. Surveillance for waterborne disease outbreaks associated with drinking water—United States, 2011–2012. *MMWR Morb. Mortal. Wkly. Rep.* 64 (31): 842–848.

Benjamin MM and Lawler DF. 2013. *Water Quality Engineering: Physical/Chemical Treatment Processes*. Hoboken, NJ: John Wiley & Sons.

Chen JP, Yang L, Wang LK, and Zhang B. 2006. Ultraviolet radiation for disinfection. In *Advanced Physicochemical Treatment Processes*, 337. Totowa, NJ: Humana Press.

Christy AC, Byrnes KR, and Settle TL. 2014. Evaluation of medicated gel as a supplement to providing acetaminophen in the drinking water of C57BL/6 mice after surgery. *J. Am. Assoc. Lab. Anim. Sci.* 53 (2): 180–184.

Clausing P and Gottschalk M. 1989. Effects of drinking water acidification, restriction of water supply and individual caging on parameters of toxicological studies in rats. *Z. Versuchstierkd.* 32 (3): 129–134.

Colborn T. 2004. Neurodevelopment and endocrine disruption. *Environ. Health Perspect.* 112 (9): 944–949.

Curran B and Smart S. 2007. Testing and analyzing animal drinking water in the vivarium. *ALN Magazine*, December 31. https://www.alnmag.com/article/2007/12/testing-and-analyzing-animal-drinking-water-vivarium.

Dvorak BI and Skipton SO. 2013. Drinking water treatment: Activated carbon filtration, 1–4. Lincoln: University of Nebraska–Lincoln Extension, Institute of Agriculture and Natural Resources.

Dychdala GR. 2001. Chlorine and chlorine compounds. In *Disinfection, Sterilization, and Preservation*, ed. SS Bloch, 135–158. Philadelphia: Lippincott Williams & Wilkins.

Edstrom Industries. 2003. Drinking water acidification. Waterford, WI: Edstrom Industries. http://www.edstrom.com/documents/?pg=1.

Fredenburg N, Engle T, and Cooper DM. 2009. Effects of alternate water sources on weight gain, blood chemistries, and food consumption in Sprague Dawley rats and ICR mice following ground transportation. Presented at the AALAS National Meeting, Denver, CO.

Gianni FJ and Willis MA. 2005. A new alternative watering system. *Lab. Anim. (N.Y.)* 34 (10): 54–55.

Hall JE, White WJ, and Lang CM. 1980. Acidification of drinking water: Its effects on selected biologic phenomena in male mice. *Lab. Anim. Sci.* 30 (4 Pt. 1): 643–651.

Hijnen W. 2010. *Quantitative Methods to Assess Capacity of Water Treatment to Eliminate Micro-Organisms. London*: IWA Publishing.

Howdeshell KL, Peterman PH, Judy BM, Taylor JA, Orazio CE, Ruhlen RL, vom Saal FS, and Welshons WV. 2003. Bisphenol A is released from used polycarbonate animal cages into water at room temperature. *Environ. Health Perspect.* 111 (9): 1180–1187.

Hunt PA, Koehler KE, Susiarjo M, Hodges CA, Ilagan A, Voigt RC, Thomas S, Thomas BF, and Hassold TJ. 2003. Bisphenol A exposure causes meiotic aneuploidy in the female mouse. *Curr. Biol.* 13 (7): 546–553.

Joseph DK, McNair J, O'Donnoghue JM, Allen ED, and DeTolla LJ. 2016. Disposable water delivery systems role in contingency planning. Presented at the 67th National Meeting of the American Association of Laboratory Science, Charlotte, NC, October 30–November 3.

Kennedy BW and Beal TS. 1991. Minerals leached into drinking water from rubber stoppers. *Lab. Anim. Sci.* 41 (3): 233–236.

Khaing ZZ, Geissler SA, Jiang S, Milman BD, Aguilar SV, Schmidt CE, and Schallert T. 2012. Assessing forelimb function after unilateral cervical spinal cord injury: Novel forelimb tasks predict lesion severity and recovery. *J. Neurotrauma* 29 (3): 488–498.

Komulainen H. 2004. Experimental cancer studies of chlorinated by-products. *Toxicology* 198 (1–3): 239–248.

Lee KP, Arnot TC, and Mattia D. 2011. A review of reverse osmosis membrane materials for desalination—Development to date and future potential. *J. Membrane Sci.* 370 (1–2): 1–22.

Lohmiller J and Lipman N. 1998. Silicon crystals in water of autoclaved glass bottles. *Contemp. Top. Lab. Anim. Sci.* 37 (1): 62–65.

Mathieu-Denoncourt J, Wallace SJ, de Solla SR, and Langlois VS. 2015. Plasticizer endocrine disruption: Highlighting developmental and reproductive effects in mammals and non-mammalian aquatic species. *Gen. Comp. Endocrinol.* 219: 74–88.

Meltzer TH. 1993. *High Purity Water Preparation for the Semiconductor, Pharmaceutical, and Power Industries*. Littleton, CO: Tall Oaks Publishing.

Molk DM, Karr-May CL, Trang ED, and Sanders GE. 2013. Sanitization of an automatic reverse-osmosis watering system: Removal of a clinically significant biofilm. *J. Am. Assoc. Lab. Anim. Sci.* 52 (2): 197–205.

National Research Council, Committee for the Update of the Guide for the Care and Use of Laboratory Animals, Institute for Laboratory Animal Research. 2011. *Guide for the Care and Use of Laboratory Animals*. Washington, DC: National Academies Press.

NTP (National Toxicology Program). 1992. NTP toxicology and carcinogenesis studies of chlorinated water (CAS Nos. 7782-50-5 and 7681-52-9) and chloraminated water (CAS No. 10599-90-3) (deionized and charcoal-filtered) in F344/N rats and B6C3F1 mice (drinking water studies). *Natl. Toxicol. Program Tech. Rep. Ser.* 392: 1–466.

Nunamaker EA, Otto KJ, Artwohl JE, and Fortman JD. 2013. Leaching of heavy metals from water bottle components into the drinking water of rodents. *J. Am. Assoc. Lab. Anim. Sci.* 52 (1): 22–27.

Overk CR, Borgia JA, and Mufson EJ. 2011. A novel approach for long-term oral drug administration in animal research. *J. Neurosci. Methods* 195 (2): 194–199.

Pilutti PM and Nemeth EJ. 2003. Technical and cost review of commercially available MF/UF membrane products. International Desalination Association (IDA) World Congress, Bahamas.

Pruet AM, O'Neal SL, Masterson NC, and Blaser CM. 2010. HydroGel™ is an adequate source of hydration for surgically modified rat models. https://clearh2o.com/images/product/Hydrogel_Rat_Model.pdf.

Richardson SD, Plewa MJ, Wagner ED, Schoeny R, and DeMarini DM. 2007. Occurrence, genotoxicity, and carcinogenicity of regulated and emerging disinfection by-products in drinking water: A review and roadmap for research. *Mutat. Res. Rev. Mutat. Res.* 636 (1–3): 178–242.

Rochester J and Bolden A. 2015. Bisphenol S and F: A systematic review and comparison of the hormonal activity of bisphenol A substitutes. *Environ. Health Perspect.* 123 (7): 643–650.

Sofi MH, Gudi R, Karumuthil-Melethil S, Perez N, Johnson BM, and Vasu C. 2014. pH of drinking water influences the composition of gut microbiome and type 1 diabetes incidence. *Diabetes* 63 (2): 632–644.

Stoughton KLM, Duan X, and Wendel EM. 2013. Reverse osmosis optimization. Washington, DC: U.S. Department of Energy. http://energy.gov/sites/prod/files/2013/10/f3/ro_optimization.pdf.

Takimoto K, Taharaguchi M, Sakai K, Takagi H, Tohya Y, and Yamada YK. 2013. Effect of hypochlorite-based disinfectants on inactivation of murine norovirus and attempt to eliminate or prevent infection in mice by addition to drinking water. *Exp. Anim.* 62 (3): 237–245.

Tchounwou PB, Yedjou CG, Patlolla AK, and Sutton DJ. 2012. Heavy metal toxicity and the environment. *EXS* 101: 133–164.

USEPA (U.S. Environmental Protection Agency). 2012. Public drinking water systems: Facts and figures. Washington, DC: USEPA. http://water.epa.gov/infrastructure/drinkingwater/pws/factoids.cfm.

USEPA (U.S. Environmental Protection Agency). 2013. Basic information about disinfection byproducts in drinking water: Total trihalomethanes, haloacetic acids, bromate, and chlorite. Washington, DC: USEPA. http://water.epa.gov/drink/contaminants/basicinformation/disinfectionbyproducts.cfm.

USEPA (U.S. Environmental Protection Agency). 2014. Drinking water contaminants. Washington, DC: USEPA. http://water.epa.gov/drink/contaminants/index.cfm.

USEPA (U.S. Environmental Protection Agency). 2015. Consumer Confidence Reports (CCR). Washington, DC: USEPA. http://cfpub.epa.gov/safewater/ccr/index.cfm.

USEPA (U.S. Environmental Protection Agency). 2016a. Brownfields. Washington, DC: USEPA. https://www.epa.gov/brownfields.

USEPA (U.S. Environmental Protection Agency). 2016b. Table of regulated drinking water contaminants. Washington, DC: USEPA. https://www.epa.gov/ground-water-and-drinking-water/table-regulated-drinking-water-contaminants.

USGS (U.S. Geological Survey). 2014. National Water-Quality Assessment (NAWQA) Program. Reston, VA: USGS. http://water.usgs.gov/nawqa/about.html.

Wolf KJ, Daft JG, Tanner SM, Hartmann R, Khafipour E, and Lorenz RG. 2014. Consumption of acidic water alters the gut microbiome and decreases the risk of diabetes in NOD mice. *J. Histochem. Cytochem.* 62 (4): 237–250.

29

Management of Research Animal Breeding Colonies

James J. Elliott, Charles T. Miller, James A. Hagarman, Stephen T. Kelley, Suzette D. Tardif, Sander O. Hacker, and Amanda Bettis

CONTENTS

Introduction

In-house breeding of animals of any species takes a deliberate effort and recognition of the costs not only in direct expenses, but also in animal numbers, space allocation, genetic monitoring, labor, and expertise. Breeding of commercially available animals is generally not recommended. Specific scientific justification should be identified before undertaking an in-house breeding program. Some factors to consider before developing breeding colonies include but are not limited to

- Are the animals available from commercial sources? If yes, why breed them in-house?
- How much space can be allocated for this colony?
- Which breeding scheme will be used—pyramid breeding regimen (foundation, pedigree expansion, or production colony) or a simple breeding of homozygotes for production?
- Are trained personnel available to manage the colony? (The most common reason for genetic variation is human error.)
- How many animals need to be generated per month or year?
- Will both sexes be used, and which genotypes will be used (homozygotes vs. heterozygotes)?
- What will be done with the offspring of the wrong genotype?
- How will costs be recovered (per diem, labor, space, or testing)?
- How will genetic drift be monitored?
- For larger species (dogs and primates), how many troops or breeders are necessary to maintain genetic diversity?
- Is cryopreservation an option to minimize the animal numbers?

From a purely cost perspective, in-house breeding programs for rodents are normally more expensive than commercial sources when calculating in all the above factors. With rodent colonies, there is also an underlying ethical question of how many animals should be produced before arriving at the necessary genotype and numbers required. There is always the question of what should be done with the colony when the funding ends or the study concludes.

In this chapter, we cover breeding management of the most common species bred at academic and private research institutions. The chapter is divided into four sections corresponding to the species as follows: rodents, macaques and baboons, marmosets, and dogs. It is not an exhaustive management review, but we have kept it as practical as possible, with links to additional resources that will be of assistance in the management of your colonies.

Mice

Introduction

Although the following information concentrates on mouse colonies, the principles are the same for other rodent models. The laboratory mouse is an important biomedical research model. Each year, increasing numbers of highly specialized and sophisticated models are created in the hopes of developing more relevant and predictive models of human disease. Indeed, with the development and application of more sophisticated techniques of gene targeting and editing, including zinc finger nucleases, TALENs, and the CRISPR/CAS9 system, the creation of new models is expected to accelerate. With the escalating number of available mouse models come corresponding increases in cost, mouse holding space, and related resources needed to manage these animals. Consequently, the art and science of managing mouse colonies efficiently is more important than ever.

A primary decision impacting many academic researchers is the sourcing method of research models; options include commercial repositories and internal breeding. Such decisions should be considered carefully, as there are financial and ethical implications for the different approaches. Indeed, we have an obligation to ensure that mouse resources are used wisely and animal usage minimized to purposeful and informative scientific experiments. In-house breeding can offer convenience, control, and cost savings if executed properly, and is often the only available approach to generate the mice of unique genotypes required for experimental cohorts. Frequently, however, the responsibility of breeding experimental animals falls to the most inexperienced or junior members of a laboratory, which can result in animal and financial waste. Although not the primary focus of this chapter, there can be significant benefits to purchasing mice from reliable suppliers or repositories; with such animals often comes a higher adherence to animal health and genetic quality standards that have significant impacts on data quality and reproducibility. In addition, when a full accounting of institutional costs (capital, overhead, labor, insurance, and direct costs) associated with internal breeding is appreciated, it has been recognized that significant institutional financial benefits can be achieved through the careful combination of external and internal animal sourcing.

The purpose of this section is to convey well-tested and effective strategies for maintaining laboratory mouse colonies in order to assist in managing colonies efficiently. There are standardized techniques and approaches that are reliable, economical, and efficient, and ensure that the mouse research produced is valuable and interpretable and minimizes the use of valuable resources (Jackson Laboratory 2009). This section provides important insights into husbandry methodologies, but they are guidelines only; every mouse facility is different, and techniques that work well at one may not work as well at other sites.

Calculating Colony Size

Calculating the Number of Breeders Needed

To properly size a mouse colony, you must first determine how many mice must be produced. To do this, a number of factors must be considered, including the following:

- The cohort size required for experiments and controls (age and sex matched)
- How often cohorts are required
- Which sex can be used
- What genotypes will be used

- The percentage of offspring of each genotype, based on the mating scheme
- The age range that can be used
- Adding some buffer to accommodate variation in productivity (assume 10%)

For example, if a researcher needs 20 homozygous (hom) male mice per week at a single age from a heterozygous (het) × het cross, he or she will need to size the colony to produce 176 mice per week. This is 20 mice/week * 2 (males only) * 4 (25% homs) * 1.1 (10% buffer).

Steady-State Colonies

To determine the number of breeders required, the following factors must be accounted for:

- Average litter size (for a good breeder, one can assume six pups per litter).
- Average number of litters that a breeding female will produce (assume four for a good breeder).
- Rotation length (assume 32 weeks). Rotation length is the time from the mating of the animal to when the breeders are retired.
- Frequency of productive matings. Assuming a 15% nonproductive (NP) rate, the productivity for an average breeder would be 1-NP, or 0.85.
- Frequency of rotation.

The colony productivity is the average number of mice produced per breeder per week. To calculate colony productivity, divide the number of litters by the rotation length, and multiply by litter size and the NP frequency (4 ÷ 32 * 6 * 0.85 = 0.6375). To determine the number of breeders needed, divide the mice needed per week by the colony productivity (176/0.6375 = 276 breeders).

Finally, calculate the number of replacement breeders that will be needed per week (number of breeders ÷ rotation length) and the number of additional breeders required to produce these additional mice (# replacements needed per week ÷ colony productivity). In the case above, het animals are produced in excess, so no additional breeding is required to provide replacements.

It is important to keep in mind a number of opportunities to minimize the number of animals being wasted in a colony. Modifying a breeding scheme to produce a higher percentage of mice with the correct genotype and using both sexes instead of one can greatly decrease the cost of producing mice. Also, using a combination of age range and cohort frequency to maximize the use of mice produced in the colony can also greatly decrease the cost of a colony.

Time Required to Expand Colonies

The average generation time for mice is 12 weeks. Assuming you start with five breeding females, to calculate the time it would take to expand to 276 breeders, the colony manager can calculate the following:

- First 12 weeks: 5 breeding females × 3 females per litter × 0.85 (assuming 15% NP rate) = 12 female breeders produced
- Second 12 weeks: 17 breeding females × 3 females per litter × 0.85 = 43 females
- Third 12 weeks: 60 breeding females × 3 females per litter × 0.85 = 153 females
- Fourth 12 weeks: 213 breeding females × 3 females per litter × 0.85 = 543 females

Therefore, it will take ~48 weeks to expand this colony to 276 breeding females and consistently produce 20 hom male mice per week.

Monitoring the Productivity of Colonies

An extremely simple and effective way to monitor colonies is to track wean per cage (or weaning per breeding unit [WPU]). Wean per cage is the number of animals weaned in a week divided by the number of breeder units. WPU can be monitored weekly to detect any change in breeder performance. A decrease in WPU can show a seasonal breeding slump or be indicative of some other problem, such as a change in the quality of feed or colony health status, or another unanticipated environmental factor, such as a change in the light cycle of the holding room. Changes in WPU allow colony managers an opportunity to increase or decrease colony size to meet the needs of investigators.

Burst Mating

If single, irregular, or rare cohorts are required, an alternative breeding strategy is to burst mate. Assume an investigator needs a single cohort of 20 male hom mice from a het × het mating. As shown above, the colony manager would still need to produce 176 mice. In order to produce numerous pregnant females at once, we can take advantage of the Whitten effect (Van Der Lee and Boot 1955; Whitten 1966). This is a phenomenon where when females are housed together and isolated from males for at least 7 days (Dalal et al. 2001), the females' estrous cycle is suppressed (Lee–Boot effect). When those females are exposed to male pheromones, estrus is induced (Whitten effect). This will maximize the number of pregnancies that will occur in the same week (~60%). To determine how many female breeders are needed, divide the number of mice required by the litter size, divide by the productivity rate, and divide by 0.6 ($176 \div 6 \div 0.85 \div 0.6 = 58$ breeding females). Consequently, if a researcher needed infrequent cohorts of 20 male hom mice, it would be more cost-effective to keep enough breeding age mice on the shelf to do periodic burst matings than to keep a colony that would produce 20 male hom mice every week. This will allow investigators to decrease colony size, reduce labor, and minimize cost.

Husbandry

Feed

According to the *Guide for the Care and Use of Laboratory Animals* (*Guide*) (NRC 2011), animals should be fed palatable, uncontaminated diets that meet their nutritional and behavioral needs at least once daily. Nutrition has a major influence on the growth and health of individual animals, and the productivity of breeding units. The National Research Council Committee on Animal Nutrition has prepared a report on the nutrient requirements of mice (NRC 1995). There are several types of diets classified by the degree of refinement of ingredients.

Natural-ingredient diets are formulated with agricultural products and by-products. Natural-ingredient diets for research mice are commercially available. Formulations for conventional and autoclavable natural-ingredient diets used successfully for many years include NIH-07 and NIH-31. The Jackson Laboratory primarily uses the natural-ingredient diet LabDiet 5K0Q (Flurkey 2009), although breeders from some strains perform better on a higher-fat diet, 5K20. If unsure, use the diet recommended for the background strain. Natural-ingredient diets are more likely to have contaminants that can affect productivity than other, more purified diets. Finally, variation in methods of processing, storing, and milling ingredients can influence the nutrient composition of natural-ingredient diets. Natural ingredients may contain low levels of naturally occurring or artificial contaminants.

Certified diets that have been assayed for contaminants are commercially available for use in select studies, such as preclinical toxicology, conducted in compliance with Food and Drug Administration (FDA) good laboratory practice standards (Code of Federal Regulations Title 21, 2009).

Purified diets are refined such that each ingredient contains a single nutrient or nutrient class; they have less nutrient concentration variability, and the potential for chemical contamination is lower. The purified diet formulations AIN-93G and AIN-93M have been developed and evaluated by a committee of the American Institute of Nutrition (AIN) as new standard reference diets (Reeves et al. 1993). These diets replaced the widely used purified diet AIN-76A (American Institute of Nutrition 1977).

Chemically defined diets contain the most elemental ingredients available, such as individual amino acids and specific sugars (NRC 1996). The latter two types of diet are more likely to be used for specific types of studies in rodents but are not commonly used because of cost, lower palatability, and a reduced shelf life.

Bedding

Rodent bedding serves to absorb moisture and provide nesting material and warmth, and enables natural behaviors, such as burrowing (Manser et al. 1997). To ensure reproducibility of experiments, the type of bedding and background microbial load should be standardized (Kraft 1980). It has been reported that mice prefer bedding of wood origin, with aspen being the favorite from 10 different commercially available bedding materials (Mulder 1975; Odynets et al. 1991; Ras et al. 2002). Size and manipulability strongly determine preference for bedding particles, with mice preferring bedding consisting of large fibrous particles (Blom et al. 1996). Hardwood has the advantage of being cost-effective and physiologically inert, but it can be irritating and is not as absorbent as other materials (Odynets et al. 1991; Burn and Mason 2005). Moisture absorbency is an important characteristic of rodent bedding for controlling bacterial growth and ammonia production. Softwood is a cost-effective option, but has the same disadvantages of hardwood while also presenting a number of problematic physiological effects (Weichbrod et al. 1988; Pelkonen and Hanninen 1997; Sanford et al. 2002; Buddaraju and Van Dyke 2003). Measured by volume, corncob is the most absorbent bedding and is resistant to ammonia buildup, but is poor nesting material (Smith et al. 2004; Burn and Mason 2005). Paper and cotton beddings are both highly absorbent, but the small particles can be irritating, particularly in the eyes of nude mice (Bazille et al. 2001; White et al. 2008). Providing bedding consisting of a mix of materials, such as aspen chip and shavings, can allow a good combination of absorbency and manipulability (Thigpen et al. 1989). The advantages and disadvantages of various types of bedding are summarized in Table 29.1.

Enrichment

Environmental enrichment is altering the living environment of an animal to provide opportunities to express more of their natural repertoire of behaviors. While there is no legal obligation to provide environmental enrichment for mice, it is a moral obligation to do everything reasonable to ensure that mice are maintained in a manner that promotes their well-being. The presence of enrichment has been shown to reduce chronic stress in mice (Gurfein et al. 2012). The best choices for enrichment are those materials that provide opportunities for typical behaviors such as burrowing, nest building, or gnawing. Bedding composed of large particles is easily manipulated by mice to provide a substrate for burrowing and nest building. Additional enrichment should be provided for singly housed mice or those maintained on a bedding type that does not promote nest building, such as corncob. Suitable materials for nest building

TABLE 29.1

Advantages and Disadvantages of Bedding Types

Bedding Type	Advantages	Disadvantages
Hardwood	Cost-effective Physiologically inert Ammonia control? (Smith et al. 2004) Preferred by rodents	Possibly irritating Less absorbent Ammonia control? (Burn and Mason 2005)
Softwood	Cost-effective	Possibly irritating Less absorbent Possible physiological effects
Corncob	Nonirritating Absorbent Ammonia control	Moderately expensive Poor nesting material
Recycled paper, cellulose chips	Absorbent	Expensive Dusty Possibly irritating
Cotton	Absorbent Ammonia control	Expensive Possibly irritating

Source: Modified from Flurkey, K., ed., *The Jackson Laboratory Handbook on Genetically Standardized Mice*, 6th ed., Jackson Laboratory, Bar Harbor, ME, 2009.

include Kimwipes, tissues without additives, and a number of commercially available nesting materials (e.g., Nestlets and Alpha Twist). Shredded paper strips allow mice to build higher-quality nests than other materials, and rats prefer paper strips as a nesting material (Manser et al. 1997, 1998; Ras et al. 2002; Hess et al. 2008). Mice that are housed individually for extended periods should also be given enrichment that encourages activity, such as climbing or hiding, including short lengths of polyvinyl chloride (PVC) pipe or commercial materials such as a Shepherd Shack, plastic "igloo," or Bio-Tunnel. Male mice housed in a group show reduced stress when given enrichment, and numerous studies have shown that enrichment reduces aggressive behavior in group-housed mice and rats (Hunt and Hambly 2006; Abramov et al. 2008; Abou-Ismail 2011; Swetter et al. 2011; Hutchinson et al. 2012; see also Van de Weerd and Baumans 1995). Notably, some studies have shown the opposite results, or no effect due to enrichment (Vestal and Schnell 1986; McGregor and Ayling 1990; Haemisch and Gartner 1994). In one example, nesting material alone promoted aggressiveness in group-housed male mice, but providing a shelter, in addition to nesting material, prevented intracage fighting (Kaliste et al. 2006).

Light Cycle

The light cycle in a mouse room is typically 12 hours of light and 12 hours of darkness or 14 hours of light and 10 hours of darkness. A longer light cycle is practically useful because it allows technicians more time to work, but it may also have beneficial effects on breeding based on the fact that a longer light cycle correlates with seasonal changes that provide optimal breeding conditions (Trainor et al. 2006; Steinman et al. 2012). Interruptions of light cycle should be avoided because of the importance of photoperiod in normal rodent physiology, affecting such things as aggression level and reproductive development (Nelson et al. 1997; Silva et al. 2010). Mice are a nocturnal species, but most testing of mice occurs during the light phase for practical reasons. For many experiments, testing mice during the dark, active phase is optimal. For this reason, some animal rooms operate on a reversed light cycle (Yang et al. 2007). In such rooms, it is important that animal care staff only enter the room during the light phase. If staff or experimenters need to enter the room during the dark cycle, they should use night vision goggles or equip the room with red lights due to the fact that murine eyes are insensitive to red light. If superovulating mice, variable light cycles will be required for certain strains.

Housing Type

The primary housing types used for laboratory mice include conventional static cages, microisolator cages, individually ventilated cages (IVCs), and flexible film plastic isolators. The choice of housing can affect the spread of airborne microorganisms, the transfer of nonairborne fomites, and the effectiveness of ventilation. Conventional shelving has the advantages that it is the least expensive and allows easy access to the animals, but provides the lowest level of protection from pathogens and potentially poor air circulation. Use of static caging may necessitate adjusted husbandry practices, including cage change frequency, to improve the environment. Microisolator cages are similar to conventional cages but have covers containing high-efficiency particulate air (HEPA) filters. Microisolators provide a higher level of pathogen protection than conventional cages but may have worse air circulation. Microisolators provide a cost-effective way to provide a higher level of pathogen protection to parts of a room. A major advantage to IVCs is increased frequency of air changes and a reduction in the ammonia levels within cages (Smith et al. 2004). Numerous studies have demonstrated detrimental health effects of increased ammonia levels on rodents (Serrano 1971; Broderson et al. 1976; Gamble and Clough 1976; Tepper et al. 1985; Berg et al. 1986; Van Winkle and Balk 1986). IVCs also typically use HEPA-filtered air that is very effective at preventing contamination of animals by other cages or room personnel (Feistenauer et al. 2014). IVCs may also be designed so that the air pressure in the cage is positive relative to the surrounding environment, providing an additional level of protection for prevention of cross-contamination between cages. Some lines of mice may not breed well in IVCs due to excessive vibration or air changes (Tsai et al. 2003). Also, animals in IVCs may have difficulty thermoregulating, and can show decreased body weight and increased anxiety-related behavioral responses compared with those housed in static caging (Bilkei-Gorzo et al. 2008; Kostomitsopoulos et al. 2012; David et al. 2013). For many bedding types, IVC housing systems allow a reduction in cage change frequency compared with static cages (every 2 weeks vs. weekly) (Reeb-Whitaker et al. 2001; Ferrecchia et al. 2014). Flexible film plastic isolators are stand-alone units with filtered air that provide pathogen protection comparable to that of IVCs while allowing a flexible solution for light mouse loads. Isolators can be pushed together to maximize floor space economy without inhibiting air circulation. An additional advantage to isolators is that a contamination will impact a small number of boxes rather than an entire room. When mouse rooms are renovated, researchers should be aware that changing the type of caging can modify the phenotype of genetic models (Logge et al. 2013, 2014).

Pair versus Trio versus Harem Mating

While the eighth edition of the *Guide* provides recommendations of minimum space for laboratory rodents, it also advises that space requirements should be assessed and modified as necessary according to performance indices (NRC 2011). For breeding units, mice are most often mated in pairs or trios or as harem matings. Pair matings are used for pedigreed stocks where lineage tracing is required, or in cases where breeding as trios is unnecessary. For example, in a strain where litters are large, having two litters in one cage could result in overcrowding and increased mortality and/or prewean discard. Trio breeding units are very common, have been used successfully at many institutions, and occur very frequently in mouse populations (Branchi 2009; Gaskill and Pritchett-Corning 2015). Pups produced in trio units show faster growth rates than young raised by single females (Sayler and Salmon 1969). Trios maximize breeding efficiency because both females will help care for the young, and are most appropriate for strains that generate small litters or are difficult to breed. Harem matings (defined as one adult male and three or four adult females) are useful when the breeding potential of each male has to be maximized. For example, SOD1G93A mice, a model for amyotrophic lateral sclerosis (ALS), are often harem mated because transgenic carrier males have a limited productive breeding life due to the progressive, degenerative neurological phenotype. Harem mating maximizes the number of offspring that can be produced from each breeder male. When harem mating is used in conventional shoebox cages, females must be removed from the breeder unit when visibly pregnant. This separation of the pregnant females makes it impossible to take advantage of postpartum estrus in females. As described below, in vitro fertilization (IVF) may provide an attractive alternative that allows many animals to be produced from a single male even more efficiently than harem mating.

Timed Pregnancy

A practical approach to obtaining rodent embryos is to mate the male and female, and then check the female early the following morning for a vaginal plug. The plug is made of coagulated secretions from the accessory sex glands of the male, and persists for 8–24 hours after mating. The plug can be visualized by lifting the female by the base of the tail and examining the vaginal opening for a whitish mass. It can also be felt with a toothpick or pipette tip. The morning of plug detection is considered gestational day 0.5.

The presence of a vaginal plug does not guarantee pregnancy. In rodents, the estrous cycle lasts approximately 4–5 days (Nelson et al. 1982). The estrous cycle is divided into four stages: proestrus, estrus, metestrus, and diestrus. Typically, only ~10% of females are in the correct stage of estrous to become pregnant. This can be improved somewhat by using the same effects described above for burst mating, the Lee–Boot and Whitten effects. The success of timed pregnancy matings may be significantly improved by identifying females in proestrus (Byers et al. 2012). This can be done visually for mice, and improves the success of timed pregnant matings to as high as 90%. An alternative method to identify mice in proestrus is by collecting and analyzing vaginal secretions (Caligioni 2009). In the proestrus stage, vaginal secretions predominantly consist of nucleated epithelial cells. Several groups have published protocols to synchronize estrous using injections of progesterone (Hasegawa et al. 2016) or a combination of cloprostenol and progesterone (Pallares and Gonzalez-Bulnes 2009). Pregnancy can be confirmed by palpation or weight gain as early as gestational day 7 (Mader et al. 2009) (Figure 29.1).

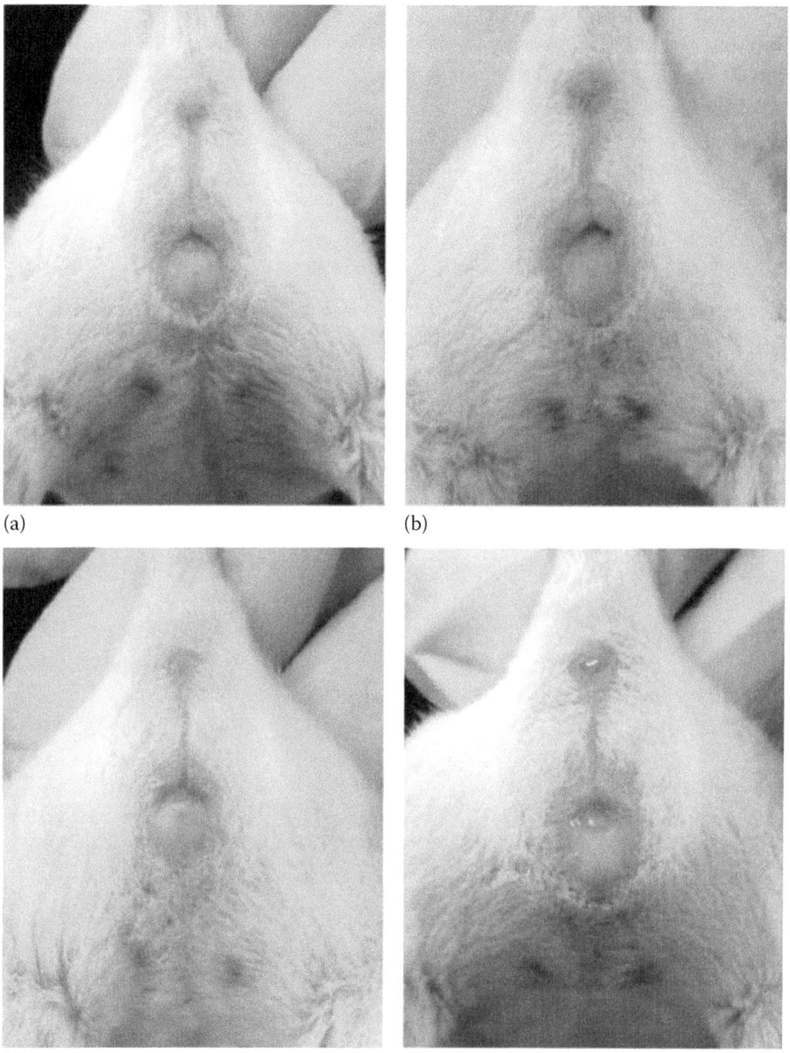

(a) (b) (c) (d)

FIGURE 29.1 **(See color insert.)** Four stages of estrous in Balb/cByJ mice. (a) Proestrus. (b) Estrus. (c) Metestrus. (d) Diestrus. (Reprinted from Byers, S.L. et al., *PLoS One*, 7, 4, 2012.)

Social Housing of Mice

According to the *Guide*, appropriate social interactions among members of the same species are essential to normal development and well-being (NRC 2011). In rodents, the effects of social deprivation appear to be critically dependent on the specific isolation conditions, in particular maternal deprivation versus peer isolation, and the age at which the isolation was experienced. At any age, social deprivation acts as a stressor. Isolation of adult rats is associated with changes in food consumption, which can affect body weight and potentially longevity (Georgsson et al. 2001). In mice, social isolation during adolescence increased voluntary ethanol intake (Lopez and Laber 2015). In adult mice, single housing has been shown to lead to lower body weight, as well as increase depressive or anxiety-like behavior (Kalliokoski et al. 2014; Demuyser et al. 2015). Also, group housing can reduce levels of anxiety and depression induced by chronic stress (Liu et al. 2013). As such, rodents should be housed in compatible same-sex or breeding groups whenever possible.

There are circumstances where mice must be housed individually. Approval for individual housing should be secured from the Institutional Animal Care and Use Committee (IACUC) before housing mice individually. Singly housed mice should be given supplemental nesting materials, as well as environmental enrichment that will encourage activity such as gnawing, climbing, or hiding. Normal reasons for individual housing include

- Particularly aggressive animals may need to be removed from group housing situations.
- Males used as studs outside of permanently maintained breeding units cannot be group housed with other males because of aggressive behavior.
- In studies involving aging, individual mice may eventually outlive their cage mates and end up living alone. Recombining these mice is a stressor that might adversely affect scientific objectives.
- Other situations may arise that necessitate individual housing and should be considered on a case-by-case basis.

Quality Control

Genetic Quality Control

There are numerous examples of genetic contamination in both commercial and private colonies of mice. The strain C57BLKS arose through the inadvertent crossing of C57BL/6J and DBA/2J mice (Naggert et al. 1995). More recently, a lab at Harvard found that the immune phenotype they were studying was due to a spontaneous mutation that created a copy number variant in the C57B6N/6Hsd substrain (Mahajan et al. 2016). These rare breeding errors will change strain characteristics substantially and compromise the reproducibility of experimental data. It is important that genetic monitoring is routinely conducted to maintain the authenticity of murine strains. The most effective way to prevent genetic contamination is for all technicians and researchers to be observant and diligent with record keeping and animal care. Simple steps to minimize the chances of strain contamination include

- Ensure adjacent strains are as phenotypically and genotypically distinct as possible, including minimizing housing colonies of the same coat color adjacent to one another
- Use different color cage cards to make each colony visually distinguishable
- Maintain diligent breeding records
- Rigorously monitor all aspects of phenotype during routine colony management
- Practice animal handling techniques that minimize the chances of an escaped animal
- Cryopreserve strains as a backup

Monitoring Genetic Background

Panels of single-nucleotide polymorphism (SNP) markers are an excellent tool for periodic monitoring of genetic background (Petkov et al. 2004a, 2004b). At the Jackson Laboratory, all breeders in foundation stock colonies are monitored using a 32-SNP panel to identify and eliminate possible contaminations before they can propagate through the colony. Pedigreed and production colonies from each strain are randomly monitored at least once per year.

Genotyping Confirmation

Strains containing mutated alleles that are maintained as hom colonies should have breeders in foundations and pedigreed colonies routinely confirmed by polymerase chain reaction (PCR) genotyping to ensure the genotype of the breeders is correct. It is also advisable to routinely confirm the genotype of mice in pool colonies that are transported from one institution to another, or even to different mouse rooms within the same institution, as much as is possible. For transgenic lines, it may be necessary to monitor transgene copy number or, as in the case of Huntington's disease models, repeat size (Zwiegers et al. 2014).

Preventing Genetic Drift

Spontaneous mutations are random and occur at a low mutation rate in most genes. Estimates of mutation rates in mammals vary widely depending on the ascertainment method, the types of genes assessed, and the calculation method. Inbred strains will acquire new mutations in the germline with each successive generation. Genetic drift is the random change in the frequency of a gene variant. Accumulated genetic drift can impact any aspect of an animal's phenotype, including breeding performance. This is especially important in small colonies, since the chance of a mutation becoming fixed is equal to its prevalence in that population (Kimura 1983). After 20 generations of intercrossing, a line is considered to be a substrain. Strategies to minimize genetic drift include cryopreservation and backcrossing. For inbred strains, once a good line is identified, embryos may be frozen and periodically used to refresh the colony in order to reduce the number of generations animals are bred and minimize genetic drift. While there is no golden rule for when to refresh, a good rule of thumb is to refresh a colony after 10 generations of breeding. Lines with mutant alleles may be maintained by continual backcrossing to the appropriate inbred background, or can also be cryopreserved using sperm or embryos. If backcrossing or cryopreserving using sperm, it is important to get inbred breeders or oocyte donors from a reliable commercial breeder with a program to maintain genetic stability (Taft et al. 2006). This will prevent the accumulation of genetic drift and ensure that the strain has a constant genetic background.

Identification of Individual Mice

There are many options to identify individual mice. Temporary forms of identification include marking the mice with a felt-tip marker, dying the fur, or clipping the fur. All these are of limited duration and also may be restricted to light-colored mice. The most common forms of permanent identification include ear notching and ear tagging. More unusual options include radio frequency identification (RFID) microchips. Each method has advantages and disadvantages. Ear notches are easily misread, and can move as the animal grows. Ear tags can fall out. Microchips implanted subcutaneously can move or be difficult to scan. Also, none of these permanent ID methods can be used on neonatal mice. For young mice, toe clipping and tattooing are reasonable alternatives, and may be used on neonates. Investigators must work with the IACUC to decide what methods are appropriate.

Breeder Selection

Breeder selection is key to preserving the phenotype of a strain. Desirable animals are those exhibiting the expected phenotype and not exhibiting undesirable traits and characteristics. For individual colonies, breeders are selected by established criteria based on the strain's breeding performance, pedigree lineages, and individual strain characteristics. New breeders come from lineages that have expected breeding characteristics. The criteria for defining normal breeding include litter frequency, litter size at birth and weaning, frequency of NP matings, development of expected phenotypes, and absence of abnormal phenotypes. For strains with segregating genotypes, select breeders that produce progeny genotypes at or near the expected Mendelian ratios. Lineages that deviate from these criteria should be eliminated and special care taken to breed away from unacceptable characteristics.

Pedigreed Colonies

Pedigreed colonies allow the tracking of each breeder pair and their progeny to their ancestors. Records are tracked by a unique pedigree number assigned to each mating in pedigree charts and on individual cage cards. The pedigree allows traceability to avoid the inadvertent fixation of unwanted mutations, especially mutations that affect reproductive performance or alter the expected phenotype. When such a mutation is suspected, it is necessary to rid the breeding colony of all copies of the mutation that may be carried as het by individuals that appear unaffected. It is advisable to maintain multiple parallel pedigreed lines of each strain derived from a common progenitor, and two or three generations of each line on the shelf. If an individual lineage produces an unexpected phenotype, animals in that line should be eliminated.

Pool Colonies

Offspring of pool colonies are pooled at wean, so any unexpected phenotype that presents among inventory in a pool colony cannot be traced back to an individual breeder pair. For this reason, pool breeders should only be one generation removed from pedigree breeders. Pool-to-pool breeding should be avoided when possible. If breeders in the pool colony have come from a different room and contain a known mutation, the genotype of the breeders should be confirmed at mating.

Record Keeping

To evaluate the breeding performance of a mouse colony, meticulously maintain accurate records and examine them regularly. The sooner a problem is detected, the sooner it can be corrected. The following habits should be cultivated:

- Investigate deviations in breeding performance immediately.
- Compare your colony's performance with that characterized by your supplier. If mutant strain data is unavailable, use data for the inbred strain.
- Keep a colony's environmental conditions stable and keep detailed records of environmental parameters.

Tips for Breeding Mouse Lines

General Information about Mice

The following list is modified from Flurkey (2009).

Number of chromosomes	20
Gestational length	18–21 days
Birth weight	0.75–1.5 g
Litter size	2–12 pups
Wean age	18–28 days
Sexual maturity	28–60 days
Adult weight	20–40 g
Estrous cycle	4–5 days
Reproductive life span (female)	6–12 months
Reproductive life span (male)	12–14 months
Life span	1.5–3 years

Optimizing Breeding Performance

Replacements. Maintain breeder units of various ages by replacing a percentage of them monthly or weekly. A colony of mixed-age breeders produces a more consistent number of pups than does a colony of even-aged breeders.

NP breeders. As a general rule, replace breeders that do not produce litters within 90 days of their birth date, if it has been 60 days since the last litter, or if they produce two or three consecutive litters that do not wean any pups.

Use young mice. Mate mice that are 6–8 weeks of age. Pairing young females with older males may also improve breeding.

Improving Breeding for Difficult Strains

Fostering

Females of some strains are poor mothers, or have a phenotype that can impact the female's ability to care for her young. In these cases, fostering is a relatively simple option. The foster mother must have a healthy litter of her own that is within 1 or 2 days of age of the fostered pups. It is helpful if there is a coat color difference between the natural pups of the foster mother and the fostered pups. The foster litter should be no larger than the natural litter. To foster, remove the foster mother from the cage. Place the foster litter in the mother's home cage and cover them with some nesting material or bedding so they acquire the foster mother's scent. Alternatively, hold the dam over the foster litter until she urinates on them. Observe the pups for a day or two to make sure the foster mother is caring for the litter. Strains that make excellent foster mothers include CD1 and FVB.

Ovarian Transplants

Some strains of mice produce normal gametes but cannot sustain a pregnancy. Other strains can breed normally, but develop a degenerative neurological phenotype that limits their reproductive life. In these cases, it can be advantageous to maintain a colony using ovarian transplants. Ovaries are removed from the donor female and transplanted into a recipient female. The donor ovaries may be cut in half so that a single donor can produce four ovarian transplant (OT)-recipient female breeders. Two key factors in determining a suitable recipient strain are histocompatibility with the donor and a coat color difference. If the recipient female's ovaries are not completely removed, some pups may come from the recipient female's gametes. Coat color differences offer a simple way to distinguish between the donor and recipient female's gametes. For strains containing mutant alleles, genotyping can also be used to verify that pups are derived from the donor ovary.

In Vitro Fertilization

IVF can be used to produce embryos for cryopreservation, create large cohorts of age-matched mice for experiments (a speed expansion), or rapidly expand a colony's size. This saves months compared with traditional breeding and only requires a single male mouse. Superovulated females are used as oocyte donors and eggs are fertilized *in vitro*. Viable embryos are transferred into pseudopregnant recipients. An additional advantage is that IVF provides an opportunity to simultaneously improve a colony's health status.

Allele-Specific Genotyping and Unwanted Alleles

Strains carrying regions flanked by loxP sequences can have the DNA between the loxP sites removed by an allele expressing *cre* recombinase. The Jackson Laboratory's Cre Repository of driver lines has more than 300 strains. Many more exist in individual labs, and they are often genotyped using generic Cre assays. When maintaining multiple lines, an accidental cross-contamination between lines that contain different Cre alleles is undetectable. It is advisable, whenever possible, to develop an allele-specific assay so genotyping results will be specific for the allele being tested.

Crossing Cre or Flp drivers with strains containing LoxP or FRT cassettes presents additional complications. A number of Cre-expressing strains have been shown to have unexpected low-level germline expression. In these cases, when crossing to a strain with a LoxP cassette, it is possible that a germline null will be created. If not specifically testing for the allele that has the LoxP cassette deleted, an investigator may be unaware that there are two distinct segregating alleles present. This can confound genotyping and experimental results. When performing this type of cross, it is advisable to design an assay for the LoxP-recombined allele and test for it as an unwanted allele.

In addition to deleted LoxP and FRT cassettes, there are a number of other common alleles that can be tested for as unwanted alleles. When importing a new strain, it is a good quality control check to assay for other common alleles using generic primers. In addition to *Cre* and *Flp*, these include commonly used genes, such as *Neo*, *GFP*, LacZ, and *luciferase*. Some commercially available high-density platforms using SNPs provide the opportunity to confirm background, as well as test for a range of unwanted contaminating alleles at a very reasonable cost.

Macaques and Baboons

Introduction

Survey results regarding the use of nonhuman primates during calendar years 2010 through 2012 indicate that the most nonhuman primates used in biomedical research and testing in the United States were from three species: *Macaca fascicularis* (cynomolgus macaque), *Macaca mulatta* (rhesus macaque), and *Macaca nemestrina* (pigtail macaque). Cynomolgus macaques were the most frequently used, followed by rhesus macaques (Lankau et al. 2014). More recent data for importation of macaques into the United States indicates that 23,003 cynomolgus and 649 rhesus, but no pigtailed, macaques were imported in 2014 (Kelley 2014). These animals were imported from China, Vietnam, Mauritius, and Cambodia. Therefore, the majority of nonhuman primates used in biomedical research and testing in the United States (i.e., cynomolgus macaques) are born and raised outside the United States.

Cynomolgus and pigtailed macaques are successfully reared in outdoor or sheltered enclosures, which lack environmental temperature support, without significant weather-related infant loss in tropical countries of origin (i.e., Mauritius, Indonesia, Vietnam, Cambodia, and Hainan Island China). Therefore, production of cynomolgus, as well as pigtailed, macaques outside the United States is attractive due to lower costs where animals that breed throughout the year can be raised in a more natural social structure in a warm environment that more closely adheres to their natural habitat.

However, there are challenges associated with production and importation of nonhuman primates from foreign sources. Many commercial airlines have significantly restricted or totally banned shipment of cargoes of nonhuman primates, making importation of smaller numbers of animals difficult and limiting shipment of these animals to charter flights. This has significantly increased the cost of shipping and stress to animals associated with handling large numbers of animals for preshipment testing, loading, and unloading, and during postarrival quarantine.

Facility construction and maintenance and animal care and management at foreign production sites are extremely variable. Before selecting foreign animal producers as sources for animals to be used in research or testing, the production sites should be audited to ensure that (1) the facilities are appropriate and well maintained, (2) the animals in the production colonies are legally obtained and shipped appropriately, (3) adequate and consistent nutrition is provided, (4) adequate sanitation practices are employed, (5) and adequate veterinary care and health monitoring are provided by appropriately trained and experienced veterinarians.

This section considers basic reproductive physiology and management considerations for the three most commonly used macaque species previously listed and baboons (*Papio* spp.). This discussion is necessarily limited in scope, and readers are directed to more detailed information on this topic located in a recent review (Tardif et al. 2012), as well as extensive numbers of publications available in the scientific literature.

Physiology

Female macaques and baboons exhibit a menstrual cycle similar to that of women, where the lining of the uterus is periodically sloughed associated with stages of the endocrine reproductive cycle (Saltzman et al. 2011). Thus, these animals are good models for studies involving reproductive biology, and menstrual cycle parameters are used for managing reproduction of these animals in the laboratory for research-associated purposes, as well as production of these animals for basic science and pharmaceutical development and testing. Female rhesus, cynomolgus, and pigtailed macaques are sexually mature at about 3.3 years of age, while female baboons (*Papio* sp.) are sexually mature at about 4 years of age. Generally, males reach sexual maturity at a slightly younger age than females, but may not be behaviorally and socially mature enough to mate successfully.

The menstrual cycle may range from approximately 26 to 30 days for macaques and 33 days for baboons. The beginning of the cycle is usually considered to be the day of menstruation onset. Menstruation may be detected overtly by observation of blood in the perineal region. Menstruation is followed by the follicular phase when the ovarian follicles are developing. The follicle phase culminates in the ovulation phase when the eggs are released from the follicles, followed by the luteal phase. The fertile period is during the late follicular phase and ovulation phase. Depending on some differences between the species and individuals, the fertile period is generally between the 10th and 14th days of the menstrual cycle.

For timed mating, where females are exposed to males for limited periods of time or for artificial insemination, a method to determine optimal fertility for individual animals is to determine the average menstrual cycle length for the previous three cycles, divide the average cycle length by two, and subtract three. The female is then exposed to the male or artificial insemination is performed for two or three subsequent days (Hendrickx and Dukelow 1995).

Female reproductive cycles may also be exhibited by changes in sex skin in the animals' perineal region. These sex skin changes are associated with cyclic hormonal changes that occur during the menstrual cycle. Rhesus macaques may demonstrate increased sex skin erythema and sometimes edema nearing the ovulatory phase of the cycle. Cynomolgus and pigtailed macaques demonstrate edema of the sex skin, but usually no marked change in skin coloration (Fortman et al. 2002). While noticeable in many animals, these sex skin changes are not consistent predictors of maximum fertility, time of ovulation, or pregnancy.

Baboons demonstrate very dramatic changes in perineal sex skin associated with the stages of the menstrual cycle, with maximum tumescence and a deep red coloration observable near ovulation. Identification of ovulatory phases of the reproductive cycle can be more precisely identified by quantitative measurement of ovarian and pituitary reproductive-associated hormones in the serum or urine (Tardif et al. 2012). Measurement of these hormones is particularly useful when employed to support artificial insemination and assisted reproductive technologies (Wolf 2004; Kubisch et al. 2006; Simerly et al. 2010).

Pregnancy Detection

In a typical macaque and baboon production setting, the most practical and quickest methods for pregnancy diagnosis are by palpation and ultrasonography. External abdominal palpation can be used to detect pregnancy by late in the first trimester. Bimanual palpation, where a finger of one hand is inserted in the rectum and the other hand supports and palpates the abdomen externally, can detect pregnancy as early as 16 days gestation when conducted by experienced individuals (Hendrickx and Dukelow 1995).

Ultrasonography is a highly reliable method for pregnancy diagnosis, as well as determining fetal viability, assessing fetal development and age, and assessing uterine health and placental condition, location, and disease (Brizzee and Dunlap 1986; Tardif et al. 2012). Pregnancy can also be determined through methods that measure the presence of chorionic gonadotropin in the serum or urine (Tardif et al. 2012); however, these methods require collecting samples from the animals, take time and laboratory equipment and support to complete, and are generally no more accurate than palpation or ultrasonography procedures that are performed by well-trained and experienced personnel. It is not unusual for females to demonstrate implantation bleeding during the first month of pregnancy. This is a normal event, but may be confused with reentry into the menstrual cycle.

Fetal loss does occur during pregnancy, with the majority of fetal mortality occurring early in pregnancy (Tardif et al. 2012). Because parturition tends to occur during the nighttime hours, animals with dystocia may not be observed until the fetus is dead or in severe distress. Additionally, dystocia may not be detected in animals housed in group enclosures or more naturalistic environments until the dam demonstrates signs of severe distress and illness. Dystocia has been attributed to many factors (Small 1982; Stockinger et al. 2011). However, certain practices, such as air transportation (Sackett 1981) and daily capture and handling (Newell-Morris et al. 1989; Tarantal and Hendrickx 1989), that would be expected to induce stress and perhaps adversely affect fetal viability do not seem to adversely affect pregnancy outcome.

Colony Management

Macaques and baboons have been produced using a variety of housing and breeding strategies, including free ranging on islands (with or without food supplementation); in large groups with multiple males and females housed in outdoor corral or field cage enclosures; in harem units with usually one male, but sometimes two, and multiple females in smaller sheltered pens or corncribs (that may or may not have some degree of environmental temperature control); and in cage housing (Kelley and Crockett 2012).

Cynomolgus, rhesus, and pigtailed macaques and baboons have menstrual cycles, ovulate, and give birth throughout the year, while rhesus macaques have a strong seasonal cycle, usually having fertile mating in the fall and giving birth in the spring or early summer. The seasonality of production is an important consideration when determining the feasibility of breeding cynomolgus or pigtailed macaques in facilities that are not environmentally controlled. Cynomolgus and pigtailed macaques are from tropical environments and do not tolerate cooler temperatures well. When infants of these species are born during the winter months in outdoor or sheltered but not environmentally controlled facilities, there may be a high infant loss.

Rhesus macaques typically do not have infants born during the cooler winter months. Therefore, these animals will produce well in outdoor or sheltered housing, without any or minimal environmental temperature control, in geographical areas with mild winters and warm weather during the spring and summer.

Baboons are more tolerant of cool temperatures and may not experience significant infant mortality if infants are born during cooler weather, as long as shelter and environmental temperature support are available.

Social Structure

In the wild, macaques and baboons live in large groups or troops comprising multiple male and female adults with progeny. Females form matriarchal hierarchies that tend to remain stable throughout multiple generations, with female progeny assuming the rank of their mothers. At maturity, males move out of their birth group and eventually join other established troops or form their own separate social groups.

When breeding in groups (either harems or multimale configurations), it is best to maintain the social structure within the groups to the extent possible to maximize production success and reduce fighting and fetal or infant mortality. Rhesus macaques tend to be most resistant to changes in social structure, with fighting and potentially severe injury or death often associated with forming new groups of adult animals, the introduction of new animals, the removal of high-ranking males or females, or other manipulation of the social hierarchy. Cynomolgus macaques and, to a lesser extent, pigtailed macaques and baboons tend to be more tolerant of changes to social structure (Williams and Bernstein 2012).

Pregnancy Successes

Pregnancy rates (number of documented pregnancies/average number of females at risk for pregnancy) for different production schemes vary considerably and range from approximately 40% to 80% (Tardif et al. 2012). Most females can produce one infant per year. For management of production colonies, an even more useful statistic is the number of progeny that survive to 1 year of age per female at risk for pregnancy, since most infant mortality occurs within a fairly short time after birth and infants that live to a year of age will likely survive to adulthood and/or for use for research or testing. If the average number of surviving infants per breeding female is 70%, you may expect 100 adult females at risk of pregnancy to produce 70 surviving progeny per year, with approximately 50% male and 50% female offspring. If the colony is self-sustaining, some offspring will have to be held for breeder replacement. The number of animals to be held for breeder replacement depends on the age of retirement of the active breeders. The prime production age tends to be from about 4 to 8 or 9 years of age for females (older for males). Primiparous females tend to be less attentive mothers, with a higher incidence of infant mortality.

Within the last couple decades, there has been an increased interest in acquiring macaques that are free of certain diseases that may affect research or are of concern for human occupational health considerations. This has led to efforts to produce specific pathogen-free (SPF) animals. The targeted agents to be eliminated or excluded for SPF status may vary but usually include *Mycobacterium tuberculosis*, *Macacine herpes virus* 1, simian immunodeficiency virus (SIV), simian retrovirus D (SRV-D), and simian T lymphotropic virus (STLV) (Morton et al. 2008). Because eliminating these agents from established breeding colonies is time-consuming and expensive, and slows production efforts, newly established breeding colonies should be started with the goal to either exclude these agents initially or rapidly eliminate infected animals from the colony.

In order to avoid introduction of excluded infectious agents, there is an interest in closing breeding colonies to new introductions from outside the institution or facility. It is paramount to maintain pedigree records and establish breeding programs to reduce inbreeding to achieve maximum heterozygosity. Pair mating and single-male harem breeding allow for parentage to be more readily recorded. Animals in large multimale and multifemale breeding groups may have to be genetically tested so that inbreeding can be avoided to the extent possible (Kanthaswamy et al. 2010).

Nutrition and Feeding

Animals should be fed a high-quality diet that has been appropriately supplemented with essential vitamins and minerals. Nutrient requirements for nonhuman primates were recently reviewed (Powers et al. 2012), along with a somewhat dated but detailed review of available information regarding nonhuman primate nutrition (NRC 2003).

Some macaque and baboon colony managers feed production animals diets with a higher protein content, approximately 25%, compared with approximately 15% for a standard diet. Group-housed animals should be offered food in a quantity and manner that allows all animals to have adequate access to food. Routine weighing of group-housed animals is a way to ensure that all animals are getting sufficient nutrition to meet their needs. Nonhuman primates are frequently provided novel food supplements (e.g., nuts, grains, fruits, and vegetables) to provide additional enrichment, as well as provide an additional source for essential vitamins (e.g., vitamin C). As with the regular diet, dietary supplements intended for enrichment should be provided in a manner that ensures that all animals can access the supplements and they do not become the primary source of nutrition or cause illness or severe social conflict (Figures 29.2 through 29.5).

FIGURE 29.2 Corn crib housing structures.

FIGURE 29.3 Corral housing for macaques.

FIGURE 29.4 Baboon indoor corral housing.

FIGURE 29.5 Indoor baboon gang cages with working chutes.

Marmosets

Marmosets are small, South American monkeys that are highly territorial and normally live in groups with only one reproductive female and one reproductive male. They are arboreal and have well-developed visual, olfactory, and auditory communication systems. In the wild, they are both a predator and a prey, and so have well-developed vigilance systems and spend much time scanning the environment for signs of danger. They cooperatively raise their offspring—fathers and older offspring will carry infants and provide them with solid food.

Housing

The temperature in marmoset housing and breeding rooms should be maintained between 22°C and 30°C, and relative humidity is recommended between 40% and 70% (Heger 1983). However, marmosets maintained indoors in drier climates may acclimate to lower humidity and develop clinical signs of a poor hair coat when exposed to humidity above 40%. The air exchange rate of 10–15 per hour is acceptable for the housing room if appropriate temperature and humidity are maintained. Sudden loud noises should be minimized.

Marmoset caging should provide ample space and diverse surfaces that promote natural behaviors. Provision of wooden perches and extensive mesh surfaces ensures that the animals can both see other animals in a room and use those surfaces for locomotion. Marmosets spend a significant amount of time clinging to vertical surfaces, including mesh cage walls. The other important aspect of the cage environment is a nest box. This should be positioned high in the cage design and serves as a resting area that gives each animal the opportunity to remove itself from view. This feature provides animals with an extra level of control over their visual interactions with others. A simple nest box can be created by cutting a hole in the side of a 1-gallon plastic container and suspending it from the side of the enclosure. More complex nest box designs can facilitate isolation and containment of animals for hands-free capture and removal from the cage.

The marmoset cage design should emphasize vertical space instead of floor space. Since marmosets spend a significant amount of time clinging to vertical surfaces, the most important dimension for marmoset cages is vertical. They spend almost all their time in quadrupedal, arboreal locomotion or vertical clinging, using perches, mesh cage surfaces, and a nest box or boxes as their primary perching areas. Mesh balconies are an additional cage design consideration. These are attached externally to cage openings and provide animals with more flexibility in terms of how other animals in a room can be observed at different times of the day. The cage floor is not commonly used for perching or other sedentary activities.

The number of breeding pairs that can be maintained in a room is variable, and depends on room size, cage design, size, and arrangement within the room (Layne 2003). Cages should initially be arranged so breeders and their offspring have the ability to break visual contact with other family groups in the room. This can be accomplished with the cage itself if the design includes solid wall or partition surfaces. Supplemental visual barriers external to the cage, such as plastic curtains, can be strategically placed within the room to facilitate visual avoidance and/or hiding among room inhabitants. After initial setup of room cages with inhabitants, it is important to maintain consistency of neighbors. A change of location of established family groups should only be considered if neighbor breeding females display signs of fascination or excessive aggression with each other that may affect breeding or infant rearing success.

Cage Sizes

All marmoset cages should contain wooden perches, as well as the nest box. The Public Health Service (PHS) guide recommends for Group 1 primates (up to 1.5 kg) 2.1 ft^2 in floor space and 30 in. in height for a pair of animals—that is, 5.25 ft^3 for a pair of marmosets or 2.62 ft^3 per animal. The goal of this space requirement is to provide enough space for the maintenance of normal postures and locomotion.

Marmosets measure approximately 13 in. from head to tip of tail. Their normal sitting posture is approximately 11 in. from top of head to tip of tail. As defined above, the PHS-recommended cage dimensions provide a clearance of around 12 in. from the tip of the animal's tail to the floor, and 6 in. from the top of the animal's head to the ceiling, and adequate room for the animal to practice normal behaviors, including quadrupedal movement, leaping, and vertical clinging.

Table 29.2 provides marmoset cage dimensions, with cubic footage that we have used—both total and per animal. Our method of calculating cage size relies more on cubic footage than square footage for providing more vertical height.

Social Housing

Breeding Groups

Breeding groups are maintained as one breeding female, one breeding male and multiple litters of their offspring. Offspring may be maintained in their natal group well into adulthood and older offspring only removed as required by either limitations on breeding cage size or signs of unacceptable levels of agonistic behavior within the groups. In an ideal situation, animals are retained in their natal groups until they are required for breeding or study.

Nonbreeding Pairs

Marmosets normally live in groups with only one reproductive male and female, with the remainder of the group composed of that pair's offspring. Marmosets of both sexes are highly territorial and extremely

TABLE 29.2

Marmoset Housing Recommendations

Housing Type (No. of Animals)	Floor Square Footage	Height (in.)	Cubic Footage	Cubic Footage per Animal
Breeding groups (up to 8)	7.47	46.5	28.95	3.62
Pair housing (2)	2.85–5.24	57.5	13.66–25.11	6.83–12.55
Single housing (1)	2.85–5.24	27.5–29.0	6.53–12.66	6.53–12.66

aggressive toward unfamiliar individuals of the same sex. The only published studies purporting to successfully pair-house unrelated marmosets of the same sex used adolescents as one member of the pair (Majolo et al. 2003). This strategy is less desirable, so adolescents should be maintained with their natal group. Therefore, when animals must be removed from their natal group and are not placed directly into breeding, attempt one of the two following pair-housing strategies when possible.

Same-Sex Siblings

If there are same-sexed siblings available, they can sometimes be removed from the natal group together and housed as a pair. These pairs must be routinely monitored for signs of extreme aggression or submission during daily husbandry activities, as their stability is highly variable.

Female: Vasectomized Male

If there are males available that are known to not be required at any point in the future for breeding or reproductive studies, those males may be vasectomized and housed with a female.

Single Housing

It is often not possible to enact a non-breeding-pair strategy when an animal is removed from its natal group. In that case, the animals are housed singly in a room with other marmosets. If marmosets are singly housed, they should have visual, auditory, and olfactory communication with at least two other animals in a room. It has been documented that the behavior and physiology of singly housed marmosets is affected by their singly housed neighbors in a way similar to that of group-housed animals (Tardif et al. 1994).

Feeding and Nutrition

Common marmosets are small, approximately 350–450 g as adults, with a relatively short gastrointestinal tract compared with other nonhuman primate species. They consume a high-energy diet, which must be supplemented with vitamin D3 in captivity. All members of a family group commonly have *ad libitum* access to what is presented as a daily ration to the cage, so there usually is no difference in diet for the breeding pair of animals. There are commercially available canned, extruded pellet and biscuit base diets available for marmosets and other New World monkeys that provide complete nutritional requirements for the species. In addition, purified or semipurified diets are also available that facilitate consistency in nutritional components and presentation. Common marmosets are omnivorous and opportunistic consumers, so periodic addition of a variety of environmentally enriching food supplements, for example, fruits, raisins, and mealworms, is a recommended feeding practice. Water is provided *ad libitum* in multiple bottles with sipper tubes attached to the cage.

Breeding Strategy

Common marmosets will breed when housed as mated pairs of sexually mature animals. Sexual maturity in males occurs from 11–13 months of age. The age of onset of sexual maturity in females is similar; however, conception and production of live offspring by females usually starts between 2 and 2.5 years of age in captive colonies (Layne and Power 2003; Tardif et al. 2003). The recommended minimum age for pairing is 18 months old. Marmosets do not show distinct, overt clinical signs of estrus, and sexual behavior is not a good indicator of ovulation, as it may be observed when the female is not ovulating. They routinely ovulate multiple ova per cycle. The estrous cycle is 28 days. The average gestation length is 143–144 days. Commercially available human or Old World nonhuman primate laboratory pregnancy determination tests are not reliable in marmosets. Early confirmation of pregnancy is possible through urine testing for a sustained increase of ovarian metabolites (Heger 1983). Ultrasound is useful for the detection of growing fetuses between 60 and 95 days of gestation. Litter size ranges from one to four; the average litter size will vary for individual colonies.

Common marmosets exhibit a fertile postpartum estrus within 10–20 days of parturition. The median interbirth interval is reported to be 154–162 days (Layne 2003; Tardif et al. 2003). A breeding pair will become a family unit or colony as offspring age and are retained with the original sire and dam. This housing strategy can be maintained indefinitely. Behaviorally dominant females will suppress ovulation of submissive cage mates or nearby females using tactile, visual, and olfactory signals. In turn, the suppressed, cohoused females, as well as the related male offspring, will assist with infant care, including carrying the young animals and offering them solid food.

Parturition usually occurs during the dark phase of the light cycle, and new offspring are first observed during morning health checks and feeding. Females will lactate for 65–90 days postpartum. Milk production will decrease significantly after day 60. Offspring will be carried by the dam or other cage members from birth to approximately 3 weeks old, when they start to venture off the adult animals. Carrying typically ceases altogether by 8 weeks of age. Offspring will start weaning onto solid food at approximately 4 weeks of age through active and passive acquisition (i.e., stealing and begging from adult cage mates).

Common marmosets produce large litters and have shorter interbirth intervals than other research-related species of nonhuman primates, so their reproductive potential is high. However, individual breeding females are reported to have the expectation of five or fewer successful litters within their life span (Tardif et al. 1994). This is due to a number of factors, the most common being a high rate of pregnancy loss and high infant mortality. The overall lifetime production of a breeding female in captivity can be expected to be 1.5–3.0 viable young per breeding year.

Clinical Issues

Given the appropriate housing and nutritional conditions, marmosets will provide rewarding production in a research setting, and individual family groups can be expected to grow in number over time. Dystocia can occur, and caesarean section can be performed successfully with appropriate surgical techniques. As mentioned previously, the two most common production issues are early pregnancy loss and infant deaths. The etiology of pregnancy loss is not clearly understood and likely multifactorial. Routine clinical health assessment of the dam will ensure that she is physically and physiologically capable of supporting pregnancy and postparturient care of the young. A stable physical and social environment, to include minimizing noise disturbance, regular timing of routine husbandry procedures, and "neighborhood" consistency within a housing room, will help minimize external stressors that may affect pregnancy maintenance.

Once born, all infants are predisposed to mortality from a number of factors. Among primary considerations are poor parenting and diminished infant condition, that is, less able or unable to cling to or move about on the dam or transfer between the dam and other adults in the cage. The most important factor associated with infant mortality is increased litter size (three). Large litters, with more than two infants, have high neonatal mortality. Provision of supplemental care by the animal care staff for weak or "extra" infants is labor-intensive but can be successful. The management of litters larger than two is dependent on the situation at the time of the birth and is ultimately the decision of the experienced colony management staff, but the following basic steps outline a common decision-making scheme:

1. *Assessment of infants' condition.* Infants who are small (<24 g) or show signs of significant weakness (cannot cling to an adult without repeatedly falling off) are extremely unlikely to be successfully reared, so these infants may be removed and euthanized shortly following birth to minimize suffering and perhaps increase the chances of successful rearing of the remaining litter.
2. *Fostering.* If all infants are judged to be in good condition and there is another dam in the colony that is nursing one infant and has given birth within 3 weeks, one of the supernumerary infants may be removed and placed with the group with one infant. This process generally involves placing the infant in the nest box and allowing the foster dam and sire access to it. This process is successful in the majority of cases.

3. *Hand rearing.* Partial or full hand earing of supernumerary infants can be successful. Those management decisions are based on production needs balanced with the acknowledgment that these infants may be phenotypically different from fully marmoset-reared animals, particularly in terms of the impact of early life nutrition. Hand rearing takes one of two forms:
   a. *Rotational rearing.* In rotational rearing, one infant per day may be removed from the group and fed formula. In these cases, the infant is best placed in a cage or incubator immediately adjacent to its home cage. The infant is returned to the group at the end of the day, and a different infant removed on the following day. In this way, each infant has access to marmoset milk for at least part of its rearing.
   b. *Full nursery rearing.* For full nursery rearing, an infant may be removed from the group and formula fed. In these cases, the infant is again best placed in a cage or incubator immediately adjacent to its home cage or adjacent to the cage of the group into which it will be fostered upon weaning. The infant is generally returned to a social group by 3 months of age.
4. *Observation and euthanasia.* If, for management reasons, hand rearing any of the infants is not chosen, then the group should be closely observed and weak infants removed and humanely euthanized in order to reduce suffering. If an infant is found alive, but on the floor in the cage, it should be immediately removed and its condition assessed. If the infant is of acceptable weight and strong enough to cling, it can be warmed and then placed into the nest box. The group can then be observed 1–2 hours later to see if the infant has been retrieved and is clinging to a carrier. Frequent observations of the group at shorter intervals may actually interfere with the parents' response to the infant, so leave the group alone for a time sufficient to let the group settle and retrieve the infant. If the infant is not retrieved or if the infant is retrieved but is again found on the floor of the cage (likely indicating insufficient strength), then the infant should be immediately removed and euthanized.

The other common clinical issue that directly affects infant marmosets is "harassment" by adults in the cage, which is frequently associated with transfer from one adult to another. Harassment behaviors are considered a normal part of cooperative infant care. They can include rubbing the infant on different parts of the cage and/or nipping of the hands, feet, and tail during transfer. The affected animals routinely undergo this behavior with little ill effects. However, exuberant tail biting is clinically observable via blood-tinged fur on the tip of a tail that may be reduced to a nub at the base of the rump over time. Routine observation during daily health checks is warranted; however, clinical intervention is usually not necessary (Figure 29.6).

FIGURE 29.6 Representation of typical sitting and vertical postures in marmosets.

Dogs

Introduction

Whether to maintain a canine breeding colony in a research setting is often dependent on the research needs of the investigators at the institution. There are several Class A producers of high-quality dogs for research use in the United States and other countries. These producers have animals that are genetically well defined, have detailed health monitoring and health status records, and can generally supply dogs of various ages and sexes as required. It is usually not feasible (due to available physical plant resources and overall cost) to maintain dog breeding colonies unless there is need for certain genetic lines to study spontaneous diseases, such as muscular dystrophy and cancer (Shelton and Engvall 2005; Shearin and Ostrander 2010; Rowel et al. 2011; Alvarez 2014; Cekanova and Rathore 2014; McGreevy et al. 2015). The number of animals necessary per year to fulfill research requirements will dictate how many bitches will be necessary. Tracking pedigrees is obviously necessary to prevent inbreeding. Programs to determine the inbreeding coefficient should be used, and the purchase of new males may be necessary to increase genetic diversity.

The physiology, veterinary, and management considerations required to produce dogs in a research environment are very similar to those used for the production of dogs by breed fanciers, hobbyists, and commercial producers of Class A dogs for research. This section is necessarily brief, and an extensive review of canine reproduction is beyond the scope of this effort. The reader is encouraged to review subject-specific material for more information (Feldman and Nelson 2004; Kustritz 2010; Greer 2014).

Physiology

Canine bitches have a monoestrus reproductive cycle with an ovulatory clinical estrus cycle occurring twice a year. The estrus cycle is divided in four stages (Kustritz 2010): proestrus (approximately 9 days, follicular stage), estrus (approximately 9 days, fertile stage with ovulation), diestrus (approximately 51 days, coincident with pregnancy and whelping if pregnant), and anestrus (approximately 4 months). The stage of the bitch's reproductive cycle can readily be detected by examining cells collected during a vaginal smear or flush as follows.

During proestrus, as estrogen levels rise, leading to capillary breakage and leakage, the vaginal smear will change over 4–7 days from noncornified intermediate cells to an increasing number of anuclear (cornified) squamous cells, neutrophils, and red blood cells. The bitch will attract male dogs but be unreceptive. A blood-tinged vaginal discharge may be present with accompanying moderate vulvar enlargement.

During estrus, the bitch becomes receptive to the male, the vaginal discharge decreases, and the vaginal smear will have nearly 100% anuclear squamous or cornified cells, superficial cells, and diminishing numbers of red cells. The average estrus period is 9 days (3 days to 3 weeks). During estrus, bitches behaviorally exhibit a "standing heat," which coincides with their most fertile period, when the bitch will readily solicit and stand for mounting and intromission by the male. If natural breeding is not possible, artificial insemination can be easily accomplished in females when they are in standing heat, also identified by the maximum numbers of cornified cells in the vaginal smear. Dogs ovulate 5–7 days prior to the onset of diestrus (7–9 days post–luteinizing hormone [LH] surge). The LH surge occurs with the initial increase in progesterone; thus, timing of ovulation can be estimated by measuring either hormone. Ovulation timing should be done in conjunction with vaginal smears to determine the beginning of estrus (70% cornified cells), followed by progesterone tests every 48 hours to detect the rise in progesterone (2–3 ng/mL), which coincides with the LH surge triggering ovulation. Breeding should occur between 2 and 7 days of the LH surge. It takes 1–3 days for the oocytes to mature, and then their life span is 2–3 days; thus, the fertile period is from day 3 to day 6 or 7 post-LH surge.

During diestrus, the vaginal smear will have more than 50% intermediate cells and some cornified cells, but transitioning to noncornified cells with neutrophils present as diestrus progresses. During anestrus, the vaginal smear will have small numbers of parabasal cells and intermediate cells, with or without neutrophils. The diestrus–anestrus period can last from 4.5 to 10 months, with an average of about 7 months.

It is not unusual for bitches to demonstrate a false pregnancy (pseudocyesis) during the diestrus phase of the estrus cycle, with mammary gland development, lactation, and nest-building behavior being demonstrated. This is a normal physiological event that can involve veterinary intervention to treat discomfort associated with marked mammary gland enlargement.

Pregnancy can generally be detected by external abdominal palpation of the bitch about 28 days after breeding. The developing embryos and adjacent structures form a series of oval firm swellings in the early gravid uterus approximately 2 in. in length. Later in pregnancy, these structures become diffusely enlarged and more difficult to discern by external abdominal palpation. However, canine fetal skeletons become calcified by about 42 days of gestation and can be detected radiographically. Ultrasound is also a very practical method for pregnancy diagnosis, with detection of the embryonic vesicles by approximately gestation days 25–28 and detection of the developing fetuses throughout the pregnancy.

Relaxin tests are commercially available and detect pregnancy in the bitch by measuring levels of the hormone. Relaxin is produced by the developing placenta following implantation of the embryo, and can be detected in the blood in most bitches as early as 22–27 days postbreeding. The level of relaxin remains elevated throughout gestation and declines rapidly following the end of the pregnancy. Relaxin can reliably differentiate between false and true pregnancies since no actual placental development occurs in pseudopregnancy.

Pregnant bitches that are nearing parturition should be housed separately from other dogs and provided nesting boxes and nesting material. Many bitches become agitated or very protective of the nest area if they are in visual contact with other dogs within the same kennel. Therefore, it is beneficial to house bitches with solid-sided kennels and without visual contact of other dogs. Impending parturition can be predicted by a drop in body temperature in the bitch to below 100°F before giving birth. As parturition approaches, bitches may be restless and nervous, engage in nesting behavior, and demonstrate panting, mild tachycardia, and increased respiratory rate.

The bitch is usually recumbent during actual delivery. The time interval between the delivery of each pup is usually less than an hour, with the bitch removing the fetal membranes and licking each pup as it is born. During whelping, human interaction should be minimal, to prevent maternal stress and cannibalism of the puppies. However, if bitches are in labor for more than 5 hours without delivering a pup or if there are more than 2 hours between delivery of subsequent pups, or pups are born dead, veterinary intervention should be considered. Oxytocin may be indicated in cases of uterine inertia (0.25–0.5 IU/kg intramuscularly [IM]) up to three times at 15- to 30-minute intervals. Oxytocin must only be used in cases of nonobstructed dystocia.

Puppies are born with eyes and ears closed and unable to micturate and defecate without stimulus from the mother (or artificial stimulation if puppies are being hand reared). Generally, the eyes open at approximately 12 days of age and ears become patent between 12 and 20 days of age. Solid food can be introduced at 4–6 weeks of age, and puppies can be weaned between 6 and 8 weeks of age.

Generally, bitches are most productive up to about 4–5 years of age, and after this time, the anestrus interval tends to increase. After about 8–9 years of age, reproductive complications increase, rendering older bitches less likely to contribute to colony production (Figures 29.7 through 29.10).

Breeding Colony Management

For discussion purposes, we use a colony of dogs with an inherited X-linked disorder (muscular dystrophy) as an example of colony management, but the principles relate to other traits as well. X-linked disorders lead to an increased incidence in males; therefore, to maintain the colony, affected males are bred to carrier females, which produce an expected ratio of 1:1 affected to nonaffected per litter (25% male affected, 25% female [homozygote] affected, 25% obligate female carriers, 25% normal males).

The mean litter size averages 7 pups (range 1–13), with an average of 2.5 litters per dam (range 1–7) produced over the course of a carrier's lifetime. With this condition, an average of 2.25 *affected* pups per litter *survive* beyond 10 days following birth. (The mean litter size for smaller-breed dogs is expected to be approximately 57% lower than for larger breeds.) Increased inbreeding will also contribute to increased pup mortality.

Conception rates are ~85%. Relaxin (pregnancy) tests are run at 30 days following the first day of diestrus, and litters are typically due 57 days from the first day of diestrus (27 days from pregnancy confirmation). Carriers are not removed from the colony at a set age but are removed based on the regularity of their estrus cycle, rate of pregnancy, litter size, and pup survival. With these factors in mind, most carriers continue to be bred until they are 5–7 years of age.

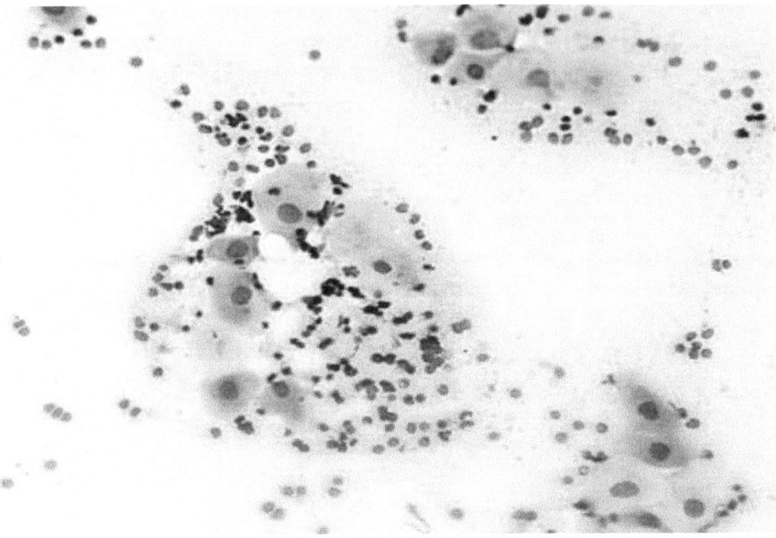

FIGURE 29.7 **(See color insert.)** Proestrus with neutrophils, red blood cells, and parabasal cells. (From R. Bowen, Vaginal Cytology, http://www.vivo.colostate.edu/hbooks/pathphys/reprod/vc/index.html.)

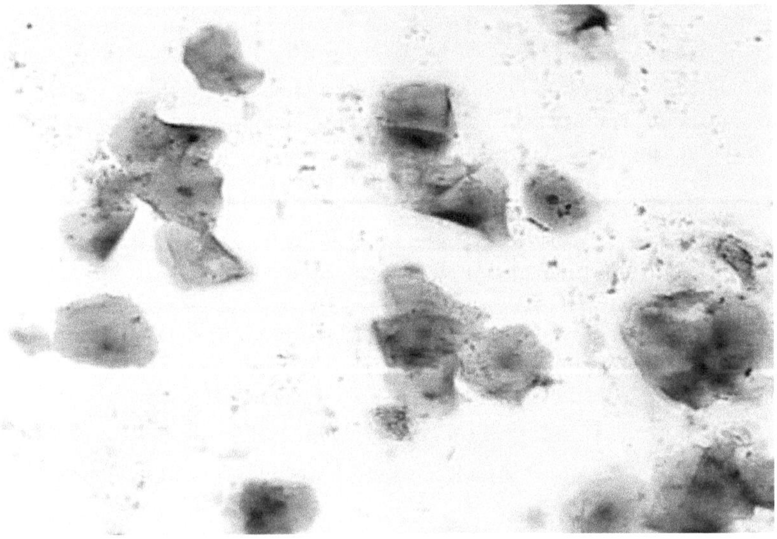

FIGURE 29.8 **(See color insert.)** Estrus with anucleate cornified cells. (From R. Bowen, Vaginal Cytology, http://www.vivo.colostate.edu/hbooks/pathphys/reprod/vc/index.html.)

Inbreeding increases the morbidity of the disease during the neonatal period, leading to increased mortality. For this reason, the inbreeding coefficient must be monitored closely in determining sire–dam breeding pairs. However, inevitably, as long as the colony remains closed, there will be a long-term increase in inbreeding. To combat this problem, periodically (every 3–5 years) introduction of new dogs into the colony or outside semen collection and artificial insemination must be considered.

To calculate the size of the breeding colony needed, put the (*surviving*) affected pup production in a formula that equates to the number of carriers per year needed (based on carrier female × affected male):

2.25 = no. of surviving affected pups per litter

0.83 = average breedings/carrier/year

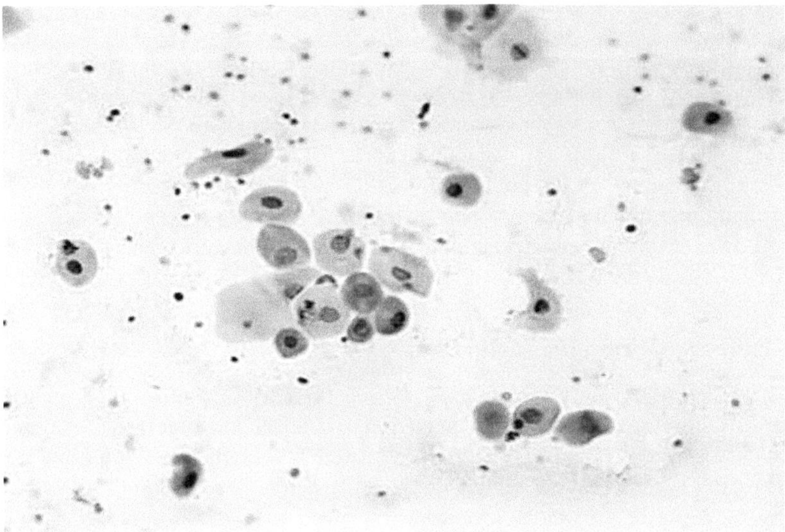

FIGURE 29.9 **(See color insert.)** Diestrus with superficial and parabasal cells. (From R. Bowen, Vaginal Cytology, http://www.vivo.colostate.edu/hbooks/pathphys/reprod/vc/index.html.)

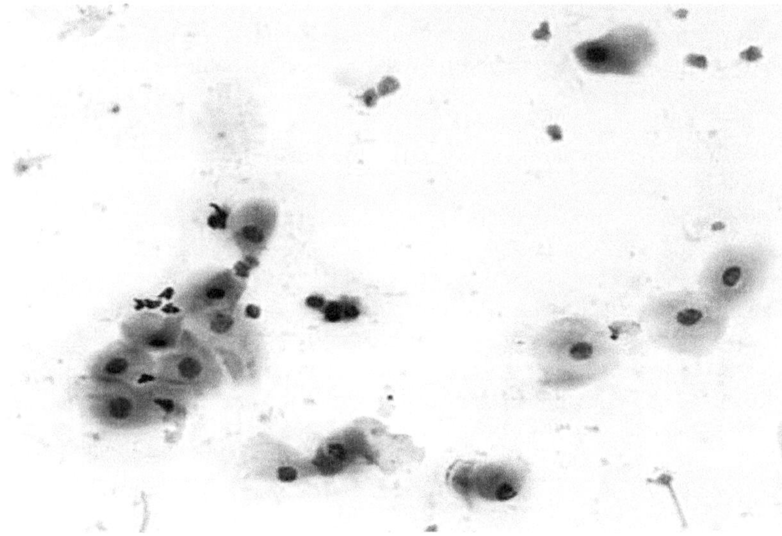

FIGURE 29.10 **(See color insert.)** Anestrus with predominantly parabasal cells and a few neutrophils. (From R. Bowen, Vaginal Cytology, http://www.vivo.colostate.edu/hbooks/pathphys/reprod/vc/index.html.)

2.25 pups/litter × 0.83 breedings/year × 0.85 pregnancies (conception rate) = litters/estrous cycle = 1.59 surviving affected pups per carrier housed annually

The ratio of affected breeding males needed to carrier females is ~1:3. Accordingly, for each affected individual required for a study, there is a demand on the breeding population of ~63% of a carrier female and 20% of an affected (stud) male. To put it another way, this means that for each Golden Retriever muscular dystrophy (GRMD) pup born, we have traditionally anticipated that we would need to budget 0.66 carrier and 0.20 breeding males. Or, based on the history of our production, we must maintain 1.5 carriers per affected dog and 0.5 breeding males per affected dog to be produced.

Breeding

Carrier bitches are monitored three times weekly (Monday, Wednesday, Friday) for estrus activity beginning at 6 months of age. When outward signs of proestrus are noted (vulvar edema and swelling, and bloody discharge), daily vaginal swabs and smears are taken to confirm the stage of estrus by cytology. Assays of serum progesterone levels are sometimes performed on select bitches to determine the optimal day for breeding. Depending on the physical condition of the males, breeding can be done naturally or through artificial insemination. Semen quality (motility, morphology, and concentration) should be evaluated in advance of their breeding. Breeding should and can be performed every other day through the first day of diestrus for an average of five matings or inseminations per bitch in estrus.

Pregnancy Determination and Parturition

At approximately 30 days postbreeding, relaxin tests can be used to confirm pregnancy. In addition to serum relaxin tests, ultrasound may be done on bitches to confirm pup viability. The anticipated parturition date is calculated as day 57 postdiestrus for pregnant bitches. One week prior to the estimated due date, recording daily body temperature and serum progesterone levels (collected ~1–3 times prior to parturition) can be performed to predict imminent whelping.

Neonatal Care

Affected pups may be weak at birth and require nutritional supplementation for extended periods. Colostrum from other dogs may be harvested using a human breast pump and frozen so that it can be used as a supplement for weaker pups that are pushed aside by more vigorous littermates and/or do not show satisfactory weight gain. Pups should be weighed soon after birth and, on average, four times daily for the first 2 weeks to monitor progress. Other observations of pup development may be righting response, whereby a pup is placed on its back and allowed to right itself; rooting; latching onto teats; and suckling ability of the pup when placed near the dam. Any evidence of weakness of these responses or failure to thrive should be noted and intervention planned. In addition to the aforementioned physical responses, any evidence of respiratory distress, abdominal distension, and/or significant weight loss (i.e., sustained lack of weight gain or loss of 20 g over a 24-hour period) should be annotated.

Those puppies perceived not to be gaining adequate weight should be supplemented by hand feeding 4–6 mL/100 g body weight of the bitch's milk or a commercially available milk replacement (Nurturall-C and/or Esbilac) at 4-hour intervals.

Puppies are allowed to nurse naturally until weaned at 6 weeks of age. They can be introduced to a gruel mixture of milled puppy kibble, 2nd Step (Esbilac) milk powder, and water at approximately 3 weeks of age and gradually weaned.

REFERENCES

Abou-Ismail UA, Mahboub HD. 2011. The effects of enriching laboratory cages using various physical structures on multiple measures of welfare in singly-housed rats. *Lab Anim* July; 45(3):145–153.

Abramov U, Puussaar T, Raud S, Kurrikoff K, Vasar E. 2008. Behavioural differences between C57BL/6 and 129S6/SvEv strains are reinforced by environmental enrichment. *Neurosci Lett* 443:223–227.

Alvarez CE. 2014. Naturally occurring cancers in dogs: Insights for translational genetics and medicine. *ILAR J* 55:16–45.

American Institute of Nutrition. 1977. Report of the American Institute of Nutrition ad hoc committee on standards for nutritional studies. *J Nutr* 107:1340–1348.

Bazille PG, Walden SD, Koniar BL, Gunther R. 2001. Commercial cotton nesting material as a predisposing factor for conjunctivitis in athymic nude mice. *Lab Anim* (NY) 30:40–42.

Berg RW, Buckingham KW, Stewart RL. 1986. Etiologic factors in diaper dermatitis: The role of urine. *Pediatr Dermatol* 3(2):102–106.

Bilkei-Gorzo A et al. 2008. Environmental modulation of anxiety-related neuronal activity and behaviors. *Behav Brain Res* 186(2):289–292.

Blom HJM, Van Tintelen G, Van Vorstenbosch CJ, Baumans V, Beynen AC. 1996. Preferences of mice and rats for types of bedding material. *Lab Anim* 30:234–244.

Branchi I. 2009. The mouse communal nest: Investigating the epigenetic influences of the early social environment on brain and behavior development. *Neurosci Biobehav Rev* 33:551–559.

Brizzee KR, Dunlap WP. 1986. Growth. In *Comparative Primate Biology*, ed. WR Dukelow, J Erwin, 363–413. New York: Alan R. Liss.

Broderson JR, Lindsey J, Crawford J. 1976. The role of environmental ammonia in respiratory mycoplasmosis of rats. *Am J Pathol* 85:115–130.

Buddaraju AKV, Van Dyke RW. 2003. Effect of animal bedding on rat liver endosome acidification. *Comp Med* 53:616–621.

Burn CC, Mason GJ. 2005. Absorbencies of six different rodent beddings: Commercially advertised absorbencies are potentially misleading. *Lab Anim* January; 39(1):68–74.

Byers SL, Wiles MV, Dunn SL, Taft RA. 2012. Mouse estrous cycle identification tool and images. *PLoS One* 7:4.

Caligioni C. 2009. Assessing reproductive status/stages in mice. *Curr Protoc Neurosci* Jul Appendix 41:1–11.

Cekanova M, Rathore K. 2014. Animal models and therapeutic molecular targets for cancer: Utility and limitations. *Drug Dev Ther* 8:1911–1922.

Dalal SJ, Estep JS, Valentin-Bon IE, Jerse AE. 2001. Standardization of the Whitten Effect to induce susceptibility to Neisseria gonorrhoeae in female mice. *Contemp Top Lab Anim Sci* 40:13–17.

David JM, Chatziioannou AF, Taschereau R, Wang H, Stout DB. 2013. The hidden cost of housing practices: Using noninvasive imaging to quantify the metabolic demands of chronic cold stress of laboratory mice. *Comp Med* October; 63(5):386–391.

Demuyser T, Bentea E, Deneyer L, Albertini G, Massie A, Smolders I. 2016. Disruption of the HPA-axis through corticosterone-release pellets induces robust depressive-like behavior and reduced BDNF levels in mice. *Neurosci Lett* 626:119–125.

Feistenauer S et al. 2014. Influence of 5 different caging types and the use of cage-changing stations on mouse allergen exposure. *J Am Assoc Lab Anim Sci* 53(4):356–363.

Feldman EC, Nelson R. 2004. *Canine and Feline Endocrinology and Reproduction*. St. Louis, MO: Saunders/Elsevier.

Ferrecchia CE, Jensen K, Van Andel R. 2014. Intracage ammonia levels in static and individually ventilated cages housing C57BL/6 mice on 4 bedding substrates. *J Am Assoc Lab Anim Sci* 53(2):146–151.

Flurkey K, ed. 2009. *The Jackson Laboratory Handbook on Genetically Standardized Mice*. 6th ed. Bar Harbor, ME: Jackson Laboratory.

Fortman JD, Hewett TA, Bennett, BT. 2002. *The Laboratory Nonhuman Primate*, 24–31. New York: CRC Press.

Gamble MR, Clough G. 1976. Ammonia build-up in animal boxes and its effect on rat tracheal epithelium. *Lab Anim* 10:93–104.

Georgsson L, Barrett J, Gietzen D. 2001. The effects of group housing and relative weight on feeding behaviour in rats. *Scand J Lab Anim Sci* 28:201–209.

Gaskill BN, Pritchett-Corning KR. 2015. The effect of cage space on behavior and reproduction in Crl:CD1(ICR) and C57BL/6NCrl laboratory mice. *PLoS One* 10:e0127875.

Greer ML. 2014. *Canine Reproduction and Neonatology*. Jackson, WY: Teton Newmedia.

Gurfein BT et al. 2012. The calm mouse: An animal model of stress reduction. *Mol Med* 18:606–617.

Haemisch A, Gartner K. 1994. The cage design affects intermale aggression in small groups of male laboratory mice: Strain specific consequences on social organization, and endocrine activations in two inbred strains (DBA/2J and CBA/J). *J Exp Anim Sci* 1994 August; 36(4–5):101–116.

Hasegawa A, Mochida K, Inoue H, Noda Y, Endo T, Watanabe G, Ogura A. 2016. High-yield superovulation in adult mice by anti-inhibin serum treatment combined with estrous cycle synchronization. *Biol Reprod* 94(1):21.

Heger HW, Neubert D. 1983. Timing of ovulation and implantation in the common marmoset, Callithrix jacchus, by monitoring of estrogens and 6 beta-hydroxypregnanolone in urine. *Arch Toxicol* September; 54(1):41–52.

Hunt C, Hambly C. 2006. Faecal corticosterone concentrations indicate that separately housed male mice are not more stressed than group housed males. *Physiol Behav* 87:519–526.

Hutchinson EK, Avery AC, Vandeqoude S. 2012. Environmental enrichment during rearing alters corticosterone levels, thymocyte numbers, and aggression in female BALB/c mice. *J Am Assoc, Lab Anim Sci* January; 51(1):18–24.

Hendrickx AG, Dukelow WR. 1995. Breeding. In *Nonhuman Primates in Biomedical Research: Biology and Management*, ed. TB Bennett, CR Abee, R Hanrickson, 335–374. Vol. 1. New York: Academic Press.

Hess SE, Rohr S, Dufour BD, Gaskill BN, Pajor EA, Garner JP. 2008. Home improvement: C57BL/6J mice given more naturalistic nesting materials build better nests. *J Am Assoc Lab Anim Sci* 47:25–31.

Jackson Laboratory. 2009. *Breeding Strategies for Maintaining Colonies of Laboratory Mice: A Jackson Laboratory Resource Manual*. Bar Harbor, ME: Jackson Laboratory.

Kaliste EK, Mering SM, Huuskonen HK. 2006. Environmental modification and agonistic behavior in NIH/S male mice: Nesting material enhances fighting but shelters prevent it. *Comp Med* 56:202–208.

Kalliokoski O, Teilmann AC, Jacobsen KR, Abelson KSP, Hau J. 2014. The lonely mouse—Single housing affects serotonergic signaling integrity measured by 8-OH-DPAT-induced hypothermia in male mice. *PLoS ONE* 9(12):e111065.

Kanthaswamy S, Capitano JP, Dubay CJ. 2010. Resources for genetic management and genomics research on nonhuman primates at the National Primate Research Centers (NPRCs). *J Med Primatol* 38(Suppl 1):17–23.

Kelley S. 2014. From Freedom of Information Act request to Fish and Wildlife Service for data regarding importation of *Macaca fascicularis*, *Macaca mulatta*, and *Macaca nemestrina* into the United States during calendar year 2014.

Kelley ST, Crockett CM. 2012. Laboratory housing of nonhuman primates. In *Nonhuman Primates in Biomedical Research: Biology and Management*, ed. CR Abee, K Mansfield, S Tardif, T Morris, 251–268. Vol. 1. New York: Academic Press, Elsevier.

Kimura M. 1983. *The Neutral Theory of Molecular Evolution*. Cambridge, MA: Cambridge University Press.

Kostomitsopoulos N et al. 2012. The effects of different types of individually ventilated caging systems on growing male mice. *Lab Anim* 41(7):192–197.

Kraft LM. 1980. The manufacture, shipping and receiving, and quality control of rodent bedding materials. *Lab Anim Sci* 30:366–376.

Kubisch HM, Gagliardi C, Williams VM, Ribka EP, Ratterree MS. 2006. In vitro fertilization in the pigtailed macaque (*Macaca nemestrina*). *Theriogenology* 66:749–754.

Kustritz MVR. 2010. *Clinical Canine and Feline Reproduction: Evidence Based Answers*. Ames, IA: Wiley-Blackwell Publications.

Lankau EW, Turner PV, Mullen RJ, Galland GG. 2014. Use of nonhuman primates in North America. *J Am Assoc Lab Anim Sci* 53:278–282.

Layne DG, Power RA. 2003. Husbandry, handling and nutrition for marmosets. *Comp Med* 53:351–359.

Liu Z, Wu R, Tai F, Jia R, Wei B, Yang X, Zhang X, Jia R. 2013. Effects of group housing on stress induced emotional and neuroendocrine alterations. *Brain Res* March 28; 1502:71–80.

Logge W, Kingham J, Karl T. 2013. Behavioural consequences of IVC cages on male and female C57BL/6J mice. *Neuroscience*, May; 237:285–293.

Logge W, Kingham J, Karl T. 2014. Do individually ventilated cage systems generate a problem for genetic mouse model research? *Genes Brain Behav* September; 13(7):713–720.

Lopez MF, Laber K. 2015. Impact of social isolation and enriched environment during adolescence on voluntary ethanol intake and anxiety in C57BL/6J mice. *Physiol Behav* September 1; 148:151–156.

Mader SL, Libal NL, Pritchett-Corning K, Yang R, Murphy SJ. 2009. Refining timed pregnancies in two strains of genetically engineered mice. *Lab Anim* 38(9):305–310.

Mahajan V, Ezana D, Mattoo H, Viswanadham V, Varki A, Morris R, Pillai S. 2016. Striking immune phenotypes in diverse gene-targeted mice are driven by a copy number variant originating from a commercially available C57BL/6 strain. *Cell Rep* May 31; 15(9):1901–1909.

Manser CE, Broom DM, Overend P, Morris TM. 1997. Operant studies to determine the strength of preference in laboratory rats for nest boxes and nest materials. *Lab Anim* 32:36–41.

Manser CE, Broom DM, Overend P, Morris TM. 1998. Investigations into the preferences of laboratory rats for nest boxes and nesting materials. *Lab Anim* 32:23–35.

Majolo B, Buchanan-Smith HM, Morris K. 2003. Factors affecting the successful pairing of unfamiliar common marmoset (Callithrix jacchus) females: Preliminary results. *Anim Welf* August; 12(3):327–337.

McGreevy JW, Hakim CH, McIntosh MA, Duan D. 2015. Animal models of Duchenne muscular dystrophy: From basic mechanisms to gene therapy. *Dis Model Mech* 8:195–213.

McGregor PK, Ayling SJ. 1990. Varied cages result in more aggression in male CFLP mice. *Appl Anim Behav Sci* 26:277–281.

Morton WR, Agy MB, Capurenio SV, Grant RF. 2008. Specific pathogen-free macaques: Definition, history, and current production. *ILAR J* 49:137–144.

Mulder, JB. 1975. Bedding preferences of pregnant laboratory-reared mice. *Behavior Research Methods & Instrumentation*. 7:21. doi:10.3758/BF03201283

Naggert JK, Mu JL, Frankel W, Bailey DW, Paigeri B. 1995. Genomic analysis of the C57BL/Ks mouse strain. *Mann Genome* February; 6(2):131–133.

Nelson JF, Felicio LS, Randall PK, Sims C, Finch CE. 1982. A longitudinal study of estrous cyclicity in aging C57BL/6J mice: I. cycle frequency, length and vaginal cytology. *Biol Reprod* 27:327–339.

Nelson RJ et al. 1997. The effects of photoperiod and food intake on reproductive development in male deer mice (*Peromyscus maniculatus*). *Physiol Behav* 62(5):945–950.

Newell-Morris LL, Farhenbach CE, Sackett GP. 1989. Prenatal psychological stress, dermatoglyphic asymmetry and pregnancy outcome in the pigtailed macaque (*Macaca nemestrina*). *Biol Neonate* 56:61–75.

NRC [National Research Council (US) Subcommittee on Laboratory Animal Nutrition]. 1995. *Nutrient Requirements of Laboratory Animals: Fourth Revised Edition*. Washington (DC): National Academies Press (US). https://www.nap.edu/catalog/4758/nutrient-requirements-of-laboratory-animals-fourth-revised-edition-1995?gclid=CKfa9fazmtMCFYW6wAodNpEIaw

NRC [National Research Council]. 1996. *Laboratory Animal Management: Rodents*. Washington (DC): National Academy Press. https://www.nap.edu/catalog/2119/rodents?gclid=CIy5rZe0mtMCFcK3wAod13IKQg

NRC (National Research Council). 2003. *Nutrient Requirements of Nonhuman Primates*. Washington, DC: National Academy Press.

NRC (National Research Council). 2011. *Guide for the Care and Use of Laboratory Animals*. 8th ed. Washington, DC: National Academies Press.

Odynets A, Simonova O, Koshunov A, Zaitsev I, Verreva A, Gnilomedova L, Rudzish R. 1991. Beddings for laboratory animals: Criteria for biological evaluation. *Laboratornye Zhyvotnye* 1:70–76.

Pallares P, Gonzalez-Bulnes A. 2009. A new method for induction and synchronization of oestrus and fertile ovulations in mice by using exogenous hormones. *Lab Anim* 43(3):295–299.

Pelkonen KH, Hänninen OO, 1997. Cytotoxicity and biotransformation inducing activity of rodent beddings: A global survey using the Hepa-1 assay. *Toxicology* Sep 26; 122(1–2):73–80.

Petkov PM, Ding Y, Cassell MA, Zhang W, Wagner G, Sargent EE, Asquith S, Crew V, Johnson KA, Robinson P, Scott VE, Wiles MV. 2004a. An efficient SNP system for mouse genome scanning and elucidating strain relationships. *Genome Res* Sep; 14(9):1906–1911.

Petkov PM, Cassell MA, Sargent EE, Donnelly CJ, Robinson P, Crew V, Asquith S, Haar RV, Wiles MV. 2004b. Development of a SNP genotyping panel for genetic monitoring of the laboratory mouse. *Genomics* May; 83(5):902–911.

Powers ML, Toddes B, Koutsos L. 2012. Nutrient requirements and dietary husbandry principles for captive nonhuman primates. In *Nonhuman Primates in Biomedical Research: Biology and Management*, ed. CR Abee, K Mansfield, S Tardif, T Morris, 269–286. Vol. 1. New York: Academic Press, Elsevier.

Ras T, van de Ven M, Patterson-Kane EG, Nelson K. 2002. Rats' preferences for corn versus wood-based bedding and nesting materials. *Lab Anim* 36:420–425.

Reeb-Whitaker CK et al. 2001. The impact of reduced frequency of cage changes on the health of mice housed in ventilated cages. *Lab Anim* 35(1):58–73.

Reeves PG, Rossow KL, Lindlauf J. 1993. Development and testing of the AIN-93 purified diets for rodents: Results on growth, kidney calcification and bone mineralization in rats and mice. *J Nutr* 123:1923–1931.

Rowel JL, McCarthy D, Alvaraz CE. 2011. Dog models of naturally occurring cancer. *Trends Mol Med* 17:380–388.

Sackett GP. 1981. Pregnancy outcomes following jet transport stress in nonhuman primates. *J Med Primatol* 10:149–154.

Saltzman W, Tardif SD, Rutherford JN. 2011. Hormone and reproductive cycles in nonhuman primates. In *Hormones and Reproduction in Vertebrates*, ed. DO Norris, KH Lopez, 291–328. Vol. 5. London: Elsevier.

Sanford AN, Clark SE, Talham G, Sidelsky MG, Coffin SE. 2002. Influence of bedding type on mucosal immune responses. *Comp Med* 52(5):429–432.

Sayler A, Salmon M. 1969. Communal nursing in mice: Influence of multiple mothers on the growth of the young. *Science* 164:1309–1310.

Serrano LJ. 1971. Carbon dioxide and ammonia in mouse cages: Effects of cage covers, population, and activity. *Lab Anim Sci* 21:75–85.

Shearin AL, Ostrander E. 2010. Leading the way: Canine models of genomics and disease. *Dis Model Mech* 3:27–34.

Shelton GD, Engvall E. 2005. Canine and feline models of human inherited muscle disease. *Neuromuscul Disord* 15:127–138.

Silva AL et al. 2010. Effects of photoperiod and experience on aggressive behavior in female California mice. *Behav Brain Res* 208(2):528–534.

Simerly CR et al. 2010. Assisted reproductive technologies (ART) with baboons generate live offspring: A nonhuman primate model for ART and reproductive sciences. *Reprod Sci* 10:917–930.

Small MF. 1982. Reproductive failure in macaques. *Am J Primatol* 2:137–147.

Smith E et al. 2004. Evaluation of cage micro-environment of mice housed on various types of bedding materials. *Contemp Top Lab Anim Sci* 43(4):12–17.

Steinman MQ, Knight JA, Trainor BC. Effects of photoperiod and food restriction on the reproductive physiology of female California mice. *Gen Comp Endocrinol* May 1;176(3):391–399.

Stockinger DE et al. 2011. Risk factors for dystocia in pigtailed macaques (*Macaca nemestrina*). *Comp Med* 61:170–175.

Swetter BJ, Karpiak CP, Cannon JT. 2011. Separating the effects of shelter from additional cage enhancements for group-housed BALB/cJ mice. *Neurosci Lett* 495(3):205–209.

Taft RA, Davisson M, Wiles MV. 2006. Know thy mouse. *Trends Genet* 22:649–653.

Tarantal AF, Hendrickx AG. 1989. Evaluation of the bioeffects of prenatal ultrasound exposure in the cynomolgus macaque (*Macaca fascicularis*). *Teratology* 39:137–147.

Tardif S, Carville A, Elmore D, Williams LE, Rice K. 2012. Reproduction and breeding of nonhuman primates. In *Nonhuman Primates in Biomedical Research: Biology and Management*, ed. CR Abee, K Mansfield, S Tardif, T Morris, 197–249. New York: Academic Press, Elsevier.

Tardif SD, Smucny Da, Abbott DH, Mansfield K, Schultz-Darken N, Yamamoto ME. 2003. Reproduction in captive common marmosets (*Callithrix jachus*). *Comp Med* August; 53(4):364–368.

Tardif SD, Hyde K, Digby L. Evidence for suppression of ovulation in singly-housed female common marmosets (Callithrix jacchus). *Lab. Primate Newsl.* 1994;33:1–4.

Tepper JS, Weiss B, Wood RW. 1985. Alterations in behavior produced by inhaled ozone or ammonia. *Fundam Appl Toxicol* 5:1110–1118.

Thigpen JE, Lebetkin EH, Dawes ML, Clark JL, Langley CL, Amy HL, Crawford D. 1989. A standard procedure for measuring rodent bedding particle size and dust content. *Lab Anim Sci* 39:60–62.

Trainor BC, Martin LB, Greiwe KM, Kuhlman JR, Nelson RJ. 2006. Social and photoperiod effects on reproduction in five species of Peromyscus. *Gen Comp Endocrinol* 148:252–259.

Tsai PP, Selzer HD, Hedrich HJ, Hackbarth H. 2003. Are the effects of different enrichment designs on the physiology and behaviour of DBA/2 mice consistent? *Lab Anim* October; 37(4):314–327.

Van de Weerd HA, Van Loo PLP, Van Zutphen LFM, Koolhaas JM, Baumans V. 1997. Nesting material as environmental enrichment has no adverse effects on behavior and physiology of laboratory mice. *Physiol Behav* 62:1019–1028.

Van der Lee S, Boot LM. 1955. Spontaneous pseudo-pregnancy in mice. *Acta Physiol Pharmacol Neerl* 4:442.

Van Winkle TJ, Balk MW. 1986. Spontaneous corneal opacities in laboratory mice. *Lab Anim Sci* 36:248–255.

Vestal BM, Schnell GD. 1986. Influence of environmental complexity and space on social interactions of mice (Mus musculus and Peromyscus leucopus). *J Comp Psychol* 100:143–154.

Weichbrod RH, Hall JE, Simmonds RC, Cisar CF. 1986. Selecting bedding material. *Lab Anim* 15:25–29.

White WJ, Hawk CT, Vasbinder MA. 2008. The use of laboratory animals in toxicology research. In *Principles and Methods in Toxicology*, 5th ed, ed. Hays AW, 1055–1101. Boca Raton, FL: CRC Press.

Williams LE, Bernstein IS. 2012. Study of nonhuman primate social behavior. In *Nonhuman Primates in Biomedical Research: Biology and Management*, ed. CR Abee, K Mansfield, S Tardif, T Morris, 131–147. New York: Academic Press, Elsevier.

Wolf DP. 2004. Assisted reproductive technologies in rhesus macaques. *Reprod Biol Endocrinol* 2:37.

Yang M et al. 2007. Social approach behaviors are similar on conventional versus reverse lighting cycles, and in replications across. *Front Behav Neurosci* 1:1.

Zwiegers P, Lee G, Shaw CA. 2014. Reduction in hSOD1 copy number significantly impacts ALS phenotype presentation in G37R (line 29) mice: Implications for the assessment of putative therapeutic agents. *J Negat Results Biomed* August; 8:13–14.

30

Managing Husbandry Programs Involving Experimental Hazards

James R. Swearengen, Rebecca K. Holt, and Ronald L. Bowman

CONTENTS

Introduction

Performing husbandry for animals exposed to potential hazards presents many unique challenges to creating and maintaining a safe work environment. While this chapter focuses largely on hazards of biological origin, it also discusses hazards of a nonbiological origin that can be common in an animal research environment, or may be of new or recent interest. For the purpose of consistency, the terms *biosafety* and *containment* may be used when discussing these nonbiological hazards, with the understanding that they are referring to the provision of safe practices and the containment of hazards to prevent inadvertent exposure to people or the environment.

The first step in ensuring a safe work environment is developing a thorough understanding of the concepts of biosafety and containment within all levels of an organization, including the animal facility manager, husbandry staff, veterinary staff, facility maintenance, investigative staff, and senior levels of management (e.g., laboratory director, chief executive officer, and institutional official [IO]). Knowledge of these concepts by all these key positions is critical to help ensure that everyone understands the need for adequate resourcing, both human and fiscal, in order to work safely with the hazards present. While ultimately every person is responsible for their own safety, they must be provided the knowledge and tools in order to work safely and responsibly.

Managing an animal facility, and the husbandry staff therein, when hazards are present requires an in-depth knowledge of the principles of biosafety and how they are applied to the provision of husbandry that also results in the highest level of welfare attainable for each species for which care is provided. The wide array of hazards that may be used in animal research and all the available mitigations for them cannot be completely covered in this chapter, but the authors' intent is to provide a basic understanding of the concerns and approaches of working with hazardous materials in animals, and to provide useful references that can bring additional and more detailed insights should the need arise. While focusing on animal husbandry programs, this chapter provides an overview of some fundamental areas, including the principles of biosafety, the key aspects of several hazards, and a variety of applicable administrative controls, after which the authors share some lessons learned in the management of husbandry programs involving various scenarios based on their experience in these areas.

Basic Principles of Biosafety

Overview of Regulations and Guidelines

The primary guidance on the safe handling and containment of biological hazards in the United States is jointly published by the Centers for Disease Control and Prevention (CDC) and the National Institutes of Health (NIH) (CDC-NIH 2009). This publication, *Biosafety in Microbiological and Biomedical Laboratories* (BMBL), is generally considered an advisory document, but in certain circumstances in the United States, these guidelines are referenced in federal law. The BMBL was first referenced in the 1997 enactment of the Laboratory Registration Select Agent Transfer Program regulation (CFR 1996), which required all facilities shipping or receiving biological select agents and toxins (BSAT) to register with the CDC and comply with the BMBL (Richmond et al. 2003). Subsequent legislation applying to the use of BSAT has been passed in response to growing concerns about its use for nefarious purposes and now applies to all aspects of BSAT possession and use in the United States. The CDC and the U.S. Department of Agriculture (USDA) Animal and Plant Health Inspection Service (APHIS) review biological infectious agents and toxins to determine their effects if misused, and then classify them as BSAT if they have the potential to pose a severe threat to human, animal, or plant health, or to animal and plant products. Once an infectious organism or toxin is designated by the CDC or APHIS as a BSAT, it then becomes subject to all the applicable federal regulations (APHIS-CDC 2015). The BMBL is also considered a reference resource by AAALAC International. The designation of the BMBL as a reference resource by AAALAC International results in its use as a standard by which animal care and use programs utilizing infectious microorganisms and hazardous biological materials are evaluated during accreditation site visits, regardless of their status as a BSAT.

There are excellent international guidelines that provide additional insights to biosafety in animal facilities and are very useful resources. The *Canadian Biosafety Standards and Guidelines* (CBSG) are published by the Public Health Agency of Canada and the Canadian Food Inspection Agency and provide excellent recommendations on handling and housing animals exposed to human and animal pathogens and toxins (PHAC 2013). An updated version of the CBSG, entitled the *Canadian Biosafety Standard* (CBS), came into force on December 1, 2015, and replaced the CBSG for regulatory purposes in Canada (PHAC 2015). The Belgium Scientific Institute of Public Health has published guidance on biosafety in animal facilities that covers many practical areas of concern for facility managers, such as facility design and construction, animal housing, traffic patterns, transport of animals exposed to infectious agents, personal protection, training and education, and emergency plans (Van Vaerenbergh et al. 2011). In addition to a few other country-level biosafety guidelines that are available through the International Federation of Biosafety Association's website, a globally developed biosafety guideline has been published by the World Health Organization (WHO 2004). Information regarding regulations and guidelines pertaining to nonbiological hazards are provided later in the chapter under the specific hazard sections.

Biosafety Levels

Four biosafety levels have been adopted essentially worldwide, and while some minor differences may be found geographically, the basic tenets are the same. Starting at the lowest biosafety level, each subsequent level builds on the standards described for the previous levels. This approach to describing increased levels of protection to personnel and the environment as the biosafety level increases requires the reader to have a solid understanding of not only the level of interest, but also all the preceding levels.

The BMBL addresses a combination of laboratory practices and techniques, safety equipment, and laboratory facilities for each of the four biosafety levels. The BMBL clearly delineates between *in vitro* and *in vivo* work involving infectious organisms and biological toxins, as evidenced by its separate sections for each. One section describes biosafety level criteria for laboratories, and the other section describes vertebrate animal biosafety level (ABSL) criteria for vivarium research facilities. The biosafety levels for activities involving infectious disease work with vertebrate animals are designated as ABSLs 1–4 (CDC-NIH 2009). A summary of the general criteria for assigning a particular infectious microorganism or toxin to a particular biosafety level and the associated practices, safety equipment, and facility guidelines for each ABSL is provided in Table 30.1.

The BMBL does describe additional biosafety levels that specifically address activities using high-consequence agricultural pathogens. High-consequence agricultural pathogens are infectious organisms that, if introduced into the United States, would have serious economic impact due to effects on both agricultural production and international trade of agricultural products. Due to this concern, the standards focus primarily on reducing the risk of the agent escaping outside the containment envelope, where it could impact animal or plant health. When using high-consequence agricultural pathogens in large or loose-housed animals, in which the room serves as the means of primary containment, there is an additional biosafety level designated as BSL-3-Agriculture (BSL-3-Ag). ABSL-3Ag requires very special facility construction standards that design the laboratory or animal room itself to serve as the primary barrier to prevent release of infectious agents into the environment. Also, there may be situations where work with some high-consequence agricultural pathogens can be conducted in small animals that can be maintained in primary containment devices within the animal room, and the room no longer serves as the primary barrier. In this situation, the work may qualify to be performed at ABSL-3 with additional enhancement unique to agriculture (ABSL-3 Enhanced) (CDC-NIH 2009).

Facility Design and Construction

There are two primary concerns when considering the design and construction of an animal facility being built with the intended purpose of working with any experimental hazard. The first is the safety of the people, and the second is the protection of the surrounding environment. While these concepts may sound relatively simple, the construction of these specialized types of facilities has grown to be

TABLE 30.1

Recommended Animal Biosafety Levels for Activities in Which Experimentally or Naturally Infected Vertebrate Animals Are Used

ABSL	Agents	Practices	Safety Equipment (Primary Barriers)	Facilities (Secondary Barriers)
1	Not known to consistently cause disease in healthy adults	Standard animal care and management practices, including appropriate medical surveillance programs	As required for normal care of each species • PPE: Laboratory coats and gloves; eye and face protection, as needed	Standard animal facility • No recirculation of exhaust air • Directional airflow recommended • Hand washing sink is available
2	• Associated with human disease • Hazard: Percutaneous exposure, ingestion, mucous membrane exposure	ABSL-1 practices plus • Limited access • Biohazard warning signs • Sharps precautions • Biosafety manual • Decontamination of all infectious wastes and of animal cages prior to washing	ABSL-1 equipment plus primary barriers • Containment equipment appropriate for animal species • PPE: Laboratory coats and gloves; face, eye, and respiratory protection, as needed	ABSL-1 facility plus • Autoclave available • Hand washing sink available • Mechanical cage washer recommended • Negative airflow into animal and procedure room recommended
3	Indigenous or exotic agents that may cause serious or potentially lethal disease through the inhalation route of exposure	ABSL-2 practices plus • Controlled access • Decontamination of clothing before laundering • Cages decontaminated before bedding removed • Disinfectant footbath as needed	ABSL-2 equipment plus • Containment equipment for housing animals and cage dumping activities • Class I, II, or III BSCs available for manipulative procedures (inoculation and necropsy) that might create infectious aerosols • PPEs: Appropriate respiratory protection	ABSL-2 facility plus • Physical separation from access corridors • Self-closing, double-door access • Sealed penetrations • Sealed windows • Autoclave available in facility • Mechanical cage washer used • Entry through anteroom or air lock • Negative airflow into animal and procedure rooms • Hand washing sink near exit of animal or procedure room
4	• Dangerous/exotic agents that pose a high risk of aerosol infections that are frequently fatal, for which there are no vaccines or treatments • Agents with a close or identical antigenic relationship to an agent requiring BSL-4 until data is available to redesignate the level • Related agents with unknown risk of transmission	ABSL-3 practices plus • Entrance through change room where personal clothing is removed and laboratory clothing is put on; shower on exiting • All wastes are decontaminated before removal from the facility	ABSL-3 equipment plus • Maximum containment equipment (i.e., Class III BSC or partial containment equipment in combination with full-body, air-supplied, positive-pressure personnel suit) used for all procedures and activities	ABSL-3 facility plus • Separate building or isolated zone • Dedicated supply and exhaust, vacuum, and decontamination systems • Other requirements outlined in the source text

Source: Centers for Disease Control and Prevention–National Institutes of Health, *Biosafety in Microbiological and Biomedical Laboratories*, ed. L.C. Chosewood and D.E. Wilson, Government Printing Office, Washington, DC, 2009.

a very complicated process that must involve experienced people from multiple professions; some of these include architects, engineers, contractors, health and safety professionals, facility managers, and veterinarians.

There are several key design and construction points to consider from the facility management and animal husbandry perspectives. The most common deficiency seen by one of the authors is the lack of adequate sealing of animal rooms. Inherent to their design, animal rooms have many penetrations in order to supply the necessary utilities (e.g., water, power, and sewer), and advances in biotelemetry have increased the number of potential penetrations into the animal room envelope in order to provide the necessary wiring for transmitting information being collected from the research subjects. Careful attention to detail must be given to ensure that penetrations are appropriately sealed in order to prevent escape of hazards, as well as prevent entry of pests.

Another relatively common concern seen with animal facilities using hazardous materials is maintaining interior surfaces in a manner that allows effective sanitation. Building materials and surfaces should be highly durable due to the significant loads placed on floors and frequent damage to walls and door frames from frequent moving of heavy caging systems and other equipment. While this type of concern may also be associated with conventional animal facilities, the ability to maintain floors and walls in a hazardous environment becomes much more complicated and the consequences of a construction- or maintenance-related failure much greater. The more complicated logistics of performing maintenance in hazardous areas should be taken into consideration when evaluating the personnel resources required for maintaining these facilities, and a cooperative relationship between animal facility managers and facility maintenance staff is essential. Educating facility maintenance staff of the importance of timely responses and the serious consequences that can occur due to lapses in maintenance may be necessary, especially when facilities that will be utilizing new hazards are being planned and constructed.

While it may not be possible to anticipate every species that will be used when designing an animal facility, planning for flexibility in the animal room design is most desirable, particularly if future research needs could change. Also, the heating, ventilation, and air-conditioning (HVAC) systems should be flexible enough to provide the appropriate environment for a variety of species. As an animal facility manager, it should be understood that even the best planning may not accommodate utilization of every species. Even the best HVAC system can struggle with adjusting for every animal species. For example, feathers and fluff from poultry with a low stocking density resulted in clogged rough filters and prefilters within 18–24 hours in a containment facility (Copps 2005). There are several excellent references that provide detailed information on many aspects of planning, constructing, operating, and managing animal facilities in which hazards are used (Frasier and Talka 2005; NIH 2008; Hessler and Britz 2009; USDA 2012).

Physical Security

With an increasingly aggressive and destructive animal rights movement, physical security has become a paramount issue in nearly all facilities performing animal research, whether or not experimental hazards are present. More sophisticated entry control procedures, such as electronic card readers, have become commonplace for most vivaria. The growing concern of the use of hazardous materials for nefarious purposes by rogue individuals or groups has led to significantly increased physical security requirements, especially where highly hazardous materials (e.g., BSAT, radioactive, and chemical) are used or maintained. The use of hazardous materials should be carefully assessed from a security perspective through the utilization of a site-specific threat risk assessment performed by qualified security professionals and hazard-specific experts.

Examples of both exterior and interior security measures for a facility housing BSAT may include vehicular and pedestrian guard stations, vehicle access barriers, security fencing, exterior remote video monitoring, a 24-hour staffed security station for building or area entry, closed-circuit video monitoring, motion detectors, emergency alarms, and biometric access controls (Jaax 2005). A close relationship with local law enforcement agencies that involves the conduct of joint response drills is necessary to ensure careful coordination and the safety of all personnel involved should a security event occur that requires entry into the animal facility. Education of internal and external security or law enforcement personnel about the risks and required risk mitigation strategies will go a long way in developing the trust and confidence necessary to mount a timely and effective response.

In addition to the heightened physical security requirements resulting from the BSAT regulations and federal oversight, the use of BSAT in animal research brings with it new responsibilities for husbandry staff. Both the USDA and the CDC expect animals exposed to infectious select agents to be accounted for with the same rigor as the agent itself, and to be included as part of the select agent inventory. This means that after any animal is exposed to an infectious select agent, it must be tracked in an inventory system that accounts for each animal from exposure until it can be verified that the agent is no longer present. Consequently, animals must be tracked all the way through euthanasia and ultimately carcass decontamination and disposal. Similarly, tissues obtained from animals exposed to infectious select agents must be tracked as well. Husbandry staff must pay extra attention to animal census, movement within the facility, and final disposition of these animals (and their tissues) exposed to infectious select agents, as the consequences of losing track of a single animal or tissue sample can be significant. Animals exposed to infectious select agents (and their tissues) whose location and/or disposition for which cannot be accounted must be reported to the appropriate Federal Select Agent Program authority (i.e., CDC or USDA) as a loss or theft of a BSAT. This brings significant repercussions involving reporting, investigations, and potential personnel actions. The CDC cites one exception to this requirement, in which animals exposed to select toxins need not be tracked in this manner since toxins bind almost immediately to tissue and are not available for anyone to readily obtain from the animal and utilize them for nefarious purposes.

Due to the increased emphasis on physical security, from both regulatory and animal rights extremism perspectives, a comprehensive approach for controlling access should be implemented. In addition to staff who access the facility on a regular basis, strong access controls should also be in place for visitors and other nonstaff personnel, some of whom may only have occasional or one-time access. This could include visiting scientists, student workers, representatives or repairmen from equipment manufacturers, external tradespersons performing facility repairs or renovations, and vendors. Having a single point of contact that is responsible for granting access to animal housing areas is an effective method for ensuring consistency in how established security procedures and criteria are applied. See Chapter 19 in this textbook for additional information on providing security and protecting research programs.

Practices and Techniques

Planning and Preexecution Activities

Good planning and preparation are key elements in managing husbandry activities for a project using experimental hazards. While many of the planning and preexecution activities are identical for research conducted in noncontainment and containment facilities, the result of a delay in research can have a much bigger impact in containment facilities due to the much higher cost to maintain and operate them. To avoid unnecessary project delays and make certain all required resources and processes are in place prior to the project initiation, animal care and veterinary staff personnel should be involved in the initial stages and throughout the project's protocol planning and risk assessment development processes. In this way, early consideration can be given to such things as the specific animal sources, caging, equipment, supplies, and procedures that may be required. For example, the major laboratory animal suppliers will usually have the most commonly used animal model species and strains available in small to moderate quantities; however, if larger quantities are required, they may need to take significant time to adjust breeding capacities to accommodate. If the project requires the use of nonhuman primates, considerable time may be necessary to make arrangements for a high-quality source of animals that have received appropriate testing to ensure that they meet acceptable health standards and are free of diseases that could negatively impact the research.

Equipment such as specialized caging systems designed for working with hazardous materials can have very long lead times for acquisition, with manufacturers often requiring 3–6 months for turnaround, depending on the quantities needed. Modifications to the manufacturer's standard caging product may be required to accommodate facility specifications, such as door openings or autoclave size clearances. Specialized accessory components may need to be designed and fabricated (e.g., mounting brackets for telemetry monitoring equipment). It is important that the design of this caging, including any modifications or newly fabricated components, is critically reviewed to ensure it is made of durable materials that can be easily sanitized and are free of sharp edges and/or pinch points that could compromise animal welfare or personal protective equipment (PPE), potentially posing exposure risks for workers to the experimental hazardous agents in use. Supplies such as the animal bedding materials may need to be investigated and evaluated ahead of time to ensure that they are appropriate for the type of caging used, can be safely changed in a containment environment, and minimize the potential for worker exposure through aerosolization of chemical or biological experimental hazards.

An ongoing and routine facility and equipment maintenance and repair program should be in place. Animal holding rooms, caging, and equipment, especially in containment facilities, are routinely exposed to harsh chemicals (e.g., bleach, quaternary ammonia compounds, vaporized hydrogen peroxide, formaldehyde-based compounds, corrosive alkaline, and acidic cage wash cleaners) and extreme temperatures and pressures (e.g., autoclaving and cage wash cycles) to ensure appropriate sanitation and decontamination. Over time, these conditions and processes can take their toll. Animal room walls, ceilings, floors, and fixtures (e.g., doors, sinks, and wall guards) should be routinely inspected for damage, peeling paint, and so forth. All wall and ceiling penetrations for such things as electric receptacles, data lines, light fixtures, telemetry wiring, and supply and exhaust air ducts should also be inspected routinely, ensuring that they are and remain properly sealed so that the secondary barrier envelope is maintained. It is often very difficult to make these types of repairs while the project is ongoing, so it is important to plan, schedule, and execute these activities prior to study initiation when possible. Many organizations have adopted the practice of annual shutdowns for all or parts of their animal facilities, to facilitate the performance of routine maintenance procedures.

Animal caging and equipment need to be inspected to ensure ongoing operational integrity prior to being put into use. This includes such things as inspecting and replacing cage-top filter material, making certain the seals and gaskets on containment caging are in good repair and seat properly and ensuring that equipment casters, door closures, cage docking, and locking mechanisms function smoothly. For equipment like biological safety cabinets (BSCs) and containment caging blower units that need to be serviced and certified within set time frames (CDC-NIH 2009), it may be advisable to plan for these procedures around study timelines when possible, even if this means they are accomplished earlier than required to avoid having them performed while the study is ongoing.

Each individual involved in the conduct of the project should have his or her specific occupational health and safety plan reviewed, updated, and adjusted if necessary, according to the potential exposure hazards of the project. If the project involves the use of biohazards for which vaccines are offered, it is prudent to perform this review early in the project planning process since some vaccines may require multiple administrations over time before they are considered to be protective. Applicable standard operating procedures (SOPs) should be reviewed and, if necessary, developed or modified, to ensure that they address project-required husbandry procedures.

Personnel training competencies need to be reviewed with respect to specific project-required procedures. If necessary, training animals should be procured for new or refresher training.

Finally, mock runs of specific procedures or groups of procedures that are dependent on each other should be considered and conducted in noncontainment settings. Regardless of how much time has been taken to consider and envision how project procedures will be accomplished, nothing can take the place of a mock run to help point out process deficiencies. These simulations should mimic actual project procedures as closely as possible and can be very valuable in fleshing out timing, logistical, and resource considerations.

Animal Housing and Husbandry

Appropriate laboratory animal caging is designed to provide a suitable microenvironment for the species being housed, allowing for appropriate temperature, humidity, ventilation, noise and vibration levels, and space (NRC 2011). Several vendors manufacture specialized caging systems that provide an additional layer of protection from hazards and which are often referred to as primary containment caging. Multiple types and levels of primary containment caging systems are available that serve as a barrier between the internal and external cage environments, preventing hazardous agents to which the animals have been exposed from escaping the cage and providing protection for the worker and other animals. A risk assessment can be conducted to determine the most appropriate type of primary containment caging for the project procedures and the experimental hazards being used. Many of the different types of primary containment caging described below are complicated pieces of equipment with multiple component parts. Husbandry personnel need to be thoroughly familiar with their specific caging systems. SOPs should be adopted to ensure there is a complete inspection performed on all components of the units prior to putting them into use to make certain they are in good repair and able to maintain operational integrity. Husbandry procedures in a containment setting and using primary containment caging are much more labor-intensive than conventional husbandry procedures, and this should be taken into account when planning and scheduling husbandry activities. Below are brief descriptions of the more common types of primary containment caging.

1. Static filter-top cages: Primarily used for smaller animals, static filter-top cages are shoe box caging with a filter integrated into the cage top. Air passively diffuses through the filter, and around the edges of the cage top in unsealed types of cage systems, between the inside and outside of the cage. The diffusion of air around the edges of the cage top still maintain containment, similarly to how a petri dish contains bacteria or viruses when being cultured. It is important to note that because of the limited airflow in this type of caging, humidity, CO_2, and ammonia levels increase much quicker than in open-top caging or the other types of containment caging described below (Lipman 1999).
2. Rodent individually ventilated caging (IVC): Providing more effective containment than the static filter-top system, IVC systems are produced by multiple caging manufacturers in several configurations (Maher and Young 2007). Although there are some differences in ventilation methods between different manufacturers of rodent IVC racks, in general, motorized fan units attached to the IVC rack provide air through supply plenums to each individual cage and pull air through ports that are sealed to the cage when it is appropriately docked to the rack and into an exhaust plenum. Exhaust air is high-efficiency particulate air (HEPA) filtered before reentering the animal holding room or can be directly tied into the containment area's building exhaust. When in containment mode, the air exhaust levels exceed the supply levels, providing a negative differential air pressure in the cage. This configuration ventilates and isolates each individual cage from every other cage. The HEPA filters remove 99.97% of particulate matter greater than 0.3 microns in size, effectively providing containment of animal allergens, fungi, bacteria, and viruses from animal workers and animals in other cages. The supply and exhaust fan motors of these units should be interlocked so that if one motor fails, the other automatically shuts down, placing the cages in a static isolator mode and avoiding a reversal of the desired cage differential air pressure. There have been multiple studies demonstrating the airflow effectiveness of IVC units at maintaining dry cages and minimizing ammonia and CO_2 levels (Hasenau et al. 1993; Reeb-Whitaker et al. 2001; Baumans et al. 2002), potentially allowing for increased time between cage changes, if an Institutional Animal Care and Use Committee (IACUC)–approved exception is granted. With most IVC, the cage lids remain attached after removal from the rack (Maher and Young 2007), and so serve as microisolator units. This allows the cages to be taken from the rack and transferred directly to appropriate primary barrier equipment for cage changing, without fear of breaking containment during transport. The use of IVC has become common in ABSL-2 through ABSL-4 containment facilities.
3. Large animal primary containment caging: Generally used for larger laboratory animal species, such as rabbits, ferrets, and nonhuman primates, this primary containment caging usually consists of stainless steel frames with glass or clear plastic windows that accept removable stainless steel cages. Once the cages are in place, the individual compartments can be closed and sealed. As the name suggests, primary containment is provided by maintaining a negative

relative air pressure inside the individual compartments. This is accomplished by an exhaust blower unit connected on the outside of the rack. Air is pulled from the room, through filtered ports on each of the cage compartments, through the individual cages, out a centralized exhaust plenum, and through a HEPA filter housed in the blower unit, prior to being expelled back into the room or directly into the building exhaust system.

4. Isolators: Available in solid-walled or flexible-walled versions, isolators can serve as an effective containment barrier for housing animal caging. Air supply to and from the isolator is HEPA filtered. Personnel access the caging through portholes fitted with sleeves and gloves. Any materials exiting the isolator enclosure must be sealed and passed through a chemical dunk tank system or an air lock chamber that can be decontaminated between studies.
5. Negative-pressure, flexible-walled, free-standing containment enclosures: These mass air displacement systems are composed of a solid skeletal structure with flexible plastic walls. Similar to the negative-air-pressure primary containment caging noted above, exhaust blowers mounted on or adjacent to the enclosure pull air out of the unit and through HEPA filtration, prior to being exhausted back into the room. This provides a negative-air-pressure environment within the enclosure. A significant advantage of these systems is that they can be designed in an infinite variety of sizes and configurations. Multiple types of caging can be housed within these enclosures. When working with large animal species such as sheep, goats, and swine, for which standard biocontainment caging is not routinely available, these enclosures can be used to surround conventional caging and pens, providing an additional level of containment.
6. Isolation cubicles: Animal isolation cubicles are oftentimes referred to as "Illinois cubicles" in reference to their original description at the University of Illinois at Chicago. These types of units are typically of solid structure, either built in to the facility or added as free-standing units postconstruction. They are solid structures that are located inside a secondary room. They are usually only large enough to hold one rack of animals and have either hinged doors or vertical sliding doors with clear polycarbonate or safety glass panels that allow full viewing of the cubicle and its contents. An excellent review of isolation cubicles and the advantages and disadvantages of these types of units is available (Hessler and Britz 2009).

Sanitation procedures represent a significant portion of animal husbandry activities. To better understand animal husbandry sanitation requirements in a containment setting, the distinction between cleaning, disinfection, decontamination, and sterilization should be understood.

- Cleaning is considered the removal of gross contamination (e.g., animal waste and debris).
- Disinfection is the reduction or elimination of unacceptable concentrations of microorganisms (NRC 2011).
- Decontamination is the reduction of microbial contamination so that disease transmission is eliminated (safe to handle).
- Sterilization is rendering an item free of all living microorganisms or viruses (BMBL, Appendix B, CDC-NIH 2009).

Performing sanitation procedures within a containment facility adds a significant level of complexity compared with similar procedures in a conventional setting. Since cage or rack washer facilities are usually located outside of the containment envelope, accommodations need to be made to decontaminate cages, animal waste, and so forth, before leaving containment. Cage and bedding changing procedures for rodents are usually performed within a BSC or other appropriate primary containment device. Soiled bedding is preferably left in the cage to reduce the risk of aerosolizing the hazardous experimental agent. Dirty caging and bedding is decontaminated (e.g., autoclaved) prior to being transported to the cage washing facilities, where the bedding can be dumped and caging sanitized using conventional means. When using larger primary containment caging for rabbits, ferrets, nonhuman primates, and so forth, additional steps are required. Hand cleaning is normally required to remove organic and inorganic debris prior to autoclaving or performing gaseous decontamination.

During the conduct of the study, secondary enclosure spaces (e.g., animal holding rooms, anterooms, and corridors) are usually sanitized with chemical disinfectant solutions (e.g., quaternary ammonia, bleach, and chlorine dioxide). Between studies, these areas are often decontaminated with gaseous or vapor disinfectants, such as formaldehyde-paraformaldehyde gas, chlorine dioxide gas, or vaporized hydrogen peroxide. The most appropriate disinfectants should be determined based on a risk assessment. Information on the different classes of disinfectants and selection criteria for use are presented in Appendix B of the BMBL (CDC-NIH 2009). Before a sanitization plan is finalized, validation studies should be performed to ensure their effectiveness for the equipment and facility-specific environment.

Animal Handling

Handling animals that have been exposed to experimental hazards may pose significant risks, due to the potential for the transmission of the hazard to the animal handler through various routes, such as bites, scratches, aerosol transmissions, or mucous membrane exposures. All procedures involving the manipulation of either infectious agents or tissues from infected animals or the generation of aerosols should be conducted within biosafety cabinets (BSCs) or other physical containment devices when practical (CDC-NIH 2009). PPE, which should be used in combination with these containment devices, is defined by risk assessment of the required activities being performed. Procedures on smaller animals, such as rodents, can usually be performed in a BSC; however, due to space limitations, working with larger animals in a BSC can be impractical. In these cases, a combination of alternate containment devices and/or appropriate PPE should be employed. Procedures and equipment that minimize the potential for exposure, such as the use of forceps and/or bite- and puncture-resistant gloves for transferring rodents during cage changing activities, should be employed. When working with larger animals that pose a greater risk from bites or scratches, such as rabbits and nonhuman primates, chemical restraint (anesthesia) should be used whenever possible. Physical restraint devices, such as squeeze cage mechanisms or species-specific restrainers, must be considered.

Waste Disposal

Procedures must be in place to handle the accumulating hazardous waste generated from contaminated PPE, animal carcasses, soiled animal bedding, and procedural materials, such as sharps, absorbent towels, and other products. The regulations governing the disposal of hazardous wastes are complicated and involve multiple levels of federal, state, and local requirements. The primary federal agencies that govern hazardous waste disposal include (1) the Environmental Protection Agency (EPA) through the Resource Conservation and Recovery Act (RCRA) (this act gives the EPA authority to govern hazardous waste identification, classification, generation, management, and disposal) (CFR 2015a), (2) the Occupational Safety and Health Administration (OSHA) (CFR 2012a), and (3) the Department of Transportation (DOT) through the Hazardous Materials Regulations that govern the transportation of hazardous materials in all modes of transportation (CFR 2015b).

The types of wastes generated from different institutions vary, depending on the experimental programs conducted and the specific activities performed. It is important to clearly identify your operation's specific waste streams to determine the proper decontamination and disposal methodologies. An individual familiar with the institution's waste streams, and well versed in all the federal, state, and local regulations applicable to those waste streams, is required. Many institutions make use of commercial hazardous waste disposal companies. These companies are familiar with the applicable regulatory requirements and can serve as an excellent training resource for helping to prepare the institution's policies for internal hazardous waste handling and clearly defining waste streams.

Safety Equipment

The BMBL defines safety equipment as BSCs, enclosed containers, and other engineering controls designed to remove or minimize exposures to hazardous biological materials. These are considered primary barriers. Additional safety equipment can include PPE, such as gloves, coats, gowns, shoe covers, boots, respirators, face shields, safety glasses, or goggles. This PPE is commonly used in combination with BSCs and other primary containment devices.

Primary Barriers

1. Biological safety cabinets: BSCs are one of the principal devices used to contain biohazards and are considered to be a primary barrier. Some types of BSCs are appropriate for use with small amounts of volatile toxic chemicals and radionucleotides; however, the use of a chemical fume hood may be more appropriate when greater than small amounts of toxic chemicals or radionucleotides are in use. The operational integrity of a BSC must be validated before it is placed in service and after it has been repaired or relocated. Relocation may break the HEPA filter seals or otherwise damage the filters or the cabinet. Each BSC should be tested and certified at least annually to ensure continued, proper operation (BMBL, Appendix A, Section VII, CDC-NIH 2009). On-site field testing of each BSC must be performed by experienced, qualified personnel. This testing must meet the appropriate National Sanitation Foundation/American National Standards Institute (NSF/ANSI) Standard 49-2014 for field certification. Personnel should be trained in the proper operation and appropriate techniques to be used before performing procedures in a BSC. For example, the effectiveness of a BSC's containment air curtain can be dramatically disrupted by rapid or sweeping arm movements, so a slow and deliberate technique is important. Walking or working behind a person using a BSC should be avoided because the resulting air movement can interfere with the inward protective curtain of air at the BSC opening (Fontes 2008). There are three classes of BSCs (I, II, and III) and multiple types within Classes I and II. The BMBL provides detailed descriptions of each, including tables indicating the protection provided by each BSC class and comparisons of the different classes for face velocity, airflow patterns, and applications (BMBL, Appendix A, CDC-NIH 2009). The most appropriate BSC for a particular task is determined by the project or activity risk assessment.
2. Transfer containers and carts: Frequently during the conduct of a study, animals may need to be transferred from one location to another. If the animals have been exposed to a hazardous agent, they will likely require containment during the transfer process. As previously mentioned, filtered containers like filter-top cages can be used for this purpose. In addition, transfer carts, specific for this purpose, are available. Some of these transfer carts have battery-powered ventilation systems and HEPA-filtered exhaust air, so that cages housed on the cart shelves are maintained in negative-air-pressure containment. Other transfer carts are mobile glove box isolators. These carts also have battery-powered ventilation systems with HEPA-filtered supply and exhaust air and require that animal rooms, procedure rooms, or other isolators have matching transfer ports with which the cart can dock.
3. Downdraft tables: Animal manipulations, such as dosing, bleeding, cage changing, and necropsy or tissue collection procedures, can usually be performed in a BSC for small animals; however, performing these procedures on larger animals, such as nonhuman primates, may not be practical due to space limitations. For these purposes, an alternate piece of primary containment equipment, the downdraft table, is frequently used. As the name implies, downdraft tables draw room air down into the work surface and away from the worker to help minimize aerosol exposures. Exhaust air is HEPA filtered and returned to the room or directly into the building exhaust system. Appropriate PPE to be worn in combination with the use of a downdraft table is determined through risk assessment. Downdraft tables used for necropsy procedures are usually fixed tables with a water supply and drainage, located in a room dedicated to necropsy procedures. Portable downdraft tables are also available for other types of large animal procedures that cannot practically be performed in a BSC. These portable tables function similarly to the fixed downdraft tables but without a water supply and with the HEPA-filtered exhaust air returned to the room.

Personal Protective Equipment

Based on the agent, selection of the proper PPE should be determined by conducting a thorough risk assessment. Personnel must also have time to learn how to wear and maintain specific PPE properly, including proper donning procedures, and safe doffing procedures for contaminated PPE. OSHA requires employers to provide PPE, such as gloves, gowns, eyewear, and masks or respirators, to create barriers that protect skin, clothing, mucous membranes, and the respiratory tract from hazardous agents. PPE is the last level of protection to prevent worker exposures to hazardous materials. PPE should not be used in place of primary barrier safety equipment, such as containment caging or a BSC, but as an adjunct to these safety engineering controls whenever possible. The most appropriate PPE should be determined by risk assessment, considering the hazards being used, the activities being performed, and the engineering controls in place. The PPE requirements for a particular area or activity should never be written in a general way, such as "Wear appropriate PPE," but should be specified and must be consistent with the hazard and required levels of containment (Fontes 2008).

1. Outer garments: Lab coats, coveralls, hair and head coverings, and shoe coverings come in a wide variety of styles and materials providing different levels of protection from different types of hazardous materials. The most appropriate style and material should be chosen according to the potential exposure risks.
2. Hand protection: Protective gloves are routinely worn to minimize skin contact with hazardous materials. Bite- and puncture-resistant gloves, gauntlets, and sleeve coverings can also be worn to reduce the risk of animal bites and scratches. Disposable gloves are available in various materials (e.g., latex, nitrile, and vinyl). The appropriate type of disposable glove material is determined by risk assessment, considering the materials with which the glove will come in contact. For example, latex gloves provide a poor barrier to petroleum-based chemicals and solvents; nitrile gloves provide a better barrier for these materials. Nondisposable gloves are also available to protect against such things as chemicals and harmful temperatures.
3. Respiratory protection: The federal agency responsible for approving and certifying respirators for use is the National Institute for Occupational Safety and Health (NIOSH). NIOSH defines a respirator as "a personal equipment device that is worn on the face, covers at least the nose and mouth, and is used to reduce the wearer's risk of inhaling hazardous airborne particles (including dust particles and infectious agents), gases or vapors." It is important to note that surgical face masks are not considered respirators. Face masks can be useful in preventing large solid or liquid hazards from entering the nose or mouth and can prevent the user from touching his or her nose or mouth, but they do not provide adequate protection from aerosolized hazards. The need for and the type of respiratory protection required is another important part of the risk assessment analysis. If respirator use is required, the institution must have a respiratory protection program with a dedicated program administrator (CFR 2007). The program includes the requirement for periodic respiratory protection training and medical clearance for wearing a respirator. In addition, fit testing is required for respirators that require a tight seal to be effective. It is important to note that facial hair can compromise a tight seal, in which case an alternate type of respirator is required. The respiratory protection most commonly used in biocontainment environments is described below:
    a. Disposable air-purifying respirators (e.g., N95, N99, and N100): These are particulate-filtering face piece respirators. When the respirator is properly fitted, the user inhales and air is drawn through the respirator filter material. These respirators are available in multiple types: N (not resistant to oil), R (somewhat resistant to oil), and P (strongly resistant to oil). N95, N99, and N100 respirators, as well as their R and P equivalents, have air-purifying efficiencies for solid and liquid particles of 95%, 99%, and 99.97%, respectively. All these respirators need a tight seal to be effective, and so fit testing is required.
    b. Reusable air-purifying respirators: These are half-face and full-face elastomeric respirators that have attached replaceable filter cartridges that can be catered to the situational need.

Similar to the disposables, when the user inhales, air is drawn through the filter cartridges. Fit testing is also required for these units.

c. Powered air-purifying respirators (PAPRs): These respirators use a battery-powered motorized blower unit to draw room air through a HEPA filter and push the filtered air into a head enclosure that has an integrated clear plastic face shield. Since there is a positive air pressure in the head enclosure, PAPRs do not require the tight seal that is necessary for the other air-purifying respirators, and so fit testing is not required.

Examples of Hazards

Biological

In the broadest sense, biological hazards occur whenever people, plants, or animals may be exposed to infectious microbes, or the toxins that are produced by these microbes, and illness or death results from this exposure (Rasco and Bledsoe 2005). These microbes typically fall into one of five categories: bacteria, viruses, fungi, protists, and archaea (NIH 2015). By far the largest categories comprising biological hazards belong to bacteria, viruses, fungi, and protists.

It was in 1876 that Robert Koch first recognized that microorganisms can cause disease when he took blood from *Bacillus anthracis*–infected cattle and used it to infect healthy cattle. Regardless of whether the microbe is studied because of its disease-producing capabilities or its link to chronic disease, working with these biological agents presents unique hazards to husbandry staff. Additionally, since many of the diseases caused by these microbes are easily transmitted by fomite, vector, aerosol, or accidental inoculation, these hazards can also present added risks to the research animals (Coelho and García Díez 2015).

It is not within the scope of this chapter to provide an exhaustive list of all potential biological agents; for a more in-depth treatment of this topic, by agent name, the reader is referred to the BMBL (CDC-NIH 2009), the NIAID's biodefense Category A, B, and C pathogens (NIAID 2015), and the American Public Health Association's *Control of Communicable Diseases Manual* (APHA 2014).

Another area of research in which biological hazards may be encountered is xenotransplantation using human tissues. In addition to the more historical use of immunodeficient rodents to study the growth and treatment of relatively well-defined tumor cell lines of human origin, many hospital-associated research programs are obtaining tumors and other tissues from hospital patients to identify potentially efficacious treatments using rodent models. It is generally recommended that tissue or cell lines of human origin be screened for the presence of human pathogens. In all cases, a risk assessment should be conducted to determine the appropriate biosafety level procedures and practices to be implemented in order for husbandry staff to safely handle the animals and any potentially contaminated bedding, cages, accessories, and so forth.

The safety of both personnel and research animals is of paramount importance when working with microorganisms or biological toxins that cause disease. Another important consideration is the need to proactively address how accidental exposures will be handled. Practicing the steps involved in treating and reporting an exposure will be key to successful implementation of an organization's emergency plan. Invariably, these sorts of mishaps occur late on a Friday afternoon or Sunday morning when the emergency contact is out of town, so having alternate contacts is an important consideration. Developing a formal, written emergency action plan to handle both medical exposures to biological hazards and evacuations and rescues is a critical component of an occupational health and safety program (OHSP), especially when working with infectious agents and biological toxins (CFR 2002).

Chemical

Chemical exposures can come from a variety of areas within a biocontainment laboratory. With the advent of smaller, more efficient automated analyzers for chemistry or toxicology testing, it is no longer a rare event to have this equipment in a containment area. Unfortunately, common features that may serve as hazards include the sharp-ended sample needles that become contaminated with a toxic chemical and can easily puncture skin, resulting in a potential exposure. High-velocity robotic arms or centrifuges associated with this equipment can also serve as routes of aerosol exposure, not only for the chemicals associated with the analyzers, but also for any infectious materials contained in the samples. Fortunately, newer models come with protective covers that can mitigate these exposures.

Standard disinfectants in containment laboratories, such as bleach, Micro-Chem™ (quaternary compound), and Virkon™ (peroxygen compound), constitute a daily chemical hazard faced by husbandry staff. Appropriate PPE should be worn whenever handling these compounds to avoid potential cutaneous and mucous membranes exposures, especially when concentrated forms of disinfectants are being handled. It should also be noted that many disinfectants can also be a slip hazard when applied to floors. The CDC maintains a website that provides information about chemical emergencies (http://emergency.cdc.gov/chemical/lab.asp), as well as other guidelines for safe laboratory work practices (CDC 2012).

Radiation

Although the Cold War era purportedly ended in the 1990s, the fear of radiological hazards associated with a deliberate or accidental release of radiation endures as an existential threat. With the advent of nuclear medicine in the 1950s and the use of radiation (radioisotopes) to diagnose and treat cancers, scientific efforts have expanded. These efforts include a focus on developing mechanisms to assess radiation injury; early, preclinical radiation countermeasures; and cutting-edge medical treatments utilizing novel radiopharmaceuticals. The lead for regulatory activities governing nuclear materials, reactors, security, and radioactive waste is the U.S. Nuclear Regulatory Commission. Its website links to much useful information for personnel involved with the study of radioactive materials. Specifically, for the medical, industrial, and academic uses of nuclear materials, a good starting point would be http://www.nrc.gov/materials/medical.html.

In developing a radiobiological research program, it is not likely the reactors or other sources of ionizing radiation will be colocated in the vivarium, so it is important to ensure that reactor staff, who might only have an ancillary association with the research animals, are nevertheless trained and covered by the institute's occupation health and safety program specific to the animal models under study. Reactor or irradiator staff are routinely trained in the safe operation of the reactor and associated equipment, but they may not be aware of the need to wear PPE to prevent direct exposure to animal allergens or bite and scratch hazards from the animal or its caging. The use of PPE may not be practiced on a routine basis if radiation studies are only performed on a sporadic basis. The need for tuberculosis (TB) tests and measles titers to minimize infected humans transmitting disease to susceptible nonhuman primates, and the need for zoonoses training in the event personnel are bitten or scratched by an animal or pinched by a contaminated piece of equipment are other considerations for the staff in a radiobiological research facility.

Another consideration is that vivaria HVAC systems are usually designed to circulate air within the animal facility and do not share unfiltered air with administrative areas. Conversely, radiation exposure areas may share their air with other areas of the building. Thus, if animals will be waiting in a prestage area, or even within the exposure area, the air circulation pathway may inadvertently expose both the animals and the susceptible human populations to each other. Thought should be given to ensuring that transport carts are utilized to minimize animal populations from coming in contact with unintended personnel as they are moved to, and wait in, the radiation exposure facility.

Husbandry staff working with animals exposed to radiation represent another population requiring close oversight. Although this population may be well versed in the various animal species and their routine husbandry requirements, they may not, at least initially, be as well versed with safe handling of radioisotopes, depleted uranium pellets, or other radiation hazards to which animals may be exposed. Close scrutiny must be paid during protocol development to ensure that these radiological hazards are addressed as they apply to contaminated animal feed and soiled bedding. One frequently overlooked aspect of radiation research may be the need to store both liquid and solid waste streams, including carcasses, for prolonged periods before they can be safely disposed of in the appropriate manner. To the uninitiated, the extra storage of accumulating contaminated waste may represent the unplanned, albeit temporary, loss of valuable vivarium square footage. In addition to the usual IACUC review, these protocols should receive additional scrutiny by trained members of the radiation safety committee.

As with other hazards associated with research, radiological emergencies may arise. Steps should be employed to preemptively address accidents and exposures before they occur. Of value would be the creation of radiobiological advisory teams with expertise in the handling and mitigation of radiation hazards. With proper training, these teams serve as a valuable adjunct to safety and occupational health departments, capable of bringing more precise subject matter expertise to the research program.

Recombinant DNA and Synthetic Nucleic Acid Molecules

The first U.S. patent for recombinant DNA technology was in 1974, followed 8 years later by the licensing of the first drug produced via recombinant DNA technology: biosynthetic human insulin (BHI), in 1982, made by inserting human genetic material into bacteria and allowing them to efficiently produce large quantities of the BHI. The processes that led to these developments are referred to as recombinant DNA technology: novel DNA created by the combination of two or more strands of DNA that may have originated from different species and combine in such a way that the genetically modified organism may never have existed in nature before. Scientists working with these recombinant techniques quickly realized that there was the potential for undesirable, unpredictable, or even hazardous side effects. This realization led to the International Conference on Recombinant DNA in February 1975, from which the U.S. government issued the *NIH Guidelines for Research Involving Recombinant or Synthetic Nucleic Acid Molecules* (*NIH Guidelines*) (Berg and Mertz 2010). These guidelines launched the concept of establishing institutional biosafety committees to provide the review and oversight of research utilizing recombinant or synthetic techniques (Hackney et al. 2012). The guidelines also provide the framework researchers must follow when designing gene therapy experiments.

These types of experiments are initially carried out at the more stringent ABSL-2 or ABSL-3 (Feldman 2003) before reducing the husbandry practices to ABSL-1 if a biosafety evaluation, with thorough risk assessments, indicates that the vectors are not environmentally persistent and do not amplify, and no known pathogenicity in humans is uncovered (Reuter et al. 2012). In most cases, recombinant organisms can be handled at the same ABSL as the wild-type recipient, while poorly defined DNA sequences from donor organisms, which might increase the virulence of the recipient organism, should be handled at higher ABSLs. Additional information on the *NIH Guidelines* can be obtained from the NIH Office of Biotechnology Activities at http://osp.od.nih.gov/office-biotechnology-activities/biosafety/nih-guidelines.

Carcinogens

Carcinogens are substances or exposures that can change a cell's DNA, either directly or indirectly, eventually causing cancer. The cancer-causing exposures can be long-term or short-term, and the doses can be high or low. Conservatively, there are more than 340 substances with the potential to cause human cancers. Extrapolation allows these human carcinogens to generally be considered animal carcinogens as well. Three of the frontline agencies developing and tracking carcinogens are the International Agency for Research on Cancer (IARC) (http://www.iarc.fr), the U.S. National Toxicology Program (NTP) http://ntp.niehs.nih.gov), and the EPA (http://www.epa.gov). The IARC began as an entity of the WHO, while NTP was formed from several different government agencies, to include the NIH, CDC, and the Food and Drug Administration (FDA). The IARC's major objective is to identify causes of cancer, while the NTP is primarily known for its periodic updates to its *Report on Carcinogens.*

The Integrated Risk Information System (ISIS) (http://www.epa.gov/iris) is an electronic database maintained by the EPA to inform the public of the human health effects from environmental substances. Each of these agencies has a different set of standards to classify these compounds, generally along the lines of carcinogenic, possibly carcinogenic, unclassifiable, and probably not carcinogenic. Ancillary federal agencies that also play a role in carcinogenic agent tracking are the CDC's NIOSH (http://www.cdc.gov/niosh) and the National Cancer Institute (http://www.cancer.gov). Preventing exposure of husbandry staff to carcinogens is critical, and steps must be taken to prevent exposure from all routes, including cutaneous, mucous membrane, ingestion, and inhalation. Working closely with research investigators and environmental health and safety (EHS) staff to develop safe husbandry procedures is essential.

Nanoparticles

Nanoparticle technology has been known since at least the ninth century, when skilled craft workers utilized a technique of applying a metallic film to the surface of pots to get a glittering effect. The technology continues today with the advent of book pages infused with silver and copper nanoparticles that filter out disease-causing bacteria in water teeming with raw sewage that can be deployed during environmental disasters or third world hot spots to alleviate shortages of potable water (Brink 2015). The benefits of this technology continues with the discovery that cone snail venom delivers analgesia 1000 times more potent than morphine but is degraded by the blood–brain barrier (BBB) unless it is hidden in a nanocontainer that can bypass the BBB (Anand et al. 2015). Interestingly, it was Michael Faraday, in 1857, who is credited with the first scientific investigations utilizing these ultrafine particles (Heiligtag and Niederberger 2013).

Once the scientific investigations began, both medical and environmental safety considerations of nanotechnology started to receive attention. As the body of evidence grows that nanomaterials can cause adverse health effects, the research also demonstrates that inhalation is a significant route of exposure (NIOSH 2012). A 2006 study by Elder demonstrated the ease with which ultrafine particles of inhaled manganese oxide particles (as might be expected to be generated from, e.g., arc welding) are easily taken up by the olfactory neuronal pathway directly to CNS tissue (Elder et al. 2006). This olfactory neuronal pathway has also been demonstrated in nonhuman primates and is likely to be a route in humans too.

The National Nanotechnology Initiative (NNI) was established in 2000 to coordinate federal research and development efforts and promote competitiveness of the United States in the nanotechnology field (http://www.nano.gov/about-nni/what). This initiative is now a federal program for science, engineering, and technology research and development for nanoscale projects. The NNI serves as the communication hub for all federal agencies engaged in nanoresearch. Out of this initiative eventually grew the Nanotechnology Research Center in 2004, a component of NIOSH, the lead agency in the U.S. government's NNI. As with most areas in the safety arena, risk management is the critical component to working with nanoparticles. Exposure assessment is the first element in mitigating problems, but exposure control is no less important. Control mechanisms include elimination, substitution, isolation, engineering controls, administrative controls, and PPE. A useful document to develop risk management strategies for work with nanoparticles is NIOSH's "General Safe Practices for Working with Engineered Nanomaterials in Research Laboratories," available at the CDC website (http://www.cdc.gov/niosh; publication No. 2012-147). Additional guidance and free publications are available at https://www.cdc.gov/niosh/topics/nanotech/pubs.html.

Another source of recent guidance on improving the safety of nanoparticle research is the Scientific Committee on Emerging and Newly Identified Health Risks (SCENIHR). Its website is http://ec.europa.eu/health/scientific_committees/emerging/index_en.htm. At the behest of the Council of the European Union in 2005, the SCENIHR drafted a document titled "The Appropriateness of Existing Methodologies to Assess the Potential Risks Associated with Engineered and Adventitious Products of Nanotechnologies," which is available at http://ec.europa.eu/health/ph_risk/documents/synth_report.pdf.

Administrative Controls

Hazard Identification and Risk Assessment

Husbandry staff working with animals exposed to experimental hazards are on the front line of research, and the hazards to which they may be exposed must be effectively identified and the appropriate level of risk assigned to those activities. Laboratory directors and principal investigators have primary responsibility for ensuring that an effective risk assessment is performed (CDC-NIH 2009). Many other entities within an organization share that responsibility.

Hazard identification is a critical aspect of the risk assessment process and is the first step in protecting the health and safety of husbandry staff. While the principal investigator is usually the best source for identifying experimental hazards and the subsequent mitigation strategies for them, the institutional EHS office should also be integrally involved in identifying potential hazards in the work environment. Other groups with in-depth knowledge of the day-to-day activities in the animal facility may provide important insight as well. Involving the husbandry staff, veterinarians, facility managers, veterinary technicians, and scientific staff will help to ensure that all aspects of the animal care and use program are evaluated.

It is important that hazard identification and the associated risk assessment be ongoing processes that involve individuals qualified to assess dangers associated with the work being performed, and that commensurate safeguards are implemented in a timely manner (NRC 2011). While routine walk-throughs of animal holding areas by EHS professionals are a very useful tool for identifying hazards (NRC 1997), the identification of hazards before they are introduced is a much more effective approach. Most institutions require the description of any experimental hazards used with animals to be included in the animal use protocol that is reviewed by the IACUC. While the safety office and/or appropriate safety committee may also be responsible for reviewing experimental hazards used with animals, the inclusion of a safety professional on the IACUC is very useful in ensuring that hazards that are planned for use in the animal facility are identified early. This is especially helpful where the use of experimental hazards in animals is relatively common.

The identification of hazards is an integral step in the ability to perform an effective risk assessment. Risk assessment is a dynamic process that involves the evaluation of multiple variables. In its basic form, two major areas are evaluated: (1) the likelihood of an injury or illness occurring from the hazards, and probable consequences of an exposure (CDC-NIH 2009), and (2) an individual's susceptibility to illness or injury. The intensity, duration, and frequency of exposure to the hazard are important in determining the level of risk involved (NRC 2011). Similarly, a change in an individual's susceptibility to illness or injury can affect the level of risk. For example, the development of a condition that results in a decrease in his or her immune response, which then would increase his or her susceptibility to a hazard, may be a low risk to an immunocompetent individual. All these factors must be reviewed on an ongoing basis, and especially when changes in either the hazard or an individual's health status occur.

While the risk assessment process can be subjective and there is no standard approach, in the authors' opinion the risk assessment process is critical when husbandry staff are working with experimental hazards, and it should be a structured and documented process. The ultimate goal of the hazard identification and risk assessment processes is to ensure that all hazards to which husbandry staff may be potentially exposed are identified, the risk to them has been thoroughly assessed, staff are informed of the hazards and have received appropriate training on them, and effective mitigations are in place to protect the staff.

Reporting and Communicating Adverse Events

A culture that encourages husbandry staff to report any adverse event should be implemented and sustained. Timely reporting of injuries or illnesses is vital for not only providing the appropriate medical treatment to staff, but also identifying trends that could affect other staff members in the future. Reporting of potential exposure to hazards, injuries, or accidents is extremely important for many reasons and is usually required by law. In the United States, employers are required to keep records of occupational deaths, injuries, and illnesses using OSHA's 300 form (CFR 1960). The practice of collecting and reviewing "near-miss reports" is considered a proactive approach to identifying potential issues before they become an actual hazard. While not usually required from a regulatory standpoint, collecting information on human error or mechanical malfunctions that "almost" result in an exposure or injury can provide important insights that can help guide a preventative approach to health and safety. In the world of infectious disease hazards, the line between a potential exposure, an exposure, and a reportable illness may not be as clearly defined as for some other hazards, with many of these infectious agents producing general and nonspecific flu-like symptoms. In order to provide additional safeguards in these situations, most institutions require husbandry staff working with animals exposed to infectious organisms to report any fever they have over a certain temperature to a designated health professional. This practice is instituted in order to minimize the possibility of a work-related exposure being missed.

Communication of Hazards

Employers have a responsibility to communicate potential hazards in the workplace to their employees. For example, OSHA's Hazard Communication Standard (HCS) (CFR 2015c), with recent updates effective in June 2015, requires that all employers with hazardous chemicals in their workplace prepare and implement a written hazard communication program. The employer must ensure that all hazardous chemical containers are appropriately labeled and that employees are provided access to the safety data sheets (SDSs) for these chemicals. In addition, it is required that an effective training program be conducted for all employees having a potential for exposure.

OSHA's Bloodborne Pathogens Standard (CFR 2001), as modified to include the Needlestick Safety and Prevention Act, prescribes safeguards to protect workers who could be reasonably anticipated to contact blood or other potentially infectious material, such as unfixed human tissues and certain body fluids, against the health hazards caused by blood-borne pathogens. These pathogens include, but are not limited to, hepatitis B virus (HBV), hepatitis C virus (HCV), and human immunodeficiency virus (HIV). The standard requires employers to establish an exposure control plan, provide worker training with annual updates, and use labels and signs to communicate hazards. In addition, the standard requires the institution to provide vaccination when appropriate and identify appropriate hazard control practices, such as the use of sharps disposal containers and self-sheathing needles.

As a part of a required occupational health program, the Department of Health and Human Services (HHS) regulation (CFR 2012b) requires that all personnel approved for access to Tier 1 BSAT be provided with the following information:

- The risk and associated health hazards for working with the Tier 1 BSAT
- Typical signs and symptoms of the diseases
- The available pre- and postexposure resources for treatment
- Whom to contact and what to do in an emergency
- Policies for immediately reporting and documenting all potential occupational exposures

As noted in the BMBL (CDC-NIH 2009), signs must be posted at the entrances to areas containing hazardous infectious agents. These signs should include the universal biohazard symbol, agent-specific information, entry and exit procedures (e.g., PPE requirements), and information for contacting responsible personnel.

As we have stressed throughout this chapter, the preparation of thorough formal risk assessments is essential for establishing practices to mitigate exposure risks to experimental hazards. These assessments serve as a primary source for the communication of hazards to husbandry staff.

Occupational Health and Safety Programs

Employers are responsible for providing safe and healthful working conditions for their employees. (OSHA, Occupational Safety and Health Act, 1970). An effective OHSP that meets all federal, state, and local regulations is critical in maintaining a safe and healthy workplace. The OHSP should consider all aspects of the institution's research program, including personnel, research activities, facilities, hazardous materials, and animal species. In addition, the OHSP should include coordination with members of the research, animal care and use, occupational health and safety, and administrative groups (Swearengen and Carpenter 2015). Some of the major references available when preparing an OHSP for husbandry staff include the *Guide for the Care and Use of Laboratory Animals* (NRC 2011); the National Research Council's *Occupational Health and Safety in the Care and Use of Research Animals* (NRC 1997); the BMBL (CDC-NIH 2009), which serves as a reference with respect to working with infectious agents and toxins; and the *NIH Guidelines* (NIH 2013). For specific guidance when working with nonhuman primates, the National Research Council's publication *Occupational Health and Safety in the Care and Use of Nonhuman Primates* (NRC 2003) is available. The four major components of an occupational health program include preplacement medical evaluations, vaccines, periodic medical evaluations, and medical support for occupational illnesses and injuries (CDC-NIH 2009).

1. Preplacement medical examinations: Individuals with a potential for exposure to hazardous agents should be enrolled in an OHSP and should receive preplacement medical evaluations (CDC-NIH 2009). The OHSP health care providers should be well versed in the hazard risks found throughout the organization's work environment and knowledgeable of the potential hazards encountered by each individual in the program. As a part of the medical evaluation, to adequately assess the individual's fitness to perform specific activities, the health care professional should evaluate any ongoing and previous medical conditions, medications being taken, and records of previous immunizations.
2. Vaccines: When available, commercial vaccines for infectious agents to which an individual could potentially be exposed should be made available. Husbandry staff are routinely vaccinated against tetanus (NRC 1997). If working with susceptible species, preexposure immunization should be offered for specific agents such as rabies virus. If working with human blood, tissues, or cell lines, HBV vaccines should be available (NRC 2011). When commercial vaccines do not exist for a biohazardous agent that represents a potential exposure risk and an investigational new drug (IND) status vaccine for the agent is available, the IND vaccine should be offered, if determined appropriate by the risk assessment. General vaccination and vaccine-specific recommendations are provided in the BMBL and the CDC ACIP (CDC 2011a).
3. Periodic medical evaluations: As part of the OHSP for individuals with the potential for being exposed to hazardous agents, routine periodic medical evaluations may be given. Medical clearances may be required in specific circumstances (e.g., respirator usage). Based on the assessment of risks and program requirements, how often medical evaluations are performed may vary. Methods for performing the evaluations can vary. Some aspects of the evaluations can be accomplished through using questionnaires and some through physical evaluations. These determinations should be made considering the health of the individual and the level of risk present. If any changes occur to an individual's health status between medical evaluations, it is extremely important that these changes are reported to the health care provider so that the individual's specific occupational health and safety plan can be updated and it can be determined if his or her activities can be performed safely. For individuals that have a substantial exposure risk to infectious agents, it may be appropriate to offer periodic laboratory testing to detect preclinical evidence of an occupationally acquired infection (CDC-NIH 2009).
4. Medical support for occupational illnesses and injuries: A key element of an OHSP is to have protocols and procedures in place for addressing potential exposures to hazardous agents. To ensure appropriate and timely response to an exposure, exposure-specific protocols should be readily available. These protocols should describe appropriate first aid, options for postexposure prophylaxis, recommended diagnostic tests, and expert medical evaluation sources (CDC 2013). Identifying potential exposures is not always easy. Signs indicative of an exposure may not present until much later and can mimic those associated with common respiratory diseases. It is important to build a workplace culture in which individuals are comfortable contacting the OHSP health provider when they have any signs that could be associated with exposure to any of the hazardous agents in their work area. Additional information about occupational health and safety in animal care and use programs can be found in Chapter 14 of this text.

Training and Competency Determination

Introduction

To minimize biological hazards, personnel must be well trained and competent in performing the procedures, well in advance of the start of the protocol. There must be sufficient time to train to competence, not just to familiarization. There are distinct differences between personnel who are trained and those who are determined competent. Competency is measurable, and while it includes evidence of knowledge, skill, and abilities, it also should include judgment and self-criticism (CDC 2011b). Not understanding the difference between training and competency is an often overlooked aspect of working with hazards and is a contributing factor in lab-acquired infections.

Animal research will continue to be an invaluable and necessary component of scientific investigations as society seeks answers to the current gaps in knowledge. A moral and ethical commitment arises from each research protocol that utilizes animals in support of this mission. These commitments are to ensure that all animals are used in the most humane manner possible, scientific advancements and human health benefits are maximized, and animals are not arbitrarily wasted on poorly designed scientific studies. In order to meet this obligation, a research facility must first develop and then maintain a robust, flexible, conscientious, and legally observant animal care and use program. To meet the minimum requirements, a research facility must have the wherewithal to evaluate this training program, and this is most effectively done with the support of an IACUC, an engaged IO, and committed cooperation from the staff both receiving and providing the training. These three components, when functioning appropriately, have the ability to identify and resolve problems before animal welfare is compromised, science is adversely affected, or instructional quality is rendered irrelevant. Although these components each have their own guiding principles of laws, regulations, and policies to assist in decision making, they also have overlapping functions that provide balance to the research program.

In addition to the three components, an actively engaged program will also avail itself of outside entities to provide additional oversight. AAALAC International and the CDC are two such bodies that can provide this additional level of outside review, especially for biocontainment environments.

Prior to outside entities performing additional oversight, a strong program will have developed strong internal measures of assessment, especially for training. Once a procedure is developed and validated, training can be created to replicate the validated procedure. When training begins, it is important to measure the success of the teaching and, finally, to have a program in place for periodic refresher training. These crucial steps must be carried out at every level of biosafety containment, not just in noncontainment scenarios, hoping that the procedures extrapolate to biocontainment. From a safety perspective, training is a major component due to the highly hazardous agents and toxins used and the difficult working conditions under which many of the tasks must be performed to minimize risk as much as possible. Introducing animals into these hazardous conditions only serves to introduce additional risk into the equation, so the personnel responsible for this work must be highly trained and exceptionally proficient, but still able to conduct the required procedures in a safe and humane manner. The only way for this proficiency to grow is through extensive training, competency determination, maintenance of skills through constant practice, and acquisition of additional skills when required by protocol obligations. When designing the training program, care must be taken to ensure that the training complies with federal law, applicable accreditation standards, and industry best practices.

Training Program

Although individual training programs can vary widely, effective programs have similar characteristics: (1) didactic training whereby personnel first read, and become familiar with, SOPs, safety manuals, disaster planning, SDSs, and facility policies; (2) one-on-one or small group orientation to the area where work will be performed; (3) task-specific training in a mock environment; (4) task-specific training in an actual containment environment; and (5) periodic refresher training and assessment of skills.

Basic Safety Principles

The didactic training should have a heavy emphasis on the principles of safely working in a containment environment before introducing animals into the training scenarios. It is important that personnel demonstrate a thorough knowledge of the following areas before proceeding with the hands-on training using live animals: facility-specific safe laboratory operations, blood-borne pathogens training, chemical hygiene, laboratory-specific SOPs, and an introductory overview of the agents in use, especially the clinical signs and symptoms expected in the event of an exposure.

Containment and Work Area Orientation

Each research facility should establish a hierarchy of trainers. These trainers are responsible for familiarizing personnel to the safe work habits expected of all employees in both the noncontainment and containment areas. At a minimum, the basic orientation should include the following topics:

- Entry and exit procedures for both regular and emergency conditions, making sure to include training on the various alarms the research facility may have (e.g., fire, low oxygen, and HVAC)
- Safe disposal of generated laboratory wastes
- Donning and doffing procedures for PPE for each area and the rationale for the use of specific PPE for each area
- Procedures to follow in the event of PPE failure and how to handle potential exposures
- How to handle medical emergencies in noncontainment and containment environments
- How to handle laboratory spills
- How to contact key personnel and key departments during an emergency
- How to handle all the above issues both during and after regular working hours, including weekends and holidays

Task-Specific Competency

Any training involving animals must be covered by an IACUC-approved protocol. In most instances, the research models and techniques used in a particular organization may stay constant, so the protocol may be written broadly enough to cover the spectrum of anticipated needs. Addenda describing novel procedures or species must be approved in advance of starting work and must be in accordance with requirements of the individual IACUC.

The training workshops should be tailored to meet the basic needs for husbandry and handling first before delving into the more complicated procedures, such as phlebotomy, anesthesia, surgery, and identification of pain and distress. Again, didactic training on the SOPs relevant to each species and technique should be familiar ground for the trainee before commencing with the hands-on training. Ideally, a test of knowledge should be required for both the didactic and hands-on portion so that learning is not just a passive receipt of information. A thorough training program will also have a mechanism to document both the didactic and hands-on training classes.

Maintenance of newly acquired skills is as important as the initial training. The system for judging competency should be responsive to the needs of the researcher but flexible enough to accommodate various levels of training personnel may bring to the research program. However, above all, training should be consistent to avoid personnel being confused while performing various techniques in biocontainment. Refresher training can be on a regular schedule, at the behest of managers or coworkers, or individuals may self-identify their additional training needs.

A key consideration for any training program is that trainers are involved early in the project planning. Unfortunately, all too often, training competencies are an afterthought. Long after the IACUC protocol is approved and the animals have been purchased, husbandry staff and veterinary technicians are given their marching orders for the imminent start of a protocol. If the protocol describes new techniques, there will be little time to adequately develop, practice, and document these competencies before the protocol starts. To avoid the temptation to begin protocol support in the absence of adequate time for training, always ensure that there is advanced planning by the key trainers or mentors and plenty of opportunities for refresher training in instances where significant time has elapsed from the training date to the implementation of the project. Additional information about education and training can be found in Chapter 12 of this text.

Mentorship Program

Mentors are those with not only exceptional experience and knowledge, but also, more importantly, an ability to successfully impart that knowledge and skill to the less experienced and less knowledgeable.

Mentorship training should occur first in a regular animal environment, followed by a simulated containment environment. Once a suitable proficiency has been attained by the trainee, the mentor will supervise the trainee in an active containment environment. Following extensive training, the mentor should verify the competency level of the trainee in all basic operational areas within the purview of the research facility's mission utilizing an institute-wide training documentation program. For consistency and for verification purposes, it is necessary that all training events, both didactic and hands-on, simulated and real, be documented in a format that is easily accessible. While a research organization may wish for zero mishaps, they will, nevertheless, occur. When the inevitable mishap does occur, a proactive safety program should immediately begin an investigation to identify the cause of the mishap. The idea is not to criticize or lay blame but, ideally, to get all parties involved in developing mitigators to prevent a recurrence and then making sure, institute-wide, that all staff can learn from the example. In instances where training has not been properly documented, is incomplete, or is outdated, it is not possible to fully gauge how much of the mishap may have been due to poor training or some other issue. This makes it difficult to thoroughly evaluate the mishap and judge where best to institute process improvements.

Lessons Learned

1. One of the most common design flaws seen from a husbandry perspective in containment facilities is the lack of janitorial support facilities in common use areas (e.g., hallways). Since containment facilities are extremely expensive to build per square foot, these types of support areas usually found in conventional animal facilities are frequently cut as part of the value engineering process. Having designated areas to fill and empty mop buckets without lifting them to a sink is important for a number of reasons. Working in a containment facility usually requires several layers of PPE, including respiratory protection that may include particulate respirators, PAPRs, and in ABSL-4, body-encapsulating suits with an attached air supply. In all these cases, the additional levels of PPE can make physical effort and visual perception difficult. Adding this to the ergonomic stresses, the risk of personal injury or spilling or splashing contaminated water in common use areas increases dramatically. Adding a simple janitorial area with a water supply that can easily reach a mop bucket remaining on the floor and curbed dumping station can greatly reduce the likelihood of injury and inadvertent contamination of common use areas where drains may not be present.
2. Decontamination of carcasses and equipment associated with an animal care program is one of the most underdesigned areas in containment facilities. A common mistake made in designing animal facilities is purchasing autoclaves that prove to be too small for the needs of the program and/or not anticipating the level of biomass and other materials that need to be processed in order to use the animal space efficiently. Deciding on the types of animals, caging, and equipment that may be used can help inform better decisions on the size of the autoclaves or other decontamination equipment that will be needed. Deciding on autoclaves with the anticipation that caging will be purchased to fit the autoclave can be a disaster. If larger animals, such as rabbits, nonhuman primates, or even larger species, are expected to be used, one might even consider alternative methods of decontamination, such as alkaline hydrolysis. Steam autoclaves, even large ones, can be limited in the amount of biomass that can be effectively decontaminated at one time. The ability to efficiently decontaminate carcasses and equipment between studies can have huge impacts on timelines for scheduling studies due to the long cycle times required for running autoclave cycles at various levels of containment versus the number of carcasses and equipment that have to be decontaminated and removed from the containment envelope before another study can begin. While air locks may be helpful in decontaminating equipment, there is still a significant turnaround time for using gas or vapor phase types of decontamination processes.
3. Incorporating sufficient electrical capacity in biocontainment animal holding rooms is an important design consideration. Containment caging systems, unlike conventional caging,

incorporate electrically powered blower units to provide appropriate airflows within the caging. Some containment caging will require separate supply and exhaust blowers for one caging unit. Primary barrier devices, such as cage changing stations or downdraft tables, may be necessary to support the work being performed. Electronic balances, computer equipment, and many other specialized pieces of equipment, which require an electric supply, may be required. Retrofitting an operational containment area with additional electric capacity can be a difficult, expensive, and time-consuming proposition. It is important to have an individual knowledgeable of the potential caging and equipment requirements for containment animal holding spaces involved in the design process. Consideration should be given to the number of electric outlets, which are required to be on emergency backup power; their individual capacity; their location in the room; and the potential for the room's total electric capacity. If all the electrical outlets are not tied into the emergency electrical power supply, then consideration should be given to ensuring the strategic placement of the limited number of emergency supply outlets throughout the room.

4. Most experts consider the unpredictability of the animal to be a major risk factor for potential exposure to hazardous materials that are being used as part of an animal research study. While this is true, the caging in which an exposed animal is housed is another major risk factor that is oftentimes overlooked. New animal caging is being designed and existing caging improved on an ongoing basis, especially in the growing field of research using hazardous materials. When ordering expensive caging and housing systems for use when hazards are anticipated, the authors have found it extremely important to order only one of the units initially, or borrow a unit from another organization that may have them, so that extensive evaluation can occur before large numbers are procured. Evaluating the unit for the presence of sharp edges or corners and pinch points is critical to reduce the risk of PPE tears, and skin cuts or abrasions. Also, it is invaluable to have experienced husbandry staff evaluate the unit for maintenance requirements and the ease of performing routine maintenance that could be required while units are still in an environment where hazardous materials are present. A thorough test of these aspects of a new caging system can identify many potential issues that can be collaboratively redesigned with the manufacturer to improve safety before a large expenditure is made.
5. As noted in the "Animal Housing and Husbandry" section of this chapter, routine cage cleaning procedures for larger animals (e.g., rabbits) that are housed in containment caging are often performed by hand. Rabbits are particularly worrisome to handle in containment due to their ability to bite quickly and scratch with their powerful hind legs. To avoid removing the rabbit from the cage and the potential hazards with handling a conscious animal that has been exposed to hazardous agents, the authors have found that a simple cage partitioning tool can be easily fabricated and safely used. The tool is a single sheet of aluminum bent into an L-shape with handles affixed on both of the external sides of the L, to allow the handler to easily maneuver the device. The length, width, and height of the tool are proportional to the internal size of the cage and the animal so that when the ends of the L meet the back and one side of the cage, it forms a rectangular compartment that restrains the rabbit into one corner. Once the rabbit is partitioned into one corner of the cage, the husbandry technician can hand clean and rinse the remainder of the cage and then maneuver the animal and partition to the other corner to complete the hand-cleaning process.
6. Modern large animal primary containment cages are constructed with consideration for the environmental and behavioral needs of the animals, as well as the ability to monitor the animals without opening the containment envelope of the caging system. This includes the incorporation of large windows made from either tempered glass or high-strength clear plastic polycarbonate material, such as Lexan™. The authors have found that with the long autoclave cycles required for ABSL-3 and ABSL-4, the tempered glass option works better. Institutions using these caging systems have noted that the clear polycarbonate windows become clouded after being subjected to multiple autoclave cycles out of their ABSL-3 and ABSL-4 containment areas, but no issues have been noted with the tempered glass.

Summary

Managing any animal husbandry program in a research environment is inherently a complex process that requires in-depth knowledge of both the activities and risks involved. Adding experimental hazards to the animal research equation brings a new level of complexity, of which a facility manager must fully understand in order to provide a safe environment and implement safe work practices for the husbandry staff. In addition to understanding the principles of working safely with known hazards, a facility manager should endeavor to stay abreast of new information relevant to the activities of the husbandry staff. Effective communication with research staff and key safety officials within the organization is essential for identifying new hazards well before they are introduced, so appropriate mitigations can be thoroughly evaluated and implemented. Due to the ever-evolving landscape in many of the hazards discussed in this chapter, it is critical for facility managers to also remain current on a wide variety of relevant topics, such as updates to applicable regulations and guidelines, improvements in facility design and construction, changes in husbandry practices and techniques, new types of caging systems and safety equipment, and advances in establishing administrative controls. While a successful approach to managing hazards in an animal facility requires the involvement and cooperation of many organizational entities, the facility manager plays a key role as an advocate for the safety of the husbandry staff.

REFERENCES

Anand, P., O'Neil, P.A., Lin, E., Trevor, D., and M. Holford. 2015. Tailored delivery of analgesic ziconotide across a blood brain barrier model using viral nanocontainers. *Sci Rep* 5:12497.

APHA (American Public Health Association). 2014. *Control of Communicable Diseases Manual*, ed. D.L. Heyman. Washington, DC: American Public Health Association.

APHIS-CDC (Animal and Plant Health Inspection Service–Centers for Disease Control and Prevention). 2015. Select agents and toxins list. Atlanta, GA: CDC. http://www.selectagents.gov/SelectAgentsandToxinsList.html (accessed August 6, 2015).

Baumans, V., Schlingmann, F., Vonck, M., and H.A. Van Lith. 2002. Individually ventilated cages: Beneficial to mice and men? *Contemp Top Lab Anim Sci* 41(1):9–13.

Berg, P., and J.E. Mertz. 2010. Personal reflections on the origins and emergence of recombinant DNA technology. *Genetics* 184(1):9–17.

Brink, S. 2015. If you spill water on this book, that's a good thing. http://www.npr.org/sections/goatsandsoda/2015/08/19/432854450/this-is-one-book-that-you-want-to-get-all-wet (accessed October 20, 2015).

CDC (Centers for Disease Control and Prevention). 2011a. National Center for Immunization and Respiratory Diseases. General recommendations on immunizations: Recommendations of the Advisory Committee on Immunization Practices (ACIP). *Morb Mortal Wkly Rep* 60(2):1–64.

CDC (Centers for Disease Control and Prevention). 2011b. Guidelines for biosafety laboratory competency. *Morb Mortal Wkly Rep* 60(Suppl):1–23. http://www.cdc.gov/mmwr/pdf/other/su6002.pdf (accessed March 3, 2016).

CDC (Centers for Disease Control and Prevention). 2012. Guidelines for safe work practices in human and animal medical diagnostic laboratories. *Morb Mortal Wkly Rep* 61:1–102. http://www.cdc.gov/mmwr/pdf/other/su6101.pdf (accessed October 20, 2015).

CDC (Centers for Disease Control and Prevention). 2013. Occupational health program guidance document for working with tier 1 select agents and toxins. 7 CFR Part 331, 9 CFR Part 121, 42 CFR Part 73. July 5. Atlanta, GA: CDC.

CDC-NIH (Centers for Disease Control and Prevention–National Institutes of Health). 2009. *Biosafety in Microbiological and Biomedical Laboratories*, ed. L.C. Chosewood and D.E. Wilson. Washington, DC: Government Printing Office.

CFR (Code of Federal Regulations). 1960. Title 29, Labor; Part 1960, Basic program elements for federal employees OHSA; Subpart I, Recordkeeping and reporting requirements. Washington, DC: Office of the Federal Register.

CFR (Code of Federal Regulations). 1996. 42 CFR Part 72, Additional requirements for facilities transferring or receiving select agents: Final rule, 55190–55200. October 24.

CFR (Code of Federal Regulations). 2001. 29 CFR 1910.1030, OSHA, quick reference guide to the bloodborne pathogens standard. https://www.osha.gov/SLTC/bloodbornepathogens/bloodborne_quickref.html (accessed September 10, 2015).

CFR (Code of Federal Regulations). 2002. 29 CFR 1910.38, Occupational Safety and Health Administration (OSHA), emergency egress. https://www.osha.gov/pls/oshaweb/owadisp.show_document?p_table=STANDARDS&p_id=9726 (accessed October 20, 2015).

CFR (Code of Federal Regulations). 2007. 29 CFR 1910.134, Occupational Safety and Health Administration (OSHA), personal protective equipment, respiratory protection. http://www.ecfr.gov/cgi-bin/text-idx?SID=cb457c7c1dbed4ad58a56ca3a1b82efb&mc=true&node=se29.5.1910_1134&rgn=div8 (accessed September 23, 2015).

CFR (Code of Federal Regulations). 2012a. 29 CFR 1910.1030, Occupational Safety and Health Standards: Toxic and hazardous substances. https://www.osha.gov/pls/oshaweb/owadisp.show_document?p_table=STANDARDS&p_id=10051 (accessed October 23, 2015).

CFR (Code of Federal Regulations). 2012b. 42 CFR Part 73, Section 12(d), Department of Health and Human Services (HHS), select agents and toxins, biosafety. http://www.ecfr.gov/cgi-bin/text-idx?SID=cb8059285690e4b4102e02e7857c5602&mc=true&node=se42.1.73_112&rgn=div8 (accessed September 10, 2015).

CFR (Code of Federal Regulations). 2015a. 40 CFR 260.1, Environmental Protection Agency, hazardous waste management system: General. http://www.ecfr.gov/cgi-bin/text-idx?SID=f1a9f760d8d8b9e0f1e30083740d7a48&mc=true&tpl=/ecfrbrowse/Title40/40cfrv26_02.tpl#0 (accessed October 23, 2015).

CFR (Code of Federal Regulations). 2015b. 49 CFR Parts 100–107, Department of Transportation, Hazardous materials and oil transportation. http://www.ecfr.gov/cgi-bin/text-idx?SID=f1a9f760d8d8b9e0f1e30083740d7a48&mc=true&tpl=/ecfrbrowse/Title49/49CIsubchapA.tpl (accessed October 21, 2015).

CFR (Code of Federal Regulations). 2015c. 29 CFR 1910.1200, Occupational Safety and Health Standards, hazard communication. http://www.ecfr.gov/cgi-bin/text-idx?SID=fb73a58e5e3db66791d7c8ba56751592&mc=true&node=se29.6.1910_11200&rgn=div8 (accessed September 9, 2015).

Coelho, A.C., and J. García Díez. 2015. Biological risks and laboratory-acquired infections: A reality that cannot be ignored in health biotechnology. *Front Bioeng Biotechnol* 3(56):1–10.

Copps, C. 2005. Issues related to the use of animals in biocontainment research facilities. *ILAR J* 46(1):35–43.

Elder, A., Gelein, R., Silva, V. et al. 2006. Translocation of inhaled ultrafine manganese oxide particles to the central nervous system. *Environ Health Perspect* 114(8):1172–1178.

Feldman, S.H. 2003. Components of gene therapy experimentation that contribute to relative risk. *Comp Med* 53(2):147–158.

Fontes, B. 2008. Institutional responsibilities in contamination control for research animals and in occupational health and safety for animal handlers. *ILAR J* 49(3):326–337.

Frasier, D.F., and J. Talka. 2005. Facility design considerations for select agent animal research. *ILAR J* 46(1):23–33.

Hackney, R.W., Myatt, T.A., Gilbert, K.M., Caruso, R.R., and S.L. Simon. 2012. Current trends in institutional biosafety committee practices. *Appl Biosafety* 17(1).

Hasenau, J.J., Baggs, R.B., and A.L. Kraus. 1993. Microenvironments in microisolation cages using BALB/c and CD-1 mice. *Contemp Top Lab Anim Sci* 33(1):11–16.

Heiligtag, F.J., and M. Niederberger. 2013. The fascinating world of nanoparticle research. *Materials Today* 16(7–8):262–271.

Hessler, J.R., and W.R. Britz. 2009. Animal isolation cubicles. In *Planning and Designing Research Animal Facilities*, ed. J.R. Hessler and N.D.M. Lehner, 151–172. San Diego: Academic Press.

Jaax, J. 2005. Administrative issues related to infectious disease research in the age of bioterrorism. *ILAR J* 46(1):8–14.

Lipman, N.S. 1999. Isolator rodent caging systems (state of the art); a critical review. *Contemp Top Lab Anim Sci* 38:9–17.

Maher, J., and D.O. Young. 2007. How to isolate a rodent; biocontainment caging options. *Lab Anim* 36(2):42–48.

NIAID (National Institute of Allergy and Infectious Diseases). 2015. NIAID emerging infectious disease/pathogens. Bethesda, MD: NIAID. http://www.niaid.nih.gov/topics/biodefenserelated/biodefense/pages/cata.aspx (accessed October 20, 2015).

NIH (National Institutes of Health). 2008. *Design Requirements Manual for Biomedical Laboratories and Animal Research Facilities.* Bethesda, MD: NIH. http://orf.od.nih.gov/PoliciesAndGuidelines/BiomedicalandAnimalResearchFacilitiesDesignPoliciesandGuidelines/Documents/Design%20Requirements%20Manual/NIH%20Design%20Requirements%20Manual%20ver%205-13.pdf (accessed August 7, 2015).

NIH (National Institutes of Health). 2015. Understanding microbes in sickness and in health. Bethesda, MD: NIH. http://www.niaid.nih.gov/topics/microbes/Documents/microbesbook.pdf (accessed September 9, 2015).

NIH (National Institutes of Health), Office of Biotechnology Activities. 2013. *Guidelines for Research Involving Recombinant or Synthetic Nucleic Acid Molecules.* Washington, DC: U.S. Department of Health and Human Services. http://osp.od.nih.gov/office-biotechnology-activities/biosafety/nih-guidelines (accessed September 30, 2015).

NIOSH (National Institute for Occupational Safety and Health). 2012. General safe practices for working with engineered nanomaterials in research laboratories. DHHS (NIOSH) Publication No. 2012-147. Atlanta, GA: NIOSH. http://www.cdc.gov/niosh/docs/2012-147 (accessed March 21, 2016).

NRC (National Research Council). 1997. *Occupational Health and Safety in the Care and Use of Research Animals*, 25. Washington, DC: National Academies Press.

NRC (National Research Council). 2003. *Occupational Health and Safety in the Care and Use of Nonhuman Primates.* Washington, DC: National Academies Press.

NRC (National Research Council). 2011. *Guide for the Care and Use of Laboratory Animals*, 18–19. 8th ed. Washington, DC: National Academies Press.

PHAC (Public Health Agency of Canada). 2013. *Canadian Biosafety Standards and Guidelines.* 1st ed. Ottawa: PHAC. http://canadianbiosafetystandards.collaboration.gc.ca/cbsg-nldcb/index-eng.php (accessed August 6, 2015).

PHAC (Public Health Agency of Canada). 2015. *Canadian Biosafety Standard.* 2nd ed. Ottawa: PHAC. http://canadianbiosafetystandards.collaboration.gc.ca/cbs-ncb/index-eng.php (accessed August 6, 2015).

Rasco, B.A., and G.E. Bledsoe. 2005. *Bioterrorism and Food Safety.* Boca Raton, FL: CRC Press/Taylor & Francis.

Reeb-Whitaker, C.K., Paigen, B., Beamer, W.G., Bronson, R.T., Churchill, G.A., Schweitzer, I.B., and D.D. Myers. 2001. The impact of reduced frequency of cage changes on the health of mice housed in ventilated cages. *Lab Anim* 35:58–73.

Reuter, J.D., Fang, X., Ly, C.S., Suter, K.K., and D. Gibbs. 2012. Assessment of hazard risk associated with the intravenous use of viral vectors in rodents. *Comp Med* 62(5):361–370.

Richmond, J.Y., Hill R.H., Weyent R.S., Nesby-O'Dell S.L., and P.E. Vinson. 2003. What's hot in animal biosafety? *ILAR J* 44(1):20–27.

Swearengen, J.R., and C.B. Carpenter. 2015. Working safely with experimental animals exposed to biohazards. In *Laboratory Animal Medicine*, ed. L.C. Anderson, G. Otto, K.R. Pritchett-Corning, M.T. Whary, and J.G. Fox, 1297–1311. San Diego: Academic Press.

USDA (U.S. Department of Agriculture). 2012. *Animal Research Service (ARS) Facilities Design Standards.* Washington, DC: USDA. http://www.afm.ars.usda.gov/ppweb/pdf/242-01m.pdf (accessed August 10, 2015).

Van Vaerenbergh, B., Van Droogenbroeck, C., Do Thi, C.D. et al. 2011. Biosafety in laboratory animal facilities: A practical approach. Brussels: WIV-ISP. http://www.biosafety.be/CU/PDF/LabAnimFacilities_SBB_2011_2505_47.pdf (accessed August 6, 2015).

WHO (World Health Organization). 2004. *Laboratory Biosafety Manual.* 3rd ed. Geneva: WHO. http://www.who.int/csr/resources/publications/biosafety/WHO_CDS_CSR_LYO_2004_11/en/ (accessed August 10, 2015).

Section VIII

Animal Health and Care

31

Veterinary Care

James M. Raber, Marek Niekrasz, Jan Linkenhoker, and Kathy A. Perdue

CONTENTS

Background

Veterinary care is a critical component of an animal care and use program, as the health and well-being of animals used in research, testing, and teaching are essential to humane and reproducible science. Although the veterinarian has primary responsibility for the well-being and clinical care of the animals, veterinary care is a team event, which spans the life of each animal and encompasses all aspects of their care and use.

The regulatory requirement for adequate and timely veterinary care and oversight is delineated in the standards developed by many countries (e.g., United States and Canada) or state unions (e.g., European Union) (European Commission 2007; Zurlo et al. 2009; CIOMS/ICLAS 2012; Bayne et al. 2014). It is the goal of program management to ensure that the provision of veterinary care is not only timely and of high quality, but also sufficiently dynamic to meet the needs of the research, testing, and/or teaching being conducted. A program of adequate veterinary care must be provided regardless of the number of animals used or the size of the research program and sponsoring institution.

Roles and Responsibilities

Although specific responsibilities vary, every individual in the animal care and use program shares the responsibility to ensure the health and well-being of the animals used in support of the program. Clear and timely communication among individuals at all program levels is imperative to ensuring that this duty is met. The specific responsibilities related to veterinary care are as follows.

Institutional Official

The institutional official (IO) bears responsibility for the entire program, which includes indirect responsibility for veterinary care. The official must ensure that the veterinary care program meets the highest quality and most ethical standards as directed by its country, state, and local laws, policies, and regulations. He or she must ensure the availability of resources (e.g., space, personnel, and funds) for both the routine operation of the veterinary care program and unforeseen circumstances (e.g., disease outbreaks and disasters). It is the IO's responsibility to ensure that the attending veterinarian is trained, qualified in laboratory animal care, knowledgeable and experienced. In addition, IO must make certain that the attending veterinarian, or other official designated to manage the program, has the authority to manage the program and that the veterinary care provided is both accessible and sufficient to meet the requirements of the program (AWR 9 CFR §1.1; CCAC 2008, Item 7; European Parliament and the Council of the European Union 2010). Clear, concise, and timely communication between the IO, attending veterinarian, and institutional animal care and use committee (IACUC) is critical to ensuring the overall program direction and animal well-being.

Institutional Animal Care and Use Committee

The IACUC or its equivalent is responsible for the assessment and oversight of all components of an institute's program, including the veterinary care program. The committee plays a key role in supporting the attending veterinarian in ensuring the health and well-being of the animals used, in addition to serving as a line of communication throughout the program.

Scientist

The scientist is responsible not only for the use of animals, but also for their well-being. It is highly recommended, and required in some countries (Bayne et al. 2014), that a scientist consult and/or collaborate with the veterinarian in both the design and execution of all projects using animals. It is important that the veterinarian be provided with a clear understanding of how the experimental manipulations may affect an animal's physical, physiologic, or behavioral appearance or profile. Anticipated changes to an animal's behavior, physiology, or appearance should be clearly outlined in the investigator's animal study proposal. Other information that is critical for the veterinary care team to understand includes, but is not limited to, (1) unique phenotypes related to each experimental model, line, strain, or species (e.g., diabetes and neurological behaviors); (2) experimental and humane endpoint criteria; (3) limitations to intervention strategies related to pain, stress, or distress; (4) limitations to medical treatments (e.g., steroids and antibiotics); and (5) limitations to palliative intervention strategies (e.g., special food, bedding, enrichment, and social housing). Because unforeseen complications may evolve throughout the course of a study and life span of an animal, continued communication with the veterinarian and veterinary care team is critical to ensure the well-being of the animals.

The scientist is also responsible for ensuring that all animal procedures and surgeries are conducted by trained, skilled personnel in accordance with his or her approved animal study protocol, which should include all current standards of veterinary care. In addition, the scientist is responsible for the safe use of hazardous agents and creating a safe work environment for his or her research team, including those who care for the animals.

Veterinarian

Regulatory Requirements

In the United States, the Animal Welfare Act is promulgated by the U.S. Department of Agriculture (USDA) as outlined in Animal Welfare Regulations (AWR) published in the Code of Federal Regulations (CFR) (AWAR 2013), Title 9, Animals and Animal Products, Subchapter A, Animal Welfare. The AWR defines the *attending veterinarian* as follows: "**Attending Veterinarian** means a person who has graduated from a veterinary school accredited by the American Veterinary Medical Association's Council on Education, or has completed the American Veterinary Medical Association's Education Commission for Foreign Veterinary Graduates certification program (https://www.avma.org), or has received equivalent formal education as determined by the Administrator; has received training and/or experience in the care and management of the species being attended; and who has direct or delegated authority for activities involving animals at a facility subject to the jurisdiction of the Secretary" (AWR 9 CFR §1.1). The quality or quantity of the required training or experience has not been specified by the USDA.

Program requirements have also been put in place by the European Union related to the designation of an advisory veterinarian with expertise on laboratory animal medicine for each animal breeder, supplier, and user (European Parliament and the Council of the European Union 2010, Article 25). But here again, like the USDA, the regulations do not specify the nature of the required expertise. The Canadian Council on Animal Care's policy statement for senior program administrators responsible for animal care and use programs (CCAC 2008, 7.2) states that veterinarians providing clinical services and/or compliance oversight must have the experience and expertise necessary to evaluate the health and welfare of each species used in the context of the work being conducted by the institution.

In most situations, the veterinarian overseeing the care of research animals has completed several years of postdoctoral training under the oversight of an experienced veterinarian. Credentialing by a certifying organization (e.g., the American College of Laboratory Animal Medicine [ACLAM]) is a common way of providing evidence of a veterinarian's knowledge and experience. Several countries have established colleges that set criteria and standards for obtaining board certification in the laboratory animal medicine specialty (i.e., Europe, Japan, and Korea) (http://www.iaclam.org/about.html). Additional information can be obtained from the International Council for Laboratory Animal Science (ICLAS) (http://iclas.org/) and the International Association of Colleges of Laboratory Animal Medicine (IACLAM) (http://www.iaclam.org/).

Although veterinary practice licensure in the state, country, province, or region may not be required to provide care to laboratory animals, licensure is commonly required to procure various pharmaceuticals and controlled drugs and sign health certificates. Veterinarians and institutions should review their local requirements for licensure and/or accreditation, which often vary by country, state, province, or region.

Veterinary Care Responsibilities

The attending veterinarian has ultimate responsibility for the day-to-day health and well-being of all animals in the program. The range and scope of these responsibilities are outlined in the *Guide for the Care and Use of Laboratory Animals* (*Guide*) (NRC 2011), as well as several position (AAALAC 2016; ACLAM 2016) and policy statements (CCAC 2008; European Parliament and the Council of the European Union 2010; USDA 2016). The size of the institutional program and nature of the research being conducted will determine whether a full-time, part-time, or consulting veterinarian is required. When the veterinarian is less than full-time, an explanation of his or her role and a prearranged schedule of when he or she will visit the animal facility should be available in writing (AWR 9 CFR §2.33 a.1). Regardless of the time commitment, the veterinarian's accountability remains the same. The definition of what is determined to be "adequate veterinary care" may vary between jurisdictions or countries. Adequate veterinary care guidelines commonly define the requirement for veterinary access to the animals and their records, the frequency of veterinary visits and observations, and the provisions for providing timely and appropriate medical care (ACLAM 2016).

A veterinarian is a required member of the IACUC and should have specialized training and/or experience in the field of laboratory animal medicine and science. The veterinarian serves as the subject area expert for issues ranging from animal husbandry to the identification and alleviation of pain and distress. The scientist should take advantage of the veterinarian's expertise during protocol development and consult with him or her throughout the protocol review and implementation process. Many IACUCs require that the institutional veterinarian be consulted on the proper use of anesthetics, analgesics, and euthanasia agents. At a minimum, the veterinarian should be involved in the review and approval of the final protocol.

The veterinarian works closely with the IACUC to facilitate the postapproval monitoring of approved animal study protocols, as well as the daily monitoring of all animal care and veterinary care program operations. The veterinarian plays a key role in the identification and prevention of occupational health and zoonotic issues. He or she must work closely with program managers and safety specialists in the development of standard operating procedures (SOPs) to mitigate or remove occupational health risk factors. The veterinarian has a role in training the investigative and animal care staff, as well as clinical support personnel. The training should include, but not be limited to, animal procurement, transportation, identification, handling, husbandry, preventive medical care, veterinary care, chemical sedation and anesthesia, sterile and aseptic surgical techniques, analgesia, euthanasia, and recognition of species-specific signs of pain or distress.

The experienced laboratory animal veterinarian is an important member of the management team and should be included in discussions of issues that involve the animal care program and animal holding facilities. Facility construction, caging purchases, investigator-purchased equipment, and so forth, can directly or indirectly impact animal health and well-being or increase disease spread risk.

Animal Facility Manager, Office Manager, and Administrative Staff

Like the veterinarian, the animal facility manager has delegated accountability for daily oversight of the facility's animal care and use program. The manager works closely with the veterinarian to ensure the health and well-being of all animals. The office manager and administrative staff oversee administrative functions (e.g., procurement and staffing) that are critical to keeping the program operating in an efficient and productive manner. Responsibilities of the facility manager, office manager, and administrative staff that directly impact the health and well-being of all animals include, but are not limited to, (1) facility access; (2) monitoring animal importation, exportation, and procurement; (3) oversight of routine animal health and behavior evaluations; (4) ensuring execution of established facility sentinel programs; (5) ensuring availability of required supplies and pharmaceuticals; and (6) oversight of the management of animal health records. In addition to the above, the facility manager works with facility engineers to ensure the integrity of facility mechanical and support systems (e.g., heating, ventilation, and air-conditioning [HVAC] and environmental control systems).

Veterinary and Laboratory Animal Technician or Technologist

Today, there is a movement to rename the veterinary and laboratory animal technician or technologist as a "veterinary nurse" (Hyde 2016; Kennedy 2016; http://www.bvna.org.uk/publications/veterinary-nursing-journal). The veterinary and laboratory animal technicians and technologists work closely with the veterinarian to provide nursing and diagnostic support to ensure the health and well-being of all animals in the facility or program. In addition to their health care responsibilities, the veterinary support staff commonly executes the facility's sentinel programs, monitors and orders required supplies and pharmaceuticals, manages animal health records, and may assist scientists with technical aspects of their research. Having additional specific training in the biology, husbandry, health, surgical care, and medical treatment of animals makes the veterinary technician or technologist a real asset. Veterinary technicians commonly hold an associate's or bachelor's degree in veterinary technology and may be designated as a "Registered" (RVT), "Certified" (CVT), or "Licensed" (LVT) Veterinary Technician. Individuals receiving their training through the U.S. military receive a 91T or 68T certification in veterinary technology. As technicians gain experience and training in laboratory animal science, they may choose to become certified by the American Association for Laboratory Animal Science (AALAS) as an assistant laboratory animal technician (ALAT), a laboratory animal technician (LAT), or a laboratory animal technologist (LATG) in the United States.

Animal Care Specialist or Husbandry Technician

The animal care specialist or husbandry technician is one of the most important contributors to the veterinary team. They are the "eyes and ears" of the program, being the first line of defense in detecting problems with an animal's health and well-being. The specialist is trained in laboratory animal husbandry and is the daily caretaker for all animals under their oversight. As a result of their daily observations, animal care specialists gain extensive knowledge of the normal physiology and behavior for each animal and are skilled at ascertaining what is "normal" for the animals under their care. They are trained on the recognition of subtle departures from normal, which may indicate the onset of a problem or disease. Upon identifying a problem, it is the animal care specialist's responsibility to report the change in the animal's condition to the technician or veterinarian for further evaluation. The specialists' contribution is also critical to the execution of many research protocols, as they are often responsible for the provision of protocol-approved special diets, food or fluid restriction, and the administration of study or therapeutic agents (e.g., analgesics).

Veterinary Care: Impact of Facility Design, Medical Support Areas, and Equipment

The requirement for medical facilities and equipment will vary with the species and research to be accommodated in a program. Program size, complexity, and location and dissemination of key program elements also play a role in determining the design of program spaces (Howard and Foucher 2008), the equipment required, the need to duplicate resources, and the amount of storage space required for normal operations and emergencies. As a general rule, rodent and aquatic facilities require less specialized veterinary medical support features than larger species, such as carnivores, nonhuman primates (NHPs), rabbits, and other nonrodent mammals. The requirements for specialized facility design features and specific equipment related to veterinary care may also vary with the country, province, or region in which the work will be conducted.

Several components of facility design (e.g., differential air pressures and surface composition) have a direct impact on a program's ability to exclude or limit the dissemination of pathogens (Hrapkiewicz et al. 2013). Attention must also be paid to personnel and equipment flow patterns, sanitization schedules, and chemical agents used (Shek et al. 2015). Pathogens do not move on their own, but rather attach themselves to airborne particles (e.g., dust), fomites, and surfaces (Brachman 1996). Therefore, if the movement, accumulation, and spread of dust are controlled, and surfaces are kept clean and sanitized, the transmission of most pathogens can be controlled. Being nonporous, free of cracks and crazing, nonabsorbent, and resistant to chemicals facilitates the ability of facility surfaces to be cleaned and sanitized (Leverage and Roberts 2009).

Since personnel, supplies, and equipment can also serve as fomites (NRC 1991; Becker et al. 2007; Clifford and Watson 2008; Watson 2013) for the transport of potential pathogens throughout the facility, these factors should also be taken into consideration when designing a facility. Whenever possible, the design should facilitate the movement of personnel and equipment from "clean" or low-risk areas (e.g., clean cage wash, food and bedding storage areas, and pathogen-free animal holding areas) to "dirty" or areas of higher risk (e.g., dirty cage wash, and isolation and quarantine areas) (Hessler 1991) rather than the reverse. In addition, attention should be given to the personal protective equipment (PPE) required in various areas of the facility to allow for the presence of gowning areas, as well as areas for the storage and disposal of PPE.

It is common to have procedure rooms or laboratories that are used to support both research and veterinary and technical procedures. Appropriate measures and SOPs must be in place to minimize the transmission of pathogens between animals while they are in these areas and exposed to common equipment. This includes, but is not limited to, surfaces where animals are used, restraint devices, scales, anesthetic machines, and equipment for behavioral studies. Policies should be developed to clearly define when animals must be handled in biosafety cabinets (BSCs) or other equipment designed to prevent the dissemination of pathogens or biological agents.

Experimental Testing and Procedure Rooms

It is important to determine if the two-way movement of animals will be permitted between the housing facility and research laboratory. If the movement of animals is restricted, the design of the animal facility should include space for the experimental manipulation of the animals. However, from a veterinary health perspective, testing and procedure rooms constitute a risk factor for the spread of pathogens from people to animals and between research animals (Koszdin and DiGiacomo 2002; Hrapkiewicz et al. 2013; Shek et al. 2015). Here again, attention to the selection of equipment, surfaces, and design of the area to facilitate cleaning and sanitation will help prevent transmission of pathogens. Closed cabinets and storage drawers help to prevent the collection of dust on porous, hard-to-sanitize supplies and materials. Consideration should be given to the presence of running water, a refrigerator, a freezer, a fireproof safety cabinet, a BSC (e.g., BSL-2 and BSL-3), a chemical fume hood, and CO_2 euthanasia chambers within the procedure area. If anesthesia will be conducted in the area, the need for oxygen and anesthetic gas scavenging (e.g., downdraft tables, snorkel scavenging systems, and passive vacuum systems) should be evaluated. If necropsies or whole-animal perfusions will be conducted, adequate ventilation systems must be present. The installation of a commercially available flushing downdraft necropsy table and systems for the collection of excess fixative to prevent contamination of public water and sewer systems should also be evaluated. Where animals will be handled on open countertops, room air pressure is commonly neutral or negative to the outside corridor. All personnel using the rooms must be educated on disease spread and their responsibility in maintaining clean and sanitary working conditions.

Medical Treatment Facilities

Ideally, an area for the treatment of sick animals should be included in the design of the facility. The size of the program, species being used, and type of research being conducted will play a role in the size and design of the treatment area. A dedicated treatment and holding area, separate from common procedure and holding rooms, may be warranted for the treatment of animals with potentially contagious conditions. Treatment areas should be easily cleaned and sanitized as outlined above for experimental procedure areas. Treatment areas should contain the equipment and supplies required to treat animals without contaminating other areas of the facility or leaving an animal unmonitored to collect needed supplies.

Diagnostic Facilities

Imaging Facilities

Like many veterinary practices around the country, it is becoming common for laboratory animal facilities to have in-house medical imaging capabilities. The requirement for and extent of in-house imaging is determined by the size of the program, the species being used, and the kinds of research being conducted. Where in-house facilities are not possible or limited, provisions should be made to obtain the needed support from outside programs.

Today, in-house imaging capabilities may include radiography, ultrasound, positron emission tomography/computed tomography (PET/CT), x-ray reconstruction of moving morphology (XROMM), magnetic resonance imaging (MRI), fluorescence imaging, and nuclear medicine. Because the imaging equipment and facility can function as a fomite to disseminate pathogens, SOPs must take into account the ease and thoroughness of routine sanitation, as well as the flow patterns of both animals and personnel. Specialized programs must be in place for training personnel who perform or attend imaging procedures, and depending on the potential exposure risk, personnel exposure monitoring may be required. Personnel protective devices such as lead aprons and gloves should be stored in a manner that maintains the integrity and functionality of the material. Personnel protective garments should be routinely evaluated to ensure their functionality. The process consists of visually and manually inspecting the garments for wrinkles, cracks, crazes, or other deteriorations, followed by testing the items using radiography. There are commercial companies that will certify the functionality of lead aprons and gloves. The purchase of preventive maintenance agreements for imaging equipment is often beneficial to keep equipment in good repair and ensure safe operation.

Depending on the research being conducted and animal models used, imaging facilities may also require additional procedural areas for animal preparation and clinical support. This is of particular importance in clinical emergencies or where invasive procedures are required for the introduction of tracers or dyes.

Clinical Pathology Facilities

Diagnostics (e.g., clinical chemistry, parasitology, bacteriology, and serology) can be provided in-house or through outside programs. In many situations, a combination of the two approaches is the best solution. In-house diagnostics offer fast results, but care must be exercised to confirm that the test results are valid for the species being tested. Reagents and equipment must be maintained in accordance with the manufacturers' instructions. An appropriate quality assurance program must also be in place to certify the validity of the test results. Personnel must be trained in sample handling, equipment operation, maintenance, and identification and resolution of problems. At a minimum, the clinical pathology area should be equipped with running water, a refrigerator, a freezer, and sufficient space for sample handling, equipment operation, and storage of supplies. Inclusion of a fireproof cabinet, BSC, chemical fume hood, and CO_2 euthanasia chambers within the area may be useful.

Necropsy Facilities

A dedicated necropsy area is advantageous for the containment of potential pathogens or known agents requiring a higher biosafety level. Containment and sanitization are key functional components of this area. As with most areas within an animal facility, it is critical that surfaces should be smooth, impervious to water, and free of cracks, crazed areas, and ledges. Specially designed flushing necropsy tables address all these requirements. Many necropsy tables are designed to also serve in downdraft mode with directional airflow to facilitate the containment of airborne particles. If the table will be used for perfusion of tissues with formalin or other fixatives, the table can be modified to collect the unused fixative as chemical waste. Necropsy rooms must be equipped with running water, soap, and paper towel dispensers. Necropsy equipment and instruments should be dedicated for that area. SOPs should be developed that outline the required PPE, traffic flow patterns, and decontamination procedures. Consideration should be given for the disposal of medical or pathological waste, as well as the dedicated refrigeration and storage of unneeded tissue and cadavers prior to disposal. Because infectious agents may be the cause of an animal's death, it is critical that the necropsy area be isolated and the personnel traffic patterns controlled to prevent cross-contamination. Unless separated from the main animal facility, the room air pressure must be negative to the outside corridor.

Pharmacy Facilities

The size and complexity of the pharmacy will depend on the size of the program, species being utilized, and research being conducted. Whether the pharmacy is a lockable cabinet or drawer in a procedure or treatment area or a dedicated room, the requirements are the same. The pharmacy must provide an environment that protects the integrity of the pharmaceuticals and allows for their security. Drugs that are governmentally regulated (i.e., controlled substances) must be stored in a securely locked, substantially constructed cabinet. These substances must be kept in a secured location that is accessible only to a minimum number of authorized individuals. Larger quantities of controlled substances may require storage in a safe. Smaller quantities of controlled substances can be stored in procedure and treatment areas, in a double-lock and double-door narcotic cabinet or other secured lockbox, for which the keys or combinations are available to only authorized personnel. Some factors that must be considered when evaluating the storage of controlled substances include (1) the regulations applicable to the storage location, (2) the number of individuals requiring access, (3) the required location of the drugs, (4) the presence of alarm systems, (5) the quantity to be stored, and (6) the past history of thefts or diversions (U.S. Department of Justice 2016).

Surgical Facilities

The facilities required to conduct rodent and aquatic survival surgery are less extensive than those that are required for other animals. For most rodents and aquatics, surgery may be performed in a laboratory or facility procedure room as long as the area being used is dedicated at the time for that purpose and appropriately managed to minimize contamination from other activities within the room during the surgery. The chosen area should be free of clutter, easily sanitized, and in a low-traffic and low-noise area of the room.

The size of the program, species to be used, and nature of the procedures being conducted will all play a role in determining the size and complexity of the surgical facility required for other species. For most species other than rodents and aquatics, a defined facility is required. At the minimum, three distinctive functional rooms or areas are needed: (1) an animal preparation area, (2) a surgeon preparation area, and (3) a dedicated surgery area (NRC 2011). Other functional areas that are commonly present include a dedicated recovery room, storage space, and instrument preparation areas. The surgical suite should provide easy access to emergency response equipment (e.g., defibrillator) and supplies.

Surgical suites are often maintained under positive air pressure to help minimize the movement of potentially contaminated airborne particles into the dedicated surgery room (NRC 2011; Perkins and Lipman 2014). While nonsurgical use is discouraged, if a surgical facility is used for other purposes, thorough decontamination is required prior to reuse for surgery (NRC 2011; Perkins and Lipman 2014).

Postoperative Recovery and Intensive Care Facilities

A dedicated area for postoperative recovery is not routinely required for most rodent and aquatic species, although these species still require close monitoring and support after surgical procedures. A dedicated area in which the animal can be closely monitored and supported throughout the postoperative period is required for other animals (NRC 2011). In many facilities, this area also supports animals that require intensive care. Intensive care areas are commonly equipped with specialized caging that is designed to provide a supportive environment for the animal and, in some cases, also prevent self-injury during the postoperative recovery period. Intensive care caging often allows for the provision of oxygen, fluid, medical, and thermal support to the recovering or convalescent animal. The area should be designed to support the electrical requirements of the various monitors, infusion pumps, and other equipment. The area is commonly stocked with supplies, pharmaceuticals, and equipment needed for treating medical emergencies.

Isolation and Quarantine Facilities

Although isolation and quarantine facilities are covered in detail in Chapter 30, it is important to understand their importance to the veterinary care program. Animals are quarantined when they are known to carry or may potentially carry a contagious organism that could adversely impact the health of other animals (Carty 2008). Animals are isolated for a variety of veterinary and research reasons. For example, pathogen-free animals are routinely isolated to verify or protect their "clean" health status. This is common practice with immunodeficient animals, such as nude and SCID mice. The amount of isolation and quarantine space required by a program will depend on the program's need to obtain animals from sources that may not meet the pathogen-free status established for the facility (Rehg and Toth 1988; Roberts and Andrews 2008) or the research being conducted. It is critical that appropriate SOPs be developed that define biocontainment practices, procedures, personnel traffic patterns, and PPE to prevent contamination and cross-contamination of critical areas within a facility. Some programs may require personnel to enter isolation and quarantine areas last or take a shower before entering or leaving the area. Facilities of this nature will require the proximity of a locker room and shower. Anterooms provide a convenient area to don or remove required PPE when entering or leaving the area. Personnel working in isolation and quarantine areas must be highly trained in disease transmission and containment to prevent cross-contamination. Isolation and quarantine facilities should have tightly controlled access and often require dedicated supplies, equipment, and autoclave support. The autoclave should be of an appropriate size and be in close proximity to the isolation or quarantine area. Some facilities are designed for a double-door, "pass-through" autoclave to process material into or out of the area. In general, the room air pressure of isolation and quarantine areas is kept negative to the outside corridor when housing animals that potentially harbor pathogens (NIH/ORF 2008; Huerkamp and Pullium 2009; Lipman et al. 2015). When isolating animals to protect their clean health status (immunodeficient animals, etc.), the room air pressure is often maintained positive to the outside corridor (Kowalski et al. 2002; Lipman et al. 2015).

Veterinary Programs

Animal Health Status

Establishment of the Required Animal Health Status

Using healthy animals is a foremost consideration in conducting sound research and generating quality data. Health status requirements will vary depending on the nature of the planned research (Baker 2003; Desrosiers 1997) and must be determined before the animals are acquired. In addition, it also must be determined if the program has the ability to maintain the health status once the animals are received. Advances in facility design features, health monitoring techniques, and specialized shipping containers allow vendors to meet a wide range of research demands. Institutionally managed breeding programs may also contribute to achieving and maintaining the appropriate animal health status. The researcher must make sure the animal procurement staff understands any nonstandard health requirements essential for the research (e.g., need for immunocompromised animals in tumor studies, and *Helicobacter* species–free mice in gastrointestinal [GI] research). It is important that research staff discuss any health requirements with the vivarium's management and veterinary staff prior to the acquisition of the animals. They can help the investigator determine if the vivarium has the capability to maintain the desired animal health status, as well as meet the other requirements of the research.

Research Requirements

There is no universal menu of research requirements. Each research project requires consideration of parameters unique to the given research. For instance, using immunocompromised animals may be appropriate for tumor studies but be quite inappropriate for infectious disease studies where a robust immune response is required. Clear communication between the researcher and the veterinary staff is paramount and should be initiated by the former. In the ideal world, every aspect of a research project should be understood by all parties involved. At a minimum, the veterinary staff needs to be familiar with the species to be used and understand the aspects of the research project that may potentially impact the health and well-being of the animals involved. Such aspects can vary from simple parameters, such as gender, weight, age, and coat color, to more complex aspects, such as strain and specific body condition (e.g., lean vs. obese). Unique features, such as physical and physiological fitness, size and functional capacity of select organs (e.g., liver and prostate in males), ability of the host immune system to fight off challenges (e.g., microorganism infection and foreign organ or tissue tolerance or rejection), spontaneous mutations, and propensity for developing endocrine imbalances (e.g., diabetes), may also be important factors to take into account.

Species Considerations

Rodent Colonies Establishment of the appropriate health status for a rodent colony or facility is dependent on many factors. First is the potential impact of animal health status on the research to be conducted. Some studies have strict requirements for germ-free animals, whereas other studies are less susceptible to the effects of pathogens. In studies requiring pathogen-free animals, further questions must be asked to determine which organisms may adversely impact the research. The term *specific pathogen-free* (SPF) is commonly used in laboratory animal science to describe animals free of a defined specific list of pathogens. The term *SPF* itself does not indicate that an animal is free of particular pathogens of interest to the receiving facility or researcher. Each vendor or facility supplying animals must be asked which pathogens they exclude from their animals. It is the receiving program's responsibility to ensure that the vendor's list of organisms they test for includes the pathogens of importance to their facility and research. In addition, the vendor or supplying facility should be asked how the current health status of their animals is determined and how the status is ensured throughout their production program, as well as during shipment of the animals to the institution.

The second factor that must be considered is what health status is optimal for the general health and well-being of the receiving colony or facility. Ideally, most program managers and veterinarians would like their animals to be free of all viral, bacterial, and parasitic pathogens, but in reality, there is a direct relationship between the colony's desired health status and the cost of doing business within a facility. Generally, the "cleaner" the desired colony health status, the higher the cost of maintaining the colony becomes. The higher cost results from several factors, including the need for (1) specialized caging and equipment, (2) increased levels of PPE, and (3) more intensified health surveillance programs. There is also a cost to the research, in that clean facilities often must not allow animals to return to the home facility after being tested or studied in another facility or laboratory. In addition, maintaining a clean facility may limit the ability to import animals from other collaborating research groups or vendors. Therefore, careful consideration must be given to the pathogen exclusion list developed for the colony or facility (Mähler et al. 2014; Shek et al. 2015). Programs must consider not only the health impact on the colony and the research, but also the final cost to maintain the desired health status (Kowalski et al. 2002).

Maintaining animals of a specific health status can require specialized barriers designed to exclude pathogens (Nicklas et al. 2015). Therefore, a third factor to consider is the available infrastructure and personnel to support the desired health status (e.g., flexible film isolators, microisolator caging, BSCs, and autoclaves), as well as the availability of funds to maintain the program.

Nonhuman Primates (New and Old World) At a minimum, NHPs need to pass a physical examination and undergo a period of quarantine to determine the presence of zoonotic agents. A 31-day quarantine is required at the port of entry into the United States (42 CFR §71.53) to safeguard against the introduction of tuberculosis-infected animals. During the quarantine, a thorough physical examination and clinical workup is conducted by a qualified veterinarian and the animal is administered a series of three intradermal tuberculosis tests. Depending on the future destination of the animal, additional tests and requirements may include combinations of the following:

- Dental exam
- Thoracic and abdominal radiology
- Hematology and serum biochemistry
- Internal and external parasitology
- Bacterial cultures: Fecal (*Shigella* sp., *Salmonella* sp., *Campylobacter* sp., and *Yersinia* sp.), nasal, oropharyngeal, tracheobronchial
- Virology: Macaque profiles—*Macacine herpesvirus* 1 (McHV-1; formerly herpes B), measles, retroviruses (Tardif et al. 2012); baboon SA8; and *Herpesvirus tamarinus* (marmosets)
- Urinalysis
- Vaccination status: Measles, rabies, and tetanus
- Neurological assessment: Hand and eye coordination and motor functions
- Psychological assessment: Basic (e.g., shy vs. aggressive, willingness to work with humans, advanced capacity for and ease of learning, memory, and suitability for complex research tasks)
- Tests focused on specific criteria required for a particular field of research: Specific phenotypes (e.g., obese, diabetic, or blind), congenital abnormalities (e.g., dwarfism, scoliosis, or polydactyly), or special conditions (e.g., amputees or geriatric)
- Genetic profiles

When available, information on the origin of the animal (e.g., wild or captive bred, country of origin), may be valuable. Additional information that may prove valuable is the animals' past social housing, medical, surgical, or research history.

There are several well-established NHP vendors offering a variety of species, ages, and sizes of monkeys with a specified health status, for example macaques free of retroviruses and McHV-1. While some institutions have adopted policies requiring their researchers to work exclusively with zoonosis-free animals, other institutions have created carefully crafted procedures for working successfully with populations of macaques known to be infected with zoonotic organisms (Mansfield 2005) or that convert to a positive status during a study. Programs must understand that even SPF animals, believed to be free of McHV-1, have been known to seroconvert and shed the virus (Weigler 1992). The only definitive, but not practical, method to ensure an animal does not carry the virus is postmortem testing. Therefore, starting with serologically negative animals can be helpful to prevent human exposures, but appropriate SOPs, including the use of PPE (see Chapter 14), must still be implemented to work safely with these animals.

Carnivores Most U.S. institutions use purpose-bred (Class A) dogs. A USDA Class A license is issued to dealers who sell animals that were bred and raised at their facility. A Class B license is issued to dealers who buy and sell warm-blooded animals that were not born and raised on their property. At the U.S. Congress's request, the National Academy of Sciences (NAS) assessed whether there is a scientific need for National Institutes of Health (NIH) grant recipients to purchase dogs and cats from random source (Class B) dealers. In May 2009, the National Research Council (NRC) released its report "Scientific and Humane Issues in the Use of Random Source Dogs and Cats in Research." (NRC 2009b). The report concluded that there was not. Moreover, in response to congressional concern, the NIH has since advised its grant recipients that it is phasing out the practice (NOT-OD-11-055). Effective October 1, 2014, the NIH no longer funds research on dogs procured from pounds, breeders, and other so-called random sources (http://www.news.sciencemag.org/news/2014/10/nih-ends-funding-experiments-using-random-source-dogs).

Class A dogs, in addition to being clinically healthy, are typically inoculated with core vaccines (e.g., canine hepatitis/adenovirus type 2, canine distemper, rabies, and canine parvovirus virus) and noncore vaccines for canine parainfluenza virus (which some protocols still consider core), *Bordetella bronchiseptica*, borreliosis, and leptospirosis (Welborn et al. 2011). Class A dogs are bred in closed colonies, which are tested for external and internal parasites and checked for apparent physical and physiological abnormalities (Nemzek et al. 2015).

Use of Class A and B cats is scrutinized using the same criteria as dogs. Class A cats, like dogs, derive from operations geared toward generating non-pet, but socialized, healthy animals. Core vaccines include feline panleukopenia, feline rhinotracheitis, feline calicivirus, and rabies virus (AAFP 2013). Noncore vaccines include feline leukemia virus, feline infectious peritonitis virus, *Bordetella*, and *Chlamydia*. Such animals are also free of external and internal parasites and apparent physical and physiological abnormalities. *Giardia* vaccination is optional.

The domestic or European ferret (*Mustella putorius furo*) has unique applications for its use in research. Ferrets should only be procured from reputable research vendors with colonies of known health status. Unless restricted by research, it is recommended that ferrets be fully immunized against rabies and canine distemper virus (AFA 2006).

Lagomorphs Conventional health status rabbits are prone to secondary bacterial infections caused mainly by *Pasteurella multocida*, which can prematurely terminate research projects (Deeb et al. 1990). There are a number of commercial vendors generating rabbits that are SPF for *Pasteurella*. These animals are significantly healthier than their conventional counterparts, allowing for successful completion of longer-term research projects. When housing *Pasteurella*-free animals in the same facility with animals known to harbor the bacterium, care must be taken to establish appropriate traffic patterns and PPE requirements to ensure that the *Pasteurella*-free animals are not inadvertently contaminated with the agent. Investigator requirements will determine the rabbit breed to be selected. For example, both albino and pigmented breeds are available, and some ocular work requires the use of pigmented eyes.

Agricultural Animals Agricultural species include pigs and small ruminants. These species require special enclosures because of their size and rapid growth, and their care may be labor-intensive (FASS 2010). Pigs are social, large animals that are very popular in teaching labs and are robust enough to be used in chronic and invasive projects. Both conventional and SPF animals are available, and the demands of contemporary research have created a plethora of genetically selected breeds (mini- and micropigs), as well as transgenic hybrids (Swindle et al. 2012).

Small ruminants (sheep and goats) have an expanding use in the research setting, although they may harbor some zoonotic concerns, such as Orf (parapoxvirus) and Q fever (*Coxiella burnetii*) (Underwood et al. 2015). Animals free of Q fever are now commercially available. For more details, see Chapter 23.

Amphibians Amphibians have long been utilized in scientific research and education as models for a variety of developmental and physiological processes, largely due to their unique ability to undergo metamorphosis and to regenerate limbs in some species. Their embryos have been used to evaluate the effects of toxins, mutagens, and teratogens. They have short generation times and genetic constructs desirable in transgenic and knockout technology, and also in genetic and genomic research. They are useful due to their sensitivity to climatic and habitat changes and environmental contamination. Amphibians begin life as aquatic larva and, through the process of metamorphosis, emerge as terrestrial adults. While many species of amphibians do not strictly adhere to this developmental pattern, they remain the only vertebrate class with such a unique adaptation (Zug et al. 2001; O'Rourke 2007). *Xenopus laevis* is the most commonly used species in the laboratory due to their ease in embryo manipulation and applicability to early pregnancy testing and developmental studies. There are well-established U.S. vendors (e.g., NASCO [https://www.enasco.com/page/xen_care]) that provide commercially raised animals with a known health status.

With the exception of African clawed frogs (*Xenopus* spp.) (https://www.enasco.com/page/xen_care and http://www.xenopus.com), amphibians are not available from research purpose-bred commercial breeding colonies. Bullfrogs (*Rana catesbeiana*) and other amphibian species may be purchased from pet suppliers, or must be collected from the wild by the investigator, or obtained from other researchers, zoos, or agencies, such as the U.S. Fish and Wildlife Service. It is essential to replicate their natural environments in order to breed them in a research setting. Unlike domesticated mammals, amphibians are wild animals and the husbandry methods used must take this into account (Browne et al. 2007). The welfare of the animals must have the highest priority in the design of animal rooms, tanks, and tank furnishings.

Fish The most commonly used fish in research is zebrafish (*Danio rerio*). It is an attractive alternative to mammalian species for the following reasons: (1) it is less sentient than mammals, (2) it has ease of reproduction and fast-growing and developing offspring, (3) it can tolerate higher doses of chemical mutagenesis than rodents, (4) transparent embryos and larvae permit noninvasive imaging strategies when following genetic manipulation or pharmacological treatment, (5) it offers insights into human disease due to functional homology with mammals, and (6) it is easy to replicate the natural environment of zebrafish, which reduces stress and its impact on experiments. There are a number of websites with information on sales, tutorials, and genetics (e.g., http://www.zfic.org/linksindex.html). Zebrafish are available from established vendors (e.g., Zebrafish International Resource Center of the University of Oregon). Zebrafish embryos are commonly washed with hypochlorite solutions prior to their introduction into an animal facility to help remove unwanted pathogens.

Birds A large number of avian species are used in laboratory animal science. The range spans many domestic species (e.g., chickens, ducks, geese, pigeons, doves, zebra finches, Japanese quail, and parakeets) and nondomestic species (e.g., crows, sparrows, and hawks) (Fair et al. 2010). Many species of birds are obtainable from commercial sources, whereas wild birds are often used solely in field studies. Avian models have advanced our understanding of developmental biology, aging, immunology, endocrinology, and genetics, as well as other aspects of medicine and science. Many avian species are highly social and should be socially housed whenever possible (Hawkins et al. 2003). When birds must be brought from the wild and maintained in the laboratory animal facility, they should ideally be housed separately from other species and, if possible, provided with conditions that approximate their natural habitat (Hawkins 2001). Along with temperature and humidity, consideration should be given to the type of food, perches, and cover provided, as well as the manner in which water is offered (NRC 1977). Both domestic and wild birds can be carriers of zoonotic diseases (e.g., chlamydiosis, cryptococcosis, histoplasmosis, psittacosis, and salmonellosis), as well as several infectious and parasitic agents that can adversely impact other birds (e.g., Newcastle disease, Marek's disease, and coccidiosis) (Baer et al. 2015; Patterson and Fee 2015; Taylor et al. 2016).

Other Considerations The desired health status of acquired animals may vary depending on the short- and long-term goals of each project. For instance, the animals needed for an acute project may require no or only a basic physical examination and short quarantine period, whereas those for lengthy projects, in most cases, require a thorough clinical evaluation and evidence of robust health during a period of quarantine. Similarly, the demand for clinical robustness will be much higher for invasive than noninvasive studies.

Acquiring Animals from Commercial Sources

The quality and health status of animals procured from vendors can vary greatly depending on the size and experience of the vendor. A program's first priority should be to maintain its own biosecurity, protecting against the introduction of unwanted pathogens. Unfortunately, under most circumstances, it is neither possible nor cost-effective to test or rederive every animal entering a facility to ensure its health status, or to breed all animals in-house. Therefore, a relationship must be established with reputable vendors to provide healthy, pathogen-defined animals.

An ideal start to procuring any animal is to find out which vendors can meet the program's requirements by talking to the animal facility veterinarian, manager, or other investigators with animals in the same facility. Additional sources of information include the materials and methods sections of published research similar to that being conducted in your program, laboratory animal association publications, and websites such as the AALAS Laboratory Animal Science Buyers Guide (http://laboratory animalsciencebuyersguide.com/) or the LabAnimal® Europe and Asia Pacific Buyers Guides (http://www.labanimaleurope.eu/buyers_guide/), as well as regional repositories. Many large vendors post their current health status online via their websites. Once potential vendors have been identified, inviting them to present a seminar on their company's products and services, including their health surveillance program and quality assurance guarantees, can be helpful. In some situations, conducting a site visit to the vendor's production facility may also be beneficial.

When establishing a long-term commitment with a vendor, developing a contract detailing the relationship between both parties should be considered. A contract can be used to detail a program's requirement for availability of specific products, custom-designed surveillance panels, shipping specifications, product returns, reimbursements, and replacements. In addition, a contract can be used to detail the vendor's responsibility to notify the program in a timely manner of any change in the health status of animals in its facility.

Acquiring Animals from Noncommercial Sources

With the advent of transgenic technology, it has become commonplace to acquire animals from noncommercial sources, such as universities or other research institutions. Noncommercial sources can vary widely in their health status and surveillance programs (Pritchett-Corning et al. 2009). When procuring animals from noncommercial sources, an assessment must be made of the risk they pose to the procuring program and how the program can mitigate that risk. It is helpful to ascertain the current and past health status history of each source. At a minimum, source colony information should include an overview of their husbandry program, use of PPE, disease barriers (e.g., use of microisolator caging and BSCs), colony health surveillance program, policies for the introduction of new animals, and colony pathogen history. It is not uncommon to request that the shipping facility provide a year's worth of relevant health surveillance testing data prior to approval of the animals for shipping. Regardless of the shipping colonies' current health status or history, some very clean facilities may require that all mice and rats coming from noncommercial sources be rederived into the facility using embryo rederivation techniques. Alternatively, other programs have a strict quarantine and testing policy to screen animals being received from noncommercial sources. Animals obtained from the wild should be quarantined (CDC, African Rodent Importation Ban, http://www.cdc.gov/poxvirus/monkeypox/african-ban.html) and thoroughly tested for zoonotic diseases, as well as detrimental organisms known to infect closely related species (42 CFR §71.56; Karesh 2005; Hutson et al. 2015).

Animal Transportation and Stress (Shipping Containers, Temperature, Food and Water, Conveyance, and Personnel)

Specific parameters that must be met during transportation to ensure the well-being of a species have been set by many countries and the International Air Transport Association (IATA) (Live Animal Regulations, http://www.iata.org/publications/store/pages/live-animals-regulation.aspx). Using properly equipped, clean vehicles in conjunction with trained and experienced personnel throughout the transportation process is essential. Appropriately sized shipping containers, control of environmental factors such as ambient temperature, provision of food and water, in-transit safety precautions, and minimization of stressful events are all factors that must be considered when transporting animals in order to preserve their health and well-being (NRC 2006). Animals traveling long distances may experience fear, anxiety, and metabolic imbalances and require time to recover once they are received (Syversen et al. 2008). It is recommended that research institutions adopt policies to protect animals from being used immediately upon receipt, without an acclimation period. The lack of an acclimation period may confound the data or make many procedures (e.g., anesthesia) unsafe for the animal. For most species, a 48- to 72-hour minimum acclimation period following transportation is considered good practice (NRC 2006). However, recent evidence indicates that up to 2 weeks may be needed for some species to recover from the stress of transportation (Ochi et al. 2016).

Animal Receipt, Examination, and Acclimation

Newly arrived animals should be received by trained and experienced personnel in a designated area that is protected from extreme heat or cold, sufficiently illuminated, clean, uncluttered, dry, and quiet. Animals appearing sick or injured at the time of receipt, as well as animals arriving in damaged shipping containers, should be isolated for veterinary assessment prior to entering the colony. Everyone involved in the animal receipt process should be notified in a timely fashion and be able to devote adequate time and effort to the task of safe and expedient placement of the new animals in their new enclosures. Ideally, cage cards and/or animal records should be prepared ahead of time. The enclosures should be supplied with water promptly, and when possible, it should be established that new animals are drinking before they are placed on an automatic watering system. Feeding animals immediately after receipt is not as critical and should be carefully considered. Food consumption should be controlled and monitored to avoid GI problems, such as emesis and diarrhea, or other health issues. Animals should be observed for a period of time and allowed to adjust to their new environment, especially if they appear anxious or aggressive, or are vocalizing.

Some species (e.g., cats, rabbits, and marmosets) and individual animals are sensitive to abrupt changes in their diet and may refuse to eat upon receipt, or display signs of GI problems, such as diarrhea or bloat. These situations may require knowledge of the originating facility's diet prior to the arrival of the animals and a careful program of transition from one diet to another.

Monitoring the Colony Health Status

Once the health status of a colony has been established, routine monitoring is essential to preserve that status. A comprehensive monitoring plan may include a variety of measures, such as daily health checks, routine physical exams, testing, health surveillance programs, and environmental monitoring. Postoperative observations and assessment of unanticipated or abnormal research-related effects and outcomes can also provide important indicators that a colony's health status has been compromised. Ideally, an initial response plan to deviations from the chosen health status should be in place before a problem occurs.

Daily Health Checks

Daily health checks are a critical way to quickly detect changes in the colony health status. These observations are mandated in the U.S. AWR (9 CFR §2.40 b.3) and the *Guide* (NRC 2011), as well as some research funding entities in the United States and Europe. The number and timing of health checks will depend on the nature of the facility, species, and research being conducted. For example, infectious disease research animals may require multiple checks each day at specific intervals, whereas breeding colony animals may require less frequent observations with more flexible timing. In some situations, it may be advantageous to observe animals when they are most active, immediately prior to, during, and after they are fed.

Daily health checks are typically performed by individuals designated by the attending veterinarian who are trained and qualified to recognize the normal behavior and appearance of the animals under their care and identify abnormalities. Because the animal care specialists observe these animals each day, they are uniquely suited to quickly recognize any subtle changes in an animal's posture, activity level, behavior, physical appearance, respiratory patterns, bodily functions, and use of its environment, including enrichment items (i.e., untouched favorite treats or decreased nest quality), regardless of the species under observation (i.e., birds, fish, amphibians, small mammals, or NHPs). Abnormalities should be documented and promptly reported to a veterinarian. It is also beneficial to provide staff with scoring sheets, potential symptoms, phenotypic differences, and pictures of common ailments seen in the research setting.

Sufficient time must be allotted for the care staff to conduct a thorough visual health check of each animal. This should be done a minimum of once a day, and preferably twice daily. For example, cages or fish tanks containing complex enrichment strategies or testing apparatus can complicate the visual assessment of the animals. In these situations, programs may require each cage to be removed from the rack daily to ensure visual assessment of all animals, or in the case of aquatics, the tank to be visible from various angles. Another key time to assess animals is during cage or tank changing. While changing cages or tanks, care staff should be attentive to both physical and behavioral signs exhibited by animals, regardless of the species, that may indicate the need for further assessment by a veterinary technician or technologist or veterinarian. Anomalous findings include, but are not limited to, changes in body conformation or physical appearance, guarding of a body part, abnormal respiration, and abnormal behavior (e.g., aggression or hiding).

When animals are housed in stacked cages, it may be helpful to equip care staff with flashlights to facilitate visualization of the animals housed in the lower cages. The use of binoculars can help visualize and assess group-housed animals in both outdoor and indoor facilities.

Postprocedural Observations and Assessment of Protocol-Related Effects

Postoperative observations are not only a way to determine if analgesia and surgical techniques are adequate; they also provide a means of assessing colony health. Changes in health status can be manifested through postoperative or postprocedural complications. For example, opportunistic pathogens may only become evident after an animal experiences a stressor such as surgery (Carty 2008). Close monitoring for abnormalities in respiratory, enteric, integumental, and other systems after surgery can alert staff to changes in the colony health status. In a similar fashion, protocol-related procedures and results may also point to problems in colony health (Gulani et al. 2016). However, one must distinguish whether these protocol-related effects are simply due to study procedures or genotypes, or are due to an unwanted pathogen in the colony. For example, weight loss could be a sequela of the experiment or a new pathogen in the colony. Abnormal performance in rodent behavioral tests could indicate a response to research interventions or new subclinical disease. Adverse anesthetic responses may be related to the protocol but could also be due to subclinical respiratory disease. When complications are observed after a surgical or other experimental procedure, the veterinarian and scientist should work together to determine whether these complications are due to changes in colony health status or to the research procedures involved.

Routine Physical Exams

Findings from recurring complete physical examinations and testing can indicate changes in colony health. Regular exams involve monitoring for abnormalities that may not be evident at visual cage- or kennel-side observations. Examples of these abnormalities may include decreased body condition, masses, enlarged lymph nodes, and periodontal disease, which could indicate the presence of tuberculosis, *Pasteurella* abscesses, shigellosis, or even McHV, in a NHP colony. Routine physicals for larger species should be performed no less than annually to determine viral, bacterial, parasitic, hematological, dental, and overall health status. The nature of the facility, the required health status, and local or national regulations will ultimately determine what tests are performed and how frequently. For smaller species (e.g., mice, rats, and fish), individual routine physical examinations are not regularly performed unless dictated by the research being conducted.

Testing and Health Surveillance Programs

The Federation of European Laboratory Animal Science Association (FELASA) has issued useful guidelines (Weber et al. 1999; Rehbinder et al. 1998, 2000; Voipio et al. 2008; Mähler et al. 2014) for monitoring the health of both large and small animals used in research. For large animals, the recommendations generally involve direct testing of research animals. Because the size of rodent colonies and modern husbandry techniques typically preclude individual examination and testing, colony surveillance testing is an efficient way to screen for many pathogens. If sentinel animals are used and tested for seroconversion, care must be taken to ensure that the animals are immunocompetent, relatively young, and capable of mounting immune responses to pathogens (Besselsen et al. 2000; Mähler et al. 2014). Sentinel animals are housed within the animal colony and, in most cases, receive dirty bedding from other colony cohorts.

In cases where pathogens are not reliably transferred by dirty bedding (Artwohl et al. 1994; Cundiff et al. 1995; Compton et al. 2004; Perdue et al. 2008; Henderson et al. 2013), sentinels may be cohabitated with research animals. Sentinel animals are typically exposed to the colony cohorts for 6–8 weeks to allow for exposure, infection, and seroconversion and then tested by a variety of methods. FELASA recommends sentinel testing a minimum of every 3 months. Testing can include serology; polymerase chain reaction (PCR) testing of fecal material, hair and anal swabs, or intestinal contents; microscopic fecal and tape test examinations; bacterial cultures of a multitude of samples; histology; and gross necropsy. As a supplement to sentinel testing, research animals (typically those no longer needed for study) can be tested by the same methods to detect those organisms that may not be transferred via dirty bedding or that the sentinels may have cleared postexposure (Clarke and Perdue 2004). Alternatively, with investigator permission, survival testing that uses blood, fecal, and swab samples can be performed. Biological samples such as bone marrow and cells may also be tested before use in animals to detect unwanted pathogens in the originating colony.

Environmental Monitoring

Testing the animals' macro- and microenvironments is an alternative or supplementary way to monitor colony health. Testing methods should be sensitive, specific, and able to detect a wide range of pathogens. Microbiological testing can be accomplished by RODAC™ (Replicate Organism Detection and Counting) or ATP-based systems (Ednie et al. 1998). RODAC is inexpensive, uses quick sampling methods, and offers quantifiable data. However, it only identifies viable aerobic bacteria and fungi that can be cultured and will not identify parasites or viral pathogens. This testing also requires several days of incubation before results are available. ATP-based systems use bioluminescence to indicate levels of ATP in live or dead organic material. These systems are especially useful in detecting and quantifying organic material on many substrates (walls, floors, and doors) and assessing the efficacy of sanitation practices (Turner et al. 2010). Unfortunately, these systems do not specify what organic material and pathogens are present and whether they are alive or dead. In addition, some disinfectants may alter the results (Turner et al. 2010). These systems also require that facilities establish their own standards for what values are acceptable and what values require action. Setting action cutoff values can be challenging and may require correlating ATP testing results with RODAC results.

Use of PCR testing to examine housing units, racking systems, supply and exhaust vents, behavioral equipment, and other items in the environment is becoming more popular as a way to monitor colony health, and incorporating its use into a surveillance program can reduce or eliminate the number of sentinel animals used (Henderson and Clifford 2013; Jensen et al. 2013; Compton and Macy 2015; Manuel et al. 2016). Food, water, and bedding can be sampled for pathogens as well (Rice et al. 2013). PCR testing is sensitive, relatively inexpensive when compared with purchasing, housing, and caring for sentinel animals; and available for a large number of organisms. Its exclusive use for health monitoring eliminates introducing sentinel animals into closed colonies. In some facilities, PCR testing has replaced rodent sentinel testing or is used in conjunction with a reduced sentinel testing program. However, care must be taken to ensure that PCR testing plans are appropriate for the type of caging system and filters used in the program. Some caging systems have been demonstrated to isolate the organisms to the animal's cage, preventing contamination of and pathogen detection in the rack plenums and ducts (Henderson and Clifford 2013; Henderson et al. 2013). Therefore, a PCR testing program must also take into account how the pathogen spreads and persists in the environment.

Potential problems with PCR testing are (1) the interpretation of negative results and (2) the potential for false-positive results. Failure to collect the sample from the contaminated area of the equipment can lead to false-negative results, whereas false positives can be the result of nucleic acids remaining on previously contaminated equipment or in the environment (Leblanc et al. 2014). Ensuring that contaminated equipment and supplies are free of all traces of lingering nucleic acids can also be problematic.

Monitoring temperature, relative humidity, and air pressures for each housing room is important and required by some regulations (NRC 2011). This monitoring is critical to health status, as temperature and humidity can affect the growth of pathogens, and differential air pressures can affect the integrity of the separation of barrier rooms, surgical suites, rooms in which biohazardous agents are being studied, or known dirty areas from other colony rooms. Monitoring can be accomplished at the room level or by building automated environmental control systems that alarm when measurements fall outside set parameters. In some cases, both types of monitoring are performed to allow for redundancy in the event of a system malfunction. The NIH Office of Laboratory Animal Welfare (OLAW) strongly encourages institutions to use electronic technology for environmental monitoring, and the *Guide* (NRC 2011) advises the use of systems with automated alarms.

Response Plan to Health Status Deviations

Appropriate contingency plans should be prepared in advance and be ready to implement when unwanted pathogens are detected in a colony. Advance preparations can include cryopreservation of important rodent lines, maintaining separation of animal rooms through strict room entry order, adjustment of air movement from clean to less clean areas, the use of PPE, and keeping complete records of animal movements. Plans for health maintenance and monitoring, as previously discussed in this chapter, will also allow for early detection of a pathogen and reduce the likelihood of an outbreak spreading. The contingency plan should also include the steps to be taken when the containment barrier is breached, for example, when a rodent is found outside of its cage or is accidentally dropped on the floor.

When a positive result is obtained, one must consider (1) the nature of the pathogen, (2) how the pathogen is shed, (3) how the pathogen is spread, (4) the sensitivity and specificity of the pathogen tests, (5) prevalence of the agent, (6) how seroconversion may complicate test results, (7) how persistent the pathogen is in the environment, and (8) how treatment may affect research results. Responses should be aimed at limiting pathogen spread and eradicating the pathogen from the colony (Reuter et al. 2011). The first, and most important, step is to isolate the potentially affected animals and conduct tests to confirm the previous results. Next, notify the appropriate institutional personnel and report the disease and quarantine as required by local and national regulations. Other important management practices to consider include changing room entry and access procedures, increasing the level of sanitation and PPE, implementing special handling requirements for animals and enclosures, closing the colony to imports and exports, and ceasing breeding in some cases. If the pathogen is zoonotic, staff will need to be educated on the risks to their health. It is critical to post signage on room or enclosure doors that very clearly describes all of this information and the new required management practices.

Responses to outbreaks are dependent on the pathogen and may involve treatment, testing and culling, caesarean section, embryo transfer, and cessation of breeding. When treatment is utilized, one may need to give consideration to the current research being conducted in the affected area. Consideration should also be given as to whether research animals will naturally clear the infection or serve as reservoirs. Using diagrams and schematics to track testing and test results can be extremely valuable in managing outbreaks.

The time, cost, and energy required to stop an outbreak will vary based on the pathogen and type of animals affected. Testing and other necessary responses, such as euthanasia of animals, can tremendously impact the responding staff, both physically and emotionally. Boosting morale and being sensitive to personnel fatigue are critical to successfully ending an outbreak. In some cases, research staff may be resistant to new management practices during an outbreak, but it is important to inform them of the enormous time, cost, and loss of research data and animals that can occur if outbreaks are not addressed in an appropriate and timely manner. If there is the potential that animals from multiple research groups may be impacted, it is useful to meet with all groups involved to explain the situation and plan for resolution.

Clinical Care of Animals

Diagnosis and Treatment of Ill Animals

When physical or behavioral abnormalities are observed in research animals, it is important to determine whether the animals are experiencing a natural or experimentally induced illness and whether intervention with treatment, monitoring, or euthanasia is appropriate. Diagnostic and treatment approaches and humane and experimental endpoints must be considered when addressing the observed illness. All these considerations must be done in the context of balancing animal welfare with research objectives.

Natural or Experimentally Induced Illness?

When routine daily observations detect physical or behavioral abnormalities in a research animal or group of animals, prompt reporting to the veterinarian is critical. Unexpected findings should also be reported to the scientist, who can provide specifics about the nature and timeline of the research. The veterinarian and scientist should work together to determine whether the illness observed is caused by a natural process or the research. Ideally, issues resulting from the research (e.g., test compounds, surgical procedures, deleterious phenotypes, and infectious agents) should have been described by the scientist in the protocol and understood by the veterinarian and animal care staff prior to study initiation.

If the abnormality is related to an unanticipated phenotype, the investigator, in consultation with the veterinarian, should determine the importance of the animal and if further characterization is justified. This may include routine diagnostics, necropsy and histology, initiating a genotyping or phenotyping study (Crawley 2000), or seeking assistance from other researchers who have experience with the model. If further characterization is justified, the IACUC should be apprised of the situation and the actions being taken to further define the potential new model.

Veterinary Diagnostic Procedures

Veterinary diagnostic procedures include physical examination, serum biochemistry panels, complete blood counts, serology, bacterial cultures, urinalysis, cytology, fecal exams, PCR testing of various samples, radiography, ultrasound, MRI and CT scans, biopsies, endoscopy, exploratory surgery, histology, and necropsy. These tests are available for rodents, just as they are for larger species, so one should not overlook their value in assessing small animal cases. Many of these tests are available through commercial services, but they can be readily performed within the facility with the proper equipment, personnel training, budgets, and space. To the greatest extent possible, the diagnostic test should be chosen with consideration of the research goals. For example, sedation for radiographs would be contraindicated just before an animal is scheduled to undergo behavioral testing at a critical research time point. The veterinarian and scientist should work together to devise a diagnostic plan that meets the needs of the animal and the study.

Treatment and Monitoring Approaches

When a diagnosis is made, a plan for treatment, monitoring, or euthanasia will be established by the veterinarian and should be made in consultation with the researcher. The decision to treat, observe only, or euthanize is complex and is discussed in the next section "To Treat, Monitor, or Euthanize?" (page 751). If treatment is chosen, there are several approaches to consider. Medications can be provided in the drinking water, regular feed, or treats, or through injections, continuous intravenous delivery systems, topical applications, or transdermal patches. Supportive care can be in the form of supplemental heat, fluids, or oxygen; additional bedding or nesting material; high-calorie and high-value foods or treats; and bandaging lesions. The treatment approach should consider whether an individual or multiple animals need to be treated, what resources are available for treatment, the nature of the sick animals, and the research objectives. For example, if mice in several cages must be treated, labor costs for providing medicated water or feed to multiple cages will be less than injecting each individual mouse needing treatment. This is an important factor to consider when staffing is limited. Providing medications through feed or water will also result in less stress to the animal than injections. If the nature of an animal is such that it refuses to take medications through water, food, or treats, treatment may be limited to oral gavage or injectable routes. Research objectives may affect the treatment plan as well. For example, if only two of five mice in a cage require treatment, and treatment may interfere with the study, the two sick mice can be treated separately from healthy cage mates or removed from the study. Also, if the research requires a strict and/or measured diet, oral medications in feed and treats may pose problems to the study and necessitate an alternative dosing route.

If the needs of the animal and the research are such that only monitoring can be performed, scoring systems may be useful in determining when treatment or euthanasia will be initiated (Langford et al. 2010;

Bekkevold et al. 2013; Nunamaker et al. 2013). Scoring systems provide an objective way to determine when endpoints have been met and/or interventions are necessary. A standard scoring system lists abnormal signs that can be observed and then asks the observer to state whether a specific abnormal finding is present or absent (Figure 31.1). Scoring systems may also involve rating abnormalities using a numerical scale or based on qualitative observations. For example, arthritis in mice can be scored using millimeters of paw swelling or using subjective numbers to indicate increasing degrees of lameness, swelling, and

Date started:________ Protocol:__________ PI/User:________ Cage #:____________________

Date		Time					
ID #	Activity	EAE score*	Body condition***	Pain score***	TX confirmed**	Initials	Comments
	QAR BAR	0 1 2 3 4 5	1 2 3 4 5	1 2 3 4	Yes No		
	QAR BAR	0 1 2 3 4 5	1 2 3 4 5	1 2 3 4	Yes No		
	QAR BAR	0 1 2 3 4 5	1 2 3 4 5	1 2 3 4	Yes No		
	QAR BAR	0 1 2 3 4 5	1 2 3 4 5	1 2 3 4	Yes No		
	QAR BAR	0 1 2 3 4 5	1 2 3 4 5	1 2 3 4	Yes No		
Date		Time					
	QAR BAR	0 1 2 3 4 5	1 2 3 4 5	1 2 3 4	Yes No		
	QAR BAR	0 1 2 3 4 5	1 2 3 4 5	1 2 3 4	Yes No		
	QAR BAR	0 1 2 3 4 5	1 2 3 4 5	1 2 3 4	Yes No		
	QAR BAR	0 1 2 3 4 5	1 2 3 4 5	1 2 3 4	Yes No		
	QAR BAR	0 1 2 3 4 5	1 2 3 4 5	1 2 3 4	Yes No		
Date		Time					
	QAR BAR	0 1 2 3 4 5	1 2 3 4 5	1 2 3 4	Yes No		
	QAR BAR	0 1 2 3 4 5	1 2 3 4 5	1 2 3 4	Yes No		
	QAR BAR	0 1 2 3 4 5	1 2 3 4 5	1 2 3 4	Yes No		
	QAR BAR	0 1 2 3 4 5	1 2 3 4 5	1 2 3 4	Yes No		
	QAR BAR	0 1 2 3 4 5	1 2 3 4 5	1 2 3 4	Yes No		
Date		Time					
	QAR BAR	0 1 2 3 4 5	1 2 3 4 5	1 2 3 4	Yes No		
	QAR BAR	0 1 2 3 4 5	1 2 3 4 5	1 2 3 4	Yes No		
	QAR BAR	0 1 2 3 4 5	1 2 3 4 5	1 2 3 4	Yes No		
	QAR BAR	0 1 2 3 4 5	1 2 3 4 5	1 2 3 4	Yes No		
	QAR BAR	0 1 2 3 4 5	1 2 3 4 5	1 2 3 4	Yes No		
Date		Time					
	QAR BAR	0 1 2 3 4 5	1 2 3 4 5	1 2 3 4	Yes No		
	QAR BAR	0 1 2 3 4 5	1 2 3 4 5	1 2 3 4	Yes No		
	QAR BAR	0 1 2 3 4 5	1 2 3 4 5	1 2 3 4	Yes No		
	QAR BAR	0 1 2 3 4 5	1 2 3 4 5	1 2 3 4	Yes No		
	QAR BAR	0 1 2 3 4 5	1 2 3 4 5	1 2 3 4	Yes No		
Date		Time					
	QAR BAR	0 1 2 3 4 5	1 2 3 4 5	1 2 3 4	Yes No		
	QAR BAR	0 1 2 3 4 5	1 2 3 4 5	1 2 3 4	Yes No		
	QAR BAR	0 1 2 3 4 5	1 2 3 4 5	1 2 3 4	Yes No		
	QAR BAR	0 1 2 3 4 5	1 2 3 4 5	1 2 3 4	Yes No		
	QAR BAR	0 1 2 3 4 5	1 2 3 4 5	1 2 3 4	Yes No		
Date		Time					
	QAR BAR	0 1 2 3 4 5	1 2 3 4 5	1 2 3 4	Yes No		
	QAR BAR	0 1 2 3 4 5	1 2 3 4 5	1 2 3 4	Yes No		
	QAR BAR	0 1 2 3 4 5	1 2 3 4 5	1 2 3 4	Yes No		
	QAR BAR	0 1 2 3 4 5	1 2 3 4 5	1 2 3 4	Yes No		
	QAR BAR	0 1 2 3 4 5	1 2 3 4 5	1 2 3 4	Yes No		

EAE scoring
0–No abnormality
1–A floppy tail with mild hind limb weakness
2–A floppy tail with moderate hind limb weakness
3–Hind leg paresis, but not complete paralysis and eventually mild forelimb weakness
4–Total paralysis of hind legs, which may be associated with moderate forelimb weakness
5–Quadriplegia or pre-moribund state

* If an animal's clinical status reaches grade 4, notify vet immediately for recommendations.
** TX confirmed = food supplements and long sipper tube present.
*** Body condition and pain score system from health evaluation of experimental laboratory mice.

QAR = quiet, alert, and responsive
BAR = bright, alert, and responsive

FIGURE 31.1 EAE score sheet. (Adapted from Division of Veterinary Resources, National Institutes of Health.)

erythema. When a single number, tally of numbers, or a certain qualitative criterion reaches a predetermined score, interventions with treatment, removal from the study, or euthanasia will occur. Scoring sheets can also list humane and experimental endpoints, provide documentation of supportive care, and provide a means to evaluate complications and general health of the animals on the study. Scoring sheets should be created prior to study initiation and be specific to the study and the species. The scoring criteria should be adapted as the study progresses to incorporate unforeseen developments. Scoring sheets should be available and readily retrievable for all staff involved in the care of the animals being scored.

Two sample scoring and monitoring sheets are provided in Figures 31.1 and 31.2. Figure 31.1 includes the experimental autoimmune encephalomyelitis (EAE) scoring system that rates the degree of the disorder on a graded scale from 0 (no abnormality) to 5 (quadriplegia or premoribund state). Figure 31.2 includes an induced arthritis scoring system that rates the degree of abnormality of each paw from 1 (no erythema or swelling) to 4 (erythema and severe swelling encompassing the foot and digits). The scores for each paw are then added for a total arthritis score. Both sheets prompt the observer to indicate several things: (1) whether a mouse is bright, alert, and responsive (BAR) or quiet, alert, and responsive (QAR); (2) the score for a given mouse according to the scoring system; (3) the body condition score (Burkholder et al. 2012); (4) the pain score (Burkholder et al. 2012); and (5) whether supportive care treatments, such as food supplements and longer sipper tubes, are present in the cage. The sheets also require observer initials, provide space for any important additional observations, and provide other important instructions (e.g., notify the veterinarian immediately if a mouse reaches an EAE score of 4).

To Treat, Monitor, or Euthanize?

The decision to treat, monitor only, or euthanize is complex and necessitates good communication between veterinarians and scientists. A complete understanding by both parties of the research objectives, the species involved, the severity and prognosis of the illness, and required IACUC-established humane and experimental endpoints, along with sound professional veterinary and scientific judgment, will facilitate the decision. In addition, the scientist should describe in the protocol any limitations on the veterinary interventions that can be provided to the animals due to adverse effects on the research being conducted (e.g., use of steroids). Monitoring may be chosen when treatment would interfere with the study and the clinical condition is mild and has a favorable prognosis. If the condition is more significant and the prognosis less favorable, it may be best to treat or euthanize the animal before data is lost due to unexpected death. Euthanasia and final sample collection may be the optimal choice if treatment negates the ability to obtain scientifically useful data from the animal. Ideally, preestablished humane endpoints and intervention contraindications outlined in the study design are detailed enough to direct decisions about treatment, monitoring, or euthanasia. While the veterinarian and researcher should collaborate on decisions about whether to treat, monitor, or euthanize, the final decision in critical cases is that of the attending veterinarian.

Humane and Experimental Endpoints

Humane and experimental endpoints are vital to managing animal welfare while concurrently meeting research objectives (Demers et al. 2006). Experimental endpoints are those that relate to the timing of the study and are typically dictated solely by the science behind the research. An example of an experimental endpoint would be to terminally collect tissues from mice 4 weeks after administration of a test drug. Humane endpoints are those that determine when animals will be relieved of pain or distress by being taken off study, treated, or euthanized. These humane endpoints minimize pain and distress to the greatest extent possible without compromising the goals of the study. Ideally, humane endpoints are devised in such a way that the study ends before an animal experiences pain or distress, and also at a time when valuable research data can still be collected. Humane endpoints can be qualitative (unkempt coat, poor response to stimuli, hunched posture, or dyspnea) or quantitative (low body temperature, weight loss, decreased food or water consumption, or predetermined biomarker values) (Toth 1997). Because humane endpoints balance the needs of the animals as well as the research, they can be challenging to establish. Some useful references regarding endpoints include those created by the Organisation for

Date started:_______ Protocol:____________ PI/User:___________ Cage #:_______________________________

Date		Time					
ID #	Activity	Total arthritis score*	Body condition***	Pain score***	TX (arthritis score >8)**	Initials	Comments
	QAR BAR		1 2 3 4 5	1 2 3 4	Yes No		
	QAR BAR		1 2 3 4 5	1 2 3 4	Yes No		
	QAR BAR		1 2 3 4 5	1 2 3 4	Yes No		
	QAR BAR		1 2 3 4 5	1 2 3 4	Yes No		
	QAR BAR		1 2 3 4 5	1 2 3 4	Yes No		
Date		Time					
	QAR BAR		1 2 3 4 5	1 2 3 4	Yes No		
	QAR BAR		1 2 3 4 5	1 2 3 4	Yes No		
	QAR BAR		1 2 3 4 5	1 2 3 4	Yes No		
	QAR BAR		1 2 3 4 5	1 2 3 4	Yes No		
	QAR BAR		1 2 3 4 5	1 2 3 4	Yes No		
Date		Time					
	QAR BAR		1 2 3 4 5	1 2 3 4	Yes No		
	QAR BAR		1 2 3 4 5	1 2 3 4	Yes No		
	QAR BAR		1 2 3 4 5	1 2 3 4	Yes No		
	QAR BAR		1 2 3 4 5	1 2 3 4	Yes No		
	QAR BAR		1 2 3 4 5	1 2 3 4	Yes No		
Date		Time					
	QAR BAR		1 2 3 4 5	1 2 3 4	Yes No		
	QAR BAR		1 2 3 4 5	1 2 3 4	Yes No		
	QAR BAR		1 2 3 4 5	1 2 3 4	Yes No		
	QAR BAR		1 2 3 4 5	1 2 3 4	Yes No		
	QAR BAR		1 2 3 4 5	1 2 3 4	Yes No		
Date		Time					
	QAR BAR		1 2 3 4 5	1 2 3 4	Yes No		
	QAR BAR		1 2 3 4 5	1 2 3 4	Yes No		
	QAR BAR		1 2 3 4 5	1 2 3 4	Yes No		
	QAR BAR		1 2 3 4 5	1 2 3 4	Yes No		
	QAR BAR		1 2 3 4 5	1 2 3 4	Yes No		
Date		Time					
	QAR BAR		1 2 3 4 5	1 2 3 4	Yes No		
	QAR BAR		1 2 3 4 5	1 2 3 4	Yes No		
	QAR BAR		1 2 3 4 5	1 2 3 4	Yes No		
	QAR BAR		1 2 3 4 5	1 2 3 4	Yes No		
	QAR BAR		1 2 3 4 5	1 2 3 4	Yes No		
Date		Time					
	QAR BAR		1 2 3 4 5	1 2 3 4	Yes No		
	QAR BAR		1 2 3 4 5	1 2 3 4	Yes No		
	QAR BAR		1 2 3 4 5	1 2 3 4	Yes No		
	QAR BAR		1 2 3 4 5	1 2 3 4	Yes No		
	QAR BAR		1 2 3 4 5	1 2 3 4	Yes No		

Arthritis scoring
1: No evidence of erythema and swelling
2: Erythema and mild swelling confined to the mid-foot (tarsals) or ankle joint
3: Erythema and mild swelling extending from the ankle to the mid-foot
4: Erythema and severe swelling encompass the ankle, foot, and digits

* Note: the total score is range from 0 to 16, i.e., score 0–4 per paw for four paws.
** TX = food supplements present if mice have arthritis score >8.
*** Body condition and pain score system from health evaluation of experimental laboratory mice.

QAR = quiet, alert, and responsive
BAR = bright, alert, and responsive

FIGURE 31.2 Arthritis score sheet. (Adapted from Division of Veterinary Resources, National Institutes of Health.)

Economic Co-operation and Development (2000), the Canadian Council on Animal Care (1998), and the National Research Council (2000). Establishing humane endpoints requires collaboration between veterinarians, scientists, and compliance bodies, such as the IACUC (Stokes 2000; NRC 2011, pp. 27). This must be done prior to initiating a study, although humane endpoints may need to be adjusted during the course of a study.

If the expected outcomes are unknown and humane endpoints cannot adequately be determined prior to study initiation, a pilot study should be performed in a small number of animals, ideally under the veterinarian's supervision. Pilot studies can determine the earliest predictive signs of adverse study effects and/or impending death. Information collected from pilot studies will allow the development of endpoints that avoid causing more pain or distress than necessary to the animals, while meeting the research objectives. Pilot studies can also be a useful tool for training staff to accurately understand, anticipate, and detect the protocol-defined humane endpoints. Although animals should always be monitored for the presence of unanticipated adverse effects, this is especially true when conducting pilot studies. When observed, adverse effects should be reported to the veterinarian, investigator, and IACUC.

Once humane endpoints have been established, the lines of authority and communication regarding animal well-being must be clearly defined. It is critical to determine who will be responsible for observations, the timing and frequency of observations, and how to properly document and follow up on findings. Close monitoring of the animals will be necessary to ensure that the endpoints are correctly implemented and no further refinements are needed. Animals that are being monitored for humane endpoints need to be clearly identified. Personnel performing the monitoring must be trained to recognize pain and distress in the species and understand the research procedures and expected outcomes. Observations may need to be more frequent than usual (e.g., every 4–6 hours), rather than once each day, and should be timed to have the maximum value (when animals are most likely to reach a humane endpoint). As often as possible, studies should be scheduled so that this timing of critical observations occurs during normal work hours. Animals should be observed first at a distance with minimal disruption, followed by assessing their responsiveness to stimuli and direct handling. Data for an endpoint assessment can be obtained by telemetry, activity monitors, video recordings, or other methods that are not disruptive to the animals.

After humane endpoints have been determined and put into practice, there may be instances where it is difficult to determine whether an animal has reached a defined humane endpoint. In this case, the veterinarian and researcher should work collaboratively to come to a consensus. If an agreement cannot be reached, the attending veterinarian has the final decision as to the animal's disposition (9 CFR §2.33 a.2).

Animal Welfare Considerations

It can be challenging to balance the welfare of research animals with study objectives. This is especially true when making the decision to treat, monitor, or euthanize and when defining and adhering to endpoints. If the research or animal care staff have any animal welfare concerns during a study, there should be multiple mechanisms in place to report their concerns without fear of reprisal. A safe reporting mechanism allows veterinarians, scientists, managers, and compliance bodies to discuss and address issues as they arise. Addressing issues early helps the laboratory animal medicine and scientific community to uphold the three Rs (replacement, reduction, and refinement) (Russell and Burch 1959) of animal research and meet regulatory requirements. Discussion and resolution of issues can result in refined endpoints, reduced pain and distress, use of fewer and/or healthier animals, and better-quality science.

Normal versus Abnormal Behavior

The only way to be able to recognize "abnormal" behavior is to devote time to learning normal behavior patterns to use as a point of reference (Arnold et al. 2011; Rock et al. 2014). Animal behavior is a set of functions that reflect the animal's response to all environmental factors and challenges. These behaviors include food and shelter and mate seeking, as well as nest-making activities and performing complex tasks, such as operating equipment and using tools by NHPs. Every individual working with animals should spend time studying a species' normal behavior in different situations—resting, feeding, interacting with conspecifics and humans, posturing, and responding to easy and difficult challenges. A careful student of behavior will notice that most behaviors displayed in nonstressful situations follow certain patterns. Learning these patterns is very useful for interpreting animal health and well-being. Knowledge of behaviors displayed in response to stressful situations is also very useful for providing insights into an animal's mental state, which can help with choosing the best animal for a needed task (e.g., an aggressive or anxious animal may not be the best candidate for a task that requires a confident and calm animal).

Pain versus Distress

Pain results from potential or actual tissue damage. Pain can be considered a potent source of stress, that is, a stressor in and of itself, that can lead to distress and maladaptive behaviors (NRC 2009a). Distress is an aversive state in which an animal is unable to adapt completely to stressors and the resulting stress, leading to maladaptive behaviors, such as abnormal feeding, absence or diminution of postprandial grooming, inappropriate social interaction with conspecifics or handlers, and inefficient reproduction (NRC 2009a).

Recognition and Assessment of Pain and Distress

The nervous system processes the sensory features of tissue-damaging stimuli, such as damage quality, intensity, location, and duration. What is perceived results in behavioral and physiologic responses that are under the influence of emotional, motivational, and cognitive processes (NRC 2009a). Stress and distress are complex syndromes that are difficult to define and even harder to interpret and recognize (Wright et al. 1985). If an animal is in pain, it might also be in distress, and vice versa.

Score Sheets and Training

Biomedical research puts much emphasis on using objective parameters to avoid misinterpretation of the data. A set of data collected by one researcher can be interpreted differently by another researcher who is not as familiar with the data, science, or subject animals. If the animal is performing behavioral tasks that are critical in the research project, it is paramount that the animal's training be done in a comprehensive and logical fashion that is transparent for all research team members and leaves no room for speculations. The training should be designed in such a way that it incorporates the individual traits of the animal.

As discussed earlier in this chapter, score sheets and scoring scales can be useful tools when evaluating an animal's well-being in a nonsubjective manner, as well as when training individuals to recognize abnormalities. Our ability to recognize an animal experiencing pain or distress and assess its severity is the first step toward prevention and treatment of the problem. Recent work has demonstrated that the facial expressions of mice, rats, and rabbits can provide a rapid and reliable means of assessing the presence and severity of pain (https://www.nc3rs.org.uk/grimacescales). "Grimace scales" that equate facial changes related to narrowing of the animal's eyes and position of the whiskers and ears have been demonstrated to be reliable indicators of both the presence of pain and its intensity.

Species-Specific Signs of Pain or Distress

The identification of pain and distress in animals is a challenging task (NRC 2009a). Because verbal feedback is not an option, veterinarians and researchers must rely on their experience, understanding, and ability to detect species-specific signs of pain or distress. Critical to the assessment of pain or distress is the ability to distinguish between normal and abnormal animal behavior. This is especially true when dealing with a species that often exhibits pain and distress with only subtle behavioral changes. Animals must be monitored by trained individuals throughout a study for pain and distress as appropriate for the species, conditions, and procedures. Therefore, it is critical that the individuals assessing an animal be trained in the species' specific signs of pain and distress, as well as the potential outcomes of the research manipulations. As discussed earlier in this chapter, pain and distress scoring is one method to convert subjective animal observations into an objective system, which some have found to be helpful in assessing animal behavior.

Rodents Ailing rats and mice most commonly show decreased motor and nest-building activity, piloerection, and an ungroomed appearance, but may exhibit other signs as outlined in Table 31.1 (Kohn et al. 2007). Although decreased food and water consumption is commonly associated with pain or distress in other species, these parameters are challenging to monitor in rodents. Also, changes in body weight are not sensitive enough to be used as an indicator of acute pain or distress (NRC 2009a).

Cavies Guinea pigs in pain usually remain quiet (as opposed to stampeding and squealing when frightened but pain-free) (Kohn et al. 2007; Shomer et al. 2015). They also display behaviors similar to those of other rodents, such as rats and mice.

Rabbits Rabbits in pain can appear apprehensive, anxious, dull, or inactive and assume a hunched appearance, attempt to hide, and squeal or cry. When experiencing acute pain, they may show aggressive

TABLE 31.1

Potential Signs Associated with Pain or Distress in Rats, Mice, and Rabbits

Potential Signs	Mice	Rats	Rabbits
Decreased food and water consumption	X	X	X
Weight loss	X	X	X
Self-imposed isolation/hiding	X	X	X
Self-mutilation, gnawing at limbs	X	X	X
Rapid breathing	X	X	X
Opened-mouth breathing	X	X	X
Abdominal breathing	X	X	X
Grinding teeth		X	X
Biting/growling/aggression		X	X
Increased/decreased movement	X	X	X
Unkempt appearance (erected, matted, or dull hair coat)	X	X	X
Abnormal posture/positioning (e.g., head pressing and hunched back)	X	X	X
Restless sleep			X
Tearing (including porphyria), lack of blinking reflex		X	X
Dilated pupils			X
Muscle rigidity, lack of muscle tone	X	X	X
Dehydration/skin tenting/sunken eyes	X	X	X
Twitching, trembling, tremor	X	X	X
Vocalization (rare)	X	X	X
Redness or swelling around surgical site	X	X	X
Increased salivation	X	X	X

Source: Kohn, D. F. et al., *J. Am. Assoc. Lab. Anim. Sci.*, 46, 97–108, 2007.

behavior with increased activity, excessive scratching and licking, and exaggerated reactions to handling. With abdominal or muscular pain, they sometimes grind their teeth and salivate excessively, in addition to displaying increased respiratory rate and inappetence. Rabbits in distress might cannibalize their young and tend to be more susceptible to the "tonic immobility reflex" (aka immobility reflex, animal hypnosis, tonic immobility, playing possum, mesmerism, and dead faint). This reflexive behavior is considered to be a mechanism of defense against predators, as it renders the animal less sensitive to pain. It abolishes voluntary motor activity. Spinal reflexes are suppressed, but not abolished. In rabbits, the most susceptible species, fine muscle tremors can occur initially or be induced by stimulation of the patellar tendon reflex. Fully immobilized rabbits exhibit pronounced catalepsy with reduced muscle tone. Additional signs of possible pain or distress are outlined in Table 31.1 (Kohn et al. 2007).

Cats A general lack of well-being is an important indication of pain in cats (Robertson 2005; Epstein et al. 2015). They may be quiet and show an apprehensive facial expression, with the forehead appearing creased. They might not have an appetite and may cry, yowl, growl, or hiss if approached or made to move. The cat's posture becomes stiff and abnormal, and it tends to hide or separate itself from other cats. If the pain is located in the head or ears, the cat might tilt its head toward the affected side. When the pain is only thoracic, the head, neck, and body might be extended, while a cat with generalized pain in both the thorax and abdomen might be crouched or hunched. A cat with abdominal or back pain might stand or lie on its side with its back arched or walk with a stilted gait. Pain in one limb is usually manifested by limping or holding up of the affected limb, with no attempt to use it. Incessant licking is sometimes associated with localized pain or psychological distress. Cats in severe or chronic pain look ungroomed and behave markedly different from normal (NRC 2009a).

Dogs Dogs in pain generally appear less alert and quieter than normal and may demonstrate stiff body movements and an unwillingness to move. A dog in severe pain might lie completely still while watchful, or adopt an abnormal posture to minimize its discomfort. With less severe pain, dogs can appear restless and more alert, but have a loss of appetite, shivering, and increased respiration with panting. Spontaneous barking is unlikely. Instead, they are more likely to whimper or howl, especially if unattended, and might growl without apparent provocation. When handled, dogs can bite, scratch, or guard painful regions and be abnormally apprehensive or aggressive (NRC 2009a; Epstein et al. 2015).

Ferrets Behavioral changes are often the best indication of a ferret's well-being. These changes can include the display of unusual aggression toward other animals or their handlers, as well as hiding. Ferrets in pain often become lethargic and stop bodily grooming, resulting in a disheveled look. Ferrets rarely vocalize when they are in pain or distress, and specific signs are largely dependent on the individual animal. Animals in pain or distress can display a loss of appetite, stop drinking, and experience weight loss (Hillyer and Queensberry 1997). Additional potential signs of pain or distress may include (1) drooped or laid-back ears; (2) closed or squinty eyes; (3) clicking, grating, or grinding of the teeth; (4) drooling, salivating, or slobbering; (5) hesitant, immobile, or slow movements; (6) guarded, non-curled, or tense movements; (7) trembling; (8) crying or whimpering; and (9) guarding, avoiding, or withdrawing from touch.

Nonhuman Primates (New and Old World) NHPs show remarkably little reaction to surgical procedures or injury (NRC 1998). Some animals can hide pain and may look well until they are gravely ill or in severe pain. A NHP that appears sick should get immediate assessment and possible intervention. A NHP in pain has a general appearance of misery and dejection. It might huddle in a crouched posture with its arms across its chest and its head forward with a grimace and glassy eyes. Acute abdominal pain can be shown by facial contortions, clenching of teeth, restlessness, and shaking accompanied by grunts and moans, with food and water usually refused. A monkey in pain can also attract altered attention from its cage mates, ranging from a lack of social grooming to attack (NRC 2009a).

Sheep and Goats Sheep and goats may appear dull and depressed, hold their heads low, and show little interest in their surroundings or eating. On handling, they might react violently or adopt a rigid posture designed to immobilize the painful region, combined with grunting and grinding of teeth. Localized pain may result in persistent licking and kicking at the offending area and, when the pain is severe, vocalizing. Sheep can tolerate severe injury without overt signs of pain or distress, although abnormal changes in posture and a reluctance to move are often apparent. Goats are more likely to vocalize in response to pain. After castration or tail docking, lambs may show signs of pain by standing and lying repeatedly, wagging their tails, occasionally bleating, and displaying neck extension, dorsal lip curling, kicking, rolling, and hyperventilation (NRC 2009a; Underwood et al. 2015).

Swine Swine normally squeal and attempt to escape when handled, and pain can accentuate these reactions. When in pain, swine often show changes in social behavior, gait, and posture and, when bedding material is present, will refrain from making a bed. Adults typically become more aggressive. Squealing is characteristic when painful areas are palpated. When in severe pain, swine are often unwilling to move and might hide in bedding (NRC 2009a; Underwood et al. 2014). As swine generally have good appetites, any individuals who go off feed should be assessed for potential signs of pain or illness.

Fish Fish respond to pain differently than mammals. They exhibit a pronounced initial response to injuries or to contact with irritants, but their response to chronic stimuli might be small or absent. Fish with severe wounds, which would cause immobility in a mammal, often appear to behave normally and even resume feeding. Fish react to noxious stimuli, such as puncture with a hypodermic needle, with strong muscular movements. When exposed to a noxious environment, such as an acidic solution, they show abnormal swimming behavior and attempt to jump out of the water, their coloring becomes darker, and their opercular movements become more rapid. Such effects indicate some, perhaps considerable, distress, but it is not possible to describe the distress unequivocally as pain induced (NRC 2009a; Sneddon 2009; Smith 2014).

Birds Birds in pain can show escape reactions, vocalization, and excessive movement (Machin 2005). Small species struggle less and emit fewer distress calls than large species. Head movements increase in extent and frequency, and there can be an increase in heart and respiratory rates. Prolonged pain results in loss of appetite, inactivity, and a drooping, miserable appearance, with the eyes held partially closed, the wings held flat against the body, and the neck retracted. When a sick bird is handled, its escape reaction is often replaced by tonic immobility (see the "Rabbits" section above). Birds with limb pain avoid use of the affected limb and "guard" it from extension (NRC 2009a). Like swine, birds generally have good appetites, and any changes from the expected consumption of food should be investigated immediately.

Amphibians Acute pain in amphibians can be characterized by flinching and muscle contractions. There might be aversive movements away from the unpleasant stimulus and attempts to bite. More chronic and persistent pain might be associated with anorexia, lethargy, and weight loss, although a direct and specific association of any of these signs with pain may be sometimes challenging (NRC 2009a).

Category E Studies: Unrelieved Pain or Distress

The *Guide* states that the IACUC should consider the use of appropriate sedation, analgesia, and anesthesia for animals (NRC 2011) expected to experience pain or distress during research manipulations. The AWR (paragraph §2.36 b.7) specifically require painful procedures, for which such drugs cannot be used, to be placed in Column E of the Animal and Plant Health Inspection Service/Animal Care (APHIS/AC) annual report (AWR). Placement of study animals in the Column E category must be scientifically justified and can only be approved after careful consideration by the IACUC. Committee review must include a careful examination of the requested animal numbers with emphasis on an investigator-provided extensive search for alternatives. A smaller pilot study to harvest preliminary data may be stipulated and include oversight by a veterinarian.

Prevention and Treatment of Pain

Prevention and relief of pain and distress in animals is humane and essential in biomedical research. The IACUC has the responsibility for ensuring that all animals under their oversight are used humanely and in accordance with a number of federal regulations and policies (NRC 2011; AWR 9 CFR §2.31 d.1.i). The principal investigator (PI) and the IACUC must fully understand their legal requirements to establish both appropriate endpoints and expectable methodologies for relieving pain and distress. The individuals responsible for monitoring the animals for pain and distress must be identified and trained.

The obligation to reduce pain and distress starts with the submission, review, and approval of an animal study proposal, but does not end there. Animals must be continually monitored for pain, distress, illness, morbidity, and/or mortality during the course of a research study. If unexpected pain or distress is observed, and is more than an isolated incident, the PI must submit an amendment delineating the unexpected problem and stating his or her proposed resolution to the issue. Alternatively, as previously stated, the PI could justify the need for unrelieved pain or distress in the amendment or, in the case of regulated species, submit a Category E justification (USDA Animal Care Policy 11, https://www.aphis.usda.gov/animal_welfare/downloads/Animal%20Care%20Policy%20Manual.pdf).

Whenever more than transient pain or distress is anticipated, measures should be taken to minimize or prevent the development of pain and/or distress (Mench and Blatchford 2014). Following any intervention strategy, the animals must be closely monitored to ensure the effectiveness of the measures taken and determine if or when additional treatment will be necessary. The extent and frequency of monitoring will depend on the level of postsurgical/procedural pain and/or distress anticipated and the chosen intervention strategy.

Intervention strategies for the management of pain and distress may include the use of both pharmacological and nonpharmacological approaches. Nonpharmacological strategies may include, but are not limited to, (1) modified housing and husbandry practices, (2) dietary modifications and supplements, (3) surgical approaches, (4) desensitization and acclimation strategies, (5) acupuncture, and (6) euthanasia under the right circumstances (e.g., https://www.primatevets.org/Content/files/Public/education/NHP_Endpoint_Guidelines.pdf). The chosen strategy will vary with the species, the procedures being

TABLE 31.2

Postprocedural Pain Potential[a]

Minimal to Mild Pain[b]	Mild to Moderate Pain[c]	Moderate to Severe Pain[d]
Catheter implantation	Minor laparotomy incisions	Major laparotomy/organ incision
Tail clipping	Thyroidectomy	Thoracotomy
Ear notching	Orchidectomy	Heterotopic organ transplantation
Subcutaneous transponder placement	C-section	Vertebral procedures
Superficial tumor implantation	Hypophysectomy	Burn procedures
Orbital sinus venotomy	Thymectomy	Trauma models
Rodent embryo transfer	Embryo transfer in nonrodents	Orthopedic procedures
Multiple injections	Bone marrow collection	
Noncorneal ocular procedures	Corneal procedures	
Intracerebral electrode implantation		
Vasectomy		
Vascular access port implantation		
Craniotomy (periosteal pain)		
Superficial lymphadenectomy		

Source: Adapted from ACLAM (American College of Laboratory Animal Medicine), Guidelines for the assessment and management of pain in rodents and rabbits, ACLAM, Chester, NH, 2006.

[a] The analgesia and monitoring required may vary due to a number of factors, such as the invasiveness of the procedure, degree of tissue trauma, surgical time, skill of the surgeon, and tissues or organs involved.

[b] Postprocedural pain relief for minimal to mild pain may be adequately addressed with preemptive analgesia, tissue infiltration with a long-acting local anesthetic, or a single dose of a long-acting nonsteroidal anti-inflammatory drug (NSAID) or mixed opioid agonist–antagonist, or other agent.

[c] Postprocedural pain relief for mild to moderate pain may be adequately addressed with tissue infiltration with a long-acting local anesthetic combined with one or more doses of a long-acting NSAID and/or an opioid or other agent, in addition to preemptive analgesic administration.

[d] Postprocedural pain relief for moderate to severe pain should encompass multimodal analgesia (e.g., combining a pure opioid agonist with a NSAID or tissue infiltration with a long-acting local anesthetic).

performed, the duration of action needed, the route of administration preferred, the degree and type of analgesia required, and the research being conducted (Table 31.2) (ACLAM 2006; http://oacu.od.nih.gov/ARAC/documents/Pain_and_Distress.pdf).

When administering pharmaceuticals, it is important to understand the pharmacokinetics and duration of action of the drugs used to alleviate pain or distress. The pharmacokinetics of an agent must be taken into consideration when designing a strategy to monitor the effectiveness of the drug intervention. The strategy should include a plan to carefully monitor the animal during the period when the effectiveness of the agent is starting to wane, to determine if additional treatment is required. Resources (Fish et al. 2008; NRC 2009a) and formularies (Hawk et al. 2005; Carpenter 2012) are available today that provide extensive information on the recognition and alleviation of pain and distress in laboratory animals.

The documentation of monitoring of the animal for pain and distress is important. The identification of cages containing animals where a potentially painful or distressful procedure has been performed can prove helpful in drawing special attention to the animal during the caretaker's daily health check. A "special observation" cage card works well for this. Cages containing animals requiring more intensive monitoring should also be appropriately identified and their monitoring and/or treatments documented at either the room, cage, or animal level (e.g., room log, cage card, or medical record). This is in addition to the investigator's notations in his or her laboratory notebook. It is important that the documentation be available to all personnel monitoring the cage or animal (e.g., IACUC, veterinarians, and animal care staff).

Anesthetics, Analgesics, Tranquilizers, and Sedatives

Anesthesia is a reversible process for the purpose of producing a convenient, safe, and inexpensive means of restraint so that clinical or experimental procedures may be conducted with a minimum of pain, discomfort, distress, or toxic side effects to the patient or subject. The goal of anesthesia is to preserve normal physiologic function without confounding the experimental protocol under consideration. The Animal Welfare Act (9 CFR 2143 a.3) requires that anesthetic, analgesic, and tranquilizing drugs be utilized in accordance with currently accepted veterinary practice and produce in the individual subject a level of anesthesia, analgesia, or tranquilization appropriate for the design of the experiment. The use of anesthetics, analgesics, and tranquilizers in laboratory animals is necessary for humane and scientific reasons. The choice and use of the most appropriate drugs are matters for the attending veterinarian and professional veterinary judgment. The veterinarian must provide research personnel with guidelines, training, and advice concerning the choice and use of these drugs. If a procedure is likely to cause more than momentary pain or discomfort to the subject animal, but the use of anesthetics, analgesics, or tranquilizing agents would defeat the purpose of the procedure, a written justification of a Category E listing for unrelieved pain or distress must be submitted and approved by the IACUC.

The following are common terms related to the use of anesthetics, sedation, and tranquilization:

- *Anesthesia*: A state of controllable, reversible insensibility in which sensory perceptions and motor responses are both markedly depressed. A total loss of sensation in the whole body or one of its parts occurs.
- *General anesthesia*: Loss of consciousness in addition to loss of sensation. Ideally, general anesthesia includes hypnosis or narcosis, hyporeflexia, and analgesia.
- *Balanced anesthesia*: Surgical anesthesia produced by a combination of two or more drugs or anesthetic techniques each contributing its own pharmacological effects. Balanced anesthesia is characterized by unconsciousness, analgesia, and muscular relaxation.
- *Dissociative anesthesia*: A central nervous system (CNS) state characterized by catalepsy, profound peripheral analgesia, and altered consciousness produced by the cyclohexamine drugs (e.g., ketamine).
- *Tranquilization* (neurolepsis or ataraxia): A state of tranquilization and calmness in which the subject is relaxed, awake, and unconcerned with its surroundings. Tranquilizers act by suppression of the hypothalamus and reticular activating system.
- *Sedation*: A mild degree of CNS depression in which the subject is awake but calm. The subject can be aroused. Action is by a dose-dependent depression of the cerebral cortex.
- *Hypnosis*: Artificially induced sleep or a trance resembling sleep from which the subject can be aroused by stimuli.
- *Narcosis*: Drug-induced sedation in which the subject is oblivious to pain, with or without hypnosis.
- *Analgesia*: Loss of sensitivity to pain.
- *Neuroleptanalgesia*: Hypnosis and analgesia produced by a combination of a neuroleptic drug and an analgesic drug.
- *Catalepsy*: A state in which there is a malleable rigidity of limbs, which, if the limbs are placed in various positions, is maintained for a time. The subject is generally unresponsive to audio, visual, or minor pain stimuli.

It is beyond the scope of this chapter to discuss the wide range of agents available to the veterinarian for anesthesia. Many excellent references are available on the selection and use of anesthetics, sedatives, and tranquilizers in laboratory animal medicine and general veterinary practice (Fish et al. 2008; Gaynor and Muir 2014; Grimm et al. 2015).

The selection of the most appropriate agents and techniques are dependent on

1. Duration of and type of operation or procedure
2. Species, breed, age, and relative size of the subject
3. Physical status of the subject and concurrent medication or treatments

4. Disposition or demeanor of the animal
5. Presence of concurrent pain or distress
6. Type of facilities and help available, including level of experience and training
7. Personal knowledge, experience, and familiarity with available agents and equipment

In general, all anesthetics cause a dose-dependent depression of consciousness, as well as a depression of normal physiological homeostasis (e.g., heart rate, respiration, blood pressure, and thermoregulatory centers) (Brunson 2008; Meyer and Fish 2008). In balanced anesthesia, various drugs are administered together that synergize each other, yielding the desired outcome at lower doses than would be possible with the administration of the single drugs alone. At lower doses, most drugs are safer and demonstrate fewer adverse effects.

All general anesthetics induce predictable stages of dose-dependent anesthesia: (1) voluntary movement or excitement, (2) involuntary excitement, (3) surgical anesthesia (Planes 1–4), and (4) medullary paralysis or death. The goal of anesthetic induction is to take the patient through the first two stages as quickly as possible. This is often facilitated by the intravenous administration of the drugs. When intravenous administration is not possible, the drugs are commonly administered intramuscularly (IM) in larger species and either subcutaneously (SQ) or intraperitoneally (IP) in smaller species. Non-intramuscular injection commonly leads to a slower absorption of the drug into the circulatory system and to prolonged voluntary and involuntary excitement phases. Care must be taken to anticipate the response of the animal and prevent injury to the animal and to protect personnel from being scratched, kicked, or bitten.

Sedatives, anxiolytics, and neuromuscular blocking agents do not provide analgesia, and anesthetics that induce unconsciousness at lower doses may not produce a pain-free state. It must be understood that unconsciousness does not always equate with a state of analgesia. In addition, it is critical to understand that neuromuscular blocking agents that produce a state of total paralysis, including respiratory paralysis, are not anesthetics and do not induce unconsciousness. Therefore, neuromuscular blocking agents must never be used in an animal that is not deeply anesthetized. When monitoring an animal that is both anesthetized and paralyzed, it is critical to, at a minimum, monitor *both* the animal's heart rate and blood pressure to ensure an adequate depth of anesthesia. A rise in heart rate with a stable or increasing blood pressure can indicate the animal's perception of pain or distress and inadequate anesthesia. However, a rise in heart rate coupled with decreasing blood pressure can mean the animal is too deeply anesthetized. It has also been recommended when using neuromuscular blocking agents that the plane of anesthesia be periodically lightened until a slight increase in heart rate is observed and then returned to the deeper plane to ensure that adequate anesthesia is present.

While careful monitoring of an animal's reflexes provides a good indication that a plane of surgical anesthesia has been reached, concurrent monitoring of the animal's heart rate and blood pressure is required to determine if the animal is pain-free. As stated above, if an animal's heart rate is increasing and its blood pressure is normal or increasing, the animal is displaying a physiological response to pain. Costly equipment is not required for monitoring; manually taking a peripheral pulse and a mucous membrane capillary refill time will suffice in most situations.

In most species, reaching a surgical plane of anesthesia is accompanied by loss of a toe pinch withdrawal response; surgical anesthesia may also be accompanied by disappearance of head shaking in response to pinching of an ear (Smith and Dannerman 2008; Gaynor and Muir 2014; Grimm et al. 2015). The pattern of respiration, slow regular deep breaths, has been demonstrated to be a good indicator of the level of surgical anesthesia.

Care must be taken throughout the anesthesia and recovery periods (Hampshire and Davis 2008) to provide the animal with appropriate monitoring and nursing support. This should include the provision of a quiet environment, a warm ambient temperature, fluid support, and additional thermal support as required. Physiological monitoring is critical, but the nature of the monitoring requirement can vary by species, length and type of procedure, available equipment, and funding. The availability of precision vaporizers, pulse oximeters, and respiratory, cardiac, and blood pressure monitors increases the safety of anesthetic use.

Whenever possible, care should be taken to prevent the development of pain, rather than focusing on the treatment of established pain. Tissue injury and other noxious stimuli cause the release of chemicals that activate sensory neurons (i.e., nociceptors). Once activated, the nociceptors send electrical impulses to the brain that are interpreted as pain. Left unchecked, the chemicals set off a cascading inflammatory reaction within the tissues around the nociceptors that become abnormally sensitive. Use of preemptive analgesia, the administration of preoperative and intraoperative medication to block pain pathways and induce analgesia (Beilin et al. 2003), can diminish or prevent the development of pain.

In many situations, pain is easier to prevent than to treat. Preemptive approaches focus on inhibiting changes in the peripheral and central nervous system that contribute to heightened postprocedural pain. By taking preemptive measures to prevent the sensitization or "windup" of pain receptors and peripheral or central pain pathways, we minimize the development of pain and make it easier to manage. Benefits of preemptive analgesia include a more stable patient throughout surgery, a smoother postoperative recovery period, and a reduction in postoperative pain and distress (Gaynor and Muir 2014). The preemptive use of opioids, nonsteroidal anti-inflammatory agents, and local anesthetics forms the foundation of most preemptive interventions. In addition, agents such as α_2 adrenergic agonists (e.g., xylazine and medetomidine), N-methyl-D-aspartate (NMDA) antagonists (e.g., ketamine), and corticosteroids are also used.

The same pain prevention drugs listed above are used to treat pain once it has developed. The drugs chosen will depend on the species, age, nature, severity of the pain, and possibly the research objectives. Combinations of analgesics are often used both preemptively and to treat pain. This strategy is called balanced or multimodal analgesia. It involves the simultaneous administration of two or more drugs classes that have additive or synergistic analgesic effects. An added benefit of using two or more drugs with additive or synergistic effects is that the dosages of each agent can often be reduced, leading to fewer adverse effects and lower incidence of drug tolerance.

Use of Pharmaceutical-Grade versus Non-Pharmaceutical-Grade Compounds

A pharmaceutical-grade compound/substance (PGC) is any active or inactive drug, biologic, or reagent for which a chemical purity standard has been established by a recognized national or regional pharmacopeia (e.g., the U.S. Pharmacopeia [USP], British Pharmacopeia [BP], European Pharmacopoeia [EP], or Japanese Pharmacopeia [JP]). These standards are used by manufacturers to help ensure the products are of the appropriate chemical purity and quality, in the appropriate vehicle solution or compound, stable, safe, and efficacious (AAALAC 2015). In addition, the Food and Drug Administration (FDA) maintains a database listing of FDA-approved commercial formulations for both FDA-approved human drugs (U.S. FDA 2016b) and veterinary drugs (U.S. FDA 2016a).

In the United States, it is a requirement that compounds used for the clinical treatment of animals or to prevent, reduce, or eliminate animal pain or distress be PGCs whenever possible (USDA Policy 3—Veterinary Care, March 25, 2011). In addition, when compounds are used to accomplish the scientific aims of the study, PGCs are preferred if available and suitable. These issues pertain to all components, both active and inactive, contained in the preparation to be administered. Therefore, the vehicle used to facilitate administration of a compound is as important of a consideration as the active compound in the preparation. The NIH OLAW suggests that the IACUC, in making its evaluation, should consider factors such as grade, purity, sterility, acid–base balance, pyrogenicity, osmolality, stability, site and route of administration, compatibility of components, unwanted effects and adverse reactions, storage, and pharmacokinetics (http://grants.nih.gov/grants/olaw/faqs.htm#662). The *Guide* (NRC 2011) further states that the use of non-PGCs in laboratory animals must be described and justified in the animal use protocol and/or be covered by an IACUC policy developed for their use and approved by the IACUC. There are many useful institutional guidelines available that can be used as templates for the development of an individualized program (NIH 2013).

Medical Records

Medical records document animal care and use, and they are an essential part of any laboratory animal care program (USDA Animal Care Policy 3; Field et al. 2007). This is true for all laboratory animal species, and the following recommendations apply to both USDA-regulated and nonregulated species. Some legislation and guidelines require the use of medical records. The U.S. AWR (9 CFR §2.35) require maintenance of some records for at least 3 years and study-related records for 3 years after study completion. European Union member states are required to maintain some records for 3 years after the death of an animal and other records for 5 years (European Parliament and the Council of the European Union 2010). The institution and attending veterinarian should determine the method by which records are maintained, and professional judgment and performance standards should be used in creating facility standards for medical records. Veterinarians should be responsible for the oversight of records, although records should be readily retrievable and available for all staff involved in animal care. Records should also be reviewed during IACUC facility inspections.

Medical record entries should be dated, legibly written, and clearly state who recorded the entry. Entries are made by those administering treatments, performing procedures or observations, evaluating test and exam results, and uploading or attaching files. Entries should be complete enough to reconstruct the care and use of the animal and allow for clear communication between research and care staff. Recommendations of what to include in records are provided in Figure 31.3. Individual records are the most common, but group records and/or entries may be utilized when performing the same procedure on a group of animals.

Medical records may be on paper or in electronic systems. Electronic records are more legible, automatically include entry dates and personnel, and may be capable of quickly generating reports for easily assessing study complications, mortality rates, and colony health. They also provide an efficient mechanism for including information in large numbers of records at the same time. They can also be utilized to easily schedule treatments, testing, and research procedures. However, electronic systems are dependent on having a reliable and secure server, easy access for updating at the point of use, and payment of licensing fees, and therefore may be cost-prohibitive and unavailable without server or computer access.

- Individual or group identification
- Species
- Gender
- Age
- Weights
- Master problem list of clinical, health, behavioral, and procedural history
- Health status (including testing history and results)
- Preventive medicine history (including routine exam findings, vaccinations, deworming)
- Medical history (including exam findings, diagnostics, and treatments)
- Anesthesia records
- Surgical history and post-operative monitoring records
- Pain assessments
- Behavioral information (including socialization history)
- Special husbandry needs
- Breeding history (including pedigree/strain/line, genotype, parents, pairing for mating, outcome of mating, complications)
- Study assignment
- Experimental history
- Treatment contraindications
- Complications from procedures, surgery, or medications
- Research data useful to animal care staff (weights, scoring sheets)
- Final disposition
- Necropsy reports

FIGURE 31.3 Recommended items to include in laboratory animal medical records.

Summary

The timely provision of quality veterinary care is a critical component of *all* laboratory animal care and use programs. Quality veterinary care not only ensures the health, well-being, and welfare of the animals, but also improves the quality, quantity, and reproducibility of the research. Because a program is only as strong as the individuals staffing it, the strength of a veterinary care program is determined by the training and experience of the attending veterinarian or designated program official and his or her staff.

REFERENCES

AAALAC. 2015. Frequently asked questions, A.9. Non-pharmaceutical-grade compounds. Frederick, MD: AAALAC. http://www.aaalac.org/accreditation/faq_landing.cfm#B9 (accessed July 11, 2016).

AAALAC. 2016. Position statements: The attending veterinarian and veterinary care. Frederick, MD: AAALAC. http://www.aaalac.org/accreditation/positionstatements.cfm#vetcare.

AAFP (American Association of Feline Practitioners). 2013. AAFP Feline Vaccination Advisory Panel report: Feline vaccination guidelines. Hillsborough, NJ: AAFP. http://www.catvets.com/guidelines/practice-guidelines/feline-vaccination-guidelines (accessed July 10, 2016).

ACLAM (American College of Laboratory Animal Medicine). 2006. Guidelines for the assessment and management of pain in rodents and rabbits. Chester, NH: ACLAM. https://www.aclam.org/Content/files/files/Public/Active/position_pain-rodent-rabbit.pdf.

ACLAM (American College of Laboratory Animal Medicine). 2016. Adequate veterinary care: Guidelines prepared by the American College of Laboratory Animal Medicine (ACLAM) to assist in the formulation and evaluation of programs of veterinary care for laboratory animals. Chester, NH: ACLAM. https://www.aclam.org/education-and-training/position-statements-and-reports.

AFA (American Ferret Association). 2006. AFA ferret (*M. putorius furo*) vaccination policy. Frederick, MD: AFA. http://www.ferret.org/read/vaccinations.html (accessed July 10, 2016).

Arnold, C., Lamp, J., Lamp, O., and A. Einspanier. 2011. Behavioral tests as indicator of pain and distress in a primate endometriosis model. *J Med Primatol* 40:317–326.

Artwohl, J. E., Cera, L. M., Wright, M. F., Medina, L. V., and L. J. Kim. 1994. The efficacy of a dirty bedding sentinel system for detecting Sendai virus infection in mice: A comparison of clinical signs and seroconversion. *Lab Anim Sci* 44:73–75.

AWAR (Animal Welfare Act Regulations). 2013. Code of Federal Regulations, Volume 1, Subchapter A, Title 9. http://awic.nal.usda.gov/government-and-professional-resources/federal-laws/animal-welfare-act.

Baer, J., Lansford, R., and K. Cheng. 2015. Japanese quail as a laboratory animal model. In *Laboratory Animal Medicine*, ed. J. G. Fox, L. C. Anderson, O. Glen, K. R. Pritchett-Corning, and M. T. Whary, 623–694. 3rd ed. Amsterdam: Elsevier Press.

Baker, D. G. 2003. *Natural Pathogens of Laboratory Animals, Their Effects on Research*. Washington, DC: ASM Press.

Bayne, K., Bayvel, A. C. D, and V. Williams. 2014. Laboratory animal welfare: International issues. In *Laboratory Animal Welfare*, ed. K. Bayne and P. V. Turner, 57–69. Amsterdam: Elsevier Press.

Becker, S. D., Bennett, M., Stewart, J. P., and J. L. Hurst. 2007. Serological survey of virus infection among wild house mice (*Mus domesticus*) in the UK. *Lab Anim* 41:229–238.

Beilin, B., Bessler, H., Mayburd, E. et al. 2003. Effects of preemptive analgesia on pain and cytokine production in the postoperative period. *Anesthesiology* 98:151–155.

Bekkevold, C. M., Robertson, K. L., Reinhard, M. K., Battles, A. H., and N. E. Rowland. 2013. Dehydration parameters and standards for laboratory mice. *J Am Assoc Lab Anim Sci* 52:233–239.

Besselsen, D. G., Wagner, A. M., and J. K. Loganbill. 2000. Effect of mouse strain and age on detection of mouse parvovirus 1 by use of serologic testing and polymerase chain reaction analysis. *Comp Med* 50:498–502.

Brachman, P. S. 1996. Epidemiology. In *Medical Microbiology*, ed. S. Baron. 4th ed. Galveston: University of Texas Medical Branch at Galveston.

Browne, R. K., Odum, R. A., Herman, T., and K. Zippel. 2007. Facility design and associated services for the study of amphibians. *ILAR J* 48:188–202.

Brunson, D. B. 2008. Pharmacology of inhalation anesthetics. In *Anesthesia and Analgesia in Laboratory Animals*, ed. R. E. Fish, P. J. Dannerman, M. Brown, and A. Z. Karas, 171–182. 2nd ed. London: Academic Press.

Burkholder, T., Foltz, C., Karlsson, E., Linton, C. G., and J. Smith. 2012. Health evaluation of laboratory mice. *Curr Protoc Mouse Biol* 2:145–165.

Carpenter, J. W. 2012. *Exotic Animal Formulary*. 4th ed. St. Louis, MO: Elsevier.

Carty, A. J. 2008. Opportunistic infections of mice and rats: Jacoby and Lindsey revisited. *ILAR J* 49:272–276.

CCAC (Canadian Council on Animal Care). 1998. Guidelines on choosing appropriate endpoint in experiments using animals for research, teaching, and testing. Ottawa: CCAC.

CCAC (Canadian Council on Animal Care). 2008. Policy statement for: Senior administrators responsible for animal care and use programs. Ottawa: CCAC.

CFR (Code of Federal Regulations). *Federal Register*. https://www.archives.gov/federal-register/cfr/.

Clarke, C. L., and K. A. Perdue. 2004. Detection and clearance of *Syphacia obvelata* infection in Swiss Webster and athymic nude mice. *Contemp Top Lab Anim Sci* 43:9–13.

Clifford, C. B., and J. Watson. 2008. Old enemies, still with us after all these years. *ILAR J* 49:291–302.

Compton, S. R., Homberger, F. R., Paturzo, F. X., and J. M. Clark. 2004. Efficacy of three microbiological monitoring methods in a ventilated cage rack. *Comp Med* 54:382–392.

Compton, S. R., and J. D. Macy. 2015. Effect of cage-wash temperature on the removal of infectious agents from caging and the detection of infectious agents on the filters of animal-bedding disposal cabinets. *J Am Assoc Lab Anim Sci* 54:745–755.

Council for International Organizations of Medical Sciences and International Council for Laboratory Animal Science (CIOMS/ICLAS). 2012. International guiding principles for biomedical research involving animals. Geneva: CIOMS. http://www.cioms.ch/images/stories/CIOMS/IGP2012.pdf (accessed May 22, 2016).

Crawley, J. N. 2000. *What's Wrong with My Mouse? Behavioral Phenotyping of Transgenic and Knockout Mice*. New York: Wiley-Liss.

Cundiff, D. D., Riley, L. K., Franklin, C. L., Hook, R. R., Jr., and C. Besch-Williford. 1995. Failure of a soiled bedding sentinel system to detect cilia-associated respiratory *Bacillus* infection in rats. *Lab Anim Sci* 45:219–221.

Deeb, B. J., DiGiacomo, R. F., Bernard, B. L., and S. M. Silbernagel. 1990. *Pasteurella multocida* and *Bordetella bronchiseptica* infections in rabbits. *J Clin Microbiol* 28:70–75.

Demers, G., Griffin, G., De Vroey, G., Haywood, J. R., Zurlo, J., and M. Bédard. 2006. Animal research: Harmonization of animal care and use guidance. *Science* 312:700–701.

Desrosiers, R. C. 1997. The value of specific pathogen-free rhesus monkey breeding colonies for AIDS research. *AIDS Res Hum Retroviruses* 13:5–6. doi:10.1089/aid.1997.13.5.

Ednie, D., Wilson, R., and M. Lang. 1998. Comparison of two sanitation monitoring methods in an animal research facility. *Contemp Top Lab Anim Sci* 37:71–74.

Epstein, M., Rodan, I., Griffenhagen, G. et al. 2015. AAHA/AAFP pain management guidelines for dogs and cats. *J Am Anim Hosp Assoc* 51:67–84.

European Commission. 2007. Commission recommendation on guidelines for the accommodation and care of animals used for experimental and other scientific purposes. *Off J Eur Union* 197:1–89.

European Parliament and the Council of the European Union (European Parliament). 2010. Directive 2010/63/EU of the European Parliament and of the council of 22 September 2010 on the protection of animals used for scientific purposes. *Off J Eur Union* 276:33–79.

Fair, J., Paul, E., Jones, J., Clark, A. B., Davie, C., and G. Kaiser. 2010. *Guidelines to the Use of Wild Birds in Research*. 3rd ed. Bethesda, MD: Ornithological Council. http://www.nmnh.si.edu/BIRDNET.

FASS (Federation of Animal Science Societies). 2010. *Guide for the Care and Use of Agricultural Animals in Research and Teaching*. 3rd ed. Champaign, IL: FASS.

Field, K., Bailey, M., Foresman, L. L. et al. 2007. Medical records for animals used in research, teaching, and testing: Public statement from the American College of Laboratory Animal Medicine. *ILAR J* 48:37–41.

Fish, R. E., Danneman, P. J., Brown, M., and A. Z. Karas, eds. 2008. *Anesthesia and Analgesia in Laboratory Animals*. 2nd ed. London: Academic Press.

Gaynor, J. S., and W. W. Muir. 2014. *Handbook of Veterinary Pain Management*. 3rd ed. St. Louis, MO: Mosby Elsevier.

Grimm, K. A., Lamont, L. A., Tranquilli, W. J., Greene, S. A., and S. A. Robertson, eds. 2015. *Veterinary Anesthesia and Analgesia*. 5th ed. Ames, IA: John Wiley & Sons.

Gulani, J., Koch, A., Chappell, M. G. et al. 2016. Cercopithecine herpesvirus 9 (simian varicella virus) infection after total-body irradiation in a rhesus macaque (*Macaca mulatta*). *J Am Assoc Lab Anim Sci* 66:150–153.

Hampshire, V. A., and J. A. Davis. 2008. Postprocedural care of commonly utilized research animal subjects. In *Anesthesia and Analgesia in Laboratory Animals*, ed. R. E. Fish, P. J. Dannerman, M. Brown, and A. Z. Karas, 171–182. 2nd ed. London: Academic Press.

Hawk, C. T., Leary, S. L., and T. H. Morris. 2005. *Formulary for Laboratory Animals*. 3rd ed. Ames, IA: Blackwell Publishing.

Hawkins, P., ed. 2001. Laboratory birds: Refinements in husbandry and procedures. Fifth report of the BVAAWF/FRAME/RSPCA/UFAW Joint Working Group on Refinement. *Lab Anim* 35 (Suppl 1):1–163.

Hawkins, P., Bairlein, F., Duncan, I. et al. 2003. Future principles for housing and care of laboratory birds. Report for the Revision of the Council of Europe Convention ETS 123 Appendix A for birds. http://www.felasa.eu/about-us/library/.

Henderson, K., and C. Clifford. 2013. A guide to integrating PCR rodent infectious agent (PRIA) panels in rodent quality control programs. Wilmington, MA: Charles River Research Animal Diagnostic Services. http://www.criver.com/files/pdfs/research-models/rm_ld_r_pria_guidebook.aspx.

Henderson, K. S., Perkins, C. L., Havens, R. B. et al. 2013. Efficacy of direct detection of pathogens in naturally infected mice by using a high-density PCR array. *J Am Assoc Lab Anim Sci* 52:763–772.

Hessler, J. R. 1991. Single versus dual-corridor systems: Advantages, disadvantages, limitations, and alternatives for effective sanitation control. In *Handbook of Facilities Planning*, vol. 2, *Laboratory Animal Facilities*, ed. T. Ruys, 59–67. New York: Van Nostrand Reinhold.

Hillyer, E., and K. Queensberry. 1997. *Ferrets, Rabbits, and Rodents: Clinical Medicine and Surgery*. Philadelphia: W. B. Saunders.

Howard, H., and Y. K. Foucher. 2008. Animal-use space. In *Planning and Designing Research Animal Facilities*, ed. J. R. Hessler and N. D. M. Lehner, 203–261. Amsterdam: Elsevier Press.

Hrapkiewicz, K., Colby, L., and P. Denison. 2013. Facility, equipment, housing and management. In *Clinical Laboratory Animal Medicine: An Introduction*, 35–40. 4th ed. Ames, IA: John Wiley & Sons.

Huerkamp, M. J., and J. K. Pullium. 2009. Quarantine facilities and operations. In *Planning and Designing Research Animal Facilities*, ed. J. R. Hessler and N. D. M. Lehner, 373. Amsterdam: Elsevier Press.

Hutson, C. L., Nakazawa, Y. J., Self, J. et al. 2015. Laboratory investigations of African pouched rats (*Cricetomys gambianus*) as a potential reservoir host species for monkeypox virus. *PLoS Negl Trop Dis* 9:e0004013.

Hyde, F. 2016. Nurses' perspectives of nomenclature and understanding of the role of veterinary technicians. *NAVTA J* 42–47.

Jensen, E. S., Allen, K. P., Henderson, K. S., Szabo, A., and J. D. Thulin. 2013. PCR testing of a ventilated caging system to detect murine fur mites. *J Am Assoc Lab Anim Sci* 52:28–33.

Karesh, W. B., Cook, R. A., Bennett, E. L., and J. Newcomb. 2005. Wildlife trade and global disease emergence. *Emerg Infect Dis* 11:1000–1002.

Kennedy, B. K. 2016. What's in a title? *Lab Anim* 45:188.

Kohn, D. F., Martin, M. E., Foley, P. L. et al. 2007. Guidelines for the assessment and management of pain in rodents and rabbits. *J Am Assoc Lab Anim Sci* 46:97–108.

Koszdin, K. L., and R. F. DiGiacomo. 2002. Outbreak: Detection and investigation. *Contemp Top Lab Anim Sci* 41:18–27.

Kowalski, W. J., Bahnfleth, W. P., and D. D. Carey. 2002. Engineering control of airborne disease transmission in animal laboratories. *Contemp Top Lab Anim Sci* 41:9–17.

Langford, D. J., Bailey, A. L., Chanda, M. L. et al. 2010. Coding of facial expressions of pain in the laboratory mouse. *Nat Methods* 7:447–449.

Leblanc, M., Berry, K., Graciano, S., Becker, B., and J. D. Reuter. 2014. False-positive results after environmental pinworm PCR testing due to rhabditid nematodes in corncob bedding. *J Am Assoc Lab Anim Sci* 53:717–724.

Leverage, N., and C. R. Roberts. 2009. Finish decisions. In *Planning and Designing Research Animal Facilities*, ed. J. R. Hessler and N. D. M. Lehner, 397–408. Amsterdam: Elsevier Press.

Lipman, N., and S. Leary. 2015. Design and management of research facilities. In *Laboratory Animal Medicine*, ed. J. G. Fox, L. C Anderson, G. Otto, K. R. Pritchett-Corning, and M. T. Whary, 1543–1597. 3rd ed. Amsterdam: Elsevier/Academic Press.

Machin, K. L. 2005. Avian pain: Physiology and evaluation. *Compend Contin Educ Vet* 27:98–108.

Mähler, M., Berard, M., Feinstein, R. et al. 2014. FELASA recommendations for the health monitoring of mouse, rat, hamster, gerbil, guinea pig, and rabbit experimental units. Report of the Federation of European Laboratory Animal Science Associations (FELASA) Working Group on Animal Health accepted by the FELASA Board of Management. *Lab Anim* 48:178–192.

Mansfield, K. G. 2005. Development of specific pathogen free nonhuman primate colonies. In *The Laboratory Primate*, ed. S. Wolfe-Coote, 229–230. New York: Elsevier.

Manuel, C. A., Pugazhenthi, U., and J. K. Leszczynski. 2016. Surveillance of a ventilated rack system for *Corynebavcterium bovis* by sampling exhaust-air manifolds. *J Am Assoc Lab Anim Sci* 55(1): 58–65.

Mench, J. A., and R. A. Blatchford. 2014. Birds as laboratory animals. In *Laboratory Animal Welfare*, ed. K. Bayne and P. Turner, 280–299. Amsterdam: Elsevier Press.

Meyer, R. E., and R. E. Fish. 2008. Pharmacology of injectable anesthetics, sedatives, and tranquilizers. In *Anesthesia and Analgesia in Laboratory Animals*, ed. R. E. Fish, P. J. Dannerman, M. Brown, and A. Z. Karas, 171–182. 2nd ed. London: Academic Press.

Nemzek, J. A., Lester, P. A., Wolfe, A. M., Dysko, R. C., and D. D. Myers. 2015. Biology and diseases of dogs. In *Laboratory Animal Medicine*, 3rd edition. eds. J. G. Fox, L. C. Anderson, O. Glen, K. R. Pritchett-Corning, and M. T. Whary, 467–469. Amsterdam: Elsevier Press.

NIH (National Institutes of Health), Office of Animal Care and Use, Animal Research Advisory Committee Guidelines. 2013. Guidelines for the use of non-pharmaceutical grade compounds in laboratory animals. Bethesda, MD: NIH. http://oacu.od.nih.gov/ARAC/documents/Pharmaceutical_Compounds.pdf (accessed July 9, 2014).

NIH/ORF (National Institutes of Health, Office of Research Facilities). 2008. *Design Requirements Manual*, Chapters 2–4, Animal research facilities. https://www.orf.od.nih.gov/PoliciesAndGuidelines/BiomedicalandAnimalResearchFacilitiesDesignPoliciesandGuidelines/Pages/DesignRequirementsManualPDF.aspx (accessed July 9, 2014).

NRC (National Research Council). 1977. *Laboratory Animal Management: Wild Birds*. Washington, DC: National Academies Press.

NRC (National Research Council). 1991. *Infectious Diseases of Mice and Rats: A Report of the Institute of Laboratory Animal Resources Committee on Infectious Diseases of Mice and Rats*. Washington, DC: National Academies Press.

NRC (National Research Council). 1998. *The Psychological Well-Being of Primates*. Washington, DC: National Academies Press.

NRC (National Research Council). 2000. *Humane Endpoints for Animals Used in Biomedical Research and Testing*. Washington, DC: National Academies Press.

NRC (National Research Council). 2006. *Guidelines for the Humane Transportation of Research Animals: A Report of the Committee on Guidelines for the Humane Transportation of Laboratory Animals*. Washington, DC: National Academies Press.

NRC (National Research Council). 2009a. *Recognition and Alleviation of Pain in Laboratory Animals*. Washington, DC: National Academies Press.

NRC (National Research Council). 2009b. *Scientific and Humane Issues in the Use of Random Source Dogs and Cats in Research*. Washington, DC: National Academies Press.

NRC (National Research Council). 2011. *Guide for the Care and Use of Laboratory Animals*. Washington, DC: National Academies Press.

Nicklas, W., Keubler, L., and A. Bleich. 2015. Maintaining and monitoring the defined microbiota status of gnotobiotic rodents. *ILAR J* 56:241–249.

Nunamaker, E. A., Artwohl, J. E., Anderson, R. J., and J. D. Fortman. 2013. Endpoint refinement for total body irradiation of C57BL/6 mice. *Comp Med* 63:22–28.

Ochi, T., Yamada, A., Nafanuma, Y., Nishina, N., and H. Koyama. 2016. Effect of road transportation on the serum biochemical parameters of cynomolgus monkeys and beagle dogs. *J Vet Med Sci* 78: 889–893.

Organisation for Economic Co-operation and Development. 2000. Guidance document on the recognition, assessment, and use of clinical signs as humane endpoints for experimental animals used in safety evaluation. Paris: OECD. http://www.oecd.org/officialdocuments/publicdisplaydocumentpdf/?cote=ENV/JM/MONO(2000)7&docLanguage=En.

O'Rourke, D. P. 2007. Amphibians used in research and teaching. *ILAR J* 48:183–187.

Patterson, M., and M. Fee. 2015. Zebra finches in biomedical research. In *Laboratory Animal Medicine*, ed. J. G. Fox, L. C. Anderson, O. Glen, K. R. Pritchett-Corning, and M. T. Whary, 623–694. 3rd ed. Amsterdam: Elsevier Press.

Perdue, K. A., Copeland, M. K., Karjala, Z. et al. 2008. Suboptimal ability of dirty-bedding sentinels to detect *Spironucleus muris* in a colony of mice with genetic manipulations of the adaptive immune system. *J Am Assoc Lab Anim Sci* 47:10–17.

Perkins, S., and N. Lipman. 2014. Inspection of surgery areas. In *The IACUC Handbook*, ed. J. Silverman, M. Suckow, and S. Murthy, 603–612. Boca Raton, FL: CRC Press.

Pritchett-Corning, K. R., Cosentino, J., and C. B. Clifford. 2009. Contemporary prevalence of infectious agents in laboratory mice and rats. *Lab Anim* 43:165–173.

Rehbinder, C., Alenius, S., Bures, J. et al. 2000. Federation of European Laboratory Animal Science Associations (FELASA) recommendations for the health monitoring of experimental units of calves, sheep, and goats. Report of the FELASA Working Group on Animal Health. *Lab Anim* 34:329–350.

Rehbinder, C., Baneux, P., Forbes, D. et al. 1998. FELASA recommendations for the health monitoring of breeding colonies and experimental units of cats, dogs, and pigs. Report of the FELASA Working Group on Animal Health. *Lab Anim* 32:1–17.

Rehg, J., and L. Toth. 1988. Rodent quarantine programs: Purpose, principles, and practice. *Lab Anim Sci* 30:323–329.

Reuter, J. D., Livingston, R., and M. Leblanc. 2011. Management strategies for controlling endemic and seasonal mouse parvovirus infection in a barrier facility. *Lab Anim* 40:145–152.

Rice, K. A., Albacarys, L. K., Metcalf Pate, K. A. et al. 2013. Comparison of diagnostic methods for *Myocoptes musculinus* according to age and treatment status of mice (*Mus musculus*). *J Am Assoc Lab Anim Sci* 52:773–781.

Roberts, J. A., and K. Andrews. 2008. Nonhuman primate quarantine: Its evolution and practice. *ILAR J* 49:145–156.

Robertson, S. A. 2005. Managing pain in feline patients. *Vet Clin North Am Small Anim Prac* 35:129–146.

Rock, M. L., Karas, A. Z., Rodriguez, K. B. et al. 2014. The time-to-integrate-to-nest test as an indicator of wellbeing in laboratory mice. *J Am Assoc Lab Anim Sci* 53:24–28.

Russell, W. M. S., and R. L. Burch. 1959. *The Principles of Humane Experimental Technique*. London: Methuen Publishing.

Shek, W. R., Smith, A. L., and K. R. Pritchett-Corning. 2015. Microbiological quality control for laboratory rodents and lagomorphs. In *Laboratory Animal Medicine*, ed. J. G. Fox, L. C. Anderson, O. Glen, K. R. Pritchett-Corning, and M. T. Whary, 467–469. 3rd ed. Amsterdam: Elsevier Press.

Shomer, N., H. Holcome, H. and J. E. Harkness. 2015. Biology and diseases of guinea pigs. In *Laboratory Animal Medicine*, ed. J. G. Fox, L. C. Anderson, O. Glen, K. R. Pritchett-Corning, and M. T. Whary, 247–277. 3rd ed. Amsterdam: Elsevier Press.

Smith, J. C., and P. J. Dannerman. 2008. Monitoring of anesthesia. In *Anesthesia and Analgesia in Laboratory Animals*, ed. R. E. Fish, P. J. Dannerman, M. Brown, and A. Z. Karas, 171–182. 2nd ed. London: Academic Press.

Smith, S. A. 2014. Welfare of laboratory fishes. In *Laboratory Animal Welfare*, ed. K. Bayne and P. Turner, 301–311. Amsterdam: Elsevier Press.

Sneddon, L. U. 2009. Pain perception in fish: Indicators and endpoints. *ILAR J* 50:338–342.

Stokes, W. 2000. Reducing unrelieved pain and distress in laboratory animals using humane endpoints. *ILAR J* 4:59–61.

Swindle, M. M., Makin, A., Herron, A. J., Clubb, F. J., Jr., and K. S. Frazier. 2012. Swine as models in biomedical research and toxicology testing. *Vet Path* 49:344–356.

Syversen, E., Pineda, F. J., and J. Watson. 2008. Temperature variations recorded during interinstitutional air shipments of laboratory mice. *J Am Assoc Lab Anim Sci* 47:31–36.

Tardif, S., Carville, A., Elmore, D., Williams, L. E., and K. Rice. 2012. Reproduction and breeding of nonhuman primates. In *Nonhuman Primates in Biomedical Research*, vol. 1, *Biology and Management*, ed. C. R. Abee, K. Mansfield, S. Tardif, and T. Morris, 197–249. 2nd ed. London: Academic Press.

Taylor, D. K., Lee, V. K., and K. R. Strait. 2016. *The Laboratory Bird*. New York: CRC Press.

Toth, L. 1997. The moribund state as an experimental endpoint. *Contemp Top Lab Anim Sci* 36:44–48.

Turner, D. E., Daugherity, E. K., Altier, C. and K. J. Maurer. 2010. Efficacy and limitation of an ATP-based monitoring system. *J Am Assoc Lab Anim Sci* 49:190–195.

Underwood, W. J., Blauwiekel, R., Delano, M., Gillesby, R., Mischer, S., and A. Schoell. 2015. Biology and diseases of ruminants (sheep, goats, and cattle). In *Laboratory Animal Medicine*, ed. J. G. Fox, L. C. Anderson, O. Glen, K. R. Pritchett-Corning, and M. T. Whary, 623–694. 3rd ed. Amsterdam: Elsevier Press.

Underwood, W. J., McGlone, J. J., Swanson, J., Anderson, K. A., and R. Anthony. 2014. Agricultural animal welfare. In *Laboratory Animal Welfare*, ed. K. Bayne and P. V. Turner, 233–278. Amsterdam: Elsevier Press.

USDA (U.S. Department of Agriculture), Animal and Plant Health Inspection Service. 2016. *Animal Care: Animal Care Policy Manual*. Washington, DC: USDA. https://www.aphis.usda.gov/animal_welfare/downloads/Animal%20Care%20Policy%20Manual.pdf.

U.S. Department of Justice, Drug Enforcement Administration, Office of Diversion Control. 2016. *Controlled Substance Security Manual*. Springfield, VA: U.S. Department of Justice. http://deadiversion.usdoj.gov/pubs/manuals/sec/index.html (accessed July 9, 2016).

U.S. FDA (Food and Drug Administration). 2016a. approved animal drug products (green book). Silver Spring, MD: U.S. FDA. http://www.fda.gov/animalveterinary/products/approvedanimaldrugproducts/default.htm (accessed July 9, 2016).

U.S. FDA (Food and Drug Administration). 2016b. Orange book: Approved drug products with therapeutic equivalence evaluations. Silver Spring, MD: U.S. FDA. http://www.accessdata.fda.gov/scripts/cder/ob/default.cfm (accessed July 9, 2016).

Voipio H., Baneux P., Gomez de Segura I., Hau J., and S. Wolfensohn. 2008. Guidelines for the veterinary care of laboratory animals: Report of the FELASA/ECLAM/ESLAV Joint Working Group on Veterinary Care. *Laboratory Animals* 42:1–11.

Watson, J. 2013. Unsterilized feed as the apparent cause of a mouse parvovirus outbreak. *J Am Assoc Lab Anim Sci* 52:83–88.

Weber, H., Berge, E., Finch, J. et al. 1999. Health monitoring of nonhuman primate colonies. Recommendations of the Federation of European Laboratory Animal Science Associations (FELASA) Working Group on Nonhuman Primate Health accepted by the FELASA Board of Management, 21 November 1998. *Lab Anim* 33:S3–S18.

Weigler, B. J. 1992. Biology of B virus in macaque and human hosts: A review. *Clin Infect Dis* 14:555–567.

Welborn, L., DeVries, J., Ford, R. et al. 2011 AAHA canine vaccination guidelines. Lakewood, CO: American Animal Hospital Association. https://www.aaha.org/professional/resources/canine_vaccine.aspx (accessed July 10, 2016).

Wright, E., Marcella, K., and J. Woodson. 1985. Animal pain: Evaluation and control. *Lab Anim* 14:20–36.

Zug, G., Vitt, L., and J. Caldwell. 2001. *Herpetology: An Introductory Biology of Amphibians and Reptiles*. 2nd ed. San Diego: Academic Press.

Zurlo, J., Bayne, K., and J. MacArthur-Clark. 2009. Adequate veterinary care for animals in research: A comparison of guidelines from around the world. *ILAR J* 50:85–88.

32

Behavioral Training as Part of the Health Care Program

Steven J. Schapiro, Elizabeth R. Magden, Lisa A. Reamer, Mary Catherine Mareno, and Susan P. Lambeth

CONTENTS

Introduction: Enhancing Care

This chapter focuses on behavioral training, specifically positive reinforcement training (PRT) techniques, as it applies to health care programs for laboratory animals. Training laboratory animals to *voluntarily participate* in necessary veterinary, husbandry, and research procedures is an important refinement (refinement is one of the 3Rs [Russell and Burch 1959]) that can substantially improve the well-being, welfare, and wellness of captive animals, enhancing the quality and utility of the animals as research models, and therefore the reliability and validity of many types of research (Graham et al. 2012).

In our opinion, one of the primary goals of state-of-the-art, comprehensive training for a health care program is to provide the animals with opportunities to voluntarily participate in their own care (Schapiro et al. 2014). Examples of this might include the animal presenting a leg for venipuncture (Graham et al. 2012), placing its face in the mask of a nebulizer (Gresswell and Goodman 2011), sitting in one place for an acupuncture or laser treatment (Magden et al. 2013, 2016), and choosing which of two medications it prefers when given a choice (Schapiro et al. 2014). In these, and similar, scenarios, the animals are given opportunities to make meaningful choices concerning their health care; they can choose to sit still for acupuncture or they can walk away. One of the strongest indications that this approach is useful for the animals would be if they continued to perform the target behavior (sitting still) in the absence of external reinforcement (food or clicks), because this could suggest that the behavior is being performed to achieve internal reinforcement (the treatments make them feel better), rather than merely for food. To us, this represents the evolution of animal care to the next level, where the animals themselves are determining the efficacy of the treatments they choose to receive.

Regulatory Requirements

While there are no regulatory requirements stating that laboratory animals must be trained to voluntarily participate in husbandry, veterinary, and/or research procedures (except perhaps chimpanzees ["acquiescent animals," Institute of Medicine 2011, p. 65]), the current version of the *Guide for the Care and Use of Laboratory Animals* (NRC 2011) states, "In most cases, principles of operant conditioning may be employed during training sessions, using progressive behavioral shaping, to induce voluntary cooperation with procedures" (p. 69). The European guidelines also state, "Accustoming animals to competent and confident handling during routine husbandry and procedures reduces stress both to animals and personnel. For some species, for example dogs and non-human primates, a training programme to encourage co-operation during procedures can be beneficial to the animals, the animal care staff and the scientific programme" (Council of Europe 2010, Appendix A, Section 4.10). Although rats, mice, and other small rodents may sometimes be trained to perform experimental tasks as dependent measures in certain types of investigations, the training of experimental animals as part of the health care program is typically restricted to dogs, pigs, some nonhuman primates, and infrequently, rabbits and cats.

Refinement

As mentioned above, training animals to voluntarily participate in management and research behaviors represents an important refinement (Russell and Burch 1959) in the continuous, ongoing evolution of laboratory animal care programs. An animal that gives a voluntary blood sample (no physical or chemical restraint) in its home enclosure should be significantly less stressed than an animal that must be restrained in order to obtain a similar sample (Graham et al. 2012). It is likely that the removal of restraint stress as a potential confounder of experimental data not only is a handling refinement, but also may result in decreases in other potential confounders, especially those related to interindividual variation. This should ultimately result in a reduction (another of the 3Rs) in the number of subjects required for certain investigations, without any adverse effects on the statistical power of the experiments.

Primates Emphasized, But the Information Is Applicable to Other Species as Well

This chapter emphasizes the application of PRT techniques in relation to health care programs for nonhuman primates in research settings, because a reasonable amount of work has been conducted addressing this issue for these taxonomic groups (Anzenberger and Gossweiler 1993; Bloomsmith et al. 1994; Laule et al. 1996; McKinley et al. 2003; Schapiro et al. 2003; Laule and Whittaker 2007; Graham et al. 2012; Perlman et al. 2012; Whittaker and Laule 2012), and relatively little work has been published on other taxonomic groups living in research settings. Where appropriate and applicable, training of pigs, cats, and a few other species for health care purposes is discussed (Adams et al. 2004; Gruen et al. 2013; Lockhard et al. 2013).

Dogs represent an interesting case for this chapter. There are relatively few publications that document the training process or the results of PRT attempts with laboratory dogs, so it is difficult to discuss the formal training of laboratory dogs. However, it is quite well known that dogs are relatively easy to acclimate to many health care–related and research-related procedures in the laboratory, using PRT and similar techniques (Mikkelsen et al. 2003; Adams et al. 2004; Meunier 2006; Tornqvist et al. 2013).

Training the Animals

In many instances, *training* is simply another word for *teaching*. Trainers teach animals to perform particular tasks by providing reinforcement immediately following the performance of the task. Animals learn that they receive reinforcement when they perform the target behavior, altering (increasing) the probability that they will perform that behavior again in the future. Psychologists have been studying learning for many years, in many different ways (Domjan 2015), and for the purposes of this chapter, we focus on one type of learning, operant conditioning, with an emphasis on one type of operant conditioning, PRT. Virtually all animals learn via operant conditioning, whether it is in a formalized laboratory training session (a dog being trained to provide a blood sample or a rat pressing a lever in a Skinner box) or under naturalistic circumstances (a chimpanzee learning to crack palm nuts with stone tools or a blue jay eating a distasteful monarch butterfly). Although punishment is an important factor in discussions of many types of "natural" learning, it has no place in applied operant conditioning for laboratory animals, especially in relation to training for health care purposes.

Training for many health care–related behaviors typically involves at least two processes. The first is the training of the animals to perform the target behaviors. This is usually a fairly straightforward process, as described in the next section. A complementary process, one that is best accomplished concurrently with training, is the desensitization (pairing of a positive reward with a potentially negative situation) of the animals to the personnel, tools, and apparatus that tend to be associated with health care behaviors. These may include people, such as the veterinarian and the veterinary technician, and/or items, such as carts, syringes, lancets, glucometers, vacuum devices, and stethoscopes, which, based on previous experiences, may have negative connotations for the animals.

Additional discussions related to both basic and applied aspects of the psychology of animal (and human) learning can be found in many psychology text books, including Domjan (2015).

Brief Description of Positive Reinforcement Training Techniques

PRT is one subset of operant conditioning techniques that relies on the *stimulus–response–reinforcement* contingency. This contingency involves (1) the trainer asking the animal for a behavior using a cue (*stimulus*), (2) the animal performing the behavior (*response*), and then (3) the animal receiving a reward (*reinforcement*) for doing so. In order for PRT to be effective, all three components of the contingency must be present in the correct order, and they must occur in temporal proximity to one another (the delay between the stimulus and the response is not as important as the delay between the response and the reinforcement). The optimal delay between the response and the reinforcement is 0.5 seconds, essentially before another behavior can occur. More extended discussions of operant conditioning and PRT can be found in Skinner (1938) and Domjan (2015).

When attempting to train a complex behavior (presenting the cephalic vein for venipuncture and blood collection using a polyvinyl chloride [PVC] sleeve) (Coleman et al. 2008), one that may be comprised of multiple steps, it is important to train the animals via "shaping" of the final behavior, sequentially rewarding successive approximations of the target behavior, rather than attempting to train the animal to perform the final behavior from the beginning. For instance, to train an animal to voluntarily provide a blood sample from the cephalic vein using the sleeve (Coleman et al. 2008), the animal would be initially rewarded for placing its arm in the sleeve, and then sequentially rewarded for holding the bolt at the end of the sleeve, for allowing alcohol to be rubbed on the arm, for allowing occlusion of the vein, and so forth, and then finally, for the behaviors of inserting a needle in the vein, drawing the blood, and applying pressure to the venipuncture site.

Most PRT scenarios with laboratory animals involve the use of both a primary reinforcer and a conditioned reinforcer. Animals willingly work for primary reinforcers (i.e., food and water) because they satisfy biological needs, while conditioned reinforcers are potentially meaningless events that only become meaningful (reinforcing) through repeated pairing with a primary reinforcer. For instance, the ringing of a bell would not stimulate salivation in Pavlov's dogs until the dogs learned that the ringing of the bell reliably preceded the presentation of meat powder (discussed in Domjan 2015). Once the association between the bell and the meat powder had been learned, the dogs would routinely salivate to the sound of the bell. In PRT for laboratory animals, a click or a whistle is typically used as the conditioned reinforcer. The click is initially repeatedly paired with food, resulting in the animals' willingness to work for the click. This is especially important when trainers want to reward animals for performing the desired behavior, but cannot provide them with food quickly enough (e.g., the animals are asked to move from the front of the enclosure to the back) or at all (e.g., prior to anesthesia). In these types of scenarios, it is more important to provide the animals with precise information (the behavior was performed correctly, as signaled by the click) than it is to provide them with food.

We rarely advocate the use of negative reinforcement training (NRT) techniques, because NRT can be interpreted to teach animals what *not* to do, rather than what to do (PRT). NRT, like PRT, increases the probability that the target behavior will be performed when the stimulus is presented. A negative reinforcer is something that the animals want to avoid, such as the squeeze-back mechanism in a primate cage. Primates can be trained to come to the front of the cage to avoid the squeeze-back (NRT) or to receive food (PRT). Most trainers agree that coming to the front of the cage to receive food is more pleasant than avoiding the squeeze-back. However, recent findings suggest that there is a small subset of animals, and circumstances, in which NRT techniques can be more effective than PRT (Wergard et al. 2015). We always suggest that PRT be tried prior to attempting NRT. If, however, analyses of the temperament profiles of certain subjects suggest that PRT is less likely, and NRT is more likely, to work, then there may be limited justification to invest large amounts of time attempting to use PRT with those animals (Hannibal et al. 2013).

Once a behavior has been trained to a criterion of "success" (e.g., the animal performs the behavior four out of the five times requested), that behavior must be maintained on a regular training schedule in order to maximize the probability that the behavior will be successfully performed when requested (needed). This means that in a smoothly operating training program, animals will receive training sessions on a frequent basis, some sessions to train new behaviors, some to maintain already trained behaviors, and some to actually perform the behaviors when required.

There are three additional general training points that are important to present. First, typically, the more experience animals have with the training process, the easier it is to train them to perform additional behaviors (Reamer et al. 2014). In other words, the animal has learned to learn, and recognizes that the approach of the trainer and/or the technician (and the relevant training items and apparatus) represents the beginning of a session in which the animal is going to be (1) requested to perform particular behaviors and (2) rewarded if it successfully performs those behaviors. Second, many animals, especially nonhuman primates, can learn trained behaviors by watching other animals being trained. This is called observational learning, and chimpanzees can accelerate the rate at which they learn to (1) urinate on command or (2) place their face in a nebulizer mask, for instance, by watching other chimpanzees being trained live, or for urination behavior, even by watching recorded training sessions (Perlman et al. 2010). Finally, animals will occasionally or regularly regress, or fall below the criterion for previously successfully trained behaviors, the first few times they actually perform the target behavior "for real." This is especially true for behaviors like "present for injection," where animals may become reticent when asked to present the next time after they have received a real injection. With proper desensitization techniques, and swamping a small number of aversive occurrences (a real injection of anesthetic with no food reward) with many neutral or positive experiences (a gentle touch with a blunt needle followed by a desirable food item), animals that have regressed can be retrained to once again meet the criteria for successful training.

Brief Description of Desensitization Techniques

The training process can be facilitated and, in many cases, must be facilitated by the use of desensitization techniques (Laule et al. 2003) to teach animals to tolerate previously "scary objects or situations." These objects or situations can include highly relevant people, places, and/or apparatus that the animals associate with fear-inducing occurrences based on prior experience. For the purposes of this chapter, we separate desensitization from PRT, because with PRT, you are training for an *overt, observable response*, while with desensitization you are teaching the animal to *tolerate* the scary object or situation. As an example of a simplified version of a desensitization plan, if an animal is fearful of syringes, the animal would be reinforced (with clicks and food) when it remains calm when, sequentially, (1) it sees an "inactive" syringe (just lying on the cart in the hallway), (2) the syringe is in the room, (3) a person briefly touches the syringe, (4) a person holds the syringe, (5) the syringe is moved toward the animal, (6) the animal is touched with the capped syringe, (7) the syringe is uncapped, and so forth. Similar sequential processes would be involved in desensitizing animals to a stethoscope, transport cage, restraint device, or the veterinarian.

Limitations of Training

Although PRT techniques can be effectively employed in many circumstances with many types of animals and behaviors, PRT does not work in every circumstance. In addition to individuals of "trainable species" that are uncooperative, certain species and certain behaviors can be difficult or impossible to train. For instance, you are unlikely to be able to effectively train (1) a rat to present for a jugular blood sample (although you can train cats for a similar procedure) (Lockhard et al. 2013), (2) a mouse to present for a retro-orbital blood sample, or (3) two adult female marmosets to live compatibly.

While the understanding of animal learning and training is a science, being an effective trainer also involves considerable sensitivity. Not all subjects will work for the same reinforcers; it is important to identify reinforcers that motivate each subject. It is rarely acceptable in training circumstances to deprive animals of a resource (food or fluid) in order to increase their motivation to perform the target behaviors. This would be especially true in circumstances in which behaviors important for health care are being trained; subjects trained for health care behaviors should be voluntary participants.

Training the Trainers

Just as there are differences in animals' interest and motivation in participating in training, there are differences in caregivers' interest and aptitude for training. Patience and consistency are two extremely important traits that are characteristic of good laboratory animal trainers. An effective trainer must utilize PRT techniques properly and be extremely patient, able to deal with the regressions that are typical in most PRT programs. Additionally, trainers must be consistent with the stimuli that they provide for the animals (these are the commands that are given to the subjects) and the reinforcers that they use. In order to maximize the success of the training program, consistency must be maintained within and across trainers. Trainers must also be sensitive to circumstances in which they are potentially *being trained by the animals*, rather than vice versa (see the example in the "Urine" training section below). While we know a few things about the aspects of temperament that make certain *animals good trainees* (Coleman et al. 2005), we know considerably less about the aspects of temperament that make certain *people good trainers.*

In general, training programs seem to work well when (1) personnel whose job titles (and skill sets) identify them as trainers are responsible for *training new* behaviors and (2) personnel whose job titles include more generalized animal care responsibilities *maintain basic* behaviors that have already been trained. Such a distribution of effort can be useful for getting the most out of the people involved in the implementation of the training program.

It is important to include veterinary technicians and/or similar workers who may be involved in the performance of the final target behavior in all phases of the training process. In some facilities, veterinary technicians may be viewed negatively by the animals due to the typical animal-related duties they perform (dosing, blood sampling, restraint, etc.). The animals may be less willing to perform the target behaviors in the presence of these "scary" individuals. By involving veterinary technicians in the early, shaping phases of the training process, the vet techs become less scary, ultimately transforming them from negative stimuli to neutral, or even positive stimuli. This is likely to occur as the vet techs participate in more and more neutral and/or positive interactions with the animals and only a small proportion of "negative" interactions.

Brief Review of Studies of the Training Process

As mentioned above, most PRT procedures are based on the stimulus–response–reinforcement contingency. Several publications are available that chronicle many aspects of the training process (Bloomsmith et al. 1994; Gillis et al. 2012). Overall, laboratory animals (primarily, but not exclusively, nonhuman primates) have been successfully trained to voluntarily perform a variety of health care–related behaviors,

TABLE 32.1
Some Health Care–Related Behaviors That Have Been Trained Using PRT Techniques

Target Behavior/ Treatment	Species	Reference
Acupuncture	Chimpanzees	Magden et al. 2013
Blood pressure	Baboons	Turkkan 1990
Capillary blood sample	Chimpanzees	Reamer et al. 2014
Choice of medication	Chimpanzees	Schapiro et al. 2014
Diabetes related	Rhesus macaques, drills, chimpanzees	Priest 1991; Laule et al. 1996; Graham et al. 2012
Enrichment	Rhesus macaques, zoo animals	Baker et al. 2010; Westlund 2014
Feces collection	Rhesus macaques	Phillippi-Falkenstein and Clarke 1991
fMRI	Dogs	Berns et al. 2012, 2013
Geriatrics	Chimpanzees	Bridges et al. 2015
Implanted devices	Rhesus macaques, chimpanzees	Graham et al. 2012; Magden et al. 2016
Nebulizer	Chimpanzees	Gresswell and Goodman 2011
Oral medication	Macaques	Crouthamel and Sackett 2004
Pinworm assessment	Chimpanzees	Schapiro et al. 2005
Present for injection	Rhesus and cynomolgus macaques, chimpanzees	Perlman et al. 2004; Videan et al. 2005; Graham et al. 2012
Saliva sample	Marmosets, rhesus macaques, chimpanzees, bonobos, orangutans, dogs, pigs	Lutz et al. 2000; Cross et al. 2004; Kutsukake et al. 2009; Kaplan et al. 2012; Behringer et al. 2014; Borah et al. 2014; Decorte et al. 2014; Evans et al. 2015
Semen collection	Chimpanzees, orangutans	VandeVoort et al. 1993; Perlman et al. 2003, 2004
Shifting	Rhesus macaques, sooty mangabeys, chimpanzees	Reinhardt 1992; Luttrell et al. 1994; Bloomsmith et al. 1998; Veeder et al. 2009
Targeting/stationing	Rhesus macaques	Schapiro et al. 2001, 2003; Fernstrom et al. 2009
Ultrasound	Bonobos, snow leopards	Broder et al. 2008; Drews et al. 2011
Urine collection	Marmosets, vervets, New World primates, chimpanzees	Kelley and Bramblett 1981; Anzenberger and Gossweiler 1993; Laule et al. 1996; Smith et al. 2004; Anestis 2005; Perlman et al. 2010
Vaginal swab	Stumptailed macaques	Bunyak et al. 1982
Venipuncture	Rhesus, cynomolgus, and stumptailed macaques; chimpanzees; cats	Reinhardt and Cowley 1990; Reinhardt 1991, 1997; Schapiro et al. 2005; Lambeth et al. 2006; Coleman et al. 2008; Graham et al. 2012; Lockhard et al. 2013

including moving between enclosures (Veeder et al. 2009), presenting for injections (Perlman et al. 2004), presenting for venipuncture (Lockhard et al. 2013), presenting for capillary blood samples (Reamer et al. 2014), and stationing for acupuncture (Magden et al. 2013). Table 32.1 provides a list of additional examples and citations for health care–related behaviors trained using PRT techniques.

Health Care Issues Amenable to Training Solutions

Many health care issues are amenable to training solutions (Laule and Whittaker 2007; Whittaker and Laule 2012). In most instances, solutions to complicated health care situations can be derived from appropriate desensitization procedures and the combination of sequences of multiple, fairly basic trained behaviors. For example, an animal can be trained to voluntarily participate in a capillary blood sampling procedure (Reamer et al. 2014) by training (shaping) a sequence of behaviors, including approach, sit, stay, hold, and present a digit.

Targeting, Shifting, and Stationing

Among the most basic of behaviors to be trained for health care purposes are shifting, targeting, and stationing. *Shifting* (sometimes called gating) refers to training the animals to move from one part of their enclosure to another part on command (Bloomsmith et al. 1998; Veeder et al. 2009). This effectively allows the entire enclosure to be cleaned and works best when the animals live in "suites" of at least two "rooms." Animals can be shifted to room B when room A is being cleaned and can then return to room A so that room B can be cleaned. When training animals to shift, it is important to remember the stimulus–response–reinforcement contingency and to provide the animals with a verbal stimulus (shift) and then reinforce them (with food, e.g., all or part of their morning ration) when they move as requested. In order for play cages or exercise pens to be valuable components of animal care programs (Griffis et al. 2013), the animals must reliably shift back and forth between the home cage and the play cage.

Training animals to *target* (Figure 32.1) and *station* is also a basic requisite for a variety of health care–related tasks. Training subjects to target involves teaching the animal to touch (with hands, feet, nose, etc.) an object (the target), such as a paddle, a stick, or a piece of PVC, when the stimulus (command) is given

FIGURE 32.1 **(See color insert.)** Squirrel monkeys "targeting" (touching) the PVC tube held by the trainer.

(Schapiro et al. 2003; Fernstrom et al. 2009). The target can be either stationary or moved to various positions around the enclosure. A target that can be moved allows the trainer to position the animal at a variety of different places within the enclosure and is typically quite helpful when training animals to shift, separate, or enter transport apparatus. Training subjects to station involves teaching the animal to stay in a desired position (typically at its target) (Schapiro et al. 2001) for progressively longer durations until the desired time frame is achieved. Animals that are reliably trained to station for periods ranging from several seconds to many minutes are (1) less likely to interfere with the training of group mates, (2) less likely to act aggressively toward people, (3) typically appropriately positioned for the performance of health care behaviors (presentation of body parts, nebulizer, acupuncture, venipuncture, etc.), and perhaps most importantly, (4) demonstrating that they are willing participants in the procedures that follow. For many of the health care–related behaviors discussed in the sections that follow, it is essential that the subject is trained to station reliably.

Examinations

Animals that are reliably trained to target and station can be easily examined. Limbs, appendages, orifices, offspring, and so forth, are quite accessible (visually and tactilely) when an animal is calmly stationed and correctly positioned at a target at the interface between the animal area and the human area. Animals can be trained to present body parts and desensitized to objects and apparatus while stationed, allowing for ocular, aural, dental, and wound examinations; body temperature and pinworm assessments (Schapiro et al. 2005); and the use of noninvasive imaging techniques (including x-rays or ultrasounds).

Training Permits Better Access to the Animal

One of the primary advantages of training subjects to perform health care–related behaviors is to improve access to the animals. This can occur in two different ways. First, animals that are trained to station at the human–animal interface of their enclosure and to present body parts can be visually examined at extremely close range. Any abnormalities (injuries, hair loss, etc.) can be assessed quickly and in considerable detail. Second, trained animals can be examined *frequently without restraint* (chemical or physical), minimizing the amount of stress involved in such examinations. Wound healing can be easily assessed through repeated stationing episodes. As with most of the health care–related behaviors, animals are requested to perform examination behaviors on a regular basis, even if no particular problems are evident, to maintain high levels of compliance for those few occasions when the behavior is necessary.

Assessing Health Based on Training Performance

Training performance is another valuable tool for monitoring health in trained subjects. Animals that are reliably trained to perform particular behaviors, but suddenly stop performing reliably, may be ill or compromised in some way. Poor performance of their trained behaviors can serve as an early signal of approaching or advancing health problems. This is particularly true and useful for animals that are perceived to be near the end of their lives and are therefore on a "quality of life watch."

Quality of Life Assessments

An extremely important goal of any animal care program is to minimize or eliminate any suffering experienced by the animals. Laboratory animal care programs are designed to guarantee that animals have at least (but usually more than) an adequate quality of life. In some situations, the quality of an animal's life may be compromised due to experimental manipulations, disease, or simply advanced age. If this compromise is serious enough, a quality of life watch may be initiated for that animal (Lambeth et al. 2013). Changes in training performance can be an especially sensitive indicator of a decrease in quality of life. For instance, a chimpanzee on a quality of life watch that suddenly stops reliably performing previously trained health care–related behaviors (e.g., stations for acupuncture) should be immediately examined to determine whether the health issue has accelerated, signaling that a significant decrease in quality of life may have occurred and euthanasia may be necessary to prevent suffering.

Diagnosis

Behaviors that have been trained using PRT techniques can play an extremely important role in the diagnosis of health care–related problems. As mentioned above, the relatively straightforward behaviors of stationing and presenting body parts allow veterinarians, technicians, and researchers to closely and frequently observe the animals for wounds or illness. Animals can also be trained for a variety of types of diagnostic sample collections; this is among the most useful training applications for health care purposes. In addition to the attainment of body fluids for diagnostic purposes, animals can be trained to allow unrestrained images (e.g., x-ray) to be taken, and to allow the interrogation of implanted monitoring units.

Monitoring Units

For those animals with acknowledged health issues, monitoring units can be surgically implanted in the animals and regularly interrogated, using the associated "reading" device. For instance, implantable loop recorders, identical to those used in humans, can be used to monitor cardiac patterns in great apes (Lammey et al. 2011; Magden et al. 2016). These devices record and store deviations from normal cardiac function, with the data downloaded approximately every 3 weeks. To download the data, animals are trained to station in a position in relation to the cage mesh that allows the technician to place the reader in close proximity to the implanted recorder (chimpanzees station with their back to the cage mesh [Magden et al. 2016]) (Figure 32.2a and b). Data can typically be transferred in less than 1 minute, leaving the implanted device ready to record data for the next 3 weeks. After remaining stationed for the download, the animal receives reinforcement for having done so and then can resume its normal activity.

(a)

(b)

FIGURE 32.2 (a) Chimpanzee with an implantable loop recorder presenting its back so that the loop recorder can be "read." (b) Implantable loop recorder being read.

Sample Collection

Many different types of samples can be collected from an animal that is (1) trained to station and present body parts at the enclosure interface and (2) desensitized to the collection device or personnel (syringe, blood sleeve, lancet, swab, vet tech, etc.). These include many types of body fluids, hair samples, skin scrapings, body temperatures, and images. The cooperative handling work with macaques of Graham and colleagues (2012) is particularly relevant here.

Body Fluids

Animals can be trained to voluntarily provide a variety of body fluids for health-related purposes. Blood is obviously the most meaningful and relevant fluid to collect, but saliva, mucous (nasal discharge for assessment of MRSA) (Figure 32.3), vaginal fluid, semen, and wound discharge can also be readily obtained from trained animals.

Saliva The most common technique for collecting saliva samples is to train animals to chew on an object (dental rope, cotton swab, gauze pad, etc.) and then retrieve the object that has been salivated on from the animals. Saliva samples have been collected from marmosets (Kaplan et al. 2012), chimpanzees (Kutsukake et al. 2009), and rhesus monkeys (Lutz et al. 2000), and have been used to assess levels of cortisol or the presence of several zoonotic pathogens (Evans et al. 2015). Obviously, obtaining saliva samples from dogs involves relatively little training (Borah et al. 2014), given a dog's propensity to salivate. Pigs also can be trained to provide saliva samples (Decorte et al. 2014).

Urine Animals (marmosets [Anzenberger and Gossweiler 1993], vervets [Kelley and Bramblett 1981], and chimpanzees [Laule and Desmond 1998; Anestis 2005; Perlman et al. 2010]) can be trained to urinate on command, providing samples for assessment of pregnancy-related hormones, glucose levels, and cortisol levels. Unlike most other applications of PRT techniques, where shaping (reinforcement of successive approximations) is typically used to train the target behavior, training an animal to urinate requires that the trainer "capture" the target behavior. This means that the trainer establishes conditions in which it becomes increasingly likely that the animal will perform the target behavior (providing the animal with a lot of fluid) while repeatedly giving the animal the stimulus (the command "urinate"). When the animal finally does urinate, the reinforcer (a very high-value reinforcer the first time, a jackpot) is immediately provided. While one might think that *initially* providing the animal with large quantities of fluids might facilitate the training of urination, this can be counterproductive, as the trainer may end up training the animal to "hold it in," rather than to urinate (the longer they hold it in, the more of the

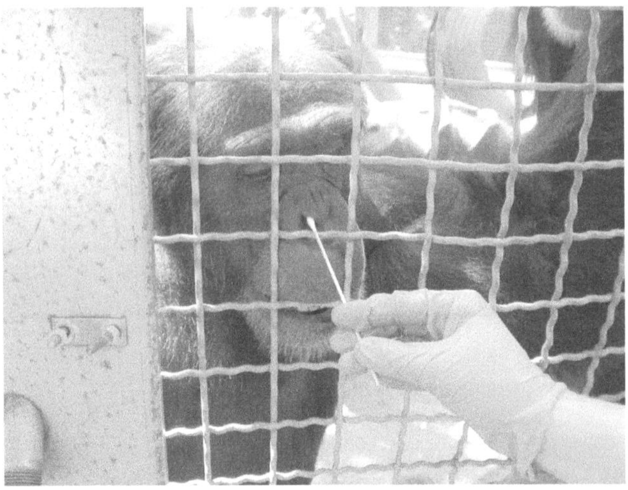

FIGURE 32.3 Chimpanzee presenting for a voluntary nasal swab.

fluid they receive). This is one (of a number of) potential circumstances in which the animals may end up training the trainer, rather than the other way around. Since midstream urine samples are usually the most desirable for analysis, some sort of apparatus (a cup and/or a wand, rather than a pan beneath the cage) (Laule et al. 1996) is typically involved in the training and collection process.

Blood The value of blood samples as diagnostic tools is well known and multifaceted. While blood samples can be obtained from restrained animals, there are many reasons to acquire voluntary blood samples from unrestrained animals, especially when these samples can be collected at "cage side." Such samples should provide data that is relatively less likely to be confounded with factors such as (1) the stress associated with manual restraint and (2) the potential for interactions with the anesthetic agents associated with chemical restraint (Schapiro et al. 2005; Lambeth et al. 2006). Voluntary blood samples can be obtained using a variety of techniques. The quantity of blood required often helps determine the blood collection technique. The same animal can be trained for multiple techniques, all of which require that considerable effort be devoted to a systematic maintenance program. Voluntary venipuncture and blood collection is one of the few behaviors that is typically trained *and maintained* by trainers, rather than by animal care technicians.

Capillary Small (80–200 μL) voluntary blood samples for the assessment of blood glucose levels and/or for use with the i-STAT system (Abbott) can be collected from animals that are trained to present a finger or toe for sampling (Figure 32.4a and b). The training for this process is straightforward; in fact, many chimpanzees among those that are trained regularly for a variety of other behaviors will perform this behavior the first time they are asked, prior to any specific training for capillary blood sampling

(a)

(b)

FIGURE 32.4 (See color insert.) (a) Using the lancet to prick the finger of a chimpanzee for a voluntary capillary blood sample. (b) Voluntary capillary blood sample from the toe of a chimpanzee for blood glucose analysis.

(Reamer et al. 2014). Training for a capillary sample typically includes sequentially reinforcing stationed animals for allowing touching of the finger or toe, cleansing with a disinfectant wipe, skin puncture with a lancet, cleansing with another wipe, gentle squeezing, and finally, collection of the blood sample. When a 200 μL sample is required, a vacuum device (Innovac Quick-Draw®) can be used to prevent clotting and/or hemolysis of the sample. When a single drop of blood is required for use in a glucometer, no vacuum device is needed.

Venous Voluntary blood samples, for analyses (complete blood counts [CBCs], chemistries, viral analyses, etc.) that require larger volumes of blood, can be collected from animals that are trained to voluntarily "enter" and remain stationary in a device that is specially designed to facilitate unrestrained blood samples (PVC sleeve, fabric sling, etc.) (Coleman et al. 2008; Stracke et al. 2011). This can be particularly important when a subject is ill and an anesthetic episode to collect a blood sample could further compromise the animal's health. In the case of the sleeve that has been successfully utilized with nonhuman primates (chimpanzees and rhesus macaques [Coleman et al. 2008]) (Figure 32.5), the animals are trained to *hold onto the bolt* at the end of the sleeve, exposing the cephalic vein for venipuncture through

FIGURE 32.5 (See color insert.) Chimpanzee providing a voluntary venous blood sample using the sleeve.

the opening cut in the sleeve. The target behavior of holding onto the bolt serves several purposes in this process: (1) it is a behavior that is incompatible with grabbing the humans, (2) it ideally positions the cephalic vein for venipuncture, (3) it indicates the animal's willingness to voluntarily provide the sample, and (4) it provides an early warning signal in the event that an animal is going to pull its arm out of the sleeve. The trainer typically touches the animal's hand during the venipuncture process, not as a method of restraint, but to assess the animal's positioning, to determine its degree of relaxation, and to provide the animal with contact comfort and reassurance during the process.

As has been mentioned for other behaviors, the training process for voluntary venipuncture using the sleeve involves both training the target behavior and desensitization of the subject to the apparatus, personnel, and process (the sleeve, the vet tech or veterinarians, the occlusion of the vein, etc.). Once trained, multiple tubes of blood using a Vacutainer® system and/or a catheter can be quickly and voluntarily obtained from subjects (Lambeth et al. 2005). Importantly, certain CBC, chemistry, and immunological parameters may differ between voluntary and nonvoluntary blood samples obtained from chimpanzees (Lambeth et al. 2005; Schapiro and Lambeth 2007).

Chimpanzees, among other species, can also be trained to voluntarily accept intravenous infusions using the same sleeve and similar training techniques (Pavonetti et al., unpublished data).

Imaging

Animals can be trained to station to facilitate the collection of images for diagnostic and research purposes. X-rays and ultrasounds are the most common diagnostic images collected. Pregnancy has been monitored via unrestrained ultrasound in a variety of zoo animals (e.g., snow leopards [Broder et al. 2008] and bonobos [Drews et al. 2011]). Extremities have been x-rayed in zoo-housed gorillas (Laule, personal communication). Dogs have been desensitized to many of the sensory aspects (noise, vibrations, tight quarters, etc.) associated with MRI scanning and have been trained to station and remain motionless in the scanner, allowing the collection of function magnetic resonance imaging (fMRI) images (Berns et al. 2012, 2013).

Treatments

PRT techniques can facilitate not only diagnostic procedures with laboratory animals, but also treatment procedures. Animals that have been successfully trained to voluntarily target and station, in the absence of restraint, can be treated in a variety of different ways, including topically, orally, via injection, with acupuncture, and/or with medicinal lasers. Preventative medicine procedures can also be performed, such as toothbrushing and skin moisturizing treatments.

There are many conceptual similarities between training animals for diagnostic and treatment behaviors: (1) voluntary participation in these behaviors is extremely important for an animal that is being diagnosed and treated because it is ill, as restraint may cause additional problems for the animal; (2) animals must be trained to perform the targeted behavior *and* desensitized to the relevant apparatus, personnel, and so forth; and (3) a comprehensive maintenance program must be established in order to maximize the probability that the target behavior will be performed when necessary.

Topical

It is extremely advantageous to have animals that are trained to present eyes, ears, mouths, digits, tails, wounds, and so forth, for topical treatment (disinfectants, sugar slurries, antibiotic creams, eye or ear drops, etc.). Socially living animals may occasionally be wounded by group mates, and individually housed animals may occasionally injure themselves; any captive animal may occasionally contract a minor infection that requires topical treatment. Having animals that are trained to voluntarily present body parts to veterinary personnel for examination and handling can facilitate the treatment of these injuries without the need for anesthesia. This can be particularly important for minor health issues among socially housed animals, as the process of anesthetizing, removing, treating, and reintroducing the animal can often create issues that are worse than the original problem. It would be less than ideal if the removal and treatment of an animal for a minor ear infection resulted in a fight upon the animal's reintroduction to the group in which the animal lost a finger or toe.

Present for Injection

There are many advantages to training animals to voluntarily present a body area for an injection (Figure 32.6). Regular subcutaneous injections for diabetic animals, intramuscular anesthetic injections for animals prior to physical examinations, vaccinations, and intramuscular antibiotic treatments for ill animals are all important applications of training animals to present for an injection. Although intuitively it may seem otherwise, training an animal to present for injection is relatively straightforward. However, animals often regress (fall below the criterion for successful training) after a real injection (especially of a dissociative anesthetic). The behavior will typically recover (return to the criterion of successful training), a process that can be significantly accelerated by an effective maintenance program (only a very small proportion of training episodes should actually result in a real injection). Typically, when an animal voluntarily presents for an injection, a positive reinforcer (food or juice) can be presented immediately after the injection, establishing a positive context for the process. Obviously, when an anesthetic is injected, food reinforcement is not possible, highlighting the importance of the conditioned reinforcer (clicker or whistle) in the training of this valuable health care–related behavior.

Present for injection is a behavior for which it is extremely important to desensitize the animals to the apparatus and personnel involved. This is especially true in situations where the veterinary technician has previously been involved in relatively few positive interactions with the animals (the presence of the vet tech may reliably indicate to the animal that something "aversive" is about to happen). It is reasonably straightforward for the trainer to train an animal to present for a "mock" injection; however, when the time comes for the real injection to be administered by the vet tech, the animal is likely to become noncooperative. Therefore, it is critical that the vet tech is involved in the early shaping and desensitization stages of present for injection training, so that the animal can learn that the vet tech is *not* a reliable indicator of an aversive interaction or outcome.

For some species, injectable antibiotic treatments are superior to oral antibiotic medications. If an animal is sick enough to require an antibiotic treatment, then the possibility exists that restraining the animal to provide the treatment may further compromise the animal's health. If the animal is trained to voluntarily present for antibiotic injections, then a significant reduction in stress can be achieved, increasing the probability of successful treatment and recovery. However, it is important to note that levels of voluntary participation for these behaviors are often lower when animals are sick.

Although not directly health care related, training animals to voluntarily present for an anesthetic injection can influence at least CBC and chemistry values in subsequently collected blood samples (Lambeth et al. 2006). Clinical parameters in venous blood samples obtained when animals voluntarily presented for an injection of anesthetic differed significantly from clinical parameters obtained when animals were nonvoluntarily anesthetized. These findings applied to both within-subject and between-subject

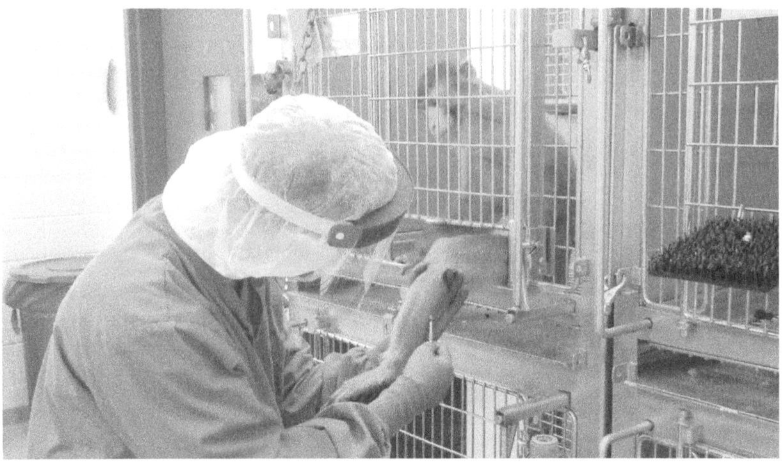

FIGURE 32.6 Rhesus monkey voluntarily presenting for an injection.

analyses of a fairly large data set (Lambeth et al. 2006), suggesting that the process of administering the anesthesia and/or the anesthesia itself could be a potential confounder in experiments that analyze parameters from venous blood samples. Voluntary presentation for anesthetic injections may minimize the confounding effects of these factors.

Nebulizer

Chimpanzees have been trained to accept the use of a nebulizer to treat respiratory problems (Gresswell and Goodman 2011; Haller et al., in preparation) (Figure 32.7). The target behavior that is trained is the placing of the animal's face (more specifically its mouth and nose) into the nebulizer mask to receive the aerosolized treatment. In essence, the animal makes a "kissy face" to position itself appropriately. Nebulizer treatments can then be administered for up to 20 minutes, several times a day. It is not uncommon for the subject's social partners to learn the behavior via observation, and even to "request" it (Haller et al., personal observation).

Acupuncture

At this point, it is important to mention again that stationing is among the most fundamental trained behaviors that underlie many of the health care–related behaviors mentioned in this chapter. Animals that are trained to position themselves appropriately at a station are demonstrating that they are willing subjects, ready to voluntarily perform any number of health care–related (and other types of) behaviors when the appropriate stimulus is presented.

Chimpanzees have been trained to allow the insertion of acupuncture needles as an adjunct therapy for the symptoms associated with osteoarthritis (Magden et al. 2013, 2016) (Figure 32.8). Animals are initially stationed and desensitized to the acupuncture needles prior to remaining still for the approximately 10-minute sessions. Initially, subjects receive primary (grapes) and conditioned (clicks) reinforcers for remaining stationed, but if the acupuncture is truly relieving their arthritis symptoms, the animals will eventually station and accept treatment without the reinforcement of clicks or grapes. This suggests that the *symptom relief* experienced as a function of the acupuncture treatment may become the primary reinforcer for the behavior, a potential example of one of the primary goals of any animal training program—the voluntary participation of the animal in its own care, because the treatment is making the animal feel better. This can be interpreted as an example of self-medication, a concept discussed in more detail below.

FIGURE 32.7 Chimpanzee voluntarily placing its face in the nebulizer mask.

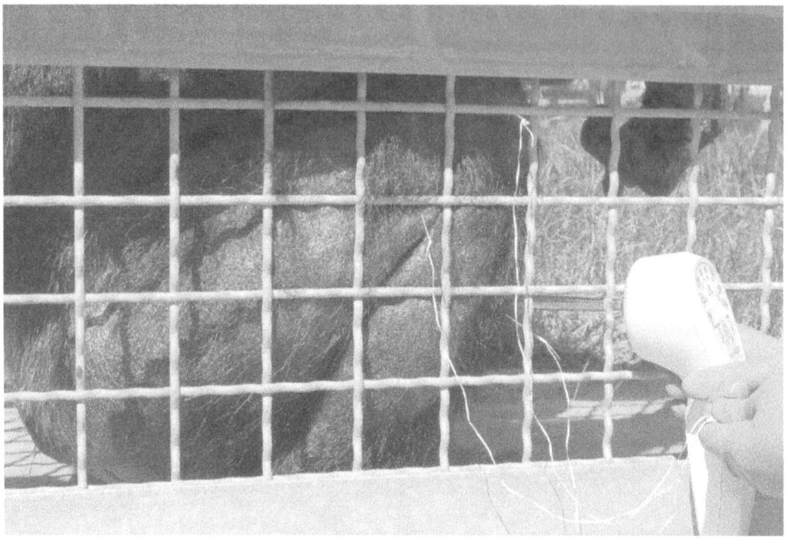

FIGURE 32.8 **(See color insert.)** Chimpanzee voluntarily presenting for a combination of acupuncture and laser therapy.

Medicinal Laser

Medicinal lasers can be used (alone or in conjunction with acupuncture) in a manner similar to that of acupuncture, as an adjunct therapy for osteoarthritis. Additionally, medicinal lasers can be used to treat injuries and/or infections. Treatment with a medicinal laser requires that the animal station and tolerate the handheld laser and its flashing light (Magden et al. 2016). One advantage of laser treatments compared with acupuncture is that animals typically have to remain stationed for much shorter periods (~2 minutes) for a laser treatment than for acupuncture (~10 minutes). External reinforcers may also become extraneous for animals receiving laser treatments when the internal reinforcement of symptom relief takes place (Schapiro et al., personal observation).

Medications: Choice Procedure

Providing animals with the opportunity to choose between two medications for treatment of an ailment is an innovative approach that can truly put the animal's health care under its own control, allowing the animal to demonstrate to care personnel which of two treatments it finds preferable. This can effectively open new lines of communication between the animals and those charged with their care and should be the next step in the continuing evolution of animal health care programs.

As an example, arthritic chimpanzees have been allowed to choose between meloxicam and ibuprofen as the treatment for their arthritis symptoms (Schapiro et al. 2014). In a small study that involved a number of necessary control conditions, chimpanzees received ibuprofen daily in colored Gatorade for 2 months, after which they received meloxicam daily in a different color of Gatorade for an identical period. The conditions were then reversed to make it an ABBA design. Finally, after a total of 8 months of being presented with one or the other medication, the animals were presented with both medications simultaneously and allowed to choose the one that they wanted. Although only a small number of subjects were involved in this study, all subjects preferred one medication (meloxicam) over the other, and perhaps more importantly, their behavior varied, depending on the medication they received or chose. In general, subjects engaged in more species-appropriate behavior when they received their preferred medication (Schapiro et al. 2014). A similar choice procedure can be used in many circumstances, for many conditions, which are typically treated by the administration of a compound, allowing for not only choices between medications, but also titration of doses. As mentioned above, such choice procedures (1) allow animals to control how they are treated, (2) open lines of communication between animals and their caregivers that can truly benefit the animals, and (3) represent the next step in the evolution of laboratory animal care.

The fact that chimpanzees self-medicate under natural conditions (Huffman 1997) is one of the reasons that these types of medication choice procedures are potentially useful for this species. If wild chimpanzees are feeling ill, they will search out the specific plant that makes them feel better and consume it and, shortly thereafter, feel better (Huffman 1997). While this may sound fairly straightforward, it represents a complicated chain of thought processes indicative of the complex cognitive abilities of chimpanzees. The choice procedure was designed with this in mind, to provide the animals with a naturalistic opportunity to use their inherent abilities to address the important situations that they are experiencing.

Data Collection

Although slightly off topic for this chapter, animals can also be trained to facilitate the collection of data that can address health care–related topics, as well as more basic and applied research questions. For instance, animals can be trained to (1) engage in affiliative activities to potentially impact immune responses (Schapiro et al. 2001), (2) solve complex problems to study the mechanisms of social learning (Hopper et al. 2015), (3) allow interspecific comparisons of trainability (Rogge et al. 2013), (4) cooperate during chair restraint (Bliss-Moreau et al. 2013), and (5) cooperate during research manipulations (Graham et al. 2012), including fMRI (Berns et al. 2013).

Cost–Benefit Analyses

Many of the benefits associated with training animals to perform health care–related behaviors have been discussed in the preceding sections. However, there are costs associated with training as well. Depending on the behavior (and the animal), a substantial amount of time may need to be invested in initially training an animal to perform a complicated behavior on command. However, overall, an effective PRT program is extremely likely to provide benefits for the people and for the animals that far outweigh these initial costs. Time and money that are invested at the beginning of a training program typically result in substantial time and money savings later in the process, when animals are trained to perform the behavior at the time that it is needed. For example, mangabeys were trained using PRT to shift between rooms in their enclosures for cleaning purposes (Veeder et al. 2009). An initial investment of 26.5 hours in the training process was recouped in just 35 days, as caregivers saved 23 minutes each time they shifted the animals (twice each day). Since those first 35 days, caregivers have experienced a savings of 46 minutes every day for several years.

In addition to straight cost–benefit analyses involving hours or dollars invested compared with hours or dollars saved, there are additional benefits that are likely to accrue as a function of working with trained animals. In certain cases, diagnostic and treatment options may only be available to animals that have been trained using PRT techniques. Specifically, the nebulizer as a treatment for respiratory issues (Gresswell and Goodman 2011) is only practical to use with animals that have been trained to insert their mouth and nose into the nebulizer mask. Anesthetizing the animals for multiple treatments each day would not be practical. Similarly, acupuncture and laser treatments (Magden et al. 2013, 2016) are only practical for animals that will voluntarily station for their 2- to 10-minute sessions.

Further benefits of training laboratory animals include better definition of the animals as biomedical models. Although not strictly a health care issue, the data collected from animals that are trained for health care–related and research-related behaviors may be "superior" to data collected from untrained animals (Lambeth et al. 2006; Graham et al. 2012). One of the easiest ways to think about this involves decreases in interindividual variation in the data obtained from subjects that are voluntarily providing samples. Each animal is trained similarly, and individual differences in stress that may exist when untrained samples are collected are likely to diminish. Voluntary samples are definitely a refinement in the way we care for laboratory animals, and if interindividual variation is significantly diminished, then a reduction in the number of subjects required for a research project may also be achieved.

Emerging Technologies

A number of emerging technologies can be utilized more efficiently with animals that have been trained. Implanted or worn devices, including telemetry systems (Lopez et al. 2014), loop recorders (Lammey et al. 2011; Magden et al. 2016), and activity monitors (Mann et al. 2005), which can be interrogated when animals are trained to station near the "reader," provide noninvasive techniques for collecting vital health (and research) data. Animals with vascular access ports can be trained to present their ports (Graham et al. 2012) and do not have to be anesthetized for sample collection or compound administration. Animals can be acclimated to wear jackets and undershirts (Kelly et al. 2014; Field et al. 2015), critical components of management systems that involve tethered catheterization. These are all examples of technologies whose utility can be enhanced by training animals to perform the behaviors that facilitate the use of these devices.

Conclusion

The use of PRT techniques to provide animals with opportunities to voluntarily participate in health care–related behaviors is an important component of laboratory animal care programs, especially those that involve nonhuman primates, dogs, and pigs. Building on relatively simple behaviors, animals can be trained to perform fairly complex behaviors that facilitate preventative medicine, as well as the diagnosis and treatment of health issues. Importantly, over time, animals that are initially trained to perform basic behaviors learn to learn, expediting the training of later, more complicated behaviors. Significant benefits accrue to people and animals when resources are invested in training animals to perform behaviors that facilitate health care, including, among numerous other behaviors, presenting for an injection, stationing for an acupuncture session, allowing a capillary blood sample, and urinating on command. Providing laboratory animals with opportunities to participate in their own care is one example of taking captive care to the next level, the goal of all progressive captive management programs.

ACKNOWLEDGMENTS

We were supported in part by Cooperative Agreement NIH U42 OD-011197. We thank the animal care staff, the behavioral management team (especially the trainers), and the executive leadership at the Keeling Center for their dedication, commitment, and support for the ongoing enhancement of the care and management programs in the nonhuman primate areas. Kristina Adams of Animal Welfare Information Center provided help with references.

REFERENCES

Adams, K. M., Navarro, A. M., Hutchinson, E. K. et al. 2004. A canine socialization and training program at the National Institutes of Health. *Lab Animal* 33(1):32–36.

Anestis, S. F. 2005. Behavioral style, dominance rank, and urinary cortisol in young chimpanzees (*Pan troglodytes*). *Behaviour* 142:1245–1268.

Anzenberger, G., and Gossweiler, H. 1993. How to obtain individual urine samples from undisturbed marmoset families. *American Journal of Primatology* 31:223–230.

Baker, K. C., Bloomsmith, M. A., Neu, K. et al. 2010. Positive reinforcement training as enrichment for singly housed rhesus macaques (*Macaca mulatta*). *Animal Welfare* 19:307–313.

Behringer, V., Stevens, J. M. G., Hohmann, G. et al. 2014. Testing the effect of medical positive reinforcement training on salivary cortisol levels in bonobos and orangutans. *PLoS One* 9:e108664.

Berns, G. S., Brooks, A. M., and Spivak, M. 2012. Functional MRI in awake unrestrained dogs. *PLoS One* 7(5):e38027.

Berns, G. S., Brooks, A., and Spivak, M. 2013. Replicability and heterogeneity of awake unrestrained canine fMRI responses. *PLoS One* 8(12):e81698.

Bliss-Moreau, E., Theil, J. H., and Moadab, G. 2013. Efficient cooperative restraint training with rhesus macaques. *Journal of Applied Animal Welfare Science* 12:98–117.

Bloomsmith, M. A., Laule, G. E., Alford, P. L. et al. 1994. Using training to moderate chimpanzee aggression during feeding. *Zoo Biology* 13:557–566.

Bloomsmith, M. A., Stone, A. M., and Laule, G. E. 1998. Positive reinforcement training to enhance the voluntary movement of group-housed chimpanzees within their enclosures. *Zoo Biology* 17:333–341.

Borah, B. M., Halter, T. J., Xie, B. et al. 2014. Kinetics of canine dental calculus crystallization: An in vitro study on the influence of inorganic components of canine saliva. *Journal of Colloid and Interface Science* 425:20–22.

Bridges, J. P., Haller, R. L., Buchl, S. J. et al. 2015. Establishing a behavioral management program for geriatric chimpanzees. *American Journal of Primatology*, 77(Suppl. 1):111.

Broder, J. M., MacFadden, A. J., Cosens, L. M. et al. 2008. Use of positive reinforcement conditioning to monitor pregnancy in an unanesthetized snow leopard (*Uncia uncia*) via transabdominal ultrasound. *Zoo Biology* 27:78–85.

Bunyak, S. C., Harvey, N. C., Rhine, R. J. et al. 1982. Venipuncture and vaginal swabbing in an enclosure occupied by a mixed-sex group of stumptailed macaques (*Macaca arctoides*). *American Journal of Primatology* 2:201–204.

Coleman, K., Pranger, L., Maier, A. et al. 2008. Training rhesus macaques for venipuncture using positive reinforcement techniques: A comparison with chimpanzees. *Journal of the American Association for Laboratory Animal Science* 47:37–41.

Coleman, K., Tully, L. A., and McMillan, J. L. 2005. Temperament correlates with training success in adult rhesus macaques. *American Journal of Primatology* 65:63–71.

Council of Europe. 2010. Directive 2010/63/EU of the European Parliament and of the council of 22 September 2010 on the protection of animals used for scientific purposes. Brussels: Council of Europe. http://register.consilium.europa.eu/pdf/en/10/st06/st06106-re01.en10.pdf.

Cross, N., Pines, M. K., and Rogers, L. J. 2004. Saliva sampling to assess cortisol levels in unrestrained common marmosets and the effect of behavioral stress. *American Journal of Primatology* 62:107–114.

Crouthamel, B., and Sackett, G. 2004. Oral medication administration: Training monkeys to take juice from a syringe. *Laboratory Primate Newsletter* 43:14–15.

Decorte, I., Van Breedam, W., Van der Stede, Y. et al. 2014. Detection of total and PRRSV-specific antibodies in oral fluids collected with different rope types from PRRSV-vaccinated and experimentally infected pigs. *BMC Veterinary Research* 10:134.

Domjan, M. 2015. *The Principles of Learning and Behavior.* 7th ed. Boston: Cengage Learning.

Drews, B., Harmann, L. M., Beehler, L. L. et al. 2011. Ultrasonographic monitoring of fetal development in unrestrained bonobos (*Pan paniscus*) at the Milwaukee County Zoo. *Zoo Biology* 30:241–253.

Evans, T. S., Barry, P. A., Gilardi, K. V. et al. 2015. Optimization of a novel non-invasive oral sampling technique for zoonotic pathogen surveillance in nonhuman primates. *PLOS Neglected Tropical Diseases* 9:e0003813.

Fernstrom, A. L., Fredlund, H., Spangberg, M. et al. 2009. Positive reinforcement training in rhesus macaques—Training progress as a result of training frequency. *American Journal of Primatology* 71:373–379.

Field, A. E., Jones, C. L., Kelly, R., III et al. 2015. Measurement of fecal corticosterone metabolites as a predictor of the habituation of rhesus macaques (*Macaca mulatta*) to jacketing. *Journal of the American Association for Laboratory Animal Science* 54:59–65.

Gillis, T. E., Janes, A. C., Kaufman, M. J. 2012. Positive reinforcement training in squirrel monkeys using clicker training. *American Journal of Primatology* 74:712–720.

Graham, M. L., Rieke, E. F., Mutch, L. A. et al. 2012. Successful implementation of cooperative handling eliminates the need for restraint in a complex non-human primate disease model. *Journal of Medical Primatology* 41:89–106.

Gresswell, C., and Goodman, G. 2011. Case study: Training a chimpanzee (*Pan troglodytes*) to use a nebulizer to aid the treatment of airsacculitis. *Zoo Biology* 30:570–578.

Griffis, C. M., Martin, A. L., Perlman, J. E. et al. 2013. Play caging benefits the behavior of singly housed laboratory rhesus macaques (*Macaca mulatta*). *Journal of the American Association for Laboratory Animal Science* 52:534–540.

Gruen, M. E., Thomson, A. E., Clary, G. P. et al. 2013. Conditioning laboratory cats to handling and transport. *Lab Animal* 42:385–389.

Hannibal, D., Minier, D., Capitanio, J. et al. 2013. Effect of temperament on the behavioral conditioning of individual rhesus monkeys (*Macaca mulatta*). *American Journal of Primatology* 75(Suppl. 1):66.

Hopper, L. M., Lambeth, S. P., Schapiro, S. J. et al. 2015. The importance of witnessed agency in chimpanzee social learning of tool use. *Behavioural Processes* 112(SI):120–129.

Huffman, M. A. 1997. Current evidence for self-medication in primates: A multidisciplinary perspective. *Yearbook of Physical Anthropology* 40:171–200.

Institute of Medicine, Committee on the Use of Chimpanzees in Biomedical and Behavioral Research. 2011. *Chimpanzees in Biomedical and Behavioral Research: Assessing the Necessity*. Washington, DC: National Academies Press.

Kaplan, G., Pines, M. K., and Rogers, L. J. 2012. Stress and stress reduction in common marmosets. *Applied Animal Behaviour Science* 137(SI):175–182.

Kelley, T. M., and Bramblett, C. A. 1981. Urine collection from vervet monkeys by instrumental conditioning. *American Journal of Primatology* 1:93–97.

Kelly, R., III, Carlson, A., and Kern, S. J. 2014. Evaluation of the use of primate undershirts as a refinement practice for jacketed rhesus macaques (*Macaca mulatta*). *Journal of the American Association for Laboratory Animal Science* 53:267–272.

Kutsukake, N., Ikeda, K., Honma, S. et al. 2009. Validation of salivary cortisol and testosterone assays in chimpanzees by liquid chromatography-tandem mass spectrometry. *American Journal of Primatology* 71:696–706.

Lambeth, S. P., Hau, J., Perlman, J. E. et al. 2006. Positive reinforcement training affects hematologic and serum chemistry values in captive chimpanzees (*Pan troglodytes*). *American Journal of Primatology* 68:245–256.

Lambeth, S. P., Perlman, J. E., Thiele, E. et al. 2005. Changes in hematology and blood chemistry parameters in captive chimpanzees (*Pan troglodytes*) as a function of blood sampling technique: Trained vs. anesthetized samples. *American Journal of Primatology* 66(Suppl. 1):182–183.

Lambeth, S. P., Schapiro, S. J., Bernacky, B. J. et al. 2013. Establishing 'quality of life' parameters using behavioural guidelines for humane euthanasia of captive non-human primates. *Animal Welfare* 22:429–435.

Lammey, M. L., Jackson, R., Ely, J. J. et al. 2011. Use of an implantable loop recorder in the investigation of arrhythmias in adult captive chimpanzees (*Pan troglodytes*). *Comparative Medicine* 61:71–75.

Laule, G., Bloomsmith, M. A., and Schapiro, S. J. 2003. The use of positive reinforcement training techniques to enhance the care, management, and welfare of primates in the laboratory. *Journal of Applied Animal Welfare Science* 6:163–173.

Laule, G., and Desmond, T. 1998. Positive reinforcement as an enrichment strategy. In *Second Nature: Environmental Enrichment for Captive Animals*, ed. D. Shepherdson, J. Mellen, and M. Hutchins, 302–313. Washington, DC: Smithsonian Institution Press.

Laule, G., Thurston, R., Alford, P. et al. 1996. Training to reliably obtain blood and urine samples from a young diabetic chimpanzee (*Pan troglodytes*). *Zoo Biology* 15:587–591.

Laule, G., and Whittaker, M. 2007. Enhancing nonhuman primate care and welfare through the use of positive reinforcement training. *Journal of Applied Animal Welfare Science* 10:31–38.

Lockhard, J., Wilson, K., and Lanman, C. 2013. The effects of operant training on blood collection for domestic cats. *Applied Animal Behaviour Science* 143:128–134.

Lopez, J. E., Vinet-Oliphant, H., Wilder, H. K. 2014. Real-time monitoring of disease progression in rhesus macaques infected with *Borrelia turicatae* by tick bite. *Journal of Infectious Diseases* 210:1639–1648.

Luttrell, L., Acker, L., Urben, M. et al. 1994. Training a large troop of rhesus macaques to cooperate during catching: Analysis of the time investment. *Animal Welfare* 3:135–140.

Lutz, C. K., Tiefenbacher, S., Jorgensen, M. J. et al. 2000. Techniques for collecting saliva from awake, unrestrained, adult monkeys for cortisol assay. *American Journal of Primatology* 52:93–99.

Magden, E. R., Haller, R., Thiele, E. et al. 2013. Acupuncture as an effective adjunct therapy for osteoarthritis in chimpanzees (*Pan troglodytes*). *Journal of the American Association for Laboratory Animal Science* 52:475–480.

Magden, E. R., Sleeper, M. M., Buchl, S. J. et al. 2016. Use of an implantable loop recorder in a chimpanzee (*Pan troglodytes*) to monitor cardiac arrhythmias and assess the effects of acupuncture and laser therapy. *Comparative Medicine* 66:52–58.

Mann, T. M., Williams, K. E., Pearce, P. C. et al. 2005. A novel method for activity monitoring in small nonhuman primates. *Laboratory Animals* 39:169–177.

McKinley, J., Buchanan-Smith, H., Bassett, L. et al. 2003. Training common marmosets (*Callithrix jacchus*) to cooperate during routine laboratory animal procedures: Ease of training and time investment. *Journal of Applied Animal Welfare Science* 6:209–220.

Meunier, L. D. 2006. Selection, acclimation, training, and preparation of dogs for the research setting. *ILAR Journal* 47:326–347.

Mikkelsen, L. F., Bentsen, H., and Gerlach, B. 2003. The restraint platform—Refinement of long term restraining of dogs for experimental procedures. *Scandinavian Journal of Laboratory Animal Science* 30:217–219.

NRC (National Research Council). 2011. *Guide for the Care and Use of Laboratory Animals*. 8th ed. Washington, DC: National Academies Press.

Perlman, J. E., Bloomsmith, M. A., Whittaker, M. A. et al. 2012. Implementing positive reinforcement animal training programs at primate laboratories. *Applied Animal Behaviour Science* 137:114–126.

Perlman, J. E., Bowsher, T. R., Braccini, S. N. et al. 2003. Using positive reinforcement training techniques to facilitate the collection of semen in chimpanzees (*Pan troglodytes*). *American Journal of Primatology* 60(Suppl. 1):77–78.

Perlman, J. E., Horner, V., Bloomsmith, M. A. et al. 2010. Positive reinforcement training, welfare, and chimpanzee social learning. In *The Chimpanzee Mind*, ed. E. V. Lonsdorf, S. R. Ross, and T. Matsuzawa, 320–331. Chicago: University of Chicago Press.

Perlman, J. E., Thiele, E., Whittaker, M. A. et al. 2004. Training chimpanzees to accept subcutaneous injections using positive reinforcement training techniques. *American Journal of Primatology* 62(Suppl. 1):96.

Phillippi-Falkenstein, K., and Clarke, M. 1991. Procedure for training corral-living rhesus monkeys for fecal and blood-sample collection. *Laboratory Animal Science* 42:83–85.

Priest, G. 1991. Training a diabetic drill (*Mandrillus leucophaeus*) to accept insulin injections and venipuncture. *Laboratory Primate Newsletter* 30(1):1–4.

Reamer, L., Malling, R., Thiele, E. et al. 2014. The use of positive reinforcement training in managing type 2 diabetes in captive chimpanzees (*Pan troglodytes*): Factors affecting initial training success. *Zoo Biology* 33:212–220.

Reinhardt, V. 1991. Training adult male rhesus monkeys to actively cooperate during in- homecage venipuncture. *Animal Technology* 42:11–17.

Reinhardt, V. 1992. Transport-cage training of caged rhesus macaques. *Animal Technology* 43:57–61.

Reinhardt, V. 1997. Training nonhuman primates to cooperate during blood collection: A review. *Animal Technology* 48:55–73.

Reinhardt, V., and Cowley, D. 1990. In-homecage blood collection from conscious stumptailed macaques. *Animal Welfare* 1:249–255.

Rogge, J., Sherenco, K., Malling, R. et al. 2013. A comparison of positive reinforcement training techniques in owl and squirrel monkeys: Time required to train to reliability. *Journal of Applied Animal Welfare Science* 16:211–220.

Russell, W. M. S., and Burch, R. L. 1959. *The Principles of Humane Experimental Technique*. London: Methuen.

Schapiro, S. J., Bloomsmith, M. A., and Laule, G. 2003. Positive reinforcement training as a technique to alter nonhuman primate behavior: Quantitative assessments of effectiveness. *Journal of Applied Animal Welfare Science* 6:175–187.

Schapiro, S. J., and Lambeth, S. P. 2007. Control, choice, and assessments of the value of behavioral management to nonhuman primates in captivity. *Journal of Applied Animal Welfare Science* 10:39–47.

Schapiro, S. J., Perlman, J. E., and Boudreau, B. A. 2001. Manipulating the affiliative interactions of group housed rhesus macaques using positive reinforcement training techniques. *American Journal of Primatology* 55:137–149.

Schapiro, S. J., Perlman, J. E., Thiele, E. et al. 2005. Training nonhuman primates to perform behaviors useful in biomedical research. *Lab Animal* 34(5):37–42.

Schapiro, S. J., Reamer, L. A., Mareno, M. C. et al. 2014. Providing chimpanzees with opportunities to voluntarily participate in their own care: Choice of medications. *American Journal of Primatology* 76 (Suppl. 1):73.

Skinner, B. F. 1938. *The Behaviour of Organisms: An Experimental Analysis.* New York: Appleton-Century-Crofts.

Smith, T. E., McCallister, J. M., Gordon, S. J. et al. 2004. Quantitative data on training New World primates to urinate. *American Journal of Primatology* 64:83–93.

Stracke, J., Bert, B., Fink, H. et al. 2011. Assessment of stress in laboratory beagle dogs constrained by a Pavlov sling. *Alternatives to Animal Experimentation* 28:317–325.

Tornqvist, H., Kujala, M. V., Somppi, S. et al. 2013. Visual event-related potentials of dogs: A non-invasive electroencephalography study. *Animal Cognition* 16:973–982.

Turkkan, J. S. 1990. New methodology for measuring blood pressure in awake baboons with use of behavioral training techniques. *Journal of Medical Primatology* 19:455–466.

VandeVoort, C. A., Neville, L. E., Tollner, T. L. et al. 1993. Noninvasive semen collection from an adult orangutan. *Zoo Biology* 12:257–265.

Veeder, C. L., Bloomsmith, M. A., McMillan, J. L. et al. 2009. Positive reinforcement training to enhance the voluntary movement of group-housed sooty mangabeys (*Cercocebus atys atys*). *Journal of the American Association for Laboratory Animal Science* 48:192–195.

Videan, E. N., Fritz, J., Murphy, J. et al. 2005. Training captive chimpanzees to cooperate for an anesthetic injection. *Lab Animal* 34(5):43–48.

Wergard, E.-M., Temrin, H., Forkman, B. et al. 2015. Training pair-housed rhesus macaques (*Macaca mulatta*) using a combination of negative and positive reinforcement. *Behavioural Processes* 113:50–59.

Westlund, K. 2014 Training is enrichment—And beyond. *Applied Animal Behaviour Science* 152:1–6.

Whittaker, M., and Laule, G. 2012. Training techniques to enhance the care and welfare of nonhuman primates. *Veterinary Clinics of North America Exotic Animal Practice* 15:445–454.

33

Managing Animal Colony Health

Kerith Luchins and George Langan

CONTENTS

Introduction

Research facilities today are frequently confronted by the challenges of adapting to new research techniques, emerging animal models, and animal research that poses potential hazards to animals and personnel. Animal program managers play a vital role in overcoming these challenges by using colony health control to ensure the protection of animal and human health while providing an appropriate environment for biomedical research, testing, and teaching. Every year, as we continue to gain more knowledge about how adventitious infections and parasites affect research animals, the field of colony health management continues to evolve—generating new methods to screen and protect animal colonies. This chapter provides information on the tools and procedures animal program managers can use to manage the animal colony health programs in their facility, specifically bioexclusion. The focus of the chapter is on rodent colonies; however, information regarding other species is included when pertinent.

More information on biosafety and biocontainment programs is available in Chapter 30 of this book and is covered in other references (CDC 2009).

Terminology

Accurately communicating objectives in an animal laboratory setting is vital, and therefore, so is a good understanding of the appropriate terminology. While terminology pertaining to managing animal colony health can sometimes be confusing, understanding the following key terms will help personnel to avoid miscommunications.

- *Isolation*: A broad term frequently utilized in the laboratory animal setting. It commonly refers to the separation of animals, personnel, and/or research materials to prevent cross-contamination of organisms and infectious agents between animals, spread of zoonotic agents, and contamination of the surrounding environment. Isolation practices can be broken down into multiple specific terms, depending on what type of isolation is required. Quarantine, biosafety, biosecurity, biocontainment, and bioexclusion are all related but distinct terms, and a careful animal program manager should avoid using them interchangeably.
- *Quarantine*: The isolation of animals possessing either unwanted or unknown health status.
- *Biosafety*: The practice of creating safe conditions in research environments by decreasing or eliminating the exposure risk of potentially hazardous biological agents to individuals and the environment (CDC 2009). This term is distinct from the others, as it typically relates to human health, not that of the animals.
- *Biosecurity*: This term has multiple definitions, depending on the discipline. Mainly, it describes the methods used to avert biological terrorism or other disease outbreaks. However, in an animal facility, the term can relate to the protection of an animal colony from contamination from known or unknown infections that may cause disease and/or lead to research variables (NRC 2011). Additionally, it relates to the protection from loss, theft, diversion, or intentional misuse of biological materials, microorganisms, and research-related information. Biosafety and biosecurity have somewhat different goals, but overall, they are generally complementary (CDC 2009; Nordmann 2010). Biosecurity practices can be further broken down into biocontainment and bioexclusion.
- *Biocontainment*: This is the specific process or engineering control that reduces the risk of external transmission and propagation of organisms that could be dangerous to human and/or animal health. It involves methods to prevent the accidental release of pathogenic organisms or agents, including bacteria, viruses, and biological toxins. Physical containment is utilized to avoid accidental occupational infections and/or release into the surrounding community.
- *Bioexclusion*: The measures taken to reduce the risk of introduction of pathogens into the animal facility. In terms of the animal colony, this involves defining a specific pathogen-free (SPF) status and the means utilized to maintain this status. This leads to the commonly used term *barrier*, which defines how a facility maintains a certain health status. Any unwanted pathogen in a facility is termed an adventitious agent.
- *Barrier*: A systematic and comprehensive program used for the prevention of pathogen introduction into the colony animals. Barriers vary widely in scope, and could relate to an individual cage or room, or an entire facility. They also consist of multiple elements, including the housing of animals with known specified pathogen status, a monitoring system to maintain this status, the design of the housing environment, and the management of the physical plant and caging environments.

Risk Assessment

Performing a risk assessment is the foundation of a good bioexclusion program. To initiate any risk assessment, the hazards or concerns need to be accurately identified and then ranked on the basis of their likelihood of occurrence and potential impact. This ranking permits the development of risk-appropriate approaches to limit the impact of the hazard, regardless of the type.

Performing a Risk Assessment

While appropriate bioexclusion practices are the cornerstone of an effective animal research facility, it is simply not practicable to exclude all pathogens. Further, unnecessary isolation and containment practices in bioexclusion programs can increase costs and personnel workload, also potentially delaying research. For these reasons, a thorough risk assessment should be made for each pathogenic agent of concern that can accurately identify the costs and benefits of exclusion. The step-by-step risk assessment process below elucidates the potential hazards in your facility and their impact to the research program. This, in turn, will allow facility leaders to develop the necessary bioexclusion practices for the agents that pose a genuine risk to the program and research (White et al. 1998; CDC 2009).

Process and Participants Involved

Prior to beginning the risk assessment process, it is important to make sure that all the necessary participants are included in the discussion. In a small facility, this may include all staff members; in a larger facility, it may be best to have a small group of individuals with subject-specific knowledge gather the necessary information for a final risk assessment. The key participants to include are the veterinarian, facility manager, diagnostic lab representative, and principal investigator or laboratory director, who can speak to the impact of agents on their research.

To conduct a risk assessment for a bioexclusion program, first rank each microorganism based on its known potential to cause disease and potential research interactions. This can be broken down further by evaluating multiple factors and how those relate to the species utilized and the research performed. For a good step-by-step discussion of performing a risk assessment for a bioexclusion program, see White et al. (1998). After evaluation, divide the microorganisms considered into three categories of risk: low, medium, and high. These categories will help determine the necessary practices when working with these organisms (White et al. 1998).

Transmission of Agents

In determining the likelihood of an agent entering the animal colony, the risk assessment needs to consider transmission of the organisms. There are several methods by which organisms can be transmitted to other animals, and the measures required to control the agents will need to correspond to the transmission pathway. For example, organisms that are transmitted via aerosol—in other words, through the spread of contaminated airborne droplets or dust—require very strict control measures to prevent spread (Shek et al. 2015). Agents may also be spread by direct contact or oral ingestion, which tend to be easier to control. Insect vectors can also be responsible for transmission of agents between animals in a facility; however, if pest control is adequate, these types of agents should be of minimal concern. A thorough review of routes of transmission is available from the Centers for Disease Control and Prevention (CDC 2009) and Whary et al. (2015).

Establishing Standards for an Animal Care and Use Program

After a thorough risk assessment, each program must establish its own standards for bioexclusion and how they will be implemented at the facility. The establishment of these standards should involve discussion with multiple individuals within the program, including facility managers, veterinarians, animal researchers, and possibly institutional oversight committees, such as the Institutional Animal Care and Use Committee (IACUC) and the Institutional Biosafety Committee (IBC). Once established, these standards should be reviewed periodically to ensure that they are still meeting the needs of the program and, where appropriate, adapt to the fluctuating needs of the research, other programmatic adjustments, and the discovery of new adventitious agents.

Defining SPF Criteria

SPF refers to an animal that is free of a list of pathogens, but otherwise has an undefined microflora (Rahija 2007). The term *SPF* alone has little meaning, as facilities often differ in the adventitious infections and parasites allowed to enter. Therefore, a critical component of any bioexclusion program is defining which adventitious infections and parasites are of concern to the facility and the steps that will be taken to prevent their entry.

The determination of whether an agent will appear on the SPF list is influenced by several factors, including, but not limited to, the potential interference of the agents with research, the ability to cause clinical disease, the need to transfer animals between facilities, and the control mechanisms needed to exclude the agents. The Federation of European Laboratory Animal Science Associations (FELASA) has developed recommendations for which infectious agents to monitor and the frequency that monitoring should be performed for mice, rats, hamsters, guinea pigs, and rabbits (Mahler Convenor et al. 2014), and this document can be helpful when developing an exclusion list. Tables 33.1 and 33.2 provide common pathogens excluded from mouse and rat barrier facilities, respectively.

In contrast to an SPF animal, a conventional animal is one that is raised in an environment that may have unknown microflora and may have an unknown disease status (Rahija 2007). These animals

TABLE 33.1

Bacteria, Viruses, and Parasites That Are Commonly Excluded from a Mouse Barrier Facility

Bacteria
Cilia-associated respiratory bacillus
Citrobacter rodentium
Clostridium piliforme
Mycoplasma pulmonis
Salmonella spp.
Viruses
Hantaviruses
K virus
Lactate dehydrogenase–elevating virus
Lymphocytic choriomeningitis virus
Minute virus of mice
Mouse adenovirus (1 and 2)
Mouse cytomegalovirus
Mouse hepatitis virus
Mouse parvovirus
Mouse polyomavirus
Mouse rotavirus
Mouse thymic virus
Mousepox virus
Pneumonia virus of mice
Reovirus type 3
Sendai virus
Theiler's murine encephalomyelitis virus
Endoparasites
Pinworms (*Syphacia* spp., *Aspiculuris tetraptera*)
Protozoa (*Giardia muris, Encephalitozoon cuniculi*)
Tapeworms (*Hymenolepis* spp.)
Ectoparasites
Mites (*Myobia musculi, Myocoptes musculinus, Radfordia affinis, Psoregates simplex*)

TABLE 33.2
Bacteria, Viruses, and Parasites That Are Commonly Excluded from a Rat Barrier Facility

Bacteria
Cilia-associated respiratory bacillus
Clostridium piliforme
Corynebacterium kutscheri
Mycoplasma pulmonis
Salmonella spp.
Viruses
Hantaviruses
Kilham rat virus
Lymphocytic choriomeningitis virus
Pneumonia virus of mice
Rat coronavirus
Rat minute virus
Rat parvovirus
Reovirus type 3
Sendai virus
Toolan's H-1 virus
Endoparasites
Pinworms (*Syphacia* spp., *Aspiculuris tetraptera*)
Protozoa (*Giardia muris, Encephalitozoon cuniculi*)
Tapeworms (*Hymenolepis* spp.)
Ectoparasites
Mites (*Myobia* spp., *Radfordia* spp.)

therefore do not necessarily need to be housed in barrier conditions. However, because unknown pathogen status can have unwanted results on research, most institutions are gradually phasing out the use of these types of colonies.

FELASA also has a working group for nonhuman primate health, which published suggestions for harmonized health management in 1998 (Weber et al. 1999). However, as these suggestions are in need of updating, more relevant publications exist that describe the SPF status for *Macaque* spp. (Morton et al. 2008). The need for SPF colonies is critical for preventing the spread of disease to human personnel, particularly as nonhuman primates harbor many zoonotic diseases (e.g., *Mycobacterium tuberculosis* and *Macacine hervpesvirus* 1). For this reason, SPF colonies usually exclude these organisms, in addition to simian immunodeficiency virus (SIV), simian type D retrovirus (SRV), and simian T cell lymphotropic/leukemia virus (STLV), as these present risks to investigators and can confound research results, especially for those studying SIV and AIDS (Morton et al. 2008).

For other animal species, health monitoring programs might not be as robust or organized; however, SPF criteria are still necessary to exclude adventitious pathogens. Restriction of these adventitious pathogens and parasites from the facility will increase research data reproducibility and reliability and decrease the risk of zoonotic transmission. FELASA has publications describing criteria for cats, dogs, and pigs (Rehbinder et al. 1998) and for small ruminants (calves, sheep, and goats) (Rehbinder et al. 2000). With the increase in zebrafish populations in biomedical research facilities, health monitoring of this species is evolving. The same principles of health monitoring apply; however, due to the aquatic environment of fish, water quality, water source, and facility design must be considered. The Zebrafish International Resource Center (ZIRC) has responded with a description of adequate health monitoring for zebrafish (Kent et al. 2009; ZIRC 2009). For all these species, a reliable SPF vendor is the first step to ensuring animals that are free of adventitious pathogens and parasites.

Maintaining SPF Status

Periodic pathogen monitoring is necessary to detect changes in the health status of an animal colony. For example, no barrier is 100% effective, and frequent monitoring is essential to assess the health status of the animals within the barrier operation. Frequent monitoring allows the detection of breaches in the barrier facility, which can then be handled promptly. Monitoring, however, can be difficult, as many of the pathogens on the colony exclusion lists do not produce overt clinical signs in affected animals and would not be detected during daily health observations. While large animal species are usually sampled and tested directly for pathogens of concern on a regular basis, for small rodent species, such as mice, rats, and hamsters, the sampling can be complicated by the small size of the animals and potential procedure-induced stress that can affect research results.

To resolve some of these issues, animal facilities often employ sentinel animals that are deliberately placed in a particular environment to detect the presence of an infectious agent. As an example, dirty bedding may be transferred from research animals to sentinel animals on a frequent basis (during cage change). Any potential pathogens should be passed on, which will be followed by colonization or infection of the sentinel animals. These sentinel animals are then tested for the pathogens on the exclusion list on a regular basis through polymerase chain reaction (PCR), serology, parasitology, and/or necropsy, followed by pathology. For rodents, one or two sentinel cages per rack allows indirect testing of the colony animals. However, soiled bedding is not adequate for the transfer of all pathogens on the exclusion list, especially airborne pathogens or ones that are present at very low levels. Soiled bedding sentinels are best utilized for pathogens with a fecal-oral transmission. Additionally, with the advent of individually ventilated caging (IVC) systems that maintain each cage in their own microenvironment, the transmission of pathogens between cages has been dramatically reduced. Thus, IVC systems are good for maintaining bioexclusion but also decrease the transmission to sentinels, making detection of pathogens difficult if they exist in the colony (Brielmeier et al. 2006). Shek (2008) provides a thorough history of rodent health monitoring strategies, along with the impact of housing modalities.

In 2013, a major refinement in the serological monitoring of sentinel rodents occurred when dried blood spot (DBS) analysis was introduced. Only a single drop (~25 μL) of whole blood is needed to test a comprehensive panel of infectious agents, making antemortem sampling easy, reducing stress to the animal, and potentially decreasing the number of sentinels utilized (DePietro et al. 2014; Myles et al. 2014). Recently, this method was further refined with the development of PCR testing for virtually all the pathogens on exclusion lists. Institutions are using PCR for confirmatory testing or to completely replace serology, as noninvasive samples such as fecal pellets, fur plucks, and oral cavity swabs are sufficient to test a comprehensive panel of infectious agents. In one study, when PCR testing of soiled bedding sentinels was compared with conventional health monitoring methods (parasitology, microbiology, and serology), both methods failed to detect the same agents, *Mycoplasma pulmonis*, *Pasteurella pneumotropica*, and *Giardia* spp. (Henderson et al. 2013).

As the yield of pathogen detection with soiled bedding sentinels can be low and the work laborious, alternatives to this technique have been developed. To increase the transmission of airborne pathogens, housing systems that exhaust all cages on the rack to the sentinel cage have been developed. The combination of using exhaust air transmission and soiled bedding sentinels allowed for the detection of intestinal flagellates, pinworms, and mouse hepatitis virus in a mouse colony; however, mouse parvovirus was not detected by either method (Brielmeier et al. 2006). Another technique uses PCR testing of the cage exhaust filter or plenum, which eliminates the need for sentinel animals, allowing a reduction in animal numbers. Filter testing was better at detecting Sendai virus and mouse hepatitis virus but was less successful at detecting mouse parvovirus in another mouse colony (Compton et al. 2004). Horizontal plenum testing can be utilized for fur mite testing, which is not reliably detected by soiled bedding sentinels (Jensen et al. 2013). As the testing methods for monitoring colony health are frequently improving, the facility manager and veterinarian will need to determine which methods will work best to detect the adventious agents and pathogens of most concern to their colony.

Colony animals can also be cohoused with sentinels, which are referred to as "direct contact" sentinels, to improve the transmission of pathogens, especially ones transmitted by aerosol, direct contact, and urine transmission. If male mice are utilized, they should be castrated to prevent fighting and pregnancy. After sufficient time, these animals are tested in the same manner as soiled bedding sentinels. In the study by Compton et al. (2004), all pathogens tested were efficiently detected by contact sentinels. However, a disadvantage of direct contact sentinels includes using many more animals, which results in increasing expense and the amount of labor required. For additional reading on colony animal sentinel programs, refer to Chapter 31.

Facility Operational Methods

Once an SPF list has been created, the next step is to delineate the ideal control measures to maintain this level of bioexclusion in the animal colony. Control measures include vendor selection, shipping, isolation or quarantine, housing, facility design, transport, and use of the animals. These control measures are used in combination with the routine monitoring program to ensure that adventitious organisms are excluded from the facility (White et al. 1998).

Operation of facilities to maintain colony health can be separated into primary and secondary levels of preventing the introduction of unwanted agents. The primary level is the first line of defense in maintaining bioexclusion. Equipment such as ventilated caging racks and biosafety cabinets or change stations to change cages is part of this primary level, along with the use of pertinent personal protective equipment (PPE), which differs based on the animal species and potential for introduction of infectious agents or pathogens. PPE such as gloves, hair bonnets, respiratory protection, boot and shoe covers, and lab coats or disposable gowns is used for coverage of exposed body surfaces, for example, face, eyes, hands, and hair, as well as exterior clothes. Although the use of PPE is an important consideration, the amount needed should be evaluated for each facility. Recent studies have shown that some PPE components do not appreciably improve and, in some cases, actually compromise attempts to maintain bioexclusion (Hickman-Davis et al. 2012; Baker et al. 2014).

Secondary levels include the design and construction of the facility, such as physical separations of functional areas and engineering controls. Examples of physical separations include doors to limit entry, anterooms to don and remove PPE, and shower-in capabilities. Engineering controls may consist of heating, ventilation, and air-conditioning (HVAC) systems and waste treatment systems (King et al. 1999; CDC 2009).

Flexibility incorporated into facility design can accommodate the needs of an ever-changing research environment, which may permit the ability to house different species (nonhuman primates vs. rodents) and animals of varying pathogen status (aerosol transmission vs. direct contact). Determining the optimum room size that will accommodate multiple species will provide the facility manager with more flexibility to use the rooms for various uses and species. Room pressurization is also important in using rooms for bioexclusion or quarantine. A room that can be changed from positive pressure, for bioexclusion, or negative pressure, for quarantine, can greatly increase the ability to accommodate different needs.

Facility surfaces should be easily sanitizable and decontaminated. Additionally, all floors, walls, and ceiling seams should be sealed to prevent incursions by feral rodents and insects. Foresight is necessary when designing a facility, as this will allow the institution to prepare for future needs. Specific information about facility design can be found in Chapter 18.

Species Differences

Bioexclusion control practices differ with the species utilized in the facility. For mice, rats, and other rodents, the control of adventitious organisms in most facilities occurs at the cage level, as these animals are housed in static microisolators or individually ventilated cages. The ventilated caging systems frequently provide high-efficiency particulate air (HEPA)—HEPA filters exclude 99.97% of particles that have a size of 0.3 μm—to each individual cage and, depending on the type of ventilated caging, may also have a HEPA filter for the air leaving the cage. For a bioexclusion program, it is preferable to house rodents in a positively pressurized cage to prevent entry of adventitious agents into the cage. Room pressurization, as discussed above, should also be considered, depending on the agents of concern, the effectiveness of the caging system at isolation, and how the animals will be managed within the room.

While all caging needs to be sanitized, the bedding, feed, water, and enrichment may need to be sterilized, depending on the level of the bioexclusion desired. This needs to be determined based on which agents are to be excluded, the health status of the animals (immunodeficient vs. immunocompetent), and the probability of the agent entering the colony via these routes.

For rodent facilities, anytime a cage needs to be opened—irrespective of whether it is for research procedures or cage change—there is a risk of compromising bioexclusion. This risk is best mitigated by handling the cage only in a Class II biological safety cabinet, laminar flow hood, or cage change station, and using an appropriate disinfectant agent on the cage and gloves of the handler. These different aspects of the animal program will need to be considered in the risk assessment process.

For larger species, such as dogs, pigs, and nonhuman primates, the bioexclusion control operates at the room level because the housing occurs in open caging or runs where contact between animals from different cages is possible. To maintain bioexclusion, air pressure differentials are set up between rooms and corridors. The building HVAC system is also utilized by controlling airflow at the facility level (e.g., moving air from the cleanest room in the barrier [surgery suite or clean cage wash] to the dirtiest room [necropsy or dirty cage wash]). In addition, cubicles can be used to isolate different groups of large animals in a single defined control area. These cubicles are each set at a negative air pressure to the room so as to ensure no cross-contamination of air between the groups of animals. However, once the door to the cubicle is open, the differential ceases to exist as it equates with the room pressure, and so only one cubicle door per room should be open at a time (White et al. 1998).

Outbreak Control

Prevention of adventitious pathogens is the key to a good bioexclusion program; however, if an unwanted adventitious pathogen or parasite is detected during colony health surveillance, the manager should address the situation quickly to decrease the chance of spread to the remainder of the colony. The urgency to address the breach and the decontamination process available will depend on the transmission characteristics of the agent detected. Additional information about specific agents can be found in Brayton et al. (2004), Clifford and Pritchett-Corning (2012), and Pritchett-Corning and Clifford (2012).

Initially, confirmation testing by a different testing method than originally utilized or other laboratory should be considered. If the positive test was found in a sentinel animal, the organism may no longer be detected in the research colony due to the time delay in health monitoring. Therefore, it can be difficult to identify the index animal. Once the pathogen has been confirmed, the suspected positive animals or room should be isolated so as to decrease the potential spread to other animals in the facility. Isolation can be established within the room, or animals can be relocated to a remote facility or room. If isolation is established in the room, modifications to the building or room function and procedural changes should be considered. If possible, room pressurization should be changed from bioexclusion to quarantine to prevent possible spread at the room level. Additional containment practices should all be considered to prevent spread of the agent as a fomite or directly on a contaminated animal. These include restricting personnel entry, limiting animal movement into and out of the room, and procedural measures, such as bagging and decontaminating dirty supplies out of the room. The procedures implemented will depend on the transmission characteristics of the agent and continued access to the animals in the room.

Once isolation has been established, the facility leadership will need to determine how to resolve the situation. Once the agent has been confirmed, the options are usually to treat, cull, or rederive the animals. The same group of individuals that participated in the risk assessment process and in the creation of the SPF criteria can be assembled to develop a plan for returning the room to SPF status. This group can help to balance the animal and research needs of the facility.

If treatment options are available, these can be pursued as long as they are feasible and will not affect the research at hand. After an adequate course of treatment, testing must be redone to ensure negative results and elimination of the organism. Some organisms can be "burnt out" by halting breeding and introductions of new animals to the colony. This option is limited to those agents that either do not persist in the environment or transmit via placental transfer. Culling the whole facility or room may be required for select organisms when this is the only acceptable method (e.g., zoonotic organisms). Culling the affected animals also may be the most efficient option when the postcontamination research value is limited and the animals are readily available from a commercial vendor. Rederivation, either by cross-fostering or embryo transfer, is the mechanism most commonly used for rodents; however, it requires technical proficiency, time, and ample funds (White et al. 1998).

Biologics

Besides live animals, biological materials, including blood products, cell lines, and other tissues derived from or passaged in rodents, are also transferred frequently between institutions. Once these biological agents are introduced into living mice, they become potential vehicles for pathogens. Consequently, pathogen testing of biologics must be done before they are introduced into live animals in the barrier facilities. This type of testing is provided by many commercial vendors using PCR technology. There has to be a fine balance between the frequency and requirement of testing so as to avoid unnecessary costs while still ensuring that the animals maintain their SPF status. As such, a risk assessment should also be completed for use of these types of materials. This type of assessment is detailed in the Peterson article (Peterson 2008).

Animal Import

Approved commercial vendors should be identified that have the same or superior health status as the barrier colony animals. However, researchers may need to use animals from nonapproved vendors, such as other biomedical research institutions. Depending on their health status, when these animals are imported into your facility, they may need to enter an import quarantine. As the exclusion list varies widely between different institutions, this decision to quarantine or not can be a difficult one. Therefore, FELASA has developed a tool where "the main objective … is to harmonize health monitoring (HM) programs (i.e., designing, sampling, monitoring, reporting and interpreting) which will help to improve knowledge about the microbiological quality of animals used in research and to meet scientific, legal, and welfare requirements" (Mahler Convenor et al. 2014). In addition, a FELASA–American Association for Laboratory Animal Science (AALAS) joint working group has been established to determine whether a common health report can be utilized for international transfers. This group has developed a customizable spreadsheet for health report formatting that can be used to exchange this information (Pritchett-Corning et al. 2014).

Once the health status of the new shipments can be ascertained, the suitability of these animals for entry into the colony must be determined. This is often done by comparing the quarantine results to established health criteria (i.e., the SPF list). Further discussion regarding evaluating and maintaining health status can be found earlier in this chapter, as well as in other references (Rehg and Toth 1998; Lipman and Homberger 2003; Roberts and Andrews 2008).

If the health statuses do not match, an import quarantine can be used to determine the pathogen status of newly arrived animals to determine if and when they can enter the colony and what actions might be needed to allow for their entry. Therefore, newly arrived animals are commonly isolated from colony animals during a quarantine period to evaluate for evidence of disease and infectious agents that are excluded from the current colony. To protect the existing colonies, the imported animals are usually placed in containment at the cage level or the room level, depending on the species and risk. Rodents are often tested or prophylactically treated for parasites (e.g., pinworms and fur mites) and tested for evidence of adventitious agents by serology, parasitology, and/or PCR. Larger animals may be tested and treated for common bacterial and parasitic infections and vaccinated if necessary. Nonhuman primates screening varies by species, but typically includes an evaluation for tuberculosis, and possible viral (e.g., *Macacine herpesvirus* 1 and retroviruses), bacterial (e.g., *Shigella* sp.), and parasitic agents. A more detailed discussion regarding primate import quarantine is available elsewhere (Kramer et al. 2012).

One refinement to the practice of live animal import quarantine, used primarily for rodents, involves the importation of cryopreserved embryos or sperm. This alleviates the need to ship live animals, which reduces stress to the animals and potential contamination during shipment. Rederivation using embryo and/or sperm, however, does not eliminate all adventitious pathogen concerns, as contamination of these tissues with mouse parvovirus (Agca et al. 2007) and endogenous ecotropic murine leukemia viruses (Hesse et al. 1999) has been documented. However, some agents, including ecto- and endoparasites, are eliminated completely, and the risk of others is greatly reduced. The testing of these tissues must be similar to the testing of biological material for contaminants (described above), which is likely to be less involved and lower in cost (Kelley 2010).

Isolation

When a few cages of rodents need to be isolated due to a disease outbreak or other situation that results in their health status being questioned, isolators can be used instead of standard caging. Many styles of isolators exist, but the most common for rodents are the semirigid and flexible film isolators. All equipment and materials entering the isolator are disinfected or sterilized to ensure that they are free of adventitious microorganisms. Isolators, which have air HEPA filtered in and out, provide bioexclusion properties provided there is no break or tear in the isolator. The use of isolators adds a lot of flexibility to a bioexclusion program (White et al. 1998).

With the increase in technological advances in research techniques and equipment, such as imaging, there has been an increase in demand for temporarily housing animals from other institutions for very short durations (e.g., 1–2 days). While some of the animals could go through the import quarantine process, some studies are time sensitive, and therefore would be invalidated if the animals were required to be quarantined. Institutions should consider how to accommodate these studies and what pathogen testing methods are necessary to protect the animal colony from adventitious agents. A containment room with the appropriate controls in place can be beneficial, as this will allow research procedures that may not usually be performed. It is important to consider how these animals of unconfirmed health status will be contained within the isolation room. Additionally, their transport and use in procedural equipment rooms, such as MRI or computed tomography (CT), will also need to be contained to prevent possible contamination of the equipment with an adventitious pathogen.

Special Considerations

Immunodeficient Animals

Immunodeficient rodents, such as athymic nude, Nod *SCID gamma*, SCID, and *Rag 1* mice strains, or any animal subjected to irradiation or immunosuppressive treatments must be protected from adventitious agents. These animals must also be protected from several other opportunistic infections that may not affect immunocompetent mice. In particular, immunodeficient mice are very susceptible to *Corynebacterium bovis*. This organism is a ubiquitous, opportunistic pathogen that typically only causes disease in immunodeficient mice, which is commonly called "scaly skin disease." Mice with this condition exhibit transient yellow-white keratin flakes adherent to the skin. These mice are usually persistently infected, and while morbidity is high, mortality is usually low (Burr et al. 2011). This is in contrast to *Pneumocystis*, which causes a chronic progressive pneumonia in immunodeficient mice and rats that can lead to death (Shek et al. 2015). To prevent opportunistic agents, immunodeficient mice are usually housed under barrier conditions, which involve IVC systems, intensive management procedures, autoclaved feed and/or water, and restricting entry to only essential individuals (Foreman et al. 2011). This will ensure their longevity as appropriate research models.

Immunodeficient large animals are more difficult to protect from opportunistic infections because the specialized caging for rodents is not readily available for larger species. In animals that will be used in immunosuppressive studies, managers should consider the use of SPF species and screening for opportunistic infections during quarantine (Wachtman and Mansfield 2008). Further protection of these species relies on practices such as isolation to limit the possible spread of opportunistic agents, good sanitation practices, and the use of PPE (Wachtman and Mansfield 2008).

Gnotobiotic Animals

Axenic, also called germ-free, animals are free of all organisms (bacteria, viruses, fungi, protozoa, and other parasitic life-forms) that are currently detectable. Gnotobiotic, also called defined flora, animals are those that have a known flora that is closely controlled (Rahija 2007). These are typically bacterial organisms that have been introduced to germ-free animals for a specific research purpose. There are several different species that have been maintained gnotobiotic; however, the mouse and swine are the most frequently used in biomedical research. The use of germ-free and gnotobiotic mice has gained renewed interest recently, as more researchers are looking at the complex interactions of the microbiome and the host, and how these interactions influence physiologic pathways in health and disease.

Gnotobiotic animal colonies require rigorous management and housing practices to maintain the health status of the colonies. These animals are typically housed in sterile semirigid or flexible film isolators with all materials (food, water, bedding, and research materials) being sterilized prior to entering the isolator. Strict attention to procedural details and frequent monitoring of autoclave performance are critical to maintaining the health status of the animals in the isolator. To maintain this level of exclusion, several steps for monitoring contamination of the isolator must occur, including daily visual inspection for damage, periodic microbiological culturing, and pathogen monitoring. The frequency of monitoring should be based on a risk assessment of the likelihood of contamination of the isolator (Rahija 2007).

Administrative Considerations

Maintaining a bioexclusion program requires a significant amount of administrative management, including the establishment of written procedures, requirements for training and documentation, and methods for ensuring compliance. Animal facilities should create programs that are clearly described and communicated, and program requirements for documentation and compliance should be straightforward and easily understood.

Access Control

Controlling access to different areas of the program allows managers to segregate animal colonies based on animal health status, as described earlier in this chapter. Access control involves several methods to segregate different areas within a facility, including signage, physical controls (lock-and-key systems, electronic door cards, and retina and fingerprint scanners), and security personnel. This allows access only to personnel who are properly trained and have completed the necessary medical clearance, prevents the spread of adventitious pathogens, and restricts unwanted entry from individuals not involved in the animal care program. Electronic control systems are also advantageous, as they provide a record of time and location of all entries and exits from the facility. The complexity of the security design depends on the size of the facility, location (city vs. rural), and type of research being conducted (NRC 2011). Security can be enhanced by providing video cameras that record entries, exits, and activities of personnel within the facility.

SOPs and Policies

Standard operating procedures (SOPs) are essential for maintaining bioexclusion in a laboratory animal research facility. All procedures—from how to quarantine import animals to regular cage change operations in a barrier facility—should be standardized to ensure uniformity. Step-by-step directions should be written by the personnel who are actually conducting the procedures to ensure that operating procedures are clear and in accordance with best practices. An active SOP program requires new personnel to be trained on SOPs, periodic revision of the SOPs and retraining, and development of new SOPs as necessary. The National Institutes of Health (NIH) FAQ website recommends that "SOPs should be reviewed by the IACUC at appropriate intervals (at least once every three years) to ensure they are up-to-date and accurate" (OLAW 2015). As SOPs are required to ensure regulatory compliance, the IACUC is responsible for oversight of the program and evaluating the effectiveness (NRC 2011). It is essential to have the SOPs readily available to all personnel, possibly including research investigators to ensure compliance. An electronic system is recommended, as it allows controlled access and the ability to easily update when revisions are necessary.

In an outbreak situation, when an adventitious pathogen has been detected, SOPs become even more important as temporary methods are put in place that might modify the standard procedures. It is necessary to ensure that all personnel who enter the facility, from the personnel that grant access to the mechanical staff, are aware of the new procedures. Social media sources, such as departmental website blogs, can be a good tool to disseminate this type of information. This allows research investigators to ask questions and voice concerns if necessary. If the outbreak occurred due to a lapse in following an established SOP, the document should be critically reviewed to ensure accuracy and retraining should occur for compliance.

Training Program

SOPs are the building blocks of any training program. Therefore, it is imperative that the training on SOPs is adequate and refreshed periodically. The training program should be tweaked to provide the necessary guidance for the different personnel who enter an animal facility. Overall, all personnel who enter the research facility must be fully trained on all aspects of a bioresearch facility. Procedures to maintain bioexclusion—such as how to don the proper PPE, follow the facility entry order, and perform decontamination procedures—are especially important for maintaining the facility health status. All initial and refresher trainings need to be documented appropriately. An electronic format is ideal to document training, as it can be easily updated and multiple users can log in to check the progress of an individual's training status.

Communication

Departmental websites are often the best method of communicating items such as the exclusion list and current SPF status of the colonies to the research staff. This allows the information to be utilized by the research staff at the institution, as well as external institutions that plan on receiving animals from your institution. If the institution has rooms with different pathogen statuses, this information should be clearly defined on the website. This list should be updated as changes are made to ensure that it is always up to date. Research staff should be familiar with the exclusion list, as there might be individuals that require a more stringent SPF status due to their research interests. For example, if a researcher is studying *Staphylococcus aureus*, he or she would not want to receive animals infected with this bacterium. Therefore, this information should be communicated to the personnel involved in animal ordering.

Summary

Maintaining animal colony health requires routine assessment of the health status of the animals. When the health status changes or new threats to the health status emerge, the manager has to know how those risks can be mitigated within the facility. A clear understanding of the capabilities of the facility design (e.g., HVAC, access control, and pressure differentials), caging systems (e.g., standard vs. ventilated), research equipment (e.g., biological safety cabinets, autoclaves, and isolators), and husbandry and management practices used to contain or exclude agents (e.g., disinfectants, SOPs, and training) is vital. Management practices such as quarantine and isolation procedures must be balanced with the needs of the research studies. This balance is best accomplished through an ongoing discussion and risk assessment process as new challenges arise.

REFERENCES

Agca, Y., B. A. Bauer, D. K. Johnson, J. K. Critser, and L. K. Riley. 2007. Detection of mouse parvovirus in Mus musculus gametes, embryos, and ovarian tissues by polymerase chain reaction assay. *Comp Med* 57 (1):51–56.

Baker, S. W., K. A. Prestia, and B. Karolewski. 2014. Using reduced personal protective equipment in an endemically infected mouse colony. *J Am Assoc Lab Anim Sci* 53 (3):273–277.

Brayton, C., M. Mähler, and W. Nicklas. 2004. Viral infections of laboratory mice. In *The Laboratory Mouse*, ed. H. J. Hedrich, 357–391. London: Academic Press.

Brielmeier, M., E. Mahabir, J. R. Needham, C. Lengger, P. Wilhelm, and J. Schmidt. 2006. Microbiological monitoring of laboratory mice and biocontainment in individually ventilated cages: A field study. *Lab Anim* 40 (3):247–260.

Burr, H. N., N. S. Lipman, J. R. White, J. Zheng, and F. R. Wolf. 2011. Strategies to prevent, treat, and provoke *Corynebacterium*-associated hyperkeratosis in athymic nude mice. *J Am Assoc Lab Anim Sci* 50 (3):378–388.

CDC (Centers for Disease Control and Prevention). 2009. *Biosafety in Microbiological and Biomedical Laboratories*. 5th ed. Washington, DC: National Institutes of Health.

Clifford, C. B., and K. R. Pritchett-Corning. 2012. Bacterial infections of laboratory mice. In *The Laboratory Mouse*, ed. H. J. Hedrich, 481–501. Boston: Academic Press.

Compton, S. R., F. R. Homberger, F. X. Paturzo, and J. M. Clark. 2004. Efficacy of three microbiological monitoring methods in a ventilated cage rack. *Comp Med* 54 (4):382–392.

DePietro, H., D. Medina, and G. Roble. 2014. Dried blood spot analysis. *ALN Magazine*, November 24.

Foreman, O., A. M. Kavirayani, S. M. Griffey, R. Reader, and L. D. Shultz. 2011. Opportunistic bacterial infections in breeding colonies of the NSG mouse strain. *Vet Pathol* 48 (2):495–499.

Henderson, K. S., C. L. Perkins, R. B. Havens, M.-J. E. Kelly, B. C. Francis, V. S. Dole, and W. R. Shek. 2013. Efficacy of direct detection of pathogens in naturally infected mice by using a high-density PCR array. *J Am Assoc Lab Anim Sci* 52 (6):763–772.

Hesse, I., A. Luz, B. Kohleisen, V. Erfle, and J. Schmidt. 1999. Prenatal transmission and pathogenicity of endogenous ecotropic murine leukemia virus Akv. *Lab Anim Sci* 49 (5):488–495.

Hickman-Davis, J. M., M. L. Nicolaus, J. M. Petty, D. M. Harrison, and V. K. Bergdall. 2012. Effectiveness of shoe covers for bioexclusion within an animal facility. *J Am Assoc Lab Anim Sci* 51 (2):181–188.

Jensen, E. S., K. P. Allen, K. S. Henderson, A. Szabo, and J. D. Thulin. 2013. PCR testing of a ventilated caging system to detect murine fur mites. *J Am Assoc Lab Anim Sci* 52 (1):28–33.

Kelley, K. 2010. Transport of mouse lines by shipment of live embryos. In *Guide to Techniques in Mouse Development, Part A: Mice, Embryos, and Cells*, ed. P. Wassarman and P. Soriano, 26–35. San Diego: Elsevier.

Kent, M. L., S. W. Feist, C. Harper, S. Hoogstraten-Miller, J. M. Law, J. M. Sánchez-Morgado, R. L. Tanguay, G. E. Sanders, J. M. Spitsbergen, and C. M. Whipps. 2009. Recommendations for control of pathogens and infectious diseases in fish research facilities. *Comp Biochem Physiol C Toxicol Pharmacol* 149 (2):240–248.

King, C. S., J. R. Hessler, and J. R. Broderson. 1999. Management of a biocontainment research facility. *Lab Anim* 28 (7):26–27, 30–32.

Kramer, J. A., E. W. Ford, and S. Capuano. 2012. Preventative medicine in nonhuman primates. In *Nonhuman Primates in Biomedical Research*, ed. C. R. Abee, K. Mansfield, S. Tardif, and T. Morris, 293–321. Boston: Academic Press.

Lipman, N. S., and F. R. Homberger. 2003. Rodent quality assurance testing: Use of sentinel animal systems. *Lab Anim* 32 (5):36–43.

Mahler Convenor, M., M. Berard, R. Feinstein, A. Gallagher, B. Illgen-Wilcke, K. Pritchett-Corning, and M. Raspa. 2014. FELASA recommendations for the health monitoring of mouse, rat, hamster, guinea pig and rabbit colonies in breeding and experimental units. *Lab Anim* 48 (3):178–192.

Morton, W. R., M. B. Agy, S. V. Capuano, and R. F. Grant. 2008. Specific pathogen-free macaques: Definition, history, and current production. *ILAR J* 49 (2):137–144.

Myles, M. H., E. K. Steffen, C. L. Besch-Williford, B. A. Bauer, R. S. Livingston, M. J. Crim, and L. K. Riley. 2014. Dried blood spot specimens for serological testing of rodents. Westbrook, ME: IDEXX BioResearch.

Nordmann, B. D. 2010. Issues in biosecurity and biosafety. *Int J Antimicrob Agents* 36 (Suppl. 1):S66–S69.

NRC (National Research Council). 2011. *Guide for the Care and Use of Laboratory Animals*. 8th ed. Washington, DC: National Academies Press.

OLAW (Office of Laboratory Animal Welfare). 2015. PHS policy on humane care and use of laboratory animals. Frequently asked questions D.14. Bethesda, MD: OLAW. http://grants.nih.gov/grants/olaw/faqs.htm#630 (accessed August 13, 2015).

Peterson, N. C. 2008. From bench to cageside: Risk assessment for rodent pathogen contamination of cells and biologics. *ILAR J* 49 (3):310–315.

Pritchett-Corning, K. R., and C. B. Clifford. 2012. Parasitic infections of laboratory mice. In *The Laboratory Mouse*, ed. H. J. Hedrich, 503–518. Boston: Academic Press.

Pritchett-Corning, K. R., J. Prins, R. Feinstein, J. Goodwin, W. Nicklas, and L. Riley. 2014. AALAS/FELASA Working Group on Health Monitoring of Rodents for Animal Transfer. *J Am Assoc Lab Anim Sci* 53 (6):633–640.

Rahija, R. J. 2007. Gnotobiotics. In *The Mouse in Biomedical Research*, ed. J. G. Fox, M. T. Davisson, F. W. Quimby, S. W. Barthold, C. E. Newcomer, and A. L. Smith, 217–233. Burlington, MA: Academic Press.

Rehbinder, C., S. Alenius, J. Bures, M. L. de las Heras, C. Greko, P. S. Kroon, and A. Gutzwiller. 2000. FELASA recommendations for the health monitoring of experimental units of calves, sheep and goats. Report of the Federation of European Laboratory Animal Science Associations (FELASA) Working Group on Animal Health. *Lab Anim* 34 (4):329–350.

Rehbinder, C., P. Baneux, D. Forbes, H. van Herck, W. Nicklas, Z. Rugaya, and G. Winkler. 1998. FELASA recommendations for the health monitoring of breeding colonies and experimental units of cats, dogs and pigs. Report of the Federation of European Laboratory Animal Science Associations (FELASA) Working Group on Animal Health. *Lab Anim* 32 (1):1–17.

Rehg, J. E., and L. A. Toth. 1998. Rodent quarantine programs: Purpose, principles, and practice. *Comp Med* 48 (5):438–447.

Roberts, J. A., and K. Andrews. 2008. Nonhuman primate quarantine: Its evolution and practice. *ILAR J* 49 (2):145–156.

Shek, W. R. 2008. Role of housing modalities on management and surveillance strategies for adventitious agents of rodents. *ILAR J* 49 (3):316–325.

Shek, W. R., A. L. Smith, and K. R. Pritchett-Corning. 2015. Microbiological quality control for laboratory rodents and lagomorphs. In *Laboratory Animal Medicine*, ed. J. G. Fox, L. C. Anderson, G. M. Otto, K. R. Pritchett-Corning, and M. T. Whary, 463–510. Boston: Academic Press.

Wachtman, L. M., and K. G. Mansfield. 2008. Opportunistic infections in immunologically compromised nonhuman primates. *ILAR J* 49 (2):191–208.

Weber, H., E. Berge, J. Finch, P. Heidt, F. J. Kaup, G. Perretta, B. Verschuere, and S. Wolfensohn. 1999. Health monitoring of nonhuman primate colonies. Recommendations of the Federation of European Laboratory Animal Science Associations (FELASA) Working Group on Nonhuman Primate Health accepted by the FELASA Board of Management. *Lab Anim* 33:S3–S18.

Whary, M. T., N. Baumgarth, J. G. Fox, and S. W. Barthold. 2015. Biology and diseases of mice. In *Laboratory Animal Medicine*, 43–149. Boston: Academic Press.

White, W. J., L. C. Anderson, J. Geistfeld, and D. G. Martin. 1998. Current strategies for controlling/eliminating opportunistic microorganisms. *ILAR J* 39 (4):291–305.

ZIRC (Zebrafish International Resource Center). 2009. Sentinel fish program. Eugene, OR: ZIRC. https://zebrafish.org/zirc/documents/protocols/pdf/health_monitoring/sentinel_fish_program.pdf (accessed March 22, 2016).

34

Surgery

Randall R. Clevenger, Jan Bernal, Michael Talcott, Teresa R. Gleason, Tracie Rindfield, and Robert F. Hoyt, Jr.

CONTENTS

Overview and Background

The role of a surgical facility manager (SFM) within a laboratory animal surgical facility, or surgery department, has changed significantly over the past 15–20 years. Historically, because of their advanced training and experience, veterinarians were usually involved in managing surgical facilities in a research environment. However, in recent years, because of increased veterinary animal care and use responsibilities, in some instances this has resulted in nonveterinarians assuming that role. In either case, a present-day SFM should have both knowledge and skills in a number of important areas to be successful. These include a comprehension of procedure scheduling, budget management and billing, equipment maintenance, procurement and inventory management, employee management, training of personnel and investigators, animal and data record maintenance, standard operating procedures (SOPs) development, and active involvement in facility maintenance and renovation projects. In addition, the SFM may even serve as a principal investigator (PI) or coinvestigator for studies. Another key to success for the SFM is to have excellent interpersonal skills. For either the veterinarian or nonveterinarian SFM, it may take a number of years through both experience and specialized training to acquire this knowledge and skills. In those facilities where the SFM is a nonveterinarian, there should be a close working relationship between the SFM and the facility veterinarian to ensure that all animal care and use activities are appropriate.

Surgical facilities or departments can vary greatly in the types of procedures performed, species of animals utilized, and support roles of the facility staff. At one end of the spectrum, the facility may be a support service for other divisions within an institution, providing only the facility, equipment, supplies, and anesthesia support and postoperative care for investigative staff that perform their own surgeries. At the other end, the facility may serve as a contract full-service provider, and handle every aspect related to a surgical study. Depending on where within this range a particular facility may fall or function, there can be a plethora of differences in required SFM expertise and equipment needed. Those facilities that perform full-service support for surgical procedures are similar in function to a human surgical and intensive care facility. In a facility where all support functions are within the same department, there is the opportunity for a close working relationship between the surgical facility staff, the investigators, and other specialized departments. Developing such relationships and then having staff members from each group involved in the presurgical planning phase can be a great benefit for the successful outcome of a surgical study (National Research Council 2011). However, when multiple other departments or facilities (magnetic resonance imaging [MRI], computed tomography [CT], etc.) or groups from outside the institution are involved in the surgical study, schedule coordination and animal transport can become more complexed for the SFM. In either case, the SFM should provide clear communication, coordination, and organization to all parties involved to help ensure successful completion of these projects. In addition, the SFM should be able to assess and allocate the resources between large numbers of requests for surgical services and organize and plan accordingly. Scheduling support staff, facility space, equipment, and time can often be quite challenging for the SFM to provide equitable resources for multiple concurrent studies. This can sometimes result in potential conflicts when there are several studies or investigators competing for these same resources or time slots. SFMs must also have good organizational, interpersonal, and communication skills, and an in-depth knowledge of the requirements for each study being supported; function as a business manager; and be a problem solver to overcome these conflicts. This chapter briefly covers key areas that may fall into the responsibilities of the SFM, and includes protocol and surgery planning, components of the surgical facility, instruments and equipment, applicable federal and state regulations, record keeping, budget management and billing, SOPs, surgical procedures and postoperative care, and training.

Protocol Planning

When any animal protocol involves a surgical procedure, it is a good idea to have the SFM involved early in the protocol development process and prior to submission to the Institutional Animal Care and Use Committee (IACUC) for review. During this preprotocol planning, the SFM can provide insight to the PI on a number of topics, including animal regulatory and welfare requirements with respect to surgical procedures, and the ethical and legal obligation for minimizing pain and distress in laboratory animals. Prior experience by the SFM can also assist the PI with refining surgical techniques and postoperative care in order to help minimize the pain and stress of the animals. This preplanning period will help identify any gaps in experience and expertise of the staff, allowing for scheduling and conducting training sessions for researchers or technical staff to address any shortcomings prior to the initiation of a new study. The topics of anesthesia and perioperative monitoring, specialized surgical techniques, and who will provide such training should specifically be addressed. This is especially true when the procedures are being performed on rodents, as investigators often now want to perform procedures on rats and mice that were previously only performed on pigs or other large animals. As such, specialized staff training in microsurgery and rodent support equipment may be needed (Hoyt et al. 2001a).

SFMs may also have developed SOPs for specific types of surgeries that can be quickly incorporated by the PI while writing the protocol. This is extremely valuable as investigators (PhD investigators, physicians, dentists, and even veterinary scientists) may lack extensive animal surgical experience or may not be familiar with species differentiations, such as anatomical differences, drug interactions or dosages, and appropriate materials and correspondingly sized equipment and supplies. Combining facility SOPs with the SFM's experiences and knowledge can greatly assist the PI with developing and writing the surgical portion of a protocol. At a minimum, the SFM should be familiarized with an upcoming surgical study, ensuring the standardization of drugs, dosages, and other species appropriate supplies are available within the facility.

When a new study involves a novel or complicated surgical technique, a pilot study may be useful to develop or perfect the procedure. In this situation, the SFM can provide valuable information to the PI based on his or her experiences with the species and techniques involved. The SFM can also provide guidance to the PI on anticipated pain and distress for a particular surgical procedure to ensure that appropriate anesthetics and analgesics are used and administered in the appropriate timeline based on the species, surgical procedures performed, and additional factors pertinent to the success of the surgical procedure or study objectives, such as age or health status (National Research Council 2009; NIH/OACU 2016a). Although a facility may have standard anesthetics and analgesics protocols that are routinely used, each study should be evaluated independently. Appropriate anesthetics and analgesics should be used and approved by the IACUC, including the type (opioid, local, nonsteroidal anti-inflammatory drugs [NSAIDs], etc.), dose requirement, administration schedule, and duration for appropriate analgesia. It should be noted that certain drugs may be identified during this preplanning phase that could be contraindicated for use on a study, as they may impact the data. As such, alternative drugs can be suggested before protocol submission and the study begins. Other, nonpharmaceutical techniques and procedures should be considered for additional relief of pain and distress, and should be included during the protocol planning phase, as modification of the postoperative care and pain assessments may be necessary. Softer or extra bedding material can be provided, and dim lights or cage coverings can provide a more soothing environment for surgical recovery. Palatable feeds or flavored liquids can help maintain hydration and nutritional support following surgery (Kohn et al. 2006). Depending on the impact that the surgical procedure may have on an animal's ability to access food or water, an IACUC-approved exception may need to be made for placement of food or water in a container on the cage floor, especially for rodents.

Another important component of the preplanning period is the SFM's role in assisting investigators with the study cost estimates, having prior knowledge of animal purchase prices and animal housing per diem charges for the intended species. He or she should also be able to provide a good estimate of the surgery and postoperative costs, the necessary support equipment and surgical supplies, and required support staffing. This can assist the PI with anticipating the overall cost of a study even if the number of animals or the species changes. An experienced facility manager, in concert with the facility veterinarian, can provide a variety of possible options to the PI during the protocol planning phase.

Surgery Planning

Following IACUC approval of an animal study protocol, detailed planning for the commencement of surgeries should proceed. It is during this time that (1) animals are identified and/or procured, undergo any quarantine requirements, and are acclimated; (2) supplies and equipment are identified and/or acquired; and (3) additional staff and investigator training is conducted in preparation for a new study. It is highly recommended for the SFM to initiate scheduling a presurgical planning meeting with all parties involved with a study (National Research Council 2011). This provides the opportunity to determine the roles of all individuals involved; convey scheduling procedures, responsibilities, and requirements; develop a plan for any required training; identify and resolve equipment requirements; and answer or address any questions, issues, or concerns.

Personnel Roles and Responsibilities

Prior to initiating surgical procedures on a new study, the responsibilities of each participant should be clearly established. These include what tasks the PI and his or her staff will handle, the responsibilities of the surgical technical staff, and the responsibilities and obligations of the SFM. These assignments may vary greatly between studies, PI groups, and even facilities, but each group or participant understanding their responsibilities, as well as being mindful of the other's responsibilities, is paramount to the success of the surgery, animal welfare, and study objectives. Knowing the roles and responsibilities in advance minimizes near misses and confusion during the start of the surgical procedures. In addition, scheduling a practice or walk-through session with all participants, prior to the first scheduled procedure, can help identify any unforeseen issues and solidify each member's role.

Surgical Supplies and Equipment

A detailed list of supplies and equipment to support each surgical experiment should be generated prior to initiating the first procedure. Such a list will identify if additional equipment is required and/or needs to be purchased. It is quite useful for the SFM to have a thorough understanding of the procurement system within the institution, including timelines and procedures required to order and receive supplies to accurately plan and schedule procedures. Should equipment need to be purchased, especially expensive capital equipment, it is important to understand that the acquisition or procurement process may have a significant impact on starting a study due to the additional time or requirements to acquire. An SFM might have an option to borrow equipment for a particular study based on personal contacts and the size and/or affiliations of the facility or institution. Having outside or network-acquired contacts can also often prove to be a valuable resource for managers. As such, it is important to explore all options. If the supply list includes items that are not normally stocked within the facility and must be specially ordered, additional lead time required to receive them must also be appreciated. It is also important to note that occasionally, specific required items may not be commercially available and need to be either custom-made or made in-house. This is especially true when dealing with small catheters to be used in rodents. The ability to evaluate their staff's knowledge and skills or utilize contacts to be able to develop or acquire these items should be in the repertoire of SFMs. It is important for a manager to learn about the surgical industry, and to have a working knowledge of suppliers for the majority of supplies required, including nonstandard items.

Scheduling of Procedures

The facility should have guidelines in place for the process of scheduling surgical procedures, and provide these requirements to each investigator during the planning and development of their study. Except for emergency procedures, surgeries should be scheduled in advance to allow for scheduling of staff, equipment, and facilities, and provide for optimum animal care and welfare. Last-minute scheduling can put undue stress and anxiety on the staff and increases the risk of inadequate and incomplete preparation. Consideration in the surgical scheduling process should be given to providing appropriate time between surgical procedures to allow for restocking of supplies, postsurgical sanitation procedures, and staff to reenergize. If multiple procedures are scheduled per day, additional consideration and planning may be necessary to have the additional required staff, surgical supplies, medications, instruments, and equipment available, highlighting the critical nature of advanced scheduling. Setting a schedule for surgical procedures must also account for any preparatory time required for presurgical activities. These routinely include medication pretreatments, preoperative animal fasting (if required), scheduling of animal transportation, specialized study-related diets, anesthetic and surgical records generation, age-based study timelines, and previous related surgical procedures. In order to provide adequate time for appropriate surgical preparation, scheduling several weeks to months in advance is usually required, depending on the facilities demand. A single point of contact actually scheduling the procedures is important for avoiding schedule conflicts, but a universal or facility-wide visible surgical calendar (web-based online scheduling system or calendar or magnetic or dry-erase board) is a valuable asset to aid in scheduling, especially when supporting multiple simultaneous studies, research units, or PIs.

Animal Transportation

Another factor to consider during the surgical planning is the housing and transportation of the animals prior to and following surgery. Ideally, the animals should be housed in close proximity to the surgical area, but this arrangement may not be available in many institutions. Extended transport time or distance can add undue stress to the animals prior to surgery. Depending on where the animals are housed, this may require additional personnel not routinely responsible for surgery-related procedures to be involved. If the animals are housed in the same facility where the surgical area is located, and the same technical staff members are performing surgical support and technical services, communication about the surgical procedures and postoperative care is simplified. However, if the animals are housed in a separate or remote facility, clear and concise communication with the appropriate personnel is critical. The housing facility and staff must be aware of the surgical schedule and related tasks, such as fasting procedures, presurgical treatment, arranged transportation pre- and postoperatively, postoperative care, and analgesic administration. It may also be necessary to arrange transportation with an additional group, and ensure that the necessary requirements are met while transporting the animals, such as appropriate caging, environmentally controlled vehicles, and separation of species. For surgical procedures involving nonhuman primates, the issues can be more complicated, as anesthesia or sedation during transportation is often necessary. During these procedures, anesthetic times can be prolonged, leading to additional potential complications. Animal transport should be arranged to minimize holding times and allow for adequate preparation by the technical staff prior to surgery. Return transportation postoperatively requires detailed communication between the surgical facility staff and the transportation and housing facility staff. This may include information about the surgical procedure, potential complications, postoperative care required, and the continuation of medications or treatments, depending on the organization. The surgical facility should have a thorough understanding of these processes and ensure that this information is being communicated.

Personnel Training

With the start of any new study or the addition of new staff, there will be a requirement for training. The training requirements will vary depending on staff experience and the procedures involved. Having qualified, experienced staff to support the surgical procedures is one of the most critical aspects of the planning process. Some institutions have dedicated staff available for the various portions of the surgical procedures, including preoperative preparation, anesthesia monitoring and support, surgical assistance, surgeons, and postoperative care. Other facilities have staff members that are trained in all these aspects but are not dedicated to these roles. They perform the surgery-related duties as part of their job descriptions. The skill set required for each of the roles directs the type of training needed to be addressed in the training program. Whether training all staff in every aspect of the surgical procedures or training individual groups on separate portions, the type of training and the amount of time required must be incorporated into the planning phase.

Surgical Facility

Requirements for the physical plant of a surgical facility are outlined in the *Guide for the Care and Use of Laboratory Animals*, and vary depending on the species used. Although an SFM may not have the opportunity to be involved in designing a facility, he or she should be aware of the requirements to effectively manage the facility. It is a good idea for the SFM to be involved in the planning and oversight of a renovation or modification of the facility as part of the team including veterinarians, technicians, researchers, engineering and architectural professionals, and facility operations personnel. Having such input from the staff who will be working in it often provides a good functional perspective. The principal components of a well-designed surgical facility are similar to those for the design of human operating rooms. "For most surgical programs, functional components of aseptic surgery include surgical support, animal preparation, surgeon's scrub, operating room, and postoperative recovery" (National Research Council 2011). All survival surgeries must be conducted under aseptic conditions in dedicated surgical suites or approved work areas. For large animal survival surgery, a dedicated surgical suite is required and separation is best achieved by physical barriers or separate rooms. Nonsurvival, minor, or emergency surgery, and rodent or nonmammal surgery can be accomplished by distance separation of the surgical space from other activities, and performed in a work area that is dedicated only to surgery during that period (Bernal et al. 2009).

When designing a new surgical suite or renovating an existing one, special attention should be given to unique requirements of the operating rooms. These include appropriate room and surgical lighting; heating, ventilation, and air-conditioning (HVAC) requirements; adequate electrical outlets and space for surgical and support equipment; medical gases, such as oxygen, medical air, or other gases to support anesthesia, surgical procedures, and equipment; anesthetic gas scavenging; and vacuum. Surgical lights should be such that they are ceiling mounted, have movable arms, generate low heat to the surgical field, and have light-beam-focusing capabilities. A backup electrical system (generator) or an alternate backup plan should be developed in the event of an electrical outage during surgical procedures in order to maintain the surgical patient.

The facility should be designed in a manner that will allow for easy disinfection and prevention of bacterial contamination. The use of mobile equipment facilitates decontamination of the operating area by allowing the room to be emptied prior to cleaning. Control of contamination and ease of cleaning should be key considerations in the design of a surgical facility (National Research Council 2011). Room lights should be recessed, flush mounted, and sealed to protect against moisture and vermin. Ceilings, walls, and other interior surfaces must be durable, smooth, impervious to moisture, and easy to clean. Floors should be made of nonslip, nonconductive materials, containing no joints or seams. Dedicated sterile surgery suite maintenance should include regular room cleaning and disinfection based on the activity and use of the room.

Dedicated operating rooms should be maintained under high-efficiency particulate air (HEPA) filtered positive-air-pressure ventilation, doors should remain closed to the greatest extent possible, and personnel movement into and out of the room should be minimized during a surgical procedure (Schonholtz 1976; Fitzgerald 1979; National Research Council 2011). The operating rooms should be positive to the other areas, including animal prep and surgeon scrub, and should have 15–20 air changes per hour (ASHE 2017). These approaches, along with the integration of HEPA-filtered supply air, can aid in the reduction of infections, by reducing the number of microbes being introduced into the rooms. The facility should implement a method to ensure that these areas are maintained at positive pressure by either a monitoring system or periodic smoke testing. In some instances, such as surgical procedures that involve the use of biohazardous or infectious agents, having a controlled system that allows conversion from positive to negative operating room pressures may be advantageous.

Animal Preparation Area

The animal preparation area should be in close proximity, but separate from the operating area. Requirements for the prep area will vary, depending on the species being used and the type of procedures. Large animal prep must be an area outside of the operating room, but rodents and nonmammals can be prepped in the same room where the surgery will occur, as long as the prep area is separate and at an appropriate distance from the surgical space to prevent contamination from hair and dander (NIH/OACU 2016b). The area should provide adequate room to work around the animal, and accommodate the equipment and supplies necessary for the species and procedures. Sufficient countertop, cabinet, and drawer storage space is needed to keep prep supplies readily available. A variety of equipment will be necessary and may include a preparation table or tub with a surface appropriate for the size of the animals; clippers for hair removal; a laryngoscope for intubation; supplemental heating, such as a circulating water blanket, forced-air warmer, or infrared warmer; patient monitoring equipment; a gas anesthesia machine; additional lighting; and ample electrical outlets to power all equipment.

Surgeon Preparation Area

Operating room personnel should change into scrub suits or similar surgical clothing in a locker room or similar area. Surgical caps, masks, shoe covers, and other protective apparel, as required by the facility, should be available and donned prior to entering the operating room for a surgery. In addition, a separate area is needed for the surgeons to scrub hands and arms immediately prior to gowning and gloving to perform surgery. The location of the scrub area should be separate from the surgical room and animal prep. Surgeons should scrub at a sink that does not require contact with their hands to start or stop the water flow, and should use accepted hand and arm surgical scrub practices. A newer alternative is now available and Food and Drug Administration (FDA) approved for use in human operating rooms. Waterless chlorhexidine- and alcohol-based hand antiseptics can be used instead of water, but should still be used outside of the operating room, in the designated surgeon prep area.

Surgical Support Area

The surgical support area is designed for cleaning and sterilization of instruments and for storage of instruments and surgical supplies. Sufficient countertop space should be available in a location conducive to assembling and wrapping instrument packs. Depending on the scope of the surgical facility, instruments may be cleaned manually using a sink, or for larger quantities, an ultrasonic instrument cleaner may be desired. Regardless of the cleaning method, instruments should be free of blood, tissue and other contaminates, and dried appropriately. Instruments should be soaked in instrument milk to lubricate and provide corrosion control. Sterilization can be accomplished with small tabletop autoclaves or cold chemical sterilizers for small quantities of instruments, or may require full-sized autoclaves when large instrument packs need to be processed. Ethylene oxide (EtO) and vaporized hydrogen peroxide (VHP) sterilizers are also effective options for surgical instruments or other items that cannot withstand the high temperatures of steam. EtO is highly flammable, and both EtO and VHP pose health risks, requiring extra safety precautions and specialized installations when using these sterilizers. Institutional safety personnel should be consulted if planning to utilize this type of sterilization. Storage space in this support area is crucial due to the intensity of the instrument and supply processing activities. Storage is necessary for sterile surgical packs and surgery supplies, extra nonsterile instruments, supplies for processing instruments and packs, and immediate use supplies (drugs, fluids, suture materials, sterile packs, drapes, gowns, gloves, etc.). Although not required, large facilities should consider a bulk storage room that may be separate from the instrument processing area, to allow the purchase of bulk supplies. Small quantities need to be stored for immediate use in the instrument area or cabinets in other rooms where they are used. A bulk storage room can also be used to store extra equipment (anesthetic machines, ventilators, patient monitors, mobile surgical lights, electrocautery units, etc.) that are not needed in the operating room for certain procedures (National Research Council 2011).

Postoperative Recovery Area

The postoperative recovery area is a critical part of the facility design necessary for the success of complex surgical studies. This area "should provide the physical environment to support the needs of the animal during the period of anesthetic and immediate postsurgical recovery and should be sited to allow adequate observation of the animal during this period" (National Research Council 2011). The equipment and space requirements will vary depending on the species and surgical procedures. To aid in successful recovery, necessary equipment and supplies should be available for care of the patient and the area should be designed with the ability to observe, evaluate, and monitor the patient. In some cases, intensive care units and additional equipment may be required, such as warming units (forced-air warmer, heat lamp, circulating warm water blanket, etc.), oxygen source, and monitoring equipment. For some procedures, 24-hour monitoring and care may be necessary. In these cases, it is critical that this possibility is addressed prior to the initiation of the surgery to ensure that the appropriate staffing and equipment is available. This could include the ability to monitor physiologic parameters and provide supplemental oxygen or supplemental heating or cooling, to maintain catheters or constant infusions, and to administer medications. Infection control is another important aspect of postoperative care, which includes the ability to keep the area, caging, and equipment clean and decontaminated at regular intervals, and especially between patients.

Adjunct Surgical Support Areas

Another functional area that can be useful for facilities that perform complicated surgical procedures is a diagnostic laboratory area to perform blood work and other testing procedures. Equipment needed is dependent on the types of testing required, but may include arterial blood gas, complete blood count, clinical serology or chemistry, packed cell volume or hematocrit, and urinalysis.

The SFM must have a thorough understanding of the requirements for each of the areas of a surgical facility, the function of each, how they relate to other areas, and what equipment may be necessary in each area. This understanding affords a manager the ability to troubleshoot facility and equipment issues, and to provide solutions or potential repairs.

Surgical Instruments and Equipment

Surgical instruments, supplies, and equipment utilized in a research setting vary based on the nature of the procedure, the species involved, and the area of the facility. Large animal procedures can generally utilize human surgical equipment, but rodents and other small animals often require specialized equipment and devices, including magnification aids (Hoyt et al. 2001a and b).

The majority of equipment can be adopted for surgery and patient monitoring of animals with equipment designed for human use. The exception relates to the use of human patient electrocardiograph (ECG) and/or oxygen saturation level (SPO_2) monitoring of mice with heart rates oftentimes >400/min, where the sampling rate of the machines is not fast enough. This equipment can be categorized into one of the following functions of a surgical program:

- Anesthesia: Administration, ventilation, and waste gas scavenging
- Patient monitoring and supportive care
- Surgical procedure: Instruments and associated tools and equipment
- Specialized or support imaging: Microsurgical, laparoscopic, and neurological
- Cleaning and sterilization

Equipment utilized in the aseptic surgical area should be dedicated to surgery and easy to clean and disinfect. Ideally, equipment in this area should be portable rather than fixed (National Research Council 2011) to facilitate cleaning and disinfection of the room on a regular schedule, defined by facility SOPs. The surgical area should be kept clear of unnecessary equipment and supplies during a procedure to allow maximum space for movement of personnel in the operating room. This practice aids in preventing contamination of the surgical site and sterile components. Surgical rooms and designated rodent surgical areas should not be used for storage of nonsurgical equipment and supplies when not in use, to minimize the introduction of contaminants. Supplies should be kept in cabinets or drawers to prevent dust and debris accumulation, and should be stored off the floor.

Anesthetic Equipment

Anesthetic equipment is necessary for the administration of agents to prevent and alleviate pain associated with a surgical procedure. The proper use of anesthetics and analgesics in surgical research programs is mandated by regulations and guidelines (Animal Welfare Act 1966; Conour et al. 2008; National Research Council 2009, 2011; NIH/OACU 2016a; PHS 2016), and is an ethical and scientific obligation. Generally, the basic configuration used for human anesthesia (i.e., oxygen source with flow meters, and inhalant gas vaporizer) can be used for most animal surgeries, including rodents. As such, oftentimes one only has to change the ventilator appropriate in size for the species being operated on should mechanical ventilation be desired. For short-duration or minor surgical procedures, injectable anesthetics may meet the requirements and necessitate minimal equipment. Longer procedures using injectable anesthetics usually require repeated injections or a constant rate infusion (CRI), which can be accomplished with precision infusion or syringe pumps. Anesthetics that work well in one species may not work as well in another, and will affect equipment needs. The use of preoperative drugs or preanesthetics can also affect the selection and dosage of general anesthetic (Kohn et al. 2006). Each of these topics needs to be addressed in consultation with the facility veterinarian, in order to determine the necessary anesthetic equipment.

Anesthetics, by their nature, cause dose-dependent physiologic changes (Kuhlman et al. 2008). These effects can be reduced by the selection of specific anesthetics and precise dosing. The use of volatile anesthetic gases can aid in the minimization of these physiological changes because they can be accurately controlled, and dosages can be adjusted quickly through the use of precision vaporizers. Each type of anesthetic gas requires a vaporizer calibrated to the specific gas being utilized. Their use increases safety and minimizes associated negative side effects by being quickly eliminated from the patient and requiring minimal metabolism. Volatile anesthetic agents can be hazardous to employees and require special safety measures. Vaporizers are usually set up with an anesthesia machine that provides pressurized oxygen or medical air that will mix with the anesthetic at specific ratios, and equipped with an anesthetic gas scavenging system. The most effective scavenging is done with a closed active system that uses vacuum or suction to pull expired gases from the system and evacuates these gases out of the facility. An alternate passive system allows waste anesthetic gases (WAGs) to pass through an activated charcoal filter canister, which traps the anesthetic and allows the remaining air to pass through. These canisters must be monitored carefully and must be weighed frequently to ensure that the maximum absorptive capacity is not exceeded and the excess gases escape into the room. It is important that the SFM ensures that all staff members are familiar with the use and safety requirements for the operation of gas anesthetic machines to protect personnel. Volatile anesthetics are especially hazardous to a developing fetus, so additional measures must be taken to minimize or eliminate exposure to personnel during pregnancy (Tannenbaum and Goldberg 1985). It is highly recommended that a WAG SOP be in place, to include that gas anesthesia machines be surveyed for leakage of volatile anesthetics on a regular basis, as may be dictated by the institution's health and safety department.

Ventilators are necessary for most open-chest surgeries, and may be recommended for many other procedures. They allow set volumes of air or oxygen and anesthetic gases to be delivered to the patient at the appropriate frequency and volume, thereby resulting in a more stable breathing pattern while under anesthesia. Ventilating animals at the appropriate tidal volume and frequency provides stable respirations and helps maintain expired CO_2 concentrations at normal physiologic levels. They may be incorporated as part of an anesthesia machine, or be a stand-alone unit that allows air or oxygen and anesthetics to be delivered through it with a flowmeter and vaporizer. The ventilator must be able to deliver precise volumes, dependent on the size of the animal. Rodent ventilators are typically the stand-alone type, whereas large animal ventilators are typically part of an integrated anesthesia machine.

Monitoring Equipment

Intraoperative patient monitoring and supportive care are essential to facilitating a successful surgical outcome. The intensity of monitoring will vary with species and procedure performed, and attention should be given to thermoregulation, cardiovascular and respiratory function, electrolyte and fluid balance, and management of pain (Kuhlman et al. 2008). Ideally, patient monitoring equipment should have the ability to monitor several of the important physiologic parameters (i.e., body temperature, cardiac and respiratory rates, cardiac elctrical activity by electrocardiogram, oxygenation, end tidal CO_2 by capnography, and blood pressure), which allows evaluation of anesthetic depth and physiologic status and stability. Most, if not all, of these parameters can be measured on larger animals using equipment made for human patients. Electrocardiography on mice using human equipment can be difficult or impossible, due to their high heart rates. Fluid replacement may also be necessary as intraoperative support therapy, depending on the duration and type of procedure and the species undergoing the surgical procedure. Postoperative pain reduction should be addressed with the selection of anesthetics and analgesics, as well as considering the use of preoperative analgesics to help minimize it. To decrease pain levels to the greatest extent possible, this should be an ongoing evaluation and adjustment process, with modifications made based on pain assessments of each previous patient in consultation with the veterinarian.

Surgical and Specialized Equipment

Additional operating room equipment, surgical instruments, and related surgical tools are necessary to perform surgical procedures. Surgical instruments are designed to perform a specific function (i.e., cutting, incising, or retracting), and will vary based on the type of surgery, species used, and research performed. They are classified by function, and names are often derived from inventors, physicians, or appearance. Most instruments are made from high-grade stainless steel, but they can also be made of titanium, especially microsurgical instruments or those used for surgical procedures conducted in an MRI facility. Surgery-related tools (drills, retractors, micromanipulators, stereotaxic equipment, etc.) should be evaluated as being appropriate for the intended species, surgical procedure, and designated research and for their ability to be appropriately sterilized for aseptic surgery.

Other surgical support equipment may include electrocautery units, infusion pumps, patient warming equipment, microsurgical and imaging equipment, and patient monitoring equipment. Maintaining normal body temperature is critical to minimize cardiovascular and respiratory disturbances caused by anesthetic and analgesic medications (Smith and Danneman 2008). This is especially important in small animals, where the high ratio of body surface to weight contributes to heat loss. Strategies for maintaining normal animal patient body temperature intraoperatively may include the use of intravenous fluid warmers, circulating warm water blankets, forced-air warming blankets, infrared warmers, electric heated pads, or even warmed fluid bags placed next to the animal. Extra care must be taken if using electrically heated pads, as they can develop hot spots and potentially cause burns. Due to this hazard, their use should be discouraged. Imaging equipment, such as radiography, fluoroscopy, MRI, x-ray CT, positron emission tomography (PET), ultrasound, or laparoscopic and microsurgical cameras, is also commonly used intraoperatively or perioperatively to support a surgical study. The larger equipment (MRI, CT, and PET) cannot be moved to the animal, so special care is needed when moving a surgical patient to the imaging areas. Other imaging equipment should be dedicated to research or animal surgery if possible. Microsurgery equipment, especially the operating microscope, should be adjusted to the surgeon and surgical procedure prior to the surgeon scrubbing, gowning, and gloving, to eliminate potential contamination from contact with a nonsterile device. Other alternatives are to use specialized sterile plastic covers for the microscope, foot controls, or sterilized hand controls, which allow the surgeon to make adjustments to the microscope during the surgery (Hoyt et al. 2001a).

The animal prep area requires some specialized equipment, including some of the same equipment used in the operating room. An anesthesia machine and waste gas scavenging equipment with or without a ventilator may be necessary, depending on anesthetics utilized and the length of time of the prep. For large animals with multiple surgical sites to prepare, and an abundance of hair, the amount of prep time could be prolonged. A method of maintaining the animal's body temperature should therefore be incorporated. If the preparation surface is stainless steel or another material that can conduct heat away from the animal, it is important to provide an insulating material between the animal and the surface. Patient monitoring equipment is also critical during anesthetic induction, as the potential of adverse effects can be elevated during this period. Monitoring parameters may vary but should typically include at least body temperature, respiratory rate, heart rate, and oxygenation saturation. A laryngoscope with various-sized blades should be available for endotracheal intubation. For intubation of rodents, specialized scope and illumination apparatus may be required. A Doppler or ultrasound machine may also be a useful aid for the percutaneous placement of catheters, especially if arterial placement is necessary in a large animal.

Cleaning and Sterilization Equipment

Instrument cleaning and sterilization is an essential component of aseptic surgery, and additional equipment and supplies are required to ensure this is done properly. If processing a large number of instruments, an ultrasonic instrument cleaner and dryer may be necessary. This will help avoid staff iatrogenic injuries, which can occur when surgical instruments are cleaned by hand, and helps ensure adequate removal of all organic material prior to sterilization. Instrument sterilization is usually accomplished using one of two common methods (physical or chemical). The physical method employs dry heat or a combination of heat, humidity, and pressure to sterilize. The chemical method employs the use of a gas or liquid sterilant. Factors contributing to the type of sterilization utilized are based on the physical characteristics of the article to be sterilized, the time required to sterilize, and the availability of the equipment (Schofield 1994; Callahan et al. 1995). All supplies and equipment should be sterilized well in advance of a scheduled procedure to prevent potential cancellation due to sterilization cycle failure.

Steam sterilizers, commonly known as autoclaves, use a physical method of sterilization utilizing heat, humidity, and pressure. High temperatures, ranging from 121°C to 134°C (250°F–270°F), at a pressure of 15–30 psi are required for sterilization to occur. Most steam sterilizers are automated with a variety of cycles available based on physical characteristics of the item, the type of wrap in which the item is contained, and the time required for completion of the cycle. The advantages of steam sterilizers are the ability to sterilize a variety of items in a readily controlled setting with multiple cycle lengths available, and sterilization occurs in a much shorter time than chemical methods. Disadvantages include the inability to properly sterilize rubber and other heat-sensitive materials, oils, powders, plastics, and electrical and/or mechanical devices. There is also limited penetration of some materials, such as fabrics, paper, and other solid materials, if thickly layered (Rutala and Weber 2008). It is important to verify the autoclave sterilization process with biological and/or temperature indicators.

Glass bead or dry heat sterilizers are another commonly used physical method of sterilization, primarily used in rodent surgery. This sterilization is an effective and convenient means of rapidly sterilizing the working surfaces of surgical instruments using a "sterile tip technique" between individual animals in sequential multiple rodent surgical procedures. When using instruments in this way, they should be sterilized via a steam or chemical sterilizer prior to the first procedure each day, and then the tips resterilized between each animal using this method. Using a single set of instruments for multiple surgeries and a glass bead sterilizer should have well-defined guidelines ensuring that the sterile technique is maintained between animals. Surgeons must confirm that the working surfaces of the instruments have cooled sufficiently before touching the animal to avoid accidently burning the tissue.

Chemical methods of sterilization include both gas and liquid agents. EtO sterilizers are noncorrosive and safe for most plastic and polyethylene materials and mechanical and electronic equipment. EtO can also permeate the lumens of catheters, flexible fiber-optic scopes, and similar devices. The disadvantages of EtO gas sterilizers are the long sterilization cycle and aeration times, and the inability to sterilize liquids and impervious packing materials. Items to be sterilized must be dry, because water will react with the EtO to form ethylene glycol. EtO sterilization is also more expensive than many other methods, and presents occupational health risks. EtO gas sterilizers have specific and strict operation and installation requirements due to the health (potential carcinogen and mutagen) and occupational (explosion) hazards. Operation of EtO sterilizers should be done in strict conformance with local and state regulations, institutional policies, and manufacturer's recommendations (Rutala and Weber 2008). VHP is another chemical-based sterilization process, with the primary advantage of having a shorter cycle than EtO gas and the ability to sterilize heat-sensitive items. Limitations of hydrogen peroxide sterilizers include incompatibility with some materials, limited penetrating ability in items with long narrow lumens, and occupational health risks (Rutala and Weber 2008). Liquids and powders cannot be processed with VHP, nylon becomes brittle, and materials must be completely dry for effective sterilization, so those that absorb moisture (i.e., paper, cotton, and cellulose) are unable to be adequately sterilized. Due to the short cycle of VHP sterilizers, the time for the vapor to completely fill small lumens is longer than the completion of the cycle. Hydrogen peroxide is used for VHP sterilization at very high concentrations and is a strong oxidant that can cause severe tissue burns and respiratory irritation, and can be toxic at these levels. Liquid chemical sterilization is a method used only for devices that can be fully immersed. Peracetic acid and glutaraldehyde can be used for specialized equipment, but they must be thoroughly rinsed and cannot be stored after sterilization, giving them limited use for general surgical sterilization.

Records and Regulations

As an SFM, record keeping and compliance with animal care and use regulations can be one of the most important and challenging aspects of the job. An abundance of federal and state laws, regulations, and guidelines dictate the types of records a surgical facility is responsible for maintaining, but the responsibilities often extend beyond the requirements. As active participants of a surgical research study, the surgical facility staff may be charged with maintaining documentation pertaining to the actual data collection and analysis. Thorough documentation of the procedure and associated events by the facility staff during the surgical study and postoperative period can be an asset to the PI, as documentation and data collected involving the surgical procedure are required for preparation of publications or addressing reviews. Records may be handwritten or entered in a digital, electronic, or computerized format and will vary depending on the type of record, study supported, or capabilities available. Complete copies of all animal records and data collected throughout the study must be maintained by the surgical facility, and then can be made available for the PI if requested. Standardized forms are valuable for the recording of pertinent information, with fields that can be filled in or items checked, to ensure that relevant information gets documented. It is imperative that the PI and staff communicate the specific types of documentation required for a procedure. It is incumbent upon the SFM to ensure that records meet the regulations and needs of the research project and PI. Accurate records for controlled substance, anesthetic, and analgesic administration are required.

Budget and Billing

It often falls on the SFM to develop and administer the facility's budget, provide cost estimates for surgical studies, establish charges, and execute billing processes. Providing accurate cost estimates requires the surgical facility staff to maintain accurate records and track costs for all supplies and equipment and the usage requirements for various types of procedures. Maintaining an adequate inventory of supplies and rotation of stock to minimize waste can impact facility costs, and should be addressed. A presurgical planning meeting (discussed above) can be an important tool in providing these estimates, by generating a detailed supply list, with input from all participants. In addition to the direct supply cost estimate, the SFM also needs to know the fixed costs of the facility and additional overhead costs. The fixed costs may be included as part of a fee paid to the institution for use of the space, or by direct payments, and may include rent, utilities, maintenance, salaries, administrative, and other space-related costs (National Center for Research Resources 2000). Overhead costs can include equipment shared across multiple studies, supply stocks (including a portion that may expire prior to use), cleaning and sterilization costs, and personnel costs. A portion of these costs should be calculated and included in the estimate for each study, along with the costs for materials, supplies, and other fees. Each facility may have different expenses that need to be recovered, and some may even be required to generate a profit.

Budget management for a surgical facility can be dependent on a variety of factors and is related to the role of the facility (support service or contract facility). A facility that provides support services for investigators within an institution may have investigators providing some supplies and equipment, but a full-contract facility may provide all required supplies and equipment. This will result in drastically different budgets and orders for supplies. The SFM should be aware of all expenditures required by the facility, the true costs associated with the surgical studies, and the income from billing for the procedures. This understanding can aid the SFM in making knowledgeable projections for future budget requirements.

Establishing charges for surgical procedures can start with the estimate and adjust for actual supplies and staff allocations based on the requirements upon completion of the first procedure. This allows for adjustment in the expense according to the actual time, supplies, and staff required. There are a variety of methods to calculate procedure charges, each with advantages and disadvantages. The most accurate way to calculate actual costs is to track every supply item utilized, and technician hours involved for each surgical procedure. This method is time-consuming and may require extra staff or time to adequately track. The least involved method is to estimate the amount of supplies and allocated staff time and approximate any additional supplies and labor and make adjustments accordingly. One method that can provide a fairly accurate procedural cost without the effort of tracking every supply item for every procedure is to track and average the cost involved in the first few procedures and use those calculations as an average across the entire study. This method allows for the opportunity to adjust if significant changes to the study occur. The SFM and technical staff must have a comprehensive understanding of what the supply costs are to evaluate the significance of the change and the cost impact to the facility. An example would be a change from an inexpensive suture to an expensive one. If these small changes become routine, this can have a tremendous impact on the facility budget.

The SFM or administrative staff should have a method in place to track surgical procedures and associated costs to enable accurate billing. The accuracy of the billing process can be achieved by involving more than one staff member and having concurrent methods of tracking procedures. For example, the surgical schedule could be utilized as one source, and maintaining a simple database as the second method, allowing comparison of the two for verification. The minimal extra effort involved in having a duplicate tracking system can easily offset efforts needed to resolve questions about inaccurate billing. A regular schedule (weekly, monthly, etc.) should be established for the billing cycle to assist in the management of the information and consistency of the billing cycle, and maintain accuracy. Depending on the size of the facility and the workload, a commercial billing software system may be necessary and may require a dedicated staff member to handle the budget and billing.

Surgical Procedures and Postoperative Care

Successful surgical procedures require appropriate presurgical planning, which has been previously addressed. Inferior planning or inadequate surgical and monitoring techniques can lead to complications during a surgical procedure, which may affect the surgical outcome and data produced, and lead to potential animal welfare concerns. Inadequate planning can affect and occur during both survival and nonsurvival surgical procedures. Surgical results are often influenced by actions taken prior to and during the procedure, such as the use of preoperative antibiotics. Prophylactic antibiotics can aid in the reduction of postoperative infections, and are commonly given for major survival surgeries, especially when body cavities are penetrated. The choice of antibiotic, along with the duration and route of administration, should be established in consultation with the researcher and the veterinarian.

The postoperative recovery time is critical, and proper assessment during this period can have a positive impact on the overall outcome of the procedure. Pain can impact the animal's welfare both physically and mentally, and proper relief of pain and distress is crucial for a positive surgical outcome. Pain is exhibited differently depending on the species, and experience and training in identification of species-specific painful behavior is necessary to ensure that pain is being relieved. Some observations, such as decreased activity, poor appetite, labored breathing, and abnormal appearances, are common signs of pain across all species, and these should be assessed during the postop period (Kohn et al. 2006; NIH/OACU 2016a). The administration of preemptive multimodal analgesics can not only aid in the reduction of postoperative pain (Corletto 2007) but also reduce the amount of anesthesia needed, shorten the overall anesthesia time, and hasten the duration of recovery. The selection of analgesics and sedatives, along with the postoperative duration of administration, should be established in consultation with the researcher and attending veterinarian (ACLAM 2016). Pain assessment for the continuation of analgesics should extend throughout the postoperative period, as pain may vary between individual patients. Variables such as the type of procedure, length of the procedure, skill level of the surgeon, species, and even an individual animal's stoicism can impact the requirement for analgesics and sedatives to ensure a comfortable surgical recovery. Complications or contraindications can occur with certain analgesics, so veterinary consultation is important prior to administration. Some complications that can occur include opioid-induced respiratory depression, nausea or ileus, NSAID toxicities in certain species (ibuprofen in dogs and cats), and local anesthetics, such as lidocaine and bupivacaine, inducing toxic effects in smaller species (Kohn et al. 2006). Studies assessing an inflammation response can be adversely impacted by the use of anti-inflammatories, and these should be avoided for those types of studies. These related topics should be part of the prestudy planning discussions so as to avoid such problems when the study is initiated. Regardless of these variables and complications, when surgical procedures are performed, all animals should be assessed for pain and analgesics should be administered as necessary.

Postoperative monitoring, documentation, and record keeping provide verification that the animal is recovering appropriately and allows for intervention if recovery is not as expected. Both objective and subjective assessments of the animal should be done at least twice daily until the animal has sufficiently recovered. For critical patients, assessments may need to be done at more frequent intervals and animals may even require around-the-clock observation. Objective assessments may include monitoring vital signs, such as body temperature, food and water intake, pulse or heart rate, and respiratory rate. Subjective assessments include evaluating the activity, appearance, and behavior of the animal. Typically, animals should resume normal posture and species-specific behavior and activity, such as walking and sitting upright, within hours of the procedure. Procedures such as spinal or thoracic surgery can delay these normal behaviors and activities and should be expected and addressed in the presurgical planning. Lameness, inactivity, and hunched postures may indicate pain or other complications and should be evaluated and referred to the veterinary staff. The fur and skin should have a healthy appearance, so animals that fail to groom or display piloerection or soiled fur should be assessed for illness or pain (National Research Council 2009). Animals that are ill or in pain may exhibit behaviors such as heightened aggression, guarding of the surgical wound, inducing self-trauma, excessive grooming, or acting more timid or fearful. An assessment of the surgical sites should be done to ensure normal wound healing and observe for signs of infection and inflammation (redness, swelling, and pain). Monitoring hydration status is also critically important to appropriate recovery, and animals may need to be supported with parental or oral fluid administration. Assessment of urine and fecal output, skin tenting, and overall condition will contribute to assessing the animal's appetite and water intake. It should be noted that animals that have had food withheld prior to surgery may require additional time to produce feces after surgery. Medications that affect gastrointestinal motility can also have implications to the postoperative monitoring and assessment based on their effects. Sutures, staples, and wound clips should be examined to ensure adequate placement and security. Nonabsorbable skin closures should be removed at the appropriate time following complete wound healing. Postoperative care might also include specialized practices, such as chest tube management or intravenous line maintenance. It is important that clear instructions be provided to the staff providing the care, and whether additional training is required should be evaluated.

Documentation of the postoperative period is required by federal regulations, and adequate record keeping ensures that proper veterinary care has been provided (Animal Welfare Act 1966). Postoperative observation, assessment, and documentation may be conducted by the research staff, animal technicians, husbandry staff, or other trained individuals. Health concerns or complications should always involve veterinary consultation. Documentation for rodent procedures may be accomplished using a variety of methods, including notations on cage cards, colored cage tags, special cards, or a logbook. Depending on the institute's IACUC, the documentation should include the animal protocol number, the surgeon, the surgery date, who monitored animal recovery, analgesics provided, and who administered them. For species covered by the U.S. Department of Agriculture (USDA) Animal Welfare Act Regulations, individual animal medical records are required for recording the postoperative recovery of the animal, medications administered, condition and behavior of the animal, physiologic parameters, and other pertinent information.

A successful surgical program must include preoperative planning; postoperative monitoring and care; consultation between the research, technical, and veterinary staff; adequate training of personnel; and compliance with the approved protocol. As such, institutions should consider establishing a system to track and monitor metrics concerning postoperative complications such as wound dehiscence, infection, or complications associated with the surgical procedure to identify training deficiencies, areas for improvement, and positive and negative trends. Surgical complications may occur with any surgical procedure and are not unexpected in an active surgical program. However, when higher incidences or trends of complications are identified, investigation into the root cause is warranted to maintain animal welfare and quality surgical support. Common causes for an increase in surgical complications are inexperienced personnel, poor training, inadequate surgical technique, inadequate postoperative observations, and faulty or malfunctioning equipment. Morbidity and mortality rates for each surgery type and/or investigator should be included in the surgical metrics within all aspects of the surgical program. Areas to also include relate to animal husbandry, specific technical and veterinary staff assisting in procedures, and types of anesthesia and analgesics utilized for specific procedures. IACUC oversight through a post-approval monitoring program can also aid in identifying these trends. This type of monitoring program provides additional data related to protocol compliance with regard to approved surgical procedures and postoperative animal care.

Surgical Facility Standard Operating Procedures

SOPs are detailed, step-by-step process descriptions intended to ensure the consistent application of institutional practices and compliance with regulations, policies, and principles (National Research Council 2011). Good laboratory practice (GLP) regulations require that facilities have written SOPs adequate to ensure the quality and integrity of data generated in studies (GLP 2011). SOPs are essential to maintain uniformity and consistency in a facility with the turnover of staff. Surgical SOPs should be composed by focusing on the scientific and medical bases for the techniques, processes, and surgical procedures in order to provide consistency and protect animal welfare. SOPs can serve as the basis for staff and investigator training. The surgical facility SOPs should also reflect the major components associated with the conduct of surgery in research (Brown et al. 1993):

- Preoperative management
- Anesthesia and analgesia management
- Intraoperative management
- Perioperative management
- Operative techniques and surgical procedures
- Postoperative management
- Facility maintenance and sanitization
- Instrument maintenance and sterilization
- Equipment use and maintenance
- Personnel responsibilities
- Record keeping

Preoperative SOPs should include a description of the presurgical evaluation process to approve or qualify an animal for surgery. For large animals, this process may include animal health record review, physical and blood work evaluations, condition determination, behavioral assessment, and any specific presurgery requirements (i.e., specific handling and restraint), special housing, imaging procedures, and antimicrobial prophylaxis. For rodents, the process should be a general health and condition assessment, along with special housing and handling requirements.

Anesthesia and analgesia SOPs should address adequate veterinary care by providing guidance to animal users on appropriate utilization of anesthetics, analgesics, and tranquilizers. They may include specific anesthetic choices and dosages for select species and procedures, and should indicate expiration and disposal methods and criteria, such as how to handle multidose vials after first being opened. Analgesia-related SOPs are based on IACUC approval for specific studies and may include specific analgesia medications, species, pain assessment, dosage, duration, storage requirements, handling instructions, and staff qualifications. Anesthesia-related SOPs should cover facility-approved anesthesia agents, indications, dosages, storage requirements or handling instructions, contraindications, species, staff qualifications, and methods of administration.

Intraoperative monitoring SOPs provide guidance and consistency in the assessment of the physiologic status of the animal throughout the procedure in order to provide optimum animal welfare with reference to maintaining a surgical plane of anesthesia, adequate pain management, and surgical success.

Perioperative management SOPs should include operating room setup and the preparation procedures for instruments, surgical supplies, animals, support staff, and surgeons.

Operative techniques and surgical procedure–related SOPs focus on proper surgical techniques, gentle tissue handling, effective hemostasis, maintenance of sufficient blood supply to tissues (systemic pressure), asepsis, accurate tissue apposition, appropriate use of monitoring equipment, and minimizing surgical time. Surgical procedure SOPs can be written for specific surgical procedures (i.e., vascular access port implantation, splenectomy, and myocardial ischemia), based on each facility's procedures, or as generalized SOPs addressing the above considerations. Surgical procedures performed as study-specific procedures may be included directly in the study protocol as opposed to an SOP, or the protocol may reference a specific facility SOP if the SOP has been approved by the IACUC. If procedures are done under GLP guidelines, any changes in the surgical procedure will need to be addressed according to GLP regulations and guidelines.

Postoperative SOPs should address three key components: recovery from anesthesia, acute postoperative care, and long-term postoperative care. From an animal welfare perspective, recovery from anesthesia is the most critical time. The SOPs should address patient monitoring and evaluation, management of physiologic parameters, specialized housing or caging during recovery, medications and treatments, and the transition from the recovery phase (i.e., normal vitals, ambulation, and return of normal behaviors). SOPs addressing the acute postoperative care should include pain assessment, appropriate relief, type and frequency of postoperative observations, surgical site observations, special housing requirements, staff qualifications, and medications and indications for treatment. Long-term postoperative care SOPs should address the type and frequency of observations, the continuation of pain assessment, staff qualifications, special housing, special diets, implant function verification, surgical site observations, and the notification process when postoperative care is completed.

Facility maintenance and sanitization SOPs should address traffic patterns with relation to contamination control, personal protective equipment (PPE) entry requirements, delineation of surgical and nonsurgical activities within the surgical suite, and monitoring of cleaning practices.

Instrument maintenance and sterilization SOPs should address the processes for cleaning, packaging, and sterilization of instruments and surgical supplies, and should include sterilization validation procedures. They may also address instrument repair, sharpening, and replacement parameters.

Equipment use and maintenance SOPs should focus on proper setup, use, maintenance, calibration, and validation of proper function and should be generated for each unique piece of equipment in the facility.

Personnel responsibility SOPs should outline unique responsibilities and requirements for staff providing surgical support and animal care within the facility, which may not be addressed in another SOP. Personnel safety practices (PPE, nonhuman primate bite and scratch kits, etc.) should also be addressed with specific SOPs.

Record-keeping SOPs should provide details on the management and maintenance of animal and facility records. They may include specific requirements for certain records, long-term storage methods and location of records, and disposal instructions.

Surgical facility SOPs should focus on the major components associated with all aspects related to the conduct of surgery and provide guidance for basic principles of analgesia and anesthesia, patient monitoring, good surgical technique, equipment operation, training and assessment of staff qualifications, and provision of optimal animal welfare.

Training

Although the responsibility for ensuring surgeons and surgical support staff are properly trained lies with the PI and the IACUC, it may fall to the SFM to help develop and/or provide the training. The SFM's role is primarily as a facilitator in the training process and includes identifying appropriate individuals to conduct training, developing or assisting with curriculum development, and monitoring the effectiveness of the training program. This includes developing or incorporating training aids and other practice methods, providing appropriate instruction and, most importantly, providing the time necessary for comprehensive training. It is essential that an institution's surgical training program be well defined and reviewed by the attending veterinarian and the IACUC.

The first step in the training process is to assess the knowledge and experience level of the trainee. In a research facility, many people of differing backgrounds may be performing surgery. This could include veterinarians, veterinary technicians, MDs, PhDs, graduate students, and research technicians.

While an MD would have extensive knowledge of human anatomy and physiology, he or she may have limited experience with the species differences that may be encountered in a research facility (Brown et al. 1993; ASR 2009). A person in a technical position may have little experience with maintaining asepsis or proper tissue handling. Therefore, training must be directed toward the needs and experience of the individual.

The curriculum of a surgical training program should progress in a logical, step-by-step manner, with each skill building on the previous one. For example, a trainee would learn instrument names, the appropriate care and use of each instrument, and the proper handling of suture and simulated tissue prior to conducting a procedure. By taking this approach, each individual must demonstrate appropriate knowledge and skill to progress to the next step, thus taking into account any previous experience as described above.

The didactic components of the curriculum are often developed within an individual institution—these may involve study modules specific to the facility or to the procedures it is most often conducting. Animal welfare regulations and species-specific anatomy and physiology are topics that lend themselves well to this approach. Managers should also take advantage of resources currently available. Many laboratory animal organizations provide training materials, free or at minimal cost, which could be incorporated into a surgical research training curriculum. Coordinators of surgical training programs can search for materials from the Animal Welfare Information Center (AWIC), the Laboratory Animal Welfare Training Exchange (LAWTE), the American Association for Laboratory Animal Science (AALAS) Learning Library, and the Academy of Surgical Research (ASR).

Wet lab training may be conducted concurrently with didactic training, but should also follow a stepped process. Hands-on training typically begins with inanimate training aids, which will allow the trainee to practice skills and techniques without the pressure of working on a live animal, but may not be required if the trainee is already a skilled surgeon and just needs to learn a new technique or learn on a different species. Inanimate training can begin with easily accessible, inexpensive items, such as oranges, bananas, chicken breasts, and pig's feet, which can be used for suture or injection training. Cadaver tissues, such as harvested arteries and veins, can be used for microsurgical training, but may have limited value as tissue properties differ from live tissue. Suture boards and simulated tissue training aids can be valuable to practice suturing, macroscopically or microscopically. Stuffed toy animals can be modified with simulated vessels made from tubing, placed under bandaging material, and then used for practicing catheter placement or blood collections. As training progresses, more sophisticated materials can be employed. Anatomic models, interactive simulators, and virtual hands-on surgical training are commercially available and could be considered. Nonsurvival training on the actual procedure is valuable, but should be limited to highly technical procedures that cannot be practiced by an alternate method. Once proficiency of a procedure is demonstrated, the trainee may proceed to survival procedures under the supervision of the trainer. As part of the process, the surgical recovery and success of the procedure must be followed to ensure competency of the individual. All training and competencies must be documented after the completion of each didactic or wet lab session.

Necessary components of a surgical training program have been well defined by the ASR (2009). This begins with training on regulatory and ethical considerations (USDA Animal Welfare Act, the *Guide for the Care and Use of Laboratory Animals*, Office of Laboratory Animal Welfare [OLAW], IACUC guidelines, FDA and Environmental Protection Agency [EPA] GLPs, and applicable national and local regulations). Presurgical planning, as described previously, needs to be part of the training program. The specifics will vary by institution, but surgical assistants and surgeons must have good foundational training on methods of analgesia, anesthesia, maintaining aseptic technique, and postoperative care. The surgical trainee will need additional training on proper surgical technique for specific species and procedures. At some institutions, the surgical staff will handle all postoperative care, but for other facilities, the animal husbandry or technical staff may take on these responsibilities. When this is the case, it is important for all individuals involved to participate in appropriate training, based on their level of involvement.

Posttraining evaluation is an ongoing process, necessary to prevent procedural drift and to continue improving techniques with advances and refinements in the field. Emerging technologies should be reviewed regularly by managers and training coordinators. Regular attendance at professional meetings such as ASR and AALAS is invaluable for learning and sharing new techniques and refinements. Professional organizations also offer courses at local and national meetings, including wet lab opportunities for hands-on training, along with certification opportunities. In addition to the AALAS certifications, which provide a good foundation for technicians, other programs for nondoctoral individuals are available with an emphasis on surgical training. The ASR certification program validates training at three different levels: anesthesia, minor surgical procedures, and major surgical procedures. Additionally, the National Association of Veterinary Technicians in America (NAVTA), the Academy of Veterinary Technicians in Anesthesia and Analgesia, and the American College of Veterinary Anesthesia and Analgesia all have specialty academies in anesthesia and research surgery.

Providing trainees the appropriate instruction, resources, and time will lay the foundation for a successful surgical department. Investing in ongoing training for existing staff will build a highly qualified, experienced team of surgeons and surgical support staff, which is critical to continued improvement in techniques and ultimately to animal welfare.

Summary

The SFM is often required to satisfy a multitude of responsibilities in the process of running the facility. He or she must have a thorough understanding of the requirements for the facility, staffing, equipment, budgeting, surgical procedures, animal care, and training. To be successful, the SFM must have a comprehensive knowledge of the functional areas of a surgical facility, their relationship to each other, and the differences between large and small animal surgery requirements. He or she serves as a mentor to facility staff, providing communication, planning, scheduling, and oversight of all facility activities. An effective SFM provides direct supervision for his or her staff and provides opportunities for career planning and advancement. The success of the surgical program is dependent on the quality and versatility of the facility management and staff.

REFERENCES

ACLAM (American College of Laboratory Animal Medicine). Guidelines for adequate veterinary care. Chester, NH: ACLAM. http://grants.nih.gov/grants/olaw/ACLAM_Adequate_Vet_Care.pdf (accessed May 3, 2016).

Animal Welfare Act. Published in the Code of Federal Regulations (CFR), Public Law 89-544, 1966, as amended.

ASHE (American Society for Healthcare Engineering). Special Departmental HVAC Issues: Operating Rooms. Chicago: ASHE. http://www.ashe.org/compliance/ec_02_05_01/01/specialdept_or.shtml (accessed April 17, 2017).

ASR (Academy of Surgical Research). Guidelines for training in surgical research with animals. *Journal of Investigative Surgery* 22:3 (2009) 218–225.

Bernal, J., M. Baldwin, T. Gleason, S. Kuhlman, G. Moore, and M. Talcott. Guidelines for rodent survival surgery. *Journal of Investigative Surgery* 22 (2009) 445–451.

Brown, M., P. T. Pearson, and F. N. Tomson. Guidelines for animal surgery in research and teaching. *American Journal of Veterinary Research* 54(9) (1993) 1544–1559.

Callahan, B. M., K. A. Hutchinson, A. L. Armstrong, and L. S. Keller. A comparison of four methods for sterilizing surgical instruments for rodent surgery. *Contemporary Topics* 34(2) (1995) 57–60.

Conour, L. A., S. T. Chou, and L. C. Anderson. Regulatory issues. In *Anesthesia and Analgesia in Laboratory Animals*, ed. R. E. Fish, M. J. Brown, P. J. Danneman, and A. Z. Karas, 569–580. 2nd ed. San Diego: Academic Press, 2008.

Corletto, F. Multimodal and balanced analgesia. *Veterinary Research Communications* 31(Suppl. 1) (2007) 59–63.

Fitzgerald, R. H. Microbiologic environment of the conventional operating room. *Arch Surg* 114 (1979) 772–775.

GLP—Good laboratory practice for nonclinical laboratory studies. Published in the Code of Federal Regulations, Title 21, Food and Drugs, Part 58 (2011) 21 CFR 58.81.

Hoyt, R. F., Jr., R. R. Clevenger, and J. A. McGehee. Introduction to microsurgery: An emerging discipline in biomedical research. *Lab Animal (N.Y.)* 30:9 (2001a) 26–35.

Hoyt, R. F., Jr., R. R. Clevenger, and J. A. McGehee. Microsurgical instrumentation and suture material. *Lab Animal (N.Y.)* 30:9 (2001b) 38–45.

Kohn, D. F., T. E. Martin, P. L. Foley, T. H. Morris, M. M. Swindle, G. A. Vogler, and S. K. Wixson. Guidelines for the assessment and management of pain in rodents and rabbits. Chester, NH: American College of Laboratory Animal Medicine, 2006.

Kuhlman, S. M., M. C. Flegal, L. K. Fox, and M. H. Murphy. Notes from the Academy of Surgical Research pages: Principles of anesthesia monitoring—Introduction. *Journal of Investigative Surgery* 21:3 (2008) 161–162.

National Center for Research Resources. *Cost Analysis and Rate Setting Manual for Animal Resource Facilities*. Bethesda, MD: National Center for Research Resources, 2000. http://grants.nih.gov/grants/policy/air/rate_setting_manual_2000.pdf.

National Research Council. *Guide for the Care and Use of Laboratory Animals*. 8th ed. Washington, DC: National Academic Press, 2011.

National Research Council, Committee on Recognition and Alleviation of Pain in Laboratory Animals. Recognition and assessment of pain. In *Recognition and Alleviation of Pain in Laboratory Animals*. Washington, DC: National Academic Press, 2009, chap. 3. http://www.ncbi.nlm.nih.gov/books/NBK32656/ (accessed May 3, 2016).

NIH/OACU (National Institutes of Health/Office of Animal Care and Use). Guidelines for pain and distress in laboratory animals: Responsibilities, recognition and alleviation. Animal Research Advisory Committee Guidelines. Bethesda, MA: NIH/OACU. https://oacu.oir.nih.gov/sites/default/files/uploads/arac-guidelines/pain_and_distress.pdf (accessed April 19, 2016a).

NIH/OACU (National Institutes of Health/Office of Animal Care and Use). Guidelines for survival rodent surgery. Animal Research Advisory Committee Guidelines. Bethesda, MA: NIH/OACU. https://oacu.oir.nih.gov/sites/default/files/uploads/arac-guidelines/rodent_surgery.pdf (accessed April 19, 2016b).

PHS (Public Health Service). Policy on humane care and use of laboratory animals. Washington, DC: National Institutes of Health, Office of Laboratory Animal Welfare, Department of Health and Human Services. http://grants.nih.gov/grants/olaw/references/PHSPolicyLabAnimals.pdf (accessed May 3, 2016).

Rutala, W., and D. Weber. Guideline for disinfection and sterilization in healthcare facilities. Atlanta, GA: Centers for Disease Control and Prevention, 2008.

Schofield, J. C. Principles of aseptic technique. In *Essentials for Animals Research: A Primer for Research Personnel*, 59–77. Washington, DC: National Agricultural Library, 1994.

Schonholtz, G. J. Maintenance of aseptic barriers in the conventional operating room. *Journal of Bone and Joint Surgery* 58 (1976) 439–445.

Smith, J. C., and P. J. Danneman. Monitoring of anesthesia. In *Anesthesia and Analgesia in Laboratory Animals*, ed. R. E. Fish, M. J. Brown, P. J. Danneman, and A. Z. Karas, 171–182. 2nd ed. San Diego: Academic Press, 2008.

Tannenbaum, T. N., and R. J. Goldberg. Exposure to anesthetic gases and reproductive outcome: A review of the epidemiologic literature. *Journal of Occupational Medicine* 27(9) (1985) 659–668.

35

Euthanasia

Christine P. Sivula and Mark A. Suckow

CONTENTS

Background and Perspective

Definition

The word *euthanasia* comes from the Greek terms *eu* (good) and *thanatos* (death). In the medical field, it is often defined as the act of painlessly killing an individual who is suffering from an incurable or painful disease. The Animal Welfare Act (AWA) and Animal Welfare Regulations (AWR), 9 CFR Part 1, §1.1 (U.S. Department of Agriculture 2002), define euthanasia as the humane destruction of an animal accomplished by a method that produces rapid unconsciousness and subsequent death without evidence of pain or distress, or a method that utilizes anesthesia produced by an agent that causes painless loss of consciousness. Although commonly used in the field of laboratory animal science to describe the killing of research animals at the end of an experiment, some assert that this is a misuse of the word (Pavlovic et al. 2011). According to this view, since animal experimentation is not in the best interest of the animal, the act of killing the animal cannot be considered euthanasia, and is instead part of the experimental process. Despite this criticism, the term *euthanasia* continues to be used in animal care and use programs in research, teaching, and testing.

Ethical Considerations

There is frequent discussion regarding the ethical considerations pertaining to the use of animals in research. Many points of view exist, from those who are vehemently opposed to those in favor, and discussions can lead to significant emotional responses (Gnadt and Leland 2002). Public perception of conditions under which animals used in research are kept is often unfavorable, and concerns regarding euthanasia of research animals are a specific area of debate on this topic. There is concern about not only the potential pain and suffering of animals while undergoing experimental procedures, but also how an animal's life ends. As biomedical research involving animals continues, society's awareness of animal welfare issues is increasing as well (Rollin 2009). According to Rollin (2009), it is important that the summation of their lives not involve fear, horror, pain, or suffering. Euthanasia of healthy research animals at the end of the experiment is a serious concern for the public (AVMA Panel on Euthanasia 2013). Recently, legislation has been passed in some states requiring that healthy dogs and cats used in research be made available for adoption rather than euthanized at the end of the experiment (Simmons 2014). While the adoption of some animals following the research project is possible, for others it is not. Often, postmortem analysis or sample collection must be performed to verify study results. Those who work in the laboratory animal field should be aware of the public's ethical concerns surrounding the use and euthanasia of research animals.

Veterinarians and technicians working in the laboratory animal field may face unique ethical challenges. Like all veterinarians, those in the laboratory field have taken an oath to use their "scientific knowledge and skills for the benefit of society through the protection of animal health and welfare, the prevention and relief of animal suffering, the conservation of animal resources, the promotion of public health, and the advancement of medical knowledge" (Nolen 2011). According to the American Veterinary Medical Association's (AVMA) Animal Welfare Principles, the responsible use of animals for human purposes, including research conducted for the benefit of both humans and animals, is consistent with the Veterinarian's Oath (AVMA 2015a). Additionally, veterinarians are expected to adhere to the AVMA's Principles of Veterinary Medical Ethics (AVMA 2015b). The first principle states that a veterinarian shall be dedicated to providing competent veterinary medical care, with compassion and respect for animal welfare and human health. Under this principle, humane euthanasia of animals is described as an ethical veterinary procedure. In some cases, however, the decision regarding if and how an animal should be euthanized may be complicated by other factors, such as the greater public good, as with euthanasia of animals used in medical research (AVMA Panel on Euthanasia 2013). Veterinarians and technicians should use their experience and knowledge to ensure the ethical care and use of research animals, promote humane euthanasia practices, and encourage refinement of procedures consistent with guidelines and regulations. They are also in a position to lend an important perspective to the ethical debate regarding animal experimentation.

Ethical concerns are frequently encountered in the laboratory animal facility and often involve animal care personnel, facility managers, veterinarians, principal investigators, and their staff (Gnadt and Leland 2002). Concerns involving endpoints for animals on study can be particularly emotional. In these cases, the veterinarian is generally called on to investigate the concern by reviewing the animal use protocol for information regarding approved endpoints. Follow-up and resolution may include discussion with the investigator or laboratory staff regarding following approved endpoints, and in some cases, Institutional Animal Care and Use Committee (IACUC) involvement may be required to resolve the issue. In all cases, the veterinarian or manager should follow up with the animal care personnel to communicate the outcome. Depending on the information discovered during investigation, this could involve a discussion about the specific study and why the approved endpoints are appropriate. Although that may not completely resolve the ethical concern for the employee, understanding why that endpoint is important to fulfill the scientific objectives of that particular study may relieve some of the concern.

Regulatory Expectations

There are several laws, regulations, and guidelines that provide guidance to the research community regarding the appropriate care and use of animals. Many of these outline specific expectations for adhering to humane endpoints and euthanasia procedures. IACUCs have the responsibility of evaluating animal care and use protocols as they relate to the welfare and use of animals within the framework of these regulations and guidelines (Committee for the Update of the Guide for the Care and Use of Laboratory Animals 2011). Special consideration should be given to protocols that involve the potential for unrelieved pain and distress, and the committee should encourage methods of replacement, refinement, and reduction and ensure a humane death (AVMA Panel on Euthanasia 2013).

Animal Welfare Act and Regulations

The AWA was initially enacted in 1966 and has been amended several times. The 1985 amendments included specific changes promoting laboratory animal welfare. Some of these changes included establishing an IACUC and the minimization of pain and distress. The importance of appropriate endpoints and euthanasia methods is highlighted in the AWR. They state that during the protocol review process, the IACUC shall determine that the proposed activities or significant changes in ongoing activities meet several requirements. One of the requirements is that animals that would otherwise experience severe or chronic pain or distress that cannot be relieved will be painlessly euthanized at the end of the procedure or, if appropriate, during the procedure. Additionally, the regulations state that methods of euthanasia used must be in accordance with the AWR definition unless a deviation is justified for scientific reasons (9 CFR Part 1, §2.31). The USDA *Animal Care Policy Manual* also addresses euthanasia of covered species, indicating that euthanasia should be in accordance with the current *AVMA Guidelines for the Euthanasia of Animals* (USDA 2016).

U.S. Government Principles for the Utilization and Care of Vertebrate Animals Used in Testing, Research, and Training

The Public Health Service (PHS) Policy on Humane Care and Use of Laboratory Animals, administered by the Office of Laboratory Animal Welfare (OLAW), implements the U.S Government Principles (Interagency Research Animal Committee 1985). There are nine principles that must be followed when U.S. government agencies perform or sponsor procedures involving the use of vertebrate animals. Like the requirements described in the AWR, Principle VI addresses appropriate humane endpoints and states that animals that would otherwise suffer severe or chronic pain or distress that cannot be relieved should be painlessly killed at the end of the procedure or, if appropriate, during the procedure. Principle VIII pertains to personnel training and indicates that all shall be appropriately qualified and experienced for conducting experiments, and that adequate arrangements should be made for in-service training, including the proper and humane care and use of laboratory animals. As euthanasia is often a component of experiments involving animals, training on proper technique when performing euthanasia procedures is also important.

Guide for the Care and Use of Laboratory Animals

The *Guide for the Care and Use of Laboratory Animals* (*Guide*) is endorsed by the Institute for Laboratory Animal Research and the National Academy of Sciences. It establishes the minimum ethical, practice, and care standards for researchers and their institutions (Committee for the Update of the Guide for the Care and Use of Laboratory Animals 2011). Regarding euthanasia of research animals, the *Guide* states that methods should be consistent with the *AVMA Guidelines for the Euthanasia of Animals*, and that it should be carried out in a manner that avoids animal distress. The *Guide* also addresses the importance of training and proficiency when performing euthanasia.

Reasons for Euthanasia

Although proper performance results in the humane death of the animal, there exist varied reasons for euthanasia. In some instances, euthanasia is performed out of concerns for the well-being of the animal, while in others, it is undertaken as part of the experimental design of studies in which the animals are being used. In either case, the reasons for euthanasia should be clearly understood by all involved personnel.

Experimental Endpoints

Many studies have included in the experimental design specified points at which animals are euthanized so that samples or other data can be collected that would otherwise not be obtainable, that is, the point at which the scientific objectives of the project are reached. In some cases, animals may need to be euthanized when the research requires that tissue samples from sites such as the heart, lung, liver, or brain be collected in a way or quantity that would not allow the animal to survive. In other cases, the research may involve surgical implantation of materials or devices that must be retrieved for further evaluation, thus necessitating euthanasia of the animal. Typically, the point at which the animal is euthanized for the purposes of harvesting tissue or implanted materials is defined by elapsed time after experimental manipulation. For example, animals might be euthanized minutes, hours, days, or longer after initial manipulation, and these time points are designed to permit the investigator to study the biological response to some manipulation over time. Interestingly, changes in tissue induced by the hypoxia associated with euthanasia has led some to argue that collection on anesthetized, but alive, animals allows superior outcome (Overmyer et al. 2015). Other experimental criteria used to define the time of euthanasia are occasionally based on other criteria related to a biological response, such as a defined tumor volume (Yapp et al. 1998).

Although not strictly an experimental endpoint, circumstances sometimes arise in which animals are euthanized because they are no longer needed for the research, for example, at the conclusion of studies that do not require euthanasia as part of the experimental design or in the case of breeding colonies that generate animals of a genotype that are not needed for the research. In the spirit of reducing the number of animals used in or produced for research, the manager should encourage oversight of breeding colonies in such a way that the number of animals of the required genotype produced closely matches the number needed. In some cases, animals that are not needed for research can be transferred to another project after approval by the IACUC to do so, assuming that the animals have not already been used in research that would preclude additional use, such as an additional survival surgical procedure. It is also possible that animals not needed for research might be adopted out as pets. In such cases, the institution should ensure that the animal is healthy and that the adopter can responsibly provide for the needs of the animal. It would be important that legal counsel for the institution review the planned process for adoption and that the policy meets all state and local regulations. Some institutions make arrangements with bird sanctuaries or zoological parks to use naïve euthanized animals as a food source.

Experimental Humane Endpoints

Sometimes, animals must be euthanized because they are ill. In some cases, this is the result of spontaneous and unexpected disease, while in other cases, illness may be associated with experimental use, age, or phenotype. In the case of the former, decisions to euthanize an animal are usually based on the professional judgment of the veterinarian and made in consultation with the research team. With respect to the latter, defined endpoints are often stated within the protocol approved by the IACUC so that personnel have a predetermined point of reference from which to make decisions when animals decline clinically. Such defined endpoints generally reflect the sort of sound veterinary judgment that guides the decision to euthanize an animal due to illness unrelated to experimentation and is based on specified clinical criteria. In any case, the overall goal is to mitigate the pain or distress of the animal that would be otherwise unrelieved.

The precise endpoints that are defined for a study may depend on the type of animal model used and the goals of the research. Identifying acceptable humane endpoints is especially important for those studies in which animals may experience significant and unrelieved pain or distress, such as those involving tumor growth, infectious disease, and toxicology. The endpoints agreed to by the IACUC, the investigator, and the veterinary team should address the needs of the animals, yet allow reasonable room for conduct of the study. It is important that all personnel involved with the care and use of the animals clearly understand up front what these endpoints are and what happens when they are reached. In addition, a plan or system that aids rapid communication between personnel and facilitates decision making as endpoint criteria are reached is essential.

It is sometimes difficult to determine the exact point at which pain or distress changes from tolerable to intolerable; thus, humane endpoints are somewhat subjective but should be based on sound professional judgment with input from the veterinary team. Some typical parameters that might be used as endpoint criteria include

1. *Level of activity.* Animals that are lethargic may be experiencing pain or distress. For example, an animal that is in pain could be unwilling to move. Similarly, animals that are severely ill due to various forms of metabolic disease may demonstrate lethargy. Many times, failure to groom will result in an unkempt appearance. In addition, vocalization can sometimes indicate the presence of pain or distress, such as the whining of a dog. Further, lack of vocalization, such as the absence of singing in a songbird, may also suggest that the animal is experiencing pain or distress.
2. *Decreased food and water consumption.* Careful attention should be paid to the amount of food and water consumed by research animals, since decreases may signal significant pain and/or distress. It is important, therefore, that the amount normally consumed be known.
3. *Body weight.* Along with decreased food and water consumption, loss in body weight can signal a decline in the animal's health. If the animal has a tumor (spontaneous or induced), body weight may not be as useful for determining health status, as the tumor can artificially alter the animal's true weight. In this case, other criteria should be considered. For example, body condition scoring has been used to evaluate health status in mice (Ullman-Cullere and Foltz 1999).
4. *Body temperature.* A persistent or precipitous rise or decrease in body temperature can indicate significant illness and can be used to define humane endpoints. In particular, hypothermia is a good indicator of impending death, and it is common to use a decrease in body temperature below a specific point as an endpoint (Toth 2000; Cates et al. 2014; Dellavalle et al. 2014).
5. *Behavior and alertness.* Listlessness, aggressiveness, or otherwise abnormal behavior can be signs of significant pain and distress in animals, and behavior and activity levels can be used as criteria in euthanasia decisions (Warren et al. 2014). As a result, it is important that personnel clearly understand what the expected normal behavior of a given species of animal is so that they may accurately note the occurrence of behaviors that are not normal.
6. *Evidence of infection.* Infection may be associated with surgical wounds or result from opportunistic microbes associated with a debilitated animal. Signs of infection vary but may include discharge from wounds, fever, and listlessness.
7. *Presence of wounds or tumors.* Unexpected traumatic wounds or nonhealing surgical wounds may be severe and merit euthanasia of the animal. Likewise, tumors that occur spontaneously or are the result of experimental manipulation may grow very large and abrade to the point of being an open wound, become necrotic, interfere with locomotion or normal function of a tissue, or significantly impact the health and well-being of the animal in other ways.
8. *Clinical pathology.* Parameters related to hematology and clinical chemistry can be used to weigh humane endpoint decisions. For example, in Ebola virus–infected nonhuman primates, extreme alterations to several commonly monitored serum parameters were related to survival outcome and could therefore be used to reasonably establish endpoint criteria, particularly when evaluated in conjunction with measurements of body temperature and activity levels (Warren et al. 2014).

The above are examples of endpoint criteria that might be used. Of course, there are others, and in all cases the specific point at which an animal would need to be euthanized should be defined before the study begins, and it should be specific. For example, it may be determined that animals on a study should be euthanized when they lose 15% or 20% of their body weight (compared with age-matched controls) or when a tumor reaches a certain measurable size. In some cases, endpoints are reached when a combination of factors are considered. In this regard, scoring systems have been developed that consider clinical observational data as a means to identify humane endpoints (Nunamaker et al. 2013; Warren et al. 2014).

Veterinary Medical Conditions

Illness that is not caused by experimental manipulation can occur in animals as a result of their genotype, or it may be spontaneous. With respect to genotype, illness may be expected in animals whose genome has been altered as a way to create an animal model of a disease, although sometimes genomic alterations result in phenotypes that are unexpected and may be deleterious to the health of the animal. In such cases, the IACUC should be informed of the unexpected phenotype-associated illness.

Similar to humans and pet animals, laboratory animals may develop spontaneous illness, such as tumors, diabetes, renal disease, and others. This is especially true in aged animals. Laboratory animals may also experience illness associated with trauma, for example, that associated with fighting.

Any animal experiencing illness should be reported to the veterinary staff. Animals with severe illness due to spontaneous conditions or phenotype should be evaluated the same as those with illness due to experimental manipulation; that is, illness that is of clinical severity associated with poor outcome should be euthanized. In this regard, it is essential to invoke the opinion of the veterinarian.

Methods of Euthanasia

The AVMA has published guidelines for the euthanasia of animals, and those guidelines are widely considered to be the authoritative document on the subject (AVMA Panel on Euthanasia 2013). The AVMA guidelines consider various methods of euthanasia as being acceptable, acceptable with conditions, or unacceptable. Acceptable methods reliably result in the humane death of the animal, methods that are acceptable with conditions are those that require some additional conditions to be met in order for humane death to reliably result, and unacceptable methods are those that are regarded as inhumane.

With respect to research animals, selection of a euthanasia method should ensure humane death, as well as meet the experimental needs of the study (Angus et al. 2008; Karmarkar et al. 2010; Hazzard et al. 2014). With respect to the former, it is common to confirm death by means of an adjunctive (secondary) method, such as induction of pneumothorax, cervical dislocation (in rodents), or exsanguination. The method chosen should also consider the impact on personnel, in terms of both safety and emotional load, as individuals may perceive and respond differently to performing or witnessing euthanasia. In addition, it is good practice to euthanize animals out of the sensory range of other animals in order to minimize distress to those nearby.

The euthanasia methods used can be categorized with respect to whether they utilize pharmacological or chemical agents versus physical methods. The methods described here are not an exhaustive list of all possible methods, but rather represent methods that are commonly used for euthanasia of research animals.

Pharmacological and Chemical Methods

Methods for euthanasia that employ either drugs or chemicals generally involve exposure of the animal to the agent via inhalation or other routes of administration, such as intravenous or, in the case of aquatic species, immersion. When such agents are used, it is advisable to use pharmaceutical-grade compounds to ensure a smooth and rapid death of the animal.

Methods Involving Inhaled Agents

Inhaled agents are generally easy to use and can be used to euthanize larger numbers of animals at a single time. This approach is very commonly used to euthanize rodents. The most common agents used include carbon dioxide (CO_2) and halogenated fluorocarbon anesthetics, such as isoflurane.

Carbon Dioxide

CO_2 is an odorless, colorless gas that rapidly induces narcosis, progressing to death when used properly. Most commonly, CO_2 is administered to animals contained within a chamber into which animals, or the home cage housing the animals, are placed. Animals placed in a chamber should not be crowded; thus, the typical chamber size accommodates rodents but not larger species. Animals to be euthanized as a group should be compatible and of the same species. A compressed gas cylinder equipped with a regulator and flowmeter is the preferred source of CO_2. Although some have historically used dry ice as a CO_2 source, it is widely regarded as unacceptable because of the inability to closely control CO_2 concentration, as well as the risk for pain and distress to the animals resulting from thermal injury.

CO_2 should be administered to chambers containing animals to be euthanized at a displacement rate of 10%–30% chamber volume per minute (AVMA Panel on Euthanasia 2013). It has been determined that this rate allows for rapid euthanasia of animals. A flowmeter attached downstream of the regulator is useful to consistently achieve flow rates that have been calculated to meet the 10%–30% displacement goal. Animals are usually left in the chamber until visible movement has ceased, and it is recommended that a method to confirm death is used. Because immature animals tolerate low oxygen levels better than adults, they require extended exposure time to CO_2 for death to result (Klaunberg et al. 2004).

Because CO_2 is heavier than atmospheric air, it is important to purge the chamber of remaining CO_2 between groups so that it does not build up to levels that would cause pain and distress to the animals (Djoufack-Momo et al. 2014). Commercially available systems exist that automatically regulate the flow of CO_2 and purge the chamber following a period that allows for death of the animals. Some such systems involve a single chamber, while others allow for multiple cages placed on a rack to be euthanized simultaneously.

Halogenated Fluorocarbon Anesthetics

The class of halogenated fluorocarbons includes several agents that are commonly used as anesthetics for surgery, including isoflurane, sevoflurane, halothane, and enflurane. Of these, isoflurane is the most commonly used. When administered as an overdose, these agents can be used for euthanasia of animals. Note that ether is sometimes used for anesthesia and euthanasia; however, its use is discouraged due to risks associated with its high flammability and explosive potential (including carcasses of animals euthanized by ether), as well as respiratory tract irritation following inadvertent inhalation by personnel.

Inhaled anesthetics are sometimes administered to rodents by means of placing an anesthetic-soaked cotton ball in a bell jar that includes a platform on which the animals may stand to avoid direct contact with the anesthetic. This method does not allow control of the anesthetic concentration, the buildup of high anesthetic concentrations may be distressful to animals, and the risk of personnel exposure to waste gas exists. The preferred method for administration of inhalant anesthetics is via precision vaporizer, a tool that carries specified concentrations of anesthetic, along with oxygen, or another carrier gas, to the animal or the chamber in which the animal is placed. The chamber should be gradually filled. Commercial systems are available that feed a specified concentration of anesthetic into a chamber at a constant rate. Any system using halogenated fluorocarbons for euthanasia should be equipped with a means to scavenge waste gas away from personnel. Another option is to initially anesthetize the animal with isoflurane and then use CO_2 for euthanasia.

In general, inhaled anesthetics are used for euthanasia of small animals weighing less than 7 kg (AVMA Panel on Euthanasia 2013). It is recommended that once voluntary movement of the animal has ceased, adjunctive procedures be followed to confirm death.

Methods Involving Noninhaled Agents

Some agents used for euthanasia are administered to animals by means other than inhalation. For the most part, such agents are administered intravenously in nonaquatic species, and via immersion in aquatic species. Although many anesthetics could be used to induce death by overdose, some of the common noninhaled agents are described here.

Euthanasia Solutions

Solutions containing compounds such as sodium pentobarbital and phenytoin are commercially available for euthanasia of animals. Although pentobarbital is classified by the U.S. Drug Enforcement Agency as a Schedule II drug, euthanasia solutions containing pentobarbital are typically classified as Schedule III drugs. These solutions require special storage and documentation requirements for their use. For purposes of euthanasia, solutions are administered as an intravenous overdose and induce profound depression of the central nervous system, resulting in deep anesthesia, followed by death. Although quick, it is important to ensure death so that recovery from very deep anesthesia does not occur.

Potassium Chloride

Solutions of potassium chloride cause cardiac arrest when administered intravenously or intracardially and have been used to induce euthanasia, typically in larger mammals such as cattle and swine. Because of the distress likely to occur in conjunction with cardiac arrest, animals need to be deeply anesthetized prior to administration of potassium chloride. Pharmaceutical-grade preparations of potassium chloride at concentrations needed for euthanasia are generally not available commercially.

Tricaine Methanesulfonate

Tricaine methanesulfonate (MS-222) is commonly used for euthanasia of aquatic species. Animals are typically immersed in solutions that are pH buffered with sodium bicarbonate. Specific concentrations used are typically 200–300 mg/L. Immersed animals progress from deep anesthesia to death; however, it is standard practice to ensure death by means of a secondary method. Pharmaceutical-grade preparations of MS-222 are available for temporary immobilization of aquatic species (U.S. Food and Drug Administration 2015).

Clove Oil

A primary component of clove oil is eugenol, a compound with anesthetic properties. Similar to MS-222, aquatic species can be euthanized by immersion in a clove oil solution, followed by a secondary method to ensure death.

For zebrafish fry, it has been demonstrated that hypothermic shock is a more effective means for euthanasia than immersion in MS-222 or eugenol, although extended exposure times are needed (Strykowski and Schech 2015).

Physical Methods of Euthanasia

A variety of physical methods can be employed for rapid euthanasia of animals, particularly small rodents. When a secondary method to ensure death is appropriate, a physical technique is most commonly employed. Some of the more common physical methods are described here. Although true for any method of euthanasia, it is particularly important for those using physical methods to be well trained and skilled. For training of individuals to properly use these methods, it is suggested that animals that have already been euthanized by other means or are deeply anesthetized be used. Although common physical methods for the euthanasia of rodents and small birds are covered here, one may wish to note that physical methods such as the captive bolt and electrocution are sometimes used for euthanasia of livestock, primarily in meat production circumstances.

Cervical Dislocation

Used primarily for euthanasia of small mammals weighing less than 200 g and small birds, the objective of cervical dislocation is to separate the first cervical vertebra from the skull, thereby disrupting essential brain function and resulting in death. Briefly, the skull is firmly held in place and the trunk of the animal quickly and firmly pulled in the reverse direction so as to separate the cervical vertebra from the skull. Devices designed for cervical dislocation of small rodents are available commercially (Stoelting Co., Wood Dale, Illinois). The method can be aesthetically displeasing to some individuals, and so consideration should be given to alternative methods when appropriate. It is preferable to use cervical dislocation for euthanasia in combination with deep anesthesia when possible.

Decapitation

Primarily used for small animals such as rodents, decapitation is sometimes used in instances in which tissues or body fluids must be recovered quickly after euthanasia. Although it is recommended that animals first be anesthetized, scientific reasons sometimes require that decapitation be performed on the unanesthetized animal. Guillotines are commercially available for the purpose of decapitation, and it is important that a program be in place for routine maintenance of the guillotine to ensure the instrument remains sharp and operational. Following use, it is important that equipment be sanitized. Neonatal rodents are sometimes euthanized in this way, and since an appropriate-size guillotine is not available, sharp scissors can be used. In all cases, it is important that the animal be immobilized so that decapitation occurs quickly and precisely, resulting in death of the animal rather than severe injury. For rodents, a disposable, flexible plastic restrainer is available that provides sufficient immobilization (DecapiCone®, Braintree Scientific, Inc., Braintree, Massachusetts). As with cervical dislocation, this method can be aesthetically displeasing to some individuals, and so consideration should be given to alternative methods when appropriate. It is preferable to use decapitation for euthanasia when combined with anesthesia. Special care needs to be exercised, due to the increased personnel hazard associated with a guillotine.

Thoracic Compression

Sometimes used for euthanasia of small mammals and birds, thoracic compression is of value in field study circumstances where it is not practical to carry equipment or compounds needed for some of the other methods of euthanasia already mentioned. Briefly, the method is performed by placing quick, firm pressure on the upper chest, resulting in respiratory compromise and relatively quick death. Although this method is advocated by some for use in field circumstances (Sikes et al. 2011), the most recent recommendation of the AVMA (2013) regards the technique as unacceptable in unanesthetized animals due to presumed distress.

Confirmation of Death

The objective of euthanasia is, of course, death of the animal, and although the methods described here are designed to achieve that objective, it is essential that death be confirmed. This is especially true of pharmacological methods. Typically, death is confirmed either through close observation of the animal or by use of adjunctive techniques. Both approaches are described here.

Observation

Very close observation can be used to confirm death of animals, but it must be undertaken with care to ensure that animals are not merely deeply anesthetized or unconscious at a point from which they might revive. Toward this end, it is helpful, in a larger animal, to listen with a stethoscope for the heartbeat. Once an extended period of time (generally10 minutes or longer) has elapsed after the euthanasia method has been performed, animals can be assessed for any movement, including respiratory movement of the chest. With enough time, the muscles of a dead animal will stiffen (rigor mortis) and observation of such can be taken as evidence of death.

Adjunctive Methods

Because confirmation of death by observation relies on the somewhat subjective skill and interpretation by personnel, the potential for error is greater compared with assurance of death by means of additional physical methods. Any of the physical euthanasia methods described earlier in this chapter can be employed after the presumptive demise of the animal to further ensure death. Typically, these methods would only be done following induction of deep anesthesia by use of a pharmacological agent. In addition to the physical methods already described, several additional techniques can be employed to ensure death:

Exsanguination

Withdrawal of a large blood volume can be used to ensure the death of the animal and prevent resuscitation. Typically, this would require removal of at least 25% of the animal's calculated blood volume.

Pithing

This method involves insertion of a sharp tool just beneath the base of the skull in an attempt to destroy the brainstem, thereby disrupting brain function essential for survival. This is sometimes performed in amphibians and requires extensive training.

Perfusion

For some studies, it is important to clear the vasculature of blood and instill a fixative, such as paraformaldehyde. This should be performed only on deeply anesthetized animals. Usually, intravascular access is established by means of a catheter or sometimes, in the case of small rodents, a needle and syringe. Saline is first administered slowly and at a volume sufficient to replace the blood. Next, the fixative is perfused in the same manner. Death quickly results from this method. It is important to seek the advice of institutional safety experts to establish practices for working with fixatives.

Personnel Training and Oversight

Training Personnel to Perform Euthanasia

The importance of proper training on euthanasia techniques is stressed in many regulatory and guidance documents (National Institutes of Health 1986; Committee for the Update of the Guide for the Care and Use of Laboratory Animals 2011; AVMA Panel on Euthanasia 2013). Proficiency in euthanasia techniques, as well as identifying humane endpoints and procedures that minimize pain and distress, promotes animal welfare and ensures the safety of personnel. Training programs should include information on the methods of euthanasia personnel are expected to utilize, depending on the species they work with, and should provide information regarding how the chosen technique induces loss of consciousness and death (AVMA Panel on Euthanasia 2013). When selecting a euthanasia method, consideration should be given to minimizing pain and fear experienced by the animal (Artwohl et al. 2006). Techniques for minimizing pain and distress may vary by species and can include euthanizing animals in their home cage, using proper methods of handling and restraint, and providing animals with a familiar and quiet environment.

Research and animal facility staff must also be trained on techniques used to verify death. This is generally accomplished by confirming cessation of vital signs, mainly the heartbeat. It should be stressed to staff during training that some animals may appear to be dead following euthanasia but may recover following disposal if death has not been confirmed. Additionally, it has been shown that neonatal rodents are more resistant to hypoxia, and a prolonged exposure to CO_2 is necessary for euthanasia by this method (Klaunberg et al. 2004; Pritchett et al. 2005). Exposure times of 10–60 minutes have been recommended for neonatal mice based on age (Pritchett et al. 2005). For these reasons, it may be necessary to require increased exposure times or include a secondary method of euthanasia, such as bilateral thoracotomy, cervical dislocation, or decapitation, following CO_2 euthanasia of neonatal rodents. In either case, the amount of time spent by staff performing euthanasia may increase, but is necessary to ensure complete and humane euthanasia in accordance with the guidelines.

Euthanasia training programs should take into account the potential for significant variation in learning styles among those who work in laboratory animal facilities. Age, cultural or socioeconomic differences, educational levels, and English as a second language may affect the way an individual processes and retains information (Kennedy 2002). Presenting information in multiple formats may be useful during training sessions. These may include verbal presentation, providing written materials, demonstrations by the trainer, and hands-on practice by trainees. This will allow the manager or trainer to confirm proficiency prior to allowing individuals to perform the procedure without observation. This is particularly important when training personnel on procedures that have a direct impact on animal welfare, such as euthanasia techniques.

Even with significant training on proper technique, performing euthanasia can be difficult for some animal facility personnel. They may find it difficult to euthanize specific animals for which they have cared or may be uneasy performing certain euthanasia methods. It is recommended that the chosen methods of euthanasia take into account the effects of the method on personnel (Artwohl et al. 2006; AVMA Panel on Euthanasia 2013). The use of CO_2 for the euthanasia of rodents may be less distressful to personnel than physical methods (Pritchett et al. 2005), such as decapitation or thoracotomy. The manager should discuss concerns regarding euthanasia practices with personnel on a case-by-case base in order to find an acceptable resolution. For instance, animal care technicians or technologists who routinely perform CO_2 euthanasia of rodents but are uncomfortable performing a secondary method per facility or institutional requirements may need to work with another member of the animal care or veterinary team to complete these procedures. While performing euthanasia is generally a job requirement of the animal care position, making it mandatory for those with a significant aversion to a specific technique may result in the loss of high-quality animal care personnel. It is recommended that managers discuss animal use protocols in detail with facility personnel, including the necessity for euthanasia (Wolfle 1985; Overhulse 2002). Principal investigators or their research staff may also be requested to speak to animal facility personnel regarding the goals of their research and the importance of the animal caretakers to their work. Being involved in larger discussions about research projects may alleviate some of their concern and facilitate their participation in euthanasia procedures, as well as instill a sense of pride in being involved in scientific discoveries that could lead to treatments for human or animal disease and ease the suffering of both (Herzog 2002).

Psychological Impact of Euthanasia on Personnel

Many animal caregivers and other animal facility personnel chose these jobs due to their affinity for animals and a desire to provide for their care and well-being. Often, bonds form between animals and those who care for them (Arluke 1996; Bayne 2002; Herzog 2002; Overhulse 2002). Being involved in the euthanasia of animals, especially those they have worked closely with, can be difficult for personnel. The killing–caring paradox describes situations in which individuals must euthanize the animals for which they provided care (Arluke 1994). Similarly, Rollin uses the term *moral stress* to describe the stress experienced by individuals who must euthanize animals for reasons that do not directly benefit the animal; it stems from a conflict between an individual's reasons for working in an animal-related job and having to euthanize the animals in their care (Rollin 1986, 2011).

Multiple studies have been conducted to evaluate the psychological impact of euthanasia on personnel from animal shelters, research facilities, and veterinary clinics. Reeve et al. (2005) sought to determine the extent of euthanasia-related strain among shelter workers and its effects on overall employee well-being. The results showed evidence that animal euthanasia was an important source of job strain for employees and was associated with increased levels of general job stress, work and family conflict, and somatic complaints. In another study, Rohlf and Bennet (2005) explored possible perpetration-induced traumatic stress (PITS) among veterinary clinic, research laboratory, and animal shelter employees whose jobs required the euthanasia of animals. PITS is similar to posttraumatic stress but differs in that those with PITS actively participate in traumatic events. They found that 11% of participants reported moderate levels of traumatic symptoms, and that PITS is a valid avenue of study in animal workers. Additionally, the results revealed that only one-quarter of the participants had received grief counseling or stress management, leading the authors to recommend that education of employees in these areas be implemented.

As some level of psychological impact is expected among those who work in animal facilities, it is important that managers acknowledge the potential stress that employees can experience due to the formation of bonds with research animals (Herzog 2002). It is also necessary for managers to recognize the impact that performing euthanasia may have on employees and provide resources for those that need additional support (Chang and Hart 2002). The American Association for Laboratory Animal Science (2015) has recommended several additional steps that managers can take to support employees involved in the euthanasia of animals:

1. Learn and recognize stressors to personnel related to euthanasia.
2. Institute an open-door policy with supervisors and administrators.
3. Provide a pleasant work environment.
4. Supply a comfortable break area.
5. Offer education relative to humane care and use and ethics.
6. Recruit investigators to conduct informational seminars for the research team highlighting the various aspects of their particular study (especially desired benefits and outcomes).
7. Recruit investigators to detail the significance of specific endpoints of the experimentation.
8. Encourage group support meetings among laboratory personnel, and enlist the aid of an outside professional to facilitate therapeutic sessions.
9. Ensure that individuals are properly trained in the procedures of euthanasia, including mechanisms of action and how each contributes to ensuring a humane death.
10. Initiate policies that do not require the technician caring for long-term animals to participate in the euthanasia of those animals.
11. Honor the request of an individual to be excused from euthanizing an animal to which he or she is particularly attached.
12. Allow homes to be found for research animals suitable for adoption (after soliciting institutional and IACUC approval).

REFERENCES

American Association for Laboratory Animal Science. 2015. Cost of caring: Human emotions in the care of laboratory animals. https://www.aalas.org/education/educational-resources/cost-of-caring#.VfM47Zc023Y (accessed September 23, 2015).

Angus, D. W., Baker, J. A., Mason, R., and Martin, I. L. 2008. The potential influence of CO2, as an agent for euthanasia, on the pharmacokinetics of basic compounds in rodents. *Drug Metab Dispos* 36:375–379.

Arluke, A. 1994. Managing emotions in an animal shelter. In *Animals and Human Society*, ed. A. Manning and J. Serpell, 145–165. New York: Routledge.

Arluke, A. 1996. Keynote speech: The well-being of animal researchers. In *The Human/Research Animal Relationship*, ed. L. Krulisch, S. Mayer, and R. Simmons, 7–19. Greenbelt, MD: Scientists Center for Animal Welfare.

Artwohl, J., Brown, P., Corning, B., and Stein, S. 2006. Report of the ACLAM Task Force on Rodent Euthanasia. *J Am Assoc Lab Anim Sci* 45:98–105.

AVMA (American Veterinary Medical Association). 2015a. AVMA animal welfare principles. Schaumburg, IL: AVMA. https://www.avma.org/KB/Policies/Pages/AVMA-Animal-Welfare-Principles.aspx (accessed September 19, 2015).

AVMA (American Veterinary Medical Association). 2015b. Principles of veterinary medical ethics of the AVMA. Schaumburg, IL: AVMA. https://www.avma.org/KB/Policies/Pages/Principles-of-Veterinary-Medical-Ethics-of-the-AVMA.aspx (accessed September 19, 2015).

AVMA (American Veterinary Medical Association) Panel on Euthanasia. 2013. *AVMA Guidelines for the Euthanasia of Animals: 2013 Edition*. Schaumburg, IL: AVMA.

Bayne, K. 2002. Development of the human-research animal bond and its impact on animal well-being. *ILAR J* 43:4–9.

Cates, C. C., McCabe, J. G., Lawson, G. W., and Couto, M. A. 2014. Core body temperature as adjunct to endpoint determination in murine median lethal dose testing of rattlesnake venom. *Comp Med* 64:440–447.

Chang, F. T., and Hart, L. A. 2002. Human-animal bonds in the laboratory: How animal behavior affects the perspectives of caregivers. *ILAR J* 43:10–18.

Committee for the Update of the Guide for the Care and Use of Laboratory Animals. 2011. *Guide for the Care and Use of Laboratory Animals*. Washington, DC: National Research Council, National Academy Press.

Dellavalle, B., Kirchhoff, J., Maretty, L., Castberg, F. C., and Kurtzhals, J. A. 2014. Implementation of minimally invasive and objective humane endpoints in the study of murine *Plasmodium* infections. *Parasitology* 4:1–7.

Djoufack-Momo, S. M., Amparan, A. A., Grunded, B., and Boivin, G. P. 2014. Evaluation of carbon dioxide dissipation within a euthanasia chamber. *J Am Assoc Lab Anim Sci* 53:404–407.

Gnadt, J. W., and Leland S. E. 2002. Bioethics and laboratory animal research. In *Management of Laboratory Animal Care and Use Programs*, ed. M. A. Suckow, F. A. Douglas, and R. H. Weichbrod, 91–115. Boca Raton, FL: CRC Press.

Hazzard, K. C., Watkins-Chow, D. E., and Garrett, L. J. 2014. Method of euthanasia influences the oocyte fertilization rate with fresh mouse sperm. *J Am Assoc Lab Anim Sci* 53:641–646.

Herzog, H. 2002. Ethical aspects of relationships between humans and research animals. *ILAR J* 43:27–31.

Interagency Research Animal Committee. 1985. U.S. government principles for the utilization and care of vertebrate animals used in testing, research, and training. Bethesda, MD: National Institutes of Health.

Karmarkar, S. W., Bottum, K. M., and Tischkau, S. A. 2010. Considerations for the use of anesthetics in neurotoxicity studies. *Comp Med* 60:256–262.

Kennedy, B. W. 2002. Management of education and training programs. In *Management of Laboratory Animal Care and Use Programs*, ed. M. A. Suckow, F. A. Douglas, and R. H. Weichbrod, 91–115. Boca Raton, FL: CRC Press.

Klaunberg, B. A., O'Malley, J., Clark, T., and Davis, J. 2004. Euthanasia of mouse fetuses and neonates. *Contemp Top* 43:29–34.

National Institutes of Health. 1986. Public Health Service policy on humane care and use of laboratory animals. Bethesda, MD: Office of Laboratory Animal Welfare. Reprint August 2015.

Nolen, S. R. 2011. Veterinarian's Oath revised to emphasize animal welfare commitment. JAVMAnews, January 1. https://www.avma.org/News/JAVMANews/Pages/x110101a.aspx (accessed July 25, 2015).

Nunamaker, E. A., Anderson, R. J., Artwohl, J. E., Lyubimov, A. V., and Fortman, J. D. 2013. Predictive observation-based endpoint criteria for mice receiving total body irradiation. *Comp Med* 63:313–322.

Overhulse, K. A. 2002. Coping with lab animal morbidity and mortality: A trainer's role. *Lab Anim* 31:39–42.

Overmyer, K. A., Thonusin, C., Qi, N. R., Burant, C. F., and Evans, C. R. 2015. Impact of anesthesia and euthanasia on metabolomics of mammalian tissues: Studies in a C57BL/6J model. *PLoS One* 10(2):e0117232.

Pavlovic, D., Spassov, A., and Lehmann, C. 2011. Euthanasia: In defense of a good, ancient word. *J Clin Res Bioeth* 2:105.

Pritchett, K., Corrow, D., Stockwell, J., and Smith, A. 2005. Euthanasia of neonatal mice with carbon dioxide. *Comp Med* 55:275–281.

Reeve, C., Rogelberg, S. G., Spitzmüller, C., and DiGiacomo, N. 2005. The caring-killing paradox: Euthanasia-related strain among animal-shelter workers. *J Appl Soc Psychol* 35:119–143.

Rohlf, V., and Bennett, P. 2005. Perpetration induced traumatic stress in persons who euthanize nonhuman animals in surgeries, animal shelters, and laboratories. *Soc Anim* 13:201–220.

Rollin, B. E. 1986. Euthanasia and moral stress. *Loss Grief Care* 1:115–126.

Rollin, B. E. 2009. Ethics and euthanasia. *Can Vet J* 50:1081–1086.

Rollin, B. E. 2011. Euthanasia, moral stress, and chronic illness in veterinary medicine. *Vet Clin North Am Small Anim Pract* 41:651–659.

Sikes, R. S., Gannon, W. L., and Animal Care and Use Committee of the American Society of Mammalogists. 2011. Guidelines of the American Society of Mammalogists for the use of wild mammals in research. *J Mammal* 92:235–253.

Simmons, A. 2014. Beagle freedom law makes history in Minnesota. *StarTribune*, May 21. http://www.startribune.com/beagle-freedom-law-makes-history-in-minnesota/260128011/ (accessed July 25, 2015).

Strykowski, J. L., and Schech, J. M. 2015. Effectiveness of recommended euthanasia methods in larval zebrafish (*Danio rerio*). *J Am Assoc Lab Anim Sci* 54:81–84.

Toth, L. A. 2000. Defining the moribund condition as an experimental endpoint for animal research. *ILAR J* 41:72–79.

Ullman-Cullere, M. H., and Foltz, C. J. 1999. Body condition scoring: A rapid and accurate method for assessing health status in mice. *Lab Anim Sci* 49:319–323.

U.S. Department of Agriculture, Animal and Plant Health Inspection Service. 2002. Animal Welfare Act and Animal Welfare Regulations. Washington, DC: U.S. Department of Agriculture.

U.S. Department of Agriculture, Animal and Plant Health Inspection Service. 2016. Veterinary care. Animal Care Policy Manual. https://www.aphis.usda.gov/animal_welfare/downloads/Animal%20Care%20Policy%20Manual.pdf.

U.S. Food and Drug Administration. 2015. ANADA 200-226 Tricaine-S—Original approval. Silver Spring, MD: U.S. Food and Drug Administration. www.fda.gov/AnimalVeterinary/Products/ApprovedAnimalDrugProducts/FOIADrugSummaries/ucm132992.htm (accessed September 23, 2015).

Warren, T. K., Trefry, J. C., Marko, S. T., Chance, T. B., Wells, J. B., Pratt, W. D., Johnson, J. C. et al. 2014. Euthanasia assessment in Ebola virus infected nonhuman primates. *Viruses* 6:4666–4682.

Wolfle, T. 1985. Laboratory animal technicians. Their role in stress reduction and human-companion animal bonding. *Vet Clin North Am Small Anim Pract* 15:449–454.

Yapp, D. T. T., Lloyd, D. K., Zhu, J., and Lehnert, S. M. 1998. The potentiation of the effect of radiation treatment by intratumoral delivery of cisplatin. *Int J Radiat Oncol Biol Phys* 42:413–420.

Section IX

Conclusion

36

Future Directions and Challenges

Joseph N. Benoit, Robert H. Weichbrod, John N. Norton, and Janet C. Garber

CONTENTS

Managing the Biomedical Research Enterprise and the Global Status of Laboratory Animal Welfare

The latter part of the twentieth century and beginning of the twenty-first century represents an age of scientific discovery, knowledge application, globalization, and social awareness that are unlike any other periods in the history of the world. The last 50 years have seen a rapid and continuous growth of the global biomedical research enterprise that includes a worldwide increase in organizations conducting research, teaching, and testing using experimental animals. Bornmann and Mutz (2015) estimated the post–World War II growth rate of global scientific output to be 8%–9% per year, which equates to a doubling of published information every 9 years. This rate of discovery is far greater than the estimated 1% of the eighteenth century and 2%–3% of the period between the two World Wars. As the research enterprise has grown, along with the speed of global communications, so has the awareness of the public with respect to new knowledge of complex biological systems, health care improvements, and oversight of research. Expectations of accountability to the public in areas of biomedical research have grown out of increased access to information, social awareness of the general public, and the lessons of history. Globalization of biomedical research has also triggered a worldwide awareness of ethics, with greater influence from the citizenry and consumers (Degeling and Johnson 2015), who are often shaping regulations designed for oversight of the research enterprise. With respect to laboratory animal welfare, there has been an increase in both the addition and amendment of laws, regulations, and guidelines designed to address public concern, promote the health and welfare of research animals, and ensure consistent practices in facilities engineering, husbandry, veterinary care, and program monitoring. Overall, the biomedical research enterprise is more attentive to the welfare of research animals than ever before. Yet parties on both sides of the discussion regarding the use of animals in research continue to debate whether the regulations and oversight are excessive or insufficient. In response, new regulations and guidelines have emerged, and in other instances, the scientific community has responded through self-imposed regulation (Degeling and Johnson 2015). The challenge is to balance the wide range of expectations facing the research community in the future.

The purpose of this chapter is to examine some of the common challenges facing the global biomedical research community, with emphasis on emerging discussions related to regulatory burden and risk mitigation. The authors also recognize that the rule of law is implemented differently around the world, and that it would be beyond the scope of this chapter to attempt to provide a meaningful review of the myriad international differences in how law is developed and enforced in different countries, and the potential challenges of harmonization (covered in Chapters 7 and 8). As such, we have approached the topic of future directions and challenges from the perspective of institutional commitment, oversight, and the regulatory environment, with the goal of helping institutional researchers and research administrators gain an understanding of the complexity of the biomedical research environment involving animal care and use.

Regulatory Challenges

King et al., in Chapter 7 of this volume, describe the oversight of animal research that exists in today's care and use environment. At the top of the list are regulations (i.e., laws, rules, and decrees), which are created by governing bodies or government agencies to address matters deemed to be reflective of the will of the citizens. Lower in the hierarchal scheme are the vast array of guidelines, best practices, industry standards, institutional policies, and accreditation standards that represent interpretations of the laws or, in some instances, implement local policies designed to extend the requirements of the law into species not covered by the law. While the benefits of government and self-regulation on animal welfare cannot be disputed, neither can the fact that administrative procedures developed to address regulations, policies, and practices are viewed as burdensome and challenging for both the scientists and the research institution. To this end, seemingly harmless administrative practices that are not required by animal welfare regulations and do little to improve animal welfare or the quality of the research become an unnecessary focus of institutional research administrators. From a researcher perspective, the collective practices, whether self-imposed or externally mandated, are viewed as a regulatory burden.

Regardless of the source, the number of regulations in the United States has grown over the past 25 years. A recent analysis of data from the U.S. Council on Government Relations (COGR) by the Federation of Societies for Experimental Biology (FASEB) has shown a dramatic increase in the cumulative number of unfunded mandates applicable to research institutions since 1991 (FASEB 2015). The number of U.S. government–promulgated new or substantially changed regulations has increased from approximately 1.5 per year in the early 1990s to approximately 5.8 per year in more recent years (2002–2014). With more regulations to attend to, many institutions have chosen to implement conservative, sometimes overzealous, practices of self-imposed administrative practices that they view as necessary to minimize institutional risks and ensure compliance. In fact, risk aversion has been suggested to be a major contributor to excessive regulation and a source of regulatory burden (Haywood and Greene 2008; Thulin et al. 2014). In addition to institutional-level policies, current data also suggests that federal agency risk aversion may be contributing to the increased regulatory burden in the United States. A recent National Academies report, *Optimizing the Nation's Investment in Academic Research* (National Academies of Sciences, Engineering, and Medicine 2016), indicated that the requirements of some federal agencies (e.g., the National Institutes of Health [NIH]) seem to strive for a zero-risk environment. In doing so, government agencies have instituted intensive oversight, monitoring, and compliance documentation processes that "require significant commitment by the institution and the investigator without any direct significant benefit for animals" (Haywood and Greene 2008).

Expanding Oversight and the Compliance Environment

Perhaps the biggest challenge, if not threat, to the biomedical and animal science research community in the United States is the varied requirements for oversight and reporting imposed by the different federal agencies that fund research. While the rules and guidance documents are developed to ensure the humane treatment of animals and to assure the public that scientific investigation is conducted in accordance with the highest ethical standards, these same regulations have placed significant documenting requirements and financial burdens on research institutions, which have had to expand administrative offices to add staff and augment associated data management tools devoted exclusively to regulatory compliance. Challenges for research institutions in areas of regulatory compliance are compounded by reductions in funding for biomedical research, as well as the development of a "culture of overcompliance" in research organizations that are attempting to avoid penalties or liabilities that may result from noncompliance. Enhanced U.S. Department of Agriculture (USDA) enforcement of the Animal Welfare Act (AWA) and revision of the USDA Animal and Plant Health Inspection Service (APHIS) inspection requirements focus additional attention to risks of noncompliance at the institutional level (American Physiological Society 2010). The tendency for risk aversion in many institutions eclipses that of rationale risk mitigation and often leads to unnecessary time-consuming practices that extend well beyond legal requirements (FASEB 2015). Concerns of the scientific community and institutional administrators that increases in regulatory burden were detracting from the work of scientists in research and discovery recently garnered the attention of the U.S. Congress. As a result, Congress tasked the National Academies of Science, Engineering, and Medicine with the examinations of federal regulations designed to support basic and applied research. In doing so, Congress requested that the report include recommendations that would (1) "assess the effectiveness of current regulations to achieve their intended purposes and modify those that are currently ineffective," (2) "decrease redundancies of effort due to different government agencies utilizing different formats and requirements for receipt of similar information," and (3) "develop new mechanisms for government agencies and academia to develop joint recommendations that best achieve regulatory intent and optimize the federal investment in research" (National Academies of Sciences, Engineering, and Medicine 2016). The findings of this report, while centered on U.S. government–funded research at U.S. universities and research centers, underscore a series of challenges facing the global research community in an increasingly "flat world" (Friedman 2005). The committee recommendation for animal research centered on three major areas: (1) feasibility of harmonization, (2) streamlining federal and agency reporting, and (3) institutional-level streamlining of self-regulation.

Identification of a need for harmonization of regulations in the United States represents a major step forward for research programs. The National Academies report on optimization uses data from the COGR and compiled by FASEB to illustrate the reporting challenges facing researchers in the United States. The data shows an increase in the number of regulatory and guidance document changes applicable to research institutions since 1991. Self-regulation, overregulation, and risk management represent major temptations and, in some instances, legal challenges for many institutions. The temptation to overregulate at the institutional level is often the result of a philosophy of zero risk rather than one of risk mitigation. Haywood and Greene (2008) examined the sources and potential consequences of self-imposed regulatory burden on animal care programs and suggested that an overzealous regulatory environment may actually alienate scientists and create an environment where researchers may attempt to (1) circumvent regulations, (2) avoid conducting research in animals, (3) bypass funding opportunities that require animal studies to address important research topics, and (4) divert institutional funds into administrative areas concerned with compliance and away from areas directly responsible for animal care.

Governmental Sources of Regulation and Legal Challenges Related to Regulatory Burden

United States

Federally funded research in the United States is awarded to research universities and research organizations by as many as 13 different government agencies. All these agencies embrace the U.S. Government Principles for the Utilization and Care of Vertebrate Animals Used in Testing, Research, and Training (U.S. Department of Health and Human Services 2015) and are required to comply with the AWA and amendments (USDA 2013). In addition to these regulations and dependent on the agency providing the funding, oversight of animal research programs may also have to be responsive to more than 20 other government-generated principles, statutes, agency policies, directives, and reference manuals. Depending on the agency, animal researchers may need to be accountable to as few as 2 (i.e., Environmental Protection Agency [EPA]) and as many as 10 (i.e., Centers for Disease Control and Prevention [CDC]) different government regulations, policies, and guidelines (Figure 36.1) (FASEB 2015). Large and complex research organizations often have research supported by multiple federal agencies, and therefore must be accountable to each agency. Complying with the rules of U.S. agencies is further complicated by the fact that interagency rules and regulations on animal welfare are inconsistent and may contradict each other. Ensuring compliance with all the aforementioned regulatory components represents one of the biggest challenges facing research administrators and researchers, as differences in federal agency regulatory requirements impose a need for extensive tracking and reporting systems. Excessive, duplicative, and specific reporting requirements place undue strain on institutional-level research administration and often result in the diversion of resources, both financial and personnel, away from conducting research to administrative support areas charged with tracking, monitoring, documenting, and reporting on compliance. Researchers are also impacted by having to devote more time to satisfying the reporting requirements and less time on the generation and dissemination of new knowledge.

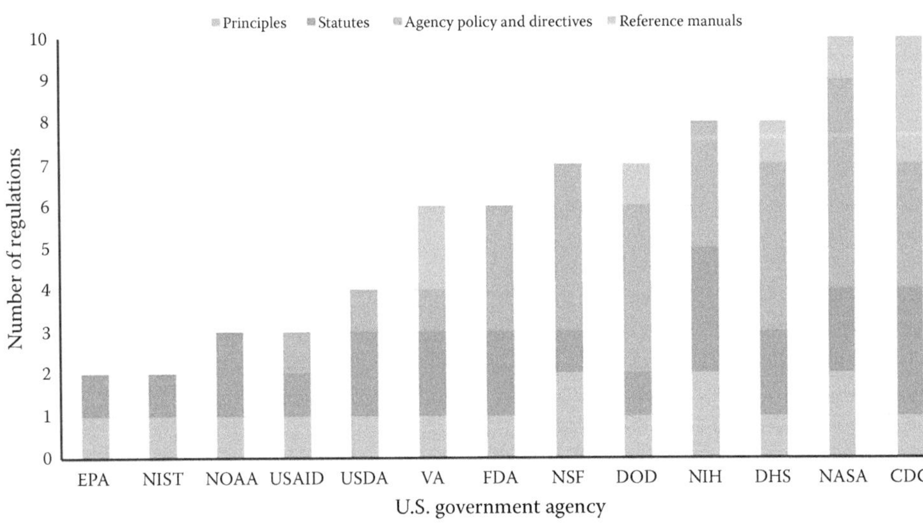

FIGURE 36.1 **(See color insert.)** Number of animal welfare regulations that different U.S. agencies require researchers to follow. NIST, National Institute of Standards and Technology; NOAA, National Oceanic and Atmospheric Administration; USAID, U.S. Agency for International Development; VA, Department of Veterans Affairs; FDA, Food and Drug Administration; NSF, National Science Foundation; DOD, Department of Defense; DHS, Department of Homeland Security. (Based on data from Dr. J. R. Haywood presented in National Academies of Sciences, Engineering, and Medicine, *Optimizing the Nation's Investment in Academic Research: A New Regulatory Framework for the 21st Century*, National Academies Press, Washington, DC, 2016.)

European Union

The governmental regulatory environment in other parts of the world parallels that observed in the United States. A recent report by Bert et al. (2016) explored some of the challenges that European Directive 2010/63/EU is placing on the scientific community in Europe. Issued as a directive rather than a regulation, 2010/63/EU sets out goals that all European Union (EU) countries must achieve by developing their own laws that are largely wrapped around the framework of the directive. The issuing of a directive by the EU provides considerable flexibility in that it allows member states to create regulations that conform to the culture and will of their people within a framework that promotes harmonization and high standards in animal welfare across the EU. However, critics of the directive claim that it is vague in some areas, which places the onus of interpretation on the member states. This could lead to discrepancies in how animal welfare regulations are managed within the EU. For example, an overzealous interpretation of the directive by some countries, particularly in areas of the directive's goal of "full replacement of procedures on live animals for scientific and educational purposes as soon as it is scientifically possible," could lead to premature halting of animal research, thereby impeding scientific research and discovery. The development and pending implementation of strict legislated limits to animal research in Italy serves as a prime example of how extreme legislation can negatively impact animal research (Abbott 2013). At present, it is difficult to predict whether disparities in regulatory burden between member states will emerge under the directive or if the goals of harmonization of animal welfare standards in Europe will be impacted by withdrawal of Britain and possibly other countries from the EU. For more details on the harmonization of standards, refer to Chapter 8 in this volume.

Other Countries

King et al., in Chapter 7 of this volume, provide a comprehensive summary of the regulations in other countries of the world, including Canada, Australia, and in Asia, South America, and Africa. Of these, the most developed oversight environments are in Australia and Canada, both of which are independent of national regulations. Emerging regulations in Asia, South America, and Africa are establishing tiered systems similar to those in other parts of the world, with attention to replacement, reduction, and refinement (the 3Rs) (Russell and Burch 1959) and ethical oversight committees playing key roles in the culture of compliance (Choe and Lee, 2014). Bayne et al. (2015) recently reported on the status of implementation of the 3Rs advanced by Russell and Burch (1959) in the worldwide arena, with particular emphasis on Brazil, China, and India. The degree to which the 3Rs are embraced and implemented in each of these three countries varies from being embedded into the scientific culture of Brazil, to steadily increasing among the scientific community of India, to aiming to align with international practices in China.

Nongovernmental Sources of Regulatory Burden

Regulatory burden in today's research environment is not only caused by government mandates and policies. It is also caused by misinterpretation or overinterpretation of rules, regulations, accreditation standards, and best practices by research institutions, ethical review boards, editorial review boards, and other members of the scientific community. Current-day ethical oversight and management of animal research is rooted in the 3Rs, as well as advances in scientific discovery, animal science, animal husbandry, facility engineering, and veterinary medicine. The benefits to animals, as well as to science, are indisputable, as pointed out by Thulin et al. (2014). Infectious diseases that were once common in research animal colonies have largely been eliminated. Greater attention to psychosocial enrichment, to a large extent, has helped to eliminate unwanted stereotypical behaviors, reduce animal injuries related to fighting in group-housed environments (Young 2003), and in some species or strains, improve breeding (Whitaker et al. 2016). The scientific community has benefitted from having fewer confounding factors introduced into sophisticated experimental design. These practices have resulted in the development of performance-based standards that are generally accepted by the public as the current standard of care for laboratory animals. Furthermore, these practices are embraced by the scientific community as a standard for pursuing and achieving scientific excellence.

However, concerns within the scientific community do arise when implementation of research oversight regulations at the institutional level results in local practices that far exceed those of the laws, regulations, and guidelines. Self-imposed regulation has been identified in recent years as a major contributor to increased regulatory burden. In a recent opinion paper, Pritt et al. (2016) described how well-intentioned assumptions made by Institutional Animal Care and Use Committees (IACUCs) and animal oversight committees create "IACUC legend," practices that become ingrained in the IACUC that go beyond the regulations and add an unnecessary administrative work burden for researchers, ethical oversight committees, and research administrative offices. Seemingly harmless administrative practices that are not required by animal welfare regulations frequently do little to improve animal welfare or the quality of the research, as they often involve only administrative office procedures. Thulin et al. (2014) suggested that programs with compliance specialists working on behalf of, but with little guidance from, the IACUC often experience a growth in self-regulation that is driven by decisions of a single person or small group of individuals that are typically not questioned by the IACUC. Over time, these procedures require an increasing effort from researchers having to respond to them, when in reality they are generally not incorporated into IACUC deliberations, institutional assurances, veterinary practices, husbandry, and facilities enhancements. Needless to say, unjustified procedural shift is rarely rewarded in the budget process, as it is difficult to justify. To this end, unauthorized implementation of self-regulation through independently acting organizationally empowered members may undermine the intent of regulations, create discord within the organization, increase financial strain on programs, and otherwise negatively impact an institution's animal care and use research programs.

One challenge is to minimize what the IACUC *does* versus what it *oversees*. For example, many IACUCs require that all personnel that will participate in a research project be listed on the written protocol proposal and qualifications to perform various procedures be described. The IACUC subsequently maintains training records and verifies that all personnel have been adequately trained. However, at some institutions, training is the responsibility of a separate group or department. Training records are retained centrally, and regularly reviewed by the training coordinator and laboratory management. In a cross-functional organization, where a technician may be asked to participate at any time in support of any research project, the requirement of listing all personnel on a research protocol may essentially be redundant and poses an unnecessary burden on the investigator and the IACUC. Rather than including technical personnel on the written protocol, the institution could take an alternate approach. Specifically, the institution could have a clearly defined training program and responsibility for record keeping, regular audits of training records (e.g., by the quality assurance unit), assurances by investigators that all personnel assigned to work on a project have been trained, and a routine monitoring program that looks at this issue. The IACUC would take on an oversight role, ensuring that the training program was functioning properly. Semiannual review of all aspects of the training program would be a component of the committee's semiannual program review. While this type of approach may not work for many programs (particularly in large academic institutions), it could be used in some smaller, centrally managed facilities.

Tracking Legislative and Rulemaking Actions in the United States

It is important for the institution to have a clear understanding of trends within the regulatory environment that can impact its program. Programmatic enhancements that are not required by laws, regulations, rules, and guidelines, or that do little to enhance animal welfare, should be avoided when possible (Haywood and Greene 2008). Understanding what is required and what is not may present a challenge to research administrators that carries significant legal implications. Any legislation that would impact regulatory compliance, access to animal models, or the availability of information may have an impact on program management because of the added costs and other burdens such laws may impose. It is important that legislation affecting research represents sound public policy, is based on evidence of need, backed by scientific data, and does not incur costs that exceed the benefits in terms of animal welfare and the quality of the science. The importance of researchers, research administrators, and members of the general public in communicating with legislators and rule makers cannot be overstated, as advocacy is key to shaping legislation to represent the will of the citizens, rather than that of a select few with special interests. Frequently, the scientific community fails to engage in advocacy before it is too late. Once laws, rules, or regulations are enacted, efforts to change them become exceedingly more difficult. In the following paragraphs, we briefly describe the processes for making laws, rules, and regulations in the United States, as an understanding of these is important for research program management.

In the United States, as many as 10,000 bills may be introduced in any given 2-year congressional session, with only about 300–500 ultimately being enacted as law. Most bills are not voted on as introduced, as they may be amended or incorporated into much larger pieces of legislation, such as an omnibus appropriations package. One example of an omnibus package with animal welfare implications is the U.S. "Farm Bill," a bundle of legislation that sets national agriculture, nutrition, conservation, and forestry policy, which is considered every 5 years (U.S. Senate 2014). While the majority of the language in the Farm Bill is typically unrelated to animal research, the potential for the inclusion of such language exists primarily because the bill must move through the House and Senate Agriculture Committees, both of which have jurisdiction over the federal AWA. Inasmuch as omnibus legislation such as the Farm Bill is considered a must-pass piece of legislation, it often serves as a vehicle for smaller pieces of legislation with special interests. A single line of text changing only one or two words in current law, that is, the AWA, can be inserted into the omnibus package as a technical amendment with the hope of being undetected as the massive 1000+-page bill moves through the legislative process. Technical amendments, although short in word count, may have enormous implications for current law.

Other approaches to passing legislation may include reintroduction of a bill by different members of Congress over many different congressional sessions. Such persistence proved to work in the research community's favor when it came to passage of the Animal Enterprise Terrorism Act (AETA) (S. 3880 [109th] 2006), which amended the federal criminal code to prohibit any person from "engaging in certain conduct for the purpose of damaging or interfering with the operations of an animal enterprise," thereby providing the U.S. Department of Justice greater authority to prosecute animal rights activists (109th U.S. Congress 2006). Educating members of Congress over a number of years allowed the bill to gain a substantial amount of support over time. In other instances, persistence alone may not be enough to ensure that legislation is passed, as evidenced by failure of special interests to catalyze passage of legislation during the last 10 congressional sessions (nearly 20 years) that would eliminate USDA Class B licenses. It is worth noting that a lack of success on the legislative front may lead special interests to achieve their goals through other means, such as through the regulatory process covered later in this chapter.

Another element that affects the odds of successful passage of legislation is whether there are similar versions of the bill in each of the two legislative bodies: one version in the U.S. House of Representatives and one version in the U.S. Senate. When the language of each of these bills is identical, the odds of passage are increased. When the language of each bill is different, a lengthier process ensues. Further, if the legislation is "jointly referred" to multiple committees in either body, the odds of passage are decreased, as this represents addition hurdles for the bill to overcome. If the legislation is "sequentially referred" to multiple committees in each body, the odds of passage become extremely low, although not impossible. Finally, if each body ultimately approves the legislation, a conference committee comprised of conferees from each legislative body enters into negotiations to reach agreement on language in a final version of the bill that will be sent to the president for either signature or veto.

U.S. Rules and Regulations

In the United States, federal regulations are created by federal agencies, boards, and commissions by a process known as rulemaking. Under the Administrative Procedure Act, federal agencies are permitted to promulgate detailed regulations from laws enacted by Congress through a public rulemaking process in which the public is invited to participate. These regulations carry the weight of law and are specifically designed to address a problem or accomplish a goal of the legislation. Once enacted, rules are published in the Code of Federal Regulations (Office of the Federal Register 2016). More information on the rule-making process is available through the Office of the Federal Register (2016). Monitoring wording of proposed legislation or rules and tracking the progress of legislation is a daunting task that individual research programs and program officers are seldom able to carry out. Researchers and institutional officers often rely on professional societies and advocacy groups, such as the National Association for Biomedical Research (NABR), to track legislation and comment on its potential impact. NABR is the only organization in the United States solely dedicated to advocating for sound public policy that recognizes the vital role that animals play in biomedical research. Staying abreast of animal welfare regulations constitutes a major challenge for animal researchers, administrators, and institutional officials. Ignorance of a law or regulation does not absolve an institution from penalty should violations occur. While there is no easy solution to navigating the animal welfare regulatory environment, there are several excellent online resources that can be consulted. These are listed in Table 36.1.

Emerging Legal Challenges Impacting Animal Care and Use Programs

Legal challenges by activist groups in areas of public disclosure, the transportation of animals, legal guardianship, and even "personhood" or legal standing for animals are developing trends that could pose a significant threat to animal research. Such tactics do not follow the traditional ethical arguments on the use of animals in research. The complexity of these issues extends far beyond the scope of this chapter, and as such, the authors have chosen to only briefly discuss them below. The reader is directed to several excellent sources that provide detailed background information on these topics (Favre 2004; Rollin 2006; Dolan 2007; Article 19 2012; Cardon et al. 2012).

Public Disclosure Laws

Seeking access to institutional documents through public disclosure laws (Freedom of Information Act [FOIA], sunshine laws, etc.) in the United States is a tactic often used by special interest groups to force research organizations to release sensitive records that can and have been used to disrupt an animal research program's operations and, in extreme cases, place animals, as well as research personnel, at risk. Originally designed to promote transparency in government, the U.S. FOIA was enacted in 1966 to instill accountability in government by granting public access to government records. In addition to U.S. federal law, every state also has a law that governs access to records in possession of the state and local governments and other public entities, such as state-supported universities. While both federal and state open records laws exempt certain categories of information from being released, there can be considerable state-to-state variability in the exemptions. For example, some state courts have held that IACUC minutes are subject to disclosure, while courts in other states have held that IACUCs are not public bodies and as such are not subject to disclosure (discussed in Cardon et al. [2012]). In public universities, open records requests could garner other sensitive information, such as detailed building maps, including vivarium spaces. In instances where electronic information is transferred to private e-mail accounts or personal electronic devices, information on home computers, smartphones, digital tablets, and so forth, could be subject to certain types of records request. Concern for the safety of employees, students, property, and research animals is real for administrators in public research institutions in the United States, particularly when the frequent requesters of biomedical research information are national animal rights groups or individuals associated with those groups (NABR 2015). In addition to the United States, laws granting access to public information are becoming more common in the global arena, as evidenced by the fact that public disclosure laws have increased from 13 countries in 1990 to nearly 100 countries in 2012 (Article 19 2012). For more information on security considerations for animal use programs, refer to Chapter 19 in this volume.

Public institutions and agencies must respond to an open records request and need to be adequately prepared to respond should a request be made. Cardon et al. (2012) provided detailed information on preparing and responding to public information requests. At the center of their discussion was the need

TABLE 36.1

Internet Sources for Animal Regulations from Different Regions of the World

Resource and URL	Description
AAALAC International www.aaalac.org	AAALAC International is a private, nonprofit organization that promotes the humane treatment of animals in science through voluntary accreditation and assessment programs. The Resources page provides links to U.S. and international regulations.
Agri-Food & Veterinary Authority (AVA) of Singapore www.ava.gov.sg	AVA has a major role in the protection of the living environment in Singapore and safeguarding the health of animals, fish, and plants.
Animal Welfare Information Center (AWIC) www.nal.usda.gov/awic	Established in 1986, AWIC is part of the USDA National Agricultural Library (2016) and serves to help the regulated community with employee training and to promote the humane care and use of animals by providing information on alternatives.
Australian and New Zealand Council for the Care of Animals in Research and Teaching (ANZCAART) www.adelaide.edu.au/ANZCCART/	ANZCCART was established in 1987 in response to particular concerns in both the wider and scientific communities about the use of animals in research and teaching. It also recognized the need to address, in a balanced and considered way, the distinctive scientific, ethical, and social issues associated with the use of animals for scientific purposes.
Canadian Council on Animal Care (CCAC) www.ccac.ca	The CCAC is the national peer review organization responsible for setting, maintaining, and overseeing the implementation of high standards for animal ethics and care in science throughout Canada.
Federation of European Laboratory Animal Science Associations www.felasa.eu	FELASA represents common interests in the furtherance of all aspects of laboratory animal science in Europe and beyond. FELASA puts the 3Rs of laboratory animal science, replacement, reduction, and refinement, center stage. FELASA advocates responsible scientific conduct with animals in the life sciences, with particular emphasis on ensuring animal welfare.
Georgetown Law Library Animal Law Research Guide http://guides.ll.georgetown.edu/AnimalLaw	This guide will assist in researching issues in animal law, animal rights, and animal welfare using resources available to Georgetown University Law Center students and faculty. Animal law topics include domestic animals, farm animals, marine mammals, wildlife, animal experimentation, and endangered species.
Government of India Ministry of Environment, Forest and Climate Change Animal Welfare Division envfor.nic.in/division/animal-welfare-division	The Animal Welfare Division is entrusted with the implementation of the provisions of the Prevention of Cruelty to Animals Act, 1960 (59 of 1960).
International Aid for Korean Animals (IAKA) Koreananimals.org	IAKA is a nonprofit organization that promotes the humane treatment of animals in South Korea.
Malaysia e-Federal Gazette www.federalgazette.agc.gov.my/	The e-Federal Gazette is the official portal for the publication of all federal legislation. Animal Welfare Act 2015 is Act 772.
Michigan State University Animal Legal & Historical Center www.animallaw.info	On this site, you will find a comprehensive repository of information about animal law, including more than 1200 full-text cases (U.S., historical, and UK), more than 1400 U.S. statutes, more than 60 topics and comprehensive explanations, legal articles on a variety of animal topics, and an international collection.
National Association for Biomedical Research (NABR) Animal Law Section www.nabranimallaw.org	The NABR Animal Law Section focuses on the potentially disruptive and rapidly growing area of animal law. Animal rights advocates are using increasingly sophisticated and coordinated legal strategies in an attempt to incrementally change our laws as they relate to animals. The potential consequences for biomedical research are not readily apparent, but it is clear that many animal rights organizations and animal rights lawyers believe research on animals should be severely restricted or prohibited completely. This website is intended to serve as a resource for those who wish to learn more about animal law. It does not endeavor to be an exhaustive survey of issues pertaining to the legal status of animals, or of the animal rights movement.

(*Continued*)

TABLE 36.1 (CONTINUED)

Internet Sources for Animal Regulations from Different Regions of the World

Resource and URL	Description
National Institutes of Health Office of Laboratory Animal Welfare (OLAW) http://grants.nih.gov/grants/olaw/	OLAW provides guidance and interpretation of the Public Health Service (PHS) Policy on Humane Care and Use of Laboratory Animals, supports educational programs, and monitors compliance with the policy by assured institutions and PHS funding components to ensure the humane care and use of animals in PHS-supported research, testing, and training, thereby contributing to the quality of PHS-supported activities.
Norecopa norecopa.no	Norecopa is Norway's National Consensus Platform for the advancement of the 3Rs in connection with animal experiments.
Philippines *Official Gazette* www.gov.ph	This website is the official journal of the Republic of the Philippines. This website, the national government portal, is updated regularly with speeches, reports, statements, press releases, and documents from the Office of the President and other departments of the Philippine government.
Taiwan Council of Agriculture http://eng.coa.gov.tw	The Council of Agriculture is the competent authority on the agricultural, forestry, fishery, animal husbandry, and food affairs in Taiwan.
Understanding Animal Research www.understandinganimalresearch.org.uk	Understanding Animal Research is a mutual society (not-for-profit organization) that explains why animals are used in medical and scientific research. It aims to achieve a broad understanding of the humane use of animals in medical, veterinary, scientific, and environmental research in the United Kingdom.
United Kingdom Home Office www.gov.uk/guidance/research-and-testing-using-animals	Research and testing using animals.

for document retention and destruction policies at the institutional level. Research administrators should fully understand their institution's policies, as well as any government requirements on document retention, archiving, and destruction. Active files relative to the animal care and use oversight program should be factual in content and devoid of extraneous information, and contain only essential information that is required for the oversight committee and research office to function. To this end, the research organization should be cognizant of what is included in the official files of the organization and of the potential impact to the institution that might arise if requested information is taken out of context and widely disseminated by the requester. Research administrators and staff should always consult their organization's record retention officers and legal counsel when they receive an open records request for information in order to ensure that (1) the requested information is not exempt from request, and (2) the information released is specific to the request. The latter is particularly important, as an organization has no obligation to provide any information beyond what is specifically requested.

Transportation of Research Animals

Pressure placed on commercial airlines by animal rights activist groups has resulted in many commercial carriers choosing to no longer carry animals that will be used for research purposes. The reduced shipping options present logistical transportation challenges, limit access to important animal models, and increase the costs associated with animal research. Disruption of the use of commercial transport for moving research animals to biomedical research organizations could significantly delay, if not halt, ongoing research projects. Although a few countries have received government support that compels airlines to transport research animals, no such regulations exist in the United States. As a result, the commercial airline has full decision-making authority on whether it will (1) carry animals and (2) refuse to carry them if they are destined for research facilities. It is likely that legal challenges to the commercial air carriers will require courts to interpret common carrier obligations as they apply to shipping live animals for research purposes. While no one can predict the outcomes of these deliberations, one can be certain that differences between countries served by any given airline will complicate and extend the discussion for years to come.

Emerging Challenge of Property versus Legal Guardianship

Legal challenges to animal research programs by lawyers and activists are testing new legal postulates that stand to grant specific legal rights to animals, including those animals that are involved in research. Leaders of this legal movement advocate the establishment of new court precedents that have the potential to significantly interfere with animal research. Unlike many earlier court cases that challenged existing laws regulating animal welfare for animals owned by a research organization, these emerging cases argue that animals are not property. Instead, they argue that animals should be placed under a legal guardianship where the guardian is held legally accountable for the welfare of the animal. Favre (2004) argues in favor of transforming property law from ownership to guardianship, which he claims would allow animals to "receive the legal respect they deserve." He further submits that the current activist approach of abolition of animal research at the level of federal laws or regulations is unrealistic, and that more achievable outcomes on the animal activist front may arise through building a foundation of change focused at the state level. Suggestions include

1. "Modifying divorce laws to require that the court decides issues relating to pets in the best interest of the animal and not on the basis of property ownership."
2. "Adopting state anticruelty laws such that criminal prosecution could be brought against guardians who violate the laws" (e.g., researchers and research organizations).
3. "Clarifying the law regarding pet adoption from a public or private agency such that the organization retains the ability to intervene on behalf of the animal. If the new owners/guardians do not fulfill their obligations toward the animal, the agency ought to be able to step in and correct the problem."

The position of gradually articulating the interests of animals into the legal system as a means to gain a stronger position for the animal rights activists (Favre 2004) could represent a significant legal challenge for both farmers and scientists. Discussions around ethical and humane treatment of research animals would be refocused toward the investigator's or research organization's ability to determine what is best for the animal. These challenges have already entered into the research animal arena with changes in policy that have resulted in the discontinuation of studies in chimpanzees and subsequent transfer of these animals to sanctuaries. In other instances, laws allowing trusts to be set up for family pets move the legal interpretation closer to legal personhood for certain species. Efforts by animal rights interests seeking to have courts recognize animals as having legal standing or personhood are on the rise. In recent years, there have been at least four such cases in the state of New York asking courts to recognize chimpanzees as having legal standing under *habeas corpus*. Although all have failed, groups appear to be committed to filing multiple test cases in courts around the country on behalf of different species until they are successful. While many of these strategies do not appear, at first glance, to affect laboratory animals, the "sum of their parts" has the potential to have an impact on life-saving medical and scientific research.

Assessing, Mitigating, and Managing Risks

Management of risk in an animal research program is a difficult task that involves a comprehensive understanding of the institution's research endeavors, the regulations governing animal welfare, and the institution's plans for the future. Bayne and Garnett (2008) describe the delicate balance among research costs, a complex legal and regulatory environment, societal issues, and the scientists' passion for discovery as obligatory if the biomedical research enterprise is to maintain the privilege to conduct research in animal models. To this end, institutions must constantly assess their animal research programs from different perspectives and do so in ways that account for the nature of the research being conducted, stakeholder expectations, and a constantly changing regulatory, oversight, and compliance environment. Assessment, mitigation, and management of risks are an essential, yet highly institution-specific, component of a contemporary animal care and use program. To meet this challenge, a research organization must have a clear understanding of its mission, be willing to accept certain risks, and be responsive to the regulatory environment in ways that do not impose excessive self-regulation. Developing a zero-risk or risk aversion philosophy has been identified by numerous sources (Haywood and Greene 2008; FASEB 2015; National Academies of Sciences, Engineering, and Medicine 2016) as a burdensome, somewhat unrealistic, and generally cost-prohibitive approach to the management of a research program. A more realistic approach is to build a risk assessment program geared toward risk mitigation, knowing that regulatory compliance can be achieved in a performance-based environment that aligns with the institutional needs. Haywood and Green (2008) suggested a commonsense three-question approach to assessing IACUC activities for regulatory burden: (1) Why do we do this? (2) Does it help the animals? (3) Can the end be achieved in a more efficient, cost-effective manner? Implementation of a performance-based program based on sound reasoning allows an institution to fully understand and address its specific animal welfare requirements without having to implement broad-sweeping burdensome practices. As a result, the institution animal research program becomes more strategic and less reactive.

A myopic approach to regulatory risk may also lead an institution and its investigators away from risk based on scientific foundation and understanding. Recent emphasis from funding agencies on the rigor of sound and unbiased design of research to ensure reproducibility underlies the importance of selecting the appropriate biological research model, whether in terms of species, sex, age, or health conditions of the animals (NIH 2016a, 2016b). Evidenced by an overdependence on male animals, preclinical studies in animals commonly neglect gender-based differences in pathophysiology and response to treatment modalities, thus leading the NIH to require grant applicants to consider and report research plans that balance the use of male and female animals (Clayton and Collins 2014). Selective reporting of research results and the pressure on researchers to publish have been cited as additional factors impacting reproducibility of research (Baker 2016). The development of the ARRIVE guidelines and agreement by many journals to endorse principles and guidelines on the reporting of animal research are appropriate avenues to enhance the rigor and reproducibility of animal research (Kilkenny et al. 2010; NIH 2016b). A balanced approach to ensuring animal welfare while promoting innovative and sound science through public education and advocacy (see Chapter 6) is warranted and best serves the concept of streamlining and achieving regulatory demands.

An important step in building a streamlined risk-managed compliant program is for the institution to have a clear understanding of trends within the regulatory environment that can impact its program. Programmatic enhancements that are not required by laws, regulations, rules, and guidelines or that do little to enhance animal welfare should be avoided when possible (Haywood and Greene 2008). Understanding what is required and what is not may present a challenge to research administrators that carries significant legal implications.

Summary and Recommendations for a Path Forward

The global legal arena for animal welfare is a complex and ever-changing one that can be somewhat unpredictable. As a result, animal research organizations are constantly having to respond to newly created or modified existing regulations and evolving guidelines. While all stakeholders in research animal welfare conversation understand and support the existence of laws and regulations that promote animal welfare and the pursuit of quality science, there is growing concern that many of the regulations create large amounts of work for administrative offices, but do little to protect the animals. As a result, less time is being spent on research, discovery, and knowledge creation.

It is interesting to look at some predictions of future challenges facing the laboratory animal community from 20 years ago. In 1995, J. Derrell Clark proposed four points for consideration in the future (Clark 1995). First, he surmised that there was very little scientific information on, or factors in, our knowledge that promote the comfort of animals in a research setting. He suggested the need for research on behavior, psychological well-being, enrichment, and effects of numerous physical and environmental factors. Second, he recognized the infancy of the concept of performance standards and the necessity to determine the animals' needs and evaluate performance, specifically the need for assessment criteria (which at the time was not well established). Third, laboratory animal scientists must conduct research in areas of animal care. And finally, there is more relevant information on care and husbandry in the literature than most are aware of. Much of the relevant data may be in studies focused on animal behavior, psychology, zoology, field studies, ethology, animal science, and psychoneuroimmunology. While Clark advocated for the mining of such data, he also cautioned that the use of published information may be difficult to interpret and even harder to apply to general situations, or put into general practice.

In the years ahead, research organizations and their team of organizational leaders, scientists, veterinarians, managers, and other animal research support staff need to work collectively together to refocus and streamline the animal research regulatory environment without compromising the intent of laws and regulations that already exist. The scientific community and agencies that fund research must continue, as well as expand, advocacy for animal research through education of the general public of the human and animal health advances that have resulted from animal research (see Chapter 6 of this volume). Harmonization of regulations and reporting requirements, as well as the sharing of information between research organizations, government agencies, oversight bodies, and accreditation groups, could alleviate some of the challenges. Assessment of oversight programs within institutions, with an eye on spotting redundancies and unnecessary office or committee practices, will also help to limit the expansion of self-imposed regulatory burden. Research administrators should understand the risks associated with their research programs and develop program-specific approaches that mitigate their identified risks, rather than implementing practices designed to avert all the possible scenarios that might occur. Research strategic plans should address goals for supporting animal research and be bold enough to strongly advocate for the significant advancement of knowledge, with particular emphasis on the 3Rs. Collectively, these and other similar maneuvers can allow research organizations to prosper and coexist in today's complex and ever-present regulatory environment.

ACKNOWLEDGMENT

The authors and editors extend their sincere appreciation to

1. Tim Allen with the U.S. Department of Agriculture's Animal Welfare Information Center for his superlative, comprehensive, and thorough literature search assistance to the writing of this chapter
2. The National Association for Biomedical Research, Washington, DC, for providing its policy expertise in the development of this chapter

REFERENCES

109th U.S. Congress. 2006. Public law 109-374, November 27. https://www.congress.gov/109/plaws/publ374/PLAW-109publ374.pdf.

Abbott, A. 2013. "Newsblog." nature.com, August 1. http://blogs.nature.com/news/2013/08/ (accessed October 26, 2016).

American Physiological Society. 2010. USDA announces enhanced Animal Welfare Act enforcement. Bethesda, MD: American Physiological Society. http://www.the-aps.org/mm/SciencePolicy/AnimalResearch/Current-Issues/USDA-Announces-Enhanced-Enforcement.html (accessed January 8, 2017).

Article 19. 2012. International standards: Right to information. April 5. https://www.article19.org/resources.php/resource/3024/en/international-standards:-right-to-information (accessed September 13, 2016).

Baker, M. 2016. Is there a reproducibility crisis? *Nature* 533: 452–454.

Bayne, K., G. S. Ramachandra, E. A. Rivera, and J. Wang. 2015. The evolution of animal welfare and the 3Rs in Brazil, China, and India. *Journal of the American Association for Laboratory Animal Science* 54 (2): 181–191.

Bayne, K. A., and N. L. Garnett. 2008. Mitigating risk, facilitating research. *ILAR Journal* 49 (4): 369–371.

Bert, B., J. Chmieleweska, A. Hensel, B. Grune, and G. Schonfelder. 2016. The animal experimentation quandry: Stuck between legislation and scientific freedom. *EMBO Reports* 17 (6): 790–792.

Bornmann, L., and R. Mutz. 2015. Growth rates of modern science: A bibliometric analysis based on the number of publications and cited references. *Journal of the Association for Information Science and Technology* 66 (11): 2215–2222.

Cardon, A. D., M. R. Bailey, and B. T. Bennett. 2012. The effect of public disclosure laws on biomedical research. *Journal of the American Association for Laboratory Animal Science* 51 (3): 306–310.

Choe, B. I., and G. H. Lee. 2014. Individual and collective responsibility to enhance regulatory compliance of the three Rs. *BMB Reports* 47 (4): 179–183.

Clark, J. D. 1995. The role of research in laboratory animal regulations and guidelines: Past, present and future. In *Current Issues and New Frontiers in Animal Research*, 7–11. Greenbelt, MD: Scientists Center for Animal Welfare.

Clayton, J. A., and F. S. Collins. 2014. NIH to balance sex in cell and animal studies. *Nature* 509: 282–283.

Degeling, C., and J. Johnson. 2015. Citizens, consumers and animals: What role do experts assign to public values in establishing animal welfare standards? *Journal of Agricultural and Environmental Ethics* 28: 961–976.

Dolan, K. 2007. *Laboratory Animal Law: Legal Control of the Use of Animals in Research*. 2nd ed. Oxford: Blackwell Publishing.

FASEB (Federation of American Societies for Experimental Biology). 2015. *Sustaining Discovery in Biological and Medical Sciences: A Discussion Framework*. Bethesda, MD: FASEB.

Favre, D. 2004. Integrating animal interests into our legal system. Social Science Research Network Electronic Paper Collection. http://ssrn.com/abstract=578570 (accessed September 13, 2016).

Friedman, T. L. 2005. *The World Is Flat: A Brief History of the Twenty-First Century*. New York: Farrar, Straus & Giroux.

Haywood, J. R., and M. Greene. 2008. Avoid an overzealous approach: A perspective on regulatory burden. *ILAR Journal* 49 (4): 426–434.

Kilkenny, C., W. J. Browne, I. C. Cuthill, M. Emerson, and D. G. Altman. 2010. Improving bioscience research reporting: The ARRIVE guidelines for reporting animal research. *PLoS Biology*. http://journals.plos.org/plosbiology/article?id=10.1371/journal.pbio.1000412 (accessed December 14, 2016).

National Academies of Sciences, Engineering, and Medicine. 2016. *Optimizing the Nation's Investment in Academic Research: A New Regulatory Framework for the 21st Century*. Washington, DC: National Academies Press.

NABR (National Association for Biomedical Research). 2015. FY14 animal rights FOIA requests. June 23. Washington, DC: NABR. http://www.nabr.org/nabr-releases-analysis-of-animal-rights-federal-foia-requests-in-fy14/ (accessed October 26, 2016).

NIH (National Institutes of Health). 2016a. Rigor and reproducibility. Bethesda, MD: NIH. https://grants.nih.gov/reproducibility/index.htm (accessed December 14, 2016).

NIH (National Institutes of Health). 2016b. Principles and guidelines for reporting preclinical research. Bethesda, MD: NIH. https://www.nih.gov/research-training/rigor-reproducibility/principles-guidelines-reporting-preclinical-research (accessed December 14, 2016).

Office of the Federal Register. 2016. www.ofr.gov (accessed October 26, 2016).

Pritt, S., J. A. McNulty, M. Greene, S. Light, and M. Brown. 2016. Decreasing institutionally imposed regulatory burden for animal research. *Lab Animal* 45 (8): 297–300.

Rollin, B. E. 2006. The regulation of animal research and the emergence of animal ethics: A conceptual history. *Theoretical Medicine and Bioethics* 27: 285–3014.

Russell, W. M. S., and R. L. Burch. 1959. *The Principles of Humane Experimental Technique*. London: Methuen & Co.

S. 3880 (109th). 2006. Animal Enterprise Terrorism Act. November 15. https://www.govtrack.us/congress/bills/109/S3880/text (accessed October 26, 2016).

Thulin, J. D., J. F. Bradfield, V. K. Bergdall, L. A. Conour, A. W. Grady, D. L. Hickman, J. N. Norton, and J. M. Wallace. 2014. The cost of self-imposed regulatory burden in animal research. *FASEB Journal* 28: 3297–3300.

USDA (U.S. Department of Agriculture). 2013. Animal Welfare Act and Animal Welfare Regulations. November. https://www.aphis.usda.gov/animal_welfare/downloads/Animal%20Care%20Blue%20Book%20-%202013%20-%20FINAL.pdf (accessed September 11, 2016).

USDA National Agricultural Library. 2016. Animal Welfare Information Center. https://www.nal.usda.gov/awic (accessed September 12, 2016).

U.S. Department of Health and Human Services, National Institutes of Health, Office of Laboratory Animal Welfare. 2015. Public Health Service policy on humane care and use of laboratory animals. Bethesda, MD: U.S. Department of Health and Human Services.

U.S. Senate. 2014. U.S. Senate Committee on Agriculture, Nutrition, & Forestry. http://www.agriculture.senate.gov/issues/farm-bill (accessed October 26, 2016).

Whitaker, Julia W., Sheryl S. Moy, Kathleen R. Pritchett-Corning, and Craig A. Fletcher. 2016. Effects of enrichment and litter parity on reproductive performance and behavior in BALB/c and 129/Sv mice. *JAALAS* 55 (4): 387–399.

Young, R. J. 2003. *Environmental Enrichment for Captive Animals*. UFAW Animal Welfare Series. Oxford: Blackwell Publishers.

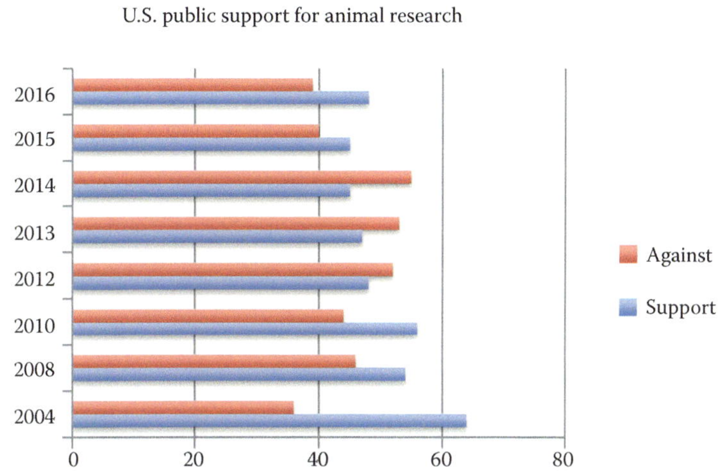

FIGURE 6.3 Surveys by the Foundation for Biomedical Research show that U.S. public support for the use of animals in medical research has declined from about 70% in 2004 to about 45% in recent years. Survey results for 2015 and 2016 included respondents who answered "unsure," 5% and 13%, respectively. Those responses are not included in the chart. (Courtesy of the Foundation for Biomedical Research.)

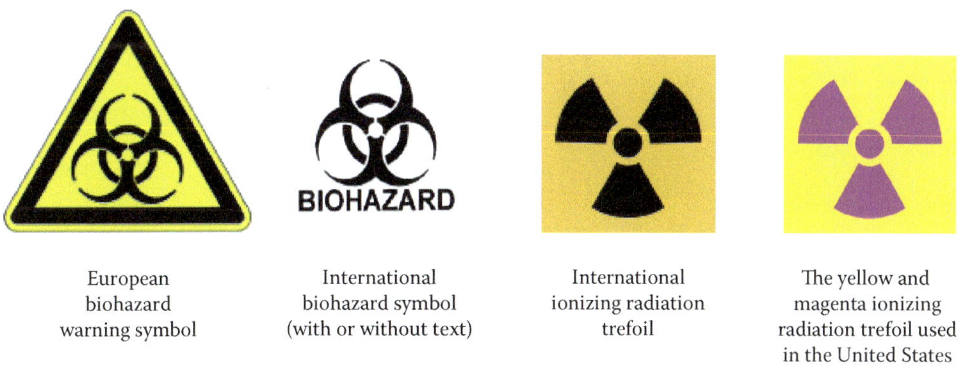

FIGURE 14.3 International warning symbols for biohazardous and radioactive materials.

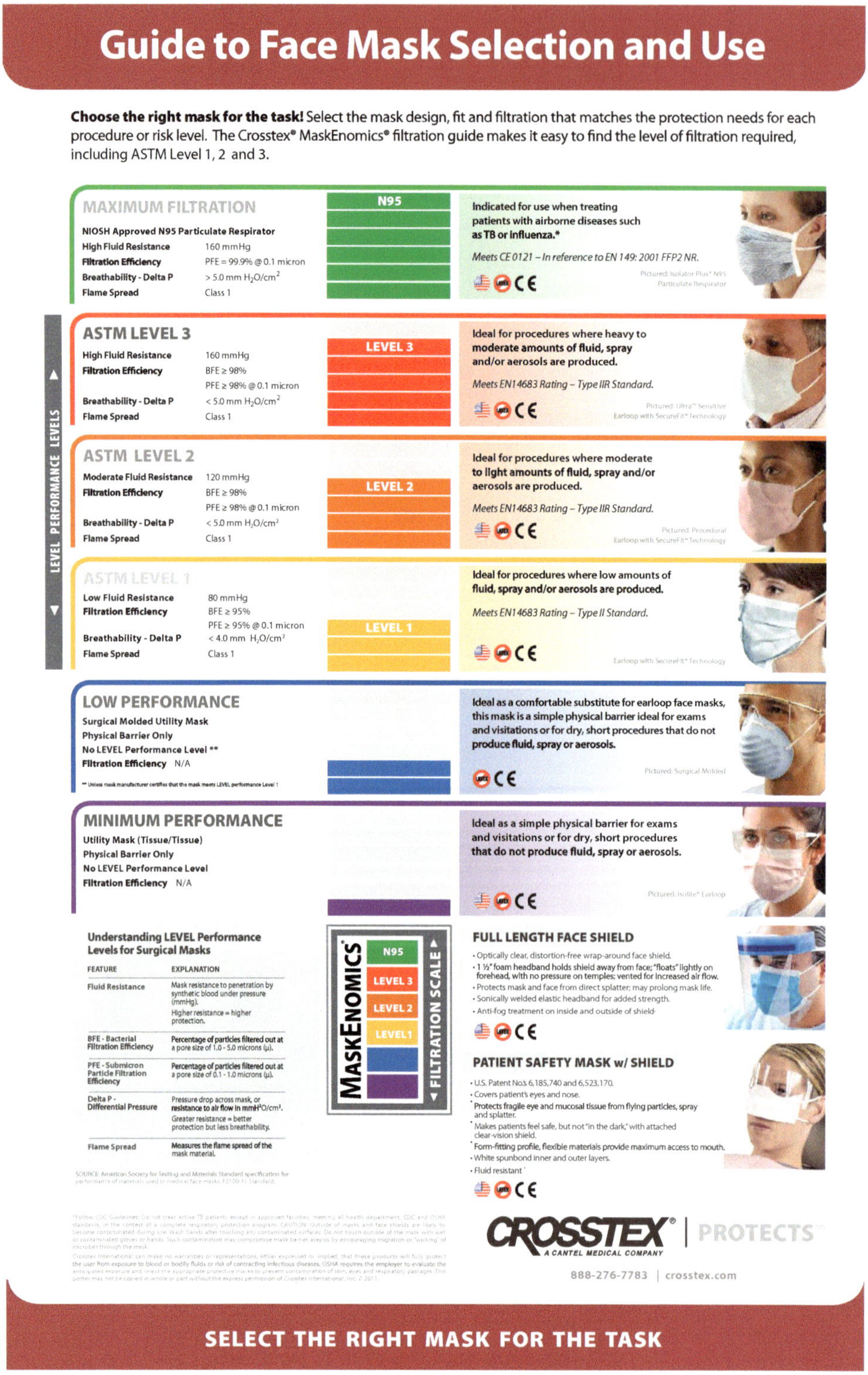

FIGURE 14.4 Guidelines for the selection and use of face masks. (From CROSSTEX, Hauppauge, NY, http://www.crosstex.com/home.asp.)

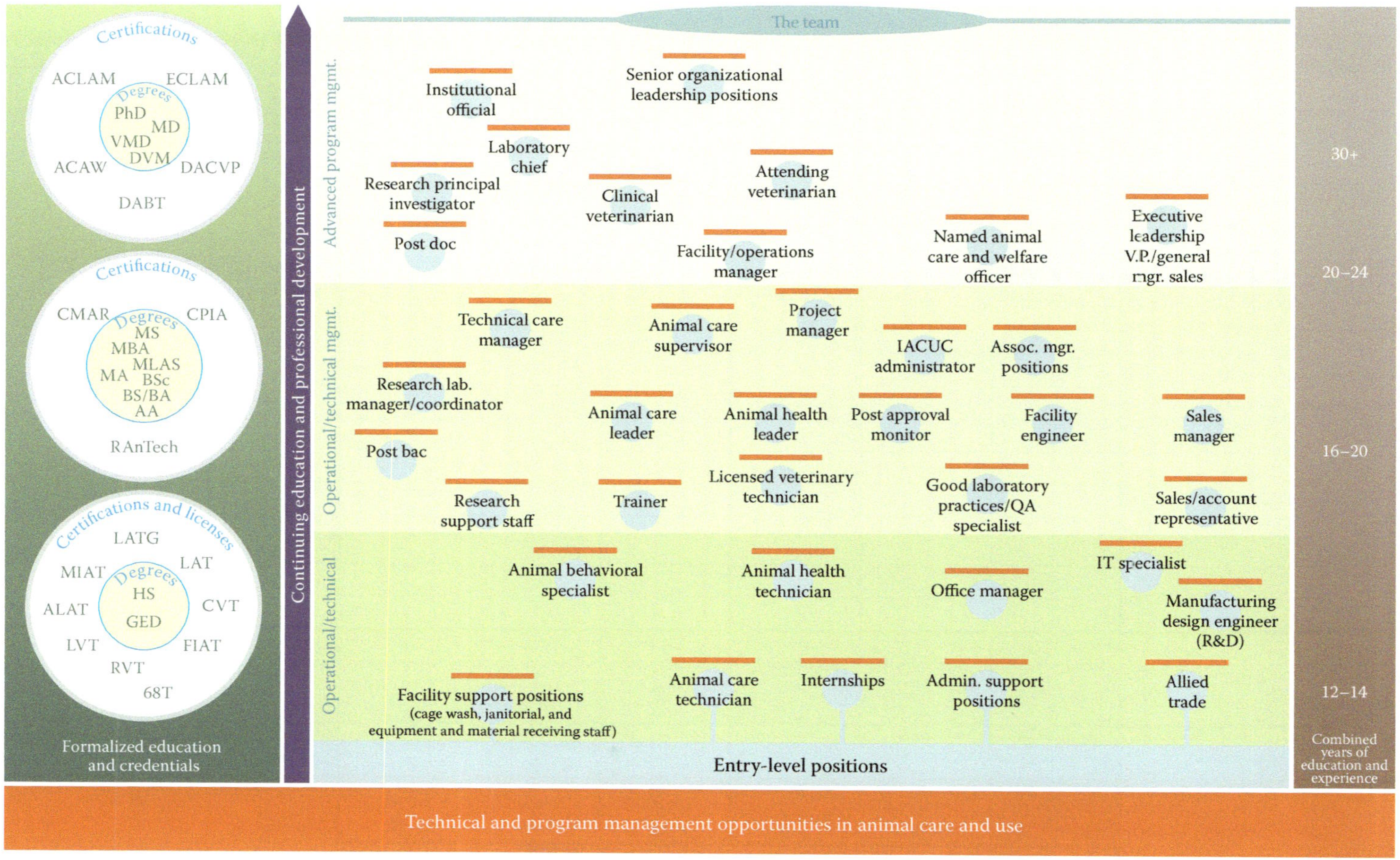

FIGURE 11.1 Example of a team that shares the responsibility to work closely together in programs of animal care and use around the world to ensure quality animal well-being and the responsible conduct of research, education, and testing with animals. This is not intended as representative of any particular organization or institution. (Developed by R. H. Weichbrod and K. McKibbin with Graphics Prints, Inc., Herndon, VA.)

Hazard and vulnerability assessment tool naturally occurring events

KAISER PERMANENTE.

Event	Probability	Severity = (magnitude – mitigation)						Risk
		Human impact	Property impact	Business impact	Preparedness	Internal response	External response	
	Likelihood this will occur	*Possibility of death or injury*	*Physical losses and damages*	*Interruption of services*	*Preplanning*	*Time, effectiveness, resources*	*Community/ mutual aid staff and supplies*	*Relative threat**
Score	*0 = N/A 1 = Low 2 = Moderate 3 = High*	*0 = N/A 1 = Low 2 = Moderate 3 = High*	*0 = N/A 1 = Low 2 = Moderate 3 = High*	*0 = N/A 1 = Low 2 = Moderate 3 = High*	*0 = N/A 1 = High 2 = Moderate 3 = Low or none*	*0 = N/A 1 = High 2 = Moderate 3 = Low or none*	*0 = N/A 1 = High 2 = Moderate 3 = Low or none*	*0–100%*
Hurricane								0%
Tornado								0%
Severe thunderstorm								0%
Snowfall								0%
Blizzard								0%
Ice storm								0%
Earthquake								0%
Tidal wave								0%
Temperature extremes								0%
Drought								0%
Flood, external								0%
Wildfire								0%
Landslide								0%
Dam inundation								
Volcano								0%

FIGURE 17.4 Example of hazard and vulnerability assessment tool. (From Kaiser Foundation Hospitals and Health Plan, Kaiser hazard vulnerability assessment (HVA), Kaiser Foundation Hospitals and Health Plan, Oakland, CA, 2001.)

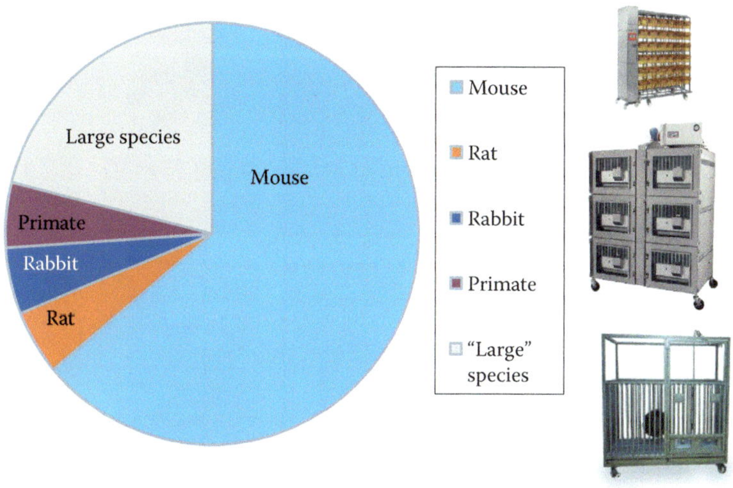

FIGURE 18.1 Relative proportion of space required by species used in research in 2006. (Figure and data compliments of HDR, Inc, Atlanta, GA)

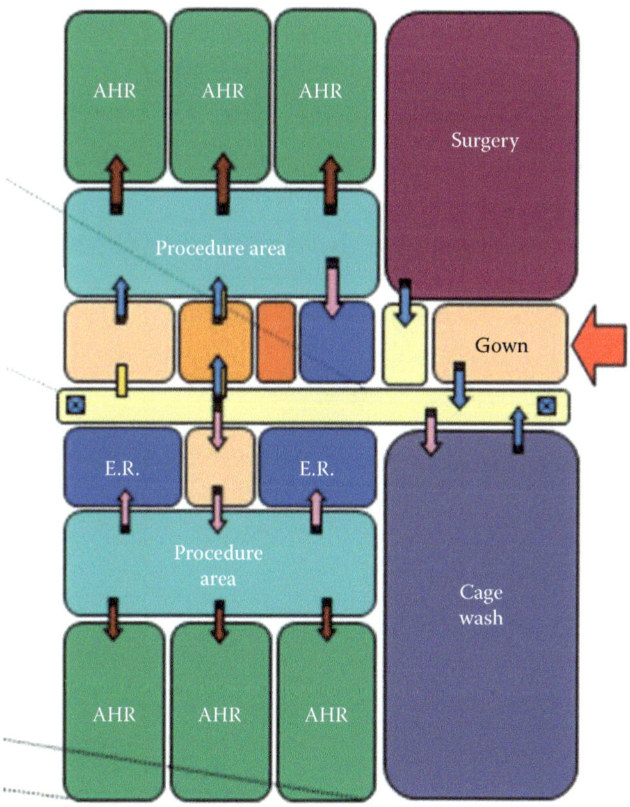

FIGURE 18.4 Modular scheme relationship and flow concept at the design stage of a construction or renovation project. AHR, animal holding room; E.R., equipment room.

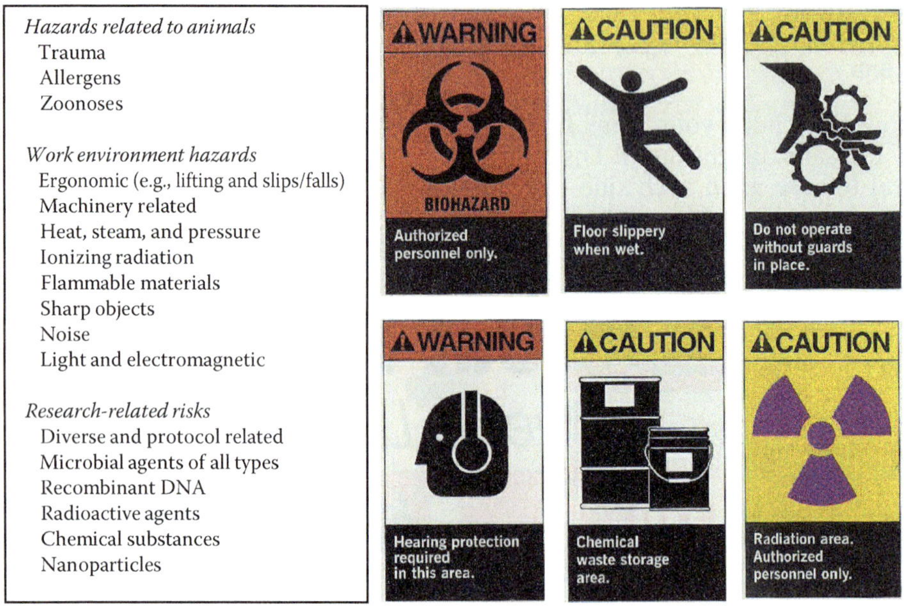

FIGURE 18.9 Presentation of various types of safety signage and a list of hazards common to animal research facilities.

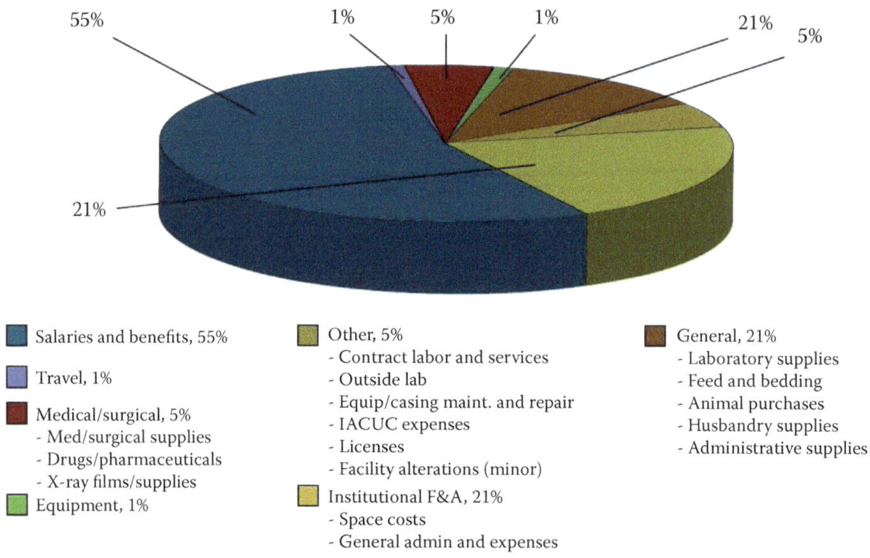

FIGURE 18.10 Representative operating costs of a functioning animal research facility. IACUC, Institutional Animal Care and Use Committee; F&A, facilities and administrative.

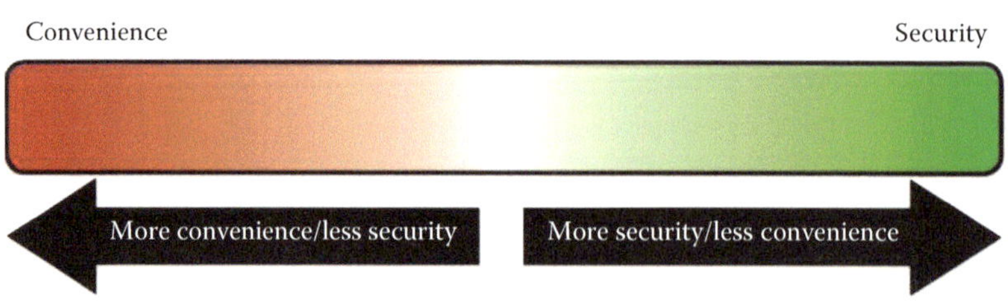

FIGURE 19.1 Relationship between convenience and security.

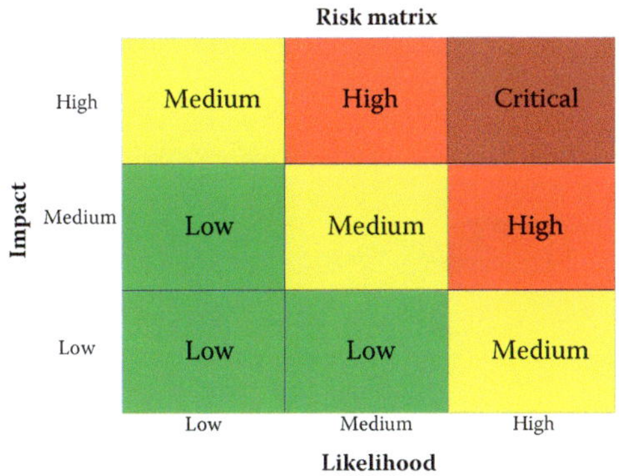

FIGURE 19.2 Risk evaluation matrix.

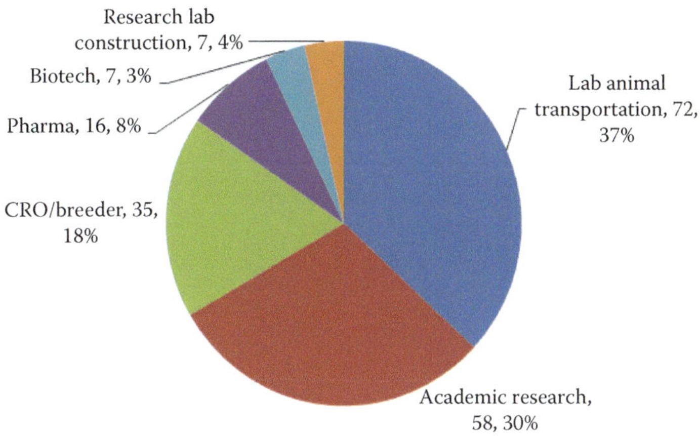

FIGURE 19.4 Breakdown of 195 direct action attacks against research-related industries from 2011 to 2015. (From INA [Information Network Associates], Data collected by INA, INA, Harrisburg, PA, 2016, www.ina-inc.com.)

FIGURE 21.1 Example of a stainless steel, mesh-bottomed caging for rats. The food bin is located on the inside of the cage. Water bottles would be placed on the outside (where the springs are located). Note the white plastic resting boards on the floors of the cage to provide the rats with a solid-bottom surface for comfort. (Photo by Deb Hickman.)

(a)

(b)

FIGURE 21.5 (a) Example of an IVC rack. There are no cages, but the rack can hold up to 160 mouse cages. The metal wall in the center of the rack provides forced air to the cages on the rack. (b) Close-up of IVC with empty bays next to the cages. On the metal wall, there are two round blue and white circles. These are the air vents that deliver and exhaust air from the cage. The silver nozzle below the air vents is an automatic watering system that provides water on demand to the animals in the cage. The yellow levers hold the cage in place, to ensure a tight seal for efficient air exchanges within the cage. (Photos by Deb Hickman.)

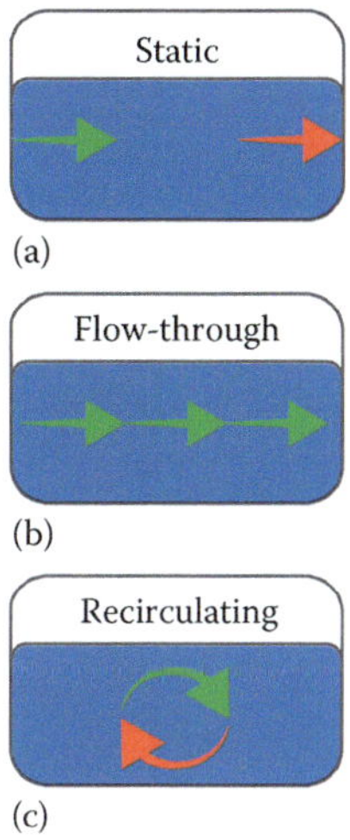

FIGURE 24.2 Three primary modes of aquaculture. (a) Static. Water is exchanged periodically. (b) Flow-through. Water flows in a unidirectional flow, removing wastes. (c) Recirculating. Clean water flows into tanks, and effluent flows out and is treated to remove wastes before flowing back into tanks.

FIGURE 25.4 Gentle misting of parrots and other species provides "showers," which promote natural grooming behaviors and feather health.

FIGURE 25.5 Venomous snakes can be housed in locking shift cages to allow husbandry with minimal risk. The red X marks the acrylic panel, which is easily removed. (Photo by D. O'Rourke.)

FIGURE 25.7 Many lizard species must have a basking site and access to UV light of the appropriate wavelength. (Photo by D. P. Baumann.)

FIGURE 25.9 Proper housing for toads includes a substrate such as coconut husks, a water source, and a PVC shelter for hiding. Note the toad at the base of the shelter buried in the substrate. (Photo by J. Cox.)

FIGURE 25.10 Provision of both an aquatic environment and a terrestrial environment is essential to species such as the tiger salamander (*Ambystoma tigrinum*) as it undergoes metamorphosis. (Photo by J. Cox.)

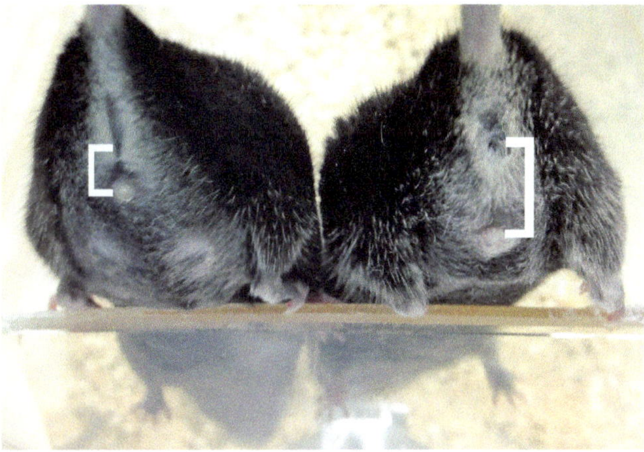

FIGURE 26.5 Sex differentiation in rodents is generally achieved by looking at anogenital distance. Anogenital distance is greater in males (right) than females (left). (Photo by Michelle Wallace-Fields.)

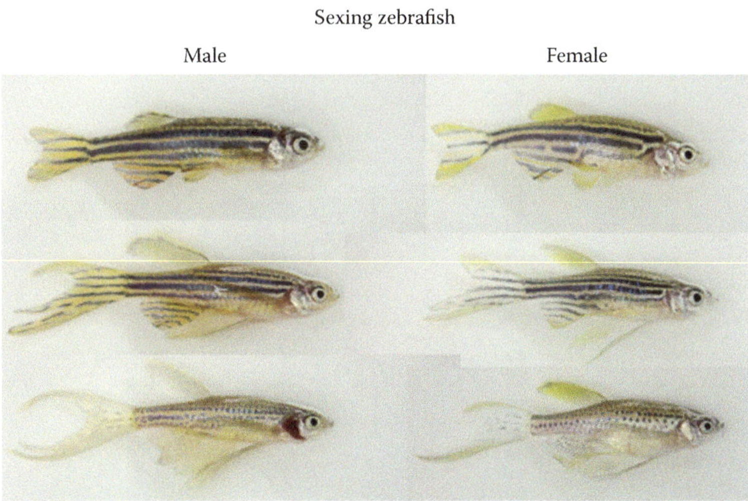

FIGURE 26.6 Sex differentiation in zebrafish. Males have a more streamlined torpedo shape; females have a rounder abdomen. On wild-type fish, females have a dorsal fin that is more yellow. Females also have a cloaca (not pictured). (Photo by Morgan Singleton.)

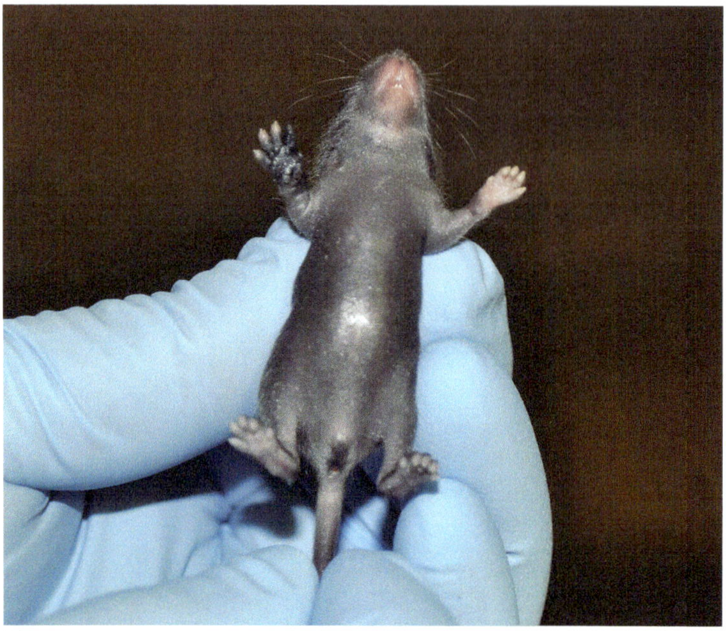

(a)

ID #	Footpad tattoo
0	○○ ○○
1	●○ ○○
2	○● ○○
3	●● ○○
4	○○ ●○
5	●○ ●○
6	○● ●○
7	●● ●○
8	○○ ○●
9	●○ ○●
10	○● ○●
11	●● ○●
12	○○ ●●
13	●○ ●●
14	○● ●●
15	●● ●●

(b)

(c)

FIGURE 26.8 (a) Neonatal animals can be identified at less than 5 days old by using footpad tattoos. (b) Up to 16 animals in a litter can be identified using this method, where (c) the right front pad is identified as 1, etc. (Photo by Michelle Wallace-Fields.)

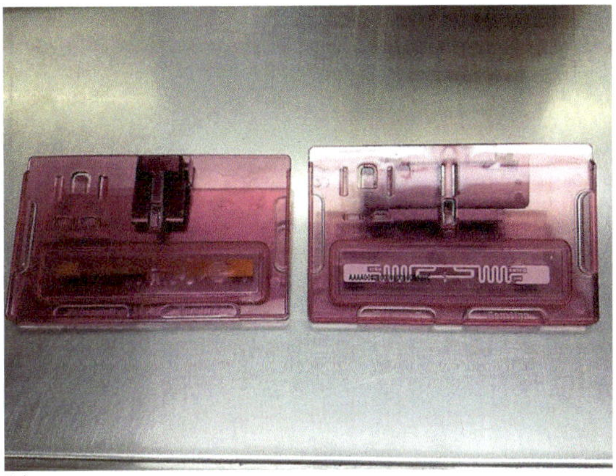

(a)

Cage Card #: 0000110963

PI: Doe, James	Requisition #: 16HSD278
Protocol Number: 123456(10)1D	Protocol Expiration Date: 10/16/2016
Strain: Hsd:ICR(CD-1) (Harlan)	Vendor: Envigo (Formerly Harlan)
Requestor: Scott, Mary	Location: RC2, 0334-RC2
Cage Type: Mouse Ventilated	DOB: 12/25/2015
Account: Doe, James - 62027731	
Special Caging:	
Comments:	

Cage Card #: 0000110963

Notes:

(b)

FIGURE 26.15 Example methods to collect census and protocol information include (a) RFID technology, which can be placed into a cage card, and (b) bar-coded cage cards. Both require special scanning devices that can be uploaded into software programs. (Photo by Michelle Wallace-Fields.)

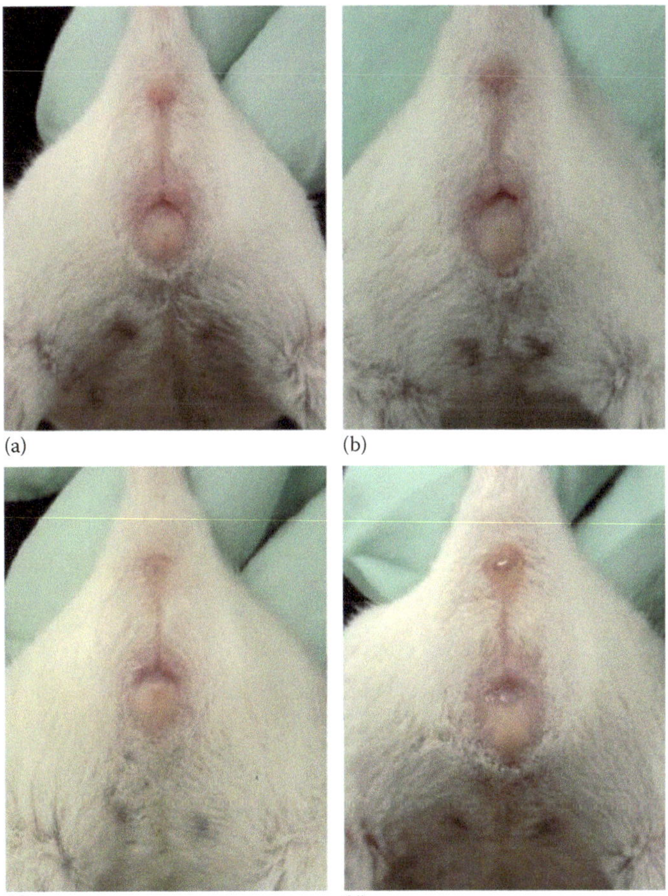

(a) (b) (c) (d)

FIGURE 29.1 Four stages of estrous in Balb/cByJ mice. (a) Proestrus. (b) Estrus. (c) Metestrus. (d) Diestrus. (Reprinted from Byers, S.L. et al., *PLoS One*, 7, 4, 2012.)

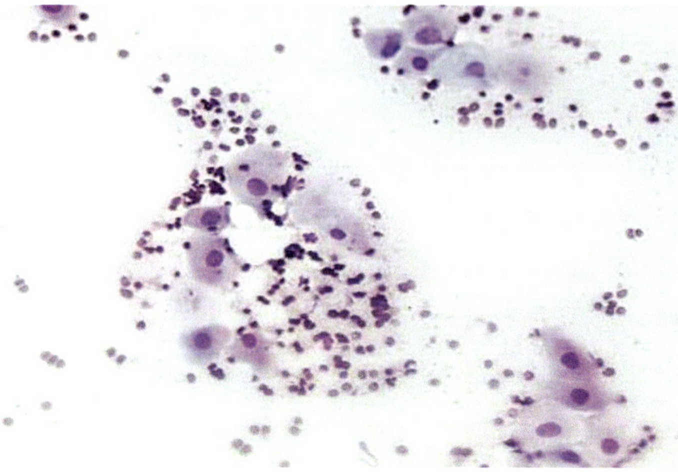

FIGURE 29.7 Proestrus with neutrophils, red blood cells, and parabasal cells. (From R. Bowen, Vaginal Cytology, http://www.vivo.colostate.edu/hbooks/pathphys/reprod/vc/index.html.)

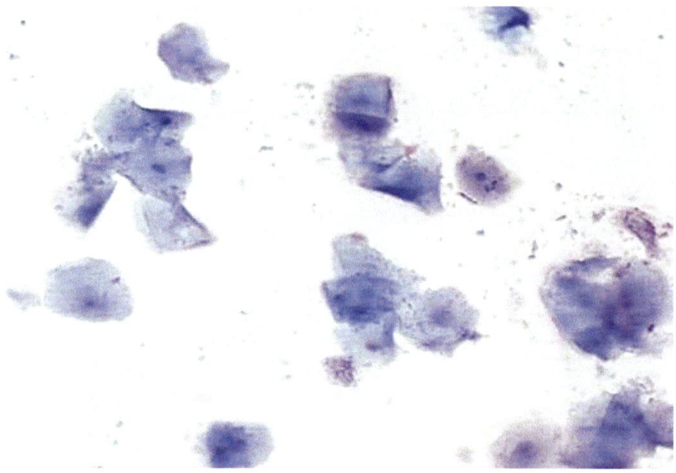

FIGURE 29.8 Estrus with anucleate cornified cells. (From R. Bowen, Vaginal Cytology, http://www.vivo.colostate.edu/hbooks/pathphys/reprod/vc/index.html.)

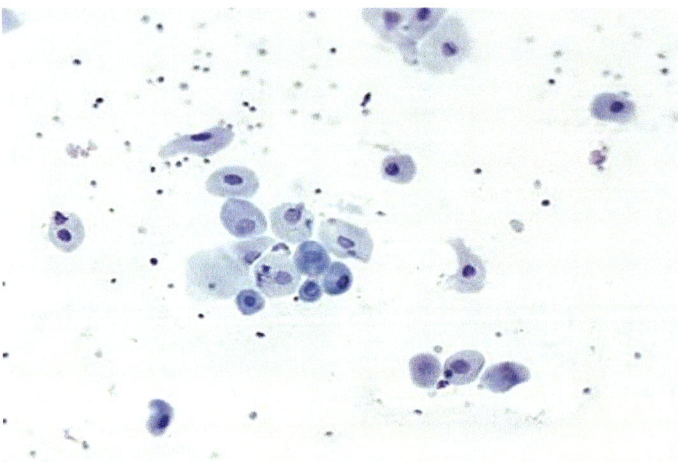

FIGURE 29.9 Diestrus with superficial and parabasal cells. (From R. Bowen, Vaginal Cytology, http://www.vivo.colostate.edu/hbooks/pathphys/reprod/vc/index.html.)

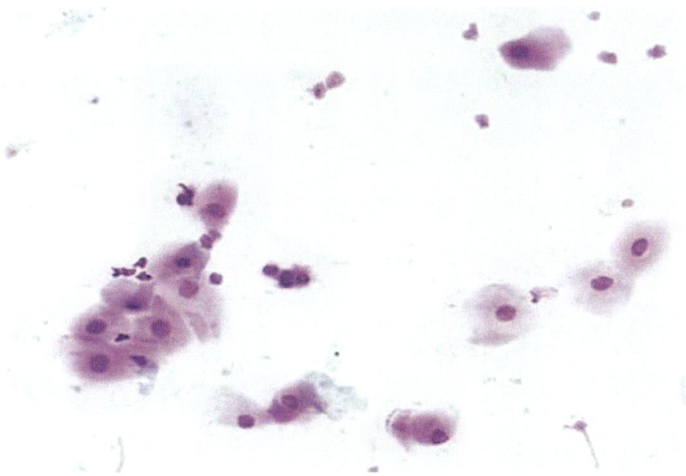

FIGURE 29.10 Anestrus with predominantly parabasal cells and a few neutrophils. (From R. Bowen, Vaginal Cytology, http://www.vivo.colostate.edu/hbooks/pathphys/reprod/vc/index.html.)

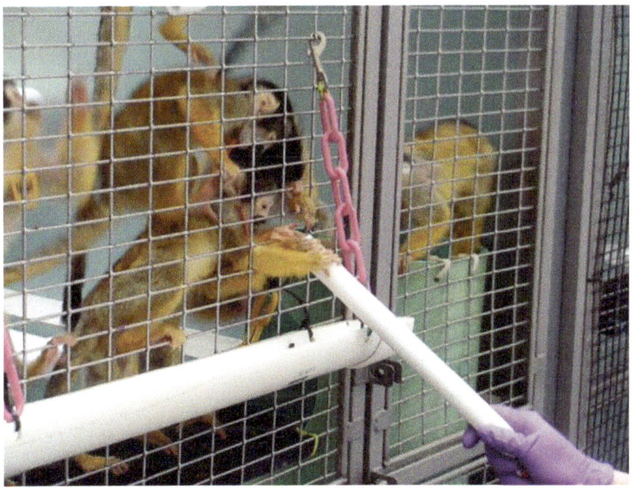

FIGURE 32.1 Squirrel monkeys "targeting" (touching) the PVC tube held by the trainer.

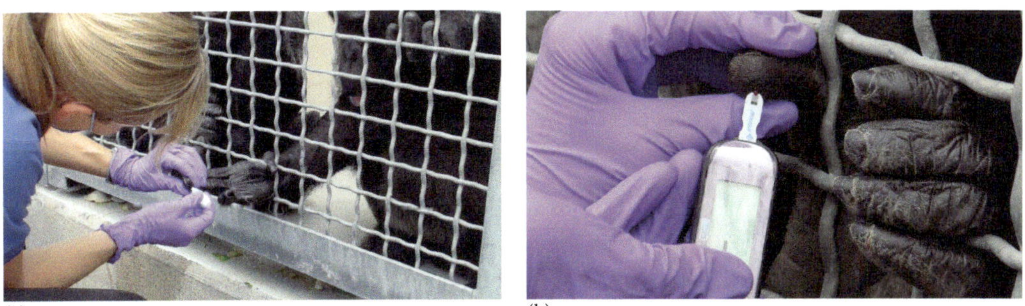

(a) (b)

FIGURE 32.4 (a) Using the lancet to prick the finger of a chimpanzee for a voluntary capillary blood sample. (b) Voluntary capillary blood sample from the toe of a chimpanzee for blood glucose analysis.

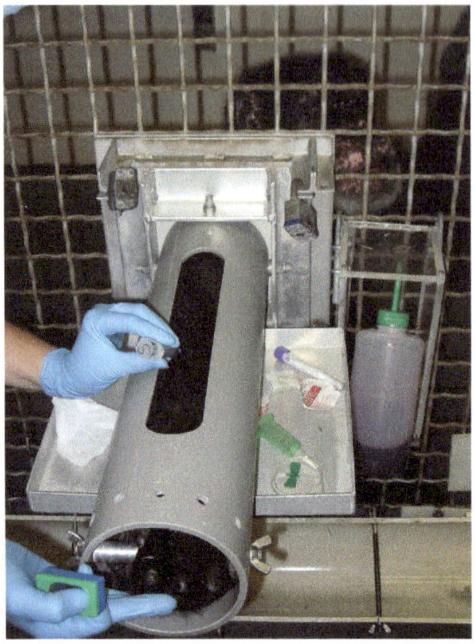

FIGURE 32.5 Chimpanzee providing a voluntary venous blood sample using the sleeve.

FIGURE 32.8 Chimpanzee voluntarily presenting for a combination of acupuncture and laser therapy.

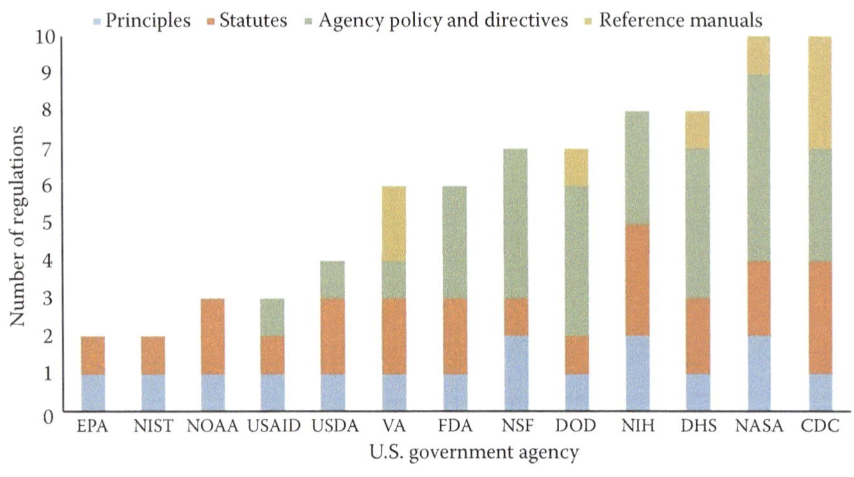

FIGURE 36.1 Number of animal welfare regulations that different U.S. agencies require researchers to follow. NIST, National Institute of Standards and Technology; NOAA, National Oceanic and Atmospheric Administration; USAID, U.S. Agency for International Development; VA, Department of Veterans Affairs; FDA, Food and Drug Administration; NSF, National Science Foundation; DOD, Department of Defense; DHS, Department of Homeland Security. (Based on data from Dr. J. R. Haywood presented in National Academies of Sciences, Engineering, and Medicine, *Optimizing the Nation's Investment in Academic Research: A New Regulatory Framework for the 21st Century*, National Academies Press, Washington, DC, 2016.)

Index

F

G

H

I

N

O

Y

Z

Bioactive Compounds and Therapeutics: Molecular Aspects, Metabolic Profiles, and Omics Studies

Bioactive Compounds and Therapeutics: Molecular Aspects, Metabolic Profiles, and Omics Studies

Topic Editors

Giovanni N. Roviello
Michele Costanzo

MDPI

Basel • Beijing • Wuhan • Barcelona • Belgrade • Novi Sad • Cluj • Manchester

Topic Editors

Giovanni N. Roviello
Institute of Biostructures and Bioimaging (IBB)
National Research Council of Italy (CNR)
Naples
Italy

Michele Costanzo
Department of Molecular Medicine and Medical Biotechnology
University of Naples Federico II
Naples
Italy

Editorial Office
MDPI AG
Grosspeteranlage 5
4052 Basel, Switzerland

This is a reprint of the Topic, published open access by the journals *Biomolecules* (ISSN 2218-273X), *Cells* (ISSN 2073-4409), *International Journal of Molecular Sciences* (ISSN 1422-0067), *Metabolites* (ISSN 2218-1989) and *Molecules* (ISSN 1420-3049), freely accessible at: www.mdpi.com/topics/LL4H8331Z3.

For citation purposes, cite each article independently as indicated on the article page online and as indicated below:

Lastname, A.A.; Lastname, B.B. Article Title. *Journal Name* **Year**, *Volume Number*, Page Range.

ISBN 978-3-7258-6570-3 (Hbk)
ISBN 978-3-7258-6571-0 (PDF)
https://doi.org/10.3390/books978-3-7258-6571-0

Contents

About the Editors

Giovanni N. Roviello

Giovanni Roviello holds a master's degree with honors in Chemistry and a Ph.D. in Biotechnology. He works as a senior researcher at the Institute of Biostructure and Bioimaging (IBB) of the Italian National Council for Research (CNR) in Naples, Italy. He has been a visiting researcher in the US (Coastal Carolina University, Gupta Science College November 2025), Germany (IMB Jena, Georg-August University, FAU University Erlangen), the UK (University of Greenwich), and Ireland (UCD Dublin), specializing in bioorganic chemistry. He has strong academic ties with AMU University of Poznan (Poland), where he lectured on nucleic acids for Ph.D. students in 2021, and with YSU (Armenia) and Geomedi University (Georgia), where he was appointed honorary professor in medicinal chemistry in 2012. Dr. Roviello is the principal investigator for CNR on various international projects, including EU- and the Royal Society (UK)-funded projects. He serves as an academic editor for several international journals and has authored more than 150 scientific papers (Scopus h-index: 33). Since 2021, he has been consistently listed among the top 2% of scientists in Stanford University's global author rankings, including the 2024 Author Database (Ioannidis, J. P. A., Updated Science-Wide Author Databases, 2025).

Michele Costanzo

Michele Costanzo is a Researcher of Biochemistry in the Department of Molecular Medicine and Medical Biotechnology at the University of Naples Federico II. He graduated with honors in Genetic Sciences and Technologies and obtained his Ph.D. in Molecular Medicine and Medical Biotechnology in 2020. He has carried out international research projects as a visiting researcher at the Proteomics facility of Necker Hospital/INSERM U24 in Paris (2017), and at the Leibniz-Institute of Analytical Sciences (ISAS) in Dortmund (2019). Dr. Costanzo's activity has focused on the development of innovative methodologies in clinical proteomics, metabolomics, and lipidomics. He has extensive experience in mass spectrometry–based technologies and in the analysis of large-scale omics datasets. His current research has a strong translational focus on rare human diseases and inborn errors of metabolism, leveraging integrative multi-omics approaches and advanced bioinformatics pipelines. He has contributed as academic editor for international peer-reviewed journals and has an extensive publication record (Scopus h-index: 23). Since 2025, he has been a member of the Directive Board of the Italian Proteomics Association.

Preface

Bioactive compounds represent one of the most dynamic and interdisciplinary frontiers in contemporary biomedical research. Their ability to modulate molecular pathways, reshape metabolic networks, and influence cellular homeostasis places them at the intersection of natural product chemistry, pharmacology, and systems biology. This reprint, developed within the Topic 'Bioactive Compounds and Therapeutics: Molecular Aspects, Metabolic Profiles, and Omics Studies', offers an integrated overview of recent scientific advances, highlighting how mechanistic studies and omics-driven approaches are transforming our understanding of therapeutic action.

More in detail, the contributions collected here span a wide spectrum of biological models and disease contexts, including metabolic dysfunction, cancer, infectious diseases, immune imbalance, as well as neurological or vascular disorders.

The volume encompasses investigations on phytochemicals, synthetic molecules, drug repurposing strategies, and innovative chemical entities, illustrating the breadth of therapeutic potential emerging from both natural and engineered bioactive structures. A recurring theme across the articles is the integration of metabolomics, proteomics, transcriptomics, and advanced biochemical assays to decode metabolic reprogramming, signaling cascades, and adaptive responses triggered by these compounds. We believe this reprint is particularly valuable for researchers, as it provides methodological insights, mechanistic frameworks, and omics-based strategies that can inspire new hypotheses and experimental designs. For clinicians, the collection highlights translational implications, emerging therapeutic candidates, and molecular evidence that may inform future precision-medicine approaches. For other readers, including students and early-career scientists, the volume serves as an accessible entry point into a rapidly evolving field, offering clear examples of how multidisciplinary research can bridge fundamental biology and therapeutic innovation.

By bringing together diverse perspectives and cutting-edge findings, this work demonstrates the importance of integrative science in identifying new drug targets, understanding disease mechanisms, and advancing the rational development of bioactive compounds. Moreover, it aims to support the scientific community in navigating the complexity of modern therapeutics and in shaping the next generation of biomedical discoveries.

Giovanni N. Roviello and Michele Costanzo
Topic Editors

International Journal of
Molecular Sciences

MDPI

ditorial

Precision Therapeutics Through Bioactive Compounds: Metabolic Reprogramming, Omics Integration, and Drug Repurposing Strategies

Michele Costanzo [1,2,*] and Giovanni N. Roviello [3,*]

1 Department of Molecular Medicine and Medical Biotechnology, University of Naples Federico II, 80131 Naples, Italy
2 CEINGE—Biotecnologie Avanzate Franco Salvatore, 80145 Naples, Italy
3 CNR Institute of Biostructures and Bioimaging, Via Tommaso De Amicis 95, 80145 Naples, Italy
* Correspondence: michele.costanzo@unina.it (M.C.); giovanni.roviello@cnr.it (G.N.R.); Tel.: +39-0812203415 (G.N.R.)

Bioactive compounds, whether derived from nature or synthetically produced, continue to redefine the landscape of therapeutic innovation. Integrating bioactive compounds with omics technologies [1,2] and metabolic profiling [3] unlocks transformative insights for precision medicine [4], drug repurposing [5], and systems-level therapeutic innovation. Remarkably, the pursuit of cutting-edge therapies for socially impactful diseases, especially neurodegenerative pathologies [6,7], centers on molecular platforms that integrate both natural and synthetic compounds, ranging from phytomolecules, amino acids, and peptides to synthetic heteroaromatic compounds [8–11]. Their ability to modulate molecular pathways, restore metabolic balance, and interact with complex biological systems positions them as key agents in the interest of precision medicine. The Topic "Bioactive Compounds and Therapeutics: Molecular Aspects, Metabolic Profiles, and Omics Studies" (https://www.mdpi.com/topics/LL4H8331Z3 accessed on 1 October 2025) brings together a diverse collection of 23 studies that explore the therapeutic potential of bioactive molecules through the lens of molecular biology, metabolic profiling, and omics technologies (Table 1). Natural products such as ursolic acid (Figure 1), rosemary extract, and hemp seed oil demonstrate promising effects on immune regulation, lipid metabolism, and adipogenesis [12,13].

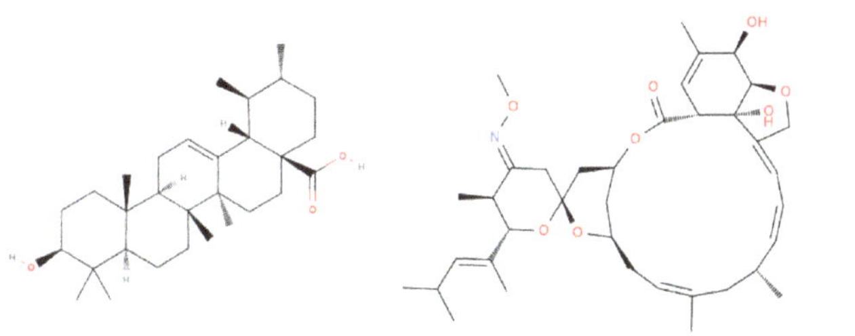

Ursolic acid **Moxidectin**

Figure 1. Structural representations of ursolic acid and moxidectin [12,13]. Moxidectin is a macrocyclic lactone with antiparasitic activity. Ursolic acid is a triterpenoid known for its anti-inflammatory and anticancer effects.

Table 1. Summary of the research collection on bioactive compounds.

Category	Examples	Therapeutic Focus
Natural Products	Ursolic acid, rosemary extract, hemp seed oil	Immune regulation, lipid metabolism, adipogenesis
Synthetic Derivatives	Halogenated boroxine, quinazoline compounds	Cancer cell autophagy, skin barrier function
Drug Repurposing	Moxidectin, GHF-201	Antimalarial activity, autophagic activation in glycogenosis
Metabolic Modulators	Methionine restriction, algal fiber-rich formulas, GPR40/GPR120 agonists	Gut health, metabolic syndrome, inflammation-linked periodontitis
Omics-Based Approaches	Glycan microarrays, metabolomic and proteomic signatures	Systems-level understanding of therapeutic mechanisms
Other Investigations	Plant stress responses, telomere biology, drug pharmacokinetics	Environmental biology, structural biology, molecular profiling
Methodologies Used	In vitro assays, in vivo models, in silico simulations	Multidisciplinary exploration from molecular insight to clinical relevance

Synthetic derivatives like halogenated boroxine [14] and novel quinazoline compounds reveal targeted actions on cancer cell autophagy and skin barrier function. The repositioning of existing drugs, exemplified by moxidectin's antimalarial activity and GHF-201's autophagic activation [15] in glycogenosis, underscores the value of drug repurposing in rare and infectious diseases. Several studies delve into the gut–brain axis [16–22], microbiota modulation, and metabolic syndrome, highlighting the systemic impact of bioactive agents. For instance, dietary methionine restriction [23] and algal fiber-rich formulas show metabolic improvements in animal models, while GPR40/GPR120 agonists [24] alleviate inflammation-linked periodontitis. The integration of omics approaches, such as glycan microarrays [25–28], metabolomic signatures in colitis [29], and proteomic insights into antibody secretion [30], provides a systems-level understanding of therapeutic mechanisms in some disease conditions. Research into bioactive compounds continues to evolve beyond traditional boundaries, encompassing not only human health but also environmental and structural biology. In particular, investigations into plant stress responses [31–33], telomere biology [34,35], and drug pharmacokinetics [36–41] exemplify this expanded scope, revealing how bioactivity influences diverse biological systems across species and contexts. These studies underscore the interconnectedness of molecular mechanisms that govern resilience, aging, and therapeutic efficacy.

Together, these insights illuminate the path from molecular understanding to clinical relevance by bridging bioactivity with omics technologies and metabolic profiling, thereby fostering a deeper understanding of how therapeutic agents can be harnessed for targeted, effective, and personalized interventions across a wide spectrum of biological systems, disease models, and therapeutic contexts, ultimately advancing precision medicine through integrative approaches that combine molecular characterization, functional analysis, and translational application.

Author Contributions: Conceptualization, M.C. and G.N.R.; writing—review and editing, M.C. and G.N.R. All authors have read and agreed to the published version of the manuscript.

Funding: This research received no external funding.

Institutional Review Board Statement: Not applicable.

Informed Consent Statement: Not applicable.

Data Availability Statement: Not applicable.

Conflicts of Interest: The authors declare no conflicts of interest.

References

1. Vailati-Riboni, M.; Palombo, V.; Loor, J.J. What are omics sciences? In *Periparturient Diseases of Dairy Cows: A Systems Biology Approach*; Springer: Berlin/Heidelberg, Germany, 2017; pp. 1–7.
2. Karczewski, K.J.; Snyder, M.P. Integrative omics for health and disease. *Nat. Rev. Genet.* **2018**, *19*, 299–310. [CrossRef]
3. Want, E.J.; Wilson, I.D.; Gika, H.; Theodoridis, G.; Plumb, R.S.; Shockcor, J.; Holmes, E.; Nicholson, J.K. Global metabolic profiling procedures for urine using UPLC–MS. *Nat. Protoc.* **2010**, *5*, 1005–1018. [CrossRef]
4. Ashley, E.A. Towards precision medicine. *Nat. Rev. Genet.* **2016**, *17*, 507–522. [CrossRef]
5. Pushpakom, S.; Iorio, F.; Eyers, P.A.; Escott, K.J.; Hopper, S.; Wells, A.; Doig, A.; Guilliams, T.; Latimer, J.; McNamee, C. Drug repurposing: Progress, challenges and recommendations. *Nat. Rev. Drug Discov.* **2019**, *18*, 41–58. [CrossRef] [PubMed]
6. Vicidomini, C.; Roviello, G.N. Therapeutic Convergence in Neurodegeneration: Natural Products, Drug Repurposing, and Biomolecular Targets. *Biomolecules* **2025**, *15*, 1333. [CrossRef] [PubMed]
7. Dugger, B.N.; Dickson, D.W. Pathology of neurodegenerative diseases. *Cold Spring Harb. Perspect. Biol.* **2017**, *9*, a028035. [CrossRef] [PubMed]
8. Simonyan, H.; Palumbo, R.; Vicidomini, C.; Scognamiglio, P.L.; Petrosyan, S.; Sahakyan, L.; Melikyan, G.; Saghyan, A.; Roviello, G.N. Binding of G-quadruplex DNA and serum albumins by synthetic non-proteinogenic amino acids: Implications for c-Myc-related anticancer activity and drug delivery. *Mol. Ther. Nucleic Acids* **2025**, *36*, 102478. [CrossRef]
9. Mittova, V.; Pirtskhalava, M.; Tsetskhladze, Z.R.; Makalatia, K.; Loladze, A.; Bebiashvili, I.; Barblishvili, T.; Gogoladze, A.; Roviello, G.N. Antioxidant Potential and Antibacterial Activities of Caucasian Endemic Plants Sempervivum transcaucasicum and Paeonia daurica subsp. mlokosewitschii Extracts and Molecular In Silico Mechanism Insights. *J. Xenobiotics* **2025**, *15*, 109. [CrossRef]
10. Ferrara, B.T.; Thompson, E.P.; Roviello, G.N.; Gale, T.F. C-Terminal Analogues of Camostat Retain TMPRSS2 Protease Inhibition: New Synthetic Directions for Antiviral Repurposing of Guanidinium-Based Drugs in Respiratory Infections. *Int. J. Mol. Sci.* **2025**, *26*, 6761. [CrossRef]
11. Roviello, G.N.; Roviello, G.; Musumeci, D.; Capasso, D.; Di Gaetano, S.; Costanzo, M.; Pedone, C. Synthesis and supramolecular assembly of 1, 3-bis (1′-uracilyl)-2-propanone. *RSC Adv.* **2014**, *4*, 28691–28698. [CrossRef]
12. Hinney, B.; Wiedermann, S.; Kaiser, W.; Krücken, J.; Joachim, A. Eprinomectin and moxidectin resistance of trichostrongyloids on a goat farm in Austria. *Pathogens* **2022**, *11*, 498. [CrossRef]
13. Mlala, S.; Oyedeji, A.O.; Gondwe, M.; Oyedeji, O.O. Ursolic acid and its derivatives as bioactive agents. *Molecules* **2019**, *24*, 2751. [CrossRef]
14. Hadzic, M.; Pojskic, L.; Lojo-Kadric, N.; Haveric, A.; Ramic, J.; Galic, B.; Haveric, S. Novel boron-containing compound, halogenated boroxine, induces selective cytotoxicity through apoptosis triggering in UT-7 leukemia. *J. Biochem. Mol. Toxicol.* **2022**, *36*, e23005. [CrossRef]
15. Mishra, K.; Sweetat, S.; Baraghithy, S.; Sprecher, U.; Marisat, M.; Bastu, S.; Glickstein, H.; Tam, J.; Rosenmann, H.; Weil, M. The Autophagic Activator GHF-201 Can Alleviate Pathology in a Mouse Model and in Patient Fibroblasts of Type III Glycogenosis. *Biomolecules* **2024**, *14*, 893. [CrossRef]
16. Bercik, P.; Collins, S.; Verdu, E. Microbes and the gut-brain axis. *Neurogastroenterol. Motil.* **2012**, *24*, 405–413. [CrossRef]
17. Chen, X.; D'Souza, R.; Hong, S.-T. The role of gut microbiota in the gut-brain axis: Current challenges and perspectives. *Protein Cell* **2013**, *4*, 403–414. [CrossRef] [PubMed]
18. Kraimi, N.; Ross, T.; Pujo, J.; De Palma, G. The gut microbiome in disorders of gut–brain interaction. *Gut Microbes* **2024**, *16*, 2360233. [CrossRef] [PubMed]
19. Mayer, E.A. Gut feelings: The emerging biology of gut–brain communication. *Nat. Rev. Neurosci.* **2011**, *12*, 453–466. [CrossRef] [PubMed]
20. Góralczyk-Bińkowska, A.; Szmajda-Krygier, D.; Kozłowska, E. The Microbiota–Gut–Brain Axis in Psychiatric Disorders. *Int. J. Mol. Sci.* **2022**, *23*, 11245. [CrossRef]
21. Doroszkiewicz, J.; Groblewska, M.; Mroczko, B. The Role of Gut Microbiota and Gut–Brain Interplay in Selected Diseases of the Central Nervous System. *Int. J. Mol. Sci.* **2021**, *22*, 10028. [CrossRef]
22. Hamamah, S.; Aghazarian, A.; Nazaryan, A.; Hajnal, A.; Covasa, M. Role of Microbiota-Gut-Brain Axis in Regulating Dopaminergic Signaling. *Biomedicines* **2022**, *10*, 436. [CrossRef]
23. Orgeron, M.L.; Stone, K.P.; Wanders, D.; Cortez, C.C.; van Nancy, T.; Gettys, T.W. The Impact of Dietary Methionine Restriction on Biomarkers of Metabolic Health. In *Glucose Homeostatis and the Pathogenesis of Diabetes Mellitus*; Elsevier: Amsterdam, The Netherlands, 2014; pp. 351–376.

24. Bianchini, G.; Nigro, C.; Sirico, A.; Novelli, R.; Prevenzano, I.; Miele, C.; Beguinot, F.; Aramini, A. A new synthetic dual agonist of GPR120/GPR40 induces GLP-1 secretion and improves glucose homeostasis in mice. *Biomed. Pharmacother.* **2021**, *139*, 111613. [CrossRef] [PubMed]
25. Rillahan, C.D.; Paulson, J.C. Glycan microarrays for decoding the glycome. *Annu. Rev. Biochem.* **2011**, *80*, 797–823. [CrossRef]
26. Kim, Y.; Hyun, J.Y.; Shin, I. Glycan microarrays from construction to applications. *Chem. Soc. Rev.* **2022**, *51*, 8276–8299. [CrossRef] [PubMed]
27. Heimburg-Molinaro, J.; Song, X.; Smith, D.F.; Cummings, R.D. Preparation and analysis of glycan microarrays. *Curr. Protoc Protein Sci.* **2011**, *64*, 12.10.1–12.10.29. [CrossRef] [PubMed]
28. Gao, C.; Wei, M.; McKitrick, T.R.; McQuillan, A.M.; Heimburg-Molinaro, J.; Cummings, R.D. Glycan microarrays as chemical tools for identifying glycan recognition by immune proteins. *Front. Chem.* **2019**, *7*, 833. [CrossRef]
29. Liu, M.; Guo, S.; Wang, L. Systematic review of metabolomic alterations in ulcerative colitis: Unveiling key metabolic signatures and pathways. *Ther. Adv. Gastroenterol.* **2024**, *17*, 17562848241239580. [CrossRef]
30. Xu, N.; Ou, J.; Si, Y.; Goh, K.Y.; Flanigan, D.D.; Han, X.; Yang, Y.; Yang, S.-T.; Zhou, L.; Liu, X.M. Proteomics insight into the production of monoclonal antibody. *Biochem. Eng. J.* **2019**, *145*, 177–185. [CrossRef]
31. Shulaev, V.; Cortes, D.; Miller, G.; Mittler, R. Metabolomics for plant stress response. *Physiol. Plant.* **2008**, *132*, 199–208. [CrossRef]
32. Sunkar, R.; Li, Y.-F.; Jagadeeswaran, G. Functions of microRNAs in plant stress responses. *Trends Plant Sci.* **2012**, *17*, 196–203. [CrossRef]
33. Mittler, R.; Zandalinas, S.I.; Fichman, Y.; Van Breusegem, F. Reactive oxygen species signalling in plant stress responses. *Nat. Rev Mol. Cell Biol.* **2022**, *23*, 663–679. [CrossRef]
34. Oeseburg, H.; De Boer, R.A.; Van Gilst, W.H.; Van Der Harst, P. Telomere biology in healthy aging and disease. *Pflügers Arch.-Eur J. Physiol.* **2010**, *459*, 259–268. [CrossRef]
35. Xu, L.; Li, S.; Stohr, B.A. The role of telomere biology in cancer. *Annu. Rev. Pathol. Mech. Dis.* **2013**, *8*, 49–78. [CrossRef]
36. Ruiz-Garcia, A.; Bermejo, M.; Moss, A.; Casabo, V.G. Pharmacokinetics in drug discovery. *J. Pharm. Sci.* **2008**, *97*, 654–690. [CrossRef] [PubMed]
37. Benedetti, M.S.; Whomsley, R.; Poggesi, I.; Cawello, W.; Mathy, F.-X.; Delporte, M.-L.; Papeleu, P.; Watelet, J.-B. Drug metabolism and pharmacokinetics. *Drug Metab. Rev.* **2009**, *41*, 344–390. [CrossRef] [PubMed]
38. Coelho, M.M.; Fernandes, C.; Remião, F.; Tiritan, M.E. Enantioselectivity in Drug Pharmacokinetics and Toxicity: Pharmacological Relevance and Analytical Methods. *Molecules* **2021**, *26*, 3113. [CrossRef] [PubMed]
39. Dobrek, L. Chronopharmacology in Therapeutic Drug Monitoring—Dependencies between the Rhythmics of Pharmacokinetic Processes and Drug Concentration in Blood. *Pharmaceutics* **2021**, *13*, 1915. [CrossRef]
40. Karaźniewicz-Łada, M.; Główka, A.K.; Mikulska, A.A.; Główka, F.K. Pharmacokinetic Drug–Drug Interactions among Antiepileptic Drugs, Including CBD, Drugs Used to Treat COVID-19 and Nutrients. *Int. J. Mol. Sci.* **2021**, *22*, 9582. [CrossRef]
41. Rombolà, L.; Scuteri, D.; Marilisa, S.; Watanabe, C.; Morrone, L.A.; Bagetta, G.; Corasaniti, M.T. Pharmacokinetic Interactions between Herbal Medicines and Drugs: Their Mechanisms and Clinical Relevance. *Life* **2020**, *10*, 106. [CrossRef]

cells

MDPI

Article

The Ivermectin Related Compound Moxidectin Can Target Apicomplexan Importin α and Limit Growth of Malarial Parasites

Sujata B. Walunj [1,2,†], Geetanjali Mishra [2], Kylie M. Wagstaff [1], Swati Patankar [2] and David A. Jans [1,*]

1 Nuclear Signaling Laboratory, Monash Biomedicine Discovery Institute, Department of Biochemistry and Molecular Biology, Monash University, Clayton, VIC 3800, Australia; sujatawlnj@gmail.com (S.B.W.); kylie.wagstaff@monash.edu (K.M.W.)

2 Molecular Parasitology Laboratory, Department of Biosciences and Bioengineering, IIT Bombay, Powai, Mumbai 400076, India; geetanjali_m0012@iitb.ac.in (G.M.); patankar@iitb.ac.in (S.P.)

* Correspondence: david.jans@monash.edu

† Current address: UCSF Helen Diller Family Comprehensive Cancer Center, San Francisco, CA 94158, USA.

Abstract: Signal-dependent transport into and out of the nucleus mediated by members of the importin (IMP) superfamily is crucial for eukaryotic function, with inhibitors targeting IMPα being of key interest as anti-infectious agents, including against the apicomplexan *Plasmodium* species and *Toxoplasma gondii*, causative agents of malaria and toxoplasmosis, respectively. We recently showed that the FDA-approved macrocyclic lactone ivermectin, as well as several other different small molecule inhibitors, can specifically bind to and inhibit *P. falciparum* and *T. gondii* IMPα functions, as well as limit parasite growth. Here we focus on the FDA-approved antiparasitic moxidectin, a structural analogue of ivermectin, for its IMPα-targeting and anti-apicomplexan properties for the first time. We use circular dichroism and intrinsic tryptophan fluorescence measurements to show that moxidectin can bind directly to apicomplexan IMPαs, thereby inhibiting their key binding functions at low μM concentrations, as well as possessing anti-parasitic activity against *P. falciparum* in culture. The results imply a class effect in terms of IMPα's ability to be targeted by macrocyclic lactone compounds. Importantly, in the face of rising global emergence of resistance to approved anti-parasitic agents, the findings highlight the potential of moxidectin and possibly other macrocyclic lactone compounds as antimalarial agents.

Keywords: *Plasmodium falciparum*; malaria; *Toxoplasma gondii*; toxoplasmosis; importins; nuclear transport inhibitors

1. Introduction

Signal-dependent transport of molecules into and out of the cell nucleus facilitated by multiple distinct members of the importin (IMP) superfamily of α and β types is central to eukaryotic cell function [1]. The best understood nuclear import pathway is that mediated by a heterodimer of IMPα and IMPβ1, where IMPα binds specifically to IMPβ1 and to the nuclear localisation signal (NLS) of cargo proteins [2–5]. IMPβ1 within the complex then mediates passage through the nuclear pore into the nucleus and subsequent release in the nucleus via distinct binding interactions with various other components of the nuclear transport machinery [1–5].

Dysregulation of nucleocytoplasmic transport can severely impact cell function [1,6–8], with the inhibition of transport through the use of small molecules a promising therapeutic strategy for cancer and infectious diseases [1,4,9]. Various inhibitors of IMPα

such as the macrocyclic lactone ivermectin [10] have been a focus of interest as antiviral agents [1,4,9,11–16], and more recently as agents to combat the apicomplexan *Plasmodium* species and *Toxoplasma gondii* [17,18], the causative agents of malaria and toxoplasmosis, respectively [19–24].

With 249 million cases and 608,000 deaths in 2022 worldwide alone, malaria is a global concern (WHO Malaria Report, 2023), with growing resistance to artemisinin-based combination therapies highlighting the need for new antiparasitic approaches [19–21]. Toxoplasmosis, caused by *T. gondii*, potentially afflicts one-third of humans asymptomatically, with immunocompromised individuals and children facing severe complications; the limitations of current therapies for toxoplasmosis, including drug resistance, emphasize the compelling need to develop new therapeutic strategies [22–25].

We recently showed that distinct inhibitors targeting IMPα can limit the growth of *P. falciparum* and *T. gondii* in culture [17,18], validating IMPα from these parasites as viable therapeutic targets [26–29]. These inhibitors included the US Food and Drug Administration (FDA)-approved anti-helminthic ivermectin [10,30], which is active against a range of different viruses, including SARS-CoV2 [9,12–16,31–33], and has been clinically investigated in an anti-malarial context [34,35]. Of a number of structural homologues of ivermectin that belong to the macrocyclic lactone class of compounds and possess anti-parasitic properties, moxidectin is also FDA-approved for human use [36,37]. Both moxidectin and ivermectin possess the ability to target glutamate-gated chloride channels [37], with moxidectin showing longer half-life/sustained release into the bloodstream that may be part of the basis of its more potent and prolonged anti-parasitic effect compared to ivermectin [38].

Although moxidectin has demonstrated efficacy against various parasites in animals [37–40], its specific activity against apicomplexan parasites, or apicomplexan IMPα, has not previously been examined. The present study addresses this directly for the first time, comparing the effects of moxidectin with those of ivermectin on IMPα from *P. falciparum* and *T. gondii*, as well as its anti-parasitic activity. Moxidectin shows a robust ability to interact with and inhibit the functions of apicomplexan IMPα, as well as limit malarial-parasite growth in culture. The results imply a class effect of macrocyclic lactones in this context, opening up the intriguing possibility of using these compounds more widely in the context of malaria and toxoplasmosis.

2. Materials and Methods

2.1. Protein Expression and Purification

P. falciparum IMPα (PfIMPα, PlasmoDB ID: Pf3D7_0812400), *T. gondii* IMPα (TgIMPα, ToxoDB ID: TGGT1_252290), *Mus musculus* IMPα (Kpna2/Rch1/pendulin/PTAC58 NCBI ID: 16647) abbreviated here to MmIMPα; MmIMPα deleted for the IMPβ1-binding domain (IBB) (MmΔIBBIMPα); and *M. musculus* IMPβ1 (MmIMPβ1, NCBI ID: 16211) GST fusion proteins were all expressed and purified as previously [14,16–18]. The His-tagged green fluorescent protein (GFP) fusion proteins T-ag-NLS-GFP and TGS1-NLS-GFP, as well as PfIMPα and TgIMPα proteins, were all purified using nickel-affinity chromatography as described [14,16–18]. GST-fusion proteins were biotinylated using the Sulfo-NHS-Biotin reagent (Pierce, Rockford, IL, USA) as previously [14,17].

2.2. Compounds

Moxidectin and ivermectin were from Sigma-Aldrich (St. Louis, MO, USA); 10 mM stock solutions were prepared in 100% DMSO.

2.3. AlphaScreen

IC_{50} analysis was performed in quadruplicate using an AlphaScreen binding assay as previously [16–18], using bacterially expressed glutathione S-transferase (GST)-C-terminally tagged PfIMPα and TgIMPα proteins [17], as well as SV40 T-ag-NLS-GFP fusion protein and MmIMPβ1 [11,41]. As previously, nickel-acceptor and streptavidin-donor beads were used [17,18]. The AlphaScreen signal was measured on a Perkin Elmer plate reader, and IC_{50} plots drawn using GraphPad Prism 9.0.2 (San Diego, CA, USA).

2.4. Circular Dichroism (CD) Spectroscopy

CD spectroscopy experiments were carried out to assess direct binding of moxidectin and ivermectin to different IMPαs as described [17,18]. CD spectra at 90.15 mg/mL protein concentration were recorded from 200–260 nm using a Jasco CD spectrometer (Jasco, Easton, MD, USA) in the absence and presence of 30 or 80 μM ivermectin or moxidectin, with percentage α-helix content estimated from the ellipticity at 222 nm as described [16,42] using the CD Multivariate secondary structure estimation analysis program.

2.5. Intrinsic Tryptophan Fluorescence Assays

Tryptophan fluorescence spectroscopy experiments were performed using a JASCO Fluorescence Spectrophotometer (JASCO, Hachioji, Tokyo, Japan) (0.5 mL quartz cuvette), with a fixed excitation wavelength of 295 nm and emission spectra collected between 310 and 400 nm with a slit width of 5 nm at 25 °C. Increasing concentrations of compounds in DMSO were incubated with the IMPαs (1 μM) in PBS for 5 min at room temperature prior to measurement. Data from three independent experiments were analysed using GraphPad Prism 10 (San Diego, CA, USA). Non-linear regression analysis was used to fit the one-site binding curves using the equation: $y = Bmax * x/(Kd + x)$, where Bmax represents the maximum binding capacity and Kd is the dissociation constant, reflecting the concentration of ligand at which half-maximal binding is achieved.

2.6. *P. falciparum* Culture

The *P. falciparum* 3D7 strain was cultured and maintained according to the standard procedures [43] with minor modifications. Cultures were maintained in RPMI 1640 (GibcoTM, Waltham, MA, USA) supplemented with 0.5% Albumax (GibcoTM), 50 mg/L hypoxanthine (Sigma-Aldrich), 2 gm/L D-glucose (Sigma-Aldrich), 2 gm/L sodium bicarbonate (Sigma-Aldrich), and 56 mg/L of gentamicin (Abbott, Chicago, IL, USA). Blood for parasite cultures was procured from volunteer blood donors with approval from the Institute Ethical Committee.

2.7. Drug Susceptibility Assays for *P. falciparum* Using HRP2 (Histidine-Rich Protein 2) ELISA

Samples from the continuous culture of *P. falciparum* were synchronised to obtain predominant rings of 0.25% parasitemia at 3% haematocrit. Compounds were made at a stock concentration of 10 mM in 100% DMSO, with the FDA-approved antimalarial dihydroartemisinin (a kind gift from IPCA Laboratories, Mumbai, India) as a control. Compound dilutions were made in RPMI 1640 to obtain the desired final test concentrations for the IC_{50} analysis.

The HRP2-sandwich ELISA assay was performed according to the standard protocol [16,17,44,45]. Briefly, the PfHRP2 protein concentration in the culture wells, indicative of the level of parasitemia, is measured using a sandwich ELISA technique. Serial 2-fold dilutions of compounds (25 μL/well) were manually dispensed into standard 96-well microculture plates (Eppendorf). Subsequently, 200 μL of *P. falciparum* culture at 0.25% parasitemia (3% haematocrit) was added to each well. The plates were then incubated for 72 h at 37 °C in 5% CO_2 in a humidified incubator. Plates were then subjected to freeze–thaw to achieve complete haemolysis. PfHRP2 levels in control and test samples were measured using a horseradish peroxidase enzyme-linked immunosorbent assay at 450 nm. GraphPad Prism 9.2.0 was used to perform non-linear regression analysis to determine the IC_{50} values [16,17].

2.8. T. gondii Tachyzoite Culture and Growth Inhibition Assay

A *T. gondii* RH strain expressing a luciferase reporter (RH-Fluc) [16] was used to assess the effect of the inhibitors on growth as described. *T. gondii* tachyzoites were maintained and cultured at 37 °C in 5% CO_2 in a humidified incubator using primary human foreskin fibroblasts (HFF, ATCC) maintained in Dulbecco's modified Eagle medium (DMEM) (GibcoTM, Waltham, MA, USA) supplemented with 3.7 g/L sodium bicarbonate and 2.38 g/L HEPES, 10% Cosmic Calf serum (HycloneTM, Logan, UT, USA) and 20 mg/L gentamicin [46,47]. Growth inhibition was monitored by luminescence [16], where 100 µL of culture medium without or with inhibitors, followed by 100 µL of DMEM containing 5000 parasites, were added to confluent monolayers of HFF cells in 96-well culture dishes (Eppendorf, Hamburg, Germany). After 48 h, 150 µL of the culture medium was discarded from each well, and 10 µL lysis buffer was added to lyse the parasites, followed by 50 µL of 2× luciferase assay reagent (Promega, Madison, WI, USA). Luminescence was then measured directly for 10 s using a Varioskan™ LUX multimode microplate reader (Thermo Fisher Scientific, Waltham, MA, USA).

2.9. MTT Assay for Host Cell Cytotoxicity

The MTT assay was used to measure the cytotoxicity of the small molecules against the HFF cells as described [48]. Briefly, freshly confluent HFF cells were incubated for 48 h in a 96-well culture plate at 37 °C in a humidified incubator (5% CO_2 atmosphere) in the presence of increasing concentrations of compounds (total volume 200 uL). A 10 µL quantity of MTT reagent (Sigma Aldrich) prepared in complete DMEM medium was then added to each well, and incubation continued for a further 3 h. 150 µL of the medium containing the MTT reagent was then removed from each well and replaced with 150 µL DMSO. Plates were then incubated room temperature in the dark for 20 min, after which absorbance for each well was measured at 570 nm. Data were analysed in GraphPad Prism 9 software.

2.10. Statistical Analysis

Non-linear regression analysis was performed in GraphPad Prism 9.2.0 (San Diego, CA, USA) to fit four parameter dose–response curves using the formula: $y = a + ((b - a)/(1 + 10^{(Log(c) - x)} \times d))$, where a and b are the minimum and maximum asymptotes respectively, c is the half-maximal inhibitory concentration value (IC_{50}), and d is the slope at the steepest part of the curve (the Hill slope).

3. Results

3.1. Moxidectin Can Block NLS Recognition by Apicomplexan IMPαs

Structurally related to ivermectin (see Figure 1a), the anti-parasitic agent moxidectin was approved for the treatment/prevention of river blindness (onchocerciasis) in humans by the FDA in 2018 [36–38]. It appears to have a longer half-life/higher efficacy against *Onchocerca volvulus* than ivermectin in animals and humans [38,49,50]. We recently showed that as for mammalian nuclear transport systems [11,12,15,16], ivermectin can block the interaction of apicomplexan IMPα proteins with NLSs as well as IMPβ1 [17]. Here, we set out for the first time to test whether moxidectin had comparable activity, performing IC_{50} analysis using an established AlphaScreen binding assay [16,17].

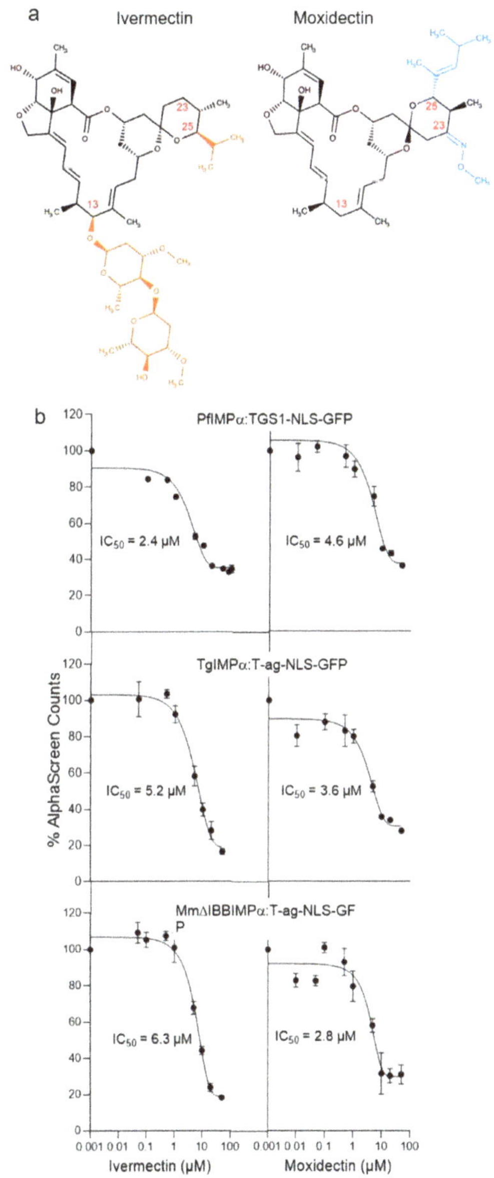

Figure 1. Close similarities of ivermectin and moxidectin in structure and IMPα inhibitory properties. (**a**). Structures of ivermectin and moxidectin are shown, with the shared chemical scaffold in black, and distinct groups highlighted in colour; an example is the C13 residue that is attached to sugar groups in the case of avermectins such as ivermectin, but is protonated in moxidectin and other members of the milbemycin family. (**b**). AlphaScreen technology was used to determine the IC_{50} values for inhibition of the binding of various IMPαs (5 nM) to NLS-containing proteins (30 nM) by ivermectin and moxidectin. Data represent the mean ± SEM for quadruplet wells from a single experiment, from a series of 3 independent experiments (see Table 1 for pooled data).

Table 1. Summary of IC_{50} analysis for inhibition of IMPα binding interactions by moxidectin and ivermectin.

	IC_{50} (μM) *	
Binding Interaction	**Ivermectin**	**Moxidectin**
PfIMPα:TGS1-NLS-GFP	5.0 ± 1.3	4.8 ± 0.4
TgIMPα:Tag-NLS-GFP	4.9 ± 0.2	3.8 ± 0.8
MmΔIBBIMPα:Tag-NLS-GFP	6.2 ± 0.2	3.6 ± 0.5
PfIMPα:MmIMPβ1	1.9 ± 0.2	3.4 ± 0.4
TgIMPα:MmIMPβ1	6.9 ± 1.2	2.7 ± 0.2
MmIMPα:MmIMPβ1	3.1 ± 0.5	4.0 ± 0.4

* Results represent the mean ± SEM (n = 3) for the values measured, as per Figures 1 and 2.

In contrast to mammalian systems where there are multiple different IMPα forms [2–4], *P. falciparum* and *T. gondii* both retain a single, essential form of IMPα, the amino acid sequences of which are 63 and 59% similar (43 and 41% identical), respectively, to that of MmIMPα (Kpna1/Rch1—see Section 2.1) used here. As previously [16–18], we first analysed the MmΔIBBIMPα (truncated) form of IMPα, which is not autoinhibited and hence can bind NLSs with high affinity in the absence of MmIMPβ1 [51,52]. We found that ivermectin inhibited MmΔIBBIMPα recognition of the well-characterised simian virus SV40 large tumour antigen (T-ag) in the context of a bacterially expressed GFP fusion protein (T-ag-NLS-GFP) at low μM concentration (Figure 1b bottom left; IC_{50} value of c. 6 μM—see Table 1), consistent with previous studies [11,16,18,41]. Strikingly, moxidectin showed similar inhibitory activity (Figure 1b bottom right) with an IC_{50} of 3–4 μM (Table 1). As previously [18], ivermectin inhibited NLS binding by PfIMPα (TGS1-NLS-GFP fusion protein) and TgIMPα (SV40:T-ag-NLS-GFP fusion protein) at low μM concentration (Figure 1b, top left and middle left panels), with IC_{50} values of c. 5 μM (Table 1). Moxidectin showed very similar ability to inhibit both apicomplexan IMPαs (Figure 1b top right/middle right panels), with comparable IC_{50} values (4–5 μM, Table 1). Clearly, moxidectin strongly resembled ivermectin in being able to inhibit NLS binding by IMPαs, including those from apicomplexans, at low μM concentrations.

Apicomplexan IMPβ1 has not been characterized as yet, but previous studies [16,18] show that apicomplexan IMPαs are able to interact with mammalian IMPβ1 with high affinity, and that ivermectin is able to inhibit this interaction through its ability to bind to IMPα to alter structure/conformation. Accordingly, we used the AlphaScreen binding assay to assess the ability of moxidectin to inhibit binding of IMPα from *Mus musculus*, *P. falciparum* and *T. gondii* to mammalian IMPβ1. Binding of full-length MmIMPα to MmIMPβ1 was inhibited by moxidectin with an IC_{50} of c. 4 μM, very similar to that of ivermectin (Figure 2, Table 1). Strikingly, it also inhibited binding of PfIMPα and TgIMPα to MmIMPβ1 with IC_{50} values of c. 3 μM, very similar to those of ivermectin (IC_{50} values of 2–7 μM; Figure 2, Table 1). Clearly, moxidectin resembles ivermectin in being able to inhibit IMPβ1 recognition by mammalian and apicomplexan IMPαs at low μM concentrations.

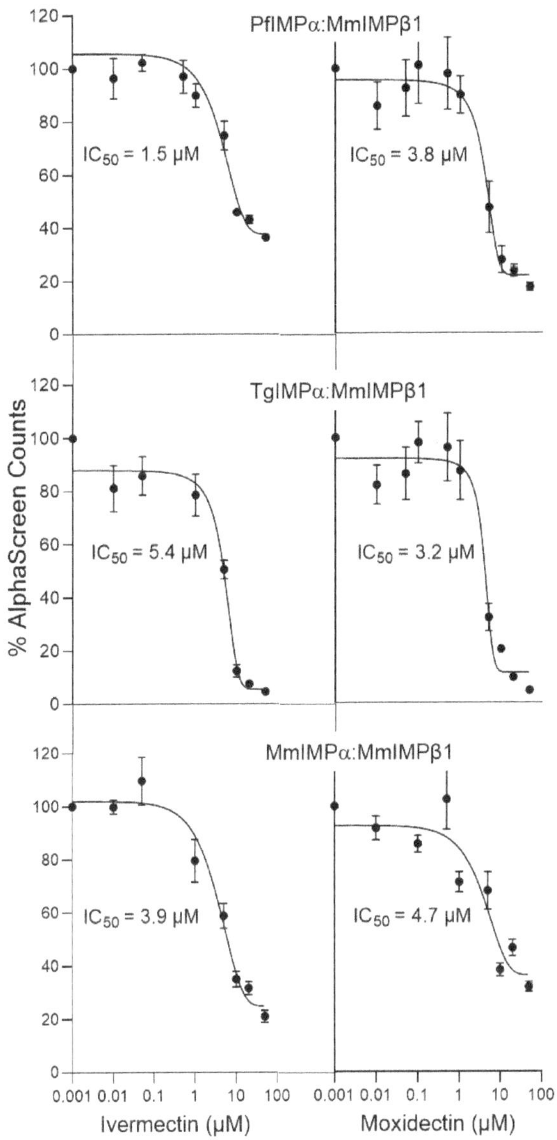

Figure 2. Like ivermectin, moxidectin can inhibit IMPα–IMPβ1 binding. AlphaScreen technology was used to determine the IC_{50} for inhibition by ivermectin and moxidectin of binding of MmIMPβ1 (30 nM) to various IMPαs (30 nM). Data represent the mean ± SEM for quadruplet wells from a single typical experiment, from a series of three independent experiments (see Table 1 for pooled data).

3.2. Moxidectin Can Bind Directly to Apicomplexan IMPα Proteins

We employed far-UV CD spectroscopy, as previously [14,18], to assess the potential of moxidectin to bind directly to IMPα proteins. The CD spectra of PfIMPα, TgIMPα, and MmIMPα all showed double minima at 208 and 222 nm (Figure 3a), consistent with the predominantly α-helical structure of IMPα (see [51]). Quantitative estimation indicated that apicomplexan IMPαs were c. 60% α-helical, compared to c. 70% for MmIMPα (Figure 3b). The CD spectra of the proteins in the presence of increasing concentrations of ivermectin and moxidectin revealed a concentration-dependent reduction in α-helicity (Figure 3a,b), indicating that, like ivermectin, moxidectin appears to be able to bind to the IMPαs directly to perturb structure.

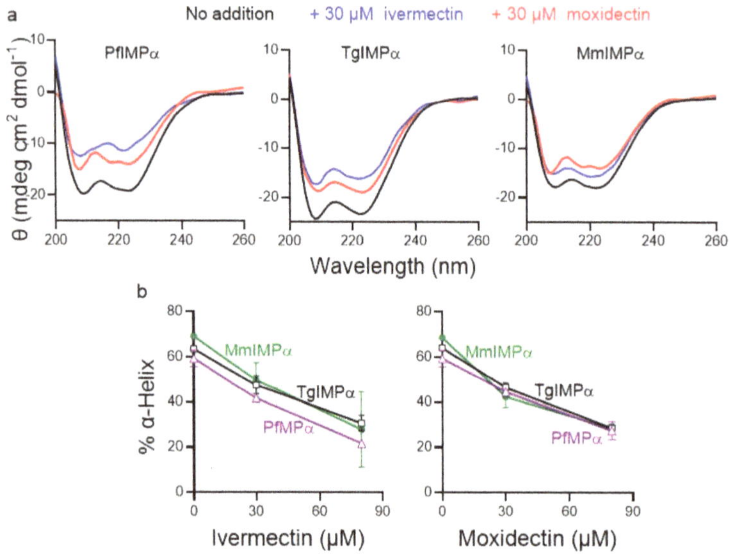

Figure 3. CD spectra for apicomplexan and mammalian IMPαs in the absence and presence of moxidectin and ivermectin. CD spectra were collected for PfIMPα, TgIMPα, and MmIMPα in the absence or presence of 30 or 80 μM ivermectin or moxidectin. (**a**). Spectra are shown from a single experiment, representative of 2 independent experiments for IMPαs without or with 30 μM ivermectin or moxidectin. Note: θ is ellipticity in mdeg cm^2 dmol^{-1}. (**b**). The α-helical content of the respective IMPαs was estimated as previously (see Section 2.4) from spectra as per Figure 3a. Results represent the mean ± SD for 2 independent experiments.

We built on these results using an intrinsic tryptophan fluorescence assay. Tryptophan residues within proteins have an intrinsic fluorescence, a change in the microenvironment of which, through ligand binding or other interactions by the protein, can alter the intrinsic fluorescence intensity [53]. We recorded tryptophan fluorescence emission spectra for the IMPαs, all of which showed a fluorescence maximum at 340 nm.

The addition of increasing concentrations of ivermectin or moxidectin resulted in a concentration-dependent decrease in fluorescence intensity at 340 nm (Figure 4a), enabling estimation of dissociation constants (Figure 4b). Moxidectin and ivermectin showed low μM IC_{50} values for mammalian IMPα (3 μM and 4 μM) and PfIMPα (6.4 μM and 7.8 μM). In contrast, IC_{50} values of moxidectin and ivermectin were higher for TgIMPα (24 μM and 19 μM). As NLS binding sites in IMPα have conserved tryptophan residues, these results suggest that moxidectin and ivermectin bind to conserved tryptophan residues on apicomplexan IMPα proteins and impact their conformation; the significance of the possibility that TgIMPα may be slightly less sensitive in this respect is not clear.

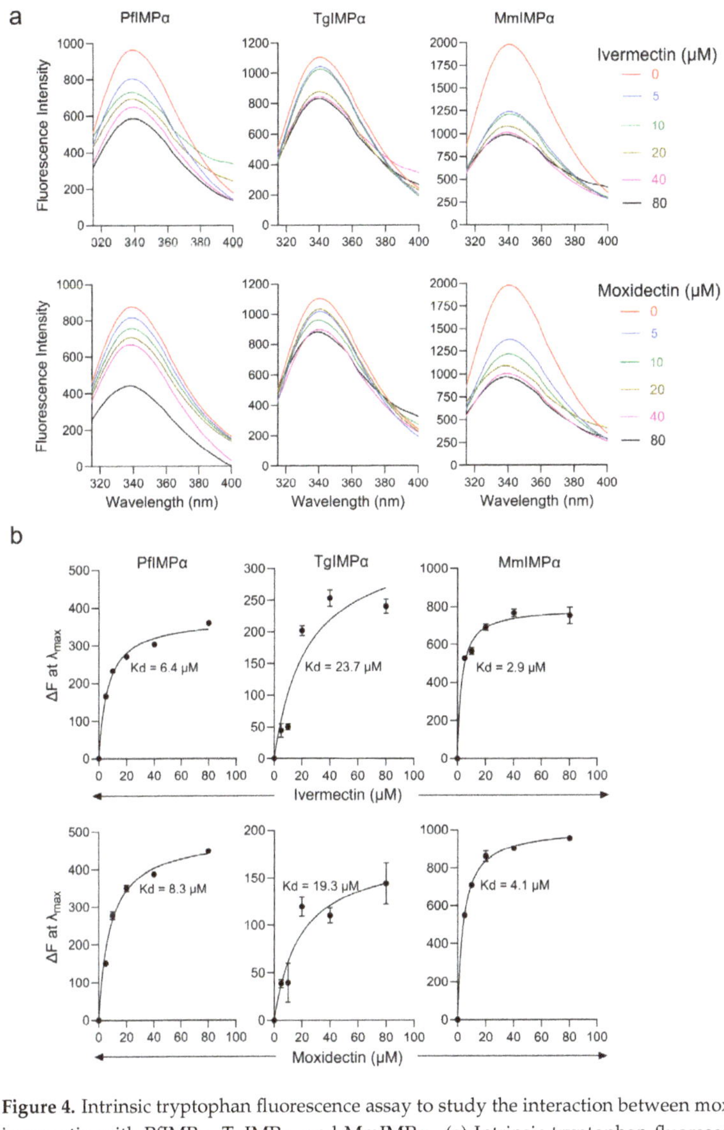

Figure 4. Intrinsic tryptophan fluorescence assay to study the interaction between moxidectin and ivermectin with PfIMPα, TgIMPα, and MmIMPα. (**a**) Intrinsic tryptophan fluorescence spectra of PfIMPα, TgIMPα, and MmIMPα were collected (excitation wavelength: 295 nm and emission from 315 nm to 400 nm) in the presence or absence of moxidectin and ivermectin in the indicated concentration ranges. (**b**). Changes in fluorescence intensity at 340 nm (λmax) from (**a**) with increasing concentrations of compound were plotted using GraphPad prism for a single typical set of measurements, representative of a series of 3 independent experiments (see Table 2).

Table 2. Summary of Kd estimations measured using intrinsic tryptophan fluorescence measurements.

Protein	Kd (μM) *	
	Ivermectin	Moxidectin
PfIMPα	6.4 ± 0.3	7.8 ± 1.2
TgIMPα	24 ± 1.4	19 ± 2.2
MmIMPα	2.9 ± 0.4	4.1 ± 0.2

* Results represent the mean ± SEM (n = 3) for the values measured as per Figure 4b.

3.3. Like Ivermectin, Moxidectin Can Limit Proliferation of *P. falciparum*

Since moxidectin appears to bind to PfIMPα and thereby inhibit NLS- and IMPβ1-binding, we carried out drug susceptibility assays for moxidectin and ivermectin against *P. falciparum* parasites using an established ELISA technique [16,17,44,45], with the clinically prescribed drug dihydroartemisinin (DHA) used as a positive control. In similar fashion to ivermectin, moxidectin showed an IC_{50} value of 1 μM against *P. falciparum* asexual stages in culture (Figure 5, Table 3), supporting the idea that moxidectin, like ivermectin, has antimalarial activity.

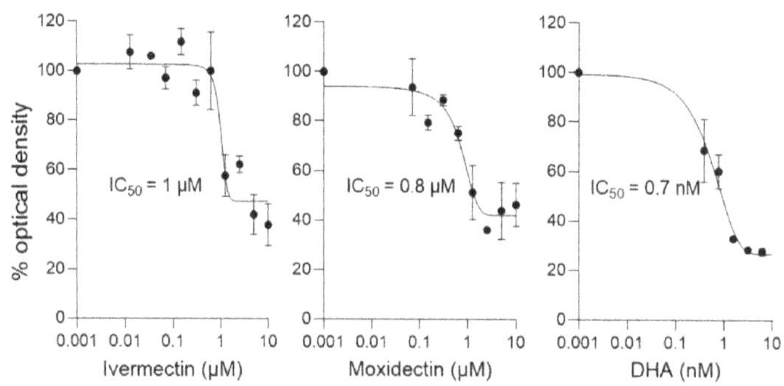

Figure 5. Ivermectin and moxidectin inhibit *P. falciparum* parasites in culture at low μM concentrations. *P. falciparum* cultures (0.25% parasitemia) were treated with increasing concentrations of the indicated compounds for 72 h, after which the HRP2-based sandwich ELISA was used to measure the HRP2 levels, determined by optical density. The results shown are from a single typical experiment performed in duplicate (SD shown), representative of a series of three independent experiments (see Table 3 for pooled data).

Table 3. Summary of the IC_{50} values for inhibition of *P. falciparum* parasites by moxidectin and ivermectin in culture.

Compound	IC_{50} (μM) *
Ivermectin	0.7 ± 0.1
Moxidectin	0.9 ± 0.1
DHA	0.001 ± 0.0005

* Results represent the mean ± SD (n = 3) for IC_{50} analysis as per Figure 5.

We also tested the effect of moxidectin on the growth of *T. gondii* tachyzoites constitutively expressing luciferase in HFF host cells [54]. IC_{50} analysis indicated values in the 30–35 μM range, comparable to that for ivermectin (see Table 4). Analysis of the potential cytotoxic effects of the compounds on the HFF cells in the infectious system using the MTT assay, however, revealed that both ivermectin and moxidectin have CC_{50} values for HFF cells of c. 10 μM (Table 4). Clearly, it cannot be excluded that the observed inhibitory effects of ivermectin/moxidectin on *T. gondii* tachyzoites in the HFF cell infectious system are attributable to cytotoxic effects on the host cells.

Table 4. Summary of IC_{50}/CC_{50} values for effects of moxidectin and ivermectin on *T. gondii* parasites in culture.

Compound	IC_{50} (μM) *	CC_{50} (μM) +
Ivermectin	29 ± 0.7	8 ± 1.8
Moxidectin	35 ± 3.1	10 ± 0.8

* Results represent the mean ± SD (n = 3). + Results represent the mean ± SD (n = 2).

4. Discussion

This study is the first to show that the macrocyclic lactone moxidectin shows promise as an inhibitor of IMPα, and accordingly has potential as an anti-parasitic agent. Through its chemical scaffold shared with ivermectin (see Figure 1a), moxidectin exhibits similar inhibitory effects on mammalian and apicomplexan IMPα proteins. Both compounds effectively inhibit NLS recognition and binding to IMPβ1 with IC_{50} values of 3–7 μM (Table 1; Figures 1 and 2); based on the fluorescence quenching and structural perturbations observed in the tryptophan fluorescence and CD spectroscopy assays here (Figures 3 and 4), conserved tryptophan residues within the IMPαs likely play a critical role in these interactions. Consistent with this, moxidectin's activity against the asexual stages of *P. falciparum* was confirmed in culture (IC_{50} of 1 μM; Figure 5). Thus, taken together our novel findings suggest that moxidectin, like ivermectin, targets/binds to conserved structural motifs in IMPα, disrupting their conformation and inhibiting key protein–protein interactions; this is likely the basis of its antimalarial activity. As it is approved by the FDA to treat helminthic infections, moxidectin's pharmacokinetic properties/safety are well documented in human studies (see [36,40,49,50]), but of course, rigorous testing of the antimalarial action of moxidectin in clinical studies, initially along the lines of those already carried out for ivermectin [34,35], will be required in the future to assess properly its true therapeutic potential against apicomplexan disease.

As mentioned, moxidectin and ivermectin share key chemical scaffolds (Figure 1a) that enable them to target glutamate-gated chloride channels in helminths/insects [54–58]. Based on the results here using CD and intrinsic tryptophan fluorescence measurements, as well as previous studies [16], it would appear that moxidectin/ivermectin's shared chemical scaffolds enable them to bind directly to specific regions of IMPα, thereby impacting key protein–protein interactions central to the nuclear transport process. Understanding the precise binding site of these compounds on apicomplexan and mammalian IMPα is crucial for the development of agents that selectively target apicomplexan without affecting mammalian host IMPα. This understanding holds the potential to pave the way for the development of a novel class of antimalarials that are selective and efficacious.

A key implication of this study is that other macrocyclic lactones related to ivermectin and moxidectin (avermectins, milbemycins etc.) may conceivably have similar properties. Although moxidectin is structurally related to ivermectin, it has distinct features (see Figure 1a and legend), which appear to be the basis of its improved stability and efficacy as an anti-parasitic compared to ivermectin in mammalian systems [38,49]. It is an intriguing possibility that other distinct avermectins/milbemycins, such as selamectin, eprinomectin, abamectin, doramectin, etc., which are currently in veterinary/agricultural use may also prove to have the same general properties. Direct experimentation is of course required to test this possibility, but in the important quest to continue to develop new strategies to combat apicomplexan disease, especially in the context of emerging resistance to existing treatments, this may well prove worthwhile. Moxidectin here showed antimalarial activity at low μM concentrations; combined with its reported longer half-life and sustained release in mammalian systems compared to ivermectin [38,49], it appears to be a more compelling candidate for further exploration in the development of anti-parasitic drugs than ivermectin. Future investigations should focus on optimising moxidectin's selectivity, exploring its efficacy in vivo, and potentially its specific targets within the parasites.

In summary, the study presented here emphasizes the exciting potential of repurposing existing drugs, such as moxidectin and ivermectin, to address the urgent need for novel and effective treatments against malaria.

Author Contributions: Conceptualisation: D.A.J., S.P., S.B.W. and K.M.W. Analysis and investigation: S.B.W. Methodology: S.B.W. and G.M. Writing original draft: S.B.W. Writing—review and editing: D.A.J., S.P. and K.M.W. Supervision: D.A.J., S.P. and K.M.W. Project administration: D.A.J. and S.P. Funding acquisition: D.A.J. and S.P. All authors have read and agreed to the published version of the manuscript.

Funding: The authors acknowledge the financial support of the National Health and Medical Research Council Australia (Senior Principal Research Fellowship APP1002486/APP1103050) to D.A.J. and the Department of Science and Technology (DST), Science and Engineering Research Board (SERB Core Research Grant CRG/2018/000129), India to S.P. S.B.W. acknowledges the support of a PhD scholarship from IITB/Monash University and support from the Department of Biotechnology, India, for project IMURA 0716 (IITB-Monash Research Academy).

Institutional Review Board Statement: Not applicable.

Informed Consent Statement: Not applicable.

Data Availability Statement: The original contributions presented in this study are included in the article. Further inquiries can be directed to the corresponding author.

Acknowledgments: The authors acknowledge IIT Bombay Hospital, staff, and volunteers for their help in the blood collection for *P. falciparum* cultures and Sanjeeva Srivastava and Prakriti Tayalia for the use of plate readers for the anti-parasitic assays.

Conflicts of Interest: The authors declare no conflicts of interest. The funders had no role in the design of the study, in the collection, analyses, or interpretation of data, in the writing of the manuscript, or in the decision to publish the results.

References

1. Kosyna, F.K.; Depping, R. Controlling the Gatekeeper: Therapeutic Targeting of Nuclear Transport. *Cells* **2018**, *7*, 221. [CrossRef] [PubMed]
2. Fulcher, A.J.; Jans, D.A. Regulation of Nucleocytoplasmic Trafficking of Viral Proteins: An Integral Role in Pathogenesis? *Biochim Biophys. Acta-Mol. Cell Res.* **2011**, *1813*, 2176–2190. [CrossRef] [PubMed]
3. Köhler, M.; Speck, C.; Christiansen, M.; Bischoff, F.R.; Prehn, S.; Haller, H.; Hartmann, E. Evidence for Distinct Substrate Specificities of Importin α Family Members in Nuclear Protein Import. *Mol. Cell. Biol.* **1999**, *16*, 7782–7791. [CrossRef]
4. Jans, D.A.; Martin, A.J.; Wagstaff, K.M. Inhibitors of Nuclear Transport. *Curr. Opin. Cell Biol.* **2019**, *58*, 50–60. [CrossRef]
5. Stewart, M. Molecular Mechanism of the Nuclear Protein Import Cycle. *Nat. Rev. Mol. Cell Biol.* **2007**, *8*, 195–208. [CrossRef]
6. Dunn, J.; McCuaig, R.D.; Tan, A.H.Y.; Tu, W.; Wu, F.; Wagstaff, K.M.; Zafar, A.; Ali, S.; Diwakar, H.; Dahlstrom, J.E.; et al Selective Targeting of Protein Kinase C (PKC)-θ Nuclear Translocation Reduces Mesenchymal Gene Signatures and Reinvigorates Dysfunctional CD8$^+$ T Cells in Immunotherapy-Resistant and Metastatic Cancers. *Cancers* **2022**, *14*, 1596. [CrossRef]
7. Gajewska, K.; Lescesen, H.; Ramialison, M.; Wagstaff, K.M.; Jans, D.A. Nuclear transporter Importin-13 plays a key role in the oxidative stress transcriptional response. *Nat. Commun.* **2021**, *12*, 5904. [CrossRef] [PubMed]
8. Zhang, S.Y.; Williamson, N.A.; Duvick, L.; Lee, A.; Orr, H.T.; Korlin-Downs, A.; Yang, P.; Mok, Y.-F.; Jans, D.A.; Bogoyevitch, M.A The ataxin-1 interactome reveals direct connection with multiple disrupted nuclear transport pathways. *Nat. Commun.* **2020** *11*, 3343. [CrossRef] [PubMed]
9. Martin, A.J.; Jans, D.A. Antivirals that target the host IMPα/β1-virus interface. *Biochem. Soc. Trans.* **2021**, *49*, 281–295. [CrossRef] [PubMed]

0. Crump, A.; Omura, S. Ivermectin, wonder drug from Japan: The human use perspective. *Proc. Jpn. Acad.* **2011**, *87*, 13–28. [CrossRef] [PubMed]
1. Wagstaff, K.M.; Sivakumaran, H.; Heaton, S.M.; Harrich, D.; Jans, D.A. Ivermectin Is a Specific Inhibitor of Importin α/β -Mediated Nuclear Import Able to Inhibit Replication of HIV-1 and Dengue Virus. *Biochem. J.* **2012**, *856*, 851–856. [CrossRef] [PubMed]
2. Caly, L.; Druce, J.D.; Catton, M.G.; Jans, D.A.; Wagstaff, K.M. The FDA-Approved Drug Ivermectin Inhibits the Replication of SARS-CoV-2 in Vitro. *Antivir. Res.* **2020**, *178*, 104787. [CrossRef]
3. Thomas, D.R.; Lundberg, L.; Pinkham, C.; Shechter, S.; Debono, A.; Baell, J.; Wagstaff, K.M.; Hick, C.A.; Kehn-hall, K.; Jans, D.A. Identification of Novel Antivirals Inhibiting Recognition of Venezuelan Equine Encephalitis Virus Capsid Protein by the Importin $\alpha/\beta 1$ Heterodimer through High-Throughput Screening. *Antivir. Res.* **2018**, *151*, 8–19. [CrossRef] [PubMed]
4. Yang, S.N.Y.; Atkinson, S.C.; Fraser, J.E.; Wang, C.; Maher, B.; Roman, N.; Forwood, J.K.; Wagstaff, K.M.; Borg, N.A.; Jans, D.A. Novel Flavivirus Antiviral That Targets The Host Nuclear Transport Importin $\alpha/\beta 1$ Heterodimer. *Cells* **2019**, *8*, 281. [CrossRef] [PubMed]
5. Shechter, S.; Thomas, D.R.; Lundberg, L.; Pinkham, C.; Lin, S.; Wagstaff, K.M.; Debono, A.; Kehn-hall, K.; Jans, D.A. Novel Inhibitors Targeting Venezuelan Equine Encephalitis Virus Capsid Protein Identified Using In Silico Structure-Based-Drug-Design. *Sci. Rep.* **2017**, *7*, 17705. [CrossRef]
6. Yang, S.N.Y.; Atkinson, S.C.; Wang, C.; Lee, A.; Bogoyevitch, M.A.; Borg, N.A.; Jans, D.A. The Broad Spectrum Antiviral Ivermectin Targets the Host Nuclear Transport Importin α/b1 Heterodimer. *Antivir. Res.* **2020**, *177*, 104760. [CrossRef]
7. Walunj, S.B.; Dias, M.M.; Kaur, C.; Wagstaff, K.M.; Dey, V.; Hick, C.; Patankar, S.; Jans, D.A. High-Throughput Screening to Identify Inhibitors of *Plasmodium falciparum* Importin α. *Cells* **2022**, *11*, 1201. [CrossRef]
8. Walunj, S.B.; Wang, C.; Wagstaff, K.M.; Patankar, S.; Jans, D.A. Conservation of Importin a Function in Apicomplexans: Ivermectin and GW5074 Target *Plasmodium falciparum* Importin a and Inhibit Parasite Growth in Culture. *Int. J. Mol. Sci.* **2022**, *23*, 13899. [CrossRef] [PubMed]
9. Greenwood, B.M.; Fidock, D.A.; Kyle, D.E.; Kappe, S.H.I.; Alonso, P.L.; Collins, F.H.; Duffy, P.E. Malaria: Progress, Perils, and Prospects for Eradication. *J. Clin. Investig.* **2008**, *118*, 1266–1276. [CrossRef]
0. Tse, E.G.; Korsik, M.; Todd, M.H. The Past, Present and Future of Antimalarial Medicines. *Malar. J.* **2019**, *18*, 93. [CrossRef]
1. Blasco, B.; Leroy, D.; Fidock, D.A. Antimalarial Drug Resistance: Linking *Plasmodium falciparum* Parasite Biology to the Clinic. *Nat. Med.* **2017**, *23*, 917–928. [CrossRef] [PubMed]
2. Montazeri, M.; Mikaeili Galeh, T.; Moosazadeh, M.; Sarvi, S.; Dodangeh, S.; Javidnia, J.; Sharif, M.; Daryani, A. The Global Serological Prevalence of *Toxoplasma gondii* in Felids during the Last Five Decades (1967–2017): A Systematic Review and Meta-Analysis. *Parasites Vectors* **2020**, *13*, 82. [CrossRef]
3. Sabin, A.B. Toxoplasmic encephalitis in children. *J. Am. Med. Assoc.* **1941**, *116*, 801–807. [CrossRef]
4. Alday, P.H.; Doggett, J.S. Drugs in Development for Toxoplasmosis: Advances, Challenges, and Current Status. *Drug Des. Dev. Ther.* **2017**, *11*, 273–293. [CrossRef] [PubMed]
5. Montazeri, M.; Mehrzadi, S.; Sharif, M.; Sarvi, S.; Tanzifi, A.; Aghayan, S.A.; Daryani, A. Drug Resistance in *Toxoplasma gondii*. *Front. Microbiol.* **2018**, *9*, 2587. [CrossRef]
6. Frankel, M.B.; Knoll, L.J. Functional Analysis of Key Nuclear Trafficking Components Reveals an Atypical Ran Network Required for Parasite Pathogenesis. *Mol. Microbiol.* **2008**, *70*, 410–420. [CrossRef] [PubMed]
7. Frankel, M.B.; Knoll, L.J. The Ins and Outs of Nuclear Trafficking: Unusual Aspects in Apicomplexan Parasites. *DNA Cell Biol.* **2009**, *28*, 277–284. [CrossRef]
8. Zhang, M.; Wang, C.; Otto, T.D.; Oberstaller, J.; Liao, X.; Adapa, S.R.; Udenze, K.; Bronner, I.F.; Casandra, D.; Mayho, M.; et al. Uncovering the Essential Genes of the Human Malaria Parasite *Plasmodium falciparum* by Saturation Mutagenesis. *Science* **2018**, *360*, eaap7847. [CrossRef] [PubMed]
9. Sidik, S.M.; Huet, D.; Ganesan, S.M.; Huynh, M.H.; Wang, T.; Nasamu, A.S.; Thiru, P.; Saeij, J.P.J.; Carruthers, V.B.; Niles, J.C.; et al. A Genome-Wide CRISPR Screen in Toxoplasma Identifies Essential Apicomplexan Genes. *Cell* **2016**, *166*, 1423–1435.e12. [CrossRef]
0. González Canga, A., Sahagún Prieto, A.M.; Diez Liébana, M.J.; Fernández Martínez, N.; Sierra Vega, M.; García Vieitez, J.J. The pharmacokinetics and interactions of ivermectin in humans—A mini-review. *AAPS J.* **2008**, *10*, 42–46. [CrossRef]
1. Jans, D.A.; Wagstaff, K.M. Ivermectin as a Broad-Spectrum Host-Directed Antiviral: The Real Deal? *Cells* **2020**, *9*, 2100. [CrossRef]
2. Ooi, E.E. Repurposing Ivermectin as an Anti-Dengue Drug. *Clin. Infect. Dis.* **2021**, *72*, e594–e595. [CrossRef] [PubMed]
3. Bryant, A.; Lawrie, T.A.; Dowswell, T.; Fordham, E.J.; Mitchell, S.; Hill, S.R.; Tham, T.C. Ivermectin for Prevention and Treatment of COVID-19 Infection: A Systematic Review, Meta-Analysis, and Trial Sequential Analysis to Inform Clinical Guidelines. *Am. J. Ther.* **2021**, *28*, e434–e460. [CrossRef]

34. Foy, B.D.; Alout, H.; Seaman, J.A.; Rao, S.; Magalhaes, T.; Wade, M.; Parikh, S.; Soma, D.; Sagna, B.; Fournet, F.; et al. Efficacy and Risk of Harms of Repeat Ivermectin Mass Drug Administrations for Control of Malaria (RIMDAMAL): A Cluster-Randomised Trial. *Lancet* **2019**, *393*, 1517–1526. [CrossRef] [PubMed]
35. Ouédraogo, A.L.; Bastiaens, G.J.H.; Tiono, A.B.; Guelbéogo, W.M.; Kobylinski, K.C.; Ouédraogo, A.; Barry, A.; Bougouma, E.C Nebie, I.; Ouattara, M.S.; et al. Efficacy and Safety of the Mosquitocidal Drug Ivermectin to Prevent Malaria Transmission after Treatment: A Double-Blind, Randomized, Clinical Trial. *Clin. Infect. Dis.* **2015**, *60*, 357–365. [CrossRef] [PubMed]
36. Sprecher, V.P.; Hofmann, D.; Savathdy, V.; Xayavong, P.; Norkhankhame, C.; Huy, R.; Khieu, V.; Sayasone, S.; Hattendorf, J.; Keiser J. Efficacy and Safety of Moxidectin Compared with Ivermectin against Strongyloides Stercoralis Infection in Adults in Laos and Cambodia: A Randomised, Double-Blind, Non-Inferiority, Phase 2b/3 Trial. *Lancet Infect. Dis.* **2023**, *24*, 196–205. [CrossRef]
37. Cobb, R.; Boeckh, A. Moxidectin: A Review of Chemistry, Pharmacokinetics and Use in Horses. *Parasit. Vectors* **2009**, *2*, S5 [CrossRef] [PubMed]
38. Prichard, R.K.; Geary, T.G. Perspectives on the Utility of Moxidectin for the Control of Parasitic Nematodes in the Face of Developing Anthelmintic Resistance. *Int. J. Parasitol. Drugs Drug Resist.* **2019**, *10*, 69–83. [CrossRef]
39. Tang, M.; Hu, X.; Wang, Y.; Yao, X.; Zhang, W.; Yu, C.; Cheng, F.; Li, J.; Fang, Q. Ivermectin, a Potential Anticancer Drug Derived from an Anti-parasitic Drug. *Pharmacol. Res.* **2021**, *163*, 105207. [CrossRef]
40. Cotreau, M.M.; Warren, S.; Ryan, J.L.; Fleckenstein, L.; Vanapalli, S.R.; Brown, K.R.; Rock, D.; Chen, C.Y.; Schwertschlag, U.S. The Anti-parasitic Moxidectin: Safety, Tolerability, and Pharmacokinetics in Humans. *J. Clin. Pharmacol.* **2003**, *43*, 1108–1115 [CrossRef]
41. Wagstaff, K.M.; Jans, D.A. Intramolecular Masking of Nuclear Localization Signals: Analysis of Importin Binding Using a Novel AlphaScreen-Based Method. *Anal. Biochem.* **2006**, *348*, 49–56. [CrossRef]
42. Oshokoya, O.O.; Roach, C.A.; Jiji, R.D. Quantification of Protein Secondary Structure Content by Multivariate Analysis of Deep-Ultraviolet Resonance Raman and Circular Dichroism Spectroscopies. *Anal. Methods* **2014**, *6*, 1691–1699. [CrossRef]
43. Trager, W.; Jensen, J.B. Human Malaria Parasites in Continuous Culture. *Science* **1976**, *193*, 673–675. [CrossRef]
44. Kurth, F.; Pongratz, P.; Bélard, S.; Mordmüller, B.; Kremsner, P.G.; Ramharter, M. In Vitro Activity of Pyronaridine against *Plasmodium falciparum* and Comparative Evaluation of Antimalarial Drug Susceptibility Assays. *Malar. J.* **2009**, *8*, 79. [CrossRef] [PubMed]
45. Noedl, H.; Bronnert, J.; Yingyuen, K.; Attlmayr, B.; Kollaritsch, H.; Fukuda, M. Simple Histidine-Rich Protein 2 Double-Site Sandwich Enzyme-Linked Immunosorbent Assay for Use in Malaria Drug Sensitivity Testing. *Antimicrob. Agents Chemother.* **2005**, *49*, 3575–3577. [CrossRef]
46. Jacot, D.; Meissner, M.; Sheiner, L.; Soldati-Favre, D.; Striepen, B. Genetic Manipulation of *Toxoplasma gondii*. In *Toxoplasma Gondii The Model Apicomplexan—Perspectives and Methods*, 2nd ed.; Academic Press: Cambridge, MA, USA, 2013; pp. 577–611. [CrossRef]
47. Striepen, B.; Soldati, D. Genetic Manipulation of *Toxoplasma gondii*. In *Toxoplasma Gondii*; Academic Press: Cambridge, MA, USA 2007; pp. 391–418. [CrossRef]
48. Riss, T.L.; Moravec, R.A.; Niles, A.L.; Duellman, S.; Benink, H.A.; Worzella, T.J.; Minor, L. Cell Viability Assays. In *Assay Guidance Manual [Internet]*; Markossian, S., Grossman, A., Arkin, M., Auld, D., Austin, C., Baell, J., Brimacombe, K., Chung, T.D.Y Coussens, N.P., Dahlin, J.L., et al., Eds.; Eli Lilly & Company and the National Center for Advancing Translational Sciences Bethesda, MD, USA, 2016.
49. Hürlimann, E.; Hofmann, D.; Keiser, J. Ivermectin and Moxidectin against Soil-Transmitted Helminth Infections. *Trends Parasitol.* **2023**, *39*, 272–284. [CrossRef]
50. Tan, B.; Opoku, N.; Attah, S.K.; Awadzi, K.; Kuesel, A.C.; Lazdins-Helds, J.; Rayner, C.; Ryg-Cornejo, V.; Sullivan, M.; Fleckenstein L. Pharmacokinetics of Oral Moxidectin in Individuals with *Onchocerca Volvulus* Infection. *PLoS Negl. Trop. Dis.* **2022**, *16*, e0010005 [CrossRef] [PubMed]
51. Kobe, B. Autoinhibition by an Internal Nuclear Localization Signal Revealed by the Crystal Structure of Mammalian Importin α *Nat. Struct. Biol.* **1999**, *6*, 388–397. [CrossRef]
52. Fanara, P.; Hodel, M.R.; Corbett, A.H.; Hodel, A.E. Quantitative Analysis of Nuclear Localization Signal (NLS)-Importin α Interaction through Fluorescence Depolarization. Evidence for auto-inhibitory regulation of NLS binding. *J. Biol. Chem.* **2000**, *275* 21218–21223. [CrossRef]
53. Sindrewicz, P.; Li, X.; Yates, E.A.; Turnbull, J.E.; Lian, L.Y.; Yu, L.G. Intrinsic Tryptophan Fluorescence Spectroscopy Reliably Determines Galectin-Ligand Interactions. *Sci. Rep.* **2019**, *9*, 11851. [CrossRef]
54. Prichard, R.; Ménez, C.; Lespine, A. Moxidectin and the Avermectins: Consanguinity but Not Identity. *Int. J. Parasitol. Drugs Drug Resist.* **2012**, *2*, 134–153. [CrossRef] [PubMed]
55. Atif, M.; Estrada-Mondragon, A.; Nguyen, B.; Lynch, J.W.; Keramidas, A. Effects of glutamate and ivermectin on single glutamate-gated chloride channels of the parasitic nematode *H. contortus*. *PLoS Pathog.* **2017**, *13*, e1006663. [CrossRef] [PubMed]
56. Hibbs, R.; Gouaux, E. Principles of activation and permeation in an anion-selective Cys-loop receptor. *Nature* **2011**, *474*, 54–60 [CrossRef] [PubMed]

57. Lynagh, T.; Lynch, J.W. A glycine residue essential for high ivermectin sensitivity in Cys-loop ion channel receptors. *Int. J. Parasitol.* **2010**, *40*, 1477–1481. [CrossRef] [PubMed]
58. Lynagh, T.; Webb, T.I.; Dixon, C.L.; Cromer, B.A.; Lynch, J.W. Molecular Determinants of Ivermectin Sensitivity at the Glycine Receptor Chloride Channel. *J. Biol. Chem.* **2011**, *286*, 43913–43924. [CrossRef] [PubMed]

International Journal of
Molecular Sciences

MDPI

Article

NBF2, an Algal Fiber-Rich Formula, Reverses Diabetic Dyslipidemia and Hyperglycemia In Vivo

Nlandu Roger Ngatu [1,*], Akram Hossain [2], Nao Maruo [3], Steeve Akumwami [4], Asadur Md. Rahman [2], Masamitsu Eitoku [3], Kanae Kanda [1], Akira Nishiyama [2], Narufumi Suganuma [3] and Tomohiro Hirao [1,*]

[1] Department of Public Health, Kagawa University Faculty of Medicine, Kagawa 761-0793, Japan; kanda.kanae@kagawa-u.ac.jp
[2] Department of Medical Pharmacology, Kagawa University Faculty of Medicine, Kagawa 761-0793, Japan; akram.hossain@kagawa-u.ac.jp (A.H.); rahman.md.asadur@kagawa-u.ac.jp (A.M.R.); nishiyama.akira@kagawa-u.ac.jp (A.N.)
[3] Department of Environmental Medicine, Kochi University School of Medicine, Nankoku 783-8505, Japan; jm-n.maruo@kochi-u.ac.jp (N.M.); meitoku@kochi-u.ac.jp (M.E.); nsuganuma@kochi-u.ac.jp (N.S.)
[4] Department of Anesthesiology, Kagawa University Faculty of Medicine, Kagawa 761-0793, Japan; s21d703@kagawa-u.ac.jp
* Correspondence: ngatu.nlandu@kagawa-u.ac.jp (N.R.N.); hirao.tomohiro@kagawa-u.ac.jp (T.H.)

Abstract: *Ulva prolifera*, known as Aonori in Japan, is an edible alga species that is mass-cultivated in Japan. Supplementation with Aonori-derived biomaterials has been reported to enhance metabolic health in previous studies. This was an experimental study that evaluated the metabolic health effects of NBF2, a formula made of algal and *junos Tanaka citrus*-derived biomaterials, on obesity and type 2 diabetes (T2DM). We used 18 obese and hyperglycemic Otsuka Long-Evans Tokushima Fatty (OLETF) rats that were assigned randomly to three groups of six animals: a high-dose NBF2 drink (20 mg/kg) group, a low-dose (10 mg/kg) NBF2 drink group and the control group that received 2 mL of tap water daily for a total of six weeks. We also used eight LETO rats as the normal control group. In addition to the glucose tolerance test (OGTT), ELISA and real-time PCR assays were performed. High-dose and lowdose NBF2 improved insulin sensitivity, as well as glycemic and lipid profiles, as compared with control rats. The OGTT showed that both NBF2 groups and LETO rats had normalized glycemia by the 90-min time-point. NBF2 up-regulated PPARα/γ-mRNA and Sirt2-mRNA gene expressions in BAT and improved the blood pressure profile. These findings suggest that the NBF2 formula, which activates PPAR-α/γ mRNA and Sirt2-mRNA, may reverse dyslipidemia and hyperglycemia in T2DM.

Keywords: algal formula; dyslipidemia; hyperglycemia; peroxisome proliferation-activated receptor; sirtuin; type 2 diabetes

1. Introduction

Drugs that are agonists of peroxisome proliferation-activated receptors (PPARs) may be effective in the management of lifestyle-related metabolic disorders such as dyslipidemia and type 2 diabetes mellitus (T2DM) [1]. The latter condition is characterized by an inadequate pancreatic beta cell response to insulin resistance. Recently, T2DM management has undergone a major conceptual change, with more emphasis on a patient-centered care approach and the safety profile of antihyperglycemic agents [1–3]. In T2DM, the loss of insulin function is reported to be closely related to adiponectin (APN) deficiency. Thus, APN replenishment in T2DM patients might reverse this condition and improve their metabolic health. APN, a protein produced by adipocytes, has been reported to stimulate fatty acid oxidation, improve insulin sensitivity and inhibit neoglucogenesis, resulting in an improvement in energy homeostasis [4]. The insulin-sensitizing effect of APN has been linked to PPAR activation [5,6]. APN also plays a role in tissue regeneration [7,8].

The up-regulation of APN production to correct its deficiency and the restoration of the function of cells involved in energy homeostasis have motivated our research. Natural compounds from *Radix astragali* [9,10] and omega-3 poly-unsaturated fatty acids have been reported to modulate APN production and improve the metabolic health markers of high-risk subjects [11]. We discovered APN-modulating biomaterials (Ngatu Bio Formulas or NBF) containing edible green marine algal species mass-cultivated in Japan, known as Aonori and Aosanori, with beneficial metabolic health effects [12,13].

This study evaluated the effects of NBF2, a formula composed of Aonori alga and *Junos citrus* peel-derived biomaterials, on metabolic health markers in obesity and T2DM model rats.

2. Results

2.1. Obesity and Diabetic Hyperglycemia Establishment in OLETF Rat Model and Effects of Daily Intake of NBF2 on Body Weight (BW) and Fasting Blood Glucose (FBG)

As shown in Figure 1, treatment with NBF2 significantly reduced weight gain in OLETF rats that received both high and low doses of NBF2 versus vehicle-treated OLETF controls ($p < 0.05$) (Figure 1A). Additionally, both low- and high-dose NBF2-treated OLETF rats had reduced fasting blood glucose (FBG) as compared with OLETF controls ($p < 0.01$), which was normalized within the third week of treatment (Figure 1B).

In this study, an oGTT was performed by administering 0.002g of glucose/1g body weight); FBG was measured at baseline (T0), 15 min (T15), 30 min (T30), 60 min (T60), and 120 min (T120). Moreover, the decrease in FBG was greater in high- ($p < 0.01$) and low-dose ($p < 0.05$) NBF2-treated rat groups, as well as in normal rats ($p < 0.001$; vs. controls) (Figure 1C).

Furthermore, only the normal (LETO) rat group had a significantly lower daily calorie intake (vs. control group; $p = 0.000$). No significant difference was observed when comparing NBF2-treated rats and the controls (Supplementary Table S1).

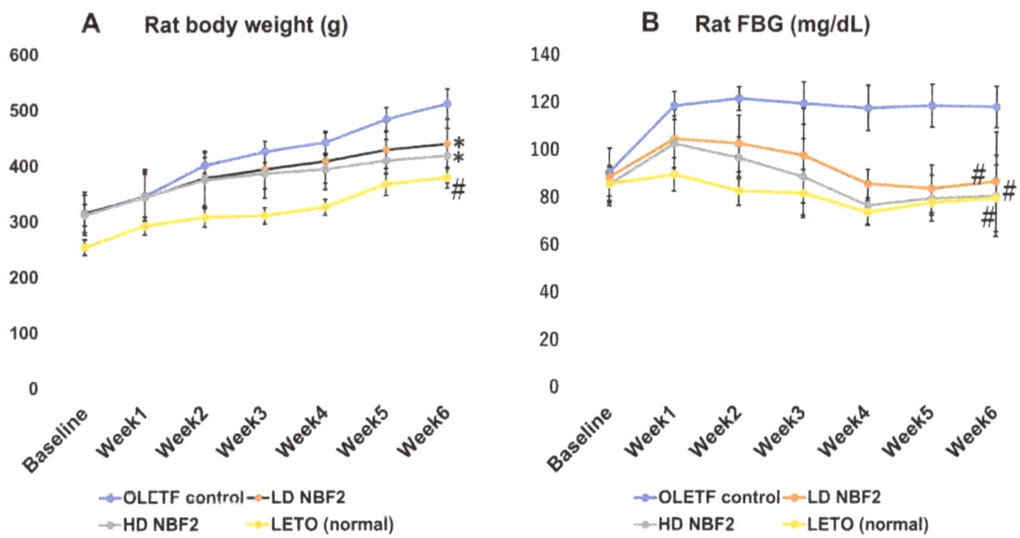

Figure 1. *Cont.*

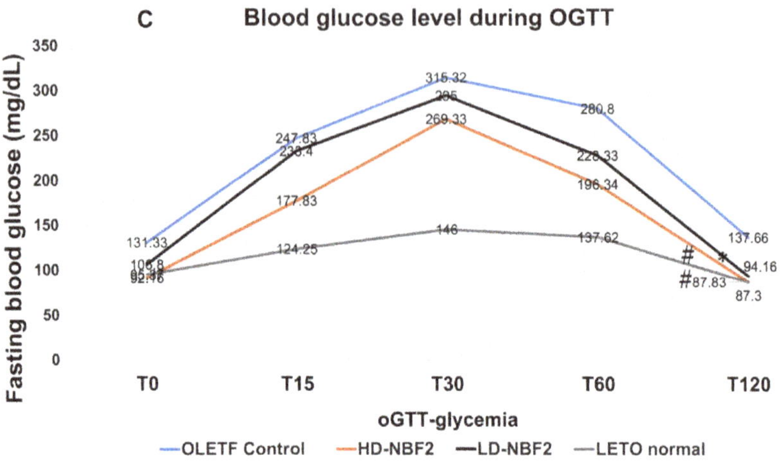

Figure 1. (**A–C**): The trend in rat body weight (**A**) and fasting blood glucose across the study period (**B**) and the blood glucose level during the glucose tolerance test (oGTT) (**C**) ($N = 26$). **Legend**: *, *p*-value less than 0.05; #, *p*-value less than 0.01 (by one-way ANOVA); OLETF, Otsuka Long-Evans Tokushima Fatty rats; LETO, Long-Evans Tokushima Otsuka rats.

2.2. *Effects of NBF2 on Plasma Cholesterol, Triglycerides, Adiponectin, Insulin Resistance and Blood Pressure Profile*

Figure 2 shows the plasma levels of cholesterols, triglycerides (TGs) and the blood pressure profile. The high-dose NBF2 group had significantly lower total cholesterol (Figure 2A), LDL-c ($p < 0.05$) (Figure 2B) and TG levels ($p < 0.01$) (Figure 2C) compared to the control group. Similarly, the low-dose NBF2 group also showed significantly reduced total cholesterol, TGs and LDL-c ($p < 0.05$). On the other hand, the HDL-c level was significantly higher in the high-dose NBF2 group (vs. control group; $p < 0.05$). Furthermore, significantly reduced blood pressure was observed in the high- and low-dose NBF2 groups for DBP and MBP compared to the control group ($p < 0.05$) (Figure 2E–G).

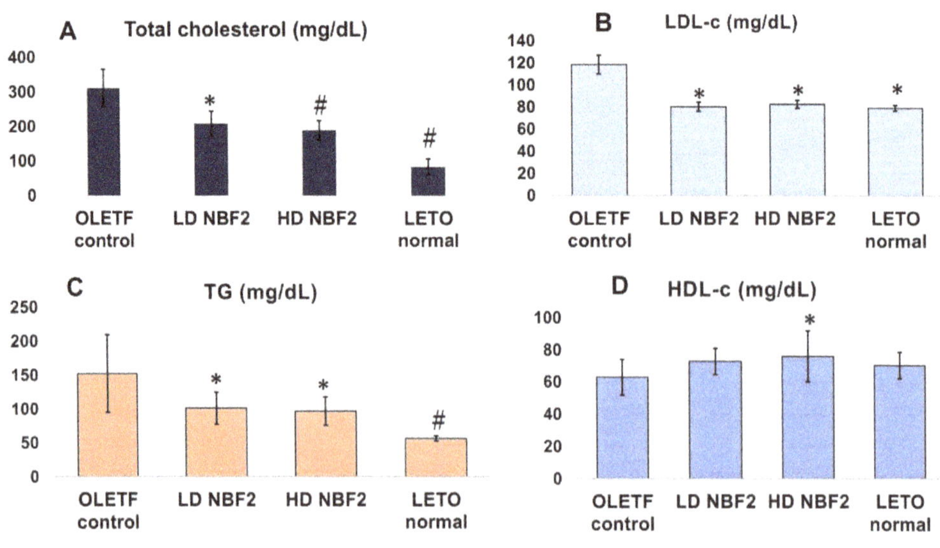

Figure 2. *Cont.*

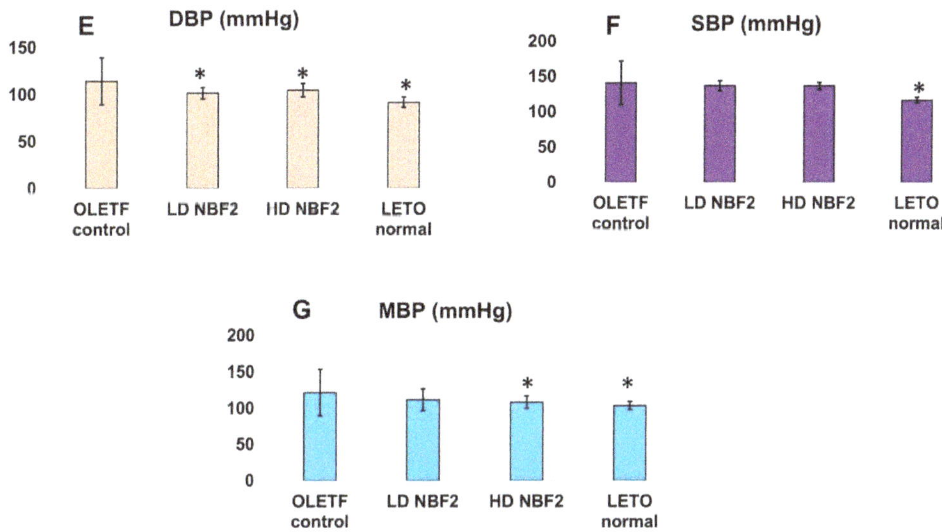

Figure 2. (**A–G**) Blood lipid and blood pressure profiles in rats ($N = 26$). **Legend**: *, p-value less than 0.05; #, p-value less than 0.01 (by one-way ANOVA); OLETF, Otsuka Long-Evans Tokushima Fatty rats; LETO, Long-Evans Tokushima Otsuka rats; T-chol, total cholesterol; LDL-c, low-density lipoprotein cholesterol; HDL-c, high-density lipoprotein cholesterol; TG, triglyceride.

ELISA assays showed significantly higher APN levels in both NBF2 groups compared to the control group ($p < 0.05$). Regarding plasma insulin, a relatively low level was observed in NBF2-treated rats but this was not significant (Figure 3A,B). Furthermore, a significant reduction in HOMA-IR was observed in the NBF2-treated groups compared to the control group (−17% versus −3% change, respectively; $p < 0.05$) (Figure 3C).

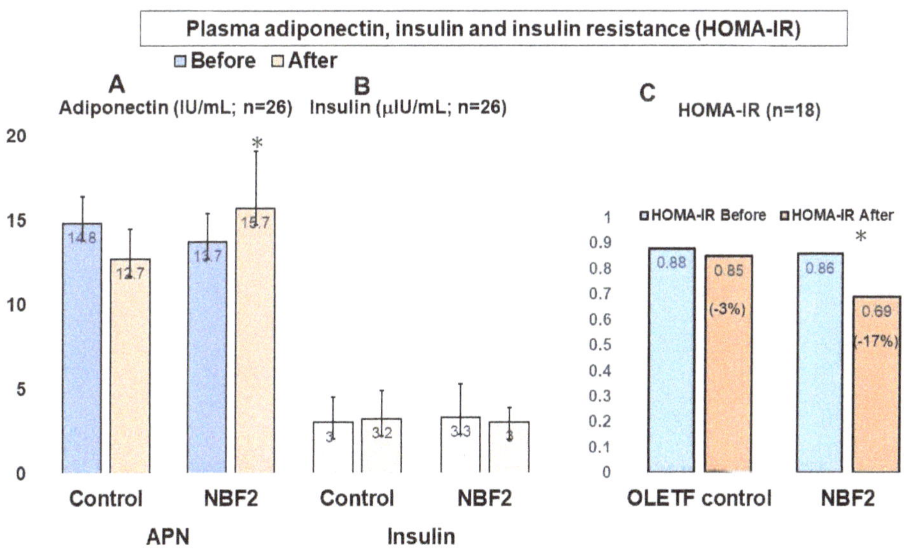

Figure 3. (**A–C**) Plasma adiponectin (**A**), insulin (**B**) and HOMA-IR (**C**) ($N = 24$). **Legend**: *, $p < 0.05$; IU, international unit; APN, adiponectin; NBF, Ngatu Bio Formula; HOMA-IR, homeostasis model assessment for insulin resistance.

3. Messenger RNA (mRNA) Gene Expression Levels of Metabolic Biomarkers

Real-time PCR assays were performed to determine the mRNA gene expression levels of metabolic biomarkers. Table 1 presents the list of primers of rat biomarkers and the corresponding mRNA gene sequences. Leptin and fatty acid synthase (FASN) are linked to the development of obesity and related metabolic disorders. It was observed that rats treated with NBF2 had significantly reduced leptin mRNA expression in brown (BAT) and white adipose tissue (WAT) and liver specimens, as compared with vehicle-treated OLETF controls ($p < 0.05$) (Figure 4A–C). Similarly, NBF2-treated rats had reduced FASN mRNA gene expression in both BAT and WAT specimens (Figure 4D), suggesting an inhibitory effect of NBF2 on fat accumulation.

Unlike FASN and leptin, sirtuin 2 (Sirt2) and the peroxisome proliferator-activated receptors alpha and gamma (PPAR-α/γ) are known to improve obesity-associated metabolic disorders and are considered therapeutic targets for diseases such as T2DM. We observed that treatment with NBF2 caused a marked increase in Sirt2-mRNA expression in the rat BAT ($p < 0.01$) and WAT ($p < 0.05$), as well as an increase in PPAR–γ and PPAR–α mRNA expressions in OLETF rat BAT as compared with the vehicle in the control rats (Figure 4G,H). The lowdose NBF2 group also had reduced APN-mRNA gene expression in BAT and APN-mRNA gene expression in WAT. ATPase sarcoplasmic/endoplasmic reticulum Ca^{2+}, also known as ATPase 2a2 or SERCA2, is an enzyme that plays an important metabolic role as a glucose transport regulator; it is down-regulated in the adipocytes of diabetic animals and humans. We found that BAT ATP2a2-mRNA expression was up-regulated in NBF2-treated rats (Figure 4M–O) (vs. control group; $p < 0.05$). Consequently, the dual PPARα/γ and Sirt2 modulatory effects of NBF2 observed in this study suggest an improvement in lipid and glucose metabolism.

Table 1. List of primers used and corresponding biomarkers' nucleic acid sequences.

Name	Primer (5′→3′)	Length	GC%	Tm	Target Gene	Amplified Size (bp)	Amplified Area
PPAR-[F]	GAGGGCACACGCTAGGAAG	19	63	60	XM_006242152.4	72	GAGGGCACACGCTAGGAAGG GCACACGCGTGCGACTTTCGG GGCCCCTGGAACTGTCCGCTA CTTCGA GTCC
PPAR-[R]	GGACTCGAAGTAGCGGACAG	20	60	60			
PPAR-[F]	GCCGCCTCAGATTTGAAAGAA	21	48	59	XM_006237009.4	142	GCCGCCTCAGATTTGAAAGAA GCTGTGAACCACTAATATCCA AGGACATTTTTGAAAACAAGG ACTACCCTTTACTGAAATTACC ATGGTTGACACAGAGATGCCA TTCTGGCCCACCAACTTCGGA ATCAGCTCTGTGGAC
PPAR-[R]	GTCCACAGAGCTGATTCCGA	20	55	60			
APN [F]	CAAGCGCTCCTGTTCCTCTT	20	55	62	XM_039087986.1	228	CAAGCGCTCCTGTTCCTCTTAA TCCTGCCCAGTCATGAAGGGA TTACTGCAACCGAAGGGCCAG GAGCTTTGGTCCCTCCACCCA AGGAAACTTGTGCAGGTTGGA TGGCAGGCATCCCAGGATATC CTGGTCACAATGGGATACCGG GCCGTGATGGCAGAGATGGCA CTCCTGGAGAGAAGGGAGAG AAGGGAGACGCAGGTGTTCT TGGTCCTAAGGGTGACCCA
APN [R]	TGGGTCACCCTTAGGACCAA	20	55	62.6			
Sirtuin 2 [F]	GACGAGCTGACCCTTGAAG	19	58	63	NM_001399630.1	148	GACGAGCTGACCCTTGAAG GAGTGACACGCTACATGCA GAGCGAGCGCTGTCGCAG GGTCATCTGTTTGGTGGGA GCTGGAATCTCCACATCCG CAGGAATCCCTGACTTCCG CTCCCCATCCACTGGCCTC TATGCAAACTTGGAGA
Sirtuin 2 [R]	TCTCCAAGTTTGCATAGAGGC	21	48	63			

Table 1. *Cont.*

Name	Primer (5′→3′)	Length	GC%	Tm	Target Gene	Amplified Size (bp)	Amplified Area
Atp2a2 [F]	TTTGGGCAGGATGAGGATG	19	53	66	NM_001110139.2	128	TTTGGGCAGGATGAGGA TGTGACATCAAAGGCTTT TACAGGGCGAGAATTTGA TGAATTAGCCCCTCAGCC CAGAGAGACGCCTGCTT AAATGCCCGTTGTTTTGCT CGAGTTGAACCTTCCCACAA
Atp2a2 [R]	TTGTGGGAAGGTTCAACTCG	20	50	65			
Fasn [F]		20	55	64	NM_017332.2	214	TCGCTCATGGGTGTGGA AGTGCGCCAGATCCTGG AACGTGAACATGATCTGG TGCTACCCATTCGTGAAG TACGGCAACTCACACTGC GGAAGCTTCAGGAAATGT CCTCCAAGGCTGGCTCAG ACACTGAGTTGGCAGCCC CCAAGTCCAAGAATGATAC ATCCCTGAAGCAGGCCCA GCTGAATCTGAGTATCCTG CTGGTGAACCCTGA
Fasn [R]		20	55	63			
Leptin [F]	GAGACCCCTGTGTCGGTTC	19	63	60	NM_013076.3	139	GAGACCCCTGTGCCGGT TCCTGTGGCTTTGGTCC TATCTGTCCTATGTTCAA GCTGTGCCTATCCACAA AGTCCAGGATGACACCA AAACCCTCATCAAGACC ATTGTCACCAGGATCAA TGACATTTCACACACGCAG
Leptin [R]	CTGCGTGTGTGAAATGTCATTG	22	45	59			
rGapdh [F]	TGATTCTACCCACGGCAAGT	20	50	64	NM_017008.4	142	TGACTCTACCCACGGCA AGTTCAACGGCACAGTC AAGGCTGAGAATGGGAA GCTGGTCATCAACGGGA AACCCATCACCATCTTCC AGGAGCGAGATCCCGCTA ACATCAAATGGGGTGATGCT
rGapdh [R]	AGCATCACCCCATTTGATGT	20	45	65			

Notes: APN, adiponectin; F, forward; R, reverse; %, percent.

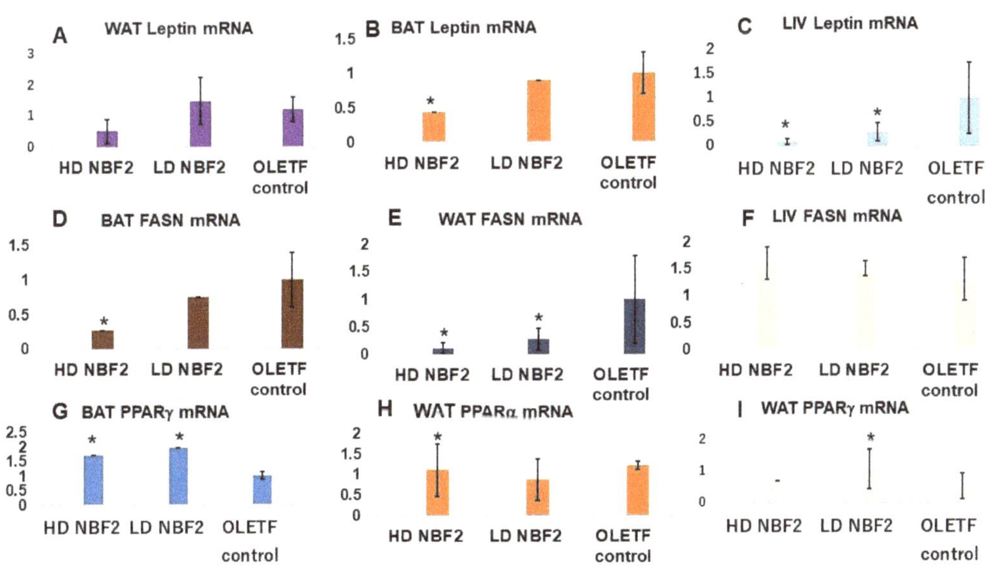

Figure 4. *Cont.*

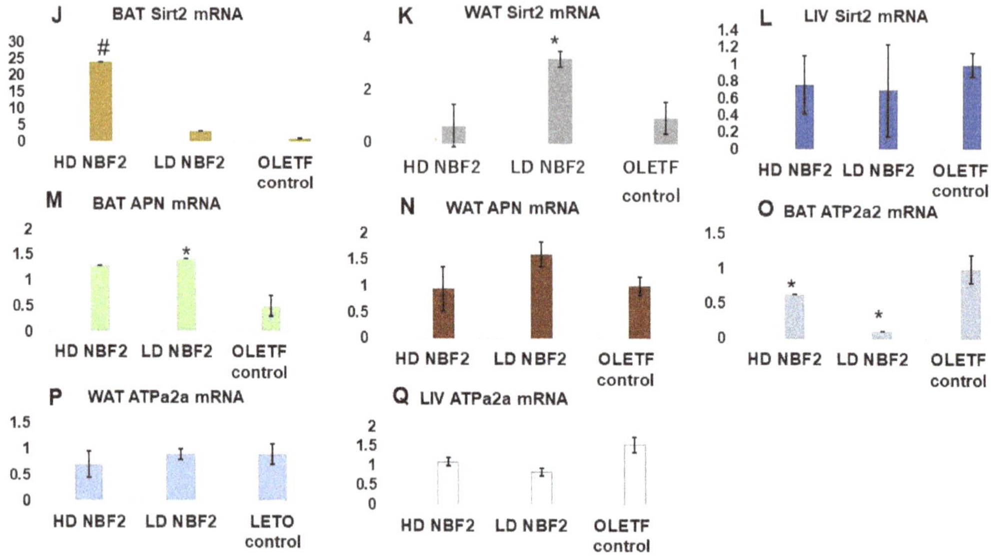

Figure 4. Messenger RNA (mRNA) gene expression of biomarkers in white/brown adipose tissue and liver biopsies (N = 20). **Legend:** *, *p*-value less than 0.05; #, *p*-value less than 0.01 (vs. OLET control; by Student's *t* test); OLETF, Otsuka Long-Evans Tokushima Fatty rats; LETO, Long-Evans Tokushima Otsuka rats; mRNA, messenger ribonucleic acid; BAT, brown adipose tissue; WAT, white adipose tissue; LIV, liver tissue; APN, adiponectin; FASN, fatty acid synthase; PPAR, peroxisome proliferator-activated receptor; Sirt2; sirtuin 2; ATPa2a or SERCA2, ATPase sarcoplasmic/endoplasmic reticulum Ca^{2+}. The unit used is the percentage (%) of the relative mRNA gene expression level for each biomarker.

In the histopathological examination, white adipose tissue (WAT) and liver sections showed the following results: Large adipocytes were mostly observed in WAT specimens from OLETF control rats (Supplementary Figure S1C,D) compared to in specimens from normal rats (Supplementary Figure S1A,B) and NBF2-treated animals (Supplementary Figure S1E,F). Regarding BAT specimens, the presence of smaller adipocytes mixed with large adipocytes was observed in OLETF controls (Supplementary Figure S1I,J), but larger ones were predominant in BAT from rats treated with NBF2 (Supplementary Figure S1K,L) and in LETO rats (Supplementary Figure S1G,H). Images of blood sinusoidal dilation and hepatic steatosis with lipid droplets were mostly observed in liver specimens from OLETF control rats (Supplementary Figure S1O,P), whereas these changes were either reduced and inexistent in liver specimens from NBF2-treated OLETF (Supplementary Figure S1Q,R) and normal rats (Supplementary Figure S1M,N).

4. Discussion

In the present experimental study, we explored the metabolic health effects of low and high (normal) doses of NBF2, a formula containing marine Aonori alga and *junos* Tanaka *citrus*-derived biomaterials, on lipid and glucose metabolism in obese and hyperglycemic OLETF rats. NBF2 treatment increased APN production.

Previously, we reported that algal biomaterial (NBF1), which represents 60% of NBF2, induced a 2- to 3-fold increase in APN in human subjects [12]. On the other hand, *junos citrus* peel, another component of NBF2 that represents 40%, is mainly composed of a potent anti-inflammatory compound, limonene [14]. Thus, the formula used in this study can be considered a potent PPAR modulator, an APN modulator and an anti-inflammatory agent that exerts beneficial effects on glucose and lipid metabolism.

In our study, NBF2 intake prevented weight gain in OLETF rats compared to the vehicle. Moreover, NBF2 could improve insulin sensitivity and also normalize blood glucose in OLETF rats to a level similar to that of lean LETO rats across the experiment. The latter finding can be explained by the increased brown fat mass, the relatively low FASN mRNA expression and the enhanced APN production in NBF2-treated OLETF rats. APN is known to ameliorate insulin resistance and glucose intolerance. By restoring insulin sensitivity through the up-regulation of APN production, NBF2 could reverse hyperglycemia. Recently, we have reported that in addition to its anti-inflammatory effect, NBF-containing algal biomaterial has a potential to improve the glycemic profile in individuals with a high cardiometabolic risk and in diabetic patients as well [13].

This study also showed an enhanced lipid metabolism and BP profile in obese and hyperglycemic rats treated with NBF2. The observed improvement in lipid profile, possibly intertwined with Sirt2 activation, may explain the slowing of weight gain and the improved BP. Recently, evidence has been accumulated suggesting that not only PPARs but also sirtuin activators can be utilized as novel therapeutic targets for metabolic disorders [15,16]. Sirt2, known as an NAD+-dependent deacetylase and which was activated by NBF2 in our study, plays an important role in terms of lipid metabolism regulation and improving liver function [17–19].

Compared to WAT, BAT has been reported to produce a larger amount of circulating APN. On the other hand, PPARs are known to induce the browning of white fat and regulate brown adipocyte thermogenic function [20,21]. The fiber-rich algal component of NBF2 has displayed a good safety profile, has exerted anti-inflammatory activity in humans in previous studies [22,23] and seems to have influenced PPAR activity in BAT in the present study. This suggests that NBF2 could be a potential natural agent with beneficial metabolic health effects. Further investigations are needed to clarify the relationship between NBF2 and PPARs.

Nonetheless, this study has some limitations. The effects of NBF2 on other metabolic markers such as glucagon-like protein-1 receptor (GLP-1R) have not been explored. Drugs that inhibit GLP-1R have recently emerged as effective therapeutic agents for T2DM, as they stimulate insulin secretion, causing a decrease in blood glucose [24].

Taken together, the findings from this study suggest that the NBF2 formula, which increased BAT, improved glucose tolerance and alleviated insulin resistance in this experiment, has a potential to reverse dyslipidemia and hyperglycemia in T2DM.

5. Materials and Methods

5.1. Animals and Intervention

The experimental protocol, as well as the animal care procedures, were carried out according to the guidelines for the care and use of animals of the Experimental Animal Center of Kagawa University (ethical approval number: 20656-1), Japan. Six-week-old male Otsuka Long-Evans Tokushima Fatty (OLETF) and genetic control Long-Evans Tokushima Otsuka (LETO) rats were purchased from Japan SLC Inc. (Shizuoka, Japan) and maintained under a controlled temperature (24 ± 2 °C) and humidity (55 ± 5%), with a 12 h light/dark cycle, at the Animal Center of Kagawa University in Japan. The rats were fed with MF chow and had access to tap water ad libitum. They were used when they reached 10 weeks of age. It has been reported that OLETF rats exhibit metabolic syndrome and prediabetes from the age of 10 weeks of age and, later on, the type 2 diabetic phase.

The NBF2-based drink was a gift from Kochi Ice Co., Ltd. (Kochi, Japan), a food processing factory located in Kochi prefecture, Japan. It is a food-based formula containing 60% ulvan-rich alga powder and 40% *Junos* Tanaka *citrus* peel powder. The chemical composition of the two biomaterials that make up the NBF2 formula is as follows:

(1) The Aonori (*Ulva prolifera*) algal biomaterial used in this study is rich in fiber (~65%), and its main bioactive component is ulvan [12], followed by protein (9~14%), omega-3 and omega-6 fatty acids (10.4~10.9%), and small amounts of flavonoids, carotenoids and minerals;

(2) The *junos citrus* peel biomaterial is rich in limonene (68~70%), other terpenoids such as γ-terpinene (11.4~12.5%), β–phellandrene (4.6~5.4%), myrcene (3.0~3.2%) and α–pinene (2.3~21.7%), and small amounts of aromatic compounds [12,15].

5.2. Experimental Protocol

In this experiment, a total of 26 rats were used, including 18 OLETF and 8 LETO (normal) rats. LETO rats were used as normal controls; prediabetic OLETF rats were randomly divided into 3 groups at week 10 of age (baseline) based on body weight and fasting blood glucose (FBG) (Figure 5):

(1) Group 1: 2 mL H_2O (vehicle)-treated OLETF rats or OLETF controls (n = 6);
(2) Group 2: 20 mg/kg/day NBF2-treated OLETF rats (n = 6);
(3) Group 3: 10 mg/kg/day NBF2-treated OLETF rats (n = 6).

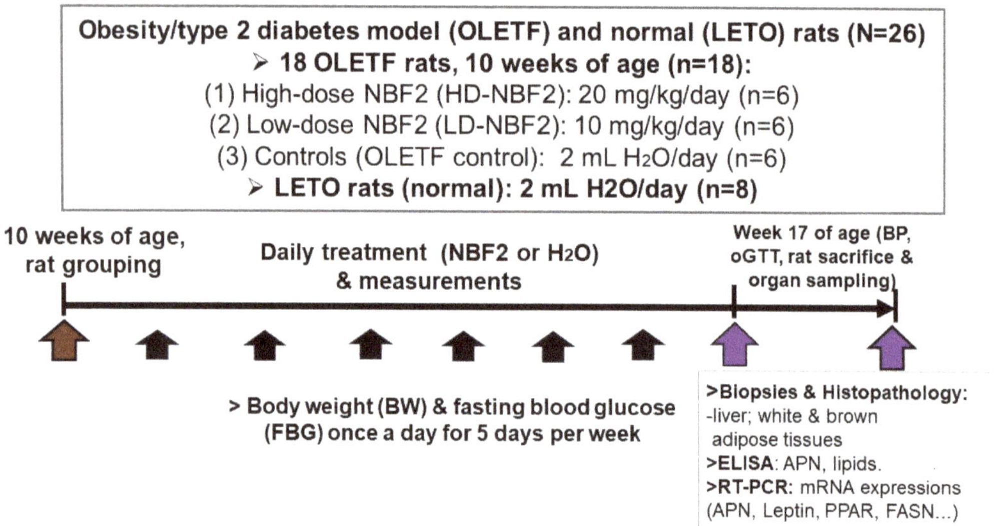

Figure 5. The experimental protocol. Figure 1 shows the procedures used in this study and the measurements performed to obtain the clinical, laboratory (ELISA, RT-PCR assays) and pathological data. **Legend**: APN, adiponectin; PPAR, peroxisome proliferator-activated receptor; FASN, fatty acid synthetase; mRNA, messenger ribonucleic acid; oGTT, glucose tolerance test. Brown arrow shows that start of experiment; purple arrows indicate the week where oGTT and biospecimen sampling were performed for bioassays, black arrows indicate the period in which daily measurements of body weight and blood glucose were carried out.

Daily calorie intake was measured on a weekly basis starting from week 10 of age; the mean value of calorie intake calculated for each group was estimated based on the energy provided by the MF chow and NBF2-containing nutrients. Furthermore, fasting blood glucose (FBG) was measured with the use of a freestyle glucose meter (Glucotest AI, Sanwa Biochemical Research Center, Kanagawa, Japan) from the tip of the rat tail once a week. Body weight was also measured on a weekly basis.

5.3. Measurements

The oral glucose tolerance test (OGTT) was carried out in duplicate within the last week of the experiment when the rats were 16 weeks old, as described previously [14]. Briefly, the rats were fasted for 12 h, and 0 min blood glucose was measured. Then, the rats were each fed 2 g/kg of 50% glucose solution by gavage; blood glucose was measured again at 30, 60 and 120 min.

Blood pressure (BP) and heart rate (HR) were measured using an automatic device (BP-98A, Softron, Tokyo, Japan), following the tail-cuff method. Briefly, a photo-elective sensor, which is attached to the animal tail by a piece of rubber tubing, detects the pulse volume oscillations through the cuff pressure and pulse waves. The pulse waves are detected as systolic BP (SBP) and mean BP (MBP); these two values are used to calculate the diastolic BP (DBP) automatically with the use of the following formula: DBP = (3 × MBP − SBP)/2. The homeostatic model assessment of insulin resistance (HOMA-IR) was calculated according to the following formula:

$$\text{HOMA-IR} = \text{glycemia (mg/dL)} \times \text{plasma insulin (UI/mL)}/405.$$

5.4. Biospecimen Sampling, Histological Examination and Real-Time PCR

On the day of animal sacrifice, the rats were fasted for 12 h and anesthetized; blood was drawn. Blood lipids were measured using a portable lipid analyzer MLA-1 (Changsha Zealson Biotech Co., Ltd., Changsha, China). Plasma adiponectin and insulin measurements were performed by using the enzyme-linked immunosorbent assay (ELISA; SRL biochemical company, Kobe, Japan). Brown adipose tissue (BAT) was collected from the scapula area, whereas white adipose tissue (WAT) was collected from the mesenteric, retroperitoneal and epididymal spaces. All necessary organs taken from the rats were weighted and stored at −80 °C prior to histological examination. The latter was performed by a pathologist. Visceral adipose tissue and liver specimens were formalin-fixed and paraffin-embedded. They were stained with hematoxylin–eosin (HE) prior to observation of the specimen with an all-in-one fluorescence microscope BZ-X800. (Keyence Corp., Osaka, Japan).

The collected adipose tissues were snap-frozen in liquid nitrogen, then stored at −80 °C until processing for RNA extraction. Real-time reverse transcription PCR (RT-PCR) was performed at the laboratory unit of the department of Environmental Medicine, Kochi University Medical School, Japan. RNA was isolated from WAT, FAT and liver tissue by using the phenol–chloroform extraction method, and cDNA was prepared. Messenger RNA (mRNA) expressions of PPAR−α, PPAR−γ, leptin and APN were analyzed in adipose tissue, whereas those of the three markers of lipid metabolism, namely fatty acid synthase (FAS), sirtuin2 (Sirt2) and ATPase sarcoplasmic/endoplasmic reticulum Ca^{2+} transporting 2 (ATP2a2), were analyzed in liver tissue by RT-PCR using ABI Prism 7000 with Power SYBR Green PCR Master Mix (Invitrogen, Tokyo, Japan) and primer sequences from Eurofins genomics (Japan). Group comparisons for mRNA gene expressions of different biomarkers were performed using the relative quantification of genes.

5.5. Ethical Considerations and Statistical Analysis of Collected Data

The study protocol was approved by the Ethical Review Board of the Experimental Animal Center of Kagawa University (approval number: 20656-1). All of the procedures were performed according to internationally adopted guidelines for the care and use of laboratory animals. Mean (SD) values were used to present continuous data. Furthermore, Student's t test was used when comparing two groups of normally distributed data related to the outcome variables. Analysis of variance (ANOVA) for repeated measures or one-way ANOVA, followed by a post hoc Tukey's test, was used when appropriate. The significance level was set at a p-value less than 0.05. Stata statistical software version 15 was used for all statistical analyses (SataCorp, College Station, TX, USA).

6. Conclusions

The present study, which is related to one of the newly discovered alga-based adiponectin-modulating formulas, NBF2, showed that its intake caused Sirt2 and dual PPAR activation, up-regulated APN, and improved glucose and lipid metabolism in a rodent model of obesity and T2DM. These findings suggest that NBF2 may be beneficial in preventing and reversing hyperlipidemia and hyperglycemia in at-risk subjects and T2DM patients.

Supplementary Materials: The following supporting information can be downloaded at: https://www.mdpi.com/article/10.3390/ijms251910828/s1.

Author Contributions: Conceptualization, N.R.N., A.H., A.M.R. and A.N.; methodology, N.R.N., T.H., K.K. and M.E.; investigation and interpretation of data, N.R.N., N.M. and N.S.; writing manuscript draft, N.R.N., S.A. and M.E.; final approval of manuscript, all authors. All authors have read and agreed to the published version of the manuscript.

Funding: No funding was received. This study was conducted with internal funds of the department of public health, Kagawa University Faculty of Medicine, Japan.

Institutional Review Board Statement: This study was conducted in accordance with the Helsinki Declaration, and the animal study protocol was approved by the Ethical Review Board of the Experimental Animal Center of Kagawa University (approval number: 20656-1).

Informed Consent Statement: Not applicable.

Data Availability Statement: The datasets used in this study are available from the corresponding authors.

Acknowledgments: We thank the manager of Kochi Ice company (Kochi, Japan), Daniel Ton-duangu from the Faculty of Medicine of Kinshasa University in D.R. Congo, Yukimi Nakamura from Takamatsu Health Center in Japan, and Joel Ngoie, President of FICE (Congo Diaspora Forum) international, for supporting this research project.

Conflicts of Interest: The authors declare no conflicts of interest.

References

1. Luo, W.; Guo, Z.; Wu, M.; Hao, C.; Hu, X.; Zhou, Z.; Yao, X.; Zhang, L.; Lui, J. Association of peroxisome proliferator-activate receptor α/δ/γ with obesity, and gene-gene interaction, in the Chinese Han population. *J. Epidemiol.* **2013**, *23*, 187–194. [CrossRef [PubMed]
2. American Diabetes Association. Pharmacologic approaches to glycemic treatment: Standards of medical care in diabetes-202 *Diabetes Care* **2021**, *44*, S111–S124. [CrossRef] [PubMed]
3. Lingvay, I. Obesity management as a primary treatment goal for type 2 diabetes: Time to reframe the conversation. *Lancet* 202 *399*, 394–405. [CrossRef] [PubMed]
4. Kadowaki, T.; Yamauchi, T.; Kubota, N.; Hara, K.; Ueki, K.; Tobe, K. Adiponectin and adiponectin receptors in insulin resistanc diabetes, and the metabolic syndrome. *J. Clin. Investig.* **2006**, *116*, 1784–1792. [CrossRef]
5. Kersten, S.; Desvergne, B.; Wahli, W. Roles of PPARs in health and disease. *Nature* **2000**, *405*, 421–424. [CrossRef]
6. Yamauchi, T.; Kamon, J.; Minokoshi, Y.; Ito, Y.; Waki, H.; Yochida, S.; Uchida, S.; Yamashita, S.; Noda, M.; Kita, S.; et a Adiponectin stimulates glucose utilization and fatty acid oxidation by activating AMP-activated protein kinase. *Nat. Med.* 200 *8*, 1288–1295. [CrossRef]
7. Karbowska, J.; Kochan, Z. Role of adiponectin in the regulation of carbohydrate and lipid metabolism. *J. Physiol. Pharmacol.* 200 *57*, 103–113. Available online: https://www.jpp.krakow.pl/journal/archive/11_06_s6/pdf/103_11_06_s6_article.pdf (accesse on 12 February 2024). [PubMed]
8. Fiaschi, T.; Magherini, F.; Gamberi, T.; Modesti, P.A.; Modesti, A. Adiponectin as a tissue regenerating hormone: More than metabolic function. *Cell Mol. Life Sci.* **2014**, *71*, 1917–1925. [CrossRef]
9. Xu, A.; Wang, H.; Hoo, R.L.C.; Sweeney, G.; Vanhoutte, P.M.; Wang, Y.; Wu, D.; Chu, W.; Qin, G.; Lam, K.S.L. Selective elevatio of adiponectin production by the natural compounds derived from medicinal herb alleviates insulin resistance and glucos intolerance in obese mice. *Endocrinology* **2009**, *150*, 625–633.
10. Hoo, R.L.C.; Wong, J.Y.L.; Qiao, C.F.; Xu, A.; Xu, H.X.; Lam, K.S.L. The effective fraction isolated from Radix astragali alleviate glucose intolerance, insulin resistance and hyperglyceridemia in diabetic db/db mice through its anti-inflammatory activity. *Nu Metab.* **2010**, *7*, 67–79. [CrossRef]
11. O'Mahony, L.L.; Matu, J.; Price, O.J.; Birch, K.M.; Ajjan, R.A.; Farrar, D.; Tapp, R.; West, D.J.; Deighton, K.; Campbell, M.I Omega-3 polyunsaturated fatty acids favourably modulate cardiometabolic biomarkers in type 2 diabetes: A meta-analysis an meta-regression of randomized controlled trials. *Cardiovasc. Diabetol.* **2018**, *17*, 98–110. [CrossRef] [PubMed]
12. Ngatu, R.N.; Ikeda, M.; Watanabe, H.; Tanaka, M.; Inoue, M.; Kanbara, S.; Nojima, S. Uncovering adiponectin replenishin property of Sujiaonori algal biomaterial in humans. *Mar. Drugs* **2017**, *15*, 32–41. [CrossRef] [PubMed]
13. Ngatu, R.N.; Ikeda, M.; Tonduangu, D.K.; Nangana, S.L.; Hirao, T. Modulatory effects of NBF1 algal-rich bioformula o adiponectin and c-reactive protein levels and its therapeutic prospects for metabolic syndrome and type 2 diabetes. *Biomedicine* **2022**, *10*, 272–2581. [CrossRef] [PubMed]

14. Hossain, A.; Yamaguchi, F.; Hirose, K.; Matsunaga, T.; Sui, L.; Hirata, Y.; Noguchi, C.; Katagi, A.; Kamitori, K.; Dong, Y.; et al. Rare sugar d-psicose prevents progression and development of diabetes in T2DM model Otsuka Long-Evans Tokushima Fatty rats. *Drug Des. Devel. Ther.* **2015**, *9*, 525–535. [CrossRef] [PubMed]
15. Watanabe, H.; Inata, Y.; Kimura, K.; Matsumoto, M.; Kaneko, S.; Kasuga, M.; Inoue, H. Sirt2 facilitates hepatic glucose uptake by deacetylating glucokinase regulatory protein. *Nat. Commun.* **2018**, *9*, 30. [CrossRef]
16. Ren, H.; Hu, F.; Wang, D.; Kang, X.; Feng, X.; Zhang, L.; Zhou, B.; Liu, S.; Yuan, G. Sirtuin 2 prevents liver steatosis and metabolic disorders by deacetylation of hepatocyte nuclear factor 4a. *Hepatology* **2021**, *74*, 723–740. [CrossRef]
17. Park, S.Y.; Chung, M.J.; Son, J.Y.; Yun, H.H.; Park, J.M.; Yim, J.H.; Jung, S.J.; Lee, S.H.; Jeong, K.S. The role of sirtuin2 in sustaining functional integrity of the liver. *Life Sci.* **2021**, *285*, 119997–120004. [CrossRef]
18. Kim, S.H.; Hong, S.H.; Park, Y.J.; Sung, J.H.; Lee, K.W.; Jung, K.; Lim, C.; Kim, J.H.; Kim, H.; Kim, H.; et al. MD001, a novel peroxisome proliferator-activated receptor α/γ agonist, improves glucose and lipid metabolism. *Sci. Rep.* **2019**, *9*, 1656–1667. [CrossRef]
19. Jain, N.; Bhansali, S.; Kurpad, A.V.; Hawkins, M.; Sharma, A.; Kaur, S.; Rastogi, A.; Bhansali, A. Effect of a dual PPAR α/γ agonist on insulin sensitivity in patients of type 2 diabetes with hyperglycemia–randomized double-blind placebo-controlled trial. *Sci. Rep.* **2019**, *9*, 19017–19026. [CrossRef]
20. Kroon, T.; Harms, M.; Mauer, S.; Ida, B.; Alexandersson, I.; Lindblom, A.; Ahnmark, A.; Nilsson, D.; Gennemark, P.; O'Mahony, G.; et al. PPARg and PPARa synergize to induce robust browning of white fat in vivo. *Mol. Metab.* **2020**, *36*, 100964. [CrossRef]
21. Shen, Y.; Su, Y.; Silva, F.J.; Weller, A.H.; Sostre-Colon, J.; Titchenell, P.M.; Steger, D.J.; Seale, P.; Soccio, P.E. Sahred PPARα/γ target genes regulate brown adipocyte thermogenic function. *Cell Rep.* **2020**, *9*, 3079–3091.e5. [CrossRef] [PubMed]
22. Gallwitz, B. Clinical perspectives on the use of the GIP/GLP-1 receptor agonist tirzepatide for the treatment of type-2 diabetes and obesity. *Front. Endocrinol.* **2022**, *13*, 1004044. [CrossRef] [PubMed]
23. Hirota, R.; Ngatu-Nlandu, R.; Nakamura, H.; Songo, H.S.; Sawamura, M. Anti-inflammatory effects of limonene from yuzu (citrus junos Tanaka) essential oil on eosinophils. *J. Food Sci.* **2010**, *75*, 87–92. [CrossRef] [PubMed]
24. Dosoky, N.S.; Setzer, W.N. Biological activities and safety of *Citrus* spp. Essential oils. *Int. J. Mol. Sci.* **2018**, *19*, 1966–1991. [CrossRef]

International Journal of
Molecular Sciences

MDPI

Article

Ursolic Acid Regulates Immune Balance, Modulates Gut Microbial Metabolism, and Improves Liver Health in Mice

Man Zhao [1,†], Yali Cui [2,†], Fengxia Wang [1], Fengyang Wu [1], Chong Li [1], Shudong Liu [1] and Baojiang Chen [1,*]

1 College of Animal Science and Technology, Hebei Agricultural University, Baoding 071051, China
2 College of Traditional Chinese Veterinary Medicine, Hebei Agricultural University, Baoding 071051, China
* Correspondence: 199051@hebau.edu.cn
† These authors contributed equally to this work.

Abstract: Ursolic acid (UA) has demonstrated significant immunomodulatory and hepatoprotective effects; however, the underlying mechanisms remain unclear. This study aims to analyze the impact of UA on the gut microbiome, metabolome, and liver transcriptome, investigate UA's role in maintaining gut immune homeostasis and liver health, and evaluate the potential contributions of gut microbes and their metabolites to these beneficial effects. Our findings indicate that UA enhances immune balance in the jejunum, fortifies intestinal barrier function, and promotes overall gut health. UA modulates the intestinal microbiota and its metabolic processes, notably increasing the abundance of beneficial bacteria such as *Odoribacter* and *Parabacteroides*, along with their metabolites, including ornithine and lactucin. Additionally, UA inhibits the expression of interleukin-1 receptor 1 (*IL1R1*) and calcium (Ca^{2+}) voltage-gated channel auxiliary subunit beta 2 (*CACNB2*) while enhancing the synthesis pathways of retinol and ascorbic acid, thereby exerting a protective influence on liver function. In summary, UA enhances intestinal immune homeostasis and promotes liver health, with these advantageous effects potentially mediated by beneficial bacteria (*Odoribacter* and *Parabacteroides*) and their metabolites (ornithine and lactucin).

Keywords: ursolic acid; gut health; liver protection; gut microbiota; metabolism; liver transcriptome

1. Introduction

Body health is closely related to gut health, relying on moderate immunity, an intact gut barrier, and a stable gut microbe. The gut harbors a diverse array of immune cells, including T helper (Th) cells and T regulatory (Treg) cells, which work together to maintain the balance of the body's immune response [1]. An overproduction of immune-related cytokines, such as interleukin (IL)-6, IL-1β, and IL-17A, can lead to intestinal cell damage and compromise the integrity of the gut barrier, consequently triggering systemic inflammation. In contrast, cytokines such as IL-10 and transforming growth factor-β (TGF-β) play a role in inhibiting inflammation [1]. Ursolic acid (UA) is a naturally occurring pentacyclic triterpene carboxylic acid found in various plants, either in free acid form or as an aglycone of triterpene saponins [2]. In recent years, there has been a surge of interest in UA, primarily concerning its pharmacological properties, which include antioxidant [3], antitumor [4], anti-inflammatory [5], and antimicrobial [2] effects. Our previous meta-analysis confirmed that UA can reduce levels of pro-inflammatory cytokines like IL-1β and IL-6 [6]. Furthermore, UA has been shown to activate the immune system and enhance the proliferation of immune cells [7]. Nevertheless, the regulatory mechanisms by which UA influences intestinal immune homeostasis and gut health remain unclear.

The gut microbiota plays a critical role in maintaining intestinal immune homeostasis and barrier function [8–10]. Research has demonstrated that gut microbes can modulate immune equilibrium through the regulation of Th17/Treg cell dynamics. For example, *Lactobacillus rhamnosus GG* is capable of modulating Th17/Treg homeostasis, thus aiding in the preservation of host intestinal health [11]. Similarly, *Bacteroides fragilis* enhances Treg cell activation by increasing *Foxp3* expression [12]. UA has exhibited significant regulatory effects on the gut microbiota [13–15]. Consequently, understanding how UA influences immune balance and gut health via the modulation of gut microbes presents a compelling area of investigation. Additionally, UA has been shown to mitigate liver damage by modulating enterohepatic axis homeostasis [13] and is recognized as a contributing factor to the hepatoprotective properties of various foods and medicinal plants [13,16]. The gut microbiota, a key component of the gut–liver axis (GLA), is also closely associated with liver health [8]. Studies indicate that *Lactobacillus* and *Bifidobacterium* can enhance metabolic processes in rats through the GLA [17]. However, the specific mechanisms through which UA exerts protective effects on the liver necessitate further exploration.

To address these questions, this study utilizes a multi-omics approach to analyze the effects of UA on the intestinal microbiota, metabolome, and liver transcriptome in mice. We evaluate the role of UA in immune balance and liver protection, elucidating the potential contributions of gut microbes and their metabolites to these processes.

2. Results

2.1. Effect of UA on Immune-Related Cytokines in Mice

As shown in Figure 1A, the body weight (BW) of mice in the UA 5 and UA 25 groups was significantly higher than that of the CON group ($p < 0.05$). Compared to the CON group, the UA 5 and UA 25 groups had significantly lower levels of interleukin (IL)-1β, IL-6, and tumor necrosis factor (TNF)-α ($p < 0.05$) and higher levels of transforming growth factor (TGF)-β and IL-10 ($p < 0.05$) (Figure 1B–G).

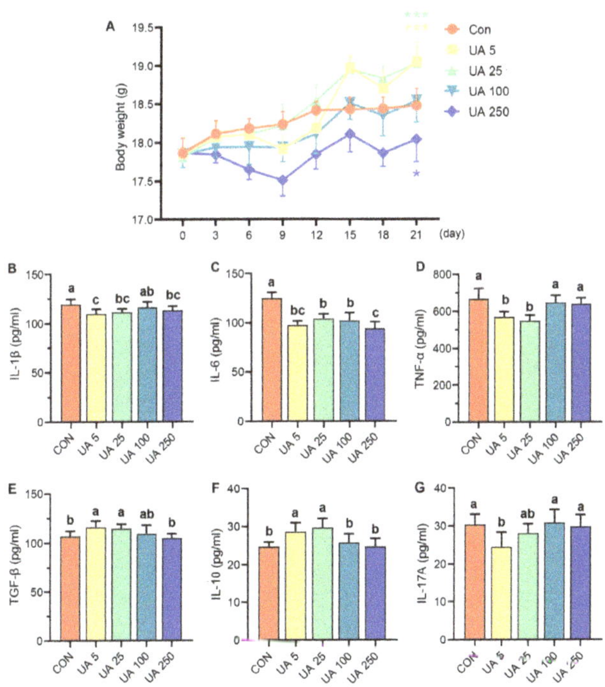

Figure 1. Effects of ursolic acid on body weight and immune-related cytokines in mice. (**A**) Body weight (*, $p < 0.05$, and ***, $p < 0.001$). (**B**) IL-1β, (**C**) IL-6, (**D**) TNF-α, (**E**) TGF-β, (**F**) IL-10, and (**G**) IL-17A cytokine levels in serum. Values with different letters (a, b, and c) differed significantly within a mice trial ($p < 0.05$). Data are shown as mean ± SD. $n = 6$.

2.2. UA Promotes Intestinal Development, Intestinal Epithelial Barrier Integrity, and Immune Balance in Mice

Mouse intestinal tissues (duodenum, jejunum, and ileum) were stained with H&E to explore the effect of UA on intestinal development in mice (Figure 2A–F). The results indicated that UA treatment (UA 5 and UA 100) significantly increased duodenal villus length (V), while (UA 5, UA 25, and UA 100) significantly increased duodenal crypt depth (C), (UA 5, UA 25, UA 100, and UA 250) significantly increased duodenal wall thickness ($p < 0.05$), and (UA 5, UA 25, UA 100, and UA 250) significantly increased duodenal crypt depth. However, there was no significant effect on the V/C ratio ($p > 0.05$). Compared to the CON group, UA treatment (UA 5, UA 25, UA 100, and UA 250) significantly increased jejunal villus length, (UA 25) crypt depth, and (UA 25) V/C and (UA 5 and UA 250) intestinal wall thickness ($p < 0.05$). UA treatment had no significant effect on the intestinal morphology (villus length, crypt depth, V/C, and intestinal wall thickness) of the ileum ($p > 0.05$). It is concluded that the effect of UA on the small intestine is mainly concentrated in the jejunum, with the optimal dose being the UA 25 group. Based on this, we further investigated the effects of optimal UA treatment (UA 25) on the intestinal barrier and immune balance in mice. The results showed that UA treatment significantly increased the mRNA expression of zonula occludens-1 (*ZO-1*) and *Occludin* in the jejunum of mice ($p < 0.05$) but had no significant effect on the mRNA expression of *Claudin1* ($p > 0.05$) (Figure 2G–I). Compared with the CON group, the UA-treated group had significantly higher forkhead box P3 (*Foxp3*) mRNA expression levels and significantly lower signal transducer and activator of transcription 3 (*Stat3*) mRNA expression levels ($p < 0.05$). UA had no significant effect on retinoid-related orphan receptor gamma t (*RORγt*) mRNA expression ($p > 0.05$) (Figure 2J–L) based on the optimal effects of the UA 25 group on body weight, serum immune factor levels, and intestinal tissue morphology, as well as beneficial effects on genes related to the intestinal barrier and immune balance. We selected the UA 25 group (UA group) for the omics study.

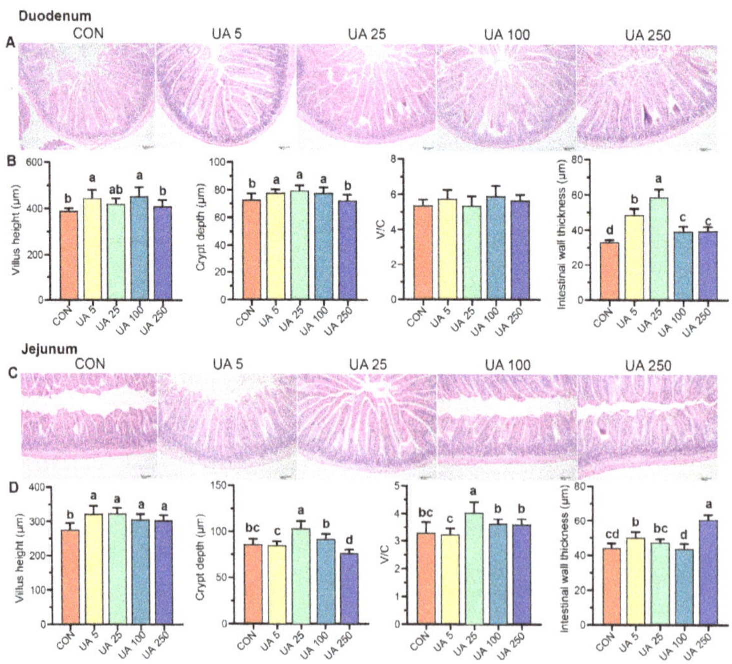

Figure 2. *Cont.*

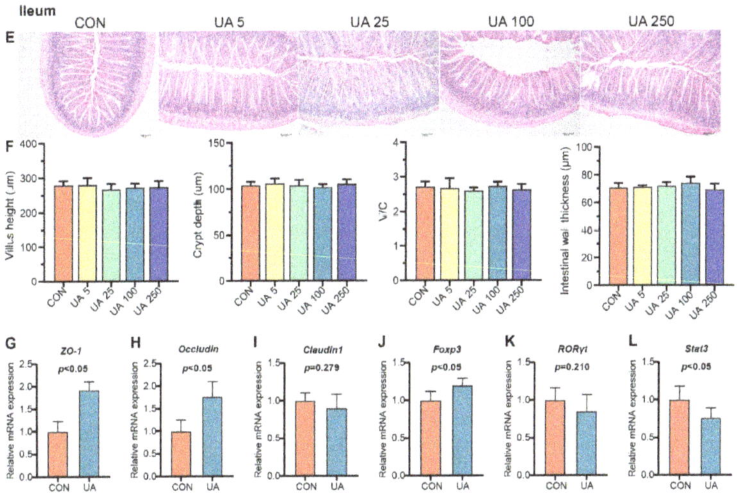

Figure 2. Effect of ursolic acid on the intestinal health of mice. Effect of ursolic acid on the intestinal morphology of the duodenum (**A**,**B**), jejunum (**C**,**D**), and ileum (**E**,**F**) in mice. Effects of ursolic acid on the expression of genes related to the intestinal barrier (**G–I**) and Th17/Treg cell-related factor (**J–L**) in the jejunum of mice. Values with different letters (a, b, c, and d) differed significantly within a mice trial ($p < 0.05$). Data are shown as mean ± SD. $n = 6$. Scale bars in (**A**,**C**,**E**) = 100 μm.

2.3. UA Changes the Composition and Structure of the Gut Microbiota in Mice

To explore whether the physiological effects of UA were related to the gut microbiota, we further investigated the effects of UA on the gut microbiota. Figure S1A–C illustrates the rank abundance curves, rarefaction curves, and Shannon curves derived from this gene sequencing. There was no significant difference in the Shannon index between the two groups ($p > 0.05$) (Figure 3A). The results of principal coordinate analysis (PCoA) indicated that the first principal component explained 42.32% of the variance between groups, and the second principal component explained 25.31% of the variance between groups. The results of the Adonis analysis demonstrated that UA treatment significantly regulated the structure of intestinal flora in mice ($p < 0.05$) (Figure 3B). Figure S1D,E presents the top 10 bacteria with relative abundance at the phylum level and the top 15 bacteria with relative abundance at the genus level. UA significantly reduced the relative abundance of *Romboutsia* and significantly increased the relative abundance of *Odoribacter*, *Coriobacteriaceae_UCG-002*, *Parabacteroides*, *Anaeroplasma*, and *Norank_f__Eggerthellaceae* ($p < 0.05$) (Figure 3C). Linear discriminant analysis (LDA) effect size (LEfSe) was employed to identify the core taxa (LDA ≥ 3, $p < 0.05$) that were most likely to explain the differences between groups. The results revealed that there were six taxa-marker bacteria in the CON group and nine taxa-marker bacteria in the UA group (Figures 3D and S1F). Notably, *Romboutsia* was further identified as the marker bacteria in the CON group, while *Odoribacter*, *Coriobacteriaceae_UCG-002*, and *Parabacteroides* were identified as the marker bacteria in the UA group.

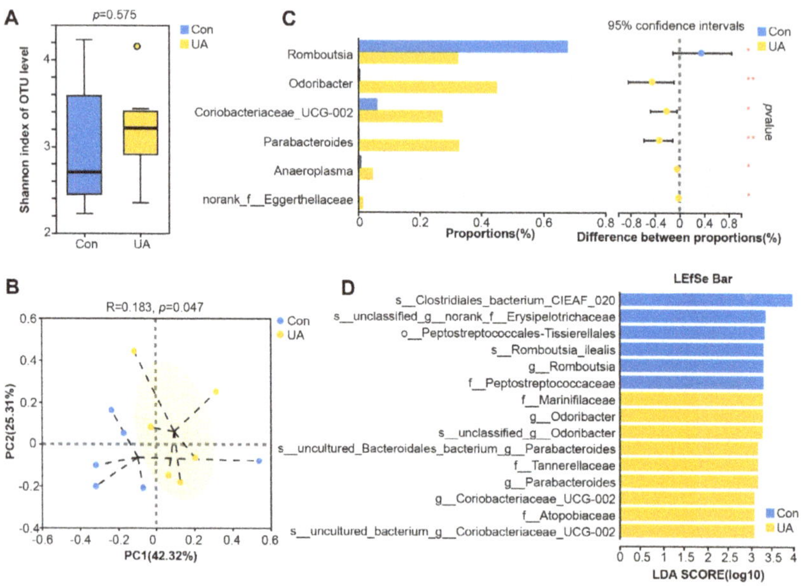

Figure 3. Effect of ursolic acid on gut microbiota in mice. (**A**) The comparison of microbiota diversity (Shannon index) and the Adonis analysis. Boxes denote the interquartile (IQR) between the first and third quartiles (25th and 75th percentiles, respectively) and the line inside denotes the median. (**B**) Principal coordinate analysis (PCoA) based on Bray–Curtis distance. (**C**) Comparison of differences in the relative abundance of bacteria at the genus level (*, $p < 0.05$, and **, $p < 0.01$). (**D**) Differentiating bacteria identified by LEfSe with LDA effect size ≥ 3. $n = 6$.

2.4. UA Also Altered Metabolism of Gut Microbiota

A total of 355 metabolites were identified in the mouse fecal metabolome. Partial least squares discriminant analysis (PLS-DA) indicated that the mice in the CON group and the UA group exhibited different metabolic characteristics (Figure S2A,B). Sixteen metabolites were significantly different between the two groups, among which indole carboxylic acid sulfate and cholic acid were significantly enriched in the CON group. Dihydroconiferyl alcohol, 5′-phosphoribosyl-N-formylglycinamide, 5-hydroperoxypent-1-enylbenzene, Asn Asn Leu Asn Val, 3-hydroxy-N^6,N^6,N^6-trimethyl-L-lysine, ornithine, lactucin, D-erythro-sphingosine C-17, dodemorph, wyerone, neoporrigenin B, Xi-4-hydroxy-4-methyl-2-cyclohexen-1-one, prolylglycine, and PA(24:0/22:6(5Z,8E,10Z,13Z,15E,19Z)-2OH(7S, 17S)) were significantly enriched in the UA group (variable importance in projection (VIP) > 1, $p < 0.05$) (Figure 4A). Kyoto Encyclopedia of Genes and Genomes (KEGG) enrichment analysis based on the 16 differential metabolites showed that three pathways were significantly enriched. It includes D-amino acid metabolism, arginine biosynthesis, and biosynthesis of various other secondary metabolites (Benjamini–Hochberg false discovery rate (FDR) < 0.05) (Figure 4B).

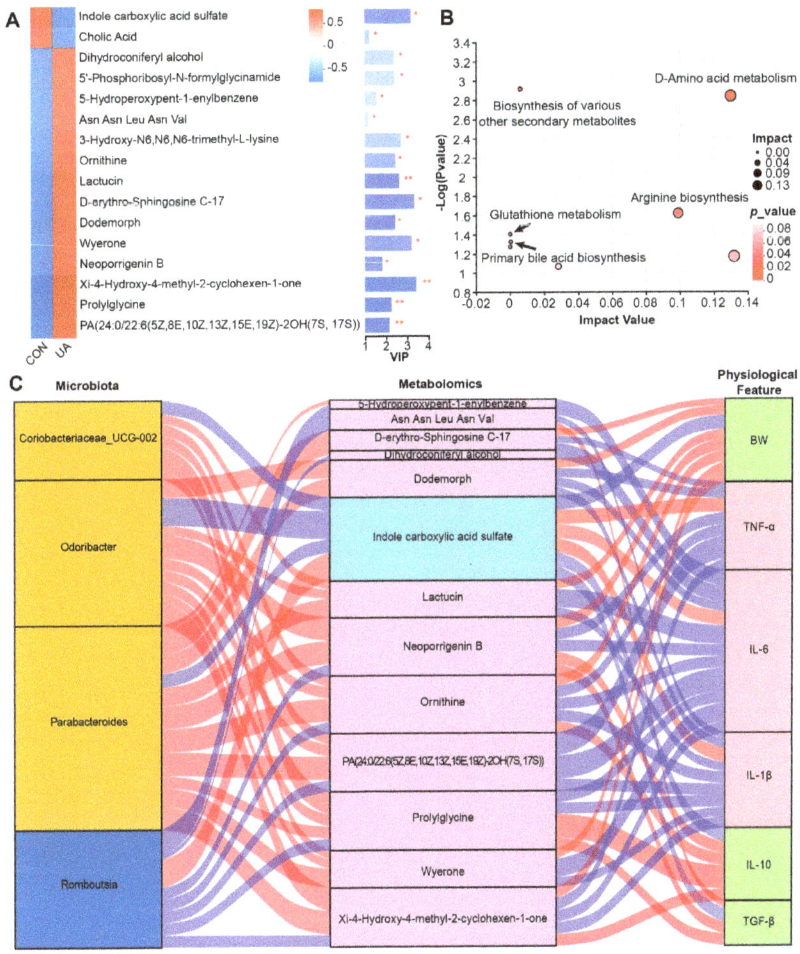

Figure 4. Effect of ursolic acid on the fecal metabolism of mice. (**A**) Heatmap of 16 different metabolites. The right side shows the variable importance in the projection (VIP) bar graph of the metabolite. The color of the bars indicates the significant difference in metabolites between the two groups of samples, and the smaller the p-value, the darker the color (*, $p < 0.05$; **, $p < 0.01$). (**B**) KEGG pathway enrichment analysis of differential metabolites. (**C**) Interrelationship between microbiota, metabolomics, and physiological features. Red connections indicate a positive correlation (Spearman correlation test, $|R| > 0.6$, $p < 0.05$,), while blue connections show correlations that were negative (Spearman correlation test, $|R| > 0.6$, $p < 0.05$). In the microbiota column, the orange stratum represents bacteria that were highly enriched in the UA group, and the stratum-colored blue was increased in the control group. In the metabolomics column, the purple stratum represents metabolites that were highly enriched in the UA group, and the stratum-colored ice blue was increased in the control group. In the physiological feature column, the green stratum represents bacteria that were highly enriched in the UA group, and the pink stratum was increased in the control group. $n = 6$.

2.5. UA Treatment Alters the Gut Microbiota and Its Metabolic Characteristics and Affects the Host Phenotype

Subsequently, we evaluated the correlation among gut microbiota, serum metabolites, and host phenotype to further explore the possible regulatory mechanisms between gut microbiota characteristics and host phenotype following UA treatment. As indicated by spearman's correlation coefficient, the four bacteria identified as marker bacteria were significantly correlated with 13 differential metabolites. These metabolites were further associated with BW, TNF-α, IL-6, IL-1β, IL-10, and TGF-β (representing the physiological features of mice) ($|R| > 0.6$, $p < 0.05$) (Figure 4C). *Odoribacter*, *Coriobacteriaceae_UCG-002*, and *Parabacteroides*, which were identified as the marker bacteria of the UA group, were positively correlated with metabolites (lactucin, ornithine, PA(24:0/22:6(5Z,8E,10Z,13Z,15E,19Z)-2OH(7S, 17S), prolylglycine, and wyerone) enriched in the UA group and were negatively correlated with Indole carboxylic acid sulfate enriched in the CON group. *Romboutsia*, which was identified as the marker bacteria in the CON group, was negatively correlated with 5-hydroperoxypent-1-enylbenzene, Asn Asn Leu Asn Val, dihydroconiferyl alcohol, neoporrigenin B, ornithine, and PA(24:0/22:6(5Z,8E,10Z,13Z,15E,19Z)-2OH(7S, 17S). Prolylglycin and Xi-4-hydroxy-4-methyl-2-cyclohexen-1-one) enriched in the UA group were positively correlated with indole carboxylic acid sulfate enriched in the CON group. Lactucin was negatively correlated with IL-6 and IL-1β and positively correlated with BW. Ornithine was negatively correlated with TNF-α, IL-6, and IL-1β, and positively correlated with BW and IL-10. In addition, there was a direct correlation between the differential bacteria and the host phenotype. *Odoribacter* and *Parabacteroides* were significantly negatively correlated with TNF-α, IL-6, and IL-1β and significantly positively correlated with BW, TGF-β, and IL-10 (Figure S3). *Coriobacteriaceae_UCG-002* was negatively correlated with IL-6 and IL-1β ($p < 0.05$) but positively correlated with IL-10 ($p < 0.05$). *Romboutsia* was significantly positively correlated with IL-6 and negatively correlated with IL-10.

2.6. UA Alters the Liver Transcriptome

To investigate the effects of UA on the mouse liver, we conducted RNA-seq analysis. A total of 29,094 genes were identified in the mouse liver. The results of the PCA analysis indicated a clear separation between the two groups of samples. The first principal component explained 48.12% of the variance between the two groups, and the second principal component explained 15.22% of the variance (Figure 5A). According to the filter threshold $|\log2FC| \geq 1$ and FDR < 0.05, we identified 636 differentially expressed genes (DEGs) in the UA vs. CON group. Compared to the CON group, there were 167 up-regulated DEGs and 469 down-regulated DEGs in the UA group (Figure 5B). In addition, the expression results of eight random mRNAs detected by qRT-PCR are presented (Figure S4A). The results of RNA-Seq (log2 fold change) and qRT-PCR (log2 fold change) were significantly positively correlated (correlation value = 0.84, $p < 0.01$) (Figure S4B). This indicates the high accuracy of both the sequencing and differential expression analysis results. Gene Ontology (GO) enrichment analysis was performed separately for down-regulated and up-regulated DEGs, and the results showed that the down-regulated genes were significantly enriched to 47 GO terms, such as cell projection organization, chemical synaptic transmission, anterograde trans-synaptic signaling, neuromuscular junction development, centrosome, and triplet codon-amino acid adaptor activity (Figure S5A). The up-regulated genes were significantly enriched in 191 GO terms, including myofilament, skeletal muscle thin filament assembly, and FATZ binding (Figure S5B).

The up-regulated and down-regulated genes were further subjected to KEGG pathway analysis to determine the functional categories. The results showed that there were no significantly enriched pathways among the down-regulated genes, but the pathway with the most enriched genes was the MAPK signaling pathway (DEGs = 11, FDR > 0.05) (Figure 5C). The top five significantly enriched pathways of up-regulated genes were retinol metabolism (DEGs = 6), ascorbate and aldarate metabolism (DEGs = 4), oxidative phosphorylation (DEGs = 7), chemical carcinogenesis-reactive oxygen species (DEGs = 9), and arginine and proline metabolism (DEGs = 5) (FDR < 0.01) (Figure 5D). Based on all DEGs, we further constructed the regulatory mechanism of UA on the ko04010, ko04668, ko00053, and ko00190 pathways (Figure 5E). Interestingly, all DEGs involved in the ko04010 and ko04668 pathways were down-regulated genes (Figure S6A), and all DEGs involved in the ko00053 (Figure S6B) and ko00190 (Figure S6C) pathways were up-regulated genes.

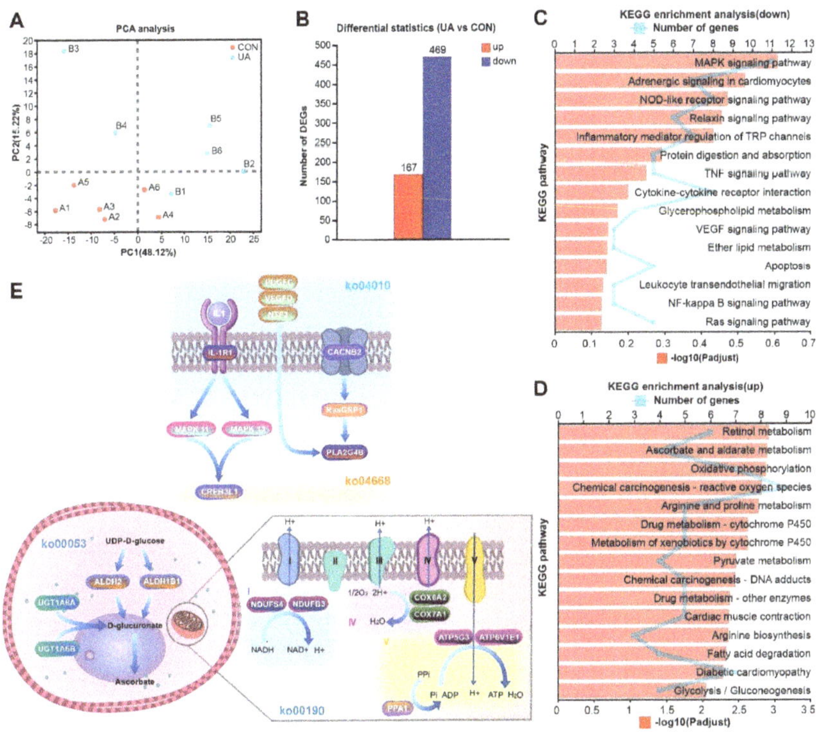

Figure 5. Effect of ursolic acid on the liver transcriptome of mice. (**A**) PCA analysis. (**B**) The number of differentially expressed genes (DEGs). KEGG enrichment pathways of down-regulated (**C**) and up-regulated (**D**) DEGs. (**E**) DEGs related to enrichment pathways (ko04010, ko04668, ko00053, and ko00190). ko04010, MAPK signaling pathway; ko04668, TNF signaling pathway; ko00053, Ascorbate and aldarate metabolism; and ko00190, Oxidative phosphorylation; I, Nicotinamide adenine dinucleotide (NADH) dehydrogenase; II, Succinate dehydrogenase; III, Cytochrome C Reductase; IV, Cytochrome c oxidase; V, ATP synthase. $n = 6$.

2.7. Multi-Omics Integrated Analysis Reveals the Possible Regulatory Mechanism of UA

Connections between bacteria, metabolites, liver genes, gut genes, and host phenotypes were established through an association study (Spearman correlation coefficient, $|R| > 0.6$, $p < 0.05$) (Figure 6). BW was positively correlated with intestinal barrier-related genes (*Occludin* and *ZO-1*) and negatively correlated with inflammation-related genes (Mitogen-activated protein kinase (*MAPK*)*13*, *MAPK11*, Interleukin-1 receptor 1 (*IL1R1*), Neurotrophic factor-3 (*NTF3*), and Calcium (Ca^{2+}) voltage-gated channel auxiliary subunit beta 2 (*CACNB2*)). The immune-related cytokines (IL-1β, IL-6, and TNF-α) were negatively correlated with the intestinal barrier-related genes (*Occludin* and *ZO-1*), and all of them were positively correlated with the inflammation-related gene (*CACNB2*). In addition, IL-1β and TNF-α exhibited a strong positive correlation with *IL1R1*, of which IL-1β had the strongest correlation with *IL1R1* (R = −0.799). *Stat3*, an immune balance-related gene, inhibited the expression of TGF-β and promoted the expression of IL-17A, while *Foxp3* facilitated the expression of TGF-β. *Odoribacter* and *Parabacteroides* were positively correlated with *Foxp3* and *Parabacteroides* was negatively correlated with *Stat3*. *Coriobacteriaceae_UCG-002*, *Odoribacter*, *Parabacteroides*, wyerone, ornithine, and lactucin could enhance intestinal barrier function by promoting the expression of *Occludin* and *ZO-1*. *Odoribacter*, wyerone, ornithine, and dodemorph showed significant negative correlations with *IL1R1* and *CACNB2*. Notably, *Odoribacter* (16 correlations: 3 negative and 13 positive), *Parabacteroides* (15 correlations: 3 negative and 12 positive), ornithine (10 correlations: 4 negative and 6 positive), wyerone (8 correlations: 4 negative and 4 positive), and lactucin (7 correlations: 4 negative and 3 positive) had the highest number of correlations, suggesting that they may serve as crucial connections in the anti-inflammatory pathway of UA.

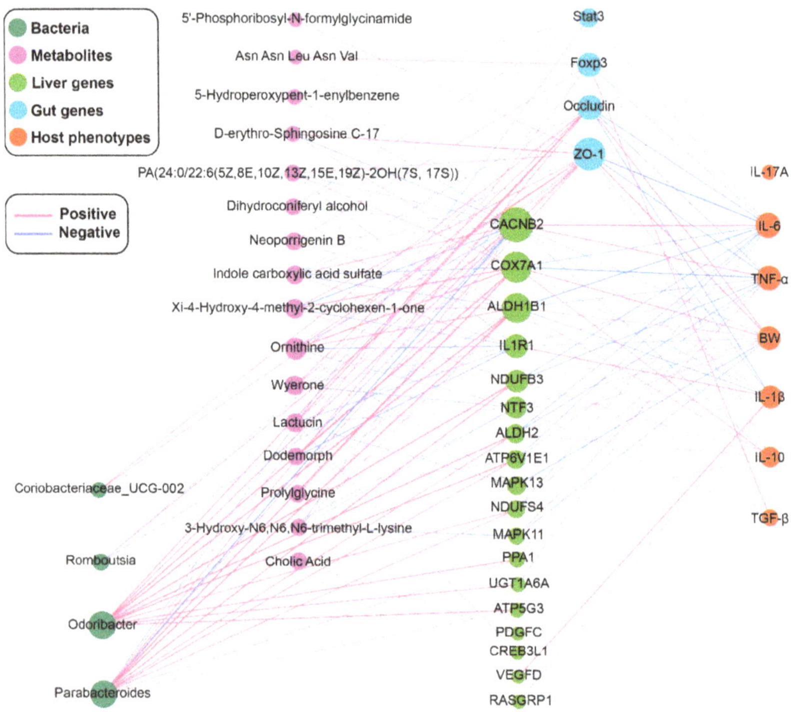

Figure 6. Spearman correlation network among the bacteria, metabolites, liver genes, gut genes, and host phenotypes (|R| > 0.6, $p < 0.05$). The width of the edges is proportional to the correlation strength. The color of the edges: red, positive; blue, and negative, $n = 6$.

3. Discussion

UA has garnered considerable attention for its array of pharmacological benefits [6,15,18,19]. In this study, the significant effects of UA in the jejunum were observed, including a notable increase in the V/C ratio and enhancement of intestinal barrier function. These findings highlight UA's role in promoting digestion and absorption as well as in improving gut health in mice. Th17 and Treg cells, which differentiate from $CD4^+$ T cells, are pivotal in the immune response and synergistically contribute to intestinal immune function [20]. The balance between Th17 and Treg cells is crucial for maintaining intestinal immune homeostasis [21]. Disruption of this balance can adversely affect nutrient digestion and absorption, thus compromising overall health [1,21,22].

IL-6-mediated Stat3 signaling stimulates the differentiation of Th17 cells by inducing the nuclear receptor RORγt, leading to the production of the inflammatory cytokine IL-17A [23]. Subsequently, IL-17A can enhance the secretion of other pro-inflammatory cytokines, including TNF-α, IL-6, and IL-1β [24]. Conversely, TGF-β promotes the transcription of Foxp3, facilitating the differentiation of Treg cells and the production of anti-inflammatory cytokines such as IL-10 and TGF-β [25]. Moreover, Foxp3 can inhibit the differentiation of Th17 cells by restricting their transcriptional potential [23]. Our results indicate that UA treatment resulted in the down-regulation of *Stat3* mRNA expression, up-regulation of *Foxp3* mRNA expression, and increased levels of TGF-β and IL-10 in mice. These findings suggest that UA can improve intestinal immune homeostasis.

The gut microbiota serves as a critical link between environmental factors and host health [26,27]. Investigating whether UA's functional role is connected to intestinal microbiota composition has become a topic of interest [13,15,28]. In our study, UA treatment significantly altered the composition and metabolites of the gut microbiota. Notably, the abundance of *Odoribacter* and *Parabacteroides* was significantly increased in the UA-treated mice. *Odoribacter*, a potential probiotic, is known to enhance mucosal immunity and maintain intestinal homeostasis [29]. Its colonization can augment Foxp3/RORγt regulatory T cells (Treg cells) and induce IL-10 production, thereby mitigating colitis in mice [30,31]. Numerous studies have highlighted the immunomodulatory effects of *Parabacteroides* [32–34], which have also been shown to promote CD4$^+$ T cell differentiation toward the CD4/Foxp3/IL-10 regulatory pathway [35,36]. Our correlation analysis demonstrated associations between *Odoribacter*, *Parabacteroides*, and key immune factors, including *Foxp3*, *Occludin*, *ZO-1*, IL-1β, IL-6, TNF-α, and IL-10. These observations suggest that UA's effects on immune balance and intestinal barrier function may be mediated by *Odoribacter* and *Parabacteroides*.

Furthermore, gut microbiota-derived ornithine has been shown to stimulate RORγt (+) IL-22 (+) cells and promote gut mucus formation, playing a vital role in maintaining intestinal homeostasis [37]. Previous research has demonstrated that lactucin can modulate gut microbiota and restore intestinal barrier function in inflammatory rat models [38]. Our findings revealed that both ornithine and lactucin were significantly enriched in UA-treated mice, indicating that these metabolites may be key factors in UA-mediated immune balance and intestinal barrier enhancement.

Additionally, our study investigates the effects of UA on liver health in mice. The MAPK signaling pathway is integral to regulating immune responses and plays a critical role in innate immunity [39]. Genes such as *IL1R1*, *MAPK11*, and *MAPK13*, expressed by immune or inflammatory cells, are crucial for inflammatory signaling [40,41]. Ras-guanine nucleotide-releasing protein 1 (RasGRP1) serves as a key regulator of the inflammatory response and is implicated in various inflammatory diseases [42]. Calcium (Ca^{2+}) voltage-gated channel auxiliary subunit beta 2 (CACNB2) may contribute to aberrant RAS-MAPK activation [43]. Our study found that *IL1R1*, *MAPK11*, *MAPK13*, *CACNB2*, and *RasGRP1* expression levels were significantly downregulated in the livers of mice in the UA group, suggesting that UA may lower the risk of liver inflammation and improve liver homeostasis. Moreover, the upregulated genes in the UA group were significantly enriched in pathways related to retinol metabolism, ascorbic acid and aldarate metabolism, and oxidative phosphorylation. Retinol, a form of vitamin A, plays an unexpected but crucial role in regulating immune responses [44]. Ascorbate, or vitamin C, acts as a potent antioxidant that supports various immune functions and enhances immune defenses [45]. Mitochondrial oxidative phosphorylation is responsible for generating the majority of ATP, which is essential for sustaining life and maintaining metabolic homeostasis [46]. These results imply that UA may enhance liver health by promoting the synthesis of retinol and ascorbic acid, thereby regulating metabolic homeostasis in the liver.

Interestingly, we observed that the *IL1R1* and *CACNB2* genes, which are associated with the MAPK signaling pathway, are broadly linked to immune-related factors, bacteria (*Odoribacter* and *Parabacteroides*), and metabolites (ornithine and lactucin). *Parabacteroides distasonis* has been shown to reduce liver damage and inflammation by inhibiting the NF-κB/MAPK pathway and activating the Nrf2 pathway, which may be related to the improvement in amino acid and bile acid metabolism in intestinal microorganisms [47]. *Parabacteroides distasonis* has a protective effect against non-alcoholic steatohepatitis (NASH) in mice by producing beneficial metabolites from dietary fiber [48]. In addition, *Odoribacter* has also been shown to be significantly negatively associated with measures of NASH progression [49]. Previous studies suggest that the liver-protective mechanism of ornithine may involve its conversion into the antioxidant glutathione and the nitric oxide precursor L-arginine, both of which are crucial for improving liver microcirculation [50,51]. Additionally, lactucin has been shown to downregulate the MAPK pathway, further enhancing liver health [52,53]. Lactucin may also enhance GLA circulation and reduce liver inflammation [38]. These observations suggest that the effects of UA on liver health may be mediated by regulating gut microbiome composition and its metabolism. However, our correlation analysis offers only indicative and predictive insights and does not definitively establish causal relationships between metabolites and hepatoprotective effects. Further studies are required to validate the relationship between these metabolites and liver health.

Our data can help us understand the function of UA in regulating immune homeostasis and improving liver health and identify the key role that gut microbiota and its metabolites may play in this process. However, we acknowledge that some limitations should be noted. We did not assess changes in gut microbial function and its association with inflammation in UA-treated mice. Further systematic studies, such as metagenomic, metatranscriptomic, or other multi-omics investigations, are needed to identify the function of gut microbiota and further explain the impact of changes in microbial metabolic function on inflammation.

In conclusion, UA can improve immune balance and enhance intestinal barrier function in mice. UA is able to increase beneficial bacteria (*Odoribacter* and *Parabacteroides*) and their metabolites (ornithin and lactucin). UA can inhibit the expression of *IL1R1* and *CACNB2* mRNA in the liver, upregulate the synthesis pathway of retinol and ascorbic acid in the liver, and promote liver health.

4. Materials and Methods

4.1. Ethics Statement

All experiments were performed in accordance with the ARRIVE guidelines and were carried out in accordance with the U.K. Animals (Scientific Procedures) Act, 1986 and associated guidelines, EU Directive 2010/63/EU for animal experiments, and approved by the Ethics Committee of Animal Experimentation of Hebei Agricultural University (Protocol 2023058).

4.2. Animals and Experimental Design

Eight-week-old female C57BL/6J mice were purchased from SPF Biotechnology Co., Ltd. (Beijing, China) and housed in pathogen-free facilities with a 12-h light and 12-h dark cycle at 22 °C.

After acclimatization to the laboratory conditions for 1 week, mice were randomly divided into 5 groups ($n = 6$ per group), including the control (CON) group, the UA 5, UA 25, UA 100, and UA 250 groups. Mice in the CON group were provided with a daily oral administration of 100 μL of PBS. Mice from the UA 5, UA 25, UA 100, and UA 250 groups received 5, 25, 100, and 250 mg/kg body weight of UA (purity $\geq$ 98%, Shenniu Pharmaceutical Co., Ltd., Dezhou, China; Figure S7) added to 100 μL PBS by daily gavage for 21 days. All group mice were administered menstruum, which was isovolumetric with 0.5% sodium carboxymethyl cellulose and 1% Tween 80. UA was dissolved in that menstruum (the menstruum and UA were administered directly into the stomach using an oral gavage needle).

Body weight was measured every three days throughout the study. On the 21st day, the mice were anesthetized with ether, the eyeballs were removed, blood samples were taken in 1.5 mL centrifuge tubes, and serum was separated by centrifugation for 15 min (3500× *g*, 4 °C), after which the mice were sacrificed. Fecal samples from all mice were collected aseptically and stored at −80 °C for microbiome and metabolomic analyses. The middle segments (0.5 cm) of the duodenum, jejunum, and ileum were flushed with PBS and fixed in a 10% formaldehyde solution for histological observation. Jejunum and liver tissue were washed with PBS, cut into slices, snap-frozen in liquid nitrogen, and stored at −80 °C for future analysis.

4.3. Enzyme-Linked Immunosorbent Assay (ELISA)

Serum samples were measured for the levels of IL-1β (MM-0040M2), IL-6 (MM-0163M2), TNF-α (MM-0132M2), TGF-β (MM-0689M2), IL-10 (MM-0176M2), and IL-17A (MM-0759M2) according to the instructions of specific ELISA kits (Jiangsu Meimian Industrial Co., Ltd., Yancheng, China).

4.4. Histological Analysis

The fixed duodenum, jejunum, and ileum samples were embedded in paraffin and stained with hematoxylin–eosin to obtain cross-sectional sections. In each section, intact and representative villi were selected, and intestinal morphology was measured. The images were evaluated using the ImageJ software 1.53j (US National Institutes of Health, Bethesda, MD, USA). Villus height (VH), crypt depth (CD), and intestinal wall thickness were measured. The ratio of villus height to crypt depth (V/C) was calculated.

4.5. Quantitative Reverse Transcription-Polymerase Chain Reaction (qRT-PCR)

Total RNA extraction and quantitative real-time PCR were performed as described previously. The extracted RNA was dissolved in RNase-free water and quantified using a NanoDrop-2000 spectrophotometer (Thermo Fisher Scientific, Waltham, MA, USA). RNA purity was evaluated by measuring the absorbance ratio at 260:280 nm, while RNA integrity was confirmed by the detection of the 18S and 28S bands after electrophoresis in 1% agarose gels. The primer sequences are shown in Table S1. The standard curve was drawn for each gene, and the linear relationship and amplification efficiency were calculated. Primers were confirmed as follows: correlation coefficient (R^2) > 0.98; the slope of the standard curve: −3–−3.5; and the amplification efficiency: 0.9–1.2. The mRNA levels of target genes were normalized to *β-actin* mRNA. Based on the cycle threshold (CT) value, the *β-actin* mRNA level was stable across the treatments in this study ($p > 0.1$) (Figure S8). The results of the relative mRNA expression of genes were calculated using the $2^{-\Delta\Delta CT}$ method [54].

4.6. 16S rRNA Gene Sequencing

Microbial genomic DNA was isolated from fecal specimens utilizing the E.Z.N.A.® soil DNA kit (Omega Bio-tek, Norcross, GA, USA). The V3-V4 hypervariable region of the bacterial 16S rRNA gene underwent amplification with the primer combinations 338F (5′-ACTCCTACGGGAGGCAGCAG-3′) and 806R (5′-GGACTACHVGGGTWTCTAAT-3′) via the T100 Thermal Cycler PCR thermocycler (BIO-RAD, Hercules, CA, USA). Subsequently, the purified amplicons were combined in equimolar proportions and subjected to paired-end sequencing on an Illumina PE300 platform (Illumina, San Diego, CA, USA) in accordance with standard operating procedures established by Majorbio Bio-Pharm Technology Co., Ltd. (Shanghai, China).

4.7. Microbiota Data Analysis

Initial sequencing data underwent quality filtering using Fastp version 0.19.6, followed by read merging using FLASH version 1.2.11. Subsequently, the sequences were optimized and grouped into operational taxonomic units (OTUs) at a 97% sequence similarity threshold utilizing UPARSE 11. The taxonomic classification of each OTU representative sequence was conducted, employing RDP Classifier version 2.13 against the Silva v138 16S rRNA gene database, with a confidence threshold set at 0.7.

Analysis of fecal microbiota data was executed on the Majorbio Cloud platform (https://cloud.majorbio.com (accessed on 15 April 2024)). Rarefaction curves and alpha diversity indices were computed using Mothur v1.30.1. Principal coordinate analysis (PCoA) and Adonis analysis, based on Bray–Curtis dissimilarity, were employed to assess the similarity among microbial communities across different samples, facilitated by the "Vegan" package. The statistical significance of non-parametric relative abundance profiles was evaluated utilizing the Mann–Whitney test. Additionally, linear discriminant analysis

(LDA) effect size (LEfSe) analysis was conducted to identify significantly enriched bacterial taxa among distinct groups (LDA score > 3, $p < 0.05$).

4.8. Liquid Chromatograph–Mass Spectrometer Analysis (LC–MS Analysis)

A solid sample weighing 50 mg underwent metabolite extraction utilizing 400 μL of a 4:1 (*v*:*v*) methanol–water extract solution. LC–MS/MS analysis was performed on a SCIEX UPLC-Triple TOF 6600 system featuring an ACQUITY HSS T3 column (100 mm × 2.1 mm i.d., 1.8 μm; Waters, Milford, MA, USA) at Majorbio Bio-Pharm Technology Co., Ltd. (Shanghai, China). Mobile phase A is 95% water + 5% acetonitrile (containing 0.1% formic acid); mobile phase B is 47.5% acetonitrile + 47.5% isopropyl alcohol + 5% water (containing 0.1% formic acid). The flow rate was 0.40 mL/min, the sample size was 10 μL, and the column temperature was 45 °C. Positive ion mode separation gradient 0–3 min and mobile phase B increased from 0% to 20%; in 3–4.5 min, mobile phase B increased from 20% to 35%; in 4.5–5 min, mobile phase B increased from 35% to 100%; at 5–6.3 min, mobile phase B remained 100%; and in 6.3–6.4 min, mobile phase B decreased from 100% to 0%. At 6.4–8 min, mobile phase B maintained 0%. Negative ion mode separation gradient: 0–1.5 min, mobile phase B from 0 to 5%; 1.5–2 min, mobile phase B increased from 5% to 10%; 2–4.5 min, mobile phase B increased from 10% to 30%; 4.5–5 min, mobile phase B increased from 30% to 100%; and at 5–6.3 min, the linearity of mobile phase B remained 100%. In 6.3–6.4 min, mobile phase B decreases from 100% to 0%. The linearity of mobile phase B remained at 0% from 6.4–8 min. Positive and negative ion scanning modes are used to collect the sample quality spectrum signal, and the quality scanning range is 50–1200 m/z. The spray gas flow rate is 50 psi, the auxiliary gas flow rate is 50 psi, the curtain gas flow rate is 35 psi, the ion source temperature is 500 °C, the positive mode ion spray voltage is set to 5500 V, the negative mode ion spray voltage is set to −4500 V, and the ion transfer tube temperature is 325 °C. Normalized collision energy is 20-40-60 eV cyclic collision energy. DDA mode was used to collect data.

Raw UHPLC–MS data were processed using Progenesis QI 2.3 software (Nonlinear Dynamics, Waters, USA) for peak detection and alignment. Metabolic features detected in at least 80% of samples were retained, and minimum imputation was applied to samples falling below the quantitation limit. The normalization of each metabolic feature was conducted by sum and quality control. Variables with relative standard deviation (RSD) exceeding 30% of QC samples were excluded, and log10 transformation was performed to obtain the final data matrix.

Conduct variance analysis on the processed data matrix file. The "ropls" R package (Version 1.6.2) was utilized for partial least squares discriminant analysis (PLS-DA). Metabolites with a variable importance in projection (VIP) value exceeding 1 and a significance level of $p < 0.05$ were deemed significantly different. Differential metabolites between the two groups were subjected to metabolic enrichment and pathway analysis based on the Kyoto Encyclopedia of Genes and Genomes (KEGG) database (http://www.genome.jp/kegg/ (accessed on 15 April 2024)) to map their biochemical pathways.

4.9. RNA Extraction, Library Construction, and Sequencing

Total RNA was isolated from the tissue using TRIzol® Reagent as per the manufacturer's protocol. Subsequently, RNA integrity was assessed using a 5300 Bioanalyser (Agilent, Santa Clara, CA, USA) and quantification was carried out using the ND-2000 (NanoDrop Technologies, Wilmington, DE, USA). RNA purification, reverse transcription, library preparation, and sequencing procedures were conducted at Shanghai Majorbio Bio-pharm Biotechnology Co., Ltd. (Shanghai, China) following the guidelines provided by Illumina (San Diego, CA, USA). The RNA-seq transcriptome library was constructed using Illumina® Stranded mRNA Prep and Ligation Kit from Illumina (San Diego, CA, USA), utilizing 1 μg of total RNA. Libraries underwent size selection for cDNA fragments of 300 bp on 2% low-range ultra-agarose, followed by PCR amplification with Phusion DNA polymerase (NEB) for 15 PCR cycles. Following quantification with Qubit 4.0, paired-end

RNA-seq libraries were sequenced using the NovaSeq 6000 sequencer (Illumina, San Diego, CA, USA) (2 × 150 bp read length).

The raw paired-end reads underwent trimming and quality control using Fastp with default parameters. Subsequently, clean reads were individually aligned to the reference genome in orientation mode employing HISAT2 software 2.2.1. The mapped reads of each sample were assembled by StringTie v2.1.4 utilizing a reference-based approach.

To identify differential expression genes (DEGs) between distinct samples, the expression level of each transcript was quantified utilizing the transcripts per million reads (TPM) method via RSEM. Differential expression analysis was conducted using DESeq2 (version 1.12.3), with DEGs meeting the criteria of $|\log2FC| \geq 1$ and false discovery rate (FDR) < 0.05 considered significantly differentially expressed genes. Additionally, functional enrichment analysis, encompassing Gene Ontology (GO) and KEGG pathways, was performed to discern which DEGs were enriched in GO terms and metabolic pathways.

4.10. Correlation Analysis

Given the extensive diversity, dimensional complexity, and discontinuous distribution of omics data, we applied the Spearman correlation examination in this investigation to investigate the associations among the bacteria, metabolites, liver genes, gut genes, and host phenotypes. A threshold of $|R| > 0.6$, with $p < 0.05$, was utilized to discern noteworthy correlations. Cytoscape version 3.9.0 was employed for network construction. Visualization of the interrelations between microbiome, metabolome, and physiological attributes was accomplished utilizing the R package "ggplot2".

4.11. Statistical Analysis

The body weight, ELISA, and histological observation data were analyzed and one-way analysis of variance (ANOVA) by Statistical Package for the Social Sciences (SPSS) version 26; the differences between means were compared by Duncan's multiple range test. Gene relative expression was compared using the *t*-test. The results were expressed as mean ± standard deviation (SD). The significance was set at $p < 0.05$.

Supplementary Materials: The supporting information can be downloaded at https://www.mdpi.com/article/10.3390/ijms251910623/s1.

Author Contributions: M.Z.: Conceptualization, Data curation, Formal analysis, Funding acquisition, Investigation, Methodology, Visualization, Writing—original draft, and Writing—review and editing. Y.C.: Data curation, Formal analysis, Methodology, and Writing—original draft. F.W. (Fengxia Wang): Data curation, Formal analysis, Methodology, and Writing—original draft. F.W. (Fengyang Wu): Data curation, Validation, and Methodology. C.L.: Data curation, Validation, and Methodology. S.L.: Data curation, Validation, and Methodology. B.C.: Conceptualization, Funding acquisition, Project administration, Resources, Supervision, and Writing—review and editing. All authors have read and agreed to the published version of the manuscript.

Funding: This research was funded by the Modern Agriculture Industry Technology System of Rabbit, China (CARS-43-B-2) and the Postgraduate Innovation Funding project of Hebei Province (CXZZBS2024075).

Institutional Review Board Statement: All experiments were performed in accordance with the ARRIVE guidelines and were carried out in accordance with the U.K. Animals (Scientific Procedures) Act, 1986, associated guidelines, EU Directive 2010/63/EU for animal experiments and approved by the Ethics Committee of Animal Experimentation of Hebei Agricultural University (Protocol 2023058).

Informed Consent Statement: Not applicable.

Data Availability Statement: The data presented in this study are openly available in the NCBI Sequence Read Archive (SRA), reference number PRJNA1115940.

Conflicts of Interest: The authors declare no conflicts of interest.

References

1. Ma, Z.; Akhtar, M.; Pan, H.; Liu, Q.; Chen, Y.; Zhou, X.; You, Y.; Shi, D.; Liu, H. Fecal microbiota transplantation improves chicke growth performance by balancing jejunal Th17/Treg cells. *Microbiome* **2023**, *11*, 137. [CrossRef] [PubMed]
2. Cargnin, S.T.; Gnoatto, S.B. Ursolic acid from apple pomace and traditional plants: A valuable triterpenoid with functiona properties. *Food Chem.* **2017**, *220*, 477–489. [CrossRef] [PubMed]
3. Fu, Y.; Liu, T.; He, S.; Zhang, Y.; Tan, Y.; Bai, Y.; Shi, J.; Deng, W.; Qiu, J.; Wang, Z.; et al. Ursolic acid reduces oxidative stres injury to ameliorate experimental autoimmune myocarditis by activating Nrf2/HO-1 signaling pathway. *Front. Pharmacol.* **202** *14*, 1189372. [CrossRef] [PubMed]
4. Gou, W.; Luo, N.; Yu, B.; Wu, H.; Wu, S.; Tian, C.; Guo, J.; Ning, H.; Bi, C.; Wei, H.; et al. Ursolic Acid Derivative UA232 Promote Tumor Cell Apoptosis by Inducing Endoplasmic Reticulum Stress and Lysosomal Dysfunction. *Int. J. Biol. Sci.* **2022**, *18*, 2639–265 [CrossRef] [PubMed]
5. Wan, Y.; Zhang, W.; Huang, C.; Jian, J.; Zhang, Y.; Liu, Q.; Chen, P.; Zhu, X. Ursolic acid alleviates Kupffer cells pyroptosis in live fibrosis by the NOX2/NLRP3 inflammasome signaling pathway. *Int. Immunopharmacol.* **2022**, *113*, 109321. [CrossRef]
6. Zhao, M.; Wu, F.; Tang, Z.; Yang, X.; Liu, Y.; Wang, F.; Chen, B. Anti-inflammatory and antioxidant activity of ursolic acid: A systematic review and meta-analysis. *Front. Pharmacol.* **2023**, *14*, 1256946. [CrossRef]
7. Zhang, B.C.; Lai, C.M.; Luo, B.Y.; Shao, J.W. Triterpenoids-templated self-assembly nanosystem for biomimetic delivery o CRISPR/Cas9 based on the synergy of TLR-2 and ICB to enhance HCC immunotherapy. *Acta Pharm. Sin. B* **2024**, *14*, 3205–321 [CrossRef]
8. Larabi, A.; Barnich, N.; Nguyen, H.T.T. New insights into the interplay between autophagy, gut microbiota and inflammator responses in IBD. *Autophagy* **2020**, *16*, 38–51. [CrossRef]
9. Singh, V.; Yeoh, B.S.; Walker, R.E.; Xiao, X.; Saha, P.; Golonka, R.M.; Cai, J.; Bretin, A.C.A.; Cheng, X.; Liu, Q.; et al. Microbiot fermentation-NLRP3 axis shapes the impact of dietary fibres on intestinal inflammation. *Gut* **2019**, *68*, 1801–1812. [CrossRef]
10. Cheng, H.; Guan, X.; Chen, D.; Ma, W. The Th17/Treg Cell Balance: A Gut Microbiota-Modulated Story. *Microorganisms* **201** *7*, 583. [CrossRef]
11. Zhang, L.L.; Xu, J.Y.; Xing, Y.; Wu, P.; Jin, Y.W.; Wei, W.; Zhao, L.; Yang, J.; Chen, G.C.; Qin, L.Q. *Lactobacillus rhamnosus* GC alleviates radiation-induced intestinal injury by modulating intestinal immunity and remodeling gut microbiota. *Microbiol. Re* **2024**, *286*, 127821. [CrossRef] [PubMed]
12. Yan, J.-B.; Luo, M.-M.; Chen, Z.-Y.; He, B.-H. The Function and Role of the Th17/Treg Cell Balance in Inflammatory Bowel Diseas *J. Immunol. Res.* **2020**, *2020*, 8813558. [CrossRef] [PubMed]
13. Yan, X.; Ren, X.; Liu, X.; Wang, Y.; Ma, J.; Song, R.; Wang, X.; Dong, Y.; Fan, Q.; Wei, J.; et al. Dietary Ursolic Acid Prevent Alcohol-Induced Liver Injury via Gut–Liver Axis Homeostasis Modulation: The Key Role of Microbiome Manipulation. *J. Agri Food Chem.* **2021**, *69*, 7074–7083. [CrossRef]
14. Sheng, Q.; Li, F.; Chen, G.; Li, J.; Li, J.; Wang, Y.; Lu, Y.; Li, Q.; Li, M.; Chai, K. Ursolic Acid Regulates Intestinal Microbiota an Inflammatory Cell Infiltration to Prevent Ulcerative Colitis. *J. Immunol. Res.* **2021**, *2021*, 6679316. [CrossRef]
15. Nie, Y.; Liu, Q.; Zhang, W.; Wan, Y.; Huang, C.; Zhu, X. Ursolic acid reverses liver fibrosis by inhibiting NOX4/NLRP inflammasome pathways and bacterial dysbiosis. *Gut Microbes* **2021**, *13*, 1972746. [CrossRef]
16. Kashyap, D.; Tuli, H.S.; Sharma, A.K. Ursolic acid (UA): A metabolite with promising therapeutic potential. *Life Sci.* **2016**, *14* 201–213. [CrossRef]
17. Chen, C.; Gao, K.; Chen, Z.; Zhang, Q.; Ke, X.; Mao, B.; Fan, Q.; Li, Y.; Chen, S. The supplementation of the multi-strain probiotic WHHPRO™ alleviates high-fat diet-induced metabolic symptoms in rats via gut-liver axis. *Front. Nutr.* **2023**, *10*, 132469 [CrossRef]
18. Li, J.; Li, N.; Yan, S.; Liu, M.; Sun, B.; Lu, Y.; Shao, Y. Ursolic acid alleviates inflammation and against diabetes-induce nephropathy through TLR4-mediated inflammatory pathway. *Mol. Med. Rep.* **2018**, *18*, 4675–4681. [CrossRef] [PubMed]
19. Li, H.; Yu, Y.; Liu, Y.; Luo, Z.; Law, B.Y.K.; Zheng, Y.; Huang, X.; Li, W. Ursolic acid enhances the antitumor effects of sorafeni associated with Mcl-1-related apoptosis and SLC7A11-dependent ferroptosis in human cancer. *Pharmacol. Res.* **2022**, *182*, 10630 [CrossRef]
20. Knochelmann, H.M.; Dwyer, C.J.; Bailey, S.R.; Amaya, S.M.; Elston, D.M.; Mazza-McCrann, J.M.; Paulos, C.M. When world collide: Th17 and Treg cells in cancer and autoimmunity. *Cell. Mol. Immunol.* **2018**, *15*, 458–469. [CrossRef]
21. Liu, Y.J.; Tang, B.; Wang, F.C.; Tang, L.; Lei, Y.Y.; Luo, Y.; Huang, S.J.; Yang, M.; Wu, L.Y.; Wang, W.; et al. Parthenolid ameliorates colon inflammation through regulating Treg/Th17 balance in a gut microbiota-dependent manner. *Theranostics* **202** *10*, 5225–5241. [CrossRef] [PubMed]
22. Zhong, Y.; Liu, W.; Xiong, Y.; Li, Y.; Wan, Q.; Zhou, W.; Zhao, H.; Xiao, Q.; Liu, D. Astragaloside IV alleviates ulcerative colitis b regulating the balance of Th17/Treg cells. *Phytomedicine* **2022**, *104*, 154287. [CrossRef] [PubMed]
23. Chung, Y.; Chang, S.H.; Martinez, G.J.; Yang, X.O.; Nurieva, R.; Kang, H.S.; Ma, L.; Watowich, S.S.; Jetten, A.M.; Tian, Q.; et a Critical regulation of early Th17 cell differentiation by interleukin-1 signaling. *Immunity* **2009**, *30*, 576–587. [CrossRef] [PubMed
24. Noack, M.; Miossec, P. Th17 and regulatory T cell balance in autoimmune and inflammatory diseases. *Autoimmun. Rev.* **2014**, *1* 668–677. [CrossRef]
25. Wang, J.; Zhao, X.; Wan, Y.Y. Intricacies of TGF-β signaling in Treg and Th17 cell biology. *Cell. Mol. Immunol.* **2023**, *20*, 1002–102 [CrossRef]

26. Kamada, N.; Seo, S.-U.; Chen, G.Y.; Núñez, G. Role of the gut microbiota in immunity and inflammatory disease. *Nat. Rev. Immunol.* **2013**, *13*, 321–335. [CrossRef]
27. Wu, Z.; Huang, S.; Li, T.; Li, N.; Han, D.; Zhang, B.; Xu, Z.Z.; Zhang, S.; Pang, J.; Wang, S.; et al. Gut microbiota from green tea polyphenol-dosed mice improves intestinal epithelial homeostasis and ameliorates experimental colitis. *Microbiome* **2021**, *9*, 184. [CrossRef]
28. Hao, W.; Kwek, E.; He, Z.; Zhu, H.; Liu, J.; Zhao, Y.; Ma, K.Y.; He, W.S.; Chen, Z.Y. Ursolic acid alleviates hypercholesterolemia and modulates the gut microbiota in hamsters. *Food Funct.* **2020**, *11*, 6091–6103. [CrossRef]
29. Zhou, J.; Li, M.; Chen, Q.; Li, X.; Chen, L.; Dong, Z.; Zhu, W.; Yang, Y.; Liu, Z.; Chen, Q. Programmable probiotics modulate inflammation and gut microbiota for inflammatory bowel disease treatment after effective oral delivery. *Nat. Commun.* **2022**, *13*, 3432. [CrossRef]
30. Lima, S.F.; Gogokhia, L.; Viladomiu, M.; Chou, L.; Putzel, G.; Jin, W.B.; Pires, S.; Guo, C.J.; Gerardin, Y.; Crawford, C.V.; et al. Transferable Immunoglobulin A–Coated *Odoribacter splanchnicus* in Responders to Fecal Microbiota Transplantation for Ulcerative Colitis Limits Colonic Inflammation. *Gastroenterology* **2021**, *162*, 166–178. [CrossRef]
31. Xing, C.; Wang, M.; Ajibade, A.A.; Tan, P.; Fu, C.; Chen, L.; Zhu, M.; Hao, Z.Z.; Chu, J.; Yu, X.; et al. Microbiota regulate innate immune signaling and protective immunity against cancer. *Cell Host Microbe* **2021**, *29*, 959–974.E7. [CrossRef] [PubMed]
32. Kverka, M.; Zakostelska, Z.; Klimesova, K.; Sokol, D.; Hudcovic, T.; Hrncir, T.; Rossmann, P.; Mrazek, J.; Kopecny, J.; Verdu, E.F.; et al. Oral administration of *Parabacteroides distasonis* antigens attenuates experimental murine colitis through modulation of immunity and microbiota composition. *Clin. Exp. Immunol.* **2011**, *163*, 250–259. [CrossRef]
33. Koh, G.Y.; Kane, A.V.; Wu, X.; Crott, J.W. *Parabacteroides distasonis* attenuates tumorigenesis, modulates inflammatory markers and promotes intestinal barrier integrity in azoxymethane-treated A/J mice. *Carcinogenesis* **2020**, *41*, 909–917. [CrossRef]
34. Sun, H.; Guo, Y.; Wang, H.; Yin, A.; Hu, J.; Yuan, T.; Zhou, S.; Xu, W.; Wei, P.; Yin, S.; et al. Gut commensal *Parabacteroides distasonis* alleviates inflammatory arthritis. *Gut* **2023**, *72*, 1664–1677. [CrossRef]
35. Cuffaro, B.; Assohoun, A.L.W.; Boutillier, D.; Súkeníková, L.; Desramaut, J.; Boudebbouze, S.; Salomé-Desnoulez, S.; Hrdý, J.; Waligora-Dupriet, A.J.; Maguin, E.; et al. In Vitro Characterization of Gut Microbiota-Derived Commensal Strains: Selection of *Parabacteroides distasonis* Strains Alleviating TNBS-Induced Colitis in Mice. *Cells* **2020**, *9*, 2104. [CrossRef] [PubMed]
36. Ezeji, J.C.; Sarikonda, D.K.; Hopperton, A.; Erkkila, H.L.; Cohen, D.E.; Martinez, S.P.; Cominelli, F.; Kuwahara, T.; Dichosa, A.E.K.; Good, C.E.; et al. *Parabacteroides distasonis*: Intriguing aerotolerant gut anaerobe with emerging antimicrobial resistance and pathogenic and probiotic roles in human health. *Gut Microbes* **2021**, *13*, 1922241. [CrossRef]
37. Qi, H.; Li, Y.; Yun, H.; Zhang, T.; Huang, Y.; Zhou, J.; Yan, H.; Wei, J.; Liu, Y.; Zhang, Z.; et al. Lactobacillus maintains healthy gut mucosa by producing L-Ornithine. *Commun. Biol.* **2019**, *2*, 171. [CrossRef] [PubMed]
38. Han, C.; Wu, X.; Zou, N.; Zhang, Y.; Yuan, J.; Gao, Y.; Chen, W.; Yao, J.; Li, C.; Hou, J.; et al. *Cichorium pumilum* Jacq Extract Inhibits LPS-Induced Inflammation via MAPK Signaling Pathway and Protects Rats from Hepatic Fibrosis Caused by Abnormalities in the Gut-Liver Axis. *Front. Pharmacol.* **2021**, *12*, 683613. [CrossRef]
39. Krishna, M.; Narang, H. The complexity of mitogen-activated protein kinases (MAPKs) made simple. *Cell. Mol. Life Sci.* **2008**, *65*, 3525–3544. [CrossRef]
40. Kyriakis, J.M.; Avruch, J. Mammalian MAPK signal transduction pathways activated by stress and inflammation: A 10-year update. *Physiol. Rev.* **2012**, *92*, 689–737. [CrossRef]
41. Petrasek, J.; Bala, S.; Csak, T.; Lippai, D.; Kodys, K.; Menashy, V.; Barrieau, M.; Min, S.Y.; Kurt-Jones, E.A.; Szabo, G. IL-1 receptor antagonist ameliorates inflammasome-dependent alcoholic steatohepatitis in mice. *J. Clin. Investig.* **2012**, *122*, 3476–3489. [CrossRef] [PubMed]
42. Wang, C.; Li, X.; Xue, B.; Yu, C.; Wang, L.; Deng, R.; Liu, H.; Chen, Z.; Zhang, Y.; Fan, S.; et al. RasGRP1 promotes the acute inflammatory response and restricts inflammation-associated cancer cell growth. *Nat. Commun.* **2022**, *13*, 7001. [CrossRef] [PubMed]
43. Pandian, V.D.; Giovannucci, D.R.; Vazquez, G.; Kumarasamy, S. CACNB2 is associated with aberrant RAS-MAPK signaling in hypertensive Dahl Salt-Sensitive rats. *Biochem. Biophys. Res. Commun.* **2019**, *513*, 760–765. [CrossRef]
44. Mora, J.R.; Iwata, M.; von Andrian, U.H. Vitamin effects on the immune system: Vitamins A and D take centre stage. *Nat. Rev. Immunol.* **2008**, *8*, 685–698. [CrossRef]
45. Carr, A.C.; Maggini, S. Vitamin C and Immune Function. *Nutrients* **2017**, *9*, 1211. [CrossRef]
46. Wilson, D.F. Oxidative phosphorylation: Regulation and role in cellular and tissue metabolism. *J. Physiol.* **2017**, *595*, 7023–7038. [CrossRef] [PubMed]
47. Wang, L.; Zheng, W.; Sun, Y.; Ren, X.; Yan, C.; Song, S.; Ai, C. Fucoidan ameliorates alcohol-induced liver injury in mice through Parabacteroides distasonis-mediated regulation of the gut-liver axis. *Int. J. Biol. Macromol.* **2024**, *279*, 135309. [CrossRef]
48. Zhao, Q.; Dai, M.Y.; Huang, R.Y.; Duan, J.Y.; Zhang, T.; Bao, W.M.; Zhang, J.Y.; Gui, S.Q.; Xia, S.M.; Dai, C.T.; et al. Parabacteroides distasonis ameliorates hepatic fibrosis potentially via modulating intestinal bile acid metabolism and hepatocyte pyroptosis in male mice. *Nat. Commun.* **2023**, *14*, 1829. [CrossRef]
49. Shu, Y.; Huang, Y.; Dong, W.; Fan, X.; Sun, Y.; Chen, G.; Zeng, X.; Ye, H. The polysaccharides from Auricularia auricula alleviate non-alcoholic fatty liver disease via modulating gut microbiota and bile acids metabolism. *Int. J. Biol. Macromol.* **2023**, *246*, 125662. [CrossRef]

50. Canbay, A.; Sowa, J.P. L-Ornithine L-Aspartate (LOLA) as a Novel Approach for Therapy of Non-alcoholic Fatty Liver Disease. *Drugs* **2019**, *79* (Suppl. 1), 39–44. [CrossRef]
51. Ding, Y.; Koda, Y.; Shashni, B.; Takeda, N.; Zhang, X.; Tanaka, N.; Nishikawa, Y.; Nagasaki, Y. An orally deliverable ornithine-based self-assembling polymer nanomedicine ameliorates hyperammonemia in acetaminophen-induced acute liver injury. *Acta Biomater.* **2023**, *168*, 515–528. [CrossRef] [PubMed]
52. Imam, K.M.S.U.; Tian, Y.; Xin, F.; Xie, Y.; Wen, B. Lactucin, a Bitter Sesquiterpene from *Cichorium intybus*, Inhibits Cancer Cell Proliferation by Downregulating the MAPK and Central Carbon Metabolism Pathway. *Molecules* **2022**, *27*, 7358. [CrossRef] [PubMed]
53. Dang, T.; Zheng, G.; Zhang, Q.; Jin, P.; Zhang, H.; Su, L.; Qin, D.; Yao, G. Sesquiterpenoids with diverse carbon skeletons from the roots of *Cichorium glandulosum* and their anti-inflammatory activities. *Fitoterapia* **2019**, *136*, 104170. [CrossRef] [PubMed]
54. Zhou, W.; Zhang, H.; Huang, L.; Sun, C.; Yue, Y.; Cao, X.; Jia, H.; Wang, C.; Gao, Y. Disulfiram with Cu^{2+} alleviates dextran sulfate sodium-induced ulcerative colitis in mice. *Theranostics* **2023**, *13*, 2879–2895. [CrossRef]

International Journal of
Molecular Sciences

MDPI

rticle

;PR40/GPR120 Agonist GW9508 Improves Metabolic ;yndrome-Exacerbated Periodontitis in Mice

anchun Li [1], Hong Yu [2], Maria F. Lopes-Virella [1] and Yan Huang [1,*]

1 Division of Endocrinology, Diabetes and Metabolic Diseases, Department of Medicine, College of Medicine, Medical University of South Carolina, Charleston, SC 29425, USA; liyya@musc.edu (Y.L.); virellam@musc.edu (M.F.L.-V.)
2 Department of Biomedical & Community Health Sciences, The James B. Edwards College of Dental Medicine, Medical University of South Carolina, Charleston, SC 29425, USA; yuho@musc.edu
* Correspondence: huangyan@musc.edu

Abstract: G protein-coupled receptor (GPR)40 and GPR120 are receptors for medium- and long-chain free fatty acids. It has been well documented that GPR40 and GPR120 activation improves metabolic syndrome (MetS) and exerts anti-inflammatory effects. Since chronic periodontitis is a common oral inflammatory disease initiated by periodontal pathogens and exacerbated by MetS, we determined if GPR40 and GPR120 activation with agonists improves MetS-associated periodontitis in animal models in this study. We induced MetS and periodontitis by high-fat diet feeding and periodontal injection of lipopolysaccharide, respectively, and treated mice with GW9508, a synthetic GPR40 and GPR120 dual agonist. We determined alveolar bone loss, osteoclast formation, and periodontal inflammation using micro-computed tomography, osteoclast staining, and histology. To understand the underlying mechanisms, we further performed studies to determine the effects of GW9508 on osteoclastogenesis and proinflammatory gene expression in vitro. Results showed that GW9508 improved metabolic parameters, including glucose, lipids, and insulin resistance. Results also showed that GW9508 improves periodontitis by reducing alveolar bone loss, osteoclastogenesis, and periodontal inflammation. Finally, in vitro studies showed that GW9508 inhibited osteoclast formation and proinflammatory gene secretion from macrophages. In conclusion, this study demonstrated for the first time that GPR40/GPR120 agonist GW9508 reduced alveolar bone loss and alleviated periodontal inflammation in mice with MetS-exacerbated periodontitis, suggesting that activating GPR40/GPR120 with agonist GW9508 is a potential anti-inflammatory approach for the treatment of MetS-associated periodontitis.

Keywords: fatty acid receptors; periodontitis; metabolic syndrome; inflammation

1. Introduction

Periodontitis, a bacteria-induced chronic inflammatory disease, is characterized by progressive destruction of tooth-supporting connective tissue and periodontal alveolar bone, leading to tooth loss [1,2]. It has been well-established that diabetes exacerbates periodontitis [3]. Interestingly, clinical studies have further demonstrated that metabolic syndrome (MetS), which is commonly associated with prediabetes, also increases the prevalence and severity of periodontitis [4–9]. MetS is defined as a cluster of cardiovascular risk factors, including obesity, high blood pressure, lipid abnormalities such as high triglycerides and low HDL, insulin resistance, and increased plasma glucose [10,11]. A recent systematic review and meta-analyses showed a dose-response relationship between the number of cardiovascular risk factors in MetS and the occurrence of periodontitis [9]. Another meta-analysis also showed that MetS is positively associated with the severity of periodontitis [12].

The findings of the relationship between MetS and periodontitis are important since 34% of the US population have MetS as contrasted to 9.3% of the US population having diabetes [13]. Given that MetS worsens periodontitis, it is important to carefully evaluate and effectively treat periodontitis in patients with MetS as patients with diabetes. Indeed, early treatment of periodontitis in MetS patients who also have prediabetes is likely to reduce the severity of periodontitis when prediabetes progresses to diabetes [14]. More importantly, patients with MetS may also have life-threatening complications such as cardiovascular disease [15], nonalcoholic fatty liver disease [16], and Alzheimer's disease [17], and increasing studies have shown that periodontitis worsens these MetS complications via periodontitis-promoted systemic inflammation [18–20]. Therefore, the development of novel and effective therapeutics for MetS-exacerbated periodontitis is essential to reduce not only tooth loss but also the severity of MetS and MetS complications.

To study the pathogenesis of MetS-associated periodontitis, our lab has developed several mouse models with both MetS and periodontitis [21–23]. In these animal models, MetS was induced by high-fat diet (HFD) feeding for 20 weeks, and periodontitis was induced by periodontal LPS injection, ligature placement around the tooth, or oral inoculation of *Porphyromonas gingivalis* [21–23]. Using these animal models, we demonstrated that MetS worsened periodontitis by increasing macrophage infiltration and the production of inflammatory cytokines in periodontal tissue and promoting osteoclastogenesis, which led to alveolar bone loss [21,22]. Obviously, MetS, as a systemic disorder, worsens periodontitis by augmenting periodontal inflammation. Hence, we hypothesized that compounds that are capable of not only improving MetS but also inhibiting tissue inflammation may have the potential to treat periodontitis in patients with MetS as an adjunctive therapy to scaling and root planning (SRP), the gold standard treatment of chronic periodontitis [24–26]. By combining anti-inflammatory therapy with SRP, the treatment of periodontitis in patients with MetS can be more effective.

In recent years, it has been well established that G protein-coupled receptor (GPR)40 and GPR120 as fatty acid (FA) receptors play important roles in not only uptaking FAs but also mediating anti-inflammatory signaling [27–29]. It has been shown that GRP40 or GPR120 activation in animal models improved inflammation-associated metabolic disorders such as diabetes and MetS [28,30]. It has also been shown that omega-3 (ω-3) FAs such as docosahexaenoic acid (DHA) and eicosapentaenoic acid (EPA) as natural agonists for both GPR40 and GPR120 [31,32] have beneficial effects on periodontitis [33,34]. To increase potency and specificity, several synthetic agonists for GPR40 or GPR120, as well as dual agonists for both GPR40 and GPR120, have been developed [35–38]. A large number of studies have shown that these agonists have therapeutic potential for many chronic diseases, including MetS and atherosclerosis, through anti-inflammatory mechanisms [32,39,40]. Since GPR40/GPR120 agonists are capable of not only improving MetS but also acting as potent anti-inflammatory agents, they may have beneficial effects on MetS-exacerbated periodontitis. However, whether activation of GPR40/GPR120 by agonists improves MetS-associated periodontitis remains unknown. In this study, we hypothesized that GPR40/GPR120 dual agonist GW9508 is capable of inhibiting the progression of periodontitis exacerbated by MetS. Our specific objectives include the following: 1. To determine the effect of GW9508 on alveolar bone loss and periodontal inflammation in a mouse model for MetS-associated periodontitis; 2. To explore the potential molecular mechanisms involved in GW9508-attenuated alveolar bone loss and periodontal inflammation.

2. Results

2.1. GW9508 Treatment Improves MetS

We first determined the effects of GW9508 on HFD-induced MetS. As shown in Figure 1, compared with control mice that were fed a low-fat diet (LFD), MetS mice that were fed HFD had increased bodyweight, fasting glucose, cholesterol, free fatty acid, insulin, and homeostatic model assessment for insulin resistance (HOMA-IR). Periodontitis (PD) induced by periodontal LPS injection had no significant effect on these metabolic parameters in LFD-fed control mice but reduced insulin and insulin resistance in HFD-fed mice. Interestingly, GW9508 treatment significantly reduced MetS-increased fasting glucose, cholesterol, and free fatty acids (FFA) in mice, suggesting that GW9508 as a GPR40/GPR120 agonist improves MetS.

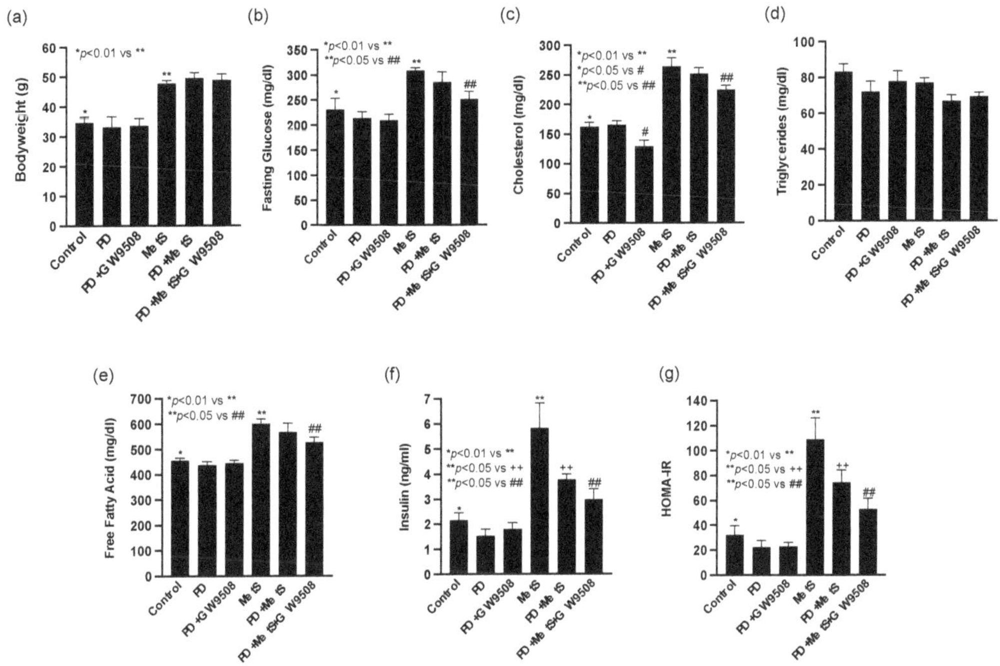

Figure 1. The effects of GW9508 on the metabolic parameters for control (LFD-fed) or MetS (HFD-fed) mice with or without LPS-induced periodontitis (PD). Eight-week-old male C57BL/6 mice were fed LFD or HFD for 20 weeks and periodontally injected with PBS or LPS in the last 4 weeks. Some of the mice were treated with GW9508 by intraperitoneal injection during the last 4 weeks. At the end of the experiment, the metabolic parameters, including bodyweight (**a**), fasting glucose (**b**), cholesterol (**c**), triglycerides (**d**), free fatty acids (**e**), insulin (**f**), and HOMA-IR (**g**) were determined. The data are presented as mean ± SD (n = 6). Group symbols: * Control, ** MetS, ++ PD+MetS, # PD+GW9508, ## PD+MetS+GW9508.

2.2. GW9508 Reduces Alveolar Bone Loss in Mice with MetS-Associated Periodontitis

At the end of the animal experiment, the mouse maxillae were dissected and scanned by micro-computed tomography (μCT) for bone volume fraction (BVF) quantification. As shown in Figure 2a and Table 1, periodontitis or MetS alone reduced BVF significantly, and the combination of periodontitis and MetS further reduced BVF. In contrast, GW9508 significantly reversed the loss of BVF in mice with both periodontitis and MetS. We also measured the distance between the cemento–enamel junction (CEJ) and alveolar bone crest (ABC) since CEJ–ABC distance is positively associated with alveolar bone loss [22]. Results showed that periodontitis or MetS significantly increased CEJ–ABC distance, and the combination of periodontitis and MetS further increased CEJ–ABC distance. Consistent with the findings from the mCT study, GW9508 significantly reduced CEJ–ABC distance in mice with both periodontitis and MetS (Figure 2b–d).

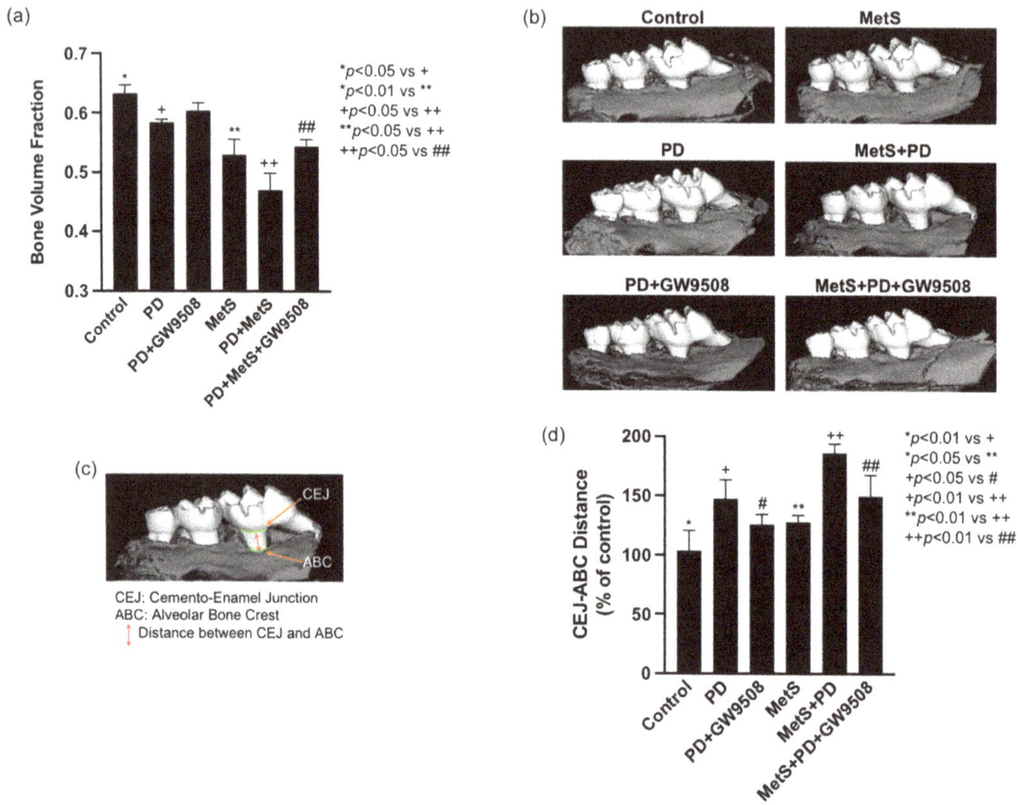

Figure 2. The effect of GW9508 on alveolar bone loss in mice with periodontitis (PD) alone or with MetS and PD. Male C57BL/6 mice were fed LFD or HFD and periodontally injected with LPS. After the treatment, the maxillae were scanned by mCT, and bone volume fraction was quantified (**a**,**b**). Additionally, the distance from CEJ to ABC was measured for the first and second molars (**b**–**d**). The data are presented as means ± SD (n = 6). Group symbols: * Control, ** MetS, + PD, ++ PD+MetS, # PD+GW9508, ## PD+MetS+GW9508.

Table 1. The effect of GW9508 on bone volume fraction (BVF) in mice with periodontitis (PD) and/or metabolic syndrome (MetS).

	BVF (Mean SD)	% BVF Reduction Compared to Control Mice	% BVF Increase by GW9508 Treatment
Control	0.632 ± 0.014	-	
PD	0.583 ± 0.005	7.75%	
PD+GW9508	0.603 ± 0.013	4.59%	3.16%
MetS	0.530 ± 0.026	16.14%	
PD+MetS	0.470 ± 0.029	25.63%	
PD+MetS+GW9508	0.544 ± 0.012	13.92%	11.71%

2.3. GW9508 Inhibits Osteoclastogenesis in Mice with MetS-Associated Periodontitis

To understand how GW9508 reduced alveolar bone loss in mice with MetS-associated periodontitis, we performed Tartrate-resistant acid phosphatase (TRAP) staining on tissue sections of the maxillae to detect osteoclasts. We focused on the area of periodontal ligaments which connects the tooth root and alveolar bone. Results showed that periodontitis or MetS increased osteoclast formation on the surface of alveolar bones, and the combination of periodontitis and MetS further increased osteoclast formation (Figure 3a,b). However, GW9508 treatment significantly reduced the number of osteoclasts in mice with periodontitis and MetS (Figure 3a,b).

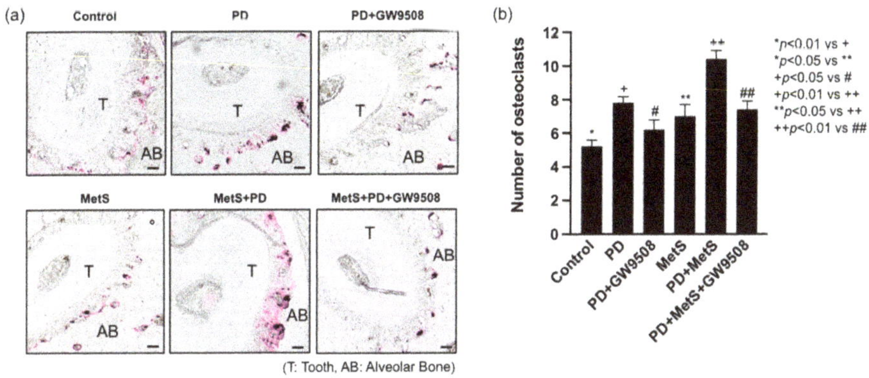

Figure 3. The effect of GW9508 on osteoclastogenesis in mice with periodontitis (PD) alone or with MetS and PD. After the maxillae were examined by mCT, as described in Figure 2, they were decalcified and sectioned. Tartrate-resistant acid phosphatase (TRAP) staining to detect osteoclasts was performed on the tissue sections. Representative tissue images from 6 groups of mice were shown (scale bar = 100 μm) (**a**), and multinucleated TRAP-positive osteoclasts per field of view were quantified and compared among the groups (**b**). The data are presented as mean ± SD (n = 6). Group symbols: * Control, + PD, # PD+GW9508, ** MetS, ++ PD+MetS, ## PD+MetS+GW9508.

2.4. GW9508 Inhibits Osteoclastogenesis Enhanced by LPS Plus Palmitic Acid (PA) In Vitro

In this study, we used a receptor activator of nuclear factor kappa-B ligand (RANKL) to induce osteoclast formation from RAW264.7 macrophages and determined the effect of GW9508 on osteoclast formation. Since we have found previously that LPS and palmitic acid (PA), a saturated fatty acid, cooperatively enhanced RANKL-induced osteoclast formation in vitro [22], we also determine the effect of GW9508 on osteoclast formation enhanced by LPS, PA, or LPS plus PA. Consistent with our previous observations [22], results showed that while either LPS or PA increased osteoclast number and area, the combination of LPS and PA further augmented osteoclast number and area (Figure 4a–c). Results also showed that GW9508 treatment significantly inhibited the osteoclast number and area enhanced by LPS plus PA in vitro (Figure 4a–c), which is consistent with the findings from our animal study shown in Figure 3.

2.5. GW9508 Reduces Periodontal Inflammation in Mice with MetS-Associated Periodontitis

Histological analysis of maxillae with a focus on the area of periodontal ligaments between the tooth root and alveolar bone was performed, and inflammatory scoring was made according to the criteria described in Materials and Methods to assess the effect of GW9508 on periodontal inflammation and bone resorption in mice. Results showed that while periodontitis increased the inflammatory scores as evidenced by increased leukocyte infiltration and bone resorption, the combination of MetS and periodontitis further increased the inflammatory score in mice (Figure 5a,b). In contrast, GW9508 treatment reduced the inflammatory scores in mice with MetS-associated periodontitis as determined by decreased leukocyte infiltration and bone resorption (Figure 5a,b).

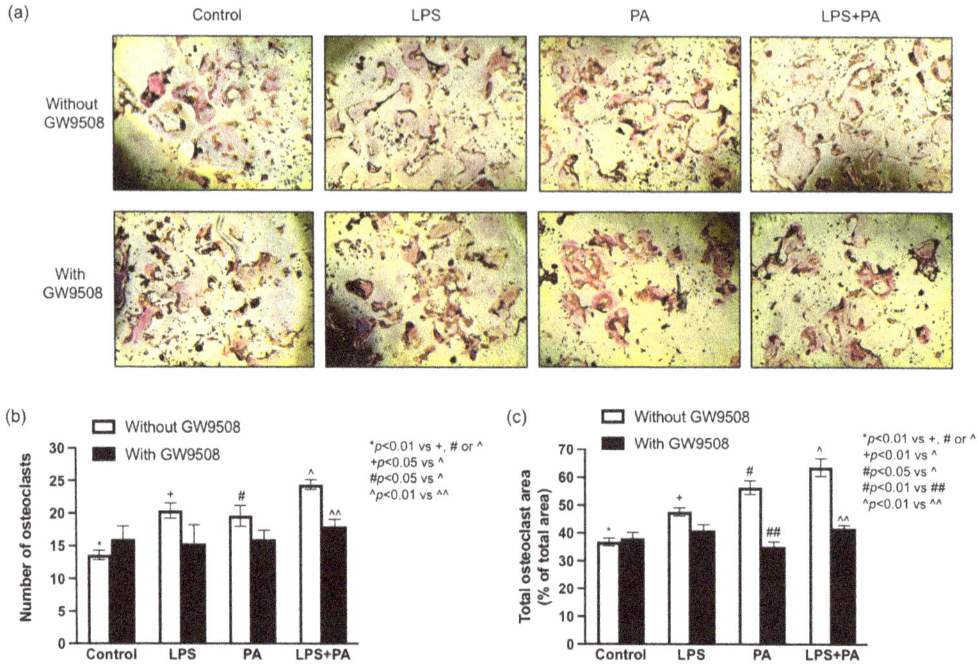

Figure 4. The effect of GW9508 on osteoclast formation stimulated by LPS, PA, or both LPS and PA. RAW264.7 macrophages were treated with 100 ng/mL of RANKL alone (control) or with RANKL plus 1 ng/mL of LPS, 100 mM of PA, or both LPS and PA for 3 days. After changing the medium, the macrophages were treated with RANKL alone or RANKL plus LPS, PA, or LPS and PA in the absence or presence of 1 mM of GW9508 for 2 days, and cells were then fixed. TRAP staining was carried out, and the osteoclasts, which were positively stained and contained more than three nuclei, were identified (**a**), counted, and compared among different groups (**b**). The total osteoclast area was also quantified as % of the total area and compared among different groups (**c**). The data are presented as mean ± SD (n = 3). Group symbols: * Control cells without GW9508 treatment, + LPS treatment without GW9508, # PA treatment without GW9508, ## PA treatment with GW9508, ^ LPS+PA without GW9508, ^^ LPS+PA with GW9508.

2.6. *GW9508 Inhibits Proinflammatory Cytokine Secretion from Macrophages In Vitro*

To explore the mechanisms by which GW9508 reduced periodontal inflammation in mice, we tested our hypothesis that GW9508 inhibits the secretion of proinflammatory cytokines from macrophages, which are immune cells known to play a pivotal role in periodontal inflammation [41]. Since LPS is released by Gram-negative bacteria that play a critical role in periodontitis, while PA is the most common saturated fatty acid in humans [42] and associated with MetS [43], we determined the effect of GW9508 on inflammatory cytokine secretion from macrophages stimulated by LPS and PA. We have shown previously that LPS and PA have a synergy on proinflammatory cytokine upregulation in macrophages [44]. Results showed that while LPS potently induced secretion of IL-6, tumor necrosis factor (TNF)a, and monocyte chemoattractant protein (MCP)-1, PA augmented the LPS-induced secretions (Figure 6). However, GW9508 significantly inhibited the cytokine secretion stimulated by LPS and PA (Figure 6).

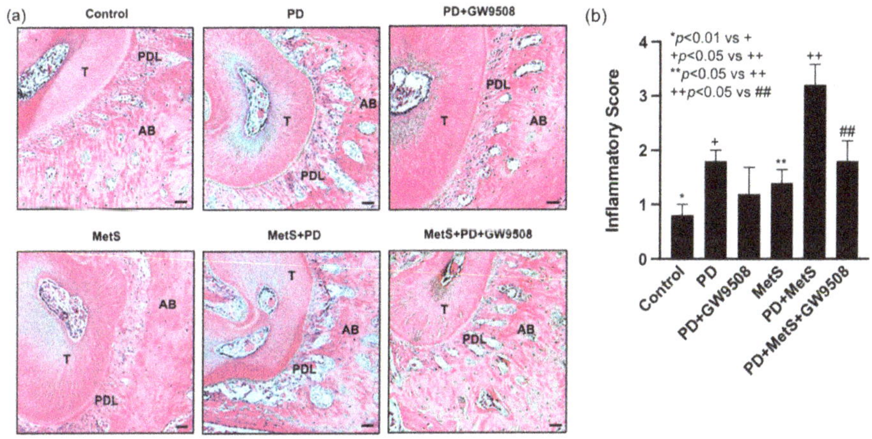

Figure 5. The effect of GW9508 on periodontal inflammation and bone resorption in mice with periodontitis (PD) or with MetS and PD. The periodontal tissue sections were stained with hematoxylin and eosin, and histologic analysis was performed. Representative photomicrographs focused on the area of periodontal ligament (PDL) along with tooth root (T) and alveolar bone (AB) were shown (scale bar = 100 μm) (**a**). The inflammatory scoring for leukocyte infiltration and bone resorption in all groups were made (**b**) according to the criteria as described in Methods. The data are presented as means ± SD ($n = 6$). Group symbols: * Control, + PD, ** MetS, ++ PD+MetS, ## PD+MetS+GW9508.

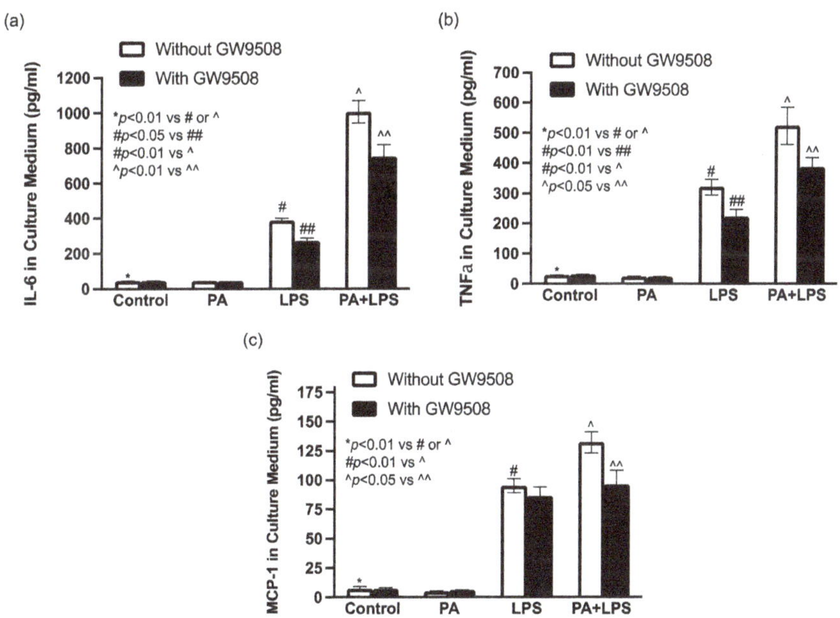

Figure 6. The effect of GW9508 on proinflammatory cytokine secretion from macrophages. RAW264.7 macrophages were treated with 1 ng/mL of LPS, 100 μM of palmitic acid (PA), or both LPS and PA in the absence or presence of 1 μM of GW9508 for 24 h. After the treatment, IL-6 (**a**), TNFa (**b**), and monocyte chemoattractant protein (MCP)-1 (**c**) in a culture medium were quantified with ELISA. The data presented as means ± SD are from one of three experiments with similar results. Group symbols: * Control cells without GW9508, # PA treatment without GW9508, ## PA treatment with GW9508, ^ LPS+PA without GW9508, ^^ LPS+PA with GW9508.

3. Discussion

GPR40 and GPR120 are G protein-coupled receptors for medium- and long-chain FAs [45,46]. It has been well documented that GPR40 is highly expressed in pancreatic β cells, and its activation enhances glucose-induced insulin secretion [47]. It has been further shown that GPR40 is involved in metabolic homeostasis [48]. In addition to pancreatic β cells, GPR40 is also expressed by many other types of cells, such as macrophages [49], osteoclasts [50], vascular endothelial cells [51,52], and hepatocytes [53], and has anti-inflammatory properties [54–58]. Park et al. showed that treatment of bone marrow-derived macrophages with GPR40 agonist TAK875 or AMG1638 suppressed the activation of the NLRP3 inflammasome by IL-1b and IL-18 via NFkB and sarco/endoplasmic reticulum Ca^{2+}-ATPase [59]. Our lab also showed recently that loss of GPR40 in mice worsened MetS-associated periodontal inflammation and that GPR40 agonists downregulated proinflammatory molecules in macrophages [23].

Similar to GPR40, GPR120 is also capable of promoting glucose homeostasis through incretin-dependent insulin secretion [60]. Furthermore, studies have shown that GPR120 is expressed by several types of immune cells, such as macrophages and $CD4^+$ T cells [61,62]. GPR120 activation suppresses proinflammatory cytokine expression and enhances anti-inflammatory cytokine expression [63]. All the above findings suggest that GPR40 or GPR120 agonists may have favorable effects on diabetes or MetS and its inflammation-related complications.

The findings that DHA and EPA, as natural dual agonists of GPR40 and GPR120, exert anti-inflammatory effects are consistent with the beneficial role of GPR40 and GPR120 in inflammatory diseases [31,64–66]. DHA and EPA have similar potencies (pEC_{50}) to GPR40 and GPR120 [32]. In addition to anti-inflammatory properties, DHA and EPA also have favorable effects on lipid metabolism and blood pressure [39]. To increase the specificity and potency, synthetic agonists for GPR40 or GPR120 and dual agonists for both GPR40 and GPR120 have been developed [37,67]. Interestingly, studies have suggested that the simultaneous activation of both GPR40 and GPR120 with a dual agonist may be more therapeutically useful than the activation of only one of them with a specific agonist. Satapati et al. reported that a dual agonist of GPR40 and GPR120 achieved superior glycemic control compared with a GPR40 selective agonist by improving insulin resistance in ob/ob mice [38], and it has been well documented that periodontitis and insulin resistance have a bidirectional relationship [68]. Furthermore, Huang et al. demonstrated that GW9508, a dual agonist of GPR40 and GPR120, has a dual effect of inhibiting osteoclastogenesis and promoting osteogenesis [69], which is closely related to periodontal health.

GW9508 is a well-characterized dual agonist with pEC_{50} of 7.32 and 5.46 for GPR40 and GPR120, respectively [70]. Studies have well documented that GW9508 is anti-inflammatory [55] and thereby exerts beneficial effects on many inflammation-related diseases, including nonalcoholic steatohepatitis [71], Alzheimer's disease [72], bacterial infection [73], and atherosclerosis [32,39,40]. In the current study, we reported for the first time that GW9508 also improved periodontitis in animal models with MetS by alleviating alveolar bone loss and periodontal inflammation.

Given that MetS exacerbates periodontitis, we first investigated the effect of GW9508 on HFD-induced MetS in this study. Results showed that GW9508 reduced metabolic parameters, including fasting glucose, cholesterol, and FFAs, clearly indicating that GW9508 improves MetS. These findings are consistent with the previous reports that GW9508 alleviates hyperglycemia [74], hyperlipidemia [71], and insulin resistance [75] in various animal models. Since it is known that MetS-related cardiovascular risk factors, including hyperglycemia, hyperlipidemia, and insulin resistance, are associated with periodontitis [21], the above results have revealed an important mechanism by which GW9508 mitigates MetS-associated periodontitis by modulating metabolic parameters.

In addition to MetS, we also investigated the effect of GW9508 on osteoclastogenesis. Results showed that GW9508 treatment strongly inhibited osteoclast formation in mice with periodontitis and MetS, indicating that GW9508 reduces alveolar bone loss by inhibiting osteoclastogenesis. To further elucidate the molecular mechanism whereby GW9508 reduced alveolar bone loss in mice with both periodontitis and MetS, we used an in vitro cell model to demonstrate that GW9508 inhibited osteoclastogenesis induced by RANKL and enhanced by LPS and PA. It is well known that LPS is increased by periodontitis [76], and PA is associated with MetS [77]. Furthermore, we demonstrated previously that LPS and PA had a cooperative effect on osteoclastogenesis [22]. Therefore, the findings from the above in vitro study indicate that the inhibition of the LPS-PA cooperation on osteoclastogenesis by GW9508 is likely a molecular mechanism involved in GW9508-reduced alveolar bone loss in mice with MetS-associated periodontitis.

In this study, we further assessed the effect of GW9508 on periodontal inflammation since it is well-known that periodontal inflammation is critically involved in osteoclastogenesis [76,78]. To test our hypothesis that GW9508 attenuated the periodontal inflammation in mice with both periodontitis and MetS, we performed histological analysis on the area of the periodontal ligament and ligament-connected tooth root and alveolar bone. Results showed that GW9508 treatment reduced leukocyte infiltration in the area in mice with both periodontitis and MetS. We also performed an in vitro study to demonstrate that GW9508 inhibited the secretion of proinflammatory cytokines such as IL-6, TNFa, and MCP-1 from macrophages stimulated by LPS and PA. Since it is well-known that proinflammatory cytokines are potent stimulators of osteoclast precursor proliferation, osteoclast formation, and bone resorption in vivo [79,80], these findings suggest that the inhibition of proinflammatory cytokine secretion by GW9508 plays a pivotal role in GW9508-reduced osteoclastogenesis and subsequent alveolar bone loss.

In the animal models for human periodontitis, the oral bacterial gavage model, periodontal bacteria or LPS inoculation model, and ligature model are commonly used mouse models [81]. All these mouse models are good models for human periodontitis [81] and were used in our research [23,82]. In the current study, we utilized the periodontal LPS inoculation model for the following reasons. We have reported previously that MetS exacerbates periodontal inflammation and bone loss in a mouse model with periodontitis induced by periodontal LPS inoculation [21]. We have also reported that LPS and PA have a synergistic stimulation on the expression of proinflammatory cytokines in macrophages [44]. In the current study, based on the above findings, we hypothesized that GW9508 attenuates the progression of periodontitis in animal models and the inhibition of the synergistic effect of LPS and PA on proinflammatory cytokine expression by GW9508 is a mechanism involved in the attenuation of periodontitis progression by GW9508. To be consistent with our hypothesis, we used the periodontal LPS inoculation model in the current study.

A limitation of this study is that only male mice were used since it is known that female C57BL/6 mice are protected against HFD-induced MetS [83]. However, clinical studies have shown that the incidence of MetS in women is similar to that in men [84]. Although it is known that periodontitis has a documented higher prevalence in men (56.4%) compared to women (38.4%) [85], the incidence of 38.4% in women is still very high. Therefore, it is important to determine the effect of GW9508 on MetS-related periodontitis in both male and female mice to understand how targeting GPR40/GPR120 affects MetS-associated periodontitis in animal models with different genders. The findings will help the development of relevant gender-based treatment options for patients with MetS-associated periodontitis. To address the effect of GW9508 on MetS-associated periodontitis in female animals, LDL receptor-deficient (LDLR −/−) mice may be used as an animal model with MetS [86,87] Wall et al. reported that female LDLR −/− mice fed an HFD developed a MetS phenotype, which is similar to that in male mice [86] These female mice had increased bodyweight, blood glucose, glucose intolerance, and lipids, including triglycerides and cholesterol.

Similar to GPR40 and GPR120, CD36, also called fatty acid translocase (FAT), is also a receptor for long-chain free fatty acids [88]. However, CD36 is different from GPR40 and GPR120 as it is a multifunctional membrane receptor involved in macrophage foam cell formation, oxidative stress, tissue degradation, and apoptosis [88]. We have shown previously that CD36 expression in periodontal tissue is upregulated by MetS in mouse models of both periodontitis and MetS and upregulated in vitro by LPS and saturated fatty acids such as PA in macrophages [89]. Since CD36 is markedly upregulated by PA, PA is likely to bind to CD36 to stimulate the expression of proinflammatory molecules such as IL-6, MCP-1, and TNFa in macrophages [90,91]. We have also shown that PA and LPS have a synergy in activating pro-inflammatory signaling and upregulating pro-inflammatory gene expression [44].

As illustrated in Figure 7, the current study showed, for the first time, that the synthetic dual GPR40/GPR120 agonist GW9508 alleviated alveolar bone loss and periodontal inflammation in mice with MetS-associated periodontitis. The investigations on the underlying mechanisms demonstrated that GW9508 inhibited osteoclastogenesis and inflammatory cytokine secretion stimulated by LPS and PA from macrophages. All these results strongly indicate that targeting GPR40/GPR120 with agonists is a potential anti-inflammatory approach in the treatment of MetS-associated periodontitis.

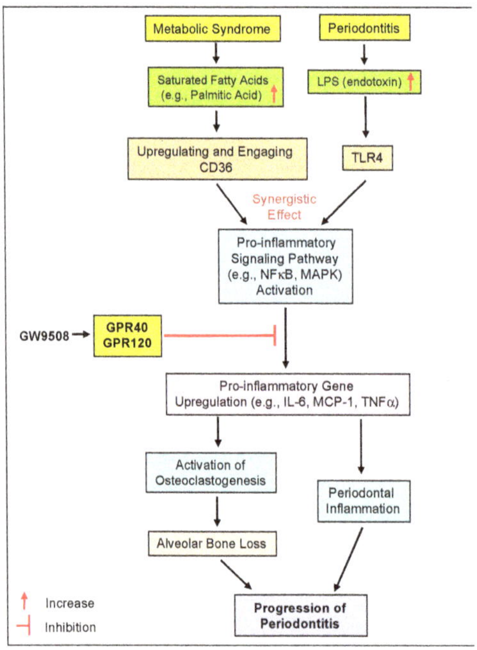

Figure 7. The potential mechanisms involved in the alleviation of alveolar bone loss and periodont inflammation by GPR40/GPR120 agonist GW9508 in mice with both MetS and periodontitis.

4. Materials and Methods

4.1. Animals

The Institutional Animal Care and Use Committee at the Medical University of South Carolina approved all experimental protocols. The study was performed in accordance with the Guidelines of ARRIVE. All mice were maintained on a 12-h light-dark cycle in a pathogen-free environment and had ad libitum access to water and food. MetS and periodontitis were induced by HFD feeding and periodontal LPS injection, respectively, and part of the mice were treated with GW9508 by intraperitoneal injection. Eight-week-old male C57BL/6 mice were randomly divided into 6 groups (n = 6): 1. Control; 2. Periodontitis; 3. Periodontitis with GW9508 treatment; 4. MetS; 5. MetS and periodontitis; 6. MetS and periodontitis with GW9508 treatment.

4.2. Induction of MetS

Mice were fed either LFD (D12450B) or HFD (D12492) (Research Diet Inc., New Brunswick, NJ, Canada) for 20 weeks. D12492 is a lard-based HFD containing 60% kcal fat, 20% kcal protein, and 20% kcal carbohydrate. For fat, it contains 254.5 g/kg fatty acid, 81.5 g/kg saturated fatty acid (SFA) and 51 g/kg PA. D12450B is a control LFD for D12492 containing 10% kcal fat, 20% kcal protein, and 70% kcal carbohydrate. For fat, it contains 43.7 g/kg fatty acid, 9.9 g/kg SFA and 1.1 g/kg PA.

4.3. Induction of Periodontitis and Treatment with GW9508

During the last 4 weeks of LFD or HFD feeding, some of the mice were injected with LPS isolated from *A.* actinomycetemcomitans (strain Y4, serotype B) [92] (0.2 mg/kg) through both left and right sides of the palatal gingiva between the maxillary 1st and 2nd molars, twice a week, for 4 weeks [93,94]. *A.* actinomycetemcomitans is one of the oral pathogens involved in chronic periodontitis [95]. For control, mice were periodontally injected with phosphate-buffered saline (PBS), the vehicle for LPS. Some of the mice were treated with GW9508 at 50 mg/kg/day by intraperitoneal injection (IP) during the last 4 weeks of LFD or HFD feeding when periodontal LPS injection was performed. The above dose of GW9508 was selected since it had been shown that treatment with GW9508 at the dose of 50 mg/kg/day through IP in C57BL/6 mice inhibited diabetes-induced neuropathy [96] and improved glucose homeostasis and insulin sensitivity [75]. The mice with IP of PBS were used as controls for GW9508 treatment.

4.4. Metabolic Measurements

Blood samples were obtained under the fasted condition, and glucose level was determined using a Precision QID glucometer (MediSense Inc., Bedford, MA, USA). Serum cholesterol was measured using the Cholestech LDX Lipid Monitoring System (Fisher Scientific, Pittsburgh, PA, USA). Serum triglycerides were measured using EnzyChrom™ Triglyceride Assay Kit (BioAssay Systems, Hayward, CA, USA). Serum FFAs were determined using the EnzyChrom™ free fatty acid kit (BioAssay systems (Hayward, CA, USA)). Serum fasting insulin was assayed using the Ultra Sensitive Insulin ELISA Kit (Crystal Chem, Inc., Downers Grove, IL, USA). Fasting whole-body insulin sensitivity was estimated with the HOMA-IR according to the formula [fasting plasma glucose (mg/dL) × fasting plasma insulin (mU/mL)]/405.

4.5. Micro-Computed Tomography (μCT), Analysis of Bone Volume Fraction (BVF), and Measurement of Cementoenamel Junction (CEJ)-Alveolar Bone Crest (ABC) Distance

Maxillae were dissected and fixed in 10% phosphate-buffered formalin for 24 h, washed with PBS, and stored in 70% ethanol. Maxillae were scanned at 55 kVp, 145 μA, 16 μm voxel resolution using Scanco Medical 40 mCT scanner (Scanco Medical, Brüttisellen, Switzerland) as described previously [97]. Three-dimensional images were generated and reconstructed for each specimen. These images were rotated with a standard orientation and threshold to discern mineralized and non-mineralized tissue. The region of interest (ROI) was indicated by the contour height of molars at the CEJ as the width and the molar cusp tips to root apices as the height. Depth was equal to the buccolingual size of the teeth plus 1.0 mm^3. BVF was calculated as the percentage of bone within the ROI using AnalyzePro software (AnalyzeDirect, Inc., Overland Park, KS, USA), https://analyzedirect.com/analyzepro-support/ (accessed on 27 August 2024). A well-trained and experienced research specialist who was blinded to the treatment groups performed the mCT analysis. Data are reported in accordance with standardized nomenclature [98]. Additionally, three-line measurements of the distance from CEJ to ABC were taken for the first and second molars, as previously described [99].

4.6. Tartrate-Resistant Acid Phosphatase (TRAP) Staining and Quantification of Osteoclasts

Formalin-fixed maxillae were decalcified in a 10% EDTA solution for 4 weeks at 4 °C. The EDTA solution was changed three times per week. The maxillae were paraffin-embedded, and 7-µm sagittal sections were prepared. TRAP staining was performed in tissue sections using a leukocyte acid phosphatase kit (Sigma Aldrich, St. Louis, MO, USA). The tissue sections were counterstained with hematoxylin after TRAP staining. Pictures were taken with a Nikon Eclipse TS-100 inverted microscope. Active osteoclasts were defined as multinucleated TRAP-positive cells in contact with the bone surface. TRAP-positive osteoclasts under the first and the second molar on the surface of the alveolar bone were counted.

4.7. Histological Study of Periodontal Inflammation

For pathological evaluation, the above tissue sections were stained with hematoxylin and eosin. The tissue inflammation and bone resorption were evaluated according to the criteria for scoring tissue inflammation and bone resorption: 0 = within normal limits; 1 = focal some leukocyte infiltration, no significant bone resorption; 2 = moderate leukocyte infiltration with mild bone resorption; 3 = severe leukocyte infiltration with moderate bone resorption; and 4 = severe leukocyte infiltration with extensive bone resorption [100].

4.8. Cell Culture

RAW 264.7 macrophages were purchased from the American Type Culture Collection (Manassas, VA, USA) and grown in DMEM (American Type Culture Collection, Manassas, VA, USA) supplemented with 10% heat-inactivated fetal calf serum (HyClone, Logan, UT, USA). The cells were maintained in a 37 °C, 90% relative humidity, 5% CO_2 environment.

4.9. PA Preparation

To prepare PA solution for cell treatments, PA (Sigma, St. Louis, MO, USA) was dissolved in 0.1 N NaOH and 70% ethanol at 70 °C to make PA with a stock concentration of 50 mM. The solution was kept at 55 °C for 10 min, mixed, and brought to room temperature.

4.10. Enzyme-Linked Immunosorbent Assay (ELISA)

Cytokines in the medium were quantified using a sandwich ELISA kit according to the protocol provided by the manufacturer (BioLegend, San Diego, CA, USA).

4.11. In Vitro Study of Osteoclast Formation

RAW264.7 macrophages were treated with 100 ng/mL of receptor activator of RANKL alone (control) or with RANKL plus 1 ng/mL of LPS, 100 mM of PA, or both LPS and PA for 3 days. The medium was then changed, and the macrophages were treated with RANKL alone or RANKL plus LPS, PA, or LPS plus PA in the absence or presence of 1 mM of GW9508 for 2 days. The RAW264.7 macrophages were treated with GW9508 during the last two days since it has been reported previously that significant changes in osteoclast differentiation induced by RANKL started on day 3 after RAW264.7 macrophages were exposed to RANKL [101]. After the treatment, cells were fixed with 4% paraformaldehyde in PBS for 10 min at room temperature. TRAP staining was carried out in accordance with the manufacturer's instructions (Sigma Aldrich, St. Louis, MO, USA). Cells were observed under light microscopy, and those containing more than three nuclei were considered osteoclasts. The means of osteoclast number per well were counted and compared among different groups. The total area of multinuclear osteoclasts was measured with ImageJ software, version 2.0.0.-rc-43/1.52n (NIH, Bethesda, MD, USA) and presented as a % of the total area per field.

4.12. Statistical Analysis

GraphPad Prism 8 (v. 8.4.3) (GraphPad Software, Inc., La Jolla, CA, USA) was used for statistical analysis. The one-way analysis of variance (ANOVA) with the post hoc test was used to determine whether there were any statistically significant differences between the means of three or more independent groups. A Student's *t*-test was performed for comparison between the two groups when data had a normal distribution. For data without normal distribution, nonparametric analysis using the Mann–Whitney test. For the in vitro study of osteoclast formation, the osteoclast numbers in three random fields per well viewed under the light microscope were counted, and the mean of the osteoclast numbers per well was calculated. The means of three wells were then compared between different groups. When an experiment was repeated, statistical analysis was also performed to compare their variances. The values were expressed as mean ± SD, and a value of $p < 0.05$ was considered significant.

Author Contributions: Conceptualization, Y.H.; methodology, Y.L. and H.Y.; supervision, Y.H. and M.F.L.-V.; analysis, Y.L., H.Y. and Y.H.; fund acquisition, Y.H.; writing—original draft preparation, Y.H.; writing—review and editing, Y.H., M.F.L.-V. and H.Y. All authors have read and agreed to the published version of the manuscript.

Funding: This work was supported by grant DE027070 funded by National Institute of Dental and Craniofacial Research (NIDCR), National Institutes of Health (NIH) of United States.

Institutional Review Board Statement: The animal study was conducted in accordance with the Declaration of Helsinki and approved by the Institutional Animal Care and Use Committee at the Medical University of South Carolina (certification no. 01154; date of approval: 13 April 2022).

Informed Consent Statement: Not applicable.

Data Availability Statement: All data generated or analyzed during the current study are available from the corresponding author upon reasonable request.

Acknowledgments: This study utilized the facilities and resources of the Center for Oral Health Research in the James B. Edwards College of Dental Medicine at the Medical University of South Carolina.

Conflicts of Interest: The authors declare no conflicts of interest.

References

Beck, J.; Garcia, R.; Heiss, G.; Vokonas, P.S.; Offenbacher, S. Periodontal disease and cardiovascular disease. *J. Periodontol.* **1996**, *67* (Suppl. S10), 1123–1137. [CrossRef]

Offenbacher, S. Periodontal diseases: Pathogenesis. *Ann. Periodontol.* **1996**, *1*, 821–878. [CrossRef] [PubMed]

Zheng, M.; Wang, C.; Ali, A.; Shih, Y.A.; Xie, Q.; Guo, C. Prevalence of periodontitis in people clinically diagnosed with diabetes mellitus: A meta-analysis of epidemiologic studies. *Acta Diabetol.* **2021**, *58*, 1307–1327. [CrossRef] [PubMed]

Katz, P.P.; Wirthlin, M.R., Jr.; Szpunar, S.M.; Selby, J.V.; Sepe, S.J.; Showstack, J.A. Epidemiology and prevention of periodontal disease in individuals with diabetes. *Diabetes Care* **1991**, *14*, 375–385. [CrossRef]

Yalda, B.; Offenbacher, S.; Collins, J.G. Diabetes as a modifier of periodontal disease expression. *Periodontology 2000* **1994**, *6*, 37–49. [CrossRef] [PubMed]

Nibali, L.; Tatarakis, N.; Needleman, I.; Tu, Y.K.; D'Aiuto, F.; Rizzo, M.; Donos, N. Clinical review: Association between metabolic syndrome and periodontitis: A systematic review and meta-analysis. *J. Clin. Endocrinol. Metab.* **2013**, *98*, 913–920. [CrossRef] [PubMed]

Gomes-Filho, I.S.; das Merces, M.C.; de Santana Passos-Soares, J.; da Cruz, S.S.; Ladeia, A.M.T.; Trindade, S.C.; de Moraes Marcilio Cerqueira, E.; Coelho, J.M.F.; Monteiro, F.M.M.; Barreto, M.L.; et al. Severity of Periodontitis and Metabolic Syndrome: Is There an Association? *J. Periodontol.* **2016**, *87*, 357–366. [CrossRef]

Gobin, R.; Tian, D.; Liu, Q.; Wang, J. Periodontal Diseases and the Risk of Metabolic Syndrome: An Updated Systematic Review and Meta-Analysis. *Front. Endocrinol.* **2020**, *11*, 336. [CrossRef]

Campos, J.R.; Martins, C.C.; Faria, S.F.S.; Carvalho, A.P.; Pereira, A.G.; Costa, F.O.; Cota, L.O.M. Association between components of metabolic syndrome and periodontitis: A systematic review and meta-analysis. *Clin. Oral Investig.* **2022**, *26*, 5557–5574. [CrossRef]

0. Grundy, S.M.; Brewer, H.B., Jr.; Cleeman, J.I.; Smith, S.C., Jr.; Lenfant, C.; National Heart, Lung, and Blood Institute (NHLBI). Definition of metabolic syndrome: Report of the National Heart, Lung, and Blood Institute/American Heart Association conference on scientific issues related to definition. *Circulation* **2004**, *109*, 433–438. [CrossRef] [PubMed]

11. Alberti, K.G.; Eckel, R.H.; Grundy, S.M.; Zimmet, P.Z.; Cleeman, J.I.; Donato, K.A.; Fruchart, J.C.; James, W.P.; Loria, C.M.; Smitl S.C., Jr.; et al. Harmonizing the metabolic syndrome: A joint interim statement of the International Diabetes Federation Tas Force on Epidemiology and Prevention; National Heart, Lung, and Blood Institute; American Heart Association; World Hear Federation; International Atherosclerosis Society; and International Association for the Study of Obesity. *Circulation* **2009**, *12* 1640–1645. [CrossRef] [PubMed]
12. Rosário-Dos-Santos, H.L.; Miranda, S.S.; Gomes-Filho, I.S.; Cruz, S.S.D.; Figueiredo, A.; Souza, E.S.; Hintz, A.M.; Loomer, P.M Passos-Soares, J.S. Periodontitis severity relationship with metabolic syndrome: A systematic review with meta-analysis. *Ora Dis.* **2023**, *29*, 2512–2520. [CrossRef]
13. Day, C. Metabolic syndrome, or What you will: Definitions and epidemiology. *Diabetes Vasc. Dis. Res.* **2007**, *4*, 32–38. [CrossRef [PubMed]
14. Srivastava, M.C.; Srivastava, R.; Verma, P.K.; Gautam, A. Metabolic syndrome and periodontal disease: An overview fo physicians. *J. Family Med. Prim. Care* **2019**, *8*, 3492–3495. [CrossRef]
15. Benetti, E.; Patel, N.S.; Collino, M. The role of PPARβ/δ in the management of metabolic syndrome and its associated cardiova cular complications. *Endocr. Metab. Immune Disord.-Drug Targets* **2011**, *11*, 273–284. [CrossRef]
16. Bogdanova, K.; Poczatkova, H.; Uherkova, L.; Riegrova, D.; Rypka, M.; Feher, J.; Marchesini, G.; Vesely, J. Non-alcoholic fatty live disease (NAFLD)—A novel common aspect of the metabolic syndrome. *Biomed. Pap.-Palacky Univ. Olomouc* **2006**, *150*, 101–10 [CrossRef] [PubMed]
17. Rojas-Gutierrez, E.; Muñoz-Arenas, G.; Treviño, S.; Espinosa, B.; Chavez, R.; Rojas, K.; Flores, G.; Díaz, A.; Guevara, J. Alzheimer' disease and metabolic syndrome: A link from oxidative stress and inflammation to neurodegeneration. *Synapse* **2017**, *71*, e2199 [CrossRef]
18. Persson, G.R.; Persson, R.E. Cardiovascular disease and periodontitis: An update on the associations and risk. *J. Clin. Periodonto* **2008**, *35* (Suppl. S8), 362–379. [CrossRef]
19. Alakhali, M.S.; Al-Maweri, S.A.; Al-Shamiri, H.M.; Al-Haddad, K.; Halboub, E. The potential association between periodontiti and non-alcoholic fatty liver disease: A systematic review. *Clin. Oral Investig.* **2018**, *22*, 2965–2974. [CrossRef]
20. Dioguardi, M.; Crincoli, V.; Laino, L.; Alovisi, M.; Sovereto, D.; Mastrangelo, F.; Russo, L.L.; Muzio, L.L. The Role of Periodontiti and Periodontal Bacteria in the Onset and Progression of Alzheimer's Disease: A Systematic Review. *J. Clin. Med.* **2020**, *9*, 49 [CrossRef]
21. Li, Y.; Lu, Z.; Zhang, X.; Yu, H.; Kirkwood, K.L.; Lopes-Virella, M.F.; Huang, Y. Metabolic syndrome exacerbates inflammatio and bone loss in periodontitis. *J. Dent. Res.* **2015**, *94*, 362–370. [CrossRef]
22. Lu, Z.; Li, Y.; Yu, H.; Lopes-Virella, M.F.; Huang, Y. High-fat diet-induced metabolic syndrome increases ligature-induced alveola bone loss in mice. *Oral Dis.* **2023**, *29*, 1312–1323. [CrossRef]
23. Li, Y.; Lu, Z.; Kirkwood, C.L.; Kirkwood, K.L.; Wank, S.A.; Li, A.J.; Lopes-Virella, M.F.; Huang, Y. GPR40 deficiency worsen metabolic syndrome-associated periodontitis in mice. *J. Periodontal Res.* **2023**, *58*, 575–587. [CrossRef] [PubMed]
24. Van Ravensteijn, M.M.; Timmerman, M.F.; Brouwer, E.A.G.; Slot, D.E. The effect of omega-3 fatty acids on active periodonta therapy: A systematic review and meta-analysis. *J. Clin. Periodontol.* **2022**, *49*, 1024–1037. [CrossRef] [PubMed]
25. Dos Santos, N.C.C.; Furukawa, M.V.; Oliveira-Cardoso, I.; Cortelli, J.R.; Feres, M.; Van Dyke, T.; Rovai, E.S. Does the use c omega-3 fatty acids as an adjunct to non-surgical periodontal therapy provide additional benefits in the treatment of periodontitis A systematic review and meta-analysis. *J. Periodontal Res.* **2022**, *57*, 435–447. [CrossRef] [PubMed]
26. Stańdo-Retecka, M.; Piatek, P.; Namiecinska, M.; Bonikowski, R.; Lewkowicz, P.; Lewkowicz, N. Clinical and microbiologica outcomes of subgingival instrumentation supplemented with high-dose omega-3 polyunsaturated fatty acids in periodonta treatment—A randomized clinical trial. *BMC Oral Health* **2023**, *23*, 290. [CrossRef] [PubMed]
27. Hara, T.; Kimura, I.; Inoue, D.; Ichimura, A.; Hirasawa, A. Free fatty acid receptors and their role in regulation of energ metabolism. *Rev. Physiol. Biochem. Pharmacol.* **2013**, *164*, 77–116. [CrossRef]
28. Miyamoto, J.; Kasubuchi, M.; Nakajima, A.; Kimura, I. Anti-Inflammatory and Insulin-Sensitizing Effects of Free Fatty Aci Receptors. *Free Fat. Acid Recept.* **2017**, *236*, 221–231. [CrossRef]
29. Talukdar, S.; Olefsky, J.M.; Osborn, O. Targeting GPR120 and other fatty acid-sensing GPCRs ameliorates insulin resistance an inflammatory diseases. *Trends Pharmacol. Sci.* **2011**, *32*, 543–550. [CrossRef]
30. Miyamoto, J.; Hasegawa, S.; Kasubuchi, M.; Ichimura, A.; Nakajima, A.; Kimura, I. Nutritional Signaling via Free Fatty Aci Receptors. *Int. J. Mol. Sci.* **2016**, *17*, 450. [CrossRef]
31. Kumari, P.; Inoue, A.; Chapman, K.; Lian, P.; Rosenbaum, D.M. Molecular mechanism of fatty acid activation of FFAR1. *Pro Natl. Acad. Sci. USA* **2023**, *120*, e2219569120. [CrossRef] [PubMed]
32. Christiansen, E.; Watterson, K.R.; Stocker, C.J.; Sokol, E.; Jenkins, L.; Simon, K.; Grundmann, M.; Petersen, R.K.; Wargent, E.T Hudson, B.D.; et al. Activity of dietary fatty acids on FFA1 and FFA4 and characterisation of pinolenic acid as a dual FFA1/FFA agonist with potential effect against metabolic diseases. *Br. J. Nutr.* **2015**, *113*, 1677–1688. [CrossRef] [PubMed]
33. Miroult, C.; Lasserre, J.; Toma, S. Effects of Omega-3 as an adjuvant in the treatment of periodontal disease: A systematic revie and meta-analysis. *Clin. Exp. Dent. Res.* **2023**, *9*, 545–556. [CrossRef]
34. Ma, J.; Kitaura, H.; Ohori, F.; Noguchi, T.; Marahleh, A.; Kinjo, R.; Kanou, K.; Ren, J.; Miura, M.; Narita, K.; et al. Generating Bor Marrow Chimeric Mouse Using GPR120 Deficient Mouse for the Study of DHA Inhibitory Effect on Osteoclast Formation an Bone Resorption. *Int. J. Mol. Sci.* **2023**, *24*, 17000. [CrossRef] [PubMed]

35. Rodrigues, D.A.; Pinheiro, P.S.M.; Ferreira, T.; Thota, S.; Fraga, C.A.M. Structural basis for the agonist action at free fatty acid receptor 1 (FFA1R or GPR40). *Chem. Biol. Drug Des.* **2018**, *91*, 668–680. [CrossRef]
36. Milligan, G.; Alvarez-Curto, E.; Watterson, K.R.; Ulven, T.; Hudson, B.D. Characterizing pharmacological ligands to study the long-chain fatty acid receptors GPR40/FFA1 and GPR120/FFA4. *Br. J. Pharmacol.* **2015**, *172*, 3254–3265. [CrossRef]
37. Bianchini, G.; Nigro, C.; Sirico, A.; Novelli, R.; Prevenzano, I.; Miele, C.; Beguinot, F.; Aramini, A. A new synthetic dual agonist of GPR120/GPR40 induces GLP-1 secretion and improves glucose homeostasis in mice. *Biomed. Pharmacother.* **2021**, *139*, 111613. [CrossRef]
38. Satapati, S.; Qian, Y.; Wu, M.S.; Petrov, A.; Dai, G.; Wang, S.P.; Zhu, Y.; Shen, X.; Muise, E.S.; Chen, Y.; et al. GPR120 suppresses adipose tissue lipolysis and synergizes with GPR40 in antidiabetic efficacy. *J. Lipid Res.* **2017**, *58*, 1561–1578. [CrossRef]
39. Liu, Y.X.; Yu, J.H.; Sun, J.H.; Ma, W.Q.; Wang, J.J.; Sun, G.J. Effects of Omega-3 Fatty Acids Supplementation on Serum Lipid Profile and Blood Pressure in Patients with Metabolic Syndrome: A Systematic Review and Meta-Analysis of Randomized Controlled Trials. *Foods* **2023**, *12*, 725. [CrossRef]
40. Suski, M.; Kiepura, A.; Wiśniewska, A.; Kuś, K.; Skałkowska, A.; Stachyra, K.; Stachowicz, A.; Gajda, M.; Korbut, R.; Olszanecki, R. Anti-atherosclerotic action of GW9508—Free fatty acid receptors activator—In apoE-knockout mice. *Pharmacol. Rep.* **2019**, *71*, 551–555. [CrossRef]
41. Wang, W.; Zheng, C.; Yang, J.; Li, B. Intersection between macrophages and periodontal pathogens in periodontitis. *J. Leukoc. Biol.* **2021**, *110*, 577–583. [CrossRef] [PubMed]
42. Carta, G.; Murru, E.; Banni, S.; Manca, C. Palmitic Acid: Physiological Role, Metabolism and Nutritional Implications. *Front. Physiol.* **2017**, *8*, 902. [CrossRef]
43. Walchuk, C.; Wang, Y.; Suh, M. The impact of EPA and DHA on ceramide lipotoxicity in the metabolic syndrome. *Br. J. Nutr.* **2021**, *125*, 863–875. [CrossRef]
44. Jin, J.; Zhang, X.; Lu, Z.; Perry, D.M.; Li, Y.; Russo, S.B.; Cowart, L.A.; Hannun, Y.A.; Huang, Y. Acid sphingomyelinase plays a key role in palmitic acid-amplified inflammatory signaling triggered by lipopolysaccharide at low concentrations in macrophages. *Am. J. Physiol. Endocrinol. Metab.* **2013**, *305*, E853–E867. [CrossRef] [PubMed]
45. Trzaskowski, B.; Latek, D.; Yuan, S.; Ghoshdastider, U.; Debinski, A.; Filipek, S. Action of molecular switches in GPCRs—Theoretical and experimental studies. *Curr. Med. Chem.* **2012**, *19*, 1090–1109. [CrossRef]
46. Itoh, Y.; Kawamata, Y.; Harada, M.; Kobayashi, M.; Fujii, R.; Fukusumi, S.; Ogi, K.; Hosoya, M.; Tanaka, Y.; Uejima, H.; et al. Free fatty acids regulate insulin secretion from pancreatic beta cells through GPR40. *Nature* **2003**, *422*, 173–176. [CrossRef]
47. Paul, A.; Nahar, S.; Nahata, P.; Sarkar, A.; Maji, A.; Samanta, A.; Karmakar, S.; Maity, T.K. Synthetic GPR40/FFAR1 agonists: An exhaustive survey on the most recent chemical classes and their structure-activity relationships. *Eur. J. Med. Chem.* **2023**, *264*, 115990. [CrossRef]
48. Al Mahri, S.; Malik, S.S.; Al Ibrahim, M.; Haji, E.; Dairi, G.; Mohammad, S. Free Fatty Acid Receptors (FFARs) in Adipose: Physiological Role and Therapeutic Outlook. *Cells* **2022**, *11*, 750. [CrossRef] [PubMed]
49. Relling, I.; Akcay, G.; Fangmann, D.; Knappe, C.; Schulte, D.M.; Hartmann, K.; Muller, N.; Turk, K.; Dempfle, A.; Franke, A.; et al. Role of wnt5a in metabolic inflammation in humans. *J. Clin. Endocrinol. Metab.* **2018**, *103*, 4253–4264. [CrossRef]
50. Cornish, J.; MacGibbon, A.; Lin, J.M.; Watson, M.; Callon, K.E.; Tong, P.C.; Dunford, J.E.; van der Does, Y.; Williams, G.A.; Grey, A.B.; et al. Modulation of osteoclastogenesis by fatty acids. *Endocrinology* **2008**, *149*, 5688–5695. [CrossRef]
51. Wang, S.; Awad, K.S.; Elinoff, J.M.; Dougherty, E.J.; Ferreyra, G.A.; Wang, J.Y.; Cai, R.; Sun, J.; Ptasinska, A.; Danner, R.L. G Protein-coupled Receptor 40 (GPR40) and Peroxisome Proliferator-activated Receptor gamma (PPARgamma): An Integrated Two-Receptor Signaling Pathway. *J. Biol. Chem.* **2015**, *290*, 19544–19557. [CrossRef] [PubMed]
52. Lu, Z.; Li, Y.; Jin, J.; Zhang, X.; Hannun, Y.A.; Huang, Y. GPR40/FFA1 and neutral sphingomyelinase are involved in palmitate-boosted inflammatory response of microvascular endothelial cells to LPS. *Atherosclerosis* **2015**, *240*, 163–173. [CrossRef] [PubMed]
53. Suh, H.N.; Huong, H.T.; Song, C.H.; Lee, J.H.; Han, H.J. Linoleic acid stimulates gluconeogenesis via Ca^{2+}/PLC, cPLA2, and PPAR pathways through GPR40 in primary cultured chicken hepatocytes. *Am. J. Physiol. Cell Physiol.* **2008**, *295*, C1518–C1527. [CrossRef]
54. Li, S.; Yu, Y.; Wang, B.; Qiao, S.; Hu, M.; Wang, H.; Fu, C.; Dong, B. Overexpression of G protein-coupled receptor 40 protects obesity-induced cardiomyopathy through the SIRT1/LKB1/AMPK pathway. *Hum. Gene Ther.* **2022**, *33*, 598–613. [CrossRef] [PubMed]
55. Xiao, J.; Cai, T.; Fang, Y.; Liu, R.; Flores, J.J.; Wang, W.; Gao, L.; Liu, Y.; Lu, Q.; Tang, L.; et al. Activation of GPR40 attenuates neuroinflammation and improves neurological function via PAK4/CREB/KDM6B pathway in an experimental GMH rat model. *J. Neuroinflamm.* **2021**, *18*, 160. [CrossRef]
56. Shen, X.; Fan, B.; Hu, X.; Luo, L.; Yan, Y.; Yang, L. Metformin Reduces Lipotoxicity-Induced Meta-Inflammation in β-Cells through the Activation of GPR40-PLC-IP3 Pathway. *J. Diabetes Res.* **2019**, *2019*, 7602427. [CrossRef]
57. Dragano, N.R.V.; Solon, C.; Ramalho, A.F.; de Moura, R.F.; Razolli, D.S.; Christiansen, E.; Azevedo, C.; Ulven, T.; Velloso, L.A. Polyunsaturated fatty acid receptors, GPR40 and GPR120, are expressed in the hypothalamus and control energy homeostasis and inflammation. *J. Neuroinflamm.* **2017**, *14*, 91. [CrossRef]
58. Gras, D.; Chanez, P.; Urbach, V.; Vachier, I.; Godard, P.; Bonnans, C. Thiazolidinediones induce proliferation of human bronchial epithelial cells through the GPR40 receptor. *Am. J. Physiol. Lung Cell Mol. Physiol.* **2009**, *296*, L970–L978. [CrossRef]

59. Park, J.; Lee, M.Y.; Seo, Y.S.; Kang, B.; Lim, S.C.; Kang, K.W. GPR40 agonist inhibits NLRP3 inflammasome activation via modulation of nuclear factor-κB and sarco/endoplasmic reticulum Ca^{2+}-ATPase. *Life Sci.* **2021**, *287*, 120127. [CrossRef]
60. Ghislain, J.; Poitout, V. Targeting lipid GPCRs to treat type 2 diabetes mellitus—Progress and challenges. *Nat. Rev. Endocrinol.* **2021**, *17*, 162–175. [CrossRef]
61. Oh, D.Y.; Talukdar, S.; Bae, E.J.; Imamura, T.; Morinaga, H.; Fan, W.; Li, P.; Lu, W.J.; Watkins, S.M.; Olefsky, J.M. GPR120 is an omega-3 fatty acid receptor mediating potent anti-inflammatory and insulin-sensitizing effects. *Cell* **2010**, *142*, 687–698. [CrossRef] [PubMed]
62. Yang, W.; Liu, H.; Xu, L.; Yu, T.; Zhao, X.; Yao, S.; Zhao, Q.; Barnes, S.; Cohn, S.M.; Dann, S.M.; et al. GPR120 Inhibits Colitis Through Regulation of $CD4^+$ T Cell Interleukin 10 Production. *Gastroenterology* **2022**, *162*, 150–165. [CrossRef] [PubMed]
63. Di Petrillo, A.; Kumar, A.; Onali, S.; Favale, A.; Fantini, M.C. GPR120/FFAR4: A Potential New Therapeutic Target for Inflammatory Bowel Disease. *Inflamm. Bowel Dis.* **2023**, *29*, 1981–1989. [CrossRef]
64. Petermann, A.B.; Reyna-Jeldes, M.; Ortega, L.; Coddou, C.; Yévenes, G.E. Roles of the Unsaturated Fatty Acid Docosahexaenoic Acid in the Central Nervous System: Molecular and Cellular Insights. *Int. J. Mol. Sci.* **2022**, *23*, 5390. [CrossRef] [PubMed]
65. Karmokar, P.F.; Moniri, N.H. Oncogenic signaling of the free-fatty acid receptors FFA1 and FFA4 in human breast carcinoma cells. *Biochem. Pharmacol.* **2022**, *206*, 115328. [CrossRef]
66. Jin, J.; Lu, Z.; Li, Y.; Cowart, L.A.; Lopes-Virella, M.F.; Huang, Y. Docosahexaenoic acid antagonizes the boosting effect of palmitic acid on LPS inflammatory signaling by inhibiting gene transcription and ceramide synthesis. *PLoS ONE* **2018**, *13*, e0193343. [CrossRef]
67. Mizuta, K.; Zhang, Y.; Mizuta, F.; Hoshijima, H.; Shiga, T.; Masaki, E.; Emala, C.W., Sr. Novel identification of the free fatty acid receptor FFAR1 that promotes contraction in airway smooth muscle. *Am. J. Physiol. Lung Cell Mol. Physiol.* **2015**, *309*, L970–L982. [CrossRef] [PubMed]
68. Islam, S.K.; Seo, M.; Lee, Y.S.; Moon, S.S. Association of periodontitis with insulin resistance, β-cell function, and impaired fasting glucose before onset of diabetes. *Endocr. J.* **2015**, *62*, 981–989. [CrossRef]
69. Huang, F.; Liu, X.; Fu, X.; Chen, Y.; Jiang, D.; Wang, T.; Hu, R.; Zou, X.; Hu, H.; Liu, C. 3D-Printed Bioactive Scaffold Loaded with GW9508 Promotes Critical-Size Bone Defect Repair by Regulating Intracellular Metabolism. *Bioengineering* **2023**, *10*, 535. [CrossRef]
70. Briscoe, C.P.; Peat, A.J.; McKeown, S.C.; Corbett, D.F.; Goetz, A.S.; Littleton, T.R.; McCoy, D.C.; Kenakin, T.P.; Andrews, J.L.; Ammala, C.; et al. Pharmacological regulation of insulin secretion in MIN6 cells through the fatty acid receptor GPR40: Identification of agonist and antagonist small molecules. *Br. J. Pharmacol.* **2006**, *148*, 619–628. [CrossRef] [PubMed]
71. Li, M.; Meng, X.; Xu, J.; Huang, X.; Li, H.; Li, G.; Wang, S.; Man, Y.; Tang, W.; Li, J. GPR40 agonist ameliorates liver X receptor-induced lipid accumulation in liver by activating AMPK pathway. *Sci. Rep.* **2016**, *6*, 25237. [CrossRef] [PubMed]
72. Gong, Y.; Li, Y.; Liu, X.; He, L. GW9508 ameliorates cognitive dysfunction via the external treatment of encephalopathy in $A\beta_{1-42}$ induced mouse model of Alzheimer's disease. *Eur. J. Pharmacol.* **2021**, *909*, 174362. [CrossRef]
73. Souza, P.R.; Walker, M.E.; Goulding, N.J.; Dalli, J.; Perretti, M.; Norling, L.V. The GPR40 Agonist GW9508 Enhances Neutrophil Function to Aid Bacterial Clearance during *E. coli* Infections. *Front. Immunol.* **2020**, *11*, 573019. [CrossRef]
74. McKillop, A.M.; Miskelly, M.G.; Moran, B.M.; Flatt, P.R. Incretins play an important role in FFA4/GPR120 regulation of glucose metabolism by GW-9508. *Life Sci.* **2023**, *318*, 121475. [CrossRef] [PubMed]
75. Ou, H.Y.; Wu, H.T.; Hung, H.C.; Yang, Y.C.; Wu, J.S.; Chang, C.J. Multiple mechanisms of GW-9508, a selective G protein-coupled receptor 40 agonist, in the regulation of glucose homeostasis and insulin sensitivity. *Am. J. Physiol. Endocrinol. Metab.* **2013**, *304*, E668–E676. [CrossRef]
76. AlQranei, M.S.; Chellaiah, M.A. Osteoclastogenesis in periodontal diseases: Possible mediators and mechanisms. *J. Oral Biosci* **2020**, *62*, 123–130. [CrossRef] [PubMed]
77. Kien, C.L. Dietary interventions for metabolic syndrome: Role of modifying dietary fats. *Curr. Diab Rep.* **2009**, *9*, 43–50. [CrossRef]
78. Baloul, S.S. Osteoclastogenesis and Osteogenesis during Tooth Movement. *Front. Oral Biol.* **2016**, *18*, 75–79. [CrossRef]
79. Zhao, B. Does TNF Promote or Restrain Osteoclastogenesis and Inflammatory Bone Resorption? *Crit. Rev. Immunol.* **2018**, *38*, 253–261. [CrossRef]
80. Lee, S.K.; Lorenzo, J. Cytokines regulating osteoclast formation and function. *Curr. Opin. Rheumatol.* **2006**, *18*, 411–418. [CrossRef]
81. Rojas, C.; García, M.P.; Polanco, A.F.; González-Osuna, L.; Sierra-Cristancho, A.; Melgar-Rodríguez, S.; Cafferata, E.A.; Vernal, R. Humanized Mouse Models for the Study of Periodontitis: An Opportunity to Elucidate Unresolved Aspects of Its Immunopathogenesis and Analyze New Immunotherapeutic Strategies. *Front. Immunol.* **2021**, *12*, 663328. [CrossRef] [PubMed]
82. Lu, Z.; Li, Y.; Chowdhury, N.; Yu, H.; Syn, W.K.; Lopes-Virella, M.; Yilmaz, Ö.; Huang, Y. The Presence of Periodontitis Exacerbates Non-Alcoholic Fatty Liver Disease via Sphingolipid Metabolism-Associated Insulin Resistance and Hepatic Inflammation in Mice with Metabolic Syndrome. *Int. J. Mol. Sci.* **2023**, *24*, 8322. [CrossRef]
83. Pettersson, U.S.; Waldén, T.B.; Carlsson, P.O.; Jansson, L.; Phillipson, M. Female mice are protected against high-fat diet induced metabolic syndrome and increase the regulatory T cell population in adipose tissue. *PLoS ONE* **2012**, *7*, e46057. [CrossRef] [PubMed]
84. Bentley-Lewis, R.; Koruda, K.; Seely, E.W. The metabolic syndrome in women. *Nat. Clin. Pract. Endocrinol. Metab.* **2007**, *3*, 696–704. [CrossRef]
85. Ioannidou, E. The Sex and Gender Intersection in Chronic Periodontitis. *Front. Public Health* **2017**, *5*, 189. [CrossRef]

6. Wall, V.Z.; Barnhart, S.; Kanter, J.E.; Kramer, F.; Shimizu-Albergine, M.; Adhikari, N.; Wight, T.N.; Hall, J.L.; Bornfeldt, K.E. Smooth muscle glucose metabolism promotes monocyte recruitment and atherosclerosis in a mouse model of metabolic syndrome. *JCI Insight* **2018**, *3*, e96544. [CrossRef] [PubMed]
7. Ferreira, D.F.; Fiamoncini, J.; Prist, I.H.; Ariga, S.K.; de Souza, H.P.; de Lima, T.M. Novel role of TLR4 in NAFLD development: Modulation of metabolic enzymes expression. *Biochim. Biophys. Acta* **2015**, *1851*, 1353–1359. [CrossRef]
8. Karunakaran, U.; Elumalai, S.; Moon, J.S.; Won, K.C. CD36 Signal Transduction in Metabolic Diseases: Novel Insights and Therapeutic Targeting. *Cells* **2021**, *10*, 1833. [CrossRef]
9. Lu, Z.; Li, Y.; Brinson, C.W.; Kirkwood, K.L.; Lopes-Virella, M.F.; Huang, Y. CD36 is upregulated in mice with periodontitis and metabolic syndrome and involved in macrophage gene upregulation by palmitate. *Oral Dis.* **2017**, *23*, 210–218. [CrossRef]
0. Kim, D.H.; Cho, Y.M.; Lee, K.H.; Jeong, S.W.; Kwon, O.J. Oleate protects macrophages from palmitate-induced apoptosis through the downregulation of CD36 expression. *Biochem. Biophys. Res. Commun.* **2017**, *488*, 477–482. [CrossRef]
1. Urso, C.J.; Zhou, H. Role of CD36 in Palmitic Acid Lipotoxicity in Neuro-2a Neuroblastoma Cells. *Biomolecules* **2021**, *11*, 1567. [CrossRef] [PubMed]
2. Yu, H.; Li, Q.; Herbert, B.; Zinna, R.; Martin, K.; Junior, C.R.; Kirkwood, K.L. Anti-inflammatory effect of MAPK phosphatase-1 local gene transfer in inflammatory bone loss. *Gene Ther.* **2011**, *18*, 344–353. [CrossRef]
3. Rogers, J.E.; Li, F.; Coatney, D.D.; Rossa, C.; Bronson, P.; Krieder, J.M.; Giannobile, W.V.; Kirkwood, K.L. Actinobacillus actinomycetemcomitans lipopolysaccharide-mediated experimental bone loss model for aggressive periodontitis. *J. Periodontol.* **2007**, *78*, 550–558. [CrossRef] [PubMed]
4. Jin, J.; Zhang, X.; Lu, Z.; Li, Y.; Lopes-Virella, M.F.; Yu, H.; Haycraft, C.J.; Li, Q.; Kirkwood, K.L.; Huang, Y. Simvastatin inhibits lipopolysaccharide-induced osteoclastogenesis and reduces alveolar bone loss in experimental periodontal disease. *J. Periodontal Res.* **2014**, *49*, 518–526. [CrossRef]
5. Silva, N.; Dutzan, N.; Hernandez, M.; Dezerega, A.; Rivera, O.; Aguillon, J.C.; Aravena, O.; Lastres, P.; Pozo, P.; Vernal, R.; et al. Characterization of progressive periodontal lesions in chronic periodontitis patients: Levels of chemokines, cytokines, matrix metalloproteinase-13, periodontal pathogens and inflammatory cells. *J. Clin. Periodontol.* **2008**, *35*, 206–214. [CrossRef]
6. Königs, V.; Pierre, S.; Schicht, M.; Welss, J.; Hahnefeld, L.; Rimola, V.; Lütjen-Drecoll, E.; Geisslinger, G.; Scholich, K. GPR40 Activation Abolishes Diabetes-Induced Painful Neuropathy by Suppressing VEGF-A Expression. *Diabetes* **2022**, *71*, 774–787. [CrossRef] [PubMed]
7. Steinkamp, H.M.; Hathaway-Schrader, J.D.; Chavez, M.B.; Aartun, J.D.; Zhang, L.; Jensen, T.; Shojaee Bakhtiari, A.; Helke, K.L.; Stumpo, D.J.; Alekseyenko, A.V.; et al. Tristetraprolin Is Required for Alveolar Bone Homeostasis. *J. Dent. Res.* **2018**, *97*, 946–953. [CrossRef]
8. Bouxsein, M.L.; Boyd, S.K.; Christiansen, B.A.; Guldberg, R.E.; Jepsen, K.J.; Muller, R. Guidelines for assessment of bone microstructure in rodents using micro-computed tomography. *J. Bone Miner. Res.* **2010**, *25*, 1468–1486. [CrossRef]
9. Park, C.H.; Abramson, Z.R.; Taba, M., Jr.; Jin, Q.; Chang, J.; Kreider, J.M.; Goldstein, S.A.; Giannobile, W.V. Three-dimensional micro-computed tomographic imaging of alveolar bone in experimental bone loss or repair. *J. Periodontol.* **2007**, *78*, 273–281. [CrossRef]
00. Jin, J.; Machado, E.R.; Yu, H.; Zhang, X.; Lu, Z.; Li, Y.; Lopes-Virella, M.F.; Kirkwood, K.L.; Huang, Y. Simvastatin inhibits LPS-induced alveolar bone loss during metabolic syndrome. *J. Dent. Res.* **2014**, *93*, 294–299. [CrossRef]
01. Pengjam, Y.; Madhyastha, H.; Madhyastha, R.; Yamaguchi, Y.; Nakajima, Y.; Maruyama, M. NF-κB pathway inhibition by anthrocyclic glycoside aloin is key event in preventing osteoclastogenesis in RAW264.7 cells. *Phytomedicine* **2016**, *23*, 417–428. [CrossRef] [PubMed]

Disclaimer/Publisher's Note: The statements, opinions and data contained in all publications are solely those of the individual author(s) and contributor(s) and not of MDPI and/or the editor(s). MDPI and/or the editor(s) disclaim responsibility for any injury to people or property resulting from any ideas, methods, instructions or products referred to in the content.

molecules

MDPI

Article

Broad-Spectrum Legionaminic Acid-Specific Antibodies in Pooled Human IgGs Revealed by Glycan Microarrays with Chemoenzymatically Synthesized Nonulosonosides

Anoopjit Singh Kooner [1], Hai Yu [1], Shani Leviatan Ben-Arye [2], Vered Padler-Karavani [2,*] and Xi Chen [1,*]

1 Department of Chemistry, University of California, Davis, CA 95616, USA; akooner@ucdavis.edu (A.S.K.); hyu@ucdavis.edu (H.Y.)

2 Department of Cell Research and Immunology, The Shmunis School of Biomedicine and Cancer Research, The George S. Wise Faculty of Life Sciences, Tel Aviv University, Tel Aviv 69978, Israel; leviata@tauex.tau.ac.il

* Correspondence: vkaravani@tauex.tau.ac.il (V.P.-K.); xiichen@ucdavis.edu (X.C.); Tel.: +972-3-640-6737 (V.P.-K.); +1-530-754-6037 (X.C.)

Abstract: The presence and the level of antibodies in human sera against bacterial glycans are indications of prior encounters with similar antigens and/or the bacteria that express them by the immune system. An increasing number of pathogenic bacteria that cause human diseases have been shown to express polysaccharides containing a bacterial nonulosonic acid called 5,7-di-*N*-acetyllegionaminic acid (Leg5,7Ac_2). To investigate the immune recognition of Leg5,7Ac_2, which is critical for the fight against bacterial infections, a highly effective chemoenzymatic synthon strategy was applied to construct a library of α2–3/6-linked Leg5,7Ac_2-glycans via their diazido-derivatives (Leg5,7diN_3-glycans) formed by efficient one-pot three-enzyme (OP3E) synthetic systems from a diazido-derivative of a six-carbon monosaccharide precursor. Glycan microarray studies using this synthetic library of a Leg5,7Ac_2-capped collection of diverse underlying glycan carriers and their matched sialoside counterparts revealed specific recognition of Leg5,7Ac_2 by human IgG antibodies pooled from thousands of healthy donors (IVIG), suggesting prior human encounters with Leg5,7Ac_2-expressing pathogenic bacteria at the population level. These biologically relevant Leg5,7Ac_2-glycans and their immune recognition assays are important tools to begin elucidating their biological roles, particularly in the context of infection and host–pathogen interactions.

Keywords: bacterial nonulosonic acid; carbohydrate; chemoenzymatic synthesis; legionaminic acid; human antibodies

1. Introduction

Human sera contain protective antibodies against a large range of carbohydrate epitopes expressed by different species [1–4]. These antibodies are developed either naturally or in response to infections, vaccinations, or other exposures to antigens. Human antibodies recognizing bacterial glycans have been revealed by profiling human sera and pooled human immunoglobulins against bacterial surface carbohydrate antigens immobilized on glycan microarrays [5]. Synthetic glycans with defined structures are important probes for these studies. Access to such synthetic glycans, especially those with complex structures that are challenging to obtain either by purification from natural sources or by chemical synthesis, is greatly facilitated by the development of efficient chemoenzymatic methods. Glycans and glycoconjugates containing nine-carbon α-keto acids called nonulosonic acids (NulOs) [6] are among the challenging synthetic targets [7]. The most well-known NulOs are sialic acids (Sias) based on a neuraminic acid (Neu) or a 2-keto-3-deoxy-nononic acid (Kdn) backbone in the deuterostome lineage of animals and in certain bacteria. In vertebrates, Sias are often presented on the termini of glycans on cell surface glycoconjugates and *N*-acetylneuraminic acid (Neu5Ac, **1**, Figure 1) is the most common and well-studied sialic acid form.

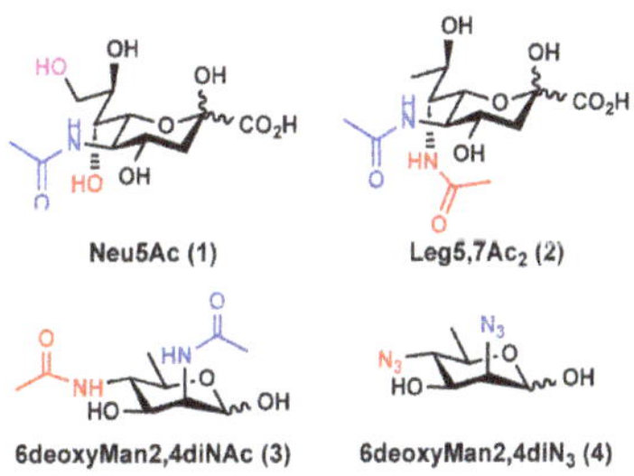

Figure 1. Structures of *N*-acetylneuraminic acid (Neu5Ac, **1**); 5,7-di-*N*-acetyllegionaminic acid (Leg5,7Ac$_2$, **2**); Leg5,7Ac$_2$ precursor 6deoxyMan2,4diNAc (**3**); and Leg5,7Ac$_2$ chemoenzymatic synthon 6deoxyMan2,4diN$_3$ (**4**).

Bacterial nonulosonic acid (NulO) 5,7-di-*N*-acetyllegionaminic acid (Leg5,7Ac$_2$, **2**) (Figure 1), or 5,7-diacetamido-3,5,7,9-tetradeoxy-D-*glycero*-D-*galacto*-non-2-ulosonic acid [6,8], is an analog of Neu5Ac (**1**). Leg5,7Ac$_2$ and Neu5Ac share an identical D-*glycero*-D-*galacto*-backbone configuration but differ for the functional groups at C7 and C9 positions. Leg5,7Ac$_2$ is the 7-acetamido-7,9-dideoxy derivative of Neu5Ac by replacing its C7-OH with an acetamido group and its C9-OH with a hydrogen atom (Figure 1) [9,10].

Chemical synthesis of Leg5,7Ac$_2$ from 2,4-diacetamido-2,4,6-trideoxy-D-mannose (6deoxyMan2,4diNAc, **3**) (Figure 1) and oxalacetic acid under basic conditions was initially reported in 2001 [11,12]. The synthetic Leg5,7Ac$_2$ was used to facilitate the identification and the correction of Leg5,7Ac$_2$-containing bacterial lipopolysaccharide (LPS) structures including those from *Legionella pneumophila* serogroup 1, a pathogen known to cause Legionnaire's disease; *Acinetobacter baumannii*, a causative agent of fatal bacteremia; and *Pseudomonas fluorescens*, *Vibrio salmonicida*, and *Vibrio alginolyticus* [11,12]. Leg5,7Ac$_2$ derivatives with variations on the stereochemistry of C4 and/or C8 and substitutions at C5 and/or C7 were also identified from bacterial sources [6,8,11,13].

The synthesis of glycans containing Leg5,7Ac$_2$ and derivatives [14] by chemical methods has been challenging [15]. The chemical synthetic strategies reported involved long routes, low overall yields, and a lack of stereochemical control for the formation of the biologically relevant α-legionaminic acid glycosyl linkage found in several bacterial polysaccharides [15]. In addition, the obtained products were limited to monosaccharides [11,12,16] or glycosides of either monosaccharides [15,17] or disaccharides [18,19].

The biosynthetic processes for Leg5,7Ac$_2$ and its cytidine 5′-monophosphate (CMP)-activated donor CMP-Leg5,7Ac$_2$ for the presumed legionaminyltransferases were reported from *Legionella pneumophila* [20] and *Campylobacter jejuni* [21] in 2008 and 2009, respectively. Both processes formed 6deoxyMan2,4diNAc (**3**, Figure 1) as a key intermediate despite the differences in using its uridine 5′-diphosphate (UDP) [20] or guanosine 5′-diphosphate (GDP) [21]-activated precursor in these two bacteria. The related enzymes had been expressed in *E. coli* for the production of Leg5,7Ac$_2$ and CMP-Leg5,7Ac$_2$ [22,23]. Recently, a metabolic labeling method using azide-modified Leg precursors led to the identification of Maf4 as a putative flagellin legionaminyltransferase from *Campylobacter jejuni* [24]. Nevertheless, information regarding legionaminyltransferases is very limited in general.

Recombinant sialyltransferases identified from bacterial and mammalian sources have been used for enzymatic and/or chemoenzymatic synthesis of Leg5,7Ac$_2$-glycans and derivatives. For example, bacterial *Pasteurella multocida* multifunctional α2–3-sialyltransferase PmST1 (previously named as tPm0188Ph [25]) [26] and *Neisseria meningitidis* MC58 α2–3-sialyltransferase [27], as well as porcine ST3Gal-I [26], were found to be efficient in forming α2–3-linked glycosides terminated with Leg5,7Ac$_2$. In comparison, bacterial *Photobacterium damselae* α2–6-sialyltransferase was less efficient in the formation of α2–6-linked glycosides terminated with Leg5,7Ac$_2$ [26] but its A235M mutant [27] had improved activity. Porcine ST3Gal-I was used to catalyze the transfer of Leg5,7Ac$_2$ from CMP-Leg5,7Ac$_2$ to a glycolipid GM1a to form a GD1a analog containing a terminal Leg5,7Ac$_2$, which was not recognized by an Neu5Ac-terminated GD1a-binding protein, myelin-associated glycoprotein (MAG or Siglec-4) [28]. Porcine ST3Gal-I and human ST6Gal-I were suitable for catalyzing the transfer of Leg5,7Ac$_2$ from CMP-Leg5,7Ac$_2$ to the O-glycans and N-glycans, respectively, on the therapeutic glycoproteins interferon-α2b and α1-antitrypsin [28]. We previously reported the chemical synthesis of the important biosynthetic intermediate 6deoxyMan2,4diNAc (**3**) and its application as an enzymatic precursor for the direct formation of Leg5,7Ac$_2$-glycosides [9] using a one-pot three-enzyme (OP3E) system containing *Pasteurella multocida* (PmAldolase), *Legionella pneumophila* CMP-Leg5,7Ac$_2$ synthetase (LpCLS), and an α2–3-sialyltransferase such as *Pasteurella multocida* multifunctional α2–3-sialyltransferase 1 (PmST1) or its M144D mutant (PmST1_M144D) with decreased sialidase and donor hydrolysis activities [10]. Nevertheless, the production of Leg5,7Ac$_2$-glycosidases by directly transferring Leg5,7Ac$_2$ from CMP-Leg5,7Ac$_2$ to appropriate acceptors using known sialyltransferases can be restricted by the types of acceptors that can be tolerated by a given sialyltransferase and the decreased efficiency in using CMP-Leg5,7Ac$_2$ as the donor substrate.

To overcome the challenges, we developed a chemoenzymatic synthon strategy for the highly efficient synthesis of Leg5,7Ac$_2$-glycosides [9]. In this strategy, a chemoenzymatic synthon 2,4-diazido-2,4,6-trideoxy-D-mannose (6deoxyMan2,4diN$_3$, **4**) (Figure 1) was chemically synthesized from commercially available D-fucose in eight steps with an overall 59.5% yield. It was a well-suited precursor for the synthesis of the corresponding 5,7-diazido-legionaminic acid (Leg5,7diN$_3$)-glycosides in 71–98% yields using a one-pot three-enzyme (OP3E) system containing PmAldolase, *Neisseria meningitidis* CMP-sialic acid synthetase (NmCSS), and an α2–3-sialyltransferase such as PmST1_M144D or an α2–6-sialyltransferase such as *Photobacterium* species α2–6-sialyltransferase (Psp2,6ST) [9]. The resulting Leg5,7diN$_3$-glycosides were conveniently converted to the corresponding target Leg5,7Ac$_2$-glycosides in 69–88% yields by treating with thioacetic acid in the presence of saturated sodium bicarbonate in aqueous solution [9]. Herein, we explore the application of the chemoenzymatic synthon 6deoxyMan2,4diN$_3$ (**4**) for the synthesis of a comprehensive library of α2–3- and α2–6-linked Leg5,7Ac$_2$-glycosides containing different underlying glycans. Glycan microarrays printed with this comprehensive collection of synthetic α2–3- and α2–6-linked Leg5,7Ac$_2$-glycosides and their related sialoside pairs demonstrate that pooled human IgGs are rich with antibodies that selectively recognize this broad spectrum of Leg5,7Ac$_2$-glycosides.

2. Results and Discussion

2.1. *Chemoenzymatic Synthesis of Leg5,7Ac$_2$-Glycosides*

Our synthetic targets were α2–3- and α2–6-linked Leg5,7Ac$_2$-glycosides containing different underlying glycans with an aglycon linker that can be used for immobilization on slides for glycan microarray assays. A propylamine aglycon was chosen as a well-suited design for the purpose. To produce the 5,7-diazido-legionaminic acid (Leg5,7diN$_3$)-glycosides efficiently from the chemoenzymatic synthon 6-deoxyMan2,4diN$_3$ (**4**) using OP3E sialyation systems [9] and allow efficient product purification, glycosides containing a hydrophobic carboxybenzyl (Cbz)-protected propylamine aglycon (ProNHCbz) [29] were chosen as sialyltransferase acceptors. Seven Cbz-tagged glycosides [30] including GalNAcαProNHCbz (Tn antigen, **5**), LacβProNHCbz (**6**), LacNAcβProNHCbz (**7**), Galβ1–3GalNAcβProNHCbz (**8**), Galβ1–3GalNAcαProNHCbz (Core 1, **9**), Galβ1–3GlcNAcβProNHCbz (Type I glycan, **10**), and Galβ1–3GlcNAcαProNHCbz (**11**) were used as sialyltransferase acceptors for OP3E sialylation reactions. The UV-detectability and the hydrophobicity of the ProNHCbz [29] in the glycoside acceptors and the resulting Leg5,7diN$_3$-glycosylated products facilitate the reaction monitoring and product purification processes.

As shown in Scheme 1, the OP3E system containing PmAldolase, NmCSS, and *Photobacterium damselae* α2–6-sialyltransferase (Pd2,6ST) or its A200Y/S232Y double mutant (Pd2,6ST_A200Y/

S232Y) [31] was used to synthesize α2–6-linked Leg5,7diN$_3$-glycosides (**12–18**) from 6deoxyMan2, 4diN$_3$ (**4**) and glycoside acceptors **5–11**. In this system, 6deoxyMan2,4diN$_3$ (**4**) reacted with pyruvate by a PmAldolase-catalyzed reaction to form Leg5,7diN$_3$, which was activated by NmCSS in the presence of cytidine 5′-triphosphate (CTP) and magnesium (Mg^{2+}) cation to form the corresponding CMP-activated sugar nucleotide CMP-Leg5,7diN$_3$ to allow the transfer of Leg5,7diN$_3$ to the sialyltransferase acceptor used in the OP3E reaction to form the desired α2–6-linked Leg5,7diN$_3$-glycoside. The target glycosides **12–18** (Table 1) were obtained with yields ranging from 70 to 99%, each was purified from the OP3E reaction with a single C18-cartridge purification process. When Galβ1–3GalNAcβProNHCbz (**8**) was used as the acceptor for Pd2,6ST in the OP3E, an undesired additional Leg5,7diN$_3$ could be added to the penultimate GalNAc moiety in the acceptor substrate, which caused complications in purifying the desired product Leg5,7diN$_3$α2–6Galβ1–3GalNAcβProNHCbz (**15**) and led to low yields. To our delight, the Pd2,6ST_A200Y/S232Y double mutant [31] was capable of adding Leg5,7diN$_3$ regio-selectively to the terminal galactose (Gal) of **8** to produce **15** in an excellent 96% yield. On the other hand, Pd2,6ST in the OP3E was efficient in synthesizing other α2–6-linked Leg5,7diN$_3$-glycoside targets (**12–14, 16–18**). Treatment of the azido-containing glycosides (**12–18**) with thioacetic acid and saturated sodium bicarbonate aqueous solution [9,30] produced the target α2–6 linked Leg5,7Ac$_2$-glycosides (**25–31**) in 65–90% yields.

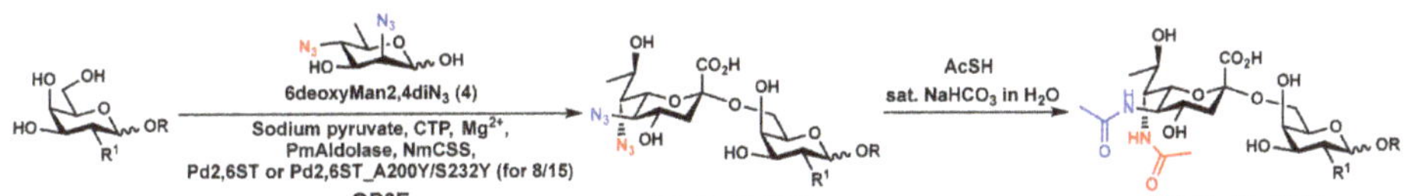

5–11 OP3E **12–18** (70–99%) **25–31** (65–90%)

Scheme 1. Chemoenzymatic synthesis of α2–6-linked Leg5,7Ac$_2$-glycosides (**25–31**) via OP3E produced α2–6-linked Leg5,7diN$_3$-glycosides (**12–18**).

Table 1. Structures of the OP3E sialyltransferase acceptors and yields of Leg5,7diN$_3$-glycosides (**12–24**) and Leg5,7Ac$_2$-glycosides (**25–37**) products.

Acceptors (5–11)	OP3E Products Leg5,7diN$_3$-glycosides (12–24)		Leg5,7Ac$_2$-glycosides (25–37)	
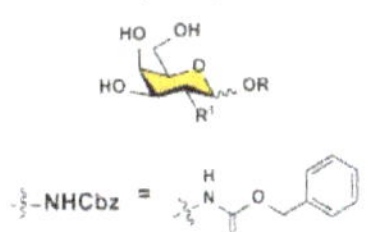	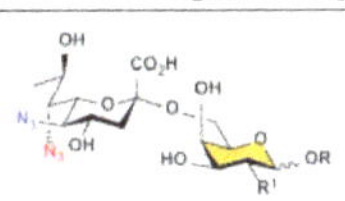 **12**–**18** (70–99%)	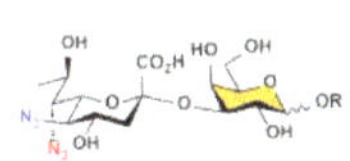 **19**–**24** (71–96%)	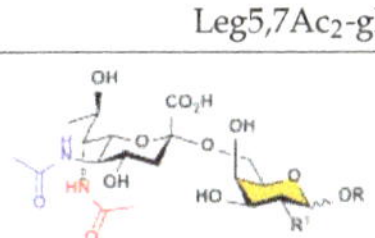 **25**–**31** (65–90%)	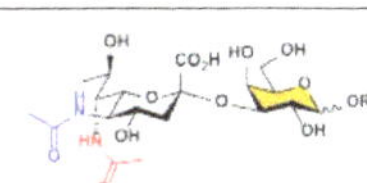 **32**–**37** (67–75%)
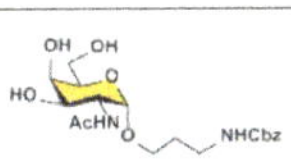 GalNAcαProNHCbz (**5**) R = αProNHCbz, R^1 = NHAc	**12** (99%)	-	**25** (65%)	-
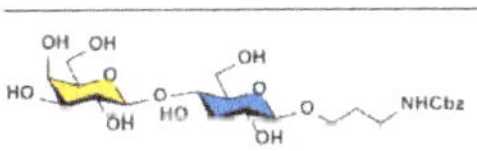 LacβProNHCbz (**6**) R = β4GlcβProNHCbz, R^1 = OH	**13** (75%)	**19** (74%)	**26** (78%)	**32** (68%)
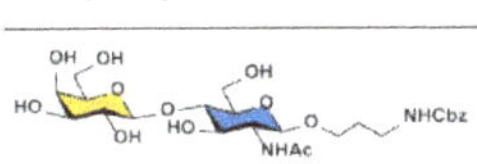 LacNAcβProNHCbz (**7**) R = β4GlcNAcβProNHCbz, R^1 = OH	**14** (70%)	**20** (71%)	**27** (68%)	**33** (67%)

Table 1. *Cont.*

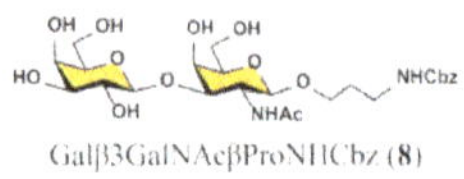 Galβ3GalNAcβProNHCbz (**8**) R = β3GalNAcβProNHCbz, R^1 = OH	**15** (93%)	**21** (96%)	**28** (90%)	**34** (67%)
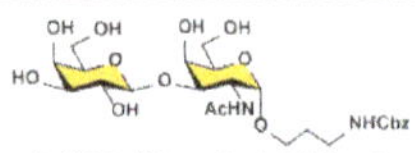 Galβ3GalNAcαProNHCbz (**9**) R = β3GalNAcαProNHCbz, R^1 = OH	**16** (77%)	**22** (77%)	**29** (68%)	**35** (69%)
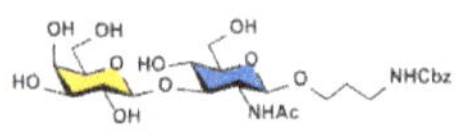 Galβ3GlcNAcβProNHCbz (**10**) R = β3GlcNAcβProNHCbz, R^1 = OH	**17** (81%)	**23** (80%)	**30** (78%)	**36** (75%)
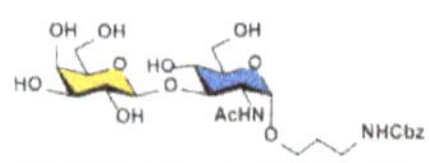 Galβ3GlcNAcαProNHCbz (**11**) R = β3GlcNAcαProNHCbz, R^1 = OH	**18** (99%)	**24** (96%)	**31** (71%)	**37** (72%)

Similarly, the OP3E system containing PmAldolase, NmCSS, and PmST1_M144D [32] was used for the efficient synthesis of α2–3-linked Leg5,7diN$_3$-glycosides (**19–24**) from galactosides **6–11**, respectively, in 71–96% yields (Scheme 2). They were readily converted to the desired α2–3-linked Leg5,7Ac$_2$-glycosides (**32–37**) in 67–75% yields using thioacetic acid in saturated sodium bicarbonate aqueous solution.

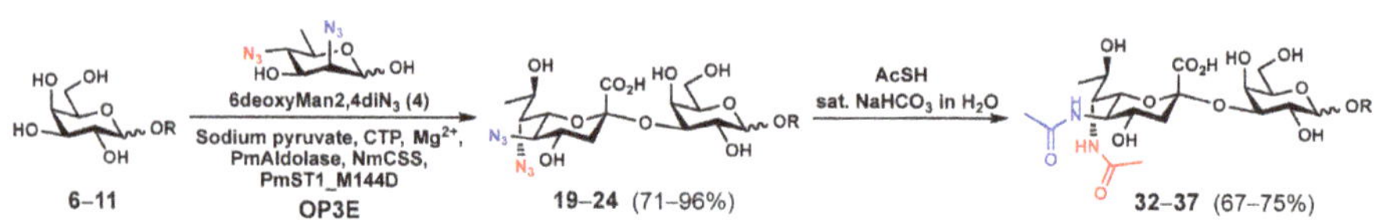

Scheme 2. Chemoenzymatic synthesis of α2–3-linked Leg5,7Ac$_2$-glycosides (**32–37**) via OP3E produced α2–3-linked Leg5,7diN$_3$-glycosides (**19–24**).

The Cbz group in Leg5,7Ac$_2$-glycosides (**25–37**) was conveniently and completely removed by catalytic hydrogenation [33] to produce Leg5,7Ac$_2$-glycosides containing a primary amino group on the propylamine aglycon, which was ready for immobilization on *N*-hydroxysuccinimide (NHS)-ester-coated [34,35] or epoxy-coated surface [36–39] for glycan microarray studies. It could also be coupled to other molecules such as proteins [40] and/or biotin [41] for various functional studies.

2.2. *Glycan Microarray Binding Studies Reveal a Broad Spectrum of Human IgG Antibodies against Leg5,7Ac$_2$-Glycosides*

A previously reported glycan microarray study using β-glycosides of a Leg5,7Ac$_2$ monosaccharide and its derivative that was chemically synthesized from D-threonine showed that human IgGs in three different samples of pooled human sera recognize the β-glycoside of Leg5,7Ac$_2$ but not its 7-amino-derivative [15]. We used glycan microarrays to characterize the immune recognition of the biologically relevant propylamine aglycon-containing Leg5,7Ac$_2$-α-glycosides obtained from **25–37** by catalytic hydrogenation to remove the Cbz group. These α-glycosides were fabricated onto epoxide-activated glass slides together with their matched pairs of sialoglycans synthesized previously, including those containing Neu5Ac [25,42], 9-*O*-acetyl-Neu5Ac (Neu5,9Ac$_2$) [40,42], *N*-glycolylneuraminic acid (Neu5Gc) [40,42,43], 9-*O*-acetyl-Neu5Gc (Neu5Gc9Ac) [40,42,43], or 7-acetamido-7-deoxy-*N*-acetylneuraminic acid (Neu5Ac7NAc) [30]. Non-sialylated glycosides were also printed as control glycan probes. The resulting glycan microarrays were used to investigate the binding profiles of several antibodies and lectins (Figures 2 and 3). Binding profiles of anti-Neu5Gc chicken immunoglobulin Y (IgY) [44], SiaFindα2–6, and *Maackia amurensis* lectin II (MAL-II) confirmed recognition of their known targets Neu5Gc-glycans, Siaα2–6-glycans, and Siaα2–3-glycans, respectively (Figure 2a). Interestingly, SiaFindα2–6 recognized all α2–6-linked glycans, including those with a terminal Leg5,7Ac$_2$ or Neu5Ac7NAc, except those linked to GalNAc that were less preferred. In contrast, MAL-II preferred sialyl T-antigen (Siaα2–3Core1). Its binding was much reduced against Neu5Ac7NAc and completely lost against terminal Leg5,7Ac$_2$ (Figure 2a). Similarly, most of the other lectins did not bind to Leg5,7Ac$_2$-glycans, except for *Lycopersicon esculentum* (tomato) lectin (LEL, TL) that recognized α2–3LacNAc glycans with terminal Sia/Leg5,7Ac$_2$ (Figure 2a). Importantly, human intravenous immunoglobulin G (IVIG; human IgG) pooled from thousands of individual donors, from two different sources (Privigen, OMRIgG-AR), showed strong binding to all Leg5,7Ac$_2$-glycans, Neu5Gc-glycans, and non-sialylated glycans, but very low binding to Neu5Ac-glycans or Neu5Ac7NAc-glycans (Figure 2a). IVIG bound most non-sialylated glycans, but the least to LacNAc (ID-45). Among sialylated glycans, IVIG showed a strong preference for Neu5Gc over Neu5Ac, which was clear by comparing binding to the matched pairs of glycans that differ only by the single additional oxygen in C5 *N*-glycolyl versus *N*-acetyl (Figure 2a). While Neu5Gc is foreign to humans, Neu5Ac is a protective self-glycan and is not normally recognized by human antibodies, as previously described [45,46]. IVIG showed stronger binding to all Leg5,7Ac$_2$-glycans compared to uncapped non-Sia underlying glycan counterparts. In contrast, there was no binding to their matched pairs of Neu5Ac7NAc-glycans, irrespective of their underlying glycan structures (Figure 2a,b). Thus, like human IgG preference toward Neu5Gc > Neu5Ac, binding of IVIG to Leg5,7Ac$_2$ > Neu5Ac7NAc occurs despite their difference in only a single oxygen atom at the C9 position in Neu5Ac7NAc that is absent in Leg5,7Ac$_2$ (Figure 2a,b). Together, these findings suggest that while Neu5Ac7NAc is less immunogenic, Leg5,7Ac$_2$ is perceived as foreign, as indicated by human IgG binding, thus likely representing secondary immune responses after previous encounters with Leg5,7Ac$_2$-expressing pathogens.

Sialic acids are known to belong to 'self-associated molecular patterns' (SAMPs) that provide a protective shield for glycoconjugates and cells [47], including from circulating antibodies [46]. To investigate human IgG immune recognition of Leg5,7Ac$_2$ within the context of 'pathogen-associated molecular patterns' (PAMPs), enzymatic cleavage of these sialic acids analogs on the arrays was evaluated by two sialidases, *Arthrobacter ureafaciens* sialidase (AUS) and *Clostridium perfringens* neuraminidase (NCP), that peel-off terminal sialic acid moieties from glycans printed on the array. Both sialidases effectively removed Neu5Ac and Neu5Gc, exposing the underlying de-capped glycans to a robust binding of IVIG anti-glycan antibodies. However, all Leg5,7Ac$_2$-glycans and Neu5Ac7NAc-glycans were found to be resistant to such cleavage by these two bacterial sialidases (Figure 3a), which correlated well with our previous results [9,48]. Despite the close similarity between Neu5Ac and Leg5,7Ac$_2$, the glycans capped with the latter become resistant to sialidase cleavage, potentially providing an immunological advantage to a human invading pathogen that would be coated with such PAMP glycans. This is further supported by recent studies of *Campylobacter jejuni*, the leading cause of foodborne bacterial gastroenteritis, revealing that expression of legionaminic acid on its bacterial flagellin serves as a virulence factor supporting enhanced biofilm formation, auto-agglutination, and host colonization [24,49]. On the other hand, pathogens also commonly use SAMPs as a molecular mimicry mechanism to evade host immunity, particularly by sialic acid sugar coating (e.g., on lipooligosaccharide; LOS) using sialyltransferases that they express. LOS sialylation shields bacteria from complement and cationic peptides and also engages inhibitory siglec receptors on host immune cells to dampen inflammatory responses [50,51].

Unlike the molecular mimicry properties of Neu5Ac used by pathogenic bacteria such as *Neisseria gonorrhoeae*, which causes sexually transmitted gonorrhea infection, incorporating Leg5,7Ac_2 instead of Neu5Ac to the LOS by *Neisseria gonorrhoeae* had been found to make the bacterium sensitive to complement and human serum [51,52]. Therefore, applying CMP-Leg5,7Ac_2, which can be used by some human pathogens to decorate their LOS with Leg5,7Ac_2, has been proposed as a novel therapeutic approach to address the global threat of multidrug-resistant bacteria [51–53]. Nevertheless, the host–pathogen interplay involving sialic acids and other nonulosonic acids with sialidases, siglecs, complement, cationic antimicrobial peptides, serum antibodies, and the complement system is complex and warrants further investigation.

a

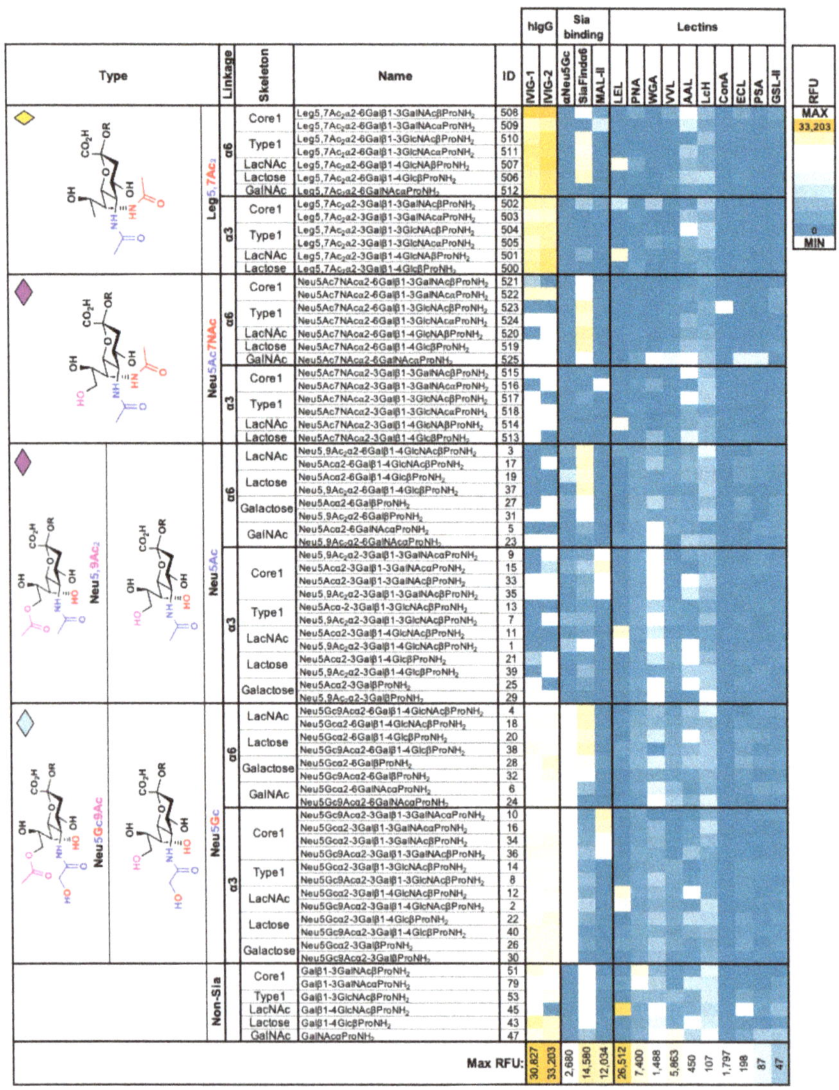

Figure 2. *Cont.*

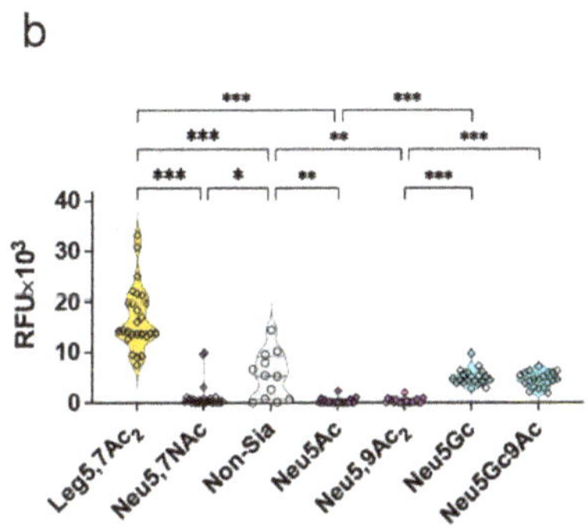

Figure 2. Glycan microarray binding studies against Leg5,7Ac$_2$-glycans and sialoglycan counterparts. (**a**) Glycan microarrays fabricated with Leg5,7Ac$_2$-glycans and sialoglycans (Neu5Gc/Neu5Ac/Neu5Ac7NAc-glycans) were examined with IVIG (IVIG-1 Privigen or IVIG-2 OMRIgG-AR; 50 µg/sub-array), anti-Neu5Gc IgY (αNeu5Gc; 1:7000), SiaFindα2-6 (0.25 µg), and 11 biotinylated plant lectins (2 µg each: MAL-II, LEL, PNA, WGA, VVL, AAL, LcH, ConA, ECL, PSA, and GSL-II) at 100 µL/subarray in blocking buffer (PBS, 1% ovalbumin; the assay buffers for plant lectins also contained divalent cations, see ESI), then detected with Cy3-anti-human IgG (0.4 ng/µL) for IVIG, Cy3-anti-chicken IgY (0.375 ng/µL) for IgY, and Cy3-sterptavidin (1.2 ng/µL) for Sia-Find and Lectins, at each 200 µL/sub-array. Slides were scanned and analyzed for Relative Fluorescence Units (RFU), which were plotted as a heatmap. (**b**) Binding of IVIG (average IVIG-1 and IVIG-2) to Leg5,7Ac$_2$-glycans compared to Neu5Ac7NAc-glycans, non-Sia glycans, and glycans terminated with Neu5Ac, Neu5,9Ac2, Neu5Gc, and Neu5Gc9Ac (mixed effects analysis, with the Geisser-Greenhouse correction and individual variances computed for each comparison using uncorrected Fisher's LSD; * $p = 0.04$, ** $p = 0.001$, *** $p < 0.001$).

To gain further insight into human IgG immune recognition of Leg5,7Ac$_2$, binding of IVIG to Leg5,7Ac$_2$-glycans on the glycan microarray was examined in the presence of selected competing glycans, including methyl α-*N*-glycolylneuraminoside (2-*O*-methyl-αNeu5Gc, or Gc2Me), glycopeptides of diverse *N*/*O*-glycans produced from mice sera digestion (mGP) [46], and the glycans that showed top preference of binding on the arrays including glycans ID-51, ID-502, ID-508 (Core1β; Leg5,7Ac$_2$α3Core1β; and Leg5,7Ac$_2$α6Core1β, respectively; Figure 3a). This analysis revealed that only αNeu5Gc (Gc2Me) did not affect IVIG binding to Leg5,7Ac$_2$-glycans, while all other glycans inhibited its binding (Figure 3b). Particularly, while IVIG binding to all Leg5,7Ac$_2$-glycans was inhibited by 50% with Core1β (ID-51) and by 80% with mGP, the competing Leg5,7Ac$_2$α3/6Core1β-glycans (ID-502/508) completely abolished it (Figure 3b,c). This suggests that human IgG is enriched with antibodies that recognize diverse Leg5,7Ac$_2$-glycans, in which the Leg5,7Ac$_2$ moiety is critical for their recognition, although with some promiscuity with the variation in the underlying glycans including Core1/Lactose/LacNAc. Based on these inhibition studies together with previous structural studies of antibodies against sialyl-Tn [54] and sialyl Lewis a (sLe^a) [55], it is plausible that the Leg5,7Ac$_2$ moiety is buried within the binding pocket of these anti-Leg5,7Ac$_2$ IgG antibodies.

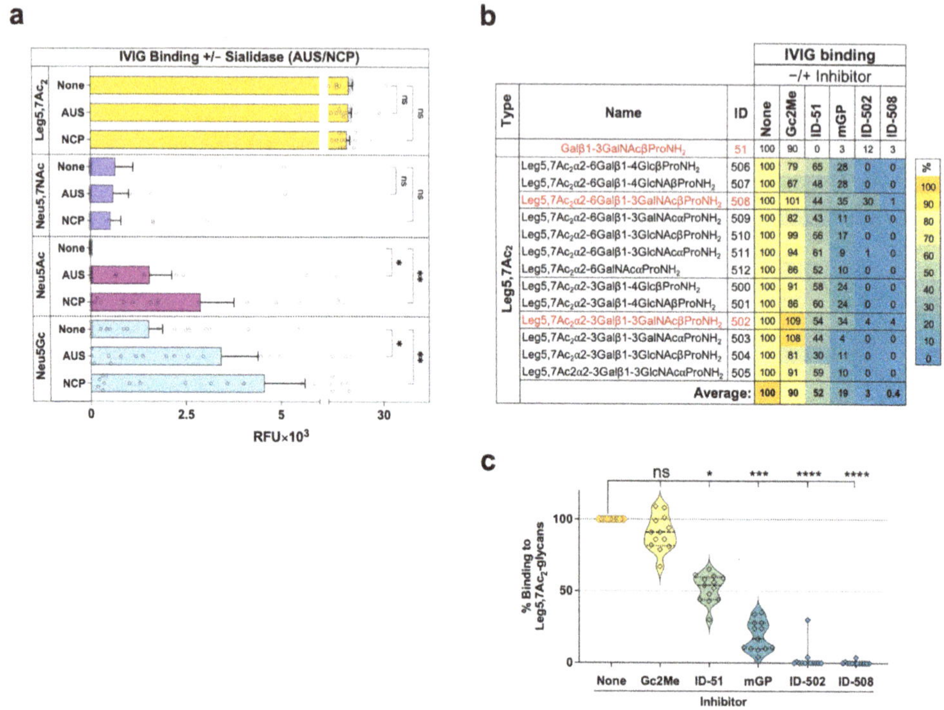

Figure 3. Characterization of IVIG binding to Leg5,7Ac$_2$-glycans. (**a**) Binding of IVIG-2 (OMRIgG-AR) to glycan microarray at 50 µg/sub-array, after pre-treatment of the arrays with/without 10 mU sialidase (*Arthrobacter ureafaciens* sialidase, AUS; *Clostridium perfringens* neuraminidase; NCP) for 2 h at 37 °C, then detected with 0.4 ng/µL Cy3-anti-human IgG. Slides were scanned, analyzed, and RFU-plotted (mixed-effects analysis, with the Geisser-Greenhouse correction, and individual variances computed for each comparison using uncorrected Fisher's LSD; ns = non-significant $p > 0.05$ ($p = 0.6132, 0.3334, 0.7287, 0.7454$ None vs AUS/NCP in Leg/Neu, respectively), * $p = 0.02$, ** $p < 0.003$). (**b**) Averaged binding of IVIG-1 and IVIG-2 to Leg5,7Ac$_2$-glycans on glycan microarrays in the presence of selected competing glycans. IVIG (IVIG-1 Privigen or IVIG-2 OMRIgG-AR, 50 µg/sub-array) were pre-incubated for 2 h on ice with various inhibitors on each sub-array, including 2-*O*-methyl-αNeu5Gc (Gc2Me; 4 mM), glycan ID-51 (Core1β, 2 mM), mouse serum glycopeptides (mGP, containing *N*/*O*-glycans [46], 0.45 mM), glycan ID-502 (Leg5,7Ac$_2$α3Core1β, 1 mM), and glycan ID-508 (Leg5,7Ac$_2$α6Core1β; 1 mM). Then, binding of the pre-complexed IVIG was examined and detected with 0.4 ng/µL Cy3-anti-human IgG. Slides were scanned, analyzed, and RFU results were plotted. Percent binding to each glycan was calculated as a ratio to binding of IVIG without inhibitor (none, 100% binding). (**c**) Statistical analysis of inhibition of IVIG binding to Leg5,7Ac$_2$-glycans of data presented in (**b**) (Friedman test with uncorrected Dunn's test; ns $p = 0.4631$, * $p = 0.0159$, *** $p = 0.0002$, and **** $p < 0.0001$).

3. Materials and Methods

3.1. Materials and Instruments

Chemicals were purchased and used without further purification. Nuclear magnetic resonance (NMR) spectra were recorded in the NMR facility of the University of California, Davis on 600 MHz and 800 MHz Bruker Avance III-NMR spectrometers and a 400 MHz Bruker Avance III HD Nanobay spectrometer (Bruker, Billerica, MA, USA). Chemical shifts are reported in parts per million (ppm) on the δ scale. High-resolution electrospray ionization (ESI) mass spectra were obtained using a Thermo Electron LTQ-

Orbitrap Hybrid mass spectrometer (Thermo Fisher Scientific, Waltham, MA, USA) at the mass spectrometry facility at the University of California, Davis. Column chromatography was performed using a CombiFlash® Rf 200i system (Teledyne ISCO, Lincoln, NE, USA) with either Redi*Sep* Rf silica columns or an ODS-SM (C18) column (51 g, 50 μm, 120 Å, Yamazen, Osaka, Japan) or manually using columns packed with silica gel 60 Å (230–400 mesh, Sorbent Technologies, Norcross, GA, USA). Thin layer chromatography (TLC) was performed on silica gel plates (Sorbent Technologies) using anisaldehyde sugar stain or 5% sulfuric acid in ethanol stain for detection. Gel filtration chromatography was performed with a column (100 cm × 2.5 cm) packed with Bio-Gel P-2 Fine resins (Bio-Rad, Hercules, CA, USA). Neu5GcaOMe (Gc2Me) was synthesized as reported previously [46]. *Pasteurella multocida* sialic acid aldolase (PmAldolase) [56], *Neisseria meningitidis* CMP-sialic acid synthetase (NmCSS) [57], *Pasteurella multocida* α2–3-sialyltransferase 1 (PmST1) [25], *Photobacterium damselae* α2–6-sialyltransferase (Pd2,6ST) [42], and Pd2,6ST_A200Y/S232Y were expressed and purified as described previously. PBSx10 was purchased from Hy-labs (Rehovot, Israel), ethanolamine from Fisher Scientific (Waltham, MA, USA), ovalbumin grade V from Sigma-Aldrich (St. Louis, MO, USA), SiaFindα2-6 from LectenzBio (Athens, GA, USA), anti-Neu5Gc IgY from BioLegend (San Diego, CA, USA), the lectins MAL-II, AAL, ConA, ECL, GSL-II, LcH, LEL, PNA, PSA, VVL, and WGA were purchased from Vector Laboratories (Newark, CA, USA), and ChromPure Human IgG whole molecule, Cy3-AffiniPure goat-anti-human IgG H+L, Cy3-sreptavidin, and Cy3-anti-chicken IgY from Jackson ImmunoResearch (West Grove, PA, USA). AlexaFlour-555-Hydraside from Invitrogen (Waltham, MA, USA). Pooled human IgG were IVIG-1 (Privigen from CSL Behring USA, King of Prussia, PA, USA) and IVIG-2 (OMRIgG-AM clinical remnants).

3.2. General Procedures for One-Pot Three-Enzyme (OP3E) Preparative-Scale Synthesis of *Leg5,7diN$_3$α2–6-Linked Nonulosonosides* (**12**–**18**)

Acceptors [30] (GalNAcαProNHCbz **5**, LacβProNHCbz **6**, LacNAcβProNHCbz **7**, Galβ1–3GalNAcβProNHCbz **8**, Galβ1–3GalNAcαProNHCbz **9**, Galβ1–3GlcNAcβProNHCbz **10**, or Galβ1–3GlcNAcαProNHCbz **11**) (0.10–0.20 mmol), 6deoxyMan2,4diN$_3$ [9] (**4**, 0.15–0.30 mmol, 1.5 eq.), sodium pyruvate (7.5 eq), and CTP (1.5 eq) were dissolved in water in a 50 mL centrifuge tube containing Tris-HCl buffer (100 mM, pH 8.5) and $MgCl_2$ (20 mM). After adding sialic acid aldolase (1–3 mg), NmCSS (0.5–1.0 mg), and Pd2,6ST (1–4 mg, for acceptors **5**–**7**, **9**–**11**) or Pd2,6ST_A200Y/S232Y28 (2 mg, for acceptor **8**), water was added to bring the final concentration of acceptor to 10 mM. The reaction mixture was incubated at 30 °C for 24–36 h. The reaction progress was monitored using TLC (ethyl acetate:methanol:water = 6:1:1, by volume, was used as the developing solvent) and mass spectrometry. The reaction mixture was diluted with the same volume of ethanol and incubated at 4 °C for 30 min. The mixture was then centrifuged and concentrated, which was purified by an automated flash chromatograph using a C18 column (acetonitrile in water gradient was used as running solvents) to obtain the desired nonulosonosides.

Leg5,7diN$_3$α2–6GalNAcαProNHCbz (**12**). A total of 110 mg, 99% yield, white amorphous solid. ^{1}H NMR (800 MHz, D_2O) δ 7.46–7.36 (m, 5H), 5.15–5.07 (m, 2H), 4.12 (dd, *J* = 11.0, 3.8 Hz, 1H), 4.02–3.81 (m, 7H), 3.78–3.62 (m, 3H), 3.58–3.52 (m, 2H), 3.46 (dt, *J* = 10.8, 5.9 Hz, 1H), 3.31–3.16 (m, 2H), 2.75 (dd, *J* = 12.5, 4.5 Hz, 1H), 2.01 (s, 3H), 1.97 (s, 3H), 1.94 (s, 3H), 1.80 (p, *J* = 6.4 Hz, 2H), 1.65 (t, *J* = 12.1 Hz, 1H), 1.15 (d, *J* = 6.3 Hz, 3H). $^{13}C\{^1H\}$ NMR (200 MHz, D_2O) δ 174.4, 174.1, 157.8, 136.5, 128.6, 128.1, 127.6, 101.4, 97.1, 72.1, 69.4, 69.0, 67.8, 67.6, 66.6, 66.1, 65.7, 65.3, 63.8, 61.4, 59.3, 49.9, 40.2, 37.5, 28.9, 22.0, 19.2. HRMS (ESI) *m/z* calculated for $C_{28}H_{39}N_8O_{13}^-$ [M − H]$^-$ 695.2642, found 695.2611.

Leg5,7diN$_3$α2–6LacβProNHCbz (**13**). A total of 57 mg, 75% yield, white amorphous solid. ^{1}H NMR (800 MHz, D_2O) δ 7.50–7.32 (m, 5H), 5.11 (s, 2H), 4.44 (d, *J* = 8.1 Hz, 1H), 4.40 (d, *J* = 7.9 Hz, 1H), 4.20–4.11 (m, 1H), 3.98–3.89 (m, 4H), 3.81–3.66 (m, 8H), 3.64 (dd, *J* = 10.0, 3.4 Hz, 1H), 3.61–3.50 (m, 6H), 3.46 (dd, *J* = 8.8, 2.4 Hz, 1H), 3.31 (t, *J* = 8.7 Hz, 1H), 3.29–3.19 (m, 2H), 2.71 (dd, *J* = 12.6, 4.8 Hz, 1H), 1.85–1.74 (m, 3H), 1.38 (d, *J* = 6.3 Hz, 3H). $^{13}C\{^1H\}$ NMR (200 MHz, D_2O) δ 173.0, 158.4, 136.6, 128.8, 128.3, 127.7, 103.3, 102.0, 100.8, 79.9, 74.6,

74.6, 73.6, 72.6, 72.4, 72.0, 70.8, 69.4, 68.3, 67.7, 66.8, 66.5, 65.8, 63.5, 63.3, 60.3, 59.3, 39.9, 37.3, 28.8, 18.9. HRMS (ESI) m/z calculated for $C_{32}H_{46}N_7O_{18}^-$ [M − H]$^-$ 816.1905, found 816.1880.

Leg5,7diN$_3$α2–6LacNAcβProNHCbz (**14**). A total of 26 mg, 70% yield, white amorphous solid. ^{1}H NMR (600 MHz, D_2O) δ 7.50–7.39 (m, 5H), 5.17–5.11 (m, 2H), 4.53 (d, *J* = 8.1 Hz, 1H), 4.42 (d, *J* = 7.9 Hz, 1H), 4.25–4.15 (m, 1H), 4.00–3.91 (m, 3H), 3.85–3.76 (m, 3H), 3.75 (s, 1H), 3.73–3.52 (m, 9H), 3.46–3.35 (m, 2H), 3.25–3.12 (m, 2H), 2.68 (dd, *J* = 12.6, 4.7 Hz, 1H), 2.06 (s, 3H), 1.83–1.71 (m, 3H), 1.40 (d, *J* = 6.3 Hz, 3H). ^{13}C{^{1}H} NMR (150 MHz, D_2O) δ 174.2, 173.0, 158.3, 136.6, 128.8, 128.4, 127.6, 103.5, 100.8, 100.6, 80.8, 74.5, 73.6, 72.5, 72.5, 72.0, 70.8, 69.2, 68.3, 67.6, 66.8, 66.3, 65.5, 63.8, 63.4, 60.4, 59.3, 55.0, 39.9, 37.3, 28.7, 22.2, 18.9. HRMS (ESI) m/z calculated for $C_{34}H_{49}N_8O_{18}^-$ [M − H]$^-$ 857.3170, found 857.3141.

Leg5,7diN$_3$α2–6Galβ1–3GalNAcβProNHCbz (**15**). A total of 73 mg, 93% yield, white amorphous solid. ^{1}H NMR (600 MHz, D_2O) δ 7.47–7.40 (m, 5H), 5.14–5.08 (m, 2H), 4.41 (d, *J* = 7.8 Hz, 1H), 4.39 (d, *J* = 7.8 Hz, 1H), 4.15–4.10 (m, 2H), 4.04–3.67 (m, 10H), 3.64–3.59 (m, 4H), 3.52–3.45 (m, 3H), 3.24–3.12 (m, 2H), 2.72 (dd, *J* = 12.6, 4.8 Hz, 1H), 2.00 (s, 3H), 1.85–1.64 (m, 3H), 1.38 (d, *J* = 6.0 Hz, 3H). ^{13}C{^{1}H} NMR (150 MHz, D_2O) δ 174.71, 173.00, 158.31, 136.67, 128.78, 128.33, 127.66, 104.59, 101.43, 100.72, 79.74, 74.85, 73.19, 72.32, 72.01, 70.47, 69.33, 68.43, 68.06, 67.46, 66.76, 66.59, 65.73, 63.59, 63.50, 61.14, 59.47, 51.20, 39.97, 37.14, 28.73, 22.20, 18.88. HRMS (ESI) m/z calculated for $C_{34}H_{49}N_8O_{18}^-$ [M − H]$^-$ 857.3170, found 857.3174.

Leg5,7diN$_3$α2–6Galβ1–3GalNAcαProNHCbz (**16**). A total of 38 mg, 77% yield, white amorphous solid. ^{1}H NMR (400 MHz, D_2O) δ 7.59–7.31 (m, 5H), 5.25–5.01 (m, 2H), 4.85 (d, *J* = 3.6 Hz, 1H), 4.45 (dd, *J* = 15.5, 7.7 Hz, 1H), 4.36–4.26 (m, 1H), 4.24–4.09 (m, 2H), 4.06–3.87 (m, 4H), 3.81–3.64 (m, 7H), 3.62–3.38 (m, 5H), 3.36–3.17 (m, 2H), 2.85–2.60 (m, 1H), 2.01 (s, 3H), 1.94–1.69 (m, 3H), 1.38 (d, *J* = 6.3 Hz, 3H). ^{13}C{^{1}H} NMR (100 MHz, D_2O) δ 174.5, 158.4, 136.7, 128.8, 128.3, 127.5, 104.7, 104.5, 97.1, 77.8, 77.4, 75.9, 74.9, 74.7, 72.5, 72.1, 70.6, 69.4, 69.0, 68.5, 67.6, 66.7, 66.4, 65.4, 63.4, 61.2, 60.9, 59.3, 48.6, 39.2, 37.6, 28.5, 22.0, 18.9. HRMS (ESI) m/z calculated for $C_{34}H_{49}N_8O_{18}^-$ [M − H]$^-$ 857.3170, found 857.3134.

Leg5,7diN$_3$α2–6Galβ1–3GlcNAcβProNHCbz (**17**). A total of 62 mg, 81% yield, white amorphous solid. ^{1}H NMR (800 MHz, D_2O) δ 7.51–7.37 (m, 5H), 5.10 (d, *J* = 5.0 Hz, 2H), 4.48 (d, *J* = 8.5 Hz, 1H), 4.35 (d, *J* = 7.8 Hz, 1H), 4.15 (dq, *J* = 8.8, 6.3 Hz, 1H), 3.97 (dd, *J* = 12.3, 2.3 Hz, 1H), 3.94–3.88 (m, 3H), 3.81–3.73 (m, 4H), 3.73–3.65 (m, 2H), 3.64–3.58 (m, 2H), 3.56 (dd, *J* = 10.4, 4.0 Hz, 1H), 3.54–3.42 (m, 5H), 3.29–3.08 (m, 2H), 2.69 (dd, *J* = 12.6, 4.8 Hz, 1H), 1.99 (s, 3H), 1.82–1.68 (m, 3H), 1.37 (d, *J* = 6.3 Hz, 3H). ^{13}C{^{1}H} NMR (200 MHz, D_2O) δ 174.6, 173.0, 158.3, 136.6, 128.8, 128.3, 127.6, 103.9, 100.9, 100.5, 84.3, 75.3, 73.5, 72.5, 72.0, 70.5, 69.3, 69.1, 68.4, 67.6, 66.8, 66.5, 65.7, 63.7, 63.5, 61.0, 59.3, 54.2, 39.9, 37.2, 28.7, 22.1, 18.9. HRMS (ESI) m/z calculated for $C_{34}H_{49}N_8O_{18}^-$ [M − H]$^-$ 857.3170, found 857.3139.

Leg5,7diN$_3$α2–6Galβ1–3GlcNAcαProNHCbz (**18**). A total of 49 mg, 99% yield, white amorphous solid. ^{1}H NMR (400 MHz, D_2O) δ 7.54–7.26 (m, 5H), 5.12 (s, 2H), 4.34 (d, *J* = 7.8 Hz, 1H), 4.22–4.10 (m, 1H), 4.06 (dd, *J* = 10.5, 3.6 Hz, 1H), 3.99–3.65 (m, 11H), 3.63–3.36 (m, 7H), 3.34–3.14 (m, 2H), 2.71 (dd, *J* = 12.7, 4.6 Hz, 1H), 2.01 (s, 3H), 1.90–1.64 (m, 3H), 1.37 (d, *J* = 6.2 Hz, 3H). ^{13}C{^{1}H} NMR (100 MHz, D_2O) δ 174.4, 173.1, 158.3, 136.6, 128.8, 128.3, 127.6, 103.8, 100.9, 96.8, 82.1, 73.5, 72.5, 71.9, 71.5, 70.6, 69.3, 68.9, 68.3, 66.7, 66.4, 65.8, 65.0, 63.7, 63.2, 60.7, 59.3, 52.2, 40.0, 37.5, 28.6, 22.0, 18.9. HRMS (ESI) m/z calculated for $C_{34}H_{49}N_8O_{18}^-$ [M − H]$^-$ 857.3170, found 857.3138.

3.3. General Procedures for One-Pot Three-Enzyme (OP3E) Preparative-Scale Synthesis of Leg5,7diN$_3$α2–3-Linked Nonulosonosides (**19**–**24**)

Acceptors [30] (LacβProNHCbz **6**, LacNAcβProNHCbz **7**, Galβ1–3GalNAcβPro NHCbz **8**, Galβ1–3GalNAcαProNHCbz **9**, Galβ1–3GlcNAcβProNHCbz **10**, or Galβ1–3GlcNAcαProNHCbz **11**) (0.10–0.20 mmol), 6deoxyMan2,4diN$_3$ [9] (**4**, 0.15–0.30 mmol, 1.5 eq.), sodium pyruvate (7.5 eq), and CTP (1.5 eq) were dissolved in water in a 50 mL centrifuge tube containing Tris-HCl buffer (100 mM, pH 8.5) and $MgCl_2$ (20 mM). After adding sialic acid aldolase (1–3 mg), NmCSS (0.5–1.0 mg), and PmST1_M144D (1–4 mg), water was added to bring the final concentration of the acceptor to 10 mM. The reaction mixture was incubated at 30 °C for 24–36 h. The reaction progress was monitored using TLC (ethyl acetate:methanol:water = 6:1:1, by volume, was used as the developing solvent) and mass spectrometry. The reaction mixture was diluted with the same volume of ethanol and incubated at 4 °C for 30 min. The mixture was then centrifuged and concentrated, which was purified by automated flash chromatography using a C18 column (CH_3CN in H_2O gradient was used as running solvents) to obtain the desired nonulosonosides.

Leg5,7diN$_3$α2–3LacβProNHCbz (**19**). A total of 56 mg, 74% yield, white amorphous solid. ^{1}H NMR (400 MHz, D_2O) δ 7.61–7.33 (m, 5H), 5.11 (s, 2H), 4.50 (d, *J* = 7.8 Hz, 1H), 4.42 (d, *J* = 8.1 Hz, 1H), 4.25–4.09 (m, 2H), 4.07–3.83 (m, 5H), 3.70–3.60 (m, 4H), 3.59–3.48 (m, 5H), 3.47–3.36 (m, 2H), 3.32–3.19 (m, 2H), 2.82–2.62 (m, 2H), 1.86–1.70 (m, 3H), 1.46–1.31 (m, 3H). ^{13}C{^{1}H} NMR (100 MHz, D_2O) δ 171.8, 158.4, 136.6, 128.8, 128.3, 127.6, 102.6, 102.2, 102.1, 78.4, 75.7, 75.1, 74.7, 74.4, 72.8, 72.2, 72.0, 69.3, 67.9, 67.7, 66.7, 65.8, 65.3, 63.5, 61.7, 61.0, 60.3, 59.3, 39.3, 37.3, 28.9, 18.8. HRMS (ESI) *m*/*z* calculated for $C_{32}H_{46}N_7O_{18}^-$ [M − H]$^-$ 816.1905, found 816.1870.

Leg5,7diN$_3$α2–3LacNAcβProNHCbz (**20**). A total of 27 mg, 71% yield, white amorphous solid. ^{1}H NMR (600 MHz, D_2O) δ 7.50–7.39 (m, 5H), 5.17–5.08 (m, 2H), 4.54 (d, *J* = 7.9 Hz, 1H), 4.47 (d, *J* = 8.3 Hz, 1H), 4.22–4.14 (m, 1H), 4.04 (dd, *J* = 9.9, 3.0 Hz, 1H), 3.99 (dd, *J* = 12.4, 2.3 Hz, 1H), 3.95–3.88 (m, 2H), 3.84 (dd, *J* = 12.5, 5.1 Hz, 1H), 3.78–3.66 (m, 8H), 3.63–3.52 (m, 4H), 3.45 (dd, *J* = 8.6, 2.2 Hz, 1H), 3.23–3.11 (m, 2H), 2.73 (dd, *J* = 12.7, 4.5 Hz, 1H), 2.02 (s, 3H), 1.91 (t, *J* = 12.2 Hz, 1H), 1.84–1.70 (m, 2H), 1.40 (t, *J* = 6.3 Hz, 3H). ^{13}C{^{1}H} NMR (150 MHz, D_2O) δ 174.5, 172.8, 158.3, 136.6, 128.8, 128.4, 127.6, 102.6, 101.1, 101.0, 78.5, 75.7, 75.1, 74.7, 72.4, 72.2, 69.4, 69.3, 67.9, 67.6, 66.6, 65.3, 63.4, 61.0, 60.1, 59.3, 55.0, 38.8, 37.2, 28.8, 22.1, 18.8. HRMS (ESI) *m*/*z* calculated for $C_{34}H_{49}N_8O_{18}^-$ [M − H]$^-$ 857.3170, found 857.3137.

Leg5,7diN$_3$α2–3Galβ1–3GalNAcβProNHCbz (**21**). A total of 23 mg, 96% yield, white amorphous solid. ^{1}H NMR (400 MHz, D_2O) δ 7.55–7.31 (m, 5H), 5.23–5.07 (m, 2H), 4.54–4.36 (m, 2H), 4.21–4.11 (m, 1H), 4.04–3.97 (m, 1H), 3.95–3.88 (m, 2H), 3.86–3.50 (m, 12H), 3.47–3.28 (m, 2H), 3.26–3.10 (m, 1H), 2.92 (d, *J* = 14.7 Hz, 2H), 2.74 (d, *J* = 12.6 Hz, 1H), 2.01 (s, 3H), 1.93–1.72 (m, 3H), 1.39 (d, *J* = 6.2 Hz, 3H). ^{13}C{^{1}H} NMR (200 MHz, D_2O) δ 174.69, 173.39, 158.34, 136.64, 128.77, 128.31, 127.60, 104.82, 104.59, 101.34, 80.20, 75.83, 74.76, 74.71, 72.09, 69.44, 68.87, 67.84, 67.58, 67.47, 66.46, 65.37, 63.30, 60.94, 60.90, 59.28, 51.10, 48.83, 39.18, 37.20, 29.56, 22.20, 18.84. HRMS (ESI) *m*/*z* calculated for $C_{34}H_{49}N_8O_{18}^-$ [M − H]$^-$ 857.3170, found 857.3135.

Leg5,7diN$_3$α2–3Galβ1–3GalNAcαProNHCbz (**22**). A total of 36 mg, 77% yield, white amorphous solid. ^{1}H NMR (400 MHz, D_2O) δ 7.50–7.37 (m, 5H), 5.19–5.07 (m, 2H), 4.85 (d, *J* = 3.7 Hz, 1H), 4.47 (d, *J* = 7.9 Hz, 1H), 4.30 (dd, *J* = 11.1, 3.6 Hz, 1H), 4.25–4.11 (m, 2H), 4.06–3.86 (m, 4H), 3.83–3.64 (m, 7H), 3.62–3.37 (m, 5H), 3.27 (q, *J* = 6.9 Hz, 2H), 2.74 (dd, *J* = 12.7, 4.7 Hz, 1H), 2.01 (s, 3H), 1.93–1.74 (m, 3H), 1.39 (d, *J* = 6.3 Hz, 3H). ^{13}C{^{1}H} NMR (100 MHz, D_2O) δ 174.5, 173.4, 158.3, 136.7, 128.8, 128.3, 127.5, 104.5, 100.4, 97.1, 77.8, 75.9, 74.7, 72.1, 70.6, 69.4, 69.0, 68.5, 67.6, 66.7, 66.4, 65.4, 65.1, 63.4, 61.2, 60.9, 48.6, 39.2, 37.6, 28.5, 22.0, 18.9. HRMS (ESI) *m*/*z* calculated for $C_{34}H_{49}N_8O_{18}^-$ [M − H]$^-$ 857.3170, found 857.3141.

Leg5,7diN$_3$α2–3Galβ1–3GlcNAcβProNHCbz (**23**). A total of 60 mg, 80% yield, white amorphous solid. ^{1}H NMR (800 MHz, D_2O) δ 7.50–7.35 (m, 5H), 5.13–5.06 (m, 2H), 4.49 (d, *J* = 8.5 Hz, 1H), 4.45 (d, *J* = 7.8 Hz, 1H), 4.19–4.08 (m, 1H), 3.99 (dd, *J* = 9.8, 3.2 Hz, 1H), 3.95–3.87 (m, 3H), 3.85–3.57 (m, 9H), 3.56–3.48 (m, 3H), 3.48–3.39 (m, 2H), 3.25–3.07 (m, 2H), 2.73 (dd, *J* = 12.6, 4.7 Hz, 1H), 2.01 (s, 3H), 1.85 (t, *J* = 12.3 Hz, 1H), 1.74 (p, *J* = 6.3 Hz, 2H), 1.37 (d, *J* = 6.4 Hz, 3H). ^{13}C{^{1}H} NMR (200 MHz, D_2O) δ 174.5, 173.3, 158.3, 136.6, 128.8, 128.3, 127.6, 103.5, 103.5, 100.8, 82.7, 75.9, 75.3, 75.1, 72.0, 69.4, 68.9, 68.7, 67.6, 67.5, 66.7, 66.5, 65.5, 63.3, 60.9, 60.7, 59.3, 54.4, 39.2, 37.2, 28.7, 22.2, 18.9. HRMS (ESI) *m*/*z* calculated for $C_{34}H_{49}N_8O_{18}^-$ [M − H]$^-$ 857.3170, found 857.3142.

Leg5,7diN$_3$α2–3Galβ1–3GlcNAcαProNHCbz (**24**). A total of 47 mg, 96% yield, white amorphous solid. ^{1}H NMR (400 MHz, D_2O) δ 7.53–7.32 (m, 5H), 5.13 (s, 2H), 4.48–4.39 (m, 1H), 4.21–4.10 (m, 1H), 4.07 (dd, *J* = 10.6, 3.5 Hz, 1H), 3.99 (dd, *J* = 9.7, 3.0 Hz, 1H), 3.93–3.65

(m, 11H), 3.63–3.37 (m, 6H), 3.35–3.14 (m, 2H), 2.74 (dd, J = 12.6, 4.6 Hz, 1H), 2.02 (s, 3H), 1.92–1.72 (m, 3H), 1.38 (d, J = 6.2 Hz, 3H). $^{13}C\{^1H\}$ NMR (100 MHz, D_2O) δ 174.3, 173.2, 158.4, 136.7, 128.8, 128.3, 127.5, 103.4, 100.6, 96.9, 81.0, 75.9, 75.0, 72.1, 71.5, 69.4, 69.1, 68.7, 67.5, 66.7, 66.5, 65.4, 65.0, 63.3, 60.9, 60.5, 59.3, 52.3, 39.2, 37.4, 28.5, 22.0, 18.9. HRMS (ESI) m/z calculated for $C_{34}H_{49}N_8O_{18}^-$ [M − H]$^-$ 857.3170, found 857.3132.

*3.4. General Procedures for Synthesizing Leg5,7Ac$_2$-Glycosides (25–37) by Converting the Azido Groups in Leg5,7diN$_3$-Glycosides (**12–24**) to NHAc Groups*

To a saturated sodium bicarbonate solution in water in a round bottom flask (100 mL), an azido-containing glycoside selected from **12–24** (10–40 mg) was added followed by drop-wise addition of 12–24 equivalents of thioacetic acid under argon at room temperature and this was stirred at 70 °C for 20 h. After completion of the reaction, the solvent was removed under vacuum. The mixture was passed through a Bio-Gel P-2 gel filtration column (water was used as an eluent). Then, the product-containing fractions were concentrated and further purified by silica gel chromatography using a mixed solvent (ethyl acetate: methanol: water = 10:1:0.1, by volume) as an eluent, followed by C18 purification (CH_3CN in H_2O gradient was used as running solvents) to obtain pure products.

Leg5,7Ac$_2$α2–6GalNAcαProNHCbz (**25**). A total of 30 mg, 65% yield, white amorphous solid. ^1H NMR (400 MHz, D_2O) δ 7.54–7.28 (m, 5H), 5.29–5.03 (m, 2H), 4.20–4.08 (m, 1H), 4.05–3.87 (m, 6H), 3.87–3.64 (m, 5H), 3.62–3.39 (m, 3H), 3.35–3.12 (m, 2H), 2.76 (dd, J = 12.6, 4.3 Hz, 1H), 2.01 (s, 3H), 1.97 (s, 3H), 1.95 (s, 3H), 1.81 (p, J = 6.3 Hz, 2H), 1.66 (t, J = 12.0 Hz, 1H), 1.15 (d, J = 6.2 Hz, 3H). $^{13}C\{^1H\}$ NMR (100 MHz, D_2O) δ 174.5, 173.9, 173.7, 173.6, 158.3, 136.6, 128.8, 128.3, 127.6, 97.0, 71.5, 69.5, 68.6, 67.6, 67.2, 66.7, 65.5, 64.2, 54.1, 54.0, 52.1, 49.9, 40.4, 37.6, 28.5, 22.1, 22.0, 21.9, 18.1. HRMS (ESI) m/z calculated for $C_{32}H_{47}N_4O_{15}^-$ [M − H]$^-$ 727.3043, found 727.3006.

Leg5,7Ac$_2$α2–6LacβProNHCbz (**26**). A total of 23 mg, 78% yield, white amorphous solid. ^1H NMR (400 MHz, D_2O) δ 7.59–7.27 (m, 5H), 5.12 (s, 2H), 4.55–4.34 (m, 2H), 4.10–3.50 (m, 18H), 3.42–3.19 (m, 3H), 2.74 (dd, J = 12.4, 4.7 Hz, 1H), 2.01 (s, 3H), 1.95 (s, 3H), 1.89–1.67 (m, 3H), 1.16 (d, J = 6.3 Hz, 3H). $^{13}C\{^1H\}$ NMR (100 MHz, D_2O) δ 173.9, 173.7, 173.4, 158.4, 136.6, 128.8, 128.3, 127.6, 103.2, 102.0, 100.3, 79.5, 74.6, 73.6, 72.8, 72.5, 72.4, 71.6, 70.7, 68.8, 68.4, 67.7, 67.1, 66.8, 63.4, 61.0, 60.2, 54.0, 52.1, 40.1, 37.3, 28.9, 22.1, 22.1, 18.1. HRMS (ESI) m/z calculated for $C_{36}H_{54}N_3O_{20}^-$ [M − H]$^-$ 848.3306, found 848.3253.

Leg5,7Ac$_2$α2–6LacNAcβProNHCbz (**27**). A total of 9 mg, 68% yield, white amorphous solid. ^1H NMR (800 MHz, D_2O) δ 7.69–7.29 (m, 5H), 5.19–5.03 (m, 2H), 4.55–4.41 (m, 2H), 4.06–3.76 (m, 9H), 3.75–3.50 (m, 10H), 3.26–3.08 (m, 2H), 2.76–2.65 (m, 1H), 2.03 (s, 3H), 1.97 (s, 3H), 1.92 (s, 3H), 1.85–1.66 (m, 3H), 1.17–1.11 (m, 3H). $^{13}C\{^1H\}$ NMR (200 MHz, D_2O) δ 174.5, 173.7, 173.6, 173.4, 158.3, 136.6, 128.8, 128.3, 127.6, 103.3, 100.8, 100.3, 80.0, 74.5, 73.5, 72.4, 72.3, 71.7, 70.7, 68.7, 68.3, 67.6, 67.1, 66.7, 63.1, 60.2, 55.1, 54.0, 52.0, 40.1, 37.2, 28.7, 22.2, 22.1, 22.0, 18.0. HRMS (ESI) m/z calculated for $C_{38}H_{57}N_4O_{20}^-$ [M − H]$^-$ 889.3572, found 889.3540.

Leg5,7Ac$_2$α2–6Galβ1–3GalNAcβProNHCbz (**28**). A total of 25 mg, 90% yield, white amorphous solid. ^1H NMR (800 MHz, D_2O) δ 7.45–7.38 (m, 5H), 5.13–5.08 (m, 2H), 4.43 (d, J = 8.0 Hz, 1H), 4.42 (d, J = 8.0 Hz, 1H), 4.19 (d, J = 3.2 Hz, 1H), 4.02–3.54 (m, 16H), 3.52–3.49 (m, 2H), 3.22–3.12 (m, 2H), 2.74 (dd, J = 12.8, 4.8 Hz, 1H), 2.00 (s, 6H), 1.94 (s, 3H), 1.76–1.72 (m, 2H), 1.62 (t, J = 4.0 Hz, 1H), 1.74 (s, 2H), 1.62 (t, J = 12.1 Hz, 1H), 1.13 (d, J = 6.4 Hz, 3H). $^{13}C\{^1H\}$ NMR (200 MHz, D_2O) δ 173.91, 173.69, 173.34, 158.34, 136.64, 128.76, 128.31, 127.61, 104.28, 101.43, 100.34, 80.36, 74.68, 73.18, 72.40, 71.63, 70.41, 68.65, 68.50, 67.73, 67.50, 67.01, 66.76, 63.64, 62.83, 61.21, 53.95, 52.10, 51.13, 40.32, 37.19, 28.72, 22.16, 22.12, 22.05, 18.01. HRMS (ESI) m/z calculated for $C_{38}H_{57}N_4O_{20}^-$ [M − H]$^-$ 889.3572, found 889.3580.

Leg5,7Ac$_2$α2–6Galβ1–3GalNAcαProNHCbz (**29**). A total of 6 mg, 68% yield, white amorphous solid. ^1H NMR (800 MHz, D_2O) δ 7.57–7.24 (m, 5H), 5.11 (q, J = 12.8 Hz, 2H), 4.42 (d, J = 7.7 Hz, 1H), 4.34–4.26 (m, 1H), 4.19–4.12 (m, 1H), 4.06–3.98 (m, 2H), 3.97–3.62 (m, 12H), 3.60–3.43 (m, 4H), 3.32–3.16 (m, 2H), 2.77 (dd, J = 12.4, 4.6 Hz, 1H), 1.94 (s, 6H), 1.92 (s, 3H), 1.84–1.75 (m, 2H), 1.71 (t, J = 12.1 Hz, 1H), 1.11 (d, J = 6.4 Hz, 3H). $^{13}C\{^1H\}$ NMR (200 MHz,

D_2O) δ 174.4, 173.9, 173.8, 173.7, 158.4, 136.7, 128.7, 128.3, 127.4, 104.4, 99.4, 97.1, 77.0, 75.6, 74.8, 71.6, 70.6, 68.9, 68.6, 67.3, 66.8, 66.7, 61.2, 60.9, 53.9, 51.9, 48.7, 40.2, 37.4, 28.4, 22.1, 22.0, 21.9, 18.0. HRMS (ESI) m/z calculated for $C_{38}H_{57}N_4O_{20}^-$ [M − H]$^-$ 889.3572, found 889.3516.

Leg5,7Ac₂α2–6Galβ1–3GlcNAcβProNHCbz (**30**). A total of 21 mg, 78% yield, white amorphous solid. ^{1}H NMR (800 MHz, D_2O) δ 7.48–7.35 (m, 5H), 5.16–5.06 (m, 2H), 4.50 (d, J = 8.5 Hz, 1H), 4.38 (d, J = 7.9 Hz, 1H), 4.06–3.96 (m, 3H), 3.95–3.87 (m, 3H), 3.86–3.77 (m, 4H), 3.75–3.68 (m, 2H), 3.66–3.60 (m, 2H), 3.59–3.52 (m, 4H), 3.51–3.46 (m, 1H), 3.25–3.10 (m, 2H), 2.72 (dd, J = 12.5, 4.7 Hz, 1H), 2.01 (s, 3H), 1.99 (s, 3H), 1.95 (s, 3H), 1.80–1.70 (m, 3H), 1.15 (d, J = 6.3 Hz, 3H).^{13}C{^{1}H} NMR (200 MHz, D_2O) δ 174.6, 173.9, 173.6, 173.4, 158.3, 136.6, 128.8, 128.3, 127.6, 104.0, 100.9, 100.5, 84.4, 75.6, 73.5, 72.5, 71.6, 70.5, 69.0, 68.7, 68.3, 67.6, 67.0, 66.8, 63.2, 60.9, 54.2, 54.0, 52.0, 40.1, 37.2, 28.7, 22.2, 22.1, 18.1. HRMS (ESI) m/z calculated for $C_{38}H_{57}N_4O_{20}^-$ [M − H]$^-$ 889.3572, found 889.3534.

Leg5,7Ac₂α2–6Galβ1–3GlcNAcαProNHCbz (**31**). A total of 17 mg, 71% yield, white amorphous solid. ^{1}H NMR (800 MHz, D_2O) δ 7.52–7.31 (m, 5H), 5.22–4.99 (m, 2H), 4.33 (d, J = 7.9 Hz, 1H), 4.13–4.05 (m, 1H), 4.04–3.78 (m, 9H), 3.76–3.67 (m, 4H), 3.62–3.43 (m, 6H), 3.33–3.17 (m, 2H), 2.70 (dd, J = 12.4, 4.7 Hz, 1H), 1.98 (s, 6H), 1.93 (s, 3H), 1.87–1.77 (m, 2H), 1.77–1.70 (m, 1H), 1.13 (d, J = 6.3 Hz, 3H). ^{13}C{^{1}H} NMR (200 MHz, D_2O) δ 174.4, 173.9, 173.6, 173.5, 158.4, 136.6, 128.8, 128.3, 127.4, 103.9, 100.1, 96.8, 82.5, 73.4, 72.5, 71.6, 71.6, 70.5, 68.8, 68.7, 68.2, 67.0, 66.7, 65.0, 63.0, 60.6, 54.0, 52.1, 52.0, 40.1, 37.4, 28.5, 22.1, 21.9, 18.0. HRMS (ESI) m/z calculated for $C_{38}H_{57}N_4O_{20}^-$ [M − H]$^-$ 889.3572, found 889.3536.

Leg5,7Ac₂α2–3LacβProNHCbz (**32**). A total of 7 mg, 68% yield, white amorphous solid. ^{1}H NMR (800 MHz, D_2O) δ 7.49–7.35 (m, 5H), 5.10 (s, 2H), 4.48 (d, J = 7.9 Hz, 1H), 4.42 (d, J = 8.0 Hz, 1H), 4.11–4.05 (m, 1H), 4.00–3.89 (m, 3H), 3.87–3.64 (m, 10H), 3.64–3.59 (m, 2H), 3.59–3.51 (m, 3H), 3.31–3.17 (m, 2H), 2.77 (dd, J = 12.4, 4.6 Hz, 1H), 1.97 (d, J = 2.0 Hz, 3H), 1.93 (s, 3H), 1.79 (p, J = 6.7 Hz, 2H), 1.73 (t, J = 12.1 Hz, 1H), 1.14 (d, J = 11.5 Hz, 3H). ^{13}C{^{1}H} NMR (200 MHz, D_2O) δ 173.9, 173.7, 169.0, 158.4, 136.6, 128.7, 128.3, 127.6, 102.6, 102.0, 95.1, 78.3, 75.3, 75.1, 74.7, 74.3, 72.7, 71.8, 69.3, 68.7, 67.7, 67.2, 66.8, 66.8, 61.0, 60.1, 58.6, 53.9, 52.0, 40.1, 37.3, 28.8, 22.1, 21.9, 18.0. HRMS (ESI) m/z calculated for $C_{36}H_{54}N_3O_{20}^-$ [M − H]$^-$ 848.3306, found 848.3271.

Leg5,7Ac₂α2–3LacNAcβProNHCbz (**33**). A total of 9 mg, 67% yield, white amorphous solid. ^{1}H NMR (800 MHz, D_2O) δ 7.41 (dq, J = 17.8, 9.0, 7.9 Hz, 5H), 5.15–5.05 (m, 2H), 4.53–4.48 (m, 1H), 4.44 (dd, J = 10.0, 8.0 Hz, 1H), 4.10 (dd, J = 9.9, 3.1 Hz, 1H), 4.00–3.92 (m, 3H), 3.90–3.78 (m, 4H), 3.74–3.52 (m, 11H), 3.26–3.07 (m, 2H), 2.77 (dd, J = 12.4, 4.6 Hz, 1H), 2.00 (s, 3H), 1.97 (d, J = 4.6 Hz, 3H), 1.92 (d, J = 2.9 Hz, 3H), 1.76–1.68 (m, 3H), 1.17–1.11 (m, 3H). ^{13}C{^{1}H} NMR (200 MHz, D_2O) δ 174.5, 174.0, 173.7, 173.7, 158.3, 136.6, 128.8, 128.3, 127.6, 103.3, 102.5, 99.5, 78.4, 75.3, 75.1, 74.7, 72.4, 71.8, 69.4, 68.7, 67.6, 67.2, 66.9, 62.4, 61.0, 60.0, 55.0, 53.9, 52.0, 40.0, 37.2, 28.7, 22.1, 22.1, 21.9, 18.0. HRMS (ESI) m/z calculated for $C_{38}H_{57}N_4O_{20}^-$ [M − H]$^-$ 889.3572, found 889.3540.

Leg5,7Ac₂α2–3Galβ1–3GalNAcβProNHCbz (**34**). A total of 5 mg, 67% yield, white amorphous solid. ^{1}H NMR (800 MHz, D_2O) δ 7.46–7.36 (m, 5H), 5.18–5.06 (m, 2H), 4.46–4.36 (m, 2H), 4.12 (d, J = 3.2 Hz, 1H), 4.06–3.98 (m, 2H), 3.96–3.86 (m, 3H), 3.82–3.53 (m, 13H), 3.28–3.09 (m, 1H), 2.90 (d, J = 46.7 Hz, 1H), 2.77 (dd, J = 12.4, 4.6 Hz, 1H), 1.96 (s, 3H), 1.92 (s, 6H), 1.85–1.65 (m, 3H), 1.11 (d, J = 6.1 Hz, 3H). ^{13}C{^{1}H} NMR (200 MHz, D_2O) δ 173.9, 173.8, 173.7, 172.9, 158.3, 136.7, 128.7, 128.3, 127.5, 104.7, 101.2, 99.3, 75.6, 74.9, 74.8, 72.0, 71.6, 68.8, 68.6, 68.0, 67.4, 67.3, 66.8, 61.0, 60.9, 60.8, 53.9, 51.9, 51.3, 40.2, 37.1, 28.6, 22.2, 22.1, 21.9, 18.0. HRMS (ESI) m/z calculated for $C_{38}H_{57}N_4O_{20}^-$ [M − H]$^-$ 889.3572, found 889.3536.

Leg5,7Ac₂α2–3Galβ1–3GalNAcαProNHCbz (**35**). A total of 13 mg, 69% yield, white amorphous solid. ^{1}H NMR (800 MHz, D_2O) δ 7.42 (dd, J = 27.7, 7.4 Hz, 5H), 5.17–5.07 (m, 2H), 4.83–4.82 (m, 1H), 4.47 (d, J = 7.9 Hz, 1H), 4.32 (dd, J = 11.1, 3.8 Hz, 1H), 4.18 (d, J = 3.2 Hz, 1H), 4.06–4.00 (m, 2H), 3.98–3.93 (m, 2H), 3.90 (d, J = 3.2 Hz, 1H), 3.86–3.79 (m, 2H), 3.77–3.68 (m, 6H), 3.60–3.52 (m, 3H), 3.51–3.43 (m, 1H), 3.33–3.16 (m, 2H), 2.79 (dd, J = 12.4, 4.5 Hz, 1H), 1.98 (s, 3H), 1.98 (s, 3H), 1.94 (s, 3H), 1.80 (h, J = 6.6 Hz, 2H), 1.73 (t, J = 12.1 Hz, 1H), 1.13 (d, J = 6.3 Hz, 3H). ^{13}C{^{1}H} NMR (200 MHz, D_2O) δ 174.4, 173.9,

173.7, 158.4, 136.7, 128.8, 128.3, 127.4, 104.4, 99.6, 97.1, 77.0, 75.7, 74.8, 71.7, 71.7, 70.6, 68.9, 68.7, 68.7, 67.3, 66.8, 66.7, 65.0, 61.2, 60.9, 53.9, 51.9, 48.7, 40.2, 37.5, 28.4, 22.1, 22.0, 21.9, 18.1. HRMS (ESI) m/z calculated for $C_{38}H_{57}N_4O_{20}^-$ $[M-H]^-$ 889.3572, found 889.3537.

Leg5,7Ac$_2$α2–3Galβ1–3GlcNAcβProNHCbz (**36**). A total of 15 mg, 75% yield, white amorphous solid. ^{1}H NMR (400 MHz, D_2O) δ 7.54–7.34 (m, 5H), 5.18–5.05 (m, 2H), 4.47 (dd, J = 20.6, 8.1 Hz, 2H), 4.07 (dd, J = 9.9, 3.0 Hz, 1H), 3.99–3.42 (m, 17H), 3.21 (q, J = 7.4 Hz, 3H), 2.80 (dd, J = 12.4, 4.5 Hz, 1H), 2.05–1.96 (m, 6H), 1.95 (s, 3H), 1.85–1.63 (m, 3H), 1.13 (d, J = 6.2 Hz, 3H). $^{13}C\{^1H\}$ NMR (100 MHz, D_2O) δ 174.3, 173.9, 173.7, 158.3, 136.6, 128.8, 128.3, 127.5, 103.6, 100.9, 100.8, 82.0, 75.7, 75.4, 75.2, 71.6, 68.8, 68.6, 68.6, 67.7, 67.4, 66.8, 61.1, 60.7, 54.5, 53.9, 51.9, 46.7, 40.2, 37.3, 28.7, 22.1, 22.1, 21.9, 18.1. HRMS (ESI) m/z calculated for $C_{38}H_{57}N_4O_{20}^-$ $[M-H]^-$ 889.3572, found 889.3532.

Leg5,7Ac$_2$α2–3Galβ1–3GlcNAcαProNHCbz (**37**). A total of 11 mg, 72% yield, white amorphous solid. ^{1}H NMR (800 MHz, D_2O) δ 7.52–7.31 (m, 5H), 5.25–5.01 (m, 2H), 4.76 (d, J = 3.6 Hz, 1H), 4.45 (d, J = 8.0 Hz, 1H), 4.12–3.99 (m, 2H), 3.97–3.64 (m, 12H), 3.63–3.49 (m, 4H), 3.48–3.40 (m, 1H), 3.33–3.14 (m, 2H), 2.77 (dd, J = 12.4, 4.5 Hz, 1H), 1.97 (s, 3H), 1.95 (s, 3H), 1.92 (s, 3H), 1.84–1.74 (m, 2H), 1.71 (t, J = 12.1 Hz, 1H), 1.11 (d, J = 6.4 Hz, 3H). $^{13}C\{^1H\}$ NMR (200 MHz, D_2O) δ 174.2, 173.9, 173.7, 158.4, 136.7, 128.7, 128.2, 127.4, 103.3, 99.4, 97.0, 79.8, 75.7, 75.0, 71.6, 71.6, 68.9, 68.6, 68.5, 67.3, 66.8, 66.7, 64.9, 61.0, 60.4, 53.9, 52.5, 51.9, 40.2, 37.3, 28.4, 22.1, 21.9, 21.9, 18.0. HRMS (ESI) m/z calculated for $C_{38}H_{57}N_4O_{20}^-$ $[M-H]^-$ 889.3572, found 889.3537.

3.5. General Procedures for Converting Cbz-Tagged Glycans to Glycosyl Propylamines

To a stirring solution of a Cbz-tagged glycoside selected from **25**–**37** (5–10 mg) in water-methanol solution (3 mL, 1:2 by volume), a catalytic amount of 10% palladium on charcoal was added to a 50 mL round bottom flask. The mixture was stirred under a hydrogen environment for 2–5 h. The solution was passed through a syringe filter. Methanol was removed from the filtrate by blowing air inside a fume hood. The residue was lyophilized and used directly for microarray assays.

3.6. Glycan Microarray Fabrication

Nano-printing of glycans microarray were fabricated as previously described [45,58]. Briefly, arrays (Version VLA1) were fabricated with NanoPrint LM-60 Microarray Printer (Arrayit) on epoxide-derivatized slides (PolyAn 2D) with 16 sub-array blocks on each slide, with each sub-array containing 340 spots arraigned in a 20 × 17 matrix. Glycoconjugates were distributed into 384-well source plates using 4 replicate wells per sample and 8 µL per well. Each glycoconjugate was diluted into 100 µM in an optimized printing buffer (300 mM phosphate buffer, pH 8.4). To monitor printing, ChromPure Human IgG whole molecule (at 20 ng/µL in PBS pH 7.4 + 10% glycerol) and AlexaFlour-555-Hydraside (at 2 ng/µL in 178 mM phosphate buffer, pH 5.5) were used for each sub-array. The arrays were printed with four 946MP3 pins (5 µm tip, 0.25 µL sample channel, ~100 µm spot diameter; Arrayit). The humidity level in the arraying chamber was maintained at about 70% during printing. Printed slides were left on the arrayer deck overnight, allowing humidity to drop to ambient levels (40–45%). Next, slides were packed, vacuum-sealed, and stored at room temperature (RT) until used.

3.7. Glycan Microarray Binding Assays

Slides were developed and analyzed as previously described [58] with some modifications. Slides were rehydrated with dH_2O and incubated for 30 min in a staining dish with 50 °C pre-warmed ethanolamine (0.05 M) in Tris-HCl (0.1 M, pH 9.0), then washed with 50 °C pre-warmed dH_2O. Slides were centrifuged at 200× g for five min then fitted with a ProPlate™ Multi-Array 16-well slide module (Grace Bio-lab) to divide into the sub-arrays (blocks). Slides were washed with PBST (0.1% Tween 20), aspirated, and blocked with 200 µL/sub-array of blocking buffer (PBS pH 7.3 + 1% w/v ovalbumin) for 1 h at RT with gentle shaking. Next, the blocking solution was aspirated and a 100 µL/sub-array of detection lectins/antibodies diluted in blocking buffer (lectins binding buffer also included divalent cations, as detailed in the ESI glycan microarray data file in Supplementary Materials) were incubated with gentle shaking for 2 h at RT.

For sialidase pretreatment, before blocking, sub-arrays were incubated with either PBS (without sialidase), 10 mU/100 µL *Arthrobacter ureafaciens* sialidase (AUS, EY Laboratories), or 10 mU/100 µL *Clostridium perfringens* neuraminidase (NCP, Rosch, Sigma, St. Louis, MO, USA), for 2 h at 37 °C with gentle shaking. Then, slides were washed twice with PBST (0.1% Tween 20), aspirated, and blocked with 200 µL/sub-array of blocking buffer (PBS pH 7.3 + 1% *w*/*v* ovalbumin) at RT for 1 h with gentle shaking.

For inhibition assays, each IVIG was pre-incubated for 2 h on ice with inhibition solution containing blocking buffer with glycans (2 mM ID#43, ID#51; or 1 mM ID#502, ID#508), glycopeptides (GP at 0.45 mM; produced from mice sera digested with Pronase) [59], 4 mM 2-*O*-methyl-α-Neu5Ac (Ac2Me), or 4 mM 2-*O*-methyl-α-Neu5Gc (Gc2Me) and only then applied to slides for 2 h. Slides were washed 5 times with PBST.

Binding was detected with Cy3-labeled secondary detection reagent diluted in PBS at 200 µL/sub-array at RT for 1 h with gentle shaking (0.4 ng/µL Cy3 AffiniPure goat-anti-human IgG (H + L), 0.375 ng/µL Cy3-anti-chicken IgY, 1.2 ng/µL Cy3-sterptavidin). Slides were washed 4 times with PBST, then with PBS (without Tween-20) for 10 min followed by removal from ProPlate™ Multi-Array slide module and immediately dipping in a staining dish with dH_2O for 10 min with shaking. Slides were then centrifuged at 200× *g* for 5 min and the dry slides were immediately scanned.

3.8. Array Slide Processing

Processed slides were scanned and analyzed as described [45,58] at 10 µm resolution with a Genepix 4000B microarray scanner (Molecular Devices, San Jose, CA, USA) using 350 gain. Image analysis was carried out with Genepix Pro 7.0 analysis software (Molecular Devices). Spots were defined as circular features with a variable radius as determined by the Genepix scanning software. Local background subtraction was performed. RFU from each spot was calculated.

3.9. Statistics

Statistical analyses were performed using GraphPad Prism 10.2.0 and described in context in the Figure legends.

4. Conclusions

We demonstrate here that the chemoenzymatic synthon strategy using 6deoxyMan2,4diN_3 (**4**) as the starting material in OP3E sialylation systems is highly efficient in constructing a comprehensive library of α2–3- and α2–6-linked Leg5,7Ac_2-glycosides via the enzymatic formation of Leg5,7diN_3, CMP-Leg5,7diN_3, and Leg5,7diN_3-glycosides intermediates, followed by chemical conversion of the azido to the acetamido groups. The hydrophobic UV-detectable Cbz group in the sialyltransferase acceptors facilitated product purification from the OP3E systems. Removal of the Cbz group in the ProNHCbz-tagged Leg5,7Ac_2-glycosides produced ProNH_2-tagged Leg5,7Ac_2-glycosides ready for immobilization on the activated glass slide surface. Glycan microarray studies using the Leg5,7Ac_2-glycosides and their structurally related sialoside pairs revealed diverse human IgG antibodies that selectively recognize the α-linked bacterial nonulosonic acid Leg5,7Ac_2. Such broad-spectrum recognition by human IgG antibodies pooled from thousands of donors (IVIG) supports previous encounters of Leg5,7Ac_2-expressing pathogens by the general human population. Overall, these resources and analytical tools are important tools for further in-depth investigation of Leg5,7Ac_2-involved bacterial physiology, virulence, pathogenicity, and host–pathogen interactions.

Supplementary Materials: The following supporting information can be downloaded at: https://www.mdpi.com/article/10.3390/molecules29163980/s1: the ^{1}H and ^{13}C NMR spectra of products **12–37** (PDF) and the glycan microarray data file (EXCEL).

Author Contributions: Conceptualization, A.S.K., H.Y., S.L.B.-A., V.P.-K., and X.C.; data curation, A.S.K., H.Y., and S.L.B.-A.; formal analysis, A.S.K., H.Y., S.L.B.-A., V.P.-K., and X.C.; funding acquisition, V.P.-K. and X.C.; investigation, A.S.K., H.Y., S.L.B.-A., V.P.-K., and X.C.; methodology, A.S.K., H.Y., S.L.B.-A., V.P.-K., and X.C.; project administration, V.P.-K. and X.C.; resources, V.P.-K. and X.C.; supervision, V.P.-K. and X.C.; writing—original draft, A.S.K., H.Y., S.L.B.-A., V.P.-K., and X.C.; writing—review and editing, A.S.K., H.Y., S.L.B.-A., V.P.-K., and X.C. All authors have read and agreed to the published version of the manuscript.

Funding: This work was financially supported by the United States (U.S.) National Institutes of Health (NIH), grant numbers R01AI130684 and R01GM141324 (to X.C.), and by the Israel Science Foundation—Israel Precision Medicine Program (ISF-IPMP no. 3042/22) and by the European Research Council Consolidator Grant ERC-2020-COG-101003021 (to V.P.-K.). The Bruker Avance-800 NMR spectrometer was funded by the U.S. National Science Foundation (NSF), grant number DBI-0722538. The Thermo Scientific Q Exactive HF Orbitrap Mass Spectrometer was purchased with a U.S. NIH, grant number S10OD025271. The contents of this publication are solely the responsibility of the authors and do not necessarily represent the official view of NIH, NSF, or ISF. The funding sponsors had no role in the design of the study; in the collection, analyses, or interpretation of data; in the wiring of the manuscript; and in the decision to publish the results.

Institutional Review Board Statement: Not applicable.

Informed Consent Statement: Not applicable.

Data Availability Statement: The original contributions presented in the study are included in the article/Supplementary Materials, further inquiries can be directed to the corresponding authors.

Conflicts of Interest: The authors declare no conflicts of interest.

References

1. Luetscher, R.N.D.; McKitrick, T.R.; Gao, C.; Mehta, A.Y.; McQuillan, A.M.; Kardish, R.; Boligan, K.F.; Song, X.; Lu, L.; Heimburg Molinaro, J.; et al. Unique repertoire of anti-carbohydrate antibodies in individual human serum. *Sci. Rep.* **2020**, *10*, 1543 [CrossRef] [PubMed]
2. Geissner, A.; Reinhardt, A.; Rademacher, C.; Johannssen, T.; Monteiro, J.; Lepenies, B.; Thepaut, M.; Fieschi, F.; Mrazkova, J Wimmerova, M.; et al. Microbe-focused glycan array screening platform. *Proc. Natl. Acad. Sci. USA* **2019**, *116*, 1958–196 [CrossRef] [PubMed]
3. Oyelaran, O.; McShane, L.M.; Dodd, L.; Gildersleeve, J.C. Profiling human serum antibodies with a carbohydrate antige microarray. *J. Proteome Res.* **2009**, *8*, 4301–4310. [CrossRef]
4. Kappler, K.; Hennet, T. Emergence and significance of carbohydrate-specific antibodies. *Genes Immun.* **2020**, *21*, 224–23 [CrossRef]
5. Marglous, S.; Brown, C.E.; Padler-Karavani, V.; Cummings, R.D.; Gildersleeve, J.C. Serum antibody screening using glycan array *Chem. Soc. Rev.* **2024**, *53*, 2603–2642. [CrossRef]
6. Lewis, A.L.; Chen, X.; Schnaar, R.L.; Varki, A. Sialic acids and other nonulosonic acids. In *Essentials of Glycobiology*, 4th ed.; Vark A., Cummings, R.D., Esko, J.D., Stanley, P., Hart, G.W., Aebi, M., Mohnen, D., Kinoshita, T., Packer, N.H., Prestegard, J.H., et a Eds.; Cold Spring Harbor Laboratory Press: Cold Spring Harbor, NY, USA, 2022; pp. 185–204.
7. Chen, X. Enabling chemoenzymatic strategies and enzymes for synthesizing sialyl glycans and sialyl glycoconjugates. *Acc. Cher Res.* **2024**, *57*, 234–246. [CrossRef] [PubMed]
8. Knirel, Y.A.; Shashkov, A.S.; Tsvetkov, Y.E.; Jansson, P.E.; Zahringer, U. 5,7-diamino-3,5,7,9-tetradeoxynon-2-ulosonic acids i bacterial glycopolymers: Chemistry and biochemistry. *Adv. Carbohydr. Chem. Biochem.* **2003**, *58*, 371–417.
9. Santra, A.; Xiao, A.; Yu, H.; Li, W.; Li, Y.; Ngo, L.; McArthur, J.B.; Chen, X. A diazido mannose analogue as a chemoenzymati synthon for synthesizing di-*N*-acetyllegionaminic acid-containing glycosides. *Angew. Chem. Int. Ed.* **2018**, *57*, 2929–293 [CrossRef]
10. McArthur, J.B.; Santra, A.; Li, W.; Kooner, A.S.; Liu, Z.; Yu, H.; Chen, X. *L. pneumophila* CMP-5,7-di-*N*-acetyllegionaminic aci synthetase (LpCLS)-involved chemoenzymatic synthesis of sialosides and analogues. *Org. Biomol. Chem.* **2020**, *18*, 738–74 [CrossRef]
11. Tsvetkov, Y.E.; Shashkov, A.S.; Knirel, Y.A.; Zahringer, U. Synthesis and identification in bacterial lipopolysaccharides c 5,7-diacetamido-3,5,7,9-tetradeoxy-D-glycero-D-galacto- and -D-glycero-D-talo-non-2-ulosonic acids. *Carbohydr. Res.* **200** *331*, 233–237. [CrossRef] [PubMed]
12. Tsvetkov, Y.E.; Shashkov, A.S.; Knirel, Y.A.; Zahringer, U. Synthesis and NMR spectroscopy of nine stereoisomeric 5,7-diacetamid 3,5,7,9-tetradeoxynon-2-ulosonic acids. *Carbohydr. Res.* **2001**, *335*, 221–243. [CrossRef] [PubMed]
13. McDonald, N.D.; Boyd, E.F. Structural and biosynthetic diversity of nonulosonic acids (NulOs) that decorate surface structures i bacteria. *Trends Microbiol.* **2021**, *29*, 142–157. [CrossRef]

14. Pradhan, K.; Kulkarni, S.S. Synthesis of nonulosonic acids. *Eur. J. Org. Chem.* **2020**, *2020*, 6819–6830. [CrossRef]
15. Matthies, S.; Stallforth, P.; Seeberger, P.H. Total synthesis of legionaminic acid as basis for serological studies. *J. Am. Chem. Soc.* **2015**, *137*, 2848–2851. [CrossRef] [PubMed]
16. Gintner, M.; Yoneda, Y.; Schmolzer, C.; Denner, C.; Kahlig, H.; Schmid, W. A versatile de novo synthesis of legionaminic acid and 4-epi-legionaminic acid starting from D-serine. *Carbohydr. Res.* **2019**, *474*, 34–42. [CrossRef]
17. Siyabalapitiya Arachchige, S.; Crich, D. Syntheses of legionaminic acid, pseudaminic acid, acetaminic acid, 8-epi-acetaminic acid, and 8-epi-legionaminic acid glycosyl donors from *N*-acetylneuraminic acid by side chain exchange. *Org. Lett.* **2022**, *24*, 2998–3002. [CrossRef] [PubMed]
18. Popik, O.; Dhakal, B.; Crich, D. Stereoselective synthesis of the equatorial glycosides of legionaminic acid. *J. Org. Chem.* **2017**, *82*, 6142–6152. [CrossRef] [PubMed]
19. Dhakal, B.; Buda, S.; Crich, D. Stereoselective synthesis of 5-epi-alpha-sialosides related to the pseudaminic acid glycosides. Reassessment of the stereoselectivity of the 5-azido-5-deacetamidosialyl thioglycosides and use of triflate as nucleophile in the Zbiral deamination of sialic acids. *J. Org. Chem.* **2016**, *81*, 10617–10630.
20. Glaze, P.A.; Watson, D.C.; Young, N.M.; Tanner, M.E. Biosynthesis of CMP-*N,N'*-diacetyllegionaminic acid from UDP-*N,N'*-diacetylbacillosamine in *Legionella pneumophila*. *Biochemistry* **2008**, *47*, 3272–3282. [CrossRef] [PubMed]
21. Schoenhofen, I.C.; Vinogradov, E.; Whitfield, D.M.; Brisson, J.R.; Logan, S.M. The CMP-legionaminic acid pathway in *Campylobacter*: Biosynthesis involving novel GDP-linked precursors. *Glycobiology* **2009**, *19*, 715–725. [CrossRef]
22. Hassan, M.I.; Lundgren, B.R.; Chaumun, M.; Whitfield, D.M.; Clark, B.; Schoenhofen, I.C.; Boddy, C.N. Total biosynthesis of legionaminic acid, a bacterial sialic acid analogue. *Angew. Chem. Int. Ed.* **2016**, *55*, 12018–12021. [CrossRef]
23. Schoenhofen, I.C.; Young, N.M.; Gilbert, M. Biosynthesis of legionaminic scid and its incorporation into glycoconjugates. *Methods Enzymol.* **2017**, *597*, 187–207.
24. Meng, X.; Boons, G.J.; Wosten, M.; Wennekes, T. Metabolic labeling of legionaminic acid in flagellin glycosylation of *Campylobacter jejuni* identifies Maf4 as a putative legionaminyl transferase. *Angew. Chem. Int. Ed.* **2021**, *60*, 24811–24816. [CrossRef] [PubMed]
25. Yu, H.; Chokhawala, H.; Karpel, R.; Yu, H.; Wu, B.; Zhang, J.; Zhang, Y.; Jia, Q.; Chen, X. A multifunctional *Pasteurella multocida* sialyltransferase: A powerful tool for the synthesis of sialoside libraries. *J. Am. Chem. Soc.* **2005**, *127*, 17618–17619. [CrossRef] [PubMed]
26. Watson, D.C.; Leclerc, S.; Wakarchuk, W.W.; Young, N.M. Enzymatic synthesis and properties of glycoconjugates with legionaminic acid as a replacement for neuraminic acid. *Glycobiology* **2011**, *21*, 99–108. [CrossRef]
27. Watson, D.C.; Wakarchuk, W.W.; Leclerc, S.; Schur, M.J.; Schoenhofen, I.C.; Young, N.M.; Gilbert, M. Sialyltransferases with enhanced legionaminic acid transferase activity for the preparation of analogs of sialoglycoconjugates. *Glycobiology* **2015**, *25*, 767–773. [CrossRef]
28. Watson, D.C.; Wakarchuk, W.W.; Gervais, C.; Durocher, Y.; Robotham, A.; Fernandes, S.M.; Schnaar, R.L.; Young, N.M.; Gilbert, M. Preparation of legionaminic acid analogs of sialo-glycoconjugates by means of mammalian sialyltransferases. *Glycoconj. J.* **2015**, *32*, 729–734. [CrossRef] [PubMed]
29. Li, W.; Ghosh, T.; Bai, Y.; Santra, A.; Xiao, A.; Chen, X. A substrate tagging and two-step enzymatic reaction strategy for large-scale synthesis of 2,7-anhydro-sialic acid. *Carbohydr. Res.* **2019**, *479*, 41–47. [CrossRef] [PubMed]
30. Kooner, A.S.; Diaz, S.; Yu, H.; Santra, A.; Varki, A.; Chen, X. Chemoenzymatic synthesis of sialosides containing 7-*N*- or 7,9-di-*N*-acetyl sialic acid as stable *O*-acetyl analogues for probing sialic acid-binding proteins. *J. Org. Chem.* **2021**, *86*, 14381–14397. [CrossRef] [PubMed]
31. Xu, Y.; Fan, Y.; Ye, J.; Wang, F.; Nie, Q.; Wang, L.; Wang, P.G.; Cao, H.; Cheng, J. Successfully engineering a bacterial sialyltransferase for regioselective α2,6-sialylation. *ACS Catal.* **2018**, *8*, 7222–7227. [CrossRef]
32. Sugiarto, G.; Lau, K.; Qu, J.; Li, Y.; Lim, S.; Mu, S.; Ames, J.B.; Fisher, A.J.; Chen, X. A sialyltransferase mutant with decreased donor hydrolysis and reduced sialidase activities for directly sialylating LewisX. *ACS Chem. Biol.* **2012**, *7*, 1232–1240. [CrossRef]
33. Maki, T.; Tsuritani, T.; Yasukata, T. A mild method for the synthesis of carbamate-protected guanidines using the Burgess reagent. *Org. Lett.* **2014**, *16*, 1868–1871. [CrossRef]
34. Song, X.; Xia, B.; Stowell, S.R.; Lasanajak, Y.; Smith, D.F.; Cummings, R.D. Novel fluorescent glycan microarray strategy reveals ligands for galectins. *Chem. Biol.* **2009**, *16*, 36–47. [CrossRef] [PubMed]
35. Blixt, O.; Head, S.; Mondala, T.; Scanlan, C.; Huflejt, M.E.; Alvarez, R.; Bryan, M.C.; Fazio, F.; Calarese, D.; Stevens, J.; et al. Printed covalent glycan array for ligand profiling of diverse glycan binding proteins. *Proc. Natl. Acad. Sci. USA* **2004**, *101*, 17033–17038. [CrossRef]
36. Park, S.; Lee, M.R.; Shin, I. Fabrication of carbohydrate chips and their use to probe protein-carbohydrate interactions. *Nat. Protoc.* **2007**, *2*, 2747–2758. [CrossRef] [PubMed]
37. Park, S.; Lee, M.R.; Shin, I. Construction of carbohydrate microarrays by using one-step, direct immobilizations of diverse unmodified glycans on solid surfaces. *Bioconjug. Chem.* **2009**, *20*, 155–162. [CrossRef] [PubMed]
38. de Boer, A.R.; Hokke, C.H.; Deelder, A.M.; Wuhrer, M. General microarray technique for immobilization and screening of natural glycans. *Anal. Chem.* **2007**, *79*, 8107–8113. [CrossRef]
39. Song, X.; Xia, B.; Lasanajak, Y.; Smith, D.F.; Cummings, R.D. Quantifiable fluorescent glycan microarrays. *Glycoconj. J.* **2008**, *25*, 15–25. [CrossRef]

40. Yu, H.; Chokhawala, H.A.; Varki, A.; Chen, X. Efficient chemoenzymatic synthesis of biotinylated human serum albumin-sialoglycoside conjugates containing *O*-acetylated sialic acids. *Org. Biomol. Chem.* **2007**, *5*, 2458–2463. [CrossRef]
41. Bochner, B.S.; Alvarez, R.A.; Mehta, P.; Bovin, N.V.; Blixt, O.; White, J.R.; Schnaar, R.L. Glycan array screening reveals a candidate ligand for Siglec-8. *J. Biol. Chem.* **2005**, *280*, 4307–4312. [CrossRef]
42. Yu, H.; Huang, S.; Chokhawala, H.; Sun, M.; Zheng, H.; Chen, X. Highly efficient chemoenzymatic synthesis of naturally occurring and non-natural alpha-2,6-linked sialosides: A *P. damsela* alpha-2,6-sialyltransferase with extremely flexible donor-substrate specificity. *Angew. Chem. Int. Ed.* **2006**, *45*, 3938–3944. [CrossRef]
43. Kooner, A.S.; Yu, H.; Chen, X. Synthesis of *N*-glycolylneuraminic scid (Neu5Gc) and its glycosides. *Front. Immunol.* **2019**, *10*, 2004. [CrossRef] [PubMed]
44. Diaz, S.L.; Padler-Karavani, V.; Ghaderi, D.; Hurtado-Ziola, N.; Yu, H.; Chen, X.; Brinkman-Van der Linden, E.C.; Varki, A.; Varki, N.M. Sensitive and specific detection of the non-human sialic scid *N*-glycolylneuraminic acid in human tissues and biotherapeutic products. *PLoS ONE* **2009**, *4*, e4241. [CrossRef] [PubMed]
45. Leviatan Ben-Arye, S.; Yu, H.; Chen, X.; Padler-Karavani, V. Profiling anti-Neu5Gc IgG in human sera with a sialoglycan microarray assay. *J. Vis. Exp.* **2017**, *125*, e56094. [CrossRef]
46. Leviatan Ben-Arye, S.; Schneider, C.; Yu, H.; Bashir, S.; Chen, X.; von Gunten, S.; Padler-Karavani, V. Differential recognition of diet-derived Neu5Gc-neoantigens on glycan microarrays by carbohydrate-specific pooled human IgG and IgA antibodies. *Bioconjug. Chem.* **2019**, *30*, 1565–1574. [CrossRef]
47. Varki, A. Since there are PAMPs and DAMPs, there must be SAMPs? Glycan "self-associated molecular patterns" dampen innate immunity, but pathogens can mimic them. *Glycobiology* **2011**, *21*, 1121–1124. [CrossRef] [PubMed]
48. Kooner, A.S.; Yuan, Y.; Yu, H.; Kang, H.; Klenow, L.; Daniels, R.; Chen, X. Sialosides containing 7-*N*-acetyl sialic acid are selective substrates for neuraminidases from influenza A viruses. *ACS Infect. Dis.* **2023**, *9*, 33–41. [CrossRef]
49. Zebian, N.; Merkx-Jacques, A.; Pittock, P.P.; Houle, S.; Dozois, C.M.; Lajoie, G.A.; Creuzenet, C. Comprehensive analysis of flagellin glycosylation in *Campylobacter jejuni* NCTC 11168 reveals incorporation of legionaminic acid and its importance for host colonization. *Glycobiology* **2016**, *26*, 386–397. [CrossRef]
50. Varki, A.; Gagneux, P. Multifarious roles of sialic acids in immunity. *Ann. N. Y. Acad. Sci.* **2012**, *1253*, 16–36. [CrossRef]
51. Gulati, S.; Schoenhofen, I.C.; Lindhout-Djukic, T.; Lewis, L.A.; Moustafa, I.Y.; Saha, S.; Zheng, B.; Nowak, N.; Rice, P.A.; Varki, A.; et al. Efficacy of antigonococcal CMP-nonulosonate therapeutics require cathelicidins. *J. Infect. Dis.* **2020**, *222*, 1641–1650. [CrossRef]
52. Gulati, S.; Schoenhofen, I.C.; Whitfield, D.M.; Cox, A.D.; Li, J.; St Michael, F.; Vinogradov, E.V.; Stupak, J.; Zheng, B.; Ohnishi, M.; et al. Utilizing CMP-sialic acid analogs to unravel *Neisseria gonorrhoeae* lipooligosaccharide-mediated complement resistance and design novel therapeutics. *PLoS Pathog.* **2015**, *11*, e1005290. [CrossRef] [PubMed]
53. Gulati, S.; Schoenhofen, I.C.; Lindhout-Djukic, T.; Schur, M.J.; Landig, C.S.; Saha, S.; Deng, L.; Lewis, L.A.; Zheng, B.; Varki, A.; et al. Therapeutic CMP-nonulosonates against multidrug-resistant *Neisseria gonorrhoeae*. *J. Immunol.* **2020**, *204*, 3283–3295. [CrossRef] [PubMed]
54. Amon, R.; Grant, O.C.; Leviatan Ben-Arye, S.; Makeneni, S.; Nivedha, A.K.; Marshanski, T.; Norn, C.; Yu, H.; Glushka, J.N.; Fleishman, S.J.; et al. A combined computational-experimental approach to define the structural origin of antibody recognition of sialyl-Tn, a tumor-associated carbohydrate antigen. *Sci. Rep.* **2018**, *8*, 10786. [CrossRef] [PubMed]
55. Borenstein-Katz, A.; Warszawski, S.; Amon, R.; Eilon, M.; Cohen-Dvashi, H.; Leviatan Ben-Arye, S.; Tasnima, N.; Yu, H.; Chen, X.; Padler-Karavani, V.; et al. Biomolecular recognition of the glycan neoantigen CA19-9 by distinct antibodies. *J. Mol. Biol.* **2021**, *433*, 167099. [CrossRef] [PubMed]
56. Li, Y.; Yu, H.; Cao, H.; Lau, K.; Muthana, S.; Tiwari, V.K.; Son, B.; Chen, X. *Pasteurella multocida* sialic acid aldolase: A promising biocatalyst. *Appl. Microbiol. Biotechnol.* **2008**, *79*, 963–970. [CrossRef] [PubMed]
57. Yu, H.; Yu, H.; Karpel, R.; Chen, X. Chemoenzymatic synthesis of CMP-sialic acid derivatives by a one-pot two-enzyme system: Comparison of substrate flexibility of three microbial CMP-sialic acid synthetases. *Bioorg. Med. Chem.* **2004**, *12*, 6427–6435. [CrossRef]
58. Padler-Karavani, V.; Song, X.; Yu, H.; Hurtado-Ziola, N.; Huang, S.; Muthana, S.; Chokhawala, H.A.; Cheng, J.; Verhagen, A.; Langereis, M.A.; et al. Cross-comparison of protein recognition of sialic acid diversity on two novel sialoglycan microarrays. *J. Biol. Chem.* **2012**, *287*, 22593–22608. [CrossRef]
59. Bashir, S.; Leviatan Ben Arye, S.; Reuven, E.M.; Yu, H.; Costa, C.; Galinanes, M.; Bottio, T.; Chen, X.; Padler-Karavani, V. Presentation mode of glycans affect recognition of human serum anti-Neu5Gc IgG antibodies. *Bioconjug. Chem.* **2019**, *30*, 161–168. [CrossRef]

biomolecules

MDPI

Article

The Autophagic Activator GHF-201 Can Alleviate Pathology in a Mouse Model and in Patient Fibroblasts of Type III Glycogenosis

Kumudesh Mishra [1], Sahar Sweetat [1], Saja Baraghithy [2], Uri Sprecher [3], Monzer Marisat [3], Sultan Bastu [4], Hava Glickstein [5], Joseph Tam [2], Hanna Rosenmann [1], Miguel Weil [3], Edoardo Malfatti [4] and Or Kakhlon [1,6,*]

1 Department of Neurology, The Agnes Ginges Center for Human Neurogenetics, Hadassah-Hebrew University Medical Center, Jerusalem 9112001, Israel; kumudeshmishra@gmail.com (K.M.); sahar.sweetat@mail.huji.ac.il (S.S.); rosenman@hadassah.org.il (H.R.)

2 Obesity and Metabolism Laboratory, Institute for Drug Research, School of Pharmacy, Faculty of Medicine, The Hebrew University of Jerusalem, Jerusalem 9112001, Israel; saja.baraghithy@mail.huji.ac.il (S.B.); yossi.tam@mail.huji.ac.il (J.T.)

3 The Shmunis School of Biomedicine and Cancer Research, The George S. Wise Faculty for Life Sciences, Sagol School of Neurosciences, Tel Aviv University, Tel Aviv 6997801, Israel; uri8sp@gmail.com (U.S.); monzermar@gmail.com (M.M.); miguelw@tauex.tau.ac.il (M.W.)

4 Centre de Reference de Maladies Neuromusculaires, UPEC—Paris Est University, IMRB INSERM U955, Team Biology of the Neuromuscular System, Faculty of Medicine, APHP Hopital Henri Mondor, 1 Rue Gustave Eiffel, 94010 Creteil, France; sultan.bastu@inserm.fr (S.B.); edoardo.malfatti@aphp.fr (E.M.)

5 Electron Microscopy Unit, The Hebrew University-Hadassah Medical School, Ein Kerem, Jerusalem 9112001, Israel; hava.glickstein@mail.huji.ac.il

6 Faculty of Medicine, Hebrew University of Jerusalem, Jerusalem 9112001, Israel

* Correspondence: ork@hadassah.org.il; Tel.: +972-546-812-380

Abstract: Glycogen storage disease type III (GSDIII) is a hereditary glycogenosis caused by deficiency of the glycogen debranching enzyme (GDE), an enzyme, encoded by *Agl*, enabling glycogen degradation by catalyzing alpha-1,4-oligosaccharide side chain transfer and alpha-1,6-glucose cleavage. GDE deficiency causes accumulation of phosphorylase-limited dextrin, leading to liver disorder followed by fatal myopathy. Here, we tested the capacity of the new autophagosomal activator GHF-201 to alleviate disease burden by clearing pathogenic glycogen surcharge in the GSDIII mouse model $Agl^{-/-}$. We used open field, grip strength, and rotarod tests for evaluating GHF-201's effects on locomotion, a biochemistry panel to quantify hematological biomarkers, indirect calorimetry to quantify in vivo metabolism, transmission electron microscopy to quantify glycogen in muscle, and fibroblast image analysis to determine cellular features affected by GHF-201. GHF-201 was able to improve all locomotion parameters and partially reversed hypoglycemia, hyperlipidemia and liver and muscle malfunction in $Agl^{-/-}$ mice. Treated mice burnt carbohydrates more efficiently and showed significant improvement of aberrant ultrastructural muscle features. In GSDIII patient fibroblasts, GHF-201 restored mitochondrial membrane polarization and corrected lysosomal swelling. In conclusion, GHF-201 is a viable candidate for treating GSDIII as it recovered a wide range of its pathologies in vivo, in vitro, and ex vivo.

Keywords: glycogen; glycogen storage disease type III; pharmacotherapy

1. Introduction

Glycogen is a branched polysaccharide with a molecular weight of nine to ten million Da. The average glycogen molecule contains about 55,000 glucose residues linked by α-1,4 (92%) and α-1,6 (8%) glycosidic bonds [1]. The synthesis of glycogen is primed by glycogenin [2] (even though aberrant forms of glycogen are also formed when glycogenin is suppressed [3]), and catalyzed by glycogen synthase (GYS) and glycogen branching enzyme (GBE). GYS links glucose residues to each other by α-1,4 glycosidic bonds to form linear chains. GBE, in turn, attaches a short linear stretch of α-1,4-linked glucose units from the outer non-reducing end of a growing glycogen chain into an α-1,6 position of a pre-existing chain, thus generating a branch. Repetitive cycles of elongation and branching generate a large (110–290 nm in liver) [4] spherical molecule with a hydrophilic surface and an increased number of reactive termini, facilitating glycogen synthesis and degradation [5,6]. Glycogen degradation, on the other hand, takes place in the cytosol and in lysosomes. In the cytosol, glycogen is degraded by glycogen phosphorylase (GP), which phosphorylytically releases outer glucose residues in the form of glucose-1-phosphate, and by glycogen debranching enzyme (GDE). GP phosphorylates outer glucose residues and releases each one as glucose-1-hosphate. GP can degrade an outer chain only until four residues from the branch point. Then, GDE is required to complete glycogenolysis. GDE catalyzes two reactions: transferring a three-glucosyl glucan to the non-reducing end of another linear strand (α-1,4-glucanotransferase reaction) and hydrolyzing the α-1-6-glycosidic bond remaining on the branch point, releasing glucose (amylo-α-1,6-glucosidase reaction, *n.b.*, glucose, rather than glucose-1-phosphate, is released [4]). In the lysosomes, glycogen degradation is catalyzed by α-glucosidase.

Glycogen is stored primarily in liver and muscle, where it, respectively, serves as a reservoir for regulating blood glucose levels and as a readily mobilizable energy reserve. Therefore, aberrations in glycogen metabolism, or glycogen storage disorders (GSDs), are primarily hepatic or muscular disorders, with two notable exceptions, APBD and LD, specifically involving the central and peripheral nervous systems. The most common disorder of glycogen metabolism is observed in diabetes [7,8], in which abnormal amounts of insulin or abnormal insulin response result in accumulation or depletion of liver glycogen. However, GSDs are normally associated with hereditary deficiencies in enzymes of glycogen and glucose metabolism causing glycogen accumulation (except for GSD type 0, which is a GYS deficiency and is therefore associated with glycogen insufficiency).

In this work, we focused on GSD type III (GSDIII, Cori disease), the third most prevalent muscle GSD (incidence of 1:100,000) after GSD types V and II [9–11], which is caused by GDE deficiency leading to accumulation of glycogen due to its reduced degradation. Phosphorylase-limit dextrin (PLD), the type of glycogen which accumulates in GSDIII, is structurally abnormal, containing shorter outer branches [12]. While phosphorylase-limit dextrin is sensitive to diastase digestion and does not form polyglucosan bodies [12], it is still associated with large vacuoles and can cause histological damage to myofibrils [13,14]. GSDIII usually starts as a liver disorder characterized by hepatomegaly, hypoglycemia, hyperlipidemia, and hyperketonemia. These result from the limited ability to breakdown glycogen to glucose, leading to excessive use of lipid oxidation as an alternative energy source with ensuing attenuation of lipid uptake by adipocytes and increase in ketone bodies as byproducts of fatty acid oxidation (FAO). Later on, myopathy pursues [15], with occasional cardiomyopathy [16], liver cirrhosis, and hepatocellular adenoma [17].

GSDIII implicates vacuolar myopathy and myofibrillar damage [14,18], also observed in GSDII [19], which shares muscle glycogen over accumulation with GSDIII (albeit lysosomal in GSDII and mostly cytoplasmic in GSDIII). We, therefore, tested the GSDIII curative capacity of GHF-201, a novel, safe compound capable of reversing glycogen accumulation and enhancing autophagy [20,21]. We show that GHF-201 is capable of improving motility and grip strength in an $Agl^{-/-}$ mouse model of GSDIII, where GDE, encoded by the murine *Agl* gene, was knocked out. At the same time, GHF-201 did not modify the deficient motor learning capacities of $Agl^{-/-}$ mice associated with a neurological deficiency not implicated in GSDIII. We further demonstrate that GHF-201 corrected hypoglycemia, hyperlipidemia, and liver and muscle damage, as well as carbohydrate over fat fuel preference. Combined with reduction of glycogen levels and muscle atrophy at the tissue level, and improvement of lysosomal and mitochondrial phenotypes at the cell level (in GSDIII patients' fibroblasts), our results strongly support that GHF-201 has a potential therapeutic capacity for treating GSDIII, either as a standalone, or as part of a combination therapy. At the moment, GSDIII has no curative alternative and patients are only managed by dietary restrictions [22]. A clinical trial for GSDIII using the UX053 gene therapy-based replacement of GDE (NCT04990388) has been initiated, but was terminated at phase I. While the gene therapy GDE replacement approach is still valid, it can also be improved by add-on therapy based on a mechanistically complementary approach. In that respect, activation of glycogen degradation by autophagy merits appraisal, especially since rapamycin-mediated activation of autophagy was indeed efficacious in a GSDIII dog model [23].

2. Materials and Methods

2.1. In Vivo Studies

In vivo studies were conducted on $Agl^{-/-}$ mice [9] and congenic C57Bl6J mice as littermate wild type controls ($Agl^{-/-}$). n = 6–10 mice/arm were used. Twice a week, 50 μL of 250 mg/kg GHF-201 in 10% DMSO was intramuscularly injected to the gastrocnemius in alternate flanks. A 10% DMSO vehicle was used as a vehicle control. Mice were treated from the age of 3 months to the age of 8 months. Animals' weight was recorded every week. A weight decrease of >10% between biweekly weightings, or >20% from treatment initiation led to removal of the animal from the study. We tested monthly moving distance and cumulative sojourn in the center of an open field, front paw grip strength using a grip strength meter (Columbus Instruments, Columbus, OH, USA), and rotarod performance (latency to fall from a rotating cylinder). The rotating cylinder (UGO Basile S.R.L., Gemonio, Italy) accelerated from 5 to 40 RPM in 99 s. Mice were tested for 3 consecutive trials with a rest period of approximately 10 min between trials.

Blood biochemistry (enzyme activities), triglycerides, and glucose were determined at the Hadassah central laboratory division. Blood was collected from the cheek pouch by submandibular bleeding. Indirect calorimetry and multi-parameter metabolic assessment (metabolic cages) were performed in Promethion Metabolic Phenotyping Systems (Sable Instruments, Las Vegas, NV, USA) as in [20]. Specifically, locomotor activity was assessed in these cages by the number of times an infrared XYZ beam array is disrupted by animal movement, while wheel running was determined by the number of wheel revolutions [24].

2.2. Ultrastructural Electron Microscopy Studies

Gastrocnemius muscle samples were fixed, embedded, sectioned, and analyzed by transmission electron microscope as described in [9]. Area percentage of glycogen dense granules in longitudinal muscle sections (10 sections/animal) was quantified in samples from n = 3 vehicle-treated WT, n = 3 vehicle-treated $Agl^{-/-}$, and n = 3 GHF-201-treated $Agl^{-/-}$ mice.

2.3. Biochemical Glycogen Quantification

Muscle glycogen was quantified biochemically using Merck's Glycogen Assay Kit (Cat. # MAK465) according to the manufacturer instructions. In brief, 10–20 mg of muscle tissues was homogenized on ice in 1 mL of 25 mM citric acid, 2.5 mg/mL NaF, pH 4.2. Samples were spun down (14,000× g, 5 min, RT) to clear the debris. Then, 10 μL of cleared supernatant was used for the colorimetric determination of glycogen using the kit. A calibration curve of 0–200 μg/mL glycogen (provided with the kit) was used for glycogen quantification. Water was used as a blank control, to which the same reagents were added. This blank reading was subtracted from the colorimetric readings of all samples and calibration standards.

2.4. High Content Analysis of GSDIII Patient Fibroblasts

Primary skin fibroblasts from 3 patients were procured from Coriell Institute (Camden, NJ, USA) and expanded in DMEM supplemented with 1% Sodium Pyruvate, 1% Penicillin-Streptomycin-Amphotericin B, 10% heat-inactivated fetal calf serum (Biological industries, Beit HaEmek, Israel), and 1% 100× non-essential amino acids solution (NEAA) at 37 °C with 5% CO_2. For live imaging, GSDIII and healthy control (HC) skin fibroblasts were seeded at 1400 cells per well and cultured at 37 °C and 5% CO_2 with or without GHF-201 (50 μM) for 24 h in specialized microscopy-grade 96-well plates (Cellvis P96-1.5H-N). Cells were then washed in HBSS, supplemented for 20 min with 1.6 mM Hoechst 33342, 0.1 mM LysoTracker Deep Red, 0.05 mM TMRE, and 0.4 mM calcein-AM Green (Thermo-Fisher Scientific, Waltham, MA, USA), and washed again with HBSS. Subsequently, images were acquired using an Operetta G1 system at 20× magnification under environmental control (37 °C and 5% CO_2) conditions. All assay parameters (including the acquisition exposure times, objective, and the analysis parameters) were kept constant for all assay repetitions. Images were analyzed by the Harmony image analysis software version 1.0. Downstream analysis included removal of wells with outlier values using GraphPad Prism (version 10.2.3)'s ROUT method (Q = 10%), feature selection by backward feature elimination, and normalization to HC levels prior to statistical analysis [25]. Multivariate analysis was performed in Python.

2.5. Statistical Analysis

Sample sizes were determined based on the effect sizes of key parameters measured (e.g., open field moving distance), as detailed in [26]. These sample sizes were sufficient to provide the acceptable power of at least 80%. Blinding in in vivo studies was obtained by encoding the animals in the different treatment groups so that the experimenters were blinded to treatment allocation. To obtain randomization, litters from mice born a few days apart were pooled and pups of similar weight were divided into vehicle- and GHF-201-treated arms. Specific statistical tests are mentioned in the relevant text and figure legends.

For the indirect calorimetry and metabolic cage analyses (Figure in Section 3.4), mean differences in metabolic parameters between wild type and vehicle-treated $Agl^{-/-}$ mice, wild type and GHF-201-treated mice and vehicle-treated and GHF-201-treated $Agl^{-/-}$ mice were calculated during light, dark, and overall (light + dark) periods. One Way ANOVA with repeated measures (and Tukey post hoc validation test) was used to determine the statistical significance of these mean differences, which are the output of repeated measures of ANOVA, for each metabolic parameter during the entire time periods analyzed (light, dark, or overall). We show in the left panels the actual time-dependent trends of the respective metabolic parameter in each treatment arm. In the right panels, we show the mean differences between the group trend averages (n = 4 mice in each group) derived from these time trends. For the food and water intake profiles, since they are cumulative, mean differences were calculated based on subtractions of the time periods end values: at 12 h for the light period, at 24 h minus 12 h for the dark period, and at 24 h for the overall periods. The statistical significance of the mean differences is shown by asterisks above the respective bars. The total energy expenditure (TEE) ANCOVA analysis was provided by the NIDDK Mouse Metabolic Phenotyping Centers (MMPC, www.mmpc.org (accessed on 1 July 2024)) using their Energy Expenditure Analysis page (http://www.mmpc.org/shared/regression.aspx (accessed on 1 July 2024)) and supported by grants DK076169 and DK115255.

3. Results

3.1. Animal Weight

GSDIII male mice presented with a higher weight throughout the treatment period as compared to WT mice ($p < 0.0001$, Two Way ANOVA with repeated measures). According to [13], GSDIII mice were lighter than WT mice, while in [27], no significant difference between the weights of GSDIII and WT mice was observed, together suggesting that such a difference might be batch-dependent. GHF-201 treatment did not significantly modify weight in GSDIII mice (Figure 1). Animal weighting was discontinued 138 d after treatment initiation due to an unexplained >10% weight drop between two weightings in all animals.

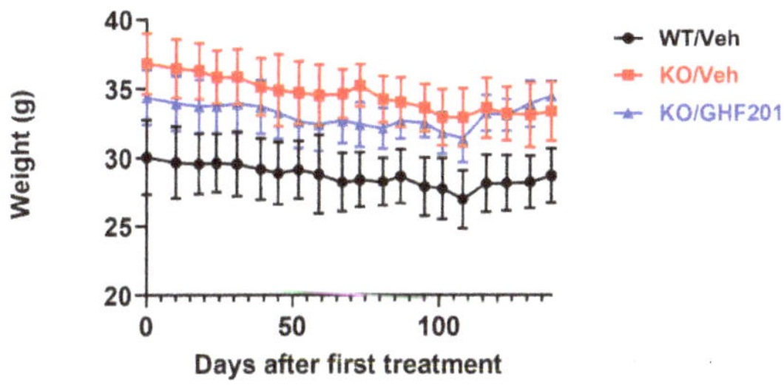

Figure 1. Weight over time in WT mice, $Agl^{-/-}$ (KO) mice injected intramuscularly biweekly with 10% DMSO solvent control, and KO mice injected in the same way with 250 mg/kg GHF-201. n = 8 WT/Veh, n = 6 KO/Veh, and n = 10 KO/GHF-201 mice were used. Throughout the period, KO mice were heavier than WT mice ($p < 0.0001$, Two Way ANOVA with repeated measures). Due to a >10% weight reduction between biweekly weightings, one mouse was removed from the KO/Veh group on day 116 after treatment initiation and one mouse was removed from the KO/GHF201 group on day 52 after treatment initiation.

3.2. Locomotor Studies

We show that GHF-201 improved open field performance at a relatively advanced stage of the disease (last time point when mice were approximately 8 months old). This improvement was demonstrated as increases in both distance traveled (Figure 2A) and sojourn in the center of the open field (Figure 2B), which indicates a more exploratory behavior associated with decreased anxiety. Sojourn in the center was corrected by GHF-201 to WT values at the age of 8 months. Interestingly, while GHF-201 significantly increased grip strength (Figure 2C) and improved open field parameters, it did not increase the latency to fall from a rotating rod in the rotarod test (Two Way ANOVA with repeated measures, $p < 0.95$, $p < 0.95$, $p < 0.97$ at 10, 47, and 108 days post-treatment initiation) and even decreased it 80 days post-treatment initiation (Two Way ANOVA with repeated measures, $p < 0.05$). Latency to fall was decreased in $Agl^{-/-}$ mice as compared to WT at all times post-treatment initiation (Two Way ANOVA with repeated measures, $p < 0.11$ (only a trend), $p < 0.03$, $p < 0.005$, and $p < 0.02$ at 10, 47, 80, and 108 days post-treatment initiation, Figure 2D). Training and motor learning capacity were, respectively, demonstrated by increased latency between runs in each test and by increased overall latency over time. As opposed to untreated and treated $Agl^{-/-}$ mice, which showed no training capacity, WT mice manifested training capacity at 80 days post-treatment initiation. This was determined by comparing the average run latencies (shown in different shades of gray, red, and blue in WT, KO/Veh, and KO/GHF-201, respectively) at each time following treatment initiation using Two Way ANOVA with repeated measures ($p < 0.04$). Motor learning capacity of WT mice was also significantly higher than that of untreated or treated $Agl^{-/-}$ mice ($p < 0.0095$, $R^2 = 0.2687$ for linear correlation of latency v time in WT mice, as compared to ($p < 0.0436$, $R^2 = 0.0732$) and ($p < 0.0691$, $R^2 = 0.0523$) for untreated and treated $Agl^{-/-}$ mice, respectively). Indeed, rotarod measures motor coordination, which is considered a neurological function of the nigrostriatal dopamine system [28,29] not implicated in GSDIII. Motor activity measured by the open field test, on the other hand, is influenced by muscle strength, which was modified by GHF-201 (Figure 2C). There are several works (e.g., [30]) which demonstrate the independence of rotarod and open field results.

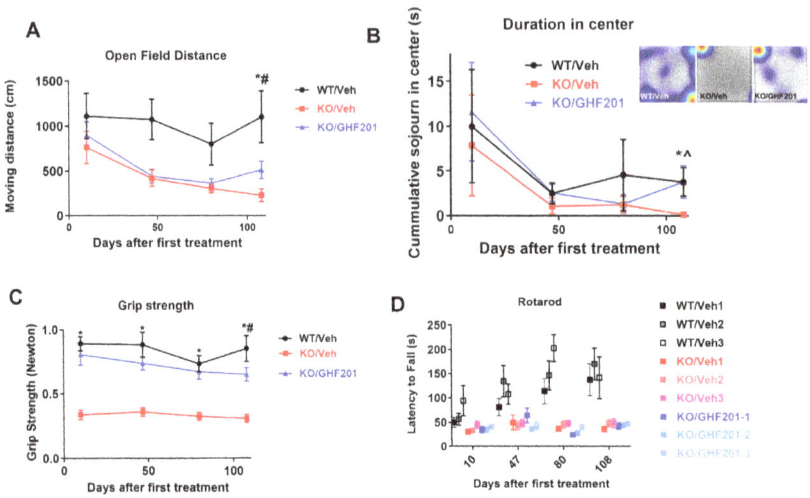

Figure 2. Movement distance (**A**), sojourn in the center of an open field arena (inset representative open field sojourn heatmaps) (**B**), grip strength (**C**), and latency to fall from a rotating rod (**D**) were measured in n = 8 $Agl^{-/-}$ mice treated with 10% DMSO vehicle (KO/Veh), n = 6 wild type mice treated with vehicle (WT/Veh), and n = 10 $Agl^{-/-}$ mice treated with GHF-201 (KO/GHF-201). In (**D**), each rotarod session included three consecutive runs (1–3), separated by 10 min pauses. Two Way ANOVA with repeated measures shows that only in (**C**) GHF-201 treatment values were higher than vehicle-treated values in KO mice throughout the period ($p < 0.05$). Linear regression analysis of overall latency over time (**D**), shows that WT animals developed a significantly higher learning capacity than KO/Veh and KO/GHF-201 mice. Training capacity, shown by increase in latency over runs, was only demonstrated in WT mice 80 days post-treatment initiation. See text for statistical analysis. *, KO/GH201 is significantly different from KO/Veh; *#, KO/GHF-201 is significantly different from both KO/Veh and WT/Veh; *^, KO/GHF-201 is significantly different only from KO/Veh and not from WT/Veh (correction effect). Difference significance determined by multiple *t*-tests with Sidak post hoc correction. All error bars represent s.e.m.

Notably, all behavioral tests had to be discontinued following the 4th test post-treatment initiation. The reason for that was that after the 4th test the weight of all mice was decreased by >10% from the last reading suggesting that GHF-201, increases muscle strength-associated mobility, but not survival, in GSDIII mice.

3.3. Blood Metabolic Panel

We tested whether GHF-201 is able to correct the hypoglycemia, hypertriglyceridemia, and liver and muscle damage as measured by blood examinations [13,31,32]. Such effects are expected from an agent, such as GHF-201, capable of inducing the catabolism of liver glycogen with an ensuing rise in blood glucose and reduction in compensatory lipid flux to the liver. Our blood metabolic panel results show that GHF-201 is able to correct hypoglycemia and hyperlipidemia and partially restore muscle (creatine kinase) and liver (alanine transferase) damage, all of which characteristically accrue with time in the $Agl^{-/-}$ GSDIII mouse model (Figure 3).

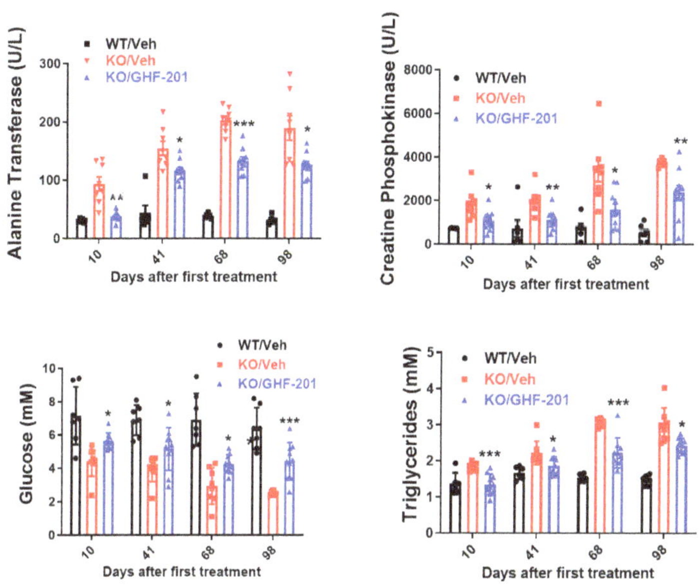

Figure 3. Blood metabolic panel based on n = 8 $Agl^{-/-}$ mice treated with 10% DMSO vehicle (KO/Veh), n = 6 wild type mice treated with vehicle (WT/Veh), and n = 10 $Agl^{-/-}$ mice treated with GHF-201, as indicated. At all time points, GHF-201 significantly reduced alanine transferase and creatine phosphokinase, demonstrating partial restoration of liver and muscle damages, respectively. Additionally, blood glucose was increased, and blood triglycerides were decreased by GHF-201 in treated $Agl^{-/-}$ mice. Statistical significance between KO/Veh and KO/GHF-201 at the different time points was determined by Two Way ANOVA with repeated measures with a Tukey post hoc test according to simple effects within rows (i.e., times). *, $0.01 < p < 0.05$; **, $0.001 < p < 0.01$; ***, $p < 0.001$.

3.4. In Vivo Metabolic Profile

The effect of GHF-201 on various in vivo metabolic parameters was determined by indirect calorimetry. Fuel preference at the whole animal level is determined by the respiratory exchange ratio (RER, the ratio of CO_2 produced to O_2 consumed). Lower RER indicates higher fat burn, while higher RER indicates higher carbohydrate burn. As our results (Figure 4(A1)) show, $Agl^{-/-}$ mice had lower RER values in the light, dark, and overall (positive mean difference in the black and red bar graphs indicates that WT/Veh levels are higher than KO/Veh and KO/GHF-201, respectively; negative mean difference indicates that they are lower). GHF-201 significantly increased RER in treated v. untreated $Agl^{-/-}$ mice in light, dark, and overall (blue bars; positive values indicate KO/Veh > KO/GHF-201, negative values indicate KO/GHF-201 > KO/Veh). However, as compared to WT mice, GHF-201 treated $Agl^{-/-}$ mice still had a significantly reduced RER in all periods (red bars). Total energy expenditure (TEE, Figure 4(A3)) was slightly reduced in $Agl^{-/-}$ mice as compared to WT mice (black bars). Like RER, TEE in $Agl^{-/-}$ mice was also significantly increased by GHF-201 in light and dark conditions and overall (blue bars), with the most significant increase ($p < 0.001$) in the light condition (*cf.* $p < 0.05$ for dark and overall). Compared to WT, TEE in GHF-201-treated $Agl^{-/-}$ mice was increased in the light but remained lower than WT in the dark and overall (red bars). Thus, GHF-201 treatment dramatically increased TEE in $Agl^{-/-}$ mice, where it was slightly reduced as compared to WT. GHF-201 correction of TEE can also be revealed by our ANCOVA analysis (Figure 4(A2),(A4)), performed on light and overall periods, where the TEE effect was the most pronounced, which also accounts for the dependence of TEE on body weight. According to the ANCOVA analysis, GHF-201 corrected weight-adjusted TEE to WT levels: the difference between WT and $Agl^{-/-}$ in light and overall conditions became insignificant when $Agl^{-/-}$ mice were treated with GHF-201. Importantly, similar to TEE correction to WT level by GHF-201, a significant difference in TEE between vehicle- and GHF-201-treated $Agl^{-/-}$ mice is also revealed by ANCOVA only under the light condition. As opposed to RER and TEE, fat oxidation (Figure 4(A5)) in $Agl^{-/-}$ mice was higher than in WT mice in all periods (black bars). GHF-201 increased fat oxidation in $Agl^{-/-}$ mice only in the light (blue bar) and maintained the higher rate of fat oxidation in $Agl^{-/-}$ as compared to WT mice (red bars). On the other hand, carbohydrate burning was compromised in $Agl^{-/-}$ mice as compared to WT with (red bars) or without (black bars) GHF-201 treatment (Figure 4(A6)). However, GHF-201 increased carbohydrate burning in $Agl^{-/-}$ mice, where it was compromised, strongly suggesting that the compound can increase glycogen catabolism. This is a therapeutic advantage as the accumulated, poorly digested, and malconstructed glycogen in GSDIII is considered a pathogenic factor. Ambulatory activity (Figure 4(B1)) was compromised in $Agl^{-/-}$ mice, as compared to WT, in all periods (black bars). This was corrected in $Agl^{-/-}$ mice by GHF-201 again only in the light (blue bars). Similar to the weight-corrected TEE (Figure 4(A2),(A4)), the ambulatory activity of $Agl^{-/-}$ mice was also rescued to WT level by GHF-201, but again only in the light period (red bar, no significant difference from WT after treatment). While no statistically significant difference between WT and $Agl^{-/-}$ mice was observed for the distance of wheel running (Figure 4(B2), black bars), GHF-201 did increase wheel running distance in $Agl^{-/-}$ mice in the light and overall periods (blue bars). This increase caused wheel running distance in $Agl^{-/-}$ mice to be significantly increased, in the light and overall periods, as compared to WT mice (red bars). Stimulation of wheel running and correction of ambulatory activity in the light (Figure 4B) are in line with open field results (Figure 2A), also acquired in the light, and with the GHF-201-mediated stimulation of carbohydrate catabolism. Together, these results suggest that GHF-201 can improve metabolic efficacy, compromised by the diseased state, at the whole animal level. Compared to WT mice, food intake (Figure 4(C1)) was not significantly modified in $Agl^{-/-}$ mice in the light and was reduced in the dark and overall (black bars). GHF-201 did not exert a significant effect on food intake, possibly excluding an increase in overall food intake which was masked by a large outlier effect of an animal in which GHF-201 reduced, rather than increased, food intake (Figure 4(C1), blue bars). Excluding this outlier, an increase in overall food uptake would have been similar to that observed in the GSDIV mouse model [20]. GHF-201 also did not reverse the reduction in food intake in $Agl^{-/-}$ mice compared to WT mice (red bars). Water uptake, on the other hand, also reduced in all periods in $Agl^{-/-}$ mice compared to WT (Figure 4(C2), black bars), was corrected by GHF-201 in $Agl^{-/-}$ mice in the dark and overall (blue bars). Nevertheless, this correction could not overcome the reduced extent of water intake in $Agl^{-/-}$ mice as compared to WT (red bars). As opposed to food intake, the effect of GHF-201 on water intake was similar in direction and in statistical significance to its effect in GSDIV mice [20]. The GHF-201-mediated increase in water intake was mainly observed in the dark (and, subsequently, overall), where activity and water (and food) consumption are normally increased in mice. A GHF-201-mediated increase in water intake with a parallel increase in fuel burning (Figure 2A) constitutes an improvement in metabolic efficiency. Please note that both food and water intake plots are cumulative and thus diverge from, rather than converge with, one another with time.

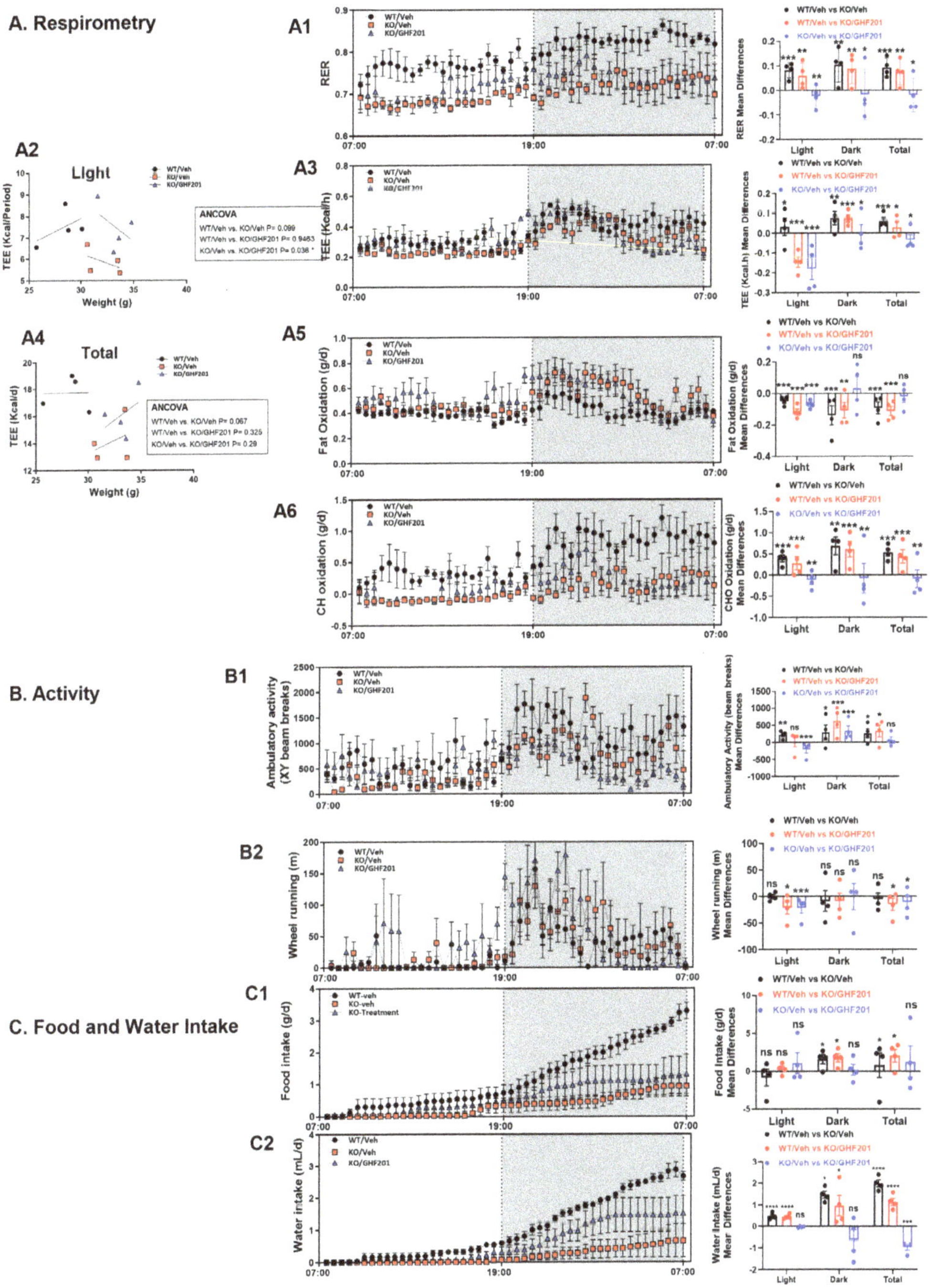

Figure 4. Indirect calorimetry and metabolic cage analysis. Following a 48 h habituation period, mice (n = 4 from each study arm indicated) were monitored over a 24 h period. Data are mean ± SEM from

eight-month-old mice. ((**A**)—Respirometry) As demonstrated by the positive mean differences, in the light, dark, and overall (total), vehicle-treated *Agl*$^{-/-}$ mice demonstrate lower respiratory exchange ratio (RER, (**A1**)), total energy expenditure (TEE, (**A2–A4**)), and carbohydrate oxidation (**A6**). As shown by the negative mean differences, fat oxidation was reduced in *Agl*$^{-/-}$ mice in light, dark, and overall as compared to WT mice (**A5**). All these parameters were increased or corrected by GHF-201 at all times, except for fat oxidation in the dark and overall (**A4**). ANCOVA results suggest that weight-dependent TEE is reduced in *Agl*$^{-/-}$ compared to WT mice in both the light condition (**A2**) and overall (**A4**), and that GHF-201 corrects this TEE reduction in *Agl*$^{-/-}$ mice in the light. ((**B**)—Activity) Ambulatory activity (**B1**) and wheel running (**B2**) were also increased (corrected), or not affected, by GHF-201 as indicated. ((**C**)—Food and Water Intake) Food intake (**C1**) was not significantly affected, and water intake ((**C2**), total cumulative values are shown in the bar graphs) was increased, mainly in the dark period, by GHF-201. Statistical significance of mean differences was determined by One Way ANOVA with repeated measures with a Tukey post hoc test. *, $0.01 < p < 0.05$; **, $0.001 < p < 0.01$; ***, $p < 0.001$; ****, $p < 0.0001$; ns, non-significant.

3.5. Muscle Glycogen Quantification

Blinded quantification by transmission EM (TEM) of gastrocnemius muscle sections showed that in muscle sections derived from GSDIII mice, a higher percentage of the area was occupied by dense glycogen granules as compared to muscle sections derived from wild type mice (Figure 5A). Myofibrils in GSDIII muscle sections were also more dissociated than wild type myofibrils due to the increased glycogen storage (Figure 5B). Both increase in glycogen granule area and myofibril dissociation were significantly corrected (even though not to wild type levels) by the GHF-201 treatment. Additional TEM images are shown in Supplementary Figure S1. These results were corroborated by a biochemical analysis of glycogen content in the same muscle sections (Figure 5C).

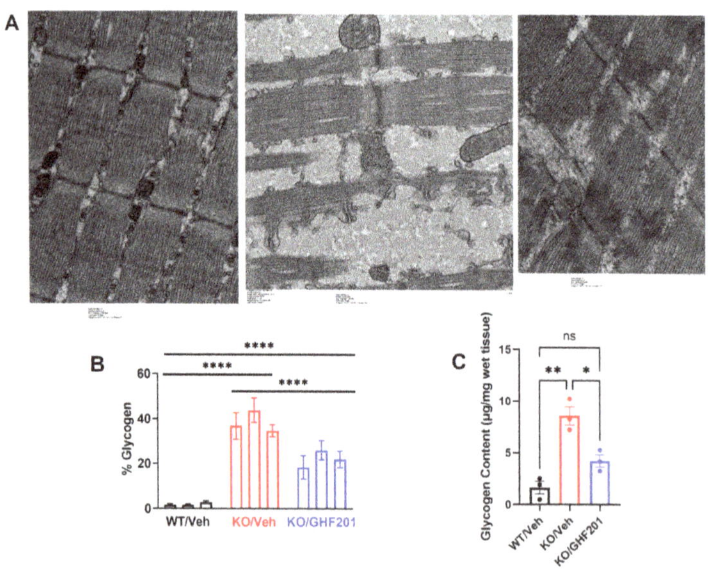

Figure 5. (**A**) Representative transmission electron microscopy images of longitudinal sections of *gastrocnemius* muscle collected from 8-month-old animals treated as indicated. Note granular glycogen material interspersing sarcomeres in the section from untreated *Agl*$^{-/-}$ mouse. Also note variable width of sarcomeres in *Agl*$^{-/-}$ mouse sample. These ultrastructural phenotypes were partially corrected in GHF-201-treated *Agl*$^{-/-}$ mouse. (**B**) Quantification of area percent of glycogen

based on analysis of TEM images from n = 3 animals. ****, significant differences between arms ($p < 0.0001$, Two Way ANOVA (factors are "animal" and "arm") with Tuckey post hoc correction). (**C**) Biochemical quantification of *Gastrocnemius* glycogen. Glycogen levels in muscle tissues collected from n = 3 mice/arm were determined by Merck's glycogen assay kit and statistically analyzed by One Way ANOVA with Tuckey post hoc correction. *, $p < 0.05$; **, $p < 0.01$, ns, $p < 0.08$.

3.6. Glycogen Levels, Lysosomal and Mitochondrial Features of GSDIII Patient-Derived Skin Fibroblasts

Consistent with the results in gastrocnemius muscle, in skin fibroblasts derived from GSDIII patients, GHF-201 also reduced glycogen levels and rendered them similar (not statistically different) to those of HC fibroblasts, as determined by periodic acid-Schiff's reagent (PAS) staining (Figure 6A). This result suggests that the mechanism of action by which GHF-201 alleviated GSDIII symptoms in mice is autolysosomal degradation of glycogen, which is known to be mediated by GHF-201 in GSDIV [20] and GSDIa [21].

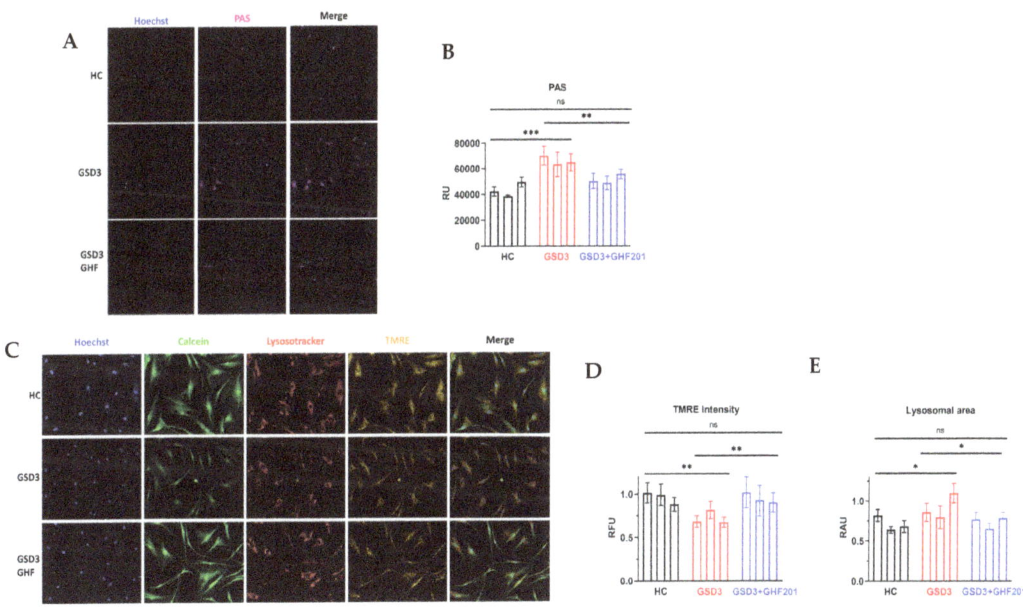

Figure 6. Skin fibroblasts from 3 GSD3 and 3 HC individuals were cultured in 96-well plates for 48 h in starvation medium without FBS and glucose. Subsequently, cells were live-stained with (**A**,**B**) PAS reagent to quantify intracellular glycogen, or (**C**–**E**) a mix of fluorescent dyes which included Hoechst, Calcein-AM, TMRE, and Lysortracker red to, respectively, stain nuclei (blue), cytoplasm (green), respiring mitochondria (yellow), and lysosomes (red). Multiple images from the live-stained cells were automatically obtained by an Operetta G1 image analyzer under environmentally controlled conditions. Shown are representative images and their quantification in relative units. Statistical significance of differences was determined by Two Way ANOVA with Tuckey post hoc correction. *, $p < 0.05$; **, $0.001 < p < 0.01$; ***, $p < 0.001$; ns, not significant.

To confirm GHF-201 mechanism of action at a cellular level, we decided to test its effects on lysosomal and mitochondrial features in fibroblasts derived from GSDIII patients. Since GHF-201 is an autophagic and catabolic activator, we anticipated that its significant effect at a cellular level will be at the lysosomal–mitochondrial axis, as shown, for instance, in GSDII [33]. Our results (Figure 6B) show that under starvation, which induces autophagic flux, GHF-201 reduced the lysosomal area, bringing it close to the area observed in HC fibroblasts. Reduced lysosomal area was also observed in other systems in healthy as compared to lysosomal impaired cells [34] and is presumably associated with increased autophagic flux and lysosomal function, which can clear overladen lysosomes from their excess content and thus reduce their size, which is increased by the lysosomal swelling this excess content causes. Mitochondrial depolarization was observed in GSDIII patient fibroblasts as compared to HC fibroblasts. This mitochondrial depolarization is probably associated with the limited glycogen degradation and carbohydrate fuel production characteristic of GSDIII (Figure 4A). In accordance with an increase in mitochondrial fueling by enhanced sutophagic catabolism, GHF-201 treatment repolarized mitochondrial membrane potential, bringing it closer to that of HC fibroblasts (Figure 6C).

4. Discussion

The long-term aim of this study was to test the capacity of GHF-201 to alleviate disease symptoms in the $Agl^{-/-}$ mouse model of GSDIII. GHF-201 specifically binds the lysosomal membrane protein LAMP1 [20]. GHF-201 was further able to re-acidify the aberrant alkaline lysosomal pH. In adult GSDIV (APBD) models, GHF-201 ameliorated the diseased state by enhancing lysosomal glycogen catabolism and autophagy.

GSDIII starts with liver involvement followed by storage of glycogen in skeletal muscle, provoking a myopathy. The disease is caused by cytoplasmic glycogen debranching enzyme (GDE) deficiency leading to accumulation of glycogen due to its reduced degradation. Glycogen accumulation in muscle is more extensive in GSDIII as compared to the other myopathy-implicating GSDs GSDII, GSDV, GSDIX, GSDX, and GSDXV [11]. In particular, human muscle biopsies from GSDIII patients show a typical and constant vacuolar myopathy, characterized by multiple and variably sized vacuoles filled with PAS-positive material [9]. Phosphorylase-limit dextrin (PLD), the type of glycogen which accumulates in GSDIII, is structurally abnormal, containing shorter outer branches [12]. While PLD is sensitive to diastase digestion and does not form polyglucosan bodies [12], it is still associated with large vacuoles and can cause histological damage to myofibrils [13,18]. In fact, as also shown by our results (Figure 5), disruption of myofibrillar structure by glycogen vacuoles is more pronounced in GSDIII than in other GSDs with prominent muscular involvement [11].

After the liver disorder, myopathy pursues [15], with occasional cardiomyopathy [16], and at later stages liver cirrhosis, and hepatocellular adenoma [17]. Interestingly, possibly due to higher overall glycogen accumulation, GSDIII manifests with excess glycogen in both cytoplasm and lysosomes, as opposed to GSDII, where excess glycogen is primarily found in lysosomes due to the aberrant function of GAA [35]. The main pathogenic factor in GSDIII, as revealed by ultrastructural analysis of muscle specimens from patients, is autophagosome accumulation. Concordantly, Laforet et al., using electron microscopy, showed the presence of large non-membrane bound sarcoplasmic deposits of normally structured glycogen as well as smaller rounded sac structures lined by a continuous double membrane containing only glycogen, corresponding to autophagosomes. A consistent SQSTM1/p62 decrease and beclin-1 increase in human muscle biopsies suggested autophagic dysregulation [9]. The latter can impede the degradation of dysfunctional organelles and proteins, and, notably, also of glycogen, in a specific pathway mediated by the starch binding domain-containing protein 1 (Stbd1), which tags glycogen for its autophagy, interacting with autophagy machinery and glycogen-related proteins, including GDE, in other domains [36]. Interestingly, among these glycogen-related proteins, GDE has the highest number of Atg8 binding motifs [36], also suggesting an important role of autophagy in GSDIII, which might take place if GDE deficiency impairs the interaction with the autophagy machinery. Moreover, dysfunctional autophagy plays a pivotal role in disrupting muscle homeostasis and causing structural damage to sarcomeres [37]. The tight link between autophagic impairment and GSDIII pathology can, thus, explain the beneficial effects of GHF-201, a general autophagic activator [20], in improving motor function (Figure 2) and skeletal muscle health (Figure 5) in a GSDIII mouse model. Other possible systemic effects of GHF-201, unrelated to autophagy enhancement, probably do not contribute to the beneficial effect of GHF-201, as it did not influence animal weight gain and histology in the long-term and behavior (Irwin test) in the short-term [20]. Even though the effect of GHF-201 on motor and behavioral features was not tested in WT mice, we assume that the beneficial effect of GHF-201 is specific to $Agl^{-/-}$ mice because PLD accumulation in this strain disrupts muscle structure (Figure 5 and Supplementary Figure S1). Therefore, partial removal of this pathogenic PLD in $Agl^{-/-}$ mice can more noticeably relieve the motor dysfunction caused by the disrupted muscle structure than in WT mice where such a disruption and consequent motor dysfunction do not exist. However, since autophagy and autophagic catabolism are generally considered cell protective and energy producing, GHF-201 might possibly improve animal health in general, even in WT mice.

Our blood biochemistry panel demonstrated that GHF-201 treatment led to improved liver and muscle function and partially corrected blood hypoglycemia and hypertriglyceridemia (Figure 3), also observed in GSDIII patients [38]. Both hypoglycemia and hyperlipidemia are expected outcomes of glycogen surcharge since more glucose is shunted from blood to intracellular glycogen and since high glycogen levels can inhibit the energy sensor AMPK [39–42], leading to activation of acetyl CoA carboxylase, the rate-limiting enzyme of fatty acid synthesis, which is phospho-inhibited by AMPK. Reduction of glycogen levels by GHF-201 (Figure 5) expectedly mitigated this hyperlipidemia. The blood lipid profile can also reflect fuel utilization. As glycogen mobilization is compromised in GSDIII and enhanced by GHF-201, we expected that GSDIII mice use lipid instead of carbohydrate fuel as a compensatory reaction. However, as suggested by the hypertriglyceridemia observed in untreated GSDIII mice (Figure 3), the mice are not capable of this metabolic compensation. Notably, as revealed by both indirect calorimetry (Figure 4) and the blood biochemistry panel (Figure 3), GHF-201 treatment enabled this metabolic compensation by boosting fat burn in the light (Figure 4), which is possibly reflected in the reduction in blood triglyceride levels (Figure 3). A similar effect of GHF-201 was observed in the GSDIV modeling $Gbe^{ys/ys}$ mice [20]. Interestingly, GHF-201-mediated improvement of fat oxidation (Figure 4A), and other in vivo metabolic effects (TEE, (Figure 4A), ambulatory activity (Figure 4B), and wheel running (Figure 2B)) were more pronounced in the light than in the dark. Since mice are nocturnal animals, in which food is mostly consumed in the dark (see also Figure 4C) and digested in the light, it can be conjectured that GHF-201, affecting fuel digestion, or autophagic substrate catabolism, exerts a larger effect in the light. Along these lines, a possible explanation for the GHF201-associated decrease of activity in the dark is that the untimely stimulation of fuel-generating catabolism in the dark (Figure 4(B1)) might lead to fuel excess known to be detrimental as it might lead to mitochondrial fragmentation in pancreatic beta cells [42] and unbalanced reactive oxygen species (ROS) production [43]. However, the fact that GHF-201 increased ambulatory activity in the dark in GSDIV modeling mice [20], while decreasing it in GSDIII modeling mice (Figure 4(B1)), calls for an explanation. One important difference between these two diseases is that in GSDIV patients, and in GSDIV-modeling mice, polyglucosan bodies (PB) formed by brancher deficiency are architecturally different from phosphorylase-limit dextrin in GSDIII formed by debrancher deficiency. The large size and unique amylopectin architecture of PB in GSDIV models causes them to clog axons, which is not the case for the aberrant glycogen (phosphorylase-limit dextrin) in GSDIII. Therefore, PB-clearance from axons and peripheral nerve by GHF-201 in a GSDIV mouse model [20] might enhance mobility by improving nerve–muscle communication. This clearance of the noxious PB from axons and neuromuscular junctions might override the putative detrimental effect of excess fuel in the dark and so might be more important for locomotion and enhance it also in the dark in GSDIV mice, as opposed to GSDIII mice.

In conclusion, we show here that GHF-201, a LAMP-1-targeting small molecule acting as a general activator of autophagic flux, can partially correct aberrant metabolic, motor, and myopathological phenotypes in the $Agl^{-/-}$ mouse model of GDE deficiency, or GSDIII, a disease implicating a particularly high glycogen burden, causing liver and myofibrillar injury.

Supplementary Materials: The following supporting information can be downloaded at https://www.mdpi.com/article/10.3390/biom14080893/s1, Figure S1: An additional representative transmission electron microscopy images of longitudinal sections of gastrocnemius muscle collected from 8-month-old animals treated as indicated.

Author Contributions: Conceptualization, O.K., M.W. and E.M.; Methodology, K.M., S.S., S.B. (Saja Baraghithy), U.S., M.M., S.B. (Sultan Bastu) and H.G.; Writing—original draft preparation, O.K.; Writing—review and editing, M.W. and E.M.; Supervision, J.T., H.R., M.W., E.M. and O.K.; funding acquisition, M.W., E.M. and O.K. All authors have read and agreed to the published version of the manuscript.

Funding: This research was funded by the Association Francaise contre les Myopathies, grant number 23116. The PhD scholarship of US was partly paid by the Yoran Institute of Genome Research and Personalized Medicine and by the Tel Aviv University Prajs-Drimmer Institute for the Development of Anti-Degenerative Drugs.

Institutional Review Board Statement: The animal study protocol was approved by the Institutional Review Board (or Ethics Committee) of THE HEBREW UNIVERSITY OF JERUSALEM (protocol code 16403 and date of approval 30 March 2022).

Informed Consent Statement: Not applicable.

Data Availability Statement: Data is contained within the article. The original contributions presented in the study are included in the article. Further inquiries can be directed to the corresponding authors.

Acknowledgments: The authors wish to thank Eran Perlson from Tel Aviv University for his help in the collection and preparation of the muscle samples and Giuseppe Ronzitti from Genethon, France, for providing the mice used in this study.

Conflicts of Interest: O.K. and M.W. are advisors to the company Golden Heart Flower (GHF) to which GHF-201 is sublicensed. The remaining authors declare that the research was conducted in the absence of any commercial or financial relationships that could be construed as a potential conflict of interest.

References

1. Illingworth, B.; Larner, J.; Cori, G.T. Structure of glycogens and amylopectins. I. Enzymatic determination of chain length. *J. Bi Chem.* **1952**, *199*, 631–640. [CrossRef]
2. Smythe, C.; Cohen, P. The discovery of glycogenin and the priming mechanism for glycogen biogenesis. *Eur. J. Biochem.* **1991**, *20* 625–631. [CrossRef] [PubMed]
3. Testoni, G.; Duran, J.; García-Rocha, M.; Vilaplana, F.; Serrano, A.L.; Sebastián, D.; López-Soldado, I.; Sullivan, M.A.; Slebe, I Vilaseca, M.; et al. Lack of Glycogenin Causes Glycogen Accumulation and Muscle Function Impairment. *Cell Metab.* **2017**, 2 256–266.e254. [CrossRef] [PubMed]
4. Adeva-Andany, M.M.; Gonzalez-Lucan, M.; Donapetry-Garcia, C.; Fernandez-Fernandez, C.; Ameneiros-Rodriguez, E. Glycoge metabolism in humans. *BBA Clin.* **2016**, *5*, 85–100. [CrossRef]
5. Prats, C.; Graham, T.E.; Shearer, J. The dynamic life of the glycogen granule. *J. Biol. Chem.* **2018**, *293*, 7089–7098. [CrossRe [PubMed]
6. Froese, D.S.; Michaeli, A.; McCorvie, T.J.; Krojer, T.; Sasi, M.; Melaev, E.; Goldblum, A.; Zatsepin, M.; Lossos, A.; Alvarez, R.; et a Structural basis of glycogen branching enzyme deficiency and pharmacologic rescue by rational peptide design. *Hum. Mol. Gen* **2015**, *24*, 5667–5676. [CrossRef] [PubMed]
7. Brereton, M.F.; Rohm, M.; Shimomura, K.; Holland, C.; Tornovsky-Babeay, S.; Dadon, D.; Iberl, M.; Chibalina, M.V.; Lee, S.; Glaser, I et al. Hyperglycaemia induces metabolic dysfunction and glycogen accumulation in pancreatic β-cells. *Nat. Commun.* **2016**, *7*, 1349 [CrossRef]
8. Ashcroft, F.M.; Rohm, M.; Clark, A.; Brereton, M.F. Is Type 2 Diabetes a Glycogen Storage Disease of Pancreatic β Cells? *Ce Metab.* **2017**, *26*, 17–23. [CrossRef]
9. Laforêt, P.; Inoue, M.; Goillot, E.; Lefeuvre, C.; Cagin, U.; Streichenberger, N.; Leonard-Louis, S.; Brochier, G.; Madelaine, A Labasse, C.; et al. Deep morphological analysis of muscle biopsies from type III glycogenesis (GSDIII), debranching enzym deficiency, revealed stereotyped vacuolar myopathy and autophagy impairment. *Acta Neuropathol. Commun.* **2019**, *7*, 16 [CrossRef]
10. Schreuder, A.; Rossi, A.; Grünert, S.; Derks, T. *Glycogen Storage Disease type III (GeneReviews® [Internet])*; University of Washingto Seattle, WA, USA, 2010.
11. Hannah, W.B.; Derks, T.G.J.; Drumm, M.L.; Grünert, S.C.; Kishnani, P.S.; Vissing, J. Glycogen storage diseases. *Nat. Rev. D Primers* **2023**, *9*, 46. [CrossRef]

12. Kishnani, P.S.; Austin, S.L.; Arn, P.; Bali, D.S.; Boney, A.; Case, L.E.; Chung, W.K.; Desai, D.M.; El-Gharbawy, A.; Haller, R.; et al. Glycogen storage disease type III diagnosis and management guidelines. *Genet. Med.* **2010**, *12*, 446–463. [CrossRef] [PubMed]
13. Vidal, P.; Pagliarani, S.; Colella, P.; Costa Verdera, H.; Jauze, L.; Gjorgjieva, M.; Puzzo, F.; Marmier, S.; Collaud, F.; Simon Sola, M.; et al. Rescue of GSDIII Phenotype with Gene Transfer Requires Liver- and Muscle-Targeted GDE Expression. *Mol. Ther.* **2018**, *26*, 890–901. [CrossRef] [PubMed]
14. DiMauro, S. *Myology*; mcGraw-Hill: New York, NY, USA, 1985; pp. 1535–1558.
15. Decostre, V.; Laforêt, P.; De Antonio, M.; Kachetel, K.; Canal, A.; Ollivier, G.; Nadaj-Pakleza, A.; Petit, F.M.; Wahbi, K.; Fayssoil, A.; et al. Long term longitudinal study of muscle function in patients with glycogen storage disease type IIIa. *Mol. Genet. Metab.* **2017**, *122*, 108–116. [CrossRef]
16. Valayannopoulos, V.; Bajolle, F.; Arnoux, J.B.; Dubois, S.; Sannier, N.; Baussan, C.; Petit, F.; Labrune, P.; Rabier, D.; Ottolenghi, C.; et al. Successful treatment of severe cardiomyopathy in glycogen storage disease type III With D,L-3-hydroxybutyrate, ketogenic and high-protein diet. *Pediatr. Res.* **2011**, *70*, 638–641. [CrossRef] [PubMed]
17. Demo, E.; Frush, D.; Gottfried, M.; Koepke, J.; Boney, A.; Bali, D.; Chen, Y.T.; Kishnani, P.S. Glycogen storage disease type III-hepatocellular carcinoma a long-term complication? *J. Hepatol.* **2007**, *46*, 492–498. [CrossRef]
18. DiMauro, S.; Hartwig, G.B.; Hays, A.; Eastwood, A.B.; Franco, R.; Olarte, M.; Chang, M.; Roses, A.D.; Fetell, M.; Schoenfeldt, R.S.; et al. Debrancher deficiency: Neuromuscular disorder in 5 adults. *Ann. Neurol.* **1979**, *5*, 422–436. [CrossRef] [PubMed]
19. Peng, S.S.; Hwu, W.L.; Lee, N.C.; Tsai, F.J.; Tsai, W.H.; Chien, Y.H. Slow, progressive myopathy in neonatally treated patients with infantile-onset Pompe disease: A muscle magnetic resonance imaging study. *Orphanet J. Rare Dis.* **2016**, *11*, 63. [CrossRef]
20. Kakhlon, O.; Vaknin, H.; Mishra, K.; D'Souza, J.; Marisat, M.; Sprecher, U.; Wald-Altman, S.; Dukhovny, A.; Raviv, Y.; Da'adoosh, B.; et al. Alleviation of a polyglucosan storage disorder by enhancement of autophagic glycogen catabolism. *EMBO Mol. Med.* **2021**, *13*, e14554. [CrossRef]
21. Sprecher, U.; D'Souza, J.; Mishra, K.; Canella Miliano, A.; Mithieux, G.; Rajas, F.; Avraham, S.; Anikster, Y.; Kakhlon, O.; Weil, M. Imprinted cell memory in glycogen storage disorder 1a. *bioRxiv* **2023**. [CrossRef]
22. Koeberl, D.D.; Koch, R.L.; Lim, J.A.; Brooks, E.D.; Arnson, B.D.; Sun, B.; Kishnani, P.S. Gene therapy for glycogen storage diseases. *J. Inherit. Metab. Dis.* **2024**, *47*, 93–118. [CrossRef]
23. Yi, H.; Brooks, E.D.; Thurberg, B.L.; Fyfe, J.C.; Kishnani, P.S.; Sun, B. Correction of glycogen storage disease type III with rapamycin in a canine model. *J. Mol. Med.* **2014**, *92*, 641–650. [CrossRef] [PubMed]
24. Higazi, A.A.; Maraga, E.; Baraghithy, S.; Udi, S.; Azar, S.; Saada, A.; Glaser, B.; Avrahami, D.; Abdeen, S.; Hamdan, Z.; et al. Characterization of metabolic alterations in the lean metabolically unhealthy alpha defensin transgenic mice. *iScience* **2024**, *27*, 108802. [CrossRef] [PubMed]
25. Caicedo, J.C.; Cooper, S.; Heigwer, F.; Warchal, S.; Qiu, P.; Molnar, C.; Vasilevich, A.S.; Barry, J.D.; Bansal, H.S.; Kraus, O.; et al. Data-analysis strategies for image-based cell profiling. *Nat. Methods* **2017**, *14*, 849–863. [CrossRef] [PubMed]
26. Charan, J.; Kantharia, N.D. How to calculate sample size in animal studies? *J. Pharmacol. Pharmacother.* **2013**, *4*, 303–306. [CrossRef]
27. Liu, K.M.; Wu, J.Y.; Chen, Y.T. Mouse model of glycogen storage disease type III. *Mol. Genet. Metab.* **2014**, *111*, 467–476. [CrossRef] [PubMed]
28. Kamel, A.S.; Mohamed, A.F.; Rabie, M.A.; Elsherbiny, M.E.; Ahmed, K.A.; Khattab, M.M.; Abdelkader, N.F. Experimental Evidence for Diiodohydroxyquinoline-Induced Neurotoxicity: Characterization of Age and Gender as Predisposing Factors. *Pharmaceuticals* **2022**, *15*, 251. [CrossRef] [PubMed]
29. Lundblad, M.; Vaudano, E.; Cenci, M.A. Cellular and behavioural effects of the adenosine A2a receptor antagonist KW-6002 in a rat model of l-DOPA-induced dyskinesia. *J. Neurochem.* **2003**, *84*, 1398–1410. [CrossRef] [PubMed]
30. Hassan, M.P.; Abdollahifar, M.A.; Aliaghaei, A.; Tabeie, F.; Vafaei-Nezhad, S.; Norouzian, M.; Abbaszadeh, H.A. Photobiomodulation therapy improved functional recovery and overexpression of interleukins-10 after contusion spinal cord injury in rats. *J. Chem. Neuroanat.* **2021**, *117*, 102010. [CrossRef]
31. Lucchiari, S.; Pagliarani, S.; Salani, S.; Filocamo, M.; Di Rocco, M.; Melis, D.; Rodolico, C.; Musumeci, O.; Toscano, A.; Bresolin, N.; et al. Hepatic and neuromuscular forms of glycogenosis type III: Nine mutations in AGL. *Hum. Mutat.* **2006**, *27*, 600–601. [CrossRef]
32. Lucchiari, S.; Santoro, D.; Pagliarani, S.; Comi, G.P. Clinical, biochemical and genetic features of glycogen debranching enzyme deficiency. *Acta Myol.* **2007**, *26*, 72–74.
33. Lim, J.A.; Li, L.; Kakhlon, O.; Myerowitz, R.; Raben, N. Defects in calcium homeostasis and mitochondria can be reversed in Pompe disease. *Autophagy* **2015**, *11*, 385–402. [CrossRef] [PubMed]
34. de Araujo, M.E.G.; Liebscher, G.; Hess, M.W.; Huber, L.A. Lysosomal size matters. *Traffic* **2020**, *21*, 60–75. [CrossRef] [PubMed]
35. Nascimbeni, A.C.; Fanin, M.; Masiero, E.; Angelini, C.; Sandri, M. The role of autophagy in the pathogenesis of glycogen storage disease type II (GSDII). *Cell Death Differ.* **2012**, *19*, 1698–1708. [CrossRef] [PubMed]
36. Koutsifeli, P.; Varma, U.; Daniels, L.J.; Annandale, M.; Li, X.; Neale, J.P.H.; Hayes, S.; Weeks, K.L.; James, S.; Delbridge, L.M.D.; et al. Glycogen-autophagy: Molecular machinery and cellular mechanisms of glycophagy. *J. Biol. Chem.* **2022**, *298*, 102093. [CrossRef] [PubMed]
37. Margeta, M. Autophagy Defects in Skeletal Myopathies. *Annu. Rev. Pathol.* **2020**, *15*, 261–285. [CrossRef] [PubMed]
38. LaBarbera, M.; Milechman, G.; Dulbecco, F. Premature coronary artery disease in a patient with glycogen storage disease III. *J. Invasive Cardiol.* **2010**, *22*, E156–E158. [PubMed]

39. Mobbs, J.I.; Koay, A.; Di Paolo, A.; Bieri, M.; Petrie, E.J.; Gorman, M.A.; Doughty, L.; Parker, M.W.; Stapleton, D.I.; Griffin, M.D.; et al. Determinants of oligosaccharide specificity of the carbohydrate-binding modules of AMP-activated protein kinase. *Biochem. J.* **2015**, *468*, 245–257. [CrossRef] [PubMed]
40. Li, X.; Wang, L.; Zhou, X.E.; Ke, J.; de Waal, P.W.; Gu, X.; Tan, M.H.; Wang, D.; Wu, D.; Xu, H.E.; et al. Structural basis of AMPK regulation by adenine nucleotides and glycogen. *Cell Res.* **2015**, *25*, 50–66, Erratum in *Cell Res.* **2015**, *25*, 398. [CrossRef] [PubMed]
41. Koay, A.; Woodcroft, B.; Petrie, E.J.; Yue, H.; Emanuelle, S.; Bieri, M.; Bailey, M.F.; Hargreaves, M.; Park, J.T.; Park, K.H.; et al. AMPK beta subunits display isoform specific affinities for carbohydrates. *FEBS Lett.* **2010**, *584*, 3499–3503. [CrossRef]
42. Alan, L.; Scorrano, L. Shaping fuel utilization by mitochondria. *Curr. Biol.* **2022**, *32*, R618–R623. [CrossRef]
43. Fisher-Wellman, K.H.; Gilliam, L.A.A.; Lin, C.T.; Cathey, B.L.; Lark, D.S.; Darrell Neufer, P. Mitochondrial glutathione depletion reveals a novel role for the pyruvate dehydrogenase complex as a key H_2O_2-emitting source under conditions of nutrient overload. *Free Radic. Biol. Med.* **2013**, *65*, 1201–1208. [CrossRef] [PubMed]

metabolites

MDPI

Review

Systematic Insights into the Relationship between the Microbiota–Gut–Brain Axis and Stroke with the Focus on Tryptophan Metabolism

Xinyu Shen [1,2] and Xiaoqin Mu [1,2,*]

[1] Genomics Research Center, Key Laboratory of Gut Microbiota and Pharmacogenomics of Heilongjiang Province, College of Pharmacy, Harbin Medical University, Harbin 150081, China; 15543547863@163.com

[2] Translational Medicine Research and Cooperation Center of Northern China, Heilongjiang Academy of Medical Sciences, Harbin 150081, China

* Correspondence: muxiaoqin@ems.hrbmu.edu.cn

Abstract: Stroke, as a serious cerebral vascular disease with high incidence and high rates of disability and mortality, has limited therapeutic options due to the narrow time window. Compelling evidence has highlighted the significance of the gut microbiota and gut–brain axis as critical regulatory factors affecting stroke. Along the microbiota–gut–brain axis, tryptophan metabolism further acquires increasing attention for its intimate association with central nervous system diseases. For the purpose of exploring the potential role of tryptophan metabolism in stroke and providing systematic insights into the intricate connection of the microbiota–gut–brain axis with the pathological procedure of stroke, this review first summarized the practical relationship between microbiota and stroke by compiling the latest case-control research. Then, the microbiota–gut–brain axis, as well as its interaction with stroke, were comprehensively elucidated on the basis of the basic anatomical structure and physiological function. Based on the crosstalk of microbiota–gut–brain, we further focused on the tryptophan metabolism from the three major metabolic pathways, namely, the kynurenine pathway, serotonin pathway, and microbial pathway, within the axis. Moreover, the effects of tryptophan metabolism on stroke were appreciated and elaborated here, which is scarcely found in other reviews. Hopefully, the systematic illustration of the mechanisms and pathways along the microbiota–gut–brain axis will inspire more translational research from metabolic perspectives, along with more attention paid to tryptophan metabolism as a promising pharmaceutical target in order to reduce the risk of stroke, mitigate the stroke progression, and ameliorate the stroke prognosis.

Keywords: stroke; gut microbiota; gut–brain axis; tryptophan metabolism

1. Introduction

As an acute cerebrovascular condition, stroke is one of the most important contributors to long-term disability and mortality, affecting millions of people each year worldwide. Ischemic stroke (IS) and hemorrhage stroke (HS) are the two main types of strokes, among which 87% are ischemic, 10% are intracerebral hemorrhage (ICH), and 3% are subarachnoid hemorrhage (SAH) according to the 2024 American Heart Association statistical update [1,2], implying the structural or functional damage to brain tissue caused by the blockage or rupture of cerebral blood vessels. Recent years have witnessed alarming growth as well as the youthful trend of stroke incidence with the prevalence of unhealthy lifestyle and eating habits. In the US, there is a projection that an additional 3.4 million adults over 18 will suffer a stroke by 2030 [1]. Although the therapy of stroke has improved during the past few decades, including intravenous thrombolysis, intra-arterial thrombectomy, recombinant tissue plasminogen activator therapy, and so on, it is limited due to the narrow therapeutic window, individual differences, as well as the potential risk [3].

In addition, dementia, cognitive impairment [4], anxiety, depression [5], and communication disorder [6] can be observed in a large amount of post-stroke cases. Thus, early prevention and intervention of stroke and understanding potential therapeutic targets are particularly essential.

Numerous studies suggested that intestinal disorders accompanied by gut microbial alteration, such as inflammatory bowel disease, irritable bowel syndrome, and constipation, have complex interactions with stroke [7–10]. Gastrointestinal disorders or gut dysbiosis, on the one hand, often appear as complications in stroke patients; on the other hand, they may increase the risk of stroke [11] or worsen cerebral infarction [12]. The gut environment influences brain function in many ways, involving the central nervous system (CNS), enteric nervous system (ENS), autonomic nervous system (ANS), neuroendocrine system (hypothalamic–pituitary–adrenal axis), and immune system [13]. Likewise, the brain fine-tunes gut activities. The intricate bidirectional communication system between the gastrointestinal tract (GIT) and the brain is called the gut–brain axis (GBA). Since gut microbiota act as an integral regulator in the GBA, the GBA was further extended to the microbiota–gut–brain axis (MGBA) [14].

A key component of the MGBA is tryptophan, an essential amino acid whose metabolism is directly or indirectly regulated by the gut microbiota [15]. Products of tryptophan metabolism, including 5-hydroxytryptamine (serotonin, 5-HT), kynurenines, indole derivatives, etc., exert profound impacts on the pathways related to MGBA [16]. Mounting evidence supports a clear correlation between stroke and Trp metabolism in the MGBA with regard to enzyme activities, metabolite level changes, and related genes [17–19]. Promisingly, tryptophan and its metabolites serve as potential biomarkers as well as therapeutic targets of neurological disorders, including stroke, endorsing high precision of diagnosis and alternative treatments.

In this review, we systematically summarize the mechanism of MGBA and comprehensively discuss the role of MGBA in stroke, with a focus on the correlation between tryptophan metabolism and the pathogenesis and progression of stroke, combining the latest knowledge so as to provide a reference for further exploration of the prevention, treatment, and rehabilitation of stroke.

2. Microbiota–Gut–Brain Axis and Stroke

2.1. Gut Microbiota and Stroke: Epidemiological Relationships

The gut microbiota refers to a community of myriad symbiotic microorganisms that populate the intestinal tract, which harbors over 1000 bacterial species containing more than 100 bacterial phyla [20]. In our bodies, almost 95% of commensal bacteria are dominated by the phyla Firmicutes and Bacteroidetes [21]. As a complex and dynamic metabolic network, the gut microbiota plays a considerable role in the host condition, affecting the immune, inflammation, neural, endocrine, and metabolism processes via GBA.

Emerging evidence elucidated the close relationship between gut dysbiosis and stroke. We looked back on and enumerated 16 case-control studies in the last 5 years (2019–2023) so as to more veritably present such connection (Table A1). By analyzing these case reports, we found that all studies reported differences in gut microbial composition between stroke patients and healthy controls through 16S ribosomal RNA gene sequencing. At the phylum level, the relative abundance of Proteobacteria was increased, while the Firmicutes were mostly reported decreased. At the family level, the amount of Enterobacteriaceae and *Lactobacillaceae* were significantly elevated, while Prevotellaceae were decreased. At the genus level, Lactobacillus, Streptococcus, Akkermansia, and Roseburia were usually upgraded, while Faecalibacterium were reduced. However, the changes in some bacteria, such as Ruminococcaceae, Bacteroidaceae, Coriobacteriaceae, Lachnospiraceae, and Bacteroidetes, remain controversial.

Beyond the altered microbial composition, the gut microbiota itself can be a potential biomarker of the outcome or severity of stroke, verified by the increased Enterobacter, Pyramidobacter, and Lachnospiraceae_UCG_001 in mild ischemic stroke patients, whereas the increased genera, such as Ruminococcaceae_UCG-002 and Christensenellaceae_R-7_group, were found in severe stroke patients [22]. Xu et al. [12] elaborated that Enterobacteriaceae was the most prevalent pathogen in both an IS cohort and mice with middle cerebral artery occlusion (MCAO). In the study of Jiang et al. [23], a positive correlation between Enterobacteriaceae and the severity of IS was found, with more abundance of Enterobacteriaceae, Faecalibacterium, Lachnospiraceae, and Ruminococcaceae but lower lactic acid bacteria of Lactobacillus in the poor outcome group. Furthermore, when inoculating *E. coli*, a common species of Enterobacteriaceae, it turned out that the genes and serum biomarkers associated with immune and inflation systems apparently upgraded, which confirmed that Enterobacteriaceae is a significant influencing factor of stroke [12]. Zheng et al. [24] elucidated that gut microbiota correlated positively with systemic inflammation in cryptogenic stroke (CS) patients and attributed the high level of lipopolysaccharide (LPS) to the elevated amount of Gram-negative bacteria, such as Escherichia–Shigella and Klebsiella of the Enterobacteriaceae family.

The gut microbiota dysbiosis could potentially lead to IS pathogenesis by modulating the host's traditional risk factors as well [25]; for example, Lactobacillaceae, Enterococcaceae, Streptococcaceae, and Enterobacteriaceae families enriched in IS patients were positively correlated with high-risk factors, including high blood pressure and ApoB/ApoA1 ratio, while negatively correlated with the preventive factor, high-density lipoprotein cholesterol (HDL-C), in the IS patients. Moreover, age, regional, and ethnic variations did make a difference in ischemic stroke when referring to the gut microbiota's taxonomic changes during the disease.

2.2. Mechanisms of the Gut–Brain Axis in Stroke

In this section, we outline the basic anatomical structure and physiology of the gut–brain axis along with the vital impact of microbiota. On that basis, we have illustrated alterations of the MGBA in stroke to provide a comprehensive understanding of the MGBA as well as its association with stroke. To be specific, this section covers four parts of the microbiota–gut–brain axis, including the barrier structure, defense system, neural network, and humoral communication. Albeit being separately recounted as follows, these parts virtually link together closely and interact with each other in the MGBA (Figure 1).

2.2.1. The Barrier Structure of the Microbiota–Gut–Brain Axis

As the gateway to the communication across the MGBA [26], the barrier of MGBA refers to the specifical cellular interfaces, which principally include the blood–brain barrier and gut epithelial barrier.

- **Gut Barrier and Its Association with Stroke**

The gut barrier can be divided into three layers: (1) The mucus layer, mainly comprising mucins secreted by goblet cells, which are also a nutritional source and adhesion sites for microbiome. (2) The gut epithelial barrier, a selectively semipermeable barrier mainly composed of absorptive enterocytes, enteroendocrine cells, goblet cells, Paneth cells, M cells, and stem cells. (3) The gut vascular barrier, which consists of microvascular endothelial cells attached by fibroblasts, enteric glia, and pericytes. All three layers jointly act as a defense for systemic circulation. To reinforce barrier integrity and restrict paracellular transport, tight junctions formed by transmembrane proteins, including claudins, Occludin, and zonula occludens (ZO), as well as adherens junctions, lie in the gut endothelium [27].

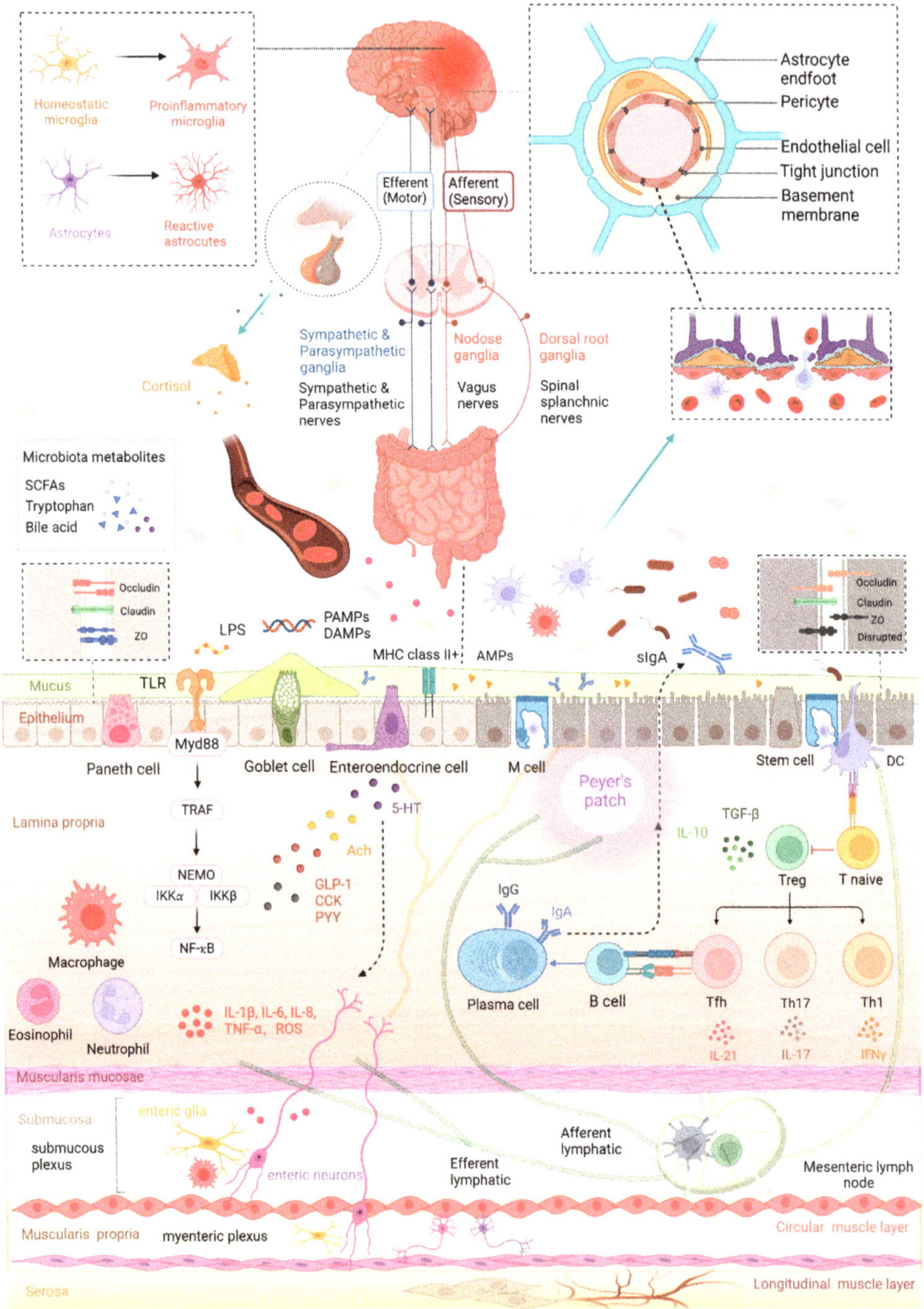

Figure 1. Crosstalk in the microbiota–gut–brain axis associated with stroke. Created with BioRender.com.

Experimental stroke mice exhibited increased gut permeability, especially vascular permeability in the jejunum and ileum, accompanied by impaired gut mucoprotein, claudin production, and goblet cells. Owing the potential mediation to the upregulated noradrenaline effect in the autonomic nervous system, the study employed β-adrenergic receptor blockers, which significantly ameliorated gut permeability and bacterial dissemination [28–30]. A recent study [31] utilizing the MCAO mice model also verified that gut cell death and apoptosis could be reversed by 6-hydroxydopamine, which denervated catecholamine peripheral neurons, and the deficiency of adrenergic receptor ADRB2 alleviated the gut permeability. The upregulated epithelial fructose metabolism gene expression, with hyperplasia of intestinal crypts and goblet cells at 24 h after stroke, was also corroborated, which can be identified as a host compensatory mechanism for the compromised gut barrier. Downregulated tight-junction (TJ) proteins, ZO-1, Occludin, and claudin-1, as well as broken TJ of the intestinal mucosa observed in ischemic stroke mice are simultaneously accompanied by the upregulation of the TNF-α-induced death receptor signaling pathway and inflammation-related proteins NF-κB, iNOS, and MPO, which were suggested to be concerned with intestinal barrier impairment [32].

Gut microbiota can influence the gut barrier by regulating the mucus secretion [33], epithelial cells' turnover [34], antimicrobial peptides' production [35], and gut permeability [36]. In that stroke often causes gut dysbiosis, several studies speculated that gut dysbiosis is a cause of damaged gut barrier in stroke, whereas the study of Kumar et al. [31] indicated that stroke-induced alterations of epithelial cells and goblet cells are independent of microbial dysbiosis, as similar changes were also observed in germ-free (GF) mice. Thus, more studies are expected to unravel the specifical role of gut microbial disruption in gut epithelial permeability related to stroke. Furthermore, the disruption of the intestinal barrier may promote translocation of gut microbiota, such as Enterococcus spp., Escherichia coli, and Morganella morganii [28], and increased plasma LBP and LPS, thus exacerbating systemic inflammation after stroke. Taken together, there is an intricate relationship between stroke, gut microbiota, and gut barrier(Figure 2).

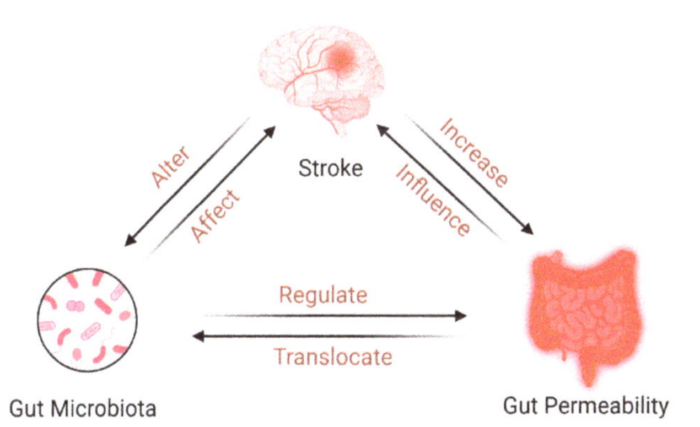

Figure 2. The relationship between stroke, gut microbiota, and gut permeability. Created with BioRender.com.

- **Blood–Brain Barrier Can Be Influenced by the Gut Microbiota**

Extensively located in capillary vessels of the brain and spinal cord parenchyma, the blood–brain barrier (BBB) mainly comprises capillary endothelial cells along with capillary basement membrane, astrocytes end-foot, pericytes, and extracellular matrix, protecting the brain tissue from the harmful substances in the systemic circulation. Similar to the gut barrier, the BBB has tight junctions as well to maintain its function [37]. Studies have reported that when stroke occurs, cerebral ischemia will induce compromised BBB [38], a key mechanism of brain injury after stroke, which can be caused by oxidative stress, overproduction of free radicals, and iron overload, while the exact molecular mechanisms remain to be explicated.

Previous studies have shown the critical influence of the gut microbiome on the BBB. Germ-free mice exhibited enhanced blood–brain barrier permeability with downgraded tight-junction proteins, Occludin and claudin-5, expression, which can be ameliorated by exposing to the microbiota of pathogen-free mice [39]. Gut dysbiosis mice also displayed decreased TJ proteins in the hippocampus [40] and an impaired cerebral endothelium, as evidenced by three weeks of an antibiotic regimen, which caused endothelial dysfunction in the brain, including an enhanced spontaneous tone as well as blunted L-NAME-mediated constrictions and ATP-mediated dilations, accompanied by lower basal eNOS activity [41]. In another study, the microbiome of spontaneously hypertensive stroke-prone rats disrupted the BBB with enhanced MMP-9 gene expression, and damaged the gut barrier with downregulated E-CAD and Muc-2 gene expression [42]. Under the circumstance of "gut leakiness", which is similar to a broken gate, it is reasonable that more bacteria and their toxic products' interactions with cells and neurons, as well as the diffusion to systemic circulation, were allowed, triggering a vast array of immune responses and inflammation with altered pathways along the GMBA. As a result, BBB permeability is affected by the above alteration, directly or indirectly rendering it more accessible to various pathogens. Thus, it can be found in various studies that changes in the gut and brain barriers are positively correlated and concomitant.

Intriguingly, clear similarities can be found in brain and gut barrier structures, which further facilitate the correlation between the gut and brain along the MGBA and implicate the amenability to be regulated by the common signals, including gut microbiota metabolites, for example, SCFAs, especially butyric acid or sodium butyrate, which exert a protective effect on both gut and brain epithelial integrity [43,44], regulating mucus secretion [45] and redistributing ZO-1, Occludin, and claudin protein expression [46–48].

2.2.2. The Defense System of the Microbiota–Gut–Brain Axis

- **The Immune System in the Gut**

As the gatekeeper [49], when threatened by a noxious substance, the intestinal immune system, the largest defense system in our body, readily delivers various strong immune responses based on its sophisticated structure [50]. At the mucous layer, molecules, including IgA, and antimicrobial peptides (AMPs) can kill bacteria, inhibit pathogens, and engage in the regulation of immune signals. Other components, such as mucin 2, zymogen granule protein 16, and IgG Fc-binding protein, have also been demonstrated as part of the innate immune mucosal defense or messengers to the host immune system [51]. Right next to the mucous layer is epithelial cells, which are equipped with cell junctions, as illustrated before, further contributing to defend against most pathogens. Beneath the epithelium lies the lamina propria, a noncellular connective tissue, which consists of a large number of immune cells [52], including monocytes or macrophages, neutrophils, eosinophils, basophils, mast cells, innate lymphoid cells (ILCs), natural killer (NK) cells, lymphocytes, and dendritic cells (DCs). These cells act to phagocyte and eliminate microorganisms and secrete anti-inflammatory factors, such as interleukin 10 (IL-10) [53], possessing anti-inflammatory effects to maintain gut homeostasis [54]. During the gut dysbiosis or stroke damage, pathogen-associated molecular patterns (PAMPs), such as LPS, peptidoglycan, and lipoteichoic acids [55], and damage-associated molecular patterns (DAMPs), including ATP, mitochondrial DNA, IL-33, etc. [56,57], can be recognized by the pattern-recognition receptors (PRRs), represented by the Toll-like receptors (TLRs) family, expressing on the surface of enterocytes or other immune cells. TLRs then interact with MyD88 through TIR domains and trigger a cascade, which leads to the translocation of NF-κB into the nucleus, regulating the expression of proinflammatory cytokines, including TNF-α, IL-1, and IL-6, in the innate and adaptive immune mediation [58].

Along with the lamina propria, Peyer's patches (PPs), mesenteric lymph nodes (MLNs), and isolated lymphoid follicles comprise the gut-associated lymphoid tissues (GALTs), the largest lymphoid in our body, which assumes the primary immune response in the gut [50,59]. Specifically in adaptive immunity, antigen is sampled by antigen-presenting cells (APCs), such as DCs, after entering through microfold cells in the follicle-associated epithelium (FAE) or through the epithelium where MHC class II+ enterocytes may also act as local APCs. Then, the processed antigens are passed on to Peyer's patches or MLNs to activate CD4+-naive T lymphocytes, which can subsequently differentiate into regulatory T cells or T helper (TH) cells with the expression of integrin $\alpha4\beta7$ and the chemokine receptor CCR9, eventually accumulating in the mucosa. B cells mature into plasma cells in the lamina propria, producing IgA to the lumen to bind the antigens, the process of which is influenced by factors such as transforming growth factor-β (TGF-β), IL-10, and signals from T cells as well as DCs [50]. Additionally, some CD8+T cells in lamina propria can be cytotoxic T-lymphocytes (CTL), while some CD4+ and CD8+T cells can be memory cells.

- **The Immune System in the Brain**

In parallel, the responsive immune cells, cytokines, and chemokines, as well as antigens, may also gain access to the bloodstream and enter the brain [60]. After stroke onset, intestinal immune cells, majorly DCs and macrophages, have been found to significantly travel to the central nervous system (CNS), where the trafficking amount is higher than other lymphoid organs [61], which aggravates the brain disease. It was thought that the existence of the BBB and the paucity of specialized antigen-presenting cells render the immune response limited in the CNS [62]. Astrocytes and microglia are two extra types of immune cells, which specifically reside in the CNS parenchyma. Astrocytes are large, ramified glial cells, most abundant in the brain, with the function of immune defense and BBB integrity protection. Microglia, the microphages considered as the earliest cells to sense antigens in the CNS [63], transit to M1/M2 proinflammatory phenotypes [64] and release proinflammatory cytokines under the pathological status, which promote inflammation and, meanwhile, facilitate the interaction with astrocytes.

During the early acute phase after stroke, DAMPs released from damaged cells interact with PRRs to initiate innate immunity, inducing the activation and differentiation of immune cells. Concomitantly, upregulation of adhesion molecules, such as intracellular adhesion molecule 1 (ICAM-1) and p-selectin, contribute to the infiltration of immune cells. The release of inflammatory mediators, including matrix metalloproteinases-9, inducible nitric oxide synthase (iNOS), and reactive oxygen species (ROS), further aggravates cells' apoptosis and the BBB disruption by damaging tight-junction proteins [65,66]. The damaged blood–brain barrier leaves the entrance open for peripheral intestinal immune cells, such as B cells and T cells, where it was reported that over one-fourth of all invading T cells were from intestinal PPs, resulting in the augmentation of neuroinflammatory responses [67].

As the immune response to stroke is a complex and multistage process [68], innate immunity plays a major role in the early acute stage. During the chronic phase, some immune cells, such as B cells, possess a dual effect in the ischemic stroke progression, with benefits related to the secretion of the anti-inflammatory cytokine IL-10, while adverse effects are mediated by autologous immunoglobulins with respect to chronic inflammation [69]. Cerebral ischemia can induce cell-mediated immunity depression, such as diminished early NK and T cell responses with reduced interferons (IFN) production by Th1 cells, all of which was found to be prevented by blocking the sympathetic nervous system [70]. The stroke-induced immunodeficiency related to spontaneously severe bacterial infections primarily contributes to the morbidity, compared to the direct brain lesions.

- **Gut Microbiota Influences the Immune System**

Moreover, microbiota exert a profound impact on gut and brain immunity along the MGBA during stroke [71]. An improved stroke outcome with increased T and B cells was observed in colonized littermates compared to the GF mice [72]. Relevant alterations in the abundant phyla, such as Actinobacteria, Bacteroidetes, and Firmicutes, will induce T cell priming, including overall increased T helper cells (CD4+) as well as polarized regulatory T cells (Foxp3+), which significantly mediates cerebral protection. Meanwhile, the microbiota influences the level of neutrophil infiltration to affect the ischemic brain. Gut dysbiosis can also cause polarization of proinflammatory Th17 (IL-17+) and Th1 (IFN-1+) cells both in the gut [67] and the ischemic brain, along with the increased Treg cells in the small intestine and less IL-17+ $\gamma\delta$ T cells in the meninges. Studies have elucidated that the gut microbiota influences such balance between local intestinal Treg cells and $\gamma\delta$T cells by inducing the differentiation of intestinal DCs, ultimately affecting the protective or damaging effects of T cells migrating to the brain after ischemic brain injury. Specifically in the brain, the function and maturation of astrocytes and microglia are regulated by microbiota as well, through the Aryl hydrocarbon receptor (AhR) signaling pathway. As numerous studies have demonstrated the role of the gut microbiota in the development, maturation, and function [73] of peripheral and central immune cells, the influence of dysbiosis under pathological conditions, especially stroke, needs more exploration. Furthermore, metabolites, such as SCFAs, tryptophan (Trp), and bile acid (BA) metabolites, or components, such as LPS and peptidoglycans, of microbiota all significantly influence the immune system [74].

2.2.3. The Neural Network of the Microbiota–Gut–Brain Axis

- **The Enteric Nervous System**

Distinguished from other peripheral organs, the gut possesses an intrinsic complete nervous system, namely the ENS, with extensive regulation of gastrointestinal motility, mucus secretory, gastrointestinal endocrine, epithelial permeability, immunologic, and inflammatory responses [75]. The ENS comprises enteric neurons and enteric glia cells, forming two main plexuses, the submucous (or Meissner's) plexus between the circular muscle layer and the muscularis mucosa, innervating the muscle motor, and the myenteric (or Auerbach's) plexus between the longitudinal and circular muscle, responsible for secretory and absorptive control. Neurons of the ENS can be classified by function into intrinsic primary afferent neurons (IPANs), interneurons, motor neurons, and intestinofugal neurons (IFN), or by morphology, mainly into Dogiel type I and type II neurons [76,77]. In general, primary transmitters in enteric neurons include acetylcholine (Ach), serotonin (5-HT), nitric oxide (NO), substance P, vasoactive intestinal peptide (VIP), and calcitonin gene-related peptide (CGRP). The myenteric neurons are mostly cholinergic or nitrergic, while submucosal neurons are mainly cholinergic neurons or VIP-expressing noncholinergic neurons [76]. Beyond the neurotransmitters, enteric neurons as well as enteric glia may also produce cytokines, such as IL-18, colony stimulatory factor 1 (CSF1), and transforming growth factor-β (TNF-β). Therein, IL-18 is necessary for the AMPs production of goblet cells to defend against pathogens, including Salmonella typhimurium [78]. TNF-β also plays an essential role in muscularis macrophages' development and maintenance, indicating the intimate association between enteric nervous systems and barriers, as well as immune function [79,80].

Loss of enteric neurons has been elucidated in the stroke (pMCAO) model, with central and peripheral galectin-3 release through a TLR4-mediated mechanism involving TAK1 and AMPK [81]. Another study also reported the myenteric neurons' loss with increased VIP-immunoreactive submucous neurons in the ileum of pMCAO mice, which simulated focal cerebral ischemia, whereas global ischemia models exhibited no significant change, implicating that different types of stroke act differently in enteric neuronal survival [82]. A recent study also uncovered that stroke induced decreased neuronal nitric oxide synthase (nNOS) in enteric myenteric neurons, leading to the injured muscle relaxation and slow gut motility. Furthermore, the impairment can be rescued by denervation of the sympathetic nervous system through 6-hydroxydopamine [83].

- **The Connections between the ENS and CNS**

The ENS is closely linked to the CNS through both afferent and efferent pathways [84]. On afferent pathways, where the gastrointestinal tract (GIT) sends sensory information to the brain, the major extrinsic nerves are the vagus nerve (VN) and spinal splanchnic sensory nerves. The vagus nerve, whose cell bodies lie in the nodose ganglia (NG), which project to the brainstem, predominate the upper gut, such as the esophagus and stomach, while spinal splanchnic sensory nerves, whose cell bodies are in the dorsal root ganglia (DRG), which project to the spinal cord dorsal horn and the dorsal column nuclei, generally fall into sympathetic and pelvic pathways responsible for the lower gut. On efferent pathways where the brain delivers the motor information to GIT, the extrinsic nerves are the sympathetic and the parasympathetic nerves. The sympathetic nerves are adrenergic with inhibitory effects, whereas the parasympathetic nerves are cholinergic with excitatory effects in the gut.

As a mixed nerve, the VN contains four principal components, including general visceral efferent, which is parallel with the GIT-distributed part of parasympathetic fibers; thus, when it refers to GIT, the parasympathetic nerves are synonymous with the VN, which has been profoundly studied. The VN connects with ENS, smooth muscle, and mucosa to innervate the gut, readily detecting various signals, including gut hormones, neurotransmitters, as well as bacterial by-products. The neuroprotective effect of VNs and vagus nerve stimulation (VNS) have been found as a promising intervention in cerebral ischemic stroke. VNS can inhibit pyroptosis with suppressed related inflammatory molecules, such as NLRP3, cleaved caspase-1, and reduced pyroptotic cells and membrane pores. It also inhibited Toll-like receptor 4/nuclear factor kappa-B expression and caused a shift from a proinflammatory phenotype in microglia with reduced NOS and TNF α, which is intricately mediated by the α7 nicotinic acetylcholine receptor (α7nAchR) [85,86].

- **Gut Microbiota Influences the Nervous System**

The gut microbiome sends significant signals in the neural network along the MGBA. Studies found that the depletion of microbiota induced activation of the gut sympathetic ganglia, which can be reversed by colonizing with SCF-producing bacteria [87]. Oral inoculation with the microbiota, such as live *Campylobacter jejuni* (*C. jejuni*), could also lead to the activated neurons bilaterally in the vagal ganglia [88], suggesting that peripheral sensory neurons send an early signal to the brain regarding pathogens or gut microbial alteration. Intrinsically, the microbiota closely interacts with the ENS. Antibiotic (Abx)-induced bacterial depletion led to the loss of enteric neurons in both submucosal and myenteric plexuses in the ileum and proximal colon, as well as reduced enteric glia in the ileal myenteric plexus, which could be recovered by microbial restoration. Furthermore, SCFs were verified to stimulate enteric neurogenesis by restoring enteric glia and neurons [89]. A regulatory circuit may also exist in the microbiome, enteric neurons, and other gut cells, including enteroendocrine cells (EECs) and immune cells. EECs can release ENS neurotransmitters, such as Ach and 5-HT, in response to altered microbiota metabolites [54]. Yan et al. revealed a relationship showing that the microbiota affects the neuronal transcriptome and structure, which subsequently tunes the generation of Treg cells lining the enteric neuron projections in the lamina propria [90]. Beyond the SCFs, tryptophan (Trp), and bile acid (BA), gut microbiota also produce and mediate other neuroactive metabolites, such as glutamate, GABA, Ach, 5-HT, and dopamine, affecting neuronal activities [91].

2.2.4. The Humoral Communication of the Microbiota–Gut–Brain Axis

- **Hypothalamic–Pituitary–Adrenal Axis and Its Association with Stroke**

The hypothalamic–pituitary–adrenal (HPA) axis is one of the principal neuroendocrine pathways and a fundamental part of the MGBA. When exposed to stressful stimuli, such as IL-6 and TNF, triggered by a disease such as stroke, the HPA axis is activated, leading to the production of glucocorticoids or cortisol, the primary stress hormones, which exert an anti-inflammatory and immunosuppressive effect on pathological tissue, along with a negative feedback on the hypothalamus and pituitary glands to control the response [92]. However, chronically elevated levels of glucocorticoids may implicate a pathological condition or underlie a bad prognosis.

A massive response associated with HPA was observed in the early stage of acute stroke, with a continuous high level of cortisol, whereas merely in the initial phase did the concentration of adrenocorticotropic hormone (ACTH) increase [93]. Cortisol levels remained high for at least seven days after stroke, as concluded by a systematic review [94], and exogenous corticosterone contributed to neuronal death with larger infarcts in MCAO mice [95]. It has been reported that post-stroke administration of metyrapone, an inhibitor of glucocorticoid synthesis, decreased the IL-6 level and the infarct size in the ischemic brain [96]. However, glucocorticoid receptor blockade was reported to cause increased blood catecholamine, while the application of prednisolone decreased β2-adrenergic receptor expression on lymphocytes in vitro, possibly reducing the sensitivity to catecholaminergic stimulation following a stroke [97]. ACTH was found to be significantly correlated with the brain lesional volume and functional outcome [93], with the effect of predicting the development of post-stroke depression as a vital and independent biomarker in a three-month follow-up study [98]. Numerous studies have identified a dysregulated HPA axis in stroke patients, and a hyper-activated HPA axis was also demonstrated in diabetic mice after stroke with an augmented inflammatory response [96]. Taken together, these studies show that the HPA axis may be a potential therapeutic target in improving stroke outcomes.

- **Gut Microbiota Influences the HPA Axis**

The gut microbiota is one of important components to activate the HPA axis. A higher HPA axis response to stress was observed in GF mice with plasma ACTH and corticosterone elevation [99]. During the inflammatory condition, bacterial translocation is also able to activate the HPA axis through the disrupted gut and brain barriers. At the genetic level, the microbiota was found to attenuate the expression of Fkbp5, a gene regulating glucocorticoid receptor sensitivity in the pituitary gland, as well as the expression of genes encoding steroidogenesis, such as MC2R and Cyp11a1 in the adrenal gland [100].

- **Gut Hormones**

As important signaling molecules within the GMBA, gut hormones spread throughout the GIT are released by enteroendocrine cells (EECs) to regulate gut motility, appetite, and maintain intestinal homeostasis. Enteroendocrine cells, which only account for 1% of gastrointestinal epithelial cells, secrete various kinds of hormones that act locally on nerve endings, enteric neurons, epithelium, and immune cells or enter the circulation to affect the CNS or other distant sites [101]. There are 12 major types of enteroendocrine cells, including enterochromaffin cells, A cells, and D cells, which secrete more than 20 hormones [102]. The main signals involved in the gut–brain crosstalk are ghrelin, polypeptide YY (PYY), glucagon-like peptide-1 (GLP-1), glucose-dependent insulinotropic polypeptide (GIP), cholecystokinin (CCK), and 5-HT [103]. EECs are equipped with a wide array of sensory receptors, which can detect microbial metabolites, such as SCFAs [104], or bacterial components, such as LPS [105]. These gut hormones may subsequently activate IPANs and extrinsic afferent nerve terminals to influence the functions of the brain [106].

3. Tryptophan Metabolism along the Microbiota–Gut–Brain Axis and Stroke

There is a growing appreciation for tryptophan metabolism along the microbiota–gut–brain axis, which has close associations with stroke. As an essential amino acid, tryptophan mainly originates from the digestion of dietary proteins in the small intestine [107]. Trp is metabolized via three major pathways, with a series of important metabolites in the gut and brain, which experience significant alterations during stroke (Figure 3), exerting a profound effect on the physiological and pathological processes.

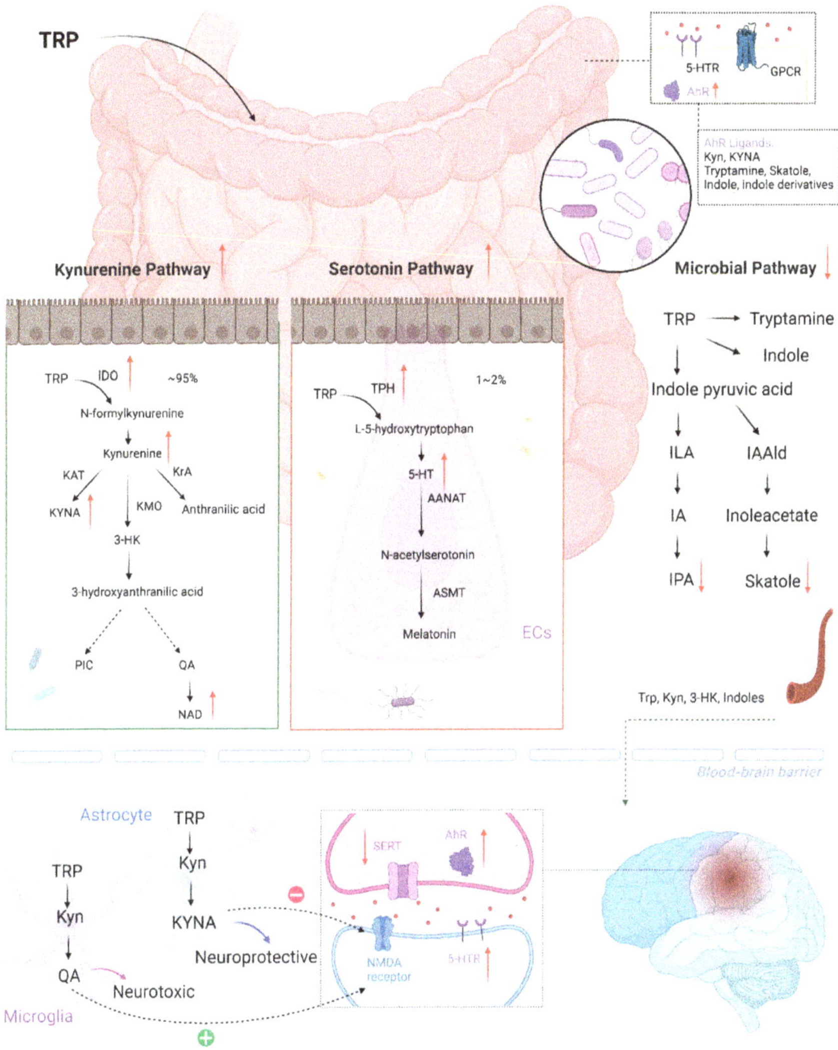

Figure 3. Tryptophan metabolism along the microbiota–gut–brain axis and its main alteration in stroke. Created with BioRender.com.

3.1. Tryptophan Metabolism Pathways along the Microbiota–Gut–Brain Axis

3.1.1. Kynurenine Pathway in the Gut and Brain

The kynurenine pathway (KP) is a key metabolic route that breaks down about 95% of free tryptophan (Trp) [15]. The process is primarily initiated by two crucial rate-limiting enzymes, indoleamine-2,3-dioxygenase (IDO) and tryptophan 2,3-dioxygenase (TDO). IDO is extensively expressed in the body, including the brain and GI tract, with two isoforms, IDO1 and IDO2. IDO1 is pivotal in the immune response and always activated by the inflammatory cytokines, especially interferon (IFN)-γ [108,109], while IDO2 is believed to be basally expressed [110]. TDO, induced by the activation of the HPA axis with increased glucocorticoid, catalyzes 90% of Trp under physiological conditions in the liver and brain [111]. With the aid of the enzyme, Trp is converted to N-formyl kynurenine, which is then catabolized to the pivotal intermediate, kynurenine (Kyn), by formamidase. AhR, a ligand-activated transcription factor that expresses on various cell types, such as T cells, neutrophils, microglia, and astrocytes, plays a master regulatory role in inflammation and immune function with the release of cytokines, including IL-6, IL-17, and IL-22 [112,113]. Kyn, a biologically active metabolite, acts as a ligand for AhR, with the functions of immunity suppression, inflammation, and carcinogenesis [114,115]. Then, Kyn can be further metabolized into three branches, which are kynurenic acid (KYNA) by kynurenine amino-transferase (KAT), 3–hydroxykynurenine (3-HK) by kynurenine-3-monoxygenase (KMO), and anthranilic acid by kynureninase A (KrA).

KYNA, an endogenous neuroprotective agent [116], can activate AhR and G protein-coupled receptor 35 (GPCR35), the activation of which may induce the inositol trisphosphate production and Ca^{2+} mobilization [117]. The 3-HK can influence the immune function by inhibiting the proliferation of CD8+ T cells and suppressing T cell responses [118,119]. Then, 3-HK converts to 3-hydroxyanthranilic acid, whose catabolism leads to the generation of quinolinic acid (QA) and picolinic acid (PIC). The N-methyl D-aspartate (NMDA) receptor is a ligand-gated cation channel, mediating the glutamate signals, with a critical role in the nervous system [120]. Overexcitation of NMDA receptors was suggested to be of great significance to the etiology of brain damage and neurodegenerative disorders, including stroke [121]. Among the major products of tryptophan, KYNA and PIC act as neuroprotective NMDA receptor antagonists, while QA is considered as a neurotoxic NMDA receptor weak agonist with proinflammatory properties [122]. Under pathophysiological conditions, QA can trigger oxidative stress and apoptosis by forming a complex with iron that contributes to ROS in different cells, including neurons and astrocytes, whose decrement, in turn, leads to lower KYNA production, thus enhancing neurological symptoms and chronic inflammation [123]. Oxidative damages to DNA induced by QA were reported to cause the overactivation of poly (ADP-ribose) polymerase (PARP) [124], an enzyme that repairs damaged DNA after oxidative injury, which would further deplete ATP and NAD [125]. Intriguingly, in mammalian cells, de novo from tryptophan is a significant biosynthesis pathway of NAD [126], which can be generated from QA as the end product of the 3-HK branch mediated by the KMO, also a rate-limiting enzyme sensitive to proinflammatory cytokines. During the inflammatory diseases, KP metabolism, as well as the production of NAD, predominantly increase, implicating the enhanced energy demand in the pathological process. Despite that QA and KYNA cannot cross the BBB, Trp, Kyn, and 3-HK can, which contribute to the accumulation and generation of neurotoxic intermediate products, including 3-HK and QA, in the brain, where microglia mainly produce QA, while astrocytes generate KYNA.

3.1.2. Serotonin Pathway in the Gut and Brain

About 1–2% of ingested Trp is metabolized in the serotonin pathway [127], which is initiated by tryptophan hydroxylase (TPH). There are two isoforms of TPH, including TPH1, which expresses in the enterochromaffin cells (ECs), and TPH2 in the CNS and ENS [111,128]. Most 5-HT is produced by enterochromaffin cells in the gut, while the rest is synthesized by neurons, leading to the distinction between mucosal 5-HT, which was related to intestinal inflammation [129], and neuronal 5-HT, which was found to promote intestinal mucosal epithelial growth and turnover, maintain myenteric neuronal density, and influence gastrointestinal motility [130,131]. Trp serves as the exclusive precursor of 5-HT, a primary monoamine neurotransmitter responsible for central neurotransmission, as well as a regulatory hormone in intestinal development and physiological activities [132–135], depending on the distinct subtypes of 5-HT receptors in various cells and tissues.

The 5-HT receptors, extensively expressed in the CNS, including the hippocampus, hypothalamus, basal ganglia, and cortex [136], are classified into seven principal classes, namely from 5-HT1 to 5-HT7, which all belong to metabotropic G-protein-coupled receptors, except for the ligand-gated ion channel 5-HT3 receptor (5-HT3R) [137]. However, 5-HT3R, which can be found on the terminals of GI vagal afferent neurons, displays an intimate involvement in gut–brain signaling [138]. In response to mechanical or chemical stimulation in the GI tract, 5-HT is released from ECs and acts locally on 5-HT3 receptors present on vagal afferent fiber, triggering smooth-muscle activity during the physiological and pathophysiological processes, such as distention, nausea, and vomiting [138,139]. Highly exposed to luminal contents, 5-HT4R, widely expressed in the intestinal epithelium as well as the ENS, specifically in myenteric neurons [140], is also vital in GI motility and epithelial secretion. Mucosal administration of 5-HT4R agonists promotes mucus secretion by goblet cells and Cl^- discharge by enterocytes [141]. Stimulation of 5-HT4R led to anti-proinflammatory, neurogenesis, and neuroprotection responses in the gut, and blocking 5-HT production induced loss of myenteric neurons and inhibition of neuronal differentiation from enteric neuronal stem cells [140]. Moreover, the agonist of the 5-HT1A receptor prevented apoptosis of intestinal epithelial cells and enteric glial cells, with reduced mucosal injury [142], and other 5-HT receptors, such as 5-HT1P and 5-HT2A, have also been verified to mediate different signals within the gut toward the CNS [143]. Additionally, the serotonin-selective reuptake transporter (SERT) exists both in the brain and intestinal epithelium, responsible for 5-HT uptake, effectively clearing serotonin from the synaptic gap and managing 5-HT concentrations [111,139]. Despite that intestinal 5-HT cannot cross the BBB, the free tryptophan that enters the circulation through the gut epithelium may cross the BBB to produce 5-HT in the brain, where some 1% of dietary tryptophan is used for the serotonin biosynthesis [132].

Another well-known product in this branch is melatonin (N-acetyl-5-methoxytryptamine), which is converted from serotonin via N-acetylserotonin through arylalkylamine N-acetyltransferase (AANAT) and acetylserotonin O-methyltransferase (ASMT). Melatonin is a hormone secreted by the pineal gland during the night, with a function in the regulation of circadian rhythm and immunity, as well as an antioxidant property [144]. Dysfunctional serotonin signaling has been widely reported in GI disorders combined with psychiatric symptoms [145].

3.1.3. Microbial Pathway

- **The Gut Microbiota's Role in Direct Trp Metabolism**

A small amount of Trp can be directly converted by the gut microbiota, primarily into indole and its derivatives. Accumulating evidence suggests that the homeostasis of microbiota–gut–brain interplay plays a central role in the regulation of Trp metabolism. Superior reviews have summarized various gut bacterial species producing different kinds of tryptophan catabolites [146,147]. Overall, phyla Actinobacteria, Firmicutes, Bacteroidetes, Proteobacteria, and Fusobacteria, as well as genera Clostridium, Burkholderia, Streptomyces, Pseudomonas, and Bacillus, are more common and enriched in tryptophan metabolism according to an in silico analysis based on bacterial genomes and gut microbiome data [148]. Specifically, some permanent-resident microbes of the gut, including the genera Clostridium, Bacteroides, as well as Escherichia coli, can metabolize Trp into indole with the expression of tryptophanase (TnA) [149,150]. Indole, viewed as a signaling molecule, affects several bacteria on quantity, virulence, drug resistance, plasmid stability, and biofilm formation, in turn [150]. It also mediates inflammation in innate immune responses [151], modulates the secretion of gut hormones, such as GLP-1, mechanically through inhibiting the voltage-gated K^+ channel, as well as weakens ATP production with NADH dehydrogenase blockage [152]. Ruminococcus gnavus and Clostridium sporogenes are known to decarboxylate Trp to tryptamine, a β-arylamine neurotransmitter, which induces the release of 5-HT by enteroendocrine cells and binds the 5-HT4 receptor to modulate colonic secretion and promote gut transition [153–155]. Along the microbial pathway, indole, tryptamine, skatole, as well as a series of indole derivatives, including indole-3-acid-acetic (IAA), indole-3-lactic acid (ILA), indole acrylic acid (IA), indole propionic acid (IPA), indole-3-acetaldehyde (IAAld), and indoxyl-3-sulfate (I3S), are ligands of AhR [156]. Typically, Bacteroides, Clostridium, Peptostreptococcus, Lactobacillus, and Bifidobacterium are producers of indole-3-lactic acid (ILA), which was suggested to significantly down-regulate the expression of IL-8 in gut epithelial cells and suppress NFκB activation in macrophages in a dose-dependent manner via activation of the AhR and nuclear factor erythroid 2-related factor 2 (Nrf2) pathway [157]. Research in human peripheral blood mononuclear cells also exhibited the anti-inflammatory and antioxidative effects, with reduced IL-6 and IL-1b and elevated IL-10 levels under the treatment of IA, which is mainly produced by Peptostreptococcus species [158]. During the inflammation induced by LPS, IPA possessed an inhibitory effect on PI3K/AKT/mTOR signaling pathway. In the study, IPA significantly enhanced the intestinal barrier through increasing claudin-1, Occludin, and ZO-1 levels, as well as facilitating the mucin secretion and the goblet cell function [159].

Indoles are known to cross the BBB and exert a profound effect on the brain [160]. Overproduction of indole in the gut increases the cerebral oxindole level. Supplementation with I3S, IPA, IAld, or tryptophanase in antibiotic-treated mice attenuated CNS inflammation via AhR signaling [161]. Equipped with the potent capacity of scavenging hydroxyl radicals and inhibiting superoxide dismutase, IPA protected primary neurons and neuroblastoma cells from oxidative damage and death induced by Aβ or hydrogen peroxide [162]. Oral administration of IPA mitigated ischemic injury, potentially by reducing DNA damage in pyramidal neurons and lipid peroxidation in ischemic hippocampal homogenates [163]. However, some of the effects can be adverse. For example, intraperitoneally administration of indoxyl sulfate can cause its further accumulation in the prefrontal cortical tissues and cerebrospinal fluid, which leads to behavioral changes and neurodegeneration, including anxiety, depression, and cognitive impairment, accompanied by impaired neuronal cell survival and neural stem cell activity, as well as disturbed expression of serotonin, corticosterone, and brain-derived neurotrophic factor (BDNF) [164].

- **Gut Microbiota Influences the Kynurenine and Serotonin Pathways**

Moreover, the gut microbiota influences the tryptophan metabolism indirectly through the kynurenine and serotonin pathways. Gene expression of the kynurenine pathway enzymes, including kynurenine aminotransferase, tryptophan 2, and 3-dioxygenase in the hippocampus, were changed in GF mice [165]. In GF pigs, tryptophan and kynurenine in the large intestine, as well as 5-HT in most organs, were significantly increased, whereas indole metabolites were generally decreased [166]. A recent study revealed relevant features of KP metabolism in Lactobacillus (L.) reuteri, with the emphasis on the qualitative differences between bacterial and mammalian metabolism. They found that kynurenine readily entered the bacterial cells and was preferentially metabolized to KYNA, whose production was positively correlated with the concentration of kynurenine and bacteria [167]. Beyond the *L. reuteri* from the phylum Firmicutes, bacteria expressing KYN pathways also include Bacteroidetes, Actinobacteria, and Proteobacteria phyla [148]. These microbiota-derived LPS were also found to stimulate colonic immune responses with the upregulation of IDO1 in the Kyn pathway, inducing Trp depletion and Kyn accumulation in the circulation [168]. Under the circumstance of inflammation, the available Trp tend to be shunted into KP rather than serotonin, which can be possibly be explained by the overaction of IDO1 [123]. When administered the antibiotics in combination with a ketogenic diet, downregulation of IDO1 and upregulation of hippocampal kynurenic acid was demonstrated [169]. A dextran sulfate sodium (DSS)-induced colitis mice model also showed increased Kyn, as well as a decreased KYNA level, with a significantly elevated Kyn/Trp ratio, and suggested a correlation with several microbial species, including Bacteroides_acidifaciens, Shigella_sonnei, Parabacteroides_distasonis, Ileibacterium_valens, Bacteroides_vulgatus, and Burkholderiales_bacterium_YL45, which may disrupt the IDO in the cortex, and Parabacteroides_distasonis, Ileibacterium_valens, and Burkholderiales_bacterium_YL45, which were indicated to regulate the KYNA level in serum directly or by affecting KAT2 [170].

Furthermore, gut microbiota metabolites are able to influence the KP as well. In a rat model of bilateral common carotid artery occlusion (BCCAO), SCFA supplementation suppressed the conversion of Trp to Kyn and restored 5-HT levels in the hippocampus and gut, inhibiting the inflammatory augmentation [171]. Butyrate treatment ameliorated serum Kyn accumulation through promoting mitochondrial functions in colonocytes as well as remodeling the gut microbiota in high-fat-diet (HFD) mice [168].

The serotonin pathway is also regulated by the gut microbiota, which plays a key role in promoting 5-HT secretion. Adult GF mice displayed deficient 5-HT concentrations in serum and plasma, with decreased expression of TPH1 in enterochromaffin cells and increased colonic 5-HT transporter SLC6A4, all of which can be reversed by the colonization of indigenous spore-forming microbes (Sp), whose metabolites, such as deoxycholate, a secondary bile acid, also raise 5-HT levels in ECs [172]. In addition, serotonergic neuronal networks were almost absent in GF mice with the alterations of 5-HT4 receptor expression [140]. Activation of 5-HT4R in GF mice may accelerate intestinal transit, concomitant with increased myenteric neurons and innervation of the colonic mucosa [140]. Orally administering gut microbiota from specific pathogen-free mice in GF mice increased 5-HT-positive endocrine cells in the upper GI and colon [173]. Probiotics, such as Bifidobacterium lactis TY-S01, maintained 5-HT in constipated mice with the upregulated mRNA expression of the 5-HT4 receptor [174], and colonization of enteropathogenic Escherichia coli boosted the 5-HT concentration by downregulating the SERT [175]. Besides, frontal cortex 5-HT exhibited negative correlations with the ratio of Firmicutes in total gut microbiota. Treatment with probiotics (fructo- and galacto-oligosaccharide) and prebiotics (Bifidobacterium longum and L. rhamnosus) diminished colonic 5-HT and TPH1 with increased SLC6A4. The decrease in intestinal 5-HT synthesis was then explained for more available tryptophan, possibly converting into 5-HT in the brain through the BBB [176]. Antibiotic-induced gut microbiota intervention was found to significantly increase the 5-HT level and the expression of its receptor 5-HT2A with the upregulation of TPH2 in the hypothalamus [177]. In alcohol-dependent mice, 5-HT content was significantly increased both in intestine and brain tissues after receiving the fecal microbiome from healthy mice [178].

It is believed that BAs activate the bile acid receptor TGR5 on ECs and intrinsic primary afferent neurons to stimulate the release of 5-HT [179], and through the TGR5/TRPA1 pathway, Bacillus subtilis was found to increase 5-HT production from ECs [180]. SFAs, such as butyrate, a positive allosteric modulator of SERT, promoted 5-HT uptake in mouse ileal organoids, and fecal concentrations of both acetate and butyrate were found negatively associated with 5-HT in serum [181]. LPS treatment prenatally increased 5-HT content in serum, the colon, and the medial prefrontal cortex (mPFC) with the increased TPH1, reduced SERT in the colon, and led to a higher serotonin 1B receptor (5-HT1BR) level in the mPFC in autism-like offspring, and they were detected to have lower mRNA expression levels of NAT and ASMT [182].

3.2. *Tryptophan Metabolism and Stroke*

Among the various abnormalities in metabolism induced by stroke, tryptophan metabolism was the most affected, according to an analysis of untargeted metabonomics in rats with stroke [183]. Furthermore, single-nucleotide polymorphisms of the genes encoding the tryptophan metabolism enzymes, such as TPH1, TPH2, KAT1, KAT2, and IDO1, were found to seriously influence the risk and pathogenesis of stroke [17]. Right before or immediately after a stroke, changes in tryptophan metabolism occurred, which were associated with the inflammatory response and oxidative stress.

Therein, the kynurenine pathway was illustrated to be activated right in the hyperacute phase (<4.5 h) of cerebral ischemia [184]. Along the activated kynurenine pathway after stroke, an increased kynurenine:tryptophan ratio and a highly significant decrease in the ratio of 3-hydroxyanthranilic acid:anthranilic acid were previously found correlated with the infarct volume [185]. IDO1 upregulation was always observed in various inflammatory states, including stroke, when the IDO1 mRNA levels increased at an early phase [19]. TDO has gained attention as well. In one study, L-Kyn levels increased in 3 h right after MCAO, with the predominant upregulation of L-Trp as well as TDO, which accounted for L-Kyn biosynthesis in neurons of the ischemic brain. Then, the Kyn activated the AhR very rapidly, which may be causally involved in brain damage at the crucial period when the lesion extended following the ischemic occlusion. Therefore, the inhibition of TDO was found to be a promising therapeutic avenue to inhibit the endogenous apoptotic and anti-neuroprotective pathways in ischemic stroke damage [186]. KYNA was demonstrated to act as a neuroprotective/scavenger/antioxidant factor, which counteracted the tissue damage and neuronal loss induced by stroke. From early phases of stroke, an overexpression of KYNA occurred in mammalian astrocytes exclusively in the infarcted regions and continued at least 21 days post-stroke via the utilization of a highly specific monoclonal antibody against KYN in transient MCAO rats [187]. The negative association between the addition of KYNA and the mortality was found in a multicenter prospective cohort study [188].

AHR is activated and overexpressed in neurons of peri-infarct, ischemic core, as well as the cortex regions in experimental MCAO mice, with a damaging role in stroke physiopathology, as revealed by the AhR agonist, which aggravated the stroke outcome. Haploinsufficiency or total deletion of AhR decreased ischemic damage and were neuroprotective. The cAMP response element (CRE)-binding protein (CREB) is a critical neuronal transcription factor, which mediates the expression of survival and antiapoptotic genes, such as neuroprotective BDNF, against ischemic injury. The increased AhR may inhibit the CREB signaling. Administration of the AhR antagonist TMF boosted the cortical levels of CREB as well as BDNF, and decreased numbers of apoptotic cells in the peri-infarct region of MCAO animals, accompanied by the downregulated proapoptotic proteins p53 and Puma as well as increased antiapoptotic Bcl-X [186]. Likewise, during the ICH, AHR overexpressed primarily in neurons in the peri-hematoma zone, which peaked at 48 h from initial insult. Subsequently, the AHR can recruit RhoA/Bax complexes, playing a detrimental role with respect to mitochondria death and free radical accumulation. Pharmacological application of TMF mitigated ROS accumulation and neuronal apoptosis in the ipsilateral hemisphere, with reduced Bax levels, Mito Sox+ cells, proapoptotic protein cleaved caspase-3, and the biomarkers of damage elicited by oxidized lipids, namely 4-HNE protein. In this setting, the kynurenine pathway was initiated immediately after stroke and kynurenine directly contributed to the AHR activation and substantial oxidative stress, with hampered neurobehavioral performance [189].

Of note, the ligands of AHR can be divided into host-derived kynurenine-based ligands, which are upregulated and detrimental after stroke with worsened microglia survival as well as poor prognosis, and microbiota-derived indole-based ligands, such as IAld and IPA, which decrease especially within the first 24 h in both the plasma and the brain after stroke, along with significant reductions in major bacterial populations involved in the regulation of Trp-derived ligands, such as Bifidobacteriales and Lactobacillales [190]. After stroke onset, activation of microglial AHR by microbial Trp-derived ligands was corroborated to suppress the expression of proinflammatory and neurotoxic genes, including TNF, IL6, IL12A, and NOS2, and boost anti-inflammatory IL10 expression. Moreover, the Trp metabolites also led to increased TGFα and downregulated VEGF-B expression via AHR in microglia, which further influenced the proinflammatory gene expression in primary human astrocytes [191]. Post-stroke treatment in GF mice with the cocktail of IPA and IAld beneficially increased microglial AHR expression associated with a higher surface expression of CD11b, MHC-II, and CD80, ameliorating antigen presentation and co-stimulation immune functions in microglia, which resulted in a reduced infarct size and neurological deficits. All these results showed the two sides of the role of AHR depending on its ligands. Herein, a balanced pool of host-derived and microbiota-derived AHR ligands was highlighted, and via post-stroke administration of indoles, stroke outcomes can be improved [190].

Along the microbial pathway, the serum IPA levels were significantly lower in mice with MCAO. And in patients with coronary artery disease (CAD), IPA was the most downregulated metabolite and decreased more evidently in stroke patients. IPA intragastrical supplementation improved stroke outcomes and the MCAO-induced alterations of the gut microbiome structure, leading to a higher abundance of the probiotic Lactobacillus and Lachnospiraceae_NK4A136_group, which was associated with the enhanced intestinal barrier function and epithelial integrity after IPA treatment. In IPA groups, immune function in MCAO mice was modulated, as proved by the reversing trend of the increased Th17 (CD4+ IL17+) and reduced Treg (CD4+ CD25+ Foxp3+) numbers in the GALT, such as PPs and MLNs, as well as decreasing CD3+ cells and increasing Tregs in the ipsilesional injured brain parenchyma, which further alleviated the neuroinflammation after stroke, with significantly decreased IL1β, IL6, and CCL2 and increased IL10 at both the mRNA and protein levels. The neuroprotective effect of IPA in stroke was further confirmed by the reduction in the activated astrocytes and decreased neuronal apoptosis in the peripheral infarction area. Moreover, the co-cultured experiments in the oxygen–glucose deprivation and reperfusion model revealed that IPA inhibited the A1 phenotype neurotoxic-reactive astrogliosis indirectly via intestinal Tregs [192]. Mechanically, a further study attributed the benefits of IPA to facilitating cholesterol efflux from macrophages to lipid-free apolipoprotein A-I (ApoA-I) through miR-142-5p/ATP-binding cassette transporter A1 (ABCA1) signaling [193]. On the contrary, another study elaborated the hypertensive effect of IPA, and intravenous infusion of IPA in rats increased heart muscle contractility, which can be reflected in the increased stroke volume [194]. In addition, indole-3-lactic acid was found significantly increased at one day after MCAO, but reduced in the chronic phase of stroke, which is potentially related to post-stroke depression [195]. Lower levels of 3-methylindole (skatole) and indoles were discovered in human plasma samples 24 h after IS [196]. In general, higher levels of indole-3-propionate were reported to be correlated with lower all-cause mortality, while indole-3-acetate and indole-3-lactate were positively correlated with fatal cerebral infarction, and indole-3-lactate levels were related to fatal stroke [197].

Higher serotonin levels were associated with the incidence of stroke, including fatal stroke [197]. In the serotonin pathway, TPH1 mRNAs was significantly upregulated in MCAO brains, with increased thrombosis and thromboembolism risk [198], and TPH1 variants were found to be more frequent in heterozygous form in the control group compared to the group of ischemic stroke patients [199], implicating a preventive and therapeutic strategy at the RNA and gene levels. Additionally, an analysis of blood samples collected from AIS patients uncovered that melatonin was significantly increased during the disease, whose synthesis may be provoked by ROS in order to defend against the oxidative injury [200], whereas 5-HT plasma concentrations did not display a distinction [19]. In patients undergoing the hyperacute phase (<4.5 h) of brain ischemia, platelet 5-HT2AR density was higher than in controls, while the platelet density of SERT was significantly lower, which coincides with the analysis of genetic polymorphisms in 834 patients with AIS and TIA, revealing that the presence of the high-expression SERT genotype is linked to a lower risk of ischemic stroke [201]. The patient group also showed temporarily increased urinary 5-HT, which had a role in promoting platelet aggregation, vasoconstriction, and ultimately, a thrombotic process. Nevertheless, these alterations decreased after one day and disappeared at three months from brain ischemia symptoms [184]. Furthermore, 5-HT1AR can also act as a serotonergic target for neuroprotection in cerebral ischemia [202]

4. Conclusions

From metabolic perspectives, emerging insights into the impacting role of the gut microbiota and gut–brain axis in cardiovascular diseases shed light on preventing and intervening in the progression of stroke, which offers a promising opportunity for translational research. A variety of effective and accessible intervention strategies have been clinically verified, encompassing probiotics, prebiotics, synbiotics, antibiotics, and diet or fecal microbiota transplantation. To harness the full potential of these interventions, a comprehensive understanding of the core pathways and mechanisms within the microbiota–gut–brain axis is essential. In this review, we systematically summarized the alterations of gut microbiota in stroke and the interactive influence between stroke and the immune, neural, and humoral pathways within the MGBA.

Tryptophan metabolism is a fundamental part of the MGBA, and its intimate correlation with CNS diseases has been widely reported. Herein, based on the profound and comprehensive perception of the MGBA, we further described the tryptophan metabolism in the kynurenine pathway, serotonin pathway, and microbial pathway. We overviewed the advances on the relationship between stroke and tryptophan metabolism, with the expectation of promoting the potentially diagnostic biomarkers and therapeutic avenues from the perspective of tryptophan metabolism along the MGBA, such as targeting the key metabolic enzymes at gene, mRNA, or protein levels, applying the antagonist or agonist of the crucial receptors, and controlling some critical metabolites. Practically, as the tryptophan index has been selected as one of the biomarkers to clarify the close and independent association between dietary patterns and stroke risk, the possibility of regulating dietary habits, including increased intakes of fresh fruits, vegetables, and fish/seafoods, in order to prevent stroke was highlighted [203]. Some specific healthy dietary patterns, such as the plant-based, Mediterranean, and Dietary Approaches to Stop Hypertension (DASH) diet patterns, are associated with elevated IPA levels as well as reduced stroke risk [204]. In addition, compounds extracted from natural products that have effects on tryptophan metabolism may provide profitable and auspicious alternatives for the treatment of stroke with few side effects and high availability [205,206].

On account of the different types and stages of strokes, sample size limitations, regional and ethnic variations, and different designs and durations, discrepancies did exist among the studies, which are expected to be refined later on. In the future, clear and precise mechanisms of gut–brain communications, how the MGBA affects stroke, and more associations between tryptophan metabolism and stroke are to be deeply explored with the utilization of advantageous technologies, such as metabolomics, proteomics, microbiome sequencing, etc. This will prompt advances in the field, shift from association to causation, and open new lines for more possibilities of clinical translatability and application, which will provide better prevention, diagnosis, and treatment of stroke, as well as ameliorative stroke prognosis and recovery.

Funding: This work was funded by the National Natural Science Foundation of China (NSFC81903631).

Institutional Review Board Statement: Not applicable.

Informed Consent Statement: Not applicable.

Data Availability Statement: No new data were created or analyzed in this study.

Conflicts of Interest: The authors declare no conflicts of interest.

Appendix A

Table A1. Major gut microbiota alterations in clinical stroke in case-control studies in the past five years.

Type of Stroke	Cohort Size	α-Diversity in Stroke	Main Changes in Gut Microbiota in Stroke	Year	Refs.
IS	30 stroke patients and 30 controls	No difference in Ace, Chao1, Simpson, and Shannon	↑*Odoribacter, Akkermansia, Ruminococcaceae e_UCG_005, norank_p_Flavobacteriaceae, norank_p_Parcubacteria, Victivallis* ↓*Anaerostipes, Ruminiclostridium_5*	2019	[22]
IS	31 stroke patients and 9 controls	No difference in Ace, Chao1, Simpson, Shannon, and observed OTUs	↑*Streptococcus * infantis, Prevotella * copri* ↓*Blautia obeum*	2019	[207]

Table A1. *Cont.*

Type of Stroke	Cohort Size	α-Diversity in Stroke	Main Changes in Gut Microbiota in Stroke	Year	Refs.
IS	79 stroke patients and 98 controls	No difference in Ace, Chao1, Simpson, Shannon, and observed OTUs	↑Proteobacteria *, Actinobacteria *, *Lactobacillus* *, *Lactococcus* *, *Escherichia/Shigella* *, *Streptococcus* *, *Collinsella* *, *Dorea* * ↓Firmicutes *, Bacteroidetes *, *Faecalibacterium* *, *Subdoligranulum* *, *Eubacterium* *, *Roseburia* *, *Lachnoclostridium* *, *Butyricicoccus* *	2020	[208]
IS, TIA, and HS	349 stroke patients and 51 controls	↓ Shannon, Simpson, and observed OTUs	↑Proteobacteria, *Escherichia/Shigella*, *Peptoniphilus*, *Ezakiella*, *Enterococcus* ↓Firmicutes, Bacteroidetes	2020	[209]
IS	20 stroke patients, 10 post-stroke, and 16 controls	↑Ace, Chao1, and observed OTUs	↑Actinobacteria *, Bacteroidaceae *, Bifidobacteriaceae *, Coriobacteriaceae *, Enterobacteriaceae, Weeksellaceae, Bacillaceae, Paenibaciiaceae, Brucellaceae, Xanthomnadaceae ↓Firmicutes *, Lachnospiraceae *, Ruminococcaceae*, Prevotellaceae*	2020	[210]
IS	28 stroke patients and 28 controls	↓Shannon and phylogenetic diversity	↑Enterobacteriaceae, Ruminococcaceae, Veillonellaceae, Lachnospiraceae ↓Bacteroidaceae, Prevotellaceae	2021	[12]
AIS	140 stroke patients and 92 controls	N/A	↑Lactobacillaceae *, Enterobacteriaceae *, Porphyromonadaceae *, *Akkermansia* * ↓Lachnospiraceae *, *Roseburia* *, *Bacteroides* *, *Faecalibacterium* *, *Blautia* *, *Anaerostipes* *	2021	[211]
AIS and AIS with T2D	150 stroke patients and 55 controls	N/A	↑Enterobacteriaceae *, Proteobacteria, Gammaproteobactera, Deltaproteobacteri, Desulfovibrionaceae, *Dorea* *, *Lactobacillus*, *Megasphaera* ↓*Lachnospira* *, *Coprococcus* *, *Prevotella*, *Faecalibacterium*, *Roseburia*	2021	[212]
IS	82 stroke patients and 82 controls	↑Shannon and observed OTUs	↑Lactobacillaceae *, Enterococcaceae *, Streptococcaceae *, Enterobacteriaceae * ↓Ruminococcaceae, Lachnospiraceae	2022	[25]
ICH	31 stroke patients and 31 controls	No difference in Chao1, Shannon, Simpson, and observed OTUs	↑Bacteroidetes, *Streptococcus*, *Bifidobacterium*, *Akkermansia*, *Lactobacillus*, *Enterococcus* ↓Firmicutes, *Faecalibacterium*, *Clostridium*, *Prevotella*, *Gemmiger*, *Blautia*	2022	[213]
AIS and AIS with PHS	17 stroke patients and 10 controls	↑Chao1 No difference in Ace, Shannon, and Simpson	↑Verrucomicrobia *, Synergistetes *, Actinobacteria, Proteobacteria, *Akkermansia* *, *Olsenella* *, *Escherichia_Shigella*, *Bifidobacterium*, *Ruminococcaceae_UCG_014* ↓Firmicutes, Bacteroidetes, *Megamonas* *, *Prevotella_9*, *Agathobacter*, *Faecalibacterium*	2022	[214]
CS	30 stroke patients and 33 controls	↑Chao1, Shannon, Ace, and observed OTUs	↑Enterobacteriaceae *, Streptococcaceae *, Lactobacillaceae *, *Escherichia–Shigella* *, *Streptococcus* *, *Lactobacillus* *, *Klebsilla* ↓Veillonellaceae *, Faecalibacterium *, Dialister, Roseburia	2022	[24]

Table A1. *Cont.*

Type of Stroke	Cohort Size	α-Diversity in Stroke	Main Changes in Gut Microbiota in Stroke	Year	Refs.
HS and IS	20 stroke patients and 20 controls	↓Chao, Ace, Shannon, Simpson, and observed OTUs	↑Fusobacteriota *, Desulfobacterota *, Akkermansiaceae, Fusobacteriota, Ruminococcaceae, Oscillospirales ↓Acidobacteriota *, *Bacteroides*	2023	[215]
AIS	90 stroke patients and 60 controls	N/A	↑Actinobacteriota, *Bacteroides*, *Parabacteroides* Faecalibacterium ↓Spirochaetes, *Faecallbacterium*, *Prevotella*, *Roseburia*, *Lachnospira*	2023	[23]
IS	10 stroke patients and 21 controls	↓Chao1 No difference in Shannon and Simpson	↑Proteobacteria *, Verrucomicrobiaceae *, Bacteroidaceae *, Streptococcaceae *, Rikenellaceae *, Enterobacteriaceae *, Christensenellacea *, *Phascolarctobacterium*, *Alistipes*, *Sutterella*, *Akkermansia*, *Bacteroides* ↓Coriobacteriaceae *, Actinobacteria, Clostridiaceae, Ruminococcaceae, *Anaerostipes* *, *Clostridiales* *	2023	[196]
AIS and AIS with HHTN	200 stroke patients and 90 controls	No difference in Ace and Shannon	↑Streptococcaceae *, Lactobacillaceae *, Marinifilaceae *, *Lactobacillus* *, *Veillonella* *, *Bacteroides*, *Bifidobacterium*, *Klebsiella* ↓*Escherichia-Shigella* *, *Streptococcus* *, *Anaerostipes* *, *Butyricicoccus* *, *Blautia*, *Faecalibacterium*	2023	[216]

H-type hypertension (HHTN), Type 2 diabetes (T2D), and phlegm-heat syndrome (PHS). Phylum, Class, Family, *Genus*, and *Species* are presented. * Available p-value < 0.05. N/A, no available data.

References

1. Martin, S.S.; Aday, A.W.; Almarzooq, Z.I.; Anderson, C.A.M.; Arora, P.; Avery, C.L.; Baker-Smith, C.M.; Barone Gibbs, B.; Beaton, A.Z.; Boehme, A.K.; et al. 2024 Heart Disease and Stroke Statistics: A Report of US and Global Data From the American Heart Association. *Circulation* **2024**, *149*, e347–e913. [CrossRef]
2. Xie, X.; Wang, L.; Dong, S.; Ge, S.; Zhu, T. Immune Regulation of the Gut-Brain Axis and Lung-Brain Axis Involved in Ischemic Stroke. *Neural Regen. Res.* **2024**, *19*, 519–528. [CrossRef]
3. Tsivgoulis, G.; Katsanos, A.H.; Sandset, E.C.; Turc, G.; Nguyen, T.N.; Bivard, A.; Fischer, U.; Khatri, P. Thrombolysis for Acute Ischaemic Stroke: Current Status and Future Perspectives. *Lancet Neurol.* **2023**, *22*, 418–429. [CrossRef] [PubMed]
4. Rost, N.S.; Brodtmann, A.; Pase, M.P.; Van Veluw, S.J.; Biffi, A.; Duering, M.; Hinman, J.D.; Dichgans, M. Post-Stroke Cognitive Impairment and Dementia. *Circ. Res.* **2022**, *130*, 1252–1271. [CrossRef] [PubMed]
5. Schöttke, H.; Giabbiconi, C.-M. Post-Stroke Depression and Post-Stroke Anxiety: Prevalence and Predictors. *Int. Psychogeriatr.* **2015**, *27*, 1805–1812. [CrossRef] [PubMed]
6. Borthwick, S. Communication Impairment in Patients Following Stroke. *Nurs. Stand.* **2012**, *26*, 35–41. [CrossRef] [PubMed]
7. Zhao, L.; Xiong, Q.; Stary, C.M.; Mahgoub, O.K.; Ye, Y.; Gu, L.; Xiong, X.; Zhu, S. Bidirectional Gut-Brain-Microbiota Axis as a Potential Link between Inflammatory Bowel Disease and Ischemic Stroke. *J. Neuroinflamm.* **2018**, *15*, 339. [CrossRef]
8. Sinagra, E.; Pellegatta, G.; Guarnotta, V.; Maida, M.; Rossi, F.; Conoscenti, G.; Pallio, S.; Alloro, R.; Raimondo, D.; Pace, F.; et al. Microbiota Gut–Brain Axis in Ischemic Stroke: A Narrative Review with a Focus about the Relationship with Inflammatory Bowel Disease. *Life* **2021**, *11*, 715. [CrossRef]
9. Tuz, A.A.; Hasenberg, A.; Hermann, D.M.; Gunzer, M.; Singh, V. Ischemic Stroke and Concomitant Gastrointestinal Complications-a Fatal Combination for Patient Recovery. *Front. Immunol.* **2022**, *13*, 1037330. [CrossRef]
10. Song, J.; Chen, W.; Ye, W. Stroke and the Risk of Gastrointestinal Disorders: A Mendelian Randomization Study. *Front. Neurol.* **2023**, *14*, 1131250. [CrossRef]
11. Roth, W.H.; Cai, A.; Zhang, C.; Chen, M.L.; Merkler, A.E.; Kamel, H. Gastrointestinal Disorders and Risk of First-Ever Ischemic Stroke. *Upd. Int. Car.* **2020**, *51*, 3577–3583. [CrossRef] [PubMed]
12. Xu, K.; Gao, X.; Xia, G.; Chen, M.; Zeng, N.; Wang, S.; You, C.; Tian, X.; Di, H.; Tang, W.; et al. Rapid Gut Dysbiosis Induced by Stroke Exacerbates Brain Infarction in Turn. *Gut* **2021**, *70*, 1486–1494. [CrossRef] [PubMed]
13. Carabotti, M.; Scirocco, A.; Maselli, M.A.; Severi, C. The Gut-Brain Axis: Interactions between Enteric Microbiota, Central and Enteric Nervous Systems. *Ann. Gastroenterol.* **2015**, *28*, 203.

4. Cryan, J.F.; O'Riordan, K.J.; Cowan, C.S.M.; Sandhu, K.V.; Bastiaanssen, T.F.S.; Boehme, M.; Codagnone, M.G.; Cussotto, S.; Fulling, C.; Golubeva, A.V.; et al. The Microbiota-Gut-Brain Axis. *Physiol. Rev.* **2019**, *99*, 1877–2013. [CrossRef] [PubMed]
5. Gao, K.; Mu, C.; Farzi, A.; Zhu, W. Tryptophan Metabolism: A Link between the Gut Microbiota and Brain. *Adv. Nutr.* **2020**, *11*, 709–723. [CrossRef] [PubMed]
6. Xue, C.; Li, G.; Zheng, Q.; Gu, X.; Shi, Q.; Su, Y.; Chu, Q.; Yuan, X.; Bao, Z.; Lu, J.; et al. Tryptophan Metabolism in Health and Disease. *Cell Metab.* **2023**, *35*, 1304–1326. [CrossRef]
7. Wigner, P.; Saluk-Bijak, J.; Synowiec, E.; Miller, E.; Sliwinski, T.; Cichon, N.; Bijak, M. Variation of Genes Encoding Tryptophan Catabolites Pathway Enzymes in Stroke. *J. Clin. Med.* **2019**, *8*, 2133. [CrossRef] [PubMed]
8. Shaw, C.; Hess, M.; Weimer, B.C. Microbial-Derived Tryptophan Metabolites and Their Role in Neurological Disease: Anthranilic Acid and Anthranilic Acid Derivatives. *Microorganisms* **2023**, *11*, 1825. [CrossRef]
9. Boros, F.A.; Maszlag-Török, R.; Szűcs, M.; Annus, Á.; Klivényi, P.; Vécsei, L. Relationships of Ischemic Stroke Occurrence and Outcome with Gene Variants Encoding Enzymes of Tryptophan Metabolism. *Biomedicines* **2021**, *9*, 1441. [CrossRef]
0. Quigley, E.M.M. Gut Bacteria in Health and Disease. *Gastroenterol. Hepatol.* **2013**, *9*, 560.
1. MetaHIT Consortium; Qin, J.; Li, R.; Raes, J.; Arumugam, M.; Burgdorf, K.S.; Manichanh, C.; Nielsen, T.; Pons, N.; Levenez, F.; et al. A Human Gut Microbial Gene Catalogue Established by Metagenomic Sequencing. *Nature* **2010**, *464*, 59–65. [CrossRef] [PubMed]
2. Li, N.; Wang, X.; Sun, C.; Wu, X.; Lu, M.; Si, Y.; Ye, X.; Wang, T.; Yu, X.; Zhao, X.; et al. Change of Intestinal Microbiota in Cerebral Ischemic Stroke Patients. *BMC Microbiol.* **2019**, *19*, 191. [CrossRef] [PubMed]
3. Jiang, Z.; Li, L.; Liu, L.; Ding, B.; Yang, Y.; He, F.; Zhang, L.; Wu, Z. Ischemic Stroke and Dysbiosis of Gut Microbiota: Changes to LPS Levels and Effects on Functional Outcomes. *Altern. Ther. Health Med.* **2023**, *29*, 284. [PubMed]
4. Zheng, Q.; Chen, Y.; Zhai, Y.; Meng, L.; Liu, H.; Tian, H.; Feng, R.; Wang, J.; Zhang, R.; Sun, K.; et al. Gut Dysbiosis Is Associated with the Severity of Cryptogenic Stroke and Enhanced Systemic Inflammatory Response. *Front. Immunol.* **2022**, *13*, 836820. [CrossRef]
5. Zhang, N.; Wang, H.; Wang, X.; Tian, M.; Tian, Y.; Li, Q.; Liang, C.; Peng, X.; Ding, J.; Wu, X.; et al. Combination Effect between Gut Microbiota and Traditional Potentially Modifiable Risk Factors for First-Ever Ischemic Stroke in Tujia, Miao and Han Populations in China. *Front. Mol. Neurosci.* **2022**, *15*, 922399. [CrossRef] [PubMed]
6. Aburto, M.R.; Cryan, J.F. Gastrointestinal and Brain Barriers: Unlocking Gates of Communication across the Microbiota–Gut–Brain Axis. *Nat. Rev. Gastroenterol. Hepatol.* **2024**, *21*, 222–247. [CrossRef] [PubMed]
7. Chelakkot, C.; Ghim, J.; Ryu, S.H. Mechanisms Regulating Intestinal Barrier Integrity and Its Pathological Implications. *Exp. Mol. Med.* **2018**, *50*, 1–9. [CrossRef]
8. Stanley, D.; Mason, L.J.; Mackin, K.E.; Srikhanta, Y.N.; Lyras, D.; Prakash, M.D.; Nurgali, K.; Venegas, A.; Hill, M.D.; Moore, R.J.; et al. Translocation and Dissemination of Commensal Bacteria in Post-Stroke Infection. *Nat. Med.* **2016**, *22*, 1277–1284. [CrossRef]
9. Houlden, A.; Goldrick, M.; Brough, D.; Vizi, E.S.; Lénárt, N.; Martinecz, B.; Roberts, I.S.; Denes, A. Brain Injury Induces Specific Changes in the Caecal Microbiota of Mice via Altered Autonomic Activity and Mucoprotein Production. *Brain Behav. Immun.* **2016**, *57*, 10–20. [CrossRef]
0. Amasheh, S.; Fromm, M.; Günzel, D. Claudins of Intestine and Nephron—A Correlation of Molecular Tight Junction Structure and Barrier Function. *Acta Physiol.* **2011**, *201*, 133–140. [CrossRef]
1. Prame Kumar, K.; McKay, L.D.; Nguyen, H.; Kaur, J.; Wilson, J.L.; Suthya, A.R.; McKeown, S.J.; Abud, H.E.; Wong, C.H.Y. Sympathetic-Mediated Intestinal Cell Death Contributes to Gut Barrier Impairment after Stroke. *Transl. Stroke Res.* **2023**. [CrossRef] [PubMed]
2. Ye, D.; Hu, Y.; Zhu, N.; Gu, W.; Long, G.; Tao, E.; Fang, M.; Jiang, M. Exploratory Investigation of Intestinal Structure and Function after Stroke in Mice. *Mediat. Inflamm.* **2021**, *2021*, 1315797. [CrossRef] [PubMed]
3. Johansson, M.E.V.; Larsson, J.M.H.; Hansson, G.C. The Two Mucus Layers of Colon Are Organized by the MUC2 Mucin, Whereas the Outer Layer Is a Legislator of Host–Microbial Interactions. *Proc. Natl. Acad. Sci. USA* **2011**, *108*, 4659–4665. [CrossRef] [PubMed]
4. Arike, L.; Seiman, A.; Van Der Post, S.; Rodriguez Piñeiro, A.M.; Ermund, A.; Schütte, A.; Bäckhed, F.; Johansson, M.E.V.; Hansson, G.C. Protein Turnover in Epithelial Cells and Mucus along the Gastrointestinal Tract Is Coordinated by the Spatial Location and Microbiota. *Cell Rep.* **2020**, *30*, 1077–1087.e3. [CrossRef]
5. Bevins, C.L.; Salzman, N.H. Paneth Cells, Antimicrobial Peptides and Maintenance of Intestinal Homeostasis. *Nat. Rev. Microbiol.* **2011**, *9*, 356–368. [CrossRef]
6. Yu, L.C.-H. Host-Microbial Interactions and Regulation of Intestinal Epithelial Barrier Function: From Physiology to Pathology. *World J. Gastrointest. Pathophysiol.* **2012**, *3*, 27. [CrossRef]
7. Lawther, B.K.; Kumar, S.; Krovvidi, H. Blood–Brain Barrier. *Contin. Educ. Anaesth. Crit. Care Pain* **2011**, *11*, 128–132. [CrossRef]
8. Kim, K.-A.; Kim, D.; Kim, J.-H.; Shin, Y.-J.; Kim, E.-S.; Akram, M.; Kim, E.-H.; Majid, A.; Baek, S.-H.; Bae, O.-N. Autophagy-Mediated Occludin Degradation Contributes to Blood–Brain Barrier Disruption during Ischemia in bEnd.3 Brain Endothelial Cells and Rat Ischemic Stroke Models. *Fluids Barriers CNS* **2020**, *17*, 21. [CrossRef]
9. Braniste, V.; Al-Asmakh, M.; Kowal, C.; Anuar, F.; Abbaspour, A.; Tóth, M.; Korecka, A.; Bakocevic, N.; Ng, L.G.; Kundu, P.; et al. The Gut Microbiota Influences Blood-Brain Barrier Permeability in Mice. *Sci. Transl. Med.* **2014**, *6*, 263ra158. [CrossRef]

40. Fröhlich, E.E.; Farzi, A.; Mayerhofer, R.; Reichmann, F.; Jačan, A.; Wagner, B.; Zinser, E.; Bordag, N.; Magnes, C.; Fröhlich, E.; et a Cognitive Impairment by Antibiotic-Induced Gut Dysbiosis: Analysis of Gut Microbiota-Brain Communication. *Brain Beha Immun.* **2016**, *56*, 140–155. [CrossRef]
41. Rustia, A.J.; Paterson, J.S.; Best, G.; Sokoya, E.M. Microbial Disruption in the Gut Promotes Cerebral Endothelial Dysfunctio *Physiol. Rep.* **2021**, *9*, e15100. [CrossRef] [PubMed]
42. Nelson, J.W.; Phillips, S.C.; Ganesh, B.P.; Petrosino, J.F.; Durgan, D.J.; Bryan, R.M. The Gut Microbiome Contributes to Blood-brai Barrier Disruption in Spontaneously Hypertensive Stroke Prone Rats. *FASEB J.* **2021**, *35*, e21201. [CrossRef] [PubMed]
43. Wang, H.; Song, W.; Wu, Q.; Gao, X.; Li, J.; Tan, C.; Zhou, H.; Zhu, J.; He, Y.; Yin, J. Fecal Transplantation from Db/Db Mic Treated with Sodium Butyrate Attenuates Ischemic Stroke Injury. *Microbiol. Spectr.* **2021**, *9*, e00042-21. [CrossRef] [PubMed]
44. Lee, J.; d'Aigle, J.; Atadja, L.; Quaicoe, V.; Honarpisheh, P.; Ganesh, B.P.; Hassan, A.; Graf, J.; Petrosino, J.; Putluri, N.; et a Gut Microbiota–Derived Short-Chain Fatty Acids Promote Poststroke Recovery in Aged Mice. *Circ. Res.* **2020**, *127*, 453–46 [CrossRef]
45. Peh, A.; Dinakis, E.; Jama, H.; Anderson, D.; Creek, D.J.; Zheng, G.; De Veer, M.; Mackay, C.R.; Zheng, T.; Kemp-Harper, B.K.; et a Dietary Fibre Reverses Adverse Post-Stroke Outcomes in Mice via Short-Chain Fatty Acids and Its Sensing Receptors GPR4 GPR43 and GPR109A. *bioRxiv* **2023**. [CrossRef]
46. Wang, H.-B.; Wang, P.-Y.; Wang, X.; Wan, Y.-L.; Liu, Y.-C. Butyrate Enhances Intestinal Epithelial Barrier Function via Up Regulation of Tight Junction Protein Claudin-1 Transcription. *Digest. Dis. Sci.* **2012**, *57*, 3126–3135. [CrossRef] [PubMed]
47. Wang, H.; Song, W.; Gao, X.; Zhu, J.; Li, J.; Wu, Q.; Tan, C.; Zhou, H.; He, Y.; Yin, J. Modulation of the Gut Microbiota of Type Diabetic Mice by Sodium Butyrate Attenuates Ischemic Stroke Injury. *Res. Sq.* **2020**. [CrossRef]
48. Zheng, L.; Kelly, C.J.; Battista, K.D.; Schaefer, R.; Lanis, J.M.; Alexeev, E.E.; Wang, R.X.; Onyiah, J.C.; Kominsky, D.J.; Colgan, S. Microbial-Derived Butyrate Promotes Epithelial Barrier Function through IL-10 Receptor–Dependent Repression of Claudin-2. *Immunol.* **2017**, *199*, 2976–2984. [CrossRef] [PubMed]
49. Powell, N.; Walker, M.M.; Talley, N.J. The Mucosal Immune System: Master Regulator of Bidirectional Gut–Brain Communication *Nat. Rev. Gastroenterol. Hepatol.* **2017**, *14*, 143–159. [CrossRef]
50. Mowat, A. McI. Anatomical Basis of Tolerance and Immunity to Intestinal Antigens. *Nat. Rev. Immunol.* **2003**, *3*, 331–34 [CrossRef]
51. Song, C.; Chai, Z.; Chen, S.; Zhang, H.; Zhang, X.; Zhou, Y. Intestinal Mucus Components and Secretion Mechanisms: What W Do and Do Not Know. *Exp. Mol. Med.* **2023**, *55*, 681–691. [CrossRef] [PubMed]
52. Honarpisheh, P.; Bryan, R.M.; McCullough, L.D. Aging Microbiota-Gut-Brain Axis in Stroke Risk and Outcome. *Circ. Res.* **202** *130*, 1112–1144. [CrossRef] [PubMed]
53. Kasarello, K.; Cudnoch-Jedrzejewska, A.; Czarzasta, K. Communication of Gut Microbiota and Brain via Immune and Neuroer docrine Signaling. *Front. Microbiol.* **2023**, *14*, 1118529. [CrossRef] [PubMed]
54. Huang, W.; Zhu, L.; Song, W.; Zhang, M.; Teng, L.; Wu, M. Crosstalk between the Gut and Brain in Ischemic Stroke: Mechanisti Insights and Therapeutic Options. *Mediat. Inflamm.* **2022**, *2022*, 6508046. [CrossRef]
55. Eletto, D.; Mentucci, F.; Voli, A.; Petrella, A.; Porta, A.; Tosco, A. Helicobacter Pylori Pathogen-Associated Molecular Pattern Friends or Foes? *Int. J. Mol. Sci.* **2022**, *23*, 3531. [CrossRef]
56. Roh, J.S.; Sohn, D.H. Damage-Associated Molecular Patterns in Inflammatory Diseases. *Immune Netw.* **2018**, *18*, e27. [CrossRe [PubMed]
57. Chaudhry, S.; Hafez, A.; Rezai Jahromi, B.; Kinfe, T.; Lamprecht, A.; Niemelä, M.; Muhammad, S. Role of Damage Associate Molecular Pattern Molecules (DAMPs) in Aneurysmal Subarachnoid Hemorrhage (aSAH). *Int. J. Mol. Sci.* **2018**, *19*, 203 [CrossRef]
58. Fadakar, K.; Dadkhahfar, S.; Esmaeili, A.; Rezaei, N. The Role of Toll-like Receptors (TLRs) in Stroke. *Rev. Neurosci.* **2014**, 2 699–712. [CrossRef]
59. Logsdon, A.F.; Erickson, M.A.; Rhea, E.M.; Salameh, T.S.; Banks, W.A. Gut Reactions: How the Blood–Brain Barrier Connects th Microbiome and the Brain. *Exp. Biol. Med.* **2018**, *243*, 159–165. [CrossRef]
60. Zhang, Y.; Wang, Z.; Peng, J.; Gerner, S.T.; Yin, S.; Jiang, Y. Gut Microbiota-Brain Interaction: An Emerging Immunotherapy fo Traumatic Brain Injury. *Exp. Neurol.* **2021**, *337*, 113585. [CrossRef]
61. Brea, D.; Poon, C.; Benakis, C.; Lubitz, G.; Murphy, M.; Iadecola, C.; Anrather, J. Stroke Affects Intestinal Immune Cell Traffickin to the Central Nervous System. *Brain Behav. Immun.* **2021**, *96*, 295–302. [CrossRef] [PubMed]
62. Waisman, A.; Liblau, R.S.; Becher, B. Innate and Adaptive Immune Responses in the CNS. *Lancet Neurol.* **2015**, *14*, 945–95 [CrossRef]
63. Reemst, K.; Noctor, S.C.; Lucassen, P.J.; Hol, E.M. The Indispensable Roles of Microglia and Astrocytes during Brain Developmen *Front. Hum. Neurosci.* **2016**, *10*, 566. [CrossRef] [PubMed]
64. Li, X.; Chen, G. CNS-Peripheral Immune Interactions in Hemorrhagic Stroke. *J. Cereb. Blood Flow Metab.* **2023**, *43*, 185–19 [CrossRef] [PubMed]
65. Kim, J.Y.; Kawabori, M.; Yenari, M.A. Innate Inflammatory Responses in Stroke: Mechanisms and Potential Therapeutic Target *Curr. Med. Chem.* **2014**, *21*, 2076–2097. [CrossRef] [PubMed]
66. Shekhar, S.; Cunningham, M.W.; Pabbidi, M.R.; Wang, S.; Booz, G.W.; Fan, F. Targeting Vascular Inflammation in Ischemic Strok Recent Developments on Novel Immunomodulatory Approaches. *Eur. J. Pharmacol.* **2018**, *833*, 531–544. [CrossRef] [PubMed]

67. Singh, V.; Roth, S.; Llovera, G.; Sadler, R.; Garzetti, D.; Stecher, B.; Dichgans, M.; Liesz, A. Microbiota Dysbiosis Controls the Neuroinflammatory Response after Stroke. *J. Neurosci.* **2016**, *36*, 7428–7440. [CrossRef] [PubMed]
68. Iadecola, C.; Buckwalter, M.S.; Anrather, J. Immune Responses to Stroke: Mechanisms, Modulation, and Therapeutic Potential. *J. Clin. Investig.* **2020**, *130*, 2777–2788. [CrossRef]
69. Wu, S.; Tabassum, S.; Payne, C.T.; Hu, H.; Gusdon, A.M.; Choi, H.A.; Ren, X.S. Updates of the Role of B-Cells in Ischemic Stroke. *Front. Cell. Neurosci.* **2024**, *18*, 1340756. [CrossRef]
70. Prass, K.; Meisel, C.; Höflich, C.; Braun, J.; Halle, E.; Wolf, T.; Ruscher, K.; Victorov, I.V.; Priller, J.; Dirnagl, U.; et al. Stroke-Induced Immunodeficiency Promotes Spontaneous Bacterial Infections and Is Mediated by Sympathetic Activation Reversal by Poststroke T Helper Cell Type 1–like Immunostimulation. *J. Exp. Med.* **2003**, *198*, 725–736. [CrossRef]
71. Pu, B.; Zhu, H.; Wei, L.; Gu, L.; Zhang, S.; Jian, Z.; Xiong, X. The Involvement of Immune Cells Between Ischemic Stroke and Gut Microbiota. *Transl. Stroke Res.* **2024**, *15*, 498–517. [CrossRef] [PubMed]
72. Singh, V.; Sadler, R.; Heindl, S.; Llovera, G.; Roth, S.; Benakis, C.; Liesz, A. The Gut Microbiome Primes a Cerebroprotective Immune Response after Stroke. *J. Cereb. Blood Flow Metab.* **2018**, *38*, 1293–1298. [CrossRef] [PubMed]
73. Grant, E.T.; Boudaud, M.; Muller, A.; Macpherson, A.J.; Desai, M.S. Maternal Diet and Gut Microbiome Composition Modulate Early-life Immune Development. *EMBO Mol. Med.* **2023**, *15*, e17241. [CrossRef]
74. Zeng, J.; Yang, K.; Nie, H.; Yuan, L.; Wang, S.; Zeng, L.; Ge, A.; Ge, J. The Mechanism of Intestinal Microbiota Regulating Immunity and Inflammation in Ischemic Stroke and the Role of Natural Botanical Active Ingredients in Regulating Intestinal Microbiota: A Review. *Biomed. Pharmacother.* **2023**, *157*, 114026. [CrossRef]
75. Furness, J.B. The Enteric Nervous System and Neurogastroenterology. *Nat. Rev. Gastroenterol. Hepatol.* **2012**, *9*, 286–294. [CrossRef] [PubMed]
76. Sharkey, K.A.; Mawe, G.M. The Enteric Nervous System. *Physiol. Rev.* **2023**, *103*, 1487–1564. [CrossRef] [PubMed]
77. Spencer, N.J.; Hu, H. Enteric Nervous System: Sensory Transduction, Neural Circuits and Gastrointestinal Motility. *Nat. Rev. Gastroenterol. Hepatol.* **2020**, *17*, 338–351. [CrossRef] [PubMed]
78. Jarret, A.; Jackson, R.; Duizer, C.; Healy, M.E.; Zhao, J.; Rone, J.M.; Bielecki, P.; Sefik, E.; Roulis, M.; Rice, T.; et al. Enteric Nervous System-Derived IL-18 Orchestrates Mucosal Barrier Immunity. *Cell* **2020**, *180*, 50–63.e12. [CrossRef]
79. Muller, P.A.; Koscsó, B.; Rajani, G.M.; Stevanovic, K.; Berres, M.-L.; Hashimoto, D.; Mortha, A.; Leboeuf, M.; Li, X.-M.; Mucida, D.; et al. Crosstalk between Muscularis Macrophages and Enteric Neurons Regulates Gastrointestinal Motility. *Cell* **2014**, *158*, 300–313. [CrossRef]
80. Viola, M.F.; Chavero-Pieres, M.; Modave, E.; Delfini, M.; Stakenborg, N.; Estévez, M.C.; Fabre, N.; Appeltans, I.; Martens, T.; Vandereyken, K.; et al. Dedicated Macrophages Organize and Maintain the Enteric Nervous System. *Nature* **2023**, *618*, 818–826. [CrossRef]
81. Cheng, X.; Boza-Serrano, A.; Turesson, M.F.; Deierborg, T.; Ekblad, E.; Voss, U. Galectin-3 Causes Enteric Neuronal Loss in Mice after Left Sided Permanent Middle Cerebral Artery Occlusion, a Model of Stroke. *Sci. Rep.* **2016**, *6*, 32893. [CrossRef] [PubMed]
82. Cheng, X.; Svensson, M.; Yang, Y.; Deierborg, T.; Ekblad, E.; Voss, U. Focal, but Not Global, Cerebral Ischaemia Causes Loss of Myenteric Neurons and Upregulation of Vasoactive Intestinal Peptide in Mouse Ileum. *Int. J. Exp. Path.* **2018**, *99*, 38–45. [CrossRef] [PubMed]
83. Kumar, K.P.; Wilson, J.L.; Nguyen, H.; McKay, L.D.; Wen, S.W.; Sepehrizadeh, T.; De Veer, M.; Rajasekhar, P.; Carbone, S.E.; Hickey, M.J.; et al. Stroke Alters the Function of Enteric Neurons to Impair Smooth Muscle Relaxation and Dysregulates Gut Transit. *JAHA* **2024**, *13*, e033279. [CrossRef]
84. Goyal, R.K. The Enteric Nervous System. *N. Engl. J. Med.* **1996**, *334*, 1106–1115. [CrossRef] [PubMed]
85. Tang, H.; Li, J.; Zhou, Q.; Li, S.; Xie, C.; Niu, L.; Ma, J.; Li, C. Vagus Nerve Stimulation Alleviated Cerebral Ischemia and Reperfusion Injury in Rats by Inhibiting Pyroptosis via A7 Nicotinic Acetylcholine Receptor. *Cell Death Discov.* **2022**, *8*, 54. [CrossRef] [PubMed]
86. Liu, Y.-L.; Wang, S.-R.; Ma, J.-X.; Yu, L.-H.; Jia, G.-W. Vagus Nerve Stimulation Is a Potential Treatment for Ischemic Stroke. *Neural Regener. Res.* **2023**, *18*, 825. [CrossRef]
87. Muller, P.A.; Schneeberger, M.; Matheis, F.; Wang, P.; Kerner, Z.; Ilanges, A.; Pellegrino, K.; Del Mármol, J.; Castro, T.B.R.; Furuichi, M.; et al. Microbiota Modulate Sympathetic Neurons via a Gut–Brain Circuit. *Nature* **2020**, *583*, 441–446. [CrossRef] [PubMed]
88. Goehler, L.E.; Gaykema, R.P.A.; Opitz, N.; Reddaway, R.; Badr, N.; Lyte, M. Activation in Vagal Afferents and Central Autonomic Pathways: Early Responses to Intestinal Infection with Campylobacter Jejuni. *Brain Behav. Immun.* **2005**, *19*, 334–344. [CrossRef] [PubMed]
89. Vicentini, F.A.; Keenan, C.M.; Wallace, L.E.; Woods, C.; Cavin, J.-B.; Flockton, A.R.; Macklin, W.B.; Belkind-Gerson, J.; Hirota, S.A.; Sharkey, K.A. Intestinal Microbiota Shapes Gut Physiology and Regulates Enteric Neurons and Glia. *Microbiome* **2021**, *9*, 210. [CrossRef]
90. Yan, Y.; Ramanan, D.; Rozenberg, M.; McGovern, K.; Rastelli, D.; Vijaykumar, B.; Yaghi, O.; Voisin, T.; Mosaheb, M.; Chiu, I.; et al. Interleukin-6 Produced by Enteric Neurons Regulates the Number and Phenotype of Microbe-Responsive Regulatory T Cells in the Gut. *Immunity* **2021**, *54*, 499–513.e5. [CrossRef]
91. Chen, Y.; Xu, J.; Chen, Y. Regulation of Neurotransmitters by the Gut Microbiota and Effects on Cognition in Neurological Disorders. *Nutrients* **2021**, *13*, 2099. [CrossRef] [PubMed]

92. Burford, N.; Webster, N.; Cruz-Topete, D. Hypothalamic-Pituitary-Adrenal Axis Modulation of Glucocorticoids in the Cardiovascular System. *Int. J. Mol. Sci.* **2017**, *18*, 2150. [CrossRef] [PubMed]
93. Fassbender, K.; Schmidt, R.; Mössner, R.; Daffertshofer, M.; Hennerici, M. Pattern of Activation of the Hypothalamic-Pituitary-Adrenal Axis in Acute Stroke. Relation to Acute Confusional State, Extent of Brain Damage, and Clinical Outcome. *Stroke* **1994**, *25*, 1105–1108. [CrossRef]
94. Barugh, A.J.; Gray, P.; Shenkin, S.D.; MacLullich, A.M.J.; Mead, G.E. Cortisol Levels and the Severity and Outcomes of Acute Stroke: A Systematic Review. *J. Neurol.* **2014**, *261*, 533–545. [CrossRef] [PubMed]
95. Sugo, N.; Hurn, P.D.; Morahan, M.B.; Hattori, K.; Traystman, R.J.; DeVries, A.C. Social Stress Exacerbates Focal Cerebral Ischemia in Mice. *Stroke* **2002**, *33*, 1660–1664. [CrossRef] [PubMed]
96. Kim, S.; Park, E.S.; Chen, P.R.; Kim, E. Dysregulated Hypothalamic–Pituitary–Adrenal Axis Is Associated With Increased Inflammation and Worse Outcomes After Ischemic Stroke in Diabetic Mice. *Front. Immunol.* **2022**, *13*, 864858. [CrossRef] [PubMed]
97. Mracsko, E.; Liesz, A.; Karcher, S.; Zorn, M.; Bari, F.; Veltkamp, R. Differential Effects of Sympathetic Nervous System and Hypothalamic–Pituitary–Adrenal Axis on Systemic Immune Cells after Severe Experimental Stroke. *Brain Behav. Immun.* **2014**, *41*, 200–209. [CrossRef] [PubMed]
98. Wang, Y.; Wang, H.; Sun, W.; Miao, J.; Liang, W.; Qiu, X.; Lan, Y.; Pan, C.; Li, G.; Zhao, X.; et al. Higher Concentration of Adrenocorticotropic Hormone Predicts Post-Stroke Depression. *Clin. Interv. Aging* **2022**, *17*, 417–427. [CrossRef]
99. Sudo, N.; Chida, Y.; Aiba, Y.; Sonoda, J.; Oyama, N.; Yu, X.; Kubo, C.; Koga, Y. Postnatal Microbial Colonization Programs the Hypothalamic–Pituitary–Adrenal System for Stress Response in Mice. *J. Physiol.* **2004**, *558*, 263–275. [CrossRef]
100. Vodička, M.; Ergang, P.; Hrnčíř, T.; Mikulecká, A.; Kvapilová, P.; Vagnerová, K.; Šestáková, B.; Fajstová, A.; Hermanová, P.; Hudcovic, T.; et al. Microbiota Affects the Expression of Genes Involved in HPA Axis Regulation and Local Metabolism of Glucocorticoids in Chronic Psychosocial Stress. *Brain Behav. Immun.* **2018**, *73*, 615–624. [CrossRef]
101. Rehfeld, J.F. A Centenary of Gastrointestinal Endocrinology. *Horm. Metab. Res.* **2004**, *36*, 735–741. [CrossRef]
102. Furness, J.B.; Rivera, L.R.; Cho, H.-J.; Bravo, D.M.; Callaghan, B. The Gut as a Sensory Organ. *Nat. Rev. Gastroenterol. Hepatol.* **2013**, *10*, 729–740. [CrossRef] [PubMed]
103. Sun, L.-J.; Li, J.-N.; Nie, Y.-Z. Gut Hormones in Microbiota-Gut-Brain Cross-Talk. *Chin. Med. J.* **2020**, *133*, 826–833. [CrossRef] [PubMed]
104. Latorre, R.; Sternini, C.; De Giorgio, R.; Greenwood-Van Meerveld, B. Enteroendocrine Cells: A Review of Their Role in Brain–Gut Communication. *Neurogastroenterol. Motil.* **2016**, *28*, 620–630. [CrossRef] [PubMed]
105. Bogunovic, M.; Davé, S.H.; Tilstra, J.S.; Chang, D.T.W.; Harpaz, N.; Xiong, H.; Mayer, L.F.; Plevy, S.E. Enteroendocrine Cells Express Functional Toll-like Receptors. *Am. J. Physiol.-Gastrointest. Liver Physiol.* **2007**, *292*, G1770–G1783. [CrossRef] [PubMed]
106. Kuwahara, A.; Matsuda, K.; Kuwahara, Y.; Asano, S.; Inui, T.; Marunaka, Y. Microbiota-Gut-Brain Axis: Enteroendocrine Cells and the Enteric Nervous System Form an Interface between the Microbiota and the Central Nervous System. *Biomed. Res.* **2020**, *41*, 199–216. [CrossRef] [PubMed]
107. Comai, S.; Bertazzo, A.; Brughera, M.; Crotti, S. Tryptophan in Health and Disease. In *Advances in Clinical Chemistry*; Elsevier: Amsterdam, The Netherlands, 2020; Volume 95, pp. 165–218. ISBN 978-0-12-821165-6.
108. Yeung, A.W.S.; Terentis, A.C.; King, N.J.C.; Thomas, S.R. Role of Indoleamine 2,3-Dioxygenase in Health and Disease. *Clin. Sci.* **2015**, *129*, 601–672. [CrossRef]
109. Huang, Y.-S.; Ogbechi, J.; Clanchy, F.I.; Williams, R.O.; Stone, T.W. IDO and Kynurenine Metabolites in Peripheral and CNS Disorders. *Front. Immunol.* **2020**, *11*, 388. [CrossRef]
110. Prendergast, G.C.; Metz, R.; Muller, A.J.; Merlo, L.M.F.; Mandik-Nayak, L. IDO2 in Immunomodulation and Autoimmune Disease. *Front. Immunol.* **2014**, *5*, 585. [CrossRef]
111. Roth, W.; Zadeh, K.; Vekariya, R.; Ge, Y.; Mohamadzadeh, M. Tryptophan Metabolism and Gut-Brain Homeostasis. *Int. J. Mol. Sci.* **2021**, *22*, 2973. [CrossRef]
112. Lamas, B.; Natividad, J.M.; Sokol, H. Aryl Hydrocarbon Receptor and Intestinal Immunity. *Mucosal Immunol.* **2018**, *11*, 1024–1038. [CrossRef] [PubMed]
113. Rothhammer, V.; Quintana, F.J. The Aryl Hydrocarbon Receptor: An Environmental Sensor Integrating Immune Responses in Health and Disease. *Nat. Rev. Immunol.* **2019**, *19*, 184–197. [CrossRef]
114. Mrštná, K.; Krčmová, L.K.; Švec, F. Advances in Kynurenine Analysis. *Clin. Chim. Acta* **2023**, *547*, 117441. [CrossRef] [PubMed]
115. Beamer, C.A.; Shepherd, D.M. Role of the Aryl Hydrocarbon Receptor (AhR) in Lung Inflammation. *Semin. Immunopathol.* **2013**, *35*, 693–704. [CrossRef] [PubMed]
116. Lőrinczi, B.; Szatmári, I. KYNA Derivatives with Modified Skeleton; Hydroxyquinolines with Potential Neuroprotective Effect. *Int. J. Mol. Sci.* **2021**, *22*, 11935. [CrossRef] [PubMed]
117. Vécsei, L.; Szalárdy, L.; Fülöp, F.; Toldi, J. Kynurenines in the CNS: Recent Advances and New Questions. *Nat. Rev. Drug Discov.* **2013**, *12*, 64–82. [CrossRef]
118. Lopez, A.; Alegre, E.; Lemaoult, J.; Carosella, E.; Gonzalez, A. Regulatory Role of Tryptophan Degradation Pathway in HLA-G Expression by Human Monocyte-Derived Dendritic Cells. *Mol. Immunol.* **2006**, *43*, 2151–2160. [CrossRef] [PubMed]

19. Weber, W.P.; Feder-Mengus, C.; Chiarugi, A.; Rosenthal, R.; Reschner, A.; Schumacher, R.; Zajac, P.; Misteli, H.; Frey, D.M.; Oertli, D.; et al. Differential Effects of the Tryptophan Metabolite3-hydroxyanthranilic Acid on the Proliferation of Human CD8$^+$ T Cells Induced by TCR Triggering or Homeostatic Cytokines. *Eur. J. Immunol.* **2006**, *36*, 296–304. [CrossRef] [PubMed]
20. Harvey-Girard, E.; Dunn, R.J. Excitatory Amino Acid Receptors of the Electrosensory System: The NR1/NR2B *N*-Methyl-D-Aspartate Receptor. *J. Neurophysiol.* **2003**, *89*, 822–832. [CrossRef]
21. Schwarcz, R.; Stone, T.W. The Kynurenine Pathway and the Brain: Challenges, Controversies and Promises. *Neuropharmacology* **2017**, *112*, 237–247. [CrossRef]
22. Pierozan, P.; Biasibetti, H.; Schmitz, F.; Ávila, H.; Parisi, M.M.; Barbe-Tuana, F.; Wyse, A.T.S.; Pessoa-Pureur, R. Quinolinic Acid Neurotoxicity: Differential Roles of Astrocytes and Microglia via FGF-2 Mediated Signaling in Redox-Linked Cytoskeletal Changes. *Biochim. Biophys. Acta (BBA)-Mol. Cell Res.* **2016**, *1863*, 3001–3014. [CrossRef] [PubMed]
23. Dehhaghi, M.; Panahi, H.K.S.; Kavyani, B.; Heng, B.; Tan, V.; Braidy, N.; Guillemin, G.J. The Role of Kynurenine Pathway and NAD$^+$ Metabolism in Myalgic Encephalomyelitis/Chronic Fatigue Syndrome. *Aging Dis.* **2022**, *13*, 698. [CrossRef] [PubMed]
24. Dehhaghi, M.; Tan, V.; Heng, B.; Braidy, N.; Mohammadipanah, F.; Guillemin, G.J. Neuroprotective Effect of Myxobacterial Extracts on Quinolinic Acid-Induced Toxicity in Primary Human Neurons. *Neurotox. Res.* **2019**, *35*, 281–290. [CrossRef] [PubMed]
25. Dehhaghi, M.; Kazemi Shariat Panahi, H.; Braidy, N.; Guillemin, G.J. Herpetosiphon Secondary Metabolites Inhibit Amyloid-β Toxicity in Human Primary Astrocytes. *J. Alzheimers Dis.* **2020**, *76*, 423–433. [CrossRef] [PubMed]
26. Zapata-Pérez, R.; Wanders, R.J.A.; Van Karnebeek, C.D.M.; Houtkooper, R.H. NAD$^+$ Homeostasis in Human Health and Disease. *EMBO Mol. Med.* **2021**, *13*, e13943. [CrossRef] [PubMed]
27. Gao, J.; Xu, K.; Liu, H.; Liu, G.; Bai, M.; Peng, C.; Li, T.; Yin, Y. Impact of the Gut Microbiota on Intestinal Immunity Mediated by Tryptophan Metabolism. *Front. Cell. Infect. Microbiol.* **2018**, *8*, 13. [CrossRef]
28. Israelyan, N.; Margolis, K.G. Serotonin as a Link between the Gut-Brain-Microbiome Axis in Autism Spectrum Disorders. *Pharmacol. Res.* **2018**, *132*, 1–6. [CrossRef]
29. Margolis, K.G.; Stevanovic, K.; Li, Z.; Yang, Q.M.; Oravecz, T.; Zambrowicz, B.; Jhaver, K.G.; Diacou, A.; Gershon, M.D. Pharmacological Reduction of Mucosal but Not Neuronal Serotonin Opposes Inflammation in Mouse Intestine. *Gut* **2014**, *63*, 928–937. [CrossRef] [PubMed]
30. Gross, E.R.; Gershon, M.D.; Margolis, K.G.; Gertsberg, Z.V.; Cowles, R.A. Neuronal Serotonin Regulates Growth of the Intestinal Mucosa in Mice. *Gastroenterology* **2012**, *143*, 408–417.e2. [CrossRef]
31. Li, Z.; Chalazonitis, A.; Huang, Y.-Y.; Mann, J.J.; Margolis, K.G.; Yang, Q.M.; Kim, D.O.; Cote, F.; Mallet, J.; Gershon, M.D. Essential Roles of Enteric Neuronal Serotonin in Gastrointestinal Motility and the Development/Survival of Enteric Dopaminergic Neurons. *J. Neurosci.* **2011**, *31*, 8998–9009. [CrossRef]
32. Wolf, D.; Klasen, M.; Eisner, P.; Zepf, F.D.; Zvyagintsev, M.; Palomero-Gallagher, N.; Weber, R.; Eisert, A.; Mathiak, K. Central Serotonin Modulates Neural Responses to Virtual Violent Actions in Emotion Regulation Networks. *Brain Struct. Funct.* **2018**, *223*, 3327–3345. [CrossRef] [PubMed]
33. Ursin, R. Serotonin and Sleep. *Sleep Med. Rev.* **2002**, *6*, 55–67. [CrossRef] [PubMed]
34. Migliarini, S.; Pacini, G.; Pelosi, B.; Lunardi, G.; Pasqualetti, M. Lack of Brain Serotonin Affects Postnatal Development and Serotonergic Neuronal Circuitry Formation. *Mol. Psychiatry* **2013**, *18*, 1106–1118. [CrossRef] [PubMed]
35. Van Kleef, E.S.B.; Gaspar, P.; Bonnin, A. Insights into the Complex Influence of 5-HT Signaling on Thalamocortical Axonal System Development. *Eur. J. Neurosci.* **2012**, *35*, 1563–1572. [CrossRef] [PubMed]
36. Filip, M.; Bader, M. Overview on 5-HT Receptors and Their Role in Physiology and Pathology of the Central Nervous System. *Pharmacol. Rep.* **2009**, *61*, 761–777. [CrossRef] [PubMed]
37. Ciranna, L. Serotonin as a Modulator of Glutamate- and GABA-Mediated Neurotransmission: Implications in Physiological Functions and in Pathology. *Curr. Neuropharmacol.* **2006**, *4*, 101–114. [CrossRef]
38. Browning, K.N. Role of Central Vagal 5-HT3 Receptors in Gastrointestinal Physiology and Pathophysiology. *Front. Neurosci.* **2015**, *9*, 413. [CrossRef] [PubMed]
39. Mawe, G.M.; Hoffman, J.M. Serotonin Signalling in the Gut—Functions, Dysfunctions and Therapeutic Targets. *Nat. Rev. Gastroenterol. Hepatol.* **2013**, *10*, 473–486. [CrossRef] [PubMed]
40. De Vadder, F.; Grasset, E.; Mannerås Holm, L.; Karsenty, G.; Macpherson, A.J.; Olofsson, L.E.; Bäckhed, F. Gut Microbiota Regulates Maturation of the Adult Enteric Nervous System via Enteric Serotonin Networks. *Proc. Natl. Acad. Sci. USA* **2018**, *115*, 6458–6463. [CrossRef] [PubMed]
41. Hoffman, J.M.; Tyler, K.; MacEachern, S.J.; Balemba, O.B.; Johnson, A.C.; Brooks, E.M.; Zhao, H.; Swain, G.M.; Moses, P.L.; Galligan, J.J.; et al. Activation of Colonic Mucosal 5-HT4 Receptors Accelerates Propulsive Motility and Inhibits Visceral Hypersensitivity. *Gastroenterology* **2012**, *142*, 844–854.e4. [CrossRef]
42. Walldorf, J.; Porzner, M.; Neumann, M.; Joodi, G.; Niess, J.H.; Von Boyen, G.; Mäder, K.; Weissbach, J.; Kleger, A.; Seufferlein, T. The Selective 5-HT1A Agonist SR57746A Protects Intestinal Epithelial Cells and Enteric Glia Cells and Promotes Mucosal Recovery in Experimental Colitis. *Inflamm. Bowel Dis.* **2022**, *28*, 423–433. [CrossRef] [PubMed]
43. Mawe, G.M.; Branchek, T.A.; Gershon, M.D. Peripheral Neural Serotonin Receptors: Identification and Characterization with Specific Antagonists and Agonists. *Proc. Natl. Acad. Sci. USA* **1986**, *83*, 9799–9803. [CrossRef] [PubMed]
44. Brzęczek, M.; Słonka, K.; Hyla-Klekot, L. Melatonin—A Pleiotropic Hormone. *Pediatr. Med. Rodz.* **2016**, *12*, 127–133. [CrossRef]

145. Navabi, S.; Gorrepati, V.S.; Yadav, S.; Chintanaboina, J.; Maher, S.; Demuth, P.; Stern, B.; Stuart, A.; Tinsley, A.; Clarke, K.; et a Influences and Impact of Anxiety and Depression in the Setting of Inflammatory Bowel Disease. *Inflamm. Bowel Dis.* **2018**, 2 2303–2308. [CrossRef] [PubMed]
146. Roager, H.M.; Licht, T.R. Microbial Tryptophan Catabolites in Health and Disease. *Nat. Commun.* **2018**, *9*, 3294. [CrossRe [PubMed]
147. Agus, A.; Planchais, J.; Sokol, H. Gut Microbiota Regulation of Tryptophan Metabolism in Health and Disease. *Cell Host Microb* **2018**, *23*, 716–724. [CrossRef] [PubMed]
148. Kaur, H.; Bose, C.; Mande, S.S. Tryptophan Metabolism by Gut Microbiome and Gut-Brain-Axis: An in Silico Analysis. *Fron Neurosci.* **2019**, *13*, 1365. [CrossRef] [PubMed]
149. Devlin, A.S.; Marcobal, A.; Dodd, D.; Nayfach, S.; Plummer, N.; Meyer, T.; Pollard, K.S.; Sonnenburg, J.L.; Fischbach, M. Modulation of a Circulating Uremic Solute via Rational Genetic Manipulation of the Gut Microbiota. *Cell Host Microbe* **2016**, 2 709–715. [CrossRef]
150. Lee, J.-H.; Lee, J. Indole as an Intercellular Signal in Microbial Communities. *FEMS Microbiol. Rev.* **2010**, *34*, 426–444. [CrossRe
151. Whitfield-Cargile, C.M.; Cohen, N.D.; Chapkin, R.S.; Weeks, B.R.; Davidson, L.A.; Goldsby, J.S.; Hunt, C.L.; Steinmeyer, S.H Menon, R.; Suchodolski, J.S.; et al. The Microbiota-Derived Metabolite Indole Decreases Mucosal Inflammation and Injury in Murine Model of NSAID Enteropathy. *Gut Microbes* **2016**, *7*, 246–261. [CrossRef]
152. Chimerel, C.; Emery, E.; Summers, D.K.; Keyser, U.; Gribble, F.M.; Reimann, F. Bacterial Metabolite Indole Modulates Increti Secretion from Intestinal Enteroendocrine L Cells. *Cell Rep.* **2014**, *9*, 1202–1208. [CrossRef]
153. Tennoune, N.; Andriamihaja, M.; Blachier, F. Production of Indole and Indole-Related Compounds by the Intestinal Microbiot and Consequences for the Host: The Good, the Bad, and the Ugly. *Microorganisms* **2022**, *10*, 930. [CrossRef] [PubMed]
154. Zhai, L.; Xiao, H.; Lin, C.; Wong, H.L.X.; Lam, Y.Y.; Gong, M.; Wu, G.; Ning, Z.; Huang, C.; Zhang, Y.; et al. Gut Microbiota-Derive Tryptamine and Phenethylamine Impair Insulin Sensitivity in Metabolic Syndrome and Irritable Bowel Syndrome. *Nat. Commu* **2023**, *14*, 4986. [CrossRef] [PubMed]
155. Williams, B.B.; Van Benschoten, A.H.; Cimermancic, P.; Donia, M.S.; Zimmermann, M.; Taketani, M.; Ishihara, A.; Kashya P.C.; Fraser, J.S.; Fischbach, M.A. Discovery and Characterization of Gut Microbiota Decarboxylases That Can Produce th Neurotransmitter Tryptamine. *Cell Host Microbe* **2014**, *16*, 495–503. [CrossRef]
156. Hubbard, T.D.; Murray, I.A.; Perdew, G.H. Indole and Tryptophan Metabolism: Endogenous and Dietary Routes to Ah Recepto Activation. *Drug Metab. Dispos.* **2015**, *43*, 1522–1535. [CrossRef] [PubMed]
157. Ehrlich, A.M.; Pacheco, A.R.; Henrick, B.M.; Taft, D.; Xu, G.; Huda, M.N.; Mishchuk, D.; Goodson, M.L.; Slupsky, C.; Barile, D et al. Indole-3-Lactic Acid Associated with Bifidobacterium-Dominated Microbiota Significantly Decreases Inflammation i Intestinal Epithelial Cells. *BMC Microbiol.* **2020**, *20*, 357. [CrossRef]
158. Wlodarska, M.; Luo, C.; Kolde, R.; d'Hennezel, E.; Annand, J.W.; Heim, C.E.; Krastel, P.; Schmitt, E.K.; Omar, A.S.; Creasey, E.A et al. Indoleacrylic Acid Produced by Commensal Peptostreptococcus Species Suppresses Inflammation. *Cell Host Microbe* **201** *22*, 25–37.e6. [CrossRef]
159. Li, J.; Zhang, L.; Wu, T.; Li, Y.; Zhou, X.; Ruan, Z. Indole-3-Propionic Acid Improved the Intestinal Barrier by Enhancing Epitheli Barrier and Mucus Barrier. *J. Agric. Food Chem.* **2021**, *69*, 1487–1495. [CrossRef] [PubMed]
160. Connell, E.; Le Gall, G.; Pontifex, M.G.; Sami, S.; Cryan, J.F.; Clarke, G.; Müller, M.; Vauzour, D. Microbial-Derived Metabolites a a Risk Factor of Age-Related Cognitive Decline and Dementia. *Mol. Neurodegener.* **2022**, *17*, 43. [CrossRef]
161. Rothhammer, V.; Mascanfroni, I.D.; Bunse, L.; Takenaka, M.C.; Kenison, J.E.; Mayo, L.; Chao, C.-C.; Patel, B.; Yan, R.; Blain, M et al. Type I Interferons and Microbial Metabolites of Tryptophan Modulate Astrocyte Activity and Central Nervous Syster Inflammation via the Aryl Hydrocarbon Receptor. *Nat. Med.* **2016**, *22*, 586–597. [CrossRef]
162. Chyan, Y.-J.; Poeggeler, B.; Omar, R.A.; Chain, D.G.; Frangione, B.; Ghiso, J.; Pappolla, M.A. Potent Neuroprotective Properti against the Alzheimer B-Amyloid by an Endogenous Melatonin-Related Indole Structure, Indole-3-Propionic Acid. *J. Biol. Che* **1999**, *274*, 21937–21942. [CrossRef]
163. Hwang, I.K.; Yoo, K.; Li, H.; Park, O.K.; Lee, C.H.; Choi, J.H.; Jeong, Y.; Lee, Y.L.; Kim, Y.; Kwon, Y.; et al. Indole-3-propion Acid Attenuates Neuronal Damage and Oxidative Stress in the Ischemic Hippocampus. *J. Neurosci. Res.* **2009**, *87*, 2126–213 [CrossRef] [PubMed]
164. Sun, C.-Y.; Li, J.-R.; Wang, Y.-Y.; Lin, S.-Y.; Ou, Y.-C.; Lin, C.-J.; Wang, J.-D.; Liao, S.-L.; Chen, C.-J. Indoxyl Sulfate Cause Behavioral Abnormality and Neurodegeneration in Mice with Unilateral Nephrectomy. *Aging* **2021**, *13*, 6681–6701. [CrossRef]
165. Moloney, G.M.; O'Leary, O.F.; Salvo-Romero, E.; Desbonnet, L.; Shanahan, F.; Dinan, T.G.; Clarke, G.; Cryan, J.F. Microbi Regulation of Hippocampal miRNA Expression: Implications for Transcription of Kynurenine Pathway Enzymes. *Behav. Bra Res.* **2017**, *334*, 50–54. [CrossRef] [PubMed]
166. Liu, B.; Yu, D.; Sun, J.; Wu, X.; Xin, Z.; Deng, B.; Fan, L.; Fu, J.; Ge, L.; Ren, W. Characterizing the Influence of Gut Microbiota o Host Tryptophan Metabolism with Germ-Free Pigs. *Anim. Nutr.* **2022**, *11*, 190–200. [CrossRef]
167. Schwarcz, R.; Foo, A.; Sathyasaikumar, K.V.; Notarangelo, F.M. The Probiotic Lactobacillus Reuteri Preferentially Synthesiz Kynurenic Acid from Kynurenine. *Int. J. Mol. Sci.* **2024**, *25*, 3679. [CrossRef] [PubMed]
168. Sun, P.; Wang, M.; Liu, Y.-X.; Li, L.; Chai, X.; Zheng, W.; Chen, S.; Zhu, X.; Zhao, S. High-Fat Diet-Disturbed Gut Microbiot Colonocyte Interactions Contribute to Dysregulating Peripheral Tryptophan-Kynurenine Metabolism. *Microbiome* **2023**, *11*, 15 [CrossRef]

169. Mu, C.; Choudhary, A.; Mayengbam, S.; Barrett, K.T.; Rho, J.M.; Shearer, J.; Scantlebury, M.H. Seizure Modulation by the Gut Microbiota and Tryptophan-Kynurenine Metabolism in an Animal Model of Infantile Spasms. *eBioMedicine* **2022**, *76*, 103833. [CrossRef]
170. Zhao, L.-P.; Wu, J.; Quan, W.; Zhou, Y.; Hong, H.; Niu, G.-Y.; Li, T.; Huang, S.-B.; Qiao, C.-M.; Zhao, W.-J.; et al. DSS-Induced Colitis Activates the Kynurenine Pathway in Serum and Brain by Affecting IDO-1 and Gut Microbiota. *Front. Immunol.* **2023**, *13*, 1089200. [CrossRef] [PubMed]
171. Xiao, W.; Li, J.; Gao, X.; Yang, H.; Su, J.; Weng, R.; Gao, Y.; Ni, W.; Gu, Y. Involvement of the Gut-Brain Axis in Vascular Depression via Tryptophan Metabolism: A Benefit of Short Chain Fatty Acids. *Exp. Neurol.* **2022**, *358*, 114225. [CrossRef]
172. Yano, J.M.; Yu, K.; Donaldson, G.P.; Shastri, G.G.; Ann, P.; Ma, L.; Nagler, C.R.; Ismagilov, R.F.; Mazmanian, S.K.; Hsiao, E.Y. Indigenous Bacteria from the Gut Microbiota Regulate Host Serotonin Biosynthesis. *Cell* **2015**, *161*, 264–276. [CrossRef] [PubMed]
173. Yang, M.; Fukui, H.; Eda, H.; Kitayama, Y.; Hara, K.; Kodani, M.; Tomita, T.; Oshima, T.; Watari, J.; Miwa, H. Involvement of Gut Microbiota in the Association between Gastrointestinal Motility and 5-HT Expression/M2 Macrophage Abundance in the Gastrointestinal Tract. *Mol. Med. Rep.* **2017**, *16*, 3482–3488. [CrossRef] [PubMed]
174. Tang, T.; Wang, J.; Jiang, Y.; Zhu, X.; Zhang, Z.; Wang, Y.; Shu, X.; Deng, Y.; Zhang, F. Bifidobacterium Lactis TY-S01 Prevents Loperamide-Induced Constipation by Modulating Gut Microbiota and Its Metabolites in Mice. *Front. Nutr.* **2022**, *9*, 890314. [CrossRef] [PubMed]
175. Xue, K.; Tao, R.; Wu, Q.; Zhang, L.; Sun, Z.; Yu, X.; Meng, J.; Bao, N.; Zhao, J. Enteropathogenic Escherichia Coli Mediates CoCrMo Particle-Induced Peri-Implant Osteolysis by Increasing Peripheral 5-HT. *Front. Cell. Infect. Microbiol.* **2022**, *11*, 796679. [CrossRef]
176. Li, H.; Wang, P.; Huang, L.; Li, P.; Zhang, D. Effects of Regulating Gut Microbiota on the Serotonin Metabolism in the Chronic Unpredictable Mild Stress Rat Model. *Neurogastroenterol. Motil.* **2019**, *31*, e13677. [CrossRef] [PubMed]
177. Li, S.; Li, X.; Wang, K.; Li, Y.; Nagaoka, K.; Li, C. Gut Microbiota Intervention Attenuates Thermogenesis in Broilers Exposed to High Temperature through Modulation of the Hypothalamic 5-HT Pathway. *J. Anim. Sci. Biotechnol.* **2023**, *14*, 159. [CrossRef] [PubMed]
178. Li, D.; Liang, W.; Zhang, W.; Huang, Z.; Liang, H.; Liu, Q. Fecal Microbiota Transplantation Repairs Intestinal Permeability and Regulates the Expression of 5-HT to Influence Alcohol-Induced Depression-like Behaviors in C57BL/6J Mice. *Front. Microbiol.* **2024**, *14*, 1241309. [CrossRef]
179. Bunnett, N.W. Neuro-humoral Signalling by Bile Acids and the TGR5 Receptor in the Gastrointestinal Tract. *J. Physiol.* **2014**, *592*, 2943–2950. [CrossRef]
180. Chen, Z.; Feng, J.; Hu, S.; Hua, Y.; Ma, S.; Fu, W.; Yang, Q.; Zhang, X. Bacillus Subtilis Promotes the Release of 5-HT to Regulate Intestinal Peristalsis in STC Mice via Bile Acid and Its Receptor TGR5 Pathway. *Dig. Dis. Sci.* **2022**, *67*, 4410–4421. [CrossRef]
181. Cai, J.; Cheung, J.; Cheung, S.W.M.; Chin, K.T.C.; Leung, R.W.K.; Lam, R.S.T.; Sharma, R.; Yiu, J.H.C.; Woo, C.W. Butyrate Acts as a Positive Allosteric Modulator of the 5-HT Transporter to Decrease Availability of 5-HT in the Ileum. *Br. J. Pharmacol.* **2024**, *181*, 1654–1670. [CrossRef]
182. Lin, F.; Wang, X.; Luo, R.; Yuan, B.; Ye, S.; Yang, T.; Xiao, L.; Chen, J. Maternal LPS Exposure Enhances the 5-HT Level in the Prefrontal Cortex of Autism-like Young Offspring. *Brain Sci.* **2023**, *13*, 958. [CrossRef] [PubMed]
183. Huang, D.; Yang, Y.; Song, W.; Jiang, C.; Zhang, Y.; Zhang, A.; Lin, Z.; Ke, X. Untargeted Metabonomic Analysis of a Cerebral Stroke Model in Rats: A Study Based on UPLC–MS/MS. *Front. Neurosci.* **2023**, *17*, 1084813. [CrossRef] [PubMed]
184. Saccaro, L.F.; Pico, F.; Chadenat, M.-L.; Richard, O.; Launay, J.-M.; Bastenaire, B.; Jullien, P.; Lambert, J.; Feuga, V.; Macquet, M.; et al. Platelet, Plasma, Urinary Tryptophan-Serotonin-Kynurenine Axis Markers in Hyperacute Brain Ischemia Patients: A Prospective Study. *Front. Neurol.* **2022**, *12*, 782317. [CrossRef]
185. Darlington, L.G.; Mackay, G.M.; Forrest, C.M.; Stoy, N.; George, C.; Stone, T.W. Altered Kynurenine Metabolism Correlates with Infarct Volume in Stroke. *Eur. J. Neurosci.* **2007**, *26*, 2211–2221. [CrossRef] [PubMed]
186. Cuartero, M.I.; Ballesteros, I.; De La Parra, J.; Harkin, A.L.; Abautret-Daly, A.; Sherwin, E.; Fernández-Salguero, P.; Corbí, Á.L.; Lizasoain, I.; Moro, M.A. L-Kynurenine/Aryl Hydrocarbon Receptor Pathway Mediates Brain Damage After Experimental Stroke. *Circulation* **2014**, *130*, 2040–2051. [CrossRef] [PubMed]
187. Mangas, A.; Yajeya, J.; González, N.; Ruiz, I.; Duleu, S.; Geffard, M.; Coveñas, R. Overexpression of Kynurenic Acid in Stroke: An Endogenous Neuroprotector? *Ann. Anat.-Anat. Anz.* **2017**, *211*, 33–38. [CrossRef]
188. Alme, K.N.; Ulvik, A.; Askim, T.; Assmus, J.; Mollnes, T.E.; Naik, M.; Næss, H.; Saltvedt, I.; Ueland, P.M.; Knapskog, A.B. Neopterin and Kynurenic Acid as Predictors of Stroke Recurrence and Mortality: A Multicentre Prospective Cohort Study on Biomarkers of Inflammation Measured Three Months after Ischemic Stroke. *BMC Neurol.* **2021**, *21*, 476. [CrossRef] [PubMed]
189. Ren, R.; Fang, Y.; Sherchan, P.; Lu, Q.; Lenahan, C.; Zhang, J.H.; Zhang, J.; Tang, J. Kynurenine/Aryl Hydrocarbon Receptor Modulates Mitochondria-Mediated Oxidative Stress and Neuronal Apoptosis in Experimental Intracerebral Hemorrhage. *Antioxid. Redox Signal.* **2022**, *37*, 1111–1129. [CrossRef]
190. Ganesh, B.P.; Peesh, P.; Blasco, M.P.; Hamamy, A.E.; Khan, R.; Guzman, G.; Honarpisheh, P.; Mohan, E.C.; Goodman, G.W.; Nguyen, J.N.; et al. Restoring a Balanced Pool of Host-Derived and Microbiota-Derived Ligands of the Aryl Hydrocarbon Receptor Is Beneficial after Stroke. *Res. Sq.* **2023**. [CrossRef]

191. Rothhammer, V.; Borucki, D.M.; Tjon, E.C.; Takenaka, M.C.; Chao, C.-C.; Ardura-Fabregat, A.; De Lima, K.A.; Gutiérrez-Vázquez, C.; Hewson, P.; Staszewski, O.; et al. Microglial Control of Astrocytes in Response to Microbial Metabolites. *Nature* **2018**, *557*, 724–728. [CrossRef]
192. Xie, Y.; Zou, X.; Han, J.; Zhang, Z.; Feng, Z.; Ouyang, Q.; Hua, S.; Liu, Z.; Li, C.; Cai, Y.; et al. Indole-3-Propionic Acid Alleviates Ischemic Brain Injury in a Mouse Middle Cerebral Artery Occlusion Model. *Exp. Neurol.* **2022**, *353*, 114081. [CrossRef] [PubMed]
193. Xue, H.; Chen, X.; Yu, C.; Deng, Y.; Zhang, Y.; Chen, S.; Chen, X.; Chen, K.; Yang, Y.; Ling, W. Gut Microbially Produced Indole-3-Propionic Acid Inhibits Atherosclerosis by Promoting Reverse Cholesterol Transport and Its Deficiency Is Causally Related to Atherosclerotic Cardiovascular Disease. *Circ. Res.* **2022**, *131*, 404–420. [CrossRef] [PubMed]
194. Konopelski, P.; Chabowski, D.; Aleksandrowicz, M.; Kozniewska, E.; Podsadni, P.; Szczepanska, A.; Ufnal, M. Indole-3-Propionic Acid, a Tryptophan-Derived Bacterial Metabolite, Increases Blood Pressure via Cardiac and Vascular Mechanisms in Rats. *Am. J. Physiol. Regul. Intgr. Comp. Physiol.* **2021**, *321*, R969–R981. [CrossRef]
195. Yoon, N.; Kim, Y.; Kim, S.-D.; Kim, M.; Jung, B.H.; Song, Y.S. Investigation of Long-Term Metabolic Alteration after Stroke in tMCAO (Transient Middle Cerebral Artery Occlusion) Mouse Model Using Metabolomics Approach. *Neurosci. Lett.* **2022**, *774*, 136492. [CrossRef] [PubMed]
196. Marsiglia, R.; Marangelo, C.; Vernocchi, P.; Scanu, M.; Pane, S.; Russo, A.; Guanziroli, E.; Del Chierico, F.; Valeriani, M.; Molteni, F.; et al. Gut Microbiota Ecological and Functional Modulation in Post-Stroke Recovery Patients: An Italian Study. *Microorganisms* **2023**, *12*, 37. [CrossRef] [PubMed]
197. Teunis, C.J.; Stroes, E.S.G.; Boekholdt, S.M.; Wareham, N.J.; Murphy, A.J.; Nieuwdorp, M.; Hazen, S.L.; Hanssen, N.M.J. Tryptophan Metabolites and Incident Cardiovascular Disease: The EPIC-Norfolk Prospective Population Study. *Atherosclerosis* **2023**, *387*, 117344. [CrossRef] [PubMed]
198. Duan, X.; Gan, J.; Xu, F.; Li, L.; Han, L.; Peng, C.; Bao, Q.; Xiao, L.; Peng, D. RNA Sequencing for Gene Expression Profiles in a Rat Model of Middle Cerebral Artery Occlusion. *Biomed. Res. Int.* **2018**, *2018*, 2465481. [CrossRef] [PubMed]
199. Tan, D.-X.; Manchester, L.C.; Sainz, R.M.; Mayo, J.C.; León, J.; Reiter, R.J. Physiological Ischemia/Reperfusion Phenomena and Their Relation to Endogenous Melatonin Production: An Hypothesis. *Endocrine* **2005**, *27*, 149–158. [CrossRef]
200. Hajsl, M.; Hlavackova, A.; Broulikova, K.; Sramek, M.; Maly, M.; Dyr, J.E.; Suttnar, J. Tryptophan Metabolism, Inflammation, and Oxidative Stress in Patients with Neurovascular Disease. *Metabolites* **2020**, *10*, 208. [CrossRef]
201. Mortensen, J.K.; Kraglund, K.L.; Johnsen, S.P.; Mors, O.; Andersen, G.; Buttenschøn, H.N. The Serotonin Transporter Gene Polymorphisms and Risk of Ischemic Stroke. *Cerebrovasc. Dis.* **2018**, *45*, 187–192. [CrossRef]
202. Aguiar, R.P.D.; Newman-Tancredi, A.; Prickaerts, J.; Oliveira, R.M.W.D. The 5-HT1A Receptor as a Serotonergic Target for Neuroprotection in Cerebral Ischemia. *Prog. Neuro-Psychopharmacol. Biol. Psychiatry* **2021**, *109*, 110210. [CrossRef]
203. Liu, D.; Tan, S.; Zhou, Z.; Gu, S.; Zuo, H. Trimethylamine N-Oxide, β-Alanine, Tryptophan Index, and Vitamin B6-Related Dietary Patterns in Association with Stroke Risk. *Nutr. Metab. Cardiovasc. Dis.* **2024**, *34*, 1179–1188. [CrossRef]
204. Bhave, V.M.; Ament, Z.; Patki, A.; Gao, Y.; Kijpaisalratana, N.; Guo, B.; Chaudhary, N.S.; Guarniz, A.G.; Gerszten, R.; Correa, A.; et al. Plasma Metabolites Link Dietary Patterns to Stroke Risk. *Ann. Neurol.* **2023**, *93*, 500–510. [CrossRef]
205. Joshi, B.; Panda, S.K.; Jouneghani, R.S.; Liu, M.; Parajuli, N.; Leyssen, P.; Neyts, J.; Luyten, W. Antibacterial, Antifungal, Antiviral, and Anthelmintic Activities of Medicinal Plants of Nepal Selected Based on Ethnobotanical Evidence. *Evid.-Based. Compl. Alt.* **2020**, *2020*, 1043471. [CrossRef]
206. Li, J.; Zhao, T.; Qiao, H.; Li, Y.; Xia, M.; Wang, X.; Liu, C.; Zheng, T.; Chen, R.; Xie, Y.; et al. Research Progress of Natural Products for the Treatment of Ischemic Stroke. *J. Integr. Neurosci.* **2022**, *21*, 14. [CrossRef] [PubMed]
207. Huang, L.; Wang, T.; Wu, Q.; Dong, X.; Shen, F.; Liu, D.; Qin, X.; Yan, L.; Wan, Q. Analysis of Microbiota in Elderly Patients with Acute Cerebral Infarction. *PeerJ* **2019**, *7*, e6928. [CrossRef] [PubMed]
208. Li, H.; Zhang, X.; Pan, D.; Liu, Y.; Yan, X.; Tang, Y.; Tao, M.; Gong, L.; Zhang, T.; Woods, C.R.; et al. Dysbiosis Characteristics of Gut Microbiota in Cerebral Infarction Patients. *Transl. Neurosci.* **2020**, *11*, 124–133. [CrossRef]
209. Haak, B.W.; Westendorp, W.F.; van Engelen, T.S.; Brands, X.; Brouwer, M.C.; Vermeij, J.D.; Hugenholtz, F.; Verhoeven, A.; Derks, R.J.; Giera, M.; et al. Disruptions of Anaerobic Gut Bacteria Are Associated with Stroke and Post-Stroke Infection: A Prospective Case–Control Study. *Transl. Stroke Res.* **2020**, *12*, 581–592. [CrossRef] [PubMed]
210. Xiang, L.; Lou, Y.; Liu, L.; Liu, Y.; Zhang, W.; Deng, J.; Guan, Y.; She, M.; You, X.; Liu, M.; et al. Gut Microbiotic Features Aiding the Diagnosis of Acute Ischemic Stroke. *Front. Cell. Infect. Microbiol.* **2020**, *10*, 587284. [CrossRef] [PubMed]
211. Tan, C.; Wu, Q.; Wang, H.; Gao, X.; Xu, R.; Cui, Z.; Zhu, J.; Zeng, X.; Zhou, H.; He, Y.; et al. Dysbiosis of Gut Microbiota and Short-chain Fatty Acids in Acute Ischemic Stroke and the Subsequent Risk for Poor Functional Outcomes. *J. Parenter. Enter. Nutr.* **2021**, *45*, 518–529. [CrossRef]
212. Chen, X.; Wu, Q.; Gao, X.; Wang, H.; Zhu, J.; Xia, G.; He, Y.; Song, W.; Xu, K. Gut Microbial Dysbiosis Associated with Type 2 Diabetes Aggravates Acute Ischemic Stroke. *mSystems* **2021**, *6*, e01304-21. [CrossRef]
213. Chen, L.; Wang, S.; Zhang, Y.; Li, Y.; Zhang, X.; Ma, J.; Zou, X.; Yao, T.; Li, S.; Chen, J.; et al. Multi-Omics Reveals Specific Host Metabolism-Microbiome Associations in Intracerebral Hemorrhage. *Front. Cell. Infect. Microbiol.* **2022**, *12*, 999627. [CrossRef]
214. Li, T.; Sun, Q.; Feng, L.; Yan, D.; Wang, B.; Li, M.; Xiong, X.; Ma, D.; Gao, Y. Uncovering the Characteristics of the Gut Microbiota in Patients with Acute Ischemic Stroke and Phlegm-Heat Syndrome. *PLoS ONE* **2022**, *17*, e0276598. [CrossRef]

15. Zhang, X.; Wang, X.; Zhao, H.; Cao, R.; Dang, Y.; Yu, B. Imbalance of Microbacterial Diversity Is Associated with Functional Prognosis of Stroke. *Neural Plast.* **2023**, *2023*, 6297653. [CrossRef]
16. Yu, S.; Chen, J.; Zhao, Y.; Liao, X.; Chen, Q.; Xie, H.; Liu, J.; Sun, J.; Zhi, S. Association Analysis of the Gut Microbiota in Predicting Outcomes for Patients with Acute Ischemic Stroke and H-Type Hypertension. *Front. Neurol.* **2023**, *14*, 1275460. [CrossRef]

biomolecules

MDPI

Article

The Rhododendron Chrysanthum Pall.s' Acetylation Modification of Rubisco Enzymes Controls Carbon Cycling to Withstand UV−B Stress

Meiqi Liu, Fushuai Gong, Wang Yu, Kun Cao, Hongwei Xu * and Xiaofu Zhou *

Jilin Provincial Key Laboratory of Plant Resource Science and Green Production, Jilin Normal University, Siping 136000, China

* Correspondence: xuhongwei@jlnu.edu.cn (H.X.); zhouxiaofu@jlnu.edu.cn (X.Z.)

Abstract: Lysine acetylation of proteins plays a critical regulatory function in plants. A few advances have been made in the study of plant acetylproteome. However, until now, there have been few data on *Rhododendron chrysanthum* Pall. (*R. chrysanthum*). We analyzed the molecular mechanisms of photosynthesis and stress resistance in *R. chrysanthum* under UV−B stress. We measured chlorophyll fluorescence parameters of *R. chrysanthum* under UV−B stress and performed a multi−omics analysis. Based on the determination of chlorophyll fluorescence parameters, *R. chrysanthum* Y(NO) (Quantum yield of non−photochemical quenching) increased under UV−B stress, indicating that the plant was damaged and photosynthesis decreased. In the analysis of acetylated proteomics data, acetylated proteins were found to be involved in a variety of biological processes. Notably, acetylated proteins were significantly enriched in the pathways of photosynthesis and carbon fixation, suggesting that lysine acetylation modifications have an important role in these activities. Our findings suggest that *R. chrysanthum* has decreased photosynthesis and impaired photosystems under UV−B stress, but NPQ shows that plants are resistant to UV−B. Acetylation proteomics revealed that up- or down-regulation of acetylation modification levels alters protein expression. Acetylation modification of key enzymes of the Calvin cycle (Rubisco, GAPDH) regulates protein expression, making Rubisco and GAPDH proteins expressed as significantly different proteins, which in turn affects the carbon fixation capacity of *R. chrysanthum*. Thus, Rubisco and GAPDH are significantly differentially expressed after acetylation modification, which affects the carbon fixation capacity and thus makes the plant resistant to UV−B stress. Lysine acetylation modification affects biological processes by regulating the expression of key enzymes in photosynthesis and carbon fixation, making plants resistant to UV−B stress.

Keywords: lysine acetylome; chlorophyll fluorescence; photosynthesis; carbon fixation; *R. chrysanthum*

1. Introduction

Rhododendron chrysanthum Pall. (*R. chrysanthum*) is a rare resource for the study of abiotic stress in plants since it grows permanently in the alpine zone of Changbai Mountain, which has strong radiation and high altitude [1]. *R. chrysanthum* has demonstrated adaptation to UV−B radiation and other abiotic stresses over the course of its lengthy growth and evolution [2].

Among the ultraviolet rays in sunlight that cause the most concern is medium−wave ultraviolet radiation (UV−B, 280–320 nm) [3]. Because of their sessile growth habit, plants are inevitably exposed to UV−B stress. Most plants that are exposed to UV−B radiation will react by producing stress [3]. According to pertinent research, UV−B stress can harm plants in a number of ways, including slowing down their growth, inhibiting photosynthesis, oxidative damage, and rupturing the integrity of vital macromolecules [4–6]. Green plants' rate of photosynthetic activity may be significantly lowered by UV−B stress [7], and parameters connected to chlorophyll fluorescence are crucial markers of photosynthetic characteristics [8]. The kale (*Brassica oleracea* L.) showed a decrease in Fv/Fm (maximum photochemical efficiency of PSII) and fluorescence transient parameters (RC/CS, ABS/CS, TR_O/CS) with UV−B stress [9]. The maximum photochemical efficiency of PSII, or the ratio of chlorophyll fluorescence parameter Fv/Fm, dramatically dropped in *Olea europaea* (L.) under long-term UV−B stress [10]. Due to its special growing environment, *R. chrysanthum* was used as the experimental material in this study. In order to ascertain the effects of UV−B radiation on photosynthesis and the ideal UV−B radiation dosage for *R. chrysanthum*, we carried out a pertinent study. We also determined the duration of UV−B radiation, which was 48 h, and found that *R. chrysanthum* photosynthesis was inhibited by UV−B but not by UV−A [2]. After UV−B treatment, we also conducted the measurement of chlorophyll fluorescence parameters, and the Fv/Fm and Fv/Fo decreased dramatically, revealing that photo-synthesis was impeded and the photosystem was injured [11].

Proteins' post-translational modifications (PTMs) can alter their activities, subcellular locations, and interaction partners, among other aspects of protein functional stability regulation [12]. One of the most common PTMs in both prokaryotes and eukaryotes is lysine acetylation (Kac) modification [13,14]. After undergoing extensive research, it was discovered that kac was involved in numerous significant processes, including the germination of seeds, the development of organs, blooming, stress response, and leaf senescence [15–20]. During the dormancy release process in poplar, acetylation modification of enzymes in the primary metabolic pathway is an important strategy for breaking the dormancy of bloom buds [21]. The acetylation of essential enzymes for starch biosynthesis to adapt to drought stress is revealed by acetylation proteomics analysis of growing wheat kernels under conditions of water deficit [22]. The OsHYPK-NatA complex plays a crucial role in coordinating rice development and stress responses, primarily through its dynamic regulation of NatA-mediated N-terminal acetylation and global protein turnaround [23]. *Sorghum bicolor* HDAC SbHDT701 enhances acetylation modifications to regulate stress response [24]. LysAc may have a major and previously unknown regulatory role in many nonhistone proteins that are involved in pathways and functions in Arabidopsis and other plants [25]. Some proteins, especially those involved in photosynthesis, glycolysis, and secondary metabolism, may be lysine-acetylated to control particular metabolic processes in tea leaves [26]. Our earlier study revealed that *R. chrysanthum's* ability to photosynthesise was hampered by UV−B radiation, but that UV−B damage to plants could be mitigated by acetylating the PSII protein. This study did not fully explore the photosynthetic carbon cycle process, which is crucial to understanding biological processes [11].

It has been investigated and determined how the *R. chrysanthum* photosystem II protein reacts to UV−B stress. But the examination initially concentrated on physiology, biochemistry, and photosystem aspects [11]. However, plants' ability to sequester carbon is also very significant. Thus, we conducted an acetylation-modified proteomics analysis of *R. chrysanthum* in order to gain a better understanding of the molecular mechanism of UV−B tolerance. This analysis offers important insights for future research on the molecular mechanism of UV−B radiation resistance.

2. Materials and Methods

2.1. Plant Material and Treatment

R. chrysanthum was kept in a climate chamber designed to mimic the conditions of a high mountain environment [2], and in the artificial climate chamber following storage, *R. chrysanthum* was continuously grown and cultured under white fluorescent lights. In the climate chamber, *R. chrysanthum* was cultivated under 50 μmol (photon) m^{-2} s^{-1} white fluorescent lights. And the following conditions were present in the artificial climate chamber: 18 °C for 14 h of light and 16 °C for 10 h of darkness. UV−B and photosynthetically active radiation (PAR) were used in this study. Two groups of *R. chrysanthum* were created: the CG group received a 48-h treatment with PAR, while the BG group received a 48-h treatment with PAR+UV−B. Each group received three biological duplicates (three repetitions for the CG group and three repetitions for the BG group) to guarantee adequate coverage. PAR radiation was performed by placing a 400 nm filter (Edward, Filter Long 2IN SQ, Barrington, NJ, USA) on the culture flasks (effective irradiance of 50 μmol m^{-2} s^{-1}), and PAR+UV−B radiation was performed by placing a 295 nm filter (Edward, Filter Long 2IN SQ, Barrington, NJ, USA) (effective irradiance of 2.3 W m^{-2}).

2.2. Measurement of Chlorophyll Fluorescence

R. chrysanthum leaves were used to evaluate the induction features of chlorophyll fluorescence using the Imaging-PAM Maxi (HeinzWalz, Effeltrich, Germany). The plants were treated for 20 min in the dark prior to measuring them. We measured the fluorescence parameters. Actual photochemical quantum yield of PSII Y(II); Maximum quantum yield of PSII (Fv′/Fm′); Electron transport rate (ETR), non-photochemical quenching coefficient (NPQ), photochemical quenching coefficient (qL), and quantum yield of uncontrolled energy dissipation of photosystem II Y(NO).

2.3. Quantitative Proteomic Study of Acetylated Modification

2.3.1. Protein Extraction

This experiment's acetylated proteomics were provided by Jingjie PTM Hangzhou Co Inc. BioLab (Hangzhou, China). Using a high−intensity ultrasonic processor (Scientz, Ningbo, China), *R. chrysanthum* was ground and pulverized in liquid nitrogen, transferred to centrifuge tubes, and sonicated three times on ice in a lysis buffer (8 M urea, 2 mM EDTA, 10 mM DTT, and 1% protease inhibitor cocktail). After centrifugation at 5500× *g* for 10 min at 4 °C, the supernatant was centrifuged for 3 min at 4 °C. Proteins in the supernatant were centrifuged for 3 min at 4 °C, and it was then precipitated with 15% TCA for 4 h at −20 °C. The rest of the sediment was washed three times with acetone after chilling in a refrigerator. Finally, the proteins were redissolved in buffer, the buffer consisted of 8 M urea, 100 mM TEAB, pH 8.0. And in order to measure the protein concentration in the supernatant, it was estimated using the 2-D Quantay Analysis Kit (Cytiva 80-6483-56).

2.3.2. Trypsin Digestion

TCA at a concentration of 20% (*m*/*v*) was slowly added to the sample to precipitate the proteins, followed by vortex mixing and incubation at 4 °C for 2 h. The precipitated proteins were collected by centrifugation at 4500× *g* for 5 min at 4 °C. Subsequently, the collected precipitated proteins were washed with pre-cooled acetone three times and allowed to dry for 1 min. The washed and dried proteins were then redissolved in 200 mM TEAB and dispersed by ultrasonication. For the first digestion, trypsin was added at a 1:50 trypsin mass ratio overnight. Subsequently, the peptides were reduced with 5 mM dithiothreitol for 60 min at 37 °C, followed by 11 mM iodoacetamide for 45 min at room temperature and in the dark. Finally, the peptides were desalted, using a Strata X SPE column.

2.3.3. Affinity Enrichment

Tryptic peptides were dissolved in NETN buffer (100 mM NaCl, 1 mM EDTA, 50 mM Tris-HCl, 0.5% NP-40, pH 8.0) to dissolve Kac-modified peptides. Pre-washed antibody beads (PTM-104, 10002D) were then added and incubated with the peptide mixture at 4 °C for an overnight period with gentle shaking. The beads were then washed twice with water and four times with NETN buffer. The use of 0.1% trifluoroacetic acid allowed the bound peptides to be released from the beads. The eluted portions were then mixed and dried under a vacuum. The resultant peptides were desalted using C18 ZipTips (ZTC18S) in accordance with the manufacturer's instructions for LC-MS/MS analysis.

2.3.4. LC-MS/MS Analysis and Database Search

Solvent A (0.1% formic acid, 2% acetonitrile/in water) was used to solubilize the tryptic peptides, which were then loaded directly onto a home-made analytical column (25-cm length, 100 μm i.d.). In a nanoElute UHPLC system (IQLAAAGABHFAPUMZZZ) running at 350 nL/min, the solvent B gradient increased from 6% to 22% (0.1% formic acid dissolved in acetonitrile) over 42 min, increased from 22% to 30% in 12 min, increased to 80% in 3 min, maintained at 80%, and then hold at 80% for 3 min. Prior to mass spectrometry analysis with timsTOF Pro, the peptides were subjected to Capillary source treatment. Then, 1.75 kV of electrospray voltage was used. At the TOF detector, precursors and fragments were analyzed using MS/MS with a scan range of 100 to 1700 m/z. Parallel accumulation serial fragmentation (PASEF) mode was used to run the timsTOF Pro. Ten PASEF-MS/MS scans were acquired per acquisition cycle, with the fragmentation of precursors with charge states 0 to 5 being the preferred range. A 24 s dynamic exclusion was set. The MaxQuant search engine (v. 1.6.6.0) was used to process the generated MS/MS data. Up to 4 missing cleavage events were made possible by using the cleavage protein trypsin/P. In the first search and the main search, the mass tolerance for precursor ions was set to 20 ppm and 20 ppm, respectively. And the mass tolerance for fragment ions was set as 20 ppm. Protein N-terminal acetylation, oxidation on Met, and acetylation on Lys were designated as variable modifications, while carbamidomethyl on Cys was designated as a fixed modification. FDR was adjusted to <1%.

2.3.5. Bioinformatics Analysis

Firstly, the average quantification of the number of multiple repetitions in the CG and BG groups was first counted separately, and then the ratio of the average of the CG and BG groups was computed. This ratio was used for final quantification. Quantitative values were log-transformed and p-values were calculated from two-tailed Student's t-tests. Significant differences were defined as protein or modified peptide $p < 0.05$, fold change (FC) ≥ 1.5. Functional annotation tools were used to identify enriched clusters of acetylated proteins.

GO enrichment analysis: protein GO annotations are classified into 3 categories: biological process, cellular component, and molecular function. The Fisher's exact test was used to determine the significance of GO enrichment of different changed proteins, and the significance was p-value < 0.05.

KEGG pathway enrichment analysis: for the enrichment analysis of the KEGG pathway, we used the KEGG database. Fisher's exact test was used to determine the significance of KEGG pathway enrichment of different changed proteins, and proteins were considered to be significant at a p-value of less than 0.05.

Protein subcellular localization: in this research, the subcellular localization of proteins is predicted using Wolfpsort(About WoLF PSORT (hgc.jp)) subcellular localization prediction software. Wolfpsort is applied to eukaryotic sequence prediction.

Protein motif analysis: by counting the pattern of amino acid sequences before and after all acetylation modification sites, a trend in the pattern of amino acid sequences within the region of the site where the modification occurs is calculated.

2.4. Modeling Acetylated Proteins Homology

To understand the structure of the protein and the level of acetylation modification sites, we used NCBI BLAST to find and compare homologous sequences. Subsequently, using SWISS-MODEL, a three-dimensional structural model of the protein was generated, labeling the acetylation sites.

2.5. Statistics and Analysis of Data

Chlorophyll fluorescence parameters measured in CG and BG groups were analyzed and plotted using SPSS Statistics 26.

3. Results

3.1. Quantitative Proteomic Analysis of Acetylated Modification in R. chrysanthum

In order to map the sites of lysine acetylation in *R. chrysanthum*, trypsin was used to break down proteins into peptides. *R. chrysanthum* was divided into two groups based on the treatments: the BG group (*R. chrysanthum* treated with PAR+UV−B for 48 h, three biological replicates) and the CG group (*R. chrysanthum* treated with PAR for 48 h, three biological replicates) groups. Then, an LC-MS/MS analysis was carried out (Supplementary Figure S1B). The majority of the acetylation-containing polypeptides were found to have lengths between 7 and 20, which is consistent with the fundamental principle of trypsin enzymatic digestion (Supplementary Figure S1C). A total of 945 lysine acetylation sites were present in 685 differentially expressed acetylated proteins (DAPs) (Supplementary Table S1). A total of 807 differentially expressed proteins (DEPs) were found using quantitative proteomics analysis; according to fold change >1.5 and $p < 0.05$, 357 of these proteins were downregulated and 450 were upregulated (Supplementary Table S2).

3.2. *Plant Carbon Assimilation in R. chrysanthum Is Impacted by Photodamage under UV−B Stress*

With the use of chlorophyll fluorescence technology, changes in leaf photosynthesis can be swiftly and precisely reflected. Chlorophyll fluorescence characteristics shift in plants exposed to UV−B stress, which in turn indicates photosynthetic damage. We determined the chlorophyll fluorescence parameters in order to investigate variations in *R. chrysanthum's* photochemical activity. For pictures of chlorophyll fluorescence, see Figure 1. The plant's NPQ is represented by the fluorescence images, which show a decline in *R. chrysanthum's* ability to protect against photodamage (Figure 1A). The plant's photosynthetic activity is reflected in qL, and under UV−B stress, *R. chrysanthum's* photosynthetic activity increases as shown by the fluorescence images (Figure 1B). Additionally, the Y(NO) image shows that the plant has photodamaged (Figure 1C). The (YII) images show a decrease in *R. chrysanthum's* actual photosynthetic rate. (Figure 1D).

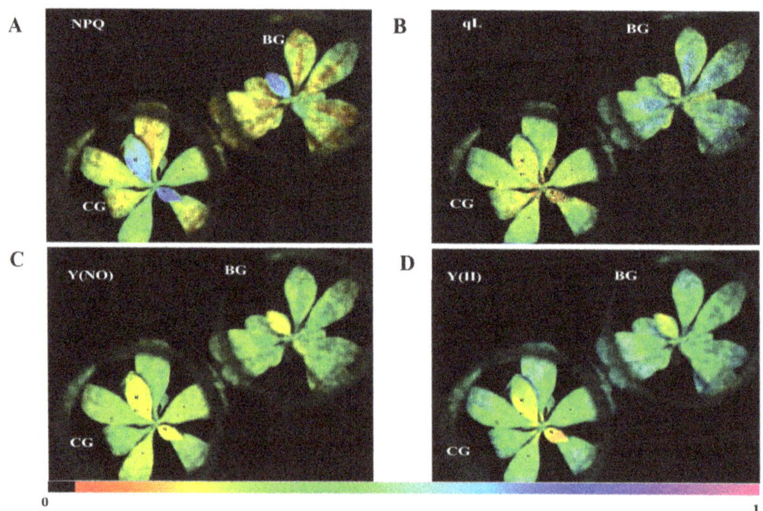

Figure 1. Imaging of real-time fluorescence of *R. chrysanthum* under UV−B stress. (**A**) average nonphotochemical quenching (NPQ); (**B**) photochemical quenching coefficient (qL)-based; (**C**) nonphotochemical quenching quantum yield Y(NO); and (**D**) actual photochemical quantum yield Y(II) for PSII. The color in the color scale of the picture indicates the size of the value, the more reddish the color indicates the smaller the value, and the more purple the color indicates the larger the value.

The rate of Fv'/Fm' significantly dropped by 9.25% under UV−B stress (Figure 2A, Table 1), suggesting that photoinhibition was taking place and damaging the reaction center of chrysanthemum photosystem II. The real light energy conversion efficiency of leaves over time is represented by Y(II) (Figure 2B, Table 1). However, Y(NO), the quantum yield of non-photochemical quenching, is a crucial marker of photodamage, and the analysis reveals a notable 15.2% increase in Y(NO) value. The plant damage caused by UV−B radiation was indicated by the significant increase in Y(NO) value (Figure 2C, Table 1). The ability of the plant to transform excess light energy into heat energy is reflected in non-photochemical quenching, which shields the plant from harm. The plant's capacity to disperse heat is indicated by NPQ. The plant's capacity to disperse heat is indicated by NPQ (Figure 2D, Table 1). Photochemical quenching reflects the photosynthetic activity of the plant, while non-photochemical quenching represents the ability of the plant to convert excess light energy into heat. Under UV−B stress, the qL of *R. chrysanthum* increased by 20.4%. This indicates that the photosynthetic activity of *R. chrysanthum* increased under UV−B stress (Figure 2E, Table 1). The rate of electron transfer can serve as an indicator of the photosynthesis's rate of electron transfer. *R. chrysanthum's* electron transfer rate (ETR) values exhibited a declining trend (Figure 2F, Table 1), suggesting that UV−B stress could have an impact on the plant. It is a crucial component of the light response curve and can be used to estimate the rate at which electron transfer energy is converted.

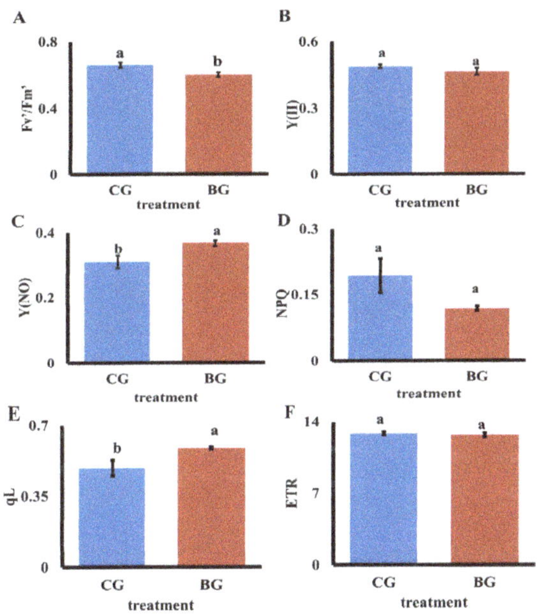

Figure 2. Chlorophyll fluorescence parameters in *R. chrysanthum*. a and b denote significant differences between PAR and PAR+UV−B ($p < 0.05$). **(A)** Maximum quantum yield of PSII (Fv′/Fm′); **(B)** Actual photochemical quantum yield of PSII [Y(II)]; **(C)** Non-photochemical quenching quantum yield [Y(NO)]; **(D)** Non-photochemical quenching coefficient (NPQ); **(E)** Photochemical quenching coefficient (qL); **(F)** Electron transfer rate (ETR). These results are the mean ± standard deviation of three samples.

Table 1. Effect of UV−B stress on chlorophyll fluorescence parameters of *R. chrysanthum*.

Group	Fv′/Fm′	Y(II)	Y(NO)	NPQ	qL	ETR
CG	0.660 ± 0.015a	0.489 ± 0.008a	0.310 ± 0.019b	0.193 ± 0.039a	0.49 ± 0.038b	12.8 ± 0.163a
BG	0.599 ± 0.014b	0.465 ± 0.016a	0.366 ± 0.009a	0.118 ± 0.006a	0.59 ± 0.005a	12.6 ± 0.205a

Note: These results represent the average ± standard deviation of 3 samples. Different lowercase letters suggest significant inter-group differences ($p < 0.05$), whereas identical lower-case characters suggest no relevant difference across groups.

3.3. *Analysis of Distribution and Motif of Acetylated Sites in Photosynthetic Carbon Cycle of R. chrysanthum*

The number of detected modification sites per protein was estimated in order to evaluate the sites' distribution in the acetylated proteins of *R. chrysanthum*. The findings suggested that 50% of proteins had only 1 site, whereas 30%, 9%, and 11% of proteins had 2, 3, or more modification sites, respectively (Figure 3A). Acetylated proteins are enriched in two processes: photosynthesis and the Calvin cycle. Only 63% of proteins had one acetylation site, whereas 12%, 5%, and 20%, respectively, of proteins had 2, 3, or more modification sites (Figure 3B). The sequence of amino acid distribution around the acetylated lysine was analyzed to comprehend the motifs in *R. chrysanthum*. Using motif-x, a total of 15 significantly enriched acetylation motifs were screened, of which the three motifs "KT", "KY", and "KS" accounted for the largest proportion. (Figure 3C, Supplementary Table S3). These results indicate that the conserved motifs and residues of the proteins in which *R. chrysanthum* undergoes acetylation are more intact and suggest that the lysine residues surrounding the basic residues in *R. chrysanthum* proteins have a greater chance of being acetylated under UV−B stress. The thermogram of compositions of amino acid around the acetylation sites showed that histidine (H) and tyrosine (Y) were apparently excessive from −4 to +4, and others, the cases in point are phenylalanine (F) and serine (S). They were highly enriched at +1 (Figure 3D, Supplementary Table S4).

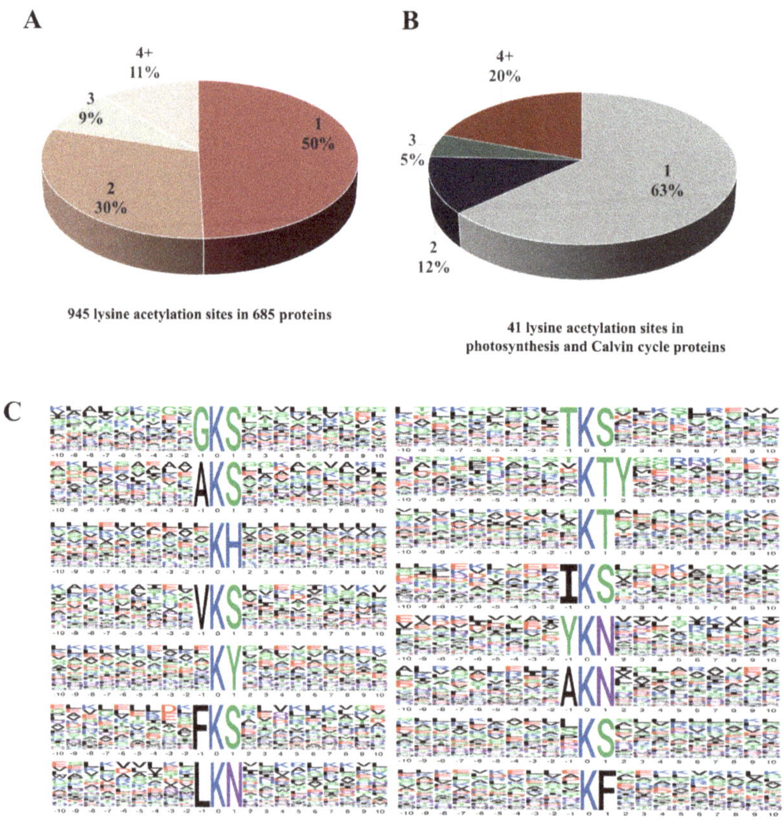

Figure 3. *Cont.*

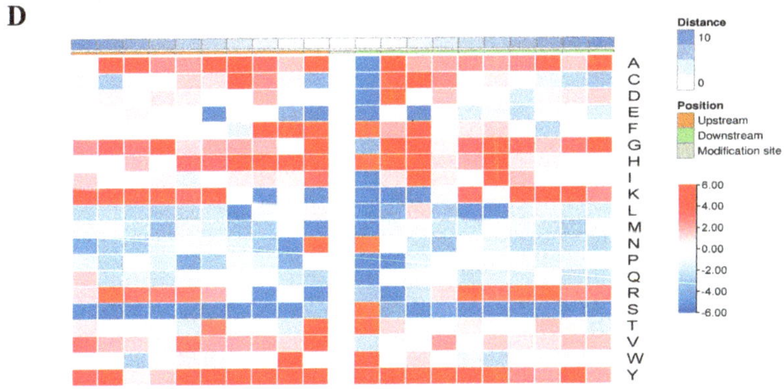

Figure 3. Properties of lysine acetylation sites in photosynthesis of *R. chrysanthum* under UV−B stress. (**A**) The number and proportion of lysine acetylation sites. (**B**) The number and percentage of lysine acetylationproteins sites in photosynthesis and Calvin cycle. (**C**) Sequence probability logos for acetylation site motifs significantly enriched for ± 10 amino acids around the lysine acetylation sites. (**D**) Clustering heatmap of amino acids up-stream and down-stream the acetylation modification site. Red color is that this amino acid is significantly enriched near the modification site, and blue is a significant decrease.

3.4. Analysis of Functional Enrichment of Acetylated Proteins and Sites in Photosynthetic Carbon Cycle of R. chrysanthum

With the aim of further investigating the acetylated proteins, they were divided into 3 main categories: biological process, molecular function, and cellular component (Supplementary Table S5). We carried out GO enrichment (Figure 4A, Supplementary Tables S6–S8). According to the cellular component analysis, *R. chrysanthum*'s acetylation-modified proteins were dramatically enriched in metabolic processes, and proteins enriched in the envelope and chloroplast were more likely to be acetylated (Figure 4A). Acetylation-modified proteins from *R. chrysanthum* were enriched in KEGG metabolic pathways, which allows a better understanding of their general function (Figure 4B, Supplementary Table S9). The findings demonstrated that some proteins in carbon metabolism undergo lysine acetylation (Supplementary Figure S2), the process of photosynthesis (Supplementary Figure S3), as well as carbon fixation in photosynthetic organisms (Supplementary Figure S4). KEGG enrichment showed that 26 acetylated proteins were acetylated in the carbon fixation of photosynthetic organisms, and 53 acetylation sites were identified. Acetylated proteins of *R. chrysanthum* in carbon fixation are localized to chloroplasts, mitochondria, and cytoplasm, as shown by subcellular localization with 17, 1, and 8 (Figure 4C).

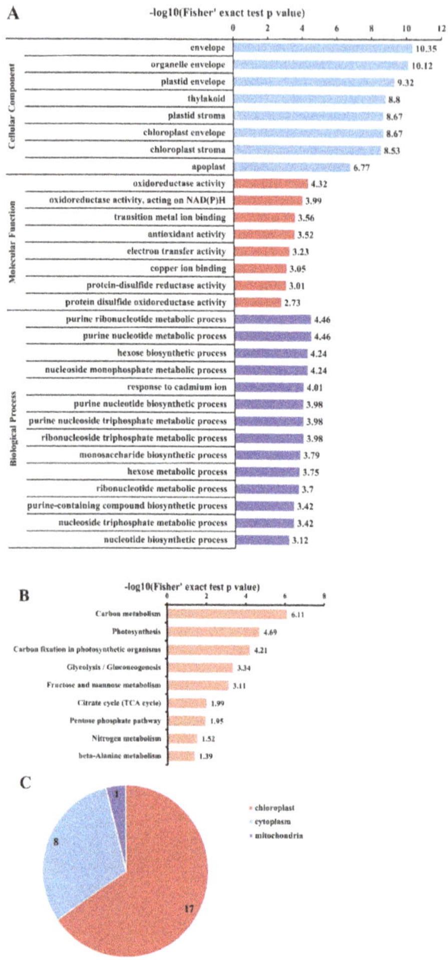

Figure 4. Enrichment analysis of the acetylated proteins in carbon fixation of *R. chrysanthum*. (**A**) The acetylated proteome is analyzed for GO enrichment. (**B**) The acetylated proteome is analyzed for KEGG pathway. (**C**) Subcellular localization of acetylated proteins of carbon fixation.

3.5. Analysis of DEPs and DAPs in Carbon Fixation in Photosynthetic Organisms of R. chrysanthum under UV−B Stress

Based on the outcomes of acetylated proteins' functional enrichment analysis, it was shown that lysine acetylation has a significant role in the pathways of photosynthesis and carbon fixation of plants. We analyzed the proteins in the carbon fixation pathway and found that the enzymes for carbon fixation under UV−B stress underwent acetylation modification with different levels of modification at different sites (Figure 5A). Carbon fixation is a reaction in photosynthesis that does not require the participation of light and involves the conversion of carbon dioxide, as well as other compounds, into glucose through a series of reactions. In the analysis of data from *R. chrysanthum* in the Calvin–Benson cycle, we found up-regulation of acetylation modification levels at the sites of metabolic enzymes Rubisco, PGK, GAPA, GAPDH, ALDO, FBP, SBPase, PRK, and TKT. In CAM, the PPC and maeB enzymes of *R. chrysanthum* were modified by acetylation under UV−B stress, and the level of modification was up-regulated. In the C4-dicarboxylic acid cycle, we also identified the acetylated protein. In this pathway, acetylation-modified MDH1, COT1, GGAT, maeB and malate dehydrogenase (E1.1.1.39) were recognized, with reduced levels of acetylation modification only at the malate dehydrogenase site (Figure 5A). We further analyzed this by quantitative proteomics and found that the only proteins that underwent acetylation modification were Rubisco and GAPDH enzymes, which showed significant differential expression of the proteins. Rubisco protein expression was up-regulated in *R. chrysanthum* after UV−B stress, which was positively correlated with the level of acetylation modification. GAPDH protein expression was down-regulated and inversely proportional to the acetylation modification level (Figure 5B, Supplementary Table S2). Taken together, these results suggest that proteins undergo acetylation modification and that the level of acetylation modification has an impact on the expression of proteins, which in turn affects the processes of photosynthesis and carbon fixation in plants in response to UV−B stress.

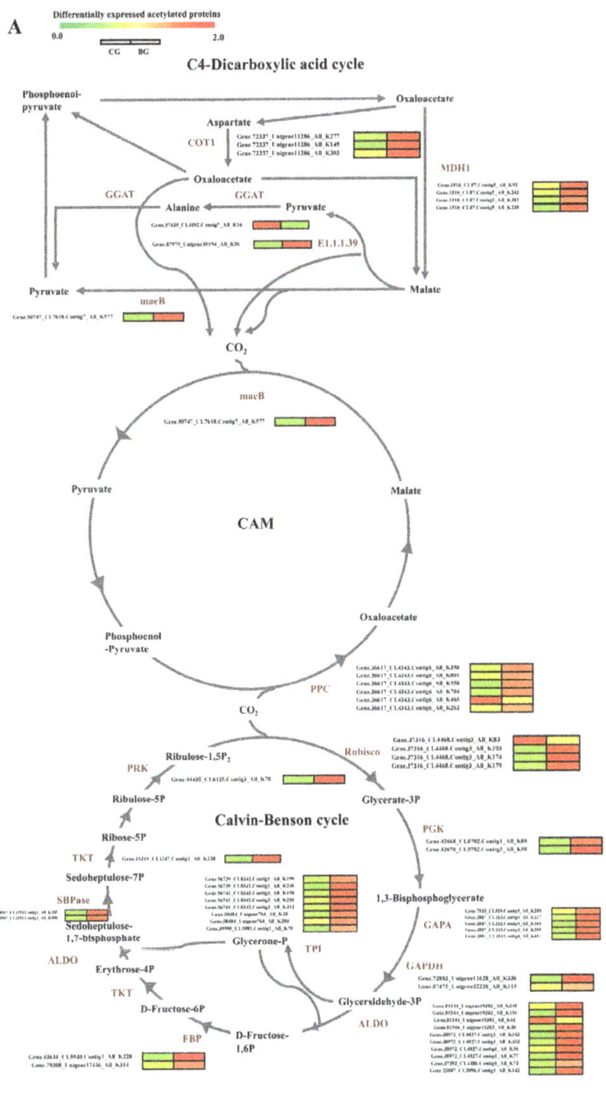

Figure 5. *Cont.*

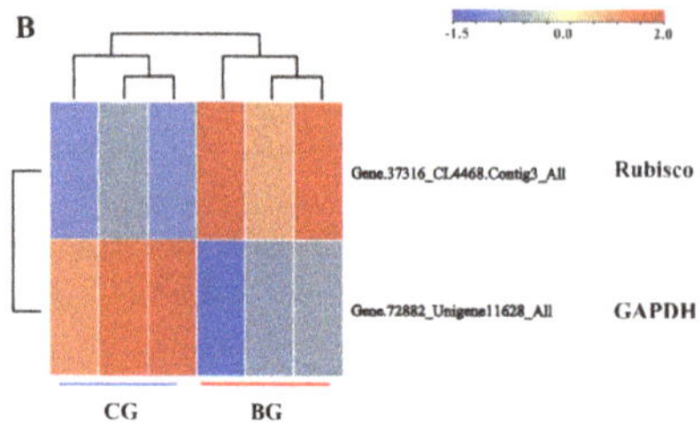

Figure 5. Working scheme of DAPs, sites and DEPs in carbon fixation in photosynthetic organisms in *R. chrysanthum*. (**A**) DAPs involved in carbon fixation in photosynthetic organisms. Rubisco: ribulose bisphosphate carboxylase/oxygenase; PGK: phosphoglycerate kinase; GAPDH: glyceraldehyde 3-phosphate dehydrogenase; GAPA: Glyceraldehyde-3-phosphate dehydrogenase B; ALDO: Fructose-bisphosphate aldolase 3; TPI: triosephosphate isomerase; FBP: fructose-1,6-bisphosphatase; TKT: Transketolase; SBPase: Sedoheputulose-1,7-bisphosphatase; PRK: phosphoribulokinase; PPC: Phosphoenolpyruvate carboxylase; maeB: NADP-dependent malic enzyme; MDH1: Malate dehydrogenase [NADP]; GGAT: Glutamate-glyoxylate aminotransferase 2; COT1: Aspartate aminotransferase; E1.1.1.39: malate dehydrogenase (decarboxylating). (**B**) Clustering heat map of DEPs in carbon fixation in photosynthetic organisms in the *R. chrysanthum* under UV−B stress.

3.6. Protein Structure Analysis of Rubisco and GAPDH Enzymes Undergoing Acetylation Modification in R. chrysanthum under UV−B Stress

The level of protein acetylation in carbon fixation and photosynthesis of *R. chrysanthum* was shown to be higher throughout the entire period of UV−B stress (Supplementary Table S10). By acetylation modification proteomic analysis, we found that acetylation modification occurred mainly for the Rubisco enzyme and the GAPDH enzyme during carbon fixation after UV−B radiation, and significant differences in expression occurred after modification. Therefore, we concluded that these two enzymes play important roles in the carbon fixation process of *R. chrysanthum* under UV−B stress. In order to understand the features and functions of Rubisco and GAPDH proteins at the molecular level, we described the hydrophobic clusters and salt bridges of the proteins. Firstly, we constructed three-dimensional models of the proteins. We used SWISS-MODEL to build statistically acceptable homology models. The acetylation sites were labelled in the 3D structure based on the information on the acetylation sites (Figure 6A). In Rubisco proteins, only the acetylation level of the Lys83 site was down-regulated, and the Lys103, Lys174, and Lys174 sites were up-regulated. In GAPDH proteins, both Lys115 and Lys336 sites were upregulated (Figure 6A(a,b)). The hydrophobic structure was analyzed, and the total area of the largest hydrophobic cluster of the GAPDH protein was 3798.7^2, with a single cluster containing 22 residues. The area of each residue is 42.7^2 and there are 89 interactions between residues (Figure 6B(a), Supplementary Table S11). In Rubisco, the total area of the largest hydrophobic cluster of the Rubisco protein was 1657.1^2, with a single cluster containing 16 residues. The area of each residue is 41.4^2 and there are 40 interactions between residues (Figure 6B(b), Supplementary Table S10). By calculating the charge separation parameters, we can obtain the fraction of charged residues (FCR) of GAPDH as 0.24 and Kappa value (K) as 0.16. The FCR of Rubisco is 0.24 and K is 0.14 (Figure 6C(a,b), Supplementary Table S11).

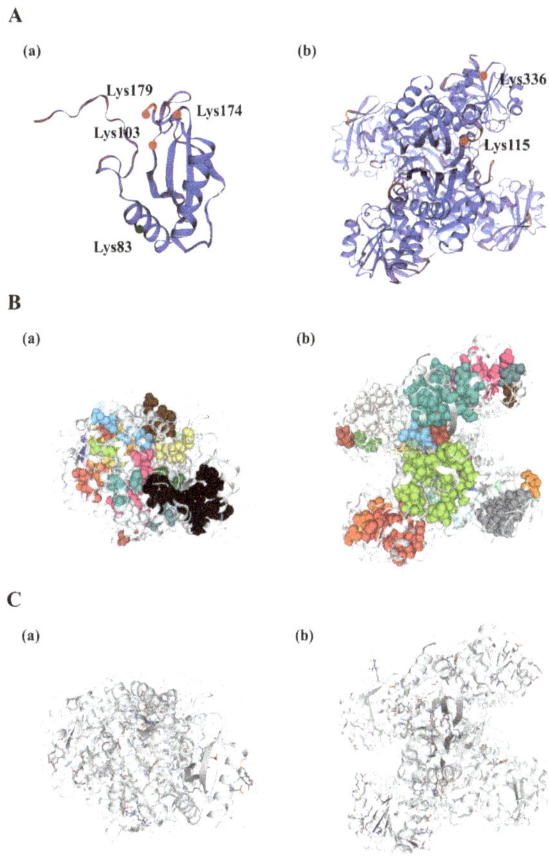

Figure 6. Three-dimensional structure, hydrophobic clusters and salt bridges of acetylation-modifying enzymes for carbon fixation in *R. chrysanthum* under UV−B stress. (**A**) Three-dimensional structure of acetylation-modifying enzymes for carbon fixation. Red is up-regulation of site acetylation modification levels, green is down-regulation of site acetylation modification levels (**B**) Hydrophobic clusters of acetylation-modifying enzymes for carbon fixation; (**C**) Salt bridges of acetylation-modifying enzymes for carbon fixation. (**a**) is Rubisco, (**b**) is GAPDH.

4. Discussion

Lysine acetylation of proteins plays an important regulatory role in plants with different biological functions. So far, acetylated proteins have been reported only in some plants, including *Arabidopsis* [27], wheat [28], poplar [21], rice [29], and pepper [30]. *R. chrysanthum* is a valuable species resource that can withstand UV−B exposure due to its unique growth environment [1]. As a result, we investigated the lysine acetylation proteomics of *R. chrysanthum* under UV−B stress, using *R. chrysanthum* as experimental material. In total, 685 acetylated proteins and 945 modification sites were found using acetylation proteomics analysis. In the natural world, photosynthesis is how plants create the organic materials needed for their own growth and development [30]. However, important enzymes involved in plant carbon fixation are somewhat impacted by abiotic stress, which has an impact on the quantity of CO_2 fixation [31]. In order to identify the mechanism underlying *R. chrysanthum*'s resistance to UV−B light, these acetylated proteins were categorized based on their GO function and examined with a focus on the acetylation of proteins during carbon fixation.

As the physiological process that provides plants with their energy, photosynthesis is crucial to the functioning of plants. Under UV−B radiation, many plants usually have obvious stress responses, resulting in damage such as inactivation of PSII reaction centers or enhanced photoinhibition [32,33]. Light plays a crucial role in photosynthesis, and the light reaction produces ATP and NADPH to drive the dark reaction. The photosynthetic system of plants utilizes light primarily through photosynthetic pigments in the membranes of the cysts [30]. Photosystem II decomposes H_2O, produces protons and releases O_2 through absorbed light energy. But, UV−B destroys PSII but has no direct effect on PSI. Nevertheless, PSI and PSII drive photosynthetic electron transport sequentially. A study demonstrates that UV−B improves chilling-light-induced PSI photoinhibition and promotes CO_2 assimilation recovery in cucumber (*Cucumis sativus* L.) [34]. Chlorophyll fluorescence parameters are useful indicators of photosynthesis. The analysis of various fluorescence parameters can reveal a number of regulatory processes within photosynthetic mechanisms. Fluorescence parameters such as Fv/Fm and Y(II) are usually reduced in plants under stress conditions [35,36]. In this study, the results of chlorophyll fluorescence parameters of *R. chrysanthum* under UV−B radiation treatment showed that Y(II) was significantly reduced compared to the control (Figure 2). This is consistent with the research results of UV−B radiation on water moss (*Fontinalis antipyretica* Hedw.) [37], indicating that under UV−B radiation stress, the PSII reaction center of plants is significantly damaged, leading to the occurrence of photoinhibition, which in turn affects photosynthetic activity. In addition, to explore the photodamage condition of *R. chrysanthum* under UV−B stress, we analyzed more chlorophyll fluorescence parameters such as NPQ, qL, ETR. Taken together, these results indicate that UV−B can cause some damage to the plant's photosynthetic system, but the plant itself responds to this damage through a series of reactions. To further investigate the molecular mechanisms of plant resistance to UV−B stress, we performed proteomic and acetylation modification analyses.

These amino acids interact and function collectively rather than separately. Since lysine's acetylation modification may have an impact on nearby amino acids, which may also play certain physiological roles in conjunction with lysine, the sequences of lysines undergoing acetylation modification and their neighboring amino acids will be shown in this study (Figure 3). More research is necessary to determine how nearby amino acids are affected by lysine acetylation modification and how lysine acetylation modification functions in conjunction with them. Hydrophobic structures and salt bridges are crucial for preserving the steric structure of proteins (Figure 6). Proteins' three-dimensional structures determine their functions, so in order to better understand how these proteins behave under UV−B stress, their salt-bridge and hydrophobic structures were further characterized after the target's protein structure was established.

Proteins ensure normal plant metabolism, and some proteins (e.g., Rubisco) play an important role in various physiological metabolic processes as biocatalytic enzymes, whose maximum absorption wavelengths fall within the wavelength range of UV−B radiation. Acetylated modification can produce a range of biochemical and physiological reactions [38]. In chloroplasts, in the Calvin cycle, GAPDH plays a crucial role by specifically binding with NADPH [39]. By shaping an invertible multi-enzyme complex with CP12, GAPDH and PRK commonly regulate and control the Calvin cycle [40]. Rubisco, a protein in the chloroplast stroma that catalyzes the carboxylation of ribulose-1,5-bisphosphate for CO_2 fixation, is an important enzyme in the Calvin cycle's first step. In plants, carbon enters the Calvin cycle in the form of carbon dioxide, so enzymes play a crucial role in the Calvin cycle process. In the study, in the Calvin cycle process, the key enzymes were acetylated in *R. chrysanthum* (Figure 5A). This is consistent with previous findings that acetylation modifications of enzymes involved in the Calvin cycle process in the face of abiotic stresses were similarly found in pepper and wheat [28,30]. Interestingly, ribulose-diphosphate carboxylase (Rubisco) contains 4 acetylation sites in *R. chrysanthum* (Supplemental Table S9).

In fact, the carbon fixation of plants is closely related to environmental factors. High UV−B radiation reduced carbon fixation rates in plants [41]. Our results suggest that under UV−B stress, plant photosystems are impaired and photosynthesis is altered, and thus the carbon assimilation process is inevitably affected to some extent. However, the level of acetylation modification affects the expression of plant proteins, which in turn adjusts the alterations that occur in the carbon assimilation process.

In our study, photosynthesis of *R. chrysanthum* was inhibited by UV−B radiation with a significant increase in qL and a significant decrease in NPQ by biochemical analyses, indicating that the photosystem was damaged. The acetylation proteome reveals that the photosynthetic proteins undergo acetylation modification, while the level of key acetylation modification enzymes in the pathway of the carbon cycle shows an upward trend (Figure 7).

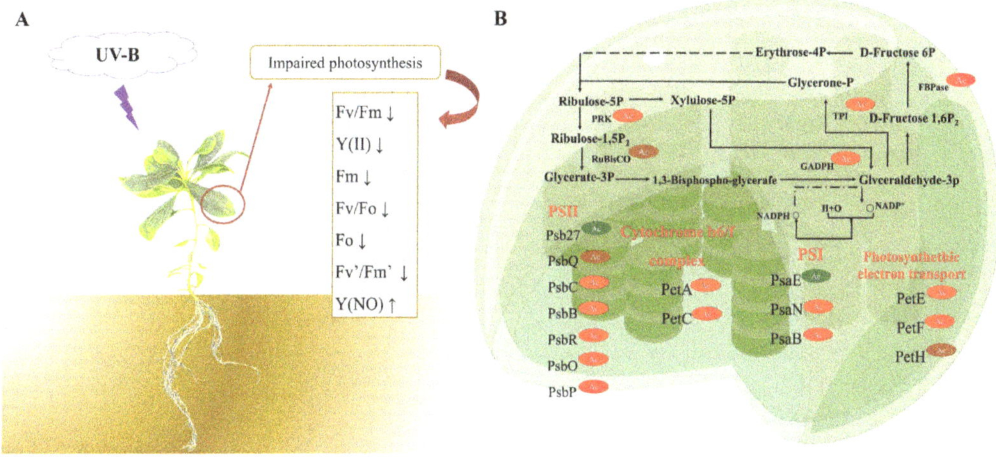

Figure 7. Mechanism diagram of *R. chrysanthum*'s photosynthesis resistance to UV−B stress. (**A**) Altered photosynthesis in *R. chrysanthum* under UV−B stress. (**B**) Schematic diagram of photosynthesis and carbon fixation acetylation of *R. chrysanthum* under UV−B stress. Upward arrows indicate elevated levels, and downward arrows indicate decreased levels.

5. Conclusions

In conclusion, we identified 685 proteins and discovered 945 acetylation modifications at their sites using acetylation proteomics data analysis. Further GO analysis revealed that these proteins are involved in a diverse set of biological processes. According to functional enrichment analysis, acetylated proteins play a role in *R. chrysanthum* photosynthesis and carbon fixation. It was discovered that acetylation modification of key Calvin cycle enzymes (Rubisco, GAPDH) modulates protein expression, causing Rubisco and GAPDH proteins to be expressed in significantly different ways, affecting *R. chrysanthum*'s carbon fixation capacity. As a result of the acetylation modification, Rubisco and GAPDH are significantly differentially expressed, affecting carbon fixation capacity and making the plant resistant to UV−B stress. This study provides valuable information about the plant's response to UV−B and lays the groundwork for future protein post-translational modification research.

Supplementary Materials: The following supporting information can be downloaded at: https://www.mdpi.com/article/10.3390/biom14060732/s1, Figure S1: Identification of lysine acetylation proteome in *Rhododendron chrysanthum*. (A) 4D-label free experimental flowchart. (B) Mass error distribution of whole identified polypeptides. (C) The length distribution of identified peptide; Figure S2. KEGG pathway enrichment analysis of the acetylated proteins in carbon metabolism. The acetylated proteins are in blue; Figure S3. KEGG pathway enrichment analysis of the acetylated proteins in photosynthesis. The acetylated proteins are in blue; Figure S4. KEGG pathway enrichment analysis of the acetylated proteins in carbon fixation in photosynthetic organisms. The acetylated proteins are in blue; Table S1. The identified acetylated sites in *Rhododendron chrysanthum*; Table S2. The identified differentially expressed proteins in *Rhododendron chrysanthum*; Table S3. Peptide motif of *R. chrysanthum*; Table S4. Frequency of the different types of amino acids around the acetylated lysine; Table S5. Go annotation detail of the acetylated proteins in *Rhododendron chrysanthum*; Table S6. Protein GO enrichment in biological process; Table S7. Protein GO enrichment in molecular function; Table S8. Protein GO enrichment in cellular component; Table S9. Protein pathway enrichment; Table S10. Differential acetylated proteins *Rhododendron chrysanthum* under UV−B stress; Table S11. Structural information on hydrophobic clusters and salt bridges in *R. chrysanthum* under UV−B stress.

Author Contributions: Writing—original draft preparation, M.L. and F.G.; writing—review and editing, H.X. and X.Z.; methodology, M.L. and W.Y.; software, K.C. and M.L.; visualization, M.L. and F.G.; supervision, H.X. and X.Z.; project administration, H.X. and X.Z. All authors have read and agreed to the published version of the manuscript.

Funding: This work was supported by the Jilin Provincial Science and Technology Development Program project (Grant numbers: YDZJ202301ZYTS517).

Institutional Review Board Statement: Not applicable.

Informed Consent Statement: Not applicable.

Data Availability Statement: The datasets generated during and/or analysed during the current study are available from the corresponding author on reasonable request.

Acknowledgments: We are grateful to Jingjie PTM Biolab (Hangzhou, China) Co., Ltd. for pro viding mass spectrometry support.

Conflicts of Interest: The authors declare no conflicts of interest.

References

1. Zhou, X.; Chen, S.; Wu, H.; Yang, Y.; Xu, H. Biochemical and proteomics analyses of antioxidant enzymes reveal the potential stress tolerance in *Rhododendron chrysanthum* Pall. *Biol. Direct.* **2017**, *12*, 10. [CrossRef] [PubMed]
2. Lyu, J.; Wang, C.; Liang, D.Y.; Liu, L.; Pandey, L.K.; Xu, H.W.; Zhou, X.F. Sensitivity of wild and domesticated Rhododendron chrysanthum to different light regime (UVA, UVB, and PAR). *Photosynthetica* **2019**, *57*, 841–849. [CrossRef]
3. Bassman, J.H. Ecosystem consequences of enhanced solar ultraviolet radiation: Secondary plant metabolites as mediators of multiple trophic interactions in terrestrial plant communities. *Photochem. Photobiol.* **2004**, *79*, 382–398. [CrossRef]
4. Knox, P.P.; Lukashev, E.P.; Gorokhov, V.V.; Grishanova, N.P.; Paschenko, V.Z. Hybrid complexes of photosynthetic reaction centers and quantum dots in various matrices: Resistance to UV irradiation and heating. *Photosynth. Res.* **2019**, *139*, 295–305. [CrossRef]
5. Widel, M.; Krzywon, A.; Gajda, K.; Skonieczna, M.; Rzeszowska-Wolny, J. Induction of bystander effects by UVA, UVB, and UVC radiation in human fibroblasts and the implication of reactive oxygen species. *Free Radic. Biol. Med.* **2014**, *68*, 278–287. [CrossRef]
6. Hideg, E.; Jansen, M.A.; Strid, A. UV−B exposure, ROS, and stress: Inseparable companions or loosely linked associates? *Trends Plant Sci.* **2013**, *18*, 107–115. [CrossRef] [PubMed]
7. Chen, Y.-E.; Su, Y.-Q.; Zhang, C.-M.; Ma, J.; Mao, H.-T.; Yang, Z.-H.; Yuan, M.; Zhang, Z.-W.; Yuan, S.; Zhang, H.-Y. Comparison of Photosynthetic Characteristics and Antioxidant Systems in Different Wheat Strains. *J. Plant Growth Regul.* **2018**, *37*, 347–359. [CrossRef]
8. Barber, J. Photosystem two. *Biochim. Biophys. Acta Bioenerg.* **1998**, *1365*, 269–277. [CrossRef]
9. Yoon, H.I.; Kim, D.; Son, J.E. Spatial and Temporal Bioactive Compound Contents and Chlorophyll Fluorescence of Kale (*Brassica oleracea* L.) Under UV−B Exposure Near Harvest Time in Controlled Environments. *Photochem. Photobiol.* **2020**, *96*, 845–852. [CrossRef]
10. Piccini, C.; Cai, G.; Dias, M.C.; Romi, M.; Longo, R.; Cantini, C. UV−B Radiation Affects Photosynthesis-Related Processes of Two Italian *Olea europaea* (L.) Varieties Differently. *Plants* **2020**, *9*, 1712. [CrossRef]
11. Liu, M.; Sun, Q.; Cao, K.; Xu, H.; Zhou, X. Acetylated Proteomics of UV−B Stress-Responsive in Photosystem II of *Rhododendron chrysanthum*. *Cells* **2023**, *12*, 478. [CrossRef] [PubMed]

2. UniProt Consortium, T. UniProt: The universal protein knowledgebase. *Nucleic Acids Res.* **2018**, *46*, 2699. [CrossRef] [PubMed]
3. Choudhary, C.; Weinert, B.T.; Nishida, Y.; Verdin, E.; Mann, M. The growing landscape of lysine acetylation links metabolism and cell signalling. *Nat. Rev. Mol. Cell Biol.* **2014**, *15*, 536–550. [CrossRef] [PubMed]
4. Narita, T.; Weinert, B.T.; Choudhary, C. Functions and mechanisms of non-histone protein acetylation. *Nat. Rev. Mol. Cell Biol.* **2019**, *20*, 156–174. [CrossRef] [PubMed]
5. Colville, A.; Alhattab, R.; Hu, M.; Labbé, H.; Xing, T.; Miki, B. Role of HD2 genes in seed germination and early seedling growth in Arabidopsis. *Plant Cell Rep.* **2011**, *30*, 1969–1979. [CrossRef]
6. Kim, W.; Latrasse, D.; Servet, C., Zhou, D X. Arabidopsis histone deacetylase HDA9 regulates flowering time through repression of AGL19. *Biochem. Biophys. Res. Commun.* **2013**, *432*, 394–398. [CrossRef] [PubMed]
7. Luo, M.; Wang, Y.Y.; Liu, X.; Yang, S.; Lu, Q.; Cui, Y.; Wu, K. HD2C interacts with HDA6 and is involved in ABA and salt stress response in Arabidopsis. *J. Exp. Bot.* **2012**, *63*, 3297–3306. [CrossRef] [PubMed]
8. Park, H.J.; Baek, D.; Cha, J.Y.; Liao, X.; Kang, S.H.; McClung, C.R.; Lee, S.Y.; Yun, D.J.; Kim, W.Y. HOS15 Interacts with the Histone Deacetylase HDA9 and the Evening Complex to Epigenetically Regulate the Floral Activator GIGANTEA. *Plant Cell* **2019**, *31*, 37–51. [CrossRef] [PubMed]
9. van Zanten, M.; Zöll, C.; Wang, Z.; Philipp, C.; Carles, A.; Li, Y.; Kornet, N.G.; Liu, Y.; Soppe, W.J. HISTONE DEACETYLASE 9 represses seedling traits in Arabidopsis thaliana dry seeds. *Plant J.* **2014**, *80*, 475–488. [CrossRef]
0. Yuan, L.; Chen, X.; Chen, H.; Wu, K.; Huang, S. Histone deacetylases HDA6 and HDA9 coordinately regulate valve cell elongation through affecting auxin signaling in Arabidopsis. *Biochem. Biophys. Res. Commun.* **2019**, *508*, 695–700. [CrossRef]
1. Liao, X.; Li, Y.; Hu, Z.; Lin, Y.; Zheng, B.; Ding, J. Poplar acetylome profiling reveals lysine acetylation dynamics in seasonal bud dormancy release. *Plant Cell Environ.* **2021**, *44*, 1830–1845. [CrossRef] [PubMed]
2. Zhu, G.R.; Yan, X.; Zhu, D.; Deng, X.; Wu, J.S.; Xia, J.; Yan, Y.M. Lysine acetylproteome profiling under water deficit reveals key acetylated proteins involved in wheat grain development and starch biosynthesis. *J. Proteom.* **2018**, *185*, 8–24. [CrossRef] [PubMed]
3. Gong, X.; Huang, Y.; Liang, Y.; Yuan, Y.; Liu, Y.; Han, T.; Li, S.; Gao, H.; Lv, B.; Huang, X.; et al. OsHYPK-mediated protein N-terminal acetylation coordinates plant development and abiotic stress responses in rice. *Mol. Plant* **2022**, *15*, 740–754. [CrossRef]
4. Du, Q.; Qu, Z.; Wang, L.; Jiang, J.; Fu, X.; Fang, Y.; Li, X.; Xie, X. Histone deacetylase SbHDT701 in Sorghum bicolor reveals functions in response to stress factors by enhancing acetylation. *Pestic. Biochem. Physiol.* **2021**, *178*, 104908. [CrossRef] [PubMed]
5. Wu, X.; Oh, M.H.; Schwarz, E.M.; Larue, C.T.; Sivaguru, M.; Imai, B.S.; Yau, P.M.; Ort, D.R.; Huber, S.C. Lysine acetylation is a widespread protein modification for diverse proteins in Arabidopsis. *Plant Physiol.* **2011**, *155*, 1769–1778. [CrossRef] [PubMed]
6. Jiang, J.; Gai, Z.; Wang, Y.; Fan, K.; Sun, L.; Wang, H.; Ding, Z. Comprehensive proteome analyses of lysine acetylation in tea leaves by sensing nitrogen nutrition. *BMC Genom.* **2018**, *19*, 840. [CrossRef]
7. König, A.C.; Hartl, M.; Boersema, P.J.; Mann, M.; Finkemeier, I. The mitochondrial lysine acetylome of *Arabidopsis*. *Mitochondrion* **2014**, *19*, 252–260. [CrossRef] [PubMed]
8. Zhang, Y.; Song, L.; Liang, W.; Mu, P.; Wang, S.; Lin, Q. Comprehensive profiling of lysine acetylproteome analysis reveals diverse functions of lysine acetylation in common wheat. *Sci. Rep.* **2016**, *6*, 21069. [CrossRef]
9. Nallamilli, B.R.; Edelmann, M.J.; Zhong, X.; Tan, F.; Mujahid, H.; Zhang, J.; Nanduri, B.; Peng, Z. Global analysis of lysine acetylation suggests the involvement of protein acetylation in diverse biological processes in rice (*Oryza sativa*). *PLoS ONE* **2014**, *9*, e89283. [CrossRef]
0. Liu, Z.; Song, J.; Miao, W.; Yang, B.; Zhang, Z.; Chen, W.; Tan, F.; Suo, H.; Dai, X.; Zou, X.; et al. Comprehensive Proteome and Lysine Acetylome Analysis Reveals the Widespread Involvement of Acetylation in Cold Resistance of Pepper (*Capsicum annuum* L.). *Front. Plant Sci.* **2021**, *12*, 730489. [CrossRef]
1. Savitch, L.V.; Leonardos, E.D.; Krol, M.; Jansson, S.; Öquist, G. Two different strategies for light utilization in photosynthesis in relation to growth and cold acclimation. *Plant Cell Environ.* **2002**, *25*, 761–771. [CrossRef]
2. Wu, J.-b.; Guan, D.-x.; Yuan, F.-h.; Zhang, X.-j. Research advances on the biological effects of elevated ultraviolet-B radiation on terrestrial plants. *J. For. Res.* **2009**, *20*, 383–390. [CrossRef]
3. Lidon, F.J.C. Micronutrients' accumulation in rice after supplemental UV−B irradiation. *J. Plant Interact.* **2012**, *7*, 19–28. [CrossRef]
4. Zhang, Z.S.; Jin, L.Q.; Li, Y.T.; Tikkanen, M.; Li, Q.M.; Ai, X.Z.; Gao, H.Y. Ultraviolet−B Radiation (UV−B) Relieves Chilling-Light-Induced PSI Photoinhibition And Accelerates The Recovery Of CO_2 Assimilation In Cucumber (*Cucumis sativus* L.) Leaves. *Sci. Rep.* **2016**, *6*, 34455. [CrossRef] [PubMed]
5. Liang, Y.; Chen, H.; Tang, M.J.; Yang, P.F.; Shen, S.H. Responses of Jatropha curcas seedlings to cold stress: Photosynthesis-related proteins and chlorophyll fluorescence characteristics. *Physiol. Plant.* **2007**, *131*, 508–517. [CrossRef]
6. Belkhodja, R.; Morales, F.; Abadia, A.; Gomez-Aparisi, J.; Abadia, J. Chlorophyll Fluorescence as a Possible Tool for Salinity Tolerance Screening in Barley (*Hordeum vulgare* L.). *Plant Physiol.* **1994**, *104*, 667–673. [CrossRef]
7. Martínez-Abaigar, J.; Núñez-Olivera, E.; Beaucourt, N.; García-Álvaro, M.A.; Tomás, R.; Arróniz, M. Different physiological responses of two aquatic bryophytes to enhanced ultraviolet−B radiation. *J. Bryol.* **2003**, *25*, 17–30. [CrossRef]
8. Ifuku, K.; Ido, K.; Sato, F. Molecular functions of PsbP and PsbQ proteins in the photosystem II supercomplex. *J. Photochem. Photobiol. B Biol.* **2011**, *104*, 158–164. [CrossRef] [PubMed]

39. McFarlane, C.R.; Shah, N.R.; Kabasakal, B.V.; Echeverria, B.; Cotton, C.A.R.; Bubeck, D.; Murray, J.W. Structural basis o light-induced redox regulation in the Calvin-Benson cycle in cyanobacteria. *Proc. Natl. Acad. Sci. USA* **2019**, *116*, 20984–2099([CrossRef]
40. Howard, T.P.; Lloyd, J.C.; Raines, C.A. Inter-species variation in the oligomeric states of the higher plant Calvin cycle enzyme glyceraldehyde-3-phosphate dehydrogenase and phosphoribulokinase. *J. Exp. Bot.* **2011**, *62*, 3799–3805. [CrossRef]
41. Martínez-Lüscher, J.; Morales, F.; Sánchez-Díaz, M.; Delrot, S.; Aguirreolea, J.; Gomès, E.; Pascual, I. Climate change condition (elevated CO_2 and temperature) and UV−B radiation affect grapevine (*Vitis vinifera* cv. Tempranillo) leaf carbon assimilatior altering fruit ripening rates. *Plant Sci.* **2015**, *236*, 168–176. [CrossRef] [PubMed]

molecules

MDPI

Article

Halogenated Boroxine $K_2[B_3O_3F_4OH]$ Modulates Metabolic Phenotype and Autophagy in Human Bladder Carcinoma 5637 Cell Line

Nikolina Elez-Burnjaković [1], Lejla Pojskić [2], Anja Haverić [2], Naida Lojo-Kadrić [2], Maida Hadžić Omanović [2], Ajla Smajlović [2], Svetoslav Kalaydjiev [3], Milka Maksimović [4], Bojan Joksimović [1,*] and Sanin Haverić [2]

1 Faculty of Medicine Foča, University of East Sarajevo, Studentska 5, 73 300 Foča, Bosnia and Herzegovina; nikolinaa85@hotmail.com

2 Institute for Genetic Engineering and Biotechnology, University of Sarajevo, Zmaja od Bosne 8, 71 000 Sarajevo, Bosnia and Herzegovina; lejla.pojskic@ingeb.unsa.ba (L.P.); anja.haveric@ingeb.unsa.ba (A.H.); naida.lojo@ingeb.unsa.ba (N.L.-K.); maida.hadzic@ingeb.unsa.ba (M.H.O.); ajla-rizvanovic@hotmail.com (A.S.); sanin.haveric@ingeb.unsa.ba (S.H.)

3 Agilent Technologies LDA UK Ltd., 5500 Lakeside, Stockport SK8 3GR, UK; svetoslav.kalaydjiev@agilent.com

4 Faculty of Science, University of Sarajevo, Zmaja od Bosne 33, 71 000 Sarajevo, Bosnia and Herzegovina; mmaksimo@yahoo.co.uk

* Correspondence: bojannjoksimovic@gmail.com; Tel.: +387-65-373-508

Abstract: Halogenated boroxine $K_2[B_3O_3F_4OH]$ (HB), an inorganic derivative of cyclic anhydride of boronic acid, is patented as a boron-containing compound with potential for the treatment of both benign and malignant skin changes. HB has effectively inhibited the growth of several carcinoma cell lines. Because of the growing interest in autophagy induction as a therapeutic approach in bladder carcinoma (BC), we aimed to assess the effects of HB on metabolic phenotype and autophagy levels in 5637 human bladder carcinoma cells (BC). Cytotoxicity was evaluated using the alamar blue assay, and the degree of autophagy was determined microscopically. Mitochondrial respiration and glycolysis were measured simultaneously. The relative expression of autophagy-related genes BECN1, P62, BCL-2, and DRAM1 was determined by real-time PCR. HB affected cell growth, while starvation significantly increased the level of autophagy in the positive control compared to the basal level of autophagy in the untreated negative control. In HB-treated cultures, the degree of autophagy was higher compared to the basal level, and metabolic phenotypes were altered; both glycolysis and oxidative phosphorylation (OXPHOS) were decreased by HB at 0.2 and 0.4 mg/mL. Gene expression was deregulated towards autophagy induction and expansion. In conclusion, HB disrupted the bioenergetic metabolism and reduced the intracellular survival potential of BC cells. Further molecular studies are needed to confirm these findings and investigate their applicative potential.

Keywords: cytotoxicity; autophagy; BC cells; gene expression; metabolic phenotype

1. Introduction

Autophagy is a dynamic process, conserved in all eukaryotes. It is responsible for the degradation of cytoplasmic content and the cell's response to stress, embryonic development, pathogens, and aging. The process is also involved in the pathology of human diseases, such as cancer, diabetes, cardiomyopathy, and neurodegenerative diseases. Among the three types of autophagy, chaperone-mediated autophagy, microautophagy, and macroautophagy, the term autophagy mainly refers to the last one (hereinafter autophagy) [1]. In normal tissues, autophagy is at a low level, and it increases in metabolic stress and in the absence of oxygen and nutrients in tumors. Many cancer types, such as bladder, lung, and colorectal [2,3], as well as melanoma [4], show increased basal autophagy and are associated with aggressiveness and a poor prognosis [5]. Metastatic tumors are highly "dependent" on autophagy, which provides the metabolic energy needed for growth. However, if extensively or inappropriately activated, autophagy can promote apoptosis (type I) or function as an alternative cell death pathway, which is called autophagic cell death (type II). Autophagy can either promote the death of cancer cells or serve as a survival mechanism against apoptotic or necrotic death induced by various anticancer treatments [6]. Given the contradictory role of autophagy during tumor initiation and progression, the therapeutic use of autophagy depends on the context and must be approached individually.

Bladder carcinoma (BC) is one of the most common malignancies, with an increasing incidence worldwide [7]. The majority of new cases are non-muscle-invasive tumors that are commonly treated by radical resection, radiotherapy, and chemotherapy. However, more than half of the superficial bladder cancers will recur, with 10–40% resulting in metastatic outcomes [8].

The non-muscle-invasive bladder carcinoma cell line 5637 has been used as an in vitro model for high-risk superficial bladder tumors with specific molecular characteristics (Bcl-2 and Bcl-X positive expression and *P53* tumor suppressor mutation) [9]. Loss of p53 tumor suppressor function, due to the P53R280T mutation [10,11], promotes the Warburg effect [11]. P53 is known to reduce the glycolysis rate by increasing the activity of fructose-2,6-bisphosphatase. This mechanism is also involved in the regulatory pathways of apoptosis [12]. This mechanism seems to increase oxidative phosphorylation. Earlier studies have confirmed increased glucose and glutamine uptake, along with pyruvate and alanine production, in 5637 cell lines in vitro [13]. Therefore, 5637 cells show inhibited apoptosis, increased cell proliferation, and chemotherapy resistance. Silencing of the *P53* mutant decreases cell viability [14]. High levels of basal autophagy in 5637 cells are confirmed by LC3 immunofluorescence staining in a nutrient-rich environment. The inhibition of basal autophagy results in apoptotic cell loss [3]. The efficacy of chemotherapy is limited because of drug resistance, while autophagy may account for the failure of chemotherapy treatment in BC. Large proportions of patients with advanced BC are resistant to chemotherapy and are characterized by specific molecular signatures [15]. The cooperative inhibition of autophagy is shown to be a vital strategy against cisplatin resistance in human BC cells [15]. To reduce chemoresistance, efforts should focus on several areas. Multi-omics sequencing and analysis can identify reliable biomarkers for predicting chemosensitivity in bladder cancer and might pinpoint the patient groups that are likely to respond to cytotoxic treatments. Targeting dysregulated pathways in chemotherapy-resistant patients could reverse resistance. Additionally, addressing other tumor microenvironment components, such as immune cells and cancer-associated fibroblasts, is crucial. Precision drug delivery using nanotechnology and antibody–drug conjugates shows promise. These strategies aim to improve response rates and increase overall and progression-free survival [16]. Meanwhile, cisplatin-based adjuvant chemotherapy remains a valid option for muscle-invasive bladder cancer [17].

Therapeutic strategies for BC may cause unfavorable side effects and complications. Therefore, autophagy induction is a promising therapeutic approach [15]. The association of the Warburg effect in cancers with tumor-related autophagy may be relevant for the further development of experimental therapeutics and cancer prevention.

In this work, we investigated the potential of dipotassium trioxohydroxytetrafluorotriborate, $K_2[B_3O_3F_4OH]$, also known as halogenated boroxine (HB), to promote autophagy in BC cell lines. HB is a derivative of the cyclic anhydride of boronic acid [18], containing four fluorine atoms substituted in a six-membered ring. Patented as a prevention and therapeutic agent for various skin changes [19,20], HB has been submitted to numerous in vitro studies [21–25]. It acts as an inhibitor of horseradish peroxidase (HRP) [26], catalase [27], superoxide dismutase [28], and carbonic anhydrases [29]. In the GR-M cell line, HB significantly downregulates the relative expression of *IGF-1*, *hTERT*, and *BCL-2* genes [24,30], suggesting its antitumor activity. At a physiological concentration of Ca^{2+}, HB inhibits the growth of tumor cells without having such an effect on normal cells [31].

The anti-tumor activity of HB in vitro has also been proven against mammary carcinoma (4T1), melanoma (B16F10), and squamous cell carcinoma (SCCVII) [32]. HB significantly inhibits tumor growth in vivo, regardless of the means of application [32]. Temporary mild-to-moderate HB dose-dependent lesions were observed in the kidneys and livers of Wistar rats after single intraperitoneal doses of HB at 25, 35, and 45 mg/kg/body weight (bw) in comparison with the control and 10 mg/kg/bw treatments. Adverse effects were not detected in a single-dose administration lower than 35 mg/kg or in repeated, nine-day dosages with 10 mg/kg/bw [33]. The median lethal dose (LD50) of HB in Sprague Dawley and Wistar rats and BALB/c mice ranges from 63 to 75 mg/kg, meaning that HB shows moderate toxicity [34].

According to the previous findings on the toxic effects of HB, we aimed to test if HB may be used as an autophagy inducer in human BC cells and its potential to modify the Warburg effect.

2. Results

2.1. Cytotoxic Effects of HB on BC Cells

The results of the alamar blue assay revealed BC cell growth inhibition of 47.16% at a 0.05 mg/mL HB concentration. HB concentrations of 0.1 and 0.2 mg/mL reduced cell growth to 65.18% and 64.43%, respectively. The inhibition percentage at 0.4 mg/mL was 53.41%. The association between the HB concentration and the percentage of cell growth inhibition was estimated by a simple linear regression calculation (Figure 1).

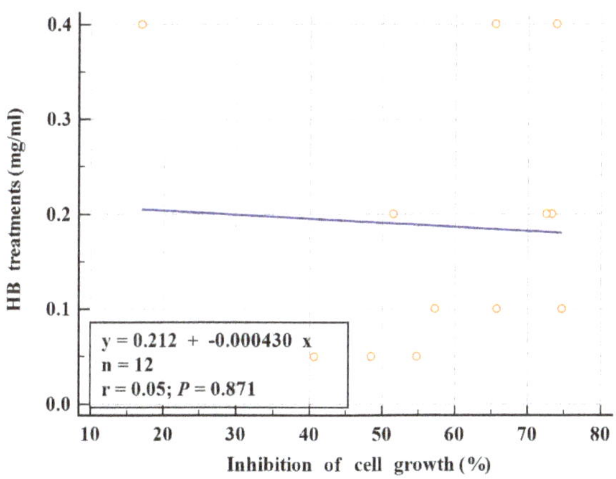

Figure 1. Association between BC cell growth and different HB treatments (simple linear regression; r = 0.05; p = 0.871). Cells were treated for 24 h after initial incubation. Cultures without test substances, set as positive control, and negative control cultures, without cells, were used to calculate the percentage of cell growth inhibition (https://tools.thermofisher.com, accessed on 15 August 2023; the alamarBlue® Assay, Thermo Fisher Scientific, Waltham, MA, USA).

2.2. Evaluation of Autophagy Induction by HB

Autophagosome induction was semi-quantified. Four categories were established according to the estimated number of autophagosome signals per cell. According to the predefined criteria, 200 cells were observed for autophagy markers (signals) for each treatment.

Compared to the negative control, which had a total of 22 cells with autophagy signals, the 0.05 mg/mL HB concentration resulted in 141 cells with autophagy signals, with most cells having 5–10 detected autophagosomes. At 0.1 and 0.2 mg/mL HB treatments, autophagy markers were detected in 166 and 148 cells, respectively. At a 0.1 mg/mL HB concentration, a higher frequency of cells with 5–10 signals was found compared to the 0.2 mg/mL HB treatment, where more cells with coupled signals were detected. At a 0.4 mg/mL HB concentration, 141 cells were positive for autophagy, with the highest frequency of cells with the coupled signals. Autophagy was successfully induced in the positive control, where 191 cells had different types of autophagy signals.

Different categories of autophagosome induction were observed in all cultures treated with HB in concentrations ranging from 0.05 to 0.4 mg/mL. Autophagy was successfully induced in the positive control, while the negative control showed low levels of basal autophagy according to the detection of autophagosome signals (Figures 2 and 3, Supplementary Materials).

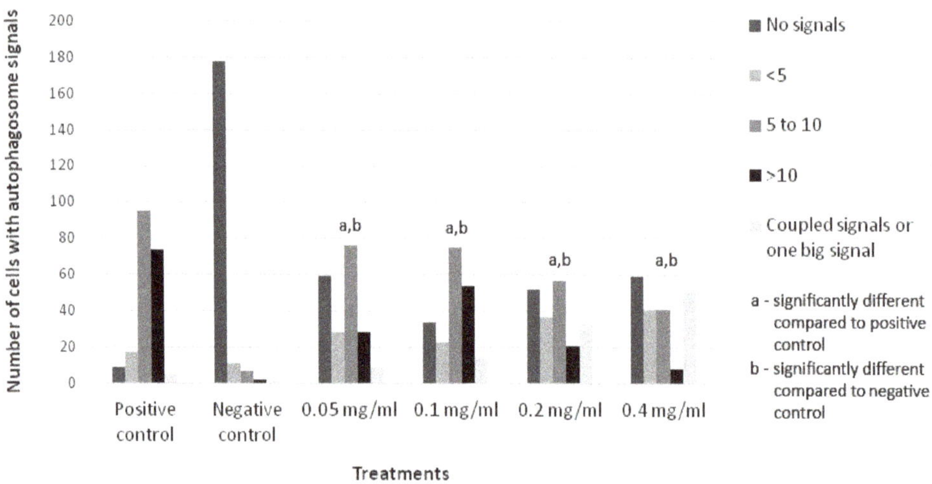

Figure 2. Distribution of frequency of detected autophagosome signals; positive (+) control—starvation-induced autophagy cells; negative (−) control—untreated cells. Four categories were established according to the estimated number of autophagosome signals per cells. According to predefined criteria, 200 cells were observed for autophagy markers (signals) for each treatment. Differences for all treatments and quantification of autophagosome signals were significant ($p < 0.05$).

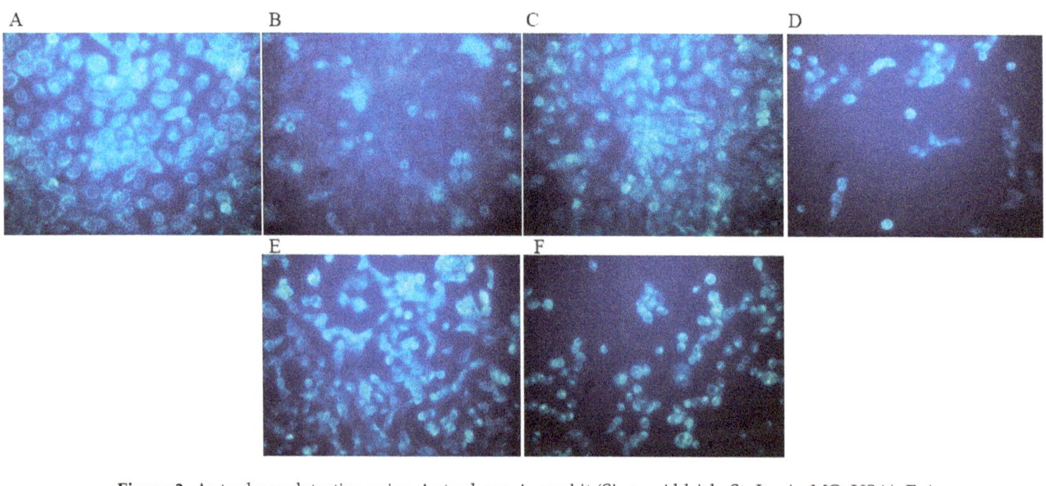

Figure 3. Autophagy detection using Autophagy Assay kit (Sigma-Aldrich, St. Louis, MO, USA). Epifluorescence detection was performed using a BX51 microscope with 400× magnification (Olympus, Tokyo, Japan); the degree of autophagy was photo-documented with a DP50 camera analyzed using View Finder Lite software version 1.0.143c (Olympus, Japan). (**A**)—positive control; (**B**)—negative control; (**C**)—0.05 mg/mL treatment; (**D**)—0.1 mg/mL treatment; (**E**)—0.2 mg/mL treatment; and (**F**)—0.4 mg/mL treatment.

Comparisons of detected autophagosome signals compared to the negative control were statistically significant in 0.05, 0.1, 0.2, and 0.4 mg/mL HB treatments ($p < 0.05$). Comparisons of autophagosome signal ratios, compared to the positive control, were statistically significant at all tested HB concentrations ($p < 0.05$).

2.3. Cellular Energy Phenotype and the Investigation of Metabolic Switching: OCR and ECAR Rates in Response to HB Treatment—Nts

The XF Cell Energy Phenotype test was performed for glycolysis and mitochondrial respiration determination. The oxygen consumption rate (OCR), as a measure of mitochondrial respiration, and the extracellular acidification rate (ECAR), as a measure of glycolysis, were determined in baseline and stressed phenotypes at 0.05, 0.1, 0.2, and 0.4 mg/mL HB concentrations in both positive and negative controls.

Significant differences between basal and stressed OCR values, tested by the Mann–Whitney test, were registered only for the positive control ($p = 0.0002$), with a decrease in the OCR values in a stressed state. ECAR values increased in a stressed state for all treatments. The differences were significant for all HB treatments except for 0.4 mg/mL ($p = 0.12$). There is a clear differentiation in OCR and ECAR values between the negative control, 0.5 and 0.1 mg/mL HB treatments on one side, the highest tested HB concentrations (0.2 and 0.4 mg/mL), and the positive control on the other side (Figure 4).

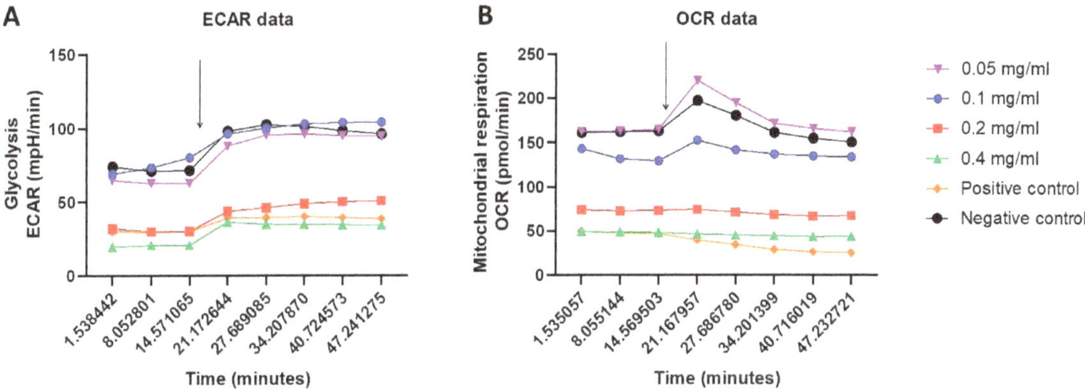

Figure 4. Phenotype Stress Test assessment of basal (three initial measurements) and stressed (five subsequent measurements) metabolic rates in BC cells seeded at 4×10^4/well: (**A**) glycolysis (ECAR) and (**B**) mitochondrial respiration (OCR) were measured on a Seahorse XF HS Mini Analyzer. ECAR—extracellular acidification rate; OCR—oxygen consumption rate. Results are presented as mean values of duplicate measurements.

When baseline and stressed OCR and ECAR values were independently compared, significantly higher values ($p < 0.05$; Kruskal–Wallis test) were found for the negative control, and lower HB concentrations (0.05 and 0.1 mg/mL) were found compared to the positive control and higher HB concentrations (0.2 and 0.4 mg/mL) (Figure 5).

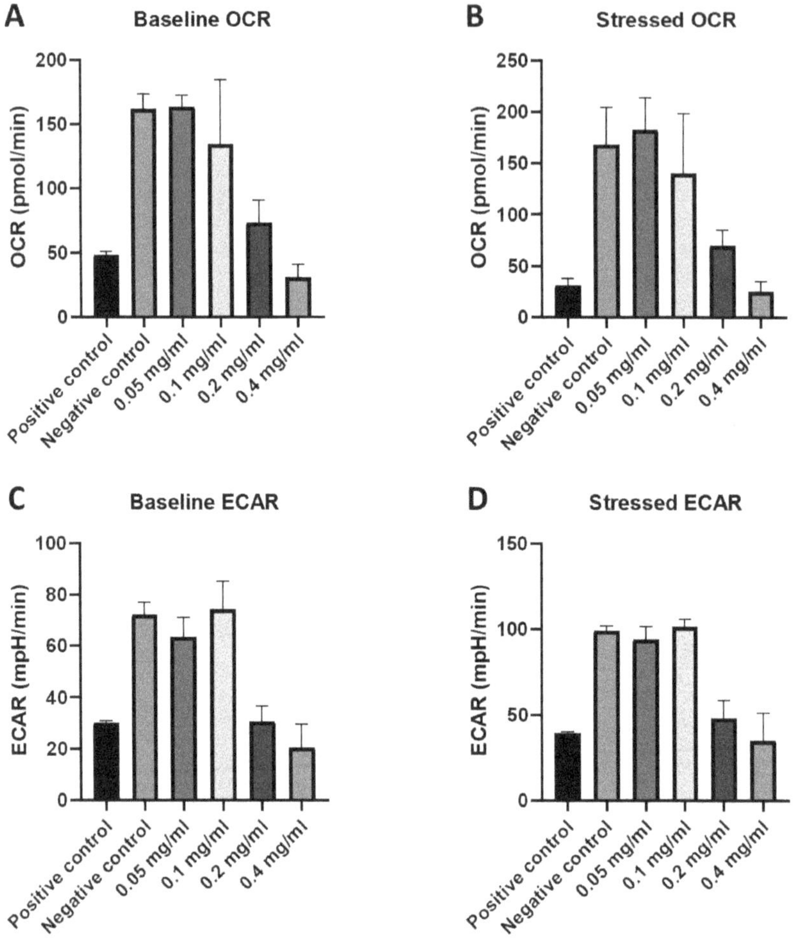

Figure 5. Phenotype Stress Test showing basal and stressed mitochondrial (**A**,**B**) and glycolytic activity (**C**,**D**) of the BC cells. Data are presented as means with SD analyzed by Kruskal–Wallis test, $p < 0.001$. ECAR—extracellular acidification rate; OCR—oxygen consumption rate.

2.4. Deregulation of Autophagy-Associate Genes by HB

The effects of different HB concentrations on cell cycle deregulation were evaluated using a relative gene expression assay for autophagy markers: *BECN-1*, *BCL-2*, *P62*, and *DRAM1*. The normalized gene expression ratios for all evaluated treatments were compared against corresponding values for negative (untreated cells) and positive controls (Figures 6 and 7).

When compared with the negative control, significant upregulation of *P62* was detected for all except for the 0.4 mg/mL HB concentration. *BECN1* was significantly upregulated in all HB treatments, with unsignificant changes at a 0.1 mg/mL HB concentration. Significant downregulation of *DRAM1* was observed in 0.1–0.4 mg/mL HB treatments compared to the positive control, while *BCL-2* was significantly downregulated at 0.1 and 0.4 mg/mL HB treatments compared to the positive control.

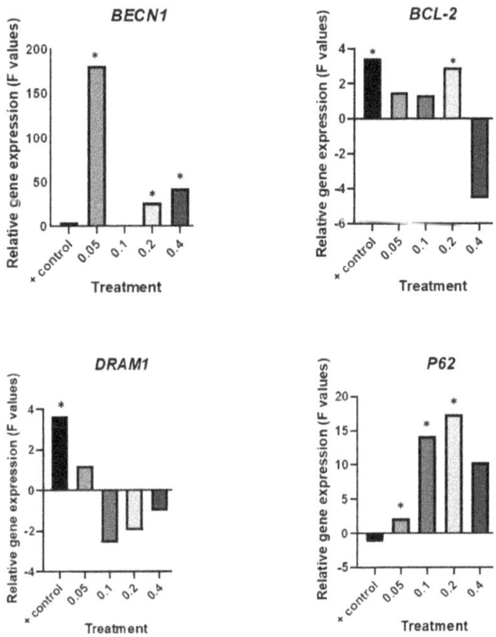

Figure 6. Results of *BECN1*, *BCL-2*, *P62*, and *DRAM1* relative gene expression analysis normalized with the *GAPDH* gene compared with HB-untreated cells (negative control). The F values are shown; * $p < 0.05$.

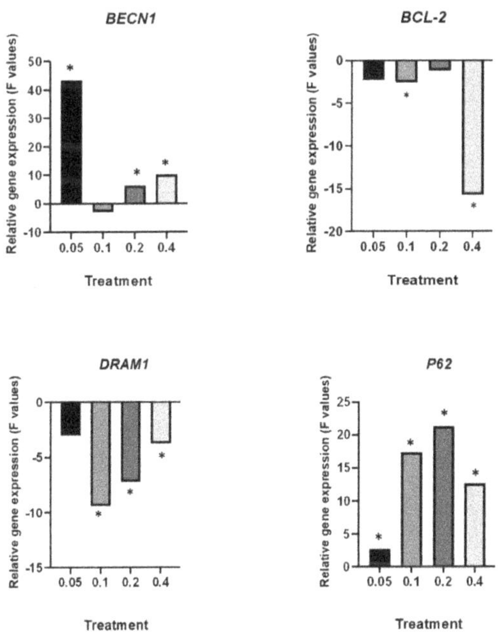

Figure 7. Results of *BECN1*, *BCL-2*, *P62*, and *DRAM1* gene expression analyses normalized with the GAPDH gene were compared with the positive control (starvation-induced autophagy cells). The F values are shown; * $p < 0.05$.

3. Discussion

Previous in vitro studies of the anti-proliferative effects of HB have revealed a selective cytotoxic effect on tumor cells and a weak cytotoxic effect on non-tumor cells [21,24,31,32,35]. In this study, we found dose-dependent HB inhibition of cell proliferation. The lowest tested HB concentration (0.05 mg/mL) slightly inhibited BC cell growth, while concentrations of 0.1, 0.2, and 0.4 mg/mL were highly cytotoxic.

Autophagy is a dynamic and complex process that is regulated by multiple molecules and has a contradictory role in tumor metabolism. In BC cells, high basal and protective autophagy against anticancer treatments [15] facilitates the avoidance of apoptotic cell death. Detected autophagosomes in cells treated with different HB concentrations (0.05–0.4 mg/mL) confirm HB-induced autophagy. The level of HB-induced autophagy was lower than in the positive control, where autophagy was successfully induced by starvation. In the negative control, a lower level of basal autophagy was registered. Therefore, HB effects on autophagy induction in BC cells, together with the inhibition of cell proliferation, indicate its high antitumor potential.

Cells possess two energy-producing pathways: mitochondrial respiration and glycolysis. Using the Seahorse XF HS Mini analyzer (Agilent, Santa Clara, CA, USA), we simultaneously measured both of these pathways in live cells, interrogating key cellular functions. Our data showed that starving BC cells (positive control) and 0.2 and 0.4 mg/mL HB treatments exhibit reduced mitochondrial respiration, indicating compromised or dysfunctional mitochondria, as compared to untreated cells and HB treatments at lower concentrations. Untreated BC cells and those treated with HB at lower concentrations showed high basal and stress glycolysis, indicating the ability to use that metabolic pathway as an energy source. These results suggest that HB disrupts glycolysis and reduces the intracellular survival potential of BC cells. Cancer cell metabolism, characterized by the Warburg effect, results in lactate production by aerobic glycolysis. Lactate production generates an acidic and hypoxic microenvironment, which promotes inflammation, tumorigenesis, invasion, and metastasis. Increased OXPHOS increases the production of reactive oxygen species (ROS) and the activation of c-JUN N-terminal kinases (JNK), resulting in autophagy induction. Hypoxia also induces the expression of a hypoxia-inducing factor (HIF-1α), which leads to the inhibition of the mammalian target of rapamycin (mTOR) and autophagy induction [36,37]. The hybrid glycolysis/OXPHOS metabolic state is suitable for the survival of invasive, aggressive tumors. It offers increased metabolic flexibility in cancer cells, which are able to utilize different nutrients in response to increased energy demands due to high proliferation and maintenance of ROS at a moderate level. It promotes genomic instability, which stimulates tumorigenesis and metastasis, and decreases susceptibility to anticancer therapy [38]. Our data strongly suggest that BC cells have a strong hybrid glycolysis/OXPHOS phenotype, which can be interrupted by HB in specific concentrations.

The HB effects on autophagy induction were considered at the gene transcription level as well. HB decreased *DRAM1* gene expression in comparison with the positive control, and it significantly increased the relative gene expression of *P62* in all applied concentrations. DRAM1 stimulates autophagosome elimination [39]. Knock-down of *DRAM1* inhibits P62 binding to the autophagosome, resulting in a decrease in autophagosome degradation. However, the same authors did not report DRAM1's role in starvation and mTOR-mediated autophagy [40]. Epi-fluorescent analyses of autophagosomes at given concentrations showed the presence of active autophagy, which is supported by increased expression of *BECN1*. Beclin 1 is responsible for the nucleation and maturation of the autophagosomes. The most significant upregulation of *BECN1* was found at 0.05 mg/mL of HB, along with a decreased number of viable cells. An increased frequency of autophagosomes and cells' degradation may also indicate a blocked autophagy flux [41]. This is supported by increased *P62* expression, which indicates higher requirements for the selective labeling of degradation contents.

One of the main mechanisms of autophagy control is the interaction between the autophagy protein beclin 1 and anti-apoptotic members of the Bcl-2 family (BCL-2, BCL-XL, and MCL-1). Under normal conditions, BCL-2 inhibits beclin 1, and during stress, they separate and further stimulate autophagy. BCL-2 is an anti-apoptotic and anti-autophagy protein [42]. Increased expression of *BCL-2* is present in various types of tumors, including bladder cancer, supporting its tumorigenesis. It is associated with a poor prognosis. In bladder cancer, inhibiting *BCL-2* expression reduces cell proliferation and increases the sensitivity of cells to chemotherapy [43]. Therefore, the regulation of *BCL-2* gene expression is the focus of anti-tumor research. Our results showed a downregulation of *BCL-2* in cells treated with HB at all tested concentrations compared to cells with starvation-induced autophagy. Because a decrease in *BCL-2* expression results in the reduced proliferation of tumor cells [43], HB should be considered a suitable agent for reducing the expression of BCL-2 in bladder tumors. Patterns of relative gene expressions, found in our study, support the hypothesis that the efficiency of HB was associated with the expression of autophagy-specific genes. It has been proven that HB at the same tested concentrations (0.2 and 0.4 mg/mL) reduces BCL-2 expression in UT-7 human leukemia cells [44]. Previously, HB has been shown to reduce *IGF-1* and *hTERT* gene expression [24]. Increased expression of both genes has been related with the poor prognosis in patients with cancer disease [45,46]. The inhibition of telomerase activity and IGF signaling, together with decreased *BCL-2* expression, are possible mechanisms explaining the reduced resistance to cell death. In our study, the results of gene expression and the detection of autophagy showed an increased presence of autophagy. It was previously explained that excessive autophagy can serve as a mechanism of cell death by promoting apoptosis or autophagy cell death due to excessive cell degradation [47]. Although autophagy allows the cell to overcome stress and maintain homeostasis [48], the constant activation of autophagy with a decrease in *BCL-2* expression, observed in BC cells treated with HB, reduced the cell's ability to survive by causing cell death.

This study has revealed HB inhibition of cell growth. A decreased percentage of BC cell growth at higher HB concentrations can be explained by the intricate interplay between extracellular nucleotides affecting the cell cycle and autophagy modulation in response to cell density, ultimately influencing cell proliferation rates [49,50], as the results of this work have suggested that HB induces autophagy in bladder carcinoma (BC) cells. Additionally, HB targeted both aerobic glycolysis and mitochondrial respiration in BC cells in vitro. By causing significant degradation, metabolic disruption, and downregulation of *BCL-2*, HB increases the cells' susceptibility to autophagy.

4. Materials and Methods

4.1. Dipotassium Trioxohydroxytetrafluorotriborate, $K_2[B_3O_3F_4OH]$, Halogenated Boroxine (HB)—Synthesis and Solution Preparation

HB is a water-soluble white powder. It was synthesized in the Laboratory for Physical Chemistry, Department of Chemistry, Faculty of Science, University of Sarajevo, as previously described [21], with 99.99% purity. The $K_2(B_3O_3F_4OH)$ stock solution was prepared by dissolving 20 mg of $K_2(B_3O_3F_4OH)$ in 1 mL of RPMI-1640 culture media.

4.2. Cell Culture Treatment

The human urinary bladder carcinoma (BC) 5637 (ATCC, Manassas, VA, USA; Cat. No. ATCC HTB-9) was obtained through collaboration with ICGEB Trieste (Trieste, Italy). Cells were grown in RPMI-1640 media (Sigma-Aldrich, St. Louis, MO, USA) with 10% fetal bovine serum (FBS) supplemented with L-glutamine and antibiotics in a humidified atmosphere of 5% CO_2 at 37 °C. After the initial 24 h, cell cultures were treated with HB at final concentrations of 0.05, 0.1, 0.2, and 0.4 mg/mL and incubated for an additional 24 h. Untreated tumor cells were used as autophagy negative controls, while for the autophagy positive controls, starvation-induced (serum-free RPMI-1640; Sigma-Aldrich, St. Louis, MO, USA) autophagy cells were used.

4.3. *Alamar Blue Assay*

Cell proliferation was assessed by an alamar blue assay (Invitrogen, Waltham, MA, USA). Each treatment was carried out in triplicate in the 96-well plate, with 10,000 cells per well seeded. Cells were treated with HB for 24 h after the initial incubation. Alamar blue dye was added in an amount equal to 10% of the culture volume 2 h prior to absorbance measuring. Cultures without test substances were set as positive assay controls, while negative assay controls did not contain cells. Medium was used as a blank. Absorbance was measured at 570 and 620 nm using a Multiscan FC plate reader (Thermo Ficher Scientific, Vantaa, Finland). The percentage of cell growth inhibition was calculated according to manufacturer instructions (https://tools.thermofisher.com, accessed on 15 August 2023; The alamarBlue® Assay).

4.4. *Autophagy Detection*

The Autophagy Assay kit (Sigma-Aldrich, St. Louis, MO, USA) was used to monitor the degree of autophagy using a fluorescent autophagosome marker that specifically binds to autophagosomes. According to the manufacturer's recommendations, 24 h after incubation of BC cells with HB, epifluorescence detection was performed using a BX51 microscope (Olympus, Japan). The degree of autophagy was photo-documented with a DP50 camera, analyzed using View Finder Lite software (Olympus, Japan), and categorized according to the number of detected signals: no signal; 0–5 signals; 5–10 signals; >10 signals; and coupled signals or one big signal. A total of 200 cells were analyzed per HB treatment or control (Figure 8).

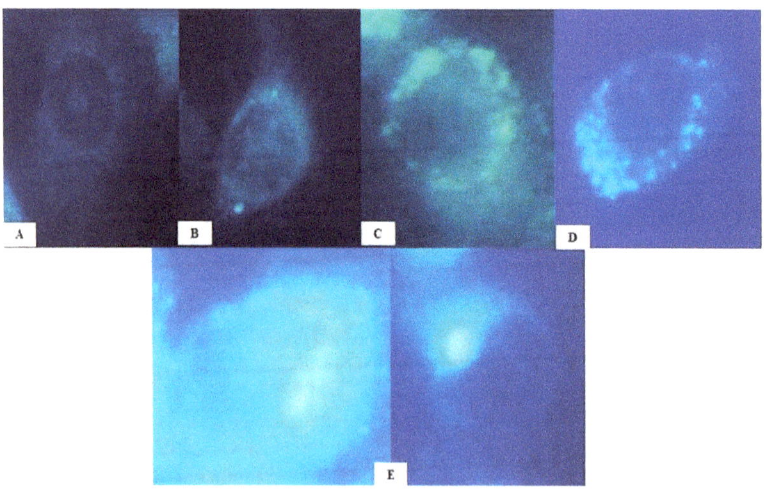

Figure 8. Evaluation of autophagy levels due to the presence and number of signals: no signal (**A**); 0–5 signals (**B**); 5–10 signals (**C**); >10 signals (**D**); and coupled signals or one big signal (**E**). Epifluorescence detection was performed using a BX51 microscope with 1000× magnification (Olympus, Tokyo, Japan).

4.5. *Oxygen Consumption Rate and Extracellular Acidification Rate*

BC cells were seeded into Seahorse XF cell culture microplates (4×10^4 cells/well) and incubated overnight. After washing the cells with Agilent Seahorse XF Base Medium, the oxygen consumption rate (OCR) and extracellular acidification rate (ECAR) were measured on a Seahorse XF HS Mini Analyzer (Agilent Technologies, Santa Clara, CA, USA). OCR in pmol/min reveals the rate of decrease in oxygen in the assay medium and, as such, is a measure of mitochondrial respiration in the cells. ECAR in mpH/minute illustrates the rate of increase in proton concentration (or decrease in pH) in the assay medium and reflets changes due to glycolysis. Alterations in metabolic function under basal and "stress" conditions were evaluated using the XFp Cell Energy Phenotype Test Kit (Cat. No. 103275-100; Agilent Technologies), according to the manufacturers' instructions. Briefly, for the measurement of basal metabolic rates, measurements of OCR and ECAR were taken at the starting assay conditions. To evaluate metabolic rates during increased energy demand, "stress" conditions were induced by the addition of a mix of oligomycin (an ATP synthase inhibitor) and the mitochondrial uncoupler carbonyl cyanide 4-[trifluoromethoxy] phenylhydrazone (FCCP). The data were analyzed using Seahorse Analytics software version 3.0.0.41 (Agilent Technologies, Santa Clara, CA, USA).

4.6. q-PCR and Gene Expression Analysis

Five cell populations used in the experiment (untreated BC cells as a negative control and four HB treatments at concentrations of 0.05, 0.1, 0.2, and 0.4 mg/mL) were subsequently analyzed. The total RNA was extracted from harvested cells using an RNA isolation kit (Nucleo Spin RNA, Macherey-Nagel, Düren, Germany). The quantity of isolated RNA was estimated by an RNA fluorometric assay using Qubit 2.0 (Life Technologies, London, UK). For cDNA synthesis, total RNA in a 40 ng/μL concentration was used. High-capacity cDNA Reverse Transcription Kits (Applied Biosystems, Foster City, CA, USA) were used for the generation of cDNA, according to the manufacturer's instructions.

Amplifications of target genes *BECN1* (beclin 1), *P62/SQSTM* (sequestosome 1), *BCL-2* (B-cell lymphoma 2), and *DRAM1* (Damage-Regulated Autophagy Modulator 1) in replicates were performed using ABI 7300 real-time PCR (Applied Biosystems, USA). The primer sequences are given in Table 1.

Table 1. Primer sequences for the target genes.

Gene	Reference Sequence (Fasta Acc Number)	Primers
BECN1	NM_003766.5	Forward: CTCCCGAGGTGAAGAGCATC Reverse: GGGGGATGAATCTGCGAGAG
P62/SQSTM	NM_003900.5	Forward: CCGTGAAGGCCTACCTTCTG Reverse: TCCTCGTCACTGGAAAAGGC
BCL-2	NM_000633.3	Forward: GGGGTCATGTGTGTGGAGAG Reverse: GAAATCAAACAGAGGCCGCA
DRAM1	NM_018370.3	Forward: TTGGTGCAGCCACGATGTAT Reverse: ACACCACAGACAAAGGCCAA

The expression of target genes was monitored and normalized with the housekeeping *GAPDH* gene in REST software, version 2 [51], which was also used to compare gene expression at the transcription level in HB treatments and the control. Subsequently, the expression ratio results of the four investigated transcripts were tested for significance ($p < 0.05$) by a randomization test.

4.7. Statistical Analysis

To determine if HB-induced autophagy is statistically significant compared to untreated samples and starvation-induced autophagy, a Chi square test was applied (PAST 3.15).

Relative gene expression analysis was evaluated using REST-384, version 2, based on a pairwise fixed reallocation randomization test [51]. The test compares the Ct values of the control and the sample (target gene).

To determine OCR and ECAR significance between baseline and stressed conditions for every treatment, a Mann–Whitney test for independent samples was applied. The Kruskal–Wallis test was applied to determine significance between baselines and stressed conditions, followed by a Conover post hoc analysis.

The Pearson correlation coefficient was calculated to correlate ECAR and OCR baselines with the stressed condition. MedCalc 18.9. software was used, and the significance level was set at 0.05.

5. Conclusions

The results of the study suggest autophagy as a cellular and molecular pathway of HB activity in BC cells. This work confirms that HB targets aerobic glycolysis and mitochondrial respiration in BC cells in vitro. Inducing massive degradation, metabolic interruption, and downregulation of *BCL-2* in BC cells HB increases their susceptibility to autophagy. Further studies are needed to elucidate signaling molecules and cellular pathways of HB in autophagy induction in different tumors. The potential of HB based anti-tumor drugs should be additionally investigated.

Supplementary Materials: The following supporting information can be downloaded at: https://www.mdpi.com/article/10.3390/molecules29122919/s1.

Author Contributions: Conceptualization, writing—original draft, formal analysis, N.E.-B.; conceptualization, funding acquisition, supervision, L.P.; data curation, writing—review and editing, A.H.; investigation, methodology, N.L.-K.; investigation, methodology, M.H.O.; methodology, data curation, A.S.; investigation, methodology, S.K.; conceptualization, supervision, M.M.; software, investigation B.J.; conceptualization, supervision, funding acquisition, writing—review and editing, S.H. All authors have read and agreed to the published version of the manuscript.

Funding: This research was funded by the Ministry of Science, Higher Education and Youth of Sarajevo Canton (Project: Cell death types—analysis, detection and classification in human tumor cultures/CELDET, number: 27-02-35-37080-8/23) and Faculty of Medicine Foča, University of East Sarajevo, Bosnia and Herzegovina.

Institutional Review Board Statement: Not applicable.

Informed Consent Statement: Not applicable.

Data Availability Statement: The data presented in this study are available in article.

Acknowledgments: The authors are grateful to Borivoj Galić, who passed away, for a huge contribution to life science and an opportunity to explore his patent, $K_2(B_3O_3F_4OH)$. The authors thank to Nataša Skoko (ICGEB, Trieste, Italy) for a donation of 5637 cell line.

Conflicts of Interest: Author Svetoslav Kalaydjiev was employed by the company Agilent Technologies LDA UK Ltd. The remaining authors declare that the research was conducted in the absence of any commercial or financial relationships that could be construed as a potential conflict of interest.

References

1. Levine, B.; Klionsky, D.J. Development by self-digestion: Molecular mechanisms and biological functions of autophagy. *Dev. Ce* **2004**, *6*, 463–477. [CrossRef]
2. Guo, J.Y.; Chen, H.Y.; Mathew, R.; Fan, J.; Strohecker, A.M.; Karsli-Uzunbas, G.; Kamphorst, J.J.; Chen, G.; Lemons, J.M.; Karantz V.; et al. Activated Ras requires autophagy to maintain oxidative metabolism and tumourigenesis. *Genes Dev.* **2011**, *25*, 460–47 [CrossRef]
3. Lin, Y.C.; Lin, J.F.; Wen, S.I.; Yang, S.C.; Tsai, T.F.; Chen, H.E.; Chou, K.Y.; Hwang, T.I. Inhibition of High Basal Level of Autophag Induces Apoptosis in Human Bladder Cancer Cells. *J. Urol.* **2016**, *195*, 1126–1135. [CrossRef]
4. Xie, X.; White, E.P.; Mehnert, J.M. Coordinate autophagy and mTOR pathway inhibition enhances cell death in melanom *PLoS ONE* **2013**, *8*, e55096. [CrossRef]
5. Ma, X.H.; Piao, S.; Wang, D.; McAfee, Q.W.; Nathanson, K.L.; Lum, J.J.; Li, L.Z.; Amaravadi, R.K. Measurements of tumour ce autophagy predict invasiveness, resistance to chemotherapy, and survival in melanoma. *Clin. Cancer Res.* **2011**, *17*, 3478–348 [CrossRef]
6. Goldsmith, J.; Levine, B.; Debnath, J. Autophagy and cancer metabolism. *Methods Enzymol.* **2014**, *542*, 25–57.
7. Antoni, S.; Ferlay, J.; Soerjomataram, I.; Znaor, A.; Jemal, A.; Bray, F. Bladder Cancer Incidence and Mortality: A Global Overvie and Recent Trends. *Eur. Urol.* **2017**, *71*, 96–108. [CrossRef]
8. Nishiyama, H.; Habuchi, T.; Watanabe, J.; Teramukai, S.; Tada, H.; Ono, Y.; Ohshima, S.; Fujimoto, K.; Hirao, Y.; Fukushima, M et al. Clinical outcome of a large-scale multi-institutional retrospective study for locally advanced bladder cancer: A surve including 1131 patients treated during 1990–2000 in Japan. *Eur. Urol.* **2004**, *45*, 176–181. [CrossRef]

9. Gazzaniga, P.; Silvestri, I.; Gradilone, A.; Scarpa, S.; Morrone, S.; Gandini, O.; Gianni, W.; Frati, L.; Agliano, A.M. Gemcitabine-induced apoptosis in 5637 cell line: An in-vitro model for high-risk superficial bladder cancer. *Anticancer Drugs* **2007**, *18*, 179–185. [CrossRef]
10. Rieger, K.M.; Little, A.F.; Swart, J.M.; Kastrinakis, W.V.; Fitzgerald, J.M.; Hess, D.T.; Libertino, J.A.; Summerhayes, I.C. Human bladder carcinoma cell lines as indicators of oncogenic change relevant to urothelial neoplastic progression. *Br. J. Cancer* **1995**, *72*, 683–690. [CrossRef]
11. Vousden, K.H.; Ryan, K.M. p53 and metabolism. *Nat. Rev. Cancer* **2009**, *9*, 691–700. [CrossRef]
12. Bensaad, K.; Tsuruta, A.; Selak, M.A.; Vidal, M.N.C.; Nakano, K.; Bartrons, R.; Gottlieb, E.; Vousden, K.H. TIGAR, a p53-inducible regulator of glycolysis and apoptosis. *Cell* **2006**, *126*, 107–120. [CrossRef]
13. Whyard, T.; Waltzer, W.C.; Waltzer, D.; Romanov, V. Metabolic alterations in bladder cancer: Applications for cancer imaging. *Exp. Cell Res.* **2016**, *341*, 77–83. [CrossRef]
14. Zhu, H.B.; Yang, K.; Xie, Y.Q.; Lin, Y.W.; Mao, Q.Q.; Xie, L.P. Silencing of mutant p53 by siRNA induces cell cycle arrest and apoptosis in human bladder cancer cells. *World J. Surg. Oncol.* **2013**, *11*, 22. [CrossRef]
15. Lin, J.F.; Lin, Y.C.; Tsai, T.F.; Chen, H.E.; Chou, K.Y.; Hwang, T.I. Cisplatin induces protective autophagy through activation of BECN1 in human bladder cancer cells. *Drug Des. Devel. Ther.* **2017**, *11*, 1517–1533. [CrossRef]
16. Liu, S.; Chen, X.; Lin, T. Emerging strategies for the improvement of chemotherapy in bladder cancer: Current knowledge and future perspectives. *J. Adv. Res.* **2022**, *39*, 187–202. [CrossRef]
17. Advanced Bladder Cancer (ABC) Meta-analysis Collaborators Group. Adjuvant Chemotherapy for Muscle-invasive Bladder Cancer: A Systematic Review and Meta-analysis of Individual Participant Data from Randomised Controlled Trials. *Eur. Urol.* **2022**, *81*, 50–61. [CrossRef]
18. Hall, D.G. *Boronic Acids*; John Wiley and Sons: New York, NY, USA, 2005.
19. Galić, B. Boroxine Composition for Removal of Skin Changes. U.S. Patent US8278289, 2 October 2012.
20. Galić, B. Removal of Skin Changes. European Patent EP1996514, 31 July 2013.
21. Haverić, S.; Haverić, A.; Bajrović, K.; Galić, B.; Maksimović, M. Effects of dipotassium trioxohydroxytetrafluorotriborate ($K_2(B_3O_3F_4OH)$) on genetic material and inhibition of cell division in human cell cultures. *Drug Chem. Toxicol.* **2011**, *34*, 250–254. [CrossRef]
22. Hadžić, M.; Haverić, S.; Haverić, A.; Galić, B. Inhibitory effects of delphinidin and luteolin on genotoxicity induced by $K_2(B_3O_3F_4OH)$ in human lymphocytes in vitro. *Biologia* **2015**, *70*, 553–558. [CrossRef]
23. Haverić, S.; Hadžić, M.; Haverić, A.; Mijanović, M.; Hadžiselimović, R.; Galić, B. Genotoxicity Evaluation of Dipotassium-Trioxohydroxytetrafluorotriborate, $K_2(B_3O_3F_4OH)$, in Human Lymphocyte Cultures and Mice Reticulocytes. *Braz. Arch. Biol. Technol.* **2016**, *59*, e16160195. [CrossRef]
24. Pojskić, L.; Haverić, S.; Lojo-Kadrić, N.; Hadžić, M.; Haverić, A.; Galić, Z.; Galić, B.; Vullo, D.; Supuran, C.T.; Miloš, M. Effects of dipotassium-trioxohydroxytetrafluorotriborate, $K_2(B_3O_3F_4OH)$, on cell viability and gene expression of common human cancer drug targets in a melanoma cell line. *J. Enzym. Inhib. Med. Chem.* **2016**, *31*, 999–1004. [CrossRef]
25. Hadžić, M.; Haverić, S.; Haverić, A.; Lojo-Kadrić, N.; Galić, B.; Ramić, J.; Pojskić, L. Bioflavonoids protect cells against halogenated boroxine-induced genotoxic damage by upregulation of hTERT expression. *Z. Naturforsch. C J. Biosci.* **2019**, *74*, 125–129. [CrossRef]
26. Ostojić, J.; Herenda, S.; Bešić, Z.; Miloš, M.; Galić, B. Advantages of an Electrochemical Method Compared to the Spectrophotometric Kinetic Study of Peroxidase Inhibition by Boroxine Derivative. *Molecules* **2017**, *22*, 1120. [CrossRef]
27. Islamović, S.; Galić, B.; Miloš, M. A study of the inhibition of catalase by dipotassium trioxohydroxytetrafluorotriborate $K_2[B_3O_3F_4OH]$. *J. Enzym. Inhib. Med. Chem.* **2014**, *29*, 744–748. [CrossRef]
28. Herenda, S.; Ostojić, J.; Hasković, E.; Hasković, D.; Miloš, M.; Galić, B. Electrochemical Investigation of the Influence of K2[B3O3F4OH] on the Activity of Immobilized Superoxide Dismutase. *Int. J. Electrochem. Sci.* **2018**, *13*, 3279–3287. [CrossRef]
29. Vullo, D.; Miloš, M.; Galić, B.; Scozzafava, A.; Supuran, C.T. Dipotassium trioxohydroxytetrafluorotriborate $K_2[B_3O_3F_4OH]$, is a potent inhibitor of human carbonic anhydrases. *J. Enzyme Inhib. Med. Chem.* **2015**, *30*, 341–344. [CrossRef]
30. Elez-Burnjaković, N.; Pojskić, L.; Haverić, A.; Lojo-Kadrić, N.; Omanović, M.H.; Ramić, J.; Smajlović, A.; Maksimović, M.; Haverić, S. New in vitro findings about halogenated boroxine cytotoxicity and deregulation of cell death-related genes in GR-M melanoma cells. *Arh. Hig. Rada Toksikol.* **2023**, *74*, 16–21. [CrossRef]
31. Ivanković, S.; Stojković, R.; Maksimović, M.; Galić, B.; Miloš, M. Impact of calcium ion on cytotoxic effect of the boroxine derivative, $K_2(B_3O_3F_4OH)$. *J. Enzym. Inhib. Med. Chem.* **2016**, *31*, 70–74. [CrossRef]
32. Ivanković, S.; Stojković, R.; Galić, Z.; Galić, B.; Ostojić, J.; Marasović, M.; Miloš, M. In vitro and in vivo antitumour activity of the halogenated boroxine dipotassium-trioxohydroxytetrafluorotriborate ($K_2(B_3O_3F_4OH)$). *J. Enzym. Inhib. Med. Chem.* **2015**, *30*, 354–359. [CrossRef]
33. Haverić, A.; Durmić-Pašić, A.; Alić, A.; Mujezinović, I.; Smajlović, A.; Ostojić, J.; Ahatović, A.; Hadžić, M.; Prašović, S.; Haverić, S.; et al. Biochemical and histomorphological findings in Swiss Wistar rats treated with potential boron-containing therapeutic—$K_2(B_3O_3F_4OH)$. *J. Trace Elem. Med. Biol.* **2020**, *62*, 126642. [CrossRef]
34. Haverić, A.; Haverić, S.; Hadžić, M.; Ezić, J.; Ćetković, T.; Galić, B. Moderate Toxicity of Potential Boron-containing Therapeutic, Dipotassium-trioxohydroxytetrafl uorotriborate—$K_2(B_3O_3F_4OH)$ in Rats and Mice. *Braz. J. Pharm. Sci.* **2023**, *59*, e21384. [CrossRef]

35. Liu, J.; Phan, E.; Saračević, O.; Burnett, M.; Lojo-Kadrić, N.; Haverić, A.; Galić, B. In vitro study of the anti-proliferative effects of dipotassium-trioxohydroxytetra fluorotriborate on the H520 non-small cell lung cancer cell line. *Genet. Appl.* **2017**, *1*, 2–7.
36. Gwangwa, M.V.; Joubert, A.M.; Visagie, M.H. Crosstalk between the Warburg effect, redox regulation and autophagy induction in tumourigenesis. *Cell Mol. Biol. Lett.* **2018**, *23*, 20. [CrossRef]
37. Nagelkerke, A.; Sweep, F.C.; Geurts-Moespot, A.; Bussink, J.; Span, P.N. Therapeutic targeting of autophagy in cancer. Part I: Molecular pathways controlling autophagy. *Semin. Cancer Biol.* **2015**, *31*, 89–98. [CrossRef]
38. Jia, D.; Park, J.H.; Jung, K.H.; Levine, H.; Kaipparettu, B.A. Elucidating the Metabolic Plasticity of Cancer: Mitochondrial Reprogramming and Hybrid Metabolic States. *Cells* **2018**, *7*, 21. [CrossRef]
39. Zhang, X.D.; Qi, L.; Wu, J.C.; Qin, Z.H. DRAM1 Regulates Autophagy Flux through Lysosomes. *PLoS ONE* **2013**, *8*, e63245. [CrossRef]
40. Galavotti, S.; Bartesaghi, S.; Faccenda, D.; Shaked-Rabi, M.; Sanzone, S.; McEvoy, A.; Dinsdale, D.; Condorelli, F.; Brandner, S.; Campanella, M.; et al. The autophagy-associated factors DRAM1 and p62 regulate cell migration and invasion in glioblastoma stem cells. *Oncogene* **2013**, *32*, 699–712. [CrossRef]
41. Vats, S.; Manjithaya, R. A reversible autophagy inhibitor blocks autophagosome-lysosome fusion by preventing Stx17 loading onto autophagosomes. *Mol. Biol. Cell* **2019**, *30*, 2283–2295. [CrossRef]
42. Pattingre, S.; Tassa, A.; Qu, X.; Garuti, R.; Liang, X.H.; Mizushima, N.; Packer, M.; Schneider, M.D.; Levine, B. Bcl-2 antiapoptotic proteins inhibit Beclin 1-dependent autophagy. *Cell* **2005**, *122*, 927–939. [CrossRef]
43. Zhou, C.; Ding, J.; Wu, Y. Resveratrol induces apoptosis of bladder cancer cells via miR21 regulation of the Akt/Bcl2 signaling pathway. *Mol. Med. Rep.* **2014**, *9*, 1467–1473. [CrossRef]
44. Hadžić, M.; Pojskić, L.; Lojo-Kadrić, N.; Haverić, A.; Ramić, J.; Galić, B.; Haverić, S. Novel boron-containing compound, halogenated boroxine, induces selective cytotoxicity through apoptosis triggering in UT-7 leukemia. *J. Biochem. Mol. Toxicol.* **2022**, *36*, e23005. [CrossRef]
45. Kucera, R.; Treskova, I.; Vrzalova, J.; Svobodova, S.; Topolcan, O.; Fuchsova, R.; Rousarova, M.; Treska, V.; Kydlicek, T. Evaluation of IGF1 serum levels in malignant melanoma and healthy subjects. *Anticancer. Res.* **2014**, *34*, 5217–5220.
46. Weroha, S.J.; Haluska, P. IGF-1 receptor inhibitors in clinical trials-early lessons. *J. Mammary Gland. Biol. Neoplasia* **2008**, *13*, 471–483. [CrossRef]
47. Shen, H.M.; Codogno, P. Autophagic cell death: Loch Ness monster or endangered species? *Autophagy* **2011**, *7*, 457–465. [CrossRef]
48. Mehrpour, M.; Esclatine, A.; Beau, I.; Codogno, P. Autophagy in health and disease. 1. Regulation and significance of autophagy: An overview. *Am. J. Physiol. Cell Physiol.* **2010**, *298*, 776–785. [CrossRef]
49. Pavel, M.; Renna, M.; Park, S.J.; Menzies, F.M.; Ricketts, T.; Fullgrabe, J.; Ashkenazi, A.; Frake, R.A.; Lombarte, A.C.; Bento, C.F.; et al. Contact inhibition controls cell survival and proliferation via YAP/TAZ-autophagy axis. *Nat. Commun.* **2018**, *9*, 2961. [CrossRef]
50. Sawa, C.; Yofu, S.; Kiriyama, K.; Sutoh, K.; Saito, T.; Kishi, S.; Gunji, M.; Inoue, Y.; Sugi, M.; Shioda, S.; et al. High concentration of extracellular nucleotides suppresses cell growth via delayed cell cycle progression in cancer and noncancer cell lines. *Heliyon* **2021**, *7*, e08318. [CrossRef]
51. Pfaffl, M.W.; Horgan, G.W.; Dempfle, L. Relative expression software tool (REST©) for group-wise comparison and statistical analysis of relative expression results in real-time PCR. *Nucleic Acids Res.* **2002**, *30*, e36. [CrossRef]

metabolites

MDPI

Article

Anaplerotic Therapy Using Triheptanoin in Two Brothers Suffering from Aconitase 2 Deficiency

Maximilian Penkl [1,*], Johannes A. Mayr [2], René G. Feichtinger [2], Ralf Reilmann [3], Otfried Debus [4], Manfred Fobker [5], Anja Penkl [1], Janine Reunert [1], Stephan Rust [1] and Thorsten Marquardt [1,*]

1 Klinik für Kinder- und Jugendmedizin, Universitätsklinikum Münster, Albert-Schweizer-Campus 1, 48149 Muenster, Germany; janine.reunert@ukmuenster.de (J.R.); stephan.rust@ukmuenster.de (S.R.)
2 Universitätsklinik für Kinder- und Jugendheilkunde, Salzburger Landeskliniken (SALK) and Paracelsus Medical University (PMU), Müllner Hauptstraße 48, 5020 Salzburg, Austria; h.mayr@salk.at (J.A.M.); r.feichtinger@salk.at (R.G.F.)
3 George-Huntington-Institut, Wilhelm-Schickard-Straße 15, 48149 Muenster, Germany; ralf.reilmann@ghi-muenster.de
4 Clemenshospital Münster, Klinik für Kinder- und Jugendmedizin, Düesbergweg 124, 48153 Muenster, Germany; o.debus@alexianer.de
5 Zentrale Einrichtung UKM Labor, Universitätsklinikum Münster, Albert-Schweitzer-Campus 1, 48149 Muenster, Germany; manfred.fobker@ukmuenster.de
* Correspondence: maximilian.penkl@gmail.com (M.P.); marquadt@uni-muenster.de (T.M.)

Abstract: Citric acid cycle deficiencies are extremely rare due to their central role in energy metabolism. The *ACO2* gene encodes the mitochondrial isoform of aconitase (aconitase 2), the second enzyme of the citric acid cycle. Approximately 100 patients with aconitase 2 deficiency have been reported with a variety of symptoms, including intellectual disability, hypotonia, optic nerve atrophy, cortical atrophy, cerebellar atrophy, and seizures. In this study, a homozygous deletion in the *ACO2* gene in two brothers with reduced aconitase 2 activity in fibroblasts has been described with symptoms including truncal hypotonia, optic atrophy, hyperopia, astigmatism, and cerebellar atrophy. In an in vivo trial, triheptanoin was used to bypass the defective aconitase 2 and fill up the citric acid cycle. Motor abilities in both patients improved.

Keywords: *ACO2* mutation; triheptanoin; citric acid cycle; anaplerotic therapy; case report

1. Introduction

The *ACO2* gene is located on chromosome 22q13.2, encoding for the mitochondrial isoform of aconitase (aconitase 2). As the second enzyme of the citric acid cycle (CAC), it is found in the mitochondrial matrix, converting citrate via cis-aconitate to isocitrate (see Figure 1). Besides being part of CAC, aconitase 2 also plays an important role in the maintenance of mitochondrial DNA (mtDNA) [1]. In *ACO2*-deficient cells, the mitochondrial respiratory capacity, the ATP turnover, the cellular ATP level [2], as well as the mtDNA copy number and mtDNA transcription, can be reduced [3]. In yeast, aconitase was found to regulate and maintain mtDNA stability. By direct binding of mtDNA, aconitase can preserve the integrity of mtDNA from the occurrence of point mutations and DNA breaks [4]. In humans, wild-type *ACO2* rescued the mtDNA depletion observed in cells from a patient with missense *ACO2* variants [1]. The maintenance of mtDNA is independent of enzymatic activity in the CAC [5]. Pathogenic variants in *ACO2* present with a neurodegenerative phenotype of variable severity [3,6], as described in over 100 cases [7]. Common symptoms are truncal hypotonia, head bobbing, and variable forms of seizures. Furthermore, patients can have ophthalmologic symptoms like strabismus, nystagmus, optic atrophy, and retinal dystrophy. Symptoms often become apparent within the first year of life [8,9]. In this study, a homozygous deletion in the *ACO2* gene in two brothers was identified. In an in vivo trial, triheptanoin, a synthetic seven-carbon fatty acid triglyceride compound that can be metabolized to succinyl-CoA, was used as an anaplerotic agent for bypassing the defective enzyme and restoring CAC function.

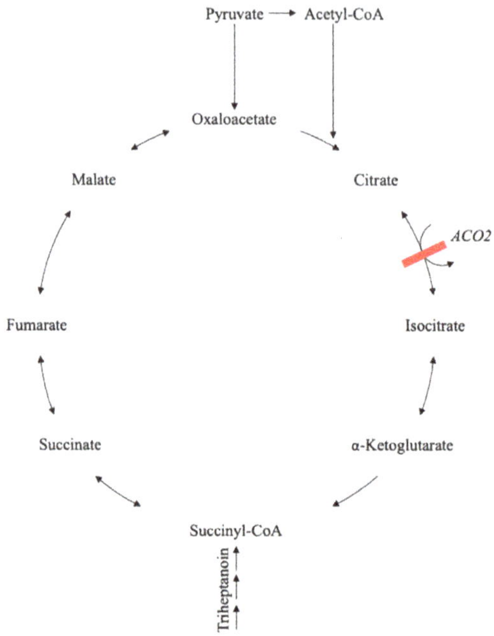

Figure 1. Scheme of the physiological and defective citric acid cycle in aconitase 2 deficiency with an overview of the CAC with its intermediates and bypassing Triheptanoin. The position of the aconitase 2 enzyme in CAC is illustrated as *ACO2*. Blocked pathway by defective *ACO2* marked with a red line. Triheptanoin bypasses aconitase 2 and feeds the CAC as succinyl-CoA.

2. Materials and Methods

2.1. Therapy with Triheptanoin

For compassionate use, triheptanoin was provided by Ultragenyx Pharmaceutical, Novato, CA, USA. Triheptanoin was prepared with banana puree or mashed potatoes. Therapy was slowly increased to 35% of daily calories coming from triheptanoin intake.

2.2. Therapy Monitoring

Standard blood tests were performed, including a complete blood count, urea, creatinine, bilirubin, transaminases, and gamma-glutamyl transferase.

Organic acid excretion in the urine was repeatedly measured for metabolic therapy monitoring.

Motor abilities were quantified by Q-Motor assessments [10], and physical therapy status was assessed with the Hammersmith Functional Motor Scale for spinal muscular atrophy (HFMS) [11] or Children's Hospital of Philadelphia Infant Test of Neuromuscular Disorders (CHOP-intended test) [12], respectively.

2.3. Isolation of Mitochondria from Cell Cultures

Primary human skin fibroblasts were used for the isolation of mitochondria. The medium was removed, and cell pellets were washed once with PBS. Cells were incubated for 3–5 min in a 10 mL trypsin solution for detachment. Cells were transferred quickly into FCS to inactivate trypsin. After centrifugation for 5 min at 1200–1300 units in 50 mL tubes, the supernatant was removed. Cell pellets were washed twice with 1 mL PBS. After centrifugation at 600× g at 4 °C for 5 min, cell pellets were weighed. Pellets were homogenized in 10 × Vol 10 mM Tris pH 7.4 with a potter 8–10 × times at maximum speed. A 1.5 M sucrose solution was immediately added to the suspension to stabilize mitochondria (volume: X = V/5; X = volume sucrose solution; V = volume resuspended cells). The suspension was mixed with a disposable plastic pipette, transferred to an Eppendorf tube, and centrifuged at 600× *g* (=1800 rpm) at 4 °C for 10 min. The supernatant, which contains the mitochondria, was centrifuged at 10,000× *g* at 4 °C for 10 min, after which the supernatant was discarded. The pellet was resuspended in 500 µL SEKT (250 mM sucrose, 2 mM EGTA, 40 mM KCl, and 20 mM Tris. pH 7.4) buffer. After centrifugation of the solution at 10,000× *g* at 4 °C for 10 min, the supernatant was discarded, and the pellet was resuspended in 1 vol of SEKT buffer (if the initial cell pellet has 50 mg, the final volume is 50 µL). The aliquoted pellets were subsequently stored at −70 °C.

2.4. Western Blotting

Isolated mitochondria from primary human skin fibroblasts were used for Western blotting. In brief, mitochondrial lysates were loaded onto a 10% SDS-PAGE gel. Electrophoresis was performed at 150 V for 35 min. Proteins were blotted onto nitrocellulose membranes (Transblot Turbo Pack; Biorad, Hercules, CA, USA) with the Transblot Turbo System from Biorad. Blocking was carried out using the Western blocking reagent (Roche, Basel, Switzerland) for 30 min at room temperature. Membranes were incubated with the following primary antibodies: mouse monoclonal anti-*ACO2* antibody overnight at 4 °C (Abcam, Cambridge, UK; ab110321) and mouse monoclonal anti-VDAC1 antibody 1 h at room temperature (Abcam, ab14734). The secondary antibody from the DAKO Envision kit was used. The Lumi-Light Western blotting substrate (Roche) was used for development.

2.5. Aconitase Activity

In order to determine aconitase activity, a commercially available kit was used (BIOXYTECH Aconitase-340; catalog number 21041). A total of 50 µg of isolated mitochondria from human skin fibroblasts were diluted in assay buffer to a final volume of 250 µL. Subsequently, 250 µL substrate, 250 µL enzyme, and 250 µL NADP were added (final reaction volume: 1 mL). Substrate, enzyme, and NADP are included in the aconitase kit. The solution was sonified for 20 s at low energy. After 15 min of preincubation at 37 °C, the change in absorbance was measured at 340 nm.

2.6. Monitoring of Plasma C7 Fatty Acids

The determination of plasma C8 and C10 fatty acid levels was performed by gas chromatography–mass spectrometry. For this purpose, 100 µL serum was mixed with 10 µL internal standard (20 µg/mL D3-Octanoic acid in methanol), 20 µL 37% HCl/H2O 1:1 (*v*/*v*), 200 µL H2O, 100 µL Methyl-t-butyl ether, and 50 µL n-hexane in a 0.5 mL Eppendorf reaction vessel and centrifuged (3 min, 10,000× *g* rpm). A total of 50 µL of the supernatant was transferred to an autosampler vial with a conical insert, and 1 µL was loaded in splitless mode onto a GC–MS (QP2010 Ultra; Shimadzu, Duisburg, Germany) equipped with a PTV injector and a capillary column (Stabilwax-DA 15 m length, 0.25 mm internal diameter, and 0.25 µm film thickness; Restek GmbH, Bad Homburg, Germany) and an autosampler AS AOC-20s. Measuring time: 58.7 min; injector temperature: 70 °C; column oven temperature program isothermal: 2 min at 40 °C; gradient 12 °C/min to 240 °C; isothermal: 40 min at 240 °C. The carrier gas was helium with a constant flow of 1 mL/min. The transfer line to the ion source was held at 260 °C. The MS detector was set to selected ion monitoring (SIM) mode as *m*/*z* 60 (C8/C10) and *m*/*z* 63 (D3-Octanoic acid, internal standard). An example of a base peak intensity chromatogram of free fatty acids is shown in Figures 2 and 3. LabSolutions software (Version 2.72; Shimadzu, Duisburg, Germany) was used for GC–MS data acquisition and processing.

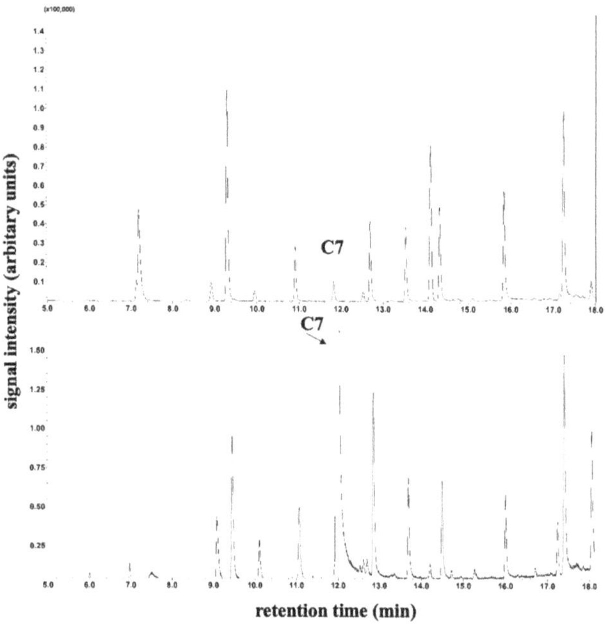

Figure 2. Determination of plasma C7 fatty acids for patient 1. Section of peak intensity chromatogram of plasma medium-chain free fatty acids by gas chromatography–single quadrupole mass spectrometry. The top image shows plasma C7 fatty acids prior to therapy, and the bottom image shows measurements during therapy.

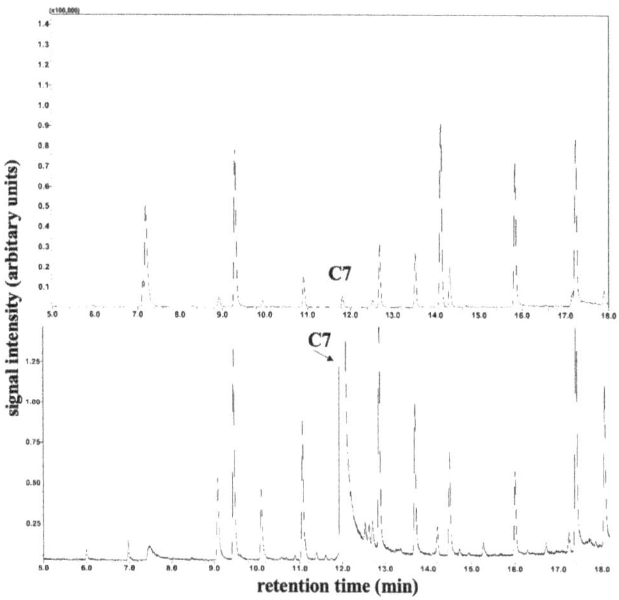

Figure 3. Determination of plasma C7 fatty acids for patient 2. Section of peak intensity chromatogram of plasma medium-chain free fatty acids by gas chromatography–single quadrupole mass spectrometry. The top image shows plasma C7 fatty acids prior to therapy, and the bottom image shows measurements during therapy.

3. Results

3.1. Clinical Report

The patients were born to a consanguineous couple from Yemen. The firstborn boy (patient 1) was born by cesarean section after an uneventful pregnancy with a birth weight of 2840 g, a length of 49 cm, and an occipitofrontal circumference of 35 cm. Apgar scores were 10/10/10.

Hexadactyly was present on both hands and was removed surgically on the third day of life.

Up until a febrile infection at the age of eight months, the development was normal. Simultaneously with the infection, the child became extraordinarily floppy, was unable to stabilize his trunk, and his head showed a head tilt. His targeted movements were atactic, and at rest, his head and body were slightly shaking. Reflexes could not be elicited clearly. His eyes exhibited a slow horizontal nystagmus. Wakefulness and interaction were not disturbed. Seizures were excluded by EEG–liquor, and MRI analysis was normal.

At the age of one year, a motor development slowdown, as well as truncal hypotonia, had progressed—even under physiotherapy. To objectify physical therapy status, 35/66 points on the Hammersmith Functional Motor Scale for spinal muscular atrophy (HFMS) [11] were calculated. In the course, infantile esotropia, downbeat nystagmus, optic atrophy, hyperopia, astigmatism, and cerebellar atrophy were diagnosed. While growing up, poor pronunciation was noticed.

At the age of eight years, a genetic analysis was performed due to the fact that the younger brother (patient 2) displayed similar symptoms.

Patient 2 was born via a planned cesarean section after an uneventful pregnancy with a birth weight of 2810 g, a length of 47 cm, and an occipitofrontal circumference of 34.8 cm. Apgar scores were 10/10/10. Development was normal until the patient was 1 month old. After a febrile infection, the boy displayed the same symptoms as the older brother. In order to objectively measure the boys' physical therapy status, the Children's Hospital of Philadelphia Infant Test of Neuromuscular Disorders (CHOP-intended test) [12] was chosen. Here, 44/64 points were calculated.

3.2. Genetic Analysis

Whole-exome sequencing identified the homozygous variant c.2335_2340delCAACAG, p.Gln779_Gln780del, i.e., deletion of the last two amino acids of the protein, for both brothers in *ACO2* (NM_001098.3). This variant was predicted to be pathogenetic by in silico simulation [13–16].

3.3. Fibroblasts Analysis

Skin fibroblasts were taken and cultured [17] from the older brother in order to quantify the residual enzyme activity and protein level of aconitase 2, with four controls also being measured for comparison purposes. In Western blot, protein levels were reduced (see Figure 4). As seen in Figure 5, aconitase 2 activity was significantly reduced to 40% compared to the four controls.

3.4. Therapy with Triheptanoin

Triheptanoin was started with a low dose and increased to a daily amount of 35% triheptanoin when the caloric intake was reached. During therapy, nausea and vomiting occurred, leading partially to the rejection of the prepared food. Nevertheless, the treatment could still be continued.

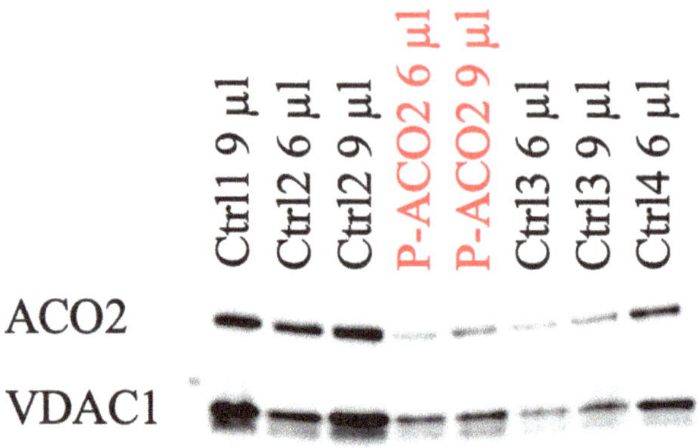

Figure 4. Western blot analysis of the *ACO2* levels of the affected individual and four controls.

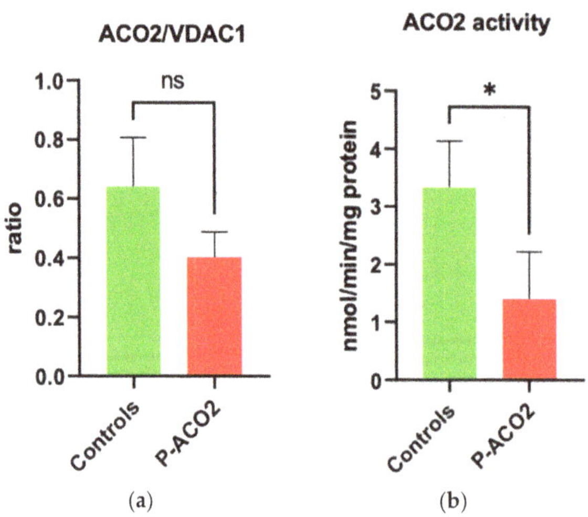

Figure 5. Determination of *ACO2* activity in the affected individual and controls. (**a**) Densitometric analysis of the *ACO2/VDAC1* ratio of the Western blot results. (**b**) Enzymatic activity of *ACO2* in the affected individual and controls. VDAC1 was used as a mitochondrial loading control. A non-parametric Mann–Whitney U test was used for statistical analysis. The enzymatic measurement was repeated twice with the same controls used in Figure 3. For both experiments, isolated mitochondria from human primary skin fibroblasts were used. For the enzymatic determination, 50 µg mitochondria per 1 mL assay volume were taken. Values present mean ± SD. (**a**): $p = 0.28$; (**b**): * $p = 0.0238$.

3.5. Motor Ability Analysis

Depending on the patient's age, either HFMS combined with Q-Motor assessments [10] or a CHOP-intended test was used for the monitoring of motor ability analysis. Prior to therapy, patient 1 scored 35/66 points at HFMS. After two months of therapy, the boy improved to 46/66 points (HFMS) and was able to maintain this level during therapy. Additionally to HFMS, Q-Motor assessments were used for patient 1. The data suggested that speeded tapping performance had improved. With a p-value < 0.05 in the Wilcoxon rank-sum test, the data are significant for the right hand and left foot (see Figures 6 and 7). For patient 2, 44/64 points in the CHOP-intended test were calculated prior to the therapy. After two months of therapy, the boy reached 60/64 points and 56/64 points after six months of therapy (see Figure 6).

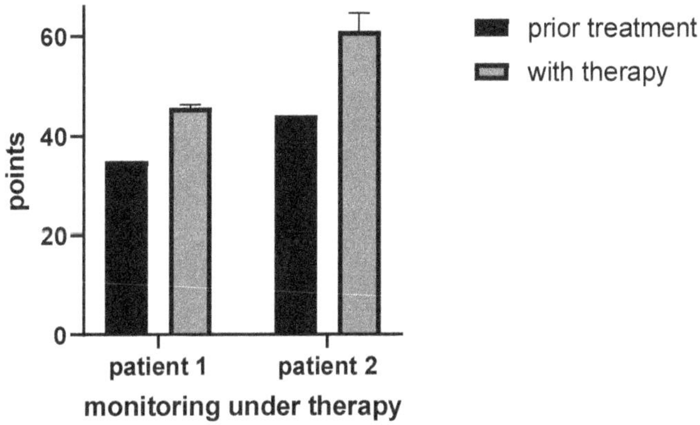

Figure 6. Monitoring of motor ability with HFMS for patient 1 and CHOP for patient 2. Patient 1: Sum of calculated points in HFMS before (n = 1) and during therapy (n = 3). Patient 2: Sum of calculated points in CHOP before (n = 1) and during therapy (n = 3).

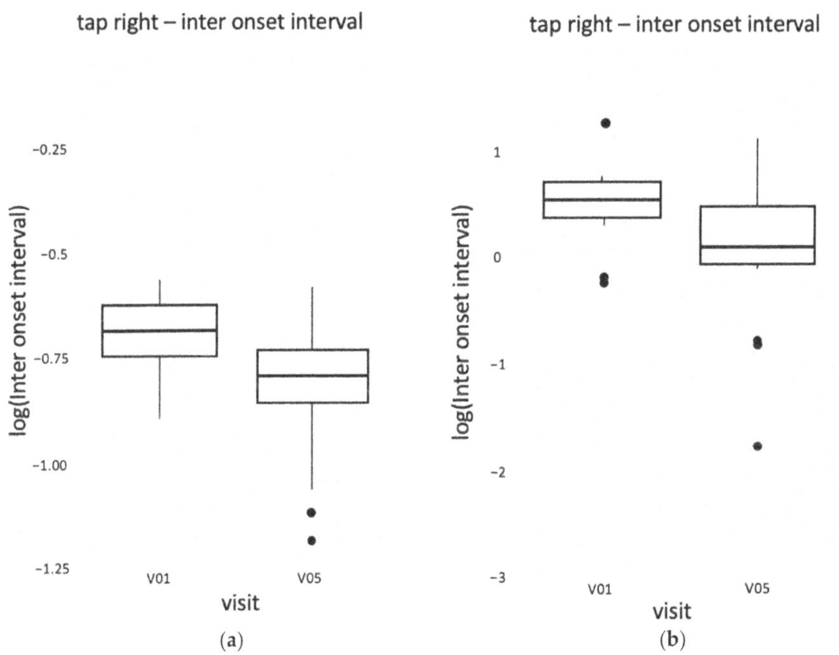

Figure 7. Q-Motor analysis of patient 1 with box plots of speeded tapping tests of the right hand and the left foot. (**a**) Speeded tapping performance of the right hand. The p-value in the Wilcoxon rank sum test is 9.813^{-9}. (**b**) Speeded tapping performance of the left foot. The p-value in the Wilcoxon rank sum test is 0.003691.

3.6. Organic Metabolites

Concurrent with the improvements in HFMS or CHOP-intended tests, respectively, an increase in CAC intermediates in urine organic acid analysis was observed during the therapy phase (see Figures 8 and 9). At the same time, the tremor of patient 1 disappeared, accompanied by improved pronunciation. Improvements in patient 2 were reduced squinting under therapy. On days with high energy demands (e.g., when affected by infection), patient 2 presented with a whole-body tremor again.

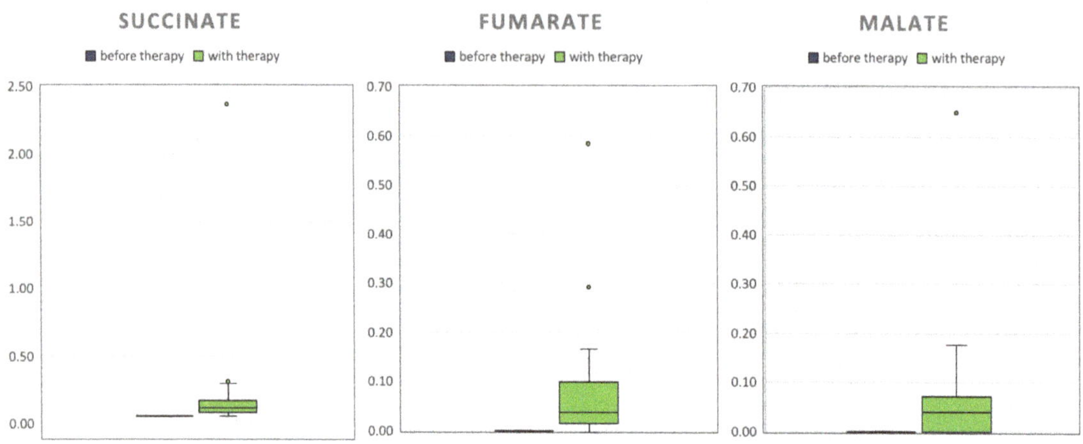

Figure 8. Progression of the measured excretory organic acids succinate, malate, and fumarate in the urine of patient 1. The organic acids fumarate and malate before therapy (blue) were not detectable ($n = 5$). With therapy (green), a higher excretion of the organic acids was detected ($n = 38$).

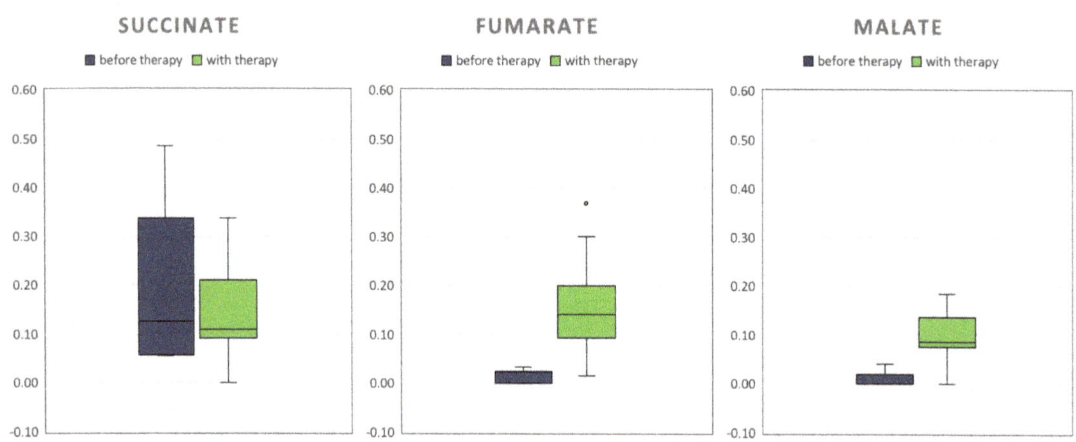

Figure 9. Progression of the measured excretory organic acids succinate, malate, and fumarate in the urine of patient 2. Five samples before therapy (blue; $n = 5$). With therapy (green), a higher excretion of the organic acids fumarate and malate was detected ($n = 38$).

3.7. C7 Fatty Acids

According to a modified method of Bartolucci et al. [18], the plasma C7 fatty acids were elevated during therapy compared to the control measured for comparison purposes (see Figures 2 and 3) by gas chromatography–mass spectrometry.

3.8. Follow-Up

After two years of therapy, the patients' parents decided to discontinue triheptanoin therapy because the success in daily life fell below the parents' expectations. During the follow-up six months after stopping the therapy, patient 1's pronunciation had deteriorated, and there was also again a marked reduction in body tension without therapy.

4. Discussion

The present study provides the first description of triheptanoin in the treatment of aconitase 2 deficiency.

Mutations in *ACO2* have been described in association with early fatal or neurodegenerative diseases. Clinical symptoms range from mildly affected patients with isolated optic atrophy to severely affected patients, including intellectual disability and seizures in early childhood. Also, hypotonia, optic nerve atrophy, cortical atrophy, and cerebellar atrophy are possibly occurring [3,7,9].

Previous studies suggested a correlation between the severity of clinical symptoms and aconitase 2 activity. In these studies, aconitase 2 activity ranged from over 60% to 5% in skin fibroblasts from patients with homozygous or compound heterozygous mutations [1,9,19]. In the case of reduced enzyme activities, a certain tissue specificity must always be considered. The degree of reduction is possibly more pronounced in other, more energy-demanding tissues than in skin fibroblasts.

Patients with higher aconitase 2 activity had a milder clinical phenotype. The same studies could show that aconitase 2 activity did not correlate with aconitase 2 protein levels [1,9,19].

While the protein level of patient 1 decreased (not significantly), there was a reduction in enzyme activity to 40%. Compared to the patients described by Metodiev et al., the phenotype is milder compared to 5% remaining activity but more severe compared to 60% remaining enzyme activity [19]. Additionally, previous investigations found a decreased mitochondrial respiratory capacity, ATP turnover, cellular ATP level, and mtDNA copy number in *ACO2*-deficient cells [1–3].

Assuming that aconitase deficiency results in insufficient energy provision, bypassing the defective aconitase 2 and increasing the mitochondrial respiratory capacity, ATP turnover, and cellular ATP level was the treatment approach. Due to the fact that aconitase 2 converts citrate to isocitrate, which cannot proceed normally in *ACO2*-deficient cells, direct isocitrate supplementation might be beneficial. Indeed, Chen et al. found an increase in cellular ATP levels in *ACO2*-deficient cells compared to controls after isocitrate supplementation in vitro [2]. Preparation of isocitrate is possible, e.g., by using yeast fermentation of sunflower oil [20]. The production of large quantities of isocitrate by this method is costly. Due to a lack of information about isocitrate uptake and metabolism in vivo, an alternative was sought. Physiologically, in CAC, isocitrate is converted to α-ketoglutarate, the next potential target for anaplerotic therapy. Direct α-ketoglutarate supplementation is possible with ornithine α-ketoglutarate; indirect supplementation is possible with a precursor of α-ketoglutarate. Longo et al. tested both ornithine α-ketoglutarate and glutamine as a precursor. They found neither ornithine α-ketoglutarate nor glutamine could increase downstream intermediates of CAC [21].

It was decided to start a trial with triheptanoin for compassionate use because it was described as having good potential in metabolic and neurodegenerative diseases [22–25]. Additionally, the application is easy to handle in homecare and has few side effects at the right dose. Triheptanoin is a synthetic seven-carbon fatty acid triglyceride compound where three C7 fatty acids (heptanoate) are esterified to one glycerol. Triheptanoin comes as an oil, which, when ingested, probably disseminates like other medium-chain triglycerides (MCT) fatty acids [26] via the venous system, mainly to the liver [27–29]. At the target cells, no active transport is required. Heptanoate diffuses through the cell and mitochondrial membrane to enter ß-oxidation by the medium-chain fatty-acid oxidase enzyme. Besides liver cells, heptanoate can be oxidized in mitochondria by most tissues [24].

Through two cycles of ß-oxidation, one molecule of heptanoate is degraded into two molecules of acetyl-CoA and one propionyl-CoA. While acetyl-CoA can enter CAC directly via citrate synthase, propionyl-CoA can be carboxylated and converted to succinyl-CoA via methylmalonyl-CoA by propionyl-CoA carboxylase, methylmalonyl-CoA epimerase, and methylmalonyl-CoA mutase. Formed succinyl-CoA can then enter CAC as an intermediate and might be used for gluconeogenetic reactions [23,27].

A second opportunity for heptanoate metabolization is to undergo just one cycle of ß-oxidation. The result is one acetyl-CoA and one remaining pentanoyl-CoA. Then, pentanoyl-CoA can be used to generate the C5-ketone bodies β-ketopentanoate and β-hydroxypentanoate. The C5-ketone bodies can cross the blood-brain barrier in order to support the brain. Subsequently, the ketone bodies are converted by the peripheral tissues to acetyl-CoA and anaplerotic propionyl-CoA [23,27]. As shown in Figure 1, triheptanoin feeds the CAC behind aconitase 2 as succinyl-CoA.

Due to the rarity of aconitase 2 deficiency, there is, to date, no specific monitoring test. For the presented patients, problems with motor mobility were the most noticeable and limiting restrictions in their daily routine. To achieve a general overview of treatment response, motor ability with Q-Motor assessments, HFMS, and CHOP-intended tests were monitored. Special monitoring with regard to, e.g., retinal dystrophy was performed in standard care and was not considered within the framework of this investigation paper.

Q-Motor assessments are usually used for neurodegenerative Huntington's disease, which presents with a variety of motor symptoms. Active cooperation of the patient is required in order to perform the tests; therefore, only patient 1 was able to take part in Q-Motor assessments.

Q-Motor assessments provide standardized measurements at different time points to avoid bias induced by examiners or study side effects. Data analysis is free from bias through blinded and automated analysis [10]. For patient 1, there was an indication of improved tapping performance in the right hand and the left foot (see Figure 7). Simultaneously with Q-Motor assessments, physical therapy status with the Hammersmith Functional Motor Scale for spinal muscular atrophy (HFMS) for patient 1 and the CHOP-intended test for patient 2 were measured. HFMS is a functional motor scale for children with spinal muscular atrophy, in particular for those with limited mobility, to provide objective information on motor ability and clinical progression [11].

The CHOP-intended test is also used to evaluate the motor skills of patients with spinal muscular atrophy, but even for infants [12].

During therapy with triheptanoin, motor abilities increased in both patients. The motor abilities of patient 1 increased from 35/66 to 46/66 points (HFMS) after two months of therapy. Patient 2 improved from 44/64 points to 56/64 points (CHOP-intended test). Even though these motor ability tests were not developed specifically for aconitase 2 deficiency and may not accurately capture all aspects of the described symptoms, this work shows that at least some of the symptomatic complex of aconitase 2 deficiency is caused by insufficient energy provision.

5. Conclusions

Pathogenic variants in *ACO2* are rare but can be found by whole-exome sequencing. When *ACO2* activity is reduced, anaplerotic triheptanoin might be beneficial. After ingestion, systemic uptake of triheptanoin has been proven by gas chromatography–mass spectrometry of the patients' plasma. By measuring the elevated levels of the excretory organic acids succinate, malate, and fumarate in the urine (in CAC behind defective *ACO2*), anaplerosis and further metabolization could be shown. Improved motor abilities, verified by Q-Motor assessments, CHOP-intended tests, and HFMS, are attributable to anaplerotic triheptanoin.

Author Contributions: Conceptualization, M.P., A.P., and T.M.; methodology, M.P., J.A.M., R.G.F., R.R., O.D., M.F., A.P., J.R., and S.R.; formal analysis, R.G.F., J.A.M., J.R., A.P., S.R., and R.R.; investigation, M.P., A.P., O.D., and T.M.; writing—original draft preparation, M.P.; writing—review and editing, T.M.; visualization, M.P.; supervision, T.M. All authors have read and agreed to the published version of the manuscript.

Funding: This research was funded by the European Joint Programme on Rare Diseases, project EJPRD19-145 GENOMIT I4695-B, financed by the Austrian Science Fund FWF to JAM.

Institutional Review Board Statement: This study was conducted in accordance with the Declaration of Helsinki and approved by the Ethics Committee of the University Clinic of Münster (UKM), Germany (31 May 2013).

Informed Consent Statement: Informed consent was obtained from all subjects involved in the study. Written, informed consent has been obtained from the patients' parents to publish this paper.

Data Availability Statement: The original contributions presented in the study are included in the article, further inquiries can be directed to the corresponding authors.

Conflicts of Interest: The authors declare no conflicts of interest.

References

1. Sadat, R.; Barca, E.; Masand, R.; Donti, T.R.; Naini, A.; De Vivo, D.C.; DiMauro, S.; Hanchard, N.A.; Graham, B.H. Functional Cellular Analyses Reveal Energy Metabolism Defect and Mitochondrial DNA Depletion in a Case of Mitochondrial Aconitase Deficiency. *Mol. Genet. Metab.* **2016**, *118*, 28–34. [CrossRef] [PubMed]
2. Chen, Y.; Cai, G.H.; Xia, B.; Wang, X.; Zhang, C.C.; Xie, B.C.; Shi, X.C.; Liu, H.; Lu, J.F.; Zhang, R.X.; et al. Mitochondrial Aconitase Controls Adipogenesis through Mediation of Cellular ATP Production. *FASEB J.* **2020**, *34*, 6688–6702. [CrossRef] [PubMed]
3. Neumann, M.A.C.; Grossmann, D.; Schimpf-Linzenbold, S.; Dayan, D.; Stingl, K.; Ben-Menachem, R.; Pines, O.; Massart, F.; Delcambre, S.; Ghelfi, J.; et al. Haploinsufficiency Due to a Novel *ACO2* Deletion Causes Mitochondrial Dysfunction in Fibroblasts from a Patient with Dominant Optic Nerve Atrophy. *Sci. Rep.* **2020**, *10*, 16736. [CrossRef] [PubMed]
4. Ciccarone, F.; Vegliante, R.; Di Leo, L.; Ciriolo, M.R. The TCA Cycle as a Bridge between Oncometabolism and DNA Transactions in Cancer. *Semin. Cancer Biol.* **2017**, *47*, 50–56. [CrossRef] [PubMed]
5. Xin, J.C.; Wang, X.; Kaufman, B.A.; Butow, R.A. Aconitase Couples Metabolic Regulation to Mitochondrial DNA Maintenance. *Science* **2005**, *307*, 714–717. [CrossRef]
6. Guehlouz, K.; Foulonneau, T.; Amati-Bonneau, P.; Charif, M.; Colin, E.; Bris, C.; Desquiret-Dumas, V.; Milea, D.; Gohier, P.; Procaccio, V.; et al. *ACO2* Clinicobiological Dataset with Extensive Phenotype Ontology Annotation. *Sci. Data* **2021**, *8*, 205. [CrossRef] [PubMed]
7. Charif, M.; Gueguen, N.; Ferré, M.; Elkarhat, Z.; Khiati, S.; LeMao, M.; Chevrollier, A.; Desquiret-Dumas, V.; Goudenège, D.; Bris, C.; et al. Dominant *ACO2* Mutations Are a Frequent Cause of Isolated Optic Atrophy. *Brain Commun.* **2021**, *3*, 205. [CrossRef] [PubMed]
8. Sharkia, R.; Wierenga, K.J.; Kessel, A.; Azem, A.; Bertini, E.; Carrozzo, R.; Torraco, A.; Goffrini, P.; Berti, C.C.; McCormick, M.E.; et al. Clinical, Radiological, and Genetic Characteristics of 16 Patients with *ACO2* Gene Defects: Delineation of an Emerging Neurometabolic Syndrome. *J. Inherit. Metab. Dis.* **2019**, *42*, 264–275. [CrossRef] [PubMed]
9. Spiegel, R.; Pines, O.; Ta-Shma, A.; Burak, E.; Shaag, A.; Halvardson, J.; Edvardson, S.; Mahajna, M.; Zenvirt, S.; Saada, A.; et al. Infantile Cerebellar-Retinal Degeneration Associated with a Mutation in Mitochondrial Aconitase, *ACO2*. *Am. J. Hum. Genet.* **2012**, *90*, 518–523. [CrossRef]
10. Reilmann, R.; Schubert, R. *Motor Outcome Measures in Huntington Disease Clinical Trials*; Elsevier B.V.: Amsterdam, The Netherladns, 2017; Volume 144, ISBN 978-0-12-801893-4.
11. Main, M.; Kairon, H.; Mercuri, E.; Muntoni, F. The Hammersmith Functional Motor Scale for Children with Spinal Muscular Atrophy: A Scale to Test Ability and Monitor Progress in Children with Limited Ambulation. *Eur. J. Paediatr. Neurol.* **2003**, *7*, 155–159. [CrossRef]
12. Glanzman, A.M.; Mazzone, E.; Main, M.; Pelliccioni, M.; Wood, J.; Swoboda, K.J.; Scott, C.; Pane, M.; Messina, S.; Bertini, E.; et al. The Children's Hospital of Philadelphia Infant Test of Neuromuscular Disorders (CHOP INTEND): Test Development and Reliability The CHOP INTEND Is a Reliable Measure of Motor Skills in Patients with SMA-I and Neuromuscular Disorders Presenting in Infancy. *Neuromuscul. Disord.* **2010**, *20*, 155–161. [CrossRef] [PubMed]
13. Schwarz, J.M.; Cooper, D.N.; Schuelke, M.; Seelow, D. Mutationtaster2: Mutation Prediction for the Deep-Sequencing Age. *Nat. Methods* **2014**, *11*, 361–362. [CrossRef] [PubMed]
14. Ng, P.C.; Henikoff, S. Predicting Deleterious Amino Acid Substitutions. *Genome Res.* **2001**, *11*, 863–874. [CrossRef]
15. Choi, Y.; Sims, G.E.; Murphy, S.; Miller, J.R.; Chan, A.P. Predicting the Functional Effect of Amino Acid Substitutions and Indels. *PLoS ONE* **2012**, *7*, e46688. [CrossRef]
16. Adzhubei, I.A.; Schmidt, S.; Peshkin, L.; Ramensky, V.E.; Gerasimova, A.; Bork, P.; Kondrashov, A.S.; Sunyaev, S.R. A Method and Server for Predicting Damaging Missense Mutations. *Nat. Methods* **2010**, *7*, 248–249. [CrossRef]
17. Kisiel, M.A.; Klar, A.S. Isolation and Culture of Human Dermal Fibroblasts. In *Skin Tissue Engineering*; Böttcher-Haberzeth, S., Biedermann, T., Eds.; Methods in Molecular Biology; Springer: New York, NY, USA, 2019; Volume 1993, pp. 71–78. ISBN 978-1-4939-9472-4.

18. Bartolucci, G.; Pallecchi, M.; Menicatti, M.; Moracci, L.; Pucciarelli, S.; Agostini, M.; Crotti, S. A Method for Assessing Plasma Free Fatty Acids from C2 to C18 and Its Application for the Early Detection of Colorectal Cancer. *J. Pharm. Biomed. Anal.* **2022**, *215*, 114762. [CrossRef]
19. Metodiev, M.D.; Gerber, S.; Hubert, L.; Delahodde, A.; Chretien, D.; Gérard, X.; Amati-Bonneau, P.; Giacomotto, M.C.; Boddaert, N.; Kaminska, A.; et al. Mutations in the Tricarboxylic Acid Cycle Enzyme, Aconitase 2, Cause Either Isolated or Syndromic Optic Neuropathy with Encephalopathy and Cerebellar Atrophy. *J. Med. Genet.* **2014**, *51*, 834–838. [CrossRef] [PubMed]
20. Heretsch, P.; Thomas, F.; Aurich, A.; Krautscheid, H.; Sicker, D.; Giannis, A. Synthesen Mit Einem Chiralen Baustein Aus Dem Citratzyklus: (2R,3S)-Isocitronensäure Aus Einer Fermentation Mit Sonnenblumenöl. *Angew. Chem.* **2008**, *120*, 1984–1986. [CrossRef]
21. Longo, N.; Price, L.B.; Gappmaier, E.; Cantor, N.L.; Ernst, S.L.; Bailey, C.; Pasquali, M. Anaplerotic Therapy in Propionic Acidemia. *Mol. Genet. Metab.* **2017**, *122*, 51–59. [CrossRef]
22. Wehbe, Z.; Tucci, S. Therapeutic Potential of Triheptanoin in Metabolic and Neurodegenerative Diseases. *J. Inherit. Metab. Dis.* **2019**, *43*, 385–391. [CrossRef]
23. Roe, C.R.; Sweetman, L.; Roe, D.S.; David, F.; Brunengraber, H. Treatment of Cardiomyopathy and Rhabdomyolysis in Long-Chain Fat Oxidation Disorders Using an Anaplerotic Odd-Chain Triglyceride. *J. Clin. Investig.* **2002**, *110*, 259–269. [CrossRef] [PubMed]
24. Mochel, F.; DeLonlay, P.; Touati, G.; Brunengraber, H.; Kinman, R.P.; Rabier, D.; Roe, C.R.; Saudubray, J.M. Pyruvate Carboxylase Deficiency: Clinical and Biochemical Response to Anaplerotic Diet Therapy. *Mol. Genet. Metab.* **2005**, *84*, 305–312. [CrossRef] [PubMed]
25. Schwarzkopf, T.M.; Koch, K.; Klein, J. Reduced Severity of Ischemic Stroke and Improvement of Mitochondrial Function after Dietary Treatment with the Anaplerotic Substance Triheptanoin. *Neuroscience* **2015**, *300*, 201–209. [CrossRef] [PubMed]
26. Lemarié, F.; Beauchamp, E.; Legrand, P.; Rioux, V. Revisiting the Metabolism and Physiological Functions of Caprylic Acid (C8:0) with Special Focus on Ghrelin Octanoylation. *Biochimie* **2016**, *120*, 40–48. [CrossRef] [PubMed]
27. Kinman, R.P.; Kasumov, T.; Jobbins, K.A.; Thomas, K.R.; Adams, J.E.; Brunengraber, L.N.; Kutz, G.; Brewer, W.U.; Roe, C.R.; Brunengraber, H. Parenteral and Enteral Metabolism of Anaplerotic Triheptanoin in Normal Rats. *Am. J. Physiol. Endocrinol. Metab.* **2006**, *291*, 860–866. [CrossRef] [PubMed]
28. Gu, L.; Zhang, G.F.; Kombu, R.S.; Allen, F.; Kutz, G.; Brewer, W.U.; Roe, C.R.; Brunengraber, H. Parenteral and Enteral Metabolism of Anaplerotic Triheptanoin in Normal Rats. II. Effects on Lipolysis, Glucose Production, and Liver Acyl-CoA Profile. *Am. J. Physiol. Endocrinol. Metab.* **2010**, *298*, 362–371. [CrossRef]
29. Aoyama, T.; Nosaka, N.; Kasai, M. Research on the Nutritional Characteristics of Medium-Chain Fatty Acids. *J. Med. Investig.* **2007**, *54*, 385–388. [CrossRef]

molecules

MDPI

Article

Hemp Seed Oil Inhibits the Adipogenicity of the Differentiation-Induced Human Mesenchymal Stem Cells through Suppressing the Cannabinoid Type 1 (CB1)

Albatul S. Almousa [1,2,*], Pandurangan Subash-Babu [1,*], Ibrahim O. Alanazi [3,4], Ali A. Alshatwi [1], Huda Alkhalaf [3], Eman Bahattab [3], Atheer Alsiyah [5] and Mohammad Alzahrani [6]

1 Department of Food Science and Nutrition, College of Food Science and Agriculture, King Saud University, P.O. Box 2460, Riyadh 11451, Saudi Arabia; alshatwi@ksu.edu.sa
2 Department of Human Nutrition, College of Home Economics, King Khalid University, P.O. Box 3236, Abha 10001, Saudi Arabia
3 The Healthy Aging Research Institute, King Abdulaziz City for Science and Technology, P.O. Box 6086, Riyadh 11442, Saudi Arabia; iaenazi@kacst.edu.sa (I.O.A.); halkhalaf@kacst.edu.sa (H.A.); ebahattab@kacst.edu.sa (E.B.)
4 Genome Research Unit, Department of Biochemistry, College of Science, King Saud University, P.O. Box 2460, Riyadh 11451, Saudi Arabia
5 The Applied Genomics Research Institute, King Abdulaziz City for Science and Technology, P.O. Box 6086, Riyadh 11442, Saudi Arabia; aalsayah@kacst.edu.sa
6 Institute of Advanced Agricultural and Food Technologies, King Abdulaziz City for Science and Technology, P.O. Box 6086, Riyadh 11442, Saudi Arabia; mzahrani@kacst.edu.sa
* Correspondence: 435203963@students.ksu.edu.sa (A.S.A.); subash@ksu.edu.sa (P.S.-B.)

Abstract: Central and peripheral mechanisms of the endocannabinoid system (ECS) favor energy intake and storage. The ECS, especially cannabidiol (CBD) receptors, controls adipocyte differentiation (hyperplasia) and lipid accumulation (hypertrophy) in adipose tissue. In white adipose tissue, cannabidiol receptor 1 (CB1) stimulation increases lipogenesis and inhibits lipolysis; in brown adipose tissue, it decreases mitochondrial thermogenesis and biogenesis. This study compared the availability of phytocannabinoids [CBD and Δ9-tetrahydrocannabinol (THC)] and polyunsaturated fatty acids [omega 3 (ω3) and omega 6 (ω6)] in different hemp seed oils (HSO). The study also examined the effect of HSO on adipocyte lipid accumulation by suppressing cannabinoid receptors in adipogenesis-stimulated human mesenchymal stem cells (hMSCs). Most importantly, Oil-Red-O′ and Nile red tests showed that HSO induced adipogenic hMSC differentiation without differentiation agents. Additionally, HSO-treated cells showed increased peroxisome proliferator-activated receptor gamma (PPARγ) mRNA expression compared to controls (hMSC). HSO reduced PPARγ mRNA expression after differentiation media (DM) treatment. After treatment with HSO, DM-hMSCs had significantly lower CB1 mRNA and protein expressions than normal hMSCs. HSO treatment also decreased transient receptor potential vanilloid 1 (TRPV1), fatty acid amide hydrolase (FAAH), and monoacylglycerol lipase (MGL) mRNAs in hMSC and DM-hMSCs. HSO treatment significantly decreased CB1, CB2, TRPV1, and G-protein-coupled receptor 55 (GPCR55) protein levels in DM-hMSC compared to hMSC in western blot analysis. In this study, HSO initiated adipogenic differentiation in hMSC without DM, but it suppressed CB1 gene and protein expression, potentially decreasing adipocyte lipid accumulation and lipogenic enzymes.

Keywords: stem cells; cannabidiol; cannabinoid receptor; adipogenesis; lipogenic enzymes

1. Introduction

The endocannabinoid system (ECS) is a complicated cell signaling system that governs many aspects of the body's physiology, including appetite, mood, pain, sensation, and immunological function. Recent developments shed light on how the ECS regulates the metabolism of adipose tissue. ECS influences various physiological processes related to adipocytes, such as phenotype determination and stimulating preadipocyte proliferation [1]. The primary components of this ECS included endocannabinoids, cannabinoid receptors and non-cannabinoid receptors [2].

The most well-known endocannabinoids are anandamide (AEA) and 2-arachidonoylglycerol (2-AG). Endocannabinoids are produced by the body basis of necessary, serving as signaling molecules that selectively bind to specific receptors. CB1 and CB2 receptors are the two main types of cannabinoids. The CB1 receptor is found primarily in the brain and central nervous system, whereas the CB2 receptor is most frequently located in the Immune system and other body tissues. The CB1 receptor was identified in adipocytes in 2003 marked the initial recognition of ECS specific role in adipocyte physiology [1]. In addition to CB1 and CB2, G-protein-coupled receptor 55 (GPCR55) and transient receptor potential vanilloid 1 (TRPV1) have been recently identified as non-cannabinoid receptors of the ECS [3,4]. Moreover, the ECS primary enzymes involved in this process are fatty acid amide hydrolase (FAAH), which breaks down anandamide, and monoacylglycerol lipase (MGL), which breaks down 2-AG [5].

There are naturally occurring substances that regulate the ECS. For example, the psychoactive Δ9-tetrahydrocannabinol (THC) derived from cannabis mimics endocannabinoid activity and activates ECS [3]. However, the non-psychoactive cannabidiol (CBD) blocks the CB_1 receptor, leading to the downregulation of the ECS [6]. Furthermore, a high intake of omega 3 (ω3) downregulates ECS in adipose tissue, heart, and liver [7], while a high intake of omega 6 (ω6) upregulates ECS, suggesting an inverse relationship between ECS tone and ω3 levels and a positive relationship between ECS tone and ω6 levels in the body. Hemp seeds, derived from the *Cannabis sativa* L. plant, have been a crucial source of food, fiber, and medicine for centuries [8]. Hemp seeds contain elevated levels of the natural CBD and a trace level of THC [9]. Additionally, hemp seed contains major essential fatty acids, including ω3 and ω6, representing the optimal ratio of ω3:ω6 (1:3) [8].

Distinct dietary bioactive compounds can impact various stages of the cell cycle. Human mesenchymal stem cells (hMSCs) may experience differentiation to new cell types, including adipocytes, in response to natural products [10]. Hemp seeds represent most compounds that alter the ECS, including polyunsaturated fatty acids (ω3 and ω6), THC, and CBD. Nevertheless, the increasing recreational use of hemp seed makes comprehending how this compound affects the ECS balance essential. Therefore, we hypothesize that hemp seed oil may have a regulatory effect on the ECS, controlling adipogenesis, fat accumulation, and hMSC differentiation to the preadipocyte stage. This research will focus on different aspects of hemp seed oil (HSO). Initially, the hemp seed oil was extracted using shell (S) and no-shell (NS) seeds with different extraction methods, and the phytoconstituents were compared to cold-pressed oil. Furthermore, we studied its ability on adipogenic differentiation among hMSCs, and the involvement of the ECS system downstream gene and protein signaling cascade molecules expression levels have been analyzed. Our results may lead to the involvement of hemp seed and its derivatives in the alteration of the ECS system and its association with lipid metabolism during the early stages of adipocyte differentiation.

2. Results

2.1. Soxhlet Extraction Method Produces More Oil and Cold-Pressed Hemp Seed Oil Has the Optimal Ratios to Carry out the Research

Due to inconsistencies in the scientific literature regarding the chemical components of hemp seeds [11], we sought to test the products used in this study to establish a more coherent understanding of the seeds and select the optimal product for the research. To this end, HSO was extracted from shelled (S) and no-shell (NS) hemp seeds following three different extraction techniques [dynamic maceration (DyM), Soxhlet (Sox), and cannabinoids extracting method (CEM)] using methanol to chloroform (3:1) solution (10 g/experiment). Extracted oils were then compared to the ready-to-use cold-pressed hemp seed oil using GC–MS and HPLC to identify the composition of lipids and cannabinoids. Table 1 shows the oil yield of each extraction method and its average ± SD. Sox extraction had the most oil yield, with an average of 3.8 mL compared to 3 mL from DyM experiments and 2.75 mL from the CEM experiments. Also, it is well noted that Sox extraction from NS hemp seeds significantly yielded more oil than DyM and CEM, S hemp seeds extraction (p-value ≤ 0.002) (Table 1).

Table 1. Hemp seed oil yield following different extraction methods.

Type of Seed	Extraction Method	Average (mL)	Average (mL) by Extraction Method
S	DyM	2.4 ± 0.39	DyM
	Sox	3.3 ± 0.50	3 mL
	CEM	2.5 ± 0.25	Sox
NS	DyM	3.6 ± 0.52 ^{d}	3.8 mL
	Sox	4.3 ± 0.19 ^{a,b,c}	CEM
	CEM	3 ± 0.12	2.75 mL

Oil was extracted from 10 g of seeds. Each value is mean ± SD Sox (NS): ^{a}—$p = 0.002$ compared to DM (S), ^{b}—$p = 0.002$ compared to CEM (S), ^{c}—$p = 0.024$ compared to CEM (NS). DM (NS): ^{d}—$p = 0.038$ compared to DM (S). DyM: dynamic maceration; Sox: Soxhlet; CEM: cannabinoid extraction method; S: shilled hemp seed; NS: not-shilled hemp seed.

GC–MS results showed that Soxhlet (Sox) experiments yield the highest amount of ω3 and ω6 compared to DyM and CEM. In addition, long-chain saturated fatty acids (LCSFA), including palmitic acid and stearic acid, were present among all extraction methods; however, Sox extraction presented both acids. Meanwhile, DyM presented 1.79% stearic acid from S hemp seeds only. On the other hand, CEM presented palmitic acid only, and it was higher in S hemp seeds compared to NS hemp seeds. Additionally, CEM yields a remarkably high amount of ω6 (60.71%) compared to a meager amount of ω3 (0.695%). Regarding CBD and THC, they were not detected by GC–MS in all samples. Table 2 displays the compounds detected by GC–MS and their percentage.

Table 2. Hemp seed extracted oil GC–MS analysis.

Oil Content	DyM		Sox		CEM		HSO
	S	NS	S	NS	S	NS	Cold-Pressed
ω3	30.65%	14.46%	28.78%	34.08% *	1.24%	0.69%	13.36%
ω6	41.98%	20.4%	64.74% *	55.02%	54.06%	60.71%	25.09%
Vitamin E	0	0	0	0	0	2.65% *	0
Palmitic acid	0	0	4.13%	6.78%	7.68% *	5.71%	5.02%
Stearic acid	1.79%	0	0.89%	2.20% *	0	0	1.87%
Pentadecanoic acid	0	0	0	0	0	0	3.80% *

* The highest among all extraction methods.

HPLC results showed a trace amount of CBD but not THC in all samples. The highest amount of CBD was found in the cold-pressed HSO (0.327 ppm) compared to Sox (S) (0.162 ppm) and Sox (NS) (0.144 ppm). Table 3 displays the amount of CBD detected by HPLC and their concentration (ppm). GC–MS chromatograms of Sox extraction of NS hemp seeds, Sox extraction of S hemp seed oil, and HPLC chromatogram of HSO (Sox extraction) are presented in Figure S1. The observed results confirmed that cold-pressed techniques possess the favorable ratio of ω3 and ω6 fatty acid, and the availability of highest CBD compared to the Sox, DyM, and CEM methods of extraction. So, further studies were carried out using cold-pressed oil.

Next, to examine the cytotoxic effect of HSO on hMSC, we seeded hMSCs at 5×10^{4} cells/well in a 96-well plate and incubated for 24 h. HSO was emulsified using tween-20 at various concentrations (0.5%, 1%, 1.5%, 2%, 2.5%, 3%, 3.5%), then mixed with medium. Later, cells were treated with MTS in accordance with the directions given by the manufacturer after 72 h. A microplate reader was used to measure the absorbance at 570 nm. MTS assay indicated that the lower range of HSO was not toxic. A higher concentration of the HSO was required to kill the hMSCs. As shown in Figure 1, the IC_{50} value for HSO-treated cells was (1.922% ± 0.342). Ultimately, two physiobiological doses of HSO (0.1% and 0.05%) were carried out through all experiments.

Table 3. CBD concentration in hemp seed extracted oils using HPLC.

DyM		Sox		CEM		HSO
S	NS	S	NS	S	NS	Cold-Pressed
0.131	0.141	0.162	0.144	0.101	0.109	0.327 *

* The highest among all extraction methods.

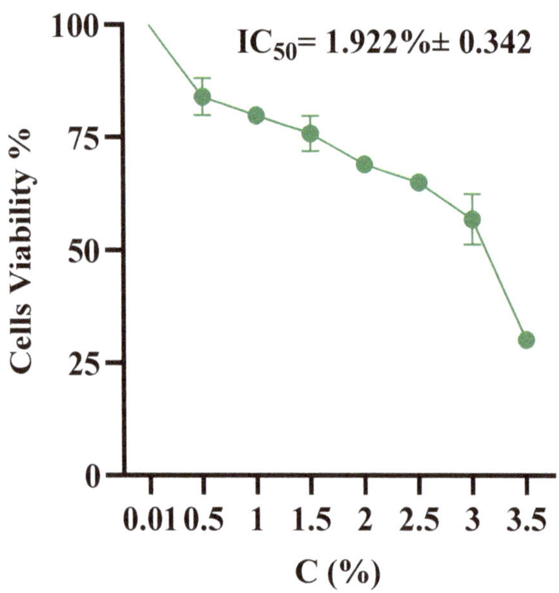

Figure 1. The IC_{50} of hMSC after 72 h incubation with HSO at various concentrations.

2.2. *HSO, CBD, and THC Differentiate hMSC into Adipocyte*

Since active ECS favors adipogenesis, we thought to study the effect of two doses of HSO on hMSC and compare the results with CBD (ECS antagonists) and THC (ECS agonists) with and without differentiation media. Thus, hMSCs were seeded in 24-well plates and allowed to reach 70% confluence. Then, cells were treated with media containing treatment only (TO) or media containing differentiation agents (DM) + treatment for 72 h. Both the groups were treated with 0.05% HSO, 0.1% HSO, 1 μM of CBD, or 1 μM THC. To confirm the adipogenic differentiation among hMSCs, we first performed an Oil Red O (ORO) and Nile red staining. As appears in Figure 2a,b (ORO) and Figure 2c,d (NR), HSO (in a dose-dependent manner), CBD, and THC started the adipogenic differentiation on hMSC with or without DM. Subsequently, the intracellular lipid content was measured by quantifying the absorbance of the released Oil Red O bound at 620 nm after adding 400 μL of 100% isopropanol overnight. As seen in Figure 2e, the ORO bound test showed that all treatment groups were significantly higher in their lipid content compared to their controls (hMSC for TO and DM-hMSC for DM).

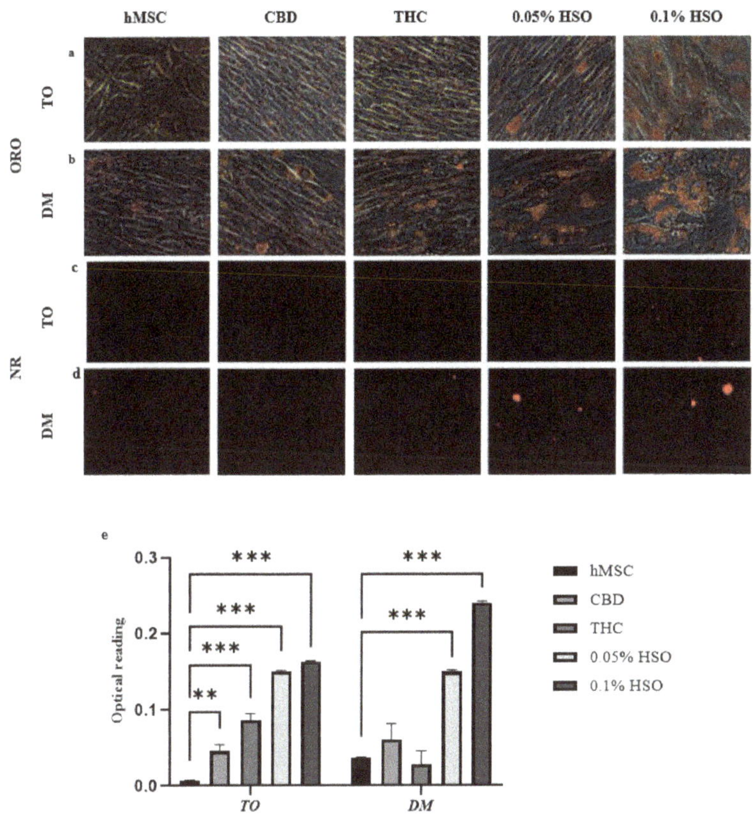

Figure 2. The morphological changes in hMSC after 72 h incubation with CBD, THC, 0.05% HSO, or 0.1% HSO in either Differentiation Media (DM) or Treatment Only (TO) using (**a**,**b**) ORO stain, (**c**,**d**) fluorescence dye—Nile Red at 50µm scale. (**e**) Intracellular lipid content represented by ORO density using 100% absolute isopropanol; the mean ± SD are shown; ANOVA as ** $p \leq 0.01$, *** $p \leq 0.001$.

Second, we sought to confirm the Oil Red O by quantifying the transcription levels of fat accumulation and adipogenic genes: peroxisome proliferator-activated receptor gamma (PPARγ) and CCAAT/enhancer-binding protein alpha (CEBPα). Subsequently, cells were harvested, cDNA was prepared and utilized for quantitative reverse transcription-polymerase chain reaction (qRT-PCR) analysis. Results showed that expression levels of these genes differed significantly between TO and DM (Figure 3). As far as PPARγ expression is concerned, Figure S1 shows the $2^{-\Delta\Delta Ct2}$ (fold change) between each group compared to the control group (hMSC). Although there was a very minimal increase among all TO groups, it was significantly higher than the control group when compared with DM groups (Figure S2). However, because the fold change of the TO groups was minimal, whereas the fold change of the DM groups was enormous in comparison to the control (hMSC), we decided to compare the results to the DM control to determine whether adding the treatments to the DM affected the results significantly. As Figure 3 shows, adding THC to the DM has increased PPARγ expression significantly by 0.573-fold ± 0.132 compared to the DM control ($p \leq 0.001$); meanwhile, adding 0.1% HSO treatment to the DM has decreased PPARγ expression significantly by 0.177-fold ± 0.130 compared to the DM control ($p \leq 0.01$). Similarly, the fold change of the CEBPα mRNA expression among DM groups was higher than TO when compared to the control hMSCs (Figure S1). Also, comparing all groups to the DM control showed that all DM treatments except THC have significantly increased the expression of CEBPα mRNA. Meanwhile, a significant decrease was seen among all TO groups ($p \leq 0.05$) (Figure 3). Interestingly, Figure S2 shows that comparing the groups (TO and DM) separately from each other resulted in the following: among the TO groups, all treatments groups (CBD, THC, 0.05% HSO and 0.1% HSO) have decreased the expression of CEBPα mRNA significantly, and the most decrease was observed among the THC treatment group by 0.9095-fold ± 0.345 ($p \leq 0.001$). However, among the DM groups, adding treatments to the DM has increased the expression of CEBPα mRNA except for the THC treatment that decreased the expression, suggesting that THC might have a direct effect on the expression of CEBPα mRNA (Figure S2).

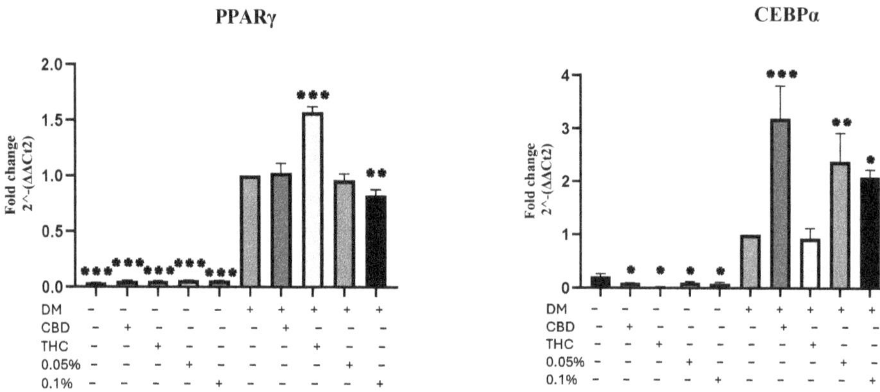

Figure 3. Gene expression. mRNA levels of adipogenic gene (PPARγ and CEBPα) were investigated by quantitative RT-PCR after 72 h of culturing in CBD, THC, 0.05% HSO, or 0.1% HSO treatments with/without DM; the mean ± SD are shown compared to the DM control; ANOVA as * $p \leq 0.05$, ** $p \leq 0.01$, *** $p \leq 0.001$.

2.3. *HSO, CBD, and THC Treatments Regulated the ECS*

Since HSO and its active compounds (CBD and THC) initiated the adipogenic differentiation, we sought to examine whether this effect was related to the activation of the ECS. Thus, we studied the trends of EC receptors (CB1, CB2, TRPV1, GPCR 55) and degrading enzymes (FAAH and MGL) in terms of their mRNAs expression upon treatment with TO or DM. As shown in Figure 4, CB1 mRNA expression compared to the control (hMSC) was significantly downregulated in all groups except for TO-CBD, and TO-0.1% HSO was decreased but not significantly. Meanwhile, CB2 mRNA expression was significantly upregulated upon the treatment with CBD in both TO and DM groups by 0.631 ± 0.212 ($p \leq 0.05$) and 0.915 ± 0.008 ($p \leq 0.01$), respectively. It is possible that CBD is a direct activator of CB2. On the other hand, THC treatment has decreased the CB2 mRNA expression in both groups (TO and DM) but not significantly. Generally, a similar pattern can be noted in the CB2 mRNA expression when comparing the TO groups to the DM groups (Figure 4). Regarding TRPV1 mRNA expression, we found that adding DM has downregulated the expression significantly in all DM groups, except DM-0.1% HSO was not significant. However, TO-0.1% HSO along with TO-CBD has significantly increased the expression of TRPV1 mRNA ($p \leq 0.01$). In terms of GPCR55 mRNA, adding DM has increased the GPCR55 mRNA expression in all DM groups but was only significant among DM control, DM-CBD, and DM-0.05% HSO ($p \leq 0.01$, $p \leq 0.001$, and $p \leq 0.01$, respectively). Interestingly, TO-CBD has also increased the expression of GPCR55 mRNA but not significantly, whereas TO-0.1% HSO has significantly increased the expression of GPCR55 mRNA ($p \leq 0.001$). As far as the EC enzymes FAAH and MGL are concerned, we found that adding DM has significantly decreased FAAH mRNA in all groups ($p \leq 0.001$)

and significantly decreased MGL mRNA in all groups except for the DM-THC group, where it was significantly increased ($p \leq 0.001$). Interestingly, THC treatment has significantly increased both enzymes compared to their controls. THC treatment has increased the expression of MGL mRNA significantly ($p \leq 0.001$) in the TO group. Furthermore, THC treatment in the TO has increased the expression of FAAH mRNA significantly ($p \leq 0.05$), suggesting that there might be a direct relationship between THC treatment and EC-degrading enzymes (FAAH and MGL).

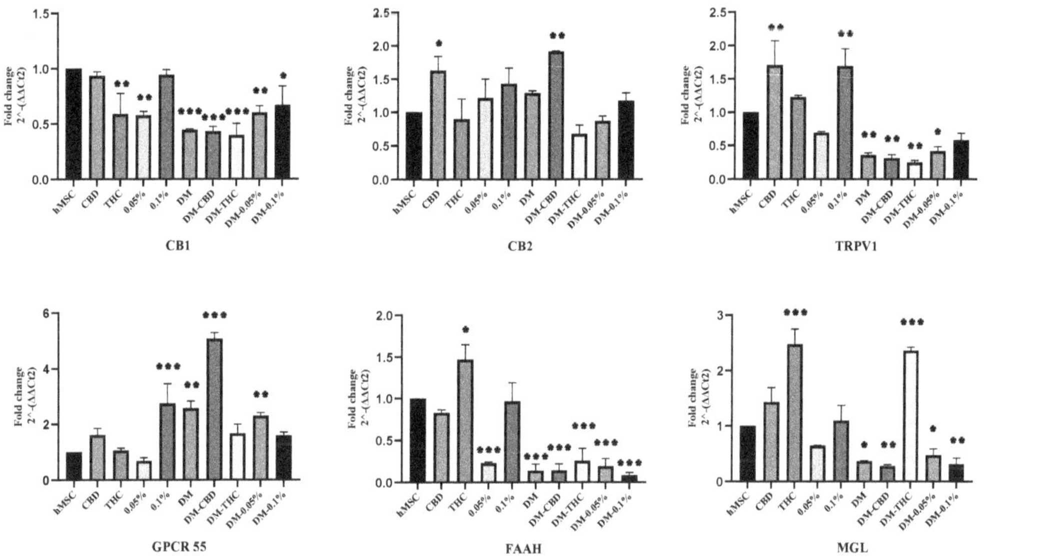

Figure 4. Gene expression. mRNA levels of ECS genes (CB1, CB2, TRPV1, GPCR55, FAAH, and MGL) were investigated by quantitative RT-PCR after 72 h of culturing with CBD, THC, 0.05% HSO, or 0.1% HSO treatment with/out DM; the mean ± SD are shown; ANOVA compared to hMSC as * $p \leq 0.05$, ** $p \leq 0.01$, *** $p \leq 0.001$.

Next, we sought to study the protein expression trends of the EC receptors (CB1, CB2, TRPV1, GPCR55). To this end, cells were harvested, and immunoblotting was performed on whole-cell lysates followed by blot quantification using densitometry scanning. As Figure 4 shows, protein expression of CB1 has confirmed the CB1 mRNA expression and was significantly decreased in all groups ($p \leq 0.001$), and the highest decrease was among the DM-treated groups. As far as CB2 protein expression is concerned, it was significantly decreased in all DM-treated groups, except DM-0.1% HSO was not significant; meanwhile, it was slightly but not significantly increased in all TO groups. It is noteworthy that CB2 protein expression was upregulated upon the treatment of 0.1% HSO when comparing each treated group to their controls. Figure 5d shows that TRPV1 protein expression has upregulated in the CBD- and THC-treated groups in the TO and DM; meanwhile, it was downregulated after the oil treatment (0.05% HSO and 0.1% HSO) in both TO and DM groups. In contrast to the mRNA result, we found that the protein level of TRPV1 in DM, DM CBD, and DM-THC was significantly higher than the control ($p \leq 0.001$). This could be due to the stableness and degradation resistance of the resulting protein even with low levels of mRNA. In addition, some proteins have longer half-lives than their mRNA, despite the rapid degradation of their mRNA [12]. As far as GPCR 55 protein levels are concerned, Figure 5e shows that the expression was notably and significantly decreased in all DM-treated groups compared to the control ($p \leq 0.001$). However, the expression of GPCR55 in DM-THC, DM-0.05% HSO, and DM-0.1% HSO was increased compared to their control (DM control). In addition, GPCR 55 protein expression was significantly increased in the TO-0.05% HSO and TO-CBD ($p \leq 0.001$).

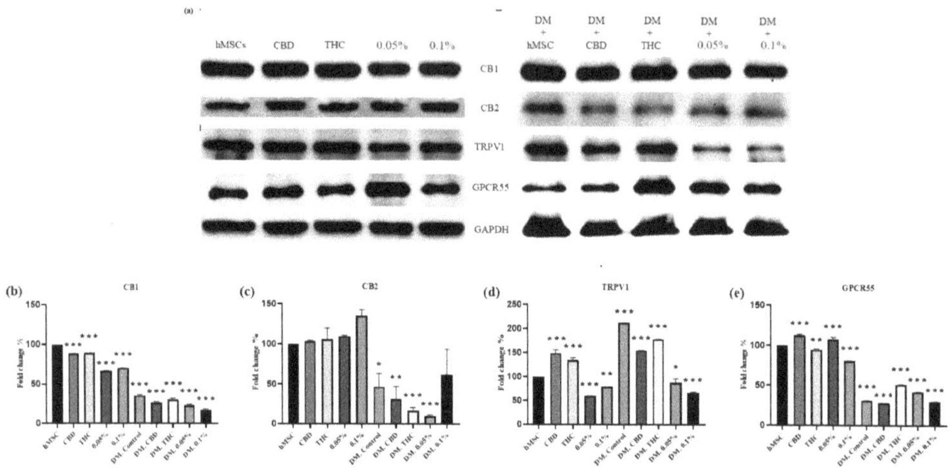

Figure 5. Western blotting of ECS receptors on hMSCs and treatments with or without DM (**a**). Densitometry analysis of CB1, CB2, TRPV1, and GPCR55 are shown in (**b–e**), respectively. The means ± SD are shown; ANOVA as * $p \leq 0.05$, ** $p \leq 0.01$, *** $p \leq 0.001$.

3. Discussion

In this investigation, we aimed to identify the role of active ECS in the effects of HSO on adipogenesis in hMSCs. Cold-pressed HSO used in this study contained around a 0.5:1 ratio of ω3:ω6 and 0.327 ppm CBD. CBD (1 μM) and THC (1 μM) treatments were used for comparison. Our present study confirmed for the first time that HSO was able to initiate adipocyte differentiation among hMSCs without the differentiation media (DM) in a dose-dependent manner (Figure 2). Additionally, HSO treatment increased the size of adipocytes in DM-treated cells (Figure 2). This might be due to the high lipid accumulation in DM-treated cells. Similarly, based on ORO results in Figure 2, we found that CBD and THC treatments in TO and DM have initiated the adipogenic differentiation compared to the control (hMSC). However, THC treatment in the DM-treated cells have decreased intracellular lipid content compared to its control (DM control) (Figure 2e). This could be attributed to the fact that the acid form of THC (THCA) creates fewer and smaller lipid droplets than rosiglitazone in hMSCs [13,14] and, as previously stated, that most of the CBD and THC were appropriately stored in hemp seed oil in the form of acidic precursor [15,16]. Furthermore, as PPAR-γ and CEBP-α are two master transcription factors that govern the differentiation and proliferation of adipocytes [17], we confirmed that the adipogenic gene CEBP-α expression level has been significantly downregulated in HSO-, CBD-, and THC-treated hMSCs compared to its control (hMSC) (Figure S2a); however, adding DM increased the CEBP-α gene expression significantly in HSO- and CBD-treated maturing adipocytes (DM-cells) compared to its control (DM-hMSC) (Figure S2b). Interestingly, we found that THC treatment has downregulated the expression of the CEBP-α mRNA in both TO- and DM-treated cells (Figure S2), suggesting a direct relationship between THC treatment and CEBP-α mRNA expression. However, THC treatment has significantly increased the expression of PPAR-γ mRNA in both TO- and DM-treated cells when compared to the control in each group (Figure S2). It is also noteworthy that all treatments in TO (HSO, CBD, and THC) have increased the expression of PPAR-γ mRNA compared to their control (hMSC). However, in the DM-treated cells, HSO treatment has decreased the PPAR-γ mRNA expression compared to the DM control (Figure S2). Both HSO treatments (0.1% HSO and 0.05% HSO) in the DM-treated cells have decreased the expression of PPAR-γ mRNA—in a dose-dependent manner—by 0.04-fold $\pm$ 0.053 and 0.176-fold $\pm$ 0.057, respectively (Table S2). This suggests that HSO might have a controlling role of lipid accumulation and adipogenesis. In addition, the ability of CBD administration to activate PPAR-γ mRNA and drive adipogenesis in both mice and hMSCs has been validated and identified such that CBD's impact might be attributed to its tendency to stimulate lipogenesis without the balance of lipolysis [18]. CBD in our study has similarly increased the expression of PPAR-γ mRNA significantly ($p > 0.05$) when in hMSCs; however, adding CBD to DM has maintained the same expression levels compared to the DM control.

We also studied the EC receptors' and enzymes' involvement in hMSC adipocyte differentiation with and without differentiation media. Upon discovery of cannabinoid receptor-1 (CB1) and its role in peripheral organs, many studies led to the discovery that CB1 receptors are directly involved in the physiological processes of energy homeostasis in adipocytes and many tissues [19,20]. CB1 receptor agonist in adipocytes and chronic activation in white adipose tissue have been found with impaired mitochondrial functions and decreased mitochondrial biogenesis [20]. In contrast, Roche et al. [19] found that activation of the CB1 receptor with the endocannabinoid 2-AG has increased the intracytoplasmic cyclic adenosine monophosphate (cAMP), which may lead to increased lipolysis and decreased lipid accumulation. Another study found that CB1 antagonist stimulates the expression of mitochondrial biogenesis enzymes, including Krebs cycle, and increased the oxidative phosphorylation potential [20]. In the current study, all treatments—especially HSO—significantly decreased CB1 mRNA expression and protein expression in all groups ($p \leq 0.001$); meanwhile, DM-treated cells showing the highest reduction. The adipogenic effect of HSO on hMSCs is evident by ORO and PPAR-γ mRNA expression—in an early stage—but might not be mediated by EC CB1 receptor. Bajzer et al. [21] confirmed that persistent CB1 antagonism has been linked to increased energy consumption, lipolysis, mitochondrial biogenesis, and activation of thermogenesis. Further, inhibition of adipocyte CB1 potentially influences lipid homeostasis in a positive way and the production of adipokines, which facilitate the lipolysis or inhibit lipid accumulation and hypertrophy [1,22].

As far as cannabinoid receptor-2 (CB2) is concerned, it was previously linked to immune system cells only. However, Roche et al. [19] were able to discover the biological functions of CB2 receptor in human pre-adipocytes and mature adipose tissue; the activation of adipocyte CB2 receptor inhibits cyclic adenosine monophosphate (cAMP) synthesis, which may lead to decreased lipolysis. In the present study, we found that CBD treatment showed strong affinity towards CB2 compared to THC. CBD treatment has significantly increased the CB2 mRNA expression in both TO- and DM-treated cells ($p \leq 0.05$, $p \leq 0.01$, respectively). On the other hand, THC treatment has decreased the CB2 mRNA expression in both TO- and DM-treated cells (not significantly). Furthermore, the CB2 protein levels have been slightly elevated in both CBD and THC treatments in the TO; however, when added to DM-treated cells, both treatments have decreased the CB2 protein levels significantly ($p \leq 0.01$, $p \leq 0.001$, respectively). Meanwhile, HSO treatment, especially 0.1% HSO, has slightly elevated the expression of CB2 mRNA in both TO- and DM-treated cells, and held this elevation even at protein level, suggesting that there are other substances in HSO that activate the CB2 receptor besides CBD, such as arachidonic acid, an $\omega 6$ PUFA that is a precursor for AEA.

Another EC receptor is TRPV1; AEA and other endocannabinoids are natural ligands for TRPV1 [23,24], and it is produced when intracellular Ca^{2+} levels increase [3]. In addition, a cross talk between CB1 and TRPV1 has been demonstrated by several studies [25–28]. CB1 has been demonstrated to either enhance or reduce TRPV1 channel activity [26]. In the present study, we found similar expression trends between TRPV1 and CB1 (Figure 4), as CBD and 0.1% HSO held the highest expression of CB1 compared to other treatment groups, it also increased the expression of TRPV1 mRNA significantly ($p \leq 0.01$). Similarly, as DM has decreased the expression of CB1 in all treatment groups significantly, it also decreased the TRPV1 mRNA expression significantly, except for 0.1% HSO it was not significant. However, this was not the case when it comes to TRPV1 protein expression, where DM-treated cells (DM control), DM-CBD, and DM-THC have significantly increased the TRPV1 protein expression. Meanwhile, HSO treatments (0.05% HSO and 0.1% HSO) have decreased its expression significantly in both TO and DM. It appears that the interaction between CB1 and TRPV1 is dependent on the amount of circulating AEA. Research has shown that TRPV1 may be activated by supplementing high dosages of AEA (1–10 M) [28,29]. In contrast, low doses of AEA (3–30 nM) decrease TRPV1 expression [30,31]. Therefore, the enzymatic production and breakdown of endogenous cannabinoids may be crucial determinants of TRPV1 activity in tissues that co-express TRPV1 and CB1 [25].

The enzymes FAAH and MGL are degrading enzymes of the intracellular endocannabinoids AEA and 2-AG, produced in response to high levels of the endocannabinoid to retain hemostasis of the active ECS [20,32]. In the present study, we found that FAAH mRNA expression was downregulated in all groups except TO-THC treatment. Significant reduction of FAAH mRNA was noted among all DM-treated groups ($p \leq 0.001$). Similar findings were observed on MGL mRNA expression. DM-treated groups have significantly reduced its expression except DM-THC, which significantly increased the expression of MGL mRNA ($p \leq 0.001$). Also, THC treatment in the TO has significantly upregulated both FAAH and MGL mRNAs ($p \leq 0.05$, $p \leq 0.001$, respectively), while CBD and HSO treatments have upregulated MGL mRNA and downregulated FAAH mRNA expression but not significantly. Furthermore, since FAAH is expressed on demand upon high levels of anandamide to retain homeostasis of the active ECS [33], our present study suggests that the ECS might not be highly activated in the differentiation process of hMSC to adipocyte. Also, the balance ratio of ω3 to ω6 in the HSO might have a role in maintaining lower endocannabinoids (AEA and 2-AG) and consequently lower the expression levels of FAAH and MGL, which are associated with low lipid accumulation. In this context, an in vitro study has demonstrated that ω3 polyunsaturated fatty acids (PUFA) have the potential to reduce the amounts of AEA and 2-AG in differentiated adipocytes from mice [34]. Conversely, arachidonic acid (ω6, PUFA) has been seen to elevate endocannabinoid levels [35].

l-α-lysophosphatidylinositol (LPI) is known as a regulator of the intracellular Ca^{2+} accumulation, and it is believed to be the only endogenous ligand of G-protein-coupled receptor-55 (GPCR55) that is presumed to be a cannabinoid receptor [36]. In addition, a positive association between obesity, type 2 diabetes, and GPCR55 expression in visceral adipose tissue (VAT) has been identified [37]. Consistently, obese patients have been identified with higher plasma levels of LPI than lean subjects [36]. In our study, we observed that all groups subjected to DM treatment exhibited a notable upregulation in the expression of GPCR55 mRNA. However, a contrasting trend was observed in the levels of GPCR55 protein, as DM treatment resulted in a significant decrease in its expression across all groups ($p \leq 0.001$). Similar findings were noted in TO-HSO (both doses 0.05% HSO and 0.1% HSO), where TO-0.1% HSO has significantly increased GPCR55 mRNA but significantly reduced the protein levels, whereas TO-0.05% HSO decreased GPCR55 mRNA but significantly increased its protein expression.

Generally, the observed discrepancy between the findings of our research and those of previous studies might perhaps be attributed to variances in the level of maturation or cell type. In our current study, endpoint measures were conducted three days after initiating adipocyte differentiation in hMSCs and treatment, whereas other investigations employed distinct cell types or conducted in vivo experiments, or achieved a fully developed state of adipocyte, often after a differentiation period of 14 days.

4. Materials and Methods

4.1. Plant Products and Chemicals

Raw hemp seeds (shelled-S and not shelled-NS-Canadian source) and cold-pressed hemp seed oil (HSO) (Moreno Valley, CA, USA) were bought from iHerb (Irvine, CA, USA). Liquid Cannabidiol (CBD) and Δ^9-tetrahydrocannabinol (THC) standards were obtained from RESTEK® (Bellefonte, PA, USA). Powder CBD was obtained from Abcam (ab120448) (Cambridge, UK). Methanol, acetonitrile, chloroform, formic acid, acetone, and isopropanol HPLC grade were supplied thankfully by the King Saud University (KSU) chemical warehouse.

4.2. Instrument

Gas Chromatography–Mass Spectrometry (GC–MS) was utilized for the analysis of extracts to determine the chemical components present and their respective concentrations. The column utilized was a DB-5MS with dimensions of 30 m in length, 0.250 mm in diameter, and 0.25 μm in thickness. The flow rate was 3 mL/min, pressure set at 10.42 psi, operating in splitless mode, with an injection volume of 3 μL. The oven temperature was 60 °C for 3 min, then increased to 100 °C at a rate of 3 °C per minute, followed by reaching 200 °C in 1 min at the same rate, and finally reaching 300 °C in 1 min at a rate of 5 °C per minute. GC–MS data were interpreted using FAME and NIST libraries. The chromatogram threshold in all experiments was set at 17.

High-Performance Liquid Chromatography: Ultraviolet/Diode Array Detector (HPLC-UV/DAD): HPLC-UV/DAD (Shimadzu, Colombia, MD, USA) was used to identify and quantify chemical compounds using a standard for each analyte (CBD and THC). Hypersil® column (100 × 4.6 mm, Thermo Scientific, Waltham, MA, USA) with a particle size of 5 μm ODS was used. The temperature was set at 30 °C, and injection volume was 10 microliters. Mobile phase solutions were A: 70% acetonitrile and B: 30% formic acid (0.1%). Samples and standards were prepared following Brighenti et al. [38]. The ratio of samples to isopropanol was 20 μL:4 mL, then vortexed for 30 s, and 2 mL was taken into an HPLC vial.

4.3. Extraction Method

Three extraction techniques (dynamic maceration-DyM, Soxhlet-Sox, and cannabinoids extraction method-CEM) were followed using methanol/chloroform solution (M/C, 3:1). DyM extraction: a modified method of Brighenti et al. [38] was followed, shelled-(S) and no-shell-(NS) hemp seeds (HS) were grounded for 30 s each. An amount of 10 g of extract(s) as initial weight was added with 40 mL of the solvent and placed on a magnetic stirrer for 4 h at room temperature. The solution was filtered using No. 1 Whatman™ filter papers (Maidstone, UK, 125 mm), and the residue was extracted twice with another 40 mL of the solvent, followed by 20 mL for 1 h 30 min and 30 min, respectively, and was placed on magnetic stirring. Finally, the filtrate was evaporated under a vacuum, and the crude extract was collected. Soxhlet-Sox extraction: a modified method of Devi and Khanam [39] was followed. An amount of 10 g of S or NS hemp seed powder was weighed and placed in a thimble (Whatman™ cellulose extraction thimbles 25 mm × 100 mm). Then, 100 mL of solvent was added to keep the solvent and HS ratio at 1:10 g/mL. The procedure was carried out for 6 h, the mixture was evaporated under a vacuum, and the crude extract was collected. CEM extraction: a modified method of Mandrioli et al. [40] was followed, whereby 10 g of S or NS hemp seed powder was mixed with 100 mL of solvent. The oscillator node was placed inside the beaker touching solvent (Physical Field) at 350 hertz and amplitude of 10 vpp and "Sin" wave shape for 30 min. Then, the beaker was placed in a sonic bath for 30 min. The process was repeated on the oscillator for 20 min and a sonic bath for 20 min. The mixture was then filtered into the evaporator flask using No. 1 Whatman™ filter papers (125 mm). Then, the mixture evaporated, and crude oil was collected. All extraction methods were performed in triplicate.

An evaporator was used at 55 °C throughout the experiments, along with a vacuum to collect the solvent-free crude extract. Then, crude extracts were collected in centrifuge tubes for further purification process. Firstly, liquid–liquid extraction (LLE) was utilized, 2 mL of crude extract was placed in a centrifuge tube along with 2 mL of H_2O and 2 mL of methanol, centrifuged at 5000 rpm for 5 min, then 1.5 mL of the solvent layer was taken into an Eppendorf tube. Secondly, a syringe filter (PTFE-0.45 μm- Thermo Scientific, Waltham, MA, USA) was used, followed by another syringe filter (Nylon-0.45 μm- Thermo Scientific,

Waltham, MA, USA), then brought to dryness using a stream of nitrogen. Finally, the residue was diluted in acetone for analysis.

4.4. Cell Culture

The hMSCs were obtained from the American Type Culture Collections (ATCC) in Manassas, VA, USA. Early passage of hMSCs (≤passage 3) were cultured in Dulbecco's Modified Eagle's Medium (DMEM, Gibco, Waltham, MA, USA). The cells were cultured in a medium containing 10% fetal bovine serum (FBS, Gibco™, Grand Island, NY, USA) and 1% penicillin–streptomycin (Gibco™, NY, USA) at 37 °C in a humidified environment with 5% CO_2. At 70% confluency, cells were maintained in either DMEM + treatment only (5 groups: control/DMEM only, CBD, THC, 0.05% HSO, and 0.1% HSO for 72 h or in a differentiation media containing Dexamethasone (1 μM), IBMX (0.5 μM), Rosiglitazone (1 μM), recombinant human insulin (167 nM) for 72 h [41] along with treatments (5 groups: DM-hMSC, DM-CBD, DM-THC, DM-0.05% HSO, and DM-0.1% HSO) for 72 h. HSO was dissolved in 1% tween-20 to ensure homogenization. CBD and THC were dissolved in methanol. Finally, condition media of DM groups were collected for further use.

4.5. Cell Viability Assay

Human mesenchymal stem cells (hMSCs) were placed in a 96-well plate with a concentration of 1×10^4 cells per well and left to incubate for 24 h. Cells were exposed to varying amounts of hemp seed oil (HSO) (0.5%, 1%, 1.5%, 2%, 2.5%, and 3%) for 72 h. HSO was dissolved in 1% tween-20 and emulsified using a sonication device at 30 °C and 40 RPM for 5 min to ensure homogenization. hMSC treated with 0.5% tween-20 was used as a control. Then, cells were treated with MTS in accordance with the directions given by the manufacturer after 72 h. A microplate reader (SPECTROstar® Nano—BMG Labtech, Ortenberg, Germany) was used to measure the absorbance at 570 nm. Three replicate wells were used for each data point throughout the experiments. Ultimately, two physiobiological doses of HSO (0.05%) and (0.1%) were used along with (1 μM) of CBD and THC for comparison.

4.6. Oil Red O and Nile Red Staining

hMSC were seeded in a 24-well plate and allowed to reach 70% confluence. Cells were treated with HSO, CBD, and THC. After 72 h of treatment, growth media was removed, and cells were gently washed using phosphate-buffered saline (PBS). Subsequently, the cells were fixed with a 10% formalin solution for a duration of 1 h at room temperature, then washed three times with deionized water. Oil Red O (ORO) solution (4:6 water to 6% Oil Red O dye in isopropanol) was applied to the cells for 10 min, then unbound Oil Red O was removed by washing with PBS. The stained lipid droplets were captured using a light microscope (ZEISS, Axiocam 506 color, Jena, Germany). Finally, bound ORO were extracted with 100% isopropanol (400 μL/well) overnight, and the concentration of intracellular ORO was assessed, which is directly proportional to lipid accumulation. The absorbance at 620 nm was measured in duplicate wells using a microplate reader.

For Nile Red (NR) staining, media from the experimental cells were discarded and washed with PBS, further NR staining was added to each well. Then, the cells were incubated at room temperature for 10 min, then washed with deionized water to remove the unbound dye before imaging. The stained lipid droplets were captured using the green filter of an inverted florescence microscope (ZEISS, Axiocam 506 color).

4.7. RNA Extraction and Real-Time Polymerase Chain Reaction (RT-PCR)

hMSCs were seeded in a 24-well plate and allowed to reach 70% confluence. Cells were treated with HSO, CBD, and THC for 72 h; then, total RNA was extracted from hMSC using Qiagen, RNEASY MINI KIT (Qiagen, Inc., Tulsa, OK, USA). After that, the concentration and purity of the extracted RNA were measured using Quickdrop (SpectraMax, QuickDrop, Framingham, MA, USA). Then, 1 μg RNA was converted to cDNA using a

High-Capacity cDNA Reverse Transcription Kit (Thermo Fisher Scientific, Vilnius, Lithuania). The QuantStudio 3 (Thermo Fisher Scientific, Waltham, MA, USA), along with SYBR® green Universal Master Mix (Life Technologies Ltd., Paisley, UK), were used to quantify the transcription levels of specific genes (PPARγ, CEBPα, CB1, CB2, TRPV1, GPCR55, FAAH, MGL). For each sample, PCR reactions were repeated twice, and the transcription levels of each gene were standardized to GAPDH (primer sequences are presented in Table S3).

4.8. Immunoblot Analysis

Cell lysates were prepared using RIPA lysis buffer and Halt™ Protease inhibitor (PI) (Thermo Fisher Scientific, USA) at a 1:25 ratio (PI:RIPA). Cell lysates were homogenized for an hour on ice, then centrifuged at 14,000 rpm for 30 min at 4 °C. The Bradford assay (Bio-Rad Protein Assay kit, Hercules, CA, USA) was used to measure protein concentration. A total protein concentration was set at 35 µg/µL for all samples. The sample buffer was prepared at a 1:19 ratio of β-mercaptoethanol:2x-laemmli-sample-buffer. Then, a 1:1 ratio of samples to sample buffer was mixed and heated at 95 °C for 5 min before uploading to the 10% SDS-polyacrylamide gel electrophoresis (PAGE). Gels were transferred to 0.2 µm polyvinylidene difluoride blotting membrane (PVDF, Amersham™ Hybond™, Cytiva, München, Germany) after electrophoresis, then blocked for 1 h with 5% skimmed milk. Next, TBST 1X buffer (50 mL of TBS 20X + 950 mL DW + 1 mL of Tween 20) was used to rinse the membrane three times. Then, the membrane was incubated overnight at 4 °C with primary monoclonal antibodies anti-CB_1 (rabbit, 1:4000), primary polyclonal antibodies anti-CB_2 (rabbit, 1:200), primary polyclonal antibodies anti-$TRPV_1$ (rabbit, 1:500), and primary polyclonal antibodies anti-GPCR55 (rabbit, 1:166) (Abcam). After that, the membrane was incubated for 2 h at room temperature with primary monoclonal antibodies anti-GAPDH (mouse, 1:10,000). The membrane was washed three times and then incubated for 1 h with anti-(rabbit/mouse) IgG secondary antibody horseradish peroxidase-conjugated (1:1000, Abcam) in 5% skim milk.

Finally, images were taken using the ChemiDoc™ Touch Imaging System (Bio-Rad, USA) after incubating the membrane in the Clarity Western ECL Substrate kit (Bio-Rad, USA). Image lab 6.1 software was used to quantify band intensities.

4.9. Statistical Analysis

The mean and standard deviation (SD) of at least duplicate experiments are used to represent all data. In addition, one-way ANOVA was used to examine the statistical significance among multiple groups, followed by the Dunnett test for multiple comparisons using GraphPad Prism 9.4.1 (Windows, GraphPad Software, San Diego, CA, USA) https://www.graphpad.com (accessed on 13 September 2022). Statistical significances were indicated as p-values ≤ 0.05.

5. Conclusions

The endocannabinoid system (ECS) is essential in promoting fat storage in animals through many mechanisms. EC tone in hMSC—including receptors and enzymes—was investigated following treatment with HSO (0.05% HSO and 0.1% HSO), CBD, and THC with and without DM during the early stages of adipogenesis. The agonism and antagonism of CB1 pose a direct effect on adipocytes and regulate energy intake and expenditure through central and peripheral pathways. In the present study, HSO effectively downregulated the CB1 mRNA, and protein expression further reduced the endogenous ligand TRPV1. Endogenous CB1 inhibition neutralizes or reduced the FAAH and MGL expressions, which play a major role in the endocannabinoid AEA and AA degradation. Although HSO independently has the capacity to initiate the adipogenic differentiation among hMSCs and upregulation of adipogenic gene PPAR-γ, HSO inhibits CB1, TRPV1, and PPAR-γ when differentiation was induced by DM. That indicates that HSO alone might have the ability to initiate the differentiation; however, it reduces the lipid accumulation and energy metabolism when differentiation is induced by DM compared to CBD or THC alone. Therefore, hemp seed might be considered as a natural agent for CB1 inhibition, to overcome adipocyte lipogenesis and obesity. Finally, we recommend an in vivo study to confirm whether HSO will continue to inhibit CB1 and define the adipocyte maturation.

Supplementary Materials: The following supporting information can be downloaded at: https://www.mdpi.com/article/10.3390/molecules29071568/s1, Figure S1: GC–MS chromatograms: (a) Sox extraction of NS hemp seeds. (b) Sox extraction of S hemp seed oil. (c) HPLC chromatogram of HSO (cold-pressed); Figure S2: Gene expression. mRNA levels of adipogenic gene (PPARγ and CEBPα) were investigated by quantitative RT-PCR after 72 h of culturing with CBD, THC, 0.05% HSO, or 0.1% HSO treatments without DM (a), or with DM (b); the mean ± SD are shown compared to their controls; ANOVA as * $p \leq 0.05$, ** $p \leq 0.01$, *** $p \leq 0.001$; Table S1: Mean fold change ($2^{-\Delta\Delta Ct2}$) of PPARγ and CEBPα mRNAs expression compared to the control (hMSC); Table S2: Mean fold change ($2^{-\Delta\Delta Ct2}$) of PPARγ and CEBPα mRNAs expression compared to the control (DM-hMSC) and Table S3: Sequences of the primers used for qRT-PCR.

Author Contributions: Conceptualization, A.S.A. and A.A.A.; data curation, A.S.A. and P.S.-B.; formal analysis, A.S.A.; funding acquisition, A.A.A.; investigation, A.S.A., E.B., A.A. and M.A.; methodology, P.S.-B., I.O.A. and H.A.; project administration, P.S.-B.; resources, A.A.A. and H.A.; software, A.S.A., P.S.-B., E.B., A.A. and M.A.; supervision, P.S.-B., I.O.A., A.A.A. and H.A.; validation, P.S.-B., I.O.A. and H.A.; visualization, A.S.A., E.B., A.A. and M.A.; writing—original draft, A.S.A. and P.S.-B.; writing—review and editing, P.S.-B., I.O.A. and H.A. All authors have read and agreed to the published version of the manuscript.

Funding: This research was funded by the Researchers Supporting Project number (RSP2024R178), King Saud University, Riyadh, Saudi Arabia.

Institutional Review Board Statement: Ethical review and approval were waived for this study as it exclusively involved the use of established human cell lines. The use of these cell lines does not involve human subjects directly, and therefore, IRB approval was not required. All experiments were conducted following standard laboratory protocols and in accordance with institutional guidelines.

Informed Consent Statement: Not applicable.

Data Availability Statement: The data presented in this study are openly available in FigShare at [https://doi.org/10.6084/m9.figshare.25218842.v1 (accessed on 1 March 2024)].

Conflicts of Interest: The authors declare no conflicts of interest. The funders had no role in the design of the study; in the collection, analyses, or interpretation of data; in the writing of the manuscript; or in the decision to publish the results.

References

1. Matias, I.; Belluomo, I.; Cota, D. The fat side of the endocannabinoid system: Role of endocannabinoids in the adipocyte. *Cannab. Cannabinoid Res.* **2016**, *1*, 176–185. [CrossRef]
2. Bermudez-Silva, F.J.; Viveros, M.P.; McPartland, J.M.; Rodriguez de Fonseca, F. The endocannabinoid system, eating behavior an energy homeostasis: The end or a new beginning? *Pharmacol. Biochem. Behav.* **2010**, *95*, 375–382. [CrossRef] [PubMed]
3. Silvestri, C.; Di Marzo, V. The endocannabinoid system in energy homeostasis and the etiopathology of metabolic disorders. *Ce Metab.* **2013**, *17*, 475–490. [CrossRef] [PubMed]
4. Tudurí, E.; Imbernon, M.; Hernández-Bautista, R.J.; Tojo, M.; Fernø, J.; Diéguez, C.; Nogueiras, R. GPR55: A new promising targ for metabolism? *J. Mol. Endocrinol.* **2017**, *58*, R191–R202. [CrossRef]
5. Bennetzen, M.F.; Wellner, N.; Ahmed, S.S.; Ahmed, S.M.; Diep, T.A.; Hansen, H.S.; Richelsen, B.; Pedersen, S.B. Investigations c the human endocannabinoid system in two subcutaneous adipose tissue depots in lean subjects and in obese subjects before an after weight loss. *Int. J. Obes.* **2011**, *35*, 1377–1384. [CrossRef]
6. De Petrocellis, L.; Ligresti, A.; Moriello, A.S.; Allarà, M.; Bisogno, T.; Petrosino, S.; Stott, C.G.; Di Marzo, V. Effects of cannabinoid and cannabinoid-enriched Cannabis extracts on TRP channels and endocannabinoid metabolic enzymes. *Br. J. Pharmacol.* **201** *163*, 1479–1494. [CrossRef] [PubMed]
7. Batetta, B.; Griinari, M.; Carta, G.; Murru, E.; Ligresti, A.; Cordeddu, L.; Giordano, E.; Sanna, F.; Bisogno, T.; Uda, S.; et a Endocannabinoids may mediate the ability of (n-3) fatty acids to reduce ectopic fat and inflammatory mediators in obese Zucke rats. *J. Nutr.* **2009**, *139*, 1495–1501. [CrossRef] [PubMed]
8. Callaway, J.C. Hempseed as a nutritional resource: An overview. *Euphytica* **2004**, *140*, 65–72. [CrossRef]
9. NHA. Hemp Resources—Hemp Facts and Statistics. The National Hemp Association. Available online: https:/ nationalhempassociation.org/hemp-resources/#Hemp%20Facts%20and%20Statistics (accessed on 8 February 2024).

10. Subash-Babu, P.; Alshatwi, A.A. Aloe-emodin inhibits adipocyte differentiation and maturation during in vitro human mesenchymal stem cell adipogenesis. *J. Biochem. Mol. Toxicol.* **2012**, *26*, 291–300. [CrossRef]
11. Mead, A. The legal status of cannabis (marijuana) and cannabidiol (CBD) under U.S. law. *Epilepsy Behav.* **2017**, *70*, 288–291. [CrossRef]
12. Greenbaum, D.; Colangelo, C.; Williams, K.; Gerstein, M. Comparing protein abundance and mRNA expression levels on a genomic scale. *Genome Biol.* **2003**, *4*, 117. [CrossRef] [PubMed]
13. Villarroel-Vicente, C.; Gutiérrez-Palomo, S.; Ferri, J.; Cortes, D.; Cabedo, N. Natural products and analogs as preventive agents for metabolic syndrome via peroxisome proliferator-activated receptors: An overview. *Eur. J. Med. Chem.* **2021**, *221*, 113535. [CrossRef] [PubMed]
14. Palomares, B.; Ruiz-Pino, F.; Garrido-Rodriguez, M.; Eugenia Prados, M.; Sánchez-Garrido, M.A.; Velasco, I.; Vazquez, M.J.; Nadal, X.; Ferreiro-Vera, C.; Morrugares, R.; et al. Tetrahydrocannabinolic acid A (THCA-A) reduces adiposity and prevents metabolic disease caused by diet-induced obesity. *Biochem. Pharmacol.* **2020**, *171*, 113693. [CrossRef] [PubMed]
15. Cardenia, V.; Toschi, T.G.; Scappini, S.; Rubino, R.C.; Rodriguez-Estrada, M.T. Development and validation of a Fast gas chromatography/mass spectrometry method for the determination of cannabinoids in *Cannabis sativa* L. *J. Food Drug Anal.* **2018**, *26*, 1283–1292. [CrossRef] [PubMed]
16. Pellegrini, M.; Marchei, E.; Pacifici, R.; Pichini, S. A rapid and simple procedure for the determination of cannabinoids in hemp food products by gas chromatography-mass spectrometry. *J. Pharm. Biomed. Anal.* **2005**, *36*, 939–946. [CrossRef]
17. Oosterveer, M.H.; Koolman, A.H.; de Boer, P.T.; Bos, T.; Bleeker, A.; van Dijk, T.H.; Bloks, V.W.; Kuipers, F.; Sauer, P.J.; van Dijk, G. Resistance to diet-induced adiposity in cannabinoid receptor-1 deficient mice is not due to impaired adipocyte function. *Nutr. Metab.* **2011**, *8*, 93. [CrossRef]
18. Chang, R.C.; Thangavelu, C.S.; Joloya, E.M.; Kuo, A.; Li, Z.; Blumberg, B. Cannabidiol promotes adipogenesis of human and mouse mesenchymal stem cells via PPARγ by inducing lipogenesis but not lipolysis. *Biochem. Pharmacol.* **2022**, *197*, 114910. [CrossRef]
19. Roche, R.; Hoareau, L.; Bes-Houtmann, S.; Gonthier, M.-P.; Laborde, C.; Baron, J.-F.; Haffaf, Y.; Cesari, M.; Festy, F. Presence of the cannabinoid receptors, CB1 and CB2, in human omental and subcutaneous adipocytes. *Histochem. Cell Biol.* **2006**, *126*, 177–187. [CrossRef]
20. Tedesco, L.; Valerio, A.; Cervino, C.; Cardile, A.; Pagano, C.; Vettor, R.; Pasquali, R.; Carruba, M.O.; Marsicano, G.; Lutz, B.; et al. Cannabinoid type 1 receptor blockade promotes mitochondrial biogenesis through endothelial nitric oxide synthase expression in white adipocytes. *Diabetes* **2008**, *57*, 2028–2036. [CrossRef]
21. Bajzer, M.; Olivieri, M.; Haas, M.K.; Pfluger, P.T.; Magrisso, I.J.; Foster, M.T.; Tschöp, M.H.; Krawczewski-Carhuatanta, K.A.; Cota, D.; Obici, S. Cannabinoid receptor 1 (CB1) antagonism enhances glucose utilisation and activates brown adipose tissue in diet-induced obese mice. *Diabetologia* **2011**, *54*, 3121–3131. [CrossRef]
22. Rakotoarivelo, V.; Sihag, J.; Flamand, N. Role of the endocannabinoid system in the adipose tissue with focus on energy metabolism. *Cells* **2021**, *10*, 1279. [CrossRef] [PubMed]
23. Caterina, M.J.; Julius, D. The vanilloid receptor: A molecular gateway to the pain pathway. *Annu. Rev. Neurosci.* **2001**, *24*, 487–517. [CrossRef] [PubMed]
24. Premkumar, L.S.; Sikand, P. TRPV1: A target for next generation analgesics. *Curr. Neuropharmacol.* **2008**, *6*, 151–163. [CrossRef]
25. Christie, S.; Wittert, G.A.; Li, H.; Page, A.J. Involvement of TRPV1 channels in energy homeostasis. *Front. Endocrinol.* **2018**, *9*, 420. [CrossRef] [PubMed]
26. Hermann, H.; De Petrocellis, L.; Bisogno, T.; Schiano Moriello, A.; Lutz, B.; Di Marzo, V. Dual effect of cannabinoid CB 1 receptor stimulation on a vanilloid VR1 receptor-mediated response. *Cell. Mol. Life Sci.* **2003**, *60*, 607–616. [CrossRef] [PubMed]
27. Matsuda, L.A.; Lolait, S.J.; Brownstein, M.J.; Young, A.C.; Bonner, T.I. Structure of a cannabinoid receptor and functional expression of the cloned cDNA. *Nature* **1990**, *346*, 561–564. [CrossRef] [PubMed]
28. Premkumar, L.S.; Ahern, G.P. Induction of vanilloid receptor channel activity by protein kinase C. *Nature* **2000**, *408*, 985–990. [CrossRef] [PubMed]
29. Vellani, V.; Mapplebeck, S.; Moriondo, A.; Davis, J.B.; McNaughton, P.A. Protein kinase C activation potentiates gating of the vanilloid receptor VR1 by capsaicin, protons, heat and anandamide. *J. Physiol.* **2001**, *534*, 813–825. [CrossRef]
30. Mahmud, A.; Santha, P.; Paule, C.C.; Nagy, I. Cannabinoid 1 receptor activation inhibits transient receptor potential vanilloid type 1 receptor-mediated cationic influx into rat cultured primary sensory neurons. *Neuroscience* **2009**, *162*, 1202–1211. [CrossRef]
31. Sántha, P.; Jenes, Á.; Somogyi, C.; Nagy, I. The endogenous cannabinoid anandamide inhibits transient receptor potential vanilloid type 1 receptor-mediated currents in rat cultured primary sensory neurons. *Acta Physiol. Hung.* **2010**, *97*, 149–158. [CrossRef]
32. Della Pietra, A.; Giniatullin, R.; Savinainen, J.R. Distinct activity of endocannabinoid-hydrolyzing enzymes MAGL and FAAH in key regions of peripheral and central nervous system implicated in migraine. *Int. J. Mol. Sci.* **2021**, *22*, 1204. [CrossRef] [PubMed]
33. Di Marzo, V.; Maccarrone, M. FAAH and anandamide: Is 2-AG really the odd one out? *Trends Pharmacol. Sci.* **2008**, *29*, 229–233. [CrossRef] [PubMed]
34. Matias, I.; Carta, G.; Murru, E.; Petrosino, S.; Banni, S.; Marzo, V.D. Effect of polyunsaturated fatty acids on endocannabinoid and N-acyl-ethanolamine levels in mouse adipocytes. *Biochim. Biophys. Acta Mol. Cell Biol. Lipids* **2008**, *1781*, 52–60. [CrossRef]
35. Koolman, A.H. *The Role of the Endocannabinoid System in the Regulation of Energy Balance;* University of Groningen: Groningen, The Netherlands, 2010.

36. Moreno-Navarrete, J.M.; Catalán, V.; Whyte, L.; Díaz-Arteaga, A.; Vázquez-Martínez, R.; Rotellar, F.; Guzmán, R.; Gómez-Ambrosi, J.; Pulido, M.R.; Russell, W.R.; et al. The L-α-lysophosphatidylinositol/GPR55 system and its potential role in human obesity. *Diabetes* **2012**, *61*, 281–291. [CrossRef]
37. Choi, J.W.; Lee, C.-W.; Chun, J. Biological roles of lysophospholipid receptors revealed by genetic null mice: An update. *Biochim. Biophys. Acta* **2008**, *1781*, 531–539. [CrossRef] [PubMed]
38. Brighenti, V.; Pellati, F.; Steinbach, M.; Maran, D.; Benvenuti, S. Development of a new extraction technique and HPLC method for the analysis of non-psychoactive cannabinoids in fibre-type *Cannabis sativa* L. (hemp). *J. Pharm. Biomed. Anal.* **2017**, *143*, 228–236. [CrossRef] [PubMed]
39. Devi, V.; Khanam, S. Comparative study of different extraction processes for hemp (*Cannabis sativa*) seed oil considering physical, chemical and industrial-scale economic aspects. *J. Clean. Prod.* **2019**, *207*, 645–657. [CrossRef]
40. Mandrioli, M.; Tura, M.; Scotti, S.; Gallina Toschi, T. Fast detection of 10 cannabinoids by RP-HPLC-UV method in *Cannabis sativa* L. *Molecules* **2019**, *24*, 2113. [CrossRef]
41. PromoCell. Preadipocyte/Adipocyte Media Instruction Manual. PromoCell. Available online: https://promocell.com/wp-content/uploads/product-information/manual/C-27436.pdf (accessed on 8 February 2024).

International Journal of
Molecular Sciences

MDPI

Article

Dietary Methionine Restriction Improves Gut Health and Alters the Plasma Metabolomic Profile in Rats by Modulating the Composition of the Gut Microbiota

Mei Yang [1,2,†], Qian Xie [1,2,†], Yintao Xiao [1,2], Minglong Xia [1,2], Jiashun Chen [1,2], Bi-E Tan [1,2,*] and Yulong Yin [2,3,*]

1 Key Laboratory of Hunan Province for the Products Quality Regulation of Livestock and Poultry, College of Animal Science and Technology, Hunan Agricultural University, Changsha 410128, China; aym1223@163.com (M.Y.)

2 Yuelushan Laboratory, Changsha 410128, China

3 Key Laboratory of Agro-Ecological Processes in Subtropical Region, National Engineering Laboratory for Pollution Control and Waste Utilization in Livestock and Poultry Production, Institute of Subtropical Agriculture, Chinese Academy of Sciences, Changsha 410125, China

* Correspondence: bietan@hunau.edu.cn (B.-E.T.); yinyulong@isa.ac.cn (Y.Y.)

† These authors contributed equally to this work.

Abstract: Dietary methionine restriction (MetR) offers an integrated set of beneficial health effects, including delaying aging, extending health span, preventing fat accumulation, and reducing oxidative stress. This study aimed to investigate whether MetR exerts entero-protective effects by modulating intestinal flora, and the effect of MetR on plasma metabolites in rats. Rats were fed diets containing 0.86% methionine (CON group) and 0.17% methionine (MetR group) for 6 weeks. Several indicators of inflammation, gut microbiota, plasma metabolites, and intestinal barrier function were measured. 16S rRNA gene sequencing was used to analyze the cecal microbiota. The MetR diet reduced the plasma and colonic inflammatory factor levels. The MetR diet significantly improved intestinal barrier function by increasing the mRNA expression of tight junction proteins, such as zonula occludens (ZO)-1, claudin-3, and claudin-5. In addition, MetR significantly increased the levels of short-chain fatty acids (SCFAs) by increasing the abundance of SCFAs-producing *Erysipclotxichaceae* and *Clostridium_sensu_stricto_1* and decreasing the abundance of pro-inflammatory bacteria *Proteobacteria* and *Escherichia-Shigella*. Furthermore, MetR reduced the plasma levels of taurochenodeoxycholate-7-sulfate, taurocholic acid, and tauro-ursodeoxycholic acid. Correlation analysis identified that colonic acetate, total colonic SCFAs, 8-acetylegelolide, collettiside I, 6-methyladenine, and cholic acid glucuronide showed a significant positive correlation with *Clostridium_sensu_stricto_1* abundance but a significant negative correlation with *Escherichia-Shigella* and *Enterococcus* abundance. MetR improved gut health and altered the plasma metabolic profile by regulating the gut microbiota in rats.

Keywords: methionine restriction; intestinal barrier; *ZO-1*; SCFAs; BAs

1. Introduction

Nutrition exerts profound effects on health, and dietary interventions are commonly used to treat metabolic disorders. Methionine is a sulfur-containing essential amino acid that is required for normal growth and development [1]. Methionine restriction (MetR) is gaining popularity due to its several health benefits, including improved stress resistance, increased longevity, and inhibition of fat deposition in experimental animals [2,3]. More specifically, both 80 and 40% MetR diets are associated with lifespan extension and metabolic health from yeast to rodents, which may be partially mediated through a reduction in oxidative stress [4]. MetR improves hepatic steatosis and delays the development of non-alcoholic steatohepatitis through antioxidant effects and alteration in lipid composition in obese mice [5,6]. In a recent study, MetR reduced liver fat content by regulating fatty acid oxidation and lipid catabolism in high-fat-fed mice [7].

The diverse and dynamic microbial community contributes to maintaining healthy gut homeostasis [8]. The integrity of the intestinal barrier, which is supported by the gut microbiota, is essential for maintaining the stability of the body [9]. Disruption and depletion of the gut microbiota may therefore affect the gut barrier and cause persistent systemic inflammation [10]. Microbiome–host interactions and microbial metabolites are associated with the capacity to produce inflammatory factors [11]. As end products of carbohydrates fermented by the gut microbiota, short-chain fatty acids (SCFAs) are involved in ameliorating host metabolic syndromes [12].

Research has highlighted the role of metabolites in mediating the interaction between the gut microbiota and the host diet. For example, bile acids (BAs) play a critical role in modulating host inflammation and the gut barrier [12,13]. In addition to their well-established role in lipid metabolism and maintaining cholesterol homeostasis, BAs also act as signaling molecules that reshape the composition of the gut microbiota through BA receptors, such as the farnesoid X receptor (FXR), which are involved in entero-protection in order to inhibit bacterial overgrowth and mucosal injury in the ileum [14].

Although previous studies have speculated that MetR exerts beneficial effects on the gut by improving epithelial barrier function in normal rats [15], the role and mechanism of the gut microbiota, and especially its metabolites, on gut health in dietary MetR intervention require further research. Therefore, we hypothesized that MetR plays a role in gut health and investigate whether the effects of MetR on plasma metabolomics are related to the modulation of the gut microbiota composition. The aim of this study was to analyze the effect of MetR diet on gut health in rats using 16S rRNA sequencing combined with plasma metabolomics. We sought to identify plasma metabolites that were associated with the MetR dietary supplements, including taurine-conjugated BAs. We also determined inflammatory cytokines, intestinal tight junction proteins, and SCFAs to investigate the effects of MetR on the plasma metabolome and to identify underlying mechanisms for maintaining gut health. The present study may contribute to the provision of new therapeutic strategies from the perspective of gut microbiota.

2. Results

2.1. *Effects of MetR on Plasma LpS, D-Lactate, DAO, and Cytokine Levels*

The plasma level of D-lactate was significantly reduced in rats in the MetR group compared to those in the CON group ($p < 0.05$) (Figure 1B). In addition, the LPS and DAO level in the plasma were not significantly different between the CON group and MetR group ($p > 0.05$, Figure 1A,C). Furthermore, MetR significantly decreased the TNF-α level relative to the control rat ($p < 0.05$) (Figure 1D). There were no significant differences in the plasma levels of IL-1β, IL-2, IL-4, IL-6, and IL-12 between the MetR and CON groups ($p > 0.05$, Figure 1D).

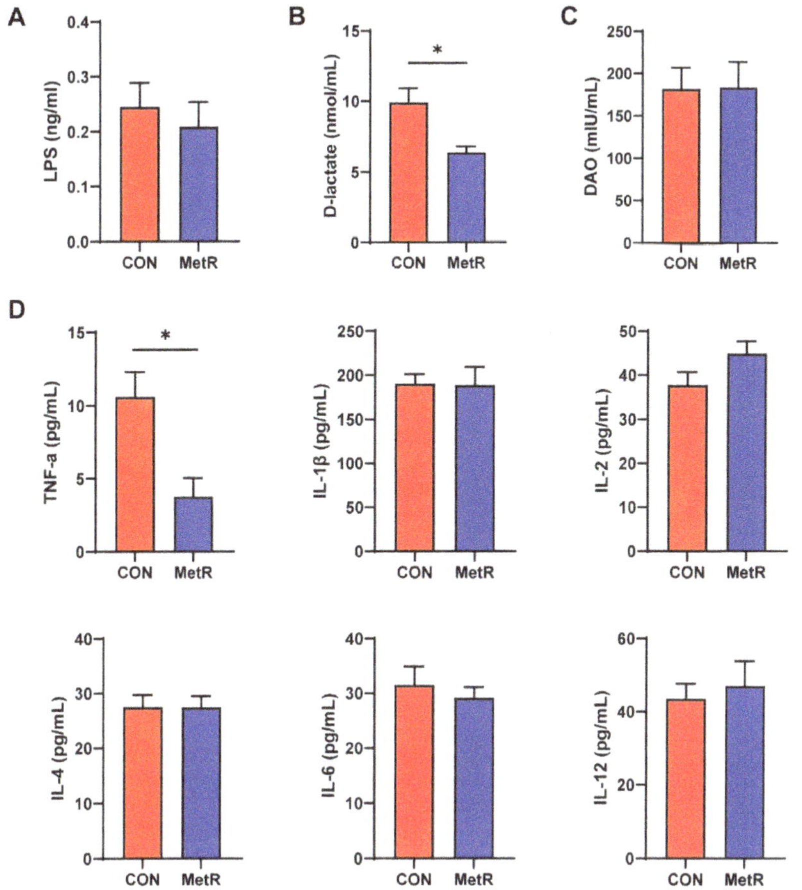

Figure 1. The plasma (**A**) LPS, (**B**) DAO, (**C**) D-lactate, (**D**) TNF-α, IL-1β, IL-2, IL-4, IL-6, and IL-12 levels in the rat. Data are expressed as the mean ± SEM for $n = 10$, * $p < 0.05$ versus CON.

2.2. Effects of MetR on Colonic Cytokine Levels, Barrier Function, and the Morphological Structure of the Colon

The cytokine levels in the colon are shown in Figure 2A. The levels of TNF-α, IL-2, IL-6, and IL-12 were significantly decreased in the MetR group compared to the CON group ($p < 0.05$). The qRT–PCR analysis revealed an increase in the abundance of zonula occludens (*ZO*)-1, claudin-3, and claudin-5 in colonic samples from rats in the MetR group relative to those in the CON group ($p < 0.05$) (Figure 2B). However, no significant changes in the mRNA expression of occludin and claudin-1 were observed in the colon of rats in the MetR group. Representative photo-micrographs of the H and E-stained and PAS-stained colon tissues are shown in Figure 3A. Goblet cell in MetR group was significantly higher than that in CON group ($p < 0.05$) (Figure 3B). Similarly, MetR significantly upregulated the protein expression of *ZO-1* compared to the CON group ($p < 0.05$) (Figure 3C,D).

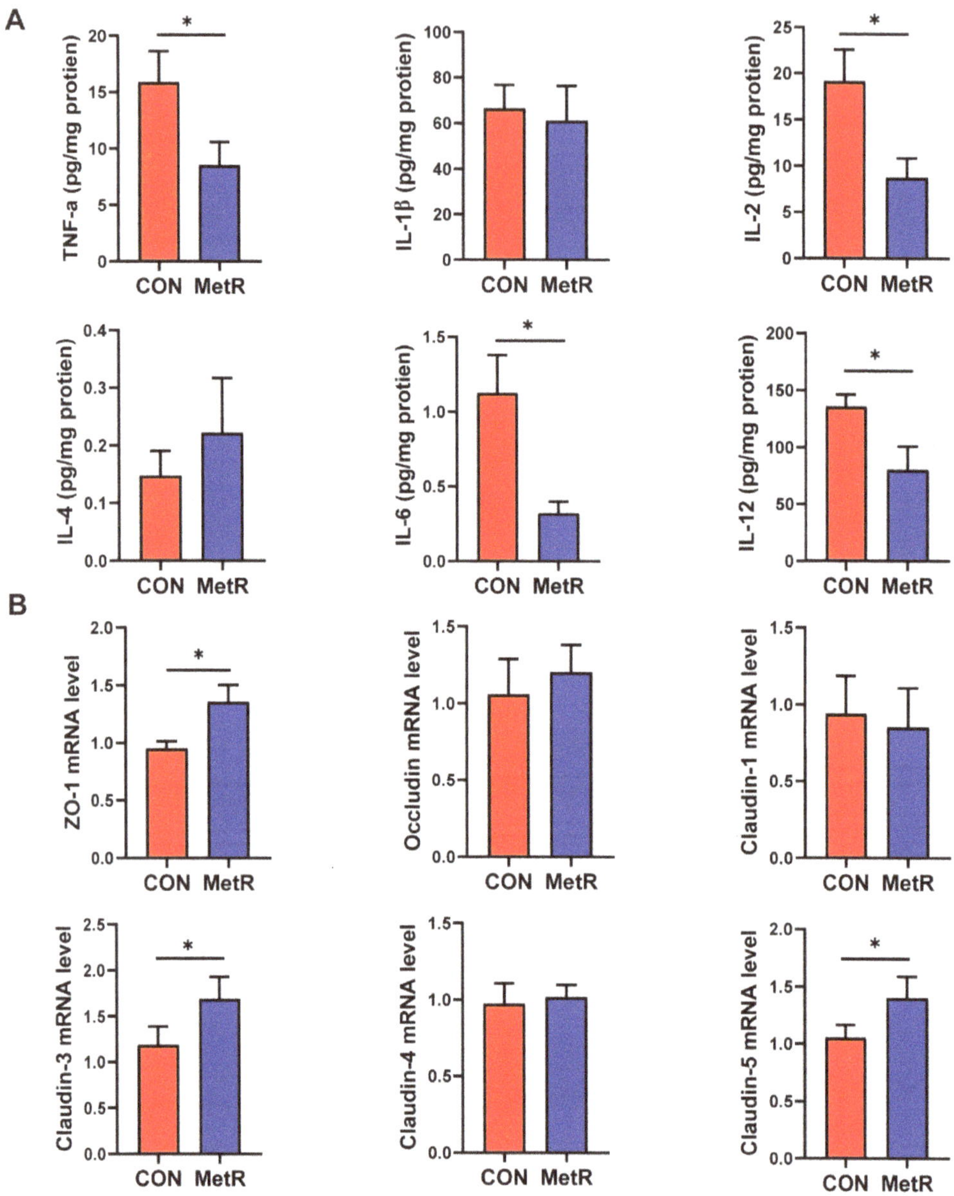

Figure 2. (**A**) Inflammatory factors and (**B**) mRNA expression levels of markers of tight junction proteins on the colon tissue. Data are expressed as the mean ± SEM for $n = 10$, * $p < 0.05$ versus CON.

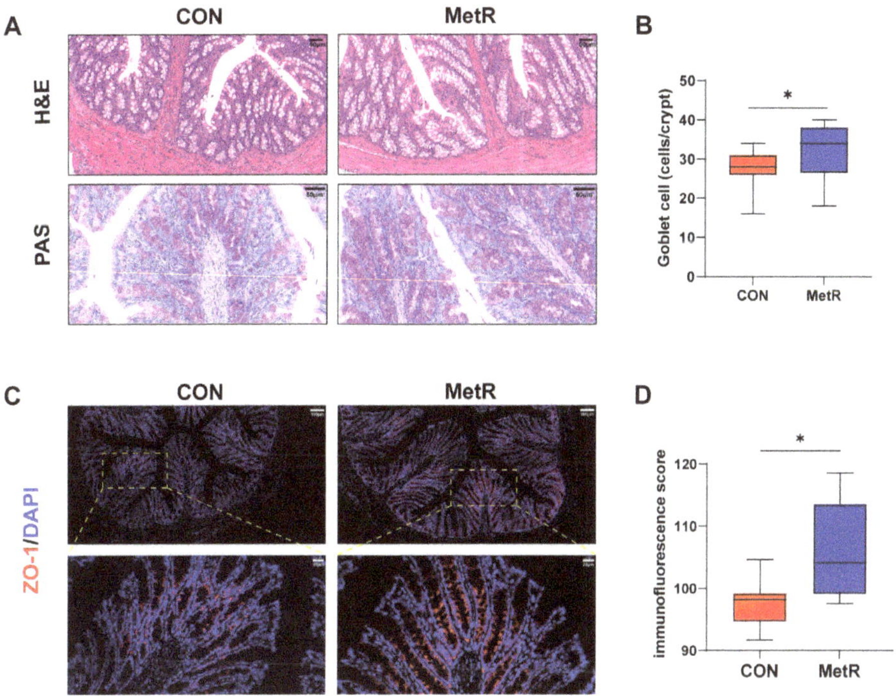

Figure 3. (**A**) Representative images of colon H and E and PAS staining. (**B**) The goblet cell numbers per crypt. (**C**) Representative immunofluorescence images of colonic sections stained with ZO-1 antibody. The dashed rectangle shows the image below with higher magnification. (**D**) Immunofluorescence score. Data are expressed as the mean ± SEM for $n = 10$, * $p < 0.05$ versus CON.

2.3. *Effects of MetR on Cecal Microbiota*

The results of the common alpha-diversity analysis (Sob index, Shannon index, Simpson index, Ace index, Chao index, and Coverage index) are shown in Figure 4A. The Simpson index of rats in the MetR group was significantly decreased relative to that of rats in the CON group ($p < 0.05$, Figure 4A). Moreover, MetR administration tended to restore the reduction of the ACE index, but the increase did not reach statistical difference ($p > 0.05$, Figure 4A). We found 296 common OTUs among two groups, while the number of OTUs, about 153, shared by MetR group was higher than that in CON group, with only 41 OTUs (Figure 4C). The Bray-Curtis distance-based PCoA and cluster tree revealed that the CON and MetR group exhibited a distinct clustering of microbiota composition (Figure 4B,D).

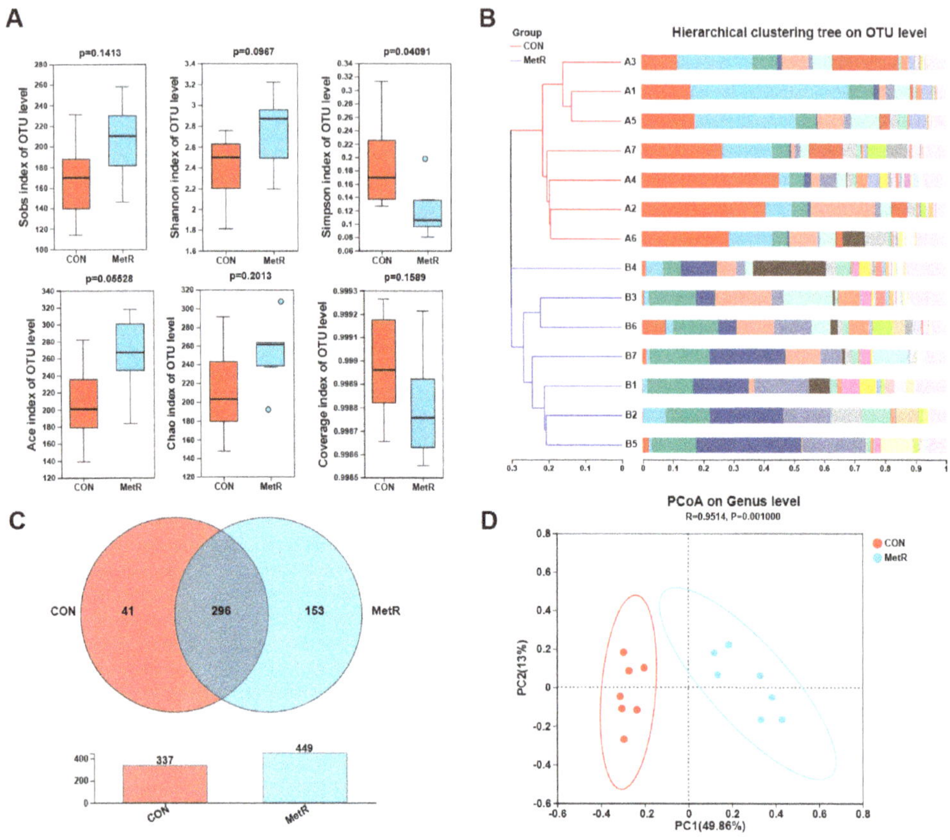

Figure 4. (**A**) Community diversity and richness among CON and MetR group. (**B**) Different colors in the figure represent samples in different groups. The degree of similarity between samples is reflected by the level of aggregation in the graph. (**C**) Venn diagram showing shared and unique OTUs among the two groups. (**D**) PCoA of the bacterial community structure among CON and MetR group. PCoA score plots showed significantly separated clusters between CON and MetR group. Data are expressed as the mean ± SEM for $n = 7$.

The bacterial sequences of cecal contents from the two experimental groups were examined at the phylum level. Based on the relative abundance of gut microbiota, the top five phyla are displayed (Figure 5A). The phylum level analysis showed that MetR feeding resulted in a significant decrease in the relative abundance of *Proteobacteria* and a significant increase in that of *Firmicutes* ($p < 0.05$). To further investigate the differences, an order-level analysis was performed. As shown in Figure 5B, the MetR diet markedly decreased the relative abundance of *Enterobacteriaceae* and *Enterococcaceae* but increased the relative abundance of *Erysipclotxichaceae* ($p < 0.05$). In addition, the MetR diet significantly decreased the relative abundance of *Escherichia-Shigella* and *Enterococcus* but increased the relative abundance of *Clostridium_sensu_stricto_1* at the genus level ($p < 0.05$) (Figure 5C).

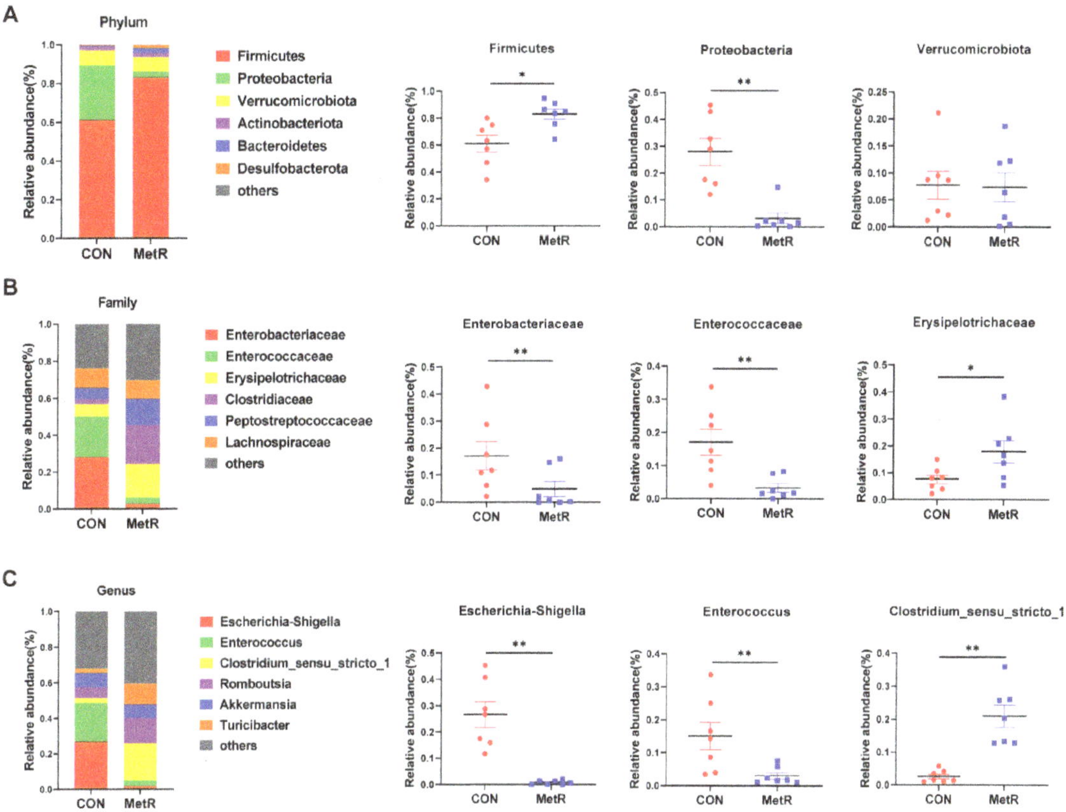

Figure 5. Microbial composition at the (**A**) phylum level, (**B**) family level, and (**C**) genus level. Each bar represents the relative abundance of each bacterial taxa of rats. Data are expressed as the mean ± SEM for $n = 7$, * $p < 0.05$, ** $p < 0.01$ versus CON.

2.4. Effects of MetR on Gut Microbiota and Cecal, Colon, and Fecal SCFA Levels

The abundances of Kyoto Encyclopedia of Genes and Genome (KEGG) pathways in each group were predicted using PICRUSt ($p < 0.01$, Figure 6A). Notably, the biosynthesis of amino acids and secondary metabolites was significantly enriched in tissues from rats in the MetR group relative to rats in the CON group (Figure 6A). In addition, other metabolic pathways, such as ABC transporters and microbial metabolism in different environments, were significantly upregulated in the CON group but downregulated in the MetR group. Figure 6B shows the levels of SCFAs in cecal, colonic, and fecal contents. The concentrations of acetate and total SCFA were significantly higher in the MetR group than in the CON group ($p < 0.05$). In particular, the concentrations of propionate and valerate in the colonic contents showed a notable increase in the MetR group compared to the CON group ($p < 0.05$, Figure 6B).

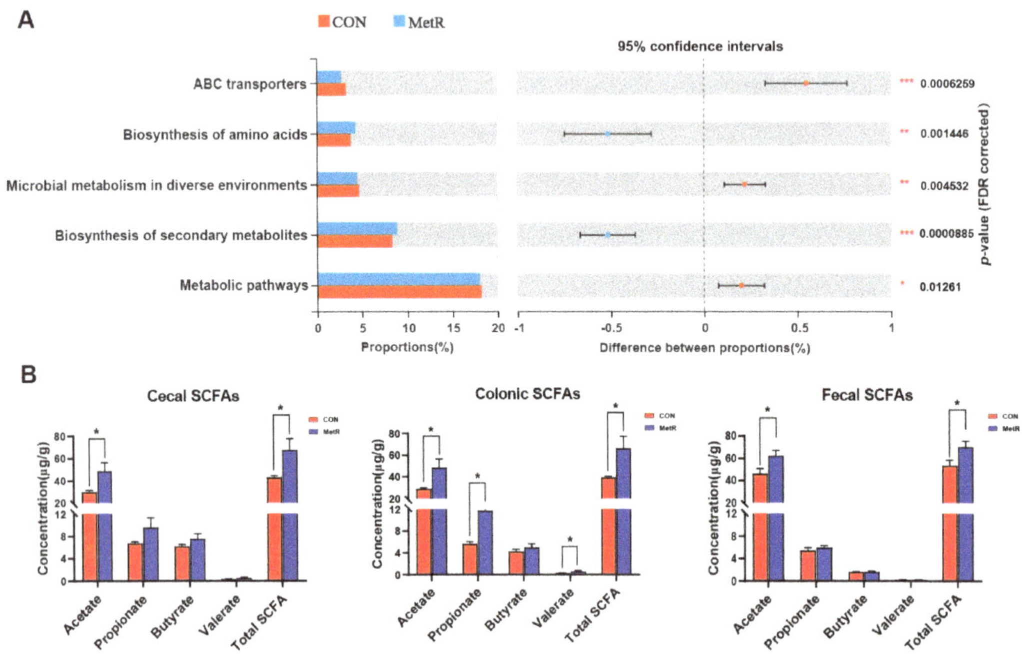

Figure 6. (**A**) Inferred metabolic functions of cecum microbiota by PICRUSt analysis among the two groups. (**B**) Concentrations of SCFAs in cecal content, colon content and fecal content. Data are expressed as the mean ± SEM for $n = 10$, * $p < 0.05$, ** $p < 0.01$, *** $p < 0.001$ versus CON.

2.5. Effects of MetR on Plasma Metabolome

A total of 149 metabolites were identified in the rat plasma, among which 61 were upregulated and 88 were downregulated (Figure 7A). A supervised OPLS-DA model (R2 = 0.996, Q2 = 0.916, $p < 0.05$) showed a robust separation of the diet groups as visualized by the score plot (Figure 7B). The KEGG databases were used to establish the metabolic pathways, which included glycine, serine, and threonine metabolism, primary bile acid biosynthesis, and bile secretion (Figure 7C). A total of 30 metabolites had VIP > 1, suggesting that they contribute to the discrimination of the groups (Figure 8A). The levels of 8-acetylegelolide, collettiside I, cholic acid glucuronide, 6-methyladenine, and cholic acid glucuronide were significantly increased in the MetR group relative to the CON group ($p < 0.01$, Figure 8B). In addition, the levels of taurochenodeoxycholate-7-sulfate, taurocholic acid (TCA), tauro-ursodeoxycholic acid (TUDCA), and 10-acetyl-3,7-dihydroxyphenoxazine were markedly reduced in the MetR group relative to the CON group ($p < 0.01$, Figure 8B).

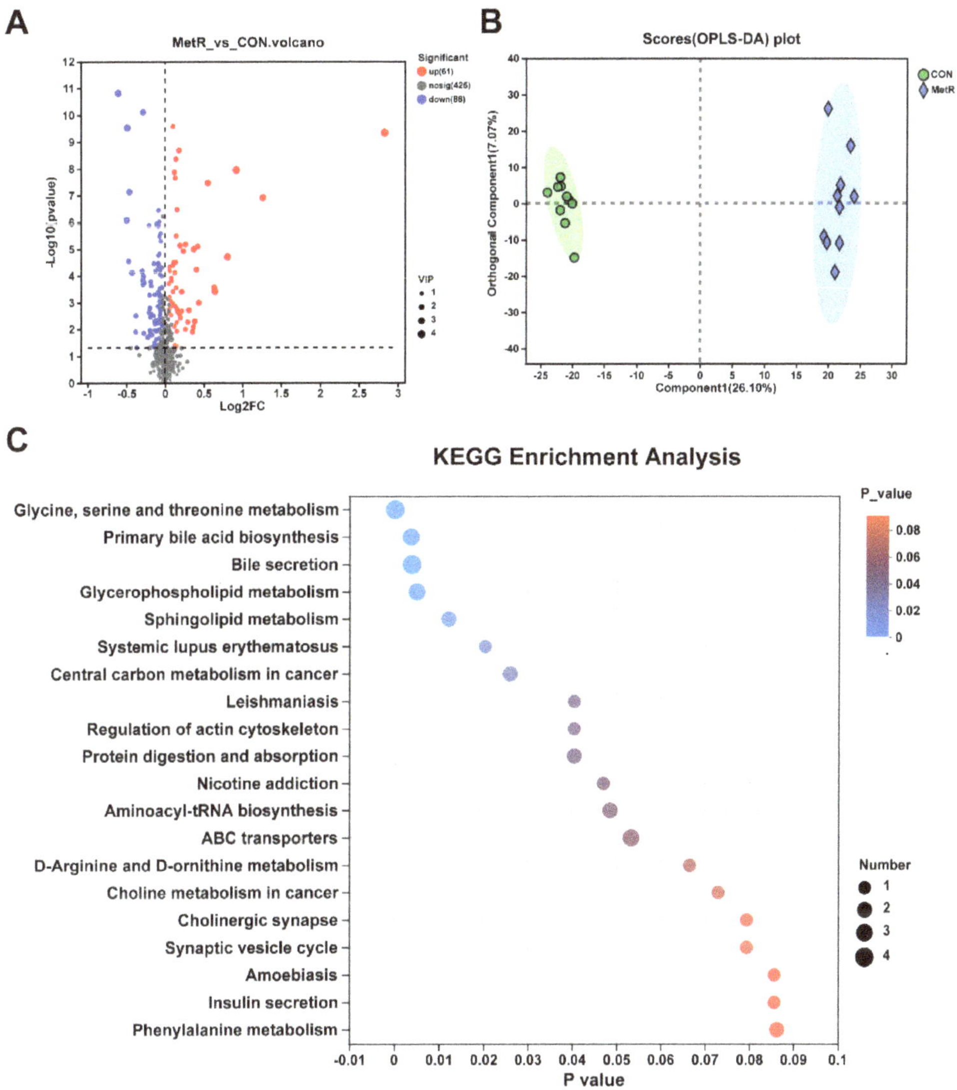

Figure 7. (**A**) Volcano plot for the CON and MetR groups, where each point represents a metabolite. Differentially expressed metabolites in combined CON and MetR (|log2 (Fold Change)| > 1 and nominal $p < 0.05$). Red dots represent upregulated metabolites; blue dots represent downregulated metabolites; and gray dots indicate non-significant differences. (**B**) Scatter plot of scores from OPLS-DA of the CON and MetR groups. (**C**) KEGG pathways enriched in the MetR group compared to the CON group.

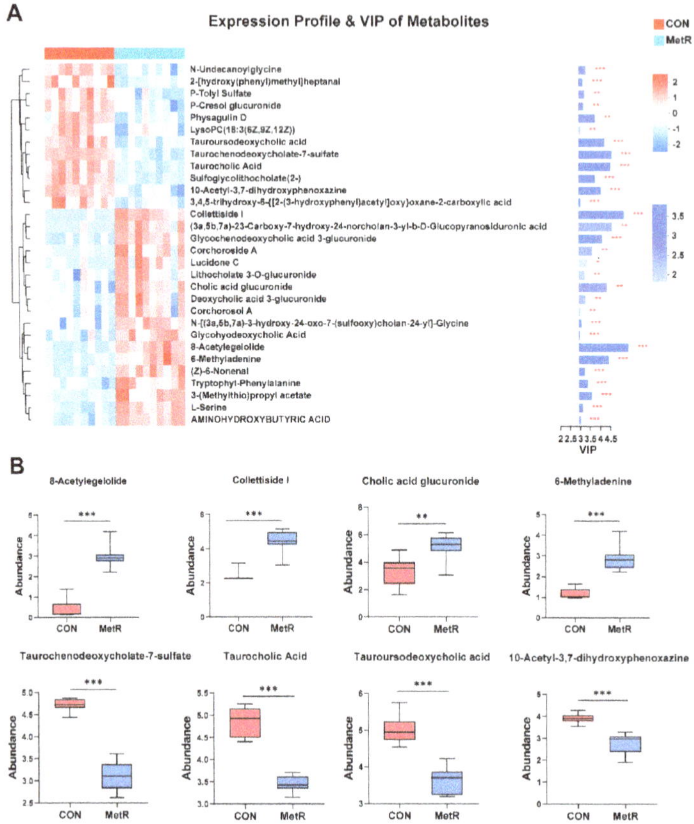

Figure 8. (**A**) Heatmap showing normalized values of 30 metabolites that were differentially abundant among the CON and MetR groups. The normalized abundance values are depicted from red to green, where red represents the highest abundance and green represents the lowest abundance. The dendrogram shows the distances of the metabolites based on their relative abundances. (**B**) The identified and change trend of the potential biomarkers of rat intervened by Wilcox test, unpaired. Data are expressed as the mean ± SEM for $n = 10$, * $p < 0.05$, ** $p < 0.01$, *** $p < 0.001$ versus CON.

2.6. Spearman's Correlation Analysis between Plasma Metabolites, Colonic SCFAs, and Gut Microbiota

The correlations between plasma metabolites, colonic SCFAs and gut microbiota were evaluated by Spearman's correlation analysis (Figure 9). The levels of colonic acetate, total colonic SCFA, 8-acetylegelolide, collettiside I, 6-methyladenine, and cholic acid glucuronide were positively correlated with the abundance of *Clostridium_sensu_stricto_1* but negatively correlated with the abundance of *Escherichia-Shigella* and *Enterococcus*. Furthermore, the levels of taurochenodeoxycholate-7-sulfate, TCA, TUDCA, and 10-acetyl-3,7-dihydroxyphenoxazine were positively correlated with the abundance of *Escherichia-Shigella* and *Enterococcus* but positively correlated with the abundance of *Clostridium_sensu_stricto_1* and *Turicibacter*.

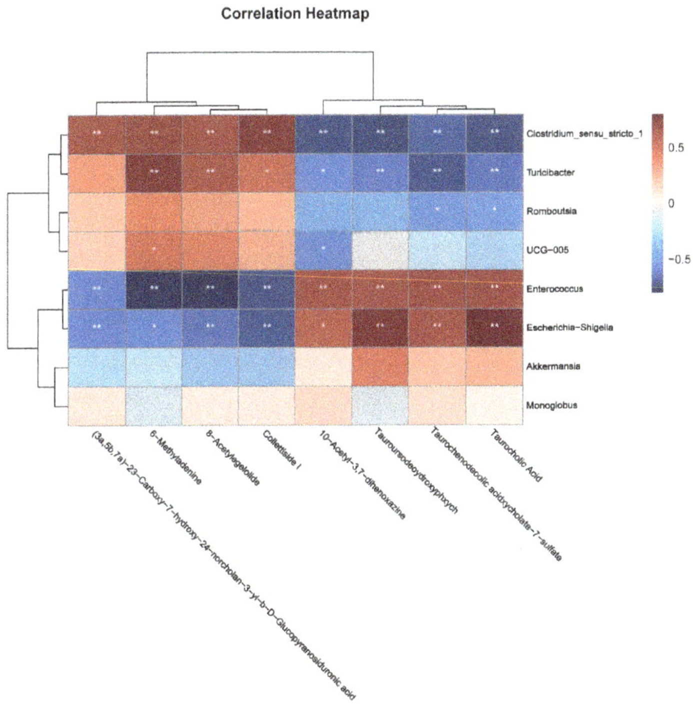

Figure 9. Heat map of the relationship between plasma metabolites, colonic SCFAs and gut microbiota. Red squares represent significant positive correlations (R > 0.5, $p < 0.05$), blue squares represent significant negative correlations (R < −0.5, $p < 0.05$), and white squares represent non-significant correlations. * $p < 0.05$, ** $p < 0.01$.

3. Discussion

There is overwhelming evidence consistently demonstrating that the MetR diet has many beneficial health effects, including reducing hepatic steatosis [6], ameliorating fat accumulation [16], preventing systemic inflammation [17], and regulating energy and protein homeostasis [7] in high-fat-fed mice. Dietary MetR prevents cognitive impairment in D-galactose-induced aging mice and extends lifespan in progeroid mice [3,18]. In this study, MetR diet increased the abundance of *Clostridium_sensu_stricto_1*, resulting in a considerable upregulation of SCFAs and downregulation of the plasma taurine-conjugated bile acids, like TCA and TUDCA. Moreover, SCFAs facilitated health benefits in the rats, including decreasing systemic inflammation and enhancing the integrity of the gut barrier.

Dietary MetR has been used as a research model for preventing and treating obesity-related metabolic diseases. The gut epithelial barrier is an essential barrier between the environment and the internal milieu of the body, limiting the passage of pathogens and toxins [19]. The disruption of the integrity of the gut barrier allows the passage of bacterial products, leading to the production of pro-inflammatory cytokines [20]. In the current study, we found that MetR reduced the levels of serum and colonic pro-inflammatory factors in rats. This finding may be related to the enhancement of tight junction proteins by MetR. The gut microbiota plays a critical role in gut growth, nutrient absorption, and host health [21]. Higher Simpson index values indicate lower community diversity [22]. The MetR group had the richest microbiota species diversity compared to the CON group, as evidenced by the reduced Simpson indices. Our results indicate that MetR selectively decreased the abundance of *Proteobacteria*. Several studies have shown that *Proteobacteria* are closely correlated with inflammatory bowel disease [23,24]. The integrity of the gut barrier plays a critical role in maintaining gut homeostasis by ensuring a complex crosstalk between gut microbes and the host [25]. Previous studies have shown that pro-inflammatory cytokines facilitate intestinal barrier damage and invasion of pathogens [26]. Ren et al. [27] showed that MetR can improve gut barrier function in aged mice by alleviating the oscillations of inflammation-related microbes. In general, an increase in the relative abundance of *Enterobacteriaceae* causes inflammatory damage to the gut epithelium barrier [28]. Our data showed that the levels of colonic D-lactate, TNF-α, IL-2, IL-6, and IL-12 in the MetR group were significantly lower than those in the CON group. Consistent with this result, MetR decreased the abundance of *Enterobacteriaceae*, *Enterococcaceae*, and *Escherichia-Shigella*; this also corresponded with an increase in *ZO-1*, claudin-3, and claudin-5 mRNA expression [29,30].

SCFAs are important energy sources for intestinal epithelial cells, allowing them to modulate the gut barrier function and host metabolism [12]. The present work revealed that MetR effectively elevated the concentration of SCFAs (acetate, propionate, and valerate) and improved enteral homeostasis. *Erysipclotxichaceae, Bacteroides, Lactobacillus*, and *Clostridium_sensu_stricto_1* consist of microbes that produce SCFAs, which are beneficial for preventing inflammation [31,32]. Moreover, *Proteobacteria* and *Escherichia-Shigella* are associated with microbiota inflammatory properties [33,34]. In this study, we found that MetR selectively increased the abundance of representative SCFAs producers (*Erysipclotxichaceae* and *Clostridium_sensu_stricto_1*), along with enhanced levels of SCFAs in the cecal, colonic, and fecal contents [35,36], indicating that MetR maintained intestinal homeostasis by promoting the growth of beneficial gut bacteria. What is more interesting is that colonic acetate and total SCFAs levels were found to positively correlate with relative *Clostridium_sensu_stricto_1* abundance. This is consistent with a previous study in which MetR increased the abundance of SCFA-producing bacteria, decreased the abundance of inflammation-related bacteria, and improved the intestinal mucosal immune barrier in a time-dependent manner [16]. Furthermore, PICRUSt indicates MetR may have mediated amino acid and secondary metabolite biosynthesis in the gut microbiota. These results suggested that MetR supplementation effectively increased the abundance of SCFA-producing bacteria and increased amino acid biosynthesis in the gut microbiota.

The metabolomic profiling revealed that eight plasma metabolites associated with the MetR diet were mainly involved in glycine–serine–threonine metabolism, primary BA biosynthesis, and bile secretion. This is consistent with the findings of Aon et al. (2020), who reported that glycine–serine–threonine metabolism influences longevity and related life-sustaining mechanisms [37]. Furthermore, BAs promote health and longevity in various organisms [38]. In our study, the levels of taurochenodeoxycholate-7-sulfate, TCA, and TUDCA were reduced and negatively correlated with the abundance of *Clostridium_sensu_stricto_1*. Excess taurine-conjugated BAs in the serum, particularly TCA, TUDCA, and tauro-chenodeoxycholic acid (TCDCA), may exacerbate the progression of non-alcoholic steatohepatitis and the severity of drug-induced liver injury [39–41]; this supports our observations in this study, indicating that the reduction in the levels of taurine-conjugated BAs might represent a means by which MetR exerts its hepato-protective effect. Moreover, TCA and TCDCA downregulate the expression of hepatic FXR and are positively associated with pro-inflammatory cytokines, such as TNF-α [42,43], supporting our observation in this study. It has been reported that bile salt hydrolase-producing bacteria such as *Clostridium_sensu_stricto_1* and *Bacteroidetes* hydrolyze taurine- and glycine-conjugated BAs to unconjugated free forms, which can be further converted to secondary BAs [40,44]. Therefore, the increased abundance of *Clostridium_sensu_stricto_1* and *Bacteroidetes* in the cecal contents may account for the depleted taurine-conjugated BAs in the MetR-fed rats.

However, the exact mechanism by which MetR affects serum BAs levels and improves gut health remains to be elucidated.

4. Methods and Materials

4.1. Animal Experiment

The animal protocols were carried out under the guidelines of animal care and were approved by the Institutional Animal Care and Use Committee of Hunan Agricultural University (approval number: CACAHU 2022-0426). Seven-week-old male Sprague–Dawley rats were purchased from Tianqin Biotechnology (Changsha, China) and were individually housed in a conventional animal facility that was maintained at 22 ± 2 °C and 55 ± 10% relative humidity with a 12-h light/dark photoperiod. After one week of acclimation, all the rats were randomly divided into two experimental diet groups: (1) the control group fed with the normal diet (CON, 0.86% methionine); (2) methionine-restricted diet group (MetR, 0.17% methionine). The composition and nutritional value of the diets are presented in Table S1 as previously described [15,45,46].

4.2. Sample Collection and Preparation

After six weeks of experimental administration, blood samples were collected from the retro-orbital of each rat under fasting conditions and centrifuged at 3500× *g* at 4 °C for 10 min. The separated plasma was then stored in a refrigerator at −80 °C. The contents of the cecum of each rat were removed, snap-frozen in liquid nitrogen, and stored at −80 °C for subsequent 16S rRNA gene sequencing analysis. The contents of the colon and the fresh feces were collected for analysis.

4.3. Enzyme-Linked Immunosorbent Assay (ELISA)

The levels of lipopolysaccharide (LPS) and D-lactate in rat plasma were determined using the ELISA kit (Jiangsu Meimian Industrial Co., Ltd., Yancheng, China). The concentrations of tumor necrosis factor-alpha (TNF-α), interleukin-1 beta (IL-1β), interleukin-2 (IL-2), interleukin-4 (IL-4), interleukin-6 (IL-6), and interleukin-12 (IL-12) were quantified using enzyme-linked immunosorbent assay kits (CSB-E11987r, CSB-E08055r, CSB-E08055r, CSB-E04635r, CSB-E04635r, CSB-E07364r, CUSABIO, Wuhan, China, https://www.cusabio.com/, 4 June 2022) according to the manufacturer's instructions. The protein concentration of the colon tissues was measured using the Omni-Easy™Instant BCA Protein Assay Kit (ZJ102, Epizyme Biotech; Shanghai; China).

4.4. Gene Expression

Total RNA was isolated from colon tissues using RNAiso Plus (Cat # 9109, Takara Biomedical Technology (Beijing) Co., Ltd., Beijing, China) according to the manufacturer's instructions. The concentration of each RNA sample was determined using Nanodrop One (Thermo Fisher Scientific, Waltham, MA, USA). Genomic DNA contamination was removed by incubating with the Evo *M-MLV* RT Kit with gDNA Clean for qPCR (AG11705) before reverse transcription. RT-qPCR was conducted using SYBR Green Premix Pro Taq HS qPCR Kit (AG11701, Accurate Biotechnology (Hunan) Co., Ltd., Changsha, China) by real-time PCR instrument (LightCycler480II, Roche, Germany). The relative expression levels of the target genes were calculated using the $2^{-\Delta\Delta Ct}$ method. GAPDH was used as the reference gene.

4.5. Histological Analysis

Colon tissues were fixed in 4% paraformaldehyde for 24 h, dehydrated using a gradient of ethanol and xylene, and finally embedded in paraffin. A microtome (Leica, RM2016, Vizna, Germany) was then used to make 4 µm sections that were subsequently dewaxed in xylene. The dewaxed sections were then rehydrated by lowering the concentration of ethanol before being stained with hematoxylin and eosin (H and E). Finally, the sections were dehydrated with gradient ethanol before dehydration with xylene. Carnoy's colonic tissues were stained with PAS according to the manufacturer's instructions (Servicebio, Wuhan, China). The sections were briefly stained with PAS dye solution for 10 to 15 min and then rinsed with distilled water. The slides were dehydrated with absolute ethyl alcohol and xylene before image acquisition on an inverted fluorescence microscope (Axio Vert A1, Zeiss, Germany).

4.6. Immunofluorescence

Paraffin sections of colon tissue of 4 μm thickness were subjected to antigen extraction with citrate buffer (pH 6.0). Tissue sections were then incubated overnight at 4 °C with primary antibodies (Abcam, UK) against ZO-1 (1:200, ab221547). Next, sections were incubated with Alexa Fluor 488 conjugated secondary antibodies (1:100, Invitrogen, Life Technologies, Carlsbad, CA, USA) for 50 min at 37 °C. Samples were observed under an inverted fluorescence microscope (Axio Vert A1, Zeiss, Germany).

4.7. 16S rRNA Gene Sequencing Analysis

Microbial genomic DNA from cecal contents was extracted using the E.Z.N.A.® Soil DNA Kit (Omega Bio-tek, Norcross, GA, USA) according to the manufacturer's instructions. Total bacterial concentrations in each sample were measured using a qPCR assay that targets the V3–V4 hypervariable region of the 16S rRNA gene from the cecal microbiota was amplified using a forward primer (338F: 5′-ACTCCTACGGGAGGCAGCAG-3′) and a reverse primer (806R: 5′-GGACTACHVGGGTWTCTAAT-3′). The sequencing libraries of bacterial 16S rRNA genes were produced using NEXTFLEX® Rapid DNA-Seq Kit (Bio Scientific, Phoenix, AZ, USA) for high throughput sequencing. The sequencing was carried out at Majorbio Bio-Pharm Technology Co., Ltd. (Shanghai, China) on an Illumina MiSeq PE300 platform. The original DNA fragments were quality filtered and merged using fastp (version 0.20.0) and Flash (version 1.2.11). The sequence reads were classified into operational taxonomic units (OTUs) at 97% identity using Uparse (version 7.1). The RDP classifier (version 2.2) was used to classify and annotate each sequence, and a comparison was conducted against the Silva 16S rRNA database (v138) using a threshold of 70%.

4.8. Plasma Metabolomics Analysis

A 50-μL aliquot of plasma in an Eppendorf tube was mixed with a pre-cooled methanol/water solution (4:1, *v*/*v*). The mixture was vortexed for 5 min and centrifuged at 12,000× *g* at 4 °C for 10 min, after which the supernatant was carefully transferred into a new Eppendorf tube and thoroughly dried using a vacuum centrifugal concentrator (SPD130P1, Thermo Fisher, Waltham, MA, USA). A volume of 200 μL of acetonitrile: water (1:1) was added to completely dissolve the dried supernatant. After centrifugation at 12,000× *g* at 4 °C for 5 min, the sample was carefully transferred into sample vials for combined ultra-high liquid chromatography–mass spectrometry (Q-Exactive Plus, Thermo Fisher, USA). The EZinfo 2.0 software was used to perform orthogonal partial least squares discriminant analysis (OPLS-DA) by analyzing each ion. All metabolite variables were scaled using the Pareto method before conducting the OPLS-DA. The model validity was evaluated using model parameters R2 and Q2, which provide information on the interpretability and predictability of the model, respectively, and avoid the risk of over-fitting. Metabolites with variable importance in the projection (VIP) > 1 and $p < 0.05$ were considered statistically significant.

4.9. SCFAs

Colonic, cecal, and fecal content samples (1 g) were mixed with ultrapure water via processes of homogenization and centrifugation. The supernatant was collected in a 10 mL Eppendorf tube, mixed with 25% metaphosphate at a ratio of 9:1 (*v*/*v*) and filtered through a 0.22 μm membrane filter. The SCFAs (acetate, propionate, butyrate, and valerate) were quantified using gas chromatography (8890, Agilent, Santa Clara, CA, USA).

4.10. Statistical Analysis

Data were analyzed using the IBM SPSS 23.0 statistical software (SPSS Inc., Chicago, IL, USA). Differences between means were tested using Student's *t*-test. Beta-diversity containing principal co-ordinates analysis (PCoA) with a Bray-Curtis diversity distance matrix were performed using R software (version 3.3.1). The different relative abundance of bacteria within groups was calculated by Wilcoxon rank-sum test. The Spearman correlation coefficients were generated using GraphPad Prism (version 9.5.0). All data are presented as mean ± SEM. $p < 0.05$ was considered statistically significant.

5. Conclusions

These results indicated that MetR treatment increased the concentration of intestinal SCFAs by remodeling the gut microbiota, as shown by the increased relative abundance of *Erysipclotxichaceae* and *Clostridium_sensu_stricto_1* and the decreased relative abundance of *Proteobacteria, Enterobacteriaceae, Enterococcaceae*, and *Escherichia-Shigella* in the cecum. The results indicate that MetR has the potential to enhance the integrity of the gut barrier and reduce the levels of taurine-conjugated BAs. These alterations in plasma metabolite levels have contributed to the beneficial effects of MetR. Therefore, MetR improved gut health and modified plasma BA levels and conjugation by regulating the gut microbiota in rats. These findings offer valuable insights that can guide the implementation of nutritional strategies aimed at promoting gut health and longevity.

Supplementary Materials: The following supporting information can be downloaded at: https://www.mdpi.com/article/10.3390/ijms25073657/s1.

Author Contributions: M.Y. and Q.X.: Conceptualization, Writing—original draft. Y.X.: Formal analysis, Methodology. M.X.: Methodology, Data curation. J.C.: Conceptualization, Project administration. B.-E.T.: Conceptualization, Supervision, Funding acquisition. Y.Y.: Conceptualization, Supervision, Writing—review and editing, Funding acquisition. All authors have read and agreed to the published version of the manuscript.

Funding: This work was supported by the National Key R&D Program (2021YFD1301004, 2021YFD1301005, 2021YFD1300401), National Natural Science Foundation of China (U20A2054, 32072745, 32102571, 32130099), Natural Science Foundation of Hunan Province (2024JJ1004) the earmarked fund for CARS (CARS-35), the Science and Technology Major Project of Yunnan Province (202202AE090032).

Institutional Review Board Statement: The animal protocols were carried out the guidelines of animal care and were approved by the Institutional Animal Care and Use Committee of Hunan Agricultural University (approval number: CACAHU 2022-0426. date of approval: 20 May 2022).

Informed Consent Statement: Not applicable.

Data Availability Statement: Data will be made available on request. Sequencing data generated in this study were deposited in Sequence Read Archive (SRA) with the SRA accession numbers PRJNA1065593.

Conflicts of Interest: The authors declare no conflicts of interest.

Abbreviations

MetR, Methionine restriction; LPS, lipopolysaccharide; DAO, diamine oxidase; TNF-α, tumor Necrosis Factor Alpha; IL-1β, interleukin-1 Inhibitors; IL-2, interleukin-2; IL-4, interleukin-4; IL-6, interleukin-6; IL-12, interleukin-12; ZO-1, zonula occludens-1; H and E, hematoxylin and eosin; SCFAs, short-chain fatty acids; OPLS-DA, orthogonal partial least squares discriminant analysis; VIP, variable importance in the projection; TCA, taurocholic acid; TUDCA, tauro-ursodeoxycholic acid; TCDCA, tauro-chenodeoxycholic acid; BA, bile acid.

References

1. Gao, X.; Sanderson, S.M.; Dai, Z.; Reid, M.A.; Cooper, D.E.; Lu, M.; Richie, J.P., Jr.; Ciccarella, A.; Calcagnotto, A.; Mikhael, P.G.; et al. Dietary methionine influences therapy in mouse cancer models and alters human metabolism. *Nature* **2019**, 572, 397–401. [CrossRef]
2. Han, L.; Wu, G.; Feng, C.; Yang, Y.; Li, B.; Ge, Y.; Jiang, Y.; Shi, Y.; Le, G. Dietary methionine restriction improves the impairment of cardiac function in middle-aged obese mice. *Food Funct.* **2020**, *11*, 1764–1778. [CrossRef]

3. Bárcena, C.; Quirós, P.M.; Durand, S.; Mayoral, P.; Rodríguez, F.; Caravia, X.M.; Mariño, G.; Garabaya, C.; Fernández-García, M.T.; Kroemer, G.; et al. Methionine Restriction Extends Lifespan in Progeroid Mice and Alters Lipid and Bile Acid Metabolism. *Cell Rep.* **2018**, *24*, 2392–2403. [CrossRef]
4. Kitada, M.; Ogura, Y.; Monno, I.; Xu, J.; Koya, D. Effect of Methionine Restriction on Aging: Its Relationship to Oxidative Stress. *Biomedicines* **2021**, *9*, 130. [CrossRef]
5. Mladenović, D.; Radosavljević, T.; Hrnčić, D.; Rasic-Markovic, A.; Stanojlović, O. The effects of dietary methionine restriction on the function and metabolic reprogramming in the liver and brain—Implications for longevity. *Rev. Neurosci.* **2019**, *30*, 581–593. [CrossRef]
6. Yang, Y.; Wang, Y.; Sun, J.; Zhang, J.; Guo, H.; Shi, Y.; Cheng, X.; Tang, X.; Le, G. Dietary methionine restriction reduces hepatic steatosis and oxidative stress in high-fat-fed mice by promoting H_2S production. *Food Funct.* **2019**, *10*, 61–77. [CrossRef]
7. Yang, Y.; Zhang, J.; Wu, G.; Sun, J.; Wang, Y.; Guo, H.; Shi, Y.; Cheng, X.; Tang, X.; Le, G. Dietary methionine restriction regulated energy and protein homeostasis by improving thyroid function in high fat diet mice. *Food Funct.* **2018**, *9*, 3718–3731. [CrossRef]
8. Gasaly, N.; de Vos, P.; Hermoso, M.A. Impact of Bacterial Metabolites on Gut Barrier Function and Host Immunity: A Focus on Bacterial Metabolism and Its Relevance for Intestinal Inflammation. *Front. Immunol.* **2021**, *12*, 658354. [CrossRef]
9. Martel, J.; Chang, S.H.; Ko, Y.F.; Hwang, T.L.; Young, J.D.; Ojcius, D.M. Gut barrier disruption and chronic disease. *Trends Endocrinol. Metab. TEM* **2022**, *33*, 247–265. [CrossRef]
10. Banerjee, S.; Sindberg, G.; Wang, F.; Meng, J.; Sharma, U.; Zhang, L.; Dauer, P.; Chen, C.; Dalluge, J.; Johnson, T.; et al. Opioid-induced gut microbial disruption and bile dysregulation leads to gut barrier compromise and sustained systemic inflammation. *Mucosal Immunol.* **2016**, *9*, 1418–1428. [CrossRef]
11. Schirmer, M.; Smeekens, S.P.; Vlamakis, H.; Jaeger, M.; Oosting, M.; Franzosa, E.A.; Ter Horst, R.; Jansen, T.; Jacobs, L.; Bonder, M.J.; et al. Linking the Human Gut Microbiome to Inflammatory Cytokine Production Capacity. *Cell* **2016**, *167*, 1125–1136.e1128. [CrossRef]
12. Martin-Gallausiaux, C.; Marinelli, L.; Blottière, H.M.; Larraufie, P.; Lapaque, N. SCFA: Mechanisms and functional importance in the gut. *Proc. Nutr. Soc.* **2021**, *80*, 37–49. [CrossRef]
13. Akhtar, M.; Chen, Y.; Ma, Z.; Zhang, X.; Shi, D.; Khan, J.A.; Liu, H. Gut microbiota-derived short chain fatty acids are potential mediators in gut inflammation. *Anim. Nutr. (Zhongguo Xu Mu Shou Yi Xue Hui)* **2022**, *8*, 350–360. [CrossRef]
14. Ma, J.; Hong, Y.; Zheng, N.; Xie, G.; Lyu, Y.; Gu, Y.; Xi, C.; Chen, L.; Wu, G.; Li, Y.; et al. Gut microbiota remodeling reverses aging-associated inflammation and dysregulation of systemic bile acid homeostasis in mice sex-specifically. *Gut Microbes* **2020**, *11*, 1450–1474. [CrossRef]
15. Ramalingam, A.; Wang, X.; Gabello, M.; Valenzano, M.C.; Soler, A.P.; Ko, A.; Morin, P.J.; Mullin, J.M. Dietary methionine restriction improves colon tight junction barrier function and alters claudin expression pattern. *Am. J. Physiol. Cell Physiol.* **2010**, *299*, C1028–C1035. [CrossRef]
16. Wu, G.; Shi, Y.; Han, L.; Feng, C.; Ge, Y.; Yu, Y.; Tang, X.; Cheng, X.; Sun, J.; Le, G.W. Dietary Methionine Restriction Ameliorated Fat Accumulation, Systemic Inflammation, and Increased Energy Metabolism by Altering Gut Microbiota in Middle-Aged Mice Administered Different Fat Diets. *J. Agric. Food Chem.* **2020**, *68*, 7745–7756. [CrossRef]
17. Yang, Y.; Zhang, Y.; Xu, Y.; Luo, T.; Ge, Y.; Jiang, Y.; Shi, Y.; Sun, J.; Le, G. Dietary methionine restriction improves the gut microbiota and reduces intestinal permeability and inflammation in high-fat-fed mice. *Food Funct.* **2019**, *10*, 5952–5968. [CrossRef]
18. Xu, Y.; Yang, Y.; Li, B.; Xie, Y.; Shi, Y.; Le, G. Dietary methionine restriction improves gut microbiota composition and prevents cognitive impairment in D-galactose-induced aging mice. *Food Funct.* **2022**, *13*, 12896–12914. [CrossRef]
19. Barbara, G.; Barbaro, M.R.; Fuschi, D.; Palombo, M.; Falangone, F.; Cremon, C.; Marasco, G.; Stanghellini, V. Inflammatory and Microbiota-Related Regulation of the Intestinal Epithelial Barrier. *Front. Nutr.* **2021**, *8*, 718356. [CrossRef]
20. Bischoff, S.C.; Barbara, G.; Buurman, W.; Ockhuizen, T.; Schulzke, J.D.; Serino, M.; Tilg, H.; Watson, A.; Wells, J.M. Intestinal permeability--a new target for disease prevention and therapy. *BMC Gastroenterol.* **2014**, *14*, 189. [CrossRef]
21. Kuziel, G.A.; Rakoff-Nahoum, S. The gut microbiome. *Curr. Biol. CB* **2022**, *32*, R257–R264. [CrossRef] [PubMed]
22. Xue, Y.; Jin, T.; Gao, C.; Li, C.; Zhou, T.; Wan, D.; Yang, M. Effects of biodegradable film mulching on bacterial diversity in soils. *Arch. Microbiol.* **2022**, *204*, 195. [CrossRef]
23. Zhao, C.; Hu, X.; Bao, L.; Wu, K.; Zhao, Y.; Xiang, K.; Li, S.; Wang, Y.; Qiu, M.; Feng, L.; et al. Gut dysbiosis induces the development of mastitis through a reduction in host anti-inflammatory enzyme activity by endotoxemia. *Microbiome* **2022**, *10*, 205. [CrossRef]
24. Yang, M.; Gu, Y.; Li, L.; Liu, T.; Song, X.; Sun, Y.; Cao, X.; Wang, B.; Jiang, K.; Cao, H. Bile Acid-Gut Microbiota Axis in Inflammatory Bowel Disease: From Bench to Bedside. *Nutrients* **2021**, *13*, 3143. [CrossRef]
25. Stolfi, C.; Maresca, C.; Monteleone, G.; Laudisi, F. Implication of Intestinal Barrier Dysfunction in Gut Dysbiosis and Diseases. *Biomedicines* **2022**, *10*, 289. [CrossRef]
26. Kaminsky, L.W.; Al-Sadi, R.; Ma, T.Y. IL-1β and the Intestinal Epithelial Tight Junction Barrier. *Front. Immunol.* **2021**, *12*, 767456 [CrossRef] [PubMed]
27. Ren, B.; Wang, L.; Mulati, A.; Liu, Y.; Liu, Z.; Liu, X. Methionine Restriction Improves Gut Barrier Function by Reshaping Diurnal Rhythms of Inflammation-Related Microbes in Aged Mice. *Front. Nutr.* **2021**, *8*, 746592. [CrossRef] [PubMed]
28. Zhai, Z.; Zhang, F.; Cao, R.; Ni, X.; Xin, Z.; Deng, J.; Wu, G.; Ren, W.; Yin, Y.; Deng, B. Cecropin A Alleviates Inflammation Through Modulating the Gut Microbiota of C57BL/6 Mice With DSS-Induced IBD. *Front. Microbiol.* **2019**, *10*, 1595. [CrossRef]

9. Jang, H.M.; Kim, J.K.; Joo, M.K.; Shin, Y.J.; Lee, K.E.; Lee, C.K.; Kim, H.J.; Kim, D.H. Enterococcus faecium and Pediococcus acidilactici deteriorate Enterobacteriaceae-induced depression and colitis in mice. *Sci. Rep.* **2022**, *12*, 9389. [CrossRef]
0. Li, S.; Guo, J.; Liu, R.; Zhang, F.; Wen, S.; Liu, Y.; Ren, W.; Zhang, X.; Shang, Y.; Gao, M.; et al. Predominance of Escherichia-Shigella in Gut Microbiome and Its Potential Correlation with Elevated Level of Plasma Tumor Necrosis Factor Alpha in Patients with Tuberculous Meningitis. *Microbiol. Spectr.* **2022**, *10*, e0192622. [CrossRef]
1. Liu, S.; Li, E.; Sun, Z.; Fu, D.; Duan, G.; Jiang, M.; Yu, Y.; Mei, L.; Yang, P.; Tang, Y.; et al. Altered gut microbiota and short chain fatty acids in Chinese children with autism spectrum disorder. *Sci. Rep.* **2019**, *9*, 287. [CrossRef]
2. Gao, X.; Du, L.; Randell, E.; Zhang, H.; Li, K.; Li, D. Effect of different phosphatidylcholines on high fat diet-induced insulin resistance in mice. *Food Funct.* **2021**, *12*, 1516–1528. [CrossRef] [PubMed]
3. Baltazar-Díaz, T.A.; González-Hernández, L.A.; Aldana-Ledesma, J.M.; Peña-Rodríguez, M.; Vega-Magaña, A.N.; Zepeda-Morales, A.S.M.; López-Roa, R.I.; Del Toro-Arreola, S.; Martínez-López, E.; Salazar-Montes, A.M.; et al. Escherichia/Shigella, SCFAs, and Metabolic Pathways-The Triad That Orchestrates Intestinal Dysbiosis in Patients with Decompensated Alcoholic Cirrhosis from Western Mexico. *Microorganisms* **2022**, *10*, 1231. [CrossRef] [PubMed]
4. Han, L.; Li, T.; Du, M.; Chang, R.; Zhan, B.; Mao, X. Beneficial Effects of Potentilla discolor Bunge Water Extract on Inflammatory Cytokines Release and Gut Microbiota in High-Fat Diet and Streptozotocin-Induced Type 2 Diabetic Mice. *Nutrients* **2019**, *11*, 670. [CrossRef] [PubMed]
5. Firrman, J.; Liu, L.; Mahalak, K.; Tanes, C.; Bittinger, K.; Tu, V.; Bobokalonov, J.; Mattei, L.; Zhang, H.; Van den Abbeele, P. The impact of environmental pH on the gut microbiota community structure and short chain fatty acid production. *FEMS Microbiol. Ecol.* **2022**, *98*, fiac038. [CrossRef] [PubMed]
6. Soriano-Lerma, A.; García-Burgos, M.; Alférez, M.J.M.; Pérez-Carrasco, V.; Sanchez-Martin, V.; Linde-Rodríguez, Á.; Ortiz-González, M.; Soriano, M.; García-Salcedo, J.A.; López-Aliaga, I. Gut microbiome-short-chain fatty acids interplay in the context of iron deficiency anaemia. *Eur. J. Nutr.* **2022**, *61*, 399–412. [CrossRef] [PubMed]
7. Aon, M.A.; Bernier, M.; Mitchell, S.J.; Di Germanio, C.; Mattison, J.A.; Ehrlich, M.R.; Colman, R.J.; Anderson, R.M.; de Cabo, R. Untangling Determinants of Enhanced Health and Lifespan through a Multi-omics Approach in Mice. *Cell Metab.* **2020**, *32*, 100–116.e104. [CrossRef] [PubMed]
8. Bárcena, C.; Valdés-Mas, R.; Mayoral, P.; Garabaya, C.; Durand, S.; Rodríguez, F.; Fernández-García, M.T.; Salazar, N.; Nogacka, A.M.; Garatachea, N.; et al. Healthspan and lifespan extension by fecal microbiota transplantation into progeroid mice. *Nat. Med.* **2019**, *25*, 1234–1242. [CrossRef]
9. Li, X.; Zhao, W.; Xiao, M.; Yu, L.; Chen, Q.; Hu, X.; Zhao, Y.; Xiong, L.; Chen, X.; Wang, X.; et al. Penthorum chinense Pursh. extract attenuates non-alcholic fatty liver disease by regulating gut microbiota and bile acid metabolism in mice. *J. Ethnopharmacol.* **2022**, *294*, 115333. [CrossRef]
0. Zhang, X.; Coker, O.O.; Chu, E.S.; Fu, K.; Lau, H.C.H.; Wang, Y.X.; Chan, A.W.H.; Wei, H.; Yang, X.; Sung, J.J.Y.; et al. Dietary cholesterol drives fatty liver-associated liver cancer by modulating gut microbiota and metabolites. *Gut* **2021**, *70*, 761–774. [CrossRef]
1. Tian, Q.; Yang, R.; Wang, Y.; Liu, J.; Wee, A.; Saxena, R.; Wang, L.; Li, M.; Liu, L.; Shan, S.; et al. A High Serum Level of Taurocholic Acid Is Correlated With the Severity and Resolution of Drug-induced Liver Injury. *Clin. Gastroenterol. Hepatol. Off. Clin. Pract. J. Am. Gastroenterol. Assoc.* **2021**, *19*, 1009–1019.e1011. [CrossRef] [PubMed]
2. Jia, W.; Xie, G.; Jia, W. Bile acid-microbiota crosstalk in gastrointestinal inflammation and carcinogenesis. *Nat. Rev. Gastroenterol. Hepatol.* **2018**, *15*, 111–128. [CrossRef]
3. Wang, L.; Ren, B.; Zhang, Q.; Chu, C.; Zhao, Z.; Wu, J.; Zhao, W.; Liu, Z.; Liu, X. Methionine restriction alleviates high-fat diet-induced obesity: Involvement of diurnal metabolism of lipids and bile acids. *Biochim. Biophys. Acta. Mol. Basis Dis.* **2020**, *1866*, 165908. [CrossRef]
4. Shen, B.; Zhou, C.; Gu, T.; Shen, Z.; Guo, Y.; Dai, W.; Liu, Y.; Zhang, J.; Lu, L.; Dong, H. Kuhuang alleviates liver fibrosis by modulating gut microbiota-mediated hepatic IFN signaling and bile acid synthesis. *Front. Pharmacol.* **2022**, *13*, 1080226. [CrossRef]
5. Maddineni, S.; Nichenametla, S.; Sinha, R.; Wilson, R.P.; Richie, J.P., Jr. Methionine restriction affects oxidative stress and glutathione-related redox pathways in the rat. *Exp. Biol. Med.* **2013**, *238*, 392–399. [CrossRef] [PubMed]
6. Tamanna, N.; Kroeker, K.; Braun, K.; Banh, S.; Treberg, J.R. The effect of short-term methionine restriction on glutathione synthetic capacity and antioxidant responses at the whole tissue and mitochondrial level in the rat liver. *Exp. Gerontol.* **2019**, *127*, 110712. [CrossRef] [PubMed]

biomolecules

MDPI

Article

Overexpression of *S30* Ribosomal Protein Leads to Transcriptional and Metabolic Changes That Affect Plant Development and Responses to Stress

Alin Finkelshtein [1], Hala Khamesa-Israelov [1] and Daniel A. Chamovitz [2,*]

1 School of Plant Sciences and Food Security, Tel Aviv University, Ramat Aviv 69978, Israel
2 The Department of Life Sciences, Ben-Gurion University of the Negev, Be'er Sheva 84105, Israel
* Correspondence: dchamovitz@bgu.ac.il

Abstract: ICT1 is an *Arabidopsis thaliana* line that overexpresses the gene encoding the S30 ribosomal subunit, leading to tolerance to exogenous indole-3-carbinol. Indole-3-carbinol (I3C) is a protective chemical formed as a breakdown of I3M in cruciferous vegetables. The overexpression of S30 in ICT1 results in transcriptional changes that prime the plant for the I3C, or biotic insult. Emerging evidence suggests that ribosomal proteins play important extra-ribosomal roles in various biochemical and developmental processes, such as transcription and stress resistance. In an attempt to elucidate the mechanism leading to I3C and stress resistance in ICT1, and using a multi-pronged approach employing transcriptomics, metabolomics, phenomics, and physiological studies, we show that overexpression of S30 leads to specific transcriptional alterations, which lead to both changes in metabolites connected to biotic and oxidative stress tolerance and, surprisingly, to photomorphogenesis.

Keywords: ribosome; indole-3-carbinol; metabolomics; photomorphogenesis; plant development; stress resistance

1. Introduction

ICT1 is an *Arabidopsis thaliana* line that overexpresses the gene encoding the *S30* ribosomal subunit, leading to tolerance to exogenous indole-3-carbinol (I3C) [1]. I3C is formed from the breakdown of indole-3-methylglucosinolate (I3M), the predominant indole glucosinolate [2], by a reaction primarily catalyzed by myrosinase [3,4], though the formation of I3M can also occur in the absence of myrosinase [5]. In Arabidopsis, I3C serves as a signaling molecule that modulates the auxin response by antagonizing auxin binding to the TIR1 receptor and other mechanisms [6–8]. Hence, exogenous I3C at high concentrations inhibits *Arabidopsis* growth and development [7].

While ICT1 was specifically tolerant to I3C and not to other toxic molecules, it was also tolerant to bacterial and fungal pathogens, as well as to oxidative stress [1]. The overexpression of *S30* in ICT1 results in transcriptional changes that prime the plant for the I3C, or biotic insult. For example, the genes involved in responses to biotic stress, immune responses, and stress responses are expressed constitutively, priming the plant to respond more quickly [1].

The ribosome, a two-subunit ribonucleoprotein complex, plays a pivotal role in protein synthesis but also has emerging roles in transcriptional regulation. In *Arabidopsis thaliana*, the large 60S subunit consists of 48 ribosomal proteins (RPs), while the small 40S subunit contains 32 RPs, including *S30* [9]. *S30* is required for ribosome assembly and protein synthesis by promoting the interaction between the tRNA anticodon stem-loop and the 18S rRNA 530 loop [10].

Mounting evidence indicates that differences in RP expression can impact functions unrelated to ribosome assembly or protein synthesis. For example, mutations and differential expression in ribosomal genes have been found to play a role in regulating normal development and responding to environmental stresses [11–14]. In an attempt to elucidate the mechanism leading to I3C and stress resistance in ICT1, and using a multi-pronged approach employing transcriptomics, metabolomics, phenomics, and physiological studies, we show that overexpression of *S30* leads to specific transcriptional alterations, which lead to both changes in metabolites connected to biotic and oxidative stress tolerance and, surprisingly, to photomorphogenesis.

2. Materials and Methods

2.1. Plant Material and Growth Assays

The *Arabidopsis* lines used in this work were all WT ecotype Columbia-0 backgrounds. Transgenic and mutant lines were previously reported: ICT1 [1], $S30^{t\text{-}DNA}$ (Salk_087911.4) [15], *tgg1/2* double mutant [16], and *cyp79B2/B3* double mutant [17]. Seeds were cultivated in Petri plates using medium containing 0.8% agar (Acumedia, Lansing, MI, USA, 7178A), half-strength Murashige and Skoog salts (MS) (Sigma, Yokohama, Japan, M5519), and 1% sucrose (Bio Lab, Rockville, MD, USA, 19220591, pH 5.7. The Petri plates were placed in chambers at 21 °C under light/dark conditions of 16 h white light at 75 lmol m^2 sec^{-1} and 8 h darkness, at 55% relative humidity. For root phenotype experiments, plates were placed vertically in the chambers. All treatments were conducted under the same day-time conditions. For experiments that included ICT1, at least 3 independent lines of ICT1 were used in each; the results are the average of all lines.

Hydroponic growth: For treatments with I3C, plants were grown hydroponically in tall petri dishes with liquid MS. ~60 seedlings were in each plate for two weeks in normal conditions (long day, 21 °C, humidity 55%). All lines were exposed for two hours to either 0.05% DMSO or 400 μM I3C (diluted in 0.05% DMSO). All samples were collected at the same time to eliminate circadian clock effects and washed three times with fresh DDW. Moreover, 10 seedlings for each sample were pooled, frozen in liquid nitrogen, and lyophilized. Indole-3-carbinol (I3C) (Sigma I7256)was dissolved in 0.05% DMSO (Merck, Rahway, NJ, USA, 1.02931) to produce a 1 M solution and stored in the dark at −20 °C.

2.2. Yield

ICT1 and WT lines were grown in pots in optimal conditions until full ripening. Seeds were collected separately from each plant and weighted. The weight was transformed into the number of seeds with the aid of a calibration curve. The average weight and number of seeds per plant were calculated.

2.3. LC-MS-MS

Sample preparation: Seedlings were grown hydroponically. Dry samples were powdered with 2 stainless-steel balls in a beat beater (1 min, MM400, Retsch, Haan, Germany). One mg of each sample was extracted with 80% methanol (Bio Lab, 1368060200) in 0.2% formic acid (500 uL), containing ethylparaben (1 ng/mL) as an internal standard (IS), by shaking (20,000 rpm, 65 °C, 15 min, Thermomixer C, Eppendorf, Hamburg, Germany). Then, the mixtures were centrifuged (14,000× *g*, 10 min), and supernatants were collected and placed in LC-MS vials. Standard curve mixtures of 0.001–10 ug/mL of I3C and DIM were prepared in 80% methanol/0.2% formic acid, containing 1 ug/mL of ethylparaben as IS.

Standards: Indole-3-carbinol (I3C) I7256, 3,3′-diindolylmethane (DIM) D9568, and ethylparaben PHR1011 (internal standard, IS) were purchased from Sigma/Merck.

LC-MS-MS analysis was performed on the Acquity UPLC system and the triple quadrupole Xevo TQ-S (both Waters). TargetLynx (Waters, Milford, MA, USA) was applied for quantitation based on standard curves.

LC parameters: Acquity BEH C18 column (2.1 × 100 mm, 1.7 μm; Waters) at 35 °C. Gradient with mobile phases A—0.01% aqueous formic acid and B—0.01% formic acid in acetonitrile: 0 min, 20%B; 3 min, 100%B; 4 min, 100%B; 4.5 min, 20%B; 7 min, 20%

B. Injection volume: 2.0 μL; flow rate: 0.3 mL/min. Retention times: 2.25 min for I3C, 2.96 min for DIM, and 2.18 min for IS.

MS/MS parameters: Electrospray ionization in positive-ion mode; desolvation temperature, 400 °C; desolvation gas flow, 600 L/hr; cone gas flow, 150 L/hr; nebulizer pressure, 7 Bar; capillary voltage, 3.56 kV; cone voltage, 44 V; with argon 0.10 mg/min as the collision gas. The MRM transitions are: 130.1 > 77.0 and 130.1 > 103.1 with collision energy (CE) 22 V in both I3C and DIM; 167.1 > 95.0 and 167.1 > 139.1, with CE 13 and 10 V, respectively, for IS.

2.4. LC-MS

Sample preparation: seedlings were grown in MS plates, with ~60 seeds in each plate, vertically standing for 16 days in normal conditions.

With and without I3C treatment, ICT1 and WT were grown hydroponically. Each tube contains 100 mg of wet weight.

Metabolic profiling of the semipolar phase was performed using the Waters ACQUITY UPLC system coupled to a Vion IMS QTof mass spectrometer (Waters Corp., MA, USA). The chromatographic separation was performed on an ACQUITY UPLC BEH C18 column (2.1 × 100 mm, i.d., 1.7 μm) (Waters Corp., MA, USA). The mobile phase A consisted of 95% water (UPLC grade) and 5% acetonitrile, with 0.1% formic acid; the mobile phase B consisted of 100% acetonitrile with 0.1% formic acid. The column was maintained at 35 °C, and the flow rate of the mobile phase was 0.3 mL*min^{-1}. Mobile phase A was initially run at 100%, and it was gradually reduced to 72% at 22 min, following a decrease to 60% at 22.5 min and 0% at 23 min. Then, mobile phase B was run at 100% until 26.5 min, and mobile phase A was set to 100% at 27 min. Finally, the column was equilibrated at 100% A for 28 min. MS parameters were as follows: the source and de-solvation temperatures were maintained at 120 °C and 350 °C, respectively. The capillary voltage was set to 2 kV, and the cone voltage was set to 40 V. Nitrogen was used as de-solvation gas and cone gas at a flow rate of 700 L*h^{-1} and 50 L*h^{-1}, respectively. The mass spectrometer was operated in full-scan HDMSE negative ionization over a mass range of 50–2000 Da. For the high energy scan function, a collision energy ramp of 20–80 eV was applied; for the low energy scan function, 5 eV was applied. Leucine-enkephalin was used as a lock-mass reference standard.

Semipolar compound identification and data analysis: LC-MS data were analyzed and processed with UNIFI (Version 1.9.4, Waters Corp., MA, USA). The putative identification of the different semipolar species was performed by comparing the accurate mass, fragmentation pattern, and ion mobility (CCS) values of theoretical structures in an in-house-generated library. The data were normalized using internal standards and sample weights.

2.5. Photomorphogenesis

WT and six independent ICT1 lines (1, 2, 28, 33, 478, and 8098) were used for the experiment. Seeds were exposed to 5 h of light, covered with aluminum foil, and grown at 21 °C for ten days. Hypocotyl length, cotyledon opening %, and angle were measured.

2.6. Abiotic Stresses

All lines were grown for two weeks and then exposed to the following conditions: 1. Salt stress: pots were irrigated with a 200 mM NaCl solution; 2. Constitutive heat (28 °C) for 3 days; 3. Intermittent high heat (44 °C) two hours a day for a week; 4. Drought: plants were irrigated with half the normal volume of water (1.5 lit/64 pots of soil); 5. UV stress: single exposure to UVb 5000 ERG for 1 min; 6. Cold stress: 4 °C during the night for a week. Untreated samples were used as controls. Plants were monitored for plant area, roundness, water content, and photosynthesis (QY max, Fv/Fm_Lss, NPQ_Lss, and RFD_Lss) using a Plant Screen^{TM} Phenotyping System (Photon Systems Instruments (PSI),

Drasov, Czech Republic). Seedlings grown in MS were transferred to PSI standard pots (one seedling per pot), and imaging was performed on days 14–31 with a 3–4-day interval.

2.7. Statistics

All the experiments were repeated at least three times. The variance between the groups that were statistically compared was similar. The samples were analyzed using a *t*-test. We considered $p < 0.05$ to be significant.

PCA is a dimension reduction analysis that transforms the original data into a new set of uncorrelated variables (principal components (PC)) that capture the maximum variance in the data. To see how many PCs can explain the most variance, we use the scree plot. In a scree plot, the point where the eigenvalues start to level off or show a bend is often referred to as the "elbow" of the plot. The number of principal components before this bend is typically considered for retention. In our case, six PCs are the most significant ones. Thus, we have chosen the combinations between them that give the best separation between samples for visualization.

3. Results

3.1. Tolerance to Exogenous I3C in ICT1 Likely Arises from Reduced Endogenous Levels of I3C

To elucidate the mechanisms leading to I3C tolerance in ICT1, we took several approaches. To investigate if the tolerance of ICT1 mutants to I3C can be attributed to reduced uptake of extracellular I3C, we conducted I3C uptake assays using LC-MS-MS. The lines used for the assay were: ICT1, WT, *S30*$^{t\text{-}DNA}$, *tgg1/2* (myrosinase-deficient), and *cyp79B2/B3* (deficient in indolic GLS synthesis).

In control samples treated with dimethyl sulfoxide (DMSO), the measured I3C content represents I3C endogenously produced by the plant. We observed similar concentrations of endogenous I3C levels in WT and *tgg1/2* and reduced concentrations in ICT1, while no detectable I3C was observed in the *cyp79B2/B3* double mutant (Figure 1a). The similar concentrations in WT and *tgg1/2* indicate that in an undamaged plant, endogenous I3C is hydrolyzed by a non-enzymatic breakdown, consistent with previously shown myrosinase-dependent and independent I3C hydrolysis pathways [5]. The reduction in endogenous levels of I3C in ICT1 correlates with the downregulation of several glucosinolate (GLS) biosynthetic genes in ICT1 (Figure 1b). The fact that *cyp79B2/B3* and ICT1 have reduced I3C levels supports the idea that downregulation of GLS genes contributes to the lower abundance of I3C in ICT1. Interestingly, while GLS synthesis is down-regulated in ICT1, this line has increased expression of the myrosinase-encoding *tgg* genes (Figure S1). Conversely, the uptake of exogenous I3C by ICT1 mutants was found to be twice as high as in WT plants (Figure 1a). Therefore, tolerance to exogenous I3C in ICT1 is not due to defects in I3C uptake.

We next tested whether the increased degradation of I3C contributes to ICT1′s tolerance of I3C. I3C is a highly unstable metabolite that forms various oligomers (Figure S2), such as the formation of 3,3′-diindolylmethane (DIM) [18]. As depicted in Figure 1c, I3C-treated samples exhibited approximately 3-times higher levels of DIM compared to I3C alone, indicating that a significant portion of absorbed I3C was indeed degraded to DIM.

As endogenous I3C levels are reduced in ICT1, and as *cyp79B2/B3* (Figure 1a) contains no I3C, we thus hypothesized that *cyp79B2/B3*, like ICT1, would also be tolerant to I3C. Using the root growth assay, as seen in Figure 2, *cyp79B2/B3* is indeed tolerant to exogenous I3C. Thus, the explanation for the tolerance to exogenous I3C in both lines is that the tolerance likely arises from reduced endogenous levels of I3C.

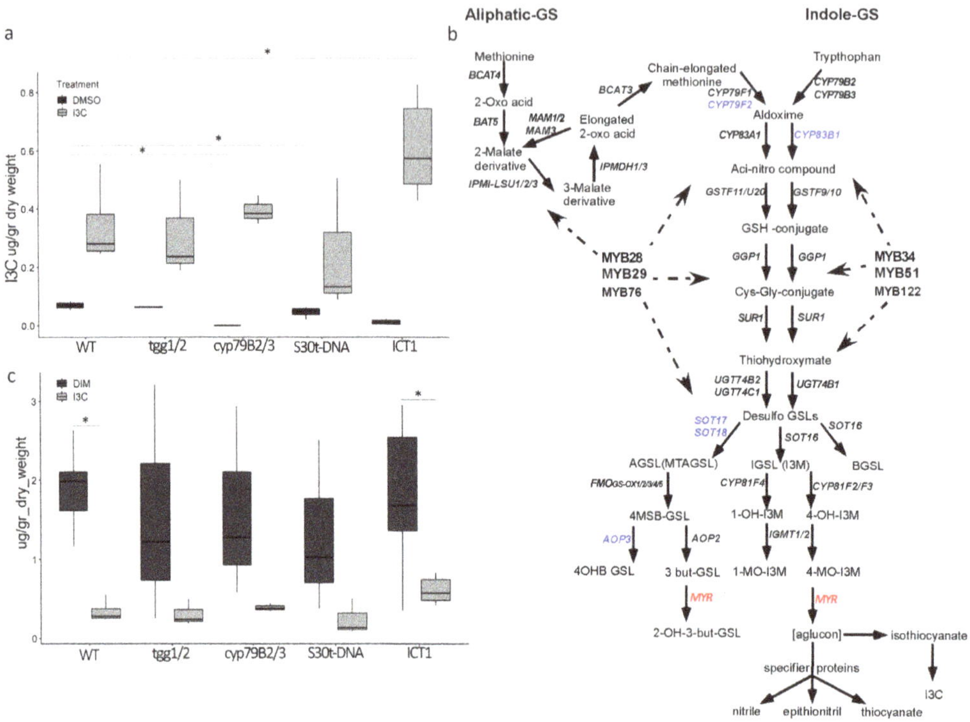

Figure 1. Uptake of exogenous I3C. LC-MS-MS results. (**a**) Total I3C content (mg/g dry weight) for each line. Each sample was treated with either 1 M DMSO or I3C 400 μM (in 1 M DMSO) for 2 h. (**b**) Glucosinolate pathway: genes marked in blue are down-regulated, and genes marked in red are up-regulated in ICT1. (**c**) Total DIM and I3C content following treatment with I3C (mg/g dry weight) for each line. * *p* value < 0.009 *t*-test. 3 > *n* > 6.

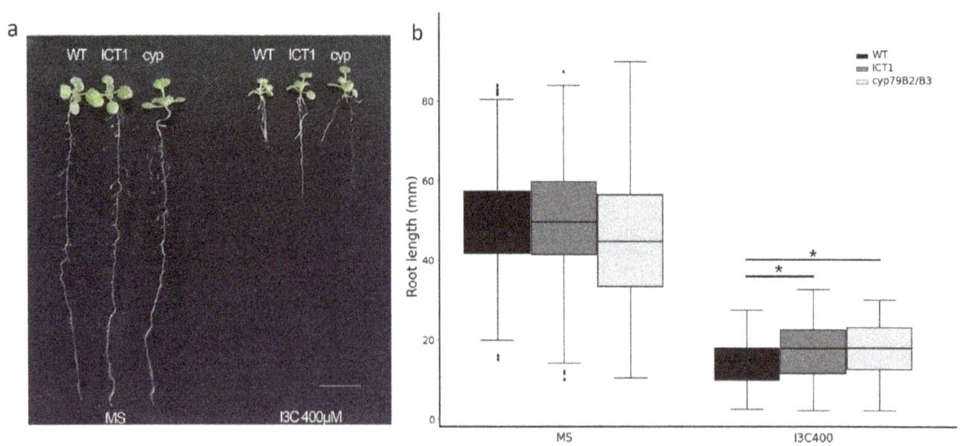

Figure 2. cyp79B2/B3 is tolerant of I3C. (**a**). Phenotypes of WT, ICT1, and cyp79B2/B3 on MS and I3C 400 μM after 14 days. Scale bar = 1 cm. (**b**) The average root length on MS and 400 μM I3C. *t*-test * *p* value < 0.05. 127 < *n* < 258. The dots over the whiskers are the outliers.

3.2. ICT1 Displays Unique Metabolomic Profiles That Correlate with Known Phenotypes

As our previous transcriptomic study revealed differences in expression of specific metabolic pathways, we sought to determine if these differences corresponded to metabolomic changes, which led us to conduct a metabolomic profiling experiment. Specifically, we investigated the profiles of GLS metabolites (GLS) and non-GLS secondary metabolites (SM) in ICT1 compared to WT and *cyp79B2/B3*. Supplementary Table S1 presents the complete list of metabolites and abundances. The principal component analysis of this data clearly differentiates between the three lines in both SM and GLS content (Figure 3a,b). In the comparison between ICT1 and WT, we identified 39 SM that are significantly different, 24 with lower relative abundance in ICT1 and 15 with higher relative abundance (Table S1). Among the 15 prevalent SMs in ICT1, some are associated with ICT1 phenotypes (Figure 3c). Notably, our previous study demonstrated that ICT1 exhibited tolerance to both oxidative and biotic stress [1]. The tolerance to oxidative stress correlates with an increase in four specific secondary metabolites—quercitrin, linarin, citric acid, and 4-O-p-coumaroylquinic acid—each known to be effective against oxidative stress [19–23]. Tolerance to both bacteria and fungi correlates with five secondary metabolites—salicylic acid, kaempferol-7-O-glucoside, linarin, quercitrin, and ascorbigen, each of which shows strong antipathogen activities [23–29].

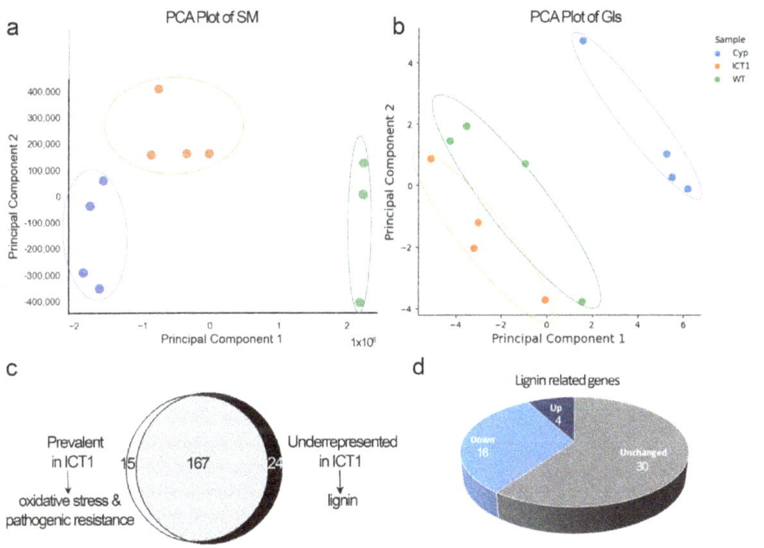

Figure 3. Secondary metabolites (SM) and glucosinolates (GLS) profiles. (**a**,**b**) Principal component analysis of the metabolite profiles of WT, ICT1, and *cyp*79B2/B3 for SM (**a**) and GLS (**b**). Each dot in PCA represents a biological replicate. In the case of SM, two dots in the WT sample overlap. (**c**) Out of 206 SM identified, 15 were overrepresented in ICT1, while 24 were underrepresented. In the overrepresented group, there are four SMs related to oxidative tolerance and eight related to pathogen tolerance. In the underrepresented group, the majority are lignin-related metabolites. (**d**) Out of 50 genes indicated in lignin synthesis, 16 (32%) are downregulated in ICT1 vs. WT [1]. The numbers of lignin-related genes are indicated on the pie chart.

Furthermore, we observed three prevalent glucosinolates in ICT1: glucobrassicin, a significant indole glucosinolate serving as a precursor for active indoles; 6-hydroxyindole-3-carboxylate hexoside isomer 1; and 2-hydroxy-2-methylpropyl glucosinolate/4-hydroxybutyl glucosinolate. Indolic glucosinolates are well-known determinants for pest and disease resistance in plants [30–32]. Consequently, the overexpression of *S30* in ICT1 results in a higher relative abundance of metabolites associated with oxidative stress and pathogen tolerance, which correlates biochemically with the observed corresponding phenotypes.

Among the secondary metabolites with reduced content in ICT1, 62.5% are lignin related (Table S1). This reduction correlates with transcriptome results [1], which showed a down-regulation in 16 out of 50 lignin biosynthetic genes (Figure 3d).

3.3. I3C Treatment Impacts the Metabolic Profiles of WT More Than That of ICT1

We further tested the differences in metabolic profiles between WT and ICT1 before and after treatment with I3C (Table S2). Using principal component analysis, we see a clear differentiation between WT vs. ICT1 for both SM and GLS profiles, whether treated or untreated with I3C. In other words, the overexpression of 30S clearly influences the metabolic profiles of ICT1, differentiating it from WT. Following treatment of WT with I3C, the metabolic profiles of both SM and GLS are clearly separated, indicating that the I3C treatment leads to clear changes in the metabolic profiles of WT. The situation for ICT1 was, though, different. Treatment of ICT1 with I3C did not cause a significant shift in SM profile, which correlates with tolerance to I3C in this mutant (Figure 4a). In other words, just as ICT1 exhibits tolerance to I3C-induced inhibition of root growth, it also exhibits tolerance to I3C-induced changes in SM profiles.

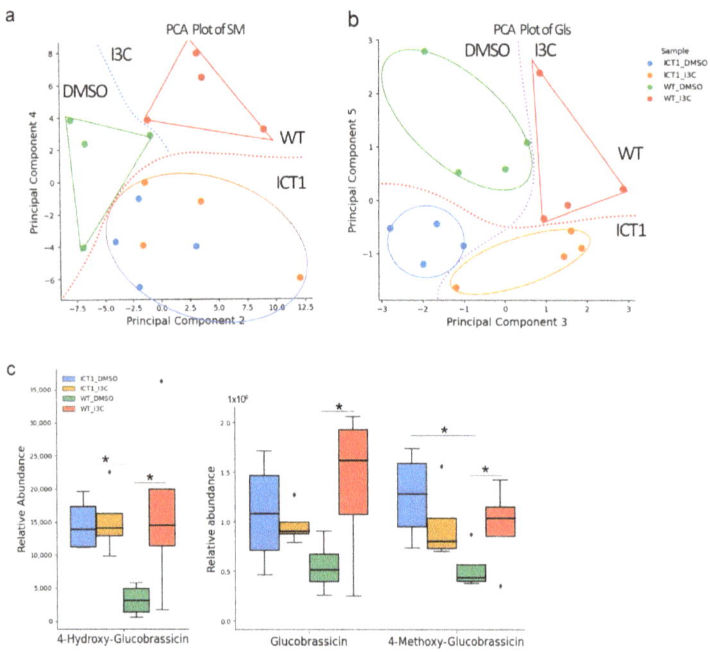

Figure 4. Metabolomic analysis of WT and ICT1 before and following I3C treatment. PCA plots of (**a**) secondary metabolites (SM) and (**b**) glucosinolates (GLS). The red lines differentiate between WT and ICT1. The blue and purple lines differentiate between I3C-treated and untreated. (**c**). Relative abundance of indolic GLS upstream of I3C. Following exposure to exogenous I3C, the levels of these three metabolites increase in WT while remaining constant in ICT1. * *t*-test p value < 0.05, $n = 4$.

However, the GLS profiles of ICT1 are influenced by I3C treatment (Figure 4b). However, we discovered that for three specific GLS, the I3C treatment has no effect on their levels in ICT1, while having increased levels in WT. These glucosinolates, glucobrassicin, 4-hydroxy glucobrassicin, and 4-methoxy glucobrassicin, are upstream of I3C production. The fact that they are induced by I3C treatment in WT but not in ICT1 also correlates with the tolerance in ICT1 (Figure 4c).

3.4. ICT1 Displays Constitutive Photomorphogenic Phenotypes

Further inspection of the transcriptome [1] revealed a group of genes related to skotomorphogenesis that are mis-regulated in ICT1, with 30 genes including Csn8 and LSH6 [33] being downregulated and 18 genes including SOB7, PCH1, CIP1, and PHOT2 [34–36] being upregulated (Figure S3). To test if ICT1 is affected by skotomorphogenesis, WT and six independent ICT1 lines were grown in the dark, and phenotypes were measured after 10 days. While in WT the cotyledons were closed and the hypocotyl was long, all ICT1 lines presented a constitutive photomorphogenesis (cop) phenotype of open cotyledons and short hypocotyl (Figure 5a–d). Thus, the misregulation of skotomorphogenesis-related genes in ICT1 suggests that ICT1 is defective in the regulation of photomorphogenesis.

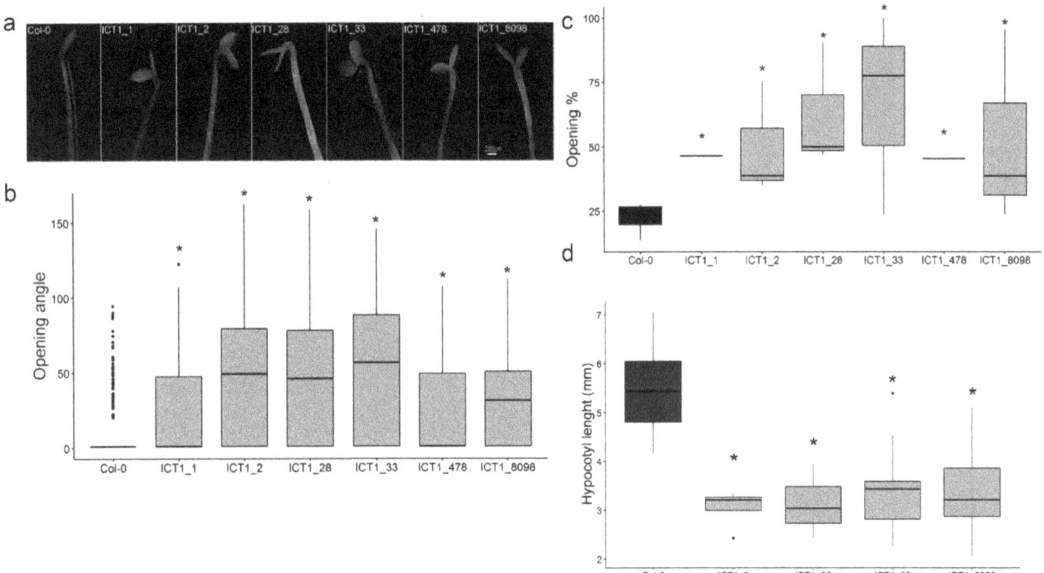

Figure 5. ICT1 shows a constitutive photomorphogenesic phenotype. (**a**) Representative photographs of cotyledons of 14-day-old, dark-grown WT and ICT1. Scale bar = 200 μm. (**b**–**d**) Quantitative measurements of angle of cotyledon opening (**b**), percentage of open cotyledons (**c**), and hypocotyl lengths (**d**). n = 55–187. * t-test p value < 0.05.

3.5. ICT1 Does Not Have Wide Resistance to Abiotic Stresses

Since ICT1 is primed in stress response pathways and showed resistance to oxidative and biotic stress [1], we wanted to see if ICT1 has a wide range of abiotic stress resistance. To test this, six additional abiotic stresses were tested: (1) salt stress; (2) constitutive 3 days of heat; (3) high heat (44 °C) 2 h a day for a week; (4) drought; (5) UV stress; and (6) cold stress. Untreated samples were used as controls. To test if the stress caused phenotypical changes in the plants such as rosette area, leaf roundness, water content, and photosynthesis (QY max, Fv/Fm_Lss, NPQ_Lss, and RFD_Lss), we have scanned all lines by phenomics from day 14 up to day 31 with a 3–4-day interval. Not only was ICT1 not tolerant to any of these stresses, it was hypersensitive to cold treatment (Figure 6). Thus, based on these results, ICT1 did not present wide resistance to abiotic stresses, further supporting the specificity of resistance to specific biotic and abiotic challenges.

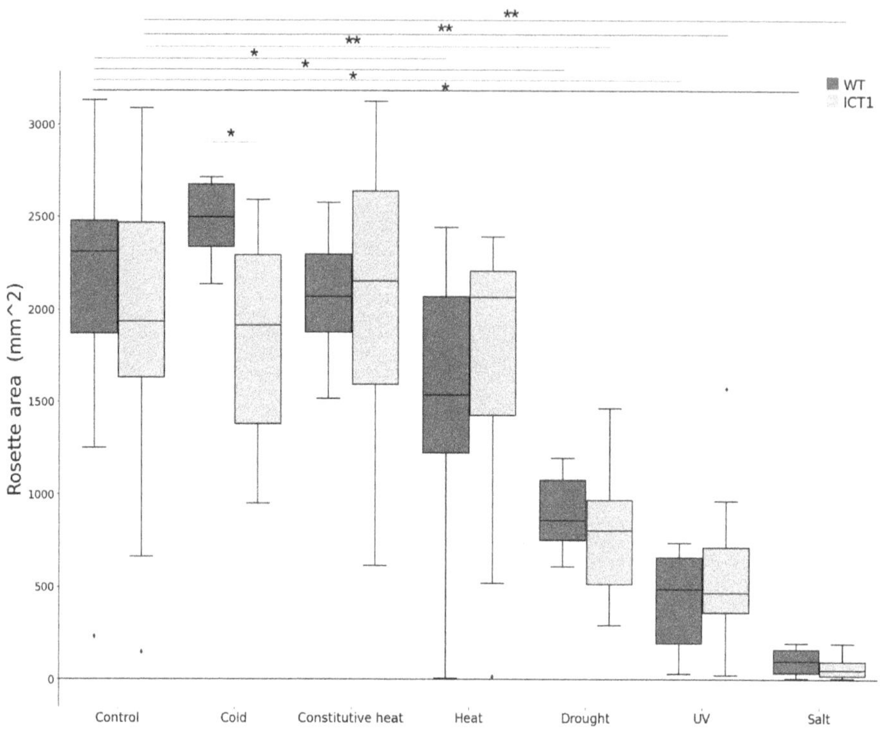

Figure 6. Differences in the rosette area following abiotic stresses. Rosette area following stresses: salt stress, constitutive 3 days of heat (28 °C), high heat (44 °C) 2 h a day for a week, drought, UV stress, and cold stress. Untreated samples were used as controls. *t*-test *p* value < 0.05 * WT, ** ICT1.

4. Discussion

Tolerance or resistance to specific toxic molecules can arise through several mechanisms, including but not limited to target site mutation [37], increased activity of efflux pumps that actively remove the specific molecule from inside the cell [38], reduced uptake of the molecule [39,40], and detoxification or degradation of the specific molecule [41–44]. We have shown here that tolerance to exogenously applied I3C in ICT1 is not due to defects in the uptake of exogenous I3C or increased degradation, but is likely due to reduced levels of endogenous I3C.

This conclusion is supported by our analysis of both ICT1 and the *cyp*79B2/B3 double mutant. *Cyp*79B2/B3 is deficient in GLS synthesis due to the knockout of two enzymes at the beginning of the indolic GLS pathway [17]. We showed that several genes involved in the GLS pathway are down-regulated in ICT1 (Figure 1b) (based on the transcriptome [1]). This deficiency in GLS synthesis in *cyp*79B2/B3 and the reduction in GLS synthesis in ICT1 lead to a lack or reduction of endogenous I3C levels in each line, respectively. Thus, as we found that *cyp*79B2/B3 is, like ICT1, also tolerant to exogenous I3C (Figure 2), we concluded that the tolerance to I3C in these lines results from reduced levels of endogenous I3C.

However, this reduction in I3C levels is likely not the only contributor to the I3C tolerant phenotype. As we previously showed, over-expression of *S30* in ICT1 leads to transcriptional priming, which primes the plant for the I3C treatment [1]. We further showed here that these specific transcriptional changes lead to both changes in metabolite profiles and development.

I3C is produced by the hydrolysis of indol-3-ylmethylglucosinolate (I3M-GS) in the presence of myrosinase [45]. As the glucosinolates and myrosinases are stored in different compartments, tissue rupture is necessary to bring them into contact [46]. Thus, I3C is produced mainly during pathogen attacks or mechanical damage [7]. Endogenous I3C found in the healthy, undamaged plant is a result of the non-enzymatic breakdown of I3M [5]. This is supported by similar levels of I3C in both WT and *tgg1/2* myrosinase-deficient mutants (Figure 1a). Interestingly, ICT1 overexpresses all functional myrosinase-encoding *Tgg* genes (Figure S1). This is likely a compensatory mechanism for the reduction in expression of GLS biosynthetic genes and consequent reduction in I3M levels.

We want to emphasize that the tolerance/resistance phenotypes detected for ICT1 are highly specific and correlate with specific transcriptional changes. In our original study, we showed that ICT1 also exhibited tolerance to oxidative stress and fungal and bacterial pathogens and that this tolerance was due to changes in gene expression patterns that primed the plants for response to the various treatments [1]. We showed here that these resistances are specific, as ICT1 was not tolerant to salt, heat, drought, UV, or cold stresses. Perhaps not unsurprisingly, we also did not detect transcriptional priming for these stress response pathways (based on the transcriptome [1]). Thus, just as we earlier showed that the tolerance in ICT1 is specific to I3C treatment and not a general tolerance response to all indoles/glucosinolates or other toxic molecules, the environmental tolerances identified were also specific to oxidative stress and pathogens and not to stresses in general. Thus, the overexpression of *S30* leads to specific transcriptional changes that lead to tolerance to abiotic and oxidative stresses, not a general overall stress response. It is important to notice that these specific tolerances carry with them a yield penalty (Figure S4).

The results from metabolic profiling coupled with transcriptome analysis (based on transcriptome [1]) further supported the hypothesis that overexpression of *S30* leads to changes in gene expression that influence specific metabolic pathways, which then lead to specific resistance phenotypes. Among the secondary metabolites enriched in ICT1, we notice metabolites associated with resistance to oxidative stress and pathogen resistance. These include quercitrin, linarin, citric acid, and 4-O-p-coumaroylquinic acid, each known to be effective against oxidative stress [19–23], and salicylic acid, kaempferol-7-O-glucoside, linarin, quercitrin, ascorbigen, and three glucosinolates, including glucobrassicin, each of which show strong antipathogen activities [23–32]. Thus, the overexpression of *S30* in ICT1 results in a higher relative abundance of metabolites associated with oxidative stress and pathogen tolerance, which provides a biochemical explanation for the observed corresponding phenotypes.

Genes related to lignin biosynthesis were also downregulated in ICT1 (Figure 3d) (based on the transcriptome [1]), and unsurprisingly, we identified a reduction in a number of lignin-related secondary metabolites (Table S1). Lignin is one of the most important secondary metabolites, which is produced by the phenylalanine/tyrosine metabolic pathway in plant cells [47,48]. However, this reduction did not correlate with any of the phenotypes tested, and indeed, a reduction in lignin would be thought to lead to an increase in pathogen or drought sensitivity. ICT1 could be a tool for further analyzing the role of lignin in plant development.

The most unexpected finding for us was that overexpression of the ribosomal gene *S30* in ICT1 affects light-mediated plant development. The misregulation of key genes (based on the transcriptome [1]), such as the down-regulation of COP9 signalosome subunit 8 (*Csn8,* also known as *Cop9*) and light-sensitive hypocotyls 6 (*LSH6*), and the up-regulation of a number of positive regulators of light signaling, including *Cop1-interating protein 1* (CIP1, [49]), photoperiodic control of hypocotyl 1 (*Pch1*), suppressor of PHYB-4 7 (*SOB7*), and phototropin 2 (PHOT2) (Figure 5), led us to test if ICT1 has a phenotype related to photomorphogenesis. Indeed, ICT1 has a constitutive photomorphogenic phenotype in the dark, including short hypocotyls and open and expanded cotyledons (Figure 6), similar, for example, to mutations in the Cop9 signalosome [33] or overexpression of positive regulators of photomorphogenesis [50]. Considering that the overexpression of *S30* interacts with I3C responses, this finding of a cop phenotype for a plant resistant to I3C is not that unexpected. Indeed, a number of studies have connected the Cop9 signalosome with hormonal as well as environmental responses [51,52], and treatment with I3C induces transcriptional changes in *Arabidopsis* similar to mutations in the CSN.

While our motivation was to further identify the mechanism of I3C tolerance in ICT1, we keep meeting the surprising findings that overexpression of a single ribosomal subunit, *S30*, triggers the misregulation of specific gene families. How this subunit mediates these responses is unclear. Indeed, the function of many individual ribosomal subunits is unclear, and in this case, we do not know if the over-expressed protein solely incorporates into the ribosome or has some mode of action in a form independent of the ribosome. Unfortunately, attempts at using antibodies or tagged proteins to answer this question were inconclusive.

Our results add to mounting evidence that differences in RP expression can impact functions unrelated to ribosome assembly or protein synthesis. For example, the constitutive expression of *L6* in rice has been linked to enhanced tolerance to salt stress (NaCl 150–200 mM) [53], while the downregulation of *S6* in cotton has resulted in reduced levels of salicylic acid and jasmonic acid, rendering cotton plants more susceptible to *V. dahliae*. On the other hand, overexpression of *S6* in *Arabidopsis thaliana* has imparted resistance to *V. dahliae* [54]. Studies employing *S. cerevisiae* also provide evidence that RP effect transcription as over-production of specific ribosomal proteins repress mutations of RNA Pol III in *S. cerevisiae* [55,56]. These and other studies indicate that individual ribosomal proteins have extra-ribosomal functions in addition to those in the translation machinery [57–60]. Future studies must elucidate the mechanism by which *S30* influences transcription.

Supplementary Materials: The following supporting information can be downloaded at: https://www.mdpi.com/article/10.3390/biom14030319/s1. Figure S1: Expression of *Tgg* functional genes in ICT1 vs. WT; Figure S2: I3C hydrolysis products; Table S1: Secondary metabolites significantly different between ICT1 and WT; Table S2: A list of significantly differing metabolites following I3C treatment; Figure S3: Transcript levels of Skoto/photo morphogenesis-related genes in ICT1 vs. WT; Figure S4: Reduced yield in ICT1.s [1,5,61–63].

Author Contributions: Conceptualization, A.F. and D.A.C.; methodology, A.F.; validation, A.F. and H.K.-I.; formal analysis, A.F. and H.K.-I.; investigation, A.F.; resources, D.A.C.; data analysis, A.F.; writing—original draft preparation, A.F. and D.A.C.; writing—review and editing, A.F. and D.A.C.; visualization, A.F.; supervision, D.A.C.; project administration, D.A.C.; funding acquisition, D.A.C. All authors have read and agreed to the published version of the manuscript.

Funding: This research was funded by the Israeli Scientific Foundation (ISF), No. 1041/18.

Institutional Review Board Statement: Not applicable.

Informed Consent Statement: Not applicable.

Data Availability Statement: The information about the transcriptomic data can be found in [1].

Conflicts of Interest: The authors declare no conflicts of interest.

Nomenclature

Glucosinolates (GLS); ribosomal proteins (RP); secondary metabolites (SM); relative abundance (RA).

References

1. Finkelshtein, A.; Khamesa, H.; Tuan, L.A.; Rabanim, M.; Chamovitz, D.A. Overexpression of the ribosomal *S30* subunit leads to indole-3-carbinol tolerance in *Arabidopsis thaliana*. *Plant J.* **2021**, *105*, 668–677. [CrossRef] [PubMed]
2. Wittstock, U.; Burow, M. Glucosinolate breakdown in Arabidopsis: Mechanism, regulation and biological significance. *Arab. Book* **2010**, *8*, e0134. [CrossRef]
3. Andréasson, E.; Bolt Jørgensen, L.; Höglund, A.S.; Rask, L.; Meijer, J. Different myrosinase and idioblast distribution in *Arabidopsis* and *Brassica napus*. *Plant Physiol.* **2001**, *127*, 1750–1763. [CrossRef] [PubMed]
4. Koroleva, O.A.; Gibson, T.M.; Cramer, R.; Stain, C. Glucosinolate-accumulating S-cells in *Arabidopsis* leaves and flower stalks undergo programmed cell death at early stages of differentiation. *Plant J.* **2010**, *64*, 456–469. [CrossRef] [PubMed]

. Kim, J.H.; Lee, B.W.; Schroeder, F.C.; Jander, G. Identification of indole glucosinolate breakdown products with antifeedant effects on *Myzus persicae* (green peach aphid). *Plant J.* **2008**, *54*, 1015–1026. [CrossRef] [PubMed]

. Katz, E.; Nisani, S.; Sela, M.; Behar, H.; Chamovitz, D.A. The effect of indole-3-carbinol on PIN1 and PIN2 in *Arabidopsis* roots. *Plant Signal. Behav.* **2015**, *10*, e1062200. [CrossRef] [PubMed]

. Katz, E.; Nisani, S.; Yadav, B.S.; Woldemariam, M.G.; Shai, B.; Obolski, U.; Ehrlich, M.; Shani, E.; Jander, G.; Chamovitz, D.A. The glucosinolate breakdown product indole-3-carbinol acts as an auxin antagonist in roots of *Arabidopsis thaliana*. *Plant J.* **2015**, *82*, 547–555. [CrossRef]

. Katz, E.; Chamovitz, D.A. Wounding of *Arabidopsis* leaves induces indole-3-carbinol-dependent autophagy in roots of *Arabidopsis thaliana*. *Plant J.* **2017**, *91*, 779–787. [CrossRef]

Barakat, A.; Szick-Miranda, K.; Chang, I.-F.; Guyot, R.; Blanc, G.; Cooke, R.; Delseny, M.; Bailey-Serres, J. The Organization of Cytoplasmic Ribosomal Protein Genes in the *Arabidopsis* Genome. *Plant Physiol.* **2001**, *127*, 398. [CrossRef]

0. Feng, Z.; Zhang, L.; Wu, Y.; Wang, L.; Xu, M.; Yang, M.; Li, Y.; Wei, G.; Chou, M. The *Rpf84* gene, encoding a ribosomal large subunit protein, RPL22, regulates symbiotic nodulation in *Robinia pseudoacacia*. *Planta* **2019**, *250*, 1897–1910. [CrossRef]

1. Nagaraj, S.; Senthil-Kumar, M.; Ramu, V.S.; Wang, K.; Mysore, K.S. Plant Ribosomal Proteins, RPL12 and RPL19, Play a Role in Nonhost Disease Resistance against Bacterial Pathogens. *Front. Plant Sci.* **2015**, *6*, 1192. [CrossRef]

2. Zhang, J.; Yuan, H.; Yang, Y.; Fish, T.; Lyi, S.M.; Thannhauser, T.W.; Zhang, L.; Li, L. Plastid ribosomal protein S5 is involved in photosynthesis, plant development, and cold stress tolerance in Arabidopsis. *J. Exp. Bot.* **2016**, *67*, 2731–2744. [CrossRef] [PubMed]

3. Ramu, V.S.; Dawane, A.; Lee, S.; Oh, S.; Lee, H.K.; Sun, L.; Senthil-Kumar, M.; Mysore, K.S. Ribosomal protein QM/RPL10 positively regulates defence and protein translation mechanisms during nonhost disease resistance. *Mol. Plant Pathol.* **2020**, *21*, 1481–1494. [CrossRef] [PubMed]

4. Zhang, H.; Pei, Y.; He, Q.; Zhu, W.; Jahangir, M.; Haq, S.U.; Khan, A.; Chen, R. Salicylic acid-related ribosomal protein CaSLP improves drought and *Pst.DC3000* tolerance in pepper. *Mol. Hortic.* **2023**, *3*, 6. [CrossRef]

5. Sessions, A.; Burke, E.; Presting, G.; Aux, G.; McElver, J.; Patton, D.; Dietrich, B.; Ho, P.; Bacwaden, J.; Ko, C.; et al. A High-Throughput *Arabidopsis* Reverse Genetics System. *Plant Cell* **2002**, *14*, 2985–2994. [CrossRef] [PubMed]

6. Barth, C.; Jander, G. Arabidopsis myrosinases *TGG1* and *TGG2* have redundant function in glucosinolate breakdown and insect defense. *Plant J.* **2006**, *46*, 549–562. [CrossRef]

7. Zhao, Y.; Hull, A.K.; Gupta, N.R.; Goss, K.A.; Alonso, J.; Ecker, J.R.; Normanly, J.; Chory, J.; Celenza, J.L. Trp-dependent auxin biosynthesis in *Arabidopsis*: Involvement of cytochrome P450s *CYP79B2* and *CYP79B3*. *Genes Dev.* **2002**, *16*, 3100–3112. [CrossRef]

8. Stresser, D.M.; Bjeldanes, L.F.; Bailey, G.S.; Williams, D.E. The anticarcinogen 3,3′-diindolylmethane is an inhibitor of cytochrome P-450. *J. Biochem. Toxicol.* **1995**, *10*, 191–201. [CrossRef]

9. Ganzon, J.G.; Chen, L.-G.; Wang, C.-C. 4-O-Caffeoylquinic acid as an antioxidant marker for mulberry leaves rich in phenolic compounds. *J. Food Drug Anal.* **2018**, *26*, 985–993. [CrossRef]

0. Mallhi, Z.I.; Rizwan, M.; Mansha, A.; Ali, Q.; Asim, S.; Ali, S.; Hussain, A.; Alrokayan, S.H.; Khan, H.A.; Alam, P.; et al. Citric Acid Enhances Plant Growth, Photosynthesis, and Phytoextraction of Lead by Alleviating the Oxidative Stress in Castor Beans. *Plants* **2019**, *8*, 525. [CrossRef]

1. Mottaghipisheh, J.; Taghrir, H.; Boveiri Dehsheikh, A.; Zomorodian, K.; Irajie, C.; Mahmoodi Sourestani, M.; Iraji, A. Linarin, a Glycosylated Flavonoid, with Potential Therapeutic Attributes: A Comprehensive Review. *Pharmaceuticals* **2021**, *14*, 1104. [CrossRef] [PubMed]

2. Tahjib-Ul-Arif, M.; Zahan, M.I.; Karim, M.M.; Imran, S.; Hunter, C.T.; Islam, M.S.; Mia, M.A.; Hannan, M.A.; Rhaman, M.S.; Hossain, M.A. Citric acid-mediated abiotic stress tolerance in plants. *Int. J. Mol. Sci.* **2021**, *22*, 7235. [CrossRef] [PubMed]

3. Chen, J.; Li, G.; Sun, C.; Peng, F.; Yu, L.; Chen, Y.; Tan, Y.; Cao, X.; Tang, Y.; Xie, X.; et al. Chemistry, pharmacokinetics, pharmacological activities, and toxicity of Quercitrin. *Phytother. Res.* **2022**, *36*, 1545–1575. [CrossRef] [PubMed]

4. Vlot, A.C.; Dempsey, D.M.A.; Klessig, D.F. Salicylic acid, a multifaceted hormone to combat disease. *Annu. Rev. Phytopathol.* **2009**, *47*, 177–206. [CrossRef] [PubMed]

5. An, C.; Mou, Z. Salicylic Acid and its Function in Plant Immunity. *J. Integr. Plant Biol.* **2011**, *53*, 412–428. [CrossRef] [PubMed]

5. Kátay, G.; Tyihák, E.; Kátay, E. Effect of ascorbigen and 1′-methylascorbigen on disease resistance of bean plants to *Uromyces phaseoli*. *Nat. Prod. Commun.* **2011**, *6*, 1934578X1100600508. [CrossRef]

7. Singh, D.; Sharma, S.; Rani, R.; Mishra, S.; Sharma, R. Kaempferol-7-O-glucoside and their antimicrobial screening isolate from Cassia renigera Wall. *Int. J. Pharm. Clin. Res.* **2011**, *3*, 30–34.

8. Klessig, D.F.; Choi, H.W.; Dempsey, D.M.A. Systemic Acquired Resistance and Salicylic Acid: Past, Present, and Future. *Mol. Plant-Microbe Interact.* **2018**, *31*, 871–888. [CrossRef]

9. Davoodi, M.; Nejad-Ebrahimi, S.; Rustaiyan, A.; Esmaeili, D. The chemical composition and antibacterial activity of a methanolic extract of *Satureja khuzistanica*. *Braz. J. Pharm. Sci.* **2022**, *58*, e19233. [CrossRef]

0. Bednarek, P.; Pislewska-Bednarek, M.; Svatos, A.; Schneider, B.; Doubsky, J.; Mansurova, M.; Humphry, M.; Consonni, C.; Panstruga, R.; Sanchez-Vallet, A.; et al. A glucosinolate metabolism pathway in living plant cells mediates broad-spectrum antifungal defense. *Science* **2009**, *323*, 101–106. [CrossRef]

31. Prince, D.C.; Rallapalli, G.; Xu, D.; Schoonbeek, H.J.; Çevik, V.; Asai, S.; Kemen, E.; Cruz-Mireles, N.; Kemen, A.; Belhaj, K.; et a *Albugo*-imposed changes to tryptophan-derived antimicrobial metabolite biosynthesis may contribute to suppression of non-hos resistance to *Phytophthora infestans* in *Arabidopsis thaliana. BMC Biol.* **2017**, *15*, 20. [CrossRef] [PubMed]
32. Nguyen, V.P.T.; Stewart, J.; Lopez, M.; Ioannou, I.; Allais, F. Glucosinolates: Natural Occurrence, Biosynthesis, Accessibilit Isolation, Structures, and Biological Activities. *Molecules* **2020**, *25*, 4537. [CrossRef] [PubMed]
33. Qin, N.; Xu, D.; Li, J.; Deng, X.W. COP9 signalosome: Discovery, conservation, activity, and function. *J. Integr. Plant Biol.* **202(** *62*, 90–103. [CrossRef] [PubMed]
34. Turk, E.M.; Fujioka, S.; Seto, H.; Shimada, Y.; Takatsuto, S.; Yoshida, S.; Wang, H.; Torres, Q.I.; Ward, J.M.; Murthy, G.; et al. *BAS* and *SOB7* act redundantly to modulate *Arabidopsis* photomorphogenesis via unique brassinosteroid inactivation mechanism *Plant J. Cell Mol. Biol.* **2005**, *42*, 23–34. [CrossRef] [PubMed]
35. Huang, H.; Yoo, C.Y.; Bindbeutel, R.; Goldsworthy, J.; Tielking, A.; Alvarez, S.; Naldrett, M.J.; Evans, B.S.; Chen, M.; Nusinov D.A. PCH1 integrates circadian and light-signaling pathways to control photoperiod-responsive growth in Arabidopsis. *eLif* **2016**, *5*, e13292. [CrossRef] [PubMed]
36. Zhao, X.; Zhao, Q.; Xu, C.; Wang, J.; Zhu, J.; Shang, B.; Zhang, X. Phot2-regulated relocation of NPH3 mediates phototropi response to high-intensity blue light in *Arabidopsis thaliana. J. Integr. Plant Biol.* **2018**, *60*, 562–577. [CrossRef] [PubMed]
37. Murphy, B.P.; Tranel, P.J. Target-Site Mutations Conferring Herbicide Resistance. *Plants* **2019**, *8*, 382. [CrossRef]
38. Pasqua, M.; Grossi, M.; Zennaro, A.; Fanelli, G.; Micheli, G.; Barras, F.; Colonna, B.; Prosseda, G. The Varied Role of Efflux Pump of the MFS Family in the Interplay of Bacteria with Animal and Plant Cells. *Microorganisms* **2019**, *7*, 285. [CrossRef]
39. Vila-Aiub, M.M.; Balbi, M.C.; Distéfano, A.J.; Fernández, L.; Hopp, E.; Yu, Q.; Powles, S.B. Glyphosate resistance in perennia *Sorghum halepense* (Johnsongrass), endowed by reduced glyphosate translocation and leaf uptake. *Pest. Manag. Sci.* **201** *68*, 430–436. [CrossRef]
40. Hu, X.; Wang, S.; Zhang, H.; Zhang, H.; Feng, S.; Qiao, K.; Lv, F.; Gong, S.; Zhou, A. Plant cadmium resistance 6 from *Sali linearistipularis* (SlPCR6) affects cadmium and copper uptake in roots of transgenic *Populus. Ecotoxicol. Environ. Saf.* **202** *245*, 114116. [CrossRef]
41. Singh, B.K.; Walker, A. Microbial degradation of organophosphorus compounds. *FEMS Microbiol. Rev.* **2006**, *30*, 428–47 [CrossRef] [PubMed]
42. Drawz, S.M.; Bonomo, R.A. Three decades of beta-lactamase inhibitors. *Clin. Microbiol. Rev.* **2010**, *23*, 160–201. [CrossRef [PubMed]
43. Xia, J.; Guo, Z.; Yang, Z.; Han, H.; Wang, S.; Xu, H.; Yang, X.; Yang, F.; Wu, Q.; Xie, W.; et al. Whitefly hijacks a plant detoxificatio gene that neutralizes plant toxins. *Cell* **2021**, *184*, 3588. [CrossRef] [PubMed]
44. Tang, Z.; Wang, H.Q.; Chen, J.; Chang, J.D.; Zhao, F.J. Molecular mechanisms underlying the toxicity and detoxification of trac metals and metalloids in plants. *J. Integr. Plant Biol.* **2023**, *65*, 570–593. [CrossRef] [PubMed]
45. Singh, A.A.; Patil, M.P.; Kang, M.-J.; Niyonizigiye, I.; Kim, G.-D. Biomedical application of Indole-3-carbinol: A mini-reviev *Phytochem. Lett.* **2021**, *41*, 49–54. [CrossRef]
46. Kliebenstein, D.J.; Kroymann, J.; Mitchell-Olds, T. The glucosinolate-myrosinase system in an ecological and evolutionary contex *Curr. Opin. Plant Biol.* **2005**, *8*, 264–271. [CrossRef] [PubMed]
47. Liu, Q.; Luo, L.; Zheng, L. Lignins: Biosynthesis and Biological Functions in Plants. *Int. J. Mol. Sci.* **2018**, *19*, 335. [CrossRef]
48. Vanholme, R.; De Meester, B.; Ralph, J.; Boerjan, W. Lignin biosynthesis and its integration into metabolism. *Curr. Opin. Biotechnc* **2019**, *56*, 230–239. [CrossRef]
49. Kang, H.; Zhang, T.-T.; Fu, L.-L.; Yao, Y.-X.; You, C.-X.; Wang, X.-F.; Hao, Y.-J. The apple MdCOP1-interacting protein 1 negativel regulates hypocotyl elongation and anthocyanin biosynthesis. *BMC Plant Biol.* **2021**, *21*, 15. [CrossRef]
50. Myers, Z.A.; Kumimoto, R.W.; Siriwardana, C.L.; Gayler, K.K.; Risinger, J.R.; Pezzetta, D.; Holt Iii, B.F. NUCLEAR FACTOR Subunit C (NF-YC) Transcription Factors Are Positive Regulators of Photomorphogenesis in *Arabidopsis thaliana. PLoS Gene* **2016**, *12*, e1006333. [CrossRef]
51. Singh, A.K.; Chamovitz, D.A. Role of Cop9 Signalosome Subunits in the Environmental and Hormonal Balance of Plan *Biomolecules* **2019**, *9*, 224. [CrossRef] [PubMed]
52. Singh, A.K.; Yadav, B.S.; Dhanapal, S.; Berliner, M.; Finkelshtein, A.; Chamovitz, D.A. CSN5A Subunit of COP9 Signalosom Temporally Buffers Response to Heat in Arabidopsis. *Biomolecules* **2019**, *9*, 805. [CrossRef] [PubMed]
53. Moin, M.; Saha, A.; Bakshi, A.; Madhav, M.S.; Kirti, P.B. Constitutive expression of Ribosomal Protein L6 modulates salt toleranc in rice transgenic plants. *Gene* **2021**, *789*, 145670. [CrossRef] [PubMed]
54. Zhu, D.; Zhang, X.; Zhou, J.; Wu, Y.; Zhang, X.; Feng, Z.; Wei, F.; Zhao, L.; Zhang, Y.; Shi, Y.; et al. Genome-Wide Analys of Ribosomal Protein GhRPS6 and Its Role in Cotton Verticillium Wilt Resistance. *Int. J. Mol. Sci.* **2021**, *22*, 1795. [CrossRe [PubMed]
55. Hermann-Le Denmat, S.; Sipiczki, M.; Thuriaux, P. Suppression of yeast RNA polymerase III mutations by the *URP2* ger encoding a protein homologous to the mammalian ribosomal protein S20. *J. Mol. Biol.* **1994**, *240*, 1–7. [PubMed]
56. Dieci, G.; Ruotolo, R.; Braglia, P.; Carles, C.; Carpentieri, A.; Amoresano, A.; Ottonello, S. Positive modulation of RNA polymeras III transcription by ribosomal proteins. *Biochem. Biophys. Res. Commun.* **2009**, *379*, 489–493. [CrossRef] [PubMed]
57. Tchórzewski, M.; Boldyreff, B.; Grankowski, N. Extraribosomal function of the acidic ribosomal P1-protein YP1alpha fror Saccharomyces cerevisiae. *Acta Biochim. Pol.* **1999**, *46*, 901–910. [CrossRef]

58. Tu, W.Y.; Huang, Y.C.; Liu, L.F.; Chang, L.H.; Tam, M.F. Rpl12p affects the transcription of the *PHO* pathway high-affinity inorganic phosphate transporters and repressible phosphatases. *Yeast* **2011**, *28*, 481–493. [CrossRef]
59. Prakash, C.; Pandey, M.; Talwar, S.; Singh, Y.; Kanojiya, S.; Pandey, A.K.; Kumar, N. Extra-ribosomal functions of *Mtb* RpsB in imparting stress resilience and drug tolerance to mycobacteria. *Biochimie* **2020**, *177*, 87–97. [CrossRef] [PubMed]
60. Kour, R.; Kim, J.; Roy, A.; Richardson, B.; Cameron, M.J.; Knott, J.G.; Mazumder, B. Loss of function of ribosomal protein L13a blocks blastocyst formation and reveals a potential nuclear role in gene expression. *FASEB J.* **2023**, *37*, e23275. [CrossRef]
61. Aggarwal, B.B.; Ichikawa, H. Molecular Targets and Anticancer Potential of Indole-3-Carbinol and Its Derivatives. *Cell Cycle* **2005**, *4*, 1201–1215. [CrossRef] [PubMed]
62. Katz, E.; Nisani, S.; Chamovitz, D.A. Indole-3-carbinol: A plant hormone combatting cancer. *F1000Research* **2018**, *7*, 689. [CrossRef] [PubMed]
63. Centofanti, F.; Buono, A.; Verboni, M.; Tomino, C.; Lucarini, S.; Duranti, A.; Pandolfi, P.P.; Novelli, G. Synthetic Methodologies and Therapeutic Potential of Indole-3-Carbinol (I3C) and Its Derivatives. *Pharmaceuticals* **2023**, *16*, 240. [CrossRef]

metabolites

MDPI

Article

Optimization of Fermentation Process for New Anti-Inflammatory Glycosylceramide Metabolite from *Aspergillus* sp.

Yung-Husan Chen [1,2,*], Qiaoqiao Zhu [1,2], Jingyi Li [1,2], Rong Yang [1,2], Jingwen Zhang [1,2], Minxin You [1,2], Lianzhong Luo [1,2] and Bingye Yang [1,2,*]

[1] Xiamen Key Laboratory of Natural Products Resources of Marine Medicine, Xiamen Medical College, Xiamen 361023, China; zhuqq@stu.njmu.edu.cn (Q.Z.); 202005440053@xmmc.edu.cn (J.L.); 201400080002@xmmc.edu.cn (R.Y.); 202102460027@xmmc.edu.cn (J.Z.); 202205480025@xmmc.edu.cn (M.Y.); llz@xmmc.edu.cn (L.L.)

[2] Fujian Provincial University Marine Biomedical Resources Engineering Research Center, Xiamen Medical College, Xiamen 361023, China

* Correspondence: cyxuan@xmmc.edu.cn (Y.-H.C.); yby@xmmc.edu.cn (B.Y.)

Abstract: A novel ceramide compound, named Aspercerebroside A (AcA), was successfully isolated from the ethyl acetate layer of the marine symbiotic fungus *Aspergillus* sp. AcA exhibited notable anti-inflammatory activity by effectively inhibiting the production of nitric oxide (NO) in RAW 264.7 cells at concentrations of 30 μg/mL and 40 μg/mL, offering a promising avenue for the treatment of inflammatory diseases. To optimize the yield of glycosylceramide (AcA), a series of techniques, including single-factor experiments, orthogonal experiments, and response surface optimization, were systematically employed to fine-tune the composition of the fermentation medium. Initially, the optimal carbon source (sucrose), nitrogen source (yeast extract powder), and the most suitable medium salinity (14 ppt) were identified through single-factor experiments. Subsequently, orthogonal experiments, employing an orthogonal table for planning and analyzing multifactor experiments, were conducted. Finally, a mathematical model, established using a Box–Behnken design, comprehensively analyzed the interactions between the various factors to determine the optimal composition of the fermentation medium. According to the model's prediction, when the sucrose concentration was set at 37.47 g/L, yeast extract powder concentration at 19.66 g/L, and medium salinity at 13.31 ppt, the predicted concentration of glycosylceramide was 171.084 μg/mL. The experimental results confirmed the model's accuracy, with the actual average concentration of glycosylceramide under these conditions measured at 171.670 μg/mL, aligning closely with the predicted value.

Keywords: *Aspergillus* sp.; glycosylceramid; orthogonal experiments; fermentation optimization; response surface optimization

1. Introduction

As valuable sources of novel pharmaceuticals and bioactive compounds, secondary metabolites have been generated by various species within the *Aspergillus* genus, and their prominence has been steadily increasing. Among the common marine fungi belonging to the *Aspergillus* genus are *Aspergillus* nidulans, *Aspergillus* oryzae, *Aspergillus* versicolor, *Aspergillus* nidurans, *Aspergillus* carneus, *Aspergillus* wentii, and others [1,2]. These fungal cells have the capacity to produce a diverse array of secondary metabolites, including alkaloids, polyketides, terpenes, peptides, and a variety of other categories [3].

Chemical analysis has revealed that ceramides consist of long-chain bases (LCBs) and long-chain fatty acid amides (FABs), linked together by an amide bond [4]. Biological analysis has shown that ceramides and their glycosylated counterparts, cerebrosides, can exhibit a wide range of biological activities, including cytotoxicity, antifungal properties, immunostimulation, and immunosuppressive effects [5]. These ceramides and cerebrosides have been extracted from various marine invertebrates, including starfish, sea anemones, sponges, corals, ascidians, and bryozoans, and from some terrestrial plants as well.

Glycosylceramides have progressively found applications in various aspects of daily life. Their potential properties for cancer prevention, immunostimulation, and anticancer effects have been demonstrated. Consequently, they have been employed as functional foods or pharmaceuticals to enhance human physiological well-being. Clinical research has highlighted age-related declines in glycosylceramide levels. In particular, glycosylceramide supplementation has shown the potential to effectively slow down the aging process of facial skin and hand skin [6,7].

Glycosylceramides were discovered to play a pivotal role as structural precursors for various sphingolipids in previous studies. They could be modified to produce hydrophilic head groups, resulting in a range of compounds, including glycosphingolipids, sphingomyelins, and glycosylceramides. These substances, identified as critical constituents of biological membranes, contributed to the formation of diverse complex sphingolipids, gangliosides, and cerebrosides within the hydrophobic backbone of the body [8]. Glycosylceramides are typically localized in specific regions of cell membranes, where membrane integrity was preserved, and they become involved in cell signaling through influencing receptor oligomerization, signal amplification, or modification of cell signals. Moreover, owing to their hydrophobic nature, glycosylceramides have been identified as essential components of the skin's stratum corneum membrane, where they aid in the establishment of the skin's permeability barrier and play a pivotal role in skin hydration [9].

The research highlights the significant role played by long-chain fatty acid amides (FABs) in skin barrier function, particularly in the formation of lipid envelopes within keratinocytes [10]. Reduced sweating ability is often experienced by individuals with atopic dermatitis (AD) due to epidermal barrier dysfunction, resulting in a marked reduction in glycosylceramides esterified with linoleic acid (LA) within the stratum corneum. This leads to sweat duct blockage and prolonged sweating latency. The elevation of glycosylceramide levels in the stratum corneum has been shown to ameliorate epidermal barrier dysfunction, reduce sweating latency, and enhance stratum corneum function [11,12].

Glycosylceramides, serving as novel and highly effective moisturizers, have found extensive application in skincare products. They effectively mitigate moisture and lipid loss in the epidermis and dermis by stabilizing glycosylceramides on cell membranes, thereby delivering exceptional moisturization [13]. Additionally, they enhance skin elasticity, resistance, and slow the aging process of the skin [14]. However, due to their unique molecular structure, glycosylceramides possess a high melting point and low solubility, making them unsuitable for direct use as components in skincare products [15]. To address this challenge, numerous cosmetics companies are actively researching and developing innovative raw materials such as glycosylceramide liposomes and glycosylceramide encapsulates. These glycosylceramide-based skincare products have gained popularity among consumers.

In this study, a fungus was isolated from *Dendronephthya* sp. on Xiongdiyu Island, Dongshan Island, Zhangzhou City, Fujian Province, and was preliminarily identified as *Aspergillus* sp. From the ethyl acetate extract of this fungus, a glycosylceramide compound named AcA was isolated. Through spectroscopy and chemical analysis, it was discovered that AcA inhibited the production of nitric oxide (NO) in RAW 264.7 macrophage cells stimulated by lipopolysaccharide (LPS) at concentrations of 30 μg/mL and 40 μg/mL, demonstrating its anti-inflammatory properties. Building upon single-factor experiments, the Design-expert 8.0.6.1 software was utilized to establish a mathematical model based on the Box–Behnken design to optimize the fermentation medium composition of this fungus. On the basis of the results, response surface methodology (RSM) was applied to optimize the optimized glycosylceramide condition. The objective was to enhance the content of glycosylceramide, particularly AcA, in the secondary metabolites of this fungal strain, thus laying the experimental groundwork for large-scale industrial fermentation production of glycosylceramides. Through the successful optimization of the fermentation medium for AcA production by Aspergillus sp., an extract rich in glycosylceramides from marine filamentous fungi was obtained.

2. Materials and Methods

2.1. Chemicals and Equipment

The chemicals and reagents used in this study included methanol, acetic acid, ethyl acetate, glucose, sucrose, glycerol, and soluble starch (purchased from Xilong Scientific Co., Ltd., Guangzhou, China); chromatographic methanol (purchased from Guangzhou Tongyuan Chemical Technology Co., Ltd., Guangzhou, China); marine saltwater (provided by Jiangxi Yantong Technology Co., Ltd., Ningdu, China); N-acetylglucosamine (purchased from Shanghai Maclin Biochemical Technology Co., Ltd., Shanghai, China); peptone, yeast extract powder, potato glucose broth dry medium, and potato glucose agar dry medium (purchased from Guangdong Huanke Microbial Science and Technology Co., Ltd., Guangzhou, China); silica gel (200–300 mesh, Qingdao Marine Chemical Co., Ltd., Qingdao, China); Sephadex LH-20 gel (purchased from Nanjing Yuanbaofeng Medical Technology Co., Ltd., Nanjing, China); a laminar flow cabinet (model SW-CJ-2FD, Suzhou Antai Air Technology Co., Ltd., Suzhou, China); an autoclave sterilizer (model GR85DR, Zhengzhou Instrument Co., Ltd., Zhengzhou, China); a biochemical incubator (model SPX-250B-Z, Shanghai Boyuan Industrial Co., Ltd. Medical Equipment Factory, Shanghai, China); an orbital shaker (model ZQZY-CGF8, Shanghai Zhichu Instrument Co., Ltd., Shanghai, China); an ultrasonic cleaning machine (model DS-080S, Dongsen Washing Technology Co., Ltd., Zhengzhou, China); an Abbott salinometer (Wuhan Zhongxia Medical College undergraduate thesis design Hongtu measurement instrument Co., Ltd., Wuhan, China); a rotary evaporator (model RE212-B, Shanghai Guigu Technology Development Co., Ltd.); and an Agilent 1260 high-performance liquid chromatography system (Agilent Technologies, Inc., Santa Clara, CA, USA). Ultraviolet (UV) spectra were obtained using a Shimadzu UV-1800 UV/vis spectrophotometer. IR data were acquired by a Bruker TENSOR27 spectrophotometer. High resolution ESIMS and MS/MS experiments were carried out on a Q-Exactive (Thermo Scientific, Waltham, MA, USA) mass spectrometer. NMR spectra were determined on a BRUKER AVANCE III 500 MHz NMR spectrometer at 298.0 K equipped with 5 mm probes (500 MHz for ^{1}H and 125 MHz for ^{13}C NMR, Bruker BioSpin, Billerica, MA, USA). The ^{1}H and ^{13}C NMR spectrum were referenced to the solvent peaks for DMSO-d6. HPLC purification was performed on a PURIFLASH® 4250 equipped with QuikSep-DAC50 series DAC-50 dynamic axial compression column system (YMC-Triart C18 S-10 μm particles, H&E Co., Ltd., Beijing, China). All solvents were purchased from Merck KGaA (Darmstadt, Germany), if not mentioned.

2.2. LC-ESI-MS/MS Analysis

One μL of AcA content was injected into the LCESI-MS/MS system (LC system: Waters CapLC; MS system: nanoACQUITY/QExactive). The flow rate of the LC system was 6 μL/min through the column system (Waters Atlantis dC18″, 3 μm, 75 μm × 50 μm column) and the flow rate eventually was reduced to 0.3 μL/min by the splitter before entering into chromatographic column system. The solvent gradient system consisted of 5–100% methanol. The solvent increased from 5% to 100% in 4 min and then equilibrated for 1 min.

Mass spectrometry was performed in +ESI ionization mode with voltage of 4.50 kV and desolvation temperature of 290 °C. Triple quadrupole detection was controlled at an accelerating voltage of 9.1 kV, cone voltage of 100 V, and impact energy of 10 eV. The scanning range of mass spectrometry was monitored between *m/z* 600 and 800. The selected reaction model (SRM) mode was used to quantify the AcA contents. The ion chromatography peak area of the product ion was compared with the standard, and then the contents of AcA in the *Aspergillus* sp. extracts were calculated. The mass spectrometer tuning and calibration standard was [Glu]-Fibrinopeptide B (Sigma Chemicals, Kewdale, WA, USA).

2.3. *Strain and Fermentation Conditions*

The marine fungus *Aspergillus* sp., isolated from the waters of Dongshan Island, Zhangzhou City, Fujian Province, is currently preserved in the Guangdong Microbial Culture Collection Center (registration number GDMCC61933). Seed culture media included potato dextrose agar dry medium (PDA) and potato dextrose broth dry medium (PDB). Fermentation media comprised carbon source single-factor fermentation medium, nitrogen source single-factor fermentation medium, and response surface-optimized fermentation medium.

2.4. *Strain Propagation*

Propagation of the strain involved transferring a 1 cm^2 single colony from a successfully passaged culture onto potato dextrose agar dry medium (PDA) and incubating at room temperature for 1–2 months under sealed conditions.

2.5. *Organic Solvent Extraction*

Sterile extraction was carried out using a mixed solvent A (ethyl acetate: methanol: acetic acid = 80:15:5). Repeated extraction was performed to obtain secondary metabolites, and rotary evaporation was used to concentrate the extract, resulting in a weight of M0 (134.59 g).

A mixed solvent B (ethyl acetate: water = 1:1) was used for liquid–liquid extraction of the concentrated extract, retaining the ethyl acetate layer. Rotary evaporation was then employed to obtain a weight of M (52.11 g).

2.6. *Silica Gel Column Chromatography*

The ethyl acetate layer concentrate (DS142-1) was passed through a normal-phase silica gel column (200–300 mesh) using the following elution conditions: 100% petroleum ether → petroleum ether: ethyl acetate = 90:1 → petroleum ether: ethyl acetate = 50:1 → petroleum ether: ethyl acetate = 30:1 → petroleum ether: ethyl acetate = 10:1 → petroleum ether: ethyl acetate = 5:1 → petroleum ether: ethyl acetate = 3:1 → petroleum ether: ethyl acetate = 1:1 → 100% petroleum ether → 100% methanol. This yielded 18 components.

A portion of DS142-1-18-2 (2.1719 g) was passed through a normal-phase flash column (300–400 mesh silica gel) using the elution conditions: dichloromethane: methanol = 36:1 → dichloromethane: methanol = 32:1 → dichloromethane: methanol = 30:1 → dichloromethane: methanol = 20:1 → 100% methanol. This resulted in five components: DS142-1-18-1 (20.1 mg), DS142-1-18-2 (17.5 mg), DS142-1-18-3 (1.0511 g), DS142-1-18-4 (200 mg), DS142-1-18-5 (880 mg).

DS142-1-18-2-3 was subjected to liquid-phase analysis and purification using a Welch Ultimate® XB-C18 column (5 μm, 10 × 250 mm) with 100% methanol as the eluent, monitored at 210 nm and 254 nm, and a flow rate of 2 mL/min. This process yielded compound 1 (2.3 mg).

2.6.1. Optimization of Culture and Fermentation Conditions

Seed Cultivation: The frozen strain suspension stored at −80 °C was gradually thawed, and a small amount of the thawed culture was spread and activated on solid PDA medium within a laminar flow hood. After incubating at 25 °C for 72 h, spores from the activated strain were collected. These spores were then inoculated into 250 mL triangular flasks containing 100 mL of PDB liquid culture medium and incubated on a light-shaking incubator at 180 rpm for 24 h to obtain the seed culture.

Seed Culture Medium Fermentation Conditions: Salt concentration in the culture medium: 15 ppt; pH: 7.2–7.4; temperature: 25 °C; shaking speed: 180 rpm; fermentation time: 24 h.

Optimization of Culture Medium Fermentation Conditions: pH: 7.2–7.4; temperature: 25 °C; shaking speed: 150 rpm; fermentation time: 72 h.

2.6.2. Sample Extraction

The fermentation broth was filtered and separated using four layers of gauze to obtain mycelial balls and clear liquid. The clear liquid was subjected to extraction with an equal volume of ethyl acetate, while the mycelial balls were subjected to ultrasound-assisted extraction 2–3 times using a mixed solution system of ethyl acetate: methanol: acetic acid = 80:15:5. After filtering and separating the mycelial balls and the extract, the extract was further extracted with ethyl acetate twice, and the combined ethyl acetate extracts were concentrated using rotary evaporation to obtain the crude extract. The weight of the crude extract was accurately measured and recorded.

Dry loading was performed using 1–1.5 times the mass of diatomaceous earth as an adsorbent, and 1–2 mL of methanol was used to dissolve the sample. The two components were mixed thoroughly to ensure that diatomaceous earth fully adsorbed the sample. After the organic reagents evaporated, the sample was dried at 45 °C for 3 h in an oven. Subsequently, a C18 SPE reverse-phase solid-phase extraction column was used for initial separation, retaining the extract under 100% methanol elution conditions. After rotary evaporation, 1 mL of methanol was used for ultrasonic dissolution, and the sample was filtered through a 0.22 μm filter membrane to obtain the final sample for HPLC analysis.

2.6.3. Determination of Glycosylceramides Content

Chromatographic Detection Conditions: Ultimate XB-C18 column: 4.6 × 250 mm, 5 μm; mobile phase: ultrapure water–methanol, flow rate: 0.8 mL/min; detection wavelength: 210 nm; injection volume: 8 μL; temperature: 26 °C; methanol elution for 20 min.

Construction of Standard Curve: Precisely weigh 10 mg of glycosylceramide standard into a 10 mL volumetric flask, dissolve it in methanol, shake it well, and bring it to a standard solution of 1 mg/mL. Subsequently, prepare a series of glycosylceramide standard solutions with concentrations of 200 μg/mL, 100 μg/mL, 75 μg/mL, 50 μg/mL, 25 μg/mL, and 10 μg/mL from the stock solution, and analyze them in triplicate using HPLC. Finally, construct a linear standard curve with peak area as the independent variable and glycosylceramides concentration as the dependent variable. The equation for the standard curve is $y = 3.5932x - 3.6172$, $R^2 = 0.9996$, where x represents peak area and y represents glycosylceramides concentration.

2.6.4. Determination of NO Content

An in vitro NO production inhibition experiment was conducted following a method reported in the literature [16]. The experiment was divided into four groups: normal group, LPS group, Ac A group (30 ug/mL, 40 ug/mL. RAW 264.7 cells (5×10^6 cells in a 96-well plate) were incubated for 6 h in a humidified incubator (37 °C, 5% CO_2). After incubation, the culture medium for the normal group was replaced with serum-containing medium. The LPS group was treated with LPS (0.2 ug/mL), and the two Ac A groups were treated with Ac A (30 ug/mL and 40 ug/mL) and LPS (0.2 ug/mL), respectively, followed by a 24 h incubation. Equivalent supernatants from RAW 264.7 cell cultures were collected and subjected to NO activity assays using a NO production inhibition assay kit (Cat NO S00215, Beyotime Biotechnology, Nantong, China).

2.6.5. Single-Factor Impact Experiments

Effect of Different Carbon Sources on glycosylceramides Production: Different carbon sources (glucose/sucrose/soluble starch/glycerol) at a concentration of 40 g/L were used as carbon sources in the fermentation medium, with 10 g/L of peptone as the nitrogen source for single-factor experiments. Each experimental group was repeated in triplicate. After fermentation, glycosylceramides were extracted from the metabolic products and analyzed by HPLC. The concentration of glycosylceramides under different carbon source conditions was determined using the standard curve. Significant differences in glycosylceramides production under different carbon source conditions were analyzed using SPSS software (analysis method: LSD, $p < 0.01$).

Effect of Different Nitrogen Sources on glycosylceramides Production: Different nitrogen sources (peptone/yeast extract/N-acetylglucosamine) at a concentration of 10 g/L were used as nitrogen sources in the fermentation medium, with 40 g/L of sucrose as the carbon source for single-factor impact experiments. Each experimental group was repeated in triplicate. After fermentation, glycosylceramides in the metabolic products were qualitatively and quantitatively determined by HPLC. The concentration of glycosylceramides under different nitrogen source conditions was determined using the standard curve. SPSS software was used to analyze the results (analysis method: LSD, as mentioned earlier), and the nitrogen source that resulted in the highest glycosylceramides content was identified.

Effect of Different Medium Salinities on glycosylceramides Production: Using 40 g/L sucrose as the carbon source and 10 g/L yeast extract as the nitrogen source, the salinity of the fermentation medium was varied to 6 ppt, 14 ppt, 22 ppt, and 30 ppt to investigate the effect of different medium salinities on glycosylceramide production by the strain. Each salinity condition was repeated in triplicate. After fermentation, combining the qualitative and quantitative results obtained by HPLC with the standard curve, the optimal medium salinity for glycosylceramide production by the strain was determined using SPSS-22.0 software.

2.6.6. Orthogonal Experimental Design

Based on the results of the single-factor experiments for various influencing factors and their levels, *Aspergillus* sp. was cultured in a fermentation medium containing 45 g/L sucrose, 8 g/L yeast extract, and an appropriate amount of marine salt. An orthogonal table L9 (34) was used according to the selected influencing factors and the number of levels.

2.6.7. Response Surface Analysis for Optimization of Fermentation Medium

Using the selected factors of sucrose, yeast extract, and medium salinity, with glycosylceramides production as the response value, a three-factor, three-level experiment was established using the Box–Behnken design in the response surface design software Design-expert 8.0.6.1. This was performed to optimize the fermentation medium formula for *Aspergillus* sp. The factors and levels for the response surface experiments are listed in Table 1.

Table 1. Factors and levels for response surface experiments.

Factor	Level		
	−1	0	1
Soluble starch g/L	35	40	45
Yeast extract g/L	8	14	20
Culture-medium salinity ppt	6	14	22

2.6.8. Model Validation

The optimal culture medium formula for *Aspergillus* sp. from marine filamentous fungi, as determined by the Box–Behnken design method, was experimentally validated to assess the reliability of the model and analyze the final optimization results.

3. Results

3.1. Structural Analysis

The crude extracts from *Aspergillus* sp. were separated by a silica gel column and purified with Sephadex LH-20 and high-performance liquid chromatography to obtain AcA. The structure of the isolated compound was established by analyzing its spectroscopic data including UV, IR, 1D (^{1}H and ^{13}C), and 2D NMR (^{1}H-^{1} H COSY, HMQC, HMBC, NOESY) and HR-ESI-MS] and comparing the obtained data with what was reported in refs [17–20].

AcA was isolated as a colorless amorphous powder. The HR-ESI-MS (positive ion mode) *m/z*: 768.56 $[M+H]^+$ and the ^{13}C-NMR data suggested a molecular formula of $C_{44}H_{81}NO_9$. The infrared spectrum of AcA showed absorption bands of hydroxy (3310 cm^{-1}) and carbonyl amide (1626 cm^{-1} and 1532 cm^{-1}) functional groups. The ^{1}H and ^{13}C-NMR spectra of AcA (Table 2) showed characteristic signals of glycoside derivatives (also known as cerebroside). These signals included anomeric protons at δ_H 4.11 (d; J = 7.75), carbon oxides (sugar portion) of H-1″ and δ_C 61.06–76.88, carbonyl carbons of H-1′ and δ_C 172.01, C-1′ (amide bond) and terminal methyl protons of δ_H 0.85 (t; J = 6.875), and methylene protons of δ_H 1.23–3.97 (m) (long-chain hydrocarbons). The glucose ^{13}C-NMR signals were δ_C 61.06 (C-6″), 70.01 (C-4″), 73.39 (C-2″), 76.54 (C-3″), 76.88 (C-5″), 103.50 (C-1″). The trans geometry of the C-8 double bond was established by the chemical shifts of the methyl groups attached to the C-9, which was δ_C 15.72 (C-18).

Table 2. ^{1}H and ^{13}C NMR Data for 1 (500, 125 MHz,DMSO-d6, TMS, δ ppm).

No.	δ_C	δ_H (J in Hz)	No.	δ_C	δ_H (J in Hz)
1″	103.50	4.11, ^{1}H, d(7.75)	1	68.60	3.97, m
2″	73.39	2.96, ^{1}H, m			3.51, dd(10.4,3.8)
3″	76.54	3.14, m	2	52.90	3.78, m
4″	70.01	3.04, m	3	70.51	3.97, m
5″	76.88	3.08, m	4	130.91	5.36, ^{1}H, dd(15.4,6.7)
6″	61.06	3.66, brdd(10.1,6.1)	5	130.91	5.67, ^{1}H, dt(15.4,6.7)
1′	172.01	—	6	27.37	1.94, m
2′	71.91	4.30, brt(4.78) t	7	32.13	1.94, m
3′	129.05	5.43, ^{1}H, dd(15.4,5.4)	8	123.5	5.09, ^{1}H, brt(6.1)
4′	130.91	5.45, ^{1}H, dt(15.4,5.4)	9	134.91	—
5′	31.63	1.94, m	10	39.10	1.94, m
6′-15′	29.00	1.24, m	11	27.29	1.23, m
16′	31.28	1.24, m	12–14	29.05	1.23, m
17′	22.08	1.24, m	15	28.69	1.23, m
18′	13.92	0.85, 3H, t(6.875)	16	22.08	1.23, m
NH	—	7.38, d(9.2)	17	13.92	0.85, ^{3}H, t(6.875)
OH-2′	—	5.77, d(5)	18	15.72	1.54, ^{3}H, s
OH-2″	—	4.98, d(4.6)	OH-3	—	4.97, d(4.2)
OH-4″	—	4.94, d(4.6)	OH-3″	—	4.92, d(5.0)
			OH-6″	—	4.52, t(5.9)

The length of the sphingoid long-chain base (LCB) and the amide-linked long-chain fatty acid base (FAB) were suggested as 17 and 20 carbons, respectively. The fragment ions at m/z 310.76 $[FAB+H]^+$ confirmed the length of the amide-linked long-chain fatty acid (FAB). Moreover, the fragment ions at m/z 280.26$[LCB+H]^+$ revealed the length of the sphingoid long-chain base (LCB) (as depicted in Figure S3). Based on the above data, the structure of AcA was defined as (2R,3E)-N-((2S,3R,4E,8E)-3-hydroxy-9-methyl-1-(β-D-glucopyranosyloxy) 2-hydroxyheptadeca-4,8-dien-2-yl))icos-3-enamide and was named Aspercerebroside A (as depicted in Figure 1).

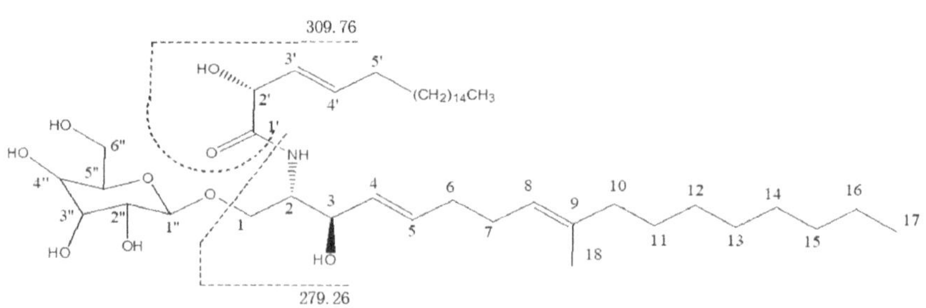

Figure 1. The chemical structural formula of Aspercerebroside A.

3.2. NO Production Assay

The inhibitory activity of the obtained aspercerebroside A on nitric oxide (NO) production was evaluated using an LPS-stimulated RAW 264.7 cell model. The NO release inhibition assay demonstrated that this compound exhibits significant anti-inflammatory effects at concentrations of 30 μg/mL and 40 μg/mL in Figure 2.

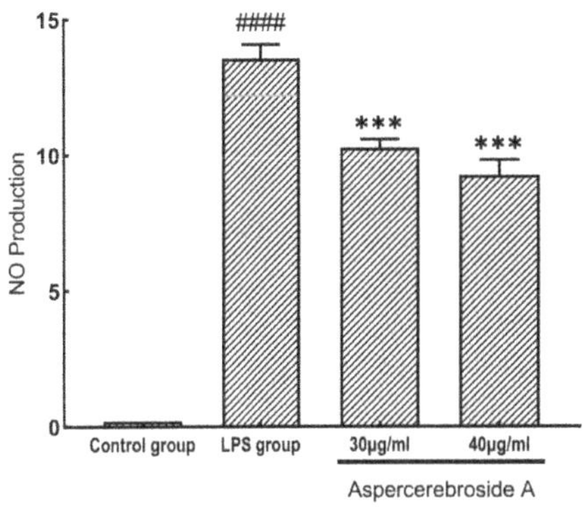

Figure 2. Inhibition of NO production by aspercerebroside A at concentrations of 30 μg/mL and 40 μg/mL in RAW 264.7 cells. Values represent the mean ± SD of three independent experiments. *** $p < 0.001$ (significant difference between the Aspercerebroside A treatment group and the LPS group). #### $p < 0.001$ (significant difference within the control group).

3.3. Single-Factor Experiments

3.3.1. Influence of Carbon Source Types on Glycosylceramide Production

Carbon sources are essential nutrients in microbial culture media. They provide the carbon skeleton for cell growth and the synthesis of cellular products, as well as the energy required for cellular activities. As shown in Figure 3, when sucrose is used as the carbon source in the fermentation medium for *Aspergillus* sp., the glycosylceramide production is the highest, with a yield of 89.32 μg/mL. Soluble starch, glucose, and glycerol have a similar impact on glycosylceramide production, which may be related to the enzymatic system utilized by this strain in glycosylceramide biosynthesis. Therefore, based on the experimental results, sucrose was selected as the optimal carbon source for the fermentation medium.

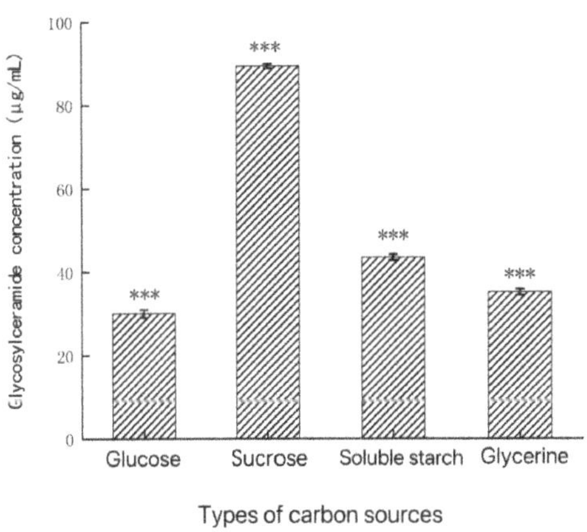

Figure 3. Influences of different carbon sources on glycosylceramide production from *Aspergillus* sp. ***: $p < 0.001$ (significant difference within the control group).

As shown in Figure 4, with the increase in sucrose concentration, glycosylceramide production first increased and then decreased. At 40 g/L, the maximum glycosylceramide production was 121 μg/mL. With the continuous increase in sucrose concentration, glycosylceramide production began to decline, which may have had a blocking effect.

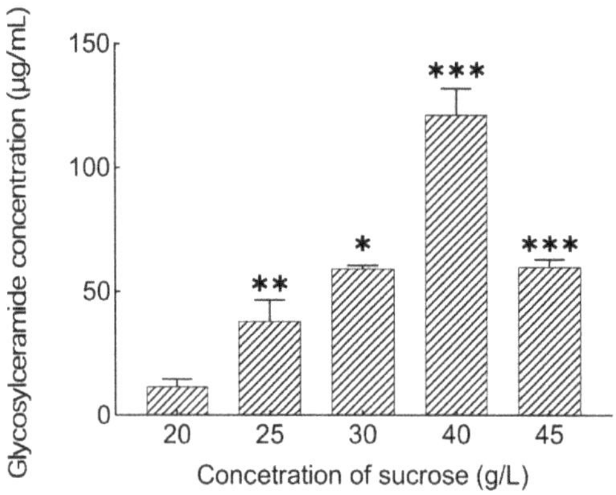

Figure 4. Influences of sucrose concentration on glycosylceramide production from *Aspergillus* sp. *: $p < 0.05$, **: $p < 0.01$, ***: $p < 0.001$ (significant difference within the control group).

3.3.2. Influence of Nitrogen Source Types on Glycosylceramide Production

Nitrogen sources are essential components in the composition of living organisms. Their primary function is to provide the necessary nitrogen elements for the synthesis of amino acids, proteins, nucleic acids, and other essential compounds. As shown in Figure 5, in the single-factor experiments with different nitrogen sources, when yeast extract was used as the nitrogen source, the glycosylceramide production was the highest, reaching 136.80 μg/mL. Next was peptones, and then the lowest production was observed with N-acetylglucosamine as the nitrogen source. Therefore, yeast extract was selected as the nitrogen source for the culture medium based on these results.

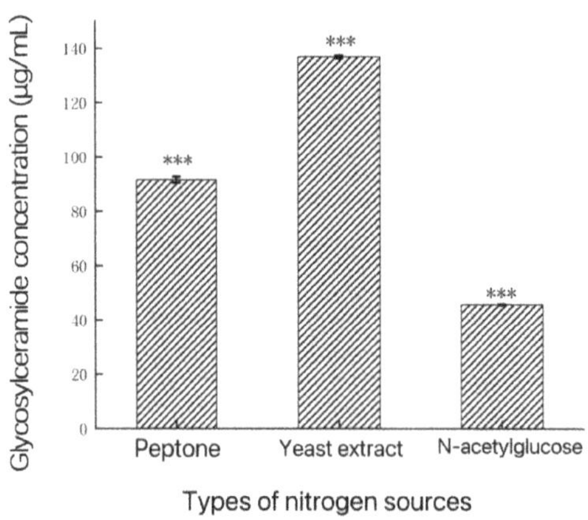

Figure 5. Influences of different nitrogen sources on glycosylceramide production from *Aspergillus* sp. ***: $p < 0.001$ (significant difference within the control group).

As shown in Figure 6, Yeast extract powder has the greatest influence on glycosylceramide yield. When the concentration of yeast extract powder is 14 g/L, the glycosylceramide yield is the highest. When the concentration of yeast extract powder continues to increase, the glycosylceramide yield is inhibited and begins to decline, Therefore, the optimal concentration is 14 g/L.

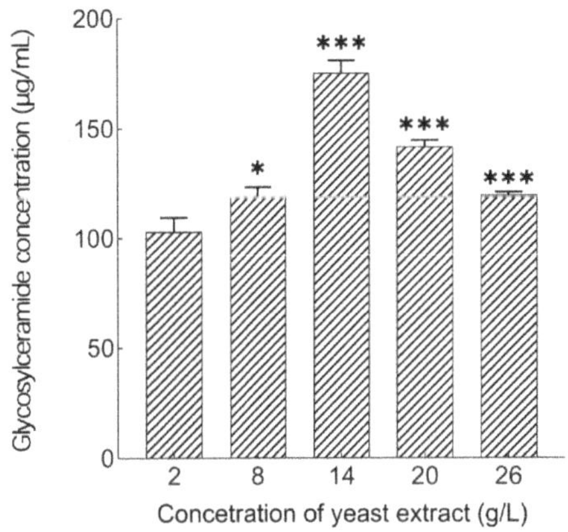

Figure 6. Influences of yeast extract concentration on glycosylceramide production from *Aspergillus* sp. *: $p < 0.05$, ***: $p < 0.001$ (significant difference within the control group).

3.3.3. Influence of Culture Medium Salinity on Glycosylceramide Production

Marine microorganisms face a unique environment characterized by high salinity, which differs from that of terrestrial microorganisms. The addition of ecological sea salt to the culture medium not only helps replicate the components of seawater but also provides essential inorganic salts for microbial growth. Changes in culture medium salinity may affect the concentration of glycosylceramides in the secondary metabolites of *Aspergillus*. sp. As shown in Figure 7, when the culture medium salinity was 14 ppt, the marine *Aspergillus* fungus produced the maximum amount of glycosylceramides, reaching 245.46 μg/mL. However, as the salinity of the culture medium increased beyond this point, the concentration of glycosylceramides gradually decreased. When the culture medium salinity reached 30 ppt, the glycosylceramide concentration was at its minimum, at only 11.47 μg/mL.

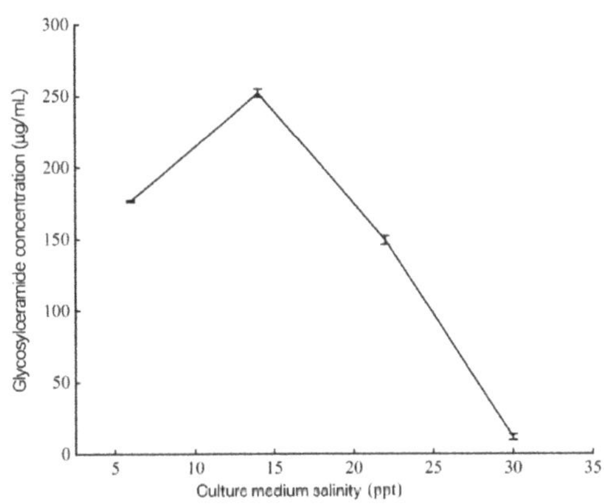

Figure 7. Influences of different nitrogen sources on glycosylceramide production from *Aspergillus* sp.

3.4. Orthogonal Array Experiment

Based on the single-factor results of various influencing factors and their levels, *Aspergillus* sp. was cultured in a fermentation medium containing 45 g/L sucrose, 8 g/L yeast extract, and an appropriate amount of ecological sea salt. An orthogonal array L9 (34) was used according to the selected influencing factors and their levels. The influencing factors and their levels to be investigated are shown in Table 3. The orthogonal experimental table L9 (34) and the experimental results are presented in Table 4, while the variance analysis data processing can be found in Table 5.

Table 3. Experimental factors of SPSS design and their levels.

Level	A (Initial pH Value)	B (Initial Salinity/ppt)	C (Temperature/°C)	D (Time/d)
1	6.2~6.4	6	20	3
2	7.2~7.4	14	25	4
3	8.2~8.4	22	30	5

Table 4. SPSS design table and experimental results.

Number	A (Initial pH Value)	B (Initial Salinity/ppt)	C (Temperature/°C)	D (Time/d)	Glycosylceramide Concentration Group 1 (μg/mL)	Glycosylceramide Concentration Group 2 (μg/mL)	Glycosylceramide Concentration Group 1 (μg/mL)
1	1	1	1	1	28.62	27.06	21.44
2	1	2	3	2	193.07	177.85	187.81
3	1	3	2	3	201.87	229	211.22
4	2	1	3	3	209.74	178.74	185.28
5	2	2	2	1	73.62	83.87	76.82
6	2	3	1	2	74.82	90.04	114.03
7	3	1	2	2	169.61	139.41	171.84
8	3	2	1	3	122.38	139.19	141.14
9	3	3	3	1	177.40	178.04	215.03
K1	1277.94	1131.96	758.72	881.9			
K2	1087.18	1195.75	1279.11	1318.48			
K3	1454.04	1491.45	1703.18	1618.78			
K1/9	141.99	125.77	84.3	97.99			
K2/9	120.80	132.86	142.12	146.5			
K3/9	161.56	165.72	189.24	179.86			
R	40.76	39.94	104.94	81.88			

Table 5. Analysis of variance.

Source of Variance	Type III Square Sum	Df	the Average Square	F Value	Significance
A	7480.994	2	3740.497	18.178	$p < 0.05$
B	8175.582	2	4087.791	19.866	$p < 0.05$
C	50,737.610	2	25,368.805	123.287	$p < 0.05$
D	30,510.160	2	15,255.080	74.137	$p < 0.05$
Error	3703.852	18	205.770		

F0.05(2,18) = 3.555, F0.01(2,18) = 6.01.

From the intuitive analysis in Table 4 using the range analysis method, it can be seen that the most significant factor affecting the production of aspercerebroside A by marine-derived *Aspergillus* sp. is temperature, and the optimal combination for aspercerebroside A production is A3B3C3D3. The K values refer to the sum of the experimental data Y for each factor at every level. For instance, in this experiment, K1 for factor A represents the sum of the nervonic acid yield for all experimental groups at pH levels ranging from 6.2 to 6.4. R value stands for the range of factors, reflecting the degree of variation among different levels. By analyzing its magnitude, it is possible to make an initial ranking of the selected factors based on their differences. K1, K2, and K3 are the sum values corresponding to different factor levels. K1/9, K2/9, and K3/9 are the means of these sum values. The R value is calculated by subtracting the smallest mean value from the largest mean value.

Table 5, utilizing the variance analysis method, shows that the order of influence on aspercerebroside A production is temperature > time > initial salinity > initial pH value. All four factors have a highly significant impact on aspercerebroside A yield. Based on the analysis of the results, the discrepancy between the order of influence of initial salinity and initial pH value observed in the range analysis and variance analysis may be due to the closeness of their effects on aspercerebroside A production. Moreover, range analysis cannot estimate the inevitable errors that occur during experiments and the measurement of the results. Therefore, the results are considered more reliable based on variance analysis.

To validate these orthogonal results, marine derived *Aspergillus* sp. was cultured under the optimized fermentation conditions from the best combination. The aspercerebroside A content obtained from this experiment was 436.86 μg/mL.

3.4.1. Response Surface Optimization Results and Analysis

Using the sucrose concentration, yeast extract concentration, and initial salinity as independent variables, a three-factor, three-level response surface experiment was conducted through the Box–Behnken design method to analyze the interactive effects among various influencing factors. The response surface results are shown in Table 6.

Table 6. Response surface design test and results.

Test Number	Level			Glycosylceramide Concentration (μg/mL)
	Sucrose Concentration (A)	Concentration of Yeast Extract Powder (B)	Salinity of Medium (C)	
1	0	−1	−1	155.6
2	0	0	0	168.4
3	1	−1	0	157.9
4	1	0	1	162.2
5	1	0	−1	160.1
6	0	1	−1	165.8
7	0	−1	1	157.7
8	−1	0	1	161.9
9	−1	−1	0	159.1
10	0	1	1	164.9
11	1	1	0	166.8
12	0	0	0	169.1
13	0	0	0	167.1
14	0	0	0	168.8
15	−1	1	0	171.3
16	−1	0	−1	163.2
17	0	0	0	169.3

The regression equation for the neuralamide concentration (Y) in relation to sucrose concentration (A), yeast extract concentration (B), and initial salinity (C), as obtained through data analysis using the Design-Expert 8.0.6.1 software, is as follows:

$$Y = 168.54 - 1.14A + 4.81B + 0.25C - 0.83AB + 0.70AC - 0.75BC - 1.96A^2 - 2.81B^2 - 4.73C^2$$

3.4.2. Variance Analysis

From Table 7, it can be observed that the *p*-value of the model is less than 0.0001, indicating that the regression equation model is highly significant. The lack-of-fit *p*-value is 0.4638, which is greater than 0.05, suggesting that errors due to random factors do not significantly affect the experimental results. The model's determination coefficient (R^2) is 0.9851, indicating that 98.51% of the response variation in neuralamide concentration is explained by the selected variables. The adjusted R^2 (R^2adj) is 0.9960, signifying that the model can explain 99.60% of the variation in neuralamide concentration response. Overall, the established model fits well and can be used to predict the concentration of neuralamide in fermentation production.

Table 7. Regression model analysis of variance.

Source of Variation	Quadratic Sum	Degree of Freedom	Mean Square	F-Value	*p*-Value
Model	360.73	9	40.08	51.53	<0.0001
Concentration of sucrose (A)	10.35	1	10.35	13.31	0.0082
Concentration of yeast extract (B)	185.28	1	185.28	238.22	<0.0001
Salinity of medium (C)	0.50	1	0.50	0.64	0.4490
AB	2.72	1	2.72	3.50	0.1035
AC	1.96	1	1.96	2.52	0.1564
BC	2.25	1	2.25	2.89	0.1328
A^2	16.13	1	16.13	20.74	0.0026
B^2	33.19	1	33.19	42.67	0.0003
C^2	94.30	1	94.30	121.24	<0.0001
Residual error	5.44	7	0.78		
Out of fit term	2.39	3	0.80	1.05	0.4638
Net error	3.05	4	0.76		
Total deviation	366.18	16			
$R^2 = 0.9851$			R^2adj = 0.9960		

Note: $p < 0.01$ indicates a very significant difference; $p < 0.05$ indicates a significant difference.

The analysis of variance (ANOVA) reveals that the first-order terms A and B have highly significant differences, indicating that the sucrose concentration and yeast extract concentration both have a highly significant impact on the concentration of neuralamide produced by the strain. The second-order terms A^2, B^2, and C^2 exhibit highly significant differences, suggesting that the relationship between neuralamide concentration and sucrose concentration, yeast extract concentration, and initial salinity is not linear. The interaction terms AB, AC, and BC all have *p*-values greater than 0.05, indicating that the pairwise interactions between the three factors are not significant. Based on the F-values, the order of significance of the three selected variables in influencing the neuralamide production concentration is as follows: yeast extract concentration (B) > sucrose concentration (A) > initial salinity (C).

3.4.3. RSM Analysis

The interaction between sucrose concentration and yeast extract concentration is depicted in Figure 8. These figures include a contour plot and a response surface plot, with the culture medium salinity being held constant at 14 ppt. The contour plot predominantly showcases an elliptical shape, underscoring the substantial interaction between sucrose concentration and yeast extract concentration in their collective impact on glycosylceramide concentration. In the 3D surface plot, it becomes apparent that the surface representing sucrose concentration undergoes more gradual changes when compared to the surface representing yeast extract concentration. This observation suggests that the yeast extract concentration exerts a more pronounced influence on glycosylceramide concentration within the strain.

Furthermore, Figure 9 displays the contour plot and response surface plot for the interaction between sucrose concentration and initial salinity when yeast extract concentration is fixed at 14 g/L. The contour plot is elliptical in shape, indicating a significant interaction between sucrose concentration and initial salinity on neuralamide concentration. In the 3D surface plot, the initial salinity surface changes more gradually compared to the sucrose concentration surface, implying that sucrose concentration has a greater influence on the neuralamide concentration produced by the strain.

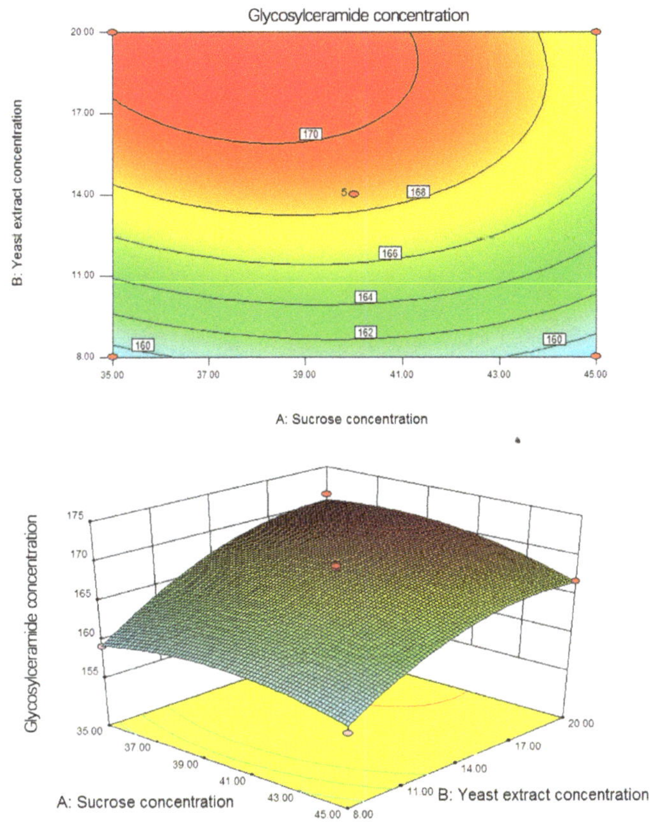

Figure 8. Contour and response surface diagram of sucrose concentration and yeast extract powder concentration on glycosylceramide concentration.

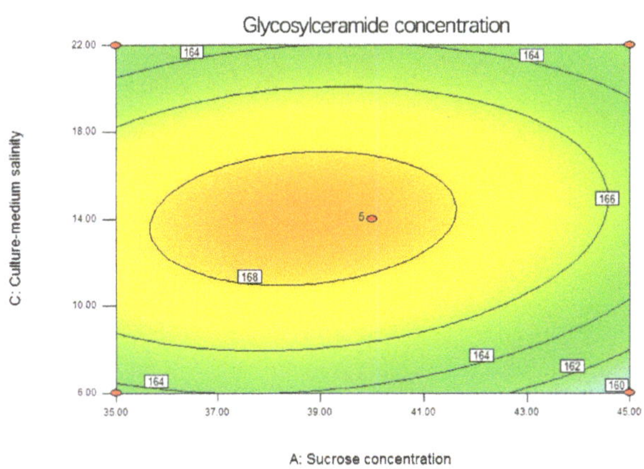

Figure 9. *Cont.*

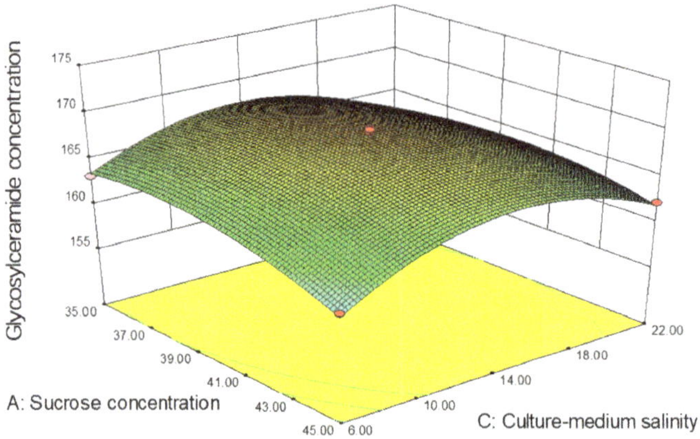

Figure 9. Contour and response surface diagram of sucrose concentration and culture-medium salinity on glycosylceramide concentration.

The influence on neuralamide concentration is influenced by the strain. Figure 10 depicts a contour plot and a response surface plot illustrating the interaction between the yeast extract concentration and initial salinity, with the sucrose concentration fixed at 40 g/L. The contour plot displays an overall elliptical shape, indicating a strong interaction between yeast extract concentration and initial salinity concerning neuralamide concentration. In the 3D surface plot, the surface representing yeast extract concentration changes more significantly compared to the initial salinity surface, suggesting that the yeast extract concentration has a notably greater impact on the glycosylceramide production of the strain than sucrose concentration.

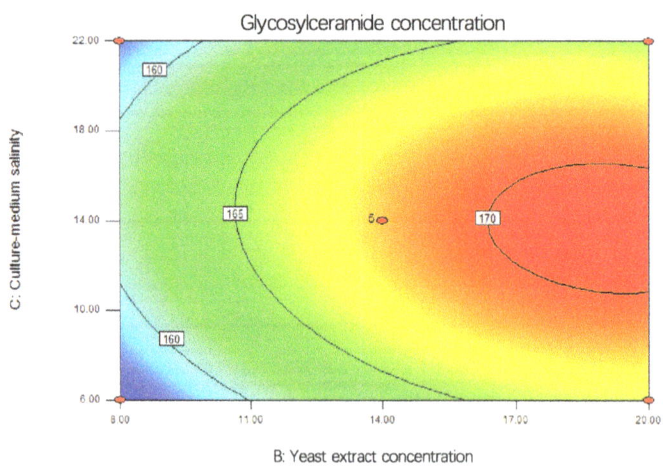

Figure 10. *Cont.*

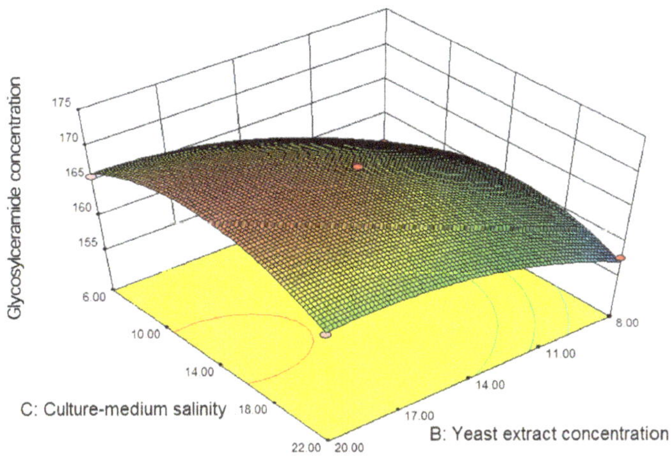

Figure 10. Contour and response surface diagram of yeast extract powder concentration and culture-medium salinity on glycosylceramide concentration.

3.5. Verification Experiment Results

Based on the response surface analysis, the predicted optimal culture medium formula for producing glycosylceramides with *Aspergillus* sp. was determined as follows: a sucrose concentration of 37.47 g/L, yeast extract concentration of 19.66 g/L, and a culture medium salinity of 13.31 ppt. Under these conditions, the glycosylceramide concentration was predicted to be 171.084 μg/mL. In practical experiments, the model parameters were adjusted to a sucrose concentration of 37.5 g/L, yeast extract concentration of 19.7 g/L, and culture medium salinity of 13.3 ppt. Three parallel repeated experiments were conducted, resulting in glycosylceramide concentrations of 172.421 μg/mL, 171.976 μg/mL, and 170.612 μg/mL, with an average value of 171.670 μg/mL. These results closely aligned with the predicted values, indicating a strong model fit and high model credibility.

4. Discussion

It is worth noting that within *Aspergillus* sp., glycosylceramide compounds falling under the glycosylceramide class have been isolated [21]. These compounds exhibit similarities to AcA, sharing a molecular weight range between 760 and 780 [22]. This similarity indicates a resemblance between these glycosylceramide variants and underscores their association with aspercerebroside A within the glycosylceramide category. Such findings further emphasize the existence of analogous glycosylceramide compounds in the *Aspergillus* sp. genus, aligning with our identification of glycosylceramide (AcA) and supporting its classification within the broader glycosylceramide spectrum.

In marine environments, microorganisms harbor abundant resources, and yet the production of potentially valuable metabolites often demonstrates low yields during initial fermentation. Hence, the optimization of microbial fermentation processes has become pivotal to enhance metabolite production and fully exploit marine microbial resources. Typically, optimizing microbial fermentation conditions involves a blend of statistical and mathematical modeling techniques, as evidenced in previous studies. For instance, Wang Cong et al. [23] successfully optimized fermentation conditions for enhancing lincomycin D production by Streptomyces parvulus OUCMDZ-2554, resulting in a remarkable 3.6-fold increase in yield. Similarly, Dong Yanjuan et al. [24] optimized the fermentation conditions for elevating xylanase production by strain YS1069, achieving a remarkable five-fold increase in xylanase activity.

However, studies focusing on enhancing glycosylceramide fermentation yields have been relatively limited. Prior research primarily centered on the applications and extraction methods of glycosylceramides, such as elevating glycosylceramide production through Ganoderma lucidum fermentation or preparing glycosylceramides through emulsion fermentation. This study bridges a gap in this domain by augmenting glycosylceramide production through the optimization of culture medium specifically designed for marine-derived microorganisms.

During the optimization of conditions by selecting the initial pH value and initial salinity as variables, it was observed that the initially fixed salinity of the culture medium would fluctuate due to subsequent pH adjustments. Throughout the fermentation process, both salinity and pH dynamically changed. Comprehensive monitoring of salinity and acidity throughout the fermentation might identify the most suitable conditions for synthesizing glycosylceramides. The experimental findings revealed that at a temperature of 20 °C, the strain's growth was notably inhibited compared to the standard temperature (25 °C), whereas at 30 °C, the growth accelerated significantly. Temperature was identified as a critical factor influencing glycosylceramide production, suggesting the potential to increase temperature moderately to enhance the strain's metabolic capacity.

Simultaneously, throughout fermentation, time displayed a direct correlation with glycosylceramide accumulation. Identifying an optimal fermentation timepoint combined with an appropriate temperature could significantly enhance fermentation capability, optimizing the cost-effectiveness of glycosylceramide production. In summary, optimizing the fermentation of the marine-derived fungus Aspergillus sp. can be achieved by enhancing its own development and nutrient synthesis through adjustments in suitable culture medium components such as inorganic salts, the number of carbon and nitrogen source types, as well as regulating the fermentation liquid's pH acidity and salinity and altering the fermentation temperature, duration, inoculum size, and speed to investigate their impact on metabolite production and content. Further exploration of the interaction between these factors could refine the fermentation conditions for optimal glycosylceramide synthesis.

5. Conclusions

This investigation explored the anti-inflammatory potential of AcA using LPS-induced RAW 264.7 cells, revealing significant inhibitory effects of Compound 1 on nitric oxide (NO) production in RAW 264.7 cells, particularly at concentrations of 30 μg/mL and 40 μg/mL. This discovery holds crucial importance in identifying potent active components derived from marine fungi.

In this study, a comprehensive investigation was conducted to determine the fermentation parameters for *Aspergillus* sp. derived from marine fungi. Initially, single-factor impact experiments were employed to explore the range and trends of carbon sources, nitrogen sources, initial pH values, salinity, temperature, and time fluctuations during fermentation. Subsequently, an orthogonal study was conducted, focusing on four influencing factors—initial pH value, initial salinity, temperature, and time—at three suitable levels, resulting in the optimal fermentation combination of A3B3C3D3. Experimental validation confirmed that the most suitable fermentation conditions for Aspergillus sp. were as follows: 45 g/L sucrose, 8 g/L yeast extract, initial salinity of 22 ppt (using ecological mineral salts to adjust fermentation medium), initial pH range of 8.2 to 8.4, and cultivation at 30 °C for 5 days, resulting in a glycosylceramide content of 436.86 μg/mL. Additionally, these optimized fermentation conditions resulting in a glycosylceramide content represent a significant finding in this research. Glycosylceramide production under such precise and enhanced conditions, previously sparsely discussed in the existing literature, emphasizes the novelty and importance of this study's outcomes.

Furthermore, employing a combination of single-factor impact experiments and response surface optimization led to the successful determination of the optimal culture medium formulation for glycosylceramide production by Aspergillus sp. derived from marine fungi. This formulation consisted of 37.5 g/L sucrose, 19.7 g/L yeast extract, and a culture medium salinity of 13.3 ppt. Experimental validation demonstrated a close alignment between actual and predicted production values, affirming the high reliability of the established regression model.

In conclusion, The discovery of these specific fermentation parameters offers a valuable contribution to the field, particularly in elucidating optimal conditions for glycosylceramide biosynthesis in *Aspergillus* sp. derived from marine fungi. It not only expands the limited literature discussing these specific fermentation conditions but also paves the way for potential applications in industrial-scale glycosylceramide production.

Supplementary Materials: The following supporting information can be downloaded at: https://www.mdpi.com/article/10.3390/metabo14020099/s1. Figure S1: ^{1}H NMR spectrum of 1, Figure S2: ^{13}C NMR spectrum of 1, Figure S3: mass spectrum of 1, Figure S4: fungal culture on YM medium, Figure S5: glycosylceramide standard curve, Figure S6: influences of different carbon sources, Figure S7: influences of different nitrogen sources, Figure S8: influences of different nitrogen sources, Table S1: response surface design test and results, Table S2: regression model analysis of variance, Figure S9: contour and response surface diagram of sucrose concentration and yeast extract powder concentration, Figure S10: contour and response surface diagram of sucrose concentration and culture-medium salinity, Figure S11: contour and response surface diagram of yeast extract powder concentration and culture-medium salinity, Figure S12: contour and response surface diagram of sucrose concentration and yeast extract powder concentration on glycosylceramide concentration, Figure S13: contour and response surface diagram of sucrose concentration and culture-medium salinity on glycosylceramide concentration, Figure S14: contour and response surface diagram of yeast extract powder concentration and culture-medium salinity on glycosylceramide concentration.

Author Contributions: Conceptualization, Y.-H.C. and B.Y. methodology, Q.Z.; software, J.L.; formal analysis, R.Y.; investigation Q.Z. and B.Y.; resources, L.L.; data curation, Q.Z. and R.Y.; writing—original draft preparation, J.Z.; writing—review and editing, M.Y.; visualization, Y.-H.C.; supervision, Y.-H.C.; project administration, B.Y.; funding acquisition, Y.-H.C. and B.Y. All authors have read and agreed to the published version of the manuscript.

Funding: This research was funded by the Xiamen Natural Science Foundation, grant number 3502Z20227311, Technology Innovation Program of Xiamen Ocean and Fisheries Development Special Funds, number 23YYZP004QCB28, 2023 Natural Science Foundation of Fujian Province Grant No. 2023J011651 and the Science and Technology Program of Fujian province, number 2023J011659.

Institutional Review Board Statement: Not applicable.

Informed Consent Statement: Not applicable.

Data Availability Statement: All data are available in the manuscript and in the Supplementary Information.

Conflicts of Interest: The authors declare no conflicts of interest.

References

Lin, S.F.; Jiang, H.; Fu, Q.Q.; Huang, S.J.; Tang, L.Y.; Li, A.; Liu, M.Q. Effects of Pulsed Light on Mycelium Growth and Conidiation in *Aspergillus oryzae*. *Fermentation* **2023**, *9*, 674. [CrossRef]

Ghazawi, K.F.; Fatani, S.A.; Mohamed, S.G.A.; Mohamed, G.A.; Ibrahim, S.R.M. *Aspergillus nidulans*—Natural Metabolites Powerhouse: Structures, Biosynthesis, Bioactivities, and Biotechnological Potential. *Fermentation* **2023**, *9*, 325. [CrossRef]

Zhuravleva, O.I.; Afiyatullov, S.S.; Denisenko, V.A.; Svetlana, P.E.; Natalya, N.S.; Pavel, S.D.; Natalya, Y.K. Secondary metabolites from a marine-derived fungus *Aspergillus carneus Blochwitz*. *Phytochemistry* **2012**, *80*, 123–131. [CrossRef] [PubMed]

Jin, W.; Rinehart, K.L.; Jares-Erijman, E.A. Ophidiacerebrosides: Cytotoxic glycosphingolipids containing a novel sphingosine from a sea star. *J. Org. Chem.* **1994**, *59*, 144–147. [CrossRef]

Tan, R.X.; Chen, J.H. The cerebrosides. *Nat. Prod. Rep.* **2003**, *20*, 509. [CrossRef] [PubMed]

Rossi, C.; Ferrante, R.; Valentinuzzi, S.; Zucchelli, M.; Buccolini, C.; Di Rado, S.; Trotta, D.; Stuppia, L.; Federici, L.; Aricò, M. Noninvasive DBS-Based Approaches to Assist Clinical Diagnosis and Treatment Monitoring of Gaucher Disease. *Biomedicines* **2023**, *11*, 2672. [CrossRef] [PubMed]

Pu, Y.F.; Zhang, W.M.; Zhong, G. Functions and Applications of Ceramides. *Cereals Oils* **2005**, *7*, 14–16.

Bruno, M.C.; Manuel, P.; Liana, C.S. Ceramide: A simple sphingolipid with unique biophysical properties. *Prog. Lipid Res.* **2014**, *54*, 53–67.

Wei, J.; Zhang, W.M.; Zhong, G. Function of ceramides in skin and application in cosmetic. *Cereals Oils* **2007**, *1*, 21–24.

10. Bakar, J.; Michael-Jubeli, R.; Libong, D.; Baillet-Guffroy, A.; Tfayli, A. Stratum corneum ceramide profiles in vitro, ex vivo, and in vivo: Characterization of the α-hydroxy double esterified ceramides. *Anal. Bioanal. Chem.* **2022**, *414*, 3675–3685. [CrossRef]
11. Stelmasiewicz, M.; Świątek, Ł.; Ludwiczuk, A. Chemical and Biological Studies of Endophytes Isolated from *Marchantia polymorpha*. *Molecules* **2023**, *28*, 2202. [CrossRef]
12. Hu, Z.; Yu, C.S. Research Progress on Skin Barrier Function. *Chin. Clin. Physician J. (Electron. Ed.)* **2013**, *7*, 3101–3103.
13. Shindo, S.; Murota, H.; Seki, T.; Mori, K.; Kaizu, K.; Nishizaka, T.; Takagi, Y.; Katayama, I. Effects of a moisturizer containing pseudo-ceramide and a eucalyptus extract on sweating function in adult atopic dermatitis: A double-blind, randomized controlled left-right comparison clinical trial. *J. Cosmet. Dermatol.* **2022**, *21*, 4503–4509. [CrossRef] [PubMed]
14. Cheawchanlertfa, P.; Chitcharoen, S.; Raethong, N.; Liu, Q.; Chumnanpuen, P.; Soommat, P.; Song, Y.; Koffas, M.; Laoteng, K.; Vongsangnak, W. Enhancing Genome-Scale Model by Integrative Exometabolome and Transcriptome: Unveiling Carbon Assimilation towards Sphingolipid Biosynthetic Capability of *Cordyceps militaris*. *J. Fungi* **2022**, *8*, 887. [CrossRef] [PubMed]
15. Bi, S.Y.; Liu, W. Influence of Ceramides on Skin Barrier Function in Atopic Dermatitis. *Chin. J. Med. Aesthet. Cosmetol.* **2015**, *21*, 317–318.
16. Wang, H.C.; He, C.F. Research Status and Application of Ceramides in the Skin. *China Surfactant Deterg. Cosmet.* **2019**, *49*, 51–5[illegible]
17. Wang, W.L.; Wang, Y.; Tao, H.W.; Tao, H.W.; Peng, X.P.; Liu, P.P.; Zhu, W.M. Cerebrosides of the Halotolerant Fungus *Alternaria raphani* Isolated from a Sea Salt Field. *J. Nat. Prod.* **2009**, *72*, 1695–1698. [CrossRef]
18. Wu, S.; Chen, Y.; Miao, C. Secondary metabolites of endophytic fungus *Xylaria* sp. YC-10 of *Azadirachta indica*. *Chem. Nat. Compd.* **2011**, *47*, 858–861. [CrossRef]
19. Hamed, A.; Abdel-Razek, A.S.; Araby, M.; Abu-Elghait, M.; El-Hosari, D.G.; Frese, M.; Soliman, H.S.M.; Stammler, H.G.; Sewald, N.; Shaaban, M. Meleagrin from marine fungus *Emericella dentata* Nq45: Crystal structure and diverse biological activity studies. *Nat. Prod. Res.* **2021**, *35*, 3830–3838. [CrossRef]
20. Boulis, A.G.; Hamed, A.A.; El-awady, M.E.; Mohanmed, A.R.; Shaaban, M. Diverse bioactive metabolites from *Penicillium* sp. MMA derived from the red sea: Structure identification and biological activity studies. *Arch. Microbiol.* **2020**, *202*, 1985–199[illegible] [CrossRef]
21. Kitagaki, H. Medical application of substances derived from non-pathogenic fungi *Aspergillus oryzae* and A. luchuensis-containing *koji*. *J. Fungi* **2021**, *7*, 243–258. [CrossRef] [PubMed]
22. Toledo, M.S.; Levery, S.B.; Straus, A.H.; Suzuki, E.; Momany, M.; Glushka, J.; Takahashi, H.K. Characterization of sphingolipids from mycopathogens: Factors correlating with expression of 2-hydroxy fatty acyl (E)-Δ3-unsaturation in cerebrosides of Paracoccidioides brasiliensis and *Aspergillus fumigatus*. *Biochemistry* **1999**, *38*, 7294–7306. [CrossRef]
23. Wang, C.; Liu, P.P.; Wang, Y.; Sun, K.L.; Zhu, W. Optimization of Fermentation Conditions for the Production of Lincomycin D by Marine-Derived Actinomycete *Streptomyces parvulus* OUCMDZ-2554. *Chin. J. Mar. Drugs* **2014**, *33*, 34–42.
24. Dong, Y.J.; Hao, J.H.; Zheng, H.F.; Sun, Y. Optimization of Enzyme Production Conditions by the Strain YS1069. *Prog. Fish. Sci.* **2016**, *37*, 154–159.

molecules

MDPI

Review

G-Quadruplexes in Human Telomere: Structures, Properties, and Applications

Yan Xu [1,*] and Makoto Komiyama [2,*]

1 Division of Chemistry, Department of Medical Sciences, Faculty of Medicine, University of Miyazaki, 5200 Kihara, Kiyotake, Miyazaki 889-1692, Japan
2 Research Center for Advanced Science and Technology (RCAST), The University of Tokyo, 4-6-1 Komaba, Meguro, Tokyo 153-8904, Japan
* Correspondence: xuyan@med.miyazaki-u.ac.jp (Y.X.); komiyama.makoto@emer.rcast.u-tokyo.ac.jp (M.K.)

Abstract: G-quadruplexes, intricate four-stranded structures composed of G-tetrads formed by four guanine bases, are prevalent in both DNA and RNA. Notably, these structures play pivotal roles in human telomeres, contributing to essential cellular functions. Additionally, the existence of DNA:RNA hybrid G-quadruplexes adds a layer of complexity to their structural diversity. This review provides a comprehensive overview of recent advancements in unraveling the intricacies of DNA and RNA G-quadruplexes within human telomeres. Detailed insights into their structural features are presented, encompassing the latest developments in chemical approaches designed to probe these G-quadruplex structures. Furthermore, this review explores the applications of G-quadruplex structures in targeting human telomeres. Finally, the manuscript outlines the imminent challenges in this evolving field, setting the stage for future investigations.

Keywords: DNA G-quadruplex; RNA G-quadruplex; DNA:RNA hybrid G-quadruplex

1. Introduction

The term telomere is derived from two Greek words: telos (meaning "end") and meros (meaning "part"). Human telomeric DNA consists of a duplex region composed of TTAGGG repeats, which terminates with a G-rich single-stranded overhang of 100–200 nucleotide (nt) length. Though telomeres are crucial for many cellular functions, they are subject to progressive decreases in length at each round of DNA replication. It has been demonstrated by many studies that human telomere DNA plays important roles in cancer and aging [1,2]. Telomeres provide the machinery to protect chromosomes from such detriments. The protective function of telomeres might depend on their state, e.g., whether they have "capping" or "uncapping" structures. Telomere DNA sequences can form G-quadruplexes with various folding topologies, as demonstrated by many structural studies [1]. As shown in Figure 1a, four guanine bases are bound to a tetrad through Hoogsteen hydrogen bondings, and these guanine tetrads stack each other into G-quadruplexes. Metal cations (e.g., Na^+ and K^+) are usually bound to the center of guanine tetrad to stabilize the whole structures. G-quadruplexes are promising for numerous approaches that were developed to target telomeres [3–5]. Several ligands for G-quadruplexes were developed, and their interactions were investigated in detail. For a long time, telomere DNA was believed not to be transcribed into RNA. However, more recent findings showed that telomere DNA is in fact transcribed into RNA (referred to as telomere RNA) in cells [6,7] These telomere RNAs exist in human cells with heterogeneous length. It was proposed that telomere RNA is associated with the end chromosome, and reveals a new level of protection of chromosomal ends [8–10]. Valuable insight into biological processes such as cancer and aging has been provided.

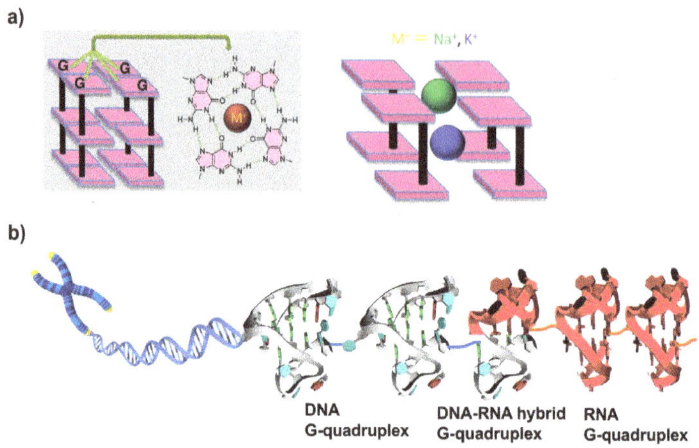

Figure 1. (**a**) Structures of G tetrads and G-quadruplexes. Cations with larger ionic radii, such as K^+ (depicted in blue), are positioned between the tetrads. In contrast, smaller cations, such as Na^+ (illustrated in green), can coordinate within the plane of the tetrads or assume an intermediate position. (**b**) Superhelix structure formed at the ends of chromosomes from telomeric DNA G-quadruplexes, DNA:RNA G-quadruplexes, and RNA G-quadruplexes. As the result, this essential portion is sufficiently protected and further functionalized.

Recently, telomere RNA was suggested to form G-quadruplexes which perform complicated functions [11–14]. With the use of probes provided by chemical approaches, telomere DNA and RNA G-quadruplex structures were detected in vitro and also in human cells [15–17]. Moreover, an DNA:RNA hybrid G-quadruplex is formed from telomeric DNA and telomeric RNA [18–22]. These results indicate that DNA G-quadruplexes, DNA:RNA hybrid G-quadruplexes, and RNA G-quadruplexes can be key players to protect the terminus of telomere (Figure 1b).

In order to deeply understand the biology of telomere, precise information on the structures of G-quadruplexes and their physicochemical properties is essential. Accordingly, the present review focuses on the structural information of DNA G-quadruplexes, RNA G-quadruplexes, and DNA:RNA hybrid G-quadruplexes in human telomere. Various chemical approaches to probe human telomere G-quadruplex structures, as well as applications of G-quadruplex structures to target human telomere, are also described. Lastly, the future aspects of these unique noncanonical structures are discussed.

2. DNA G-Quadruplexes in Human Telomere

The telomeres of human chromosome ends are composed of telomeric repeats of 5′-TTAGGG-3′ sequence. Through many structural studies on the human telomere, a large number of G-quadruplexes of different structures have been already identified, as depicted in Figure 2 [23–32]. These structural differences are primarily based on the syn/anti conformations of guanine residues and the relative orientation of the G-quartet core [33]. Thus, the topologies of DNA G-quadruplex are drastically dependent on the sequences and lengths of telomere DNA, the types of linking loops, and the nature of the associated metal cations.

For example, in NaCl solution, a 22-mer telomeric DNA d[AGGG(TTAGGG)$_3$] forms an antiparallel G-quadruplex with anti/syn guanines (Figure 2a) [34]. One loop is diagonal, and two other loops are lateral. In KCl solution, however, this sequence forms the mixtures of hybrid-1 and hybrid-2 G-quadruplexes, in which the loop types are external, lateral, and lateral (Figure 2c,d) [35–38]. In the crystal grown in the presence of K^+ ion, the same sequence adopts a completely different parallel G-quadruplex conformation having three external linking loops (Figure 2b) [39]. On the other hand, another telomeric sequence d[GGG(TTAGGG)$_3$T] forms in KCl solution an antiparallel G-quadruplex with only two G-tetrad layers (Figure 2e) [40–42]. More recently, d[(GGGTTA)$_2$GGGTTTGGG] sequence was reported to take another antiparallel structure, so-called "chair" with lateral loops (Figure 2f) [43]. Circular dichroism (CD) spectroscopy is very useful to distinguish G-quadruplex structures in terms of fundamental characteristic bands [44]. For example, the CD spectrum of human telomeric sequence d[AGGG(TTAGGG)$_3$] in 100 mM Na^+ solution exhibits a 295 nm positive band and a 265 nm negative band, which is characteristic of an antiparallel G-quartet structure [35]. In a 100 mM K^+ ion solution, however, this sequence shows a stronger positive CD band at 290 nm, with weak negative peaks near 255 and 235 nm, indicating that a mixture of hybrid G-quadruplexes is formed under these conditions.

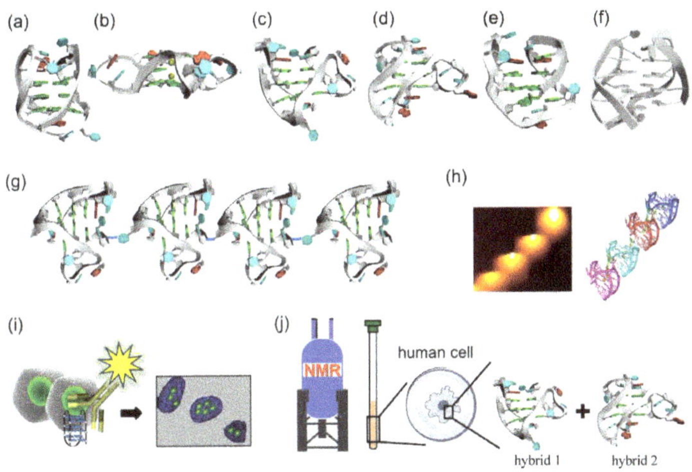

Figure 2. Variety of conformations of DNA G-quadruplexes observed under different conditions. (**a**) Antiparallel G-quadruplex from d[AGGG(TTAGGG)$_3$] formed in Na^+ solution. (**b**) A crystal structure from d[AGGG(TTAGGG)$_3$] in the presence of K^+ ions. (**c**,**d**) Hybrid-1 type and hybrid-2 type G-quadruplexes from d[AGGG(TTAGGG)$_3$] prepared in K^+ solution. (**e**) Antiparallel G-quadruplex of d[GGG(TTAGGG)$_3$T] with only two G-tetrad layers in K^+ solution. (**f**) Chair antiparallel conformation from d[(GGGTTA)$_2$GGGTTTGGG] in K^+ solution. (**g**,**h**) Higher order G-quadruplex observed by AFM. (**i**) Visualization of DNA G-quadruplex structures in cells by immunostaining with the use of structure-specific antibody. (**j**) In-cell ^{19}F NMR of d[AGGG(TTAGGG)$_3$] in Hela cells.

For the analysis of the structures of G-quadruplexes from still longer telomeric sequence, atomic force microscope (AFM) was employed (Figure 2g,h) [45]. In the AFM image, a telomere DNA (around 100-nucleotide length) forms a higher-order assembly of G-quadruplex, in which adjacent G-quadruplex conformation units are arranged in end-to-end structure. These G-quadruplex units stack each other into higher-order packing structures. Some small molecules and antibodies were also used to investigate the DNA G-quadruplexes in living cells (Figure 2i) [46,47].

The topology of G-quadruplex of telomeric DNA in living human cells has not yet been sufficiently clarified. To shed light on this problem, in-cell ^{19}F NMR experiments were performed in the environment close to the inside of human cells. ^{19}F NMR spectroscopy is suitable for this purpose, since it is characterized by (i) simplicity of spectra, (ii) high sensitivity of chemical shift to the environment, and (iii) absence of natural background signals in cells. For example, 3,5-bis(trifluoromethyl)phenyl moiety was connected to the 5′-end of telomeric DNA sequence (d[AGGG(TTAGGG)$_3$]), and this ^{19}F-labelled DNA was injected into human cells. According to in-cell ^{19}F NMR, this telomeric DNA forms the mixture of hybrid-1 and hybrid-2 G-quadruplexes in the human cells (Figure 2j) [48,49].

In *Xenopus laevis* oocytes, a hybrid-2 type G-quadruplex is formed [50]. The two-tetrad G-quadruplex in Figure 2e was further examined by ^{19}F NMR spectroscopy. Only one ^{19}F NMR signal was observed both in K^+ solution and in HeLa cells [48]. These results provide valuable information on the understanding of the structures of human telomeric DNA in living human cells and also on the design new drugs that target telomeric DNA.

Many fluorescent probes to detect DNA G-quadruplexes have been reported, and employed to image living cells [51–57]. In 2015, naturally occurring alkaloid epiberberine was found to emit fluorescence upon the binding with DNA G-quadruplex [58]. Later, various fluorescent probes were developed [59–62], and used to construct quadruplex-based analytical sensing platforms. Among these probes, a tripodal cationic fluorescent probe was featured by notable differences in fluorescence lifetime between G-quadruplexes and other DNA topologies [63]. Furthermore, its fluorescence quantum yield was greatly increased upon G-quadruplex binding. These two features allowed a clearcut differentiation of DNA G-quadruplexes from background signals in cell imaging, and a quantitative analysis was accomplished in terms of photon counts. According to NMR spectroscopy, this probe interacts with G-quadruplexes using three arms through π–π stacking. An artificial protein of 6.7 kD is useful to detect G-quadruplexes in living cells with the ChIP-Seq technique [64]. In this protein probe, two identical G4-binding domains (23-amino acid residues each) are connected with linker peptide to bind G-quadruplexes with sufficient strength. Small molecular ligands, which distinguish different conformations of G-quadruplexes each other, were also reported [65–67].

It is noteworthy that small molecules targeting G-quadruplexes are also important as potent drugs for therapy of cancer and other diseases [68–79]. For example, a small molecule, phenyl-ethenyl-quinoline, specifically binds G-quadruplex in oncogene MYC promoter, and is promising as a lead compound of anticancer drug [80]. This compound consists of phenyl and quinoline moieties linked via rather flexible ethenyl bridge, and has +1 charge on the methylated N1 atom. In addition to its direct interaction with the G-quadruplex core, this molecule additionally interacts with the 3′- and the 5′-flanking residues of target G-quadruplex that are immediately adjacent to the G-core. These interactions help this binder to distinguish the MYC promoter G-quadruplex from other G4 structures, as proposed previously [81]. Important roles of structured water in these G-quadruplex-small molecule interactions have been also indicated [82].

3. RNA G-Quadruplexes in Human Telomere

Telomere DNA is transcribed into telomeric RNA (UUAGGG repeat) which shows various biological functions [6,7]. Importantly, RNA G-quadruplexes in most cases take parallel conformation. For example, a 12-mer telomeric RNA r(UAGGGUUAGGGU) forms a parallel dimer G-quadruplex in Na^+ solution, as determined by ^{1}H-NMR (Figure 3a) [83]. Both in K^+ solution and in K^+-stabilized crystal, this 12-mer RNA also forms a G-quadruplex of the same topology [12,14]. This feature is highly in contrast with the results for DNA G-quadruplexes, in which the structures enormously depend on the conditions (Figure 2). Stereochemical constraints inherent to RNA should be one of the governing factors for this difference. Very interestingly, the formation of parallel G-quadruplex from four strands of r(UAGGGU) is accompanied by simultaneous formation of a U-tetrad [13]. As shown in Figure 3b, this U-tetrad (in red), formed through Hoogsteen bindings of four uridines at the 3′-end of each RNA strand, sits on the parallel G-quadruplex structure (in green). It is noteworthy that this U-tetrad dramatically stabilizes the whole structure of RNA assembly. In cells, this kind of U-tetrad could be also formed, and promote the functions of human telomere. In many biological regulations by RNA G-quadruplex, specific proteins bind to this noncanonical structure. For example, heterogeneous nuclear ribonucleoprotein A1 (hnRNPA1) cooperates with RNA G-quadruplex to facilitate telomere capping [84–86]. The loops of G-quadruplex are primarily responsible for the interactions with hnRNPA1.

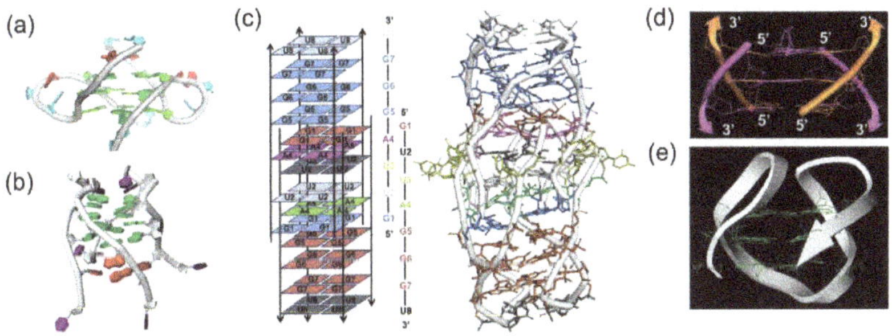

Figure 3. (**a**) Dimeric parallel telomeric RNA G-quadruplex from r(UAGGGUUAGGGU). (**b**) Tetramolecular RNA G-quadruplex from r(UAGGGU) possessing a novel U-tetrad on the bottom (in red). (**c**) RNA G-quadruplex having 14 tetrads, in which an eight-stranded helix part intercalates at the center. (**d**) Antiparallel RNA G-quadruplex from the fragments having 8-bromoguanosine. (**e**) Antiparallel RNA G-quadruplex formed by a 22-mer telomeric RNA.

Until very recently, it was believed that almost all natural RNAs form only parallel G-quadruplexes [87,88]. However, it was evidenced that an antiparallel G-quadruplex is formed in Na^+ solution from a 22-mer RNA of telomere sequence (Figure 3e) [89]. An antiparallel G-quadruplex was also obtained in K^+ solution, when 6-mer RNA (r(UAGGGU)) was modified with 8^{Br}rG (Figure 3d) [90]. On the other hand, 8-mer telomere RNA GUUAGGGU forms an unusual RNA architecture which involves 14 tetrads such as A-, U- and G-tetrad (Figure 3c) [91]. A central scaffold with eight stranded helices intercalates two symmetry subunit G-quadruplexes. This conformation is stable and is retained even under denaturing conditions. Apparently, the structures of RNA G-quadruplexes are more complicated than expected before [92].

Formation of RNA G-quadruplexes in vivo was first proposed in terms of the dimerization of two genomic RNA copies of HIV through RNA-RNA interactions [93]. This dimerization was dramatically stabilized by K^+ cation. This result, as well as other evidence, indicated the formation of an intermolecular RNA G-quadruplex, in which four guanine bases are assembled to tetrad as already known in DNA G-quadruplexes [87,94–96]. Later, RNA G-quadruplexes were found in many biological systems, and their essential roles were evidenced in many bioprocesses: regulation of mRNA translation [97–103], telomere homeostasis [104], mRNA splicing [105–107], and others [106,108]. Of course, their relevance to diseases is apparent [109,110]. In RNA viruses, G-quadruplexes in their genomes control viral replication, transcription, and other processes. For example, there exist many G-quadruplex folding sequences in the open reading frame regions of SARS-CoV-2 (the virus of COVID-19) [111], which caused a terrible pandemic from 2019 to 2023. Direct interaction of one of these G-quadruplexes with viral helicase, which is inevitable enzyme for the viral functions, was proposed.

In order to detect the formation of RNA G-quadruplex in cells, a light-switching probe was fabricated by incorporating pyrenes into both the 5′- and the 3′-ends of r(GGGUUAGGG) (Figure 4a) [17]. When G-quadruplex is formed from this RNA, the two pyrene chromophores are placed in close proximity to form their exciplexes. As the results, the maximum of fluorescence emission shifts from 400 nm (blue emission) to 480 nm (green emission). The system is employable for in-cell detection, as depicted in fluorescence microscopy images in the bottom. Furthermore, ^{19}F NMR was successfully used to investigate the structure of telomeric RNA in cells (Figure 4b) [112–117]. Telomeric RNA r(UAGGGUUAGGGU) was labeled with 3,5-bis(trifluoromethyl)phenyl moiety, and injected into *Xenopus laevis* oocytes [112]. A new ^{19}F-NMR signal was observed (the bottom spectrum), and assigned to stacking assembly of two G-quadruplex units with the use of the reference spectrum obtained in vitro (the top).

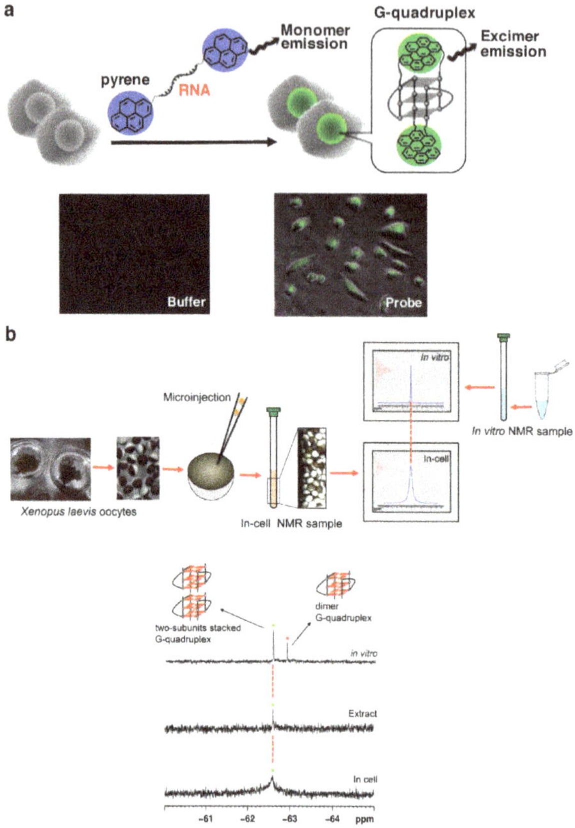

Figure 4. (**a**) A pyrene probe used to detect in-cell formation of G-quadruplex from telomeric RNA (r(GGGUUAGGG)). Fluorescence microscopy images of living cells with or without the probe are shown below. (**b**) In-cell ^{19}F NMR experiment in *Xenopus* oocytes. The spectra of (**top**) RNA telomeric G-quadruplex in vitro and (**bottom**) in *Xenopus* oocytes.

Alternatively, RNA G-quadruplexes were detected in cells by using a coumarin-hemicyanine fluorophore (QUMA-1), which was prepared through the condensation of a coumarin aldehyde and an N-methylated quinoline derivative [118]. Its fluorescence emission at 660 nm is only weak in buffer, but is significantly enhanced by RNA G-quadruplexes. This enhancement is specific to G-quadruplexes, and either single-stranded RNA or double-stranded RNA hardly affects the emission intensity. This red-emitting fluorescent probe was successfully employed for selective, continuous, and real-time visualization of RNA G-quadruplexes in living cells. It was proposed that the tight binding of the probe to RNA G-quadruplexes restricts the rotation around the methine bridge between the coumarin and the N-methylated quinoline moiety, and locks the probe in its fluorescently active conformation. In another probe PyroTASQ, four guanines are attached to a pyrene as fluorescent chromophore [119]. The intrinsic fluorescence from this probe is minimal, since the emission from the pyrene chromophore is quenched by their own guanines through intramolecular photo-induced electron transfer. Upon binding with G-quadruplex, however, this probe achieves a conformational switch to assemble its four guanines into an intramolecular G-quartet, which stacks on the analyte G-quadruplex. As a result, the electrons in the guanines are redistributed to suppress the fluorescence quenching, and the emission from the pyrene is restored. These two fluorescent probes are highly promising to sensitively detect G-quadruplexes in vivo and in vitro, since they light up only with the coexistence of G-quadruplexes and are otherwise virtually silent.

Diastereomerically pure Fe(II) metametallohelices strongly bind the G-quadruplex of human telomeric repeat-containing RNA (TERRA rG4), and stabilize this non-canonical structure [120]. Accordingly, DNA synthesis on RNA template containing four repeats of human telomeric sequence is inhibited by them. As tailor-made binders, unique aptamers towards TERRA rG4 were prepared using unnatural L-RNA units (mirror image of natural D-RNA nucleotides) [121]. Strong and specific binding of TERRA rG4 by the aptamer was evidenced. Notably, they recognized the target TERRA rG4 very strictly, and never bound other forms of constructs. Neither simple RNA hairpin nor telomeric DNA G-quadruplex was bound, and still more importantly, even other rG4s (hTERC, NRAS, miRNA149, and others) were not bound either. Cationic porphyrins (CPs) bearing an alkyl chain was used for photodynamic therapy of cancer [122]. The porphyrin derivatives preferentially bind to the RNA G-quadruplex formed in mRNA, and destroys this mRNA upon photoirradiation.

4. DNA:RNA Hybrid G-Quadruplexes in Human Telomere

The discovery of telomere RNA soon raised a crucial question of how telomeric RNA is specifically associated with telomeric DNA to accomplish various functions (chromosome-end regulation, protection, and others). An association between telomere RNA and telomere DNA through the formation of an intermolecular DNA:RNA G-quadruplex may be an answer to this problem. However, it is technically difficult to study the DNA:RNA hybrid G-quadruplex structure by traditional methods, such as NMR spectroscopy and X-ray crystallography, since the DNA G-quadruplex, RNA G-quadruplex, and DNA:RNA hybrid G-quadruplex may coexist as a mixture in solution. Thus, concrete evidence for the formation of DNA:RNA G-quadruplex is hard to obtain. Accordingly, the experiments in Figure 5a,b were designed to trap DNA:RNA hybrid G-quadruplex by click reaction and produce a snapshot of the inter-converting structures that are present in complicated solutions. Human telomeric DNA and RNA are modified with either an alkyne or an azide, and mixed each other. The detection of "azide-alkyne cycloaddition" product between the DNA and the RNA is used as the evidence for the formation of DNA:RNA hybrid G-quadruplex. In Figure 5a, an alkyne is bound to the 5′-end of RNA, and an azide is bound to the 5′-end of DNA [19]. Thus, only when the DNA:RNA hybrid G-quadruplex is formed in the reaction mixture, the alkyne and the azide are placed nearby and the click reaction between them should efficiently proceed [123]. In fact, it was clearly shown by gel-electrophoresis that DNA:RNA hybrid G-quadruplex is formed from telomere DNA and telomeric RNA in aqueous K^+ solutions [19,20,45,123]. Furthermore, this method was applied to in-cell reactions, where the DNA was labelled with an azide-bearing profluorophore (7-azidocoumarin), and incorporated into human cells together with the alkyne-bearing RNA. Direct evidence for the formation of DNA:RNA hybrid G-quadruplexes in cells was provided as fluorescent signals in the bottom images in Figure 5b [20,21,123]. Consistently, no fluorescent signals were detected in the absence of Cu catalysts for the click reactions (upper images).

With the use of ^{19}F NMR spectrum, the formation of DNA:RNA hybrid G-quadruplex in living human cells was also assessed and strongly confirmed [22]. In Figure 5c, a ^{19}F-labelled 12-mer G-rich RNA was mixed with 12-mer G-rich DNA in 1:3 ratio. Among two major peaks, the signal at −62.61 ppm was assigned to the DNA:RNA hybrid G-quadruplex (HQ). As shown in the top spectrum, this signal was clearly observed in living human cells, confirming the formation of this DNA:RNA hybrid G-quadruplex. Concurrently, the RNA G-quadruplex (RQ) was also formed.

Telomere RNA plays a crucial role in the "capping" of telomeres through the formation of telomeric DNA:RNA G-quadruplexes [20,124–126]. This argument is supported by insights from a single-molecule FRET study examining the folding patterns of telomere DNA in conjunction with telomere RNA. Notably, telomere RNA exhibits a preference for the formation of DNA:RNA hybrid G-quadruplexes at the 3′ end of telomeric DNA.

Furthermore, DNA G-quadruplexes is unfolded by a complementary C-rich sequence, but this effect is not observed for DNA:RNA hybrid G-quadruplexes. These findings underscore previous arguments that these hybrid G-quadruplexes protect the 3′ end of the telomeric DNA overhang, which is critical for the interaction with telomerase and other telomere-associated proteins. Unique contribution of RNA in reinforcing the structural integrity of telomeres has been evidenced. According to bioinformatic analysis, putative DNA:RNA hybrid G-quadruplex-forming sequences are abundant on both sides of the transcription start sites in the genome of warm-blooded animals [125]. Their formation is dependent on the negative supercoiling generated by RNA polymerases. Genome-wide occurrence of DNA:RNA hybrid G-quadruplexes has been also indicated [126]. The biological roles of DNA:RNA hybrid G-quadruplexes are gradually getting clearer.

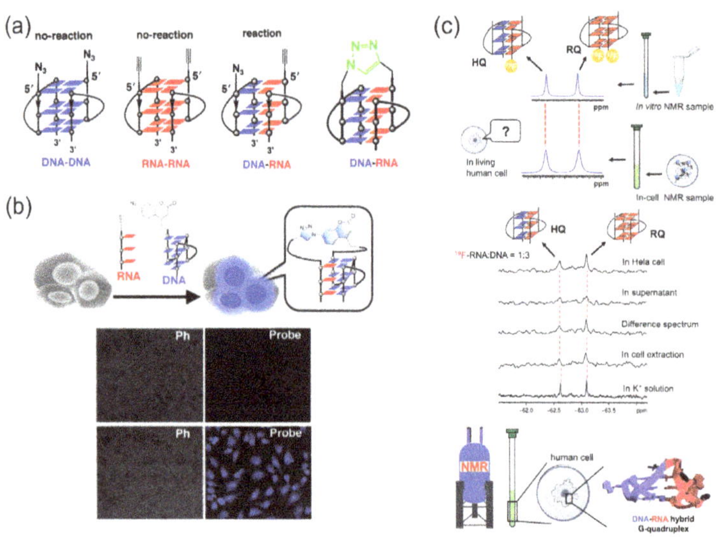

Figure 5. (**a**) Formation of DNA:RNA hybrid G-quadruplex promotes a click reaction to differentiate this hybrid G-quadruplex from DNA G-quadruplex and RNA G-quadruplex. (**b**) Fluorescence images with and without the probe in cells. Formation of DNA:RNA hybrid G-quadruplex in cells was confirmed. Upper images; without Cu catalyst for the click reaction. Bottom images; with Cu catalyst. Ph; phase-contrast imaging. (**c**) ^{19}F NMR showing the formation of DNA:RNA hybrid G-quadruplex in human cells.

5. Binding Proteins of Telomere DNA and RNA G-Quadruplexes

Notably, POT1 has been demonstrated to unravel intramolecular G-quadruplexes to facilitate its binding to telomeric DNA [127,128]. The unfolding of G-quadruplexes by POT1 also impacts the capacity of telomeric sequences to undergo extension by telomerase, at least in vitro [1]. Initially, it was shown with telomerase from three distinct species of ciliated protozoa that oligonucleotides adopting antiparallel intramolecular G-quadruplex structures do not serve as favorable substrates for telomerase [129,130]. The antiparallel and hybrid G-quadruplex structures assumed by a 4-repeat oligonucleotide in K^+ solution have been also demonstrated to be ineffective substrates for telomerase [129–134]. These G-quadruplex conformations [133,134] inhibit the binding of telomerase to the DNA, leading to a non-processive attempted extension of the DNA by telomerase [129]. POT1 can reinstate the capability of telomerase to engage with these substrates by unfolding the G-quadruplex and capturing the DNA in a linear form with a protruding tail [127]. The dynamic interplay of G-quadruplex folding and their unfolding by POT1 may thus serve as a regulatory mechanism, governing the access of telomerase to the telomere.

HnRNPA1 is an RNA binding protein which is involved in the telomere maintenance machinery. The binding of hnRNPA1 to telomere RNA with structure-dependence can regulate human telomere function. In Figure 6, the interactions of G-quadruplex RNA with hnRNPA1 were analyzed by gel electrophoresis in which RNA G-quadruplexes of different structures were employed [85,86]. It was demonstrated that the loops of G-quadruplex RNA are very important for interaction with hnRNPA1. Furthermore, a macrocyclic heptaoxazole molecule, 7OTD, and its fluorescent derivative, Cy5−7OTD, have been developed to bind to telomere RNA G-quadruplex and visualize its complex formation with hnRNPA1 [85]. In Figure 7a, the green band is for telomere RNA G-quadruplex. The red band of smaller mobility, observed after the addition of Cy5−7OTD, is assigned to the interaction of RNA G-quadruplex with Cy5−7OTD. Upon further addition of hnRNPA1, a new band appeared in the upper part of the gel, due to its binding with the RNA G-quadruplex. Furthermore, Cy5−7OTD was employed in Figure 7b to analyze the interactions of RNA telomere G-quadruplex with hnRNPA1 in cells. By incubating HeLa cells with this probe, the telomere RNA G-quadruplex was detected as red signals in fluorescence microscopy. On the other hand, hnRNPA1 in the cells was visualized by green fluorescence from anti-hnRNPA1 antibody-based probe. Quite importantly, the green fluorescence of hnRNPA1 and the red fluorescent foci of the RNA G-quadruplex notably overlapped each other. Consistently, the intensity of the red signal was reduced by RNase treatment, confirming that this signal came from the Cy5−7OTD bound to the RNA G-quadruplex.

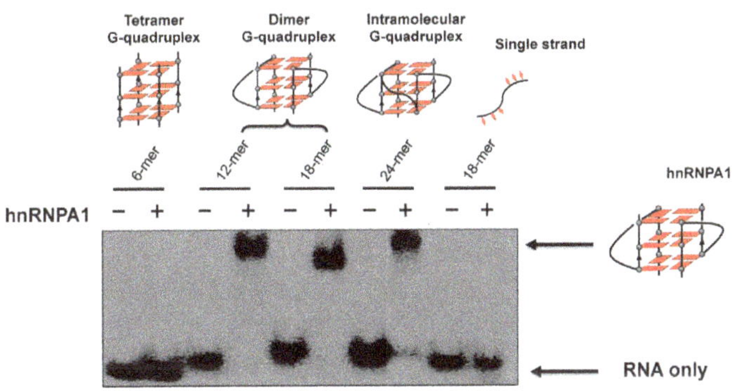

Figure 6. Binding of hnRNPA1 to telomere RNA G-quadruplexes of various structures. Only the RNA G-quadruplexes involving loops (the dimer and the intramolecular G-quadruplexes) bind hnRNPA1, and neither the tetramer RNA G-quadruplex without loops nor single-stranded RNA binds the protein.

The C-terminal region of TLS, a human telomere-binding protein, engages in the formation of a ternary complex with G-quadruplex DNA and RNA from human telomeres in vitro, specifically within the Arg-Gly-Gly domain [135]. Notably, TLS exhibits binding affinity to G-quadruplex telomere DNA within double-stranded regions and G-quadruplex RNA. This dual interaction plays a crucial role in regulating histone modifications associated with telomeres, influencing overall telomere length. The findings propose that the G-quadruplex serves as a scaffold for TLS, orchestrating the modulation of telomere length through histone modifications. In parallel, the N-terminal Gly/Arg-rich (GAR) domain of TRF2 demonstrates recognition capabilities for G-quadruplex RNA [136]. The G-quadruplex structure within telomere RNA emerges as a pivotal recognition element for the TRF2 GAR domain. The interaction between TRF2 GAR and telomere RNA is deemed essential for maintaining telomere stability, underlining its significance in the intricate regulatory mechanisms governing telomeric function.

Alternative lengthening of telomeres (ALT), a telomerase-independent mechanism for maintaining telomeres, is orchestrated primarily through break-induced replication

(BIR) [137,138]. The human protein RAD52 plays a key role in promoting ALT by facilitating the formation of D-loops. However, it is noteworthy that ALT can also occur via a RAD52-independent BIR pathway. In the absence of RAD52, telomere RNA contributes significantly to ALT activity by forming dynamic telomeric R-loops. These R-loops, in turn, augment the presence of G-quadruplex structures at telomeres. The increased G-quadruplex formation enhances ALT, even in cases where telomeric RNA is depleted. This observation suggests that G-quadruplexes operate downstream of R-loops, playing a pivotal role in promoting break-induced replication and contributing to the ALT process.

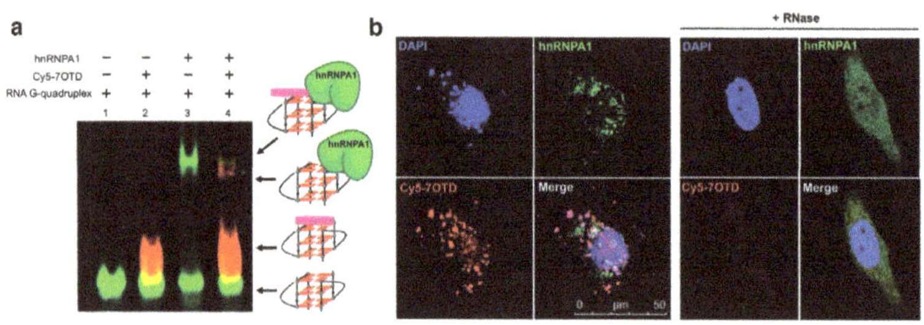

Figure 7. Analysis of the binding of hnRNP1 with RNA G-quadruplex using Cy5−7OTD. (**a**) Gel electrophoresis on the binding. Lane 1, RNA G-quadruplex visualized with fluorescein in green; lane 2, RNA G-quadruplex with Cy5−7OTD in red; lane 3, the mixture of RNA G-quadruplex and hnRNPA1 visualized by fluorescein in green; lane 4, the complex formation of RNA G-quadruplex with hnRNPA1 colored in red by Cy5− 7OTD. (**b**) Microscopy images of human cells incubated with Cy5−7OTD (red) and then with antibody of hnRNPA1 (green). DNA was stained by DAPI (blue fluorescence). In the merged panel, colocalization of RNA G-quadruplex (red) and hnRNPA1 (green) is apparent. As expected, the signals of RNA G-quadruplex were notably weakened when the cells were treated by RNase.

Telomeric RNA exhibits the ability to adopt a G-quadruplex structure and engage in interactions with the chromatin remodeler ATRX [139]. In this interaction, telomere RNA acts as a regulator, preventing ATRX from binding to subtelomeric regions. Instead, it facilitates the promotion of G-quadruplex formation in proximity to transcription start sites (TSS) that are distant from telomeres. Upon depletion of telomere RNA, there is an observed increase in ATRX binding at TSS, leading to the unwinding of G-quadruplex DNA and subsequent repression of gene expression. This unveils an epigenetic regulatory mechanism by which telomere RNA serves to sequester ATRX, thereby preserving DNA G-quadruplex structures and influencing gene expression patterns.

6. Applications of G-Quadruplexes

Unique structural restraints of G-quadruplexes are effectively employed as scaffolds for specific biological and/or chemical reactions. Freedoms of structural design of the systems are highly advantageous for these applications. In terms of the formation of DNA G-quadruplex, a chemical method to selectively cleave human telomere DNA was developed [83]. In Figure 8a, 16-mer DNA of three-repeat human telomere sequence (^{Br}GGGTTA^{Br}G^{Br}GGTTAGGGT) binds to a telomere DNA substrate through the formation of dimeric G-quadruplex. To the 5′-end of this 16-mer DNA, an ethylenediamine-*N,N,N′,N′*-tetrakis(methylenephosphonic acid) (EDTP) as metal-binding group is conjugated. Upon the addition of $Ce^{IV}(NH_4)_2(NO_3)_6$, Ce^{IV} ion (a catalyst for DNA hydrolysis) is bound to the EDTP, and placed near the phosphodiester linkage(s) in the telomere DNA substrate. As a result, only telomere DNA is picked up from reaction mixtures through G-quadruplex formation, and cleaved through hydrolytic pathway.

A similar strategy was employed to design a photo-controllable inhibitor of telomerase, which is vigorously expressed in cancer cells (Figure 8b) [140]. To the 5′-end of 6-mer DNA (TA^{Br}GGGT), a photosensitive dye psoralen (green colored) was conjugated. This photo-crosslinking reagent is a linear furanocoumarin, and selectively reacts with pyrimidine bases (T and C) for crosslinking. This psoralen-conjugated 6-mer DNA binds to telomere DNA through the formation of intermolecular G-quadruplex. Upon photo-irradiation, the psoralen crosslinked the telomere DNA with the 6-mer DNA. As the result, the duplication of telomere DNA by telomerase was sterically inhibited, triggering the death of cancer cells. These two results apparently show that G-quadruplex formation serves as an efficient approach to target human telomere for its effective transformation.

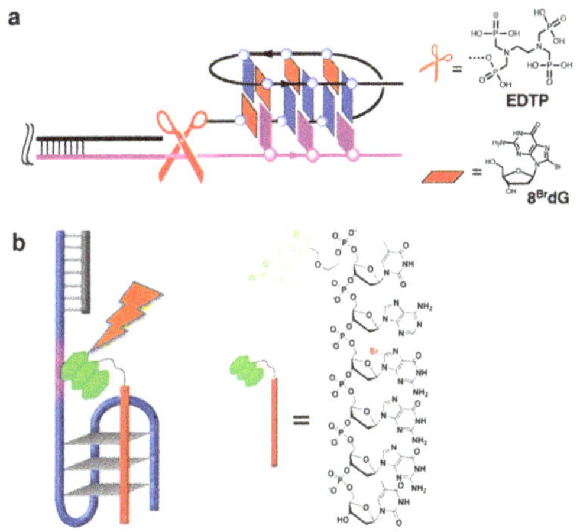

Figure 8. (**a**) Telomere DNA specific cleavage by G-quadruplex formation. The structures of ligand EDTP and $8^{Br}dG$ are shown. (**b**) Photochemical inhibition of telomerase by targeting human telomere DNA with a 6-mer DNA (red bar) through G-quadruplex formation. The 6-mer DNA bears a psoralen (green) at the 5′-end.

Alternatively, the formation of an intermolecular G-quadruplex presents a novel mechanism to regulate the efficiency of translation [141]. A G-rich sequence (5′-GGGCCCGGG-3′) was introduced to the 5′-UTR of mRNA, and a telomere related G-rich RNA (5′-GGGUUAGGG-3′) was employed as on-off controller of translation. An enhanced green fluorescent protein reporter (EGFP1) was employed. Without the addition of the short telomere RNA, the 5′-UTR of mRNA was in the normal state, and EGFP1 protein was efficiently expressed to emit green fluorescence. Upon the addition of the telomere short RNA. However, EGFP1 was hardly expressed. Apparently, the mRNA translation was inhibited by the formation of RNA G-quadruplex between the short telomere RNA and the G-rich sequence in the 5′-UTR. The short telomere RNA successfully worked as an on-off controller of translation.

7. Future Prospects and Conclusions

Telomere biology has come a long way in the past 70 years, from the observation that chromosomes require protection to the award of the 2009 Nobel Prize for the discovery of telomerase and the effects of telomere shortening on cells. Functional telomeres have been experimentally implicated in a number of molecular cell processes. Molecular components of telomeres, including terminal double- and single-stranded telomeric DNA, protein complex shelterin, and telomerase form physical states which are associated with functional capping and uncapping. In this decade in particular, structural information of DNA G-quadruplexes and RNA G-quadruplexes has been rapidly accumulated, deepening our understanding on the biological roles of telomeres. Pre-clinical experiments suggested that telomere is an encouraging target for cancer therapy. The findings of telomere RNA molecules open new doors to a better understanding of the complicated functions of telomeres. These telomere RNAs are bound by a complicated set of proteins [142]. Furthermore, telomeric RNA G-quadruplex and DNA G-quadruplex show homogeneous dimerization and mutual association. Telomere RNA, a newly emerging player, may be useful for cancer treatment, since these RNA molecules accompanies malignant transformation. Thus, telomere RNA G-quadruplex structures should be also a valuable target for anticancer agents [1]. Stabilization of telomere RNA G-quadruplex structures by small molecules can lead to telomerase inhibition. All of these results confirm that there are crucial needs to revisit the structural and functional mechanisms of telomeres.

The current status of our understanding on telomere science is schematically depicted in Figure 9. Telomere DNA, telomere RNA, and related-proteins construct TDRP body (telomeric DNA, RNA and proteins body), such as P bodies for mRNA surveillance, mRNA decay, RNA-mediated silencing and translational control [143]. Telomere capping is promoted by the binding of POT1 protein to single-strand telomere DNA [84]. Many evidences show that telomere RNAs are associated with various elements in the telomere. Through the formation of telomeric DNA:RNA G-quadruplexes, telomere RNA interacts with telomeric DNA for the "capping" of telomere. Moreover, telomeric RNA is a structural component of the telomere nucleoprotein complex, and is also bound to the telomere through interactions with Shelterin components such as TRF1 and TRF2. Three domains of TRF2 interact with telomeric RNA including the amino-terminal basic domain (GAR) [144]. Telomeric RNA further interacts with the ORC1 region that associates with TRF2 GAR. Interestingly, the TRF2 GAR has a relatively high affinity towards G-rich RNA capable of forming G-quadruplex structures, indicating that the recognition mainly depends on G-quadruplex structure, rather than a specific sequence [85]. Telomeric RNA promotes POT1 binding to telomeric DNA by removing hnRNPA1 [84]. Thus, the accumulation of telomeric RNA is helpful for the formation of TDRP body and promotes telomere capping. It was also shown that acridine ligands notably change the UUA loop conformation of RNA G-quadruplex [145]. The resultant stabilization of G-quadruplexes should facilitate the formation of TDRP bodies, in which higher-order telomere structures are satisfactorily capped to accomplish homeostatic functions.

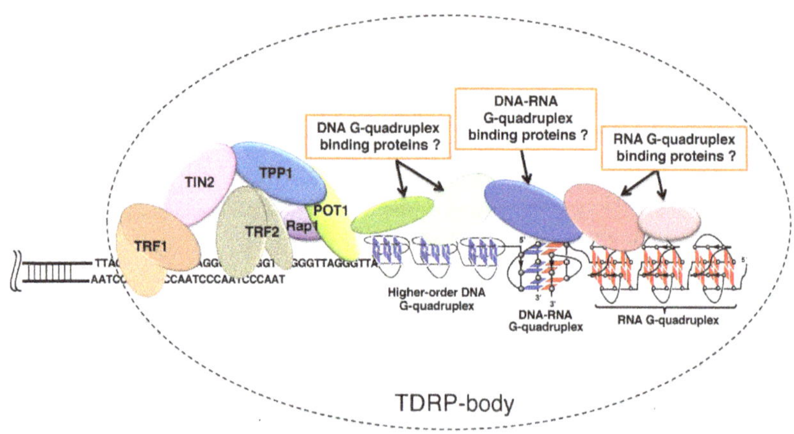

Figure 9. A schematic view of human telomere end: telomere DNA, telomere RNA, and related-proteins in the TDRP body. Proteins can recognize DNA telomere G-quadruplex, RNA telomere G-quadruplex and hybrid DNA:RNA G-quadruplex. The TDRP body at the ends of chromosomes provides a protective structure. Note that some of the contents could be speculative, and have not necessarily been firmly established.

Furthermore, it should be noted that newly discovered telomere RNA may provide a solution to "the end replication problem" in which a protruding single-stranded 3′-DNA end causes a progressive shortening of telomeric DNA at each round of DNA replication [146,147]. In normal cells, telomere RNA can act as a template for telomere elongation synthesis by reverse transcriptase. Support for this hypothesis comes from the finding that telomere RNAs are significantly downregulated in various types of human cancers, compared to normal tissues. In normal cells, the reverse transcription pathway is involved in the maintenance of telomere length, whereas telomerase is mainly responsible for telomere length maintenance in cancer cells.

Funding: This research received no external funding.

Conflicts of Interest: The authors declare no conflicts of interest.

References

1. Xu, Y. Chemistry in Human Telomere Biology: Structure, Function and Targeting of Telomere DNA/RNA. *Chem. Soc. Rev.* **2011**, *40*, 2719–2740. [CrossRef] [PubMed]
2. Hurley, L.H. DNA and its associated processes as targets for cancer therapy. *Nat. Rev. Cancer* **2002**, *2*, 188–200. [CrossRef] [PubMed]
3. Monchaud, D.; Teulade-Fichou, M.P. A hitchhiker's guide to G-quadruplex ligands. *Org. Biomol. Chem.* **2008**, *6*, 627–636. [CrossRef] [PubMed]
4. Georgiades, S.N.; Abd Karim, N.H.; Suntharalingam, K.; Vilar, R. Interaction of metal complexes with G-quadruplex DNA. *Angew. Chem. Int. Ed.* **2010**, *49*, 4020–4034. [CrossRef]
5. Ou, T.M.; Lu, Y.J.; Tan, J.H.; Huang, Z.S.; Wong, K.Y.; Gu, L.Q. G-quadruplexes: Targets in anticancer drug design. *ChemMedChem* **2008**, *3*, 690–713. [CrossRef]
6. Azzalin, C.M.; Reichenbach, P.; Khoriauli, L.; Giulotto, E.; Lingner, J. Telomeric repeat containing RNA and RNA surveillance factors at mammalian chromosome ends. *Science* **2007**, *318*, 798–801. [CrossRef]
7. Schoeftner, S.; Blasco, M.A. Developmentally regulated transcription of mammalian telomeres by DNA-dependent RNA polymerase II. *Nat. Cell Biol.* **2008**, *10*, 228–236. [CrossRef]
8. Luke, B.; Lingner, J. TERRA: Telomeric repeat-containing RNA. *EMBO J.* **2009**, *28*, 2503–2510. [CrossRef]
9. Schoeftner, S.; Blasco, M.A. A 'higher order' of telomere regulation: Telomere heterochromatin and telomeric RNAs. *EMBO J.* **2009**, *28*, 2323–2336. [CrossRef]
10. Feuerhahn, S.; Iglesias, N.; Panza, A.; Porro, A.; Lingner, L. TERRA biogenesis, turnover and implications for function. *FEBS Lett.* **2010**, *84*, 3812–3818. [CrossRef]
11. Xu, Y.; Kaminaga, K.; Komiyama, M. G-Quadruplex Formation by Human Telomeric Repeats-Containing RNA in Na^+ Solution. *J. Am. Chem. Soc.* **2008**, *130*, 11179–11184. [CrossRef]
12. Martadinata, H.; Phan, A.T. Structure of propeller-type parallel-stranded RNA G-quadruplexes, formed by human telomeric RNA sequences in K^+ solution. *J. Am. Chem. Soc.* **2009**, *131*, 2570–2578. [CrossRef] [PubMed]
13. Xu, Y.; Ishizuka, T.; Kimura, T.; Komiyama, M. A U-tetrad stabilizes human telomeric RNA G-quadruplex structure. *J. Am. Chem. Soc.* **2010**, *132*, 7231–7233. [CrossRef] [PubMed]
14. Collie, G.W.; Haider, S.M.; Neidle, S.; Parkinson, G.N. A crystallographic and modelling study of a human telomeric RNA (TERRA) quadruplex. *Nucleic Acids Res.* **2010**, *38*, 5569–5580. [CrossRef]
15. Xu, Y.; Sugiyama, H. Highly efficient photochemical 2′-deoxyribonolactone formation at the diagonal loop of a 5-iodouracil-containing antiparallel G-quartet. *J. Am. Chem. Soc.* **2004**, *126*, 6274–6279. [CrossRef] [PubMed]
16. Shrestha, P.; Jonchhe, S.; Emura, T.; Hidaka, K.; Endo, M.; Sugiyama, H.; Mao, H. Confined space facilitates G-quadruplex formation. *Nat. Nanotechnol.* **2017**, *12*, 582–588. [CrossRef] [PubMed]
17. Xu, Y.; Suzuki, Y.; Ito, K.; Komiyama, M. Telomeric repeat-containing RNA structure in living cells. *Proc. Natl. Acad. Sci. USA* **2010**, *107*, 14579–14584. [CrossRef]
18. Xu, Y.; Kimura, T.; Komiyama, M. Human telomere RNA and DNA form an intermolecular G-quadruplex. *Nucleic Acids Symp. Ser.* **2008**, *52*, 169–170. [CrossRef]
19. Xu, Y.; Suzuki, Y.; Komiyama, M. Click chemistry as an approach to probing G-quadruplex structure. *Angew. Chem. Int. Ed.* **2009**, *48*, 3281–3284. [CrossRef]
20. Xu, Y.; Ishizuka, T.; Yang, J.; Ito, K.; Katada, H.; Komiyama, M.; Hayashi, T. Oligonucleotide models of telomeric DNA and RNA form a hybrid G-quadruplex structure as a potential component of telomeres. *J. Biol. Chem.* **2012**, *287*, 41787–41796. [CrossRef]
21. Xu, Y. Recent progress in human telomere RNA structure and function. *Bioorg. Med. Chem. Lett.* **2018**, *28*, 2577–2584. [CrossRef] [PubMed]
22. Bao, H.L.; Xu, Y. Telomeric DNA-RNA-hybrid G-quadruplex exists in environmental conditions of HeLa cells. *Chem. Commun.* **2020**, *56*, 6547–6550. [CrossRef] [PubMed]
23. Guschlbauer, W.; Chantot, J.F.; Thiele, D. 4-Stranded nucleic-acid structures 25 years later -From guanosine gels to telomer DNA. *J. Biomol. Struct. Dyn.* **1990**, *8*, 491–511. [CrossRef] [PubMed]
24. Williamson, J.R. G-Quartet Structures in Telomeric DNA. *Annu. Rev. Biophys. Biomol. Struct.* **1994**, *23*, 703–730. [CrossRef] [PubMed]
25. Simonsson, T. G-quadruplex DNA structures--variations on a theme. *Biol. Chem.* **2001**, *382*, 621–628. [CrossRef] [PubMed]

26. Lane, A.N.; Chaires, J.B.; Gray, R.D.; Trent, J.O. Stability and kinetics of G-quadruplex structures. *Nucleic Acids Res.* **2008**, *36*, 5482–5515. [CrossRef] [PubMed]
27. Gilbert, D.E.; Feigon, J. Multistranded DNA structures. *Curr. Opin. Struct. Biol.* **1999**, *9*, 305–314. [CrossRef]
28. Phan, A.T.; Kuryavyi, V.; Patel, D.J. DNA architecture: From G to Z. *Curr. Opin. Struct. Biol.* **2006**, *16*, 288–298. [CrossRef]
29. Burge, S.; Parkinson, G.N.; Hazel, P.; Todd, A.K.; Neidle, S. Quadruplex DNA: Sequence, topology and structure. *Nucleic Acids Res.* **2006**, *34*, 5402–5415. [CrossRef]
30. Huppert, J.L. Four-stranded nucleic acids: Structure, function and targeting of G-quadruplexes. *Chem. Soc. Rev.* **2008**, *37*, 1375–1384. [CrossRef]
31. Mergny, J.L.; De Cian, A.; Ghelab, A.; Sacca, B.; Lacroix, L. Kinetics of tetramolecular quadruplexes. *Nucleic Acids Res.* **2005**, *33*, 81–94. [CrossRef]
32. Phan, A.T. Human telomeric G-quadruplex: Structures of DNA and RNA sequences. *FEBS J.* **2010**, *277*, 1107–1117. [CrossRef]
33. Williamson, J.R. Guanine quartets. *Curr. Opin. Struct. Biol.* **1993**, *3*, 357–362. [CrossRef]
34. Wang, Y.; Patel, D.J. Solution Structure of the Human Telomeric Repeat d[$AG_3(T_2AG_3)_3$] G-Tetraplex. *Structure* **1993**, *1*, 263–282 [CrossRef]
35. Xu, Y.; Noguchi, Y.; Sugiyama, H. The New Models of the Human Telomere d[AGGG(TTAGGG)3] in K^+ Solution. *Bioorg. Med. Chem.* **2006**, *14*, 5584–5591. [CrossRef]
36. Ambrus, A.; Chen, D.; Dai, J.; Bialis, T.; Jones, R.A.; Yang, D. Human Telomeric Sequence Forms a Hybrid-Type Intramolecular G-quadruplex Structure with Mixed Parallel/Antiparallel Strands in Potassium Solution. *Nucleic Acids Res.* **2006**, *34*, 2723–2735 [CrossRef]
37. Luu, K.N.; Phan, A.T.; Kuryavyi, V.; Lacroix, L.; Patel, D.J. Structure of the Human Telomere in K^+ Solution: An Intramolecular (3 + 1) G-quadruplex Scaffold. *J. Am. Chem. Soc.* **2006**, *128*, 9963–9970. [CrossRef]
38. Dai, J.; Carver, M.; Punchihewa, C.; Jones, R.A.; Yang, D. Structure of the Hybrid-2 Type Intramolecular Human Telomeric G-quadruplex in K^+ Solution: Insights into Structure Polymorphism of the Human Telomeric Sequence. *Nucleic Acids Res.* **2007**, *35*, 4927–4940. [CrossRef]
39. Parkinson, G.N.; Lee, M.P.; Neidle, S. Crystal Structure of Parallel Quadruplexes from Human Telomeric DNA. *Nature* **2002**, *417*, 876–880. [CrossRef] [PubMed]
40. Zhang, Z.; Dai, J.; Veliath, E.; Jones, R.A.; Yang, D. Structure of a two-G-tetrad intramolecular G-quadruplex formed by a variant human telomeric sequence in K^+ solution: Insights into the interconversion of human telomeric G-quadruplex structures. *Nucleic Acids Res.* **2010**, *38*, 1009–1021. [CrossRef] [PubMed]
41. Lim, K.W.; Amrane, S.; Bouaziz, S.; Xu, W.; Mu, Y.; Patel, D.J.; Luu, K.N.; Phan, A.T. Structure of the human telomere in K^+ solution: A stable basket-type G-quadruplex with only two G-tetrad layers. *J. Am. Chem. Soc.* **2009**, *131*, 4301–4309. [CrossRef] [PubMed]
42. Hansel, R.; Lohr, F.; Trantirek, L.; Dotsch, V. High-resolution insight into G-overhang architecture. *J. Am. Chem. Soc.* **2013**, *135*, 2816–2824. [CrossRef]
43. Liu, C.; Zhou, B.; Geng, Y.; Yan Tam, D.; Feng, R.; Miao, H.; Xu, N.; Shi, X.; You, Y.; Hong, Y.; et al. A chair-type G-quadruplex structure formed by a human telomeric variant DNA in K^+ solution. *Chem. Sci.* **2019**, *10*, 218–226. [CrossRef]
44. Shiekh, S.; Kodikara, S.G.; Balci, H. Structure, Topology, and Stability of Multiple G-quadruplexes in Long Telomeric Overhangs *J. Mol. Biol.* **2023**, *434*, 168205. [CrossRef]
45. Xu, Y.; Ishizuka, T.; Kurabayashi, K.; Komiyama, M. Consecutive Formation of G-Quadruplexes in Human Telomeric-Overhang DNA: A Protective Capping Structure for Telomere Ends. *Angew. Chem. Int. Ed.* **2009**, *48*, 7833–7836. [CrossRef]
46. Biffi, G.; Tannahill, D.; McCafferty, J.; Balasubramanian, S. Quantitative Visualization of DNA G-quadruplex Structures in Human Cells. *Nat. Chem.* **2013**, *5*, 182–186. [CrossRef]
47. Zhang, S.; Sun, H.; Wang, L.; Liu, Y.; Chen, H.; Li, Q.; Guan, A.; Liu, M.; Tang, Y. Real-Time Monitoring of DNA G-quadruplexes in Living Cells with a Small-Molecule Fluorescent Probe. *Nucleic Acids Res.* **2018**, *46*, 7522–7532. [CrossRef]
48. Bao, H.L.; Liu, H.S.; Xu, Y. Hybrid-type and two-tetrad antiparallel telomere DNA G-quadruplex structures in living human cells *Nucleic Acids Res.* **2019**, *47*, 4940–4947. [CrossRef] [PubMed]
49. Wu, G.; Chen, L.; Liu, W.; Yang, D. Molecular Recognition of the Hybrid-Type G-Quadruplexes in Human Telomeres. *Molecules* **2019**, *24*, 1578. [CrossRef] [PubMed]
50. Manna, S.; Sarkar, D.; Srivatsan, S.G. A Dual-App Nucleoside Probe Provides Structural Insights into the Human Telomeric Overhang in Live Cells. *J. Am. Chem. Soc.* **2018**, *140*, 12622–12633. [CrossRef]
51. Ma, D.; Zhang, Z.; Wang, M.; Lu, L.; Zhong, H.; Leung, C. Recent Developments in G-Quadruplex Probes. *Chem. Biol.* **2015**, *22*, 812–828. [CrossRef] [PubMed]
52. Xu, J.; Jiang, R.; He, H.; Ma, C.; Tang, Z. Recent advances on G-quadruplex for biosensing, bioimaging and cancer therapy. *Trends Anal. Chem.* **2021**, *139*, 116257. [CrossRef]
53. Chen, L.; Dickerhoff, J.; Sakai, S.; Yang, D. DNA G-Quadruplex in Human Telomeres and Oncogene Promoters: Structures Functions, and Small Molecule Targeting. *Acc. Chem. Res.* **2022**, *55*, 2628–2646. [CrossRef] [PubMed]
54. Yuan, J.H.; Shao, W.; Chen, S.B.; Huang, Z.S.; Tan, J.H. Recent advances in fluorescent probes for G-quadruplex nucleic acids *Biochem. Biophys. Res. Commun.* **2020**, *531*, 18–24. [CrossRef] [PubMed]

5. Raguseo, F.; Chowdhury, S.; Minard, A.; Antonio, M.D. Chemical-biology approaches to probe DNA and RNA G-quadruplex structures in the genome. *Chem. Commun.* **2020**, *56*, 1317–1324. [CrossRef] [PubMed]
6. Iida, K.; Suzuki, N.; Sasaki, A.; Ishida, S.; Arai, T. Development of a novel light-up probe for detection of G-quadruplexes in stress granules. *Sci. Rep.* **2022**, *12*, 12892. [CrossRef] [PubMed]
7. Holden, L.; Gkika, K.S.; Burke, C.S.; Long, C.; Keyes, T.E. Selective, Disruptive Luminescent Ru(II) Polypyridyl Probes of G-Quadruplex. *Inorg. Chem.* **2023**, *62*, 2213–2227. [CrossRef] [PubMed]
8. Zhang, L.; Liu, H.; Shao, Y.; Lin, C.; Jia, H.; Chen, G.; Yang, D.; Wang, Y. Selective Lighting Up of Epiberberine Alkaloid Fluorescence by Fluorophore-Switching Aptamer and Stoichiometric Targeting of Human Telomeric DNA G-Quadruplex Multimer. *Anal. Chem.* **2015**, *87*, 730–737. [CrossRef]
9. Chang, C.C.; Wu, J.Y.; Chien, C.W.; Wu, W.S.; Liu, H.; Kang, C.C.; Yu, L.J.; Chang, T.C. A Fluorescent Carbazole Derivative: High Sensitivity for Quadruplex DNA. *Anal. Chem.* **2003**, *75*, 6177–6183. [CrossRef]
0. Chang, C.C.; Kuo, I.C.; Ling, I.F.; Chen, C.T.; Chen, H.C.; Lou, P.J.; Lin, J.J.; Chang, T.C. Detection of Quadruplex DNA Structures in Human Telomeres by a Fluorescent Carbazole Derivative. *Anal. Chem.* **2004**, *76*, 4490–4494. [CrossRef]
1. Chang, C.C.; Chu, J.F.; Kao, F.J.; Chiu, Y.C.; Lou, P.J.; Chen, H.C.; Chang, T.C. Verification of Antiparallel G-Quadruplex Structure in Human Telomeres by Using Two-Photon Excitation Fluorescence Lifetime Imaging Microscopy of the 3,6-Bis(1-Methyl-4-Vinylpyridinium)Carbazole Diiodide Molecule. *Anal. Chem.* **2006**, *78*, 2810–2815. [CrossRef] [PubMed]
2. Liu, W.; Lin, C.; Wu, G.; Dai, J.; Chang, T.-C.; Yang, D. Structures of 1:1 and 2:1 Complexes of BMVC and MYC Promoter G-Quadruplex Reveal a Mechanism of Ligand Conformation Adjustment for G4-Recognition. *Nucleic Acids Res.* **2019**, *47*, 11931–11942. [CrossRef] [PubMed]
3. Liu, L.; Liu, W.; Wang, K.; Zhu, B.; Xia, X.; Ji, L.; Mao, Z. Quantitative Detection of G-Quadruplex DNA in Live Cells Based on Photon Counts and Complex Structure Discrimination. *Angew. Chem. Int. Ed.* **2020**, *59*, 9719–9726. [CrossRef] [PubMed]
4. Zheng, K.W.; Zhang, J.Y.; He, Y.D.; Gong, J.Y.; Wen, C.J.; Chen, J.N.; Hao, Y.H.; Zhao, Y.; Tan, Z. Detection of genomic G-quadruplexes in living cells using a small artificial protein. *Nucleic Acids Res.* **2020**, *48*, 11706–11720. [CrossRef] [PubMed]
5. Biver, T. Discriminating between Parallel, Anti-Parallel and Hybrid G-Quadruplexes: Mechanistic Details on Their Binding to Small Molecules. *Molecules* **2022**, *27*, 4165. [CrossRef] [PubMed]
6. Ma, Y.; Iida, K.; Nagasawa, K. Topologies of G-quadruplex: Biological functions and regulation by ligands. *Biochem. Biophys. Res. Commun.* **2020**, *531*, 3–17. [CrossRef] [PubMed]
7. Nagatsugi, F.; Onizuka, K. Functional G-Quadruplex Binding Molecules. *Chem. Lett.* **2020**, *49*, 771–780. [CrossRef]
8. Sun, D.; Thompson, B.; Cathers, B.E.; Salazar, M.; Kerwin, S.M.; Trent, J.O.; Jenkins, T.C.; Neidle, S.; Hurley, L.H. Inhibition of Human Telomerase by a G-Quadruplex-Interactive Compound. *J. Med. Chem.* **1997**, *40*, 2113–2116. [CrossRef]
9. Neidle, S.; Parkinson, G. Telomere Maintenance as a Target for Anticancer Drug Discovery. *Nat. Rev. Drug Discov.* **2002**, *1*, 383–393. [CrossRef]
0. Yang, D.; Okamoto, K. Structural Insights into G-Quadruplexes: Towards New Anticancer Drugs. *Future Med. Chem.* **2010**, *2*, 619–646. [CrossRef]
1. Wang, K.B.; Elsayed, M.S.A.; Wu, G.; Deng, N.; Cushman, M.; Yang, D. Indenoisoquinoline Topoisomerase Inhibitors Strongly Bind and Stabilize the MYC Promoter G-Quadruplex and Downregulate MYC. *J. Am. Chem. Soc.* **2019**, *141*, 11059–11070. [CrossRef] [PubMed]
2. Monsen, R.C. Higher-order G-quadruplexes in promoters are untapped drug targets. *Front. Chem.* **2023**, *11*, 1211512. [CrossRef] [PubMed]
3. Shu, H.; Zhang, R.; Xiao, K.; Yang, J.; Sun, X. G-Quadruplex-Binding Proteins: Promising Targets for Drug Design. *Biomolecules* **2022**, *12*, 648. [CrossRef] [PubMed]
4. Kosiol, N.; Juranek, S.; Brossart, P.; Heine, A.; Paeschke, K. G-quadruplexes: A promising target for cancer therapy. *Mol. Cancer* **2021**, *20*, 40. [CrossRef] [PubMed]
5. Teng, F.Y.; Jiang, Z.Z.; Guo, M.; Tan, X.Z.; Chen, F.; Xi, X.G.; Xu, Y. G-quadruplex DNA: A novel target for drug design. *Cell. Mol. Life Sci.* **2021**, *78*, 6557–6583. [CrossRef] [PubMed]
6. Zou, M.; Li, J.Y.; Zhang, M.J.; Li, J.H.; Huang, J.T.; You, P.D.; Liu, S.W.; Zhou, C.Q. G-quadruplex binder pyridostatin as an effective multi-target ZIKV inhibitor. *Int. J. Biol. Macromol.* **2021**, *190*, 178–188. [CrossRef]
7. Hu, X.X.; Wang, S.Q.; Gan, S.Q.; Liu, L.; Zhong, M.Q.; Jia, M.H.; Jiang, F.; Xu, Y.; Xiao, C.D.; Shen, X.C. A Small Ligand That Selectively Binds to the G-quadruplex at the Human Vascular Endothelial Growth Factor Internal Ribosomal Entry Site and Represses the Translation. *Front. Chem.* **2021**, *9*, 781198. [CrossRef]
8. Miglietta, G.; Marinello, J.; Russo, M.; Capranico, G. Ligands stimulating antitumour immunity as the next G-quadruplex challenge. *Mol. Cancer* **2022**, *21*, 180. [CrossRef]
9. Dhamodharan, V.; Pradeepkumar, P.I. Specific Recognition of Promoter G-Quadruplex DNAs by Small Molecule Ligands and Light-up Probes. *ACS Chem. Biol.* **2019**, *14*, 2102–2114. [CrossRef]
0. Dickerhoff, J.; Dai, J.; Yang, D. Structural recognition of the MYC promoter G-quadruplex by a quinoline derivative: Insights into molecular targeting of parallel G-quadruplexes. *Nucleic Acids Res.* **2021**, *49*, 5905–5915. [CrossRef]
1. Calabrese, D.R.; Chen, X.; Leon, E.C.; Gaikwad, S.M.; Phyo, Z.; Hewitt, W.M.; Alden, S.; Hilimire, T.A.; He, F.; Michalowski, A.M.; et al. Chemical and structural studies provide a mechanistic basis for recognition of the MYC G-quadruplex. *Nat. Commun.* **2018**, *9*, 4229. [CrossRef] [PubMed]

82. Neidle, S. Structured Waters Mediate Small Molecule Binding to G-Quadruplex Nucleic Acids. *Pharmaceuticals* **2022**, *15*, [CrossRef] [PubMed]
83. Xu, Y.; Suzuki, Y.; Lönnberg, T.; Komiyama, M. Human telomeric DNA sequence-specific cleaving by G-quadruplex formatio *J. Am. Chem. Soc.* **2009**, *131*, 2871–2874. [CrossRef] [PubMed]
84. Flynn, R.L.; Centore, R.C.; O'Sullivan, R.J.; Rai, R.; Tse, A.; Songyang, Z.; Chang, S.; Karlseder, J.; Zou, L. TERRA and HnRNPA Orchestrate an RPA-to-POT1 Switch on Telomeric Single-Stranded DNA. *Nature* **2011**, *471*, 532–536. [CrossRef] [PubMed]
85. Liu, X.; Ishizuka, T.; Bao, H.L.; Wada, K.; Takeda, Y.; Iida, K.; Nagasawa, K.; Yang, D.; Xu, Y. Structure-dependent Binding o hnRNPA1 to Telomere RNA. *J. Am. Chem. Soc.* **2017**, *139*, 7533–7539. [CrossRef] [PubMed]
86. Liu, X.; Xu, Y. HnRNPA1 Specifically Recognizes the Base of Nucleotide at the Loop of RNA G-Quadruplex. *Molecules* **201** *23*, 237. [CrossRef] [PubMed]
87. Agarwala, P.; Pandey, S.; Maiti, S. The tale of RNA G-quadruplex. *Org. Biomol. Chem.* **2015**, *13*, 5570–5585. [CrossRef]
88. Mendoza, O.; Porrini, M.; Salgado, G.F.; Gabelica, V.; Mergny, J.L. Orienting Tetramolecular G-Quadruplex Formation: The Ques for the Elusive RNA Antiparallel Quadruplex. *Chem. Eur. J.* **2015**, *21*, 6732–6739. [CrossRef]
89. Xiao, C.-D.; Shibata, T.; Yamamoto, Y.; Xu, Y. An intramolecular antiparallel G-quadruplex formed by human telomere RN *Chem. Commun.* **2018**, *54*, 3944–3946. [CrossRef]
90. Xiao, C.-D.; Ishizuka, T.; Xu, Y. Antiparallel RNA G-quadruplex Formed by Human Telomere RNA Containing 8-Bromoguanosin *Sci. Rep.* **2017**, *7*, 6695. [CrossRef]
91. Xiao, C.-D.; Ishizuka, T.; Zhu, X.-Q.; Li, Y.; Sugiyama, H.; Xu, Y. Unusual Topological RNA Architecture with an Eight-Strande Helical Fragment Containing A-, G-, and U-Tetrads. *J. Am. Chem. Soc.* **2017**, *139*, 2565–2568. [CrossRef] [PubMed]
92. Banco, M.T.; Ferré-D'amaeré, A.R. The emerging structural complexity of G-quadruplex RNAs. *RNA* **2021**, *27*, 390–402. [CrossRef [PubMed]
93. Sundquist, W.I.; Heaphy, S. Evidence for interstrand quadruplex formation in the dimerization of human immunodeficienc virus 1 genomic RNA. *Proc. Natl. Acad. Sci. USA* **1993**, *90*, 3393–3397. [CrossRef] [PubMed]
94. Marquet, R.; Paillart, J.C.; Skripkin, E.; Ehresmann, C.; Ehresmann, B. Dimerization of human immunodeficiency virus type RNA involves sequences located upstream of the splice donor site. *Nucleic Acids Res.* **1994**, *22*, 145–151. [CrossRef] [PubMed]
95. Piekna-Przybylska, D.; Sharma, G.; Bambara, R.A. Mechanism of HIV-1 RNA Dimerization in the Central Region of the Genom and Significance for Viral Evolution. *J. Biol. Chem.* **2013**, *288*, 24140–24150. [CrossRef] [PubMed]
96. Shen, W.; Gao, L.; Balakrishnan, M.; Bambara, R.A. A Recombination Hot Spot in HIV-1 Contains Guanosine Runs That Ca Form a G-quartet Structure and Promote Strand Transfer in Vitro. *J. Biol. Chem.* **2009**, *284*, 33883–33893. [CrossRef] [PubMed]
97. Gomez, D.; Guédin, A.; Mergny, J.-L.; Salles, B.; Riou, J.-F.; Teulade-Fichou, M.-P.; Calsou, P. A G-quadruplex structure within th 5′-UTR of TRF2 mRNA represses translation in human cells. *Nucleic Acids Res.* **2010**, *38*, 7187–7198. [CrossRef]
98. Norseen, J.; Johnson, F.B.; Lieberman, P.M. Role for G-Quadruplex RNA Binding by Epstein-Barr Virus Nuclear Antigen 1 i DNA Replication and Metaphase Chromosome Attachment. *J. Virol.* **2009**, *83*, 10336–10346. [CrossRef]
99. Wang, Q.; Liu, J.-q.; Chen, Z.; Zheng, K.-w.; Chen, C.-y.; Hao, Y.-h.; Tan, Z. G-quadruplex formation at the 3′ end of telomere DN inhibits its extension by telomerase, polymerase and unwinding by helicase. *Nucleic Acids Res.* **2011**, *39*, 6229–6237. [CrossRef]
100. Arora, A.; Dutkiewicz, M.; Scaria, V.; Hariharan, M.; Maiti, S.; Kurreck, J. Inhibition of translation in living eukaryotic cells by a RNA G-quadruplex motif. *RNA* **2008**, *14*, 1290–1296. [CrossRef]
101. Patel, D.J.; Phan, A.T.; Kuryavy, V. Human telomere, oncogenic promoter and 5′-UTR G-quadruplexes: Diverse higher orde DNA and RNA targets for cancer therapeutics. *Nucleic Acids Res.* **2007**, *35*, 7429–7455. [CrossRef] [PubMed]
102. Mullen, M.A.; Olson, K.J.; Dallaire, P.; Major, F.; Assmann, S.M.; Bevilacqua, P.C. RNA G-Quadruplexes in the model plant specie *Arabidopsis thaliana*: Prevalence and possible functional roles. *Nucleic Acids Res.* **2010**, *38*, 8149–8163. [CrossRef] [PubMed]
103. Morris, M.J.; Basu, S. An unusually stable G-quadruplex within the 5′-UTR of the MT3 matrix metalloproteinase mRNA represse translation in eukaryotic cells. *Biochemistry* **2009**, *48*, 5313–5319. [CrossRef] [PubMed]
104. Gros, J.; Guédin, A.; Mergny, J.-L.; Lacroix, L. G-Quadruplex formation interferes with P1 helix formation in the RNA componer of telomerase hTERC. *ChemBioChem* **2008**, *9*, 2075–2079. [CrossRef] [PubMed]
105. Ribeiro, M.M.; Teixeira, G.S.; Martins, L.; Marques, M.R.; de Souza, A.P.; Line, S.R.P. G-quadruplex formation enhances splicin efficiency of *PAX9* intron 1. *Hum. Genet.* **2015**, *134*, 37–44. [CrossRef]
106. Huang, H.; Zhang, J.; Harvey, S.E.; Hu, X.; Cheng, C. RNA G-quadruplex secondary structure promotes alternative splicing vi the RNA-binding protein hnRNPF. *Genes Dev.* **2023**, *31*, 2296–2309. [CrossRef]
107. Jara-Espejo, M.; Fleming, A.M.; Burrows, C.J. Potential G-Quadruplex Forming Sequences and N^6-Methyladenosine Colocalize a Human Pre-mRNA Intron Splice Sites. *ACS Chem. Biol.* **2020**, *15*, 1292–1300. [CrossRef]
108. Georgakopoulos-Soares, I.; Parada, G.E.; Wong, H.Y.; Medhi, R.; Furlan, G.; Munita, R.; Miska, E.A.; Kwok, C.K.; Hemberg, M Alternative splicing modulation by G-quadruplexes. *Nat. Commun.* **2022**, *13*, 2404. [CrossRef]
109. Asamitsu, S.; Yabuki, Y.; Ikenoshita, S.; Kawakubo, K.; Kawasaki, M.; Usuki, S.; Nakayama, Y.; Adachi, K.; Kugoh, H.; Ishii, K et al. CGG repeat RNA G-quadruplexes interact with FMRpolyG to cause neuronal dysfunction in fragile X-related tremor/ataxi syndrome. *Sci. Adv.* **2021**, *7*, eabd9440. [CrossRef]
110. Mcrae, E.K.S.; Dupas, S.J.; Booy, E.P.; Piragasam, R.S.; Fahlman, R.P.; Mckenna, S.A. An RNA guanine quadruplex regulate pathway to TRAIL-sensitization by DDX21. *RNA* **2020**, *26*, 44–57. [CrossRef]

111. Ji, D.; Juhas, M.; Tsang, C.M.; Kwok, C.K.; Li, Y.; Zhang, Y. Discovery of G-quadruplex-forming sequences in SARS-CoV-2. *Brief. Bioinform.* **2021**, *22*, 1150–1160. [CrossRef] [PubMed]
112. Bao, H.L.; Ishizuka, T.; Sakamoto, T.; Fujimoto, K.; Uechi, T.; Kenmochi, N.; Xu, Y. Investigation of higher-order RNA G-quadruplex structures in vitro and in living cells by ^{19}F NMR spectroscopy. *Nucleic Acids Res.* **2017**, *45*, 5501–5511. [CrossRef] [PubMed]
113. Ishizuka, T.; Zhao, P.Y.; Bao, H.L.; Xu, Y. A Multi-functional Guanine Derivative for Studying DNA G-quadruplex Structure. *Analyst* **2017**, *142*, 4083–4088. [CrossRef] [PubMed]
114. Ishizuka, T.; Yamashita, A.; Asada, Y.; Xu, Y. Studying DNA G-quadruplex aptamer by ^{19}F NMR. *ACS Omega* **2017**, *2*, 8843–8848. [CrossRef] [PubMed]
115. Bao, H.L.; Ishizuka, T.; Iwanami, A.; Oyoshi, T.; Xu, Y. A Simple and Sensitive ^{19}F NMR Approach for Studying the Interaction of RNA G-Quadruplex with Ligand Molecule and Protein. *ChemistrySelect* **2017**, *2*, 4170–4175. [CrossRef]
116. Bao, H.L.; Xu, Y. Investigation of higher-order RNA G-quadruplex structures in vitro and in living cells by ^{19}F NMR spectroscopy. *Nat. Protoc.* **2018**, *13*, 652–665. [CrossRef] [PubMed]
117. Bao, H.L.; Ishizuka, T.; Yamashita, A.; Furukoji, E.; Asada, Y.; Xu, Y. Improving thermodynamic stability and anticoagulant activity of thrombin binding aptamer by incorporation of 8-trifluoromethyl-2′-deoxyguanosine. *J. Med. Chem.* **2021**, *1*, 711–718. [CrossRef]
118. Chen, X.; Chen, S.; Jing Dai, J.; Yuan, J.; Ou, T.; Huang, Z.; Tan, J. Tracking the Dynamic Folding and Unfolding of RNAG-Quadruplexes in Live Cells. *Angew. Chem. Int. Ed.* **2018**, *57*, 4702–4706. [CrossRef]
119. Laguerre, A.; Stefan, L.; Larrouy, M.; Genest, D.; Novotna, J.; Pirrotta, M.; Monchaud, D. Twice-As-Smart Synthetic G-Quartet: PyroTASQ Is Both a Smart Quadruplex Ligand and a Smart Fluorescent Probe. *J. Am. Chem. Soc.* **2014**, *136*, 12406–12414. [CrossRef]
120. Malina, J.; Scott, P.; Brabec, V. Stabilization of human telomeric RNA G-quadruplex by the water-compatible optically pure and biologically-active metallohelices. *Sci. Rep.* **2020**, *10*, 14543. [CrossRef]
121. Chan, C.Y.; Kwok, C.K. Specific Binding of a d-RNA G-Quadruplex Structure with an l-RNA Aptamer. *Angew. Chem. Int. Ed.* **2020**, *59*, 5293–5297. [CrossRef] [PubMed]
122. Ferino, A.; Nicoletto, G.; D'Este, F.; Zorzet, S.; Lago, S.; Richter, S.N.; Tikhomirov, A.; Shchekotikhin, A.; Xodo, L.E. Photodynamic Therapy for ras-Driven Cancers: Targeting G-Quadruplex RNA Structures with Bifunctional Alkyl-Modified Porphyrins. *J. Med. Chem.* **2020**, *63*, 1245–1260. [CrossRef] [PubMed]
123. Xu, Y.; Suzuki, Y.; Ishizuka, T.; Xiao, C.-D.; Liu, X.; Xiao, C.-D.; Komiyama, M. Finding a Human Telomere DNA-RNA Hybrid G-Quadruplex Formed by Human Telomeric 6-mer RNA and 16-mer DNA using Click Chemistry: A Protective Structure for Telomere End. *Bioorg. Med. Chem.* **2014**, *22*, 4419–4421. [CrossRef]
124. Choi, B.E.; Lee, H.T. DNA-RNA hybrid G-quadruplex tends to form near the 3′ end of telomere overhang. *Biophys. J.* **2022**, *121*, 2962–2980. [CrossRef] [PubMed]
125. Zhang, J.; Xia, Y.; Hao, Y.; Tan, Z. DNA:RNA hybrid G-quadruplex formation upstream of transcription start site. *Sci. Rep.* **2020**, *10*, 7429. [CrossRef]
126. Ren, C.; Duan, R.; Hao, Y.; Tan, Z. Genome-Wide Formation of DNA:RNA Hybrid G-Quadruplexes in Live Yeast Cells. *bioRxiv* **2023**. [CrossRef]
127. Zaug, A.J.; Podell, E.R.; Cech, T.R. Human POT1 disrupts telomeric G-quadruplexes allowing telomerase extension in vitro. *Proc. Natl. Acad. Sci. USA* **2005**, *102*, 10864–10869. [CrossRef]
128. Wang, H.; Nora, G.J.; Ghodke, H.; Opresko, P. Single molecule studies of physiologically relevant telomereic tails reveals Pot1 mechanism for promoting G-quadruplex unfolding. *J. Biol. Chem.* **2011**, *286*, 7479–7489. [CrossRef]
129. Zahler, A.M.; Williamson, J.R.; Cech, T.R.; Prescott, D.M. Inhibition of telomerase by G-quartet DNA structures. *Nature* **1991**, *350*, 718–720. [CrossRef]
130. Oganesian, L.; Moon, I.K.; Bryan, T.M.; Jarstfer, M.B. Extension of G-quadruplex DNA by ciliate telomerase. *EMBO J.* **2006**, *25*, 1148–1159. [CrossRef]
131. Moye, A.L.; Porter, K.C.; Cohen, S.B.; Phan, T.; Zyner, K.G.; Sasaki, N.; Lovrecz, G.O.; Beck, J.L.; Bryan, T.M. Telomeric G-quadruplexes are a substrate and site of localization for human telomerase. *Nat. Commun.* **2015**, *6*, 7643. [CrossRef] [PubMed]
132. De Cian, A.; Cristofari, G.; Reichenbach, P.; De Lemos, E.; Monchaud, D.; Teulade-Fichou, M.P.; Shin-Ya, K.; Lacroix, L.; Lingner, J.; Mergny, J.L. Reevaluation of telomerase inhibition by quadruplex ligands and their mechanisms of action. *Proc. Natl. Acad. Sci. USA* **2007**, *104*, 17347–17352. [CrossRef]
133. Hwang, H.; Kreig, A.; Calvert, J.; Lormand, J.; Kwon, Y.; Daley, J.M.; Sung, P.; Opresko, P.L.; Myong, S. Telomeric overhang length determines structural dynamics and accessibility to telomerase and ALT-associated proteins. *Structure* **2014**, *22*, 842–853. [CrossRef] [PubMed]
134. Lee, H.T.; Bose, A.; Lee, C.Y.; Opresko, P.L.; Myong, S. Molecular mechanisms by which oxidative DNA damage promotes telomerase activity. *Nucleic Acids Res.* **2017**, *45*, 11752–11765. [CrossRef] [PubMed]
135. Takahama, K.; Takada, A.; Shota Tada, S.; Shimizu, M.; Sayama, K.; Kurokawa, R.; Oyoshi, T. Regulation of Telomere Length by G-Quadruplex Telomere DNA and TERRA-Binding Protein TLS/FUS. *Chem. Biol.* **2013**, *20*, 341–550. [CrossRef] [PubMed]

136. Mei, Y.; Deng, Z.; Vladimirova, O.; Gulve, N.; Johnson, F.B.; Drosopoulos, W.C.; Schildkraut, C.L.; Lieberman, P.M. TERRA G-quadruplex RNA interaction with TRF2 GAR domain is required for telomere integrity. *Sci. Rep.* **2021**, *11*, 3509. [CrossRef] [PubMed]
137. Yang, S.Y.; Emily Chang, E.Y.C.; Lim, J.; Kwan, H.H.; Monchaud, D.; Yip, S.; Stirling, P.C.; Wong, J.M.Y. G-quadruplexes mark alternative lengthening of telomeres. *NAR Cancer* **2021**, *3*, zcab031. [CrossRef]
138. Yadav, T.; Zhang, J.M.; Ouyang, J.; Leung, W.; Simoneau, A.; Zou, L. TERRA and RAD51AP1 promote alternative lengthening of telomeres through an R- to D-loop switch. *Mol. Cell* **2022**, *82*, 3985–4000. [CrossRef]
139. Tsai, R.X.; Fang, K.C.; Yang, P.C.; Hsieh, Y.H.; Chiang, I.T.; Chen, Y.; Lee, H.G.; Lee, J.T.; Chu, H.C. TERRA regulates DNA G-quadruplex formation and ATRX recruitment to chromatin. *Nucleic Acids Res.* **2022**, *50*, 12217–12234. [CrossRef]
140. Xu, Y.; Ito, K.; Komiyama, M. A 6-mer photocontrolled oligonucleotide as an effective telomerase inhibitor. *J. Am. Chem. Soc* **2010**, *132*, 631–637. [CrossRef]
141. Ito, K.; Go, S.; Komiyama, M.; Xu, Y. Inhibition of Translation by Small RNA-Stabilized mRNA Structures in Human Cells. *J. Am Chem. Soc.* **2011**, *133*, 19153–19159. [CrossRef] [PubMed]
142. de Silanes, I.L.; d'Alcontres, M.S.; Blasco, M.A. TERRA transcripts are bound by a complex array of RNA-binding proteins. *Nat Commun.* **2010**, *1*, 33. [CrossRef] [PubMed]
143. Eulalio, A.; Behm-Ansmant, I.; Izaurralde, E. P bodies: At the crossroads of post-transcriptional pathways. *Nat. Rev. Mol. Cell Biol.* **2007**, *8*, 9–22. [CrossRef] [PubMed]
144. Deng, Z.; Norseen, J.; Wiedmer, A.; Riethman, H.P.; Lieberman, M. TERRA RNA Binding to TRF2 Facilitates Heterochromatin Formation and ORC Recruitment at Telomeres. *Mol. Cell* **2009**, *35*, 403–413. [CrossRef]
145. Collie, G.W.; Sparapani, S.; Parkinson, G.N.; Neidle, S. Structural Basis of Telomeric RNA Quadruplex−Acridine Ligand Recognition. *J. Am. Chem. Soc.* **2011**, *133*, 2721–2728. [CrossRef]
146. Wynford-Thomas, D.; Kipling, D. The end-replication problem. *Nature* **1997**, *389*, 551–552. [CrossRef]
147. Bonnell, E.; Pasquier, E.; Wellinger, R.J. Telomere Replication: Solving Multiple End Replication Problems. *Front. Cell Dev. Biol* **2021**, *9*, 668171. [CrossRef]

molecules

MDPI

rticle

Absorption, Distribution, Metabolism, and Excretion of [14C]BS1801, a Selenium-Containing Drug Candidate, in Rats

Cheng Yang [1,2], Mingzhen Xue [3], Yifei He [2], Hanwei Yin [4], Chen Yang [2], Dafang Zhong [2], Huihui Zeng [4,*], Yuandong Zheng [2,*] and Xingxing Diao [1,2,*]

1 School of Chinese Materia Medica, Nanjing University of Chinese Medicine, Nanjing 210023, China; yc2550983083@163.com
2 Shanghai Institute of Materia Medica, Chinese Academy of Sciences, Shanghai 201210, China; s20-heyifei@simm.ac.cn (Y.H.); yangchen@simm.ac.cn (C.Y.); dfzhong@simm.ac.cn (D.Z.)
3 Jiangsu Key Laboratory for Functional Substances of Chinese Medicine, Nanjing University of Chinese Medicine, Nanjing 210023, China; xuemingzhen@foxmail.com
4 Shanghai Yuanxi Pharmaceutical Technology Co., Ltd., Shanghai 201203, China; yinhanwei@yxmedicine.com
* Correspondence: zenghh@bjmu.edu.cn (H.Z.); ydzheng@simm.ac.cn (Y.Z.); xxdiao@simm.ac.cn (X.D.)

Abstract: BS1801 is a selenium-containing drug candidate with potential for treating liver and lung fibrosis. To fully elucidate the biotransformation of BS1801 in animals and provide sufficient preclinical drug metabolism data for human mass balance study, the metabolism of BS1801 in rats was investigated. We used radiolabeling techniques to investigate the mass balance, tissue distribution, and metabolite identification of BS1801 in Sprague–Dawley/Long–Evans rats after a single oral dose of 100 mg/kg (100 μCi/kg) [^{14}C]BS1801: 1. The mean recovery of radioactive substances in urine and feces was 93.39% within 168 h postdose, and feces were the main excretion route. 2. Additionally, less than 1.00% of the dose was recovered from either urine or bile. 3. BS1801-related components were widely distributed throughout the body. 4. Fifteen metabolites were identified in rat plasma, urine, feces, and bile, and BS1801 was detected only in feces. 5. BS1801-M484, the methylation product obtained via a N–Se bond reduction in BS1801, was the most abundant drug-related component in plasma. The main metabolic pathways of BS1801 were reduction, amide hydrolysis, oxidation, and methylation. Overall, BS1801 was distributed throughout the body, and excreted mainly as an intact BS1801 form through feces. No differences were observed between male and female rats in distribution, metabolism, and excretion of BS1801.

Keywords: [^{14}C]BS1801; BS1801 (butaselen); tissue distribution; metabolism; mass balance; selenium

1. Introduction

The thioredoxin (Trx) system, which consists of Trx, thioredoxin reductase (TrxR), and nicotinamide adenine dinucleotide phosphate (NADPH), is a key antioxidant system that defends against oxidative stress by regulating the protein dithiol/disulfide balance through its disulfide reductase activity [1]. TrxR/Trx is involved in the development of various diseases, especially cancer [2,3]. In addition, the Trx system is essential for normal cellular function, so it is necessary to develop its inhibitors as a therapeutic strategy [4]. TrxR is a selenocysteine-containing enzyme responsible for maintaining redox homeostasis in cells and plays a key role in regulating diverse cellular redox signaling pathways [5]. However, few pharmaceutical companies have been actively involved in the development of TrxR enzyme inhibitors, and the inhibitors described in the literature and patents fall into the following main categories: metal complexes, Michael receptors, sulfur/selenium (Se)-containing compounds, and others [4]. Nevertheless, most of the disclosed inhibitors were developed based on the modification of the selenocysteine residue (the selenol group) of TrxR, which has an overall similar but relatively stronger response (nucleophilicity and metal-binding property) than that of its counterpart cysteine residue (the thiol group) [2,4,6]. Importantly, the concentration of thiols in cells is much higher than that of selenols, so the abundance of thiols compensates for their low reactivity; as a result, the inhibitor may undergo cross-reactivity with thiols [4]. Therefore, inhibitors of TrxR should be further developed.

In recent years, Se-containing compounds have attracted increasing attention. Se is an essential trace element for many biological functions and a component of selenoproteins [7]. It is generally believed that Se has multiple positive effects on human health [8,9]. A study summarizing more than 100 Se-containing compounds found that Se-containing compounds exhibit a variety of anticancer, antioxidant, and antifibrotic activities and that incorporating Se atoms into small molecules significantly increases their biological activity [6]. Different structures of Se derivatives have different efficacy [6,7]. Compared with the toxicity and adverse reactions of inorganic selenium compounds, organic selenium derivatives have attracted more attention, and many organic selenium compounds have been found to have strong anti-tumor activity [6,10,11]. Selenium-containing drugs are effective chemopreventive and chemotherapeutic agents for cancer [6,12]. Se is also ingested in the form of dietary supplements [13]. Additionally, the presence of Se affects the pharmacokinetics, metabolism, and detection of drugs. Se contains six stable isotopes with different abundances, which form characteristic peak clusters and facilitate the identification of compounds containing Se.

Ebselen (2-phenyl-1,2-benzisoselenazol-3(2H)-one), a nontoxic organoselenium compound discovered in the 1980s, was used to prevent and treat a variety of diseases because of its wide range of targets and ability to modulate a variety of biological processes [9,14,15]. The treatment of stroke in 1997 was not approved in Japan due to insufficient efficacy data [14]. However, researchers have not stopped exploring the compound, and it is currently in clinical stage III [16]. BBSKE (1,2-[bis(1,2-benzisoselenazolone-3 (2H)-ketone)]-ethane) is a heterocyclic organoselenium compound that has been reported to exhibit antitumor effects and is a novel antitumor agent [17]. Ebselen and BBSKE had a high binding rate with plasma proteins, resulting in their inability to perform pharmacokinetic studies by directly quantifying the parent drug. Several studies quantified the plasma metabolites (selenol derivatives) of ebselen or BBSKE via different sample pretreatment methods or the LC–MS method combined with other methods to assess their pharmacokinetics [18–20].

BS1801 (butaselen), a selenium-containing drug candidate, is a novel TrxR inhibitor developed by Shanghai Yuanxi Pharmaceutical Technology Co., Ltd. It is a structural analog of ebselen and BBSKE, all of which contain Se. Ebselen exerts its effects by mimicking glutathione peroxidase activity, whereas BS1801 and BBSKE exert their effects by inhibiting TrxR activity [21–23]. Although ebselen has been included in a number of clinical trials in the United States and Asia, there is no specific prospect of clinical application [24]. Studies on BBSKE have focused more on its anticancer activity, while the current drug candidate BS1801 has been proposed for the clinical treatment of fibrosis and is currently in phase I clinical trials [17,25,26].

To our knowledge, there are only two studies on the metabolism and pharmacokinetics related to BS1801. In the first study, the covalent binding product of BS1801 and human serum albumin was described, as well as the mechanism of its reduction and methylation [27]. In another study, M2 (BS1801-M484) was identified as the major metabolite of BS1801 in human plasma via mass spectrometry response and UV absorption peak area, and the pharmacokinetics of M2 were evaluated [25]. However, the mass spectrometry and UV response do not accurately reflect the abundance of metabolites since they are susceptible to structural modifications of metabolites.

Among the available studies, there were no accurate data to investigate the proportion of metabolites in BS1801, so the safety of major metabolites cannot be evaluated. At the same time, there was a lack of tissue distribution and mass balance data for BS1801 in vivo. Since the concentration of the drug in the target organ is directly related to the efficacy and toxicity, tissue distribution must be investigated in animals to further advance the research progress of BS1801 for eventual marketing [28]. A mass balance study of radiolabeled BS1801 in humans is also needed. In the draft *Clinical Pharmacology Considerations for*

Human Radiolabeled Mass Balance Studies published by the US FDA in May 2022, it was noted that the dose of human radioactive administration should be estimated based on animal research data [29]. Therefore, the radiolabeled mass balance study in animals must be performed before human-radiolabeled mass balance study.

In this study, the distribution, metabolism, and excretion of [^{14}C]BS1801(Figure 1) in rats were studied. We evaluated the overall metabolism of BS1801 in rats, providing a strong basis for human mass balance study and promoting clinical research.

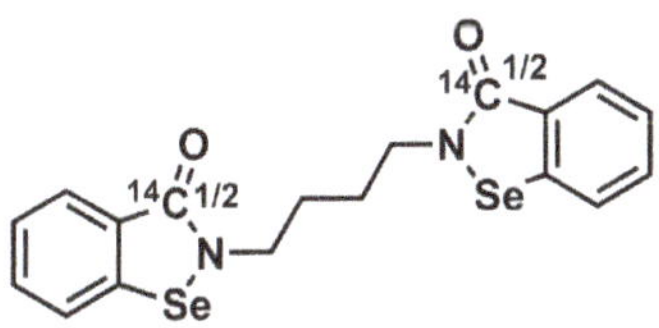

Figure 1. Chemical structure of [^{14}C]BS1801.

2. Results

2.1. Mass Balance

The mean radioactive recovery in urine and feces was 93.39% in SD rats after a single oral administration from 0 to 168 h. Among them, feces was the main excretion route, and the cumulative excretion of radioactive substances accounted for 92.77% of the dose, while urine accounted for only 0.62%. The mean radioactive recovery in urine and feces within 0–48 h was 91.57% after administration. The cumulative radioactive recoveries of urine and feces are provided in Figure 2A (detailed data are provided in Supplemental Table S2).

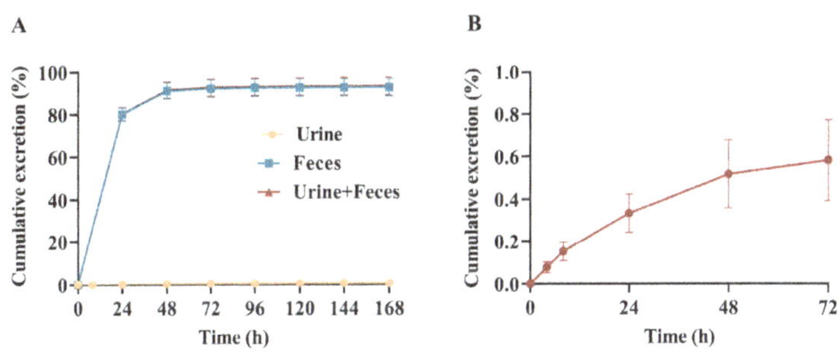

Figure 2. Mean cumulative excretion of total radioactivity (3 male and 3 female SD rats) in urine and feces (**A**) and bile (**B**) in SD rats after a single oral dose of 100 mg/kg (100 μCi/kg) [^{14}C]BS1801.

The recovery of radioactivity in bile within 72 h after a single intragastric administration in SD rats undergoing BDC was 0.58% (Figure 2B and detailed data are provided in Supplemental Table S2). The drug excreted through the feces consisted mainly of the part absorbed via the bile and the unabsorbed part, and it was clear from the biliary excretion data that BS1801 in feces was predominantly unabsorbed.

2.2. Tissue Distribution

All tissues of twenty-four LE rats (3 male and 3 female at each time point) were collected at 2, 10, 48, and 96 h after intragastric administration. The drug-related substances in male rats were mainly concentrated in the liver, stomach, large intestine, and eyes (Figure 3A). The ratios of $AUC_{0–96h}$ in these tissues to plasma $AUC_{0–96h}$ were 11.80, 10.60, 7.48, and 5.35, respectively (Supplemental Table S3). In female rats, drug-related substances were mainly concentrated in the large intestine and liver, and the $AUC_{0–96h}$ ratios of these tissues to the plasma $AUC_{0–96h}$ were 16.82 and 6.59, respectively (Figure 3B and Supplemental Table S3). The ratios of other tissues $AUC_{0–96h}$ to plasma $AUC_{0–96h}$ were lower than 5.00 postdose in all rats (Supplemental Table S3).

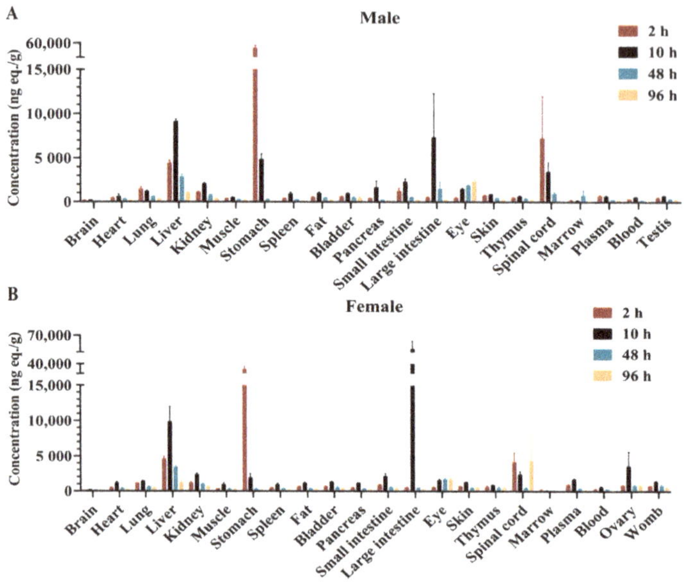

Figure 3. Tissue concentration–time distribution in male (**A**) and female (**B**) LE rats after a single oral dose of 100 mg/kg (100 μCi/kg) [^{14}C]BS1801.

In male LE rats, the T_{max} was 2 or 10 h for all tissues except the eyes and marrow. In female LE rats, the T_{max} was 2 or 10 h for all tissues except the eyes and spinal cord (Figure 3A,B).

The average ratio of blood $AUC_{0–96h}$ to plasma $AUC_{0–96h}$ was 0.73 in male rats post-dose, while that of female rats was 0.45 (Supplemental Table S3), indicating that the drug-related substances preferred to not be distributed in blood cells.

2.3. Metabolite Identification

Figure 4 presents radio-chromatograms of pooled plasma (0–72 h), urine (0–96 h), feces (0–48 h), and bile (0–48 h) for male and female rats. A total of twelve metabolites in male rats and thirteen metabolites in female rats were identified after drug administration. Supplemental Table S4 summarizes the proposed metabolic pathways, elemental composition, determined m/z ([M + H]$^+$), mass error (ppm), and characteristic fragment ions for BS1801 and its metabolites.

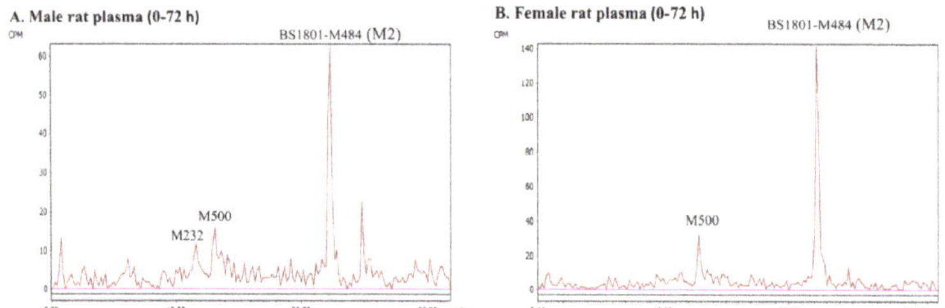

Figure 4. *Cont.*

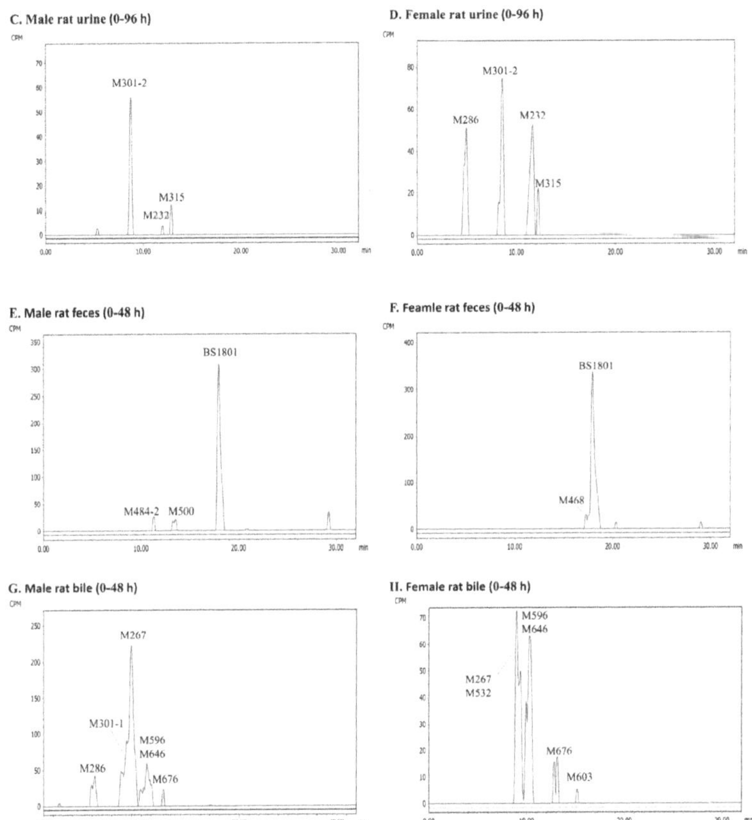

Figure 4. Representative radio-chromatograms of metabolites of BS1801 after a single oral dose of 100 mg/kg (100 μCi/kg) [^{14}C]BS1801 in male and female SD rats. (**A**) Male rat plasma (0–72 h); (**B**) female rat plasma (0–72 h); (**C**) male rat urine (0–96 h); (**D**) female rat urine (0–96 h); (**E**) male rat feces (0–48 h); (**F**) female rat feces (0–48 h); (**G**) male rat bile (0–48 h); (**H**) female rat bile (0–48 h).

2.3.1. HR-MS Analysis of BS1801 and M2 Reference Standards

BS1801. Under the experimental conditions of this study, BS1801 ($C_{18}H_{16}N_2O_2Se_2$) eluted at 17.65 min, and the molecular ion ([M + H]$^+$) obtained in the full-scan MS was m/z 452.9617. The major fragment ions of BS1801 were m/z 254.0079, 211.9608, 199.9608, and 184.9501. The fragment ions m/z 254.0079 and 199.9608 were mainly produced by C–N bond cleavage. The C–C bond breakage produced a fragment ion m/z 211.9608. The fragment ion m/z 184.9501 was produced by the amide bond breakage and N–Se bond breakage (Figure 5A).

M2. M2 was the product of BS1801 reduction and further methylation, and its retention time was 22.28 min. M2 yielded a molecular ion ([M + H]$^+$) m/z 485.0242 and its major fragment ions m/z 270.0392, 198.9656, 183.9418, and 72.0815 in full-scan MS. The cleavage of the C–N bond produced a fragment ion m/z 270.0392, and the cleavage of the amide bond produced a fragment ion m/z 198.9656. Amide bond breakage and the loss of $CH_3\cdot$ radical produced the fragment ion m/z 183.9418. The fragment ion m/z 72.0815 was produced by C–N bond breakage (Figure 5B).

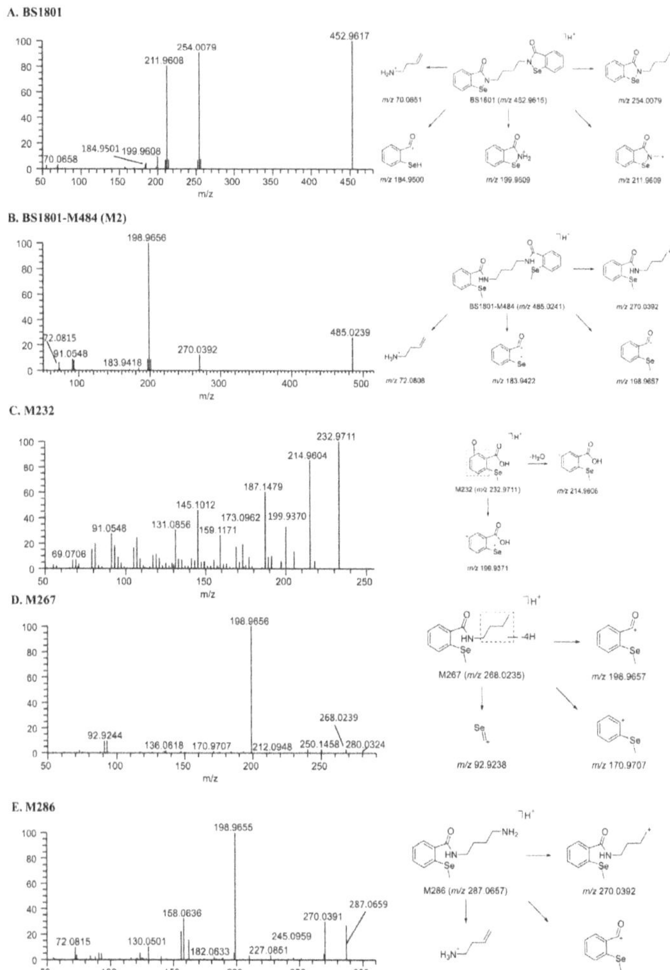

Figure 5. Mass spectra of metabolites and their proposed fragmentation profiles. (**A**) BS1801; (**B**) BS1801-M484 (M2); (**C**) M232; (**D**) M267; (**E**) M286.

2.3.2. Plasma

BS1801 was not detected in $AUC_{0\text{–}72h}$ pooled male rat plasma. Three metabolites were identified via HR-MS and radio-chromatographic peaks. Among them, M2 accounted for 65.41% of the total radioactivity and was the most abundant metabolite. M500 and M232 accounted for 20.07% and 11.59% of the total radioactivity, respectively. In female rats, BS1801 was also undetected in plasma. Only two metabolites, M2 and M500, were detected, accounting for 81.33% and 14.00% of the total plasma radioactivity, respectively (Figure 4, and Supplemental Table S5).

2.3.3. Urine

The cumulative excretion of radioactive substances in urine was only 0.62% of the dose, as shown in the mass balance study. A total of three radioactive chromatographic peaks were identified in 0–96 h pooled male rat urine sample, and the parent drug BS1801 was not detected. M232, M301-2, and M315 accounted for 0.018%, 0.41%, and 0.068% of the dose, respectively. Four radioactive peaks were identified in 0–96 h pooled female rat urine sample, and BS1801 was not detected. M232, M286, M301-2, and M315 accounted for 0.23%, 0.20%, 0.21%, and 0.046% of the dose, respectively (Figure 4, and Supplemental Table S5).

2.3.4. Feces

According to the mass balance study, feces was the main excretion route. The cumulative excretion of radioactive substances in male rats feces accounted for 92.32% of the dose, and BS1801 accounted for 76.97% of the dose. Two metabolites were identified by combining HR-MS and radio-chromatographic peaks. Among them, M500 was the major metabolite, accounting for 5.73% of the dose; M484-2 was the minor metabolite, accounting for 4.00% of the dose. The cumulative excretion of radioactive substances in the feces of female rats accounted for 93.22% of the dose, and BS1801 accounted for 81.89% of the dose. Only the metabolite M468 was identified by combining HR-MS and radio-chromatographic peaks, accounting for 7.77% of the dose (Figure 4, and Supplemental Table S5).

2.3.5. Bile

The cumulative excretion of radioactive substances in bile accounted for only 0.58% of the dose. Six metabolites were identified using HR-MS and radio-chromatographic peaks in pooled 0–48 h male rat bile. Among them, M267 was the major metabolite, accounting for 0.46% of the dose. M301-1 was the second abundant metabolite, accounting for 0.22% of the dose. Co-eluting metabolites M596 and M646 accounted for 0.18% of the dose. The contents of M286 and M676 were less than 0.1% of the dose. Six metabolites were also identified in pooled female rat bile. Among them, M267 and M532 co-eluted; M596 and M646 also co-eluted. The contents of all metabolites in bile were less than 0.1% of the dose (Figure 4, and Supplemental Table S5).

2.3.6. Description of Metabolites

BS1801 and M2. The structures of BS1801 and M2 were confirmed by comparison with the retention time and LC–MS/MS spectra of the reference standards.

M232. The metabolite of M232 (m/z 232.9713) was composed of $C_8H_8O_3Se$, which was produced by the amide hydrolysis of M484-2. Its main fragment ions were m/z 214.9604 and 199.9370. The fragment ion m/z 214.9604 was 18.0109 Da (H_2O) smaller than M232, which was obtained by the neutral loss of water from the precursor ion. The fragment ion m/z 199.9370 was produced by further losing $CH_3\cdot$ radical from the precursor ion m/z 214.9604 (Figure 5C).

M267. The elemental composition of M267 (m/z 268.0237) was $C_{12}H_{13}NOSe$. The precise molecular weight of M267 was the same as that of M286 minus an NH_5, corresponding to deamination and dehydrogenation. The main product ions were m/z 198.9656, 170.9707, and 92.9244. Among them, the product ion m/z 198.9656 was the same as the product ion m/z M286. This supports the inferred structure of M267 (Figure 5D).

M286. The elemental composition of M286 (m/z 287.0660) was $C_{12}H_{18}N_2OSe$, which was produced by *N*-dealkylation and methylation of BS1801. The main fragment ions were m/z 270.0391, 198.9655, and 72.0815. Among them, the fragment ion m/z 270.0391 was produced by the neutral loss of NH_3; fragment ion m/z 198.9655 was produced by amide bond cleavage (Figure 5E).

M301-1 and M301-2. M301-1 and M301-2 were produced by oxidative deamination and further oxidation of M286. The $[M + H]^+$ of M301-1 and M301-2 ($C_{12}H_{15}NO_3Se$) were m/z 302.0290, which was identical to the theoretical value. The main fragment ion m/z 198.9655 (the same as the fragment ions of M286) was produced by amide bond breakage, and the fragment ion m/z 183.9421 was produced by amide bond cleavage and losing $CH_3\cdot$ radical. The fragment ion m/z 86.0602 supported the modification site of M301-1. The fragment ions of M301-2 were similar to those of M301-1 (Figure 6A).

M315. Metabolite M315 (m/z 316.0447) was composed of $C_{13}H_{17}NO_3Se$, which was produced by M301 methylation. M315 produced the same fragment ions as M301 at m/z 198.9656 and 183.9426. The fragment ion m/z 284.0182 was the fragment of M315 after the neutral loss of methanol. These results indicated that M315 was a methylated product of M301 (Figure 6B).

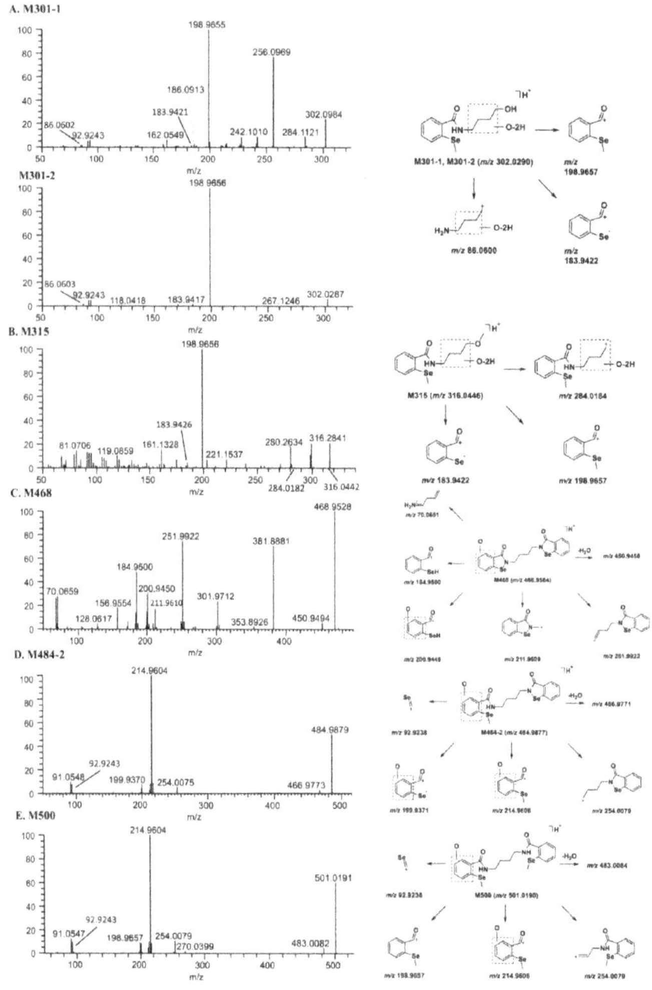

Figure 6. Mass spectra of metabolites and their proposed fragmentation profiles. (**A**) M301-1 and M301-2; (**B**) M315; (**C**) M468; (**D**) M484-2; (**E**) M500.

M468. M468 ($C_{18}H_{16}N_2O_3Se_2$), with $[M + H]^+$ at m/z 468.9552, was 15.9945 Da (O) larger than that of BS1801, suggesting that M468 was produced by the mono-oxidation of BS1801. Its major product ions m/z 211.9610, 184.9500, and 70.0659 were identical to those of BS1801. The product ion m/z 450.9494 was obtained by the neutral loss of water from the precursor ion. The difference between the product ions m/z 200.9450 and 184.9500 was 15.9950 Da (O), so it was inferred that the oxidation site was on the benzene ring (Figure 6C).

M484-2. The $[M + H]^+$ of M484-2, m/z 484.9880, was 16.0328 Da (CH_4) larger than that of M468. Its elemental composition was $C_{19}H_{20}N_2O_3Se_2$. Therefore, it was inferred that M484-2 was the product of Se methylation of M468. The fragment ion m/z 254.0075 was produced by C–N bond cleavage. Amide bond cleavage produced a fragment ion m/z 214.9604, and further losing $CH_3\cdot$ radical produced a fragment ion m/z 199.9370. The fragment ion m/z 466.9773 was obtained from the neutral loss of water from M484-2 (Figure 6D).

M500. The $[M + H]^+$ of M500, m/z 501.0192, was 15.9950 Da (O) larger than that of M2. According to the precise molecular weight, the molecular formula of M500 was $C_{20}H_{24}N_2O_3Se_2$, indicating that it was derived from the mono-oxidation of M2. The main product ions of M500 were m/z 483.0082, 254.0079, 214.9604, 198.9657, and 92.9243. The product ion m/z 483.0082 was obtained by the neutral loss of water from the precursor ion. The product ion m/z 198.9657 was similar to the product ion of M2 and was 15.9947 Da (O) smaller than the product ion m/z 214.9604, indicating that the oxidation site was on the benzene ring (Figure 6E).

M532. The $[M + H]^+$ of M532 ($C_{20}H_{24}N_2O_5Se_2$) was m/z 533.0091, which was produced by the di-oxidation of M500. The neutral loss of water from M532 produced fragment ion m/z 514.9987, and the further loss of water produced fragment ion m/z 496.9887. The breakage of the amide bond produced fragment ion m/z 230.9554, the further removal of one oxygen atom produced ion m/z 214.9605, and the removal of two oxygen atoms produced ion m/z 198.9656 (Figure 7A).

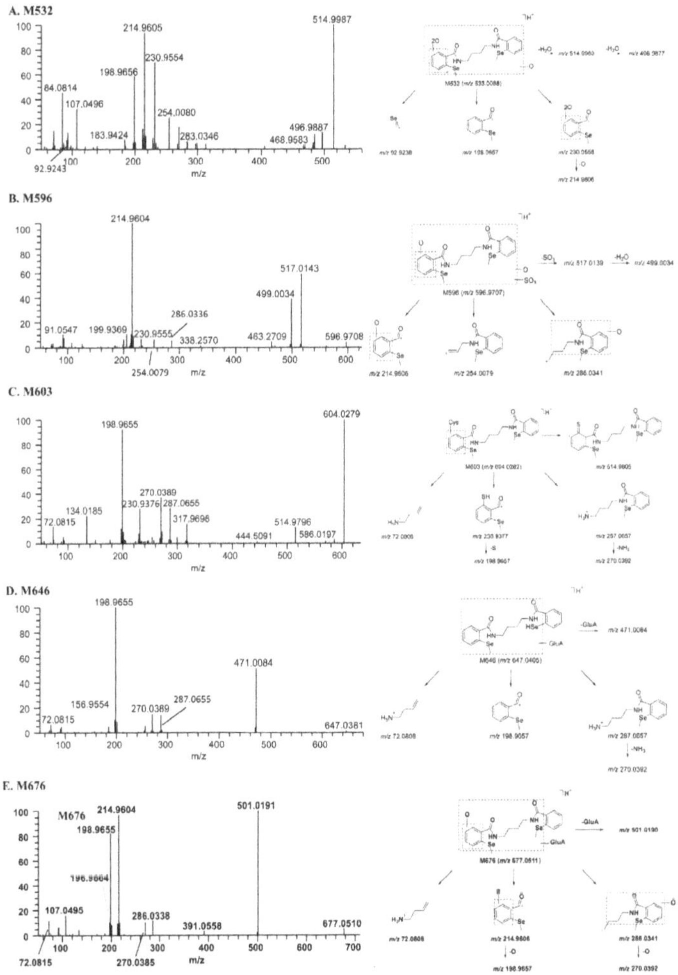

Figure 7. Mass spectra of metabolites and their proposed fragmentation profiles. (**A**) M532; (**B**) M596; (**C**) M603; (**D**) M646; (**E**) M676. Cys, cysteine conjugation; GluA, glucuronidation; SO_3, sulfation.

M596. The elemental composition of M596 (m/z 596.9709) was $C_{20}H_{24}N_2O_7Se_2S$. M596 had one more SO_4 than M500, indicating sulfation and oxidation. Its major product ions were m/z 517.0143 (the loss of SO_3), 499.0034 (the loss of SO_3 and water), 286.0336 (C–N bond breakage), 254.0079, and 214.9604. Among them, the product ions m/z 254.0079, and 214.9604 were identical to those of M500. However, the specific conjugation and oxidation sites were not determined (Figure 7B).

M603. The $[M + H]^+$ of M603 ($C_{23}H_{29}N_3O_4Se_2S$) was m/z 604.0291. Compared with M2, M603 was $C_3H_5NO_2S$ more than M2, corresponding to cysteine. The product ions m/z 270.0389, 198.9655, and 72.0815 of M603 were the same as those of M2. The product ion m/z 230.9376 was 31.9720 Da (S) more than the product ion m/z 198.9656 of M2; therefore, the binding site of M603 to cysteine was located on the benzene ring. The product ion m/z 514.9796 further supported the proposed structure of M603 (Figure 7C).

M646. The elemental composition of M646 (m/z 647.0406) was $C_{25}H_{30}N_2O_8Se_2$. It was the product of BS1801 reduction, methylation, and glucuronidation. The fragment ion m/z 471.0084 was produced by the neutral loss of glucuronic acid. The breakage of the amide bond on one side yielded the product ions m/z 287.0655 and 198.9655. The neutral loss of NH_3 (17.0266 Da) from the ion m/z 287.0655 generated the product ion m/z 270.0389 (Figure 7D).

M676. M676 (m/z 677.0506) had an elemental composition of $C_{26}H_{32}N_2O_9Se_2$, which was $C_6H_8O_6$ more than M500. After the neutral loss of a molecule of glucuronic acid ($C_6H_8O_6$), M676 yielded product ion m/z 501.0191. When one side of the C–N or amide bond was broken, product ions m/z 286.0338 or 214.9604 were obtained; the further loss of an O atom yielded product ions m/z 270.0385 and 198.9655, respectively. The product ions m/z 214.9604 and 198.9655 were identical to M500 (Figure 7E).

The main metabolic pathways of BS1801 are shown in Figure 8.

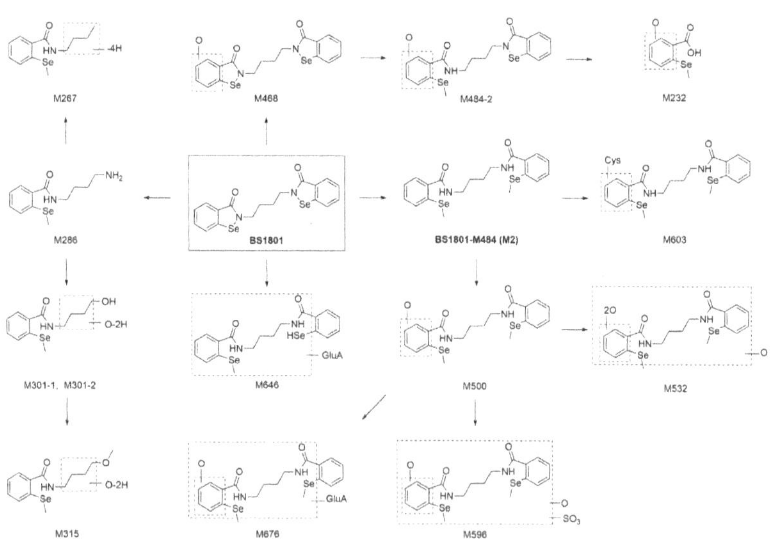

Figure 8. Metabolic pathway of BS1801 in rat. Cys, cysteine conjugation; GluA, glucuronidation; SO_3, sulfation. Bold means reference standards available.

3. Discussion

This study investigated the mass balance, tissue distribution, and metabolism of BS1801 in rats after a single oral dose of 100 mg/kg (100 μCi/kg) [^{14}C]BS1801. BS1801 and its metabolites were identified in plasma, urine, feces, and bile via an accurate mass determination and diagnostic MS fragment ions.

In mass balance studies, the results measured with radiolabeling techniques are much more accurate than those measured with non-radiolabeled compounds [30]. In this study, 91.57% of the radioactivity was excreted within 48 h after oral administration of [^{14}C]BS1801, indicating that [^{14}C]BS1801 was rapidly excreted in rats. The mean cumulative radioactivity recovery in urine and feces was 93.39% within 168 h after administration, of which 92.77% was in feces, and only 0.62% was in urine, suggesting that feces was the main excretion route. The average cumulative radioactive recovery in bile was 0.58% of the dose. The radioactive recovery in either urine or bile was less than 1% of the dose; thus, we inferred that BS1801 was not well absorbed in rats (Supplemental Table S2). However, previous studies have shown that it has very good pharmacological effect, which may indicate that a small amount of BS1801 can exert a strong therapeutic effect [26]. Therefore, further research is needed to investigate whether BS1801 absorption may be improved by altering the dosage form, etc., which can minimize the administered dose. Drug-related substances were widely distributed after the intragastric administration of [^{14}C]BS1801, and radioactivity was detected in all collected tissues. However, the overall distribution of radioactivity concentrations was low, which may be attributed to its low absorption. In male LE rats, high radioactivity concentration was detected mainly in the liver, stomach, large intestine, and eyes; in female LE rats, high radioactivity concentration was detected mainly in the large intestine and liver (Figure 3A,B). The relatively high concentration levels in the liver are favorable for the targeted treatment of liver fibrosis. The ratio of $AUC_{0–96h}$ of brain to that of plasma was the lowest in all analyzed tissues, which may be due to the blood–brain barrier limiting the entry of BS1801 into the brain. The results of mass balance and tissue distribution of [^{14}C]BS1801 in rats provide a basis for human dosimetry.

It is well known that high-resolution mass spectrometry is commonly used to quantitatively analyze compounds, but quantifying metabolite content is also crucial [31]. This study used high-resolution mass spectrometry combined with radio-chromatography to qualitatively and quantitatively analyze the metabolites associated with BS1801 simultaneously, which can provide the ratio of metabolites to dose. This will help us identify major metabolites that need further attention. This is the greatest advantage of radiolabeling technology, which can reflect the metabolite content and reveal the biotransformation pathway. Based on the high radioactive recovery of BS1801, a total of 15 metabolites were identified in plasma, urine, feces, and bile. The major metabolic pathways of BS1801 were reduction, amide hydrolysis, oxidation, and methylation, and the minor metabolic pathways were glucuronidation and sulfation (Figure 8). BS1801 was not detected in plasma, urine, and bile but was observed only in feces. Unchanged BS1801 was the major component in rat feces, accounting for more than 75.00% of the dose. M2, a product of methylation following reduction in the N–Se bond, was the major metabolite in plasma, accounting for 65.41% and 81.33% of plasma total radioactivity in male and female SD rats, respectively (Supplemental Table S5). This result is consistent with the findings obtained for unradiolabeled BS1801 in humans [25]. However, the metabolite of M2 was only present in plasma and was completely converted to other metabolites before excretion. In this study, M500 accounted for 20.07% and 14.00% of the plasma total radioactivity in male and female rats, respectively (Supplemental Table S5). This is inconsistent with a previous study in which the exposure of other metabolites was less than 5% of M2 [25]. This may result from several reasons, such as species differences or MS signal changes from metabolic modifications. Notably, it is likely that the MS response and UV peak area do not accurately reflect the metabolite content due to the influence of metabolic modifications. Additionally, we could not quantify each metabolite by using reference standards. In contrast, radiolabeling technology exhibits a unique advantage due to quantitative radioactivity unaffected by metabolite structure. M232 was a newly identified metabolite in the plasma of male rats, accounting for 11.59% of the total plasma radioactivity (Supplemental Table S5). Considering that M500 and M232 accounted for more than 10% of total radioactivity in plasma, we need to pay attention to their exposure in future human study. The advantages of the radiolabeling technique were demonstrated by comparing the results with those obtained by non-radiolabeled human plasma. Because the method can more accurately quantify the content of metabolites, some metabolites that are easily overlooked but need attention or safety assessment can be found.

BS1801 is further modified from the structure of ebselen and BBSKE. It has been reported that ebselen and BBSKE have high binding rates with albumin [19,32]. However, it is worth mentioning that when BS1801 plasma pooling was performed for protein precipitation, the recoveries of radioactivity in the supernatants of male and female rats were 105.85% and 91.23%, respectively, which demonstrated that the bottom-layer precipitation was almost free of BS1801 and its related metabolites. It has been reported that BS1801 can bind to human serum albumin rapidly and thoroughly, but the covalent binding is unstable; thus, the compound can be reduced and eventually metabolized to BS1801-M484 (M2) [25]. This explains why the bottom-layer protein precipitates in our study contain almost no radioactive substances. According to the free drug hypothesis, after compounds bind to plasma proteins, only free compounds are used for hepatic uptake through active transport or passive diffusion, and only free drugs can reach target tissues to interact with the target [29,33,34]. The properties of BS1801 covalently bound to albumin and eventually metabolized to the free product M2 are more conducive to drug efficacy. The pharmacokinetic/pharmacodynamic modeling (PK/PD modeling) of BS1801 needs to be further explored.

4. Materials and Methods

4.1. Chemicals and Reagents

BS1801 (99.5% content) and the related reference standard M2 (97.7% content) were provided by Shanghai Yuanxi Pharmaceutical Technology Co., Ltd. (Shanghai, China). [^{14}C]BS1801 (Figure 1, 99.71% content) was provided by Wuxi Beta Co., Ltd. (Wuxi, China). The liquid scintillation solution was purchased from Zhejiang Dingxin (Shaoxing, China). The alkaline liquid scintillation cocktail was obtained from RDC^{TM} (Hilladale, NJ, USA). Acetonitrile and methanol of high-performance liquid chromatography (HPLC) grade were purchased from Sigma (St. Louis, MO, USA). Formic acid was purchased from Rhawn Chemical Reagent Co., Ltd. (Shanghai, China). Ammonium acetate was purchased from Sinopharm (Shanghai, China). Ultrapure water was obtained from a Milli-Q integral water purification system (Millipore, Molsheim, France). All other reagents used were HPLC grade or analytical grade.

4.2. Animals

Sprague–Dawley (SD, *Rattus norvegicus*) rats (6–8 weeks, 200–300 g) and Long–Evans (LE, *Rattus norvegicus*) rats (6–8 weeks, 160–220 g) were purchased from Zhejiang Charles River Laboratory Animals Co., Ltd. (Jiaxing, China). The animal use license number of Shanghai Institute of Materia Medica, Chinese Academy of Sciences is SYXK (Hu) 2020-0042. The ethics committee approval number for this study is 2022-07-DXX-06, and it was valid from July 2022 to July 2023. Healthy eligible rats were housed in cages containing toys for 2–5 days in a 12 h day/night constant-temperature laboratory with free access to food and water after receipt in cages with no more than 5 rats per cage. The rats were fasted for 12 h prior to drug administration. The Animal Ethics Committee of the Shanghai Institute of Materia Medica, Chinese Academy of Science (Shanghai, China) approved the study protocol. The study design is shown in Supplementary Table S1.

4.3. Dosing Solutions and Oral Administration

BS1801 and [^{14}C]BS1801 were prepared in 0.5% sodium carboxymethyl cellulose solution at a target concentration of 10 mg/mL (10 μCi/mL). The radioactivity of [^{14}C]BS1801 was measured using a liquid scintillation counter (LSC) (Hidex, Turku, Finland) before and after administration, and a single dose of 10 mL/kg was administered orally.

4.4. Plasma Samples for Metabolite Identification

Three SD rats of each sex were used to obtain plasma samples. Approximately 200 μL of blood was collected from the jugular vein of six SD rats in a centrifuge tube containing heparin sodium before and 0.5, 2, 4, 6, 10, 12, 24, 48, 72, and 96 h after administration, followed by centrifugation (3500× *g*, 5 min, 4 °C), to obtain plasma. All samples were stored at −20 °C for analysis.

4.5. Mass Balance

Three male and three female SD rats were housed individually in metabolic cages. Urine samples were collected from each rat at predose and 0–8, 8–24, 24–48, 48–72, 72–96, 96–120, 120–144, and 144–168 h postdose. When each urine sample was collected, the metabolic cage was flushed with a small amount of water for 0–72 h and methanol/water (50:50, *v:v*) for 72–168 h, and the flushing solution was added to the urine sample for the corresponding period. Fecal samples were also collected at predose and 0–24, 24–48, 48–72, 72–96, 96–120, 120–144, and 144–168 h postdose. Urine and fecal samples were weighed and recorded during each collection period. All samples were stored at −20 °C for testing.

Bile duct cannulation (BDC) was performed in 3 male and 3 female SD rats before drug administration. After the operation, the wounds were wiped with iodophor for 2 days, and painkillers were administered appropriately. The state of the rats was observed daily, and the drug was administered only when no abnormality was observed within 2 days. Bile was collected before and 0–4, 4–8, 8–24, 24–48, and 48–72 h after dosing. The weights of the bile samples were recorded, and the samples were stored at −20 °C until testing.

4.6. Tissue Distribution

Twenty-four LE rats (12 of each sex) were used for tissue distribution studies. Three male and three female rats were sacrificed through abdominal aortic bleeding at 2, 10, 48, and 96 h after administration. Approximately 1/5 of the blood was stored frozen at −20 °C, and the remaining blood was centrifuged (3500× *g*, 5 min, 4 °C) to obtain plasma. The brain, heart, lung, liver, stomach wall, spleen, kidney, muscle, pigmented fat, ovary, womb, testis, bladder wall, pancreas, small intestine wall, large intestine wall, eye, pigmented skin, thymus, spinal cord, and marrow were also collected. The tissues were washed with saline, dried, and stored at −20 °C until testing.

4.7. Sample Preparation for Radioactivity Analysis

4.7.1. Urine, Bile, and Plasma Samples

Samples of urine (100 μL), bile (30 μL), and plasma (100 μL from the tissue distribution study) were weighed and added to 15 mL scintillation solution. Among them, urine and bile samples were duplicated, and plasma samples were in singlicate. The radioactivity was measured directly via LSC.

4.7.2. Feces, Blood, and Tissue Samples

Approximately three times the weight of methanol/water (50:50, *v:v*) was added to each fecal sample and homogenized with a homogenizer (SKGZ, Shanghai Biheng, Shanghai, China). Approximately 0.2~0.3 g of the fecal homogenate was weighed in a combustion boat, and dried overnight. Blood samples (100 μL) and tissue samples (approximately 0.1 g) were weighed into burning boats. After the samples were completely dried, they were combusted using a Biological Sample Oxidizer (OTCS-11/1, Shanghai Yixing, Shanghai, China). The $^{14}CO_2$ generated was captured with 15 mL alkaline scintillation solution; then, the radioactivity was measured using LSC.

4.8. Metabolite Radio-Profiling and Characterization

4.8.1. Sample Preparation

The plasma samples (predose and 0.5, 2, 4, 6, 10, 12, 24, 48, and 72 h postdose) were pooled according to the AUC-pooling principle [35,36]. Then, 3 mL of methanol/water

(50:50, *v:v*) was slowly added to the pooled plasma (1 mL), followed by vortexing for 1 min and sonication for 10 min twice. Then, the upper extract and the lower precipitate were separated by centrifugation (5000 rpm, 10 min). The supernatant was transferred to another new centrifuge tube. Then, 1 mL of water was added to the precipitate, vortexed for 1 min, and sonicated for 10 min. Then, 2 mL of methanol/acetonitrile (50:50, *v:v*) was added to the mixture, two vortexes were performed for 1 min, and the mixture was ultrasonicated for 10 min, followed by centrifugation (5000 rpm, 10 min). The two-times supernatant extracts were combined and dried under a nitrogen stream and finally redissolved with acetonitrile/water (20:80, *v:v*). The recoveries of extraction and reconstitution processes were measured via LSC.

Urine samples were pooled based on equal percentage according to the volume of urine from 0–8, 8–24, 24–48, 48–72, and 72–96 h, and the corresponding volumes of urine samples were mixed with the same gender. The pooled urine was vortexed and centrifuged (5000 rpm, 10 min). After centrifugation, the supernatant was dried under a nitrogen stream and then redissolved in acetonitrile/water (20:80, *v:v*). The bile samples at 0–4, 4–8, 8–24, and 24–48 h were pooled according to the same process as for urine.

The feces samples were pooled based on equal percentage according to the feces weight at 0–24 and 24–48 h for the same sex. The pooled fecal homogenate (1 g) was extracted by using the same method as for the plasma. The supernatant was then evaporated under a nitrogen stream and redissolved in 300 μL acetonitrile/water (20:80, *v:v*).

4.8.2. UHPLC-β-RAM/UV/QE MS Analysis

An ACQUITY UPLC BEH Pheny column (100 × 2.1 mm, 1.7 μm) with a Phenomenex KrudKatcher Ultra HPLC In-Line Filter guard column (2 μm, Depth Filter × 0.004 in ID) was used to perform metabolite radio-profiling and characterization on the Vanquish UHPLC system, which was combined with a QExactive Plus mass spectrometer (Thermo, Waltham, MA, USA) containing an electrospray ionization (ESI) source. The column temperature was 40 °C, and the flow rate was 0.45 mL/min. The mobile phases contained 5 mM ammonium acetate with 0.1% formic acid in water (A) and acetonitrile (B). The gradient elution procedure was as follows: 0 min, 5% B; 2 min, 10% B; 27 min, 40% B; 28 min, 95% B; 31 min, 95% B; 32 min, 5% B; and 35 min, 5% B. The eluent was monitored using a photodiode array detection with a full range of wavelengths. The mass spectrometer was operated in positive (+ESI) mode with a m/z 100–1500 Da scanning range. The optimized MS parameters were as follows: sheath gas, 35 L/min; auxiliary gas, 10 L/min; capillary temperature, 320 °C; and spray voltage, 3.8 kV. Data acquisition was performed using Xcalibur 4.1 software (Thermo, Waltham, MA, USA), and data analysis was accomplished using Compound Discoverer software 3.3 SP2 (Thermo, Waltham, MA, USA).

The extracted plasma, urine, feces, and bile samples were separated via UHPLC and detected using an online/offline radioactivity detector to obtain the corresponding radioactive metabolite profiles. Due to low radioactivity of plasma, the UHPLC eluent was collected using a UHPLC fraction collector (Thermo, Waltham, MA, USA) over time (10 s/well) on Deepwell LumaPlateTM 96-well plates. After evaporation via a vacuum centrifuge concentrator, the radioactivity was determined with an offline Sense Beta detector (1 min/well, Hidex, Turku, Finland). For the urine, feces, and bile samples, the UHPLC eluent was analyzed using an online radioactivity detector β-RAM. Radioactive metabolite profiling was integrated to obtain the radio-chromatogram with Laura 6.0.5.101 software (Lablogic, Sheffield, UK). The percentage of each metabolite in urine, feces, and bile (%dose) and the ratio of metabolites in the plasma (% area under the curve, %AUC) were calculated.

5. Conclusions

In brief, this study investigated the distribution, metabolism, and excretion of [^{14}C]BS1801 in rats. The mean recovery of radioactive substances in urine and feces was 93.39% in SD rats within 168 h postdose, and feces was the main excretion route with the cumulative excretion of radioactive substances accounting for 92.77% of the dose. Additionally, less than 1.00% of the dose was recovered from either urine or bile. The main excretion period of [^{14}C]BS1801 after administration was within 48 h, during which the mean radioactive dose recovery in urine and feces reached 91.57%. Drug-related substances were mainly distributed in the liver, stomach, large intestine, and eyes, and high concentrations in the liver facilitate targeted liver fibrosis therapy. A total of fifteen metabolites of BS1801 were detected in rats, and the main metabolic pathways were reduction, amide hydrolysis, oxidation, and methylation. Consistent with most thiol metabolites, the Se site was susceptible to methylation. BS1801-M484, the methylation product obtained by reduction in the N–Se bond in BS1801, was the most abundant drug-related component, and its radioactivity accounted for 65.41% and 81.33% of the pooled AUC_{0-72h} plasma in male and female rats, respectively. In conclusion, these data are expected to provide a strong basis for further clinical studies.

Supplementary Materials: The following are available online at https://www.mdpi.com/article/10.3390/molecules28248102/s1, Table S1: Study design. Table S2: Mean cumulative dose recovery in urine, feces, urine + feces, and bile following a single oral administration of [^{14}C]BS1801 at the dose of 100 mg/kg (100 μCi/kg) to SD rats. Table S3: AUC_{0-96h} of radioactivity in tissues and plasma of rats, and their ratio to plasma following a single oral administration of [^{14}C]BS1801 at the dose of 100 mg/kg (100 μCi/kg). Table S4: Information of BS1801and its metabolites detected in rat plasma, urine, feces, and bile by using UHPLC-Q Exactive Plus MS. Table S5: The distribution of radioactivity from BS1801 and its major metabolites in rat plasma, urine, feces, and bile and the percent of dose in urine and feces (%).

Author Contributions: Data curation, C.Y. (Cheng Yang) and Y.H.; Formal analysis, C.Y. (Cheng Yang); Funding acquisition, H.Z.; Investigation, C.Y. (Cheng Yang), M.X., Y.H. and C.Y. (Chen Yang); Methodology, H.Y., H.Z. and Y.Z.; Project administration, X.D.; Resources, X.D.; Supervision, H.Y., D.Z., Y.Z. and X.D.; Writing—original draft, C.Y. (Cheng Yang); Writing—review and editing, X.D. All authors have read and agreed to the published version of the manuscript.

Funding: This study was sponsored by Shanghai Yuanxi Medicine Corp. This work was also supported by grants from the National Natural Science Foundation of China (NO. 82104275), the National Key R&D Program of China (NO. 2022YFF1202600), the National Natural Science Foundation of China (NO. 82373938), and the Key Technologies R&D Program of Guangdong Province (NO. 2023B1111030004).

Institutional Review Board Statement: The animal study protocol was approved by the Animal Ethics Committee of the Shanghai Institute of Materia Medica, Chinese Academy of Sciences for studies involving animals. The ethics committee approval number for this study is 2022-07-DXX-06, and it was approved in July 2022.

Informed Consent Statement: Not applicable.

Data Availability Statement: The research data used to support the findings of this study are included in the article.

Conflicts of Interest: Hanwei Yin and Huihui Zeng are employees of the Shanghai Yuanxi Pharmaceutical Technology Co., Ltd. The other authors declare that they have no competing interests.

References

1. Lu, J.; Holmgren, A. The thioredoxin antioxidant system. *Free Radic. Biol. Med.* **2014**, *66*, 75–87. [CrossRef] [PubMed]
2. Zhang, J.; Li, X.; Han, X.; Liu, R.; Fang, J. Targeting the Thioredoxin System for Cancer Therapy. *Trends Pharmacol. Sci.* **2017**, *38*, 794–808. [CrossRef]
3. Ghareeb, H.; Metanis, N. The Thioredoxin System: A Promising Target for Cancer Drug Development. *Chem. Eur. J.* **2020**, *26*, 10175–10184. [CrossRef]
4. Zhang, B.; Zhang, J.; Peng, S.; Liu, R.; Li, X.; Hou, Y.; Han, X.; Fang, J. Thioredoxin reductase inhibitors: A patent review. *Expert Opin. Ther. Pat.* **2017**, *27*, 547–556. [CrossRef]
5. Chupakhin, E.; Krasavin, M. Thioredoxin reductase inhibitors: Updated patent review (2017-present). *Expert Opin. Ther. Pat.* **2021**, *31*, 745–758. [CrossRef]
6. Chuai, H.; Zhang, S.Q.; Bai, H.; Li, J.; Wang, Y.; Sun, J.; Wen, E.; Zhang, J.; Xin, M. Small molecule selenium-containing compounds: Recent development and therapeutic applications. *Eur. J. Med. Chem.* **2021**, *223*, 113621. [CrossRef] [PubMed]

7. Ye, R.H.; Huang, J.Q.; Wang, Z.X.; Chen, Y.X.; Dong, Y.L. Trace Element Selenium Effectively Alleviates Intestinal Diseases. *Int. J. Mol. Sci.* **2021**, *22*, 11708. [CrossRef]
8. Steinbrenner, H.; Speckmann, B.; Klotz, L.O. Selenoproteins: Antioxidant selenoenzymes and beyond. *Arch. Biochem. Biophys.* **2016**, *595*, 113–119. [CrossRef]
9. Barchielli, G.; Capperucci, A.; Tanini, D. The Role of Selenium in Pathologies: An Updated Review. *Antioxidants* **2022**, *11*, 251 [CrossRef]
10. Gandin, V.; Khalkar, P.; Braude, J.; Fernandes, A.P. Organic selenium compounds as potential chemotherapeutic agents for improved cancer treatment. *Free Radic. Biol. Med.* **2018**, *127*, 80–97. [CrossRef]
11. Jariwalla, R.J.; Gangapurkar, B.; Nakamura, D. Differential sensitivity of various human tumour-derived cell types to apoptosis by organic derivatives of selenium. *Br. J. Nutr.* **2008**, *101*, 182–189. [CrossRef] [PubMed]
12. Ip, C.; Dong, Y.; Ganther, H.E. New concepts in selenium chemoprevention. *Cancer Metastasis Rev.* **2002**, *21*, 281–289. [CrossRef]
13. Rayman, M.P. Selenium and human health. *Lancet* **2012**, *379*, 1256–1268. [CrossRef] [PubMed]
14. Parnham, M.J.; Sies, H. The early research and development of ebselen. *Biochem. Pharmacol.* **2013**, *86*, 1248–1253. [CrossRef]
15. Wang, J.; Wang, P.; Dong, C.; Zhao, Y.; Zhou, J.; Yuan, C.; Zou, L. Mechanisms of ebselen as a therapeutic and its pharmacology applications. *Future Med. Chem.* **2020**, *12*, 2141–2160. [CrossRef] [PubMed]
16. Sound Pharmaceuticals lnc. Ebselen. 2022. Available online: https://data.pharmacodia.com/drug#/main/drugInfo (accessed on 12 September 2023).
17. Lan, L.; Zhao, F.; Wang, Y.; Zeng, H. The mechanism of apoptosis induced by a novel thioredoxin reductase inhibitor in A549 cells: Possible involvement of nuclear factor-kappaB-dependent pathway. *Eur. J Pharmacol.* **2007**, *555*, 83–92. [CrossRef]
18. Zheng, W.; He, R.; Boada, R.; Subirana, M.A.; Ginman, T.; Ottosson, H.; Valiente, M.; Zhao, Y.; Hassan, M. A general covalent binding model between cytotoxic selenocompounds and albumin revealed by mass spectrometry and X-ray absorption spectroscopy *Sci. Rep.* **2020**, *10*, 1274. [CrossRef]
19. Zhou, H.Y.; Dou, G.F.; Meng, Z.Y.; Lou, Y.Q.; Zhang, G.L. High performance liquid chromatographic determination of 1,2-[bis(1,2-benzisoselenazolone-3(2H)-ketone)]-ethane (BBSKE), a novel organoselenium compound, in dog plasma using pre-column derivatization and its application in pharmacokinetic study. *J. Chromatogr. B* **2007**, *852*, 617–624. [CrossRef]
20. Zhou, H.Y.; Dou, G.F.; Meng, Z.Y.; Lou, Y.Q.; Zhang, G.L. Development of a rapid and sensitive liquid chromatography/tandem mass spectrometry method for the determination of 1,2-[bis(1,2-benzisoselenazolone-3(2H)-ketone)]-ethane (BBSKE), a novel anti-cancer agent in rat plasma. *Biomed. Chromatogr.* **2008**, *22*, 1123–1129. [CrossRef]
21. Sies, H. Ebselen, a Selenoorganic Compound as Glutathione-Peroxidase Mimic. *Free Radic. Biol. Med.* **1993**, *14*, 313–323. [CrossRef]
22. Zheng, X.; Ma, W.; Sun, R.; Yin, H.; Lin, F.; Liu, Y.; Xu, W.; Zeng, H. Butaselen prevents hepatocarcinogenesis and progression through inhibiting thioredoxin reductase activity. *Redox. Biol.* **2018**, *14*, 237–249. [CrossRef]
23. Zheng, X.Q.; Chen, Y.F.; Bai, M.; Liu, Y.X.; Xu, B.Y.; Sun, R.X.; Zeng, H.H. The antimetastatic effect and underlying mechanisms of thioredoxin reductase inhibitor ethaselen. *Free Radic. Biol. Med.* **2019**, *131*, 7–17. [CrossRef] [PubMed]
24. Santi, C.; Scimmi, C.; Sancineto, L. Ebselen and Analogues: Pharmacological Properties and Synthetic Strategies for Their Preparation. *Molecules* **2021**, *26*, 4230. [CrossRef] [PubMed]
25. Tian, Q.; Jiang, J.; Yin, H.; Ma, J.; Deng, G.; Zhou, J.; Zhong, D. Quantification of the major circulating metabolite of BS1801, an ebselen analog, in human plasma. *J. Pharm. Biomed. Anal.* **2022**, *212*, 114638. [CrossRef]
26. Chen, Y.; Yin, H.; Sun, J.; Zhang, G.; Zhang, Y.; Zeng, H. TrxR/Trx inhibitor butaselen ameliorates pulmonary fibrosis by suppressing NF-κB/TGF-β1/Smads signaling. *Biomed. Pharmacother.* **2023**, *169*, 115822. [CrossRef] [PubMed]
27. Tian, Q.; Jiang, J.; Yin, H.; Zhang, Y.; Li, Y.; Wu, P.; Peng, C.; Wang, Z.; Zhou, J.; Zeng, H.; et al. Investigating the Metabolic Mechanisms of Butaselen, An Ebselen Analog. *Curr. Drug Metab.* **2022**, *23*, 928–939. [CrossRef] [PubMed]
28. Lu, J.J.; Pan, Q.Q.; Zhou, J.Q.; Weng, Y.; Chen, K.L.; Shi, L.; Zhu, G.X.; Chen, C.L.; Li, L.; Geng, M.Y.; et al. Pharmacokinetics, distribution, and excretion of sodium oligomannate, a recently approved anti-Alzheimer's disease drug in China. *J. Pharm. Anal* **2022**, *12*, 145–155. [CrossRef]
29. Schulz, J.A.; Stresser, D.M.; Kalvass, J.C. Plasma protein-mediated uptake and contradictions to the free drug hypothesis: A critical review. *Drug Metab. Rev.* **2023**, *55*, 205–238. [CrossRef]
30. Zheng, Y.D.; Zhang, H.; Zhan, Y.; Bian, Y.C.; Ma, S.; Gan, H.X.; Lai, X.J.; Liu, Y.Q.; Gong, Y.C.; Liu, X.F.; et al. Pharmacokinetics, mass balance, and metabolism of [(14)C]vicagrel, a novel irreversible P2Y12 inhibitor in humans. *Acta Pharmacol. Sin.* **2021**, *42*, 1535–1546. [CrossRef]
31. Wang, L.; Li, S.; Li, J.; Cheng, Z.; Feng, Y.; Ouyang, H.; Du, Z.; Jiang, H. Comprehensive metabolic profiling of Alismatis Rhizoma triterpenes in rats based on characteristic ions and a triterpene database. *J. Pharm. Anal.* **2021**, *11*, 96–107. [CrossRef]
32. Wagner, G.; Schuch, G.; Akerboom, T.P.; Sies, H. Transport of ebselen in plasma and its transfer to binding sites in the hepatocyte *Biochem. Pharmacol.* **1994**, *48*, 1137–1144. [CrossRef]
33. Tillement, J.P.; Urien, S.; Chaumet-Riffaud, P.; Riant, P.; Bree, F.; Morin, D.; Albengres, E.; Barre, J. Blood binding and tissue uptake of drugs. Recent advances and perspectives. *Fundam. Clin. Pharmacol.* **1988**, *2*, 223–238. [CrossRef] [PubMed]
34. Hammarlund-Udenaes, M. Active-site concentrations of chemicals—Are they a better predictor of effect than plasma/organ/tissue concentrations? *Basic Clin. Pharmacol. Toxicol.* **2010**, *106*, 215–220. [CrossRef] [PubMed]
35. Hop, C.E.C.A.; Wang, Z.; Chen, Q.; Kwei, G. Plasma-Pooling Methods To Increase Throughput for in Vivo Pharmacokinetic Screening. *J. Pharm. Sci.* **1998**, *87*, 901–903. [CrossRef] [PubMed]

5. Hamilton, R.A.; Garnett, W.R.; Kline, B.J. Determination of mean valproic acid serum level by assay of a single pooled sample. *Clin. Pharmacol. Ther.* **1981**, *29*, 408–413. [CrossRef]

biomolecules

MDPI

Article

Rhododendron chrysanthum's Primary Metabolites Are Converted to Phenolics More Quickly When Exposed to UV-B Radiation

Fushuai Gong, Wang Yu, Qingpan Zeng, Jiawei Dong, Kun Cao, Hongwei Xu and Xiaofu Zhou *

Jilin Provincial Key Laboratory of Plant Resource Science and Green Production, Jilin Normal University, Siping 136000, China
* Correspondence: zhouxiaofu@jlnu.edu.cn

Abstract: The plant defense system is immediately triggered by UV-B irradiation, particularly the production of metabolites and enzymes involved in the UV-B response. Although substantial research on UV-B-related molecular responses in *Arabidopsis* has been conducted, comparatively few studies have examined the precise consequences of direct UV-B treatment on *R. chrysanthum*. The ultra-high-performance liquid chromatography tandem mass spectrometry (UPLC-MS/MS) methodology and TMT quantitative proteomics are used in this study to describe the metabolic response of *R. chrysanthum* to UV-B radiation and annotate the response mechanism of the primary metabolism and phenolic metabolism of *R. chrysanthum*. The outcomes demonstrated that following UV-B radiation, the primary metabolites (L-phenylalanine and D-lactose*) underwent considerable changes to varying degrees. This gives a solid theoretical foundation for investigating the use of precursor substances, such as phenylalanine, to aid plants in overcoming abiotic stressors. The external application of ABA produced a considerable increase in the phenolic content and improved the plants' resistance to UV-B damage. Our hypothesis is that externally applied ABA may work in concert with UV-B to facilitate the transformation of primary metabolites into phenolic compounds. This hypothesis offers a framework for investigating how ABA can increase a plant's phenolic content in order to help the plant withstand abiotic stressors. Overall, this study revealed alterations and mechanisms of primary and secondary metabolic strategies in response to UV-B radiation.

Keywords: UV-B radiation; *Rhododendron chrysanthum*; primary metabolite; phenolic compounds; ultra-performance liquid chromatography tandem mass spectrometry (UPLC-MS/MS); ABA

1. Introduction

China's Changbai Mountain are home to the medicinal properties ornamental shrub *R. chrysanthum*. The predominant abiotic stressor in the Changbai Mountains is UV-B radiation because of the area's unique alpine habitat, which includes cold temperatures, a thin atmosphere, and intense sun radiation. As a result, *R. chrysanthum* is a unique plant resource for the investigation of plant adaptability, and our earlier research has enough physiological support for the plant's great tolerance to UV-B exposure.

UV-B radiation (280–315 nm) was a natural component of sunlight before the 1980s and 1990s, when a surge in the atmosphere's hydrochlorofluorocarbon caused a severe thinning of the stratospheric ozone layer and a corresponding rapid rise in the amount of UV-B light reaching the Earth's surface [1,2]. Obviously, this has a detrimental effect on farming and the vegetation in the natural ecosystem [3]. Even while UV-B only makes up a little portion of the energy, it has a lot of energy on its own [4]. Induced by the UV resistance motif 8 (UVR8)-dependent UV-B signaling pathway, UV-B-responsive genes encourage the formation of phenolic substances such as hydroxycinnamic acid, Kaempferol-3-hexoside rhamnose, and luteolin (5,7,3′,4′-Tetrahydroxyflavone) [5–7]. Numerous studies have looked at the long-term impact of UV-B radiation on plants, but few have concentrated on the short-term consequences. Studies on the impact of short-term UV-B exposure may aid in the understanding of the UV-induced pathways of signaling as well as the modifications in compounds and enzymes that occur immediately in reaction to UV irradiation. In light of this hypothesis, it is critical to comprehend how *R. chrysanthum* metabolites adjust to increased UV-B exposure.

Phenolic compounds make up the biggest class of secondary metabolites found in plants. In addition to aiding in the creation of cell walls, these substances shield plants against diseases and UV radiation [8]. Phenolics, particularly flavonoids, are the secondary metabolites that are most important for UV protection out of the more than 8000 that exist. According to Marco Santin et al., UV-B radiation triggers the downstream activation of several phenylpropane biosynthetic and regulatory genes, which causes the levels of particular phenolic compounds to significantly increase [9]. Depending on their chemical makeup, phenolic compounds can be categorized into three classes and show a variety of structural variations in nature. Specifically, there are three types of phenolic compounds: (1) those with a C6C1 carbon dioxide in the framework such as 4-hydroxybenzoic acid and 2,5-Dihydroxybenzoic acid; (2) those with C6C3 carbon bones, such as caffeic acid, p-coumaric acid, and ferulic acid; and (3) those with a C6C3C6 carbon skeleton (flavonoids are the typical C6C3C6) [10,11]. Higher plants produce secondary metabolites through the shikimate pathway, which is also the pathway by which phenolic compounds are often generated. Phosphoenolpyruvate and erythrose 4-phosphate are the starting points of the route, whereas chorismite is its last stop [12,13]. A vital metabolite produced by the chorion is phenylalanine [14]. In plant metabolism, phenylalanine is regarded as a universal precursor of the molecules C6C1-, C6C3-, and C6C3C6 and their polymers. The phenolic compound buildup, which is controlled by abiotic stressors, has received a lot of attention in studies. A possible explanation for the large carbon flux through the phenylalanine metabolic network is the vast range of physiological functions of these substances and their significance for plant growth, development, defense, and environmental reactions [15]. There are many phenolic chemicals in *R. chrysanthum* that can be studied.

Abscisic acid (ABA), a plant hormone, is crucial for numerous physiological processes, including seed germination, dormancy, growth, and adaptation to biotic and abiotic stressors. By controlling the processes of starch breakdown, source pool transfer, and leaf senescence, ABA may encourage bud dormancy. Additionally, ABA shields plants from dry stress by encouraging the build-up of proline. It was demonstrated that *TaFDL2-1A* enhances transgenic wheat's resistance to drought by positively regulating ABA production, ABA responsiveness, and ROS scavenging [16].

Metabolomics is a rapidly developing field that plays a critical role in understanding the complicated primary and secondary metabolite variations in higher plants [17]. Modern biology is increasingly adopting and recognizing metabolomics techniques. Nuclear magnetic resonance spectroscopy (NMR), gas chromatography–mass spectrometry (GC-MS), and liquid chromatography–mass spectrometry (LC-MS) are typical analytical methods used in plant metabolomics research [18,19]. Due to recent advancements in UPLC-MS techniques, which offer greater and fewer run durations than traditional HPLC-MS techniques, metabolite analysis has become more reliable [20]. UV-B stabilizes health-promoting bioactive chemicals that are necessary for the bread-making process [21]. Additionally, the analysis of flavonoids using ultra-high-performance liquid chromatography revealed that UVR8-mediated flavonoid induction is a UV-B tolerance feature that is conserved in terrestrial plants and could be an early adaptation to life on Earth [22].

UV-B light activates a number of phenolic compound biosynthesis enzymes, including chalcone synthase (CHS, E.C. 2.3.1.74) and phenylalanine deaminase (PAL, E.C. 4.3.1.25) [23]. The enzymes PAL, cinnamate 4-hydroxylase (C4H, E.C. 1.14.13.11), and 4-coumarate coenzyme A ligase (4CL, E.C. 6.2.1.12) are engaged in the upstream phenylalanine route, which provides precursors for the downstream flavonoid metabolism pathway as well as the flavonol metabolism pathway.

In general, plants' primary defense against harmful UV-B radiation is to encourage the creation of UV-absorbing substances such phenolic compounds, flavonoids, and hydroxycinnamates [24]. The mechanisms of *R. chrysanthum*'s metabolic reactions to UV-B light are poorly understood. We investigated the impact of short-duration UV-B rays on the formation of both primary and phenolic compounds specific for primary and phenolic metabolites in *R. chrysanthum* using ultra-performance liquid chromatography–mass spectrometry (UPLC-MS) metabolomics in order to investigate in depth the reactions and mechanisms of how metabolic reprogramming reaches a new homeostasis after UV-B light.

2. Materials and Methods

2.1. Plant Material, Growing Conditions, and Treatments

Three of the nine (n = 3) genetically identical *R. chrysanthum* seedlings that were chosen were separated into three groups as an array of biological duplicates. Before receiving UV-B therapy, all seedlings were cultivated in a sophisticated artificial climate room (http://www.nbsaifu.cn/about_about/gsjj9f6.html accessed on 15 July 2023; Ningbo Saifu Experimental Instrument Co., Ningbo, China) at a temperature of 25 °C/25 °C (day/night) and a photoperiod of 14 h/10 h (day/night). One batch of seedlings was grown in a media that also contained exogenous ABA, whereas the remainder of the two groups were both cultured in regular *R. chrysanthum* medium. After four months of growth, all three groups of seedlings were at once moved to an artificial climate chamber used for this experiment that replicated an alpine environment (18 °C (light for 14 h)/16 °C (dark for 10 h)) with white light radiation of 50 mol (photons) $m^{-2}s^{-1}$ and 60% relative humidity. Group M of seedlings grown on regular *R. chrysanthum* media were the only ones to receive PAR light. Group N of seedlings grown on regular *R. chrysanthum* medium were only exposed to UV-B light [25]. For group Q, seedlings grown in media with exogenous ABA application likewise only received UV-B light. The three groups' seedlings underwent two days of artificial climate chamber incubation with 8 h of light per day.

In a simulated climate room, UV-B light (UV-B) (295–315 nm) and PAR (400–700 nm) were applied to three sets of seedlings [26]. UV intensity measurements were made by Sentry Optronics Corp. in China. The culture flasks were specifically covered during the UV-B treatment using a 295 nm long-pass filter (Filter Long 2IN SQ Filter, Edmund, NJ, USA). Above the culture flasks, 400 nm long-pass filters were installed for the PAR treatments (Edmund). According to the transmission function of the long-pass filter, the samples actually received an irradiance of 50 mol (photons) m^{-2} s^{-1} of PAR and 2.3 W m^{-2} UV-B, respectively.

2.2. Identification and Quantification of Primary Metabolites and Secondary Metabolites by UPLC-MS/MS

The following steps were followed when conducting the metabolomic analysis with Metware Biotech Inc. in Wuhan, China: At −20 °C, 1200 µL of an aqueous solution of 70% methanol was used to extract 50 mg of crushed, lyophilized powder. The supernatant was aspirated and put through a 0.22 µm membrane after being centrifuged at 12,000 rpm for 3 min. The metabolites in samples of *R. chrysanthum* were broadly quantified using a liquid chromatography–electrospray ionization tandem mass spectrometry (LC-ESI-MS/MS) technology. Following the established techniques previously described, Wuhan METWARE Biotechnology Co., Ltd. in Wuhan (Wuhan, China) employed an LC-ESI-MS/MS system (UPLC, gasket-packed UFLC CBM30A, Shimadzu, Kyoto, Japan; MS, 6500 QTRAP, Applied Biosystems, Norwalk, CT, USA), and analyses were performed on all sample extracts [27,28]. Ultrapure water with a formic acid content of 0.1% and acetonitrile with a formic acid content of 0.1% were the two mobile phases A and B, respectively. According to the gradient program, phase B was raised from 5% to 95% in 9 min, held for 1 min, and then immediately lowered to 5% and adjusted for 14 min. The column temperature was 40 °C, the mobile phase flow rate was 0.35 mL/min, and the injection volume was 2 µL.

By using electrospray ionization (ESI), metabolites were ionized. They were recognized using the Metware database's secondary fragment ion spectra. By using triple quadrupole mass spectrometry, the identified metabolites were measured through multiple reaction monitoring. The relative content of metabolites was calculated using peak regions. Variable importance (VIP) in projection scores with orthogonal partial least squares discriminant analysis (OPLS-DA) > 1 and |fold change| > 1.2 were regarded as differential metabolites (DMs). Based on the KEGG database, metabolites were enhanced and categorized.

2.3. Analysis of Metabolite Data

The statistics function prcomp in R (www.r-project.org, accessed on 15 July 2023) was used to perform unsupervised PCA (principal component analysis). Prior to unsupervised PCA, the data were unit-variance-scaled. Using the program MetaboAnalyst 5.0 (https://www.metaboanalyst.ca/, accessed on 17 July 2023), metabolite correlation network diagrams were created for the metabolites in the samples. Utilizing the KEGG Compound database (http://www.kegg.jp/kegg/compound/, accessed on 20 July 2023), identified metabolites were annotated. The annotated metabolites were then linked to the KEGG Pathway database (http://www.kegg.jp/kegg/pathway.html, accessed on 25 July 2023). The p-values for the hypergeometric test were used to establish the importance of the pathways where significantly regulated metabolites had been mapped.

2.4. Identification and Quantification of Proteins by Liquid Chromatography–Mass Spectrometry Analysis

In order to execute the proteome analysis, the following procedures were followed: protein extraction, trypsin digestion, Tandem Mass Tag (TMT) tagging, peptide fractionation, LC-MS/MS analysis, protein annotation, and functional enrichment [29].

2.5. Statistical Analysis

R (http://cran.r-project.org/, accessed on 2 July 2023) and IBM SPSS statistical software (https://www.ibm.com/analytics/spss-statistics, IBM SPSS Statistics 26, accessed on 5 July 2023) were used for the statistical analysis. Results show letters to indicate the levels of significance. A one-way analysis of variance (ANOVA) and a Pearson correlation test with an 80% threshold of significance were used to evaluate the data.

3. Results

3.1. The Primary Metabolism Reprogramming in the Presence of UV-B Stress

The repeatability of various experimental materials under UV-B exposure was ensured by inserting quality control (QC) samples into every 10 test samples during the identification of metabolites using the UPLC-MS/MS method [30]. The curves of the total ion current for metabolite detection showed high overlap when the total ion current plots (TIC plots) of various QC samples were analyzed by mass spectrometry detection (Supplementary Figures S1 and S2). This means that the retention times and peak intensities were consistent, indicating that mass spectrometry had better signal stability when detecting the same sample at various times.

To comprehend the general metabolic variations and variability between groups among the samples, Principal Component Analysis (PCA) was carried out on the specimens, including the QC samples. PCA determines whether there are variations among metabolomes by separating metabolome trends into groups. Insignificant differences were found between UV-B pressure treatment specimens of the same materials in PCA of the control, experimental, and QC samples; however, there was a clear pattern of segregation between the various groups (Figure 1a). The Pearson correlation coefficient (PCC) R was employed as an evaluation metric for the correlation of the three biological replicates. Heat maps of the correlation between the three biological replicates and the QC samples were also created. The data showed a strong positive association among the three biological replicates (Figure 1b).

The MetWare Database (MWDB), multiple reaction monitoring (MRM), and the Kyoto Encyclopedia of Genes and Genomes (KEGG) Compound Database were used in the qualitative and quantitative mass spectrometry analysis of the metabolites in each UV-B stress sample. Comprehensive targeted metabolic analysis yielded a total of 2148 metabolites, of which 615 were primary metabolites, comprising organic acids, lipids, amino acids and their derivatives, nucleotides and their derivatives, and other classes (Figure 1c). The majority of amino acids and their derivatives, such as L-phenylalanine, L-tyrosine*, etc., were upregulated in the MN group as well as in the NQ group. Of these, phenylalanine attenuates the adverse effects of abiotic stresses. While nucleotides and their derivatives were upregulated in the MN group but did not show any significant changes in the NQ group, the majority of lipid metabolites did not exhibit any appreciable changes. The same trend was also observed in the organic acids, such as 3-Dehydroshikimic acid and L-Tartaric acid. The 3-Dehydroshikimic acid substantially altered in response to UV-B exposure, indicating that it functions as a crucial intermediary product influencing the metabolic pathways involved in the biosynthesis of aromatic amino acids. The majority of the saccharides, including D-Lactose* and DMelezitose O-rhamnoside, among other classes, were downregulated in the MN group but not significantly altered or upregulated in the NQ group; 1359 different secondary metabolites were found, such as phenolic acids, flavonoids, quinones, lignans and coumarins, ellagitannins, alkaloids, terpenoids, steroids, and others (Figure 1d). Significant changes were observed in the phenolic acids and flavonoids. For example, phenolic acid metabolites such as 4-O-Galloylquinic acid, chlorogenic acid (3-O-Caffeoylquinic acid), methyl caffeate, etc. were upregulated in the MN group and also tended to be upregulated in the NQ group.

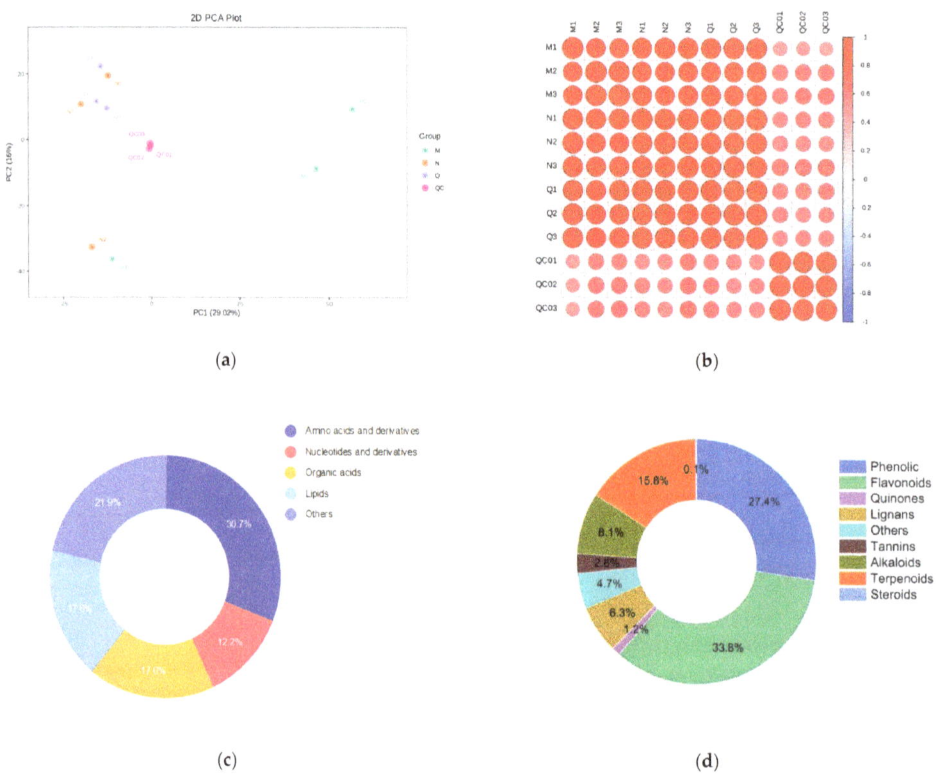

(a) (b)

(c) (d)

Figure 1. *Cont.*

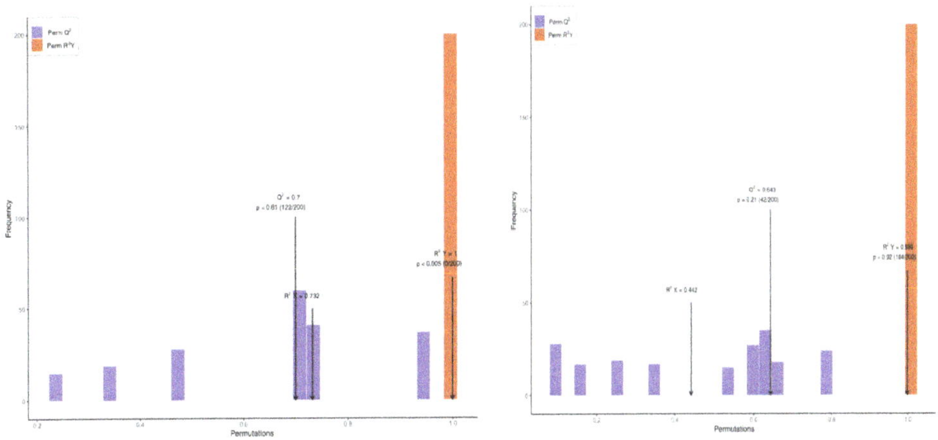

(e) (f)

Figure 1. UV-B-treated as well as externally applied ABA UV-B-treated samples and quality control (QC) samples were examined in different groups and categorized for primary and secondary metabolites. (**a**) Comparing control, UV-B-treated, and quality control (QC) samples using principal component analysis (PCA); (**b**) comparison of control, UV-B-treated, and quality control (QC) data with those from various groups; (**c**) classification and proportions of primary metabolites; (**d**) classification and proportions of secondary metabolites; (**e**,**f**) the variables that led to differences between groups were monitored and calculated using the OPLS-DA model. The explanatory rates of the X and Y matrices are denoted by R^2X and R^2Y, respectively. Q^2Y represents the model's ability to forecast. The model is more stable and reliable the closer its value is near one. Additionally, $Q^2Y > 0.5$ and $Q^2Y > 0.9$ are regarded as valid and outstanding models, respectively.

A supervised OPLS-DA model was also used to compare metabolite levels and pinpoint the factors that contributed to group disparities. The OPLS-DA model was used to compute the differences between M and N ($R^2X = 0.732$, $R^2Y = 1$, $Q^2 = 0.7$) (Figure 1e) and between N and Q ($R^2X = 0.442$, $R^2Y = 0.998$, $Q^2 = 0.643$) (Figure 1f). Both comparisons had Q^2 values that were more than 0.5, showing that the two models were steady and had distinct metabolic phenotypes in response to UV-B exposure. These findings imply that the metabolite composition of *R. chrysanthum* is significantly altered by UV-B stress.

A total of 56 significantly modified metabolites were found between N and Q (39 upregulated, 17 downregulated) (Figure 2a,c), and 203 were found between M and N (126 upregulated, 77 downregulated) (Figure 2a,b) using the thresholds of $|\log2\text{Fold Change}| \geq 1.2$ and VIP (variable importance in project, VIP) > 1. Under UV-B stress, groups M and N as well as groups N and Q only co-influenced eighteen DMs, with seventeen of these co-influencing compounds located in the upper Venn diagram (Figure 2e), including S-methyl-L-cysteine and 6-O-methylguanine, and just one in the lower Venn diagram (Figure 2f), namely, Nystose. This is due to the fact that the differentially expressed metabolites that are most frequently impacted in both Venn diagrams are those metabolites whose expression levels varied between the two groups (Figure 2d). Our study focused on the DMs of these two trends, where the first trend's DMs were those that responded favorably to UV-B stress and the second trend's DMs were those that responded adversely to UV-B stress. The differential compounds of the two trends were classified as core DMs. Amino acids and their derivatives, which are the main primary metabolites of plants and, as monomers, are crucial to the synthesis of proteins, are the most varied group of core differential metabolites. The categorization and expression data for the 18 core DMs are shown in Figure 2g, together with information about the core DMs (Supplementary Table S1), Table 1 shows information on the first seven core DMs.

3.2. *Dynamics of Phenolic Compounds under UV-B Radiation*

The screening parameters for secondary metabolites matched those for primary metabolites. The MN group of secondary metabolites had a total of 392 DMs that were upregulated and 88 which were downregulated, while the NQ group had 180 DMs that were upregulated and 136 that were downregulated (Figure 3a). According to the VIP value, Figure 3b,c show the first 15 metabolites were up- and downregulated in the MN group and the NQ group, respectively. In addition, following UV-B stress and ABA addition, cluster analysis revealed three distinct metabolite trends (Figure 3d). Naturally, we focused more on the sixth cluster. There, we discovered that the DMs that were clustered to the sixth cluster showed a growing trend in the MN group, indicating that these metabolites are crucial in the response to UV-B stress, and continued to show an upward trend in the NQ group, showing that the outwardly applied ABA helps the *R. chrysanthum* to withstand UV-B stress. The majority of the secondary DMs in cluster VI are phenolic substances. As a result, we suggest that variations in phenolic content may help to improve the fluidity of membranes and promote the protective effects of reactive oxygen species (ROS) when exposed to UV-B stress. There were statistical differences between groups in the quantity of phenolic chemicals (Figure 3e). It was discovered that *R. chrysanthum* was able to withstand UV-B stress thanks to the external application of ABA, which also caused phenolic abundance to grow greatly following UV-B stress.

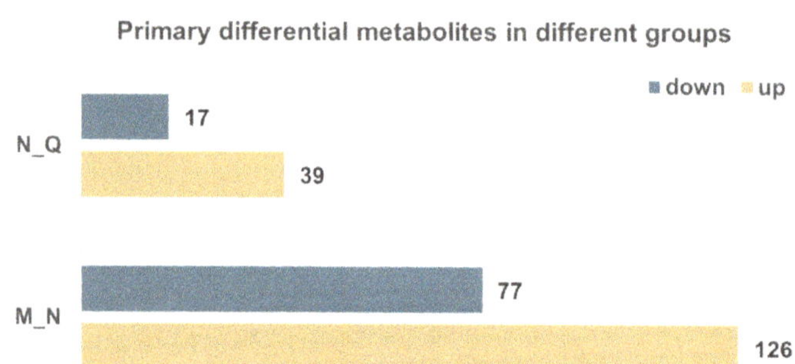

(a)

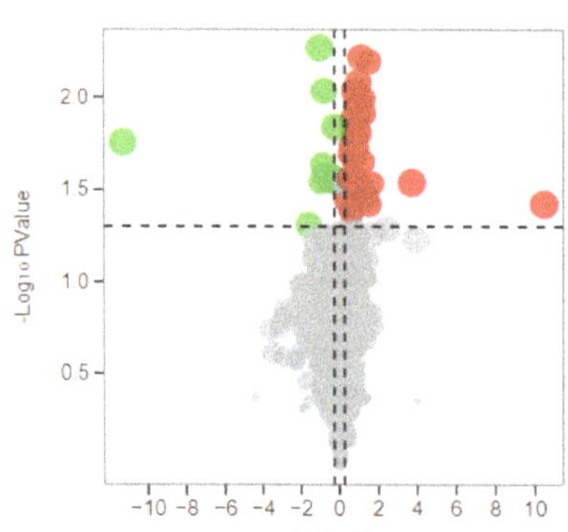

(b)

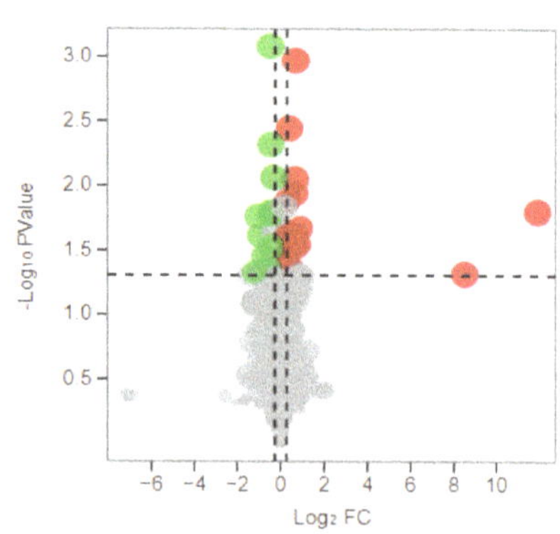

(c)

Figure 2. *Cont.*

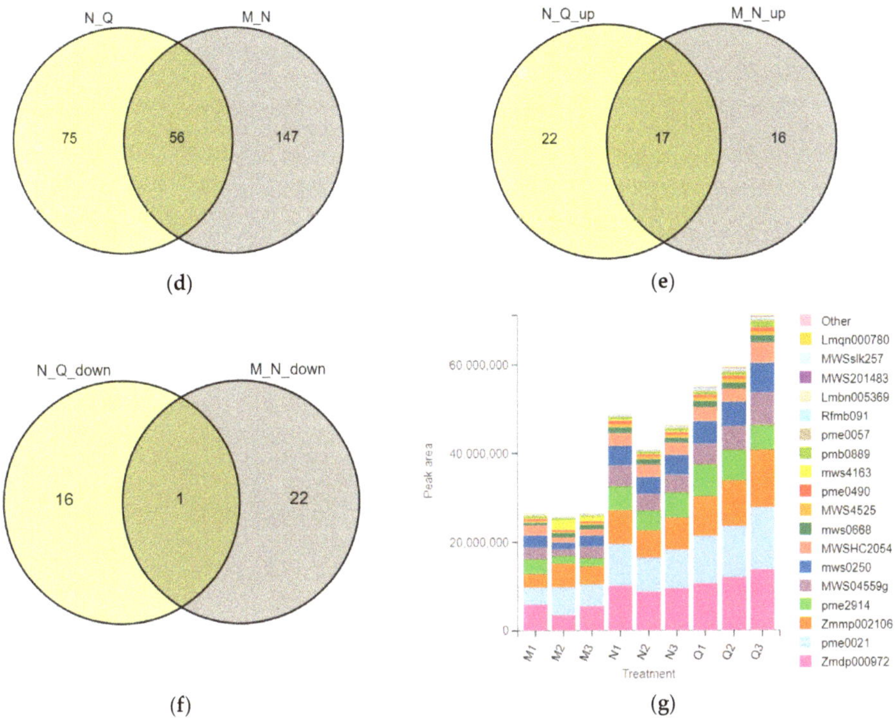

Figure 2. Analysis of the primary DMs in *R. chrysanthum* following UV-B exposure and UV-B exposure with externally administered ABA. (**a**) Statistic differences in metabolites across groups; (**b**,**c**) DMs of the M−N and N−Q groups are displayed on volcano plots, respectively. In the volcano plot, red circles represent up-regulated substances and green circles represent down-regulated substances, horizontal dashed lines indicate screening $p < 0.05$, vertical dashed lines screening $|\text{fold change}| > 1.2$; (**d**–**f**) Venn diagrams for DMs, upregulated DMs, and downregulated DMs are shown for the M−N and N−Q groups, respectively; (**g**) The three MNQ groups' core DMs are shown in bar graphs.

The ratio of the phenolic content in *R. chrysanthum* after UV-B radiation only to the phenolic content in *R. chrysanthum* without UV-B radiation as well as the ratio of the phenolic content in *R. chrysanthum* after UV-B radiation with externally applied ABA to the phenolic content in *R. chrysanthum* after UV-B radiation with no ABA applied were greater than one (Figure 4a–c). In other words, phenolics contribute to *R. chrysanthum*'s ability to fend off UV-B harm. The plant's ability to withstand UV-B stress was also improved by the effective promotion of stomatal closure by ABA when administered externally. Additionally, the abundance changes of all C6C3C6 backbone phenolics were less pronounced in the MN group than in the NQ group, indicating that this group of phenolics likely contributes significantly to *R. chrysanthum*'s UV-B protection (Figure 4c).

3.3. Network Diagram and Correlation Analysis of R. chrysanthum's Main and Phenolic Metabolite Relationships

A metabolite interaction network was carried out to further investigate how the primary and phenolic metabolites in the *R. chrysanthum* responded to UV-B exposure. To supervise the network's simplification, the Degree Filter was set to 2.0 and the Betweenness Filter was set to 1.0. Adenine is a significant "hub" in the network since it has the greatest degree and the most connections to other nodes (84). The purine metabolism pathway is the one that is reportedly damaged, according to the network diagram. With a degree of 71, L-phenylalanine is also a more significant "hub" in the network, influencing pathways such as the biosynthesis of phenylalanine, tyrosine, and tryptophan as well as the metabolic pathway of phenylalanine. Both of these pathways have *p*-values of 0.01, and L-phenylalanine is labeled in the figure with its structural formula. L-tyrosine's degree is also relatively large, 63, and plays an important role in the network, affecting pathways. In addition to the biosynthesis of phenylalanine, tyrosine, and tryptophan, there is also the tyrosine metabolic pathway, which is labeled as the L-tyrosine structural formula in the figure. Many plant secondary metabolites, such as rhodiolosides, betalains, and tocopherols (vitamin E), are derived from tyrosine.

For each category of phenolic core DMs and each category of primary core DMs in M, N, and Q, Pearson's pairwise correlation coefficient values were obtained. Significant correlations were defined as those with a *p*-value of 0.05 or below and a relationship analysis threshold of 0.8.

Table 1. The first seven key core DMs' VIP and *p*-value changes in different components under UV-B circumstances of stress.

Index	Structural Formula	Compounds	VIP (MN)	*p*-Value (MN)	VIP (NQ)	*p*-Value (NQ)
pme0490	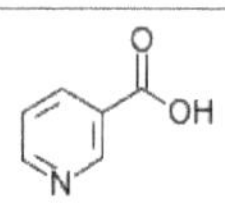 C00253	Nicotinic acid (Vitamin B3)	1.707	0.021	1.290	0.182
Zmdp000972	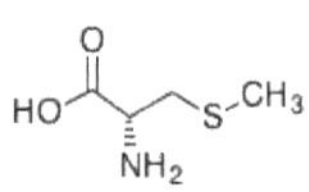 C22040	S-methyl-L-cysteine	1.683	0.010	1.610	0.078
pme2914	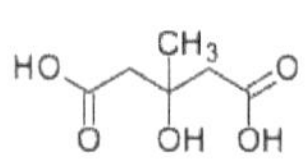 C03761	3-hydroxy-3-methylpentane-1,5-dioic acid	1.676	0.012	1.425	0.097
mws0250	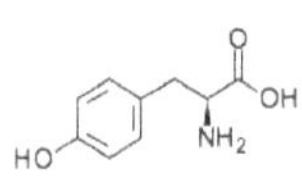 C00082	L-tyrosine*	1.655	0.013	1.627	0.079
pme0021	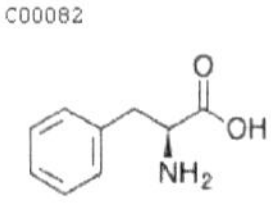 C00079	L-phenylalanine	1.655	0.017	1.694	0.057
MWS4525	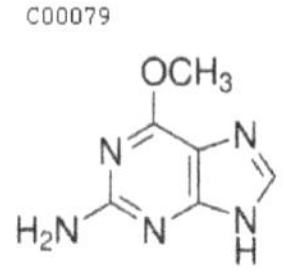	6-O-methylguanine	1.602	0.029	1.268	0.166
MWSslk257	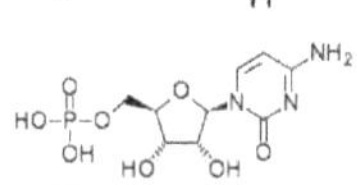 C00065	Cytidine 5′-monophosphate(Cytidylic acid)	1.595	0.112	1.268	0.192

In the correlation analysis between amino acids and their derivatives and phenolic acids, there were 57 positive correlations and no negative correlations (Figure 5b). For instance, L-phenylalanine showed a positive correlation with ferulic acid-4-O-glucoside, 3-O-Feruloylquinic acid-O-glucoside, and particularly with Trans-5-O-(p-Coumaroyl) shikimate, as well as significantly positive correlations with chlorogenic acid (3-O-Caffeoylquinic acid) and correlation coefficients of 0.95 and 0.90, respectively. It was demonstrated (Figure 5c) that there were 40 positive correlations and, once more, no negative correlations in the correlation study of amino acids and their derivatives with flavonoids. For instance, L-tyrosine* showed a strong association with 5,7,3′-Trihydroxy-4,6,5′-trimethoxyxanthone, luteolin-7-O-glucuronid e-5-O-rhamnoside, and especially with dihydromyricetin (ampelopsin), which had a correlation value of 0.86. There were 10 positive associations and no negative correlations in the correlation study of amino acids and their derivatives with lignans, alkaloids, and terpenoids (Figure 5d). For instance, N-(2-hydroxy-4-methoxyphenyl) acetamide was strongly positively associated with L-hom ocysteine, Dimethylfraxetin, Dehydrodiconiferyl alcohol, and others, with a correlation coefficient of 0.96. Combining these three correlation analyses, it was discovered that amino acids and their derivatives significantly correlated with phenolic acid compounds, and the interaction network was more complex. This finding suggests that amino acids and their derivatives, along with phenolic acids, play significant roles in the *R. chrysanthum*'s defense against UV-B stress and that the two classes of compounds can be used as markers for assessing the *R. chrysanthum*.

Again, only 25 positive correlations and no negative correlations were found in the correlation analysis of nucleotides and their derivatives with phenolic acids (Figure 5e). For instance, with correlation values of 0.92 and 0.88, respectively, adenine strongly positively linked with 2-phenylethanol, and xanthosine significantly positively associated with Trans-5-O-(p-Coumaroyl) shikimate. Only 22 positive correlations between nucleotides and their derivatives and flavonoids were found in the correlation study (Figure 5f). For instance, adenine and dihydromyricetin (ampelopsin) had a correlation of 0.91, which was significantly positive. In addition, 6-O-methylguanine and dihydromyricetin (ampelopsin) exhibited positive correlations with adenine and 13(S)-HODE, respectively. There were only seven positive connections and no negative correlations in the correlation study of nucleotides and their derivatives with lignans, alkaloids, and terpenoids (Figure 5g). For instance, adenine and N-(2-hydroxy-4-methoxyphenyl) acetamide had a correlation coefficient of 0.94 that was significantly positive. Combining the results of the three correlation analyses, it was discovered that nucleotides and their derivatives had the strongest link with phenolic acids and interacted with the most compounds, followed by flavonoids.

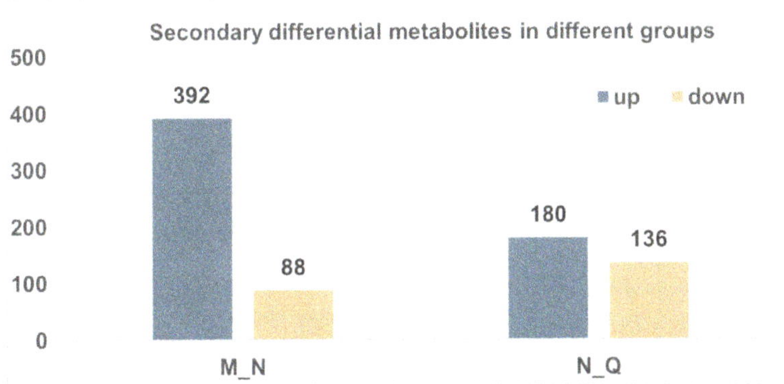

(a)

Figure 3. *Cont.*

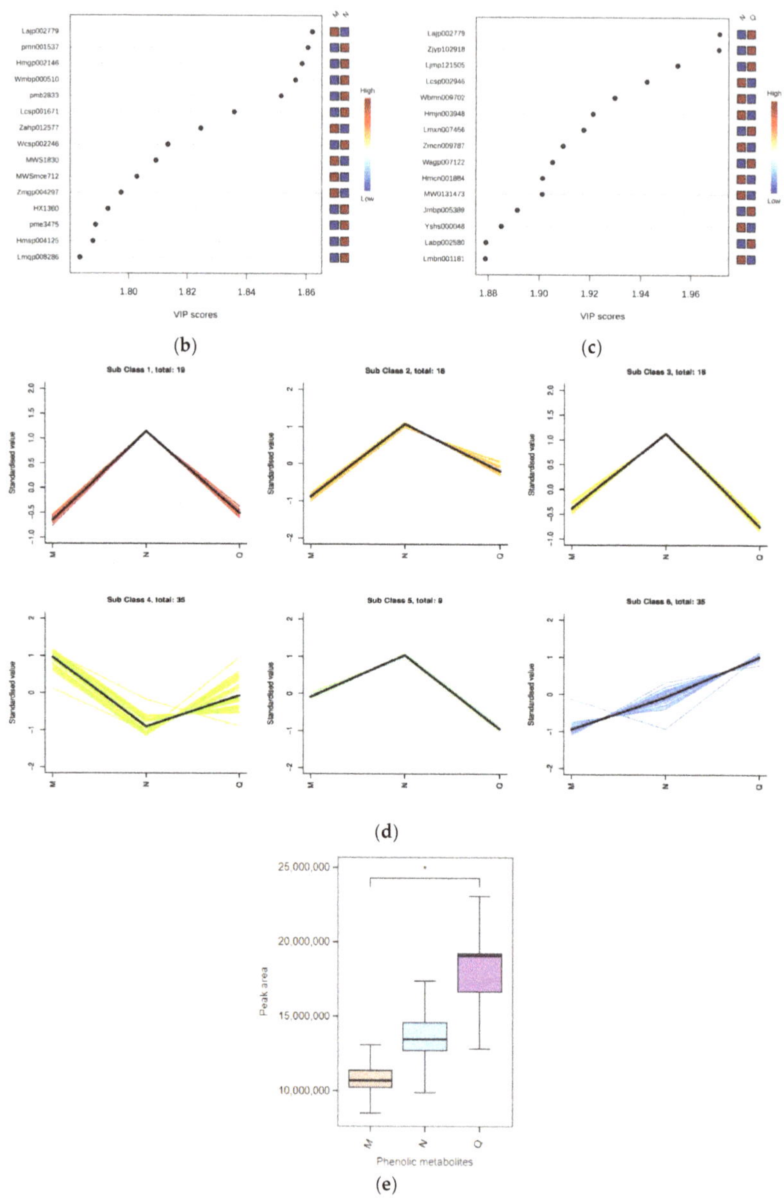

Figure 3. Secondary *R. chrysanthum* metabolites in reaction to UV-B light and UV-B radiation with externally administered ABA. (**a**) Comparative metabolite histogram statistics between different categories. (**b**,**c**) For the M−N and N−Q groups, respectively, the top 15 different metabolites ranked by VIP value of secondary differential compounds are displayed, with red representing high levels and green indicating low levels; (**d**) analyses of metabolites using the K-means clustering technique; (**e**) trends in phenolic metabolite abundance with UV-B therapy and exposure to UV-B rays with externally administered ABA are shown in a boxed line graph, * indicates $p < 0.05$.

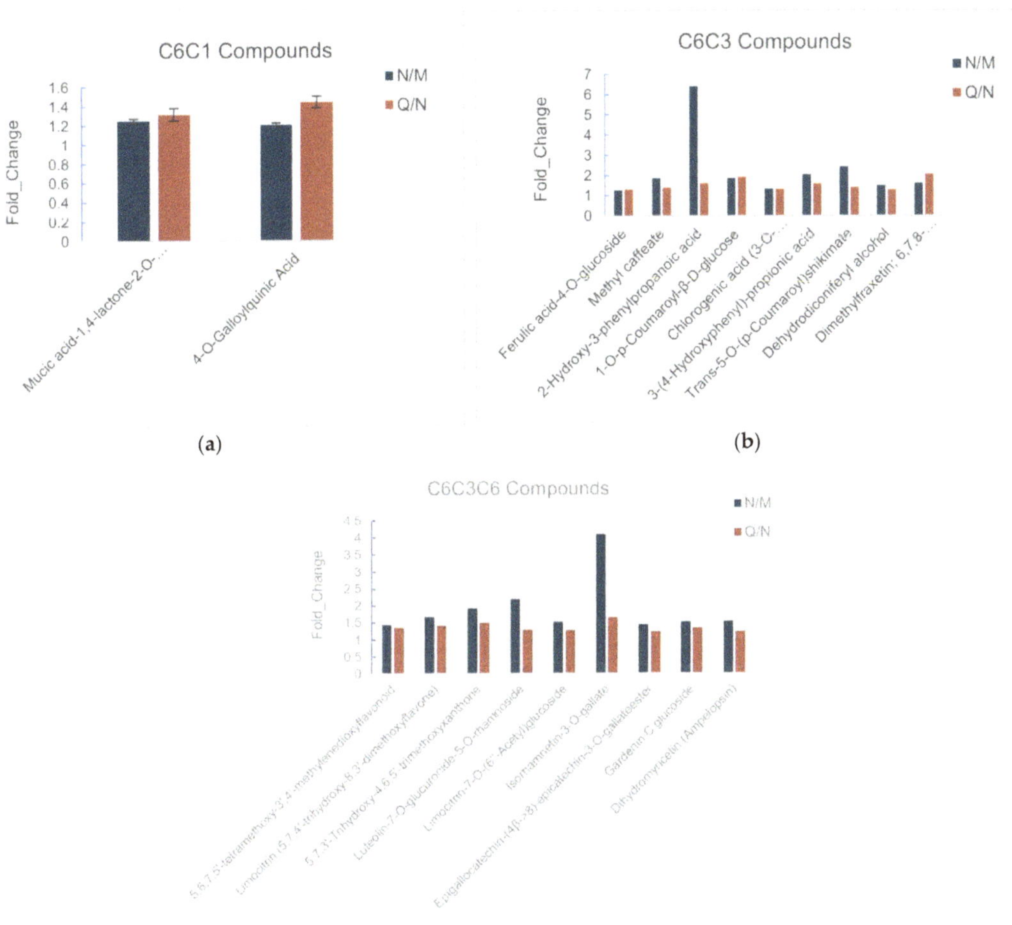

(c)

Figure 4. Statistics of phenolic and flavonoid abundance under the effects of UV-B and UV-B treatment with externally administered ABA. The proportion of each substance's expression is set as the vertical coordinate. (**a**) C6C1-carbon skeleton phenols; (**b**) phenolics with the carbon skeleton C6C3; (**c**) phenolics with the carbon skeleton C6C3C6.

Thirty-three correlations were found in the correlation analysis of vitamins, lipids, sugars, and phenolic acids (Figure 5h), twenty-nine of which were positive and four of which were negative. For instance, Nystose and methyl caffeate were negatively connected, while 13(S)-HODE and Elaidolinolenic Acid were positively correlated. Additionally, there was a substantial positive link between chlorogenic acid (3-O-caffeoylquinic acid) and nicotinic acid (Vitamin B3), with correlation coefficients of 0.88 and 0.90, respectively. Twenty-six correlations were found in the correlation analysis of vitamins, lipids, sugars, and flavonoids (Figure 5i), of which twenty-two were positive and four were negative. Specifically, dihydromyricetin (ampelopsin) and nicotinic acid (Vitamin B3) revealed a substantial positive link with a correlation of 0.94. Nicotinic acid (Vitamin B3) was positively correlated with 13(S)-HODE, Nystose was negatively correlated with luteolin-7-O-glucuronide-5-O-rhamnoside, etc. Only seven associations, four of which were positive and three of which were negative, were discovered when vitamins, lipids, and carbohydrates were correlated with lignans, alkaloids, and terpenoids (Figure 5j). It was found that 13(S)-HODE and N-(2-hydroxy-4-methoxyphenyl) acetamide had positive and negative correlations, respectively. With a correlation coefficient of 0.95, N-(2-hydroxy-4-methoxyphenyl) acetamide in particular showed a highly positive link with nicotinic acid (Vitamin B3). These three correlation analyses combined showed that there were more metabolite interactions between vitamins, lipids, and sugars in correlation with phenolic acids to help *R. chrysanthum* resist UV-B stress, indicating that phenolic acid compounds can be used as markers for *R. chrysanthum*'s reaction to UV-B stress.

When viewed collectively, the nine correlation analyses demonstrate the critical role that amino acids and their derivatives play in the network of correlation analyses as well as the complex interactions that phenolic compounds, including phenolic acids and flavonoids, exhibit with other compounds in this network. Additionally, it was shown that only sugars, lipids, and vitamins had negative correlations with other substances, which may be connected to Nystose's decreased expression in response to UV-B radiation. Together, our findings imply that in response to extreme UV-B exposure, *R. chrysanthum* mobilizes more complicated metabolite relationships to cooperate.

3.4. Building a Comprehensive Picture of the Primary and Phenolic Metabolite Networks of the R. chrysanthum

Simplified primary and phenolic compound metabolic pathways were employed to demonstrate the *R. chrysanthum*'s metabolic response to UV-B radiation in order to more clearly illustrate the linkages between the data mentioned above. Heat maps are used to display the expression levels of the relevant enzymes in the proteome in an effort to completely reveal the enzymes associated with the L-phenylalanine and flavonoid biosynthesis pathways as well as in the monolignol biosynthesis pathway (Figure 6a). Metabolite abundance and the levels of metabolite-related enzyme expression in metabolic pathways are intricately connected. Some amino acids, such as S-methyl-L-cysteine, L-phenylalanine, and L-tyrosine*, considerably elevated in response to UV-B radiation for primary metabolites. Moreover, following UV-B treatment, the concentrations of acids implicated in the tricarboxylic acid (TCA) cycle, such as fumarate, malate, and succinate, were elevated. In terms of the phenolic components, Trans-5-O-(p-Coumaroyl) shikimate and naringenin chalcone were the primary phenolic acids and flavonoids that were upregulated in *R. chrysanthum*. Furthermore, the shikimate acid pathway uses L-phenylalanine as a crucial node in the synthesis of phenolic compounds, which may explain why most phenolic compounds rise in concentration when exposed to UV-B light. The boxed line graphs show how the expression of L-phenylalanine, naringenin chalcone, and Trans-5-O-(p-Coumaroyl) shikimate changes in response to UV-B stress. Significant changes in the expression of L-phenylalanine and naringenin chalcone can be observed, indicating a potential protective effect of *R. chrysanthum* against UV-B stress.

3.5. Enrichment Pathway Analysis and Correlation between R. chrysanthum Proteomics and Metabolomics in Response to UV-B Stress

KEGG enrichment analysis of the 11 identified enzymes was performed on a total of eight metabolic pathways (Figure 7a,b). These included flavonoid biosynthesis; circadian rhythm of plant, propane, piperidine and pyridine alkaloids biosynthesis; secondary metabolites biosynthesis; metabolic pathways, flavonoids, and flavonoids biosynthesis; phenyl metabolic pathway, piperidine, and pyridine alkaloids biosynthesis; secondary metabolites biosynthesis; metabolic pathways, flavonoids, and flavonols biosynthesis; phenylalanine, tyrosine, and tryptophan biosynthesis; and amino acid biosynthesis. According to the results, L-phenylalanine, naringenin chalcone, and Trans-5-O-(p-Coumaroyl) shikimate increased in response to UV-B stress while cinnamic acid, p-Coumaric acid, and naringenin (5,7,4′-Trihydroxyflavanone) * were continually depleted. Among these, Trans-5-O-(p-Coumaroyl) shikimate, naringenin chalcone, and L-phenylalanine displayed considerably higher levels and are important nodes in the UV-B stress regulation network.

Pearson correlation studies were carried out (Figure 7c) to further evaluate the links between important enzymes and essential DMs. The hub of the entire network, CHS and CHI, exhibits a favorable connection with metabolites. The 2-hydroxy-3-phenylpropanoic acid, isorhamnetin-3-O-gallate, and 3-O-Feruloylquinic acid-O-glucosid all showed strong positive associations with CHS1 and CHS, with the least significant correlation coefficient being 0.88 and the most significant correlation coefficient being as high as 0.99. With correlation coefficients around 0.9, CHI and CHI3 demonstrated valuable positive relationships with 3-O-Feruloylquinic acid-O-glucosid, 2-hydroxy-3-phenylpropanoic acid, and L-phenylalanine. Two metabolites, 1-O-p-Coumaroyl-D-glucose and 3-hydroxy-3-methylpentane-1,5-dioic acid, are under the control of F3H. Additionally, PPA-AT favorably regulated 1-O-p-Coumaroyl-D-glucose and 2-hydroxy-3-phenylpropanoic acid. These six metabolites belonged to amino acids and their derivatives, phenolic acids, and flavonoids, indicating that CHS, CHI, F3H, and PPA-AT collectively affected primary metabolites as well as phenolic compounds against UV-B stress in the *R. chrysanthum*. This further affected phenylpropane biosynthesis and controlled the primary as well as phenolic compound metabolic pathways. The information about 11 important enzymes is shown in Table 2.

Table 2. Information about 11 important enzymes under UV-B stress.

Protein Accession	KEGG KO No.	KEGG Gene	M/N Ratio
Gene.41041_CL5299.Contig1_All	K00475	F3H; naringenin 3-dioxygenase [EC:1.14.11.9]	0.573
Gene.13909_CL1098.Contig2_All	K00660	CHS; chalcone synthase [EC:2.3.1.74]	0.48
Gene.75412_Unigene13888_All	K00660	CHS; chalcone synthase [EC:2.3.1.74]	0.546
Gene.13913_CL1098.Contig3_All	K00660	CHS; chalcone synthase [EC:2.3.1.74]	0.557
Gene.22273_CL2057.Contig1_All	K01850	E5.4.99.5; chorismate mutase [EC:5.4.99.5]	0.727
Gene.77444_Unigene15688_All	K01859	E5.5.1.6; chalcone isomerase [EC:5.5.1.6]	0.781
Gene.9527_CL706.Contig2_All	K01859	E5.5.1.6; chalcone isomerase [EC:5.5.1.6]	0.865

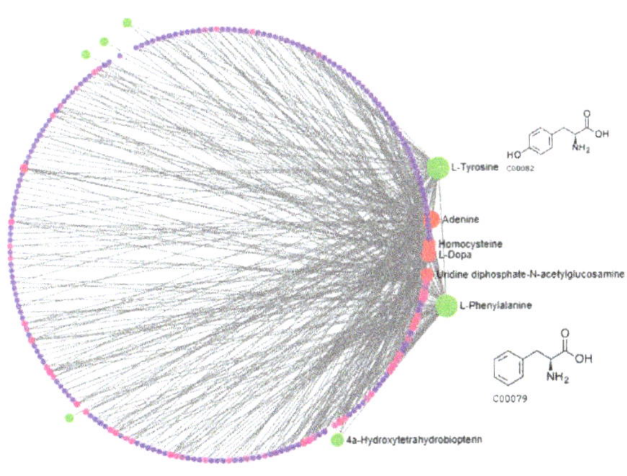

(a)

Figure 5. *Cont.*

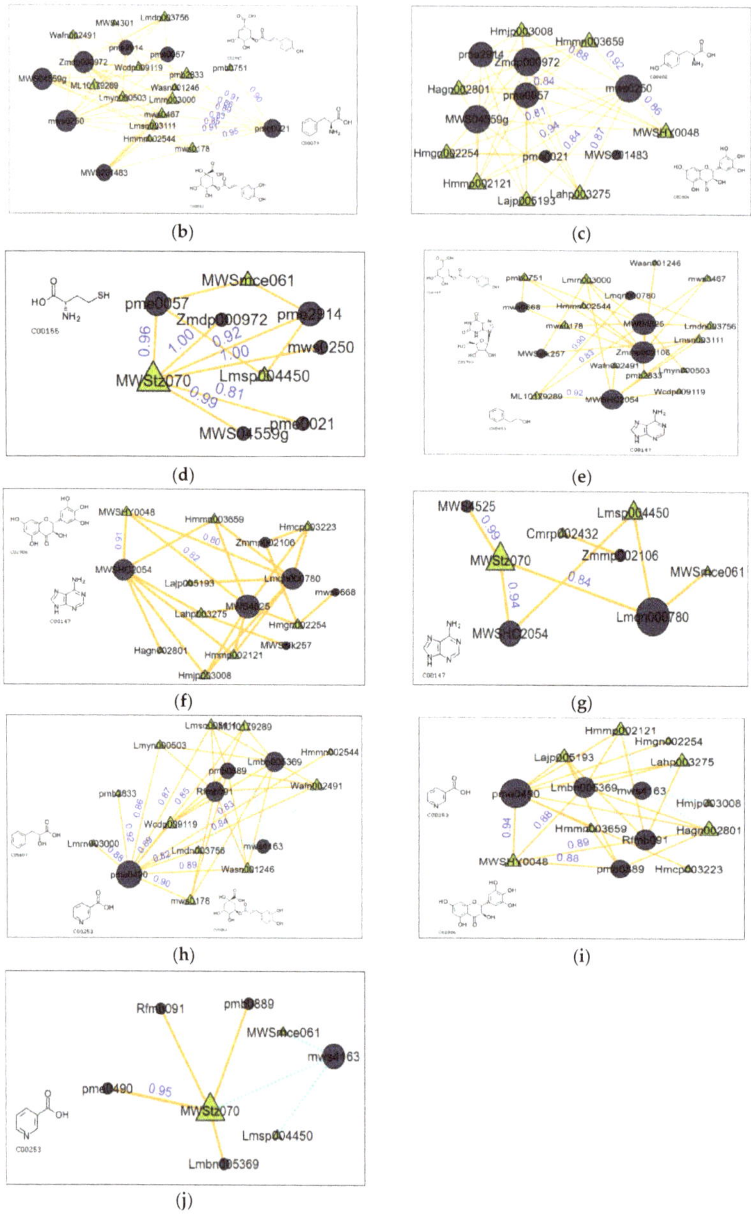

Figure 5. Network analysis and correlation of different primary and secondary metabolites linked to *R. chrysanthum*'s reaction to UV-B stress are shown. (**a**) Degrees are the number of connections to other points, and green is used to highlight the substances that are the focus of the study. (**b–d**) Correlation analysis of amino acids and their derivatives with phenolic acids, flavonoids, and lignans, alkaloids, and terpenoids, respectively. (**e–g**) Correlation analysis of nucleotides and their derivatives with phenolic acids, flavonoids, and lignans, alkaloids, terpenoids, respectively. (**h–j**) Correlation analysis of sugars, vitamins, and free fatty acids with phenolic acids, flavonoids, and lignans, alkaloids, and terpenoids, respectively.

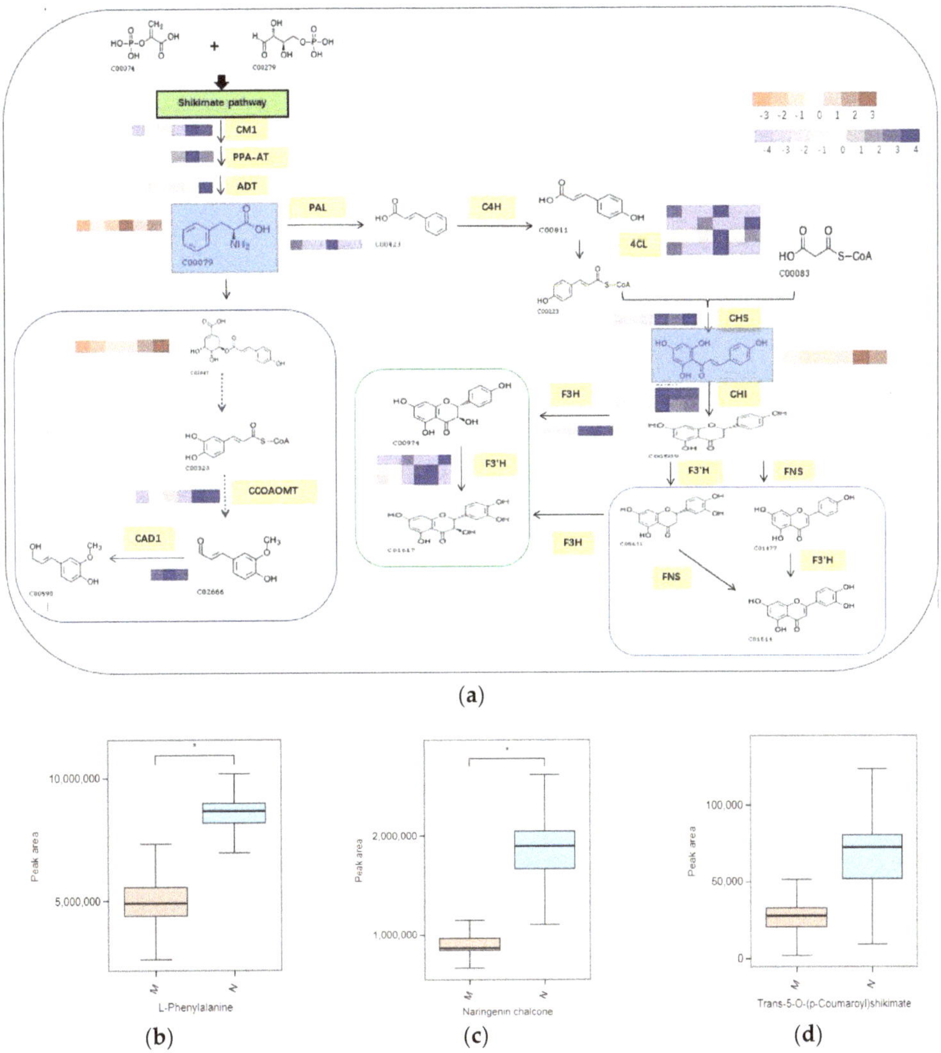

Figure 6. Visualization of dynamic pathways of primary and phenolic metabolites. (**a**) Yellow fills indicate enzymes that cause changes in the pathway, while green fills indicate metabolites with significant changes in the pathway. The flavanone metabolic pathway is represented by pink boxes, the flavonol metabolic pathway by green boxes, and the monolignol metabolic pathway by blue boxes. In the heat map, the color pink denotes that UV-B duress has almost no impact on the enzyme, the color light purple denotes that UV-B radiation inhibits the enzyme's activity, and the color dark purple denotes that UV-B radiation increases the enzyme's activity; likewise, a metabolite's light brown hue suggests that UV-B radiation is downregulating it, whereas a metabolite's dark brown color indicates that UV-B radiation is upregulating it. (**b**–**d**) L-phenylalanine, naringenin chalcone, and Trans-5-O-(p-Coumaroyl) shikimate all show significant alterations. Significant difference (* $p < 0.05$).

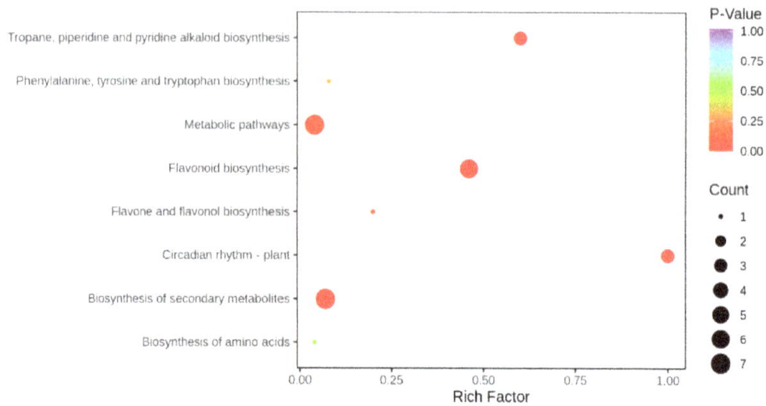

(**a**)

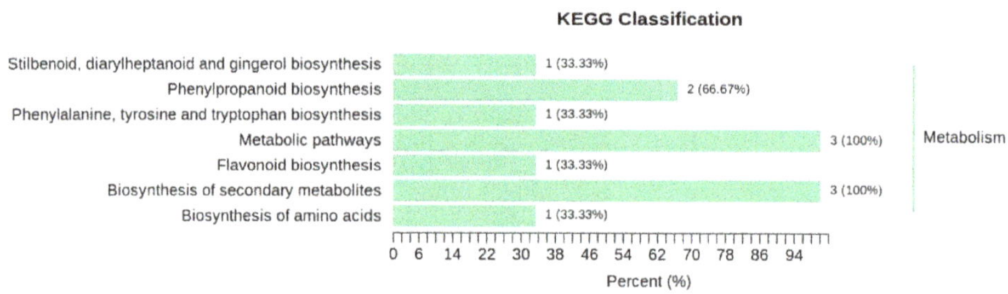

(**b**)

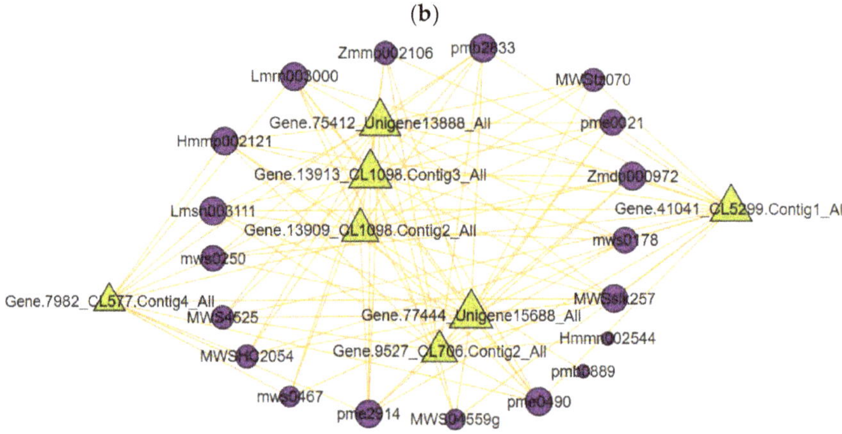

(**c**)

Figure 7. Key enzymes involved in *R. chrysanthum*'s reaction to UV-B exposure were analyzed using the KEGG pathway method, and correlations between key enzymes and metabolites were examined. (**a**,**b**) Key enzymes' KEGG bubble and bar graphs; (**c**) links between proteins and metabolomes. Proteins are represented by triangles, metabolites by circles, and positive relationships by solid yellow lines.

4. Discussion

Numerous responses of plants to elevated UV-B radiation have an impact on plant development, growth, and the accumulation of secondary metabolites [31]. According to the UPLC-MS/MS detection platform, the abundance and characterization of metabolites were investigated in this study, and *R. chrysanthum*'s response to UV-B stress was examined at the level of both primary and secondary metabolites. These analyses together revealed alterations to the metabolic profiles of various metabolites of the *R. chrysanthum* under UV-B stress and explored the reprogramming of primary and secondary metabolic responses to UV-B. In order to validate the changes in the expression of important metabolites in the pathway in response to UV-B stress, we also examined protein expression under UV-B stress based on LC-MS/MS and evaluated changes in the abundance of the key enzymes of L-phenylalanine, tyrosine, and tryptophan; the flavonol metabolism pathway, the flavonoid metabolism pathway, and the monolignol biosynthesis pathway, as well as changes in the pathways of enzyme action through the proteome.

Based on UPLC-MS/MS metabolomics, we examined specific variations in the principal metabolites of *R. chrysanthum* under UV-B radiation. Growth requires energy from sugar, and glucose is a significant soluble sugar and signaling molecule that clearly plays a fundamental role in plant stress [32]. Following UV-B radiation, the *R. chrysanthum*'s decreased glucose levels may be linked to a more pronounced transition from carbon absorption to carbon storage [31]. Additionally, larger concentrations of monosaccharides from the pentose phosphate pathway (PPP), such as D-erythrose-4-phosphate, were found in the UV-B-treated group, which may be due to the PPP pathway's higher respiratory frequency in the *R. chrysanthum*. Through the glycolytic pathway and the TCA cycle, carbohydrates like sucrose can be oxidized, and since glycolysis can result in an increase in ATP content, the plant's capacity to adapt to its surroundings under abiotic stress circumstances is unavoidably improved [33]. The fundamental metabolites that plants primarily produce are amino acids, which are crucial for protecting them from abiotic stressors and, more critically, serving as precursors to secondary metabolites that increase plant resistance to UV-B radiation. This research uses L-phenylalanine as an example, which exhibits a substantial rise under UV-B stress. The flavonoid pathway is indirectly impacted by the activity of pathways involving phenylpropanoid substances linked with amino acid metabolism, as has been shown in prior investigations, and the same results have been drawn in this paper [34].

The amount of flavanols and flavonols in *R. chrysanthum* increased in response to UV-B radiation, according to the dynamic response pattern of the phenolic compounds. According to physiological research, phenolics play a role in UV-B protection [35]. Within the phenolics, flavonoids and flavonols shield the cells as they collect in the *R. chrysanthum*'s epidermis, acting as filters and absorbing the majority of UV-B radiation [36,37]. Ryan et al. reported that UV radiation caused the production of flavonols with a significant amount of hydroxylation in the *Petunia* and the *Arabidopsis thaliana*. They hypothesized that flavonols might act on an as-yet-uncharacterized role in the UV stress reaction due to their UV-absorbing qualities and that the hydroxylation of flavonols may have a positive impact on antioxidant capacity [38]. Thus, under UV-B radiation, the biggest increases in the concentrations of flavonoid and flavonol substances such as luteolin (5,7,3′,4′-Tetrahydroxyflavone) and quercetin-3-O-rhamnoside (quercitrin) were seen. Additionally, it was found that C6C3C6 carbon compounds were accumulating at a faster rate than C6C3 carbon compounds like cinnamic acid and ferulic acid, which were shown to be accumulating at a slower rate. The greater UV-B absorption and antioxidant activity of C3C6C3 carbon compounds compared to C3C6 carbon compounds (Figure 4a–c) may be the cause of this event, which led to the incorporation of C6C3C6 carbon components at the expense of C3C6 carbon compounds.

For the *R. chrysanthum*, interaction network diagrams were created for the primary and phenolic metabolites, and correlation analysis diagrams were examined between each of the primary core difference metabolites and the secondary phenolic metabolites in order to better understand the differences in the specificity of metabolites identified between the UV-B-treated and control groups (Figure 5). The network diagram was mainly enriched with phenylalanine, tyrosine, and tryptophan biosynthesis as well as phenylalanine biosynthesis pathways, suggesting the significance of L-phenylalanine in the *R. chrysanthum*'s defense against UV-B stress; and the absence of phenolics-annotated pathways may be due to the fact that some of the detected phenolics were not annotated to the KEGG data, which can be further explored in future studies. A diagram of the primary and phenolic metabolite pathways was created (Figure 6) to help further clarify the mechanisms of primary as well as phenolic metabolites against UV-B damage. UV-B light causes a rise in TCA cycle intermediates such as fumarate, a key regulator of carbon and nitrogen interactions. After plants mobilize their stored carbon reserves to boost flux along the phenylpropane pathway, it appears highly likely that fumarate will be elevated under UV-B light [39]. The same conclusion was obtained in our experiments. Shikimic acid is the initiator of secondary metabolism in higher plants, linking sugar metabolism to polyphenol metabolism, and is an essential pathway for the biosynthesis of aromatic compounds. In plants, phenylalanine is recognized as a general precursor of C6C1-, C6C3-, and C6C3C6 compounds and their polymers such as tannins and lignin [40]. Thus, taken as a whole, these data imply that in response to UV-B, there is an "initiating" mechanism at the primary metabolic level in plant cells that involves the reprogramming of metabolism to effectively transfer carbon to aromatic amino acid precursors of the phenylpropane pathway [41]. Tryptophan and phenylalanine compete for chorismate to synthesize alkaloids and phenols, respectively [31]. Phenylalanine accumulated greater under UV-B radiation, according to the findings. This demonstrates that the increase in phenolic compounds during UV-B radiation, which is caused by the movement of carbon in primary metabolism, is the result of metabolic reprogramming. Phenylalanine, a critical node in phenolic metabolism, dramatically increased under UV-B radiation.

Proteomics was used to validate the aforementioned inferences made from metabolomics (Figure 7a–c). On the one hand, qualitative and quantitative analyses of the key enzymes in the metabolic pathway showed that the changes between the key enzymes in the pathway and the key metabolites showed consistency; specifically, the key enzymes in the metabolic pathway were significantly increased in the *R. chrysanthum* against UV-B stress, which in turn affected the increase in the expression of the related metabolites and the response of the entire pathway to UV-B stress. The findings of a correlation analysis of 11 major enzymes from mid-species with the core DMs, on the other hand, showed that the proteome and metabolome responded consistently to UV-B stress.

We also discovered that after applying ABA externally and then applying UV-B radiation, alterations in significant primary metabolites and phenolics were considerably more pronounced than they were after the standard UV-B radiation treatment. In contrast to flavonoids, which are secondary metabolites with antioxidant activity and UV absorbers, ABA enhances UV-B stress acclimation by promoting ROS scavenging, according to earlier investigations [42]. The results of the present study's results, which had a much greater flavonoid content, imply that externally administered ABA may work in concert with UV-B to encourage the conversion of primary metabolites to phenolic compounds.

5. Conclusions

Using ultra-performance liquid chromatography tandem mass spectrometry (UPLC-MS/MS) technology, we were able to determine the metabolic profiles and explore the particular metabolic pathways of *R. chrysanthum* in response to short-duration UV-B radiation. Our findings indicate that plant cells reprogrammed the L-phenylalanine biosynthesis pathway to undergo a more intensive shift from carbon absorption to carbon accumulation as a main metabolic level mechanism in response to UV-B exposure. This gives phenylalanine, a precursor substance, a solid theoretical foundation for research on its potential to assist plants in coping with abiotic stressors. Simultaneously, alterations in pertinent metabolites and enzymes within the flavonoid and monolignol biosynthesis pathways were initiated at the secondary metabolic level. In phenolics, C6C1 carbon compounds (phenolic acids) and C6C3C6 carbon compounds (flavonoids and flavonols) were upregulated at the expense of C6C3 carbon compounds in order to withstand UV-B stress. Furthermore, we discovered that *R. chrysanthum*'s resistance to UV-B was markedly improved and that phenolic abundance increased significantly following the external application of ABA; in other words, external application of ABA may encourage the conversion of primary metabolites to phenolics in concert with UV-B, thereby strengthening the plant's capacity for both UV-B uptake and antioxidant defense. In addition to offering insights for future research on the mechanism by which externally applied ABA can mitigate UV-B damage in the *R. chrysanthum*, this study advances our knowledge of how primary and secondary metabolites in this plant respond to UV-B stress.

Supplementary Materials: The following supporting information can be downloaded at: https://www.mdpi.com/article/10.3390/biom13121700/s1, Figure S1: Total ion flow diagram in positive ion mode; Figure S2: Total ion flow map in negative ion mode; Table S1: Information on 18 core differential metabolites.

Author Contributions: Writing—original draft preparation, F.G. and W.Y.; writing—review and editing, K.C., H.X. and X.Z.; methodology, F.G. and Q.Z.; software, F.G. and J.D.; visualization, F.G. and W.Y.; supervision, K.C. and X.Z.; project administration, H.X. and X.Z. All authors have read and agreed to the published version of the manuscript.

Funding: This work was supported by the Jilin Provincial Science and Technology Development Program project (Grant numbers: YDZJ202301ZYTS517).

Institutional Review Board Statement: Not applicable.

Informed Consent Statement: Not applicable.

Data Availability Statement: The datasets generated during and/or analyzed during the current study are available from the corresponding author on reasonable request.

Acknowledgments: We are grateful to Jingjie PTM Biolab (Hangzhou, China) Co., Ltd. for providing mass spectrometry support.

Conflicts of Interest: The authors declare no conflict of interest.

eferences

Hossaini, R.; Chipperfield, M.P.; Montzka, S.A.; Leeson, A.A.; Dhomse, S.S.; Pyle, J.A. The increasing threat to stratospheric ozone from dichloromethane. *Nat. Commun.* **2017**, *8*, 15962. [CrossRef]

Neale, R.E.; Barnes, P.W.; Robson, T.M.; Neale, P.J.; Williamson, C.E.; Zepp, R.G.; Wilson, S.R.; Madronich, S.; Andrady, A.L.; Heikkilä, A.M.; et al. Environmental effects of stratospheric ozone depletion, UV radiation, and interactions with climate change: UNEP Environmental Effects Assessment Panel, Update 2020. *Photochem. Photobiol. Sci.* **2021**, *20*, 1–67. [CrossRef]

Bornman, J.F.; Barnes, P.W.; Robinson, S.A.; Ballaré, C.L.; Flint, S.D.; Caldwell, M.M. Solar ultraviolet radiation and ozone depletion-driven climate change: Effects on terrestrial ecosystems. *Photochem. Photobiol. Sci.* **2015**, *14*, 88–107. [CrossRef]

Parihar, P.; Singh, S.; Singh, R.; Singh, V.P.; Prasad, S.M. Changing scenario in plant UV-B research:UV-B from a generic stressor to a specific regulator. *J. Photochem. Photobiol. B* **2015**, *153*, 334–343. [CrossRef]

Jenkins, G.I. Signal Transduction in Responses to UV-B Radiation. *Annu. Rev. Plant Biol.* **2009**, *60*, 407–431. [CrossRef]

Neugart, S.; Fiol, M.; Schreiner, M.; Rohn, S.; Zrenner, R.; Kroh, L.W.; Krumbein, A. Interaction of Moderate UV-B Exposure and Temperature on the Formation of Structurally Different Flavonol Glycosides and Hydroxycinnamic Acid Derivatives in Kale (*Brassica oleracea* var. sabellica). *J. Agric. Food. Chem.* **2014**, *62*, 4054–4062. [CrossRef] [PubMed]

Wargent, J.J.; Moore, J.P.; Roland Ennos, A.; Paul, N.D. Ultraviolet Radiation as a Limiting Factor in Leaf Expansion and Development. *Photochem. Photobiol.* **2009**, *85*, 279–286. [CrossRef] [PubMed]

Frolov, A.; Henning, A.; Böttcher, C.; Tissier, A.; Strack, D. An UPLC-MS/MS Method for the Simultaneous Identification and Quantitation of Cell Wall Phenolics in *Brassica napus* Seeds. *J. Agric. Food. Chem.* **2013**, *61*, 1219–1227. [CrossRef]

Santin, M.; Simoni, S.; Vangelisti, A.; Giordani, T.; Cavallini, A.; Mannucci, A.; Ranieri, A.; Castagna, A. Transcriptomic Analysis on the Peel of UV-B-Exposed Peach Fruit Reveals an Upregulation of Phenolic- and UVR8-Related Pathways. *Plants* **2023**, *12*, 1818. [CrossRef]

). Gómez-Caravaca, A.M.; López-Cobo, A.; Verardo, V.; Segura-Carretero, A.; Fernández-Gutiérrez, A. HPLC-DAD-q-TOF-MS as a powerful platform for the determination of phenolic and other polar compounds in the edible part of mango and its by-products (peel, seed, and seed husk). *Electrophoresis* **2016**, *37*, 1072–1084. [CrossRef]

1. Abad-García, B.; Berrueta, L.A.; Garmón-Lobato, S.; Gallo, B.; Vicente, F. A general analytical strategy for the characterization of phenolic compounds in fruit juices by high-performance liquid chromatography with diode array detection coupled to electrospray ionization and triple quadrupole mass spectrometry. *J. Chromatogr. A* **2009**, *1216*, 5398–5415. [CrossRef]

12. Maeda, H.; Dudareva, N. The Shikimate Pathway and Aromatic Amino Acid Biosynthesis in Plants. *Annu. Rev. Plant Biol.* **201** *63*, 73–105. [CrossRef]
13. Herrmann, K.M.; Weaver, L.M. The Shikimate Pathway. *Annu. Rev. Plant Physiol. Plant Mol. Biol.* **1999**, *50*, 473–503. [CrossRef]
14. Babst, B.A.; Harding, S.A.; Tsai, C.-J. Biosynthesis of Phenolic Glycosides from Phenylpropanoid and Benzenoid Precursors i *Populus*. *J. Chem. Ecol.* **2010**, *36*, 286–297. [CrossRef]
15. Marković, S.; Tošović, J. Application of Time-Dependent Density Functional and Natural Bond Orbital Theories to the UV–v Absorption Spectra of Some Phenolic Compounds. *J. Phys. Chem. A* **2015**, *119*, 9352–9362. [CrossRef]
16. Wang, B.; Li, L.; Liu, M.; Peng, D.; Wei, A.; Hou, B.; Lei, Y.; Li, X. *TaFDL2-1A* confers drought stress tolerance by promoting AB biosynthesis, ABA responses, and ROS scavenging in transgenic wheat. *Plant J.* **2022**, *112*, 722–737. [CrossRef]
17. Razzaq, A.; Sadia, B.; Raza, A.; Khalid Hameed, M.; Saleem, F. Metabolomics: A Way Forward for Crop Improvement. *Metabolite* **2019**, *9*, 303. [CrossRef]
18. Obata, T.; Fernie, A.R. The use of metabolomics to dissect plant responses to abiotic stresses. *Cell. Mol. Life Sci.* **2012**, *69*, 3225–324 [CrossRef]
19. Fiehn, O. Combining Genomics, Metabolome Analysis, and Biochemical Modelling to Understand Metabolic Networks. *Com Funct. Genom.* **2001**, *2*, 914970. [CrossRef]
20. Rogachev, I.; Aharoni, A. UPLC-MS-based metabolite analysis in tomato. *Methods Mol. Biol.* **2012**, *860*, 129–144. [CrossRef] [PubMed]
21. Klopsch, R.; Baldermann, S.; Voss, A.; Rohn, S.; Schreiner, M.; Neugart, S. Narrow-Banded UVB Affects the Stability of Secondar Plant Metabolites in Kale (*Brassica oleracea* var. sabellica) and Pea (Pisum sativum) Leaves Being Added to Lentil Flour Fortifie Bread: A Novel Approach for Producing Functional Foods. *Foods* **2019**, *8*, 427. [CrossRef]
22. Clayton, W.A.; Albert, N.W.; Thrimawithana, A.H.; McGhie, T.K.; Deroles, S.C.; Schwinn, K.E.; Warren, B.A.; McLachlan, A.R.G Bowman, J.L.; Jordan, B.R.; et al. UVR8-mediated induction of flavonoid biosynthesis for UVB tolerance is conserved between th liverwort *Marchantia polymorpha* and flowering plants. *Plant J.* **2018**, *96*, 503–517. [CrossRef]
23. Logemann, E.; Tavernaro, A.; Schulz, W.; Somssich, I.E.; Hahlbrock, K. UV light selectively coinduces supply pathways fror primary metabolism and flavonoid secondary product formation in parsley. *Proc. Natl. Acad. Sci. USA* **2000**, *97*, 1903–190 [CrossRef]
24. Hahlbrock, K.; Scheel, D. Physiology and Molecular Biology of Phenylpropanoid Metabolism. *Annu. Rev. Plant Physiol. Plant Mo Biol.* **1989**, *40*, 347–369. [CrossRef]
25. Maher, M.; Ahmad, H.; Nishawy, E.; Li, Y.; Luo, J. Novel Transcriptome Study and Detection of Metabolic Variations i UV-B-Treated Date Palm (*Phoenix dactylifera* cv. Khalas). *Int. J. Mol. Sci.* **2021**, *22*, 2564. [CrossRef]
26. Liu, Y.; Fan, H.; Dong, J.; Chen, J.; Xu, H.; Zhou, X. Phosphoproteomics of cold stress-responsive mechanisms in *Rhododendro chrysanthum*. *Mol. Biol. Rep.* **2022**, *49*, 303–312. [CrossRef] [PubMed]
27. Giovannoni, J. Tomato Multiomics Reveals Consequences of Crop Domestication and Improvement. *Cell* **2018**, *172*, 6–8. [CrossRef]
28. Chen, W.; Gong, L.; Guo, Z.; Wang, W.; Zhang, H.; Liu, X.; Yu, S.; Xiong, L.; Luo, J. A Novel Integrated Method for Large-Scal Detection, Identification, and Quantification of Widely Targeted Metabolites: Application in the Study of Rice Metabolomics. *Mc Plant* **2013**, *6*, 1769–1780. [CrossRef]
29. Zhou, X.; Chen, S.; Wu, H.; Yang, Y.; Xu, H. Biochemical and proteomics analyses of antioxidant enzymes reveal the potentia stress tolerance in *Rhododendron chrysanthum* Pall. *Biol. Direct* **2017**, *12*, 10. [CrossRef]
30. Li, Q.; Song, J. Analysis of widely targeted metabolites of the euhalophyte *Suaeda salsa* under saline conditions provides ne insights into salt tolerance and nutritional value in halophytic species. *BMC Plant Biol.* **2019**, *19*, 388. [CrossRef]
31. Liu, Y.; Liu, J.; Abozeid, A.; Wu, K.X.; Guo, X.R.; Mu, L.Q.; Tang, Z.H. UV-B Radiation Largely Promoted the Transformation c Primary Metabolites to Phenols in *Astragalus mongholicus* Seedlings. *Biomolecules* **2020**, *10*, 504. [CrossRef]
32. Yoon, J.; Cho, L.-H.; Tun, W.; Jeon, J.-S.; An, G. Sucrose signaling in higher plants. *Plant Sci.* **2021**, *302*, 110703. [CrossRef]
33. Mookerjee, S.A.; Gerencser, A.A.; Nicholls, D.G.; Brand, M.D. Quantifying intracellular rates of glycolytic and oxidative AT production and consumption using extracellular flux measurements. *J. Biol. Chem.* **2017**, *292*, 7189–7207. [CrossRef]
34. Sun, Q.; Liu, M.; Cao, K.; Xu, H.; Zhou, X. UV-B Irradiation to Amino Acids and Carbohydrate Metabolism in *Rhododendro chrysanthum* Leaves by Coupling Deep Transcriptome and Metabolome Analysis. *Plants* **2022**, *11*, 2730. [CrossRef]
35. Peer, W.A.; Murphy, A.S. Flavonoids and auxin transport: Modulators or regulators? *Trends Plant Sci.* **2007**, *12*, 556–563. [CrossRef]
36. Ryan, K.G.; Swinny, E.E.; Markham, K.R.; Winefield, C. Flavonoid gene expression and UV photoprotection in transgenic an mutant *Petunia* leaves. *Phytochemistry* **2002**, *59*, 23–32. [CrossRef] [PubMed]
37. Taylor, L.P.; Grotewold, E. Flavonoids as developmental regulators. *Curr. Opin. Plant Biol.* **2005**, *8*, 317–323. [CrossRef]
38. Ryan, K.G.; Swinny, E.E.; Winefield, C.; Markham, K.R. Flavonoids and UV Photoprotection in Arabidopsis Mutants. *Z. Fü Naturforschung C* **2001**, *56*, 745–754. [CrossRef]
39. Gibon, Y.; Pyl, E.-T.; Sulpice, R.; Lunn, J.E.; HÖHne, M.; GÜNther, M.; Stitt, M. Adjustment of growth, starch turnover, protei content and central metabolism to a decrease of the carbon supply when *Arabidopsis* is grown in very short photoperiods. *Plan Cell Environ.* **2009**, *32*, 859–874. [CrossRef]
40. Robbins, M.P. Biochemistry of plant secondary metabolism. Annual Plant Reviews, Volume 2. Edited by Michael Wink. *Eur. Plant Pathol.* **2000**, *106*, 487. [CrossRef]

41. Kusano, M.; Fukushima, A.; Redestig, H.; Saito, K. Metabolomic approaches toward understanding nitrogen metabolism in plants. *J. Exp. Bot.* **2011**, *62*, 1439–1453. [CrossRef]
42. Hewage, K.A.H.; Yang, J.-F.; Wang, D.; Hao, G.-F.; Yang, G.-F.; Zhu, J.-K. Chemical Manipulation of Abscisic Acid Signaling: A New Approach to Abiotic and Biotic Stress Management in Agriculture. *Adv. Sci.* **2020**, *7*, 2001265. [CrossRef]

metabolites

MDPI

Article

Unravelling Effects of Rosemary (*Rosmarinus officinalis* L.) Extract on Hepatic Fat Accumulation and Plasma Lipid Profile in Rats Fed a High-Fat Western-Style Diet

Sidsel Madsen [1], Steffen Yde Bak [2], Christian Clement Yde [2], Henrik Max Jensen [2], Tine Ahrendt Knudsen [2], Cecilie Bæch-Laursen [3], Jens Juul Holst [3], Christoffer Laustsen [4] and Mette Skou Hedemann [1,*]

1 Department of Animal and Veterinary Sciences, Aarhus University, Blichers Allé 20, DK-8830 Tjele, Denmark
2 IFF—Nutrition Biosciences Aps, Edwin Rahrs Vej 38, DK-8220 Brabrand, Denmark; steffen.yde.bak@iff.com (S.Y.B.); christian-clement.yde@iff.com (C.C.Y.); henrik.max.jensen@iff.com (H.M.J.); tine.ahrendt.knudsen@iff.com (T.A.K.)
3 Department of Biomedical Sciences and Novo Nordisk Foundation, Center for Basic Metabolic Research, Faculty of Health and Medical Sciences, University of Copenhagen, Blegdamsvej 3, DK-2200 Copenhagen, Denmark; cecilie.b@sund.ku.dk (C.B.-L.); jjholst@sund.ku.dk (J.J.H.)
4 The MR Research Centre, Department of Clinical Medicine, Aarhus University, Palle Juul-Jensens Boulevard 99, DK-8200 Aarhus, Denmark; cl@clin.au.dk
* Correspondence: mette.hedemann@anivet.au.dk

Abstract: The objective of the study was to investigate the preventive effect on obesity-related conditions of rosemary (*Rosmarinus officinalis* L.) extract (RE) in young, healthy rats fed a high-fat Western-style diet to complement the existing knowledge gap concerning the anti-obesity effects of RE in vivo. Sprague Dawley rats (71.3 ± 0.46 g) were fed a high-fat Western-style diet (WD) or WD containing either 1 g/kg feed or 4 g/kg feed RE for six weeks. A group fed standard chow served as a negative control. The treatments did not affect body weight; however, the liver fat percentage was reduced in rats fed RE, and NMR analyses of liver tissue indicated that total cholesterol and triglycerides in the liver were reduced. In plasma, HDL cholesterol was increased while triglycerides were decreased. Rats fed high RE had significantly increased fasting plasma concentrations of Glucagon-like peptide-1 (GLP-1). Proteomics analyses of liver tissue showed that RE increased enzymes involved in fatty acid oxidation, possibly associated with the higher fasting GLP-1 levels, which may explain the improvement of the overall lipid profile and hepatic fat accumulation. Furthermore, high levels of succinic acid in the cecal content of RE-treated animals suggested a modulation of the microbiota composition. In conclusion, our results suggest that RE may alleviate the effects of consuming a high-fat diet through increased GLP-1 secretion and changes in microbiota composition.

Keywords: obesity; metabolomics; proteomics; polyphenol; *Rosmarinus officinalis*

1. Introduction

Worldwide, the prevalence of obesity has increased markedly over the last decades, and obesity is now defined as a pandemic. Obesity is linked to a sedentary lifestyle combined with a high intake of energy-dense food products and is associated with an increased risk of numerous comorbidities including type 2 diabetes, certain cancers, cardiovascular diseases, asthma, gallbladder disease, osteoarthritis, and chronic back pain, resulting in higher overall mortality [1].

In recent years, the interest in natural, bioactive compounds with anti-obesity effects has increased. Polyphenols are a group of compounds found in many fruits, vegetables, and cereals [2]. Polyphenols have been studied in relation to human health, and regular consumption is thought to be associated with a decreased risk of several chronic diseases [3,4]. Obesity is one of the conditions, where polyphenols appear to have a preventive effect.

Different polyphenols have shown the ability to modulate hormonal appetite regulation [5], inhibit the activity of digestive enzymes [6], inhibit fatty acid synthesis [7], or increase fatty acid oxidation [8], all of which are considered expedient in attenuating the development of obesity.

Rosemary (*Rosmarinus officinalis* L.) is an aromatic, evergreen plant containing high concentrations of polyphenols, with carnosic acid and rosmarinic acid being the most abundant bioactive compounds [9]. In medical use, rosemary has been shown to be beneficial in several conditions, including obesity, metabolic syndrome, and type 2 diabetes, and potential mechanisms of action include an ability to limit absorption of lipids [10] and inhibit hepatic fat accumulation and adipocyte hypertrophy [11].

Recent in vitro studies showed that rosemary extract (RE) stimulated the secretion of the satiety-inducing hormone Glucagon-like peptide-1 (GLP-1) from immortalized GLP-1 producing cell lines [12]. GLP-1 is a peptide hormone secreted from intestinal L-cells in response to nutrients in the intestinal lumen. Following secretion, GLP-1 is rapidly inactivated by enzymes in plasma and on cell surfaces, resulting in a half-life of intact GLP-1 of only one to two minutes in humans [13]. Upon entering the blood-stream, GLP-1 is circulated to several target organs, where its metabolic functions include stimulation of insulin secretion, induction of satiety, inhibition of hepatic gluconeogenesis, stimulation of glycogen synthesis and glucose oxidation in muscle tissue, and stimulation of free fatty acid (FFA) synthesis and glucose uptake in adipose tissue, although these effects may be secondary to the actions on the endocrine pancreas or the effects on body weight [14–17].

Dyslipidemia is often seen in obese patients and includes reduced levels of circulating high-density lipoproteins (HDL) and enhanced levels of low-density lipoproteins (LDL) and triglycerides [18]. Obesity is associated with liver diseases, such as non-alcoholic fatty liver disease (NAFLD). NAFLD is characterized by excessive intrahepatic triglyceride levels and may be accompanied by excessive levels of FFAs and cholesterol in the plasma. It is often associated with alterations in whole-body metabolic homeostasis [19]. Consumption of polyphenols has been associated with the prevention and treatment of NAFLD [20], and the mechanisms may involve the effects of microbial gut metabolites as well as stimulation of gut-derived hormones [21].

We hypothesized that RE reduces feed intake in rats and inhibits diet-induced weight gain through increased fasting GLP-1 levels. The study was carried out in young, non-obese rats with the aim to investigate if daily consumption of a high-fat diet with RE over a period of six weeks influenced weight gain. We studied metabolic parameters related to obesity and used untargeted metabolomics and proteomics to decipher the mode of action.

2. Materials and Methods

2.1. Plant Material and Extract

The RE (IFF-Nutrient Bioscience ApS, Brabrand, Denmark) used in the present experiment was a dried powder from a water/ethanol extract of rosemary leaves, containing min. 65% carnosic acid and min. 70% carnosic acid + carnosol.

2.2. Animals

In vivo study: The study included 64 male Sprague Dawley rats obtained from Taconic Biosciences (Ejby, Denmark) with a starting weight of 71.34 g (±0.46 SEM). The rats were divided into pairs of two rats (32 pairs) and the pairs were divided into four experimental groups (8 pairs per group). Each group was fed one of four pelleted experimental diets; standard rodent chow (Altromin 1324, Altromin GmbH, Lage, Germany) (ST), Western-style diet (D12079B; 40% fat, 43% carbohydrates, 17% protein; Research Diets, New Brunswick, NJ, USA) (WD), WD supplemented with 100 mg rosemary extract (RE) pr. 100 g feed (WD+low-RE), or WD supplemented with 400 mg RE pr. 100 g feed (WD+high-RE). The dose of RE added was based on a previous experiment where reduced weight gain was observed in mice fed a high-fat diet supplemented with RE [22]. The rats were expected to eat 10–20 g feed/day which corresponds to daily intakes of 40–80 mg of RE in rats fed the highest level of RE. The chemical composition of the diets was determined as previously described [23] and is shown in Supplementary Table S1.

Animals were housed in pairs in clear plastic boxes (38 × 60 cm) bedded with shavings and followed a 12:12 h light-dark cycle, at 25 °C and 60% humidity, with ad libitum access to water and feed during the entire experiment. Rats were individually marked to enable measurement of individual body weight (BW). BW gain and feed intake were measured twice weekly (see supplementary methods). The experiment was conducted over 46 days.

2.3. Sample Collection

At the end of the experiment, and after six hours of food deprivation to allow for the normalization of metabolic markers, all rats were euthanized by an intraperitoneal injection of 0.5 mL pentobarbital (200 mg/mL) with lidocaine (20 mg/mL). From each box, one rat was euthanized and kept intact at −20 °C for whole body composition analysis, while the second rat in the box was used for tissue and blood collection. When death was ensured, blood was collected from the heart into EDTA-coated tubes containing 1% (*V*/*V*) dipeptidylpeptidase-4 (DPP-4) inhibitor (Merck KGaA, Darmstad, Germany). Blood samples were immediately centrifuged at 2000× *g* for 12 min at 4 °C before plasma was stored at −80 °C until further analysis. The abdominal cavity was opened, and the liver was dissected out, weighed, and snap-frozen in liquid nitrogen. Samples were collected from three segments of the small intestine (5 cm distal to the pylorus sphincter, a mid-section, and 5 cm proximal to the ileocaecal valve). The samples were washed thoroughly in saline before being placed in RNAlater®(ThermoFisher Scientific, Waltham, MA, USA), and further treated according to the manufacturer's protocol. All tissue samples were stored at −80 °C until further analysis. The caecum was dissected free and cecal tissue and content were weighed. Content was collected and stored at −80 °C until further analysis.

2.4. Studies on Isolated Perfused Rat Small Intestine

Male Wistar rats (~250 g) were obtained from Janvier (Saint Berthevin Cedex, France) and housed with ad libitum access to standard chow and water, following a 12:12 h light:dark cycle. Rats were acclimatized for a least a week before the experiments. On the day of the experiment, rats were transferred to our perfusion facility and anesthetized, and the entire small intestine was isolated and perfused as described in detail in the supplementary methods and elsewhere [24,25].

2.5. Total Body Fat Content

Body composition data were acquired using a 9.4T pre-clinical MR scanner (Agilent, Yarnton, UK) equipped with a 72 mm proton volume coil (RAPID Biomedical GmbH, Rimpar, Germany). Full body proton NMR spectra from each animal were Fourier Transformed and spectrally analyzed by fitting the water and the fat peaks using a general linear model approach that was developed in-house (MATLAB, The MathWorks, Inc., Natick, MA, USA). Subsequently, the fitted peak areas were used to calculate the fat percentage by division by the water component and fat component signal areas.

2.6. Fasting Total GLP-1 Plasma Concentrations

Fasting total GLP-1 concentrations were measured using an in-house radioimmunoassay, as previously described [26], based on an antiserum directed against the amidated C-terminus of GLP-1. The antibody reacts equally with intact amidated GLP-1 (7–36 amide), the primary metabolite (9–36 amide), and truncated sequences produced by mid-site cleavage (e.g., by NEP 24.11). Therefore, the concentrations represent the total GLP-1 concentration. I125-labeled GLP-1 7-36NH2 (a kind gift from Novo Nordisk A/S) was used as a tracer. The experimental detection limit was 1 pmol/L, and the coefficient of variation was <6% at 20 pM.

2.7. RNA Extraction and Gene Expression

Total RNA was extracted from the whole wall of the small intestine using a NucleoSpin RNA kit (Macherey-Nagel GmbH & Co., Ltd., KG, Duren, Germany), in accordance with the manufacturer's protocol. RNA concentration and purity were evaluated using a NanoDrop spectrophotometer (ND-1000, NanoDrop Technologies, Wilmington, DE, USA). A High-Capacity cDNA Reverse Transcriptase Kit (Life Technologies, Taastrup, Denmark) was used for cDNA synthesis. The qPCR of glyceraldehyde-3-phosphate dehydrogenase (GAPDH), β-actin, and proglucagon (GCG) genes was run in 384-well plates with a 1:4 ratio of cDNA and TaqMan Master Mix (Applied Biosystems, Foster, CA, USA), according to the manufacturer's protocol. The analyses were performed in triplicates for the housekeeping genes and in duplicates for the target genes, under standard amplification conditions determined for the Applied Biosystem ViiA 7 Real-Time PCR system (Life Technologies, Taastrup, Denmark). Expression levels were first normalized to the expression level in the most proximal sample in the respective rats, and values were then normalized to the control group fed the standard diet. For the interaction between the treatment and the tissue, normalization to the most proximal segment for the standard diet treatment was used.

2.8. Plasma Metabolic Markers

Plasma cholesterol, HDL, LDL, and triglycerides were quantified by standard procedures (Siemens Diagnostics® Clinical Methods for ADVIA 1800). FFAs were determined using the Wako, NEFA C ACS-ACOD assay method. All analyses were performed using an autoanalyzer, ADVIA 1800 Chemistry System (Siemens Medical Solutions, Tarrytown, NY 10591, USA).

2.9. Analysis of Hepatic Fat Percentage

Liver samples were homogenized in twice the amount of methanol (w/v) using an Ultra-Turrax homogenizer (IKA Labortechnik, Staufen, Germany) while placed in an ice bath. Of this homogenate, 600 mg was mixed with 1.0 mL water, 1.5 mL methanol, and 1.0 mL chloroform, before being shaken for one minute. Water (1.0 mL) and chloroform (2.0 mL) were added to the mixture, following shaking for one minute. Phase separation was accelerated by centrifugation at 2000× g for 5 min. The lower chloroform phase (2 mL) was collected and dried, and the residue was weighed for determination of fat content.

2.10. Proteomics Analysis of Liver Tissue

Approximately 100–200 mg of tissue was excised from the left lateral lobe. Frozen tissue was covered with 1 mL lysis buffer (5% SDS in 100 mM triethylammonium bicarbonate, TEAB, pH 7.5). Tissue homogenization was done using a ball mill at 25 Hz for 10 min (Retsch GmbH, Haan, Germany). Homogenates were centrifuged for 10 min at 14,000× g. The supernatants were heated to 95 °C for 10 min to fully denature proteins. DNA was sheared using a Bioruptor Pico (Diagenode Diagnostics, Liege, Belgium) high-energy water bath sonicator system using 10 cycles of 30 s sonication followed by a 30 s break. The temperature was set to 5 °C throughout the entire process. Protein concentrations were estimated in triplicates using a BCA micro assay kit (Thermo Scientific Pierce, Rockford, IL, USA).

Liver lysates were digested into peptides using the S-trap™ micro digestion protocol [27] using a trypsin-to-protein ratio of 1:30. Peptide elutes were dried down and resuspended in 50 µL 0.1% TFA. Concentrations of the peptides were estimated using a micro-BCA™ protein assay kit (Thermo Scientific Pierce, Rockford, IL, USA) and samples were adjusted to equal concentrations. A QC sample (mixture of all samples) was prepared by taking 10 µL from all samples and mixing well. Data acquisition and analysis were performed as described in the supplementary methods.

Data Acquisition: Nano LC−MS/MS analyses were performed using an UltiMate™ 3000 RSLCnano System (Thermo Fisher Scientific, Waltham, MA, USA) interfaced to a Q Exactive™ HF Hybrid Quadrupole-Orbitrap™ Mass Spectrometer (Thermo Fisher Scientific, Waltham,

MA, USA). Samples were dissolved in 0.1% TFA and loaded onto a 20 mm nanoViper Trap Column (Acclaim™ PepMap™ 100 C18, 3μm particle size, and an i.d. 0.075mm) connected to a 400 mm analytical column (PepSep, ReproSil 3 μm C18 beads, pore diameter 120Å and an i.d. 0.075 mm). Separation was performed at a flow rate of 300 nL/min using a 40 min gradient of 0−41% Solvent B (100% ACN, 0.1% FA) into the Nanospray Flex Ion Source (Thermo Scientific). The Q Exactive HF instrument was operated in a data-dependent MS/MS using HCD fragmentation. The peptide masses were measured by the Orbitrap (MS scans were obtained with a resolution of 60,000 (FWHM) at m/z 200). The top 12 most intense ions were selected and subjected to fragmentation. Ions were isolated by the quadrupole using a 1.2 Da width isolation window. Fragment spectra were recorded in the Orbitrap with a resolution of 30,000 (FWHM) at m/z 200. Dynamic exclusion was enabled with an exclusion duration of 30 s, and an exclusion mass tolerance width of ±10 ppm relative to masses on the list. Samples were analyzed in a randomized order with a QC sample after every fifth sample and with five QC samples at the beginning of the sample sequence.

Data analysis: The LC−MS/MS data were processed (smoothing, background subtraction, and centroiding) using Proteome Discoverer (Version 2.2, Thermo Scientific). The processed LC−MS/MS data were subjected to database searching against a UniProt Rattus Norvegicus database containing 29,944 sequences using an in-house Mascot server. Trypsin was chosen as an enzyme with a maximum of two missed cleavages allowed. S-Carbamidomethyl cysteine was defined as a fixed modification and oxidation of methionine as a variable modification. The MS/MS results were searched with a peptide ion mass tolerance of ±10 ppm and a fragment ion mass tolerance of ±0.8 Da. A percolator [28] was used for calculating the false discovery rate. Only peptides that were identified as rank 1 peptides and with a confidence value of 1% ($q < 0.01$) were considered for further analysis.

All raw files were imported into Expressionist v.12.0.9 (Genedata, Basel, Schwitzerland) for data analysis. Imported files were noise filtered using a chemical noise subtraction. Data were RT aligned using a pairwise alignment, filtered, and smoothed before peak detection, based on volumes. Detected peaks were isotopically clustered and singletons were filtered out. Peak clusters were matched with Proteome Discoverer identifications and peptides were grouped based on protein identifications. Proteins were quantified based on the three most intense peptides' Hi-3 [29]. Quantitative results were exported into Analyst v.12.0.9 (Genedata) for normalization, statistical filtering, and testing. Principle Component Analysis was performed using the GeneData Analyst module. A total of 1570 proteins were identified and quantified and within these, 999 proteins were quantified with a minimum of two peptides per protein. The average finding percentage of the 999 proteins throughout all 44 runs was 99.3%, giving an almost complete data matrix. Data were normalized using an intensity drift normalization, where intensities from QC samples are used to correct for drift in the LC-MS/MS performance. Using an ANOVA, we found 247 proteins significantly changing between one of the four diets with an overall FDR of 5%. Regulated proteins were manually annotated.

2.11. NMR Spectroscopy

A dual-phase methanol/chloroform/water (1:1:1) extraction protocol was performed on the left lateral lobe liver samples (weight 259 ± 94 mg), as previously described [30]. The NMR measurements were performed on a 600 MHz Bruker Avance III spectrometer operating at a frequency of 600.13 MHz for 1H nucleus (Bruker Biospins, Rheinstetten, Germany) and equipped with a 5mm TCI CryoProbe. Polar extracts: 1H NMR spectra were recorded at 298 K using a 1D Noesy pulse sequence with pre-saturation. A total of 256 scans collected into 64 K data points were acquired with a spectral width of 20.02 ppm, a recycle delay of 5 s, and an acquisition time of 2.73 s. Non-polar extracts: 1H NMR spectra were obtained at 298 K using a single pulse sequence with a 30° flip angle. A total of 128 scans collected into 64K data points were acquired with a spectral width of 20.02 ppm, a recycle delay of 3 s, and an acquisition time of 2.73 s. The anomeric glycogen H-1 signal at 5.4 ppm in the polar extracts was manually quantified and normalized to the weight of the liver for each rat.

Hepatic lipid signals in the non-polar phase were mainly tentatively assigned according to Fathi et al. [31]. Prior to data analysis, the residual water signal was excluded from the "Polar" spectra, and all 1H NMR spectra were normalized to the weight of each sample. Baseline correction by distribution-based classification [32], alignment using icoshift version 1.3.1 [33], and binning according to an optimized bucketing algorithm [34] were performed for both data sets. Multivariate data analysis was applied using Principal Component Analysis (PCA) on Pareto-scaled data using PLS Toolbox 8.7 (eigenvector Research, USA) in Matlab R2018b (The MathWorks, Inc., Natick, MA, USA) to detect clustering behavior.

2.12. Metabolomic Analysis of Plasma and Cecal Contents

LC-MS was used to analyze the plasma and cecal metabolome in an untargeted approach. Plasma samples (150 μL) were placed in the wells of a 96-well plate and deproteinized with 450 μL ACN containing internal standards (*p*-chlorophenylalanine and glycocholic acid (Glycine-1 13C), final concentration 0.01 mg/mL) and processed as previously described [35]. Cecal contents were freeze-dried and milled prior to analysis. Cecal content (50 mg) was mixed with 500 μL 80% ACN with added internal standards as described above. The samples were vortexed for 20 min at room temperature and centrifuged for 10 min at 10,000× *g* and 4 °C. The supernatant was transferred to a 96-square well, 0.45 μM polypropylene filter plate and aspirated into a new plate. The filtered supernatant was treated as described for the plasma samples. UPLC-MS analysis was performed using a Dionex UltiMate 3000 (Dionex, Sunnyvale, CA, USA) ultra-high pressure liquid chromatography system coupled with an Impact HD Quadrupole Time-of-Flight (QTOF) mass spectrometer (Bruker Daltonics GmbH, Bremen, Germany) operating in positive electrospray ionization mode (ESI+) and negative electrospray ionization mode (ESI-) using the instrumental parameters previously described [35,36]. The chromatographic behavior of the samples, pre-processing, and identification of metabolites was assessed as previously described [35].

2.13. SCFA Analysis

Short-chain fatty acid (SCFA) concentrations were measured as previously described [37], with some modifications as described in the supplementary methods.

2.14. Statistical Analyses

Differences in fasting plasma GLP-1 concentrations, metabolic markers in plasma, total and hepatic fat accumulation, weight, and SCFA were assessed by One-way ANOVA with Tukey multiple-comparison correction. Differences in the presence of liver proteins were assessed by One-way ANOVA using Storey-Tibshriani correction. $p < 0.05$ was considered statistically significant whereas a p-value between 0.05 and 0.10 was considered as an indication of a trend towards significance.

Gene expression data were analyzed using a normal mixed model:

$$Yij = \mu + \alpha i + \upsilon j + \varepsilon ij,$$

where Yij is the gene expression; μ is the overall mean; αi is the effect of treatment (i = ST, WD, WD+low-RE, WD+high-RE); υj is the random effect of a block (j = 1, 2); and εij is the residual error component. The data are presented as least square means with a 95% confidence interval. Effects were considered significant when $p < 0.05$, whereas $0.05 \leq p \leq 0.10$ represented a tendency.

For perfusion studies, GLP-1 total outputs were calculated by multiplying perfusion flow (7.5 mL/min) with the GLP-1 concentration in the perfusion effluents (measured in pmol/L). Hormone outputs (fmol/min) are presented as means ± SEM. To test for the statistical significance of responses, total GLP-1 outputs were calculated by summing up the outputs during the entire period of stimulus administrations (15 min intra-luminal/intra-vascular RE administration) and were compared to respective total GLP-1 outputs at baseline conditions. Baseline outputs were calculated over a similar number of samples with half of the samples taken immediately before stimulus administration and the other half coming from the terminal part of the subsequent baseline period. Statistical calculations were performed using GraphPad Prism 7 software (La Jolla, CA, USA), employing One-way ANOVA for repeated measurements followed by the Bonferroni multiple comparison test or student t-test as indicated in the figure legends. $p < 0.05$ was considered significant. Prior to testing, the D'Agostino-Pearson omnibus normality test was performed to confirm the Gaussian distribution of the data.

3. Results

3.1. Body Weight and Feed Intake

As presented in Figure 1A, during the six-week experimental period, a clear effect of time ($p < 0.0001$) was observed for body weight, while no overall effect of treatments was observed ($p = 0.10$). The treatments influenced the 24-h feed intake significantly ($p = 0.006$; Figure 1B). When determining the daily weight gain pr. energy intake (g/kcal), an overall effect of treatment was observed ($p < 0.0001$; Figure 1C). When comparing this for the different treatments, a significant difference was seen on days 3–17; however, no overall effect of treatment was observed after day 13.

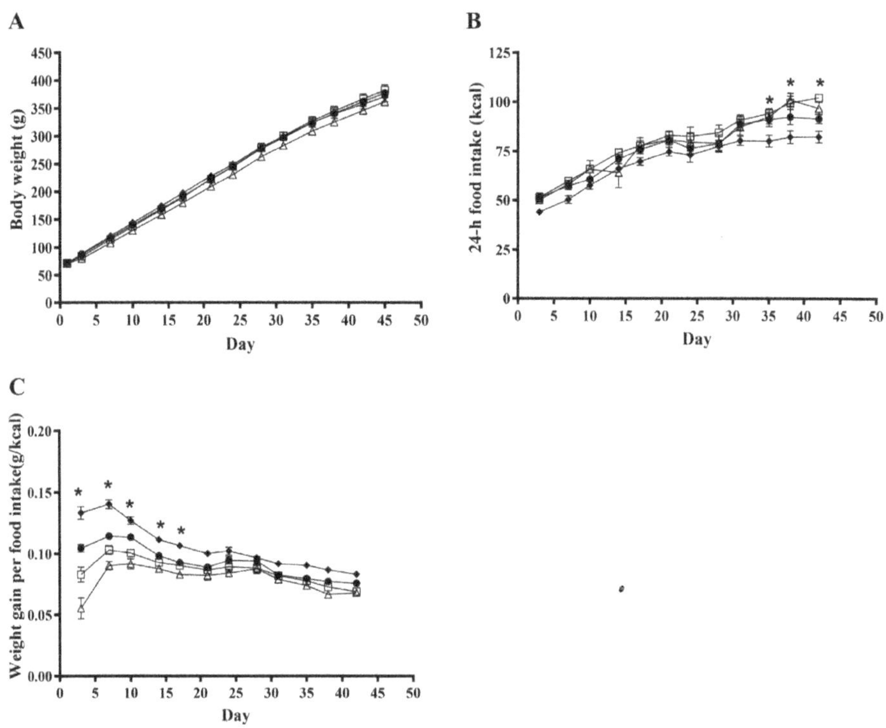

Figure 1. Effects of rosemary extract on body weight, $n = 16$ (**A**), 24-h feed intake, $n = 8$ (**B**), and weight gain per energy intake, $n = 8$ (**C**), in rats fed a ST (♦), WD (•), WD+low-RE (□), or WD+high-RE (Δ) over 46 days. Results are presented as mean ± SEM. * indicates a significant difference ($p < 0.05$) between two or more treatments: (**B**) ST and WD+low-RE differ significantly at days 35, 38, and 42, and ST and WD+high-RE differ significantly at day 38. (**C**) All four treatments differ at day 3. In addition, WD and WD+high-RE differ at day 7, and ST and WD+high-RE differ at days 7, 10, 14, and 17. ST, standard diet; WD, Western-style diet; WD+low-RE, WD supplemented with low concentration of rosemary extract; WD+high-RE, WD supplemented with high concentration of rosemary extract.

3.2. Liver Weight and Fat Accumulation in the Liver and Whole Body

Mean total body fat percentages (Figure 2A) following six weeks of treatment were 14.1% (ST), 17.3% (WD), 18.9% (WD+low-RE), and 15.0% (WD+high-RE), and no difference between the treatments was observed ($p = 0.18$). Rats treated with WD had significantly higher liver fat percentages compared to ST treatment ($p < 0.0001$). Treatment with WD+high-RE strongly decreased the fat mass in the liver (Figure 2B), with an average reduction of 46% compared to the WD. The mean weights of wet liver tissue (Figure 2C), expressed as a percentage of total body weight, were 3.99% (ST), 4.50% (WD), 5.26% (WD+low-RE), and 5.16% (WD+high-RE). ST treatment resulted in the lowest relative liver weight, it increased when feeding WD, and the addition of RE to the diet increased the relative liver weight further.

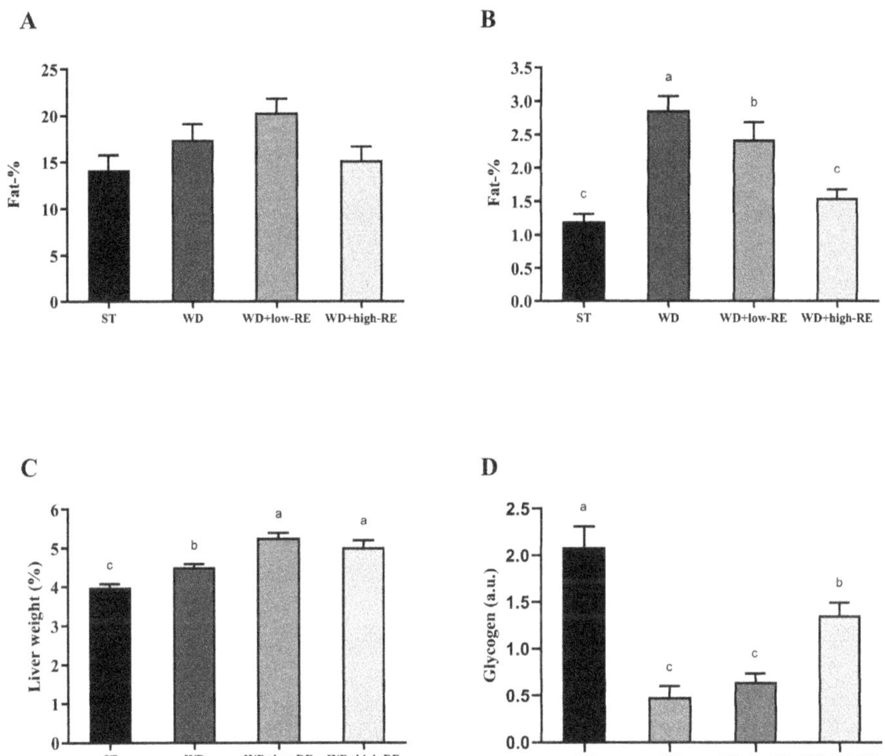

Figure 2. Effects of rosemary extract on (**A**) total body fat percentage, (**B**) liver fat percentage, (**C**) liver weight as a percentage of total body weight, and (**D**) hepatic glycogen (a.u) in rats fed a standard diet (ST), Western-style diet (WD), or a WD supplemented with either low (WD+low-RE) or high (WD+high-RE) concentrations of rosemary extract, measured at day 46. $n = 8$. Results are presented as mean ± SEM. Means without common letters differ significantly ($p < 0.05$).

3.3. Plasma Metabolic Markers

An overall effect of treatment was observed for total cholesterol ($p = 0.0008$; Figure 3A). Thus, feeding WD, WD+low-RE, and WD+high-RE caused a significantly higher level of total plasma cholesterol, compared to ST, but no significant differences were observed between the three. For mean plasma levels of LDL-cholesterol (Figure 3B), only WD+low-RE treatment resulted in a significantly higher level than ST, whereas WD+high-RE was similar to ST and WD was intermediate. Analysis of mean plasma levels of HDL-cholesterol

(Figure 3C) showed that WD+low-RE and WD+high-RE treatment caused significantly higher plasma HDL levels than ST treatment. An effect of treatment on the plasma level of triglycerides was observed ($p = 0.0002$; Figure 3D). The mean plasma levels of plasma triglycerides were significantly higher with WD and WD+low-RE compared to the ST, whereas levels were intermediate in rats fed WD+high-RE. Plasma FFA levels (Figure 3E) were significantly higher for WD compared to ST ($p = 0.005$), whereas WD-low-RE and WD+high-RE were intermediate.

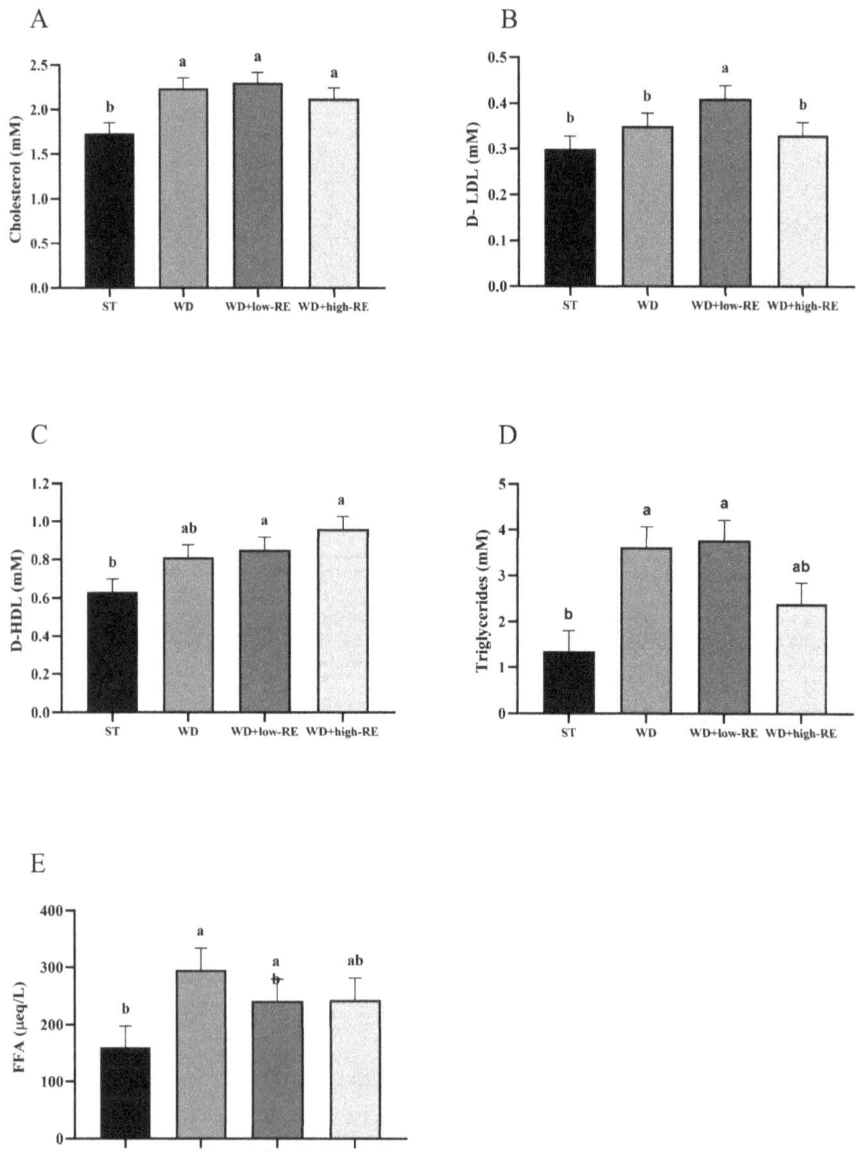

Figure 3. Effects of rosemary extract on plasma levels of (**A**) total cholesterol (mM), (**B**) low-density lipoprotein (D-LDL) (mM), (**C**) high-density lipoprotein (D-HDL) (mM), (**D**) triglycerides (mM), and

(E) free fatty acids (FFA) (μeq/L) in rats fed a standard diet (ST), Western-style diet (WD), or a WD supplemented with either low (WD+low-RE) or high (WD+high-RE) concentrations of rosemary extract. Measured at day 46. $n = 8$. Results are presented as mean ± SEM. Means without common letters differ significantly ($p < 0.05$).

3.4. Fasting GLP-1 Plasma Concentrations and Proglucagon Expression

An effect of treatment was observed for the fasting plasma GLP-1 concentration ($p = 0.01$, Figure 4A). Thus, fasting GLP-1 plasma concentrations were increased by WD+high-RE treatment compared to ST and WD treatment, but concentrations were not different from the WD+low-RE group and also did not differ from ST and WD.

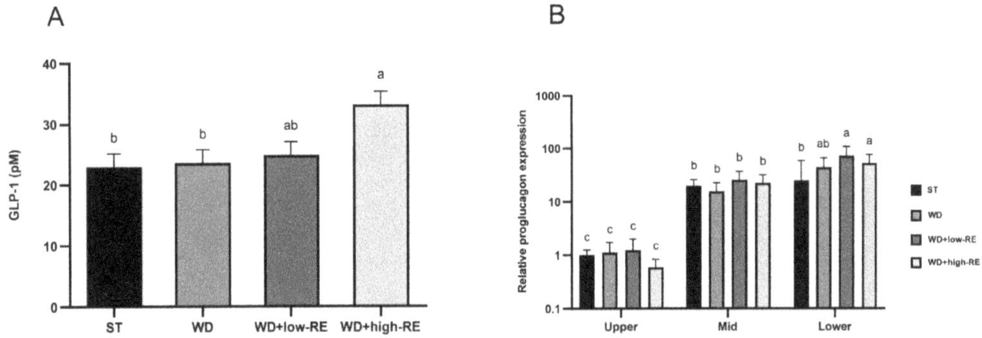

Figure 4. Effects of rosemary extract on (**A**) fasting levels of glucagon-like peptide-1 (GLP-1) (pM) and (**B**) relative proglucagon expression for the interaction between segment and treatment, normalized to ST treatment in upper small intestine (SI) in rats fed a standard diet (ST), Western-style diet (WD), or a WD supplemented with either low (WD+low-RE) or high (WD+high-RE) concentrations of rosemary extract. Measured at day 46. $n = 8$. Results on GLP-1 are presented as mean ± SEM and results on relative proglucagon expression are presented as means with 95% confidence intervals. Means without common letters differ significantly ($p < 0.05$).

An interaction between treatment and segment was observed for relative proglucagon expression ($p = 0.03$; Figure 4B). The interactive effect was observed in the lower SI where the relative expression was higher in rats fed WD+low-RE and WD+high-RE compared to rats fed ST. For segment-specific changes in relative proglucagon gene expression, the mid-part and the distal part of the small intestine were 21-fold and 48-fold higher than the proximal part, respectively.

3.5. Acute Effects of RE on GLP-1 Secretion from Isolated Perfused Rat Small Intestine

The baseline GLP-1 output was (mean ± SEM) 171 ± 11.2 fmol/min. GLP-1 output was neither affected by intra-luminal RE infusion ($p > 0.99$ between total output at 15 min baseline and 15 min stimulation), nor by intra-arterial RE infusion ($p = 0.97$ between preceding total baseline output and 15 min of stimulation) (n = 6, Figure 5). Confirming responsiveness of the perfused preparation, intra-arterial taurodeoxycholic acid (100 μM, a conjugated secondary bile acid) instantly and reversibly increased secretion by a factor of 2–3 during 15 min of stimulation ($p < 0.05$ compared to preceding output at baseline).

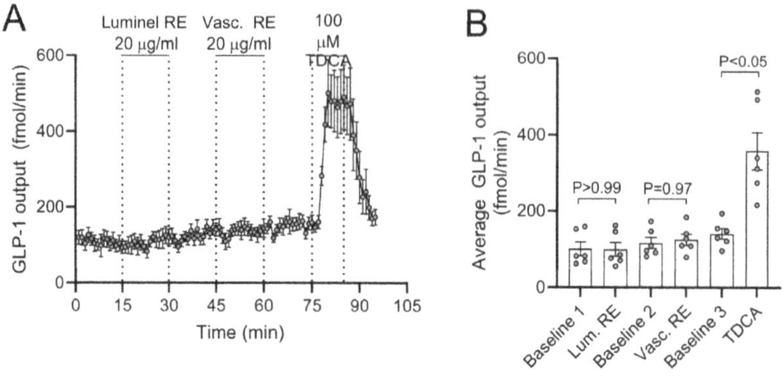

Figure 5. Luminal and vascular rosemary extract (RE)-stimulated glucagon-like peptide-1 (GLP-1) secretion. (**A**) GLP-1 output and (**B**) average GLP-1 output are shown as means ± SEM in response to two subsequent stimulations with 20 μg/mL RE. Intra-luminal infusion of 100μM taurodeoxycholice acid (TDCA) was included at the end of all experiments as a positive control.

3.6. Proteomic Analysis of Liver Tissue

In this study, we used label-free LC-MS/MS-based proteomics. A principal components analysis (PCA) score plot of the liver proteins is shown in Figure 6. Principal component (PC) 2 shows the second largest variation (10.5%) in the dataset and, in this case, separates the different diets.

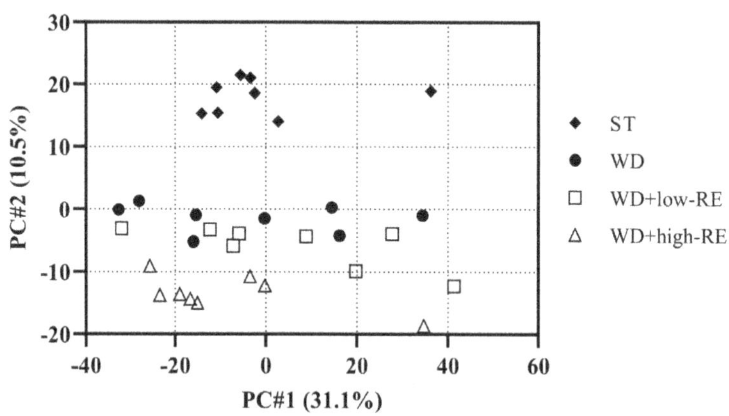

Figure 6. Principal components analysis score plot of liver proteins of rats fed ST (♦), WD (•), WD+low-RE (□), or WD+high-RE (Δ), over 46 days. ST, standard diet; WD, Western-style diet; WD+low-RE, WD supplemented with a low concentration of rosemary extract; WD+high-RE, WD supplemented with a high concentration of rosemary extract.

A number of proteins involved in lipid metabolism and oxidation of lipids were found to be affected in WD-fed animals compared to ST-fed animals. The proteins found related to fatty acid metabolism and oxidation and regulated in comparison to the ST diet are shown in Table 1. Interestingly, a number of these proteins are regulated differentially in response to the RE supplement. The hydroxyacyl-coenzyme A dehydrogenase and all-trans-retinol 13,14-reductase (RETSAT) were upregulated by WD whereas, in rats supplemented with RE, these proteins were not significantly altered compared to ST.

Table 1. Gene and protein name of quantified proteins involved in fatty acid metabolism. Fold changes between the different diets in relation to rats fed a standard diet (ST), Western-style diet (WD), or a WD supplemented with either low (low-RE) or high (high-RE) concentrations of rosemary extract, measured at day 46. $n = 8$.

Protein Name	Gene	WD/ST	WD+Low-RE/ ST	WD+High-RE/ ST
Arylacetamide deacetylase	*AADAC*	1.7	1.6	1.7
ATP-binding cassette sub-family D member 3	*ABCD3*	1.2	1.3	1.5
Short/branched chain specific acyl-CoA dehydrogenase, mitochondrial	*ACADS*	0.5	0.6	0.5
Aldehyde oxidase 1	*AOX1*	1.2	1.2	1.3
CDGSH iron-sulfur domain-containing protein 1	*CISD1*	0.9	0.8	0.8
2,4-dienoyl-CoA reductase, mitochondrial	*DECR1*	2.0	1.6	1.8
Enoyl-CoA delta isomerase 1, mitochondrial	*ECI1*	2.1	1.8	2.2
Hydroxyacyl-coenzyme A dehydrogenase, mitochondrial	*HADH*	1.3	1.0	1.0
Trifunctional enzyme subunit alpha, mitochondrial	*HADHA*	1.4	1.3	1.5
Trifunctional enzyme subunit beta, mitochondrial	*HADHB*	1.3	1.0	1.3
Hydroxymethylglutaryl-CoA synthase, mitochondrial	*HMGCS2*	1.4	1.2	1.2
Corticosteroid 11-beta-dehydrogenase isozyme 1	*HSD11B1*	0.5	0.3	0.5
Estradiol 17-beta-dehydrogenase 2	*HSD17B2*	1.2	1.4	1.6
MICOS complex subunit Mic60 (Fragment)	*IMMT*	0.8	1.2	1.7
Enoyl-[acyl-carrier-protein] reductase, mitochondrial	*MECR*	1.6	2.2	4.1
Macrophage migration inhibitory factor	*MIF*	1.1	1.3	1.6
5′-AMP-activated protein kinase catalytic subunit alpha-2	*PRKAA2*	1.4	1.2	1.5
All-trans-retinol 13,14-reductase	*RETSAT*	2.4	1.4	1.2
Stearoyl-coenzyme A desaturase 1	*SCD1*	9.8	10.1	7.6

The mitochondrial protein Enoyl-[acyl-carrier-protein] reductase and the protein macrophage migration inhibitory factor (MIF) increased in the WD+high-RE diet compared to ST, whereas WD only enhanced the abundance slightly. Changes were also observed in the mitochondrial protein MICOS complex subunit Mic60 which increased by 1.7-fold in the WD+high-RE diet compared to ST. Stearoyl-CoA desaturase 1 (SCD1) was found to be the most regulated protein in response to WD as the protein was 9.8-fold upregulated in the WD samples compared to ST samples. In WD+high-RE treatment, this increase was slightly lower at 7.6-fold.

Several members of UDP-glucuronosyltransferases and glutathione reductases and glutathione S-transferases belonging to the Xenobiotic pathway phase II response were also found to be regulated in response to a high-RE diet (Table S2).

3.7. NMR Analysis of Liver Tissue

A PCA model on the non-polar liver extracts resulted in a clear separation of the treatments along PC1 (Figure 7A). The WD+low-RE treatment could not be separated from WD in PC1, whereas WD+high-RE showed a shift towards the lipid profile of the ST-fed rats. The PC1 loadings for the hepatic free and total cholesterol could be ranked according to treatments: WD and WD+low-RE > WD+high-RE > ST (Figure 7B). The same pattern was also shown for glycerol backbone signals, which could indicate less hepatic triglyceride in WD+high-RE rats compared to WD rats. The PC1 loadings in Figure 7B of the fatty acyl chain signals have in general both positive and negative loadings for the same functional group (e.g., (CH2)2 at 1.24–1.37 ppm). This could be seen as minor differences in the lipid signals when comparing different treatments. Moreover, the unsaturated fatty acid broad peak from 5.23 to 5.38 ppm shows mainly negatively valued loadings, which means that livers from WD+high-RE rats have a higher prevalence of unsaturated fatty acids than livers from WD and WD+low-RE rats.

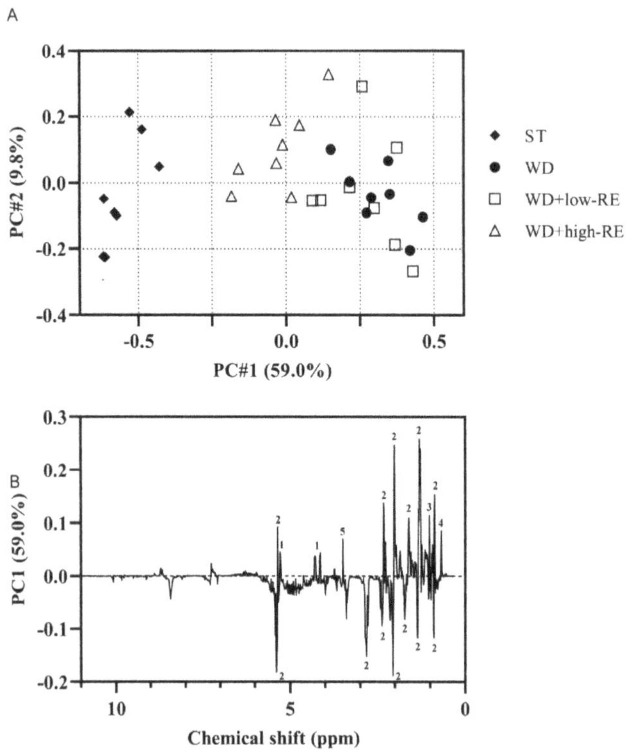

Figure 7. (**A**): Principal components analysis score plot of the non-polar extracts from the liver of rats fed ST (♦), WD (•), WD+low-RE (□), or WD+high-RE (Δ), over 46 days. ST, standard diet; WD, Western-style diet; WD+low-RE, WD supplemented with a low concentration of rosemary extract; WD+high-RE, WD supplemented with a high concentration of rosemary extract. (**B**): PC1 loadings plot of non-polar extracts. Assignments: 1, glycerol backbone; 2, fatty acyl chain; 3, free cholesterol; 4, total cholesterol; 5, residual methanol.

The hepatic glycogen levels were quantified in the polar liver extracts by NMR spectroscopy (Figure 2D). A significant treatment effect was found for glycogen hepatic levels with elevated levels in ST and WD+high-RE compared to WD and WD+low-RE rats.

3.8. LC-MS Metabolomics of Plasma and Caecum Content

The PCA scores plot of cecal digesta samples analyzed in negative mode revealed four separated clusters based on treatment (Figure 8A). Likewise, the PCA score plot of plasma (Figure 8B) in negative mode showed a clear separation between rats fed ST and WD. The separation between WD and WD with added RE was less clear; however, a ranking of the samples was evident.

In cecal content, the ST-treated rats differed from WD-, WD+low-RE-, and WD+high-RE-treated rats by having high levels of metabolites linked to whole grain intake, e.g., enterolactone and dihydroferulic acid (Figure S2, Tables S2 and S3). Moreover, ST-treated rats had higher levels of certain lipid metabolites, indicating different lipid sources in ST and WD. In general, the treatment with WD, WD+low-RE, and WD+high-RE caused high levels of putatively identified bile acid (BA) metabolites (Figure S3). However, for many of these BA metabolites, the concentration was decreasing with the increase in RE concentration. Cholic acid, a primary BA, appears to be the only identified BA metabolite increasing with RE concentration. Many metabolites were seen increasing with RE concentration, most of which were connected to rosemary. These include carnosol, carnosic acid, and glucuronides of these.

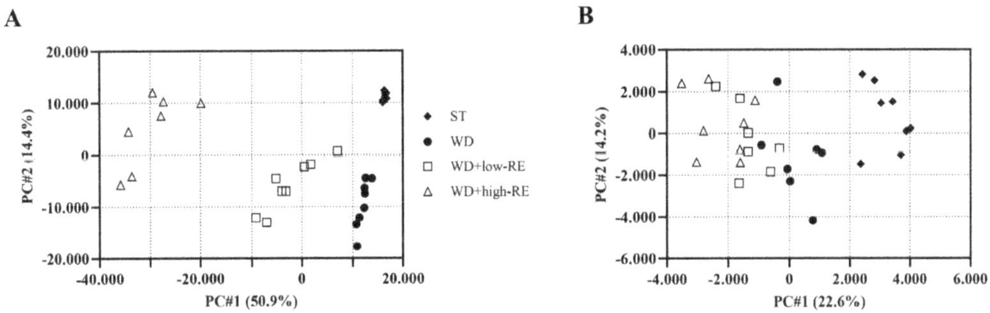

Figure 8. Principal components analysis score plot of the metabolome, acquired in negative mode, of cecal content (**A**) and plasma (**B**) of rats fed ST (♦), WD (•), WD+low-RE (□), or WD+high-RE (Δ), over 46 days. ST, standard diet; WD, Western-style diet; WD+low-RE, WD supplemented with a low concentration of rosemary extract; WD+high-RE, WD supplemented with a high concentration of rosemary extract.

In plasma, many RE-derived metabolites, including carnosol, carnosic acid, and rosmadial, were found in WD+low-RE- and WD+high-RE-treated rats, with a clear positive correlation to concentration in the diets (Figure S6, Tables S4 and S5). Furthermore, lysophosphatidylcholines comprised a major part of the metabolites differing in the plasma metabolome of rats fed the experimental diets.

3.9. Short-Chain Fatty Acids

ST-fed rats had significantly ($p < 0.0001$) more content in the caecum than rats fed WD, who on the other hand had less cecal content than rats fed WD supplemented with RE (Table 2). The total amount of carboxylic acids in the caecum was significantly higher ($p < 0.0001$) in ST-treated rats compared to treatment with WD, WD+low-RE, and WD+high-RE. The same trend was seen in the pool size of acetic, propionic, and butyric acid. In contrast, the pool size of succinic acid was significantly lower in ST- and WD-treated rats compared to treatment with WD+low-RE and WD+high-RE.

Table 2. Weight of cecal content and pool size of total and individual carboxylic acids in caecum content in rats fed a standard diet (ST), Western-style diet (WD), or a WD supplemented with either low (low-RE) or high (high-RE) concentrations of rosemary extract, measured at day 46. $n = 8$. Mean values with 95% confidence intervals.

	ST	WD	WD+Low-RE	WD+High-RE	*p*-Value
Cecal content (g)	4.64 ^{a} [3.69–5.83]	1.79 ^{c} [1.42–2.25]	2.44 ^{b} [1.94–3.07]	2.62 ^{b} [2.01–3.41]	<0.0001
Carboxylic acids, pool size (μmol)					
Total carboxylic acids	555 ^{a} [445–692]	84 ^{c} [67–104]	125 ^{b} [100–155]	128 ^{b} [96–169]	<0.0001
Acetic acid	302 ^{a} [241–379]	52.3 ^{b} [41.7–65.5]	69.0 ^{b} [55.1–86.5]	69.5 ^{b} [52.3–92.5]	<0.0001
Propionic acid	58.9 ^{a} [44.7–77.5]	13.2 ^{b} [10.0–17.3]	15.5 ^{b} [11.8–20.4]	12.7 ^{b} [9.0–18.0]	<0.0001
Butyric acid	165 ^{a} [128–214]	6.75 ^{b} [5.21–8.74]	3.15 ^{c} [2.43–4.08]	1.72 ^{d} [1.27–2.32]	<0.0001
Succinic acid	4.03 ^{b} [2.09–7.75]	2.45 ^{b} [1.41–4.26]	31.6 ^{a} [18.9–53.1]	21.8 ^{a} [12.0–39.7]	<0.0001

$^{a, b, c, d}$ Mean values within a row with unlike superscript letter were significantly different ($p < 0.05$).

4. Discussion

Over the six-week intervention, we did not observe an effect on body weight after supplementing a Western-style diet high in fat with RE, despite an elevated fasting total GLP-1 concentration, and hence our hypothesis was rejected. In a previous study [10], daily administration of carnosic acid (20 mg/kg BW/day), a polyphenol found in high concentrations in rosemary, to six-week-old mice fed a high-fat diet ad libitum for 14 days was sufficient to reduce weight gain significantly. For comparison, the WD+high-RE rats consumed up to approximately 150 mg RE per kg body weight per day, corresponding to 98 mg of carnosic acid per kg body weight per day. Thus, the lack of effect of the treatment on body weight in the present experiment is not likely to be due to insufficient dosing. Furthermore, adult, obese mice on a high-fat diet supplemented with 200 mg/kg body weight of RE had a significantly lower weight gain after 35 days of treatment compared to those who did not receive RE [38]. The lack of effect on BW observed in the present study may be explained by the choice of animal and strain. In a study with rats with a start weight of approximately 100 g, feeding the rats either a standard chow or a high-fat diet did not induce any diet-related differences in body weight over a period of six weeks in Sprague Dawley rats, while in Long-Evans rats, the animals exposed to a high-fat diet gained significantly more weight [39]. In addition, for outbred Sprague Dawley rats fed a diet high in fat and energy, only about half will develop obesity, while the other half is resistant to diet-induced obesity and will gain weight at the same pace as standard chow-fed controls [40].

We showed that feeding WD+high-RE for six weeks significantly increased fasting total GLP-1 plasma concentrations. This is not supported by our studies on isolated perfused rat small intestine, where intra-luminal or intra-arterial RE infusion did not stimulate the secretion of intestinal GLP-1. However, our previous in vitro studies showed augmented secretion of GLP-1 from STC-1 and HuTu-80 cells when stimulated with RE [12]. The discrepancy between the long-term and acute effects of RE on GLP-1 secretion points towards an adaptive effect of RE, as previously seen for nutritional modulation of GLP-1 secretion [41]. It is well documented that RE is a potent inhibitor of the primary enzyme catalyzing GLP-1 degradation, DPP-4 [42]. Our metabolomics data demonstrated that several RE-derived metabolites, including carnosol, carnosic acid, and rosmadial, were present in plasma which could possibly lead to inhibition of DPP-4 and thereby a higher concentration of active GLP-1. Confirmation of this would require measurement of active GLP-1. We observed a higher proglucagon gene expression after treatment with WD+low-RE and WD+high-RE, but this does not correlate with the increased fasting levels after WD+high-RE only. The expression increased progressively from the proximal duodenum to the distal ileum, which corresponds to the tissue concentrations of GLP-1 in rats [43], in agreement with the distribution of GLP-1-secreting cells.

GLP-1 has been associated with an increased activity of peroxisome proliferator-activated receptor (PPAR)-α, and PPARγ [44] and polyphenols have been shown to modulate PPARα expression when added to high-fat diets [45]. PPARα and PPARγ are both known to downregulate the expression of SCD1 and RETSAT [44,46,47]. Our proteomics data show upregulation of SCD1 in the WD compared to ST, which is in accordance with previous studies, showing that SCD1 activity is upregulated in diets high in saturated fatty acids [48]. Interestingly, this upregulation was significantly lower in WD+high-RE compared to WD. High SCD1 activity is associated with hypertriglyceridemia and reduced levels of HDL cholesterol [49]. Consequently, a reduction in the upregulation of the high-fat diet-induced SCD1 expression by RE may contribute to lower concentrations of TG through SCD1. Our data also show RETSAT to be upregulated by WD, but this protein was no longer significantly upregulated when the rats received RE supplements. In humans, hepatic RETSAT expression correlates with serum TGs and steatosis, and depletion of RETSAT in obese mice lowers hepatic and circulating TGs [50]. Consequently, the effect of RE on RETSAT may contribute to the reduction in TG. Several studies have associated lower expression of SCD1 and RETSAT with a favorable lipid profile in plasma and liver tissue, such as decreased plasma triglycerides and increased plasma HDL, lower total liver fat, triglycerides, and cholesterol in the liver, and a higher proportion of unsaturated fatty acids in the liver [47,49,51,52]. The levels of MIF in the liver were increased by RE. MIF has a hepatoprotective role in in vivo models of NAFLD [53]. Hence, it is possible that RE contributed to lower hepatic fat accumulation through its upregulation of MIF.

The observed increase in fasting GLP-1 concentrations might cause an increase in fatty acid β-oxidation and a decrease in fatty acid synthesis, resulting in the reduction in relative fat mass in the liver. Animal studies have reported a decrease in hepatic fat accumulation upon treatment with the GLP-1 receptor agonist exenatide [54,55]. It has also been concluded that rats on a high-fat diet treated with exenatide have increased levels of transcripts of factors involved in fatty acid β-oxidation, resulting in a reduction in total hepatic lipid accumulation [56]. Others have found an increased plasma GLP-1 concentration to be associated with a reduction in mRNA and protein expression of transcription factors required for cholesterol biosynthesis and enzymes that catalyze fatty acid synthesis [57]. Repeating our study protocol in animals deficient in GLP-1 signaling will be required to ultimately test this hypothesis.

Our data showed that rats consuming a high-fat diet for six weeks increased hepatic fat percentage by 2.4-fold, but RE inhibited the accumulation of fat in the liver, resulting in a fat percentage only 1.29-fold higher than ST for WD+high-RE. This finding agrees with that of others who have found polyphenol-rich extracts capable of attenuating steatosis in human and animal models [58,59]. The observations on fat in liver tissue are consistent with our NMR data, where the lipid profile of the liver showed that consumption of a Western-style diet for six weeks resulted in higher levels of cholesterol and triglycerides, and that the present fatty acids were more often saturated fatty acids than in a standard diet. Interestingly, by supplementing the diet with high-RE, the overall lipid profile in the liver shifted to resemble a mix of standard and Western-style diets, with less free and total cholesterol, less triglycerides, and more unsaturated fatty acids than the Western-style diet. This is consistent with previous studies in mice showing that RE can reduce triglyceride levels in the liver by 39% [38]. The rats fed WD supplemented with RE had a relatively increased liver weight. This is most probably associated with the enzyme induction seen due to the xenobiotic metabolism taking place in the liver [60]. This is supported by the increase in several proteins involved in metabolic detoxification pathways, as well as the observed glucuronidation of RE-derived metabolites.

In our study, we did not see any differences in total body fat mass between ST and the other treatments, even though they were based on a high-fat diet. Interestingly, our WD+high-RE treated rats showed decreased levels of plasma triglycerides and increased levels of HDL, demonstrating a positive effect of continuous consumption of rosemary extract. This contradicts the study by Harach et el. [38], where a RE concentration corresponding to half of the concentration in our WD+high-RE did not lower the plasma triglyceride concentration, although hepatic triglyceride concentration was reduced. We observed higher total cholesterol in plasma in rats fed a diet based on high fat but did not see an effect of RE. A meta-analysis that included 750 participants evaluated the effect of polyphenol-rich cinnamon on blood lipid concentrations and concluded that supplementation with this polyphenol source reduced plasma triglycerides and total cholesterol but did not have any effect on LDL and HDL [61]. In another study, 25 healthy participants received a polyphenol-rich pine bark extract for six weeks, resulting in decreased LDL and increased HDL [62].

Our metabolomics data showed that the concentration of numerous putative BA metabolites in cecal content was inversely proportional to the RE concentration. In contrast to the present study, several studies have shown that polyphenols induce increases in BA excretion [63]. In the present study, the excretion of cholic acid, a primary BA, was increased with increasing addition of RE to the diet. However, the excretion of deoxycholic acid, a secondary BA, and numerous putative BA metabolites were decreased with increasing RE in the diet. Dietary fat stimulates the secretion of BA into the intestinal lumen which promotes changes in the microbiota that increases the conversion of primary to secondary

BA [64]. In the present study, this effect was counteracted by the addition of RE which is in agreement with previous studies [65,66], and it has been suggested that polyphenols may regulate the profile and concentration of BA [65] through modulation of the profile of the microflora [67].

Changing the diet from ST to WD and supplementing the high-fat diet with RE caused noticeable changes in the pool of carboxylic acids found in the cecal content, indicating that the microbiota composition was modulated. The considerably lower microbial activity observed when feeding WD is due to the lower content of substrate (dietary fiber) for the microflora in WD. However, the pool of succinic acid increased approximately 10-fold when adding RE to WD and this is in accordance with a previous study showing that the addition of some polyphenols to a high-fat diet increased the concentration of succinic acid in cecal content [68]. It has been shown that RE increases *Prevotella* groups [69], and *Prevotella copri* has been recognized as a succinate producer [70]. Succinate contributes to improved plasma glucose through the activation of intestinal gluconeogenesis [71] and increased glycogen storage in the liver [72], as also seen in our study.

5. Conclusions

We did not observe an increase in body weight or total body fat percentage in young, healthy rats during the experimental period when the animals were fed a high-fat diet. However, the treatment with the high-fat diet-induced dyslipidemia manifested as a higher hepatic fat percentage, higher cholesterol and triglyceride concentrations in the plasma and liver, and higher LDL in plasma which was ameliorated when supplementing 4 g RE per kg feed to the diet. Furthermore, RE increased plasma HDL. Our proteomics data suggest this is accomplished through the upregulation of proteins involved in fatty acid metabolism. Altogether, our study strongly indicates that RE improves the lipid profile in animals before the onset of classical signs of obesity, perhaps associated with an increase in fasting GLP-1 or possibly through the upregulation of PPARα and PPARγ. The study indicates that the effect of RE is exerted both via bioavailable metabolites that exert their effect systemically and through gut microbiota-associated metabotypes.

Supplementary Materials: The following supporting information can be downloaded at: https://www.mdpi.com/article/10.3390/metabo13090974/s1, Supplementary methods; Figure S1: Heat maps of metabolites discriminating between the ST and the WD diet identified and annotated in cecal content of rats and being dominant in the (A) ST diet or (B) WD diet, respectively; Figure S2: Heat maps of metabolites discriminating between the WD and the WD+high-RE diet identified and annotated in caecal content of rats and being dominant in the (A) WD diet or (B) WD+high-RE diet, respectively; Figure S3: Heat maps of metabolites discriminating between the ST and the WD diet identified and annotated in plasma of rats and being dominant in the (a) ST diet or (b) WD diet, respectively; Figure S4: Heat maps of metabolites discriminating between the WD and the WD+high-RE diet identified and annotated in plasma of rats and being dominant in the (a) WD diet or (b) WD+high-RE diet, respectively; Table S1: Ingredient composition and chemical composition1 of standard diet (ST), Western style diet (WD), or a WD supplemented with either low (low-RE) or high (high-RE) concentrations of rosemary extract; Supplementary Table S2. Gene and protein name of quantified proteins involved in metabolic detoxification pathways; Supplementary Table S3. List of metabolic features discriminating between rats fed ST and WD in caecal content; Supplementary Table S4. List of metabolic features discriminating between rats fed WD and WD+high-RE in caecal content; Supplementary Table S5. List of metabolic features discriminating between rats fed ST and WD in plasma; Supplementary Table S6. List of metabolic features discriminating between rats fed WD and WD+high-RE in plasma.

Author Contributions: Conceptualization: S.M., H.M.J. and M.S.H.; Data curation: S.M., C.C.Y., S.Y.B., C.B.-L., C.L. and M.S.H.; Formal analysis: S.M., C.C.Y., S.Y.B. and M.S.H.; Funding acquisition; T.A.K., H.M.J. and M.S.H.; Investigation: S.M. and M.S.H.; Methodology: S.M., C.C.Y., S.Y.B., M.S.H., C.B.-L., J.J.H. and C.L.; Project administration: T.A.K.; Supervision: M.S.H.; Writing—original draft: S.M.; Writing—review & editing: all authors. All authors have read and agreed to the published version of the manuscript.

Funding: This research was funded by Innovation Found Denmark, grant number 5158-00011B.

Institutional Review Board Statement: The study on the 64 intact rats was conducted with permission from the Danish Animal Experiments Inspectorate (2018-15-0201-01454). The study complied with the guidelines of the Danish Ministry of Justice, Act no. 474 af 15 May 2014, concerning the care and use of animals in research as stipulated in the executive order no. 12 of 7 January 2016, and under consideration of the Arrive Guidelines [73]. The intestinal perfusion studies were conducted with permission from the Danish Animal Experiments Inspectorate (Perfusions: 2018-15-0201-01397) and the local ethical committee in accordance with the guidelines of Danish legislation governing animal experimentation (1987) and the National Institutes of Health (publication number 85-23) and the European Convention for the Protection of Vertebrate Animals used for Experimental and other Scientific Purposes (Council of Europe No 123, Strasbourg 1985).

Informed Consent Statement: Not applicable.

Data Availability Statement: None of the data were deposited in an official repository. The data that support the study findings are available from the authors upon request.

Acknowledgments: Karin Kvorning Jensen and Emilie Sommerlund, Department of Animal and Veterinary Sciences, Aarhus University are gratefully acknowledged for rearing and handling of the animals. Kasper V. Poulsen, Anette M. Pedersen, and Thomas Rebsdorf are gratefully acknowledged for their excellent technical assistance.

Conflicts of Interest: The authors declare no conflict of interest.

eferences

1. Guh, D.P.; Zhang, W.; Bansback, N.; Amarsi, Z.; Birmingham, C.L.; Anis, A.H. The incidence of co-morbidities related to obesity and overweight: A systematic review and meta-analysis. *BMC Public Health* **2009**, *9*, 88. [CrossRef] [PubMed]
2. Williamson, G. The role of polyphenols in modern nutrition. *Nutr. Bull.* **2017**, *42*, 226–235. [CrossRef] [PubMed]
3. Arts, I.C.W.; Hollman, P.C.H. Polyphenols and disease risk in epidemiologic studies. *Am. J. Clin. Nutr.* **2005**, *81*, 317s–325s. [CrossRef]
4. Scalbert, A.; Manach, C.; Morand, C.; Remesy, C.; Jimenez, L. Dietary polyphenols and the prevention of diseases. *Crit. Rev. Food Sci. Nutr.* **2005**, *45*, 287–306. [CrossRef] [PubMed]
5. Panickar, K.S. Effects of dietary polyphenols on neuroregulatory factors and pathways that mediate food intake and energy regulation in obesity. *Mol. Nutr. Food Res.* **2013**, *57*, 34–47. [CrossRef] [PubMed]
6. Gu, Y.Y.; Hurst, W.J.; Stuart, D.A.; Lambert, J.D. Inhibition of Key Digestive Enzymes by Cocoa Extracts and Procyanidins. *J. Agric. Food Chem.* **2011**, *59*, 5305–5311. [CrossRef] [PubMed]
7. Matsui, N.; Ito, R.; Nishimura, E.; Yoshikawa, M.; Kato, M.; Kamei, M.; Shibata, H.; Matsumoto, I.; Abe, K.; Hashizume, S. Ingested cocoa can prevent high-fat diet-induced obesity by regulating the expression of genes for fatty acid metabolism. *Nutrition* **2005**, *21*, 594–601. [CrossRef] [PubMed]
8. Shimoda, H.; Tanaka, J.; Kikuchi, M.; Fukuda, T.; Ito, H.; Hatano, T.; Yoshidao, T. Effect of Polyphenol-Rich Extract from Walnut on Diet-Induced Hypertriglyceridemia in Mice via Enhancement of Fatty Acid Oxidation in the Liver. *J. Agric. Food Chem.* **2009**, *57*, 1786–1792. [CrossRef]
9. Erkan, N.; Ayranci, G.; Ayranci, E. Antioxidant activities of rosemary (*Rosmarinus Officinalis* L.) extract, blackseed (*Nigella sativa* L.) essential oil, carnosic acid, rosmarinic acid and sesamol. *Food Chem.* **2008**, *110*, 76–82. [CrossRef]
10. Ibarra, A.; Cases, J.; Roller, M.; Chiralt-Boix, A.; Coussaert, A.; Ripoll, C. Carnosic acid-rich rosemary (*Rosmarinus officinalis* L.) leaf extract limits weight gain and improves cholesterol levels and glycaemia in mice on a high-fat diet. *Br. J. Nutr.* **2011**, *106*, 1182–1189. [CrossRef]
11. Ninomiya, K.; Matsuda, H.; Shimoda, H.; Nishida, N.; Kasajima, N.; Yoshino, T.; Morikawa, T.; Yoshikawa, M. Carnosic acid, a new class of lipid absorption inhibitor from sage. *Bioorg. Med. Chem. Lett.* **2004**, *14*, 1943–1946. [CrossRef] [PubMed]
12. Yue, Y.; Madsen, S.; Hedemann, M.S.; Knudsen, K.E.B.; Sparsø, F.V.; Laursen, A.; Jensen, H.M.; Knudsen, T.A.; Purup, S. Effect of food ingredients on glucagon-like peptide-1 secretion in STC-1 and HuTu-80 cells. *Int. J. Food Sci. Technol.* **2019**, *54*, 3149–3155. [CrossRef]
13. Vilsbøll, T.; Christensen, M.; Junker, A.E.; Knop, F.K.; Gluud, L.L. Effects of glucagon-like peptide-1 receptor agonists on weight loss: Systematic review and meta analyses of randomised controlled trials. *BMJ* **2012**, *344*, d7771. [CrossRef] [PubMed]
14. Gribble, F.M.; Reimann, F. Metabolic Messengers: Glucagon-like peptide 1. *Nat. Metab.* **2021**, *3*, 142–148. [CrossRef] [PubMed]
15. Knauf, C.; Cani, P.D.; Perrin, C.; Iglesias, M.A.; Maury, J.F.; Bernard, E.; Benhamed, F.; Gremeaux, T.; Drucker, D.J.; Kahn, C.R.; et al. Brain glucagon-like peptide-1 increases insulin secretion and muscle insulin resistance to favor hepatic glycogen storage. *J. Clin. Investig.* **2005**, *115*, 3554–3563. [CrossRef] [PubMed]
16. Green, C.J.; Henriksen, T.I.; Pedersen, B.K.; Solomon, T.P.J. Glucagon like peptide-1-induced glucose metabolism in differentiated human muscle satellite cells is attenuated by hyperglycemia. *PLoS ONE* **2012**, *7*, e44284. [CrossRef] [PubMed]

17. Farr, S.; Taher, J.; Adeli, K. Glucagon-like peptide-1 as a key regulator of lipid and lipoprotein metabolism in fasting an postprandial states. *Cardiovasc. Hematol. Disord. Drug Targets* **2014**, *14*, 126–136. [CrossRef] [PubMed]
18. Howard, B.V.; Ruotolo, G.; Robbins, D.C. Obesity and dyslipidemia. *Endocrinol. Metab. Clin. North Am.* **2003**, *32*, 855–86 [CrossRef]
19. Fabbrini, E.; Sullivan, S.; Klein, S. Obesity and nonalcoholic fatty liver disease: Biochemical, metabolic, and clinical implication *Hepatology* **2010**, *51*, 679–689. [CrossRef]
20. Abenavoli, L.; Larussa, T.; Corea, A.; Procopio, A.C.; Boccuto, L.; Dallio, M.; Federico, A.; Luzza, F. Dietary Polyphenols an Non-Alcoholic Fatty Liver Disease. *Nutrients* **2021**, *13*, 494. [CrossRef]
21. Wang, Z.; Zeng, M.; Wang, Z.; Qin, F.; Chen, J.; He, Z. Dietary Polyphenols to Combat Nonalcoholic Fatty Liver Disease via th Gut–Brain–Liver Axis: A Review of Possible Mechanisms. *J. Agric. Food Chem.* **2021**, *69*, 3585–3600. [CrossRef] [PubMed]
22. Zhao, Y.; Sedighi, R.; Wang, P.; Chen, H.; Zhu, Y.; Sang, S. Carnosic Acid as a Major Bioactive Component in Rosemary Extra Ameliorates High-Fat-Diet-Induced Obesity and Metabolic Syndrome in Mice. *J. Agric. Food Chem.* **2015**, *63*, 4843–4852. [CrossRe [PubMed]
23. Hedemann, M.S.; Hermansen, K.; Pedersen, S.; Bach Knudsen, K.E. Resistant Starch but Not Enzymatically Modified Waxy Maiz Delays Development of Diabetes in Zucker Diabetic Fatty Rats. *J. Nutr.* **2017**, *147*, 825–834. [CrossRef] [PubMed]
24. Kuhre, R.E.; Frost, C.R.; Svendsen, B.; Holst, J.J. Molecular Mechanisms of Glucose-Stimulated GLP-1 Secretion From Perfuse Rat Small Intestine. *Diabetes* **2015**, *64*, 370–382. [CrossRef] [PubMed]
25. Kuhre, R.E.; Holst, J.J. Mechanisms Underlying Gut Hormone Secretion Using the Isolated Perfused Rat Small Intestine. *J. V Exp.* **2019**, *144*, e58533. [CrossRef]
26. Ørskov, C.; Rabenhøj, L.; Wettergren, A.; Kofod, H.; Holst, J.J. Tissue and plasma concentrations of amidated and glycine-extende glucagon-like peptide I in humans. *Diabetes* **1994**, *43*, 535–539. [CrossRef] [PubMed]
27. Zougman, A.; Selby, P.J.; Banks, R.E. Suspension trapping (STrap) sample preparation method for bottom-up proteomics analysi *Proteomics* **2014**, *14*, 1006-1000. [CrossRef]
28. Käll, L.; Canterbury, J.D.; Weston, J.; Noble, W.S.; MacCoss, M.J. Semi-supervised learning for peptide identification from shotgu proteomics datasets. *Nat. Methods* **2007**, *4*, 923–925. [CrossRef]
29. Silva, J.C.; Gorenstein, M.V.; Li, G.-Z.; Vissers, J.P.C.; Geromanos, S.J. Absolute Quantification of Proteins by LCMSE: A Virtue o Parallel ms Acquisition *S. *Mol. Cell. Proteom.* **2006**, *5*, 144–156. [CrossRef]
30. Yde, C.C.; Clausen, M.R.; Ditlev, D.B.; Lillefosse, H.; Madsen, L.; Kristiansen, K.; Liaset, B.; Bertram, H.C. Multi-block PC and multi-compartmental study of the metabolic responses to intake of hydrolysed versus intact casein in C57BL/6J mice b NMR-based metabolomics. *Metabolomics* **2014**, *10*, 938–949. [CrossRef]
31. Fathi, F.; Brun, A.; Rott, K.H.; Falco Cobra, P.; Tonelli, M.; Eghbalnia, H.R.; Caviedes-Vidal, E.; Karasov, W.H.; Markley, J. NMR-Based Identification of Metabolites in Polar and Non-Polar Extracts of Avian Liver. *Metabolites* **2017**, *7*, 61. [CrossRe [PubMed]
32. Wang, K.-C.; Wang, S.-Y.; Kuo, C.-h.; Tseng, Y.J. Distribution-Based Classification Method for Baseline Correction of Metabolom 1D Proton Nuclear Magnetic Resonance Spectra. *Anal. Chem.* **2013**, *85*, 1231–1239. [CrossRef] [PubMed]
33. Savorani, F.; Tomasi, G.; Engelsen, S.B. icoshift: A versatile tool for the rapid alignment of 1D NMR spectra. *J. Magn. Reson.* **201** *202*, 190–202. [CrossRef] [PubMed]
34. Sousa, S.A.A.; Magalhães, A.; Ferreira, M.M.C. Optimized bucketing for NMR spectra: Three case studies. *Chemom. Intell. La Syst.* **2013**, *122*, 93–102. [CrossRef]
35. Curtasu, M.V.; Tafintseva, V.; Bendiks, Z.A.; Marco, M.L.; Kohler, A.; Xu, Y.; Nørskov, N.P.; Nygaard Lærke, H.; Bach Knudse K.E.; Hedemann, M.S. Obesity-Related Metabolome and Gut Microbiota Profiles of Juvenile Göttingen Minipigs—Long-Terr Intake of Fructose and Resistant Starch. *Metabolites* **2020**, *10*, 456. [CrossRef]
36. Curtasu, M.V.; Knudsen, K.E.B.; Callesen, H.; Purup, S.; Stagsted, J.; Hedemann, M.S. Obesity Development in a Miniatu Yucatan Pig Model: A Multi-compartmental Metabolomics Study on Cloned and Normal Pigs Fed Restricted or Ad Libitur High-Energy Diets. *J. Proteome Res.* **2019**, *18*, 30–47. [CrossRef] [PubMed]
37. Jensen, M.T.; Cox, R.P.; Jensen, B.B. Microbial-Production of Skatole in the Hind Gut of Pigs Given Different Diets and Its Relatio to Skatole Deposition in Backfat. *Anim. Sci.* **1995**, *61*, 293–304. [CrossRef]
38. Harach, T.; Aprikian, O.; Monnard, I.; Moulin, J.; Membrez, M.; Beolor, J.C.; Raab, T.; Mace, K.; Darimont, C. Rosemary (*Rosmarinu officinalis* L.) Leaf Extract Limits Weight Gain and Liver Steatosis in Mice Fed a High-Fat Diet. *Planta Med.* **2010**, *76*, 566–57 [CrossRef]
39. Song, Z.B.; Xie, W.R.; Chen, S.S.; Strong, J.A.; Print, M.S.; Wang, J.I.; Shareef, A.F.; Ulrich-Lai, Y.M.; Zhang, J.M. High-fat die increases pain behaviors in rats with or without obesity. *Sci. Rep.* **2017**, *7*, 10350. [CrossRef]
40. Levin, B.E.; DunnMeynell, A.A.; Balkan, B.; Keesey, R.E. Selective breeding for diet-induced obesity and resistance in Spragu Dawley rats. *Am. J. Physiol-Regul. Integr. Comp. Physiol.* **1997**, *273*, R725–R730. [CrossRef]
41. Bodnaruc, A.M.; Prud'homme, D.; Blanchet, R.; Giroux, I. Nutritional modulation of endogenous glucagon-like peptide- secretion: A review. *Nutr. Metab.* **2016**, *13*, 92. [CrossRef] [PubMed]
42. Bower, A.M.; Hernandez, L.M.R.; Berhow, M.A.; de Mejia, E.G. Bioactive Compounds from Culinary Herbs Inhibit a Molecula Target for Type 2 Diabetes Management, Dipeptidyl Peptidase IV. *J. Agric. Food Chem.* **2014**, *62*, 6147–6158. [CrossRef] [PubMed

3. Wewer Albrechtsen, N.J.; Kuhre, R.E.; Toräng, S.; Holst, J.J. The intestinal distribution pattern of appetite- and glucose regulatory peptides in mice, rats and pigs. *BMC Res. Notes* **2016**, *9*, 1–6. [CrossRef] [PubMed]
4. Patel, V.; Joharapurkar, A.; Kshirsagar, S.; Sutariya, B.; Patel, M.; Patel, H.; Pandey, D.; Patel, D.; Bahekar, R.; Jain, M. Central administration of coagonist of GLP-1 and glucagon receptors improves dyslipidemia. *Biomed. Pharmacother.* **2018**, *98*, 364–371. [CrossRef] [PubMed]
5. Domínguez-Avila, J.A.; González-Aguilar, G.A.; Alvarez-Parrilla, E.; De la Rosa, L.A. Modulation of PPAR Expression and Activity in Response to Polyphenolic Compounds in High Fat Diets. *Int. J. Mol. Sci.* **2016**, *17*, 1002. [CrossRef] [PubMed]
6. Yao-Borengasser, A.; Rassouli, N.; Varma, V.; Bodles, A.M.; Rasouli, N.; Unal, R.; Phanavanh, B.; Ranganathan, G.; McGehee, R.E., Jr.; Kern, P.A. Stearoyl-coenzyme A desaturase 1 gene expression increases after pioglitazone treatment and is associated with peroxisomal proliferator-activated receptor-gamma responsiveness. *J. Clin. Endocrinol. Metab.* **2008**, *93*, 4431–4439. [CrossRef] [PubMed]
7. Schupp, M.; Lefterova, M.I.; Janke, J.; Leitner, K.; Cristancho, A.G.; Mullican, S.E.; Qatanani, M.; Szwergold, N.; Steger, D.J.; Curtin, J.C.; et al. Retinol saturase promotes adipogenesis and is downregulated in obesity. *Proc. Natl. Acad. Sci. USA* **2009**, *106*, 1105–1110. [CrossRef] [PubMed]
8. Sampath, H.; Miyazaki, M.; Dobrzyn, A.; Ntambi, J. Stearoyl CoA desaturase-1 mediates the pro-lipogenic effects of dietary saturated fat. *FASEB J.* **2007**, *21*, A109. [CrossRef]
9. Chu, K.; Miyazaki, M.; Man, W.C.; Ntambi, J.M. Stearoyl-coenzyme A desaturase 1 deficiency protects against hypertriglyceridemia and increases plasma high-density lipoprotein cholesterol induced by liver X receptor activation. *Mol. Cell. Biol.* **2006**, *26*, 6786–6798. [CrossRef]
0. Heidenreich, S.; Witte, N.; Weber, P.; Goehring, I.; Tolkachov, A.; von Loeffelholz, C.; Docke, S.; Bauer, M.; Stockmann, M.; Pfeiffer, A.F.H.; et al. Retinol saturase coordinates liver metabolism by regulating ChREBP activity. *Nat. Commun.* **2017**, *8*, 384. [CrossRef]
1. Silbernagel, G.; Kovarova, M.; Cegan, A.; Machann, J.; Schick, F.; Lehmann, R.; Haring, H.U.; Stefan, N.; Schleicher, E.; Fritsche, A.; et al. High hepatic SCD1 activity is associated with low liver fat content in healthy subjects under a lipogenic diet. *J. Clin. Endocrinol. Metab.* **2012**, *97*, E2288–E2292. [CrossRef] [PubMed]
2. Miyazaki, M.; Kim, Y.C.; Gray-Keller, M.P.; Attie, A.D.; Ntambi, J.M. The biosynthesis of hepatic cholesterol esters and triglycerides is impaired in mice with a disruption of the gene for stearoyl-CoA desaturase 1. *J. Biol. Chem.* **2000**, *275*, 30132–30138. [CrossRef] [PubMed]
3. Heinrichs, D.; Berres, M.L.; Coeuru, M.; Knauel, M.; Nellen, A.; Fischer, P.; Philippeit, C.; Bucala, R.; Trautwein, C.; Wasmuth, H.E.; et al. Protective role of macrophage migration inhibitory factor in nonalcoholic steatohepatitis. *FASEB J.* **2014**, *28*, 5136–5147. [CrossRef] [PubMed]
4. Ding, X.K.; Saxena, N.K.; Lin, S.B.; Gupta, N.; Anania, F.A. Exendin-4, a glucagon-like protein-1 (GLP-1) receptor agonist, reverses hepatic steatosis in ob/ob mice. *Hepatology* **2006**, *43*, 173–181. [CrossRef] [PubMed]
5. Trevaskis, J.L.; Griffin, P.S.; Wittmer, C.; Neuschwander-Tetri, B.A.; Brunt, E.M.; Dolman, C.S.; Erickson, M.R.; Napora, J.; Parkes, D.G.; Roth, J.D. Glucagon-like peptide-1 receptor agonism improves metabolic, biochemical, and histopathological indices of nonalcoholic steatohepatitis in mice. *Am. J. Physiol-Gastrointest. Liver Physiol.* **2012**, *302*, G762–G772. [CrossRef] [PubMed]
6. Svegliati-Baroni, G.; Saccomanno, S.; Rychlicki, C.; Agostinelli, L.; De Minicis, S.; Candelaresi, C.; Faraci, G.; Pacetti, D.; Vivarelli, M.; Nicolini, D.; et al. Glucagon-like peptide-1 receptor activation stimulates hepatic lipid oxidation and restores hepatic signalling alteration induced by a high-fat diet in nonalcoholic steatohepatitis. *Liver Int* **2011**, *31*, 1285–1297. [CrossRef] [PubMed]
7. Khound, R.; Taher, J.; Baker, C.; Adeli, K.; Su, Q. GLP-1 Elicits an Intrinsic Gut-Liver Metabolic Signal to Ameliorate Diet-Induced VLDL Overproduction and Insulin Resistance. *Arterioscler. Thromb. Vasc. Biol.* **2017**, *37*, 2252–2259. [CrossRef]
8. Chang, H.C.; Peng, C.H.; Yeh, D.M.; Kao, E.S.; Wang, C.J. Hibiscus sabdariffa extract inhibits obesity and fat accumulation, and improves liver steatosis in humans. *Food Funct.* **2014**, *5*, 734–739. [CrossRef]
9. Gan, L.; Meng, Z.J.; Xiong, R.B.; Guo, J.Q.; Lu, X.C.; Zheng, Z.W.; Deng, Y.P.; Luo, B.D.; Zou, F.; Li, H. Green tea polyphenol epigallocatechin-3-gallate ameliorates insulin resistance in non-alcoholic fatty liver disease mice. *Acta Pharmacol. Sin.* **2015**, *36*, 597–605. [CrossRef]
0. Maronpot, R.R.; Yoshizawa, K.; Nyska, A.; Harada, T.; Flake, G.; Mueller, G.; Singh, B.; Ward, J.M. Hepatic Enzyme Induction: Histopathology. *Toxicol. Pathol.* **2010**, *38*, 776–795. [CrossRef]
1. Maierean, S.M.; Serban, M.C.; Sahebkar, A.; Ursoniu, S.; Serban, A.; Penson, P.; Banach, M.; Meta-anal, L.B.P. The effects of cinnamon supplementation on blood lipid concentrations: A systematic review and meta-analysis. *J. Clin. Lipidol.* **2017**, *11*, 1393–1406. [CrossRef] [PubMed]
2. Devaraj, S.; Vega-Lopez, S.; Kaul, N.; Schonlau, F.; Rohdewald, P.; Jialal, I. Supplementation with a pine bark extract rich in polyphenols increases plasma antioxidant capacity and alters the plasma lipoprotein profile. *Lipids* **2002**, *37*, 931–934. [CrossRef] [PubMed]
3. Chambers, K.F.; Day, P.E.; Aboufarrag, H.T.; Kroon, P.A. Polyphenol Effects on Cholesterol Metabolism via Bile Acid Biosynthesis, CYP7A1: A Review. *Nutrients* **2019**, *11*, 2588. [CrossRef] [PubMed]
4. Ridlon, J.M.; Kang, D.-J.; Hylemon, P.B. Bile salt biotransformations by human intestinal bacteria. *J. Lipid Res.* **2006**, *47*, 241–259. [CrossRef]

65. Fotschki, B.; Juśkiewicz, J.; Jurgoński, A.; Rigby, N.; Sójka, M.; Kołodziejczyk, K.; Mackie, A.; Zduńczyk, Z. Raspberry pomace alters cecal microbial activity and reduces secondary bile acids in rats fed a high-fat diet. *J. Nutr. Biochem.* **2017**, *46*, 13–20. [CrossRef] [PubMed]
66. Holscher, H.D.; Guetterman, H.M.; Swanson, K.S.; An, R.; Matthan, N.R.; Lichtenstein, A.H.; Novotny, J.A.; Baer, D.J. Walnut Consumption Alters the Gastrointestinal Microbiota, Microbially Derived Secondary Bile Acids, and Health Markers in Healthy Adults: A Randomized Controlled Trial. *J. Nutr.* **2018**, *148*, 861–867. [CrossRef] [PubMed]
67. Han, Y.; Haraguchi, T.; Iwanaga, S.; Tomotake, H.; Okazaki, Y.; Mineo, S.; Moriyama, A.; Inoue, J.; Kato, N. Consumption of Some Polyphenols Reduces Fecal Deoxycholic Acid and Lithocholic Acid, the Secondary Bile Acids of Risk Factors of Colon Cancer. *J. Agric. Food Chem.* **2009**, *57*, 8587–8590. [CrossRef]
68. Haraguchi, T.; Kayashima, T.; Okazaki, Y.; Inoue, J.; Mineo, S.; Matsubara, K.; Sakaguchi, E.; Yanaka, N.; Kato, N. Cecal Succinate Elevated by Some Dietary Polyphenols May Inhibit Colon Cancer Cell Proliferation and Angiogenesis. *J. Agric. Food Chem.* **2014**, *62*, 5589–5594. [CrossRef]
69. Romo-Vaquero, M.; Selma, M.-V.; Larrosa, M.; Obiol, M.; García-Villalba, R.; González-Barrio, R.; Issaly, N.; Flanagan, J.; Roller, M.; Tomás-Barberán, F.A.; et al. A rosemary extract rich in carnosic acid selectively modulates caecum microbiota and inhibits β-glucosidase activity, altering fiber and short chain fatty acids fecal excretion in lean and obese female rats. *PLoS ONE* **2014**, *9*, e94687. [CrossRef]
70. Fernández-Veledo, S.; Vendrell, J. Gut microbiota-derived succinate: Friend or foe in human metabolic diseases? *Rev. Endocr. Metab. Disord.* **2019**, *20*, 439–447. [CrossRef]
71. De Vadder, F.; Kovatcheva-Datchary, P.; Zitoun, C.; Duchampt, A.; Bäckhed, F.; Mithieux, G. Microbiota-Produced Succinate Improves Glucose Homeostasis via Intestinal Gluconeogenesis. *Cell Metab.* **2016**, *24*, 151–157. [CrossRef]
72. Kovatcheva-Datchary, P.; Nilsson, A.; Akrami, R.; Lee, Y.S.; De Vadder, F.; Arora, T.; Hallen, A.; Martens, E.; Björck, I.; Bäckhed, F. Dietary Fiber-Induced Improvement in Glucose Metabolism Is Associated with Increased Abundance of Prevotella. *Cell Metab.* **2015**, *22*, 971–982. [CrossRef]
73. Percie du Sert, N.; Ahluwalia, A.; Alam, S.; Avey, M.T.; Baker, M.; Browne, W.J.; Clark, A.; Cuthill, I.C.; Dirnagl, U.; Emerson, M.; et al. Reporting animal research: Explanation and elaboration for the ARRIVE guidelines 2.0. *PLoS Biol.* **2020**, *18*, e3000411. [CrossRef]

molecules

Article

Novel Quinazoline Derivative Induces Differentiation of Keratinocytes and Enhances Skin Barrier Functions against Th2 Cytokine-Mediated Signaling

ukyung Park [1,†], Huddar Srigouri [2,†] and Dongwon Kim [1,3,*]

1 Department of Energy and Biotechnology, Graduate School, Dongseo University, Busan 47011, Republic of Korea
2 Drug Information Platform Center, Korea Research Institute of Chemical Technology, Daejeon 34114, Republic of Korea; srigouri@krict.re.kr
3 Department of Bio-Pharmaceutical Engineering, College of Bio-Health Convergence, Dongseo University, Busan 47011, Republic of Korea
* Correspondence: dwkim@dongseo.ac.kr; Tel.: +82-051-320-1972
† These authors contributed equally to this work.

Abstract: Atopic dermatitis (AD) is a common inflammatory skin disease characterized by pruritic lesions and skin barrier dysfunction. In this study, we evaluated the effect of a quinazoline derivative, SH-340, on TSLP expression and signaling in human primary keratinocytes. Our results demonstrated that SH-340 significantly increased factors for differentiation and skin barrier function including KRT1, KRT2, KRT10, IVL, LOR, CLDN1, OVOL1, and FLG, whereas it inhibited TSLP expression in a dose-dependent manner, both at the mRNA and protein levels. Furthermore, SH-340 was found to inhibit the phosphorylation of STAT6, a downstream signaling molecule of IL-4 and IL-13, in keratinocytes. These findings suggest that SH-340 may suppress TSLP expression by inhibiting the IL-4/IL-13-STAT6 signaling pathway. Finally, SH-340 may potentially contribute to both the alleviation of inflammation and the restoration of skin barrier function.

Keywords: atopic dermatitis; quinazoline; filaggrin; TSLP; STAT6

1. Introduction

Quinazolines are a class of heterocyclic compounds that have been extensively studied in the field of medicinal chemistry due to their wide range of biological activities. These compounds have been found to exhibit a variety of pharmacological activities, including anticancer, antimicrobial, and anti-inflammatory activities [1,2]. Previous studies support the biological activities of quinazolines in skin research. Studies have demonstrated that quinazolines show potent anti-inflammatory effects by inhibiting pro-inflammatory cytokines such as tumor necrosis factor-alpha (TNF-α), interleukin (IL)-1 beta, and IL-6 [3]. In addition, quinazolines inhibit the proliferation of skin cells and modulate the immune response in the skin [4,5]. Additionally, quinazolines have shown promising effects in promoting wound healing by stimulating fibroblast proliferation, collagen synthesis, and angiogenesis [6]. Furthermore, quinazolines have shown notable antioxidant activity, effectively scavenging reactive oxygen species (ROS) and protecting skin cells from oxidative stress-induced damage [7]. Some quinazoline derivatives have potential antimicrobial effects against common skin pathogens including *Staphylococcus aureus* [5,8]. Collectively, these findings highlight the significant potential of quinazolines in various aspects of skin research associated with inflammation modulation, wound healing, antioxidant protection, and antimicrobial applications.

Atopic dermatitis (AD) is characterized by chronic inflammation and a defective skin barrier, which allows for the penetration of irritants and allergens [9–11]. In AD patients, the skin barrier is weakened due to defects in the filaggrin (FLG) gene, which codes for a protein that helps to maintain the structural integrity of the stratum corneum, the outermost layer of the skin [10,12]. Overall, the regulation of the FLG gene involves a complex interplay of signaling mechanisms, transcription factors, cytokines, and genetic variations [11–13]. Accumulating evidence strongly supports that T help type 2 (Th2)-mediated cytokines (IL-4 and IL-13) have been implicated in the regulation of FLG gene expression by activating JAK (Janus kinase)-STAT signaling pathway [10]. Thus, current therapeutic strategies for AD aim to restore and maintain skin barrier function by inducing FLG expression, as well as reducing inflammation and pruritus. JAK inhibitors work by blocking the activity of JAKs, which leads to the activation of immune cells and the production of pro-inflammatory cytokines.

Several JAK inhibitors are currently being investigated for the treatment of AD, including baricitinib, upadacitinib, and abrocitinib [14,15]. Clinical studies have demonstrated that these JAK inhibitors can improve the signs and symptoms of AD, including reducing itching, redness, and skin lesions [16,17]. Recently, a quinazoline derivative called JTE-052 (delgocitinib) has been shown to reduce inflammation and improve skin barrier function in a mouse model of atopic dermatitis [18]. JTE-052 was found to inhibit the production of pro-inflammatory cytokines and chemokines such as TNF-α, IL-1β, IL-6, and CCL17 in the skin. It also decreased the infiltration of inflammatory cells such as eosinophils and mast cells into the skin. In addition, JTE-052 was found to increase the expression of proteins that are involved in maintaining the integrity of the skin barrier, such as loricrin (LOR), involucrin (IVL), and FLG.

Despite the existence of a quinazoline-based drug for AD, the development of new quinazoline derivatives with improved efficacy and fewer side effects is required to enhance the existing treatments for AD. Given the limited efficacy of current treatments for atopic dermatitis, we wanted to explore novel compounds to develop more effective and targeted therapies. Here, we synthesized a quinazoline derivative SH-340, the synthesis method is shown in Figure 1, refs. [19–23] N-(3-chloro-4-fluorophenyl)-7-((4-methyl piperazin-1-yl)methyl)-7,8-dihydro [1,4]dioxino [2,3-g]quinazolin-4-amine), and investigated biological activities in human primary keratinocytes using in vitro AD model.

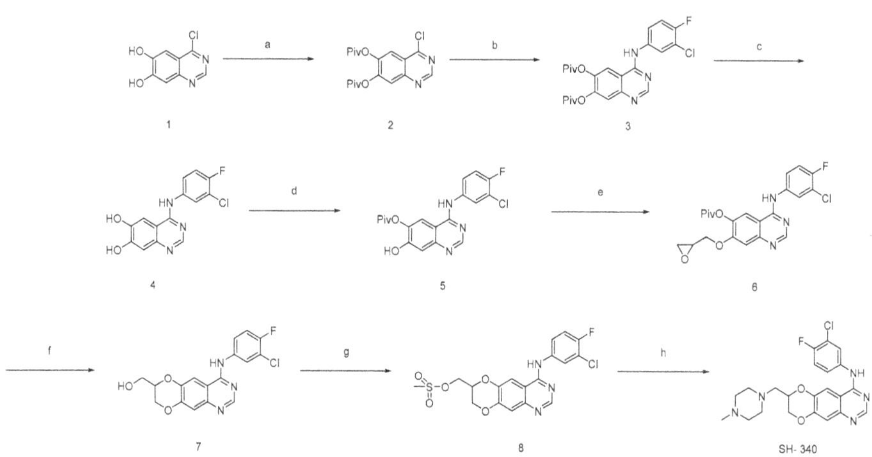

Figure 1. Chemical structure and synthesis of SH-340. 1. Reagents and condition: (**a**) Piv-Cl, TEA, DCM, −10 °C to r.t; (**b**) 3-chloro-4-fluoroaniline, IPA reflux 12 h; (**c**) 7N NH_3 in MeOH, 0–23 °C 12 h; (**d**) Piv-Cl, Et_3N, THF −40 °C 2 h, (regioselective acylation gives 5 major product) (**e**) oxiran-2-ylmethyl 4-nitrobenzenesulfonate, K_2CO_3, DMF, 0 °C-rt, 12 h; (**f**) K_2CO_3, MeOH 12 h, rt; (**g**) Et_3N, MsCl, THF 0–23 °C, 16 h; and (**h**) 1-methylpiperazine, Et_3N, DMF 85 °C 24 h.

2. Results

2.1. Synthesis and Functionalization of Novel Quinazoline Compound SH-340 via Multi-Step Reactions

Commercially purchased diol (1) was subsequently protected with pivaloyl groups. We found that the use of pivaloyl groups over acetyl-protecting groups was essential in aiding the purification of the subsequent chloride (2), and the substitution of aromatic halides with nitrogen nucleophiles, 3-chloro-2-fluoroaniline, was achieved in excellent yield in i-PrOH at reflux. Deprotection of (3) via ammonolysis afforded the stable diol (4), Finally, regioselective protection of (5) was achieved using pivaloyl chloride in the presence of TEA at −40 °C to afford the key intermediate, and its regiochemistry was discussed in chemical synthesis 4.5; compound (5) was treated with oxiran-2-yl-methyl-3-nitrobenzenesulfonate to give corresponding epoxide (6), which transforms the ring opening of epoxides, leads to cyclization, and gives free hydroxy group (7); mesylation of the resultant free hydroxy group with MsCl/TEA gives (8); and subsequent displacement of mesylate with 1-methylpiperazine affords desired novel quinazoline compound SH-340.

2.2. SH-340 Inhibits Proliferation of Keratinocytes

To determine the appropriate concentration of primary keratinocytes, we initially checked the cytotoxicity of SH-340 (Figure 2) at different concentrations. Cell viability results show that SH-340 has significant cytotoxicity from 0.5 to 25 μM (Figure 2C). Although keratinocytes cultured in keratinocyte growth media (KGM) maintain their stemness, they are allowed to differentiate in the basal media (KBM) without growth factors. When normal human epidermal keratinocytes (NHEKs) were treated with SH-340, we found that their growth is slow and the morphology shows a differentiated cell type with the increased cell size (Figure 2A,B). Given that the cytotoxicity assay measures the metabolic activities of mitochondria, our results indicate that SH-340 may inhibit proliferation and induce differentiation.

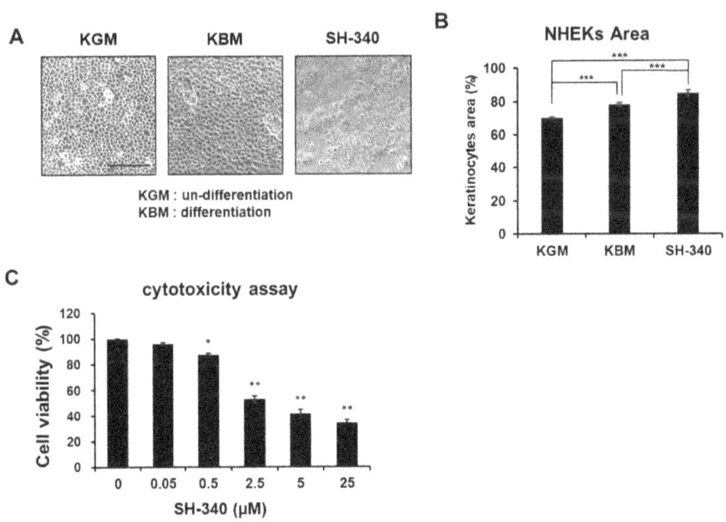

Figure 2. SH-340 inhibits the proliferation of keratinocytes. The morphology (**A**) and quantified cell area (**B**) of keratinocytes cultured in KGM, KBM, and KBM + SH-340 (0.5 μM). (**C**) Cell viability at different concentrations of SH-340. KGM indicates undifferentiated keratinocytes cultured in keratinocyte growth media, and KBM indicates differentiated keratinocytes cultured in keratinocyte basal media without growth factors. Scale bar: 250 μm. DMSO 1%: apoptotic positive control. ($n = 3$, *** $p < 0.001$, ** $p < 0.01$, * $p < 0.05$).

2.3. SH-340 Induces Differentiation of Keratinocytes

To examine whether SH-340 stimulates the differentiation of keratinocytes, we checked differentiated markers in NHEKs. NHEKs treated with SH-340 (0.5 μM) show remarkably increased gene expression of differentiation markers such as KRT1, KRT2, KRT2, and Ovo Like Transcriptional Repressor 1 (OVOL1) (Figure 3A). We also checked the expression of IVL, LOR, and Cloudin1 (CLDN1), and FLG is increased by SH-340 (Figure 2A). We confirmed that the protein expression of KRT1, KRT10, and FLG is increased in a dose-dependent manner in NHEK treated with SH-340 (Figure 3B). These results suggest that SH-340 is a potent material that can improve skin barrier function by effectively inducing NHEK differentiation.

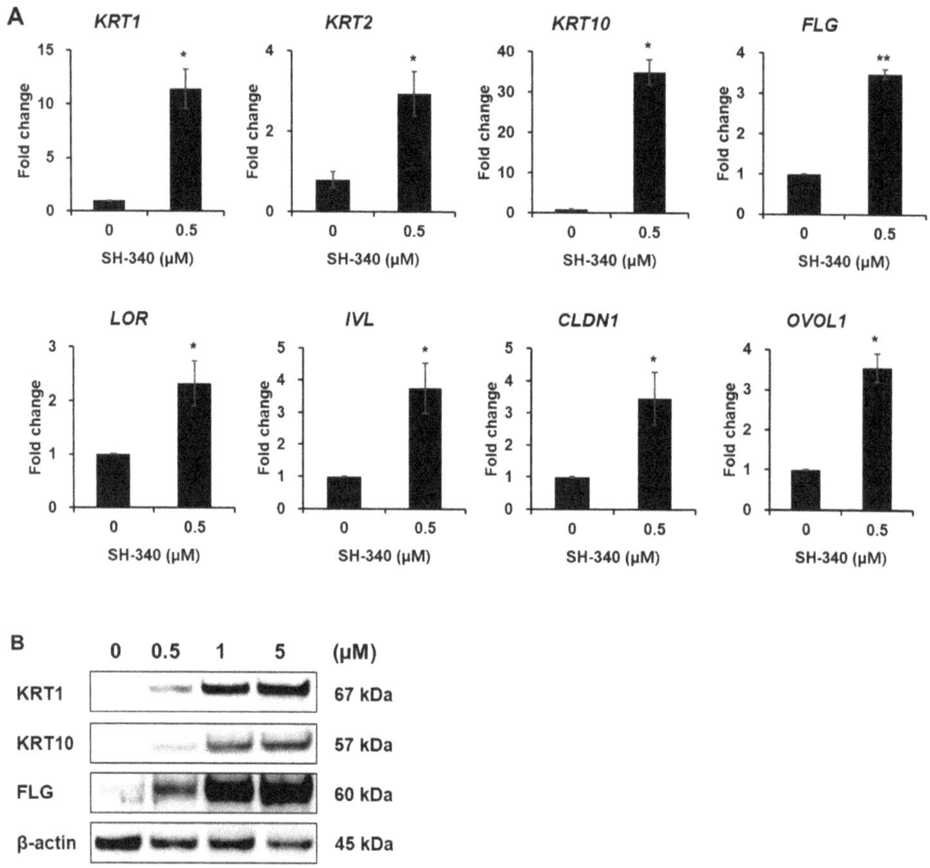

Figure 3. SH-340 induces the differentiation of keratinocytes. (**A**) Quantified real-time PCR results of gene expression for keratinocyte differentiation. (**B**) Western blot analysis of KRT1, KRT10, and FLG in keratinocytes treated with different concentrations of SH-340. β-actin was used as a loading control (n = 4–5, ** $p < 0.01$, * $p < 0.05$).

2.4. SH-340 Inhibits IL-4/IL-13-Induced STAT6 Signaling

AD is characterized by overexpression of Th2 cytokines interleukin-4 (IL-4) and IL-13, which leads to defective keratinocyte differentiation [24]. Strong research evidence proved that Th2-mediated IL-4/IL-13 inhibits gene expression of FLG by activating the JAK-STAT pathway, resulting in disruption of the skin barrier [14,25]. Thus, we attempted to verify if SH-340 regulates the IL-4/IL-13-induced signaling pathway. We identified that SH-340 induces KRT1, KRT10, and FLG protein expression (Figure 4A). Moreover, SH-340 significantly restores their expression in the presence of IL-4/IL-13 (Figure 4A). Moreover, SH-340 showed a 45% inhibition of IL-4/IL-13-induced STAT6 phosphorylation in NHEKs (Figure 4B,C). Thus, these results indicate that SH-340 has the potential to counteract the negative effects of IL-4 and IL-13 on keratinocyte differentiation and skin barrier function.

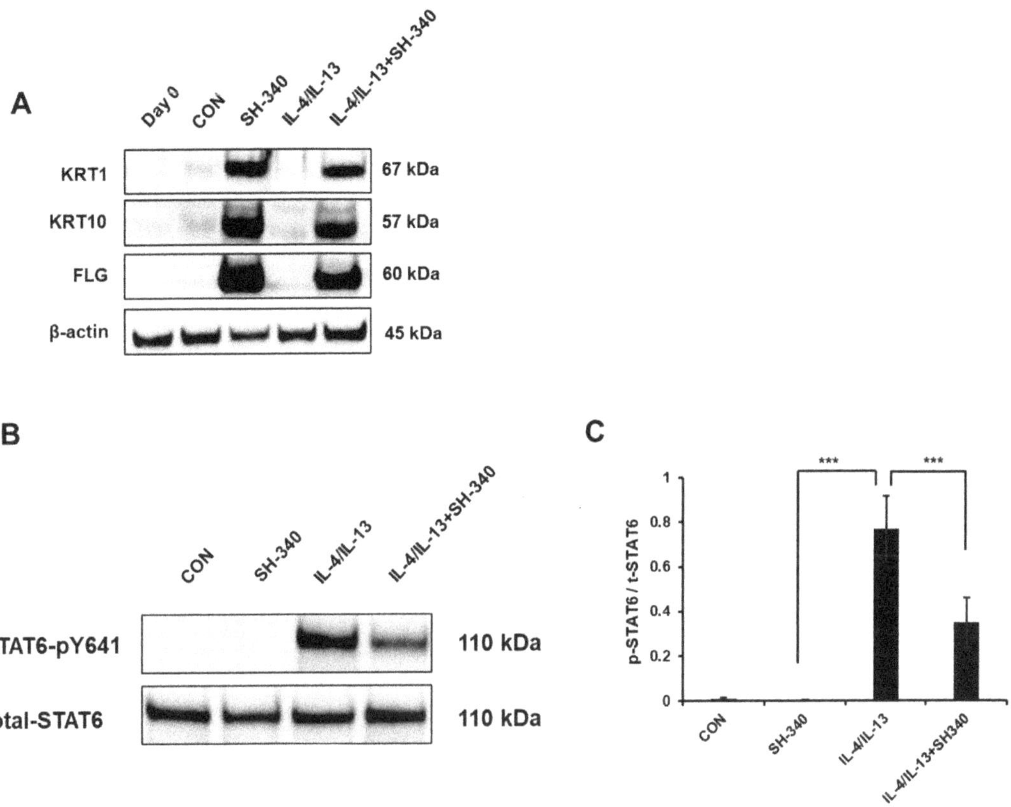

Figure 4. SH-340 inhibits Th2 cytokine-mediated signaling. (**A**) SH-340 restores KRT1, KRT10, and FLG expression in keratinocytes with IL-4 and IL-13. (**B**) Analysis of STAT6 phosphorylation in keratinocytes. SH-340 inhibits and significantly inhibits STAT6 phosphorylation induced by IL-4 and IL-13. (**C**) Quantified results of STAT6 phosphorylation ($n = 6$, *** $p < 0.001$).

2.5. SH-340 Inhibits TSLP Expression in NHEK

The presence of IL-4 and IL-13 in skin lesions of atopic dermatitis is associated with the stimulation of thymic stromal lymphopoietin (TSLP) production in keratinocytes upon exposure to external substances [26]. Previous studies demonstrated that toll-like receptor 3 (TLR3) activation induces the expression of TSLP in human keratinocytes [27]. In addition, damaged tissues release non-coding double-stranded RNAs (dsRNAs) to activate TLR3, resulting in enhanced wound healing and tissue regeneration [28,29]. Given the repeated scratching that occurs in the skin of patients with AD, we were interested in investigating whether exposure of keratinocytes to a synthetic dsRNA (polyinosinic-polycytidylic acid; poly I:C), IL-4, and IL-13 can lead to the production of TSLP. Compared to the control group, poly I:C induced TSLP expression in keratinocytes (Figure 5). Moreover, keratinocytes exposed to poly I:C, IL-4, and IL-13 produce TSLP more than poly I:C alone (Figure 5). However, SH-340 significantly suppressed TSLP production in keratinocytes treated with poly I:C, IL-4, and IL-13 (Figure 5), suggesting that SH-340 may have the potential to alleviate pruritus caused by AD.

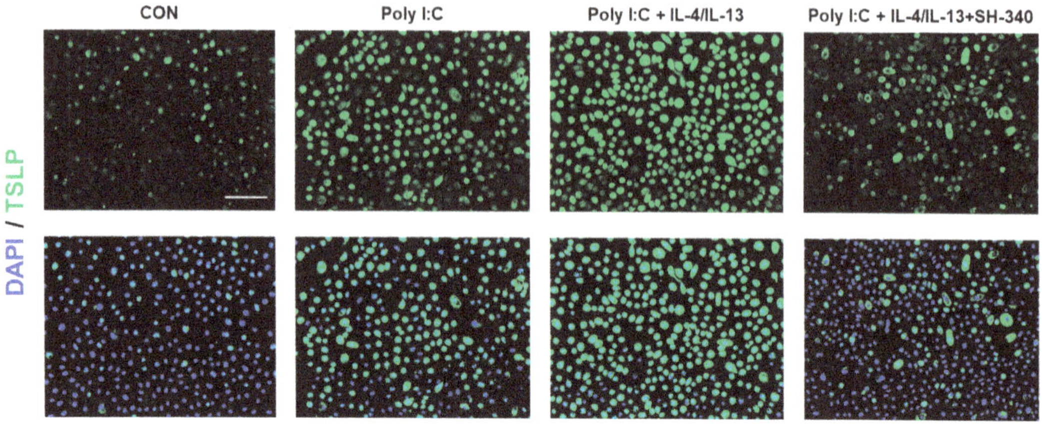

Figure 5. SH-340 inhibits TSLP expression induced by poly I:C or poly I:C with IL-4/IL-13 treatment. Green indicates TSLP-positive keratinocytes. Blue indicates nuclear counterstaining. The scale bar is 125 μm ($n = 3$).

3. Discussion

Quinazoline derivatives have been studied and used in various research areas, including cancer research, neuropharmacology, inflammatory diseases, and infectious diseases. Especially, JTE-052 is a quinazoline-based drug currently being studied for the treatment of inflammatory skin diseases such as AD [18]. JTE-052 is a potent JAK inhibitor, which has been extensively studied in the context of various inflammatory disorders. Despite the existence of a few quinazoline-based drugs for AD, there are some significant reasons why we still need to study and develop quinazoline derivatives, for example, limited efficacy or safety concerns. Thus, studying and understanding the mechanisms of action of quinazoline derivatives could help researchers identify new targets and develop novel treatments for various diseases. Here, we investigated a new quinazoline molecule, SH-340, to demonstrate its potential as a therapeutic agent for AD. First, SH-340 at 0.5 μM inhibits proliferation and induces differentiation, although a high concentration of SH-340 (from 2.5 μM) showed cytotoxicity. When keratinocytes transition from basal cells to more differentiated cells, they normally change morphology, gene expression, and protein production [30,31]. The increase in cell size during keratinocyte differentiation is accompanied by the synthesis and deposition of structural proteins, which contribute to the development of a protective skin barrier. SH-340 inhibits proliferation and increases cell size, which indicates that keratinocytes may undergo a differentiation process in the presence of SH-340. We also confirmed this by analyzing differentiation factors and found that SH-340 induces keratinocyte differentiation by increasing the gene expression for initial and terminal differentiation and skin barrier functions. Second, SH-340 inhibits IL-4/IL-13-induced STAT6 signaling, which plays a key role in the pathogenesis of AD. Third, we found that SH-340 inhibits TSLP expression in NHEK, suggesting that it may attenuate the symptoms of AD by reducing the activation of dendritic cells and Th2 cells.

The potent inhibition of TSLP by SH-340 represents a significant breakthrough in immunological and therapeutic research. Since TSLP plays a pivotal role in initiating and perpetuating allergic responses, it can be a crucial target for allergic diseases such as asthma, eczema, and allergic rhinitis. The discovery that SH-340 can effectively inhibit TSLP opens up new avenues for designing innovative treatments that could attenuate allergic reactions. From a scientific standpoint, this finding deepens our comprehension of the intricate pathways underlying allergic responses, potentially paving the way for further insights into immunomodulation strategies. Clinically, SH-340′s ability to suppress TSLP suggests a potential breakthrough in treating a range of allergic diseases, offering a novel therapeutic angle that could complement or even surpass existing approaches. Although JTE-052's JAK inhibitory activity positions it as a promising therapeutic agent in various inflammatory conditions, by targeting JAKs, key mediators of cytokine signaling, JTE-052 broadly affects multiple cytokine pathways, leading to the suppression of a wide range of inflammatory responses. However, when considering its impact on TSLP, the downstream consequences of JAK inhibition need to be evaluated. While TSLP expression is influenced by JAK-STAT signaling, JTE-052 has not been studied to demonstrate its effect on TSLP inhibition. Thus, the specific regulation of the IL-4/IL-13-STAT6 pathway by SH-340 offers a targeted approach to addressing TSLP-associated inflammation. By focusing on this specific pathway, SH-340's inhibition of TSLP expression represents a precise strategy for curbing the initial triggers of AD. This targeted approach not only reduces the risk of unintended downstream effects on unrelated cytokine networks but also highlights the importance of TSLP in the context of AD. Furthermore, as TSLP influences autoimmune diseases and certain skin conditions, SH-340 might hold promise in addressing these disorders. Moreover, the compound's mechanism of action could be harnessed for enhancing the efficacy of cancer immunotherapies by modulating immune responses. This versatility positions SH-340 as a versatile tool for exploring new therapeutic directions across diverse medical fields, making it a significant asset in drug development and clinical research.

Taken together, our results suggest that SH-340 has potential as a novel therapeutic agent for the treatment of AD. By targeting keratinocyte differentiation, skin barrier functions, and TSLP expression, SH-340 may offer a multi-faceted approach to the treatment of AD, which could address both the inflammatory and barrier dysfunction aspects of the disease. Despite these results, the effect of SH-340 on enhancing skin barrier function is reported for the first time in this study, and the exact mechanism is not yet known. Therefore, further studies are needed to fully elucidate the mechanism of action of SH-340 and to evaluate its efficacy and safety in vivo.

4. Materials and Methods

4.1. Chemical Synthesis

General methods reagents, solvents, and starting products were acquired from commercial sources. When indicated, the reaction products were purified by "flash" chromatography on silica gel (35–70 μm) with the indicated solvent system. HRMS analysis was conducted in EI mode, 70 eV, resolution 5000 using a JEOL JMS-700 high-resolution mass spectrometer (MS), LC-Ms were analyzed by ESI^+ represented as m/z $(M + H)^+$, NMR spectra were recorded in $CDCl_3$ at 300 MHz, 400 MHz, 500 MHz (1H), and 101 MHz (13C), and chemical shifts are reported in δ values downfield from TMS or relative to residual chloroform (7.26 ppm, 77.0 ppm) or residual DMSO (2.50 ppm) as an internal standard. Data are reported in the following manner: chemical shift, multiplicity, coupling constant (J) in hertz (Hz), and integrated intensity and assignment (when possible). Multiplicities are reported using the following abbreviations: s, singlet; d, doublet; dd, doublet of doublets; t, triplet; q, quadruplet; m, multiplet; br s, broad signal; and app, apparent. Assignments and stereochemical determinations are given only when they are derived from definitive dimensional NMR experiments (g-HSQC) (Supplementary Materials).

4.2. Synthesis of 4-Chloroquinazoline-6,7-diyl bis(2,2-dimethylpropanoate) (2)

To a stirred suspension of commercially available 4-chloroquinazoline-6,7-diol (2.0 g, 10.17 mmol) and TEA (4.9 mL, 35.59 mmol) in dichloromethane (20 mL), trimethylacetyl chloride (Pivolyl chloride) (2.7 mL, 22.38 mmol) was added dropwise. The resulting suspension was stirred at room temperature for 2 h, distilled of excess DCM, diluted with EtOAc (500 mL), and washed with water (5 × 20 mL). The organic phase was concentrated to dryness and the residue triturated with water, and the resulting precipitate was collected by filtration, washed with water (5 × 20 mL), and dried to a constant weight to afford the title compound (2.6 g, 91%), LCMS: ESI^+ m/z 364.82 $(M + H)^+$, ^1H NMR (300 MHz, Chloroform-d) δ 9.05 (s, 1H), 8.06 (s, 1H), 7.91 (s, 1H), 1.43 (d, J = 4.0 Hz, 18H).

4.3. *Synthesis of 4-((3-Chloro-4-fluorophenyl)amino)quinazoline-6,7-diyl bis(2,2-dimethyl propanoate)* (**3**)

A mixture of 4-chloroquinazoline-6,7-diyl bis(2,2-dimethylpropanoate) (1 g, 2.741 mmol) in *i*-PrOH (10.9 mL) was treated with 3-chloro-4-fluoroaniline (1.5 eq) and stirred at 95 °C for 12 h. When TLC showed completion of the reaction, it cooled to room temperature and evaporated to dryness. And the residue was suspended several times in diethyl ether; solidification was done and obtained a pure yellow solid compound after filtration. Obtained yield (1.2 g, 93%) LCMS: ESI$^+$ m/z 474.5 (M + H)$^+$, ^{1}H NMR (300 MHz, Methanol-d_4) δ 8.89 (s, 1H), 8.64 (s, 1H), 8.00 (dd, J = 6.6, 2.6 Hz, 1H), 7.88 (s, 1H), 7.71 (ddd, J = 8.9, 4.2, 2.7 Hz, 1H), 7.41 (t, J = 8.9 Hz, 1H), 1.44 (d, J = 9.0 Hz, 18H).

4.4. *Synthesis of 4-((3-Chloro-4-fluorophenyl)amino)quinazoline-6,7-diol* (**4**)

A stirred slurry of (1.0 g, 2.11 mmol) was treated at 0 °C, with 7 M solution of NH_3 in MeOH (7 mL); the mixture was stirred at 0 °C for 15 min, and then at 23 °C for 4.5 h. The mixture was evaporated, and the residue was suspended in water (40 mL), stirred overnight, and filtered. The residue was washed with water (50 mL), DCM (4 × 50 mL), Et_2O (2 × 50 mL), and dried in a desiccator to afford the title compound (0.587 g, 91%), LCMS: ESI$^+$ m/z 306.7 (M + H)$^+$, ^{1}H NMR (300 MHz, DMSO-d_6) δ 9.54 (s, 1H), 8.43 (s, 1H), 8.21 (dd, J = 6.9, 2.5 Hz, 1H), 7.87–7.72 (m, 1H), 7.41 (t, J = 9.1 Hz, 1H), 7.12 (s, 1H).

4.5. *Synthesis of 4-((3-Chloro-4-fluorophenyl)amino)-7-hydroxyquinazolin-6-yl pivalate* (**5**)

A stirred suspension of 4-((3-chloro-4-fluorophenyl)amino)quinazoline-6,7-diol (1.4 g, 4.57 mmol) in DMF (15 mL) was treated with Et_3N (1.9 mL, 13.71 mmol), cooled to −40 °C, and treated dropwise with Piv_2O (0.672 mL, 5.5 mmol). The mixture was stirred at −40 °C for 1 h, after which the cooling bath was removed, and stirring was continued for 2.5 h. The reaction mixture was diluted with DCM (50 mL), washed with 10% citric acid (2 × 10 mL), dried over (Na_2SO_4), filtered, and evaporated flash column (DCM: EtOAc 1:1) to afford solid (beige–yellow solid). Yield obtained (0.510 g, 30%) LCMS: ESI$^+$ m/z 390.9 (M + H)$^+$, ^{1}H NMR (500 MHz, DMSO-d_6) δ 11.04 (s, 1H), 9.66 (s, 1H), 8.54 (s, 1H), 8.20 (d, J = 5.5 Hz, 2H), 7.82 (dq, J = 9.7, 3.4, 2.9 Hz, 1H), 7.44 (t, J = 9.1 Hz, 1H), 7.18 (s, 1H), 1.38 (s, 9H).

The regiochemistry of the 6-OPiv-7-OH intermediate 6 was determined by NMR experiments. In comparison, to the parent bis-phenol 4, the proton at C-5 shifted 0.5 ppm downfield (as opposed to the C-8 proton which did not move) and a NOE was observed between the C-5 proton and the aniline N–H.

4.6. *Synthesis of 4-((3-Chloro-4-fluorophenyl)amino)-7-(oxiran-2-ylmethoxy)quinazolin-6-yl pivalate* (**6**)

For the synthesis, the following instructions should be followed: Dissolve 4-((3-chloro-4-fluorophenyl)amino)-7-hydroxyquinazolin-6-yl pivalate (0.180 g, 0.461 mmol) in N, N-dimethylformamide (2 mL) then add K_2CO_3 at 0 °C and stir for 10 min, then add oxiran-2-ylmethyl 4-nitrobenzenesulfonate (1.1 eq) and stir at 0 °C, for 30 min more, and bring RM to room temperature; continue stirring for 6 h. When TLC shows completion of the reaction, extract RM with EtOAc (50 mL) and wash with water (3 × 20 mL), then wash with a brine solution to remove excess DMF. This organic phase was concentrated and used in the next step without purification. Obtained yield, 0.140 mg, 68%. LCMS: ESI$^+$ m/z 446.9 (M + H)$^+$, ^{1}H NMR (400 MHz, Methanol-d_4) δ 8.53 (s, 1H), 8.10 (s, 1H), 8.06 (dd, J = 6.7, 2.5 Hz, 1H), 7.69 (dt, J = 8.8, 3.3 Hz, 1H), 7.29–7.22 (m, 2H), 4.52 (dd, J = 11.2, 2.1 Hz, 1H), 4.00 (dd, J = 11.2, 6.6 Hz, 1H), 3.37 (dd, J = 6.5, 4.0 Hz, 1H), 2.91 (t, J = 4.6 Hz, 1H), 2.78 (dd, J = 4.9, 2.6 Hz, 1H), 1.45 (s, 9H).

4.7. *Synthesis of (4-((3-Chloro-4-fluorophenyl)amino)-7,8-dihydro-[1,4]dioxino[2,3-g]quinazolin-7-yl) methanol* (**7**)

For the synthesis, the following instructions should be followed: Dissolve 4-((3-chloro-4-fluorophenyl)amino)-7-(oxiran-2-ylmethoxy)quinazolin-6-yl pivalate (0.230 g, 0.515 mmol) in MeOH (5 mL) then add K_2CO_3 (0.142 g, 1.0 mmol) and stir at rt for 12 h. When TLC shows completion of the reaction, evaporate RM to dryness, add H_2O, adjust pH to 7 by NH_4Cl, extract RM with EtOAc (50 mL) and wash with water (3 × 20 mL), and evaporate organic layer to dryness. There is no need to purify, by column, it can be directly used for the next step. Obtained yield (140 mg, 75%). LCMS: ESI^+ m/z 362.85 $(M + H)^+$.

*4.8. Synthesis of (4-((3-Chloro-4-fluorophenyl)amino)-7,8-dihydro-[1,4]dioxino[2,3-g]quinazolin-7-yl) methyl methanesulfonate (**8**)*

A solution of (4-((3-chloro-4-fluorophenyl)amino)-7,8-dihydro-[1,4]dioxino[2,3-g] quinazolin-7-yl)methanol (0.160 g, 0.442 mmol) in THF (10 mL) was treated with Et_3N (1.5 eq), cooled to 0 °C, and treated dropwise with MsCl (1.3 eq). The mixture was stirred at rt for 16 h, cooled to 0 °C, treated with $NaHCO_3$ (pH-7), extracted with EtOAc (50 mL), and washed with water (3 × 20 mL). The organic layer was evaporated to dryness. There was no need to purify, by column, it was directly used for the next step. Obtained yield (215 mg, 100%) LCMS: ESI^+ m/z 440.1 $(M + H)^+$.

4.9. Synthesis of N-(3-Chloro-4-fluorophenyl)-7-((4-methylpiperazin-1-yl)methyl)-7,8-dihydro-[1,4]dioxino[2,3-g]quinazolin-4-amine (SH-340)

Nucleophilic substitution of quinazolinyl mesylates with secondary amines: A mixture of quinazolinyl mesylate (150 mg, 0.341 mmol) in DMF (5 mL) was treated with 1-methylpiperazine (<10 eq) and Et_3N (<10 eq), and the mixture was stirred at 85 °C for 24 h; the mixture was cooled to rt, residue was dissolved in EtOAc (50 mL) and washed with water (3 × 20 mL), excess DMF was removed by brine wash, the organic layer evaporated, and purification was conducted by 10% Mc: MeOH. Obtained yield (82 mg, 54%) LCMS: ESI^+ m/z 444.5 $(M + H)^+$ HRMS exact mass observed m/z 443.1523 with int 100%, ^{1}H NMR (300 MHz, Chloroform-*d*) δ 8.63 (s, 1H), 8.02 (dd, J = 6.5, 2.4 Hz, 1H), 7.62–7.54 (m, 1H), 7.48 (s, 1H), 7.35 (s, 1H), 7.15 (t, J = 8.8 Hz, 1H), 4.48–4.38 (m, 2H), 4.13 (dd, J = 11.7, 8.0 Hz, 1H), 2.73 (ddt, J = 24.0, 15.8, 8.3 Hz, 10H), 2.42 (s, 3H). ^{13}C NMR (101 MHz, $CDCl_3$) δ 171.95, 156.27, 153.44, 149.04, 146.45, 143.61, 135.33, 123.77, 121.31, 121.24, 116.68, 116.46, 113.95, 110.14, 106.36, 71.61, 67.05, 62.59, 58.81, 58.14, 55.08, 54.98, 54.67, 53.57, 51.01, 46.02, 45.85, 45.56, 45.18, 41.68, 41.47.

4.10. Culture of Human Keratinocytes

Normal human epidermal keratinocytes (NHEK) were commercially provided from Lonza. NHEKs were grown in KGM media by adding 1 mL of KGM SingleQuots™ supplements to 500 mL of KGM basal medium as per the manufacturer's instructions. One vial of NHEKs frozen stock was thawed in a 37 °C water bath. Once thawed, the cells were transferred to a 15 mL conical tube containing 10 mL of KGM. After centrifugation at 3000 rpm for 5 min, the supernatant was removed and the cell pellet was resuspended in 2 mL of KGM. NHEKs were plated in a sterile culture flask at a density of 2×10^5 cells/cm^2 and incubated in a humidified CO_2 incubator at 37 °C and 5% CO_2. Media was replaced with fresh KGM every other day. Once NHEKs reached about 70–80% confluency, they were split using 0.025% trypsin-EDTA solution for a subculture. For all experiments, cells were used in passages 3–6.

4.11. Cytotoxicity Assay

The cytotoxicity of the extract was evaluated using the Cell Counting Kit-8 (CCK-8, Dojindo, Kumamoto, Japan) assay. Normal human epidermal keratinocytes (NHEK) were seeded at a density of 1×10^4 cells/well in a 96-well plate and cultured for 24 h. After replacing the medium, the cells were treated with various concentrations (1, 2, 5, and 10 μg/mL) of the extract and incubated for 48 h in standard cell culture conditions. Next, the culture medium was supplemented with the CCK-8 solution, and the cells were further incubated for 3 h in a CO_2 incubator. Absorbance readings were taken at 450 nm using an

INNO-M microplate spectrophotometer (LTEK). The cell viability at each concentration was determined using the provided equation.

$$Cell\ Viability\ (\%) = \frac{A_{sample}}{A_{control}} \times 100$$

4.12. RNA Isolation and qRT-PCR

NHEKs were seeded at a concentration of 7×10^4 cells/well in a 12-well plate and incubated for 24 h. After the incubation period, the cells were treated with the extract at a concentration of 1 μg/mL and further incubated for 72 h. Total RNA was extracted from the cells using the Quick-RNA Micro Prep Kit (ZYMO Research, Irvine, CA, USA), following the instructions provided by the manufacturer. The concentration and purity of the isolated RNA were assessed using a Jenway Genova Nano spectrophotometer (Cole-Parmer Ltd., Vernon Hills, IL, USA). For cDNA synthesis, the high-capacity cDNA Kit (ThermoFisher Scientific, Waltham, MA, USA) was utilized, with 1–2 μg of total RNA being used as the starting material for the reaction. For real-time quantitative polymerase chain reaction (qRT-PCR), the synthesized cDNA was prepared using a QuantStudio 3 instrument (Applied Biosystems, Carlsbad, CA, USA) and TaqMan Fast Advanced Master Mix (ThermoFisher Scientific). Specific target probes (ThermoFisher Scientific) were used, including KRT1 (Hs00196158_m1), KRT2 (Hs00166294_m1), KRT9 (Hs00413861_m1), IVL (Hs00846307_s1), FLG (Hs00856927_g1), CLDN1 (Hs00221623_m1), and TSLP (Hs00263639_m1). To normalize the results, the ribosomal protein lateral stalk subunit P0 (RPLP0, Hs00420895) was used as a loading control. The qRT-PCR conditions involved an initial incubation at 50 °C for 5 min, followed by denaturation at 95 °C for 20 s, annealing at 95 °C for 3 s, and extension at 60 °C for 20 s. A total of 40 cycles were performed. The amplified gene expression was quantitatively analyzed using the ddCt method [32] with the assistance of QuantStudio™ Design & Analysis Software (version 1.5.1; Applied Biosystems).

4.13. Western Blot

NHEKs were plated in a 6-well plate at a density of 1×10^5 cells/well and incubated for 24 h. Following the incubation period, the cells were treated with SH-340 (0.5 μM), Recombinant Human IL-4 (10 ng/mL), and IL-13 (10 ng/mL) obtained from PeproTech (Cranbury, NJ, USA) for 96 h. Subsequently, the cells were collected using a scraper, and protein lysates were prepared using RIPA lysis buffer (ThermoFisher Scientific) supplemented with protease inhibitors (GenDepot, Katy, TX, USA) and phosphatase inhibitors (GenDepot). Proteins were first extracted using RIPA buffer (#89900, ThermoFisher Scientific). NHEKs were first homogenized in ice-cold RIPA buffer containing protease and phosphatase inhibitors by sonication with 125 W and 20 kHz conditions 10 times in 2 s. The lysates were then incubated on ice for 30 min to allow complete cell lysis. After centrifugation at the maximum speed for 10 min at 4 °C, the supernatant containing the extracted proteins was collected and stored at −80 °C until further use. The protein concentration was determined using a BCA (bicinchoninic acid) assay (#23225, ThermoFisher Scientific) followed by the manual. For the experiment, protein samples were separated by electrophoresis on a polyacrylamide gel (10%) (#1610158, Bio-Rad). The separated proteins were then transferred onto the PVDF membrane (#10600021, Cytiva, Marlborough, MA, USA). The membrane was then blocked with non-fat dry milk in TBS including 0.1% Tween 20 (TBS-T) to prevent non-specific binding of the primary antibody. The primary antibody (KRT1: #sc-376224, KRT10: #sc-23877, FLG: #sc-80609, Santacruz biotechnology) was then incubated with the membrane overnight at 4 °C. After washing the membrane using TBS-T three times, a secondary antibody conjugated with horseradish peroxidase (HRP) (#31430, Invitrogen, Carlsbad, CA, USA) was added, and the membrane was incubated for 1 h at room temperature. The presence of TSLP was visualized by adding a chemiluminescent substrate using a chemiluminescent imaging system. The intensity of the signal can be quantified using densitometry. Beta-actin (#sc-47778, Santacruz biotechnology) was used as a reference protein to normalize the levels of the protein of interest.

4.14. Measuring of STAT6 Phosphorylation

Cells were plated in a 6-well plate with a density of 2×10^5 cells/well and incubated for 24 h. Following the incubation, the cells were exposed to varying concentrations of SH-340 (0.5 μM) for a duration of 2 h. After the pre-treatment, the cells were subjected to IL-4 and IL-13 at a concentration of 10 ng/mL for 30 min to induce phosphorylation. Phosphorylation levels were assessed using Western blotting, employing phospho-STAT6 antibody (1:2000, BD Biosciences, San Jose, CA, USA) and total STAT6 antibody (1:2000, Cell Signaling, Danvers, MA, USA) through the same technique.

4.15. Immunostaining

To detect thymic stromal lymphopoietin (TSLP) in NHEKs, cells were seeded onto chamber slides (#154461PK, ThermoFisher Scientific). Then cells were fixed with 4% paraformaldehyde for 10 min and permeabilized with PBS including 0.4% Triton X-100 for 15 min. After washing with PBS three times, cells were blocked with PBS containing 10% normal goat serum for 1 h at room temperature. After blocking, cells were incubated with a primary TSLP antibody (1:100 dilution; Rabbit polyclonal, Cat# A13134, ABclonal, Woburn, MA, USA) in a blocking buffer overnight at 4 °C. The cells were then washed with PBS and incubated with a fluorescently labeled secondary antibody (goat anti-rabbit IgG, 1:1000 dilution, RSA1245, BioActs, Incheon, South Korea) for 1 h at room temperature. The cells were then counterstained with a nuclear stain, such as 4′,6-diamidino-2-phenylindole (DAPI) (ab104139, Abcam, Boston, MA, USA), and mounted on glass slides with a mounting medium. The cells were visualized using a fluorescence microscope (EVOS M5000 Imaging System, Invitrogen) and images were stored as JPEG files.

4.16. Analysis of Cell Size

To analyze the average cell size, six random regions were selected from the cell images of each group, and the area of the cell was measured using Image J software. The average values of the measured areas from three images per group were calculated, and statistical analysis was performed using Student's *t*-test.

4.17. Statistics

The experiments were conducted at least three times, and the findings were presented as the mean ± standard error of the mean (SEM). Statistical significance was evaluated using Student's *t*-test for comparing two groups and one-way ANOVA for comparing more than two groups with a significance level of $p < 0.05$.

5. Conclusions

In summary, the quinazoline derivatives SH-340 shows promising potential as a therapeutic for AD. Its ability to address keratinocyte differentiation, skin barrier functions, and TSLP inhibition holds implications not only for AD but also for broader allergic diseases like asthma and eczema. Additionally, SH-340's innovative TSLP modulation offers a targeted approach to curb AD triggers without affecting unrelated pathways. Its versatility extends to autoimmune diseases and cancer immunotherapies, positioning SH-340 as a versatile tool in diverse medical fields. However, further investigation is needed to unravel its mechanisms fully, validate efficacy, and ensure safety. This study underscores SH-340's transformative potential in dermatology and beyond, redefining AD treatment and immunotherapeutic strategies.

Supplementary Materials: The following supporting information can be downloaded at: https://www.mdpi.com/article/10.3390/molecules28166119/s1, Figure S1: LC-MS analysis data, Figure S2: SH340 HRMS analysis report, Figure S3: SH-340 1H-NMR (300 MHz Chloroform), Figure S4: SH-340 13C-NMR (300 MHz Chloroform); Table S1: SH-340 LC-MS: Analytical equipment and conditions.

Author Contributions: Conceptualization, D.K.; methodology, D.K., Y.P. and H.S.; validation, Y.P. and H.S.; formal analysis, Y.P. and H.S.; investigation, Y.P. and H.S.; data curation, D.K., Y.P. and H.S.; writing—original draft preparation, Y.P. and H.S.; writing—review and editing, D.K., Y.P. and H.S.; visualization, Y.P. and H.S.; supervision, D.K.; project administration, D.K.; funding acquisition, D.K. All authors have read and agreed to the published version of the manuscript.

Funding: This study received support from the National Research Foundation of Korea (Grant No. 2020R1C1C1005206).

Institutional Review Board Statement: Not applicable.

Informed Consent Statement: Not applicable.

Data Availability Statement: Not applicable.

Acknowledgments: The chemical library used in this study was kindly provided by Korea Chemical Bank (www.chembank.org (accessed on 1 June 2021)) of Korea Research Institute of Chemical Technology.

Conflicts of Interest: The authors declare no conflict of interest.

Sample Availability: Not applicable.

References

1. Alsibaee, A.M.; Al-Yousef, H.M.; Al-Salem, H.S. Quinazolinones, the winning horse in drug discovery. *Molecules* **2023**, *28*, 978. [CrossRef] [PubMed]
2. Li, Z.; Zhao, L.; Bian, Y.; Li, Y.; Qu, J.; Song, F. The antibacterial activity of quinazoline and quinazolinone hybrids. *Curr. Top. Med. Chem.* **2022**, *22*, 1035–1044. [CrossRef] [PubMed]
3. Lv, L.; Maimaitiming, M.; Huang, Y.; Yang, J.; Chen, S.; Sun, Y.; Zhang, X.; Li, X.; Xue, C.; Wang, P.; et al. Discovery of quinazolin-4(3h)-one derivatives as novel ache inhibitors with anti-inflammatory activities. *Eur. J. Med. Chem.* **2023**, *254*, 115346. [CrossRef]
4. Powell, T.J.; Ben-Bassat, H.; Klein, B.Y.; Chen, H.; Shenoy, N.; Mccollough, J.; Narog, B.; Gazit, A.; Harzstark, Z.; Chaouat, M.; et al. Growth inhibition of psoriatic keratinocytes by quinazoline tyrosine kinase inhibitors. *Br. J. Dermatol.* **1999**, *141*, 802–810. [CrossRef] [PubMed]
5. Bouley, R.; Ding, D.; Peng, Z.; Bastian, M.; Lastochkin, E.; Song, W.; Suckow, M.A.; Schroeder, V.A.; Wolter, W.R.; Mobashery, S.; et al. Structure–activity relationship for the 4(3h)-quinazolinone antibacterials. *J. Med. Chem.* **2016**, *59*, 5011–5021. [CrossRef] [PubMed]
6. Bagheri, E.; Saremi, K.; Hajiaghaalipour, F.; Faraj, L.F.; Ali, M.H.; Abdulla, A.M.; Khaing, L.S.; Salehen, N.A. Synthesis of novel derivatives of quinazoline schiff base compound promotes epithelial wound healing. *Curr. Pharm. Des.* **2018**, *24*, 1395–1404. [CrossRef] [PubMed]
7. Perveen, S.; Saad, M.S.; Khan, M.K.; Choudhary, I.M. Antioxidant and ros inhibitory activities of heterocyclic 2-aryl-4(3h)-quinazolinone derivatives. *Lett. Drug Des. Discov.* **2021**, *18*, 806–815. [CrossRef]
8. Malasala, S.; Ahmad, M.N.; Akunuri, R.; Shukla, M.; Kaul, G.; Dasgupta, A.; Madhavi, Y.V.; Chopra, S.; Nanduri, S. Synthesis and evaluation of new quinazoline-benzimidazole hybrids as potent anti-microbial agents against multidrug resistant staphylococcus aureus and mycobacterium tuberculosis. *Eur. J. Med. Chem.* **2021**, *212*, 112996. [CrossRef]
9. Mortlock, R.D.; Ma, E.C.; Cohen, J.M.; Damsky, W. Assessment of treatment-relevant immune biomarkers in psoriasis and atopic dermatitis: Toward personalized medicine in dermatology. *J. Investig. Dermatol.* **2023**, *143*, 1412–1422. [CrossRef]
10. Drislane, C.; Irvine, A.D. The role of filaggrin in atopic dermatitis and allergic disease. *Ann. Allergy Asthma Immunol.* **2020**, *124*, 36–43. [CrossRef]
11. Ong, P.Y. New insights in the pathogenesis of atopic dermatitis. *Pediatr. Res.* **2014**, *75*, 171–175. [CrossRef] [PubMed]
12. Cohen, I.; Birnbaum, R.Y.; Leibson, K.; Taube, R.; Sivan, S.; Birk, O.S. Znf750 is expressed in differentiated keratinocytes and regulates epidermal late differentiation genes. *PLoS ONE* **2012**, *7*, e42628. [CrossRef] [PubMed]
13. Furue, M. Regulation of filaggrin, loricrin, and involucrin by il-4, il-13, il-17a, il-22, ahr, and nrf2: Pathogenic implications in atopic dermatitis. *Int. J. Mol. Sci.* **2020**, *21*, 5382. [CrossRef] [PubMed]
14. Huang, I.H.; Chung, W.H.; Wu, P.C.; Chen, C.B. Jak-stat signaling pathway in the pathogenesis of atopic dermatitis: An updated review. *Front. Immunol.* **2022**, *13*, 1068260. [CrossRef]
15. Renert-Yuval, Y.; Guttman-Yassky, E. New treatments for atopic dermatitis targeting beyond il-4/il-13 cytokines. *Ann. Allergy Asthma Immunol.* **2020**, *124*, 28–35. [CrossRef] [PubMed]

6. Paller, A.S.; Ladizinski, B.; Mendes-Bastos, P.; Siegfried, E.; Soong, W.; Prajapati, V.H.; Lio, P.; Thyssen, J.P.; Simpson, E.L.; Platt, A.M.; et al. Efficacy and safety of upadacitinib treatment in adolescents with moderate-to-severe atopic dermatitis: Analysis of the measure up 1, measure up 2, and ad up randomized clinical trials. *JAMA Dermatol.* **2023**, *159*, 526–535. [CrossRef]
7. Blauvelt, A.; Szepietowski, J.C.; Papp, K.; Simpson, E.L.; Silverberg, J.I.; Kim, B.S.; Kwatra, S.G.; Kuligowski, M.E.; Venturanza, M.E.; Wei, S.; et al. Itch-free state in patients with atopic dermatitis treated with ruxolitinib cream: A pooled analysis from two randomized phase 3 studies. *J. Am. Acad. Dermatol.* **2023**, *88*, 651–653. [CrossRef] [PubMed]
8. Amano, W.; Nakajima, S.; Kunugi, H.; Numata, Y.; Kitoh, A.; Egawa, G.; Dainichi, T.; Honda, T.; Otsuka, A.; Kimoto, Y.; et al. The janus kinase inhibitor jte-052 improves skin barrier function through suppressing signal transducer and activator of transcription 3 signaling. *J. Allergy Clin. Immunol.* **2015**, *136*, 667–677.e7. [CrossRef]
9. Lee, J.Y.; Park, Y.K.; Seo, S.H.; Yang, B.S.; Park, H.; Lee, Y.S. 7-substituted-[1,4]dioxano[2,3-g]quinazolines as inhibitors of epidermal growth factor receptor kinase. *Arch. Pharm.* **2002**, *335*, 487–494. [CrossRef]
0. Hassan, H.M.A.; Denetiu, I.; Khan, S.A.; Rehan, M.; Sakkaf, K.; Gauthaman, K. Synthesis and biological evaluation of novel triazolyl 4-anilinoquinazolines as anticancer agents. *Med. Chem. Res.* **2019**, *28*, 1766–1772. [CrossRef]
1. Vollono, L.; Falconi, M.; Gaziano, R.; Iacovelli, F.; Dika, E.; Terracciano, C.; Bianchi, L.; Campione, E. Potential of curcumin in skin disorders. *Nutrients* **2019**, *11*, 2169. [PubMed]
2. Harris, C.S.; Hennequin, L.F.; Willerval, O. Three-point variation of a gefinitib quinazoline core. *Tetrahedron Lett.* **2009**, *50*, 1600–1602. [CrossRef]
3. Harris, C.S.; Hennequin, L.F.; Kettle, J.G.; Willerval, O.A. Selective alkylation of a 6,7-dihydroxyquinazoline. *Tetrahedron Lett.* **2005**, *46*, 7715–7719. [CrossRef]
4. Gittler, J.K.; Shemer, A.; Suárez-Fariñas, M.; Fuentes-Duculan, J.; Gulewicz, K.J.; Wang, C.Q.; Mitsui, H.; Cardinale, I.; de Guzman Strong, C.; Krueger, J.G.; et al. Progressive activation of t(h)2/t(h)22 cytokines and selective epidermal proteins characterizes acute and chronic atopic dermatitis. *J. Allergy Clin. Immunol.* **2012**, *130*, 1344–1354. [CrossRef] [PubMed]
5. Hu, X.; Li, J.; Fu, M.; Zhao, X.; Wang, W. The jak/stat signaling pathway: From bench to clinic. *Signal Transduct. Target. Ther.* **2021**, *6*, 402. [CrossRef] [PubMed]
6. Soumelis, V.; Reche, P.A.; Kanzler, H.; Yuan, W.; Edward, G.; Homey, B.; Gilliet, M.; Ho, S.; Antonenko, S.; Lauerma, A.; et al. Human epithelial cells trigger dendritic cell mediated allergic inflammation by producing tslp. *Nat. Immunol.* **2002**, *3*, 673–680. [CrossRef] [PubMed]
7. Vu, A.T.; Chen, X.; Xie, Y.; Kamijo, S.; Ushio, H.; Kawasaki, J.; Hara, M.; Ikeda, S.; Okumura, K.; Ogawa, H.; et al. Extracellular double-stranded rna induces tslp via an endosomal acidification- and nf-κb-dependent pathway in human keratinocytes. *J. Investig. Dermatol.* **2011**, *131*, 2205–2212. [CrossRef] [PubMed]
8. Kim, D.; Chen, R.; Sheu, M.; Kim, N.; Kim, S.; Islam, N.; Wier, E.M.; Wang, G.; Li, A.; Park, A.; et al. Noncoding dsrna induces retinoic acid synthesis to stimulate hair follicle regeneration via tlr3. *Nat. Commun.* **2019**, *10*, 2811. [CrossRef]
9. Nelson, A.M.; Reddy, S.K.; Ratliff, T.S.; Hossain, M.Z.; Katseff, A.S.; Zhu, A.S.; Chang, E.; Resnik, S.R.; Page, C.; Kim, D.; et al. Dsrna released by tissue damage activates tlr3 to drive skin regeneration. *Cell Stem Cell* **2015**, *17*, 139–151. [CrossRef]
0. Ali, D.; Alhattab, D.; Jafar, H.; Alzubide, M.; Sharar, N.; Bdour, S.; Awidi, A. Differential marker expression between keratinocyte stem cells and their progeny generated from a single colony. *Int. J. Mol. Sci.* **2021**, *22*, 10810. [CrossRef]
1. Finnegan, A.; Cho, R.J.; Luu, A.; Harirchian, P.; Lee, J.; Cheng, J.B.; Song, J.S. Single-cell transcriptomics reveals spatial and temporal turnover of keratinocyte differentiation regulators. *Front. Genet.* **2019**, *10*, 775. [CrossRef]
2. Livak, K.J.; Schmittgen, T.D. Analysis of relative gene expression data using real-time quantitative pcr and the 2(-delta delta c(t)) method. *Methods* **2001**, *25*, 402–408. [CrossRef]

molecules

MDPI

Review

Research Progress of Takeda G Protein-Coupled Receptor 5 in Metabolic Syndrome

Xianmei Gou [1], Lin Qin [1], Di Wu [1], Jian Xie [1], Yanliu Lu [1,2], Qianru Zhang [1,*] and Yuqi He [1,*]

1 Guizhou Engineering Research Center of Industrial Key-Technology for Dendrobium Nobile, Zunyi Medical University, Zunyi 563000, China; gxm8777@outlook.com (X.G.)
2 Joint International Research Laboratory of Ethnomedicine of Ministry of Education, Zunyi Medical University, Zunyi 563000, China
* Correspondence: zhangqianru@zmu.edu.cn (Q.Z.); yqhe2016@zmu.edu.cn (Y.H.)

Abstract: Bile acids are acknowledged as signaling molecules involved in metabolic syndrome. The Takeda G protein-coupled receptor 5 (TGR5) functions as a significant bile acid receptor. The accumulated evidence suggests that TGR5 involves lipid homeostasis, glucose metabolism, and inflammation regulation. In line with this, recent preclinical studies also demonstrate that TGR5 plays a significant role in the generation and progression of metabolic syndrome, encompassing type 2 diabetes mellitus, obesity, atherosclerosis, and non-alcoholic fatty liver disease (NAFLD). In this review, we discuss the role of TGR5 in metabolic syndrome, illustrating the underlying mechanisms and therapeutic targets.

Keywords: bile acids; TGR5; Takeda G protein-coupled receptor 5; metabolic syndrome

1. Introduction

Metabolic syndrome (MS) refers to a group of metabolic dysregulations characterized by central obesity, hyperglycemia, insulin resistance, and hypertension. Individuals with MS have an increased risk of cardiovascular diseases and type 2 diabetes mellitus (T2DM) [1,2]. MS has become an increasing global public health problem, with the worldwide prevalence of metabolic syndrome predicted to encompass one-fourth of the world's population, equating to an excess of one billion people worldwide being affected by metabolic syndrome [3,4]. It is most significant for lowering blood glucose and lipid levels of MS individuals in the therapeutic and preventive aspects of cardiovascular diseases.

Takeda G protein-coupled receptor 5 (TGR5, GPBAR1, M-BAR, or GPR131) is a bile acid membrane receptor and a member of the G protein-coupled receptor subfamily. TGR5 can influence bile acid composition and enterohepatic circulation [5]. Accumulated evidence has demonstrated that TGR5 plays special roles in lipid and glucose metabolism, energy homeostasis, and immune and anti-inflammatory regulation [6–9]. TGR5 is widely expressed in the liver, intestine, pancreas, central nervous system, and immune cells [10]. TGR5 activation by bile acids can influence lipid, glucose, and energy metabolism (Figure 1). Hepatic TGR5 acts as a suppressor of lipid accumulation in the liver by regulating the bile acids composition and enterohepatic circulation [11,12]. Mouse models with MS have illustrated that the activation of TGR5 could promote insulin secretion and insulin sensitivity [13,14]. TGR5 is also involved in neuronal regulation in the central system to influence food intake and body weight [15]. Studies using tissue-specific mice have revealed that TGR5 could increase energy expenditure in skeletal muscle and adipose tissue [16,17]. TGR5 also involves in the pathogenesis of atherosclerosis by regulating the secretion of vasodilators and vasoconstrictors [18,19] (Figure 1).

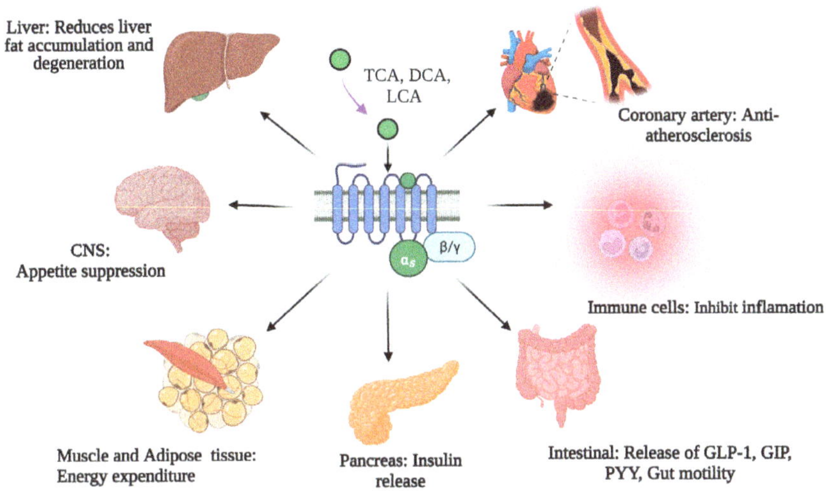

Figure 1. The role of TGR5 in metabolic syndrome in different cells and tissues. TCA, taurocholic acid; DCA, deoxycholic acid; LCA, Lithocholic acid; CNS, center nervous system; NAFLD, non-alcoholic fatty liver disease; NASH, non-alcoholic steatohepatitis; GLP-1, glucagon-like peptide-1; GIP, gastric inhibitory polypeptide; PYY, Peptide YY. This figure was created by Biorender (https://app.biorender.com/, accessed on 9 June 2023).

Bile acids (BAs) have been acknowledged as pivotal signal factors involving diverse biological functions including metabolic homeostasis and the immune response, mainly through activating two bile acid receptors: TGR5 and farnesoid X receptor (FXR). Recently, hundreds of research articles about TGR5 and MS have been collected in SCIE and PubMed up to 2023. However, the approved drugs targeting TGR5 for the treatment of MS are still limited. This review aims to map the underlying mechanisms of TGR5 and its potential as a drug target in the treatment of MS.

2. TGR5

The encoded sequence of the TGR5 gene is evolutionarily conserved in humans, rats, and other mammals and is located on human chromosome 2q35 and mouse chromosome 1c3 [6,20]. TGR5 gene expression is densely found in human tissues and cells, including the brain, liver, endocrine glands, skeletal muscles, spleen, lymph nodes, lung, placenta, nerve cells, and adipocytes. In the liver, TGR5 is highly expressed in bile duct cells, activated liver stem cells, sinusoidal endothelial cells, Kupffer cells, and NKT cells [10,21].

There are seven transmembrane domains (TMH1-7) of TGR5, including three extracellular loops that bind to ligands (ECL1-3) and three intracellular loops responsible for transmitting signals (ICL1-3) [22]. TGR5 is activated by bile acids in the order of LCA > DCA > CDCA > CA [23]. Bile acids and TRG5 agonists interact with TGR5 in the binding pocket and deliver a G-protein complex with α_s, β, and γ subunits [24]. GDP is subsequently released from the G protein complex and replaced by GTP, inducing the separation of G-protein α_s with β-γ dimers. G-protein α_s activates intracellular adenylyl cyclase (AC), resulting in ATP to cAMP conversion and protein kinase A (PKA) activation, which further induce downstream signaling involving bile acid synthesis, insulin sensitivity, atherosclerosis development, energy expenditure, and inflammation [25,26].

3. TGR5 and Bile Acids

BAs, including primary and secondary types, are synthesized from cholesterol and other precursors within hepatocytes. BAs have been acknowledged as metabolic messengers that could regulate their own metabolism and immune system by targeting FXR and TGR5 [27,28]. A disruption of the bile acids pool component has been described in MS, including T2DM, insulin-resistant, and non-alcoholic hepatitis [29–31]. Cholesterol 12α hydroxylase (CYP8B1) is an important enzyme in the synthesis of bile acids, particularly with the amount of 12α-hydroxy bile acids. TGR5 activation decreases the expression of CYP8B1 and reduces the ratio of 12α-hydroxylated bile acids to non-12α-hydroxylated bile acids in the bile acid pool. Mcgavgan et al. [11] found that reduced circulating bile acid profile hydrophobicity was related to a TGR5-dependent expression of the CYP8B1 protein. Meanwhile, the TGR5 agonist (INT-777) could downregulate hepatic Cyp8b1 expression [32]. The levels of Cyp8b1 and hydrophobic bile acids, especially 12α-hydroxy bile acids, have been upregulated in tgr5$^{-/-}$ mice and significantly reduced in tgr5$^{+/+}$ mice [33,34]. Thus, selective under-regulation of the expression level of hepatic Cyp8b1 may be a mechanism by which TGR5 regulates the bile acid profiles (Figure 2).

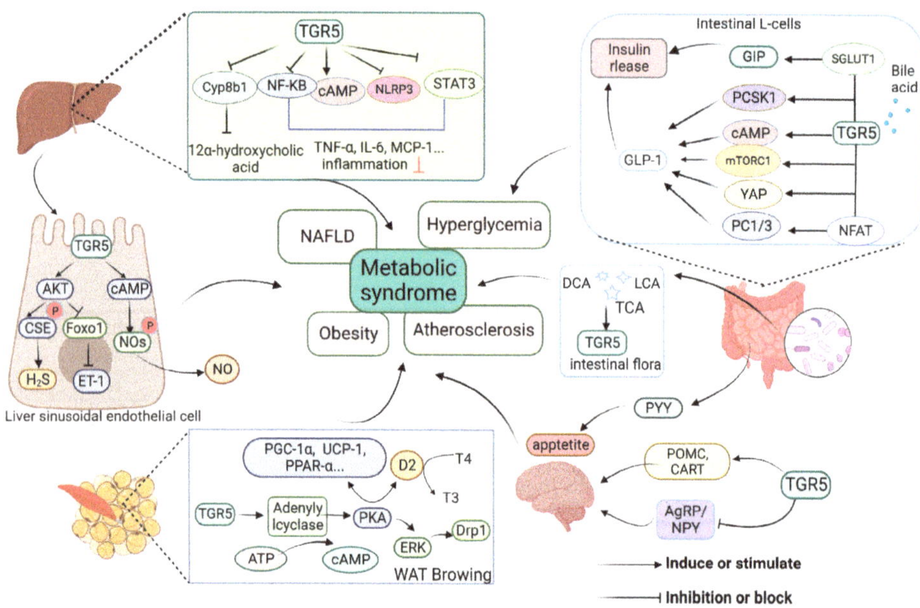

Figure 2. Review on the mechanisms of TGR5 in metabolic syndrome. TGR5 signaling is involved in the regulation of bile acid synthesis, vascular factor function [11,18], amelioration of inflammatory diseases by inhibiting inflammation [30,35,36], modulation of endocrine peptide release to ameliorate glucose disorders [17,37,38], appetite suppression, and promotion of energy expenditure [39,40]. TGR5 is also indirectly activated by regulating intestinal flora [41,42]. TGR5, takeda G protein-coupled receptor 5; TCA, taurocholic acid; DCA, deoxycholic acid; LCA, lithocholic acid; CYP8B1, cholesterol 12α hydroxylase; cAMP, cyclic adenosine monophosphate; PKA, protein kinase A; GLP-1, glucagon-like peptide-1; PC1/3, prohormone convertase 1/3; PCSK1, proprotein convertase subtilisin/kexin type 1; mTOR, mammalian target of rapamycin; NFAT, nuclear factor of activated T cells; YAP, yes associated protein; GIP, gastric inhibitory polypeptide; SGLT1, sodium glucose cotransporter 1; PYY, peptide YY; TNF-α, tumor necrosis factor-α; IL-6, interleukin-6; NLRP3, NOD-like receptor domain associated protein 3; WAT, white adipose tissue; ERK, extracellular regulated protein kinases; Drp1,

dynamin-related protein 1; D2, deiodinase 2; T4, thyroxine; T3, 3,5,3′-triiodothyronine; Ucp1, uncoupling protein 1; PGC-1α, PPAR-γ coactivator-1α; POMC, pro-opiomelanocortin; CART, cocaine-amphetamine-regulated transcript; YAgRP/NPY, neuropeptide Y/agouti-related peptide; MCP-1, monocyte chemoattractant protein-1; NOS, nitric oxide synthase; NO, nitric oxide; STAT3, signal transducer and activator of transcription 3; CSE, cystathionine γ-lyase; FoxO1, forkhead box protein O1; ET-1, endothelin-1. This figure was created by Biorender (https://app.biorender.com/, accessed on 11 June 2023).

4. TGR5 and T2DM

TGR5 is an essential receptor in maintaining the balance of glucose levels in the body and controlling insulin sensitivity. It was suggested that TGR5$^{-/-}$ mice display impaired glucose tolerance and insulin resistance compared to their WT counterparts, whereas TGR5-overexpressing mice showed good tolerance to glucose [9]. Meanwhile, the administration of TGR5 agonists was shown to modulate glucose levels in high-fat diet (HFD) mice, further demonstrating that TGR5 plays an active role in the disease process of T2DM [43].

4.1. Promote GLP-1 Secretion

Several studies suggest that TGR5 activation elevates glucagon-like peptide-1 (GLP-1) secretory capacity and expands the L-cell population (Figure 2). The intestinal L cells are the main synthesizing and releasing cells of GLP-1, and TGR5 sits in the basal layer of L cells [44]. GLP-1 is an important peptide hormone that binds to the GLP-1 receptor in pancreatic β cells; its functions include stimulating insulin release, delaying gastric emptying, enhancing satiety, and inhibiting glucagon secretion [45].

On the one hand, TGR5 was found to induce GLP-1 production and regulate glucose metabolism through different metabolic pathways. In STC-1 cells, handling of bile acid-activated TGR5 stimulated GLP-1 secretion through the cAMP pathway [13]. Prohormone convertase 1/3 (PC1/3), processing and modifying GLP-1, is an important enzyme for the production and secretion of GLP-1. It has been found that altered intestinal bile acid composition induced an elevation of the TGR5 agonist, taurocholic acid (TCA), which stimulated the nuclear factor of activated T cells (NFAT). Then, PC1/3 binds to NFAT binding, promotes PC1/3 gene expression, and induces GLP-1 secretion, thereby improving glucose tolerance [46]. Meanwhile, proprotein convertase subtilisin/kexin type 1 (PCSK1) encoding the PC1/3 is important for GLP-1 synthesis and can affect its secretion. It was shown that the administration of oligofructose (OFS) to wild-type mice improved glucose tolerance compared to TGR5$^{-/-}$ mice. The mechanism was as follows: OFS altered the composition of the intestinal BAs, particularly 6α-hydroxy bile acids, increased TGR5 activity, and promoted PCSK1 expression and, subsequently, GLP-1 release [47]. mTORC1 is a complex of mammalian target of rapamycin (mTOR). Several studies have shown that ileal TGR5–mTOR1 signaling engages in GLP-1 secretion in L cells. The cellular experiment showed that deoxycholic acid (DCA) activated the TGR5–mTOR1 signaling pathway, phosphorylated mTOR, S6, and S6 kinases (S6Ks), and induced GLP-1 synthesis and release. However, the effect of DCA-induced GLP-1 synthesis is abolished when siRNA is knocked down of mTORC1 or TGR5 [48].

On the other hand, TGR5 could expand the L-cell population indirectly to increase GLP-1 production. Ginsenoside K has been found to activate TGR5 by elevating lithocholic acid (LCA) and DCA content. This, in turn, activates the RhoA/ROCK/YAP signaling pathway, increasing the abundance of L-cells, and ultimately enhancing GLP-1 release [49]. When TGR5 is knocked down, yes-associated protein (YAP) exhibits low expression and loses its glucose regulatory function [50]. Therefore, TGR5 is a promising target for use against T2DM.

FXR and TGR5 are essential receptors in the feedback of bile acids synthesis and en-terohepatic circulation [51]. Moreover, these two receptors are co-expressed in the ileum and colon [52]. In intestinal endocrine L cells, FXR and TGR5 could stimulate GLP-1 secretion, which contributes to increasing insulin secretion to improve insulin sensitivity, respectively. A study using FXR/TGR5-knockout mice has illustrated that FXR and TGR5 may coordinately regulate GLP-1 production and secretion [53]. Further research on the roles of FXR and TGR5 in metabolic syndrome is still needed.

4.2. Promote GIP Secretion

A glucose-dependent gastric inhibitory polypeptide (GIP) increases insulin secretion and maintains glucose homeostasis in the intestine (Figure 2). Upon food ingestion, intestinal cells release GIP into the bloodstream, where it binds to the GIP receptor (GIPR) on pancreatic β-cells. This activation of GIPR leads to increased cAMP levels and proinsulin gene transcription, which ultimately promotes insulin secretion and regulates the survival and reproduction of insulin-independent pancreatic β cells [37]. Zhu et al. [54] found that saponins derived from *Camellia Sinensis* seeds could stimulate GIP secretion in mouse and STC-1 cells via sodium glucose cotransporter 1 (SGLT1) and TGR5. However, TGR5 siRNA was able to counteract the increase in GIP expression, revealing that TGR5 participates in the regulation of GIP expression for the first time. Furthermore, network pharmacology analysis showed that PCSK1 was related to SGLT1, TGR5, and GIP [54]. Interestingly, the downregulation of PCSK1 inhibits GIP and GLP-1 production in STC-1 cells [55]. Thus, the TGR5–PCSK1 axis is likely critical for the simultaneous induction of GLP-1 and GIP secretion, and in vivo experiments are still needed to verify the involved mechanisms further. The TGR5–GIP signaling pathway offers a promising avenue for creating novel hypoglycemic medications.

4.3. Promote PYY Secretion

Peptide YY (PYY) is a gastrointestinal hormone released by L cells in the ileum and colon. PYY could slow gastric emptying, delay gastrointestinal transport, and suppress appetite through the hypothalamic circuits [56]. Bile acids promote the secretion of PYY by activating the TGR5 receptor in L cells (Figure 2). In vitro studies have shown that the TGR5 agonist DCA induces the release of PYY in a dose-dependent manner [23]. However, this effect was not observed in TGR5 knockout mice [57]. In a clinical study, the administration of DCA and TCA can simulate a rapid and noticeable increase in plasma PYY. In contrast, applying bile acids with a low binding affinity to TGR5 cannot induce PYY secretion significantly [58]. Of mechanistic relevance was the demonstration that activation of TGR5 stimulates the production of Gα and cAMP to release PYY via the Epac/PLC-ε-CA^{2+} pathway in a cAMP-dependent manner [59]. The findings above demonstrate that TGR5 signaling is associated with bile acid-induced PYY secretion.

4.4. Increase Glucose Metabolism in Skeletal Muscle

TGR5 is highly expressed in skeletal muscle, the largest organ of glucose handling in humans. Skeletal muscle also increases glucose clearance and improves insulin resistance to regulate body glucose homeostasis, in addition to promoting energy metabolism (Figure 2). One TGR5 agonist, known as 4-Phenoxynicotinamide (MN6), could enhance insulin-stimulated glucose transport via the cAMP/PKA pathway, thereby ameliorating insulin resistance and glucose tolerance in skeletal muscle in HFD mice and C2C12 myotubes [60]. Furthermore, TGR5 has been proven to alleviate glucose homeostasis and age-related glucose disorders in skeletal muscle. Sasaki et al. [17] found that the activation of the glycolytic pathway in skeletal muscle-specific TGR5 transgenic mice led to increased glucose utilization and clearance in HFD mice. Moreover, observation shows that old transgenic mice exhibit increased the glucose clearance and strength of the skeletal muscles. More importantly, the activation of TGR5 increases a higher muscle mass, which is beneficial in improving diabetes. Sasaki et al. [21] showed that exercise-induced unfolded protein response (UPR) and activating transcription factor 6α (ATF6α) increase the expression of TGR5, which, in turn, promotes the differentiation of skeletal muscle cells and muscle hypertrophy in mice. Moreover, Sund et al. [61] found that TGR5 strengthens skeletal muscle regeneration and myogenic process through the AKT/mTOR/FoXO3 signaling pathway in C2C12 cells. This research reveals that activating TGR5 in skeletal muscle effectively improves glucose metabolism disorder, opening new avenues for drug development to treat diabetes.

4.5. Inhibit Inflammation

Chronic inflammation is a crucial initiating factor of insulin resistance. TGR5 signaling becomes a promising approach for treating T2DM by interfering with the inflammatory response (Figure 2). Several studies have shown a high expression of TGR5 in monocytes, macrophages, and immune organs such as the spleen. The bile acid–TGR5 pathway inhibits the secretion of various cytokines, including tumor necrosis factor-α (TNF-α) and interleukin (IL)-1α, IL-1β, IL-6, and IL-8 [62].

In Kupffer cells, TGR5 activation reduces cytokine expression induced by LPS and inhibits NF-κB-dependent inflammatory responses, improving glucose tolerance [63]. According to a study, activation of TGR5 reduces chemokines expression in macrophages via the AKT–mTOR pathway [64]. This pathway involves the phosphorylation of AKT and mTORC1, which induces the expression of 4E-binding proteins (4E-BP) and CCAAT/enhancer binding protein beta (C/EBPβ) isoform liver-inhibitory protein (LIP). Therefore, the TGR5–AKT–mTOR-LIP pathway is a newly discovered pathway for macrophages to promote anti-diabetic effects [64]. Moreover, abnormal glucose metabolism is closely associated with regulating NOD-like receptor domain-associated protein 3 (NLRP3) inflammatory vesicles. TGR5 activation induces PKA kinase, leading to the ubiquitination of NLRP3. Therefore, the TGR5–cAMP–PKA axis inhibits the formation of NLRP3 inflammatory vesicles to attenuate inflammatory response, glucose tolerance, and insulin resistance induced by a high-fat diet [65]. It also estimated that the inflammation and insulin resistance of chimeric mice simultaneous upregulated after that specifically removing TGR5 from macrophages and bone marrow [26]. According to these studies, TGR5 signaling could prevent hepatic insulin resistance development and progression, ultimately improving systemic glucose levels.

5. TGR5 and Obesity

5.1. Promote Energy Consumption and Fat Browning

Obesity is mainly induced by an imbalance between energy intake and expenditure, ultimately resulting in increased fat storage in adipose tissue. Adipose tissue can be divided into white adipose tissue (WAT) and brown adipose tissue (BAT). WAT primarily stores energy, while BAT contains more mitochondria and plays a key role in energy consumption. Therefore, an increase in BAT is conducive to improving obesity. TGR5 is a crucial target for regulating mitochondrial function and promoting adipose tissue remodeling (Figure 2).

BAs haven been acknowledged as metabolic regulators affecting energy expenditure via TGR5 stimulation. In brown adipose tissue and skeletal muscle, the interaction of BA and TGR5 promotes the activation of type 2 deiodinase (Dio2), which induces the expression of the crucial metabolic genes including uncoupling protein 1 (UCP1) and peroxisome proliferator-activated receptor-gamma coactivator-1 alpha (PGC-1α) in mitochondria [66–68]. Dio2 converts inactive thyroxine (T4) to active 3,5,3′-triiodothyronine (T3), inducing the activation of UCP1 [16,69]. Mitochondrial UCP1 diverts the electron gradient through proton leak and induces uncoupled respiration [70,71]. PGC–1α co-activates proliferator-activated receptor γ (PPARγ) and fine-tunes mitochondrial biogenesis, oxidative metabolism, and thermogenesis [72]. The thermogenic activity of BA supplementation was disappeared in Tgr5-knockout mice that were fed a high-fat diet [73]. BA–TGR5–Dio2 signaling promotes thermogenic activity instead of ATP production. In addition, an in-vivo study demonstrated administration of TGR5-activated bile acid, which can transform white adipose tissue into beige adipose tissue and promote mitochondrial fission via the extracellular regulated protein kinases/dynamin-related protein 1 (ERK/DRP1) pathway [70].

5.2. Increase Intestinal Peristalsis

TGR5 is involved in bile acids-induced intestinal peristalsis and significantly prevents central obesity caused by constipation. Studies have demonstrated that TGR5 agonists can enhance the intestinal motility of mice, while the absence of TGR5 results in a longer colonic transit time and decreased frequency of defecation in mice [50]. Activated TGR5 could stimulate the secretions of peristaltic mediators, including 5-hydroxytryptamine (5-HT) and calcitonin gene-related peptide (CGRP) in enterochromaffin cells (ECs), while the effect was lost in TGR5$^{-/-}$ mice [74]. In vitro and in vivo experiments have demonstrated that TGR5 activation can enhance the production of 5-HT from EC cells. This improvement in constipation is achieved by activating the transient receptor potential ankyrin 1 (TRPA1) and the PLC-γ1/PIP2 signaling pathway [75].

5.3. Regulate Hypothalamic Neurons

TGR5 signaling in the hypothalamus is an important factor in protecting against obesity caused by HFD. Hypothalamic bile acid level is reduced in diet-induced obese (DIO) mice compared to the control group [15]. However, when DIO mice are administered with bile acids or central TGR5 agonists, their body weight and fat mass are reduced by activating the sympathetic nervous system, thereby regulating energy homeostasis. Conversely, when hypothalamic TGR5 expression in the medio basal hypothalamus is downregulated or deleted, it contributes to obesity by weakening sympathetic nerve activity [15]. Together, hypothalamic TGR5 is an important receptor for anti-obesity via the supplementation of bile acid, which provides a basis for the development of BA–TGR5 as a pathway for anti-obesity drugs.

Findings from a study of the central neurons involved in TGR5 regulation of food intake to improve obesity (Figure 2). TGR5 not only promotes the expression of anorexia neurons to increase satiety signals but also inhibits the expression of hunger-signaling neurons to suppress hunger signals, thereby reversing obesity. A study has shown that DCA and cholecystokinin octapeptide (CCK8) act synergistically to stimulate pro-opiomelanocortin (POMC) and cocaine–amphetamine-regulated transcript (CART) neurons via the TGR5 receptor of the valgus nerve. Then, it delivers satiety signals to the hypothalamus, reducing food intake [76]. Chen et al. [77] investigated the mechanism of bile acids releasing POMC in mouse hypothalamus GT1-7 cells. The results showed that phosphorylation of the signal transducer and activator of transcription 3 (p-STAT3), phosphorylation of threonine kinase (p-AKT), suppressor of cytokine signaling (SOCS3), and TGR5 expression were significantly increased. This led to the conclusion that bile acid may regulate anorexigenic neuropeptides through p-AKT and p-STAT3–SOCS3 signaling. In the other study, TGR5 activation inhibits the secretion of hypothalamic appetite-promoting neuropeptide Y/agouti-related peptide (AgRP/NPY) neurons via the Rho/ROCK/Actin-remodeling pathway, which, in turn, suppresses hunger signaling to reduce food intake. In contrast, the deletion of TGR5 results in a notable rise in food consumption and encourages the onset of obesity [78]. In addition to POMC, CART, and AgRP/NPY neurons, other neurons still play a role in regulating body weight through TGR5 and related mechanisms. Further exploration is needed to understand their involvement fully.

6. TGR5 and Atherosclerosis

6.1. Regulate Inflammation

The development and progression of atherosclerosis are promoted by dyslipidemia and chronic inflammation, with TGR5 being associated with the pathogenesis of atherosclerosis. When macrophages remove the modified forms of LDL-C from the vascular wall, it eventually forms foam cells and produces local inflammatory factors and chemokines, causing chronic inflammation in the vascular wall and promoting lipid accumulation in the cells [24]. It has been shown that activation of TGR5 blunts the production of a cluster of differentiation 36 (CD36), scavenger receptor A (SR-A), monocyte chemoattractant protein-1 (MCP-1), and chemokine ligand-5 (CCL5) through the cAMP–NF-κB signaling pathway [79], which reduces the uptake of oxidized LDL (Ox-LDL) and the formation of foam cells, thereby inhibiting the progression of atherosclerosis. In the atherosclerosis model of $LDL^{-/-}$ mice, activation of TGR5 reduces atheromatous plaques by inhibiting inflammation and lipid accumulation. However, when the Tgr5 gene was knocked out, there was no inhibitory effect on atherosclerosis [79]. This suggests that the activation of TGR5 plays a crucial role in preventing atherosclerosis by modulating the immune system.

6.2. *Promote NO Release*

Nitric oxide (NO) is an essential molecule in preventing atherosclerosis. It has multiple beneficial effects, including vasodilation and increased blood flow. Moreover, NO reduces the adhesion of monocytes to vascular endothelial cells, thus preventing the formation of atheromatous plaque. Activation of TGR5 may offer a novel therapeutic strategy for treating atherosclerosis. When bile acid binds to TGR5 and triggers the expression of cAMP in liver sinusoidal endothelial cells, this process results in an elevation of intracellular Ca^{2+} levels, leading to the activation of nitric oxide synthase (NOS) phosphorylation and subsequent release of nitric oxide (NO) (Figure 2) [80].

TGR5 is activated by endogenous bile acids, with TCA being the most effective, followed by LCA, DCA, chenodeoxycholic acid (CDCA), and cholic acids (CA). A study revealed that individuals with atherosclerosis had significantly lower levels of LCA in their serum. This reduction level of LCA could impede TGR5 on anti-inflammatory in macrophages, leading to increasing plaque formation. Thus, LCA may be a marker for the clinic diagnosis of coronary atherosclerosis [19].

7. TGR5 and NAFLD

Prior research has demonstrated that TGR5 protects against several liver diseases. Non-alcoholic fatty liver disease (NAFLD) is a common complication of obesity and T2DM. NAFLD has the potential to advance to non-alcoholic steatohepatitis (NASH), liver fibrosis, hepatocellular injury, cirrhosis, and hepatocellular carcinoma, because of lipid buildup in the liver (Figure 2). The TGR5 receptor may play a role in the regulation of NAFLD/NASH, as indicated by a notable decrease in TGR5 expression observed in the liver tissues of both NASH patients and NASH animals [35,65]. Meanwhile, $TGR5^{-/-}$ HFD mice showed lipid accumulation and increased hepatic fat energy [11]. Furthermore, the impact of TGR5 on lipid accumulation may vary depending on the gender of the individual. Research conducted by Vassileva et al. [9] on DIO mice revealed that male $Tgr5^{-/-}$ mice showed higher levels of hepatic steatosis than female $Tgr5^{-/-}$ mice, highlighting the regulatory role of TGR5 in this process.

7.1. *Improve Insulin Resistance*

The development of NAFLD is significantly influenced by insulin resistance in both the liver and adipose tissue [81]. The TGR5 signaling pathway is critical in regulating intestinal GLP-1 secretion and improving insulin resistance. According to Lambert et al. [82], GLP-1 has been shown to improve insulin sensitivity and reduce hepatic glucose production in the animal model of NAFLD, which inhibits the development of liver steatosis, NASH, and liver fibrosis. A clinical study has further demonstrated that treatment with liraglutide, a GLP-1 agonist, can hinder the histological deterioration of NAFLD. This is achieved by reducing body weight and fasting blood glucose, increasing insulin sensitivity and inhibiting de novo lipogenesis and inflammation [83]. In addition, activation of TGR5 releases Glucagon-like peptide-2 (GLP-2), which possesses anti-inflammatory properties and accelerates liver regeneration after partial hepatectomy [84,85]. Thus TGR5 is expected to be a novel target that improves NAFLD/NASH via increasing the release of GLP-1 as well as GLP-2.

7.2. Inhibit Inflammation

Monocytes and macrophages are important sources of inflammatory factors in the liver, playing a crucial role in the development of NASH and hepatocellular carcinoma [86]. The TGR5 receptor improves NASH in various immune cells by inhibiting inflammation (Figure 2).

Activation of TGR5 in macrophages has been shown to effectively inhibit inflammatory responses through the toll-like receptor 4 (TLR4)/NF-κB pathway [87]. Additionally, TGR5 activation reduces the binding activity of NF–κB–p65 DNA, inhibiting the NF–κB pathway and decreasing TNF-α secretion in CD14 macrophages both in vitro and in vivo [88]. Moreover, TGR5 agonists increase the phosphorylation of PKA and induce ubiquitination of NLRP3 inflammasome, which inhibits NLRP3-mediated M1 macrophage polarization, thereby alleviating liver steatosis and inflammation in NASH [89]. Activation of TGR5 has been observed to inhibit lipopolysaccharide (LPS)-induced inflammatory factors production in Kupffer cells and THP-1 cells of TGR5 overexpressed. Additionally, TGR5 activation has been found to reduce phosphorylated level of the signal transducer and activator of transcription 3 (STAT3), resulting in a hepatoprotective effect [90]. Conversely, LPS-induced $TGR5^{-/-}$ mice show severe liver injury, hepatocyte apoptosis, and inflammation and cause elevated serum alanine and aspartate aminotransferase [91]. The above studies show that TGR5 signaling pathways inhibit the occurrence and development of NASH by regulating inflammatory factors, providing a new idea for the treatment of NASH.

7.3. Ameliorate Effect on Fibrin

TGR5 agonists ameliorate liver fibrosis by inducing apoptosis, inhibiting inflammation, and improving blood circulation.

LCA has been identified as the effective natural agonist of TGR5, which plays a crucial role in regulating metabolism and inhibiting inflammation. A recent study has investigated the effects of LCA on immune cells and hematopoietic stem cells (HSCs) in liver fibrosis. According to the findings, LCA could hinder the activity of transforming growth factor beta (TGF-β) and glycolysis while enhancing oxidative phosphorylation. This leads to the differentiation of macrophages towards the M2 type and prevents their differentiation into the M1 type [92]. Meanwhile, LCA increases the recruitment of NK cells and decreases NKT cell activation, ameliorating liver fibrosis in mice induced by carbon tetrachloride (CCL4) [92]. BAR501, a TGR5 agonist, has been shown to reverse liver and vascular damage in mouse models of NASH in a TGR5-dependent manner. BAR501 treatment induces the browning of white adipose tissue and inhibits the development of aortic intima–media thickness [93]. Additionally, it reduces the expression of specific inflammatory genes and adhesion molecules in the aorta, such as IL-6, TNF-a, vascular cell adhesion molecule (VCAM), intercellular adhesion molecule-1 (ICAM-1), and endothelial selection [93].

The activation of TGR5 in liver sinusoidal endothelial cells (LSECs) promotes vasodilators secretion and inhibition of vasoconstrictors release (Figure 2). The former is achieved through the induction of endothelial nitric oxide synthase (eNOS) phosphorylation, which increases nitric oxide (NO) production [18]. Similarly, TGR5 activation increases cystathionine γ-lyase (CSE) activity via Akt-induced serine phosphorylation, increasing H_2S production. Moreover, the activation of Akt leads to the reduced transcription of forkhead box protein O1 (FoxO1), inhibiting the expression of vasoconstrictor endothelin-1 (ET-1) [94]. These findings indicate that increased TGR5 signaling has the potential to improve liver function by decreasing lipid buildup in the liver, reducing fibrosis and inflammation.

8. Gut Microbiota Regulates TGR5

The interaction between gut flora and the TGR5 receptor has become a current research hotspot. Accumulating evidence shows that gut flora indirectly activates the TGR5 pathway by metabolizing the composition of the bile acid profile, facilitating the effect on a series of MS. In contrast, TGR5 deficiency causes gut microbiota imbalance and aggravates liver injury.

Studies have demonstrated that knocking out hypoxia-inducible factor 2α (HIF-2α) in mice can effectively reduce obesity by decreasing gut lactate levels (Figure 2). This reduction leads to a decrease in the frequency of *Bacteroides vulgatus* and an increase in the abundance of *Ruminococcus torques*. These changes result in higher TCA and DCA and TGR5 activation levels. Activation of TGR5 enhances the thermogenic capacity of white adipose tissue by increasing the expression of UCP-1 and creatine kinase mitochondrial 2 (CKMT2) [44]. A recent study has shown that flaxseed can alleviate NASH by reducing the abundance of *Firmicutes* and increasing the abundance of *Bifidobacterium* and *Actinobacteria*. Additionally, it can increase the proportion of LCA and DCA, which can inhibit liver inflammation and protect against HFD-induced hepatic steatosis via the TGR5–NF-κB signaling pathway [95]. Dietary supplementation with probiotics can enhance insulin sensitivity and stimulate adipose tissue browning, ultimately resulting in weight loss. The mechanism increases the abundance of *Bacteroides* that further induce the TGR5–GLP-1 pathway and inhibit intestinal dipeptidyl peptidase-4 (DDP4) [96]. Meanwhile, a clinical study has also shown that probiotics consumption improves metabolic parameters such as lowering cholesterol, improving intestinal function and increasing insulin sensitivity in obese and T2DM patients [41].

9. TGR5 Agonists

TGR5 activation could improve insulin resistance, increase adipose tissue thermogenesis and energy expenditure, and inhibit inflammation. However, the approved drug targeting TGR5 is limited in the treatment of MS. Bile acid analogs and non-bile acid regulators exhibit therapeutic potential for the treatment of metabolic disorders, including obesity, diabetes, and NAFLD. Ursodeoxycholic acid (UDCA) is a secondary bile acid used as a first-line drug in the treatment of primary biliary cholangitis [97]. A clinical study has shown that administering low doses of UDCA can enhance GLP-1 secretion through TGR5 signaling in diabetic patients. A combination of sitagliptin and UDCA is in phase-IV clinical trials [98]. The steroid agonist 6α-ethyl-23(S)-methylcholic acid (INT-777), specific to TGR5, has been found to promote thermogenesis and energy metabolism in adipose tissue [99]. Furthermore, it also stimulates the secretion of GLP-1 and inhibits macrophage-mediated inflammation and lipid accumulation [32], which can lead to improved insulin sensitivity and reduced hepatic steatosis [100]. Recent studies have shown that INT-777 inhibits neuroinflammation and improves nerve injury [101,102]. Meanwhile, in vivo studies using mouse model with NAFLD also validated the hypoglycemic effects of INT-777 [103].

RDX8940 is a highly effective and selective TGR5 agonist, which could promote the secretion of GLP-1, GLP-2, and PYY in mouse intestinal L cells with minimal systemic side effects. Unlike other agonists, it does not inhibit gallbladder emptying in mice, and it is also non-toxic to the heart, liver, and pancreas. Furthermore, RDX8940 improves hepatic steatosis and insulin sensitivity in insulin-resistant mouse models, which has the potential to treat NASH and NAFLD [14]. A-Ionone is an aromatic diterpenoid compound with the same TGR5-activating effect as CDCA and is used in food and cosmetics with a high safety coefficient. The administration of 0.2% α-Ionone to mice fed with HFD led to a significant rise in the expression of thermogenic genes, such as PGC-1α, UCP-1, and Dio2, in their brown adipose tissue. It inhibits weight gain in DIO mice by activating TGR5 compared to normal HFD mice. It is suggested that α-Ionone has shown potential as a TGR5 agonist for preventing obesity [104]. To improve TGR5 agonists for treating metabolic diseases, Zhao et al. [105] developed a new series of 1-benzyl–1H-imidazole–5-carboxyamide derivatives. The most potent compounds, 19d and 19e, activated TGR5 better than the control drugs INT-777 and LCA while exhibiting good FXR selectivity. In vivo studies also showed significant hypoglycemic effects.

Nevertheless, systemic TGR5 activation can cause some unpleasant adverse effects including excessive gallbladder filling and insufficient emptying [106,107]. TGR5 agonists have also been associated with "bile-acid-related" itching and analgesia [108]. Animal studies have illustrated that structurally diverse TGR5 agonists reduce vascular tone and blood pressure, resulting in reflex tachycardia and inotropy [109].

10. Conclusions

In summary, TGR5 activation influences glucolipid metabolism, energy metabolism, inflammatory response, and endogenous signaling molecules expression through multiple signal transduction pathways. It has excellent potential effects in treating T2DM, obesity, atherosclerosis, and NAFLD, which is a key direction and hot area for drug research and development at home and abroad. However, the activity of TGR5 has been demonstrated in animal and cellular assays and needs to be further validated and observed in clinical trials. A comprehensive understanding of the potential mechanisms of TGR5 in metabolic syndrome can serve as a theoretical foundation and reference value for future TGR5-specific targeting in the treatment of metabolic syndrome.

Author Contributions: Conceptualization, Q.Z., L.Q. and X.G.; writing—original draft preparation, X.G.; writing—review and editing, L.Q. and D.W.; visualization, J.X. and Y.L.; funding acquisition, Y.H. All authors have read and agreed to the published version of the manuscript.

Funding: The study was funded by the National Nature Science Foundation of China (82260843, 82060649, 82160740); Zunyi City of China (ZSKHHZ [2021]290, QKHPTRC [2020]-16); and the Guizhou Engineering Research Center of Industrial Key-technology for Dendrobium Nobile (QJJ [2022]048, QJJ [2022]006).

Institutional Review Board Statement: Not applicable.

Informed Consent Statement: Not applicable.

Data Availability Statement: Not applicable.

Conflicts of Interest: The authors declare there are no competing interest.

Sample Availability: Not applicable.

References

1. Hu, C.; Jia, W. Multi-omics profiling: The way toward precision medicine in metabolic diseases. *J. Mol. Cell Biol.* **2021**, *13*, 576–593 [CrossRef] [PubMed]
2. Ambroselli, D.; Masciulli, F.; Romano, E.; Catanzaro, G.; Besharat, Z.M.; Massari, M.C.; Ferretti, E.; Migliaccio, S.; Izzo, L.; Ritieni, A.; et al. New Advances in Metabolic Syndrome, from Prevention to Treatment: The Role of Diet and Food. *Nutrients* **2023**, *15*, 640. [CrossRef] [PubMed]
3. Saklayen, M.G. The Global Epidemic of the Metabolic Syndrome. *Curr. Hypertens. Rep.* **2018**, *20*, 1–8. [CrossRef] [PubMed]
4. Vrdoljak, J.; Kumric, M.; Vilovic, M.; Martinovic, D.; Rogosic, V.; Borovac, J.A.; Kurir, T.T.; Bozic, J. Can Fasting Curb the Metabolic Syndrome Epidemic? *Nutrients* **2022**, *14*, 456. [CrossRef] [PubMed]
5. Chiang, J.Y.L. Bile Acid Metabolism and Signaling. *Compr. Physiol.* **2013**, *3*, 1191–1212. [CrossRef]
6. Duboc, H.; Taché, Y.; Hofmann, A.F. The bile acid TGR5 membrane receptor: From basic research to clinical application. *Dig. Liver Dis.* **2014**, *46*, 302–312. [CrossRef]
7. van Nierop, F.S.; Scheltema, M.J.; Eggink, H.M.; Pols, T.W.; Sonne, D.P.; Knop, F.K.; Soeters, M.R. Clinical relevance of the bile acid receptor TGR5 in metabolism. *Lancet Diabetes Endocrinol.* **2016**, *5*, 224–233. [CrossRef]
8. Ferrell, J.M.; Pathak, P.; Boehme, S.; Gilliland, T.; Chiang, J.Y.L. Deficiency of Both Farnesoid X Receptor and Takeda G Protein-Coupled Receptor 5 Exacerbated Liver Fibrosis in Mice. *Hepatology* **2019**, *70*, 955–970. [CrossRef]
9. Vassileva, G.; Hu, W.; Hoos, L.; Tetzloff, G.; Yang, S.; Liu, L.; Kang, L.; Davis, H.R.; Hedrick, J.A.; Lan, H.; et al. Gender-dependent effect of Gpbar1 genetic deletion on the metabolic profiles of diet-induced obese mice. *J. Endocrinol.* **2010**, *205*, 225–232. [CrossRef]
10. Holter, M.M.; Chirikjian, M.K.; Govani, V.N.; Cummings, B.P. TGR5 Signaling in Hepatic Metabolic Health. *Nutrients* **2020**, *12*, 2598. [CrossRef]
11. McGavigan, A.K.; Garibay, D.; Henseler, Z.M.; Chen, J.; Bettaieb, A.; Haj, F.G.; Ley, R.E.; Chouinard, M.L.; Cummings, B.P. TGR5 contributes to glucoregulatory improvements after vertical sleeve gastrectomy in mice. *Gut* **2015**, *66*, 226–234. [CrossRef] [PubMed]
12. Chevre, R.; Castaño, D.; Chua, T.; Corlianò, M.; Patankar, J.V.; Sng, L.; Sim, L.; Juin, T.L.; Carissimo, G.; Ng, L.F.P.; et al. Therapeutic modulation of the bile acid pool by *Cyp8b1* knockdown protects against nonalcoholic fatty liver disease in mice. *FASEB J.* **2018**, *32*, 3792–3802. [CrossRef] [PubMed]
13. Maczewsky, J.; Kaiser, J.; Gresch, A.; Gerst, F.; Düfer, M.; Krippeit-Drews, P.; Drews, G. TGR5 Activation Promotes Stimulus-Secretion Coupling of Pancreatic β-Cells via a PKA-Dependent Pathway. *Diabetes* **2018**, *68*, 324–336. [CrossRef] [PubMed]

4. Finn, P.D.; Rodriguez, D.; Kohler, J.; Jiang, Z.; Wan, S.; Blanco, E.; King, A.J.; Chen, T.; Bell, N.; Dragoli, D.; et al. Intestinal TGR5 agonism improves hepatic steatosis and insulin sensitivity in Western diet-fed mice. *Am. J. Physiol. Liver Physiol.* **2019**, *316*, G412–G424. [CrossRef]
5. Castellanos-Jankiewicz, A.; Guzmán-Quevedo, O.; Fénelon, V.S.; Zizzari, P.; Quarta, C.; Bellocchio, L.; Tailleux, A.; Charton, J.; Fernandois, D.; Henricsson, M.; et al. Hypothalamic bile acid-TGR5 signaling protects from obesity. *Cell Metab.* **2021**, *33*, 1483–1492.e10. [CrossRef]
6. Drigo, R.A.; Fonseca, T.L.; Werneck-De-Castro, J.P.S.; Bianco, A.C. Role of the type 2 iodothyronine deiodinase (D2) in the control of thyroid hormone signaling. *Biochim. Biophys. Acta* **2013**, *1830*, 3956–3964. [CrossRef]
7. Sasaki, T.; Watanabe, Y.; Kuboyama, A.; Oikawa, A.; Shimizu, M.; Yamauchi, Y.; Sato, R. Muscle-specific TGR5 overexpression improves glucose clearance in glucose-intolerant mice. *J. Biol. Chem.* **2021**, *296*, 100131. [CrossRef]
8. Kida, T.; Tsubosaka, Y.; Hori, M.; Ozaki, H.; Murata, T. Bile Acid Receptor TGR5 Agonism Induces NO Production and Reduces Monocyte Adhesion in Vascular Endothelial Cells. *Arter. Thromb. Vasc. Biol.* **2013**, *33*, 1663–1669. [CrossRef]
9. Duboc, H.; Aelion, H.; Rainteau, D.; Rajca, S.; Sokol, H.; Humbert, L.; Farabos, D.; Coffin, B.; Weber, S.; Porcher, R.; et al. Crosstalk between the hepatologist and the cardiologist: A future place for the lithocholic acid as a coronary atheroma risk factor? *Hepatology* **2012**, *56*, 2426. [CrossRef]
0. Tiwari, A.; Maiti, P. TGR5: An emerging bile acid G-protein-coupled receptor target for the potential treatment of metabolic disorders. *Drug Discov. Today* **2009**, *14*, 523–530. [CrossRef]
1. Sasaki, T.; Kuboyama, A.; Mita, M.; Murata, S.; Shimizu, M.; Inoue, J.; Mori, K.; Sato, R. The exercise-inducible bile acid receptor Tgr5 improves skeletal muscle function in mice. *J. Biol. Chem.* **2018**, *293*, 10322–10332. [CrossRef]
2. Hov, J.R.; Keitel, V.; Laerdahl, J.K.; Spomer, L.; Ellinghaus, E.; ElSharawy, A.; Melum, E.; Boberg, K.M.; Manke, T.; Balschun, T.; et al. Mutational Characterization of the Bile Acid Receptor TGR5 in Primary Sclerosing Cholangitis. *PLoS ONE* **2010**, *5*, e12403. [CrossRef]
3. Kuhre, R.E.; Wewer Albrechtsen, N.J.; Larsen, O.; Jepsen, S.L.; Balk-Møller, E.; Andersen, D.B.; Deacon, C.F.; Schoonjans, K.; Reimann, F.; Gribble, F.M.; et al. Bile acids are important direct and indirect regulators of the secretion of appetite- and metabolism-regulating hormones from the gut and pancreas. *Mol. Metab.* **2018**, *11*, 84–95. [CrossRef]
4. Kawamata, Y.; Fujii, R.; Hosoya, M.; Harada, M.; Yoshida, H.; Miwa, M.; Fukusumi, S.; Habata, Y.; Itoh, T.; Shintani, Y.; et al. A G Protein-coupled Receptor Responsive to Bile Acids. *J. Biol. Chem.* **2003**, *278*, 9435–9440. [CrossRef] [PubMed]
5. Qi, Y.C.; Li, P.F. Progress in the study of TGR5 receptor-mediated signaling pathway. *Chin. J. Pharmacol.* **2020**, *36*, 153–157.
6. Herbert, M.R.; Siegel, D.L.; Staszewski, L.; Cayanan, C.; Banerjee, U.; Dhamija, S.; Anderson, J.; Fan, A.; Wang, L.; Rix, P.; et al. Synthesis and SAR of 2-aryl-3-aminomethylquinolines as agonists of the bile acid receptor TGR5. *Bioorg. Med. Chem. Lett.* **2010**, *20*, 5718–5721. [CrossRef] [PubMed]
7. Perino, A.; Demagny, H.; Velazquez-Villegas, L.A.; Schoonjans, K. Molecular physiology of bile acid signaling in health, disease, and aging. *Physiol. Rev.* **2021**, *101*, 683–731. [CrossRef]
8. Wang, X.X.; Xie, C.; Libby, A.E.; Ranjit, S.; Levi, J.; Myakala, K.; Bhasin, K.; Jones, B.A.; Orlicky, D.J.; Takahashi, S.; et al. The role of FXR and TGR5 in reversing and preventing progression of Western diet–induced hepatic steatosis, inflammation, and fibrosis in mice. *J. Biol. Chem.* **2022**, *298*, 102530. [CrossRef] [PubMed]
9. Li, T.; Francl, J.M.; Boehme, S.; Ochoa, A.; Zhang, Y.; Klaassen, C.D.; Erickson, S.K.; Chiang, J.Y. Glucose and insulin induction of bile acid synthesis: Mechanisms and implication in diabetes and obesity. *J Biol Chem.* **2012**, *287*, 1861–1873. [CrossRef] [PubMed]
0. Xie, G.; Jiang, R.; Wang, X.; Liu, P.; Zhao, A.; Wu, Y.; Huang, F.; Liu, Z.; Rajani, C.; Zheng, X.; et al. Conjugated secondary 12α-hydroxylated bile acids promote liver fibrogenesis. *Ebiomedicine* **2021**, *66*, 103290. [CrossRef]
1. Ticho, A.L.; Malhotra, P.; Dudeja, P.K.; Gill, R.K.; Alrefai, W.A. Intestinal Absorption of Bile Acids in Health and Disease. *Compr. Physiol.* **2019**, *10*, 21–56. [CrossRef]
2. Pathak, P.; Liu, H.; Boehme, S.; Xie, C.; Krausz, K.W.; Gonzalez, F.; Chiang, J.Y.L. Farnesoid X receptor induces Takeda G-protein receptor 5 cross-talk to regulate bile acid synthesis and hepatic metabolism. *J. Biol. Chem.* **2017**, *292*, 11055–11069. [CrossRef]
3. Xie, C.; Huang, W.; Young, R.L.; Jones, K.L.; Horowitz, M.; Rayner, C.K.; Wu, T. Role of Bile Acids in the Regulation of Food Intake, and Their Dysregulation in Metabolic Disease. *Nutrients* **2021**, *13*, 1104. [CrossRef]
4. Donepudi, A.C.; Boehme, S.; Li, F.; Chiang, J.Y. G-protein-coupled bile acid receptor plays a key role in bile acid metabolism and fasting-induced hepatic steatosis in mice. *Hepatology* **2017**, *65*, 813–827. [CrossRef]
5. Gillard, J.; Clerbaux, L.-A.; Nachit, M.; Sempoux, C.; Staels, B.; Bindels, L.B.; Tailleux, A.; Leclercq, I.A. Bile acids contribute to the development of non-alcoholic steatohepatitis in mice. *JHEP Rep.* **2022**, *4*, 100387. [CrossRef]
6. Qi, Y.; Duan, G.; Wei, D.; Zhao, C.; Ma, Y. The Bile Acid Membrane Receptor TGR5 in Cancer: Friend or Foe? *Molecules* **2022**, *27*, 5292. [CrossRef]
7. Scheen, A.J. Dual GIP/GLP-1 receptor agonists: New advances for treating type-2 diabetes. *Ann. D'endocrinologie* **2023**, *84*, 316–321. [CrossRef]
8. Ballantyne, G.H.; Longo, W.E.; Savoca, P.E.; Adrian, T.E.; Vukasin, A.P.; Bilchik, A.J.; Sussman, J.; Modlin, I.M. Deoxycholate-stimulated release of peptide YY from the isolated perfused rabbit left colon. *Am. J. Physiol. Liver Physiol.* **1989**, *257*, G715–G724. [CrossRef]
9. Saito, M.; Matsushita, M.; Yoneshiro, T.; Okamatsu-Ogura, Y. Brown Adipose Tissue, Diet-Induced Thermogenesis, and Thermogenic Food Ingredients: From Mice to Men. *Front. Endocrinol.* **2020**, *11*, 222. [CrossRef]

40. Svensson, P.-A.; Olsson, M.; Andersson-Assarsson, J.C.; Taube, M.; Pereira, M.J.; Froguel, P.; Jacobson, P. The TGR5 gene i expressed in human subcutaneous adipose tissue and is associated with obesity, weight loss and resting metabolic rate. *Biochen Biophys. Res. Commun.* **2013**, *433*, 563–566. [CrossRef]
41. Depommier, C.; Everard, A.; Druart, C.; Plovier, H.; Van Hul, M.; Vieira-Silva, S.; Falony, G.; Raes, J.; Maiter, D Delzenne, N.M.; et al. Supplementation with Akkermansia muciniphila in overweight and obese human volunteers: proof-of-concept exploratory study. *Nat. Med.* **2019**, *25*, 1096–1103. [CrossRef]
42. Wu, Q.; Liang, X.; Wang, K.; Lin, J.; Wang, X.; Wang, P.; Zhang, Y.; Nie, Q.; Liu, H.; Zhang, Z.; et al. Intestinal hypoxia-inducibl factor 2α regulates lactate levels to shape the gut microbiome and alter thermogenesis. *Cell Metab.* **2021**, *33*, 1988–2003.e [CrossRef] [PubMed]
43. Shen, H.; Ding, L.; Baig, M.; Tian, J.; Wang, Y.; Huang, W. Improving glucose and lipids metabolism: Drug development based o bile acid related targets. *Cell Stress* **2021**, *5*, 1–18. [CrossRef] [PubMed]
44. Bouzas, C.; Pastor, R.; Garcia, S.; Monserrat-Mesquida, M.; Martínez-González, M.; Salas-Salvadó, J.; Corella, D.; Goday, A Martínez, J.A.; Alonso-Gómez, M.; et al. Comparative effects of glucagon-like peptide-1 receptors agonists, 4-dipeptidyl peptidas inhibitors, and metformin on metabolic syndrome. *Biomed. Pharmacother.* **2023**, *161*, 114561. [CrossRef] [PubMed]
45. Christiansen, C.B.; Trammell, S.A.J.; Wewer Albrechtsen, N.J.; Schoonjans, K.; Albrechtsen, R.; Gillum, M.; Kuhre, R.E Holst, J.J. Bile acids drive colonic secretion of glucagon-like-peptide 1 and peptide-YY in rodents. *Am. J. Physiol. Live Physiol.* **2019**, *316*, G574–G584. [CrossRef]
46. Morimoto, K.; Watanabe, M.; Sugizaki, T.; Irie, J.-I.; Itoh, H. Intestinal Bile Acid Composition Modulates Prohormone Convertas 1/3 (PC1/3) Expression and Consequent GLP-1 Production in Male Mice. *Endocrinology* **2016**, *157*, 1071–1081. [CrossRef]
47. Makki, K.; Brolin, H.; Petersen, N.; Henricsson, M.; Christensen, D.P.; Khan, M.T.; Wahlström, A.; Bergh, P.-O.; Tremaroli, V Schoonjans, K.; et al. 6α-hydroxylated bile acids mediate TGR5 signalling to improve glucose metabolism upon dietary fibe supplementation in mice. *Gut* **2022**, *72*, 314–324. [CrossRef]
48. Zhai, H.; Li, Z.; Peng, M.; Huang, Z.; Qin, T.; Chen, L.; Li, H.; Zhang, H.; Zhang, W.; Xu, G. Takeda G Protein-Coupled Recepto 5-Mechanistic Target of Rapamycin Complex 1 Signaling Contributes to the Increment of Glucagon-Like Peptide-1 Productio after Roux-en-Y Gastric Bypass. *Ebiomedicine* **2018**, *32*, 201–214. [CrossRef]
49. Tian, F.; Xu, W.; Chen, L.; Chen, T.; Feng, X.; Chen, J.; Wei, D.; Huang, Q. Ginsenoside compound K increases glucagon-lik peptide-1 release and L-cell abundance in db/db mice through TGR5/YAP signaling. *Int. Immunopharmacol.* **2022**, *113*, 10940 [CrossRef]
50. Sorrentino, G.; Perino, A.; Yildiz, E.; El Alam, G.; Sleiman, M.B.; Gioiello, A.; Pellicciari, R.; Schoonjans, K. Bile Acids Signal vi TGR5 to Activate Intestinal Stem Cells and Epithelial Regeneration. *Gastroenterology* **2020**, *159*, 956–968.e8. [CrossRef]
51. Chen, M.-J.; Liu, C.; Wan, Y.; Yang, L.; Jiang, S.; Qian, D.-W.; Duan, J.-A. Enterohepatic circulation of bile acids and their emergin roles on glucolipid metabolism. *Steroids* **2020**, *165*, 108757. [CrossRef]
52. Chiang, J.Y.; Pathak, P.; Liu, H.; Donepudi, A.; Ferrell, J.; Boehme, S. Intestinal Farnesoid X Receptor and Takeda G Protein Coupl Receptor 5 Signaling in Metabolic Regulation. *Dig. Dis.* **2017**, *35*, 241–245. [CrossRef]
53. Zhai, Z.; Niu, K.; Liu, H.; Lin, C.; Tu, Y.; Liu, Y.; Cai, L.; Ouyang, K.; Liu, J. Policosanol alleviates hepatic lipid accumulatio by regulating bile acids metabolism in C57BL6/mice through AMPK–FXR–TGR5 cross-talk. *J. Food Sci.* **2021**, *86*, 5466–547 [CrossRef]
54. Zhu, H.; Wang, K.; Chen, S.; Kang, J.; Guo, N.; Chen, H.; Liu, J.; Wu, Y.; He, P.; Tu, Y.; et al. Saponins from *Camellia sinensis* Seed Stimulate GIP Secretion in Mice and STC-1 Cells via SGLT1 and TGR. *Nutrients* **2022**, *14*, 3413. [CrossRef]
55. Zhang, Y.; Huang, S.; Li, P.; Chen, Q.; Li, Y.; Zhou, Y.; Wang, L.; Kang, M.; Zhang, B.; Yang, B.; et al. Pancreatic cancer-derive exosomes suppress the production of GIP and GLP-1 from STC-1 cells in vitro by down-regulating the PCSK1/3. *Cancer Let* **2018**, *431*, 190–200. [CrossRef]
56. Wu, Y.; He, H.; Cheng, Z.; Bai, Y.; Ma, X. The Role of Neuropeptide Y and Peptide YY in the Development of Obesity via Gut-brai Axis. *Curr. Protein Pept. Sci.* **2019**, *20*, 750–758. [CrossRef]
57. Yang, H.; Liu, H.; Jiao, Y.W.; Qian, J. Roux-en-Y Gastrointestinal Bypass Promotes Activation of TGR5 and Peptide YY. *Endoc Metab. Immune Disord. Drug Targets* **2020**, *20*, 1262–1267. [CrossRef]
58. Guida, C.; Stephen, S.; Watson, M.; Dempster, N.; Larraufie, P.; Marjot, T.; Cargill, T.; Rickers, L.; Pavlides, M.; Tomlinson, J.; et a PYY plays a key role in the resolution of diabetes following bariatric surgery in humans. *Ebiomedicine* **2019**, *40*, 67–76. [CrossRef
59. Bala, V.; Rajagopal, S.; Kumar, D.P.; Nalli, A.D.; Mahavadi, S.; Sanyal, A.J.; Grider, J.R.; Murthy, K.S. Release of GLP-1 and PYY i response to the activation of G protein-coupled bile acid receptor TGR5 is mediated by Epac/PLC-Îµ pathway and modulated b endogenous H2S. *Front. Physiol.* **2014**, *5*, 420. [CrossRef]
60. Huang, S.; Ma, S.; Ning, M.; Yang, W.; Ye, Y.; Zhang, L.; Shen, J.; Leng, Y. TGR5 agonist ameliorates insulin resistance in skeleta muscles and improves glucose homeostasis in diabetic mice. *Metabolism* **2019**, *99*, 45–56. [CrossRef]
61. Sun, L.; Li, F.; Tan, W.; Zhao, W.; Li, Y.; Zhu, X.; Gao, P.; Shu, G.; Wang, S.; Jiang, Q.; et al. Lithocholic acid promotes skeleta muscle regeneration through the TGR5 receptor. *Acta Biochim. Biophys. Sin.* **2023**, *55*, 51–61. [CrossRef]
62. Sahakyan, G.; Vejux, A.; Sahakyan, N. The Role of Oxidative Stress-Mediated Inflammation in the Development of T2DM-Induce Diabetic Nephropathy: Possible Preventive Action of Tannins and Other Oligomeric Polyphenols. *Molecules* **2022**, *27*, 903 [CrossRef]

63. Xiao, H.; Sun, X.; Liu, R.; Chen, Z.; Lin, Z.; Yang, Y.; Zhang, M.; Liu, P.; Quan, S.; Huang, H. Gentiopicroside activates the bile acid receptor Gpbar1 (TGR5) to repress NF-kappaB pathway and ameliorate diabetic nephropathy. *Pharmacol. Res.* **2019**, *151*, 104559. [CrossRef]
64. Perino, A.; Pols, T.W.H.; Nomura, M.; Stein, S.; Pellicciari, R.; Schoonjans, K. TGR5 reduces macrophage migration through mTOR-induced C/EBPβ differential translation. *J. Clin. Investig.* **2014**, *124*, 5424–5436. [CrossRef]
65. Guo, C.; Xie, S.; Chi, Z.; Zhang, J.; Liu, Y.; Zhang, L.; Zheng, M.; Zhang, X.; Xia, D.; Ke, Y.; et al. Bile Acids Control Inflammation and Metabolic Disorder through Inhibition of NLRP3 Inflammasome. *Immunity* **2016**, *45*, 802–816. [CrossRef]
66. Watanabe, M.; Houten, S.M.; Mataki, C.; Christoffolete, M.A.; Kim, B.W.; Sato, H.; Messaddeq, N.; Harney, J.W.; Ezaki, O.; Kodama, T.; et al. Bile acids induce energy expenditure by promoting intracellular thyroid hormone activation. *Nature* **2006**, *439*, 484–489. [CrossRef]
67. de Jesus, L.A.; Carvalho, S.D.; Ribeiro, M.O.; Schneider, M.; Kim, S.-W.; Harney, J.W.; Larsen, P.R.; Bianco, A.C. The type 2 iodothyronine deiodinase is essential for adaptive thermogenesis in brown adipose tissue. *J. Clin. Investig.* **2001**, *108*, 1379–1385. [CrossRef]
68. Gereben, B.; Zavacki, A.M.; Ribich, S.; Kim, B.W.; Huang, S.A.; Simonides, W.S.; Zeld, A.; Bianco, A.C. Cellular and Molecular Basis of Deiodinase-Regulated Thyroid Hormone Signaling. *Endocr. Rev.* **2008**, *29*, 898–938. [CrossRef]
69. Fonseca, T.L.; Russo, S.C.; Luongo, C.; Salvatore, D.; Bianco, A.C. Inactivation of Type 3 Deiodinase Results in Life-long Changes in the Brown Adipose Tissue Transcriptome in the Male Mouse. *Endocrinology* **2022**, *5*, 1–10. [CrossRef]
70. Velazquez-Villegas, L.A.; Perino, A.; Lemos, V.; Zietak, M.; Nomura, M.; Pols, T.W.H.; Schoonjans, K. TGR5 signalling promotes mitochondrial fission and beige remodelling of white adipose tissue. *Nat. Commun.* **2018**, *9*, 1–13. [CrossRef]
71. Krauss, S.; Zhang, C.-Y.; Lowell, B.B. The mitochondrial uncoupling-protein homologues. *Nat. Rev. Mol. Cell Biol.* **2005**, *6*, 248–261. [CrossRef]
72. Jeong, J.H.; Cho, S.; Pak, Y.K. Sterol-independent repression of low density lipoprotein receptor promoter by peroxisome proliferator activated receptor γ coactivator-1α (PGC-1α). *Exp. Mol. Med.* **2009**, *41*, 406–416. [CrossRef]
73. Fan, M.; Wang, Y.; Jin, L.; Fang, Z.; Peng, J.; Tu, J.; Liu, Y.; Zhang, E.; Xu, S.; Liu, X.; et al. Bile Acid–Mediated Activation of Brown Fat Protects From Alcohol-Induced Steatosis and Liver Injury in Mice. *Cell. Mol. Gastroenterol. Hepatol.* **2021**, *13*, 809–826. [CrossRef]
74. Lund, M.L.; Egerod, K.L.; Engelstoft, M.S.; Dmytriyeva, O.; Theodorsson, E.; Patel, B.A.; Schwartz, T.W. Enterochromaffin 5-HT cells—A major target for GLP-1 and gut microbial metabolites. *Mol. Metab.* **2018**, *11*, 70–83. [CrossRef]
75. Zhan, Y.; Wen, Y.; Zhang, L.-L.; Shen, X.-L.; Chen, X.-H.; Wu, X.-H.; Tang, X.-G. Paeoniflorin Improved Constipation in the Loperamide-Induced Rat Model via TGR5/TRPA1 Signaling-Mediated 5-Hydroxytryptamine Secretion. *Evid.-Based Complement Altern. Med.* **2021**, *2021*, 6076293. [CrossRef]
76. Wu, X.; Li, J.-Y.; Lee, A.; Lu, Y.-X.; Zhou, S.-Y.; Owyang, C. Satiety induced by bile acids is mediated via vagal afferent pathways. *J. Clin. Investig.* **2020**, *5*, e132400. [CrossRef] [PubMed]
77. Chen, C.; Zhou, Y.; Huang, R.; Wang, M.; Li, Y.; Li, J. Bile acids regulate anorexigenic neuropeptide through p-STAT3-SOCS3 signaling in mouse hypothalamic cells. *Nan Fang Yi Ke Da Xue Xue Bao* **2020**, *40*, 1001–1007. [PubMed]
78. Perino, A.; Velázquez-Villegas, L.A.; Bresciani, N.; Sun, Y.; Huang, Q.; Fénelon, V.S.; Castellanos-Jankiewicz, A.; Zizzari, P.; Bruschetta, G.; Jin, S.; et al. Central anorexigenic actions of bile acids are mediated by TGR. *Nat. Metab.* **2021**, *3*, 595–603. [CrossRef] [PubMed]
79. Pols, T.W.; Nomura, M.; Harach, T.; Sasso, G.L.; Oosterveer, M.H.; Thomas, C.; Rizzo, G.; Gioiello, A.; Adorini, L.; Pellicciari, R.; et al. TGR5 Activation Inhibits Atherosclerosis by Reducing Macrophage Inflammation and Lipid Loading. *Cell Metab.* **2011**, *14*, 747–757. [CrossRef]
80. Guizoni, D.M.; Vettorazzi, J.F.; Carneiro, E.M.; Davel, A.P. Modulation of endothelium-derived nitric oxide production and activity by taurine and taurine-conjugated bile acids. *Nitric Oxide* **2019**, *94*, 48–53. [CrossRef]
81. Hashimoto, E.; Taniai, M.; Tokushige, K. Characteristics and diagnosis of NAFLD/NASH. *J. Gastroenterol. Hepatol.* **2013**, *28*, 64–70. [CrossRef] [PubMed]
82. Lambert, J.E.; Ramos-Roman, M.A.; Browning, J.D.; Parks, E.J. Increased De Novo Lipogenesis Is a Distinct Characteristic of Individuals With Nonalcoholic Fatty Liver Disease. *Gastroenterology* **2014**, *146*, 726–735. [CrossRef]
83. Cegla, J. Liraglutide safety and efficacy in patients with non-alcoholic steatohepatitis (LEAN): A multicentre, double-blind, randomized, placebo-controlled phase 2 study. *Ann. Clin. Biochem. Int. J. Biochem. Lab. Med.* **2016**, *53*, 518. [CrossRef] [PubMed]
84. El-Jamal, N.; Erdual, E.; Neunlist, M.; Koriche, D.; Dubuquoy, C.; Maggiotto, F.; Chevalier, J.; Berrebi, D.; Dubuquoy, L.; Boulanger, E.; et al. Glugacon-like peptide-2: Broad receptor expression, limited therapeutic effect on intestinal inflammation and novel role in liver regeneration. *Am. J. Physiol. Gastrointest. Liver Physiol.* **2014**, *307*, G274–G285. [CrossRef]
85. Ivory, C.P.A.; Wallace, L.E.; McCafferty, D.-M.; Sigalet, D.L. Interleukin-10-independent anti-inflammatory actions of glucagon-like peptide. *Am. J. Physiol. Liver Physiol.* **2008**, *295*, G1202–G1210. [CrossRef]
86. Baffy, G. Kupffer cells in non-alcoholic fatty liver disease: The emerging view. *J. Hepatol.* **2009**, *51*, 212–223. [CrossRef]
87. Mobraten, K.; Haugbro, T.; Karlstrom, E.; Kleiveland, C.R.; Lea, T. Activation of the bile acid receptor TGR5 enhances LPS-induced inflammatory responses in a human monocytic cell line. *J. Recept. Signal Transduct.* **2015**, *35*, 402–409. [CrossRef]
88. Yang, H.; Luo, F.; Wei, Y.; Jiao, Y.; Qian, J.; Chen, S.; Gong, Y.; Tang, L. TGR5 protects against cholestatic liver disease via suppressing the NF-κB pathway and activating the Nrf2/HO-1 pathway. *Ann. Transl. Med.* **2021**, *9*, 1158. [CrossRef]

89. Shi, Y.; Su, W.; Zhang, L.; Shi, C.; Zhou, J.; Wang, P.; Wang, H.; Shi, X.; Wei, S.; Wang, Q.; et al. TGR5 Regulates Macrophage Inflammation in Nonalcoholic Steatohepatitis by Modulating NLRP3 Inflammasome Activation. *Front. Immunol.* **2020**, *11*, 609060. [CrossRef]
90. Zhao, L.; Zhang, H.; Liu, X.; Xue, S.; Chen, D.; Zou, J.; Jiang, H. TGR5 deficiency activates antitumor immunity in non-small cell lung cancer via restraining M2 macrophage polarization. *Acta Pharm. Sin. B* **2022**, *12*, 787–800. [CrossRef]
91. Rao, J.; Yang, C.; Yang, S.; Lu, H.; Hu, Y.; Lu, L.; Cheng, F.; Wang, X. Deficiency of TGR5 exacerbates immune-mediated cholestatic hepatic injury by stabilizing the β-catenin destruction complex. *Int. Immunol.* **2020**, *32*, 321–334. [CrossRef] [PubMed]
92. Shao, J.; Ge, T.; Tang, C.; Wang, G.; Pang, L.; Chen, Z. Synergistic anti-inflammatory effect of gut microbiota and lithocholic acid on liver fibrosis. *Inflamm. Res.* **2022**, *71*, 1389–1401. [CrossRef]
93. Carino, A.; Marchianò, S.; Biagioli, M.; Bucci, M.; Vellecco, V.; Brancaleone, V.; Fiorucci, C.; Zampella, A.; Monti, M.C.; Distrutti, E.; et al. Agonism for the bile acid receptor GPBAR1 reverses liver and vascular damage in a mouse model of steatohepatitis. *FASEB J.* **2019**, *33*, 2809–2822. [CrossRef]
94. Häussinger, D.; Keitel, V. Role of TGR5 (GPBAR1) in Liver Disease. *Semin. Liver Dis.* **2018**, *38*, 333–339. [CrossRef]
95. Yang, C.; Wan, M.; Xu, D.; Pan, D.; Xia, H.; Yang, L.; Sun, G. Flaxseed Powder Attenuates Non-Alcoholic Steatohepatitis via Modulation of Gut Microbiota and Bile Acid Metabolism through Gut–Liver Axis. *Int. J. Mol. Sci.* **2021**, *22*, 10858. [CrossRef]
96. Yang, J.-Y.; Lee, Y.-S.; Kim, Y.; Lee, S.-H.; Ryu, S.; Fukuda, S.; Hase, K.; Yang, C.-S.; Lim, H.; Kim, M.-S.; et al. Gut commensal Bacteroides acidifaciens prevents obesity and improves insulin sensitivity in mice. *Mucosal Immunol.* **2017**, *10*, 104–116. [CrossRef] [PubMed]
97. Suraweera, D.; Rahal, H.; Jimenez, M.; Viramontes, M.; Choi, G.; Saab, S. Treatment of primary biliary cholangitis ursodeoxycholic acid non-responders: A systematic review. *Liver Int.* **2017**, *37*, 1877–1886. [CrossRef] [PubMed]
98. Murakami, M.; Une, N.; Nishizawa, M.; Suzuki, S.; Ito, H.; Horiuchi, T. Incretin secretion stimulated by ursodeoxycholic acid in healthy subjects. *Springerplus* **2013**, *2*, 20. [CrossRef] [PubMed]
99. Pellicciari, R.; Gioiello, A.; Macchiarulo, A.; Thomas, C.; Rosatelli, E.; Natalini, B.; Sardella, R.; Pruzanski, M.; Roda, A.; Pastorini, E.; et al. Discovery of 6α-Ethyl-23(*S*)-methylcholic Acid (*S*-EMCA, INT-777) as a Potent and Selective Agonist for the TGR5 Receptor, a Novel Target for Diabesity. *J. Med. Chem.* **2009**, *52*, 7958–7961. [CrossRef]
100. Thomas, C.; Gioiello, A.; Noriega, L.; Strehle, A.; Oury, J.; Rizzo, G.; Macchiarulo, A.; Yamamoto, H.; Mataki, C.; Pruzanski, M.; et al. TGR5-Mediated Bile Acid Sensing Controls Glucose Homeostasis. *Cell Metab.* **2009**, *10*, 167–177. [CrossRef]
101. Liu, X.; Guan, J.; Wu, Z.; Xu, L.; Sun, C. The TGR5 Agonist INT-777 Promotes Peripheral Nerve Regeneration by Activating cAMP-dependent Protein Kinase A in Schwann Cells. *Mol. Neurobiol.* **2023**, *60*, 1901–1913. [CrossRef]
102. Huang, R.; Gao, Y.; Chen, J.; Duan, Q.; He, P.; Zhang, J.; Huang, H.; Zhang, Q.; Ma, G.; Zhang, Y.; et al. TGR5 Agonist INT-777 Alleviates Inflammatory Neurodegeneration in Parkinson's Disease Mouse Model by Modulating Mitochondrial Dynamics in Microglia. *Neuroscience* **2022**, *490*, 100–119. [CrossRef]
103. de Oliveira, M.C.; Gilglioni, E.H.; de Boer, B.A.; Runge, J.H.; de Waart, D.R.; Salgueiro, C.L.; Ishii-Iwamoto, E.L.; Elferink, R.P.O.; Gaemers, I.C. Bile acid receptor agonists INT747 and INT777 decrease oestrogen deficiency-related postmenopausal obesity and hepatic steatosis in mice. *Biochim. Biophys. Acta* **2016**, *1862*, 2054–2062. [CrossRef]
104. Sasaki, T.; Ikari, N.; Hashimoto, S.; Sato, R. Identification of α-ionone, nootkatone, and their derivatives as TGR5 agonists. *Biochem. Biophys. Res. Commun.* **2023**, *653*, 147–152. [CrossRef] [PubMed]
105. Zhao, S.; Li, X.; Wang, L.; Peng, W.; Ye, W.; Li, W.; Wang, Y.-D.; Chen, W.-D. Design, synthesis and evaluation of 1-benzyl-1H-imidazole-5-carboxamide derivatives as potent TGR5 agonists. *Bioorganic Med. Chem.* **2020**, *32*, 115972. [CrossRef] [PubMed]
106. Vassileva, G.; Golovko, A.; Markowitz, L.; Abbondanzo, S.J.; Zeng, M.; Yang, S.; Hoos, L.; Tetzloff, G.; Levitan, D.; Murgolo, N.J.; et al. Targeted deletion of Gpbar1 protects mice from cholesterol gallstone formation. *Biochem. J.* **2006**, *398*, 423–430. [CrossRef]
107. Piotrowski, D.W.; Futatsugi, K.; Warmus, J.S.; Orr, S.T.M.; Freeman-Cook, K.D.; Londregan, A.T.; Wei, L.; Jennings, S.M.; Herr, M.; Coffey, S.B.; et al. Identification of Tetrahydropyrido[4,3-*d*]pyrimidine Amides as a New Class of Orally Bioavailable TGR5 Agonists. *ACS Med. Chem. Lett.* **2012**, *4*, 63–68. [CrossRef]
108. Alemi, F.; Kwon, E.; Poole, D.P.; Lieu, T.; Lyo, V.; Cattaruzza, F.; Cevikbas, F.; Steinhoff, M.; Nassini, R.; Materazzi, S.; et al. The TGR5 receptor mediates bile acid–induced itch and analgesia. *J. Clin. Investig.* **2013**, *123*, 1513–1530. [CrossRef] [PubMed]
109. Fryer, R.M.; Ng, K.J.; Mazurek, S.G.N.; Patnaude, L.; Skow, D.J.; Muthukumarana, A.; Gilpin, K.E.; Dinallo, R.M.; Kuzmich, D.; Lord, J.; et al. G Protein–Coupled Bile Acid Receptor 1 Stimulation Mediates Arterial Vasodilation through a K_{Ca}1.1 (BK_{Ca})–Dependent Mechanism. *Experiment* **2014**, *348*, 421–431. [CrossRef] [PubMed]

metabolites

MDPI

Article

Mucosal Metabolomic Signatures in Chronic Colitis: Novel Insights into the Pathophysiology of Inflammatory Bowel Disease

Nathan Calzadilla [1], Aisha Qazi [2], Anchal Sharma [2], Kai Mongan [2], Shane Comiskey [2], Jahnavi Manne [2], Alvin G. Youkhana [2], Sonam Khanna [2], Seema Saksena [2,3], Pradeep K. Dudeja [2,3], Waddah A. Alrefai [2,3] and Ravinder K. Gill [2,3,*]

1 Department of Biomedical Engineering, University of Illinois Chicago, Chicago, IL 60612, USA; ncalza3@uic.edu

2 Division of Gastroenterology & Hepatology, University of Illinois Chicago, Chicago, IL 60612, USA; aqazi7@uic.edu (A.Q.); anchals@uic.edu (A.S.); kmongan@neomed.edu (K.M.); shanec@uic.edu (S.C.); jmanne2@uic.edu (J.M.); ayoukh4@uic.edu (A.G.Y.); skhann5@uic.edu (S.K.); saksena@uic.edu (S.S.); pkdudeja@uic.edu (P.K.D.); walrefai@uic.edu (W.A.A.)

3 Jesse Brown VA Medical Center, Chicago, IL 60612, USA

* Correspondence: rgill@uic.edu; Tel.: +1-(312)-569-7463

Abstract: Inflammatory bowel diseases (IBD) involve complex interactions among genetic factors, aberrant immune activation, and gut microbial dysbiosis. While metabolomic studies have focused on feces and serum, fewer investigations have examined the intestinal mucosa despite its crucial role in metabolite absorption and transport. The goals of this study were twofold: to test the hypothesis that gut microbial dysbiosis from chronic intestinal inflammation leads to mucosal metabolic alterations suitable for therapeutic targeting, and to address gaps in metabolomic studies of intestinal inflammation that have overlooked the mucosal metabolome. The chronic DSS colitis was induced for five weeks in 7–9-week-old wild-type C57BL/6J male mice followed by microbial profiling with targeted 16srRNA sequencing service. Mucosal metabolite measurements were performed by Metabolon (Morrisville, NC). The data were analyzed using the bioinformatic tools Pathview, MetOrigin, and Metaboanalyst. The novel findings demonstrated increases in several host- and microbe-derived purine, pyrimidine, endocannabinoid, and ceramide metabolites in colitis. Origin analysis revealed that microbial-related tryptophan metabolites kynurenine, anthranilate, 5-hydroxyindoleacetate, and C-glycosyltryptophan were significantly increased in colon mucosa during chronic inflammation and strongly correlated with disease activity. These findings offer new insights into the pathophysiology of IBD and provide novel potential targets for microbial-based therapeutics.

Keywords: chronic colitis; mucosal inflammation; host–microbe interactions; mucosal metabolomics

1. Introduction

Inflammatory bowel diseases (IBD) are chronic, debilitating illnesses with two major subtypes: ulcerative colitis and Crohn's disease. Between 1990 and 2017, the number of affected individuals rose from 3.7 million to 6.8 million, with the highest prevalence rates in North America [1]. In addition to diarrhea, abdominal pain, ulcers, and rectal bleeding, IBD carries an increased risk of colon cancer along with several extraintestinal manifestations [2]. Ulcerative colitis is characterized by mucosal inflammation localized to the colon, while Crohn's disease is characterized by full thickness inflammation anywhere along the gastrointestinal tract with skip lesions [3]. While there are many therapeutic options, treatment is complex, not curative, and carries significant side effects [4]. Moreover, patients with refractory disease often require surgery and possibly parenteral nutrition. Further complicating the clinical picture are the various factors in the pathophysiology of

IBD, including host genetic factors, intestinal barrier disruption, immune dysregulation, gut microbial dysbiosis, and altered metabolite pools [5]. This has led to several groups utilizing metabolomic and integrated omic tools to better understand the pathophysiology and progression of IBD [6–15].

For instance, meta-analysis of gene expression data in IBD patients revealed that ileal and colonic disease have distinct molecular patterns that are not adequately considered with clinical diagnosis and management [16]. These approaches have also been successfully applied to predict patient responses to anti-integrin therapy for the implementation of more efficacious combinatorial treatments [17]. Several studies have also performed microbiota profiling and metabolomics in the serum and feces of IBD patients to detect novel disease signatures [18,19]. Furthermore, human studies provide tremendous value in their translational potential, but due to the myriad of confounding variables in clinical studies, it is much more difficult to reach mechanistic conclusions. Therefore, animal models of IBD are often used to mimic human disease and supplement human studies [20]. Additionally, animal models allow for more uniformity and the ability to capture a disease process at a specific timepoint. For instance, one such study performed untargeted mucosal and serum metabolomics in a model of acute DSS colitis and found that alterations in tryptophan metabolism affected AhR-mediated IL-10 signaling [21]. Moreover, these high throughput approaches have also been used to identify novel drivers of intestinal inflammation such as proteins involved in the actin cytoskeleton and beneficial dietary compounds from human studies with further investigation using animal models to gain mechanistic insight [22,23]. However, there is a need for such studies in additional models that are more reflective of the chronically relapsing and remitting nature of IBD.

Here we performed untargeted colonic mucosal metabolomic profiling in a murine model of chronic DSS colitis that resembles IBD. The objectives of this study were to understand gut microbial changes during chronic colitis, identify differentially altered microbial metabolites, and uncover altered pathways during chronic intestinal inflammation for future study. This study was driven by the hypothesis that chronic colonic inflammation drives unique changes in the mucosal metabolome reflective of IBD due to inflammation-induced gut microbial dysbiosis and disturbed intestinal architecture.

2. Materials and Methods

2.1. Animals

All animal studies were performed in accordance with institutional guidelines and regulations and approved by the Animal Care Committees of the University of Illinois at Chicago and the Jesse Brown Veterans Affairs Medical Center (Approval Code: ACC 21-078). Wild-type C57/BL6 mice were purchased from Jackson Laboratory (Bar Harbour, ME, USA). Autoclaved polypropylene cages with corncob bedding were used to house the mice. The mice were provided free access to food and water (Teklad Irradiated LM-485 Mouse/Rat Diet 7912, Envigo, Indianapolis, IN, USA) under a 12 h light/dark cycle. Male mice aged 7–9 weeks were provided either water alone (control group) or water supplemented with 2.5% *w*/*v* dextran sulfate sodium (DSS) (colitis group) for 5 weeks to induce chronic colitis. DSS was administered to mice for three cycles (one week/cycle), alternating with water only between the cycles (DSS was administered during weeks 1, 3, and 5, while water was administered during weeks 2 and 4) [24].

2.2. Metabolomic Analysis

Samples were processed and analyzed by Metabolon Inc. (Durham, NC, USA) as previously described [25]. Additional details are provided below.

2.2.1. Sample Preparation

Hamilton Company's automated MicroLab STAR® system was used to prepare samples. For quality control purposes, several recovery standards were added prior to the first step in the extraction process. Proteins were precipitated with methanol for 2 min with vigorous shaking, followed by centrifugation for 5 min at 700× *g* to dissociate small molecules bound to protein or trapped in the precipitated protein matrix and recover chemically diverse metabolites (Glen Mills GenoGrinder 2000). The resulting extract was divided into five fractions: two for analysis using two separate reverse phases (RP)/UPLC-MS/MS methods with positive ion mode electrospray ionization (ESI), one for analysis using RP/UPLC-MS/MS with negative ion mode ESI, one for analysis using HILIC/UPLC-MS/MS with negative ion mode ESI, and one for backup. To remove the organic solvent, samples were briefly placed on a TurboVap® (Zymark; Hopkinton, MA, USA). Before preparing for analysis, the sample extracts were dried overnight under nitrogen gas [25].

2.2.2. Ultrahigh Performance Liquid Chromatography–Tandem Mass Spectroscopy (UPLC-MS/MS)

Waters ACQUITY ultra-performance liquid chromatography (UPLC) and a Thermo Scientific Q-Exactive high resolution/accurate mass spectrometer interfaced with a heated electrospray ionization (HESI-II) source and Orbitrap mass analyzer operated at 35,000 mass resolution were used in all methods. The sample extract was dried before being reconstituted in solvents suitable for each of the four methods. To ensure injection and chromatographic consistency, each reconstitution solvent contained a series of standards at fixed concentrations. One aliquot was analyzed under acidic positive ion conditions, which were chromatographically tailored for more hydrophilic substances. In this procedure, the extract was gradient eluted from a C18 column (Waters UPLC BEH C18-2.1 × 100 mm, 1.7 m) using water and methanol containing 0.05% perfluoropentanoic acid (PFPA) and 0.1% formic acid (FA). Another aliquot was examined under acidic positive ion conditions that had been chromatographically tuned for more hydrophobic substances. The extract was gradient-eluted from the same aforementioned C18 column using methanol, acetonitrile, water, 0.05% PFPA, and 0.01% FA and was operated at a higher total organic content in this method. A separate dedicated C18 column was used to evaluate another aliquot under basic negative ion optimized conditions. The basic extracts were gradient-eluted from the column with methanol and water containing 6.5 mM ammonium bicarbonate at pH 8. After being eluted from an HILIC column (Waters UPLC BEH Amide 2.1 × 150 mm, 1.7 m) using a gradient of water and acetonitrile with 10 mM ammonium formate, pH 10.8, the fourth aliquot was subjected to negative ionization analysis. Dynamic exclusion was used to alternate between the MS analysis and data-dependent MSn scans. The scan range varied slightly between methods, but it was between 70 and 1000 m/z. The raw data files were archived and extracted in the manner described below [25].

2.2.3. Compound Identification and Metabolite Quantification

Using Metabolon's equipment and software (Durham, NC, USA), raw data were retrieved, peaks were found, and quality control was performed. Compounds were discovered by comparing them to purified standards in the library. The library was built on authenticated standards and included retention time/index (RI), mass-to-charge ratio (m/z), and chromatographic data (including MS/MS spectrum data). Compound identification was based on the retention index within a limited RI window of the proposed identification, an accurate mass match to the library within 10 ppm, and the MS/MS forward and reverse scores between experimental data and authenticated standards. Peaks were measured using the area-under-the-curve method and normalized to total protein as determined using the Bradford assay [25].

2.3. Bioinformatic Analysis

Heatmaps were generated for statistically significant differential metabolites using ComplexHeatMap in R [26,27]. Origin analysis was performed utilizing the online MetOrigin interface [28]. Briefly, a list of significantly altered metabolites with Kyoto Encyclopedia of Genes and Genomes (KEGG) and Human Metabolome Database (HMBD) IDs were uploaded, and we proceeded with simple analysis. Enrichment analysis and sPLSDA analysis were performed using Metaboanalyst [29].

2.4. Myeloperoxidase Activity

With some minor modifications, the Krawisz [30] method was used to measure myeloperoxidase activity (MPO) in the distal colon. MPO activity is expressed as units per milligram of tissue after being normalized to the amount of tissue.

2.5. Hematoxylin and Eosin Staining

The intestines were sampled, and they were quickly rinsed in PBS. The tissues were immediately snap-frozen in liquid nitrogen after being embedded in OCT media. Hematoxylin and eosin (H&E) staining was performed according to the manufacturer's protocol (ScyTek Laboratories, Logan, UT, USA) on longitudinal sections that were 5 μm thick, as previously described [31].

2.6. Microbial Sampling and 16S rRNA Sequencing

Cecal contents were collected and immediately snap-frozen in liquid nitrogen. Samples were sent to Zymo Research (Irvine, CA, USA) for 16S rRNA sequencing. Amplicon sequence variant (ASV) analysis was performed with DADA2. Linear discriminant analysis (LDA) was performed with bioBakery [32,33].

2.7. Statistical Analysis

For metabolite abundances, Welch's two-sample *t*-test was used to assess the significance and set at a level of $p < 0.05$. A false discovery rate (FDR) method was also applied with a set $q < 0.1$. Welch's two-sample *t*-test was used to assess significance for all other comparisons.

3. Results

3.1. Gut Microbial Dysbiosis Is Associated with Chronic DSS Treatment

Mucosal inflammation is a key feature in IBD [34]. Optimal treatment options focusing on complete healing of the mucosa are warranted. To understand the key mucosal alterations occurring during chronic intestinal inflammation, we utilized the chronic DSS colitis to model the relapsing and remitting nature of human IBD. Wild-type mice were administered alternating cycles of 2.5% DSS in drinking water and water alone for five weeks to induce chronic colitis (Figure 1A). We first assessed the severity of experimentally induced inflammation. As shown in Figure 1B and Supplementary Figure S1, colon length was significantly decreased in the group treated with DSS, suggestive of intestinal inflammation. The average colon length decreased from 6.3 cm to 5.0 cm during chronic DSS treatment ($p < 0.0001$). Additionally, the colon weight-to-length ratio, a marker used to determine diarrhea, was significantly increased in the group treated with chronic DSS ($p < 0.0001$) (Figure 1C). Given that another key aspect of colitis is neutrophil infiltration, the severity of inflammation was further determined by measuring myeloperoxidase (MPO) activity in colonic mucosa from the distal colon. Figure 1D showed that the average MPO activity increased from 0.495 units/mg of tissue to 6.163 units/mg of tissue ($p = 0.0333$). Further, qPCR demonstrated increases in the mRNA expression of proinflammatory cytokines IL-1β ($p = 0.0067$) (Figure 1E) and CXCL1 ($p = 0.0224$) (Figure 1F). Finally, representative histological analysis revealed architectural distortion, active inflammation, and a greater than 25% ulcerated area in the distal colon during chronic DSS colitis compared to controls (Figure 1G).

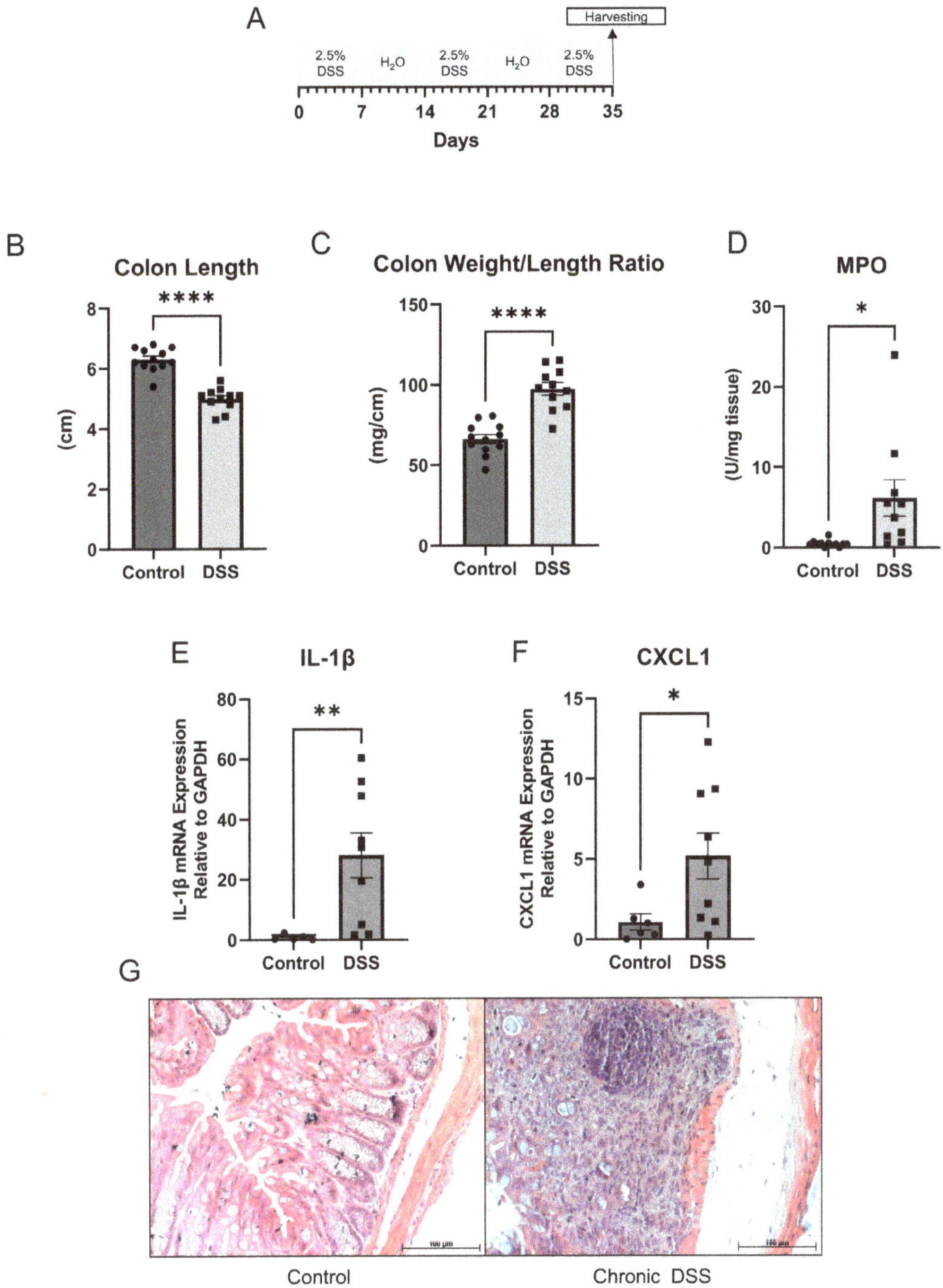

Figure 1. (**A**) Schematic for study design. (**B**) Colon length (cm) (**C**) Colon weight/length ratio (mg/cm) (**D**) Myeloperoxidase (MPO) activity units per mg of tissue (control $n = 12$; dextran sulfate sodium (DSS) $n = 11$). (**E**,**F**) Quantitative PCR for IL-1β and CXCL1 in distal colon mucosa (control $n = 5$; DSS $n = 9$). (**G**) Representative hematoxylin and eosin staining of control and chronic DSS mice. (Welch t-test; * $p < 0.05$, ** $p < 0.01$, **** $p < 0.0001$).

Gut microbial dysbiosis, another hallmark of chronic colitis, was further examined [35]. 16S rRNA sequencing was performed on the cecal contents, followed by LEfSe analysis to identify differentially altered microbes. The cladogram depicted in Figure 2A shows the various taxonomic levels of microbes associated with chronic colitis and control groups and their taxonomic relationship to each other from LEfSe (linear discriminant analysis effect size) analysis. A tabular summary of these taxa are depicted with their effect size and p-value in Supplemental Table S1. Additionally, changes observed in this study that are consistent with IBD are depicted in Supplemental Table S2 [36–42]. Figure 2B depicts the top 20 taxa associated with the control and DSS groups, and Supplemental Figures S2 and S3 depict all taxa. Of note, the phyla Verrucomicrobia, Bacteroidetes, and Actinobacteria were significantly associated with chronic DSS colitis. Furthermore, alpha diversity was assessed using several indices, including the total number of observed species, Chao1, Fisher, Shannon Diversity, and Simpson index. Overall, the data demonstrated a loss of alpha diversity within samples of mice with chronic colitis (Figure 2D). Specifically, the overall species observed dropped from 226.7 in the control group to 189.5 in the colitis group (p = 0.0001). Additionally, the Chao1, Fisher, Simpson, and Shannon diversity indices, which consider both the number of species and the abundance of each species, decreased from 226.8 in the control group to 189.5 during chronic colitis (p = 0.0001), 35.84 to 29.03 (p = 0.0001), 0.9819 to 0.9696 (p = 0.0002), and 6.629 to 6.117 ($p < 0.0001$), respectively. As expected, chronic colitis led to both a decrease in the overall number of species and the distribution of species. Furthermore, beta diversity was assessed using the weighted UniFrac method [43], which accounts for the presence of organisms, their abundance, and their taxonomic relationship to each other. The analysis showed that each sample's microbial communities clustered according to their treatment group (Figure 2E). Furthermore, a hallmark of microbial changes in IBD is a loss of short-chain fatty acid-producing bacteria [44]. There was a decrease in the relative abundance of the genus *Roseburia* ($p < 0.001$) (Figure 2C) and several *Roseburia* species (Supplemental Figure S4), which were identified as biomarkers for the control groups through LEfSe analysis. There have also been reported increases in the genus Lactobacillus in IBD cohorts [37], which were also found in the chronic DSS samples (Figure 2C, p = 0.0070). Further, we found increases in *Turicibacter sanguinis* ($p < 0.001$) and *Erysipelatoclostridium* species ($p < 0.001$), which also correspond well with previous reports on human cohorts with IBD (Figure 2C) [36]. Overall, our findings showed that chronic DSS treatment in mice resulted in chronic colitis phenotype with associated dysbiosis, providing an adequate model to further explore the mucosal changes that occur in the colon during on-again, off-again inflammatory insults.

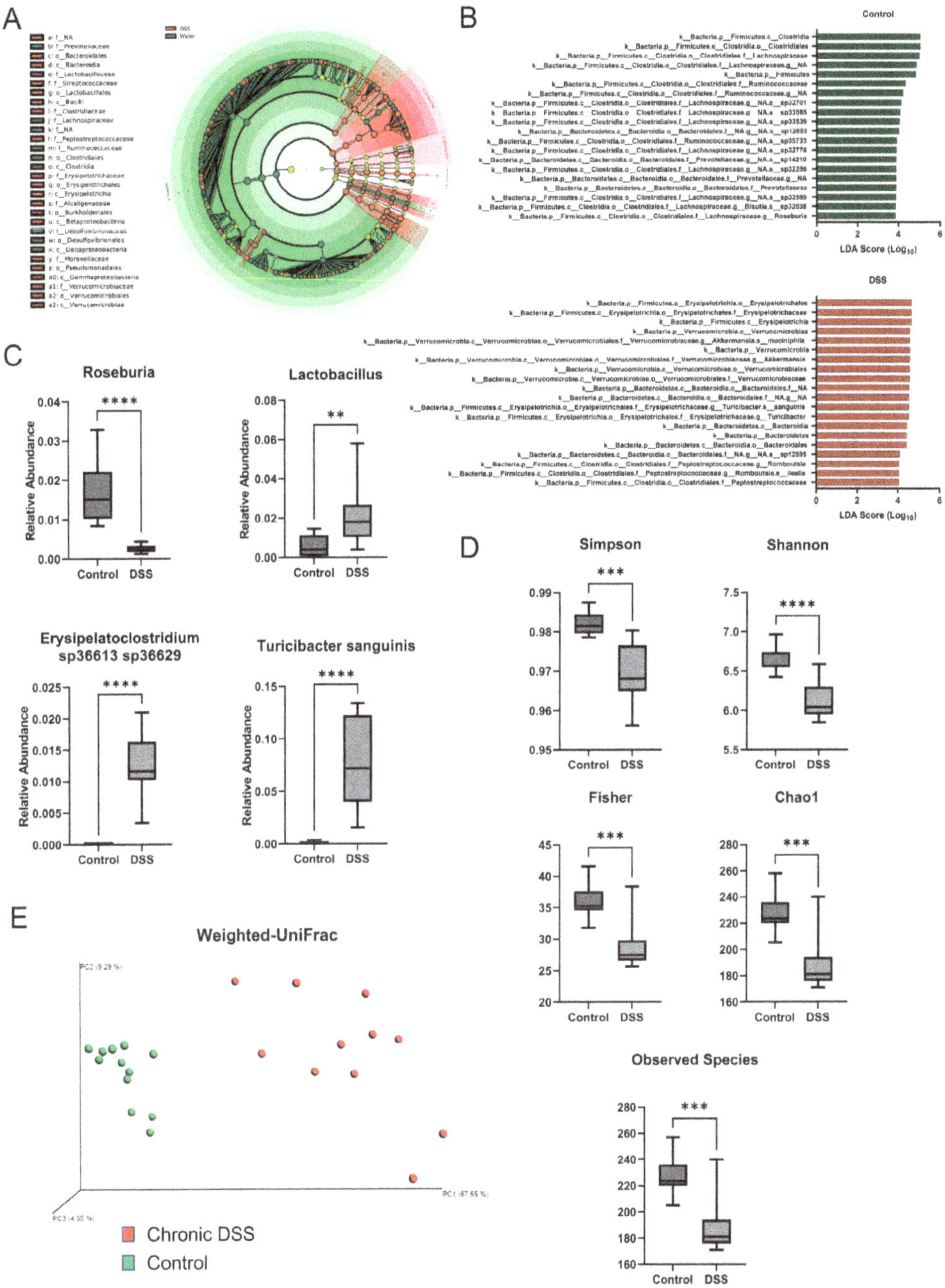

Figure 2. (**A**) Cladogram highlighting taxa associated with control water and dextran sulfate sodium (DSS) treatments based on linear discriminant analysis effect size. (**B**) Top 20 taxa associated with control group based on linear discriminant analysis effect size (LEfSe) analysis (**top**) and top 20 taxa associated with DSS group based on LEfSe analysis (**bottom**). (**C**) Boxplots of relative abundances for various taxa associated with inflammatory bowel diseases (IBD) ($p < 0.05$, Welch t-test; control $n = 12$; DSS $n = 11$). (**D**) Boxplots for alpha diversity measures. (**E**) Weighted UniFrac beta diversity plot. (Welch t-test; ** $p < 0.01$, *** $p < 0.001$, **** $p < 0.0001$).

3.2. *Chronic Colitis Leads to Distinct Metabolomic Profile in the Colonic Mucosa*

Given that functional changes in the gut microbiome are manifested through metabolic alterations, untargeted metabolomic analysis was next performed to take an unbiased approach to discovering novel changes in the intestinal mucosa during chronic inflammation. The analysis revealed a unique metabolic signature in the colon mucosa after chronic colitis. Overall, there were 314 differentially altered metabolites ($p < 0.05$, $q < 0.1$), with 80 being decreased and 234 metabolites increased in the colonic mucosa of mice with chronic colitis (Figure 3A, Supplemental Figure S5). Of the 314, the largest proportion of altered metabolites was involved in lipid metabolism, with the next largest group involved in amino acid metabolism. The same trend was observed for both the increased and decreased metabolites when compared to the control (Figure 3B). Additionally, sparse partial least squares discriminant analysis (sPLSDA) revealed two distinct clusters when comparing the colon mucosa metabolome of control mice to those with chronic colitis (Figure 3C). This assessment was followed up using enrichment analysis, which considers how many metabolites are altered within a given Kyoto Encyclopedia of Genes and Genomes (KEGG) pathway (Figure 3D). Of note, several nucleotide and amino acid metabolism pathways were among the most enriched.

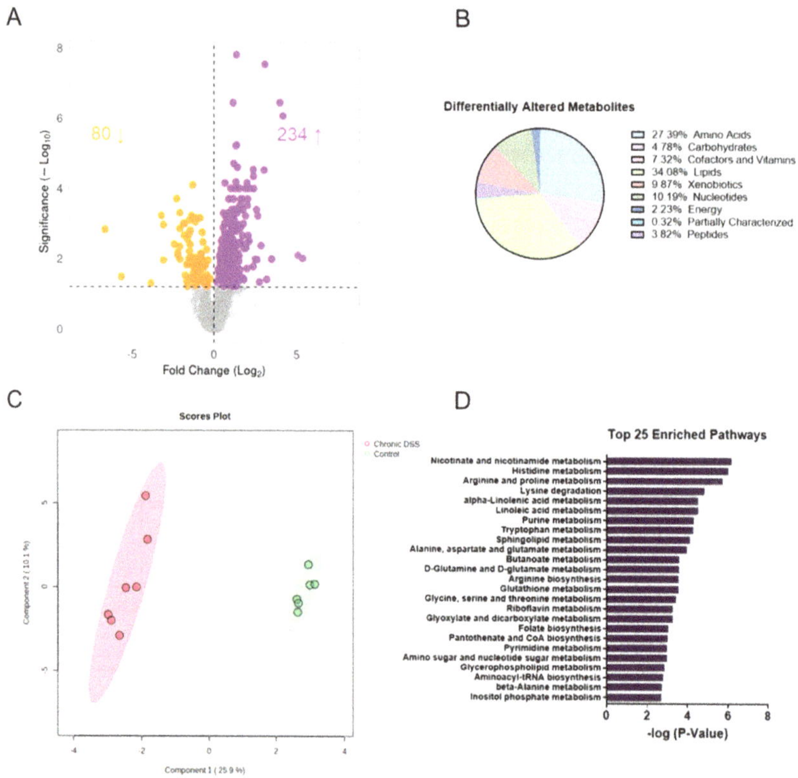

Figure 3. (**A**) Volcano plot (the log2 of the fold change of metabolites in wild-type (WT) dextran sulfate sodium (DSS) compared to WT was plotted against the −log10 of the p-value) where the up-facing arrow indicates the number of significantly metabolites increased in abundance and the down-facing arrow indicates the number of decreased and (**B**) pie chart showing the proportion of differentially altered metabolites by metabolite subtype. (**C**) Sparse partial least squares discriminant analysis (sPLSDA) plots generated in Metaboanalyst using the Statistical Analysis function. (**D**) Top 25 enriched pathways in mucosa of control vs. DSS treated mice.

3.3. *Chronic Colitis Leads to Increased Mucosal Localization of Inflammatory Metabolites and Impaired Enterohepatic Circulation*

Our investigations revealed striking differences in the mucosal abundances of several metabolites that are implicated in inflammatory processes such as polyamines, amino acid metabolites, eicosanoids, and TCA cycle intermediates. Figure 4A highlights the distributions of the significantly altered inflammatory mediators in the mucosa during chronic DSS colitis. It is worth highlighting the vast increase in mucosal itaconate levels of nearly 18-fold ($p < 0.001$) during chronic inflammation, as it has been shown to have anti-inflammatory properties in models for sepsis, autoimmune disease, and reperfusion injury [45]. Furthermore, kynurenine, a tryptophan metabolite known to be upregulated during inflammation, was increased nearly 16-fold ($p < 0.001$) in the mucosa after chronic DSS treatment. Additionally, our data revealed several interesting changes in metabolites involved in enterohepatic circulation (Figure 4B). For instance, the secondary bile acids deoxycholate and lithocholate, which are directly influenced by gut microbiota, are significantly decreased during chronic colitis ($p < 0.05$). Moreover, heme and its degradation products bilirubin, biliverdin, and protoporphyrin IX were significantly increased in the chronically inflamed mucosa, suggesting increased synthesis and/or absorption of erythrocyte byproducts.

3.4. *Contribution of Microbial Mucosal Metabolomic Profile during Chronic Colitis*

An unbiased approach was utilized to examine the potential link between gut microbial dysbiosis and the altered metabolic profile in the colonic mucosa. The bioinformatic tool Met-Origin was utilized to aid in classifying metabolites as related to microbes or co-metabolized by both the host and microbes. This analysis revealed a multitude of metabolites that were altered, at least in part, through changes in gut microbial composition (Figure 5). Our analysis showed 23 microbial-related metabolites differentially altered in chronic colitis (Figure 5A). Interestingly, fucose was increased in the colon mucosa of mice with chronic colitis. Additionally, several differentially altered metabolites were significantly associated with disease severity assessed by MPO levels (Figure 5B, Supplemental Figure S6). Our analysis revealed that in addition to several co-metabolized nucleotides highlighted in Figure 5, there were increases in heme-related metabolites protoporphyrin IX and biliverdin in the colonic mucosa of mice with chronic colitis (Figure 5C) that are also highlighted in Figure 4B in more detail. This may suggest increased heme breakdown and compensatory synthesis in response to chronic intestinal inflammation. Ultimately, this analysis allows for a better understanding of the contribution of gut microbial dysbiosis to the metabolomic alterations in the mucosa of mice with chronic colitis.

3.5. *Chronic Colitis Alters Mucosal Dietary, Nucleotide, Amino Acid, and Lipid Metabolism*

Food and plant components: Differentially altered metabolites were subsequently organized by their metabolic pathway. For instance, there were several alterations in dietary and plant-derived metabolites in the colonic mucosa of mice with chronic colitis (Figure 6A). Stachydrine and ergothioneine, for instance, are both increased in the mucosa during chronic colitis. Given that these are dietary metabolites, our findings suggest altered colonic nutrient absorption in response to inflammation. Our analysis also revealed alterations in several nucleotide metabolism pathways.

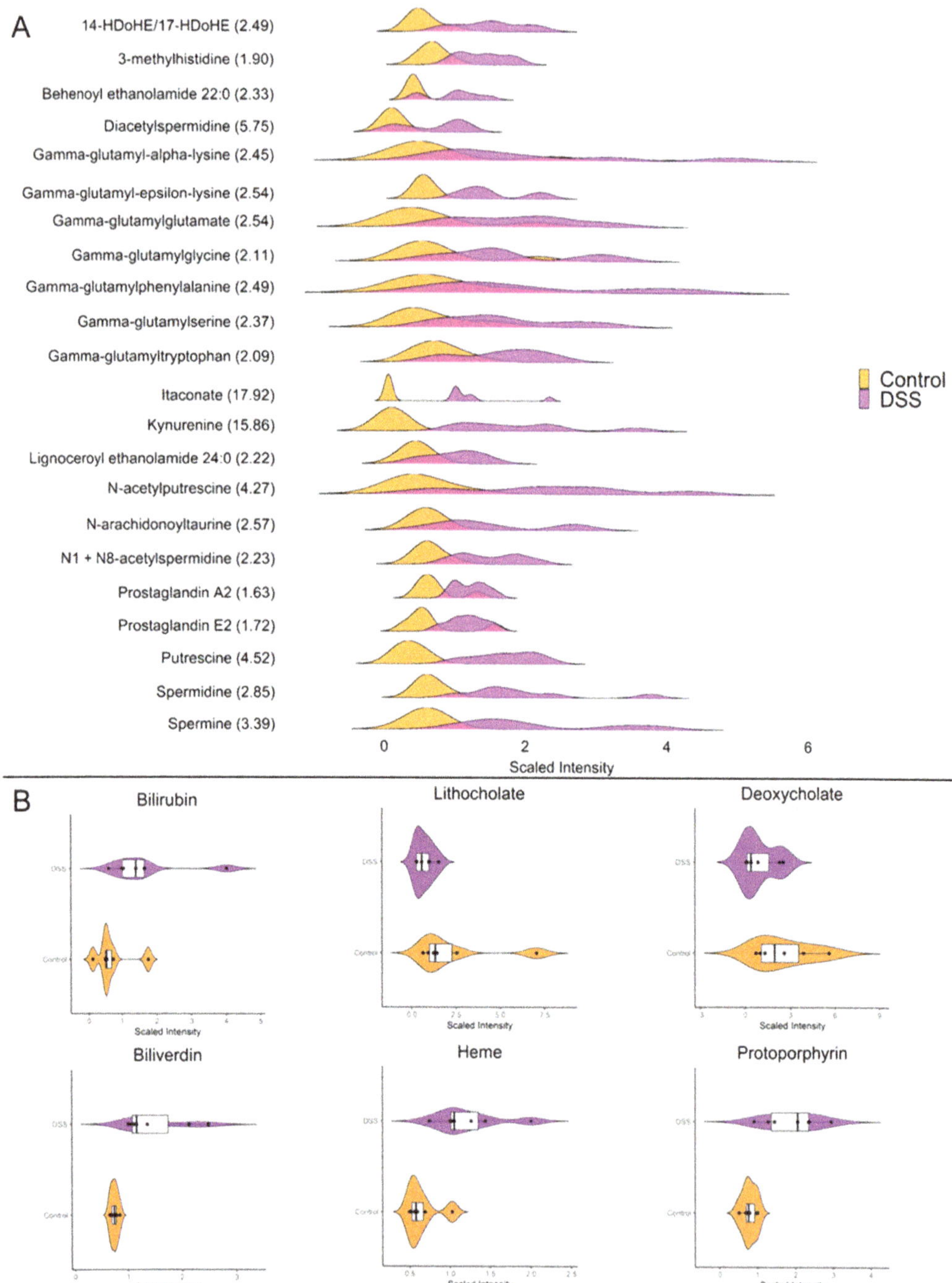

Figure 4. (**A**) Ridgeline plot showing the distributions of differentially altered metabolites with known roles as inflammatory mediators ($p < 0.05$, $q < 0.01$, Welch t-test; control $n = 6$; DSS $n = 7$). The values within parenthesis indicate the fold change between DSS and control. (**B**) Raincloud plots showing abundance and distributions of differentially altered enterohepatic metabolites ($p < 0.05$, $q < 0.01$, Welch t-test; control $n = 6$; DSS $n = 7$).

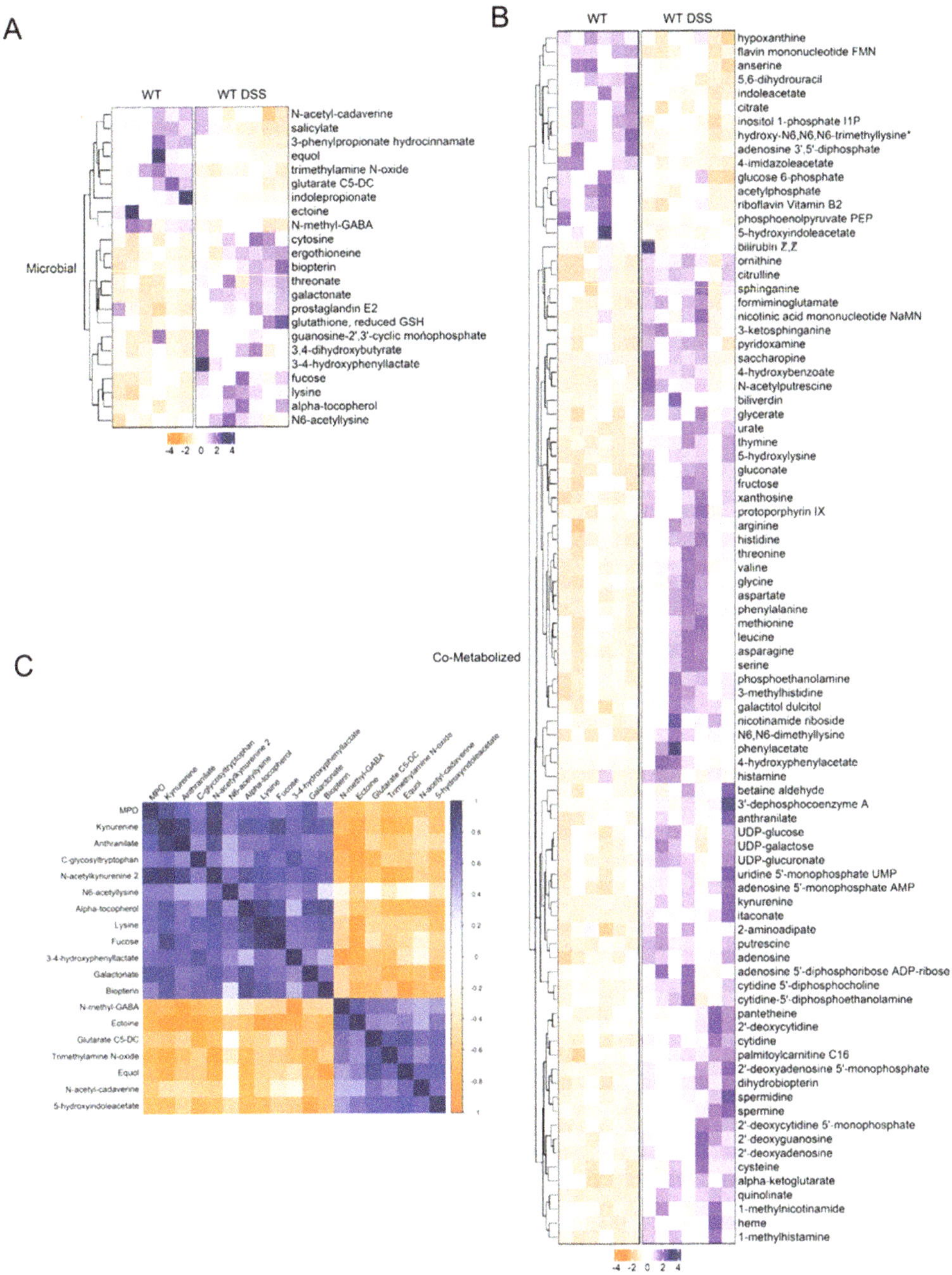

Figure 5. (**A**) Heatmap of microbial-related and (**B**) co-metabolized differentially altered metabolites between control and chronic colitis mice ($p < 0.05$, $q < 0.1$, Welch t-test; control $n = 6$; DSS $n = 7$. (**C**) Correlation plot showing Spearman correlation coefficients between microbial metabolites and MPO activity that were statistically significant ($p < 0.05$). * indicates compounds with lack of reference standard acquisition but their identity is based upon orthogonal information.

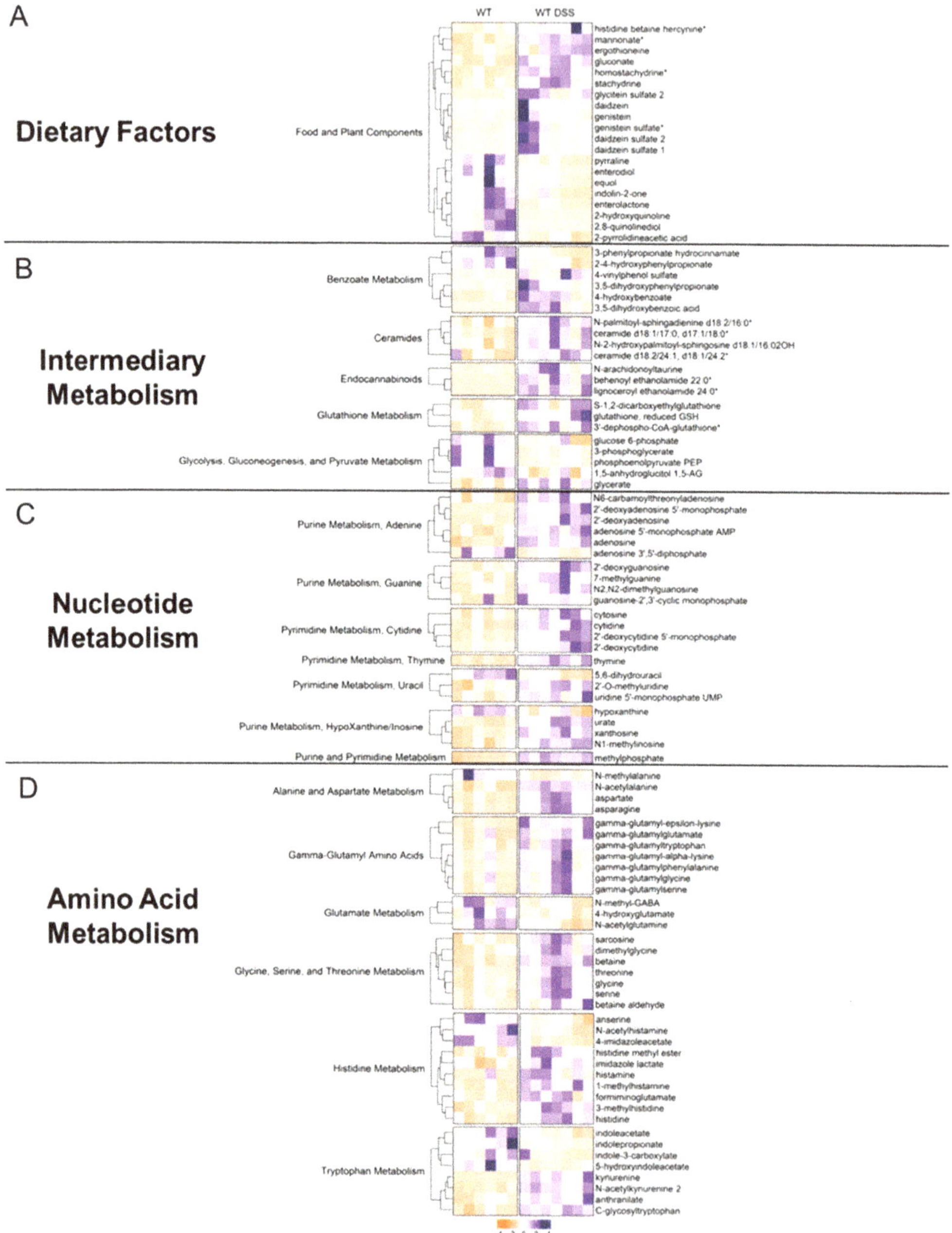

Figure 6. Heatmaps of differentially altered metabolites ($p < 0.05$, $q < 0.1$, Welch t-test) in (**A**) dietary factors, (**B**) intermediary metabolism, (**C**) nucleotide metabolism, and (**D**) amino acid metabolism. Heatmap was scaled from −4 to 4 (control = 6; DSS $n = 7$). * indicates compounds with lack of reference standard acquisition but their identity is based upon orthogonal information.

Intermediary metabolism: Several interesting increases in other classes of metabolites were also found. For instance, endocannabinoids (Figure 6B), a class of compounds newly emerging as a possible therapeutic target in IBD, showed general increases in the mucosa of mice with chronic colitis. Other interesting findings included increases in ceramides. Furthermore, the mucosa of mice with chronic colitis demonstrated decreased glycolytic intermediates such as glucose 6-phosphate, phosphoenolpyruvate, and 3-phosphoglycerate, suggesting decreased glycolytic activity (Supplemental Figure S7).

Nucleotide metabolism: Generally, there were increases in metabolites involved in adenine, guanine, cytidine, thymine, uracil, and inosine metabolism (Figure 6C, Supplemental Figure S8) in the colon mucosa of chronic colitis mice, suggesting increased nucleotide production due to increased epithelial cell turnover. For instance, adenosine was increased in the mucosa during chronic colitis. Interestingly, adenosine receptor agonists have been shown to mitigate the effects of DSS-induced colitis through NF-κB signaling. Another interesting finding was the increase in methylated nucleotide metabolites such as N1-methylinosine, 2′-O-methyluridine, and N2, N2-dimethylguanosine, suggesting changes in genetic regulation.

Amino acid metabolism: Our data showed that gamma-glutamyl amino acids were increased during chronic colitis (Figure 6D, Supplemental Figure S9). Interestingly, decreases in microbial indole tryptophan metabolites and increases in co-metabolized kynurenine were also found, revealing alterations in tryptophan metabolism within intestinal mucosa during inflammation. Increases in the amino acids glycine, serine, threonine, histidine, aspartate, and asparagine were also identified, suggesting altered transport and/or metabolism in the intestinal mucosa during chronic inflammation. This analysis revealed several findings corroborated in human cohorts of IBD and other IBD models while also describing novel metabolomic findings in the colonic mucosa during chronic colitis.

4. Discussion

In this study, it was hypothesized that chronic colonic inflammation results in a unique microbially influenced mucosal metabolomic profile, providing new insight into the pathophysiology of inflammatory bowel diseases (IBD). This study investigated the gut microbial composition and mucosal metabolome of mice with induced chronic colitis. This work found that chronic colitis resulted in increased myeloperoxidase activity in the colon, as well as increased levels of proinflammatory cytokines IL-1β and CXCL1. Gut microbial analysis revealed expected losses in alpha diversity, two distinct beta diversity clusters, and microbial composition changes consistent with IBD patients. Furthermore, mucosal metabolomics revealed 328 differentially altered metabolites with several enriched metabolic pathways. These changes were alterations in microbial-related metabolites, food and plant components, nucleotides, amino acids, and lipids.

4.1. Facts and Perspectives

Gut microbial dysbiosis is a critical factor in the pathophysiology of IBD [35]. In this regard, several groups have investigated changes in the gut microbiome in IBD patients and in animal models for IBD with one characteristic feature being a loss of alpha diversity [46,47], a measure for the total number of species as well as the abundance of such species, which were observed in the present study. Additionally, beta diversity, which has also been consistently affected by IBD [47], was also found to be affected by induced chronic colitis in the current study. Furthermore, several characteristic microbial changes have been highlighted in the chosen disease model that have also been reported in IBD patients. The genus Roseburia was identified as a characteristic feature of untreated control mice and was significantly decreased in abundance in mice with chronic colitis. Roseburia are known to produce short-chain fatty acids, which have been shown to decrease in abundance in IBD, and administration of Roseburia intestinalis supernatant has been shown to alleviate induce colitis [48–50]. Although there was not a statistically significant decrease in the mucosal levels of the SCFA butyrate, there was a trend towards a decrease in the chronically inflamed mucosa. This could be due to SCFAs serving as the key fuel for colonocytes and thus being rapidly metabolized. There was also an increase in the genus

Lactobacillus, Turicibacter sanguinis, and an Erysipleatoclostridium species, which have also been observed in IBD patients [36,37,39].

While it was not surprising to observe distinct mucosal metabolic profiles during chronic colitis, there were several findings that have been corroborated in IBD patients parallel to novel discoveries that can be used for further mechanistic inquiries. For example, decreases in indole metabolites such as indoleacetate, indolepropionate (IPA), and indole-3-carboxylate have been shown in the serum of subjects with ulcerative colitis with IPA serving as a marker for disease remission [51]. Interestingly, our data show that the decreases in these indole metabolites are occuring in the intestinal mucosa during inflammation. This suggests that the decreases in the serum of those with IBD is due to decreased microbial production and not increased transport in mucosal epithelial or immune cells. Collectively, these findings can be exploited for the benefit of patients with IBD through microbial-based therapy with cocktails of microbes capable of synthesizing beneficial indole metabolites from tryptophan. Regarding dietary factors, increases in ergothioneine and stachydrine were found in the inflamed mucosa. Stachydrine has been shown to have anti-inflammatory properties in primary human chondrocytes isolated from subjects with osteoarthritis that were treated with IL-1β [52]. Ergothioneine was previously shown to also demonstrate protective effects against DSS-induced colitis by decreasing MPO activity levels and improving intestinal barrier function [53]. Taken together, our findings suggest a compensatory increase in the absorption of anti-inflammatory metabolites stachydrine and ergothioneine in response to mucosal inflammation. This suggests that increasing epithelial absorption of stachydrine and ergothioneine could be used as a potential therapeutic approach in IBD, possibly by increasing the activity of the ergothioneine transporter [54]. Increases in mucosal ceramides, which have been associated with increased inflammation in the mucosa of subjects with IBD, were found [55]. It was also found that the microbial-derived metabolite fucose, a sugar known to participate in the formation of Lewis antigens, was increased in the colon mucosa of mice with chronic colitis. This is particularly notable, as Lewis antigens have been associated with ulcerative colitis [56]. In fact, IgG and IgM reactivity has been shown to be increased towards fucosylated oligosaccharides in the serum of patients with Crohn's disease, implicating fucose-carrying Bacteroides species in intestinal inflammation [57]. In fact, our analysis identified the order Bacteroidales as a key distinguishing feature for mice with chronic colitis. Considering these findings, therapeutics targeting fucose degradation could serve as a novel strategy for inflammation in IBD.

The present study also identified several altered metabolic pathways in response to chronic colitis. There were decreased levels of glycolytic intermediates, which is important to note, as a previous study found that patients with IBD had antibodies that recognized several glycolytic enzymes as antigens [58]. One such enzyme was phosphoglycerate mutase, the enzyme responsible for conversion of glycerate-3P to glycerate-2P. Our rendering of the KEGG glycolysis pathway demonstrated decreased levels of glycerate-2P in the intestinal mucosa during chronic colitis, highlighting the need to further explore the implications of disrupted glycolysis. Our data also highlight novel increases in endocannabinoids in the colon mucosa during chronic inflammation. This is an important finding, as the role of the endocannabinoid system in IBD is an emerging area of interest [59]. Moreover, this work found increases in several ceramides, a class of sphingolipids, which are emerging as a new area of investigation in IBD [55]. Ultimately, this study provides a comprehensive report of alterations in mucosal metabolism during chronic inflammation in a relevant model that highlights microbial contributions, novel pathways for future investigation, and identifies newer therapeutic targets for intestinal inflammation.

This work characterized the gut microbial alterations associated with chronic inflammation and further elucidated the impacts of those changes by analyzing differentially altered microbial-related metabolites. For instance, these studies revealed increases in heme degradation products protoporphyrin IX and biliverdin within the chronically inflamed colon mucosa. Simultaneously, we found that *Turicibacter sanguinis*, a bacterial species that synthesizes biliverdin through heme oxygenase [60], and the family of bacteria *Erysipelotrichaceae*, which are known to utilize protoporphyrin IX as a substrate [61], were both increased in the intestinal lumens of mice with chronic colitis. Our study yielded significant findings indicating elevated levels of various purine, pyrimidine, endocannabinoid, and ceramide metabolites derived from both the host and the gut microbiota in colitis with the potential to both serve as disease biomarkers and as areas of future mechanistic investigation. Notably, our analysis of the origin of these metabolites revealed a microbial contribution, particularly in the case of tryptophan metabolites, including kynurenine, anthranilate, 5-hydroxyindoleacetate, and C-glycosyltryptophan. These microbial-related tryptophan metabolites displayed substantial increases within the colon mucosa during chronic inflammation and exhibited strong correlations with disease activity. Consequently, these findings provide opportunity to develop targeted therapies for harnessing the beneficial metabolites through engineered microbes that can break the proinflammatory metabolites or shunt them towards the synthesis of beneficial compounds. For example, kynurenine aminotransferase-expressing microbes can be utilized to convert increased kynurenine into kynurenate [62]. These findings provide valuable insights into the underlying pathophysiology of IBD and present promising new targets for the development of microbial-based therapeutics.

4.2. Strengths of the Study

This study utilized a preclinical translational model for IBD in the form of chronic DSS colitis that mimics remitting and relapsing inflammation observed in human disease. Furthermore, while several metabolomic studies in IBD have focused on less invasive biological compartments such as serum and feces, the present study rigorously examined the mucosal metabolome—an understudied tissue compartment. Moreover, this work utilized a discovery approach with the integration of gut microbiota and mucosal metabolites. This resulted in the identification of differentially altered microbial metabolites resulting from gut microbial dysbiosis that can be utilized as potential biomarkers and targeted with microbial-based therapeutics. The limitations of our study include that metabolomic profiling and microbial sequencing were performed on two distinct sets of mice. However, the approaches utilized in this work were able to identify microbial contributions to microbial-related metabolic changes with MetOrigin. Furthermore, our study used only male mice and should be repeated in female mice at various stages of their estrous cycles, with those stages taken under consideration as a variable. While there are not strong differences in the incidence of IBD between males and females, there are some slight differences that require future study [63]. For instance, females tend to have a higher risk of CD while males tend to have a higher risk of UC. Future studies can focus on further exploring the mechanistic processes involved in the disruption of identified metabolic pathways by separating the mucosa into epithelial cell fractions and lamina propria fractions to better understand which changes are coming from epithelial cells or immune cells specifically.

5. Conclusions

In summary, our studies provide a comprehensive view into the metabolomic alterations in the intestinal mucosa during chronic inflammation that adds to the growing number of studies utilizing metabolomics to gain new insight into disease [64,65]. This study identified novel mucosal metabolites that could be used as potential inflammatory biomarkers and that should warrant further investigation such as disturbances in glycolytic metabolism, endocannabinoids, and ceramides. Finally, this work identified several potential therapeutic targets: increasing microbial production of indole metabolites, shunting kynurenine pathway, targeting fucose degradation, and increasing epithelial cell ergothioneine and stachydrine influx.

Supplementary Materials: The following supporting information can be downloaded at: https://www.mdpi.com/article/10.3390/metabo13070873/s1, Figure S1: Representative colon pictures; Figure S2: LEfSe analysis of taxa for control mice; Figure S3: LEfSe analysis of taxa for chronic DSS mice; Figure S4: Relative abundance of *Roseburia* species; Figure S5: Differentially altered mucosal metabolites; Figure S6: Associations between microbial metabolites and MPO; Figure S7: Metabolic alterations in glycolysis/gluconeogenesis; Figure S8: Metabolic alterations in pyrimidine and purine metabolism; Figure S9: Metabolic alterations in amino acid metabolism; Table S1: Notable changes in gut microbiota determine by LEfSe; Table S2: Overlapping microbiota changes with human IBD; Data S1: Gut microbial composition across samples; Data S2: Colon mucosa metabolomics.

Author Contributions: Conceptualization, W.A.A. and R.K.G.; Formal analysis, N.C.; Funding acquisition, N.C., S.S., P.K.D., W.A.A. and R.K.G.; Investigation, N.C., A.Q., A.S., K.M., S.C., J.M., A.G.Y. and S.K.; Methodology, N.C., A.Q. and A.S.; Writing—original draft, N.C.; Writing—review and editing, N.C., S.S., P.K.D., W.A.A. and R.K.G. All authors have read and agreed to the published version of the manuscript.

Funding: These studies were supported by the NIDDK grants R01 DK098170 (RKG); the Crohn's and Colitis Foundation CCFA 581138 (RKG), and the Department of Veterans Affairs I01BX000152 (WAA), I01BX002867 (SS), VA Merit Award I01BX002011 (PKD), VA Senior Research Career Scientist Award 1IK6BX005242 (PKD), VA BCCMA Award I01BX005862 (PKD), NIH NIDDK grant R01 DK092441 and R01 DK054016 (PKD), and F30AA030222 (NC), the Veterans Affairs Research Career Scientist Award 1IK6BX005243 (WAA).

Institutional Review Board Statement: All animal studies were performed in accordance with institutional guidelines and regulations and approved by the Animal Care Committees of the University of Illinois at Chicago and the Jesse Brown Veterans Affairs Medical Center (Approval Code: ACC 21-078).

Informed Consent Statement: Not applicable.

Data Availability Statement: All data is publicly available as part of supplemental data.

Conflicts of Interest: The authors declare no conflict of interest.

References

1. Kaplan, G.G.; Ng, S.C. Understanding and Preventing the Global Increase of Inflammatory Bowel Disease. *Gastroenterology* **2017**, *152*, 313–321.e2. [CrossRef] [PubMed]
2. Guan, Q. A Comprehensive Review and Update on the Pathogenesis of Inflammatory Bowel Disease. *J. Immunol. Res.* **2019**, *2019*, 7247238. [CrossRef] [PubMed]
3. Gohil, K.; Carramusa, B. Ulcerative Colitis and Crohn's Disease. *P T* **2014**, *39*, 576–577. [PubMed]
4. Al-Bawardy, B.; Shivashankar, R.; Proctor, D.D. Novel and Emerging Therapies for Inflammatory Bowel Disease. *Front. Pharmacol.* **2021**, *12*, 651415. [CrossRef]
5. Chang, J.T. Pathophysiology of Inflammatory Bowel Diseases. *N. Engl. J. Med.* **2020**, *383*, 2652–2664. [CrossRef]
6. Lloyd-Price, J.; Arze, C.; Ananthakrishnan, A.N.; Schirmer, M.; Avila-Pacheco, J.; Poon, T.W.; Andrews, E.; Ajami, N.J.; Bonham, K.S.; Brislawn, C.J.; et al. Multi-Omics of the Gut Microbial Ecosystem in Inflammatory Bowel Diseases. *Nature* **2019**, *569*, 655–662. [CrossRef]
7. Franzosa, E.A.; Sirota-Madi, A.; Avila-Pacheco, J.; Fornelos, N.; Haiser, H.J.; Reinker, S.; Vatanen, T.; Hall, A.B.; Mallick, H.; McIver, L.J.; et al. Gut Microbiome Structure and Metabolic Activity in Inflammatory Bowel Disease. *Nat. Microbiol.* **2019**, *4*, 293–305. [CrossRef]
8. Zheng, L.; Wen, X.-L.; Duan, S.-L. Role of Metabolites Derived from Gut Microbiota in Inflammatory Bowel Disease. *World J. Clin. Cases* **2022**, *10*, 2660–2677. [CrossRef]
9. Aldars-García, L.; Gisbert, J.P.; Chaparro, M. Metabolomics Insights into Inflammatory Bowel Disease: A Comprehensive Review. *Pharmaceuticals* **2021**, *14*, 1190. [CrossRef]
10. Kolho, K.-L.; Pessia, A.; Jaakkola, T.; de Vos, W.M.; Velagapudi, V. Faecal and Serum Metabolomics in Paediatric Inflammatory Bowel Disease. *J. Crohn's Colitis* **2017**, *11*, 321–334. [CrossRef]
11. Le Gall, G.; Noor, S.O.; Ridgway, K.; Scovell, L.; Jamieson, C.; Johnson, I.T.; Colquhoun, I.J.; Kemsley, E.K.; Narbad, A. Metabolomics of Fecal Extracts Detects Altered Metabolic Activity of Gut Microbiota in Ulcerative Colitis and Irritable Bowel Syndrome. *J. Proteome Res.* **2011**, *10*, 4208–4218. [CrossRef]
12. Bjerrum, J.T.; Wang, Y.; Hao, F.; Coskun, M.; Ludwig, C.; Günther, U.; Nielsen, O.H. Metabonomics of Human Fecal Extracts Characterize Ulcerative Colitis, Crohn's Disease and Healthy Individuals. *Metabolomics* **2015**, *11*, 122–133. [CrossRef]

3. Manzella, C.R.; Jayawardena, D.; Pagani, W.; Li, Y.; Alrefai, W.A.; Bauer, J.; Jung, B.; Weber, C.R.; Gill, R.K. Serum Serotonin Differentiates Between Disease Activity States in Crohn's Patients. *Inflamm. Bowel Dis.* **2020**, *26*, 1607–1618. [CrossRef] [PubMed]
4. De Preter, V.; Machiels, K.; Joossens, M.; Arijs, I.; Matthys, C.; Vermeire, S.; Rutgeerts, P.; Verbeke, K. Faecal Metabolite Profiling Identifies Medium-Chain Fatty Acids as Discriminating Compounds in IBD. *Gut* **2015**, *64*, 447–458. [CrossRef] [PubMed]
5. Jansson, J.; Willing, B.; Lucio, M.; Fekete, A.; Dicksved, J.; Halfvarson, J.; Tysk, C.; Schmitt-Kopplin, P. Metabolomics Reveals Metabolic Biomarkers of Crohn's Disease. *PLoS ONE* **2009**, *4*, e6386. [CrossRef] [PubMed]
6. Perez, K.; Ngollo, M.; Rabinowitz, K.; Hammoudi, N.; Seksik, P.; Xavier, R.J.; Daly, M.J.; Dotan, I.; Le Bourhis, L.; Allez, M. Meta-Analysis of IBD Gut Samples Gene Expression Identifies Specific Markers of Ileal and Colonic Diseases. *Inflamm. Bowel Dis.* **2022**, *28*, 775–782. [CrossRef]
7. Abreu, M.T.; Davies, J.M.; Quintero, M.A.; Delmas, A.; Diaz, S.; Martinez, C.D.; Venables, T.; Reich, A.; Crynen, G.; Deshpande, A.R.; et al. Transcriptional Behavior of Regulatory T Cells Predicts IBD Patient Responses to Vedolizumab Therapy. *Inflamm. Bowel Dis.* **2022**, *28*, 1800–1812. [CrossRef] [PubMed]
8. Gallagher, K.; Catesson, A.; Griffin, J.L.; Holmes, E.; Williams, H.R.T. Metabolomic Analysis in Inflammatory Bowel Disease: A Systematic Review. *J. Crohn's Colitis* **2021**, *15*, 813–826. [CrossRef]
9. Aldars-García, L.; Chaparro, M.; Gisbert, J.P. Systematic Review: The Gut Microbiome and Its Potential Clinical Application in Inflammatory Bowel Disease. *Microorganisms* **2021**, *9*, 977. [CrossRef]
0. Baydi, Z.; Limami, Y.; Khalki, L.; Zaid, N.; Naya, A.; Mtairag, E.M.; Oudghiri, M.; Zaid, Y. An Update of Research Animal Models of Inflammatory Bowel Disease. *Sci. World J.* **2021**, *2021*, 7479540. [CrossRef]
1. Lanis, J.M.; Alexeev, E.E.; Curtis, V.F.; Kitzenberg, D.A.; Kao, D.J.; Battista, K.D.; Gerich, M.E.; Glover, L.E.; Kominsky, D.J.; Colgan, S.P. Tryptophan Metabolite Activation of the Aryl Hydrocarbon Receptor Regulates IL-10 Receptor Expression on Intestinal Epithelia. *Mucosal. Immunol.* **2017**, *10*, 1133–1144. [CrossRef]
2. Chen, Y.; Mai, Q.; Chen, Z.; Lin, T.; Cai, Y.; Han, J.; Wang, Y.; Zhang, M.; Tan, S.; Wu, Z.; et al. Dietary Palmitoleic Acid Reprograms Gut Microbiota and Improves Biological Therapy against Colitis. *Gut Microbes* **2023**, *15*, 2211501. [CrossRef]
3. Kang, S.; Kim, J.; Park, A.; Koh, M.; Shin, W.; Park, G.; Lee, T.A.; Kim, H.J.; Han, H.; Kim, Y.; et al. TRIM40 Is a Pathogenic Driver of Inflammatory Bowel Disease Subverting Intestinal Barrier Integrity. *Nat. Commun.* **2023**, *14*, 700. [CrossRef] [PubMed]
4. Chassaing, B.; Aitken, J.D.; Malleshappa, M.; Vijay-Kumar, M. Dextran Sulfate Sodium (DSS)-Induced Colitis in Mice. *Curr. Protoc. Immunol.* **2014**, *104*, 15.25.1–15.25.14. [CrossRef]
5. Guo, L.; Milburn, M.V.; Ryals, J.A.; Lonergan, S.C.; Mitchell, M.W.; Wulff, J.E.; Alexander, D.C.; Evans, A.M.; Bridgewater, B.; Miller, L.; et al. Plasma Metabolomic Profiles Enhance Precision Medicine for Volunteers of Normal Health. *Proc. Natl. Acad. Sci. USA* **2015**, *112*, E4901–E4910. [CrossRef] [PubMed]
6. Gu, Z. Complex Heatmap Visualization. *iMeta* **2022**, *1*, e43. [CrossRef]
7. Gu, Z.; Eils, R.; Schlesner, M. Complex Heatmaps Reveal Patterns and Correlations in Multidimensional Genomic Data. *Bioinformatics* **2016**, *32*, 2847–2849. [CrossRef] [PubMed]
8. Yu, G.; Xu, C.; Zhang, D.; Ju, F.; Ni, Y. MetOrigin: Discriminating the Origins of Microbial Metabolites for Integrative Analysis of the Gut Microbiome and Metabolome. *iMeta* **2022**, *1*, e10. [CrossRef]
9. Xia, J.; Psychogios, N.; Young, N.; Wishart, D.S. MetaboAnalyst: A Web Server for Metabolomic Data Analysis and Interpretation. *Nucleic Acids Res.* **2009**, *37*, W652–W660. [CrossRef]
0. Krawisz, J.E.; Sharon, P.; Stenson, W.F. Quantitative Assay for Acute Intestinal Inflammation Based on Myeloperoxidase Activity. Assessment of Inflammation in Rat and Hamster Models. *Gastroenterology* **1984**, *87*, 1344–1350. [CrossRef]
1. Kumar, A.; Anbazhagan, A.N.; Coffing, H.; Chatterjee, I.; Priyamvada, S.; Gujral, T.; Saksena, S.; Gill, R.K.; Alrefai, W.A.; Borthakur, A.; et al. Lactobacillus Acidophilus Counteracts Inhibition of NHE3 and DRA Expression and Alleviates Diarrheal Phenotype in Mice Infected with Citrobacter Rodentium. *Am. J. Physiol. Gastrointest. Liver Physiol.* **2016**, *311*, G817–G826. [CrossRef] [PubMed]
2. McIver, L.J.; Abu-Ali, G.; Franzosa, E.A.; Schwager, R.; Morgan, X.C.; Waldron, L.; Segata, N.; Huttenhower, C. BioBakery: A Meta'omic Analysis Environment. *Bioinformatics* **2018**, *34*, 1235–1237. [CrossRef] [PubMed]
3. Beghini, F.; McIver, L.J.; Blanco-Míguez, A.; Dubois, L.; Asnicar, F.; Maharjan, S.; Mailyan, A.; Manghi, P.; Scholz, M.; Thomas, A.M.; et al. Integrating Taxonomic, Functional, and Strain-Level Profiling of Diverse Microbial Communities with BioBakery 3. *eLife* **2021**, *10*, e65088. [CrossRef] [PubMed]
4. Ahmad, R.; Sorrell, M.F.; Batra, S.K.; Dhawan, P.; Singh, A.B. Gut Permeability and Mucosal Inflammation: Bad, Good or Context Dependent. *Mucosal. Immunol.* **2017**, *10*, 307–317. [CrossRef]
5. Santana, P.T.; Rosas, S.L.B.; Ribeiro, B.E.; Marinho, Y.; de Souza, H.S.P. Dysbiosis in Inflammatory Bowel Disease: Pathogenic Role and Potential Therapeutic Targets. *Int. J. Mol. Sci.* **2022**, *23*, 3464. [CrossRef]
6. Qiu, Z.; Yang, H.; Rong, L.; Ding, W.; Chen, J.; Zhong, L. Targeted Metagenome Based Analyses Show Gut Microbial Diversity of Inflammatory Bowel Disease Patients. *Indian J. Microbiol.* **2017**, *57*, 307–315. [CrossRef]
7. Wang, W.; Chen, L.; Zhou, R.; Wang, X.; Song, L.; Huang, S.; Wang, G.; Xia, B. Increased Proportions of Bifidobacterium and the Lactobacillus Group and Loss of Butyrate-Producing Bacteria in Inflammatory Bowel Disease. *J. Clin. Microbiol.* **2014**, *52*, 398–406. [CrossRef]
8. Glover, J.S.; Browning, B.D.; Ticer, T.D.; Engevik, A.C.; Engevik, M.A. Acinetobacter Calcoaceticus Is Well Adapted to Withstand Intestinal Stressors and Modulate the Gut Epithelium. *Front. Physiol.* **2022**, *13*, 880024. [CrossRef]

39. Facchin, S.; Vitulo, N.; Calgaro, M.; Buda, A.; Romualdi, C.; Pohl, D.; Perini, B.; Lorenzon, G.; Marinelli, C.; D'Incà, R et al. Microbiota Changes Induced by Microencapsulated Sodium Butyrate in Patients with Inflammatory Bowel Disease *Neurogastroenterol. Motil.* **2020**, *32*, e13914. [CrossRef]
40. Dahal, R.H.; Kim, S.; Kim, Y.K.; Kim, E.S.; Kim, J. Insight into Gut Dysbiosis of Patients with Inflammatory Bowel Disease and Ischemic Colitis. *Front. Microbiol.* **2023**, *14*, 1174832. [CrossRef]
41. Sasaki, K.; Inoue, J.; Sasaki, D.; Hoshi, N.; Shirai, T.; Fukuda, I.; Azuma, T.; Kondo, A.; Osawa, R. Construction of a Model Culture System of Human Colonic Microbiota to Detect Decreased Lachnospiraceae Abundance and Butyrogenesis in the Feces of Ulcerative Colitis Patients. *Biotechnol. J.* **2019**, *14*, e1800555. [CrossRef] [PubMed]
42. Liu, S.; Zhao, W.; Lan, P.; Mou, X. The Microbiome in Inflammatory Bowel Diseases: From Pathogenesis to Therapy. *Protein Cell* **2021**, *12*, 331–345. [CrossRef] [PubMed]
43. Chang, Q.; Luan, Y.; Sun, F. Variance Adjusted Weighted UniFrac: A Powerful Beta Diversity Measure for Comparing Communities Based on Phylogeny. *BMC Bioinform.* **2011**, *12*, 118. [CrossRef] [PubMed]
44. Parada Venegas, D.; De la Fuente, M.K.; Landskron, G.; González, M.J.; Quera, R.; Dijkstra, G.; Harmsen, H.J.M.; Faber, K.N Hermoso, M.A. Short Chain Fatty Acids (SCFAs)-Mediated Gut Epithelial and Immune Regulation and Its Relevance for Inflammatory Bowel Diseases. *Front. Immunol.* **2019**, *10*, 277. [CrossRef]
45. Peace, C.G.; O'Neill, L.A. The Role of Itaconate in Host Defense and Inflammation. *J. Clin. Investig.* **2022**, *132*, e148548. [CrossRef
46. Clooney, A.G.; Eckenberger, J.; Laserna-Mendieta, E.; Sexton, K.A.; Bernstein, M.T.; Vagianos, K.; Sargent, M.; Ryan, F.J.; Moran C.; Sheehan, D.; et al. Ranking Microbiome Variance in Inflammatory Bowel Disease: A Large Longitudinal Intercontinental Study. *Gut* **2021**, *70*, 499–510. [CrossRef]
47. Abdel-Rahman, L.I.H.; Morgan, X.C. Searching for a Consensus Among Inflammatory Bowel Disease Studies: A Systematic Meta-Analysis. *Inflamm. Bowel Dis.* **2023**, *29*, 125–139. [CrossRef]
48. Nie, K.; Ma, K.; Luo, W.; Shen, Z.; Yang, Z.; Xiao, M.; Tong, T.; Yang, Y.; Wang, X. Roseburia Intestinalis: A Beneficial Gut Organism from the Discoveries in Genus and Species. *Front. Cell. Infect. Microbiol.* **2021**, *11*, 757718. [CrossRef]
49. Luo, W.; Shen, Z.; Deng, M.; Li, X.; Tan, B.; Xiao, M.; Wu, S.; Yang, Z.; Zhu, C.; Tian, L.; et al. Roseburia Intestinalis Supernatant Ameliorates Colitis Induced in Mice by Regulating the Immune Response. *Mol. Med. Rep.* **2019**, *20*, 1007–1016. [CrossRef]
50. Tilg, H.; Danese, S. Roseburia Hominis: A Novel Guilty Player in Ulcerative Colitis Pathogenesis? *Gut* **2014**, *63*, 1204–120 [CrossRef]
51. Alexeev, E.E.; Lanis, J.M.; Kao, D.J.; Campbell, E.L.; Kelly, C.J.; Battista, K.D.; Gerich, M.E.; Jenkins, B.R.; Walk, S.T.; Kominsky D.J.; et al. Microbiota-Derived Indole Metabolites Promote Human and Murine Intestinal Homeostasis through Regulation of Interleukin-10 Receptor. *Am. J. Pathol.* **2018**, *188*, 1183–1194. [CrossRef] [PubMed]
52. Wu, H.; Zhang, M.; Li, W.; Zhu, S.; Zhang, D. Stachydrine Attenuates IL-1β-Induced Inflammatory Response in Osteoarthritis Chondrocytes through the NF-KB Signaling Pathway. *Chem. Biol. Interact.* **2020**, *326*, 109136. [CrossRef] [PubMed]
53. Gao, Y.; Zhou, B.; Zhang, H.; Chen, L.; Wang, X.; Chen, H.; Zhou, L. L-Ergothioneine Exhibits Protective Effects against Dextran Sulfate Sodium-Induced Colitis in Mice. *ACS Omega* **2022**, *7*, 21554–21565. [CrossRef]
54. Gründemann, D.; Harlfinger, S.; Golz, S.; Geerts, A.; Lazar, A.; Berkels, R.; Jung, N.; Rubbert, A.; Schömig, E. Discovery of the Ergothioneine Transporter. *Proc. Natl. Acad. Sci. USA* **2005**, *102*, 5256–5261. [CrossRef] [PubMed]
55. Bryan, P.-F.; Karla, C.; Edgar Alejandro, M.-T.; Sara Elva, E.-P.; Gemma, F.; Luz, C. Sphingolipids as Mediators in the Crosstalk between Microbiota and Intestinal Cells: Implications for Inflammatory Bowel Disease. *Mediat. Inflamm.* **2016**, *2016*, 989014 [CrossRef]
56. Hu, D.; Zhang, D.; Zheng, S.; Guo, M.; Lin, X.; Jiang, Y. Association of Ulcerative Colitis with FUT2 and FUT3 Polymorphisms in Patients from Southeast China. *PLoS ONE* **2016**, *11*, e0146557. [CrossRef]
57. Kappler, K.; Lasanajak, Y.; Smith, D.F.; Opitz, L.; Hennet, T. Increased Antibody Response to Fucosylated Oligosaccharides and Fucose-Carrying Bacteroides Species in Crohn's Disease. *Front. Microbiol.* **2020**, *11*, 1553. [CrossRef]
58. Vermeulen, N.; Vermeire, S.; Arijs, I.; Michiels, G.; Ballet, V.; Derua, R.; Waelkens, E.; Van Lommel, L.; Schuit, F.; Rutgeerts, P.; et al. Seroreactivity against Glycolytic Enzymes in Inflammatory Bowel Disease. *Inflamm. Bowel Dis.* **2011**, *17*, 557–564. [CrossRef]
59. Hryhorowicz, S.; Kaczmarek-Ryś, M.; Zielińska, A.; Scott, R.J.; Słomski, R.; Pławski, A. Endocannabinoid System as a Promising Therapeutic Target in Inflammatory Bowel Disease—A Systematic Review. *Front. Immunol.* **2021**, *12*, 790803. [CrossRef]
60. Maki, J.J.; Nielsen, D.W.; Looft, T. Complete Genome Sequence and Annotation for Turicibacter Sanguinis MOL361T (DSM 14220 *Microbiol Resour Announc* **2020**, *9*, e00475-20. [CrossRef]
61. Lim, S.; Rhee, M.-S.; Chang, D.-H.; Kim, B.-C. Whole-Genome Sequence of Erysipelothrix Larvae LV19T (=KCTC 33523T), a Useful Strain for Arsenic Detoxification, from the Larval Gut of the Rhinoceros Beetle, Trypoxylus Dichotomus. *J. Biotechnol.* **201** *223*, 40–41. [CrossRef] [PubMed]
62. Chen, L.-M.; Bao, C.-H.; Wu, Y.; Liang, S.-H.; Wang, D.; Wu, L.-Y.; Huang, Y.; Liu, H.-R.; Wu, H.-G. Tryptophan-Kynurenine Metabolism: A Link between the Gut and Brain for Depression in Inflammatory Bowel Disease. *J. Neuroinflamm.* **2021**, *18*, 13 [CrossRef] [PubMed]
63. Rustgi, S.D.; Kayal, M.; Shah, S.C. Sex-Based Differences in Inflammatory Bowel Diseases: A Review. *Therap. Adv. Gastroentero* **2020**, *13*, 1756284820915043. [CrossRef]

64. Sochal, M.; Ditmer, M.; Binienda, A.; Gabryelska, A.; Białasiewicz, P.; Talar-Wojnarowska, R.; Fichna, J.; Małecka-Wojciesko, E. Relation between Selected Sleep Parameters, Depression, Anti-Tumor Necrosis Factor Therapy, and the Brain-Derived Neurotrophic Factor Pathway in Inflammatory Bowel Disease. *Metabolites* **2023**, *13*, 450. [CrossRef]
65. Jo, J.-K.; Seo, S.-H.; Park, S.-E.; Kim, H.-W.; Kim, E.-J.; Kim, J.-S.; Pyo, J.-Y.; Cho, K.-M.; Kwon, S.-J.; Park, D.-H.; et al. Gut Microbiome and Metabolome Profiles Associated with High-Fat Diet in Mice. *Metabolites* **2021**, *11*, 482. [CrossRef] [PubMed]

cells

MDPI

Article

Identification of Small Molecules Affecting the Secretion of Therapeutic Antibodies with the Retention Using Selective Hook (RUSH) System

Mathilde Coulet [1,2,3], Sylvie Lachkar [2,3], Marion Leduc [2,3], Marc Trombe [1], Zelia Gouveia [4], Franck Perez [4], Oliver Kepp [2,3], Guido Kroemer [2,3,5,*,†] and Stéphane Basmaciogullari [1,*,†]

1 Sanofi R&D, 94400 Vitry-sur-Seine, France
2 Metabolomics and Cell Biology Platforms, Gustave Roussy Cancer Center, 94800 Villejuif, France
3 Centre de Recherche des Cordeliers, Equipe Labellisée par la Ligue Contre le Cancer, Université de Paris Cité, Sorbonne Université, Inserm U1138, Institut Universitaire de France, 75006 Paris, France
4 Institut Curie, PSL Research University, Sorbonne Université, CNRS, UMR 144, 26 rue d'Ulm, 75005 Paris, France
5 Department of Biology, Institut du Cancer Paris CARPEM, Hôpital Européen Georges Pompidou, AP-HP, 75015 Paris, France
* Correspondence: kroemer@orange.fr (G.K.); stephane.basmaciogullari@sanofi.com (S.B.)
† These authors contributed equally to this work.

Abstract: Unlocking cell secretion capacity is of paramount interest for the pharmaceutical industry focused on biologics. Here, we leveraged retention using a selective hook (RUSH) system for the identification of human osteosarcoma U2OS cell secretion modulators, through automated, high-throughput screening of small compound libraries. We created a U2OS cell line which co-expresses a variant of streptavidin addressed to the lumen-facing membrane of the endoplasmic reticulum (ER) and a recombinant anti-PD-L1 antibody. The heavy chain of the antibody was modified at its C-terminus, to which a furin cleavage site, a green fluorescent protein (GFP), and a streptavidin binding peptide (SBP) were added. We show that the U2OS cell line stably expresses the streptavidin hook and the recombinant antibody bait, which is retained in the ER through the streptavidin–SBP interaction. We further document that the addition of biotin to the culture medium triggers the antibody release from the ER, its trafficking through the Golgi where the GFP-SBP moiety is clipped off, and eventually its release in the extra cellular space, with specific antigen-binding properties. The use of this clone in screening campaigns led to the identification of lycorine as a secretion enhancer, and nigericin and tyrphostin AG-879 as secretion inhibitors. Altogether, our data support the utility of this approach for the identification of agents that could be used to improve recombinant production yields and also for a better understanding of the regulatory mechanism at work in the conventional secretion pathway.

Keywords: RUSH assay; antibody; high-throughput screening; protein secretion

1. Introduction

Secretion is a fundamental cell-biology process that enables the release of molecules in the extracellular environment. This capacity is necessary to fulfill a multitude of needs, and the released molecules can play roles in growth, chemotactism, homeostasis, immunity, and cell–cell signaling, to mention a few [1]. Eukaryotic cells have a functional network of interconnected organelles, enabling the synthesis and the controlled transport of a variety of proteins along multiple routes. In particular, the conventional secretory pathway allows for the exocytosis of proteins synthetized in the endoplasmic reticulum (ER) to the extracellular environment. Upon completion of mRNA translation and protein translocation to the lumen of the ER, the newly formed proteins are sorted in specialized membrane structures that bud coat protein complex II (COPII)-coated vesicles. These vesicles are transported along the microtubule network towards the Golgi apparatus, where fusion with the *cis*-Golgi takes place, involving the interaction of soluble N-ethylmaleimide-sensitive-factor attachment receptors (SNAREs) present on both the vesicles and *cis*-Golgi membranes. Proteins undergo additional post-translational modifications as they reach the *trans*-Golgi face from which the clathrin-coated vesicles bud. These vesicles transport their payload to the plasma membrane, where SNARE-mediated fusion occurs, and the proteins are eventually released into the extracellular space [2]. In addition to the structural components that make the trafficking of vesicles from the ER to the plasma membrane possible, many regulatory mechanisms act as quality control during this process, such as the unfolded protein response (UPR) within the ER, where misfolded proteins are diverted from the secretory pathway, in order to ensure the secretion of functional proteins. Defects in protein folding and secretion can lead to various pathologies, such as cancer, and neurodegenerative, metabolic, and inflammatory diseases, highlighting the importance of better understanding the underlying cellular mechanisms, which would pave the way for the identification of novel therapeutic molecules [3–6].

An understanding of the secretion mechanisms is also of paramount interest for the pharmaceutical industry, which has started a transition from the all-chemical era to biologics. Indeed, the proportion of drugs made from living organisms has greatly increased in the past decades, today representing one quarter of all approved therapies; half of them being for therapeutic antibodies [7]. Such antibodies are produced after transfection of suitable cells with specific heavy and light chain-encoding vectors, cell culture upscaling, and antibody purification from the cell culture supernatant [7]. Several directions have been investigated with the final goal of optimizing titers of recombinant proteins, and these include cell host selection, vector design, and medium formulation, together with upstream and downstream process optimization. Small-scale experiments are thus conducted in development labs, before intermediate and final upscaling of the production process in bioreactors for clinical or commercial manufacturing [8]. Besides these canonical optimization efforts, little has been reported in the engineering literature regarding the secretion mechanism itself, from the protein assembly in the ER down to its secretion in the extracellular environment [9]. The identification of modulators of protein secretion is an efficient strategy for enriching the available knowledge on the key elements of the secretory pathway. Among the inhibitors reported in the literature, several small molecules have enabled highlighting the role of the cytoskeleton in protein secretion. Brefeldin A is known for its capacity to impact the anterograde transport of vesicles from the ER to the Golgi, while nocodazole causes depolymerization of microtubules, and cytochalasin B destabilizes the actin network [10,11]. In addition, monensin was shown to block the transport of the protein within the Golgi apparatus [12]. All of these compounds significantly decrease protein secretion and have made it possible to study the key molecular mechanisms of the conventional secretory pathway. Other approaches have also allowed the identification of practices or compounds capable of increasing protein secretion. For example, cell cycle arrest has been correlated with enhanced protein production. Although the exact mechanisms remain to be understood, this strategy is used in the industry to increase production yields. Cell cycle arrest can be achieved through genetic engineering, such as in cell cycle regulatory gene over-expression; through the induction of the hypothermic response; or through the use of small molecules, such as sodium butyrate [13,14]. The importance of chaperone proteins for the correct folding of nascent proteins and secretion has also been widely documented, and their over-expression obtained through genetic manipulation resulted in enhanced protein secretion in several studies [2,15,16]. For example, targeting the UPR via the upregulation of genes involved in this pathway or using ER stress-inducing small molecules resulted in increased cell-specific productivity, confirming the importance of this response [17]. Finally, the conventional secretory pathway was successfully activated through strategies involving the proteins required for the correct addressing of vesicles, such as SNAREs [18], ceramide transferase protein (CERT) [19], and Munc18b [2], or proteins participating in the dynamics of the cytoskeleton, including cofilin [20], actin alpha cardiac muscle 1 (ACTC1), and guanosine triphosphate GTPase-activating protein (TAGAP) [21,22].

In line with these studies, we aimed to identify new modulators of protein secretion by leveraging retention using a selective hook (RUSH) system in a medium-throughput drug library screening approach. The RUSH system was first described by Boncompain et al. [23] and is based on the retention of the protein of interest in a specific subcellular location, until its triggered release in the secretory pathway. This model is particularly useful in a kinetics-based screening approach, where events need to be synchronized. In this work, the RUSH system was adapted to study the secretion of a modified anti-PDL1 antibody and the screening of small compound libraries. A GFP moiety was added to the C-terminal portion of the antibody heavy chain, to allow secretion monitoring with video microscopy. In addition, a furin-cleavable sequence was inserted between the Fc domain of the antibody and the GFP moiety. This ensured the processing of the fusion protein by Golgi-resident furins [24] and the release of a native antibody in the extracellular space that was directly usable in functional assays. Our screen identified inhibitors and activators of antibody secretion, which were further validated in an antigen binding assay.

2. Materials and Methods

2.1. Cell Culture, Chemicals, and Antibodies

Human osteosarcoma U2OS cells obtained from the Kroemer lab were maintained at 37 °C in a humidified incubator with 5% CO_2 in Dulbecco's modified Eagle medium (DMEM) supplemented with 10% fetal bovine serum (FBS), 1% penicillin-streptomycin, and 1% HEPES (Gibco, Carlsbad, CA, USA). CHO-K1 (CCL-61) cells were obtained from the ATCC and cultured at 37 °C in a humidified incubator with 5% CO_2 in F-12 Ham's nutrient mixture (Sigma, Saint-Louis, MO, USA) supplemented with 10% FBS and 2 mM L-Glutamine (Gibco). PD-L1-expressing CHO cells (Sanofi property) were cultured in similar conditions in medium supplemented with 7.5 µg/mL puromycin (Gibco). Prestwick, the BML-2840 ICCB Known Bioactives, and the Autophagy Compounds libraries were purchased from Prestwick Chemicals (Illkirch, France), Enzo Life Sciences (Farmingdale, NY, USA), and TargetMol (Boston, MA, USA), respectively. Anti-streptavidin (Santa Cruz Biotechnology, Dallas, TX, USA), anti-calreticulin (Abcam, Cambridge, UK), anti-SBP (Santa Cruz Biotechnology), anti- human heavy and light IgG chains (Invitrogen), and the anti-GFP (Cell Signaling, Danvers, MA, USA) antibodies were used at 1:500, 1:200, 1:500, 1:5000, and 1:1000 dilutions, respectively, in Western blotting and immunofluorescence experiments. The reference anti-human PD-L1 antibody used in FACS experiments was produced at Sanofi labs from cells transfected with plasmid constructs based on the published Atezolizumab heavy and light chain sequences available on the IMGT website (imgt.org, accessed on 12 May 2023) [25]. The secondary Alexa Fluor 568- and 488-conjugated anti-mouse and rabbit IgG antibodies (Life Technologies, Carlsbad, CA, USA) were used at 1:1000 and 1:500 dilutions, respectively, in immunofluorescence and FACS experiments. The secondary HRP-conjugated goat anti-mouse and rabbit IgG (Southern Biotech, Birmingham, AL, USA) antibodies were used at a 1:5000 dilution in the Western blotting experiments. The anti-human kappa light chain antibody (Thermo Fisher Scientific, Waltham, MA, USA) was used at a 1:4000 dilution in ELISA.

2.2. Constructs Used for the RUSH Antibody Secretion System

The streptavidin-KDEL plasmid was described in Zhao et al. [26]. The amino acid sequences of the heavy and light chains of the anti-PD-L1 Atezolizumab antibody available on IMGT were used to generate nucleic acid sequences adapted to expression in human cells. Sequences encoding a furin-cleavable site (AVSKERSKRSP), a linker $(G_4S)_2$, enhanced green fluorescent protein (eGPF), and a streptavidin binding protein (SBP) were added in that order to the 3′ end of the anti-PD-L1 heavy chain coding sequence. Anti-PD-L1 heavy and light chain constructs were manufactured by GeneArt (Ratisbonne, Germany) and provided in a shuttle vector. The heavy and light chain sequences were then cloned into the multiple cloning sites 2 and 1, respectively, of the pVITRO1 plasmid (Invivogen, San Diego, CA, USA) using a NEBuilder assembly kit (New England Biolabs, Ipswich, MA, USA). The integrity of the construct, hereafter referred to as anti-PD-L1 IgG RUSH plasmid, was confirmed by sequencing (Eurofins, Luxemburg).

2.3. Establishment of the RUSH-Antibody Cell Line

U2OS cells were transfected with the streptavidin-KDEL plasmid using Lipofectamine 2000 (Thermo Fisher Scientific) and selected in medium supplemented with G418 (400 μg/mL, Gibco-Invitrogen) for 2 weeks. Limiting dilution in 96-well plates was used for single cell isolation from the selected pool. Clones were amplified and selected based on streptavidin immunofluorescence. Clone #BD6 showing the highest streptavidin expression was further expanded in medium supplemented with G418 (200 μg/mL) and transfected with the anti-PD-L1 IgG RUSH plasmid using Lipofectamine 2000 and selected in medium supplemented with G418 (200 μg/mL) and hygromycin (100 μg/mL, Santa Cruz Biotechnology) for 2 weeks. Limiting dilution in 96-well plates was used for single cell isolation from clone #BD6-derived selected pool. Clones were amplified and selected based on GFP fluorescence, and clone #7 was selected for further characterization and screening campaigns.

2.4. Immunofluorescence

U2OS cells were seeded in black 96-well imaging plates (Greiner bio one, Kremsmünster, Austria). After 24 h, the medium was removed, cells were rinsed with 37 °C pre-heated PBS and fixed in PBS supplemented with 4% paraformaldehyde (PFA, Sigma) and 1 μg/mL Hoechst (Life Technologies, Carlsbad, CA, USA) for 20 min under agitation at room temperature (RT). Cells were rinsed with PBS before addition of a 5% FBS/0.3% Triton X-100/PBS permeabilizing solution for 30 min under agitation at RT. Cells were rinsed with PBS and incubated overnight under agitation at 4 °C with the primary antibody in 1% BSA/PBS. Cells were washed with PBS and incubated for 45 min under agitation at RT with the secondary antibody in 1% BSA/PBS. Finally, cells were washed and processed for image acquisition using a IXM XL BioImager (Molecular Devices, Sunnyvale, CA, USA).

2.5. Antibody Expression and Secretion Characterization by Western Blot

U2OS clone #7 cells were seeded in T-175 flasks and kept for 24 h at 37 °C in a humidified incubator with 5% CO_2. The next day, the supernatant was removed; cells were rinsed with 37 °C pre-heated PBS; and new serum-free medium containing no biotin, 40 μM biotin, or 40 μM biotin together with 100 μM of furin inhibitor I (Sigma) was added to the flask for 4 h. At the end of the incubation time, cells were harvested and lysed in RIPA buffer (Thermo Fisher Scientific) supplemented with protease and phosphatase inhibitors (Roche, Basel, Switzerland). Additionally, cell culture supernatants were collected and concentrated ~100-fold in a Amicon 50 kDa filter unit (Sigma) according to the manufacturer's instructions. Cell lysate and concentrated supernatant protein concentrations were determined using a BCA™ protein assay kit (Thermo Fisher Scientific). Next, 10 μg of total protein content from lysates and supernatants was resolved on NuPAGE™ 4–12% Bis-Tris gels (Invitrogen, Carlsbad, CA, USA) under reducing conditions and transferred to a nitrocellulose membrane (BioRad, Hercules, CA, USA) using a BioRad system. The membranes were blocked for 1 h with 0.01% Tween-20/5% non-fat dry milk/TBS, and the primary antibodies were added overnight under agitation at 4 °C. The membranes were rinsed with 0.01% Tween-20/5% non-fat dry milk/TBS and incubated for 1 h with secondary antibodies under agitation at 4 °C. The membranes were then washed 3 times for 5 min with 0.01% Tween-20/TBS, and the peroxidase activity was evaluated with Amersham ECL Primer Western Blotting Detection Reagent (GE Healthcare, Little Chalfont, UK) on ImageQuant LAS4000 (GE Healthcare).

2.6. Flow Cytometric Validation of Antibody Reactivity

U2OS clone #7 cell culture supernatants were prepared as described above and used to assess the binding of the RUSH antibody to PD-L1-expressing cells through flow cytometry. Three million CHO-K1 and CHO PD-L1 cells were harvested, rinsed with PBS, and incubated for 30 min at 4 °C under agitation in the dark with either 150 μL of the recombinant Atezolizumab prepared in Sanofi research labs or concentrated U2OS clone #7 cell culture supernatant. Cells were then rinsed with PBS and secondarily stained for 30 min at 4 °C under agitation in the dark with an Alexa Fluor 647-coupled anti-human IgG antibody, before flow cytometry analysis of 100 μL solution on a MACSQuant (Miltenyi Biotec, Bergisch Gladbach, Germany).

2.7. High-Throughput Compound Screening for Modulators of Protein Secretion

U2OS clone #7 cells were seeded in black 384-well imaging plates (Greiner bio one), at 2000 cells/well, and incubated for 24 h at 37 °C in a humidified incubator with 5% CO_2. Cells were treated on the next day with compounds from either the Prestwick library at 20 μM or 10 μM, the ICCB Known Bioactives library at $1/100^e$ or $1/1000^e$, or the autophagy compound library at 10 μM or 1 μM for 4 h. At the end of the incubation time, biotin was added at 40 μM for 1 h (untreated cells and DMSO-treated cells were used as a control). Cells were then fixed as described above and processed for subsequent automated image acquisition using a robot-assisted IXM XL BioImager (Molecular Devices, Sunnyvale, CA, USA) equipped with a Sola light source (Lumencor, Beaverton, OR, USA), adapted excitation and emission filters (Semrock, Rochester, NY, USA), and a 16-bit monochrome sCMOS PCO.edge 5.5 camera (PCO, Kelheim, Germany). A 20 X PlanAPO objective (Nikon, Tokyo, Japan) was used to acquire a minimum of 4 view fields in each well. The acquired images were processed using the open-access R software (https://www.r-project.org, accessed on 12 May 2023) with the use of the freely available packages EBImage (available on the Bioconductor repository https://www.bioconductor.org, accessed on 12 May 2023) and RBioFormats (https://github.com/aoles/RBioFormats, accessed on 12 May 2023), and the custom packages MetaxpR. (https://github.com/asauvat/MetaxpR, accessed on 12 May 2023) and MorphR (https://github.com/kroemerlab/MorphR, accessed on 12 May 2023). Nuclei were first detected using the Hoechst 33342 signal and used as a marker to detect cytoplasmic areas based on GFP signal, allowing for the evaluation of cytoplasmic GFP intensity. Data were subsequently extracted and statistically evaluated using R (https://www.r-project.org, accessed on 12 May 2023). Data were normalized using negative and positive controls, reduced, and transformed into a Z-score. The experiments were repeated manually at low scale with selected drugs, to confirm their effects on secretion; in this second set of assays, U2OS clone #7 cells incubated with secretion inhibitors were fixed and stained with Hoechst as described above, for imaging and viability determination based on cell counts.

2.8. ELISA

U2OS clone #7 cells were seeded in 6-well plates and kept for 24 h at 37 °C in a humidified incubator with 5% CO_2. The next day, the supernatants were removed, cells were rinsed with 37 °C pre-heated PBS, and new medium containing the chemical treatment was added for 4 h. At the end of the incubation time, no biotin or 40 μM biotin was added to the flask for 30 min (inhibitory selected molecules) or 1 h (activator selected molecules). Then supernatants were collected and used for the ELISA assay, as follows: MaxiSorp 96-well plates (Thermo Fisher Scientific) were coated with 0.5 μg/mL hFc-human-PD-L1 (R&D Systems, Minneapolis, MN, USA) in PBS for 5.5 h at RT. Coating solution was then removed and 2% BSA/PBS blocking solution was added overnight at 4 °C. The next day, wells were washed 3 times with 0.5% BSA/0.05% Tween-20/PBS solution. Supernatants were added after a 1/3 dilution in 0.5% BSA/0.05% Tween-20/PBS, for 2.3 h at RT. Wells were washed 3 times with 0.5% BSA/0.05% Tween-20/PBS solution. Anti-human kappa light chain HRP-coupled antibody was added diluted in 0.5% BSA/0.05% Tween-20/PBS at 1/4000[e] for 2.3 h at RT. Wells were washed 3 times with 0.5% BSA/0.05% Tween-20/PBS solution, before incubation with a substrate reagents pack (R&D Systems) for 15 min in the dark. Then, 50 μL of STOP solution (R&D Systems) was added to each well and the absorbance in each well was measured at 450 nm, and 635 nm for background subtraction, using a VICTOR™ X4 (PerkinElmer, Waltham, MA, USA).

2.9. Statistics

Unless otherwise specified, data are reported as the mean ± SD of two replicates in a minimum of three independent experiments. Statistical significance was assessed using a one-sided Mann–Whitney U-test.

3. Results

3.1. Design of an Antibody RUSH System

The RUSH system is composed of a hook that is streptavidin and a bait that is a protein fused to a streptavidin-binding peptide (SBP). When these two molecules are co-expressed, the bait is sequestered by the hook, due to the high-affinity streptavidin/SBP interaction (K_d in the range of 10^{-8} to 10^{-9} Mol) [27], and released upon addition of biotin, which outcompetes SBP due to its higher affinity for streptavidin (K_d in the range 10^{-14} Mol) [28]. Here, we adapted this cell-based assay to monitor the secretion kinetics of an antibody (the bait) and screen small compound libraries to identify secretion modulators. As antibodies are secreted through the conventional pathway, the streptavidin hook was fused to a KDEL peptide to ensure its retention in the endoplasmic reticulum (ER) through the interaction with the KDEL receptors present in the ER [29]. The antibody was also modified to fit the purpose of the assay. A polycistronic vector was used that encodes both the heavy and light (kappa) chains of a human PD-L1-specific antibody [30]. The C-terminus of the heavy chain was fused to a $(Gly_4Ser)_2$ linker, a furin-cleavable site, the green fluorescent protein (GFP), and SBP (Figure 1B). We reasoned that the fusion molecule (hereafter called RUSH antibody) would assemble into mature homodimers upon protein synthesis and remain in the ER due to its interaction with the streptavidin KDEL hook, until addition of biotin would trigger its release and maturation in the Golgi through its trafficking to the extracellular space (Figure 1A).

The hook and bait expressing clone was obtained using U2OS cell transfection with the hook-encoding construct and the selection a clone with high hook expression levels, in order to ensure efficient retention of the bait and minimize leakiness in the absence of biotin. As shown in Figure 1C, Clone #BD6 staining with streptavidin-specific antibodies revealed the colocalization of the hook with the prototypic ER-sessile protein calreticulin [31]. Clone #BD6 was further transfected with the RUSH antibody encoding vector, followed by single cell cloning and the selection of the hook/bait clone used in this study (clone #7). This clone was selected based on two criteria: (i) a bright GFP-dependent fluorescent signal, indicating high abundance of the RUSH antibody; and (ii) discrete localization of the GFP signal in the perinuclear area, indicating ER retention (Figure 2A). Fluorescence video-microscopic observation of cells from clone #7 revealed that GFP remained in the same subcellular compartment over the 2 h of acquisition. On the contrary, the addition of biotin resulted in the relocation of the GFP signal towards discrete puncta that corresponded to the Golgi apparatus within 40 to 120 min, with a complete cytoplasmic GFP signal decay within 4 h, most likely due to the release of the fusion protein into the extracellular space (Figure 2A). The GFP signal co-localized with streptavidin at baseline and the addition of biotin did not modify this pattern, confirming the retention of the hook in the ER, regardless of the presence or absence of biotin in the medium. However, the GFP/streptavidin co-localization was lost 4 h after addition of biotin (Figure 2B). These results confirmed the expression of the bait and its retention in the ER through the hook, and suggest its progressive release in the extracellular space upon addition of biotin.

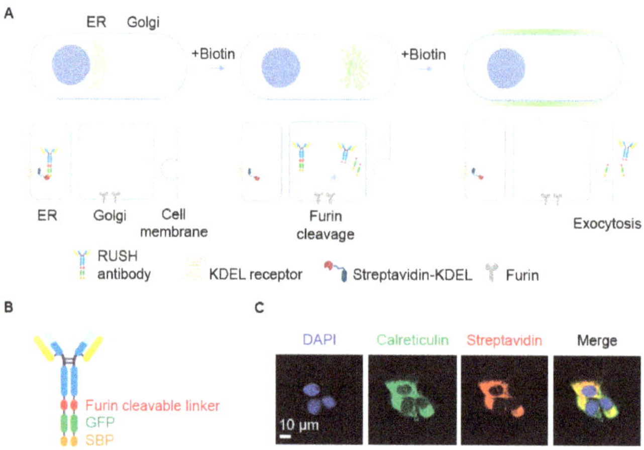

Figure 1. Design of the RUSH antibody secretion system (**A**) Principle of the RUSH assay for antibody secretion. The fusion protein (in this case, the antibody fused to the streptavidin binding domain, SBD, separated by a linker that contains a furin cleavage site) is retained in the ER lumen via the interaction of SBP with a streptavidin-KDEL hook. Upon addition of biotin, the interaction between the hook and the fusion protein is competitively disrupted, and the fusion protein is hence free to follow the classical secretory pathway to the Golgi, where the resident furin protease cleaves the linker between SBD and the antibody. The resulting parts are then secreted in the extracellular environment. The GFP moiety enables following the process with fluorescence microscopy. (**B**) Schematic representation of the custom anti-PD-L1 RUSH antibody, a fusion protein composed of an anti-PD-L1 IgG modified in its C-terminal end by the addition of a furin cleavable linker, a GFP, and an SBP sequence. (**C**) Immunostaining of fixed cells showing the colocalization of the streptavidin-KDEL hook with calreticulin in the ER. Pearson's correlation coefficient (PCC) = 0.69. Surface overlap coefficient (SOC) = 0.45.

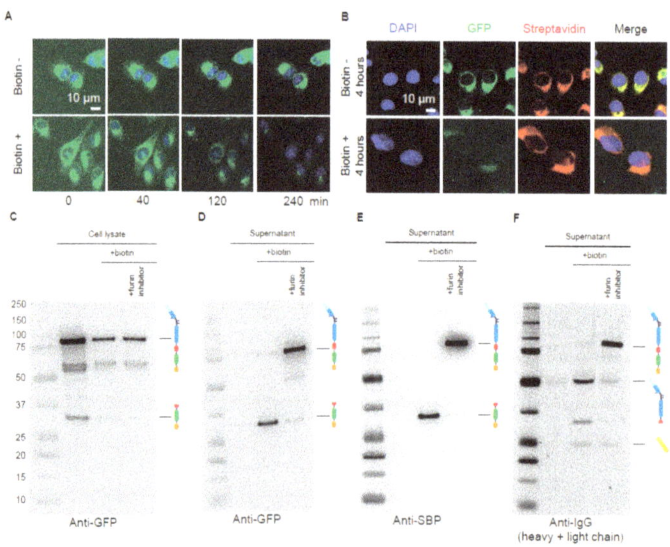

Figure 2. Characterization of the anti-PD-L1 antibody RUSH system (**A**) Time lapse microscopy of clone #7 cells incubated in the absence or in presence of 40 μM biotin, fixed and stained with DAPI.

The secretion rate could be deducted from the decreasing intracellular GFP intensity upon addition of biotin to the cell cultures. (**B**) Immunostaining of fixed cells incubated for 4 h in the absence or in the presence of biotin showing the colocalization of the streptavidin-KDEL hook with the GFP-tagged antibody in the absence of biotin and the loss of colocalization 4 h post biotin addition. PCC = 0.51 in the absence of biotin, and PCC = 0.29 in the presence of biotin. (**C–F**) Western blot of cell lysates (**C**) or concentrated supernatants (**D–F**) of clone #7 incubated for 4 h in the absence of biotin, in the presence of 40 μM biotin, or in the presence of 40 μM biotin together with 100 μM furin inhibitor I. Membranes were probed with the indicated antibodies to detect proteins that contain the GFP moiety (**C**,**D**), the SBP moiety (**E**), or epitopes from the immunoglobulin heavy and light chains (**F**). Molecular mass standards (in kDa) are indicated on the left, in panel (**D**). The structures of the proteins detected by immunoblot are indicated based on the scheme shown in Figure 1B (blue, antibody heavy chain; yellow, antibody light chain; green, GFP; orange, SBP; red shape, linker with furin cleavage site; truncated red shape, linker residues left after cleavage by furin). Results are representative of at least three independent experiments.

3.2. *Biochemical Validation of the Antibody RUSH System*

In the next step, we proceeded to the biochemical validation of clone #7, before its use in a screening campaign. For this aim, we characterized the molecular species synthesized and released by clone #7 cultured with or without biotin and a furin protease inhibitor. As can be seen in Figure 2C, the GFP-specific antibodies used in the Western blotting experiments revealed the presence of a major molecular species of ~80 kDa and two minor species of ~55 and ~35 kDa in the lysate of cells grown in the absence of biotin. The 80 kDa mass was compatible with the full-length heavy chain of the RUSH antibody (theoretical mass of 84 kDa), while the 35 kDa mass was compatible with the GFP-SBP C-terminal fragment of the heavy chain, which was released upon cleavage by furin (theoretical mass of 32 kDa). A similar analysis performed on the cell culture supernatant with GFP, SBP, and human IgG- specific antibodies only revealed trace amounts of recombinant protein in the absence of biotin (Figure 2D–F). This is consistent with the minor processing observed in cell lysates and therefore the efficient retention of the RUSH antibody in pre-Golgi intracellular compartments, as documented through the immunofluorescence (Figure 2A,B).

When cells were cultured in the presence of biotin, the Western blotting pattern in the cell lysate was unchanged, except that the intensity of the signal was lower, which was concomitant with the detection of recombinant species in the cell culture supernatant (Figure 2C–F). This demonstrates that the intracytoplasmic GFP decay observed via immunofluorescence in Figure 2A,B was a direct consequence of the protein release in the extracellular space. Moreover, the molecular masses of the material detected in the supernatant suggest an almost full processing of the released RUSH antibody. This was further confirmed by the comparable patterns obtained in cell lysates and in the cell culture supernatant when cells were grown in the presence of furin inhibitor. It can therefore be concluded that the RUSH antibody was released through the Golgi network and processed upon addition of biotin.

Taken together, these findings demonstrated that clone #7 allowed for biotin-stimulated release, as well as for the furin-dependent proteolytic maturation, of the RUSH antibody.

3.3. *Immunological Validation of the Antibody RUSH System*

Besides the mechanistic validation of clone #7, in terms of the antibody production, retention, maturation, and release upon addition of biotin, we also addressed the PD-L1 binding properties of the secreted RUSH antibody. We further investigated the RUSH antibody binding properties to PD-L1 in its native conformation when it was presented at the plasma membrane. For this aim, clone #7 was incubated with biotin and the cell culture supernatant was assayed for binding on PD-L1^{pos} and PD-L1^{neg} CHO cells by FACS. As can be seen in Figure 3A,B, native recombinant anti-PD-L1 binding to PD-L1^{pos} (CHO-PDL1) but not PD-L1^{neg} CHO (CHO-K1) cells was documented, which validated the use of such a system for further characterization of the RUSH antibody. Similarly, the supernatant of clone #7 cultured in the presence of biotin demonstrated binding activity to CHO-PDL1 cells but not to CHO-K1 cells (Figure 3C,D), indicating that the released RUSH antibody conserved its antigen-binding properties, despite the presence of extra ammino-acids at its C-terminus, left after processing with furins. Of note, PD-L1 binding was also detectable in the supernatant of clone #7 cultured in the presence of biotin and a furin protease inhibitor (Figure 3E,F). In these conditions, binding was detected through both the secondary staining antibody and the GFP moiety of the unprocessed RUSH antibody, which was not detectable with the native antibody (Figure 3B) and only very slightly detected in the case of the supernatant of clone #7 cultured with biotin only (Figure 3D). Altogether, these observations indicate that clone #7 was fit for both screening campaigns and further antibody characterization in PD-L1-binding assays.

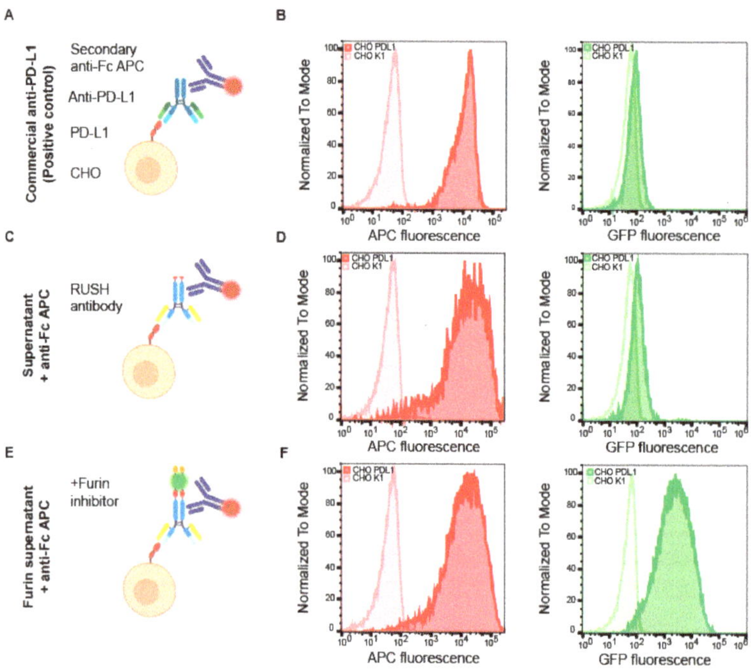

Figure 3. Characterization of the binding capacity of the anti-PD-L1 RUSH antibody released through the biotin-activated RUSH system. (**A**,**C**,**E**) Schematic representation of the antibody binding to PD-L1^{pos} cells and detection by flow cytometry. (**B**) FACS histogram plot of native recombinant and purified anti-PD-L1 antibody. (**D**,**F**) FACS histogram plots of processed and unprocessed RUSH anti-PD-L1 antibody released by clone #7 in the presence of biotin and in the absence or presence of furin inhibitor, respectively. Dark colors, binding to PD-L1^{pos} CHO cells (CHO-PDL1); light colors, binding to PD-L1^{neg} parental CHO cells (CHO-K1). Antibody binding was detected by means of an indirect immunofluorescence assay with a secondary anti-Fc APC conjugated (red) or by direct measurement of GFP fluorescence.

3.4. Pharmacological Screen for Modulators of Biotin-Induced Antibody Release

We next sought to use clone #7 in screening campaigns for the identification of drugs capable of enhancing or inhibiting antibody secretion. For this, clone #7 was cultured in 384 well plates and treated for 4 h with drugs from chemical libraries, followed by the addition of biotin for 1 h and the determination of the intra-cytoplasmic GFP signal (Figure 4A). Cell counting was facilitated by nuclear counterstaining with Hoechst 33324, in order to assess the viability and distinguish drug effects on the secretion kinetics from direct cytotoxicity, and to select relevant drugs for further characterization. Two independent screening campaigns were performed, to test a total of 2300 drugs at 2 different concentrations (i.e., a low concentration and a 10-fold higher concentration, referred to as "low" and "high" doses, as described in the Section 2).

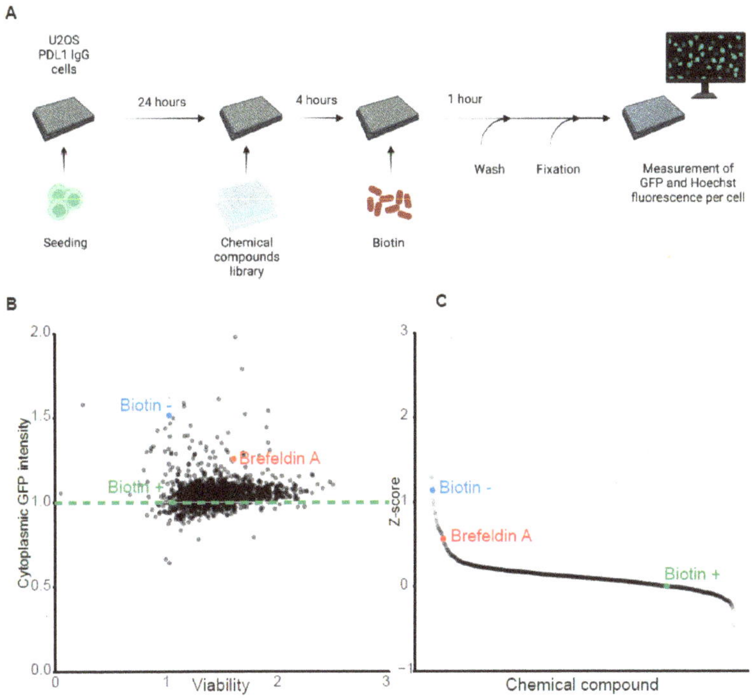

Figure 4. High-throughput screening of chemical compound libraries for the identification of modulators of protein secretion. (**A**) Screening workflow. U2OS clone #7 was plated in 384-well plates and incubated for 24 h, before treatment with each compound of the Prestwick, ICCB Known Bioactives, or autophagy compounds at two concentrations: low and high (with a 10-fold difference). Biotin was added after 4 h, and cells were washed and fixed 1 h after addition of biotin and processed for automated image acquisition of the intracellular Hoechst 33324 and GFP fluorescence. (**B**) Dot plot representation of drugs based on their impact on GFP release (cytoplasmic GFP signal) and nucleus integrity (viability). (**C**) Z-score ranking of the cytoplasmic GFP intensity after treatment. The results of the Prestwick library screening at a low drug concentration were plotted. Brefeldin A was used as a positive control for the inhibition of secretion. The compounds causing a cytoplasmic GFP intensity below that observed for the biotin-only treated cells were potential enhancers of protein secretion. Drugs that induced a GFP fluorescence level higher than that measured in the presence of biotin only were potential inhibitors of protein secretion.

The biotin-induced decrease of the cellular GFP signal was influenced by a minority of compounds from various chemical libraries, as indicated through the plotting of primary data (Figure 4B) or Z-score analyses (Figure 4C). As an internal control, brefeldin A, which is an inhibitor of conventional Golgi-dependent protein secretion [32], inhibited the biotin-induced decrease of the GFP signal (Figure 4B,C). Plotting of the Z-scores obtained at the low and high concentrations for each compound (Figure 5A,B) led to the identification of several compounds that consistently decreased the fluorescent signal, therefore acting as secretion activators. In order to select candidates to be further characterized, the 20 best activators identified at either the low or high concentration were identified, and among these, activators working at both concentrations were shortlisted. In the Prestwick library, balsalazide sodium (no longer commercially available), methenamine, pyrithyldione, cycloheximide, digoxigenin, verteporfin, and ropinirole fulfilled these criteria (Figure 5A). Verteporfin was eliminated due to its phototoxicity that leads to GFP bleaching [33]. In a combined screen involving compounds contained in two additional libraries (ICCB Known Bioactives and TargetMol Autophagy), several additional compounds (anisomycin, emetine, and pinocembrine) were active at both low and high concentrations. Additionally, although lycorine was not among the 20 most active compounds at high concentration, it was more efficient than any other compound at low concentration and was therefore selected (Figure 5B).

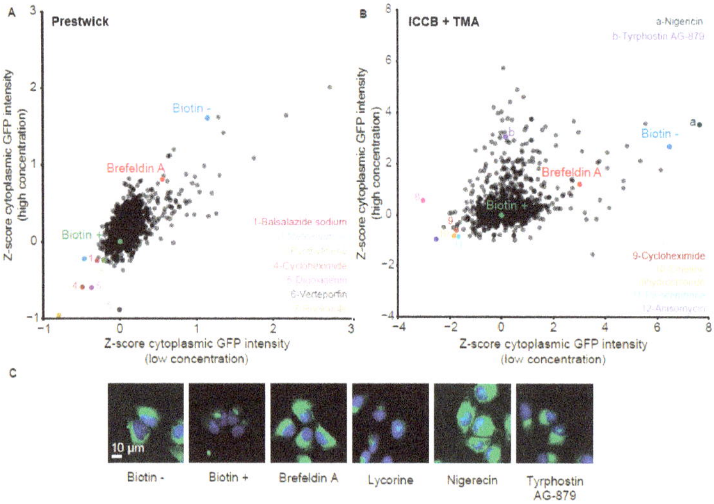

Figure 5. Selection of potential modulators of protein secretion. (**A**,**B**) Dot plot representation of drug efficiency based on the Z-scores measured at low and high drug concentration ((**A**) Prestwick library; (**B**) ICCB Known Bioactives and TargetMol Autophagy libraries). Compounds selected for further characterization are listed and highlighted in color. (**C**) Clone #7 was treated with the indicated compounds, fixed and analyzed as described in Figure 4A legend.

Several drugs were also identified that inhibited the biotin-induced cytoplasmic GFP signal decay, therefore acting as potential secretion inhibitors, among which nigericin and tyrphostin AG-879 were selected for further characterization (Figure 5B) because they had not been identified previously by Zhao et al. [26].

3.5. Validation of Stimulators and Inhibitors of Biotin-Induced Antibody Release

Medium- and high-throughput screenings often yield false positive and negative hits, calling for independent validation in low-throughput experiments with orthogonal technologies. In a first round of experiments, we evaluated the effects of shortlisted compounds with potential antibody secretion enhancer activity. The compounds identified in the screening campaigns were retested on clone #7 at a low concentration (Figure 6A) or a ten-fold higher concentration (Figure 6B) for their ability to increase antibody secretion. This experiment showed that all selected hits were active at the highest concentration, which confirmed the results of the screening (Figure 6B). We further characterized these hits in an PD-L1 binding ELISA. In this system, sodium butyrate induced a two-fold increase of RUSH antibody release upon biotin addition, as expected from its effect on protein expression [34,35]. Among the drugs tested, only lycorine (tested at both 1 and 50 µM) was considered to be active, although its effect was milder than that of sodium butyrate, while all the other compounds failed to display any antibody secretion stimulatory activity (Figure 6B).

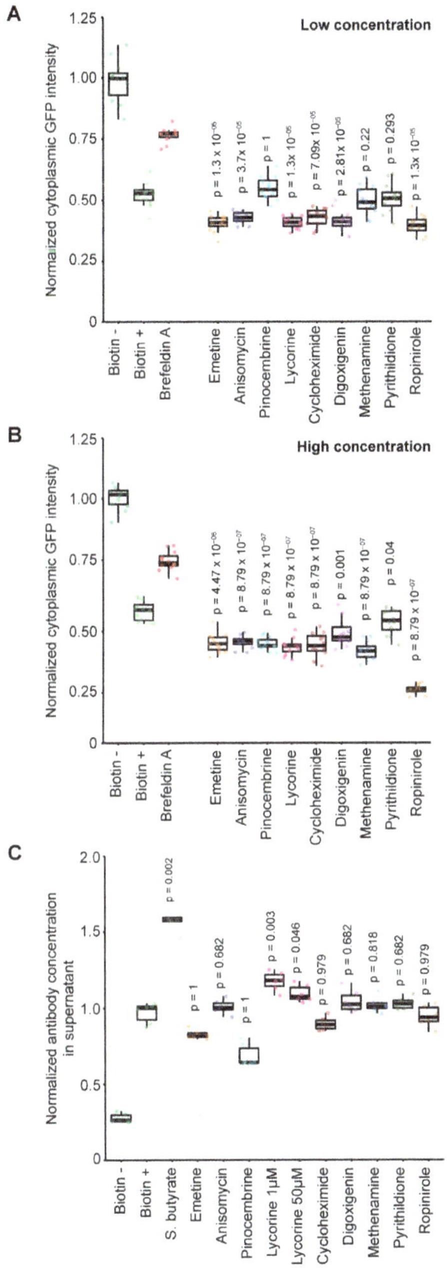

Figure 6. Validation of enhancers of protein secretion. (**A**,**B**) Clone #7 cells were treated with the indicated compounds and analyzed as indicated in the legend of Figure 4A. Drugs were used at low (**A**) or high concentration (**B**), as indicated in the text, and the post-treatment intracytoplasmic fluorescence was normalized to that measured in mock-treated cells (no biotin). (**C**) ELISA assay on immobilized PD-L1. Clone #7 cells were treated with the indicated compounds, and supernatants were assayed for PD-L1 binding. Sodium butyrate was used as a protein secretion enhancer positive control.

Similarly, potential secretion inhibitors identified during the screening campaigns were also tested in an ad hoc assay. Both nigericin and tyrphostin AG-879 (used at 50 µg/mL) turned out to be as potent as brefeldin A in suppressing antibody secretion (Figure 7A), without affecting cellular viability (Figure 7B). Cell imaging further confirmed the secretion activator effect of lycorine and the inhibitory effect of nigericin and tyrphostin AG-879 (Figure 5C).

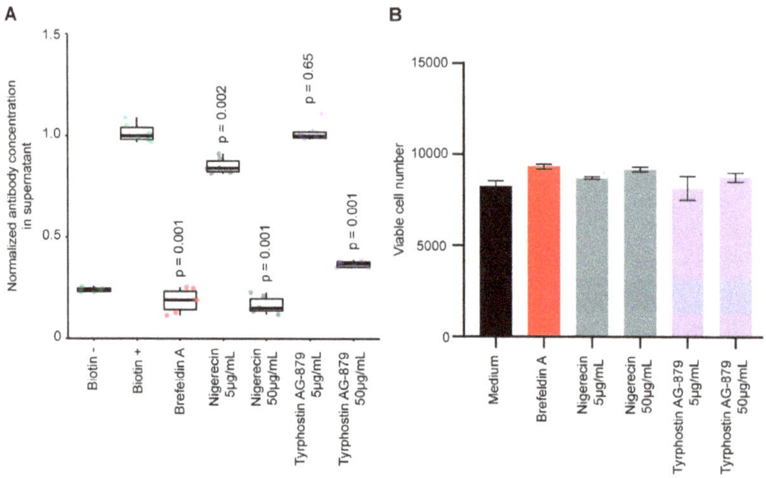

Figure 7. Validation of inhibitors of protein secretion. (**A**) ELISA assay on immobilized PD-L1. Clone #7 cells were treated with the indicated compounds and supernatants were assayed for PD-L1 binding. Brefeldin A was used as a protein secretion inhibitor positive control. (**B**) Cells treated in A were subjected to a toxicity assay. Hoechst 33324 fluorescence was measured, which enabled counting the adherent living cells in each well.

4. Discussion

In the present study, we provide evidence that the RUSH system can be adapted to the controlled release of functional antibodies (exemplified by the immunotherapeutic anti-PD-L1 antibody Atezolizumab) from cultured U2OS cells via the addition of biotin. Using this antibody compatible RUSH system, small chemical compounds libraries were successfully screened using a high-throughput fluorescence imaging system, for the identification of modulators of conventional antibody secretion. Here, we report one novel activator and two novel inhibitors of antibody secretion validated using enzyme-linked immunosorbent assay (ELISA) antibody titration.

The RUSH system offers a cost-effective, high-throughput, high-content, and screening-compatible alternative to complementary assays such as ELISA. Thus, expressed in firmly adherent human osteosarcoma U2OS cells, the RUSH system becomes a valuable tool for the image-based kinetic assessment of protein trafficking. Of note, U2OS cells are neither antibody producing nor specialized secretory cells, which necessitates validation in physiologically relevant systems. In previous publications, variations of the RUSH system were generated by fusing different proteins, such as calreticulin [36], giantin [37], transferrin receptor [38], and tumor necrosis factor [37] to SBP and a reporter protein, either GFP, which can be directly visualized by fluorescence microscopy, or other non-fluorescent proteins tracked by alternative methods. The particularity of these model cargos is their monomeric nature, and no large homo-dimeric proteins such as antibodies have ever been studied in the RUSH system, in order to document whether secretion can be triggered and whether antibodies maintain antigen-binding properties upon release after long-term immobilization in the ER lumen. These questions needed to be addressed prior to using the RUSH system for the identification of antibody secretion modulators in screening campaigns.

Setting up a RUSH system requires the addition of SBP to the antibody for reversible ER targeting and a fluorescent protein (e.g., GFP) for automatized video microscopy monitoring of antibody secretion. As highlighted by Haas et al. and Luria et al., this represents a challenge for antibodies, since immunoglobulins and GFP do not normally originate from the same cellular compartment, and the addition of foreign sequences to antibodies has been shown to interfere with their secretion and functionality [39,40]. The mechanism of IgG heavy and light chain assembly prior to mature antibody secretion poses another challenge. Indeed, while antibody heavy chains require assembly with light chains for secretion, a free light chain might be secreted. We therefore opted to retain the RUSH antibody through its heavy chain, in order to confidently monitor the secretion of fully assembled antibodies [41–44]. With these constraints considered, the RUSH machinery was added to the C-terminus of the anti-PD-L1 heavy chain. A furin cleavage site was also added between the heavy chain and the RUSH construct for the removal of the SBP and GFP sequences past the Golgi network, which might have affected the functions of the antibody otherwise.

Our data showed the efficient detection and retention of the antibody-GFP fusion in the ER of clone #7, and its release and maturation upon addition of biotin in the cell culture medium. ELISA and cell-based assays also demonstrated the specific binding of the released antibody to its PD-L1 target. The fact that the majority of secreted molecules were processed by furins suggests that the secretory machinery of clone #7 was not saturated, which makes this system relevant for the identification of secretion modulators. Of note, the uncleaved full-length antibody retained its specificity and was detectable through its GFP moiety. This has also been described by others and confirms the potential interest of such constructs for live imaging of cells, organs, or organisms [40].

Using this antibody-compatible RUSH system, we screened chemical libraries to identify small molecules that would modulate antibody secretion. The conditions were adapted to identify inhibitors of protein secretion and compare the results with previous screening campaigns performed with a distinct cargo [26], but also with activators. Among the 31 inhibitory compounds of the Prestwick library identified by Zhao et al., 13 were also identified in our screen, due to their ability to increase intracytoplasmic GFP intensity either at high or low concentration, suggesting consistency between the two approaches, despite the use of different clones and cargos. Among the two unique compounds that caused the retention of the RUSH antibody in the cytoplasm, tyrphostin AG-879, which is characterized as a tyrosine kinase inhibitor (IC_{50} = 10 μM), could be validated as an inhibitor of antibody secretion when it was used at a concentration of 50 μg/mL. The identification of a kinase inhibitor is consistent with the involvement of various kinases that act on the Golgi and that are responsible for protein maturation, secretion, and triage to the degradation pathways [45]. This is also consistent with the need for phosphorylation of effectors of the secretory pathway, such as the eukaryotic initiation factor 2 (eiF2), resulting in the decrease of the UPR and hence of the secretory capacity of the cells [46]. Nigericin was also confirmed as a secretion inhibitor when used at a concentration of 50 μg/mL (but not 5 μg/mL). This potassium ionophore might interfere with bioenergetic functions, pH regulation, and ion homeostasis, and induce pleiotropic cellular effects, leading to a decreased secretion potential [47].

Our screening campaigns also led to the identification of candidate molecules that might increase antibody secretion in the biotin-stimulated RUSH system, because they decreased the intracytoplasmic GFP fluorescence signal, among which eight compounds (anisomycin, cycloheximide, digoxigenin, lycorine, methenamine, ropinirole, pinocembrine, and pyrithyldione) were selected for further characterization by ELISA. This assay measures the RUSH antibody binding to its immobilized PD-L1 target and was used to measure antibody titers in cells treated with candidate molecules. Unexpectedly, among these eight compounds that consistently decreased intracytoplasmic GFP fluorescence, which was interpreted as accelerated antibody release, only lycorine was shown to increase RUSH antibody titers in treated cell culture supernatant. These discrepancies might be attributable to the nature of the read out of both assays, with one being the measurement of intracytoplasmic GFP fluorescence intensity decrease and the second being the antibody titration in the cell culture supernatant. As a possible explanation, compounds selected based on the screening results might have had an unsuspected negative impact on the affinity of the RUSH antibody/antigen interaction, therefore masking their direct impact on antibody secretion. On the contrary, the fact that lycorine was validated in both assays and in several independent experiments strongly supports its stimulatory effect on antibody secretion in our system.

Lycorine likely has multiple effects on cell biology, with several cellular targets and a proposed utility for rather diverse indications, including cancer [48], fungal infection [49], and Alzheimer's disease [50]; however, no effect on secretion has been published to the best of our knowledge. Further investigations will be required to decipher its mode of action as an activator, as well as that of the inhibitors identified in this work.

In future applications, the screening approach presented in this study could be upscaled to larger pharmacological screening campaigns at the industrial level, for the discovery of novel molecules with secretion modulatory effects. Potential clinical applications would involve the direct use of the identified compounds to modulate protein secretion in related diseases, or the identification of their cellular target for subsequent development of therapeutic solutions [5]. In an industrial context, the RUSH system adapted to producer cell lines in liquid culture could be seen as a tool to control the secretion of recombinant therapeutics proteins of interest with toxic properties once secreted in the culture supernatant, resulting in cell growth inhibition or even cell death. This would represent a valuable alternative to the inducible transcription systems that were initially explored for the decoupling of cell growth and product formation, which proved challenging to develop [51].

5. Conclusions

Overall, while the addition of lycorine in bioreactors might hamper drug substance purification in industrial processes, deciphering its mechanism of action would be of great interest. A direct action of lycorine on cellular targets through genetic engineering could unlock the secretory potential of cell hosts. Such hosts would bring added value to the field of recombinant therapeutic protein production and to the pharmaceutical industry.

Author Contributions: Conceptualization, M.C., O.K., S.B. and G.K.; investigation, M.C., S.L. and M.L. software, M.L.; resources, M.T., Z.G. and F.P.; writing—original draft preparation, M.C., S.B. and G.K.; writing—review and editing, M.C., O.K., S.B. and G.K.; supervision, O.K., S.B. and G.K. All authors have read and agreed to the published version of the manuscript.

Funding: This work was primarily supported by a research contract involving Sanofi and Institut Gustave Roussy. M.C. is supported by a grant from the French ANRT agency (CIFRE No 2020/0106). O.K. receives funding from the Institut National du Cancer (INCa) and the DIM ELICIT initiative of the Ile de France. GK is supported by the Ligue contre le Cancer (équipe labellisée); Agence National de la Recherche (ANR)—Projets blancs; AMMICa US23/CNRS UMS3655; Association pour la recherche sur le cancer (ARC); Cancéropôle Ile-de-France; European Research Council Advanced Investigator Grand "ICD-Cancer", Fondation pour la Recherche Médicale (FRM); a donation by Elior; Equipex Onco-Pheno-Screen; European Joint Programme on Rare Diseases (EJPRD); European Research Council (ICD-Cancer), European Union Horizon 2020 Projects Oncobiome and Crimson; Fondation Carrefour; Institut National du Cancer (INCa); Institut Universitaire de France; LabEx Immuno-Oncology (ANR-18-IDEX-0001); a Cancer Research ASPIRE Award from the Mark Foundation; the RHU Immunolife; Seerave Foundation; SIRIC Stratified Oncology Cell DNA Repair and Tumor Immune Elimination (SOCRATE); and SIRIC Cancer Research and Personalized Medicine (CARPEM). This study contributes to the IdEx Université de Paris ANR-18-IDEX-0001. The work in FP lab was supported by LabEx CelTisPhyBio (ANR-10-LBX-0038 part of the IDEX PSL no. ANR-10-IDEX-0001-02, by the ANR grants ANR-22-CE14-0066-01, ANR-20-CE14-0017-02,

ANR-19-CE13-0006-03, ANR-19-CE13-0002-03) and by the Fondation pour la Recherche Médicale (EQU201903007925).

Institutional Review Board Statement: Not applicable. This study did not involve humans or animals.

Informed Consent Statement: Not applicable. This study did not involve humans.

Data Availability Statement: The data presented in this study are available on request from the corresponding authors. The data are not publicly available since they are being used for filing a patent.

Acknowledgments: We acknowledge Christine DeMaria and Henry Lin at Sanofi for a critical review of the manuscript. Figures 1A,B, 2C–F, 3A,C,E and 4A were created with BioRender.com, accessed on 31 March 2023.

Conflicts of Interest: G.K. received research support from Sanofi. M.C., M.T. and S.B. are full-time employees of Sanofi and may hold shares and/or stock options in the company. Outside of this work, G.K. has held research contracts with Daiichi Sankyo, Eleor, Kaleido, Lytix Pharma, PharmaMar, Osasuna Therapeutics, Samsara Therapeutics, Tollys, and Vascage. G.K. is on the Board of Directors of the Bristol Myers Squibb Foundation France. G.K. is a scientific co-founder of everImmune, Osasuna Therapeutics, Samsara Therapeutics, and Therafast Bio. G.K. is in the scientific advisory boards of Hevolution, Institut Servier, and Longevity Vision Funds. G.K. is the inventor of patents covering therapeutic targeting of aging, cancer, cystic fibrosis, and metabolic disorders. G.K.'s wife, Laurence Zitvogel, has held research contracts with Glaxo Smyth Kline, Incyte, Lytix, Kaleido, Innovate Pharma, Daiichi Sankyo, Pilege, Merus, Transgene, 9 m, Tusk, and Roche; was on the on the Board of Directors of Transgene; is a cofounder of everImmune; and holds patents covering the treatment of cancer and the therapeutic manipulation of the microbiota. G.K.'s brother, Romano Kroemer, was an employee of Sanofi and now consults for Boehringer-Ingelheim. O.K. is a scientific co-founder of Samsara Therapeutics. Z.G. and F.P. were employed by the start-up Honing Biosciences until May 2023. Sanofi and Inserm have filed a patent application (not yet published) covering the methods for secretion modulation reported here. M.C., O.K., S.B. and G.K. are inventors.

References

1. Viotti, C. ER to Golgi-Dependent Protein Secretion: The Conventional Pathway. *Methods Mol. Biol.* **2016**, *1459*, 3–29. [CrossRef]
2. Torres, M.; Hussain, H.; Dickson, A.J. The secretory pathway—The key for unlocking the potential of Chinese hamster ovary cell factories for manufacturing therapeutic proteins. *Crit. Rev. Biotechnol.* **2022**, *43*, 628–645. [CrossRef]
3. Jena, B.P. Molecular machinery and mechanism of cell secretion. *Exp. Biol. Med.* **2005**, *230*, 307–319. [CrossRef] [PubMed]
4. Luesch, H.; Paavilainen, V.O. Natural products as modulators of eukaryotic protein secretion. *Nat. Prod. Rep.* **2020**, *37*, 717–736. [CrossRef] [PubMed]
5. Wang, M.; Kaufman, R.J. Protein misfolding in the endoplasmic reticulum as a conduit to human disease. *Nature* **2016**, *529*, 326–335. [CrossRef]
6. Dejeans, N.; Manié, S.; Hetz, C.; Bard, F.; Hupp, T.; Agostinis, P.; Samali, A.; Chevet, E. Addicted to secrete—Novel concepts and targets in cancer therapy. *Trends Mol. Med.* **2014**, *20*, 242–250. [CrossRef]
7. Walsh, G.; Walsh, E. Biopharmaceutical benchmarks 2022. *Nat. Biotechnol.* **2022**, *40*, 1722–1760. [CrossRef] [PubMed]
8. Wurm, F.M. Production of recombinant protein therapeutics in cultivated mammalian cells. *Nat. Biotechnol.* **2004**, *22*, 1393–1398. [CrossRef]
9. Chiu, M.L.; Goulet, D.R.; Teplyakov, A.; Gilliland, G.L. Antibody Structure and Function: The Basis for Engineering Therapeutics. *Antibodies* **2019**, *8*, 55. [CrossRef]
10. Pelham, H.R. Multiple targets for brefeldin A. *Cell* **1991**, *67*, 449–451. [CrossRef]
11. Badr, C.E.; Hewett, J.W.; Breakefield, X.O.; Tannous, B.A. A highly sensitive assay for monitoring the secretory pathway and ER stress. *PLoS ONE* **2007**, *2*, e571. [CrossRef]
12. Aowicki, D.; Huczyński, A. Structure and antimicrobial properties of monensin A and its derivatives: Summary of the achievements. *BioMed Res. Int.* **2013**, *2013*, 742149. [CrossRef]
13. Fussenegger, M.; Mazur, X.; Bailey, J.E. A novel cytostatic process enhances the productivity of Chinese hamster ovary cells. *Biotechnol. Bioeng.* **1997**, *55*, 927–939. [CrossRef]
14. Sunley, K.; Butler, M. Strategies for the enhancement of recombinant protein production from mammalian cells by growth arrest. *Biotechnol. Adv.* **2010**, *28*, 385–394. [CrossRef] [PubMed]
15. Chung, J.Y.; Lim, S.W.; Hong, Y.J.; Hwang, S.O.; Lee, G.M. Effect of doxycycline-regulated calnexin and calreticulin expression on specific thrombopoietin productivity of recombinant Chinese hamster ovary cells. *Biotechnol. Bioeng.* **2004**, *85*, 539–546. [CrossRef]

16. Berger, A.; Le Fourn, V.; Masternak, J.; Regamey, A.; Bodenmann, I.; Girod, P.A.; Mermod, N. Overexpression of transcription factor Foxa1 and target genes remediate therapeutic protein production bottlenecks in Chinese hamster ovary cells. *Biotechnol. Bioeng.* **2020**, *117*, 1101–1116. [CrossRef]
17. Chandrawanshi, V.; Kulkarni, R.; Prabhu, A.; Mehra, S. Enhancing titers and productivity of rCHO clones with a combination of an optimized fed-batch process and ER-stress adaptation. *J. Biotechnol.* **2020**, *311*, 49–58. [CrossRef] [PubMed]
18. Peng, R.W.; Abellan, E.; Fussenegger, M. Differential effect of exocytic SNAREs on the production of recombinant proteins in mammalian cells. *Biotechnol. Bioeng.* **2011**, *108*, 611–620. [CrossRef]
19. Rahimpour, A.; Vaziri, B.; Moazzami, R.; Nematollahi, L.; Barkhordari, F.; Kokabee, L.; Adeli, A.; Mahboudi, F. Engineering the cellular protein secretory pathway for enhancement of recombinant tissue plasminogen activator expression in Chinese hamster ovary cells: Effects of CERT and XBP1s genes. *J. Microbiol. Biotechnol.* **2013**, *23*, 1116–1122. [CrossRef]
20. Hammond, S.; Lee, K.H. RNA interference of cofilin in Chinese hamster ovary cells improves recombinant protein productivity. *Biotechnol. Bioeng.* **2012**, *109*, 528–535. [CrossRef]
21. Pourcel, L.; Buron, F.; Garcia, F.; Delaloix, M.S.; Le Fourn, V.; Girod, P.A.; Mermod, N. Transient vitamin B5 starving improves mammalian cell homeostasis and protein production. *Metab. Eng.* **2020**, *60*, 77–86. [CrossRef]
22. Pourcel, L.; Buron, F.; Arib, G.; Le Fourn, V.; Regamey, A.; Bodenmann, I.; Girod, P.A.; Mermod, N. Influence of cytoskeleton organization on recombinant protein expression by CHO cells. *Biotechnol. Bioeng.* **2020**, *117*, 1117–1126. [CrossRef]
23. Boncompain, G.; Divoux, S.; Gareil, N.; de Forges, H.; Lescure, A.; Latreche, L.; Mercanti, V.; Jollivet, F.; Raposo, G.; Perez, F. Synchronization of secretory protein traffic in populations of cells. *Nat. Methods* **2012**, *9*, 493–498. [CrossRef]
24. Shapiro, J.; Sciaky, N.; Lee, J.; Bosshart, H.; Angeletti, R.H.; Bonifacino, J.S. Localization of endogenous furin in cultured cell lines. *J. Histochem. Cytochem. Off. J. Histochem. Soc.* **1997**, *45*, 3–12. [CrossRef]
25. Lefranc, M.P.; Ehrenmann, F.; Kossida, S.; Giudicelli, V.; Duroux, P. Use of IMGT(®) Databases and Tools for Antibody Engineering and Humanization. *Methods Mol. Biol.* **2018**, *1827*, 35–69. [CrossRef] [PubMed]
26. Zhao, L.; Liu, P.; Boncompain, G.; Loos, F.; Lachkar, S.; Bezu, L.; Chen, G.; Zhou, H.; Perez, F.; Kepp, O.; et al. Identification of pharmacological inhibitors of conventional protein secretion. *Sci. Rep.* **2018**, *8*, 14966. [CrossRef] [PubMed]
27. Barrette-Ng, I.H.; Wu, S.C.; Tjia, W.M.; Wong, S.L.; Ng, K.K. The structure of the SBP-Tag-streptavidin complex reveals a novel helical scaffold bridging binding pockets on separate subunits. *Acta Crystallographica. Sect. D Biol. Crystallogr.* **2013**, *69*, 879–887. [CrossRef]
28. Delgadillo, R.F.; Mueser, T.C.; Zaleta-Rivera, K.; Carnes, K.A.; González-Valdez, J.; Parkhurst, L.J. Detailed characterization of the solution kinetics and thermodynamics of biotin, biocytin and HABA binding to avidin and streptavidin. *PLoS ONE* **2019**, *14*, e0204194. [CrossRef] [PubMed]
29. Capitani, M.; Sallese, M. The KDEL receptor: New functions for an old protein. *FEBS Lett.* **2009**, *583*, 3863–3871. [CrossRef] [PubMed]
30. Zhang, F.; Qi, X.; Wang, X.; Wei, D.; Wu, J.; Feng, L.; Cai, H.; Wang, Y.; Zeng, N.; Xu, T.; et al. Structural basis of the therapeutic anti-PD-L1 antibody atezolizumab. *Oncotarget* **2017**, *8*, 90215–90224. [CrossRef]
31. Nomura, R.; Orii, M.; Senda, T. Calreticulin-2 is localized in the lumen of the endoplasmic reticulum but is not a Ca^{2+}-binding protein. *Histochem. Cell Biol.* **2011**, *135*, 531–538. [CrossRef]
32. Misumi, Y.; Misumi, Y.; Miki, K.; Takatsuki, A.; Tamura, G.; Ikehara, Y. Novel blockade by brefeldin A of intracellular transport of secretory proteins in cultured rat hepatocytes. *J. Biol. Chem.* **1986**, *261*, 11398–11403. [CrossRef] [PubMed]
33. Coulet, M.; (Sanofi R&D, Vitry-sur-Seine, France). Personal communication, 2023.
34. Jiang, Z.; Sharfstein, S.T. Sodium butyrate stimulates monoclonal antibody over-expression in CHO cells by improving gene accessibility. *Biotechnol. Bioeng.* **2008**, *100*, 189–194. [CrossRef]
35. Mimura, Y.; Lund, J.; Church, S.; Dong, S.; Li, J.; Goodall, M.; Jefferis, R. Butyrate increases production of human chimeric IgG in CHO-K1 cells whilst maintaining function and glycoform profile. *J. Immunol. Methods* **2001**, *247*, 205–216. [CrossRef] [PubMed]
36. Liu, P.; Zhao, L.; Loos, F.; Marty, C.; Xie, W.; Martins, I.; Lachkar, S.; Qu, B.; Waeckel-Énée, E.; Plo, I.; et al. Immunosuppression by Mutated Calreticulin Released from Malignant Cells. *Mol. Cell* **2020**, *77*, 748–760.e9. [CrossRef]
37. Fourriere, L.; Divoux, S.; Roceri, M.; Perez, F.; Boncompain, G. Microtubule-independent secretion requires functional maturation of Golgi elements. *J. Cell Sci.* **2016**, *129*, 3238–3250. [CrossRef] [PubMed]
38. Deffieu, M.S.; Cesonyte, I.; Delalande, F.; Boncompain, G.; Dorobantu, C.; Song, E.; Lucansky, V.; Hirschler, A.; Cianferani, S.; Perez, F.; et al. Rab7-harboring vesicles are carriers of the transferrin receptor through the biosynthetic secretory pathway. *Sci. Adv.* **2021**, *7*, eaba7803. [CrossRef]
39. Haas, A.K.; Mayer, K.; Brinkmann, U. Generation of fluorescent IgG fusion proteins in mammalian cells. *Methods Mol. Biol.* **2012**, *901*, 265–276. [CrossRef] [PubMed]
40. Luria, Y.; Raichlin, D.; Benhar, I. Fluorescent IgG fusion proteins made in *E. coli*. *MAbs* **2012**, *4*, 373–384. [CrossRef]
41. Leitzgen, K.; Knittler, M.R.; Haas, I.G. Assembly of immunoglobulin light chains as a prerequisite for secretion. A model for oligomerization-dependent subunit folding. *J. Biol. Chem.* **1997**, *272*, 3117–3123. [CrossRef] [PubMed]
42. Hammond, C.; Helenius, A. Quality control in the secretory pathway. *Curr. Opin. Cell Biol.* **1995**, *7*, 523–529. [CrossRef] [PubMed]
43. Dul, J.L.; Aviel, S.; Melnick, J.; Argon, Y. Ig light chains are secreted predominantly as monomers. *J. Immunol.* **1996**, *157*, 2969–2975. [CrossRef] [PubMed]

4. Hopper, J.E.; Papagiannes, E. Evidence by radioimmunoassay that mitogen-activated human blood mononuclear cells secrete significant amounts of light chain Ig unassociated with heavy chain. *Cell. Immunol.* **1986**, *101*, 122–131. [CrossRef]
5. Sreelatha, A.; Kinch, L.N.; Tagliabracci, V.S. The secretory pathway kinases. *Biochim. Et Biophys. Acta* **2015**, *1854*, 1687–1693. [CrossRef]
6. Harding, H.P.; Zhang, Y.; Zeng, H.; Novoa, I.; Lu, P.D.; Calfon, M.; Sadri, N.; Yun, C.; Popko, B.; Paules, R.; et al. An integrated stress response regulates amino acid metabolism and resistance to oxidative stress. *Mol. Cell* **2003**, *11*, 619–633. [CrossRef]
7. Gao, G.; Liu, F.; Xu, Z.; Wan, D.; Han, Y.; Kuang, Y.; Wang, Q.; Zhi, Q. Evidence of nigericin as a potential therapeutic candidate for cancers: A review. *Biomed. Pharmacother.* **2021**, *137*, 111262. [CrossRef] [PubMed]
8. Roy, M.; Liang, L.; Xiao, X.; Feng, P.; Ye, M.; Liu, J. Lycorine: A prospective natural lead for anticancer drug discovery. *Biomed. Pharmacother.* **2018**, *107*, 615–624. [CrossRef]
9. Zhao, S.; Guo, Y.; Wang, Q.; An, B. Antifungal effects of lycorine on Botrytis cinerea and possible mechanisms. *Biotechnol. Lett.* **2021**, *43*, 1503–1512. [CrossRef]
0. Kola, A.; Lamponi, S.; Currò, F.; Valensin, D. A Comparative Study between Lycorine and Galantamine Abilities to Interact with AMYLOID β and Reduce In Vitro Neurotoxicity. *Int. J. Mol. Sci.* **2023**, *24*, 2500. [CrossRef]
1. Li, F.; Vijayasankaran, N.; Shen, A.Y.; Kiss, R.; Amanullah, A. Cell culture processes for monoclonal antibody production. *MAbs* **2010**, *2*, 466–479. [CrossRef]

cells

MDPI

Article

Ginsentide TP1 Protects Hypoxia-Induced Dysfunction and ER Stress-Linked Apoptosis

Bamaprasad Dutta [1], Shining Loo [1,2], Antony Kam [1,3], Siu Kwan Sze [1,4] and James P. Tam [1,*]

1 School of Biological Sciences, Nanyang Technological University, Singapore 637551, Singapore; bama0001@e.ntu.edu.sg (B.D.); shining.loo@xjtlu.edu.cn (S.L.); antony.kam@xjtlu.edu.cn (A.K.); nsze@brocku.ca (S.K.S.)
2 Academy of Pharmacy, Xi'an Jiaotong-Liverpool University, Suzhou 215123, China
3 Department of Biological Sciences, Xi'an Jiaotong-Liverpool University, Suzhou 215123, China
4 Department of Health Sciences, Brock University, Niagara Region, St. Catharines, ON L2S 3A1, Canada
* Correspondence: jptam@ntu.edu.sg; Tel.: +65-6316-2833

Abstract: Hypoxia-induced vascular endothelial dysfunction (VED) is a significant contributor to several severe human diseases, including heart disease, stroke, dementia, and cancer. However, current treatment options for VED are limited due to the lack of understanding of the underlying disease mechanisms and therapeutic leads. We recently discovered a heat-stable microprotein in ginseng, called ginsentide TP1, that has been shown to reduce vascular dysfunction in cardiovascular disease models. In this study, we use a combination of functional assays and quantitative pulsed SILAC proteomics to identify new proteins synthesized in hypoxia and to show that ginsentide TP1 provides protection for human endothelial cells against hypoxia and ER stress. Consistent with the reported findings, we also found that hypoxia activates various pathways related to endothelium activation and monocyte adhesion, which in turn, impairs nitric oxide (NO) synthase activity, reduces the bioavailability of NO, and increases the production of reactive oxygen species that contribute to VED. Additionally, hypoxia triggers endoplasmic reticulum stress and initiates apoptotic signaling pathways associated with cardiovascular pathology. Treatment with ginsentide TP1 reduced surface adhesion molecule expression, prevented activation of the endothelium and leukocyte adhesion, restored protein hemostasis, and reduced ER stress to protect against hypoxia-induced cell death. Ginsentide TP1 also restored NO signaling and bioavailability, reduced oxidative stress, and protected endothelial cells from endothelium dysfunction. In conclusion, this study shows that the molecular pathogenesis of VED induced by hypoxia can be mitigated by treatment with ginsentide TP1, which could be one of the key bioactive compounds responsible for the "cure-all" effect of ginseng. This research may lead to the development of new therapies for cardiovascular disorders.

Keywords: vascular endothelial dysfunction; hypoxia; ginsentide; ER stress; unfolded protein response; antioxidation; anti-inflammation; anti-apoptosis; pulsed SILAC; quantitative proteomics

1. Introduction

Hypoxia-induced vascular endothelial dysfunction (VED) plays a key role in the pathogenesis of multiple human disorders, including heart disease, stroke, dementia, and diabetic complications; however, the molecular mediators of VED largely remain unknown. Clinical data suggest that hypoxia induced by sleep apnea or high-altitude conditions is strongly associated with VED and may increase the risk of cardiovascular disorders and heart failure [1–6]. Models of pulmonary vasoconstriction studied under low-oxygen conditions also suggest that hypoxia strongly influences the endothelium via connexin-40-mediated effects on the endothelial membrane potential [7]. However, hypoxia also modulates the functions of P2X and P2Y cell surface receptors, which stimulate the release of nitric oxide (NO) and hyperpolarizing factors [8]. In addition, VED restricts blood flow and oxygen supply to tissues and endothelium, forming a vicious cycle and amplifying the hypoxia-induced VED. Acting through these underlying mechanisms, hypoxia can trigger proinflammatory endothelial responses, including the upregulation of cell adhesion molecules that increase macrophage infiltration [9], which is a known risk factor for adverse cardiovascular events [1,4–6]. Therapeutic targeting of hypoxia-VED will break this vicious cycle and limit the progression of vascular pathology underlying major human diseases.

Under normal physiological conditions, NO generation maintains the quiescent state of the endothelium [10]. In contrast, disrupting NO signaling triggers endothelial activation and the expression of cytokines, chemokines, and adhesion molecules that promote leukocyte adhesion and inflammation [10,11]. During oxidative stress, NO reacts with superoxide anions to generate peroxynitrite, which reduces the function of endothelial nitric oxide synthases (eNOSs) and limits the production of NO to promote VED [10,12,13]. Hypoxia also induces endoplasmic reticulum (ER) stress, which triggers proapoptotic signaling pathways that are implicated in cardiovascular disease [14,15]. In contrast, the inhibition of ER stress-induced endothelial apoptosis is known to prevent VED and reduce the cardiovascular complications of diabetes [16]. Therefore, ER stress represents a critical component of VED and a priority target for therapeutic interventions; however, methods for limiting ER stress in the hypoxic endothelium have not been identified.

Ginseng plants belong to the *Panax* (panacea) genus and were named for their "cure-all" medicinal effects. For many centuries, ginseng has been used as a herbal medicine to treat various disorders, including heart disease and diabetes [17,18]. Currently, identified bioactive compounds are primarily derived from small-molecule metabolites. Moreover, the molecular basis underlying the beneficial effects of ginseng on VED in human patients has not been clarified. Studies conducted in our laboratory led to the identification of a novel family of cysteine-rich peptides (CRPs) or microproteins from ginseng plants designated as ginsentides [19]. To date, 14 ginsentides (TP1–14) have been identified from three common and medicinally valuable ginseng species, i.e., *Panax ginseng*, *Panax notoginseng*, and *Panax quinquefolius*. [19]. All ginsentides are 31–33 amino acids long and contain a conserved 8-cysteine motif and four disulfide linkages [19]. The prototypic member ginsentide TP1 (hereafter given as TP1) is the most abundant CRP found in *P. ginseng* and *P. notoginseng*. The 31-residue TP1 is both cysteine- and glycine-rich (Figure 1A). The eight cysteine residues in TP1 form novel disulfide connectivity of Cys I–IV, II–VI, III–VII, and V–VIII, resulting in a pseudocyclic knot in which the first and last residues are cystine forming disulfides with a knotted and multicyclic structure (Figure 1A,B) [19]. This unique disulfide connectivity provides the structural rigidity to TP1 and braces it into a hyperstable and compact microprotein that is resistant to heat, acid, and both exopeptidase and endopeptidase degradation [19–22].

TP1 and other members of ginsentides (TP2–14) form a new CRP family. They contain eight cysteine residues and a cysteine motif similar to heveins found in *Hevea brasiliensis* and hevein-like peptides, but they can be distinguished from heveins by a distinctly different disulfide connectivity, biosynthesis, and the absence of a chitin-binding domain which is essential for the host-defense functions of heveins and hevein-like peptides [19–24]. Recently, we showed that ginsentides TP1, TP3, and TP8 modulate multiple biological systems to exert potent effects on vascular biology [25].

Here, we show that TP1 protects endothelial cells (ECs) from hypoxia-induced and stress-linked dysfunctions underlying VED pathogens by synthesizing new proteins critical to their adaptive responses; we used pulsed SILAC (pSILAC) quantitative proteomics to pulse/chase de novo protein synthesis in ECs subjected to low-oxygen stress with or without TP1. Guided by the pSILAC results, we further investigated how TP1 protects ECs from hypoxia stress using functional assays. Using these approaches, we show that TP1 protects human ECs from hypoxia-induced VED by preventing ER stress-mediated apoptotic signaling and by increasing cell survival. We also compared TP1 with ginsenosides, a family of well-studied bioactive metabolites from ginseng. Our findings showed that ginsenosides did not exert protective effects against hypoxia-induced VED when compared to the novel ginsentides.

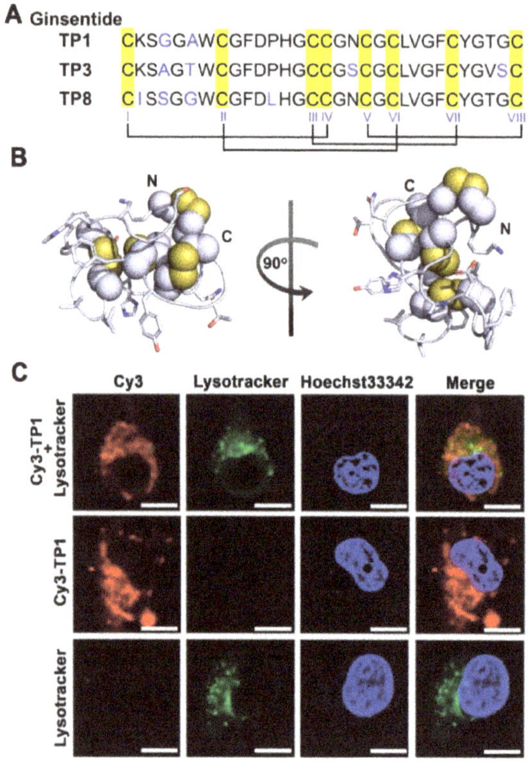

Figure 1. Ginsentides: cysteine- and glycine-rich cell-penetrating peptides from ginseng. (**A**) Primary sequence and disulfide connectivity (Cys I–IV, Cys II–VI, Cys III–VII, and Cys V–VIII) of ginsentides TP1, TP3, and TP8. (**B**) Cartoon representation of the structure of the pseudocyclic TP1 (PDB code: 2ML7). The unusual disulfide connectivity between cysteine residues forms a pseudocyclic knot, in which the first and last residues are cysteines. (**C**) Images of HUVEC-CS cells after incubation with 1 μM Cy3–TP1, captured using live-cell confocal microscopy at 37 °C. Cy3–TP1 is shown in red, whereas lysosomes and nuclei are counterstained with LysoTracker (green) and Hoechst 33342 (blue), respectively. Scale bar: 5 μm.

2. Materials and Methods

2.1. Reagents

All chemicals and reagents were purchased from Sigma-Aldrich (St. Louis, MO, USA) unless otherwise indicated. Antibodies against GAPDH (6C5) and actin (C-2) were purchased from Santa Cruz Biotechnology (Santa Cruz, CA, USA), whereas anti-LC3B (L7543) antibody was purchased from Sigma-Aldrich. Anti-eNOS (D9A5L), anti-phospho-eNOS (Ser1177), anti-integrin αV (D2N5H), anti-integrin β1 (D6S1W), anti-integrin β3, HRP-linked anti-mouse IgG, and HRP-linked anti-rabbit IgG antibodies were obtained from Cell Signaling Technology (Danvers, MA, USA). Clarity Max Western ECL Blotting Substrate (1705062) was purchased from Bio-Rad Laboratories (Milan, Italy). The cell lines, including HUVEC-CS (ATCC CRL-2873), H9c2 (ATCC CRL-1446), C2C12 (ATCC CRL-1772), A7r5 (ATCC CRL-1444) and THP-1 (ATCC TIB-202) were acquired from ATCC.

2.2. Extraction and Purification of TP1 and Other Ginsentides

Peptide extraction was performed using the method described by Tam et al. [19]. Briefly, dried *P. ginseng* flowers were pulverized and blended with Milli-Q water (100 g/L). The extract was then centrifuged, and the obtained supernatant was filtered through a

0.22 μm membrane (Thermo Fisher Scientific, Waltham, MA, USA). Purification was performed using multiple rounds of reversed-phase high-performance liquid chromatography (RP-HPLC), which was conducted using a C18 column (250 × 22 mm; 5 μm particle size; Phenomenex, Torrance, CA, USA) at a flow rate of 5 mL/min with a linear gradient of 1%/min of 10–80% buffer B [0.1% trifluoroacetic acid (TFA) in acetonitrile (ACN)] and buffer A (0.1% TFA in HPLC water). Final purification was performed via analytical RP-HPLC using a C18 column (250 × 4.6 mm; 5 μm particle size; Phenomenex) at a flow rate of 1 mL/min using the same gradient. MALDI-TOF mass spectrometry was used to identify the peptide-containing fractions. Ginsentide TP3 and TP8 were isolated from dried *P. ginseng* seeds and dried *P. quinquefolius* flowers, respectively, and purified using the same method.

2.3. Cell Culture

HUVEC-CS cells, rat cardiomyoblast cells (H9c2), rat aorta thoracic/smooth muscle cells (A7r5), and mouse myoblast cells (C2C12) were maintained in DMEM supplemented with 10% FBS at 37 °C in 5% CO_2. THP1 cells were maintained in an RPMI medium supplemented with 10% FBS at 37 °C in 5% CO_2.

2.4. Confocal Microscopic-Based Cellular Uptake Analyses

Cy3–TP1 was used to study the cellular uptake of TP1 in HUVEC-CS cells. To visualize the intracellular distribution of Cy3–TP1, cells were grown on an 8-well chamber slide (Ibidi, Gräfelfing, Germany) overnight before they were incubated with 1 μM Cy3–TP1 at 37 °C for 1 h and counterstained with LysoTracker and Hoechst 333241. Postincubation, the medium was drained, and the cells were washed gently with PBS. Fresh medium was added prior to imaging. Slides were observed and imaged using a Zeiss LSM 710 confocal microscope.

2.5. Pulsed SILAC Labeling

For pSILAC labeling, 2×10^6 HUVEC-CS cells were seeded onto 10 cm Petri dishes and cultured overnight at 37 °C in 5% CO_2. The supernatant was aspirated the following day and replaced with SILAC medium containing $^{13}C_6$ L-arginine and $^{13}C_6$ and $^{15}N_2$ L-lysine together with 10% SILAC-compatible dialyzed FBS. The cells were then supplemented with 20 μM TP1 or PBS vehicle control, after which the dishes were transferred into hypoxia chambers from which the air was removed by flashing the chambers with hypoxic gas (95% N_2, 5% CO_2, and <0.1% O_2) for 10 min at a constant flow rate of 15 L/min. The chambers were then sealed and incubated at 37 °C for 24 h. Control cells were instead maintained under standard culture conditions (i.e., normoxia: 21% O_2 and 5% CO_2) at 37 °C for the 24 h incubation period. Subsequently, the cells were trypsinized and washed three times with ice-cold PBS before they were stored at −80 °C until they were used in proteomic experiments. The entire experimental protocol was performed in triplicate, as summarized in Figure S2B.

2.6. Protein Extraction and Tryptic Digestion

The cell pellets obtained from each experimental condition were suspended in lysis buffer (8 M urea in 100 mM ammonium bicarbonate supplemented with a protease inhibitor cocktail) and sonicated at 30% amplitude for 30 s. The cell lysates were clarified via centrifugation at 20,000× *g* for 10 min, and the protein content of the lysates was quantified using the Bradford protein assay. For each sample, 500 μg of protein was diluted with 100 mM ammonium bicarbonate buffer to decrease the urea concentration to <1 M. The samples were then reduced using 10 mM dithiothreitol at 37 °C for 2 h, after which they were alkylated with 55 mM iodoacetamide at room temperature for 45 min in the dark. Finally, the samples were digested via incubation with trypsin (with an enzyme:protein ratio of 1:50) overnight at 37 °C, after which the reaction was stopped by adding formic acid to a final concentration of 0.5%. Tryptic peptides were then desalted using Sep-Pak C18 cartridges (Waters, Milford, MA, USA), vacuum dried, and stored for HPLC fractionation.

2.7. Reverse-Phase Chromatography Fractionation and Liquid Chromatography with Tandem Mass Spectrometry

Tryptic peptides were fractionated using reverse-phase chromatography separation performed at pH 8. Briefly, the desalted tryptic peptides were reconstituted in buffer A (0.02% NH_4OH in water) and fractionated using an X-Bridge C18 column (4.6 × 200 mm; 5 µm particle size; 130 Å pore size; Waters) at a flow rate of 1 mL/min in an HPLC unit (Prominence, Shimadzu, Kyoto, Japan). The gradient was as follows: 100–95% buffer A for 3 min; 5–35% buffer B (80% ACN and 0.02% NH_4OH) for 40 min; 35–70% buffer B for 12 min; and 70% buffer B for 5 min. The 280 nm wavelength was selected for chromatogram recording, and 64 fractions were collected within a 65 min period. The collected fractions were combined into 13 pools using a concatenated pooling strategy, after which they were dried via vacuum centrifugation. The samples were then subjected to liquid chromatography with tandem mass spectrometry (LC–MS/MS) analysis using a Q Exactive mass spectrometer coupled with an online Dionex Ultimate 3000 RSLC nano-LC system (Thermo Fisher Scientific). Peptide samples were separated using a Dionex EASY-spray column (PepMap C18; 3 µm; 100 Å) and sprayed through an EASY nanospray source at a voltage of 1.5 kV. The range 350–1600 m/z was used for the full MS scan with a resolution of 70,000 at m/z 200. The maximum ion accumulation time was set to 100 ms, and dynamic exclusion was set to 30 s. The top 10 ions above a threshold of 1000 counts were selected for high-energy collision dissociation fragmentation using 28% normalized collision energy with a maximum ion accumulation time of 120 ms as a fragmentation parameter. The other parameters were set as follows: an AGC setting of 1×10^6 for the full MS scan and 2×10^5 for the MS2 scan, an isolation width of 2 Da for the MS2 scan, the exclusion of single and unassigned charged ions from the MS/MS scan, and an underfill ratio of 0.1%. All raw data were acquired using Xcaliber 2.2 (Thermo Scientific, Waltham, MA, USA). Samples from triplicate experiments were run as technical duplicates.

2.8. Mass Spectrometric Data Analysis and Bioinformatics

Protein identification and quantification were performed using MaxQuant version 1.5.2.8 (Max Planck Institute of Biochemistry, Planegg, Germany). The parameters used were as follows: carbamidomethyl at cysteine as a static modification; N-term acetyl, methionine oxidation, and asparagine and glutamine deamidation as dynamic modifications; trypsin/P and a maximum of two missed cleavages as digestion parameters; and a 30-ppm precursor mass and 0.5 Da fragment mass tolerance. The UniProt Knowledgebase of human proteins (downloaded on 25 July 2016, including 70,849 sequences and 23,964,784 residues) was used as a search database with a false discovery rate cutoff of <0.01% both at the peptide and protein levels. For SILAC quantification, Arg6 and Lys8 were set as labeling residues. Searched data were extracted in text format for further analysis. Technical duplicates were searched in combination, and statistical analysis was conducted on data collected in triplicate experiments. The online bioinformatics tool GOrilla was used for bioinformatics analyses [26].

2.9. LDH Assay

A LDH release–based cytotoxicity assay was conducted using a CytoSelect LDH Cytotoxicity Assay Kit (CBA-241, Cell Biolabs, Inc., San Diego, CA, USA). Briefly, HUVEC-CS cells were cultured with or without 100 µM TP1 for 24 h at 37 °C in 5% CO_2. PBS was used as vehicle control, and 1% Triton X-100 was used as a positive control, i.e., as a cell death and membrane-damaging agent. The culture medium was collected, transferred, and mixed with the LDH assay reagent at a ratio of 9:1. The reaction mixture was then incubated at 37 °C for 2 h. Quantification was performed calorimetrically using a wavelength of

450 nm. The assay was conducted in triplicate, and statistical analyses and graphing were performed using GraphPad Prism (version 6.01).

2.10. MTT Assay

Cells were cultured in DMEM with or without 20 μM TP1 (or dose maintained) under normoxic and hypoxic conditions for 24 h, after which MTT reagent (final concentration of 0.5 mg/mL) was added, and the cells were incubated for a further 2 h at 37 °C. The culture medium was then aspirated, and formazan crystals were dissolved in dimethyl sulfoxide to conduct colorimetric quantification at 570 nm using a microplate reader (Tecan Magellan, Männedorf, Switzerland) with a reference wavelength of 630 nm. Experiments were conducted in triplicate, and the resultant data were used for statistical analysis. For ginsentide TP3 and TP8 and ginsenoside Rb1 and Rg1 treatments, cells were treated with 20 μM of the respective compounds for 24 h under normoxic and hypoxic conditions, respectively, with PBS used as vehicle control. Cell viability was assessed using a MTT assay, as described.

2.11. Monocyte–Epithelium Adhesion Assay

HUVEC-CS cells were seeded in 6-well plates and grown as a monolayer in DMEM supplemented with 20 μM TP1 (or the PBS control) under normoxic or hypoxic conditions for 24 h. The ginsenosides Rb1 and Rg1 were used as reference controls at 20 μM. THP-1 cells were cultured in RPMI medium supplemented with 10% FBS. Immediately prior to their use in adhesion assays, the THP-1 cells were harvested, washed three times with PBS, and labeled with 10 μM carboxyfluorescein succinimidyl ester (CFSE) for 10 min at 37 °C. The cell culture medium was aspirated from the HUVEC-CS monolayers, which were then washed with a serum-free RPMI medium. CFSE-labeled THP1 cells were then seeded onto the HUVEC-CS monolayer at a density of 2×10^5 cells/well in serum-free RPMI medium and incubated for 90 min at 37 °C. After incubation, the culture medium was drained, and the HUVEC-CS monolayer was washed three times with PBS to remove nonadherent leukocytes. Images of adherent leukocytes were then captured using the green fluorescence channel of a Nikon ECLIPSE Ti-S inverted microscope system (Nikon, Yokosuka, Kanagawa, Japan). Cells were quantified via cell counting, and statistical analyses were performed using data collected in triplicate experiments.

2.12. NO Detection

Intercellular NO levels were measured using a diaminofluorescein-2 diacetate (DAF-2DA)–based assay. ECs were cultured under various experimental conditions for 24 h before the assay began. For hypoxic cultures, the cells were also treated with 100 μM $CoCl_2$. First, the culture medium was aspirated, after which the cells were washed with Hank's Balanced Salt Solution. The cells were incubated with 1 μM DAF-2DA for 60 min at 37 °C, after which the medium was refreshed, and images of labeled cells were captured using the green fluorescence channel of the Nikon ECLIPSE Ti-S inverted microscope system. Fluorescence intensity was quantified by analyzing the labeled cells via ImageJ. The experiment was performed in triplicate.

2.13. Western Blot Analysis

First, 35 μg of each protein sample was resolved on a 10% acrylamide gel and transferred onto a PVDF membrane. Immunoblotting was then performed using anti-protein antibodies, and detection was achieved using an ECL system (Millipore, Burlington, MA, USA).

2.14. Cell Cycle Analysis

Cells cultured under different experimental conditions were harvested via trypsinization and washed with ice-cold multivalent cation-free PBS. The cells were then fixed in ethanol at −20 °C for 24 h and stained with 0.5 mg/mL propidium iodide for 15 min at 37 °C. DNA content was measured using a BD FACSCalibur flow cytometer (BD Biosciences, San Jose, CA, USA), and data analysis was performed using Weasel (version 3.3.2).

2.15. qRT-PCR

RNA was extracted from cells using RNA isolation kits (Macherey-Nagel, Duren, Germany) according to the 'manufacturer's protocol. RevertAid H Minus First Strand cDNA Synthesis Kits (Thermo Fisher Scientific) were used to synthesize first strand cDNA from 2 µg of total RNA. The primer sequences used in this study are listed in Supplementary Table S1. The primers used to assess ER stress pathway–related genes were designed as described by Oslowski et al. [27]. Real-time PCR amplification was performed using a CFX Connect Real-Time PCR Detection System (Bio-Rad Laboratories, Singapore) with KAPA SYBR FAST qPCR Kits (KAPA Biosystems, Boston, MA, USA). The cycling temperatures were as follows: enzyme activation at 95 °C for 3 min; denaturation at 95 °C for 10 s; annealing and extension at 60 °C for 30 s. Data were normalized against the expression level of 18S ribosomal RNA in each sample. The experiment was performed in triplicate.

2.16. Apoptosis Detection

Apoptotic cells were detected using an Annexin V Apoptosis Detection Kit with PI (BioLegend Inc., San Diego, CA, USA). HUVEC-CS cells were cultured in DMEM with or without 20 µM TP1 under normoxic or hypoxic conditions for 24 h. In some experiments, the apoptosis of HUVEC-CS cells was induced using 0.2 µg/mL doxorubicin treatment for 24 h. After incubation, the cells were stained with FITC-conjugated annexin V and propidium iodide according to the manufacturer's protocol. Apoptotic cells were quantified using the BD FACSCalibur flow cytometry system (BD Biosciences).

2.17. ROS Detection

Intracellular ROS activity was measured using a DCFDA-based fluorogenic assay. Briefly, ECs were grown as a monolayer for 24 h under various experimental conditions, washed with serum-free medium, and stained with 1 µM DCFDA for 60 min at 37 °C. The dye-containing medium was then replaced with a fresh medium, and images of DCFDA-labeled cells were captured using the green fluorescence channel of the Nikon ECLIPSE Ti-S inverted microscope system. Quantification of fluorescence intensity was performed by analyzing labeled cells in ImageJ. Statistical analyses were performed using data from triplicate experiments.

2.18. Immunostaining

Thioflavin-T staining was used to detect UPR activity. Cells were grown on coverslips under various experimental conditions for 24 h, washed with PBS, and fixed with 4% paraformaldehyde for 20 min at room temperature. The fixed cells were then permeabilized for 5 min with 0.2% Triton X-100 at room temperature, stained with 1 µg/mL DAPI for 5 min, and counterstained with 5 µM thioflavin-T for 10 min. Coverslips were then mounted on glass slides with Clarion mounting medium, and images were captured using the Nikon ECLIPSE Ti-S inverted microscope system. Quantification of fluorescence intensity was performed by analyzing labeled cells in ImageJ. The experiment was performed in triplicate.

2.19. Aggregated Protein Isolation

Cells were cultured with or without TP1 under normoxic or hypoxic conditions for 24 h, after which aggregated proteins were isolated from 10×10^6 cells per condition. Briefly, cells were lysed in lysis buffer [50 mM Tris–HCl (pH 7.6), 150 mM NaCl, 1 mM EDTA, 1% NP-40, 0.1% SDS, 0.5% SDC, and protease inhibitor cocktail] with sonication at 4 °C. Cell debris was removed via centrifugation at $3000\times g$. Aggregated proteins were isolated from precleared lysates via ultracentrifugation at $112{,}000\times g$ and 4 °C for 1 h. Aggregates were washed once with lysis buffer and dissolved in a modified lysis buffer containing 5% ammonium hydroxide. Bradford quantification was used to measure the protein content. The experiment was performed in triplicate.

2.20. Statistical Analyses

Statistical analysis was performed using GraphPad Version 6.01 software (GraphPad Software, Inc., La Jolla, CA, USA). All data used for the analysis were obtained from individual experimental triplicates or as specified in the experiments. For data analysis, a one-way analysis of variance (ANOVA) with a Newman–Keuls posthoc test was conducted. Additionally, a two-way ANOVA with Šidák's multiple comparisons test was considered for the groups with two independent variables. The data are presented as the mean ± standard deviation (SD) or standard error of the mean (SEM). A significance level of $p < 0.05$ was considered statistically significant.

3. Results

3.1. TP1 Is a Hyperstable Nontoxic and Non-Membranolytic Cell-Penetrating Peptide

TP1 is a positively charged and hydrophobic peptide. To determine whether it interacts with intracellular proteins to protect against hypoxia-induced dysfunction, we site-specifically labeled TP1 with Cy3 at its lysine side chain. The cellular uptake of Cy3-labeled TP1 (Cy3–TP1) in live cells was observed using confocal microscopy to minimize artifacts in localizing labeled peptides under fluorescence microscopy [28]. The live-cell images of human umbilical vein/vascular endothelium (HUVEC-CS) cells after incubation with 1 μM Cy3–TP1 for 1 h revealed that Cy3–TP1 was internalized and distributed throughout the cytoplasm, whereas LysoTracker counterstaining showed that Cy3–TP1 also accumulated in lysosomes (Figure 1C) due its internalization through energy-dependent endocytosis as observed in our previous studies [25].

To investigate the toxicity of TP1, we used a lactate dehydrogenase (LDH) release assay and an MTT assay–based cell viability assay to assess multiple cell lines, including HUVEC-CS, H9c2, A7r5, and C2C12 cells. The LDH release assay revealed that treatment with TP1 up to 100 μM produced no significant cytotoxicity or cell-membrane-damaging effects on HUVEC-CS cells (Figure S2A). In agreement with the LDH assay results, the MTT-based cell viability assay also indicated that HUVEC-CS cell viability was not altered following TP1 treatment up to 100 μM (Figure S2B–D). The MTT-based cell viability assay also confirmed the nontoxic effect of TP1 at the effective dosage of 20 μM in multiple cell lines used in subsequent assays (Figure S2D). In addition, the MTT-based assay indicated that TP1 has no significant effect on cell proliferation, which was validated via a flowcytometry–based cell cycle assessment (Figure S2C). In agreement with our previous in vivo toxicity study, which did not show acute toxicity effects, e.g., cytotoxic, hemolytic, and immunogenic effects [19], the present results reflect the comparable nontoxic effect of TP1. Thus, we conclude that TP1 does not exert membranolytic, cytotoxic, or mitogenic effects at a dosage of up to 100 μM. The dose–response of TP1 was investigated within a range of 0–60 μM (Figure S2E), and a dose of 20 μM was chosen for subsequent studies.

3.2. TP1 Alters Hypoxia-Modulated Endothelial Proteome

To study hypoxia-induced pathophysiology in ECs, we used a pSILAC proteomic method to pulse/chase the incorporation of stable isotope-labeled lysine and arginine into newly synthesized proteins [29]. Briefly, HUVEC-CS cells were cultured for 24 h in SILAC medium containing $^{13}C_6$ L-arginine and $^{13}C_6$ and $^{15}N_2$ L-lysine in the presence or absence of 20 μM TP1 [or the phosphate-buffered saline (PBS)–only control] under normal oxygenation or hypoxic conditions (Figure S3A,B). After a 24 h incubation, the cells were lysed, and their protein content was extracted for LC-MS/MS proteomic analysis, in which 7614 proteins were identified with high confidence (3 experimental replicates and a false discovery rate of <1%), including 7101 proteins identified based on more than one component peptide (Supplementary dataset). According to Pearson's correlation analysis, a linear relationship ($R^2 > 0.9$) existed between replicates (Figure S4), indicating the high reproducibility and reliability of the experiments. All statistical analyses were conducted using data from triplicate experiments, *p*-values were calculated using 'Student's heteroscedastic *t*-test with a two-tailed distribution, and volcano plots were used to determine data distribution and cutoff levels (Figure S5A–C). In total, 2802 proteins were significantly enriched under hypoxia, either in the presence or absence of TP1, and the relative abundance of the newly synthesized proteins was subjected to hierarchical clustering (Figure 2A). Cutoffs for differential protein synthesis were determined using $\log_2$ ratio distributions: ratio values of <0.76 [log2 (Ratio) < −0.4] and >1.32 [log2 (Ratio) > 0.4] for reduced and induced proteins, respectively (Figure S5A–C). Using these cutoff values, we found that 148 and 431 proteins were induced and suppressed, respectively, under hypoxic conditions compared to normoxic conditions. We also found that TP1 treatment induced and repressed a further 39 and 172 distinct proteins, respectively, during hypoxia.

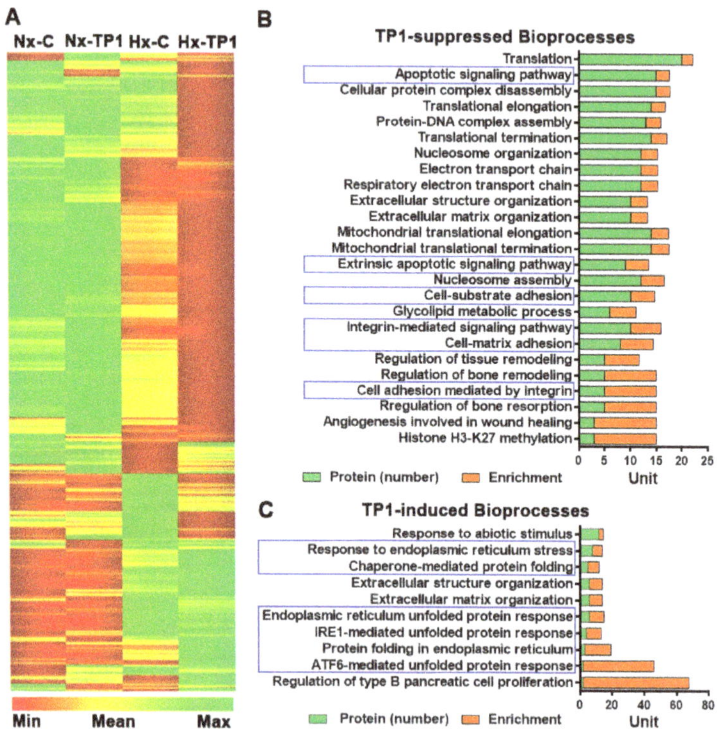

Figure 2. TP1-mediated differential protein synthesis in HUVEC-CS cells under normoxic and hypoxic conditions. (**A**) Heat map representing the relative abundance of newly synthesized proteins under the indicated experimental conditions. HUVEC-CS cells were treated with 20 μM TP1 for 24 h under normoxic or hypoxic conditions, respectively, and PBS was used as the vehicle control. A SILAC-based protein labeling approach was used to label the newly synthesized proteins, which were later quantified using a LC-MS/MS-based quantitative proteomic technique. Cluster analysis of the identified proteins was performed using the online bioinformatics tool Gene Pattern (http://genepattern.broadinstitute.org, accessed on 20 November 2017) with hierarchical clustering and Pearson's correlation options. Vehicle control under normoxic (Nx-C) or hypoxic (Hx-C) conditions and TP1 treatment under normoxic (Nx-TP1) or hypoxic (Hx-TP1) conditions are shown. (**B**,**C**) Bioinformatics analysis of TP1-regulated bioprocesses in hypoxic HUVEC-CS cells. Figures represent TP1-suppressed bioprocesses (**B**) and TP1-induced bioprocesses (**C**) during 24 h of hypoxic stress.

3.3. TP1 Limits Hypoxia-Induced Endothelial Activation and THP-1 Cell Adhesion

We performed functional analyses of the proteins modulated by low-oxygen stress with or without TP1 treatment. Data analysis performed using the online bioinformatics tool GOrilla (http://cbl-gorilla.cs.technion.ac.il, accessed on 30 December 2016) [26] revealed that hypoxia induced the expression of exocytosis regulators, key mediators of leukocyte activation/inflammatory responses and various metabolic intermediaries (Figure S6A–C). Importantly, TP1-treatment significantly reduced the synthesis of proteins involved in VED pathogenesis, including integrin-mediated cell adhesion molecules, cell-matrix/substrate adhesion regulators, and apoptotic signaling pathways (Figure 2B). Additionally, hypoxia stimulates endothelial activation and promotes inflammatory responses, which are significantly reduced by treatment with TP1. Indeed, TP1 treatment also induced the synthesis of proteins associated with cellular responses to ER stress, including the ATF6-mediated unfolded protein response (UPR), IRE1-mediated UPR, and chaperone-mediated protein folding (Figure 2B,C). Taken together, TP1 decreases hypoxia-induced ER stress in the human endothelium.

Bioinformatics analysis of our pSILAC data indicated that hypoxia stress-activated inflammatory response pathways were active in ECs (Figure S6A–C), leading to leukocyte activation and recruitment. Therefore, we used quantitative PCR to determine the expression patterns of leukocyte adhesion receptors during hypoxia. Our finding showed that the expression level of *ICAM1* mRNA increased 9-fold and the expression of *ALCAM*, *L1CAM*, and *VCAM1* mRNAs was also substantially upregulated (Figure 3A). Analysis of proteomic data showed that ICAM1 protein synthesis was upregulated in hypoxic ECs relative to normoxic cells (Figure 3B). In addition, cell adhesion assays using THP-1 human monocytic cells showed that THP-1 binding to hypoxic HUVEC-CS cells increased 2-fold (Figure 3C,D). Proteomic data analysis also revealed that TP1 treatment significantly decreased the expression of ICAM1 and ALCAM at the mRNA and protein levels in hypoxic ECs (Figure 3A,B) and reduced the transcription of L1CAM and VCAM1 (Figure 3A). Integrins are key signaling molecules involved in the regulation of various bioprocesses such as the activation of inflammatory responses [30]. The activation of integrin subsets, including $\alpha v\beta 3$ and $\alpha 5\beta 1$, is considered to trigger inflammatory responses through the activation of proinflammatory gene expression [31–33]. In the present study, we found that integrin-subtype-$\beta 3$ synthesis was induced by hypoxia (Figure S7A), which also significantly upregulated total integrin levels, including the levels of integrin subtypes α_v, β_1, and β_3 in ECs (Figure S7B,C). Proteomic data analysis revealed that TP1-treatment suppressed the synthesis of critical integrin subtypes in hypoxic ECs (Figure S7B,C) and significantly downregulated the total expression levels of integrin subtypes α_v, β_1 and β_3 in hypoxic ECs (Figure S7B,C). Therefore, the TP1-mediated suppression of integrin expression might prevent the activation of integrin subtypes, including $\alpha v\beta 3$ and $\alpha 5\beta 1$, which could result in the deactivation of proinflammatory gene expression [31–33], including the expression of *ALCAM*, *ICAM1*, *L1CAM*, and *VCAM1*, in hypoxic ECs. Accordingly, THP1 cell adhesion to hypoxic ECs was decreased 2.9-fold following TP1 treatment (Figure 3C,D), suggesting the high potential of TP1 affecting VED pathogenesis in vitro. In contrast, treatment with two major bioactive ginseng metabolites, ginsenosides Rb1 and Rg1, did not affect hypoxia-mediated proinflammatory gene activation (Figure 3A), nor did these ginseng metabolites prevent hypoxia-induced leukocyte adhesion to ECs (Figure 3C,D). Thus, unlike TP1, ginsenosides Rb1 and Rg1 do not have any protective impacts against hypoxia-induced endothelium activation.

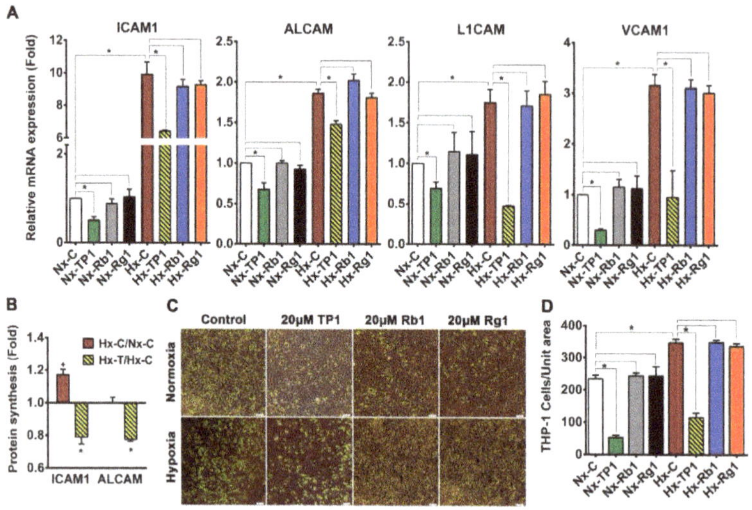

Figure 3. Ginsentide TP1 inhibits THP-1 cell adhesion to human vascular endothelial cells. (**A**) Relative mRNA expression of adhesion receptors in control and treated HUVEC-CS cells cultured under normoxic or hypoxic conditions. Data means ± SEM, and statistical significance was determined using triplicate biological replicates. Vehicle control cells were cultured for 24 h under normoxic (Nx-C) or hypoxic (Hx-C) conditions and TP1/Rb1/Rg1-treated cells were cultured for 24 h under normoxic (Nx-TP1/Nx-Rb1/Nx-Rg1) or hypoxic (Hx-TP1/Hx-Rb1/Hx-Rg1) conditions, respectively. * $p < 0.05$. (**B**) Fold-change in the newly synthesized cell surface adhesion receptors ICAM1 and ALCAM under hypoxic conditions relative to normoxic conditions (Hx-C/Nx-C) or in TP1-treated hypoxic cells compared with vehicle-treated hypoxic cells (Hx-TP1/Hx-C). Data are means ± standard error of the mean (SEM), and statistical significance was calculated based on triplicate experiments. + $p < 0.05$ respect to Nx-C and * $p < 0.05$ respect to Hx-C. (**C**) Differential binding of human monocytic THP1 cells to a HUVEC-CS monolayer cultured under normoxic or hypoxic conditions in the presence of TP1/Rb1/Rg1 or the PBS vehicle-only control. Carboxyfluorescein succinimidyl ester (CFSE)-labeled THP-1 cells are shown in green. Scale bar: 100 μm. (**D**) THP-1 adhesion was quantified via cell counting, and statistical significance was determined using triplicate experiments.

3.4. TP1 Promotes EC Survival by Antagonizing Apoptotic Signaling during Hypoxia

Given our findings that TP1 protects ECs against hypoxia-induced oxidative stress and ER stress, we assessed whether TP1 increases cell viability under hypoxic conditions. Briefly, ECs were cultured in Dulbecco's modified Eagle medium (DMEM) containing 10% fetal bovine serum (FBS) with or without TP1 and then incubated in a normoxic or hypoxic environment for 24 h before cell survival was determined using an MTT assay. These experiments revealed that TP1 treatment increased EC viability by 20% under hypoxic conditions (Figure 4A,B). Importantly, comparable protective effects were also observed in other TP1-treated cell types, including rat heart myoblast H9c2 cells (32%), rat aorta thoracic/smooth muscle A7r5 cells (31%), and mouse myoblast C2C12 cells (19%) (Figure 4A). We also tested two other ginsentide members, TP3 (*P. ginseng* seed) and TP8 (*P. quinquefolius* flower), which differ from TP1 by one and three amino acid residues, respectively (Figure 1A). Like TP1, TP3, and TP8 also exerted similar protective effects against hypoxic cell death in ECs (Figure 4B). Collectively, these results suggest that TP1 and its closely related homologs TP3 and TP8 provide various cell types with protection against hypoxia-induced death.

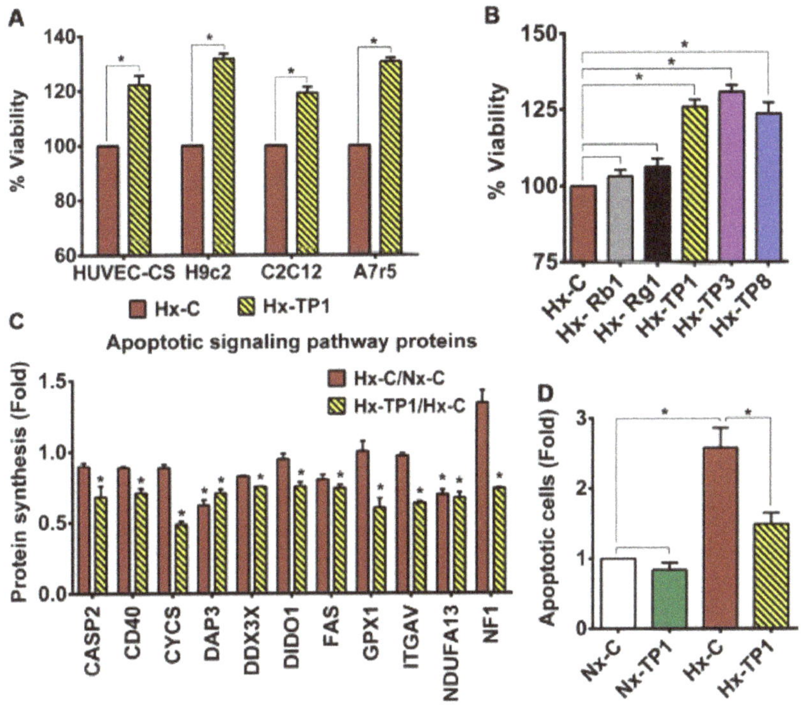

Figure 4. Ginsentide TP1 prevents hypoxic stress-mediated cell death. (**A**) Viability of hypoxic cells measured using an MTT assay after treatment with TP1 or vehicle control. Relative survival is expressed as mean values and statistical significance was calculated using three independent experimental replicates. (**B**) Effects of ginsentides, i.e., TP1, TP3, and TP8, and ginsenosides, i.e., Rb1 and Rg1, on the viability of hypoxic HUVEC-CS cells. Viability was measured using an MTT assay after treatment with 20 μM ginsenoside (Rb1 or Rg1), 20 μM ginsentide (TP1, TP3, or TP8), or PBS (vehicle control). Relative survival is expressed as mean values ± standard error of the mean (SEM) from three independent experimental replicates. * $p < 0.05$. (**C**) Relative change in the expression of apoptotic signaling pathway proteins after 24 h of hypoxic culture in the presence of TP1. Statistical significance was calculated using data from triplicate experiments. (**D**) Effect of TP1 on hypoxia-induced apoptosis in HUVEC-CS cells. Relative quantification was based on annexin V and propidium Iodide (PI) staining used to identify apoptotic cells. Statistical significance was calculated using three independent experimental replicates. Data are means ± SEM for triplicate experiments. * $p < 0.05$.

Hypoxia is known to induce cell cycle arrest in the G0/G1 phase and impair mitochondrial function, leading to the apoptosis of ECs and progression to VED [34]. Our proteomic data analysis showed that TP1 could suppress the synthesis of apoptosis signaling proteins in hypoxic ECs (Figure 4C); therefore, we investigated the antiapoptotic effects of TP1 via the flow cytometric analysis of annexin V and propidium iodide (PI) staining. We observed a marked increase in the proportion of apoptotic and necrotic HUVEC-CS cells cultured under hypoxic conditions; however, treatment with TP1 was sufficient to abolish this increase (Figure 4D and Figure S8A,B). TP1 also conferred significant protection against cell death in an alternative doxorubicin-induced apoptosis model (Figure S8C,D). This result is significant because doxorubicin is a leading drug against certain types of cancer, and TP1 exerts antiapoptotic effects to prevent doxorubicin-induced cell death during hypoxia.

We also investigated the effects of ginsenosides on the viability of ECs during hypoxic stress but found no significant effects of Rb1 or Rg1 treatment on cell survivability

(Figure 4B). Thus, unlike the tested ginsentides, we found these two ginsenosides have no protective effects against hypoxia-mediated cell death.

3.5. TP1 Treatment Rescues ECs from Hypoxia-Induced ER Stress

The disruption of ER homeostasis during hypoxia leads to the accumulation of misfolded/unfolded proteins in the ER lumen, resulting in ER stress. Hypoxia-induced ER stress initiates the unfolded protein response (UPR), which promotes cellular adaptation to stress conditions and promotes the apoptosis of cells that fail to restore normal function [35,36]. To better understand the mechanism underlying the promotion of hypoxic cell survival by TP1, we measured the mRNA expression levels of the UPR marker XBP1. We found that the spliced isoform was upregulated 20-fold in hypoxic HUVEC-CS cells compared to normoxic HUVEC-CS cells (Figure S9). Under these conditions, according to our proteomic data analysis, TP1 induced the synthesis of several ER stress sensor (ATF6/IRE1)–regulated downstream proadoptive UPR genes, including those encoding chaperone proteins and autophagy proteins, in hypoxic ECs (Figure 5A,B). Consistent with these proteomic findings, we also detected the upregulation of spliced-XPB1 mRNA in TP1-treated hypoxic ECs (Figure S9) as well as the enhanced expression of ATF6 transcripts and multiple downstream UPR genes, e.g., *BiP* (*HSPA5*), *GRP94* (*HSP90B1*), and *EDEM* (Figure 5C). Our quantitative reverse transcription–polymerase chain reaction (qRT-PCR) analyses also indicated that TP1-treatment significantly downregulated the expression of multiple genes in the PERK–eIF2a–ATF4 apoptosis pathway, including *ATF4*, *CHOP*, *DR5*, and *TRB3*, in hypoxic cells (Figure 5D). Conversely, this qRT–PCR analysis showed that the reference compounds, ginsenosides Rb1 and Rg1, had no significant effect on the mRNA expression of ER stress-related genes (Figure S10A,B).

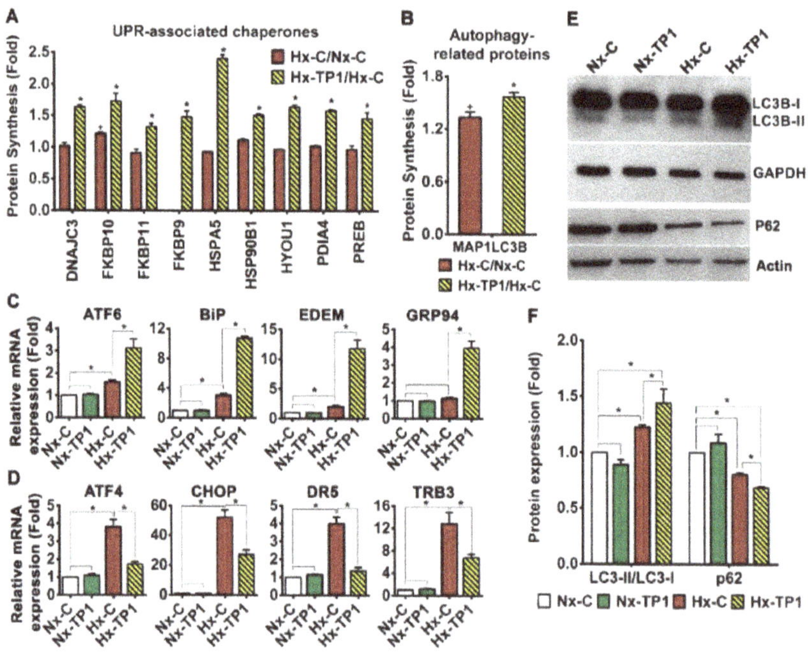

Figure 5. Ginsentide TP1 rescues endothelial cells from hypoxia-induced ER stress. (**A**,**B**) Relative expression levels of UPR-associated chaperone proteins (**A**) and autophagy-related proteins (**B**) under hypoxic conditions compared with normoxic conditions (Hx-C/Nx-C) or in TP1-treated hypoxic cells relative to PBS-treated hypoxic cells (Hx-TP1/Hx-C). Data are means ± standard error of the mean (SEM), and statistical significance was calculated using triplicate experiments. Significance

+ $p < 0.05$ respect to Nx-C and * $p < 0.05$ respect to Hx-C. (**C**,**D**) Relative mRNA expression of ER stress-related genes in HUVEC-CS cells cultured under various conditions. Data are means ± SEM, and statistical significance was calculated using triplicate biological replicates. (**E**) Western blot images of autophagy markers, including LC3B-I and II and P62, under the indicated experimental conditions. (**F**) Relative quantification of the LC3B-II/LC3B-I ratio and relative expression of P62 under the given experimental conditions (quantified based on Western blot band intensity). Data are means ± SEM, and statistical significance was calculated using triplicate biological replicates. Vehicle control cells were cultured for 24 h under normoxic (Nx-C) or hypoxic (Hx-C) conditions and TP1-treated cells were cultured for 24 h under normoxic (Nx-TP1) or hypoxic (Hx-TP1) conditions. * $p < 0.05$.

Autophagy is an essential cell process, a principal component of the integrated stress response [37], and part of the ER stress-mediated proadoptive UPR response [35,36]. Indeed, Western blot analysis revealed that the ratio of LC3B-II/LC3B-I increased significantly upon TP1 treatment under hypoxic conditions (Figure 5E,F), indicating an increase in autophagosome formation [37]. TP1-treatment also increased P62 degradation in hypoxic ECs and resulted in a lower level of P62 in these cells (Figure 5E,F), further confirming the activation of autophagy. Overall, our Western blot results indicated that TP1-induced selective autophagy occurred in ginsentide-treated hypoxic cells. Thus, TP1 promotes the clearance of accumulated proteins and suppresses apoptotic signaling to prevent the ER-stress-induced death of hypoxic cells. Indeed, when we measured protein aggregation using thioflavin-T staining and fluorescence microscopic analysis, we found a 1.6-fold increase in protein accumulation in hypoxic HUVEC-CS cells, which was effectively reversed by TP1 treatment (Figure 6A,B). The protective effect of TP1 against hypoxia was also confirmed using alternative cell lines, i.e., H9c2, A7r5, and C2C12, in which the reduction in thioflavin-T staining upon TP1-treatment was comparable to that in HUVEC-CS cells (Figure S11A,B). Consistent with the fluorescence-based analyses, the ultracentrifugation and quantitation of protein aggregates confirmed that protein accumulation in HUVEC-CS cells during hypoxic stress was markedly reduced by TP1 treatment, suggesting that TP1 is an efficient stimulator of the UPR (Figure 6C).

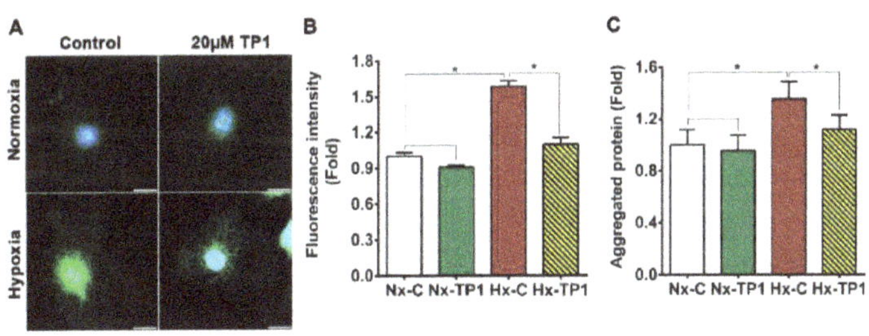

Figure 6. TP1-mediated recovery of ER/protein homeostasis in hypoxic endothelial cells. (**A**) Thioflavin-T (ThT) staining of protein aggregates in HUVEC-CS cells cultured under the indicated conditions. Aggregated proteins were stained with ThT (green), and nuclei were counterstained with DAPI (blue). Scale bar: 20 μm. (**B**) Relative quantification of protein aggregation in cells subjected to the indicated experimental conditions. Protein aggregation was quantified by calculating the average fluorescence intensity of >100 individual cells, and statistical analysis was performed using four experimental replicates. (**C**) Relative abundance of aggregated proteins isolated from HUVEC-CS cells cultured under different conditions. Data are means ± standard error of the mean, and statistical significance was calculated using triplicate experiments. Vehicle control cells were cultured for 24 h under normoxic (Nx-C) or hypoxic (Hx-C) conditions and TP1-treated cells were cultured for 24 h under normoxic (Nx-TP1) or hypoxic (Hx-TP1) conditions. * $p < 0.05$.

3.6. TP1 Restores NO Production and Reduces Reactive Oxygen Species Generation in Hypoxic ECs

Under normal physiological conditions, NO maintains the quiescent state of the endothelium, while disruption of NO signaling can trigger the endothelial expression of mediators that promote leukocyte adhesion and inflammation [38]. Under oxidative stress conditions, the reaction of NO with superoxide anions generates peroxynitrite, which significantly impairs NO production by nitric oxide synthases, thereby promoting endothelial activation and progression to VED [38–40]. Given our finding that TP1 limits endothelial activation under hypoxic conditions, we assessed whether TP1 acts via effects on the generation of NO and reactive oxygen species (ROS). Using a fluorogenic staining approach, i.e., a 2′,7′-dichlorofluorescein diacetate (DCFDA) assay, we found that intracellular ROS levels were higher in hypoxic ECs than normoxic cells; however, TP1-treatment reduced ROS levels 1.4-fold in hypoxic ECs (Figure 7A,B). Similarly, a 1.3-fold reduction in intracellular NO was detected in hypoxic ECs, but this deficit was effectively reversed by TP1 treatment (Figure 7B). Accordingly, when the expression of eNOS was assessed using Western blotting, protein levels that were reduced under hypoxic conditions were partially restored by TP1 treatment (Figure 7C). Consistent with these data, hypoxic stress was found to induce a 1.3-fold decrease in eNOS phosphorylation at residue Ser^{1177}, which is known to modify enzyme activity under both physiological and pathological conditions [41,42]; however, TP1-treatment reversed this deficiency in eNOS phosphorylation (Figure 7C). Taken together, our data indicate that TP1 limits ROS generation and maintains NO bioavailability to prevent endothelial activation under hypoxic conditions. Bioactive ginsenosides are known to counteract redox regulation, promote NO production, and prevent ROS generation; thus, they protect cells against oxidative damage [25,43–45]. Consistent with these reported effects, we found that treatment with the ginsenosides Rb1 and Rg1 reduced the intracellular ROS levels of both normoxic and hypoxic ECs, comparable with the impact of TP1 treatment (Figure S12A,B).

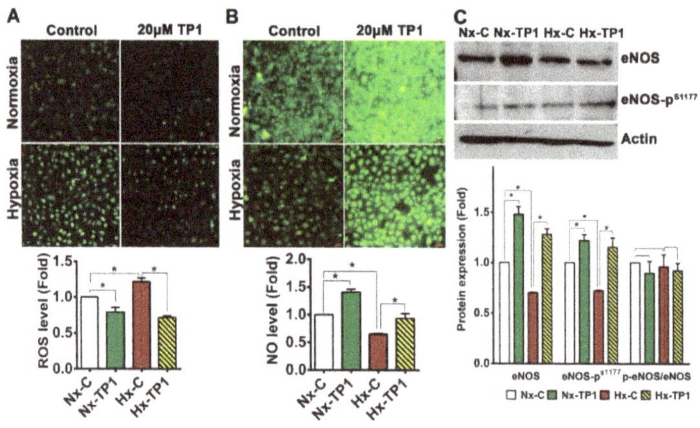

Figure 7. Ginsentide TP1 reduces intracellular reactive oxygen species (ROS) generation and restores nitric oxide (NO) generation in endothelial cells. (**A**) Effect of TP1 on ROS levels in HUVEC-CS cells cultured under different experimental conditions. ROS activity was measured by (DCFDA)-based fluorogenic assay. Scale bar: 25 µm. Relative quantification was performed using the relative fluorescence intensity. Statistical significance was calculated using five biological replicates. (**B**) HUVEC-CS cells were cultured with or without TP1 for 24 h under normoxic or hypoxic conditions, and NO levels were measured via DAF-2DA staining. Scale bar: 25 µm. Relative quantification of NO was based on the fluorescence intensity in each image, and statistical significance was calculated using four biological replicates. (**C**) Relative expression of eNOS and Ser^{1177}-phosphorylated eNOS (eNOS-p^{S1177}) under the indicated experimental conditions (quantified based on Western blot band intensity). Data are shown as means ± standard error of the mean, and statistical significance was calculated using biological triplicates. Vehicle control cells were cultured for 24 h under normoxic (Nx-C) or hypoxic (Hx-C) conditions and TP1-treated cells were cultured for 24 h under normoxic (Nx-TP1) or hypoxic (Hx-TP1) conditions. * $p < 0.05$.

4. Discussion

Ginseng, the most valuable herb in traditional medicines, is prescribed to treat a wide range of disorders, including cardiovascular disease [17,18,46]. Thus far, ginsenosides, which belong to the small-molecule metabolites of the triterpene saponin family, have received the most attention as the active compounds to account for the ethnomedicinal effects of ginseng [46,47]. Despite an abundance of ginseng research, ginsenosides alone cannot adequately explain the "cure-all" effect of ginseng in its ethnomedicinal uses [46,47]. Recently, we reported that the cell-penetrating cystine-dense microprotein TP1 coordinates multiple systems to exert anti-stress and cardiovascular benefits [25]. This report shows that ginsentide TP1 differing from the two key ginsenosides, Rb1 and Rg1, plays an important adaptogenic role in maintaining protein homeostasis in hypoxic cells and suppressing ER-stress-induced apoptosis.

Ginsentides have two distinct advantages over most unconstrained peptide biologics of comparable size. First, ginsentides, containing about 25% cysteine, are highly cystine-braced, rendering them structurally compact and highly resistant to proteolytic degradation. Second, ginsentide TP1, the representative and most abundant ginsentide, is a cell-penetrating peptide and can target intracellular protein–protein interactions. Additionally, TP1 does not exhibit any toxic or membrane-damaging effects. In comparison, ginsenosides have poor bioavailability, thereby limiting their clinical applications [48]. TP1's cell-penetrating property can also be found in certain highly cystine-braced microproteins, such as the plant-derived 8C-hevein-like peptides [20–24,28]. Thus, hyperstability and cell-penetrating properties of ginsentides are desirable attributes for drug development.

The vascular endothelium is comprised of a monolayer of cells that maintain hemostasis by regulating vascular tone, cell adhesion, smooth muscle proliferation, and tissue inflammation. The dysregulation of these mechanisms leads to VED, which can result in systemic inflammatory pathology, adverse thrombotic events, and vascular tissue remodeling. A link between hypoxia and VED has been established, as well as their consequences in relation to cardiovascular disorders [1–6]. Because preventing hypoxia-linked VED will be beneficial for treating cardiovascular disorders, we established a pulsed SILAC–based quantitative proteomic approach to determine the effects of TP1 on hypoxia-linked VED. We profiled the newly synthesized proteome of normoxic and hypoxic endothelium cells with or without TP1 treatment and analyzed the differentially expressed newly synthesized proteins. Our findings showed that hypoxia and TP1 treatment alter the expression profile of newly synthesized proteins in endothelium cells. Gene ontology analysis revealed that hypoxia promotes the expression of endothelium dysfunction–linked pathway proteins. In addition, TP1 treatment reduced the expression of VED pathogenesis-linked pathway proteins, including integrins, cell adhesion molecules, and apoptotic signaling proteins. In this regard, we showed that pulsed SILAC-based quantitative proteomics is useful in the functional characterization of cell-penetrating and cystine-dense microproteins such as ginsentides.

Hypoxia promotes inflammatory signaling [9] in ECs, resulting in the increased surface expression of adhesion molecules, including ICAM1 and VCAM1, which promote leukocyte/platelet binding and progression to VED [49,50]. Integrin complexes, including $\alpha v\beta 3$, are involved in the regulation of surface adhesion molecules [31–33] and the TP1-mediated suppression of integrin αv, $\beta 1$, and $\beta 3$ expression, which lead to decreased surface molecule expression levels in hypoxic ECs and reduced THP-1 cell adhesion to these cells. In the present study, hypoxic ECs enhanced the expression of adhesion molecules and increased leukocyte binding by 2-fold. In contrast, hypoxic ECs treated with TP1 exhibited reduced adhesion molecule expression at both the mRNA and protein levels. Accordingly, leukocyte binding to hypoxic ECs decreased 2.9-fold in the presence of TP1, indicating the potent effects of ginsentides on the activation of the endothelium [51,52]. In contrast, ginsenosides Rb1 and Rg1, two bioactive components of ginseng, displayed no notable effects on the surface molecule expression and THP-1 cell adhesion of ECs.

Cellular stress, including hypoxia, can cause defects in protein folding that lead to protein aggregation in the ER lumen and induce ER stress [53,54], which is a crucial component of endothelial dysfunction and cardiovascular disease [15,55,56]. The eNOS/NO pathway plays a critical role in maintaining the intracellular redox balance and protecting against ER stress in ECs [57]. Hence, the disruption of this pathway results in oxidative stress and proinflammatory responses that promote VED [15,58]. One mechanism underlying the protection of hypoxic cells against ER stress is the evolutionarily conserved UPR, which first attempts to restore protein homeostasis and later triggers apoptosis if recovery is unsuccessful [35,36,59]. The UPR plays a vital role in determining the fate of stressed cells. In the current study, hypoxic HUVEC-CS cells exhibited the 18-fold upregulation of spliced XBP1 mRNA, an archetypal marker of UPR activation [60]. Indeed, low-oxygen conditions also promoted the accumulation of unfolded proteins in multiple cell types, including ECs, myocytes, cardiomyocytes, and vascular smooth muscle cells, and the significant upregulation of the cell death signaling genes *ATF4*, *CHOP*, *DR5*, and *TRB3*. However, treatment with TP1 increased the hypoxic EC expression of anti-aggregation ERED genes and molecular chaperones that support protein folding, including *DNAJC3*, *HSP90B1*, *HSPA5*, and *HYOU1*, which help restore protein homeostasis and reduce ER stress in hypoxic ECs. In addition, TP1 promoted selective autophagy as a part of the proadaptive UPR response [36,37], thereby enhancing protein clearance and restoring protein homeostasis in hypoxic ECs. Our proteomic results revealed that the expression of apoptosis signaling proteins and cell death rates were significantly reduced following TP1 treatment in hypoxic ECs, myocytes, cardiomyocytes, and vascular smooth muscle cells. Taken together, our data indicate that TP1 exerts potent effects on the UPR pathway and reduces ER-stress-induced apoptotic signaling, thereby increasing overall EC survival under hypoxic conditions (Figure 8). The ability of TP1 to protect ECs from stress-induced apoptosis could be exploited clinically to treat various human disorders, particularly hypoxia-associated cardiovascular disorders.

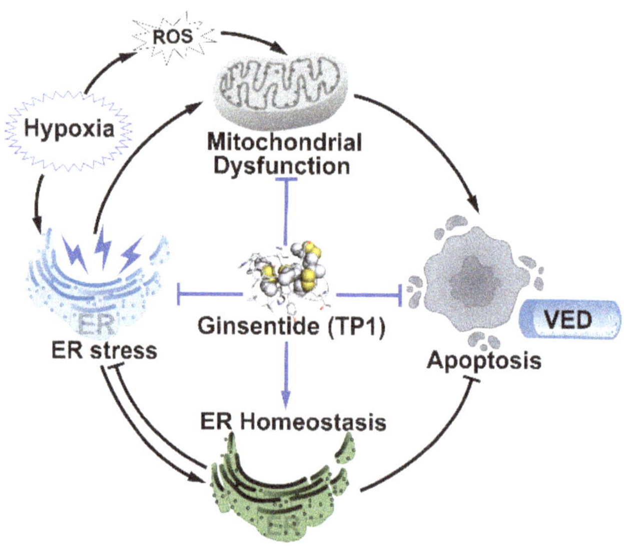

Figure 8. Proposed model for TP1-mediated protective effects against hypoxic stress in the endothelium. Endoplasmic reticulum (ER) stress initiates the unfolded protein response (UPR) as a part of the integrated stress response. UPR promotes both cellular adaptations to stress conditions and the apoptosis of cells that fail to restore normal function. Adaptive UPR prevents ER stress by restoring ER/protein homeostasis. In contrast, prolonged ER stress induces mitochondrial dysfunction and promotes apoptotic cell death. Hypoxia-induced ER stress in endothelium cells initiates proadaptive responses. Prolonged hypoxia leads to prolonged ER stress, which is proposed to cause the failure

of proadaptive responses and ensure apoptotic cell death, leading to endothelium dysfunction. Ginsentide TP1 is thought to minimize the effect of ER stress in hypoxic endothelial cells by restoring ER/protein homeostasis via the activation of proadaptive responses and suppression of apoptotic cell death.

Hypoxia has been shown to impair the activity of the NO biosynthetic enzyme eNOS via multiple mechanisms, including the subduction of protein phosphorylation at Ser^{1177} [41,42,61,62], which leads to the disruption of NO-mediated protective pathways and endothelial activation by redox signaling [61,63]. Loss of eNOS function reduces NO production and increases superoxide radical generation in hypoxic ECs, leading to endothelium activation. However, in the present study, we found a significant decrease in eNOS expression and NO levels, in parallel with elevated ROS production, in hypoxic ECs. Importantly, TP1 treatment reversed this hypoxia-induced effect in hypoxic ECs, increasing eNOS expression and phosphorylation at Ser^{1177}, which led to the recovery of NO synthesis. The TP1-mediated restoration of the redox balance in hypoxic cells via the eNOS/NO pathway may prevent endothelium dysfunction [25,51,52]. Notably, the tested ginsenosides Rb1 and Rg1 exerted a comparable effect [25], i.e., they restored the redox balance by increasing the NO level and reducing the ROS load in hypoxic ECs.

5. Conclusions

Hypoxia impairs essential cellular functions to promote endothelial cell activation and progression to VED. Low oxygen conditions also induce ER stress, which activates apoptotic signaling in hypoxic endothelial cells and multiple other cell lineages affected by cardiovascular pathology. Ginsentide TP1 rescues protein homeostasis in hypoxic cells and suppresses ER-stress-induced apoptotic signals to improve the survival of multiple cardiovascular cell types. TP1 also restores NO bioavailability and reduces intercellular ROS in hypoxic endothelial cells via restorative effects on the NO/eNOS pathway, which also protects against hypoxia-induced VED. Our results provide strong support that ginsentides are unexplored microproteins responsible for the ethnomedicinal benefits of ginseng in cardiovascular diseases. Additionally, ginsentides could provide therapeutic leads useful in treating patients at risk of VED and associated cardiovascular complications. However, to ascertain the efficacy of TP1 as a therapeutic, further in vivo studies would be necessary.

Supplementary Materials: The following supporting information can be downloaded at: https://www.mdpi.com/article/10.3390/cells12101401/s1, Supplementary Data S1: Figure S1: HPLC and mass spectrometry profiles of purified ginsentide TP1 extracted from *Panax ginseng* flowers; Figure S2: Toxicity and dose–response curve of ginsentide TP1; Figure S3: Schematic representation of the experimental workflow of pulsed SILAC labelling and proteomics; Figure S4: Linear regression plots representing the correlation among triplicate experiments under different experimental; Figure S5: Statistical analysis of proteomic data; Figure S6: Hypoxia-induced biological processes in endothelial cells (ECs); Figure S7: Ginsentide TP1 suppresses integrin expression in hypoxic endothelial cells; Figure S8: TP1 prevents apoptosis in HUVEC-CS; Figure S9: Effect of TP1 on the hypoxia-induced unfolded protein response (UPR) in endothelial cells; Figure S10: Effects of ginsenosides on hypoxia-induced ER stress-related gene expression in ECs; Figure S11: TP1-mediated recovery of ER/protein homeostasis in hypoxic cardiovascular cell types; Figure S12: Effect of ginsenoside treatment on the intracellular ROS load in ECs; Table S1: List of genes and corresponding primers used in quantitative RT–PCR. Supplementary Data S2: Table S2: List of all proteins identified; and Table S3: List of proteins identified with multiple peptides.

Author Contributions: B.D. involved in designing and performing experiments, analyzing data, and writing the manuscript; S.L. and A.K. performed peptide extraction, purification, and confocal microscopic-based cellular uptake studies; S.K.S. conceived and oversaw the proteomics analysis and edited the manuscript; J.P.T. conceived the idea and supervised and edited the manuscript. All authors have read and agreed to the published version of the manuscript.

Funding: This study was supported in part by the Nanyang Technological University Internal 702 Funding to Synzymes and Natural Products (SYNC) and the AcRF Tier 3 funding (MOE2016703 T3-1-003).

Institutional Review Board Statement: Not applicable.

Informed Consent Statement: Not applicable.

Data Availability Statement: The mass spectrometry proteomics raw data files, along with the Proteome Discoverer search data file (including protein summary), have been deposited to the ProteomeXchange Consortium (http://proteomecentral.proteomexchange.org, accessed on 18 January 2023) via the jPOSTrepo (Japan ProteOme STandard Repository) partner repository [64] with the dataset identifier PXD039938 for ProteomeXchange and JPST001996 for jPOST. The data preview page access URL is https://repository.jpostdb.org/preview/32803372563e3629fe4684 (Access key 8294).

Conflicts of Interest: The authors declare that they have no potential conflict of interest.

References

1. Hoyos, C.M.; Melehan, K.L.; Liu, P.Y.; Grunstein, R.R.; Phillips, C.L. Does obstructive sleep apnea cause endothelial dysfunction? A critical review of the literature. *Sleep Med. Rev.* **2015**, *20*, 15–26. [CrossRef] [PubMed]
2. Wilkins, M.R.; Ghofrani, H.A.; Weissmann, N.; Aldashev, A.; Zhao, L. Pathophysiology and treatment of high-altitude pulmonary vascular disease. *Circulation* **2015**, *131*, 582–590. [CrossRef] [PubMed]
3. Costanzo, M.R.; Khayat, R.; Ponikowski, P.; Augostini, R.; Stellbrink, C.; Mianulli, M.; Abraham, W.T. Mechanisms and clinical consequences of untreated central sleep apnea in heart failure. *J. Am. Coll. Cardiol.* **2015**, *65*, 72–84. [CrossRef] [PubMed]
4. Chen, J.; Lin, S.; Zeng, Y. An Update on Obstructive Sleep Apnea for Atherosclerosis: Mechanism, Diagnosis, and Treatment. *Front. Cardiovasc. Med.* **2021**, *8*, 647071. [CrossRef]
5. Haranczyk, M.; Konieczynska, M.; Plazak, W. Endothelial dysfunction in obstructive sleep apnea patients. *Sleep Breath* **2022**, *26*, 231–242. [CrossRef]
6. Peracaula, M.; Torres, D.; Poyatos, P.; Luque, N.; Rojas, E.; Obrador, A.; Orriols, R.; Tura-Ceide, O. Endothelial Dysfunction and Cardiovascular Risk in Obstructive Sleep Apnea: A Review Article. *Life* **2022**, *12*, 537. [CrossRef]
7. Wang, L.; Yin, J.; Nickles, H.T.; Ranke, H.; Tabuchi, A.; Hoffmann, J.; Tabeling, C.; Barbosa-Sicard, E.; Chanson, M.; Kwak, B.R.; et al. Hypoxic pulmonary vasoconstriction requires connexin 40-mediated endothelial signal conduction. *J. Clin. Investig.* **2012**, *122*, 4218–4230. [CrossRef]
8. Burnstock, G.; Ralevic, V. Purinergic signaling and blood vessels in health and disease. *Pharmacol. Rev.* **2014**, *66*, 102–192. [CrossRef]
9. Erlinge, D.; Burnstock, G. P2 receptors in cardiovascular regulation and disease. *Purinergic Signal.* **2008**, *4*, 1–20. [CrossRef]
10. Cai, H.; Harrison, D.G. Endothelial dysfunction in cardiovascular diseases: The role of oxidant stress. *Circ. Res.* **2000**, *87*, 840–844. [CrossRef]
11. Pober, J.S. Endothelial activation: Intracellular signaling pathways. *Arthritis Res. Ther.* **2002**, *4* (Suppl. S3), S109. [CrossRef] [PubMed]
12. Handy, D.E.; Loscalzo, J. Responses to reductive stress in the cardiovascular system. *Free. Radic. Biol. Med.* **2017**, *109*, 114–124. [CrossRef] [PubMed]
13. Szocs, K. Endothelial dysfunction and reactive oxygen species production in ischemia/reperfusion and nitrate tolerance. *Gen. Physiol. Biophys.* **2004**, *23*, 265–295. [PubMed]
14. Bourdier, G.; Flore, P.; Sanchez, H.; Pepin, J.L.; Belaidi, E.; Arnaud, C. High-intensity training reduces intermittent hypoxia-induced ER stress and myocardial infarct size. *Am. J. Physiol. Heart Circ. Physiol.* **2016**, *310*, H279–H289. [CrossRef]
15. Battson, M.L.; Lee, D.M.; Gentile, C.L. Endoplasmic reticulum stress and the development of endothelial dysfunction. *Am. J. Physiol. Heart Circ. Physiol.* **2017**, *312*, H355–H367. [CrossRef]
16. Choi, Y.Y.; Kim, S.; Han, J.H.; Nam, D.H.; Park, K.M.; Kim, S.Y.; Woo, C.H. CHOP deficiency inhibits methylglyoxal-induced endothelial dysfunction. *Biochem. Biophys. Res. Commun.* **2016**, *480*, 362–368. [CrossRef]
17. Wee, J.J.; Mee Park, K.; Chung, A.S. Biological Activities of Ginseng and Its Application to Human Health. In *Herbal Medicine: Biomolecular and Clinical Aspects*, 2nd ed.; Benzie, I.F.F., Wachtel-Galor, S., Eds.; CRC Press/Taylor & Francis LLC: Boca Raton, FL, USA, 2011.
18. Wang, H.; Peng, D.; Xie, J. Ginseng leaf-stem: Bioactive constituents and pharmacological functions. *Chin. Med.* **2009**, *4*, 20. [CrossRef]
19. Tam, J.P.; Nguyen, G.K.T.; Loo, S.; Wang, S.; Yang, D.; Kam, A. Ginsentides: Cysteine and Glycine-rich Peptides from the Ginseng Family with Unusual Disulfide Connectivity. *Sci. Rep.* **2018**, *8*, 16201. [CrossRef]
20. Dutta, B.; Huang, J.; To, J.; Tam, J.P. LIR Motif-Containing Hyperdisulfide beta-Ginkgotide is Cytoprotective, Adaptogenic, and Scaffold-Ready. *Molecules* **2019**, *24*, 2417. [CrossRef]

1. Loo, S.; Kam, A.; Li, B.B.; Feng, N.; Wang, X.; Tam, J.P. Discovery of Hyperstable Noncanonical Plant-Derived Epidermal Growth Factor Receptor Agonist and Analogs. *J. Med. Chem.* **2021**, *64*, 7746–7759. [CrossRef]
2. Tam, J.P.; Wang, S.; Wong, K.H.; Tan, W.L. Antimicrobial Peptides from Plants. *Pharmaceuticals* **2015**, *8*, 711–757. [CrossRef] [PubMed]
3. Loo, S.; Tay, S.V.; Kam, A.; Lee, W.; Tam, J.P. Hololectin Interdomain Linker Determines Asparaginyl Endopeptidase-Mediated Maturation of Antifungal Hevein-Like Peptides in Oats. *Front. Plant Sci.* **2022**, *13*, 899740. [CrossRef] [PubMed]
4. Loo, S.; Tay, S.V.; Kam, A.; Tang, F.; Fan, J.S.; Yang, D.; Tam, J.P. Anti-Fungal Hevein-like Peptides Biosynthesized from Quinoa Cleavable Hololectins. *Molecules* **2021**, *26*, 5909. [CrossRef]
5. Loo, S.; Kam, A.; Dutta, B.; Zhang, X.; Feng, N.; Sze, S.K.; Liu, C.-F.; Wang, X.; Tam, J.P. Decoding the Cure-all Effects of Ginseng. *bioRxiv* **2023**. [CrossRef]
6. Eden, E.; Navon, R.; Steinfeld, I.; Lipson, D.; Yakhini, Z. GOrilla: A tool for discovery and visualization of enriched GO terms in ranked gene lists. *BMC Bioinform.* **2009**, *10*, 48. [CrossRef] [PubMed]
7. Oslowski, C.M.; Urano, F. Measuring ER stress and the unfolded protein response using mammalian tissue culture system. *Methods Enzym.* **2011**, *490*, 71–92. [CrossRef]
8. Kam, A.; Loo, S.; Dutta, B.; Sze, S.K.; Tam, J.P. Plant-derived mitochondria-targeting cysteine-rich peptide modulates cellular bioenergetics. *J. Biol. Chem.* **2019**, *294*, 4000–4011. [CrossRef] [PubMed]
9. Park, J.E.; Tse, S.W.; Xue, G.; Assisi, C.; Maqueda, A.S.; Ramon, G.P.X.; Low, J.K.; Kon, O.L.; Tay, C.Y.; Tam, J.P.; et al. Pulsed SILAC-based proteomic analysis unveils hypoxia- and serum starvation-induced de novo protein synthesis with PHD finger protein 14 (PHF14) as a hypoxia sensitive epigenetic regulator in cell cycle progression. *Oncotarget* **2019**, *10*, 2136–2150. [CrossRef]
0. Finney, A.C.; Stokes, K.Y.; Pattillo, C.B.; Orr, A.W. Integrin signaling in atherosclerosis. *Cell. Mol. Life Sci. CMLS* **2017**, *74*, 2263–2282. [CrossRef]
1. Chen, J.; Green, J.; Yurdagul, A., Jr.; Albert, P.; McInnis, M.C.; Orr, A.W. alphavbeta3 Integrins Mediate Flow-Induced NF-kappaB Activation, Proinflammatory Gene Expression, and Early Atherogenic Inflammation. *Am. J. Pathol.* **2015**, *185*, 2575–2589. [CrossRef]
2. Akbar, N.; Digby, J.E.; Cahill, T.J.; Tavare, A.N.; Corbin, A.L.; Saluja, S.; Dawkins, S.; Edgar, L.; Rawlings, N.; Ziberna, K.; et al. Endothelium-derived extracellular vesicles promote splenic monocyte mobilization in myocardial infarction. *JCI Insight* **2017**, *2*, e93344. [CrossRef] [PubMed]
3. Roebuck, K.A.; Finnegan, A. Regulation of intercellular adhesion molecule-1 (CD54) gene expression. *J. Leukoc. Biol.* **1999**, *66*, 876–888. [CrossRef] [PubMed]
4. Tang, X.; Luo, Y.X.; Chen, H.Z.; Liu, D.P. Mitochondria, endothelial cell function, and vascular diseases. *Front. Physiol.* **2014**, *5*, 175. [CrossRef] [PubMed]
5. Xu, C.; Bailly-Maitre, B.; Reed, J.C. Endoplasmic reticulum stress: Cell life and death decisions. *J. Clin. Investig.* **2005**, *115*, 2656–2664. [CrossRef]
6. Rozpedek, W.; Pytel, D.; Mucha, B.; Leszczynska, H.; Diehl, J.A.; Majsterek, I. The Role of the PERK/eIF2alpha/ATF4/CHOP Signaling Pathway in Tumor Progression During Endoplasmic Reticulum Stress. *Curr. Mol. Med.* **2016**, *16*, 533–544. [CrossRef]
7. Kroemer, G.; Marino, G.; Levine, B. Autophagy and the integrated stress response. *Mol. Cell* **2010**, *40*, 280–293. [CrossRef]
8. Tousoulis, D.; Kampoli, A.M.; Tentolouris, C.; Papageorgiou, N.; Stefanadis, C. The role of nitric oxide on endothelial function. *Curr. Vasc. Pharmacol.* **2012**, *10*, 4–18. [CrossRef]
9. Montezano, A.C.; Touyz, R.M. Reactive oxygen species and endothelial function-role of nitric oxide synthase uncoupling and Nox family nicotinamide adenine dinucleotide phosphate oxidases. *Basic Clin. Pharmacol. Toxicol.* **2012**, *110*, 87–94. [CrossRef]
0. Norouzirad, R.; Gonzalez-Muniesa, P.; Ghasemi, A. Hypoxia in Obesity and Diabetes: Potential Therapeutic Effects of Hyperoxia and Nitrate. *Oxid. Med. Cell. Longev.* **2017**, *2017*, 5350267. [CrossRef]
1. Michell, B.J.; Griffiths, J.E.; Mitchelhill, K.I.; Rodriguez-Crespo, I.; Tiganis, T.; Bozinovski, S.; de Montellano, P.R.; Kemp, B.E.; Pearson, R.B. The Akt kinase signals directly to endothelial nitric oxide synthase. *Curr. Biol.* **1999**, *9*, 845–848. [CrossRef]
2. Kolluru, G.K.; Siamwala, J.H.; Chatterjee, S. eNOS phosphorylation in health and disease. *Biochimie* **2010**, *92*, 1186–1198. [CrossRef]
3. Mohanan, P.; Subramaniyam, S.; Mathiyalagan, R.; Yang, D.C. Molecular signaling of ginsenosides Rb1, Rg1, and Rg3 and their mode of actions. *J. Ginseng Res.* **2018**, *42*, 123–132. [CrossRef] [PubMed]
4. Xu, M.; Ma, Q.; Fan, C.; Chen, X.; Zhang, H.; Tang, M. Ginsenosides Rb1 and Rg1 Protect Primary Cultured Astrocytes against Oxygen-Glucose Deprivation/Reoxygenation-Induced Injury via Improving Mitochondrial Function. *Int. J. Mol. Sci.* **2019**, *20*, 6086. [CrossRef] [PubMed]
5. Rajput, S.A.; Shaukat, A.; Rajput, I.R.; Kamboh, A.A.; Iqbal, Z.; Saeed, M.; Akhtar, R.W.; Shah, S.A.H.; Raza, M.A.; El Askary, A.; et al. Ginsenoside Rb1 prevents deoxynivalenol-induced immune injury via alleviating oxidative stress and apoptosis in mice. *Ecotoxicol. Environ. Saf.* **2021**, *220*, 112333. [CrossRef] [PubMed]
6. Ratan, Z.A.; Haidere, M.F.; Hong, Y.H.; Park, S.H.; Lee, J.O.; Lee, J.; Cho, J.Y. Pharmacological potential of ginseng and its major component ginsenosides. *J. Ginseng Res.* **2021**, *45*, 199–210. [CrossRef]
7. Choi, S.H.; Jung, S.W.; Lee, B.H.; Kim, H.J.; Hwang, S.H.; Kim, H.K.; Nah, S.Y. Ginseng pharmacology: A new paradigm based on gintonin-lysophosphatidic acid receptor interactions. *Front. Pharmacol.* **2015**, *6*, 245. [CrossRef]
8. Leung, K.W.; Wong, A.S. Pharmacology of ginsenosides: A literature review. *Chin. Med.* **2010**, *5*, 20. [CrossRef]

49. Schaefer, A.; van Duijn, T.J.; Majolee, J.; Burridge, K.; Hordijk, P.L. Endothelial CD2AP Binds the Receptor ICAM-1 To Contro Mechanosignaling, Leukocyte Adhesion, and the Route of Leukocyte Diapedesis In Vitro. *J. Immunol.* **2017**, *198*, 4823–483([CrossRef]
50. Anbarasan, C.; Bavanilatha, M.; Latchumanadhas, K.; Ajit Mullasari, S. ICAM-1 molecular mechanism and genome wide SNP' association studies. *Indian Heart J.* **2015**, *67*, 282–287. [CrossRef]
51. De Caterina, R.; Libby, P.; Peng, H.B.; Thannickal, V.J.; Rajavashisth, T.B.; Gimbrone, M.A., Jr.; Shin, W.S.; Liao, J.K. Nitric oxid decreases cytokine-induced endothelial activation. Nitric oxide selectively reduces endothelial expression of adhesion molecule and proinflammatory cytokines. *J. Clin. Investig.* **1995**, *96*, 60–68. [CrossRef]
52. Li, S.; Xu, J.; Yao, W.; Li, H.; Liu, Q.; Xiao, F.; Irwin, M.G.; Xia, Z.; Ruan, W. Sevoflurane pretreatment attenuates TNF-alpha induced human endothelial cell dysfunction through activating eNOS/NO pathway. *Biochem. Biophys. Res. Commun.* **2015**, *46(* 879–886. [CrossRef] [PubMed]
53. Badiola, N.; Penas, C.; Minano-Molina, A.; Barneda-Zahonero, B.; Fado, R.; Sanchez-Opazo, G.; Comella, J.X.; Sabria, J.; Zhu, C Blomgren, K.; et al. Induction of ER stress in response to oxygen-glucose deprivation of cortical cultures involves the activation o the PERK and IRE-1 pathways and of caspase-12. *Cell Death Dis.* **2011**, *2*, e149. [CrossRef] [PubMed]
54. Thuerauf, D.J.; Marcinko, M.; Gude, N.; Rubio, M.; Sussman, M.A.; Glembotski, C.C. Activation of the unfolded protein respons in infarcted mouse heart and hypoxic cultured cardiac myocytes. *Circ. Res.* **2006**, *99*, 275–282. [CrossRef] [PubMed]
55. Minamino, T.; Kitakaze, M. ER stress in cardiovascular disease. *J. Mol. Cell. Cardiol.* **2010**, *48*, 1105–1110. [CrossRef] [PubMed]
56. Cimellaro, A.; Perticone, M.; Fiorentino, T.V.; Sciacqua, A.; Hribal, M.L. Role of endoplasmic reticulum stress in endothelia dysfunction. *Nutr. Metab. Cardiovasc. Dis. NMCD* **2016**, *26*, 863–871. [CrossRef]
57. Takaki, A.; Morikawa, K.; Tsutsui, M.; Murayama, Y.; Tekes, E.; Yamagishi, H.; Ohashi, J.; Yada, T.; Yanagihara, N.; Shimokawa H. Crucial role of nitric oxide synthases system in endothelium-dependent hyperpolarization in mice. *J. Exp. Med.* **2008**, *20* 2053–2063. [CrossRef]
58. Wu, S.; Gao, X.; Yang, S.; Meng, M.; Yang, X.; Ge, B. The role of endoplasmic reticulum stress in endothelial dysfunction induce by homocysteine thiolactone. *Fundam. Clin. Pharmacol.* **2015**, *29*, 252–259. [CrossRef]
59. Oslowski, C.M.; Urano, F. The binary switch between life and death of endoplasmic reticulum-stressed beta cells. *Curr. Opir Endocrinol. Diabetes Obes.* **2010**, *17*, 107–112. [CrossRef]
60. Chen, X.; Iliopoulos, D.; Zhang, Q.; Tang, Q.; Greenblatt, M.B.; Hatziapostolou, M.; Lim, E.; Tam, W.L.; Ni, M.; Chen, Y.; et a XBP1 promotes triple-negative breast cancer by controlling the HIF1alpha pathway. *Nature* **2014**, *508*, 103–107. [CrossRef]
61. De Pascali, F.; Hemann, C.; Samons, K.; Chen, C.A.; Zweier, J.L. Hypoxia and reoxygenation induce endothelial nitric oxid synthase uncoupling in endothelial cells through tetrahydrobiopterin depletion and S-glutathionylation. *Biochemistry* **2014**, *5* 3679–3688. [CrossRef]
62. Fish, J.E.; Yan, M.S.; Matouk, C.C.; St Bernard, R.; Ho, J.J.; Gavryushova, A.; Srivastava, D.; Marsden, P.A. Hypoxic repression o endothelial nitric-oxide synthase transcription is coupled with eviction of promoter histones. *J. Biol. Chem.* **2010**, *285*, 810–82([CrossRef] [PubMed]
63. Widmer, R.J.; Lerman, A. Endothelial dysfunction and cardiovascular disease. *Glob. Cardiol. Sci. Pract.* **2014**, *2014*, 291–30 [CrossRef] [PubMed]
64. Okuda, S.; Watanabe, Y.; Moriya, Y.; Kawano, S.; Yamamoto, T.; Matsumoto, M.; Takami, T.; Kobayashi, D.; Araki, N.; Yoshizawa A.C.; et al. jPOSTrepo: An international standard data repository for proteomes. *Nucleic Acids Res.* **2017**, *45*, D1107–D111 [CrossRef] [PubMed]

MDPI AG
Grosspeteranlage 5
4052 Basel
Switzerland
Tel.: +41 61 683 77 34

MDPI Books Editorial Office
E-mail: books@mdpi.com
www.mdpi.com/books

www.ingramcontent.com/pod-product-compliance
Lightning Source LLC
Chambersburg PA
CBHW041938260926
51850CB00018B/254
9783725865703